Handbook of Biochemical Kinetics

Handbook of Biochemical Kinetics

Daniel L. Purich
R. Donald Allison

Department of Biochemistry
and Molecular Biology

University of Florida
College of Medicine

Gainesville, Florida

 **ACADEMIC PRESS**

San Diego London Boston New York Sydney Tokyo Toronto

Academic Press
A Division of Harcourt Brace & Company
525 B Street, Suite 1900, San Diego, CA 92101-4495
http://www.apnet.com

Academic Press
24–28 Oval Road, London NW1 7DX
http://www.hbuk.co.uk/ap/

Library of Congress Catalog Card Number: 99-63958
International Standard Book Number: 0-12-568048-1

Printed in the United States of America
99 00 01 02 03 04 MM 9 8 7 6 5 4 3 2 1

Table of Contents

Preface

The biotic world is doubtlessly the best known example of what Nobelist Murray Gell-Mann has termed "complex adaptive systems"—a name given to those systems possessing the innate capacity to learn and evolve by utilizing acquired information. Those familiar with living systems cannot but marvel at each cell's ability to grow, to sense, to communicate, to cooperate, to move, to proliferate, to die and, even then, to yield opportunity to succeeding cells. If we dare speak of vitalism, especially as the new millennium is eager to dawn, we only do so to recognize that homeostatic mechanisms endow cells with such remarkable resilience that early investigators mistook homeostasis as a persuasive indication that life is self-determining and beyond the laws of chemistry and physics. A shared goal of modern molecular life scientists is to understand the mechanisms and interactions responsible for homeostasis. One approach for analyzing the mechanics of complex systems is to determine the chronology of discrete steps within the overall process—a pursuit called "kinetics." This strategy allows an investigator to assess the structural and energetic determinants of transitions from one step to the next. By identifying voids in the time-line, one considers the possibility of other likely intermediates and ultimately identifies all elementary reactions of a mechanism.

Kinetics is an analytical approach deeply rooted in chemistry and physics, and biochemists have intuitively and inventively honed the tools of chemists and physicists for experiments on biological processes. Biochemical kinetics first began to flourish in enzymology—a field which has gainfully exploited advances in physical organic chemistry, structural chemistry, and spectroscopy in order to dissect the individual steps comprising enzyme mechanisms. No apology is offered, nor should any be required, for our strong emphasis on chemical kinetics and enzyme kinetics. Scientists working within these disciplines have enjoyed unparalleled success in dissecting complex multistage processes. Mechanisms are tools for assessing current knowledge and for designing better experiments. As working models, mechanisms offer the virtues of simplicity, precision, and generativity. "Simplicity" arises from the symbolic representation of the interactions among the minimal number of components needed to account for all observed properties of a system. "Precision" emerges by considering how rival models have nonisomorphic features (i.e., testable differences) that distinguish one from another. "Generativity" results from the recombining of a model's constituent elements to admit new findings, to predict new properties, and to stimulate additional rounds of experiment. For chemical and enzyme kineticists, the goal of this recursive enterprise is to determine a mechanism (a) that accounts for responses to changes in each component's concentration, (b) that explains the detailed time-evolution of all chemical events, (c) that defines the concentration and structure of transient intermediates, (d) that makes sense of relevant changes in positional properties (i.e., conformation, configuration, and/or physical location), and (e) that reconciles the thermodynamics of all reactions steps and transitions. In this respect, the rigor of chemical and enzyme kinetics teaches us all how best to invent new approaches that appropriately balance theory and experiment.

The inspiration for this HANDBOOK stemmed from our shared interest in teaching students about the logical and systematic investigation of enzyme catalysis and metabolic control. We began twenty-five years ago with the teaching of graduate-level courses ("Chemical Aspects of Biological Systems"; "Enzyme Kinetics and Mechanism") at the University of California Santa Barbara as well as a course entitled "Enzyme Kinetics" at the Cornell University Medical College. More recently, we have taught undergraduate students ("A Survey of Biochemistry and Molecular Biology") as well as graduate students ("Advanced Metabolism"; "Physical Biochemistry and Structural Biology"; "Dynamic Processes in the Molecular Life Sciences") here at the University of Florida. Our lectures have included material on the theory and practice of steady-state kinetics, rapid reaction kinetics, isotope-exchange kinetics, inhibitor design, equilibrium and kinetic isotope effects, protein oligomerization and polymerization kinetics, pulse-chase kinetics, transport kinetics, biomineralization kinetics, as well as ligand binding, cooperativity, and allostery. Because no existing text covered the bulk of these topics, we resorted to developing an extensive set of lecture notes—an activity that encouraged us to consider writing what we initially had envisioned as a short textbook on biochemical kinetics.

What also became clear was that, before and during any detailed consideration of a molecular process, teachers must always take pains to explain the associated terminology adequately. In 1789, the French chemist Antoine Lavoisier aptly declared: "Every branch of physical science must consist of the series of facts that are the objects of the science, the ideas that represent these facts, and the words by which these ideas are expressed. And, as ideas are preserved and communicated by means of words, it necessarily follows that we cannot improve the science without improving language or nomenclature." We recognized that there was no published resource to help students come to grips with the far-ranging terminology of biochemical kinetics. Furthermore, as the distinction between scientific disciplines becomes blurred by what may be called "the interdisciplinary imperative," students and practicing scientists from other disciplines will require a reliable sourcebook that explains terminology. Far too much time is wasted when students trace a finger over many pages of a textbook, only to find a partial definition for a sought-after term. Moreover, as more bioscientists come from countries not using English as a working language, there is an even greater need for a reliable and clearly written sourcebook of definitions.

A dictionary format became an appealing possibility for our Handbook, but we also wished to treat many terms in greater depth than found in any dictionary. This led us to adopt the present word-list format which in many

respects resembles the "Micropaedia" section of the Encyclopaedia Britannica. One loses the seamless organization that can be realized in multichapter expositions that systematically develop a series of topics. We have accordingly attempted to mitigate this problem by including longer tracts on absorption and fluorescence spectroscopy, biomineralization, chemical kinetics, enzyme kinetics, Hill and Scatchard treatments, ligand binding cooperativity, kinetic isotope effects, and protein polymerization. Likewise, we have extensively inserted cross-references at appropriate locations within many entries. One intrinsic advantage of a mini-encyclopedia, however, is that in subsequent printings we should be able to make corrections and additions, or even deletions of an entire term, without upsetting the overall format. We also felt that readers should be encouraged to consult the most authoritative sources on particular topics. For this reason, we developed a collection of over 5000 literature references and, in many cases, our citations credit the original papers on a given topic. We have included the names of nearly 1000 enzymes, along with chemical reactions, EC numbers, and, in many instances, their biochemical and catalytic properties. The references cited are not intended to be comprehensive; rather, they serve to guide the reader to further interesting and helpful reading on subjects we have discussed. Where possible, at least one reference is included to provide information on assay protocols for the listed enzyme. We also urge readers to use the Wordfinder (included at the back of the Handbook) to take fullest advantage of the text and reference material. The nearly 8000 entries in the Wordfinder represent all listed source words as well as other subheadings, keywords, or synonyms. Each entry is immediately followed by the recommended source entries, and many source words are also cross-referenced to guide the reader to other related source words.

One of us (D.L.P.) has been a member of the Methods in Enzymology family of editors for well over two decades. The volumes in this series on "Enzyme Kinetics and Mechanism" have become a standard for those interested in biological catalysis. As the form of this book began to emerge, we quickly recognized that Methods in Enzymology could serve as an additional source for annotations on recommended theories and practices for kinetic studies on a wide range of topics. The Handbook contains nearly 6000 Methods in Enzymology citations, and we have indicated the topic, volume, and beginning page for each at the foot of many of the source words. We trust that users of our Handbook will benefit from improved access to the first 280 volumes of METHODS IN ENZYMOLOGY.

For the derivations presented in this Handbook, we have assumed that the reader is familiar with the fundamentals of differential and integral calculus. To those who are loathe to engage in the rigor of mathematics, we say "Take heart!" The successful study of kinetics requires only that students work out a considerable number of problems which are both theoretical and numerical in character. We are reminded that Mithridates VI, the Grecian king of Pontus, is said to have acquired a tolerance to poison by taking gradually increasing doses. To aid those seeking their own intellectual mithridate (i.e., acquired antidote), we provide scores of step-by-step derivations and practical advice on how to derive particular rate expressions. Likewise, we have included detailed protocols for H. J. Fromm's systematic "theory-of-graphs" method as well as W. W. Cleland's net reaction rate method. We are also greatly indebted to Dr. Charles Y. Huang for permitting us to include entire tracts from his chapter (which originally appeared in Volume 63 of METHODS IN ENZYMOLOGY) on the derivation of initial velocity and isotope exchange rate equations. We immediately recognized how daunting the task would be to attempt to surpass Dr. Huang's truly outstanding treatment.

The success of our HANDBOOK OF BIOCHEMICAL KINETICS can only be judged by those using this manual for some period of time. We have come to recognize that we could not possibly represent all of the topics falling within the realm of biochemical kinetics—and certainly not within a first edition. We are also certain that, despite a determined effort to cover the terminology of chemical and enzyme kinetics, we have still overlooked some important issues. We had also hoped to include additional kinetic techniques applied in pharmacokinetics, cell biology, electrophysiology, and metabolic control analysis. Time constraints also prevented our developing mathematical sections on Laplace transforms, vector algebra, distribution functions, and especially statistics. Eventually, we aspire to create a compact disk version of this Handbook, appropriately presented as hyperlinked text; that same CD should have room for selected problems/exercises along with step-by-step solutions, as well as down-loadable programs for kinetic simulation, algorithms for symbolic derivation of rate equations, molecular dynamics and related modeling techniques, and tried-and-true statistical methods. We also hope that our readers will not hesitate to advise us of shortcomings, missed terms, as well as techniques meriting definition, mention, or further explanation. We shall be forever grateful for such guidance.

We thank our students and colleagues for reading earlier drafts of our manuscript, and we are especially grateful to both Shirley Light and Dolores Wright of Academic Press for their insights, help, and thorough editing of the text. We also thank Academic Press for allowing us to incorporate the numerous annotations to METHODS IN ENZYMOLOGY.

Finally, as first- and second-generation disciples of Professor Herbert J. Fromm, we dedicate this book to him, in recognition of his germinal and indelible contributions to the field of enzyme kinetics and mechanism.

Daniel L. Purich

R. Donald Allison

January, 1999

Abbreviations & Symbols

Roman Letters and Symbols

A	Molecule in the ground state Acceptor molecule (in fluorescence)
A*	Molecule in the excited state
A	SI symbol for absorbance (unitless) SI symbol for Helmholtz energy (J) SI symbol for the pre-exponential term in Arrhenius equation $(\text{mol}^{-1}\text{m}^3)^{n-1}\ \text{s}^{-1}$
ΔA	Change in absorbancy
%A	Percent absorption of light $(100 - \%T)$
[A]	Concentration of A
A, B, C, ...	Substrate A, B, C, ...
$A, B, C, ...$	Coulombic contributions to the potential energy of interaction Moments of inertia of transition-state complex
A_{ij}	Amplitude of kinetic decay
Å	Angstrom unit $(10^{-10}\ \text{m})$
a_i	Thermodynamic activity of species i
a_o	Bohr radius
B	Generalized base
B	$e^2\beta/2\varepsilon kT$ or $[e^3/(\varepsilon kT)^{2/3}](2\pi N_o/1000)^{1/2}$, a constant in the Debye-Hückel limiting law
$B_{i,j}$	Second virial coefficient for the mutual interactions of species i and j
Bq	Becquerel (unit of radioactivity = 1 disintegration per second)
C	Coulomb

C or c	Molar concentration (M or moles/L)
C	SI symbol for heat capacity (J K^{-1})
$\Delta C_p{}^\circ$	Constant pressure standard heat capacity per mole
$\hat{C}_i$ or $\hat{c}_i$	Weight concentration of the ith species
c or Ci	Symbol for Curie (old unit of radioactivity)
c_o	SI unit for speed of light in a vacuum; (value = 2.998×10^8 m s^{-1})
c	SI unit for speed of light in a medium (m s^{-1}) Concentration
D	SI symbol for debye (unitless) Donor molecule (in fluorescence)
D	SI symbol for translational diffusion constant (m^2 s^{-1}) Spectroscopic energy of dissociation of a diatomic molecule in the Morse equation
$\Delta D_o{}^\circ$	Difference in dissociation energies of products and reactants measured from zero-point energies
Da	Dalton
D_{rot}	Rotational diffusion constant
$d_{20,\text{w}}$	Density extrapolated to 20°C, water
d	Density Collision diameter
E	Free or uncomplexed enzyme General symbol for enzyme Effector molecule

E	Energy	$\Delta G°$	Standard Gibbs free energy change per mole
	Initial kinetic energy of relative motion of reactants	$\Delta^{\ddagger}G°$	Standard Gibbs free energy of activation
$\mathbf{E}$	Electric vector of light		
E_a	Activation energy	g	Gram
$[E_o]$ or E_o	Total enzyme concentration	g_e	Degeneracy of the lower state
Eq or eq	Equivalent	g_u	Degeneracy of the upper state
	Equilibrium	H	Henry (unit of self-inductance and mutual inductance)
$E°$	Standard electromotive force		
E_s	Energy of molecule in excited singlet state	H	Enthalpy
			Hamiltonian
E_{tor}	Torsional potential energy	H_{local}	Magnetic field strength at nucleus of a molecule
e	Exponential function		
e	Charge on an electron (value = 1.602×10^{-19} coulombs)	$\Delta H°$	Standard enthalpy change per mole
		$\Delta^{\ddagger}H°$	Standard enthalpy of activation
	Equatorial position of a substituent on a molecule	$\mathbf{H}$	Magnetic field
e'	Pseudo-equatorial position of a substituent on a molecule	$\mathbf{H}_{res}$	Magnetic field intensity at which resonance takes place
F	Modified enzyme form in ping pong mechanisms	Hz	Hertz (unit of frequency cycles per second)
$\mathbf{F}$	Force	h	Planck's constant = 6.626×10^{-34} J·sec or 6.626×10^{-27} erg·sec
F	Free energy (archaic)		
	Fluorescence	$\hbar$	$h/2\pi = 1.055 \times 10^{-34}$ J·sec or 1.055×10^{-27} erg·sec
	Rotational-vibrational energy distribution function		
	Momentum distribution function	I	Nuclear spin quantum number
			Inhibitor
$\mathscr{F}$	Faraday	I	Intensity of radiation
f	Fugacity		Ionic strength
	Number of sites for acceptor on ligand (so-called "ligand valence")		Light transmittance
	Oscillator strength	I_{50} or $I_{0.5}$	Inhibitor yielding 50% inhibition or 0.5 the uninhibited rate
	Translational frictional coefficient		
f	Fractional attainment of isotopic equilibrium	$I(\lambda)$	Intensity of light at wavelength λ
		$I(\lambda)_f$	Intensity of emitted light at wavelength λ
F_{rel}	Fluorescence$_{sample}$/Fluorescence$_{standard}$		
G	Gibbs free energy	i	Square root of (-1)
	Gravitational constant	J	Joule

J	Nuclear magnetic resonance coupling coefficient
	Flux density (units = particles area^{-1} time^{-1})
j	Apparent order of a binding reaction
K	Symbol for Kelvin
K or K_{eq}	Macroscopic equilibrium constant
K_a	Acid dissociation constant
$K_A, K_B, \ldots$	Dissociation constant for ligand A, B, C, ...
K_{ap}	Apparent equilibrium constant
K_D	Dissociation constant
K_F	Formation constant (synonym of association constant)
	Dissociation constant for ligand F for an allosteric protein
K_i	Macroscopic inhibition constant
	Macroscopic ionization constant
$K_{ia}, K_{ib}, \ldots$	Dissociation constants in enzyme kinetics
K_R	Dissociation constant for ligand that binds to R-state of allosteric protein
K_S	Equilibrium constant for dissociation of ES complex
K_T	Dissociation constant for ligand that binds to T-state of an allosteric protein
K_w	The constant equal to the product of $[H^+]$ (or, $[H_3O^+]$) and $[OH^-]$ in an aqueous solution
$K_1, K_2, K_3, \ldots$	Stepwise binding or dissociation constants for successive attachments of ligand to an oligomeric receptor
k or k_B	Boltzmann constant
k	Rate constant
	Microscopic equilibrium constant
k_{cat}	Catalytic constant; turnover number
k_{cat}/K_m	Specificity constant

k_d	Intrinsic dissociation constant (reciprocal of $k_{i,j}$)
k_{iH}, k_{iD}, k_{iT}	A rate constant for isotopic isomers containing H, D, or T
k_H/k_D	Kinetic isotope effect
$k_{i,j}$	Intrinsic association or binding constant (reciprocal of k_d) for interaction between sites on species i and j
KE	Kinetic energy
L	Liter
L	Avogadro's number
L	Angular momentum
$\mathscr{L}$	Allosteric constant equal to $[T_o]/[R_o]$
M	Molecular weight
	Molar
$M^{\ddagger}$	Transition-state complex
M	Magnetization
$\overline{M_n}$	Number average molecular weight
M_r	Relative molecular mass
$\overline{M_w}$	Weight average molecular weight
m	Meter
m_e	Mass of electron at rest value = 9.1094×10^{-28} g
N	Newton (unit of force)
N_o	Avogadro's number = 6.0221×10^{23} mol^{-1}
n	Refractive index
	Number of moles
n → π^*	Electronic transition
n → σ^*	Electronic transition
n_H or n_{Hill}	Hill coefficient
P	Generalized symbol for product
P or p	Pressure
pK_a	$-\log_{10} K_a$
pO_2	Oxygen partial pressure

$(pO_2)_{0.5}$ Oxygen partial pressure at 0.5 saturation

Q   Coulomb (unit of electrostatic charge)

Q Heat absorbed by a defined system

Q_{CO_2} Amount CO_2 released by tissue

Q_{syn} Synergism quotient

q Quantum yield

q_o Unquenched quantum yield

R Universal gas constant
Electric resistance
Gross rate of isotopic exchange

R_∞ Rydberg constant

$\overline{R}$ Fraction of allosteric protein in the R-state

R_G Radius of gyration

r Radius
Distance of separation

$\boldsymbol{r}$ Polymer end-to-end vector

S   Svedberg unit (10^{-13} s)

S_A Partial molal entropy

S_A' Unitary part of the partial molal entropy

$\Delta S°$ Standard entropy change

$\Delta^{\ddagger}S°$ Standard entropy of activation

$\boldsymbol{S}$ Scattering vector

S_1 Singlet state

s   Second (unit of time)

s Sedimentation coefficient
Equilibrium constant for helix growth

$s_{20,w}$ Sedimentation coefficient corrected to 20°C, water

$\hat{s}$ Unit vector along scattered radiation

T Temperature

$\%T$ Percent transmission of light

T_m Melting temperature

T_1 Longitudinal relaxation time

T_2 Transverse relaxation time

t Time

$t_{1/2}$ Half-life

U Internal energy

V Volume

$\Delta V^{\ddagger}$ Volume of activation

V_h Hydrated volume

V_m or V_{max} Maximal velocity

$V_{m,f}$ or $V_{max,f}$ Maximal velocity in the forward direction

$V_{m,r}$ or $V_{max,r}$ Maximal velocity in the reverse direction

V/K Ratio of V_{max} to K_m

v Speed
Initial velocity of enzyme-catalyzed reaction
Vibrational frequency

$\dot{X}$ dX/dt

$(\overline{X_i})$ Equilibrium concentration of substance X_i

$\Delta(X_i)$ Difference between temporal and equilibrium concentration of X_i

z   Charge on a macromolecule or ion in units of e

Greek Letters and Symbols

α Degree of association
Alpha particle
Electric polarizability
Reduced concentration ($[F]/K_F$) for allosteric protein

α_H Hill coefficient

β Reduced concentration ($[I]/K_I$) for allosteric protein

β_e Bohr magneton

Γ	Surface concentration (mol m^{-2})
	Parameter affecting relaxation amplitude
γ	Reduced concentration ([A]/K_A) for allosteric protein
δ	Phase shift
	Chemical shift in nuclear magnetic resonance
∂	Torque
ε	Molar absorptivity
	Dielectric constant
$\Delta\varepsilon$	Ellipticity in circular dichroism
η	Solution viscosity
η_o	Solvent viscosity
$[\eta]$	Intrinsic viscosity
θ_i	Fraction of ligand saturation of ith site
$[\theta]$	Molar ellipticity
κ	Transmission coefficient for transition state
	Inverse screening length
λ	Wavelength
	Kinetic decay time
μ_i	Chemical potential of ith species per mole
$\mu_i{}^o$	Standard chemical potential per mole
μ_m	Magnetic moment
μ	Ionic strength
	Electric dipole moment operator
	Reduced mass, $\mu = m_A m_B/(m_A + m_B)$
ν	Frequency
$\overline{\nu}$	Fractional saturation of ligand binding sites
Ξ	Grand partition function in the Wyman treatment
Π	Product algorithm
	Osmotic pressure
ρ	Density (mass per unit volume)

$\rho(r)$	Electron density
Σ	Summation algorithm
τ	Relaxation time (t)
	Lag time (s)
Φ	Phi relation in enzyme kinetics
	Electrical potential
ϕ	Quantum yield (unitless)
χ	Mole fraction of component I
Ω	Solid angle
$\Omega_{n,i}$	Statistical factor for ligand-i binding at n sites on a macromolecule
ω	Angular momentum
	Circular frequency (Hz)
	Ionic strength (archaic)
ω_o	Larmor frequency (Hz)
ω	Angular velocity (rad s^{-1} or s^{-1})

Mathematical Symbols

α,β,γ	Directional angles
f'	First derivative
f''	Second derivative
∂	Partial derivative, Jacobian
$\int$	Integral
$<>$	Average
$<\|>$	Overlap interval
$<\|\|>$	Expectation value integral
$*$	Superscript designating radioactive substance
	Superscript designating excited state
	Subscript designating complex conjugate
$\ddagger$	Superscript for transition state
$(\partial x/\partial t)_y$	Partial differential of x with respect to time at constant y
∇	Vector differential or gradient, $$\mathbf{i}\frac{\partial}{\partial x} + \mathbf{j}\frac{\partial}{\partial y} + \mathbf{k}\frac{\partial}{\partial z}$$

∇^2	Second derivative operator, $\dfrac{\partial^2}{\partial x^2} + \dfrac{\partial^2}{\partial y^2} + \dfrac{\partial^2}{\partial z^2}$
Δ	Constant time interval
( )	Activity of a solute
(,)	Open interval
[ ]	Concentration of a solute
[,]	Closed interval
∞	Infinite dilution, typically as a subscript

Multiples/Submultiples

10^{12}	Tera (symbol = T)
10^{9}	Giga (symbol = G)
10^{6}	Mega (symbol = M)
10^{3}	Kilo (symbol = k)
10^{-1}	Deci (symbol = d)
10^{-2}	Centi (symbol = c)
10^{-3}	Milli (symbol = m)
10^{-6}	Micro (symbol = m)
10^{-9}	Nano (symbol = n)
10^{-12}	Pico (symbol = p)
10^{-15}	Femto (symbol = f)
10^{-18}	Atto (symbol = a)
10^{-21}	Zepto (symbol = z)

Biochemical Abbreviations

A	Adenine Alanine or alanyl
aa	Amino acid
aaRS	Aminoacyl-tRNA
ACAT	Acyl-CoA:cholesterol acyltransferase
ACES	N-(2-Acetamido)-2-aminoethanesulfonic acid
ACh	Acetylcholine

ACP	Acyl carrier protein
ADA	Adenosine deaminase
Ade	Adenine
ADH	Alcohol dehydrogenase
ADP	Adenosine 5′-diphosphate
Ala	Alanine or alanyl
ALA	δ-Aminolevulinic acid or δ-aminolevulinate
AMP	Adenosine 5′-monophosphate
amu	Atomic mass unit (1.66×10^{-27} kg or 1.66×10^{-24} g)
Arg	Arginyl or arginyl
Asn	Asparagine or asparaginyl
Asp	Aspartic acid, aspartate, or aspartyl
Asx	Aspartate + asparagine or aspartyl + asparaginyl
ATCase	Aspartate transcarbamoylase
ATP	Adenosine 5′-triphosphate
B	Aspartate + asparagine (or aspartyl + asparaginyl)
BES	N,N-Bis(2-hydroxyethyl)-2-aminoethanesulfonic acid
Bi	Two-substrate enzyme system
bp	Base pair
BPG	D-2,3-Bisphosphoglycerate
BPTI	Bovine pancreatic trypsin inhibitor
Bq	Becquerel
C	Cytosine Cysteine or cysteinyl
CaM	Calmodulin
cAMP	Cyclic AMP
CAP	Catabolite gene activating protein
cAPK	Protein kinase A (or cyclic AMP-stimulated protein kinase)
CAPS	3-(Cyclohexylamino)propanesulfonic acid

Cbz-	Benzyloxycarbonyl-		EDTA	Ethylenediaminetetraacetic acid or its conjugate base
cDNA	Complimentary strand DNA		EF	Elongation factor
CDP	Cytidine 5′-diphosphate		EGTA	Ethylene glycol bis(β-aminoethyl ether)-N,N,N',N'-tetraacetic acid or its conjugate base
CHES	3-(Cyclohexylamino)ethanesulfonic acid		EPPS	N-2-Hydroxyethylpiperazinepropane-sulfonic acid (also known as HEPPS)
Chl	Chlorophyll		EPR or ESR	Electron paramagnetic resonance or Electron spin resonance
CM	Carboxymethyl			
cmc	Critical micelle concentration		F	Phenylalanine or phenylalanyl
CMP	Cytidine 5′-monophosphate		FAD	Oxidized flavin adenine dinucleotide
CoA	Coenzyme A		FADH·	Radical form of reduced flavin adenine dinucleotide
CoASH	Coenzyme A			
CoQ	Coenzyme Q		FADH$_2$	Reduced flavin adenine dinucleotide
CTP	Cytidine 5′-phosphate		FBP	Fructose 1,6-bisphosphate
Cys	Cysteine or cysteinyl		Fd	Ferredoxin
D	Dalton Aspartic acid, aspartate, or aspartyl		fMet	N-Formylmethionine
d	Deoxy		FMN	Flavin mononucleotide
dd	Dideoxy		F1P	Fructose 1-phosphate
DEAE	Diethylaminoethyl		F6P	Fructose 6-phosphate
DFP	Diisopropyl fluorophosphate		G	Guanine Glycine or glycyl
DG	sn-1,2-Diacylglycerol			
DHAP	Dihydroxyacetone phosphate		GABA	γ-Aminobutyric acid
DHF	Dihydrofolate		Gal	Galactose
DHFR	Dihydrofolate reductase		GalNAc	N-Acetylglucosamine
DMF	Dimethylformamide		GAP	Glyceraldehyde 3-phosphate
DMS	Dimethyl sulfate		GDP	Guanosine 5′-diphosphate
DMSO	Dimethylsulfoxide		Gla	4-Carboxyglutamic acid or 4-carboxyglutamyl
DNP	2,4-Dinitrophenyl			
Dol	Dolichol		Glc	Glucose
L-DOPA	L-3,4-Dihydroxyphenylalanine		Gln	Glutamine or glutaminyl
DPN$^+$	see recommended abbreviation NAD$^+$		Glu	Glutamic acid, glutamate, or glutamyl
DSC	Differential scanning calorimetry		Gly	Glycine or glycyl
E	Glutamic acid, glutamate, or glutamyl		GMP	Guanosine 5′-monophosphate

G1P	Glucose 1-phosphate		IR	Infrared
G6P	Glucose 6-phosphate		IS	Insertion sequence
GSH	Glutathione (sometimes referred to as reduced glutathione)		ITP	Inosine 5′-triphosphate
			K	Lysine or lysyl
GSSG	Glutathione disulfide (sometimes referred to as oxidized glutathione)		kb	Kilobase pair
			kD or kDa	Kilodalton
GTP	Guanosine 5′-triphosphate		KF	Klenow factor
H	Histidine or histidyl		L	Leucine or leucyl
HA	Hemagglutinin		LCAT	Lecithin:cholesterol acyl transferase
Hb	Hemoglobin		LDH	Lactate dehydrogenase
HbA	Adult hemoglobin		LDL	Low density lipoprotein
HbCO	Carbon monoxide hemoglobin		Leu	Leucine or leucyl
HbO$_2$	Oxyhemoglobin		Lys	Lysine or lysyl
HbS	Sickle cell hemoglobin		M	Methionine or methionyl
HbF	Fetal hemoglobin		MALDI-MS	Matrix-assisted laser desorption ionization-mass spectroscopy
HDL	High density lipoprotein			
HEPES	N-2-Hydroxyethylpiperazine-N'-2-ethanesulfonic acid (also written as Hepes)		Man	Mannose
			Mb	Myoglobin
			MbCO	Carbon monoxide myoglobin
HEPPS	See EPPS		MbO$_2$	Oxymyoglobin
HGPRT	Hypoxanthine:guanine phosphoribosyltransferase		MES	2-(N-Morpholino)ethanesulfonic acid
			Met	Methionine or methionyl
His	Histidine or histidyl		MetHb	Methemoglobin (or, Fe(III)Hb)
HMG-CoA	β-Hydroxymethylglutaryl-CoA		MetMb	Metmyoglobin (or, Fe(III)Mb)
hnRNA	Heterogenous nuclear RNA		MOPS	3-(N-Morpholino)propanesulfonic acid
hsp	Heat shock protein		MS	Mass spectroscopy/spectrometry
Hyp	Hydroxyproline		N	Asparagine or asparaginyl
I	Isoleucine or isoleucyl			
IDL	Intermediate density lipoprotein		NAD$^+$	Oxidized nicotinamide adenine dinucleotide
IF	Initiation factor		NADH	Reduced nicotinamide adenine dinucleotide
IgG	Immunoglobulin			
Ile	Isoleucine or isoleucyl		NAG	N-Acetylglucosamine
IMP	Inosine 5′-monophosphate		NAM	N-Acetylmuramic acid
IPTG	Isopropylthiogalactoside		NANA	N-Acetylneuraminic acid

NMN	Nicotinamide mononucleotide	Q_{CO_2}	The amount (in microliters) of CO_2 given off (under standard conditions of pressure and temperature) per milligram of tissue per hour
NMR	Nuclear magnetic resonance		
NOESY	Nuclear Overhauser effect spectroscopy		
NTP	Nucleoside 5'-triphosphate	QELS	Quasi-elastic laser light scattering
P	Proline or prolyl	QH_2	Ubiqinol
P (or p)	Phosphate	Quad	Four substrate enzyme system
PEP	Phosphoenolpyruvate	R	Arginine or arginyl
PFK	Phosphofructokinase	r	Ribo
PG	Prostaglandin	R5P	Ribose 5-phosphate
2PG	2-Phosphoglycerate	RPC	Reverse phase chromatography
3PG	3-Phosphoglycerate	RT	Reverse transcriptase
PGI	Phosphoglucoisomerase	RTK	Receptor tyrosine kinase
PGM	Phosphoglucomutase	$Ru1,5P_2$	Ribulose 1,5-bisphosphate
Phe	Phenylalanine or phenylalanyl	Ru5P	Ribulose 5-phosphate
PIP_2	Phosphatidylinositol 4,5-bisphosphate	S	Serine or seryl Svedberg constant
PIPES	Piperazine-N,N'-bis(2-ethanesulfonic acid)	SAM	S-Adenosylmethionine
		Ser	Serine or seryl
PK	Pyruvate kinase	T	Threonine or threonyl Thymine
PKA	Protein kinase A (or cyclic AMP-stimulated protein kinase)	TAPS	Tris(hydroxymethyl)methylamino-propanesulfonic acid
PKC	Protein kinase C	TCA	Tricarboxylic acid
PKU	Phenylketonuria	Ter	Three-substrate enzyme system
PLP	Pyridoxal 5-phosphate	TES	N-Tris(hydroxymethyl)methyl-2-aminoethanesulfonic acid
Pol	DNA polymerase		
PP_i	Inorganic pyrophosphate	THF	Tetrahydrofolate
Pro	Proline or prolyl	Thr	Threonine or threonyl
PrP	Prion protein	TIM or TPI	Triose-phosphate isomerase
PRPP	5-Phosphoribosyl-α-pyrophosphate	TPP	Thiamin pyrophosphate (or thiamin diphosphate)
PS	Photosystem	Tris	Tris(hydroxymethyl)aminomethane
Q	Glutamine or glutaminyl Ubiqinone (Coenzyme Q or CoQ)	TS	Thymidylate synthase Transition state
		TX	Transition state intermediate

Tyr	Tyrosine or tyrosyl	Xaa	Unspecified amino acid or amino acyl residue
U	Uridine		
Uni	One-substrate enzyme system	XAFS	X-ray analysis for structure
V	Valine or valyl	Y	Tyrosine or tyrosyl
Val	Valine or valyl	YADH	Yeast alcohol dehydrogenase
X	Nonstandard or unknown amino acid or amino acyl	Z	Glutamate + glutamine or glutamyl + glutaminyl

Abbreviated Binding Schemes

The following diagrams indicate the binding interactions for enzyme kinetic mechanisms. To conserve space, the notation used in this Handbook is a compact version of the diagrams first introduced by Cleland[1]. His diagram for the Ordered Uni Bi Mechanism is as follows:

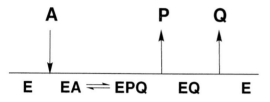

Throughout this handbook, we have used the following single-line, compact notation:

$$_E\underline{\textbf{A} \downarrow (\textbf{EA} \leftrightarrow \textbf{EPQ}) \downarrow \textbf{P (EQ)} \downarrow \textbf{Q}}_E$$

This convention offers the advantage that it can be readily reproduced on virtually all word processors and typesetting devices without creating any special artwork. The enzyme surface is represented by the underline, in this case preceded by a subscript E to indicate the unbound (or free) enzyme before any substrate addition and after all products desorb. An arrow pointing to the line indicates binding, and the reader should understand that reversible binding is taken for granted. Moreover, unlike the original Cleland diagram, product release always is indicated by a downward arrow, because this systematic usage emphasizes the symmetry of certain mechanisms. The symbol "**A** ↓" indicates that substrate A adds; the symbol "**A** or **B** ↓ ↓", "**A** or **B** or **C** ↓ ↓ ↓", etc., indicates *random* addition of two or three substrates, respectively. Interconversions of enzyme-bound reactants can be reversible (↔) or irreversible (→).

Taking the Ordered Uni Bi mechanism as an example, we can consider several additional possibilities:

EA-to-EX and EX-to-EPQ reversible:

$$_E\underline{\textbf{A} \downarrow (\textbf{EA} \leftrightarrow \textbf{EX} \leftrightarrow \textbf{EPQ}) \downarrow \textbf{P (EQ)} \downarrow \textbf{Q}}_E$$

EA-to-EX reversible and EX-to-EPQ irreversible:

$$_E\underline{\textbf{A} \downarrow (\textbf{EA} \leftrightarrow \textbf{EX} \rightarrow \textbf{EPQ}) \downarrow \textbf{P (EQ)} \downarrow \textbf{Q}}_E$$

EA-to-EX irreversible and EX-to-EPQ irreversible:

$$_E\underline{\textbf{A} \downarrow (\textbf{EA} \rightarrow \textbf{EX} \rightarrow \textbf{EPQ}) \downarrow \textbf{P (EQ)} \downarrow \textbf{Q}}_E$$

For the case of so-called iso mechanisms, the compact diagrams are as follows:

$$_E\underline{\textbf{A} \downarrow (\textbf{EA} \leftrightarrow \textbf{EX} \rightarrow \textbf{EPQ}) \downarrow \textbf{P} \downarrow \textbf{Q}}_{F \leftrightarrow E}$$

where F↔E represents the reversible isomerization step.

Other examples of one-substrate and two-substrate kinetic mechanisms include:

Ordered Uni Bi Mechanism

$$_E\underline{\textbf{A} \downarrow (\textbf{EA} \leftrightarrow \textbf{EPQ}) \downarrow \textbf{P (EQ)} \downarrow \textbf{Q}}_E$$

Random Uni Bi Mechanism

$$_E\underline{\textbf{A} \downarrow (\textbf{EA} \leftrightarrow \textbf{EPQ}) \downarrow \downarrow \textbf{P or Q}}_E$$

Random Bi Uni Mechanism

$$_E\underline{\textbf{A or B} \downarrow \downarrow (\textbf{EAB} \leftrightarrow \textbf{EP}) \downarrow \textbf{P}}_E$$

Ordered Bi Uni Mechanism

$$_E\underline{\textbf{A} \downarrow (\textbf{EA}) \textbf{B} \downarrow (\textbf{EAB} \leftrightarrow \textbf{EP}) \downarrow \textbf{P}}_E$$

Ordered Bi Bi Mechanism

$$_E\underline{\textbf{A} \downarrow (\textbf{EA}) \textbf{B} \downarrow (\textbf{EAB} \leftrightarrow \textbf{EPQ}) \downarrow \textbf{P (EQ)} \downarrow \textbf{Q}}_E$$

Ordered Bi Bi Theorell-Chance Mechanism

$$_E\underline{\textbf{A} \downarrow (\textbf{EA}) \textbf{B} \downarrow \downarrow \textbf{P (EQ)} \downarrow \textbf{Q}}_E$$

Ping Pong Bi Bi Mechanism

$$\underset{E}{\underline{\quad A\downarrow \ (EA\leftrightarrow FP) \ \downarrow P \ (F) \ B\downarrow \ (FB\leftrightarrow EQ) \ \downarrow Q}}_{E}$$

[1] W. W. Cleland (1963) *Biochem. Biophys. Acta.* **67**, 104.

Random Bi Bi Mechanism

$$\underset{E}{\underline{\quad A\ or\ B \ \downarrow\downarrow \ (EAB\leftrightarrow EPQ\ ) \ \downarrow\downarrow \ P\ or\ Q}}_{E}$$

Source Words

A

A, B, C, . . ./P, Q, R, . . .

Symbols for substrates and products, respectively, in multisubstrate enzyme-catalyzed reactions. In all ordered reaction mechanisms, A represents the first substrate to bind, B is the second, *etc.*, whereas P denotes the first product to be released, Q represents the second, *etc.* **See** *Cleland Nomenclature*

AB INITIO MOLECULAR-ORBITAL CALCULATIONS

A method of molecular-orbital calculations for determining bonding characteristics and other structural information about a wide variety of compounds and molecular configurations, including those that may not be directly observable (*e.g.*, transition state configurations with partial bonds). Although *ab initio* calculations are typically applied to systems with a small number of atoms, these computationally intensive calculations can be helpful in providing insights about the enzyme-catalyzed reactions. Related methods, known as semiempirical methods, use simplifying assumptions in the calculations and are determined more quickly than standard *ab initio* methods.

W. J. Hehre, L. Radom, P. von R. Schleyer & J. A. Pople (1986) *Ab Initio Molecular Orbital Theory*, Wiley, New York.
T. Clark (1985) *A Handbook of Computational Chemistry*, Wiley, New York.
W. G. Richards & D. L. Cooper (1983) *Ab Initio Molecular Orbital Calculations for Chemists*, 2nd ed., Oxford Press, Oxford.
W. Thiel (1988) *Tetrahedron* **44**, 7393.

ABM-1 & ABM-2 SEQUENCES IN ACTIN-BASED MOTORS

Consensus docking sites[1] for actin-based motility, defined by the oligoproline modules in *Listeria monocytogenes* ActA surface protein and human platelet vasodilator-stimulated phosphoprotein (VASP). Analysis of known actin regulatory proteins led to the identification of distinct A̲ctin-B̲ased M̲otility (or A̲ctin-B̲ased-M̲otor) homology sequences:

ABM-1: (D/E)FPPPPX(D/E) [where X = P or T]

ABM-2: XPPPPP [where X = G, A, L, P, or S]

Actin-based motility involves a cascade of binding interactions designed to assemble actin regulatory proteins into functional locomotory units. *Listeria* ActA surface protein contains a series of nearly identical EFPPPP TDE-type oligoproline sequences for binding vasodilator-stimulated phosphoprotein (VASP). The latter, a tetrameric protein with 20-24 GPPPPP docking sites, binds numerous molecules of profilin, a 15 kDa regulatory protein known to promote actin filament assembly[2]. Laine *et al.*[3] recently demonstrated that proteolysis of the focal contact component vinculin unmasks an ActA homologue for actin-based *Shigella* motility. The ABM-1 sequence (PDFPPPPPDL) is located at or near the C-terminus of the p90 proteolytic fragment of vinculin. Unmasking of this site serves as a molecular switch that initiates assembly of an actin-based motility complex containing VASP and profilin. Another focal adhesion protein zyxin[4] contains several ABM-1 homology sequences that are also functionally active in reorganizing the actin cytoskeletal network.

[1]D. L. Purich & F. S. Southwick (l997) *Biochem. Biophys. Res. Comm.* **231**, 686.
[2]F. S. Southwick & D. L. Purich (1996) *New Engl. J. Med.* **334**, 770.
[3]R. O. Laine, W. Zeile, F. Kang, D. L. Purich & F. S. Southwick (1997) *J. Cell Biol.* **138**, 1255.
[4]R. M. Golsteyn, M. C. Beckerle, T. Koay & E. Friedrich (1997) *J. Cell Sci.* **110**, 1893.

ABORTIVE COMPLEXES

Nonproductive reversible complexes of an enzyme with various substrates and/or products. The International Union of Biochemistry[1] distinguishes dead-end complex from abortive complex, and the latter term is regarded as a synonym for "nonproductive complex". Complexes that fail to undergo further reactions along the catalytic pathway are called dead-end complexes, and the reactions producing them are called dead-end reactions.

Some ambiguity exists in the literature regarding the usage of "abortive complex". For example, the term has been used to describe the nonproductive complex formed between an enzyme and a competitive inhibitor[2] or to describe that inhibition of depolymerases resulting from shifted registration of the substrate within the enzyme's set of subsites[2,3]. Still others have used the term interchangeably with dead-end complexes[4]. The term abortive complexes formation is treated as a special case of dead-end complexation and is restricted to nonproductive complexes involving the binding of substrate(s) and/or product(s) to one or more enzyme forms. Thus, nonproductive complexes that culminate in substrate inhibition are abortive complexes. For a discussion concerning formation of EB and EP complexes in rapid-equilibrium ordered Bi Bi reactions, see the section on the Frieden Dilemma. Enzyme-substrate-product complexes that often form with multisubstrate enzymes are also abortive complexes.

Early product inhibition studies of *Aerobacter aerogenes* ribitol dehydrogenase[5] demonstrated the formation of the E-NAD$^+$-D-ribulose and E-NADH-D-ribitol complexes. In the lactate dehydrogenase reaction, the E-NADH-lactate and E-NAD$^+$-pyruvate complexes are stable[6], and determination of the K_d values indicates that the E-NAD$^+$-pyruvate ternary complex is physiologically relevant[7]. Abortive complexes have been reported for a wide variety of enzymes. Isotope exchange at equilibrium is used to identify the E-NADH-malate abortive with bovine heart malate dehydrogenase[7]. Creatine kinase forms an E-MgADP-creatine complex[8]. Inhibition at high substrate-product concentrations may arise from factors other than abortive complexes; for example, the inhibition observed in an equilibrium exchange experiment may be related to high ionic strength of reaction solutions[9].

Wong and Hanes[10] pointed out that equilibrium exchange studies can be useful in detecting the presence of abortive species. Although abortive complexes can complicate exchange kinetic behavior, the Wedler-Boyer protocol[11] minimizes the influence of abortives on equilibrium exchange studies.

Different abortives may be formed with alternative products or substrates. Such procedures can be useful in helping to distinguish Theorell-Chance mechanisms from ordered systems with abortive complexes[12]. In the case of lactate dehydrogenase, the E-pyruvate-NAD$^+$ and E-lactate-NADH abortive complexes may play a regulatory roles in aerobic *versus* anaerobic metabolism.

Computer simulations[13] also point to the regulatory potential of these non-productive complexes. *See* Deadend Complexes; Inhibition; Nonproductive Complexes; Product Inhibition; Substrate Inhibition; Isotope Trapping; Isotope Exchange at Equilibrium; Enzyme Regulation

[1] International Union of Biochemistry (1982) *Eur. J. Biochem.* **28**, 281.
[2] M. Dixon & E. C. Webb (1979) *Enzymes*, 3rd ed., Academic Press, New York.
[3] J. D. Allen (1979) *Meth. Enzymol.* **64**, 248.
[4] H. J. Fromm (1975) *Initial Rate Enzyme Kinetics*, Springer-Verlag, New York.
[5] H. J. Fromm & D. R. Nelson (1962) *J. Biol. Chem.* **231**, 215.
[6] H. J. Fromm (1963) *J. Biol. Chem.* **238**, 2938.
[7] H. Gutfreund, R. Cantwell, C. H. McMurray, R. S. Criddle & G. Hathaway (1968) *Biochem. J.* **106**, 683.
[8] E. Silverstein & G. Sulebele (1969) *Biochemistry* **8**, 2543.
[9] J. F. Morrison & W. W. Cleland (1966) *J. Biol. Chem.* **241**, 673.
[10] J. T.-F. Wong & C. S. Hanes (1964) *Nature* **203**, 492.
[11] F. C. Wedler & P. D. Boyer (1972) *J. Biol. Chem.* **247**, 984.
[12] C. C. Wratten & W. W. Cleland (1965) *Biochemistry* **4**, 2442.
[13] D. L. Purich & H. J. Fromm (1972) *Curr. Topics in Cell. Reg.* **6**, 131.

Selected entries from *Methods in Enzymology* [vol, page(s)]:
Formation, **63**, 43, 419-424, 432-436; chymotrypsin, **63**, 205; isotope exchange, **64**, 32, 33, 39-45; isotope trapping, **64**, 58; limitation, **63**, 432-436; multiple, one-substrate system, **63**, 473, 474; pH effects, **63**, 205; practical aspects, **63**, 477-480; substrate inhibition, **63**, 500, 501; two-substrate system, **63**, 474-478; in product inhibition studies, **249**, 188-189, 193, 199-200, 205; identification of, **249**, 188-189, 202, 206, 208-209.

ABSCISSA

The *x*-coordinate axis for a graph of Cartesian coordinates [*x*,*y*] or [*x*, *f*(*x*)] or the *x*-value for any [*x*,*y*] ordered pair. This corresponds to the [*Substrate Concentration*]-axis in *v* versus [S] plots or the 1/[*Substrate Concentration*]-axis in so-called double-reciprocal or Lineweaver-Burk plots.

ABSOLUTE CONFIGURATION

A method for designating the stereoisomeric configuration of a chiral carbon atom within a molecular entity. The designation D was arbitrarily assigned to (+)-glyceraldehyde, and (−)-glyceraldehyde was assigned the label

L. Compounds that can be derived from L-glyceraldehyde without inversion reactions of the chiral center are likewise designated L- (and, their mirror images, D-). X-ray crystallographic studies later showed that D-glyceraldehyde had the configuration shown below[1]. D- and L-Alanine are also depicted.

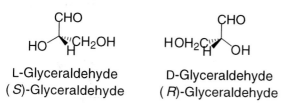

L-Glyceraldehyde
(*S*)-Glyceraldehyde

D-Glyceraldehyde
(*R*)-Glyceraldehyde

The DL-system is in common use with respect to amino acids and sugars, but the Cahn-Ingold-Prelog system (the *RS*-system) is more systematic and should be used.

The literature is replete with reports failing to specify the stereochemistry of certain reactions, and one must often infer the enantiomer. **See** *Configuration; Cahn-Ingold-Prelog System; Corn Rule; Diastereomers; Enantiomers; (R/S)-Convention*

[1]J. M. Bijuoet, A. F. Peerdeman & A. J. van Bommel (1951) *Nature* **168**, 271.
[2]W. Klyne & J. Burkingham (1978) *Atlas of Stereochemistry*, 2nd ed., vol. **2**, Oxford Univ. Press, New York.
[3]J. Jacques, C. Gros, S. Bourcier, M. J. Brienne & J. Toullec (1977) *Absolute Configurations*, Georg Thieme Publ., Stuttgart.

ABSOLUTE TEMPERATURE

A temperature measured on an absolute temperature scale (*i.e.*, a scale in which zero degrees is equivalent to absolute zero). In the Kelvin scale, the degree unit is the kelvin, abbreviated as K; it does not have the superscript o used to indicate degree as on the Celsius scale. K has the same magnitude as degree Celsius (°C).

ABSOLUTE UNCERTAINTY

The uncertainty in measured values expressed in units of the measurement. For example, a reaction velocity of 10.2 M/min is presumed to be valid to a tenth, and the absolute uncertainty is 0.1 M/min. **See** *Relative Uncertainty*

ABSOLUTE ZERO

The temperature at thermal energy of random motion of molecular entities of a system in thermal equilibrium is zero. This temperature is equal to −273.15°C or −459.67°F. Note that even at absolute zero, chemical bonds still retain zero point energy.

ABSORBANCE

A quantitative measure of photon absorption by a molecule, expressed as the $\log_{10}$ of the ratio of the radiant intensity I_o of light transmitted through a reference sample to the light I transmitted through the solution [*i.e.*, $A = \log(I_o/I)$]. Out-moded terms for absorbance such as optical density, extinction, and absorbancy should be abandoned.

The International Union of Pure and Applied Chemistry recommends that the definition should now be based on the ratio of the radiant power of incident radiation (P_o) to the radiant power of transmitted radiation (P). Thus, $A = \log(P_o/P) = \log T^{-1}$. In solution, P_o would refer to the radiant power of light transmitted through the reference sample. T is referred to as the transmittance. If natural logarithms are used, the quantity, symbolized by B, is referred to as the Napierian absorbance. Thus, $B = \ln(P_o/P)$. The definition assumes that light reflection and light scattering are negligible. If not, the appropriate term for $\log(P_o/P)$ is "attenuance." **See** *Beer-Lambert Law; Absorption Coefficient; Absorption Spectroscopy*

ABSORBED DOSE

1. The quantity of absorbed energy absorbed per unit mass of a substance, object, or organism in an irradiated medium. Symbolized by D, the SI unit is the gray (Gy; joules per kilogram). The unit rad is also commonly used (1 rad = 0.01 Gy). 2. The amount of substance (*e.g.*, pharmaceutical) absorbed by an organism or cell.

ABSORPTION

1. The process of being taken up or becoming a part of another body. The unrelated, but often confused term "adsorption" describes a particular type of surface phenomenon. **See also** *Adsorption*. 2. The process of transfer of energy (*e.g.*, from an electromagnetic field) to a structure or entity (*e.g.*, a molecular entity). 3. The process of transport of a substance into a cell.

ABSORPTION COEFFICIENT

The log-base$_{10}$ attenuance (or absorbance) (*i.e.*, D or A) divided by the optical pathlength (l). This coefficient, symbolized by a, is thus equal to $l^{-1}\log(P_o/P)$. **See** *Beer-Lambert Law; Absorbance; Molar Absorption Coefficient; Absorption Spectroscopy*

ABSORPTION SPECTROSCOPY

Absorption spectroscopy is widely used to follow the course of enzyme-catalyzed reactions. Absorbance measurements should be made under conditions that permit use of the Beer-Lambert Law:

$$A = \log (I_o/I) = \varepsilon cl$$

where A is the absorbance (a dimensionless parameter), I_0 and I are the incident radiant intensity and the transmitted radiant intensity, ε is the molar (base$_{10}$) absorption coefficient, c is the molar concentration of the absorbing species, and l is the absorption pathlength (*i.e.*, the distance through solution that light must pass). [Note: IUPAC[1] now favors $A = \log(P_o/P)$ where P_o and P are the incident radiant power and the transmitted radiant power, respectively.]

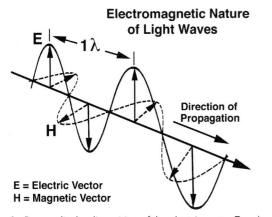

Electromagnetic Nature of Light Waves

E = Electric Vector
H = Magnetic Vector

Figure 1. Perpendicular disposition of the electric vector **E** and magnetic vector **H** of a light wave traveling from its source in the direction of propagation shown by the arrow. Note that electromagnetic radiation interacts with molecules in two ways: (a) in absorption, the energy of a photon is absorbed by an electron (hence, the familiar term electronic absorbance spectrum) when the direction of the electric vector is aligned with the transition dipole of the molecule; (b) in light scatterring, only the direction of propagation is changed, and very little, if any, energy is lost.

Practical Considerations. Typical absorption assay methods utilize ultraviolet (UV) or visible (vis) wavelengths. With most spectrophotometers, the measured absorbance should be less than 1.2 to obtain a strictly linear relationship (*i.e.*, to obey the Beer-Lambert Law). Nonlinear *A versus c* plots can result from micelle formation, sample turbidity, the presence of stray light (see below), bubble formation, stacking of aromatic chromophores, and even the presence of fine cotton strands from tissue used to clean the faces of cuvettes. One is well advised to confirm the linearity of absorbance with respect to product (or substrate) concentration under the exact assay conditions to be employed in

rate experiments. Prior centrifugation or filtration may be needed to reduce light scattering or turbidity.

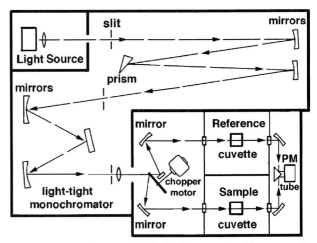

Figure 2. Design features of a double-beam UV/visible spectrophotometer. Note that rays of light pass through a set of slits as they enter the light-tight monochromator. Rotation of the prism determines the wavelength of dispersed light that passes on to the sample compartment. The chopper-motor rotates a beam-splitter that allows half of the in-coming light to travel to the reference compartment, while reflecting the other half of the in-coming light to the sample compartment.

Instruments with double monochromator configurations, or equivalent multi-pass configurations, can greatly reduce stray light (which is any radiation of wavelength other than that of the columnated light beam). Absorbance in the presence of stray light can be expressed as:

$$A = \log [(I_o + I_s)/(I + I_s)]$$

where I_s is the stray light intensity. The larger the value of I_s (relative to I_o or I), the greater the error in concentration or rate measurements[2].

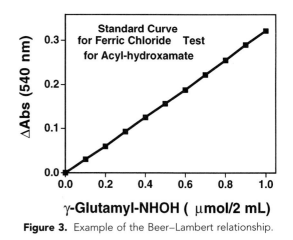

Figure 3. Example of the Beer–Lambert relationship.

The Beer-Lambert relationship is additive (*i.e.*, the absorption of light by one chemical species is unaffected by the presence of other species, irrespective of whether those other species absorb light at the same wavelength). Thus, $A = \Sigma(\varepsilon_i c_i l)$. The greater the difference in the molar absorption coefficients between the substrate(s) and the product(s), the larger the change in absorbance with time and the greater the ease in velocity determination. This statement assumes that the two chemical species do not interact with each other.

Note that the Beer-Lambert relationship does not require one to monitor a reaction at the wavelength maximum value (λ_{max}) of either the substrate or the product. All other factors being equal, one should chose that wavelength yielding the greatest $\Delta\varepsilon$ value. For example, AMP has a λ_{max} value at 259 nm at a pH value of 7, whereas IMP has a λ_{max} value of 248.5 nm. Yet, the AMP deaminase reaction is measured best at 265 nm, the wavelength affording the largest change in ε. Likewise, one can use a wavelength other than 340 nm to assay NADH or NADPH, but one should avoid unnecessary loss of signal-to-noise by working as close to the wavelength yielding maximal absorption.

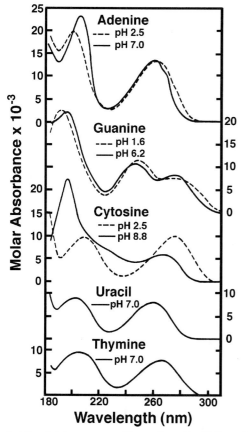

Figure 4. Ultraviolet spectrum of bases found in DNA and RNA.

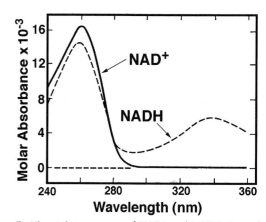

Figure 5. Ultraviolet spectrum of NAD$^+$ and NADH. Note that the absorption band centered at 340 nm serves as a valuable way to assay many dehydrogenases as well as other enzymes that form products that can be coupled to NAD$^+$ reduction or NADH oxidation.

Occasionally, one can increase the $\Delta\varepsilon$ by utilizing alternative substrates. For example, 3-acetyl-NAD$^+$ or thio-NAD$^+$ can often be used with NAD$^+$-dependent dehydrogenases. Note however that an alternative substrate may change the kinetic mechanism, as compared to that observed with the naturally occurring substrate. Alternative substrates are of particular value when the normal substrate(s) and product(s) do not efficiently absorb UV or visible light. For example, many p-nitroaniline or p-nitrophenyl derivatives have proved to be quite useful in enzyme assays because they exhibit intense absorption around 410 nm.

Ideally, other components in the reaction mixture should not absorb significantly at the monitored wavelength. In addition, colored impurities should be removed. For example, commercial imidazole, a commonly used buffer, contains a yellow impurity that can be easily removed upon recrystallization from ethyl acetate.

General Principles. Light absorption is quantized. The energy change associated with an electronic transition occurs within the ultraviolet (UV) or visible (vis) spectrum. Visible light absorption corresponds to low-energy electronic transitions, such as those observed with certain transition metal ions or with molecules having conjugated double bonds. The near ultraviolet (200 and 400 nm) corresponds to electronic transitions in molecular entities with smaller conjugated systems (*e.g.*, 1,3,5-hexatriene, ATP, *etc.*). Isolated double bonds, such as those within peptide bonds, absorb light around 200-210 nm. Dioxygen, carbon dioxide, and water can absorb light of wavelength less than 180 nm, and spectroscopy in the far-UV typically requires a vacuum, hence the term "vacuum UV" for low-wavelength light.

Sharp absorption bands are typically not observed in UV and visible absorption spectra of liquid samples. This is the consequence of the presence of the vibrational and rotational fine structure that become superimposed on the potential energy surfaces of the electronic transitions. Fine structure in UV/vis absorption spectra can be detected for samples in vapor phase or in nonpolar solvents.

The intensity of light absorption is governed by a number of factors that determine the transition probability (*i.e.*, the probability of interaction between the radiant energy and the electronic system). This probability is proportional to the square of the transition moment and is thus related to the electronic charge distribution in the molecular entity. Hence, absorption bands (with $\varepsilon > 10,000$ cm^{-1} M^{-1}) suggest that the transition is accompanied by a large change in the transition moment. Intensity is also affected by the polarity of the excited state and the target area of the absorbing system. Transitions associated with UV-visible absorption spectroscopy consist of an electron in a filled molecular orbital being excited to the next higher energy orbital (an antibonding orbital). Although many exceptions are known, the relative transition energies roughly are: $\sigma \rightarrow \sigma^* > n \rightarrow \sigma^* > n \rightarrow \pi^* > \pi \rightarrow \pi^*$.

$\sigma \rightarrow \sigma^*$ Transitions. These transitions typically occur between 120 and 220 nm (*i.e.*, in the far-UV). The λ_{max} value for typical $\sigma \rightarrow \sigma^*$ transitions of carbon-carbon or carbon-hydrogen bonds is usually around 150 nm. For example, the λ_{max} value for ethane is 135 nm.

$n \rightarrow \sigma^*$ Transitions. These transitions typically occur at wavelengths greater than that needed for $\sigma \rightarrow \sigma^*$ transitions. Roughly, $n \rightarrow \sigma^*$ transitions involving $-\ddot{O}-$ occur with wavelengths at about 185 nm, with $-\ddot{N}-$ and $-\ddot{S}-$ at about 195 nm, and with carbonyls at about 190 nm. Examples of λ_{max} values (and ε values) of $n \rightarrow \sigma^*$ transitions include: water ($\lambda_{max} = 167$ nm with $\varepsilon = 7000$), methanol ($\lambda_{max} = 183$ nm with $\varepsilon = 500$), acetone ($\lambda_{max} = 188$ nm with $\varepsilon = 1860$), and methyl iodide ($\lambda_{max} = 259$ nm with $\varepsilon = 400$).

$n \rightarrow \pi^*$ Transitions. These are "forbidden" transitions according to symmetry rules, but molecular vibrations

allow these transitions to occur, albeit with low intensities. Nonbonding electrons of carbonyl groups will often have $n \rightarrow \pi^*$ transition λ_{max} values of around 300 nm. Some examples include acetone ($\lambda_{max} = 279$ nm, $\varepsilon = 15$), acetophenone (319 nm, $\varepsilon = 50$), thiourea (a shoulder at 291 nm, $\varepsilon = 71$), and acetic acid (204 nm, $\varepsilon = 41$). The $n \rightarrow \pi^*$ transition can also be detected in optical rotatory dispersion measurements. Moreover, $n \rightarrow \pi^*$ transitions often exhibit a blue shift in polar solvents or environments.

$\pi \rightarrow \pi^*$ Transitions. Typically occurring at wavelengths in the near-UV, transitions of this type are the most commonly utilized spectral signals in kinetic and structural studies.

[1]IUPAC (1988) *Pure and Appl. Chem.* **60**, 1055.

[2]R. D. Allison & D. L. Purich (1979) *Meth. Enzymol.* **63**, 3.

[3]C. R. Cantor & P. R. Schimmel (1980) *Biophysical Chemistry*, part II, pp. 344-408, Freeman, San Francisco.

[4]R. P. Bauman (1962) *Absorption Spectroscopy*, Wiley, New York.

[5]H. H. Jaffe & M. Orchin (1962) *Theory and Application of Ultraviolet Spectroscopy*, Wiley, New York.

232, 389; spectrometer for, **232**, 392-401; performance characteristics, **232**, 389-390.

ABSORPTIVITY

A parameter in spectroscopy and photochemistry, equal to the absorptance $(1 - (P/P_0))$ divided by the optical pathlength (l) of the sample containing the absorbing agent. Thus, it equals $(1 - (P/P_0))/l$ where P_0 is the radiant power of light being transmitted through a reference sample, and P is the radiant power being transmitted through the solution. The Commission on Photochemistry does not recommend the use of this term. *See Absorbance; Absorption Coefficient; Beer-Lambert Law; Absorption Spectroscopy*

ABSTRACTION REACTION

Any chemical process in which one reactant removes an atom (neutral or charged) from the other reacting entity. An example is the generation of a free radical by the action of an initiator on another molecule. If abstraction takes place at a chiral carbon, racemization is almost always observed in nonenzymic processes. On the other hand, enzymes frequently abstract and reattach atoms or groups of atoms in a fashion that maintains stereochemistry.

ACCELERATION

In physics, the time rate of change of motional velocity resulting from changes in a body's speed and/or direction. In biochemistry, acceleration refers to an increased rate of a chemical reaction in the presence of an enzyme or other catalyst. *See Catalytic Rate Enhancement; Catalytic Proficiency; Efficiency Function*

ACCRETION

Solute or particulate accumulation onto an aggregated phase (or solid state) that grows together by the addition of material at the periphery. Both cohesive and adhesive forces are thought to be driving forces in accretion. Sea shells and kidney stones are also known to form as layers of crystallites and amorphous components by accretion of external substances.

ACCURACY

The closeness or proximity of a measured value to the true value for a quantity being measured. Unless the magnitude of a quantity is specified by a formal SI definition, one typically uses reference standards to establish the accepted true value for a given quantity. *See also Precision*

ACETALDEHYDE DEHYDROGENASE (ACETYLATING)

This enzyme [EC 1.2.1.10] catalyzes the oxidation of acetaldehyde in the presence of NAD^+ and coenzyme A to form acetyl-CoA + NADH + H^+. Other aldehyde substrates include glycolaldehyde, propanal, and butanal.

E. R. Stadtman & R. M. Burton (1955) *Meth. Enzymol.* **1**, 222 and 518.
F. B. Rudolph, D. L. Purich & H. J. Fromm (1968) *J. Biol. Chem.* **243**, 5539.

2-(ACETAMIDOMETHYLENE)SUCCINATE HYDROLASE

This enzyme [EC 3.5.1.29] catalyzes the hydrolysis of 2-(acetamidomethylene)-succinate to yield acetate, succinate semialdehyde, carbon dioxide, and ammonia.

R. W. Burg (1970) *Meth. Enzymol.* **18(A)**, 634.

ACETATE KINASE

This phosphotransferase [EC 2.7.2.1] catalyzes the thermodynamically favored phosphorylation of ADP to form ATP (*i.e.*, K_{eq} = [ATP][acetate]/{[acetyl phosphate][ADP]} = 3000). GDP is also an effective phosphoryl group acceptor. This enzyme is easily cold-denatured, and one must use glycerol to maintain full catalytic activity. Initial kinetic evidence, as well as borohydride reduction experiments, suggested the formation of an enzyme-bound acyl-phosphate intermediate, but later kinetic and stereochemical[1] data indicate that the kinetic mechanism is sequential and that there is direct in-line phosphoryl transfer. Incidental generation of a metaphosphate anion during catalysis may explain the formation of an enzyme-bound acyl-phosphate. Acetate kinase is ideally suited for the regeneration of ATP or GTP from ADP or GDP, respectively.

[1]P. A. Frey (1992) *The Enzymes* **20**, 160.

Selected entries from *Methods in Enzymology* [vol, page(s)]:
Acetate assay with, **3**, 269; activation, **44**, 889; activity assay, **44**, 893, 894; alternative substrates, **87**, 11; bridge-to-nonbridge transfer, **87**, 19-20, 226, 232; chiral phosphoryl-ATP, **87**, 211, 258, 300; cold denaturation, **63**, 9; cysteine residues, **44**, 887-889; equilibrium constant, **63**, 5; exchange properties, **64**, 9, 39, **87**, 5, 18, 656; hydroxylaminolysis, **87**, 18; immobilization, **44**, 891, 892; inhibitor, **63**, 398; initial rate kinetics, **87**, 18; metal-ion bind-

ing, **63**, 275-278; metaphosphate, **87**, 12, 20; metaphosphate synthesis and, **6**, 262-263; nucleoside diphosphate kinase activity; **63**, 8; phosphorylation potential, **55**, 237; pH stability profile, **87**, 18; promoting microtubule assembly, **85**, 417-419; purine nucleoside diphosphate kinase activity, **63**, 8; regenerating GTP from GDP, **85**, 417; ribulose-5-phosphate 4-epimerase and, **5**, 253-254; *Veillonella alcalescens* acetate kinase [ATP formation assay, **71**, 312; hydroxamate assay, **71**, 311; properties, **71**, 315; stability to heat, **71**, 313; stimulation by succinate, 71, 316; substrate specificity, **71**, 316; xylulose-5-phosphate 3-epimerase and, **5**, 250-251; xylulose-5-phosphate phosphoketolase and, **5**, 26; purine inhibitor, **63**, 398; metal-ion binding, **63**, 275-278; phosphorothioates, **87**, 200, 205, 226, 232, 258; from *Bacillus stearothermophilus*; assay, **90**, 179; properties, **90**, 183; purification, **90**, 180; in acetyl phosphate and acetyl-CoA determination, **122**, 44; and hexokinase, in glucose 6-phosphate production, **136**, 52; dihydroxyacetone phosphate synthesis with, **136**, 277; glucose 6-phosphate synthesis with, **136**, 279; *sn*-glycerol 3-phosphate synthesis with, **136**, 276; in pyruvic acid phosphoroclastic system, **243**, 96, 99.

ACETATE KINASE (PYROPHOSPHATE)

This enzyme [EC 2.7.2.12] converts acetate and pyrophosphate to form acetyl phosphate and orthophosphate.

H. G. Wood, W. E. O'Brien & G. Michaels (1977) *Adv. Enzymol.* **45**, 8555.

ACETAZOLAMIDE

A diuretic agent (5-acetamido-1,3,4-thiadiazole-2-sulfonamide) that acts as a potent noncompetitive inhibitor (K_i 10^{-8} M) of carbonic anhydrase.

ACETOACETATE DECARBOXYLASE

This enzyme [EC 4.1.1.4] catalyzes the decarboxylation of acetoacetate to form acetone and carbon dioxide.

M. H. O'Leary (1992) *The Enzymes*, 3rd ed., **20**, 235.
D. J. Creighton & N. S. R. K. Murthy (1990) *The Enzymes*, 3rd ed., **19**, 323.
D. S. Sigman & G. Mooser (1975) *Ann. Rev. Biochem.* **44**, 889.
I. Fridovich (1972) *The Enzymes*, 3rd ed., **6**, 255.

ACETOLACTATE SYNTHASE

This enzyme [EC 4.1.3.18] catalyzes the reversible carboxylation of 2-acetolactate with carbon dioxide to form two pyruvate ions. Thiamin pyrophosphate is a required cofactor.

B. A. Palfey & V. Massey (1998) *Comprehensive Biological Catalysis: A Mechanistic Reference* 3, 83.
R. L. Schowen (1998) *Comprehensive Biological Catalysis: A Mechanistic Reference* 2, 217.
J. V. Schloss & M. S. Hixon (1998) *Comprehensive Biological Catalysis: A Mechanistic Reference* 2, 43.
A. Schellenberger, G. Hübner & H. Neef (1997) *Meth. Enzymol.* **279**, 131.

J. S. Holt, S. B. Powles & J. A. M. Holtum (1993) *Ann. Rev. Plant Physiol. Plant Mol. Biol.* **44**, 203.
R. Kluger (1992) *The Enzymes*, 3rd ed., **20**, 271.
J. H. Jackson (1988) *Meth. Enzymol.* **166**, 230.

ACETYLCHOLINESTERASE

This enzyme [EC 3.1.1.7], also known as true cholinesterase, choline esterase I, and cholinesterase, catalyzes the hydrolysis of acetylcholine to produce choline and acetate. The enzyme will also act on a number of acetate esters as well as catalyze some transacetylations.

D. M. Quinn & S. R. Feaster (1998) *Comprehensive Biological Catalysis: A Mechanistic Reference* 1, 455.
H. Okuda (1991) *A Study of Enzymes* 2, 563.
T. L. Rosenberry (1975) *Adv. Enzymol.* **43**, 103.
H. C. Froede & I. B. Wilson (1971) *The Enzymes*, 3rd ed., **5**, 87.
L. T. Potter (1971) *Meth. Enzymol.* **17(B)**, 778.

Selected entries from *Methods in Enzymology* [vol, page(s)]:
Acetylthiocholine as substrate, **251**, 101-102; assay by ESR, **251**, 102-105; inhibitors, **251**, 103; modification by symmetrical disulfide radical, **251**, 100; thioester substrate, **248**, 16; transition state and multisubstrate analogues, **249**, 305; enzyme receptor, similarity to collagen, **245**, 3.

ACETYL-CoA (or, ACETYL COENZYME A)

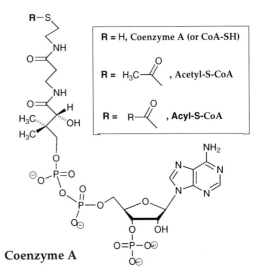

R = H, Coenzyme A (or CoA-SH)

R = H₃C—C(=O)—, Acetyl-S-CoA

R = R—C(=O)—, Acyl-S-CoA

Coenzyme A

As the principal thiolester of intermediary metabolism, acetyl coenzyme A is involved in two-carbon biosynthetic and degradative steps. An essential component is the vitamin pantithenic acid, which provides the sulfur atom for the thiolester formation.

Selected entries from *Methods in Enzymology* [vol, page(s)]:
Assay, **1**, 611; **3**, 935-938; **63**, 33; separation by HPLC, **72**, 45; extraction from tissues, **13**, 439; formation of, **1**, 486, 518, 585; **5**, 466; free energy of hydrolysis, **1**, 694; substrate for the following enzymes [acetyl-coenzyme A acyl carrier protein transacylase, **14**, 50; acetyl-coenzyme A carboxylase, **14**, 3, 9; acetyl-coenzyme A synthetase, **13**, 375; *N*-acetyltransferase, **17B**, 805; aminoacetone

synthase, **17B**, 585; carnitine acetyltransferase, **13**, 387-389; **14**, 613; choline acetyltransferase, **17B**, 780, 788, 798; citrate synthase, **13**, 3, 4, 8, 9, 11, 12, 15-16, 19-20, 22, 25; **14**, 617; fatty acid synthase, **14**, 17, 22, 33, 40.

ACETYL-CoA *C*-ACETYLTRANSFERASE (or, THIOLASE)

This enzyme [EC 2.3.1.9], also known as thiolase, transfers an acetyl group from one acetyl-CoA molecule to another to form free coenzyme A and acetoacetyl-CoA.

D. J. Creighton & N. S. R. K. Murthy (1990) *The Enzymes*, 3rd ed., **19**, 323.
U. Gehring & F. Lynen (1972) *The Enzymes*, 3rd ed., **7**, 391.

ACETYL-CoA *C*-ACYLTRANSFERASE

This enzyme [EC 2.3.1.16], also known as 3-ketoacyl-CoA thiolase, transfers an acyl group from an acyl-CoA to acetyl-CoA to form free coenzyme A and 3-oxoacyl-CoA.

J. V. Schloss & M. S. Hixon (1998) *Comprehensive Biological Catalysis: A Mechanistic Reference* **2**, 43.

ACETYL-CoA:ACP TRANSACYLASE

This enzyme [EC 2.3.1.38], also referred to as acetyl-CoA:[acyl-carrier protein] *S*-acetyltransferase, transfers an acetyl group from one acetyl-CoA to an acyl-carrier-protein (ACP) to form free coenzyme A and the acetyl-[acyl-carrier-protein]. *See also* Fatty Acid Synthase

S. J. Wakil & J. K. Stoops (1983) *The Enzymes*, 3rd ed., **16**, 3.
P. R. Vagelos (1973) *The Enzymes*, 3rd ed., **8**, 155.
A. W. Alberts, P. W. Majerus & P. R. Vagelos (1969) *Meth. Enzymol.* **14**, 50.

ACETYL-CoA CARBOXYLASE

This enzyme [EC 6.4.1.2] catalyzes the reaction of acetyl-CoA, bicarbonate, and ATP to form malonyl-CoA, orthophosphate, and ADP. The plant enzyme will also act on propionyl-CoA and butanoyl-CoA. The enzyme will also catalyze certain transcarboxylations and it requires biotin as a cofactor.

J. N. Earnhardt & D. N. Silverman (1998) *Comprehensive Biological Catalysis: A Mechanistic Reference* **1**, 495.
J. V. Schloss & M. S. Hixon (1998) *Comprehensive Biological Catalysis: A Mechanistic Reference* **2**, 43.
K.-H. Kim (1997) *Ann. Rev. Nutr.* **17**, 77.
S. B. Ohlrogge & J. G. Jaworski (1997) *Ann. Rev. Plant Physiol. Plant Mol. Biol.* **48**, 109.
R. W. Brownsey & R. M. Denton (1987) *The Enzymes*, 3rd ed., **18**, 123.
K. Bloch (1977) *Adv. Enzymol.* **45**, 1.
A. W. Alberts & P. R. Vagelos (1972) *The Enzymes*, 3rd ed., **6**, 37.

ACETYL-CoA SYNTHETASE

This enzyme [EC 6.2.1.1], also referred to as acetate-CoA ligase or acetate thiokinase, catalyzes the reaction of acetate, coenzyme A, and ATP to form acetyl-CoA, AMP, and pyrophosphate. The enzyme will also utilize propanoate and propenoate as substrates.

L. A. Kleczkowski (1994) *Ann. Rev. Plant Physiol. Plant Mol. Biol.* **45**, 339.
J. C. Londesborough & L. T. Webster, Jr. (1974) *The Enzymes*, 3rd ed., **10**, 469.
Selected entries from *Methods in Enzymology* [vol, page(s)]: Adenosine 5'-*O*-(1-thiotriphosphate), **87**, 224, 230-231; bridge-nonbridge oxygens, **87**, 251-253; kinetics, **87**, 355; mechanism, **87**, 251, 355; NMR, **87**, 251-253; stereochemistry **87**, 206, 212, 224, 230-233, 251-253.

ACETYLENE MONOCARBOXYLATE HYDRATASE

This enzyme [EC 4.2.1.71] adds water to propynoate to form 3-hydroxypropenoate. The enzyme will also act on 3-butynoate to form acetoacetate.

J. V. Schloss & M. S. Hixon (1998) *Comprehensive Biological Catalysis: A Mechanistic Reference* **2**, 43.

N-ACETYLGALACTOSAMINE-4-SULFATE SULFATASE

This enzyme [EC 3.1.6.12] acts on 4-sulfate groups of the *N*-acetylgalactosamine 4-sulfate moieties in chondroitin sulfate and dermatan sulfate.

H. Kresse & J. Glössl (1987) *Adv. Enzymol.* **60**, 217.

N-ACETYLGALACTOSAMINE-6-SULFATE SULFATASE

This enzyme [EC 3.1.6.4] acts on 6-sulfate groups of the *N*-acetylgalactosamine 6-sulfate moieties in chondroitin sulfate and the galactose 6-sulfate groups in keratan sulfate.

H. Kresse & J. Glössl (1987) *Adv. Enzymol.* **60**, 217.

N-ACETYLGALACTOSAMINIDE SIALYLTRANSFERASE

This enzyme [EC 2.4.99.3] catalyzes the reaction of a glycano-1,3-(*N*-acetylgalactosaminyl)-glycoprotein and CMP-*N*-acetylneuraminate to produce CMP and the glycano-(2,6-α-*N*-acetylneuraminyl)-(*N*-acetylgalactosaminyl)-glycoprotein.

T. A. Beyer, J. E. Sadler, J. I. Rearick, J. C. Paulson & R. L. Hill (1981) *Adv. Enzymol.* **52**, 23.

N-ACETYLGLUCOSAMINE KINASE

This enzyme [EC 2.7.1.59] catalyzes the phosphorylation by ATP of N-acetylglucosamine to generate ADP and N-acetylglucosamine 6-phosphate. The bacterial enzyme is also reported to act on glucose as well.

S. S. Barkulis (1966) *Meth. Enzymol.* **9**, 415.

N-ACETYLGLUCOSAMINE-6-PHOSPHATE 2-EPIMERASE

This enzyme catalyzes the epimerization at the 2-position of N-acetylglucosamine 6-phosphate. **See also** *N-Acylglucosamine-6-phosphate 2-Epimerase*

M. E. Tanner & G. L. Kenyon (1998) *Comprehensive Biological Catalysis: A Mechanistic Reference* **2**, 7.

N-ACETYLGLUCOSAMINE-6-SULFATE SULFATASE

This enzyme [EC 3.1.6.14] catalyzes the hydrolysis of the 6-sulfate moieties of the N-acetylglucosamine 6-sulfate subunits of heparan sulfate and keratan sulfate. It has been suggested that this enzyme might be identical to N-sulfoglucosamine-6-sulfatase.

H. Kresse & J. Glössl (1987) *Adv. Enzymol.* **60**, 217.

α-N-ACETYLGLUCOSAMINIDASE

This enzyme [EC 3.2.1.50] catalyzes hydrolysis of terminal nonreducing N-acetylglucosamine residues in N-acetyl-α-glucosaminides.

H. Kresse & J. Glössl (1987) *Adv. Enzymol.* **60**, 217.

β-N-ACETYLGLUCOSAMINIDASE

This enzyme, reportedly catalyzing the hydrolysis of terminal, nonreducing N-acetyl-β-glucosamine moieties in chitobiose and higher analogs, is now a deleted EC entry [EC 3.2.1.30].

P. M. Dey & E. del Campillo (1984) *Adv. Enzymol.* **56**, 141.

N-ACETYLGLUTAMATE SYNTHASE

This enzyme [EC 2.3.1.1], also referred to as amino-acid N-acetyltransferase and acetyl-CoA : glutamate N-acetyltransferase, catalyzes the reaction of acetyl-CoA with glutamate to form coenzyme A and N-acetylglutamate. The enzyme will also acts on aspartate and, more slowly, with some other amino acids. The mammalian enzyme is activated by L-arginine. **See also** *Glutamate Acetyltransferase*

S. G. Powers-Lee (1985) *Meth. Enzymol.* **113**, 27.
T. Sonoda & M. Tatibana (1983) *J. Biol. Chem.* **258**, 9839.
H. J. Vogel & R. H. Vogel (1974) *Adv. Enzymol.* **40**, 65.

N-ACETYL-γ-GLUTAMYL-PHOSPHATE REDUCTASE

This enzyme [EC 1.2.1.38], also known as N-acetylglutamate semialdehyde dehydrogenase and NAGSA dehydrogenase, catalyzes the reaction of N-acetylglutamate 5-semialdehyde with NADP$^+$ and phosphate to generate N-acetyl-5-glutamyl phosphate and NADPH.

H. J. Vogel & R. H. Vogel (1974) *Adv. Enzymol.* **40**, 65.

β-N-ACETYLHEXOSAMINIDASE

This enzyme [EC 3.2.1.52], also referred to as β-hexosaminidase and N-acetyl-β-glucosaminidase, catalyzes the hydrolysis of terminal nonreducing N-acetylhexosamine residues in N-acetyl-β-hexosaminides. N-Acetylglucosides and N-acetylgalactosides are substrates.

H. Kresse & J. Glössl (1987) *Adv. Enzymol.* **60**, 217.
H. M. Flowers & N. Sharon (1979) *Adv. Enzymol.* **48**, 29.

O-ACETYLHOMOSERINE (THIOL)-LYASE

This enzyme [EC 4.2.99.10], also referred to as O-acetylhomoserine sulfhydrylase, catalyzes the reaction of O-acetylhomoserine with methanethiol to generate methionine and acetate. The enzyme can also act on other thiols or H$_2$S, producing homocysteine or thioethers. The enzyme isolated from baker's yeast will also catalyze the reaction exhibited by O-acetylserine (thiol)-lyase [EC 4.2.99.8], albeit more slowly.

S. Yamagata (1987) *Meth. Enzymol.* **143**, 465.
I. Shiio & H. Ozaki (1987) *Meth. Enzymol.* **143**, 470.

N-ACETYLNEURAMINATE LYASE

This enzyme [EC 4.1.3.3], also known as N-acetylneuraminate aldolase, will convert N-acetylneuraminate to N-acetylmannosamine and pyruvate. The enzyme will also act on N-glycoloylneuraminate and on O-acetylated sialic acids, other than O^4-acetylated derivatives.

K. N. Allen (1998) *Comprehensive Biological Catalysis: A Mechanistic Reference* **2**, 135.
W. A. Wood (1972) *The Enzymes*, 3rd ed., **7**, 281.

N^2-ACETYLORNITHINE AMINOTRANSFERASE

This enzyme [EC 2.6.1.11] catalyzes the pyridoxal-phosphate-dependent reaction of 2-acetylornithine with α-ketoglutarate to produce *N*-acetylglutamate 5-semialdehyde and glutamate.

H. J. Vogel & R. H. Vogel (1974) *Adv. Enzymol.* **40**, 65.
A. E. Braunstein (1973) *The Enzymes*, 3rd ed., **9**, 379.
H. J. Vogel & E. E. Jones (1970) *Meth. Enzymol.* **17(A)**, 260.

N²-ACETYLORNITHINE DEACETYLASE

This enzyme [EC 3.5.1.16], also known as acetylornithinase and *N*-acetylornithinase, catalyzes the reaction of water with N^2-acetylornithine to produce acetate and ornithine. The enzyme also catalyzes the hydrolysis of *N*-acetylmethionine.

H. J. Vogel & R. H. Vogel (1974) *Adv. Enzymol.* **40**, 65.

O-ACETYLSERINE (THIOL)-LYASE

This enzyme [EC 4.2.99.8], also known as cysteine synthase and *O*-acetylserine sulfhydrylase, catalyzes the pyridoxal-phosphate-dependent reaction of H_2S with O^3-acetylserine to produce cysteine and acetate. Some alkyl thiols, cyanide, pyrazole, and some other heterocyclic compounds can also act as acceptors.

T. Nagasawa & H. Yamada (1987) *Meth. Enzymol.* **143**, 474.
A. E. Martell (1982) *Adv. Enzymol.* **53**, 163.
L. Davis & D. E. Metzler (1972) *The Enzymes*, 3rd ed., **7**, 33.

ACHIRAL

The absence of chirality. Achiral molecules have an internal plane or point of symmetry. A molecular configuration is said to be achiral when it is superimposable on its mirror image. **See** *Chirality*

ACID

A substance that liberates protons as a consequence of its dissolution or dissociation. **See** *Brønsted Theory; Lewis Acid; Lewis Base*

ACID-BASE EQUILIBRIUM CONSTANTS

The following table contains p*K* values for a selected group of biochemically important substances. The associated thermodynamic parameters and the original literature references were presented by Edsall and Wyman[1].

pK_a Values for Selected Substances at 25°C

Substance	pK_a
Hydroxyproline (pK_1)	1.82
Proline (pK_1)	1.97
Aspartic acid (pK_1)	2.00
Threonine (pK_1)	2.10
Phosphoric acid (pK_1)	2.15
Serine (pK_1)	2.19
Chloroacetic acid	2.28
Glycine (pK_1)	2.35
Alanine (pK_1)	2.35
Formic acid	3.75
Lactic acid	3.86
Aspartic acid (pK_2)	3.91
Succinic acid (pK_1)	4.21
Acetic acid	4.76
Propionic acid	4.87
Succinic acid (pK_2)	5.63
Carbonic acid (apparent pK_1)	6.35
Glucose 1-phosphate (pK_2)	6.50
Phosphoric acid (pK_2)	7.20
Threonine (pK_2)	9.10
Serine (pK_2)	9.21
Ammonium ion	9.25
Ethanolammonium ion	9.50
Hydroxyproline (pK_2)	9.66
Glycine (pK_2)	9.78
Trimethylammonium ion	9.79
Alanine (pK_2)	9.87
Aspartic acid (pK_3)	10.00
Carbonic acid (pK_2)	10.33
Methylammonium ion	10.62
Proline (pK_2)	10.64
Dimethylammonium ion	10.77
Water (pK_w)	13.997

[1]J. T. Edsall & J. Wyman (1958) *Biophysical Chemistry*, pp. 452, Academic Press, New York.

ACID CATALYSIS

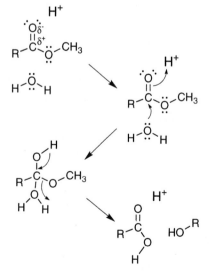

Acid Catalysis of Ester Hydrolysis

Any process for which the rate of reaction is accelerated through the participation of an acid as a catalyst. **See** *General Acid Catalysis; Specific Acid Catalysis*

ACIDITY

1. The tendency for a Brønsted acid to act as a proton donor, expressed in terms of the compound's dissociation constant in water. 2. The term also refers to the tendency to form a Lewis adduct, as measured by a dissociation constant.

With reference to a solvent, this term is usually restricted to Brønsted acids. If the solvent is water, the pH value of the solution is a good measure of the proton-donating ability of the solvent, provided that the concentration of the solute is not too high. For concentrated solutions or for mixtures of solvents, the acidity of the solvent is best indicated by use of an acidity function. *See Degree of Dissociation; Henderson-Hasselbalch Equation; Acid-Base Equilibrium Constants; Brønsted Theory; Lewis Acid; Acidity Function; Leveling Effect*

ACIDITY FUNCTION

A thermodynamic measure of the proton-donating or proton-accepting ability of a solvent system (or closely related thermodynamic property such as the ability of a solvent system to form Lewis adducts)[1-3].

There are many types of acidity scales: depending on the nature of the indicator, of conjugate bases with a -1 charge, of acids that form stable carbocations, *etc.* Perhaps the best known acidity function is the Hammett acidity function, H_o, which is used for concentrated acidic solvents having a high dielectric constant[4,5]. For any solvent (including a mixture of solvents in which the relative proportions are specified), H_o is defined to be $pK_{BHW+} - \log([BH^+]/[B])$. The value of H_o is measured using a weak indicator base whose pK value in water is known. Two common indicators are the aniline derivatives, *o*-nitroanilinium ion and 2,4-nitroanilinium ion, having pK_{BHW+} values of -0.29 and -4.53, respectively, in water. The $[BH^+]/[B]$ ratio for one indicator is measured, usually spectrophotometrically, in the given solvent. Knowledge of a substance's pK in water (*i.e.*, pK_{BHW+}) allows the investigator to calculate H_o. Once H_o is known, pK_a values can be determined for any other acid-base pair.

The value of h_o is defined as $a_{H^+} f_I/f_{HI^+}$ where a_{H^+} is the chemical activity of the proton, and f_I and f_{HI^+} are the activity coefficients of the indicator. H_o and h_o are related by $H_o = -\log h_o$. For dilute solutions, $H_o = pH$. This is the situation for most biochemical reactions. The subscript is a reference to the net charge on the base. Hence, H_- is the corresponding acidity function for mononegatively charged bases in equilibrium with neutral acids.

The Hammett acidity function only applies to acidic solvents having a high dielectric constant, further requiring that the f_I/f_{HI^+} ratio be independent of the nature of the indicator. Thus, the Hammett acidity function is applicable for uncharged indicator bases that are aniline derivatives. As mentioned above, other acidity functions have been suggested. No single formulation has been developed that satisfies different solvent systems or types of bases.

Bunnett and Olsen[6-8] used a different approach and derived the equation:

$$\log([SH^+]/[S]) + H_o = \Phi(H_o + \log[H^+]) + pK_{SH+}$$

in which S is a base that can be protonated by an acidic solvent. Plotting $\log([SH^+]/[S]) + H_o$ versus $H_o + \log[H^+]$ will result in a reasonably linear line having a slope of Φ. The method uses a linear free energy relationship to address the problems of defining basicity for weak organic bases in solutions of moderate concentrations of mineral acids. The value and sign of Φ is a measure of the response of the acid-base equilibria to changes in acid concentration. An analogous equation for kinetic data is

$$\log k_\psi + H_o = \Phi(H_o + \log[H^+]) + \log k_2^o$$

where k_ψ is the pseudo-first-order rate constant for a reaction in acidic solution and k_2^o is the corresponding constant at infinite dilution in water. Here Φ is suggested to represent the response of the chemical reaction rate to changes in the acid concentration. Attempts have been made to apply this method to basic media. More-O'Ferrell[9] suggested that the slope may be related semi-quantitatively to the structure of the transition state.

A related treatment of the acidity function considers changes in acidity of the solvent on an acid-base equilibria[10]. With this procedure, an equilibrium is chosen as reference, having an equilibrium constant K'_o for the reference reaction in a reference solvent and K' for the reference reaction in the particular medium under study. The reaction under study has the corresponding equilibria of K_o (in the reference solvent) and K (in the particular medium). Thus,

$$\log (K'/K_o') = m^* \log (K/K_o)$$

where the slope m^* corresponds to $1 - \Phi$ in the Bunnett-Olsen treatment. Bunnett[11] has also plotted $\log k_\psi + H_o$ vs. $\log a_{water}$ where a_{water} is the activity of water and has indicated that the slope of the line, w, suggests certain possibilities for the chemical reaction mechanism in moderately concentrated aqueous acids. For example, if w is between -2.5 and zero, then water does not participate in the formation of the transition state. If w is between 1.2 and 3.3, water participates as a nucleophile in the rate-determining step. Long and Bakule[12] have criticized this approach. **See also** Degree of Dissociation; Henderson-Hasselbalch Equation; Acid-Base Equilibrium Constants; Bunnet-Olsen Equations; Cox-Yeats Treatment

[1]C. H. Rochester (1970) *Acidity Functions*, Academic Press, New York.
[2]J. March (1985) *Advanced Organic Chemistry*, Wiley, New York.
[3]J. Hine (1962) *Physical Organic Chemistry*, McGraw-Hill, New York.
[4]L. Zucker & L. P. Hammett (1939) *J. Amer. Chem. Soc.* **61**, 2791.
[5]L. P. Hammett (1970) *Physical Organic Chemistry*. McGraw-Hill, New York.
[6]J. F. Bunnett & F. P. Olsen (1966) *Can. J. Chem.* **44**, 1899 and 1917.
[7]J. F. Bunnett, R. L. McDonald & F. P. Olsen (1974) *J. Amer. Chem. Soc.* **96**, 2855.
[8]V. Lucchini, G. Modena, G. Scorrano, R. A. Cox & Y. Yates (1982) *J. Am. Chem. Soc.* **104**, 1958.
[9]R. A. More O'Ferrall (1972) *J. Chem. Soc., Perkin Trans.* **2**, 976.
[10]A. Bagno, G. Scorrano & R. A. More O'Ferrall (1987) *Rev. Chem. Interm.* **7**, 313.
[11]J. F. Bunnett (1961) *J. Amer. Chem. Soc.* **83**, 4956.
[12]F. A. Long & R. Bakule (1963) *J. Amer. Chem. Soc.* **85**, 2313.

ACID-LABILE SULFIDES

The bridging sulfur atoms in iron-sulfur proteins are often referred to as acid-labile sulfides, because treatment of such proteins with acids generates H_2S.

ACID PHOSPHATASE

This enzyme [EC 3.1.3.2], also referred to as acid phosphomonoesterase, phosphomonoesterase, and glycerophosphatase, catalyzes the hydrolysis of an orthophosphoric monoester to generate an alcohol and orthophosphate. The enzyme, which has a wide specificity, will also catalyze transphosphorylations.

M. Cohn (1982) *Ann. Rev. Biophys. Bioeng.* **11**, 23.
V. P. Hollander (1971) *The Enzymes*, 3rd ed., **4**, 449.

ACONITASE

This [4Fe-4S] cluster-containing enzyme [EC 4.2.1.3], also known as citrate hydro-lyase and aconitate hydratase, will act on citrate to generate *cis*-aconitate ((Z)-prop-1-ene 1,2,3-tricarboxylate) and water. The enzyme will also catalyze the conversion of isocitrate into *cis*-aconitate.

V. E. Anderson (1998) *Comprehensive Biological Catalysis: A Mechanistic Reference* **2**, 115.
B. G. Fox (1998) *Comprehensive Biological Catalysis: A Mechanistic Reference* **3**, 261.
J. V. Schloss & M. S. Hixon (1998) *Comprehensive Biological Catalysis: A Mechanistic Reference* **2**, 43.
H. Lauble, M. C. Kennedy, H. Beinert & D. C. Stout (1992)*Biochemistry* **31**, 2735.
L. Zheng, M. C. Kennedy, H. Beinert & H. Zalkin (1992) *J. Biol. Chem.* **267**, 7895.
J. B. Howard & D. C. Rees (1991) *Adv. Protein Chem.* **42**, 199.
P. A. Srere (1975) *Adv. Enzymol.* **43**, 57.
J. P. Glusker (1971) *The Enzymes*, 3rd ed., **5**, 413.

ACONITATE DECARBOXYLASE

This enzyme [EC 4.1.1.6] catalyzes the conversion of *cis*-aconitate to itaconate (or, 2-methylenesuccinate) and carbon dioxide.

J. V. Schloss & M. S. Hixon (1998) *Comprehensive Biological Catalysis:A Mechanistic Reference* **2**, 43.
R. Bentley (1962) *Meth. Enzymol.* **5**, 593.

ACONITATE Δ-ISOMERASE

This enzyme [EC 5.3.3.7] catalyzes the interconversion of *trans*-aconitate to *cis*-aconitate, the reaction reportedly to occur by an allelic rearrangement.

D. J. Creighton & N. S. R. K. Murthy (1990) *The Enzymes*, 3rd ed., **19**, 323.

ACROSIN

This enzyme [EC 3.4.21.10] catalyzes the hydrolysis of Arg-Xaa and Lys-Xaa peptide bonds. The enzyme belongs to the peptidase family S1 and is inhibited by naturally occurring trypsin inhibitors.

J. S. Bond & P. E. Butler (1987) *Ann. Rev. Biochem.* **56**, 333.

ACTIN ASSEMBLY ASSAYS

The 43 kDa actin monomer (often termed globular actin or G-actin) polymerizes to form filamentous (or F-) actin, a process that can be measured by a number of biochemical and biophysical techniques[1]. Actin self-assembly obeys the kinetics of nucleated polymerization[2], and the stages of polymerization include nucleation, elongation, and polymer length redistribution. Cooper and Pollard[1] have presented the following table indicating the advantages and disadvantages of each of these techniques.

direct assessment of the length of suitably fixed and contrast-stained actin filaments. (7) *DNase inhibition* of actin polymerization takes advantage of the observation that DNase I preferentially binds to actin monomers with sufficient affinity to block any polymerization of DNase-actin complex. (8) *Pelleting* relies on the much higher sedimentation coefficient of actin filaments as compared to monomeric actin. (9) *Millipore filtration assays* allow one to rapidly separate monomeric and polymeric actin using filter disks with 0.45 μm pores.

Table I

Methods to Measure Actin Polymerization

Method	Signal α[polymer]	Signal : noise ratio	Sensitivity to length	Shear rate	Native actin	Expense	Sample size (ml)
1. Capillary viscometry	$\pm$	High	Yes	High	Yes	Low	≥ 0.6
2. ΔOD_{232}	Yes	Moderate	No	0	Yes	High	≥ 0.6
3. Flow birefringence	$\pm$	High	Yes	Variable	Yes	High	≥ 1
4. Fluorescence							
a. NBD-NEM-actin	Yes	Moderate	No	0	No	High	≥ 0.5
b. Pyrene-actin	Yes	High	No	0	No	High	≥ 0.5
5. Light scattering	Yes	High	No	0	Yes	High	≥ 1
6. Electron microscopy	Yes	High	Yes	0	Yes	High	< 0.1
7. DNase inhibition	Yes	High	No	0	Yes	Moderate	< 0.1
8. Pelleting	Yes	High	Yes	Moderate	Yes	Moderate	0.2
9. Millipore filtration	Unknown	Moderate	Yes	High	Yes	Low	< 0.1

(1) *Capillary viscometry* is based on the higher viscosity of F-actin compared to G-actin, and one determines the time needed for a solution to pass through the orifice of a glass capillary viscometer. (2) *Difference spectroscopy* at 232 nm can be utilized to estimate the amount of F-actin in a solution, but one must exercise care to correct for light scattering contributions to the apparent optical density change (ΔOD_{232}). (3) *Flow birefringence* relies on the alignment of actin filaments with the direction of flow imparted by the rotation of one of two concentric cylinders containing the filaments in the annular space between the cyclinders. (4) *Fluorescence* changes in an extrinsic chromophore (NBD or pyrene) covalently attached to actin monomers at cysteine-373 in the actin monomer. (5) *Light scattering* can be used to study polymerization because the filaments scatter light much more intensely than individual actin subunits. The 90° light scattering intensity is directly related to polymer weight concentration. (6) *Electron microscopy* can provide a

Also see *Protein Polymerization; Self-Assembly Mechanisms.*

[1] J. A. Cooper & T. D. Pollard (1982) *Meth. Enzymol.* **85**, 182.
[2] F. Oosawa & S. Asakura (1975) *Thermodynamics of Protein Polymerization*, Academic Press, New York.

ACTIN ASSEMBLY KINETICS

The self-assembly of actin filaments occurs through sequential head-to-tail polymerization, a process that passes through phases called nucleation, elongation, and establishment of the monomer-polymer equilibrium, polymer length redistribution, and treadmilling[1-3]. The last two are properties of assembled filaments. Monomeric actin is called G-actin (or globular actin), and polymeric actin is termed F-actin (or filamentous actin). The most extensively studied actins are those obtained in high abundance from rabbit muscle and *Acanthamoeba castellani*. Many features of actin polymerization are analogous to tubulin, and the reader may wish to consult that entry. **See** *Microtubule Assembly Kinetics.*

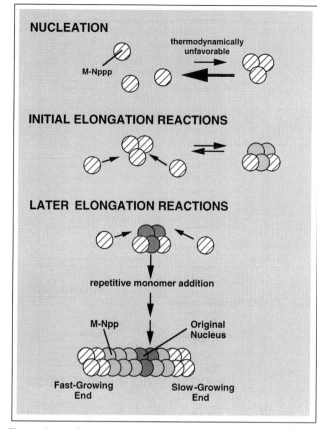

NUCLEATION

INITIAL ELONGATION REACTIONS

LATER ELONGATION REACTIONS

Figure 1. Nucleation and growth of actin filaments. Nucleation is shown here as a thermodynamically unfavored process, which in the presence of sufficient actin-ATP will undergo initial elongation to form small filament structures that subsequently elongate with rate constants that do not depend on filament length. Elongation proceeds until the monomeric actin (or G-actin) concentration equals the critical concentration for actin assembly.

Nucleation is the most thermodynamically unfavorable step (Fig. 1) in which three actin·ATP molecules must come together to form a polymerization nucleus. Only when the actin·ATP concentration is sufficiently high can these unstable nuclei persist at sufficient concentrations for polymerization to occur. Uncoordinated spontaneous nucleation must be suppressed in certain regions of the cytoplasm, and actin monomer-sequestering proteins may limit nucleation by decreasing the available concentration of actin·ATP monomers. The ADP·actin complex appears to be a potent inhibitor of nucleation. The nucleation process cannot persist for long if elongation is occurring. Indeed, if A_{total} is the total monomer concentration at the beginning of elongation, then the rate of spontaneous nucleation will be proportional to the third power of A_{total}:

$$Rate_{Nucleation} = k_N [A_{total}]^3$$

and a 20% drop in A_{total} should reduce the rate of nucleation by a factor of two:

$$Rate_{0.8}/Rate_{1.0} = k_N [0.8A_{total}]^3/k_N [A_{total}]^3$$
$$= (0.8)^3[A_{total}]^3/[A_{total}]^3 = 0.5$$

This effect will be even greater if there is any ADP-actin in the total pool of actin monomers, especially if ADP-actin inhibits nucleation.

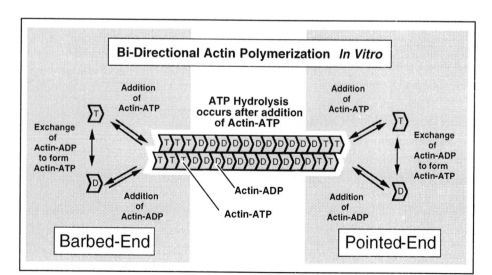

Figure 2. Scheme for the addition and loss of actin-ATP or actin-ADP from the barbed and pointed ends of an actin filament. The barbed end is the faster growing and more stable end of an actin filament. While the exchange of actin-ADP with ATP to yield actin-ATP and ADP is shown here as a spontaneous process, the actin regulatory protein profilin greatly accelerates the exchange process. Note also that hydrolysis is thought to occur after (and not coincident with) addition of actin-ATP at either end.

Elongation is the repetitive addition reactions of actin·ATP (Fig. 2). The actin-bound ATP is hydrolyzed during/after monomer addition to filaments, forming polymer-bound ADP and releasing orthophosphate. The kinetics of elongation conform to that predicted by the following rate law:

$$dCp/dt = k_+ N[A_{total}] - k_{-1} N$$

where Cp is the concentration of polymerized actin, k_+ is the bimolecular rate constant (equal to the sum of the on-rate constants for actin monomer addition to both ends of an actin filament, N is the number concentration of polymer ends that are capable of reacting with actin monomers, $[A_{total}]$ is the total monomer concentration, and k_{-1} is the rate constant for monomer dissociation. If N is constant during the elongation phase, the observed rate process will fit a simple first-order decay curve. Elongation continues until the elongation "on"-rate (*i.e.*, $k_+[actin]$) is exactly balanced by the "off"-rate (k_-). This condition defines the critical actin concentration ($[actin]_{critical} = k_-/k_+$), a parameter which represents the concentration of actin that coexists with assembled filaments (Fig. 3, top).

Because the onset of monomer-polymer equilibrium can occur before the filaments achieve their own equilibrium concentration behavior, these filaments will undergo polymer length redistribution. This is a slow process *in vitro* that in many respects resembles crystallization (**See** *Ostwald Ripening*).

Treadmilling is the net opposite-end assembly/disassembly process first described by Wegner's study[4] of actin polymer dynamics. He recognized that there was no reason to believe that the release of free energy of ATP hydrolysis need be identical for addition reactions occurring on opposite filament ends. The so-called (+)-end and (−)-end have different rate and equilibrium constants: $K^+ = k^+_{off}/k^+_{on}$ and $K^- = k^-_{off}/k^-_{on}$, such that $[actin]_{critical} = k_{off}/k_{on} = (k^+_{off} + k^-_{off})/(k^+_{on} + k^-_{on})$. The macroscopic actin critical concentration obeys the following inequality:

$$k^+_{off}/k^+_{on} < [actin]_{critical} < k^-_{off}/k^-_{on}$$

As the name implies, actin treadmilling does not involve any net increase in the amount of polymerized actin; instead, when ATP is present, monomers are spontane-

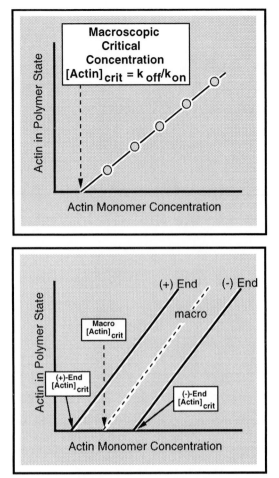

Figure 3. Critical concentration behavior of actin self-assembly. For the top diagram depicting the macroscopic critical concentration curve, one determines the total amount of polymerized actin by methods that measure the sum of addition and release processes occurring at both ends. Examples of such methods are sedimentation, light scattering, fluorescence assays with pyrene-labeled actin, and viscosity measurements. For the bottom curves, the polymerization behavior is typically determined by fluorescence assays conducted under conditions where one of the ends is blocked by the presence of molecules such as gelsolin (a barbed-end capping protein) or spectrin-band 4.1-actin (a complex prepared from erythrocyte membranes, such that only barbed-end growth occurs). Note further that the barbed end (or (+)-end) has a lower critical concentration than the pointed end (or (−)-end). This differential stabilization requires the occurrence of ATP hydrolysis to supply the free energy that drives subunit addition to the (+)-end at the expense of the subunit loss from the (−)-end.

ously released from the less stable (−)-ends, and monomers are taken up by the more stable (+)-ends (Fig. 3). This results in a flux of monomers that can be quantitatively assessed by the inclusion of $[^3H]ATP$ or $[^{14}C]ATP$ which exchanges with unlabeled actin in the monomeric pool and upon polymerization is trapped in filaments as actin-ADP.

Although investigations of the *in vitro* polymerization process have provided a theoretical underpinning for probing intracellular actin dynamics, there are obvious voids in our understanding. Actin-based motility in many cell types reaches a rate of 1 μm/second, corresponding to the steady-state addition/loss of 50 to 100 monomers per filament per second. This is much faster than any *in vitro* kinetic study, and the mechanism of assembly may differ substantially from the model studies. For example, although exchange of ATP for ADP on actin is not a rate-limiting reaction *in vitro*, this may not be true in the cell. Intracellular F-actin formation is also highly localized, occurring within discrete polymerization zones established by assembly of actin-based motor complexes from a battery of regulatory proteins. These include capping proteins, severing proteins, sequestering proteins, bundling and cross-linking proteins, actin-related proteins, as well as a growing list of actin-based motor complex components. Their interactions are controlled by binding interactions (**See** *Actin Filament Capping Protein, Actin Filament Severing Protein, Actin Filament Bundling/Cross-Linking Protein, and ABM-1 & ABM-2 Sequences in Actin-Based Motors*). In addition, actin assembly is controlled by extracellular cues transmitted to the cell's interior by focal adhesion and adherens junction proteins.

[1]T. P. Stossel (1993) *Science* **260**, 1086.
[2]M. F. Carlier (1989) *Int. Rev. Cytol.* **115**, 139.
[3]T. D. Pollard & J. A. Cooper (1986) *Annual Rev. Biochem.* **55**, 987.
[4]A. Wegner (1982) *J. Mol. Biol.* **161**, 607.

Selected entries from *Methods in Enzymology* [vol, page(s)]:
Staining with fluorochrome-conjugated phalloidin, **194**, 729; binding activity of ponticulin, **196**, 58; crosslinking in agonist-stimulated cells, assay, **196**, 486; depolymerization [actin-binding protein effects, **215**, 74; myosin effects, **215**, 74; dilution effect on platelet cytoskeleton, **215**, 76]; expression in *Escherichia coli*, **196**, 368 [growth conditions for, **196**, 378; solubility, effect of bacterial lysis, **196**, 380; vector-related variability, **196**, 370]; fluorescein labeling, **196**, 50; actin-gelsolin complex [assay, **215**, 94; isolation from platelets, **215**, 89]; globular, purification from *Dictyostelium discoideum*, **196**, 89; isoforms, separation, **196**, 105; monomeric columns with, preparation, **196**, 310; nucleation activity [agonist-stimulated, assay of inhibitor, **196**, 495; measurement, **196**, 493]; platelet [characterization, **215**, 58; purification, **215**, 58, 66; recombination with actin-binding protein and α-actinin, **215**, 73]; polymerization in agonist-stimulated cells, assay, **196**, 486; preparation, **196**, 402; actin-profilin complex, isolation from cell extracts, **196**, 97; purification, **196**, 50, 74; separation from profilin:actin, **196**, 115; skeletal muscle, labeling with pyrene, **196**, 138; sliding over myosin-coated surfaces, assay, **196**, 399; stored, rejuvenation by recycling, **196**, 403.

Actin filaments: affinity column [preparation, **196**, 49; properties, **196**, 305]; assay [in cells, **196**, 486; in platelet lysates, **215**, 54]; associated proteins, identification, **215**, 50; attachment to column matrix, **196**, 309; chromatography, **196**, 312; crosslinking in cells, measurement, **196**, 491; fluorescent labeling, **196**, 403; imaging, **196**, 408; length measurement, **196**, 416; proteins binding, affinity chromatography, **196**, 303.

Actin-binding proteins: ABP-50, purification, **196**, 78; ABP-120, purification, **196**, 79; ABP-240 purification, **196**, 76; effect on actin depolymerization, **215**, 74; extraction, **196**, 311; isolation from *Dictyostelium discoideum*, **196**, 70; platelet-derived actin binding proteins [characterization, **215**, 58; purification, **215**, 58, 64; recombination with actin, **215**, 73]; 30-kDa *Dictyostelium discoideum* actin-crosslinking protein [assays, **196**, 91; preparation, **196**, 84]; actin-depolymerizing factor [assay, **196**, 132]; DNase assay, **196**, 136; platelet-derived a-actinin [characterization, **215**, 58; purification, **215**, 58, 70; recombination with actin, **215**, 73].

ACTIN ATPase REACTION (Apparent Irreversibility)

The pathway of ATP hydrolysis associated with rabbit muscle actin polymerization was investigated using an assay of intermediate ^{18}O-positional isotope exchange reactions. Under a variety of conditions that influence the rate and extent of F-actin self-assembly, ATP hydrolysis proceeded without any evidence of multiple reversals that are characteristic of any reversible phosphoanhydride-bond cleavage[1]. These results are in harmony with published findings[2] indicating that GTP hydrolysis during tubulin polymerization also fails to display any evidence of intermediate exchange reactions.

[1]M. F. Carlier, D. Pantaloni, J. A. Evans, P. K. Lambooy, E. D. Korn & M. R. Webb (1988) *FEBS Lett.* **235**, 211.
[2]J. M. Angelastro & D. L. Purich (1990) *Eur. J. Biochem.* **191**, 507.

ACTIN-BASED PATHOGEN MOTILITY

Actin-based motility involves the highly regulated assembly/disassembly of actin filaments in the cytoplasm nearest the peripheral membrane. The energetics and polarity of ATP-dependent filament assembly produce the expansive force and directionality for peripheral membrane protrusion during ameboid motion[1]. Two bacterial systems bypass the complexities of signal transduction cascades as well as membrane geometry: *Listeria monocytogenes*, a gram-positive rod that causes severe meningitis and bacteremia, and *Shigella flexneri*, a gram-negative rod that is a leading cause of bacillary dysentery. *Listeria* produces proteins known as internalins to induce phagocytosis[2,3], and *Shigella* uses the surface proteins, Ipa B and C[4]. Although most bacteria are trapped and killed in phagolysosomes, ingested *Listeria* and *Shigella* readily escape by producing hemolysins that disrupt the phagolysosomal membrane[5-7]. After entering the host cytoplasm, they grow readily, doubling every hour. These pathogens usurp host contractile proteins, allowing them to mimic the activated membrane state that stimulates localized actin filament assembly for their own motility. Then an initial halo of actin filaments surrounding the bacteria is remodeled over several hours, thereby confining actin assembly to one bacterial pole[8,9]. They begin to locomote and to travel throughout the cytoplasm, eventually reaching the peripheral membrane, where growth of their actin-rich tails forces the membrane to protrude outwardly, forming filopods. Bacteria-laden filopods are ingested by adjacent cells, and the life-cycle begins anew. In this manner, both *Listeria* and *Shigella* evade humoral immunity and antibiotics that poorly penetrate host cells[10].

Kinetics of Bacterial Motility. Examination of *Listeria*-infected host cells has provided many important clues about pathogen motility. Renal tubular epithelial PtK2 cells spread and flatten when grown on glass coverslips; after infection, extracellular bacteria are washed away, and addition of gentamicin (10 μg/mL) prevents extracellular bacterial growth. This antibiotic penetrates cells poorly and does not impair proliferation of intracellular bacteria[8,11,12]. Within 3-4 hours postinfection[13], *Listeria* begin moving at rates of 0.1 to 0.45 μm/sec. As they move, phase-dense tails form, and those bacteria moving more rapidly have longer tails. These tails contain numerous actin filaments, as indicated by staining with fluorescently conjugated phalloidin, a mushroom toxin that specifically binds to actin filaments. Microinjection of live, infected cells with rhodamine-conjugated actin monomers allows the kinetic analysis of actin assembly and disassembly in these tails. More rapid motility is always matched by more rapid actin assembly. By contrast, the rate of actin disassembly is constant ($t_{1/2}$ = 40–60 sec) and does not depend on bacterial speed[13,14]. Time-lapse video also revealed that the actin filament tails are not dragged along by the bacterium; rather, these tails remain fixed in the cytoplasm, and the progressive addition of actin subunits near the bacterium allows the pathogen to move forward. Labeled monomers are observed to add only at the junction between the bacterium and the actin tail[13]. Thus, new actin filament assembly takes place where the bacterium contacts the tail. These filaments become bundled by α-actinin, and they also become linked to the existing cy-

toskeletal network to form a stable platform. This prevents backward movement of the bacterium as actin filaments at the bacteria-tail junction lengthen; filament elongation propels each bacterium through the cytoplasm in the direction opposite to the tail. These experiments constitute the first clear demonstration that actin filaments can produce force and actively generate movement.

The site of motor assembly lies directly behind the moving bacterium in what has been termed a polymerization zone. *Listeria* assembles its actin-based motor by concentrating host cell proteins that stimulate actin assembly. Transposon mutagenesis[15,16] demonstrated that a single 67 kDa protein, ActA, is needed for the formation of actin rocket tails as well as for the spread of bacteria from cell to cell. The ability to cause disease in mice was also markedly impaired (the LD$_{50}$ increasing from 10^4 to 10^8 organisms) in ActA-deficient *Listeria*. ActA has a series oligoproline FEFPPPPTDE repeats. Microinjection[17] of CFEFPPPPTDE, a synthetic peptide analogue of the second oligoproline repeat into *Listeria*-infected PtK2 cells competitively blocks *Listeria* actin-based motility. Time-lapse video phase microscopy revealed that within 30 sec after introducing the peptide, *Listeria* ceased all further motility. Over this same time-frame, the phase-dense tails also disappeared. At intracellular peptide concentration as low as 50 nM, *Listeria* actin-based motility was totally arrested[17]. Microinjection of a combinatorial library of peptides containing the very same amino acids, but in random order, was completely without effect at 60 μM. Similarly, two SH3 peptides failed to inhibit even at concentrations of 60 μM. These SH3 sequences contain intervening positively charged residues rather than negatively charged residues of the ActA peptide. Little or no inhibition was observable when 10 μM poly(L-proline) was microinjected. These experiments demonstrate the specificity of ActA peptide interactions in *Listeria* motility.

Chakrabarty *et al.*[18] discovered that the host cell protein vasodilator-stimulated phosphoprotein (VASP) concentrates at the back of motile *Listeria* at the bacteria-tail junction. VASP docks onto the surface of *Listeria* by binding to the oligoproline sequences of ActA concentrated on the rearward pole of moving bacteria. In uninfected host cells, VASP is normally found in focal contacts, regions that allow cells to adhere to the substratum and to process extracellular stimuli[19]. Focal contacts also contain high concentrations of actin filaments as well as other actin-binding proteins, including α-actinin, vinculin, zyxin, paxillin, and talin. Microinjection of the FEFPPPPTDE peptide into uninfected cells dissociates VASP from these focal contacts, suggesting the existence of an ActA-like VASP binding protein in host cells[18].

By binding profilin, a 15.5 kDa actin-regulatory protein, VASP acts as an adapter protein to link profilin to the surface of *Listeria*.[19] Each 45 kDa VASP monomer contains a series of 5–6 oligoproline Gly-Pro-Pro-Pro-Pro-Pro repeats (abbreviated GP_5) that bind to profilin[20]. The native VASP tetramer therefore contains 20–24 GP_5 sites for profilin binding. Immunofluorescence microscopy demonstrated that profilin concentrates at the back of *Listeria* in the very same region where VASP is found[21]. Microinjection of $GP_5GP_5GP_5$ blocks *Listeria* actin-based motility by competitively inhibiting profilin binding to VASP[20].

BASIC FEATURES OF THE ACTIN-BASED MOTOR COMPLEX. These experiments support the following model for *Listeria* actin-based motility. The *Listeria* surface protein ActA uses its oligoproline sequences (FEFPPPPTDE) to bind the host cell protein VASP. VASP in turn deploys its own larger set of GPPPPP sequences to bind profilin.

This system allows one ActA molecule to concentrate up to 96 profilin molecules, and, *Listeria* may concentrate as much as 0.5–1 mM profilin on its surface[20]. Profilin markedly potentiates growth of actin filaments[22], and highly concentrated profilin in the polymerization zone would be expected to result in explosive actin filament assembly, thus generating the forces required for *Listeria* intracellular movement.

IDENTIFICATION OF HOST CELL ActA HOMOLOGUES. Sansonetti's group first demonstrated that the 120 kDa bacterial surface protein IcsA is the only protein required for the generation of the actin-based motor in *Shigella*[23]. Comparisons of the derived amino acid sequences however demonstrate no homology between the ActA protein and IcsA, which contains no oligoproline sites. IcsA attracts a host cell protein as an ActA homologue (*i.e.*, the host cell protein must function in place of ActA by providing oligoproline sequences suitable for high-affinity binding of VASP). Zeile *et al.*[12] microinjected the *Listeria* ActA peptide into *Shigella*-infected cells to show that 80 nM FEFPPPPTDE sequence also blocked *Shigella* actin-based motility. Furthermore, immunofluorescence experiments with anti-VASP antibody demonstrated that VASP became localized on the rearward pole of motile *Shigella*[18]. Microinjection of the GP5 peptide into *Shigella*-infected cells also blocked profilin binding to VASP, and was effective in arresting *Shigella* motility at 10–12 μM.

That motile *Shigella* concentrate an ActA homologue on their surface provided a way to identify a proteolytic vinculin fragment as the homologue. Laine *et al.*[24] used synthetic FEFPPPPTDE as an immunogen to generate a polyclonal antibody (Ab). Microinjection of this antibody blocked *Shigella* motility in infected cells, and the Ab was localized to the bacterial tails. Western blots of

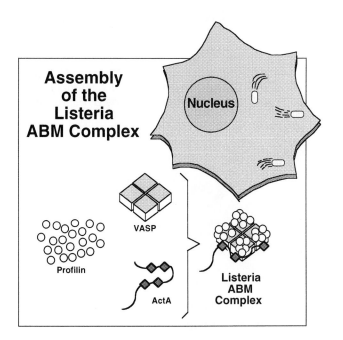

The *Listeria* Actin-Based Motility complex. The bacterial surface protein ActA contains four oligoproline repeats (designated as dark rectangles) that bind vasodilator-stimulated phosphoprotein VASP, a tetrameric actin-regulatory protein in the cytoplasm of the host cell. Each VASP tetramer contains 20–24 profilin-binding sites of the type GPPPPP, allowing high concentrations of profilin to bind onto the surface of *Listeria*. This creates a polymerization zone that enhances production and polymerization of ATP-actin monomers. Profilin probably ushers ATP-actin monomers to actin filaments growing at the bacterium–actin tail junction.

platelet extracts revealed a single, strongly cross-reactive polypeptide (M_r = 90 kDa; pI = 6.5) isolated by 2-dimensional IEF/SDS-PAGE. Microsequencing of five lysC protease fragments matched corresponding stretches in the head domain of vinculin, an actin regulatory protein localized in focal adhesion contacts[24]. The carboxy-terminus of the 90 kDa head region contains a DFPPPPPDLE sequence that is virtually identical to those in ActA. The anti-ActA Ab failed to react with full-length vinculin, binding preferentially to its 90 kDa fragment in cell extracts. This was confirmed using the p90 head domain (generated *in vitro* by thermolysin treatment) and full-length vinculin, indicating that this antigenic site was masked in the intact molecule. Thus, despite the harsh conditions of SDS-PAGE, refolding must occur upon processing the blot, and the ActA-like sequence is masked, rendering it incapable of attracting VASP until the head region is freed from the tail[24]. Vinculin p90 accumulates in *Shigella*-infected PtK2 cells over the 1–3 hour period after initiating infection, but no p90 was observed in uninfected cells[24]. Separate microinjections of intact vinculin and p90 head fragment into *Shigella*-infected cells demonstrated that 120 nM vinculin was completely without effect on the rate of *Shigella* locomotion, but 120 nM p90 dramatically accelerated the velocity of the *Shigella* intracellular movement, increasing mean velocities by a factor of three[24]. These experiments suggest a variation on the *Listeria* model: (1) *Shigella* infection leads to vinculin proteolysis; (2) an ActA-like oligoproline sequence is unmasked in the vinculin head region; (3) IcsA binds the vinculin head domain, and the exposed ActA-like sequence binds VASP; (4) the latter in turn binds profilin onto the rearward pole of the bacterium; and (5) the high concentration of profilin within this polymerization zone stimulates actin filament assembly. Two phylogenetically unrelated bacteria have adopted similar subcellular mechanisms that permit them to spread from cell to cell. Both usurp host components to assemble actin-based motor complexes that are structurally and functionally related. *Listeria* uses ActA to attract VASP which then concentrates profilin. **See** *ABM-1 and ABM-2 Sequences in Actin-Based Motors*

[1]T. P. Stossel (1994) *Sci. Am.* **271** (September), 54.
[2]J. L. Gaillard, P. Berche, C. Frehel, E. Gouin & P. Cossart (1991) *Cell* **65**, 1127.
[3]S. Dramsi, I. Biswas, E. Maguin, L. Braun, P. Mastroeni & P. Cossart (1995) *Mol. Microbiol.* **16**, 251.

[4]R. Menard, M. C. Prevost, P. Gounon, P. Sansonetti & C. Dehio (1996) *Proc. Natl. Acad. Sci. U. S. A.* **93**, 1254.
[5]S. Kathariou, P. Metz, H. Hof & W. Goebel (1987) *J. Bacteriol.* **169**, 1291.
[6]D. A. Portnoy, P. S. Jacks & D. J. Hinrichs (1988) *J. Exp. Med.* **167**, 1459.
[7]S. Barzu, Z. Benjelloun-Touimi, A. Phalipon, P. Sansonetti & C. Parsot (1997) *Infect. Immun.* **65**, 1599.
[8]G. A. Dabiri, J. M. Sanger, D. A. Portnoy & F. S. Southwick (1990) *Proc. Natl. Acad. Sci. U. S. A.* **87**, 6068.
[9]L. G. Tilney & D. A. Portnoy (1989) *J. Cell. Biol.* **109**, 1597.
[10]F. S. Southwick & D. L. Purich (1996) *New Engl. J. Med.* **334**, 770.
[11]E. A. Havell (1986) *Infect. Immun.* **54**, 787.
[12]W. L. Zeile, D. L. Purich & F. S. Southwick (1996) *J. Cell Biol.* **133**, 49.
[13]J. M. Sanger, J. W. Sanger & F. S. Southwick (1992) *Infect. Immun.* **60**, 3609.
[14]J. A. Theriot, T. J. Mitchison, L. G. Tilney & D. A. Portnoy (1992) *Nature* **357**, 257.
[15]E. Domann, J. Wehland, M. Rohde, S. Pistor, M. Hartl, W. Goebel, M. Leimeister-Wachter, M. Wuenscher & T. Chakraborty (1992) *EMBO J.* **11**, 1981.
[16]C. Kocks, E. Gouin, M. Tabouret, P. Berche, H. Ohayon & P. Cossart (1992) *Cell* **68**, 521.
[17]F. S. Southwick & D. L. Purich (1994) *Proc. Natl. Acad. Sci. U. S. A.* **91**, 5168.
[18]T. Chakrabarty, F. Ebel & E. Domann (1995) *EMBO J.* **14**, 1314.
[19]M. Reinhard, M. Halbrugge, U. Scheer, C. Wiegand, B. M. Jockusch & U. Walter (1992) *EMBO J.* **11**, 2063.
[20]F. Kang, R. O. Laine, M. R. Bubb, F. S. Southwick & D. L. Purich (1997) *Biochemistry* **36**, 8384.
[21]J. A. Theriot, J. Rosenblatt, D. A. Portnoy, P. J. Goldschmidt-Clermont & T. J. Mitchison (1994) *Cell* **76**, 505.
[22]D. Pantaloni & M. F. Carlier (1993) *Cell* **75**, 1007.
[23]M. L. Bernardini, J. Mounier, H. d'Hauteville, M. Coquis-Rondon & P. J. Sansonetti (1989) *Proc. Natl. Acad. Sci. U. S. A.* **86**, 3867.
[24]R. O. Laine, W. Zeile, F. Kang, D. L. Purich & F. S. Southwick (1997) *J. Cell Biol.* **138**, 1255.

ACTIN FILAMENT CAPPING PROTEIN

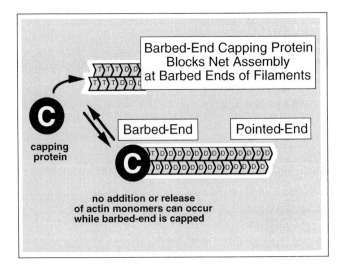

A cytoskeletal regulatory protein that binds to one end (usually the (+)-end or so-called barbed end) of an actin filament, thereby blocking gain or loss of actin monomers. The affinity of a capping protein can be readily determined as the concentration of capping protein needed to inhibit the initial rate of dilution-induced disassembly of polymer containing pyrenyl-actin. The latter is a fluorescently tagged form of actin that gives a direct report of assembly indicator. Barbed-end capping proteins include Cap G, Cap Z, capping protein, as well as the actin capping/severing proteins below.

If a capping protein binds in a rapid equilibrium manner to the end of an actin filament, the capped filament (a) will no longer permit polymerization; (b) will thereby decrease the number of growing ends during elongation; and (c) should even retard subunit loss from the capped end if the sample is added to a actin-free solution containing a capping protein concentration sufficient to maintain the overall fraction of filaments that are capped. If $[N]$ equals the number concentration of uncapped filaments, and if $[N_{capped}]$ equals the number concentration of capped filaments, then in the presence of capping protein Y, the dissociation equilibrium constant for capping is given by

$$K_d = [Y_{free}][N]/[N_{capped}]$$

Because the total number concentration $[N_{total}]$ is the sum of $[N] + [N_{capped}]$, we can combine this relationship with the equilibrium equation to obtain:

$$[N] = [N_{total}] \{K_d /([N_{capped}] + K_d)\}$$

We then see that the rate of elongation in the presence of a capping protein will be given by the following rate law:

$$d[A_{poly}]/dt = k_+[A_{total}] [N_{total}] \{K_d/([N_{capped}] + K_d)\}$$
$$- k_{-1}N$$

Because the concentration of capped filaments is a denominator term, we immediately recognize that the elongation rate will be retarded by capping.

Note that if a capping protein binds to monomeric actin, the capping protein will also be a monomer-sequestering agent. A good example of such behavior is profilin. **See also** ABM-1 & ABM-2 Sequences in Actin-Based Motors; Actin-Based Bacterial Motility; Actin Assembly Kinetics

T. P. Stossel (1993) Science **260**, 1086.
T. P. Stossel (1994) Sci. Am. **271**(September), 54.

ACTIN FILAMENT CROSS-LINKING PROTEIN/BUNDLING

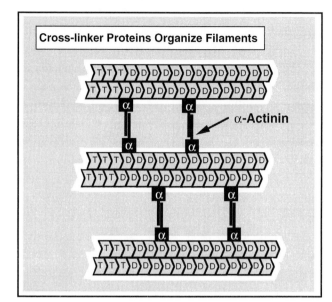

A cytoskeletal regulatory protein that contains pairs or binding regions that attaching laterally to actin filament and resulting in cross-linking and/or bundling. These include α-actinin, ABP-30, ABP-120, ABP-240, spectrin, dematin, fascin, fimbrin, MARCKS, and dystrophin. Many of these proteins are antiparallel dimers of monomeric units that each contain a single actin filament binding site. **See also** ABM-1 & ABM-2 Sequences in Actin-Based Motors; Actin-Based Bacterial Motility; Actin Assembly Kinetics

T. P. Stossel (1993) Science **260**, 1086.
T. P. Stossel (1994) Sci. Am. **271**(September), 54.

ACTIN FILAMENT GROWTH
(Polymerization Zone Model)
A model for localized, polar growth of actin filaments during lamellipod protrusion, filopod formation, or actin-based motility of intracellular pathogens such as *Listeria monocytogenes*, *Shigella flexneri*, or *vaccinia*. The polymerization zone is characterized by the presence of docking sequences that serve to concentrate or cluster filament growth-promoting factors, including but not limited to the actin regulatory component known as profilin. The cardinal feature of filament growth within the polymerization zone is that the concentration of polymerizable actin-ATP far exceeds that present in the bulk phase of the cytoplasm.

Actin filaments grow rapidly within cells, and the clearest evidence of this rapid growth is the ability of the cell's leading edge to move at rates of 0.5 to 1 micrometer per second. Likewise, actin-based motility of *Listeria* and *Shigella* can attain rates of nearly 0.5 micrometers per second. Because microfilaments contain about 360 actin monomers per micrometer of length, a motility rate of 0.5 to 1 micrometer per second corresponds to an apparent first-order rate constant (*i.e.*, $k_{apparent} = k_{on}$ [Actin–ATP]) of about 180–360 s^{-1}. The bimolecular rate constant k_{on} for actin-ATP addition to the barbed end has a nominal value of 2–3×10^6 M^{-1} s^{-1}. Therefore, one can estimate that 60–180 μM actin–ATP (a value equal to $k_{apparent}$/k_{on} or 180–360 s^{-1}/2–3×10^6 M^{-1} s^{-1} = 0.06–0.18 mM) would be needed to sustain a filament growth rate of 0.5 to 1 μm/s. This local actin-ATP concentration range greatly exceeds the 0.25–0.3 μM concentration of G-actin thought to occur in most cells.

How then may one account for this discrepancy? We can first dispense with any trivial explanation suggesting that the value of k_{on} is unreasonably low, because bimolecular rate constants for protein addition to filaments typically have an upper bound of 10^7 M^{-1} s^{-1}. And although the total intracellular actin concentration often lies in the 200–300 μM concentration range, very little is present as uncomplexed actin–ATP monomer. In fact, cells typically have about 10–30% of total actin present as filaments, and the remainder is present as thymosin-β4–actin complex (60–70%), profilin-actin complex (10–20%), and substantially less than 1% as uncomplexed actin–ATP monomers. The most attractive idea is that regions of very active growth are apt to contain high local concentrations of actin–ATP that can readily mobilized to fulfill the needs of rapidly elongating filaments. Kang *et al.*[1] recently proposed one such model that relies on a hierarchy of oligoproline sequences (**See** *ABM-1 & ABM-2 Sequences In Actin-Based Motors*) to localize profilin (or the profilin–actin complex) at the sites of rapid filament assembly. As shown in Fig. 1, this cascade of binding interactions can result in the concentration of profilin and/or the profilin–actin complex within a polymerization zone defined by the activated cluster region. Host cells are also likely to utilize an analogous strategy for placing

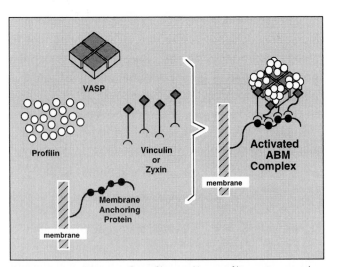

Figure 1. Localization of profilin and/or profilin–actin complex into an activated cluster. VASP, Vasodilator-stimulated phosphoprotein.

profilin and/or profilin–actin complex into an activated complex. This cluster creates a polymerization zone where the ability of profilin to facilitate ATP exchange with actin–ADP ensures that high levels of actin–ATP will be available for barbed-end growth of nearby actin filaments (see Fig. 2). Because actin–ADP is much less efficient in actin assembly, conversion to actin–ATP complex favors rapid elongation of filaments. Note also that depolymerization occurs exclusively at the pointed ends, and the much lower concentration of untethered profilin is likely to greatly limit the nucleotide exchange rate. The polymerization zone may also limit the ability of capping proteins (*e.g.*, Cap G, Cap Z, and gelsolin) to bind to the barbed end, an action that reduces the rate of filament assembly.

Other components, such as a-actinin and actin-related proteins (or ARPs), may also be recruited to the polymerization zone, whereas depolymerizing factors such as ADF and cofilin are apt to be selectively localized in the depolymerization zone. Note: Agents that disrupt the formation of activated ABM clusters will also suppress the rate of assembly in the polymerization zone. **See** *ABM-1 & ABM-2 Sequences In Actin-Based Motors*

[1]F. Kang, R. O. Laine, M. R. Bubb, F. S. Southwick & D. L. Purich (1997) *Biochemistry* **36**, 8384.

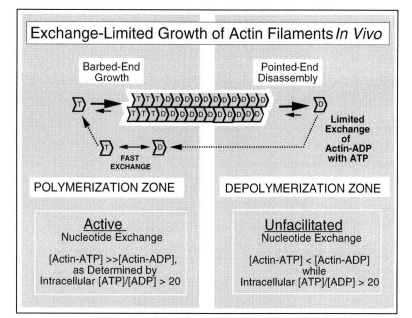

Figure 2. Characteristics of adenine nucleotide exchange that define actin polymerization and depolymerization zones within cells.

ACTIN FILAMENT LATERAL BINDING PROTEIN

A cytoskeletal regulatory protein that attaches laterally to actin filament without resulting in filament cross-linking and/or bundling. These include troponin, tropomyosin, calponin, tropomodulin, adducin, caldesmon, and hisactophilin. *See also ABM-1 & ABM-2 Sequences in Actin-Based Motors; Actin-Based Bacterial Motility; Actin Assembly Kinetics*

T. P. Stossel (1993) *Science* **260**, 1086.
T. P. Stossel (1994) *Sci. Am.* **271**(September), 54.

ACTIN FILAMENT SEVERING PROTEIN

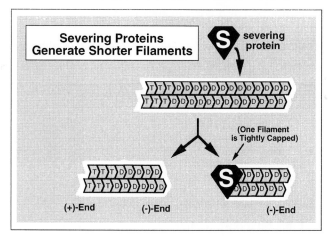

A cytoskeletal regulatory protein that binds to and cleaves actin filaments. These proteins do not catalyze filament cleavage; instead, they act stoichiometrically by remaining bound to one of the filament ends generated in the severing reaction. The prototypical severing protein is gelsolin, but other severing proteins include fragmin, villin, and ADF. Addition of a severing protein increases the number of polymer ends, and the rate of dilution-induced depolymerization rate will increase as more severing protein is added. Note also that many severing proteins form tightly bound caps on the so-called barbed end of one of the filaments produced by the severing process. A good example is gelsolin whose dissociation constant for the barbed end lies in the sub-picomolar range. *See also ABM-1 & ABM-2 Sequences in Actin-Based Motors; Actin-Based Bacterial Motility; Actin Assembly Kinetics*

T. P. Stossel (1993) *Science* **260**, 1086.
T. P. Stossel (1994) *Sci. Am.* **271**(September), 54.

ACTIN FILAMENTS (Mechanical Properties)

The mechanical properties of actin filament networks depend on the manner in which actin monomer is prepared and stored, as well as how they are polymerized conditions. Differences in mechanical properties are not the consequence of using two different types of forced oscillatory rheometers. Xu *et al.*[1] found that filaments assembled in EGTA and Mg^{2+} from fresh, gel-filtered ATP-actin monomer (1 mg/mL) have an elastic storage modulus (G') of approximately 1 Pa at a deformation frequency of 0.1–1 Hz. G' is slightly higher when actin is polymerized in the presence of K^+, Ca^{2+}, and Mg^{2+}. Actin monomer storage in the absence of frequent buffer changes to maintain ATP and reduced dithiothreitol can result in changes that increase the G' of filaments by a factor of 10 or more. The authors also stress that frozen storage can preserve the properties of monomeric actin, as long as one takes steps to prevent protein denaturation or aggregation due to freezing or thawing.

[1]J. Xu, W. H. Schwarz, J. A. Kas, T. P. Stossel, P. A. Janmey & T. D. Pollard (1998) *Biophys. J.* **74**, 2731.

ACTIN MONOMER SEQUESTERING PROTEIN

A cytoskeletal regulatory protein that alters the distribution of monomeric and polymeric actin by sequestering monomers in a form that does not engage in polymer elongation reactions. These sequestering proteins do not directly affect the rate constants of actin polymerization/depolymerization or the magnitude of the critical concentration. The presence of a sequestering protein reduces the concentration of actin monomers that are polymerization-dependent, thereby altering the rate and extent of actin polymerization. Moreover, addition of sequestering protein to a solution containing actin polymer and monomer in an equilibrium (or steady-state) distribution results in net depolymerization, depending on the concentration and affinity of sequestering protein for actin monomers. Thymosin $\beta4$ is the most abundant actin sequestering protein in nonmuscle cells, followed by profilin. Other sequestering proteins include actobindin, profilin, DNase I, ABP-50, and vitamin D binding protein.

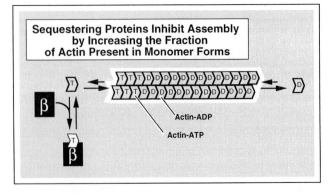

Sequestering Proteins Inhibit Assembly by Increasing the Fraction of Actin Present in Monomer Forms

Actin-ADP
Actin-ATP

The effect of monomer sequestration on actin assembly can be described as follows. ition to filaments, forming

polymer-bound ADP and releasing orthophosphate. The kinetics of elongation conform to that predicted by the following rate law:

$$d[A_{poly}]/dt = k_+N[A_{total}] - k_{-1}N$$

where $[A_{poly}]$ is the concentration of polymerized actin, k_+ is the bimolecular rate constant (equal to the sum of the on-rate constants for actin monomer addition to both ends of an actin filament), N is the number concentration of polymer ends that are capable of reacting with actin monomers, and $[A_{total}]$ is the total monomer concentration. Upon actin monomer sequestration by protein X, we must consider an additional form of actin, namely AX:

$$K_d = [A_{free}][X_{free}]/[AX]$$

where K_d is the dissociation constant of complex AX, and $[A_{free}]$, $[X_{free}]$, and $[AX]$ are the respective concentrations of free or uncomplexed actin monomer, free or uncomplexed sequestering protein, and actin·X complex. By conservation, we can write the following relationship:

$$[AX] = [A_{total}] - [A_{free}]$$

which upon substitution into the above expression for K_d yields

$$[A_{free}] = K_d [A_{total}]/([X_{free}] + K_d)$$

Sequestered monomers bound to the sequestering protein cannot bind to an actin filament, and only free actin can participate in elongation. Thus,

$$d[A_{poly}]/dt = k_+N[A_{total}] \{K_d/([X_{free}] + K_d)\} - k_{-1}N$$

This rate law predicts that the rate of polymerization will be suppressed by the reduction in the free actin concentration as a consequence of AX formation.

In the absence of a sequestering protein, actin polymerization will proceed to equilibrium or it will reach a steady-state extent of polymerization, at which point $d[A_{poly}]/dt = 0$. $[A]_\infty$ is the critical concentration, equal to k_-/k_+. In the presence of a sequestering protein, $[A_{total}] = [A_{free}] + [A_{poly}] + [AX]$, and after polymerization reaches equilibrium, the unpolymerized actin concentration $[A]_\infty'$ will equal $\{[AX] + [A_{free}]\}$. Hence,

$$[A]_\infty'/[A]_\infty = \{[AX] + [A_{free}]\}/[A_{free}]$$

and this relationship can be rearranged into the following equation:

$$[A]_\infty' = [A]_\infty \{1 + [X_{free}]/K_d\}$$

This equation clearly demonstrates that the observed critical concentration will be increased by the presence of sequestering protein. In this respect, addition of a sequestering protein should reduce *both* the rate and the extent of polymerization. **See also** *ABM-1 & ABM-2 Sequences in Actin-Based Motors; Actin-Based Bacterial Motility; Actin Assembly Kinetics*

T. P. Stossel (1993) *Science* **260**, 1086.
T. P. Stossel (1994) *Sci. Am.* **271**(September), 54.

ACTINOMETER

An instrument or suitable chemical system that allows one to determine the number of photons (or number of quanta) absorbed in a beam of electromagnetic radiation.

ACTION

The multiplicative product of work (unit = Joule) and time (unit = sec). The alternative term is *impulse*.

ACTION POTENTIAL

The transient change in the transmembrane potential upon excitation. An action potential cycle consists of a transient depolarization of the cell membrane of an excitable cell (such as a neuron) as a result of increased permeability of ions across the membrane, followed by repolarization, hyperpolarization, and finally a return to the resting potential. This cycle typically lasts 1–2 milliseconds and travels along the axon from the cell body (or, axon hillock) to the axonal terminus at a rate of 1–100 meters per second. **See** *Membrane Potential*

ACTION SPECTRUM

A plot of the photoresponse of a biological or chemical process (normalized with respect to the number of incident photons) as a function of wavelength or energy of radiation. Diverse biological phenomena have been investigated by determining action spectra[1]. For example, action spectroscopy was used to demonstrate (a) that about 10 absorbed photons are sufficient to yield one molecule of dioxygen by photosynthesis; (b) that about 10^{10} photons are needed for hemolysis of red blood cells; (c) that only one photon is needed to dissociate carboxyhemoglobin into deoxyHb and CO. The method requires the use of an photomultiplier and amplifier suitable for measuring light intensity as well as a reference standard to calibrate the system. **See also** *Spectral Responsivity; Spectral Effectiveness; Quantum Yield; Efficiency Spectrum; Excitation Spectrum*

[1] R. B. Setlow & E. C. Pollard (1962) *Molecular Biophysics*, pp. 267–351, Addison-Wesley, Reading, MA.

ACTIVATION

The increase in energy content of an atom, ion, or molecular entity or the process that makes an atom, ion, or molecular entity more active or reactive. In enzymology, activation often refers to processes that result in increased enzyme activity. For example, increasing temperature often can have a positive effect on enzyme activity (**See** *Arrhenius Equation*). Other examples of enzyme activation include: (1) proteolysis of zymogens; (2) alterations in ionic strength; (3) alterations due to pH changes; (4) activation in cooperative systems; (5) lipid or membrane interface activation; (6) metal ion effects; (7) autocatalysis; and (8) covalent modification.

Dixon and Webb[1] present an extensive consideration of activation mechanisms involving the reversible binding of an activator (less often termed an "agonist") to the enzyme. Nonessential activation refers to enzyme-dependent processes that can convert substrate(s) to product(s) in the absence of the activator, albeit at a slower rate. Essential activators are molecular entities that are required by the enzyme in the catalysis of a reaction. In a sense, essential activators are similar to second (or third) substrates, albeit they are not converted to products. An example of an essential activator might be an enzyme that requires the binding of a metal ion for catalysis to proceed. Below are a few cases of essential activation.

Essential Activation in a Uni Uni Mechanism–Type I (Activator Binds First). Consider the following reaction scheme in which an activator (A) binds to the free enzyme (E) prior to the binding of the substrate (S):

$$\underline{{}_{E}A \downarrow S \downarrow (ESA \leftrightarrow EPA) \downarrow P \downarrow A_{E}}$$

For the rapid equilibrium case, all binding steps are fast relative to the ESA-to-EPA interconversion step. Thus, $K_A = [E][A]/[EA]$, $K_S = [EA][S]/[ESA]$, and $K_P = [EA][P]/[EPA]$. The initial rate expression ($[P] 0$) for this case is

$$v = V_{max,f}[A][S]/\{K_S(K_A + [A]) + [A][S]\}$$

where $V_{max,f} = k_5[E_{total}]$ where k_5 is the forward rate constant for the ESA → EPA interconversion. Note that, if [A] is saturating (i.e., [A] >> K_A), then the equation reduces to the normal Michaelis-Menten expression. A double-reciprocal plot (1/v versus 1/[A]) at different, nonsaturating levels of the activator will yield a series of straight lines having a common intersection point on the vertical axis (at $1/v = 1/V_{max,f}$).

The steady-state expression for this scheme (where [P] ≈ 0 and the forward rate constants are k_1, k_3, k_5, and k_7 for the E + A → EA, the EA + → ESA, the ESA → EPA, and the EPA → EA + P steps, respectively, and k_2, k_4, k_6, and k_8 are the corresponding rate constants for the reverse reactions of those steps, respectively) is the same general expression as presented for the rapid equilibrium case except where $V_{max,f} = k_5k_7[E_{total}]/(k_5 + k_6 + k_7)$, $K_A = k_2/k_1$, and $K_S = (k_4(k_6 + k_7) + k_5k_7)/(k_3(k_5 + k_6 + k_7))$.

Essential Activation in a Uni Uni Mechanism—Type II (Activator Binds Second). In this scheme, the essential activator can only bind to the enzyme-substrate or enzyme-product binary complexes:

$$\underset{E}{\underline{\quad S\downarrow\ A\downarrow (ESA \leftrightarrow EPA)\downarrow A\ \downarrow P\quad}}_{E}$$

Such a mechanism is a form of substrate-induced activation. If all of the binding steps are rapid relative to the ESA-to-EPA interconversion step, the initial-rate rapid-equilibrium equation for this scheme is

$$v = V_{max,f}[A][S]/\{K_AK_S + K_A[S] + [A][S]\}$$

where $K_S = [E][S]/[ES]$, $K_A = [ES][A]/[ESA]$, and $V_{max,f} = k_5[E_{total}]$ in which k_5 is the forward rate constant for the ESA → EPA conversion. Note the difference in this expression from the rapid equilibrium expression for the previous Type I scheme. In the earlier scheme, there is a $K_S[A]$ term in the denominator whereas there is a $K_A[S]$ term in this scheme. Hence, a double-reciprocal plot (1/v versus 1/[S]) will yield a series of straight lines intersecting at a common point not on the vertical axis. This intersection point will have coordinates of $1/[S] = -1/K_S$ and $1/v = 1/V_{max,f}$. Hence, the rapid equilibrium Type II scheme should be readily distinguishable from the previous scheme.

The steady-state expression for this same scheme (in which the forward rate constants are k_1, k_3, k_5, k_7, and

k_9 for the E + S → ES, the ES + A → ESA, the ESA → EPA, the EPA → EP + A steps, and EP → E + P respectively, and k_2, k_4, k_6, k_8, and k_{10} are the corresponding rate constants for the reverse reactions of those steps, respectively), in the absence of product is $v = V_{max,f}[A][S]/\{K_SK_A + K_S[A] + K_A[S] + [A][S] + [A]^2[S]/K_{IA}\}$ where $V_{max,f} = k_3k_5k_7k_9[E_{total}]/\Phi$, $K_S = (k_2k_4k_6k_8 + k_3k_5k_7k_9)/(k_1\Phi)$, $K_A = k_9(k_4(k_6 + k_7) + k_5k_7)/\Phi$, $1/K_{IA} = k_3k_8(k_5 + k_6)/\Phi$, and $\Phi = k_4k_6k_8 + k_3(k_9(k_6 + k_7) + k_5(k_7 + k_9))$. A double-reciprocal plot of such a steady-state mechanism, at varying, constant concentrations of the activator, will yield a series of straight lines. A replot of the slopes (slopes vs. 1/[A]) will provide a value for K_A. However, a replot of the vertical intercepts (intercepts vs. 1/[A]) will be nonlinear (curving up at elevated levels of the activator), depending on the relative magnitude of the squared term.

Essential Activation in a Uni Uni Mechanism—Type III (Activator and Substrate Bind Randomly).

$$\underset{E}{\underline{\quad A\ or\ S\downarrow\downarrow\ (ESA \leftrightarrow EPA)\ \downarrow\downarrow P\ orA\quad}}_{E}$$

where "A or P double arrows down" indicates random binding in two successive steps. The rapid equilibrium expression for this scheme, in which the sole rate-determining step is the ESA-to-EPA interconversion, is

$$v = V_{max,f}[A]/[S]/\{K_{iS}K_A + K_A[S] + K_S[A] + [A][S]\}$$

where $V_{max,f} = k_9[E_{total}]$, $K_{iS} = [E][S]/[ES]$, $K_A = [ES][A]/[ESA]$, $K_S = [EA][S]/[ESA]$, and k_9 is the forward rate constant for the ESA → EPA conversion. Double-reciprocal plots (1/v versus 1/[S] or 1/[A]) will yield a series of straight lines. In the 1/v versus 1/[S] plot, the intercept replot will provide a value for K_A and $V_{max,f}$. The intercept replot of the 1/v versus 1/[A] data will provide a value for K_S. The slope replots will then yield K_{iS}. In the standard double-reciprocal primary plots (i.e., 1/v vs. 1/[S]), the lines will intersect at a common point having coordinates of $1/[S] = -1/K_{iS}$ and $1/v = (1/V_{max,f})(1 - (K_S/K_{iS}))$. Hence, the point-of-intersection will be in the second quadrant if $K_S >> K_{iS}$, in the third quadrant if $K_S << K_{iS}$, and on the horizontal axis if $K_S = K_{iS}$.

The steady-state expression for v for this scheme is rather complex, containing squared terms in [S] and in [A] in the numerator as well as terms containing $[S]^2$, $[A]^2$, $[S]^2[A]$, $[S][A]^2$, $[A]^3$, $[S]^2[A]^2$, and $[S][A]^3$ in the denominator.

NONESSENTIAL ACTIVATION. In these cases, if the nonessental activator can bind to the free enzyme, the resulting equations are very similar in format to kinetic expressions for partial inhibition. Consider the following scheme for a Uni Uni mechanism in which a nonessential activator (A) can bind to the free enzyme. Thus, E + S ⇌ ES ⇌ EP ⇌ E + P and E + A ⇌ EA is followed by EA + S ⇌ ESA ⇌ EPA ⇌ EA + P. In addition, one can have ES + A ⇌ ESA as well as EP + A ⇌ EPA. Considering the rapid equilibrium case; *i.e.*, where all binding steps are rapid relative to the interconversion steps (ES ⇌ EP and ESA ⇌ EPA) and can be described by dissociation constants. Thus, $K_S = [E][S]/[ES]$, $K_A = [E][A]/[EA]$, $K_{AS} = [EA][S]/[ESA]$, $K_P = [E][P]/[EP]$, and $K_{AP} = [EA][P]/[EPA]$. In this scheme, the resulting rate expression will have the same general format irrespective of whether the activator can bind to the ES or EP complexes. Under initial rate conditions (*i.e.*, $[P] \approx 0$) and letting $k_{11} = \alpha k_3$ (where k_3 is the forward rate constant for the ES → EP conversion and k_{11} is the corresponding rate constant for the ESA → EPA conversion), and remembering that $\alpha > 1.0$ for an activator, then $v = V_{\max,f}[S](1 + (\alpha[AK_S/(K_A K_{AS}))) \div \{K_S + [S] + ([A]K_S/K_A)(1 + [S]/K_{AS})\}\}$ where $V_{\max,f} = k_3[\mathrm{E_{total}}]$. Note that, if $[A] = 0$, the equation reduces to the standard Michaelis-Menten equation. Plotting protocols are available that allow one to graphically determine the various kinetic parameters of this, and related, systems.

[1]M. Dixon & E. C. Webb (1979) *Enzymes*, 3rd ed., Academic Press, New York.

ACTIVATION, CALMODULIN-DEPENDENT

Upon binding calcium ions, the small acidic protein known as calmodulin can activate enzymes by binding to a wide variety of proteins containing calmodulin-binding domains. Such proteins include cAMP phosphodiesterase, calmodulin-dependent nitric oxide synthase, calmodulin kinases, the plasma membrane calcium pump, calcineurin, and calmodulin-dependent inositol-(1,4,5)-trisphosphate 3-kinase. *See also Activation; Autoinhibition*

P. James, T. Vorherr & E. Carafoli (1995) *Trends in Biochem. Sci.* **20**, 38.

ACTIVATOR

A substance, agent, or factor (other than a catalyst) the presence of which will increase the rate of a catalyzed reaction. A substance that activates an enzyme-catalyzed reaction by binding to the enzyme is often referred to as an enzyme activator. *See Activation; Allosteric Effector; Linked Functions; London-Steck Plot*

ACTIVE SITE

The topologically defined region(s) on an enzyme responsible for the binding of substrate(s), coenzymes, metal ions, and protons that directly participate in the chemical transformation catalyzed by an enzyme, ribozyme, or catalytic antibody. Active sites need not be part of the same protein subunit, and covalently bound intermediates may interact with several regions on different subunits of a multisubunit enzyme complex. *See Lambda (Λ) Isomers of Metal Ion-Nucleotide Complexes; Lock and Key Model of Enzyme Action; Low-Barrier Hydrogen Bonds: Role in Catalysis; Yaga-Ozawa Plot; Yonetani-Theorell Plot; Induced-Fit Model; Allosteric Interaction*

ACTIVE-SITE TITRATION

The reaction of a chemically active substance with a group located in an enzyme's active site, such that the stoichiometry of active sites can be determined analytically. *See Acetylcholinesterase; Affinity Labeling; Diisopropylfluorophosphate*

Selected entries from *Methods in Enzymology* [vol, page(s)]: Collagenase, **248**, 100–101, 502–503; DNA polymerase, **249**, 47; gelatinase, **248**, 97, 100–101, 474, 502–503; glycyl endopeptidase, **244**, 543–544; Kex2 protease, **244**, 158; matrix metalloproteinases, **248**, 502–503; myeloblastin, **244**, 63; peptidases, **248**, 85–101; stromelysin, **248**, 100–101, 460–461, 502–503; trypsinlike enzymes, **248**, 14.

ACTIVE TRANSPORT

Any process whereby a substance is transported from a region of low concentration to a region of higher concentration by coupling the otherwise unfavorable $\Delta G_{\mathrm{diffusion}}$ with the highly favorable $\Delta G_{\mathrm{hydrolysis}}$ for ATP conversion to ADP and orthophosphate. Other metabolites displaying high group transfer potential can replace ATP in the hydrolase-type reaction, as long as sufficient free energy is liberated. Contrary to generally held ideas that active transport requires a membrane, one can also have active transport in a single compartment, such as the cytoplasm. The only condition that must be fulfilled is that the transport occurs against a concentration gradient. **See** *Membrane Transporters; Ion Pumps;* $\Delta G_{diffusion}$

ACTIVITY

1. With respect to an enzyme, the rate of substrate-to-product conversion catalyzed by an enzyme under a given set of conditions, either measured by the amount of substance (*e.g.*, micromoles) converted per unit time or by concentration change (*e.g.*, millimolarity) per unit time. **See** *Specific Activity; Turnover Number.* 2. Referring to the measure of a property of a biomolecule, pharmaceutical, procedure, *etc.*, with respect to the response that substance or procedure produces. 3. **See** *Optical Activity.* 4. The amount of radioactive substance (or number of atoms) that disintegrates per unit time. **See** *Specific Activity.* 5. A unitless thermodynamic parameter which is used in place of concentration to correct for nonideality of gases or of solutions. The absolute activity of a substance B, symbolized by λ_B, is related to the chemical potential of B (symbolized by μ_B) by the relationship $\mu_B = RT \ln \lambda_B$ where R is the universal gas constant and T is the absolute temperature. The ratio of the absolute activity of some substance B to some absolute activity for some reference state, $\lambda_B^{\ominus}$, is referred to as the relative activity (usually simply called "activity"). The relative activity is symbolized by a and is defined by the relationship $a_B = \lambda_B/\lambda_B^{\ominus} = e^{(\mu_B \mu_B^{\ominus})/RT}$. If the states are referenced with respect to Raoult's Law, the relative activity is symbolized by a. However, if referenced with respect to Henry's Law, the activities can have a molality, a concentration, or a mole fraction basis and are symbolized by a_m, a_c, and a_x, respectively. **See also** *Activity Coefficient; Chemical Potential; Fugacity; Turnover Number; Latent Activity; Biomineralization*

ACTIVITY COEFFICIENT

A unitless correction factor that relates the relative activity of a substance to the quantity of the substance in a mixture. Activity coefficients are frequently determined by emf (electromotive force) or freezing-point depression measurements. At infinite dilution, the activity coefficient equals 1.00. Activity coefficients for electrolytes can vary significantly depending upon the concentration of the electrolyte. Activity coefficients can exceed values of 1.00. For example, a 4.0 molal HCl solution has a coefficient of 1.76 and a 4.0 molal LiCl has a value of

1.46. The magnitude of the coefficient reflects the electric charge distribution of the ionic species. A 0.1 molal solution of $Al_2(SO_4)_3$ has an activity coefficient of only 0.035. It should also be noted that, in dilute solutions, activity coefficients of electrolytes decrease in magnitude with increasing concentration. A minimum is reached and the coefficient then increases with concentration. **See** *Activity; Debye-Hückel Law; Biomineralization*

ACTOMYOSIN ATPase

The ATPase enzyme activity of actomyosin has been assigned the classification number EC 3.6.1.32. Myosin catalyzes the hydrolysis of ATP to ADP and orthophosphate. In the absence of actin, myosin is a more feeble ATPase. **See also** *Myosin*

A. Oplatka (1997) *Crit. Rev. Biochem. Mol. Biol.* **32**, 307.

D. A. Schafer & J. A. Cooper (1995) *Ann. Rev. Cell Dev. Biol.* **11**, 497.

P. J. McLaughlin & A. G. Weeds (1995) *Ann. Rev. Biophys. Biomol. Struct.* **24**, 643.

J. Condeelis (1993) *Ann. Rev. Cell Biol.* **9**, 411.

F. Solomon (1991) *Ann. Rev. Cell Biol.* **7**, 633.

J. M. Squire (1990) *Molecular Mechanisms of Muscular Contraction*, CRC Press, Boca Raton.

W. P. Jencks (1980) *Adv. Enzymol.* **51**, 75.

ACYL–ACYL-CARRIER-PROTEIN Δ^9-DESATURASE

This enzyme catalyzes the NADPH- and dioxygen-dependent insertion of *cis* double bonds into the methylene region of fatty acyl structures covalently attached to the phosphopantetheine portion of an acyl carrier protein.

B. G. Fox (1998) *Comprehensive Biological Catalysis: A Mechanistic Reference* **3**, 261.

I. Nishida & N. Murata (1996) *Ann. Rev. Plant Physiol. Plant Mol. Biol.* **47**, 541.

ACYLAL INTERMEDIATE

An enzymatic reaction intermediate formed by acylation of an acetal hydroxyl group. Such an intermediate is thought to occur in a number of reactions involving carbohydrates. The sucrose phosphorylase reaction is thought to proceed by way of an acyl-glucosyl intermediate.

ACYLAMINOACYL-PEPTIDASE

This enzyme [EC 3.4.19.1], also known as acylamino-acid releasing enzyme and *N*-acylpeptide hydrolase, catalyzes the hydrolysis of an acylaminoacyl-peptide to generate an acylamino acid and the free peptide. Catalysis is most efficient when the P1 site is occupied by a seryl, alanyl, or methionyl residue. The enzyme is a poor catalyst if P1 is a glycyl, tyrosyl, aspartyl, asparaginyl, or prolyl residue. The EC number accounts for a group of similar enzymes that liberate *N*-acetyl or *N*-formyl amino acids from proteins and peptides.

W. M. Jones, A. Scaloni & J. M. Manning (1994) *Meth. Enzymol.* **244**, 227.

ACYL-CoA:CHOLESTEROL *O*-ACYLTRANSFERASE

This enzyme [EC 2.3.1.26], also known as sterol *O*-acyltransferase, sterol-ester synthase, and cholesterol acyltransferase, catalyzes the reaction of an acyl-coenzyme A derivative with cholesterol to produce coenzyme A and the cholesterol ester. The animal enzyme is highly specific for transfer of acyl groups having a single *cis* double bond at C9.

T.-Y. Chang, C. C. Y. Chang & D. Cheng (1997) *Ann. Rev. Biochem.* **66**, 613.
T.-Y. Chang & G. M. Doolittle (1983) *The Enzymes*, 3rd ed., **16**, 523.

ACYL-CoA DEHYDROGENASES

A number of enzymes given this name catalyze the formation of a 2,3-dehydroacyl-coenzyme A derivative from the acyl-CoA substrate. Acyl-CoA dehydrogenase ($NADP^+$) [EC 1.3.1.8], also known as 2-enoyl-CoA reductase, utilizes $NADP^+$. The liver enzyme acts on enoyl-CoA substrates having a carbon chain length of 4 to 16 carbons, with an optimum activity on 2-hexenoyl-CoA. In *E. coli*, both *cis*-specific and *trans*-specific enzymes exist (EC 1.3.1.37, *cis*-2-enoyl-CoA reductase (NADPH), and EC 1.3.1.38, *trans*-2-enoyl-CoA reductase (NADPH)). Medium-chain acyl-CoA dehydrogenase [EC 1.3.99.3] utilizes an electron-transferring flavoprotein (requiring FAD) instead of $NADP^+$. ***See also*** *Butyryl-CoA Dehydrogenase; Long-Chain Acyl-CoA Dehydrogenase*

B. A. Palfey & V. Massey (1998) *Comprehensive Biological Catalysis: A Mechanistic Reference* **3**, 83.
M. A. Ator & P. R. Ortez de Montellano (1990) *The Enzymes*, 3rd ed. **19**, 213.

ACYL-CoA DESATURASE

This enzyme [EC 1.14.99.5], also known as stearoyl-CoA desaturase, fatty acid desaturase, and Δ^9-desaturase, will catalyze the reaction of stearoyl-CoA with a hydrogen donor and dioxygen to produce oleoyl-CoA, water, and the oxidized factor. The rat liver enzyme is an enzyme system utilizing cytochrome b_5 and cytochrome b_5 reductase. ***See also*** *Desaturases*

I. Nishida & N. Murata (1996) *Ann. Rev. Plant Physiol. Plant Mol. Biol.* **47**, 541.

ACYL-CoA:RETINOL *O*-ACYLTRANSFERASE

This enzyme [EC 2.3.1.76], also referred to as retinol fatty-acyltransferase, catalyzes the reaction of an acyl-CoA derivative with retinol to generate coenzyme A and the retinyl ester. The CoA derivative can be palmitoyl-CoA or other long-chain fatty-acyl derivatives of coenzyme A.

R. Blomhoff, M. H. Green & K. R. Norum (1992) *Ann. Rev. Nutr.* **12**, 37.
J. C. Saari & D. L. Bredberg (1990) *Meth. Enzymol.* **190**, 156.
A. C. Ross (1990) *Meth. Enzymol.* **189**, 442.

ACYL ENZYME INTERMEDIATES

An enzyme reaction intermediate (Enz—O—C(O)R or Enz—S—C(O)R), formed by a carboxyl group transfer (*e.g.*, from a peptide bond or ester) to a hydroxyl or thiol group of an active-site amino acyl residue of the enzyme. Such intermediates are formed in reactions catalyzed by serine proteases[1–4], transglutaminase[5], and formylglycinamide ribonucleotide amidotransferase[6]. Acyl-enzyme intermediates often can be isolated at low temperatures, low pH, or a combination of both[4]. For acyl-seryl derivatives, deacylation at a pH value of 2 is about 10^5-fold slower than at the optimal pH[4]. A primary isotope effect can frequently be observed with a ^{13}C-labeled substrate. If an amide substrate is used, it is possible that a secondary isotope effect may be observed as well[7]. ***See also*** *Active Site Titration; Serpins (Inhibitory Mechanism)*

[1] A. L. Fink & M. H. Geeves (1979) *Meth. Enzymol.* **63**, 336.
[2] P. Douzou (1977) *Cryoenzymology: An Introduction.*, Academic Press, New York.
[3] A. L. Fink & G. A. Petsko (1981) *Adv. Enzymol.* **52**, 177.
[4] R. J. Coll, P. D. Compton & A. L. Fink (1982) *Meth. Enzymol.* **87**, 66.
[5] J. E. Folk (1982) *Meth. Enzymol.* **87**, 36.
[6] J. M. Buchanan (1982) *Meth. Enzymol.* **87**, 76.
[7] W. W. Cleland (1982) *Meth. Enzymol.* **87**, 625.

N-ACYLGLUCOSAMINE 2-EPIMERASE

This enzyme [EC 5.1.3.8], also known as *N*-acetylglucosamine 2-epimerase, catalyzes the epimerization of *N*-acylglucosamine to *N*-acylmannosamine. ATP is reported to be required in this reaction.

M. E. Tanner & G. L. Kenyon (1998) *Comprehensive Biological Catalysis: A Mechanistic Reference* **2**, 7.
E. Adams (1976) *Adv. Enzymol.* **44**, 69.
L. Glaser (1972) *The Enzymes*, 3rd ed., **6**, 355.

N-ACYLGLUCOSAMINE-6-PHOSPHATE 2-EPIMERASE

This enzyme [EC 5.1.3.9], also known as *N*-acetylglucosamine-6-phosphate 2-epimerase, catalyzes the interconversion of *N*-acylglucosamine 6-phosphate to *N*-acylmannosamine 6-phosphate.

M. E. Tanner & G. L. Kenyon (1998) *Comprehensive Biological Catalysis: A Mechanistic Reference* **2**, 7.
E. Adams (1976) *Adv. Enzymol.* **44**, 69.
L. Glaser (1972) *The Enzymes*, 3rd ed., **6**, 355.

1-ACYLGLYCEROL-3-PHOSPHATE ACYLTRANSFERASE

This enzyme catalyzes the reaction of an acyl-CoA derivative with 1-acyl-*sn*-glycerol 3-phosphate to generate coenzyme A and 1,2-diacyl-*sn*-glycerol 3-phosphate. The animal enzyme is reported to be specific for the transfer of unsaturated fatty acyl groups. Interestingly, the acyl-[acyl-carrier-protein] can also act as an acyl donor.

R. M. Bell & R. A. Coleman (1983) *The Enzymes*, 3rd ed., **16**, 87.
J. D. Esko & C. R. H. Raetz (1983) *The Enzymes*, 3rd ed., **16**, 207.

ACYLGLYCERONE-PHOSPHATE REDUCTASE

This enzyme [EC 1.1.1.101], also known as palmitoyldihydroxyacetone-phosphate reductase, catalyzes the reaction of 1-palmitoylglycerol 3-phosphate and NADP$^+$ to produce palmitoylglycerone phosphate and NADPH. The enzyme can also utilize as substrates alkylglycerone 3-phosphate and alkylglycerol 3-phosphate.

A. K. Hajra, S. C. Datta & M. K. Ghosh (1992) *Meth. Enzymol.* **209**, 402.

1-ACYLGLYCERO-3-PHOSPHOCHOLINE O-ACYLTRANSFERASE

This enzyme [EC 2.3.1.23], also called lysolecithin acyltransferase and lysophosphatidylcholine acyltransferase, catalyzes the reaction of an acyl-CoA derivative with 1-acyl-*sn*-glycero-3-phosphocholine to yield coenzyme A and 1,2-diacyl-*sn*-glycero-3-phosphocholine. The enzyme preferentially acts on unsaturated acyl-CoA derivatives, but 1-acyl-*sn*-glycero-3-phosphoinositol can also act as the acceptor.

P. C. Choy, P. G. Tardi & J. J. Mukherjee (1992) *Meth. Enzymol.* **209**, 90.
J. D. Esko & C. R. H. Raetz (1983) *The Enzymes*, 3rd ed., **16**, 207.

ACYL-LYSINE DEACYLASE

This enzyme [EC 3.5.1.17] catalyzes the hydrolysis of N^6-acyllysine to generate a fatty acid anion and lysine.

I. Chibata, T. Ishikawa & T. Tosa (1970) *Meth. Enzymol.* **19**, 756.

N-ACYLMANNOSAMINE KINASE

This enzyme [EC 2.7.1.60] catalyzes the ATP-dependent phosphorylation of *N*-acylmannosamine to yield ADP and *N*-acylmannosamine 6-phosphate. The enzyme can act on both the acetyl and the glycolyl derivatives.

W. Kundig & S. Roseman (1966) *Meth. Enzymol.* **8**, 195.

N-ACYLNEURAMINATE CYTIDYLYLTRANSFERASE

This enzyme [EC 2.7.7.43], also referred to as CMP-*N*-acetylneuraminic acid synthetase and CMP-sialate synthase, catalyzes the conversion of CTP and *N*-acylneuraminate to yield pyrophosphate and CMP-*N*-acylneuraminate. The protein will act on both the *N*-acetyl and *N*-glycolyl derivatives.

E. L. Kean & S. Roseman (1966) *Meth. Enzymol.* **8**, 208.

N-ACYLNEURAMINATE 9-PHOSPHATASE

This enzyme [EC 3.1.3.29] catalyzes the hydrolysis of *N*-acylneuraminate 9-phosphate to produce *N*-acylneuraminate and phosphate.

G. W. Jourdian, A. Swanson, D. Watson & S. Roseman (1966) *Meth. Enzymol.* **8**, 205.

N-ACYLNEURAMINATE 9-PHOSPHATE SYNTHASE

This enzyme [EC 4.1.3.20] catalyzes the reaction of *N*-acylneuraminate 9-phosphate with phosphate to produce *N*-acetylmannosamine 6-phosphate, phosphoenolpyruvate, and water. The protein will act on *N*-glycoloyl and *N*-acetyl derivatives.

D. Watson, G. W. Jourdian & S. Roseman (1966) *Meth. Enzymol.* **8**, 201.

ACYLPHOSPHATASE

This enzyme [EC 3.6.1.7] catalyzes the hydrolysis of an acyl phosphate to yield a fatty acid anion and orthophosphate.

I. Harary (1963) *Meth. Enzymol.* **6**, 324.

ACYL-PHOSPHATE–HEXOSE PHOSPHOTRANSFERASE

This enzyme [EC 2.7.1.61] catalyzes the reaction of an acyl phosphate with a hexose to produce an acid and hexose phosphate. If the sugar is D-glucose or D-mannose, phosphorylation is on O6. If the sugar is D-fructose, phosphorylation is on O1 or O6.

R. L. Andersin & M. Y. Karmel (1966) *Meth. Enzymol.* **9**, 392.

ACYL-PHOSPHATE INTERMEDIATE

An enzymatic reaction intermediate formed by phosphoryl transfer to a carboxyl group on an enzyme. Acyl-phosphates are structurally analogous to acid anhydrides ($R-CO-O-CO-R'$), and they are thermodynamically less stable than either of the two phosphoanhydride bonds in ATP. This is evident by the fact that the acetate kinase reaction (ADP + acetyl-phosphate = ATP + acetate) favors ATP formation with an equilibrium constant of about 3,000. Acetyl-phosphate can be chemically synthesized by reacting orthophosphate with acetic anhydride.

Acyl-phosphates can participate in so-called in-line phosphoryl transfer reactions, and depending on reaction conditions, acyl-phosphate compounds can form a metaphosphate anion:

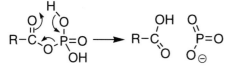

Metaphosphate Formation from Acyl-P

Acyl-phosphates also display an unusual "U"-shaped pH profile for the rate of hydrolysis (see Fig. 1). This reflects the fact that acyl-phosphate compounds are susceptible to both acid- and base-catalyzed hydrolysis, and the shoulder at less acidic pH values reflects the slightly faster hydrolysis of the monoanion, compared to the dianion. Observation of a "U"-shape pH-rate profile for a previously uncharacterized phosphoryl-enzyme intermediate supports the inference that an acyl-phosphate has formed.

Acyl-phosphate compounds also undergo hydroxaminolysis by hydroxylamine ($HO–NH_2$). Initial attack occurs by the oxygen atom of hydroxylamine, and the O-linked product reacts with a second molecule of hydroxylamine to yield the more stable N-acylated acid hydroxamate.

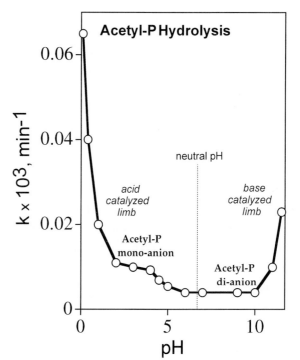

Figure 1. Acid- and base-catalyzed hydrolysis of acetyl-phosphate. From G. Di Sabato & Jencks (1961) *J. Am. Chem. Soc.* **83**, 4400; reproduced with permission of the authors and the American Chemical Society.

If an acyl-phosphate intermediate involves a side-chain carboxyl of glutamyl or aspartyl residue on the enzyme, the corresponding hydroxamate should undergo dinitrophenylation with Sanger's reagent, and Lössen rearrangement should decarboxylate the aspartate or glutamate. Likewise, an acyl-phosphate intermediate involving a side-chain carboxyl of glutamyl or aspartyl residue on the enzyme should undergo nucleophilic reduction by sodium [³H]borohydride to form the the corresponding ω-hydroxy-α-amino acids after acid hydrolysis of the protein. [Note: If an enzyme-catalyzed reaction involves a thiol-ester intermediate, these activated carboxyl compounds are also susceptible to the nucleophilic reactions cited above.] ***See*** *D-Alanine-D-Alanine Ligase; Mapping Substrate Interactions Using Substrate Data*

ACYL-SERINE INTERMEDIATE

An enzymatic reaction intermediate formed by carboxyl group transfer to a hydroxyl group of a serine residue within the active site.

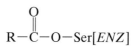

Such an intermediate is known to be formed in reactions catalyzed by trypsin, chymotrypsin, thrombin, other enzymes of the blood-clotting cascade (except angiotensin-converting enzyme, which is an aspartic protease). An acyl-serine intermediate is also formed in the acetylcholinesterase reaction. The active site serine of this enzyme and the serine proteases can be alkylated by diisopropyl-fluorophosphate. **See also** *Active Site Titration*

ADAIR CONSTANTS

The thermodynamic association (or dissociation) constants used in the Adair equation for a ligand binding at sites in a multisite protein. The term "Adair constants" originally referred only to the four constants for the reversible binding of dioxygen to hemoglobin. **See** *Adair Equation*

ADAIR EQUATION

An equation presented by Gilbert Adair in 1924 to describe the binding of molecular oxygen to hemoglobin. Let the microsccopic dissociation constant for the binding of O_2 to Hb be represented by k_1, the microscopic dissociation constant for O_2 binding to the $Hb(O_2)$ complex be represented by k_2, and k_3 and k_4 are the corresponding microscopic dissociation constants for the third and fourth O_2, respectively. Thus, in terms of macroscopic dissociation constants, $k_1 = 4K_1$, $k_2 = (3/2)K_2$, $k_3 = (2/3)K_3$, and $k_4 = (1/4)K_4$, in which the coefficients are statistical factors derived from the number of available binding sites (and sites from which an already-bound ligand can leave). If Y_S represents the fractional saturation of the ligand binding (*i.e.*, fraction of occupied sites divided by total concentration of all sites), then

$$Y_S = \{([O_2]/k_1) + (3[O_2]^2/(k_1 k_2)) + (3[O_2]^3/(k_1 k_2 k_3)) + ([O_2]^4/(k_1 k_2 k_3 k_4))\} \div \{1 + (4[O_2]/k_1) + (6[O_2]^2/(k_1 k_2)) + (4[O_2]^3/(k_1 k_2 k_3)) + ([O_2]^4/(k_1 k_2 k_3 k_4))\}$$

Analogous equations can be derived for the binding of other ligands to multisite proteins.

If the microscopic dissociation constants are not equal, then the Adair equation predicts a nonhyperbolic (*i.e.*, cooperative) curve. Values for k_1 and k_4 can be obtained by studying ligand binding at very low and very high ligand concentrations. Values for k_2 and k_3 can be ob-

tained from computer fitting protocols or by using a Hill plot.

The general Adair equation for the binding of a ligand X to a multisite protein where K_i represents the thermodynamic macroscopic association constant for the *i*th site and where n is the total number of sites is

$$nY_S = \left\{ \sum_{i=1}^{n} (i[X]^i \prod_{j=1}^{i} (1/K_j)) \right\} \div \left\{ 1 + \sum_{i=1}^{n} ([X]^i \prod_{j=1}^{i} (1/K_j)) \right\}$$

Note that the Adair equation does not provide a reason for why identical sites would have different dissociation (or association) constants. **See** *Allosterism; Cooperativity; Koshland-Nemethy-Filmer Model*

Selected entries from *Methods in Enzymology* [vol, page(s)]: Apparent fractional saturation, **232**, 615-618; global parameters, **232**, 625-626, 631-632; overview, **232**, 606-611; oxygen electrode data, treatment, **232**, 611-615; hemoglobin dimer-tetramer equilibrium in, **232**, 597-606; to oxygen equilibrium curve of hemoglobin; binding scheme, **232**, 560-563; importance of bottom and top data, **232**, 569-572; conversion of absorbance values to oxygen saturation values, **232**, 566-567; errors of best-fit Adair constant values, **232**, 572-574; extrapolation procedure, **232**, 566-567; nonlinear least-squares method, **232**, 565-566; statistical weighting, **232**, 567-569.

ADDITION REACTION

A general type of chemical reaction between two compounds, A and B, such that there is a net reduction in bond multiplicity (*e.g.*, addition of a compound across a carbon-carbon double bond such that the product has lost this π-bond). An example is the hydration of a double bond, such as that observed in the conversion of fumarate to malate by fumarase. Addition reactions can also occur with strained ring structures that, in some respects, resemble double bonds (*e.g.*, cyclopropyl derivatives or certain epoxides). A special case of a hydro-alkenyl addition is the conversion of 2,3-oxidosqualene to dammaradienol or in the conversion of squalene to lanosterol. Reactions in which new moieties are linked to adjacent atoms (as is the case in the hydration of fumarate) are often referred to as 1,2-addition reactions. If the atoms that contain newly linked moieties are not adjacent (as is often the case with conjugated reactants), then the reaction is often referred to as a 1,*n*-addition reaction in which *n* is the numbered atom distant from 1 (*e.g.*, 1,4-addition reaction). In general, addition reactions can take place via electrophilic addition, nucleophilic addition, free-radical addition, or via simultaneous or pericyclic addition.

ADDITIVITY PRINCIPLE

The principle that different structural domains, moieties, or features of a molecular substance contribute separately and additively to a property of a substance. In 1840, G. H. Hess introduced the Law of Constant Heat Summation, a relation that allows one to calculate the heat of a reaction from collected measurements of seemingly different reactions, as long as the summation of a series of reactions yields the same overall chemical reaction as the one of interest. Thermodynamic additivity requires that if two components, A and B, contribute independently to some process, then the total change in free energy (or enthalpy or entropy) is the sum of components, $\Delta G = \Delta G_A + \Delta G_B$. In view of its broad use in examining chemical and physical principles, Benson[1] has even offered the view that additivity is the "fourth" law of thermodynamics.

This principle has been utilized to assess intrinsic subsite binding energies for enzymes that have substrate binding subsites and exosites topologically distant from the reactive site (e.g., polysaccharide depolymerases[2] and proteinases).

Dill[3] recently discussed the merits and limitations of models that assume thermodynamic additivity and independence (of energy types, of neighbor interactions, of conformational freedom, of monomer contact pairing frequencies, etc.). He states that biological molecules may achieve stability in the face of thermal uncertainty, as polymers do, by compounding many small interactions; this summing can stump modelers because application of the additivity principle leads to accumulated error. Entropies and free energy may not be additive to describe weak interactions that are ensembles of states. He concludes that additivity principles appear to be few and limited in scope in biochemistry.

[1]S. W. Benson (1976) Thermochemical Kinetics: Methods for Estimation of Thermochemical Data and Rate Parameters, 2nd ed., Wiley, New York.
[2]J. D. Allen (1979) Meth. Enzymol. **64**, 248.
[3]K. A. Dill (1997) J. Biol. Chem. **272**, 701.

ADDUCT

A distinct, chemical species that is formed by direct combination between two or more molecules such that no structural changes have occurred "within" the original interacting chemical species. The resulting interactions between the species are often covalent or coordinate covalent. Examples include Lewis adducts and Meisenheimer adducts. The term "complex" usually refers to a looser association between interacting species. Intramolecular adducts are also possible. *See* Complex; Lewis Adduct; Meisenheimer Adduct

ADENINE DEAMINASE

This enzyme [EC 3.5.4.2], also called adenase, adenine aminohydrolase, and adenine aminase, catalyzes the hydrolysis of adenine to hypoxanthine and ammonia.

C. L. Zielke & C. H. Suelter (1971) The Enzymes, 3rd ed., **4**, 47.

ADENINE NUCLEOTIDE TRANSLOCASE

A nucleotide transporter (located in the outer mitochondrial membrane) that mediates one-for-one translocation/exchange of cytosolic ADP for mitochondrial ATP. This translocase is potently inhibited by atractyloside and bonkregic acid.

M. Klingenberg, E. Winkler & S.-G. Huang (1995) Meth. Enzymol. **260**, 369.
M. Klingenberg (1991) A Study of Enzymes **2**, 367.
P. V. Vignais, M. R. Block, F. Boulay, G. Brandolin, P. Dalbon & G. J. M. Lauquin (1985) Structure & Properties of Cell Membranes **2**, 139.

Selected entries from *Methods in Enzymology* [vol, page(s)]:
Abundance, **260**, 370; atractyloside loading, **260**, 376-377, 379; exchange kinetics, **260**, 377-379; fluorescence probes, **260**, 384-385; inhibitors, **260**, 371-372; nucleotide binding, **260**, 370-371; purification; carboxyatractylate loading, **260**, 373; gel filtration, **260**, 374-375; hydroxyapatite chromatography, **260**, 374; solubilization, **260**, 373; quantitation, **260**, 379; reconstitution into phospholipid vesicles, **260**, 375-376; bovine heart protein, **260**, 376; yeast protein, **260**, 376-377; structure, **260**, 371; translocation intermediates in mitochondrial import, **260**, 270-271; and uncoupling protein, comparison, **260**, 370.

ADENINE PHOSPHORIBOSYLTRANSFERASE

This enzyme [EC 2.4.2.7], also referred to as AMP pyrophosphorylase and transphosphoribosidase, catalyzes the reaction of AMP and pyrophosphate (or, diphosphate) to generate adenine and 5-phospho-α-ribose 1-diphosphate. In the reverse reaction, 5-amino-4-imidazolecarboxyamide can replace adenine.

J. G. Flaks (1963) Meth. Enzymol. **6**, 136.

ADENOSINE DEAMINASE

This enzyme [EC 3.5.4.4], also known as adenosine aminohydrolase, catalyzes the hydrolysis of adenosine to yield inosine and ammonia.

A. Radzicka & R. Wolfenden (1995) *Meth. Enzymol.* **249**, 284.
D. K. Wilson & F. A. Quiocho (1994) *Nature Struct. Biol.* **1**, 691.
J. J. Villafranca & T. Nowak (1992) *The Enzymes*, 3rd ed., **20**, 63.
D. K. Wilson, F. B. Rudolph & F. A. Quiocho (1991) *Science* **252**, 1278.
L. F. Thompson & J. E. Seegmiller (1980) *Adv. Enzymol.* **51**, 167.
C. L. Zielke & C. H. Suelter (1971) *The Enzymes*, 3rd ed., **4**, 47.

ADENOSINE KINASE

This enzyme [EC 2.7.1.20] catalyzes the ATP-dependent phosphorylation of adenosine to generate AMP and ADP. 2-Aminoadenosine can also act as a substrate for this enzyme.

P. A. Frey (1992) *The Enzymes*, 3rd ed., **20**, 141.
P. A. Frey (1989) *Adv. Enzymol.* **62**, 119.
E. P. Anderson (1973) *The Enzymes*, 3rd ed., **9**, 49.

S-ADENOSYLHOMOCYSTEINE LYASE

This enzyme [EC 3.3.1.1], also known as S-adenosyl-homocysteinase, catalyzes the hydrolysis of S-adenosyl-homocysteine to yield adenosine and homocysteine.

A. R. Clarke & T. R. Dafforn (1998) *Comprehensive Biological Catalysis: A Mechanistic Reference* **3**, 1.
M. A. Ator & P. R. Ortez de Montellano (1990) *The Enzymes*, 3rd ed. **19**, 213.
P. K. Chiang (1987) *Meth. Enzymol.* **143**, 377.
A. Guranowski & H. Jakubowski (1987) *Meth. Enzymol.* **143**, 430.

S-ADENOSYLHOMOCYSTEINE NUCLEOSIDASE

This enzyme [EC 3.2.2.9] catalyzes the hydrolysis of S-adenosylhomocysteine to generate adenine and S-ribosylhomocysteine. The enzyme will also act on 5'-methylthioadenosine to give adenine and 5-methyl-thioribose. ***See also*** *Methylthioadenosine Nucleoside Hydrolase*

S-ADENOSYL-L-METHIONINE

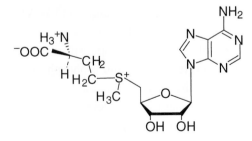

A sulfonium derivative, abbreviated AdoMet or SAM, that is primarily generated by the action of methionine adenosyltransferase (ATP + L-methionine + H_2O → SAM + P_i + pyrophosphate). SAM is a major methylat-

ing agent in the cell (***See*** *Methyl Transfer Reaction*). SAM has six chiral centers: four in the ribose moiety, the α-carbon of methionine, and the sulfonium sulfur. The sulfur in the biologically active form of SAM has the $S(+)$ configuration. This center readily undergoes epimerization; at a pH value of 7.5 and at 37°C, the half-life is about 12 hours[1]. Even the solid halide salt of SAM will undergo 12% decomposition within a week at 3°C. Commercial SAM is subject to as much as 10% loss per day at 25°C; therefore, care must be exercised whenever studying SAM-dependent systems. This lability is not significant in biological systems, because the turnover time for SAM biosynthesis and consumption is short (*e.g.*, less than 25 minutes in human and rat liver). Protein binding of SAM may also improve its *in vivo* stability[1]. Kinetic studies of SAM-dependent methyltransferases should use freshly prepared SAM. If solutions are to be stored, they should be kept for no longer than one month, preferably at −20°C under acidic conditions (typically below pH 2.5).

[1]S. E. Wu, W. P. Huskey, R. T. Borchardt & R. L. Schowen (1984) *J. Amer. Chem. Soc.* **106**, 5762.

S-ADENOSYLMETHIONINE DECARBOXYLASE

This enzyme [EC 4.1.1.50] catalyzes the pyruvate-dependent conversion of S-adenosylmethionine to (5-deoxy-5-adenosyl)(3-aminopropyl)methylsulfonium salt and carbon dioxide.

M. L. Hackert & A. E. Pegg (1998) *Comprehensive Biological Catalysis: A Mechanistic Reference* **2**, 201.
P. D. van Poelje & E. E. Snell (1990) *Ann. Rev. Biochem.* **59**, 29.
M. H. Stipanuk (1986) *Ann. Rev. Nutr.* **6**, 179.
C. W. Tabor & H. Tabor (1984) *Adv. Enzymol.* **56**, 251.
F. Schlenk (1983) *Adv. Enzymol.* **54**, 195.
E. A. Boeker & E. E. Snell (1972) *The Enzymes*, 3rd ed., **6**, 217.

S-ADENOSYLMETHIONINE HYDROLASE

This enzyme [EC 3.3.1.2], also referred to as S-adenosyl-methionine cleaving enzyme and methylmethionine-sulfonium-salt hydrolase, catalyzes the hydrolysis of S-adenosylmethionine to produce methylthioadenosine and homoserine. The enzyme will also convert methyl-methionine sulfonium salt to dimethyl sulfide and homoserine.

M. L. Gefter (1971) *Meth. Enzymol.* **17B**, 406.

ADENYLATE CYCLASE
(or, ADENYLYL CYCLASE)

This enzyme [EC 4.6.1.1], also known as adenylyl cyclase and 3′,5′-cyclic AMP synthetase, catalyzes the conversion of ATP to 3′,5′-cyclic AMP and pyrophosphate. The enzyme requires pyruvate as a cofactor and will also utilize dATP as a substrate (thereby producing 3′,5′-cyclic dAMP). In the presence of NAD(P)+-arginine ADP-ribosyltransferase, this enzyme is activated by covalent modification.

R. Taussig & A. G. Gilman (1995) *J. Biol. Chem.* **270**, 1.

K. A. Mintzer & J. Field (1995) *Meth. Enzymol.* **255**, 468.

P. A. Frey (1992) *The Enzymes*, 3rd ed., **20**, 141.

J. M. Stadel, A. De Lean & R. J. Lefkowitz (1982) *Adv. Enzymol.* **53**, 1.

ADENYLATE ISOPENTENYLTRANSFERASE

This enzyme [EC 2.5.1.27], also called 2-isopentenyl-diphosphate:AMP Δ^2-isopentenyltransferase and cytokinin synthase, catalyzes the reaction of AMP with Δ^2-isopentenyl diphosphate to yield pyrophosphate and N^6-(Δ^2-isopentenyl)adenosine 5′-monophosphate.

R. A. Gibbs (1998) *Comprehensive Biological Catalysis: A Mechanistic Reference* **1**, 31.

D. S. Letham & L. M. S. Palmi (1983) *Ann. Rev. Plant Physiol.* **34**, 163.

ADENYLATE KINASE

This enzyme [EC 2.7.4.3], also known as myokinase, catalyzes the reversible reaction of MgATP with AMP to produce MgADP and ADP. Inorganic triphosphate can also act as substrate with this enzyme. ***See*** *Energy Charge; Metal Ions in Nucleotide-Dependent Reactions*

M.-D. Tsai, R.-T. Jiang, T. Dahnke & Z. Shi (1995) *Meth. Enzymol.* **249**, 425.

L. A. Kleczkowski (1994) *Ann. Rev. Plant Physiol. Plant Mol. Biol.* **45**, 339.

J. J. Villafranca & T. Nowak (1992) *The Enzymes*, 3rd ed. **20**, 63.

M. Hamada, H. Takenaka, M. Sumida & S. A. Kuby (1991) *A Study of Enzymes* **2**, 403.

P. A. Frey (1989) *Adv. Enzymol.* **62**, 119.

A. S. Mildvan & D. C. Fry (1987) *Adv. Enzymol.* **59**, 241.

L. Noda (1973) *The Enzymes*, 3rd ed., **8**, 279.

Selected entries from *Methods in Enzymology* [vol, page(s)]:
As frequent contaminant in enzyme preparations, **64**, 24; a contaminant in phosphotransferases, **63**, 7; coupled assay for, **63**, 6, 32; inhibition, **63**, 7, 398, 401, 483; isotope exchange properties, **64**, 8; mass action ratio, **63**, 18; mechanism, **63**, 18; substrate, metal-ion complex, **63**, 259; AMP phosphorylation, **238**, 34-35; bifurcation analysis, **240**, 809-812; kinetics, **240**, 809-810, 812; NMR spectroscopy, nonresonant effects, **239**, 77; in photoactivatable probe synthesis, **237**, 88-89, 93; substrates, **240**, 810; P¹,P⁵-di(adenosine-5′)-pentaphosphate inhibition of adenylate kinase, **63**, 7, 401, 483; P¹,P⁴-di(adenosine-5′)-tetraphosphate inhibition

of adenylate kinase, **63**, 7, 401, 483; phosphorus stereospecificity [toward adenosine 5′-monothiophosphate, **249**, 426-427 (enhancement with R97M mutant, **249**, 432-434; reversal with R44M mutant, **249**, 431-432; wild-type, confirmation, **249**, 429-431); toward adenosine 5-(1-thiotriphosphate), **249**, 433 (perturbation with T23A mutant, **249**, 437-440; relaxation with R128A mutant, **249**, 435-437); at AMP site, **249**, 426-427; at ATP site, **249**, 426-427; demonstration, 249, 426; manipulation by site-directed mutagenesis, **249**, 425-443 (active site conformations and, **249**, 441-442; kinetic experiments, **249**, 428; methods, **249**, 428-429; microscopic rates and, **249**, 440-441; procedures, **249**, 429-433; results interpretation, **249**, 440-442; at P α of MgATP, **249**, 433-440; wild-type and site-directed mutant enzymes, comparison, **249**, 440-441].

ADENYLATE KINASE INHIBITORS

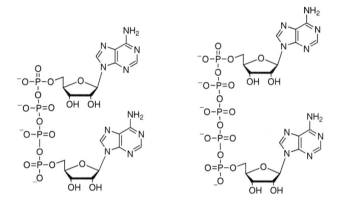

P¹,P⁴-di-(Adenosine-5')-P₄   **P¹,P⁵-di-(Adenosine-5')-P₅**

Because many cells maintain ATP, ADP, and AMP concentrations at or near the mass action ratio of the adenylate kinase reaction, the cellular content of this enzyme is often quite high. A consequence of such abundance is that, even after extensive purification, many proteins and enzymes contain traces of adenylate kinase activity. The presence of this kinase can confound the quantitative analysis of processes that either require ADP or are carried out in the presence of both ATP and AMP. Furthermore, the equilibrium of any reaction producing ADP may be altered if adenylate kinase activity is present. To minimize the effect of adenylate kinase, one can utilize the bisubstrate geometrical analogues Ap4A and Ap5A to occupy simultaneously both substrate binding pockets of this kinase[1,2]. Typical inhibitory concentrations are 0.4 and 0.2 mM, respectively. Of course, as is the case for the use of any inhibitor, one must always determine whether Ap4A or Ap5A has a direct effect on a particular reaction under examination. For example, Powers *et al.*[3] studied the effect of a series of α,ω-di-(adenosine 5′)-polyphosphates (*e.g.*, Ap$_n$A, where n =

2, 3, 4, 5 and 6) on carbamyl phosphate synthetase. Only Ap_5A was found to be an effective inhibitor of the overall reaction as well as two partial reactions catalyzed by the enzyme (namely, HCO_3^--dependent ATP hydrolysis and ATP synthesis from carbamyl-P and ADP).

[1]D. L. Purich & H. J. Fromm (1972) *Biochim. Biophys. Acta* **276**, 563.
[2]G. E. Leinhard & I. I. Secemski (1973) *J. Biol. Chem.* **248**, 1121.
[3]S. G. Powers, O. W. Griffith & A. Meister (1977) *J. Biol. Chem.* **252**, 3558.

Selected entries from *Methods in Enzymology* [vol, page(s)]:
Adenylate kinase contamination in in phosphotransferases, **63**, 7; as contaminant enzyme preparations, **64**, 24; P¹,P⁴-di(adenosine-5')-tetraphosphate and P¹,P⁵-di(adenosine-5')-pentaphosphate inhibition of adenylate kinase, **63**, 7, 401, 483.

ADENYLOSUCCINATE LYASE

This enzyme [EC 4.3.2.2], also referred to as adenylosuccinase, catalyzes the conversion of N^6-(1,2-dicarboxyethyl)AMP (or, adenylosuccinate) to yield fumarate and AMP. The enzyme will also convert 1-(5-phosphoribosyl)-4-(*N*-succinocarboxamide)-5-aminoimidazole to 5'-phosphoribosyl-5-amino-4-imidazolecarboxamide.

S. Ratner (1972) *The Enzymes*, 3rd ed., **7**, 167.

ADENYLOSUCCINATE SYNTHETASE

Adenylosuccinate synthetase[1-3] (AMPSase; EC 6.3.4.4) catalyzes the following reversible reaction:

$$\text{IMP} + \text{Aspartate} + \text{MgGTP}^{2-} = \text{AMP-Succinate} + \text{MgGDP}^{1-} + \text{P}_i$$

Aside from the natural nucleophile, aspartate, the only other agent that can attack the 6-carbon of IMP in a highly specific manner is NH_2OH.

As the first committed step in the biosynthesis of AMP from IMP, AMPSase plays a central role in *de novo* purine nucleotide biosynthesis. A 6-phosphoryl-IMP intermediate appears to be formed during catalysis, and kinetic studies of *E. coli* AMPSase demonstrated that the substrates bind to the enzyme active sites randomly. With mammalian AMPSase, aspartate exhibits preferred binding to the E·GTP·IMP complex rather than to the free enzyme. Other kinetic data support the inference that Mg·aspartate complex formation occurs within the adenylosuccinate synthetase active site and that such a complex may be an important factor in the activation of the protonated amino group of aspartate, enhancement of the enzyme's binding affinity, and its specificity for aspartate.

[1]R. Honzatko & H. J. Fromm (1998) *Adv. Enzymol.* **73**, in press.
[2]B. F. Cooper & F. B. Rudolph (1995) *Meth. Enzymol.* **249**, 188.
[3]F. M. Raushel & J. J. Villafranca (1988) *Crit. Rev. Biochem.* **23**, 1.

Selected entries from *Methods in Enzymology* [vol, page(s)]:
Equilibrium isotope exchange study of kinetic mechanism, **249**, 466; site-directed mutagenesis of *Escherichia coli* enzyme, **249**, 93; positional isotope exchange studies, **249**, 423; product inhibition studies of three substrates:three products reactions, **249**, 207-208.

ADENYLYLSULFATE—AMMONIA ADENYLYLTRANSFERASE

This enzyme [EC 2.7.7.51] catalyzes reaction of adenylylsulfate and ammonia to form adenosine 5'-phosphoramidate and sulfate.

A. Schmidt & K. Jäger (1992) *Ann. Rev. Plant Physiol. Plant Mol. Biol.* **43**, 325.
H. Frankhauser, J. A. Schiff, L. J. Garber & T. Saidha (1987) *Meth. Enzymol.* **143**, 354.

ADENYLYLSULFATE KINASE

This enzyme [EC 2.7.1.25], also referred to as APS kinase, catalyzes the reaction of ATP and adenylylsulfate to generate ADP and 3'-phosphoadenylylsulfate.

T. S. Leyh (1993) *Crit. Rev. Biochem. Mol. Biol.* **28**, 515.
A. Schmidt & K. Jäger (1992) *Ann. Rev. Plant Physiol. Plant Mol. Biol.* **43**, 325.
J. A. Schiff & T. Saidha (1987) *Meth. Enzymol.* **143**, 329.
H. D. Peck, Jr. (1974) *The Enzymes*, 3rd ed., **10**, 651.

ADENYLYLSULFATE REDUCTASE

This enzyme [EC 1.8.99.2] catalyzes the reaction of AMP, sulfite, and some acceptors (for example, methyl viologen) to yield adenylylsulfate and the reduced acceptor. Both FAD and iron are cofactors.

J. Lampreia, A. S. Pereira & J. J. G. Moura (1994) *Meth. Enzymol.* **243**, 241.
C. Dahl, N. Speich & H. G. Trüper (1994) *Meth. Enzymol.* **243**, 331.
C. Dahl & H. G. Trüper (1994) *Meth. Enzymol.* **243**, 400.
Y. Hatefi & D. L. Stiggall (1976) *The Enzymes*, 3rd ed., **13**, 175.
G. Palmer (1975) *The Enzymes*, 3rd ed.,**12**, 1.

ADENYLYLSULFATE SULFOHYDROLASE

This enzyme [EC 3.6.2.1], also called adenylylsulfatase, catalyzes the hydrolysis of adenylylsulfate to generate AMP and sulfate.

A. Schmidt & K. Jäger (1992) *Ann. Rev. Plant Physiol. Plant Mol. Biol.* **43**, 325.

A. B. Roy (1971) *The Enzymes*, 3rd ed., **5**, 1.

ADIABATIC PROCESS

1. Any process in which heat neither enters nor leaves the system. Thus, $q = 0$ or, for a differential process, $dq = 0$. If there is a reversible change in volume with such a system, $dU = -PdV$ where U is the internal energy, P is the pressure, and V is the volume. Any reversible adiabatic change is isentropic. (For an irreversible adiabatic change, $dU = -P_{ex}dV$ where P_{ex} is the external pressure.) An example of such a process is the adiabatic expansion of a gas, resulting in the cooling of the gas. The rapid release of vapor from an aerosol can is thus accompanied by a drop in temperature since the release is so fast there is little time for heat to be transferred. An adiabatic compression of a gas will result in the heating of the gas. An example can be provided by the use of a hand pump when inflating a basketball or bicycle tire. The act of pumping, by compressing the gas, is a near adiabatic process. Often, this increase in temperature is translated to the needle valve and can be felt by the individual doing the pumping. [Note that, for an ideal gas undergoing adiabatic expansion or compression, $PV\gamma = K$ where K is a constant and γ is the ratio of the heat capacities, C_P/C_V (provided that the two heat capacities are constants). Recall that, for an isothermal, nonadiabatic process, PV = constant.] 2. In chemical reactions, one normally considers that the reaction path is continuous over a potential energy surface. Such a reaction is said to be an adiabatic reaction. If a reaction pathway moves from one potential energy surface to another, it is often termed a nonadiabatic (or diabatic) reaction. The use of the term in these instances is not the same as its use in thermodynamics. Nonadiabatic processes can occur when the movement of electrons is not rapid relative to the change of a nuclear position. Under these conditions, the Born-Oppenheimer approximation is not valid and one can have the crossing of potential energy surfaces. *See Isothermal Process; Heat Capacity*

ADP-RIBOSYLATION

Covalent modification of a protein by the linkage of an ADP-ribosyl moiety to the protein. The resulting product typically exhibits altered kinetic and/or regulatory properties. ADP-ribosyltransferases catalyze the transfer of the ADP-ribosyl group of NAD^+ to a protein acceptor, producing the modified protein and free nicotinamide. The reaction scheme catalyzed by the cholera toxin A subunit and a rabbit muscle protein has been determined to be random[1]. The amino acyl residues that have been reported to have been modified in these reactions include a cysteinyl residue (*e.g.*, via pertussis toxin[2]), an arginyl residue (via cholera toxin A), an asparaginyl residue (via botulinum toxin), and a modified histidyl residue (via diphtheria toxin). In addition, the subunit structure appears to have a role in substrate specificity[13].

The widespread occurrence of ADP-ribosyltransferases[3-7] has generated considerable interest in assessing the role of ADP-ribosylation in cellular metabolism. Regulatory proteins associated with this modification event have also been identified. A small G-protein referred to as ARF (<u>A</u>DP-<u>r</u>ibosylation <u>f</u>actor) has been shown to enhance this covalent modification[8-10]. ARF actually represents a family of proteins, both cytosolic and membrane-bound. Interestingly, ARF has been shown to have a crucial role in vesicle transport from the Golgi[11,12]. When ARF-GDP binds to a developing vesicle, a GDP|GTP exchange occurs, and the coatomer begins to form. After the vesicle pinches off, the GTP is hydrolyzed and the coatomer and ARF are released, allowing the vesicle to fuse with the target membrane.

[1]J. S.-A. Larew, J. E. Peterson & D. J. Graves (1991) *J. Biol. Chem.* **266**, 52.

[2]A. G. Gilman (1987) *Ann. Rev. Biochem.* **56**, 615.

[3]M. D. Brightwell, C. E. Leech, M. K. O'Farrell, W. J. D. Whish & S. Shall (1975) *Biochem. J.* **147**, 119.

[4]J. Moss, S. J. Stanley & P. A. Watkins (1980) *J. Biol. Chem.* **255**, 5838.

[5]J. Moss & S. J. Stanley (1981) *Proc. Natl. Acad. Sci. U. S. A.* **78**, 4809.

[6]J. Moss & S. J. Stanley (1981) *J. Biol. Chem.* **256**, 7830.

[7]G. Soman, J. R. Mickelson, C. F. Louis & D. J. Graves (1984) *Biochem. Biophys. Res. Commun.* **120**, 973.

[8]R. A. Kahn & A. G. Gilman (1986) *J. Biol. Chem.* **261**, 7906.

[9]L. Monoco, J. J. Murtaugh, K. B. Newman, S.-C. Tsai, J. Moss & M. Vaughn (1990) *Proc. Natl. Acad. Sci. U. S. A.* **87**, 2206.

[10]M. Tsuchiya, S. R. Price, M. S. Nightingale, J. Moss & M. Vaughn (1989) *Biochemistry* **28**, 9668.

[11]J. E. Rothman (1994) *Nature* **372**, 55.

[12]J. E. Rothman and L. Orci (1996) *Sci. Amer.* **274** (3), 70.

[13]D. J. Graves, B. L. Martin & J. H. Wang (1994) *Co- and Posttranslational Modification of Proteins*, Oxford Univ. Press, New York.

Selected entries from *Methods in Enzymology* [vol, page(s)]:
By bacterial toxin, **235**, 617-632; by cholera toxin [catalysis, **237**, 45; assay, **235**, 642-647; G protein as subunit, **237**, 48]; by pertussis toxin, **237**, 24-26, 71, 132-133; G-protein subunits, **237**, 236-238; time-course of transducin ADP-ribosylation, **237**, 77-79, 91, 93-94.

ADRENODOXIN

A [2Fe-2S] ferredoxin participating in electron transfer reactions from NADPH to cytochrome P-450 within the adrenal gland.

ADSORPTION

The process of adhesion of molecular entities to surfaces of solids or liquids with which the entities are coming into contact. For example, adsorption chromatography utilities different adsorption properties to separate molecules. **See also** Absorption; Biomineralization; Micellar Catalysis; Langmuir Isotherm

AEQUORIN

A Ca^{2+}-binding protein isolated from jellyfish (*Aequorea* sp.) that is frequently used as a chemiluminescent calcium indicator.[1,2] Aequorin contains a hydrophobic prosthetic group, coelenterazine. Investigators have been able to express the transfected gene recombinantly in a number of cells, and upon addition of the cofactor, one can measure intracellular calcium concentration. **See** *Fura-2*

[1]E. R. Ridgway & C. C. Ashley (1967) *Biochem. Biophys. Res. Commun.* **29**, 229.
[2]A. Azzi & B. Chance (1969) *Biochim. Biophys. Acta* **189**, 141.

Selected entries from *Methods in Enzymology* [vol, page(s)]: Calibration of calcium binding, **260**, 425-427; luminescence measurement, **260**, 424-425; mitochondria-targeted hybrid protein [calcium quantitation, **260**, 418, 422, 424-428; stable expression, **260**, 418-421; transient expression, **260**, 420; intracellular localization, **260**, 421-425; reconstitution with coelenterazine, **260**, 422-422]; structure, **260**, 418.

AFFINITY LABELING

The covalent modification of an enzyme, most frequently at a reactive amino acid residue located within the active site. Affinity labeling is usually achieved by employing a substrate or coenzyme analogue bearing a reactive group that undergoes chemical, or photochemical, reaction with a nearby nucleophilic or electrophilic group located at the active center. Classical examples of affinity labeling reagents include: (a) tosyl-L-phenylalanine chloromethyl ketone (abbreviated TPCK) which binds into the aromatic substrate binding pocket of chymotrypsin and subsequently undergoes a specific reaction with histidine-57 to inactivate the enzyme; (b) diisopropylphosphofluoridate (abbreviated DIFP or DFP) which binds into the acetylcholine site of acetylcholinesterase and subsequently inactivates the enzyme by attaching to the active site seryl residue with expulsion of fluoride ion; and (c) bromohydroxyacetone phosphate which inactivates triose-phosphate isomerase by esterification of the active site glutamate-165.

The basic kinetic scheme involves site-directed binding (hence, affinity) of a chemically or photochemically reactive substance I, followed by first-order conversion to a covalent adduct:

$$E + I \overset{K_i}{\rightleftharpoons} EI \overset{k}{\rightarrow} E - I$$

The rate law for covalent modification shows saturation kinetic behavior if the second step is rate-determining:

$$v_{\text{inactivation}} = -d[E]/dt = k[E][I]/(K_i + [I])$$

This equation will describe the loss of enzyme activity, if the affinity label is binding at the active site. Moreover, reversibly bound ligands capable of occupying the same site as the affinity label will competitively inhibit the rate of enzyme inactivation.

Baker[1] was among the earliest investigators to promote the study of active site-directed irreversible inhibitors as probes of enzyme active sites and as potential pharmacologic agents. He reasoned that while different isozymes may contain similar reactive groups in the active site, they may have unique sites outside the active site. His strategy was to locate reactive substituents in a way that promotes reactivity with residues lying outside the active site, thereby achieving what he termed "exo-alkylation". His goal was to inactivate certain isozymes that may be required by cancerous cells, but not their untransformed counterparts.

Plapp[2] has presented a valuable account of the use of affinity labeling to identify essential amino acid residues involved in enzymic catalysis, to map the general topological features of the active site, and to gain insights about the most likely mechanism of catalysis. He also discusses what one must consider in designing active site-directed reagents (a) to achieve specificity or, at least, selectivity and (b) to chose among the many options for the covalent chemistry (*e.g.*, alkylation, acylation, or photoinsertion). Issues related to the evaluation of affinity labels (such as kinetics of inactivation, reactivity, substrate protection, and specificity) are also discussed.

Colman[3] has also produced a definitive account of site-specific modification of enzymes, and her chapter is particularly instructive about the range and utility of reaction types that can be gainfully exploited in affinity labeling experiments.

As in the case of enzymic catalysis, one can have one protonation state of the enzyme, say EH, that reacts with the affinity label and two other states, say E^- and EH_2^+, that do not.

$$EH_2^+$$
$$\updownarrow$$
$$EH + I \overset{K_i}{\rightleftharpoons} EHI \overset{k}{\to} EH-I$$
$$\updownarrow$$
$$E^-$$

In such a case, a Michaelis pH function becomes useful in describing the pH profile for inactivation. Starting with the conservation of enzyme equation:

$$[E_o] = [EH] + [E^-] + [EH_2^+]$$

or

$$[E_o]/[EH] = 1 + [E^-]/[EH] + [EH_2^+]/[EH]$$

or

$$[E_o]/[EH] = 1 + K_1/[H^+] + [H^+]/K_2$$

where $K_1 = [H^+][E^-]/[EH]$ and $K_2 = [H^+][EH]/[EH_2^+]$. If $v_{inactivation} = k[EHI]$, then one can rearrange the expression to get an equation that predicts a bell-shaped profile for the rate of inactivation as a function of pH. Likewise, the state of protonation of the affinity reagent can affect the kinetics of enzyme inactivation by affinity labels. Brocklehurst[4] has given a useful account of two-protonic-state electrophiles as probes of enzyme catalytic mechanism.

[1]B. R. Baker (1967) *Design of Active-Site Directed Irreversible Enzyme Inhibitors*, Wiley, New York.
[2]B. V. Plapp (1982) *Meth. Enzymol.* **87**, 469.
[3]R. F. Colman (1990) *The Enzymes*, 3rd. ed., **19**, 283.
[4]K. Brocklehurst (1982) *Meth. Enzymol.* **87**, 427.

Selected entries from *Methods in Enzymology* [vol, page(s)]:
General Methodology: An overview, **46**, 3; application of affinity labeling for studying structure and function of enzymes, **87**, 469; transition state analogs as potential affinity labeling reagents, **46**, 15; mechanism-based irreversible enzyme inhibitors, **46**, 28; syncatalytic enzyme modification: characteristic features and differentiation from affinity labeling, **46**, 41; paracatalytic enzyme modification by oxidation of enzyme-substrate carbanion intermediates, **46**, 48; catalytic competence: a direct criterion for affinity labeling, **46**, 54; differential labeling: a general technique for selective modification of binding sites, **46**, 59; photoaffinity la-

beling, **46**, 69; design of exo affinity labeling reagents, **46**, 115; haloketones as affinity labeling reagents, **46**, 130; haloacetyl derivatives, **46**, 153; acetylenic irreversible enzyme inhibitors, **46**, 158; organic isothiocyanates as affinity labels, **46**, 164; photo-cross-linking of protein-nucleic acid complexes, **46**, 168; affinity labeling of multicomponent systems, **46**, 180.

Enzymes: Reaction of serine proteases with halomethyl ketones, **46**, 197; reaction of serine proteases with aza-amino acid and aza-peptide derivatives, **46**, 208; active site-directed inhibition with substrates producing carbonium ions: chymotrypsin, **46**, 216; peptide aldehydes: potent inhibitors of serine and cysteine proteases, **46**, 220; carboxypeptidases A and B, **46**, 225; N-substituted arginine chloromethyl ketones, **46**, 229; renin, **46**, 235; nucleotide and nucleic acid systems; adenosine derivatives for dehydrogenases and kinases, **46**, 240; alcohol dehydrogenases, **46**, 249; arylazido nucleotide analogs in a photoaffinity approach to receptor site labeling, **46**, 259; aromatic thioethers of purine nucleotides, **46**, 289; N^6-o- and p-fluorobenzoyladenosine 5'-triphosphates, **46**, 295; 6-chloropurine ribonucleoside 5'-phosphate, **46**, 299; carboxylic-phosphoric anhydrides isosteric with adenine nucleotides, **46**, 302; active-site labeling of thymidylate synthetase with 5-fluoro-2'-deoxyuridylate, **46**, 307; inert Co(III) complexes as reagents for nucleotide binding sites, **46**, 312; the active site of ribonucleoside disphosphate reductase, **46**, 321; adenosine deaminase, **46**, 327; direct photoaffinity labeling with cyclic nucleotides, **46**, 335; adenosine 3',5'-cyclic monophosphate binding sites, **46**, 339; 5-formyl-UTP for DNA-dependent RNA polymerase, **46**, 346; DNA-dependent RNA polymerase, **46**, 353; staphylococcal nuclease, **46**, 358.

Carbohydrate Systems: Carbohydrate binding sites, **46**, 362; glucosidases, **46**, 368; glycidol phosphates and 1,2-anhydrohexitol 6-phosphates, **46**, 381; β-galactosidase, **46**, 398; lysozyme, **46**, 403.

Amino Acid Systems: Glutamine binding sites, **46**, 414; labeling of the active site of L-aspartate β-decarboxylase with β-chloro-L-alanine, **46**, 427; active site of L-asparaginase: reaction with diazo-4-oxonorvaline, **46**, 432; labeling of serum prealbumin with N-bromoacetyl-L-thyroxine, **46**, 435; a pyridoxamine phosphate derivative, **46**, 441.

Steroid Systems: Labeling of steroid systems, **46**, 447; irreversible inhibitors of Δ^5-3-ketosteroid isomerase: acetylenic and allenic 3-oxo-5,10-secosteroids, **46**, 461; labeling of Δ^5-3-ketosteroid isomerase by photoexcited steroid ketones, **46**, 469.

Antibodies: Affinity labeling of antibody combining sites as illustrated by anti-dinitrophenyl antibodies, **46**, 479; p-azobenzenearsonate antibody, **46**, 492; affinity cross-linking of heavy and light chains, **46**, 501; bivalent affinity labeling haptens in the formation of model immune complexes, **46**, 505; DNP-based diazoketones and azides, **46**, 508; labeling of antilactose antibody, **46**, 516.

Other Proteins: The ouabain-binding site on (Na^+/K^+)-adenosine-5'-triphosphatase, **46**, 523; penicillin isocyanates for β-lactamase, **46**, 531; active site-directed addition of a small group to an enzyme: the ethylation of luciferin, **46**, 537; mandelate racemase, **46**, 541; dimethylpyrazole carboxamidine and related derivatives, **46**, 548; labeling of catechol O-methyltransferase with N-haloacetyl derivatives, **46**, 554; affinity labeling of binding sites in proteins by sensitized photooxidation, **46**, 561; bromocolchicine as a label for tubulin, **46**, 567.

Receptors and Transport Systems: Affinity labeling of receptors, **46**, 572; nicotinic acetylcholine receptors, **46**, 582; β-adrenergic receptors, **46**, 591; opiate receptors, **46**, 601; amino acid transport proteins, **46**, 607; the biotin transport system, **46**, 613.

Nucleic Acids and Ribosomes: Affinity labeling of ribosomal functional sites, **46**, 621; photoaffinity labeling of 23S RNA in ribosomes, **46**, 637; photoaffinity labels for nucleic acids, **46**, 644; identification of GDP binding sites with 4-azidophenyl-GDP, **46**, 649; photoactivated GTP analogs, **46**, 656; a photoactivated analog of streptomycin, **46**, 660; haloacylated streptomycin and puromycin analogs, **46**, 662; chemically reactive oligonucleotides, **46**, 669; aromatic ketone derivatives of aminoacyl-tRNA as photoaffinity labels for ribosomes, **46**, 676; photoaffinity-probe-modified tRNA for the analysis of ribosomal binding sites, **46**, 683; analogs of chloramphenicol and their application to labeling ribosomes, **46**, 702; a homologous series of photoreactive peptidyl-tRNAs for probing the ribosomal peptidyltransferase center, **46**, 707; photoaffinity labeling of ribosomes with the unmodified ligands puromycin and initiation factor 3, **46**, 711.

AGARITINE γ-GLUTAMYLTRANSFERASE

This enzyme [EC 2.3.2.9] catalyzes the reaction of agaritine with an acceptor substrate to generate 4-hydroxymethylphenylhydrazine and the γ-glutamyl-acceptor. Examples of substrate acceptors include 4-hydroxyaniline, cyclohexylamine, 1-naphthylhydrazine, and similar compounds. The enzyme will also be the catalyst in the hydrolysis of agaritine. **See also** γ-Glutamyl Transpeptidase

B. Levenberg (1970) *Meth. Enzymol.* **17A**, 877.

AGMATINE DEIMINASE

This enzyme [EC 3.5.3.12], also known as agmatine iminohydrolase, catalyzes the hydrolysis of agmatine to produce *N*-carbamoylputrescine and ammonia. The plant enzyme also catalyzes the reactions of EC 2.1.3.3 (ornithine carbamoyltransferase), EC 2.1.3.6 (putrescine carbamoyltransferase) and EC 2.7.2.2 (carbamate kinase), thereby functioning as a putrescine synthase, converting agmatine and ornithine into putrescine and citrulline, respectively.

T. A. Smith (1985) *Ann. Rev. Plant Physiol.* **36**, 117.
C. W. Tabor & H. Tabor (1984) *Ann. Rev. Biochem.* **53**, 749.

AGONIST

Any chemical substance (*i.e.*, hormone, drug, metabolite, *etc.*) that associates with a receptor binding site and produces a biological response through such binding. The occupancy assumption of agonist action states that the magnitude of the bioeffect is determined by the number of receptors occupied by agonist. Agonists may differ with respect to the affinity of their interactions with a

receptor; however, once occupied, the receptor is assumed to make the same contribution to the overall response in a manner that is independent of the bound agonist. If two different agonists are present at concentrations allowing each to occupy the same fraction of receptor sites, and if the response is different at this same level of receptor occupancy, then one uses the term efficacy to describe the biological effectiveness of each agonist-receptor complex. **See** *Receptor-Agonist Action*

D-ALANINE:D-ALANINE LIGASE

This enzyme [EC 6.3.2.4], also known as D-alanylalanine synthetase, catalyzes an essential reaction for growth of certain enterococci, and failure to synthesize this dipeptide prevents the cross-linking of the underlying peptidoglycan structure of the bacterial cell wall. The resulting aminoacyl-D-Ala-D-Ala strand is the target of vancomycin binding, such that this antibiotic arrests peptidoglycan cross-linking and blocks cell-wall synthesis.

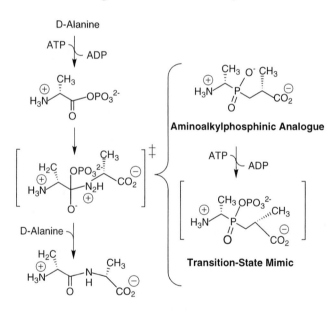

Mechanism of D-Ala-D-Ala Synthesis and Inhibition

The kinetic reaction mechanism appears to be random, and for the reaction to proceed, all substrates must reside as a E·D-Ala·D-Ala·MgATP quaternary complex. Except for its activation of an α-carboxylate to form a peptide bond, the enzyme's mechanism appears to be completely analogous to that catalyzed by glutamine synthetase, which forms a γ-glutamyl-phosphate intermediate. There is strong evidence for the participation of an acyl-phosphate intermediate which then suffers attack by the amino group of a second D-Ala molecule. Indeed, a phosphinate dipeptide analogue[1] is converted by enzymatic phosphorylation from low-affinity inhibitor to extremely tightly bound analogue of the ligase's reaction intermediates. [1(S)-Aminoethyl][2-carboxy-2(R)-methyl-1-ethyl]phosphinate is an ATP-dependent, slow-binding inhibitor of the D-Ala:D-Ala ligase from *Salmonella typhimurium*, and the enzyme-inhibitor complex (after ATP-dependent phosphorylation) has a half-life of 17 days at 37°C. The mechanism of inhibition is analogous to that of glutamine synthetase by methionine sulfoximine and phosphinothricin.

McDermott *et al.*[2] used rotational resonance, a then newly developed solid-state NMR method, for structural studies of an inhibited complex formed by reaction of D-alanine:D-alanine ligase, ATP, and the aminoalkyl dipeptide analogue. The measured NMR coupling properties indicate that the two species are bridged in a P—O—P linkage, with a P—P through-space distance of 2.7 ± 0.2 Å. This work unambiguously demonstrated that the inactivation mechanism involves phosphorylation of enzyme-bound inhibitor by ATP to form a phosphoryl-phosphinate adduct[3].

Vancomycin-resistant enterococci produce a mutant enzyme[4] (designated, VanA) that substitutes α-hydroxy acids in place of D-alanine. Because these so-called depsipeptides fail to bind vancomycin, but still function in the essential cell wall cross-linking reactions, these mutant bacteria are resistant to vancomycin antibiosis. Fan *et al.*[5] determined the ligase structure (co-crystallized with an S,R-methylphosphinate analogue and ATP) by X-ray diffraction to a resolution of 2.3 Å. The authors suggest a catalytic mechanism in which a helix dipole and a hydrogen-bonded triad of tyrosine, serine, and glutamic acid facilitate the binding and deprotonation steps. A different triad exists in a D-alanine:D-lactate ligase (VanA) present in vancomycin-resistant enterococci. Another D-Ala:D-Ala-ligase-related enzyme has the capacity to make D-alanyl-D-serine[6]. Alignment of the deduced amino acid sequences with those of other related enzymes from gram-negative and gram-positive bacteria revealed the presence of four distinct sequence patterns in the putative substrate-binding sites, each correlating with specificity to a particular substrate.

[1] K. Duncan & C. T. Walsh (1988) *Biochemistry* **27**, 3709.
[2] A. E. McDermott, F. Creuzet, R. G. Griffin, L. E. Zawadzke, Q. Z. Ye & C. T. Walsh (1990) *Biochemistry* **29**, 5767.
[3] C. T. Walsh (1993) *Science* **261**, 308 and *Science* **262**, 164.
[4] T. D. Bugg, S. Dutka-Malen, M. Arthur, P. Courvalin, & C. T. Walsh (1991) *Biochemistry* **30**, 2017.
[5] C. Fan, P. C. Moews, C. T. Walsh & J. R. Knox (1994) *Science* **266**, 439.
[6] S. Evers, B. Casadewall, M. Charles, S. Dutka-Malen, M. Galimand & P. Courvalin (1996) *Mol. Evol.* **42**, 706.

ALANINE AMINOTRANSFERASE

This enzyme [EC 2.6.1.2], also known as glutamic-pyruvic transaminase and glutamic-alanine transaminase, catalyzes the pyridoxal-phosphate-dependent reaction of alanine with 2-ketoglutarate, resulting on the production of pyruvate and glutamate. 2-Aminobutanoate will also react, albeit slowly. There is another alanine aminotransferase [EC 2.6.1.12], better known as alanine-oxo-acid aminotransferase, which catalyzes the pyridoxal-phosphate-dependent reaction of alanine and a 2-keto acid to generate pyruvate and an amino acid. *See also Alanine:Glyoxylate Aminotransferase*

A. J. L. Cooper (1985) *Meth. Enzymol.* **113**, 69.
A. E. Braunstein (1973) *The Enzymes*, 3rd ed., **9**, 379.

D-ALANINE AMINOTRANSFERASE

This enzyme [EC 2.6.1.21], also known as D-aspartate aminotransferase, D-amino acid aminotransferase, and D-amino acid transaminase, catalyzes the reversible pyridoxal-phosphate-dependent reaction of D-alanine with α-ketoglutarate to yield pyruvate and D-glutamate. The enzyme will also utilize as substrates the D-stereoisomers of leucine, aspartate, glutamate, aminobutyrate, norvaline, and asparagine. *See* D-*Amino Acid Aminotransferase*

A. E. Braunstein (1973) *The Enzymes*, 3rd ed., **9**, 379.

β-ALANINE AMINOTRANSFERASE

This enzyme [EC 2.6.1.18], also known as β-alanine-pyruvate aminotransferase, catalyzes the reversible pyridoxal-phosphate-dependent reaction of β-alanine with pyruvate to generate 3-oxopropanoate and alanine.

K. Yonaha, S. Toyama & K. Soda (1987) *Meth. Enzymol.* **143**, 500.
O. W. Griffith (1986) *Ann. Rev. Biochem.* **55**, 855.
A. E. Braunstein (1973) *The Enzymes*, 3rd ed., **9**, 379.

ALANINE DEHYDROGENASE

This enzyme [EC 1.4.1.1] catalyzes the reaction of alanine with water and NAD$^+$ to produce pyruvate, ammonia, and NADH.

A. R. Clarke & T. R. Dafforn (1998) *Comprehensive Biological Catalysis: A Mechanistic Reference* **3**, 1.

N. M. W. Brunhuber & J. S. Blanchard (1994) *Crit. Rev. Biochem. Mol. Biol.* **29**, 415.

A. Yoshida & E. Freese (1970) *Meth. Enzymol.* **17A**, 176.

ALANINE:GLYOXYLATE AMINOTRANSFERASE

This enzyme [EC 2.6.1.44] catalyzes the reversible, pyridoxal-phosphate-dependent reaction of alanine with glyoxylate to generate pyruvate and glycine. One component of the animal enzyme can utilize 2-oxobutanoate as a substrate instead of glyoxylate. A second component of the enzyme can also catalyze the reaction of alanine with 3-hydroxypyruvate to produce pyruvate and serine.

See Serine:Pyruvate Aminotransferase

A. E. Braunstein (1973) *The Enzymes*, 3rd ed., **9**, 379.

D-ALANINE HYDROXYMETHYLTRANSFERASE

This enzyme [EC 2.1.2.7], also called 2-methylserine hydroxymethyltransferase, catalyzes the reaction of D-alanine with water and 5,10-methylenetetrahydrofolate to produce tetrahydrofolate and 2-methylserine. The enzyme can also use 2-hydroxymethylserine as a substrate.

E. W. Miles (1971) *Meth. Enzymol.* **17B**, 341.

ALANINE RACEMASE

This enzyme [EC 5.1.1.1] catalyzes the pyridoxal-phosphate-dependent interconversion of D-alanine with L-alanine.

M. E. Tanner & G. L. Kenyon (1998) *Comprehensive Biological Catalysis: A Mechanistic Reference* **2**, 7.

R. A. John (1998) *Comprehensive Biological Catalysis: A Mechanistic Reference* **2**, 173.

M. A. Ator & P. R. Ortez de Montellano (1990) *The Enzymes*, 3rd ed., **19**, 213.

A. E. Martell (1982) *Adv. Enzymol.* **53**, 163.

E. Adams (1976) *Adv. Enzymol.* **44**, 69.

E. Adams (1972) *The Enzymes*, 3rd ed., **6**, 479.

β-ALANOPINE DEHYDROGENASE

This enzyme [EC 1.5.1.17] catalyzes the reversible reaction of 2,2'-iminodipropanoate with water and NAD^+ to produce alanine, pyruvate and NADH. In the reverse reaction, alanine can be replaced as a substrate by cysteine, serine, or threonine. Glycine acts very slowly as a substrate (that is, the reaction catalyzed by strombine dehydrogenase).

J. Thompson & S. P. F. Miller (1991) *Adv. Enzymol.* **64**, 317.

D-ALANYL-D-ALANINE PEPTIDASE

This enzyme [EC 3.4.16.4], also known as serine-type D-alanyl-D-alanine carboxypeptidase, catalyzes the hydrolysis of D-alanyl-D-alanine to yield two D-alanine. This enzyme comprises a group of membrane-bound, bacterial enzymes of the peptidase family S11. They are distinct from the zinc D-alanyl-D-alanine carboxypeptidase [EC 3.4.17.14]. The enzyme also hydrolyzes the D-alanyl-D-alanine peptide bond in the polypeptide of the cell wall. In addition, the enzyme will also catalyze the transpeptidation of peptidyl-alanyl moieties that are N-acetyl-substituents of D-alanine. The protein is inhibited by β-lactam antibiotics, which acylate the active-site seryl residue.

M. Jamin, J.-M. Wilkin & J.-M. Frère (1995) *Essays in Biochem.* **29**, 1.

B. Granier, M. Jamin, M. Adam, M. Galleni, B. Lakaye, W. Zorzi, J. Grandchamps, J.-m. Wilkin, C. Fraipont, B. Joris, C. Duez, M. Nguyen-Distèche, J. Coyette, M. Leyh-Bouille, J. Dusart, L. Christiaens, J.-M. Frère & J.-M. Guysen (1994) *Meth. Enzymol.* **244**, 249.

β-ALANYL-CoA AMMONIA-LYASE

This enzyme [EC 4.3.1.6] catalyzes the conversion of β-alanyl-CoA to acrylyl-CoA and ammonia.

P. R. Vagelos (1962) *Meth. Enzymol.* **5**, 587.

O-ALANYLPHOSPHATIDYLGLYCEROL SYNTHASE

This enzyme [EC 2.3.2.11], also referred to as alanyl-tRNA:phosphatidylglycerol alanyltransferase, catalyzes the reaction of alanyl-tRNA and phosphatidylglycerol to generate tRNA and O^3-alanyl-O^1-phosphatidylglycerol.

R. L. Soffer (1974) *Adv. Enzymol.* **40**, 91.

ALBERY DIAGRAM

A graphical method for assessing the most likely transition-state geometry of a particular reaction. Albery[1] employed this diagram to evaluate the contributions of different components of a reaction to the formation of the transition state[1].

Consider the case of a simple S_N2 reaction in which the bond order for the breaking of the C—X bond in the reactant (where X is the leaving group), B_{CX}, varies from 1.0 to 0.0 and the bond order for the formation of the N—C bond (where N is the nucleophile), B_{NC}, varies from 0.0 to 1.0. The Albery diagram is a rectangular map (in this case, plotting B_{NC} on the horizontal axis, varying from 0.0 to 1.0, and B_{CX} on the vertical axis, going from

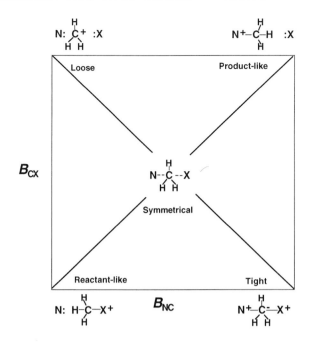

1.0 to 0.0) with the unreacted reactants in the lower left-hand corner (where $B_{CX} = 1.0$ and $B_{NC} = 0.0$) and the products in the upper right-hand corner ($B_{CX} = 0.0$ and $B_{NC} = 1.0$). The symmetric transition state lies at the center of the diagram. The lower right-hand corner represents tight transition-state structure in which the entering and leaving groups both have a large bond order associated with the reaction center. The upper left-hand corner represents a loose transition-state structure in which neither moiety exhibits any significant bonding interaction with the reaction center.

A perfectly symmetrical reaction would be represented by a diagonal running from the lower left-hand corner to the upper right-hand corner. The ''looseness'' and ''tightness'' of the leaving and entering group are represented by the other diagonal. Note that any transition state for a particular reaction can be represented by a point on this diagram.

The reader should note such a diagram is neither a free-energy diagram nor a potential-energy surface.

[1]W. J. Albery (1993) *Adv. Phys. Org. Chem.* **29**, 139.

ALBUMIN

The abundance and ease of purification made bovine serum albumin (BSA) an early standard in protein chemistry, and BSA is widely used as protein standard in biuret, Lowry, and Bradford assays as well as a molecular weight

marker. BSA is also a nutrient in cell culture, an aid in antibody production, and a component that is frequently used for protein and enzyme stabilization. When added to a enzyme reaction mixture at about 1-2 mg/mL, BSA often stabilizes enzymes. BSA is also added to counter the effects of enzyme dilution and to prevent inactivation of the enzyme due to adsorption onto the walls of a reaction vessel. BSA binds a wide variety of substances, including substrates, cofactors, or other components of the reaction mixture. In such cases, BSA may inhibit enzyme activity, and one should exercise care in its use.

ALCOHOL DEHYDROGENASE

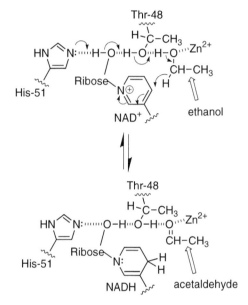

**Hydride Transfer in the
Alcohol Dehydrogenase Reaction**

This zinc metalloenzyme [EC 1.1.1.1 and EC 1.1.1.2] catalyzes the reversible oxidation of a broad spectrum of alcohol substrates and reduction of aldehyde substrates, usually with NAD$^+$ as a coenzyme. The yeast and horse liver enzymes are probably the most extensively characterized oxidoreductases with respect to the reaction mechanism. Only one of two zinc ions is catalytically important, and the general mechanistic properties of the yeast and liver enzymes are similar, but not identical. Alcohol dehydrogenase can be regarded as a model enzyme system for the exploration of hydrogen kinetic isotope effects.

B. V. Plapp (1995) *Meth. Enzymol.* **249**, 91.
M. W. W. Adams & A. Kletzin (1996) *Adv. Protein Chem.* **48**, 101.
B. J. Bahnson & J. P. Klinman (1995) *Meth. Enzymol.* **249**, 373.

W. W. Cleland (1995) *Meth. Enzymol.* **249**, 341.

T. A. Hansen & C. M. H. Hensgens (1994) *Meth. Enzymol.* **243**, 17.

N. J. Oppenheimer & A. L. Handlon (1992) *The Enzymes*, 3rd ed., **20**, 453.

C.-I. Brändén, H. Jörnvall, H. Eklund & B. Furugren (1975) *The Enzymes*, 3rd ed., **11**, 103.

K. Dalziel (1975) *The Enzymes*, 3rd ed., **11**, 1.

Selected entries from *Methods in Enzymology* [vol, page(s)]:
Abortive complex formation, **63**, 420; affinity chromatography, of NAD, **66**, 43-48; assay, of NAD, **66**, 41, 42, 80; binding, to agarose-bound nucleotides, **66**, 196-208; coupled assay, **63**, 34; cryoenzymology, **63**, 338; equilibrium perturbation, **64**, 120, 123; half-site reactivity, **64**, 184; inhibitor, **63**, 398, 408; isotope effect, **63**, 110; isotope exchange, **64**, 8, 28, 29; mechanism, **63**, 51; product inhibition, **63**, 436; reaction intermediates, stopped-flow absorbance studies, **61**, 319; substrate effect, **63**, 494; substrate inhibition, **63**, 507; affinity partitioning with dye ligands, **228**, 132-135; chromogenic substrates, **246**, 177-178, 189-190; cloning from bacteria, mutant complementation, **258**, 217-220; free sulfhydryl groups, determination by ESR, **251**, 97-98; horse liver ADH [active site, **249**, 94; isotope effects on oxidation of benzyl alcohol, **249**, 383-386; rapid scanning stopped-flow spectroscopy (alcohol oxidation, **246**, 184, 186-187, 189-190; aldehyde reduction, **246**, 184; cobalt substitution for zinc, **246**, 181-183; experimental design, **246**, 183-184; isobutyramide binding, **246**, 183; elucidation of mechanism, 246, 191, 193; pyrazole inhibition, **246**, 183; structural assignment of intermediates, **246**, 190); reaction, hydrogen tunneling in, **249**, 390-393, 396-397; site-directed mutants (altered pH dependencies, **249**, 110-111; catalytic efficiency, **249**, 104-106; steady-state kinetic analysis, **249**, 101-104; transient kinetic analysis, **249**, 108-109); structure, **249**, 94]; human liver, site-directed mutants, acid-base catalysis, **249**, 117-118; liquid-liquid partition chromatography, **228**, 195, 197; mechanism, **246**, 178-179, 191, 193; metal ligands, **246**, 182-183; modification by symmetrical disulfide radical, **251**, 99; *Desulfovibrio gigas* NAD-dependent ADH, **243**, 17-18; NAD$^+$ photoaffinity labeling, **237**, 72; secondary structure analysis, 246, 514; catalytic efficiency, **249**, 104-107; transition state and multisubstrate analogs, **249**, 304; yeast ADH [reaction (hydrogen tunneling in, **249**, 383-390, 396-397; isotope effects, **249**, 383); site-directed mutants (altered pH dependencies, **249**, 110-111; catalytic efficiency, **249**, 104-106; steady-state kinetic analysis, **249**, 101-104.

ALCOHOL DEHYDROGENASE (QUINOLINE-DEPENDENT)

This enzyme [EC 1.1.99.8], also referred to as alcohol dehydrogenase (acceptor) and methanol dehydrogenase, catalyzes the oxidation-reduction reaction of a primary alcohol with an acceptor to generate an aldehyde and the reduced acceptor. The cofactor for this enzyme is pyrroloquinoline quinone (PQQ). A wide variety of primary alcohols can act as the substrate. ***See also*** *Alcohol Dehydrogenase*

C. Anthony (1998) *Comprehensive Biological Catalysis: A Mechanistic Reference* 3, 155.

B. J. Bahnson & J. P. Klinman (1995) *Meth. Enzymol.* **249**, 373.

B. W. Groen & J. A. Duine (1990) *Meth. Enzymol.* **188**, 33.

ALCOHOL OXIDASE

This enzyme [EC 1.1.3.13], also known as methanol oxidase and AOX, catalyzes the reaction of a primary alcohol with dioxygen to generate an aldehyde and hydrogen peroxide. The enzyme utilizes FAD as the coenzyme and lower primary alcohols and unsaturated alcohols as the substrate. However, branched-chain and secondary alcohols are not acted upon by this enzyme.

I. J. van der Klei, L. V. Bystrykh & W. Harder (1990) *Meth. Enzymol.* **188**, 420.

ALCOHOL SULFOTRANSFERASE

This enzyme [EC 2.8.2.2], also referred to as hydroxysteroid sulfotransferase, catalyzes the reaction of 3′-phosphoadenylylsulfate with an alcohol to produce adenosine 3′,5′-bisphosphate and an alkyl sulfate. The alcohols that can act as substrates include aliphatic alcohols, ascorbate, chloramphenicol, ephedrine, hydroxysteroids, and other primary and secondary alcohols. However, phenolic steroids will not serve as substrates (such alcohols can be acted upon by steroid sulfotransferases).

J. D. Gregory (1962) *Meth. Enzymol.* **5**, 977.

ALDEHYDE DEHYDROGENASE

A number of enzymes have been referred to as aldehyde dehydrogenase. Aldehyde dehydrogenase (NAD$^+$) [EC 1.2.1.3] catalyzes the reaction of an aldehyde with NAD$^+$ and water to form a carboxylic acid and NADH. The enzyme is capable of acting on a wide variety of aldehydes. It will also oxidize D-glucuronolactone to form D-glucarate. Aldehyde dehydrogenase (NADP$^+$) [EC 1.2.1.4] will also catalyze the oxidation of an aldehyde to an acid except in this case it uses NADP$^+$ as the coenzyme. Aldehyde dehydrogenase (NAD(P)$^+$) [EC 1.2.1.5] can use either NAD$^+$ or NADP$^+$ as the coenzyme. Aldehyde dehydrogenase (acceptor) [EC 1.2.99.3], better known as aldehyde dehydrogenase (pyrroloquinoline-quinone), catalyzes the oxidation of an aldehyde to an acid using PQQ. This enzyme can utilize a wide variety of aldehydes as substrates including straight-chain aldehydes containing as many as ten carbon atoms, aromatic aldehydes, glyoxylate, and glyceraldehyde.

N. J. Oppenheimer & A. L. Handlon (1992) *The Enzymes*, 3rd ed., **20**, 453.

J. C. Murrell & W. Ashraf (1990) *Meth. Enzymol.* **188**, 26.

M. M. Attwood (1990) *Meth. Enzymol.* **188**, 314.

T.-K. Li (1977) *Adv. Enzymol.* **45**, 427.

ALDEHYDE HYDRATION

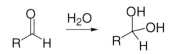

In aqueous solutions, aldehydes [RHC=O] undergo general acid-catalyzed addition of water to yield the hydrate [RHC(OH)$_2$], and the equilibrium position lies in favor of the hydrate. Jencks[1] summarized the most likely mechanism for the hydration reaction.

[1]W. P. Jencks (1969) *Catalysis in Chemistry and Enzymology*, p. 211, McGraw-Hill, New York. [Also available as Jencks, W. P. (1986) *Catalysis in Chemistry and Enzymology*, Dover Publications, New York.]

ALDEHYDE OXIDASE

This enzyme [EC 1.2.3.1] catalyzes the reaction of an aldehyde with water and dioxygen to produce a carboxylic acid and hydrogen peroxide. The enzyme uses both heme and molybdenum as cofactors. In addition, the enzyme can also catalyze the oxidation of quinoline and pyridine derivatives. In some systems this enzyme may be identical with xanthine oxidase.

P. E. Baugh, D. Collison, C. D. Garner & J. A. Joule (1998) *Comprehensive Biological Catalysis: A Mechanistic Reference* 3, 377.
J. G. Ferry (1992) *Crit. Rev. Biochem. Mol. Biol.* 27, 473.
R. C. Bray (1980) *Adv. Enzymol.* 51, 107.
R. C. Bray (1975) *The Enzymes*, 3rd ed., 12, 299.
G. Palmer (1975) *The Enzymes*, 3rd ed., 12, 1.

ALDEHYDE OXIDOREDUCTASE

A family of enzymes that catalyze the molybdenum- or tungsten-cofactor dependent conversion of aldehydes to carboxylates, often using ferredoxin. *See also* Aldose Reductase

C. Kisker, H. Schindelin & D. C. Rees (1997) *Ann. Rev. Biochem.* 66, 233.
M. W. W. Adams & A. Kletzin (1996) *Adv. Protein Chem.* 48, 101.
J. J. G. Moura & B. A. S. Barata (1994) *Meth. Enzymol.* 243, 24.
J. G. Ferry (1992) *Crit. Rev. Biochem. Mol. Biol.* 27, 473.

ALDOLASE

Fructose-1,6-bisphosphate aldolase [EC 4.1.2.13] catalyzes the reversible cleavage of D-fructose 1,6-bisphosphate to dihydroxyacetone phosphate and D-glyceraldehyde 3-phosphate. Interestingly, the $\Delta G^{\ominus'}$ value for this reaction is +23.9 kJ/mol, and under standard conditions the reaction is highly endergonic. Although appropriate for gluconeogenesis, it might appear that glycolysis would be hindered by this value. Nevertheless, in the direction of glycolysis, under typical intracellular concentrations of reactants and products, the ΔG value is about −1.3 kJ/mol, and the cleavage reaction is thermodynamically favored.

Aldolases have been classified into two different groups: Class I aldolases are not inhibited by chelating agents such as EDTA and an intermediate can be trapped with borohydride treatment. This class of enzymes proceed via a mechanism involving covalent catalysis. Class II aldolases are inhibited by EDTA and a covalent intermediate with the enzyme is not formed. This class of enzymes, which require the presence of a metal ion such as Zn^{2+} or Fe^{2+}, proceed via a mechanism involving metal ion catalysis. Some organisms express both classes of aldolases. In Class II aldolases, a divalent cation (*e.g.*, Fe^{2+} or Zn^{2+}) stabilizes the enolate intermediate or forms a metal ion-sugar complex that acts as an electrophile. These aldolases occur in bacteria, fungi, and algae.

K. N. Allen (1998) *Comprehensive Biological Catalysis: A Mechanistic Reference* 2, 135.
J. V. Schloss & M. S. Hixon (1998) *Comprehensive Biological Catalysis: A Mechanistic Reference* 2, 43.
J. A. Littlechild & H. C. Watson (1993)*Trends Biochem. Sci.* 18, 36.
R. Kluger (1992) *The Enzymes*, 3rd ed., 20, 271.
D. J. Hupe (1991) *A Study of Enzymes* 2, 485.
D. J. Creighton & N. S. R. K. Murthy (1990) *The Enzymes*, 3rd ed., 19, 323.
B. L. Horecker, O. Tsolas & C. Y. Lai (1972) *The Enzymes*, 3rd ed., 7, 213.

ALDOL CONDENSATION

The reversible reaction (usually base catalyzed) of two carbonyl compounds[1,2] (*i.e.*, two aldehydes, two ketones, or one of each), in which at least one compound has an α-hydrogen, to produce a β-hydroxy carbonyl compound.

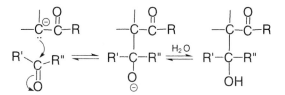

Originally, the term "aldol condensation" referred specifically to the reaction of an aldehyde (having an α-hydrogen) with an aldehyde/ketone to form a β-hydroxy aldehyde (the aldol). The reverse reaction is often referred to as a retrograde aldol reaction, a retro-aldol condensation (or reaction), or an aldol cleavage. March[3] categorizes aldol condensations into five classes. The first is condensation between two identical aldehydes (obviously, both aldehydes have to contain α-hydrogens). The reaction equilibrium lies to the right[4]. The second class is the condensation reaction between two identical ketones. In this case, the equilibrium is to the left. However, if the product is removed as the reaction progresses, then the synthesis can be successful. The third category is the condensation reaction between two different aldehydes, in which case a mixture of products will be obtained. If only one of the aldehydes has an α-hydrogen, then only two aldol products are possible. In many cases, the crossed product is the main product (the crossed aldol condensation is sometimes referred to as the Claisen-Schmidt reaction; see also category five). March's fourth category is the condensation between two different ketones, which is seldom attempted in the laboratory. The fifth and final category is the condensation between an aldehyde and a ketone. If there is no α-hydrogen on the aldehyde (*e.g.*, an aromatic aldehyde), usually only a single product (the β-hydroxy ketone or the subsequent α,β-unsaturated ketone) is obtained since there is little competition with the ketone condensing with another ketone.

Beyond the classical glycolytic enzyme (**See** *Aldolase*), numerous enzymes catalyze aldol and aldol-like condensation/cleavage reactions. Among them are the following: (1) Sedoheptulose-1,7-bisphosphate aldolase, which converts sedoheptulose 1,7-bisphosphate to erythrose 4-phosphate and dihydroxyacetone phosphate. (2) Transaldolase, which catalyzes reactions with D-erythrose 4-phosphate and D-fructose 6-phosphate as substrates. As in the case of fructose-1,6-bisphosphate aldolase, this enzyme uses a ε-amino side-chain to form a Schiff base intermediate. In this case, however, the triose phosphate moiety is not released but is transferred to the other aldose (in this case, the aldotetrose). (3) Phospho-2-keto-3-deoxygluconate aldolase, an enzyme in the Entner-Doudoroff pathway, that catalyzes the cleavage 6-phospho-2-keto-3-deoxy-D-gluconate to form pyruvate and D-glyceraldehyde 3-phosphate. (4) 3-Deoxy-D-*manno*-octulosonate aldolase, which facilitates the reaction of pyruvate with D-arabinose to form 3-deoxy-D-*manno*-octulosonate, an important component of bacterial cell walls. (5) 2-Oxo-4-hydroxyglutarate aldolase, which converts 2-oxo-4-hydroxyglutarate to pyruvate and glyoxylate. (6) Deoxyribose-phosphate aldolase, which converts 2-deoxy-D-ribose 5-phosphate

to D-glyceraldehyde 3-phosphate and acetaldehyde. (7) δ-Aminolevulinate dehydratase (also called porphobilinogen synthase), which catalyzes the second step in protoporphyrin biosynthesis (two 5-aminolevulinates = porphobilinogen + $2H_2O$). There are at least three distinct steps in this enzyme-catalyzed reaction: an aldol condensation, a dehydration, and a cyclic imine formation.

[1]Z. G. Hajos (1979) *Carbon-Carbon Bond Formation*, vol. **1**, p. 1, Marcel Dekker, New York.
[2]A. T. Nielsen and W.J. Houlihan (1968) *Org. React.* **16**, 1.
[3]J. March (1992) *Advanced Organic Chemistry*, 4th ed., p. 937, Wiley, New York.

ALDOSE 1-EPIMERASE

This enzyme [EC 5.1.3.3], also known as mutarotase, catalyzes the epimerization of the hemiacetal carbon atom of aldoses (thus, anomerization). Hence, α-D-glucose is reversibly converted to β-D-glucose. Other sugars can act as substrates (*e.g.*, L-arabinose, D-xylose, D-galactose, maltose, and lactose).

M. E. Tanner (1998) *Comprehensive Biological Catalysis: A Mechanistic Reference* **1**, 208.
E. Adams (1976) *Adv. Enzymol.* **44**, 69.

ALDOSE REDUCTASE

This enzyme [EC 1.1.1.21], also known as aldehyde reductase and polyol dehydrogenase (NADP+), catalyzes the reaction of an alditol with NAD(P)+ to generate an aldose and NAD(P)H. The enzyme exhibits a broad specificity for the alditol.

J. G. Ferry (1992) *Crit. Rev. Biochem. Mol. Biol.* **27**, 473.

ALGINATE LYASE

This enzyme [EC 4.2.2.3], also referred to as poly(β-D-mannuronate) lyase and poly(mana) alginate lyase, catalyzes the eliminative cleavage of polysaccharides containing β-D-mannuronate residues to give oligosaccharides with 4-deoxy-α-L-*erythro*-hex-4-enopyranuronosyl groups at their ends.

J. Preiss (1966) *Meth. Enzymol.* **8**, 641.

ALKALINE PHOSPHATASE

This enzyme [EC 3.1.3.1], also known as alkaline phosphomonoesterase, phosphomonoesterase, and glycerophosphatase, catalyzes the hydrolysis of many orthophosphoric monoesters (the substrate specificity is quite wide) to generate an alcohol and orthophosphate. The enzyme, which utilizes zinc and magnesium ions as cofactors, has a high pH optimum. In addition to this phosphatase activity, the enzyme can also catalyze certain transphosphorylations. In some systems, the enzyme can also act on diphosphate (pyrophosphate) (hence, a pyrophosphatase activity).

A. C. Hengge (1998) *Comprehensive Biological Catalysis: A Mechanistic Reference* **1**, 517.
J. A. Gerlt (1992) *The Enzymes*, 3rd ed., **20**, 95.
J. E. Coleman (1992) *Ann. Rev. Biophys. Biomol. Struct.* **21**, 441.
B. L. Vallee & A. Galdes (1984) *Adv. Enzymol.* **56**, 283.
H. W. Wyckoff, M. Handschumacher, K. Murthy & J. M. Sowadski (1983) *Adv. Enzymol.* **55**, 453.
M. Cohn & G. H. Reed (1982) *Ann. Rev. Biochem.* **51**, 365.
W. W. Cleland (1977) *Adv. Enzymol.* **45**, 373.
T. W. Reid & I. B. Wilson (1971) *The Enzymes*, 3rd ed., **4**, 373.
H. N. Fernley (1971) *The Enzymes*, 3rd ed., **4**, 417.

ALKANE HYDROXYLASE

This enzyme [EC 1.14.15.3], also known as alkane 1-monooxygenase, lauric acid ω-hydroxylase, ω-hydroxylase, and fatty acid ω-hydroxylase, catalyzes the reaction of octane with dioxygen and reduced rubredoxin to produce 1-octanol, water, and oxidized rubredoxin. The enzyme can also hydroxylate fatty acids in the ω-position. In some systems this enzyme is a heme-thiolated protein (P-450).

B. G. Fox (1998) *Comprehensive Biological Catalysis: A Mechanistic Reference* **3**, 261.

ALKENYLGLYCEROPHOSPHOCHOLINE HYDROLASE

This enzyme [EC 3.3.2.2], also known as lysoplasmalogenase, catalyzes the hydrolysis of 1-(1-alkenyl)-*sn*-glycero-3-phosphocholine to generate an aldehyde and *sn*-glycero-3-phosphocholine.

M. S. Jurkowitz-Alexander & L. A. Horrocks (1991) *Meth. Enzymol.* **197**, 483.

ALKENYLGLYCEROPHOSPHOETHANOL-AMINE HYDROLASE

This enzyme [EC 3.3.2.5], also known as lysoplasmalogenase, catalyzes the hydrolysis of 1-(1-alkenyl)-*sn*-glycero-3-phosphoethanolamine to yield an aldehyde and *sn*-glycero-3-phosphoethanolamine.

M. S. Jurkowitz-Alexander & L. A. Horrocks (1991) *Meth. Enzymol.* **197**, 483.

1-ALKYL-2-ACETYLGLYCEROL CHOLINE PHOSPHOTRANSFERASE

This enzyme [EC 2.7.8.16] was previously thought to catalyze the reaction of CDP-choline with a 1-alkyl-2-acetyl-*sn*-glycerol to generate CMP and 1-alkyl-2-acetyl-*sn*-glycero-3-phosphocholine. It is now deleted as an EC entry.

T.-c. Lee & F. Snyder (1992) *Meth. Enzymol.* **209**, 279.

1-ALKYL-2-ACETYLGLYCEROPHOSPHOCHOLINE ESTERASE

This enzyme [EC 3.1.1.47], also known as platelet-activating factor acetylhydrolase, 2-acetyl-1-alkyl-glycerophosphocholine esterase, and LDL-associated phospholipase A_2, catalyzes the hydrolysis of 2-acetyl-1-alkyl-*sn*-glycero-3-phosphocholine to yield 1-alkyl-*sn*-glycero-3-phosphocholine and acetate.

D. M. Stafforini, T. M. McIntyre, G. A. Zimmerman & S. M. Prescott (1997) *J. Biol. Chem.* **272**, 17895.
D. M. Stafforini, S. M. Prescott & T. M. McIntyre (1991) *Meth. Enzymol.* **197**, 411.

S-ALKYLCYSTEINE LYASE

This enzyme [EC 4.4.1.6] catalyzes the pyridoxal-phosphate-dependent hydrolysis of an *S*-alkyl-L-cysteine to generate an alkyl thiol, ammonia, and pyruvate. The reaction is an α,β-elimination. In yeast, the enzyme may be identical to cystathionine β-lyase. **See also** *Alliin Lyase*

Y. Nishizuka (1971) *Meth. Enzymol.* **17B**, 470.

ALKYLDIHYDROXYACETONE PHOSPHATE SYNTHASE

This enzyme [EC 2.5.1.26], also known as alkylglycerone-phosphate synthase, catalyzes the reaction of 1-acylglycerone 3-phosphate with a long-chain alcohol to produce 1-alkylglycerone 3-phosphate and a long-chain acid anion. In this reaction, the ester-linked fatty acid of the substrate is removed and replaced with a long-chain alcohol in an ether linkage.

A. Brown & F. Snyder (1992) *Meth. Enzymol.* **209**, 377.

ALKYLGLYCEROL KINASE

This enzyme [EC 2.7.1.93] catalyzes the reaction of ATP with 1-*O*-alkyl-*sn*-glycerol to generate ADP and 1-*O*-alkyl-*sn*-glycerol 3-phosphate.

F. Snyder (1992) *Meth. Enzymol.* **209**, 211.

1-ALKYLGLYCEROPHOSPHOCHOLINE *O*-ACYLTRANSFERASE

This enzyme [EC 2.3.1.63] catalyzes the reaction of an acyl-CoA and 1-alkyl-*sn*-glycero-3-phosphocholine to yield coenzyme A and 1-alkyl-2-acyl-*sn*-glycero-3-phosphocholine. In some systems, this enzyme may be identical to 1-acylglycerophosphocholine acyltransferase [EC 2.3.1.23].

P. C. Choy & C. R. McMaster (1992) *Meth. Enzymol.* **209**, 86.

1-ALKYL-2-LYSOGLYCERO-3-PHOSPHOCHOLINE *O*-ACETYLTRANSFERASE

This enzyme [EC 2.3.1.67], also referred to as 1-alkylglycerophosphocholine acetyltransferase, catalyzes the reaction of acetyl-CoA with 1-alkyl-*sn*-glycero-3-phosphocholine to produce coenzyme A and 1-alkyl-2-acetyl-*sn*-glycero-3-phosphocholine.

T.-c. Lee, D. S. Vallari & F. Snyder (1992) *Meth. Enzymol.* **209**, 396.

ALLANTOICASE

This enzyme [EC 3.5.3.4], also called allantoate amidinohydrolase, catalyzes the hydrolysis of allantoate to produce (−)-ureidoglycolate and urea. The enzyme can also catalyze the hydrolysis of (+)-ureidoglycolate to generate glyoxylate and urea.

N. E. Tolbert (1981) *Ann. Rev. Biochem.* **50**, 133.

ALLANTOINASE

This enzyme [EC 3.5.2.5] catalyzes the hydrolysis of allantoin to generate allantoate.

T. G. Cooper (1984) *Adv. Enzymol.* **56**, 91.
N. E. Tolbert (1981) *Ann. Rev. Biochem.* **50**, 133.

ALLENE-OXIDE CYCLASE

This enzyme [EC 5.3.99.6] catalyzes the conversion of (9*Z*)-(13*S*)-12,13-epoxyoctadeca-9,11,15-trienoate to (15*Z*)-12-oxophyto-10,15-dienoate. The allene oxides which are formed by the action of allene-oxide synthase [EC 4.2.1.92] are converted into cyclopentenone derivatives.

G. Sembdner & B. Parthier (1993) *Ann. Rev. Plant Physiol. Plant Mol. Biol.* **44**, 569.

ALLENE-OXIDE SYNTHASE

This enzyme [EC 4.2.1.92], also called hydroperoxide dehydratase and hydroperoxide isomerase, catalyzes the conversion of (9*Z*,11*E*,14*Z*)-(13*S*)-hydroperoxyoctadeca-(9,11,14)-trienoate to (9*Z*)-(13*S*)-12,13-epoxyoctadeca-9,11-dienoate and water. The enzyme uses a heme-thiolate cofactor and utilizes a number of unsaturated fatty-acid hydroperoxides as substrates, forming the corresponding allene oxides.

R. A. Creelman & J. E. Mullet (1997) *Ann. Rev. Plant Physiol. Plant Mol. Biol.* **48**, 355.
A. R. Brash & W. Song (1996) *Meth. Enzymol.* **272**, 250.

ALLIIN LYASE

This enzyme [EC 4.4.1.4], also known as alliinase and cysteine sulfoxide lyase, catalyzes the conversion of an *S*-alkyl-L-cysteine *S*-oxide to an alkyl sulfenate and 2-aminoacrylate. The enzyme requires pyridoxal phosphate.

A. E. Braunstein & E. V. Goryachenkova (1984) *Adv. Enzymol.* **56**, 1.
L. Davis & D. E. Metzler (1972) *The Enzymes*, 3rd ed., **7**, 33.

ALLOPHANATE HYDROLASE

This enzyme [EC 3.5.1.54] catalyzes the hydrolysis of urea-1-carboxylate to yield 2 CO_2 and 2 ammonia. Interestingly, the enzyme isolated from yeast (but not the corresponding enzyme from green algae) will also catalyze the reaction of urea with ATP and CO_2 to produce ADP, orthophosphate, 2 ammonia, and 2 CO_2.

T. G. Cooper (1984) *Adv. Enzymol.* **56**, 91.

ALLOSE KINASE

This enzyme [EC 2.7.1.55], also referred to as allokinase, catalyzes the reaction of ATP with D-allose to yield ADP and D-allose 6-phosphate.

F. J. Simpson & L. N. Gibbins (1966) *Meth. Enzymol.* **9**, 412.

ALLOSTERY

Linked-function mechanisms for cooperative binding interaction of metabolites and/or drugs, based on the presence of two or more different conformational states of the protein or receptor. ***See*** *Adair Equation; Cooperative Ligand Binding; Hemoglobin; Hill Equation & Plot; Koshland-Némethy-Filmer Model; Monod-Wyman-Changeux Model; Negative Cooperativity; Positive Cooperativity*

ALLOTOPIC EFFECT

Interference in receptor-agonist interactions arising from the binding of an antagonist at a separate binding site.

While the antagonist site is topologically distinct from the agonist binding site, the two are linked by alterations in receptor conformation. The analogous behavior is commonly called allosterism for regulatory enzymes, and the prefix "allo" (Greek for "another") emphasizes the involvement of spatially separate binding sites.

ALLYLIC SUBSTITUTION REACTION

A reaction of an allylic system ($CH_2=CH-CH_2-$) in which substitution occurs at the carbon adjacent to the double bond (*e.g.*, at position one if the double bond is between positions 2 and 3). The incoming group may become attached to the same atom that contained the leaving group or it may become attached to the "relative position" 3, with the double bond shifting to the position between carbons one and two.

ALPHA (α)

1. First item in a series (for example, the first carbon linked to a carboxyl group). 2. Abbreviation for alpha particle. 3. Symbol for angle of optical rotation. 4. Symbol for degree of dissociation. 5. Symbol for electric polarizability of a molecule. 6. Often with a subscript number (i) the coefficient of $[A]^i$ in the numerator of the generalized rate expression. 7. Symbol for "is proportional to." 8. Symbol for Napierian absorption coefficient. 9. In brackets, symbol for specific optical rotation.

ALPHA-ADDITION (α-ADDITION)

The reversal of α-elimination; an addition reaction between two or more reacting chemical species in which two new chemical bonds are formed on the same atom.

ALPHA CLEAVAGE (α-CLEAVAGE)

A homolytic cleavage of a bond associated with an atom or group that is bonded to a specified group. An example of an alpha-cleavage would be a Norrish type I cleavage.

Comm. on Photochem. (1988) *Pure and Appl. Chem.* **60**, 1055.

ALPHA EFFECT

Enhanced nucleophilicity of an attacking nucleophile attributable to one or more unshared pairs of electrons on an atom that lies immediately adjacent to the nucleophilic group[1-4]. The increased nucleophilicity of hydrazines and hydroxylamines is due to an alpha effect, as is the increased nucleophilicity of HO_2^- compared to HO^- in solution. This effect on nuclephile reactivity can

be significant for substitution reactions at a carbonyl group, at an unsaturated carbon, or in reactions of a nucleophile with a carbocation. The effect is largest when the transition state displays considerable bond formation. Substitution reactions at a saturated carbon generally tend to exhibit little or no alpha effect.

As is also true for ambident anions, substances exhibiting alpha effects in their reactions consistently deviate from the anticipated structure-reactivity correlations known for simple nucleophiles.

One explanation for the alpha effect is ground-state destabilization: Repulsive electronic interactions between the alpha atom's lone-pair and the nucleophile occur in the ground-state, and such destabilization is expected to be relieved as a covalent bond is forming in the transition-state of a nucleophilic substitution reaction. Reduced solvation in molecules exhibiting the alpha effect may also play a role in the increased nucleophilicity[5]; for example, OH_2^- displays no effect in the gas phase, but a substantial effect is observed in solution[6]. Another factor may be the ability of the alpha lone-pair to stabilize any partially positive group formed in the transition state.

[1] A. P. Grekov & V. Y. Veselov (1978) *Russ. Chem. Rev.* **47**, 631.
[2] N. J. Fina & J. O. Edwards (1973) *Int. J. Chem. Kinet.* **5**, 1.
[3] S. Wolfe, D. J. Mitchell, H. B. Schlegel, C. Minot & O. Eisenstein (1982) *Tet. Lett.* **23**, 615.
[4] S. Hoz & E. Buncel (1985) *Isr. J. Chem.* **26**, 313.
[5] D. Herschlag & W. P. Jencks (1990) *J. Amer. Chem. Soc.* **112**, 1951.
[6] C. H. DePuy, E. W. Della, J. Filley, J. J. Grabowski & V. M. Bierbaum (1983) *J. Amer. Chem. Soc.* **105**, 2481.

ALPHA-ELIMINATION (α-ELIMINATION)

A reaction involving the loss or elimination of two substituents attached to a common atom (usually carbon). α-Elimination is often observed in the synthesis of certain carbenes and nitrenes. **See also** Carbene

ALPHA EXPULSION

Any reaction characterized by the expulsion or loss of a group covalently bound to the carbon atom lying immediately adjacent (hence the designation "alpha") to a chromophore. After photoexcitation, the group that is expelled either contains an odd number of electrons or is an anionic species.

ALPHA PARTICLE

A decay product (corresponding to the nucleus of a helium atom) formed in certain nuclear disintegration reactions. These are the least energetic particles formed during radioactive decay, and their motion can be stopped by a single thickness of paper.

ALTERNATIVE HYPOTHESIS

A statistical term for a hypothesis that a particular set of data is at odds with a given null hypothesis. *See Statistics (A Primer)*

ALTERNATIVE PRODUCT INHIBITION

Rate experiments that are typically carried out in the presence of different concentrations of an alternative product (or product analog) while using the normal substrates[1,2]. This approach can be particularly useful when the normal product cannot be used because it is unstable, insoluble, or ineffective (the latter indicated by a very high K_i value). Moreover, the normal product may be consumed as an essential substrate in a coupled assay system[2] for the primary enzyme. Fromm and Zewe used the alternative product inhibition approach in their study of hexokinase[3]. Wratten and Cleland later applied this procedure to exclude the Theorell-Chance mechanism for liver alcohol dehydrogenase[4]. *See Abortive Complexes*

[1]H. J. Fromm (1975) *Initial Rate Enzyme Kinetics*, Springer-Verlag, New York.
[2]F. B. Rudolph (1979) *Meth. Enzymol.* **63**, 411.
[3]H. J. Fromm & V. Zewe (1962) *J. Biol. Chem.* **237**, 3027.
[4]C. C. Wratten & W. W. Cleland (1965) *Biochemistry* **4**, 2442.

ALTERNATIVE SUBSTRATES

Any compounds (other than the physiologically relevant substrate) that can serve as substrates for a particular enzyme. Alternative substrates compete with the natural substrate and with each other for access to the enzyme's active site. Thus, if one is utilizing an assay that measures the production of the true substrate, then the presence of the alternative substrate will result in competitive inhibition relative to the true substrate.

GENERAL PROPERTIES. Because enzymes rarely exhibit absolute substrate specificity, their use of alternative substrates affords the opportunity to investigate the nature of enzymic catalysis under conditions where the reaction's kinetic and thermodynamic parameter may reveal more about rate-limiting steps or other mechanistic aspects. An excellent example of an enzyme with many alternative substrates is bovine brain hexokinase (Table I).

Table I
Substrate Specificity of Bovine Brain Hexokinase

Substrate	Michaelis Constant	Relative Rate
Glucose	0.05 mM	1.0
Fructose	1.6 mM	1.5
Mannose	0.07 mM	0.4
2-Deoxyglucose	0.27 mM	1.0
1,5-Anhydro-D-glucitol[a]	10 mM	1.0
Glucosamine	0.8 mM	0.6
Galactose	100 mM	0.02

[a]Note: This cyclic ether analogue of glucose is conformationally analogous to the β-D-glucopyranose, except it lacks the C-1 hydroxyl group.

MULTISUBSTRATE SYSTEMS. Wong and Hanes[1,2] were probably among the first to suggest that alternative substrates may be useful in mechanistic studies. Fromm's laboratory was the first to use and extend the theory of alternative substrate inhibition to address specific questions about multisubstrate enzyme kinetic mechanisms[3-5]. Huang demonstrated the advantages of a constant ratio approach[6-8] when dealing with alternative substrate kinetics.

There are three basic methods for carrying out alternative substrate inhibition studies. In the first, the investigator seeks to observe numerical changes in the coefficients of the double-reciprocal form of the enzyme rate expression in the presence and absence of the alternative substrate. For some mechanisms, only certain coefficients will be altered[2]. This method requires extremely accurate estimates of the magnitudes of the coefficients and should always be supplemented with other kinetic probes[6].

In the second and most commonly used method, the investigator studies the alterations in the patterns of the initial rate data (usually graphically presented as double-reciprocal plots). In these studies of multisubstrate and multiproduct enzyme-catalyzed reactions, an investigator can measure the rate of the reaction by either by observing any increase in the concentration of a common product (*i.e.*, a product formed by both the normal substrate and the alternative substrate) or by detecting an increase in the concentration of the product not formed from the alternative substrate. Each procedure results in its own characteristic set of rate expressions and will produce different patterns in the double-reciprocal plots (1/*v* vs. 1/[A] at different, constant concentrations of the alternative substrate). The common-product approach, although very useful, will often produce nonlinearity in the double-reciprocal plots[4,6,9]. Huang has suggested that a constant-ratio approach, in which the substrate and the alternative substrate are varied in a constant ratio, can further assist an investigator in distinguishing between mechanisms[6-8].

Whenever using alternative substrate inhibition procedures, the investigator must demonstrate that initial rate conditions remain valid throughout the course of the experiment. This is particularly true of the other substrate(s) in multisubstrate enzymes. Because both the substrate under study and its analog are present in the reaction mixtures, the other cosubstrates will be depleted faster. This should always be a consideration in the design of the experiment.

ENERGETICS. While isozymes are apt to have different energies of activation, even under the same assay conditions, an enzyme acting on different substrates can in some circumstances exhibit the same energy of activation[10,11]. Yeast sucrase, for example, has an energy of activation of 46 kJ·mol^{-1} (or 11.0 kcal·mol^{-1}) for both sucrose and raffinose[11-13]. The rate-determining step in the enzyme-catalyzed reaction may differ with alternative substrates, and this may be reflected in the observed energy of activation. Likewise, if the rate-determining step changes with protein modification, assay conditions, or through site-directed mutagenesis, Arrhenius plots should reflect those changes. An example is the myosin ATPase which exhibits biphasicity in the Arrhenius plot with ITP as a substrate, but a typical linear Arrhenius plot with ATP as the substrate. Levy, Sharon, and Koshland[14] suggested that this may be the result of the 6-amino group on ATP interacting with some functional moiety on the protein, thereby producing an enzyme-substrate complex insensitive to temperature change. *See* Competitive Inhibitor; Abortive Complexes; Map-

ping Substrate Interactions Using Substrate Data; Membrane Transport; Energy of Activation; Q_{10}; Arrhenius Equation; van't Hoff Relationship

[1]J. T. Wong & C. S. Hanes (1962) Can. J. Biochem. Physiol. **40**, 763.
[2]J. T. F. Wong (1975) Kinetics of Enzyme Mechanisms, Academic Press, New York.
[3]V. Zewe, H. J. Fromm & R. Fabino (1964) J. Biol. Chem. **239**, 1625.
[4]H. J. Fromm (1964) Biochim. Biophys. Acta **81**, 413.
[5]F. B. Rudolph & H. J. Fromm (1971) Arch. Biochem. Biophys. **147**, 515.
[6]C. Y. Huang (1979) Meth. Enzymol. **63**, 486.
[7]C. Y. Huang & S. Kaufman J. Biol. Chem. **248**, 4242.
[8]C. Y. Huang (1977) Arch. Biochem. Biophys. **184**, 488.
[9]I. G. Darvey (1976) Mol. Cell. Biochem. **11**, 3.
[10]M. Dixon & E. C. Webb, Enzymes, 3rd ed., Academic Press, New York (1979).
[11]I. W. Sizer (1943) Adv. Enzymol. **3**, 35.
[12]I. W. Sizer (1938) Enzymologia **4**, 215.
[13]I. W. Sizer (1937) J. Cellular and Comp. Physiol. **10**, 61.
[14]H. M. Levy, N. Sharon & D. E. Koshland, Jr. (1959) Biochim. Biophys. Acta **33**, 288.

ALTRONATE DEHYDRATASE

This enzyme [EC 4.2.1.7] catalyzes the conversion of D-altronate to 2-dehydro-3-deoxy-D-gluconate and water.

J. Robert-Baudouy, J. Jimeno-Abendano & F. Stoeber (1982) Meth. Enzymol. **90**, 288.
W. A. Wood (1971) The Enzymes, 3rd ed., **5**, 573.

AMADORI REARRANGEMENT

A reaction in which *N*-glycosides of aldoses are converted into *N*-glycosides of the corresponding ketoses *via* acid or base catalysis. In tryptophan biosynthesis, the conversion of *N*-(5'-phosphoribosyl)anthranilate to enol-1-*o*-carboxyphenylamino-1-deoxyribulose phosphate involves an Amadori rearrangement. Similarly, in the biosynthesis of folic acid, riboflavin, and the dimethylbenzimidazole group of vitamin B$_{12}$, an Amadori rearrangement occurs at an early step. In histidine biosynthesis, the rearrangement occurs in the interconversion of N^1-5'-phosphoribosylformimino-5-aminoimidazole-4-carboxamide ribonucleotide to N^1-5'-phosphoribulosylformimino-5-aminoimidazole-4-carboxamide ribonucleotide. Amadori rearrangements can also occur during the glycation of proteins.

[1]M. Amadori (1925) Atti. Accad. Nazl. Lincei **2**(6), 337.
[2]M. Amadori (1929) Atti. Accad. Nazl. Lincei **9**(6), 68 and 226.
[3]G. Hodge (1955) Adv. Carbohyd. Chem. **10**, 169.
[4]R. U. Lemieux (1964) in Molecular Rearrangements (P. de Mayo, ed.), p. 753, Wiley, New York.
[5]J. E. Hodge & B. E. Fisher (1963) Methods in Carbohyd. Chem. **2**, 99.

AMBIDENT NUCLEOPHILE

A term describing a chemical species containing two distinct, yet strongly interacting, reactive centers at which a covalent bond can be formed. These centers are positioned so that a reaction at one center either stops (or greatly retards) a reaction at the other center. The more basic group in the ambident anion preferentially reacts with the polarizable group under nucleophilic attack. Ambident anions consistently deviate from anticipated structure-activity correlations known for substances containing only a single reactive group.

Examples of such molecules include conjugated nucleophiles such as the enolate anion. Such nucleophiles have potentially two attacking atoms (in the case of the enolate anion, the oxygen or the α-carbon); reaction conditions affect which will be the more prevalent species. Other examples include the cyanide (CN^-) and the nitrite (NO_2^-) ions.

Molecules having two reactive centers that are noninteracting or only weakly interacting, such as the carboxylate anions in malonate or succinate, are not considered to be ambident. The term "bifunctional" should be used to describe these compounds. If a molecule contains more than two interacting centers, the terms "polydent" or "multident" should be used. Examples of such nucleophiles include anions derived from malonic esters, β-keto esters, β-diketones, as well as phenoxide ions.

ω-AMIDASE

This enzyme [EC 3.5.1.3] catalyzes the hydrolysis of a monoamide of a dicarboxylic acid to generate a dicarboxylate and ammonia. The enzyme can utilize as a substrate glutaramate, succinamate, and their corresponding 2-keto derivatives.

A. J. L. Cooper, T. E. Duffy & A. Meister (1985) Meth. Enzymol. 113, 350.
A. J. L. Cooper & A. Meister (1976) Crit. Rev. Biochem. 4, 281.

AMIDOPHOSPHORIBOSYLTRANSFERASE

This enzyme [EC 2.4.2.14], also known as glutamine phosphoribosyl-pyrophosphate amidotransferase, catalyzes the reaction of glutamine with 5-phospho-α-D-ribose 1-diphosphate and water to produce 5-phospho-β-D-ribosylamine, diphosphate (or, pyrophosphate), and glutamate.

G. Davies, M. L. Sinnott & S. G. Withers (1998) Comprehensive Biological Catalysis: A Mechanistic Reference 1, 119.
H. Zalkin (1993) Adv. Enzymol. 66, 203.
H. Zalkin (1985) Meth. Enzymol. 113, 264.
W. D. L. Musick (1981) Crit. Rev. Biochem. 11, 1.
W. N. Keley & J. B. Wyngaarden (1974) Adv. Enzymol. 41, 1.
D. E. Koshland, Jr. & A. Levitzki (1974) The Enzymes, 3rd ed. 10, 539.

AMINE OXIDASES

This group of enzymes catalyzes the oxidation of amines. Amine oxidase [EC 1.4.3.4], a flavin-containing enzyme (also known as monoamine oxidase, tyramine oxidase, tyraminase, or adrenalin oxidase) catalyzes the reaction of an organic amine (i.e., $R—CH_2—NH_2$) with dioxygen and water to yield an aldehyde, ammonia, and hydrogen peroxide. The enzyme utilizes FAD as a cofactor and readily acts on primary amines. Secondary and tertiary amines can also serve as substrates.

Amine oxidase [EC 1.4.3.6] (also known as diamine oxidase, diamino oxhydrase, and histaminase) catalyzes the reaction of an organic amine (i.e., $R—CH_2—NH_2$) with dioxygen and water to yield an aldehyde, ammonia, and hydrogen peroxide. Cofactors are a copper ion and 6-hydroxy-DOPA.

C. Anthony (1998) Comprehensive Biological Catalysis: A Mechanistic Reference 3, 155.
A. Messerschmidt (1998) Comprehensive Biological Catalysis: A Mechanistic Reference 3, 401.
C. Hartmann & W. S. McIntire (1997) Meth. Enzymol. 280, 98.
J. P. Klinman (1996) J. Biol. Chem. 271, 27189.
J. M. Janes & J. P. Klinman (1995) Meth. Enzymol. 258, 20.
B. J. Bahnson & J. P. Klinman (1995) Meth. Enzymol. 249, 373.
E. J. Brush & J. W. Kozarich (1992) The Enzymes, 3rd ed., 20, 317.
M. A. Ator & P. R. Ortez de Montellano (1990) The Enzymes, 3rd ed., 19, 213.
J. A. Duine, J. Frank & J. A. Jongejan (1987) Adv. Enzymol. 59, 169.
V. Massey & P. Hemmerich (1975) The Enzymes, 3rd ed., 12, 191.
B. G. Malmström, L.-E. Andréasson & B. Reinhammer (1975) The Enzymes, 3rd ed., 12, 507.

AMINE SULFOTRANSFERASE

This enzyme [EC 2.8.2.3], also known as arylamine sulfotransferase, catalyzes the reaction of 3'-phosphoadenylylsulfate with an arylamine to produce adenosine 3',5'-bisphosphate and an arylsulfamate. Primary and secondary amines which can act as substrates include aniline, 2-naphthylamine, cyclohexylamine, and octylamine.

S. G. Ramaswamy & W. B. Jakoby (1987) Meth. Enzymol. 143, 201.

D-AMINO ACID AMINOTRANSFERASE

This enzyme catalyzes the reaction of a D-amino acid and and an α-keto acid to produce a new D-amino acid and a new α-keto acid. This enzyme activity is often catalyzed by D-alanine aminotransferase [EC 2.6.1.21]. *See D-Alanine Aminotransferase*

R. A. John (1998) in *Comprehensive Biological Catalysis: A Mechanistic Reference* **2**, 173.

H. Hayashi, H. Wada, T. Yoshimura, N. Esaki & K. Soda (1990) *Ann. Rev. Biochem.* **59**, 87.

C. T. Walsh (1984) *Ann. Rev. Biochem.* **53**, 493.

D-AMINO-ACID DEHYDROGENASE

This enzyme [EC 1.4.99.1] catalyzes the oxidation-reduction reaction of a D-amino acid with an acceptor and water to generate an α-keto acid, ammonia, and the reduced acceptor. The enzyme utilizes FAD as a cofactor. Most D-amino acids, with the noted exceptions of D-aspartate and D-glutamate, can be used as substrates.

K. Tsukada (1971) *Meth. Enzymol.* **17B**, 623.

L-AMINO-ACID DEHYDROGENASE

This enzyme [EC 1.4.1.5] catalyzes the oxidation-reduction reaction of an L-amino acid with NAD^+ and water to produce an α-keto acid, ammonia, and NADH. The amino acids used as substrates are the aliphatic amino acids.

N. M. W. Brunhuber & J. S. Blanchard (1994) *Crit. Rev. Biochem. Mol. Biol.* **29**, 415.

D-AMINO ACID OXIDASE

This FAD-dependent enzyme [EC 1.4.3.3] catalyzes the reaction of a D-amino acid with dioxygen and water to generate an α-keto acid, ammonia, and hydrogen peroxide. Substrate specificity is wide and includes glycine.

B. A. Palfey & V. Massey (1998) *Comprehensive Biological Catalysis: A Mechanistic Reference* **3**, 83.

K. Yagi (1991) *A Study of Enzymes* **2**, 271.

G. A. Hamilton (1985) *Adv. Enzymol.* **57**, 85.

H. J. Bright & D. J. T. Porter (1975) *The Enzymes*, 3rd ed., **12**, 421.

L-AMINO-ACID OXIDASE

This FAD-dependent enzyme [EC 1.4.3.2] catalyzes the reaction of an L-amino acid with dioxygen and water to generate an α-keto acid, ammonia, and hydrogen peroxide.

B. A. Palfey & V. Massey (1998) *Comprehensive Biological Catalysis: A Mechanistic Reference* **3**, 83.

H. J. Bright & D. J. T. Porter (1975) *The Enzymes*, 3rd ed., **12**, 421.

ω-AMINO ACID : PYRUVATE AMINOTRANSFERASE

This enzyme catalyzes the reaction of an ω-amino acid with pyruvate to produce L-alanine and an aldehyde-containing carboxylic acid. *See Lysine 6-Aminotransferase; Ornithine Aminotransferase*

K. Yonaha, S. Toyama & K. Soda (1987) *Meth. Enzymol.* **143**, 500.

AMINO ACID RACEMASE

This enzyme [EC 5.1.1.10] catalyzes the interconversion of L- and D-amino acids, using pyridoxal phosphate.

M. E. Tanner & G. L. Kenyon (1998) *Comprehensive Biological Catalysis: A Mechanistic Reference* **2**, 7.

E. Adams (1972) *The Enzymes*, 3rd ed., **6**, 479.

AMINO ACID TURNOVER KINETICS

Wolfe[1] has presented an excellent description of the systematic application of stable and radioactive isotope tracers in determining the kinetics of leucine metabolism and other amino acids in living systems.

[1]R. R. Wolfe (1992) *Radioactive and Stable Isotope Tracers in Biomedicine*, p. 357, Wiley-Liss, New York.

AMINOACYLASE

This enzyme [EC 3.5.1.14] (also referred to as histozyme, hippuricase, benzamidase, dehydropeptidase II, aminoacylase I, and acylase I) catalyzes the hydrolysis of an *N*-acyl-L-amino acid to yield a fatty acid anion and an L-amino acid. The enzyme has a wide specificity for the amino acid derivative. It will also catalyze the hydrolysis of dehydropeptides.

J. P. Greenstein (1955) *Meth. Enzymol.* **2**, 109.

AMINOACYL-tRNA HYDROLASE

This enzyme [EC 3.1.1.29], also called peptidyl-tRNA hydrolase, catalyzes the hydrolysis of an *N*-substituted aminoacyl-tRNA to produce an *N*-substituted amino acid and tRNA.

P. Yot, D. Paulin & F. Chapeville (1971) *Meth. Enzymol.* **20**, 194.

AMINOACYL-tRNA SYNTHETASES

These enzymes [EC 6.1.1.x] catalyze the reaction of a specific amino acid with the corresponding tRNA and ATP to generate AMP, diphosphate (or, pyrophosphate) and the aminoacyl-tRNA derivative.

E. A. First (1998) *Comprehensive Biological Catalysis: A Mechanistic Reference* **1**, 573.

D. C. H. Yang (1996) *Curr. Top. Cell. Reg.* **34**, 101.
C. W. Carter, Jr. (1993) *Ann. Rev. Biochem.* **62**, 715.
D. D. Buechter & P. Schimmel (1993) *Crit. Rev. Biochem. Mol. Biol.* **28**, 309.
M. Mirande (1991) *Prog. Nucleic Acid Res. Mol. Biol.* **40**, 95.
P. R. Schimmel (1979) *Adv. Enzymol.* **49**, 187.
L. L. Kisselev & O. O. Favorova (1974) *Adv. Enzymol.* **40**, 141.
D. Söll & P. R. Schimmel (1974) *The Enzymes*, 3rd ed., **10**, 489.

2-AMINOADIPATE AMINOTRANSFERASE

This enzyme [EC 2.6.1.39] catalyzes the reversible reaction of 2-aminoadipate with 2-oxoglutarate (or, α-ketoglutarate) to generate 2-oxoadipate and glutamate. The enzyme requires pyridoxal phosphate.

H. P. Broquist (1971) *Meth. Enzymol.* **17B**, 119.

2-AMINOADIPATE 6-SEMIALDEHYDE DEHYDROGENASE

This enzyme [EC 1.2.1.31], also known as α-aminoadipate reductase, catalyzes the reaction of L-2-aminoadipate 6-semialdehyde with $NADP^+$ and water to generate L-2-aminoadipate and NADPH.

V. W. Rodwell (1971) *Meth. Enzymol.* **17(B)**, 188.

4-AMINOBUTYRATE AMINOTRANSFERASE

This enzyme [EC 2.6.1.19] catalyzes the reversible reaction of γ-aminobutyrate with α-ketoglutarate to yield succinate semialdehyde and glutamate. A number of enzyme preparations have been reported to also use β-alanine, 5-aminopentanoate, and (R,S)-3-amino-2-methylpropanoate as substrates.

R. A. John (1998) *Comprehensive Biological Catalysis: A Mechanistic Reference* **2**, 173.
R. B. Silverman (1995) *Meth. Enzymol.* **249**, 240.
M. A. Ator & P. R. Ortez de Montellano (1990) *The Enzymes*, 3rd ed., **9**, 213.
A. J. L. Cooper (1985) *Meth. Enzymol.* **113**, 80.
A. E. Braunstein (1973) *The Enzymes*, 3rd ed., **9**, 379.

α-AMINO-ε-CAPROLACTAM RACEMASE

This enzyme [EC 5.1.1.15], also referred to as 2-aminohexano-6-lactam racemase, catalyzes the reversible interconversion of the L- and D-stereoisomers of 2-aminohexano-6-lactam. The enzyme, which utilizes pyridoxal phosphate, will also catalyze the interconversion of 2-aminopentano-5-lactam and 2-amino-3-mercaptohexano-6-lactam. The enzyme exhibits a minor aminotransferase activity with certain α-amino acids.

M. E. Tanner & G. L. Kenyon (1998) *Comprehensive Biological Catalysis: A Mechanistic Reference* **2**, 7.
H. Hayashi, H. Wada, T. Yoshimura, N. Esaki & K. Soda (1990) *Ann. Rev. Biochem.* **59**, 87.

1-AMINOCYCLOPROPANE-1-CARBOXYLATE SYNTHASE

This enzyme [EC 4.4.1.14] catalyzes the conversion of S-adenosyl-L-methionine to 1-aminocyclopropane-1-carboxylate methylthioadenosine via an α,γ-elimination. The required cofactor for this enzyme is pyridoxal phosphate.

H. Kende (1993) *Ann. Rev. Plant Physiol. Plant Mol. Biol.* **44**, 283.
D. O. Adams & S. F. Yang (1987) *Meth. Enzymol.* **143**, 426.

2-AMINO-4-HYDROXY-6-HYDROXY-METHYLDIHYDROPTERIDINE PYROPHOSPHOKINASE

This enzyme [EC 2.7.6.3], also known as 6-hydroxymethyl-7,8-dihydropterin pyrophosphokinase and 7,8-dihydro-6-hydroxymethylpterin pyrophosphokinase, catalyzes the reaction of ATP with 2-amino-4-hydroxy-6-hydroxymethyl-7,8-dihydropteridine to generate AMP and 2-amino-7,8-dihydro-4-hydroxy-6-(diphosphooxymethyl)pteridine.

R. L. Switzer (1974) *The Enzymes*, 3rd ed., **10**, 607.

β-AMINOISOBUTYRATE AMINOTRANSFERASE

This enzyme [EC 2.6.1.40], also known as (R)-3-amino-2-methylpropionate–pyruvate aminotransferase, catalyzes the reversible reaction of (R)-3-amino-2-methylpropanoate with pyruvate to generate 2-methyl-3-oxopropanoate and alanine.

O. W. Griffith (1986) *Ann. Rev. Biochem.* **55**, 855.

5-AMINOLEVULINATE AMINOTRANSFERASE

This pyridoxal phosphate-dependent enzyme [EC 2.6.1.43] catalyzes the reversible reaction of 5-aminolevulinate with pyruvate to produce 4,5-dioxopentanoate and alanine.

A. E. Braunstein (1973) *The Enzymes*, 3rd ed., **9**, 379.

5-AMINOLEVULINATE DEHYDRATASE

This enzyme [EC 4.2.1.24], also known as porphobilinogen synthase, catalyzes the reaction of two molecules of

5-aminolevulinate to generate porphobilinogen and two water. This enzyme is a zinc-dependent system.

K. N. Allen (1998) *Comprehensive Biological Catalysis: A Mechanistic Reference* **2**, 135.
P. M. Jordan & P. N. B. Gibbs (1985) *Biochem. J.* **227**, 1015.
B. L. Vallee & A. Galdes (1984) *Adv. Enzymol.* **56**, 283.
S. Granick & S. I. Beale (1978) *Adv. Enzymol.* **46**, 33.
D. Shemin (1972) *The Enzymes*, 3rd ed., **7**, 323.

5-AMINOLEVULINATE SYNTHASE

This enzyme [EC 2.3.1.37] catalyzes the reaction of succinyl-CoA with glycine to yield 5-aminolevulinate, coenzyme A, and carbon dioxide. Pyridoxal phosphate is used as a cofactor in this reaction. In mammals, the enzyme isolated from erythrocytes is genetically distinct from that in other tissues.

P. M. Shoolingin-Jordan, J. E. LeLean & A. J. Lloyd (1997) *Meth. Enzymol.* **281**, 309.
S. Granick & S. I. Beale (1978) *Adv. Enzymol.* **46**, 33.
P. M. Jordan & D. Shemin (1972) *The Enzymes*, 3rd ed., **7**, 339.

AMINOMALONATE DECARBOXYLASE

This enzymatic activity is now included under aspartate β-decarboxylase.

E. A. Boeker & E. E. Snell (1972)*The Enzymes*, 3rd ed., **6**, 217.

AMINOPEPTIDASE

Enzymes with this designation [EC 3.4.11.x] are α-aminoacylpeptide hydrolases. **See also** *specific aminopeptidase*

R. J. DeLange & E. L. Smith (1971) *The Enzymes*, 3rd ed., **3**, 81.

AMINOPEPTIDASE P

This enzyme [EC 3.4.11.9] (also known as Xaa–Pro aminopeptidase, X–Pro aminopeptidase, proline aminopeptidase, and aminoacylproline aminopeptidase) catalyzes the hydrolysis of a peptide bond at the *N*-terminus of a peptide provided that the *N*-terminal amino acyl residue is linked to a prolyl residue by that peptide bond. The enzyme will also act on dipeptides and tripeptides with that same restriction. Either manganese or cobalt is needed as a cofactor. This enzyme appears to be a membrane-bound system in both mammalian and bacterial cells. The protein belongs to the peptidase family M24B.

A. Yaron & A. Berger (1970) *Meth. Enzymol.* **19**, 521.

AMOUNT-OF-SUBSTANCE

A chemical term (symbolized by n, often with a subscript denoting the specific type of substance) denoting the quantity of atoms, molecules, ions, radicals, or electrons present. The SI unit is the mole. The amount-of-substance divided by the total volume is the concentration. **See** *Concentration*

AMP DEAMINASE

This enzyme [EC 3.5.4.6], also known as AMP aminohydrolase and adenylic acid deaminase, catalyzes the hydrolysis of AMP to yield IMP and ammonia.

C. L. Zielke & C. H. Suelter (1971) *The Enzymes*, 3rd ed., **4**, 47.

AMPHIPHILIC

Referring to a chemical species having two distinct domains with opposing solubility properties (*e.g.*, a charged domain and an apolar or hydrophobic domain). Such molecules often readily form micelles when present in dilute aqueous solutions.

AMPHIPROTIC

A term applied to solvents, such as water and ethanol, that are both protogenic (*i.e.*, release protons) and protophilic (*i.e.*, bind protons). **See** *Protogenic*

AMPHOTERIC

Referring to chemical species possessing the ability to exhibit either acidic or basic properties (*e.g.*, amino acids). This property actually depends on the medium. For example, sulfuric acid is an acid, when studied in water; however, it becomes amphoteric when studied in superacids.

AMPLIFICATION

The multiplicative enhancement of a signal, leading to an output of much greater magnitude through the relay action of one or more transducers. Concerned about how the dark-adapted eye senses single-photon events, Wald[1] was among the first to recognize that only certain cyclic reaction schemes display the capacity to amplify an initially weak signal. The diagram shown below illustrates (A) one-to-one stoichiometric processes that proceed without amplification and (B) a sequence of enzymatic activities arranged as cascades that produce robust amplification.

(A) Without amplification

stimulus

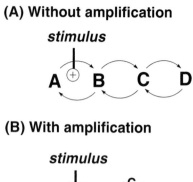

(B) With amplification

stimulus

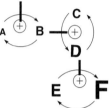

Hypothetical hierarchy of reactions: (A) those that fail to yield amplification and (B) those that can achieve any level of amplification, depending only on the number of cycles and the kinetic parameters of the active enzymes formed by successive conversions of proenzyme into active enzymes.

The classical example is blood clotting, where successive steps involving enzyme-catalyzed proteolysis converts an inactive (or weakly active) proenzyme into its highly active form. Although unknown at the time of Wald's classical report, kinase-type and nucleotidyltransferase-type reactions (*See* Enzyme Cascade Kinetics) are frequently the source of biological signal transduction and amplification.

[1]G. Wald (1965) *Science* **150**, 1028.

AMPLITUDE

1. The maximum distance between the value of a periodic function (*i.e.*, a function with repeated values for $f(x)$ for all integer multiples of a constant displacement or increment along the independent variable axis) and the function's mean value. 2. A term used in classical mechanics to define the magnitude of the maximum displacement of a body experiencing an oscillatory motion. 3. A term used in relaxation kinetics to indicate the magnitude of displacement of a chemical reaction.

AMP NUCLEOSIDASE

This enzyme [EC 3.2.2.4] catalyzes the hydrolysis of AMP to yield adenine and D-ribose 5-phosphate.

G. Davies, M. L. Sinnott & S. G. Withers (1998) *Comprehensive Biological Catalysis: A Mechanistic Reference* **1**, 119.
M. H. O'Leary (1989) *Ann. Rev. Biochem.* **58**, 377.

AMYLASES

Amylases are glycosidases that hydrolyze *O*-glucosyl bonds in glucans. α-Amylase [EC 3.2.1.1], also referred to as 1,4-α-D-glucan glucanohydrolase, catalyzes the endohydrolysis of a 1,4-α-D-glucosidic bond in a polysaccharide or oligosaccharide (there must be at least three glucosyl units in the oligosaccharide). Starch, glycogen, and related polysaccharides and oligosaccharides will serve as substrates. The endoglycosidic linkages are hydrolyzed in a random manner. The glucosyl group that has the newly generated anomeric carbon will be released with that anomeric carbon in the α-configuration. β-Amylase [EC 3.2.1.2], also known as saccharogen amylase and 1,4-α-D-glucan maltohydrolase, catalyzes the hydrolysis of 1,4-α-glucosidic linkages in polysaccharides near the nonreducing end of the polysaccharide chain such that successive maltose units are released from that nonreducing end. Starch, glycogen, and related polysaccharides and oligosaccharides can serve as substrates, all generating β-maltose by an inversion. *See* Limit Dextran

G. Davies, M. L. Sinnott & S. G. Withers (1998) *Comprehensive Biological Catalysis: A Mechanistic Reference* **1**, 119.
M. W. Bauer, S. B. Halio & R. M. Kelly (1996) *Adv. Protein Chem.* **48**, 271.
E. H. Van Beers, H. A. Büller, R. J. Grand, A. W. C. Einerhand & J. Dekker (1995) *Crit. Rev. Biochem. Mol. Biol.* **30**, 197.
J. Lehmann & M. Schmidt-Schuchardt (1994) *Meth. Enzymol.* **247**, 265.
G. Mooser (1992) *The Enzymes*, 3rd ed., **20**, 187.
I. S. Pretorius, M. G. Lambrechts & J. Marmur (1991) *Crit. Rev. Biochem. Mol. Biol.* **26**, 53.
J. D. Allen (1980) *Meth. Enzymol.* **64**, 248.
J. A. Thoma, J. E. Spradlin & S. Dygert (1971) *The Enzymes*, 3rd ed., **5**, 115.
T. Takagi, H. Toda & T. Isemura (1971) *The Enzymes*, 3rd ed., **5**, 235.

AMYLOSUCRASE

This enzyme [EC 2.4.1.4], also referred to as sucrose-glucan glucosyltransferase, catalyzes the reaction of sucrose with a glucan (specifically, $[(1,4)\text{-}\alpha\text{-D-glucosyl}]_n$) to yield D-fructose and a longer glucan (that is, $[(1,4)\text{-}\alpha\text{-D-glucosyl}]_{n+1}$).

R. J. Hehre (1955) *Meth. Enzymol.* **1**, 178.

ANALOG COMPUTER SIMULATION

An approximate numerical integration corresponding to the time evolution of a chemical rate process, as achieved by an electronic circuit operating as a computer by manipulating continuous physical variables. The electronic analog computer uses voltages and/or currents in circuits of resistors, capacitors, switches, and operational amplifiers to carry out analogous mathematical functions that simulate the time dependence of the process being modeled. Inherent limitations on the precision of circuitry components led kineticists to abandon analog computers and to favor the much greater reliability of digital computer methods. **See also** *Computer Simulation; KINSIM/FITSIM*

ANAPLEROTIC

Referring to reactions, pathways, or processes that replenish or add to intermediates of a metabolic cycle (usually the tricarboxylic cycle). The process itself is referred to as anaplerosis.

ANATION

The replacement of one or more water ligands within a coordination complex (or coordination entity) by an anion.

ANCHIMERIC ASSISTANCE

The well-known facilitation[1,2] of nucleophilic displacement reactions (particularly solvolytic reactions) resulting from a suitably positioned neighboring group. This functional group may help in directing the nucleophile to the electrophilic center or it may increase reactivity by stabilizing a reaction intermediate. Jencks[3] provides the following example of the importance of intramolecular assistance: Hydrolysis of tetramethylsuccanilic acid at pH 5 has a half-time of 30 minutes, whereas the hydrolysis of acetanilide at the same pH has a half-time of around 300 years.

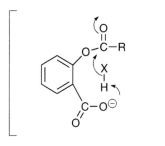

Anchimeric assistance may also explain slight changes in mechanism resulting from the presence of a neighboring group. One likely case is the acid-catalyzed hydrolysis of phenylglycosides. Raftery's group[4] found that the secondary kinetic isotope effect (k_H/k_D) was 1.13 for acid hydrolysis of phenyl-4-O-(2-acetamido-2-deoxy-β-D-glu-

copyranosyl)-β-D-glucopyranoside, but k_H/k_D was later determined to be only 1.08 for phenyl-4-O-(2-acetamido-2-deoxy-β-D-glucopyranosyl)-2-acetamido-2-deoxy-β-D-glucopyranoside[5]. Such a result suggests that the kinetics for forming the likely oxocarbonium ion intermediate are influenced by the neighboring acetamido group. Interestingly, the value of k_H/k_D for lyzozyme-catalyzed hydrolysis are 1.11 and 1.14, respectively, suggesting that substitution of the acetamido group for a hydrogen has little effect on the enzymic process. **See also** *Neighboring Group Mechanism*

[1]C. K. Ingold (1953) *Structure and Mechanism in Organic Chemistry*, p. 511 Cornell University Press, Ithaca.
[2]S. Winstein & E. Grunwald (1948) *J. Am. Chem. Soc.* **70**, 828.
[3]W. P. Jencks (1969) *Catalysis in Chemistry and Enzymology*, p. 211, McGraw-Hill, New York. [Also available as Jencks, W. P. (1986) *Catalysis in Chemistry and Enzymology*, Dover Publications, New York]
[4]F. W. Dahlquist, T. Rand-Meir & M. A. Raftery (1969) *Biochemistry* **8**, 4214.
[5]L. H. Mohr, L. E. H. Smith & M. A. Raftery (1973) *Arch. Biochem. Biophys.* **159**, 505.

ANCHOR PRINCIPLE

A concept suggesting that the relatively large size of coenzymes and other substrates is likely to assist in the proper positioning of the reaction center on the enzyme. This anchoring is attended by a corresponding loss of translational and rotational motion. The binding energy includes the energy required for achieving the proper orientation. The binding of well-anchored large substrates may initiate the conformational reorganization of the enzyme's active site. If these factors are enhanced in the transition state, so too will they favorably contribute to catalysis.

[1]W. P. Jencks (1975) *Adv. Enzymol.* **43**, 219.

ANGIOTENSIN CONVERTING ENZYME

This zinc-dependent enzyme [EC 3.4.15.1] (also known as dipeptidyl carboxypeptidase I, dipeptidyl-dipeptidase A, kininase II, peptidase P, and carboxycathepsin) catalyzes the release of a *C*-terminal dipeptide at a neutral pH. The enzyme will also act on bradykinin. The presence of prolyl residues in angiotensin I and in bradykinin results in only single dipeptides being released due to the activity of this enzyme, a protein which belongs to the peptidase M2 family. The enzyme is a glycoprotein, generally membrane-bound, that is chloride ion-dependent.

P. Corval, T. A. Williams & F. Soubrier (1995) *Meth. Enzymol.* **248**, 283.

A. A. Patchett & E. H. Cordes (1985) *Adv. Enzymol.* **57**, 1.

B. L. Vallee & A. Galdes (1984) *Adv. Enzymol.* **56**, 283.

M. A. Ondetti & D. W. Cushman (1984) *Crit. Rev. Biochem.* **16**, 381.

ANGULAR MOMENTUM

A measure of the rotational momentum of a body. For circular motion, the angular momentum equals the product of a body's mass and its angular velocity.

ANION EFFECTS ON pH-RATE DATA

As noted by Dixon and Webb[1], the dependence of enzyme activity can be significantly altered by the presence of anions[2]. For example, the presence of monovalent anions shifts the pH_{opt} of salivary α-amylase to a more alkaline pH value[3]. In addition, Cl^- and Br^- clearly can have an activating role with this enzyme as well. Another example is fumarase[4,5]. This enzyme is activated by several divalent and trivalent anions. Monovalent anions have either little effect or are inhibitory. There is also a shift in the pH_{opt} to a more alkaline pH value whether the anion is activating or inhibiting. This behavior led a number of investigators to suggest that these anions alter the magnitude of the ionization constant of the enzyme-substrate complex, affecting the alkaline side of the pH-rate profile, with little or no effect on the dissociation constant governing the "acid" limb of the curve. Chloride ions behave differently: they show inhibitory effects on the "acid" limb of the curve and exhibit little influence on the other dissociation constant. Nevertheless, in both scenarios, there is a shift to a higher pH_{opt}. Anions can also effect k_{cat} without altering the magnitude of dissociation constants[6].

Alberty[7] analyzed the anion effect on pH-rate data. He first considered a one-substrate, one-product enzyme-catalyzed reaction in which all binding interactions were rapid equilibrium phenomena. He obtained rate expressions for effects on V_{max} and K_m, thereby demonstrating how an anion might alter a pH-rate profile. He also considered how anions may act as competitive inhibitors. The effect of anions on alcohol dehydrogenase has also been investigated[8]. Chloride ions appear to affect the on- and off-rate constants for NAD^+ and NADH binding. **See also** *pH Studies; Activation; Optimum pH*

[1] M. Dixon & E. C. Webb (1979) *Enzymes*, 3rd ed., p. 395, Academic Press, New York.

[2] S. W. Cole (1904) *J. Physiol.* **30**, 202.

[3] K. Myrbäck (1926) *Hoppe-Seyler's Zeit. Physiol. Chem.* **159**, 1.

[4] P. J. G. Mann & B. Woolf (1930) *Biochem. J.* **24**, 427.

[5] V. Massey (1953) *Biochem. J.* **53**, 67.

[6] E. C. Webb & P. F. W. Morrow (1959) *Biochem. J.* **73**, 7.

[7] R. A. Alberty (1954) *J. Amer. Chem. Soc.* **76**, 2494.

[8] H. Theorell, A. P. Nygaard & R. Bonnischsen (1955) *Acta Chem. Scand.* **9**, 1148.

ANISOTROPIC

A term used with respect to a medium or a substance having one or more physical properties that have different values depending upon the direction with which that property is measured. **See also** *Isotropic*

ANNULATION (and ANNELATION)

A chemical transformation in which a ring structure is fused to a molecular entity by way of two newly formed bonds. Some investigators distinguish annulation (the formation of a ring from acyclic precursors) from annelation (the fusion of an additional ring to a ring already in existence).

ANOMER

One of the diastereomers of sugars (and sugar derivatives) differing only in the configuration about the internal hemiacetal or hemiketal carbon (*i.e.*, the so-called anomeric carbon)[1].

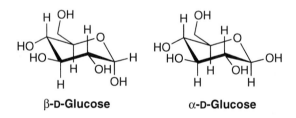

β-**D-Glucose**   α-**D-Glucose**

Anomers are a special class of epimers. The two anomeric forms are designated by α and β prefixes placed immediately before the configurational symbol (*e.g.*, α-D-glucose and β-L-mannose). The α anomeric configuration is defined as that configuration that has the same configuration as the reference carbon (in the cases above, C5) in the Fischer projection. The more dextrorotatory of the α/β-pair of D-sugar anomers is written as "α-D-," and the other is designated as "β-D-." The more levorotatory of the α/β-pair of L-sugar anomers is written as "α-L-," and its anomer is written as "β-L-. **See** *Epimers; Anomeric Specificity; Diastereomers*

[1] J. F. Robyt (1998) *Essentials of Carbohydrate Chemistry*, p. 33, Springer, New York.

ANOMERIC EFFECT

An observed effect of certain substituents on the conformation and stability of glycosides. In the case of α- and β-glucosides, when the substituent attached to the anomeric position of D-glucose is a nonpolar alkyl group, such a substituent will favor the equatorial position (*i.e.*, β). However, if the substituent is polar, the α-position (*i.e.*, axial) is favored. Thus with polar moieties, α-glucosides exhibit greater stability than β-glucosides.

A. J. Kirby (1983) *The Anomeric Effect and Related Stereoelectronic Effects at Oxygen*, Springer-Verlag, New York.

ANTAGONIST

Any substance that blocks agonist binding and/or inhibits an agonist-induced biological response. Many antagonists are structural analogues of the corresponding agonists, and molecular mimicry permits antagonists to bind at a receptor binding site in a manner that blocks or alters the typical agonist-induced response. Antagonists can sometimes also interfere with downstream steps in signal-transduction cascades triggered by an agonist. Some antagonists show no obvious structural similarity to the agonist, and antagonist binding to the receptor is thought to convert the occupied receptor to a conformation which blocks agonist binding.

ANTARAFACIAL & SUPRAFACIAL MIGRATIONS

Intramolecular antarafacial and suprafacial reactions are associated with a sigmatropic rearrangement. In such reactions, a σ-bond, adjacent to one or more π-systems, migrates to a new position with the concomitant reorganization of the π-system. For example, if a hydrogen migrates along one face of the π-system, it is designated a suprafacial migration. If the hydrogen migrates across the π-system, it is designated an antarafacial migration. Orbital symmetry rules[1] can identify which migrations are forbidden and those that are allowed. For example, the rules predict that antarafacial thermal [1,3] sigmatropic rearrangements are allowed, whereas suprafacial migrations are forbidden. However, in photochemical reactions, the suprafacial pathway is allowed, but the antarafacial migration is forbidden. For [1,5] rearrangements, the thermal process is suprafacial whereas the photochemical path is antarafacial. In general, for [1,*j*] migrations, if *j* is of the form $4n + 1$, where n is an integer, suprafacial migrations are thermally allowed.

If *j* is of the form $4n - 1$, suprafacial migrations are photochemically allowed. For antarafacial migrations, the restrictions are reversed. ***See also*** *Antarafacial & Suprafacial Reactions*

[1] R. B. Woodward & R. Hoffman (1970) *The Conservation of Orbital Symmetry*, Academic Press, New York.

ANTARAFACIAL & SUPRAFACIAL REACTIONS

Spatially distinct reactions occurring when a molecule or molecular fragment can simultaneously undergo two distinct changes in bonding at a common center or centers. If the reaction occurs at opposite faces of the portion of the molecule undergoing change, the reaction is designated antarafacial. If the changes occur at the same face of the molecule or molecular fragment, the reaction is designated suprafacial. The distinction between the two processes is obvious when considering reactions at a *p*-orbital or with a conjugated π-electron system[1,2]. ***See also*** *Antarafacial & Suprafacial Migrations*

[1] IUPAC (1979) *Pure Appl. Chem.* **51**, 1725.
[2] T. L. Gilchrist & R. C. Storr, (1979) *Organic Reactions and Orbital Symmetry*, Cambridge Univ. Press, Cambridge.

ANTHOCYANIDIN SYNTHASE

This enzyme catalyzes the conversion of *cis*-leucoanthocyanidins to anthocyanidins.

A. G. Prescott & P. John (1996) *Ann. Rev. Plant Physiol. Plant Mol. Biol.* **47**, 245.

ANTHRANILATE 1,2-DIOXYGENASE

This enzyme [EC 1.14.12.1], also known as anthranilate hydroxylase, decarboxylating, catalyzes the reaction of anthranilate with NAD(P)H, dioxygen, and two water molecules to produce catechol, carbon dioxide, $NAD(P)^+$, and ammonia. The enzyme requires an iron ion as a cofactor.

J. V. Schloss & M. S. Hixon (1998) *Comprehensive Biological Catalysis: A Mechanistic Reference* **2**, 43.
O. Hayaishi, M. Nozaki & M. T. Abbott (1975) *The Enzymes*, 3rd ed., **12**, 119.

ANTHRANILATE HYDROXYLASE

This enzyme [EC 1.14.13.35], also referred to as anthranilate 3-monooxygenase (deaminating) and anthranilate 2,3-dioxygenase (a transferred entry, formerly EC 1.14.12.2), catalyzes the reaction of anthranilate with NADPH and dioxygen to generate 2,3-dihydroxybenzoate, NADP⁺, and ammonia. The enzyme isolated from *Aspergillus niger* is an iron protein, whereas the yeast *Trichosporon cutaneum* enzyme utilizes FAD as a cofactor.

B. A. Palfey & V. Massey (1998) *Comprehensive Biological Catalysis: A Mechanistic Reference* **3**, 83.
J. V. Schloss & M. S. Hixon (1998) *Comprehensive Biological Catalysis: A Mechanistic Reference* **2**, 43.
O. Hayaishi, M. Nozaki & M. T. Abbott (1975) *The Enzymes*, 3rd ed., **12**, 119.

ANTHRANILATE PHOSPHORIBOSYL-TRANSFERASE

This enzyme [EC 2.4.2.18], also referred to as phosphoribosyl-anthranilate pyrophosphorylase, catalyzes the reaction of anthranilate with phosphoribosylpyrophosphate to produce *N*-5′-phosphoribosylanthranilate and pyrophosphate. In certain species, this enzyme is part of a multifunctional protein, together with one or more other components of the system for the biosynthesis of tryptophan (*i.e.*, indole-3-glycerol-phosphate synthase, anthranilate synthase, tryptophan synthase, and phosphoribosylanthranilate isomerase).

R. Bentley (1990) *Crit. Rev. Biochem. Mol. Biol.* **25**, 307.
R. Bauerle, J. Hess & S. French (1987) *Meth. Enzymol.* **142**, 366.
W. D. L. Musick (1981) *Crit. Rev. Biochem.* **11**, 1.

ANTHRANILATE SYNTHASE

This enzyme [EC 4.1.3.27] catalyzes the reaction of chorismate with glutamine to generate anthranilate, pyruvate, and glutamate. In certain species, this enzyme is part of a multifunctional protein together with one or more other components of the system for the biosynthesis of tryptophan (*i.e.*, indole-3-glycerol-phosphate synthase, anthranilate phosphoribosyltransferase, tryptophan synthase, and phosphoribosylanthranilate isomerase). The anthranilate synthase that is present in these complexes has been reported to be able to utilize either glutamine or ammonia as the nitrogen source. However, it has also been reported that when anthranilate synthase is separated from this complex, only ammonia can serve as a substrate.

B. Bartel (1997) *Ann. Rev. Plant Physiol. Plant Mol. Biol.* **48**, 51.
H. Zalkin (1993) *Adv. Enzymol.* **66**, 203.
R. Bentley (1990) *Crit. Rev. Biochem. Mol. Biol.* **25**, 307.
I. P. Crawford (1987) *Meth. Enzymol.* **142**, 300.
H. Zalkin (1985) *Meth. Enzymol.* **113**, 287.
D. E. Koshland, Jr., & A. Levitzki (1974) *The Enzymes*, 3rd ed., **10**, 539.
J. M. Buchanan (1973) *Adv. Enzymol.* **39**, 91.
H. Zalkin (1973) *Adv. Enzymol.* **38**, 1.

ANTI

A prefix that designates stereochemical relationships in molecular entities by referring to opposite sides of a reference plane, as contrasted with the prefix "*syn*" which refers to the same side of the reference plane. For example, two substituents are said to be in an *anti* arrangement if they are attached to atoms that are joined by a single bond and if the dihedral angle formed by the substituents is greater than 90°.

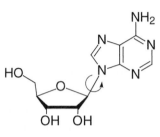

For substituents attached to adjacent atoms of a carbon-carbon double bond, the terms *cis* and *trans* or *E* and *Z* are used instead of *anti* and *syn*. This is also true of oximes and related molecules as well as to cyclic structures. The terms *anti* and *syn* are also used in the context of chemical reactions or transformations, designating the relative orientation of substituent addition or elimination.

ANTIAROMATICITY

A property associated with compounds that are destabilized by a closed loop of electrons. Antiaromatic compounds are typically planar and contain 4*n* electrons, where *n* is a positive integer, in overlapping parallel *p* orbitals. These compounds also have a paramagnetic ring current. Thus, protons on the outside of the ring will exhibit an upfield NMR chemical shift. *See* Aromatic

R. Breslow (1973) *Acc. Chem. Res.* **6**, 393.
F. Sondheimer (1972) *Acc. Chem. Res.* **5**, 81.

ANTI-ARRHENIUS BEHAVIOR

Reaction behavior that fails to agree with the Arrhenius temperature dependence. Enzyme-catalyzed reactions have been reported to display anti-Arrhenius behavior. As the temperature rises, the temperature dependence of the reaction rate does exhibit a normal Arrhenius pattern; however, at elevated temperatures, the reaction velocity falls off as a result of denaturation.

A reaction for which the mechanism changes in a temperature-dependent manner may also exhibit anti-Arrhenius behavior. *See* Arrhenius Equation; Arrhenius Plot, Nonlinear

ANTIBODY-HAPTEN INTERACTIONS

Specific, rapid, and high-affinity recognition of antigens by antibodies is essential for the host's immune system to respond to invasion by foreign cells and pathogens. Pecht and Lancet[1] have presented a cogent analysis of antibody interactions with haptens, and the interested reader will find their examination of the kinetics and thermodynamics to be particularly lucid. Briefly, for the simple one-step binding scheme,

$$\text{Antibody} + \text{Hapten} = \text{Complex}$$

their examination of the literature indicated the following interesting information about the magnitudes for the forward and reverse rate constants: (a) k_{on} varies over a range of three orders of magnitude (from 0.005 to 5×10^8 M^{-1}s^{-1}); (b) k_{off} varies over a range of eight powers of ten (from 3×10^{-5} s^{-1} to 6×10^3 s^{-1}); (c) rigorous treatment of the kinetics requires an additional isomerization step typical of most binding reactions. The latter is suggested, because the limiting rate of diffusion places an upper bound on the magnitude of the equilibrium constant for a one-step binding mechanism. Thus, in a one-step reaction mechanism, the value of k_{off} must be particularly important in models that attempt to explain wide variations in binding affinity that are achieved with different haptens as ligands.

A general two-step ligand-induced isomerization mechanism, as described by Eigen[2], can be written as:

$$A + B \underset{k_{-1}}{\overset{k_{+1}}{\rightleftharpoons}} AB \underset{k_{-2}}{\overset{k_{+2}}{\rightleftharpoons}} C .$$

His treatment puts the values for the first-step rate constants as

$$k_{+1} = (4\pi N/1000)\{(D_A + D_B)/R_{AB}\}$$

and

$$k_{-1} = 3(D_A + D_B)/R_{AB}^2$$

where N equals Avogadro's number, D_A and D_B are the diffusion constants, and R_{AB} is the so-called encounter distance. Pecht and Lancet[1] estimate the latter to be 3

to 15 Å, and for a hapten of molecular weight ~400, they use a diffusivity D_B of about 5×10^{-5} cm^2/s and an R_{AB} value of 7.5 Å. Taking a typical diffusivity for an immunoglobulin ($D_A = 3 \times 10^{-9}$ cm^2/sec), the values of k_{+1} and k_{-1} turn out to be 3×10^9 M^{-1}s^{-1} and 3×10^9 s^{-1}, suggesting that the equilibrium constant K_1 for the weakly interacting, encounter-controlled first step is around 1 M. The second step involves the hapten-induced conformational change from species AB to form species C. For the encounter step, Pecht and Lancet[1] suggest that k_{-1} (as well as the $\Delta H_{-1}^{\ddagger}$ and $\Delta S_{-1}^{\ddagger}$) may be considered to be constant for virtually all hapten-immunoglobulin reactions. On the other hand, values of k_{+1} and $\Delta S_{+1}^{\ddagger}$ fall within a range that depends on the geometry of ligand approach to its receptor site. For the isomerization step, unimolecular rate constants range from 10 to 100 per second, and the slowness of the transition is suggestive of long-range conformational changes that involve more than a few residues contacting each other in the complex.

[1] I. Pecht & D. Lancet (1977) in *Chemical Relaxation in Molecular Biology* (I. Pecht & R. Rigler, eds.) pp. 2-3, Springer-Verlag, Berlin.
[2] M. Eigen (1974) in *Quantum Statistical Mechanics in the Natural Sciences* (S. L. Minz & S. M. Wiedermayer, eds.) pp. 37-61, Plenum Press, New York.

Selected entries from *Methods in Enzymology* [vol, page(s)]:
Basic Principles of Antigen-Antibody Reactions: Proteins and polypeptides as antigens, **70**, 49; the experimental induction of antibodies to nucleic acids, **70**, 70; the preparation of antigenic hapten-carrier conjugates: a survey, **70**, 85; production of reagent antibodies, **70**, 104; preparation of Fab fragments from IgGs of different animal species, **70**, 142; use of carbodiimides in the preparation of immunizing conjugates, **70**, 151; use of glutaraldehyde as a coupling agent for proteins and peptides, **70**, 159; immunochemical analysis by antigen-antibody precipitation in gels, **70**, 166

Methods for the Detection of Antigens/Antibodies: Equilibrium and kinetic inhibition assays based upon fluorescence polarization, **70**, 3; fluorescence excitation transfer immunoassay (FETI), **70**, 28; indirect quenching fluoroimmunoassay, **70**, 60; the homogeneous substrate-labeled fluorescent immunoassay, **70**, 79; fluorescence immunoassays using plane surface solid phases (FIAPS), **70**, 87.

ANTIBONDING ORBITAL

The combination of two atomic orbitals to form a bond between two elements actually produces two molecular orbitals having different energies. The molecular orbital with the lower energy is referred to as the bonding orbital; the other orbital tends to push atoms apart and is termed the antibonding orbital. *See* Molecular Orbital; Bonding Orbital; Orbital

ANTICOOPERATIVITY

Another descriptor used in place of "negative cooperativity" to describe ligand binding interactions. Anticooperativity results whenever partial ligand saturation of a cooperative system results in reduced binding, either as a consequence of changes in the apparent ligand binding constants or from the reduction in the concentration of sites available for ligand binding. **See** Ligand Binding Cooperativity; Negative Cooperativity

ANTIPORT

A membrane protein or protein complex that is responsible for the simultaneous transport of two molecular entities or ions across the membrane, each entity traveling in opposite directions. **See** Uniport; Symport; Membrane Transport

APICOPHILICITY

The preference of a substituent to adopt an apical position in associative mechanisms for nucleophilic reactions on phosphorus. This property is discussed in detail by Benkovic and Shray[1], and the interested reader will also wish to consult Westheimer[2].

[1]S. J. Benkovic & K. J. Shray (1978) in *Transition States of Biochemical Processes* (R. D. Gandour & R. L. Schowen, eds.) p. 496, Plenum Press, New York.
[2]F. H. Westheimer (1968) *Acc. Chem. Res.* **1**, 70.

APOACTIVATOR

Any regulatory protein that stimulates transcription from one or more genes in the presence of a coactivator molecule.

APROTIC

A term, usually referring to a solvent, describing a compound which act neither as a proton donor nor a proton acceptor. Examples of polar aprotic solvents include dimethylformamide, dimethylsulfoxide, acetone, acetonitrile, sulfur dioxide, and hexamethylphosphoramide. Examples of nonpolar aprotic solvents include benzene and carbon tetrachloride. Studies of reactions in protic and aprotic solvents have demonstrated the importance of solvation on reactants, leaving groups, and transition states. Degrees of nucleophilicity as well as acidity are different in aprotic solvents. For example, small, negatively charged nucleophiles react more readily in polar aprotic solvents. It should also be noted that extremely strong acids or bases may still protonate or deprotonate compounds normally considered aprotic.

C. Reichardt (1988) *Solvents and Solvent Effects in Organic Chemistry*, 2nd ed., VCH, New York.
G. W. Klumpp (1982) *Reactivity in Organic Chemistry*, Wiley, New York.
T. W. Bentley & P. von R. Schleyer (1977) *Adv. Phys. Org. Chem.* **14**, 1.
F. G. Bordwell & D. L. Hughes (1981) *J. Org. Chem.* **46**, 3570.

A *PRIORI* PROBABILITY

A statistical method wherein the probability is assigned to an event or parameter prior to any testing.

APYRASE

This calcium ion-dependent enzyme [EC 3.6.1.5] (also known as ATP-diphosphatase, adenosine diphosphatase, and ATP-diphosphohydrolase) catalyzes the hydrolysis (with two H_2O) of ATP to generate AMP and two orthophosphate ions. The enzyme will also utilize ADP as a substrate as well as other nucleoside triphosphates and diphosphates.

P. S. Krishnan (1955) *Meth. Enzymol.* **2**, 591.

AQUACOBALAMIN REDUCTASE

Aquacobalamin reductase (NADH) [EC 1.6.99.8] catalyzes the reaction of two aquacob(III)alamin with NADH to form 2 cob(II)alamin and NAD^+. Aquacobalamin reductase (NADPH) [EC 1.6.99.11] catalyzes the exact same reaction, albeit with NADPH. In addition, this enzyme can utilize hydroxycobalamin as a substrate, but not cyanocobalamin.

F. Watanabe & Y. Nakano (1997) *Meth. Enzymol.* **281**, 289.

AQUAMOLALITY

A concentration scale for solutes in aqueous solutions, equal to moles of solute/55.51 mol water. It is frequently used in studies of solvent isotope effects. As pointed out by Schowen and Schowen[1], the choice of standard states can change the sign for the free energy of transfer of a species from one solvent to another, even from HOH and DOD. The commonly used concentration scales are molarity, mole fraction, aquamolality, and molality. Free energies tend to be nearly the same on all but the molality scale, on which they are about 63 cal mol^{-1} more positive at 298 K than on the first three scales. The interested reader should consult Table I of Schowen and Schowen[1] for a compilation of the relative stabilities of various molecules in isotopic waters at 298 K. **See** *Concentration*

[1]K. B. Schowen & R. L. Schowen (1982) *Meth. Enzymol.* **87**, 551.

AQUATION

The process by which one or more water molecules are incorporated into another molecular entity until no other atoms, ions, and groups or moieties are further displaced. In the case of metal ions, aquation reflects the ability of nonbonded electrons on water oxygens to form coordinate covalent bonds with the metal ion. **See also** *Hydration; Water*

α-L-ARABINOFURANOSIDASE

This enzyme [EC 3.2.1.55], also called arabinosidase, catalyzes the hydrolysis of terminal nonreducing α-L-arabinofuranoside residues in α-L-arabinosides. Actual substrates include α-L-arabinofuranosides, α-L-arabinans containing (1,3)- and/or (1,5)-linkages, arabinoxylans, and arabinogalactans. It should be noted that some β-galactosidases and β-D-fucosidases will also hydrolyze α-L-arabinosides.

G. Davies, M. L. Sinnott & S. G. Withers (1998) *Comprehensive Biological Catalysis: A Mechanistic Reference* **1**, 119.

D-ARABINOKINASE

This enzyme [EC 2.7.1.54] catalyzes the reaction of ATP with D-arabinose to generate ADP and D-arabinose 5-phosphate.

W. A. Volk (1966) *Meth. Enzymol.* **9**, 442.

ARABINONATE DEHYDRATASE

Arabinonate dehydratase [EC 4.2.1.5] catalyzes the conversion of D-arabinonate to 2-dehydro-3-deoxy-D-arabinonate and water whereas L-arabinonate dehydratase [EC 4.2.1.25] catalyzes the conversion of L-arabinonate to 2-dehydro-3-deoxy-L-arabinonate and water.

W. A. Wood (1971) *The Enzymes*, 3rd ed., **5**, 573.

ARABINOSE ISOMERASE

Arabinose isomerase [EC 5.3.1.3] catalyzes the interconversion of D-arabinose and D-ribulose. The enzyme will also utilize L-fucose and more slowly utilize L-galactose and D-altrose as substrates. L-Arabinose isomerase [EC 5.3.1.4] catalyzes the interconversion of L-arabinose and L-ribulose.

D. J. Creighton & N. S. R. K. Murthy (1990) *The Enzymes*, 3rd ed., **19**, 323.
I. A. Rose (1975) *Adv. Enzymol.* **43**, 491.
E. A. Noltmann (1972) *The Enzymes*, 3rd ed., **6**, 271.

ARABINOSE-5-PHOSPHATE ISOMERASE

This enzyme [EC 5.3.1.13] catalyzes the interconversion of D-arabinose 5-phosphate to D-ribulose 5-phosphate.

E. A. Noltmann (1972) *The Enzymes*, 3rd ed., **6**, 271.

ARACHIDONYL-CoA SYNTHETASE

This enzyme [EC 6.2.1.15], also known as arachidonate:CoA ligase, catalyzes the reaction of arachidonate with ATP and coenzyme A to generate arachidonyl-CoA, AMP, and pyrophosphate (or, diphosphate). The enzyme can also use 8,11,14-icosatrienoate as a substrate, but not the other long-chain fatty acids. It should be noted that this enzyme is not identical to long-chain acyl-CoA synthetase [EC 6.2.1.3].

M. Laposata (1990) *Meth. Enzymol.* **187**, 237.

ARGINASE

This enzyme [EC 3.5.3.1], also referred to as arginine amidinase and canavanase, catalyzes the hydrolysis of arginine resulting in the formation of urea and ornithine. The enzyme, which uses a manganese ion as a cofactor, will also utilize α-N-substituted L-arginines and canavanine as substrates.

J. E. Penner-Hahn (1998) *Comprehensive Biological Catalysis: A Mechanistic Reference* **3**, 439.
S. Ratner (1973) *Adv. Enzymol.* **39**, 1.

ARGININE DECARBOXYLASE

This enzyme [EC 4.1.1.19] uses pyridoxal phosphate as a cofactor in catalyzing the conversion of arginine to carbon dioxide and agmatine.

T. Hashimoto & Y. Yamada (1994) *Ann. Rev. Plant Physiol. Plant Mol. Biol.* **45**, 257.
M. H. O'Leary (1980) *Meth. Enzymol.* **64**, 83.
E. A. Boeker & E. E. Snell (1972) *The Enzymes*, 3rd ed., **6**, 217.

ARGININE DEIMINASE

This enzyme [EC 3.5.3.6], also known as arginine dihydrolase, catalyzes the hydrolysis of arginine (with canavanine as a substrate as well) to produce citrulline and ammonia.

J. Thompson & S. P. F. Miller (1991) *Adv. Enzymol.* **64**, 317.

ARGININE KINASE

This enzyme [EC 2.7.3.3] catalyzes the reaction of arginine with ATP to produce ADP and *N*-phosphoarginine.

J. F. Morrison (1973) *The Enzymes*, 3rd ed., **8**, 457.

ARGININE 2-MONOOXYGENASE

This enzyme [EC 1.13.12.1] catalyzes the reaction of arginine with dioxygen to generate 4-guanidobutanamide, carbon dioxide, and water. The enzyme can also utilize canavanine and homoarginine as substrates.

S. Yamamoto & Y. Ishimura (1991) *A Study of Enzymes* **2**, 315.
V. Massey & P. Hemmerich (1975) *The Enzymes*, 3rd ed., **12**, 191.

ARGININE RACEMASE

This enzyme [EC 5.1.1.9] catalyzes the interconversion of L- and D-arginine. The enzyme requires pyridoxal phosphate as a cofactor.

M. E. Tanner & G. L. Kenyon (1998) *Comprehensive Biological Catalysis: A Mechanistic Reference* **2**, 7.
E. Adams (1976) *Adv. Enzymol.* **44**, 69.
E. Adams (1972) *The Enzymes*, 3rd ed., **6**, 479.

ARGININOSUCCINATE LYASE

This enzyme [EC 4.3.2.1], also known as arginosuccinase, catalyzes the conversion of *N*-argininosuccinate to fumarate and arginine.

V. E. Anderson (1998) *Comprehensive Biological Catalysis: A Mechanistic Reference* **2**, 115.
J. V. Schloss & M. S. Hixon (1998) *Comprehensive Biological Catalysis: A Mechanistic Reference* **2**, 43.
L. S. Mullins & F. M. Raushel (1995) *Meth. Enzymol.* **249**, 398.
H. J. Vogel & R. H. Vogel (1974) *Adv. Enzymol.* **40**, 65–90.
S. Ratner (1973) *Adv. Enzymol.* **39**, 1.
S. Ratner (1972) *The Enzymes*, 3rd ed., **7**, 167.

ARGININOSUCCINATE SYNTHETASE

This enzyme [EC 6.3.4.5], also referred to as citrulline:aspartate ligase, catalyzes the reaction of citrulline with aspartate and ATP to produce argininosuccinate, AMP, and pyrophosphate (or, diphosphate).

F. M. Raushel & J. J. Villafranca (1988) *Crit. Rev. Biochem.* **23**, 1.
H. J. Vogel & R. H. Vogel (1974) *Adv. Enzymol.* **40**, 65.
S. Ratner (1973) *Adv. Enzymol.* **39**, 1.

ARISTOLOCHENE SYNTHASE

This enzyme [EC 2.5.1.40] catalyzes the conversion of *trans,trans*-farnesyl diphosphate (or, *trans,trans*-farnesyl pyrophosphate) to aristolochene and diphosphate (or, pyrophosphate). The enzyme-catalyzed reaction proceeds through an initial internal transfer of the farnesyl group to the C10 position resulting in the monocyclic intermediate germacrene A. This is followed by further cyclization and an internal methyl transfer converting the intermediate into aristolochene.

R. A. Gibbs (1998) *Comprehensive Biological Catalysis: A Mechanistic Reference* **1**, 31.

AROMATIC

1. Referring to a substance that behaves chemically like benzene. 2. Referring to a molecule that exhibits an ability to sustain an induced ring current. With few exceptions, protons covalently linked to an aromatic ring exhibit a downfield chemical shift in an NMR experiment as compared to protons that are not associated with an aromatic ring. Because of resonance stabilization, aromatic compounds are especially stable, and they exhibit greater stability than that of any specific canonical form or so-called contributing form (*e.g.*, any of the well known Kekulé or Dewar structures of benzene). Most commonly, aromatic molecules have cyclically delocalized π electrons, and the number of π electrons obeys Hückel's rule. In addition, aromatic rings will exhibit equal, or near equal, bond lengths, planarity, and the ability to undergo aromatic substitution reactions. **See** *Antiaromaticity; Hückel's Rule*

D. Lloyd (1989) *The Chemistry of Conjugated Cyclic Compounds*, Wiley, New York.
P. J. Garratt (1986) *Aromaticity*, Wiley, New York.
K. Jug & A. M. Koster (1991) *J. Phys. Org. Chem.* **4**, 163.
Z. Zhou & R. G. Parr (1989) *J. Amer. Chem. Soc.* **111**, 7371.
A. R. Katritzky, P. Barczynski, G. Musumarra, D. Pisano & M. Szafran (1989) *J. Amer. Chem. Soc.* **111**, 7.

AROMATIC-AMINO-ACID AMINOTRANSFERASE

This enzyme [EC 2.6.1.57] catalyzes the reversible reaction of an aromatic amino acid with α-ketoglutarate to generate an aromatic oxo acid and glutamate. Pyridoxal phosphate is a required cofactor. Methionine can also act as a weak substrate, substituting for the aromatic amino acid. Oxaloacetate substitutes for α-ketoglutarate.

J. DiRuggiero & F. T. Robb (1996) *Adv. Protein Chem.* **48**, 311.
C. Mavrides (1987) *Meth. Enzymol.* **142**, 253.
A. J. L. Cooper (1985) *Meth. Enzymol.* **113**, 73.

AROMATIC-AMINO-ACID DECARBOXYLASE

This enzyme [EC 4.1.1.28] (also referred to as DOPA decarboxylase, tryptophan decarboxylase, and hydroxytryptophan decarboxylase) catalyzes the conversion of tryptophan to tryptamine and carbon dioxide. Pyridoxal phosphate is a required cofactor for this enzyme. The enzyme can also utilize 5-hydroxytryptophan and dihydroxyphenylalanine as substrates.

T. Hashimoto & Y. Yamada (1994) *Ann. Rev. Plant Physiol. Plant Mol. Biol.* **45**, 257.
M. A. Ator & P. R. Ortez de Montellano (1990) *The Enzymes*, 3rd ed., **19**, 213.
T. L. Sourkes (1987) *Meth. Enzymol.* **142**, 170.
E. A. Boeker & E. E. Snell (1972) *The Enzymes*, 3rd ed., **6**, 217.

AROMATIC-AMINO-ACID:GLYOXYLATE AMINOTRANSFERASE

This enzyme [EC 2.6.1.60] catalyzes the reversible reaction of an aromatic amino acid with glyoxylate to generate an aromatic oxo acid and glycine. Aromatic amino acids that can act as substrates include phenylalanine, kynurenine, tyrosine, and histidine. Pyruvate and hydroxypyruvate can act as glyoxylate substitutes as well.

T. Takada & T. Noguchi (1987) *Meth. Enzymol.* **142**, 273.

ARRHENIUS CONSTANT

A dimensionless number equal to the activation energy of a reaction (E_a) divided by the product of the universal gas constant (R) and the absolute temperature, T: thus, $E_a/(RT)$.

ARRHENIUS EQUATION & PLOT

A relationship[1] describing the temperature dependence of an individual rate constant:

$$(\partial \ln k/\partial T)_P = E_a/RT^2$$

where E_a is the energy of activation for the reaction having the rate constant k at constant pressure P, R is the universal gas constant, and T is the absolute temperature. The International Organization for Standardization (ISO), the International Union of Pure and Applied Chemistry (IUPAC), and the International Union of Pure and Applied Physics (IUPAP) have recommended that E_a or E, written in italic type[2], be the symbol for activation energy. Nevertheless, E_{act}, E^*, $E^{\ddagger}$, and $E^{\#}$ are often used (in the older literature, the symbol μ was also common).

If one assumes that E_a is not temperature-dependent, integration of the above differential equation yields:

$$\ln k = -\frac{E_a}{RT} + \text{Constant}$$

Letting the constant of integration equal $\ln A$, the above equation becomes

$$k = Ae^{-E_a/RT}$$

This is the well-known Arrhenius equation in which A is referred to as the frequency factor or pre-exponential factor.

A plot of $\ln k$ against the reciprocal of the absolute temperature (an Arrhenius plot) will produce a straight line having a slope of $-E_a/R$. The frequency factor can be obtained from the vertical intercept, $\ln A$. The Arrhenius relationship has been demonstrated to be valid in a large number of cases (for example, colchicine-induced GTPase activity of tubulin[3] or the binding of *N*-acetyl-phenylalanyl-tRNA[Phe] to ribosomes[4]). In practice, the Arrhenius equation is only a good approximation of the temperature dependence of the rate constant, a point which will be addressed below.

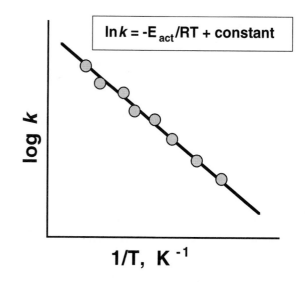

Arrhenius recognized that for molecules to react they must attain a certain critical energy, E_a. On the basis of collision theory, the rate of reaction is equal to the number of collisions per unit time (the frequency factor) multiplied by the fraction of collisions that results in a reaction. This relationship was first developed from the kinetic theory of gases[5,6]. For a bimolecular reaction, the bimolecular rate constant, k, can be expressed as

$$k = Ze^{-E_a/RT}$$

where $e^{-E_a/RT}$ is the Boltzmann factor representing the probability that a collision will be effective and Z is the number of collisions between reactants per second in $1\ cm^3$.

Effect of Inhibitors on Arrhenius Plots. The presence of inhibitors can lead to nonlinear Arrhenius plots as well as an increase in E_a. For example, jack bean urease exhibits a linear Arrhenius plot in the absence of an inhibitor, but a curved plot in the presence of 34 mM sodium sulfite[7].

Effect of Solvent on Arrhenius Plots. If water is a substrate, then the presence of an organic solvent, which may disrupt the structure and/or orientation of water, may alter the Arrhenius plot. For example, a linear plot is seen with fumarate hydratase in the presence of 10% methanol. However, the plot is biphasic in the presence of 10% ethanol[8]. *See Boltzmann Distribution; Collision Theory; Temperature Dependency, Transition-State Theory; Energy of Activation; Q_{10}*

[1]S. Arrhenius (1889) *Zeit. für Physik. Chem.* **4**, 226.
[2]I. Mills (1988) *Quantities, Units, and Symbols in Physical Chemistry*, Blackwell Scientific Publ., Oxford.
[3]B. Perez-Ramirez & S. N. Timasheff (1994) *Biochemistry* **33**, 6262.
[4]S. Schilling-Bartetzko, A. Bartetzko & K. H. Nierhaus (1992) *J. Biol. Chem.* **267**, 4703.
[5]W. C. McC. Lewis (1918) *J. Chem. Soc.* **113**, 471.
[6]M. Trautz (1916) *Zeit. Anorg. Chem.* **96**, 1.
[7]G. B. Kistiakowsky & R. Lumry (1949) *J. Amer. Chem. Soc.* **71**, 2006.
[8]V. Massey (1953) *Biochem. J.* **53**, 72.

ARRHENIUS PLOTS (NONLINEAR)

Arrhenius plots of enzyme-catalyzed reactions occasionally exhibit curvilinearity[1] and appear to have two straight-line segments intersecting each other at an angle. While suggestive of a change in mechanism signaled by a break in the activation energy at some specific transition temperature, Kistiakowsky and Lumry[2] have pointed out that in many of those cases for which two intersecting lines are drawn, intersecting at a "transition temperature," the data are better described by a smooth curve. Dixon and Webb[3] offered the following summary of possible explanations for nonlinearity in Arrhenius plots. These include: (1) a change in the properties of the solvent with temperature; (2) the occurrence of two or more parallel or competing reactions (*e.g.*, catalyzed by isozymes), each with different energies of activation;

(3) the presence of two successive reactions with different energies of activation; (4) the enzyme has two (or more) different forms, each with its own activation energy; (5) a reversible inactivation of the enzyme; or (6) a discontinuity affecting the forward reaction only.

Finally, yet another issue enters into the interpretation of nonlinear Arrhenius plots of enzyme-catalyzed reactions. As is seen in the examples above, one typically plots $\ln V_{max}$ (or, $\ln k_{cat}$) *versus* the reciprocal absolute temperature. This protocol is certainly valid for rapid equilibrium enzymes whose rate-determining step does not change throughout the temperature range studied (and, in addition, remains rapid equilibrium throughout this range). However, for steady-state enzymes, other factors can influence the interpretation of the nonlinear data. For example, for an ordered two-substrate, two-product reaction, k_{cat} is equal to $k_5 k_7/(k_5 + k_7)$ in which k_5 and k_7 are the off-rate constants for the two products. If these two rate constants have a different temperature dependency (*e.g.*, $k_5 > k_7$ at one temperature but not at another temperature), then a nonlinear Arrhenius plot may result. *See Arrhenius Equation; Q_{10}; Transition-State Theory; van't Hoff Relationship*

[1]K. J. Laidler & B. F. Peterman (1979) *Meth. Enzymol.* **63**, 183.
[2]G. B. Kistiakowski & R. Lumry (1949) *J. Amer. Chem. Soc.* **71**, 2006.
[3]M. Dixon & E. C. Webb (1979) *Enzymes*, 3rd ed., Academic Press, New York.

ARSENOLYSIS

Arsenate-dependent "hydrolysis" of a covalent bond. Arsenate dianion ($HAsO_4^{2-}$) is a structural analogue of orthophosphate dianion (HPO_4^{2-}) and often sacts as a phosphate analogue in many enzyme-catalyzed reaction utilizing phosphate as a substrate. (Note that the pK_a values for arsenic acid (at 18°C) are 2.25, 6.77, and 11.60 whereas the pK_a values for H_3PO_4 (at 25°C) are 2.15, 7.20, and 12.33.) The most significant fact is that arsenate esters (*i.e.*, $R-O-AsO_3^{2-}$) are kinetically and thermodynamically unstable in aqueous solutions. When formed on and released from the enzyme's surface, they rapidly hydrolyze to form ROH and $HAsO_4^{2-}$. By contrast, phosphate esters are far more kinetically stable.

An example of an enzyme that exhibits arsenolysisis is sucrose phosphorylase[1], an enzyme that catalyzes the reaction:

$$\text{Sucrose} + \text{Orthophosphate} = \text{D-Fructose}$$
$$+ \text{D-Glucose 1-phosphate}$$

In the presence of arsenate (and absence of phosphate), the enzyme will exhibit a sucrase activity. The enzyme also catalyzes arsenolysis of D-glucose 1-phosphate. In the absence of the enzyme, there is very little hydrolysis (or arsenolysis) of D-glucose 1-phosphate. In the presence of the enzyme and arsenate, however, D-glucose 1-phosphate rapidly converts to D-glucose and orthophosphate. This observation supports the formation of a glucosyl-enzyme intermediate. When arsenate is present in significant amounts, the arsenate ester is formed and is almost immediately hydrolyzed to D-glucose and arsenate. Interestingly, maltose phosphorylase and cellobiose phosphorylase, which are both able to form D-glucose 1-phosphate as a product, do not facilitate arsenolysis. This is one of many lines of evidence indicating that the latter enzymes have different mechanisms than that of sucrose phosphorylase. Other enzyme reactions that undergo arsenolysis include glutamine synthetase, tubulin:tyrosine ligase (ADP-forming), and D-alanyl-D-alanine synthetase.

[1]J. Mieyal & R. Abeles (1972) *The Enzymes*, 3rd ed., **7**, 515.

ARYLALKYLAMINE N-ACETYLTRANSFERASE

This enzyme [EC 2.3.1.87], also known as serotonin acetyltransferase and serotonin acetylase, catalyzes the reaction of acetyl-CoA and arylalkylamine to generate coenzyme A and N-acetylarylalkylamine. The enzyme exhibits a rather narrow specificity toward other arylalkylamines. This enzyme is distinct from arylamine acetyltransferase.

M. A. Namboodisi, R. Dubbels & D. C. Klein (1987) *Meth. Enzymol.* **142**, 583.

ARYLAMINE N-ACETYLTRANSFERASE

This enzyme [EC 2.3.1.5], also known as acetyl-CoA: arylamine N-acetyltransferase and arylamine acetylase, catalyzes the reaction of acetyl-CoA with an arylamine to produce coenzyme A and an N-acetylarylamine. This enzyme exhibits a low specificity with respect to the aromatic amine substrate. In fact, even serotonin can serve as a substrate. The enzyme has also been reported to catalyze acetyl-transfer reactions between arylamines without the use of coenzyme A.

ARYLFORMAMIDASE

This enzyme [EC 3.5.1.9], also called kynurenine formamidase and formylkynureninase, catalyzes the hydrolysis of N-formylkynurenine to yield formate and kynurenine. The enzyme will also use other aromatic formylamines as substrates.

E. Katz, D. Brown & M. J. M. Hitchcock (1987) *Meth. Enzymol.* **142**, 225.

ARYLSULFATASES

Arylsulfatase [EC 3.1.6.1], also known simply as sulfatase, catalyzes the hydrolysis of a phenol sulfate, thereby producing a phenol and sulfate. This enzyme classification represents a collection of enzymes with rather similar specificities. (1) Steryl-sulfatase [EC 3.1.6.2], also referred to as arylsulfatase C and steroid sulfatase, catalyzes the hydrolysis of 3-β-hydroxyandrost-5-en-17-one 3-sulfate to 3-β-hydroxyandrost-5-en-17-one and sulfate. The enzyme utilizes other steryl sulfates as substrates. (2) Cerebroside-sulfatase [EC 3.1.6.8], or arylsulfatase A, catalyzes the hydrolysis of a cerebroside 3-sulfate to yield a cerebroside and sulfate. The enzyme will also hydrolyze the galactose 3-sulfate bond present in a number of lipids. In addition, the enzyme will also hydrolyze ascorbate 2-sulfate and other phenol sulfates.

G. Lowe (1998) *Comprehensive Biological Catalysis: A Mechanistic Reference* **1**, 627.
E. Conzelmann & K. Sandhoff (1987) *Adv. Enzymol.* **60**, 89.
A. B. Roy (1987) *Meth. Enzymol.* **143**, 207.
Y.-T. Li & S.-C. Li (1983) *The Enzymes*, 3rd ed., **16**, 427.
A. B. Roy (1971) *The Enzymes*, 3rd ed., **5**, 1.
R. G. Nicholls & A. B. Roy (1971) *The Enzymes*, 3rd ed., **5**, 21.

ARYLSULFOTRANSFERASE

This enzyme [EC 2.8.2.1], also known as phenol sulfotransferase and sulfokinase, catalyzes the reaction of 3'-phosphoadenylylsulfate with a phenol to yield adenosine 3',5'-bisphosphate and an aryl sulfate. The enzyme can utilize a number of aromatic compounds as substrates.

G. Lowe (1998) *Comprehensive Biological Catalysis: A Mechanistic Reference* **1**, 627.
J. V. Schloss & M. S. Hixon (1998) *Comprehensive Biological Catalysis: A Mechanistic Reference* **2**, 43.
S. G. Ramaswamy & W. B. Jakoby (1987) *Meth. Enzymol.* **143**, 201.

ASCORBATE–CYTOCHROME b_5 REDUCTASE

This enzyme [EC 1.10.2.1] catalyzes the reaction of ascorbate with ferricytochrome b_5 to yield monodehydroascorbate and ferrocytochrome b_5.

J. V. Schloss & M. S. Hixon (1998) *Comprehensive Biological Catalysis: A Mechanistic Reference* **2**, 43.

ASCORBATE OXIDASE

This enzyme [EC 1.10.3.3], also called ascorbase, catalyzes the reaction of two ascorbate molecules with dioxygen to produce two dehydroascorbate molecules and two water molecules. The enzyme utilizes a copper ion as a cofactor.

A. Messerschmidt (1998) *Comprehensive Biological Catalysis: A Mechanistic Reference* **3**, 401.
J. V. Schloss & M. S. Hixon (1998) *Comprehensive Biological Catalysis: A Mechanistic Reference* **2**, 43.
E. T. Adman (1991) *Adv. Protein Chem.* **42**, 145.
B. G. Malmström, L.-E. Andréasson & B. Reinhammer (1975) *The Enzymes*, 3rd ed., **12**, 507.

ASCORBATE PEROXIDASE

This enzyme [EC 1.11.1.11] catalyzes the reaction of ascorbate with hydrogen peroxide to produce dehydroascorbate and two water molecules.

H. B. Dunford (1998) *Comprehensive Biological Catalysis: A Mechanistic Reference* **3**, 195.

ASPARAGINASE

This enzyme [EC 3.5.1.1] catalyzes the hydrolysis of asparagine to ammonia and aspartate.

J. C. Wriston, Jr. (1985) *Meth. Enzymol.* **113**, 608.
J. C. Wriston, Jr. & T. O. Yellin (1973) *Adv. Enzymol.* **39**, 185.
J. C. Wriston, Jr. (1971) *The Enzymes*, 3rd ed., **4**, 101.

ASPARAGINE AMINOTRANSFERASE

This vitamin B_6-dependent enzyme [EC 2.6.1.14], also referred to as asparagine–oxo-acid aminotransferase, catalyzes the reversible reaction of asparagine and a 2-oxo acid to yield 2-oxosuccinamate and an amino acid.

A. J. L. Cooper & A. Meister (1985) *Meth. Enzymol.* **113**, 602.

ASPARAGINE SYNTHETASE

Asparagine synthetase (glutamine-hydrolyzing) [EC 6.3.5.4] catalyzes the reaction of aspartate with ATP and glutamine to produce asparagine, AMP, diphosphate (or, pyrophosphate), and glutamate. The enzyme will also hydrolyze glutamine to glutamate and ammonia in the absence of the other substrates. Asparagine synthetase (ammonia-using) [EC 6.3.1.1], also known as aspartate—ammonia ligase, catalyzes the reaction of aspartate with ammonia and ATP to produce asparagine, AMP, and diphosphate (or, pyrophosphate). Asparagine synthetase (ammonia-using, ADP-forming) is an enzyme [EC 6.3.1.4] reported to catalyze the reaction of aspartate with ammonia and ATP to produce asparagine, ADP, and orthophosphate.

N. Richards & S. M. Schuster (1998) *Adv. Enzymol.* **72**, 145.
H.-M. Lam, K. T. Coschigamo, I. C. Oliveira, R. Melo-Oliveira & G. M. Coruzzi (1996) *Ann. Rev. Plant Physiol. Plant Mol. Biol.* **47**, 569.
H. Zalkin (1993) *Adv. Enzymol.* **66**, 203.
A. Meister (1974) *The Enzymes*, 3rd ed., **10**, 561.
J. M. Buchanan (1973) *Adv. Enzymol.* **39**, 91.

ASPARTATE AMINOTRANSFERASE

This enzyme [EC 2.6.1.1] (also known as transaminase A, glutamic:oxaloacetic transaminase, and glutamic:aspartic transaminase) catalyzes the reversible reaction of aspartate with α-ketoglutarate to produce oxaloacetate and glutamate. Pyridoxal phosphate is a required cofactor. The enzyme has a relatively broad specificity, and tyrosine, phenylalanine, and tryptophan can all serve as substrates.

R. A. John (1998) *Comprehensive Biological Catalysis: A Mechanistic Reference* **2**, 173.
J. DiRuggiero & F. T. Robb (1996) *Adv. Protein Chem.* **48**, 311.
H.-M. Lam, K. T. Coschigamo, I. C. Oliveira, R. Melo-Oliveira & G. M. Coruzzi (1996) *Ann. Rev. Plant Physiol. Plant Mol. Biol.* **47**, 569.
M. Martinez-Carrion, A. Artigues, A. Berezov, M. L. Bianconi, A. M. Reyes & A. Iriarte (1995) *Meth. Enzymol.* **259**, 590.
M. A. Ator & P. R. Ortez de Montellano (1990) *The Enzymes*, 3rd ed., **19**, 213.
D. J. Creighton & N. S. R. K. Murthy (1990) *The Enzymes*, 3rd ed., **19**, 323.
C. Mavrides (1987) *Meth. Enzymol.* **142**, 253-267.
T. Yagi, H. Kagamiyama, M. Nozaki & K. Soda (1985) *Meth. Enzymol.* **113**, 83.
A. E. Martell (1982) *Adv. Enzymol.* **53**, 163.
D. E. Metzler (1979) *Adv. Enzymol.* **50**, 1.
A. E. Braunstein (1973) *The Enzymes*, 3rd ed., **9**, 379.

ASPARTATE AMMONIA-LYASE

This enzyme [EC 4.3.1.1], also known as aspartase and fumaric aminase, catalyzes the conversion of aspartate to fumarate and ammonia.

V. E. Anderson (1998) *Comprehensive Biological Catalysis: A Mechanistic Reference* **2**, 115.
J. V. Schloss & M. S. Hixon (1998) *Comprehensive Biological Catalysis: A Mechanistic Reference* **2**, 43.
M. Tokushige (1985) *Meth. Enzymol.* **113**, 618.
K. R. Hanson & E. A. Havir (1972) *The Enzymes*, 3rd ed., **7**, 75.

ASPARTATE CARBAMOYLTRANSFERASE

The prototypical allosteric regulatory protein isolated from *Escherichia coli* (molecular weight 310,000). The enzyme [EC 2.1.3.2], also known as aspartate transcarbamylase and carbamylaspartotranskinase, catalyzes the first committed step (*i.e.*, the transfer of a carbamoyl

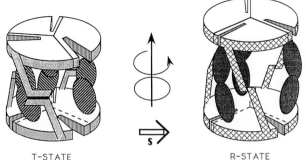

T-STATE   R-STATE

Structural model for conformational changes in ATCase, based on the X-ray crystallographic investigations of Lipscomb and colleagues. Note that substrates are thought to enter through a channel, and allosteric effectors alter channel accessibility.

group from carbamoyl phosphate to the α-amino group of aspartate with the liberation of phosphate) in pyrimidine biosynthesis. This enzyme (ATCase) can be dissociated by mercurials into two catalytic trimers and three regulatory dimers. ATCase is feedback-inhibited by the pyrimidine nucleotide biosynthetic end-product CTP and is activated by ATP.

N. M. Allewell & V. J. LiCata (1995) *Meth. Enzymol.* **259**, 608.
W. W. Cleland (1995) *Meth. Enzymol.* **249**, 341.
W. N. Lipscomb (1994) *Adv. Enzymol.* **68**, 67.
E. Lolis & G. A. Petsko (1990) *Ann. Rev. Biochem.* **59**, 597.
E. R. Kantrowitz & W. N. Lipscomb (1988) *Science* **241**, 669.
J. S. Brabson, M. R. Maurizi & R. L. Switzer (1985) *Meth. Enzymol.* **113**, 627.
B. L. Vallee & A. Galdes (1984) *Adv. Enzymol.* **56**, 283.
G. R. Jacobson & G. R. Stark (1973) *The Enzymes*, 3rd ed., **9**, 225.

Selected entries from *Methods in Enzymology* [*vol*, page(s)]: Activation, **64**, 177; concerted allosteric model, **64**, 173; isotope exchange properties, **64**, 10; negative cooperativity, **64**, 189; rapid relaxation measurement, **64**, 188, 189; sequential model, **64**, 173; site-directed mutagenesis, **202**, 717, 725; structure-function analysis, **202**, 694; supersecondary structures, genetic exchange within domains, **202**, 704; allosteric mechanism, **259**, 614, 627; binding site number, **259**, 615-616; concentration, **259**, 615; differential scanning calorimetry [assembly effects, **259**, 624-625; buffer sensitivity, **259**, 625; ligation effects, **259**, 625; mutation effects, **259**, 626]; electrostatic interactions, **259**, 626-628; *Escherichia coli* ATCase [cooperativity in (allosteric structures and model testing, **249**, 554-555; experimental evaluation, **249**, 548-554; heterotropic, **249**, 552-553; homotropic, **249**, 551-552; mutational analysis, **249**, 554; structural model, **249**, 549-551); intersubunit ligand binding sites, **249**, 559-560; site-directed mutants, acid-base catalysis, **249**, 115-116; ligand binding [assays, **259**, 615-616; cooperativity, detection, **259**, 616-617; dissociation constants, **259**, 616; enthalpy, **259**, 618-620; entropy, **259**, 618-621; free energy determination, **259**, 623; nucleotide triphosphates, **259**, 619-620, 624; sites, interaction distances, **249**, 561-562; subunit interactions, **259**, 610, 622-624]; linkage analysis, **259**, 614, 621-624, 627; mechanism, **249**, 362-363 [contribution to protein structure knowledge, **259**, 612; equilibrium isotope exchange investigation, **249**, 470; molecular modeling, **259**, 612; thermodynamic definition by state functions, **259**, 611, 628]; modifier action, equilibrium isotope exchange investigation, **249**, 471-472; pK values, sensitivity to side-chain positions, **259**, 627-628; reac-

tion catalyzed, **259**, 609; regulation, **259**, 609-610; reversible denaturation, **259**, 626; site-directed mutants, equilibrium isotope exchange investigations, **249**, 474-478; structure, **259**, 609; subunit interactions, free energy, **259**, 613, 617-618, 620-621, 623.

ASPARTATE α-DECARBOXYLASE

This enzyme [EC 4.1.1.11] catalyzes the conversion of aspartate to β-alanine and carbon dioxide. Pyruvate is a cofactor for the *E. coli* enzyme.

M. L. Hackert & A. E. Pegg (1998) *Comprehensive Biological Catalysis: A Mechanistic Reference* **2**, 201.
J. M. Williamson (1985) *Meth. Enzymol.* **113**, 589.

ASPARTATE β-DECARBOXYLASE

This enzyme [EC 4.1.1.12], also known as desulfinase, catalyzes the conversion of aspartate to alanine and carbon dioxide. Pyridoxal phosphate is a required cofactor. The enzyme will also catalyze the decarboxylation of aminomalonate as well as the desulfination of 3-sulfinoalanine to sulfite and alanine.

A. E. Braunstein (1973) *The Enzymes*, 3rd ed., **9**, 379.
E. A. Boeker & E. E. Snell (1972) *The Enzymes*, 3rd ed., **6**, 217.

ASPARTATE KINASE

Aspartate kinase [EC 2.7.2.4], also known as aspartokinase, catalyzes the reaction of aspartate with ATP to produce 4-phosphoaspartate and ADP. The enzyme isolated from *E. coli* is a multifunctional protein, also exhibiting the ability to catalyze the reaction of homoserine with $NAD(P)^+$ to produce aspartate 4-semialdehyde and $NAD(P)H$ (that is, the activity of homoserine dehydrogenase, EC 1.1.1.3).

I. Saint-Gerons, C. Parsot, M. M. Zakin, O. Bârzy & G. N. Cohen (1988) *Crit. Rev. Biochem.* **23**, S1.
G. N. Cohen (1985) *Meth. Enzymol.* **113**, 596.
P. Truffa-Bachi (1973) *The Enzymes*, 3rd ed., **8**, 509.

D-ASPARTATE OXIDASE

This enzyme [EC 1.4.3.1] catalyzes the reaction of D-aspartate with water and dioxygen to generate oxaloacetate, ammonia, and hydrogen peroxide. FAD is the cofactor for this enzyme.

G. A. Hamilton (1985) *Adv. Enzymol.* **57**, 85.
M. Dixon (1970) *Meth. Enzymol.* **17A**, 713.

ASPARTATE RACEMASE

This enzyme [EC 5.1.1.13] catalyzes the interconversion of L- and D-aspartate. The enzyme will also utilize alanine as a substrate, albeit at only half the reported rates of aspartate.

M. E. Tanner & G. L. Kenyon (1998) *Comprehensive Biological Catalysis: A Mechanistic Reference* **2**, 7.

ASPARTATE-SEMIALDEHYDE DEHYDROGENASE

This enzyme [EC 1.2.1.11] catalyzes the reaction of aspartate-4-semialdehyde with orthophosphate and NADP$^+$ to produce 4-aspartyl phosphate and NADPH.

I. Saint-Gerons, C. Parsot, M. M. Zakin, O. Bârzy & G. N. Cohen (1988) *Crit. Rev. Biochem.* **23**, S1.
G. N. Cohen (1985) *Meth. Enzymol.* **113**, 600.
G. D. Hegeman, G. N. Cohen & R. Morgan (1970) *Meth. Enzymol.* **17A**, 708.

ASPARTIC PROTEASES

The aspartic proteinases are a subset of peptide hydrolases [EC 3.4.23.x] that have a participating active-site carboxyl group and a pH optimum below 5. The microbial aspartic proteinases [formerly represented by EC 3.4.23.6] represents a large group of enzymes isolated from microorganisms with varying characteristics. For example, aspergillopepsin I [EC 3.4.23.18] (also known as awamorin and trypsinogen kinase) is a peptidase, a member of peptidase family A1, isolated from several *Aspergillus* species (imperfect fungi). It catalyzes the hydrolysis of proteins with broad specificity. The enzyme generally favors hydrophobic residues in P_1 and P_1' sites. Its ability to accept a lysyl residue in P_1 site leads to trypsinogen activation. The enzyme will not clot milk. However, endothiapepsin [EC 3.4.23.22] (also known as *Endothia* aspartic proteinase) does clot milk. This enzyme, isolated from the ascomycete *Endothia parasitica* and a member of the peptidase family A1 as well, has a broad specificity similar to that of pepsin A. This enzyme prefers hydrophobic residues at P_1 and P_1'. *See also specific enzyme*

T. D. Meek (1998) *Comprehensive Biological Catalysis: A Mechanistic Reference* **1**, 327.
N. D. Rawlings & A. J. Barrett (1995) *Meth. Enzymol.* **248**, 105.
E. Lolis & G. A. Petsko (1990) *Ann. Rev. Biochem.* **59**, 597.
Selected entries from *Methods in Enzymology* [vol, page(s)]:
Active site, structure, **241**, 214; catalytic mechanism, **241**, 223-224; crystal structure, **241**, 214, 216; comparative studies with HIV protease [catalytic properties, **241**, 205-224; evolutionary relationships, **241**, 196-197; screening for HIV-1 protease inhibitors, **241**, 318-321; structure, **241**, 254-257, 280; substrate specificity, **241**, 255, 283].

β-ASPARTYL-N-ACETYLGLUCOSAMINIDASE

This enzyme [EC 3.2.2.11], also referred to as 1-aspartamido-β-N-acetylglucosamine amidohydrolase, catalyzes the hydrolysis of 1-β-aspartyl-N-acetylglucosaminylamine to yield N-acetylglucosamine and asparagine.

E. Conzelmann & K. Sandhoff (1987) *Adv. Enzymol.* **60**, 89.
E. H. Eylar & M. Murakami (1966) *Meth. Enzymol.* **8**, 597.

β-ASPARTYLDIPEPTIDASE

This enzyme [EC 3.4.19.5], now known as β-aspartyl-peptidase, is a mammalian cytosolic protein that catalyzes the hydrolysis of a β-linked aspartyl residue from the N-terminus of a polypeptide. Other isopeptide linkages (*e.g.*, a γ-glutamyl linkage) are not hydrolyzed by this enzyme.

E. E. Haley (1970) *Meth. Enzymol.* **19**, 730 and 737.

ASPARTYL-tRNA SYNTHETASE

This enzyme [EC 6.1.1.12], also known as aspartate:tRNA ligase, catalyzes the reaction of aspartate with ATP and tRNA^{Asp} to generate aspartyl-tRNA^{Asp}, AMP, and diphosphate (or, pyrophosphate). *See also Aminoacyl-RNA Synthetases*

E. A. First (1998) *Comprehensive Biological Catalysis: A Mechanistic Reference* **1**, 573.
J. Cavarelli, G. Eriani, B. Rees, M. Ruff, M. Boeglin, A. Mitschler, F. Martin, J. Gangloff, J.-C. Thierry & D. Moras (1994) *EMBO J.* **13**, 327.
P. Schimmel (1987) *Ann. Rev. Biochem.* **56**, 125.

ASSOCIATION

The combining of two or more substances or molecular entities to yield a single substance or molecular entity, a process that involves either covalent or noncovalent bonding. Included in this definition is the formation of ion pairs from free ions, the noncovalent aggregation of monomers to form polymeric structures or complexes, as well as colligation. The opposite of association is dissociation.

ASSOCIATION CONSTANT

The equilibrium constant (often symbolized by K_a) describing the association of two or more compounds, ions, or atoms into a new compound or complex. For example, the association constant, or binding constant, for oxygen binding to myoglobin is [myoglobin–O_2]/{[myoglobin][O_2]}, whereas the average association constant for hapten-antibody reaction is [hapten-bound antibody]/{[free antibody] [free hapten]}. The reciprocal of the association constant is the dissociation constant. *See Dissociation Constant; Stability Constant*

ASTACIN

This enzyme [EC 3.4.24.21] is a zinc-dependent digestive endopeptidase from the cardia of the crayfish *Astacus fluviatilis*. It catalyzes the hydrolysis of peptide bonds in substrates having five or more amino acids.

W. L. Mock (1998) *Comprehensive Biological Catalysis: A Mechanistic Reference* **1**, 425.
W. Stöcker & R. Zwilling (1995) *Meth. Enzymol.* **248**, 305.

ASYMMETRIC INDUCTION

The process by which a stereochemically inactive center is converted to a specific stereoisomeric form. In most cases, the reacting center is prochiral. Such processes can occur with reactions involving an optically active reagent, solvent, or catalyst (*e.g.*, an enzyme). The reaction produced by such a process is referred to as an enantioselective reaction[1-3]. In principle, use of circularly polarized light in photochemical reactions of achiral reactants might also exhibit asymmetric induction. However, reported enantioselectivities in these cases[4,5] have been very small.

[1]J. D. Morrison (1985) *Asymmetric Synthesis*, vol. **5**, Academic Press, New York.
[2]K. Tomioka (1990) *Synthesis*, 541.
[3]G. Consiglio and R. Waymouth (1989) *Chem. Rev.* **89**, 257.
[4]O. Buchardt (1974) *Angew Chem. Int. Ed. Engl.* **13**, 179.
[5]L. D. Barron (1986) *J. Amer. Chem. Soc.* **108**, 5539.

ASYMMETRY PARAMETER

A parameter, η, used for describing nonsymmetric fields in nuclear quadrupole resonance spectroscopy. It is defined as $\eta = (q_{xx} - q_{yy})/q_{zz}$ in which q_{xx}, q_{yy} and q_{zz} are the components of the field gradient q (which is the second derivative of the time-averaged electric potential) along the x-, y- and z-axes. By convention, q_{zz} refers to the largest field gradient, q_{yy} to the next largest, and q_{xx} to the smallest when all three values are different.

M. W. G. de Bolster (1997) *Pure Appl. Chem.* **69**, 1251.

ASYMPTOTE

A straight line whose perpendicular distance from a curve becomes progressively smaller as the distance from the origin at [0,0] becomes greater. For example, in a plot of velocity *versus* [Substrate Concentration] for an enzyme-catalyzed reaction, the asymptote reaches the maximal velocity when the enzyme molecules become saturated with substrate.

ATMOSPHERE

1. A unit of pressure, symbolized by atm, equal to 760 mmHg or 101,325 pascals (or, newtons per square meter) or 1.01325 bar. At sea level, atmospheric pressure roughly fluctuates around a value of one atm. 2. The air surrounding the Earth. The composition of dry atmospheric air, in percent volume is: N_2, 78.084; O_2, 20.946; Ar, 0.934; CO_2, 0.033; and trace amounts of Ne, He, Kr, Xe, and CH_4. The density of atmospheric air at sea level is 1220 g/m^3, while at an altitude of 10 km, the density is 425 g/m^3.

ATOM PERCENT EXCESS

A measure of the amount of a stable isotopic label that exceeds its natural abundance in unlabeled tracee. This is most directly accomplished using an ion ratio mass spectrometer to measure the ratio of ion currents for isotopomers such as $^{12}CO_2$ at mass 44 and $^{13}CO_2$ at mass 45. From the difference between the ion current ratio for a sample (r_{sample}) and the ion current ratio for a reference gas ($r_{reference}$), the atom percent excess (APE) can be estimated. **See** *Tracer/Tracee Ratio; Compartmental Analysis; Isotope Exchange Kinetics*

ATOMIC MASS UNIT

A unit of mass (abbreviated by a.m.u. or amu) equal to 1/12-th the mass of the carbon atom of mass number 12 ($^{12}_6C$). Thus, 1 a.m.u. $= m_u = (1/12)(^{12}_6C$ mass$) = 1.6605402 \times 10^{-27}$ kg. Hence, carbon-12 has a mass of exactly 12.000000 a.m.u.

ATOMIC ORBITAL

The wavefunction of an electron associated with an atomic nucleus. The orbital is typically depicted as a three-dimensional electron density cloud. If an electron's azimuthal quantum number (l) is zero, then the atomic orbital is called an *s* orbital and the electron density graph is spherically symmetric[1]. If l is one, there are three spatially distinct orbitals, all referred to as *p* orbitals, having a dumb-bell shape with a node in the center where the probability of finding the electron is extremely small. (Note: For relativistic considerations[2], the probability of an electron residing at the node cannot be zero.) Electrons having a quantum number l equal to two are associated with *d* orbitals.

[1]P. S. C. Matthews (1986) *Quantum Chemistry of Atoms and Molecules*, Cambridge Univ. Press, Cambridge.
[2]R. E. Powell (1968) *J. Chem. Ed.* **45**, 558.

ATOMIZATION

The conversion of a molecule into its constituent atoms.

ATP

Adenosine 5'-triphosphate, the major energy-storing nucleotide within living organisms. The standard Gibbs free energy of hydrolysis is -30.5 kJ/mol.

Selected entries from *Methods in Enzymology* [**vol**, page(s)]:
[32]P-labeled [detection, **238**, 40, 48, 50-51; disposal, **238**, 53; half-life, **238**, 40; α-phosphate labeling specificity, **238**, 39-40; quality of preparations, **238**, 40]; regeneration systems, **238**, 34-35; stability, **238**, 352-354; tritiated [assay advantages, **238**, 38; detection, **238**, 39, 48, 50-51; disposal, **238**, 52-53; half-life, **238**, 38; stability of label, **238**, 39].

ATPase, Ca²⁺

This class of enzymes [EC 3.6.1.38] includes calcium pumps that catalyze hydrolysis of ATP, producing ADP and orthophosphate; this hydrolysis is coupled to the transport of Ca^{2+} ions.

F. Wuytack & L. Raeymaekers (1992) *J. Bioenerg. Biomembr.* **24**, 285.
W. P. Jencks (1992) *Biochem. Soc. Trans.* **20**, 555.
G. Inesi, D. Lewis, D. Nikic, A. Hussain & M. E. Kirtley (1992) *Adv. Enzymol.* **65**, 185.
W. P. Jencks (1980) *Adv. Enzymol.* **51**, 75.
W. Hasselbach (1974) *The Enzymes*, 3rd ed., **10**, 431.

ATPase, F₁Fₒ

See *F₁Fₒ ATPase; Binding Change Mechanism; CF₁CFₒ ATPase*

ATPase, H⁺/K⁺

This class of enzymes [EC 3.6.1.36] (also known as the hydrogen/potassium-exchanging ATPase, the potassium-transporting ATPase, proton pump, and the gastric H⁺/K⁺ ATPase) catalyzes the hydrolysis of ATP to ADP and orthophosphate, coupled with the exchange of H⁺ and K⁺ ions. The gastric mucosal enzyme has been the best characterized.

M. Besanèon, J. M. Shin, F. Mercier, K. Munson, E. Rabon, S. Hersey & G. Sachs (1992) *Acta Physiol. Scand.* **146**, 77.
G. Sachs, M. Besanèon, J. M. Shin, F. Mercier, K. Munson & S. Hersey (1992) *J. Bioenerg. Biomembr.* **24**, 301.
Q. Al-Awqati (1986) *Ann. Rev. Cell Biol.* **2**, 179.
C. Tanford (1983) *Ann. Rev. Biochem.* **52**, 379.

ATPase, Na⁺/K⁺

Nerve stimulation results in a net influx of sodium ions, and normal conditions are restored by the outward transport of sodium ions against an electrochemical gradient. While several earlier workers had identified ATPases in the sheath of giant squid axons, it was Skou[1,2] who first connected the sodium, potassium ATPase [EC 3.6.1.37] with the ion flux of neurons. This discovery culminated

in his sharing the Nobel Prize in 1997, some forty years after his seminal work, with Boyer and Walker who were cited for their studies on the mechanism and structure of ATP synthase. Later studies by Kanazawa *et al.*[3] measured the rates of formation and decomposition of the phosphorylated intermediate in the reaction catalyzed by sodium, potassium ATPase.

[1]J. C. Skou (1957) *Biochim. Biophys. Acta* **23**, 394.
[2]J. C. Skou (1960) *Biochim. Biophys. Acta* **42**, 6.
[3]T. Kanazawa, M. Saito & Y. Tonomura (1970) *J. Biochem. (Tokyo)* **67**, 693.

ATPase (Stereochemistry of Phosphoryl Transfer)

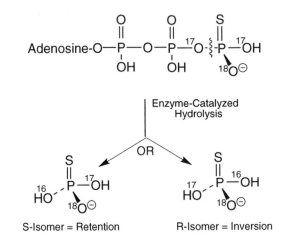

Stereochemical Course of ATPase Reactions

[¹⁶O,¹⁷O,¹⁸O,S]ATPγS provides a powerful means for answering a fundamental question in ATPase catalysis, namely whether a particular ATPase is transiently phosphorylated during the catalytic reaction cycle. Formation and hydrolysis of an E-P intermediate requires two in-line nucleophilic substitution reactions, each suffering inversion of stereochemical configuration at tht the phosphorus atom of the phosphoryl group. This results in overall conservation of stereochemical configuration. However, when an ATPase catalyzes direct transfer to water, there will be an inversion of configuration.

M. R. Webb (1982) *Meth. Enzymol.* **87**, 301.

ATP CITRATE LYASE

This enzyme [EC 4.1.3.8], also known as ATP–citrate (*pro-S-*)-lyase and citrate cleavage enzyme, catalyzes the reaction of citrate with ATP and coenzyme A to yield oxaloacetate, acetyl-CoA, ADP, and orthophosphate.

The multiprotein enzyme complex can be dissociated into smaller components of which two are identical with citryl-CoA lyase and citrate:CoA ligase.

D. J. Creighton & N. S. R. K. Murthy (1990) *The Enzymes*, 3rd ed., **19**, 323.
P. A. Srere (1975) *Adv. Enzymol.* **43**, 57.
L. B. Spector (1972) *The Enzymes*, 3rd ed., **7**, 357.

ATP:GLUCOSE-1-PHOSPHATE ADENYLYLTRANSFERASE

This enzyme [EC 2.7.7.27], also known as ADP–glucose synthase and ADP–glucose pyrophosphorylase, catalyzes the reaction of α-D-glucose 1-phosphate with ATP to produce ADP–glucose and pyrophosphate.

A. M. Smith, K. Denyer & C. Martin (1997) *Ann. Rev. Plant Physiol. Plant Mol. Biol.* **48**, 67.
J. Preiss (1973) *The Enzymes*, 3rd ed., **8**, 73.

ATP & GTP DEPLETION

Because ATP and GTP are required in many biosynthetic and mechanochemical reactions, the ability to manipulate their concentrations is particularly important in many kinetic studies. The three enzymes most commonly employed for this purpose are hexokinase, phosphofructokinase, and alkaline phosphatase. Each of these enzymes offers advantages and disadvantages, and one must consider the design requirements for any particular experiment before using one of these to deplete ATP or GTP.

Yeast hexokinase (1–2 international units per mL reaction solution) acting for 20 min in the presence of 10 mM D-glucose has proven to be quite effective in depleting ATP at levels not exceeding 1 mM. At pH 7 in the presence of 1–2 mM uncomplexed magnesium ion, the equilibrium constant for the hexokinase reaction is about 1500; thus, one can anticipate substantial conversion of l mM ATP, as indicated by the following equation:

$$\frac{[\text{ADP}]}{[\text{ATP}]} = 1500 \frac{[\text{D-Glucose}]}{[\text{D-Glucose 6-P}]}$$

Using the expression $\{x/(1\text{mM} - x) = 1500\ (\sim 10\text{ mM})/x\}$, one can solve the quadratic equation or iteratively estimate the value of x that satisfies the equation. This would place x at about 0.99994, indicating that 99.99% of the ATP would be depleted if the hexokinase reaction equilibrium is reached.

As written, the above expression helps one to recognize that hexokinase can be used to maintain the [ADP]/ [ATP] concentration ratio at any desired value, even in the presence of a slow ATPase activity by the enzyme under investigation. By knowing the equilibrium constant for the hexokinase reaction at the pH for a particular experiment, one can merely use D-glucose and D-glucose 6-phosphate in the ratio that will yield the desired [ADP]/[ATP] in the presence of sufficient hexokinase to reach equilibrium.

GTP is also a substrate for yeast hexokinase, but phosphofructokinase (PFK) acts on GTP more efficiently than hexokinase. Typically, one uses 1–2 international units PFK per milliliter of reaction solution (for 20 min in the presence of 5 mM D-fructose 6-phosphate) to deplete GTP initially present at concentrations up to 1 mM.

Because alkaline phosphatase converts ATP or GTP to their respective nucleosides, use of alkaline phosphatase to deplete ATP or GTP should be reserved for those cases where this does not present a problem[1]. Furthermore, alkaline phosphatase is potently inhibited by orthophosphate; so higher than would be anticipated amounts of enzyme are required for efficiently depleting ATP or GTP.

[1]D. L. Purich & R. K. MacNeal (1978) *FEBS Lett.* **96**, 83.

ATP PHOSPHORIBOSYLTRANSFERASE

This enzyme [EC 2.4.2.17], also known as phosphoribosyl-ATP pyrophosphorylase, catalyzes the reaction of ATP with 5-phospho-α-D-ribose 1-diphosphate to generate 1-(5-phospho-D-ribosyl)-ATP and pyrophosphate.

W. D. L. Musick (1981) *Crit. Rev. Biochem.* **11**, 1.
R. G. Martin, M. A. Berberich, B. N. Ames, W. W. Davis, R. F. Goldberger & J. D. Yourno (1971) *Meth. Enzymol.* **17B**, 3.

ATP, THERMODYNAMICS OF HYDROLYSIS

Hydrolysis of the terminal (or, β–γ) phosphoanhydride bond of ATP is the primary thermodynamic driving force in metabolism. Written as a biochemical equation at a specified pH, ATP hydrolysis can be represented as:

$$\text{ATP} + \text{H}_2\text{O} = \text{ADP} + \text{P}_i$$

The apparent equilibrium constant, $K'_{(\text{T,P,pH})}$ for this process is :

$$K'_{(\text{T,P,pH})} = [\text{ADP}]\,[\text{P}_i]\,/\{[\text{ATP}]c^\circ\}$$

where $c°$ is the standard state concentration of 1 M. Biochemists can often advantageously describe biochemically relevant equations in terms of reactants at specified pH and concentrations of free metal ions that are bound by reactant species. The relevant equilibrium constant is written as K'. A reconceptualization of the thermodynamics of biochemical processes, based on equilibria written in terms of reactants at a specified pH and concentrations of free metal ions bound by reactant species[1,2].

In chemical thermodynamics, temperature and pressure are specified and a system is defined in terms of species, using the change in the Gibbs free energy:

$$dG = -SdT + VdP + \sum_{i=1}^{N} \mu_i dn_i$$

where G, S, T, V, and P are the Gibbs free energy, entropy, temperature, volume, and pressure. The terms μ_i, n_i, and N are the chemical potential of species i, the amount of species i, and the number of species in the system.

In the new biochemical thermodynamic approach, the various protonated species of a metabolite are treated as pseudoisomeric forms having the same transformed chemical potential μ_i at a specified pH, such that they are collected in the term n_i' (which is equal to Σn_i).

$$(dG)_{pH} = -S' dT + VdP + \sum_{i=1}^{N'} \mu_i' dn_i'$$

[1] R. A. Alberty (1994) *Biochim. Biophys. Acta* **1207**, 1.
[2] R. A. Alberty (1996) *Eur. J. Biochem.* **240**, 1. (A useful account of the IUPAC-IUBMB recommendations for nomenclature and tables in biochemical thermodynamics)

ATROPISOMERS

Isomers that lack internal planes of symmetry or points of symmetry (hence, are chiral) due to a prevention (or significant hindrance) to rotation about a single bond.

M. Oki (1983) *Top. Stereochem.* **14**, 1.

ATROXASE

This zinc-dependent nonhemorrhagic endopeptidase [EC 3.4.24.43] isolated from the venom of the rattlesnake *Crotalus atrox* is a peptidase family M12B member that catalyzes the hydrolysis of the His[5]—Leu[6], Ser[9]—His[10],

His[10]—Leu[11], Ala[14]—Leu[15], and the Tyr[16]—Leu[17] bonds in the insulin B chain. It will also act on fibrinogen.

J. B. Bjarnason & J. W. Fox (1995) *Meth. Enzymol.* **248**, 345.

ATTENUANCE FILTER

A substance or device (synonymous with neutral density filter) that reduces the radiant power of incident radiation by a constant factor over all wavelengths within the instrument's operating range. **See also** Filter

ATTO-

A prefix, symbolized by a, used in submultiples of units and corresponding to a value of 10^{-18}.

AUTOCATALYSIS

The self-activating process for converting an inactive catalyst, I, to its active form, A. Such processes typically display a discernible lag-phase followed by accelerated conversion of I to A. Zymogen activation can display autocatalysis, as exemplified by the conversion of pepsinogen to active pepsin at low pH. Many growth processes can also be described via an autocatalysis curve.

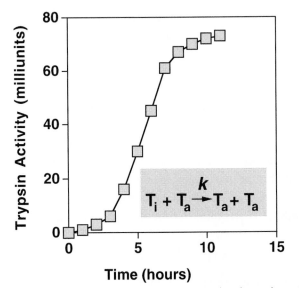

The time-course of an autocatalytic process. Note that during the early stages the amount of active catalyst is low, but changes dramatically as a function of time.

The process is a special case of a second-order autocalytic reaction in which a product or reaction intermediate functions as a catalyst. In the early stages of the reaction, the rate-of-reaction of autocatalytic processes increases with the time from its initial value, as seen in the plot above. If a reaction of A to P is autocatalytic such that the rate expression is

$$v = -\frac{d[A]}{dt} = k[A][P]$$

where t is time and k is a second-order rate constant, then the integrated rate expression is

$$\frac{1}{[A_o] + [P_o]} \ln \frac{[A_o][P]}{[P_o][A]} = kt$$

Chain reactions and explosions are also autocatalytic processes. **See also** Bifurcation Theory; Prion Plaque Formation

AUTOINHIBITION

A term used to describe a characteristic of certain enzymes that have intermolecular interactions which result in the inhibition of the enzyme's catalytic activity[1]. For example, the calmodulin-binding domain of the human plasma-membrane Ca^{2+} pump overlaps with an autoinhibitory domain[2,3]. In the absence of calmodulin and phosphorylation by protein kinase C, the Ca^{2+} pump is inhibited. Binding of calmodulin to the calmodulin-binding domain results in activation of the pump. A second such case is the flavin-containing heme protein nitric oxide synthase[4] (NOS) which generates NO· in neurons from L-arginine, NADPH, and O_2. The enzyme autoinactivates upon reacting with steady-state concentrations of NO· to form a ferrous-nitrosyl complex during aerobic catalysis. Other examples of autoinhibition include calmodulin-dependent cAMP phosphodiesterase[5] and smooth-muscle myosin light-chain kinase[6]. In a sense, all enzymes that can undergo activation by intermolecular effectors display the phenomena of autoinhibition. Activation is simply a depression of an inhibitory or partially inhibitory aspect of the protein. However, the term autoinhibition remains quite useful in describing domains in such proteins as well as the conformational changes and binding equilibria associated with these phenomena. **See also** Activation

[1]P. James, T. Vorherr & E. Carafoli (1995) *Trends in Biochem. Sci.* **20**, 38.

[2]M. Zurini, J. Krebs, J. T. Penniston & E. Carafoli (1984) *J. Biol. Chem.* **259**, 618.

[3]P. James, T. Vorherr, J. Krebs, A. Morelli, G. Castello, D. J. McCormick, J. T. Penniston, A. DeFlora & E. Carafoli (1989) *J. Biol. Chem.* **264**, 8289.

[4]H. M. Abu-Soud, J. Wang, D. L. Rousseau, J. M. Fukuto, L. J. Ignarro & D. J. Stuehr (1995) *J. Biol. Chem.* **270**, 22997.

[5]C. B. Klee (1980) in *Protein Phosphorylation and Bioregulation* (eds., G. Thomas, E. J. Podesta & J. Gardon), p. 61, Karger.

[6]A. M. Edelman, K. Takio, D. K. Blumental, R. S. Hansen, K. A. Walsh, K. Titani & E. G. Krebs (1985) *J. Biol. Chem.* **260**, 11275.

AUTONOMOUS CATALYTIC DOMAIN

A catalytic function or region of a multienzyme system that is associated with a domain that is separate and distinct from the domain(s) responsible for the other catalytic function(s).

Nomenclature Comm., IUB (1989) *Eur. J. Biochem.* **185**, 485.

AUTOPHAGY

Intracellular digestion occurring in living cells in which regions of the cytoplasm and/or cell components are digested. Many authors consider autophagy to include the normal processes occurring within the lysosome. Heterophagy then refers to intracellular digestion of exogenous materials brought into the cell by pinocytosis and phagocytosis. Some authors make a distinction between autophagy and autolysis, whereas others consider autolysis as a form of autophagy. In this context, autolysis is a process that occurs specifically in dying or dead cells. It is programmed and controlled cell destruction.

AUTOPHOSPHORYLATION

The self-phosphorylation process catalyzed by many protein kinases as part of the regulatory mechanism for their own activation. Because true autophosphorylation is a unimolecular reaction involving enzyme both as catalyst and phosphoryl acceptor, the fraction of autophosphorylated enzyme at any time after addition of ATP (or another phosphoryl donor) will be independent of the initial concentration of the enzyme. This criterion was first applied to the autophosphorylation of cardiac muscle cyclic AMP-stimulated protein kinase[1], now designated protein kinase A (PKA). At a fixed concentration of $MgATP^{2-}$, the fraction of autophosphorylated protein will follow the first-order rate laws, $[A]/[A_o] = e^{-kt}$, where k is a first-order rate constant.

[1]J. A. Todhunter & D. L. Purich (1977) *Biochim. Biophys. Acta* **485**, 87.

AUTOPROTOLYSIS

Any process in which a molecule acts as an acid in a reaction with another identical molecule acting as a base. Autoprotolytic reactions can only occur with amphiprotic molecular entities.

The best known example of autoprotolysis is the ionization of water:

$$2H_2O = H_3O^+ + OH^-$$

The equilibrium constant for this reaction is given by the expression

$$K_{eq} = [H_3O^+][OH^-]/[H_2O]^2$$

Because the activity (or concentration) values for H_3O^+ and OH^- are much smaller than the corresponding value for H_2O (*i.e.*, the solvent), the solvent concentration remains unchanged. (The density of pure water at 25°C is about 0.997 g/ml. Hence, the concentration of pure water at 25°C is (0.997 g/mL)(1000 mL/L)(18.0153 g/mol) = 55.3 M. This value is clearly larger than $[H_3O^+]$ and $[OH^-]$ and the value of $[H_2O]^2$ can be regarded as a constant.) Thus, the expression for the autolysis of water (or, the ionization of water) can be simplified to $K_{eq}[H_2O]^2 = K_w = [H_3O^+][OH^-]$. This value is equal to 10^{-14} M^2 at 25°C. *See Temperature Effects on pK_w*

Autoprotolysis constants exist for any amphiprotic solvent and can be determined from electric conductivity studies of the solvent and solutions. A few examples[1] include:

Solvent	Autoprotolysis Constant[a]
H_2O	$[H_3O^+][OH^-] = 10^{-14}$
NH_3	$[NH_4^+][NH_2^-] = 10^{-33}$
H_2SO_4	$[H_3SO_4^+][HSO_4^-] = 2 \times 10^{-4}$
HCOOH	$[HC(OH)_2^+][HCOO^-] = 6 \times 10^{-7}$
CH_3COOH	$[CH_3C(OH)_2^+][CH_3COO^-] = 10^{-13}$
CH_3OH	$[CH_3OH_2^+][CH_3O^-] = 2 \times 10^{-17}$
CH_3CH_2OH	$[CH_3CH_2OH_2^+][CH_3CH_2O^-] = 3 \times 10^{-20}$

[a]All values are those measured at 25°C, except for NH_3 (−33°C).

[1]L. Pauling (1970) *General Chemistry*, 3rd ed., Freeman, San Francisco (Dover republication, 1988).

AUTOPROTOLYSIS CONSTANT

The product of the molar concentrations (or, more accurately, the activities) of the species produced as a result of autoprotolysis. The autoprotolysis constant for water is K_w, equal to $[H_3O^+][OH^-]$, or 1.0×10^{-14} at 25°C. It is a temperature-dependent constant, increasing with increasing temperature (at 37°C, $K_w = 2.51 \times 10^{-14}$). *See Autoprotolysis*

AVOGADRO'S NUMBER

1. The number of atoms, symbolized by either L or N_A, in 0.012 kg of pure carbon-12, equivalent to 6.0221367 $\times 10^{23}$ per mole. 2. The number of molecules in one gram-molecular weight (1 mole) of any compound.

AZEOTROPE

For a solution or mixture of two or more distinct liquid components, an azeotrope is that composition (typically measured in mole fractions or percent weight and referred to as the azeotropic solution) with which there is either a maximum point (a negative azeotrope) or a minimum point (a positive azeotrope) in a boiling point *versus* composition diagram at constant pressure.

AZIDO PHOTOAFFINITY REAGENTS

Any member of a class of substances that generate a highly reactive nitrene species upon absorption of photic energy.

Selected entries from *Methods in Enzymology* [**vol**, page(s)]: 2-Azidoadenosine synthesis, **237**, 84; ^{32}P-labeled 2-azido-ADP-ribose, intra- and intermolecular transfer from G protein a subunit carboxy terminus, **237**, 95-99; 2-azido-AMP [nonradioactive, synthesis, **237**, 85, 93; ^{32}P-labeled, chemical synthesis, **237**, 82-86; in preparation of 2-azido-[^{32}P]NAD+, **237**, 82]; 4-azidoanilido-GTP [G protein labeling, **237**, 100-110; preparation, **237**, 101-103; structure, **237**, 101; tubulin labeling, **237**, 106-107]; P3-(4-azidoanilido)-P1-5'-guanosine triphosphate **195**, 282; 2-azido-ATP [^{32}P-labeled, enzymatic synthesis, **237**, 87-89, 92; in preparation of 2-azido-[^{32}P]NAD+, **237**, 82]; 3'-(p-azido-m-iodophenylacetyl)-2',5'-dideoxy-adenosine [adenylyl cyclase inactivation, **238**, 68-71; photocoupling, **238**, 67-68; radioactive labeling, **238**, 68; solubility, **238**, 68; synthesis, **238**, 65]; 3'-(p-azido-m-iodophenylbutyryl)-2',5'-dideoxy-adenosine [adenylyl cyclase inactivation, **238**, 68-71; photocoupling, **238**, 67-68; radioactive labeling, **238**, 68; solubility, **238**, 68; synthesis, **238**, 65]; 2-azido-NAD [as G-protein structure probe, **237**, 71; determination of kinetic constants, **237**, 91-92; nonradioactive, synthesis, **237**, 86-87; ^{32}P-labeled (enzyme cleavage analysis, **237**, 91; synthesis, **237**, 81-95]]; 5-azido-UDPglucose [^{32}P-labeled, applications in glycobiology, **230**, 334-339; synthesis, **230**, 330-334]; 5-azido-UDPglucuronic acid [^{32}P-labeled, applications in glycobiology, **230**, 334-339; synthesis, **230**, 330-334]; 4'-azidowarfarin, **206**, 57-60

AZURIN

A bacterial electron-transfer protein, dark blue in color, that contains a type 1 copper site.

B

BACTERIAL LEADER PEPTIDASE I

This enzyme [EC 3.4.21.89], also known as signal peptidase I and phage-procoat-leader peptidase, catalyzes the hydrolysis of N-terminal leader sequences from secreted and periplasmic protein precursors. It acts on a single bond Ala–Ala in the m-13 phage procoat protein and creates the signal (leader) peptide and coat protein. It is a member of the peptidase family S26 but is unaffected by inhibitors of most serine peptidases.

W. R. Tschantz & R. E. Dalbey (1994) *Meth. Enzymol.* **244**, 285.

BAIT-AND-SWITCH MECHANISM

A strategy for raising of catalytic antibodies for a desired reaction type that employs a hapten as "bait" to induce the production of antibodies that are most likely to possess certain side-chain residues that may be important in binding and/or catalysis. One then "switches" to the substrate.

R. A. Lerner & S. J. Benkovic (1988) *Bioessays* **9**, 107.
K. D. Janda, M. I. Weinhouse, D. M. Schloeder, R. A. Lerner & S. J. Benkovic (1990) *J. Amer. Chem. Soc.* **112**, 1274.

BAND-PASS FILTER

A substance or device which allows the transmission of electromagnetic radiation of a specified wavelength range and does not allow transmission of radiation outside that range. *See also* *Filter; Cut-off Filter*

BASAL RATE

The residual or unstimulated activity of (a) an enzyme reaction, (b) a series of reactions, or (c) the energy metabolism of an individual organism. Although one most frequently considers basal reaction rates in enzyme kinetics, any unfacilitated transport or leakage of ions and/or metabolites across a membrane barrier can also be considered a basal activity in studies of transporter kinetics.

BASAL METABOLIC RATE (or B.M.R.). An index of metabolic activity of an individual organism, usually measured by the rate of oxygen consumption while in a resting, nonsleeping state. The basal metabolic rate can also be determined as the rate of heat evolution in a state of resting and without recent consumption of foodstuffs. Because temperature control is of vital importance to mammals, the basal metabolic rate is roughly proportional to body surface area. A young healthy adult male (mass 70 kg) typically has a basal metabolic rate of 300–350 kJ/hour, corresponding to about 70–80 watts.

ENZYME KINETICS. Most enzyme-catalyzed reactions proceed at rates that are much faster than their nonenzymatic counterparts. Nonetheless, there are a number of conditions that account for basal rates for biochemical reactions in the absence, or apparent absence, of the enzyme under consideration:

(a) Some reactions proceed at significant uncatalyzed rates, including the hydration of carbon dioxide and the formation or decomposition of certain cyclic lactones. In these cases, one must determine the basal rate (in the absence of added enzyme) over the same substrate concentration range used for enzymatic rate measurements. The difference (*i.e.*, $v_{enzymatic} - v_{nonenzymatic}$) must then be used in any determination of enzyme rate behavior. Note that nonenzymatic processes can be catalyzed by certain buffers and/or metal ions. The judicious choice of components in an assay solution can often reduce the basal reaction rate.

(b) A substrate or other component in the reaction assay may be contaminated with the enzyme of interest or some other enzyme that affects the accurate determination of enzyme reaction rate. This can be a serious problem when carrying out investigations of posttranslational modification reactions involving protein substrates. Care must be exercised to ascertain whether the protein substrate has any endogenous enzymatic activity. This often can be accomplished merely by omitting the primary enzyme and determining whether any posttranslational modification has occurred. (One should also confirm that the apparent kinetics of posttranslational modification are unaffected by the denaturation or proteolysis of the protein substrate)

(c) A coupled enzyme assay may introduce trace amounts of the enzyme of interest as a contaminant in one or more of the auxilliary enzymes added to the reaction.

(d) A nucleotide-regenerating system may likewise introduce trace amounts of the enzyme of interest as a contaminant.

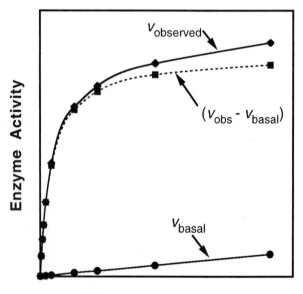

Effector Concentration

Response of an enzyme reaction in which the basal rate is proportional to the concentration of an effector. In this case, unless the basal rate is subtracted point-by-point from each of the observed velocities, the system will never achieve saturation.

ENZYME ACTIVATION. To characterize enzyme activation quantitatively, one must often correct for the presence of any basal reaction rate or basal activity in the unstimulated state. Typically, one uses ($v_{stimulated} - v_{basal}$) or ($v_{stimulated} - v_{basal}$)/($v_{maximally-stimulated} - v_{basal}$) for this purpose in studies of enzyme activation.

BASE CATALYSIS

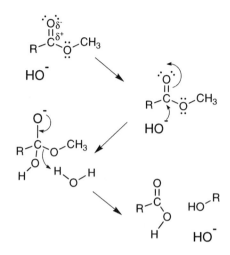

Base Catalysis of Ester Hydrolysis

Any chemical reaction that is accelerated through the catalyic participation of base. **See** *Acidity; Brønsted Relation; General Acid Catalysis; General Base Catalysis*

BASE PAIRING

The hydrogen-bond pairing interactions between pyrimidine and purine bases in DNA and RNA. **See** *DNA Breathing*

Selected entries from *Methods in Enzymology* [vol, page(s)]:
Catalysts of proton exchange [concentration in base pair lifetime determination, **261**, 405-406; efficiency, **261**, 388; external proton acceptor, **261**, 389-391; intrinsic catalysis, **261**, 390-391; mechanism, **261**, 386; pK values, **261**, 387; selection, **261**, 404-405]; kinetics of opening and proton exchange [mechanism of proton exchange, **261**, 385-386, 388-389; pH dependence, **261**, 385; time scale of exchange, **261**, 391, 393]; proton exchange measurement by NMR [base pair lifetime determination, **261**, 404-406; B-DNA, **261**, 406-407; B'-DNA, **261**, 408, 410-411; cross-relaxation effects, **261**, 396-397; cytidine imino proton exchange, **261**, 411-413; data processing, **261**, 401, 403; dissociation constant determination, **261**, 403-404; imino proton chemical shift, **261**, 404; luzopeptin-DNA complex, **261**, 411; magnetization transfer from water protons, **261**, 393-395, 399; proton/deuterium exchange in real time, **261**, 391-393; relaxation rate increments, **261**, 395, 401; sample preparation, **261**, 391-392; solvent signal suppression, **261**, 398-399].

BASE STACKING

The diffusion-controlled bonding interactions between successive heterocyclic bases in oligo- and polynucleotides that arise from extended, planar π-orbital interactions. The resulting structures usually exhibit a 3.4 Å plane-to-plane distance within the same chain. While the stabilization arising from base stacking is small (ΔG −1 kcal/mol), the additive effect of many such interactions can profoundly influence nucleic acid structure as well as interactions with enzymes that use nucleic acids as substrates. Generally speaking, purine-purine base stacking is more energetically favored.

BASICITY OF CARBONYL OXYGENS

As noted elsewhere in this Handbook, carbonyl groups are polarized: the carbon atom has a partial positive charge, whereas the carbonyl oxygen displays partial negative charge character. In fact, the carbonyl oxygen has weakly basic properties; hence, protonated carbonyls are usually strong acids, as evidenced by the conjugate acids of acetamide (pK_a 0), water (pK_a −1.7), methanol (pK_a −2), acetic acid (pK_a −6), ethylacetate (pK_a −6 to −7), and acetaldehyde (pK_a −8). The protonated form of a carbonyl oxygen is often called an oxonium salt, and such species do indeed factor into the mechanisms of acid-catalyzed reactions involving carbonyl-containing compounds.

BATHOCHROMISM

A shift (also known as a "red shift") in a substance's electronic absorption spectrum toward longer wavelengths, as a consequence of a substituent, solvent, environment, or other effect. The opposite of a bathochromic shift is referred to as a hypsochromic shift.

BECQUEREL

The SI unit of measure for radioactivity. One becquerel (Bq) is equal to one disintegration per second (dps), thereby corresponding to about 2.703×10^{-11} curie.

BEER-LAMBERT LAW

A function relating the absorbance of light relative to concentration and thickness of a sample. The absorbance of light is directly proportional to the concentration of the absorbing chromophore multiplied by the thickness of the substance, containing the chromophore, through which the light is being transmitted. Thus, $A = \varepsilon bc$ where A is the absorbance, ε is the molar extinction coefficient (which varies with wavelength), b is the thickness of the

solution, and c is the concentration. **See** *Absorption Spectroscopy*

BENZENE 1,2-DIOXYGENASE

This enzyme [EC 1.14.12.3], also known as benzene hydroxylase, catalyzes the NADH-linked reaction of benzene and dioxygen to produce *cis*-1,2-dihydrobenzene-1,2-diol and NAD$^+$. This multiprotein complex contains an iron-sulfur flavoprotein, an iron-sulfur oxygenase, and ferredoxin. **See also** *Phthalate Dioxygenase*

B. G. Fox (1998) *Comprehensive Biological Catalysis: A Mechanistic Reference* **3**, 261.
P. J. Geary, J. R. Mason & C. L. Joannou (1990) *Meth. Enzymol.* **188**, 52.
O. Hayaishi, M. Nozaki & M. T. Abbott (1975) *The Enzymes*, 3rd ed., **12**, 119.

BENZOATE-CoA LIGASE

This enzyme [EC 6.2.1.25] catalyzes the reaction of benzoate with ATP and coenzyme A to produce benzoyl-CoA, AMP, and pyrophosphate (or, diphosphate). The enzyme will also use 2-, 3- and 4-fluorobenzoate as substrate. The corresponding chlorobenzoates will all serve as substrates, albeit less effectively.

J. Gibson, J. F. Geissler & C. S. Harwood (1990) *Meth. Enzymol.* **188**, 154.

BENZOATE 1,2-DIOXYGENASE

Benzoate 1,2-dioxygenase [EC 1.14.12.10], also called benzoate hydroxylase, catalyzes the reaction of benzoate with dioxygen and NADH to generate catechol, carbon dioxide, and NAD$^+$. This is a multiprotein system which contains a reductase which is an iron-sulfur flavoprotein (FAD) and an iron-sulfur oxygenase.

B. G. Fox (1998) *Comprehensive Biological Catalysis: A Mechanistic Reference* **3**, 261.
J. V. Schloss & M. S. Hixon (1998) *Comprehensive Biological Catalysis: A Mechanistic Reference* **2**, 43.
O. Hayaishi, M. Nozaki & M. T. Abbott (1975) *The Enzymes*, 3rd ed., **12**, 119.

BENZOYLFORMATE DECARBOXYLASE

This enzyme [EC 4.1.1.7] catalyzes the conversion of benzoylformate to benzaldehyde and carbon dioxide. Thiamin pyrophosphate is a cofactor.

M. H. O'Leary (1989) *Ann. Rev. Biochem.* **58**, 377.

BERBERINE BRIDGE ENZYME

This enzyme [EC 1.5.3.9], also called reticuline oxidase and tetrahydroprotoberberine synthase, catalyzes the reaction of (S)-reticuline with dioxygen to generate (S)-scoulerine and hydrogen peroxide. FAD is the cofactor for this enzyme. The reaction product is a precursor of protopine, protoberberine, and benzophenanthridine-alkaloid biosynthesis in plants.

T. Hashimoto & Y. Yamada (1994) *Ann. Rev. Plant Physiol. Plant Mol. Biol.* **45**, 257.

BETA (β)

1. The second letter in the Greek alphabet; hence, used to denote the second item in a series (for example, the second methylene carbon from the carboxyl group of a fatty acid). 2. Symbol for the coefficient of $[B]^i$ in the denominator of a generalized rate expression. 3. Symbol for reciprocal temperature parameter, $\beta = 1/kT$. 4. Symbol for pressure coefficient, $\beta = (\partial p / \partial T)_v$. 5. Symbol for depth of penetration of light (Napierian). 6. β^-, Symbol for electron. 7. β^+, Symbol for positron.

BETAINE–ALDEHYDE DEHYDROGENASE

This enzyme [EC 1.2.1.8] catalyzes the reaction of betaine aldehyde with NAD^+ and water to produce betaine and NADH.

D. Rhodes & A. D. Hanson (1993) *Ann. Rev. Plant Physiol. Plant Mol. Biol.* **44**, 357.

BETAINE–HOMOCYSTEINE S-METHYLTRANSFERASE

This enzyme [EC 2.1.1.5] catalyzes the reaction of trimethylammonioacetate with homocysteine to yield dimethylglycine and methionine.

W. E. Skiba, M. S. Wells, J. H. Mangam & W. M. Awad, Jr. (1987) *Meth. Enzymol.* **143**, 384.

BIFUNCTIONAL CATALYSIS

A catalytic process (often involving proton transfer) in which both functional groups of a bifunctional chemical species participate in the rate-controlling step. In such systems, the catalytic coefficient is larger than would be expected were only one functional group present. Bifunctional catalysis is not the same as two different catalysts acting in concert.

BIFUNCTIONAL ENZYME

An enzyme possessing the capacity to catalyze two mechanistically different chemical reactions. Examples include the fructose-6-phosphate-2-kinase/fructose-2,6-bisphosphatase, glutathionylspermidine synthetase/amidase, and the *Escherichia coli* glutamine synthetase cascade's UTase/UR enzyme reactions.

C. H. Lin, D. S. Kwon, J. M. Bollinger, Jr., & C. T. Walsh (1997) *Biochemistry* **36**, 14930.

BIFURCATION THEORY

A theoretical framework for considering how the behavior of dynamical systems change as some parameter of the system is altered. Poincaré first applied the term bifurcation for the "splitting" of asymptotic states of a dynamical system. A bifurcation is a period-doubling, -quadrupling, *etc.*, that precede the onset of chaos and represent the sudden appearance of a qualitatively different behavior as some parameter is varied. Bifurcations come in four basic varieties: flip bifurcations, fold bifurcations, pitchfork bifurcations, and transcritical bifurcations. In principle, bifurcation theory allows one to understand qualitative changes of a system change to, or from, an equilibrium, periodic, or chaotic state.

Selected entries from *Methods in Enzymology* [vol, page(s)]: Biological applications, **240**, 781-782, 791; development, **240**, 784-790; enzyme kinetic applications [hysteresis, **240**, 782-783; autocatalysis, **240**, 799–803; substrate cycling, **240**, 803-806]; detection with Jacobian eigenvalues, **240**, 795; algorithms, **240**, 793-795 [AUTO computer program, **240**, 795-798, 815; DSTOOL computer program, **240**, 796-797; GEPASI computer program, **240**, 796; LOCBIF computer program, **240**, 796-797]; Newton's method, **240**, 792-794; numerical continuation, **240**, 792-797; model preparation, **240**, 797-798.

BILIVERDIN REDUCTASE

This enzyme [EC 1.3.1.24] catalyzes the reversible reaction of bilirubin with $NAD(P)^+$ to generate biliverdin and NAD(P)H.

B. F. Cooper & F. B. Rudolph (1995) *Meth. Enzymol.* **249**, 188.

BIMOLECULAR

Bimolecular reactions are elementary reactions involving two distinct entities that combine to form an activated complex. For reactions in solution, the solvent contributes to the reaction's molecularity only when it is a reactant of the system. Bimolecular reactions are usually second order, but it is important to stress that some second order reactions need not be bimolecular.

Bimolecular rate constants typically have units of M^{-1} s^{-1}. Strict compliance to SI units for bimolecular rate constants requires expressing concentrations in the SI system as $mol \cdot m^{-3}$, but concentrations are more frequently given as $mol \cdot L^{-1}$ (or, M^{-1} or $mol \cdot dm^{-3}$). When gas-phase reactions are studied (and even in the case of gases in solution), "concentrations" are often reported in terms of partial pressures rather than molarity. Nevertheless, it is usually preferable to convert such units to true concentration units when able.

Bimolecular processes are very common in biological systems. The binding of a hormone to a receptor is a bimolecular reaction, as is substrate and inhibitor binding to an enzyme. The term "bimolecular mechanism" applies to those reactions having a rate-limiting step that is bimolecular. **See** *Chemical Kinetics; Molecularity; Reaction Order; Elementary Reaction; Transition-State Theory*

BINARY COMPLEX

A noncovalent complex between two molecules. Binary complex often refers to an enzyme-substrate complex, designated ES in single-substrate reactions or as EA or EB in certain multisubstrate enzyme-catalyzed reactions. **See** *Michaelis Complex*

BINDING CHANGE MECHANISM

The $F_o F_1$-ATPase mechanism (a) for coupling chemiosmotic gradient free energy of proton translocation to the synthesis of ATP or (b) for coupling the free energy of ATP hydolysis to create transmembrane proton gradients. While translocation of protons is the function of F_o component, the formation or cleavage of ATP's $\beta - \gamma$ phosphoanhydride bond is catalyzed by the $\alpha_3 \beta_3$ F_1 component. When associated into a functional proton-translocating ATP synthase/hydrolase, F_o and F_1 can mediate either of the processes mentioned above.

The binding change mechanism for the trimeric F_1 component resembles the action of a Wankle (or rotary) engine used in some automobiles. F_1's "cylinders" are made up of three identical polypeptide chains, and each protomeric unit is in a different conformation (designated L for "Loose", T for "Tight", and O for "Open"). ADP and P_i bind to the L-site. Then, an energy-driven step occurs that successively converts $L \cdot ADP \cdot P_i$ to $T \cdot ATP$ and $T \cdot ATP$ to $O + ATP$:

$$L + ADP + P_i = L \cdot ADP \cdot P_i$$

$$L \cdot ADP \cdot P_i \rightarrow T \cdot ATP + HOH$$

$$T \cdot ATP \rightarrow O + ATP$$

These three conformational states are driven by the mechanical rotation of the $\alpha_3 \beta_3$ F_1 trimer around an axis connecting F_o to F_1. Again by analogy to the rotary engine, the L-, T-, and O-states represent fueling, igniting, and discharging steps.

The concept of rotational catalysis[1] by ATP synthase is based on: (a) ^{32}P and ^{18}O exchange rate data attesting to strong cooperativity with sequential participation of several catalytic sites; (b) P_i and ATP ^{18}O-isotopomer distributions indicating that all catalytic sites exhibit identical catalysis; and (c) that catalysis is strongly influenced by the γ-subunit whose primary structure was not likely to account for spatially similar interactions with the β-subunits[2,3]. The model was found to be compatible with the 2.8 Å resolution structure of bovine heart mitochondrial F_1-ATPase[4].

This radical mechanism was without precedent in biochemistry, and direct microscopical observation was reported in 1997 by Noji *et al.*[5] Rotation of γ-subunit of thermophilic F_1-ATPase was observed directly with an epifluorescence microscope. The enzyme was immobilized on a nickel-coated coverslip through a His-tag introduced to the N-termini of the β-subunit, and the enzyme was attached to a fluorescently labeled actin filament. The latter served like a hand of a clock for detecting rotary motions in their observations. The authors describe the motor as follows: "a central rotor of radius approximately 1 nm, formed by its γ-subunit, turns in a stator barrel of radius approximately 5 nm formed by three α- and three β-subunits. F_1-ATPase, together with the membrane-embedded proton-conducting unit F_o, forms the H^+-ATP synthase that reversibly couples transmembrane proton flow to ATP synthesis/hydrolysis in respiring and photosynthetic cells." In the presence of ATP, the filament rotated for more than 100 revolutions in an anticlockwise direction when viewed from the "membrane" side. The rotary torque produced reached more than 40 pNewtons nm^{-1} under high load. As emphasized by Noji *et al.*[5], cells employ many linear motors (or chemical-to-mechanical transducers), including dynein, kinesin, myosin, elongation factors, and RNA polymerase, to move along and exert force on a filamentous structures. With the exception of bacterial flagellum's rotary motor, no other rotary motor had been demonstrated prior to this work on the F_oF_1-ATPase.

The binding change mechanism represents an intellectual triumph in kinetic and mechanistic enzymology, and Paul Boyer was awarded the 1997 Nobel Prize in Chemistry for creating an entirely new way of analyzing energy transduction mechanisms. He shared the prize with John Walker, whose group determined the atomic level structure of F_1, and Jens Skou who first identified the Na^+,K^+-ATPase responsible for establishing the sodium/potassium gradients in neurons.

[1]P. D. Boyer (1997) *Annu. Rev. Biochem.* **66**, 717.
[2]P. D. Boyer & W. E. Kohlbrenner (1981) in *Energy Coupling in Photosynthesis* (R. Selman & S. Selman-Reiner, eds.) pp. 231-240, Elsevier Biomedical, Amsterdam.
[3]P. D. Boyer (1983) in *Biochemistry of Metabolic Processes* (B. L. F. Lennon, F. W. Stratman & R. N. Zalten, eds.) pp. 465-477, Elsevier, Amsterdam.
[4]J. P. Abrahams, A. G. Leslie, R. Lutter & J. E. Walker (1994) *Nature* **370**, 621.
[5]H. Noji, R. Yasuda, M. Yoshida & K. Kinoshita, Jr. (1997) *Nature* **386**, 299.

BINDING INTERACTION

Any noncovalent association guided by electrostatics and complementarity, in which a smaller solute (known as a ligand) binds to a site on its larger molecular weight binding partner (frequently called a protein, enzyme, or receptor). Although kinetics is the major focus of this handbook, binding is a fundamental step in nearly all kinetic mechanisms. Detailed discussions of equilibrium ligand binding interactions lie beyond the scope of this book, and interested readers are encouraged to consult Winzor and Sawyer[1] or Wyman and Gill[2]. **See** *Allosteric Interaction; Binding Isotherm; Biosensor; Cooperative Ligand Binding; Equilibrium Constant; Equilibrium Dialysis; Hummel-Dreyer Method; Linked Function Theory; Klotz Plot; Macroscopic Constant; Microscopic Constant; Molecular Crowding; Scatchard Plot*

[1]D. J. Winzor & W. H. Sawyer (1995) *Quantitative Characterization of Ligand Binding*, p. 168, Wiley, New York.
[2]J. Wyman & S. J. Gill (1990) *Binding and Linkage: Functional Chemistry of Biological Macromolecules*, p. 330, University Science Books, Mill Valley, CA.

BINDING SITE

A topologically defined site on a macromolecular entity that interacts with another molecular entity (commonly referred to as a ligand) *via* electrostatic interactions, hydrophobic interactions, π-orbital overlapping, hydrogen bonding, van der Waals forces, coordination, or even covalent bonding.

BINOMIAL THEOREM

A mathematical theorem defining the form of the polynomial corresponding to $(a + b)^n$, where n is a positive integer. The coefficients for each term of such a polynomial are given by Pascal's triangle. **See** *Pascal's Triangle*

BIOAVAILABILITY

The pharmacokinetic characteristics describing how the active drug in a particular formulation is absorbed and retained in the bloodstream or a organ. Typically determined parameters are (1) the circulating concentration of a drug at certain time intervals after its specified route of administration (*i.e.*, oral, intravenous, *etc.*), and (2) the duration of the drug's persistence in blood. Because differences in the pharmaceutical formulation of the same drug can greatly influence the chemical and physical determinants of bioavailability, any reliable study of bioavailability must be conducted with the formulated drug, and not the pure drug itself. Formulations exhibiting identical bioavailability are said to be bioequivalent.

BIOCATALYTIC ELECTRODE

Any molecule-selective electrode that incorporates an enzyme-catalyzed reaction to transduce the concentration of an analyte into an electrochemical signal. Such electrodes fall under the general category of biosensors, and they rely on the catalytic properties of a covalently-attached or matrix-enclosed enzyme for their high selectivity and responsiveness. Voltammetric sensors containing glucose oxidase (reaction catalyzed: glucose + $O_2 \rightarrow$ glucono-1,5-lactone + H_2O_2) have proved to be commercially successful; the hydrogen peroxide product can be oxidized to molecular oxygen, thereby generating several electrons (reaction: $H_2O_2 + 2OH^- \rightarrow H_2O + 2e^-$). **See also** *Biosensor*

BIOCHEMICAL NOMENCLATURE: RECOMMENDATIONS

The following are the exact literature references dealing with the recommended nomenclature for biochemical compounds and their associated enzyme-catalyzed reactions.

GENERAL:

Abbreviations and symbols for chemical names of special interest in biological chemistry
  J. Biol. Chem. (1966) **241**, 527-533
Abbreviations and symbols: a compilation
  Eur. J. Biochem. (1977) **74**, 1-6
Trivial names of miscellaneous compounds of importance in biochemistry
  J. Biol. Chem. (1966) **241**, 2987-2994
Citation of bibliographic references in biochemical journals
  J. Biol. Chem. (1973) **248**, 7279-7280
Fundamental stereochemistry
  J. Org. Chem. (1970) **25**, 2849-2867
  Final version may be found in Pure Appl. Chem. (1976) **45**, 11-30.

AMINO ACIDS, PEPTIDES & PROTEINS:

Recommended nomenclature and symbolism for amino acids and peptides
  J. Biol. Chem. (1985) **260**, 14-42
  Biochemistry (1975) **14**, 449-462
Abbreviations and symbols for the description of the conformation of polypeptide chains
  J. Biol. Chem. (1970) **245**, 6489-6497
Nomenclature of iron-sulfur proteins
  Eur. J. Biochem. (1979) **93**, 427-430
  Corrections: *Eur. J. Biochem.* (1979) **102**, 315
Nomenclature of peptide hormones
  J. Biol. Chem. (1975) **250**, 3215-3216
Nomenclature of human immunoglobulins
  Eur. J. Biochem. (1974) **45**, 5-6
Recommended nomenclature of glycoproteins, glycopeptides, and peptidoglycans
  J. Biol. Chem. (1987) **262**, 13-18
Recommended nomenclature of electron-transfer proteins
  Eur. J. Biochem. (1991) **200**, 599-611

BIOTHERMODYNAMICS:

Recommendations for the measurement and presentation of biochemical equilibrium data
  J. Biol. Chem. (1976) **251**, 6879-6885
Recommendations for the presentation of thermodynamic and related data in biology
  Eur. J. Biochem. (1985) **153**, 429-434

CARBOHYDRATES:

Nomenclature of Carbohydrates (Recommendations 1996)
  available at: http://www.qmw.ac.uksim;ugca000/iupac/2carb
Tentative rules for carbohydrate nomenclature. Part 1
  J. Biol. Chem. (1972) **247**, 613-634
  Corrections, *Eur. J. Biochem.* (1972) **25**, 4
Conformational nomenclature for five- and six-membered ring forms of monosaccharides and their derivatives
  Eur. J. Biochem. (1980) **111**, 295-298
Nomenclature of unsaturated monosaccharides
  Eur. J. Biochem. (1981) **119**, 1-3
Nomenclature of branched-chain monosaccharides
  Eur. J. Biochem. (1981) **119**, 5-8
Abbreviated terminology of oligosaccharide chains
  J. Biol. Chem. (1982) **257**, 3347-3351
Polysaccharide nomenclature
  J. Biol. Chem. (1982) **257**, 3352-3354
Symbols for specifying the conformation of polysaccharide chains
  Eur. J. Biochem. (1983) **131**, 5-7

CAROTENOIDS:

Tentative rules for the nomenclature of carotenoids
  J. Biol. Chem. (1972) **247**, 2633-2643
  Revision, *Biochemistry* (1975) **14**, 1803
  Pure Appl. Chem. (1975) **41**, 407-431

CORRINOIDS:

Nomenclature of corrinoids
  Biochemistry (1974) **13**, 1555-1560

CYCLITOLS:

The nomenclature of cyclitols. Recommendations
  Eur. J. Biochem. (1975) **57**, 1-7

ENZYMES:

Enzyme Nomenclature. Recommendations
  (1992) Academic Press, New York
Nomenclature of multiple forms of enzymes
  J. Biol. Chem. (1977) **252**, 5939-5941
Catalytic activity
  Units of enzyme activity
    Eur. J. Biochem. (1979) **97**, 319-320
  Symbolism and terminology in enzyme kinetics
    Eur. J. Biochem. (1982) **128**, 281-291

FOLIC ACID:
Nomenclature and symbols for folic acid and related
  compounds
  Eur. J. Biochem. **2**, 5-6 (1967)

LABELED COMPOUNDS:
  Eur. J. Biochem. (1978) **86**, 9-25
  Eur. J. Biochem. (1979) **102**, 315-316
  Final version: *Pure Appl. Chem.* (1979) **51**, 353-380.

LIPIDS:
The nomenclature of lipids. Recommendations
  Lipids (1977) **12**, 455-468

NUCLEOTIDES AND NUCLEIC ACIDS:
Abbreviations and symbols for nucleic acids, polynucleo-
  tides and their constituents
  J. Biol. Chem. (1970) **245**, 5171-5176
  Corrections, *J. Biol. Chem.* (1971) **246**, 4894
Abbreviations and symbols for the description of confor-
  mations of polynucleotide chains
  Eur. J. Biochem. (1983) **131**, 9-15
Nomenclature for incompletely specified bases in nucleic
  acid sequences
  J. Biol. Chem. (1986) **261**, 13-17
Nomenclature of junctions and branch points in nucleic
  acids
  Eur. J. Biochem. (1995) **230**, 1-2

PHOSPHORUS:
Nomenclature of phosphorus-containing compounds of
  biochemical importance
  Proc. Natl. Acad. Sci. U.S.A. (1977) **74**, 2222-2230

QUINONES:
Nomenclature of quinones with isoprenoid side chains
  Eur. J. Biochem. (1975) **53**, 15-18

RETINOIDS:
Nomenclature of retinoids
  J. Biol. Chem. (1983) **258**, 5329-5333

STEROIDS:
Nomenclature of steroids. Revised tentative rules
  Biochemistry (1969) **8**, 2227-2242
  Amendments (1971) and corrections
  Biochemistry (1971) **10**, 4994-4995

Definitive rules for steroids
  Pure Appl. Chem. (1972) **31**, 285-322.

TETRAPYRROLES:
The nomenclature of tetrapyrroles
  Eur. J. Biochem. (1980) **108**, 1-30

TOCOPHEROLS:
Nomenclature of tocopherols and related compounds
  Eur. J. Biochem. (1982) **123**, 473-475

VITAMIN B_6
Nomenclature for vitamin B_6 and related compounds
  Eur. J. Biochem. (1973) **40**, 325-327

VITAMIN D
Nomenclature of vitamin D
  Eur. J. Biochem. (1982) **124**, 223-227

BIOCHEMICAL SELF-ASSEMBLY

Caspar and Klug[1] distinguished two fundamental types
of self-assembly processes: (a) *True self-assembly*—a series of reactions relying on the propensity of subunits to
condense and form assembled structures strictly as a
result of the information encoded in the architecture
of the components. (b) *Template-directed assembly*—a
process depending on the presence of a separate template that imparts structural constraints on the pathway
for constructing the final assembled structure.

True self-assembly is observed in the formation of many
oligomeric proteins. Indeed, Friedman and Beychok[2] reviewed efforts to define the subunit assembly and reconstitution pathways in multisubunit proteins, and all of
the several dozen examples cited in their review represent true self-assembly. Polymeric species are also
formed by true self-assembly, and the G-actin to F-actin
transition is an excellent example[3]. By contrast, there
are strong indications that ribosomal RNA species play
a central role in specifying the pathway to and the structure of ribosome particles. And it is interesting to note
that the assembly of the tobacco mosaic virus (TMV)
appears to be a two-step hybrid mechanism: the coat
protein subunits first combine to form 34-subunit disks
by true self-assembly from monomeric and trimeric com-

ponents; then, the assembly of the virus particle is directed by single-stranded RNA at a specific site in the sequence of the polynucleotide.

The cardinal feature of template-free self-assembly is the ability of the protomers to engage in nucleation, a series of cooperative reactions leading to the formation of polymerization nuclei. These nuclei are thermodynamically unstable, and the high molecularity of nucleation hinders the process. Once formed, however, such nuclei can promote further assembly, because the rate constants for the subsequent protomer addition steps often favor elongation. These rate constants, of course, need not be identical because the first few steps beyond nucleation may be necessary to create the regular lattice for elongation reactions to become independent of polymer length. In this respect, the nuclei represent self-assembly intermediates with unfavorable equilibria leading to them and far more favorable equilibria beyond them. As noted earlier, pure tubulin contains all of the structural information required to form microtubules, and there has been considerable interest in characterizing the initiation process. *See* *Actin Assembly Kinetics; Hemoglobin S Polymerization; Microtubule Assembly Kinetics*

[1]D. L. Caspar & A. Klug (1962) *Cold Spring Harbor Symp. Quant. Biol.* **27**, 1.
[2]F. K. Friedman & S. Beychok (1979) *Annu. Rev. Biochem.* **48**, 217.
[3]F. Oosawa & M. Kasai (1962) *J. Mol. Biol.* **4**, 10.

BIOCONJUGATE

Any biologically produced molecular entity or species having substituents derived from different origins or pathways. For example, glutathione reacts with many arene oxides, and the bioconjugate product is then processed for excretion. Bioconjugates are often more water-soluble, and they are typically more readily compartmentalized and/or metabolized.

BIOCONVERSION

The biological transformation of one molecular entity or substance to form another.

BIOLEACHING

Any biological process, usually involving microorganisms, that extracts substances, typically metals, from ores or soil.

BIOMASS

Any material(s) produced by the growth of a microorganism(s), a plant(s), or an animal(s).

BIOMIMETIC

Any process or procedure designed to imitate one or more natural chemical processes. A molecular entity may also be biomimetic by mimicking the structure and/or function of a biological substance.

BIOMINERALIZATION

Any process in which inorganic cations and inorganic/organic anions combine at concentrations exceeding their respective solubility product(s) to create crystalline or paracrystalline structures.

INTRODUCTORY REMARKS. The most common biominerals include calcium phosphate, calcium carbonate, and calcium oxalate. While these substances readily form regular crystals and paracrystals, living systems rely on biospecific binding interactions to achieve exquisite morphological control over the biomineralization process. The complex and varied crystal habits adopted by marine organisms demonstrate how stereospecific intervention can lead to an astonishing variety of shell forms through the directed deposition of the principal ingredient, calcium carbonate. In higher organisms, the diverse morphologies of endoskeleton, exoskeleton, and teeth are achieved (a) by modulating interactions with cell membranes, proteins, and other inorganic and organic electrolytes, and (b) by controlling concentration/supersaturation of mineral salts through the action of chelating agents, ion pumps, and energy metabolism. Biomineralization also represents an efficient method for removing excess cations, particularly calcium ion, from biological fluids. Within the urinary tract, crystallization of calcium oxalate and calcium phosphate occurs opportunistically and quite freely, and these events probably represent an efficient means for efficient excretion of low-sububility substances. Renal stone disease apparently arises when crystallites assemble into flow-obstructing stones[1-3]. While a comprehensive treatment of biomineralization lies far beyond the bounds of this handbook, there is a set of physicochemical events that underlie all of the above-mentioned processes, especially as regards those solution and surface phenomena controlling nucleation and growth phases. This entry relies on examples using calcium oxalate monohydrate, the major component in idiopathic renal biomineralization.

SOLUTION BEHAVIOR. Biomineralization is dominated by physical chemical considerations[1,2], and we begin with a discussion of real electrolyte solutions in which the concentration of a substance exceeds its thermodynamically defined solubility. In such a case, the presence of a coexisting crystal surface will lead to crystal growth.

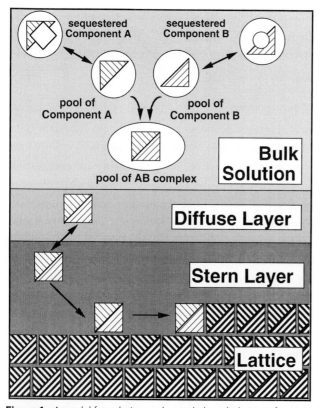

Figure 1. A model for solution and crystal phase behavior of uncomplexed and complexed components (shown here as light and dark triangles, respectively).

On the left-hand side of Fig. 1 are the various mineral-forming components in their complexed and uncomplexed forms. They can directly bind to the crystal growth sites or they can combine to form the crystallization monomers (half-filled squares). These processes can be blocked competitively by the presence of other substances that form nonproductive complexes, thereby depleting the concentrations of precursors through mass action. The diagram also shows a second phase that helps to explain the nature of oriented diffusion and subsequent adsorption of the monomers. This so-called "dou-

ble-layer region" includes the "diffuse layer" (or "Gouy-Chapman layer") and the "Stern layer". The double-layer model allows one to rationalize the forces that monomers experience within the gradient of the electric field created by the surface. This can result in attractive or repulsive diffusion of the monomer to or from the crystal surface. Upon approaching more closely to the surface, each monomer enters the Stern layer and encounters a combination of local electrostatic and chemical forces. Upon arriving at the crystal lattice surface, the monomer undergoes two-dimensional diffusion on the crystal lattice until it is incorporated into the growing crystal. These growth sites are depicted as discontinuities in the crystal surface, such as a step from one lattice layer to the next.

Once formed, crystals can interact with or adhere to other components. Adhesion can occur directly between crystals to form larger structures which may act as the nidus for further mineralization. Crystals can also interact with metabolites and/or tissues to form such structures. These interactions can and do alter the stereochemical course of crystal assembly, and molecular recognition probably accounts for the striations and intricate geometries adopted by biomineralized structures. The role and composition of metabolites comprising the matrix may be useful to understanding the local and large-scale structure of biominerals.

LIQUID-PHASE BEHAVIOR. The liquid phase contains dissolved substances and contacts the solid phase. For our purposes, the "liquid phase" is used synonymously with "aqueous phase", and all processes discussed in this section take place in aqueous solutions. The dissolved monomers of the solid phase are formed in equilibrium with their uncomplexed components. Such components may be uncomplexed ions (which are charged atoms or molecules) free in solution or ionic complexes in equilibrium with dissociated ions. Concentrations of the uncomplexed ions, therefore, depend upon the concentrations of all chemical substances competing for binding interactions with them. Each complexation reaction is defined by either a solution equilibrium constant:

$$K_a = [\text{XY Complex}]/\{[\text{Component X}][\text{Component Y}]\}$$

or by a solubility product constant

$$K_{sp} = [\text{Component A}][\text{Component B}]$$

Accordingly, if a particular ion undergoes complexation with a number of other substances, then the free (or uncomplexed) concentration is defined as the total analytical concentration of the ion *minus* the total concentrations of all complexed forms of that ion. In other words, the total amount of an ion in solution is distributed into its ionic form (uncomplexed form) and the other complexed forms. In some cases, mineral salts are so weakly associated in solution that their uncomplexed ionic forms predominate; such is the case with sodium chloride. We use the term "speciation" to designate the detailed distribution of a substance's ionic and complexed species in solution. Such an account must deal with the equilibria interconnecting such species, and the magnitude of their corresponding thermodynamic equilibrium constants quantitatively define each of their concentrations. To achieve a detailed description of a substance's speciation, one must explicitly know all of the possible complexation reactions that the biomineral ions undergo, and one must have good values for the stability constant (synonymous with "formation constant" or "association constant" in this presentation) for each complex. The magnitude of a stability constant is expressed in reciprocal molar units (M^{-1}) and depends on the temperature and ionic strength of the solution. The reader may recall that ionic strength may be defined as one-half the sum of a series of terms, each corresponding to the concentration of a particular species multiplied by the square of its charge. **See** *Ionic Strength, Solubility Product*

So far, we have expressed the number of ions per unit volume by concentration. Physical chemists prefer to use "activity" to characterize the behavior of solutes in solution. As a solution becomes more concentrated and as the ionic strength increases, ions behave as if there were fewer of them present than would be indicated by their analytical concentrations. The activity of a component is related to its concentration by a proportionality constant known as an "activity coefficient". Considerations regarding ionic activity become particularly important, when fluids are quite concentrated. An extension of this treatment is the use of the "activity product" to

represent the mathematical product of the activities for a set of ions.

Solubility is the measure of the amount of a solid that can be held in solution, and this is achieved when a solution remains in contact with the solid for a sufficient period to ensure equilibration (saturation). Solubility is typically expressed as an activity product which is generally called the "solubility product constant". A more useful way of expressing the degree of saturation is provided by the "relative supersaturation," which can be regarded as the activity product of a solid in solution divided by the solubility product constant. This ratio defines how much of a solid is in solution compared to how much should remain dissolved under stable, equilibrium conditions. Solutions can be made supersaturated with respect to a particular solid by combining solutions of the constituent ions of the solid, or by lowering the temperature of a saturated solution. Usually, the temperature dependence of saturation is not very high, so that lowering the temperature is a more gentle means of supersaturating a solution. Any amount of solute that exceeds the solubility product will over time reach equilibrium by precipitating from solution. Relative supersaturation is actually the driving force for nucleation and crystal growth.

THE DOUBLE LAYER. On a microscopic level, boundaries between solids and liquids (also referred to as "interfaces") are complicated transition zones (see Fig. 2). The physical theories for interfacial behavior[1,2] are different from those for regular solutions, and this is especially true for crystals in aqueous environments.

The surfaces of these crystals often possess characteristic "surface charge" which influences the composition of the solution near the surface. As noted earlier, the double-layer theory divides this region into two parts: (1) a layer near the surface that has a excess of ions whose electric charge is opposite to that of the surface itself (because opposite charges attract); and (2) an outer layer or region containing ions that are oppositely charged relative to that of the first (*i.e.*, the outer layer has the same charge as the surface). The double-layer also maintains overall electric neutrality. The innermost layer is called the Stern layer. Ions and molecules within this layer experience both electrostatic interaction with and

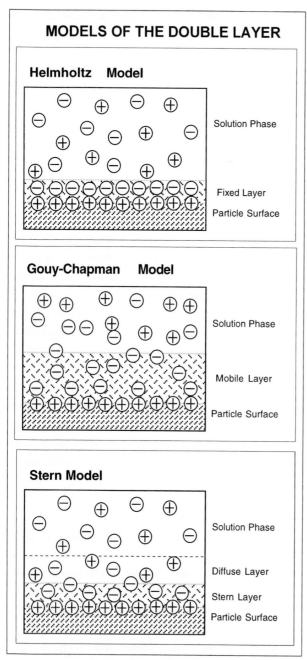

Figure 2. The ionic nature of the interactions at an interface of a electrically charged surface and an aqueous solution containing ionizable solutes. The oldest model, first proposed by Helmholtz, unsuccessfully treats the interface as though it were composed of rigidly held, oppositely charged ions, lined up next to each other in a manner akin to a parallel-plate capacitor. The Gouy-Chapman model was based on the idea that one must consider thermal motions that disturb ionic mobility and interactions at the surface. This model is characterized by the presence of a mobile phase of counterions immediately above a charged surface. The most realistic treatment, proposed by Stern in 1924, accounts for adsorption and exchange of counterions within a layer that extends from the interface for a distance equal to several ionic diameters. The electrical properties of double layers also allow one to treat colloidal interactions, in the same manner as mutual repulsion can explain the stability of so-called lyophobic sols. (Note: To avoid unnecessary duplication, only a positively charged surface is considered.)

chemical attachment to the surface. The composition of this layer gives rise to many of the distinctive behaviors of particles in solution, especially influencing aggregation. A very important parameter characterizing the Stern layer is the Debye-Hückel length, a parameter which expresses the thickness of this layer and strongly depends on solution ionic strength. The Gouy-Chapman layer is the outer or *diffuse* layer. Ions and molecules within this layer are not bound, yet they are influenced by the electrostatic properties of the Stern layer. While it is meaningful to consider the amounts of chemical species in the Stern layer, the Gouy-Chapman layer is best thought of as possessing only a total charge. *Oriented diffusion* is a term coined for the influence that solutes may experience in different ways by the charge structure of the double-layer region. This influence could amount to a facilitation or obstacle with regard to the participation of certain substances in surface processes.

Some ions and other molecular substances attach to surfaces under the influence of electrostatic and chemical forces. This process is known as *adsorption*, and those substances adhering to the surface are called "adsorbates." Various theories treat adsorption as a hyperbolic binding curve (**See** *Langmuir Isotherm*) which relates the substance's surface concentration to its bulk solution concentration. An isotherm describes the gross or macroscopically observable behavior of adsorbates, whereas the Stern theory is a way of describing adsorption at a molecular level, while still taking into account the specific properties of the adsorbates and the surfaces. Surface-adsorbed ions and molecules are usually not rigidly attached, and they often can migrate along the surface to other points of attraction where they may dwell for a time. This process is called "surface diffusion." The interface may be pictured as an active region where molecules and ions approach and attach, detach and diffuse, or detach and return to the bulk solution. Some substances may anchor themselves more or less irreversibly to the surface, and this appears to be a particularly important viewpoint for larger biomolecules which may have multiple points of attachment, so-called *polypodal attachment*. At any instant, such molecules may release one or two points of attachment, but releasing all points of attachment simultaneously is less and less likely with every additional site of attachment. Naturally, some of the substances attached to the surface are those of which the solid itself is composed, and various ionization states or complexes of these substances may still interact with the surface without leading to crystal growth. The distinction between incorporated and adsorbed species must be kept in mind. Incorporation into the solid leads to net crystal growth. Likewise, during dissolution, some of the exposed solid may leave, thereby returning to the solution. Other lattices may have just the right structure to permit other unrelated solid lattices to be formed. When a crystal of one composition gives rise to a crystal of another composition in this way, it is usually called *epitaxy* or *epitaxial growth*, but the term *oriented overgrowth* may also be seen.

For a solution supersaturated with solute, crystallization cannot take place in the absence of a higher-order process known as *nucleation*. In one sense, nucleation refers to any process affording a route for a substance from one physical phase to another, usually in the direction

of greater order. Examples are the slow formation of bubbles in carbonated beverages or the condensation of water vapor into droplets. Of greatest interest to biomineralization is the spontaneous appearance of microscopic crystals from supersaturated solutions.

IONIC SPECIATION. Ions interact continually in aqueous solution. Ions are complexed with water molecules. Even when we say that a certain ion is uncomplexed, the fact is that the ion is still complexed, in this case with water molecules. Association constants (also known in the literature as stability or formation constants) allow one to quantitate the extent to which an ion is complexed with any particular substance in solution. They also allow comparisons of the relative affinity of different complexing agents for a particular chemical substance. Speciation is a chain of linked binding functions (see Fig. 3). Such diagrams show the relative concentrations of the various complexes in solution, and the reversible equilibria existing between these pools are shown by the arrows.

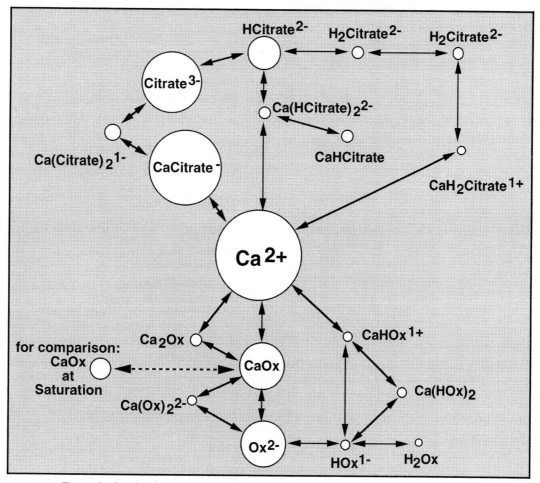

Figure 3. Graphical representation of ionic and complexed species as a "pool diagram."

The linkage between all the circles through various pathways is a graphical display of the fact that a change in the concentration of any complex or component on the diagram will be transmitted to all other circles on the diagram. For example, if citrate is reduced, the calcium which was bound as calcium citrate complex will be released, and this will increase the available calcium; increased available calcium will, in turn, compete with other cations for phosphate, sulfate, and so on. The net result will be an increased concentration of all other calcium complexes, a reduction of all other cation complexes, and an increase in the uncomplexed cations. Because of the number of complexes in the scheme, the effect of small changes in any one constituent is buffered. The key to the physical success of the scheme is that all the reactions in it are reversible; that is, though the complexes are stable relative to the component activities, if those activities decrease, there is nothing preventing the complexes from breaking down into their components. Irreversible reactions change the way this scheme works; such reactions could be the irreversible binding of a cation or anion, reactions of complexing agents with extremely high affinities, or, as we shall consider in the next section, the precipitation of insoluble salts.

SOLUBILITY PRODUCT. For ions that combine and form an insoluble phase, we use an alternative form of an equilibrium constant known as the solubility product. Consider the following reaction:

$$\text{Cation} + \text{Anion} = \text{Solid Phase}$$

The ion solubility product (K_{sp}) is given by the expression,

$$K_{sp} = [\text{Anion}][\text{Cation}]/1$$

where "1" in the denominator designates the assigned activity of the solid phase, and is usually not written. The law of mass action requires that ions in solution will remain completely soluble so long as the product of their concentrations is less than or equal to their corresponding solubility product. Whenever the product of their concentrations exceeds the solubility product, the solution will be initially supersaturated, and if precipitation occurs, the system will reach thermodynamic equilibrium only when the product of their concentrations is once

again less than or equal to their corresponding solubility product.

RELATIVE SUPERSATURATION. The free energy that drives nucleation and crystal growth is directly related to a substance's relative supersaturation (RS), and the latter is a unitless parameter equal to $[X]_{instantaneous}/[X]_{saturation}$:

$$\Delta G = -RT \ln [X]_{instantaneous}/[X]_{saturation} = -RT \ln(\text{RS})$$

The higher the relative supersaturation, the more likely nucleation becomes, and the faster crystal growth proceeds. Molecules or ions will remain dissolved provided that the conditions are energetically favorable. However, all molecules above a certain threshold solution activity will remain in, or become part of, the solid phase. Molecules are in equilibrium between the solid state and the dissolved state. The extent to which the equilibrium balance favors the dissolved phase indicates the degree of solubility.

Once formed, the solid phase will grow until the activity of the complex from which the solid phase is formed reaches its equilibrium value.

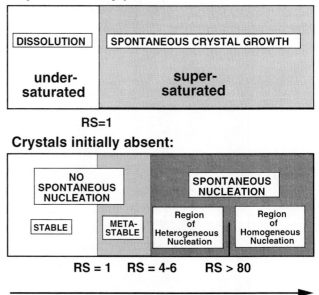

Figure 4. Phase diagram for crystallization as a function of relative supersaturation.

These solid phases are connected to the components in Fig. 4, with which they are in reversible equilibrium. For example, if magnesium ion were added to a complex solution containing solid calcium oxalate monohydrate (COM), the magnesium would compete with calcium for an increased share of the oxalate; this would reduce the amount of the calcium oxalate complex, and finally a small amount of calcium oxalate solid would dissolve to restore the complex concentration to its equilibrium value. In urine, this picture must be extended to account for the molecular substances that coat crystals and reduce access of the solution to the surface; coated crystals do not redissolve readily.

SURFACE PHENOMENA. Assuming the presence of crystal surfaces in living systems, adsorption is the process most actively affecting their nature. Adsorbates alter the chemical, physical, and electrical properties of surfaces; by doing so, they affect crystal growth and aggregation. Whereas one might tend to identify a solid by its bulk composition, it is the surface composition that interacts with substances in the solution. Again using COM (calcium oxalate monohydrate) as our example, the bulk composition of this stone salt will be one calcium ion to one oxalate ion to one water molecule. The surface composition depends on how the crystal is "sliced" (i.e., which of the different crystal faces are stable in solution). These different faces have different ratios of calcium to oxalate, and because these ratios need not be strictly one-to-one, different faces can actually have different charges. Each face of the crystal will have a specific surface charge. It is often impractical to try to view the charging of ions at such a microscopic level, but the overall charging of the crystal is experimentally determined in "Zeta potential" measurements using a microscope to follow the electrophoresis of crystals.

NUCLEATION. Nucleation creates a new phase that is organizationally more related to the crystal lattice than to the monomeric species that undergoes crystallization. This process permits solutions that are of high relative supersaturation to crystallize and thereby reach equilibrium between liquid and solid phases[1,2,4]. Nucleation occurs when the local concentration of components that will comprise the solid phase exceeds a threshold level as a result of short-range concentration fluctuations in the bulk solution. In this respect, the kinetics of nucle-ation reflect the frequency and duration of such fluctuations.

To understand nucleation, consider five different conditions that arise as the relative supersaturation is increased in the initial absence or presence of crystals. At RS values greater than one, the solution is supersaturated and will undergo spontaneous crystal growth until unit saturation occurs. One can identify so-called "stable" and "metastable" states which do not undergo spontaneous nucleation despite the fact that the metastable condition is somewhat supersaturated. At higher relative supersaturation (say RS 4–6 for the case of calcium oxalate monohydrate), one can observe spontaneous nucleation. This region of spontaneous nucleation can be further divided conceptually into two regions, based on the magnitude of the liquid-solid interfacial tension (σ). Heterogeneous nucleation allows other substances to reduce the metastable limit and is characterized by a σ value of approximately 30 erg/cm^2. On the other hand, homogeneous nucleation requires considerably higher relative supersaturation (hence, higher σ values) to allow local concentration fluctuations to be the primary route for forming nuclei. Homogeneous nucleation can occur slowly, even at the relative supersaturation values commonly associated with heterogeneous nucleation.

The rate at which the concentration of supersaturated components are depleted through crystal growth depends on the concentration of nuclei. The inhibitory action of various agents reflects a lowering of the nuclei concentration either through inhibition of nuclei formation or through promotion of nuclei dissolution, and promoters of nucleation must act to either increase the rate of nucleation or decrease the loss of nuclei from the solution. Inhibition or promotion of nucleation must be analyzed under conditions where the relative supersaturation remains the same. These definitions do not include mechanisms which could be explained as resulting merely from changes in the relative supersaturation, as would be the case for calcium ion and oxalate complexing agents.

The induction time tau for the precipitation process, where $\tau = gC^n$. C is the square root of the molar ion concentration product of the precipitating salt, and g and n are empirical constants. In this respect, n represents the apparent molecularity for nucleation; for calcium oxalate at 25°C, these investigators obtained values for g and n of 1.03×10^{-7} S and -3.33, respectively.

GROWTH & AGGREGATION. After nucleation occurs, crystal growth ensues until the RS approaches unity. During this growth, crystals can remain as single growing units or they can aggregate to form larger particles. In this regard, one should distinguish between crystal size distributions and particle size distributions; the former arise from growth of crystallites in the absence of aggregation, and the latter reflect a summation of individual crystal growth as well as that arising from aggregation.

Aggregation can be regarded as a higher-order process in which particles in aqueous media associate and form extended structures. Charged surfaces on particles can interact to bond them together, forming the larger structures or aggregates. Understanding the composition of the Stern layer is therefore an essential basis to a detailed account of the aggregation behavior of suspended particles. Aggregation is strongly dependent on the hydrodynamic forces in the medium, and it is divided broadly into two categories: (i) orthokinetic aggregation occurs in the presence of stirring, and (ii) perikinetic aggregation takes place in the effective absence of stirring. When particles aggregate so tightly that they merge together, they are said to be *sintered*. Aggregation might be effected also by substances that can cross-link surfaces by a sort of gluing phenomenon. Interactions leading to aggregation are: electrostatic attraction; van der Waals interactions; liquid bridging; capillarity; viscous binder effects; and solid bridging.

[1]B. Finlayson (1978) *Kidney Internat.* **13**, 344.
[2]C. M. Brown & D. L. Purich (1992) in *Disorders of Bone and Mineral Metabolism* (F. L. Coe & M. J. Flavus, eds.) pp. 613-624, Raven Press, New York.
[3]W. G. Robertson & M. Peacock (1985) in *Urolithiasis: Etiology-Diagnosis* (H.-J. Schneider, ed.) p. 183, Springer-Verlag, Berlin.
[4]J. A. Christiansen & A. E. Nielsen (1951) *Acta Chim. Scand.* **5**, 673.

BIOREACTOR

An open or closed reaction system or device designed to manipulate physical parameters (such as heat/mass transfer, pH, reactant concentration(s), *etc.*) as a means for regulating biocatalysis in order to maximize product

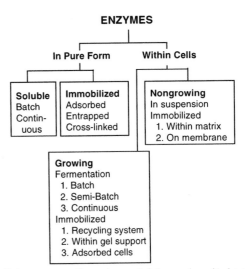

Range of bioreactor configurations gainfully employed in biotechnology and agri-business.

output and to minimize process costs. **See** *Immobilized Enzymes*

[1]C. L. Cooney (1984) in *Biotechnology & Biological Frontiers* (P. H. Abelson, ed.) pp. 242-253, American Association for the Advancement of Science, Washington.
[2]R. A. Messing (1975) *Immobilized Enzymes for Industrial Reactors*, Academic Press, New York.

BIOSENSOR

A highly sensitive detection system that relies on surface plasmon resonance (SPR) to obtain real time monitoring of biochemical interactions without requiring stable/radioactive isotopic labels or absorption/fluorescence reporter groups. The optical phenomenon known as SPR occurs when surface plasmon waves are excited at a metal/liquid interface. Gold is an excellent conductor as a result of its "gaseous cloud" of free electrons, and this cloud of ions can be a form of plasma. These free electrons in the detector's gold foil surface are remarkably sensitive to changes in the electrostatic interactions occurring in the liquid phase as a result of biospecific binding interactions. Moreover, these free electrons determine the refractive index near the foil's surface. In the BIACORE® probe, SPR is detected using the gold film at the sensor probe tip. "White" light traversing the probe's optical fiber undergoes total internal reflection at the surface interfaces, but SPR reduces the reflected light intensity at certain wavelengths. Binding interactions alter the refractive index close to the surface, such that the wavelength of the reflected light intensity minimum shifts. The BIACORE® probe allows the instrument to sense the spectral output of the reflected light. Because nearly all proteins have very similar specific refractive indices, SPR is a mass detector that does not depend on the nature of the interacting species. The surface of a sensor probe is coated with a carboxymethylated-dextran matrix that permits covalent attachment of a detecting molecule. An intrinsic advantage of this methodology is that extremely small volumes of material are needed because all of the binding interactions occur in an ultrathin layer. *See also* Biocatalytic Electrode

Selected entries from *Methods in Enzymology* [vol, page(s)]: Theory, **240**, 323-324; instrumentation [BIACORE, **240**, 323-324; data output, **240**, 325]; data analysis [computer program, **240**, 330; linearity of models, **240**, 343-345; linear least-squares, **240**, 325, 343, 345-346, 348; nonlinear least-squares, **240**, 325-326, 343, 345-346, 348]; ligand immobilization, **240**, 329; ligate preparation, **240**, 329-330; phase data [association, **240**, 330-332; dissociation, **240**, 333-334]; rate constants [association, **240**, 340-342, 348-349; dissociation, **240**, 334-340, 348-349]; rate equations, **240**, 326-328; reproducibility in parameter determinations, **240**, 342-343; kinetic analysis of macromolecular interactions by surface plasmon resonance biosensors [theory, **295**, 269-270; experimental design, **295**, 271-277; data analysis, **295**, 277; global analysis of data, **295**, 279; CLAMP biosensor data analysis program, **295**, 279-285; testing reaction models experimentally, **295**, 285; establishing parameter confidence, **295**, 286; interpreting results, **295**, 290-293].

BIOTIN AND DERIVATIVES

Biotin[1] is a vitamin that was first recognized as an agent for preventing dermatitis and paralysis in rats fed large quantities of uncooked egg white (which contains abundant quantities of the protein avidin). The vitamin is covalently attached to an ε-amino group of an active site lysyl residue in biotin-dependent enzymes. Carboxyphosphate is thought to be formed prior to carboxyl group transfer to biotin. The N-1' position of biotin then acts as a carboxyl group carrier involved in carboxylation reactions.

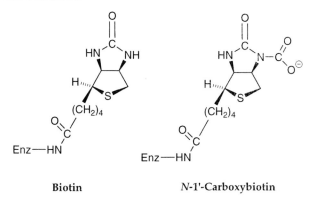

Biotin   *N*-1'-Carboxybiotin

Beyond its fundamental role in metabolism, the high affinity of biotin binding to avidin (K_d 10^{-15} M) and streptavidin has made biotin labeling especially valuable in cell biology[1]. A protein of interest may be covalently labeled with suitably derivatized, commercially available biotin. For example, one can use *N*-hydroxysuccinimide or *p*-nitrophenyl derivatives of biotin to react with amino groups on the target protein.[2] The protein's location within cells can be detected by using streptavidin along with a fluorescently labeled anti-streptavidin antibody.

[1]For a comprehensive discussion of the reactions and properties of vitamins and cofactors, readers are urged to consult *Biochemistry: The Chemical Reactions of Living Cells* (2nd ed.) by D. E. Metzler (2000) Academic Press, Orlando. This text and reference book contains a valuable compilation of literature citations.
[2]The Pierce Chemical Company (Rockford, Illinois) freely distributes *Avidin-Biotin Chemistry: A Handbook* to interested researchers.

Selected entries from *Methods in Enzymology* [vol, page(s)]: Isotopic dilution assay for biotin use of [^{14}C]biotin, **62**, 279; use of [^{3}H]biotin, **62**, 284; rapid fluorometric assay for avidin and biotin, **62**, 287; microassay of avidin, **62**, 290; radioimmunoassay for chicken avidin, **62**, 292; the synthesis and use of spin-labeled analogs of biotin in the study of avidin, **62**, 295; the avidin-biotin complex in affinity cytochemistry, **62**, 308; egg yolk biotin-binding protein: assay and purification, **62**, 316; antibodies that bind biotin and inhibit biotin-containing enzymes, **62**, 319; microbiological biosynthesis of biotin, **62**, 326; preparation of pimeloyl-coenzyme A, **62**, 339; 7,8-diaminopelargonic acid aminotransferase, **62**, 342; dethiobiotin synthetase, **62**, 348; synthesis and use of specifically tritiated dethiobiotin in the study of biotin biosynthesis by *Aspergillus niger*, **62**, 353; the biotin transport system in yeast, **62**, 371; enzymic reduction of D-biotin D-sulfoxide to D-biotin, **62**, 379; isolation and characterization of D-allobisnorbiotin, **62**, 385; analysis of microbial biotin proteins, **62**, 390; identification of biocytin, **62**, 398.

BIOTIN HOLOCARBOXYLASE SYNTHETASE

This enzyme [EC 6.3.4.10], also known as biotin—[propionyl-CoA-carboxylase] ligase and holocarboxylase synthetase, catalyzes the reaction of biotin with ATP and apo-[propanoyl-CoA:carbon-dioxide ligase (ADP-forming)] to produce AMP, pyrophosphate, and propanoyl-CoA:carbon-dioxide ligase (ADP-forming).

Y. Suzuki & K. Narisawa (1997) *Meth. Enzymol.* **279**, 386.
K. Dakshinamurti & J. Chauhan (1988) *Ann. Rev. Nutr.* **8**, 211.
N. H. Goss & H. G. Wood (1984) *Meth. Enzymol.* **107**, 261.

BIRADICAL

A molecule, molecular species, or entity containing two unpaired electrons on different atoms (*e.g.*, R—$\dot{C}$H—CH$_2$CH$_2$—R—$\dot{C}$H—R); also referred to as diradicals. If the unpaired electrons are widely separated, the entity behaves spectrally like two doublets. If the electrons are close enough for interaction, they are referred to as triplets. *See also* Carbene; Free Radicals

W. T. Borden (1982) *Diradicals*, Wiley, New York.
L. J. Johnston & J. C. Scaiano (1989) *Chem. Rev.* **89**, 521.
C. Doubleday, Jr., N. J. Turro & J. Wang (1989) *Acc. Chem. Res.* **22**, 199.

BISPHOSPHOGLYCERATE MUTASE

This enzyme [EC 5.4.2.4] (also known as bisphosphoglycerate synthase, diphosphoglycerate mutase, and glycerate phosphomutase) catalyzes the reversible interconversion of 3-phospho-D-glyceroyl phosphate to 2,3-bisphospho-D-glycerate. The "cofactor" for this enzyme is 3-phosphoglyceroyl phosphate. In the reaction scheme, the enzyme is phosphorylated by this "cofactor" to yield a phosphoenzyme and 3-phosphoglycerate. The latter compound is then rephosphorylated by the enzyme to yield 2,3-bisphosphoglycerate, but this reaction is slowed down by dissociation of 3-phosphoglycerate from the enzyme, which is therefore more active in the presence of added 3-phosphoglycerate.

L. A. Fothergill-Gilmore & H. C. Watson (1989) *Adv. Enzymol.* **62**, 227.
Z. B. Rose (1982) *Meth. Enzymol.* **87**, 42-51.
Z. B. Rose (1980) *Adv. Enzymol.* **51**, 211-253.
W. J. Ray, Jr., & E. J. Peck, Jr. (1972) *The Enzymes*, 3rd ed., **6**, 407.

BI TER REACTION

An enzyme-catalyzed reaction in which there are two substrates and three reaction products.

BI UNI REACTION

An enzyme-catalyzed reaction involving two substrates and one product. There are two basic Bi Uni mechanisms (not considering reactions containing abortive complexes or those catagorized as Iso mechanisms). These mechanisms are the ordered Bi Uni scheme, in which the two substrates bind in a specific order, and the random Bi Uni mechanism, in which either substrate can bind first. Each of these mechanisms can be either rapid equilibrium or steady-state systems.

"BLACK BOX"

A term typically used in steady-state enzyme kinetics to acknowledge that there exist many kinetic models where

measurements of the changes in enzyme velocity as a function of substrate or effector concentration cannot provide any mechanistic information about the number or nature of isomerizations occurring within the so-called central complexes. The fact is that steady-state enzyme kinetics can provide useful tools for a very limited repertoire of mechanistic studies, and one must quickly proceed to experiments that employ other methods, such as isotope exchange, stopped-flow, rapid mix/quench, temperature-jump, rapid scan spectroscopy, kinetic isotope effects, transition-state inhibitor design, and NMR as well as any of the other environmentally sensitive spectroscopic methods.

A quote from W. P. Jencks' prefatory chapter[1] in the 1997 *Annual Review of Biochemistry* says it all rather succinctly. Referring to his twenty years investigating the ATP-dependent sarcoplasmic reticulum calcium pump, he remarked: "We studied this reaction cycle by using rapid mixing experiments to determine the individual steps of the reaction. I had previously decided that any conclusion that was reached from measurements of steady-state kinetics is wrong. This could be an overreaction. However, I think this is a good place to start, especially when one is examining complicated reactions."

[1]W. P. Jencks (1997) *Ann. Rev. Biochem.* **66**, 1.

BLEACHING

A term used in photochemistry and photomicroscopy to indicate the loss of absorption or emission intensity.

BLEOMYCIN

A glycopeptide molecule, often abbreviated BLM, that can serve as a metal chelating ligand. The bleomycins comprise a family of related antibiotics. The Fe(III) complex of bleomycin is an antitumor agent. Bleomycin's activity is associated with single-strand DNA cleavage through the generation of hydroxyl free radicals.

BLUE COPPER PROTEIN

An electron-transfer protein containing a type-1 copper site that manifests itself by a strong visible absorption spectum. Another characteristic attributable to copper coordination by a cysteinyl sulfur is the EPR signal that displays an unusually small hyperfine coupling to the copper nucleus.

BOHR

An atomic unit of length used in quantum mechanical calculations of electronic wavefunctions. It is symbolized by a_o and is equivalent to the Bohr radius, the radius of the smallest orbit of the least energetic electron in a Bohr hydrogen atom. The bohr is equal to $\alpha/(4\pi R_\infty)$, where α is the fine-structure constant, π is the ratio of the circumference of a circle to its diameter, and R_∞ is the Rydberg constant. The parameter α includes $\hbar$, as well as the electron's rest mass and elementary charge, and the permittivity of a vacuum. One bohr equals $5.29177249 \times 10^{-11}$ meter (or, about 0.529 angstroms).

BOHR EFFECT

Hemoglobin's linked function behavior of cooperative oxygen binding that results in reduced oxygen binding as the pH of a red cell or hemoglobin solution is lowered. Protonation of hemoglobin stabilizes the deoxygenated form, thereby reducing the ability of this heme protein to bind oxygen. In a reciprocal manner, oxygen binding must likewise alter the acid/base properties of hemoglobin. *See also* Linked Functions; Adair Equation; Hemoglobin

Selected entries from *Methods in Enzymology* [vol, page(s)]: Description, **231**, 97; **232**, 130; mechanism, **232**, 97-99; role of hemoglobin A histidyl residues, **232**, 130-133, 138; role of hemoglobin chloride-binding site, **231**, 225-226; ^{1}H NMR analysis, **232**, 97; role of hemoglobin side-chain carboxyl groups, **231**, 246-247; anion dependence, **259**, 524, 529, 534-535; characteristics, **259**, 16-18, 513; cryoelectrophoresis determination, **259**, 475-477; dissociation of individual proton-binding sites, **259**, 515, 531-534; electrostatic energy of proton binding, **259**, 526-527; energetics calculation, limitations of algorithms, **259**, 536-538; free energy [calculation from titration curves, **259**, 517-519; salt and pH dependence, **259**, 527]; measurement, **259**, 475-477, 520; microscopic proton-binding isotherms, **259**, 520-521, 535-536; proton binding site, average charge computation, **259**, 525-526; proton titration curves, **259**, 514-517, 523-524, 530-534.

BOLTZMANN DISTRIBUTION

A quantitative function[1] describing how large collections of electrons, atoms, or molecules are distributed in terms of their occupancy of quantum mechanically allowed energy states (*i.e.*, electronic, vibrational, rotational, and translational), based on energy differences defined by solution of the Schrödinger equation. At any given temperature, the Boltzmann distribution can be used to determine the average energy of the entire collection of molecules—a fact that helped determine an absolute temperature scale. This distribution also helps one recognize how chemical reactivity is related to the fraction of molecules (within an ensemble) having sufficient energy to exceed some energy barrier to reactivity.

Consider a collection of particles with energies ε_1, ε_2, ε_3, . . ., ε_i, each corresponding to a single quantum state, starting with the lowest lying state and proceedith to the ith state. The number of particles in each state can be written symbolically as N_0, N_1, N_2, N_3, . . . N_i, The Boltzmann distribution function relating the relative occupancy of two states is written as:

$$\frac{N_i}{N_0} = \exp[-(\varepsilon_i - \varepsilon_0)/k_B T] = \exp(-\Delta\varepsilon_i)/k_B T$$

where k_B is the Boltzmann constant (where k_B equals R/N_0, the gas constant R divided by Avogadro's number). If more than one state has the same energy, then the number of such states represents the degeneracy (symbolized by g). The number of particles in g degenerate states relative to the number in the lowest lying quantum state can be determined using the following expression:

$$\frac{N_i}{N_0} = \frac{g_i}{g_0} \exp(-\Delta\varepsilon_i)/k_B T$$

The exponential nature of this function results in very small values of (N_i/N_0) whenever $\Delta\varepsilon_i$ is much greater than $k_B T$.

The Boltzmann distribution helps us understand how intermediates can become trapped in energy wells between successive transition states in a multistep reaction mechanism. This behavior forms the basis for a special field of enzymology known as cryoenzymology. By appropriate choice of water-miscible solvents, enzymes can be studied at ultra-low temperatures where the rates of interconversion of enzyme species can be greatly retarded. *See Cryoenzymology*

[1]Derivation of the Boltzmann distribution function is based on statistical mechanical considerations and requires use of Stirling's approximation and Lagrange's method of undetermined multipliers to arrive at the basic equation, $(N_i/N_0) = (g_i/g_0)\exp[-\beta \Delta\varepsilon_i]$. The exponential term β defines the temperature scale of the Boltzmann function and can be shown to equal $1/k_B T$. In classical mechanics, this distribution is defined by giving values for the coordinates and momenta for each particle in three-coordinate space and the lin-

ear moment, ρ_x, ρ_y, and ρ_z (each equal to the particles mass multiplied by the velocity component in the x, y, and z directions). The probability that a particle lies in the intervals x to $x + dx$, y to $y + dy$, and z to $z + dz$, and the probability lies correspondingly between ρ_x and $\rho_x + d\rho_x$, ρ_y and $\rho_y + d\rho_y$, and ρ_z and $\rho_z + d\rho_z$, can be written as: $P = e^{-W/k_B T}\, dx\, dy\, dz\, d\rho_x\, d\rho_y\, d\rho_z$, where W is the energy and k_B the Boltzmann constant, and T the absolute temperature. The probability of a particle having energy w above the ground state can likewise be written as: $N_i/N_o = g_i e^{-W/k_B T}$, where g_i takes into account the number of possible states and any energetic degeneracy. Note that N_i is the number of molecules having energy W, and N_o is the total number of molecules or particles. Further note that the ratio N_i/N_o is the probability of occurrence of molecule N_i in a total ensemble N_o. An analogous treatment can be obtained by defining the relative probabilities of various quantum states of a system in equilibrium.

BOLUS

A large quantity of substance delivered or added to an organism or system, all at once. A single administration of a large amount of a drug is often described as a "bolus dose".

BOLUS TRACER INJECTION

A bolus injection initially leads to an increase in the tracer/tracee ratio within a compartment, and this ratio then undergoes a first-order decline based on the rate constant for metabolite elimination[1].

[1] R. R. Wolfe (1992) *Radioactive and Stable Isotope Tracers in Biomedicine*, Wiley-Liss, New York.

BONDING ORBITAL

A molecular orbital having a lower energy level than the atomic orbitals from which the bonding orbital is formed. Such an orbital can contain two electrons, and their presence results in a strong bond when the overlap of the atomic orbitals is large. Two overlapping atomic orbitals combine to yield one low-energy bonding orbital (designated σ) and one high-energy antibonding orbital (designated σ^*). Two paired electrons are sufficient to fill the s orbital, and any additional electrons must occupy the high energy σ^* orbital where, rather than stabilizing the bond, they lead to repulsion between the atoms.

BOND NUMBER

The number of electron-pair bonds shared by two atoms. In acetone, for example, the bond number for the carbon and oxygen is two, whereas the bond number is one for carbon bonded to hydrogen.

BOND ORDER

An index of the degree of bonding between two atoms in a molecule. A normal single bond is given a bond order of one. In valence-bond theory, the bond order for a particular bond is calculated as the weighted sum of all the canonical forms of the molecule. For example, if benzene were completely represented by two Kekulé resonance forms, then one would predict that a carbon-carbon bond in benzene has a bond order of 1.5. However, the three so-called Dewar structures also contribute to the structure of benzene[1]. The carbon-carbon bond order for benzene using valence-bond theory is thus determined to be 1.463. Hence, a carbon-carbon bond in benzene is not halfway between a single and a double bond, but actually somewhat less[2,3]. Bond orders are calculated differently in molecular orbital theory[1,2]. The order is determined from the weights of the atomic orbitals in each molecular orbital. With this method, the carbon-carbon bond order in benzene is 1.667.

[1] A. Pullman & B. Pullman (1958) *Prog. Org. Chem.* **4**, 31.
[2] D. Clarkson, C. A. Coulson & T. H. Goodwin (1963) *Tetrahedron* **19**, 2153.
[3] W. C. Herndon & C. Parkanyi (1976) *J. Chem. Ed.* **53**, 689.

BOND VALENCE SUM ANALYSIS

An empirical method for correlating the oxidation state of a metal ion with the coordination geometry and the bond lengths[1]. Bond valence sum analysis has been used in characterizing the structural features of vanadium-dependent haloperoxidases[2,3].

[1] W. T. Liu & H. H. Thorp (1993) *Inorg. Chem.* **32**, 4102.
[2] A. Butler & M. J. Clague (1995) *Mechanistic Bioinorganic Chemistry* (H. H. Thorp & V. L. Pecoraro, eds.), pp. 329-349, ACS, Washington, DC.
[3] C. J. Carrano, M. Mohan & S. M. Holmes (1994) *Inorg. Chem.* **33**, 646.

BORN-OPPENHEIMER APPROXIMATION

An approximation stating that the motion of nuclei in ordinary molecular vibrations is slow relative to the motions of electrons. Thus, the nuclei can be held in fixed positions when doing calculations of electronic states. Such an assumption is useful in determining potential energy surfaces and is central in studying the quantum mechanical properties of molecules. ***See also*** Adiabatic Photoreaction; Diabatic Photoreaction

BORNYL PYROPHOSPHATE SYNTHASE

This enzyme [EC 5.5.1.8], also known as geranyl-diphosphate cyclase, catalyzes the conversion of geranyl diphosphate to (+)-bornyl diphosphate.

R. A. Gibbs (1998) *Comprehensive Biological Catalysis: A Mechanistic Reference* **1**, 31.

F. M. Raushel & J. J. Villafranca (1988) *Crit. Rev. Biochem.* **23**, 1.

BOROHYDRIDE REDUCTION

Sodium borohydride ($NaBH_4$) and its isotopic forms (NaB^2H_4 or NaB^3H_4) are impressively useful reducing agents[1-4]. $NaBH_4$ exhibits much greater selectivity than lithium aluminum hydride ($LiAlH_4$), and normally reduces ketones and aldehydes to the corresponding alcohols without acting on carbon-carbon double bonds, nitriles, esters, nitro groups, *etc.* However, conditions can be altered (*e.g.*, the presence of transition metal salts) to increase the reducing power of $NaBH_4$. For example, alkenes and alkynes can be reduced with $NaBH_4$ provided the unsaturated bonds are complexed with salts such as $FeCl_2$ or $CoBr_2$. Amides are not reduced by $NaBH_4$, unless other reagents are present. In general, $NaBH_4$ will reduce aldehydes and ketones to alcohols, acyl halides to alcohols (or to aldehydes) under some conditions, and act on Schiff bases, imines, and hydrazones. This nucleophilic reductant will also convert *N*-alkylimino esters ($RC(OR')=NR''$) to secondary amines.

In biochemistry, metal hydrides such as $NaBH_4$ have been widely used in synthesis. For example, $NaBH_4$ has been used in the preparation of alkyl cobalamins from cyanocobalamin[5,6], and in the synthesis of the chiral [β,γ-$^{17}O;\gamma^{17}O,^{18}O$]ATP$\gamma$S from the 5'-aldehyde of adenosine[7,8]. Lithium triethylborohydride (also known as "Super Hydride") has been used in the preparation of chiral acetate and other chiral methyl groups[9,10].

$NaBH_4$ has greatly facilitated the chemical trapping of covalent intermediates that are formed in enzyme-catalyzed reactions. The metal hydrides have played a significant role in characterizing the mechanism of enzyme processes (Table I). Interestingly, borohydride treatment can occasionally be used to exclude certain mechanistic possibilities. Treating galactose-1-phosphate uridylyltransferase[11] with NaB^3H_4 in the presence of substrate, for example, failed to result in radiolabel incorporation. This observation, when taken in the context of other experimental findings, indicated the very low probability of an acyl phosphate intermediate as a catalytic feature of this enzymic process.

[1] A. Hajos (1979) *Complex Hydrides*, Elsevier, New York.
[2] H. O. House (1972) *Modern Synthetic Reactions*, 2nd ed., pp. 49-71, Benjamin, New York.
[3] O. H. Wheeler (1966) in *The Chemistry of the Carbonyl Group* (S. Patai, ed.), Part 1, pp. 507-566, Interscience, New York.
[4] H. C. Brown (1972) *Boranes in Organic Chemistry*, Cornell Univ. Press, Ithaca, New York.
[5] T. C. Stadtman (1971) *Science* **171**, 859.
[6] F. Wagner (1966) *Ann. Rev. Biochem.* **35**, Part I, 405.
[7] M. R. Webb & D. R. Trentham (1981) *J. Biol. Chem.* **256**, 4884.
[8] M. R. Webb (1982) *Meth. Enzymol.* **87**, 301.
[9] M. Kajiwara, S.-F. Lee, A. I. Scott, M. Akhtar, C. R. Jones & P. M. Jordan (1978) *J. Chem. Soc., Chem. Comm.*, p. 967.
[10] H. G. Floss (1982) *Meth. Enzymol.* **87**, 126.
[11] J. C. Speck, P. T. Rowley & B. L. Horecker (1963) *J. Amer. Chem. Soc.* **85**, 1012.
[12] I. Fridovich (1972) *The Enzymes*, 3rd ed., **6**, 255.
[13] J. A. Todhunter & D. L. Purich (1975) *J. Biol. Chem.* **250**, 3505.
[14] R. Lane & E. Snell (1976) *Biochemistry* **15**, 4175, 4180.
[15] R. Wickner, C. Tabor & H. Tabor (1970) *J. Biol. Chem.* **245**, 2132.
[16] S. Shapiro & D. Dennis (1965) *Biochemistry* **4**, 2283.
[17] J. Butler, W. Alworth & M. Nugent (1974) *J. Amer. Chem. Soc.* **96**, 1617.
[18] J. A. Todhunter, K. B. Reichel & D. L. Purich (1976) *Arch. Biochem. Biophys.* **174**, 120.
[19] D. S. Hodgins & R. H. Abeles (1967) *J. Biol. Chem.* **242**, 5158.
[20] P. A. Frey, L.-J. Wong, K.-F. Sheu & S.-L. Yang (1982) *Meth. Enzymol.* **87**, 20.

Table I

Borohydride Reduction as a Probe of Enzyme Reaction Intermediates

Enzyme Name	Reaction Type	Intermediate
Fructose-1,6-bisphosphate aldolase[12]	Class I Aldolases	Imine
Acetoacetate decarboxylase[13]	Decarboxylase	Imine
Glutamine synthetase[14]	Synthetase	Acyl-phosphate
Histidine decarboxylase[15]	Decarboxylase	Dehydroalanine
S-Adenosylmethionine decarboxylase[16]	Decarboxylase	Dehydroalanine
Lactate racemase[17]	Racemase	Lactyl thiolester
5'-Dehydroquinate dehydrase[18]	Dehydrase	Imine
Acetate kinase[19]	Phosphotransferase	Acyl-phosphate (?)
D-Proline reductase[20]	Reductase	Dehydroalanine

BOSE-EINSTEIN CONDENSATE

A newly discovered, highly organized state of matter in which clusters of 20–30 component atoms are magnetically contained and adiabatically cooled to within 2–3×10^{-9} K of absolute zero. At this point, the motions of the contained atoms are overcome by very weak cohesive forces of the Bose-Einstein condensate. While of no apparent relevance to biochemical kinetics, the Bose-Einstein condensate represents one of the most perfect forms of self-assembly, inasmuch as all atoms within the condensate share identical Schrödinger wave equations.

BRANCHED-CHAIN-AMINO-ACID AMINOTRANSFERASE

This enzyme [EC 2.6.1.42], also referred to as transaminase B, catalyzes the reversible reaction of leucine with α-ketoglutarate (or, 2-oxoglutarate) to produce 4-methyl-2-oxopentanoate and glutamate. The pyridoxal-phosphate-dependent enzyme will also utilize isoleucine and valine as substrates. However, this enzyme is distinct from that of valine:pyruvate aminotransferase [EC 2.6.1.66]. **See also** Leucine Aminotransferase

T. K. Korpela (1988) *Meth. Enzymol.* **166**, 269.
R. Kido (1988) *Meth. Enzymol.* **166**, 275.
A. J. L. Cooper (1985) *Meth. Enzymol.* **113**, 71.
A. E. Braunstein (1973) *The Enzymes*, 3rd ed., **9**, 379.

BRANCHED-CHAIN α-KETO ACID DEHYDROGENASE COMPLEX

This enzyme complex [EC 1.2.4.4], also known as 3-methyl-2-oxobutanoate dehydrogenase (lipoamide) and 2-oxoisovalerate dehydrogenase, catalyzes the reaction of 3-methyl-2-oxobutanoate with lipoamide to produce S-(2-methylpropanoyl)dihydrolipoamide and carbon dioxide. Thiamin pyrophosphate is a required cofactor. The complex also can utilize (S)-3-methyl-2-oxopentanoate and 4-methyl-2-oxopentanoate as substrates. The complex contains branched-chain α-keto acid decarboxylase, dihydrolipoyl acyltransferase, and dihydrolipoamide dehydrogenase [EC 1.8.1.4].

D. T. Chuang & R. P. Cox (1988) *Meth. Enzymol.* **166**, 135.
D. T. Chuang (1988) *Meth. Enzymol.* **166**, 146.
D. J. Danmer & S. C. Heffelfinger (1988) *Meth. Enzymol.* **166**, 298.
P. J. Randle, P. A. Patston & J. Espinal (1987) *The Enzymes*, 3rd ed., **18**, 97.
G. W. Goodwin, B. Zhang, R. Paxton & R. A. Harris (1988) *Meth. Enzymol.* **166**, 189.

BRANCHED-CHAIN α-KETO ACID DEHYDROGENASE KINASE

This enzyme phosphorylates branched-chain α-keto acid dehydrogenase using ATP.

J. Espanal, M. Beggs & P. J. Randle (1988) *Meth. Enzymol.* **166**, 166.
R. Paxton (1988) *Meth. Enzymol.* **166**, 313-320.
P. J. Randle, P. A. Patston & J. Espinal (1987) *The Enzymes*, 3rd ed., **18**, 97.

BRANCHING ENZYME

This enzyme [EC 2.4.1.18] (also known as 1,4-α-glucan branching enzyme, glycogen branching enzyme, amylo-(1,4→1,6)transglucosidase, and amylo-(1,4→1,6)transglycosylase) catalyzes the formation of 1,6-glucosidic linkages of glycogen by transferring a segment of a 1,4-α-glucan to a C6 hydroxyl of a similar glucan chain. The enzyme also converts amylose into amylopectin. The actual name given to the enzyme depends upon the product. If the product is glycogen, then glycogen branching enzyme should be used. If the product is amylopectin, then amylopectin branching enzyme is used. The latter has frequently been termed Q-enzyme.

O. Nelson & D. Pan (1995) *Ann. Rev. Plant Physiol. Plant Mol. Biol.* **46**, 475.
G. Mooser (1992) *The Enzymes*, 3rd ed., **20**, 187.

BRANCHPOINT

Any biochemical pathway intermediate that can proceed along more than one route in a network of metabolic reactions.

BREATHING

The continual loss and regain of local or global structure of a macromolecule. In terms of DNA structure, for example, breathing results from the breaking and remaking of hydrogen bonds that stabilize the double helix. The rate and extent of DNA breathing can often be assessed by tritium isotope exchange, because intact hydrogen bonds retain the isotope. In proteins, breathing also occurs, but lifetimes range over the nanosecond-to-hour time-scale.

BRIDGED CARBOCATION

A carbocation in which a group or moiety bridges two or more potential carbenium centers (such that there are alternative Lewis structures having different carbenium centers).

BRIDGE-TO-NONBRIDGE OXYGEN SCRAMBLING

Midelfort and Rose[1] made a major advance in mechanistic enzymology through their development of an isotope scrambling method for the detection of transient [Enz·ADP·P-X] formation from [^{18}O]ATP in ATP-coupled enzyme reactions. This ingenious method takes advantage of the torsional symmetry of the newly terminal phosphoryl oxygen atoms in the enzyme-bound ADP that is held near the newly formed acyl-phosphate intermediate of glutamine synthetase. ATP, specifically labeled with an ^{18}O atom in the β–γ bridge oxygen, was incubated with enzyme, and reversible cleavage of the P_β—O—P_γ phosphoanhydride bond was detected by the appearance of ^{18}O in the β-nonbridge oxygens of the ATP pool. This requires formation and reversal of the phosphoryl transfer steps leading to formation of enzyme·ADP·acyl-phosphate, such that the resynthesized ATP is free to depart from the enzyme and to mix with the pool of [β–γ bridge-labeled ^{18}O]ATP. The Midelfort and Rose[1] experiments with sheep brain and *Escherichia coli* glutamine synthetases showed that cleavage of ATP to form enzyme-bound ADP and P—X requires glutamate.

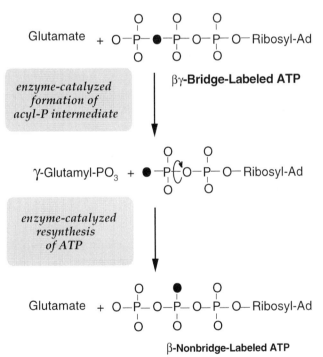

Bridge-to-nonbridge positional isotope exchange catalyzed by glutamine synthetase. Note that ATP (labeled with oxygen-18 in the bridge position) is incubated with enzyme and glutamate. When formation of the acyl-phosphate occurs, the oxygen atoms on the terminal phosphoryl of ADP become torsio-symmetric. With rotation about the $\alpha\beta$-single bond and subsequent resynthesis of ATP, the oxygen-18 atom scrambles into the nonbridge position, and this can be detected in subsequent analysis of the ATP.

The exchange catalyzed by the *E. coli* enzyme with glutamate occurs in the absence of ammonia and is partially inhibited by added ammonium chloride, as would be expected if the exchange reflects the mechanistic pathway for glutamine synthesis. Their positional isotope exchange results provided compelling kinetic evidence for a two-step mechanism wherein phosphoryl transfer from ATP to glutamate to form the acyl-phosphate intermediate precedes reaction with ammonia.

[1]C. F. Midelfort & I. A. Rose (1976) *J. Biol. Chem.* **251**, 5881.

BRIGGS-HALDANE EQUATION

The derivation of an enzyme initial-rate expression utilizing the steady-state assumption: *i.e.*, that the rate of change in the concentration of any enzyme form is much less than the rate of change of the product concentration or the substrate concentration. In such cases, the rate of change of the concentration of any enzyme form is set equal to zero. For a single-substrate, single-product, enzyme-catalyzed reaction, this equation (in the absense of significant product concentrations) is $v = V_{max}[S]/(K_m + [S])$ where v is the initial rate, V_{max} is the maximum velocity for the forward reaction, [S] is the initial substrate concentration, and K_m is the Michaelis constant. *See* Enzyme Kinetics; Steady-State Assumption; Uni Uni Mechanism

BRØNSTED PLOT

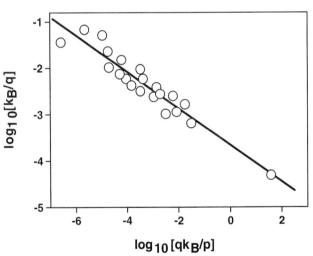

Brønsted plot for base catalysis of the mutarotation of glucose.

A graphical depiction of $\log(k_a/p)$ plotted as a function of $\log(qK_a/p)$, where k_a is the rate constant for general acid catalysis, K_a is the acid ionization constant, p is the number of equivalent protons on the acid, and q is the number of equivalent positions where a proton can be accepted in the conjugate base. An analogous Brønsted plot is used for analyzing general base catalysis: $\log(k_b/q)$ is plotted as a function of $\log(pK_b/q)$.

BRØNSTED RELATION

A mathematical expression[1-6] relating the reaction rate constant to the strength of the acid or base participating in the reaction. This relationship, more commonly known as the Brønsted Catalysis Law, is an often used principle for analyzing systems undergoing general acid or general base catalysis. IUPAC[7] recommends against the use of the term Brønsted Catalysis Law because the relationship has been demonstrated for a number of reactions (*e.g.*, certain proton transfer reactions) that are not catalyzed.

The Brønsted acid relation can be written as:

$$\frac{k_{HA}}{p} = G\left(\frac{qK_{HA}}{p}\right)^{\alpha}$$

where α and G are constants for a given reaction, solvent, and temperature (in which α is known as the Brønsted constant), k_{HA} is the rate constant for general acid catalysis (or the rate constant for an uncatalyzed reaction dependent on the concentration of HA [*i.e.*, the acid]), K_{HA} is the acid dissociation constant, and p and q are statistical factors. The number of equivalent protons in the acid HA is p, and the number of equivalent basic sites in the conjugate base A$^-$ is q. Rate data can be plotted in the form $\log(k_{HA}/p) = \log G + \alpha \log(K_{HA}q/p)$. It should be noted that the chosen values for p and q should always be explicitly indicated by the authors. In addition, if one is applying transition-state theory, the rate constant should be corrected by using symmetry numbers for the transition state and the reactants. Thus, for general acid catalysis, k_{HA} is multiplied by $\sigma^{\ddagger}/(\sigma_{HA}\sigma_r)$. Unfortunately, establishing a good estimate for this particular correction factor is frequently challenging[8-10].

The corresponding Brønsted base relation is given by:

$$\frac{k_A}{q} = G\left(\frac{qK_{HA}}{p}\right)^{-\beta}$$

where β is a Brønsted constant characteristic of the reaction, solvent, and temperature, and k_A is the rate

constant for general base catalysis (or, for uncatalyzed reactions, the rate constant dependent on the concentration of the base). Thus, $\log(k_A/q) = \log G - \beta \log (K_{HA}q/p)$.

For systems that undergo general acid and general base catalysis, one might expect that there would be a direct relationship between the observed rate constant and the strength of the acid or base. Brønsted's proposal[1,2] in the mid-1920's was, in fact, the first linear free energy relationship. According to these relationships, a plot of the logarithm of the catalytic rate constant *versus* the logarithm of the dissociation constant for a number of acids (or bases) will yield a straight line if the system exhibits general acid (or general base) catalysis. The slope of this line provides the value for the Brønsted constant, α or β. Brønsted constants typically range between zero and one and are often considered to be an estimate of the extent of proton transfer in the transition state. A value of zero strongly suggests that the transition state resembles the reactants. On the other hand, the transition state will most likely resemble the products, when the Brønsted constant has a value near unity. Nevertheless, there are exceptions to this rule-of-thumb. In general, a will have a value of one when the rate constant is independent of the strength of the catalytic acid; the value is zero if the reaction is catalyzed by the solvent. Still others have suggested that Brønsted theory does not provide insight about the extent of proton transfer in the transition state[11-14].

There are a number of limitations on the Brønsted relationship. First of all, the relation holds only for similar types of acids (or bases). For example, carboxylic acids may have a different α values compared to sulfonic acids or phenols. Because charge, and likewise solvation, can greatly influence the reaction rate, deviations of net charge from one catalyst to another can also influence Brønsted plots. Another limitation on this relationship relates to temperature. Reaction rates and the corresponding dissociation constants for the acids must all be measured at the same temperature (and, most rigorously, in the same solvent). For some systems, this may prove infeasible. A third limitation is that the reaction must indeed be subject to general acid (or base) catalysis. For certain catalysts, deviations from a linear relationship may indicate other modes of action beyond general acid/

base catalysis (*e.g.*, the occurrence of nucleophilic reactions). Other deviations of catalysts from a Brønsted line may be due to slow proton transfer processes and/or inefficient general acid/base catalysis (*e.g.*, by carbon acids). Such deviations may arise as a consequence of resonance stabilization or structural reorganization of the resulting anion. Differential solvation of catalysts may also cause deviations from a Brønsted line.

As mentioned above, the Brønsted constant is an attempt to measure the relative stabilization of the transition state by different acids. However, values less than zero and greater than one have also been reported, possibly due to intermolecular effects on the transition state[15,16].

The linear part of the Brønsted relationship represents a segment of a curved line. The theory for the deviations from linearity was provided by Marcus[11,12,17].

Rowlett and Silverman[18] used a Brønsted plot to examine the interaction of external buffers with human carbonic anhydrase II. The buffers act as proton acceptors in the removal of the proton generated by the enzyme-catalyzed reaction. The Brønsted plot displays a plateau at a value of about 10^9 $M^{-1}s^{-1}$ for the catalytic rate constant, which is close to the rate for proton transfer with small molecules. This plot indicated that the donor group on the enzyme has a pK_a value of about 7.6. All of the data are "consistent with proton transfer directly between the active site and buffer or between a proton shuttle group on the enzyme of pK_a near 7 and buffer"[18]. Using site-directed mutants of human carbonic anhydrase III having pK_a values for zinc-bound water ranging between 5 and 9, Silverman *et al.*[19] were able to demonstrate a Brønsted relationship and apply Marcus rate theory. The results suggest a "facile proton transfer" having a small intrinsic energy barrier of about 6.3 kJ/mol. The larger overall energy barrier of about 42 kJ/mol suggest that other processes, such as solvent reorganization, conformational changes, *etc.*, accompany the proton transfer.

[1] J. N. Brønsted K. Pedersen (1924) *Zeit. Phys. Chem.* **A108**, 185.
[2] J. N. Brønsted (1928) *Chem. Rev.* **5**, 322.
[3] G. W. Klumpp (1982) *Reactivity in Organic Chemistry*, pp. 167-179, Wiley, New York.
[4] R. P. Bell (1978) in *Correlation Analysis in Chemistry: Recent Advances* (eds., N. B. Chapman & J. Shorter) pp. 55-84, Plenum Press, New York.
[5] J. W. Moore & R. G. Pearson (1981) *Kinetics and Mechanism*, Wiley, New York.
[6] W. P. Jencks (1969) *Catalysis in Chemistry and Enzymology*, McGraw-Hill, New York. [Also available as Jencks, W. P. (1986) *Catalysis in Chemistry and Enzymology*, Dover Publications, New York].
[7] IUPAC (1979) *Pure and Appl. Chem.* **51**, 1725.
[8] S. W. Benson (1958) *J. Amer. Chem. Soc.* **80**, 5151.
[9] E. Pollak & P. Pechukos (1978) *J. Amer. Chem. Soc.* **100**, 2984.
[10] D. R. Coulson, E. Pollak & P. Pechukos (1978) *J. Amer. Chem. Soc.* **100**, 2992.
[11] R. A. Marcus (1968) *J. Phys. Chem.* **72**, 891.
[12] R. A. Marcus (1969) *J. Amer. Chem. Soc.* **91**, 7224.
[13] F. G. Bordwell & W. J. Boyle, Jr. (1971) *J. Amer. Chem. Soc.* **93**, 511.
[14] F. G. Bordwell & W. J. Boyle, Jr. (1972) *J. Amer. Chem. Soc.* **94**, 3901.
[15] A. J. Kresge, H. L. Chen, Y. Chiang, E. Murrill, M. A. Payne & D. S. Sagatys (1971) *J. Amer. Chem. Soc.* **93**, 413.
[16] A. J. Kresge (1970) *J. Amer. Chem. Soc.* **92**, 3210.
[17] J. Hine (1971) *J. Amer. Chem. Soc.* **93**, 3703.
[18] R. S. Rowlett & D. N. Silverman (1982) *J. Amer. Chem. Soc.* **104**, 6737.
[19] D. N. Silverman, C. Tu, X. Chen, S. M. Tanhauser, A. J. Kresge & P. J. Laipis (1993) *Biochemistry* **32**, 10757.

BRØNSTED THEORY

An acid/base theory stating (a) that an acid is any substance that tends to donate or release protons (also called hydrogen ions) to a base, and (b) that a base is a substance that accepts or removes protons from an acid. Examples of Brønsted acids include H_3O^+, H_2O, CH_3COOH, and HSO_4^-. Examples of Brønsted bases include H_2O, OH^-, CH_3COO^-, and SO_4^{2-}.

BUBBLES

Those experimentalists who use spectrophotometry or spectrofluorimetry to measure rates of biochemical reactions should always be mindful that "bubble clearance" frequently displays first-order kinetics. This applies to bubbles adhering to the inside wall of the cuvette as well as bubbles released from solution itself. The presence of bubbles within a cuvette may introduce artifactual kinetic behavior resulting (a) from refractive index differences between the gas trapped in the bubbles and that of the test solution, and (b) from the high reflectance of the air/water interface surrounding some bubbles.

Slow warming of a cold viscous solution is a particularly effective way of generating bubbles. Should this prove to be the case, buffers can be degassed by centrifugation at 400-500 *g* for 10 min or by briefly warming the cold solution (obviously minus any protein constituents) to 60-70°C for 5 min. Centrifugation offers the advantage that dust particles, another source of light scattering, can also frequently be removed from the buffer. Protein solutions often tend to froth, even upon careful mixing. This can lead to denaturation of any proteins passing into the thin bubble wall, a region where exposure to air and surface tension can wreak havoc on many proteins. Avoid shaking the cuvette after the addition of a protein or enzyme; repeated, delicate inversion of the cuvette, capped with a small strip of Parafilm, is usually sufficient to achieve adequate mixing without introducing bubbles.

BUFFER

Any substance that stabilizes the concentration of one of its dissociated species, most commonly a proton (or more properly a hydronium ion). The free concentration of a metal ion may also be buffered by use of a metal ion-chelator complex, such as Ca·EGTA or K·Mg·EDTA.

Selected entries from *Methods in Enzymology* [vol, page(s)]: Buffer capacity, **63**, 4; choice, **63**, 19, 20, 285; metal ion chelation effects, **63**, 225, 226, 287, 298, 299; dielectric constant effect on pK, **63**, 226; dilution, **63**, 20; equilibrium constant effects, **63**, 18; heavy water, **63**, 226, 227; ionic strength effects, **63**, 226, 227; metal ion stability constant measurements, **63**, 298, 299; theory, **182**, 24; broad-range, overview, **182**, 30; broad-range, preparation, **104**, 410; metal ions and buffers [control of, **140**, 409; detection of catalytic metals, **186**, 125]; buffers: [solubilization of membrane proteins, choice, **104**, 310; **182**, 257; for cell permeabilization studies, ion selection, **192**, 297; for chemical cross-linking of histones, **170**, 553; for chromatofocusing proteins in different pH ranges, **182**, 384; for NMR studies, **104**, 412-414; for plant tissue homogenization, **182**, 179; for yeast extract preparation, **182**, 167; for preparation of extracts from higher eukaryotes, **182**, 196; for protein cleavage, **182**, 620; for reversed-phase HPLC, **193**, 404; volatile buffer removal, **182**, 29]; inhibitory effects: [of cacodylate, **9**, 669; of citrate, **9**, 620, 669; of 2,4,6-collidine, **9**, 722; of glycylglycine, **9**, 722; of imidazole, **9**, 722; of maleate, **9**, 722; of phosphate, **9**, 669; of Tris, **9**, 209, 345, 591, 722].

BUFFER CAPACITY

A measurement of the ability of a buffer system to limit the change in pH of a solution upon the addition of an increment of strong base[1-3]. It is the reciprocal of the slope of the pH-neutralization curve[2]. Consider the simple equilibrium, $HA \rightleftharpoons H^+ + A^-$ where $K_a' = [H^+][A^-]/[HA]$ in which K_a' is a practical dissociation constant determined under conditions of constant ionic strength. In such systems the practical pK_a' is equal to the pH of solution when there are equal concentrations of the two buffer species. Since the total concentration of the two

species remains constant (*i.e.*, $c = [HA] + [A^-]$) and charges have to be balanced (*i.e.*, the cation concentrations are equal to the anion concentrations: *e.g.*, $[H^+] + [Na^+] + \ldots = [OH^-] + [A^-] + \ldots$) then an expression for the buffer capacity, β, can be derived: $\beta = d[B]/dpH = (\log e)^{-1} [(K_a'c [H^+]/(K_a' + [H^+])^2) + [H^+] + K_w/[H^+]]$ where [B] is the concentration of strong base added and K_w is the ion product of water ($= [H^+][OH^-]$). The second and third terms in the expression become important factors in strongly acidic or strongly basic solutions, respectively. Between pH values of about 3 and 11, these two terms are negligible. (However, this "rule-of-thumb" fails if the buffer concentration is very low.) Note that the buffer capacity is never zero since water will also contribute, albeit usually as a minor contributor. *See* Pseudo Buffers

[1] D. D. Van Slyke (1922) *J. Biol. Chem.* **52**, 525.
[2] D. D. Perrin & B. Dempsey (1974) *Buffers for pH and Metal Ion Control*, Chapman and Hall, London.
[3] R. G. Bates (1964) *Determination of pH, Theory and Practice*, Wiley, New York.

BUFFER EFFECTS ON ENZYMATIC PROTON TRANSFER

Beyond their control of bulk solvent pH, buffers can occasionally become an integral component in enzymatic catalysis. Carbonic anhydrase presents an instructive case where the catalytic efficiency is so great ($k_{cat} > 10^6$ s^{-1}) that proton transfer becomes rate-limiting. The rate was found to depend on the concentration of the protonated form of buffers in the solution. Indeed, Silverman and Tu[1] adduced the first convincing evidence for the role of buffer in carbonic anhydrase catalysis through their observation of a imidazole buffer-dependent enhancement in equilibrium exchanges of oxygen isotope between carbon dioxide and water. The effect is strictly on k_{cat}, and k_{cat}/V_{max} is unaffected because the latter is described by a rate law that doesn't include proton transfer steps.

[1] D. N. Silverman & C. K. Tu (1975) *J. Amer. Chem. Soc.* **97**, 2263.

BUFFER SELECTION

A number of criteria are crucial in the selection of an appropriate buffer to be utilized in an enzyme-kinetic study. Included in these criteria are:

1. Appropriate pK_a Value. The pK_a value of the buffer should be within one pH unit of the pH value chosen for the study; preferably closer. For example, Tris should not be used as a buffer if the system is to be studied at pH 6.5. It should also be recalled that the pK_a of a buffer changes with temperature. For example, the dpK_a/dT value for Tris is -0.028. If the assay solution is prepared and stored at one temperature, whereas the enzyme is assayed at another temperature, the buffer may very well not be appropriate under both temperature conditions.

2. Stability. The buffer should be stable under the conditions of the biochemical experiment or enzyme assay.

3. Minimal Interactions with Enzyme. The buffer should not be an effector of the enzyme system under study. It is not unusual to expect buffers containing carboxyl groups, phosphoryl groups, etc., to bind to the enzyme and possibly to alter the enzyme activity. In the initial series of studies, an entire set of buffers should be examined. Buffers yielding maximal activity should be examined (under saturating substrate concentrations) at several buffer concentrations (at constant ionic strength). Buffer dilution should have minimal effect on enzyme activity. These studies are also useful in identifying potential classes of molecules that may act as potential activators or inhibitors. In addition, the investigator should supplement the study by looking at a number of different counterions.

4. Minimal Interactions with Assay Components. If a buffer binds to or reacts with a component of the system, it should be avoided. For example, borate buffers can complex with alcohols and carbohydrates and great care should be exercised when utilizing these compounds. Some arsenate-based buffers have oxidizing potential with respect to thiols. Buffers should also have a low binding capacity with respect to divalent cations.

5. Effects with Respect to Assay System. The buffer should not affect the physical procedure being utilized. For example, if proton NMR studies are being conducted, the buffer should not contain ^1H. If a substance (e.g., product, substrate, or protein) has to be isolated at a later step, it may be necessary to use a volatile buffer. If a spectroscopic technique is being utilized (e.g., UV absorption spectroscopy), the buffer (both protonated and deprotonated) should be transparent at the wavelength being utilized.

BUNDLING PROTEIN

Any of the cytoskeletal proteins that cross-link cytoskeletal filaments into colinear arrays, such as the action of α-actinin in promoting the formation of actin stress fibers. Bundling proteins typically contain pairs of binding sites for attachment to cytoskeletal filaments.

BUNNETT-OLSEN EQUATIONS

Certain expressions describing a solvent acidity function, where S is a base that is protonated by an aqueous mineral acid solution. The equations describe a linear free-energy relationship between $\log([SH^+]/[S]) + H_o$ and $H_o + \log[H^+]$, where H_o is Hammett's acidity function and where $H_o + \log[H^+]$ represents the activity function $\log(\gamma_S\gamma_{H^+}/\gamma_{SH^+})$ for the nitroaniline reference bases to build H_o. Thus, $\log([SH^+]/[S]) - \log[H^+] = (\phi - 1)(H_o + \log[H^+]) + pK_{SH^+}$ and $\log([SH^+]/[S]) + H_o = \phi(H_o + \log[H^+]) + pK_{SH^+}$. Here, ϕ represents the response of the $S + H^+ \rightleftharpoons SH^+$ equilibrium to changes in the acid concentration. A corresponding equation can be written for kinetic data: $\log k_\psi + H_o = \phi(H_o + \log[H^+]) + \log k_2^o$ where k_ψ is the pseudo-first-order rate constant for the reaction of a weak base in an acidic solution and k_2^o is the second-order rate constant at infinite dilution in water. In this case, ϕ represents the response of the reaction rate to changing acid concentrations. **See also** *Acidity Function*

J. F. Burnett & F. P. Olsen (1966) *Can. J. Chem.* **44**, 1899.

J. F. Burnett, R. L. McDonald & F. P. Olsen (1974) *J. Amer. Chem. Soc.* **96**, 2855.

BURST KINETICS

Bursts in product formation can occur when an enzyme is first combined with its substrate(s), depending on the nature of the kinetic mechanism, the relative magnitudes of the rate constants for each step, as well as the relative concentrations of active enzyme and substrate(s). This is especially apparent when one uses fast reaction kinetic techniques and when the chromophoric product is released in a fast step, which is then followed by a slower release of the second product. This is depicted below, where product formation is indicated by the appearance of a chromophoric product.

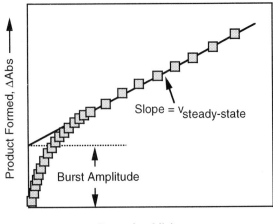

The classical example is chymotrypsin hydrolysis[1] of *p*-nitrophenyl ethylcarbonate or *p*-nitrophenyl acetate. In both cases, the first product released is *p*-nitrophenol, and the acyl-enzyme intermediate then breaks down at a slower rate. Because the first product absorbs visible light so strongly, product formation is readily detectable. In this diagram, the instantaneous velocity (*i.e.*, the line drawn tangent to the curve at any chosen moment) is highest at the time of mixing enzyme and substrate and then quickly falls off to a linear production rate characteristic of steady-state kinetic behavior. Because substrate depletion is relatively insignificant over the period shown in the figure, any change in reaction rate directly reflects the change in the concentration of free enzyme available to react with substrate. Moreover, extrapolation of the tangent line during the steady-state phase back to time zero gives the concentration change during the burst phase. This represents the concentration of active enzyme that was present at time zero.

A chemically realistic representation of the overall catalytic process can be written as follows:

$$E + S = E \cdot P_1 \cdot P_2 = E \cdot P_2 + P_1$$

$$E \cdot P_2 = E + P_2$$

In this case, P_1 is the chromophoric product that is released prior to the slower conversion of EP_2 to form P_2,

thereby completing the reaction cycle. For the purposes of understanding the kinetics of bursts, we can simplify the above scheme to allow for formation of EP_2, followed by the slower release of P_2:

$$E + S \xrightarrow{k_1'} E \cdot P_2 + P_1$$

$$E \cdot P_2 \xrightarrow{k_2'} E + P_2$$

As indicated by Fersht[2,3], the magnitude of the burst equals $[E_{active}]\{k_1'/(k_1' + k_2')\}^2$, which reduces to just the concentration of active enzyme $[E_{active}]$ if k_2' is much smaller than k_1'. Aside from providing valuable mechanistic information, burst kinetic experiments can be gainfully employed to estimate the fraction of active enzyme (*i.e.*, $[E_{active}]/[E_{total}]$) for an enzyme preparation.

[1]B. S. Hartley & B. A. Kilby (1954) *Biochem. J.* **56**, 288.
[2]A. Fersht (1977) *Enzyme Structure and Mechanism*, pp. 108-109 & 175-176, Freeman, Reading.
[3]A. R. Fersht (1999) *Structure and Mechanism in Protein Science: A Guide to Enzyme Catalysis and Protein Folding*, pp. 650, W. H. Freeman, New York.

γ-BUTYROBETAINE HYDROXYLASE

This enzyme [EC 1.14.11.1], also known as *γ*-butyrobetaine, 2-ketoglutarate dioxygenase, catalyzes the reaction of 4-trimethylammoniobutanoate with *δ*-ketoglutarate (or, 2-oxoglutarate) and dioxygen to yield 3-hydroxy-4-trimethylammoniobutanoate, succinate, and carbon dioxide. Both iron ions and ascorbate are needed as cofactors.

B. G. Fox (1998) *Comprehensive Biological Catalysis: A Mechanistic Reference* **3**, 261.
O. Hayaishi, M. Nozaki & M. T. Abbott (1975) *The Enzymes*, 3rd ed., **12**, 119.

BUTYRYLCHOLINE ESTERASE

This enzyme [EC 3.1.1.8] (also known as cholinesterase, pseudocholinesterase, acylcholine acylhydrolase, non-specific cholinesterase, and benzoylcholinesterase) catalyzes the hydrolysis of an acylcholine to generate choline and a carboxylic acid anion. A variety of choline esters and a few other compounds can serve as substrates.

D. M. Quinn & S. R. Feaster (1998) *Comprehensive Biological Catalysis: A Mechanistic Reference* **1**, 455.

H. Okuda (1991) *A Study of Enzymes,* **2**, 563.

BUTYRYL-CoA DEHYDROGENASE

This enzyme [EC 1.3.99.2], also referred to as short-chain acyl-CoA dehydrogenase and unsaturated acyl-CoA reductase, catalyzes the reaction of butanoyl-CoA and an electron-transferring flavoprotein to produce 2-butenoyl-CoA and the reduced electron-transferring flavoprotein. The enzyme, which uses FAD, forms a complex with the electron-transferring flavoprotein and electron-transferring flavoprotein dehydrogenase [EC 1.5.5.1] that will reduce ubiquinone and other acceptors.

B. A. Palfey & V. Massey (1998) *Comprehensive Biological Catalysis: A Mechanistic Reference* **3**, 83.

M. A. Ator & P. R. Ortez de Montellano (1990) *The Enzymes*, 3rd ed., **19**, 213.

C

CAFFEINE

A naturally occurring substance (1,3,7-trimethylxanthine) that competitively inhibits 3′,5′-cyclic AMP phosphodiesterase and acts as a powerful central nervous system stimulant. The inhibitory effect can be reversed at high cAMP concentrations. A related substance, theophylline (also known as aminophylline) has a similar mode of action.

CAGE

An aggregate collection of molecules surrounding another molecular entity. A solvent cage forms by electrostatic interactions around various ions in solution.

CAGE COMPOUND

1. Clathrate. 2. Any molecular entity that is nonplanar and often polycyclic that encloses a cavity. 3. Photosensitive precursors that generate chemically reactive species upon irradiation with a laser or flash-lamp. *See Photoreactive Caged Compounds*

IUPAC (1979) *Pure and Appl. Chem.* **51**, 1725.

CAGED ATP

A chemically substituted ATP molecule (such as adenosine-5′-triphospho-1-(2-nitrophenyl)ethanol) that is not capable of participating in ATP-dependent hydrolase/phosphotransferase reactions until the protecting group is photochemically removed. Excitation with UV light at a wavelength of 360 nm immediately photolyzes caged ATP to produce ATP, thereby permitting the kinetic analysis of processes occurring on time-scales as short a 1-2 milliseconds.

CAHN-INGOLD-PRELOG SYSTEM

A self-consistent set of rules (also referred to as the *RS*-system) that is used to designate the three-dimensional configuration of chiral molecules[1]. This system replaces the DL absolute configuration scheme. In publishing experimental results using chiral substrates, the *RS*-designation should always be provided. If the substrate under consideration is an amino acid or sugar, then the DL absolute configuration may be used. However, one should recall that there is no direct correlation between the two systems. For example, L-alanine (or, (+)-alanine) is equivalent to (*S*)-alanine whereas L-cysteine (or, (−)-cysteine) corresponds to (*R*)-cysteine. The individual rules for the Cahn-Ingold-Prelog system can be found in the references below or in any basic organic chemistry text. The important rules are: (1) List substituents directly attached to a chiral carbon in decreasing atomic number. (2) If two or more atoms are directly linked to the chiral carbon, then the atomic number of the atom(s) linked to those atoms determines the ranking. Thus, in $(CH_3)_2CH-CHCl-CH_2SH$, the $-CH_2SH$ group takes precedence over the $CH(CH_3)_2$ group because sulfur has a higher atomic number than carbon. (3) Double bonds are counted as two single bonds, and triple bonds count as three single bonds. (4) When dealing with different isotopes of the same atom, the heavier isotope takes precedence over the lighter one. Thus, tritium takes precedence over hydrogen. (5) All atoms except hydrogen are considered to have a valence of four; thus, $-NH^{\oplus}(CH_3)_2$ takes precedence over $-N(CH_3)_2$.

IUPAC (1976) *Pure and Appl. Chem.* **45**, 13.
R. S. Cahn, C. K. Ingold & V. Prelog (1966) *Angew. Chem. Int. Ed. Engl.* **5**, 385.
V. Prelog & G. Helmchen (1982) *Angew. Chem. Int. Ed. Engl.* **21**, 567.

CALCINEURIN

A calmodulin-binding phosphoprotein phosphatase, found at highest abundance in brain tissue, that removes phosphoryl groups from phosphoseryl and phosphothreonyl residues.

B. G. Fox (1998) *Comprehensive Biological Catalysis: A Mechanistic Reference* **3**, 261.

C. B. Klee, G. F. Draetta & M. J. Hubbard (1988) *Adv. Enzymol.* **61**, 149.

C. B. Klee, M. H. Krinks, A. S. Manalan, P. Cohen & A. A. Stewart (1983) *Meth. Enzymol.* **102**, 227.

CALCIUM-45 (^{45}Ca)

A β-emitting (0.257 MeV) radioactive calcium nuclide with a decay half-life of 163 days. The following decay data indicate the time (days), followed by the fraction of original amount at the specified time: 0.0, 1.000; 5, 0.979; 10, 0.958; 15, 0.938; 20, 0.918; 25, 0.898; 30, 0.879; 35, 0.862; 40, 0.844; 45, 0.826; 50, 0.808; 100, 0.654; 150, 0.528; 200, 0.427; 250, 0.345; 300, 0.279; 350, 0.226; 400, 0.183; 450, 0.148; 500, 0.119; 550, 0.096; 600, 0.078.

Special precautions: Bone is the critical organ in terms of dose. The effective *biological* half-life in humans is nearly 20 years. Double gloving is recommended when working with this radionuclide.

CALCIUM/CALMODULIN-DEPENDENT PROTEIN KINASE

This enzyme [EC 2.7.1.123], also referred to as calcium/calmodulin-dependent protein kinase type II, and microtubule-associated protein MAP2 kinase, catalyzes the reaction of ATP with a protein to produce ADP and an *O*-phosphoprotein. The enzyme requires calcium ions and calmodulin. Proteins that can serve as substrates include vimentin, synapsin, glycogen synthase, the myosin light-chains, and the microtubule-associated tau protein. This enzyme is distinct from myosin light-chain kinase [EC 2.7.1.117], caldesmon kinase [EC 2.7.1.120], and tau-protein kinase [EC 2.7.1.135].

P. I. Hanson & H. Schulman (1992) *Ann. Rev. Biochem.* **61**, 559.

S. I. Walaas & P. Greengard (1987) *The Enzymes*, 3rd ed., **18**, 285.

J. T. Stull, M. H. Nunnally & C. H. Michnaff (1986) *The Enzymes*, 3rd ed., **17**, 113.

CALCIUM ION

This Group IIA (or Group 2) element (atomic symbol, Ca; atomic number, 20; atomic weight, 40.078; electronic configuration = $1s^2 2s^2 2p^6 3s^2 3p^6 4s^2$) loses both $4s^2$ electrons to form a divalent cation of 0.99Å ionic radius. Ionic calcium combines readily with oxygen ligands (chiefly water, phosphates, polyphosphates, and carboxylates) to form stable metal ion complexes. ^{41}Ca undergoes nuclear disintegration by electron capture with a decay energy of 0.421 MeV and a half-life of 103,000 years. ^{45}Ca is a beta particle emitter with a decay energy of 0.257 MeV and a half-life of 162.7 days. The high degree of spatial and temporal precision occurring during calcium ion control of cellular processes suggests that there are efficient mechanisms for localized control release and sequestration of calcium ion. This is possible through the highly coordinated interplay of calcium ion channels, stores, and oscillations.

Selected entries from *Methods in Enzymology* [vol, page(s)]: Chelation, **238**, 74, 76, 297; buffers [for analysis of exocytosis, **221**, 132; preparation, **219**, 186; modulation of cytosolic buffering capacity with quin2, **221**, 159]; fluorescence assay, **240**, 724-725, 740-742; fluorescence imaging, **225**, 531; **238**, 303-304, 322-325, 334-335; free intracellular levels after bacterial invasion, **236**, 482-489; free calcium in solutions for membrane fusion analysis, calculation and control, **221**, 149; homeostasis mechanisms, **238**, 80; hormonal elevation, **238**, 79; inositol phosphate effect on release, **238**, 207; determination of cytosolic levels [computer methods, **238**, 73-75; with fura-2, **238**, 73, 146; with indo-1, **238**, 298, 316-317; with quin-2, **238**, 297]; hormone effects, **238**, 79; ionomycin effects, **238**, 79; membrane depolarization effects, **238**, 80; NMR determination of calcium binding to proteins, **227**, 115; ^{45}Ca, **238**, 153-154; free ion determination, **238**, 139-140; spectrofluorometric assay, **236**, 485; calcium-binding proteins, **228**, 248-253; chelators, **236**, 473-475.

CALCIUM ION INDICATOR DYES

Tsien's development[1,2] of calcium ion sensors in the form of fluorescent chelator/indicator molecules has allowed major advances regarding regulatory mechanisms based on calcium ion. It is now both feasible and routine to measure [Ca^{2+}] directly in single living cells at high spatial resolution, using a digital imaging microscope and the highly fluorescent Ca^{2+}-sensitive dye known as fura-2.

Kao and Tsien[3] studied the Ca^{2+}-binding kinetics of fura-2 and azo-1 by temperature-jump relaxation methods. In 140 mM KCl at 20°C, the respective association and dissociation rate constants for fura-2 were 6×10^8 $M^{-1}s^{-1}$ and 97 s^{-1}; these kinetic properties were insensitive to hydrogen ion concentration over the pH range from 7.4 to 8.4. Azo-1 was studied in 140 mM KCl: At 10°C, azo-1 exhibited respective association and dissociation rate constants of 1.4×10^8 $M^{-1}s^{-1}$ and 780 s^{-1}; at 20°C, these values were 3.99×10^8 $M^{-1}s^{-1}$ and 1200 s^{-1}. Thus, fura-2 and azo-1 permit rapid monitoring changes in intracellular calcium ion concentration. ***See** Fura-2*

Tsien[4] recently introduced a simple technique for loading cells with Ca^{2+}-selective chelators that are rendered temporarily membrane permeable by masking their four carboxylates as esters which then hydrolyze once inside the cells and unmask the active chelator.

Ca^{2+} activation of processes at cell membranes (such as those involved in contraction, secretion, and neurotransmitter release) require *in situ* or *in vitro* Ca^{2+} concentrations that are typically 10—100 times higher than those measured in intact cells during stimulation. This discrepancy might be reconciled if the local cell membrane $[Ca^{2+}]$ is very different from that in the rest of the cell. Soluble Ca^{2+} indicators indicate spatially averaged cytoplasmic $[Ca^{2+}]$ and cannot resolve fluctuations in near-membrane Ca^{2+} signals. However, the new Ca^{2+} indicator FFP18 allows one to monitor near-membrane $[Ca^{2+}]$ selectively[5]. Images of the intracellular distribution of FFP18 show that >65% is located on or near the plasma membrane. Recorded $[Ca^{2+}]$ transients using FFP18 during membrane depolarization-induced Ca^{2+} influx show that near-membrane $[Ca^{2+}]$ rises faster and reaches micromolar levels earlier than cytoplasmic $[Ca^{2+}]$; the latter of which only reaches 0.1–0.2 μM. High-speed digital imaging revealed that the near-membrane $[Ca^{2+}]$ rises within 20 msec, peaks at 50–100 msec, and then subsides[5].

See also Calcium Ion; Aequorin; Fura-2

[1] R. Y. Tsien (1989) *Annu. Rev. Neurosci.* **12**, 227.
[2] R. W. Tsien & R. Y. Tsien (1990) *Annu. Rev. Cell. Biol.* **6**, 715.
[3] J. P. Kao & R. Y. Tsien (1988) *Biophys. J.* **53**, 635.
[4] R. Y. Tsien (1981) *Nature* **290**, 527.
[5] E. F. Etter, A. Minta, M. Poenie & F. S. Fay (1996) *Proc. Natl. Acad. Sci. U.S.A.* **93**, 5368.

CALMODULIN

A regulatory protein initially discovered as an activator of cyclic nucleotide phosphodiesterase (or 3′,5′-cyclic nucleotide 5′-PDE, EC 3.1.4.17). Kinetic studies[1] on the activation of cyclic nucleotide PDE as a function of calmodulin and Ca^{2+} concentrations were analyzed by what is likely to be a general approach for characterizing activation mechanisms. The method accounts for the various interactions between phosphodiesterase and different calmodulin-Ca^{2+} complexes. All four Ca^{2+} must be bound to calmodulin for the protein to form an activated complex with PDE. The authors identified several regulatory advantages of engaging the four Ca^{2+} sites on calmodulin: (1) The activation of PDE as a function of Ca^{2+} is highly cooperative (if the PDE·calmodulin·$(Ca^{2+})_4$ complex is the dominant active species), and the

system behaves as an "on/off" switch for PDE activation. (2) At normal intracellular Ca^{2+} levels (<0.1 μM), PDE and calmodulin fail to complex, thereby allowing the distribution of calmodulin complexes with its various target enzymes to be reshuffled after resequestration of Ca^{2+} following each surge. (3) The affinity between the enzyme and the fully liganded calmodulin is 10^4-10^5 times that observed in the absence of Ca^{2+}. Thus, a very sharp increase in affinity is achievable through a 10- to 20-times increase in the affinity of Ca^{2+} for the enzyme-calmodulin complex in each of the four binding steps.

[1] C. Y. Huang, V. Chau, P. B. Chock, J. H. Wang & R. K. Sharma (1981) *Proc. Natl. Acad. Sci. U.S.A.* **78**, 871.

Selected entries from *Methods in Enzymology* [vol, page(s)]: Calmodulin purification and fluorescent labeling, **102**, 1; assay of calmodulin by Ca^{2+}-dependent phosphodiesterase, **102**, 39; myosin light chain phosphorylation in smooth muscle and nonmuscle cells as a probe of calmodulin function, **102**, 62; spectroscopic analyses of calmodulin and its interactions, **102**, 82; Ca^{2+} binding to calmodulin, **102**, 135; preparation of fluorescent labeled calmodulins, **102**, 148; techniques for measuring the interaction of drugs with calmodulin, **102**, 171; detection of calmodulin-binding polypeptides separated in SDS-polyacrylamide gels by a sensitive [^{125}I]calmodulin gel overlay assay, **102**, 204; use of calmodulin affinity chromatography for purification of specific calmodulin-dependent enzymes, **102**, 210; chemical approaches to the calmodulin system, **102**, 296; ^{13}C chemical shift, **239**, 369; contamination in creatine phosphokinase, **238**, 76; removal of endogenous calmodulin, **238**, 77; heteronuclear relaxation studies, **239**, 564; hydrophobic affinity partitioning, **228**, 257; myosin light-chain kinase peptide complex, **239**, 664, 682, 708; regulation of adenylyl cyclase, **238**, 77; as retroviral protease substrate, **241**, 290; 3D ^{1}H-detected [^{13}C-^{1}H] long-range correlation spectrum, **239**, 97.

CALOMEL ELECTRODE

A reference electrode composed of metallic mercury and solid mercurous chloride. The redox half-reaction is:

$$Hg_2Cl_2 + 2e^- = 2Hg_{(solid)} + 2Cl^-$$

The potential of this electrode only depends on the chloride ion concentration which is strictly controlled by the saturated potassium chloride solution in the electrode. The $E°$ value for the above reaction is 0.26808 volts (at 273.15 K and a pressure of 1 atm (or, 101325 Pa).

CALORIMETRY

The measurement of the amount of heat released or absorbed in a reaction; also included in this category are determinations of heat capacities, latent heats, and caloric values of fuels. *See also* Differential Scanning Calorimetry

Selected entries from *Methods in Enzymology* [vol, page(s)]: Calorimetry: Accuracy, **240**, 98-99; algorithm for nonlinear least-squares fit, **240**, 94; enthalpy change for ligand binding, **240**, 91-92, 96-99; entropy change for ligand binding, **240**, 93; free energy of ligand binding, **240**, 93-94; free ligand concentration calculation, **240**, 93; heat of binding, **240**, 92; heat capacity, analysis, **240**, 96-99; parameter estimation, **240**, 95; sequential analysis, **240**, 96-99; simultaneous analysis, **240**, 96-99; titration experiments, **240**, 91

Differential Scanning Calorimetry: Baseline estimation [effect on parameters, **240**, 542-543, 548-549; importance, **240**, 540; polynomial interpolation, **240**, 540-541, 549, 567; proportional method, **240**, 541-542, 547-548, 567]; changes in solvent accessible surface areas, **240**, 519-520, 528; cooperativity measurement error, **240**, 642; data simulation, **240**, 537-539; effect of ligand binding, **240**, 533; enthalpy determination, **240**, 504, 506-507, 513-514, 534-537; entropy determination, **240**, 506-507, 514-517, 535-536, 539; folding/unfolding partition function determination, **240**, 505, 508-509, 532, 535; F-statistics, **240**, 554-559, 562; gel-liquid crystalline phase transition, **240**, 571-572; Gibbs free energy determination, **240**, 503, 518, 521, 535-536; heat capacity determination, **240**, 506, 510-513, 531, 534, 546-547; industrial applications, **240**, 531, 568; α-lactalbumin molten globule state, **240**, 522, 528-530; noise effect on parameters, **240**, 539-540, 562, 565; two-dimensional, guanidinium hydrochloride thermal stability surfaces [accuracy, **240**, 566-567; methods, **240**, 537-543; multiple folding state fitting, **240**, 543-566].

CALPAIN

This enzyme [EC 3.4.22.17] is an intracellular, nonlysosomal member of the peptidase family C2. The enzyme catalyzes the calcium ion-dependent hydrolysis of peptide bonds with preference for Tyr–Xaa, Met–Xaa, or Arg–Xaa with a leucyl or valyl residue at the P_2 position. There are two main types of calpain. One has a high calcium sensitivity in the micromolar range and is called μ-calpain or calpain I. The other calpain has a low calcium sensitivity in the millimolar range and is called m-calpain or calpain II. Forms of calpain exhibiting intermediate calcium sensitivity also exist.

E. Shaw (1990) *Adv. Enzymol.* **63**, 271.
J. S. Bond & P. E. Butler (1987) *Ann. Rev. Biochem.* **56**, 333.

CAMPHOR 5-MONOOXYGENASE

This enzyme [EC 1.14.15.1], also known as camphor 5-*exo*-methylene hydroxylase, and cytochrome P450-cam, catalyzes the reaction of (+)-camphor with putidaredoxin and dioxygen to generate (+)-*exo*-5-hydroxy-camphor, oxidized putidaredoxin, and water. A heme-thiolate acts as a cofactor. The enzyme can also utilize (−)-camphor as a substrate, and 1,2A-campholide will result in the formation of 5-*exo*-hydroxy-1,2-campholide.

V. Ullrich & W. Duppel (1975) *The Enzymes*, 3rd ed., **12**, 253.

cAMP PHOSPHODIESTERASE

This enzyme [EC 3.1.4.17] catalyzes the hydrolysis of 3′,5′-cAMP or other 3′,5′cyclic nucleotides (cNMP) to form AMP or NMP. The mammalian enzyme is activated by calcium ions in the presence of calmodulin.

E. Degerman, P. Belfrage & V. Manganiello (1997) *J. Biol. Chem.* **272**, 6823.
J. A. Gerlt (1992) *The Enzymes*, 3rd ed., **20**, 95.

cAMP-STIMULATED PROTEIN KINASE (PKA)

This protein kinase (known as protein kinase A or PK-A) has an R_2C_2 quaternary structure that binds 3′,5′-cAMP at its dimeric regulatory (R) subunit with resultant release of two catalytic (C) subunits. The free energy of hydrolysis of the cyclic nucleotide activator is large (ΔG −13 kcal/mol) and allows the 3′,5′-cAMP to be virtually irreversibly converted to AMP by the action of a specific phosphodiesterase. This protein kinase, originally discovered by the Nobelists Edwin Krebs and Edward Fischer, is now considered to be the prototype for over two thousand members of the protein kinase superfamily.

D. Bossemeyer, R. A. Engh, V. Kinzel, H. Ponstingl & R. Huber (1993) *EMBO J.* **12**, 849.
L. N. Johnson & D. Barford (1993) *Ann. Rev. Biophys. Biomol. Struct.* **22**, 199.
J. Larner (1990) *Adv. Enzymol.* **63**, 173.
S. I. Walaas & P. Greengard (1987) *The Enzymes*, 3rd ed., **18**, 285.
S. J. Beebe & J. D. Corbin (1986) *The Enzymes*, 3rd ed., **17**, 43.
D. A. Walsh & E. G. Krebs (1973) *The Enzymes*, 3rd ed., **8**, 555.

CANNIZZARO REACTION

The conproportionation reaction of aromatic aldehydes or aldehydes lacking α-hydrogens under alkaline conditions, yielding an alcohol and an acid.

S. Cannizaro (1853) *Ann.* **88**, 129.
T. A. Geissman (1944) *Org. React.* **2**, 94.

CARBAMOYL-PHOSPHATE SYNTHETASE I AND II

Carbamoyl-phosphate synthetase (ammonia) [EC 6.3.4.16], also known as carbamoyl-phosphate synthetase I, catalyzes the reaction of two molecules of ATP with carbon dioxide, ammonia, and water to produce two molecules of ADP, orthophosphate, and carbamoyl phosphate. Carbamoyl-phosphate synthetase (glutamine-hydrolyzing) [EC 6.3.5.5], also known as carbamoyl-phosphate synthetase II, catalyzes the reaction of two molecules of ATP with carbon dioxide, glutamine, and water to produce two molecules of ADP, orthophosphate, glutamate, and carbamoyl phosphate.

J. N. Earnhardt & D. N. Silverman (1998) *Comprehensive Biological Catalysis: A Mechanistic Reference* **1**, 495.

L. S. Mullins & F. M. Raushel (1995) *Meth. Enzymol.* **249**, 398.

H. Zalkin (1993) *Adv. Enzymol.* **66**, 203.

A. Meister (1989) *Adv. Enzymol.* **62**, 315.

D. S. Kaseman & A. Meister (1985) *Meth. Enzymol.* **113**, 305.

I. A. Rose (1979) *Adv. Enzymol.* **50**, 361.

D. E. Koshland, Jr., & A. Levitzki (1974) *The Enzymes*, 3rd ed., **10**, 539.

S. Ratner (1973) *Adv. Enzymol.* **39**, 1.

J. M. Buchanan (1973) *Adv. Enzymol.* **39**, 91.

N-CARBAMOYLPUTRESCINE AMIDOHYDROLASE

This enzyme [EC 3.5.1.53], also known as *N*-carbamoylputrescine amidase, catalyzes the hydrolysis of *N*-carbamoylputrescine to produce putrescine, carbon dioxide, and ammonia.

T. A. Smith (1985) *Ann. Rev. Plant Physiol.* **36**, 117.

CARBANION

An anion species in which an even number of electrons and a significant negative charge remain located on at least one carbon atom.

CARBENE

A highly reactive compound containing a neutral divalent carbon with two nonbonding electrons (*i.e.*, $:CH_2$ or a substitution derivative). The nonbonding electrons can have parallel spins (triplet state) or antiparallel spins (singlet state). The parent species, $:CH_2$, is also known as methylene. A number of carbene derivatives have been used as photoaffinity labels of proteins. Irradiation of 3′-*O*-(4-benzoyl)benzoyl-ATP will cause 70% inactivation of mitochondrial F_1-ATPase[1].

[1]N. Williams & P. S. Coleman (1982) *J. Biol. Chem.* **257**, 2834.

CARBENIUM CENTER

The carbon atom (within a carbenium ion) on which the excess positive charge may be considered to be largely concentrated.

IUPAC (1979) *Pure and Appl. Chem.* **51**, 1725.

CARBENIUM ION

A term for a positively-charged trivalent carbon atom (a carbocation). R_3C^+ is a carbenium ion. The term "carbenium ion" is intended to replace the term "carbonium ion," but the habits of practicing chemists persist. *See* *Carbocation; Carbonium Ion; Carbene*

CARBOCATION

A cation in which significant positive charge is located on at least one carbon atom, itself having an even number of electrons. Both carbenium ions and pentavalent species (such as the methanonium ion, CH_5^+) are carbocations. However, radical cations and carbynium ions are not considered to be carbocations. *See* *Carbenium Ion; Bridged Carbocation; Carbonium Ion*

CARBOHYDRATE MODIFICATION REACTIONS

Because carbohydrates are so frequently used as substrates in kinetic studies of enzymes and metabolic pathways, we refer the reader to the following topics in Robyt's excellent account[1] of chemical reactions used to modify carbohydrates: formation of carbohydrate esters, pp. 77-81; sulfonic acid esters, pp. 81-83; ethers [methyl, p. 83; trityl, pp. 83-84; benzyl, pp. 84-85; trialkyl silyl, p. 85]; acetals and ketals, pp. 85-92; modifications at C-1 [reduction of aldehydes and ketones, pp. 92-93; reduction of thioacetals, p. 93; oxidation, pp. 93-94; chain elongation, pp. 94-98; chain length reduction, pp. 98-99; substitution at the reducing carbon atom, pp. 99-103; formation of glycosides, pp. 103-105; formation of glycosidic linkages between monosaccharide residues, 105-108]; modifications at C-2, pp. 108-113; modifications at C-3, pp. 113-120; modifications at C-4, pp. 121-124; modifications at C-5, pp. 125-128; modifications at C-6 in hexopyranoses, pp. 128-134.

[1]J. F. Robyt (1998) *Essentials of Carbohydrate Chemistry*, Springer, New York.

CARBON-14 (^{14}C)

A radioactive β-emitting (0.156 MeV) carbon nuclide. The decay half-life is 5715 years.

Special precautions: Bone concentrates carbonates and is the critical organ in terms of dose. The effective *biological* half-life in humans is around 10 days. Double gloving is recommended when working with this radionuclide.

CARBON DIOXIDE

Carbon dioxide is a symmetrical, linear triatomic molecule ($O=C=O$) with a zero dipole moment. The carbon-to-hydrogen bond distances are about 1.16Å, which is about 0.06Å shorter than typical carbonyl double bonds. This shorter bond length was interpreted by Pauling[1] to indicate that greater resonance stabilization occurs with CO_2 than with aldehydes, ketones, or amides. When combined with water, carbonic acid (H_2CO_3) forms, and depending on the pH of the solution, carbonic acid loses one or two protons to form bicarbonate and carbonate, respectively. The various thermodynamic parameters[2] of these reactions are shown in Table I.

Table I

Thermodynamics of CO_2 Hydration and Ionization at 298 K

Reaction	pK	ΔG (kcal/mol)	ΔH (kcal/mol)	ΔS (e.u.)
$K_{hydration} = [H_2CO_3]/[CO_2]$	2.59	3.53	1.13	−8
$K_{carbonic\ acid} = [H^+][HCO_3^-]/[H_2CO_3]$	3.77	5.17	1.01	−14
$K_1 = [H^+][HCO_3^-]/\{[CO_2] + [H_2CO_3]\}$	6.35	8.67	2.24	−21.6
$K_2 = [H^+][CO_3^{2-}]/[HCO_3^-]$	10.3	14.1	3.60	−35.2

Edsall[3] has presented a cogent analysis of the various rates of hydration and dehydration, including the ability of oxygen acids to catalyze these reactions. **See** Oxygen, Oxides & Oxo Anions

[1]L. Pauling (1960) The Nature of the Chemical Bond (3rd ed.), pp. 267-269, Cornell Univ. Press, Ithaca.

[2]J. T. Edsall & J. Wyman (1958) Biophysical Chemistry, vol. I, chap. 10, Academic Press, New York.

[3]J. T. Edsall (1969) in CO₂: Chemical, Biochemical, and Physiological Aspects (R. E. Forster, J. T. Edsall, A. B. Otis & F. J. W. Roughton, eds.), pp. 15-27, NASA Publication SP-188.

Selected entries from Methods in Enzymology [vol, page(s)]: As anesthetic, **77**, 115; binding equilibria analysis, **76**, 487-511; by chemical-chromatographic method, **76**, 487-496; breath tests, **77**, 3-6; carbonic anhydrase, **64**, 120; carbon isotope effect, **64**, 86-95; collection of ¹³C-labeled carbon dioxide, **77**, 9; contamination, **64**, 93, 102; distillation, **64**, 91; effect on oxygen equilibrium curve for hemoglobin, **76**, 446, 468-470; equilibrium isotope effect, **64**, 109; exhalation, **77**, 3-9; pharmacokinetics, **77**, 8, 9; fixation [measurement in mutants, **69**, 16, 17, 20-23; photosynthetic mutants, **69**, 9; release of products, **69**, 608, 609]; formation, **77**, 3-6; controlled supply, **69**, 567-569; measurement, **69**, 563; mass spectrometer calibration, **69**, 554; measurement, **64**, 92; **69**, 563, 564; comparison of techniques, **77**, 6-8; reaction with α-amino and ε-amino groups, **76**, 488-489, 495, 504, 505; resistance to diffusion, **69**, 453; solubility, **64**, 93-94; in hemoglobin solutions, **76**, 494; standard curve, **69**, 564-566; nitrogen and water removal, **64**, 91.

CARBONIC ANHYDRASE

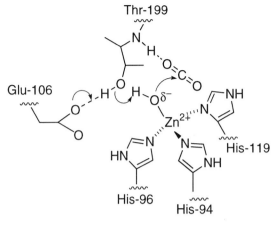

Carbonic Anhydrase-II Catalysis of H₂O Attack on CO₂

This enzyme [EC 4.2.1.1], also referred to as carbonate dehydratase, is a zinc-dependent enzyme that catalyzes the reaction of carbon dioxide with water to form carbonic acid (or, of bicarbonate and a proton). **See also** Proton Transfer in Aqueous Solution; Manometric Assay Methods; Marcus Rate Theory

J. N. Earnhardt & D. N. Silverman (1998) Comprehensive Biological Catalysis: A Mechanistic Reference **1**, 483.

D. N. Silverman (1995) Meth. Enzymol. **249**, 479.

W. S. Sly & P. Y. Hu (1995) Ann. Rev. Biochem. **64**, 375.

M. R. Badger & G. D. Price (1994) Ann. Rev. Plant Physiol. Plant Mol. Biol. **45**, 369.

D. W. Christianson (1991) Adv. Protein Chem. **42**, 281.

B. L. Vallee & A. Galdes (1984) Adv. Enzymol. **56**, 283.

D. N. Silverman & S. H. Vincent (1983) Crit. Rev. Biochem. **14**, 207.

D. N. Silverman (1982) Meth. Enzymol. **87**, 732.

Y. Pocker & S. Sarkanen (1978) Adv. Enzymol. **47**, 149.

S. Lindskog, L. E. Henderson, K. K. Kannan, A. Liljas, P. O. Nyman & B. Strandberg (1971) The Enzymes, 3rd ed., **5**, 587.

CARBONIUM ION

1. A carbocationic species in which there is at least one pentavalent carbon atom (e.g., CH_5^+). 2. Traditional name for chemical species that are now referred to as carbenium ions. Considerable confusion exists in the literature with this term for carbocations. The "-onium" suffix usually refers to a higher covalency when compared to the neutral atom; thus, CH_5^+ would be a true carbonium ion (in terms of the first definition). Additional ambiguity results when the term "ethyl carbonium ion" is used to describe both $CH_3CH_2^+$ and to $R-CH_2CH_2^+$. For these reasons, the terms carbocation or carbenium ion are now preferred.

CARBON MONOXIDE

The properties of this odorless and extremely poisonous diatomic gas are described under *Oxygen, Oxides & Oxo Anions.*

Selected entries from *Methods in Enzymology* [vol, page(s)]: Hemoglobin binding, stepwise CO combination and dissociation rate constants, double mixing experiment, **232**, 432-436; infrared spectra, **232**, 140-151; interaction with partially liganded hemoglobin intermediates, double mixing experiments, **232**, 432-445; rebinding to hemoglobin [after photodissociation, **232**, 72-73, 416; after photolysis, **232**, 80-83, 416; carboxyhemoglobin Fe-C-O configuration, **232**, 280-281; optical spectroscopy, **232**, 62-63; structural relaxation dynamics studies, **232**, 345, 348].

CARBON MONOXIDE DEHYDROGENASE

This enzyme [EC 1.2.99.2], also known as acetyl-CoA synthase, catalyzes the reaction of carbon monoxide with water and an acceptor to produce carbon dioxide and the reduced acceptor. The cofactors of this enzyme include nickel and zinc ions as well as non-heme iron. Methyl viologen can act as the acceptor substrate. The enzyme is isolated from *Clostridium* sp. Interestingly, it also catalyzes an exchange reaction of carbon between C1 of acetyl-CoA and carbon monoxide. The protein participates in the synthesis of acetyl-CoA from carbon dioxide and hydrogen in the organisms.

B. T. Golding & W. Buckel (1998) *Comprehensive Biological Catalysis: A Mechanistic Reference* **3**, 239.
P. E. Baugh, D. Collison, C. D. Garner & J. A. Joule (1998) *Comprehensive Biological Catalysis: A Mechanistic Reference* **3**, 377.

N^5-(CARBOXYETHYL)ORNITHINE SYNTHASE

This enzyme [EC 1.5.1.24], also referred to as N^5-(1-L-carboxyethyl)-L-ornithine:NADP$^+$ oxidoreductase, reversibly catalyzes the reaction of ornithine with pyruvate and NADPH to produce N^5-(1-carboxyethyl)ornithine, NADP$^+$, and water. Lysine can also serve as a substrate, acting on N^6, albeit not as effectively as ornithine.

J. Thompson & J. A. Donkersloot (1992) *Ann. Rev. Biochem.* **61**, 517.
J. Thompson & S. P. F. Miller (1991) *Adv. Enzymol.* **64**, 317.

CARBOXYLESTERASES

These enzymes [EC 3.1.1.1] (also referred to as ali-esterase, B-esterase, monobutyrase, cocaine esterase, methylbutyrase, and procaine esterase) catalyze the hydrolysis of a carboxylic ester to yield a carboxylate anion and an alcohol. They exhibit a broad specificity, even acting on vitamin A esters.

K. Krisch (1971) *The Enzymes*, 3rd ed., **5**, 43.

CARBOXYPEPTIDASE A

This zinc-dependent enzyme [EC 3.4.17.1], a member of the peptidase family M14, catalyzes the hydrolysis of peptide bonds at the C-terminus of polypeptides. Little hydrolytic action occurs if the C-terminal amino acid is aspartate, glutamate, arginine, lysine, or proline. Carboxypeptidase A is formed from a precursor protein, procarboxypeptidase A.

J. E. Coleman (1992) *Ann. Rev. Biochem.* **61**, 897.
J. J. Villafranca & T. Nowak (1992) *The Enzymes*, 3rd ed., **20**, 63.
D. W. Christianson (1991) *Adv. Protein Chem.* **42**, 281.
J. A. Hartsuck & W. N. Lipscomb (1971) *The Enzymes*, 3rd ed., **3**, 1.

CARBOXYPEPTIDASE B

This zinc-dependent enzyme [EC 3.4.17.2], a member of the peptidase family M14, catalyzes the hydrolysis of peptide bonds located at the C-terminus of a polypeptide. The enzyme prefers C-terminal lysine or arginine residues.

J. E. Folk (1971) *The Enzymes*, 3rd ed., **3**, 57.

CARBOXYPEPTIDASE C

This enzyme [EC 3.4.16.5] (also known as serine-type carboxypeptidase I, cathepsin A, carboxypeptidase Y, and lysosomal protective protein) is a member of the peptidase family S10 and catalyzes the hydrolysis of the peptide bond, with broad specificity, located at the C-terminus of a polypeptide. The pH optimum ranges from 4.5 to 6.0. The enzyme is irreversibly inhibited by diisopropyl fluorophosphate and is sensitive to thiol-blocking reagents.

S. J. Remington & K. Breddan (1994) *Meth. Enzymol.* **244**, 231.

CARBOXYPEPTIDASE D

This enzyme [EC 3.4.16.6], also known as carboxypeptidase KEX1 and carboxypeptidase S1, is a member of the peptidase family S10 and catalyzes the hydrolysis of a peptide bond located at the C-terminus of a polypeptide with a preference for a C-terminal arginine or lysine residue. This broadly specific enzyme is inhibited by the diisopropyl fluorophosphate and thiol-blocking reagents. The pH optimum ranges from 4.5 to 6.0.

S. J. Remington & K. Breddan (1994) *Meth. Enzymol.* **244**, 231.

CARBOXYPEPTIDASE M

This enzyme [EC 3.4.17.12], which is a member of peptidase family M14, catalyzes the hydrolysis of a peptide bond at the C-terminus of a polypeptide provided that the C-terminal amino acid is an arginine or a lysine residue. The protein is a membrane-bound enzyme which has a pH optimum at neutral pH. This enzyme is distinct from arginine carboxypeptidase (carboxypeptidase N) [EC 3.4.17.3] and from carboxypeptidase H [EC 3.4.17.10].

F. Tan, P. A. Deddish & R. A. Skidgel (1995) *Meth. Enzymol.* **248**, 663.

CARBOXYPEPTIDASE N

This enzyme [EC 3.4.17.3] (also referred to as lysine carboxypeptidase, arginine carboxypeptidase, kininase I, or anaphylatoxin inactivator) is a zinc-dependent member of peptidase family M14. The enzyme hydrolyzes the peptide bond at the C-terminus provided that the C-terminal amino acid is either arginine or lysine. The enzyme inactivates bradykinin and anaphylatoxins in blood plasma.

R. A. Skidgel (1995) *Meth. Enzymol.* **248**, 653.

CARBOXYPEPTIDASE T

This enzyme [EC 3.4.17.18], a member of peptidase family M14, catalyzes the hydrolysis of a peptide bond at the C-terminus of a polypeptide provided that the amino acid has a C-terminal amino acid that is either hydrophobic or positively charged.

V. M. Stepanov (1995) *Meth. Enzymol.* **248**, 675.

CARDIOLIPIN SYNTHASE

The following are recent reviews on the molecular and physical properties of this enzyme.

I. Shibuya & S. Hiraoka (1992) *Meth. Enzymol.* **209**, 321.
J. D. Esko & C. R. H. Raetz (1983) *The Enzymes*, 3rd ed., **16**, 207.

CARNITINE

$(CH_3)_3N^+-CH_2-CH(OH)-CH_2-COO^-$, an acyl-transfer metabolite, found in high abundance in skeletal muscle, liver, and yeast. *O*-Acylcarnitine, is a high-energy ester like its structurally related *O*-acetylcholine, $(CH_3)_3N^+-CH_2-CH_2-O-C(=O)-CH_3$.

Selected entries from *Methods in Enzymology* [vol, page(s)]:
Radioisotopic assay of acetylcarnitine and acetyl-CoA, **123**, 259; short-chain acylcarnitines: identification and quantitation, **123**, 264; purification and assay of carnitine acyltransferases, **123**, 276; determination of the specific activity of long-chain acylcarnitine esters, **123**, 284; synthesis of carnitine precursors and related compounds, **123**, 290; synthesis of radioactive (−)-carnitine from γ-aminobutyrate, **123**, 297; S-adenosylmethionine:ε-N-L-lysine methyltransferase, **123**, 303

CARNITINE *O*-ACETYLTRANSFERASE

This enzyme [EC 2.3.1.7], also known as carnitine acetylase, catalyzes the reversible reaction of acetyl-CoA with carnitine to yield coenzyme A and *O*-acetylcarnitine. The enzyme can also use propanoyl-CoA and butanoyl-CoA as substrates.

L. L. Bieber & S. Farrell (1983) *The Enzymes*, 3rd ed., **16**, 627.
K. F. Tipton & H. B. F. Dixon (1979) *Meth. Enzymol.* **63**, 183.

CARNITINE *O*-OCTANOYLTRANSFERASE

This enzyme [EC 2.3.1.137] catalyzes the reversible reaction of octanoyl-CoA with carnitine to yield coenzyme A and *O*-octanoylcarnitine. The enzyme utilizes a range of acyl-CoA derivatives as substrates, with optimal activity reported with C_6 or C_8 acyl groups.

L. L. Bieber & S. Farrell (1983) *The Enzymes*, 3rd ed., **16**, 627.

CARNITINE *O*-PALMITOYLTRANSFERASE

This enzyme [EC 2.3.1.21] catalyzes the reversible reaction of palmitoyl-CoA with carnitine to yield coenzyme A and *O*-palmitoylcarnitine. The enzyme exhibits a broad specificity toward acyl-CoA substrates, covering a range from C_8 to C_{18}, but optimal activity is observed with palmitoyl-CoA.

L. L. Bieber & S. Farrell (1983) *The Enzymes*, 3rd ed., **16**, 627.

CARNOSINASE

This enzyme [EC 3.4.13.3] (also referred to as Xaa–His dipeptidase, X–His dipeptidase, aminoacylhistidine dipeptidase, and homocarnosinase), is a zinc-dependent dipeptidase that catalyzes the hydrolysis of Xaa–His dipeptides. Carnosine, homocarnosine, and anserine are preferred substrates for this mammalian cytosolic enzyme. Other aminoacylhistidine dipeptides are weaker substrates (including homoanserine). The enzyme is activated by thiols and inhibited by metal-chelating agents.

O. W. Griffith (1986) *Ann. Rev. Biochem.* **55**, 855.

CARNOSINE SYNTHETASE

This enzyme [EC 6.3.2.11] catalyzes the reaction of histidine with β-alanine and ATP to produce carnosine, AMP, and pyrophosphate.

O. W. Griffith (1986) *Ann. Rev. Biochem.* **55**, 855.
G. D. Kalyankar & A. Meister (1971) *Meth. Enzymol.* **17B**, 102.

CARNOT CYCLE

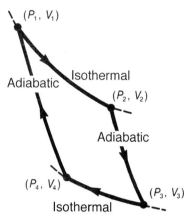

A hypothetical cycle for achieving reversible work, typically consisting of a sequence of operations: (1) isothermal expansion of an ideal gas at a temperature T_2; (2) adiabatic expansion from T_2 to T_1; (3) isothermal compression at temperature T_1; and (4) adiabatic compression from T_1 to T_2. This cycle represents the action of an ideal heat engine, one exhibiting maximum thermal efficiency. Inferences drawn from thermodynamic consideration of Carnot cycles have advanced our understanding about the thermodynamics of chemical systems. **See** Carnot's Theorem; Efficiency; Thermodynamics

CARNOT'S THEOREM

A principle stating that for any engine working between the same two temperatures, maximum efficiency will occur by a engine working *reversibly* between those same two temperatures. Thus, all reversible engines have the same efficiency between the same temperatures and that efficiency is dependent only on those temperatures and not on the nature of the substance being acted upon. **See** Efficiency; Thermodynamics, Laws of; Carnot Cycle

β-CAROTENE (or, β-CAROTENOID) 15,15′-DIOXYGENASE

This enzyme [EC 1.13.11.21] catalyzes the reaction of β-carotene with dioxygen to produce two retinal molecules. Both iron ions and bile salts are required cofactors.

L. Villard-Mackintosh & C. J. Bates (1993) *Meth. Enzymol.* **214**, 168.
M. R. Lakshman & C. Okoh (1993) *Meth. Enzymol.* **214**, 256.
J. A. Olson & M. R. Lakshman (1990) *Meth. Enzymol.* **189**, 425.
O. Hayaishi, M. Nozaki & M. T. Abbott (1975) *The Enzymes*, 3rd ed., **12**, 119.

CASBENE SYNTHASE

This enzyme [EC 4.6.1.7] catalyzes the conversion of geranylgeranyl diphosphate to the antifungal diterpene casbene and pyrophosphate (or, diphosphate). The enzyme was first isolated from the castor bean.

R. A. Gibbs (1998) *Comprehensive Biological Catalysis: A Mechanistic Reference* **1**, 31.

CASPASES (or, CYSTEINYL ASPARTATE-SPECIFIC PROTEINASES)

Caspase-1 [EC 3.4.22.36] (also known as interleukin-1β converting enzyme and interleukin-1β convertase precursor) is a member of the peptidase family C14. It catalyzes the hydrolysis of the Asp[116]–Ala[117] and Asp[27]–Gly[28] in the precursor protein, resulting in the release of interleukin-1β. The enzyme will also hydrolyze the small peptide, Ac-TyrValAlaAsp—NHMEC.

D. W. Nicholson & N. A. Thornberry (1997) *Trends Biochem. Sci.* **22**, 299.

CATALASE

This enzyme [EC 1.11.1.6] catalyzes the conversion of hydrogen peroxide to dioxygen and two water molecules. Both heme and manganese ions are used as cofactors. Several organic substances, *e.g.* ethanol, can act as the hydrogen donor. A manganese protein containing Mn(III) in the resting state is often called pseudocatalase.

H. B. Dunford (1998) *Comprehensive Biological Catalysis: A Mechanistic Reference* **3**, 195.
J. E. Penner-Hahn (1998) *Comprehensive Biological Catalysis: A Mechanistic Reference* **3**, 439.
D. C. Rees & D. Farrelly (1990) *The Enzymes*, 3rd ed., **19**, 37.
M. Ueda, S. Mozaffar & A. Tanaka (1990) *Meth. Enzymol.* **188**, 463.
G. R. Schonbaum & B. Chance (1976) *The Enzymes*, 3rd ed., **13**, 363.

CATALYST

A substance that accelerates a chemical reaction but does not become consumed, generated, or permanently changed by such reaction. Thus, a catalyst does not alter the overall stoichiometric expression for the reaction or the overall equilibrium constant. The enhanced reactivity produced by a catalyst is referred to as catalysis.

An important characteristic of a catalyst's action is that the mechanism of the reaction is altered in a manner that allows for a lower activation energy. In a number of nonenzymatic examples, a reactant or product can act as a catalyst as well, and the definition must be altered to include substances that appear in the overall rate expression with a power higher than the corresponding stoichiometry number. One may also describe certain substances as catalysts even when, upon accelerating the reaction, they are destroyed by the process. In fact, nearly every catalyst for hydrogenation of organic molecules becomes progressively more and more poisoned with successive rounds of catalysis. In biology, we recognize that bacterial luciferases catalyze only a single turnover. Likewise, actin and tubulin respectively accelerate ATP and GTP hydrolysis in a single catalytic round associated with polymerization.

At very high concentrations, the enzyme can alter the equilibrium constant. If K_{eq} is calculated by determining the equilibrium concentrations of all free products and reactants, and if the products and reactants have different affinities for the free enzyme, then high $[E_{tot}]$ favors formation of significant amounts of EA and EP, and this may cause an apparent shift in K_{eq}. In such instances, the enzyme is now a stoichiometric participant in the reaction, and the true equilibrium constant has to take this into account.

CATALYST EFFECT ON EQUILIBRIUM CONSTANT

The cardinal feature of catalysis is that the equilibrium constant of a chemical reaction is unaffected by the presence of a catalyst. This is true as long as the concentration of the catalyst is insignificant relative to the concentration(s) of the least abundant reactant(s) or product(s). Thus for an uncatalyzed reaction (A $\rightleftharpoons$ B) with rate constants k_+ and k_- (for the forward and reverse rates, respectively), the equilibrium constant $K_{eq} = k_+/k_-$. In the presence of catalyst, the rate constants are increased, say to xk_+ and yk_-, and the new equilibrium constant $K_{eq}' = xk_+/yk_- = (x/y)(k_+/k_-) = (x/y)K_{eq}$. Because the rate enhancement in the presence of enzyme will always be the same for forward and reverse reactions, $x = y$, such that K_{eq}' must still equal K_{eq}.

CATALYTIC ANTIBODIES

Antibodies possessing catalytic activity as a result of their selection using transition state mimics as immunogenic or antigenic agents. Antibody catalysis takes advantage of the molecular diversity of the immune system and current knowledge of mechanistic enzymology and physical organic chemistry[1], allowing production of novel catalysts for reactions previously lacking biological

catalysis. As pointed out by Schultz and Lerner[1], systematic mechanistic and structural investigations of catalytic antibodies should provide insights about enzymic catalysis and the evolution of catalytic function.

The concept of catalytic antibodies was suggested succinctly by Jencks[2]. "If complementarity between the active site and the transition state contributes significantly to enzymatic catalysis, it should be possible to synthesize an enzyme by constructing such an active site. One way to do this is to prepare an antibody to a haptenic group which resembles the transition state of a given reaction. The combining sites of such antibodies should be complementary to the transition state and should cause an acceleration by forcing bound substrates to resemble the transition state."

Many of the 60 known reactions catalyzed by monoclonal antibodies involve kinetically favored reactions (e.g., ester hydrolysis), but abzymes can also speed up kinetically disfavored reactions. Stewart and Benkovic[3] applied transition-state theory to analyze the scope and limitations of antibody catalysis quantitatively. They found the observed rate accelerations can be predicted from the ratio of equilibrium binding constants of the reaction substrate and the transition-state analogue used to raise the antibody. This approach permitted them to rationalize product selectivity displayed in antibody catalysis of disfavored reactions, to predict the degree of rate acceleration that catalytic antibodies may ultimately afford, and to highlight some differences between the way that they and enzymes catalyze reactions.

The nature of antibody catalysis remains to be elucidated, and antibodies will not reach the efficiency of enzymes until they can emulate the conformational changes, acid/base, redox, and/or nucleophilic/electrophilic reactivities of catalytic residues along the entire reaction coordinate. It is worthy of note that Hollfelder et al.[4] recently demonstrated that serum albumins catalyze the eliminative ring-opening of a benzoisoxazole at rates that are similar to those observed with catalytic antibodies[5]. They suggest that formal general base catalysis contributes only modestly to the efficiency of both systems, and they favor the view that the antibody catalysis may be enhanced in some cases by nonspecific medium effects.

Antibody Aldolases: An Illustrative Example. Catalytic antibodies can be prepared by reactive immunization[6,7], a method that selects antibody catalysts *in vivo* in terms of their ability to catalyze a particular chemical reaction. As pointed out by Barbas *et al.*[8], the protocol differs from earlier methods relying on immunization with transition-state mimics that are as inert as possible so that resulting antibodies can react with target substrates that are in their native state. In the reactive immunization technique, a reactive antigen is designed so that a desired chemical reaction can take place within the antibody binding pocket during its induction. These efforts allowed Barbas *et al.*[8] to prepare antibody aldolases that displayed typical Michaelis-Menten kinetics, and Michaelis constants for substrates in cross-aldol reactions ranged upward from 1 mM to 1 M. The acceptor aldehydes displayed K_m values in the 0.02–0.50 mM range. Some of the observed catalytic rate enhancements are shown in Fig. 1.

Barbas *et al.*[8] offered the following mechanisms shown in Fig. 2 for the Class 1 aldolase reaction and Ab38C2 or Ab 33F12 antibody aldolase reactions, where R is 4-isobutyramidobenzyl or *n*-butyl. Both mechanisms involve a chemically reactive lysyl ε-amino group,

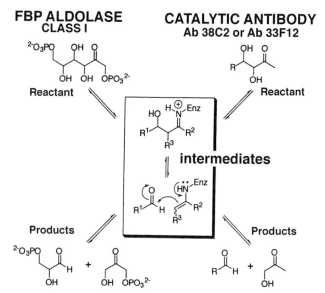

Figure 1. Kinetic parameters for the selection of antibody-catalyzed aldol and *retro*-aldol reactions, reflecting the biocatalyst's ability to accept substrates that differ clearly with respect to their molecular geometry. No background reaction was observed for the self-condensation of cyclopentanone. The indicated value for cyclopentanone addition to pentanal was estimated using the published k_{uncat} value of $2.28 \times 10^{-7}\ \mathrm{M^{-1}\,s^{-1}}$ for the aldol addition of acetone to an aldehyde. Reproduced with permission of the authors and the American Association for the Advancement of Science.

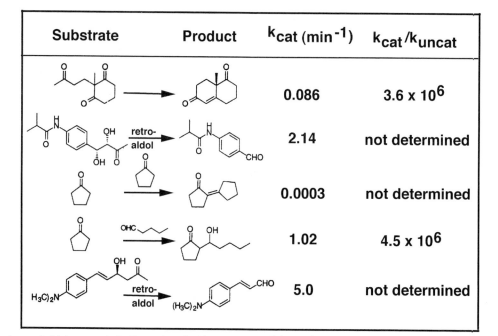

Substrate	Product	k_{cat} (min^{-1})	k_{cat}/k_{uncat}
		0.086	3.6×10^6
		2.14	not determined
		0.0003	not determined
		1.02	4.5×10^6
		5.0	not determined

Figure 2. Mechanisms of aldolase catalysis. Reproduced with permission of the authors and the American Association for the Advancement of Science.

and Barbas *et al.*[8] used binding studies with pyridoxal as well as aldolase substrates to demonstrate that Ab33F12 contains such a residue. From spectrophotometric studies of the pH dependence of the catalytic antibody's reaction with 3-methyl-2,4-pentanedione, the observed titration curve for enamine formation yielded a pK_a value of 5.5 to 6.0. This is in the range originally observed by Westheimer's group for the perturbed lysyl ε-amino group in the acetoacetate decarboxylase reaction, and Barbas *et al.*[8] provided X-ray crystal structural evidence that the Fab' fragment has what they refer to as a promiscuous hydrophobic pocket surrounding a structurally unusual lysyl residue.

[1] P. G. Schultz & R. A. Lerner (1995) *Science* **269**, 1835.
[2] W. P. Jencks (1969) *Catalysis in Chemistry and Enzymology*, p. 288, McGraw-Hill, New York. [Also available as Jencks, W. P. (1986) *Catalysis in Chemistry and Enzymology*, Dover Publications, New York].
[3] J. D Stewart & S. J. Benkovic (1995) *Nature* **375**, 388.
[4] F. Hollfelder, A. J. Kirby & D. S. Tawfik (1996) *Nature* **383**, 60.
[5] S. N. Thorn, R. G. Daniels, M. T. Auditor & D. Hilvert (1995) *Nature* **373**, 228.
[6] J. Wagner, R. A. Lerner & C. F. Barbas III (1995) *Science* **270**, 1797.
[7] P. Wirsching, J. A. Ashley, C.-H. Lo, K. D. Janda & R. A. Lerner (1995) *Science* **270**, 1775.
[8] C. F. Barbas III, A. Heine, G. Zhong, T. Hoffmann, S. Gramatikova, R. Björnestedt, B. List, J. Anderson, E. A. Stura, I. A. Wilson & R. A. Lerner (1997) *Science* **278**, 2085.

Selected entries from *Methods in Enzymology* [vol, page(s)]: Design, **178**, 551; immunoassay, **178**, 542; production, **178**, 531; purification, **178**, 543 ; substrates and enzymatic assay, **178**, 544; derivatization with spectroscopic probe, **178**, 567; ester cleavage assays, **178**, 565; fluorescence quenching binding assay, **178**, 567; substrates, synthesis, **178**, 564; generation [for acyl-group transfer, **203**, 342; for β-elimination reactions, **203**, 330; for Claisen rearrangements, **203**, 339; for Diels-Alder reactions, **203**, 340; from existing antibodies, **203**, 347; for hydrolysis of esters, carbonates, and amide bonds, **203**, 333; for stereospecific hydrolysis of unactivated esters, **203**, 337; for thymine dimer cleavage, **203**, 333]; strategies, **203**, 329; thermodynamics of catalytic mechanism, compared to enzymes, **259**, 685.

CATALYTIC CONSTANT

A proportionality factor for the dependency of a particular reaction on the concentration of a particular catalyst. The rate of a reaction, v, is equal to the sum of the rates of the uncatalyzed reaction (having a proportionality constant (or, rate constant), k_o) and the rate of the catalyzed reaction. Thus,

$$v = (k_0 + \Sigma k_i [C_i]^{n_i})[A]^{\alpha}[B]^{\beta} \ldots$$

where $[C_i]$ is the concentration of one of a set of i catalysts, k_i is the catalytic constant for catalyst i, n_i is the partial order of the reaction with respect to catalyst i

(usually $n_i = 1$), [A], [B], *etc.*, are the concentrations of the reactants, and α, β, *etc.*, are the partial order of the reaction with respect to each substrate. Thus, k_i is an $(\alpha + \beta + \ldots + 1)$th rate constant. *See* Turnover Number

IUPAC (1979) *Pure and Appl. Chem.* **51**, 1725.

CATALYTIC DOMAIN

That part of a polypeptide chain(s) possessing catalytic function. The catalytic domain may comprise more than one structural domain[1], and a multienzyme polypeptide contains at least two types of such domains. *See* Multienzyme Polypeptide; Autonomous Catalytic Domain

[1] Nomenclature Committee, IUB (1989) *Eur. J. Biochem.* **185**, 485.

CATALYTIC ENHANCEMENT

The ratio of the rate of an enzyme- (or ribozyme-) catalyzed reaction to the rate of the reaction in the absence of catalysis. This ratio equals k_{cat}/k_{non}, where k_{cat} is the turnover number and k_{non} is the noncatalyzed rate constant. *See* Catalytic Proficiency

CATALYTIC PROFICIENCY

A quantitative measure[1] of an enzyme's ability to lower the activation barrier for the reaction of a substrate in solution. Catalytic proficiency (a unitless parameter) equals the enzyme-catalyzed reaction rate constant (expressed as k_{cat}/K_m) divided by the rate constant (k_{non}) for the noncatalyzed reference reaction.

In the case of orotic acid, nonenzymatic decarboxylation proceeds with a half-time ($t_{1/2}$) of about 2.45×10^{15} s near pH 7 at room temperature, as indicated by reactions in quartz tubes at elevated temperatures[1]. Orotidine 5'-phosphate decarboxylase thus appears to be an extremely proficient enzyme which enhances the reaction rate by a factor of 10^{17}. They estimate the transition state form of the substrate has a dissociation constant that is less than 5×10^{-24} M.

Radzicka and Wolfenden[1] also report other catalytic proficiencies: staphylococcal nuclease, 5.9×10^{19}; adenosine deaminase, 7.8×10^{16}; AMP nucleosidase, 5.0×10^{16}; cytidine deaminase, 9.1×10^{15}, carboxypeptidase A, 2.2×10^{15}; ketosterol isomerase, 1.8×10^{15}; triose-phosphate isomerase, 5.6×10^{13}; chorismate mutase, 4.2×10^{13}; carbonic anhydrase, 9.2×10^{8}; prolylpeptidyl isomerase, or cyclophilin, 5.3×10^8. Estimated values of catalytic proficiencies represent upper limits on the dissociation of the enzyme-substrate complex in the transition state for cases where (a) the mechanisms of enzymatic and noncatalyzed reactions are not the same or (b) the enzyme reaction rate may be limited by a step other than bond-making or bond-breaking in substrates, such as the desorption of product(s)[1-2].

Mandelate racemase, another pertinent example, catalyzes the kinetically and thermodynamically unfavorable α-carbon proton abstraction. Bearne and Wolfenden[3] measured deuterium incorporation rates into the α-position of mandelate and the rate of (R)-mandelate racemization upon incubation at elevated temperatures. From an Arrhenius plot, they obtained a $\Delta G^{\ddagger}$ for racemization and deuterium exchange rate was estimated to be around 35 kcal/mol at 25°C under neutral conditions. The magnitude of the latter indicated mandelate racemase achieves the remarkable rate enhancement of 1.7×10^{15}, and a level of transition state affinity ($K_{TX} = 2 \times 10^{-19}$ M). These investigators also estimated the effective concentrations of the catalytic side chains in the native protein: for Lys-166, the effective concentration was 622 M; for His-297, they obtained a value 3×10^3 M; and for Glu-317, the value was 3×10^5 M. The authors state that their observations are consistent with the idea that general acid-general base catalysis is efficient mode of catalysis when enzyme's structure is optimally complementary with their substrates in the transition-state. *See Reference Reaction; Catalytic Enhancement*

[1]A. Radzicka & R. Wolfenden (1995) *Science* **267**, 90.
[2]R. Wolfenden (1976) *Ann. Rev. Biophys. Bioeng.* **5**, 271.
[3]S. L. Bearne & R. Wolfenden (1997) *Biochemistry* **36**,1646.

CATALYTIC RNA

Enzymatically active ribonucleic acid segments (some of which are known as ribozymes) with the capacity to catalyze RNA self-splicing[1] or peptide bond formation[2]. The overall catalytic rate enhancement is around 10^{11}.

The hammerhead ribozyme was the first identified motif to catalyze sequence-specific self-cleavage in plant virus satellite RNA molecules (Fig. 1). This motif facilitates magnesium ion-dependent transesterification with a turnover rate of about 1 s^{-1}.

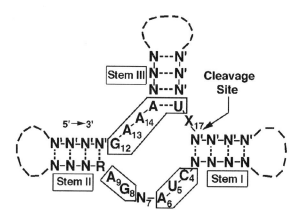

Figure 1. Consensus nucleotide sequence and stem–loop structure of hammerhead ribozyme.

In the microorganism *Tetrahymena*, group I intron splicing is accomplished in the catalytic cycle shown in Fig. 2. The intron first binds guanosine or one of its 5′-phosphorylated derivatives which is then used as a nucleophile in cleaving the 5′ splice site.

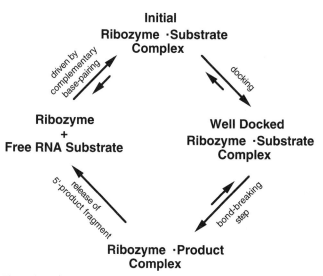

Figure 2. The catalytic reaction cycle of the *Tetrahymena* ribozyme (E), showing the binding and docking reactions (leading to the formation of E·S complex), followed by a bond cleavage breaking step (the rate constant for which is k_{chem}) and release of the 5′-fragment in the multiturnover steps (rate constant equals k_{mt}).

Ribozyme catalysis of phosphoryl transfer in appears to be promoted by destabilization of the ground state[3]. It also seems clear that magnesium ion complexation and hydrogen bonding interactions stabilize the negative charge that develops on the leaving group during entry of the nucleophile. This transesterification reaction is mechanistically analogous to that used in the mRNA spliceosome as well as in other DNA topoisomerase and

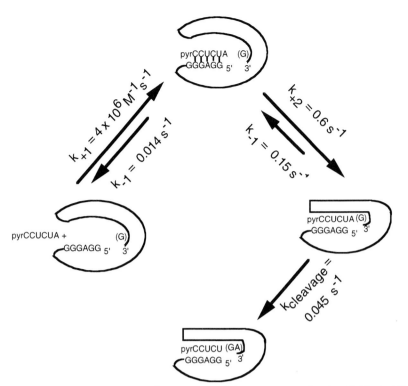

Figure 3. Minimal mechanism for pyrCCUCUA binding in the presence of the acceptor molecule pG. Vertical lines represent base-pairing interactions, and the boldface dots show those nucleotides that require 2'-OH groups needed for optimal binding. (From Turner et al.[5] with permission of the authors and the publisher Springer-Verlag.)

transposition reactions. The true catalytic nature of the ribozyme was demonstrated by the discovery that the RNA component of RNase P catalytically processed tRNA precursors[4].

As shown by Turner et al.[5], fluorescence experiments using a 5'-pyrenylated oligonucleotides have aided the determination of rate constants and equilibrium constants that define (a) the initial base-pairing step in substrate binding, (b) the so-called docking step that reflects a substrate-induced conformational step, and (c) the bond cleavage step *per se*. The scheme shown in Fig. 3 represents a beautiful example of Koshland's induced-fit model at work in ribozyme action.

The self-splicing reaction of group II introns requires the folding of intronic RNA to form the catalytically active unit. The most likely chemical pathway during the splicing reaction is shown in Fig. 4. The reaction mechanism commences with proton abstraction from the ribose 2'-OH, and the 2'-alkoxide then attacks the phosphorus, forming a pentavalent intermediate. In the second step, rate-limiting breaking of the P—O bond may

be assisted by acid catalysis. Two magnesium ions are thought to play key roles in this process.

Containing only around 30 nucleotides in their catalytic core, the so-called hammerhead ribozymes are the smallest catalytic RNA's. They display Michaelis-Menten kinetics in their action on substrates: k_{cat} 1–3 × 10^{-2} s^{-1};

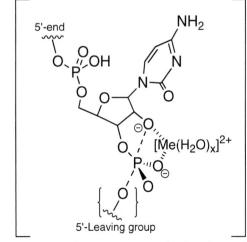

Figure 4. Transesterification reaction mechanism of group II introns. Nucleophilic attack leads to the pentavalent phosphorus intermediate which subsequently decomposes to yield the cleavage products.

K_m values 20–200 nM. Product release is fast, such that the rate-determining step is phosphodiester bond scission. Values for k_{cat}/K_m fall in the $1–4 \times 10^7$ M^{-1} s^{-1} range.

In their lucid comparison of mechanisms of ribozyme and protein RNase action, Narlikar and Herschlag[6] recently described the current state of RNA catalysis. Of particular relevance to catalysis is whether RNA catalysts share the same power exhibited by proteins to perturb the microenvironment of active-site relative to the aqueous solution. Proteins have a good mix of side-chain residues that can lower or raise the dielectric constant of regions within the active site. The authors argue that RNA is likely to be less effective in creating a low effective dielectric constant within the active site, and they suggest that the larger size of RNA catalysts may be related the need to be rigid to preferentially stabilize the transition state. They also believe that this bears directly on Jencks' observations about the role of intrinsic binding energy in enzymic catalysis (i.e., the energy derived from binding interactions, while not strongly manifested in the ground-state enzyme·substrate complex, is profoundly engaged in transition state stabilization)[7].

The bacterial ribosome is composed of three RNA components (5S, comprising 120 nucleotides; 16S, 1542 nucleotides; and 23S, 2904 nucleotides) along with some 55 protein components. Nitta et al.[2] demonstrated that the 23S ribosomal RNA (rRNA) catalyzes the peptide bond reaction without any requirement for ribosomal proteins. They also individually synthesized the six domains of the 23S rRNA, and after combining these components, peptide bond formation could be demonstrated. By omitting or including various combinations of the six domains, Nitta et al.[2] also showed that activity could be reconstituted by domain V only, when present at a concentration that was ten-times higher than that of intact 23S ribosomal RNA. This important finding demonstrates that RNA can catalyze peptide bond synthesis (i.e., C–N ligation) in addition to the well-known phosphotransfer reactions of ribozymes.

[1] A. J. Zaug & T. R. Cech (1982) Nucl. Acids Res. 10, 2823.
[2] I. Nitta, Y. Kamada, H. Noda, T. Ueda & K. Watanabe (1998) Science 281, 666.
[3] T. R. Cech & D. Hershlag (1996) Catalytic RNA (F. Eckstein & D. M. J. Lilley, eds.) p. 4, Springer-Verlag, Berlin.
[4] C. Guerrier-Takada, K. Gardiner, T. Marsh, N. Pace & S. Altman (1983) Cell 35, 849.

[5] D. H. Turner, Y. Li, M. Fountain, L. Profenno & P. C. Bevilacqua (1996) Catalytic RNA (F. Eckstein & D. M. J. Lilley, eds.) p. 23, Springer-Verlag, Berlin.
[6] G. J. Narlikar & D. Herschlag (1997) Ann. Rev. Biochem. 66, 19.
[7] W. P. Jencks (1975) Adv. Enzymol. 43, 219.

CATASTROPHE THEORY

A formal mathematical approach for explaining the occurrence of sudden or abrupt changes in the behavior of a system. These abrupt changes occur at frequencies that often are dictated by one or more rate constants describing the interconversion of metastable and unstable states. The process is said to be stochastic if the frequency of transition is controlled by a random variable.

The self-assembly of microtubules is attended by abrupt switching between states of constant growth and shrinkage, and this process is termed "dynamic instability." The growth-to-shrinkage transitions occur catastrophically, and the reverse transition, known as rescue, returns the microtubules to a growth phase. These growth-to-shrinkage transitions had been viewed as random events occurring in proportion to the microtubule number concentration, thus following first-order kinetics. Odde et al.[1] recently applied a probabilistic analysis of data (based on direct microscopical observation of individual microtubules in vitro) in terms of the distribution of growth times. They observed that the slower growing and biologically inactive ends (termed ($-$)-ends) obeyed first-order catastrophe kinetics, but the faster growing and biologically active ($+$)-ends did not. The kinetic behavior at ($+$)-ends led them to infer that elongating ($+$)-ends of microtubules have an effective frequency of catastrophe that depends on the duration of growth prior to their experiencing a growth-to-shrinkage transition. While initially low, this frequency rises asymptotically to a limiting value. The authors conclude that the dynamic instability of microtubules is more complex than previous treatments would suggest.

[1] D. J. Odde, L. Cassimeris & H. M. Buettner (1995) Biophys. J. 69, 796.

CATECHOL 1,2-DIOXYGENASE

This enzyme [EC 1.13.11.1] catalyzes the reaction of catechol with dioxygen to produce cis,cis-muconate. Iron ions are used as a cofactor. Among its physiological roles, catechol 1,2-dioxygenase participates in the metabolism of nitro-aromatic compounds by a strain of *Pseudomonas putida.*

B. G. Fox (1998) *Comprehensive Biological Catalysis: A Mechanistic Reference* **3**, 261.
J. V. Schloss & M. S. Hixon (1998) *Comprehensive Biological Catalysis: A Mechanistic Reference* **2**, 43.
S. Yamamoto & Y. Ishimura (1991) *A Study of Enzymes* **2**, 315.
K.-L. Ngai, E. L. Neidle & L. N. Ornston (1990) *Meth. Enzymol.* **188**, 122.
O. Hayaishi, M. Nozaki & M. T. Abbott (1975) *The Enzymes*, 3rd ed., **12**, 119.

CATECHOL 2,3-DIOXYGENASE

This iron-dependent enzyme [EC 1.13.11.2], also called metapyrocatechase, catalyzes the reaction of catechol with dioxygen to produce 2-hydroxymuconate semialdehyde. The enzyme from *Alcaligenes* sp. strain O-1 reportedly catalyzes the reaction of 3-sulfocatechol with dioxygen and water to produce (2*E*,4*Z*)-2-hydroxymuconate and bisulfite.

B. G. Fox (1998) *Comprehensive Biological Catalysis: A Mechanistic Reference* **3**, 261.
S. Yamamoto & Y. Ishimura (1991) *A Study of Enzymes* **2**, 315.
M. A. Ator & P. R. Ortez de Montellano (1990) *The Enzymes*, 3rd ed., **19**, 213.
I. A. Kataeva & L. A. Golovleva (1990) *Meth. Enzymol.* **188**, 115.
O. Hayaishi, M. Nozaki & M. T. Abbott (1975) *The Enzymes*, 3rd ed., **12**, 119.

CATECHOL *O*-METHYLTRANSFERASE

This enzyme [EC 2.1.1.6] catalyzes the reaction of catechol with *S*-adenosylmethionine to produce *S*-adenosylhomocysteine and guaiacol (or, *o*-methoxyphenol).

F. Takusagawa, M. Fujioka, A. Spies & R. L. Schowen (1998) *Comprehensive Biological Catalysis: A Mechanistic Reference* **1**, 1.
T. Lotta, J. Vidgren, C. Tilgmann, I. Ulmanen, K. Melén, I. Julkunen & J. Taskinen (1995) *Biochemistry* **34**, 4202.
X. Cheng (1995) *Ann. Rev. Biophys. Biomol. Struct.* **24**, 293.
J. Vidgren, L. A. Svenson & A. Liljas (1994) *Nature* **368**, 354.

CATECHOL OXIDASE

These copper ion-dependent enzymes [EC 1.10.3.1] (also referred to as diphenol oxidases, *O*-diphenolase, phenolases, polyphenol oxidases, or tyrosinases) catalyze the reaction of two catechol molecules with dioxygen to produce two 1,2-benzoquinone and two water. A variety of substituted catechols can act as substrates. Many of the enzymes listed under this classification also catalyze a monophenol monooxygenase activity [*i.e.*, EC 1.14.18.1].

See also *Monophenol Monooxygenase; Tyrosine Monooxygenase*

N. H. Horowitz, M. Fling & G. Horn (1970) *Meth. Enzymol.* **17A**, 615.

CATENANE

A molecular entity consisting of two or more polymeric rings interlinked together but not covalently bonded to each other. Mitochondrial DNA are known to form catenated structures consisting of two interlinked nucleic acid circles (*i.e.*, a [2]catenane).

CATENARY MODEL FOR COMPARTMENT ANALYSIS

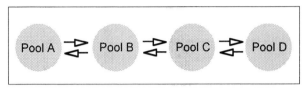

A linear series of metabolite compartments used in mathematical algorithms for modeling the kinetic behavior of metabolite and drug turnover.

CATHEPSIN B

This lysosomal enzyme [EC 3.4.22.1], also known as cathepsin B1, is a member of the peptidase family C1. The catalyzed reaction is the hydrolysis of peptide binds with a broad specificity. The enzyme prefers the ArgArg—Xaa bond in small peptide substrates (thus distinguishing this enzyme from cathepsin L). The enzyme also exhibits a peptidyl-dipeptidase activity, releasing C-terminal dipeptides from larger polypeptides.

H. Kirschke & B. Wiederanders (1994) *Meth. Enzymol.* **244**, 500.
H. Scholze & E. Tannich (1994) *Meth. Enzymol.* **244**, 512.

CATHEPSIN D

This lysosomal endopeptidase [EC 3.4.23.5] is similar to pepsin A, except that the specificity is narrower and will not hydrolyze the Gln^4—His^5 peptide bond in the B chain of insulin. The enzyme is a member of the peptidase family A1.

T. D. Meek (1998) *Comprehensive Biological Catalysis: A Mechanistic Reference* **1**, 327.
J. S. Bond & P. E. Butler (1987) *Ann. Rev. Biochem.* **56**, 333.

CATHEPSIN E

This enzyme [EC 3.4.23.34], also known as slow-moving proteinase and erythrocyte membrane aspartic proteinase, is similar to cathepsin D, albeit with a slightly broader specificity.

T. D. Meek (1998) *Comprehensive Biological Catalysis: A Mechanistic Reference* **1**, 327.
T. Kageyama (1995) *Meth. Enzymol.* **248**, 120.

CATHEPSIN G

This endopeptidase [EC 3.4.21.20], a member of the peptidase family S1, has substrate specificity similar to that of chymotrypsin C.

J. S. Bond & P. E. Butler (1987) *Ann. Rev. Biochem.* **56**, 333.

CATHEPSIN H

This mammalian lysosomal endopeptidase [EC 3.4.22.16] is also known as aleurain, cathepsin B3, cathepsin BA, and benzoylarginine:naphthylamide hydrolase. A member of the peptidase family C1, the enzyme also acts with an aminopeptidase activity, preferring Arg—Xaa peptide bonds.

H. Kirschke & B. Wiederanders (1994) *Meth. Enzymol.* **244**, 500.

CATHEPSIN L

This peptidase family C1 enzyme [EC 3.4.22.15] is an lysosomal endopeptidase with specificity akin to papain. Cathepsin L displays a higher activity toward protein substrates than does cathepsin B.

H. Kirschke & B. Wiederanders (1994) *Meth. Enzymol.* **244**, 500.

CATHEPSIN S

This peptidase family C1 enzyme [EC 3.4.22.27] is a lysosomal cysteinyl-dependent endopeptidase with substrate specificity similar to cathepsin L.

H. Kirschke & B. Wiederanders (1994) *Meth. Enzymol.* **244**, 500.

CATHEPSIN T

This protease [EC 3.4.22.24] interconverts the three forms of tyrosine aminotransferase [EC 2.6.1.5].

H. C. Pitot & E. Gohda (1987) *Meth. Enzymol.* **142**, 279.

CATIONIC INTERMEDIATE ANALOGUES

Positively charged reaction intermediates generated in a wide variety of enzyme-catalyzed reactions. Analogues containing a protonated amino group in a position where a carbocation is generated during catalysis have proved to be excellent tightly bound enzyme inhibitors[1]. For example, 1-amino sugars inhibit a number of glycosidases[2], and aza-substrate derivatives can serve as strong inhibitors.

[1]A. Radzicka & R. Wolfenden (1995) *Meth. Enzymol.* **249**, 284.
[2]H. L. Levine, R. S. Brody & F. H. Westheimer (1980) *Biochemistry* **19**, 4993.

CATION π-INTERACTIONS

A relatively strong, noncovalent binding interaction of a cation to the π face of an aromatic structure. This interaction can be considered to result from an electrostatic attraction between a positive charge and the quadrupole moment of the aromatic ring. Dougherty[1] suggests that one can also view the side chains of phenylalanine, tyrosine, and tryptophan as candidates for these interactions.

[1]D. A. Dougherty (1996) *Science* **271**, 163.

CDP-4-DEHYDRO-6-DEOXYGLUCOSE REDUCTASE

This enzyme [EC 1.17.1.1], also known as CDP-4-keto-6-deoxyglucose reductase, catalyzes the reversible reaction of CDP-4-dehydro-3,6-dideoxy-D-glucose with $NAD(P)^+$ and water to produce CDP-4-dehydro-6-deoxy-D-glucose and NAD(P)H.

S. Matsuhashi & J. L. Strominger (1966) *Meth. Enzymol.* **8**, 310.

CDP-DIACYLGLYCEROL:*sn*-GLYCEROL-3-PHOSPHATE 3-PHOSPHATIDYLTRANSFERASE

This enzyme [EC 2.7.8.5] (also referred to as phosphatidylglycerophosphate synthase, glycerophosphate phosphatidyltransferase, and 3-phosphatidyl-1′-glycerol-3′-phosphate synthase) catalyzes the reaction of CDP-diacylglycerol with glycerol 3-phosphate to produce CMP and 3-(3-phosphatidyl)glycerol 1-phosphate.

W. Dowhan (1992) *Meth. Enzymol.* **209**, 313.
R. A. Pieringer (1983) *The Enzymes*, 3rd ed., **16**, 255.
J. D. Esko & C. R. H. Raetz (1983) *The Enzymes*, 3rd ed., **16**, 207.
T. S. Moore, Jr. (1982) *Ann. Rev. Plant Physiol.* **33**, 235.

CDP-GLUCOSE 4,6-DEHYDRATASE

This enzyme [EC 4.2.1.45] catalyzes the conversion of CDP-glucose to CDP-4-dehydro-6-deoxy-D-glucose and water. The enzyme uses NAD^+ as a cofactor.

L. Glaser & H. Zarkowsky (1971) *The Enzymes*, 3rd ed., **5**, 465.

CELLOBIOSE 2-EPIMERASE

This enzyme [EC 5.1.3.11] catalyzes the interconversion of cellobiose to D-glucosyl-D-mannose.

M. E. Tanner & G. L. Kenyon (1998) *Comprehensive Biological Catalysis: A Mechanistic Reference* **2**, 7.
E. Adams (1976) *Adv. Enzymol.* **44**, 69.
L. Glaser (1972) *The Enzymes*, 3rd ed., **6**, 355.

CELLOBIOSE PHOSPHORYLASE

This enzyme [EC 2.4.1.20] catalyzes the reaction of cellobiose with orthophosphate to produce α-D-glucose 1-phosphate and D-glucose.

J. J. Mieyal & R. H. Abeles (1972) *The Enzymes*, 3rd ed., **7**, 515.
M. Cohn (1961) *The Enzymes*, 2nd ed., **5**, 179.
M. Doudoroff (1961) *The Enzymes*, 2nd ed., **5**, 229.

CELL PROLIFERATION KINETICS

When cells divide or undergo binary fission under batch culture conditions, the proliferation process often can be depicted as follows:

$$2^0 N_o \rightarrow 2^1 N_o \rightarrow 2^2 N_o \rightarrow \ldots \rightarrow 2^i N_o \rightarrow \ldots \rightarrow 2^n N_o$$

where $2^i N_o$ describes the algebraic product of 2^i (the number of doublings) and N_o (the original number of cells). In the most ideal case, every cell will sustain its viability and proliferative potential. Even so, such a collection of otherwise identical cells will not divide synchronously for any substantial period, because access to nutrients as well as geometrical considerations will affect each living cell in slightly different ways. If the number of doublings ν equals the elapsed time t divided by the doubling time t_d (*i.e.*, $\nu = t/t_d$), the factor f_t for the fold-increase in cell number can be written as:

$$f_t = N_t/N_o = 2^\nu$$

This equation describes a geometric growth process which can be transformed into the following logarithmic expression:

$$\ln N_t - \ln N_o = 0.693 t/t_d$$

Note that this equation will only apply during the initially exponential growth phase. As the number of cells becomes much greater, the daughter cells compete for limiting quantities of one or more nutrients. This will reduce the efficiency of cell division, and the proliferation curve will reach a plateau that is often called the stationary growth phase.

CELL PROLIFERATION (Thymidine Incorporation Assay)

Measurements of the rate and extent of cell proliferation are essential for understanding the effects of cell growth factors and inhibitors. In the absence of aneuploidy, the DNA content is an invariant characteristic of each particular cell type, and typically the incorporation of tritiated thymidine into DNA is a useful measure of cell proliferation. Such measurements start with the incubation of the cells of interest with [³H]thymidine, preferably ethanol-free because the presence of ethanol can alter the redox state of cells, and ethanol is also toxic to cells. One centrifuges the cells at low relative centrifugal field for 10–20 min, a procedure that works best with cells in suspension, although adherent cells can be trypsinized to release them from culture dishes without any confounding effect on the thymidine incorporation measurement. After resuspending cells in phosphate-buffered saline and subsequent recentrifugation to pellet the cells, one uses 10% (w/v) trichloroacetic acid to release low-molecular-weight nucleosides and nucleotides from the acid-precipitated DNA which can be collected on filter disks that are then further extracted with absolute ethanol. A nonradioactive assay can also be accomplished by replacing labeled thymidine with 5-bromo-2'-deoxyuridine (BrdU) which is subsequently detected with a BrdU-specific monoclonal antibody. This approach works best if cells are exposed simultaneously to [³H]thymidine (or BrdU) and 5-fluorodeoxyuridine which blocks the thymidylate synthase reaction, thereby reducing any competition by endogenous thymidine in the DNA incorporation step.

CELLULASE

Also referred to as endo-1,4-β-glucanase and carboxymethyl cellulase, this enzyme [EC 3.2.1.4] catalyzes the endohydrolysis of 1,4-β-D-glucosidic linkages in cellulose. The enzyme also catalyzes the hydrolysis of 1,4-linkages in β-D-glucans also containing 1,3-linkages.

G. Davies, M. L. Sinnott & S. G. Withers (1998) *Comprehensive Biological Catalysis: A Mechanistic Reference* **1**, 119-208.
P. Béguin & M. Lemaire (1996) *Crit. Rev. Biochem. Mol. Biol.* **31**, 201.
M. W. Bauer, S. B. Halio & R. M. Kelly (1996) *Adv. Protein Chem.* **48**, 271.
S. C. Fry (1995) *Ann. Rev. Plant Physiol. Plant Mol. Biol.* **46**, 497.
D. R. Whitaker (1971) *The Enzymes*, 3rd ed., **5**, 273.

CELLULOSE POLYSULFATASE

This enzyme [EC 3.1.6.7] catalyzes the hydrolysis of the 2- and 3-sulfate groups of the polysulfates of cellulose and charonin.

A. B. Roy (1971) *The Enzymes*, 3rd ed., **5**, 1.

CELLULOSE SYNTHASE

Also referred to as UDP-glucose-β-D-glucan glucosyltransferase and UDP-glucose-cellulose glucosyltransferase, cellulose synthase (UDP-forming) [EC 2.4.1.12] catalyzes the reaction of UDP-glucose with [(1,4)-β-D-glucosyl]$_n$ to produce UDP and [(1,4)-β-D-glucosyl]$_{(n+1)}$. This enzyme participates in the biosynthesis of cellulose. Cellulose synthase (GDP-forming) [EC 2.4.1.29] catalyzes the reaction of GDP-glucose with [(1,4)-β-D-glucosyl]$_n$ to produce GDP and [(1,4)-β-D-glucosyl]$_{(n+1)}$. This enzyme also participates in the biosynthesis of cellulose.

G. Davies, M. L. Sinnott & S. G. Withers (1998) *Comprehensive Biological Catalysis: A Mechanistic Reference* **1**, 119.

CELSIUS TEMPERATURE

The temperature, symbolized by θ or t and measured in units of degree Celsius (symbolized by °C), in which $\theta = T - 273.15$ (in which T is the temperature in kelvin) where the triple point of water is set at 273.16 K (or, 0.01°C). The conversion of Celsius temperature to Fahrenheit temperature (°F) is given by °F = (9/5)(°C) + 32.

CENNAMO TREATMENT

A protocol[1-3] for assessing the behavior of enzymes having one or more isomerization steps. The method facilitates determination of K_{iip}, a kinetic parameter associated with the enzyme interconversion. **See** *Iso Mechanisms*

[1]C. Cennamo (1968) *J. Theort. Biol.* **21**, 260.
[2]C. Cennamo (1969) *J. Theort. Biol.* **23**, 53.
[3]K. L. Rebholz & D. B. Northup (1995) *Meth. Enzymol.* **249**, 211.

CENTRAL ATOM

The centrally located atom in a coordination entity to which other atoms or group of atoms (generally referred to as ligands) binds.

[1]M. W. G. de Bolster (1997) *Pure Appl. Chem.* **69**, 1251.

CERAMIDASE

This enzyme [EC 3.5.1.23], also known as acylsphingosine deacylase, catalyzes the hydrolysis of an *N*-acylsphingosine to produce a fatty acid anion and sphingosine.

J. D. Esko & C. R. H. Raetz (1983) *The Enzymes*, 3rd ed., **16**, 207.
Y. Kishimoto (1983) *The Enzymes*, 3rd ed., **16**, 357.

CERAMIDE CHOLINEPHOSPHOTRANSFERASE

This enzyme [EC 2.7.8.3] catalyzes the reaction of CDP-choline with *N*-acylsphingosine to generate CMP and a sphingomyelin.

E. P. Kennedy (1962) *Meth. Enzymol.* **5**, 486.

CEREBROSIDE SULFATASE

This enzyme [EC 3.1.6.8], also known as arylsulfatase A, catalyzes the hydrolysis of a cerebroside 3-sulfate to produce a cerebroside and sulfate. The enzyme will also catalyze analogous reactions on the galactose 3-sulfate residues in a number of lipids as well as on ascorbate 2-sulfate and many phenol sulfates.

A. B. Roy (1971) *The Enzymes*, 3rd ed., **5**, 1.

CERULOPLASMIN

A blue, copper-containing glycoprotein present in mammalian blood plasma and containing type 1, type 2, and type 3 copper centers. The type 2 and type 3 copper centers are close together, forming a trinuclear copper cluster. Ceruloplasmin has an important role in the transport and storage of copper ions. Thus, it participates in the metabolism of copper-containing enzymes.

CF$_1$CF$_o$-ATPase

H$^+$-Transporting ATP synthase [EC 3.6.1.34] in plants, also referred to as chloroplast ATPase and CF$_1$CF$_o$-ATPase, catalyzes the hydrolysis of ATP to produce ADP and orthophosphate. When coupled with proton transport the reverse reaction results in the synthesis of ATP by this multisubunit complex. CF$_1$, isolated from the rest of the membrane-bound complex, retains the ATPase activity but not the proton-translocating activity.

B. J. Barkla & O. Pantoja (1996) *Ann. Rev. Plant Physiol. Plant Mol. Biol.* **47**, 159.
M. R. Sussman (1994) *Ann. Rev. Plant Physiol. Plant Mol. Biol.* **45**, 211.
L. A. Kleczkowski (1994) *Ann. Rev. Plant Physiol. Plant Mol. Biol.* **45**, 339.
B. Rubinstein & D. G. Luster (1993) *Ann. Rev. Plant Physiol. Plant Mol. Biol.* **44**, 131.
R. Serrano (1989) *Ann. Rev. Plant Physiol. Plant Mol. Biol.* **40**, 61.
H. S. Penefsky (1974) *The Enzymes*, 3rd ed., **10**, 375.

cGMP-DEPENDENT PROTEIN KINASE

This general class of ATP-dependent protein kinases (abbreviated cGPK) catalyzes 3′,5′-cyclic-GMP-stimulated phosphorylation of protein substrates. cGPK's are single-chain kinases containing two copies of the cyclic nucleotide-binding domain in their N-terminal regions. The nucleotide specificity of cAMP-protein kinases and cGMP-protein kinases can be traced to the conserved region of β-barrel 7: a threonyl residue is invariant in cGPK, and an alanyl residue is invariant in most cAPK's. One preferred substrate is vasodilator-stimulated phosphoprotein (VASP), an adapter in actin-based motility mechanisms. (**See** *Actin-Based Pathogen Motility*) VASP's Ser-239 appears to be the preferred cGMP-dependent protein kinase phosphorylation site in intact cells and platelets.

A. M. Edelman, D. K. Blumenthal & E. G. Krebs (1987) *Ann. Rev. Biochem.* **56**, 567.
S. J. Beebe & J. D. Corbin (1986) *The Enzymes*, 3rd ed., **17**, 43.
A. C. Nairn, H. C. Hemmings, Jr., & P. Greengard (1985) *Ann. Rev. Biochem.* **54**, 931.
D. A. Flockhart & J. D. Corbin (1982) *Crit. Rev. Biochem.* **12**, 133.
E. G. Krebs & J. A. Beavo (1979) *Ann. Rev. Biochem.* **48**, 923.

cGMP PHOSPHODIESTERASE

This membrane-bound homodimeric cyclic nucleotide phosphodiesterase (EC 3.1.4.17) catalyzes the conversion of 3',5'-cyclic-GMP to form 5'-GMP.

A. Tar, T. D. Ting & Y.-K. Ho (1994) *Meth. Enzymol.* **238**, 3.
N. D. Goldberg & M. K. Haddox (1977) *Ann. Rev. Biochem.* **46**, 823.

CHA METHOD

A method for deriving enzyme-rate expressions combining both rapid equilibrium and steady-state procedures first illustrated by Cha[1]. With this method, demonstrated by Fromm[2] and Huang[3], a different rate expression will be obtained depending on which steps are chosen to be in rapid equilibrium and which steps are not. **See** *Enzyme Kinetic Derivations; Turnover Number*

[1]S. Cha (1988) *J. Biol. Chem.* **243**, 820.
[2]H. J. Fromm (1972) *Initial Rate Enzyme Kinetics*, Springer-Verlag, New York.
[3]C. Y. Huang (1979) *Meth. Enzymol.* **63**, 54.

CHAIN PROPAGATION

The step that regenerates a reactive species, thereby permitting a chain reaction to proceed.

CHAIN REACTION

A reaction having a mechanism utilizing reactive intermediates (chain carriers) that participate in a cycle of steps, allowing reactive species to be regenerated after each cycle (*i.e.*, chain propagation steps). The chemical kinetic expression first given independently by Christiansen[1], Herzfeld[2], and Polanyi[3], has the same general form of the empirical rate expression provided by Bodenstein

and Lind[4]. **See also** *Chain Transfer; Initiation; Termination*

[1]J. A. Christiansen (1919) *K. Dan. Vidensk. Selsk., Mat.-Fys. Medd.* **1**, 14.
[2]K. F. Herzfeld (1919) *Ann. Phys.* **59**, 635.
[3]M. Polanyi (1920) *Zeit. Elektrochem.* **26**, 49.
[4]M. Bodenstein & S. C. Lind (1907) *Zeit. Phys. Chem.* **57**, 168.

CHAIN RULE

A procedure in calculus for changing variables in a differential function, such that:

$$dy/dx = dy/da + da/dx$$

as long as y and a are differentiable functions of a and x, respectively.

CHALCONE ISOMERASE

This enzyme [EC 5.5.1.6], also known as chalcone:flavonone isomerase, catalyzes the interconversion of a chalcone to a flavoanone.

J. V. Schloss & M. S. Hixon (1998) *Comprehensive Biological Catalysis: A Mechanistic Reference* **2**, 43.
R. A. Dixon, P. M. Dey & C. J. Lamb (1983) *Adv. Enzymol.* **55**, 1.

CHALCONE SYNTHASE

This enzyme [EC 2.3.1.74] (also known as naringenin-chalcone synthase, flavonone synthase, and 6'-deoxychalcone synthase) catalyzes the reaction of three malonyl-CoA with 4-coumaroyl-CoA to produce four coenzyme A, three carbon dioxide, and naringeninchalcone. If both NADH and a particular reductase is also present, the final product is 6'-deoxychalcone.

R. A. Dixon, P. M. Dey & C. J. Lamb (1983) *Adv. Enzymol.* **55**, 1.

CHANGE IN MECHANISM

For any chemical or physical process, several different mechanisms may compete with each other, depending on the energy that is available to the system. A change in one or more reaction conditions can alter the degree to which competing pathways contribute to the kinetics of the overall process. The resulting shift from one pathway to another is referred to as a change in mechanism. An instructive, but as yet unproven, example might apply in the reaction catalyzed by hen egg white lysozyme. The enzyme is believed to operate via an oxocarbonium ion intermediate stabilized by neighboring carboxylate groups at the enzyme's active site. If another set of reaction conditions were to lead to the formation of an acylal intermediate (containing a covalent bond between one of the carboxyl groups and the C-1 position of the sugar), then we would say that a change in mechanism has occurred.

CHANNEL ACCESS

Control of enzymic activity arising from the modulated access of substrates to a channel leading to the active site. Such a scheme was suggested for aspartate carbamoyltransferase[1] which has its complement of active sites located on the interior surface of the complex comprised of catalytic and regulatory subunits. Nonetheless, isotope exchange studies of this enzyme suggest that this form of enzyme regulation does not apply in the case of aspartate transcarbamoylase[2,3].

[1]S. G. Warren, B. F. Edwards, D. R. Evans, D. C. Wiley & W. N. Lipscomb (1973) *Proc. Natl. Acad. Sci. U.S.A.* **70**, 1117.
[2]F. C. Wedler & F. J. Gasser (1974) *Arch. Biochem. Biophys.* **163**, 69.
[3]F. C. Wedler & W. H. Shalongo (1982) *Meth. Enzymol.* **87**, 647.

CHAOTROPIC AGENTS

Agents which, upon addition to an aqueous solution, increase the solubility of nonpolar entities. These ions (*e.g.*, Li^+, Mg^{2+}, Ca^{2+}, Ba^{2+}, I^-, ClO_4^-, SCN^-, and the guanidinium cation) are thought to promote some proteins to undergo denaturation by disrupting hydrogen-bonding interactions. Thus, in enzyme-catalyzed reactions, one should always be concerned about the chaotropic nature of various ionic species. The major intracellular monovalent cations (*e.g.*, potassium ion, arginine guanidinium ion, histidine imidazolium ion, and lysine ε-ammonium ion) are chaotropes. ***See*** *Hofmeister Series; Kosmotropes; Ion Interactions with Water*

CHAPERONES (Chaperonins)

Proteins that assist in the folding or translocation of other proteins[1]. These specialized proteins possess the ability (a) to inhibit incorrect folding and/or aggregation of newly biosynthesized proteins during and following translation; (b) to target or direct denatured or oxidatively damaged proteins to lysosomes where they undergo degradation; and (c) to facilate transmembrane translocation of targeted proteins.

Chaperonins frequently carry out their functions through ATP hydrolysis-dependent steps, and the kinetic properties of these reactions suggest that the equilibrium constants for the protein binding or releasing reactions are modulated by thermodynamic coupling to the free energy of ATP hydrolysis. The heat shock cognate or Hsc70 protein is among the most highly conserved proteins in nature. The amino-terminal domain of this protein contains an ATP binding/hydrolysis function, and the carboxyl-terminus is thought to bear the polypeptide binding site protein. Indeed, the observed rate of the dissociation of denatured proteins from Hsc70 exceeds the rate of ATP hydrolysis has been taken as evidence that ATP binding, and not ATP hydrolysis itself, somehow facilitates the dissociation reaction[2,3]. Ha and McKay[4] recently provided evidence for an induced conformational change that follows ATP binding.

How molecular chaperones facilitate protein folding by rapidly binding non-native proteins of very different sequence and function is a question of central significance. Perrett *et al.*[5] studied this process by determining the effect of ionic strength on the refolding of barnase (a bacterial RNase) on GroEL as well as on the thermal denaturation of the enzyme in the presence of chaperonins GroEL and SecB. Both chaperones bind barnase in its denatured state and lower the melting temperature (or, T_m) value. Refolding of barnase in the presence of GroEL is obviously multiphasic: (1) the slowest phase corresponds to the refolding of a singly bound molecule of barnase complexed to GroEL; and (2) the fastest phase involves association of barnase and GroEL. When Perrett *et al.*[1] used high GroEL-to-barnase ratios and an ionic strength less than 0.2 M, the fast phase matched the observed rate of binding. The association rate for barnase and GroEL was highly dependent on ionic strength; with a high ionic strength >0.6 M, most barnase molecules refolded free in solution, thereby escaping the binding interaction with GroEL. Their findings led them to suggest that ionic interaction between barnase and GroEL occurs initially and transiently; then hydrophobic binding appears to occur. Such a mechanism affords the opportunity for diffusion-controlled association, followed by slow dissociation of unfolded polypeptide.

While hydrogen exchange kinetics have been used to infer important features of protein folding pathways, Clarke *et al.*[6] recently advanced the argument that hydrogen exchange does not represent a short cut for studying protein-folding pathways, because the technique does not allow one to distinguish paths involving intermediates from those that are unrelated side reactions. Studies of barnase and chymotrypsin inhibitor-2 indicate no obvious relationship between hydrogen exchange at equilibrium and their folding pathways[7].

[1]M. J. Gething & J. Sambrook (1992) *Nature* **355**, 33.
[2]D. R. Palleros, K. L. Reid, L. Shi, W. J. Welch & A. L. Fink (1993) *Nature* **365**, 664.
[3]K. Prasad, J. Heuser, E. Eisenberg & L. Greene (1994) *J. Biol. Chem.* **269**, 6931.
[4]J. H. Ha & D. B. McKay (1995) *Biochemistry* **34**, 11635.
[5]S. Perrett, R. Zahn, G. Stenberg & A. R. Fersht (1997) *J. Mol. Biol.* **269**, 892.
[6]J. Clarke, L. S. Itzhaki & A. R. Fersht (1997) *Trends Biochem. Sci.* **22**, 284.
[7]A. R. Fersht (1999) *Structure and Mechanism in Protein Science: A Guide to Enzyme Catalysis and Protein Folding*, pp. 650, W. H. Freeman, New York.

CHARGE DENSITY

The quantity (symbolized by ρ) of electrical charge per unit volume, equal to Q/V, with SI units of coulombs per cubic meter. More technically, this definition refers to the volume charge density. ***See*** *Surface Charge Density*

CHARGED INTERFACES

When particles or large molecules make contact with water or an aqueous solution, the polarity of the solvent promotes the formation of an electrically charged interface. The accumulation of charge can result from at least three mechanisms: (a) ionization of acid and/or base groups on the particle's surface; (b) the adsorption of anions, cations, ampholytes, and/or protons; and (c) dissolution of ion-pairs that are discrete subunits of the crystalline particle, such as calcium·oxalate and calcium·phosphate complexes that are building blocks of kidney stone and bone crystal, respectively. The electric charging of the surface also influences how other solutes, ions, and water molecules are attracted to that surface. These interactions and the random thermal motion of ionic and polar solvent molecules establishes a diffuse part of what is termed the electric double layer, with the surface being the other part of this double layer.

The Gouy-Chapman model describes the properties of the diffuse region of the double-layer. This intuitive model assumes that counterions are point charges that obey a Boltzmann distribution, with highest concentration nearest the oppositely charged flat surface. The polar solvent is assumed to have the same dielectric constant within the diffuse region. The effective surface potential depends on the charge density of the surface and the ionic properties of the aqueous medium and its solutes. A more realistic case allows hydrated ions only to approach the surface at a finite distance, thereby defining what is called the Stern layer—so named to honor the model's originator. ***See*** *Biomineralization*

D. J. Shaw (1980) *Introduction to Colloid and Surface Chemistry*, 3rd ed., Butterworth & Co., London.

CHARGE RELAY SYSTEM

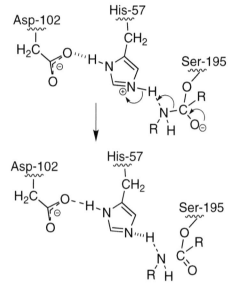

Charge Relay in Chymotrypsin Catalysis

A term used to describe the catalytic action of a triad of side-chain residues in the so-called serine proteases and serine esterases. In the case of chymotrypsin, the β-carboxylate of Asp-102 tugs on the proton bound to the nitrogen at position-1 of the imidazole group of His-57; this is thought to increase the basicity of the other ring nitrogen (at position 3) which deprotonates the hydroxyl of Ser-195 as it attacks the carbonyl of the peptide bond. (One difficulty with this mechanistic proposal is that it remains unclear how His-57 is a sufficiently strong base to abstract a proton to generate an incipient alkoxide from the serine hydroxyl group.)

T. A. Steitz & R. G. Shulman (1982) *Ann. Rev. Biophys. Bioeng.* **11**, 419.
D. R. Corey & C. S. Craik (1992) *J. Am. Chem. Soc.* **114**, 1784.

CHARGE-TRANSFER COMPLEX

Any aggregate (containing two or more molecules, molecular entities, or functional groups) in which an electric charge is transferred from a donor to an acceptor.

M. W. G. de Bolster (1997) *Pure Appl. Chem.* **69**, 1251.

CHARGE-TRANSFER TRANSITION

An electronic transition in which a significant electric charge is transferred from an electron donor to an electron acceptor. A charge-transfer complex (also given the nonstandard appellation ''electron donor-acceptor (EDA) complex'')[1,2] is a ground-state complex exhibiting a charge-transfer absorption spectrum (in general, a spectrum that is not the same as the sum of the spectra of the individual molecular entities)[3-6]. Such complexes are usually colored.

Examples of such charge-transfer complexes include the complex formed between a metal ion and a π orbital of a double bond or an aromatic system[7,8], the complex formed between polynitro aromatics (such as picric acid)[9] and other π-orbital-containing molecules, and complexes of I_2 and Br_2 with amines, ketones, aromatics, etc.[10] Phenols and quinones also form charge-transfer complexes[11].

The term ''charge-transfer'' is not intended to imply that full transfer of charge is necessary for the complexation to occur.

[1]IUPAC (1979) Pure and Appl. Chem. **51**, 1725.
[2]Comm. on Photochem (1988) Pure and Appl. Chem. **60**, 1055.
[3]R. Foster (1969) Organic Charge-Transfer Complexes, Academic Press, New York.
[4]R. Foster (1976) Chem. Berich. **12**, 18.
[5]O. K. Poleshchuk & Y. K. Maksyutin (1976) Russ. Chem. Rev. **45**, 1077.
[6]D. V. Banthorpe (1970) Chem. Rev. **70**, 295.
[7]S. D. Ittel & J. A. Ibers (1976) Advan. Organomet. Chem. **14**, 33.
[8]F. R. Hartley (1973) Chem. Rev. **73**, 163.
[9]V. P. Parini (1962) Russ. Chem. Rev. **31**, 408.
[10]O. Hassel & C. Romming (1962) Q. Rev. Chem. Soc. **16**, 1.
[11]R. Foster & M. I. Foreman (1974) in The Chemistry of the Quinonoid Compounds (ed., S. Patai), pt. 1, pp. 257-333, Wiley, New York.

CHELATION

The formation of two or more bonds between a single central atom (often a metal ion) and a chemical species, termed a multidentate ligand. The complex formed is referred to as a chelate. One naturally occurring example is the $MgATP^{2-}$ complex. The terms bidentate, tridentate, etc., refer to the number of sites on the chelating ligand that are bond to the central atom. EDTA and EGTA have four carboxylic acid groups to form a tetradentate ligand, and many crown ethers are polydentate in terms of their ion binding character. **See also** Sequestration; Cryptand

CHELEX

A weak cation chelating resin [matrix: styrene divinylbenzene; functional group: $R-CH_2N(CH_2COO^-)_2$] that is highly selective for divalent cations and can be used to prepare metal-depleted biochemical solutions and cell culture media. The resin has the following properties: (a) operating pH range of 4–14; (b) chemically stable from pH 0 to 14; (c) minimal wet capacity of 0.4 mEq/mL; (d) nominal density of 0.65 g/mL; (e) wet bead size ranges of 75–150, 150–300, and 300–1200 mm; and (f) approximate molecular weight exclusion of 3,500.

A typical batch procedure involves direct addition of one part Chelex to four parts buffer solution (weight/volume) with subsequent filtration to remove the resin. This treatment will readily remove iron, copper, zinc, calcium, and magnesium ions. Optimal chelation for iron and calcium ions occurs at pH 5–6.4; for copper, zinc, and magnesium ions, the optimal pH range is 7.4–8.0.

Proteins, especially those with higher isoelectric points, will bind to Chelex, and care must be exercised to minimize protein binding whenever Chelex is used to ''demetallate'' a solution containing protein. In such cases, one should operationally determine the least amount of Chelex and time of exposure needed to remove metal ions without significantly reducing the protein concentration. Chelex may even be used to study the effects of cations on cells in culture. Again, one must determine the most appropriate pH and resin/specimen to minimize loss of other essential nutrients and/or cofactors.

Selected entries from *Methods in Enzymology* [vol, page(s)]:
Affinity for calcium ion, **57**, 302, 307; arsenazo III purification, **56**, 318; assay of nucleoside phosphorylase, **51**, 539; for preparation of metal-free buffers, **54**, 476; removal of calcium ions, **57**, 301, 302; of paramagnetic contaminants, **54**, 154; for removing iron contamination, **233**, 85-86, 123; DNA preparation and, **29**, 376, 377, 380, 384, 404; methionyl tRNA and, **20**, 190; in deferration of laboratory media, **235**, 327; for iron contamination removal, **233**, 85-86, 123.

CHEMICAL BOND

Pauling[1] offered the following definition of a chemical bond: ''. . .there is a chemical bond between two atoms or groups of atoms. . . (if) forces acting between them are such as to lead to the formation of an aggregate with sufficient stability to make it convenient for the chemist to consider it as an independent molecular species.'' Chemical bonds include ionic bonds, coordinate covalent bonds, covalent bonds, and the fractional bonds of solid metals. Weak dipole-dipole interactions are not considered as forming a chemical bond.

[1]L. Pauling (1960) *The Nature of the Chemical Bond*, 3rd ed., p. 10, Cornell University Press, Ithaca.

CHEMICAL EXCHANGE EFFECT ON NMR SPECTRUM

The electromagnetic environment surrounding a nucleus within molecule A may change if A undergoes a reversible chemical reaction to form molecule B, especially if the chemical transformation involves electrons that are sufficiently near the nucleus of interest. The same can be said of a nucleus within a ligand molecule undergoing reversible binding to another molecule, such as a protein; the nucleus will then experience two different electromagnetic fields in free and bound forms of the ligand, say A and B. At first thought, one might surmise that the NMR spectrum will always exhibit two resonances: one for A and another for B at chemical shifts δ_A and δ_B. The fact is that the observed NMR spectrum (see below) will depend on the time-scale of the chemical exchange between A and B relative to

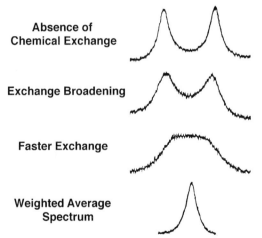

Effect of chemical exchange on NMR resonance spectra.

the NMR relaxation rate. If exchange is very slow, two sharp resonances, or "lines" in the parliance of NMR spectroscopy, will be observed. At the other extreme where A↔B exchange is very fast, the single sharp line observed in the spectrum will reflect a time-averaging of the electronic field effects of both A and B. At intermediate rates of exchange, one or two exchange-broadened lines will be observed, depending on the relative magnitude of rate constants for exchange and NMR relaxation.

CHEMICAL FLUX

The change in concentration of a molecular entity (being transformed) per unit time and usually symbolized by ϕ. This change in concentration (*i.e.*, dc_i/dt) occurs in one direction only and applies to the progress of a reaction step (or sequence of steps) in a complex scheme that may even involve a set of parallel reactions. The term chemical flux can also refer to the progress of a chemical reaction(s) in one direction while that system is at equilibrium (*i.e.*, via isotope exchange at equilibrium).

IUPAC (1979) *Pure and Appl. Chem.* **51**, 1725.

CHEMICAL KINETICS

GENERAL INTRODUCTION. When conducted with appropriate rigor and design, kinetic studies provide the opportunity to draw strong inferences about how reacting molecules combine with each other. One can also deduce how electrons and atoms within molecules rearrange geometrically and/or electronically as they approach the transition state. The validity of such inferences and deductions can also enable one to manipulate reaction conditions, thereby altering the outcome in terms of yields and types of products formed. Likewise, mechanistic information can be used to organize ideas about a large family of related reactions, and the inferential skills of all chemists ultimately stem from systematic studies of reaction kinetics.

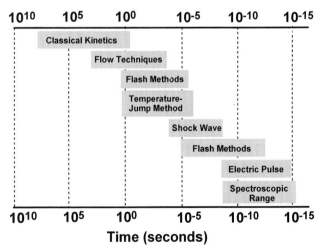

Figure 1. Techniques for determining the rates of chemical and physical processes occurring on the time-scales indicated as gray zones.

All kinetic studies begin with empirical rate equations accounting for the dependence of reaction velocity (v)

or reactant concentrations, with k as the temperature-dependent constant of proportionality:

$$v = k \cdot f$$

where f can be regarded as a combination of reactant concentration(s) and a partition function having the properties of a well-defined mathematical system. The goal of kinetic mechanism studies is to identify all reaction schemes that reasonably account for the mathematical form of the empirical rate equation. One then develops other experiments to test which of the rival mechanistic schemes is self-consistent, allowing for a direct correspondence of observed rate behavior over a defined range of experimental conditions with that predicted by the rate equation. The ingenuity of the kineticist rests with his/her ability to create and to execute rate experiments that test the reasonableness of rival mechanisms. One approach is to chose a set of reaction conditions (taking into account such factors as pressure, temperature, solvent polarity, ionic strength, pH, *etc.*) that introduces unique and distinguishable features into the derived rate equations (or rate laws) for each rival mechanism. This amounts to a perturbation method wherein seemingly identical kinetic behavior under one set of conditions is altered or perturbed experimentally to divulge divergent behavior that is diagnostic with respect to a choice of reaction schemes.

Kineticists often apply a variety of mechanistic probes to avoid being duped into reaching an inappropriate conclusion based on limited information. Even so, one cannot test all possible sets of reaction conditions, and the "proof by exclusion" method allows one only to discard mechanisms that fail to be consistent with a given set of experimental results. Kinetic studies are therefore said to lack the ability to prove any reaction mechanism; they only have the power to disprove a particular mechanistic scheme. In mathematics, one is content with rigorous proofs achieved by exclusion or by *reductio ad absurdum* arguments. In fact, *all* science, not just kinetics, is based on the tenet that conclusions always remain open to question and can be overturned by reliable evidence to the contrary. Kineticists create mechanisms which are "logic boxes"—strong enough to persuade themselves and others of the chemical reasonableness of their findings. This hallmark of good science is hypothesis-driven research, and mechanisms are hypotheses that unify seemingly diverse properties of nature.

SOLUTION BEHAVIOR. Virtually all biochemical reactions occur in water, and one should first consider some aspects of rate processes in aqueous solutions. From X-ray dispersion studies, one gets a sense that water displays considerable local order. (For further consideration, **See** *Water*). The hydrogen bonding capacity of water greatly influences the structure and reactivity of components present in an aqueous solution. In a condensed solution phase, ions and molecules move about less randomly than they would in the gas phase; this results from the fact that these ions and molecules are influenced by electronic and van der Waals interactions with other adjacent solvent and solute molecules. Translation diffusion is dependent both on the mean free path and electrostatic interactions, both of which are far more restricted in solutions. Any moving ion experiences drag exerted by neighboring ions and solvent. A variety of arguments suggest that while root-mean square velocities of molecules in solution and gas phases are somewhat comparable, collisions occur much more frequently in solution. In aqueous solutions, the local ordering of H_2O molecules creates a shell, or solvent cage, around solutes. Once two reactants do collide into each other, a solvent cage surrounds and holds them in an "encounter complex," a transient and vaguely understood configuration of reactants and solvent that favors chemical reaction between encountering reactants. So while some ions and certainly protons can react at high rates, aqueous solute reactivity is said to be encounter-controlled, meaning that reactants must first succeed in forming an encounter complex. Accordingly, one may correctly anticipate that a more viscous medium retards the frequency of such encounters. This can be illustrated as follows with the water molecules in the solvent cages indicated in parentheses, along with subscripts to emphasize that the number of solvent molecules need not be the same for each solvent cage:

$$(H_2O)_q + B(H_2O)_r \rightleftharpoons [(H_2O)_q A \cdot \cdot B(H_2O)_r]$$
collision

$$[(H_2O)_q A \cdot \cdot B(H_2O)_r] \rightleftharpoons [(H_2O)_s A \cdot \cdot B(H_2O)_t]$$
encounter

$$[(H_2O)_s A \cdot \cdot B(H_2O)_t] \rightleftharpoons [(AB(H_2O)_u]$$
complex formation

One's intuition also leads to the notion that the lifetime of an encounter complex must be of importance in promoting reactivity of ions and solutes.

Transition-state theory (*see below*) indicates that solvation can accelerate or retard the rates of a chemical reaction. In all cases, solvation lowers the potential energy of reactants. The single most important issue is the stability of the solvated ground state reactants in solution relative to that of the reactants in the transition state. Moreover, solvent interactions alter the equilibrium constants for many reactions, as might be inferred from the magnitude of $\Delta H_{\text{hydration}}$ for ions and even neutral solutes. Ionic solutes make substantial demands on the solvating properties of water, and this is especially true for ions of low ionic radius. Capable of acting both as an acid or a base, water is an especially good solvent for anions and cations. Furthermore, the extensive hydrogen network affects the ability of water to solvate molecules. During solvation, bonds between water molecules must break to create a cavity into which the solute may be placed. The ease with which such a cavity can form is greatly influenced by the surface tension of the solvent. Lower values for the latter permit cavity formation at lower energies.

Proton transfer in aqueous solutions can also be quite rapid (**See** *Water*). Protons also readily penetrate hydrogen-bonded complexes, and their transfer is sometimes availed by quantum mechanical tunneling. The ability of protons to move rapidly in hydrogen-bonded regions may also explain why proton binding and release are rarely rate-limiting in enzymic reactions. Contrary to intuition, electron transfer is not faster than proton transfer. Electrons do not exist as mobile charge carriers in water. Close proximity of electron donor and acceptor can overcome this limitation, and redox enzymes probably promote electron transfer by altering water structure or excluding water molecules altogether.

REACTION ORDER & MOLECULARITY. In the context of reaction rate equations, one can identify systems obeying the following mathematical forms:

Zero Order $v = k_0$

First Order $v = k_1[A]$

Second Order $v = k_2[A]^2$ or $v = k[A][B]$

Third Order $v = k_3[A]^3$ or $v = k[A]^2[B]$
       or $v = k[A][B][C]$

where the respective units of k_0, k_1, k_2, and k_3 are $M \cdot s^{-1}$, s^{-1}, $M^{-1} \cdot s^{-1}$, and $M^{-2} \cdot s^{-1}$. These expressions and the units of their rate constants all express reaction rate as a molarity change per unit time (*i.e.*, $-d[X]/dt$). Note also that, while we list a third-order reaction above, no such processes are known to occur for chemical reactions in the liquid phase.

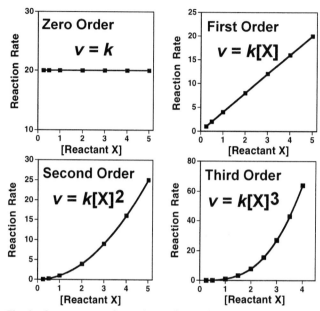

Fig. 2. Concentration dependency of zero-, first-, second-, and third-order chemical reactions.

The functional nature of some rate equations requires that rate constants occasionally can be of a mixed order, (*i.e.*, they can have nonintegral dimensions with respect to reactant molarity). Whether the reaction order is an integer or a fractional value, its value n is obtained from the slope of a plot of reaction rate *versus* $\log_{10}$ [Reactant].

Another term used to describe rate processes is "molecularity," which can be defined as an integer indicating the molecular stoichiometry of an elementary reaction, which is a one-step reaction. Collision theory treats molecularity in terms of the number of molecules (or atoms, if one or more of the reacting entities are single atoms) involved in a simple collisional process that ultimately leads to product formation. Transition-state theory considers molecularity as the number of molecules (or entities) that are used to form the activated complex. For reactions in solution, solvent molecules are counted in the molecularity, only if they enter into the overall process and not when they merely exert an environmental or solvent effect.

Again, the molecularity of a reaction is always an integer and only applies to elementary reactions. Such is not always the case for the order of a reaction. The distinction between molecularity and order can also be stated as follows: molecularity is the theoretical description of an elementary process; reaction order refers to the entire empirically derived rate expression (which is a set of elementary reactions) for the complete reaction. Usually a bimolecular reaction is second order; however, the converse need not always be true. Thus, unimolecular, bimolecular, and termolecular reactions refer to elementary reactions involving one, two, or three entities that combine to form an activated complex.

The rate constant (sometimes called the specific reaction rate) is commonly designated by k. The SI unit of time is the "second" (symbolized by s). Thus, unimolecular rate constants are typically expressed in s^{-1}, and unimolecular processes are by definition concentration-independent reactions. A slight difficulty arises regarding SI units and bi- and termolecular rate constants. Concentrations in the SI system would be mol per cubic meter, but in chemistry concentrations are expressed in $mol \cdot dm^{-3}$ (or more commonly $mol \cdot L^{-1}$ or simply M^{-1}). Thus, a bimolecular rate constant typically has units of $M^{-1}s^{-1}$, whereas a termolecular rate constant is expressed with units of $M^{-2}s^{-1}$.

FIRST-ORDER PROCESSES. In a first-order reaction, the rate is directly proportional to the concentration of reactant, as indicated by the following rate law:

$$v = -d[A]/dt = k[A]$$

Rearranging this expression allows one to separate variables to form:

$$-d[A]/[A] = kdt$$

where the concentration of A at time t is $[A]_t$ and that originally present is $[A]_o$. Integration of the above equation yields:

$$[A]_t = [A]_o \exp(-kt)$$

The half-life of a first-order process (i.e., the time $t_{1/2}$ required for $[A]_o$ to decay to $0.5[A]_o$) is given as:

$$\ln 2 = 0.693 = kt_{1/2} \quad \text{or} \quad \log 2 = (kt_{1/2})/(2.30)$$

These equations tell us that at successive time intervals equal to $n \times t_{1/2}$ where $n = 1, 2, 3$, etc.), the concentration

of A will fall to 0.50, 0.25, 0.125, etc., of the original amount of A (Fig. 3).

First-order kinetic behavior is observed throughout Nature. One familiar example is radioactive decay:

$$^{14}C \rightarrow {}^{14}N + \beta^-\ t_{1/2} = 5715 \text{ years}$$

$$^{3}H \rightarrow {}^{3}He + \beta^-\ t_{1/2} = 12.32 \text{ years}$$

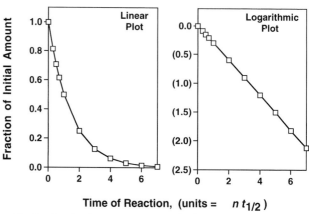

Fig. 3. Graphical representations of a first-order rate process.

First-order reaction kinetics is frequently observed in organic chemistry in the form of an S_N1 reaction, indicating it is a first-order nucleophilic substitution type. An example is the solvolysis of *tert*-butylbromide at alkaline pH to form *tert*-butanol and bromide ion. The reaction probably proceeds in two steps:

$$\overset{slowest}{A} \rightarrow \overset{fast}{B} \rightarrow C$$

The first-order nature of this reaction reminds us that rate-limiting hydroxide attack on *tert*-butylbromide is not a feature of this S_N1 reaction.

Another instance where first-order kinetics applies is the conversion of reversible enzyme-substrate complex (ES) to regenerate free enzyme (E) plus product (P) as part of the Michaelis-Menten scheme:

$$E + S = ES \rightarrow E + P$$

Other reactions may exhibit the sequential first-order conversion of one reactant, followed by a second first-order process:

$$\overset{k_1}{A} \rightarrow \overset{k_2}{B} \rightarrow C$$

These are called series first-order reactions, and they are commonly analyzed using numerical integration, as shown in Fig. 4.

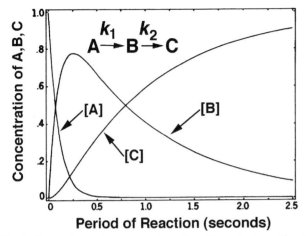

Fig. 4. Series first-order process A → B → C, beginning with rate constants $k_1 = 10$ s^{-1} and $k_2 = 1$ s^{-1}. [Reactant A (concentration = 1 mM), intermediate species B (concentration = 0 mM), and product C (concentration = 1 mM)].

BIMOLECULAR RATE PROCESSES. Many chemical processes involve the rate-limiting reaction of two components:

$$A + B \xrightarrow{k} C$$

where k has units equal to M^{-1} s^{-1}.

1. General Rate Equation. The differential equation for a bimolecular rate process can be written as:

$$v = \mathrm{d}x/\mathrm{d}t = k[A][B] = k(a - x)(b - x)$$

where a equals the initial reactant A concentration; b equals the initial reactant B concentration; x equals the amount of product at time t; $(a - x)$ equals the concentration of A at time t; and $(b - x)$ equals the concentration of B at time t. After separating variables, we get:

$$\mathrm{d}x/\{(a - x)(b - x)\} = k\mathrm{d}t$$

After integration by the method of partial fractions, the result is:

$$\frac{\ln(a - x) - \ln(b - x)}{a - x} = kt + \text{Constant}$$

where at $t = 0$ and $x = 0$, the constant of integration

equals $\ln\{(a/b)/(a - b)\}$. This yields the integrated second-order rate law ($a \neq b$):

$$\frac{1}{(a - b)} \ln \frac{b(a - x)}{a(b - x)} = kt$$

2. Pseudo-First-Order Behavior. In some cases, one reactant may be present in great excess over the other, and the last two equations can be simplified to their pseudo-first-order form. In Fig. 5, we illustrate the time course of a bimolecular reaction between equivalent initial concentrations of A and B.

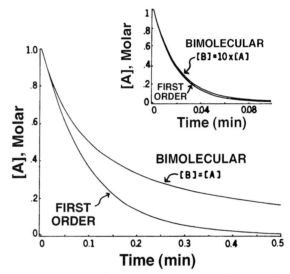

Fig. 5. Divergence in plots of the concentration of reactant A as a function of time for first-order and bimolecular rate processes (in the latter cases [A] = [B]). The inset shows what occurs in the pseudo-first-order condition (*i.e.*, a condition where a ten-times molar excess of reactant B allows the conversion of reactant A in a nearly perfect first-order process).

The observant reader will notice that the rate of conversion of A is initially the same as the first-order process shown. However, as A and B decrease, the bimolecular reaction rate slows appreciably, and the decrease in A concentration tails off more slowly than the first-order process. As illustrated in the inset, however, when $[B_o] = 10 \times [A_o]$, the second-order and first-order reactions are virtually indistinguishable. In fact, plots of $\ln[A_o]$ *versus* time look virtually linear (*i.e.*, pseudo-first-order) when $[B_o] \geq 3[A_o]$.

3. Effect of Electrostatic Interactions on the Bimolecular Rate Constant. The bimolecular rate equation presented above does not account for the effect of electrostatic interactions on reactivity of ionic molecules. Brønsted and Bjerrum, among others, recognized that the behavior of ionic reactants will depend on ionic strength and not simply on the concentrations of each reacting component. These electrostatic effects must be factored into kinetic and thermodynamic expressions for transition-state behavior. The usual scheme, written here as:

$$X_1 + X_2 \rightleftharpoons (X_1 \cdots X_2)^{\ddagger} \rightarrow \text{Products}$$

where reactants X_1 and X_2 first form the transition-state species $(X_1 \cdots X_2)^{\ddagger}$, must be amended for ionic reactants of charge z_1 and z_2.

$$X_1^{z_1} + X_2^{z_2} \rightleftharpoons (X_1^{z_1} \cdots X_2^{z_2})^{\ddagger} \rightarrow \text{Products}$$

In both of these schemes, the parentheses are used to indicate the chemical activities, and not simply the respective concentrations of each species. One can write an equilibrium relationship for the reactants and transition-state complex:

$$K^{\ddagger} = \frac{(X_1^{z_1} \cdots X_2^{z_2})}{(X_1^{z_1})(X_2^{z_2})} = \frac{[X_1^{z_1} \cdots X_2^{z_2}]^{\ddagger}}{[X_1^{z_1}][X_2^{z_2}]} \times \frac{\gamma_{12}^{\ddagger}}{\gamma_1 \gamma_2}$$

where the parentheses indicate activities, the brackets represent concentrations, and the Greek letter γ_i symbolizes the respective activity coefficients. As previously written, the bimolecular reaction rate is:

$$-\frac{dc_1}{dt} = k_{12} c_1 c_2$$

From transition-state theory, the bimolecular rate constant k_{12} can be written as:

$$k_{12} = \frac{\nu c^{\ddagger}}{c_1 c_2} = \frac{k_B T}{h} \times \frac{c^{\ddagger}}{c_1 c_2} = \frac{k_B T}{h} \times K^{\ddagger} \times \frac{\gamma^{\ddagger}}{\gamma_1 \gamma_2}$$

This equation contains the activity coefficients γ_1, γ_2, and $\gamma^{\ddagger}$. Recall from the Debye-Hückel treatment of ionic interactions in dilute solutions that the magnitude of these coefficients shows the following dependence on ionic strength μ for a solution of electrolytes:

$$\log \gamma_+ = -0.5091 \, z_+^2 (\mu)^{1/2}$$

$$\log \gamma_- = -0.5091 \, z_-^2 (\mu)^{1/2}$$

Therefore, because the bimolecular rate constant deals with two components,

$$k_{12} = \log \left(\frac{k_B T}{h} K^{\ddagger} \right) + 1.0182 \, z_1 z_2 (\mu)^{1/2}$$

[Recall that $\mu = (1/2) \sum_{i=1}^{n} c_i z_i^2$, where c_i is the concentration of the ith ion, z_i is the charge, and the summation extends over all n ions within the solution.] The equation for k_{12} predicts that for two ionic reactants of the same charge the reaction rate will increace with ionic strength. This behavior is an immediate consequence of the reduced electrical repulsion of ions of like charge when present in a more concentrated electrolyte solution. Such behavior is often called ionic screening. If two reactants are of opposite ionic charge, their mutually attractive interactions will be favored at low ionic strength, but disfavored at higher ionic strength. Thus, for oppositely charged reactants, the observed reaction rate will decrease as the ionic strength increases.

Experimenters would do well to avoid any unnecessary changes in the ionic composition of reaction samples within a series of experiments. If possible, chose a standard set of reaction conditions, because one cannot readily correct data from one set of experimental conditions in any reliable manner that reveals the reactivity under a different set of conditions. Maintenance of ionic strength and solvent composition is desirable, and correction to constant ionic strength often effectively minimizes or eliminates electrostatic effects. Even so, remember that Debye-Hückel theory only applies to reasonably dilute electrolyte solutions. Another important fact is that ion effects and solvent effects on the activity coefficients of polar transition states may be more significant than more modest effects on reactants.

EXPERIMENTAL CONSIDERATIONS. Accurate analytical measurement of reactant depletion or product accumulation is of central importance in kinetics. Usually, the experimenter develops an assay based on product accumulation, especially when no product or product-like analyte is initially present. In this situation, $\Delta P = [P_t] - [P_{\text{initial}}]$, where $[P_{\text{initial}}]$ is most often negligible. This contrasts with the greater uncertainty in measuring a drop in substrate concentration ($\Delta S = [S_t] - [S_{\text{initial}}]$); here, the difference of two larger, nearly equal values challenges the precision or reproducibility of rate measurements. The analytical method must not introduce complications arising from lack of sensitivity or any confounding effects such as slow response of the measuring instrument. For example, one should avoid spectrophotometric assays where the molar absorption coefficient is unavoidably small; this situation can lead to great imprecision. Likewise, the long response time of some pH and ion-selective electrodes may overlap with processes having a similar time scale for reaction. *See Absorption Spectroscopy, Fluorescence Spectroscopy*

1. Progress Curve Analysis. Determination of the rate constants for a first-order or bimolecular rate process can be achieved by graphical analysis. In the first case, one plots $\ln\{([A_o] - [A_t])/([A_o] - [A_\infty])\}$ *versus* time, where $[A_o]$, $[A_t]$, and $[A_\infty]$ are the respective reactant concentrations at time zero, time t, and infinity. [Note: Many reactions are not irreversible, and one cannot assume that $[A_\infty]$ will be zero; likewise, there may be a baseline correction in the analytical method, and this will frequently show up in the value of $[A_\infty]$. The slope equals $-k$. For determining rate constants of a bimolecular process, a logarithmic plot is employed to obtain a linear graphical transformation of the reaction progress curve. In those cases where the so-called "infinity value" cannot be determined, one may apply the Guggenheim method. In this protocol, values of the physical property (*e.g.*, pressure, absorbance, *etc.*) are measured at various times (*i.e.*, property values, λ_1, λ_2, λ_3, ..., are obtained at times t_1, t_2, t_3, ...). Then, values of the same physical property (λ_1', λ_2', λ_3', ...) are obtained at times $t_1 + \Delta$, $t_2 + \Delta$, $t_3 + \Delta$, ..., in which Δ is a constant time interval. Typically, for chemical reactions, Δ is a value between $t_{1/2}$ and $3t_{1/2}$. For λ_n and l_n', $\ln(\lambda_n' - \lambda_n) = -kt_n +$ constant where λ_n is the measurement of the physical property at t_n, λ_n' is the property measurement at $t_n + \Delta$, k is the first-order rate constant, and the constant is $\ln[(\lambda_\infty - \lambda_o)(1 - e^{-k\Delta})]$. Thus, a plot of $\ln(\lambda_n' - \lambda_n)$ *versus* time will yield a straight line having a slope of $-k$.

2. Initial Rate Assumption. The entire reaction progress curve, or at least a substantial portion of it, is typically required to accurately determine the rate constant for a first-order or second-order reaction. Nonetheless, one can frequently estimate the rate constant by measuring the velocity over a brief period (known as the initial rate phase) where only a small amount of reactant is consumed. This leads to a straight-line reaction progress curve (*see* Fig. 6) which is drawn as a tangent to the initial reaction velocity.

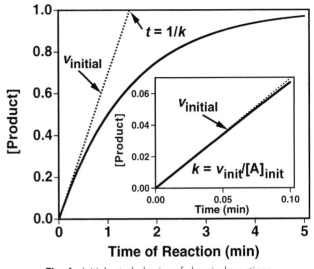

Fig. 6. Initial rate behavior of chemical reactions.

Then, k equals the observed reaction rate divided by the initial reactant concentration (*i.e.*, $k = v_{initial}/[A_{initial}]$). This method is most useful when one has an assay method that is sufficiently sensitive to ensure that only a small fraction, say 3–5%, of the reactant is depleted during the rate measurements. Typically, this is satisfactorily achieved with a UV-visible spectrophotometer, a fluorescence spectrometer, or a radioactively labeled reactant. The initial rate method is extremely convenient, and the preponderance of enzyme rate data has been obtained by initial rate measurements. Finally, one should note that the initial rate method can yield erroneous results if the initial reactant concentration is in doubt. This is not true for the plots of $\ln\{([A_o] - [A_t]/[A_o] - [A_\infty])\}$ *versus* reaction time because one is considering the fraction of reactant A remaining.

PRINCIPLE OF DETAILED BALANCE. For chemical reactions at thermodynamic equilibrium, the likelihood of any particular molecular configuration strictly depends on the energy of that configuration. This is a requirement of microscopic reversibility, a principle rooted in classical equilibrium thermodynamics and equally well grounded in statistical mechanics and quantum mechanics. One may demonstrate that for chemical reactions at equilibrium, or rigorously at steady-state, the number of molecules passing from configurations "a" to "b" must be equivalently balanced by the number of molecules converting from configurations "b" to "a." Another way of stating this principle is to say that should there exist a pathway from configuration "a" to configuration "b," a pathway must likewise exist for transition from "b" to "a." Each step must be balanced by forward and reverse reaction steps. This Principle of Detailed Balancing can be most useful in efforts to construct kinetic mechanisms for chemical reactions and exchange processes.

Consider the following reversible chemical reaction between reactant A and product B:

$$A \underset{k_{-1}}{\overset{k_1}{\rightleftharpoons}} B$$

for which the rate of reaction at equilibrium can be written as:

$$\frac{d[A]}{dt} = -k_1[A] + k_{-1} = 0$$

Suppose for the sake of argument that there is an alternative path in which A passes through an intermediate form X on its way to B. This could be schematically drawn as follows:

$$A \overset{k_{ax}}{\rightarrow} X \overset{k_{xb}}{\rightarrow} B \quad \text{and} \quad B \overset{k_{ba}}{\rightarrow} A$$

One may wonder whether this cyclic scheme can provide a pathway to maintain A and B in equilibrium through the intermediate X. In fact, such a mechanism would violate the Principle of Detailed Balancing, because at equilibrium the rate of any particular reaction step must be exactly equal to (*i.e.*, balanced by) the rate of the exact reverse of that reaction step. In the case presented here, we are considering collections of molecules on a large or macroscopic scale. Were we to consider the same process on the molecular scale, the corresponding rule is known as the Principle of Microscopic Reversibility, a condition that can be rigorously proven on the basis of quantum mechanical expressions for transition probabilities. This is intuitively reasonable on the basis of transition state theory: molecules following a given path between reactant and product will naturally pass over the barrier of lowest potential energy. It also follows that one cannot propose any reaction mechanism that defies the principles of detailed balancing and microscopic reversibility.

TRANSITION STATE THEORY. Arrhenius and van't Hoff independently proposed that the rate constant for a chemical reaction must depend on the absolute temperature T, thereby transforming the empirical relationship:

$$k = Ae^{-B/T}$$

(where A and B are suitably chosen constants) into a related form:

$$k = Ae^{-\Delta E/RT}$$

where R is the universal gas constant. This equation is widely known as the Arrhenius Law, and the pre-exponential term is known as the frequency factor (with the same units as k).

The constant A, in many respects, is analogous to the highest achievable magnitude of the rate constant k, and the $e^{-\Delta E/RT}$ term defines the fraction of molecules having sufficient energy to react. If $\Delta E \approx 0$, then $e^{-\Delta E/RT} \approx e^0$, which equals unity; thus, $k \approx A$. On the other hand, if ΔE is very large, such that $E \approx 20RT$, then $e^{-\Delta E/RT} \approx e^{-20}$, which is approximately 2×10^{-9}; therefore, k is extremely small in this example. This behavior obeys the Boltzmann principle which states that the fraction of collisions with energy exceeding a threshold value of ΔE is $e^{-\Delta E/RT}$. This fraction, also known as the Boltzmann factor, is greater: (a) as the absolute temperature T becomes greater or (b) as ΔE becomes smaller.

The Arrhenius equation can be rearranged to yield the following expression:

$$\ln k = \ln A - (\Delta E/RT)$$

or

$$\log_{10} k = \log_{10} A - (\Delta E/2.30259RT)$$

If a process obeys the Arrhenius law, then a plot of $[\log_{10} k]$ *versus* T^{-1} should yield a straight line with a slope of $-(\Delta E/2.30259R)$. This indicates that ΔE represents a barrier to reactivity, perhaps related to organizational events that reflect the conversion of the reactant to an activated species.

Eyring and Polanyi extended the Arrhenius concept concerning the nature of barriers to chemical reactivity. They considered the concept of a transition state which can be defined as the highest energy species or configuration in a chemical reaction. Transition-state theory now forms the basis of uncatalyzed and catalyzed chemical reactions, whether homogeneous, heterogeneous, or interfacial processes (see below). The simplest schematic representation is:

$$A + B \rightleftharpoons X^{\ddagger} \rightarrow \text{Product(s)}$$

where $X^{\ddagger}$ is the activated transition-state (TS) complex in the transition state. We deliberately treat the system as the simplest two-step process here (Fig. 7). [Another reasonable approach would be to involve an initial step forming AB before forming $X^{\ddagger}$.]

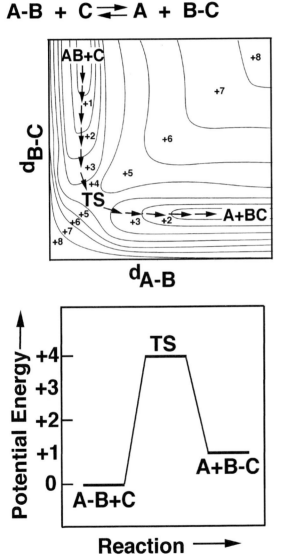

A-B + C $\rightleftharpoons$ A + B-C

Fig. 7. Energetics of a bimolecular rate process. *Top*: Representation of the potential energy surface along coordinate axes corresponding to the interatomic distance of B-to-C and A-to-B, where incremental displacements along the potential energy axis are shown as a series of isoenergetic lines (each marked by arbitrarily chosen numbers to indicate increased energy of the transition-state (TS) intermediate relative to the reactants). *Bottom*: Typical reaction coordinate diagram for a bimolecular group transfer reaction.

Written as an equilibrium dissociation constant, $K^{\ddagger}$ equals $[A][B]/[X^{\ddagger}]$; and if we rearrange this expression (by substituting $\Delta G^{\ddagger} = -nRT \ln K^{\ddagger}$ in the form: $\exp(-\Delta G^{\ddagger}/nRT) = K^{\ddagger}$), we get:

$$[X^{\ddagger}] = [A][B]e^{-\Delta G^{\ddagger}/nRT}$$

This expression can be combined with the equation, $v = k_2[A][B] = k^{\ddagger}[X^{\ddagger}]$ where $k^{\ddagger}$ is a measure of the frequency with which product is formed from $X^{\ddagger}$. In this respect,

$$k_2 = k^{\ddagger}e^{-\Delta G^{\ddagger}/nRT}$$

This simplified approach is analogous to the more rigorous absolute rate treatment. The important conclusion is that the bimolecular rate constant is related to the magnitude of the barrier that must be surmounted to reach the transition state. Note that there is no activation barrier (*i.e.*, that $\Delta G^{\ddagger} = 0$) in cases where no chemical bond is broken prior to chemical reaction. One example is the combination of free radicals. (In other cases where electrons and hydrogen ions can undergo quantum mechanical tunneling, the width of the reaction barrier becomes more important than the height.)

The Eyring activated-complex (or transition-state) treatment relates the observed rate constant k to $K^{\ddagger}$ multiplied by the frequency factor $k_B T/h$, where k_B is the Boltzmann constant, T is the absolute temperature, and h is Planck's constant:

$$k = K^{\ddagger}(k_B T/h)$$

At 298 K, $k_B T/h \approx 6 \times 10^{12}$ s^{-1}, and when multiplied by $K^{\ddagger}$ which is an association equilibrium constant (with units of molarity^{-1}), the units M^{-1}s^{-1} of k correspond to those of a bimolecular rate constant.

DIFFUSION-CONTROLLED PROCESSES. Even reactions having very small activation energies are not instantaneous; they still require their reactants to diffuse through the liquid before encountering each other. The rate of these so-called "diffusion-controlled reactions" depends solely on the time needed to form an encounter complex. The motions of molecules in solutions are greatly influenced by the potential fields exerted by their neighbors. Thus, the potential energy of individual molecules varies. There are also rapid collisions with molecules making up the solvent cage around a molecule as it proceeds to move by a series of discontinuous and haphazard displacements. Two molecules in solution that become neighbors may even tend to collide with each other a number of times (often referred to as an encounter) before they either react or separate. Particles experiencing random or Brownian motion can be described by applying Fick's First Law and the Einstein diffusion relation, yielding an estimate of the upper limit of the bimolecular rate constant k:

$$k = 4\pi d_{AB}(D_A + D_B)L$$

where d_{AB} is the hard-sphere collision diameter discussed in collision theory, D_A and D_B are the diffusion coefficients for A and B, and L is Avogadro's number. This result was first obtained by Smoluchowski by assuming there is no intermolecular potential energy between the reacting molecules. With nominal values of d_{AB} at 5×10^{-8} cm, and $(D_A + D_B)$ at 10^{-5} cm$^2\cdot$sec^{-1}, and by adjusting for appropriate units, the above equation shows that k is equal to about 7×10^9 M$^{-1}\cdot$sec^{-1}. The corresponding upper limit value for a bimolecular rate constant in the gas phase is about 10^{11} M$^{-1}\cdot$sec^{-1}. Thus in solutions, bimolecular rate constants cannot exceed 10^9–10^{10} M$^{-1}\cdot$sec^{-1} because the rate of diffusion then becomes a limiting factor.

Recall that the diffusion coefficient of a molecule will decrease with increasing viscosity of the solvent. Thus, the rate of encounter complex formation will decrease in a viscous medium. Since viscosity is itself temperature dependent, such encounters in solution will have their own activation energy.

The above expression represents the maximum rate for an encounter in a solution. However, most reactions have an energy barrier and/or strict steric requirements. Typically, many collisions occur before a reaction ensues. One must also consider the duration of an encounter and the frequency of vibrational collision during an encounter. The duration of an encounter, τ_{AB}, between two molecules A and B can be estimated to be

$$\tau_{AB} = \lambda^2/6D$$

In this expression, $D = D_A + D_B$, the sum of the diffusion coefficients for A and B, and λ^2 is the mean square displacement needed to terminate an encounter. The

value of λ can be set equal to the diameter of a solvent molecule or set equal to $2d_{AB}$. The probability of a reaction occurring during an encounter is

$$P_r = \frac{\nu p e^{-LE/RT}}{\nu p e^{-LE/RT} + 1/\tau_{AB}}$$

where ν is the frequency of collision between A and B during an encounter and p is the steric factor. Thus, the collision theory bimolecular rate constant in solution becomes

$$k = (2/3)\pi d_{AB}\lambda^2 \nu p e^{-LE/RT}L$$

If the magnitude of τ_{AB}^{-1} is small relative to the magnitude of $\nu p e^{-LE/RT}$, then the reaction is effectively diffusion-controlled. **See** *Arrhenius Equation; Transition-State Theory; Fick's Laws*

CATALYSIS. Any condition promoting X‡ formation will tend to speed up the reaction rate, and catalysts are thought to accomplish rate enhancement chiefly by stabilizing the transition state. Shown in Fig. 8 is an enzyme-catalyzed process in which reactant S (more commonly called substrate in enzymology) combines with enzyme to form an enzyme-substrate complex. This complex leads to formation of the transition state complex EX‡ which may proceed to form enzyme-product complex. The catalytic reaction cycle is then completed by the release of product P, whereupon the uncombined enzyme returns to its original state.

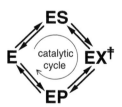

Fig. 8. Enzyme reaction cycle. In each catalytic round, the enzyme returns to the same chemical form.

In this case, there are three barriers to surmount: (1) the activated complex X_1^* which precedes ES formation; (2) the least stable configuration $X_2^\ddagger$, corresponding to the transition state; and (3) the activated complex X_3^* which precedes EP breakdown to form enzyme plus product (Fig. 9).

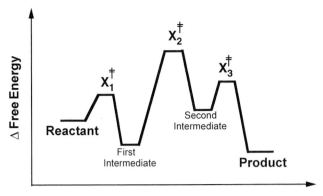

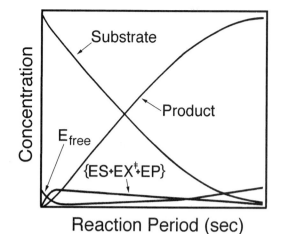

Fig. 9. Reaction coordinate diagram for a catalyzed reaction involving two intermediate species and three transition states. The highest lying species $X_2^{\ddagger}$ corresponds to the transition state (TS) of the reaction.

Fig. 10. Time-course of an enzyme-catalyzed reaction, illustrating time-dependent changes in all reactant concentrations.

Catalysts are very effective at stabilizing the transition state, and enzymatic reactions often proceed at rates that are 10^7–10^{11} times faster than their uncatalyzed counterparts. Because a catalyst returns to the same chemical form after each catalytic round, the catalyst cannot alter the equilibrium constant of a reaction. Thus, an enzyme must accelerate the $S \rightarrow P$ interconversion by the equivalent factor representing the acceleration of $P \rightarrow S$.

Although enzyme-catalyzed reactions are described in many other entries in this Handbook, some mention of the time-evolution of an enzymatic process should be considered here. Shown in Fig. 10 is an representation of a typical reaction progress curve. A rapid rise in the concentration of reactant-bound species {ES + E‡ + EP} occurs in the pre-steady-state phase. Then, the total concentration of ES + E‡ + EP remains relatively stable during the steady-state phase, the period over which product formation is relatively constant. As substrate becomes depleted, the driving force for maintaining the concentrations of ES and E‡ weakens, and although not shown here, [EP] becomes the dominant reactant-bound species as product accumulates.

In terms of enzyme catalysis, the following factors are likely to influence the magnitude of the rate enhancement in enzymatic processes: (a) proximity and orientation effects; (b) electrostatic complementarity of the enzyme's active site with respect to the reactant's stabilized transition state configuration; (c) enzyme-bound metal ions that serve as template, that alter pK's of catalytic groups, that facilitate nucleophilic attack, and that have

special redox properties; (d) fine tuning of rate constants that optimize the throughput of reactants.

We use the term homogeneous catalysis for those cases occurring in a single phase, gaseous or liquid. Most enzymes are homogeneous catalysts. The term heterogeneous catalysis refers to catalysis of multiphase chemical reactions, and this form usually occurs at the interface of the different phases. Lipolysis, for instance, is catalyzed by lipases acting at the interface between aqueous and lipid phases. Another example, this time from bioorganic chemistry, is the so-called micellar catalysis of phosphate monoester hydrolysis in two-phase systems composed of aqueous and tertiary cetylammonium micelles.

Chemical reactivity is also facilitated by what has been called acid-base catalysis. Because most nucleophilic, electrophilic, and redox reactions in water are strongly influenced by pH, the fundamental nature of acid-base catalysis becomes obvious. Indeed, systematic investigation of bioorganic and enzymic reactions requires an understanding of the salient features of acid-base catalysis. This field began in the early nineteenth century with Kirchoff's studies of the hydrolysis of starch into glucose, as well as Wilhemy's work on acid-catalyzed inversion of cane sugar. Research over the past fifty years has proven the overall significance of acid-base catalysis in biotic systems. ***See*** *Brønsted Theory; pH Kinetics*

RELAXATION KINETICS. The term "relaxation" describes the readjustment of a system from one state of dynamic equilibrium to a new state in response to the abrupt addition or removal of energy to or from the system. The relaxation time refers to the period that must elapse for an exponentially decreasing variable to fall from its original value $X_{initial}$ to $X_{initial}/e$ (or approximately $0.368 X_{initial}$). In 1866, J. C. Maxwell was the first to use the term "time of relaxation" to represent the period needed for the elastic force of a fluid to decay to X/e, where X is the initial value of the imposed force. Although Einstein was the first to treat the relaxation kinetics of a dissociating gas subjected to adiabatic compression and dilatation in a sound wave, many other chemical reactions in the solution phase also exhibit relaxation behavior.

1. BRIEF DESCRIPTION OF RELAXATION METHODS.

Chemical relaxation theory and the ingenious methods for applying relaxation theory to rapid chemical reactions were developed largely through the work of Nobelists Eigen, Porter, and Norrish. In the temperature-jump method, energy is added to a system by the electrical discharge of a capacitor which results in Ohmic heating that arises as a consequence of the frictional drag on ions that are suddenly induced to move toward their respective electrodes. This technique (see below) is widely utilized to determine the kinetics of elementary reactions that make up a multistage chemical reaction mechanism.

In the pressure-jump technique, a system reaches equilibrium at an initially pressurized condition; then, the sudden rupture of a foil diaphragm causes rapid depressurization. The system responds to a loss of energy (*i.e.*, for a gas, any increase in volume will increase the mean free path between reactants, thereby lowering its energy; likewise, if conversion of reactant to product has a non-zero volume change, the chemical system will be perturbed by a sudden change in pressure). Pressure effects can also be induced by shock tubes that create a sudden compression wave that passes through a reacting medium. In pulsed radiolysis and flash protolysis, the sudden absorption of high energy radiation rapidly perturbs a system that was previously in dynamic equilibrium. In concentration correlation analysis, the chaotic movement of molecules leads to an increase or decrease in the statistically average number of molecules in a volume element, and the system relaxes in response to the perturbation from thermodynamic equilibrium. Hammes has edited the definitive treatment on the investigation of elementary reaction steps and very fast reactions in solution.

2. BASIC KINETIC BEHAVIOR.

Relaxation experiments are usually carried out with systems that are fairly close to equilibrium, and as such they can be treated by linear differential equations. If a perturbation leads to a larger displacement from equilibrium, the mathematics becomes intractable, and this is certainly the case if multiple step systems are substantially displaced from equilibrium (Fig. 11).

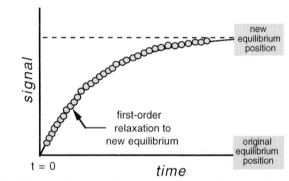

Chemical Relaxation

Fig. 11. Relaxation behavior of a chemical reaction where a sudden change from the original equilibrium position results in an increase in some signal that is proportional to the change in reactant concentration. Note that the process is first order.

Case 1. Unimolecular Isomerization: Consider a simple chemical reaction between species A and B at dynamic equilibrium:

$$A \underset{k_{-1}}{\overset{k_1}{\rightleftharpoons}} B$$

If we let the quantity a equal the total concentration of A and B (*i.e.*, $a = [A] + [B]$), and if we designate the concentration of B at any time to be x, then we can write the following differential equation:

$$dx/dt = k_1(a - x) - k_{-1}x$$

which at equilibrium (designated below by the subscript e) becomes:

$$dx/dt = 0 = k_1(a - x_e) - k_{-1}x_e$$

and by solving for $k_1 a$,

$$k_1 a = (k_1 + k_{-1})x_e$$

By analogy, we can express the displacement of x from its equilibrium concentration x_e as follows,

$$\Delta x = x - x_e$$

Thus, $x = \Delta x + x_e$, and the rate of change of Δx can be rewritten as

$$d(\Delta x)/dt = k_1 a - k_1 \Delta x - k_1 x_e - k_{-1} \Delta x - k_{-1} x_e$$

Now, substituting the previously expression $k_1 a = (k_1 + k_{-1})x_e$, we get:

$$d(\Delta x)/dt =$$

$$k_1 x_e + k_{-1} x_e - k_1 \Delta x - k_1 x_e - k_{-1} \Delta x - k_{-1} x_e$$

which upon canceling terms of opposite sign yields the final rate law:

$$d(\Delta x)/dt = -(k_1 + k_{-1})\Delta x$$

Integration of this differential equation yields the following equation:

$$\Delta x = \Delta x_o \exp(-t/\tau)$$

where τ equals $(k_1 + k_{-1})$.

Case 2. Unimolecular Forward/Bimolecular Reverse: Consider a simple chemical reaction between species A and B at dynamic equilibrium:

$$A \underset{k_{-1}}{\overset{k_1}{\rightleftharpoons}} B + C$$

Again, we let a equal the total concentration of A and B (i.e., $a = [A] + [B]$), and x is the concentration of B or C. We can write the following differential equation for the system at equilibrium:

$$dx/dt = 0 = k_1(a - x_e) - k_{-1}x^2$$

Using the expression $\Delta x = x - x_e$, we get

$$d(\Delta x)/dt = k_1(a - \Delta x - x_e) - k_{-1}(\Delta x + x_e)^2$$

Because $(\Delta x)^2$ is very small for a system near equilibrium, then

$$d(\Delta x)/dt = k_1 a - k_1 \Delta x - k_1 x_e - k_{-1} x_e^2$$

Now, by first substituting the previous expression $k_1 a = (k_1 x_e + k_{-1} x_e^2)$ and then canceling terms of opposite sign yields the final rate law:

$$d(\Delta x)/dt = -(k_1 + 2k_{-1}x_e)\Delta x$$

Integration yields the following equation:

$$\Delta x = \Delta x_o e^{-t/\tau}$$

where τ equals $(k_1 + 2k_{-1}x_e)$. Note that $(A_e + B_e)$ make separate additive contributions from the forward and reverse reaction. It is possible to arrange to study the system at several different values of x_e, such that a plot of τ^{-1} *versus* x_e has a slope of $2k_{-1}$ and an intercept of k_1.

3. IMPORTANT ONE-STEP REACTION SYSTEMS.

Without further derivation, we now summarize the relaxation kinetic parameters for the most frequently encountered one-step reactions.

Process	Relaxation Time
$A \underset{k_{-1}}{\overset{k_1}{\rightleftharpoons}} B$	$\tau^{-1} = k_1 + k_{-1}$
$A + B \underset{k_{-1}}{\overset{k_1}{\rightleftharpoons}} C$	$\tau^{-1} = k_1(A_e + B_e) + k_{-1}$
$2A \underset{k_{-1}}{\overset{k_1}{\rightleftharpoons}} A_2$	$\tau^{-1} = 4k_1 A_e + k_{-1}$
$A + B \underset{k_{-1}}{\overset{k_1}{\rightleftharpoons}} C + D$	$\tau^{-1} = k_1(A_e + B_e) + k_{-1}(C_e + D_e)$
$A + C \underset{k_{-1}}{\overset{k_1}{\rightleftharpoons}} C + B$	$\tau^{-1} = k_1 C_e + k_{-1} C_e$ [where C is a catalyst]
$A + B \underset{k_{-1}}{\overset{k_1}{\rightleftharpoons}} C$	$\tau^{-1} = k_1 B_e + k_{-1}$ [where B is buffered]

4. TEMPERATURE-JUMP METHOD.

This rapid kinetic technique is frequently used to study the behavior of transient processes. The equilibrium position of most biochemical reactions is temperature-dependent, and a chemical equilibrium can be perturbed by suddenly increasing the reaction temperature. Moreover, for those reactions exhibiting little or no temperature dependence, one can link the formation of one reactant or product to another temperature-dependent process. For example, a reaction that produces or consumes a proton can be carried out in the presence of a buffer that is strongly temperature dependent. As noted earlier, Joule heating is accomplished by discharging a high-voltage capacitor through an aqueous electrolyte solution confined in the reaction cuvette. This compartment is like other cuvettes used in absorption or fluorescence spectrophotometry, insofar as it contains optically transparent windows. Typically, the optical path for the cuvette is around 1 cm. However, as shown below, the temperature-jump cell must also contain the high-voltage electrode and the grounding electrode, suitably positioned to maximize the heating within a short period. Moreover, the electrodes must be free of any sharp edges that might permit unwanted sparking of the applied high voltage. For experiments with costly biochemical components, the reaction volume can be as little as 0.2 mL, and voltages of about 10 kV are required to obtain a 10°C temperature change in about 10 microseconds. If this results in a 5–10% displacement of a 1 mM reactant, then the corresponding 50–100 mM change in reactant concentration can easily be detected by absorption or fluorescence spectroscopy. If we consider an isomerization reaction with k_1 and k_{-1} each equal to 500 s^{-1}, then τ will be about 1 millisecond, a relaxation period that would be well separated from the 3-10 microsecond period needed for Joule heating.

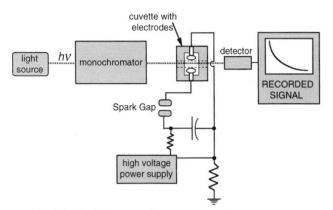

Fig. 12. Block diagram of a temperature-jump apparatus.

The schematic shown in Fig. 12 illustrates the general components that make up the temperature-jump apparatus. A spark-gap triggers the discharge of electrical energy stored within a capacitor, and discharge imparts a substantial electric field on the sample. As the principal charge carriers, small ions present in the reaction solution begin to migrate rapidly toward an oppositely charged pole; the acceleration of these ions must necessarily be balanced by frictional forces of the solvent, leading to the consequential release of energy as heat. Because the ions are present throughout the sample, this sudden heating occurs relatively uniformly throughout the sample. The speed with which the capacitor discharges tends to be the major factor limiting the time required for Joule heating. To assure rapid and uniform heating, large chemically inert, gold-plated electrodes are placed in the sample compartment and make direct and intimate contact with the reaction solution. The temperature change is experienced by the reaction system

as a change in rate constants, such that the system undergoes chemical relaxation (Fig. 13). Usually, absorbance or fluorescence signals are recorded on an oscilloscope or some other data acquisition device, and these can be analyzed to obtain values for the rate constants. Other optical techniques, such as polarimetry and light scattering, have been employed. The temperature-jump technique can also be combined with rapid mixing devices to increase the types of chemical processes that can be observed.

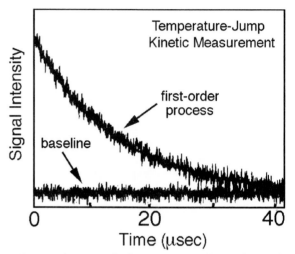

Fig. 13. Typical tracing of a kinetic process observed through the use of the temperature-jump method.

Temperature-jump devices provide satisfactory signal-to-noise characteristics (see Fig. 13). This is accomplished by using bright light sources such as xenon or mercury vapor sources as well as quartz halide lamps. The choice of photomultiplier depends on the method of detection, with fluorescence measurements requiring a higher quality PM tube than necessary for detecting changes in UV-visible absorbance.

J. W. Moore & R. G. Pearson (1981) *Kinetics and Mechanism*, Wiley, New York.

G. G. Hammes (1974) *Investigation of Rates and Mechanisms of Reactions*, Part II, vol. **6** in *Techniques of Chemistry*, Wiley-Interscience, New York.

G. G. Hammes & P. R. Schimmel (1970) *The Enzymes*, 3rd ed., **2**, 67.

C. A. Fierke & G. G. Hammes (1995) *Meth. Enzymol.* **249**,1.

CHEMICAL POTENTIAL

The thermodynamic parameter μ_i which J. Willard Gibbs defined as the change in the free energy (now termed the Gibbs free energy change or ΔG) with a change in the amount of substance in moles. For component B,

$$\mu_B = (\partial G / \partial n_B)_{T,P,nj \neq B}$$

If component B is an ionic species,

$$\mu_B = RT \ln a_B + \mu_B^{\ominus}$$

where R and T are the universal gas constant and absolute temperature, respectively, and a_B is the activity of component B (*i.e.*, $a_B = \gamma_B m_B$, with γ_B defined as the ion activity coefficient and m_B expressed as the molarity of B). Hence,

$$\mu_B = \mu_B(\text{ideal}) + \mu_B(\text{electric})$$

or

$$\mu_B = [RT \ln a_B + \mu_B^{\ominus}] + RT \ln \gamma_B$$

where $\ln \gamma_B = z_i^2 e_i^2 b / 8 \pi \varepsilon_o k_B T$ in the well known Debye-Hückel electrostatic model. **See** *Chemical Thermodynamics, Debye-Hückel Model, Gibbs Free Energy.*

CHEMICAL RESCUE

The capacity of an acidic/basic solute to replace a functional catalytic group, usually located within or near an enzyme's active site, that has been removed by random mutation or site-directed mutagenesis[1]. In their studies on aspartate aminotransferase, Toney and Kirsch[1] used site-directed mutagenesis to replace Lys-258, the endogenous general base, with an alanine residue. Catalytic activity was then restored to this catalytically inactive mutant by providing exogenous amines. Eleven amines were studied to generate a Brønsted correlation ($\beta = 0.4$) for the transamination of cysteine sulfinate. These investigators concluded that localized mutagenesis allows one to apply classical Brønsted analysis of transition-state structure to enzyme-catalyzed reactions. Another good example is the use of exogenous amines of a site-directed mutant of ribulose-1,5-bisphosphate carboxylase/oxygenase, also lacking a key lysyl residue[2]. Harpel *et al.*[2] demonstrated that ethylamine enhances the carboxylation rate of K329A by about 80–fold and strengthens complexation of 2-carboxyarabinitol 1,5-bisphosphate. Rescue of K329A follows an apparent Brønsted relationship with a β of 1, suggesting complete protonation of amine in the rescued transition state. Rate saturation with respect to amine concentration and the different steric preferences for amines between K329A

and K329C suggest that the amines bind to the enzyme in the position voided by the mutation.

[1]M. D.Toney & J. F. Kirsch (1989) *Science* **243**, 1485.
[2]M. R. Harpel & F. C. Hartman (1994) *Biochemistry* **33**, 5553.

CHEMICAL SHIFT

The displacement in the magnetic resonance frequency of a nucleus as a consequence of the electronic environment in which the nucleus resides. Because moving electrons generate their own magnetic fields, a nucleus surrounded by these electrons experiences an effective field, H_{eff}, which is defined by $(1 - \alpha)H_o$, where α is the so-called screening constant and H_o is the applied magnetic field. A chemical shift is typically reported as a dimensionless displacement (units = parts per million, or simply ppm) from a reference standard. If the magnetic field is varied while the radio frequency ν is held constant, then the chemical shift (ppm) equals $\{[H_{\text{sample}} - H_{\text{reference}}]/H_{\text{reference}}\} \times 10^6$. If the radio frequency is varied at constant magnetic field, the chemical shift (ppm) equals $\{[\nu_{\text{sample}} - \nu_{\text{reference}}]/n_{\text{reference}}\} \times 10^6$. Tetramethylsilane is frequently used as a proton chemical shift reference compound in organic chemistry, because it yields a single sharp NMR signal in a region well removed from the resonance lines of most other protons.

CHEMISORPTION

Any adsorptive process or interaction involving significant electron sharing or transfer. Bond strengths typically fall in the 1–5 eV range per chemisorbed molecule, corresponding to about 100 to 500 kJ per mole. As opposed to physisorption, chemisorption is more specific, and the strength of binding much stronger. When chemisorption involves electron interactions with transition metals, the stronger interactions probably arise from the localized and highly directional *d*-orbital interactions.

CHEMOSELECTIVITY

The preferential chemical reaction of a reagent with one of two or more different functional groups. For example, sodium borohydride exhibits greater chemoselectivity as a reducing agent than does lithium aluminum hydride, because the latter reacts with a wider spectrum of substances.

IUPAC (1979) *Pure and Appl. Chem.* **51**, 1725.

CHI (χ)

1. Symbol for the bond angle beween the C2 (that is, α) carbon and the adjacent "side-chain" carbon of an amino acid in a peptide or protein. 2. Symbol for surface electric potential. 3. A parameter associated with a distribution in statistics, more commonly referred to as chi-squared distributions (χ^2-distribution). Chi-square is a sum of terms in which each term is a quotient obtained by dividing the square of the difference between the observed and the theoretical value of a quantity by the theoretical value.

CHIRALITY

The condition of handedness or lack of superimposability of molecules differing only with respect to the stereochemical arrangement of identical substituents around a tetrahedral center. In 1893, Lord Kelvin succinctly defined chirality: "I call any geometrical figure, or any group of points, *chiral*, and say that it has *chirality*, if its image in a plane mirror, ideally realized, cannot be brought to coincide with itself."

Because proteins are made up of L-amino acids, they exhibit chirality in their structures, lacking planes or points of symmetry. Proteins also can exhibit chirality in their interactions with other chiral molecules as well as prochiral centers in other molecules. This latter point is beautifully illustrated by fumarase's catalysis of the dehydration of L-malate, a molecule containing two seemingly equivalent hydrogen atoms:

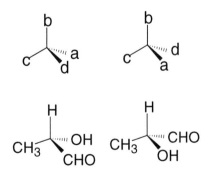

Nonsuperimposable Structures

The hydrogen in the *pro-R* position[1] is abstracted by the enzyme, and the *pro-S* hydrogen remains unaffected. In the reverse reaction, a single deuteron from D_2O is added during the hydration of fumarate, and that deuteron always occupies the *pro-R* position. The stereospecificity of fumarase is completely attributable to its ability to bind substrates in a manner that permits a base on the enzyme's surface to add or abstract only the *pro-R* hydrogen.

[1]Note that the *pro-R* and *pro-S* hydrogens in L-malate can be readily discerned by mentally replacing either hydrogen by a deuterium atom and then applying the *RS* or Cahn-Ingold-Prelog system for defining *R* and *S* stereoisomers.

CHIRALITY ANALYSIS OF ENZYME REACTIONS

A method used to study the stereochemical course of an enzyme-catalyzed reaction involving the use of functional moieties (of the normal substrate) that are not typically chiral. Consider, for example, a prochiral methylene group ($R—CH_2—R'$). Because the methylene group is prochiral, an enzyme is usually able to distinguish between the two hydrogen atoms[1], but this is not true for an achiral methyl group ($R—CH_3$), unless specifically isotopically substituted. Nevertheless, the stereochemical course of an enzyme acting on an achiral center can often be analyzed by first converting the group into a chiral entity (*e.g.*, for a methyl group into $R—CHDT$) and then treating with the enzyme, followed by chirality analysis of the substrate and/or product[2-7]. Synthesis of the stereospecifically pure enantiomer necessitates the use of a series of stereospecific reactions of which the steric course has been well established.

If a chiral group is generated in the system to be studied, then chirality analysis of that group typically entails the removal of that moiety in a defined set of reactions such that the stereochemistry is not altered or the reactions have a known stereochemical course. For example, chirality analysis of chiral methyl groups typically involves conversion of the methyl groups into acetic acid.

Various applications of chirality analysis can be found in refs. 5, 6, and 7. **See** *Stereospecificity; Resolution*

[1]A. G. Ogston (1948) *Nature* **162**, 963.
[2]J. W. Cornforth, J. W. Redmond, H. Eggerer, W. Buckel & C. Gutschow (1969) *Nature* **221**, 1212.
[3]J. Luthy, J. Retey & D. Arigoni (1969) *Nature* **221**, 1213.

[4]J. W. Cornforth, J. W. Redmond, H. Eggerer, W. Buckel & C. Gutschow (1970) *Eur. J. Biochem.* **14**, 1.

[5]J. W. Cornforth (1973) *Chem. Soc. Rev.* **2**, 1.

[6]H. G. Floss & M.-D. Tsai (1979) *Adv. Enzymol.* **50**, 243.

[7]H. G. Floss (1982) *Meth. Enzymol.* **87**, 126.

CHIRALITY PROBES

Stereochemical probes of the specificity of substrates, products, and effectors in enzyme-catalyzed reactions, receptor-ligand interactions, nucleic acid-ligand interactions, *etc.* Most chirality probe studies attempt to address the stereospecificity of the substrates or ligands or even allosteric effectors. However, upon use of specific kinetic probes, isotopic labeling of achiral centers, chromium- or cobalt-nucleotide complexes, *etc.*, other stereospecific characteristics can be identified, all of which will assist in the delineation of the kinetic mechanism as well as the active-site topology. A few examples of chirality probes include:

(1) Substrate/Product/Effector/Ligand Stereospecificity. Assessing the specificity for particular stereoisomers as substrates, products or effectors (surprisingly, effectors such as enzyme activators, rarely have their stereospecificities reported). In such studies, the investigators should clearly state the degree of chiral purity which is present and how it was determined.

(2) Anomeric Stereospecificity. For enzymes utilizing sugar or sugar derivatives as substrates or effectors, in addition to assessing issues of sugar specificity, it should be recalled that most sugars exist in two anomeric configurations. In many cases, the anomeric interconversion rate is slow and an investigator can assess the anomeric specificity for a given system[1]. For example, D-glucose oxidase is specific for the β isomer[2]. For systems with faster anomerization rates, both substrate analog and kinetic methods are available[1].

(3) Chiral Methyl or Methylene Groups. Upon following a specific set of reactions, utilizing isotopically labeled reagents, chiral methylene groups (*e.g.*, *R*-CHD—RO′) or chiral methyl groups (*e.g.*, R—CHDT) can be substituted or incorporated into substrates. With such labeled compounds, stereochemical issues not apparent with the unlabeled substrate can be readily addressed[3].

(4) Hydride Transfer in NAD⁺- and NADP⁺-Dependent Enzymes. The transfer of the hydride ion in redox reaction of NAD^+- and $NADP^+$-dependent enzymes can occur either to the *re*- or the *si*-face of the pyridine ring of the coenzyme[4,5]. Such stereochemistry is crucial in the characterization of these enzymes. The same enzymes from different sources can express different stereospecificities. For example, *E. coli* $NAD(P)^+$ transhydrogenase expressed one form of stereospecificity[6] whereas the *Pseudomonas aeruginosa* enzyme catalyzes the identical reaction with the other NAD^+ form[7].

(5) Chromium(III) and Cobalt(III) Nucleotide Complexes. Coordination isomers of Cr(III)- and Co(III)-nucleotides can be separated and used to test for specificity. Since Cr(III) is diamagnetic, EPR and NMR experiments can supplement the chirality probes[8,9].

(6) Phosphorothioate Nucleotides Analogues. The stereochemical course of of nucleotide-utilizing enzymes can be studies using a nucleotide analog in which a non-bridging oxygen at the α- or β-phosphorus of nucleoside 5′-triphosphate has been replaced by a sulfur atom. These phosphorothioate analogues have proven very useful in characterizing an enzyme-catalyzed reaction by observing whether inversion or retention has occurred[10].

(7) Chiral Phosphates Utilizing Oxygen Isotopes as Nucleotide Analogues. Chiral phosphoryl groups in nucleotides can be achieved using various oxygen isotopes[11,12]. Chiral [$^{16}O,^{17}O,^{18}O$]phosphoric monoesters have the additional advantage in that they are chemically equivalent to the naturally occurring substrate and can also be used in a number of NMR probes[13].

[1]S. J. Benkovic (1979) *Meth. Enzymol.* **63**, 370.

[2]D. Keilin & E. F. Hartree (1952) *Biochem. J.* **50**, 341.

[3]H. G. Floss (1982) *Meth. Enzymol.* **87**, 126.

[4]K. You, L. J. Arnold, Jr., W. S. Allison & N. O. Kaplan (1978) *Trends in Biochem. Sci.* **3**, 265.

[5]K. You (1982) *Meth. Enzymol.* **87**, 101.

[6]J. B. Hoek, J. Rydström & B. Hojeberg (1974) *Biochem. Biophys. Acta* **333**, 237.

[7]D. D. Louie & N. O. Kaplan (1970) *J. Biol. Chem.* **245**, 5691.

[8]W. W. Cleland (1982) *Meth. Enzymol.* **87**, 159.

[9]J. J. Villafranca (1982) *Meth. Enzymol.* **87**, 180.

[10]F. Eckstein, P. J. Romaniuk & B. A. Connolly (1982) *Meth. Enzymol.* **87**, 197.

[11]P. A. Frey, J. P. Richard, H. Ho, R. S. Brody, R. D. Sammons & K. Sheu (1982) *Meth. Enzymol.* **87**, 213.

[12]S. L. Buchwald, D. E. Hansen, A. Hassett & J. R. Knowles (1982) *Meth. Enzymol.* **87**, 279.

[13]M.-D. Tsai (1982) *Meth. Enzymol.* **87**, 235.

CHIRAL METHYL GROUPS

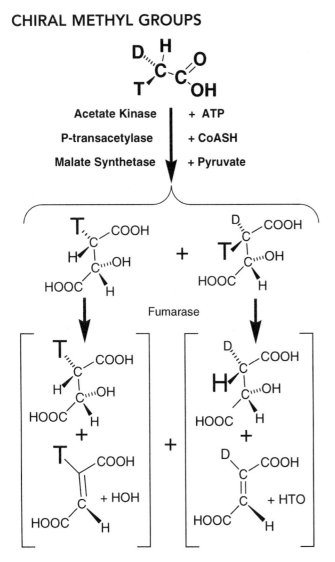

Stereochemically defined, isotopically prepared methyl groups used to determine the stereochemical course of methyltransferases[1,2]. One typically utilizes [(S)-$^3H,^2H,^1H$]CH$_3$X or [(R)-$^3H,^2H,^1H$]CH$_3$X where X is the remaining part of the substrate. This process cannot be observed when all three hydrogen atoms are ^1H, because they are equivalent and stereochemically indistinguishable.

[1]K. R. Hanson & I. A. Rose (1975) *Accts. Chem. Res.* **8**, 1.
[2]H. G. Floss (1982) *Meth. Enzymol.* **87**, 126.

CHI-SQUARE DISTRIBUTION

A statistical term referring to a monoparametric distribution used to obtain confidence intervals for the variance of a normally distributed random variable. The so-called chi-square (χ^2) test is a protocol for comparing the goodness of fit of observed and theoretical frequency distributions.

CHITINASE

This enzyme [EC 3.2.1.14] (also referred to as chitodextrinase, 1,4-β-poly-N-acetylglucosaminidase, and poly-β-glucosaminidase) catalyzes the hydrolysis of the 1,4-β-linkages of N-acetyl-D-glucosamine polymers of chitin. It should be noted that some chitinases will also display the activity observed with lysozyme [EC 3.2.1.17].

G. Davies, M. L. Sinnott & S. G. Withers (1998) *Comprehensive Biological Catalysis: A Mechanistic Reference* **1**, 119.
E. Cabib (1987) *Adv. Enzymol.* **59**, 59.
T. Imoto, L. N. Johnson, A. C. T. North, D. C. Phillips & J. A. Rupley (1972) *The Enzymes*, 3rd ed., **7**, 665.

CHITIN DEACETYLASE

This enzyme [EC 3.5.1.41] catalyzes the hydrolysis of chitin (or, more specifically, the N-acetamido groups of N-acetyl-D-glucosamine residues in chitin) to produce chitosan and acetate.

E. Cabib (1987) *Adv. Enzymol.* **59**, 59.

CHITIN SYNTHASE

This enzyme [EC 2.4.1.16], also known as chitin-UDP N-acetylglucosaminyltransferase, catalyzes the reaction of UDP-N-acetyl-D-glucosamine with [(1,4)-(N-acetyl-β-D-glucosaminyl)]$_n$ to produce UDP and [(1,4)-(N-acetyl-β-D-glucosaminyl)]$_{(n+1)}$. The enzyme ultimately converts UDP-N-acetyl-D-glucosamine into chitin and UDP.

E. Cabib (1987) *Adv. Enzymol.* **59**, 59.

CHLORAMPHENICOL ACETYLTRANSFERASE

This enzyme [EC 2.3.1.28] catalyzes the reaction of acetyl-CoA with chloramphenicol to produce coenzyme A and chloramphenicol 3-acetate.

W. V. Shaw & A. G. W. Leslie (1991) *Ann. Rev. Biophys. Biophys. Chem.* **20**, 363.
W. V. Shaw (1983) *Crit. Rev. Biochem.* **14**, 1.

CHLOROPEROXIDASE

This enzyme [EC 1.11.1.10], also called chloride peroxidase, catalyzes the reaction of hydrogen peroxide with two RH and two Cl$^-$ to produce two R—Cl and two water molecules. A heme group is one of the cofactors. This enzyme can also catalyze bromination and iodination, but not fluorination.

H. B. Dunford (1998) *Comprehensive Biological Catalysis: A Mechanistic Reference* **3**, 195.
A. Butler (1998) *Comprehensive Biological Catalysis: A Mechanistic Reference* **3**, 427.
M. A. Ator & P. R. Ortez de Montellano (1990) *The Enzymes*, 3rd ed., **19**, 213.

CHLOROPHYLLASE

This enzyme [EC 3.1.1.14] catalyzes the hydrolysis of chlorophyll to produce phytol and chlorophyllide. The enzyme has also been reported to catalyze chlorophyllide transfer reactions (for example, in converting chlorophyll to methylchlorophyllide).

S. Granick & S. I. Beale (1978) *Adv. Enzymol.* **46**, 33.

CHOLERA TOXIN

An ADP-ribosyltransferase that permanently activates the G_S regulatory protein in the adenylate cyclase pathway. Because ADP-ribosylated $G_S \cdot GTP$ complex cannot be converted to its metabolic inactive form, the adenylate cyclase remains in its activated state. The following are recent reviews on the molecular and physical properties of this ADP-ribosylating enzyme.

J. Moss, R. S. Hawn, S.-C. Tsai, C. F. Welsh, F.-J. S. Lee, S. R. Price & M. Vaughan (1994) *Meth. Enzymol.* **237**, 44.
J. Moss, S.-C. Tsai & M. Vaughan (1994) *Meth. Enzymol.* **235**, 640.
N. J. Oppenheimer & A. L. Handlon (1992) *The Enzymes*, 3rd ed., **20**, 453.
G. S. Kopf & M. J. Woolkalis (1991) *Meth. Enzymol.* **195**, 257.
J. Moss & M. Vaughan (1988) *Adv. Enzymol.* **61**, 303.

CHOLESTERYL ESTER HYDROLASE

This enzyme [EC 3.1.1.13] (also known as cholesterol esterase, sterol esterase, cholesterol ester synthase, and triterpenol esterase) catalyzes the hydrolysis of a steryl ester to produce a sterol and a fatty acid anion. This class represents a group of enzymes exhibiting broad specificity. They act on esters of sterols and long-chain fatty acids, and may also bring about the esterification of sterols. These enzymes are typically activated by bile salts. *See also* Esterases

D. P. Hajjar (1994) *Adv. Enzymol.* **69**, 45.
M. A. Wells & N. A. DiRenzo (1983) *The Enzymes*, 3rd ed., **16**, 113.

CHOLINE *O*-ACETYLTRANSFERASE

This enzyme [EC 2.3.1.6], also known as choline acetylase, catalyzes the reaction of acetyl-CoA with choline to produce coenzyme A and *O*-acetylcholine. The enzyme can also utilize propionyl-CoA as a substrate, albeit as a weaker reactant.

S. H. Zeisel (1981) *Ann. Rev. Nutr.* **1**, 95.
H. G. Mautner (1976) *Crit. Rev. Biochem.* **4**, 341.
D. S. Sigman & G. Mooser (1975) *Ann. Rev. Biochem.* **44**, 889.
L. T. Potter (1971) *Meth. Enzymol.* **17(B)**, 778.
A. K. Prince (1971) *Meth. Enzymol.* **17(B)**, 788.
L. T. Potter & V. A. S. Glover (1971) *Meth. Enzymol.* **17(B)**, 798.

CHOLINE DEHYDROGENASE

This enzyme [EC 1.1.99.1] is PQQ-dependent and catalyzes the reaction of choline with the acceptor to produce betaine aldehyde and the reduced acceptor.

S. H. Zeisel (1981) *Ann. Rev. Nutr.* **1**, 95.
Y. Hatefi & D. L. Stiggall (1976) *The Enzymes*, 3rd ed., **13**, 175.

CHOLINE KINASE

This enzyme [EC 2.7.1.32] catalyzes the phosphoryl transfer from ATP with choline to produce ADP and *O*-phosphocholine. Ethanolamine and its methyl and ethyl derivatives can also serve as substrates.

C. Kent (1995) *Ann. Rev. Biochem.* **64**, 315.
S. H. Zeisel & J. K. Blusztajn (1994) *Ann. Rev. Nutr.* **14**, 269.
K. Ishidate & Y. Nakazawa (1992) *Meth. Enzymol.* **209**, 121.
J. D. Esko & C. R. H. Raetz (1983) *The Enzymes*, 3rd ed., **16**, 207.

CHOLINE OXIDASE

This enzyme [EC 1.1.3.17] catalyzes the reaction of choline with dioxygen to produce betaine aldehyde and hydrogen peroxide. FAD is a cofactor. The enzyme has also been reported to be able to oxidize betaine aldehyde to betaine.

J. H. Quastel (1955) *Meth. Enzymol.* **1**, 674.

CHOLINE-PHOSPHATE CYTIDYLYLTRANSFERASE

This enzyme [EC 2.7.7.15], also referred to as phosphorylcholine transferase, catalyzes the reaction of CTP with choline phosphate to produce pyrophosphate (or, diphosphate) and CDP-choline.

C. Kent (1995) *Ann. Rev. Biochem.* **64**, 315.
S. H. Zeisel & J. K. Blusztajn (1994) *Ann. Rev. Nutr.* **14**, 269.
Y. Kishimoto (1983) *The Enzymes*, 3rd ed., **16**, 357.

CHOLINEPHOSPHOTRANSFERASE

This enzyme [EC 2.7.8.2] (also known as CDP-choline:1,2-diacylglycerol cholinephosphotransferase, diacylglycerol cholinephosphotransferase, phosphorylcholine:glyceride transferase, alkylacylglycerol cholinephosphotransferase, and 1-alkyl-2-acetylglycerol cholinephosphotransferase) catalyzes the reaction of CDP-choline with 1,2-diacylglycerol to produce CMP and a phosphatidylcholine. 1-Alkyl-2-acylglycerol derivatives can also serve as substrates.

S. H. Zeisel & J. K. Blusztajn (1994) *Ann. Rev. Nutr.* **14**, 269.
R. B. Cornell (1992) *Meth. Enzymol.* **209**, 267.
R. H. Hjelmstad & R. M. Bell (1992) *Meth. Enzymol.* **209**, 272.

CHOLINE SULFATASE

This enzyme [EC 3.1.6.6] catalyzes the hydrolysis of choline sulfate to produce choline and sulfate.

A. B. Roy (1971) *The Enzymes*, 3rd ed., **5**, 1.

CHOLINE SULFOTRANSFERASE

This enzyme [EC 2.8.2.6] catalyzes the reaction of 3'-phosphoadenylylsulfate with choline to produce adenosine 3',5'-bisphosphate and choline sulfate.

J. D. Gregory (1962) *Meth. Enzymol.* **5**, 977.

CHOLOYL-CoA SYNTHETASE

This enzyme [EC 6.2.1.7], also known as cholate:CoA ligase, catalyzes the reaction of ATP with cholate and coenzyme A to produce AMP, pyrophosphate (or, diphosphate) and choloyl-CoA.

W. H. Elliott (1962) *Meth. Enzymol.* **5**, 473.

CHONDROITIN 4-SULFOTRANSFERASE

This enzyme [EC 2.8.2.5] catalyzes the reaction of chondroitin with 3'-phosphoadenylylsulfate to produce adenosine 3',5'-bisphosphate and chondroitin 4'-sulfate. Specifically, the sulfation is at the 4-position of *N*-acetylgalactosamine residues of chondroitin. This enzyme is distinct from that of chondroitin 6-sulfotransferase [EC 2.8.2.17].

S. Suzuki (1966) *Meth. Enzymol.* **8**, 496.

CHONDROSULFATASES

Chondro-4-sulfatase [EC 3.1.6.9] catalyzes the hydrolysis of 4-deoxy-β-D-gluc-4-enuronosyl-(1,4)-*N*-acetyl-D-galactosamine 4-sulfate to yield 4-deoxy-β-D-gluc-4-enuronosyl-(1,4)-*N*-acetyl-D-galactosamine and sulfate. The saturated analog will also act as a substrate but not higher oligosaccharides. Nor will 6-sulfate derivatives be hydrolyzed by this enzyme. Chondro-6-sulfatase [EC 3.1.6.10] catalyzes the hydrolysis of 4-deoxy-β-D-gluc-4-enuronosyl-(1,4)-*N*-acetyl-D-galactosamine 6-sulfate to yield 4-deoxy-β-D-gluc-4-enuronosyl-(1,4)-*N*-acetyl-D-galactosamine and sulfate. The saturated analog will also act as a substrate but not higher oligosaccharides. Nor will 4-sulfate derivatives be hydrolyzed by this enzyme.

A. B. Roy (1971) *The Enzymes*, 3rd ed., **5**, 1.

CHORISMATE MUTASE

This enzyme [EC 5.4.99.5] catalyzes the interconversion of chorismate and prephenate.

J. V. Schloss & M. S. Hixon (1998) *Comprehensive Biological Catalysis: A Mechanistic Reference* **2**, 43.
R. Bentley (1990) *Crit. Rev. Biochem. Mol. Biol.* **25**, 307.
B. E. Davidson (1987) *Meth. Enzymol.* **142**, 432-439.

CHORISMATE SYNTHASE

This enzyme [EC 4.6.1.4], also known as 5-enolpyruvylshikimate-3-phosphate phospholyase, catalyzes the conversion of 5-*O*-(1-carboxyvinyl)-3-phosphoshikimate to chorismate and orthophosphate. It should be noted that shikimate is numbered so that the double bond is between C1 and C2. However, some of the early reports on this enzyme numbered shikimate in the reverse direction.

B. A. Palfey & V. Massey (1998) *Comprehensive Biological Catalysis: A Mechanistic Reference* **3**, 83.
R. Bentley (1990) *Crit. Rev. Biochem. Mol. Biol.* **25**, 307.
F. H. Gaertner (1987) *Meth. Enzymol.* **142**, 362.

CHROMIUM-51 (^{51}Cr)

A γ-emitting (0.320 MeV) radioactive chromium nuclide having a decay half-life is 27.70 days. The following decay data indicate the time (days), followed by the fraction of original amount at the specified time: 0.0, 1.000; 1, 0.975; 2, 0.951; 3, 0.928; 4, 0.905; 5, 0.882; 6, 0.861; 7, 0.839; 8, 0.819; 9, 0.798; 10, 0.779; 20, 0.606; 30, 0.472; 40, 0.368; 50, 0.286; 60, 0.223; 70, 0.174; 80, 0.135; 90, 0.105; 100, 0.082; 110, 0.064; 120, 0.050.

Special precautions: Lung is the critical organ in terms of inhaled dose. The effective *biological* half-life in humans is around 2 years. Store behind lead shielding. Follow institutional regulations regarding (a) type and location of dosimeters; and (b) whole body counting if significant uptake is suspected.

CHROMIUM-NUCLEOTIDE COMPLEXES

Exchange-inert complexes of Cr(III) with nucleotide ligands are very stable toward hydrolysis. Such complexes have proven to be extremely useful as chirality probes in that different coordination isomers can be prepared and separated[1]. These nucleotide complexes have also proved useful as dead-end inhibitors of enzyme-catalyzed reactions. Because Cr(III) is paramagnetic, distances can be measured by measuring the effects of Cr(III) on the NMR signals of nearby atoms when the Cr(III)-nucleotide complex binds to the surface of a macromolecule[2]. **See** *Exchange-Inert Complexes*

[1]W. W. Cleland (1982) *Meth. Enzymol.* **87**, 159.
[2]J. J. Villafranca (1982) *Meth. Enzymol.* **87**, 180.

CHROMOGENIC SUBSTRATE

A synthetically derivatized substrate designed to undergo a change in absorption and/or fluorescence spectrum upon its enzymatic conversion to product. Chromogenic substrates provide valuable assays for enzymes that otherwise fail to produce a spectral change, especially phosphotransferases, amide bond synthases, isomerases, and hydrolases.

Probably the most familiar example is *p*-nitrophenylphosphate which upon treatment with alkaline phosphatase generates orthophosphate and *p*-nitrophenol. The latter is an intensely yellow substance with maximal absorbance at 410–412 nm. The continuous spectrophotometric rate assay of trypsin[1] uses the chromogenic substrate *N*-α-benzoyl-DL-arginine-*p*-nitroanilide and upon hydrolysis produces *p*-nitroaniline (λ_{max} 405 nm). β-*N*-Acetylglucosaminidase can be similarly assayed by using *p*-nitrophenyl-*N*-acetyl-β-D-glucosaminide[2,3]. In other cases, the continuous spectral assay can be achieved by employing a chromogenic reagent that reacts with a reaction product. For instance, consider the chloramphenicol acetyltransferase reaction:

CoAS-Ac + Chloramphenicol
$$\rightarrow \text{Chloramphenicol3-acetate} + \text{CoSH}$$

CoASH + 5,5'-Dithiobis(2-nitrobenzoate)
$$\rightarrow \text{CoAS-(5-thio-2-nitrobenzoate)}$$
$$+ \text{5-Thionitrobenzoate}$$

The reaction velocity is measured by the formation of thionitrobenzoate ion, a species that absorbs strongly at 412 nm.

The time course of protease reactions can be monitored continuously by using fluorogenic peptide substrates. One type contains a fluorescent substituent (termed a donor) and a second aromatic group (called an acceptor), itself incapable of fluorescing. Through Förster resonance energy transfer (or FRET), the photic energy from the donor's excited state is transferred to the acceptor, resulting in partial or complete quenching of fluorescence. Upon cleavage by the protease, the donor and acceptor are released from each other, and because the efficiency of FRET falls off as the inverse sixth power of the donor-to-acceptor distance, bond scission greatly

decreases the ability of the acceptor to quench the donor's fluorescence. In such instances, one observes an increase in the sample's fluorescence that can be used to evaluate the kinetics of the enzymatic reaction[4,5].

A severe limitation of chromogenic substrates is the presence of one or more reporter groups whose bulk and electrostatic properties can greatly alter the nature of the enzyme's interaction with its substrate. This would be especially significant when the chromogenic substituent contributes an atom near the reaction center (*i.e.*, the actual site of cleavage). In the example given above for alkaline phosphatase, the presence of a nitrophenyl group may even change important aspects of the bond making/breaking mechanism. A chromogenic substrate can also exhibit subsite interactions that are not at all representative of a protease's interactions with its natural substrate. To avoid such circumstances, the experimenter should consider resorting to the use of two substances: the chromogenic substrate and the second, a nonchromogenic substrate whose interactions with the enzyme are of primary interest. The latter can be treated as an alternative substrate inhibitor of the chromogenic substrate reaction. In this way, one can infer how the nonchromogenic substrate interacts with the substrate. This works best if the enzyme is noncooperative; otherwise, one runs the risk of site-site interactions that reflect site occupancy by both substrates. In the same way, an enzyme's interactions with nonsubstrates can be treated by using it as a competitive inhibitor relative to the chromogenic substrate. While these approaches require additional kinetic characterization, the experimenter is richly rewarded by a much better picture of the enzyme's interaction with the substrate, and not the chromogenic substituent. ***See*** *Absorption Spectrsoscopy; Fluorescence Spectroscopy.*

[1] B. Kassell (1970) *Meth. Enzymol.* **19**, 844.
[2] O. A. Bessey, O. H. Lowry & M. J. Brock (1946) *J. Biol. Chem.* **164**, 321.
[3] S.-C. Li, S & Y.-T. Li (1970) *J. Biol. Chem.* **245**, 5153.
[4] R. M. Silverstein (1975) *Anal. Biochem.* **63**, 281.
[5] J. H. Hash (1975) *Meth. Enzymol.* **43**, 737.

CHROMOPHORE

That atom, group, or moiety within a molecular entity which is primarily responsible for the electronic transition resulting in the spectral band under consideration. ***See*** *Absorption Spectrsoscopy*

CHYMASE

This enzyme [EC 3.4.21.39], also referred to as mast cell protease I and skeletal muscle (SK) protease, is an endopeptidase that has been isolated from mast cell granules. It belongs to the peptidase family S1 and catalyzes the hydrolysis of peptide bonds, preferring Phe–Xaa > Tyr–Xaa > Trp–Xaa > Leu–Xaa.

J. S. Bond & P. E. Butler (1987) *Ann. Rev. Biochem.* **56**, 333.

CHYMOPAPAIN

This enzyme [EC 3.4.22.6], also known as papaya proteinase II, is a member of the peptidase family C1. It is the major endopeptidase of papaya (*Carica papaya*) latex. It has a specificity similar to that of papain. In addition, there are a number of chromatographic forms of the enzyme.

D. K. Kunimitsu & K. T. Yasunobu (1970) *Meth. Enzymol.* **19**, 244.
D. M. Greenberg (1955) *Meth. Enzymol.* **2**, 54.

CHYMOSIN

This enzyme [EC 3.4.23.4], also known as rennin, is a member of the peptidase family A1, exhibiting a broad specificity similar to that of pepsin A. It is a neonatal gastric enzyme with high milk clotting (note that chymosin clots milk by catalyzing the hydrolysis of a single peptide bond in κ-casein [Phe[105]–Met[106]]) and weak general proteolytic activity. The protein is formed from the precursor, prochymosin. The enzyme has also been found among mammals with a postnatal uptake of immunoglobins.

D. R. Davies (1990) *Ann. Rev. Biophys. Biophys. Chem.* **19**, 189.

CHYMOTRYPSIN

This enzyme [EC 3.4.21.1], also known as α-chymotrypsin, is an endopeptidase belonging to the peptidase family S1. It catalyzes the hydrolysis of peptide bonds with the preference for Tyr–Xaa, Trp–Xaa, Phe–Xaa, and Leu–Xaa.

C. W. Wharton (1998) *Comprehensive Biological Catalysis: A Mechanistic Reference* **1**, 345.
M. A. Ator & P. R. Ortez de Montellano (1990) *The Enzymes*, 3rd ed., **19**, 213.
E. T. Kaiser, D. S. Scott & S. E. Rokita (1985) *Ann. Rev. Biochem.* **54**, 565.
R. J. Coll, P. D. Compton & A. L. Fink (1982) *Meth. Enzymol.* **87**, 66.
J. S. Fruton (1982) *Adv. Enzymol.* **53**, 239.
D. M. Blow (1971) *The Enzymes*, 3rd ed., **3**, 185.
G. P. Hess (1971) *The Enzymes*, 3rd ed., **3**, 213.

Selected entries from *Methods in Enzymology* [vol, page(s)]:
Active site titration: [by 2-hydroxy-5-nitro-α-toluenesulfonic acid sultone, **19**, 6-14; by *p*-nitrophenyl ester substrates, **19**, 14-20; by rapidly reversible, covalently bound substrates, **19**, 14-20; by slowly reversible, covalently bound inhibitors, **19**, 6-14]; assay, **11**, 235; α-bromoacetylaminoacyl derivatives, **11**, 679; intramolecular transformation of, **11**, 679; buried carboxyl groups in, **11**, 25, 618; derivative formed with 2-bromoacetamido-4-nitrophenol, difference spectra of, 11, 869, 870; photolysis of, **11**, 679; diethylphosphonyl, reactivation with *N*-phenylnicotinohydroxamic acid, **11**, 701; dimerization, **26**, 197; **27**, 402, 438; infrared spectrum, **26**, 468; phenylmethane-sulfonyl, reactivation of, **11**, 709; reaction [with 2-bromoacetamido-4-nitrophenol, **11**, 868; with α-bromoacetanilide, **11**, 564; with α-bromoacetophenone, **11**, 564; with α-bromo-*N*-benzylacetamide, **11**, 564; with α-bromo-*N*-ethylacetamide, **11**, 564; with dansyl chloride, **11**, 858, 859; with diisopropyl fluorophosphate, **11**, 694-695; with phenylmethanesulfonyl fluoride, 11, 708; with TPCK, **11**, 678]; specificity of, **11**, 222; stock solution preparation, **19**, 11; conversion to anhydrochymotrypsin, **11**, 709, 711; abortive complex, **63**, 205; activation energy, **63**, 243-245; **64**, 226; adsorption, onto glass, **61**, 77; bovine [action on β chain of oxidized insulin, **80**, 649; amino acid composition, **80**, 650; hydrolysis of low molecular weight synthetic substrates, **80**, 731; inhibition, **80**, 801; N-terminal sequence, **80**, 631; substrate specificity, **80**, 603; burst, **64**, 199; chemical modification, **63**, 219; competitive labeling, **63**, 219; conformational change, **64**, 224, 225; cryoenzymology, **63**, 338; dielectric constant, effects, **63**, 212; hysteresis, **64**, 217, 220; infrared spectrum, **63**, 212; inhibition, 63, 180; **80**, 771; inhibition by eglin, **80**, 812; inhibition by a 1-proteinase inhibitor, **80**, 763; interaction parameters, preferential, **61**, 48; kinetic parameters, **63**, 237, 238, 244; kinetics, calorimetric determination, **61**, 264; low-frequency dynamics, **61**, 443-445; low-temperature dynamics, **61**, 325, 326; microcalorimetric studies of ligand binding, **61**, 303; molecular form in normal serum, **74**, 273-275; negative cooperativity, **64**, 220; nitrogen isotope effect, **64**, 95; nuclear magnetic resonance properties, **63**, 212; partial specific volumes, **61**, 48; p*K* values, perturbed, **63**, 210, 220; progress curve analysis, **63**, 180; calorimetry of protonation, **61**, 265; radioimmunoassay, **74**, 273; reaction with a 2-antiplasmin, **80**, 404, 406; similarity to cathepsin G, **80**, 564; similarity to thrombin, **80**, 294, 295; spectrokinetic probe, **61**, 323; structure-function, **80**, 733; zymogen conformational change, **64**, 225; active-site titration, **248**, 87-89, 100-101; carboxy-terminal sequence analysis, **240**, 705, 708, 710, 712; active site residues, **244**, 23-24; members, **244**, 22-37; family [biological role, **244**, 23; catalytic serine codon, **244**, 30-31; catalytic unit, **244**, 23, 25; cleavage site specificity, **244**, 114; domains, **244**, 25, 29-30; evolution, **244**, 30-32; gene, **244**, 29-30; members, **244**, 22-23, 26-27; processing, **244**, 23; signal peptide, **244**, 23, 25; species distribution, **244**, 26-28, 31]; fluorimetric assay, **248**, 20-21; phosphonate inhibitors, **244**, 434; preparation, **235**, 566; site-directed mutants and acid-base catalysis, **249**, 114; structure, **240**, 703, 705; substrates [fluorogenic, **248**, 19; thioester, **248**, 10; transition state and multisubstrate analogs, **249**, 306; zymography in nondissociating gels with copolymerized substrates, **235**, 588-589.

CHYMOTRYPSIN C

This enzyme [EC 3.4.21.2], a member of the peptidase family S1, is an endopeptidase that exhibits a preference for hydrolysis of peptide bonds at Leu—Xaa, Tyr—Xaa, Phe—Xaa, Met—Xaa, Trp—Xaa, Gln—Xaa, and Asn—Xaa. The enzyme reacts more readily with Tos—

Leu—CH_2Cl than Tos—Phe—CH_2Cl in contrast to chymotrypsin (note: chymotypsin A and chymotypsin B were later found to be homologous and now are listed with EC 3.4.21.1).

J. E. Folk (1970) *Meth. Enzymol.* **19**, 109.

CHYMOTRYPSIN-LIKE SERINE PROTEASE SUPERFAMILY

Perona and Craik[1] recently described the serine proteases possessing a chymotrypsin-like fold. These enzymes have an almost identical fold made up of two β-barrels, with the catalytic Ser195, His57, and Asp102 residues at the interface of the two domains. To bring about accurate positioning of the nucleophilic Ser195, each enzyme contains five enzyme-substrate hydrogen bonds that serve to juxtapose the scissile peptide bond adjacent to the catalytic Ser—His couple. Many specificity-controlling structural determinants reside on surface loops, suggesting that rapid and varied evolutionary divergence can take place, while still conserving the overall tertiary fold.

[1]J. J. Perona & C. S. Craik (1997) *J. Biol. Chem.* **272**, 29987.

trans-CINNAMATE 4-MONOOXYGENASE

This enzyme [EC 1.14.13.11], also known as cinnamate 4-hydroxylase, catalyzes the reaction of *trans*-cinnamate with NADPH and dioxygen to generate 4-hydroxycinnamate, $NADP^+$, and water. The enzyme, which uses a heme-thiolate as a cofactor, can also replace NADPH with NADH (however, the reaction will proceed slower).

I. Raskin (1992) *Ann. Rev. Plant Physiol. Plant Mol. Biol.* **43**, 439.
R. A. Dixon, P. M. Dey & C. J. Lamb (1983) *Adv. Enzymol.* **55**, 1.

CIRCADIAN RHYTHM

Any biological process characterized by or occurring repeatedly with a 24-hour periodicity. Beyond their complex spatial structures, living organisms display remarkable time structures that integate complex data sets comprised of daily and seasonal cues. These include the day-night cycle, length of day, the spectrum and intensity of solar illumination, lunar cycle and consequential changes in its luminosity, tidal rhythms, seasonal changes in the flow and content of rivers, *etc.* These changes are sensed as changes in photic and thermal energy which alter an organism's biochemical, physiological and behavioral characteristics. In *Drosophila*, the gene known as period (*PER*) is required for the circadian rhythms, and oscillations in levels of its mRNA and protein in

brain may strongly influence rhythmicity. A homologous mouse gene (*CLOCK*) encodes a novel transcription factor that is likely to control periodicity by changes in DNA binding, protein dimerization, and transcriptional activation, and CLOCK protein shares a motif with PERIOD, suggesting that the circadian clock mechanism is evolutionarily conserved.

On the basis of studies with *Drosophila* and *Neurospora* circadian clocks, such oscillatory behavior appears to involve cycling gene expression. Nishida[1] summarized recent progress using molecular biological approaches to elucidate the circadian clock mechanism. He also described the importance of transcription factors in providing clues about the nature of what may be a common mechanism of circadian clock in the divergent species.

Controlled chaos may also factor into the generation of rhythmic behavior in living systems. A recently proposed model[2] describes the central circadian oscillator as a chaotic attractor. Limit cycle mechanisms have been previously offered to explain circadian clocks and related phenomena, but they are limited to a single stable periodic behavior. In contrast, a chaotic attractor can generate rich dynamic behavior. Attractive features of such a model include versatility of period selection as well as use of control elements of the type already well known for metabolic circuitry.

[1]N. Ishida (1995) *Neurosci. Res.* **23**, 231.
[2]A. L. Lloyd & D. Lloyd (1993) *Biosystems* **29**, 77.

CIRCE EFFECT

Utilization of strong attractive forces to lure a substrate into an active site and to force the substrate to undergo "extraordinary transformation of form and structure.[1]" Distortion of the substrate occurs as a consequence of favorable electrostatic interactions within and immediately surrounding the enzyme's active site. The Latin name *Circe* harkens back to the Greek mythological sorceress *Kirke* who transformed Odysseus's fellow sailors into swine, but Odysseus later succeeded in forcing her to return them to their previously human forms.

[1]W. P. Jencks (1975) *Adv. Enzymol.* **43**, 219.

CIRCULAR DICHROISM SPECTROSCOPY

A spectroscopic technique that measures the differential absorption of left and right circularly polarized light as a measure of electronic interactions arising from asymmetric (or chiral) as well as dissymmetric molecular groups. This method also allows one to detect conformational changes in biopolymers.

Selected entries from *Methods in Enzymology* [vol, page(s)]: General principles, **76**, 262-265; of hemoglobins, **76**, 262-275; globin effects, **76**, 271; information derived, **76**, 267-275 [in far UV, **76**, 267-270; in near UV, **76**, 270-272; in visible region, **76**, 272-275]; instrumentation, **76**, 266-267; of single crystals, **76**, 206-207; absorbance measurement, **240**, 637; aromatic residues, **232**, 248; hemoglobin, **232**, 248-257; near-UV region, **232**, 248, 252-257; secondary structure analysis, **232**, 248; Soret region, **232**, 257-260; visible region, **232**, 248-249, 260-263; monitoring of protein unfolding, **240**, 615; protein concentration needed for measurement, **240**, 632, 635; structural analysis correlated with X-ray crystallography, **232**, 247-248; in far-UV region, **232**, 7-8; in near-UV region, **232**, 7; ribonuclease A unfolding, **240**, 623-628; source code for two-state protein folding model, **240**, 619, 644-645; standard deviation determination, **240**, 621; staphylococcal nuclease A unfolding analysis, **240**, 628, 631-632; stationary CD, **232**, 249-263; time-resolved CD [apparatus for millisecond time range, **232**, 263; for nanosecond and microsecond time scale, **232**, 263-264; experimental kinetic data, **232**, 264-266]; high-pressure experiments, **259**, 369-370; hybrid oligonucleotide strand characterization, **246**, 31-33; instrumentation [calibration, **246**, 23-24, 41; design, **246**, 40-42 ; manufacturers, **246**, 41; stopped-flow, **246**, 61; light characteristics, **246**, 35-36]; low-density lipoprotein oxidation analysis, **233**, 428-430; multiple scanning, **246**, 43; nucleic acids [chromophores, **246**, 62; duplex stoichiometry, **246**, 4, 6, 13-15, 35; ionic strength effects, **246**, 63; ligand binding, **246**, 67-68; protein interactions, **246**, 6, 68-71; secondary structure, **246**, 4, 6, 13-15, 35, 62-67; temperature effects, **246**, 63-64]; photoacoustic detection, **246**, 60; proteins [baseline, **259**, 500; chromophores, **246**, 43-44; concentration for measurement, **240**, 632, 635; folding, **246**, 61; ligand interactions, 246, 55-58; secondary structure analysis, **246**, 5, 13-15, 35; **259**, 499-500 (accuracy of determination, **246**, 513; a helix, **246**, 44-45; analysis methods, **246**, 49-52; β sheet, **246**, 45-46; β turns, **246**, 46-47; helix-coil transition, **246**, 47-49; membrane proteins, **246**, 58-60; random coil, **246**, 47; sources of error, **246**, 54; wavelengths for data collection, **246**, 53-54]; structural analysis [correlation with X-ray crystallography, **232**, 247-248; in far-UV region, **232**, 7-8; in near-UV region, **232**, 7; tertiary structure analysis, **246**, 54-55; unfolding studies, **240**, 615; **259**, 498-500, 506-509; vibrational studies, **246**, 515-516; wavelength monitoring, **240**, 635]; reaction kinetics, **246**, 14; RNA [codon-anticodon interactions, **261**, 285, 287; duplex formation monitoring, **261**, 284; duplex stoichiometry, **246**, 24-28; ion concentration effects, **261**, 283-284; temperature effects, **261**, 283-284; wavelength sensitivity (backbone alterations, **261**, 283; base stacking, **261**, 283); sample cells, **246**, 42; sample requirements [buffers, **246**, 42; chirality, **246**, 14, 34; concentration, **246**, 42]; semiquinone intermediate in amine oxidase, **258**, 87; sensitivity, **258**, 274; theory, **258**, 274 [electric dipole transition moment, **246**, 39-40; magnetic dipole transition moment, **246**, 39-40; rotational strength, **246**, 39-40, 93-94]; transition metals [coordination geometry, **246**, 5, 16; spin state, **246**, 5]; units of measurement [ellipticity, **246**, 24, 36-37; molar circular dichroism, **246**, 24, 36]; vibrational circular dichroism; [instrumentation, **246**, 37; protein conformation determination, **246**, 52-53].

CITRAMALATE LYASE

The enzyme [EC 4.1.3.22] catalyzes the conversion of (3*S*)-citramalate to acetate and pyruvate. It should be noted that the enzyme complex can be dissociated into subunits. Two of those components are citramalate CoA-transferase [EC 2.8.3.11] and citramalyl-CoA lyase [EC 4.1.3.25].

D. J. Creighton & N. S. R. K. Murthy (1990) *The Enzymes*, 3rd ed., **19**, 323.

CITRAMALYL-CoA LYASE

This enzyme [EC 4.1.3.25] catalyzes the conversion of (3*S*)-citramalyl-CoA to acetyl-CoA and pyruvate. The (3*S*)-citramalyl thioacyl-carrier protein can also be utilized as a substrate. This enzyme has been reported to be a component of citramalate lyase [EC 4.1.3.22].

H. A. Lardy (1969) *Meth. Enzymol.* **13**, 314.

CITRATE LYASE

This enzyme [EC 4.1.3.6] (also known as citrate (*pro-3S*)-lyase, citrase, citratase, citritase, citridesmolase, and citrate aldolase) catalyzes the conversion of citrate to acetate and oxaloacetate. Citrate lyase can be dissociated into subunits, two components of which are identical with citrate CoA-transferase [EC 2.8.3.10] and citryl-CoA lyase [EC 4.1.3.34].

D. J. Creighton & N. S. R. K. Murthy (1990) *The Enzymes*, 3rd ed., **19**, 323.
P. A. Srere (1975) *Adv. Enzymol.* **43**, 57.
L. B. Spector (1972) *The Enzymes*, 3rd ed., **7**, 357.

CITRATE SYNTHASE

Citrate (*si*)-synthase [EC 4.1.3.7] (also known as citrate condensing enzyme, citrogenase, and oxaloacetate transacetase) catalyzes the reaction of acetyl-CoA with oxaloacetate and water to produce citrate and coenzyme A (in which acetyl-CoA $\longleftrightarrow$ (*pro-3S*)-CH$_2$COO$^-$). Citrate (*re*)-synthase [EC 4.1.3.28] catalyzes the same reaction, albeit with the opposite stereochemistry (thus, acetyl-CoA $\longleftrightarrow$ (*pro-3R*)-CH$_2$COO$^-$).

J. V. Schloss & M. S. Hixon (1998) *Comprehensive Biological Catalysis: A Mechanistic Reference* **2**, 43.
S. J. Remington (1992) *Curr. Top. Cell. Reg.* **33**, 209.
R. F. Colman (1990) *The Enzymes*, 3rd ed., **19**, 283.
D. J. Creighton & N. S. R. K. Murthy (1990) *The Enzymes*, 3rd ed., **19**, 323.
G. Müller-Kraft & W. Babel (1990) *Meth. Enzymol.* **188**, 350.
P. A. Srere (1975) *Adv. Enzymol.* **43**, 57.
L. B. Spector (1972) *The Enzymes*, 3rd ed., **7**, 357.

CLAISEN CONDENSATION

A base-catalyzed reaction of esters (or thioesters) containing an α-hydrogen with another ester (in some cases another molecule of the same ester) to produce a β-keto ester and an alcohol or thiol. A number of enzymes catalyze Claisen condensations, including thiolase, malate synthase, citrate synthase, and hydroxymethylglutaryl-CoA synthase.

CLATHRATE

A compound having sites or inclusion spaces for completely enclosing so-called guests, molecular entities that fit into the above-mentioned sites or spaces. A synonym for clathrate is cage compound. A clathrate compound can be formed with hydroquinone as the host and methanol, CO_2, argon, or SO_2 as the guest[1]. Water can also serve as a host in clathrate compounds[2]. **See** *Cage Compound; Guest; Host Molecule*

[1]D. D. MacNicol, J. J. McKendrick & D. R. Wilson (1978) *Chem. Soc. Rev.* **7**, 65.
[2]G . H. Cady (1983) *J. Chem. Ed.* **60**, 915.

CLELAND NOMENCLATURE

A system for describing kinetic mechanisms for enzyme-catalyzed reactions[1,2]. Reactants (*i.e.*, substrates) are symbolized by the letters A, B, C, D, *etc.*, whereas products are designated by P, Q, R, S, *etc.* Reaction schemes are also identified by the number of substrates and products utilized (*i.e.*, Uni (for one), Bi (two), Ter (three; occasionally Tri), Quad (four), Quin (five), *etc.* Thus, a two-substrate, three-product enzyme-catalyzed reaction would be a Bi Ter system. In addition, reaction schemes are identified by the pattern of substrate addition to the enzyme's active site as well as the release of products. For a two-substrate, one-product scheme in which either substrate can bind to the free enzyme, the enzyme scheme is designated a random Bi Uni mechanism. If the substrates bind in a distinct order (note that, in such cases, A binds before B; for ordered multiproduct release, P is released prior to Q, *etc.*), the scheme would be ordered Bi Uni. If the binding scheme is different than the release of product, then that information should also be provided; for example, a two-substrate, two-product reaction in which the substrates bind to the enzyme in an ordered fashion whereas the products are released randomly would be designated ordered on, random off Bi Bi scheme. If one or more Theorell-Chance steps are present, that information is also given (*e.g.*, ordered Bi Bi-(Theorell-Chance)), with the prefixes included if there is more than one Theorell-Chance step.

The schemes provided as examples above are often referred to as being "sequential" in that all substrates must bind before any product is released. However, in other well-characterized examples, a product is released prior to the binding of one or more substrates. For example, in a two-substrate, two-product scheme, substrate A could bind to the free enzyme (E) to form an EA binary complex. A chemical "event" then occurs within the complex resulting in the formation of the first product (P) and a modified form of the enzyme (note: the initial free enzyme is designated by the letter E whereas the modified enzyme (or other isomerizational forms of the enzyme) are designated by the letters F, G, H, *etc.*). After P is released, the second substrate (B) binds to F to form the FB. Another chemical event occurs, the second product (Q) is formed and the original form of the enzyme (E) is regenerated. Such a mechanism is described as a ping pong Bi Bi scheme in Cleland's nomenclature. Ping pong schemes can also occur in systems having more than two substrates and/or products. For example, consider the three-substrate, three-product reaction scheme in which the first substrate binds to the free enzyme resulting in the formation of the first product followed by the ordered binding and release of the remaining substrates and products. Such a scheme is referred to as an ordered Uni Uni Bi Bi ping pong mechanism.

If a stable enzyme form isomerizes and these isomerization steps are a part of the reaction pathway, the term Iso is provided. For a Uni Uni mechanism in which an isomerization step occurs [*e.g.*, the reaction sequence: E + A $\longleftrightarrow$ EA; EA $\longleftrightarrow$ FP; FP $\longleftrightarrow$ F + P; and F $\longleftrightarrow$ E] is designated an Iso Uni Uni scheme. If more than one isomerization event occurs, then the prefixes di-, tri-, *etc.*, are utilized with Iso. Note that this aspect of the nomenclature does not refer to isomerization steps which may or may not occur between central complexes.

Complexes of enzyme, substrates, products, inhibitors, *etc.*, are often designated as being binary, ternary, quaternary, *etc.*, depending on the number of entities present in the complex. For example, EAB would be a ternary complex. Central complexes are those transient complexes that generate products (or substrates in the reverse reaction) or which isomerize to those forms which can generate products. Thus, in an ordered Bi Bi reaction scheme, the enzyme can exist in five forms: E, EA, EAB,

EPQ, and EQ. The central complexes are EAB and EPQ. The binary complexes EA and EQ are often referred to as Michaelis complexes in that they are generated by simple binding events; but no "chemistry" occurs until one or more other reactants bind to the active site. Note that central complexes can only participate in unimolecular events whereas Michaelis complexes can participate in both unimolecular and bimolecular events.

In Cleland nomenclature, the initial velocity and individual rate constants are designated by lower case italicized letters (*e.g.*, v, k, k_2, *etc.*). Dissociation, Michaelis, and equilibrium constants utilize an upper case italicized K with the appropriate unitalicized lower case subscript. For example, the equilibrium constant would be symbolized by K_{eq} whereas the Michaelis constant for substrate B would be designated by K_b. Dissociation constants for a Michaelis complex contain a subscript i and a letter for the dissociating ligand (*e.g.*, for the EA binary complex, the dissociation constant would be K_{ia}). Maximum velocities are designated by a capital italicized V, usually with a subscript 1 or 2 depending on whether the forward or reverse reaction is referred to. (If the numerical subscript is not provided, the forward reaction is assumed. In most cases, the unitalicized subscript "max" is also provided.)

Occasionally rate expressions are described as 1/1, 2/1, *etc.*, functions, referring to the maximum power of the substrate concentration in the numerator (N) and denominator (D). For example, consider the case of the steady-state random Bi Uni mechanism. The reciprocal form of the rate expression (at constant [B]) has the general form of $1/v = (\phi_o + \phi_{DA}[A] + \phi_{DA2}[A]^2)/(\phi_{NA}[A] + \phi_{NA2}[A]^2)$ where the ϕ values are collections of rate constants. If both the numerator and denominator of this reciprocal form of the rate expression are divided by the substrate concentration raised to the highest power in which it appears (in this case, $[A]^2$), then the numerator has a term in $1/[A]^2$ (as well as $1/[A]$ and $1/[A]^0$) whereas the denominator has terms in $1/[A]$ and $1/[A]^0$. Thus, this rate expression is a 2/1 function. **See** *Multisubstrate Mechanisms*

[1]W. W. Cleland (1963) *Biochim. Biophys. Acta* **67**, 104.
[2]W. W. Cleland (1963) *Biochim. Biophys. Acta* **67**, 173.

CLOSED SYSTEM

Any system (chemical, biological, mechanical, *etc.*) that does not permit the transfer of mass with its surrounding environment. Nonetheless, in closed systems, heat or energy can pass between the system and the surroundings. **See** *Isolated System; Open System*

CLOSTRIPAIN

This enzyme [EC 3.4.22.8], also referred to as clostridiopeptidase B, is an endopeptidase belonging to the peptidase family C11 that has been isolated from the bacterium *Clostridium histolyticum*. It catalyzes the hydrolysis of peptide bonds with a preference for the Arg—Xaa bond (including the Arg—Pro bond), but not the Lys—Xaa peptide bond. This calcium ion-activated enzyme is inactivated by EDTA.

E. Shaw (1990) *Adv. Enzymol.* **63**, 271.
K. Morihara (1974) *Adv. Enzymol.* **41**, 179.
W. M. Mitchell & W. F. Harrington (1971) *The Enzymes*, 3rd ed., **3**, 699.

C.M.C.

Abbreviation for critical micelle concentration. **See** *Micelle Formation*

CMP-*N*-ACETYLNEURAMINATE-*β*-GALACTOSIDE SIALYLTRANSFERASES

CMP-*N*-acetylneuraminate-*β*-galactoside α-2,6-sialyltransferase [EC 2.4.99.1], also referred to as *β*-galactosamide α-2,6-sialyltransferase and *β*-galactoside α-2,6-sialyltransferase, catalyzes the reaction of CMP-*N*-acetylneuraminate with *β*-D-galactosyl-1,4-acetyl-*β*-D-glucosamine to produce CMP and α-*N*-acetylneuraminyl-2,6-*β*-D-galactosyl-1,4-*N*-acetyl-*β*-D-glucosamine. Note that the terminal *β*-D-galactosyl residue of the oligosaccharide of glycoproteins, as well as lactose, can act as a substrate as well.

CMP-*N*-acetylneuraminate-*β*-galactoside α-2,3-sialyltransferase [EC 2.4.99.4], also referred to as *β*-galactoside α-2,3-sialyltransferase, catalyzes the reaction of CMP-*N*-acetylneuraminate with *β*-D-galactosyl-1,3-*N*-acetyl-α-D-galactosaminyl-R to produce CMP and α-*N*-acetylneuraminyl-2,3-*β*-D-galactosyl-1,3-*N*-acetyl-α-D-galactosaminyl-R. In the acceptor substrate, R can be H, a threonyl or seryl side chain in a glycoprotein, or a glycolipid. Lactose can also be utilized as a substrate. In some systems, this enzyme may be identical with CMP-

N-acetylneuraminate:monosialoganglioside sialyltransferase [EC 2.4.99.2].

D. M. Carlson & G. W. Jourdian (1966) *Meth. Enzymol.* **8**, 358.

COACTIVATOR

A molecule that functions in conjunction with a protein apo-activator. The classical example is cAMP, the coactivator of the CAP protein in bacteria.

COBALAMIN ADENOSYLTRANSFERASE

This enzyme [EC 2.5.1.17], also referred to as cob(I)-alamin adenosyltransferase or aquacob(I)alamin adenosyltransferase, catalyzes the reaction of cob(I)alamin with ATP and water to produce adenosylcobalamin, orthophosphate, and pyrophosphate (or, diphosphate). A cofactor for this enzyme is manganese ion.

S. H. Mudd (1973) *The Enzymes*, 3rd ed., **8**, 121.

COBALAMINS AND COBAMIDES (B₁₂)
Adenosylcobalamin

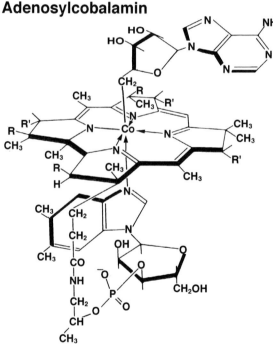

Members of the vitamin B_{12} or corrinoid group, containing a highly substituted tetrapyrrole ring which donates four nitrogen atoms to form coordinate covalent linkages with a centrally positioned hexacoordinate cobalt ion.

Selected entries from *Methods in Enzymology* [vol, page(s)]:
Spectroscopic analysis of vitamin B_{12} derivatives, **67**, 5; preparation of cryptofluorescent analogs of cobalamin coenzymes, **67**, 12; polarography of compounds of the vitamin B_{12} series in dimethylformamide, **67**, 20; comparison of serum vitamin B_{12} (co-balamin) determination by two isotope-dilution methods and by *Euglena* assay, **67**, 24; measurement of tissue vitamin B_{12} by radioisotopic competitive inhibition assay and quantitation of tissue cobalamin fractions, **67**, 31; ribosome-associated vitamin B_{12} adenosylating enzyme of *Lactobacillus leichmannii*, **67**, 41; immobilized derivatives of vitamin B_{12} coenzyme and its analogs, **67**, 57; solubilization of the receptor for intrinsic factor-B_{12} complex from guinea pig intestinal mucosa, **67**, 67; properties of proteins that bind vitamin B_{12} in subcellular fractions of rat liver, **67**, 72; isolation and characterization of vitamin B_{12}-binding proteins from human fluids, **67**, 80; biosynthesis of transcobalamin II, **67**, 89; competitive binding radioassays for vitamin B_{12} in biological fluids or solid tissues, **67**, 99; determination of cobalamins in biological material, **123**, 3; analysis of cobalamin coenzymes and other corrinoids by high-performance liquid chromatography, **123**, 14; isolation of native and proteolytically derived ileal receptor for intrinsic factor-cobalamin, **123**, 23; purification of B_{12}-binding proteins using a photodissociative affinity matrix, **123**, 28; solid-phase immunoassay for human transcobalamin II and detection of the secretory protein in cultured human cells, **123**, 36.

COBALT NUCLEOTIDES

Exchange-inert complexes of Co(III) with nucleotides that have proven to be extremely useful as chirality probes because the different coordination isomers are stable and can be prepared and separated[1]. In addition, these nucleotides can be used as dead-end inhibitors of enzyme-catalyzed reactions and, since Co(III) is diamagnetic, a number of spectroscopic protocols can be utilized[2]. ***See*** *Exchange-Inert Complexes; Chromium-Nucleotide Complexes; Metal Ion-Nucleotide Interactions*

[1]W. W. Cleland (1982) *Meth. Enzymol.* **87**, 159.
[2]J. J. Villafranca (1982) *Meth. Enzymol.* **87**, 180.

COCOONASES

Insect cocoonases are now classified under trypsin's EC number, [EC 3.4.21.4].

J. H. Law, P. E. Dunn & K. J. Kramer (1977) *Adv. Enzymol.* **45**, 389.

COENZYME Q (Ubiquinone)

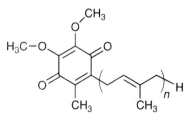

Coenzyme Q (Ubiquinone)

A polyprenylated benzoquinone derivative that is a key component in mitochondrial electron transport, and the reduced form is a hydroquinone. Coenzyme Q is not essential in humans.

CO-ION

An ion of like electric charge. Antonym: Counterion.

COLLAGENASES

(1) Microbial collagenase [EC 3.4.24.3] (also known as *Clostridium histolyticum* collagenase, clostridiopeptidase A, collagenase A, and collagenase I) is a member of the peptidase family M9. This enzyme catalyzes the hydrolysis of peptide bonds in native collagen in the triple helical region at Xaa—Gly bonds. (2) Widely distributed in vertebrates, the zinc-dependent interstitial collagenase [EC 3.4.24.7], also referred to as vertebrate collagenase and matrix metalloproteinase 1, is a member of the peptidase family M10B. It catalyzes the hydrolysis of a particular peptide bond in the native collagen triple helix. (3) Neutrophil collagenase [EC 3.4.24.34], also called matrix metalloproteinase 8, is a zinc-/calcium-dependent endopeptidase and is a member of the peptidase family M10B. (4) Hypodermin C [EC 3.4.21.49], also known as *Hypoderma* collagenase, is an endopeptidase that catalyzes the hydrolysis of peptide bonds in a number of proteins including native collagen. ***See also*** *Gelatinase*

M. Dioszegi, P. Cannon & H. E. Van Wart (1995) *Meth. Enzymol.* **248**, 413.

H. Tschesche (1995) *Meth. Enzymol.* **248**, 431.

J. D. Grubb (1994) *Meth. Enzymol.* **235**, 594.

B. L. Vallee & A. Galdes (1984) *Adv. Enzymol.* **56**, 283.

R. Kuttan & A. N. Radhakrishnan (1973) *Adv. Enzymol.* **37**, 273.

S. Seifter & E. Harper (1971) *The Enzymes*, 3rd ed., **3**, 649.

COLLISION CROSS-SECTION

A measure of the cross-sectional area swept through by a particle or chemical entity; this cross section influences collisions with another molecular or ionic entity. ***See*** *Collision Theory*

J. W. Moore & R. G. Pearson (1981) *Kinetics and Mechanism*, Wiley, New York.

COLLISION THEORY

A gas-phase reactivity model that assumes molecules react as a result of the collision of reactant molecules. The basic idea is that the kinetic energy of the impacting molecules exceeds the activation energy required for reaction. Classical mechanics is used to estimate the fraction of the collisions with enough energy to allow re-actant molecules to convert to product. A deficiency in the model is that steric or orientation factors are ignored, despite the fact that quantum mechanical considerations may require them. Furthermore, energy contributions from rotational and vibrational degrees of freedom may factor into the activation energy term in collision theory.

COLLOID SCIENCE

The systematic investigation of the chemical and physical properties of large molecules and particles having at least one dimension in the 1 nm to 1 mm range. Colloidal suspensions exhibit special properties not typically observed with smaller molecules and/or ions. In many cases, colloids are incapable of achieving a true state of dissolution, existing instead as dispersions. Their constituent macromolecules also develop an electric double layer at their junction with an aqueous electrolyte solution, and the distribution of ionic charge near their surface is not homogeneous. Colloids are often aggregates formed from both regular and irregular bonding patterns among their constituent atoms, molecules, or particles. The term "sol" describes an apparently uniform, or at least consistent, dispersion within a liquid or gaseous medium. The term "gel" refers to a disperse colloidal suspension of particles or macromolecules linked by chemical bonds or physisorption. In the sol state, colloids are characterized by substantial right angle light scattering (a phenomenon termed the Tyndall effect in honor of John Tyndall, although this property was probably first noted years earlier by Michael Faraday in his work on colloidal gold).

The nature of the forces maintaining a colloidal suspension can be largely regarded as electrostatic. In some instances, the component substance of a colloid can be brought to a state of dispersion that possesses such great stability that, however one may try, the original components can no longer be made to re-emerge from the dispersion. Such dispersions are said to be irreversible. Other colloidal suspensions (termed reversible) are readily returned to their undispersed condition.

Emulsions are colloidal suspensions produced by combining two liquids, and their dispersed state is usually stabilized by yet another component, termed an emulsifying agent. Foams are gaseous colloidal suspensions formed by combining two substances (of which one is typically a gas), and again the stability of a foam is aided by a third and minor component (in this case, a foaming agent).

Doubtlessly, Nature's most astonishing colloidal suspension is protoplasm, a dispersion of biological macromolecules (chiefly proteins) as well as numerous hydrophilic, lipophilic, and amphipathic low-molecular-weight metabolites. The cytoskeleton, various vesicles and even glycogen particles form the underlying structure of protoplasm, which is constantly being remodeled locally through the involvement of ATP-dependent molecular motors as well as numerous regulatory proteins. The viscosity and consistency of protoplasm changes in gel-sol transitions that are associated with cellular motility.

COLUMBAMINE *O*-METHYLTRANSFERASE

This enzyme [EC 2.1.1.118] catalyzes the reaction of *S*-adenosylmethionine with columbamine to generate *S*-adenosylhomocysteine and palmatine. The product of this reaction is a protoberberine alkaloid that occurs in a wide variety of plants. The enzyme is distinct from that of 8-hydroxyquercetin 8-*O*-methyltransferase.

T. Hashimoto & Y. Yamada (1994) *Ann. Rev. Plant Physiol. Plant Mol. Biol.* **45**, 257.

COMBINATORIAL LIBRARIES & MOLECULAR RECOGNITION

A method for simultaneous synthesis of a large number of peptides and/or ligands on solid supports for the discovery and/or development of high-affinity ligands that can be used as enzyme inhibitors or drugs. Initial work[1] centered on the identification of protease inhibitors using peptide libraries, but the sophistication of solution- and solid-phase chemistry has progressed to consider nonpeptidic ligands that may exhibit even higher affinity complexation with target proteins. Lam[2] describes five general approaches using combinatorial peptide library: (a) biological libraries; (b) spatially addressable parallel solid-phase or solution-phase libraries; (c) synthetic library methods requiring deconvolution; (d) the 'one-bead/one-compound' library method; and (e) synthetic library methods using affinity chromatography selection. In some cases, synthetic combinatorial libraries allow the investigator to probe enzyme/receptor interactions with an extensive library of so-called "pharmacophore" motifs rather than a library of compounds[3]. Brenner *et al.*[4] have described how the diversity of chemical synthe-

sis and the power of genetics can be linked to provide a powerful, versatile method for drug screening. They described a process of alternating parallel combinatorial synthesis for encoding individual members of a large combinatorial library, such that each is marked with unique nucleotide sequences. After the chemical entity is bound to a target enzyme/receptor, the genetic tag is then amplified by replication and used to enrich the bound ligand by serial hybridization to a subset of the library. The nature of the chemical structure bound to the receptor is readily decoded by sequencing the nucleotide tag.

The method is illustrated by the work of Bastos *et al.*[5] who synthesized two sets of noncleavable peptide-inhibitor libraries to map the subsites for substrate/inhibitor interactions with human heart chymase. A chymotrypsin-like enzyme, chymase converts angiotensin I to angiotensin II. As is customary for describing protease subsites, P1, P2, *etc.*, are numbered toward the N-terminal direction from the scissile bond, whereas P′1, P′2, *etc.*, are numbered in sequence toward the C-terminal direction from the scissile bond. Their first library consisted of peptides with 3-fluorobenzylpyruvamides in the P1 position. The P′1 and P′2 positions were varied to contain each of the 20 natural amino acids and P′3 was always an arginine. The second library consisted of peptides with Phe ketoamides at P1, Gly in P′1, and benzyloxycarbonyl (Z)-Ile in P4. The P2 and P3 positions again were varied to contain each of the naturally occurring amino acids, except for Cys and Met. The peptides of both libraries are attached to a solid support (pins). The peptide interactions with enzyme were evaluated by immersing the "pins" in a enzyme solution and then determining the amount of absorbed enzyme. The libraries selected the best and worst inhibitors within each group of peptides and provided an approximate ranking of other peptides according to K_i. The authors found that Z—Ile—Glu—Pro—Phe—CO_2Me and (F)—Phe—CO—Glu—Asp—ArgOMe appeared to be the best inhibitors of chymase in this collection of peptide inhibitors. They then synthesized the peptides and found K_i values were 1 nM and 1 mM, respectively. The respective K_i values for chymotrypsin were 10 nM and 100 mM.

Feng *et al.*[6] used combinatorial synthesis to discover several ligands containing nonpeptide binding elements to the Src SH3 domain. Their encoded library used compounds of the form Cap—M_1—M_2—M_3—PLPPLP, in which the Cap and M_i's are composed of a diverse set of organic monomers. The PLPPLP portion provided a structural bias directing the nonpeptide fragment Cap—M_1—M_2—M_3 to the SH3 specificity pocket. Fifteen ligands were selected from > 1.1 million distinct compounds. The solution structures of the Src SH3 domain complexed with two ligands containing nonpeptide elements selected from the library were determined by multidimensional NMR spectroscopy. The nonpeptide moieties of the ligands interact with the specificity pocket of Src SH3 domain differently from peptides complexed with SH3 domains. Structural information about the ligands was used to design various homologues, and their affinities for the SH3 domain were measured. These investigators demonstrated the efficacy of cycles of protein structure-based combinatorial chemistry, followed by structure determination of the few highest-affinity ligands in defining the basis of molecular recognition by Src SH3 domain.

[1]K. S. Lam, S. E. Salmon, E. M. Hersh, V. J. Hruby , W. M. Kazmierski & R. J. Knapp (1991) *Nature* **354**, 82 (publ. errata: *Nature* **358**, 434 (1992)).
[2]K. S. Lam (1997) *Anticancer Drug Des.* **12**, 145.
[3]N. F. Sepetov, V. Krchnak, M. Stankova, S. Wad, K. S. Lam & M. Lebl (1995) *Proc. Natl. Acad. Sci. U.S.A.* **92**, 5426.
[4]S. Brenner & R. A. Lerner (1992) *Proc. Natl. Acad. Sci. U.S.A.* **89**, 5381.
[5]M. Bastos, N. J. Maeji & R. H. Abeles (1995) *Proc. Natl. Acad. Sci. U.S.A.* **92**, 6738.
[6]S. Feng, T. M. Kapoor, F. Shirai, A. P. Combs & S. L. Schreiber (1996) *Chem. Biol.* **3**, 661.

COMMITMENT FACTORS IN SOLVENT ISOTOPE EFFECTS

An approach to studying transition states in enzyme-catalyzed reactions using solvent isotope effects. In this treatment, very useful in isotope effect experiments, the relative rates of contributing steps in a multistep reaction are grouped into a fraction referred to as the commitment factor[1].

[1]K. B. Schowen & R. L. Schowen (1982) *Meth. Enzymol.* **87**, 551.

COMMITMENT-TO-CATALYSIS

A unitless ratio of rate constants that serves as an index or measure of an enzyme's ability to promote catalysis as opposed to releasing its substrate before catalysis can occur. Consider the following scheme:

$$E + S \underset{k_2}{\overset{k_1}{\rightleftharpoons}} ES \overset{k_3}{\rightarrow} E + P$$

In this scheme, $V_{max} = k_3[E_{tot}]$, $K_m = (k_2 + k_3)/k_1$, and the commitment to catalysis equals $k_3/(k_2 + k_3)$. (a) If k_3 is much less than k_2, then K_m becomes the equilibrium or dissociation constant K_{ia} ($= k_2/k_1$), indicating that E and S react rapidly and reversibly to form the ES complex. In this case, the commitment-to-catalysis is very low because ES has a much greater tendency to dissociate rather than to convert S to form P. (b) If k_3 equals k_2, then $K_m = (k_2 + k_3)/k_1 = 2k_2/k_1 = 2K_{ia}$, and the commitment-to-catalysis equals 0.5. This means that the ES complex has an identical probability to dissociate to E + S or to convert to E plus P. (c) If k_3 is much greater than k_2, the commitment-to-catalysis is high and approaches unity. Whenever ES is formed, product formation would then be highly favored. ***See also*** *Rapid Equilibrium Assumption; Kinetic Isotope Effect; Isotope Trapping*

COMPARTMENTAL ANALYSIS

A kinetic approach for determining the pools accessible to a tracee (*i.e.*, a metabolite, drug, or other substance) within a living organism. The basic strategy entails the following: (a) administration of a tracer, in the same chemical form as the tracee or as a metabolic precursor; (b) determination of changes in the specific activity (*i.e.*, [mol tracer]/[mol tracer + mol tracee]) in biological samples using radioactive or stable isotopes; (c) development of compartmental model(s) consistent with the time evolution of observed changes in specific activity, including rate constants and pool sizes for each tracee compartment; and (d) systematic experimentation to adduce sufficient information for discriminating rival models, along with due consideration of any statistical uncertainty in the data.

COMPARTMENTALIZATION

The division of a cell into different regions or domains containing different components and physiological functions. Compartmentalization is a means for more effective regulation arising from the partitioning of a cell into regions with different enzymatic and regulatory components. For example, fatty acid catabolism occurs within the mitochondria whereas fatty acid biosynthesis occurs in the cytoplasm. Hence, biomembranes often serve an important role in the regulation of metabolic pathways by forming compartments as well as serving as a region of communication between different compartments. However, other types of compartmentalization can also occur. For example, proteins embedded in a biomembrane or proteins that are a part of a larger cytoplasmic aggregation of other proteins, can often be considered to be compartmentalized since communication to and from those proteins may limited by nondiffusion processes.

COMPENSATION EFFECT

Some reactions are characterized by straight-line plots of $T\Delta S^{\ddagger}$ *versus* $\Delta H^{\ddagger}$ having a slope of approximately one, where this linearity results from compensatory, or off-setting, changes of $\Delta H^{\ddagger}$ and $T\Delta S^{\ddagger}$. For this reason, the change in the Gibbs energy of activation, $\Delta G^{\ddagger} = \Delta H^{\ddagger} - T\Delta S^{\ddagger}$, is a better description of the variation in the reaction than either $\Delta H^{\ddagger}$ or $T\Delta S^{\ddagger}$ alone[1,2]. **See also** *Isokinetic Relationship*

[1]K. A. Connor (1991) *Chemical Kinetics*, VCH, New York.
[2]IUPAC (1994) *Pure and Appl. Chem.* **66**, 1077.

COMPETITION PLOT

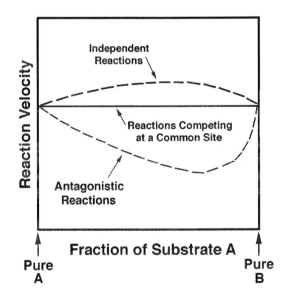

A plotting procedure[1] that serves to test whether two reactions occur at the same active site. One makes a plot of total rate *versus* p, where p varies from 0 to 1 and specifies the concentrations $(1 - p)A_0$ and pB_0 of two substrates in terms of reference concentrations A_0 and B_0. If the two substrates react at the same active site,

the competition plot gives a horizontal straight line, *i.e.*, the total rate is independent of p. Independent reactions at two separate sites give a curve with a maximum. Separate reactions with cross-inhibition generate curves with either maxima or minima according to whether the Michaelis constants of the two substrates are smaller or larger than their inhibition constants in the other reactions. Ambiguous results can sometimes arise, but the authors present experimental strategies around such difficulties.

[1]C. Chevillard, M. L. Cardenas & A. Cornish-Bowden (1993) *Biochem. J.* **289**, 599.

COMPETITIVE BINDING ASSAY

An assay protocol in which a macromolecule or binder (*i.e.*, the entity which is capable of binding a particular substance or ligand) competes for labeled and unlabeled ligand. Once binding has occurred, bound and free ligand are separated and quantitated. The bound and unbound ratios can be related to known standards. Examples of this procedure, also occasionally referred to as displacement analysis and saturation analysis, include enzyme-linked immunosorbent assays (ELISAs) and radioimmunoassays. Some investigators also use the term to describe binding assays involving an unlabeled ligand and a structurally similar ligand (*e.g.*, a competitive inhibitor) that binds to the same site on the macromolecule.

COMPETITIVE INHIBITION

Any reduction in catalytic activity by a substance binding to the same form of a enzyme (or receptor) to which a substrate (or ligand) binds. This binding is said to be competitive, because inhibitor I and substrate S cannot bind simultaneously (*i.e.*, their binding is mutually exclusive). Note: Substrate and inhibitor need not act at the same topological site on the enzyme in order to exhibit mutually exclusive binding.

For a one-substrate enzyme-catalyzed reaction, the following scheme describes the action of a competitive inhibitor:

$$E + S \rightleftharpoons EX \rightarrow E + P$$

$$E + I \rightleftharpoons EI$$

where $K_i = [E][I]/[EI]$, assuming rapid and reversible binding. The resulting rate law can be written in double-reciprocal form:

$$1/v = (1/V_{max}) + (K_m/(V_{max}[S]))\{1 + [I]/K_i\}$$

This equation has a $(1/v)$ axis intercept value of $(1/V_{max})$ and a slope of $(K_m/V_{max})\{1 + [I]/K_i\}$. Hence, all lines intersect at a common point on the $(1/v)$ axis. Plotting the slope values *versus* $[I]$ yields a secondary plot (or replot) whose slope is $K_m/(V_{max}K_i)$ and horizontal intercept is $-(1/K_i)$.

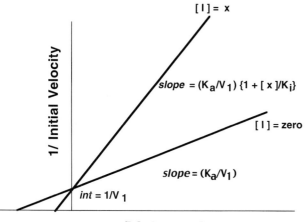

Fromm[1] first demonstrated how competitive inhibitors can be employed to distinguish the order of substrate binding for multisubstrate enzyme mechanisms. Each competitive inhibitor, with respect to one substrate, displays distinctive pattern(s) relative to the other substrate(s)[2,3].

[1]H. J. Fromm & V. Zewe (1962) *J. Biol. Chem.* **237**, 3027.
[2]H. J. Fromm (1979) *Meth. Enzymol.* **63**, 467.
[3]H. J. Fromm (1995) *Meth. Enzymol.* **249**, 123.

Selected entries from *Methods in Enzymology* [vol, page(s)]: Theory, **63**, 468-477 [one-substrate, **63**, 468-474; two-substrate, **63**, 474-478; three-substrate, **63**, 477, 479]; practical aspects, **63**, 477-485; limitation, **63**, 485, 486; complications, **63**, 406-408; computer program, **63**, 114, 115, 126, 127, 137; constant ratio to substrate, **63**, 10; coupled enzyme assay, **63**, 40; cryoenzymology, **63**, 345; equation, **63**, 10, 468-470; metal-nucleotide complex, **63**, 268, 269, 292, 328, 329; pH effects, **63**, 214; product effects, **63**, 412, 413; statistical analysis, **63**, 110; substrate binding order, **63**, 467-486; tight-binding, **63**, 445.

COMPETITIVE INHIBITORS
(Multisubstrate Enzyme Systems)

Fromm and Zewe[1] originally suggested that competitive inhibitors could aid an investigator in distinguishing be-

tween various enzyme kinetic mechanisms. The procedure[2,3] utilizes a competitive inhibitor for one of the substrates in a multisubstrate enzyme-catalyzed reaction and determines whether that same inhibitor is competitive, noncompetitive, or uncompetitive with respect to the other substrate(s) of the enzyme. For example, if an investigator had a competitive inhibitor for one of the substrates of a non-ping pong bisubstrate enzyme and that inhibitor was uncompetitive with respect to the other substrate, the researcher could eliminate the random binding mechanism (Table I).

Table I

Use of Competitive Inhibitors to Distinguish Various Bi-Reactant Systems[a]

Mechanism	1/v versus 1/[A]	1/v versus 1/[B]
Ordered Bi Uni	C	N
	U	C
Ordered Bi Bi'	C	N
	U	C
Ping Pong Bi Bi	C	U
	U	C
Random Bi Uni	C	N
	N	C
Random Bi Bi	C	N
	N	C

[a]C = competitive inhibition pattern; N = noncompetitive inhibition pattern; and U = uncompetitive inhibition pattern.

The various ter-reactant enzyme systems can also be discriminated by the use of competitive inhibitors (see Table II). Fromm[4] has also presented a cogent argument for exercising special care in choosing the concentrations of the nonvaried substrates in experiments on reactions involving three substrates.

While requiring the availability of competitive inhibitors for each of the substrates, Fromm's use of competitive inhibitors to distinguish multisubstrate enzyme kinetic pathways represents the most powerful initial rate method. **See** *Alternative Substrate Inhibition*

[1]H. J. Fromm and V. Zewe (1962) *J. Biol. Chem.* **237**, 3027.
[2]H. J. Fromm (1975) *Initial Rate Enzyme Kinetics*, pp. 94-110, Springer-Verlag, New York.
[3]H. J. Fromm (1979) *Meth. Enzymol.* **63**, 467.
[4]H. J. Fromm (1967) *Biochim. Biophys. Acta* **139**, 221.

Table II. Use of Competitive Inhibitors to Distinguish Various Ter-Reactant Enzyme Systems[a]

Mechanism	1/v versus 1/[A]	1/v versus 1/[B]	1/v versus 1/[C]
Ordered Ter Ter	C	N	N
	U	C	N[b]
	U	U	C
Random Ter Ter (rapid equilibrium)	C	N	N
	N	C	N
	N	N	C
Random AB, Ordered C (rapid equilibrium)	C	N	C[c]
	N	C	C[d]
	U	U	C
Ordered A, Random BC (rapid equilibrium)	C	N	N
	U	C	N
	U	N	C
Random AC, Ordered B (rapid equilibrium)	C	N	N
	N	C	N
	N	N	C
Hexa Uni Ping Pong	C	U	U
	U	C	U
	U	U	C
Ordered Bi Uni Uni Bi Ping Pong	C	N[e]	U
	U	C	U
	U	U	C
Ordered Uni Uni Bi Bi Ping Pong	C	U[e]	U
	U	C	N[f]
	U	U	C
Random Bi Uni Uni Bi Ping Pong	C	N	U
	N	C	U
	U	U	C
Random Uni Uni Bi Bi Ping Pong	C	U	U
	U	C	N
	U	N	C

[a]C = competitive inhibition pattern; N = noncompetitive inhibition pattern; and U = uncompetitive inhibition pattern.
[b]This plot will be nonlinear if EAI binds C to form an EAIC complex.
[c]This plot will be noncompetitive if EIB binds C to form an EIBC complex.
[d]This plot will be noncompetitive if EIA binds C to form an EIAC complex.
[e]This plot will be nonlinear if EI binds B to form an EIB complex.
[f]This plot will be nonlinear if E'I binds C to form an E'IC complex.

COMPLEX REACTION

A reaction or set of reactions that consists of several elementary reactions. **See also** *Elementary Reaction; Primitive Change; Parallel Reactions; Series Reactions*

COMPONENT

1. A chemically distinct substance or species from which a phase or constituent of a system can be made. For example, for a system consisting of ice floating in water, there are three constituents: liquid water, ice, and water vapor. The number of components can exceed the number of phases. For example, D-glucose dissolved in water in a sealed ampule results is a system having one phase, albeit containing two components. **See also** *Constituent* 2. One of the coordinates or elements of a vector. 3. A vector term that is a part of a vector sum or vector product (*i.e.*, in which the resultant is a vector). 4. With respect to any vectoral quantity, a component in any direction is the representation of the vector by the projection of that vector onto the line in the specified direction. For example, in the two-dimensional Cartesian coordinate system, for the vector going from the coordinate point (1,1) to point (2,2), the component of that vector along the *x*-axis runs from point (1,0) to point (2,0). 5. In statistics, a variable in a multivariate distribution.

COMPOSITE RATE CONSTANTS

A rate constant describing the reaction rate over a series of transition states, where if written as the sum of reciprocal equilibrium and rate constants, it describes the difference in free energy between the *n*th transition state and the original reactant[1]. An excellent example is provided in Fig. 1 which is taken from Fisher *et al.*[1]

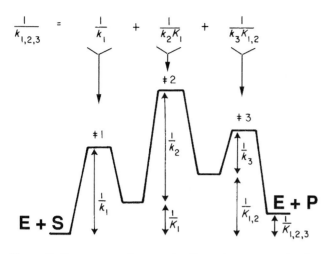

$$\frac{1}{k_{1,2,3}} = \frac{1}{k_1} + \frac{1}{k_2 K_1} + \frac{1}{k_3 K_{1,2}}$$

Figure 1. Free energy diagram of a three-step process, showing each term in the expression for the composite rate constant $k_{1,2,3}$ in the forward direction. Reproduced with permission of the authors and the American Chemical Society.

[1]L. M. Fisher, W. J. Albery & J. R. Knowles (1986) *Biochemistry* **25**, 2529.

COMPOSITE REACTION

A chemical reaction whose rate expression includes rate constants from more than one elementary reaction. Examples of composite reactions include parallel reactions and stepwise reactions.

COMPRESSION

A potential factor affecting the catalytic power of enzyme-catalyzed reactions in which the enzyme forces the attacking moiety into the valence shell of the entity under "attack." Compression is thought to be manifested in the transition state complex. The results of kinetic isotope effect measurements have suggested that compression events factor in the catechol methyltransferase reaction[1].

[1]F. Takusagawa, M. Fujioka, A. Spies & R. L. Schowen (1998) *Comprehensive Biological Catalysis: A Mechanistic Reference* **1**, 1.

COMPROPORTIONATION

Reverse of disproportionation. *See* Disproportionation

COMPUTER ALGORITHMS & SOFTWARE

The widespread availability of digital computing has greatly facilitated data acquisition and analysis. No single text, and least of all a manual on the general applications of biochemical kinetics, can describe all the computing methods at the disposal of molecular life scientists. For this reason, readers are encouraged to consult specialty monographs dealing with a specific kinetic technique.

See also KINSIM/FITSIM, SAAM, Analog Computer, Meta-Model, etc.

Selected entries from *Methods in Enzymology* [vol, page(s)]: AGIRE, **249**, 79-81, 225-226; AMBER, **240**, 416, 428-429, 433; **243**, 562, 567-590; **261**, 5, 6, 28-29, 40, 56, 94, 292, 603; AMBER/OPLS, **243**, 562; AUTO, **240**, 795-798, 803, 805, 812, 815; CAPP, **239**, 315; CHARMM, **240**, 451; **243**, 562, 567, 569-571, 573, 578; CHEOPS, **261**, 23; Connolly Molecular Surface program, **243**, 604; CURVEFIT function, **232**, 143, 151, 155, 159-160; DelPhi program, **243**, 606; for determination of metal ion-nucleotide complex dissociation constants, **249**, 182, 184-188; Dials&Windows program, **261**, 39, 43, 124; DIANA, **239**, 304, 429; **240**, 416, 435-436; DISMAN, **239**, 429; DOCK, **241**, 355-358; DSPP, **243**, 602; ECEPP/2, **243**, 562; EM algorithm, **240**, 178-180; ENZFITTER, **248**, 68-70, 75-79, 98-99; **249**, 525; for estimation of cooperativity parameters, **249**, 525-526; EZFIT, **249**, 525; FELIX, **240**, 450; FIRM, **240**, 427, 432-433, 435, 437; FITSIM, **240**, 312-322; FLOG, **241**, 359, 361-363; FLUSYS, **248**, 20, 91; GAMESS, **243**, 563; GAUSSIAN, **243**, 563; GRAFIT, **249**, 525; GROMOS, **243**, 562, 606; GROW, **241**, 358; HABAS, **239**, 422; HONDO, **243**, 563; IGOR, **240**, 330; Insight/Discover program, **243**, 561, 595; ISOBI, **249**, 449, 455-457, 478; ISOBI-HS, **249**, 456-459; ISOCALC4, **249**, 349-350, 365-368; ISOCALC5, **249**, 350, 366, 369-373; ISOCOOP, **249**, 455-457, 471; ISOMOD, **249**, 455-457, 471; ISOTER, **249**, 449, 459-461, 464-465; JMP, **249**, 525; KALEIDOGRAPH, **249**, 525; KINSIM, **240**, 312-322; **249**, 461; LIGAND, **249**, 525; LINSHA, **261**, 22-24; MARDIGRAS, **239**, 420-421, 425, 427-428; **240**, 427-428; **261**, 10, 12-16, 74, 94, 601-603; Marquardt algorithm, **246**, 696; MATHEMATICA, **249**, 525; MERLOT, **232**, 22; for molecular modeling, **243**, 560-561; Monte Carlo methods (*See* Monte Carlo Methods); MOPAC, **243**, 563; MOPAC ESP, **243**, 564; for moving average interpolation, **235**, 32; Nelder-Mead simplex algorithm, **240**, 188 Newton algorithm, **246**, 696-697; for probit analysis, **235**, 31; for protein NMR spectral analysis, **239**, 289, 308; PSEUROT, **261**, 255; QUANTA, **243**, 561; **261**, 204; Relaxation Analysis Program, **261**, 444-445; SHAKE, **261**, 29, 37, 57; SIMNOE, **261**, 78-79; SPARKY, **261**, 8; SPHINX, **261**, 22-24, 596; Statistical Analysis System, **235**, 31, 39; STEREO-SEARCH, **239**, 422; for structure-based computational chemistry calculations, **243**, 601-606; ULTRAFIT, **249**, 525; WESDYN, **241**, 186-187; **261**, 139; XPLOR, **232**, 22; **239**, 304, 473; **261**, 205, 292, 603.

CONCAVE-DOWN

A term used to describe the behavior of a curvilinear function by specifying that, over a given interval, the first derivative increases, reaches a local or global maximum, and then decreases.

CONCAVE-UP

A term used to describe the behavior of a curvilinear function by specifying that, over a given interval, the first derivative decreases, reaches a local or global minimum, and then increases.

CONCENTRATION

The amount of substance, typically measured in moles, per unit volume of solution (commonly measured in dm^{-3} [*i.e.*, liters, L]). Thus, amount-of-substance concentration, usually referred to simply as concentration and symbolized by c (often with a subscript denoting the species), can be described by $c_B = n_B/V$ where n_B is the amount-of-substance B and V is the total volume of the solution. Since the density of a solution will vary with temperature, the concentration c_B will be temperature-dependent. Hence, the temperature should always be reported. The SI unit is mol·dm^{-3} or molarity.

Concentration is the most common means for describing the composition of a solution in biochemistry. Enzyme kinetic expressions are typically expressed in these concentration units. Unless otherwise noted, this is the method used throughout this text. Nevertheless, other methods for describing compositions are utilized. For example, mole fractions are often used in Job plots. Gases in solution are commonly measured in terms of partial pressures. Below is a brief description of a few of these other conventions or methods.

Mole fraction, often symbolized by x or X followed by a subscript denoting the entity, represents the amount of a component divided by the total amount of all components. Thus, the mole fraction of component B of a solution, x_B, is equal to $n_B/\Sigma n_i$ where n_B is the amount of substance B and Σn_i is the total amount of all substances in solution. In biochemical systems, usually the solvent is disregarded in determining mole fractions. The mole fraction, a dimensionless number expressed in decimal fractions or percentages, is temperature-independent and is a useful description for solutions in theoretical studies and in physical biochemistry.

Molality, symbolized by m and usually followed by a subscript denoting the component, is also often used in physical biochemistry and is equal to the amount of substance per unit mass of solvent (*e.g.*, $m_B = n_B/$ (*mass*$_\text{solvent}$)). The SI unit for molality, another temperature-independent quantity, is mol·kg^{-1}. A quantity related to molality, yet not as widely used in physical studies, is the volume molality, symbolized by m''. The volume molality is equal to the amount of substance per unit volume of solvent (recall that concentration was equal to amount of substance per unit volume of solution).

As mentioned above, gases in solution are often measured in partial pressures. Recall that Dalton's Law of Partial Pressures states that the total pressure is the sum of the pressures partially exerted by all of the component, noninteracting gases. For one gaseous component in equilibrium with the same component dissolved in a liquid, the partial pressure of that gas in solution is the pressure that gas would exert in the atmosphere. It is symbolized by either p or P followed by the specific gas (*e.g.*, either P_{O_2} or pO_2). For example, typical blood P_{O_2} levels range between 75 and 100 torr (or, 10.0 to 13.3 kPa since the pascal (Pa) is an SI unit).

In clinical studies, compositions are often described in terms of mass of substance per unit volume. For example, normal adult blood glucose levels typically range between 70 and 110 mg/dl. While these numbers have great utility in clinical studies, they are less useful in assessing the action of the metabolite with other proteins and systems. Thus, it is valuable to provide the concentration term as well in these reports.

Protein concentrations are often reported in terms of mass per unit volume (*e.g.*, mg/ml) or in units of activity per volume. Care should be exercised when using these forms of concentrations. To have valid steady-state kinetics, the total enzyme concentration must be much less than the substrate concentration. Hence, one should always be cognizant of the true concentration of active enzyme in terms of molarity.

If the molecular weight is known for each component of the solution, then mole fractions (in which the solvent is considered one of the components) and molality can be mathematically interconverted. For example, if a solution contains two components, A and B, in which x_A, x_B, m_A, m_B, M_A, and M_B represent the mole fractions, the molalities, and the molecular weights of A and B, respectively, then $x_B = m_B/[(1000/M_A) + m_B]$.

Similarly, concentration and mole fractions can be related. If ρ is the density of the solution then, for a solution containing two components A and B, $x_B = c_B/\{[(1000\rho - c_B M_B/M_A] + c_B\}$. In dilute aqueous solutions, molarity is approximately equal to molality. **See** *Concentration Range Selection*

CONCENTRATION CORRELATION ANALYSIS

A chemical relaxation technique that measures the magnitude and time dependence of fluctuations in the concentrations of reactants[1-3]. If a system is at thermodynamic equilibrium, individual reactant and product molecules within a volume element will undergo excursions from the homogeneous concentration behavior expected on the basis of exactly matching forward and reverse reaction rates. The magnitudes of such excursions, their frequency of occurrence, and the rates of their dissipation are rich sources of dynamic information on the underlying chemical and physical processes. The experimental techniques and theory used in concentration correlation analysis provide rate constants, molecular transport coefficients, and equilibrium constants. Magde[4] has provided a particularly lucid description of concentration correlation analysis. **See** *Correlation Function*

[1] Y. Chen (1973) *J. Chem. Phys.* **59**, 5810.
[2] Y. Chen & T. L. Hill (1973) *Biophys. J.* **13**, 1276.
[3] G. Feher & M. Weissman (1973) *Proc. Natl. Acad. Sci. U.S.A.* **70**, 870.
[4] D. Magde (1977) in *Chemical Relaxation in Molecular Biology* (I. Pecht & R. Rigler, eds.), pp. 2-3, Springer-Verlag, Berlin.

CONCENTRATION-RESPONSE CURVE

A plot of the biological response produced by a drug, typically rising from a basal value (frequently zero at zero drug concentration) to a maximal response as the drug concentration is increased to a point where the extent of drug complexation with its receptor is no longer concentration-dependent. Such curves are typically plotted as the fraction of maximal response *versus* the logarithm (base$_{10}$) of drug concentration.

CONCERTED FEEDBACK INHIBITION

The action of two (or more) feedback modifiers that must be present simultaneously for any inhibition to occur. An illustrative case is the ordered addition of two effectors (designated as X and Y) in the following reaction scheme:

$$
\begin{array}{ccccc}
E + S & \rightleftharpoons & ES & \rightarrow & E + P \\
\updownarrow & & \updownarrow & & \\
EX + S & \rightleftharpoons & EXS & \rightarrow & EX + P \\
\updownarrow & & & & \\
EXY & & & &
\end{array}
$$

This mechanism involves the ordered addition of inhibitors, such that X must bind before Y can. As a result, the following are essential properties: (a) the presence of only X along with substrate S has no effect of enzyme reaction rate, because X does not affect substrate binding or the rate of ES breakdown; (b) the presence of only Y along with S is without effect, because Y cannot bind in the absence of X; and (c) inhibition will take place only when X and Y are both present, thereby allowing inactive EXY complex to accumulate.

CONCERTED PROCESS

Two or more primitive chemical/physical changes occurring within the same elementary reaction step. Typically, the simultaneous progress of these changes is thermodynamically more favorable than a set of successive changes. **See** *Primitive Changes; Elementary Reactions; Synchronous*

IUPAC (1979) *Pure and Appl. Chem.* **51**, 1725.

CONDENSATION REACTION

A reaction occurring between two or more distinct molecular entities (or, between different sites within the same molecular entity), resulting in the formation of a main product with the concomitant release of a small molecule, most commonly water. In most chemical condensations, the reaction usually consists of an addition reaction, followed by an elimination reaction.

CONFIDENCE INTERVAL

A term used in statistics to indicate the range of an interval within which the true value for a parameter lies, given a specified confidence level.

CONFIDENCE LEVEL

A term used in statistics to indicate the reliability of a measured result. A confidence level of 90% indicates that there is a 10% chance that the result is not reliable.

CONFIGURATION, DETERMINATION OF

There are a number of methods for determining the configuration of an optically active molecular entity[1,2]. Among the more common methods are:

1. Conversion of the particular compound of unknown configuration to a compound of known configuration, without altering the three-dimensional arrangement of substituents about the chiral center.

2. Conversion of the particular compound of unknown configuration to a compound of known configuration with the alternation of configuration about the chiral center, provided that the mechanism is known. For example, S_N2 mechanisms proceed with inversion at chiral carbons.

3. Use of biochemical agents such as enzymes. For example, if an enzyme has been reported to bind, or act on, only the R-isomers of a series of substrates then, if the enzyme acts on the unknown, it is likely, although not always, that the unknown also has the R-configuration.

4. In certain cases, the sign and extent of the optical rotation can aid in assigning the configuration. Often, with a homologous series of molecules, the optical rotation will change in the same direction as the particular entity increases in size. Hence, if the configurations of a number (not just one) are known, the unknown configuration of molecule that fits within that series may be determined.

5. The X-ray method of Bijvoet et al.[3] can determine the configuration directly.

6. The use of optical rotary dispersion and circular dichroism also provide means for determining configurations.

7. Using techniques associated with a symmetric or stereoselective synthesis, a particular compound can be prepared, based on knowledge of the stereochemical course of the synthesis.

[1]H. B. Kagan (1977) *Determination of Configuration by Chemical Means* (vol. **3** of Stereochemistry), Georg Thieme Publishers, Stuttgart.

[2]J. H. Brewster (1972) in *Elucidation of Organic Structures by Physical and Chemical Methods* (eds., K. W. Bentley & G. W. Kirby), vol. **4** of Techniques of Chemistry (ed., A. Weissberger), part 3, pp. 1-249, Wiley, New York.

[3]J. M. Bijvoet, A. F. Peerdeman & A. J. van Bommel (1951) *Nature* **168**, 271.

[4]H. B. Kagen (1977) *Determination of Configurations by Dipole Moments, CD, or ORD* (vol. **2** of Stereochemistry), Georg Thieme Publishers, Stuttgart.

[5]H. M. Smith (1983) *Chem. Rev.* **83**, 359.

CONFORMATIONAL CHANGES IN PROTEINS

Enzyme active sites and receptors rarely interact with ligands without some attendant change in conformation, and the ability to detect and quantify a conformational change lies at the heart of contemporary biochemical kinetics. *See* Induced Fit Model; Fluorescence Spectroscopy; Linked Functions; Hemoglobin; Cooperativity

Selected entries from *Methods in Enzymology* [vol, page(s)]: 2-Bromoacetamido-4-nitrophenol as a reporter group, **11**, 866; dansyl chloride (1-dimethylaminonaphthalene-5-sulfonyl chloride) as fluorescent reporter group, **11**, 857; difference spectroscopy, **11**, 748; environmentally sensitive groups attached to proteins, **11**, 856; fluorescence measurements, **11**, 776; hydrogen exchange, **11**, 734; immunological techniques, **11**, 917; proteolytic susceptibility, **11**, 905; microcalorimetry of protein conformational changes, **61**, 261, 287; detection of conformational changes in proteins by low temperature-rapid flow analysis, **61**, 318; investigation by [near-ultraviolet circular dichroism, **61**, 339; time-resolved fluorescence measurements, **61**, 378; low frequency vibrations and the dynamics of proteins and polypeptides, **61**, 425; carbon-13 nuclear magnetic resonance, **61**, 458]; enzymes, **61**, 259-335; **63**, 9; activation energetics, **64**, 225, 226; calcium binding, **64**, 387; calorimetry, **61**, 266; attending enzyme binding at interface, **64**, 388-392; fluorescence probes, **69**, 518, 519; hexokinase burst, **64**, 199; hysteresis, **64**, 194; isotope trapping, **64**, 58; lipase, **64**, 344, 348; lipid, **64**, 385, 388; low-temperature dynamics, **61**, 318-335; phospholipase A₂, **64**, 348, 387, 388; ribonuclease, **64**, 220; specificity, reversal, **64**, 344; structural basis, **64**, 223-225; interactions, biological systems, **61**, 24, 25; spectroscopic studies, **61**, 337-549; studies with site-specific anti-cytochrome c antibodies, **74**, 257-262; temperature effects, **63**, 239, 348; surface pressure effects, **64**, 387.

CONFORMATIONAL EQUILIBRIA, MACROMOLECULAR

Any of the so-called *all-or-none* transitions between ordered structure and random coil for biological macromolecules. These highly cooperative processes can be defined by an equilibrium constant written as: K = (fraction of molecules in random coil)/(fraction of molecules in ordered structure). Such transitions occur over very narrow temperature ranges, and the midpoint is usually given the name "the melting temperature" (T_m). The high degree of cooperativity in this two-state transition indicates that any break in local order results in loss of global order. This type of treatment can be used to characterize the conformational transitions of DNA as a function of temperature. They also apply to the heat-induced denaturation of proteins.

CONFORMATIONS

Three-dimensional arrangements in space of atoms and groups in a molecular entity that are interconvertible by free rotations about chemical bonds[1,2]. Specific conformations are often called conformers or rotamers. Typically, conformers tend to interconvert rapidly.

[1]E. L. Eliel, N. L. Allinger, S. J. Angyal & G. A. Morrison (1965) *Conformational Analysis* Interscience, New York.
[2]J. Dale (1976) *Top. Stereochem.* **9**, 199.

CONSERVATION OF CHARGE, LAW OF

The condition satisfied whenever the total net charge of any closed system (*e.g.*, a chemical reaction) remains constant.

CONSERVATION OF ENERGY, LAW OF

The condition satisfied whenever the total energy of any closed system is constant. Thus, for a chemical reaction there is no loss or gain of energy in a closed system; however, energy may be transformed from one form into another. **See** *Conservation of Mass and Energy; Thermodynamics*

CONSTANT RATIO METHODS

1. A procedure used to assist in identifying sequential mechanisms when the double-reciprocal plots exhibit parallel lines[1-4]. In some cases, bireactant mechanism can have various collections of rate constants that result in so-called parallel line kinetics, even though the mechanism is not ping pong. However, if the concentrations of A and B are kept in constant ratio with respect to each other, a sequential mechanism in a $1/v$ *vs.* $1/[A]$ plot would be nonlinear (since in the denominator the last term of the double-reciprocal form of the rate expression contains $[A]^2$: for example, for the steady-state ordered Bi Bi reaction scheme in which $[B] = \alpha[A]$, the double-reciprocal rate expression becomes $1/v = (1/V_{max}) + (1/(V_{max}[A]))(K_b/\alpha + K_a) + K_{ia}K_b/(\alpha V_{max}[A]^2))$, whereas a ping pong mechanism would still be linear (that rate expression not containing a squared term). However, the nonlinearity is still dependent on the magnitude of this last term. The influence of this squared term can be increased by altering the ratio of A to B. Thus, in such studies it is advisable to conduct at least two sets of experiments: one in which $[A] \gg [B]$ and the other in which $[B] \gg [A]$. If substrate inhibition is a factor, however, the concentration of the substrate has to be maintained below the concentration producing inhibition.

2. A similar procedure can be used in product inhibition studies for enzymes having three or more products. Different mechanisms can be distinguished if certain products are held in a constant ratio.

3. Constant ratio methods are also employed to maintain mass action ratios in measurements of isotope exchange at equilibrium.

4. An initial-rate method introduced by Fromm[3] to discriminate between rival three-substrate enzyme kinetic mechanisms. **See** *Fromm Method for Ternary Systems.*

[1]E. Garces & W. W. Cleland (1969) *Biochemistry* **8**, 633.
[2]W. W. Cleland (1970) *The Enzymes*, 3rd ed., **2**, 1.
[3]H. J. Fromm (1975) *Initial Rate Enzyme Kinetics*, Springer-Verlag, New York.
[4]D. L. Purich (1992) *Meth. Enzymol.* **87**, 3.

CONSTITUENT

A phase of a particular chemical or physical system. For ice floating in water, there are three constituents of the system: liquid water, ice, and air. However, there are only two components, water and air. **See also** *Component*

CONTINUOUS-FLOW DEVICE

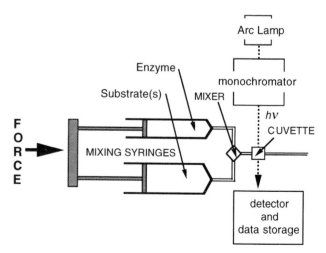

A rapid mixing device, introduced nearly eighty years ago by Hartridge and Roughton, to mix two or more reactants (mixing time $\approx$ 10-50 μsec) and to utilize a constant flow rate through a loop that then passes through the observation port of a flow cuvette. The length of the loop and flow rate determine the "age" of the sample examined in the cuvette. The chief advantage of the method is that the sample is essentially time-invariant, and high quality spectral data can be obtained in the continuous-flow mode. The chief disadvantage is the consumption of high volumes of reactants. *See Stopped-Flow Technique; Rapid Mix-Quench Method*

CONTINUOUS *versus* STOPPED-TIME ASSAYS

The velocity of an enzyme-catalyzed reaction can be measured either by a continuous assay or by a stopped-time protocol. Whenever possible, the continuous measurement of a velocity (*e.g.*, the increase or decrease in absorbance *vs.* time) should be utilized. In stopped-time assays, the investigator must demonstrate that the reaction is completely terminated at the specified point in time and that products are readily and quantitatively separated from substrates. In addition, one must show that the system is under initial rate conditions. Thus, at least three or four different time points should be chosen. Stopped-time assays also require an assay blank (for $t = 0$). In this blank, typically the quenching conditions are applied prior to the initiation step. Whenever practicable, replicate kinetic analyses should be done, even with continuous assay protocols. *See Enzyme Assay Methods; Basal Rate*

CONTRADICTION

A lack of agreement or consistency in some experimental finding or theoretical argument. In the case of kinetic studies, one can take advantage of contradictions to eliminate rival schemes as long as one is confident of both the experimental results and the design on which they are based. Accordingly, one seeks to rule out potential explanations or rival mechanisms by seeking to find contradictions between theory and experiment.

In his influential philosophical essay "What is Life?," Schrödinger[1] quotes Miguel de Unamuno: "*Si un hombre nunca se contradice, será porque nunca dice nada*" (Translated: "Should a man never contradict himself, the reason must be that he virtually never says anything at all.") Philosophers have argued that valuable scientific publications must address points of contradiction con-

cerning earlier work and that they will raise new contradictions provoking continued inquiry.

[1]E. Schrödinger (1956) *What is Life? and Other Scientific Essays*, Doubleday, Garden City, New York.

CONTRAST AGENT

Any paramagnetic (or ferromagnetic) metal complex or particle that decreases the relaxation times of nuclei detected in an image.

M. W. G. de Bolster (1997) *Pure Appl. Chem.* **69**, 1251.

CONTROL

A set of experimental conditions for which the outcome of measurement or observation is well defined on the basis of previous experience. A positive control describes a set of conditions known to result in a well characterized behavior, whereas a negative control confirms that a factor previously known to be without effect continues to be without effect.

For example, one may decide to alter the order of a series of rate assays with respect to substrate concentration (*i.e.*, in one case starting from low to high substrate concentration and then again from high to low substrate concentration). This would allow one to identify any effect of substrate concentration on the time-dependent inactivation of an enzyme. Without such a control, the experimenter might overlook an important effect of the substrate's effect on enzyme stability. Futhermore, depending upon the extent of enzyme inactivation and the order of rate assays with respect to substrate concentration, the observed rate saturation behavior could have the appearance of a cooperative system. Ideally, one can use a random number table to assign the order in which a set of measurements is obtained. In this case, the experimenter decides on a "randomization rule" (such as choosing every third entry from left to right in a table of random digits) and uses a pencil eraser to blindly locate a starting place in the table. After a random number is paired with each experimental condition, one can execute the experiment in an order determined by the magnitude of the random number in ascending or descending order. *See Statistics (A Primer)*

CONTROL COEFFICIENT

An index (symbolized by C_i^X) of the magnitude of change (or output) X exhibited by a system variable in response to a certain perturbation (or input):

$$C_i^X = \frac{\partial X}{\partial v_i} \cdot \frac{v_i}{X} = \frac{\partial \ln X_i}{\partial \ln v_i}$$

where X is the system variable, and v_i is steady-state rate of ith enzyme-catalyzed reaction. For metabolic pathways, a system variable would be a flux or a concentration of a given reactant, intermediate, or product.

COOPERATIVITY

Any process in which an earlier event affects the likelihood of occurrence of some subsequent event(s). For enzyme-catalyzed reactions, cooperativity is observed in the regulation of metabolic pathways by enzymes expressing non-hyperbolic rate-saturation behavior. The most widely accepted explanations for cooperativity relate equilibrium and kinetic models for symmetry-conserving and ligand-induced conformational models[1-3], but mnemonic models can account for some aspects of hysteretic behavior. **See** *Adair Equation; Allosterism; Hemoglobin; Hill Equation and Plot; Hysteresis; Koshland-Neméthy-Filmer-Model; Ligand Binding Analysis; Monod-Wyman-Changeux Model; Negative Cooperativity; Positive Cooperativity; Scatchard Equation and Plot*

[1]K. E. Neet (1980) *Meth. Enzymol.* **64**, 139.
[2]K. E. Neet & G. R. Ainslie, Jr. (1980) *Meth. Enzymol.* **64**, 192.
[3]K. E. Neet (1995) *Meth. Enzymol.* **249**, 519.

Selected entries from *Methods in Enzymology* [vol, page(s)]: Adair model, **249**, 526-528; analysis [graphical representation in, **249**, 520, 522-525; mutational, **249**, 554]; definition, **249**, 519-520; apparent in [bisubstrate mechanism, **64**, 162, 163; concerted symmetrical model, **64**, 167, 168]; catalytic effect, **64**, 174-176; comparative aspects, **64**, 171-174; complex, **64**, 140, 141, 181-185; concerted Monod-Wyman-Changeux model [assumptions, **64**, 166, 166; **249**, 528-530, 533-534, 546-549, 553-554; binding model (exclusive, **64**, 166, 172; nonexclusive, **64**, 166, 172); catalytic effect, **64**, 174, 175; comparison with other model, **64**, 171-174; constant, **64**, 154; equilibrium expression, **64**, 166, 167; hysteresis, **64**, 214, 215; modifier action, **64**, 167; negative, **64**, 172-175]; concerted-symmetrical model, **64**, 165-168; constants, **64**, 153-159; crystal lattice as allosteric effector stabilizing quaternary conformation, **232**, 15; curvature, reciprocal plot, **64**, 142-144; data analysis, **64**, 150, 151; equilibrium model, **64**, 141, 151-153, 159-164, 185-189; estimation, **64**, 142-153; false, **63**, 6, 7; gamma coefficient, **64**, 142, 143, 149-151; heterotropic interaction, **64**, 140, 141; Hill model, **64**, 144-147, 150, 151; homotropic interaction, **64**, 140; hysteretic systems, **64**, 163, 164, 192-226; induced-fit model, **64**, 185-189; isotope exchange, **64**, 4, 17; kinetic aspects, **64**, 141, 159-164; kinetic mechanisms, **249**, 520-521, 539-546 [steady-state random enzyme, **249**, 539-541]; Koshland-Nemethy-Filmer model, **249**, 530-534, 548 Kurganov method, **64**, 148; least squares method, **64**, 156-159; lipase-colipase system, **64**, 379; metal ions, **63**, 266; multisite, **64**, 160-162; negative, **64**, 140, 181-183; **249**, 521, 540 [graphical represen-

tation, **249**, 523-524 chymotrypsin, **64**, 220; concerted model, **64**, 172, 173; double-reciprocal plot, **64**, 142-144; half-site reactivity, **64**, 183-185; Hill plot, **64**, 144-147, 150-153; hysteresis, **64**, 164, 195, 196, 205, 206, 210, 212, 214; maximal velocity determination, **64**, 145; mixed, **64**, 181-183; parameters describing allosteric enzymes [computer programs for, **249**, 525-526; estimation by curve fitting, **249**, 525-526]; product effect, **64**, 177; protomer-oligomer equilibrium, **64**, 180; Scatchard plot, **64**, 147, 148; sensitivity index, **64**, 302; sequential model, **64**, 170, 175]; nomenclature, **64**, 159, 160; positive, **64**, 139, 140; **249**, 521, 540; protomer-oligomer equilibrium, **64**, 178-181; proton binding, **63**, 195-198, 232; PRPP:ATP phosphoribosyltransferase, **63**, 9; R_S, **64**, 147, 149, 150; R_V, **64**, 143, 144, 149-151; rate equation, **64**, 151-153; $S_{0.5}$, **64**, 146, 147, 150, 152; Scatchard plot, **64**, 145, 147, 148, 150; sequential model, **64**, 168-171; site-site interactions, **64**, 164-178; Sturgill-Biltonen method, **64**, 148; thermodynamic analysis, from graphs, **249**, 557; V system, **64**, 168.

COORDINATE-COVALENT BOND

A covalent bond formed when the two bonding electrons originate from or are said to be donated by the same atom.

COORDINATION

1. A covalent bond, also called a coordinate-covalent bond, in which the two electrons associated with that bond have come from only one of the two atoms or groups linked by that bond. While such bonds are often written as A: → B, the nature of this bond is similar to that of a covalent bond with significant dipolar character.

2. The process of formation of a coordinate-covalent bond; the reverse of unimolecular heterolysis. 3. The number of ligands surrounding a central atom. **See also** *Dipolar Bond*

COORDINATION NUMBERS & GEOMETRIES

1. The number of atoms, groups, molecules, or ions surrounding a particular atom or ion in a crystal or complex. 2. The number of atoms, groups, or ions directly linked to a particular atom or ion in a crystal or complex.

Pauling[1] offers the following distinction: the ligancy or coordination number of the central atom is the number of atoms bonded to that central atom, whereas its covalence is given by the number of covalent bonds formed by the central atom. The distinction becomes evident when double bonds join a central atom to a binding partner.

COORDINATION NUMBERS & GEOMETRIES

Coordination Number 2:

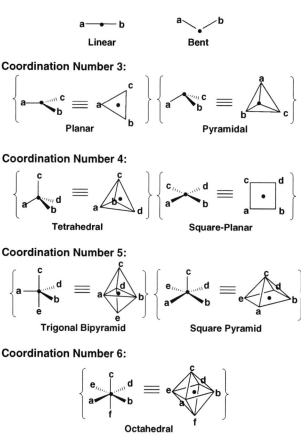

Coordination Number 3:

Planar **Pyramidal**

Coordination Number 4:

Tetrahedral **Square-Planar**

Coordination Number 5:

Trigonal Bipyramid **Square Pyramid**

Coordination Number 6:

Octahedral

Note: The symbol • represents the central atom in the complex.

The following are descriptions and examples of complexes of metal ions having coordination numbers of two to six.

Coordination Number Two. Although rarely found in nature, complexes of Cu(I), Ag(I), Au(I) and Hg(I) adopt linear coordination geometries, such as that observed with [H_3N—Ag—NH_3].

Coordination Number Three. These include planar complex ions such as boric acid $B(OH)_3$ and out-of-plane pyramidal complexes (including $SnCl_3^{1-}$).

Coordination Number Four. These complex ions either have (a) a square planar geometry with the four ligands at each corner, such that the metal ion lies in-plane at the center, or (b) a tetrahedral complex where the centrally located metal ion has four ligands arranged as the hydrogen atoms in methane. Square planar complexes of

Cu(II), Ni(II), and Pt(II) which have eight d-orbital electrons, forming four electron pairs filling all d orbitals (but not the $d_{x^2-y^2}$ orbital) and hybridizing to create four dsp^2 orbitals extending in-plane from the central metal ion. A good example is the antineoplastic agent cis-platin, or cis-[Pt(II)Cl_2(NH_3)$_2$] which has two chloride ions next to each (and hence cis) and two ammonia groups likewise situated next to each other. The $trans$-complex has alternating ligands, with no two alike next to each other. Examples of tetrahedral complexes are the rubredoxins which use four sulfhydryl groups to ligate the central iron atom. In this case, however, the bond angles are not all 109.5°, as observed in methane.

Coordination Number Five. Although much less frequently observed, two geometries have coordination numbers of five. The first has two apical ligands situated above and below a plane that contains three trigonally disposed ligands around it. [This pentavalent or trigonal bipyramid geometry is analogous to that achieved in phosphoryl transfer reactions, where the incoming nucleophile and the leaving group (*i.e.*, exiphile) are in apical positions, and the three other oxygen atoms are arranged trigonally and in-plane with respect to the phosphorus atom.] The second, so-called square-pyramid geometry resembles a five-sided Egyptian pyramid with a central metal ion and four ligands situated in the plane of its base, and the fifth ligand occupying an above-plane or apical position. Such a five-cordinate complex is that adopted by the high-spin Fe(II) atom in deoxyhemoglobin, except that the fifth ligand, a histidine imadazole nitrogen atom, pulls the iron out-of-plane by about 0.75 Å.

Coordination Number Six. These are among the most frequently observed complexes. All six ligands are arranged in a hexacoordinate structure with octahedral (*i.e.*, eight-sided) geometry. A good example is O_2 bound as the sixth ligand in oxygenated hemoglobin. Likewise, the structure of cyanocobalamin (vitamin B_{12}) is octahedral, with four in-plane porphyrin ring nitrogens, and cyanide and a nitrogen supplied by the vitamin's dimethylbenzimidazole substituent. In living systems, the bound cyanide is hydrolyzed, allowing water to become the sixth ligand.

[1]L. Pauling (1960) *The Nature of the Chemical Bond*, 3rd ed., p. 63, Cornell University Press, Ithaca.

COPROPORPHYRINOGEN OXIDASE

This enzyme [EC 1.3.3.3] (also referred to as coproporphyrinogenase, coproporphyrinogen-III oxidase, and coprogen oxidase) catalyzes the reaction of coproporphyrinogen-III with dioxygen to generate protoporphyrinogen-IX and two carbon dioxide molecules. Iron ions are required as cofactors.

T. Yoshinaga (1997) *Meth. Enzymol.* **281**, 355.
S. Granick & S. I. Beale (1978) *Adv. Enzymol.* **46**, 33.

CORNFORTH CHIRAL METHYL STEREOCHEMICAL METHOD

This procedure[1,2] involves the enzymatic conversion of the chiral acetate ($C(HDT)COO^-$), obtained from experiments involving chiral methyl groups[2], to labeled malate. The enzymes used in this procedure include acetate kinase, phosphotransacetylase, malate synthase, and fumarase.

[1] J. W. Cornforth, J. W. Redmond, H. Eggerer, W. Buckel & C. Gutschow (1970) *Eur. J. Biochem.* **14**, 1.
[2] H. G. Floss (1982) *Meth. Enzymol.* **87**, 126.

"CORN" RULE

The mnemonic for visualizing the correct three-dimensional representation of L-amino acids. As an individual sights along the H—C_α bond, with the hydrogen facing the person, the three remaining substituents should spell CO-R-N (for CO [*i.e.*, the carbonyl], R for the side chain, and N for the amino nitrogen) if the amino acid is an L-amino acid.

J. S. Richardson (1981) *Adv. Protein Chem.* **34**, 171.

CORPHIN

Corphin is the F-430 cofactor found in methyl-coenzyme M reductase, a nickel-containing enzyme that participates in the conversion of carbon dioxide to methane in methanogenic bacteria. The nickel ion in F-430 is coordinated by the tetrahydrocorphin ligand, which contains structural elements of both porphyrins and corrins.

M. W. G. de Bolster (1997) *Pure Appl. Chem.* **69**, 1251.

CORRELATION COEFFICIENT

A statistical parameter, often symbolized by r, that indicates the degree to which two variables have a linear relationship. The value of r will range between -1 and $+1$ in which a value of 1.00 indicates a perfect positive relationship between the two variables.

CORRELATION DIAGRAM

A depiction of how molecular orbitals are formed from atomic orbitals of separated atoms and/or how these molecular orbitals correlate with atomic orbitals of the atoms united as the nuclei come together to form a molecular species. Such diagrams are especially useful in rationalizing the symmetry properties of orbitals.

CORRELATION FUNCTION

A mathematical expression that defines how an ensemble average fluctuates with time. An example is the analysis of Brownian motion, where one may seek to understand the nature of the frictional force. If one considers τ as a time interval that is very small on the macroscopic scale, but large on the microscopic scale, then

$$\tau \gg \tau^*$$

where τ^* is the correlation time which will be on the same time-scale as the average period of fluctuations of the force $F(t)$. One can also say that τ^* is a direct measure of the relaxation time needed for the degrees of freedom responsible for the force to return to equilibrium after being suddenly disturbed. Using this formalism, one can ask for a certain time t_1 what the probability is that the force $F(t)$ assumes a value between $F(t_1)$ and $F(t_1) + dF(t_1)$. Likewise, one can evaluate the joint probability that at a time t_1 the force lies between $F(t_1)$ and $F(t_1) + dF(t_1)$ and that at t_2 the force lies between $F(t_2)$ and $F(t_2) + dF(t_2)$.

Correlation functions are powerful tools in statistical physics, and in the above example they permit one to examine the behavior of a fluctuating system from a reference time back to previous times. Such fluctuations can occur in the concentration of two (or more) interconverting chemical species in dynamic equilibrium, and the technique of concentration correlation analysis permits one to determine the forward and reverse rate constants for their interconversion. **See** *Concentration Correlation Analysis*

CORRELATION SPECTROSCOPY

A technique, also known as fluctuation correlation spectroscopy, that uses a correlation algorithm to detect similarity of a fluctuating signal occurring over a period of time. This property aids in the kinetic characterization of fluctuations in light scattering, or other electromagnetic properties, from a small number of particles or molecules present in the small volume element that is sampled by an incident beam of light. Correlation spectroscopy allows an investigator to characterize aspects of the local nature of the system being studied.

Application of the correlation function shows how similar two signals are with respect to each other and indicates how long the two signals remain similar as one is shifted in a time domain relative to the other. Correlating a signal with itself is called autocorrelation. Different sorts of signals have distinctly different autocorrelation functions: (a) Random noise is said to be uncorrelated, meaning that noise is only instantaneously similar to itself. Upon applying a correlation function, any shift along the time axis fails to yield any correlation at all, and correlation function of random noise with itself is a single sharp spike at a time shift of zero. (b) Periodic signals go in and out of phase, as one is shifted with respect to the other. This means that such signals will show strong correlation at any shift allowing the peaks to coincide. The autocorrelation function of a periodic signal is also a periodic, with a time-period that is identical to that of the original signal. Signals that are of short duration can only remain self-similar for small time shifts, so their autocorrelation functions are short. *See Correlation Function; Concentration Correlation Analysis*

Selected entries from *Methods in Enzymology* [vol, page(s)]:
Acquisition of frequency-discriminated spectrum, **239**, 162-166, 170; sensitivity, **239**, 169-173; constant-time, **239**, 23-26; double-quantum filtered, **239**, 236; gradient pulse experiments, **239**, 185-189; protein structural information, **239**, 377-379; pulse sequence and coherence transfer pathway, **239**, 148-149; paramagnetic metalloprotein, **239**, 494-497; data recording, SWAT method, **239**, 166-169, 172; line shapes, effects of gradient pulses, **239**, 162-166; identification of protein amino acid resonances, **232**, 100; cyclosporin A, **239**, 240-241.

CORRELATION TIME

The average time (symbolized τ_C) between molecular collisions, taking into account rotational, dipolar, and electron spin lattice relaxation rates.

Selected entries from *Methods in Enzymology* [vol, page(s)]:
Anisotropy effects, **261**, 427-430; determination by dynamic laser light scattering (quasi-elastic light scattering), **261**, 432-433; determination for nucleic acids by NMR [accuracy, **261**, 432-433; algorithms, **261**, 11-13, 425, 430; carbon-13 relaxation, **261**, 11-12, 422-426, 431, 434-435; cross-relaxation rates, **261**, 419-422, 435; error sources, **261**, 430-432; phosphorus-31 relaxation, **261**, 426-427, 431; proton relaxation, **261**, 51, 418-422; relaxation matrix calculations, **261**, 12]; deuterium solvent viscosity effects, **261**, 433; effect

on NOE, **261**, 155, 413-414; length dependence in B-DNA, **261**, 432-433; rotational (*See Rotational Correlation Time*); spectral density function, **261**, 471-473; temperature effects, **261**, 431, 433-434.

CORTISONE β-REDUCTASE

This enzyme [EC 1.3.1.3] catalyzes the reversible reaction of 4,5-β-dihydrocortisone with NADP$^+$ to produce cortisone and NADPH.

G. M. Tomkins (1962) *Meth. Enzymol.* **5**, 499.

COSMOGENIC DATING

Any geochronometric method for estimating the age of objects based upon the generation of radioactive isotopes by cosmic radiation, followed by isotopic incorporation into the biosphere/geosphere, and their subsequent first-order decay with release of radiation and/or accumulation of daughter isotopes. These methods take advantage of the lack of any dependence of the decay rate on temperature, pressure, pH, or other physical parameters. *See Radiocarbon Dating*

COULOMB

The SI unit for electric charge (symbolized by C) and equal to the amount of electricity transferred by a current of one ampere in one second.

COULOMB'S LAW

The relationship describing the interaction, in terms of mutual force (F), between two electrostatic point charges: $F = Q_1Q_2/(4\pi\varepsilon d^2)$ where Q_1 and Q_2 are the respective point charges, d is the distance separating them, and ε is the permittivity of the medium.

COULOMETRIC TITRATION

An analytical technique based upon the constant-current electrochemical generation of a redox-active titrant that can quantitatively react with a redox-active analyte. This technique takes advantage of the high precision with which the instrumentation controls electric current and time to produce standardized solutions of titrant.

4-COUMAROYL-CoA SYNTHETASE

This enzyme [EC 6.2.1.12] (also referred to as hydroxycinnamoyl-CoA ligase and 4-coumarate:CoA ligase) catalyzes the reaction of 4-coumarate with ATP and coenzyme A to yield 4-coumaroyl-CoA, AMP, and pyrophosphate (or, diphosphate).

R. A. Dixon, P. M. Dey & C. J. Lamb (1983) *Adv. Enzymol.* **55**, 1.

COUNTERION

An ion of opposite electric charge. Antonym: Co-ion.

COUNTING STATISTICS

While radioactive decay is itself a random process, the Gaussian distribution function fails to account for probability relationships describing rates of radioactive decay[1]. Instead, appropriate statistical analysis of scintillation counting data relies on the use of the Poisson probability distribution function:

$$P = (\mu^{x_i}/x_i!) \exp{(-\mu)}$$

where P is the frequency of occurrence of a given count x_i, and μ is the mean for a large data set. If N is the number of counts over a certain time interval, then the relative standard deviation ($\sigma_{m,rel}$) can be written as follows:

$$\sigma_{m,rel} = \sigma_m/N = N^{-1/2}$$

because σ_m equals $N^{1/2}$. The standard deviation in the counting rate $r = dN/dt$, or simply N/t, one can show that the standard deviation in r is σ_r which is equal to $(r/t)^{1/2}$. Likewise, $\sigma_{r,rel}$ is given by the expression $(rt)^{-1/2}$. The confidence interval for a measurement can be defined as the limits around a measured quantity within which the true mean can be expected to fall with a stated probability. If the measured standard deviation closely approximates the true standard deviation, the confidence limit for the counting rate is the mean rate $\pm z\sigma$, and a tenfold decrease in the relative uncertainty necessitates approximately a hundred fold greater number of counts. A good rule of thumb is to collect 10,000 counts (above the accumulated background count) to achieve a precision of around one percent. **See** Statistics (A Primer)

[1]G. Friedlander, J. W. Kennedy, E. S. Macias & J. M. Miller (1981) *Nuclear and Radiochemistry*, 3rd ed., chap. 9, Wiley, New York.

COUPLED ENZYME ASSAYS

A protocol for continuous enzyme assay that involves one or more auxiliary enzymes to convert a product of the primary reaction in a second or auxiliary reaction that produces a change in absorbance or fluorescence. As noted below, coupled enzyme assays, while convenient, are fraught with experimental limitations that must be overcome in order to obtain valid initial velocity data.

Primary enzyme reaction: A → B

Auxiliary enzyme reaction: B → C

McClure[1] was among the first to treat the kinetics of coupled enzyme assays, and others have discussed essential aspects of experimental design.[2,3] Rudolph *et al.*[3] provided an extensive list of coupled-enzyme assays for selected enzyme reaction products, and several examples are shown in Table I.

Among the practical considerations for utilizing coupled enzyme protocols are the following:

(1) The components of the coupling system should neither inhibit nor activate the primary enzyme. Moreover, care must be exercized to ascertain that the auxiliary enzyme(s) is not contaminated with other minor enzyme activities capable of influencing the primary enzymatic activity. The results from any coupled enzyme assay must exactly match the results obtained with other valid initial rate assays to ensure that the presence of the auxiliary system in no way affects the activity of the primary enzyme. This is typically accomplished by comparing data obtained from the coupled assay with stopped-time assay results to ensure that similar results are obtained.

(2) Sufficient coupling enzymes must be provided. Various treatments for determining the necessary concentration of auxiliary enzyme are available[2]. Good rules-of-thumb are (a) that further addition of auxiliary enzyme should not change the rate of the primary enzyme reaction, (b) that doubling the concentration of the primary enzyme results in an exact doubling of observed reaction rate. These conditions should be confirmed at the highest and lowest concentrations of the primary reaction substrate(s) if the coupled assay system is to be used to generate a set of reaction velocity data as a function of those same substrate concentrations.

(3) Investigators should always determine that the assays are linear and that the lag times are reasonable (the amplitude change before the onset of linearity should rarely exceed 1–2% of total product formed if the reaction proceeds to equilibrium).

(4) When an investigator studies the effects of other substances (*e.g.*, competitive inhibitors, *etc.*) on the en-

Table I

Auxiliary Enzyme Assay Methods for Initial Rate Analysis

Product Measured	Species Monitored	Coupling Enzymes	Examples
ATP	NADH Disappearance	Hexokinase and Glucose-6-P Dehydrogenase	Adenylate Kinase, Creatine Kinase[a]
ADP	NADH Disappearance	Pyruvate Kinase and Lactate Dehydrogenase	Fructokinase,[b] γ-Glutamylcysteine Synthetase,[c] Mannokinase,[d] ATPase[e,f]
ADP	NADH Disappearance	Pyruvate Kinase, Lactate Dehydrogenase, and 3′-Nucleotidase	Adenosine-5′-phosphosulfate Kinase[g]
AMP	NADH Disappearance	Adenylate Kinase	Adenosine Phosphotransferase,[h] GMP Synthetase[i]
GDP	NADH Disappearance	Pyruvate Kinase and Lactate Dehydrogenase	Adenylosuccinate Synthetase[j]
GMP	NADH Disappearance	GMP Kinase, Pyruvate Kinase, and Lactate Dehydrogenase	Hypoxanthine-Guanine Phosphoribosyltransferase[k]
UDP	NADH Disappearance	Pyruvate Kinase and Lactate Dehydrogenase	Glycogen Synthetase[l]
Fructose 6-P	NADH Formation	Glucosephosphate Isomerase, Glucose-6-P Dehydrogenase	Fructokinase[b]
Galactose	NADH Formation	Galactose Dehydrogenase	Lactase[m]
D-Glucose	O_2 Consumption	Glucose Oxidase and Catalase	Aldose-1-epimerase[n]
3-Phosphoglycerate	NADH Disappearance	Phosphoglyceromutase, Enolase, Pyruvate Kinase, and Lactate Dehydrogenase	Phosphoglycerate Kinase[o]
Pyruvate	NADH Disappearance	Lactate Dehydrogenase	Neuraminidase[p]
Acetyl-CoA	p-Nitroaniline Disappearance	Arylamine Acetyltransferase	Acetyl-CoA Synthetase, ATP-Citrate Lyase[q]
Fumarate	NADH Formation	Fumarase, Malate Dehydrogenase	Argininosuccinate Lyase[r]
Oxaloacetate	NADH Disappearance	Malate Dehydrogenase	Phosphoenolpyruvate Carboxy-kinase[s]
Aspartate	NADH Disappearance	Aspartate Aminotransferase, Malate Dehydrogenase	Asparaginase[t]

[a]L. T. Oliver, *Biochem. J.* **61**, 116 (1955).
[b]B. Sabater, J. Sebastián, and C. Asensio, *Biochim. Biophys. Acta* **284**, 414 (1972).
[c]W. B. Rathbun and H. D. Gilbert, *Anal. Biochem.* **54**, 153 (1973).
[d]B. Sabater, J. Sebastián, and C. Asensio, *Biochim. Biophys. Acta* **284**, 406 (1972).
[e]B. C. Monk and G. M. Kellerman, *Anal. Biochem.* **73**, 187 (1976).
[f]D. J. Horgan, R. K. Tume, and R. P. Newbold, *Anal. Biochem.* **48**, 147 (1972).
[g]J. N. Burnell and F. R. Whatley, *Anal. Biochem.* **68**, 281 (1975).
[h]W. D. Park, M. E. Tischler, R. B. Dunlop, and R. R. Fisher, *Anal. Biochem.* **54**, 495 (1973).
[i]T. Spector, R. L. Miller, J. A. Fyfe, and T. A. Krenitsky, *Biochim. Biophys. Acta* **370**, 585 (1974).
[j]T. Spector and R. L. Miller, *Biochim. Biophys. Acta* **445**, 509 (1976).
[k]A. Giacomello and C. Salerno, *Anal. Biochem.* **79**, 263 (1977).
[l]J. S. Passonneau and D. A. Rottenburg, *Anal. Biochem.* **51**, 528 (1973).
[m]N. Asp and A. Dahlqvist, *Anal. Biochem.* **47**, 527 (1972).
[n]M. K. Weibel, *Anal. Biochem.* **70**, 489 (1976).
[o]M. Ali and Y. S. Brownstone, *Biochim. Biophys. Acta* **445**, 74 (1976).
[p]D. N. Ziegler and H. D. Hutchinson, *Appl. Microbiol.* **23**, 1060 (1972).
[q]G. Hoffman, L. Weiss, and O. H. Weiland, *Anal. Biochem.* **84**, 441 (1978).
[r]J. F. Sherwin and S. Natelson, *Clin. Chem.* **21**, 230 (1975).
[s]R. J. Hansen, H. Hinz, and H. Holzer, *Anal. Biochem.* **74**, 576 (1976).
[t]H. N. Jayaram, P. A. Cooney, S. Jayaram, and L. Rosenblum, *Anal. Biochem.* **59**, 327 (1974).

zyme activity, those other substances should not affect the coupling system. An activator may accelerate the primary enzyme reaction to a point where the auxiliary system becomes limiting, and the presence of inhibitors, metal ion chelators, and other metabolites may reduce the effectiveness of the auxiliary enzyme. In both cases, increasing the amount of auxiliary enzymes may be essential for maintaining a valid coupled enzyme assay.

Even under seemingly ideal conditions, the steady-state concentration of the first reaction product may exceed the inhibitory constant for the binding of that product to the primary enzyme. In such cases, the linearity of the coupled assay can be misleading, and the investigator must validate the coupled enzyme kinetic data by direct comparison with the results obtained by another technique such as a stopped-time radiometric assay. This reference assay must be sufficiently sensitive to permit activity assays under conditions that avoid significant accumulation of product.

Easterby[4] proposed a generalized theory of the transition time for sequential enzyme reactions where the steady-state production of product is preceded by a lag period or transition time during which the intermediates of the sequence are accumulating. He found that if a steady state is eventually reached, the magnitude of this lag may be calculated, even when the differentiation equations describing the process have no analytical solution. The calculation may be made for simple systems in which the enzymes obey Michaelis-Menten kinetics or for more complex pathways in which intermediates act as modifiers of the enzymes. The transition time associated with each intermediate in the sequence is given by the ratio of the appropriate steady-state intermediate concentration to the steady-state flux. The theory is also applicable to the transition between steady states produced by flux changes. Application of the theory to coupled enzyme assays makes it possible to define the minimum requirements for successful operation of a coupled assay. The theory can be extended to deal with sequences in which the enzyme concentration exceeds substrate concentration.

Yang and Schulz[5] also formulated a treatment of coupled enzyme reaction kinetics that does not assume an irreversible first reaction. The validity of their theory is confirmed by a model system consisting of enoyl-CoA hydratase (EC 4.2.1.17) and 3-hydroxyacyl-CoA dehydrogenase (EC 1.1.1.35) with 2,4-decadienoyl coenzyme A as a substrate. Unlike the conventional theory, their approach was found to be indispensible for coupled enzyme systems characterized by a first reaction with a small equilibrium constant and/or wherein the coupling enzyme concentration is higher than that of the intermediate. Equations based on their theory can allow one to calculate steady-state velocities of coupled enzyme reactions and to predict the time course of coupled enzyme reactions during the pre-steady state.

Occasionally, one may also wish to use an auxiliary enzyme not as an assay system but strictly as a means for maintaining the steady-state concentration of a primary reactant in a multisubstrate reaction system. For instance, acetate kinase (and its substrate acetyl phosphate) or creatine kinase (and its substrate creatine phosphate) can be utilized to regenerate the concentration of ATP, a substrate of the primary enzyme. **See** *ATP/GTP Regenerating Systems*

[1]W. R. McClure (1969) *Biochemistry* **8**, 2782.
[2]A. Storer & A. Cornish-Bowden (1974) *Biochem. J.* **141**, 205.
[3]F. B. Rudolph, B. W. Baugher & R. S. Beissner (1979) *Meth. Enzymol.* **63**, 22.
[4]J. S. Easterby (1981) *Biochem. J.* **199**, 155.
[5]S. Y. Yang & H. Schulz (1987) *Biochemistry* **26**, 5579.
Selected entries from *Methods in Enzymology* [vol, page(s)]:
Analysis, **63**, 22-30; assumptions, **63**, 23; dehydrogenase, **63**, 31, 32; fluorescence assay, **63**, 37; isozymes, **63**, 30; kinase, **63**, 5, 6; kinetic theory, **63**, 22-30; lag time, **63**, 23-29; levels, **63**, 23, 26, 28, 37-39; model, **63**, 22-30; pH effects, **63**, 225; practical aspects, **63**, 30-39; precautions, **63**, 39-41; progress curve, **63**, 177.

COVALENT BOND

A strongly stabilizing interaction between two atoms in which certain of their atomic orbitals overlap, thereby resulting in a region of high electron density. This sharing of two electrons in the orbital by the two bonded atoms describes a molecular orbital.

COX-YEATS EQUATION

A modification of the Bunnett-Olsen equation concerned with solvent acidity in which $\log([SH^+]/[S]) - \log[H^+] = m^*X + pK_{SH^+}$ where [S] and [SH$^+$] are the solvent and protonated solvent concentrations, and X is the activity function $\log[(\gamma_S\gamma_{H^+}/\gamma_{SH^+})]$ for an arbitrary reference base. In practice, $X = -(H_o + \log[H^+])$, called the excess acidity (where H_o is the Hammett acidity function, $m^* = 1 - \phi$, and ϕ represents the response of the $S + H^+ \rightleftharpoons SH^+$ equilibrium to changes in the acid concentration). **See** *Acidity Function; Bunnett-Olsen Equation*

R. A. Cox & K. Yates (1978) *J. Amer. Chem. Soc.* **100**, 3861.
R. A. Cox & K. Yates (1981) *Can. J. Chem.* **59**, 2116.
V. Lucchini, G. Modena, G. Scorrano, R. A. Cox & K. Yates (1982) *J. Amer. Chem. Soc.* **104**, 1958.

CRATIC FREE ENERGY

A term introduced by Gurney[1] to describe the contribution to the total free energy of a solution and to free energy changes arising as a result of mixing. The cratic free energy can be regarded as the entropy of mixing, and this contribution to the free energy of binding must be taken into account when attempting to relate the energetics of subsite binding in multisite enzymes with microscopic thermodynamic parameters as well as in the thermodynamics of micelle formation.

[1]R. W. Gurney (1953) *Ionic Processes in Solution*, McGraw-Hill, New York.

CREATINE KINASE

This muscle phosphotransferase (EC 2.7.3.2) catalyzes the reversible rephosphorylation of ADP to form ATP (*i.e.*, K_{eq} = [ATP][Creatine]/([Creatine phosphate][ADP]) = 30). In resting muscle, creatine phosphate is synthesized at the expense of abundant stores of ATP; intracellular creatine phosphate stores often reach 50–60 mM. If ATP is suddenly depleted by muscle contraction, its product ADP is immediately converted back into ATP by the reverse of the creatine kinase reaction. Depending on the pH at which the enzyme is studied, the kinetic reaction can be either rapid equilibrium random or rapid equilibrium ordered. *N*-Ethylglycocyamine can also act as a substrate.

J. J. Villafranca & T. Nowak (1992) *The Enzymes*, 3rd ed., **20**, 63.
G. L. Kenyon & G. H. Reed (1983) *Adv. Enzymol.* **54**, 367.
D. L. Purich & R. D. Allison (1980) *Meth. Enzymol.* **64**, 3.
D. C. Watts (1973) *The Enzymes*, 3rd ed., **8**, 384-455.

Selected entries from *Methods in Enzymology* [vol, page(s)]:
Abortive complex formation, **63**, 432; adenylate cyclase assay, **79**, 165; coupled enzyme assay, **63**, 32; inhibition, **63**, 463 [competitive, **63**, 292; tight-binding, **63**, 463]; mechanism, **64**, 7, 23, 32, 33, 39; metal ion activation, **63**, 258, 263, 264, 275-278, 292, 327-329; (2'-5')-oligoadenylate synthetase assay, **79**, 157, 158; pH effect, **63**, 280-283; creatine kinase isoenzymes [clinical significance, **74**, 198-199; hybridization, **74**, 201; normal plasma levels, **74**, 199; properties, **74**, 198; purification, **74**, 200-202]; radioimmunoassay, **74**, 198-209; ATP-regeneration system, **238**, 34-35; contamination with calmodulin, **238**, 76; use in photoactivatable probe synthesis, **237**, 88-89, 93.

CRITICAL CONCENTRATION

The threshold concentration of monomer that must be exceeded for any observable polymer formation in a self-assembling system. In the context of Oosawa's condensation-equilibrium model for protein polymerization, the cooperativity of nucleation and the intrinsic thermodynamic instability of nuclei contribute to the sudden onset of polymer formation as the monomer concentration reaches and exceeds the critical concentration. Condensation-equilibrium processes that exhibit critical concentration behavior *in vitro* include F-actin formation from G-actin, microtubule self-assembly from tubulin, and fibril formation from amyloid β protein. Critical concentration behavior will also occur in indefinite isodesmic polymerization reactions that involve a stable template. One example is the elongation of microtubules from centrosomes, basal bodies, or axonemes.

The condensation equilibrium model allows us to consider the equilibrium between unpolymerized protein X and its polymeric form P,

$$X + P_n \rightleftharpoons P_{n+1}$$

where n and $n + 1$ indicate the total number of assembled units of protein. In this process, the rate of monomer addition is governed by the rate constant k_{on}, and the rate of monomer release depends on the magnitude of k_{off}. If both rate constants are taken to be independent of the length of the polymer, then the kinetics for monomer addition/release from a large collection of polymers of varying lengths will be:

$$\frac{d[X]}{dt} = k_{on}[X] \sum_{i=1}^{m} [P_i] - k_{off} \sum_{i=1}^{m+1} [P_{i+1}]$$

The summations are made over all polymer lengths from one to a maximal length m and typically represents the polymer weight concentration; *i.e.*, the total amount of monomer present as polymeric species. Thus the two summations are alternate expressions of [P], the approximate concentration of polymer ends. At equilibrium, $dX/dt = 0$, and the dissociation constant ($K_d = k_{off}/k_{on}$) reduces to [X], which is the concentration of monomer that coexists with polymer at equilibrium. We chose to use the infinity sign as a subscript to distinguish this equilibrium concentration of monomer (*i.e.*, $[M]_\infty$) from that concentration [M] that may be present at any other extent of reaction.

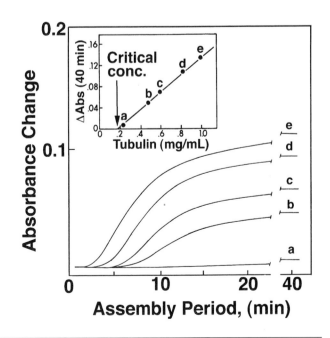

We can consider three situations: (a) if $[M]_\infty > [M]_\infty$, there will be a net increase in polymerization; (b) if $[M] < [M]_\infty$, there will be net depolymerization; and (c) when $[M] = [M]_\infty$, the monomer and polymer are in equilibrium. The last case does not imply a static condition; rather, monomer addition and loss can constantly occur, but the total polymer weight concentration will always remain unchanged. In head-to-tail polymerizations, like those of actin and tubulin assembly *in vitro*, each of the two polymer ends can interact with monomer; the critical concentration $[M]_\infty$ then equals $(k_{on}^+ + k_{off}^-)/(k_{on}^+ + k_{on}^-)$, where the supercripts indicate the plus and minus ends of the assembled polymers. ***See*** *Actin Assembly Kinetics; Microtubule Assembly Kinetics; Endwise Depolymerization*

CRITICAL MICELLE CONCENTRATION

The threshold concentration above which a free amphiphile will self-assemble and spontaneously form micelles. Below this concentration, no micelles are detectable. Note that micelle formation is not equivalent to a phase separation[1]. For most systems, the concentration range associated with the monomer-to-micelle transition is fairly narrow, and c.m.c. values can be obtained from graphical procedures. The symbol for c.m.c. is c_m. Whenever c.m.c. values are reported, the method for the determination should be clearly stated, because different physical techniques may be more or less sensitive to changes in the amphiphile's aggregation state. ***See*** *Micelle Formation*

[1]C. Tanford (1973) *The Hydrophobic Effect*, Wiley-Interscience, New York.

CRITICAL MICELLE TEMPERATURE

The temperature, abbreviated c.m.t., at which a detergent/solvent system or a lipid/solvent system passes from a hydrated crystalline state to an isotropic micellar solution. For a number of lipids, the c.m.t. is below the freezing point of the solvent. The Krafft point, T_k, is the c.m.t. at the critical micelle concentration.

CROSS-CONJUGATION

A form of p orbital overlap in certain types of molecules in which two groups of moieties are not conjugated with each other yet both are conjugated with a third.

An example of a cross-conjugated system would be $C=C-C(=O)-C=C$.

N. F. Phelan & M. Orchin (1968) *J. Chem. Educ.* **45**, 633.

CROSSOVER THEOREM

A principle in metabolic regulation that allows one to identify the inhibited step within a metabolic pathway as that reaction for which the concentrations of reactants and products rise and fall, respectively, from their steady-state values when an inhibitor is introduced. In the context of the electron transfer chain, the crossover-point refers to that reaction step demarking the transition from more reduced to more oxidized respiratory enzymes.

CROSS-REACTION

The binding interaction between an antibody (or antiserum) and a substance that was not the original immunogen employed in the antibody's production. The interaction results from the presence of one or more binding determinants (called epitopes), and at least one of two cases holds true: (1) the substance has an epitope that mimicks the conformation and/or electrostatic properties of the original epitope; or (2) the original immunogen was contaminated with another substance giving rise to the cross-reactivity. Because each antibody has its own well-defined epitope binding site, the second case above is unlikely with monoclonal antibodies. Should a monoclonal Ab show cross-reactivity, the first case is the most reasonable explanation.

CROTONASE SUPERFAMILY

Members of this superfamily catalyze coenzyme A thioester reactions requiring a stabilized oxyanion intermediate. Babbit and Gerlt[1] point out that an enolate anion is generated by abstraction of the α-proton of the thioester in reactions, resulting in β-addition or elimination of water (crotonase and carnitine racemase), 1,3-proton transfer (3,2-*trans*-enoyl-CoA isomerase), and carbon-carbon bond formation (naphthoate synthase, intramolecular addition of the enolate anion to a carboxyl group). An enolate anion is also generated by nucleophilic aromatic addition in the reaction catalyzed by 4-chlorobenzoyl-CoA dehalogenase. ***See*** *Enoyl-CoA Hydratase*

[1]P. C. Babbit & J. A. Gerlt (1997) *J. Biol. Chem.* **272**, 30591.

CRYOENZYMOLOGY

The application of very low temperatures to detect, to thermally trap, and to characterize intermediates in enzyme-catalyzed reactions[1-5]. This is made possible by the fact that each individual, elementary step in a reaction pathway has its own activation energy. Lowering the temperature reduces the fraction of molecules that can react in certain steps, thereby permitting otherwise reactive species to accumulate.

Because cryosolvents must be used in studies of biochemical reactions in water, it is important to recall that the dielectric constant of a solution increases with decreasing temperature. Fink and Geeves[1] describe the following steps: (1) preliminary tests to identify possible cryosolvent(s); (2) determination of the effect of cosolvent on the catalytic properties; (3) determination of the effect of cosolvent on the structural properties; (4) determination of the effect of subzero temperature on the catalytic properties; (5) determination of the effect of subzero temperature on the structural properties; (6) detection of intermediates by initiating catalytic reaction at subzero temperature; (7) kinetic, thermodynamic, and spectral characterization of detected intermediates; (8) correlation of low-temperature findings with those under "normal" conditions; and (9) structural studies on trapped intermediates.

[1]A. L. Fink & M. A. Geeves (1979) *Meth. Enzymol.* **63**, 336.
[2]A. L. Fink (1977) *Acc. Chem. Res.* **10**, 233.
[3]P. Douzou (1977) *Cryobiochemistry*, Academic Press, New York.
[4]P. Douzou (1977) *Adv. Enzymol.* **45**, 157.
[5]R. J. Coll, P. D. Compton & A. L. Fink (1982) *Meth. Enzymol.* **87**, 66.

Selected entries from *Methods in Enzymology* [vol, page(s)]: Theory, **63**, 340-352; measurement, **63**, 365; cryosolvent [catalytic effect, **63**, 344-346; choice, **63**, 341-343; dielectric constant, **63**, 354; electrolyte solubility, **63**, 355, 356; enzyme stability, **63**, 344; pH measurements, **63**, 357, 358; preparation, **63**, 358-361; viscosity effects, **63**, 358]; intermediate detection, **63**, 349, 350; mixing techniques, **63**, 361, 362; rapid reaction techniques, **63**, 367-369; temperature control, **63**, 363-367; temperature effect on catalysis, **63**, 348, 349; temperature effect on enzyme structure, **63**, 348.

CRYPTANDS

Monocyclic or polycyclic molecules containing binding sites for one or more so-called guest molecules or ions. Examples include the crown ethers as well as numerous macrocycles containing other electronegative atoms besides oxygen. ***See*** *Inclusion Complexes; Host-Guest Interactions; Cyclodextrins*

J.-M. Lehn (1988) *Angew. Chem. Int., Ed. Eng.* **27**, 90.
P.G. Potvin & J.-M. Lehn (1987) *Prog. Macrocycl. Chem.* **3**, 167.
W. Kiggen & F. Vögtle (1987) *Prog. Macrocycl. Chem.* **3**, 309.

CRYPTATE EFFECT

An effect arising from the binding of a ligand to a cryptand: namely, high stability, high selectivity, slow exchange rates, and shielding of the substrate or guest from the environment. One simple example of a cryptate effect is the binding of NH_4^+ to a protonated polyamine cryptand. The pK_a value of the NH_4^+-cryptate is about six pH units higher than that of free NH_4^+.

J.-M. Lehn (1988) *Angew Chem. Int. Ed. Engl.* **27**, 89.

CRYPTIC CATALYSIS

A kinetic behavior first proposed by Wedler and Boyer[1] based on their studies of modifier effects on bacterial glutamine synthetase. They noted that particular effectors inhibited certain equilibrium exchanges, while other enzymic exchange reactions remained largely unaffected. They suggested that isotope-exchange-at-equilibrium experiments with the appropriate metabolic modifier may divulge characteristics of the catalytic process not typically observable in typical initial-rate studies (hence, use of the term cryptic catalysis). Nonetheless, before one attributes such behavior to cryptic catalysis, great care should be exercised whenever the effector can also act as an alternative substrate or when impurities are present (*e.g.*, the presence of GTP in the GDP stock solutions or *vice verca*). ***See*** *Isotope Exchange at Equilibrium*

[1]F. C. Wedler & P. D. Boyer (1972) *J. Biol. Chem.* **247**, 993.

CRYPTIC STEREOCHEMISTRY

The stereochemical course of a reaction at an achiral center which can be only observed by utilizing isotopic substitution (*i.e.*, a procedure for converting the achiral group into a chiral moiety).

K. R. Hanson & I. A. Rose (1975) *Acc. Chem. Res.* **8**, 81.
H. G. Floss (1982) *Meth. Enzymol.* **87**, 126.

CRYSTAL FIELD SPLITTING

Loss of degeneracy of energy levels of molecular entities due to lower site symmetry created by a crystalline environment; this term is not synonymous with ligand field splitting[1]. ***See*** *Ligand Field Splitting*

[1]Comm. on Photochem. (1988), *Pure and Appl. Chem.* **60**, 1055.

CRYSTAL FIELD THEORY

A systematic approach for rationalizing the properties of coordination entities by assuming that ligands interact with the central metal atom solely by electrostatics, with ligands as negative point charges surrounding an electropositive central atom. No covalent bonding is assumed to occur. Crystal field splitting refers to separation of a transition metal ion's five degenerate d orbitals into energetically different orbitals containing paired or unpaired electrons.

CTP:ETHANOLAMINEPHOSPHATE CYTIDYLYLTRANSFERASE

This enzyme [EC 2.7.7.14], also referred to as ethanolamine-phosphate cytidylyltransferase and phosphorylethanolamine transferase, catalyzes the reaction of CTP with ethanolamine phosphate to produce CDP-ethanolamine and pyrophosphate.

L. M. Tijburg, P. S. Vermeulen & L. M. G. van Golde (1992) *Meth. Enzymol.* **209**, 258.

CTP SYNTHETASE

This enzyme [EC 6.3.4.2], also referred to as UTP:ammonia ligase, catalyzes the reaction of UTP with ATP and ammonia (or glutamine) to produce CTP, ATP, and orthophosphate (and glutamate if glutamine had been used as the nitrogen source).

H. Zalkin (1993) *Adv. Enzymol.* **66**, 203.
H. Zalkin (1985) *Meth. Enzymol.* **113**, 282.
D. E. Koshland, Jr., & A. Levitzki (1974) *The Enzymes*, 3rd ed., **10**, 539.
J. M. Buchanan (1973) *Adv. Enzymol.* **39**, 91.

CUMULATIVE DISTRIBUTION FUNCTION

A distribution function for a random variable X defined as:

$$F(x) = P(X \le x)$$

where x is any real number and $F(x)$ is a function. The distribution function is obtained from the probability function by noting that:

$$F(x) = P(X \le x) = \Sigma f(u)$$

where the sum is taken over all values of u for which u is less than or equal to x.

CUMULATIVE FEEDBACK INHIBITION

A proposal for the regulation of bacterial glutamine synthetase by end products containing nitrogen atoms derived from the amido group of glutamine[1,2]. Such regulatory behavior would allow feedback inhibitors to throttle the enzyme's catalytic activity without completely arresting glutamine biosynthesis, unless all of the end products reached saturating concentrations. Each feedback inhibitor was believed to bind at its own topologically distinct regulatory site on the synthetase, and this incorrect model has become a standard entry in general biochemistry textbooks. Dahlquist & Purich[3] demonstrated that no special sites are involved, and X-ray crystallography has verified their suggestion that many ligands bind directly at the active site[4]. Purich[5] recently presented a critical analysis of the cumulative inhibition of glutamine synthetase.

[1]C. A. Woolfolk & E. R. Stadtman (1964) *Biochem. Biophys. Res. Commun.* **17**, 313.
[2]C. A. Woolfolk & E. R. Stadtman (1967) *Arch. Biochem. Biophys.* **118**, 736.
[3]F. W. Dahlquist & D. L. Purich (1975) *Biochemistry* **14**, 1980.
[4]S. H Liaw, C. Pan & D. S. Eisenberg (1993) *Proc Natl. Acad. Sci. U.S.A.* **90**, 4996.
[5]D. L. Purich (1998) *Adv. Enzymol.* **72**, 9.

CURIE

(1) A unit of measurement of radioactivity; one curie (Ci) is equal to 3.70×10^{10} disintegrations per second (dps) or 3.70×10^{10} becquerels (Bq). (2) Formerly, the amount of radioactivity equal to the amount of radon in equilibrium with one gram of radium.

CURVE FITTING

Any statistical procedure that defines a set of experimental data in terms of a mathematical function. The most common curve-fitting protocol is the least squares method (also known as analysis of covariance).

β-CYANOALANINE SYNTHASE

This enzyme [EC 4.4.1.9] catalyzes the reaction of cysteine with cyanide to produce H_2S and 3-cyanoalanine.

A. E. Braunstein & E. V. Goryachenkova (1984) *Adv. Enzymol.* **56**, 1.

3′,5′-CYCLIC AMP

This prototypical second messenger exerts excitatory control in cells principally by stimulating protein phosphorylation through the activation of kinase cascades.

Selected entries from *Methods in Enzymology* [vol, page(s)]: Radioimmunoassay, **73**, 109; 79, 163, 164; binding assays, **65**, 859, 860; **237**, 394-397; protein binding assays, **238**, 90; ^{3}H-labeled cAMP [formation from [^{3}H]adenine, **195**, 22; chromatographic separation, **195**, 27]; ^{32}P-labeled [synthesis, **195**, 35; purification, **195**, 40]; radio-immunoassay assays with tritiated adenine, **238**, 91-92; binding, GTP inhibition, **237**, 399-400; isotope dilution assays, **237**, 390-394; phospholipase C stimulation, **238**, 212-213; purification, **238**, 45-47, 49-52, 92-94; second messenger responses, **237**, 398-399.

3′,5′-CYCLIC-AMP PHOSPHODIESTERASE

This enzyme [EC 3.1.4.17] catalyzes the hydrolysis of a nucleoside 3′,5′-cyclic phosphate to produce a nucleoside 5′-phosphate. The enzyme can utilize 3′,5′-cyclic AMP, 3′,5′-cyclic dAMP, 3′,5′-cyclic IMP, 3′,5′-cyclic GMP, and 3′,5′-cyclic CMP as substrates.

CYCLODEXTRIN

A host-type molecule (also called cycloamyloses and Schardinger dextrins) consisting of six (α-cyclodextrin), seven (β-cyclodextrin), or eight (γ-cyclodextrin) glucosyl residues interlinked in a large ring by glycosidic bonds. The structure formed is a hollow, truncated cone. The interior portion of the cone is less polar than the exterior and the cyclodextrins (CDs) can act as hosts to many different kinds of guests. Covalently modified cyclodextrins have served as models for enzyme catalysis[1-7]. In addition, if cyclodextrins are stacked, they can act as channel-type complexes.

For additional information on cyclodextrins, reader should consult the indicated pages in Robyt[8]: properties of the α-, β-, γ-, δ-, and ε-cyclodextrins, p. 248; solubilities of α-, β-, and γ-cyclodextrins, p. 249; cyclic β-1→2-glucans, pp. 251-252; cyclic β-1→6- and β-1→3-glucans, pp. 252-254; cyclomaltodextrins, pp. 254-255; cycloalterotetraose, pp. 255-256; cycloinulodextrin, p. 256; synthesis of cyclodextrins, pp. 256-258; and macrocyclic maltodextrins, p. 258.

[1]M. L. Bender & M. Komiyama (1978) *Cyclodextrin Chemistry,* Springer, New York.

[2]W. Saenger (1984) in *Inclusion Compounds* (J. L. Atwood, J. E. Davies & D. D. MacNicol, eds.), vol. **2**, pp. 231-259, Academic Press, New York.

[3]R. Bergeron (1984) in *Inclusion Compounds* (J. L. Atwood, J. E. Davies & D. D. MacNicol, eds.), vol. **3**, pp. 391-443, Academic Press, New York.

[4]I. Tabushi (1984) in *Inclusion Compounds* (J. L. Atwood, J. E. Davies & D. D. MacNicol, eds.), vol. **3**, pp. 445-471, Academic Press, New York.

[5]R. Breslow (1984) in *Inclusion Compounds* (J. L. Atwood, J. E. Davies & D. D. MacNicol, eds.), vol. **3**, pp. 473-508, Academic Press, New York.

[6]A. P. Croft & R. A. Bartsch (1983)*Tetrahedron* **39**, 1417.

[7]I. Tabushi & Y. Kuroda (1983)*Adv. Catal.* **32**, 417.

[8]J. F. Robyt (1998) *Essentials of Carbohydrate Chemistry,* Springer, New York.

CYCLODEXTRIN GLYCOSYLTRANSFERASE

Cyclodextrins are formed in the degradation of starch and dextran by the action of cyclodextrin/glucanosyltransferases [EC 2.4.1.19]. These enzymes also catalyze so-called acceptor reactions in which the cyclodextrin ring is opened and an acceptor molecule (e.g., glucose) is added to the reducing end of the maltodextrin chain.

J. F. Robyt (1998) *Essentials of Carbohydrate Chemistry,* Springer, New York.

I. S. Pretorius, M. G. Lambrechts & J. Marmur (1991) *Crit. Rev. Biochem. Mol. Biol.* **26**, 53.

M. Vihinen & P. Mäntsälä (1989) *Crit. Rev. Biochem. Mol. Biol.* **24**, 329.

CYCLOEUCALENOL–OBTUSIFOLIOL ISOMERASE

This enzyme [EC 5.5.1.9], also called cycloeucalenol cycloisomerase, catalyzes the interconversion of cycloeucalenol to obtusifoliol.

P. Benveniste (1986) *Ann. Rev. Plant Physiol.* **37**, 275.

CYCLOHEXADIENYL DEHYDROGENASE

This enzyme [EC 1.3.1.43], also referred to as arogenate dehydrogenase and pretyrosine dehydrogenase, catalyzes the reaction of arogenate with NAD^+ to produce tyrosine, NADH, and carbon dioxide. Both prephenate and D-prephenyllactate can act as alternative substrates.

R. Bentley (1990) *Crit. Rev. Biochem. Mol. Biol.* **25**, 307.

R. Fischer & R. Jensen (1987) *Meth. Enzymol.* **142**, 488.

CYCLOHEXANONE MONOOXYGENASE

This enzyme [EC 1.14.13.22] catalyzes the reaction of cyclohexanone with NADPH and dioxygen to produce 6-hexanolide, $NADP^+$, and water. FAD is used as a cofactor. A number of other cyclic ketones can serve as substrates.

B. A. Palfey & V. Massey (1998) *Comprehensive Biological Catalysis: A Mechanistic Reference* **3**, 83.
M. A. Ator & P. R. Ortez de Montellano (1990) *The Enzymes*, 3rd ed., **19**, 213.

CYCLOPENTANONE 1,2-MONOOXYGENASE

This enzyme [EC 1.14.13.16] catalyzes the reaction of cyclopentanone with NADPH and dioxygen to produce 5-valerolactone, NADP$^+$, and water [EC 1.14.13.16].

P. W. Trudgill (1990) *Meth. Enzymol.* **188**, 77.

CYSTATHIONINE β-LYASE

This pyridoxal-phosphate-dependent enzyme [EC 4.4.1.8], also referred to as β-cystathionase and cystine lyase, catalyzes the hydrolysis of cystathionine to yield homocysteine, pyruvate, and ammonia.

J. Giovanelli (1987) *Meth. Enzymol.* **143**, 443.
L. Davis & D. E. Metzler (1972) *The Enzymes*, 3rd ed., **7**, 33.

CYSTATHIONINE γ-LYASE

This pyridoxal-phosphate-dependent enzyme [EC 4.4.1.1] (also referred to as homoserine deaminase, homoserine dehydratase, γ-cystathionase, cystine desulfhydrase, cysteine desulfhydrase, and cystathionase) catalyzes the hydrolysis of cystathionine to produce cysteine, ammonia, and α-ketobutanoate (or, 2-oxobutanoate).

T. Nagasawa, H. Kanzaki & H. Yamada (1987) *Meth. Enzymol.* **143**, 486.
A. E. Braunstein & E. V. Goryachenkova (1984) *Adv. Enzymol.* **56**, 1.
A. E. Martell (1982) *Adv. Enzymol.* **53**, 163.
L. Davis & D. E. Metzler (1972) *The Enzymes*, 3rd ed., **7**, 33.

CYSTATHIONINE β-SYNTHASE

This pyridoxal-phosphate-dependent enzyme [EC 4.2.1.22] (also known as serine sulfhydrase, β-thionase, and methylcysteine synthase) catalyzes the reaction of homocysteine with serine to produce cystathionine and water.

J. P. Kraus (1987) *Meth. Enzymol.* **143**, 388.
A. E. Braunstein & E. V. Goryachenkova (1984) *Adv. Enzymol.* **56**, 1.
L. Davis & D. E. Metzler (1972) *The Enzymes*, 3rd ed. **7**, 33.

CYSTEAMINE DIOXYGENASE

This enzyme [EC 1.13.11.19] catalyzes the reaction of cysteamine with dioxygen to produce hypotaurine. The protein uses iron ions as cofactors.

R. B. Richerson & D. M. Ziegler (1987) *Meth. Enzymol.* **143**, 410.
A. J. L. Cooper (1983) *Ann. Rev. Biochem.* **52**, 187.
O. Hayaishi, M. Nozaki & M. T. Abbott (1975) *The Enzymes*, 3rd ed., **12**, 119.

CYSTEINE *S*-CONJUGATE *N*-ACETYLTRANSFERASE

This enzyme [EC 2.3.1.80] catalyzes the reaction of acetyl-CoA with an *S*-substituted L-cysteine to produce coenzyme A and an *S*-substituted *N*-acetyl-L-cysteine.

M. W. Duffel & W. B. Jakoby (1985) *Meth. Enzymol.* **113**, 516.

CYSTEINE *S*-CONJUGATE β-LYASE

This pyridoxal-phosphate-dependent enzyme [EC 4.4.1.13] catalyzes the conversion of an *S*-substituted cysteine (that is, RS—CH$_2$—CH(NH$_3^+$)COO$^-$) to RSH, ammonia, and pyruvate. **See also** S-Substituted Cysteine Sulfoxide Lyase

J. L. Stevens & W. B. Jakoby (1985) *Meth. Enzymol.* **113**, 510.

D-CYSTEINE DESULFHYDRASE

This enzyme [EC 4.4.1.15] catalyzes the hydrolysis of D-cysteine to produce pyruvate, ammonia, and H$_2$S.

A. Schmidt (1987) *Meth. Enzymol.* **143**, 449.

CYSTEINE DIOXYGENASE

This enzyme [EC 1.13.11.20] catalyzes the reaction of cysteine with dioxygen to yield 3-sulfinoalanine. Both iron ions and NAD(P)H are used as cofactors.

K. Yamaguchi & Y. Hosokawa (1987) *Meth. Enzymol.* **143**, 395.
O. Hayaishi, M. Nozaki & M. T. Abbott (1975) *The Enzymes*, 3rd ed. **12**, 119.

CYSTEINE LYASE

This pyridoxal-phosphate-dependent enzyme [EC 4.4.1.10] catalyzes the reaction cysteine with sulfite to produce cysteate and H$_2$S. The enzyme can also catalyze the reaction of two cysteines (thereby producing lanthionine) as well as other alkyl thiols as substrates.

A. E. Braunstein & E. V. Goryachenkova (1984) *Adv. Enzymol.* **56**, 1.
L. Davis & D. E. Metzler (1972) *The Enzymes*, 3rd ed., **7**, 33.

CYSTEINE SULFINATE DECARBOXYLASE

This pyridoxal-phosphate-dependent enzyme [EC 4.1.1.29], also referred to as sulfinoalanine decarboxylase, catalyzes the conversion of 3-sulfino-L-alanine to hypotaurine and carbon dioxide. The enzyme can also utilize L-cysteate as a substrate.

C. L. Weinstein & O. W. Griffith (1987) *Meth. Enzymol.* **143**, 404.
E. A. Boeker & E. E. Snell (1972) *The Enzymes*, 3rd ed., **6**, 217.

CYTIDINE DEAMINASE

This enzyme [EC 3.5.4.5], also called cytidine aminohydrolase, catalyzes the hydrolysis of cytidine to produce uridine and ammonia.

A. Radzicka & R. Wolfenden (1995) *Meth. Enzymol.* **249**, 284.

CYTIDYLATE KINASE

This enzyme [EC 2.7.4.14], also called deoxycytidylate kinase, catalyzes the reaction of CMP with ATP to form CDP and ADP, as well as similar reactions with dCMP, UMP, and dUMP.

E. P. Anderson (1973) *The Enzymes*, 3rd ed., **9**, 49.

CYTOCHROME c

This heme-dependent peripheral membrane protein is a member of the electron-transport chain, and shuttles electrons between components of Complexes III and IV.

H. B. Gray & J. R. Winkler (1996) *Ann. Rev. Biochem.* **65**, 537.
R. J. Maier (1996) *Adv. Protein Chem.* **48**, 35.
G. R. Moore & G. W. Pettigrew (1990) *Cytochromes c. Evolutionary, Structural and Physiochemical Aspects*, Springer-Verlag, New York.
D. C. Rees & D. Farrelly (1990) *The Enzymes*, 3rd ed., **19**, 37.
A. A. DiSpirito (1990) *Meth. Enzymol.* **188**, 289.
T. E. Meyer & M. D. Kamen (1982) *Adv. Protein Chem.* **35**, 105.
R. E. Dickerson & R. Timkovich (1975) *The Enzymes*, 3rd ed., **11**, 397.

Selected entries from *Methods in Enzymology* [vol, page(s)]:
Acid-denatured, conformation, **232**, 3; acid-induced refolding, **232**, 5-6; properties, **234**, 342-343; synthesis, **234**, 339-342; as chemiluminescence catalyst for luminol HPLC hydroperoxide assay, **233**, 325-327; engineered metal-binding sites, in metal affinity partitioning, **228**, 178-179; amino acid composition, **243**, 159-160; amino acid sequence, **243**, 163-165; domain structure, **243**, 164-165; heme-binding sites, **243**, 163-165; heme ligands, **243**, 164-165; histidine residues, **243**, 163-165; properties, **243**, 158-162; redox properties, **243**, 160161; spectral properties, **243**, 158-159; spin state, **243**, 161; HETCOR experiment with low-spin iron(III), **239**, 500; metal affinity partitioning, **228**, 173; reduction [NADPH oxidase assay with, **233**, 223-225; by superoxide, **233**, 155]; superfamily, **243**, 104-105.

CYTOCHROME c₃ HYDROGENASE

This enzyme [EC 1.12.2.1], also called hydrogenase, catalyzes the reaction of H_2 with two ferricytochrome c_3 to produce two H^+ and two ferrocytochrome c_3. The enzyme uses iron ions as well as a cofactor. Methylene blue and other acceptors can be reduced with H_2 with this system. ***See also*** *Hydrogenase; Hydrogen Dehydrogenase*

M. A. Halcrow (1998) *Comprehensive Biological Catalysis: A Mechanistic Reference* **3**, 359.

CYTOCHROME c OXIDASE

This enzyme [EC 1.9.3.1] (also referred to as cytochrome aa_3 and cytochrome oxidase) catalyzes the reaction of four ferrocytochrome c with dioxygen to produce four ferricytochrome c and two water molecules. This protein also contains copper ions as cofactors.

H. B. Dunford (1998) *Comprehensive Biological Catalysis: A Mechanistic Reference* **3**, 195.
A. Messerschmidt (1998) *Comprehensive Biological Catalysis: A Mechanistic Reference* **3**, 401.
H. B. Gray & J. R. Winkler (1996) *Ann. Rev. Biochem.* **65**, 537.
R. J. Maier (1996) *Adv. Protein Chem.* **48**, 35.
R. O. Poyton, B. Goehring, M. Droste, K. A. Sevarino, L. A. Allen & X.-J. Zhao (1995) *Meth. Enzymol.* **260**, 97.
R. A. Capaldi, M. F. Marusich & J.-W. Taanman (1995) *Meth. Enzymol.* **280**, 117.
S. M. Musser, M. H. B. Stowell & S. I. Chan (1995) *Adv. Enzymol.* **71**, 79.
B. L. Trumpower & R. B. Gennis (1994) *Ann. Rev. Biochem.* **63**, 675.
E. T. Adman (1991) *Adv. Protein Chem.* **42**, 145.
W. S. Caughey, W. J. Wallace, J. A. Volpe & S. Yoshikawa (1976) *The Enzymes*, 3rd ed., **13**, 299.

CYTOCHROME c PEROXIDASE

This enzyme [EC 1.11.1.5], having a heme derivative as a cofactor, catalyzes the reaction of two ferrocytochrome c with hydrogen peroxide to produce two ferricytochrome c and two water molecules.

H. B. Dunford (1998) *Comprehensive Biological Catalysis: A Mechanistic Reference* **3**, 195.
J. S. Zhou & B. M. Hoffman (1994) *Science* **265**, 1693.
H. Pelletier & J. Kraut (1992) *Science* **258**, 1748.
D. C. Rees & D. Farrelly (1990) *The Enzymes*, 3rd ed., **19**, 37.
T. E. Meyer & M. D. Kamen (1982) *Adv. Protein Chem.* **35**, 105.
T. L. Poulos & J. Kraut (1980) *J. Biol. Chem.* **255**, 10322.
T. Yonetani (1976) *The Enzymes*, 3rd ed., **13**, 345.

CYTOCHROME P-450

A superfamily of heme-dependent monooxygenases that utilize molecular oxygen and NADPH. These enzymes are localized in the endoplasmic reticulum, and are often used as a marker for microsomal fractions obtained upon homogenization of cells. Especially abundant in liver, cytochrome P-450 enzymes play a major role in detoxification. They also catalyze the first step in ω-oxidation of medium- and long-chain fatty acids.

H. B. Dunford (1998) *Comprehensive Biological Catalysis: A Mechanistic Reference* **3**, 195.

Y. Watanabe & J. T. Groves (1992) *The Enzymes*, 3rd ed., **20**, 405.

F. P. Guengerich (1991) *J. Biol. Chem.* **266**, 10019.

D. C. Rees & D. Farrelly (1990) *The Enzymes*, 3rd ed., **19**, 37.

M. A. Ator & P. R. Ortez de Montellano (1990) *The Enzymes*, 3rd ed., **19**, 213.

S. D. Black & M. J. Coon (1987) *Adv. Enzymol.* **60**, 35.

P. Ortiz de Montellano, ed., (1986) *Cytochrome P-450: Structure, Mechanism, and Biochemistry*, Plenum, New York.

M. J. Coon & D. R. Koop (1983) *The Enzymes*, 3rd ed., **16**, 645.

I. C. Gunsalus & S. G. Sligar (1978) *Adv. Enzymol.* **47**, 1.

V. Ullrich & W. Duppel (1975) *The Enzymes*, 3rd ed., **12**, 253.

Selected entries from *Methods in Enzymology* [vol, page(s)]:
Activity in ethanol oxidation, **233**, 118; in hydroxyethyl radical formation; analysis with reconstituted vesicles, **233**, 127; characterization, **233**, 123-125; monooxygenase activity, **231**, 574-575; reductase [hemoglobin-catalyzed reactions with, **231**, 573-574; oxygen concentration and, **231**, 579; pH dependence, **231**, 580; reaction mixtures, **231**, 578; oxygen content, measurement, **231**, 587-588; reductase concentration and, **231**, 580; time dependence, **231**, 578; preparation, **231**, 577].

CYTOCHROME P-450 REDUCTASE

This enzyme [EC 1.6.2.4] (also referred to as NADPH:ferrihemoprotein reductase, NADPH:cytochrome P450 reductase, TPNH$_2$ cytochrome c reductase, and ferrihemoprotein P450 reductase) catalyzes the reaction of NADPH with two ferricytochrome to produce NADP$^+$ and two ferrocytochrome. The protein requires FMN and FAD. In addition, it also catalyzes the reduction of heme-thiolate-dependent monooxygenases (*e.g.*, the unspecific monooxygenase [EC 1.14.14.1]) and it is a part of the microsomal hydroxylating system. This reductase also is capable of reducing cytochrome b_5 and cytochrome c.

B. A. Palfey & V. Massey (1998) *Comprehensive Biological Catalysis: A Mechanistic Reference* **3**, 83.

β-(9-CYTOKININ)-ALANINE SYNTHASE

This enzyme [EC 4.2.99.13], also known as lupinic acid synthase, catalyzes the reaction of O-acetyl-L-serine with zeatin to produce lupinate and acetate. Zeatin is N^6-(4-hydroxy-3-methyl-butyl-*trans*-2-enylamino)purine. A number of other N^6-substituted purines can function as substrates for this enzyme.

D. S. Letham & L. M. S. Palmi (1983) *Ann. Rev. Plant Physiol.* **34**, 163.

CYTOKININ 7β-GLUCOSYLTRANSFERASE

This enzyme [EC 2.4.1.118], also known as UDP-glucose-zeatin 7-glucosyltransferase, catalyzes the reaction of UDP-glucose with an N^6-alkylaminopurine to produce UDP and an N^6-alkylaminopurine-7-β-D-glucoside. The enzyme can act on a number of N^6-substituted adenine derivatives (*e.g.*, zeatin and N^6-benzylaminopurine). However, it has been reported that N^6-benzyladenine does not serve as a substrate. Depending on the substrate, a 9-β-D-glucoside may be the product formed.

D. S. Letham & L. M. S. Palmi (1983) *Ann. Rev. Plant Physiol.* **34**, 163.

D

DAKIN-WEST REACTION

A chemical reaction in which the carboxyl group of an α-amino acid is replaced by an acyl group (*i.e.*, an acyl-decarboxylation). In this reaction, an α-amino acid is reacted with an anhydride in the presence of pyridine and the resulting product is an *N*-acylated ketone. In certain instances, *N*-substituted amino acids will also act as reactants.

H. D. Dakin & R. West (1928) *J. Biol. Chem.* **78**, 745.
G. L. Buchanan (1988) *Chem.Soc. Rev.* **17**, 91.

DARVEY PLOT

A graphical procedure for characterizing isomerization mechanisms[1,2]. The protocol uses data from product inhibition, and $1/[v_{[P]=0} - v_{[P]0}]$ is plotted *versus* $1/[P]$ at various constant concentrations of the substrate (where $v_{P=0}$ is the initial velocity in the absence of product and $v_{[P]0}$ is the initial velocity in the presence of product). Secondary and ternary replots allows one to characterize the mechanism[1,2]. This procedure requires very accurate estimation of initial rates.

[1]I. G. Darvey (1972) *Biochem. J.* **128**, 383.
[2]K. L. Rebholz & D. B. Northrop (1995) *Meth. Enzymol.* **249**, 211.

DAVIES EQUATION

An equation for estimating how the activity coefficient of an ion in dilute solutions is influenced by ionic strength. ***See*** *Debye-Hückel treatment*

C. W. Davies (1962) *Ion Association*, Butterworth, London.

dCMP DEAMINASE

This enzyme [EC 3.5.4.12], also called deoxycytidylate deaminase, catalyzes the hydrolysis of dCMP to produce

dUMP and ammonia. A number of 5-substituted dCMPs can also serve as substrates.

P. Reichard (1988) *Ann. Rev. Biochem.* **57**, 349.

DEAD-END INHIBITION

Decreased enzymatic activity observed when complexation of substance with an enzyme (or a specific enzyme form in a kinetic mechanism) will not permit catalytic turnover to form product(s), unless that substance dissociates.

DEAD-END INHIBITION (CLELAND'S RULES)

Cleland[1,2] described the following rules for reversible inhibition patterns observed in double-reciprocal plots of initial rate behavior.

Rule 1. Upon obtaining a double-reciprocal plot of $1/v$ *vs.* $1/[A]$ (where [A] is the initial substrate concentration and v is the initial velocity) at varying concentrations of the inhibitor (I), if the vertical intercept varies with the concentration of the reversible inhibitor, then the inhibitor can bind to an enzyme form that does not bind the varied substrate. For example, for the simple Uni Uni mechanism (E + A $\rightleftharpoons$ EX $\rightleftharpoons$ E + P), a noncompetitive or uncompetitive inhibitor (both of which exhibit changes in the vertical intercept at varying concentrations of the inhibitor), I binds to EX, a form of the enzyme that does not bind free A. In such cases, saturation with the varied substrate will not completely reverse the inhibition.

If the inhibitor can bind more than once to the enzyme form that doesn't bind the varied substrate (*e.g.*, forming EXII in the example above), then a secondary replot (vertical intercepts *vs.* [I]) will be nonlinear, exhibiting curvature that is concave up. If multiple binding does not occur, then the secondary plot will be linear.

Rule 2. For a plot of $1/v$ vs. $1/[A]$ at varying concentrations of reversible inhibitor I, if the slope of the lines varies with [I], then (a) the inhibitor either binds to the same enzyme form to which the varied substrate binds or (b) the inhibitor binds to an enzyme form capable of altering the concentration of another enzyme form that reacts with the varied substrate. (Note that in case (b), there must be a reversible reaction between these two enzyme forms.) Example of these cases can be seen with the steady-state ordered Bi Bi reaction scheme having the reciprocal rate expression $1/v = (1/V_{max}) + (K_a + (V_{max} [A])) + (K_b/(V_{max} [B])) + (K_{ia}K_b/(V_{max}[A][B]))$, where V_{max} is the maximum velocity in the forward direction, K_a is the Michaelis constant for the substrate A, K_b is the Michaelis constant for the substrate B, and K_{ia} is the dissociation constant for A. A competitive inhibitor of A, in a $1/v$ vs. $1/[A]$ plot, will exhibit a change in the slope at different concentrations of I. Note also that the slope will vary with [I] in the $1/v$ vs. $1/[B]$ plot. Here, the inhibitor binds to the free enzyme, which is an enzyme form reversibly connected to the enzyme form binding B (*i.e.*, EA).

Note that this rule requires reversible steps between the appropriate enzyme forms. Fromm[3] has pointed out that, for product inhibition with saturating levels of one of the substrates, an effectively irreversible step may exist (*e.g.*, saturating B in the ordered Bi Bi example above: thus, EA + B $\rightleftharpoons$ EAB is essentially irreversible). In this case there would be no slope effect of the product P in the $1/v$ vs. $1/[A]$ plot (at saturating B).

Rule 3. Both Rules 1 and 2 are applied if the reversible inhibitor binds to more than one enzyme form. Secondary plots can be nonlinear. *See* Abortive Complexes

[1]W. W. Cleland (1963) *Biochim. Biophys. Acta* **67**, 188.
[2]K. M. Plowman (1972) *Enzyme Kinetics*, McGraw-Hill, New York.
[3]H. J. Fromm (1975) *Initial Rate Enzyme Kinetics*, Springer-Verlag, New York.

DEBRANCHING ENZYME

4α-Glucanotransferase [EC 2.4.1.25] (also referred to as disproportionating enzyme, dextrin glycosyltransferase, oligo-1,4$\rightarrow$1,4-glucantransferase, and amylomaltase) catalyzes the transfer of a segment of a (1,4)-α-D-glucan to a new 4-position in an acceptor which may be glucose or a (1,4)-α-D-glucan. This enzymatic activity is also a portion of the mammalian and yeast glycogen debranching system (with an amylo-1,6-glucosidase activity). Amylo-1,6-glucosidase [EC 3.2.1.33], also referred to as dextrin 6-α-D-glucosidase, catalyzes the endohydrolysis of 1,6-α-D-glucoside linkages at points of branching in the polysaccharide chains of 1,4-linked α-D-glucose residues. As mentioned above, in mammals and yeast this enzyme is linked to a glycosyltransferase similar to 4α-glucanotransferase [EC 2.4.1.25]. Together, these two activities constitute the glycogen debranching system.

G. Davies, M. L. Sinnott & S. G. Withers (1998) *Comprehensive Biological Catalysis: A Mechanistic Reference* **1**, 119.
E. Y. C. Lee & W. J. Whelan (1971) *The Enzymes*, 3rd ed., **5**, 191.

DEBYE

The SI unit for the measurement of dipole moments (symbolized by D) equal to 10^{-18} esu·cm or 3.33564×10^{-30} coulomb-meter.

DEBYE-HÜCKEL TREATMENT

A mathematical approach that attempts to explain how the solution behavior of electrolytes depends on their ionic charge and the overall ionic strength of the solution. This treatment allows one to estimate the mean activity coefficients that relate the analytical concentration of an electrolyte to its solution activity. We also can come to understand how the activity coefficient behaves as a function of ionic strength.

Significance of Activity Coefficients. While we typically focus our attention on the analytical concentration of reactant(s) and product(s) for a given chemical process, the thermodynamic concept of equilibrium depends on the chemical potential of a species. This is shown by the following relationship

$$dG = \Sigma \mu_i dn_i$$

where μ_i is the chemical potential and dn_i is the change in moles of substance i added to (+ sign) or removed from (− sign) a solution. Let us consider making a solution by adding the reactants for the following equilibrium

$$aA + bB = cC + dD$$

We initially add reactants A and B in only minute amounts (say Aε and Bε) or by adding products C and

D, again only minute amounts (say $C\varepsilon$ and $D\varepsilon$). In such a case, $dG_{forward}$ and $dG_{reverse}$ can be written as follows

$$dG_{forward} = \mu_A A\varepsilon + \mu_B B\varepsilon$$

$$dG_{reverse} = \mu_C C\varepsilon + \mu_D D\varepsilon$$

Equilibrium is attained when $dG_{forward}$ and $dG_{reverse}$ are equal, and

$$\mu_A A\varepsilon + \mu_B B\varepsilon = \mu_C C\varepsilon + \mu_D D\varepsilon$$

We used small amounts of reactants and products to assure that we reached equilibrium, and we now cancel the subscripts in each term to arrive at

$$\mu_A A + \mu_B B = \mu_C C + \mu_D D$$

We can then write out the chemical potential μ_i as

$$\mu_i = \mu_i^{\circ} + RT \ln[A_i/A_i^{\circ}]$$

where μ_i° is the standard chemical potential of substance i, R is the universal gas constant in calories/mole or Joules/mole, T is the absolute temperature in units K, A_i is the activity (or the *effective concentration*) of the substance, and A_i° is its activity under standard conditions. In the explanation given above, we added infinitesmal amounts of reactants and products to the solution, but we need to relate activity A_i to concentration [A], because in practice we most often work at higher concentrations. The activity coefficient γ_i is a proportionality factor which changes under different solution conditions, such that

$$A_i = \gamma_i[A]$$

To understand why γ_i depends on solution conditions, we must recognize that the behavior of solutes in solution depends upon the presence of other similar and/or dissimilar solutes, and electrolytes (*i.e.*, charged solutes) are especially affected by the presence of all ionic species in solution. Unless we can account for these effects on the value of α for each substance, we cannot know the effective concentration of a substance at any particular analytical concentration, and we cannot comprehend the thermodynamic properties of these substances in solution. The Debye-Hückel treatment offers us a means for estimating α_i.

The Debye Hückel Limiting Law. Although beyond the scope of this Handbook, the derivation of this quantitative relationship rests on the following simplifying assumptions: (a) Electrolytes are assumed to be completely dissociated. The ions are taken to have a spherical charge distribution, such that they can be treated as if all their electric charge is localized as a point charge. (b) Interactions between charged solute species are dominated by electrostatic attraction/repulsion, as dictated by Coulomb's Law ($F = q_1q_2/Dr^2$), where F is the force of attraction/repulsion, q_1 and q_2 are the charges expressed in electrostatic units, r is the distance between the ions which are considered to be point charges, and D is the dielectric constant of the medium. (c) The long-range interactions are considered to occur in a uniform medium, and the dielectric constant and viscosity of pure water are used in this model. (d) No contribution is made by short-range forces such as ion-dipole interactions, ion pair formation, and van der Waals attractions. (e) Thermal motions of ions disrupt any oriented interionic attractions in a manner that can be defined by the Boltzmann distribution law.

For simple monovalent cations and anions, the Debye-Hückel limiting law can be written as follows:

$$-\log \gamma_i = Az_i^2(I)^{1/2}$$

where γ_i is the activity coefficient, z_i is the charge, I is the ionic strength ($I = (1/2)\Sigma c_i z_i^2$). The proportionality constant A is written as:

$$A = (1/2.303)(e^2/2Dk_BT)(8\pi e^2 N/1000Dk_BT)^{1/2}$$

where N is Avogadro's number, k_B is the Boltzmann constant, and T is the absolute temperature. The value of A for water is 0.512, such that

$$-\log \gamma_i = 0.512z_i^2(I)^{1/2}$$

By introducing the mean activity coefficient (γ)

$$(\gamma)^{(z_M+z_N)} = (\gamma_M)^{z_N}(\gamma_N)^{z_M}$$

Taking the negative logarithm of each side, we obtain

$$-(z_M + z_N)\log \gamma = -z_N \log \gamma_M - z_M\log \gamma_N$$

Substituting the Debye-Hückel limiting law into each term on the right-hand side of the equation, and then rearranging, we get:

$$-\log \gamma = 0.512 z_M z_N I^{1/2}$$

where one uses the absolute values of z_M and z_N.

To illustrate how this equation is used, we can calculate the value of γ for a solution of 0.1 F HCl. [Note that formality (F) is the number of gram formula weights per liter solution.] We distinguish formality and molarity: the first indicates explicitly how the solution was prepared, and the second specifies the concentration of actual species in solution. As to the value of ionic strength, we can write the following:

$$I = 1/2 \sum c_i z_i^2 = 0.5 \{(0.1)(1)^2 + (0.1)(-1)^2\}$$
$$= 0.5\{0.2\} = 0.1$$

$$-\log \gamma = 0.512 z_M z_N I^{1/2} = 0.512(1)(1)(0.1)^{1/2} = 0.16$$

Taking the antilogarithm of each side, we get γ equal to 0.69, and this value is to be compared to a measured activity of 0.80.

The Extended Debye-Hückel Equation. This exercise reminds us that the Debye-Hückel limiting law is not sufficiently accurate for most physicochemical studies. To estimate the calculated activity coefficient more accurately, one must consider the fact that ions are not point charges. To the contrary, ions are of finite size relative to the distance over which the ions interact electrostatically. This brings us to the extended Debye-Hückel equation:

$$-\log \gamma = 0.512 z_M z_N \{I^{1/2}[1 + 50.3 \, (DT)^{1/2} \, a \, I^{1/2}]$$

For pure water at 298 K, D equals 78.5, such that

$$-\log \gamma = 0.512 z_M z_N \{\mu^{1/2} \, [1 + 0.328 \, a \, I^{1/2}]$$

where the adjustable parameter a is the radius of the hydrated ion in angstrom units.

Dependence of Single-Ion Activity Coefficients on Ionic Strength

Ionic Species	Ionic Strength of Solution		
	$\mu = 0.01$	$\mu = 0.05$	$\mu = 0.10$
H^+	0.91	0.86	0.83
Na^+	0.90	0.82	0.76
Cl^-	0.90	0.81	0.76
Mg^{2+}	0.69	0.51	0.45
Ca^{2+}	0.69	0.49	0.41

DEFINITE INTEGRAL

An integral Y that can be evaluated between two defined (or definite) limits $x = a$ and $x = b$, and written as:

$$Y = \int_{x=a}^{x=b} f(x) \, dx$$

Definite integrals often can be graphically evaluated as so-called Riemann sums by partitioning the area under the curve into rectangles, the horizontal width of which

corresponds to intervals along the x-axis. Because the area of each partition can be determined and summed, one can obtain the overall area.

$$Y = \int_{x=a}^{x=b} f(x) \, dx = \lim_{n \to \infty} \sum_{i=1}^{n} f\left(\frac{x_i + x_{i-1}}{2}\right) \cdot \Delta x_i$$

In this expression, we use so-called Δ-notation to indicate increments along the horizontal axis, such that the area of each partition is $f(x_i) \cdot (\Delta x_i)$, where $f(x_i)$ is the value of the function at the midpoint $(x_i + x_{i-1})/2$ of the interval $\Delta x_i = (x_i - x_{i-1})$.

DEGREE OF ACTIVATION

A dimensionless quantity, symbolized by ε_a, used to assess the extent of activation by a particular compound on the initial rate of an enzyme-catalyzed reaction. The degree of activation is equal to $(v_a - v_o)/v_o$ where v_a is the reaction rate in the presence of the activator and v_o is the initial rate in the absence of the activator. Whenever ε_a values are reported, the activator concentration and initial rate conditions have to be reported as well. The degree of activation, a useful parameter in the initial stages of an investigation, does not address issues related to the mechanism of activation. **See** *Activation; Basal Rate*

DEGREE OF DISSOCIATION

A dimensionless parameter (usually symbolized as α) indicating the fraction of monoprotic acid present as its dissociated form. This parameter can be defined quantitatively as follows:

$$\alpha = \frac{[\text{conjugate base}]}{[\text{total acid}]} = \frac{[\text{conjugate base}]}{[\text{acid HA}] + [\text{conjugate base}]}$$

If c is the initial or total concentration of an an acid HA, then the concentrations of A^- and H^+ are each equal to αc, where α lies in the interval $[0 < \alpha < 1]$. Likewise, the equilibrium concentration of HA must be $(1 - \alpha)c$. Thus,

$$K_{\text{dissociation}} = \frac{\alpha^2 c^2}{(1 - \alpha)c} = \frac{\alpha^2 c}{(1 - \alpha)}$$

We can consider several important cases. (1) While α can approach unity, the degree of dissociation can never be complete (*i.e.*, $\alpha \neq 1$); otherwise, the system would no longer be at equilibrium. This is a reasonable expectation for any equilibrium process, and in the case of an acid, there must be some HA present, however low its concentration. (2) Under the condition where $\alpha \ll 1$, the above equation becomes:

$$\alpha^2 = \frac{K_{\text{dissociation}}}{c} \quad \text{or} \quad \alpha = \sqrt{\frac{K_{\text{dissociation}}}{c}}$$

Because $K_{\text{dissociation}}$ is an equilibrium dissociation constant, this expression reminds us that the degree of dissociation will depend on the total concentration of HA; α will get smaller as c is raised, and α will get larger as c is lowered.

In most cases, $K_{\text{dissociation}}$ is written as K_a or K_A, and one may take logarithms of the dissociation constant to obtain the familiar Henderson-Hasselbalch equation:

$$\text{pH} = pK_a + \log \frac{[A^-]}{[\text{HA}]} = pK_a + \log \frac{\alpha}{1 - \alpha}$$

If pK_a is known, this equation permits one to calculate α at any pH. **See** Henderson-Hasselbalch Equation

DEGREE OF INHIBITION

A dimensionless quantity used to assess the extent of inhibition by a particular compound at a specific concentration on the initial rate of an enzyme-catalyzed reaction. The degree of inhibition, symbolized by ε_i, is equal to $(v_o - v_i)/v_o$ where v_o in the reaction rate in the absence of the inhibitor and v_i is the rate in the presence of the inhibitor. Whenever ε_i values are reported, the inhibitor concentration and initial rate conditions have to be provided as well. The degree of inhibition is a useful parameter in the early stages of an investigation. However, it does not address the type of inhibition nor provide much information on the dissociation constant for the inhibitor. **See** Inhibition (as well as specific type of inhibition)

DEGREES OF FREEDOM

1. The number of independent variables required to specify the state of a mechanical or thermodynamic system. Degrees of freedom arise from the possible motions of molecules or particles in a system. (The term "generalized coordinates" is also used in physics to designate the minimal number of coordinates needed to specify the state of a mechanical system.) 2. The number of independent or unrestricted random variables constituting a statistic. **See** Statistics (A Primer)

The heat capacity of a substance depends on the extent to which thermal energy can be absorbed and stored in the form of molecular motions. The kinetic theory of gases was first used to define the contributions of degrees of freedom to the heat capacity. Naumann[1] concluded that a monoatomic ideal gas possesses no thermal energy other than that of translation. Elementary kinetic theory predicts that the contribution equals $3R/2$ for the constant volume heat capacity (C_V) or $5R/2$ for the constant pressure heat capacity (C_P). R is the universal gas constant. A diatomic molecule has five degrees of freedom: three translational motions (along x, y, and z axes) as well as separate rotational motions (one about each of the two mutually perpendicular axes). One must also consider the vibrational motion of the two atoms as they periodically move toward and away from one another, and this contribution to the heat capacity was first formulated by Einstein[2]. Polyatomic molecules are slightly more complicated. For carbon dioxide, each of the three atoms has three degrees of freedom, yielding nine degrees of freedom for the $O{=}C{=}O$ molecule: three correspond to the motions of the molecule along the x, y, and z axes; two to rotations in the xz and xy planes; and four to vibrations that are most frequently called the two symmetric and asymmetric stretching modes and the two bending modes.

In chemical reaction kinetics, there may be thermodynamic contributions arising from any change(s) in the degrees of freedom of the atoms involved in bond-breaking and bond-making steps. Coupled vibratory motions can also be a factor: whenever the frequencies of two uncoupled vibrators are sufficiently close to each other, introduction of a physical connection between the vibrators will result in a frequency shift for the coupled system.

[1]A. Naumann (1867) Ann. Chem. Pharm. **142**, 265.
[2]A. Einstein (1907) Ann. Physik **22**, 180.

DEHYDROALANINE

A specialized amino acid residue that serves as an essenstial electrophilic center in several enzymatic reactions, including those catalyzed by L-phenylalanine ammonia lyase (Reaction: L-phenylalanine $\rightarrow$ trans-cinnamate + NH_3) and L-histidine ammonia lyase (Reaction: L-histidine $\rightarrow$ urocanate + NH_3). The former facilitates the elimination of ammonia and the pro-S hydrogen of phenylanine, and the initial step is nucleophilic attack of phenylalanine's amino group on the methenyl group of a dehydroalanyl residue.

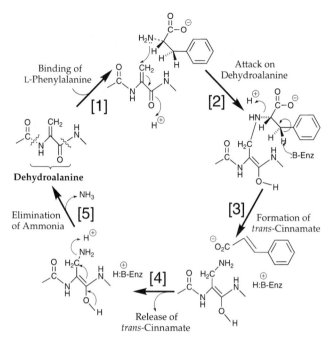

The dehydroalanyl is thought to be formed via expulsion of the elements of water from an active site serine of an inactive precursor form of these lyases. **See also** *Borohydride Reduction*

DEHYDROASCORBATE REDUCTASE

This enzyme [EC 1.6.5.4], also known as monodehydroascorbate reductase (NADH), catalyzes the reaction of NADH with two monodehydroascorbate molecules to produce NAD$^+$ and two molecules of ascorbate.

E. Maellaro, B. Del Bello, L. Sugherini, M. Comporti & A. F. Casini (1997) *Meth. Enzymol.* **279**, 30.

7-DEHYDROCHOLESTEROL REDUCTASE

This enzyme [EC 1.3.1.21] catalyzes the reaction of cholesterol with NADP$^+$ to yield cholesta-5,7-dien-3-β-ol and NADPH.

M. E. Dempsey (1969) *Meth. Enzymol.* **15**, 501.

2-DEHYDRO-3-DEOXY-D-GLUCONATE 5-DEHYDROGENASE

This enzyme [EC 1.1.1.127], also known as 2-keto-3-deoxygluconate dehydrogenase, catalyzes the reaction of 2-dehydro-3-deoxy-D-gluconate with NAD$^+$ to produce (4S)-4,6-dihydroxy-2,5-dioxohexanoate and NADH. The enzyme isolated from *Pseudomonas* can utilize NADP$^+$ as the coenzyme whereas the enzymes isolated

from *Erwinia chrysanthemi* and *E. coli* are more specific for NAD$^+$.

J. Preiss (1966) *Meth. Enzymol.* **9**, 203.

2-DEHYDRO-3-DEOXY-D-GLUCONATE 6-DEHYDROGENASE

This enzyme [EC 1.1.1.126], also known as 2-keto-3-deoxy-D-gluconate dehydrogenase, catalyzes the reaction of 2-dehydro-3-deoxy-D-gluconate with NADP$^+$ to produce (4S,5S)-4,5-dihydroxy-2,6-dioxohexanoate and NADPH.

J. Preiss (1966) *Meth. Enzymol.* **9**, 206.

2-DEHYDRO-3-DEOXYGLUCONOKINASE

This enzyme [EC 2.7.1.45], also known as 2-keto-3-deoxygluconokinase and 3-deoxy-2-oxo-D-gluconate kinase, catalyzes the reaction of ATP with 2-dehydro-3-deoxy-D-gluconate to produce ADP and 6-phospho-2-dehydro-3-deoxy-D-gluconate.

G. Ashwell (1962) *Meth. Enzymol.* **5**, 190.

DEHYDROGLUCONOKINASE

This enzyme [EC 2.7.1.13], also known as ketogluconokinase, catalyzes the reaction of 2-dehydro-D-gluconate with ATP to produce ADP and 6-phospho-2-dehydro-D-gluconate.

W. A. Wood & E. W. Frampton (1962) *Meth. Enzymol.* **5**, 291.

3-DEHYDROQUINATE DEHYDRATASE

This enzyme [EC 4.2.1.10] catalyzes the reaction of 3-dehydroquinate to produce 3-dehydroshikimate and water.

K. N. Allen (1998) *Comprehensive Biological Catalysis: A Mechanistic Reference* **2**, 135.
R. Bentley (1990) *Crit. Rev. Biochem. Mol. Biol.* **25**, 307.
S. Chaudhuri, K. Duncan & J. R. Coggins (1987) *Meth. Enzymol.* **142**, 320.

3-DEHYDROQUINATE SYNTHASE

This enzyme [EC 4.6.1.3] catalyzes the conversion of 3-deoxy-*arabino*-heptulosonate 7-phosphate to form 3-dehydroquinate and orthophosphate. The enzyme requires cobalt and the hydrogen atoms located on C7 of the substrate are retained on C2 of the product.

A. R. Clarke & T. R. Dafforn (1998) *Comprehensive Biological Catalysis: A Mechanistic Reference* **3**, 1.
J. V. Schloss & M. S. Hixon (1998) *Comprehensive Biological Catalysis: A Mechanistic Reference* **2**, 43.
R. Bentley (1990) *Crit. Rev. Biochem. Mol. Biol.* **25**, 307.

3-DEHYDROSPHINGANINE REDUCTASE

This enzyme [EC 1.1.1.102] catalyzes the reversible reaction of sphinganine with $NADP^+$ to produce 3-dehydrosphinganine and NADPH.

A. H. Merrill, Jr., & E. Wang (1992) *Meth. Enzymol.* **209**, 427.

DEINHIBITION

Any ligand-induced process that nullifies the inhibitory effect of an enzyme-bound inhibitor. In such cases, the ligand is said to deinhibit the system. Such behavior is distinct from activation, because increased activity is not observed when the deinhibiting ligand interacts with the enzyme in the absence of the inhibitor.

An excellent example of deinhibition is the reversal of glucose 6-phosphate inhibition of brain hexokinase (type I isozyme)[1]. As the pacemaker of glycolysis in brain tissue, hexokinase exhibits unique regulatory properties in that physiological levels of phosphate relieve potent inhibition by the reaction product, glucose 6-phosphate. On the other hand, orthophosphate is without effect on brain hexokinase in the absence of glucose 6-phosphate. This behavior contributes in part to the action of orthophosphate in accelerating glycolysis, a process widely known as the Crabtree Effect.

[1]D. L. Purich, F. B. Rudolph & H. J. Fromm (1973) *Advan. Enzymol.* **39**, 249.

DELOCALIZATION

A chemical concept that applies to π-bonding in conjugated systems. Electrons in conjugated π bonds are not restricted to just two atoms. Each bond of conjugated systems has a partial double bond character, reflected by the bond order for a given bond.

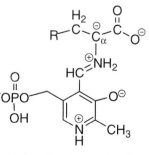

Delocalized α-Carbanion

With conjugated systems having a net charge, delocalization of the charge also occurs. The situation is analogous for unpaired electrons in conjugated radicals.

DELTA (Δ, δ)

Δ: 1. Symbol for application of heat in a reaction (*e.g.*, $A \overset{\Delta}{\to} P$). 2. Symbol for change. 3. In organic structures, used to denote the presence of a double bond, usually with a superscript number to indicate position. 4. Symbol for mass excess.

δ: 1. Fourth letter in the Greek alphabet; hence, used to denote the fourth item in a series (for example, the fourth methylene carbon in a fatty acid). 2. Symbol for thickness. 3. Symbol for chemical shift in NMR.

DEOXYADENOSINE KINASE

This enzyme [EC 2.7.1.76] catalyzes the reaction of ATP with deoxyadenosine to produce ADP and dAMP. The enzyme will also utilize deoxyguanosine as a substrate. In some systems, this enzyme may be identical with deoxycytidine kinase.

E. P. Anderson (1973) *The Enzymes*, 3rd ed., **9**, 49.

DEOXYADENYLATE KINASE

This enzyme [EC 2.7.4.11] catalyzes the reaction of ATP with dAMP to produce ADP and dADP. The enzyme can also use AMP as a substrate (thus producing another ADP as a product; hence, an activity identical to that of adenylate kinase).

E. P. Anderson (1973) *The Enzymes*, 3rd ed., **9**, 49.

DEOXYCYTIDINE KINASE

This enzyme [EC 2.7.1.74] catalyzes the reaction of a nucleoside triphosphate with deoxycytidine to produce a nucleoside diphosphate and dCMP. This enzyme can use any nucleoside triphosphate (except dCTP) as the phosphate-donor substrate and it can phosphorylate cytosine arabinoside as well.

E. P. Anderson (1973) *The Enzymes*, 3rd ed., **9**, 49.

DEOXYCYTIDYLATE HYDROXYMETHYLTRANSFERASE

This enzyme [EC 3.6.1.12], also known as dCTP pyrophosphatase, catalyzes the hydrolysis of dCTP to produce dCMP and pyrophosphate (or, diphosphate). The enzyme can also utilize dCDP as a substrate, producing CMP and orthophosphate.

J. I. Rader & F. M. Huennekens (1973) *The Enzymes*, 3rd ed., **9**, 197.

3-DEOXY-D-*manno*-OCTULOSONATE ALDOLASE

This enzyme [EC 4.1.2.23] catalyzes the reversible conversion of 3-deoxy-D-*manno*-octulosonate to pyruvate and D-arabinose.

W. A. Wood (1972) *The Enzymes*, 3rd ed., **7**, 281.

3-DEOXYOCTULOSONATE 8-PHOSPHATE SYNTHASE

This enzyme [EC 4.1.2.16] (also known as phospho-2-dehydro-3-deoxyoctonate aldolase, phospho-2-keto-3-deoxyoctonate aldolase, and 3-deoxy-D-*manno*-octulosonic acid 8-phosphate synthetase) catalyzes the reaction of 2-dehydro-3-deoxy-D-octonate 8-phosphate and orthophosphate to produce phosphoenolpyruvate, D-arabinose 5-phosphate, and water.

J. V. Schloss & M. S. Hixon (1998) *Comprehensive Biological Catalysis: A Mechanistic Reference* **2**, 43.
D. H. Levin & E. Racker (1966) *Meth. Enzymol.* **8**, 216.

(DEOXY)NUCLEOSIDE MONOPHOSPHATE KINASE

This enzyme [EC 2.7.4.13] catalyzes the reaction of ATP with a deoxynucleoside phosphate to produce ADP and a deoxynucleoside diphosphate. The enzyme can also utilize dATP as the phosphorylating substrate as well.

M. J. Bessman (1963) *Meth. Enzymol.* **6**, 166.

DEOXYRIBONUCLEASES

Enzymes catalyzing cleavage of DNA, including: endodeoxyribonucleases that generate 5'-phosphomonoesters [EC 3.1.21.x], endodeoxyribonucleases that produce products other than 5'-phosphomonoesters [EC 3.1.22.x], site-specific endodeoxyribonucleases acting on altered bases [EC 3.1.25.x], and exodeoxyribonucleases producing 5'-phosphomonoesters [EC 3.1.11.x]. A few examples are:

(1) Deoxyribonuclease I [EC 3.1.21.1], also known as pancreatic DNase and thymonuclease, catalyzes the endonucleolytic cleavage of DNA, preferring dsDNA, to a 5'-phosphodinucleotide and 5'-phosphooligonucleotide end products.

(2) Deoxyribonuclease IV (phage T₄-induced) [EC 3.1.21.2], also called endodeoxyribonuclease IV (phage T₄-induced) and endonuclease IV, catalyzes the endonucleolytic cleavage of DNA, preferring single-stranded, to 5'-phosphooligonucleotide end products.

(3) Type I site-specific deoxyribonuclease [EC 3.1.21.3], also referred to as type I restriction enzyme, catalyzes the endonucleolytic cleavage of DNA to give random, double-stranded fragments with terminal 5'-phosphates. ATP (or dATP) is simultaneously hydrolyzed to ADP and orthophosphate. This classification represents a large group of enzymes that have an absolute requirement for ATP (or dATP) and *S*-adenosylmethionine. They recognize specific short DNA sequences and hydrolyze bonds at sites remote from the recognition sequence.

(4) Type II site-specific deoxyribonuclease [EC 3.1.21.4], also referred to as type II restriction enzyme, catalyzes the endonucleolytic cleavage of DNA to give specific, double-stranded fragments with terminal 5'-phosphates. Magnesium ions are required as cofactors.

(5) Type III site-specific deoxyribonuclease [EC 3.1.21.5], also referred to as type III restriction enzyme, catalyzes the endonucleolytic cleavage of DNA to give specific, double-stranded fragments with terminal 5'-phosphates. This class of enzymes has an absolute requirement for ATP but does not hydrolyze the ATP. *S*-Adenosylmethionine stimulates the reaction, but is not absolutely required. These enzymes recognize specific short DNA sequences and hydrolyze bonds that are a short distance away from the recognition sequence.

(6) Deoxyribonuclease II [EC 3.1.22.1], also referred to as pancreatic DNase II, catalyzes the endonucleolytic hydrolysis of DNA (preferring double-stranded DNA) to produce 3'-phosphomononucleotide and 3'-phosphooligonucleotide end products.

(7) Deoxyribonuclease V [EC 3.1.22.3], also known as endodeoxyribonuclease V, catalyzes the endonucleolytic cleavage of DNA at apurinic or apyrimidinic sites to products with a 3'-phosphate.

(8) Crossover junction endoribonuclease [EC 3.1.22.4], also referred to as Holliday junction nuclease, catalyzes the endonucleolytic hydrolysis of a bond(s) in DNA at a junction such as a reciprocal single-stranded crossover between two homologous DNA duplexes (Holliday junction).

(9) Deoxyribonuclease X [EC 3.1.22.5] catalyzes the endonucleolytic cleavage of supercoiled plasma DNA to linear DNA duplexes. The enzyme exhibits a preference for supercoiled DNA.

(10) Deoxyribonuclease (pyrimidine dimer) [EC 3.1.25.1] catalyzes the endonucleolytic hydrolysis of a bond in DNA near pyrimidine dimers to generate products with 5'-phosphates. The enzyme acts on damaged strands of DNA, 5' from the damaged site.

(11) DNA-(apurinic or apyrimidinic site) lyase [EC 4.2.99.18, formerly EC 3.1.25.2] acts on the $C-O-P$ bond 3' to the apurinic or apyrimidinic site in DNA. This bond is broken by a β-elimination reaction, leaving a 3'-terminal unsaturated sugar and a product with a terminal 5'-phosphate. Note that this "nicking" of the phosphodiester bond is a lyase-type reaction, not hydrolysis.

(12) Exodeoxyribonuclease I [EC 3.1.11.1] catalyzes the hydrolysis of bonds in DNA (preferring single-stranded DNA), acting progressively in a 3'- to 5'-direction, releasing 5'-phosphomononucleotides.

(13) Exodeoxyribonuclease III [EC 3.1.11.2] catalyzes the hydrolysis of bonds in DNA (preferring double-stranded DNA), acting progressively in a 3'- to 5'-direction, releasing 5'-phosphomononucleotides.

(14) Exodeoxyribonuclease (lambda-induced) [EC 3.1.11.3], also called lambda exonuclease, catalyzes the hydrolysis of double-stranded DNA, acting progressively in a 5'- to 3'-direction and releasing 5'-phosphomononucleotides. This enzyme does not attack single-strand breaks.

(15) Exodeoxyribonuclease (phage SP_3-induced) [EC 3.1.11.4] catalyzes exonucleolytic cleavage of DNA (preferring single-stranded) in the 5'- to 3'-direction to yield 5'-phosphodinucleotides.

(16) Exodeoxyribonuclease V [EC 3.1.11.5] catalyzes the exonucleolytic cleavage of double-stranded DNA in the presence of ATP, in either the 5'- to 3'- or the 3'- to 5'-direction to yield 5'-phosphooligonucleotides. Thus, this enzyme exhibits a DNA-dependent ATPase activity.

(17) Exodeoxyribonuclease VII [EC 3.1.11.6] catalyzes the exonucleolytic cleavage of DNA (preferring single-stranded) in either the 5'- to 3'- or the 3'- to 5'-direction to yield 5'-phosphomononucleotides.

N. H. Williams (1998) *Comprehensive Biological Catalysis: A Mechanistic Reference* **1**, 543.
J. A. Gerlt (1992) *The Enzymes*, 3rd ed., 20, 95.
Two entire volumes dedicated to DNases: *The Enzymes*, 3rd ed.,**14** and **21**.
I. R. Lehman (1971) *The Enzymes*, 3rd ed., **4**, 251.
G. Bernardi (1971) *The Enzymes*, 3rd ed., **4**, 271.
M. Laskowski, Sr., (1971) *The Enzymes*, 3rd ed., **4**, 289.

2-DEOXYRIBOSE-5-PHOSPHATE ALDOLASE

This enzyme [EC 4.1.2.4], also referred to as phosphodeoxyriboaldolase and deoxyriboaldolase, catalyzes the reversible conversion of 2-deoxy-D-ribose 5-phosphate to D-glyceraldehyde 3-phosphate and acetaldehyde.

K. N. Allen (1998) *Comprehensive Biological Catalysis: A Mechanistic Reference* **2**, 135.
D. S. Feingold & P. A. Hoffee (1972) *The Enzymes*, 3rd ed., **7**, 303.

DEOXYTHYMIDINE KINASE

This enzyme [EC 2.7.1.21], better known as thymidine kinase, catalyzes the reaction of ATP with thymidine to produce ADP and thymidine 5'-phosphate. Deoxyuridine can serve as a substrate and dGTP can substitute for ATP. The deoxypyrimidine kinase complex induced by herpes simplex virus catalyzes this reaction as well as those of AMP:thymidine kinase, ADP:thymidine kinase, and dTMP kinase.

E. P. Anderson (1973) *The Enzymes*, 3rd ed., **9**, 49.

DEPHOSPHO-CoA KINASE

This enzyme [EC 2.7.1.24] catalyzes the reaction of ATP with dephospho-CoA to produce ADP and coenzyme A.

G. W. E. Plaut, C. M. Smith & W. L. Alworth (1974) *Ann. Rev. Biochem.* **43**, 899.
Y. Abiko (1970) *Meth. Enzymol.* **18A**, 358.

DEPOLARIZATION

1. Reduction, neutralization, or change in direction of polarity. 2. Transient reduction in the membrane potential. *See* Action Potential; Hyperpolarization

DEPOLYMERIZATION

The spontaneous or enzyme-catalyzed breakdown (or disassembly) of a biopolymer into its component subunits.

Biopolymers are typically formed from subunits through the formation of covalent and/or noncovalent bonds. Examples of the covalently linked biopolymers include proteins formed from amino acids, nucleic acids formed from ribo- or deoxyribonucleotides, and polysaccharides formed from monosaccharides. These are depolymerized by hydrolases which can add and release from the biopolymers during each hydrolytic cycle or remain associated with the polymer for several or many covalent bond-breaking cycles. Biopolymers formed by noncovalent forces include oligomeric proteins [such as hemoglobin ($\alpha_2\beta_2$), tubulin ($\alpha\beta$), and lactate dehydrogenase (M_4, M_3H_1, M_3H_1, M_2H_2, M_1H_3, and H_4 tetramers formed from muscle (M) and heart (H) polypeptide chains) or polymeric structures [such as microtubules, filamentous actin, and intermediate filaments]. These depolymerize spontaneously in response to changes in pH, temperature or as the result of the presence/absence of other solutes (metal ions, metabolites, or other inorganic anions). **See also** *Processivity; Dynamic Instability; Treadmilling; Polymerization; Depolymerization, End-Wise*

DEPOLYMERIZATION, END-WISE

Linear supramolecular polymers (such as those formed from actin, tubulin, and tobacco mosaic virus coat protein) are formed as products of efficient entropy-driven self-assembly[1,2], and the dynamics of depolymerization can be quantitatively analyzed[3,4] if the following conditions hold true: (a) the polymer undergoes stepwise release of subunits in a series first-order manner; (b) the off-rate constant for subunit release is independent of polymer length over the course of depolymerization; (c) the rate of polymerization is effectively zero under the conditions of the experiment; (d) turbidity or some other physically determinable parameter is a measure of the polymer weight concentration and is independent of changes in polymer length distribution; and (e) the concentrations of polymers of various lengths can be estimated by electron microscopy or some other suitably accurate method. For microtubules, these conditions can be met by isothermal dilution of the total tubulin concentration to below its critical concentration[5], by cold-induced depolymerization[5] after lowering the sample temperature to 5–6°C, or by calcium ion induced disassembly[6]. Rapid isothermal dilution can be accomplished within 1–3 seconds after dilution using the mixer shown in Fig. 1.

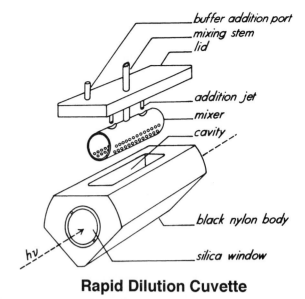

Rapid Dilution Cuvette

Figure 1. Mixing device for microtubule depolymerization kinetic studies requiring prompt dilution while minimizing shearing forces that may alter the polymer length distribution.

A typical fit of the theoretical depolymerization kinetics to the experimentally determined rate behavior for dilution-induced depolymerization is illustrated in Fig. 2

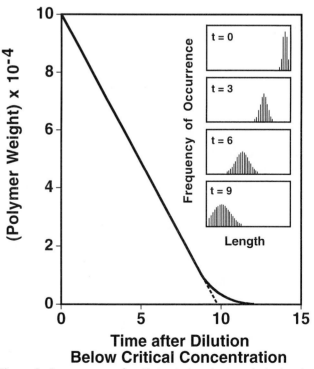

Figure 2. Progress curve for dilution-induced microtubule depolymerization. *Inset:* Polymer length distribution prior to dilution-induced disassembly. The data points are experimentally determined, and the solid line is based on the theoretical treatment[3,4].

for microtubules composed of tubulin and microtubule-associated proteins. The theoretical depolymerization curve (solid line) was obtained by taking into account the polymer length distribution for the microtubule sample immediately preceding depolymerization.

The generality of the end-wise depolymerization kinetic model is indicated by the comparison of the observed and predicted time-courses of cold-induced microtubule disassembly (Fig. 3). *See Self-Assembly Protein Polymerization*

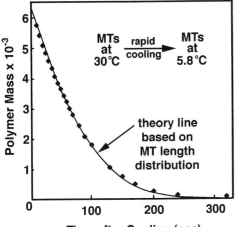

Figure 3. Progress curve for cold-induced depolymerization after chilling microtubules (assembled at 30°C using 2.1 mg/ml tubulin and 0.4 mg/ml microtubule-associated proteins). The data points are experimentally determined, and the solid line is based on the theoretical treatment[3,4].

[1]M. A. Lauffer (1975) *Entropy-Driven Processes in Biology*, Springer-Verlag, Berlin.
[2]F. Oosawa and M. Kasai (1962) *J. Mol. Biol.* **4**, 10.
[3]D. Kristofferson, T. L. Karr & D. L. Purich (1980) *J. Biol. Chem.* **255**, 8567.
[4]D. L. Purich, T. L. Karr & D. Kristofferson (1980) *Meth. Enzymol.* **85**, 439.
[5]T. L. Karr, D. Kristofferson & D. L. Purich (1980) *J. Biol. Chem.* **255**, 8560.
[6]T. L. Karr, D. Kristofferson & D. L. Purich (1980) *J. Biol. Chem.* **255**, 11853.

DESATURASES, FATTY ACYL-CoA

There are several fatty acyl-CoA desaturases. Stearoyl-CoA desaturase [EC 1.14.99.5] (also referred to as acyl-CoA desaturase, fatty acid desaturase, and Δ^9-desaturase) catalyzes the reaction of stearoyl-CoA with dioxygen and a hydrogen donor (AH_2) to produce oleoyl-CoA, two water molecules, and A. The enzyme requires iron as a cofactor. Linoleoyl-CoA desaturase [EC 1.14.99.25], also referred to as Δ^6-desaturase, cata-

lyzes the reaction of linoleoyl-CoA with dioxygen and AH_2 to produce γ-linolenoyl-CoA, A, and two water molecules.

I. Nishida & N. Murata (1996) *Ann. Rev. Plant Physiol. Plant Mol. Biol.* **47**, 541.
P. W. Holloway (1983) *The Enzymes*, 3rd ed., **16**, 63.

DESOLVATION

A potential factor for enhancing the effectiveness of catalyzed reactions. The relative importance of this factor will depend on the polarity of the substrate(s), transition state, and reaction product(s). The energy associated with desolvation of substrates must be compensated for by the binding interactions between the substrates and the enzyme.

DETHIOBIOTIN SYNTHASE

This enzyme [EC 6.3.3.3] catalyzes the reaction of ATP with 7,8-diaminononanoate and carbon dioxide to produce ADP, orthophosphate, and dethiobiotin. CTP can be used as a substrate as well although it is not as effective as ATP.

G. Schneider & Y. Lindqvist (1997) *Meth. Enzymol.* **279**, 376.

DEXTRANASE

This enzyme [EC 3.2.1.11], also known as α-1,6-glucan-6-glucanohydrolase, catalyzes the endohydrolysis of 1,6-α-D-glucosidic linkages in dextran.

J.-C. Janson & J. Porath (1966) *Meth. Enzymol.* **8**, 615.

DEXTRANSUCRASE

This enzyme [EC 2.4.1.5], also known as sucrose 6-glucosyltransferase, catalyzes the reaction of sucrose with $[(1,6)-\alpha$-D-glucosyl]$_n$ to produce fructose and $[(1,6)-\alpha$-D-glucosyl]$_{(n+1)}$.

G. Davies, M. L. Sinnott & S. G. Withers (1998) *Comprehensive Biological Catalysis: A Mechanistic Reference* **1**,119.
G. Mooser (1992) *The Enzymes*, 3rd ed., **20**, 187.

DEUTERIUM

An isotope of hydrogen having a nucleus (referred to as the deuteron) consisting of one proton and one neutron. Deuterium is a stable isotope (symbolized by ^2H or D) having an atomic weight of 2.0140 amu and a natural abundance of 0.015% relative to all hydrogen isotopes.

DEVIATION

The extent of variation within a series of replicate determinations of some experimental parameter. Let a_1, a_2, and a_3 represent several values of replicate measurements, and let a represent the arithmetic mean (or average). The deviations Δ_1, Δ_2, and Δ_3 are $(a_1 - a)$, $(a_2 - a)$, and $(a_3 - a)$, respectively. These quantities may be greater than, less than, or equal to zero, and the sign indicates the direction of deviation.

DEWAR STRUCTURES

Canonical forms of benzene that are calculated to contribute about 22% to the resonance stabilization of benzene. Such resonance structures have no separate physical reality or independent existence. For the case of benzene, the two Kekulé structures with alternating double bonds (i.e., "cyclohexatriene" structures) contribute equally and predominantly to the resonance hybrid structure. A dotted circle is often used to indicate the resonance-stabilized bonding of benzene. Nonetheless, the most frequently appearing structures of benzene are the two Kekulé structures. **See** Kekulé Structures

A. Pullman & B. Pullman (1958) Prog. Org. Chem. **4**, 31.

DIACYLGLYCEROL O-ACYLTRANSFERASE

This enzyme [EC 2.3.1.20], also known as diglyceride acyltransferase, catalyzes the reaction of an acyl-CoA with 1,2-diacylglycerol to produce coenzyme A and a triacylglycerol. The acyl-CoA derivative can be palmitoyl-CoA or other long-chain acyl-CoA compounds.

R. A. Coleman (1992) Meth. Enzymol. **209**, 98.
R. M. Bell & R. A. Coleman (1983) The Enzymes, 3rd ed., **16**, 87.

1,2-DIACYLGLYCEROL 3-β-GALACTOSYLTRANSFERASE

This enzyme [EC 2.4.1.46], also referred to as UDP-galactose:1,2-diacylglycerol 3-β-galactosyltransferase, catalyzes the reaction of UDP-galactose with a 1,2-diacylglycerol to produce UDP and a 3-β-D-galactosyl-1,2-diacylglycerol.

J. Browse & C. Somerville (1991) Ann. Rev. Plant Physiol. Plant Mol. Biol. **42**, 467.

DIACYLGLYCEROL KINASE

This enzyme [EC 2.7.1.107], also known as diglyceride kinase, catalyzes the reaction of ATP with a 1,2-diacylglycerol to produce ADP and a 1,2-diacylglycerol 3-phosphate.

J. P. Walsh & R. M. Bell (1992) Meth. Enzymol. **209**, 153.
J. D. Esko & C. R. H. Raetz (1983) The Enzymes, 3rd ed., **16**, 207.
R. A. Pieringer (1983) The Enzymes, 3rd ed., **16**, 255.

DIAMINE AMINOTRANSFERASE

This enzyme [EC 2.6.1.29] catalyzes the reaction of an α,ω-diamine with α-ketoglutarate (or, 2-oxoglutarate) to generate an ω-aminoaldehyde and glutamate.

A. E. Braunstein (1973) The Enzymes, 3rd ed., **9**, 379.

DIAMINOHEXANOATE DEHYDROGENASE

This enzyme [EC 1.4.1.11], better referred to as L-erythro-3,5-diaminohexanoate dehydrogenase, catalyzes the reaction of L-erythro-3,5-diaminohexanoate with NAD^+ and water to produce (S)-5-amino-3-oxohexanoate, ammonia, and NADH.

N. M. W. Brunhuber & J. S. Blanchard (1994) Crit. Rev. Biochem. Mol. Biol. **29**, 415.

2,4-DIAMINOPENTANOATE DEHYDROGENASE

This enzyme [EC 1.4.1.12] catalyzes the reaction of 2,4-diaminopentanoate with $NAD(P)^+$ and water to produce 2-amino-4-oxopentanoate, ammonia, and NAD(P)H. The enzyme can also utilize 2,5-diaminohexanoate as a substrate (although not as effectively as the substrate mentioned above) forming 2-amino-5-oxohexanoate, which then cyclizes nonenzymatically to form 1-pyrroline-2-methyl-5-carboxylate.

N. M. W. Brunhuber & J. S. Blanchard (1994) Crit. Rev. Biochem. Mol. Biol. **29**, 415.

DIAMINOPIMELATE DECARBOXYLASE

This enzyme [EC 4.1.1.20] is a pyridoxal-phosphate-dependent protein that catalyzes the conversion of meso-2,6-diaminoheptanedioate to lysine and carbon dioxide.

E. A. Boeker & E. E. Snell (1972) The Enzymes, 3rd ed., **6**, 217.

DIAMINOPIMELATE DEHYDROGENASE

This enzyme [EC 1.4.1.16] catalyzes the reaction of meso-2,6-diaminoheptanedioate with $NADP^+$ and water to produce L-2-amino-6-oxoheptanedioate, ammonia, and NADPH.

N. M. W. Brunhuber & J. S. Blanchard (1994) Crit. Rev. Biochem. Mol. Biol. **29**, 415.

DIAMINOPIMELATE EPIMERASE

This enzyme [EC 5.1.1.7] catalyzes the interconversion of LL-2,6-diaminoheptanedioate to *meso*-diaminoheptanedioate.

M. E. Tanner & G. L. Kenyon (1998) *Comprehensive Biological Catalysis: A Mechanistic Reference* **2**, 7.
E. Adams (1972) *The Enzymes*, 3rd ed., **6**, 479.

DIASTEREOMERS

Stereoisomers with more than one chiral center and which are not mirror images of each other; hence, stereoisomers that are not enantiomers of each other. For example, L-threonine and D-threonine are an enantiomeric pair whereas L-threonine and D-allothreonine are a diastereomeric pair (as is L-threonine and L-allothreonine). Diastereomers will have similar physical, chemical, and spectral properties but those properties will *not* be identical. If n is the number of chiral centers, then the maximum number of stereoisomers will be equal to 2^n. However, the actual number for a given set of isomers may be less than 2^n due to the presence of meso forms. **See** *Enantiomer; Epimer; Meso Form*

DIATHERMIC WALL

A wall that separates systems but allows them to come into thermal equilibrium; necessary for isothermal processes. **See** *Adiabatic Wall*

DICHLOROMETHANE DEHALOGENASE

This enzyme [EC 4.5.1.3] catalyzes the hydrolysis of dichloromethane to generate formaldehyde and two HCl. Glutathione is required as a cofactor.

T. Leisinger & D. Kohlert-Staub (1990) *Meth. Enzymol.* **188**, 355.

DIELECTRIC CONSTANT

A measure of the ease of dissociation of ion pairs (or point charges) dissolved (or embedded) in a dielectric solvent. The force of ionic attraction or repulsion can be treated in accordance with Coulomb's inverse-square law, such that $F = q_1 q_2 / Dr^2$, where the numerator is the product of the ionic charges q_1 and q_2, r is the charge-to-charge distance, and D is the dielectric constant. D is a dimensionless constant that has a characteristic value for each dielectric substance or insulator. Suppose we are dealing with a pair of oppositely charged ions; the attraction will be correspondingly lessened in a medium where D is large. The following are dielectric constants for several liquids at 20°C: hexane, 1.87; benzene, 2.28; diethyl ether, 4.33; chloroform, 5.05; acetone, 21.40; ethanol, 24.0; water, 80; and hydrogen cyanide, 116.0. Water has a high dielectric constant, whereas the value of D for benzene is much lower. In the former case, the electric field around each ion is effectively lowered because water is easily polarized; the induced dipoles of the water molecules are aligned so as to yield an overall dipole moment that greatly reduces the field strength. Benzene is not so easily polarized, and the interionic force of attraction of oppositely charged ions is much greater. **See also** *Debye-Hückel Treatment*

The reader should also recall that the dielectric constant of the medium greatly influences chemical reactivity. This is especially true for nucleophilic substitution, where electric charge plays an essential role in all steps of the reaction.

DIETHYL PYROCARBONATE

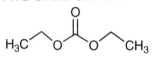

A potent enzyme inhibitor (abbreviated DEP) that acts by ethoxyformylation of proteins, usually at histidine residues. DEP is an irreversible inhibitor of ribonuclease, and rinsing glassware with a 0.1% (weight/volume) DEP solution is recommended to render glassware nuclease-free. Aqueous solutions must be freshly prepared for maximal effectiveness, because DEP will hydrolyze in 6–12 hours at neutral pH.

Selected entries from *Methods in Enzymology* [vol, page(s)]:
Inhibitory properties, **68**, 212; inactivator, of ribonucleases, **65**, 681; lipase modification, **64**, 390; nuclease inactivation, **79**, 63; ribonuclease inactivation, **79**, 52, 112-113, 267.

DIFFERENTIAL SCANNING CALORIMETRY (DSC)

A thermochemical method that simultaneously measures differences in heat flow into a test substance and a reference substance (whose thermochemical properties are already well characterized) as both are subjected to programmed temperature ramping of the otherwise thermally isolated sample holder. The advantage of differential scanning calorimetry is a kinetic technique that allows one to record differences in heat absorption directly rather than measuring the total heat evolved/

absorbed by a sample of interest. Moreover, one can make measurements during the course of chemical reactions or as certain polymerization reactions proceed. Power-compensated DSC involves heating of the sample and reference substances by separate heaters, and one determines the power (usually in milliwatts) on each heater as the temperatures of sample and reference are kept equal. Heat flux DSC actually detects the differential heat flow from chromel discs lying beneath pans containing the sample and reference as the temperature is ramped. Instead of measuring the power of each heater (or furnace), heat flux DSC relies on high sensitivity thermocouples to generate the differential signal.

Selected entries from *Methods in Enzymology* [vol, page(s)]: Aspartate transcarbamylase [assembly effects, **259**, 624-625; buffer sensitivity, **259**, 625; ligation effects, **259**, 625; mutation effects, **259**, 626]; baseline estimation [effect on parameters, **240**, 542-543, 548-549; importance of, **240**, 540; polynomial interpolation, **240**, 540-541, 549, 567; proportional method for, **240**, 541-542, 547-548, 567]; baseline subtraction and partial molar heat capacity, **259**, 151; changes in solvent accessible surface areas, **240**, 519-520, 528; characterization of membrane phase transition, **250**, 179-181; cooperativity measurement error, **240**, 642; data simulation, **240**, 537-539; enthalpy determination, **240**, 504, 506-507, 513-514, 534-537; entropy determination, **240**, 506-507, 514-517, 535-536, 539; excess heat capacity function [deconvolution analysis, **259**, 156-158; estimation, **259**, 152; statistical thermodynamic definition, **259**, 152-154]; folding/unfolding partition function determination, **240**, 505, 508-509, 532, 535; F-statistics, **240**, 554-559, 562; gel-liquid crystalline phase transition, **240**, 571-572; Gibbs free energy determination, **240**, 503, 518, 521, 535-536; heat capacity determination, **240**, 506, 510-513, 531, 534, 546-547; industrial application, **240**, 531, 568; α-lactalbumin molten globule state, **240**, 522, 528-530; ligand binding effects, **240**, 533; lipid-protein interaction analysis, **259**, 605-607; noise effect on parameters, **240**, 539-540, 562, 565; oligonucleotide thermodynamic parameters [entropy change, **259**, 237, 239; excess heat capacity, **259**, 234-235; free energy change, **259**, 155, 237, 239; van't Hoff enthalpy, **259**, 235-237]; protein measurement [enthalpy, **259**, 155-156, 168, 488, 595; entropy, **259**, 155-156, 168; heat capacity (absolute heat capacity, **259**, 145, 150-151, 154-155; change on unfolding, **259**, 165; conformational analysis, **259**, 163-165; curve-fitting, **259**, 164; hydration contribution, **259**, 158, 161-162, 164-165; noncovalent interactions, contribution, **259**, 158, 161, 163, 165; prediction, **259**, 167; primary heat capacity, **259**, 158-159; protonizable groups, contribution, **259**, 162-163; unfolded state, **259**, 162; transition temperature, **259**, 156]; protein unfolding, monitoring, **259**, 502, 601; RNA structure determination [data analysis, **259**, 289-290; sensitivity, **259**, 284]; two-dimensional, guanidinium hydrochloride thermal stability surfaces [accuracy, **240**, 566-567; methods, **240**, 537-543; multiple folding state fitting, **240**, 543-566].

DIFFERENTIATION

A procedure employed in calculus to describe the rate of change of a function with respect to one or more variables. The following are differentials for functions (where x is a variable and A is a constant) frequently encountered in chemistry.

$$dA/dx = 0$$

$$dx/dx = 1$$

$$dAx/dx = A \, dx/dx = A$$

$$d(e^x)/dx = e^x$$

$$d(e^{Ax})/dx = Ae^{Ax}$$

$$d(\ln x)/dx = 1/x$$

$$d(A \ln x)/dx = A/x$$

$$d[f(x) + g(x)]/dx = d[f(x)]/dx + d[g(x)]/dx$$

$$d[f(x)g(x)]/dx = g(x)d[f(x)]/dx + f(x)d[g(x)]/dx$$

$$d[f(x)/g(x)]/dx = \{g(x)d[f(x)/dx] - f(x)d[g(x)]/dx\}/g(x)^2$$
$$= [1/g(x)]df(x)/dx$$

DIFFUSION

The process by which random molecular motions result in a net flow of molecules from regions of high concentration to those of low concentration.

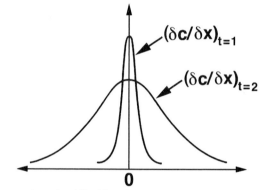

Concentration of a diffusible substance at times t_1 and t_2. Note that the concentration gradient decreases over a period of time.

Diffusing molecules achieve a rate of flow J that is proportional to the concentration gradient ($\partial c/\partial x$), characterized by a proportionality constant, D (also known as the diffusivity or diffusion coefficient):

$$J = -D\frac{\partial c}{\partial x}$$

The negative sign designates that the direction of flow occurs spontaneously from the region of higher concentration. The diffusion coefficient D represents the amount of solute that diffuses across a 1 cm² area per sec under the influence of a concentration gradient of 1 g/cm². D is characteristic for each type of solute molecule, and its value takes into account both molecular mass and shape. ***See also*** *Transport Processes; Fick's Laws*

DIFFUSION COEFFICIENT/CONSTANT

A parameter indicating the mass of material that will diffuse across a unit area in unit time under a concentration gradient of unity. The diffusion coefficient, D, is the proportionality constant in Fick's Laws of Diffusion. **See** *Fick's Laws*

Diffusion Constants of Biomolecules

Particle	Diffusion Coefficient (cm^2/s) $\times 10^7$	Molecular Mass (Daltons)
Hemoglobin	6.3	67,000
Catalase	4.1	250,000
Urease	3.5	470,000
Tobacco Mosaic Virus	0.5	31,000,000

DIFFUSION CONTROL

Most treatments encountered in discussions of collision theory primarily are concerned with reactions in the gas phase. However, most of the reactions in chemistry and biochemistry occur in solutions. In solutions, the molecules are moving in the potential field of their neighbors rather than freely as in the gas phase. Thus, the potential energy varies and holes in the solvation shell permits displacement of the molecule from its original position. There are also rapid collisions with molecules that make up the solvent cage as the molecule makes its series of discontinuous displacements. Two molecules in solution that become neighbors will tend to collide a number of times (often referred to as an encounter) before they separate (or before they react).

Fick's laws describe the interactions or encounters between noninteracting particles experiencing random, Brownian motion. Collisions in solution are diffusion-controlled. As is discussed in most physical chemistry texts[1,2], by applying Fick's First Law and the Einstein diffusion relation, the upper limit of the bimolecular rate constant k would be equal to

$$k = 4\pi d_{AB}(D_A + D_B)L$$

where d_{AB} is the hard-sphere collision diameter discussed in collision theory, D_A and D_B are the diffusion coefficients for A and B, L is Avogadro's number, and there is an assumption that there is no intermolecular potential energy between the reacting molecules (this result was first obtained by Smoluchowski[3,4]). Using typical values of d_{AB} (5×10^{-8} cm), $D_A + D_B$ (10^{-5} cm²·sec⁻¹), and adjusting for appropriate units, the above equation shows that k is equal to about 7×10^9

$M^{-1}\cdot sec^{-1}$. The corresponding upper limit value for a bimolecular rate constant in the gas phase is about 10^{11} $M^{-1}\cdot sec^{-1}$. Thus in solutions, bimolecular rate constants cannot exceed 10^9-10^{10} $M^{-1}\cdot sec^{-1}$ since diffusion control takes over from collision control.

It should also be recalled that the diffusion coefficient of a molecule will decrease with increasing viscosity of the solvent. Thus, as might be expected with cytosols or lipid bilayers, a viscous medium will slow down the rate of encounters. Since viscosity is itself temperature dependent, such encounters in solution will have their own activation energy.

The above expression represents the maximum rate for an encounter in a solution. However, most reactions have an energy barrier and/or strict steric requirements. Typically, many collisions occur before a reaction ensues. One must also consider the duration of an encounter and the frequency of vibrational collision during an encounter.

The duration of an encounter, τ_{AB}, between two molecules A and B can be estimated[1] to be

$$\tau_{AB} = \frac{\overline{\lambda}^2}{6D}$$

where $D = D_A + D_B$, the sum of the diffusion coefficients for A and B, and $\overline{\lambda}^2$ is the mean square displacement needed to break off an encounter. The value of λ can be set equal to the diameter of a solvent molecule or set equal to $2d_{AB}$.

The probability of a reaction occurring during an encounter[1] is

$$P_r = \frac{vp \exp(-LE/RT)}{vp \exp(-LE/RT) + \tau_{AB}^{-1}}$$

where v is the frequency of collision between A and B during an encounter and p is the steric factor. Thus, the collision theory bimolecular rate constant in solution becomes

$$k = \frac{2\pi d_{AB}\lambda^2 vpL \exp(-LE/RT)}{3}$$

If τ_{AB}^{-1} is small relative to $vpe^{-LE/RT}$, then the reaction is diffusion controlled. If not, then $k = 4\pi d_{AB}DLP_r$. This is the liquid-phase equivalent of the equation for the bimolecular constant derived in the gas phase with the steric factor p included. Collision theory predicts that Arrhenius frequency factors will be slightly larger in solution than in the gas phase for bimolecular reactions. **See** Arrhenius Equation; Transition-State Theory; Fick's Laws

[1] J. W. Moore & R. G. Pearson (1981) Kinetics and Mechanism, Wiley, New York.
[2] W. J. Moore (1972) Physical Chemistry, Prentice-Hall, Englewood Cliffs, New Jersey.
[3] M. V. Smoluchowski (1916) Phys. Zeit. **17**, 557.
[4] M. V. Smoluchowski (1916) Phys. Zeit. **17**, 583.

DIFFUSION-CONTROLLED CRYSTAL GROWTH

For an aqueous suspension of crystals to grow, the solute must (a) make its way to the surface by diffusion, (b) undergo desolvation, and (c) insert itself into the lattice structure. The first step involves establishment of a stationary diffusional concentration field around each particle. The elementary step for diffusion has an activation energy ($\Delta G^{\ddagger}$), and a molecule or ion changes its position with a frequency of $(k_B T/h)\exp[-\Delta G_D^{\ddagger}/k_B T]$. Einstein's treatment of Brownian motion indicates that a displacement of λ will occur within a time τ if λ equals the square root of $2D\tau$. Thus, the rate constant for change of position equal to one ionic diameter d will be

$$\tau^{-1} = \frac{2D}{d^2} = (k_B T/h)\exp(-\Delta G^{\ddagger}/k_B T)$$

Nielsen[1] treats the solute concentration $c(\rho,t)$ in the space surrounding a particle as a function of distance (radius ρ from the center) and time t, and the diffusion process obeys the following equations for dc/dt and j (defined as the amount of solute diffusing in the direction of the solute concentration gradient per unit area of a cross section lying perpendicular to the gradient):

$$j = Ddc/d\rho$$

$$dc/dt = D[d^2c/d\rho^2 + (2/\rho)dc/d\rho]$$

The following boundary conditions apply: (a) $c(\rho,0) = c_{\text{bulk}}$, the concentration at "infinite distance" or bulk concentration; (b) $c(r,t) = c_{\text{solubility}}$, the solute concentration at saturation; and (c) $c(\infty,t) = c_{\text{bulk}}$. The solution (assuming constant r and c_{bulk}) is

$$c = s + (c_{\text{bulk}} - s)\ \{1 - (r/\rho)\ (1 - \text{erf}[(\rho - r)/(2Dt)^{1/2}])\}$$

which, when t is infinity (i.e., the system becomes stationary), reduces to

$$c = s + (c_{\text{bulk}} - s)\ \{1 - (r/\rho)\}$$

Nielsen[2] obtained the approximate solution for the time evolution of the radius:

$$r = [2Dv(c_{\text{bulk}} - s)t/q]^{1/2}$$

where q is a function of $v(c_{\text{bulk}} - s)$. This equation also predicts that diffusion-controlled crystal growth depends on solution viscosity which is one determinative factor in the magnitude of D. **See** Biomineralization

[1] A. E. Nielsen (1964) Kinetics of Precipitation, pp. 135-138, Pergamon, Oxford.
[2] A. E. Nielsen (1961) J. Phys. Chem. **65**, 46.

DIFFUSION-LIMITED REACTION

A reaction whose rate is limited (or controlled) only by the speed with which reactants diffuse to each other. For a ligand binding to a protein, the bimolecular rate constant for diffusion-limited association is around 10^9 $M^{-1}s^{-1}$. The enzyme acetylcholinesterase has an apparent on-rate constant of 1.6×10^8 $M^{-1}s^{-1}$ with its natural cationic substrate acetylcholine, and the on-rate constant of about 6×10^8 with acetylselenoylcholine and about 2.5×10^8 with acetylthiocholine. On the other hand, neutral substrates such as phenylacetate and isoamylacetate have bimolecular rate constants of 6×10^5 and 1.6×10^7 $M^{-1}s^{-1}$. This illustrates the degree to which electrostatic interactions can change a reaction from a diffusion-limited process to one that involves other types of protein-ligand interactions. **See** Chemical Kinetics; Diffusion-Limited Reaction

DIFFUSION OF LIGAND TO RECEPTOR

The random-motion process by which a ligand approaches and penetrates a tissue surface: first, by passing down a concentration gradient by three-dimensional Brownian motion in the bulk solution; then, by reaching the surface where the dimensionality of random excursions reduces it to a two-dimensional process; and finally, by docking at the receptor site[1-3]. Once even weakly associated within the tissue surface, receptor electrostatics and mobility probably work together to guide the ligand to its binding site. Moreover, the kinetics of ligand diffusion to its receptor(s) is likely to greatly influence the time dependence of metabolite and drug action.

[1]H. C. Berg & E. M. Purcell (1977) *Biophys. J.* **20**, 193.
[2]J. Crank (1975) *The Mathematics of Diffusion*, 2nd ed., Clarendon Press, Oxford.
[3]T. Kenakin (1993) *Pharmacologic Analysis of Drug-Receptor Interaction*, 2nd ed., pp. 20-24, Raven Press, New York.

DIFFUSION OF MOLECULES INTO A PORE

As illustrated by experiments on mediated uptake of proteins[1] through the nuclear pores of frog oocyte nuclei, molecular diffusion through large pores depends on the pore's cross-sectional area and the size and shape of the diffusing molecule[1-3]. As the dimension of the molecule approaches that of the pore, movement into and through the pore will become hindered, because molecules must enter without hitting the rim of the pore. Assuming no bulk solvent flow and pure diffusion of a rigid sphere through a pore, the pore's effective cross-sectional area, $A_{effective}$, will be reduced by the radius of the entering molecule. For a sphere entering a cylindrical pore,

$$A_{effective} = A_o(r - a)2/K_1r^2$$

where A_o is the actual cross-sectional area, a is the sphere's radius, and K_1 is the drag coefficient; the latter of which is a complex function of a/r.

[1]C. M. Feldherr, R. J. Cohen & J. A. Ogburn (1983) *J. Cell Biol.* **96**, 1486.
[2]W. L. Haberman & R. M. Sayre (1958) Motion of Rigid and Fluid Spheres in Stationary and Moving Liquids Inside Cylindical Tubes, *David Taylor Model Basin Report No. 1143*, U. S. Navy, Washington, D. C.
[3]P. L. Paine & P. Scherr (1975) *Biophys. J.* **15**, 1087.

DIFFUSION OF OXYGEN WITHIN RED CELLS

As pointed by Schroeder and Holmquist[1], the rate of oxygen transport within an erythrocyte depends on several parameters: the partial pressure of O_2 at the inner surface of the red cell membrane; the solubility of molecular oxygen, which also depends on the hemoglobin concentration (typically, 34% *w/v*), the fractional saturation of heme sites with oxygen, the kinetics of oxygen binding and release steps, the self-diffusion constant of hemoglobin, and the shape and volume of the red cell itself. Using the Einstein relation, $t = x^2/2D$, Schroeder and Holmquist[1] estimated the approximate time needed for diffusion of an O_2 molecule to the red cell's interior by using values of 1 micrometer for the half-thickness x of a red cell and 7.3×10^{-6} cm^2/s for the diffusivity of molecular oxygen. This puts the diffusion time at about 0.0007 s, which is about 100 times faster than the 0.058 s period required for hemoglobin ($D_{Hb} = 1.75 \times 10^{-7}$ cm^2/s) to diffuse a similar distance at 37°C. Because half-life for oxygen dissociation from hemoglobin is 0.0078 s, each oxygen molecule must successively "hitch a ride" on 7–8 hemoglobin molecules on the way to the red cell's center. Note also that the transfer of oxygen does not involve what Scholander[2] termed a "bucket brigade" wherein diffusing hemoglobin molecules hand off oxygen molecules to each other. Theoretical and experimental grounds[3] offer no support for direct bimolecular transfer of Hb-bound O_2 to another hemoglobin.

[1]W. A. Schroeder & W. R. Holmquist (1968) in *Structural Chemistry and Biology* (A. Rich and N. Davidson, eds.), pp. 245-246, W. H. Freeman, San Francisco.
[2]P. F. Scholander (1960) *Science* **131**, 505.
[3]J. B. Wittenberg (1966) *J. Biol. Chem.* **241**, 104.

DIFFUSION TIME

The period of time t required for a molecule with a diffusion coefficient D (units of cm^2 s^{-1}) to diffuse a mean square distance x^2, as given by the expression

$$t = \frac{x^2}{2D}$$

For example, a molecule with a diffusivity of 10^{-5} cm^2 s^{-1} requires about 0.5 seconds to diffuse about 10 micrometers (or 10^{-1} cm), which corresponds to the dimension of many cells.

DIFFUSION WITHIN CELLULAR/ EXTRACELLULAR SPACES

Free diffusion of molecules in solution is characteristically a haphazard process with net directionality determined only by solute gradients and diffusion coefficients. Within cellular and extracellular spaces, however, diffusion can be strongly influenced by noncovalent interactions of solvent and solute molecules with membranes as well as the cellular and extracellular matrix[1-3]. Channels and orifices can also alter the movement of solute and solvent molecules. These interactions can greatly alter the magnitude of the diffusion coefficient for a molecule from its isotropic value D in water to apparent diffusion coefficient D^* (which often can be directionally resolved into D_x^*, D_y^*, and D_z^*). The parameter λ, known as the tortuosity, equals $(D/D^*)^{1/2}$. In principle, λ has x, y, and z components that need not be equal if there is any anisotropy in the local electrical fields or porosity of the matrix.

Diffusional anisotropy can be measured by many procedures. Fluorescence microscopy[1] is an especially powerful technique that permits one to use video microscopy to follow the redistribution of fluorescently tagged molecules introduced into specific cellular or extracellular spaces. With the availability of microinjection methods, one can initiate such studies by placing a bolus dose of tagged molecules with considerable spatial precision. In NMR microscopy, one takes advantage of the fact that NMR signal intensities depend in part on translational self-diffusion, and any field gradient affecting transverse magnetization prior to or during signal acquisition will result in loss of signal strength[4-6]. An example of the NMR approach is provided by the work of Inglis *et al.*[4] on the diffusional anisotropy of water molecules in excised rat spinal cord. Anisotropic diffusion coefficients were measured in perfused-fixed rat cervical spinal cord in a 300 MHz imaging spectrometer by using apparent diffusion tensor (ADT) imaging. The method has an in-plane resolution of 10 μm, sufficient to discern diffusional characteristics in defined histologic regions (see the table below).

Anisotropic Diffusion Coefficients for Water Protons in Rat Cervical Spinal Cord[4]

Anatomical Region	D_{xx}	D_{yy}	D_{zz}
Dorsal funiculus	0.026[a]	0.033	0.209
Ventrolateral funiculus	0.039	0.051	0.194
Ventral funiculus	0.027	0.037	0.187
Dorsolateral funiculus	0.049	0.045	0.195
Dorsal horn	0.144	0.163	0.183
Ventral horn	0.181	0.157	0.150
Gray commissure	0.178	0.097	0.124

[a]All anisotropic apparent diffusion coefficients are expressed in units of 10^{-3} mm^2 s^{-1}. The z-axis runs parallel to the spinal cord.

[1]C. Nicholson & E. Sykova (1998) *Trends Neuro. Sci.* **21**, 207.
[2]H. D. Lux & E. Neher (1973) *Exp. Brain Res.* **17**, 190.
[3]C. Nicholson & L. Tao (1993) *Biophys. J.* **65**, 2277.
[4]B. A. Inglis, L. Yang, E. D. Wirth, D. Plant & T. H. Mareci (1997) *Magn. Reson. Imag.* **15**, 441.
[5]H. C. Torrey (1955) *Phys. Rev.* **104**, 563.
[6]P. T. Callaghan (1991) *Principles of Nuclear Magnetic Resonance Microscopy*, pp. 157-169 and pp. 201-208, Clarendon Press, Oxford.

DIHYDROFOLATE REDUCTASE

This enzyme [EC 1.5.1.3], also called tetrahydrofolate dehydrogenase, catalyzes the reversible reaction of 7,8-dihydrofolate with NADPH to produce 5,6,7,8-tetrahydrofolate and NADP$^+$. The enzyme isolated from mammals and some microorganisms can also slowly catalyze the reduction of folate to 5,6,7,8-tetrahydrofolate. **See also** *Dihydrofolate Reductase-Thymidylate Synthase*

A. R. Clarke & T. R. Dafforn (1998) *Comprehensive Biological Catalysis: A Mechanistic Reference* **3**, 1.
R. L. Blakley (1995) *Adv. Enzymol.* **70**, 23.
H. Eisenberg, M. Mevarech & G. Zaccai (1992) *Adv. Protein Chem.* **43**, 1.
K. A. Johnson & S. J. Benkovic (1990) *The Enzymes*, 3rd. ed., **19**, 159.
R. F. Colman (1990) *The Enzymes*, 3rd ed., **19**, 283.
J. Kraut & D. A. Matthews (1987) *Biological Macromolecules & Assemblies* (F. A. Jurnak & A. McPherson, eds.), pp. 1-71, Wiley, New York.

Selected entries from *Methods in Enzymology* [vol, page(s)]: Cryoenzymology, **63**, 338; inhibitor, **63**, 396, 399; tight-binding inhibitor, **63**, 437, 462, 466; hysteresis, **240**, 318; deuteration for protein-ligand interaction study, **239**, 684-685; mechanism, **240**, 317-318; NADPH off-rate, **240**, 318; pre-steady-state burst, **240**, 318; unfolding, reaction coordinate diagram, **202**, 117; catalysis, transient kinetics, **249**, 23-27; *Escherichia coli* DHFR [hysteresis, structural basis, **249**, 545; kinetic mechanism, derivation, **249**, 25-27; ligand binding, association and dissociation rate constants, **249**, 24-27; site-directed mutants, altered pH dependencies, **249**, 111-112; transient kinetics, **249**, 23-27 (inhibitor binding, **249**, 23-27; site-directed mutant, **249**, 108-109)]; hysteresis, **240**, 318; mechanism, **240**, 317-318; mitochondrial import of fusion protein, **260**, 243-245, 248, 271-272; NADPH off-rate, **240**, 318; pre-steady-state burst, **240**, 318; reaction catalyzed, **249**, 23; transition state and multisubstrate analogs, **249**, 304

DIHYDROFOLATE SYNTHETASE

This enzyme [EC 6.3.2.12] catalyzes the reaction of ATP with dihydropteroate and glutamate to generate ADP, orthophosphate, and dihydrofolate.

G. M. Brown & J. M. Williamson (1982) *Adv. Biochem.* **53**, 345.

DIHYDROLIPOAMIDE DEHYDROGENASE

This enzyme [EC 1.8.1.4] (also known as lipoamide reductase (NADH), the E3 component of α-ketoacid dehydrogenase complexes, lipoyl dehydrogenase, dihydrolipoyl dehydrogenase, and diaphorase) requires FAD and catalyzes the reaction of dihydrolipoamide with NAD$^+$ to produce lipoamide and NADH. The protein is also a component of the pyruvate dehydrogenase complex and the α-ketoglutarate (or, 2-oxoglutarate) dehydrogenase complex. **See also** *Pyruvate Dehydrogenase*

A. R. Clarke & T. R. Dafforn (1998) *Comprehensive Biological Catalysis: A Mechanistic Reference* **3**, 1.
B. A. Palfey & V. Massey (1998) *Comprehensive Biological Catalysis: A Mechanistic Reference* **3**, 83.
M. S. Patel, N. N. Vettakkorumakankav & T.-C. Liu (1995) *Meth. Enzymol.* **252**, 186.
C. H. Williams, Jr. (1976) *The Enzymes*, 3rd ed., **13**, 89.

DIHYDRONEOPTERIN ALDOLASE

This enzyme [EC 4.1.2.25] catalyzes the conversion of 2-amino-4-hydroxy-6-(D-*erythro*-1,2,3-trihydroxypropyl)-7,8-dihydropteridine to 2-amino-4-hydroxy-6-hydroxymethyl-7,8-dihydropteridine and glycolaldehyde.

G. M. Brown & J. M. Williamson (1982) *Adv. Biochem.* **53**, 345.

DIHYDRONEOPTERIN TRIPHOSPHATE PYROPHOSPHOHYDROLASE

The following is a recent review on the molecular and physical properties of this enzyme.

G. M. Brown & J. M. Williamson (1982) *Adv. Biochem.* **53**, 345.

DIHYDROOROTASE

This enzyme [EC 3.5.2.3], also called carbamoylaspartic dehydrase, catalyzes the hydrolysis of (*S*)-dihydroorotate to yield *N*-carbamoyl-L-aspartate.

M. E. Jones (1980) *Ann. Rev. Biochem.* **49**, 253.

DIHYDROOROTATE DEHYDROGENASE

This enzyme [EC 1.3.99.11] catalyzes the reaction of (*S*)-dihydroorotate with an acceptor substrate to produce orotate and the reduced acceptor. Both iron and zinc ions are needed as cofactors. Acceptor substrates include 2,6-dichloroindophenol, 1,10-phenanthroline, and dioxygen (although dioxygen isn't as effective as the first two).

M. E. Jones (1980) *Ann. Rev. Biochem.* **49**, 253.
G. Palmer (1975) *The Enzymes*, 3rd ed., **12**, 1.

DIHYDROOROTATE OXIDASE

This enzyme [EC 1.3.3.1] catalyzes the reaction of (*S*)-dihydroorotate with dioxygen to produce orotate and hydrogen peroxide. The enzyme requires FAD and FMN as cofactors. Ferricyanide can also serve as a substrate as well.

M. H. O'Leary (1989) *Ann. Rev. Biochem.* **58**, 377.

DIHYDROPTERIDINE REDUCTASE

This enzyme [EC 1.6.99.7] catalyzes the reaction of NAD(P)H with 6,7-dihydropteridine (that is, the quinoid form of dihydropteridine) to produce NAD(P)$^+$ and 5,6,7,8-tetrahydropteridine. The enzyme is not identical with dihydrofolate reductase.

A. R. Clarke & T. R. Dafforn (1998) *Comprehensive Biological Catalysis: A Mechanistic Reference* **3**, 1.
S. Kaufman (1987) *Meth. Enzymol.* **142**, 97.
H. Hasegawa & N. Nakanishi (1987) *Meth. Enzymol.* **142**, 103.

F. A. Firgaira, R. G. H. Cotton, I. Jennings & D. M. Danks (1987) *Meth. Enzymol.* **142**, 116.
K. G. Scrimgeour & S. Cheema-Dhadli (1987) *Meth. Enzymol.* **142**, 127.

DIHYDROPTEROATE SYNTHASE

This enzyme [EC 2.5.1.15], also known as dihydropteroate pyrophosphorylase, catalyzes the reaction of 2-amino-4-hydroxy-6-hydroxymethyl-7,8-dihydropteridine diphosphate with 4-aminobenzoate to produce pyrophosphate and dihydropteroate.

G. M. Brown & J. M. Williamson (1982) *Adv. Enzymol.* **53**, 345.

DIHYDROPYRIMIDINASE

This enzyme [EC 3.5.2.2], also called hydantoinase, catalyzes the hydrolysis of 5,6-dihydrouracil to produce 3-ureidopropionate. The enzyme can also utilize dihydrothymine and hydantoin as substrates.

O. W. Griffith (1986) *Ann. Rev. Biochem.* **55**, 855.

DIHYDROURACIL DEHYDROGENASE

Dihydrouracil dehydrogenase (NAD$^+$) [EC 1.3.1.1] catalyzes the reaction of 5,6-dihydrouracil with NAD$^+$ to produce uracil and NADH. Dihydrouracil dehydrogenase (NADP$^+$) [EC 1.3.1.2], also called dihydropyrimidine dehydrogenase (NADP$^+$) and dihydrothymine dehydrogenase, catalyzes the reaction of 5,6-dihydrouracil with NADP$^+$ to produce uracil and NADPH. The enzyme can also use dihydrothymine as a substrate.

O. W. Griffith (1986) *Ann. Rev. Biochem.* **55**, 855.

DIHYDROXYACETONE PHOSPHATE ACYLTRANSFERASE

This enzyme [EC 2.3.1.42], also known as glycerone-phosphate *O*-acyltransferase, catalyzes the reaction of an acyl-CoA with dihydroxyacetone phosphate (or, glycerone phosphate) to produce coenzyme A and an acyldihydroxyacetone phosphate (or, an acylglycerone phosphate). The acyl-CoA derivatives of palmitate, stearate, and oleate can all be utilized as substrates, with highest activity observed with palmitoyl-CoA.

C. Kent (1995) *Ann. Rev. Biochem.* **64**, 315.
K. O. Webber & A. K. Hajra (1992) *Meth. Enzymol.* **209**, 92.
R. M. Bell & R. A. Coleman (1983) *The Enzymes*, 3rd ed., **16**, 87.
J. D. Esko & C. R. H. Raetz (1983) *The Enzymes*, 3rd ed., **16**, 207.

DIHYDROXYACETONE SYNTHASE

This enzyme catalyzes the reaction of formaldehyde with xylulose 5-phosphate to produce dihydroxyacetone and glyceraldehyde 3-phosphate.

L. V. Dystrykh, W. de Koning & W. Harder (1990) *Meth. Enzymol.* **188**, 435.

DIHYDROXY-ACID DEHYDRATASE

This enzyme [EC 4.2.1.9] catalyzes the conversion of 2,3-dihydroxy-3-methylbutanoate to 3-methyl-2-oxobutanoate and water.

B. G. Fox (1998) *Comprehensive Biological Catalysis: A Mechanistic Reference* **3**, 261.
K. Kiritani & R. P. Wagner (1970) *Meth. Enzymol.* **17A**, 755.

2,3-DIHYDROXYBIPHENYL 1,2-DIOXYGENASE

This enzyme [EC 1.13.11.39], also referred to as biphenyl-2,3-diol 1,2-dioxygenase, catalyzes the reaction of biphenyl-2,3-diol with dioxygen to produce 2-hydroxy-6-oxo-6-phenylhexa-2,4-dienoate and water. The enzyme can also use 3-isopropylcatechol as a substrate, forming 7-methyl-2-hydroxy-6-oxoocta-2,4-dienoate as a product. This protein is not identical with catechol 2,3-dioxygenase.

B. G. Fox (1998) *Comprehensive Biological Catalysis: A Mechanistic Reference* **3**, 261.

DIHYDROXYFUMARATE DECARBOXYLASE

This enzyme [EC 4.1.1.54] catalyzes the conversion of dihydroxyfumarate to tartronate semialdehyde and carbon dioxide.

J. V. Schloss & M. S. Hixon (1998) *Comprehensive Biological Catalysis: A Mechanistic Reference* **2**, 43.

2,3-DIHYDROXYINDOLE 2,3-DIOXYGENASE

This enzyme [EC 1.13.11.23] catalyzes the reaction of 2,3-dihydroxyindole with dioxygen to produce anthranilate and carbon dioxide.

O. Hayaishi, M. Nozaki & M. T. Abbott (1975) *The Enzymes*, 3rd ed., **12**, 119.

3,4-DIHYDROXYPHENYLACETATE 2,3-DIOXYGENASE

This enzyme [EC 1.13.11.15], also called homoprotocatechuate 2,3-dioxygenase, is an iron-dependent enzyme that catalyzes the reaction of 3,4-dihydroxyphenylacetate with dioxygen to produce 2-hydroxy-5-carboxymethylmuconate semialdehyde.

J. E. Penner-Hahn (1998) *Comprehensive Biological Catalysis: A Mechanistic Reference* **3**, 439.
O. Hayaishi, M. Nozaki & M. T. Abbott (1975) *The Enzymes*, 3rd ed., **12**, 119.

DIHYDROXYPHTHALATE DECARBOXYLASE

This enzyme [EC 4.1.1.55] catalyzes the conversion of 4,5-dihydroxyphthalate to 3,4-dihydroxybenzoate and carbon dioxide.

J. V. Schloss & M. S. Hixon (1998) *Comprehensive Biological Catalysis: A Mechanistic Reference* **2**, 43.

2,5-DIHYDROXYPYRIDINE 5,6-DIOXYGENASE

This enzyme [EC 1.13.11.9] is an iron-dependent protein that catalyzes the reaction of 2,5-dihydroxypyridine with dioxygen to produce maleamate and formate.

O. Hayaishi, M. Nozaki & M. T. Abbott (1975) *The Enzymes*, 3rd ed., **12**, 119.

2,3-DIHYDROXY-9,10-SECOANDROSTA-1,3,5(10)-TRIENE-9,17-DIONE 4,5-DIOXYGENASE

This enzyme [EC 1.13.11.25], also referred to as steroid 4,5-dioxygenase and 3-alkylcatechol 2,3-dioxygenase, is an iron-dependent enzyme that catalyzes the reaction of 3,4-dihydroxy-9,10-secoandrosta-1,3,5(10)-triene-9,17-dione with dioxygen to produce 3-hydroxy-5,9,17-trioxo-4,5:9,10-disecoandrosta-1(10),2-dien-4-oate. The enzyme can also use 3-isopropylcatechol and 3-*tert*-butyl-5-methylcatechol as substrates.

O. Hayaishi, M. Nozaki & M. T. Abbott (1975) *The Enzymes*, 3rd ed., **12**, 119.

DIIODOPHENYLPYRUVATE REDUCTASE

This enzyme [EC 1.1.1.96] catalyzes the reversible reaction of β-(3,5-diiodo-4-hydroxyphenyl)lactate with NAD^+ to produce β-(3,5-diiodo-4-hydroxyphenyl)pyruvate and NADH. The substrates for this enzyme must contain an aromatic ring with a pyruvate side chain (or lactate), the most active substrates being halogenated derivatives. Potential substrates with hydroxyl or amino groups in the 3 or 5 position are inactive.

V. G. Zannoni (1970) *Meth. Enzymol.* **17A**, 665.

DIIODOTYROSINE AMINOTRANSFERASE

This pyridoxal-phosphate-dependent enzyme [EC 2.6.1.24] catalyzes the reversible reaction of 3,5-diiodotyrosine with α-ketoglutarate (or, 2-oxoglutarate) to generate 3,5-diiodo-4-hydroxyphenylpyruvate and glutamate. Also acting as substrates are the 3,5-dichloro, 3,5-dibromo, and the 3-iodo derivatives of tyrosine, as well as thyroxine and triiodothyronine.

A. E. Braunstein (1973) *The Enzymes*, 3rd ed., **9**, 379.

DIISOPROPYL FLUOROPHOSPHATE

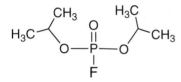

A potent irreversible inhibitor (abbreviated DFP) of many serine proteinases and serine esterases (especially acetylcholinesterase). This substance is EXTREMELY POISONOUS, but the vapor state can be minimized by using dry, water-miscible solvents such as 2-propanol[1]. Aqueous solutions become inactivated by hydrolysis, but solutions made with dry 2-propanol are stable at $-20°C$ for many months.

Enzyme inactivation by DFP and related organophosphorus compounds (diethyl fluorophosphate, dipropyl fluorophosphate, isopropylmethyl fluorophosphate, diethyl-4-nitrophenylphosphate) can be treated as a bimolecular process:

$$EH + P\text{–}X \rightarrow E\text{–}P + X^- + H^+$$

where the enzyme engages in nucleophilic attack on the electrophilic phosphorus atom. Bimolecular rate constants[1] for inhibition of acetylcholinesterase, chymotrypsin, and trypsin are 4.6×10^4, 1.5×10^4, and 9.2×10^2 M^{-1} min^{-1}, respectively.

[1]Cohen, R. A. Osterman & F. Berends (1967) *Meth. Enzymol.* **11**, 686.

Selected entries from *Methods in Enzymology* [vol, page(s)]:
Types of organophosphorus inhibitors, **11**, 686-688; toxicity hazards, **11**, 688; purity and analysis, **11**, 688; solutions of organophosphorus compounds, **11**, 689; estimation of specific radioactivity of organophosphorus compounds, **11**, 689-690; method for estimating phosphorus content, **11**, 691; reactions with enzymes, **11**, 691-701 [rate constants, **11**, 692; phosphorylation of chymotrypsin, **11**, 694-696; identification of phosphoryl and phosphonyl peptides, **11**, 697]; inhibition of serine peptidases, **244**, 7-8, 432; mechanism of inactivation, **244**, 432-433; phosphorylserine bond formation, **244**, 433; stability inhibitor, **244**, 432; toxicity, **244**, 7-8, 432.

DILUTION EFFECTS

The reduction in concentration of reactants, enzymes, and solute molecules can provide important information about kinetic systems. For example, one can readily differentiate a first-order process from a second-order process by testing whether the period required to reduce a reactant concentration to 50% of its initial value depends on dilution. First-order processes and intramolecular processes should not exhibit any effect on rate by diluting a reactant. In terms of enzyme-catalyzed processes, the Michaelis-Menten equation requires that the initial reaction velocity depends strictly on the concentration of active catalyst. Dilution can also be used to induce dissociation of molecular complexes or to promote depolymerization of certain polymers (such as F-actin and microtubules).

Selected entries from *Methods in Enzymology* [vol, page(s)]:
Dilution of enzyme samples, **63**, 10; lipolysis substrate effect, **64**, 361, 362; dilution jump kinetic assay, **74**, 14-19, 28; dilution method [for dissociation equilibria, **61**, 65-96; continuous dilution cuvette, **61**, 78-96; data analysis, **61**, 74, 75; equations, **61**, 70-74; errors, **61**, 76-78; experimental procedures, **61**, 69, 70; merits, **61**, 75, 76; theory, **61**, 68, 69

DIMERIZATION

A reaction between two identical molecular entities of chemical species to give a new species. Dimerization need not require covalent bond formation. Formation of a complex between two polypeptide chains or formation of an excimer are processes that do not require covalent bond formation. If the two polypeptide chains are identical, the complex is usually referred to as a homodimer; otherwise, the resulting complex is referred to as heterodimer.

DIMETHYLALLYL*TRANS*TRANSFERASE

This enzyme [EC 2.5.1.1] (also referred to as prenyltransferase and geranyl-diphosphate synthase) catalyzes the reaction of dimethylallyl diphosphate and isopentenyl diphosphate to produce geranyl diphosphate and pyrophosphate (or, diphosphate). The enzyme will not accept larger prenyl diphosphates as substrates.

H. Kleinig (1989) *Ann. Rev. Plant Physiol. Plant Mol. Biol.* **40**, 39.

DIMETHYLANILINE MONOOXYGENASE

This enzyme [EC 1.14.13.8], also referred to as microsomal flavin-containing monooxygenase and dimethylaniline oxidase, catalyzes the reaction of N,N-dimethylaniline with NADPH and dioxygen to produce N,N-dimethylaniline N-oxide, NADP$^+$, and water. The enzyme uses FAD and can utilize a number of dialkylarylamines as substrates.

B. A. Palfey & V. Massey (1998) *Comprehensive Biological Catalysis: A Mechanistic Reference* **3**, 83.

F. P. Guengerich (1990) *Crit. Rev. Biochem. Mol. Biol.* **25**, 97.

DIMINISHED ISOTOPE EFFECT

The value corresponding to the observed isotope effect, expressed as the ratio: $[(k/k') - 1]$, as a consequence of other rate-contributing steps[1].

[1]D. B. Northrop (1982) *Meth. Enzymol.* **87**, 607.

DIOL DEHYDRASE

The following are recent reviews on the molecular and physical properties of this enzyme.

J. A. Stubbe (1989) *Ann. Rev. Biochem.* **58**, 257.

B. M. Babior & J. S. Krouwer (1979) *Crit. Rev. Biochem.* **6**, 35.

R. H. Abeles (1971) *The Enzymes*, 3rd ed., **5**, 481.

DIPEPTIDASES

There are many dipeptidases [EC 3.4.13.x]. Cytosol nonspecific dipeptidase [EC 3.4.13.18] (also referred to as peptidase A, glycylglycine dipeptidase, glycylleucine dipeptidase, and N^2-β-alanylarginine dipeptidase) catalyzes the hydrolysis of dipeptides. Membrane dipeptidase [EC 3.1.13.19] (also known as microsomal dipeptidase, renal dipeptidase, and dehydropeptidase I) is a zinc-dependent enzyme (a member of the peptidase family M19) that also catalyzes the hydrolysis of dipeptides.

S. S. Tate (1985) *Meth. Enzymol.* **113**, 471.

R. J. DeLange & E. L. Smith (1971) *The Enzymes*, 3rd ed., **3**, 81.

DIPEPTIDYL-PEPTIDASE I

This enzyme [EC 3.4.14.1], also called cathepsin C and cathepsin J, catalyzes the hydrolysis of a peptide bond resulting in the release of an N-terminal dipeptide, XaaXbb–Xcc, except when Xaa is an arginyl or a lysyl residue, or Xbb or Xcc is a prolyl residue. This enzyme, a member of the peptidase family C1, is a Cl$^-$-dependent lysosomal cysteine-type peptidase.

M. J. Mycek (1970) *Meth. Enzymol.* **19**, 285.

DIPEPTIDYL-PEPTIDASE IV

This enzyme [EC 3.4.14.5] (also referred to as dipeptidyl aminopeptidase IV, Xaa–Pro-dipeptidylaminopeptidase, Gly–Pro naphthylamidase, and postproline dipeptidyl aminopeptidase IV) catalyzes the hydrolysis of a peptide bond in a protein such that there is a release of an N-terminal dipeptide, XaaXbb–Xcc, preferentially when Xbb is a prolyl residue and provided Xcc is neither a prolyl nor a hydroxyprolyl residue. This protein, a member of the peptidase family S9B, is a membrane-bound serine-type peptidase isolated from mammalian tissue.

Y. Ikehara, S. Ogata & Y. Misumi (1994) *Meth. Enzymol.* **244**, 215.

DIPHOSPHOMEVALONATE DECARBOXYLASE

This enzyme [EC 4.1.1.33], also known as mevalonate pyrophosphate decarboxylase, catalyzes the reaction of ATP with (R)-5-diphosphomevalonate to produce ADP, orthophosphate, carbon dioxide, and isopentenyl diphosphate.

DIPHTHERIA TOXIN

This bacterial ADPRtransferase inactivates host cell ribosomal elongation factor eEF-2 by NAD$^+$-dependent covalent modification of diphthamide, an acceptor-modified histidyl residue, 2-[3-carboxyamido-3-(trimethylammonio)propyl]histidine.

L. Passador & W. Iglewski (1994) *Meth. Enzymol.* **235**, 617.

N. J. Oppenheimer & A. L. Handlon (1992) *The Enzymes*, 3rd ed., **20**, 453.

C.-Y. Lai (1986) *Adv. Enzymol.* **58**, 99.

DIPOLAR BOND

A bond formed by coordination of two neutral molecules or moieties. IUPAC now favors use of this term over various synonyms, such as coordinate covalent bond.

IUPAC (1979) *Pure and Appl. Chem.* **51**, 1725.

DIPOLE-DIPOLE INTERACTION

An interaction, either intermolecular or intramolecular, between molecules, groups, or bonds having a permanent electric dipole moment. The distance and relative orientation between the two dipoles governs the strength of this interaction. ***See also*** *van der Waals Forces*

DIPOLE MOMENT

A vectorial entity (usually symbolized by μ) having a magnitude equal to Qr where Q represents the magnitude of either charge, r is the distance separating them.

The dipole moment is a fundamental property of a molecule (or any dipole unit) in which two opposite charges are separated by a distance[1,2]. This entity is commonly measured in debye units (symbolized by D), equal to 3.33564×10^{-30} coulomb·meters, in SI units). Since the net dipole moment of a molecule is equal to the vectorial sum of the individual bond moments, the dipole moment provides valuable information on the structure and electrical properties of that molecule. The dipole moment can be determined by use of the Debye equation for total polarization. Examples of dipole moments (in the gas phase) are water (1.854 D), ammonia (1.471 D), nitromethane (3.46 D), imidazole (3.8 D), toluene (0.375 D), and pyrimidine (2.334 D). Even symmetrical molecules will have a small, but measurable dipole moment, due to centrifugal distortion effects. Methane[3,4,] for example, has a value of about 5.4×10^{-6} D.

Dipole Moments (μ) of Selected Molecules

Molecule	μ	Molecule	μ
Oxygen, O_2 (gas)	0	Ethyl acetate (gas)	1.76
Methane (gas)	0	Methyl bromide (gas)	1.80
Carbon dioxide (gas)	0	Water, H_2O (liquid)	1.85
Trimethylamine (gas)	0.62	Methyl chloride (gas)	1.86
Dimethylamine (gas)	1.02	Formic acid (in dioxane)	2.07
Hydrogen chloride (gas)	1.08	Pyridine (in benzene)	2.21
Methylamine (gas)	1.33	Acetaldehyde (gas)	2.68
Phenol (gas)	1.40	Acetone (gas)	2.84
Ammonia (gas)	1.46	Nitromethane (gas)	3.50
Methyl iodide (gas)	1.59	Acetamide (in benzene)	3.63
n-Propyl alcohol (gas)	1.66	Urea (in dioxane)	4.56

[1]P. Debye (1945) *Polar Molecules*, Dover, New York.
[2]O. Exmer (1975) *Dipole Moments in Organic Chemistry*, Georg Thieme, Stuttgart.
[3]I. Ozier(1971) *Phys. Rev. Lett.* **27**, 1329.
[4]A. Rosenberg, I. Ozier & A. K. Kudian (1972) *J. Chem. Phys.* **57**, 568.

DIRECT-LINEAR PLOT

A graphical method for analyzing initial rate enzyme kinetic data that avoids some of the weighting and statistical problems associated with the other graphical procedures[1-3]. The method offers the advantage that it is relatively insensitive to aberrant observations or data points. In a Cartesian-coordinate system, a substrate concentration is marked along the horizontal axis (*i.e.*, the *x* axis) to the left of the origin. The initial velocity (*v*) observed with that substrate concentration is placed on the vertical axis (*i.e.*, the *y* axis) above the origin. A line is then drawn connecting those two points. The process is repeated for every substrate concentration and initial velocity pair.

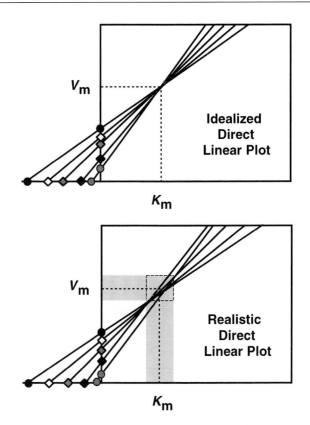

With perfect data, there will be a common intersection point at coordinates (K_a, V_{max}) in the first quadrant. However, for real experimental data, there will be a number of intersection points. The maximum number of intersection points will be $(n/2)(n - 1)$ where *n* is the number of lines. If different sets of lines have a common intersection point, then that point is treated with weighting. The estimate of the K_a value will be the median value for the *x*-coordinate of these intersection points. Correspondingly, the median value of the *y*-coordinate of these intersection points provides the value for V_{max}.

The method is not as useful with multisubstrate systems; however, Cornish-Bowden and Endrenyi[4] have presented a robust regression method to treat data with more parameters.

[1]R. Eisenthal & A. Cornish-Bowden (1974) *Biochem. J.* **139**, 715.
[2]A. Cornish-Bowden & R. Eisenthal (1974) *Biochem. J.* **139**, 721.
[3]A. Cornish-Bowden & R. Eisenthal (1978) *Biochim. Biophys. Acta* **523**, 268.
[4]A. Cornish-Bowden & L. Endrenyi (1986) *Biochem. J.* **234**, 21.

DISPROPORTIONATION

Any reaction in which one molecular entity reacts with an identical molecular entity, transferring an electron, atom or group from one reactant to the other:

$$A + A \rightarrow A' + A''$$

The reverse of disproportionation is referred to as comproportionation. A special case of disproportionation (or dismutation) is radical disproportionation which serves as a termination process in the fate of free radicals. For example, $2\ CH_3CH_2\cdot \rightarrow CH_3CH_3 + CH_2{=}CH_2$.

DISSOCIATION

The taking apart or separation of a single substance or molecular entity into two or more other substances or different molecular entities. Dissociation includes heterolytic and homolytic reactions, the release of free ions from ion pairs, the depolymerization process of a polymer consisting of smaller subunits, *etc.*

DISSOCIATION CONSTANT

The equilibrium constant (commonly symbolized K_d) describing the dissociation of a compound into two or more compounds, ions, or atoms. For the reaction A + B = A–B, the dissociation constant is [A][B]/[A–B]. The dissociation constant is the reciprocal of the association constant and the negative logarithm of the K_d is the pK_d. The dissociation constant for a Brønsted acid (commonly symbolized K_a) is [H$^+$][A$^-$]/[HA], where HA is the undissociated acid and A$^-$ represents the conjugate base. In aqueous solutions, water participates in this equilibrium. However, since the concentration of water is large and essentially constant, it has been incorporated into the constant K_a. The negative logarithm is referred to as the pK_a for the acid.

The dissociation constant for a base (symbolized K_b) equals [B$^+$][OH$^-$]/[BOH], where BOH represents the undissociated base. As mentioned above, in aqueous solutions, the concentration of water is considered high and constant and is a portion of K_b. The "dissociation constant" (or, autoprotolysis constant) of water (symbolized K_w) equals [H$^+$][OH$^-$] = 10^{-14} at 25°C. At 30°C, $K_w = 10^{-13.8}$ and at 37°C, $K_w = 10^{-13.6}$. The negative logarithm is pK_w. **See** *Association Constant*

DISSOCIATION CONSTANTS
(Metal-Nucleotide Complexes)

O'Sullivan and Smithers[1] describe various protocols for determining dissociation constants (or, stability constants) of metal ion-nucleotide complexes. Morrison and Cleland[2,3] have presented a kinetic method that has

proven to be quite useful. In this procedure the nucleotide competes for Mg^{2+} and another metal ion which binds to the nucleotide more tightly than Mg^{2+}. An enzyme is chosen that utilizes the Mg^{2+}-nucleotide complex as a substrate but which does not use the other metal ion-nucleotide complex. In addition, it is assumed that the metal ion under study (as well as Mg^{2+}) does not form an insoluble precipitate at the pH value of the studies and the concentration of the free metal ion. Upon determining the K_m value for MgATP, the K_d value for the MgATP complex, and the K_i value for the metal ion-nucleotide complex with the enzyme, all under the same set of experimental conditions, the K_d for the metal ion-nucleotide complex can be readily obtained. A computer program has been provided to assist in K_d determinations[3].

[1]W. J. O'Sullivan & G. W. Smithers (1979) *Meth. Enzymol.* **63**, 294.
[2]J. F. Morrison & W. W. Cleland (1980) *Biochemistry* **19**, 3127.
[3]W. W. Cleland (1995) *Meth. Enzymol.* **249**, 181.

DISSOCIATION CONSTANTS
(Temperature Effects)

A temperature change can have an effect on the degree of ionization of various moieties on a protein as well as on substrates and effectors. Since proton dissociation constants (*i.e.*, K_a values) are thermodynamic parameters, a change in temperature can result in a significant alteration of the pH-activity curve[1].

Investigators have often used pH studies to aid in the identification of active-site residues. However, care must be exercised in this regard since observed pK values may refer to macroscopic parameters rather than microscopic values and pK values of groups can be greatly affected by the immediate environment. Because of these influences, it may be difficult to identify specific moieties responsible for binding and/or catalysis.

As was demonstrated in the discussion for the van't Hoff relationship:

$$\Delta H = -2.303RT^2 dpK_a/dT$$

(Note that care should be exercised here. Changes in enthalpies are determined from equilibrium constants. The K_a values used in pH studies are dissociation constants and not association constants.) Plotting pK_a *vs.* the reciprocal of the absolute temperature will provide a value for ΔH. An early study using this procedure was reported by Gutfreund[2] with trypsin. Gutfreund reported a pK_a value of 6.25 at 25°C and a ΔH value of 7000 cal·mol^{-1} (29 kJ·mol^{-1}), concluding that the data agree with the participation of an imidazole moiety at the active site. In another early study (although only two temperatures were studied), Massey and Alberty[3] concluded that the ΔH of ionization was less than 4500 cal·mol^{-1} (19 kJ·mol^{-1}) in the inhibition studies of fumarate hydratase with thiocyanate, suggesting the participation of a carboxyl group. However, Knowles[4] has pointed out that the factors that are sufficient to perturb a pK_a can also affect changes in enthalpy and/or entropy.

An investigator should also be concerned about the buffer since the pK_a of a buffer can also vary with temperature[5-7]. The pH of a buffer solution will vary with temperature depending on the variance of the activity coefficient terms and the pK_a of the buffer.

The temperature-dependent change in pK_a for some compounds can be significant. For instance, the ΔpK_a for Tris (tris(hydroxymethyl)aminomethane) between 4°C and 37°C is 0.924. It is therefore advisable to verify the pH of the solution throughout the temperature range studied when this buffer is used. Ellis and Morrison[8] have discussed the practical pK_a value (written as pK_a^*), referring to the pK_a value under chosen experimental conditions. They present basic programs for calculating pK_a^* values at different temperatures and ionic strengths. **See** *Buffers; pH Effects; van't Hoff Relationship*

[1]K. F. Tipton & H. B. F. Dixon (1979) *Meth. Enzymol.* **63**, 183.
[2]H. Gutfreund (1955) *Trans. Faraday Soc.* **51**, 441.
[3]V. Massey & R. A. Alberty (1954) *Biochim. Biophys. Acta* **13**, 354.
[4]J. R. Knowles (1976) *CRC Crit. Rev. Biochem.* **4**, 165.
[5]D. D. Perrin & B. Dempsey (1974) *Buffers and pH and Metal Ion Control*, Chapman and Hall, London.
[6]D. D. Perrin, B. Dempsey & E. P. Serjeant (1981) *pKa Prediction for Organic Acids and Bases.* Chapman and Hall, London.
[7]D. D. Perrin (1964) *Austral. J. Chem.* **17**, 484.
[8]K. J. Ellis & J. F. Morrison (1982) *Meth. Enzymol.* **87**, 405.

DISSOCIATION KINETICS

Measurement of the rates of dissociation of enzyme-substrate and protein-ligand complexes, usually promoted by dilution. The utility of dissociation kinetics is well illustrated in the report of Dunn and Raftery[1] who examined the kinetics of [³H]acetylcholine and [³H]suberyldicholine dissociation from the membrane-bound acetylcholine receptor. Each agonist binds to two high-affinity sites on the receptor, and the equilibrium dissociation constant is around 15 nM. Upon triggering dissociation of [³H]acetylcholine from the receptor complex by dilution, a monophasic process is observed with an apparent rate of 0.023 ± 0.010 s^{-1}. When μM levels of acetylcholine, carbamylcholine, or suberyldicholine was included in the dilution buffer, this rate increased about 5-fold. This accelerating effect, which is not expected in simple independent binding processes, occurred even when the two high-affinity sites were initially saturated with the radiolabeled ligand. The authors inferred that there is likely to be an additional site (or subsite) for agonist with a K_d in the μM range. Yet, over the 0 to 20 μM range, no additional sites for [³H]acetylcholine were detectable. These findings led Dunn and Raftery[1] to propose that each high-affinity site has two mutually exclusive subsites, designated A and B. With [³H]acetylcholine initially occupying the A site, occupancy of site B by unlabeled ligand reduces the affinity for site A and accelerates the dissociation of the radiolabeled ligand. Their other dissociation kinetic measurements using a large bis-quaternary agonist, [³H]suberyldicholine, provided an additional clue about the properties of these proposed subsites. Its dissociation rate was similar to that observed with acetylcholine (0.028 ± 0.012 s^{-1}), but this rate was only weakly influenced by the presence of unlabeled ligands. They suggest that suberyldicholine may cross-link the two subsites or sterically occlude the second site.

[1]S. M. Dunn & M. A. Raftery (1997) *Biochemistry* **36**, 3846.

DISTORTION, FREE ENERGY OF

Enzymes are thought to achieve their enormous rate accelerations by distorting the substrate into their most reactive electronic configuration and conformation (**See** *Entropy Trap Model*). Accordingly, there is great interest in estimating the free energy of distortion. One way to treat this problem was considered by Schowen[1] who defines $\Delta G_{distortion}$ as the total work needed (a) to convert the substrate from its standard-reaction state to the transition state, including that work for breaking away of any extra covalently attached groups, desolvation, *etc.*, and (b) to convert the enzyme from its substrate-free form into exactly the enzyme structure achieved in the transition state, again including protein conformational changes, desolvation of the active site, *etc.* This "thought" experiment provides a way to estimate the energy changes from the substrate's and enzyme's ground states to their so-called poised structures (*i.e.*, their respective structures in the transition state). Schowen also defines the vertical binding energy as the net change in free energy that is realized when one combines the poised structure of the substrate with the poised structure of the enzyme.

[1] R. L. Schowen (1978) in *Transition States of Biochemical Processes*, p. 88 (R. D. Gandour & R. L. Schowen, eds.), Plenum, New York.

DISTRIBUTION

The set of possible values of a random variable, most typically presented as a plot of frequency of occurrence on the vertical axis *versus* magnitude of the variable on the horizontal axis. *See Boltzmann Distribution Law*

DITHIONITE

A nonbiological reductant ($S_2O_4^{2-}$) that has proven to be of immense value in converting uncomplexed and porphyrin-bound Fe(III) to the +2 oxidation state with the concomitant formation of two molecules of SO_2 gas. Dithionite also reacts with heme-bound oxygen to produce deoxyhemoglobin, and treatment of intact red blood cells with dithionite can induce sickling in cells containing hemoglobin S. Dithionite also reduces NAD^+ to NADH.

Selected entries from *Methods in Enzymology* [vol, page(s)]:
Use in cytochrome assays, **69**, 181, 182; ferredoxin-nitrite reductase assay, **69**, 256; flavodoxins, **69**, 405, 406; interference, with analysis of nitrogenase, **69**, 755, 756; iron-sulfur center reduction, **69**, 241, 243, 246; oxygen reaction, **76**, 637-638; oxygen removal, **76**, 267, 293; oxyhemoglobin reaction, **76**, 637-639; preparation, **76**, 58-59, 791-792; reaction center isolation, **69**, 175; induction of sickling in hemoglobin S, **76**, 788; use in ligand-binding studies, **76**, 427; to reduce methemoglobin, **76**, 11-12.

DITHIOTHREITOL

A reagent introduced by W. W. Cleland (hence the alternative cognomen "Cleland's Reagent") to take advantage of the ability of 1,4-dithiothreitol to react with disulfide cross-linked protein (P–S–S–P) to form 2 P–SH plus the oxidized six-membered cyclic disulfide form of the reagent. This reagent and its partner, dithioerythritol, are frequently added (at 0.5 to 2.0 mM concentrations) to cell-free extracts to prevent adventitious oxidation of protein sulfhydryl groups during purification.

Selected entries from *Methods in Enzymology* [vol, page(s)]:
Absorption maxima of disulfide bond, **251**, 55; fatty acylation of

purified proteins, **250**, 462; oxidation potential, **251**, 20; protective effect against protease inactivation, **241**, 111-112; protein disulfide reduction, **233**, 398, 405; reaction with protein sulfhydryls, **251**, 361; reaction with superoxide, **251**, 84; redox potential, **252**, 179; reduction of disulfide bond, rate, **251**, 167, 171, 173; glutathione disulfide, **251**, 171; immunoglobulin, **251**, 171; measurement, **251**, 169-170; papain, **251**, 170-172; trypsinogen, **251**, 170-171; thiyl radical, **251**, 35.

DIXON PLOT

A plot of the reciprocal of the initial velocity as a function of inhibitor concentrations at different fixed concentrations of substrate[1]. For a competitive inhibitor, the lines passing through the data will intersect at a common point in the second quadrant (whose coordinates are the ordered pair ($-K_{is}$, $1/V_{max}$). Here, K_{is} is the dissociation constant for the competitive inhibitor and V_{max} is the maximum velocity.

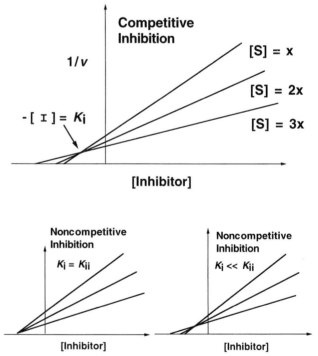

The intersecting line pattern observed for competitive inhibition (top plot) can be distinguished from the action of a noncompetitive inhibitor characterized by $K_i = K_{ii}$. However, if K_i is much less than K_{ii}, the observed pattern cannot be distinguished.

For a noncompetitive inhibitor[2], the intersection point occurs at coordinates ($-K_{is}$, $(1/V_{max})(1 - [K_{is}/K_{ii}])$), where K_{ii} is the dissociation constant of the inhibitor with respect to the enzyme-substrate complex. Thus, a noncompetitive inhibitor can have an intersection point in the second quadrant (when $K_{is} < K_{ii}$), on the horizontal axis (when $K_{is} = K_{ii}$), or in the third quadrant (when $K_{is} > K_{ii}$). For this reason, a Dixon plot should *never* be used for the purpose of distinguishing modes of inhibitor action[3]. Standard double-reciprocal plots (*i.e.*, $1/v$ *versus* $1/[A]$ at varying concentrations of the inhibitor) offer a more reliable means.

[1]M. Dixon (1953) *Biochem. J.* **55**, 170.
[2]D. L. Purich & H. J. Fromm (1972) *Biochem. Biophys. Acta*, **268**, 1.
[3]Note: Uncompetitive inhibitors display parallel lines in Dixon plots.

DIXON-WEBB PLOTS

A plot used in pH studies of enzyme activity: a plot of log V_{max} *vs.* pH or a plot of log (V_{max}/K_m) *vs.* pH. These plots are useful when the pK values of the enzyme are separated by at least 3.5 pH units. *See* pH Studies

M. Dixon & E. C. Webb (1979) *Enzymes*, 2nd ed., Longmans, Green, New York.
K. F. Tipton & H. B. F. Dixon (1979) *Meth. Enzymol.* **63**, 183.

DNA (ADENINE-N^6)-METHYLTRANSFERASE

This enzyme [EC 2.1.1.72] catalyzes the reaction of *S*-adenosyl-L-methionine with an adenine residue in DNA to produce *S*-adenosyl-L-homocysteine and a 6-methylaminopurine residue in DNA.

F. Takusagawa, M. Fujioka, A. Spies & R. L. Schowen (1998) *Comprehensive Biological Catalysis: A Mechanistic Reference* **1**, 1.
I. Ahmad & D. N. Rao (1996) *Crit. Rev. Biochem. Mol. Biol.* **31**, 361.
D. Landry, J. M. Barsomian, G. R. Feehery & G. G. Wilson (1992) *Meth. Enzymol.* **216**, 244.

DNA BREATHING

The continual loss and regain of local and global structure in DNA that results from the breaking and remaking of hydrogen bonds between A and T as well as G and C base pairs in double-stranded DNA. The ability of these base pairs within DNA to break and remake their hydrogen bonds is essential for DNA replication and transcription, and this so-called ''breathing'' process must also factor in the recognition, flexibility, and structure of DNA. Base-pair opening is most frequently studied by following the exchange of protons from imino groups with water using nuclear magnetic resonance measurements based on line broadening, longitudinal relaxation, and magnetization transfer from water. At room temperature, base-pair lifetimes occur on the 10–40 ms time-scale.

DNA (CYTOSINE-5-)-METHYLTRANSFERASE

DNA (cytosine-5-)-methyltransferase [EC 2.1.1.37] catalyzes the reaction of *S*-adenosyl-L-methionine with DNA to produce *S*-adenosyl-L-homocysteine and DNA containing a 5-methylcytosine residue. Site-specific DNA methyltransferase (cytosine-specific) [EC 2.1.1.73] catalyzes the reaction of *S*-adenosyl-L-methionine with DNA containing a cytosine to produce *S*-adenosyl-L-homocysteine and DNA containing a 5-methylcytosine.

I. Ahmad & D. N. Rao (1996) *Crit. Rev. Biochem. Mol. Biol.* **31**, 361.
X. Cheng (1995) *Ann. Rev. Biophys. Biomol. Struct.* **24**, 293.
S. Kumar, X. Cheng, S. Klimasauskas, S. Mi, J. Posfai, R. J. Roberts & G. G. Wilson (1994) *Nucleic Acid Res.* **22**, 1.
D. Landry, J. M. Barsomian, G. R. Feehery & G. G. Wilson (1992) *Meth. Enzymol.* **216**, 244.
S. Hattman (1981) *The Enzymes*, 3rd ed., **14**, 517.
S. J. Kerr & E. Borek (1973) *The Enzymes*, 3rd ed., **9**, 167.

DNA GLYCOSYLASES

DNA α-glucosyltransferase [EC 2.4.1.26] catalyzes the transfer of an α-D-glucosyl residue from UDPglucose to a hydroxymethylcytosine residue in DNA. DNA β-glucosyltransferase [EC 2.4.1.27] catalyzes the transfer of a β-D-glucosyl residue from UDPglucose to a hydroxymethylcytosine residue in DNA. Glucosyl-DNA β-glucosyltransferase [EC 2.4.1.28] catalyzes the transfer of a β-D-glucosyl residue from UDPglucose to a glucosylhydroxymethylcytosine residue in DNA.

G. Davies, M. L. Sinnott & S. G. Withers (1998) *Comprehensive Biological Catalysis: A Mechanistic Reference* **1**, 119.
L. D. Samson (1992) *Essays in Biochem.* **27**, 69.
B. Weiss & L. Grossman (1987) *Adv. Enzymol.* **60**, 1.
E. C. Friedberg, T. Bonura, E. H. Radany & J. D. Love (1981) *The Enzymes*, 3rd ed., **14**, 251.
B. K. Duncan (1981) *The Enzymes*, 3rd ed., **14**, 565.

DNA GYRASES & DNA REVERSE GYRASE

This entry contains a set of enzymes for which the terms gyrase and reverse gyrase have been used. DNA topoisomerase (ATP-hydrolyzing) [EC 5.99.1.3] (also referred to as DNA topoisomerase II, DNA gyrase, and Type II DNA topoisomerase) catalyzes the ATP-dependent breakage, passage, and rejoining of double-stranded DNA. The enzyme can introduce negative superhelical turns into double-stranded circular DNA. One domain of the protein has a nicking-closing activity, whereas another unit catalyzes the supertwisting and hydrolysis of ATP. *See also* DNA Topoisomerases

R. A. Grayling, K. Sandman & J. N. Reeve (1996) *Adv. Protein Chem.* **48**, 437.

D. B. Wigley (1995) *Ann. Rev. Biophys. Biomol. Struct.* **24**, 185.
R. J. Reece & A. Maxwell (1991) *Crit. Rev. Biochem. Mol. Biol.* **26**, 335.
M. Gellert (1981) *The Enzymes*, 3rd ed., **14**, 345.

DNA LIGASES

DNA ligase (ATP) [EC 6.5.1.1] catalyzes the reaction of ATP with (deoxyribonucleotide)$_n$ and (deoxyribonucleotide)$_m$ to produce AMP, pyrophosphate, and (deoxyribonucleotide)$_{(n+m)}$. This activity results in the formation of a phosphodiester at the site of a single-strand break in duplex DNA. RNA can also act as substrate to some extent.

DNA ligase (NAD$^+$) [EC 6.5.1.2] (also referred to as polydeoxyribonucleotide synthase (NAD$^+$), polynucleotide ligase (NAD$^+$), DNA repair enzyme, and DNA joinase) catalyzes the reaction of NAD$^+$ with (deoxyribonucleotide)$_n$ and (deoxyribonucleotide)$_m$ to produce AMP, nicotinamide nucleotide, and (deoxyribonucleotide)$_{(n+m)}$. This forms a phosphodiester at the site of a single-strand break in duplex DNA. RNA can also act as substrate to some extent.

A. R. Clarke & T. R. Dafforn (1998) *Comprehensive Biological Catalysis: A Mechanistic Reference* **3**, 1.
B. F. Cooper & F. B. Rudolph (1995) *Meth. Enzymol.* **249**, 188.
P. A. Frey (1992) *The Enzymes*, 3rd ed., **20**, 141.
M. J. Engler & C. C. Richardson (1982) *The Enzymes*, 3rd ed., **15**, 3.
I. R. Lehman (1974) *The Enzymes*, 3rd ed., **10**, 237.

DNA METHYLTRANSFERASES

Methylation of certain bases within DNA molecules plays a major role in gene expression, as is evidenced by the transcriptional silencing of X chromosome genes through methylation. The following are recent reviews on the molecular and physical properties of these enzymes [EC 2.1.1.x]. **See also** *specific methyltransferase*

F. Takusagawa, M. Fujioka, A. Spies & R. L. Schowen (1998) *Comprehensive Biological Catalysis: A Mechanistic Reference* **1**, 1.
I. Ahmad & D. N. Rao (1996) *Crit. Rev. Biochem. Mol. Biol.* **31**, 361.
D. Landry, J. M. Barsomian, G. R. Feehery & G. G. Wilson (1992) *Meth. Enzymol.* **216**, 244.
R. L. P. Adams (1990) *Biochem. J.* **265**, 309.
R. L. P. Adams & R. H. Burdon (1982) *Crit. Rev. Biochem.* **13**, 349.
S. Hattman (1981) *The Enzymes*, 3rd ed., **14**, 517.
S. J. Kerr & E. Borek (1973) *The Enzymes*, 3rd ed., **9**, 167.

DNA OXIDATIVE DAMAGE AND REPAIR

Henle and Linn[1] recently reviewed the formation, prevention, and repair of DNA damage by iron/hydrogen peroxide (so-called Fenton oxidants). Berlett and Stadt-man[2] also discussed the corresponding pathways by which oxidatively modified forms of proteins accumulate during aging, oxidative stress, and in some pathological conditions.

[1] E. S. Henle & S. Linn (1997) *J. Biol. Chem.* **272**, 19095.
[2] B. S. Berlett & E. R. Stadtman (1997) *J. Biol. Chem.* **272**, 20313.

DNA PHOTOLYASE

This enzyme [EC 4.1.99.3] catalyzes the reactivation by light of irradiated DNA by converting cyclobutadipyrimidine residues in DNA to two pyrimidine residues in DNA. This is a flavoprotein containing a second chromophore group.

B. A. Palfey & V. Massey (1998) *Comprehensive Biological Catalysis: A Mechanistic Reference* **3**, 83.
S.-T. Kim, P. F. Heelis & A. Sancar (1995) *Meth. Enzymol.* **258**, 319.
A. Sancar (1994) *Biochemistry* **33**, 2.
E. J. Brush & J.W. Kozarich (1992) *The Enzymes*, 3rd ed., **20**, 317.
B. M. Sutherland (1981) *The Enzymes*, 3rd ed. **14**, 482.

DNA POLYMERASES

DNA-directed DNA polymerases [EC 2.7.7.7], also called DNA nucleotidyltransferases (DNA-directed), are enzymes that catalyze the DNA template-directed extension of the 3'-end of a nucleic acid strand one nucleotide at a time. Thus, n deoxynucleoside triphosphates produce n pyrophosphate (or, diphosphate) ions and DNA$_n$. This enzyme cannot initiate the synthesis of a polymeric chain *de novo*; it requires a primer which may be DNA or RNA. RNA-directed DNA polymerases [EC 2.7.7.49], also referred to as reverse transcriptases, DNA nucleotidyltransferases (RNA-directed), and revertases, are enzymes that catalyze the RNA template-directed extension of the 3'-end of a nucleic acid strand one nucleotide at a time. Thus, n deoxynucleoside triphosphates produce n pyrophosphate (or, diphosphate) ions and DNA$_n$. As was the case above, this enzyme cannot initiate the synthesis of a polymeric chain *de novo*; it requires a primer which may be DNA or RNA.

N. H. Williams (1998) *Comprehensive Biological Catalysis: A Mechanistic Reference* **1**, 543.
F. B. Perler, S. Kumar & H. Kong (1996) *Adv. Protein Chem.* **48**, 377.
Z. Kelman & M. O'Donnell (1995) *Ann. Rev. Biochem.* **64**, 171.
C. S. McHenry (1991) *J. Biol. Chem.* **266**, 19127.
P. A. Frey (1989) *Adv. Enzymol.* **62**, 119.
V. Mizrahi & S. J. Benkovic (1988) *Adv. Enzymol.* **61**, 437.
B. L. Vallee & A. Galdes (1984) *Adv. Enzymol.* **56**, 283.
See also chapters in *The Enzymes*, 3rd. ed., vols. **10**, **14**, **19**, and **20**.

DNA RENATURATION KINETICS

Upon mixing, complementary DNA sequences will tend to recombine and form a double helix in a process commonly called DNA renaturation. Treatment of DNA renaturation kinetics is greatly simplified by assuming: (a) that the rate behavior can be treated as a bimolecular process limited by the chains finding proper registration of their respective hydrogen bonding partners on complementary strands; (b) that, after initial chain recombination, there is a much more rapid zippering process for hydrogen bonding and formation of the double helix; (c) that a collection of complementary strands does not contain inhibitors (*i.e.*, components of slightly mismatched that retard the reassociation of genuinely complementary strands of DNA. When these conditions hold, the following scheme will describe how complementary sequences A and B will recombine:

$$A + B \xrightarrow{k_{+1}} H$$

By expressing the molar concentrations of each sequence A and B and helix H, we can obtain the following rate expression describes this process:

$$\frac{d[H_i]}{dt} = k_{+1}[A][B]$$

As noted by Bloomfield, Crothers, and Tinoco[1], however, there is value in re-expressing the molar concentrations of each strand in terms of the number of sequence repeats, where a total double-helix length of T base pairs, broken into fragments of length L, may contain n_i copies of repeat sequence-i. In this way, one can designate the concentration of strands having sequence-i as $[A]_i$, and $[B]_i$, which in turn can be defined with respect to their nucleotide content:

$$[A]_i = (n_i/2T)[A]_N \text{ and } [B]_i = (n_i/2T)[B]_N$$

where the factor "2" in the denominator changes base pairs into bases, and the subscript N indicates that concentration is expressed in terms of nucleotides. Under these conditions,

$$\frac{d[H_i]_N}{dt} = 2L \frac{d[H_i]}{dt} = k_{+1} 2L \left(\frac{n_i}{2T}\right)^2 = \frac{k_{+1}L}{2} \left(\frac{n_i}{T}\right)^2$$

This expression can also be written as follows:

$$\frac{d[H_i]_N}{dt} = \frac{k_N}{2} \left(\frac{n_i}{T}\right)^2 C_N$$

where C_N is the total single-strand nucleotide concentration, and k_N equals $k_{+1} n_i / T$.

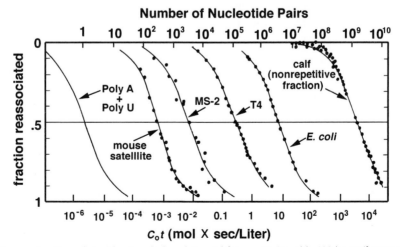

Number of Nucleotide Pairs

Figure 1. Progress curves for renaturation of double-stranded nucleic acid fragments (roughly 400 bases/fragment). From Britten and Kohne[2] with permission.

$C_o t$ Curve Analysis. By making a few simplifying assumptions (including that L remains constant), Britten and Kohne[2] developed a straightforward kinetic approach to demonstrate the high frequency of occurrence of repeated sequences in DNA from the genomes of higher organisms. To analyze the time-course of DNA renaturation, they rearranged the integrated rate equation:

$$\left(\frac{1}{C_N} - \frac{1}{C_o}\right) = \frac{k_N t}{2}$$

(where C_o is the total nucleotide concentration) to obtain the following final kinetic expression:

$$f = \frac{1}{1 + 0.5\, k_N C_o t}$$

Because the time-courses for renaturation are presented in terms of $C_o t$, and not time, the progress curves shown in Fig. 1 are frequently called $C_o t$, curves. Here, the parameter f is the fraction of total nucleotides in the helical form (*i.e.*, $f = C_N/C_o$), and when $f = 0.5$, k_N can be evaluated readily as $C_o t/2$.

[1]V. A. Bloomfield, D. M. Crothers & I. Tinoco, Jr. (1974) *Physical Chemistry of Nucleic Acids*, pp. 361-367 Harper & Row, New York.
[2]R. J. Britten & D. E. Kohne (1968) *Science* **161**, 529.

DNA TOPOISOMERASES

DNA topoisomerase [EC 5.99.1.2] (also referred to as DNA topoisomerase I, relaxing enzyme, untwisting enzyme, swivelase, type I DNA topoisomerase, nicking-closing enzyme, and ω-protein) catalyzes the ATP-independent breakage of single-stranded DNA, followed by the passage and rejoining of that DNA. The enzyme catalyzes the conversion of one topological isomer of DNA into another (*e.g.*, the relaxation of superhelical turns in DNA, the interconversion of simple and knotted rings of single-stranded DNA, and the intertwisting of single-stranded rings of complementary sequences. DNA topoisomerase (ATP-hydrolyzing) [EC 5.99.1.3] (also referred to as DNA topoisomerase II, DNA gyrase, and Type II DNA topoisomerase) catalyzes the ATP-dependent breakage, passage, and rejoining of double-stranded DNA. The enzyme can introduce negative superhelical turns into double-stranded circular DNA. One unit of the protein has a nicking-closing activity whereas

another unit catalyzes the supertwisting and hydrolysis of ATP.

J. C. Wang (1996) *Ann. Rev. Biochem.* **65**, 635.
R. A. Grayling, K. Sandman & J. N. Reeve (1996) *Adv. Protein Chem.* **48**, 437.
R. J. Reese & A. Maxwell (1991) *Crit. Rev. Biochem. Mol. Biol.* **26**, 335.
A. Maxwell & M. Gellert (1986) *Adv. Prot. Chem.* **38**, 69.
S. A. Wasserman & N. R. Cozzarelli (1986) *Science* **232**, 951.
J. C. Wang (1981) *The Enzymes*, 3rd ed., **14**, 331.
M. Gellert (1981) *The Enzymes*, 3rd ed., **14**, 345.
H. A. Nash (1981) *The Enzymes*, 3rd ed., **14**, 471.

DNA UNWINDING (Kinetic Model For Small DNA)

Frictional resistance of the viscous medium is thought to be the chief energy barrier determining the unwinding rate[1]. Longer polymers experience much greater friction, and the time for untwisting of DNA was thought to be proportional to at least the second power of length. Nonetheless, Cohen and Crothers[2] demonstrated that even relatively small DNA molecules (with masses of 65,000–2,300,000 Daltons) exhibit frictionally limited unwinding. The rate of unwinding is also influenced by the rate of breaking hydrogen bonds.

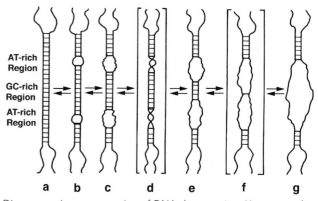

Diagrammatic representation of DNA denaturation. Upon perturbation, the first "melting" away of structure occurs within AT-rich regions, and further perturbation causes this local disorder to grow, such that GC-rich regions are the last to become disordered. The DNA unwinding reaction is limited by viscous resistance.

DNA unwinding in the above diagram[3] occurs initially in the AT-rich regions of the DNA duplex (indicated as "a" above), and these local regions of disorder (shown in "c" and "d") require some twisting motions (shown in "d"). As these areas grow ("e"), one reaches the point of nearly complete loss of local order ("f"), followed by formation of a large loop ("g"). As would be expected, the reverse rate constants are much lower for larger loops, because achieving proper registration of complementary bases pairs may lead to a higher activation energy.

Cohen and Crothers[2] also present an excellent account of their use of the Guggenheim method for extracting rate constants from incomplete segments of decay processes, in which either the starting- and/or end-points of the spectral decay are not well defined. *See also* Chemical Kinetics (1. Progress Curve Analysis)

[1]D. M. Crothers (1969) *Acc. Chem. Res.* **2**, 225.
[2]R. J. Cohen & D. M. Crothers (1971) *J. Mol. Biol.* **61**, 525.
[3]We are indebted to our colleague, Professor Robert J. Cohen, for the original drawing on which this diagram is based.

DOCKING METHODS

Computer-based techniques for locating and analyzing ligand binding sites on macromolecules. These methods must take into account the steric and electrostatic properties of ligand and receptor as well as ligand-induced conformational changes.

Selected entries from *Methods in Enzymology* [**vol**, page(s)]:
DOCK computer program, **241**, 355-358; applications to HIV protease, **241**, 355-358, 362-370; automated FLOG approach, **241**, 359, 361-365, 367-370; MINDEX flexibase, **241**, 362-370; ligand-receptor matching and evaluation, **241**, 361-362; ligand representation, **241**, 359-360; ligand selection, **241**, 366-370; receptor definition, **241**, 365-366; receptor representation, 241, 360-361; search results, **241**, 362.

DOMAIN

A topologically separate or autonomous subregion of a protein, usually encoded by a contiguous sequence. Such regions are often stable to limited proteolysis, allowing them to be excised from the polypeptide chain without significant loss of structure or functional binding of ligands. Most domains in multidomain proteins form from continuous segments and adopt the same structural class.

Domains may be regarded as the basic units for the structure, function and evolution of proteins, but the definition of a domain remains fuzzy. They are most often treated as compact or connected areas that are apparent from a visual inspection of protein models. To avoid subjectivity and ambiguities of visual inspection, computer algorithms have been developed to localize domains. Rashin[1] offered an alternative interpretation: domains are stable globular fragments, generated in biochemical experiments that refold autonomously and retain specific functions. He proposed a method for localiz-

ing these globular fragments on the basis of surface area measurements. Islam *et al.*[2] used an algorithm based on interresidue contacts to identify domains in proteins. The results of applying their algorithm agreed with those derived by visual inspection for nearly four-fifths of the 284 nonredundant chains that they considered. They also examined the relationship between accessible surface area and molecular weight for domains and chains. Rigid domains, defined by Nichols *et al.*[3] as a local region of tertiary structure common to two or more different protein conformations, can be identified numerically from atomic coordinates by finding sets of residues, such that the distance between any two residues within the set belonging to one conformation is the same as the distance between the two structurally equivalent residues within the set belonging to any other conformation. They described two algorithms that employ the difference-distance matrix to search directly for rigid domains from atomic coordinates.

[1]A. A. Rashin (1981) *Nature* **291**, 85.
[2]S. A. Islam, J. Luo & M. J. Sternberg (1995) *Protein Engineering* **8**, **513**.
[3]W. L. Nichols, G. D. Rose, L. F. Ten Eyck & B. H. Zimm (1995) *Proteins* **23**, 38.

DOMAIN SWAPPING

A mechanism[1] for explaining how oligomeric and polymeric forms of a protein or polypeptide can arise by three-dimensional swapping of internal domains in monomers. The basic idea is that a protein's ternary structure is stabilized by the contacts between its domains, and under certain circumstances, the domains of several polypeptide chains may bind in a way that gives rise to nonphysiologic oligomers, polymers, and aggregates.

The following diagram, adapted from that first presented by Bennett *et al.*[1], describes a postulated pathway for evolution of a protein dimer from single-domain proteins. The scheme begins with the fusion of two single-domain polypeptides and proceeds through the evolution of interdomain contacts, and in the case of enzymes, development of an active site. These same interdomain contacts can also stabilize formation of a domain-swapped dimer which then undergoes further evolution into a present-day dimer.

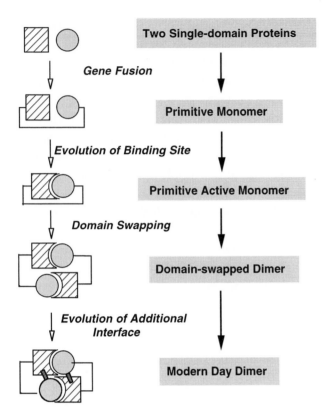

Two Single-domain Proteins

Gene Fusion

Primitive Monomer

Evolution of Binding Site

Primitive Active Monomer

Domain Swapping

Domain-swapped Dimer

Evolution of Additional Interface

Modern Day Dimer

As illustrated in the diagram below, domain swapping can also result in indefinite polymerization to form linear supramolecular structures. These may correspond to present-day polymers of proteins such as microtubules, or they may represent abnormal structures, like the straight and paired-helical filaments in the neurofibrillary tangles observed in the brain tissue of those afflicted with Alzheimer's disease.

Polymerization as a Consequence of Domain Swapping

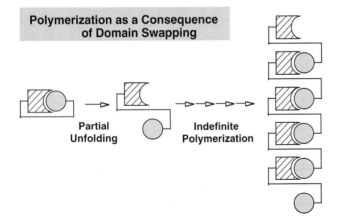

Partial Unfolding   Indefinite Polymerization

The idea of domain swapping traces its origin to the work of Crestfield *et al.*[2] who noted that a mixture of His-12 alkylated RNase monomer and His-119 alkylated RNase monomer could recover catalytic activity through the formation of a dimeric species. As pointed out by Bennett *et al.*[1], this is an excellent example of protein complementation[3], wherein two different protein molecules, each lacking an essential protein motif, can combine to form a functional molecule.

[1] M. J. Bennett, M. P. Schlunegger & D. Eisenberg (1995) *Protein Science* **4**, 2455.
[2] A. M. Crestfield, W. H. Stein & S. Moore (1963) *J. Biol. Chem.* **238**, 2421.
[3] I. Zabin & M. R. Villarejo (1975) *Annu. Rev. Biochem.* **44**, 295.

DONNAN EQUILIBRIUM

The equilibrium (also known as the Donnan effect) established across a semipermeable membrane or the equivalent of such a membrane (such as a solid ion-exchanger) across which one or more charged substances, often a protein, cannot diffuse. Diffusible anions and cations are distributed on the two sides of the membrane, such that the sum of concentrations (in dilute solutions) of diffusible and nondiffusible anions on either side of the membrane equals the sum of concentrations of diffusible and nondiffusible cations. Thus, the diffusible ions will be asymmetrically distributed across the membrane and a Donnan potential develops.

DOPAMINE β-MONOOXYGENASE

This enzyme [EC 1.14.17.1], also known as dopamine β-hydroxylase, is a copper-dependent system catalyzing the reaction of 3,4-dihydroxyphenethylamine with ascorbate and dioxygen to produce noradrenaline, dehydroascorbate, and water. The enzyme is stimulated by fumarate.

M. A. Ator & P. R. Ortez de Montellano (1990) *The Enzymes*, 3rd ed., **19**, 213.
F. P. Guengerich (1990) *Crit. Rev. Biochem. Mol. Biol.* **25**, 97.
T. Ljones (1987) *Meth. Enzymol.* **142**, 596.
R. P. Frigon (1987) *Meth. Enzymol.* **142**, 603.
R. M. Weppelman (1987) *Meth. Enzymol.* **142**, 608.
S. M. Miller & J. P. Klinman (1982) *Meth. Enzymol.* **87**, 711.
V. Ullrich & W. Duppel (1975) *The Enzymes*, 3rd ed., **12**, 253.

DOSE

1. A measured quantity of a substance or solution given to an organism or system at one time. 2. The energy or amount of photons absorbed by an object or substance per unit area or unit volume in a given exposure time; the word is often preceded by the type of light (*e.g.*, UV dose). 3. Fluence; the energy or amount of photons received by an object or substance per unit area or unit volume during a given exposure time.

DOSE, LETHAL

The dose of a substrate which is likely to cause death in an organism. It is symbolized by LD and is usually supplemented with a subscript number indicating the percentage of test animals that died: the absolute lethal dose is LD_{100}, the median lethal dose is LD_{50}, and the minimal lethal dose is LD_{05}. These values will vary with the type of test animal and the route of administration; hence, that information should always be provided.

DOUBLE-RECIPROCAL PLOT

A linear reciprocal transformation of a function of the form $[f(\alpha) = \alpha/(1 + \alpha)]$, such that $1/f(\alpha)$ is plotted on the vertical axis and $1/\alpha$ is plotted on the horizontal axis. In the case of one-substrate enzyme kinetics, the hyperbolic function is:

$$v = V_{max}[S]/(K_m + [S])$$

and the double-reciprocal form[1] is as follows:

$$1/v = (1/V_{max}) + (K_m/(V_{max}[S]))$$

Because any linear equation can be expressed in terms of its slope (m) and vertical intercept (b), such that $y = mx + b$, the double-reciprocal form of the Michaelis-Menten equation yields a slope of K_m/V_{max}, a vertical-axis intercept of $1/V_{max}$, and a horizontal-axis intercept of $-1/K_m$.

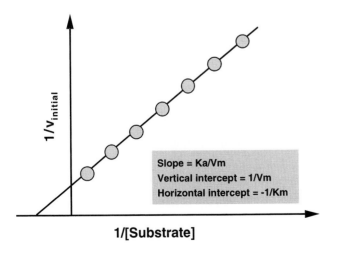

Slope = Ka/Vm
Vertical intercept = 1/Vm
Horizontal intercept = -1/Km

1/[Substrate]

This graphical representation of initial rate data for enzyme-catalyzed reactions is also referred to as the Lineweaver-Burk plot[2] or a Woolf-Lineweaver-Burk plot[1,2]. Double-reciprocal plots are also used for Bi-, Ter-, and higher reactant systems. Its broad usage and ease of interpretation have made the double-reciprocal plot the most common graphical representation of initial-rate data. However, investigators[3,4] have identified problems with the double-reciprocal plot, particularly in terms of a greater variation in statistical weighting compared to other graphical representations. Nevertheless, Fromm has reviewed practical reasons for favoring double-reciprocal plots[5,6], and he has also demonstrated that data weighting and computer methods can eliminate or minimize the problems encountered by earlier workers[7].

[1]B. Woolf quoted in J. B. S. Haldane & K. G. Stern, *Allgemeine Chemie der Enzyme*, S. 119, Steinkopff, Dresden-Leipzig.
[2]H. Lineweaver & D. Burk (1934) *J. Amer. Chem. Soc.* **56**, 658.
[3]G. N. Wilkinson (1961) *Biochem. J.* **80**, 324.
[4]J. E. Dowd & D. S. Riggs (1965) *J. Biol. Chem.* **240**, 863.
[5]H. J. Fromm (1975) *Initial Rate Enzyme Kinetics*, Springer-Verlag, New York.
[6]F. B. Rudolph & H. J. Fromm (1979) *Meth. Enzymol.* **63**, 138.
[7]D. Siano, J. W. Zyskind & H. J. Fromm (1975) *Arch. Biochem. Biophys.* **170**, 587.

DRUG ABSORPTION/ELIMINATION (First-Order Model)

The time evolution of plasma drug concentration can be treated by a simple model that assumes first-order absorption and first-order elimination of the drug. If $[D_o]$ is the effective concentration of the drug dose, appearance of drug in the plasma will obey the relationship:

$$[D] = [D_o]\exp(-k_a t)$$

where [D] is the instantaneous plasma drug concentration at time t and k_a is the rate constant for absorption. Letting [X] represent the amount of drug absorbed into the circulation,

$$d[X]/dt = k_a[D] - k_e[X]$$

where k_e is the first-order rate constant for elimination of the drug from plasma. Separating variables, we get

$$d[X]/dt + k_e[X] = k_a[D]$$

and substituting $[D_o]\exp(-k_a t)$ for [D] then yields

$$d[X]/dt + k_e[X] = k_a[D_o]\exp(-k_a t)$$

Integration of this rate expression yields

$$[X] = \{k_a/(k_e - k_a)\}[D_o] \{\exp(-k_a t) - \exp(-k_e t)\}$$

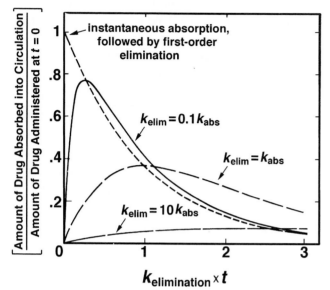

Time-courses of drug absorption and elimination, showing how the relative magnitudes of k_{abs} and k_{elim} alter the amount of drug absorbed into the circulation relative to the amount of drug administered initially.

As shown above, the plasma drug concentration usually exhibits a biphasic rise and fall that are characteristic of all two-step series first-order processes. When k_e is zero, there will only be first-order absorption without any elimination, and when $k_a \gg k_e$, drug absorption will be virtually instantaneous, followed by first-order elimination.

DRUG ACTION (Operational Model)

An alternative model to the classical drug-receptor occupation treatment for drug action[1,2]. The tissue response as a function of agonist concentration can be described by the equation:

$$E_a/E_m = [AR]/[R_t] = [A_{free}]/\{[A_{free}] + K_A\}$$

where E_a is the response, E_m is the maximal response, [A] is the agonist concentration, [R_t] is the total concentration of receptor, K_A is the dissociation constant for the AR complex, and K_E is the value of [AR] that evokes half-maximal tissue response.

[1]J. W. Black & P. Leff (1983) *Proc. Roy. Soc. Lond. [Biol.]* **220**, 141.
[2]T. Kenakin (1993) *Pharmacologic Analysis of Drug-Receptor Interaction*, 2nd ed., pp. 20-24, Raven Press, New York.

DRUG RECEPTOR

A specific drug binding site that transduces site occupancy into a biological response. Although originally proposed as receptor sites located on the outer surface of a target cell membrane, drug receptors can reside at any location within a cell. They typically display a high degree of stereochemical recognition, such that nonspecific drug binding can be compensated for by using a pharmacologically inactive, stereoisomeric form of the active drug. Many specific drug receptors also can be classified into subtypes which are products of highly related genes, alternative mRNA splicing, unique post-translational covalent modification, or even characteristic subcellular localization.

DYAD SYMMETRY

A molecular structural property wherein a rotation of 180° leads to the same superimposable structure. Because proteins are made up of L-amino acid residues, neither a single subunit nor a multisubunit protein can possess a plane or point of symmetry. However, proteins do exhibit dyad symmetry. This structural property allows many DNA-binding proteins to make equivalent contacts to the two antiparallel chains of the nucleic acid.

DYNAMIC EQUILIBRIUM

The state of relentless interconversion of reactant(s) and product(s) at thermodynamic equilibrium. This fundamental concept that reactant(s) and product(s) are ceaselessly shuttling forth and back was adduced on the basis of kinetic observations, years before the field of chemical thermodynamics had undergone its modern transformation. Wilhemy[1] in 1850 used polarimetry to demonstrate that the rate of acid-catalyzed hydrolysis of sucrose was linearly dependent on the concentration of sugar. A few years later, Berthelot and St. Gilles[2,3] reached the same conclusion in their studies of ethyl acetate hydrolysis. These observations led Guldberg and Waage[4,5] to postulate that chemical equilibria are dynamic, realizing as they did that the reaction rate in each direction depended on the solution concentration (an intensive variable), and not the amount (an extensive variable) of reactant(s) and product(s).

[1]L. Wilhelmy (1850) *Ann. Physik u. Chem., Poggendorff's* **81**, 413 and 499.
[2]M. Berthelot & P. De St. Gilles (1862) *Ann. Chim. Phys.* [3]**65**, 385.
[3]M. Berthelot (1862) *Ann. Chim. Phys.* [3]**66**, 110.
[4]C. M. Guldberg & P. Waage (1867) *Études sur les affinités chimiques*, Brøgger and Christie, Christiania (Oslo).
[5]C. M. Guldberg & P. Waage (1879) *J. prakt. Chem.* [2]**19**, 69.

DYNAMICS

The branch of physics concerned with matter in motion as well as the forces associated with those motions, as contrasted with its interchangeable usage for the kinetics of processes in chemistry and biochemistry.

DYNEIN

This family of ATP-dependent motors [EC 3.6.1.33] that forms ADP and orthophosphate during its energy-dependent translocation along the surface of microtubules. Unique "cargo" sites on dynein molecules allow for the specific transport of cellular organelles and other macromolecular components.

S. Khan & M. Sheetz (1997) *Ann. Rev. Biochem.* **66**, 785.
R. Vallee & E. L. F. Holzbauer (1994) *Ann. Rev. Cell Dev. Biol.* **10**, 339.
K. A. Johnson (1992) *The Enzymes*, 3rd ed., **20**, 1.
I. R. Gibbons (1988) *J. Biol. Chem.* **263**, 15837.

E

e

A special irrational number known in mathematics as the Euler number in honor of the prolific Swiss mathematician Leonhard Euler (1707-1783). This number is the least upper bound of the set of all numbers

$$\left(\frac{n+1}{n}\right)^n$$

where n is a positive integer. (A rational approximation of this number to ten significant figures is 2.718281828.)

The number e emerges from the exponential function, which is the sum of the exponential series, such that

$$e^x = \exp x = \sum_{n=0}^{\infty} \frac{z^n}{n!}$$
$$= 1 + z + \frac{z^2}{2} + \frac{z^3}{6} + \frac{z^4}{24} + \cdots$$

When z equals unity, we get the number e, which can be expressed in several equivalent ways:

$$e = \sum_{n}^{\infty} \frac{1}{n!} = 1 + 1 + \frac{1}{2} + \frac{1}{6} + \frac{1}{24} + \cdots$$
$$= \lim_{n \to \infty} \left(\frac{n+1}{n}\right)^n$$

This number forms the basis of the natural (or Naperian) logarithmic scale. For each positive number x, there is unique number y, such $x = e^y$, and y is called the natural logarithm of x. We can therefore write

$$y = \log_e x = \ln x$$

With this definition, one should immediately recognize that for every number x, $e^{\ln x} = x$, where $x > 0$, such that

$\ln e^x = x$. This leads to the following properties of natural logarithms: (1) $\ln x \cdot y = \ln x + \ln y$; (2) $\ln (x/y) = \ln x - \ln y$; (3) $\ln x^c = c \ln x$; (4) $\ln x = \ln y$, if and only if $x = y$; and (5) $\ln x < \ln y$, if and only if $x < y$.

The number e is a special number that allows one to describe the exponential laws of growth and decay. If $f(t) > 0$ is the amount of a substance present at time t, then the rate of change of f at time t is proportional to $f(t)$ and can be written as: $f'(t) = k f(t)$ for some constant k and every time t in some interval. We integrate this differential equation as follows:

$$\int_{t=0}^{t=t} \frac{f'(t)}{f(t)} \, dt = \int_0^t k \, dt$$

Changing variables by allowing $u = f(t)$ and $du = f'(t) \, dt$ allows one rewrite the above integral as:

$$\int_{f(0)}^{f(t)} \frac{1}{u} \, du = \int_0^t k \, dt \text{ or } \ln u \Big|_{f(0)}^{f(t)} = kt \Big|_0^t$$

Therefore, $\ln f(t) - \ln f(0) = kt$; $\ln \{f(t)/f(0)\} = kt$; $f(t)/f(0) = e^{kt}$; and $f(t) = f(0)e^{kt}$. For growth, the exponent is positive [i.e., $f(t) = f(0)e^{kt}$]; for decay, the exponent is negative [i.e., $f(t) = f(0)e^{-kt}$].

The number e also appears in considerations of another form of growth, namely compound interest (i.e., the interest that accumulates when a principal P earns, and thereby grows from, successive payments at a fixed interest rate of i percent). The total principal after n periods (years in the case of many financial investments) can be determined by the following expression:

$$P_{\text{total}} = P\{1 + (i/100)\}^n$$

If the interest is paid m times per period n, the total principal will be greater:

$$P_{\text{total}} = P\{1 + (i/100nm)\}^{nm}$$

Furthermore, if the interest payments are instantaneous (*i.e.*, as *n* approaches infinity), growth in the total principal is governed by the following expression:

$$P_{\text{total}} = P \times \exp(i/100)$$

The principle of compound interest also applies to many biological processes, where an inherited incremental advantage of one organism allows it to gain greater and greater advantage over a competitor. For example, suppose a graduate student accomplishes 10% more per year than another student; then after five years, his/her total accomplishment will be $\{1 + (10/100)\}^5$ or 1.61 greater than the other student. (This moral is instructive!)

Another special property of e is that the derivative of e^x exactly equals e^x; thus, this is an function that also happens to be its own derivative!

EADIE-HOFSTEE PLOT

A graphical procedure[1-4] for plotting enzyme initial rate data as *v versus v*/[S] (also known as the Woolf-Augustinsson-Hofstee plot), where the initial velocity *v* is the so-called (*y*)-axis variable and *v* divided by the initial substrate concentration [S] is the so-called (*x*)-axis variable, such that the vertical-axis intercept equals V_{max}, the slope equals $-K_{\text{m}}$, and the horizontal-axis intercept is $V_{\text{max}}/K_{\text{m}}$.

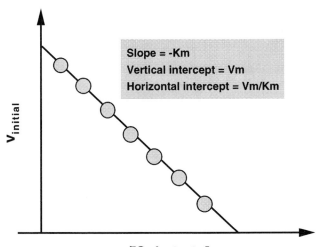

Slope = -Km
Vertical intercept = Vm
Horizontal intercept = Vm/Km

v_{initial}

v_{initial} /[Substrate]

This transformation of the Michaelis-Menten equation for a one-substrate enzyme can be written as:

$$v = -K_{\text{m}}(v/[\text{S}]) + V_{\text{max}}$$

Others have plotted the data by changing the horizontal and vertical axes: thus, *v*/[S] *versus v* (also known as the Eadie-Scatchard plot). In this case, the slope is $-1/K_{\text{m}}$, the vertical intercept is now $V_{\text{max}}/K_{\text{m}}$, and the horizontal intercept is V_{max} (thus, $v/[\text{S}] = -(v/K_{\text{m}}) + (V_{\text{max}}/K_{\text{m}})$).

[1]B. Woolf (1932), cited by J. B. S. Haldane & K. G. Stern, *Allgemeine Chemie der Enzyme*, p. 119, Steinkopff, Dresden-Leipzig.
[2]G. S. Eadie (1942) *J. Biol. Chem.* **146**, 85.
[3]K.-B. Augustinsson (1948) *Acta Physiol. Scand.* **15**, Suppl., 52.
[4]G. Scatchard (1949) *Ann. N. Y. Acad. Sci.* **51**, 660.

EC$_{50}$

The molarity of an effector or agonist that produces 50% of the maximal effect of that effector or agonist.

EDITING

Any process leading to the elimination or correction of errors in molecular structure, particularly as it applies to the sequence of monomeric units in nucleic acids and proteins as well as the synthesis of aminoacyl-tRNA's.

EDITING MECHANISMS

Intrinsic enzymatic activities of enzymes that minimize or prevent erroneous introduction of closely related structural analogues of metabolites into biopolymers. Hopfield[1] has discussed the general need and likely strategies for kinetic editing or proofreading. Fersht's group[2,3] carried out an exhaustive analysis of aminoacyl-tRNA synthetase editing. Mischarged aminoacylated-tRNA's are recognized and hydrolyzed to prevent their inclusion in polypeptides by translation. DNA and RNA polymerases also use editing to excise incorrectly incorporated deoxyribotide and ribotide subunits. Yarus[4] presented a cogent analysis showing that enzymic editing (or proofreading) requires (1) a branched pathway that diverts and recycles any incorrectly produced product; (2) constrained use of substrate specificity; (c) consumption of energy; and (d) a compromise between accuracy and yield. His report also treats the ways in which a branched Michaelis-Menten-type enzyme might produce useful proofreading properties.

[1]J. J. Hopfield (1974) *Proc. Natl. Acad. Sci. U.S.A.* **71**, 4135.
[2]A. R. Fersht, R. S. Mulvey & G. L. E. Koch (1975) *Biochemistry* **14**, 13.
[3]A. R. Fersht (1999) *Structure and Mechanism in Protein Science: A Guide to Enzyme Catalysis and Protein Folding*, pp. 650, W. H. Freeman, New York.
[4]M. Yarus (1992) *Trends Biochem. Sci.* **17**, 130.

EDTA

Ethylenediamine-N,N,N',N'-tetraacetic acid, an important chelating agent and metal ion buffer: $(HOOCCH_2)_2NCH_2CH_2N(CH_2COOH)_2$. The pK_a values of EDTA (at an ionic strength of 0.1 and a temperature of 20°C) are 2.0, 2.67, 6.16, and 10.26. Thus, at a pH range of 4 to 5 most of the EDTA in an aqueous solution is present as the divalent anion whereas between pH 7 to 9, the trivalent anion is the more prevalent species. However, the formation of a metal ion-EDTA complex will result in the displacement of the remaining protons from EDTA. Thus, $M^{n+} + H_2(EDTA)^{2-} \rightleftharpoons M(EDTA)^{(n-4)+} + 2H^+$ between pH 4 and 5 and $M^{n+} + H(EDTA)^{3-} \rightleftharpoons M(EDTA)^{(n-4)+} + H^+$ between pH 7 and 9. Hence, the tetravalent anion of EDTA is the species of importance in equilibrium calculations.

Metal ion complex formation is typically measured via stability, or formation, constants. Thus, for $M^{n+} + EDTA^{4-} \rightleftharpoons M(EDTA)^{n-4)+}$, $K_{M(EDTA)} = [M(EDTA)^{(n-4)+}]/([M^{n+}][EDTA^{4-}])$. A number of values for $\log K_{M(EDTA)}$ at 25°C and an ionic strength approaching zero are provided in the table below:

Formation Constants for Selected Metal-EDTA Complexes

Metal Ion	$\log K_{M(EDTA)}$	Metal Ion	$\log K_{M(EDTA)}$
Co(III)	36	Zn	16.4
V(III)	25.9	Co(II)	16.31
Fe(III)	24.23	Al	16.13
Cr(III)	23	Fe(II)	14.33
Hg(II)	21.80	Mn(II)	13.8
Cu(II)	18.7	V(II)	12.7
Ni	18.56	Ca	11.0
Pb(II)	18.3	Mg	8.64
V(V)	18.05	Ag	7.32
V(IV)	18.0	Na(I)	1.7

EFFECTIVE DOSE

The dose, symbolized by ED, producing a desired effect by a substance. The symbol is usually followed by a subscript number indicating the dose that has the desired effect on a certain percentage of the tested population. The median effective dose is ED_{50}.

EFFECTIVE MOLARITY

The ratio of a first-order rate constant for an intramolecular reaction (involving two functional groups or moieties within the same molecular entity) to the second-order rate constant of the analogous intermolecular elementary reaction[1]. Effective molarity has units of concentration. While the highest effective concentration in an aqueous solution was thought for many years to be 55.5 M (corresponding to that of pure water), Kirby[2] has tabulated much higher effective concentrations. These include values of 3,700 M for cyclic thiolester formation from 4-thiolbutyrate; 200 M for dithiolane ring formation with 1,4-dithioerythritol; and 190,000 M for cyclic anhydride formation from succinic acid. **See also** Intramolecular Catalysis.

[1]IUPAC (1979) Pure and Appl. Chem. **51**, 1725.
[2]A. J. Kirby (1980) Adv. Phys. Org. Chem. **17**, 183.

EFFECTOR

A general term applying to substances or agents that elevate or depress the efficient action of a catalyst, transporter, or receptor. Both activators and inhibitors are considered to be effectors of enzyme action and transport, and the analogous terms for receptors are agonists and antagonists. In many instances, the effector and modifier are used synonymously, but the Nomenclature Committee[1] of the International Union of Biochemistry suggests that effector be restricted to those substances producing a significant physiological effect. The term modifier applies to substances that are artificially added to in vitro systems. **See also** Modifier; Activator; Inhibitor

[1]International Union of Biochemistry (1982) Eur. J. Biochem. **128**, 281.

EFFICIENCY FUNCTION

A quantitative expression developed by Albery and Knowles[1] to describe the effectiveness of a catalyst in accelerating a chemical reaction. The function, which depends on magnitude of the rate constants describing individual steps in the reaction, reaches a limiting value of unity when the reaction rate is controlled by diffusion. For the interconversion of dihydroxacetone phosphate and glyceraldehyde 3-phosphate, the efficiency function equals 2.5×10^{-11} for a simple carboxylate catalyst in a nonenzymic process and 0.6 for the enzyme-catalyzed process. Albery and Knowles[1] suggest that evolution has produced a nearly perfect catalyst in the form of triosephosphate isomerase. **See** Reaction Coordinate Diagram

[1]W. J. Albery & J. R. Knowles (1977) Biochemistry **15**, 5631.

EF-HAND

A motif (found in more than 160 different calcium-binding proteins) consisting of two alpha helices, designated "E" and "F," that are joined by a Ca^{2+}-binding loop. Functional EF-hands always occur in pairs. The EF-hand family includes calmodulin, troponin C, myosin regulatory light chain, parvalbumin, calbindins, α-actinin, calcineurin B subunit, calpain, and fimbrin. A striking feature of the EF-hand family is the ability to modulate the activity of a number of enzymes. EF-hands have been identified inferentially in numerous calcium ion binding proteins based upon similarity classifications of amino acid sequence and in some cases confirmed directly by crystal structures.

EIGENFUNCTION

Any solution to one or more differential equations for which there are only certain allowed parameters.

EIGEN-TAMM MECHANISM FOR METAL ION COMPLEXATION

A kinetic description of rapid metal ion complexation reactions. Desolvation of the primary coordination sphere of the metal ion is treated as the rate-limiting step in forming its inner-sphere coordination complex with a ligand.

M. Eigen & K. Tamm (1962) *Z. Elektrochem.* **66**, 107.

EINSTEIN

A unit used principally in photochemistry to designate the energy equivalent to an Avogadro's number of quanta (*i.e.*, 6.0221367×10^{23} quanta).

EINSTEIN-SUTHERLAND EQUATION

The quantitative expression ($D = k_B T/f$) describing the relationship between the diffusion constant D of a particle and its frictional coefficient f, where k_B and T are the Boltzmann constant and the absolute temperature, respectively. **See** *Diffusion*

ELASTASE

This enzyme catalyzes the hydrolysis of proteins, including elastin, with preferential cleavage at Ala–Xaa. The following are reviews on the molecular and physical properties of this enzyme [EC 3.4.21.36 (pancreatic) and EC 3.4.21.37 (leukocyte)].

L. Rust, C. R. Messing & B. H. Iglewski (1994) *Meth. Enzymol.* **235**, 554.
E. Lolis & G. A. Petsko (1990) *Ann. Rev. Biochem.* **59**, 597.
M. A. Ator & P. R. Ortez de Montellano (1990) *The Enzymes*, 3rd ed., **19**, 213.
L. Robert & W. Hornebeck, eds. (1989) *Elastin and Elastases*, CRC Press, Boca Raton.
J. S. Bond & P. E. Butler (1987) *Ann. Rev. Biochem.* **56**, 333.
E. T. Kaiser, D. S. Scott & S. E. Rokita (1985) *Ann. Rev. Biochem.* **54**, 565.
T. A. Steitz & R. G. Schulman (1982) *Ann. Rev. Biophys. Bioeng.* **11**, 419.
R. J. Coll, P. D. Compton & A. L. Fink (1982) *Meth. Enzymol.* **87**, 66.
B. S. Hartley & D. M. Shotton (1971) *The Enzymes*, 3rd ed., **3**, 323.

ELASTICITY

A metabolic control parameter (symbolized by ε) that provides a measure of the direction and magnitude of a change in enzyme (or pathway) velocity as a function of a change in the concentration x of substance X acting as a substrate or effector[1].

$$\varepsilon \equiv (\partial \ln v / \partial \ln x) \equiv (x \partial v / v \partial x)$$

When ε equals zero, v shows no functional dependence on the concentration of X. If ε is positive, v increases as the concentration X increases; if ε is negative, v decreases as a increases.

For a lucid account of the kinetics of multi-enzyme systems, the reader should consult Cornish-Bowden[1] who defines such related parameters as flux control coefficients, summation relationships, and response coefficients.

[1]A. Cornish-Bowden (1996) *Fundamentals of Enzyme Kinetics*, Portland Press, London.

ELASTICITY COEFFICIENT

The partial derivative of the velocity of a metabolic step or pathway with respect to the concentration of a substrate, product, and/or effector.

ELECTRIC CHARGE

The positive or negative charge (commonly symbolized by Q or q) on a molecule, radical, or particle resulting from the deficient or excess accumulation of electrons. Electric charge need not be an integer value. The SI unit "coulomb" (abbreviated C) equals the quantity of electricity transferred by an electric current of 1 ampere over the period of 1 second.

ELECTRIC POTENTIAL

A quantity (commonly symbolized by V) for the work needed to bring a unit positive charge to that point in space from an infinite distance. Thus, $V = dw/dQ$ where w is the work and Q is the electric charge. The SI unit for electric potential is the volt (V). The electric potential difference, also measured in volts and symbolized by U, ΔV, or $\Delta \Phi$, is equal to the difference in potential between two points (i.e., $U = V_2 - V_1$) as measured by the work needed to transfer a unit positive charge from one point to the other. *See also* Electromotive Force

ELECTROACTIVE SPECIES

Any chemical compound that can be oxidized or reduced in a manner that generates a flow of electrons from a solution to an electrode.

ELECTROCATALYSIS

A term used in electrochemistry to describe facilitated electron transfer, resulting in an increased rate of half-cell reactions at electrode surfaces.

ELECTRODE KINETICS

The systematic representation of the time dependence of chemical reactions occurring at an electrode surface. These reactions can be analyzed as a succession of events: (a) diffusion of reactants within the bulk solution to the more condensed layer of solution near the electrode surface; (b) penetration of that layer to achieve adsorption on the electrode's surface; (c) electron transfer to (i.e., reduction) or from (i.e., oxidation) the adsorbed reactant(s); product desorption, penetration of the condensed layer, and diffusion into the bulk solution.

K. J. Laidler (1970) *J. Chem. Education* **47**, 600.

ELECTROFUGE

If the leaving group in a chemical reaction does not leave with the bonding electron pair, that leaving group is referred to as an electrofuge. The tendency of such a moiety to leave is referred to as electrofugicity. *See also* Electrophile; Nucleofuge

ELECTROMOTIVE FORCE

The amount of energy obtained from an electrical source per unit of electric charge. The SI unit for electromotive force (commonly symbolized by E) is the volt or V. The electromotive force is equal to $\int (F/Q) \cdot ds$ where F is the force, Q is the electric charge, and s is the position vector. *See also* Electric Potential

ELECTRON

A fundamental elementary particle possessing a negative charge of $1.60217733 \times 10^{-9}$ coulomb (the elementary charge) and a rest mass, m_e, of $9.1093897 \times 10^{-31}$ kg (or, 0.51099906 MeV).

ELECTRON ACCEPTOR

1. A substance which will accept or interact with an electron. 2. A Lewis acid (Note, however, that IUPAC now discourages the use of this term in this manner.)

IUPAC (1979) *Pure and Appl. Chem.* **51**, 1725.

ELECTRON AFFINITY

The change in enthalpy due to the association of an additional electron to a particular atom or molecular entity. This physical quantity is usually measured in the gas phase.

IUPAC (1979) *Pure and Appl. Chem.* **51**, 1725.

ELECTRON DENSITY

The electron probability distribution in a molecular entity. If $P(x,y,z)$ is the electron density (in Cartesian coordinates) at coordinates (x,y,z), then $P(x,y,z)\, dx\, dy\, dz$ represents the probability of finding an electron in the volume element $dx\, dy\, dz$ at coordinates (x,y,z). *See also* Charge Density

ELECTRON DENSITY FUNCTION

An approximate quantum mechanical expression[1,2] that allows one to calculate the electrostatic surface potential around atoms, radicals, ions, and molecules by assuming that the ground-state electron density uniquely specifies the Hamiltonian of the system and thereby all the properties of the ground state. This approach greatly facilitates computational schemes for exact calculation of the ground-state energy and electron density of orbitals.

[1]P. Hohenberg & W. Kohn (1964) *Phys. Rev. B* **136**, 864.
[2]W. Kohn & L. J. Sham (1965) *Phys. Rev. A* **140**, 1133.

ELECTRON DONOR

A substance or molecular entity that can transfer (or donate) an electron to another chemical species. Note:

IUPAC has discouraged the use of this term as a synonym for Lewis base.

IUPAC (1979) *Pure and Appl. Chem.* **51**, 1725.

ELECTRONEGATIVITY

A measure of the ability of an atom within a molecule to attract bonding electrons toward itself[1,2]. For a bond between two atoms of different electronegativities, the electron molecular orbital cloud is not symmetric, and the atom with the higher electronegativity attracts the larger proportion of the cloud. The most popular quantitative description was presented by Pauling[3], who based his scale on bond dissociation energies (measured in kcal per mol).

[1]K. D. Sen & C. K. Jorgensen (1987) *Electronegativity*, Springer, New York.
[2]S. S. Batsanov (1968) *Russ. Chem. Rev.* **37**, 332.
[3]L. Pauling (1960) *The Nature of the Chemical Bond*, 3rd ed., Cornell Univ. Press, Ithaca.

ELECTRONEUTRALITY PRINCIPLE

A conceptual framework, first proposed by Pauling[1,2], for writing the correct structural formulas for polyatomic molecules and ions. The electroneutrality principle requires that the electronic structure of stable molecules is such that the electric charge on each atom is close to zero (*i.e.*, electroneutrality), and varies between at most $+1$ and -1. Application of this principle requires knowledge of electronegativities as well as a general understanding of the relationship between ionic character of a bond and the electronegativity difference of the bonded atoms.

One classical example that applies the electroneutrality principle is the electronic structure of carbon monoxide, a diatomic molecule with a very small dipole moment of 0.110 debye. The only electronic structure that satisfies the octet rule for CO is $:C{\equiv}O:$, a structure that corresponds to C^- and O^+, if the shared electron pairs are equally devided by the two atoms. Pauling showed that the electronegativity difference of 1.0 would correspond to about 22% partial ionic character for each bond, and to charges of $C^{0.34-}$ and $O^{0.34+}$. A second possible electronic structure, $:C{=}\ddot{O}:$, does not complete the octet for carbon. The partial ionic character of the bonds corresponds to $C^{0.44+}$ and $O^{0.44-}$. If these two structures contribute equally to a hybrid molecular structure, the resultant electronic charge on each atom would be very small.

Thus, in agreement with the elecroneutrality principle and the very small dipole moment, one can reasonably write the electron configuration for CO as the resonance hybrid: $\{:C{\equiv}O: \leftrightarrow :C{=}\ddot{O}:\}$.

[1]L. Pauling (1948) *J. Chem. Soc.*, 1461.
[2]L. Pauling (1948) *General Chemistry*, Freeman, San Francisco (also, the 1988 Dover republication of the third edition of Pauling's classic text (1970) *General Chemistry*, pp. 192-194).

ELECTRON-NUCLEAR DOUBLE RESONANCE (ENDOR)

A magnetic resonance spectroscopic technique used for detect hyperfine interactions between electrons and nuclear spins. In its continuous-wave mode, the ESR signal intensity is measured as radio frequency is applied. In pulsed mode, pulses of radio frequency energy are applied, and the ESR signal is detected as a spin-echo. In either case, enhanced EPR signal strength occurs when the radio frequency is in resonance with the coupled nuclei.

ELECTRON SINK

A term describing that region of a molecule having sufficient electrophilic character, such that it can capture or trap an electron during the course of a reaction.

ELECTRON SPIN RESONANCE

A condition arising when a chemical species containing an odd or unpaired electron is placed in a magnetic field (typically around 3000 gauss or 0.3 tesla), such that the components (corresponding to $+h/4\pi$ and $-h/4\pi$) of spin angular momentum of the electron are quantized along the direction of the applied magnetic field. In such cases, the difference in energy levels corresponds to that observed in the microwave region of the electromagnetic spectrum (*e.g.*, 10^{-23} joules is equivalent to a microwave frequency of about 0.3 cm^{-1}). When positioned in a magnetic field, a resonance effect is observed when the sample is irradiated by microwaves of energy equal to the aforementioned energy difference. The following band designations are used: L (1.1 gigahertz or GHz), S (3.0 GHz), X (9.5 GHz), K (22.0 GHz), and Q (35.0 GHz). Free radicals and other biological substances, such as nitric oxide (with its 15 electrons) and manganous ions, can be gainfully examined by ESR methods. Because the energy difference is relatively small, ESR spectra are frequently sensitive to the environment in which the unpaired electron resides. Synthetic moieties containing free radicals can often be attached to macromolecules or membranes to use the ESR spectral properties to investigate mechanistic issues.

ELECTRON TRANSFER

The exchange of an electron between two molecular entities or between two sites within the same molecular entity. *See* *Marcus Equation; Electrode Kinetics*

ELECTRON TRANSFER REACTIONS

Rees and Farrelly[1] have presented an instructive general treatment of the following electron transfer reaction mechanism:

$$A_o + B_r \underset{k_{-1}}{\overset{k_1}{\rightleftharpoons}} A_oB_r \underset{k_{-2}}{\overset{k_2}{\rightleftharpoons}} A_{cr}B_o \underset{k_{-3}}{\overset{k_3}{\rightleftharpoons}} A_r + B_o$$

where the three successive equilibria are association of reactants, electron transfer, and product dissociation; each has an assigned formation constant K_1, K_2, and K_3 equal to k_i/k_{-i}. Treating the system in a manner akin to a steady-state one-substrate/two-intermediate enzyme mechanism with $k_{-3}[A_r][B_o] = 0$, the reaction velocity v is given as follows:

$$v = k_{obs}[A_o][B_r]$$
$$= \{k_1 k_2 k_3/(k_{-1}k_{-2} + k_{-1}k_3 + k_2 k_3)\}[A_o][B_r]$$

They describe three cases: (a) rate-limiting electron transfer, in which event, $v = k_{-2}K_1[A_o][B_r]$; (b) rate-limiting substrate association, thereby yielding $v = k_1[A_o][B_r]$; and pre-equilibration of the first two steps with rate-limiting product release, giving $v = k_3 K_1 K_2[A_o][B_r]$. They note that k_{obs} never gives k_2 alone, and one would need to know which of the aforementioned cases applies in order to get a good estimate of k_2. The reader is encouraged to consult their assessment of the occasions to which each case is apt to pertain. Rees and Farrelly[1] also give an account of how Marcus theory and quantum electronic theories might apply to electron transfer reactions.

[1]D. C. Rees & D. Farrelly (1990) *The Enzymes*, 3rd. ed., **20**, 37.

ELECTRON TRANSPORT

The transfer of one or more electrons within and/or between molecules or other electroactive species. Unlike classical mechanics in which a system must surmount any potential energy barrier between two states, quantum mechanics allows certain electron transfer systems to tunnel or leak through the barrier without surmounting it. Tunneling can occur when the de Broglie wavelength of small particles (particularly electrons, hydrogen atoms, as well as protons and hydride ions) is comparable to the width of the potential energy barrier. In these cases, the Uncertainty Principle allows for the probability of the electron residing on either side of the barrier.

Selected entries from *Methods in Enzymology* [**vol**, page(s)]:
Electron-transport chain [components, **69**, 205, 206; sites of inhibition, **69**, 676, 677]; chloroplast [autoxidizable carriers, **69**, 416, 417; DBMIB, **69**, 422, 423; dichlorophenolindophenol and related carriers, **69**, 418; ferricyanide, **69**, 417, 418; isolated, **69**, 85, 86; NADP/ferredoxin, **69**, 416; substituted benzoquinones, **69**, 422; inhibitors, chemical modifiers, **69**, 503, 504; measurement, **69**, 653, 654; mutations, **69**, 16, 19; reverse, **69**, 630, 631; acid-base-driven, **69**, 639-641; luminescence, **69**, 639, 640; properties, **69**, 640, 641; reduction of coenzyme Q, **69**, 639; ATP-driven, **69**, 632-639; nitrate reductase sequence, **69**, 278, 279.

ELECTROPHILE

A reagent or reactant that is preferentially attracted to a region of high electron density within a particular molecular entity (or, in the case of intramolecular processes, within the same molecular entity). Typically, an electrophile bonds by accepting both bonding electrons for the reaction partner (termed the nucleophile). The term electrophile is also used with respect to certain polar radicals exhibiting high reactivities with sites of high electron density. **See also** *Electrophilicity; Nucleophile*

ELECTROPHILIC CATALYSIS

Catalysis arising from the participation of a Lewis acid in catalyzed or uncatalyzed chemical reactions.

IUPAC (1979) *Pure and Appl. Chem.* **51**, 1725.

ELECTROPHILICITY

1. The relative reactivity expressed by an electrophile. Electrophilicity is measured by the relative rate constants for a particular reaction of different electrophiles for a common substrate. 2. The property of being electrophilic. **See also** *Electrophile; Nucleophilicity*

IUPAC (1979) *Pure and Appl. Chem.* **51**, 1725.

ELECTROPHILIC SUBSTITUTION REACTION

A reaction in which an electrophile participates in heterolytic substitution of another molecular entity that supplies both of the bonding electrons. In the case of aromatic electrophilic substitution (AES), one electrophile (typically a proton) is substituted by another electron-deficient species. AES reactions include halogenation (which is often catalyzed by the presence of a Lewis acid salt such as ferric chloride or aluminum chloride), nitration, and so-called Friedel-Crafts acylation and alkylation reactions. On the basis of the extensive literature on AES reactions, one can readily rationalize how this process leads to the synthesis of many substituted aromatic compounds. This is accomplished by considering how the transition states structurally resemble the carbonium ion intermediates in an AES reaction.

ELECTROSTATIC BOND

A bond resulting from attractive electric interactions between chemical species that include, but are not limited to, ions. In this regard, polarized species such as carbonyl and amino groups can engage in electrostatic interactions that give rise to hydrogen bonds.

ELECTROSTATIC SURFACE POTENTIAL

A measure of the electric charge density and distribution along the surface of molecules. The surface potential is approximately determined by the laws of classical electrostatics or more rigorously by quantum mechanics. The magnitude of the attractive or repulsive energy (U) between two point charges (q_1 and q_2) separated at a distance r is equal to (kq_1q_2/Dr), where k is 9×10^9 joulemeter/coulomb^2 and D is the dielectric constant of the medium. To estimate the electrostatic surface potential, calculations are made for the attractive or repulsive interactions experienced by a probe proton passing above the molecule's surface at a fixed distance.

Selected entries from *Methods in Enzymology* [vol, page(s)]:
Computation, **240**, 648-649, 652-653; electrostatic potential difference effect on protein partitioning away from isoelectric point, **228**, 229; salt effects, **240**, 653-654; hemoglobin, **240**, 656-657; ion binding site determination, **240**, 648; static accessibility parameter algorithm, **240**, 648-649; Tanford-Kirkwood model, **240**, 648, 654.

ELECTROSTRICTION

1. The deformation of a dielectric substance or body in an applied electric field. 2. The orientation of solvent water molecules interacting with ions or charge surfaces of molecules. Proteins contain many ionic groups on their surface, and each group has the capacity to attract and deform nearby water molecules. These tightly bound water molecules experience an electrostrictive effect of about -16 Å^3 per charged group, and electrostriction slightly increases the density of bound water. Thus, a protein containing around 50 charged groups on its surface will increase the number of water molecules in its hydration shell by about 25, amounting to a density increase of 1–2%. Moreover, when a peptide bond is hydrolyzed, there is most frequently a decrease in the volume of the aqueous medium; this occurs when the newly formed carboxylate and protonated amino groups contract the surrounding water. Volume changes are directly determined by the use of a dilatometer which is comprised of the glass bulb of a pycnometer whose tapered glass cap is replaced by a tightly fitting cap with a vertical bore connected to a calibrated capillary.

Because the motions of bound water are restricted, electrostriction results in a considerable reduction in entropy. A consequence is that ionization in a solution reduces the entropy; by contrast, neutralization of two aquated counterions leads to a release of bound water and a rise in entropy.

ELECTROTAXIS

The directed movement or locomotion of an organism placed within an electric field.

ELECTROVALENT BOND

A less frequently used synonym for ionic bond.

ELEMENTARY CHARGE

The magnitude of electric charge on an electron or proton that gives rise to their mutually attractive interaction. The charge on the elementary particles is referred to as elementary charge, is symbolized by e, and has a value of $1.60217733 \times 10^{-19}$ coulombs.

ELEMENTARY REACTION

A reaction that takes place on a molecular scale in a single step following an individual collision or other elementary process and in which no stable intermediate need be postulated and no simpler reaction can be suggested.

ELIMINAND

An alternative term for the leaving group in an elimination reaction.

ELONGATION (With No Change in Polymer Number Concentration)

Although the self-assembly of polymeric structures can involve nucleation and elongation steps (**See** *Actin Assembly Kinetics; Microtubule Assembly Kinetics*), one can simplify the assembly process through what is known as "seeded" assembly. At an initial monomer concentration [M], "seeded" assembly is induced by the addition of pre-assembled polymeric structures; consequently the polymer number concentration must remain constant. The rate of monomer incorporation into indefinite length polymers can be written as follows:

$$-d[M]/dt = k_+[M] - k_-$$

where k_+ corresponds to the product of the bimolecular rate constant for elongation and the concentration of elongating seeds, and k_- is likewise the product of the unimolecular rate constant for monomer dissociation and the concentration of elongating seeds. Multiplying the right-hand side of the equation by k_+/k_+, we obtain

$$-d[M]/dt = k_+[M_o] - k_+k_-/k_+ = k_+\{[M_o] - [M_\infty]\}$$

Here, the ratio k_-/k_+ defines the critical monomer concentration [M], the point below which no polymerization can occur. Integration of $-d[M]dt = k_+\{[M_o] - [M_\infty]\}$ yields the expression:

$$[M] = \{[M_o] - [M_\infty]\}\exp[-k_+ t]$$

which defines a single exponential decay from the initial monomer concentration $\{[M] - [M_\infty]\}$ to the equilibrium monomer concentration $[M_\infty]$.

If an experimenter wishes to study actin filament elongation, for example, there are three ways to induce the elongation process: (a) by adding short preformed filaments that have unblocked barbed and pointed ends (thereby yielding bidirectional elongation); (b) by introducing gelsolin-capped actin filaments (thereby permitting elongation strictly at the pointed end); or (c) by supplying red cell spectrin-4.1 complex (thereby inducing barbed-end elongation). In the case of tubulin polymerization, one may employ: (a) short pre-assembled microtubule seeds to induce bidirectional elongation; (b) flagellar axonemes to initiate morphologically distinguishable bidirectional growth; or (c) centrosomes to promote plus-end growth of microtubules. When bidirectional growth takes place, k_+ represents the sum of the on-rate constants for both ends.

EMBRYO

A subcritical aggregate having fewer subunit components than a nucleus. When this term is applied in the kinetics of precipitation, n refers to the number of subunits in a particle and n^* defines the number of subunits in a particle of critical size. This definition avoids confusion by distinguishing between subcritical ($n < n^*$ subunits), critical ($n = n^*$ subunits), and supercritical ($n > n^*$ subunits) particle sizes. If a nucleus is defined as containing n n^* subunits, then an embryo contains n n^* subunits. Note that in this treatment, we are not using a phase-transition description to describe nucleation, and we are focusing on the smallest step in the process that leads to further aggregation.

EMISSION

1. Radiative deactivation of an excited state of a molecular entity. 2. The release of a nucleon (usually an electron, neutron, or photon) during radioactive decay.

EMISSION SPECTRUM

The emitted spectral radiant power (or exitance) or the emitted spectral photon irradiance (or exitance) plotted as a function of the frequency, wavenumber, or wavelength. The corrected emission spectrum has been corrected for wavelength-dependent variations in the equipment response. **See also** Action Spectrum; Excitation Spectrum

ENANTIOSELECTIVE REACTION

A reaction in which an optically inactive compound (or achiral center of an optically active molecule) is selectively converted to a specific enantiomer (or chiral center). **See** Asymmetric Induction

ENCOUNTER COMPLEX

Any weakly attractive, short-lived complex that is typically formed as an intermediate in a reaction mechanism. When there are only two such molecular entities engaged in the formation of a particular encounter complex, that complex is often called an encounter pair.

IUPAC (1979) Pure and Appl. Chem. **51**, 1725.

ENCOUNTER-CONTROLLED RATE

A reaction velocity equal to the rate of encounter of reacting molecular entities (also known as diffusion-controlled rate). For a bimolecular reaction in aqueous solutions at 25°C, the corresponding second-order rate constant for the encounter-controlled rate is typically about 10^{10} M^{-1}s^{-1}. **See** Diffusion Control for Bimolecular Collisions in Solution

ENDERGONIC PROCESS

A change or reaction for which the Gibbs free energy (ΔG) becomes more positive. (The origin of this compound term is derived from the Greek prefix *end*- which is a contraction of *endo* and the Greek root *ergon*, standing for work.)

ENDO-α (or β)-N-ACETYLGALACTOSAMINIDASE

A number of enzyme have been reported to catalyze the hydrolysis of glycosidic linkages in glycoproteins at N-acetylgalactosamine residues, exhibiting specificities at either the α- or the β-linkage. Enzymes that are reported to act on both α- and β-linkages have to be clearly characterized as to their specificity.

R. J. Staneloni & L. F. Leloir (1982) Crit. Rev. Biochem. **12**, 289.
H. M. Flowers & N. Sharon (1979) Adv. Enzymol. **48**, 29.

ENDO-β-N-ACETYLGLUCOSAMINIDASE

This family of enzymes [EC 3.2.1.96] catalyzes the endo-hydrolysis of the di-N-acetylchitobiosyl unit in high-mannose glycopeptides and [Man(GlcNAc)₂]Asn-containing glycoproteins, releasing the rest of the oligosaccharide intact, while leaving a single N-acetyl-D-glucosamine residue on the protein.

E. Conzelmann & K. Sandhoff (1987) *Adv. Enzymol.* **60**, 89.
H. M. Flowers & N. Sharon (1979) *Adv. Enzymol.* **48**, 29.

ENDOENZYME

Any enzyme catalyzing a hydrolytic reaction that occurs within a polymeric substrate molecule. Those enzymes acting at the terminus of a polymeric substrate molecule are called exoenzymes.

ENDO-β-GALACTOSIDASE

A family of enzymes that catalyzes the endohydrolysis of β-galactosides. They include the endogalactosidase [EC 3.2.1.102] that hydrolyzes the 1,4-β-D-galactosidic linkages in blood groups A and B and the endohydrolase [EC 3.2.1.103] that acts on the 1,4-β-D-galactosidic linkages in keratan sulfate.

R. J. Staneloni & L. F. Leloir (1982) *Crit. Rev. Biochem.* **12**, 289.
H. M. Flowers & N. Sharon (1979) *Adv. Enzymol.* **48**, 29.

ENDONUCLEASE

The following reviews describe the molecular and physical properties of this broad class of enzymes that catalyze the endohydrolysis of deoxyribonucleic acids and ribonucleic acids. The class includes deoxyribonuclease II [EC 3.1.22.1], *Aspergillus* deoxyribonuclease K₁ [EC 3.1.22.2], deoxyribonuclease V [EC 3.1.22.3], crossover junction endoribonuclease [EC 3.1.22.4], and deoxyribonuclease X [EC 3.1.22.5]. **See also** *Deoxyribonucleases; Restriction Enzymes; Ribonucleases*

N. H. Williams (1998) *Comprehensive Biological Catalysis: A Mechanistic Reference* **1**, 543.
V. E. Anderson (1998) *Comprehensive Biological Catalysis: A Mechanistic Reference* **2**, 115.
A. S. Bhagwat (1992) *Meth. Enzymol.* **216**, 199.
I. R. Lehman (1981) *The Enzymes*, 3rd ed., **14**, 193.
E. C. Friedberg, T. Bonura, E. H. Radany & J. D. Love (1981) *The Enzymes*, 3rd ed., **14**, 251.
L. Grossman, A. Braun, R. Feldberg & I. Mahler (1975) *Ann. Rev. Biochem.* **44**, 19.

ENDONUCLEASE III

This enzyme (also known as DNA-(apurinic or apyrimidinic site) lyase, AP endonuclease class I, *E. coli* endonuclease III, Phage-T4 UV endonuclease, and *Micrococcus luteus* UV endonuclease) catalyzes the cleavage of the phosphodiester bond in a lyase-type reaction, not hydro-

lysis. This action leaves a 3′-terminal unsaturated sugar and a product with a terminal 5′-phosphate. This family of enzymes was previously listed under EC 3.1.25.2.

B. Demple & L. Harrison (1994) *Ann. Rev. Biochem.* **63**, 915.
E. C. Friedberg, T. Bonura, E. H. Radany & J. D. Love (1981) *The Enzymes*, 3rd ed., **14**, 251.

ENDONUCLEASE IV

This enzyme [EC 3.1.21.2], also known as endodeoxyribonuclease IV (phage T4-induced), catalyzes the endonucleolytic cleavage of DNA to 5′-phosphooligonucleotide end products. The enzyme exhibits a preference for single-stranded DNA.

B. Demple & L. Harrison (1994) *Ann. Rev. Biochem.* **63**, 915.
E. C. Friedberg, T. Bonura, E. H. Radany & J. D. Love (1981) *The Enzymes*, 3rd ed., **14**, 251.

ENDONUCLEASE V (PYRIMIDINE DIMER)

The enzyme [EC 3.1.25.1] acts on a damaged strand (at a position that is 5′ from the damaged site) and catalyzes the endonucleolytic cleavage near pyrimidine dimers to products with 5′-phosphate.

A. Sancar & G. B. Sancar (1988) *Ann. Rev. Biochem.* **57**, 29.
T. Lindahl (1982) *Ann. Rev. Biochem.* **51**, 61.
E. C. Friedberg, T. Bonura, E. H. Radany & J. D. Love (1981) *The Enzymes*, 3rd ed., **14**, 251.

ENDONUCLEASE VII

This apurinic/apyrimidinic endonuclease from *E. coli* catalyzes the hydrolysis of single-stranded polydeoxyribonucleotides and DNA that contains depyrimidinic sites.

E. C. Friedberg, T. Bonura, E. H. Radany & J. D. Love (1981) *The Enzymes*, 3rd ed., **14**, 251.

ENDO-OLIGOPEPTIDASE A

This zinc-dependent enzyme [EC 3.4.24.15], also referred to as thimet oligopeptidase and soluble metalloendopeptidase, catalyzes the hydrolysis of peptide bonds with a preferential cleavage at positions with hydrophobic residues at P1, P2, and P3′ and a small amino acid residue at P1′. Substrates for this enzyme contain five to fifteen amino acid residues.

E. Shaw (1990) *Adv. Enzymol.* **63**, 271.

ENDOPEPTIDASE Clp

This enzyme [EC 3.4.21.92], also referred to as endopeptidase Ti and caseinolytic protease, catalyzes the hydrolysis of peptide bonds in small peptides and proteins in the presence of ATP and magnesium ion. The protein usually used in assays of this enzyme is α-casein. If ATP is absent, only pentapeptides or smaller are hydrolyzed.

M. R. Maurizi, M. W. Thompson, S. K. Singh & S.-H. Kim (1994) *Meth. Enzymol.* **244**, 314.

ENDOPOLYGALACTURONIDASE

This cell wall hydrolase catalyzes the hydrolysis of α-(1→4)-galacturonide linkages.

R. L. Fischer & A. B. Bennett (1991) *Ann. Rev. Plant Physiol. Plant Mol. Biol.* **42**, 675.

ENDOPOLYPHOSPHATASE

This enzyme [EC 3.6.1.10] (also known as polyphosphate depolymerase, metaphosphatase, and polyphosphatase) catalyzes the hydrolysis of polyphosphate to yield oligophosphate products containing four or five phosphate residues.

H. G. Wood & J. E. Clark (1988) *Ann. Rev. Biochem.* **57**, 235.

ENDOTHERMIC PROCESS

A change in a system or chemical reaction for which there is an absorption of heat (*i.e.*, the process requires heat to proceed). In such systems, ΔH is a positive value (where H is the enthalpy). ***See also*** *Enthalpy; Exothermic; Endogonic; Endoergic*

ENDOTHIAPEPSIN

This aspartic proteinase [EC 3.4.23.22], from the ascomycete *Endothia parasitica*, catalyzes the hydrolysis of proteins with broad specificity similar to that of pepsin A, with preferential action on substrates containing hydrophobic residues at P1 and P1′.

D. R. Davies (1990) *Ann. Rev. Biophys. Biophys. Chem.* **19**, 189.

END PRODUCT

The final product(s) generated by a sequence of enzyme-catalyzed steps within a metabolic pathway. End products frequently act as potent inhibitors of the first committed step in a metabolic pathway leading to their synthesis. End products are also known to accelerate the rate of another metabolic pathway leading to a different set of end products. The highly coordinated biosynthesis of adenine and guanine nucleotides provides examples of negative and positive effects of end products of each pathway.

END-TO-END ANNEALING

The reattachment of two supramolecular polymers, such as that observed when a high concentration of sheared microtubules is permitted to incubate *in vitro*. End-to-end annealing of microtubules is an unlikely process in living cells, because (a) the viscosity of the cytoplasm is apt to reduce greatly the tumbling of these polymers and (b) the so-called minus ends of the microtubules are usually firmly attached to microtubule-organizing centers.

ENE REACTION

A reaction occurring between a molecular entity possessing both a double bond and an allylic hydrogen (known as the ene) and a molecule with a multiple bond (or enophile).

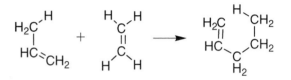

Ene Reaction of Propene with Ethene

In an ene reaction, there is a transfer of the allylic hydrogen and subsequent reorganization of the bonding. An example of an ene reaction is the addition of ethene (the enophile) to propene (the ene), thereby producing 1-pentene:

$$H_2C^*{=}CH{-}CH_2D + H_2C{=}C^{\#}H_2$$
$$\rightarrow H_2C{=}CH{-}C^*H_2{-}CH_2C^{\#}H_2D$$

Superscript symbols are provided to facilitate an understanding of the reaction. The reverse of an ene reaction is often referred to as the retro-ene reaction.

IUPAC (1979) *Pure and Appl. Chem.* **51**, 1725.

ENERGY BARRIER

1. In a reaction, the threshold energy that must be exceeded for molecules to become activated toward chemical reaction. 2. In conformational isomerizations, the energy needed to interconvert stable conformers. 3. More generally, the energy needed to transform, activate, or interconvert a particular molecular entity(ies).

ENERGY CHARGE (or, ADENYLATE ENERGY CHARGE)

An index of the phosphoanhydride (*i.e.*, P—O—P) bond content of the adenine nucleotides of a cell, based on a hypothetical model[1-3] that attempts to explain the metabolic basis for control of ATP utilization and regeneration. Later studies[4,5] demonstrated that the energy charge model is overly simplistic and that its principles are unlikely to constitute a useful model for the control of energy metabolism within biological systems.

The energy charge (or adenylate energy charge) is estimated by substituting the measured ATP, ADP, and AMP concentrations into the following expression:

$$\text{Energy Charge} = \frac{[\text{ATP}] + 0.5[\text{ADP}]}{[\text{ATP}] + [\text{ADP}] + [\text{AMP}]}$$

The energy charge quotient has a value of unity (or, 1.00) when only ATP is present and a value of zero when only AMP is present. Thus, the adenine nucleotide system is said to be fully charged at EC = 1 and fully discharged at EC = 0. At intermediate values, the adenine nucleotides are interconverted by adenylate kinase, and their concentrations are constrained by the adenylate kinase mass action ratio:

$$K_{eq} = [\text{ADP}]^2/\{[\text{ATP}][\text{AMP}]\}$$

The adenylate energy charge model asserts that ATP-utilizing enzymes (called U-systems) and ATP-regenerating enzymes (called R-systems) constitute a homeostatic mechanism for buffering the energy charge in a range of 0.75–0.85. Accordingly, a drop in the cellular energy charge would be expected to reduce the velocity of ATP-utilizing reactions, while simultaneously stimulating ATP-replenishing processes.

The model has numerous short-comings[4,5]. Chiefly, it fails to account for the fact that U- and R-enzymes frequently have other phosphoryl-acceptor and -donor substrates, and the model does not explain how fluctuations in these cosubstrate concentrations may alter the so-called responsiveness of enzymes. Moreover, the concentration of orthophosphate is wholly omitted from the model, despite the fact that key regulatory steps in ATP utilization and regeneration are exquisitely sensitive to changes in orthophosphate concentration. The energy-charge model also fails to consider magnesium ion complexion by nucleotides or the effects of adenine nucleotide compartmentation in living cells. Moreover, many signal transduction processes involve U-systems (*e.g.*, ATP-dependent protein kinases) to stimulate enzymes in other so-called R-systems by downstream activation of ATP-regenerating enzymes. This would suggest that the initial hormone- or effector-stimulated kinase is itself unlikely to obey the predicted energy charge behavior of a U-system.

In view of these serious flaws, as well as other technical difficulties[4-6], the energy charge model is misleadingly simplistic in attempting to interpret the intracellular behavior of regulatory enzymes utilizing or replenishing ATP. In limited circumstances and with the understanding that the energy-charge hypothesis is flawed, the EC quotient may still be helpful in discussing the broad homeostatic features of ATP-utilizing and ATP-replenishing reaction.

[1]D. E. Atkinson (1968) *Biochemistry* **11**, 4030.
[2]D. Atkinson (1977) *Trends Biochem. Sci.* **2**, N 198.
[3]D. E. Atkinson & A. G. Chapman (1979) *Meth. Enzymol.* **55**, 229.
[4]D. L. Purich & H. J. Fromm (1972) *J. Biol. Chem.* **247**, 249.
[5]D. L. Purich & H. J. Fromm (1973) *J. Biol. Chem.* **248**, 461.
[6]H. Fromm (1977) *Trends Biochem. Sci.* **2**, N 198.

ENERGY OF ACTIVATION

Any energy change (from the ground state to the transition state in a chemical reaction) that accounts (a) for the presence of a reaction barrier and (b) for the temperature dependence of a rate constant. This quantity is symbolized by E_a and may also be referred to as the Arrhenius energy of activation and activation energy.

An elementary reaction where radicals (e.g., A· + B· → A–B) recombine may not involve an energy barrier and would therefore have no energy of activation.

Occasionally, a negative activation energy is reported. For example, an E_a value of $-13.0\,\text{kJ·mol}^{-1}$ was obtained for tRNA[Phe] binding to the E-site of poly(U)programmed ribosomes containing unlabeled tRNA[Phe] at its P-site[1]. In such circumstances, one must be especially wary, because there are several reasons why an erroneously negative E_a value is obtained:

(a) The actual experimental design was flawed, and the rate parameter was not correctly measured. If the true activation energy is a very small number, experimental inaccuracy or imprecision may be especially problematic. One should conduct an error analysis that appropriately evaluates the overall effect of imprecise measurement on the derived parameter.

(b) The observed process is not governed by a simple bimolecular reaction and involves at least one endothermic equilibration with a large $-\Delta H$ value. An example is the reaction of nitric oxide with molecular oxygen ($2 \cdot NO + O_2 \rightarrow 2 \cdot NO_2$) which was originally described by the third-order rate law: ($-d[\cdot NO]/dt = 2k_{obs}$ $[\cdot NO]^2[O_2]$)[2,3]. The actual process is more likely to involve a fast equilibrium that is followed by a slower bimolecular step ($\cdot NO + O_2 \rightleftharpoons NO_3$; $NO_3 + \cdot NO \rightarrow 2 \cdot NO_2$)[4,5]. The observed rate constant, k_{obs}, is the product of a bimolecular rate constant, k, for the second step and an equilibrium constant, K, for the formation of NO_3. Thus, the observed activation energy, E_{obs}, is the sum of the E_a for the second step and ΔH for the equilibration step (i.e., $E_{obs} = E_a + \Delta H$). Because the equilibration step has a particularly negative ΔH value, a negative activation energy is observed[5].

(c) A much less likely explanation is that a process thought to be bimolecular is actually trimolecular. Application of transition-state theory to a trimolecular reaction is known to result in an abnormal activation energy[6].
See Arrhenius Equation; Transition-State Theory; Q_{10}

[1]S. Schilling-Bartetzko, A. Bartetzko & K. H. Nierhaus (1992) J. Biol. Chem. **267**, 4704.
[2]M. Boderstein (1918) Zeit. Elektrochem. **24**, 183.
[3]M. Boderstein (1922) Zeit. Phys. Chem. **100**, 68.
[4]R. G. Pearson (1976) Symmetry Rules for Chemical Reactions, Wiley-Interscience, New York.
[5]J. W. Moore & R. G. Pearson (1981) Kinetics and Mechanism, 3rd ed., Wiley, New York.
[6]H. Gershinowitz & H. Eyring (1935) J. Amer. Chem. Soc. **57**, 985.

ENERGY TRANSFER

1. The process by which a molecular entity absorbs light and a phenomenon originates from the excited state of another molecular entity. 2. The process by which an excited state of one molecular entity is deactivated to a lower-lying state upon the transfer of energy to a second molecular entity, thus raising that new entity to a higher energy state. If the two entitites involved are different moieties of the same molecular entity, the process is referred to as an intramolecular energy transfer. **See**

also Förster Resonance Excitation Transfer (FRET); Radiative Energy Transfer; Spectral Overlap; Fluorescence

Comm. on Photochem. (1988) Pure and Appl. Chem. **60**, 1055.

ENHANCER

Any DNA sequence that binds protein factors and thereby stimulates transcription. This stimulatory effect may occur over an appreciable distance from the enhancer's site. Enhancers act in either strand orientation, and they can be found either upstream or downstream with respect to the promoter.

ENOLASE

Enolase [EC 4.2.1.11] catalyzes the interconversion of 2-phosphoglycerate and phosphoenolpyruvate in a stepwise mechanism. Proton abstraction and at least one additional step (e.g., hydroxide loss or product release) may limit the overall reaction rate.

R. Jaenicke, H. Schurig, N. Beaucamp & R. Ostendorp (1996) Adv. Protein Chem. **48**, 181.
J. J. Villafranca & T. Nowak (1992) The Enzymes, 3rd ed., **20**, 63.
L. Leboida & B. Stec (1991) Biochemistry **30**, 2817.
F. Wold (1971) The Enzymes, 3rd ed., **5**, 499.

ENOLASE SUPERFAMILY

Members of this superfamily catalyze at least eleven different chemical reactions[1], including racemization, epimerization, and both syn and anti β-elimination reactions involving water, ammonia, or an intramolecular carboxylate group as a leaving group. Babbit and Gerlt[1] point out that reactions catalyzed by members of the enolase superfamily are initiated by a common partial reaction, metal-assisted, general base-catalyzed abstraction of the α-proton of a carboxylate anion to generate a stabilized enolate anion intermediate. They further state that the fate of the intermediate (protonation in the case of racemization and epimerization reactions and vinylogous β-elimination in the others) must be determined by the different functional groups that "surround" the intermediate in the active site.

[1]P. C. Babbit & J. A. Gerlt (1997) J. Biol. Chem. **272**, 30591.

ENOLIZATION

Any process that produces an enol, including the formation of enolate anions. Enols (i.e., entities containing the moiety $HO-C(R_1)=C(R_2R_3)$) appear as intermediates in a wide variety of enzyme-catalyzed reactions, and Rose[1] has presented the following diagram to describe the various categories of stepwise processes involving enol intermediates.

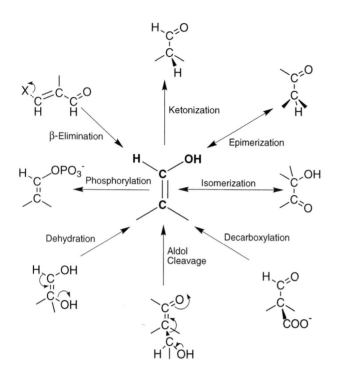

Enzyme Reaction Types Involving Enol Intermediates

A number of methods[1] can assist in identifying and characterizing enol intermediates (as well as eneamine and carbanion intermediates) in enzyme-catalyzed reactions. These include: (1) proton isotope exchange; (2) oxidation of the intermediate; (3) coupled elimination; (4) spectrophotometric methods; (5) use of transition-state inhibitors; (6) use of suicide inhibitors; (7) isolation of the enol; and (8) destructive analysis.

[1] I. A. Rose (1982) *Meth. Enzymol.* **87**, 84.

5-*Enol*PYRUVYLSHIKIMATE-3-PHOSPHATE SYNTHASE

This enzyme [EC 2.5.1.19] catalyzes the reaction of phosphoenolpyruvate and 3-phosphoshikimate, yielding orthophosphate and 5-*O*-(1-carboxyvinyl)-3-phosphoshikimate.

J. S. Holt, S. B. Powles & J. A. M. Holtum (1993) *Ann. Rev. Plant Physiol. Plant Mol. Biol.* **44**, 203.
K. A. Johnson (1992) *The Enzymes*, 3rd ed., **20**, 1.
R. Bentley (1990) *Crit. Rev. Biochem. Mol. Biol.* **25**, 307.

ENOYL-ACP REDUCTASE

This family of enzymes catalyzes the reaction of an acyl–[acyl-carrier-protein] with NAD$^+$ [EC 1.3.1.9] or with NADP$^+$ [EC 1.3.1.10] to produce 2,3-dehydroacyl–[acyl-carrier-protein] and NADH or NADPH. **See also** *Fatty Acid Synthetase*

J. Browse & C. Somerville (1991) *Ann. Rev. Plant Physiol. Plant Mol. Biol.* **42**, 467.
S. J. Wakil & J. K. Stoops (1983) *The Enzymes*, 3rd ed., **16**, 3.

ENOYL-CoA HYDRATASE or (CROTONASE)

This enzyme catalyzes the addition of water to *trans*-2(or 3)-enoyl-CoA to yield (3*S*)-3-hydroxyacyl-CoA. The enzyme will act on *cis* derivatives, generating the (3*R*)-3-hydroxyacyl-CoA product.

The following is a classical review on the molecular and physical properties of this enzyme [EC 4.2.1.17].

R. L. Hill & J. W. Teipel (1971) *The Enzymes*, 3rd ed., **5**, 539.

ENOYLPYRUVATE

Although enolpyruvate has been long regarded as a pyruvate kinase reaction intermediate, direct chemical trapping and/or isolation remained elusive. Seeholzer *et al.*[1] recently developed a method that distinguishes enolpyruvate, based on its reaction with bromine in acid, followed by derivatization of the bromopyruvate with thionitrobenzoate. The thioether derivative (10 pmol) could be reliably quantitated, allowing their determination of the internal equilibria, including the E·ATP·enolpyruvate complex. Phosphoenolpyruvate also reacts with bromine to yield bromopyruvate, and to quantitate enolpyruvate specifically in a background of phosphoenolpyruvate, the authors took advantage of the much greater acid stability of phosphoenolpyruvate.

[1] S. H. Seeholzer, A. Jaworowski & I. A. Rose (1991) *Biochemistry* **30**, 727.

ENTATIC STATE

A hypothetical conformational state[1] defined by a geometrically and electronically strained site within an enzyme thought to facilitate the conversion of an enzyme-substrate complex to the transition state. Vallee and Williams[1] defined the entatic state as an abnormal condition of localized strain transmitted by relief of compression or other steric clashes elsewhere in the enzyme. They suggested that catalytic rate enhancements arise from the heightened reactivity of catalytic group(s) that have experienced unimolecular activation. Williams[2] presented additional ideas on how metalloenzymes may utilize such a mechanism to assist in the conversion of enzyme-bound substrate to the reaction's transition state.

[1]B. L. Vallee & R. J. P. Williams (1968) *Proc. Natl. Acad. Sci. U.S.A.* **59**, 498.

[2]R. J. P Williams (1971) *Cold Spring Harbor Symp. Quant. Biol.* **36**, 53.

ENTERING GROUP

The atom, ion, or group of a molecular entity that approaches and forms a covalent or coordinate covalent bond with the substrate during a reaction. Any attacking nucleophile is an entering group.

ENTEROPEPTIDASE

This enzyme [EC 3.4.21.9], also known as enterokinase, activates trypsinogen by catalyzing the cleavage of Lys^6–Ile^7 bond in trypsinogen.

J. S. Bond & P. E. Butler (1987) *Ann. Rev. Biochem.* **56**, 333.

ENTEROTOXIN

There are a large number of proteins that share this designation. Perhaps the best known is the heat-labile enterotoxin from *E. coli* that catalyzes the ADP-ribosylation of a number of proteins.

L. Passador & W. Iglewski (1994) *Meth. Enzymol.* **235**, 617.
C.-Y. Lai (1986) *Adv. Enzymol.* **58**, 99.

ENTHALPY

A thermodynamic quantity, symbolized by H, and equal to $U + PV$, where U is the thermal energy, P is pressure, and V is volume. If the pressure is constant and no other work is done other than $P \Delta V$ work, then the increase in enthalpy is equal to the heat absorbed at constant pressure ($\Delta H = q_P$). If $\Delta H < 0$, the process is regarded as exothermic; if $\Delta H > 0$, the process is regarded as endothermic. ***See also*** *Innate Thermodynamic Quantities*

Selected entries from *Methods in Enzymology* [vol, page(s)]: Antibody-antigen interactions, **247**, 295-296; binding enthalpy [ligand concentration effect, **259**, 212; limitations on interpretation, **259**, 211-212; measurement by calorimetry, **259**, 196, 209-210; pH effect, **259**, 212; protein isomerization effects, **259**, 214-218; range and ligand interactions, **259**, 212; temperature effect, **259**, 212, 214-221; ternary complex formation, **259**, 213, 217-218]; calculation, **247**, 290-291; coupled equilibria, driving of processes, **259**, 708-709; differential scanning calorimetric determination, **240**, 504, 506-507, 513-514, 531, 534-537; enthalpy-entropy compensation in titration calorimetry, **247**, 300-302; errors in isothermal equilibria studies, **259**, 629; ionization enthalpies of buff-ers, **240**, 518; motive enthalpy, temperature dependence, **259**, 641-642; protein analysis by differential scanning calorimetry, **259**, 155-156, 168; protonation effects, **240**, 517-518; spectroelectrochemical determination, **246**, 711; van't Hoff analysis [oligonucleotide (differential scanning calorimetry, **259**, 235-237; melting curves, **259**, 225-234); protein, **259**, 149-150, 196, 618-620]; van't Hoff enthalpy, **240**, 507-508, 528, 531.

ENTHALPY-DRIVEN REACTION

Any reaction or process for which the enthalpy change is large and negative, such that the ΔH term greatly dominates the $-T\Delta S$ term in the Gibbs equation ($\Delta G = \Delta H - T\Delta S$), and acts thereby as the thermodynamic driving force for that reaction or process.

ENTHALPY OF ACTIVATION

The standard enthalpy difference between reactant(s) of a reaction and the activated complex in the transition state at the same temperature and pressure. It is symbolized by $\Delta H^{\ddagger}$ and is equal to $(E_a - RT)$, where E_a is the energy of activation, R is the molar gas constant, and T is the absolute temperature (provided that all non-first-order rate constants are expressed in temperature-independent concentration units, such as molarity, and are measured at a fixed temperature and pressure). Formally, this quantity is the enthalpy of activation at constant pressure. ***See*** *Transition-State Theory (Thermodynamics); Transition-State Theory; Gibbs Free Energy of Activation; Entropy of Activation; Volume of Activation*

ENTROPY

A fundamental thermodynamic state function (symbolized by S), and as such, not dependent on the path by which a particular state is reached. For a reversible process, the differential change in entropy, dS, is equal to the amount of energy absorbed by the system, dq, divided by the absolute temperature, T. Thus,

$$\Delta S = \int_A^B dq/T$$

The SI units for entropy is joules per kelvin. If irreversibility occurs at any point in the process under consideration, then $dS > dq/T$. The entropy of an isolated system undergoing an irreversible process always increases (in other words, for any isolated system $\Delta S \geq 0$).

In statistical mechanics[1], entropy of a system with respect to a particular state is related to the probability, W, of a system being in that state: *i.e.*, $S = k_B \ln W + b$ where b is a constant and k_B is the Boltzmann constant. Hence, entropy represents the degree of disorder within a system. *See also* Gibbs Free Energy; Enthalpy; Thermodynamics

[1]R. Dickerson (1969) *Molecular Thermodynamics*, p. 452, W. A. Benjamin, New York. (This is a remarkably lucid treatment of thermodynamics.)

Selected entries from *Methods in Enzymology* [vol, page(s)]:
Amino acid side chain contribution, **240**, 516; approximate entropy [biological application, **240**, 71; calculation, **240**, 74-75; computer implementation, **240**, 78, 88; data points, **240**, 75-76, 88; data processing, **240**, 73; development, **240**, 71; family of statistics, **240**, 76; filters, **240**, 74-77, 88; implicit noise filtering, **240**, 79; length of compared runs, **240**, 74-75; model independence, **240**, 79; noise effects, **240**, 82, 88; outlier sensitivity, **240**, 78-79, 88; quantification of data regularity, **240**, 70, 74-75, 82; relationship to other approaches (feature recognition algorithms, **240**, 80-81; Kolmogorov-Sinai entropy, **240**, 71, 76, 81-83; moment statistics, **240**, 80; phase space plots, **240**, 83-84; power spectra, **240**, 83-84); relative consistency, **240**, 77; statistical validity, **240**, 79-80]; Kolmogorov-Sinai entropy [comparison with approximate entropy, **240**, 71, 76, 81-83; limitations, **240**, 81; noise effects, **240**, 81]; coupled equilibria, driving of processes, **259**, 708-709; differential scanning calorimetric determination, **240**, 506-507, 514, 516-517, 535-536, 539; errors in isothermal equilibria studies, **259**, 629, 639-640, 709; gas, **259**, 562-563; ligand binding-induced change, **240**, 93; maximum entropy method [applications (anisotropy decay, **240**, 301-310; distribution of protein conformational states, **240**, 269-273; frequency domain fluorescence, **240**, 273-286; time-correlated single-photon counting, **240**, 290-310); classical approach (automatic noise scaling, **240**, 267-268; formula, **240**, 266-267; inference about reconstruction, **240**, 267; positive-negative distributions, **240**, 268); computer implementation, **240**, 268-269, 290; derivation from Bayesian probability theory, **240**, 264-265; historic maximum entropy, **240**, 269-270; Laplace transform handling, **240**, 262, 287; in multidimensional NMR, **239**, 257-259, 265; properties, **240**, 310; roots, **240**, 263-264]; mixing, **259**, 709, 715; motive, crystalline solids, **259**, 639-641; peptide backbone contribution, **240**, 516-517; protein analysis by differential scanning calorimetry, **259**, 155-156, 168; range and ligand interactions, **259**, 212; single-particle translational partition function, **259**, 709; solvent reorganization, **259**, 561-564; spectroelectrochemical determination, **246**, 710-711; thermal [crystalline solids, **259**, 639-641; derivation of change, **259**, 631; water, **259**, 561-562].

ENTROPY AND ENTHALPY OF ACTIVATION (Enzymatic)

Determination of values for $\Delta H^{\ddagger}$ and $\Delta S^{\ddagger}$ allows one the opportunity to further understand the action of enzyme catalysis, provided that these quantities can be unambiguously evaluated. High catalytic efficiency is often attributed to $\Delta S^{\ddagger}$ effects. The magnitude and sign of $\Delta S^{\ddagger}$ can provide information on the nature of the activated complexes. Both positive and negative entropies of activation have been reported: as pointed out by Laidler[1,2], Fromm[3], and Jencks[4], many factors can affect $\Delta S^{\ddagger}$ values, including a reversible conformational change in the enzyme during or after substrate binding. In addition, whenever a substrate interacts with an enzyme, both

substrate and enzyme are "lost" to the solvent as separate solute species. Laidler[2,5] referred to this contribution as the "unmixing" of the solute species with the solvent and affecting $\Delta S^{\ddagger}$, whereas Jencks[4] attributed this to the entropy of dilution. Other factors affecting the entropy of activation include the orientation of the substrates in the transition state relative to the unassociated molecules, the relative hydrophobicity of both substrate and protein, the net change in number of hydrogen bonds between substrate and enzyme, the loss of translational and possibly rotational degrees of freedom upon the binding of the substrate[6,7] (if the substrate has lost degrees of freedom, then the enzyme-substrate complex must have gained degrees of freedom[8]), the release of water molecules from the enzymes active site upon binding of the substrate, the relative orientation of solvent molecules in the transition state and in the solvent cage around the substrate molecule, the molecularity of the reaction[3] (the net volume of activation is expected to be negative for a bimolecular reaction), and electrostatic effects.

The physical meaning for the entropy of activation is often overinterpreted, and care should be exercised in analyzing multistep processes that may not attain thermodynamic equilibrium during each catalytic reaction cycle. If there is a temperature-dependent change in the rate-determining step for a reaction, reliable interpretations of $\Delta S^{\ddagger}$ may be exceedingly difficult to achieve. This will also be true if there is a change in the heat capacity of the solvent or a temperature-dependent change in protein structure.

Hammes[9] suggested that "energy compensation" has a role in enzyme-catalyzed reactions, such that there is synergism between bond cleavage and bond formation.
See *Transition-State Theory; Transition-State Theory (Thermodynamics); Transition-State Theory in Solutions; Entropy of Activation; Volume of Activation*

[1]K. J. Laidler & P. S. Bunting (1973) *The Chemical Kinetics of Enzyme Action*, 2nd ed., Clarendon Press, Oxford.
[2]K. J. Laidler (1955) *Discussion Faraday Soc.* **20**, 83.
[3]H. J. Fromm (1975) *Initial Rate Enzyme Kinetics*, Springer-Verlag, New York.
[4]W. P. Jencks (1969) *Catalysis in Chemistry and Enzymology*, McGraw-Hill, New York. [Also available as Jencks, W. P. (1986) *Catalysis in Chemistry and Enzymology*, Dover Publications, New York].
[5]M. L. Barnard & K. J. Laidler (1952) *J. Amer. Chem. Soc.* **74**, 6099.
[6]P. D. Bartlett & R. R. Hiatt (1958) *J. Amer. Chem. Soc.* **80**, 1398.

[7] E. G. Foster, A. C. Pope & F. Daniels (1947) *J. Amer. Chem. Soc.* **69**, 1893.

[8] F. H. Westheimer (1962) *Adv. Enzymol.* **24**, 441.

[9] G. G. Hammes (1964) *Nature* **204**, 342.

ENTROPY OF ACTIVATION

The standard entropy difference between the reactant(s) of a reaction and the activated complex of the transition state, at the same temperature and pressure. Entropy of activation is symbolized by either $\Delta S^{\ddagger}$ or $\Delta^{\ddagger} S^{\ominus}$ and is equal to $(\Delta H^{\ddagger} - \Delta G^{\ddagger})/T$ where $\Delta H^{\ddagger}$ is the enthalpy of activation, $\Delta G^{\ddagger}$ is the Gibbs free energy of activation, and T is the absolute temperature (provided that all rate constants other than first-order are expressed in temperature-independent concentration units such as molarity). Technically, this quantity is the entropy of activation at constant pressure, and from this value, the entropy of activation at constant volume can be deduced. *See* Transition-State Theory (Thermodynamics); Gibbs Free Energy of Activation; Enthalpy of Activation; Volume of Activation; Entropy and Enthalpy of Activation (Enzymatic)

ENTROPY TRAP MODEL

The concept that catalytic rate enhancement stems from an enzyme's ability to freeze the substrate into a particularly reactive configuration, which in the absence of an enzyme would only be present at vanishingly low concentration. In this model, reacting groups within an enzyme's active site are brought together in precisely the correct positions that maximize reactivity. The ability to achieve such high precision in orienting reactive groups occurs at a price, namely a very large loss of entropy: hence the name "entropy trap". Jencks[1] has estimated that rate enhancements of 10^8 to 10^{12} might be realistically accessible for two reactants at 0.01 M. Such precise binding requires restricted motion of reactant on the enzyme, a thermodynamically unfavored circumstance that most probably reduces the observed affinity of an enzyme for its substrate. Jencks has also emphasized that this reduced affinity stems from strain needed to force the substrate to bind at an active site, which is itself designed to be complementary to the transition state, and not the substrate's ground state.

[1] W. P. Jencks (1975) *Adv. Enzymol.* **43**, 219.

ENZYME

Any biological substance acting as a catalyst. This expanded definition now includes proteins classically defined as enzymes, other proteins that facilitate otherwise rate-limiting exchanges of macromolecule-bound effector molecules, catalytic antibodies designed in the laboratory, as well as ribozymes which are catalysts made up of ribonucleic acid.

The Enzyme Commission of the International Union of Biochemistry devised a scheme for enzyme classification based upon the types of reactions catalyzed. There are six main classes: the oxidoreductases (all enzymes that catalyze oxidation-reduction reactions, such as dehydrogenases, reductases, oxidases, transhydrogenases, peroxidases, desaturases, dismutases, and a few synthases); the transferases (those enzymes that catalyze the transfer of some moiety, such as a methyl or acetyl group, from one compound to another); the hydrolases which catalyze the hydrolytic cleavage of a bond; lyases, those enzymes that catalyze the cleavage of a bond by elimination, producing double bonds, or, conversely, adding groups to double bonds; the isomerases which catalyze configurational or structural changes within a molecule; and the ligases (and synthetases) that catalyze bond formation with the concomitant hydrolysis of a pyrophosphate bond in ATP or similar molecule. *See* Enzyme Commission Nomenclature

ENZYME CASCADE KINETICS

The rate behavior of linear and multicyclic systems of enzymes arranged such that each enzyme in a step or cycle is itself a substrate for an enzyme-catalyzed covalent modification reaction. The systematic investigation of the kinetics of these cascade systems promises to provide insights about the design features accounting for the signal amplification and information processing achieved by such hierarchies of enzymes with other enzymes serving as their substrates.

Covalent interconversion of enzymes is well established as a fundamental theme in metabolic regulation[1]. The prototypic reversible interconverting systems include the sequence of phosphorylation/dephosphorylation steps in the activation of mammalian glycogen phosphorylase and pyruvate dehydrogenase[2] as well as the nucleotidylation/denucleotidylation using UTP and ATP in the bacterial glutamine synthetase cascade[3] (see Fig. 1.).

Covalent Interconversion of Bacterial Glutamine Synthetase

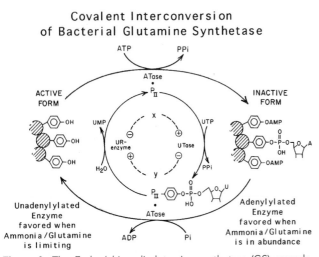

Figure 1. The *Escherichia coli* glutamine synthetase (GS) cascade, indicating the metabolic regulation of glutamine synthesis by multiple effector molecules that directly reflect the status of nitrogen metabolism in the bacterium. In this scheme, GS is the metabolically active form of glutamine synthetase, and GS-AMP is the metabolically inactive adenylylated enzyme. The same converting enzyme, designated ATase for adenylyltransferase, catalyzes adenylylation and deadenylylation to form GS-AMP and GS, respectively. As shown in the diagram, however, those metabolites serving as positive effectors for adenylylation behave as negative effectors for deadenylylation, and *vice versa*. Even more striking is the involvement of a bifunctional enzyme UR/UT which serves to release UMP from P_{II}-UMP or to uridylate tyrosinyl residues on P_{II}. The dotted lines from P_{II} and P_{II}-UMP indicate the ability of these forms of the regulatory protein to direct the ATase's action on glutamine synthetase.

Unidirectional activation of the intrinsic and extrinsic pathways of the blood clotting cascade exemplify a linear sequence of proteolysis steps that amplify an initial signal into a massive, well-controlled biological response. In this respect, enzyme cascades are not merely amplification systems; the multiplicity of enzymes involved in a cascade allows for a multiplicity of effector binding sites, thereby enhancing a cascade's responsiveness to changes in a regulatory input signal[3]. This is illustrated by the numerous calcium ion binding sites in the blood clotting cascade, and a small change in the concentration of this divalent cation elicits a major change in the rapidity with which fibrin is generated in the last step of this sequence of converting proteolytic zymogens into their catalytically active forms. A schematic representation of a generalized multicyclic cascade is shown in Fig. 2, where the increasing number of vertical arrows indicate an increase in catalytic activity of the enzymes.

Cascade of Enzyme-Catalyzed Interconversion Reactions

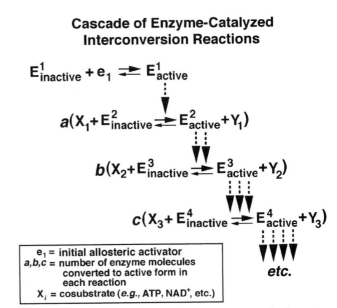

Figure 2. A multistep enzyme interconversion cascade illustrating the sequential modification and conversion of target enzymes from one state of activity to another. Although written here as a cascade of increasing catalytic activity, enzyme-catalyzed covalent modification can either activate or inhibit target enzymes, depending on the particular system under study.

Accordingly, the kinetic properties of cascades involving "converter" and "target" enzymes[5] becomes of paramount significance in any consideration of covalently interconverting enzyme systems. Stadtman and Chock[4–6] led the way in developing a steady-state treatment for analyzing such cascades. They have offered the following emerging concepts regarding the behavior of enzyme cascades:

1. Covalent interconversion of an enzyme is a dynamic process, and the specific activity of the target enzyme is determined by coupling of two opposing, dynamic cascades. This results in a steady-state distribution of active and inactive species of the interconvertible enzyme. The target enzyme's activity smoothly and continuously responds to feedback signals that alter the efficiency of the enzymes catalyzing the interconversion steps as well as the ability of the target enzyme to be an effective substrate for the interconverting enzymes.

2. A minimum of three enzymes are typically involved in each interconvertible enzyme cycle that can integrate many allosteric effectors into a single output, the specific activity of the target enzyme in the cascades.

3. Cyclic cascades can achieve great signal amplification, and interconvertible enzymes can respond to effectors at concentrations considerably below their respective dissociation constants.

4. Cyclic cascades can propagate a sigmoidal dose-response curve showing how an increasing concentration of allosteric effector manifests itself in activity changes of the target enzyme.

Future kinetic work on the time evolution of cascades, rather than steady-state rate behavior, is likely to reveal how time-dependent changes in the responsiveness are related to the kinetic properties of individual components. Presently, no multicyclic cascade has been rigorously experimentally analyzed, either by the steady-state approach or through the use of an integrated rate equation method. An effort in this direction is the recent work of Huang and Ferrell[7] on the mitogen-activated protein kinase cascade. These investigators found that this cascade shows a sharp stimulus-response curve that may be fundamentally important in the function of those molecular switches controlling such high-level processes as mitogenesis.

[1]E. G. Krebs (1994) *Trends Biochem. Sci.* **19**, 439.
[2]F. Huchs, D. D. Randall, T. E. Roche, M. W. Burgett, J. W. Pelley & L. J. Reed (1977) *Arch. Biochem Biophys.* **151**, 328.
[3]S. P. Adler, D. L. Purich & E. R. Stadtman (1975) *J. Biol. Chem.* **250**, 6264.
[4]P. B. Chock & E. R. Stadtman (1980) *Meth. Enzymol.* **64**, 297.
[5]E. R. Stadtman & P. B. Chock (1977) *Proc. Natl. Acad. Sci. U.S.A.* **74**, 2761.
[6]P. B. Chock & E. R. Stadtman (1977) *Proc. Natl. Acad. Sci. U.S.A.* **74**, 2766.
[7]C.-Y. F. Huang & J. E. Ferrell (1996) *Proc. Natl. Acad. Sci. U.S.A.* **93**, 10078.

ENZYME COMMISSION NOMENCLATURE

The International Union of Biochemistry and Molecular Biology (IUBMB) system for rational naming and categorizing of enzymes into six major classes: [1] oxidoreductases which catalyze reactions involving the loss or gain of electrons by substrates; [2] transferases which catalyze the transfer of reactive groups from a donor substrate to an acceptor substrate; [3] hydrolases which catalyze the introduction of the elements of water into a substrate; [4] lyases which catalyze elimination of a group from a substrate with formation of a double bond or addition of a group to a double bond; [5] isomerases which catalyze intramolecular rearrangements; and [6] ligases which catalyze the joining of molecules by the formation of covalent bonds.

The EC number contains four numerical elements each separated by points. For example, alcohol dehydrogenase is assigned the number EC 1.1.1.1. The first numerical element (the furthest one to the left) identifies which of the six main classes the enzyme belongs. For alcohol dehydrogenase, this class is the oxidoreductases. The second element identifies the subclass; for EC 1.1.x.x, this refers to oxidoreductases acting on the CH—OH group of donors. The third element identifies the sub-sub-class; for EC 1.1.1.x this refers to oxidoreductases that use NAD^+ or $NADP^+$ as the acceptor. The final sub-sub-sub-class is unique for that enzyme-catalyzed reaction.

1. X. X.X	OXIDOREDUCTASES (Acceptor = Oxidant; Donor = Reductant)
1. 1. X.X	Acting on the CH—OH group of donors.
1. 1. 1.X	Acceptor: NAD^+ or $NADP^+$.
1. 1. 2.X	Acceptor: a cytochrome.
1. 1. 3.X	Acceptor: oxygen.
1. 1. 4.X	Acceptor: a disulfide.
1. 1. 5.X	Acceptor: a quinone or similar compound.
1. 1.99.X	Acceptor: other.
1. 2. X.X	Acting on the aldehyde or oxo group of donors.
1. 2. 1.X	Acceptor: NAD^+ or $NADP^+$.
1. 2. 2.X	Acceptor: a cytochrome.
1. 2. 3.X	Acceptor: oxygen.
1. 2. 4.X	Acceptor: a disulfide.
1. 2. 7.X	Acceptor: an iron-sulfur protein.
1. 2.99.X	Acceptor: other.
1. 3. X.X	Acting on the CH—CH group of donors.
1. 3. 1.X	Acceptor: NAD^+ or $NADP^+$.
1. 3. 2.X	Acceptor: a cytochrome.
1. 3. 3.X	Acceptor: oxygen.
1. 3. 5.X	Acceptor: a quinone or related compound.
1. 3. 7.X	Acceptor: an iron-sulfur protein.
1. 3.99.X	Acceptor: other.
1. 4. X.X	Acting on the CH—NH_2 group of donors.

1. 4. 1.X	Acceptor: NAD$^+$ or NADP$^+$.	
1. 4. 2.X	Acceptor: a cytochrome.	
1. 4. 3.X	Acceptor: oxygen.	
1. 4. 4.X	Acceptor: a disulfide.	
1. 4. 7.X	Acceptor: an iron-sulfur protein.	
1. 4.99.X	Acceptor: other.	
1. 5. X.X	Acting on the CH—NH group of donors.	
1. 5. 1.X	Acceptor: NAD$^+$ or NADP$^+$.	
1. 5. 3.X	Acceptor: oxygen.	
1. 5. 4.X	Acceptor: a disulfide.	
1. 5. 5.X	Acceptor: a quinone or similar compound.	
1. 5.99.X	Acceptor: other.	
1. 6. X.X	Acting on NADH or NADPH.	
1. 6. 1.X	Acceptor: NAD$^+$ or NADP$^+$.	
1. 6. 2.X	Acceptor: a cytochrome.	
1. 6. 4.X	Acceptor: a disulfide.	
1. 6. 5.X	Acceptor: a quinone or similar compound.	
1. 6. 6.X	Acceptor: a nitrogenous group.	
1. 6. 8.X	Acceptor: a flavin.	
1. 6.99.X	Acceptor: other.	
1. 7. X.X	Acting on other nitrogenous compounds as donors.	
1. 7. 2.X	Acceptor: a cytochrome.	
1. 7. 3.X	Acceptor: oxygen.	
1. 7. 7.X	Acceptor: an iron-sulfur protein.	
1. 7.99.X	Acceptor: other.	
1. 8. X.X	Acting on a sulfur group of donors.	
1. 8. 1.X	Acceptor: NAD$^+$ or NADP$^+$.	
1. 8. 2.X	Acceptor: a cytochrome.	
1. 8. 3.X	Acceptor: oxygen.	
1. 8. 4.X	Acceptor: a disulfide.	
1. 8. 5.X	Acceptor: a quinone or similar compound.	
1. 8. 7.X	Acceptor: an iron-sulfur protein.	
1. 8.99.X	Acceptor: other.	
1. 9. X.X	Acting on a heme group of donors.	
1. 9. 3.X	Acceptor: oxygen.	
1. 9. 6.X	Acceptor: a nitrogenous group.	
1. 9.99.X	Acceptor: other.	
1.10. X.X	Acting on diphenols and related substances as donors.	
1.10. 1.X	Acceptor: NAD$^+$ or NADP$^+$.	
1.10. 2.X	Acceptor: a cytochrome.	
1.10. 3.X	Acceptor: oxygen.	
1.10.99.X	Acceptor: other.	

1.11. X.X	Acting on a peroxide as acceptor (peroxidases).
1.12. X.X	Acting on hydrogen as donor.
1.12. 1.X	Acceptor: NAD$^+$ or NADP$^+$.
1.12. 2.X	Acceptor: a cytochrome.
1.12.99.X	Acceptor: other.
1.13. X.X	Acting on single donors with incorporation of molecular oxygen (oxygenases).
1.13.11.X	Acceptor: incorporation of two atoms of oxygen.
1.13.12.X	Acceptor: incorporation of one atom of oxygen.
1.13.99.X	Miscellaneous (requires further characterization).
1.14. X.X	Acting on paired donors with incorporation of molecular oxygen.
1.14.11.X	Donor: 2-oxoglutarate as one donor, and incorporation of one atom each of oxygen into both donors.
1.14.12.X	Donor: NADH or NADPH, and incorporation of two atoms of oxygen into one donor.
1.14.13.X	Donor: NADH or NADPH, and incorporation of one atom of oxygen.
1.14.14.X	Donor: reduced flavin or flavoprotein, and incorporation of one atom of oxygen.
1.14.15.X	Donor: a reduced iron-sulfur protein, and incorporation of one atom of oxygen.
1.14.16.X	Donor: reduced pteridine, and incorporation of one atom of oxygen.
1.14.17.X	Donor: ascorbate, and incorporation of one atom of oxygen.
1.14.18.X	Donor: another compound, and incorporation of one atom of oxygen.
1.14.99.X	Miscellaneous (requires further characterization).
1.15. X.X	Acting on superoxide radicals.
1.16. X.X	Oxidizing metal ions.
1.16. 1.X	Acceptor: NAD$^+$ or NADP$^+$.
1.16. 3.X	Acceptor: oxygen.

1.17. X.X	Acting on —CH$_2$ groups.
1.17. 1.X	Acceptor: NAD$^+$ or NADP$^+$.
1.17. 3.X	Acceptor: oxygen.
1.17. 4.X	Acceptor: a disulfide.
1.17.99.X	Acceptor: other.
1.18. X.X	Acting on reduced ferredoxin as donor.
1.18. 1.X	Acceptor: NAD$^+$ or NADP$^+$.
1.18. 6.X	Acceptor: dinitrogen.
1.18.99.X	Acceptor: H$^+$.
1.19. X.X	Acting on reduced flavodoxin as donor.
1.19. 6.X	Acceptor: dinitrogen.
	Other oxidoreductases.
1.97. X.X	
2. X. X.X	TRANSFERASES.
2. 1. X.X	Transferring one-carbon groups.
2. 1. 1.X	Methyltransferases.
2. 1. 2.X	Hydroxymethyl-, formyl- and related transferases.
2. 1. 3.X	Carboxyl- and carbamoyltransferases.
2. 1. 4.X	Amidinotransferases.
2. 2. X.X	Transferring aldehyde or ketone residues.
2. 3. X.X	Acyltransferases.
2. 3. 1.X	Acyltransferases.
2. 3. 2.X	Aminoacyltransferases.
2. 4. X.X	Glycosyltransferases.
2. 4. 1.X	Hexosyltransferases.
2. 4. 2.X	Pentosyltransferases.
2. 4.99.X	Transferring other glycosyl groups.
2. 5. X.X	Transferring alkyl or aryl groups, other than methyl groups.
2. 6. X.X	Transferring nitrogenous groups.
2. 6. 1.X	Transaminases (aminotransferases).
2. 6. 3.X	Oximinotransferases.
2. 6.99.X	Transferring other nitrogenous groups.
2. 7. X.X	Transferring phosphorus-containing groups.
2. 7. 1.X	Phosphotransferases with an alcohol group as acceptor (nucleophile).
2. 7. 2.X	Phosphotransferases with a carboxyl group as acceptor (nucleophile).
2. 7. 3.X	Phosphotransferases with a nitrogenous group as acceptor (nucleophile).

2. 7. 4.X	Phosphotransferases with a phosphate group as acceptor (nucleophile).
2. 7. 6.X	Diphosphotransferases.
2. 7. 7.X	Nucleotidyltransferases.
2. 7. 8.X	Transferases for other substituted phosphate groups.
2. 7. 9.X	Phosphotransferases with paired acceptors.
2. 8. X.X	Transferring sulfur-containing groups.
2. 8. 1.X	Sulfurtransferases.
2. 8. 2.X	Sulfotransferases.
2. 8. 3.X	CoA-transferases.
2. 9. X.X	Transferring selenium-containing groups.
3. X. X.X	HYDROLASES.
3. 1. X.X	Acting on ester bonds.
3. 1. 1.X	Carboxylic ester hydrolases.
3. 1. 2.X	Thiolester hydrolases.
3. 1. 3.X	Phosphoric monoester hydrolases.
3. 1. 4.X	Phosphoric diester hydrolases.
3. 1. 5.X	Triphosphoric monoester hydrolases.
3. 1. 6.X	Sulfuric ester hydrolases.
3. 1. 7.X	Diphosphoric monoester hydrolases.
3. 1. 8.X	Phosphoric triester hydrolases.
3. 1.11.X	Exodeoxyribonucleases producing 5′-phosphomonoesters.
3. 1.13.X	Exoribonucleases producing 5′-phosphomonoesters.
3. 1.14.X	Exoribonucleases producing other than 5′-phosphomonoesters.
3. 1.15.X	Exonucleases active with either ribo- or deoxyribonucleic acids and producing 5′-phosphomonoesters.
3. 1.16.X	Exonucleases active with either ribo- or deoxyribonucleic acids and producing other than 5′-phosphomonoesters.
3. 1.21.X	Endodeoxyribonucleases producing 5′-phosphomonoesters.
3. 1.22.X	Endodeoxyribonucleases producing other than 5′-phosphomonoesters.
3. 1.25.X	Site-specific endodeoxyribonucleases specific for altered bases.

3. 1.26.X	Endoribonucleases producing 5′-phosphomonoesters.
3. 1.27.X	Endoribonucleases producing other than 5′-phosphomonoesters.
3. 1.30.X	Endonucleases active with either ribo- or deoxyribonucleic acids and producing 5′-phosphomonoesters.
3. 1.31.X	Endonucleases active with either ribo- or deoxyribonucleic acids and producing other than 5′-phosphomonoesters.
3. 2. X.X	Glycosidases.
3. 2. 1.X	Hydrolysing O-glycosyl compounds.
3. 2. 2.X	Hydrolysing N-glycosyl compounds.
3. 2. 3.X	Hydrolysing S-glycosyl compounds.
3. 3. X.X	Acting on ether bonds.
3. 3. 1.X	Thioether hydrolases.
3. 3. 2.X	Ether hydrolases.
3. 4. X.X	Acting on peptide bonds (peptide hydrolases).
3. 4.11.X	Aminopeptidases.
3. 4.13.X	Dipeptidases.
3. 4.14.X	Dipeptidyl-peptidases and tripeptidyl-peptidases.
3. 4.15.X	Peptidyl-dipeptidases.
3. 4.16.X	Serine-type carboxypeptidases.
3. 4.17.X	Metallocarboxypeptidases.
3. 4.18.X	Cysteine-type carboxypeptidases.
3. 4.19.X	Omega peptidases.
3. 4.21.X	Serine endopeptidases.
3. 4.22.X	Cysteine endopeptidases.
3. 4.23.X	Aspartic endopeptidases.
3. 4.24.X	Metalloendopeptidases.
3. 4.99.X	Endopeptidases of unknown catalytic mechanism.
3. 5. X.X	Acting on carbon-nitrogen bonds, other than peptide bonds.
3. 5. 1.X	In linear amides.
3. 5. 2.X	In cyclic amides.
3. 5. 3.X	In linear amidines.
3. 5. 4.X	In cyclic amidines.
3. 5. 5.X	In nitriles.
3. 5.99.X	In other compounds.
3. 6. X.X	Acting on acid anhydrides.
3. 6. 1.X	In phosphorus-containing anhydrides.
3. 6. 2.X	In sulfonyl-containing anhydrides.

3. 7. X.X	Acting on carbon-carbon bonds.
3. 7. 1.X	In ketonic substances.
3. 8. X.X	Acting on halide bonds.
3. 8. 1.X	In C-halide compounds.
3. 9. X.X	Acting on phosphorus-nitrogen bonds.
3.10. X.X	Acting on sulfur-nitrogen bonds.
3.11. X.X	Acting on carbon-phosphorus bonds.
3.12. X.X	Acting on sulfur-sulfur bonds.
4. X. X.X	LYASES.
4. 1. X.X	Carbon-carbon lyases.
4. 1. 1.X	Carboxy-lyases.
4. 1. 2.X	Aldehyde-lyases.
4. 1. 3.X	Oxo-acid-lyases.
4. 1.99.X	Other carbon-carbon lyases.
4. 2. X.X	Carbon-oxygen lyases.
4. 2. 1.X	Hydro-lyases.
4. 2. 2.X	Acting on polysaccharides.
4. 2.99.X	Other carbon-oxygen lyases.
4. 3. X.X	Carbon-nitrogen lyases.
4. 3. 1.X	Ammonia-lyases.
4. 3. 2.X	Amidine-lyases.
4. 3. 3.X	Amine-lyases.
4. 3.99.X	Other carbon-nitrogen-lyases.
4. 4. X.X	Carbon-sulfur lyases.
4. 5. X.X	Carbon-halide lyases.
4. 6. X.X	Phosphorus-oxygen lyases.
4.99. X.X	Other lyases.
5. X. X.X	ISOMERASES.
5. 1. X.X	Racemases and epimerases.
5. 1. 1.X	Acting on amino acids and derivatives.
5. 1. 2.X	Acting on hydroxy acids and derivatives.
5. 1. 3.X	Acting on carbohydrates and derivatives.
5. 1.99.X	Acting on other compounds.
5. 2. X.X	cis-trans-Isomerases.
5. 3. X.X	Intramolecular oxidoreductases.
5. 3. 1.X	Interconverting aldoses and ketoses.
5. 3. 2.X	Interconverting keto- and enol-groups.
5. 3. 3.X	Transposing C=C groups.
5. 3. 4.X	Transposing S−S bonds.
5. 3.99.X	Other intramolecular oxidoreductases.

5. 4. X.X	Intramolecular transferases (mutases).	
5. 4. 1.X		Transferring acyl groups.
5. 4. 2.X		Phosphotransferases (phosphomutases).
5. 4. 3.X		Transferring amino groups.
5. 4.99.X		Transferring other groups.
5. 5. X.X	Intramolecular lyases.	
5.99. X.X	Other isomerases.	

6. X. X.X	LIGASES.	
6. 1.X.X	Forming carbon—oxygen bonds.	
6. 1. 1.X		Ligases forming aminoacyl-tRNA and related compounds.
6. 2. X.X	Forming carbon—sulfur bonds.	
6. 2. 1.X		Acid—thiol ligases.
6. 3. X.X	Forming carbon—nitrogen bonds.	
6. 3. 1.X		Acid—ammonia (or amine) ligases (amide synthases).
6. 3. 2.X		Acid—amino-acid ligases (peptide synthases).
6. 3. 3.X		Cyclo-ligases.
6. 3. 4.X		Other carbon—nitrogen ligases.
6. 3. 5.X		Carbon—nitrogen ligases with glutamine as amido-*N*-donor.
6. 4. X.X	Forming carbon—carbon bonds.	
6. 5. X X	Forming phosphoric ester bonds.	

ENZYME CONCENTRATION EFFECTS

In steady-state kinetic studies, the total concentration of the enzyme should be much less than the concentration of the substrate(s), product(s), and effector(s); typically, by at least a thousandfold. When this condition is not true, the steady-state condition will not be valid and other methods, such as global analysis, have to be utilized to analyze the kinetic data.

In standard kinetic studies, the initial velocity (v) should be directly proportional to the total enzyme concentration. Indeed, this plot should be one of the first items analyzed in the kinetic characterization of an enzyme. The enzyme concentration range in this plot should span from below to above the concentration to be used in the inital rate studies (and, ideally, the line should go through the origin). When curvature is observed in such plots, there are several potential explanations[1,2].

1. A toxic impurity present in the reaction mixture (but not in the stock enzyme solution) may result in upward curvature; at high concentrations of enzyme, a larger amount of enzyme remains active.

2. A dissociable activator (or coenzyme) present in the enzyme solution may result in upward curvature; as the enzyme concentration increases, so also does the concentration of the activator.

3. Upward curvature may occur when the enzyme undergoes self-association or aggregation into a more active form; such a process is favored at elevated concentrations.

4. If the enzyme under consideration is assayed via a coupling enzyme(s) system, downward curvature in the plot will be observed when the primary enzyme fails to limit the observed velocity. In such instances, it would be necessary to increase the amount of coupling enzymes such that they remain in excess under all conditions. This problem is often encountered in manometric assays, where the rate of diffusion of gas into the liquid begins to limit the enzyme reaction rate.

5. Elevated concentrations of enzyme can also affect the observed rate if those concentrations interfere with the physical methods of observing the reaction velocity (*e.g.*, increased light scattering at high [E_{total}] in spectrophotometric assays; increased absorbance by the protein; *etc.*)

6. Nonlinearity can also be observed if increased amounts of the stock enzyme solution alter some other reaction parameter (*e.g.*, pH, ionic strength, *etc.*). These issues are easily assessed and remedied.

7. If a coenzyme or activator is present in the reaction mixture at relatively low concentrations, then addition of increasing concentrations of total enzyme to the reaction mixture may reduce the coenzyme or activator to suboptimal levels. In such cases, a downward-oriented curve may be seen in the v *vs.* [E_{total}] plot.

8. Downward curvature can also be seen if steady-state and initial rate conditions are not observed (*e.g.*, the product concentration approaches inhibitory levels or substrate depletion occurs).

9. If a reversible inhibitor is present in the stock enzyme solution, then as more enzyme is added to the reaction, more inhibitor is also simultaneously introduced, and this can lead to progressively less activity in such *v versus* [E_total] plots.

10. If the enzyme undergoes aggregation or a conformational change at elevated concentrations that results in reduced activity, the *v versus* [E_total] plots will curve downward at elevated enzyme concentrations.

11. If the enzyme is unstable and steadily degrades in the stock solution, nonlinearity may be observed in plots of initial velocity *v versus* [E_total]. This can be corrected by using a standardized enzyme activity assay periodically during the course of the kinetic experiment to determine the amount of lost activity. The likelihood for enzyme instability arises whenever the kinetic experiment requires one to transfer the enzyme from its storage buffer to that used in the experiment.

[1]M. Dixon & E. C. Webb (1979) *Enzymes*, 3rd ed., Academic Press, New York.
[2]D. L. Purich & R. D. Allison (1980) *Meth. Enzymol.* **63**, 3.
Selected entries from *Methods in Enzymology* [vol, page(s)]:
Dilution of enzyme samples, **63**, 10; lipolysis substrate effect, **64**, 361, 362; dilution jump kinetic assay, **74**, 14-19, 28; dilution method [for dissociation equilibria, **61**, 65-96; continuous dilution cuvette, **61**, 78-96; data analysis, **61**, 74, 75; equations, **61**, 70-74; errors, **61**, 76-78; experimental procedures, **61**, 69, 70; merits, **61**, 75, 76; theory, **61**, 68, 69.

ENZYME DEPLETION (Apparent Noncompetitive Inhibition)

Enzyme inhibition by an extremely tight-binding inhibitor[1]. When the substrate(s), *regardless of the detailed mode of inhibition*, has (have) a negligible effect on the formation of enzyme-inhibitor (E-I) complex, the net result is depletion (*i.e.*, the removal of enzyme by the inhibitor from the reaction). The observed kinetic pattern is identical to the simple noncompetitive inhibition case; the substrate and the inhibitor do not affect each other's binding, because only V_{max} is changed due to reduced enzyme concentration, while K_m remains unaltered.

An example of enzyme depletion is the ribonuclease inhibitor isolated from human placenta by Blackburn, Wilson & Moore[2]. This protein forms a 1:1 complex with bovine pancreatic RNase A and is a noncompetitive inhibitor of the pancreatic enzyme, with a K_i of 3×10^{-10} M.

The rate equation to be used for kinetic analysis of enzyme depletion is that for *simple noncompetitive* inhibition. If the Henderson equation or similar types are not employed, keep in mind that the inhibitor concentration [I] is the *free* inhibitor concentration. Determination of K_i may not be feasible if the rate assay is insensitive and requires an enzyme concentration much greater than K_i. Alternatively, K_i may be obtained by measuring the on-off rate constants of the E·I complex, provided the rate constants for any conformationl change steps involved are also known.

[1]The authors thank Dr. Charles Y. Huang for contributing this entry.
[2]P. Blackburn, G. Wilson & S. Moore (1977) *J. Biol. Chem.* **252**, 5904.

ENZYME ENERGETICS (The Case of Proline Racemase)

Although triosephosphate isomerase[1] is the enzyme that has been used in the most widely known systematic investigation of catalysis, every student interested in using kinetics to explore catalysis should also school herself or himself on the reaction mechanism and energetics of proline racemase. This bacterial enzyme efficiently interconverts L-proline to D-proline, and Rudnick and Abeles[2] showed that the enzyme is a homologous dimer containing a single active site made up of residues contributed from both subunits. They also demonstrated that catalysis proceeds by a two-base mechanism, and proton abstraction alternatively occurs from one or the other of the two faces of the proline ring. Two enzyme forms, designated E_1 and E_2, each have one protonated base and one unprotonated base, between which there appears to be relatively little or no proton exchange (Fig. 1).

The research team headed by Albery and Knowles presented a series of seven papers[3-9] on the dynamics of the catalytic process. First, they investigated the reaction kinetics over a range of substrate concentration[3]. While the standard free energy changes for a series of intermediate steps and the associated transition states are fixed for a particular enzyme acting on a given substrate, the actual distribution of enzyme among substrate-bound and substrate-free species is sensitive to the degree of enzyme saturation by substrate. Accordingly, Fisher *et*

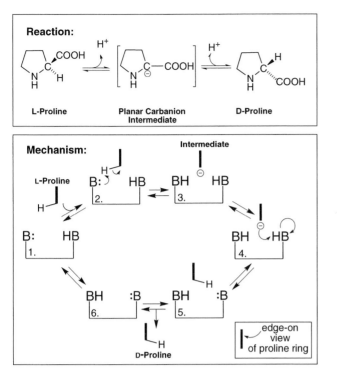

Figure 1. Reaction and mechanism of proline racemase.

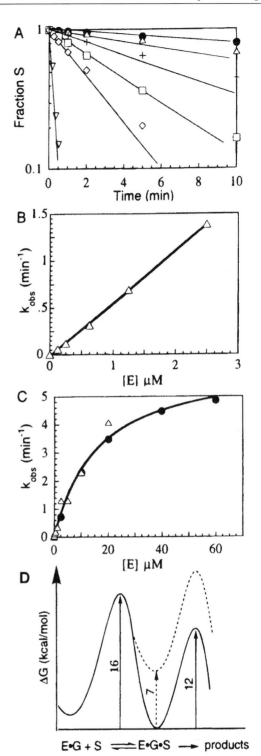

al.[3] considered three different substrate concentration regimes, which they termed: (a) *undersaturation*, a condition where the rate-limiting free energy difference is that between the lowest energy free enzyme form and the highest energy bound transition state; (b) *saturation*, a situation where the reaction proceeds from a bound intermediate over a bound transition state or from the free enzyme form over a free transition state; and (c) *oversaturation*, a condition where reactivity is determined by the rate of conversion from the lowest bound enzyme form over the highest free enzyme transition state.

Over these substrate concentration changes, kinetic measurements can be made to determine rate constants and other parameters that define the dynamic properties of segments along the reaction coordinate are shown in Fig. 2. The reaction catalyzed follows the scheme:

$$E_1 \overset{S}{\rightleftarrows} EX_1^{\ddagger} \rightleftarrows E_1S \rightleftarrows EX_2^{\ddagger} \rightleftarrows E_1P$$

$$\rightleftarrows EX_3^{\ddagger} \overset{P}{\rightleftarrows} E_2 \rightleftarrows EX_4^{\ddagger} \rightleftarrows E_1$$

From left to right, these forms are represented in Fig. 1 as E_1, 1, E_1S, 2, E_2P, 3, E_2, 4, and back to E_1. Note that different segments of the reaction cycle manifest

Figure 2. Free energy diagrams illustrating the kinetic regimes of undersaturation, saturation, and oversaturation. The states that determine the reaction rate observed under these conditions are marked as heavy bars. The changes for interversion of free enzyme forms E_1 and E_2 are shown by the dotted lines. Reproduced with permission of the authors and the American Chemical Society.

their kinetic properties under conditions of undersaturation, saturation and oversaturation. In the second paper, Fisher et al.[4] used the tracer perturbation technique under oversaturation conditions. This clever enzyme kinetic technique, introduced by Britton and co-workers[10,11], permits one to measure the equilibrium distribution of enzyme-bound substrate(s), intermediate(s), and product(s). Radiolabeled substrate or product is initially permitted to react with enzyme for sufficient time to equilibrate. At thermodynamic equilibrium, all the different enzyme-bound species will be present at concentrations reflecting their stability relative to each other. One can then add a large excess of unlabeled substrate. Any unbound *or* newly released radiolabel will mix with this large unlabeled pool of substrate or product, where it will undergo substantial reduction in radiospecific activity. This dilution effectively reduces or eliminates any detectible recycling of released radiolabel. One can then sample the system as a function of time, and the rates and amounts of labeled substrate and labeled product depend on the kinetics of synthesis and/or interconversion of bound radiolabel to substrate and product, which upon release is unlikely rebind. One can also evaluate the "peak-switch" concentration c_p, a parameter that reveals the point of substrate saturation at which the rate-determining step shifts from substrate-to-product interconversion and then depends on the $[E_2 \leftrightarrow EX_4^{\ddagger} \leftrightarrow E_1]$ conversion. Fisher et al.[5] then determined the rates, fractionation factors, and buffer catalysis in the oversaturated region to explore the interconversion of these two forms of free enzyme. Using equimolar substrate and product that was specifically deuterated at the 2-position for some or all of one enantiomer, competitive deuterium wash-out experiments were carried out after the reaction mixture reached chemical and isotopic equilibrium in the presence of enzyme. The authors indicate that their experimental results were most compatible with a stepwise reaction for the interconversion of the free enzyme forms, a process in which a proton is abstracted from a bound water molecule to give a reaction intermediate bearing a hydroxide ion bound to the diprotonated form of the enzyme. Other fractionation factors provided additional information on (1) the catalytic reaction's concertedness[6]; (2) the nature of essential catalytic groups[7]; and (3) the interconveresion of the two monoprotonated enzyme forms E_1 and E_2[8]. These efforts culminated in the construction of what is probably the most

detailed free energy profiles for an enzyme catalyzed reaction[9,12]. This profile is shown in Fig. 3, in which the transition state has an enzyme-proline carbanion sandwiched between the two catalytic thiols.

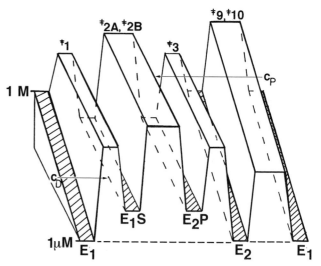

Figure 3. Computed three-dimensional free energy diagram for proline racemase: x-axis, the reaction coordinate; y-axis, free energy; and z-axis, substrate saturation. At the front of the diagram, [S] = [P] = 1 μM, where the enzyme is unsaturated; at the back of the diagram, [S] = [P] = 1 M, where the enzyme is oversaturated. Reproduced from reference 9 with permission of the authors and the American Chemical Society.

[1] J. R. Knowles & W. J. Albery (1977) *Acc. Chem. Res.* **85**, 2047.
[2] G. Rudnick & R. H. Abeles (1975) *Biochemistry* **14**, 4515.
[3] L. M. Fisher, W. J. Albery & J. R. Knowles (1986) *Biochemistry* **25**, 2529.
[4] L. M. Fisher, W. J. Albery & J. R. Knowles (1986) *Biochemistry* **25**, 2538.
[5] L. M. Fisher, J. G. Belasco, T. W. Bruice, W. J. Albery & J. R. Knowles (1986) *Biochemistry* **25**, 2543.
[6] J. G. Belasco, W. J. Albery & J. R. Knowles (1986) *Biochemistry* **25**, 2552.
[7] J. G. Belasco, T. W. Bruice, W. J. Albery & J. R. Knowles (1986) *Biochemistry* **25**, 2558.
[8] J. G. Belasco, T. W. Bruice, L. M. Fisher, W. J. Albery & J. R. Knowles (1986) *Biochemistry* **25**, 2564.
[9] W. J. Albery & J. R. Knowles (1986) *Biochemistry* **25**, 2572.
[10] H. G. Britton & J. B. Clarke (1972) *Biochem. J.* **110**, 161.
[11] H. G. Britton & J. B. Clarke (1972) *Biochem. J.* **130**, 397.
[12] W. J. Albery & J. R. Knowles (1986) *J. Theoret. Biol.* **124**, 173.

ENZYME EQUIVALENT WEIGHT

The weight (or more correctly, the mass) of a protein expressed in grams per mole of active sites. Not all oligomeric proteins, even some with identical subunits, have a number of active sites equal to the number of subunits. Enzyme normality (*i.e.*, the concentration of enzyme active sites) is typically determined by active site titration with an active-site-directed irreversible inhibitor. This is illustrated by the use of ^{32}P-labeled diisopropylfluorophosphate[1] to determine the enzyme equivalent weight of acetylcholinesterase. Alternatively, one can use so-called "burst" kinetics using stoichiometric quantities of enzyme and substrate.

[1]H. O. Michel & S. Krop (1951) *J. Biol. Chem.* **190**, 119.

ENZYME INHIBITION

In addition to the entries on specific types of reversible and irreversible inhibitors, one may also wish to consult the following selected entries from *Methods in Enzymology* [vol, page(s)].

Competitive, **249**, 123, 146, 190 [partial, **249**, 124; progress curve equations for, **249**, 176, 180; for three-substrate systems, **249**, 133, 136]; competitive-uncompetitive, **249**, 138; concave-up hyperbolic, **249**, 143; dead-end, **249**, 124 [for bireactant kinetic mechanism determination, **249**, 130-133; definition of kinetic constants, **249**, 220-221; effects on enzyme progress curves, nonlinear regression analysis, **249**, 71-72; inhibition constant evaluation, **249**, 134-135; kinetic analysis with, **249**, 123-143; one-substrate systems, **249**, 124-126; unireactant systems, theory, **249**, 124-130]; hyperbolic, **249**, 145 [identification, **249**, 162]; irreversible types, **249**, 240 (**See also** *Mechanism-Based Enzyme Inhibition*); kinetic models [with constant concentration of some components, **249**, 165-166; differential and conservation equation sets, **249**, 166-168; experimental data, processing, **249**, 168-169; introduction of appropriate kinetic mechanism, **249**, 163; introduction of simplifying assumptions, **249**, 163-166; parameters for, **249**, 166-170; progress curve equations, derivation, **249**, 168; with rapid equilibrium in some steps, **249**, 164-165; rate equations, derivation, **249**, 168; steady-state assumptions, **249**, 163-164; linear [identification, **249**, 162; uncompetitive, one-substrate systems, **249**, 128-129]; noncompetitive, **249**, 146, 190-191 [one-substrate systems, **249**, 126-128; progress curve equations for, **249**, 176, 180; pure, **249**, 146 (progress curve equations for, **249**, 175-178, 180); nonlinear, **249**, 162 [competitive, **249**, 138; noncompetitive, **249**, 138; one-substrate systems, **249**, 129-130]; parabolic, **249**, 138, 192-193 [one-substrate systems, **249**, 130]; parameters, **249**, 166 [determination, **249**, 160-170 (slow, tight-binding inhibition, 249, 159-160); reliability, **249**, 169-170]; product inhibition (**See** *Product Inhibition*); slow, tight-binding [definition, **249**, 149; inhibition parameters, determination, **249**, 159-160; isomerization and, kinetic models, **249**, 160; kinetic models, **249**, 158-160]; slow-binding [definition, **249**, 149; inhibition parameters, **249**, 170-172; isomerization and, kinetic models, **249**, 155-158; kinetic models, **249**, 155-158, 170-172; rapid equilibrium in some steps, **249**, 164-165; time dependence, **249**, 149]; tight-binding [characteristics, **249**, 148-149; consequences, **249**, 148; definition, **249**, 148-149; hyperbolic, **249**, 173-174; identification, **249**, 148-149, 162; inhibition parameters, **249**, 173-178; kinetic models, **249**, 149-153, 173-178; progress curve equations for, **249**, 174-178; reaction conditions, **249**, 148]; uncompetitive, **249**, 146, 190-192 [progress curve equations for, **249**, 176, 180].

ENZYME-LINKED IMMUNOSORBENT ASSAYS (ELISA)

Any immunoassay that is based on the binding of soluble enzyme-linked antibody to an surface-immobilized antigen, such that the amount of antibody-antigen complexation is quantitatively measured by the activity of the linked enzyme. The major advantage offered by ELISA methodologies is that there is no need for radioactivity, because the high catalytic power of the covalently linked enzyme gives a high signal-to-noise ratio in these measurements. Various strategies may be employed in immobilizing the antigen, but the technique requires that the binding of enzyme-linked antibody to antigen in no way reduces the enzyme's activity. In some competitive ELISA techniques, the enzyme-linked antibody is reacted with a soluble antigen that forms a complex and can no longer react with the surface-immobilized antigen.

Selected entries from *Methods in Enzymology* [vol, page(s)]: Detection and quantification, **74**, 608-616; effect of antigen/antibody ratio, **74**, 613, 614; applications, **74**, 614-615; controls, **74**, 612; data analysis, **74**, 612-613; immunoglobulin G [conjugation to glucose oxidase, **74**, 610; heat aggregation, **74**, 609-610; heat denaturation, **74**, 610; immobilization, **74**, 610]; precision, **74**, 612, 614; principle, **74**, 608-609; procedure, **74**, 611-612 [adsorption of endogenous rheumatoid factor, **74**, 611; enzyme activity assay, **74**, 612; enzyme-immunoglobulin conjugate binding, **74**, 611-612; sample competitive binding, **74**, 611]; reference serum, preparation, **74**, 610-611; comparative results, 74, 611, 614; sample storage, **74**, 611; sensitivity, **74**, 612; specificity, **74**, 613]; leukocyte interferon, **79**, 595; solid phase, **73**, 483, 523-550, 656, 666 [classification, **73**, 483-486; conjugate; (preparation, **73**, 147-166; purification, **73**, 163, 166-175; stability, **73**, 162, 163); enzyme selection, **73**, 148, 657-659; factors affecting, **73**, 158, 159, 506-511; kinetic analysis, **73**, 521, 522; magnetic method, **73**, 471-482, 663, 664 [for immunoglobulin E (enzyme conjugation, **73**, 475; immobilization of protein, **73**, 474, 475; reagents, **73**, 474; specific direct method, **73**, 477-479 indirect method, **73**, 479-481; total bridged avidin-biotin method, **73**, 481, 482 sandwich method, **73**, 477, 478; magnetic materials, **73**, 472-474)]; protein A as tracer, **73**, 186-191; conjugation procedure, **73**, 176-191; substrate choice, **73**, 657-659.

ENZYME MECHANISM

A systematic representation of the bond-breaking and bond-making chemical steps that explains: (a) the solvent and electrostatic factors affecting ligand binding and desorption steps; (b) the order of substrate binding and product release; (c) steric factors influencing an enzyme's chiral recognition and disposition of substrates, intermediates, products, and effectors; (d) the configurations of all atoms of the enzyme's active site and reactants during the entire reaction cycle; (e) the energetics and rates affecting the accumulation and interconversion of intermediates; (f) the structure and properties of the transition state; and (g) the activation or inhibition of enzyme catalysis by regulatory molecules.

In principle, no reaction mechanism can ever be completely understood until one achieves a complete quantum mechanical determination of the state-to-state differential scattering cross section. Enzyme catalysis involves far too many atoms for successful application of such a detailed quantum mechanical approach, even with the availability of supercomputers. This does not imply that quantum mechanics is without value in characterizing enzyme catalysis. Quantum mechanics allows one to comprehend aspects of chemical reactivity by considering the detailed properties of substrate(s), likely intermediates, and reaction products. One may also apply quantum mechanical to analyze the behavior of a limited number of atoms comprising a restricted region surrounding the reaction center. In any case, the inability to attain a complete quantum mechanical treatment in no way minimizes the quality of critical thinking required to conduct rigorous mechanistic studies on enzymes. In view of the celerity of enzyme processes, great ingenuity and experimental skill is required to investigate intermediates that accumulate at such low concentrations during the reaction cycle.

Selected entries from *Methods in Enzymology* [vol, page(s)]: Acid-base catalysis [with site-directed mutants, **249**, 110-118; altered pH dependencies, **249**, 110]; commitment to [in determination of intrinsic isotope effects, **249**, 343, 347-349; in interfacial catalysis, **249**, 598-599; equilibrium isotope exchange in, **249**, 443-479; hydrogen tunneling in, **249**, 373-397]; interfacial [competitive inhibitors, kinetic characterization, **249**, 604-605; equilibrium parameters, **249**, 587-594; forward commitment to, **249**, 598-599; interpretation, **249**, 578-586 (constraining variables for high processivity, **249**, 582-586; kinetic variables at interface, **249**, 581-582); kinetic basis for, **249**, 567-614; kinetic parameters, **249**, 594-599; uses, **249**, 599-605; Michaelis-Menten formalism, **249**, 570, 578, 581-582; in scooting mode, **249**, 581, 583, 585-586, 600-603, 613]; site-directed mutagenesis studies, **249**, 91-119 [active site essential residues, **249**, 91-93; catalysis evolution, **249**, 92; catalytic efficiency, **249**, 92, 104-107; role of specific interactions or residues, **249**, 91-96; transition state theory, **249**, 284-288].

ENZYME PURITY

The availability of completely pure enzyme affords the opportunity (a) to investigate mechanistic and regulatory properties in the absence of contaminating enzyme activities, and (b) to determine those kinetic parameters requiring an accurate knowledge of the enzyme concentration. Nevertheless, most kinetic studies do not require pure enzyme. What is required in that there be no (or, at least, minimal) contaminating activities that interfere with the initial-rate assay. A contaminating protein may act on the substrate(s), product(s), or effector(s). Hence,

each of the central reaction constituents should be examined separately[1]. One method for doing this is to incubate the substrate, product, or effector with the enzyme preparation (under the reaction conditions) in which a required constituent of the reaction (*e.g.*, one of the substrates in a multisubstrate reaction) is absent. The presence of the enzyme preparation should not result in a change in the concentration of a substrate, product, inhibitor, effector, or any other constituent (*e.g.*, buffer) in the absence of a necessary constituent for the reaction (at least, not in the time frame of the experiment).

In a number of cases there may be a contaminating enzyme present which acts on one or more of the substrates, products, or effectors of the system under study. It may be necessary to include in the reaction mixture a specific inhibitor for that contaminating activity. For example, adenylate kinase is often present in preparations of a number of phosphotransferases. It is often advantageous, in such instances, to include a specific inhibitor of adenylate kinase (*e.g.*, P^1,P^4-di(adenosine-5′)-tetraphosphate or P^1,P^5-di(adenosine-5′)-pentaphosphate).[2,3] If an inhibitor of the contaminating activity is added as an additional constituent of the reaction mixture, the investigator should demonstrate that the inhibitor is not an effector of the enzyme under study.

It should be recalled that, in some instances, the contaminating activity may be an intrinsic property of the enzyme. For example, sucrose phosphorylase exhibits a transglycosylation reaction[4], γ-glutamyl transpeptidase has a glutaminase activity[5], *E. coli* acetate kinase has a purine nucleoside-5′-phosphate kinase activity[6], and hexokinase has a weak ATPase activity[7]. In order for a system to be thoroughly characterized, each of these other activities should be studied as well. However, it may not be easily demonstrated that the same polypeptide (or polypeptide complex) is responsible for both the central reaction and the side reaction(s); or, that both activities have a common active site. Protocols to address this issue include: (1) inactivation studies demonstrating parallel losses of activities; (2) successive proteolysis studies in which activities are lost at the same or different steps; and (3) site-directed mutagenesis studies identifying both activities with the same protein and loss of both or separate activities with the modification of particular amino acid residues.

Isozymes are also a common presence in enzyme preparations and they can often be detected via polyacrylamide gel electrophoresis. The detected presence of isozymes may result in the need for further purification steps and the kinetic characterization of each isozyme. It may be necessary to use nondenaturing electrophoretic procedures to separate the different isozymes. *See Isozymes; Enzyme Concentration*

[1] R. D. Allison & D. L. Purich (1979) *Meth. Enzymol.* **63**, 3.

[2] D. L. Purich & H. J. Fromm (1972) *Biochim. Biophys. Acta* **276**, 563.

[3] G. E. Lienhard & I .I. Secemski (1973) *J. Biol. Chem.* **248**, 1121.

[4] J. J. Mieyal & R. H. Abeles (1972) *The Enzymes*, 3rd ed., **7**, 515.

[5] R. D. Allison (1985) *Meth. Enzymol.* **113**, 419.

[6] B. C. Webb, J. A. Todhunter & D. L. Purich (1976) *Arch. Biochem. Biophys.* **173**, 282.

[7] D. L. Purich, H. J. Fromm & F. B. Rudolph (1973) *Adv. in Enzymol.* **39**, 249.

Selected entries from *Methods in Enzymology* [vol, page(s)]:
Why Purify Enzymes?, **182**, 1; strategies and considerations for protein purifications, **182**, 9; rethinking your purification procedure, **182**, 779.

General Methods for Handling Proteins and Enzymes: Buffers,**104**, 404; **182**, 24; preparation of high-purity laboratory water, **104**, 391; measuring enzyme activity, **182**, 38; quantitating protein, **182**, 50; protein assay for dilute solutions, **104**, 415; concentrating proteins and removing solutes, **182**, 68; maintaining protein stability, **182**, 83; vectors, hosts, and strategies for overproducing proteins in *Escherichia coli* , **182**, 93; overexpressing proteins in eukaryotes, **182**, 112.

Extracts and Subcellular Fractionation: Preparing extracts from prokaryotes, **182**, 147; preparing yeast extracts, **182**, 154; preparing extracts from plants, **182**, 174; preparing extracts from higher eukaryotes, **182**, 194; isolating subcellular organelles, **182**, 203; preparing membrane fractions, **182**, 225.

Solubilization Procedures; Detergents: An overview, **182**, 239; solubilization of functional membrane proteins, **104**, 305; solubilization of native membrane proteins, **182**, 253; solubilization of protein aggregates,**182**, 264; removal of detergents from membrane proteins, **104**, 318; **182**, 277.

Purification Procedures: Precipitation techniques, **182**, 285; protein precipitation with polyethylene glycol, **104**, 351; **182**, 301; ion exchange methods, **182**, 309; gel filtration, **182**, 317; hydroxyapatite columns, **182**, 329; hydrophobic chromatography, **182**, 339; chromatography on immobilized reactive dyes, **182**, 343; affinity chromatography, **104**, 3; **182**, 357, 371; immobilization of affinity ligands with organic sulfonyl chlorides, **104**, 56; hydrophobic chromatography, **104**, 69; affinity chromatography on immobilized dyes, **104**, 97; displacement chromatography of proteins, **104**, 113; chromatofocusing, **182**, 380; HPLC, **182**, 392; high-performance size exclusion chromatography, **104**, 154; high-performance ion-exchange chromatography, **104**, 170; reversed-phase HPLC, **104**, 190; high-performance liquid affinity chromatography, **104**, 212; care of HPLC columns, **104**, 133; optimal pH conditions for ion exchangers on macroporous supports, **104**, 223.

Electrophoretic Methods: Systems for polyacrylamide gel electrophoresis, **104**, 237; preparative isoelectric focusing,**104**, 256; gel protein stains: silver stain, **104**, 441; gel protein stains: glycoproteins, **104**, 447; gel protein stains: phosphoproteins, **104**, 451; Western blots, **104**, 455; fluorography for the detection of radioactivity in gels, **104**, 460; affinity electrophoresis, **104**, 275; preparative isotachophoresis, **104**, 281; enzyme localization in gels, **104**, 416; one-dimensional gel electrophoresis, **182**, 425; two-dimensional polyacrylamide gel electrophoresis,**182**, 441; isoelectric focusing,**182**, 459; gel staining techniques, **182**, 477; elution of protein from gels, **182**, 488.

Specialized Purification Procedures: Purification of membrane proteins, **182**, 499; purification of integral membrane proteins, **104**, 329; reconstitution of membrane proteins, **104**, 340; purification of DNA-binding proteins by site-specific DNA affinity chromatography, **182**, 521; purification of glycoproteins, **182**, 529; purification of multienzyme complexes, **182**, 539.

Characterization of Purified Proteins: Assessing purity, **182**, 555; determining size, molecular weight, and presence of subunits, **182**, 566; amino acid analysis, **182**, 587; limited N-terminal sequence analysis, **182**, 602; peptide mapping, **182**, 613; analysis for protein modifications and nonprotein cofactors, **182**, 626; protein crystallization, **182**, 646.

Immunological Procedures: Immunosorbent separations, **104**, 381; preparing polyclonal antibodies, **182**, 663; preparing monoclonal antibodies,**182**, 670; protein blotting and immunodetection, **182**, 679; immunoprecipitating proteins, **182**, 688; immunoassays, **182**, 700.

Additional Techniques: Radiolabeling of proteins, **182**, 721; using purified protein to clone its gene, **182**, 738; computer analysis of protein structure, **182**, 751.

ENZYME RATE EQUATIONS
(1. The Basics)

The time course and kinetics of an enzyme-catalyzed reaction can be evaluated by a variety of procedures, but the initial rate method is by far most widely applied. The initial reaction phase refers to a period of time over which initial substrate concentration remains constant and for which the accumulated product has no measurable effect. This phase is characterized by a linear accumulation of product with respect to the elapsed time after mixing enzyme and substrate. The slope of a line drawn tangent to the initial progress curve has a slope of $\Delta[P]/\Delta t$ or $\{[P]_t - [P]_{initial}\}/\Delta t$, where $[P]$ is the product concentration (initial and at time t), and this represents the reaction rate. Note that this fits with the accepted convention of expressing the rate of a chemical reaction as the change in molarity of a substrate or product, normalized with respect to the stoichiometry of the reaction. A doubling of the reaction time during the initial rate phase should result in a doubling of product formed.

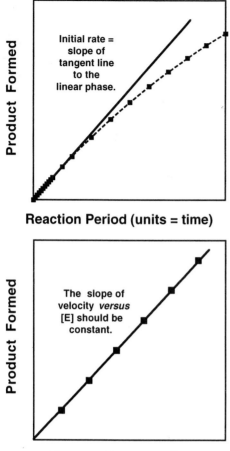

Reaction Period (units = time)

Initial rate = slope of tangent line to the linear phase.

Enzyme Concentration

The slope of velocity *versus* [E] should be constant.

The relationship between reaction velocity and enzyme concentration (in the absence of self-association of the enzyme) should also be adjusted such that reaction rate is linearly related to catalyst concentration, $[E_{total}]$. Initial rates typically fail to obtain if $[E_{total}] = 0.01 \, [A]_{initial}$ where $[A]_{initial}$ is the initial substrate concentration. As a general rule, the substrate concentration will not have changed more than 5–10% of its value over the initial rate phase of the reaction. This rule-of-thumb applies only to thermodynamically favorable reactions, and investigators are well advised to limit substrate consumption to well below 5%.

It was Henri who first proposed that enzyme catalysis depended on the formation of a transient complex of enzyme and substrate, followed by the breakdown (*i.e.*, chemical conversion) of bound substrate into product. Nonetheless, credit for derivation of the rate expression for the initial rate phase of one-substrate enzyme-catalyzed reactions is given to Michaelis and Menten. Both treatments gave the same general result:

$$v = V_{max}[S]/\{K_m + [S]\}$$

where v is the initial reaction velocity, V_{max} is the maximal reaction velocity, K_m is the Michaelis constant, and [S] is the initial substrate concentration.

DERIVATION OF THE MICHAELIS-MENTEN EQUATION. To obtain the Michaelis-Menten equation (that is, the rapid equilibrium expression) for a one-substrate enzyme-catalyzed reaction, one must make a series of assumptions: (a) that all experimental conditions (*e.g.*, pH, temperature, ionic strength) remain constant over the course of the experiment; (b) that the enzyme, substrate, and product are all stable under the assay conditions over the time course of the experiment; (c) that at least one binary central complex forms (*e.g.*, EX); (d) that E and S react rapidly and reversibly to form this complex; (e) that the binding of S to E can be described by an equilibrium dissociation constant, $K_s = k_2/k_1$, where k_1 is the forward bimolecular rate constant for S binding to E and k_2 is the unimolecular rate constant for the release of S from the central complex; (f) that the conversion of EX to E + P, having the unimolecular rate constant k_3, is slow and irreversible (or, if reversible, the low concentration of [P] over the time course of the experiment precludes the reverse reaction); (g) that the total enzyme concentration $[E_{tot}]$ is the sum of uncomplexed enzyme concentration, [E], and the enzyme-substrate central complex (*i.e.*, conservation of matter requires that $[E_{total}] = [E] + [EX]$); (h) that the velocity, v, of the reaction, equal to d[P]/dt, equals [EX] multiplied by the unimolecular rate constant k_3; (i) that the concentration of substrate is essentially unchanged during the initial velocity period (*i.e.*, $\Delta[S] \ll [S_{initial}]$); and (j) that the total substrate concentration, $[S_{total}]$, far exceeds the total enzyme concentration, $[E_{total}]$, such that $[S_{total}] = [S]$, where [S] is the uncomplexed substrate concentration.

We can then begin the derivation by considering two independent equations:

$$K_s = [E][S]/[EX]$$

where K_s is the dissociation constant for S and

$$[E_{total}] = [E] + [EX]$$

This first equation can be rewritten as $[E] = K_s [EX]/[S]$ and substituted for [E] in the second equation to yield

$$[E_{total}] = (K_s[EX]/[S]) + [EX]$$

Dividing both sides by [EX], we get

$$[E_{total}]/[EX] = (K_s/[S]) + 1$$

Then, by multiplying the left side by k_3/k_3 (note that $v = k_3[EX]$ and defining the maximum velocity, V_{max}, as being equal to $k_3[E_{total}]$), we get:

$$v = \frac{V_{max}[S]}{K_s + [S]}$$

The above expression represents the equation of a hyperbola (i.e., $f(x) = ax/(b + x)$) and a plot of the initial velocity as a function of [S] will result in a rectangular hyperbola. The limiting value of the reaction rate at saturating substrate concentration is termed V_{max}. The K_m is obtained by determining the substrate concentration yielding one-half maximal velocity. At low [S], the reaction rate is first-order with respect to substrate concentration, but at saturating substrate concentrations, the reaction is said to display zero-order kinetics. Another way for representing the Michaelis-Menten equation is by using the double-reciprocal (or Lineweaver-Burk) transformation:

$$1/v = (1/V_{max}) + (K_s/V_{max})(1/[S])$$

This method is widely used because it provides linear transformation of the hyperbolic function describing the rate saturation process. Double-reciprocal plots can be reasonably accurate if rate data can be obtained over a reasonable range of saturation, say from 0.3 V_{max} to 0.8 V_{max}.

STEADY STATE TREATMENT. While the Michaelis-Menten model requires the rapid equilibrium formation of ES complex prior to catalysis, there are many enzymes which do not exhibit such rate behavior. Accordingly, Briggs and Haldane considered the case where the enzyme and substrate obey the steady state assumption, which states that during the course of a reaction there will be a period over which the concentrations of various enzyme species will appear to be time-invariant (i.e., $d[EX]/dt \cong 0$). Such an assumption then provides that

$$d[EX]/dt = k_1[E][S] - k_2[EX] - k_3[EX] \cong 0$$

where we write a positive sign in front of all rates leading to EX and a minus sign in front of terms going away from EX. Transposing and collecting like terms gives

$$k_1[E][S] = k_2[EX] + k_3[EX] = (k_2 + k_3)[EX]$$

or

$$[E] = \{(k_2 + k_3)/k_1[S]\}[EX]$$

We can now substitute our new expression for [E] into the same conservation equation ($[E_{total}] = [E] + [EX]$) as used earlier to get

$$[E_{total}]/[EX] = 1 + (k_2 + k_3)/(k_1[S])$$

and, by multiplying each enzyme species by k_3, we arrive at

$$V_{max}/v = 1 + (k_2 + k_3)/(k_1[S])$$

(This means that we again use $k_3[EX]$ to represent v). The term $(k_2 + k_3)/k_1$ has the same units as the constant K_s obtained in the rapid equilibrium treatment. In honor of Leonor Michaelis, we use the designation K_m (or Michaelis constant) to represent this quotient, and the steady state rate law reduces to:

$$V_{max}/v = 1 + K_m/[S]$$

This result is identical in overall form to that obtained by Michaelis and Menten, and both equations may be used to deal with one-substrate enzymic processes. In fact, without detailed knowledge of the magnitudes k_1, k_2, and k_3, one would be unable to discern a difference between the two equations.

[The reader should also readily recognize that the same equation is obtained if we use the conservation equation along with the following differential equation:

$$d[E]/dt = k_2[EX] + k_3[EX] - k_1[E][S] \cong 0$$

This fact reminds us that in making the steady state assumption, we are treating both E and EX as time-invariant forms.]

THE MICHAELIS-MENTEN EQUATION AS A LIMITING CASE OF THE STEADY STATE EQUATION. To achieve a rapid equilibrium between E and EX, $k_1[S]$ and k_2 must each be much greater than k_3. [Note: the rate constant k_1 is a bimolecular rate constant with units of molarity^{-1} seconds^{-1}, and we must use $k_1[S]$

to compare its magnitude to either k_2 or k_3.] Thus, returning to K_m, we see that when $k_2 \gg k_3$,

$$K_m = (k_2 + k_3)/k_1 \cong k_2/k_1 = K_s$$

Moreover, the rapid equilibrium assumption used in obtaining the Michaelis-Menten equation requires that

$$\text{Rate}_{\text{Substrate Binding}} = k_1[\text{E}][\text{S}] \gg k_3[\text{EX}]$$
$$= \text{Rate}_{\text{Product Formation}}$$

EFFECT OF ADDITIONAL "CENTRAL COMPLEX" SPECIES ON THE GENERAL FORM OF THE STEADY STATE RATE EQUATION. Up to now, we have actually considered a chemically unrealistic model for enzyme catalysis in that we have assumed that a single enzyme-bound species, namely EX, accounts for the catalytic process. We now treat a more reasonable representation of the kinetic mechanism

$$\text{E} + \text{S} \underset{k_2}{\overset{k_1}{\rightleftharpoons}} \text{ES} \underset{k_4}{\overset{k_3}{\rightleftharpoons}} \text{EP} \underset{k_6}{\overset{k_5}{\rightleftharpoons}} \text{E} + \text{P}$$

by using an additional constant k_5 for the breakdown of EP, the enzyme·product complex, and the rate constants for the ES-to-EP interconversion. We will specify that $v = k_5[\text{EP}] - k_6[\text{P}]$. Again, if we take $[\text{P}] \cong 0$, then $v = k_5[\text{EP}]$ and the rate law can be derived by using the conservation equation (now containing EP as an additional species), along with two linear differential equations to solve for the three different enzyme forms:

$$[\text{E}_{\text{total}}] = [\text{E}] + [\text{ES}] + [\text{EP}]$$

$$d[\text{ES}]/dt = k_1[\text{E}][\text{S}] + k_4[\text{EP}] - k_2[\text{ES}] - k_3[\text{ES}] \cong 0$$

$$d[\text{EP}]/dt = k_3[\text{ES}] + k_6[\text{E}][\text{P}] - k_4[\text{EP}] - k_5[\text{EP}] \cong 0$$

This last differential equation is further simplified since we are considering the case in which $[\text{P}] \cong 0$: thus, $d[\text{EP}]/dt = k_3[\text{ES}] - (k_4 + k_5)[\text{EP}] \cong 0$. We need only to use the differential equations to obtain expressions for $[\text{E}]$ and $[\text{ES}]$ in terms of $[\text{EP}]$ (since the initial rate is directly proportional to $[\text{EP}]$: $v = k_5[\text{EP}]$) and then substitute into the first equation to obtain

$$[\text{E}_{\text{total}}] = [\text{E}] + [\text{ES}] + [\text{EP}]$$
$$= \left(\frac{k_2(k_4 + k_5) + k_3k_5}{k_1k_3[\text{S}]} + \frac{k_4 + k_5}{k_3} + 1 \right)[\text{EP}]$$

which is substituted into the denominator of $v = k_5[\text{EP}][\text{E}_{\text{total}}]/[\text{E}_{\text{total}}]$. The $[\text{EP}]$ terms thus cancel and the equation rearranges into

$$v = \frac{k_1k_3k_5[\text{S}][\text{E}_{\text{total}}]}{k_2(k_4 + k_5) + k_3k_5 + k_1(k_3 + k_4 + k_5)[\text{S}]}$$

By dividing the numerator and denominator by $k_1(k_3 + k_4 + k_5)$, the coefficient of the substrate concentration term in the denominator, this equation can be converted to the same general form as before:

$$v = V_{\text{max}}[\text{S}]/([\text{S}] + K_m)$$

where $V_{\text{max}} = k_3k_5[\text{E}_{\text{total}}]/(k_3 + k_4 + k_5)$ and $K_m = \{k_2(k_4 + k_5) + k_3k_5\}/\{k_1(k_3 + k_4 + k_5)\}$. In this case, if $k_3 \ll k_5$ and $k_4 \ll k_5$, the rate limiting step becomes the breakdown of ES in the forward direction, and the rate expression reduces to the steady state derivation for the result we obtained earlier with the single EX intermediate.

It is worth repeating that in the absence of knowledge concerning the magnitudes of individual rate constants, the two-intermediate case is indistinguishable from the single intermediate example derived previously. Thus, initial rate steady-state kinetics alone tells us nothing about the rate processes occurring after substrate binding and before product release.

DERIVATION OF MORE COMPLICATED RATE EQUATIONS. So far, the rate equations that describe one-substrate enzyme systems have been fairly simple, and the usual algebraic manipulations of substitution and/or addition of simultaneous equations have permitted us to obtain the pertinent rate law. When the number of steps increases and especially when there are branched pathways involved, these manual methods become cumbersome, and more systematic procedures are required. The next two sections should allow the reader to develop a working knowledge of effective methods for obtaining multisubstrate enzyme rate expressions.

ENZYME RATE EQUATIONS
(2. Derivation of Initial Velocity Equations*)

A rate equation for an enzymic reaction is a mathematical expression that depicts the process in terms of rate constants and reactant concentrations. It serves as a link between the experimentally observed kinetic behavior and a plausible model or mechanism. The characteristics of the rate equation permit tests to be designed to verify the mechanism. Conversely, the experimental observations provide clues to what the mechanism may be, hence, what form the rate expression shall take.

Derivation of rate equations is an integral part of the effective usage of kinetics as a tool. Novel mechanisms must be described by new equations, and familiar ones often need to be modified to account for minor deviations from the expected pattern. The mathematical manipulations involved in deriving initial velocity or isotope exchange-rate laws are in general quite straightforward, but can be tedious. It is the purpose of this entry, therefore, to present the currently available methods with emphasis on the more convenient ones.

The derivation of initial velocity equations invariably entails certain assumptions. In fact, these assumptions are often conditions that must be fulfilled for the equations to be valid. Initial velocity is defined as the reaction rate at the early phase of enzymic catalysis during which the formation of product is linear with respect to time. This linear phase is achieved when the enzyme and substrate intermediates reach a steady state or quasi-equilibrium. Other assumptions basic to the derivation of initial rate equations are as follows:

1. The enzyme and the substrate form a complex.

2. The substrate concentration is much greater than the enzyme concentration, so that the free substrate concentration is equivalent to the total concentration. This condition further requires that the amount of product formed is small, such that the reverse reaction or product inhibition is negligible.

3. During the reaction, constant pH, temperature, and ionic strength are maintained.

STEADY-STATE TREATMENT. During the steady state, the concentrations of various enzyme intermediates are essentially unchanged; that is, the rate of formation of a given intermediate is equal to its rate of disappearance. This assumption was first introduced to the derivation of enzyme kinetic equations by Briggs and Haldane[1].

To derive a rate equation, the first step is to write a reaction mechanism. The nomenclature used by Fromm[2] will be adopted here with the exception that rate constants in the forward and reverse directions will be denoted by positive and negative subscripts. For example, the simplest one substrate—one product reaction can be written as:

$$E + A \underset{k_{-1}}{\overset{k_1}{\rightleftharpoons}} EA \overset{k_2}{\longrightarrow} E + P \qquad (1)$$

Since both the k_{-1} and k_2 steps (or branches) lead from EA to E, the two branches, as has been shown by Volkenstein and Goldstein[3], can be combined into a single branch. This simplification procedure will be used whenever feasible.

$$E \underset{k_{-1}+k_2}{\overset{k_1 A}{\rightleftharpoons}} EA$$

The initial rate is given by

$$v = d[P]/dt = k_2[EA]$$

Applying the steady-state assumption, we have

$$d[EA]/dt = k_1[A][E] - (k_{-1} + k_2)[EA] = 0 \qquad (2)$$

To obtain an expression for [EA], the enzyme conservation equation

$$\text{Total enzyme} = [E_o] = [E] + [EA] \qquad (3)$$

is required. Substitution of $[E] = ([E_o] - [EA])$ into Eq. (2) yields

$$[ES] = \frac{[E_o][A]}{\{(k_{-1} + k_2)/k_1\} + [A]}$$

and

$$v = k_2[EA] = \frac{k_2[E_o][A]}{\{(k_{-1} + k_2)/k_1\} + [A]}$$
$$= \frac{V_1[A]}{K_m + [A]} \qquad (4)$$

where V_1 is the maximum velocity in the forward direction and K_m is the Michaelis constant.

It should be noted that the validity of the steady-state method does not depend on the assumption $d[EA]/dt = 0$. Without setting Eq. (2) equal to zero, one can obtain the following expression from Eqs. (2) and (3):

$$[EA] = \frac{k_1[A][E_o] - d[EA]/dt}{k_1[A] + k_{-1} + k_2}$$

Wong[4] has pointed out that the steady-state approximation only requires that $d[EA]/dt$ be small compared with $k_1[A][E_o]$. In the early phase of the reaction, if $[A] \gg [E_o]$, the rate of change of $[EA]$ due to diminishing $[A]$ will be relatively slow. It is clear that the validity of steady state is intimately tied to the condition of high substrate to enzyme ratio.

The Determinant Method. For a mechanism involving several enzyme-containing species, derivation of the rate equation can be done by solving the simultaneous algebraic equations by the determinant method. Consider the mechanism described by Eq. (1) with the addition of an EP intermediate.

$$E \underset{k_{-1}}{\overset{k_1[A]}{\rightleftharpoons}} EA \underset{k_{-2}}{\overset{k_2}{\rightleftharpoons}} EP \overset{k_3}{\longrightarrow} A + P \quad (5)$$

The three simultaneous equations are given in the following form:

$$
\begin{array}{cccc}
& E & EA & EP \\
d[E]/dt = & \begin{vmatrix} -k_1[A] & k_{-1} & k_3 \\ k_1[A] & -(k_{-1}+k_2) & k_{-2} \\ 0 & k_2 & -(k_{-2}+k_3) \end{vmatrix} & & = 0 \\
d[EA]/dt = & & & = 0 \\
d[EP]/dt = & & & = 0
\end{array}
$$

The determinant, or distribution term for [E], for example, can be calculated from the coefficients listed above, after deleting the E column. For a mechanism of n intermediates, only $n - 1$ equations are needed. Thus, by leaving out the $d[EP]/dt$ row, we can write

$$[E] = \begin{vmatrix} k_{-1} & k_3 \\ -(k_{-1}+k_2) & k_{-2} \end{vmatrix} = k_{-1}k_{-2} + k_3(k_{-1}+k_2)$$

If the $d[E]/dt$ row is omitted instead, we have

$$[E] = \begin{vmatrix} -(k_{-1}+k_2) & k_{-2} \\ k_2 & -(k_{-2}+k_3) \end{vmatrix}$$
$$= k_{-1}(k_{-2}+k_3) + k_2(k_{-2}+k_3) - k_2 k_{-2}$$
$$= k_{-1}k_{-2} + k_3(k_{-1}+k_2)$$

Note that deletion of different equations often leads to different amounts of algebraic manipulations. Application of the same operations to [EA] and [EP] yields

$$[EA] = k_1(k_{-2}+k_3)[A]$$

$$[EP] = k_1 k_2[A]$$

The rate equation is readily obtained as

$$\frac{v}{[E_o]} = \frac{k_3[EP]}{[E] + [EA] + [EP]}$$
$$= \frac{k_1 k_2 k_3[A]}{k_{-1}k_{-2} + k_3(k_{-1}+k_2) + k_1(k_{-2}+k_3)[A] + k_1 k_2[A]}$$

or

$$v = \frac{k_2 k_3[E_o][A]/(k_2 + k_{-2} + k_3)}{\{[k_{-1}k_{-2} + k_3(k_{-1}+k_2)]/[k_1(k_2 + k_{-2} + k_3)]\} + [A]}$$
$$= \frac{V_1[A]}{K_m + [A]} \quad (6)$$

Equation (6) is identical in form with Eq. (4). In fact, if $k_3 \gg k_2, k_{-2}$, Eq. (6) reduces to Eq. (4). Although Eq. (5) is a more realistic mechanism compared with Eq. (1), especially when the rapid-equilibrium treatment is applied to the reversible reaction, the information obtainable from initial-rate studies of such unireactant system remains nevertheless the same: V_1 and K_m. This serves to justify the simplification used by the kineticist; that is, the elimination of certain intermediates to maintain brevity of the rate equation (provided the mathematical form is unaltered). Thus, the *forward* reaction of an ordered Bi Bi mechanism is generally written as diagrammed below.

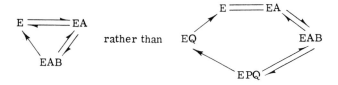

This use of the determinant method for complex enzyme mechanisms is time-consuming because of the stepwise expansion and the large number of positive and negative terms that must be canceled. It is quite useful, however, in computer-assisted derivation of rate equations[5].

The King and Altman Method. King and Altman[6] developed a systematic approach for deriving steady-state rate equations, which has contributed to the advance of enzyme kinetics. The first step of this method is to draw an *enclosed* geometric figure with each enzyme form as one of the corners. Equation (5), for instance, can be rewritten as:

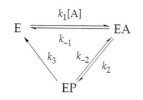

The second step is to draw all the possible patterns that connect all the enzyme species without forming a loop. For a mechanism with n enzyme species, or a figure with n corners, each pattern should contain $n - 1$ lines. The number of valid patterns for any single-loop mechanism is equal to the number of enzyme forms. Thus, there are three patterns for the triangle shown above:

The determinant for a given enzyme species is obtained as the summation of the product of the rate constants and concentration factors associated with all the branches in the patterns *leading toward* this particular enzyme species. The same patterns are used for each species, albeit the direction in which they are read will vary. However, when an irreversible step is present, *e.g.*, the $EP \rightarrow E$ step, some patterns become invalid for certain enzyme forms.

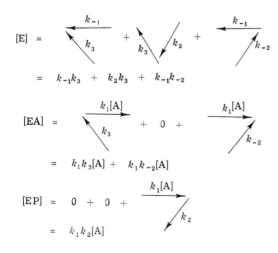

The rate equation is obtained as

$$\frac{v}{[E_o]} = \frac{k_3[EP]}{[E] + [EA] + [EP]}$$

where [E], [EA], and [EP] are the determinants for E, EA, and EP, respectively.

The presence of an enzyme intermediate(s) that is not part of a loop will not affect the number of King-Altman patterns. For instance, the addition of a competitive inhibitor, I, to the above system will result in the same number of patterns.

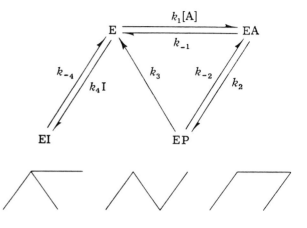

The additional $E \rightleftharpoons EI$ branch is present in *all* the diagrams. Thus, in calculating the number of valid King-Altman patterns, only the *closed loops* need be considered. The determinants of E, EA, EAB, and EI can be obtained by the method just described:

$$[E] = k_{-4}(k_{-1}k_3 + k_2k_3 + k_{-1}k_{-2})$$

$$[EA] = k_{-4}(k_1k_3[A] + k_1k_{-2}[A])$$

$$[EP] = k_{-4}(k_1k_2[A])$$

$$[EI] = k_4[I](k_{-1}k_3 + k_2k_3 + k_{-1}k_{-2})$$

It is more convenient, however, to treat this case by first considering only the loop portion (ignoring the additional $E \rightleftharpoons EI$ step for the time being).

$$[E] = k_{-1}k_3 + k_2k_3 + k_{-1}k_{-2}$$

$$[EA] = k_1(k_{-2} + k_3)[A]$$

$$[EP] = k_1k_2[A]$$

The determinant for EI is then obtained as

$$[EI] = [E]k_4[I]/k_{-4}$$
$$= (k_{-1}k_3 + k_2k_3 + k_{-1}k_{-2})k_4[I]/k_{-4}$$
$$= (k_{-1}k_3 + k_2k_3 + k_{-1}k_{-2})[I]/K_i$$

where $K_i = k_{-4}/k_4$.

The King-Altman method is most convenient for single-loop mechanisms. In practice, there is no need to write down the patterns. One can use an object; say, a paper clip, to block one branch of the loop, write down the appropriate term for each enzyme species, then repeat the process until every branch in the loop has been blocked once.

For more complex mechanisms having alternate pathways that form several closed loops, the precise number of valid King-Altman patterns must be calculated to avoid omission of terms. To illustrate the various situations that may occur, let us consider Scheme 1.

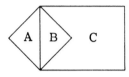

Scheme 1

The total number of patterns with $n - 1$ lines is given by

$$\frac{m!}{(n - 1)!(m - n + 1)!}$$

where m = the number of lines in the complete geometric figure. In the above scheme, $m = 8$ and $n = 6$, and the total number of patterns with 5 lines is

$$\frac{8!}{5!\,3!} = \frac{(8 \times 7 \times 6 \times 5 \times 4 \times 3 \times 2 \times 1)}{(5 \times 4 \times 3 \times 2 \times 1)(3 \times 2 \times 1)} = 56$$

This number, however, includes patterns that contain the following loops, which must be subtracted from the total:

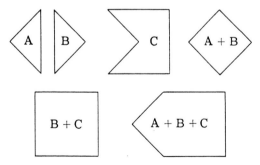

The number of patterns for a loop with r lines is given by

$$\frac{(m - r)!}{(n - 1 - r)!\,(m - n + 1)!}$$

According to this equation, for loops A and B, $r = 3$, we have 10 patterns each; for loops A + B and B + C,

$r = 4$, we have 4 patterns each; and for loops C and A + B + C, $r = 5$ (note that $0! = 1$), we have 1 pattern each. The total number of loop-containing patterns to be subtracted is 30. One of the patterns, however, occurred three times in the above calculations, but should be discarded only once. This pattern involves both loop A and loop B (solid lines indicate the loop that gives rise to this pattern).

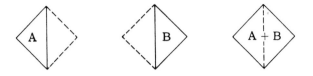

Thus, the total loop-containing patterns is 28, and the total number of valid patterns is $56 - 28 = 28$.

The 28 5-lined patterns are shown below.

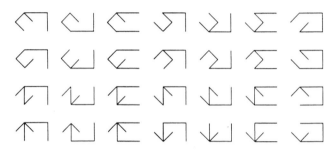

The conventional way of computing the valid King-Altman patterns is rather tedious. A set of formulas developed by the author allows the calculation of the desired number in a very short time. Each of these formulas is applicable to a particular geometric arrangement. For any figure consisting of three subfigures arrayed in sequence like the one shown in Scheme 1, the general formula for calculating the exact number of the valid King-Altman pattern, π, is

$$\pi = a \cdot b \cdot c - (l_{AB}^{2} \cdot c + l_{BC}^{2} \cdot a)$$

where a, b, and c = the number of lines in subfigures A, B, and C; l_{AB} and l_{BC} = the number of lines in the common boundaries between A and B, and B and C, respectively. For $a = 3$, $b = 3$, $c = 5$, $l_{AB} = 1$, and $l_{BC} = 2$ (Scheme 1), we have

$$\begin{aligned} \pi &= 3 \times 3 \times 5 - (1^{2} \times 5 + 2^{2} \times 3) \\ &= 45 - (5 + 12) \\ &= 28 \end{aligned}$$

In the case of two subfigures A and B sharing a common boundary as shown in Scheme 2

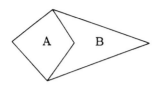

Scheme 2

the formula is given by

$$\pi = a \cdot b - l_{AB}^2$$
$$= 4 \times 4 - 2^2$$
$$= 12$$

Formulas for calculation of up to four subfigures in every possible geometric arrangement have been established.

The Method of Volkenstein and Goldstein. Volkenstein and Goldstein[3] have applied the theory of graphs to the derivation of rate equations. Their approach has three main features: the use of an auxiliary "node," the "compression" of a path into a point, and the addition of parallel branches. These can be best explained by an example (Scheme 3).

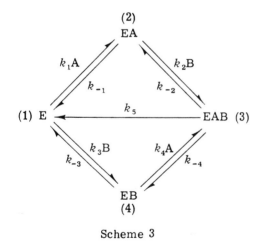

Scheme 3

Each enzyme-containing species is assigned a number and referred to as a node.

Suppose we want to calculate the determinant for EA (node 2). First, we choose another node, say node 3, as the auxiliary node (a reference starting point). The choice of the auxiliary node is arbitrary; it will not affect the outcome of the derivation, but may affect the amount of work involved.

All the possible pathways (flow patterns) leading from (3) to (2) are then written (marked by solid branches).

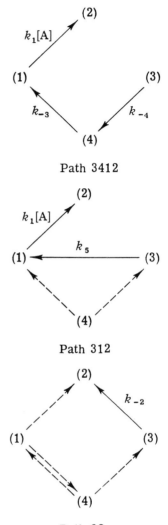

The nodes not included in the pathways retain the branches leading *away* from them (dashed branches). Since path 3412 flows through all the nodes, it is one of the terms of the determinant with a path value of $k_1 k_{-3} k_{-4}[A]$. Path 312 ($= k_1 k_5[A]$) and path 32 ($= k_{-2}$) are now compressed into points.

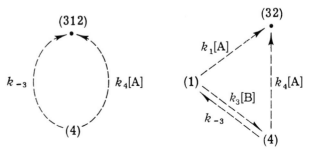

The parallel branches leading from (4) to the compressed point (312) can be added together to yield

(312)

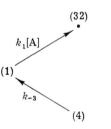

$(k_{-3} + k_4[A])$

(4)

and the expression for this part is $(P312)(k_{-3} + k_4[A]) = k_1k_5[A](k_{-3} + k_4[A])$. The part containing point (32) can be treated by selecting a secondary auxiliary node, say node (4), and repeating the procedure described at the onset.

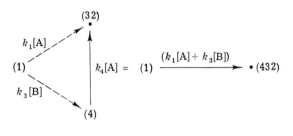

and

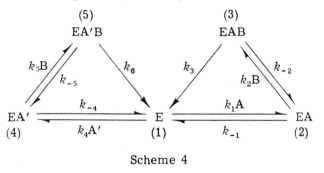

The contribution from this part is $(P432)(k_1[A] + k_3[B]) = k_{-2}k_4[A](k_1[A] + k_3[B])$ and $(P32)(k_1k_{-3}[A]) = k_1k_{-2}k_{-3}[A]$. Adding the terms together, we obtain the determinant for EA

$$[EA] = k_1k_{-3}k_{-4}[A] + k_1k_5[A](k_{-3} + k_4[A])$$
$$+ k_{-2}k_4[A](k_1[A] + k_3[B]) + k_1k_{-2}k_{-3}[A]$$
$$= k_1k_{-3}(k_{-2} + k_{-4} + k_5)[A] + k_{-2}k_3k_4[A][B]$$
$$+ k_1k_4(k_{-2} + k_5)[A]^2$$

The determinants for E, EB, and EAB can be obtained in a similar fashion. The complete rate equation is given by

$$\frac{v}{[E_o]} = \frac{k_5[EAB]}{[E] + [EA] + [EB] + [EAB]}$$

Rate equations for more complex mechanisms can be derived by repeating the procedure described above as many times as necessary. The choice of the auxiliary

point becomes important for reaction schemes containing several loops. The process is analogous to deciding which row (equation) should be omitted from the matrix in the determinant method. In general, one should choose, by inspection of the geometric structure of the mechanism, a node such that, if one removes from the figure the auxiliary node and the node whose determinant is desired, the remaining nodes do not form a closed loop. In addition, one should select a node situated in a symmetrical position with respect to the desired node. For instance, node (4) is a better choice as an auxiliary node for the calculation of the determinant for node (2). Node (3) was chosen for the sole purpose of illustrating the use of secondary auxiliary nodes.

The Systematic Approach. The systematic approach for deriving rate equations was first devised by Fromm[7] based on certain concepts advanced by Volkenstein and Goldstein[3]. Its underlying principles, however, are more akin to the graphic method of King and Altman[6]. The procedure to be described here[8] is a modified method that includes the contributions from the aforementioned workers and from Wong and Hanes[9].

Let us use as an example the ordered Bi Bi mechanism, in which an alternative substrate, A', for the first substrate, A, is present (Scheme 4).

Scheme 4

Each enzyme-containing species is assigned a node number as previously described in the Volkenstein and Goldstein method. For each node, a node value is written, which is simply the summation of all branch values (rate constant and concentration factor) leading *away* from the node[7]:

$$(1) = k_1[A] + k_4[A']$$
$$(2) = k_{-1} + k_2[B]$$
$$(3) = k_{-2} + k_3$$
$$(4) = k_{-4} + k_5[B]$$
$$(5) = k_{-5} + k_6$$

The determinant of a given enzyme species is equal to the noncyclic terms generated by multiplying together all the node values, excluding its own[9]. For example,

$$[E] = (2)(3)(4)(5)$$
$$= (k_{-1} + \underline{k_2[B]})(\underline{k_{-2}} + k_3)(k_{-4} + \underline{k_5[B]})(\underline{k_{-5}} + k_6)$$

and

$$[EAB] = (1)(2)(4)(5)$$
$$= (k_1[A] + \underline{k_4[A']})(k_{-1} + k_2[B])$$
$$(k_{-4} + \underline{k_5[B]})(\underline{k_{-5}} + k_6)$$

The cyclic terms—terms that contain the products of reversible steps (underlined: $k_1[A] \cdot k_{-1}$, $k_2[B] \cdot k_{-2}$, $k_4[A'] \cdot k_{-4}$, and $k_5[B] \cdot k_{-5}$; readily identified by their subscripts differing only in positive and negative signs)—or the product of a closed loop [underscored by dashed lines; e.g., $k_4[A'] \cdot k_5[B] \cdot k_6$, constituting the (1) (4) (5) loop] are to be deleted to obtain the correct expression. The "reversible-step" terms can be eliminated during expansion:

$$[E] = (k_{-1} + \underline{k_2[B]})(\underline{k_{-2}} + k_3)(k_{-4} + \underline{k_5[B]})(\underline{k_{-5}} + k_6)$$

$$= [(k_{-1} + \underline{k_2[B]}) \frac{\underline{k_{-2}}}{X}$$

$$+ (k_{-1} + k_2[B])k_3][k_{-4} + \underline{k_5[B]}) \frac{\underline{k_{-5}}}{X}$$

$$+ (k_{-4} + k_5[B])k_6]$$

$$= k_{-1}k_{-4}(k_{-2} + k_3)(k_{-5} + k_6)$$

$$+ [k_2k_3k_{-4}(k_{-5} + k_6) + k_{-1}k_5k_6(k_{-2} + k_3)][B]$$

$$+ k_2k_3k_5k_6[B]^2$$

Note that the canceled terms are marked by an X. Also note that $k_2[B]$ and $k_5[B]$, instead of k_{-2} and k_{-5}, are crossed out because the products $k_{-1}k_{-2}$ and $k_{-4}k_{-5}$ are not "cyclic."

The "loop terms" can be eliminated prior to expansion by a "branching" technique. From the expression

$$[EAB] = (1) (2) (4) (5)$$

a quick glance at Scheme 4 will reveal that (1) (4) (5) form a closed loop. Thus, the "one-branch" approach of

Fromm[7] is first applied. With this approach, the determinant of an enzyme species, e.g., EAB, is obtained as the summation of the products of the nearest branch values leading to it (for EAB, there is only one nearest branch, EA (R) EAB or 23) and the remaining node values

$$[EAB] = (23) (1) (4) (5)$$

Similarly

$$[EA] = (12) (3) (4) (5) + (32) (1) (4) (5)$$

Note that the (1) (4) (5) loop is not eliminated by the one-branch approach. We now apply the "consecutive-branch" technique[8]

$$[EAB] = (23) (12) (4) (5)$$
$$[EA] = (12) (3) (4) (5) + (32)X$$

The procedure for using this technique is as follows: (a) Only the *loop-containing* terms require further branching; e.g., since (3) (4) (5) do not form a loop, the term (12) (3) (4) (5) remains unchanged. *Unnecessary consecutive-branching may result in omission of terms.* (b) For the loop-containing terms, the first branch(es) is followed by its nearest branches and its remaining nodes not involved in these branches; e.g., the 23 branch is followed by the 12 branch. This is done by inspection of the reaction scheme. (c) In the case of (32) (1) (4) (5), since there is no branch leading from other nodes to (3), the whole term is deleted (marked by X).

When all the loops have been removed, the resultant expression is expanded to obtain the desired determinant. Thus,

$$[EAB] = k_2[B] \cdot k_1[A](k_{-4} + \underline{k_5[B]})(k_{-5} + k_6)$$
$$= k_1k_2k_{-4}(k_{-5} + k_6)[A][B]$$
$$+ k_1k_2k_5k_6[A][B]^2$$

$$[EA] = k_1[A] (k_{-2} + k_3)(k_{-4} + \underline{k_5[B]})(k_{-5} + k_6)$$

$$= k_1k_{-4}(k_{-2} + k_3)(k_{-5} + k_6)[A] + k_1k_5k_6$$

$$(k_{-2} + k_3)[A][B]$$

Note that the product of reversible steps, $k_5[B] \cdot k_{-5}$, is canceled in the expanding process, as has been previously described.

The determinant of EA′ and EA′B can be obtained by the same approach, and the complete rate equation is expressed as $v/[E_o]$. For the example given here, if the common product P is measured, we have

$$\frac{v}{[E_o]} = \frac{k_3[EAB] + k_6[EA'B]}{[E] + [EA] + [EAB] + [EA'] + [EA'B]}$$

The first rule for using the systematic approach, broadly stated, is as follows:

Rule 1: The determinant of a given enzyme-containing species is equal to the product of the node values of the other enzyme species, minus the reversible-step terms. When the nodes form one or more closed loops, apply the one-branch approach. Apply the consecutive-branch approach to any *remaining* loop-containing terms until all loops are eliminated.

It should be noted that unnecessary application of the one-branch approach may lead to needless algebraic manipulations. Furthermore, the branching technique often results in "redundant terms" that must be searched out and deleted. As an example, let us write the determinant for E in Scheme 4 by the one-branch method:

$$[E] = (21)(3)(4)(5) + (31)(2)(4)(5) + (41)(2)(3)(5)$$
$$+ (51)(2)(3)(4)$$
$$= k_{-1}(k_{-2} + k_3)(k_{-4} + k_5[B])(k_{-5} + k_6)$$
$$+ k_3(k_{-1} + k_2[B])(k_{-4} + k_5[B])(k_{-5} + k_6)$$
$$+ k_{-4}(k_{-1} + k_2[B])(k_{-2} + k_3)(k_{-5} + k_6)$$
$$+ k_6(k_{-1} + k_2[B])(k_{-2} + k_3)(k_{-4} + k_5[B])$$

Expansion of the above equation will generate many redundant terms; *e.g.*, $k_{-1}k_3k_{-4}k_6$ can be found in all four terms above, but only one is needed. These redundant terms can be eliminated by using Rule 2.

Rule 2: The nearest branch values cannot appear in *subsequent* node values. When the consecutive-branch approach is used, the *product* of the consecutive branches cannot appear in *subsequent* terms.

Consequently, the 21 term, k_{-1}, is crossed out from all subsequent terms; the 31 term. k_3, from all subsequent node (3) terms, but not from the node (3) term preceding it; and the 41 term, k_{-4}, from the subsequent (4) term:

$$[E] = k_{-1}(k_{-2} + k_3)(k_{-4} + k_5[B])(k_{-5} + k_6)$$
$$+ k_3(\underset{X}{k_{-1}} + k_2[B])(k_{-4} + k_5[B])(k_{-5} + k_6)$$
$$+ k_{-4}(\underset{X}{k_{-1}} + \underset{X}{k_2[B]})(\underset{X}{k_{-2} + k_3})(k_{-5} + k_6)$$
$$+ k_6(\underset{X}{k_{-1}} + \underset{X}{k_2[B]})(\underset{X}{k_{-2} + k_3})(\underset{X X}{k_{-4} + k_5[B]}) \Big\} \text{Canceled}$$

After elimination of the redundant terms, the last two terms are canceled because they all contain the $k_2[B] \cdot k_{-2}$ reversible-step product. The remaining terms are identical with the expression obtained from $[E] = (2)(3)(4)(5)$, demonstrating the fact that unnecessary branching may lead to wasteful algebraic exercise.

There are situations where certain redundant terms are difficult to eliminate. The following example serves to illustrate a procedure useful for the complicated cases. Consider the hypothetical mechanism shown in Scheme 5.

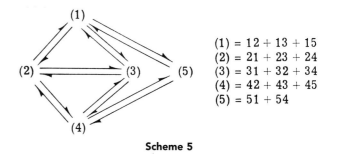

(1) = 12 + 13 + 15
(2) = 21 + 23 + 24
(3) = 31 + 32 + 34
(4) = 42 + 43 + 45
(5) = 51 + 54

Scheme 5

The numerical notation used for Scheme 5 has several advantages. (a) The tedium of writing down all the rate contants and concentration factors is avoided. The constants can be substituted into the final expression (after canceling all the loop terms, redundant terms, and reversible-step terms) to obtain the desired determinant. (b) When two or more steps are assigned the same rate constant, *e.g.*, k_1 and k_{-1}, this method prevents the cancelation of $k_1 \cdot k_{-1}$ product formed from different steps. (c) The reversible-step terms can still be easily identified and eliminated by their numbers, *e.g.*, $12 \cdot 21$, $35 \cdot 53$, *etc.*

Suppose we want to calculate the determinant for (5), we can write $D_5 = (1)(2)(3)(4)$. Applying the consecutive-branch technique, we have

$$D_5 = (15)\,(2)\,(3)\,(4) + (45)\,(1)\,(2)\,(3)$$
$$= 15\,[21\,(3)\,(4) + 31\,(2)\,(4)] + 45\,[24\,(1)\,(3)$$
$$+\ 34\,(1)\,(2)]$$
$$= 15\,[21\,(31 + 32 + 34)(42 + 43 + 45)$$
$$+\ 31\,(\underline{21} + 23 + 24)(42 + 43 + 45)]$$
$$+\ 45\,[24\,(12 + 13 + \underline{15})(\underline{31} + 32 + 34)$$
$$+\ 34\,(12 + 13 + \underline{15})(\underline{21} + 23 + \underline{24})]$$

Note that within the same brackets, the redundant terms (underlined by solid lines) are readily removed, but the products of consecutive branch terms, $15 \cdot 21$ and $15 \cdot 31$ (underlined by dashed lines) within the second set of brackets cannot be canceled without further manipulation. One can either eliminate them during further expansion of the equation or repeat the consecutive-branch approach until *all* the consecutive branches end with a branch leading away from a *common* node. This "common-node consecutive branching" approach is always valid—even when all the loop-containing terms have been eliminated. Using this approach, we can write

$$D_5 = 15\,[21\,(3)\,(4) + 31\,(2)\,(4)] + 45\,[24\,(1)\,(3)$$
$$+\ 34\,(1)\,(2)]$$
$$= 15\,\{21\,(3)\,(4) + 31\,[23\,(4) + 43 \cdot 24]\}$$
$$+\ 45\,\{24\,(1)\,(3) + 34\,[23\,(1) + 13 \cdot 21]\}$$

In the above operation, we chose node (2) as the common node; we could have chosen node (3) and obtained the same determinant. The principle involved here is analogous to the "auxiliary node" used by Volkenstein and Goldstein. It is not routinely used because the extra operations are not needed under most circumstances.

Comparison of Different Steady-State Methods. For relatively simple mechanisms, all the diagrammatic and systematic procedures illustrated in the foregoing sections are quite convenient. The King-Altman method is best suited for single-loop mechanisms, but becomes laborious for more complex cases with five or more enzyme forms because of the work involved in the calculation and drawing of valid patterns. With multiloop reaction schemes involving four to five enzyme species, the systematic approach requires the least effort, especially

when irreversible steps are present, since it does away with pattern drawing. When the number of enzyme forms reaches six or more in a mechanism with several alternate pathways, all the manual methods become tedious owing to the sheer number of terms involved.

THE RAPID-EQUILIBRIUM TREATMENT. The first rate equation for an enzyme-catalyzed reaction was derived by Henri and by Michaelis and Menten, based on the rapid-equilibrium concept. With this treatment it is assumed that there is a slow catalytic conversion step and the combination and dissociation of enzyme and substrate are relatively fast, such that they reach a state of quasi-equilibrium or rapid equilibrium.

The derivation of initial-velocity equations for any rapid equilibrium system is quite simple. When the equilibrium relationships among various enzyme-substrate complexes are defined, the rate equation can be written simply by inspection. Consider the one-substrate system

$$E + A \underset{K_a}{\rightleftharpoons} EA \xrightarrow{k_2}$$

$$[EA] = \{[E][A]\}/K_a, \quad v = k_2[EA]$$

We can write

$$\frac{v}{[E_o]} = \frac{k_2[EA]}{[E] + [EA]} = \frac{k_2[E][A]/K_a}{[E] + [E][A]/K_a}$$

$$= \frac{k_2([A]/K_a)}{1 + [A]/K_a} = \frac{k_2[A]}{(K_a + [A])}$$

It is clear that there is no need to write down [E], and one can obtain the rate expression by replacing [E] with 1. Thus, for the Random Bi Bi mechanism shown below

$$E + A \underset{K_{ia}}{\rightleftharpoons} EA \qquad EA + B \underset{K_b}{\rightleftharpoons} EAB \xrightarrow{k_5} E + P$$
$$E + B \underset{K_{ib}}{\rightleftharpoons} EB \qquad EB + A \underset{K_a}{\rightleftharpoons} EAB \xrightarrow{k_5} E + P$$

we can quickly write

$$\frac{v}{[E_o]} = \frac{k_5[A][B]/(K_{ia}K_b)}{1 + [A]/K_{ia} + [B]/K_{ib} + [A][B]/K_{ia}K_b}$$

Note that although two equations describe the formation of EAB, only one of them is used (because $K_{ia}K_b = K_{ib}K_a$).

In using the equilibrium treatment, one should bear in mind that the rate laws so obtained are generally different in form from those derived by the steady-state assumption for the same mechanism.

THE COMBINED EQUILIBRIUM AND STEADY-STATE TREATMENT. There are a number of reasons why a rate equation should be derived by the combined equilibrium and steady-state approach. First, the experimentally observed kinetic patterns necessitate such a treatment. For example, several enzymic reactions have been proposed to proceed by the rapid-equilibrium random mechanism in one direction, but by the ordered pathway in the other. Second, steady-state treatment of complex mechanisms often results in equations that contain many higher-order terms. It is at times necessary to simplify the equation to bring it down to a manageable size and to reveal the basic kinetic properties of the mechanism.

The procedure to be described here was originally developed by Cha[10]. The basic principle of his approach is to treat the rapid-equilibrium segment as though it were a single enzyme species at steady state with the other species. Let us consider the hybrid Rapid-Equilibrium Random-Ordered Bi Bi system:

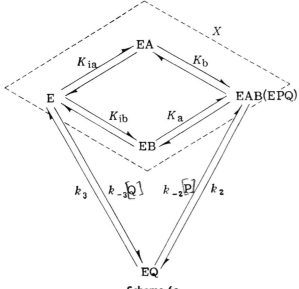

Scheme 6a

The area enclosed by dashed lines in Scheme 6a is the rapid-equilibrium random segment. The segment is called X, and Scheme 6a is reduced to a basic figure of X and EQ connected by two pathways (Scheme 6b):

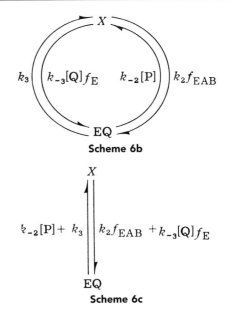

Scheme 6b

Scheme 6c

Scheme 6b can be further condensed into one line (Scheme 6c) according to the principle of "addition of parallel branches." The rate constants k_2 and $k_{-3}[Q]$ are multiplied by the fractional concentration of the enzyme species within the equilibrium segment that participates in that pathway, f_{EAB} and f_E.

$$f_{EAB} = \frac{[EAB]}{[E] + [EA] + [EB] + [EAB]}$$

$$= \frac{[A][B]/K_{ia}K_b}{1 + [A]/K_{ia} + [B]/K_{ib} + [A][B]/K_{ia}K_b}$$

$$f_E = \frac{1}{1 + [A]/K_{ia} + [B]/K_{ib} + [A][B]/K_{ia}K_b}$$

From Scheme 6c, we have

$$[EQ] = k_2 f_{EAB} + k_{-3}[Q]f_E$$

$$[X] = k_{-2}[P] + k_3$$

Thus, we obtain the rate equation

$$\frac{v}{[E_o]} = \frac{k_2 f_{EAB}[X] - k_{-2}[P][EQ]}{[X] + [EQ]}$$

$$= \frac{k_2 k_3 f_{EAB} - k_{-2}k_{-3}[P][Q]f_E}{k_{-2}[P] + k_3 + k_2 f_{EAB} + k_{-3}[Q]f_E}$$

$$= \frac{k_2 k_3([A][B]/K_{ia}K_b - k_{-2}k_{-3}[P][Q]}{(k_{-2}[P] + k_3)(1 + [A]/K_{ia} + [B]/K_{ib} + [A][B]/K_{ia}K_b) + k_2([A][B]/K_{ia}K_b) + k_{-3}[Q]}$$

This equation reveals atypical product inhibition patterns for a random mechanism: P is noncompetitive with both A and B; Q is competitive with both A and B. Whenever abnormal product inhibition patterns are observed, therefore, partial equilibrium treatment of the mechanism may be considered.

In the case of a rate-limiting step *within* a rapid equilibrium segment, it is necessary to include such a rate-limiting step in the velocity equation. Scheme 7a serves as an example for this type of mechanism:

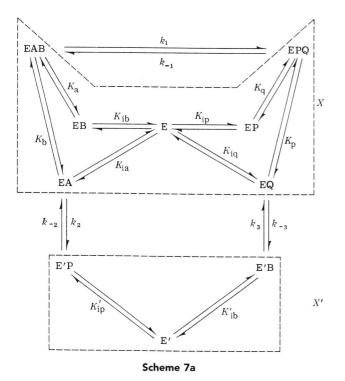

Scheme 7a

The two rapid-equilibrium segments X and X' are again indicated by areas enclosed by dashed lines. Within X, there is a rate-limiting step involving the interconversion of EAB and EPQ. By treating X and X' as though they were two enzyme species, adding parallel branches together, and multiplying the rate constants with appropriate fractional enzyme concentrations, we arrive at Scheme 7b.

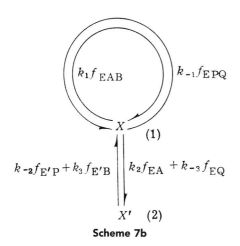

Scheme 7b

To obtain the determinants of X and X', the internal rate-limiting step in X is not included since it leads both away from and toward X. This can be readily shown by applying the systematic method described in the derivation of steady-state equations. Recall that the node values for X and X' are the summation of branch values leading *away* from them, as follows:

$$(1) = k_2 f_{EA} + k_{-3} f_{EQ}$$
$$(2) = k_{-2} f_{E'P} + k_3 f_{E'B}$$

The determinants for X and X' are defined as the product of other node values excluding their own: that is, $X = (2) = k_{-2} f_{E'P} + k_3 f_{E'B}$ and $X' = (1) = k_2 f_{EA} + k_{-3} f_{EQ}$.

The fractional enzyme concentrations are obtained as their fractions in the appropriate equilibrium segments:

$$f_{EA} = \frac{[A]/K_{ia}}{\begin{array}{c}1 + [A]/K_{ia} + [B]/K_{ib} + [A][B]/K_{ia}K_b \\ + [P]/K_{ip} + [Q]/K_{iq} + [P][Q]/K_{ip}K_q\end{array}}$$

$$f_{EQ} = \frac{[Q]/K_{iq}}{\begin{array}{c}1 + [A]/K_{ia} + [B]/K_{ib} + [A][B]/K_{ia}K_b \\ + [P]/K_{ip} + [Q]/K_{iq} + [P][Q]/K_{ip}K_q\end{array}}$$

$$f_{EAB} = \frac{[A][B]/K_{ia}K_b}{\begin{array}{c}1 + [A]/K_{ia} + [B]/K_{ib} + [A][B]/K_{ia}K_b \\ + [P]/K_{ip} + [Q]/K_{iq} + [P][Q]/K_{ip}K_q\end{array}}$$

$$f_{EPQ} = \frac{[P][Q]/K_{ip}K_q}{\begin{array}{c}1 + [A]/K_{ia} + [B]/K_{ib} + [A][B]/K_{ia}K_b \\ + [P]/K_{ip} + [Q]/K_{iq} + [P][Q]/K_{ip}K_q\end{array}}$$

$$f_{E'B} = \frac{[B]/K_{ib}'}{1 + [B]/K_{ib}' + [P]/K_{ip}'}$$

$$f_{E'P} = \frac{[P]/K_{ip}'}{1 + [B]/K_{ib}' + [P]/K_{ip}'}$$

In writing an expression for the initial velocity for Scheme 7b, however, one must include the internal rate-limiting step. Thus, we have

$$v = (k_1 f_{EAB} - k_{-1} f_{EPQ})X + k_2 f_{EA}X - k_{-2} f_{E'P}X'$$

Note that in steady-state treatment, any of the pathways can be used to write the velocity expression. Either one of the two pathways linking X and X' (see Scheme 7a) will yield the same expression:

$$k_2 f_{EA} X - k_{-2} f_{EP} X' = k_2 f_{EA}(k_{-2} f_{E'P} + k_3 f_{E'B})$$
$$- k_{-2} f_{E'P}(k_2 f_{EA} + k_{-3} f_{EQ})$$
$$= k_2 f_{EA} k_3 f_{E'B} - k_{-2} f_{E'P} k_{-3} f_{EQ}$$

$$k_3 f_{E'B} X' - k_{-3} f_{EQ} X = k_3 f_{E'B}(k_2 f_{EA} + k_{-3} f_{EQ})$$
$$- k_{-3} f_{EQ}(k_{-2} f_{E'P} + k_3 f_{E'B})$$
$$= k_2 f_{EA} k_3 f_{E'B} - k_{-2} f_{E'P} k_{-3} f_{EQ}$$

The complete rate equation is given by

$$\frac{v}{[E_o]} = \frac{(k_1 f_{EAB} - k_{-1} f_{EPQ})X + k_2 f_{EA} X - k_{-2} f_{E'P} X'}{X + X'}$$

$$= \frac{\begin{array}{c}(k_1 f_{EAB} - k_{-1} f_{EPQ})(k_{-2} f_{E'P} + k_3 f_{E'B}) \\ + k_2 f_{EA} k_3 f_{E'B} - k_{-2} f_{E'P} k_{-3} f_{EQ}\end{array}}{k_{-2} f_{E'P} + k_3 f_{E'B} + k_2 f_{EA} + k_{-3} f_{EQ}}$$

For mechanisms involving three or more rapid-equilibrium segments, once the segments are properly represented as "nodes" in a scheme, the rate equation can be obtained by the usual "systematic approach." For example, consider the case of one substrate—one product reaction in which a modifier M is in rapid equilibrium with E, EA, and EP.

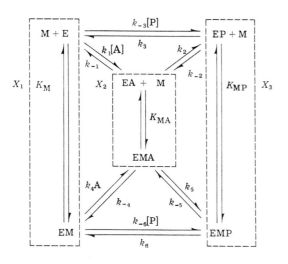

Treating X_1, X_2, and X_3 as nodes and adding parallel branches together, we obtain

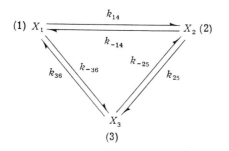

(3)

where $k_{14} = k_1[A]f_E + k_4[A]f_{EM}$, $k_{-14} = k_{-1}f_{EA} + k_{-4}f_{EMA}$, $k_{25} = k_2 f_{EA} + k_5 f_{EMA}$, $k_{-25} = k_{-2}f_{EP} + k_{-5}f_{EMP}$, $k_{36} = k_3 f_{EP} + k_6 f_{EMP}$, and $k_{-36} = k_{-3}[P]f_E + k_{-6}[P]f_{EM}$. Note that $v = k_{36}X_3 - k_{-36}X_1$.

CONCLUDING REMARKS. In this entry, the derivation of initial-velocity equations under steady-state, rapid-equilibrium, and the hybrid rapid-equilibrium and steady-state conditions has been covered. Derivation of initial velocity equation for the quasi-equilibrium case is quite straightforward once the equilibrium relationships among various enzyme-containing species are defined. The combined rapid-equilibrium and steady-state treatment can be reduced to the steady-state method by treating the equilibrium segments as though they were enzyme intermediates.

Albeit several methods for deriving steady-state rate equations have been described, one has to be proficient at only one of them since, for a given reaction scheme, they all lead to the same equation. The method to be recommended here is the *systematic* approach. It requires the least amount of time and work, especially when irreversible steps are present in the reaction scheme. Furthermore, no pattern drawing is needed in this approach. The initial setup is done by systematic inspection of the reaction diagram, and the risk of omission of terms (due to omission of certain patterns) is minimized. It should be emphasized that the validity of a rate equation depends also on whether the reaction diagram is properly constructed; that is, whether the number and types of enzyme species involved, the pathways linking these species, the sequence of reaction steps, *etc.*, are carefully considered and appropriately arranged. While the procedures presented in this entry allow one to obtain correct algebraic expressions, the rate equations so obtained are only as valid as the assumptions made in constructing the reaction scheme. **See** *Fromm's Method For Deriving Enzyme Rate Equations; Net Rate Constant Method*

[1]G. E. Briggs & J. B. S. Haldane (1925) *Biochem. J.* **19**, 338.
[2]H. J. Fromm (1979) *Meth. Enzymol.* **63**, 42.
[3]M. V. Volkenstein & B. N. Goldstein (1966) *Biochim. Biophys. Acta* **115**, 471.

*This section, authored by Charles Y. Huang, originally appeared in *Methods in Enzymology*, volume 63, as the first part of Chapter [4], pp. 54-75 ("Derivation of Initial Velocity and Isotope Exchange Rate Equations"). All equations and references in this entry are numbered sequentially as they originally appeared.

[4]J. T. Wong (1975) *Kinetics of Enzyme Mechanisms*, Academic Press, New York.
[5]H. J. Fromm (1979) *Meth. Enzymol.* **63**, 84.
[6]E. L. King & C. Altmann (1956) *J. Phys. Chem.* **60**, 1375.
[7]H. J. Fromm (1970) *Biochem. Biophys. Res. Commun.* **40**, 692.
[8]C. Y. Huang (1978) *Fed. Proc., Fed. Am. Soc. Exp. Biol.* **37**, 1423.
[9]J. T. Wong & C. S. Hanes (1962) *Can. J. Biochem. Physiol.* **40**, 763.
[10]S. Cha (1968) *J. Biol. Chem.* **243**, 820.

ENZYME RATE EQUATIONS (3. Derivation of Isotope Exchange Rate Equations*)

Isotope-exchange rate equations can be classified into two types: exchange at equilibrium and at steady state. The theory and technique of equilibrium exchange was pioneered by Boyer[1]. Most applications of isotope-exchange methods to enzyme systems have been of this type. Under equilibrium conditions there are no net chemical changes; whereas under steady-state conditions there is net catalysis. Although the derivation of these two types of rate equations differs in the assumptions involved, an equation derived by the steady-state approach can be readily converted into one for equilibrium exchange. Therefore, procedures intended for steady-state exchange are equally applicable to the derivation of equilibrium-exchange rate equations.

EQUATIONS FOR EXCHANGES AT EQUILIBRIUM. The derivation of equilibrium isotope-exchange equations for enzymic reactions was first formulated by Boyer[1]. Others have subsequently contributed to its development. Yagil and Hoberman[2] and Flossdorf and Kula[3] have devised generalized approaches that treat the flux of a label in a chemical reaction in a way analogous to the flow of charge in an electrical circuit. In this approach, (a) the equilibrium velocity of a reaction proceeding through n parallel steps is equal to the sum of the n individual rates; (b) when proceeding through n consecutive steps, the reciprocal of equilibrium velocity is equal to the sum of the n reciprocal. To demonstrate the use of this method, let us consider the B (R) P exchange in an Ordered Bi Bi mechanism (Scheme 1).

$$E \underset{k_{-1}}{\overset{k_1[A]}{\rightleftharpoons}} EA \underset{k_{-2}}{\overset{k_2[B^*]}{\rightleftharpoons}} EAB^* \underset{k_{-3}}{\overset{k_3}{\rightleftharpoons}} EP^*Q \underset{k_{-4}[P]}{\overset{k_4}{\rightleftharpoons}} EQ \underset{k_{-5}[Q]}{\overset{k_5}{\rightleftharpoons}} E$$

Scheme 1

The asterisks mark the labeled species. Since the system is at equilibrium and the exchange involves only three

steps, we can write a new scheme in the direction of isotopic flux:

$$EA \xrightarrow{k_2[B^*]} EAB^* \xrightarrow{k_3} EP^*Q \xrightarrow{k_4} EQ + P^*$$

The reverse steps are not shown because they are included in the equilibrium relationships to be substituted into the equation later.

Using the rule governing consecutive steps, we have

$$\frac{1}{v_{B\to P}^*} = \frac{1}{k_2[B^*][EA]} + \frac{1}{k_3[EAB^*]} + \frac{1}{k_4[EP^*Q]} \quad (1)$$

From the equilibrium relationships

$$[EAB^*] = \frac{k_2[B^*][EA]}{k_{-2}}$$

$$[EP^*Q] = \frac{k_3[EAB^*]}{k_{-3}} = \frac{k_2 k_3[B^*][EA]}{k_{-2} k_{-3}}$$

we can substitute the expressions of $[EAB^*]$ and $[EP^*Q]$ into Eq. (1) to obtain

$$\frac{1}{v_{B\to P}^*} = \frac{k_3 k_4 + k_{-2}(k_{-3} + k_4)}{k_2 k_3 k_4[B^*][EA]} \quad (2)$$

or

$$\frac{v_{B\to P}^*}{[E_o]} = \frac{k_2 k_3 k_4[B^*][EA]/[E_o]}{k_3 k_4 + k_{-2}(k_{-3} + k_4)} \quad (3)$$

where

$$\frac{[EA]}{[E_o]} = \frac{[EA]}{[E] + [EA] + [EAB] + [EPQ] + [EQ]}$$

Hence,

$$\frac{v_{B\to P}^*}{[E_o]} = \frac{k_1 k_2 k_3 k_4[A][B^*]}{k_{-1}[k_3 k_4 + k_{-2}(k_{-3} + k_4)]}$$
$$\scriptstyle [1 + k_1[A]/k_{-1} + k_1 k_2[A][B]/k_{-1}k_{-2}$$
$$\scriptstyle + k_{-5}[Q]/k_5 + k_{-4}k_{-5}[P][Q]/k_4 k_5]$$

The derivation of exchange-rate equations for mechanisms with branched pathways requires the rules governing consecutive and parallel steps. Consider as an example the A → P exchange in the Random Bi Bi mechanism (Scheme 2):

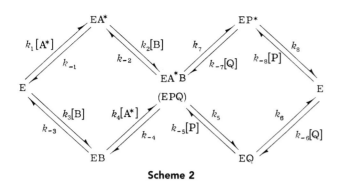

Scheme 2

Again, only the flux in one direction need be considered, and the steps not involved in the flux are ignored (Scheme 3).

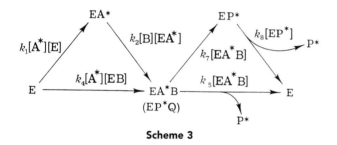

Scheme 3

The first step is to obtain the expressions for the E → EA* → EA*B and EA*B → EP* → E pathways. Let us call them k_{12} and k_{78}:

$$\frac{1}{k_{12}} = \frac{1}{k_1[A^*][E]} + \frac{1}{k_2[B][EA]};$$

$$k_{12} = \frac{k_1[A^*][E] \cdot k_2[B][EA]}{k_1[A^*][E] + k_2[B][EA]}$$

$$\frac{1}{k_{78}} = \frac{1}{k_7[EA^*B]} + \frac{1}{k_8[EP^*]};$$

$$k_{78} = \frac{k_7[EA^*B] \cdot k_8[EP^*]}{k_7[EA^*B] + k_8[EP^*]}$$

Scheme 3 can now be represented by Scheme 4.

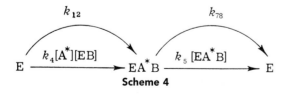

Scheme 4

Since parallel steps are additive, we have

$$E \xrightarrow{k_4[A^*][EB] + k_{12}} EA^*B \xrightarrow{k_5[EA^*B] + k_{78}} E$$

and finally

$$\frac{1}{v^*_{A \to P}} = \frac{1}{k_4[A^*][EB] + k_{12}} + \frac{1}{k_5[EA^*B] + k_{78}} \quad (4a)$$

The exchange-rate equation is obtained by substituting the expressions for k_{12}, k_{78}, and the equilibrium expressions for the enzyme intermediates into Eq. (4a).

From the following equilibrium relationships

$$[EA^*] = k_1[A^*][E]/k_{-1}$$

$$[EB] = k_3[B][E]/k_{-3}$$

$$[EA^*B] = k_4[A^*]k_3[B][E]/k_{-4}k_{-3}$$

$$[EP^*] = k_7[EA^*B]/k_{-7}[Q]$$

and

$$k_1[A]k_2[B] = k_3[A]k_4[B], \qquad k_{-1}k_{-2} = k_{-3}k_{-4}$$

it can be shown that

$$\frac{1}{v^*_{A \to P}} = \frac{1}{k_4[A^*][EB] + \{k_1[A^*][E] \cdot k_2[B][EA^*]\}/\{k_1[A^*][E] + k_2[B][EA^*]\}}$$

$$= \frac{1}{k_5[EA^*B] + \{k_7[EA^*B] \cdot k_8[EP^*]\}/\{k_7[EA^*B] + k_8[EP^*]\}}$$

$$= \frac{1}{[E]} \cdot \frac{\left\{ \begin{array}{l} (k_{-1} + k_2[B])(k_{-7}[Q] + k_8)(k_{-4} + k_5) \\ + k_7 k_8(k_{-1} + k_2[B]) \\ + k_{-1}k_{-2}(k_{-7}[Q] + k_8) \end{array} \right\}}{\left\{ \begin{array}{l} \{k_1[A]k_2[B] + k_4[A]k_3[B]\} \\ \cdot (k_{-1} + k_2[B])/k_{-3}\} \\ \cdot \{k_5(k_{-7}[Q] + k_8) + k_8 k_7\} \end{array} \right\}}$$

DERIVATION BY THE STEADY-STATE METHOD. Britton[4] first derived isotope flux equations under steady-state rather than equilibrium conditions. To illustrate his procedure, we shall again use Scheme 1, the B → P exchange in Ordered Bi Bi mechanism, as an example, so that the results can be compared (Scheme 1a).

$$EA \underset{}{\overset{k_2[B^*][EA]}{\rightleftharpoons}} EAB^* \underset{k_{-3}}{\overset{k_3}{\rightleftharpoons}} EP^*Q \overset{k_4}{\longrightarrow} EQ + P^*$$

$$(1) \qquad\qquad (2) \qquad\qquad (3) \qquad\qquad (4)$$

Scheme 1a

Note that (a) the reverse steps are needed for steady-state treatment; (b) the k_{-4} step is not shown because the initial rate of exchange is being measured; and (c) only the unlabeled enzyme concentration is included in the derivation because the concentration factors of labeled enzyme forms will cancel out during the derivation. Also, each enzyme form is assigned a number for reference purposes.

The procedure is to calculate the B → P or 1 → 4 isotope transfer in a stepwise manner using partition theory:

$$\text{Flux } 1 \to 2 = k_2[B^*][EA]$$

$$\text{Flux } 1 \to 3 = (1 \to 2) \cdot \frac{(2 \to 3)}{(2 \to 3) + (2 \to 1)}$$

$$= \frac{k_2[B^*][EA]k_3}{k_3 + k_{-2}}$$

$$\text{Flux } 1 \to 4 = (1 \to 3) \cdot \frac{(3 \to 4)}{(3 \to 4) + (3 \to 1)} \qquad (5)$$

$$= \frac{k_2[B^*][EA]k_3}{k_3 + k_{-2}} \cdot \frac{k_4}{k_4 + \{k_{-2}k_{-3}/(k_3 + k_{-2})\}}$$

$$= \frac{k_2k_3k_4[B^*][EA]}{k_4(k_3 + k_{-2}) + k_{-2}k_{-3}}$$

$$= v^*_{B \to P}$$

Note that Eq. (5) is identical with Eq. (2). More recently, Cleland[5] developed an approach that is more convenient. His procedure starts with the release of labeled product and works backwards as follows:

$$\text{Flux } 3 \to 4 = k_4$$

$$\text{Flux } 2 \to 4 = (2 \to 3) \cdot \frac{(3 \to 4)}{(3 \to 4) + (3 \to 2)}$$

$$= \frac{k_3k_4}{k_4 + k_{-3}}$$

$$\text{Flux } 1 \to 4 = (1 \to 2) \cdot \frac{(2 \to 4)}{(2 \to 4) + (2 \to 1)}$$

$$= \frac{k_2[B^*][EA]k_3k_4/(k_4 + k_{-3})}{\{k_3k_4/(k_4 + k_{-3})\} + k_{-2}}$$

$$= \frac{k_2k_3k_4[B^*][EA]}{k_3k_4 + k_{-2}(k_4 + k_{-3})}$$

These procedures, however, are more suitable for deriving exchange-rate equations for mechanisms without branched pathways. For more complex mechanisms, the schematic method of Cleland[6] based on the approach of King and Altman is quite convenient. This method can be adapted to the systematic approach and combined with the deletion of steps linking unlabeled species previously described to further reduce the amount of work involved[7]. Consider the A → P exchange in the Random Bi Bi mechanism (cf. Scheme 2; the figure has been redrawn in a folded geometric form) shown in Scheme 5.

The pathways connecting the unlabeled enzymes species, E ⇌ EB and E ⇌ EQ, as has been shown in Scheme 3, can be eliminated; and the EB ⇌ EAB and EAB → EQ steps can be directly linked to E (Scheme 5a).

By combining the k_{-4} and k_5 branches, Scheme 5a can be further reduced to Scheme 5b.

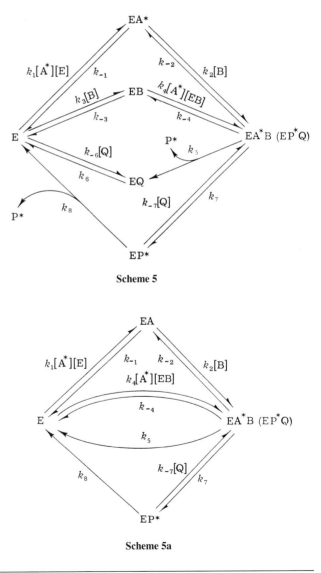

Scheme 5

Scheme 5a

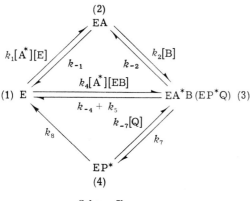

Scheme 5b

Now we apply the systematic method to Scheme 5b by assigning node numbers to the enzyme species and write down the node values:

$$(1) = k_1[A^*][E] + k_4[A^*][EB]$$

$$(2) = k_{-1} + k_2[B]$$

$$(3) = k_{-2} + k_{-4} + k_5 + k_7$$

$$(4) = k_{-7}[Q] + k_8$$

We then use Cleland's procedure to write the expression for the initial rate of formation of labeled P from A.

$$v^* = d[P^*]/dt = k_5[EA^*B] + k_8[EP^*]$$

$$= \frac{k_5 N_{EA^*B} + k_8 N_{EP^*}}{D}$$

where N_{EA^*B}, N_{EP^*}, and D are the "determinants" of EA*B, EP*, and E in Scheme 5b. The N terms vary with mechanism, depending on the enzyme species from which the labeled product is released, but the D term is always obtained as the "determinant" of E.

Thus, we can write

$$N_{EA^*B} = (1)(2)(4) = \{k_1[A^*][E] + k_4[A^*][EB]\}$$
$$(k_{-1} + k_2[B])(k_{-7}[Q] + k_8)$$
$$= k_1[A^*][E](k_2[B])(k_{-7}[Q] + k_8)$$
$$+ k_4[A^*][EB](k_{-1} + k_2[B])(k_{-7}[Q] + k_8)$$

$$N_{EP^*} = (34)(1)(2) = k_7\{k_1[A^*][E]$$
$$+ k_4[A^*][EB]\}(k_{-1} + k_2[B])$$
$$= k_7 \cdot k_1[A^*][E] \cdot k_2[B]$$
$$+ k_7 k_4[A^*][EB](k_{-1} + k_2[B])$$

$$D = (2)(3)(4) = (k_{-1} + k_2[B])$$
$$(k_{-2} + k_{-4} + k_5 + k_7)(k_{-7}[Q] + k_8)$$
$$= (k_{-4} + k_5)(k_{-1} + k_2[B])(k_{-7}[Q] + k_8)$$
$$+ k_{-1}k_{-2}(k_{-7}[Q] + k_8) + k_7 k_8(k_{-1} + k_2[B])$$

The complete exchange equation is obtained as

$$\frac{v^*}{[E_o]} = \frac{(k_5 N_{EA^*B} + k_8 N_{EP^*})/D}{[E] + [EA] + [EB] + [EAB] + [EP] + [EQ]}$$

It should be noted that [E], [EA], [EB], [EAB], [EP], and [EQ] are now the *normal* determinants of E, EA, EB, EAB, EP, and EQ obtained from the steady-state treatment of the intact reaction scheme. If the equilibrium-exchange rate equation is desired, [E], [EA], [EAB], [EP], and [EQ] should be obtained from the equilibrium relationships.

The following derivations demonstrate that an isotope exchange rate equation obtained by the steady-state treatment can be converted to one of exchange at equilibrium:

$$v^* = \frac{\left\{\begin{array}{c} k_5\{k_1[A^*]k_2[B](k_{-7}[Q] + k_8)[E] + k_4[A^*] \\ \cdot(k_{-1} + k_2[B])(k_{-7}[Q] + k_8[EB]\} \\ + k_8\{k_7 k_1[A^*]k_2[B][E] \\ + k_7 k_4[A^*](k_{-1} + k_2[B])[EB]\} \end{array}\right\}}{\left\{\begin{array}{c} (k_{-4} + k_5)(k_{-1} + k_2[B])(k_{-7}[Q] + k_8) \\ + k_{-1}k_{-2}(k_{-7}[Q] + k_8) + k_7 k_8(k_{-1} + k_2[B]) \end{array}\right\}}$$

Substitution of $[EB] = k_3[B][E]/k_{-3}$ into the above expression yields

$$v^* = \frac{[k_5(k_{-7}[Q] + k_8) + k_8 k_7][k_1[A^*]k_2[B] + k_4[A^*]k_3[B](k_{-1} + k_2[B])/k_{-3}][E]}{(k_{-4} + k_5)(k_{-1} + k_2[B])(k_{-7}[Q] + k_8) + k_{-1}k_{-2}(k_{-7}[Q] + k_8) + k_7 k_8(k_{-1} + k_2[B])}$$

which is identical with Eq. (4b) derived by the equilibrium treatment.

The systematic method is equally convenient for the derivation of rate equations for simple mechanisms. Scheme 1, for example, can be redrawn as an enclosed figure after deleting the pathways between unlabeled enzyme forms.

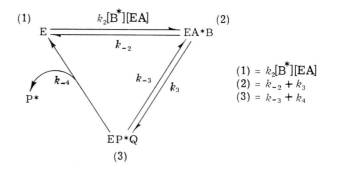

$$(1) = k_2[B^*][EA]$$
$$(2) = k_{-2} + k_3$$
$$(3) = k_{-3} + k_4$$

$$v_{B \to P}^* = \frac{k_4 N_{EP^*Q}}{D}$$

$$N_{EP^*Q} = (1)(2) = k_2[B^*][EA](k_{-2} + k_3)$$
$$= k_2 k_3[B^*][EA]$$

$$D = (2)(3) = (k_{-2} + k_3)(k_{-3} + k_4)$$
$$= k_{-2}(k_{-3} + k_4) + k_3 k_4$$

Thus

$$v^* = \frac{k_2 k_3 k_4[B^*][EA]}{k_{-2}(k_{-3} + k_4) + k_3 k_4}$$

[1]P. D. Boyer (1959) *Arch. Biochem. Biophys.* **82**, 387.
[2]G. Yagil & H. D. Hoberman (1969) *Biochemistry* **8**, 352.
[3]J. Flossdorf & M. Kula (1972) *Eur. J. Biochem.* **30**, 325.
[4]H. G. Britton (1964) *J. Physiol. (London)* **170**, 1.
[5]W. W. Cleland (1975) *Biochemistry* **14**, 3220.
[6]W. W. Cleland (1967) *Ann. Rev. Biochem.* **36**, 77.
[7]C. Y. Huang (1978) *Fed. Proc., Fed. Am. Soc. Exp. Biol.* **37**, 1423.

*This entry, authored by Charles Y. Huang, originally appeared in *Methods in Enzymology*, volume 63, as the second part of Article [4], pp. 76—84 ("Derivation of Initial Velocity and Isotope Exchange Rate Equations").

ENZYME STABILITY

In most kinetic investigations, one *assumes* the enzyme remains stable over the course of the measurement. When this is the case, corrective measures must be taken to obtain valid kinetic data. A useful test for any enzyme system is to plot enzyme activity *versus* time. This is readily accomplished by using a standardized assay (usually at optimal or saturating substrate concentrations) to measure the enzyme's specific activity periodically during the course of some experiment. This approach may fail to detect a reduction in activity characterized by lower affinity for substrate; however, use of a subsaturating substrate concentration in a time-course study will reveal this behavior.

Efforts should be made to stabilize an enzyme's activity. Certain agents (such as glycerol, ammonium ions, boric acid, polyethylene glycol, and even talcum powder or bentonite clay) have proven widely to be effective enzyme stabilizers. For multisubstrate enzymes, inclusion of one particular substrate with the enzyme (in the absence of other substrates or cofactors) often stabilizes an enzyme's catalytic activity. Such a substrate may also assist in "unlocking" the enzyme from a particularly inactive conformational form. In addition to substrates, other ligands and effectors (including reaction products, effectors, and even reversible inhibitors) can often exert a stabilizing influence. Some investigators have successfully employed an inhibitor as a stabilizing agent in the concentrated stock solution of enzyme; when a small aliquot is added to the reaction mixture, the inhibitor's concentration is diluted, such that the inhibitory effect is undectable[1].

Some interactions between macromolecules are relatively slow, and lags in the reaction are often minimized by preincubation. For multisubstrate enzymes acting on a protein substrate, one may need to incubate the enzyme and protein substrate together before initiating the reaction. An example is bovine brain tubulin:tyrosine ligase[2], and the observed lag is completely eliminated when the ligase and tubulin are incubated together prior to initiating the reaction.

Stabilization may also be achieved by altering reaction conditions (*e.g.*, pH, temperature, ionic strength, *etc.*). In other instances, loss of activity may attend the adsorption of the enzyme to the walls of the reaction container (*e.g.*, glass or plastic can strongly interact with proteins). Using a different container or simply including a "spectator" protein (*e.g.*, albumin) may increase the degree of stability.

The procedure adopted for isolating an enzyme can often influence enzyme stability. If the preparation procedure involves heat treatment or use of organic solvent, the enzyme may become "locked" into a less-active conformational form. The experimenter should consider using another preparative method.

Many enzymes (and protein substrates) require free thiol residues for proper conformation and/or activity. In this case, including a thiol reagent (such as 2-mercaptoethanol, dithiothreitol, dithioerythritol, or cysteine) may maintain activity.

Enzyme dilution often adversely influences enzyme stability. The presence of another protein (*e.g.*, 1-2 mg/mL serum albumin) may afford greater stability. When any protein is added to a system for stabilizing purposes, one should be certain that there is no effect on other substrates, products, or effectors in the assay mixture. Remember that some commercial preparations of bovine serum albumin, unless specifically "defatted" by extraction with an organic solvent, may contain fatty acids and other fat-soluble factors, Bovine serum albumin also binds ATP, and this can be a problem if the ATP concentration is varied in kinetic measurements conducted in the presence of BSA.

[1]R. D. Allison & D. L. Purich (1979) *Meth. Enzymol.* **63**, 3.
[2]N. L. Deans, R. D. Allison & D. L. Purich (1992) *Biochem. J.* **286**, 243.

ENZYME-SUBSTRATE COMPLEX
Any binary, ternary, or quaternary complex formed upon the association of a substrate with the free enzyme or other enzyme form; such a complex is often called the Michaelis complex.

EPIMERIZATION
The stereochemical isomerization or structural rearrangement resulting in the interconversion of epimers. For example, the conversion of β-D-glucose to β-D-galactose involves epimerization at the C-4 carbon atom, and the epimerization at the C-2 of β-D-glucose results in the synthesis of β-D-mannose.

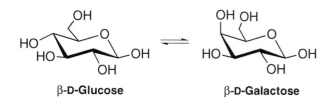

β-D-Glucose   **β-D-Galactose**

EPITAXY
Growth of a crystal containing a different substance atop a pre-existing crystal. The occurrence of epitaxy suggests that the presence of the first crystal promotes a second, otherwise disfavored crystallization process. Epitaxy need not require the any intimate geometrical fit between the new and old solid phases, surface tension changes or weak adhesive forces can also be responsible. One theory of urinary stone disease posits that calcium oxalate monohydrate crystals do not spontaneously occur by homogeneous nucleation; rather, epitaxy is thought to take place at sites of mineralized calcium phosphate known as Randall's plaque.

EPOXIDE HYDROLASE
This enzyme [EC 3.3.2.3], also known as epoxide hydratase and arene-oxide hydratase, catalyzes the hydrolysis

of a variety of epoxides and arene oxides to yield the corresponding glycol.

R. N. Armstrong (1987) *Crit. Rev. Biochem.* **22**, 39.

trans-EPOXYSUCCINATE HYDROLASE
This enzyme [EC 3.3.2.4], also known as *trans*-epoxysuccinate hydratase and tartrate epoxidase, catalyzes the hydrolysis of *trans*-2,3-epoxysuccinate to yield *meso*-tartrate.

W. B. Jakoby & T. A. Fjellstedt (1972) *The Enzymes*, 3rd ed., **7**, 199.

EPSILON (ε)
1. The fifth letter in the Greek alphabet; hence, used to denote the fifth in a series (for example, the fifth methylene carbon in a fatty acid). 2. Symbol for molar absorption coefficient or extinction coefficient. 3. Symbol for permittivity (ε_0 refers to permittivity of a vacuum; ε_r refers to relative permittivity). 4. ε_a, Symbol for degree of activation (IUB (1982) *Eur. J. Biochem.* **128**, 281). 5. ε_i, Symbol for degree of inhibition. 6. Symbol for efficiency. 7. Symbol for linear strain. 8. Symbol for emittance.

EQUATIONS OF STATE
Thermodynamic expressions for the functional interdependence of a number of physical chemical parameters (usually, pressure, volume, temperature, and amount) of a particular substance. While equations of state have been developed for gases, liquids, and solids, the theories are most advanced for gaseous systems.

EQUI-EFFECTIVE DOSE/MOLARITY RATIO
The ratio (EDR) of two doses or the ratio (EMR) of two molarities of test and reference substances that are isoactive in their effectiveness as an agonist or antagonist.

EQUILIBRIUM
1. A condition of balance in a chemical system, at which no further change in reactant and product concentrations occurs. 2. If actions occurring within the reaction result in no net change in the reactant and product concentrations, the system is said to be in dynamic equilibrium. A chemical equilibrium is a dynamic equilibrium when the reaction rate in the forward direction is balanced by the rate in the reverse direction. The potential energy

of a system reaches a minimum when a system is at equilibrium. If the potential energy is at a maximum, the system's state is said to be unstable or metastable.

EQUILIBRIUM CONSTANT

A ratio equal to the product of the concentrations of all products at equilbrium divided by the product of the concentrations of all reactants. The equilibrium constant is most commonly symbolized by K_{eq}.

At equilibrium, there is no net change in the concentrations of any reactants or products, and the rate of the forward reaction equals the rate of the reverse reaction. For a general reaction,

$$\nu_A A + \nu_B B + \cdots = \nu_P P + \nu_Q Q + \cdots$$

then

$$K_{eq} = ([P]_{eq}^{\nu_P}[Q]_{eq}^{\nu_Q} \cdots)/\{[A]_{eq}^{\nu_A}[B]_{eq}^{\nu_B} \cdots\}$$

where $[A]_{eq}$, $[B]_{eq}$, *etc.*, represent the concentrations of the substances at equilibrium. The negative logarithm of the equilibrium constant is pK_{eq}.

The equilibrium constant for a dissociation reaction is usually referred to as the dissociation constant (K_d), and the products of that dissociation are then written in the numerator. For reactions not at equilbrium, the value for the product of the concentrations of all products divided by the product of the concentrations of all reactants is frequently referred to as the mass action ratio. *See* Association Constant; Dissociation Constant; Mass Action Ratio

EQUILIBRIUM CONSTANT DETERMINATIONS

A number of methods can be utilized in determining the equilibrium position of chemical reactions:

1. Starting with substrates only (or products only), add sufficient enzyme to equilibrate the reaction and periodically sample the reaction mixture until substrate and product concentrations reach a time-invariant value. The difficulty with this method is that product inhibition can greatly retard the approach to equilibrium. In addition, if the enzyme is unstable, denaturation may occur before equilibration has in fact occurred. In such instances, a false K_{eq} will be obtained. Likewise, if a proton is con-

sumed or produced in the reaction, the pH of the reaction mixture may change. A good practice is to approach the equilibrium from both reaction directions; in such cases, the identical K_{eq} value must be obtained.

2. Applying the Haldane relation to obtain an equilibrium constant from initial rate kinetics. Because of the inherently greater uncertainty in determinations of rate parameters, this method often proves to be unreliable! *See* Haldane Relation

3. Another effective method of determining the equilibrium constant is to prepare a series of reaction samples in which all substrates and products are present at concentrations that would be expected to correspond roughly to the equilibrium constant. Once enzyme is added to mixtures having slightly different mass-action ratios, the investigator can plot these changes *versus* the initially chosen mass-action ratio. If the ratio is lower than the equilibrium constant, there will be net conversion to product (that is, measuring D[P], possibly using labeled products and/or substrates). If the ratio is greater than the equilibrium constant, there will be net conversion to reactant. The mass action ratio that yields a zero deviation must exactly match the equilibrium constant.

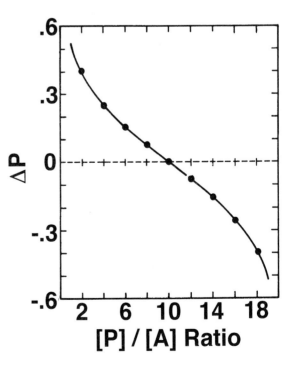

For multisubstrate, multiproduct enzymes, one also has the opportunity to vary specific substrate:product pairs[2].

For example, consider a two-substrate, two-product reaction (A + B = P + Q). In a series of solutions, the concentrations of A and P can be set at several different absolute values such that the ratio is the same in all mixtures. The ratio of the other substrate:product pair ([B]/[Q]) can now be varied such that [P][Q]/([A][B]) varies about the estimated equilibrium constant. Once enzyme is added, deviations in one or more substrates or products are then measured after a sufficient incubation period.

4. The use of a doubly-labeled system is also ideal for demonstrating equilibration has occurred. For example, if ^{14}C-labeled substrate is mixed with ^{3}H-labeled product in the presence of enzyme, when equilibration has been reached, the [3]H : [14]C ratio in substrate and product must be equal.

With each of these methods, the investigator should carefully note the amount of enzyme used. At very high enzyme concentrations, there is always the possibility a contaminating enzyme activity may alter the measurements. High enzyme concentrations can also lead to selective binding of a reaction component (*e.g.*, metal ion), thereby displacing the K_{eq}. A good procedure is to determine the equilibrium constant at several enzyme concentrations. The value of K_{eq} should be independent of [E_{total}]. If this is not the case, the investigator has an opportunity to determine the basis for the perturbation.

The equilibrium constant of an enzyme-catalyzed reaction can depend greatly on reaction conditions. Because most substrates, products, and effectors are ionic species, the concentration and activity of each species is usually pH-dependent. This is particularly true for nucleotide-dependent enzymes which utilize substrates having pK_a values near the pH value of the reaction. For example, both ATP^{4-} and $HATP^{3-}$ may be the nucleotide substrate for a phosphotransferase, albeit with different K_m values. Thus, the equilibrium constant with ATP^{4-} may be significantly different than that of $HATP^{3-}$. In addition, most phosphotransferases do not utilize free nucleotides as the substrate but use the metal ion complexes. Both ATP^{4-} and $HATP^{3-}$ have different stability constants for Mg^{2+}. If the buffer (or any other constituent of the reaction mixture) also binds the metal ion, the buffer (or that other constituent) can also alter the observed equilibrium constant[1,2].

Whenever reporting equilibrium constants, detailed information concerning the reaction conditions should always be indicated. Alberty[3] has also presented an important review of biochemical thermodynamics in which he discusses the apparent equilibrium constant for biochemical reactions (K') in terms of sums of reactant species.

[1]R. D. Allison & D. L. Purich (1979) *Meth. Enzymol.* **63**, 3.
[2]D. L. Purich & R. D. Allison (1980) *Meth. Enzymol.* **64**, 3.
[3]R. A. Alberty (1994) *Biochim. Biophys. Acta* **1207**, 1.

EQUILIBRIUM DIALYSIS

An equilibrium binding technique that allows one to measure the net accrual of macromolecule-bound ligand in one of two compartments separated by a semipermeable membrane.

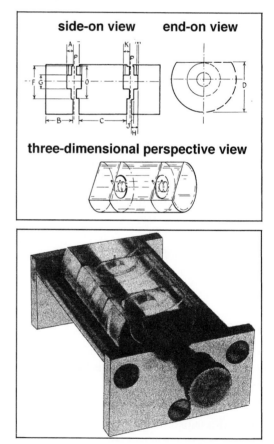

Miniaturized dialysis cells for quantitative analysis of equilibrium ligand binding. Protein (20 mL) is placed in a compartment located on one side of a semipermeable membrane, and radiolabeled ligand is added to the second 20 mL-compartment. Equilibration is typically achieved within 90 min using these miniaturized cells. Equal volume aliquots are removed from both compartments, and the net accrual of radiolabel in the protein sample indicates the amount of bound ligand. (From P. T. Englund, J. A. Huberman, T. M. Jovin & A. Kornberg (1969) *J. Biol. Chem.* **244**, 3038, with permission of the authors and the American Society for Biochemistry & Molecular Biology.)

A now classical example of the power of equilibrium dialysis is the determination by Englund *et al.*[1] that DNA polymerase binds all four nucleoside 5'-triphosphate substrates at the same subsite in the enzyme's catalytic center. ***See also*** *Womack-Colowick Technique*

[1]P. T. Englund, J. A. Huberman, T. M. Jovin & A. Kornberg (1969) *J. Biol. Chem.* **244**, 3038.

Selected entries from ***Methods in Enzymology*** [vol, page(s)]:
Theory, **76**, 561-562; determination, of association constant, **70**, 30; nonheme ligand studies, **76**, 535, 539-541; binding-parameter evaluation, **76**, 539-540; using bromthymol blue, **76**, 540-541; rapid rate of organophosphate-hemoglobin interaction, **76**, 559-577; binding assay, **76**, 567-570; buffers, **76**, 567; radioactive ligand synthesis, **76**, 562-567.

EQUILIBRIUM ENZYMES IN METABOLIC PATHWAYS

A commonly held belief is that, for an enzyme reaction within a metabolic pathway, a large excess of catalytic capacity relative to a pathway's metabolic flux ensures that a given step is at or near thermodynamic equilibrium. Brooks[1] recently treated the kinetic behavior of reaction schemes one might judge to be at equilibrium, and he showed that individual steps can remain far from equilibrium, even at a high ratio of an enzyme's flux to a pathway's steady-state flux. His calculations indicate that whether a reaction is near equilibrium depends on (a) the overall flux through the enzyme locus and (b) the kinetic parameters of the other enzymes in the pathway.

[1]S. P. Brooks (1996) *Biochem. Cell Biol.* **74**, 411.

EQUILIBRIUM ISOTOPE EFFECT

A nonunity ratio (sometimes called a thermodynamic isotope effect) of the equilibrium constants (K_{light}/K_{heavy}) for two reactions differing only in the isotopic composition at one or more positions of their otherwise chemically identical substances[1,2]. If the equilibrium isotope effect is attributable to a covalent bond making/breaking, then the effect is often referred to as a *primary* equilibrium isotope effect. If isotopic substitution at a position other than the scissile bond results in an equilibrium isotope effect, the term *secondary* equilibrium istope effect is used.

Since potential energy surfaces of isotopic molecules are nearly identical, equilibrium isotope effects can only arise from the effect of isotopic mass on the nuclear motions of the reactants and products. Thus the ratio can be expressed in terms of partition functions for nuclear

motion. In the case of light atoms at moderate temperatures (*e.g.*, protium *vs.* deuterium), the equilibrium isotope effect is governed primarily by zero-point energy differences. ***See*** *Kinetic Isotope Effects*

[1]IUPAC (1979) *Pure and Appl. Chem.* **51**, 1725.
[2]M. Wolfsberg (1972) *Acc. Chem. Res.* **5**, 225.

EQUILIBRIUM PERTURBATION METHOD

The observed perturbation of a dynamic chemical process from its equilibrium position when isotopic substitution on a substrate or product changes the initial rate of conversion of substrate to product or the initial rate of conversion of product to substrate[1–3].

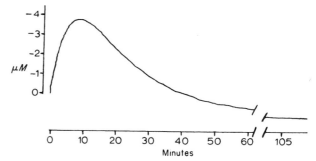

Equilibrium perturbation[2] with malic enzyme added to a pH 7.1 solution containing 0.419 mM malate-2-D, 0.079 mM NADPH, 20 mM potassium bicarbonate (3.8 mM carbon dioxide), 20 mM magnesium sulfate, 0.102 mM NADP+, and 3.83 mM pyruvate. NADPH initially disappears within the first 10 min, followed by its reappearance over the ensuing 60 min period.

The initially observed perturbation (or equilibrium isotope effect) will disappear as isotopic scrambling (or mixing) subsequently occurs. This is illustrated above for the malic enzyme; note the initial displacement of the equilibrium toward malate, followed by readjustment of the system to its final equilibrium position. The magnitude of the initial perturbation provides information on the occurrence of kinetic isotope effects and the nature of the rate-limiting step in an enzymatic process. ***See also*** *Kinetic Isotope Effect; Solvent Isotope Effect*

[1]M. I. Schimerlik, J. E. Rife & W. W. Cleland (1975) *Biochemistry* **14**, 5347.
[2]W. W. Cleland (1980) *Meth. Enzymol.* **64**, 104.
[3]W. W. Cleland (1982) *Meth. Enzymol.* **87**, 641.

EQUILIBRIUM THERMODYNAMICS (Measurable Quantities)

Thermodynamic studies[1] of systems at equilibrium permit one to gain insights regarding mechanisms of chemical and biochemical reactions. Thermodynamic considerations and inferences are founded on the experimenter's ability to evaluate accurately certain measurable quantities, mainly: heat capacities; latent heats, heats of mixing, heats of chemical reaction; equilibrium constants for reactions (including analogous electrochemical analogues of the chemical equilibrium constant); vapor pressures; solubilities; coefficients of compressibility and expansion; and volume changes attending a chemical reaction.

[1]K. Denbigh (1968) *The Principles of Chemical Equilibrium*, 2nd ed., pp. 494, Cambridge University Press, London.

ERG

A unit of energy, work, or heat defined as the equivalent of a force of one dyne acting through a distance of one centimeter. A joule is equal to 10^7 ergs.

ERGODIC

A descriptor for a process that applies when every sizable sample represents the whole. The term stems from statistical mathematics and refers to any system which, after a sufficiently long time of wandering from one state to another, returns to states that are similar to those previously encountered. Assuming that system under study is ergodic is a standard practice in modern dynamics, statistical mechanics, and atomic theory. For example, chemical reaction dynamics frequently assumes that vibrational modes are well coupled (*i.e.*, ergodic) and redistribute energy rapidly with respect to the course of the reaction. Most large dynamical systems are assumed to exhibit ergodic dynamics, but in small systems, the free interchange of energy between degrees of freedom need not always occur.

ERGODIC PROCESS

A principle in statistical physics holding that, for each state within a system, there is an equal probability of occurrence; for this reason, a system can rapidly pass through all such states.

ERGOTHIONASE

This enzyme catalyzes the conversion of L-ergothioneine to 2-thiolurocanic acid and trimethylamine.

J. B. Wolff (1971) *Meth. Enzymol.* **17B**, 105.

ESTERASES

This is a broad class of enzymes that catalyze the hydrolysis of esters; some of these enzymes are quite specific. *See also* Cholesteryl Ester Hydrolase; specific esterase

H. Okuda (1991) *A Study of Enzymes* **2**, 563.
K. Krisch (1971) *The Enzymes*, 3rd ed., **5**, 43.

ESTER HYDROLYSIS MECHANISMS

Esters can be hydrolyzed nonenzymatically in acidic and alkaline solutions. In both cases, a tetrahedral intermediate is indicated on the basis of the measured rates of hydrolysis and the rates of oxygen isotope exchange.

C. K. Ingold (1953) *Structure and Mechanism in Organic Chemistry*, Cornell University Press, Ithaca.

ETA (η)

1. Symbol for viscosity. 2. Symbol for the efficiency of a step. 3. The seventh letter of the Greek alphabet; hence, used to denote the seventh in a series (for example, the seventh methylene carbon in a fatty acid). 4. Symbol for overpotential.

ETHANOLAMINE AMMONIA-LYASE

This cobalamin-dependent enzyme [EC 4.3.1.7] catalyzes the interconversion of ethanolamine to form acetaldehyde and ammonia.

P. A. Frey & G. H. Reed (1993) *Adv. Enzymol.* **66**, 1.
J. A. Stubbe (1989) *Ann. Rev. Biochem.* **58**, 257.
B. M. Babior & J. S. Krouwer (1979) *Crit. Rev. Biochem.* **6**, 35.
T. C. Stadtman (1972) *The Enzymes*, 3rd ed., **6**, 539.

ETHANOLAMINE KINASE

This enzyme [EC 2.7.1.82] catalyzes the ATP-dependent phosphorylation of ethanolamine to yield *O*-phosphoethanolamine and ADP.

D. Rhodes & A. D. Hanson (1993) *Ann. Rev. Plant Physiol. Plant Mol. Biol.* **44**, 357.
K. Ishidate & Y. Nakazawa (1992) *Meth. Enzymol.* **209**, 121.
J. D. Esko & C. R. H. Raetz (1983) *The Enzymes*, 3rd ed., **16**, 207.

ETHANOLAMINE *N*-METHYLTRANSFERASE

This enzyme catalyzes the methylation of ethanolamine to produce *N*-methylethanolamine.

D. Rhodes & A. D. Hanson (1993) *Ann. Rev. Plant Physiol. Plant Mol. Biol.* **44**, 357.

ETHANOLAMINEPHOSPHOTRANSFERASE

This enzyme [EC 2.7.8.1] converts CDP-ethanolamine and 1,2-diacylglycerol into CMP and phosphatidylethanolamine.

R. H. Hjelmstad & R. M. Bell (1992) *Meth. Enzymol.* **209**, 272.

ETHANOLAMINE—SERINE BASE-EXCHANGE ENZYME

This enzyme catalyzes the reaction of a phospholipid (for example, phosphatidylserine) with ethanolamine to produce phosphatidylethanolamine and the free base (*i.e.*, the amine-containing metabolite serine), thereby preserving the phosphodiester linkage. Ethanolamine can be replaced with serine, choline, monomethyletha-nolamine, and dimethylethanolamine.

J. N. Kanfer (1992) *Meth. Enzymol.* **209**, 341.

ETIOCHOLANOLONE SULFATASE

This enzyme catalyzes the hydrolysis of etiocholanolone sulfate to produce etiocholanolone and sulfate.

A. B. Roy (1971) *The Enzymes*, 3rd ed., **5**, 1.

EXCHANGE-INERT METAL-NUCLEOTIDE COMPLEXES

A metal-nucleotide complex[1,2] that exhibits low rates of ligand exchange as a result of substituting higher oxidation state metal ions with ionic radii nearly equal to the naturally bound metal ion. Such compounds can be prepared with chromium(III), cobalt(III), and rhodium(III) in place of magnesium or calcium ion. Because these exchange-inert complexes can be resolved into their various optically active isomers, they have proven to be powerful mechanistic probes, particularly for kinases, NTPases, and nucleotidyl transferases. In the case of Cr(III) coordination complexes with the two phosphates of ATP or ADP, the second phosphate becomes chiral, and the screw sense must be specified to describe the three-dimensional configuration of atoms.

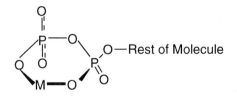

Λ Isomer

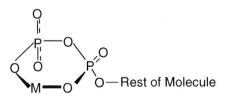

Δ Isomer

In this case, the reference axis is drawn through the metal ion in a fashion perpendicular to the chelate ring. The skew line defining the helical sense (L for left-hand, and D for right-hand) is the bond from the chelate ring to the rest of the molecule. The L and D complexes are diastereomeric and can be separated from each other for direct stereochemical studies of their individual binding and inhibitory properties.

[1]W. W. Cleland (1982) *Meth. Enzymol.* **87**, 159.
[2]J. J. Villafranca (1982) *Meth. Enzymol.* **87**, 180.

EXCITED STATE

A state of higher energy experienced by a molecular entity, relative to its ground state. In photic processes, the excited state can be achieved by absorption of photons, by heating, or by extreme pressurization.

EXERGONIC

An adjective describing a process, change, or reaction in which the change in Gibbs free energy (ΔG) is a negative value. If a compound has a negative change in the standard Gibb's free energy of formation, the substance is referred to as an exergonic compound.

EXOCYTOSIS

The release of intracellular constituents by fusion of an intracellular vesicle with the plasma membrane, thereby exposing the contents to the outer surface.

EXOENZYME

Any enzyme acting *at the terminus* of a polymeric substrate molecule. By contrast, the term endoenzyme refers to an enzyme catalyzing a hydrolytic reaction that occurs *within* a polymeric substrate molecule.

EXOERGIC

An adjective describing a process or reaction of nuclear systems which produces heat.

EXO-MALTOHEXAOHYDROLASE

This enzyme [EC 3.2.1.98] removes successive maltohexaose residues from the nonreducing chain ends of 1,4-α-D-glucosidic linkages in amylaceous polysaccharides.

M. Vihinen & P. Mäntsälä (1989) *Crit. Rev. Biochem. Mol. Biol.* **24**, 329.

EXO-MALTOTETRAOHYDROLASE

This enzyme [EC 3.2.1.60] removes successive maltotetraose residues from the nonreducing chain ends by hydrolysis of 1,4-α-D-glucosidic linkages in amylaceous polysaccharides.

M. Vihinen & P. Mäntsälä (1989) *Crit. Rev. Biochem. Mol. Biol.* **24**, 329.

EXONUCLEASE

Any enzyme that catalyzes breakage of a phosphodiester linkage at one or the other end of a polynucleotide chain, resulting in the release of single nucleotides or small oligonucleotides. *See specific exonuclease*

B. Weiss (1981) *The Enzymes*, 3rd ed., **14**, 203.
K. M. T. Muskavitch & S. Linn (1981) *The Enzymes*, 3rd ed., **14**, 233.
L. Grossman, A. Braun, R. Feldberg & I. Mahler (1975) *Ann. Rev. Biochem.* **44**, 19.

EXONUCLEASE III

This enzyme [EC 3.1.11.2] degrades double-stranded DNA progressively in a 3'- to 5'-direction, thereby releasing 5'-phosphomononucleotides.

B. Demple & L. Harrison (1994) *Ann. Rev. Biochem.* **63**, 915.
B. Weiss (1981) *The Enzymes*, 3rd ed., **14**, 203.
E. C. Friedberg, T. Bonura, E. H. Radany & J. D. Love (1981) *The Enzymes*, 3rd ed., **14**, 251.

EXONUCLEASE V

This enzyme [EC 3.1.11.5] catalyzes 5'- to 3'- or 3'- to 5'-exonucleolytic cleavage (when ATP is present), yielding 5'-phosphooligonucleotides.

M. M. Cox & I. R. Lehman (1987) *Ann. Rev. Biochem.* **56**, 229.
K. M. T. Muskavitch & S. Linn (1981) *The Enzymes*, 3rd ed., **14**, 233.

EXOPOLYGALACTURONISIDASE

This enzyme [EC 3.2.1.82] catalyzes the hydrolysis of pectic acid from the nonreducing end, thereby releasing digalacturonate.

R. L. Fischer & A. B. Bennett (1991) *Ann. Rev. Plant Physiol. Plant Mol. Biol.* **42**, 675.

EXOPOLYPHOSPHATASES

These enzymes [EC 3.6.1.11] catalyze the sequential hydrolysis of orthophosphate from polyphosphates.

H. G. Wood & J. E. Clark (1988) *Ann. Rev. Biochem.* **57**, 235.

EXO-SITE

A topologically distinct region on an enzyme that recognizes and binds specific molecules or structures and is far removed from the enzyme's active site. An example is the fibrinogen recognition site on thrombin. This site is spatially distinct from the site for cleavage of the scissile bond of fibrinogen.

J. W. Fenton (1981) *Ann. N.Y. Acad. Sci.* **370**, 468.
M. T. Stubbs & W. Bode (1995) *Trends in Biochem. Sci.* **20**, 23.

EXOTHERMIC

An adjective describing a process, change, or reaction that releases heat. If pressure is constant and no work other than $P\Delta V$ work is done, then $\Delta H = q_P$ where H is the enthalpy and q_P is the heat released at constant pressure. Thus, in such systems, exothermic systems will have negative ΔH values. Since volume changes in chemical and biochemical reactions in solution are small, typically $\Delta H = \Delta U$ in such systems in which U is the internal energy. A compound which has a negative standard enthalpy of formation ($\Delta_f H^{\ominus}$) is referred to as an exothermic compound. An example would be ethanol in which $\Delta_f H^{\ominus}$ (298.15) is -287 kJ/mol. *See also* Endothermic; Exergonic; Exoergic

EXOTOXIN A

Exotoxin A is an NAD$^+$-dependent ADP-ribosylating protein from *Pseudomonas aeruginosa*.

L. Passador & W. Iglewski (1994) *Meth. Enzymol.* **235**, 617.
C.-T. Lai (1986) *Adv. Enzymol.* **58**, 99.

EXPERIMENTAL DESIGN OF INITIAL RATE ENZYME ASSAYS

The basic experimental design for enzyme kinetic assays is directed primarily by the focus of the particular study. If the objective of the assay is related to issues of clinical biochemistry or to enzyme purification, then design protocols are mainly concerned with optimizing each of the assay conditions (*e.g.*, substrate and effector concentrations, pH, temperature, ionic strength, metal ion levels, *etc.*). In such studies, the investigator is often addressing issues related to minimizing interfering agents, to increase the assay sensitivity such that low levels of enzymatic activity can be identified, and to provide an easy, straightforward, cheap, and reproducible protocol. However, the enzyme kineticist is typically working with suboptimal conditions in addressing issues of enzyme-rate behavior, mechanism, protein-ligand and protein-protein interactions, as well as regulation. It is through such detailed analyses that true optimal conditions can be identified.

1. Perhaps of first concern in determining the overall design of a particular assay is the actual method used for product identification (or for substrate depletion) per unit time. Many different methods have been utilized (*e.g.*, radiometric, spectrophotometric, fluorometric, pH-stat, polarimetric, *etc.*) No matter which method is used, the product has to be clearly identified (or substrate, if substrate depletion is being measured). With stopped-time assays, it may be necessary to separate product(s) from substrate(s) prior to determination of the amounts of the metabolite(s) present (as well as demonstration that product(s) and substrate(s) are truly separated). If so, the investigator should be able to demonstrate that the assay procedure clearly measures true initial rates (see below). Closely related to these issues are concerns about purity (**See** *Substrate Purity; Enzyme Purity; Water Purity, etc.*) and stability (**See** *Substrate Stability; Enzyme Stability, etc.*). If the components of the assay mixture are not stable over the time course of the experiment (or, if certain side reactions occur), then corrections have to be made in analyzing the rate behavior.

2. The assay protocol should measure true initial rates (**See** *Initial Rate Condition*). For most systems, this represents a time period in which less than ten percent of the substrate concentration has undergone conversion. However, if a reaction is not significantly favored thermodynamically or if product inhibition is particularly potent, then a much smaller percentage of substrate conversion may be needed such that true initial rate conditions are obtained. Addition of an auxiliary enzyme system may prove necessary to avoid product accumulation. **See** *Coupled Enzyme Assays*

3. Preincubation of the assay mixture is a crucial step in kinetic investigations. If the reaction is to be initiated by the addition of enzyme, then the reaction sample (minus the enzyme) should be incubated under the reaction conditions (particularly temperature) for several minutes. Whenever possible, the initiating agent (*e.g.*, the enzyme) should also undergo thermal equilibration; however, this may not be possible if the agent is relatively

unstable. In some systems, the substrate may be more unstable than the enzyme. In these cases, it may be better to initiate the reaction with the substrate. Nevertheless, whichever means is utilized to preincubate and to initiate the reaction, the volumes of both should be clearly stated in reports. In addition, the volume of the initiating agent should be the smaller of the two volumes, particularly if it is not possible to preincubate the initiatior at the assay temperature.

4. In kinetic studies, it is best to eliminate any lag phases that may alter the initial velocity behavior of an enzyme. As mentioned above, preincubation is an important factor in this issue. Another is mixing. When the initiating agent is added to the reaction mixture, the entire solution should be thoroughly mixed such that true steady-state conditions can be attained (**See** *Mixing Time*). In some instances, the presence of a magnetic stir bar within the reaction vessel can be quite advantageous.

5. Investigators should always be concerned in steady-state investigations that the conditions for the steady-state (*e.g.*, the assumptions in the mathematical analysis of the system) are present. For example, the total enzyme concentration should be significantly less than the concentration of the substrate or any effector (*e.g.*, a noncompetitive inhibitor). In most cases, if the concentration of the substrate(s) is 10^3-fold are greater than that of the enzyme, the steady-state relationship should hold. If not, global analysis and analysis of the total progress curve may be required. A good check on this is to look at the inital velocity of the reaction at different enzyme concentrations (such a plot should always be made in the initial studies of the kinetic mechanism). This plot should be linear, with the line intersecting the origin, in the range of the concentration of the enzyme to be utilized. Nonlinearity can be due to a number of reasons (*e.g.*, polymerization of the enzyme at higher concentration may result in a more or less active enzyme form). In such cases, it may be necessary to study the kinetics of the system under more than one condition. (**See also** *Enzyme Purity*). In addition to the *v* vs. [E_{total}] plot, a progress curve (*i.e.*, [P] *vs.* time) should be made. With single point assays (*i.e.*, periodic sampling of a reaction mixture), the time points of the aliquots taken should be within the initial linear phase of this progress curve. When using continuous assay methods (*e.g.*, observing

the change in spectrophotometric absorbance with time), the initial velocity is obtained by measuring the slope of the progress curve. This slope is also measured in a time frame in which the progress curve is still linear (as mentioned above, typically less than 10% of substrate conversion). **See** *Continuous vs. Single-Point Assays; Progress Curve; Initial Rate Condition*

6. In single-point assays where the reaction is completely terminated upon addition of some quenching agent or condition, care should be exercised to demonstrate that the reaction has indeed been terminated and that the quenching process is fast relative to the reaction velocity. If product is being measured then the product should be stable under the quenching condition (or, at least stable enough under the time frame in which the product concentration is being measured). **See** *Quenching*

EXPONENTIAL

1. A term used to describe a function or expression containing powers or exponents (*e.g.*, y^x). 2. Referring to any function or equation containing or described by the exponential function, e^x or $(2.718281828. . .)^x$. **See** *e*

EXPONENTIAL BREAKDOWN

1. The disintegration of a substance in a first-order manner. 2. A breakdown of the Swain-Schaad relationship in kinetic isotope effect studies[1], usually as a consequence of tunneling or kinetic complexity.

[1]B. J. Bahnson & J. P. Klinman (1995) *Meth. Enzymol.* **249**, 373.

EXTENSIVE PROPERTY

Any property that depends on the concentration or amount of substance present. Examples of extensive properties include mass and internal energy.

EXTENT OF REACTION

An elementary description of the course of any chemical reaction by the amount of substance used up or generated in that reaction. Consider the general reaction:

$$\nu_1 A + \nu_2 B + . . . = \nu_n P + \nu_{n+1} Q + . . .$$

where A, B, . . ., represent the reactants, P, Q, . . ., represent the products, and the corresponding stoichiometry numbers are represented by ν_i. Thus, the extent of reaction, symbolized by ξ, can be defined by $n_x = n_{x_0} \nu_x \xi$, where n_x is the amount of substance X (*i.e.*, A or B or P, *etc.*), n_{x_0} is the amount of X at the beginning of the reaction, and ν_X is the corresponding stoichiometry number (with the convention that the number is positive for products and negative for reactants). It should be noted that the amount of any substance that has reacted, ν_i, can be represented as $dn_i = \nu_i d\xi$. The SI unit for the extent of reaction is the mole. **See** *Rate of Conversion; Rate of Reaction; Velocity*

EXTINCTION COEFFICIENT

An alternative term for the *molar absorption coefficient* or *molar absorptivity*. **See** *Absorption Spectroscopy*

EXTRINSIC FACTOR X ACTIVATING COMPLEX

This complex catalyzes the activation of factor IX or factor X as a part of the overall coagulation cascade.

L. Lorand & K. G. Mann, eds. (1993) *Meth. Enzymol.*, vols. **222** and **223**.
K. G. Mann, R. J. Jenny & S. Krishnaswamy (1988) *Ann. Rev. Biochem.* **57**, 915.

EYRING EQUATION

An equation for the bimolecular rate constant, k, obtained from transition-state theory. This constant is directly proportional to the equilibrium constant between reactants and the activated complex as well as the absolute temperature: $k = (RT/Lh)K^{\ddagger}$. **See** *Transition-State Theory*

E/Z CONVENTION

Because the *cis-trans* system for designating configurational isomers can frequently lead to confusion, the *E/Z* convention is the recommended IUPAC system. The two groups attached to each end of a double bond are given priority numbers in exactly the same way that applies to the *R/S* system for specifying stereochemistry. Higher priority is given to the substituent atom having the higher atomic number; in cases where the atomic number is the same, higher priority is given to the heavier isotopomer; and in those cases where there still remains a possibility for ambiguity, one assigns priority based on the atomic number (then mass) at the first point of difference on two substituents. The so-called *Z*-isomer has the two groups of higher priority (attached to the two carbon atoms of the double bond) residing on the same side of the molecule, whereas the *E*-isomer has the two groups of higher priority residing on the opposite side of the molecule. (*Z* represents the German word *zusammen* for "together," whereas *E* represents the German word *entgagen* for "opposite.")

F

F-430

A tetrapyrrole (related to porphyrins and corrins) containing a nickel ion. This cofactor, corphin, is a crucial component of methyl-coenzyme M reductase, a bacterial enzyme participating in the formation of methane.

FACILITATED TRANSPORT

Carrier-mediated passage of a molecular entity across a membrane (or other barrier). Facilitated transport follows saturation kinetics: i.e, the rate of transport at elevated concentrations of the transportable substrate reaches a maximum that reflects the concentration of carriers/transporters. In this respect, the kinetics resemble the Michaelis-Menten behavior of enzyme-catalyzed reactions. Facilitated diffusion systems are often stereospecific, and they are subject to competitive inhibition. Facilitated transport systems are also distinguished from active transport systems which work against a concentration barrier and require a source of free energy. Simple diffusion often occurs in parallel to facilitated diffusion, and one must correct facilitated transport for the basal rate. This is usually evident when a plot of transport rate *versus* substrate concentration reaches a limiting nonzero rate at saturating substrate[1]. While the term "passive transport" has been used synonymously with facilitated transport, others have suggested that this term may be confused with or mistaken for simple diffusion. **See** Membrane Transport Kinetics

[1]W. D. Stein (1989) *Meth. Enzymol.* **171**, 23.

FARADAY'S CONSTANT

The amount of electricity required to reduce one equivalent of a monovalent ion. The Faraday, symbolized by F or $\mathcal{F}$, is equal to 96,485.309 coulombs per mole. **See** Nernst Equation

FARADAY'S LAWS

1. The amount of an electrolyte decomposed by an electric current is prportional to the size of that current. 2. In a solution containing several electrolytes and through which a current is passing, the amounts of each substance decomposed is proportional to their chemical equivalents.

FARNESYL-DIPHOSPHATE FARNESYLTRANSFERASE

This enzyme [EC 2.5.1.21], also known as farnesyltransferase, presqualene di-diphosphate synthase, and squalene synthase, catalyzes the condensation of two molecules of farnesyl diphosphate to form presqualene diphosphate and diphosphate (or, pyrophosphate). The entire enzyme complex catalyzes the NADPH-dependent reduction of presqualene diphosphate to yield squalene.

E. D. Beytía & J. W. Porter (1976) *Ann. Rev. Biochem.* **45**, 113.

FARNESYL*TRANS*TRANSFERASE

This enzyme [EC 2.5.1.29], also known as geranylgeranyl-diphosphate synthase, catalyzes the condensation of *trans,trans*-farnesyl diphosphate and isopentenyl diphosphate to yield geranylgeranyl diphosphate and pyrophosphate.

H. Kleinig (1989) *Ann. Rev. Plant Physiol. Plant Mol. Biol.* **40**, 39.

FATTY ACID SYNTHASE

This enzyme complex [EC 2.3.1.85] catalyzes the conversion of acetyl-CoA, n moles malonyl-CoA, and $2n$ moles of NADPH to yield long-chain fatty acids, plus $(n+1)$

moles coenzyme A, n moles of carbon dioxide, and $2n$ moles of NADP$^+$.

S. B. Ohlrogge & J. G. Jaworski (1997) *Ann. Rev. Plant Physiol. Plant Mol. Biol.* **48**, 109.

S. Smith (1994) *FASEB J.* **8**, 1248.

L. A. Kleczkowski (1994) *Ann. Rev. Plant Physiol. Plant Mol. Biol.* **45**, 339.

S.-I. Chang & G. G. Hammes (1990) *Acc. Chem. Res.* **23**, 363.

S. J. Wakil & J. K. Stoops (1983) *The Enzymes*, 3rd ed., **16**, 3.

K. Bloch (1977) *Adv. Enzymol.* **45**, 1.

FEEDBACK INHIBITION

A form of regulation in a metabolic pathway in which an end product (or even an intermediate in the pathway) binds to and inhibits an enzyme which catalyzes an earlier reaction in that same pathway. For example, consider the metabolic scheme $A \rightarrow B \rightarrow C \rightarrow D$ in which the steps are respectively catalyzed by the enzymes E_1, E_2, and E_3. Feedback inhibition would be seen when elevated concentrations of D (or C) inhibited E_1.

When more than one end product in branched pathways can bind to and inhibit a enzyme catalyzing an earlier step in the pathway, a number of possible forms of interaction may occur[1] (**See also** *Multiple Inhibition*). These include:

Synergistic Feedback Inhibition: Also referred to as cooperative feedback inhibition; in this scheme, binding of one inhibitor is not mutually exclusive of the other feedback inhibitor(s). For example, if one feedback inhibitor (at a particular concentration) resulted in a certain degree of inhibition, while, at another concentration, the other inhibitor exhibited a different degree of inhibition, synergistic feedback inhibition is present when the inhibition is greater than would be predicted when both inhibitors are present at those same concentrations. Various procedures are available to deal with such cases (*e.g.*, **See** *Yonetani-Theorell Plot; Yagi-Ozawa Plot*). The first step in the biosynthesis of the purine nucleotides, catalyzed by PRPP-glutamine amidotransferase, is reported to be an example of synergistic feedback inhibition[2,3].

Concerted Feedback Inhibition: Concerted feedback inhibition, also called multivalent feedback inhibition, is the limiting case of synergistic feedback inhibition. In this case neither end product alone will inhibit the target enzyme. However, when both are present, inhibition is observed.

Cumulative Feedback Inhibition: In cumulative feedback inhibition, the end products can inhibit the reaction of the target enzyme separately. Many textbooks erroneously indicate that the cumulative feedback inhibition of *E. coli* glutamine synthetase involves separate regulatory sites for each feedback inhibitor. **See** *Cumulative Feedback Inhibition*

Additive Feedback Inhibition: This term is restricted to cases in which the end product inhibitors bind to the target enzyme in a mutually exclusive manner and the degrees of inhibition are additive. Extreme care should be exercised when using this term; in most cases, the inhibition is really cumulative feedback inhibition.

[1] I. H. Segel (1975) *Enzyme Kinetics*, Wiley, New York.

[2] C. T. Caskey, D. M. Ashton & J. B. Wyngaarden (1964) *J. Biol. Chem.* **239**, 2570.

[3] D. P. Nierlich & B. Magasanik (1965) *J. Biol. Chem.* **240**, 358.

FEED-FORWARD ACTIVATION

A regulatory effect observed in many biochemical pathways; an enzyme, which catalyzes a step (typically a late step) in a pathway, is activated by elevated levels of a precursor of a substrate for that enzyme. A possible example of feed-forward activation may be the action of elevated levels of fructose 1,6-bisphosphate on pyruvate kinase.

FeMo-COFACTOR

A crucial inorganic structure found in the FeMo protein of molybdenum-dependent nitrogenase. This cluster contains Fe, Mo, and S in a 7:1:9 ratio. There are two cuboidal subunits, Fe_4S_3 and $MoFe_3S_3$, bridged by three S^{2-} ions and "anchored" to the protein by a histidyl residue linked to the Mo atom and by a cysteinyl residue linked to one of the iron ions of the Fe_4S_3 domain of the cluster. The Mo atom at the periphery of the molecule is six-coordinated. It has three sulfido ligands, the histidyl imidazole-nitrogen as a ligand, and two oxygen atoms from an (R)-homocitrate molecule.

FENTON REACTION

1. A ferrous ion-dependent reaction in which the highly reactive hydroxyl radical is generated from hydrogen peroxide: $Fe^{2+} + H_2O_2 \rightarrow Fe^{3+} + HO\cdot + OH^-$. This reaction possibly proceeds via an oxoiron(IV) intermediate. The addition of an additional reducing agent such as ascorbate leads to a cycle which can increase the

damage to biological molecules. **See also** *Haber-Weiss Reaction; Udenfriend's Reagent*

2. Originally, the Fenton reaction was specifically the oxidation of α-hydroxy acids to α-keto acids using Fenton's reagent, a fresh mixture of hydrogen peroxide and ferrous salts (usually ferrous sulfate)[1-3]. The procedure will also convert 1,2-glycols to hydroxy aldehydes. Fenton's reagent can also be used to hydroxylate aromatic rings. In these reactions, biaryls are typically found as side products. **See also** *Udenfriend's Reagent*

[1]H. J. H. Fenton (1893) *Proc. Chem. Soc.* **9**, 113.
[2]H. J. H. Fenton (1894) *J. Chem. Soc.* **65**, 899.
[3]C. Walling (1975) *Acc. Chem. Res.* **8**, 125.

FERREDOXIN

This redox protein participates in several enzyme-catalyzed reactions, including: glutamate synthase (ferredoxin) [EC 1.4.7.1]; ferredoxin-nitrite reductase [EC 1.7.7.1]; ferredoxin-nitrate reductase [EC 1.7.7.2], sulfite reductase (ferredoxin) [EC 1.8.7.1]; ferredoxin-NADP$^+$ reductase [EC 1.18.1.2]; ferredoxin-NAD$^+$ reductase [EC 1.18.1.3]; and NAD$^+$-dinitrogen-reductase ADP-D-ribosyltransferase [EC 2.4.2.37]. **See also** *Resonance Raman Spectroscopy*

M. W. W. Adams & A. Kletzin (1996) *Adv. Protein Chem.* **48**, 101.
H. Cheng & J. L. Markley (1995) *Ann. Rev. Biophys. Biomol. Struct.* **24**, 209.
P. Schürmann (1995) *Meth. Enzymol.* **252**, 274.
J. J. G. Moura, A. L. Maredo & P. N. Palma (1994) *Meth. Enzymol.* **243**, 165.
H. Eisenberg, M. Mevarech & G. Zaccai (1992) *Adv. Protein Chem.* **43**, 1.
J. B. Howard & D. C. Rees (1991) *Adv. Protein Chem.* **42**, 199.

FERREDOXIN:NADP$^+$ REDUCTASE

This FAD-enzyme [EC 1.18.1.2], also known as adrenodoxin reductase, catalyzes the reaction of reduced ferredoxin and NADP$^+$, yielding oxidized ferredoxin and NADPH.

G. Palmer (1975) *The Enzymes*, 3rd ed., **12**, 1.

FERROCHELATASE

This enzyme [EC 4.99.1.1], also known as protoheme ferro-lyase and heme synthetase, catalyzes the reaction of protoporphyrin with Fe(II) to yield protoheme and two protons.

V. M. Sellers & H. A. Dailey (1997) *Meth. Enzymol.* **281**, 378.
S. Granick & S. I. Beale (1978) *Adv. Enzymol.* **46**, 33.

FERROMAGNETIC

A large, positive magnetically susceptible substance that is capable of being magnetized by a weak magnetic field. If the material retains a high percentage of its magnetization after the applied magnetic field has been removed, that material is said to be "hard." If most of the magnetization is lost, the substance is designated "soft." Major ferromagnetic elements include iron, cobalt, and nickel. Above a certain temperature, known as the Curie point, ferromagnetic materials become paramagnetic. Below that temperature, an increasing applied magnetic field will result in increasing magnetization.

Ferrimagnetism is a special form of antiferromagnetism. Ferrites (an example of which is $Fe_2O_3 \cdot XO$ where X is a divalent metal such as Co, Ni, Zn, or Mn) exhibit both ferromagnetism and ferrimagnetism. The presence of two ions in the solid of an antiferrimagnetic substance will result in antiparallel magnetic dipole moments of adjacent ions in which the difference of the moments is unequal (or in which the number of magnetic moments in one direction is greater than that in the opposite direction). It is possible, by the selection of an appropriate ion, to design ferrimagnetic substances with specific magnetizations.

FERROXIDASE

This copper-dependent enzyme [EC 1.16.3.1], also known as ceruloplasmin, converts $4\,Fe(II) + 4\,H^+ + O_2$ into $4\,Fe(III) + 2\,H_2O$.

E. Frieden & H. S. Hsieh (1976) *Adv. Enzymol.* **44**, 187.
B. G. Malmström, L.-E. Andréasson & B. Reinhammer (1975) *The Enzymes*, 3rd ed., **12**, 507.

[2Fe–2S] CLUSTER

A designation for a two iron, two labile-sulfur cluster present in a number of iron-sulfur proteins. The [2Fe–2S] center consists of two iron atoms bridged by two sulfur atoms (the sulfido-bridges). The oxidation levels of the clusters are usually indicated by adding the charges on the iron and sulfide atoms (*e.g.*, [2Fe–2S]$^{2+}$).

[4Fe–4S] CLUSTER

A designation for a four iron, four labile-sulfur cluster present in a number of iron-sulfur proteins. Possible oxidation levels of this type of cluster include [4Fe–4S]$^{3+}$, [4Fe–4S]$^{2+}$, and [4Fe–4S]$^+$. Note that the individual iron atoms in a particular cluster do not have to have the same oxidation state. For example, there are two [4Fe–4S] clusters in ferredoxin and, in oxidized ferredoxin, there is one Fe(II) and three Fe(III) atoms in each cluster.

FICIN

The following articles describe some of the molecular and physical properties of this enzyme [EC 3.4.22.3] which catalyzes the hydrolysis of peptide bonds and exhibits a broad specificity (at Lys–, Ala–, Tyr–, Gly–, Asn–, Leu–, and Val–).

E. Shaw (1990) *Adv. Enzymol.* **63**, 271.
J. S. Fruton (1982) *Adv. Enzymol.* **53**, 239.
A. N. Glazer & E. L. Smith (1971) *The Enzymes*, 3rd ed., **3**, 502.

FICK'S LAWS OF DIFFUSION

1. The direction of diffusion or movement by solutes is always from a region of higher concentration to one of lower concentration. In addition, the diffusive flux, J_A, of a solute A across a plane at x is equal to $-D(\partial c_A/\partial x)$ where D is the diffusion coefficient and $\partial c_A/\partial x$ is the concentration gradient of A at x. 2. The increase of the concentration of A with time, $\partial c_A/\partial t$, is equal to $D(\partial^2 c_A/\partial x^2)$ where $\partial^2 c_A/\partial x^2$ is the change in the concentration gradient. **See also** *Diffusion; Transport Processes*

FIELD EFFECT

The polarization of a chemical bond in a molecule through space or through solvent molecules, by a remote pole or dipole, an effect that is dependent on the geometry of the molecule. Functional groups are often ranked as being either electron-withdrawing or electron-donating (with respect to hydrogen). Examples of electron-donating groups include O^-, COO^-, CH_3, CH_2R, CHR_2, CR_3, D, as well as atoms of low electronegativity. Examples of electron-withdrawing groups include COOH, COOR, COR, the halogens, OH, SH, SR, OR, NH_3^+, NR_3^+, and $CH{=}CR_2$. In general, sp carbons have more electron-withdrawing effects than do sp^2 carbons which, in turn, have a greater effect than sp^3 carbons. Field effects, also known as direct effects, decrease with distance and orientation and can be influenced by the solvent. **See** *Inductive Effect*

J. D. Roberts & W. T. Moreland, Jr. (1953) *J. Amer. Chem. Soc.* **75**, 2167.
E. Ceppi, W. Eckhardt & C. A. Grob (1973) *Tet. Lett.* **1973**, 3627.

FIRST DERIVATIVE TEST

A procedure used in calculus to identify local maxima and minima in a continuous and differentiable function. The function at some point x in an interval (a,b) is a local maximum if the first derivative, $df(x)/dx$, is greater than zero over the interval to the left of x, is zero at x, and is less than zero over the interval to the right of x. Likewise, the function at some point x in an interval (a,b) is a local minimum if the first derivative, $df(x)/dx$, is less than zero over the interval to the left of x, is zero at x, and is greater than zero over the interval to the right of x.

FIRST-ORDER REACTION KINETICS

A reaction is considered first order if it agrees with the rate expression $v = -d[A]/dt = k[A]$, where v is the reaction rate, t represents time, and k is the first-order rate constant having units of reciprocal time. If $[A_o]$ represents the concentration of A at $t = 0$, and x the concentration of A at time t, then another method of expressing the rate law is $dx/dt = k([A_o] - x)$. Integration of the expression yields $\ln[A] = -kt + \ln[A_o] = \ln([A_o] - x)$ or $[A] = [A_o]e^{-kt}$. A plot of $\ln[A]$ vs. t will yield a straight line having a slope of $-k$. Alternatively, a value for k can be obtained from a determination of the reaction half-life, $t_{1/2}$. From the equation above, it is directly seen that the first-order rate constant is inversely proportional to $t_{1/2}$: *i.e.*, $k = (\ln 2)/t_{1/2}$. As many authors have pointed out[1,2], true first-order rate constants are easily obtained from half-life measurements. Note that the half-life of a first-order reaction is independent of the initial reactant concentration. Thus, all that is necessary is to measure any quantity (such as spectric, radiometric, volumetric, *etc.*) that changes with the concentration of A in such a way that the rate of change in the measured quantity has the same half-life.

There are many examples of first-order reactions: dissociation from a complex, decompositions, isomerizations, *etc.* The decomposition of gaseous nitrogen pentoxide ($2N_2O_5 \rightarrow 4NO_2 + O_2$) was determined to be first order ($-d[N_2O_5]/dt = k[N_2O_5]$) as is the release of product from an enzyme-product complex (EP $\rightarrow$ E + P). In a single-substrate, enzyme-catalyzed reaction in which the substrate concentration is much less than the Michaelis constant (*i.e.*, $[S] \ll K_m$) the reaction is said to be first-order since the Michaelis-Menten equation reduces to

$v = V_{max} [S]/K_m$. A reaction of higher order is called pseudo-first-order if all but one of the reactants are high in concentration and do not change appreciably in concentration over the time course of the reaction. In such cases, these concentrations can be treated as constants. **See** *Order of Reaction; Half-Life; Second-Order Reaction; Zero-Order Reaction; Molecularity; Michaelis-Menten Equation; Chemical Kinetics*

[1] J. W. Moore & R. G. Pearson (1981) *Kinetics and Mechanism*, Wiley, New York.
[2] W. P. Jencks (1969) *Catalysis in Chemistry and Enzymology*, McGraw-Hill, New York. [Also available as Jencks, W. P. (1986) *Catalysis in Chemistry and Enzymology*, Dover Publications, New York].

FISCHER PROJECTION

A method for depicting the geometry of one or more asymmetric tetrahedral centers, where each is repre-

In this way, the Fischer projection of central carbon of D-(+)-glyceraldehyde [also known as *R*-(+)-glyceraldehyde] has the hydroxyl group projecting to the right and the hydrogen atom pointed to the left. Likewise, the Fischer projection of D-glucose in the open chain form has the C1 aldehyde group at the top (pointing vertically) and the C6 hydroxymethyl group at the bottom (again vertically); the hydroxyl groups at C2, C3, C4, and C5 respectively point right, left, right, and right. **See** *R/S Convention for Asymmetric Centers*

FITSIM COMPUTER PROGRAM

Digital computer software that allows an investigator to determine the magnitude of rate constants for individual steps within a chemical reaction mechanism. **See** *KINSIM/FITSIM*

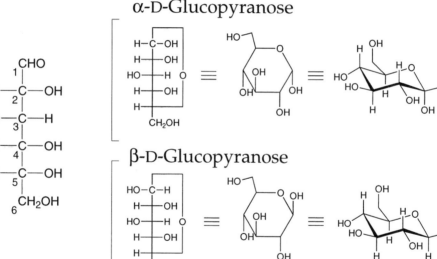

α-D-Glucopyranose

β-D-Glucopyranose

Fischer Projection   Tollens Projection   Haworth Projection   Chair Projection

sented as the point of intersection of the vertical and horizontal lines of a cross. The horizontal lines that extend leftward and rightward represent bonds to those substituents extending out and toward the observer (*i.e.*, above the plane of the paper). The vertical lines that extend upward and downward represent bonds to two other substituents extending backward and behind the plane of the paper. For carbohydrates and alkanes, the convention is that the carbon chain is positioned vertically.

Selected entries from *Methods in Enzymology* [vol, page(s)]:
Data analysis flow chart, **240**, 314-315; data point number requirements, **240**, 314; determination of enzyme kinetic parameters: multisubstrate, **240**, 316-319; single substrate, **240**, 314-316; enzyme mechanism testing, **240**, 322; evaluation of binding processes, **240**, 319321; file transfer protocol site, **240**, 312; instructions for use, **240**, 312-313.

FLAGELLAR ROTATION

The mechanochemical rotatory motion of bacterial flagella, driven by electrochemical proton gradients across the peripheral membrane. Each complete turn requires a coupling to the downhill transport of about a thousand protons. *Escherichia coli* can rotate their flagella at 10 Hertz (*i.e.*, cycles/s).

S. C. Schuster & S. Khan (1994) *Annu. Rev. Biophys. Biomol. Struct.* **23**, 509.

FLASH PHOTOLYSIS

A fast reaction technique that employs sudden photoactivation or photolysis to initiate or alter a chemical reaction system. This sudden perturbation creates a nonequilibrium situation, allowing one to determine the time course of the relaxation of a chemical reaction system back to equilibrium.

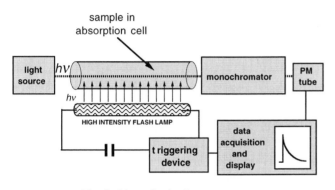

Flash Photolysis Apparatus

Light absorption is usually quite fast (time scale = 1–10 femtoseconds), and various physical measurements can be used to characterize the properties of intermediates that are formed along the reaction coordinate. This strategy was introduced by Porter who later shared the Nobel Prize in Chemistry with Eigen and Norrish for their germinal contributions to fast reaction kinetics. **See** *Chemical Kinetics*

G. Porter (1950) *Proc. Roy Soc.* **A200**, 284.

G. Porter (1960) *Zeit. Elektrochem.* **64**, 59.

G. Porter (1974) in *Investigation of Rates and Mechanisms of Reactions*, Part II, 3rd ed. (G. G. Hammes, ed.) vol. **VI**, pp. 367-462, Wiley-Interscience, New York.

Selected entries from *Methods in Enzymology* [vol, page(s)]: Apparatus, **232**, 89; as central kinetic problem, **232**, 293; electron-transfer reactions, rate measurements, **232**, 89-91; for ligand binding cooperativity measurement, **231**, 147, 149; and ligand rebinding after photodissociation, **232**, 72-73; distribution of states and, **232**, 82; double distribution and, **232**, 82-83; kinetic simulations, **232**, 82; ligand for, **232**, 80-82; photolysis pulse length for, **232**, 79-80; comparison to pulse radiolysis, **233**, 3.

FLAVANONE 3β-HYDROXYLASE

This enzyme [EC 1.14.11.9] uses iron and ascorbate as cofactors in a reaction that converts naringenin, 2-oxoglutarate and dioxygen into dihydrokaempferol, succinate, and carbon dioxide.

A. G. Prescott & P. John (1996) *Ann. Rev. Plant Physiol. Plant Mol. Biol.* **47**, 245.

FLAVINS AND DERIVATIVES

This group includes the coenzyme forms of water-soluble vitamin B₂ or riboflavin. Synthesis occurs by initial cyclohydrolase action on the guanine ring of GTP and subsequent steps lead to the synthesis of the isoalloxazine ring structure (see structures below).

Selected entries from *Methods in Enzymology* [vol, page(s)]: Determination of FMN and FAD by fluorescence titration with apoflavodoxin, **66**, 217; purification of flavin-adenine dinucleotide and coenzyme A on *p*-acetoxymercurianiline-agarose, **66**, 221; a convenient biosynthetic method for the preparation of radioactive flavin nucleotides using *Clostridium kluyveri*, **66**, 227; isolation, chemical synthesis, and properties of roseoflavin, **66**, 235; isolation, synthesis, and properties of 8-hydroxyflavins, **66**, 241; structure, properties and determination of covalently bound flavins, **66**, 253; a two-step chemical synthesis of lumiflavin, **66**, 265; syntheses of 5-deazaflavins, **66**, 267; preparation, characterization, and coenzymic properties of 5-carba-5-deaza and 1-

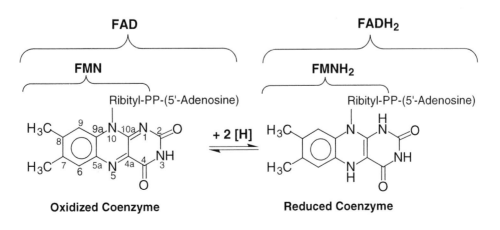

carba-1-deaza analogs of riboflavin, FMN, and FAD, **66**, 277; preparation of flavin 5′-phosphates using immobilized flavokinase, **66**, 287; flavin 1,N^6-ethenoadenine dinucleotide, **66**, 290; flavin suicide inhibitor adducts, **66**, 294; GTP cyclohydrolase II from *Escherichia coli*, **66**, 303; riboflavin synthetase from *Eremothecium ashbyii* and a salvage pathway of the by-product in the enzyme reaction, **66**, 307; continuous fluorescence assay, partial purification, and properties of flavokinase from *Megasphaera elsdenii*, **66**, 323; riboflavin α-glucoside-synthesizing enzyme from pig liver, **66**, 327; isolation and identification of schizoflavins, **66**, 333; flavin affinity chromatography, **66**, 338; activity staining for flavoprotein oxidases, **66**, 344; temperature-difference spectra of flavins and flavoproteins, **66**, 350; fluorescence and optical characteristics of reduced flavins and flavoproteins, **66**, 360; time-resolved fluorescence on flavins and flavoproteins, **66**, 373; application of nuclear magnetic resonance and photochemically induced dynamic nuclear polarization to free and protein-bound flavins, **66**, 385; laser fluorescence techniques, **66**, 416.

FLAVONE SYNTHASE

This enzyme, a P-450 monooxygenase, catalyzes the NADPH-dependent conversion of flavanones to flavones.

A. G. Prescott & P. John (1996) *Ann. Rev. Plant Physiol. Plant Mol. Biol.* **47**, 245.

FLAVONOL SYNTHASE

This enzyme catalyzes the insertion of a double bond in dihydroflavonols.

A. G. Prescott & P. John (1996) *Ann. Rev. Plant Physiol. Plant Mol. Biol.* **47**, 245.

FLUCTUATIONS

Uncertain shifts or swings in the position, motion, or energy of particles or molecules. Fluctuations about some mean value of position, motion, or energy occur for individual particles/molecules, but these microscopic events are not typically sensed when averaged over a large population of particles/molecules. The statistical treatment of fluctuations in what is called statistical mechanics permits one to derive statistical analogues of entropy and free energy. Furthermore, the equilibrium state can be well understood on the basis of statistical mechanics. One can also use the concept of local fluctuations to account for molecular light scattering.

FLUORESCENCE
(1. General Aspects)

Fluorescence has become one of the most powerful tools in the molecular life sciences[1,2]. This is understandable because most biological molecules (particularly macromolecules) absorb light over the 200–800 nm spectral range. Many biomolecules are intensely fluorescent, and even nonfluorescent substances can be investigated with respect to the nature of their interactions by using other fluorescent molecules. Nonfluorescent molecules can be extrinsically labeled with a fluorescent reporter group that can be bound (a) noncovalently at one or more adsorption sites on the macromolecule or (b) covalently through linkages formed by reaction with nucleophiles or electrophiles present on the macromolecule of interest. Indeed, anyone familiar with fluorescence spectroscopy (and microscopy) has good reason to marvel at the variety of labeling procedures at one's disposal. Although only cottage industries a decade ago, specialty companies have grown rapidly in efforts to satisfy the burgeoning needs of spectroscopists and microscopists using fluorescent compounds.

Because fluorescence spectroscopy has become such a robust field of investigation, we cannot possibly cover all of the relevant techniques in this series of entries on fluorescence. For this reason, we will consider the basics of fluorescence, the design of a research fluorimeter, and illustrative cases of ligand binding measurements. Space constraints preclude any discussion of time-dependent fluorescence phenomena including the following topics: multiexponential decay processes, time-domain lifetime measurements, frequency-domain lifetime measurements, static and dynamic quenching, fluorescence anisotropy, energy transfer and distance determinations, as well as excited state reactions. For a more thorough treatment of these and other topics, the interested reader should consult reference 2 as well as the sets of *Methods in Enzymology* entries on fluorescence applications.

BASICS OF ABSORPTION AND FLUORESCENCE. When light interacts with a substance, the incident photons typically collide elastically (in which case Rayleigh light scattering occurs) or they may be absorbed. In simplest terms, the quantum nature of light absorption reflects the fact that there are discrete energy differences (or steps) between successive electronic states, vibrational states, and even rotational states. Another basic principle is that an absorbed photon possessing more energy than that needed for an electronic transition will lead to the occupancy of higher vibrational and rotational energy states. In these cases, the excess energy is dissipated quickly as the molecule reaches the lowest-lying excited state. Furthermore, photon absorption is

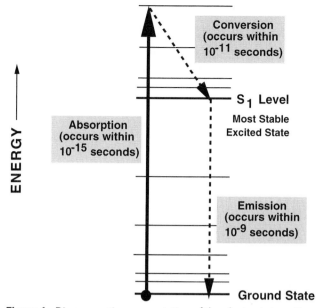

Conversion
(occurs within
10^{-11} seconds)

S$_1$ Level
Most Stable
Excited State

Absorption
(occurs within
10^{-15} seconds)

Emission
(occurs within
10^{-9} seconds)

Ground State

Figure 1. Diagrammatic representation of the electronic transitions taking place within a molecule after photon absorption (indicated here as a boldface upwardly pointed arrow). Excitation is followed by conversion to a less energetic excited state from which emission takes place.

virtually instantaneous (requiring only about 10^{-15} s), and this time is too short (by a factor of 100 to 1000) for changes to occur in the position of nuclei in the substance. (This so-called "vertical transition" is frequently regarded to be a consequence of the Franck-Condon Principle.) The lower probability of transitions to the lowest singlet state makes this transition slower (around 0.01 nanoseconds). Furthermore, the even lower probability of transition to the ground state renders the lifetime of the lowest lying excited state to be in the nanosecond or longer time range.

An absorbed photon's energy can be also be dissipated thermally (*i.e.*, as heat, when all of the excitation energy is converted into vibrational and rotational energy). Such transitions yield no photon in the ultraviolet or visible spectrum, and they are most commonly termed radiationless transitions. Another way in which a molecule can leave its excited state is for the molecule to lose only part of its energy as vibrational and/or rotational energy, thereby "dropping" down or relaxing to the lowest lying excited state (designated S$_1$ in Fig. 1 above). After this conversion, the S$_1$ singlet state may lose its energy in several ways: (a) by resonance energy transfer of its remaining energy to another molecule (termed an acceptor), and this molecule can the act as a quencher (by releasing its newly gotten energy thermally) or as a

fluorescer (by releasing a photon in the UV/visible range, though often at a longer wavelength than would have been produced by photon loss from the S$_1$ state of the donor molecule; (b) by colliding into another molecule which accepts some or all of the energy (a process called collisional quenching); (c) by fluorescence which is often called a "prompt" emission because it occurs in the nanosecond time range; or (d) by a so-called intersystem crossing to form a more stable triplet state which itself can lose the energy as phosphorescence (often termed "delayed" emission) or again by quenching.

Radiationless transitions have an associated rate constant ($k_{radiationless}$); intersystem crossings have a corresponding rate constant ($k_{crossing}$); and fluorescence is characterized by its own rate constant, designated here as $k_{fluorescence}$. [Each of these competing events are first-order relaxation processes. **See** *Chemical Kinetics: (9. Relaxation Kinetics)*] A photo-excited molecule rarely re-emits every photon by fluorescence, and the quantum yield is the ratio of the number of photons produced by fluorescence to the number of photons originally absorbed. Accordingly, the quantum yield ϕ is formally equal to $k_{fluorescence}/[k_{fluorescence} + k_{crossing} + k_{radiationless}]$. As we shall see later, the quantum yield of a substance can be greatly influenced by the fluorophore's immediate environment. Likewise, shifts in the emission spectrum can also help one to detect changes in environment, and the physicochemical basis of such changes can often be inferred.

We should also note that there are other ways in which substances can luminesce (*i.e.*, produce light). One of the most common nonfluorescent mechanisms is called chemiluminescence. Chemical reactions can produce light when radicals (*i.e.*, atoms or molecules with reactive unpaired electrons) combine to form a covalent bond. Heating of molecules can also cause them to display chemiluminescence. And, of course, there are biological processes that are accompanied by chemiluminescence. The most familiar case is the luciferase reaction associated with fire flies and luminescent marine creatures. Peroxidase reactions also produce faint luminescence.

DESIGN FEATURES OF THE SPECTROPHO-TOFLUORIMETER. A sample's fluorescence is almost always dispersed radially, and in principle much greater sensitivity could be achieved by positioning a battery of photodetectors in a spherical arrangement to capture more of the emitted light from the sample. The design (Fig. 2) of the right-angle fluorimeter, however, provides a convenient means (a) for minimizing effects of stray light (*See* *Absorption Spectroscopy*); (b) for making polarization measurements; and (c) for sampling the same region within the cuvette. Collimating slits are mechanical devices that can be controlled manually or by servo circuitry to allow preset or adjustable apertures for light entering and leaving the sample compartment. Slits ensure that the experimenter can manipulate the intensity and purity of light reaching the sample and detector. A larger slit allows more light to pass, but this light may not be uniformly of the same wavelength. A smaller slit width helps to assure that only light within a certain wavelength range will reach the sample and detector, but this diminishes light intensity and likewise lowers instrument sensitivity. A good research fluorimeter allows one to control the intensity and wavelength of the exciting beam of light as well as the photomultiplier wavelength and slit widths of the emitted light.

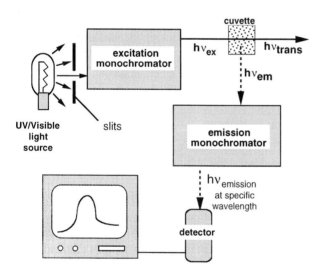

Figure 2. General design features of a fluorescence spectrophotometer.

Fluorimeters use a photomultiplier (or PM) tube that is positioned perpendicular to the incident light beam to detect fluorescence emission by a sample. The photomultiplier has a light-sensitive surface (the photocathode) which is maintained at a negative potential of about 950–1500 volts. So-called photoelectrons are released from photocathode upon light absorption, and they are

amplified by a succession of dynodes, culminating in a pulse of current at the anode. There are two basic ways to acquire fluorescence data, namely through continuous monitoring of the emission photocurrent or by single photon-counting. In the continuous or analog mode, one detects a time-averaged photocurrent, and this requires highly stable amplifiers and power supplies to avoid error. (Most modern instruments use an additional photodetector to monitor fluctuations in incident light intensity resulting from changes in the power output of the light source. Through the use of a partially silvered mirror to divert a fraction (say 5–10%) of the exciting light, this third detector contemporaneously samples the excitation beam and improves sensitivity by automatically correcting the emission photocurrent or photon count for fluctuations in incident light intensity.) In the photon-counting mode, single or individual pulses are registered upon arriving at the anode. This technique increases sensitivity, and any spurious thermally induced loss of electrons from the photocathode can be greatly reduced by using cryotechniques. (Modern photon-counting instruments have Peltier devices to keep the PM tube at $-30°$ to $-40°C$.) One difficulty with single photon counting is called coincidence, which is the simultaneous (or nearly so) arrival of two pulses that are registered as one event. The probability of coincidence increases as the number of photons per second increases, and one should keep counting rates below 1 MHz to avoid this problem.

Unlike absorption spectroscopy, the measurement of fluorescence can yield unreliable data if the experimenter is unaware of intrinsic limitations of fluorescence. To appreciate these issues, we adopt the following "thought" experiment recommended by Laurence[1]. We place a fluorescence cuvette containing an adequate volume of sample within the instrument located in a dark room, leaving the light-tight lid off the sample compartment. (In actual practice, this might require one to defeat any electronic interlocking mechanism used to protect the PM tube from exposure to bright light.) We can now view the cuvette from above, noting that this perpendicular geometry matches that of the pencil of emitted light that reaches the photomultiplier itself. Using the wavelength control on the excitation monochromator, we can scan the wavelength from longer to shorter wavelength. A sharp onset of fluorescence should be sensed by the viewer's eye, and the wavelength yielding maximal output of emitted light can be determined. If there is no large particle light scattering or turbidity in the sample, the color of the fluorescent light should not depend on the wavelength of the exciting light. Another reason why more than one color of fluorescence might occur is that the sample contains more than one fluorescent compound. The viewer should also take note of the intensity of light from the front face (*i.e.*, the surface closest to the exciting light beam) of the cuvette to the back face. The fluorescence should be of uniform intensity throughout the cuvette. If the fluorescence is brightest at the illuminated (or front) face, the solution is probably too concentrated, and insufficient light is able to reach the molecules nearer the back face. Because the emission beam is measured from the light near the center of a cuvette, extremely high concentrations of the sample will have the paradoxical effect of reducing the fluorescence detected by the photomultiplier tube. Diluting the sample should create a condition of uniform light intensity. Another reason why one might not observe uniform light intensity from the front to back face of the cuvette is the presence of a quenching agent. As noted earlier, a quenching agent can accept energy from the photo-excited sample molecule, and release of this energy by vibrational and rotational modes will reduce the intensity of fluorescence.

Even if another type of molecule does not quench the fluorescence, it may still absorb light and reduce the number of photons available for exciting the fluorophore[2]. Figure 3 reminds us that the volume element measured experimentally is located at the center of the sample cuvette. A fraction of incident light is observed over the distance x_1, and the sample only absorbs the light traveling from x_1 to x_2. Thus, the presence of one or more light-absorbing species in the cuvette reduces the photic energy available to excite the volume element that is actually measured. This latter phenomenon is called an "inner-filter" effect, because the presence of another absorbing molecule acts as though it is a filter present within the sample. If the absorbance values of the sample solution are determined at the excitation and emission wavelengths, it is possible to correct for inner-filter effects. A useful equation[2] is:

$$F_{\text{corrected}} = F_{\text{observed}} 10^{\left\{\frac{(Absorbance_{\text{excitation}} - Absorbance_{\text{emission}})}{2}\right\}}$$

The measured absorbances are divide by two, because the midpoint of a 1 cm cuvette is located at a path length of 0.5. Figure 4 shows the observed and corrected fluorescence for a hypothetical compound having appreciable absorbance at the excitation and emission wavelengths.

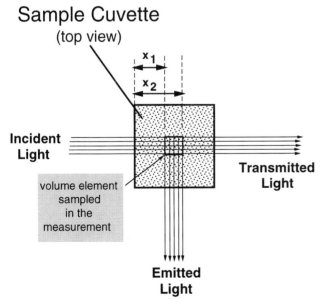

Figure 3. Illustration of the origin of inner filter effects arising by absorption of some exciting photons by one or more substances having an UV-visible absorbance spectrum in the wavelength region of light in the excitation beam.

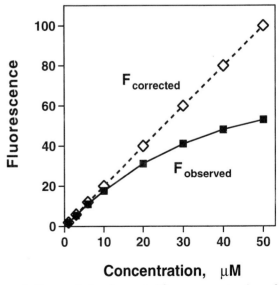

Figure 4. Uncorrected and corrected fluorescence signal as a function of the concentration of a hypothetical fluorophore having significant absorbance at the excitation wavelength (in this case $\varepsilon_{\text{excitation}} = 10,000$ M^{-1}cm^{-1}) as well as at the emission wavelength ($\varepsilon_{\text{emission}} = 10,000$ M^{-1}cm^{-1}).

Notice that the fluorescence intensity is directly proportional to the optical density only over a narrow range. Use of the above equation corrects for this apparent loss in fluorescence intensity.

INTRINSIC AND EXTRINSIC FLUORESCENCE.

Intrinsic fluorescence refers to the fluorescence of the macromolecule itself, and in the case of proteins this typically involves emission from tyrosinyl and tryptophanyl residues, with the latter dominating if excitation is carried out at 280 nm. The distance for tyrosine-to-tryptophan resonance energy transfer is approximately 14 Å, suggesting that this mode of tyrosine fluorescence quenching should occur efficiently in most proteins. Moreover, tyrosine fluorescence is quenched whenever nearby bases (such as carboxylate anions) accept the phenolic proton of tyrosine during the excited state lifetime. To examine tryptophan fluorescence only, one typically excites at 295 nm, where tyrosine weakly absorbs. [Note: While the phenolate ion of tyrosine absorbs around 293 nm, its high pK_a of 10–11 in proteins typically renders its concentration too low to be of practical concern.] The tryptophan emission is maximal at 340–350 nm, depending on the local environment around this intrinsic fluorophore.

Extrinsic fluorescence is used whenever the natural fluorescence of a macromolecule is inadequate for accurate fluorescence measurement. In this case, one can attach a fluorescent reporter group by using the reactive isocyanate or isothiocyanate derivatives of fluorescein or rhodamine, two intensely fluorescent molecules. One can covalently also label a protein's α- and ε-amino groups with dansyl chloride (i.e., N,N-dimethylaminonaphthalenesulfonyl chloride). Another useful reagent is 8-anilino-1-naphthalenesulfonic acid (abbreviated ANS). This compound is bound noncovalently by hydrophobic interactions; in aqueous solutions, ANS is only very fluorescent, but upon binding within an apolar environment, the quantum yield of ANS becomes about 100 times greater.

We should note that there are other intrinsic fluorophores that can be used in binding studies. NADH and NADPH are both highly fluorescent coenzymes. So too are riboflavin and FAD, but flavoproteins are not typically fluorescent. With the exception of the Y base in tRNA, nucleic acids are nonfluorescent.

FLUORESCENCE MEASUREMENTS OF LIGAND BINDING.

In principle, ligand binding may either enhance or quench the intrinsic or extrinsic fluorescence of its macromolecular receptor; or it may change the polarization of the fluorescence emission (see below).

The tandemly repeated Gly-Pro-Pro-Pro-Pro-Pro modules within platelet vasodilator stimulated phosphoprotein (VASP) are thought to serve as docking sites for the actin regulatory protein profilin[3]. Profilin is a ubiquitous protein involved in rearrangement of the actin cytoskeleton. An unusual property of profilin is its high binding affinity for poly(L-proline). The following fluorescence titration data illustrate how the binding of synthetic GPPPPP-triple repeat is attended by an enhancement in the intrinsic fluorescence of profilin[2] (Fig. 5). Note that the fluorescence enhancement reaches a limiting value that is indicative of saturation of available sites.

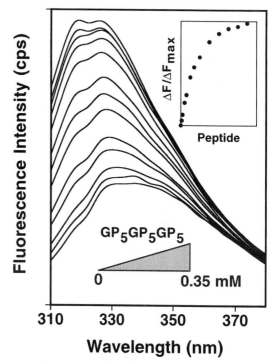

Figure 5. Enhancement of profilin fluorescence in the presence of the oligoproline peptide (GPPPPP)₃.

This is more evident in the replot of the extent of enhancement versus the concentration of the proline-rich peptide, and a dissociation constant of about 85 μM was obtained. Fluorescence studies with [His133Ser]-profilin, a site-directed mutant previously shown to be defective in binding poly(L-proline), exhibits little or no evidence of saturable GP₅GP₅GP₅ binding. On the basis of NMR experiments, the binding of oligoproline peptides to profilin appears to cover one or several tryptophan residues on the protein. Poly(L-proline) binds at the hydrophobic interface between profilin's N- and C-terminal helices and the upper face of its antiparallel β-sheet. As few as six contiguous prolines were deemed to be sufficient for binding profilin.

The binding of 8-anilino-1-naphthalenesulfonate (ANS) to ciliary dynein ATPase resulted in a marked increase in the dye's fluorescence intensity, accompanied by a blue shift in the observed fluorescence emission maximum[4] (Fig. 6). While dynein has 37 ± 3 dye binding

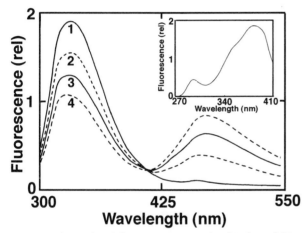

Figure 6. Enhanced ANS fluorescence attending binding of this extrinsic environmentally sensitive probe to sites on sea urchin sperm tail dynein.

sites, ANS binding does not alter enzyme activity. The fluorescence properties of the enzyme-dye complex were used as a conformational probe of dynein binding interactions with substrates and effectors. The fluorescence of the dynein-ANS complex was increased by a number of substrates, including ATP, GTP, and UTP. The transfer of excitation energy from dynein chromophores to adsorbed ANS was also investigated. Dynein appears to undergo a localized conformational change in its interaction with ATP. Native dynein was also found to be conformationally different from heat-activated or NEM-modified enzyme as evidenced by the emission and excitation spectra of the various enzyme-ANS complexes.

[1]D. J. R. Laurence (1957) *Meth. Enzymol.* **4**, 174.

[2]J. R. Lakowicz (1983) *Principles of Fluorescence Spectroscopy*, pp. 496, Plenum Press, New York.

[3]F. Kang, R. O. Laine, M. R. Bubb, F. S. Southwick & D. L. Purich (1997) *Biochemistry* **36**, 8384.

[4]A. C. Saucier, S. Mariotti, S. A. Anderson & D. L. Purich (1985) *Biochemistry* **24**, 7581.

Selected entries from *Methods in Enzymology* [vol, page(s)]:
Energy level diagram, **246**, 336; excitation transfer, **246**, 688; expected intensity, **246**, 684; front-face, **232**, 231-246; hemoglobin adducts [cryogenic methods, **231**, 677-681; room-temperature methods, **231**, 674-676]; high-pressure experiments [hemoglobin subunit dissociation, **232**, 42-55; interpretation, **232**, 50-55]; inner filter effect, **246**, 698; instrumentation, **231**, 677-679; lipid-protein interactions, **259**, 604, 607; low-temperature spectroscopy, **246**, 160-163; measurement by ODMR, **246**, 611-612, 620, 625, 629; multilinear modeling, **246**, 684-685, 688, 700; photosynthetic complexes, **246**, 6; picosecond, **232**, 244-245; protein unfolding, **259**, 496-498, 506-509, 595, 598-603; sensitivity, **232**, 42; **246**, 2, 362; synchronous scanning, hemoglobin adducts, **231**, 674-676; time-resolved [*See Fluorescence Spectroscopy (Time-Resolved)*], time scale, **232**, 42.

FLUORESCENCE (2. Time-Resolved Methods)

Time-resolved fluorescence spectroscopy allows one to measure the rate constants of the processes directly influencing fluorescence lifetimes. This information permits one to infer details about the microenvironment of the fluorophore, thereby elucidating the nature of a macromolecule's internal motions.

One can employ linearly polarized light to excite selectively those fluorophores that are in a particular orientation. The difference between excitation and emitted light polarization changes whenever fluorophores rotate during the period of time between excitation and emission. The magnitude of depolarization can be measured, and one can therefore deduce the fluorophore's rotational relaxation kinetics. Extrinsic fluorescence probes are especially useful here, because the proper choice of their fluorescence lifetime will greatly improve the measurement of rotational relaxation rates. One can also determine the freedom of motion of the probe relative to the rotational diffusion properties of the macromolecule to which it is attached. When held rigidly by the macromolecule, the depolarization of a probe's fluorescence is dominated by the the motion of the macromolecule.

Selected entries from *Methods in Enzymology* [vol, page(s)]:,
Applications, **246**, 335 [immunoassay, **246**, 343-344; nucleic acids, **246**, 344-345; photoreceptors, **246**, 341-343; protein conformation, **246**, 339-340; protein-membrane interactions, **246**, 340-341; two-dimensional imaging, **246**, 345]; energy level diagram, **246**, 336; excited state decay kinetics, **246**, 337-338; in-

terprobe distance determination, **246**, 320-321, 340; phase modulation method, **246**, 299, 345, 359 [data analysis, **246**, 361; detectors, **246**, 360; dynamic range, **246**, 360-361; excitation sources, **246**, 359-360; modulators, **246**, 360; sensitivity, **246**, 361]; pulse methods, **246**, 298-299, 345 [data deconvolution, **246**, 346-347; detectors, **246**, 352-356; global analysis, **246**, 349; lifetime analysis, **246**, 347-348; light sources, **246**, 351-352; scattering sample, **246**, 346; streak camera method, **246**, 356-357; time-correlated single-photon counting, **246**, 349-351; time shift parameter, **246**, 347; up-conversion technique, **246**, 357-359; wavelength dependence of detector response, **246**, 346-347]; resolution of fluorophore-binding extent, **246**, 299-300; rotational diffusion, **246**, 9.

FLUORESCENCE (3. Resonance Energy Transfer)

Fluorescence resonance energy transfer (or FRET) occurs if the absorbed energy of one fluorophore (termed the donor or D) is not emitted, but is instead transferred to a second molecule (termed an acceptor or A). The latter may act either as a fluorescer or as a quencher [*See Fluorescence (1) General Aspects*]. Changes in the conformation of a macromolecule can bring an acceptor molecule closer to or farther from a donor fluorophore, and the efficiency of dipole-dipole resonance energy transfer falls off as the inverse sixth power of the A-to-D distance. Thus, FRET serves as a molecular ruler in the 10–80 Å range. The most common ways for measuring FRET are as (a) a decrease in the emission intensity of the donor; (b) an increase in the emission intensity of the acceptor; or (c) a decreased lifetime for the donor.

Selected entries from *Methods in Enzymology* [**vol**, page(s)]: Analysis of GTP-binding/GTPase cycle of G protein, **237**, 411-412; applications, **240**, 216-217, 247; **246**, 301-302 [diffusion rates, **246**, 303; distance of closest approach, **246**, 303; DNA (Holliday junctions, **246**, 325-326; hybridization, **246**, 324; structure, **246**, 322-324); dye development, **246**, 303, 328; reaction kinetics, **246**, 18, 302-303, 322]; computer programs for testing, **240**, 243-247; conformational distribution determination, **240**, 247-253; decay evaluation [donor fluorescence decay, **240**, 230-234, 249-250, 252; exponential approximation of exact theoretical decay, **240**, 222-229; linked systems, **240**, 234-237, 249-253; randomly distributed fluorophores, **240**, 237-243]; diffusion coefficient determination, **240**, 248, 250-251; diffusion-enhanced FRET, **246**, 326-328; distance measurement [accuracy, **246**, 330; effect of dye orientation, **246**, 305, 312-313; limitations, **246**, 300, 302, 305, 328, 330]; donor-acceptor pairs, as protein-protein interaction readout, **237**, 419-420; dye labeling [effect on macromolecular structure, **246**, 317; extrinsic fluorophores, **246**, 314-316; intrinsic fluorophores, **246**, 313-314; site-specific attachment, **246**, 316-317]; dye selection, **246**, 18, 314-316; effect of distance between pairs, **240**, 218-219, 251-252; effect on donor fluorescence lifetime, **246**, 302, 306; energy level diagram, **246**, 311; energy transfer equations, **246**, 306-307, 309-310; fluorescence decay functions, **240**, 222-223; fluorescence measurements, **237**, 420-422; Hamiltonian of donor-acceptor interaction,

246, 310; intramolecular fluorescence quenching mediated by, **241**, 54-56, 73-75 ; lanthanides, **246**, 330-334; measurement techniques, **237**, 420-422 [controls, **246**, 322; decrease in donor intensity, **246**, 317-318; donor lifetimes, **246**, 320-321, 340; donor photobleaching rates, **246**, 321-322; sensitized emission of acceptor, **246**, 318-320]; polarization measurements, **246**, 317-318; rate of inducing transitions, **246**, 310-311; theory, **246**, 10, 18, 306-313; time evolution for concentration of pairs, **240**, 239-240; transfer rate for dipole-dipole interaction, **240**, 216; uniformity of pair distribution, **240**, 237-239.

FLUORESCENCE CORRELATION SPECTROSCOPY (FCS)

A special case of fluctuation correlation, wherein the laser-induced fluorescence from a very small probe volume (defined by a laser beam focused by a water-immersion microscope objective to an open focal light cell) is autocorrelated in time[1-3]. Random particle motion leads to fluctuations in the number of molecules within the volume element and likewise to fluctuations in the emitted light intensity. Fluctuation analysis offers a noninvasive determination of molecular dynamics in the single molecule range and can be carried out yielding chemical reaction constants or diffusion coefficients. Modest changes in the diffusion coefficient can be determined by measuring of the average decay time of the induced fluorescence light pulses. Fluorescence correlation spectroscopy was first carried out by Magde *et al.*[1], and there are numerous applications[4]. An important application is the determination of thermodynamic and association/dissociation rate constants for intermolecular reactions such as hybridization and renaturation processes between complementary DNA or RNA strands[5] as well as antibody-antigen interactions.

[1] D. Magde, E. L. Elson & W. W. Webb (1972) *Phys. Rev. Lett.* **29**, 705.
[2] E. L. Elson & D. Magde (1974) *Biopolymers* **13**, 1.
[3] D. Magde, E. L. Elson & W. W. Webb (1974) *Biopolymers* **13**, 29.
[4] M. Eigen & R. Rigler (1994) *Proc. Natl. Acad. Sci. U.S.A.* **91**, 5740.
[5] P. Schwille, F. Oehlenschläger & N. Walter (1996) *Biochemistry* **35**, 10182.

FLUORESCENCE DECAY

The process by which the energy of an excited atom, molecule, or molecular species is lost as a consequence of photon release as fluorescence. *See Fluorescence*

Selected entries from *Methods in Enzymology* [**vol**, page(s)]: Analysis, **240**, 290-310; anisotropic, **240**, 301-310; effect of material diffusion, **240**, 219, 221; evaluation of donor fluorescence decay, **240**, 230-234; optimal length of time step, **240**, 224-229; exponential approximation of exact theoretical decay, **240**, 222-229; linked systems, **240**, 234-237; measurement techniques

FLUORESCENCE DEPOLARIZATION

A phenomenon that occurs when a collection of molecules is excited by plane-polarized light, such that only those molecules whose transition dipoles are appropriately aligned will absorb energy. If these excited molecules remain rigidly fixed during excitation and emission, all emitted photons will have their electric vectors lying in the same plane. However, if rotational diffusion occurs during the fluorescence lifetime, then such rotational mobility will reduce the fluorescence light intensity, as sampled by a second polarizer (called the analyzer).

FLUORESCENCE LIFETIME IMAGING MICROSCOPY

A microscopical technique that differentiates molecules on the basis of their respective fluorescence lifetimes.

FLUORESCENCE RECOVERY AFTER PHOTOBLEACHING

A spectroscopic technique (often abbreviated FRAP) that relies on highly localized photobleaching within specific cell regions to detect and/or measure (a) the lateral diffusibility of molecules within membranes, (b) the diffusion and replacement of molecules within assembled structures, such as the cytoskeleton, and (c) related diffusion and exchange of substances within the cell's aqueous phases. The recovery (or redistribution) phase that follows the photobleaching event provides valuable information about the rates of diffusion and photolabel replacement into the photobleach target area, and one makes the implicit assumption that photobleaching does not alter the normal rates of diffusion and replacement that occur continuously within cells.

Wolf[1] has written a detailed chapter on how one goes about the task of designing, building, and using a FRAP instrument. After presenting an overview of the technique, he offers advice on: the choice of laser and microscope; methods for splitting, attenuating, and steering the laser microbeam; pertinent information on photomultipliers and output devices; and the operation of the instrument. Elson and Qian[2] have presented a cogent account of how one interprets FRAP data in terms of the underlying molecular interactions.

[1]D. E. Wolfe (1989) *Meth. Cell Biol.* **30**, 271.
[2]E. L. Elson & H. Qian (1989) *Meth. Cell Biol.* **30**, 307.

FLUOROACETATE/FLUOROCITRATE

Potent metabolic inhibitors of the citric acid cycle. Fluoroacetate ($F-CH_2COO^-$) must first be converted to fluoroacetyl-S-CoA (by acetyl-CoA synthetase) and thence to fluorocitrate (by citrate synthase) before it can act as a potent metabolic inhibitor of the aconitase reaction as well as citrate transport. Submicromolar concentrations of $(-)$-*erythro*-fluorocitrate can irreversibly inhibit citrate uptake by isolated brain mitochondria.

FLUX

1. A rate of transfer of entities, particles, fluids through a given point, surface, or pathway. For example, the different pathways for a particular enzyme-catalyzed reaction will have a different flux through each of those pathways. **See also** *Chemical Flux*. 2. A measure of the power associated with a particular quantity. **See** *Radiant Energy Flux; Radiant Power*. 3. A measure of the strength of a particular field of force (*e.g.*, magnetic flux).

In recognition of the changing nature of virtually all phenomena observable in nature, Heraclitus (*c*. 500 B.C.) is attributed to be the first to assert[1]: "All is flux, nothing stays still." In this repect, Heraclitus may be rightfully recognized as the first kineticist.

[1]Plato (*c*. 360 B.C.) *Cratylus* and Diogenes Laertius (*c*. 250 A.D.) *Lives of the Philosophers*, book IX, sec. 8.

FLUX CONTROL IN METABOLIC PATHWAYS

The regulatory control of the rate of passage of metabolites through a biochemical pathway. This active area of research includes characterization of allosteric activation and inhibition, gene induction and repression, hysteretic interactions, and posttranslational modifications of all the proteins participating in a given metabolic pathway. For example in glycolysis, flux control includes regulation of glucose influx, the interactions of fructose 2,6-bisphosphate, feedforward activation, and feedback inhibition. Interestingly, studies of yeast fermentation rates, in which the rate-limiting enzymes of glycolysis are overexpressed, have not given any significantly positive results[1]. Clearly, there are other aspects of flux control in this well-known pathway that need further elucidation. The recent reports that a subunit of the trehalose synthase complex participates in the control of the flux through yeast glycolysis suggests some unique aspects of regulatory controls in fungi and possibly other organisms[2]. Interestingly, under heat-shock conditions, yeast glucose and trehalose concentrations increase whereas levels of hexose 6-phosphate and fructose 1,6-bisphosphate fall[3]. These observations hint at an interesting mechanism to restrict passage of metabolites through a pathway since trehalose 6-phosphate is an inhibitor of yeast hexokinase.

[1] I. Schaaf, J. Heinisch & F. K. Zimmermann (1989) *Yeast* **5**, 285.
[2] J. M. Thevelein & S. Hohmann (1995) *Trends in Biochem. Sci.* **20**, 3.
[3] M. J. Neves & J. Francois (1992) *Biochem. J.* **288**, 859.

FLUX-GENERATING STEP

Any enzymic reaction that supplies substrate to a metabolic pathway. For all subsequent steps to maintain their steady-state concentrations, flux-generating reactions must exhibit zero-order kinetics.

FLUX RATIO

The ratio of two unidirectional fluxes of substances passing through some boundary or membrane.

FMN ADENYLYLTRANSFERASE

This enzyme [EC 2.7.7.2], also known as FAD synthase and FAD pyrophosphorylase, catalyzes the ATP-dependent transfer of an adenylyl group to FMN to yield FAD and diphosphate.

D. B. McCormick, M. Oka, D. M. Bowers-Komro, Y. Yanada & H. A. Hartman (1997) *Meth. Enzymol.* **280**, 407.

F_oF_1-ATPase (and Related Proton-Transporting ATPases)

This membrane-bound, multi-subunit enzyme [EC 3.6.1.34 and EC 3.6.1.35], also known as mitochondrial ATPase, chloroplast ATPase, coupling factors F_o ("o" standing for oligomycin-sensitive component), F_1 and CF_1, catalyzes the hydrolysis/synthesis of ATP. In mitochondria, chloroplasts and bacteria ATP synthesis is coupled with transport of protons. When located within sealed vesicles, ATPase activity remains latent, but hydrolase activity can be observed upon addition of a proton ionophore whose shuttling of bound protons forth and back across the membrane collapses the transmembrane proton gradient. The phosphoryl transfer step proceeds as a single displacement reaction that is attended by inversion of stereochemical configuration at the phosphorus atom in the transferred phosphoryl group (*i.e.*, no phosphoenzyme intermediate is formed). *See* Binding Change Mechanism

P. D. Boyer (1997) *Ann. Rev. Biochem.* **66**, 717.
J. H. Hurley (1996) *Ann. Rev. Biophys. & Biomol. Struct.* **25**, 137.
R. H. P. Law, S. Manon, R. J. Devenish & P. Nagley (1995) *Meth. Enzymol.* **260**, 133.
J. E. Walker, I. R. Collinson, M. J. Van Raaij & M. J. Runswick (1995) *Meth. Enzymol.* **260**, 163.
J. P. Abrahams, A. G. W. Leslie, R. Lutter & J. E. Walker (1994) *Nature* **370**, 621.
H. S. Penefsky & R. L. Cross (1991) *Adv. Enzymol.* **64**, 173.
W. P. Jencks (1980) *Adv. Enzymol.* **51**, 75.
H. S. Penefsky (1974) *The Enzymes*, 3rd ed., **10**, 375.

FOLYLPOLY-γ-GLUTAMATE SYNTHASE

This enzyme [EC 6.3.2.17], which is also known as folylpoly-γ-glutamate synthetase, catalyzes the sequential ATP-dependent addition of glutamyl groups onto tetrahydrofolyl-$[Glu]_n$ to yield tetrahydrofolyl-$[Glu]_{n+1}$ plus ADP and orthophosphate.

I. Atkinson, T. Garrow, A. Brenner & B. Shane (1997) *Meth. Enzymol.* **281**, 134.
H. C. Imeson & E. A. Cossins (1997) *Meth. Enzymol.* **281**, 141.

FOOTPRINTING

Any technique designed to characterize binding interactions by determining the accessibility of the backbone of macromolecules to cleavage or modification reactions. For nucleic acid interactions, footprinting was originally accomplished by changes in phosphodiester accessibility to DNase I, but numerous chemical and enzymatic methods continue to be elaborated.

FORCE

A vector quantity that changes the state of rest or motion of matter. The force, $\boldsymbol{F}$, is equal to the change in momentum, ρ, with respect to time (*i.e.*, $\boldsymbol{F} = d\rho/dt$). If the mass, m, of the entity is constant, the force is that which produces an acceleration, $\mathbf{a}$ (also a vector quantity), of that entity (*i.e.*, $\boldsymbol{F} = m\mathbf{a}$). The SI unit for force is the newton.

FORCE EFFECTS ON MOLECULAR MOTORS

Exertion of a mechanical force is essential for the functioning of cells and tissues. Various ATP/GTP hydrolysis-dependent motors play key roles in generating force and accomplishing useful work. Examples include the ATP-dependent polymerization of actin monomers, the GTP-dependent polymerization of tubulin dimers, the microtubule-associated dynein and kinesin motors, the ATP-dependent cross-bridging of myosin with actin filaments, various steps in mitosis and meiosis, ATP-dependent ciliary and flagellar dynein motors, the energy-driven bacterial flagellum rotor, and the F_oF_1 ATPase rotor. Within their natural context, cells also exhibit a balance of outwardly and inwardly directed forces, and Ingber[1] recently reviewed tensegrity, a term relating to the structural and organizational interconnectedness of the cytoskeleton with other cellular and extracellular elements. Tensegrity embodies the physical mechanisms that translate mechanical forces into a biochemical response and the integration of the resulting signals with those generated by growth factors and extracellular matrix.

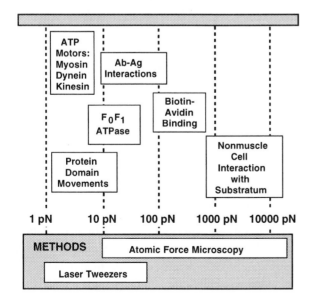

The advent of laser tweezer and atomic force microscope technology has afforded the means for learning more about transduction of chemical and mechanical energy through the measurement of forces generated on a molecular scale. Khan and Scheetz[2] recently considered various mechanisms for force effects on enzymes, broadly defined to include certain slow NTPases such as the motors mentioned above and the assembly-induced ATP and GTP hydrolases (see Figure). Their review also provides several useful insights regarding the nature of force-producing steps in mechanochemical processes.

[1]D. E. Ingber (1997) *Annu. Rev. Physiol.* **59**, 575.
[2]S. Khan & M. P. Sheetz (1997) *Annu. Rev. Biochem.* **66**, 785.

FORCING FUNCTION

A mathematical function used in fast-reaction kinetics to describe how a perturbation of definable strength and duration leads to a change in the kinetic parameters from an initial condition or state to a final state preceding or overlapping with the ensuing chemical relaxation process under investigation.

A chemical system at equilibrium can be changed by a perturbing force applied in the form of a change in pressure, temperature, or electric field strength. The experimenter seeks to convert the chemical system from one set of rate constants before perturbation to another set of rate constants afterward:

$$\underset{k_{-1}}{\overset{k_{+1}}{A \rightleftharpoons B}} \overset{\text{PERTURBATION}}{\underset{f(t)}{\ggggggggggggggggg}} \underset{k_{-1}'}{\overset{k_{+1}'}{A \rightleftharpoons B}}$$

The integrated form of the forcing function $f(t)$ describes the time evolution of the system's transition between these two states (shown in the following plots as the dashed lines; the adjustment of the chemical system in response to the altered rate constants is shown as the solid lines).

The plot shown below is realistic, because the forcing function and the relaxation process are overlapping on a temporal basis. These considerations can be formalized, as described for temporal changes of the concentration(s)[1,2] of the reacting species or in terms of the advancements[3] for each of the individual reaction step(s). Knowledge of the forcing function allows one to obtain corrected chemical relaxation data essential for appro-

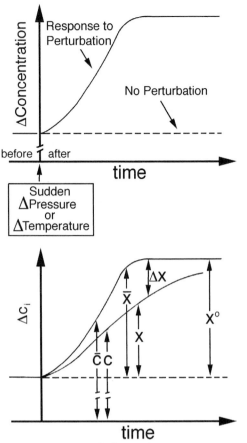

A schematic representation of a forcing function (upper curve) and a relaxation process (lower curve). The system, originally at equilibrium, responds to changes in rate constants and equilibrium constants imposed by the forcing function. For a more complete account of forcing functions, see references 1 and 2 below. Based on a diagram from Ilgenfritz[2].

priate fitting to theoretical models for various potential reaction mechanisms.

[1]M. Eigen & L. DeMaeyer (1973) in *Techniques of Chemistry*, vol. VI (Part 2), pp. 63-146, Wiley, New York.
[2]G. Ilgenfritz (1977) in *Chemical Relaxation in Molecular Biology* (I. Pecht & R. Rigler, eds.), pp. 2-3, Springer-Verlag, Berlin.
[3]P. R. Schimmel (1971) *J. Chem. Phys.* **54**, 4136.

FORMALDEHYDE DEHYDROGENASE

This enzyme [EC 1.2.1.46] converts formaldehyde and NAD^+ and water into formate and NADH. A similarly named enzyme [EC 1.2.1.1] converts formaldehyde and glutathione in the presence of NAD^+ into *S*-formylglutathione plus NADH. *See also* *Aldehyde Dehydrogenase*

M. M. Attwood (1990) *Meth. Enzymol.* **188**, 314.
K. T. Douglas (1987) *Adv. Enzymol.* **59**, 103.

FORMALITY

The number of gram formula weights of solute per liter of solution (designated by the unit Formal or F), as distinguished from molarity (M) which explicitly indicates the concentration of a molecular species. Though often confused and used interchangeably, formality defines how a particular solution is formulated (*i.e.*, prepared), whereas molarity defines what occurs after the solute is combined with its solvent. For example, a 0.01 F solution of lactic acid, when adjusted to a pH value corresponding to its pK_a, is actually 0.005 M lactic acid (the Brønsted acid) and 0.005 M lactate anion (the conjugate base). Most solutions are labeled "Molar" or M, whereas the experimenter should technically mark them as "Formal" or F. In electrochemistry, one frequently uses formality. For example, the standard hydrochloric acid solution in the glass pH electrode is 1 F. Likewise, "formal potential" is the potential of an electrode half-reaction if both reductant and oxidant concentrations are set at 1 F.

FORMAMIDOPYRIMIDINE GLYCOSIDASE

This enzyme catalyzes the release of formamidopyrimidines from methylated, alkali-treated DNA. The enzyme will also act on unmethylated formamidopyrimidine derivatives of both adenine and guanine. *See also* *Formamidomethyluracil-DNA Glycosidase*

B. Demple & L. Harrison (1994) *Ann. Rev. Biochem.* **63**, 915.
T. Lindahl (1982) *Ann. Rev. Biochem.* **51**, 61.
B. K. Duncan (1981) *The Enzymes*, 3rd ed., **14**, 565.

FORMATE DEHYDROGENASE

Enzymes that catalyze the oxidation/reduction reaction converting formate to carbon dioxide utilizing NAD^+ [EC 1.2.1.2], cytochrome b_1 [EC 1.2.2.1], cytochrome c-553 [EC 1.2.2.3], or $NADP^+$ [EC 1.2.1.43] as a redox cofactor.

T. C. Stadtman (1996) *Ann. Rev. Biochem.* **65**, 83.
W. W. Cleland (1995) *Meth. Enzymol.* **249**, 341.
D. R. Jollie & J. D. Lipscomb (1990) *Meth. Enzymol.* **188**, 331.
R. C. Bray (1980) *Adv. Enzymol.* **51**, 107.
T. C. Stadtman (1979) *Adv. Enzymol.* **48**, 1.
K. Dalziel (1975) *The Enzymes*, 3rd ed., **11**, 1.

FORMATION CONSTANT

An association constant (units = M^{-1}) for an equilibrium binding reaction of molecules, most frequently the binding interactions of a macromolecule ligand (which can be a proton, substrate, effector, anion, or cation such as a metal ion). Concentrations of the products of an association or formation reactions appear in the numerator of formation constants. Macromolecules and ligands may undergo pH-dependent changes in charge, and many binding reactions are often also characterized by changes in entropy and/or enthalpy. Thus, formation constants frequently depend on such variables as pH, temperature, and pressure. *See also Ligand Binding; Metal Ion Complexation; Stability Constant; Biomineralization*

FORMIMINOGLUTAMASE

This enzyme [EC 3.5.3.8] catalyzes the hydrolysis of formamide and L-glutamate from *N*-formimino-L-glutamate.

B. Magasanik, E. Kaminskas & J. Kimhi (1971) *Meth. Enzymol.* **17B**, 57.

FORMIMINOGLUTAMATE DEIMINASE

This enzyme [EC 3.5.3.13] catalyzes the hydrolytic deamination of *N*-formimino-L-glutamate to yield *N*-formyl-L-glutamate and ammonia.

R. B. Wickner & H. Tabor (1971) *Meth. Enzymol.* **17B**, 80.

5-FORMIMINOTETRAHYDROFOLATE CYCLODEAMINASE

This enzyme [EC 4.3.1.4] converts 5-formiminotetrahydrofolate to 5,10-methenyltetrahydrofolate and ammonia.

J. I. Rader & F. M. Huennekens (1973) *The Enzymes*, 3rd ed., **9**, 197.

FORMYLGLYCINAMIDE RIBONUCLEOTIDE AMIDOTRANSFERASE

This enzyme [EC 6.3.5.3], also known as phosphoribosylformylglycinamidine synthase, catalyzes the ATP-dependent synthesis of formylglycinamidine ribonucleotide (FGAM) from formylglycinamide ribonucleotide (FGAR) and glutamine, yielding ADP, orthophosphate, and glutamate as products[1-3]. The enzyme is inhibited by 6-diazo-5-oxo-L-norleucine, as is true for many other glutamine-dependent enzymes. Buchanan[3] has described the covalent chemistry of substrates and antimetabolites.

[1]J. M. Buchanan (1973) *Adv. Enzymol.* **39**, 91.
[2]D. E. Koshland, Jr., & A. Levitzki (1974) *The Enzymes*, 3rd ed., **10**, 539.
[3]J. M. Buchanan (1982) *Meth. Enzymol.* **87**, 76.

FORMYLGLYCINAMIDINE SYNTHETASE

This enzyme converts 5′-phosphoribosylformylglycinamide in the presence of ATP, L-glutamine, and water into 5′-phosphoribosylformylglycinamidine, L-glutamate, ADP, and orthophosphate.

H. Zalkin (1993) *Adv. Enzymol.* **66**, 203.

FORMYLMETHANOFURAN DEHYDROGENASE

This enzyme [EC 1.2.99.5] converts formylmethanofuran, in the presence of water and a suitable acceptor, into methanofuran, carbon dioxide, and reduced acceptor.

J. G. Ferry (1992) *Crit. Rev. Biochem. Mol. Biol.* **27**, 473.

FORMYLMETHANOFURAN:5,6,7,8-TETRAHYDROMETHANOPTERIN *N*-FORMYLTRANSFERASE

This enzyme [EC 2.3.1.101] catalyzes the reaction of *N*-formylmethanofuran with 5,6,7,8-tetrahydromethanopterin to yield methanofuran and 5-formyl-5,6,7,8-tetrahydromethanopterin.

J. G. Ferry (1992) *Crit. Rev. Biochem. Mol. Biol.* **27**, 473.

10-FORMYLTETRAHYDROFOLATE DEFORMYLASE

This enzyme [EC 3.5.1.10], also known as formyltetrahydrofolate amidohydrolase and formyl-FH_4 hydrolase, catalyzes the hydrolysis of 10-formyltetrahydrofolate to yield formate and tetrahydrofolate.

J. I. Rader & F. M. Huennekens (1973) *The Enzymes*, 3rd ed., **9**, 197.

10-FORMYLTETRAHYDROFOLATE DEHYDROGENASE

This enzyme [EC 1.5.1.6] catalyzes the conversion of 10-formyltetrahydrofolate in the presence of $NADP^+$ and water to yield tetrahydrofolate, NADPH, and carbon dioxide.

V. Schirch (1997) *Meth. Enzymol.* **281**, 146.

FORMYLTETRAHYDROFOLATE HYDROLASE

This enzyme catalyzes the hydrolysis of N^{10}-formyltetrahydrofolate to produce tetrahydrofolate and formate.

H. Zalkin (1997) *Meth. Enzymol.* **281**, 214.

10-FORMYLTETRAHYDROFOLATE SYNTHETASE

This enzyme [EC 6.3.4.3], whose recommended EC name is formate-tetrahydrofolate ligase, catalyzes the ATP-dependent conversion of formate and tetrahydrofolate to yield ADP, phosphate, and 10-formyltetrahydrofolate.

V. Schirch (1997) *Meth. Enzymol.* **281**, 146.
R. E. MacKenzie (1997) *Meth. Enzymol.* **281**, 171.
J. I. Rader & F. M. Huennekens (1973) *The Enzymes*, 3rd ed., **9**, 197.

FOSTER-NIEMANN PLOTS

A plot of an enzyme progress curve with respect to product inhibition. This plot is particularly useful with regard to enzymes with isomerization steps in which the product acts as a noncompetitive inhibitor. In the Foster-Niemann plot[1,2], $[P]/t$ is plotted as a function of $(1/t) \ln(1 - [P]/[P_\infty])$. As discussed by Orsi and Tipton[3] in their review of progress curves, if product inhibition is competitive (as it would be in a standard Uni Uni mechanism), the Foster-Niemann plot would be linear (note that the integrated form of the Uni Uni mechanism, in the presence of product, is $[P]/t = [((V_{max,f}/K_a) + (V_{max,r}/K_p))/(K_a^{-1} - K_p^{-1})] + \{(1 + ([A_o]/K_p))/(K_a^{-1} - K_p^{-1})) + ([A_o]/(1 + K_{eq}))\}(1/t) \ln(([P_\infty] - [P])/[P_\infty])$ where $V_{max,f}$ and $V_{max,r}$ are the maximum velocities in the forward and reverse reaction, respectively, K_a and K_p are the Michaelis constants for A and P, respectively, K_{eq} is the reaction equilibrium constant, $[P]$ is the concentration of product at time t, and $[P_\infty]$ is the final concentration of product.) However, the Foster-Niemann plot will be curved if either: (1) there is an error in zero time measurements[4] (hence, it behooves the investigator to be precise as to the starting time and conditions of the reaction), or (2) product inhibition is noncompetitive (as is the case with isomerization mechanisms such as the Iso Uni Uni mechanism). As reviewed by Rebholz and Northrop[2], these plots are valuable in that progress curves using both high and low concentrations of substrate can be placed on a normalized scale, it is usually readily apparent if the lines are curved, and the goodness-of-fit of the data can be easily compared with other integrated expressions. **See** *Iso Mechanisms*

[1]R. J. Foster & C. Niemann (1953) *Proc. Natl. Acad. Sci. U.S.A.* **39**, 999.
[2]K. L. Rebholz & D. B. Northrop (1995) *Meth. Enzymol.* **249**, 211.
[3]B. A. Orsi & K. F. Tipton (1979) *Meth. Enzymol.* **63**, 159.
[4]M. Taraszka (1962) Ph.D. Thesis, Univ. of Wisconsin, Madison.

FOURIER TRANSFORM

A mathematical procedure for expressing any single-valued periodic function as the infinite series summation of a succession of sine waveforms of higher order, such that $f(x)$, a function with a period of 2π, can be treated as a trigonometric series:

$$f(x) = a_o + \sum_{n=1}^{\infty} (a_n \cos nx + b_n \sin nx)$$

In addition to this capacity to deconvolute any periodic function, one may likewise reconstruct the original function.

Selected entries from *Methods in Enzymology* [vol, page(s)]: **Application in fluorescence, 240**, 734, 736, 757; convolution, **240**, 490-491; in NMR [discrete transform, **239**, 319-322; inverse transform, **239**, 208, 259; multinuclear multidimensional NMR, **239**, 71-73; shift theorem, **239**, 210; time-domain shape functions, **239**, 208-209]; FT infrared spectroscopy [iron-coordinated CO, in difference spectrum of photolyzed carbonmonoxymyoglobin, **232**, 186-187; for fatty acyl ester determination in small cell samples, **233**, 311-313; myoglobin conformational substrates, **232**, 186-187].

Applications in infrared spectroscopy: A and B bands, iron-coordinated CO, **232**, 186-187; application to allosteric mechanisms, **249**, 566; bacteriorhodopsin, **246**, 9, 380-381; caged compounds, **246**, 6, 520-521; DNA [base pair formation, **246**, 506; conformation, **246**, 506-507; denaturation thermodynamics, **246**, 506; ligand interactions, **246**, 6, 507; sample requirements, **246**, 506]; fatty acyl ester determination in small cell samples, **233**, 311-313; instrumentation [energy throughput, **246**, 503-504; interferometer, **246**, 504; multiplex advantage, **246**, 503-504; lipids [anesthetic perturbations, **246**, 509; disorder in membranes, **246**, 508-510; head groups, **246**, 510; membrane models, **246**, 507-508; phase transitions, **246**, 508-509; protein interactions, **246**, 510]; myoglobin conformational substrates, **232**, 186-187; photosystem II tyrosyl radicals, **258**, 316-318; polarization, protein order in membranes, **246**, 7; proteins [amide I band, **246**, 512-513; amide II mode, **246**, 512; derivative spectra, **246**, 513; proton exchange rates, **246**, 512; secondary structure, accuracy, **246**, 513-514; water interference in spectra, **246**, 138, 380, 387, 503, 513]; reaction-induced infrared difference spectroscopy [bacteriorhodopsin, **246**, 519-520; chromoproteins, **246**, 517-519]; trigger [electron transfer, **246**, 522; light, **246**, 517-522; requirements, **246**, 517]; sample requirements, **246**, 526; sensitivity, **246**, 380-381, 517; time-resolved spectroscopy [rapid-scan technique, **246**, 381, 523; step-scan method, **246**, 381, 523; stroboscopic method, **246**, 381, 523].

FOURIER TRANSFORM INFRARED/ PHOTOACOUSTIC SPECTROSCOPY TO ASSESS SECONDARY STRUCTURE

Fourier transform infrared/photoacoustic spectroscopy (FT-IR/PAS) can be used to evaluate the secondary structure of proteins, as demonstrated by experiments on concanavalin A, hemoglobin, lysozyme, and trypsin, four proteins having different distributions of secondary structures. Secondary structure assessed on the basis of FT-IR/PAS agreed well with that determined by X-ray diffraction, CD, and traditional FT-IR methods. This technique requires smaller amounts of sample, and the secondary structure can be evaluated with as little as 0.5 μg protein.

S. Luo, C. Y. Huang, J. F. McClelland & D. J. Graves (1994) *Anal. Biochem.* **216**, 67.

FRACTAL

A mathematically definable structure which exhibits the property of always appearing to have the same morphology, even when the observer "endlessly" enlarges portions of it. In general, fractals have three features: heterogeneity, self-similarity, and the absence of a well-defined scale of length. Fractals have become important concepts in modern nonlinear dynamics. *See Chaos Theory*

B. B. Mandelbrot (1983) *The Fractal Geometry of Nature*, Freeman, New York.

B. B. Mandelbrot (1984) *J. Stat. Phys.* **34**, 895.

Selected entries from *Methods in Enzymology* [vol, page(s)]: Background literature, **210**, 638; in biological data, **210**, 648; definition, **210**, 665; identification in biochemical reactions, **210**, 651; related terminology, **210**, 655; for time scaling in biochemical networks, **210**, 636; noise scaling, **210**, 643; time history analysis, **210**, 643.

FRACTAL REACTION KINETICS

Reactions occurring at interfaces of two or more physical phases (namely those processes referred to as heterogeneous reactions) as well as reactions occurring in an understirred solution often do not obey the kinetic theory that applies to reactions taking place in well-stirred solutions. These diffusion-controlled reactions typically exhibit geometric constraints. A good example is crystal growth, a process occurring at the junction between liquid and crystal phases; crystal growth kinetics reflect a complex manifold of processes, including the diffusion of molecules from the bulk solution phase into the Gouy-Chapman layer (*i.e.*, bulk diffusion), diffusion of molecules into the Stern layer, two-dimensional surface transport, and capture by the crystal lattice. The converse (*i.e.*, dissolution) is likewise the culmination of a series of diffusion-limited steps. Fractals can be used to treat the kinetics at interfaces as well as percolation clusters[1], and such approaches allow one to deal with fractal orders for some elementary reactions, self-ordering and self-unmixing of reactants, as well as rate coefficients that exhibit temporal memory. Fractal network dynamics[2]

can also be applied to the special kinetic behavior of random structures that exhibit fractal geometry.

[1] R. Kopelman (1988) *Science* **241**, 1620.
[2] R. Orbach (1986) *Science* **231**, 814.

FRACTIONATION FACTOR

A ratio defining the isotopic distribution of two isotopes equilibrated between two different chemical species. If X, followed by a subscript, represents the mole fraction of an isotope (denoted by that same subscript), then the fractionation factor, often symbolized by ϕ, with respect to chemical species A and B is $(X_1/X_2)_A/(X_1/X_2)_B$. Fractionation factors can also refer to different sites, A and B, within the same chemical species. As an example, the deuterium solvent fractionation factor, used in studying solvent isotope effects, is $\phi = (X_D/X_H)_{solute}/(X_D/X_H)_{solvent}$.

IUPAC (1979) *Pure and Appl. Chem.* **51**, 1725.

FRAGMENTATION

1. The heterolytic cleavage of a molecular entity of the form $R_1-R_2-R_3-R_4-X \rightarrow R_1-R_2^+ + R_3=R_4 + :X^-$.
2. The conversion of a radical to a diamagnetic molecule and a smaller radical: *e.g.*, $R_1-R_2-X\cdot \rightarrow R_1=R_2 + X\cdot$.
3. The conversion of a radical ion to an ion and a separate free radical.

IUPAC (1979) *Pure and Appl. Chem.* **51**, 1725.

FRANCK-CONDON PRINCIPLE

The most likely electronic transition will occur without changes in the positions of the nuclei (*e.g.*, little change in the distance between atoms) in the molecular entity and its environment. Such a state is known as a Franck-Condon state, and the transition is referred to as a vertical transition. In such transitions, the intensity of the vibronic transition is proportional to the square of the overlap interval between the vibrational wavefunctions of the two states. *See Fluorescence; Jablonski Diagram*

Comm. on Photochem. (1988) *Pure and Appl. Chem.* **60**, 1055.

FREE RADICAL

A molecular entity having an unpaired electron. Free radicals are usually short-lived and are highly reactive. Examples include methyl radical ($\cdot CH_3$), hydroxyl radical ($\cdot OH$), nitric oxide ($NO\cdot$), hydroperoxy radical ($HOO\cdot$), and the sodium atom (Na). Note that, since there is at least one unpaired electron, a free radical has a net magnetic moment, is paramagnetic, and can yield an electron spin resonance spectrum.

The unpaired electron is usually depicted as a dot adjacent to the symbol of the atom having the highest spin density. Some investigators have suggested that the word "free" not be so widely used, favoring instead the restricted use of the term "free radical" for indicating those radicals not forming parts of radical pairs. Charged free radicals are more commonly called radical anions or radical cations. σ-Radicals are those molecular entities in which the unpaired electron occupies an orbital having considerable s character whereas π-radicals are those molecular entities with an unpaired electron in an orbital with p character. **See** *Radical; Biradical; Transient Chemical Species; Persistent*

FREQUENCY

1. The number of times that an event or a periodic function repeats during a unit value of an independent variable, usually time. 2. For any periodic motion, the frequency, symbolized by ν, is the number of repetitions (*e.g.*, number of revolutions or cycles or oscillations of an electromagnetic wave) of some process occurring in a unit period of time. The SI unit for frequency, reciprocal second, is often referred to as a hertz (Hz). However, the term hertz should only be used in reference to cycles per second and not for radial (circular) frequency or angular velocity symbolized by $\varphi\,(=2\pi\nu)$ and having SI units of rad·s^{-1}.

FRIEDEN DILEMMA

A potential limitation encountered when one seeks to characterize the kinetic binding order of certain rapid equilibrium enzyme-catalyzed reactions containing specific abortive complexes. Frieden[1] pointed out that initial rate kinetics alone were limited in the ability to distinguish a rapid equilibrium random Bi Bi mechanism from a rapid equilibrium ordered Bi Bi mechanism if the ordered mechanism could also form the EB and EP abortive complexes. Isotope exchange at equilibrium experiments[2] would also be ineffective. However, such a dilemma would be a problem only for those rapid equilibrium enzymes having k_{cat} values less than 30–50 sec^{-1}.[2] For those rapid equilibrium systems in which k_{cat} is small, Frieden's dilemma necessitates the use of procedures other than standard initial rate kinetics.

[1]C. Frieden (1976) *Biochem. Biophys. Res. Commun.* **68**, 914.
[2]D. L. Purich, R. D. Allison & J. A. Todhunter (1977) *Biochem. Biophys. Res. Commun.* **77**, 753.

FRIEDEN'S METHOD FOR TER-REACTANT ENZYMIC KINETICS

Frieden[1] presented a protocol for the kinetic analysis of enzymes having three substrates (A, B, and C). In this procedure, one substrate is held constant at a high concentration (*i.e.*, greater than its corresponding Michaelis constant), albeit nonsaturating. The system is then treated as a pseudo-two-substrate enzyme. Thus as an example, C would be held at a high, nonsaturating, and constant concentration while initial rates are measured at different concentrations of A (the concentrations of B being varied at different constant levels as well). The identical procedure is then followed with high, constant, nonsaturating levels of B and then of A.

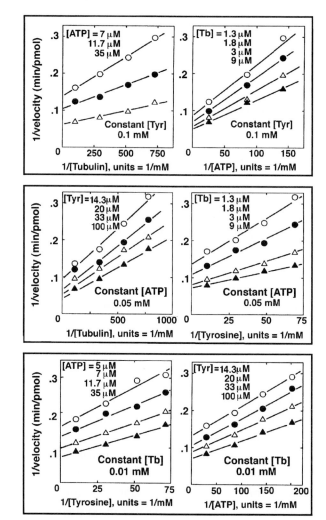

Initial rate patterns for bovine brain tubulin:tyrosine ligase (Reaction: detyrosinated tubulin (Tb) + L-tyrosine + ATP = tyrosinated tubulin + ADP + orthophosphate). The results of each set of experiments are shown as a single frame containing the same data points plotted as a function of one or the other varied substrate. The nonvaried substrate concentration is shown in the lower right-hand corner of every double-reciprocal plot. (From ref. 2.)

Hence, a ter-reactant system will result in six double-reciprocal plots (or Eadie-Hofstee plots, *etc.*). Fromm[3] has pointed out the importance of keeping the third substrate to nonsaturating conditions. Under saturating conditions, parallel lines might result, even though the enzyme is not ping pong in nature. If parallel lines are observed in ter-reactant (or higher) systems, it is strongly urged to repeat the kinetic study at a lower concentration of the unvaried substrate, in order to verify the parallel-lined behavior. The Frieden method is easily extended to quad- and higher systems. **See** *Fromm Method; Substrate Concentration Range*

[1]C. Frieden (1959) *J. Biol. Chem.* **234**, 2891.
[2]N. L. Deans, R. D. Allison, & D. L. Purich (1992) *Biochem. J.* **286**, 243.
[3]H. J. Fromm (1975) *Initial Rate Enzyme Kinetics*, Springer-Verlag, New York.

FROMM'S METHOD FOR DERIVING ENZYME RATE EQUATIONS

A useful procedure for deriving steady-state rate expressions for enzyme-catalyzed reactions[1,2]. Although not as commonly used as the King and Altman method, it is far more convenient (and less error-prone) when attempting to obtain expressions for complicated reaction schemes. One of its values is that the approach is very systematic and straightforward. The systematic nature of the procedure can be illustrated by the derivation of the steady-state ordered Bi Bi reaction.

$$E + A \underset{k_2}{\overset{k_1}{\rightleftharpoons}} EA$$

$$EA + B \underset{k_4}{\overset{k_3}{\rightleftharpoons}} EXY$$

$$EXY \underset{k_6}{\overset{k_5}{\rightleftharpoons}} EQ + P$$

$$EQ \underset{k_8}{\overset{k_7}{\rightleftharpoons}} E + Q$$

STEP ONE: Using just the different enzyme forms, write out the reaction scheme, such that a closed geometric pattern is generated. In this case (E) occurs at the start and end of the catalytic reaction cycle, and we get:

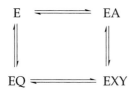

STEP TWO: Number each enzyme form with a circled number, and indicate each arrow with the corresponding unimolecular or bimolecular rate constant. For bimolecular rate constants, the corresponding reactant species (*i.e.*, substrate or product) should also be identified with the corresponding arrow. [To avoid errors by misplacing a minus sign, it is recommended that only positive numbers be used as subscripts for the rate constants (preferably odd numbers for forward reactions, even numbers for reverse reactions).] Thus, for our ordered Bi Bi scheme:

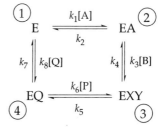

STEP THREE: Each circled number (*i.e.*, each enzyme form or enzyme-containing species) is characterized by one or more arrows leading away from that species. List these numbers in parentheses (the same numbers that were circled in step two) equal to a sum of those rate constants (including the substrate or product associated with bimolecular steps). Thus,

$$(1) = k_1A + k_8Q$$

$$(2) = k_2 + k_3B$$

$$(3) = k_4 + k_5$$

$$(4) = k_6P + k_7$$

Note that the letters represent concentrations (*i.e.*, [A], [B], [P], and [Q]).

STEP FOUR: For each enzyme form, write down all of the shortest one-step paths contributing to the formation of that enzyme form (including the rate constant and substrate or product for bimolecular steps, with each arrow). Thus, for our example:

for 1 we have $2 \xrightarrow{k_2} 1$ and $4 \xrightarrow{k_7} 1$

for 2 we have $1 \xrightarrow{k_1A} 2$ and $3 \xrightarrow{k_4} 2$

for 3 we have $2 \xrightarrow{k_3B} 3$ and $4 \xrightarrow{k_6P} 3$

for 4 we have $1 \xrightarrow{k_8Q} 4$ and $3 \xrightarrow{k_5} 4$

STEP FIVE: Adjacent to each one-step route (identified in the previous step) are written (in parentheses) the numbers of the enzyme species not appearing in that particular one-step path. These "collections" of one-step terms (*i.e.*, the "one-branch" approach of Fromm), are summed together to form the determinant for that enzyme species. Thus:

$$① = 2 \rightarrow 1(3)(4) + 4 \rightarrow 1(2)(3) = E$$

$$② = 1 \rightarrow 2(3)(4) + 3 \rightarrow 2(1)(4) = EA$$

$$③ = 2 \rightarrow 3(1)(4) + 4 \rightarrow 3(1)(2) = EXY$$

$$④ = 1 \rightarrow 4(2)(3) + 3 \rightarrow 4(1)(2) = EQ$$

(One may also opt to use Huang's "consecutive-branch" approach.[3,4])

STEP SIX: Substitute appropriate rate constants for the arrows (from step four) and the numbers in parentheses with the expressions obtained from step three. Thus,

$$[E] = ① = k_2(k_4 + k_5)(k_6P + k_7)$$
$$+ k_7(k_2 + k_3B)(k_4 + k_5)$$

$$[EA] = ② = k_1A(k_4 + k_5)(k_6P + k_7)$$
$$+ k_4(k_1A + k_8Q)(k_6P + k_7)$$

$$[EXY] = ③ = k_3B(k_1A + k_8Q)(k_6P + k_7)$$
$$+ k_6P(k_1A + k_8Q)(k_2 + k_3B)$$

$$[EQ] = ④ = k_8Q(k_2 + k_3B)(k_4 + k_5)$$
$$+ k_5(k_1A + k_8Q)(k_2 + k_3B)$$

STEP SEVEN: Expand each term in step six.

$$[E] = k_2k_4k_6P + k_2k_4k_7 + k_2k_5k_6P + k_2k_5k_7$$
$$+ k_2k_4k_7 + k_2k_5k_7 + k_3k_4k_7B + k_3k_5k_7B$$

$$[EA] = k_1k_4k_6AP + k_1k_4k_7A + k_1k_5k_6AP$$
$$+ k_1k_5k_7A + k_1k_4k_6AP + k_1k_4k_7A$$
$$+ k_4k_6k_8PQ + k_4k_7k_8Q$$

$$[EXY] = k_1k_3k_6ABP + k_1k_3k_7AB + k_3k_6k_8BPQ$$
$$+ k_3k_7k_8BQ + k_1k_2k_6AP + k_1k_3k_6ABP$$
$$+ k_2k_6k_8PQ + k_3k_6k_8BPQ$$

$$[EQ] = k_2k_4k_8Q + k_2k_5k_8Q + k_3k_4k_8BQ$$
$$+ k_3k_5k_8BQ + k_1k_2k_5A + k_1k_3k_5AB$$
$$+ k_2k_5k_8Q + k_3k_5k_8BQ$$

STEP EIGHT: Eliminate all redundant terms in the determinant for each enzyme form. For example, in the expanded expression for [E], the $k_2k_4k_7$ terms appear twice. One of these is eliminated. The same procedure is followed for each enzyme form.

STEP NINE: Eliminate all "forbidden" terms appearing in any determinant. These are terms containing products of two rate constants associated with the same reversible step. For example, in the expression for [E], $k_3k_4k_7B$ and $k_2k_5k_6P$ are forbidden terms in that k_3k_4 and k_5k_6, respectively, (*i.e.*, cyclic terms) are present.

STEP TEN: Eliminate any closed loop terms. There are no closed loop terms in the example given above (if there were it would either contain the product $k_1k_3k_5k_7$ or $k_2k_4k_6k_8$). Care should be exercised here. Although closed loop terms are rare, not considering them in enzyme rate derivations can lead to incorrect expressions and, thus, inaccurate and erroneous predictions of enzyme rate behavior. A good check is to write out all of the rate-constant products that would constitute a closed loop and then check each enzyme determinant to see if any are present. If so, they are eliminated. Huang's modification[3,4] of Fromm's systematic approach also addresses this issue of closed loops.

After each of these elimination steps, the final determinants for the enzyme forms of the steady-state ordered Bi Bi reaction scheme are:

$$[E] = k_2k_7(k_4 + k_5) + k_3k_5k_7[B] + k_2k_4k_6[P]$$

$$[EA] = k_1k_7(k_4 + k_5)[A] + k_1k_4k_6[A][P]$$
$$+ k_4k_6k_8[P][Q]$$

$$[EXY] = k_1k_3k_7[A][B] + k_2k_6k_8[P][Q]$$
$$+ k_1k_3k_6[A][B][P] + k_3k_6k_8[B][P][Q]$$

$$[EQ] = k_1k_3k_5[A][B] + k_3k_5k_8[B][Q]$$
$$+ k_2k_8(k_4 + k_5)[Q]$$

Since $[E_{total}] = [E] + [EA] + [EXY] + [EQ]$, the relative proportion of each enzyme form during the steady-state conditions is given by the ratio of the determinant of the specific enzyme form to that of the total enzyme: *e.g.*, $[E]/[E_{total}]$.

Under the typical condition of initial rate studies in which [P] and [Q] are effectively zero (or, are sufficiently below their respective K_i values to be kinetically insignificant), the terms in the determinant containing [P] and [Q] are eliminated (obviously, this would not be the case in product inhibition studies). The velocity of the enzyme catalyzed reaction is given by $d[Q]/dt = v = k_7[EQ] - k_8[E][Q]$ which reduces to $k_7[EQ]$ since $[Q] \approx 0$. Substituting the determinant for $[EQ]$ and multiplying by unity in the form of $[E_{total}]/[E_{total}] = [E_{total}]/\{[E] + [EA] + [EXY] + [EQ]\}$ in which the denominator is the sum of all the determinants obtained in the derivation, one obtains an expression for the steady-state rate of the reaction in terms of all the appropriate rate constants:

$$\nu = \frac{k_1k_3k_5k_7[A][B][E_{total}]}{\{k_2k_7(k_4 + k_5) + k_1k_7(k_4 + k_5)[A] + k_3k_5k_7[B] + k_1k_3(k_5 + k_7)[A][B]\}}$$

If the standard form of the rate expression is to be used (*i.e.*, with Michaelis constants, maximum velocities, *etc.*) (**See** *Cleland Nomenclature*), then both the numerator and denominator are divided by the coefficient of the $[A][B]$ term appearing in the denominator (*i.e.*, divide by the coefficient of the term containing all of the substrates utilized in the reaction raised to the power equal to the stoichiometry number of each substrate). Thus, the ordered Bi Bi expression becomes

$$\nu = \frac{V_{max}[A][B]}{K_{ia}K_b + K_b[A] + K_a[B] + [A][B]}$$

in which V_{max} is the maximum forward reaction velocity equal to $k_5k_7[E_{total}]/(k_5 + k_7)$, K_{ia} is the dissociation constant for substrate A and is equal to k_2/k_1, K_a is the Michaelis constant for A and is equal to $k_5k_7/\{k_1(k_5 + k_7)\}$, and K_b is the Michaelis constant for B and is equal to $k_7(k_4 + k_5)/\{k_3(k_5 + k_7)\}$. **See** *Enzyme Rate Equations (2. Derivation of Initial Velocity Equations)*

[1]H. J. Fromm (1970) *Biochem. Biophys. Res. Commun.* **40**, 692.
[2]H. J. Fromm (1975) *Initial Rate Enzyme Kinetics*, Springer-Verlag, New York.
[3]C. Y. Huang (1978) *Fed. Proc., Fed. Am. Soc. Exp. Biol.* **37**, 1423.
[4]C. Y. Huang (1979) *Meth. Enzymol.* **63**, 54.

FROMM'S METHOD FOR TER-REACTANT ENZYMIC KINETICS

A procedure to simplify the experimental method in the kinetic analysis of three-substrate, enzyme-catalyzed reactions[1-4]. In this method, the concentration of one substrate is varied while the other two substrates are kept in a constant ratio and in which the individual concentrations of these two substrates are in the neighborhood of their respective Michaelis constants. The experi-

ment is then repeated a number of times utilizing different absolute concentrations of the two substrates, albeit maintaining the same ratio. The same procedure is then used such that each substrate will be the varied reactant.

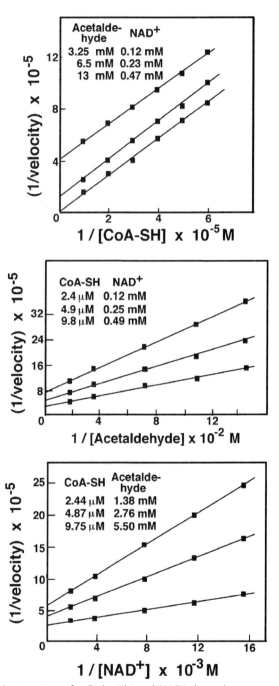

Initial rate patterns for *Escherichia coli* NAD$^+$-dependent coenzyme A-linked aldehyde dehydrogenase (Reaction: NAD$^+$ + CoA-SH + acetaldehyde = NADH + acetyl-S-CoA + H$^+$). The results of each of three experiments are shown as a single double-reciprocal plot, and the nonvaried substrate concentrations for each curve are indicated above the data points.

With this protocol, ping pong reactions will yield at least one plot containing parallel lines. However, it is difficult with this procedure to obtain values for a number of the kinetic parameters (e.g., Michaelis constants) for the reaction. An example of a system studied with this procedure is provided by E. coli CoA-linked aldehyde dehydrogenase[2]. **See** Frieden Protocol

[1]H. J. Fromm (1967) Biochim. Biophys. Acta **139**, 221.
[2]F. B. Rudolph, D. L. Purich & H. J. Fromm (1968) J. Biol. Chem. **243**, 5539.
[3]F. B. Rudolph & H. J. Fromm (1969) J. Biol. Chem. **244**, 3832.
[4]H. J. Fromm (1975) Initial Rate Enzyme Kinetics, Springer-Verlag, New York.

FRONTIER ORBITALS

The highest-energy occupied molecular orbital (HOMO) and the lowest-energy unoccupied molecular orbital (LUMO) of a particular molecular entity. The HOMO can be completely or partially filled whereas the LUMO can be completely or partially vacant. The frontier orbital method allows one to interpret reaction behavior by studying HOMO and LUMO overlaps between reacting entities.

K. Fukai (1971) Acc. Chem. Res. **4**, 57.
I. Fleming (1976) Frontier Orbitals and Organic Chemical Reactions, Wiley, New York.
K. Fukai (1982) Angew. Chem. Int. Ed. Engl. **21**, 801.

FROST-SCHWEMER METHOD

A procedure for analyzing competitive, consecutive second-order reactions[1-3] of the type:

$$A + B \xrightarrow{k_1} C + E$$

$$A + C \xrightarrow{k_2} D + E$$

This method is of use in thiol-disulfide interchanges and when using two-protonic-state electrophiles[4].

[1]A. A. Frost & W. C. Schwemer (1952) J. Amer. Chem. Soc. **74**, 1268.
[2]W. C. Schwemer & A. A. Frost (1951) J. Amer. Chem. Soc. **73**, 4541.
[3]J. W. Moore & R. G. Pearson (1981) Kinetics and Mechanism, 3rd ed., Wiley, New York.
[4]K. Brocklehurst (1982) Meth. Enzymol. **87**, 427.

1,2-β-FRUCTAN:1,2-β-FRUCTAN 1-β-FRUCTOSYLTRANSFERASE

This enzyme [EC 2.4.1.100] catalyzes fructosyl transfer between (1,2)-β-D-fructosyl_m and (1,2)-β-D-fructosyl_n to yield (1,2)-β-D-fructosyl_{m-1} plus (1,2)-β-D-fructosyl_{n+1}.

C. J. Pollock & A. J. Cairns (1991) Ann. Rev. Plant Physiol. Plant Mol. Biol. **42**, 77.

FRUCTAN β-FRUCTOSIDASE

This enzyme [EC 3.2.1.80] catalyzes the hydrolysis of terminal nonreducing 2,1- and 2,6-linked β-D-fructofuranose residues in fructans.

C. J. Pollock & A. J. Cairns (1991) Ann. Rev. Plant Physiol. Plant Mol. Biol. **42**, 77.

FRUCTOKINASE

This enzyme [EC 2.7.1.4] catalyzes the ATP-dependent phosphorylation of fructose to form fructose 6-phosphate and ADP.

W. W. Cleland (1977) Adv. Enzymol. **45**, 373.

FRUCTOSE-1,6-BISPHOSPHATASE

This hydrolase [EC 3.1.3.11] catalyzes conversion of D-fructose 1,6-bisphosphate to D-fructose 6-phosphate and orthophosphate.

H. C. Huppe & D. H. Turpin (1994) Ann. Rev. Plant Physiol. Plant Mol. Biol. **45**, 577.
S. J. Pilkis & T. H. Claus (1991) Ann. Rev. Nutr. **11**, 465.
S. J. Pilkis, T. H. Claus, P. D. Kountz & M. R. El-Maghrabi (1987) The Enzymes, 3rd ed., **18**, 3.
E. Van Schaffingen (1987) Adv. Enzymol. **59**, 315.
S. Pontremoli & B. L. Horecker (1971) The Enzymes, 3rd ed., **4**, 611.

FRUCTOSE-2,6-BISPHOSPHATASE

This hydrolase [EC 3.1.3.46] catalyzes the conversion of D-fructose 2,6-bisphosphate to yield D-fructose 6-phosphate and orthophosphate. **See also** 6-Phosphofructo-2-kinase

S. J. Pilkis, T. H. Claus, I. J. Kurland & A. J. Lange (1995) Ann. Rev. Biochem. **64**, 799.
K. Uyeda (1991) A Study of Enzymes **2**, 445.
S. J. Pilkis, T. H. Claus, P. D. Kountz & M. R. El-Maghrabi (1987) The Enzymes, 3rd ed., **18**, 3.
E. Van Schaffingen (1987) Adv. Enzymol. **59**, 315.

D-FUCONATE DEHYDRATASE

This enzyme [EC 4.2.1.67] converts D-fuconate to 2-dehydro-3-deoxy-D-fuconate along with the release of water.

W. A. Wood (1971) The Enzymes, 3rd ed., **5**, 573.

FUCOSE-1-PHOSPHATE GUANYLYLTRANSFERASE

This enzyme [EC 2.7.7.30] catalyzes the conversion of GTP and L-fucose 1-phosphate to yield GDP-L-fucose and pyrophosphate (or, diphosphate).

E. Adams (1976) *Adv. Enzymol.* **44**, 69.

α-L-FUCOSIDASE

This enzyme [EC 3.2.1.51] catalyzes the release of alcohols from α-L-fucosides to yield α-L-fucose.

S. C. Fry (1995) *Ann. Rev. Plant Physiol. Plant Mol. Biol.* **46**, 497.
E. Conzelmann & K. Sandhoff (1987) *Adv. Enzymol.* **60**, 89.
P. M. Dey & E. del Campillo (1984) *Adv. Enzymol.* **56**, 141.
H. M. Flowers & N. Sharon (1979) *Adv. Enzymol.* **48**, 29.

FUCOSYLTRANSFERASES

This class of enzymes includes galactoside 3(4)-L-fucosyltransferase, [EC 2.4.1.65]; glycoprotein 6-α-L-fucosyltransferase [EC 2.4.1.68], galactoside 2-α-L-fucosyltransferase [EC 2.4.1.69], and galactoside 3-fucosyltransferase [EC 2.4.1.152].

Y. Kishimoto (1983) *The Enzymes*, 3rd ed., **16**, 357.
K. A. Presper & E. C. Heath (1983) *The Enzymes*, 3rd ed., **16**, 449.
R. J. Staneloni & L. F. Leloir (1982) *Crit. Rev. Biochem.* **12**, 289.
T. A. Beyer, J. E. Sadler, J. I. Rearick, J. C. Paulson & R. L. Hill (1981) *Adv. Enzymol.* **52**, 23.

FUCULOSE-1-PHOSPHATE ALDOLASE

This enzyme [EC 4.1.2.17] catalyzes the aldol cleavage of L-fuculose 1-phosphate to yield glycerone phosphate and (*S*)-lactaldehyde.

D. S. Feingold & P. A. Hoffee (1972) *The Enzymes*, 3rd ed., **7**, 303.

FUMARASE

This hydratase [EC 4.2.1.2] catalyzes the reversible dehydration of (*S*)-malate to form fumarate and water.

I. A. Rose (1995) *Meth. Enzymol.* **249**, 315.
D. J. Creighton & N. S. R. K. Murthy (1990) *The Enzymes*, 3rd ed., **19**, 323.
D. J. T. Porter & H. J. Bright (1980) *J. Biol. Chem.* **255**, 4772.
W. W. Cleland (1977) *Adv. Enzymol.* **45**, 373.
R. L. Hill & J. W. Teipel (1971) *The Enzymes*, 3rd ed., **5**, 539.

FUMARATE REDUCTASE

This oxidoreductase catalyzes the reaction of succinate and NAD^+ to yield fumarate, NADH, and a proton.

K. M. Noll (1995) *Meth. Enzymol.* **251**, 470.

FURA-2

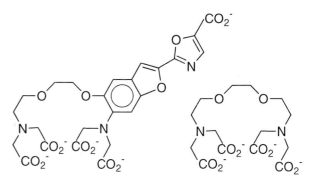

Structures of FURA-2 (left) and EGTA (right)

A calcium ion indicator dye (based on the structure of the chelator EGTA) that exhibits a strong fluorescence at 385 nm and can be used to measure changes in intracellular Ca^{2+} concentration. The approximate dissociation constant for the Ca^{2+}-Fura-2 complex is 0.1 μM, depending on cellular ion composition and pH. An esterified derivative of Fura-2 readily crosses the peripheral membrane of many cells and, after hydrolysis, the release of Fura-2 permits calcium ion measurements within cells. *See Calcium Ion Indicator Dyes; Metal Ion Complexation*

R. Y. Tsien (1988) *Trends in Neurosciences* **11**, 419.

Selected entries from *Methods in Enzymology* [vol, page(s)]: Cellular calcium determination, **238**, 73, 146, 298; calibration, **260**, 342-343; cell loading, **238**, 299; ratio fluorescence application, **238**, 298; Fura-2/AM cell loading for assay of free intracellular Ca^{2+}, **236**, 484; in assay of agonist-promoted calcium levels inside sperm, **225**, 150.

FURIN

This calcium-activated enzyme [EC 3.4.21.75] catalyzes the hydrolysis of peptide bonds in protein precursors that results in the release of mature proteins from their proproteins by hydrolysis of ArgXaaYaaArg—Zaa bonds, where Xaa can be any amino acid and Yaa is an arginyl or a lysyl residue. Albumin, complement component C3, and von Willebrand factor are thus released from their respective precursors. Furin is a member of the peptidase family S8.

C. W. Wharton (1998) *Comprehensive Biological Catalysis: A Mechanistic Reference* **1**, 345.
K. Nakayama (1994) *Meth. Enzymol.* **244**, 167.

G

$\Delta G°$ (Relation to K_{eq})

The Gibbs equation relates the change in a reaction's Gibbs standard free energy ($\Delta G°$) to its equilibrium constant:

$$\Delta G° = -RT(\ln K_{eq})$$

where R is the universal gas constant, T is the absolute temperature, and K_{eq} is the equilibrium constant under standard conditions. This fundamental equation allows one to interpret the thermodynamic properties of reactions and to apply the Additivity Principle to assess the equilibrium constants of other reactions. Because this equation is of such importance, two brief derivations of the relationship are presented here.

BASIC DERIVATION. Consider a chemical reaction of the following general description:

$$aA + bB \rightleftharpoons pP + qQ$$

where the lower case letters (a, b, c, and d) are stoichiometric coefficients for each reactant. Now, the concentration of A at a particular time can be greater than, less than, or equal to its equilibrium concentration; the same is true for B, P, and Q. The free energy of each reactant can be written as:

Free energy of a moles of A $= aG_A = \Delta G_A° + aRT\ln A$

Free energy of b moles of B $= bG_B = \Delta G_B° + bRT\ln B$

Free energy of p moles of P $= pG_P = \Delta G_P° + pRT\ln P$

Free energy of q moles of Q $= qG_Q = \Delta G_Q° + qRT\ln Q$

The free energy change for the reaction can now be written as shown in the following equations:

$$\begin{aligned}\Delta G &= G_{\text{products}} - G_{\text{reactants}}\\ &= pG_P + qG_Q - aG_A - bG_B\\ &= (\Delta G_P° + pRT\ln P) + (\Delta G_Q° + qRT\ln Q)\\ &\quad - (\Delta G_A° + aRT\ln A) - (\Delta G_B° + bRT\ln B)\end{aligned}$$

or

$$\begin{aligned}\Delta G &= [\Delta G_P° + \Delta G_Q° - (\Delta G_A° + \Delta G_B°)]\\ &\quad + RT\ln(P^pQ^q/[A^aB^b])\end{aligned}$$

When the reaction system attains equilibrium, ΔG will equal zero, and

$$\Delta G° + RT\ln(P^pQ^q/[A^aB^b]) = 0$$

or

$$\Delta G° = -RT\ln(P^pQ^q/[A^aB^b]) = -RT\ln K_{eq}$$

If the stoichiometric coefficients are each equal to one, then we arrive at the expression

$$\Delta G° = -RT\ln(PQ/(AB)) = -RT\ln K_{eq}$$

Note also that the standard free energy change ($\Delta G°$) has units of joules per mole, whereas the free energy change (ΔG) has units of joules.

A SLIGHTLY MORE RIGOROUS APPROACH. In the above description, the system was treated as though the reactants were ideal gases. A better approach is to account for the free energy of the system as a function of added reactants, and for solutions, the most technically correct action is to use partial molal quantities. Because free energy, $G(T, P, n_1, n_2, \ldots)$, is a state function depending on temperature, pressure, and the moles of each reactant, one can write the total differential of the free energy, and this will include terms of the form:

$$\mu_i = (\partial G/\partial n_i)_{P,T,n_{j(j \neq i)}}$$

Thus, the total differential of the free energy function is

$$dG = \mu_A dn_A + \mu_B dn_B + \mu_P dn_P + \mu_Q dn_Q$$

During the course of a chemical reaction, the number of moles of a particular reactant will change in a manner consistent with the stoichiometric relationships defined by the stoichiometric coefficients (a, b, p, and q):

$$dn_B/dn_A = b/a \text{ or } dn_B = (b/a)dn_A$$

$$dn_P/dn_A = -p/a \text{ or } dn_P = (-p/a)dn_A$$

$$dn_Q/dn_A = -q/a \text{ or } dn_Q = (-q/a)dn_A$$

The chemical potential of each component can likewise be written as

$$\mu_A = \mu_A{}^\circ + RT(A)$$

$$\mu_B = \mu_B{}^\circ + RT(B)$$

$$\mu_P = \mu_P{}^\circ + RT(P)$$

$$\mu_Q = \mu_Q{}^\circ + RT(Q)$$

where the parentheses denote chemical activity not molar concentration. At thermodynamic equilibrium, $dG = 0$, and the sum of the chemical potentials of the substrates equals the sum of the chemical potentials of the products:

$$\mu_A dn_A + \mu_B dn_B = \mu_P dn_P + \mu_Q dn_Q$$

Substituting each of these terms by the quantities in the seven previous equations, we get:

$$(p\mu_P + q\mu_Q - a\mu_A - b\mu_B) = -RT \ln(P^p Q^q/[A^a B^b])$$

or

$$\Delta G^\circ = -RT \ln K_{eq}$$

See *Activity Coefficients; Additivity Principle; Biochemical Thermodynamics; Chemical Potential; Equilibrium Constants; Hess's Law; Innate Thermodynamic Quantities; Molecular Crowding; Thermodynamics, Laws of; Thermodynamic Cycle; Thermodynamic Equations of State*

ΔΔG AS AN INDEX OF COOPERATIVITY

The linkage of binding energy at multiple ligand sites is commonly referred to as cooperativity, and the free energy of interaction between such sites (*i.e.*, $\Delta\Delta G$) serves as a useful index of ligand binding cooperativity. A stringent definition of cooperativity can be stated in terms of the free energy changes associated with the relative magnitudes of the respective ligand binding constants. If the change in the free energy change is zero, then binding is strictly noncooperative. If the ligand affinity for one site is raised by binding to the other site, then $\Delta\Delta G$ will be positive, and this is termed positive cooperativity. By contrast, if the ligand affinity for one site is lowered by binding to the other site, then $\Delta\Delta G$ will be negative, and this is termed negative cooperativity. This thermodynamic approach is more rigorous than the use of the Hill coefficient.

GALACTARATE DEHYDRATASE

This enzyme [EC 4.2.1.42] catalyzes the conversion of D-galactarate to 5-dehydro-4-deoxy-D-glucarate and water.

W. A. Wood (1971) *The Enzymes*, 3rd ed., **5**, 573.

GALACTITOL 2-DEHYDROGENASE

This enzyme [EC 1.1.1.16] catalyzes the reaction of galactitol with NAD^+ to produce D-tagatose and NADH. The enzyme will also catalyze the conversion of other alditols containing an L-*threo* configuration adjacent to a primary alcohol group.

D. R. D. Shaw (1962) *Meth. Enzymol.* **5**, 323.

GALACTOKINASE

This enzyme [EC 2.7.1.6] catalyzes the reaction of ATP with D-galactose to produce ADP and D-galactose 1-phosphate. D-Galactosamine can also act as the substrate.

M. R. Heinrich & S. M. Howard (1966) *Meth. Enzymol.* **9**, 407.
D. B. Wilson & D. S. Hogness (1966) *Meth. Enzymol.* **8**, 229.
E. S. Maxwell, K. Kurahashi & H. M. Kalckar (1962) *Meth. Enzymol.* **5**, 174.
L. F. Leloir & R. E. Trucco (1955) *Meth. Enzymol.* **1**, 290.

GALACTONATE DEHYDRATASE

This enzyme [EC 4.2.1.6] catalyzes the conversion of D-galactonate to yield 2-dehydro-3-deoxy-D-galactonate and water.

W. A. Wood (1971) *The Enzymes*, 3rd ed., **5**, 573.

GALACTOSE 1-DEHYDROGENASE

This enzyme [EC 1.1.1.48] catalyzes the reaction of D-galactose with NAD^+ to produce D-galactono-1,4-lactone and NADH.

K. Wallenfels & G. Kurz (1966) *Meth. Enzymol.* **9**, 112.
M. Doudoroff (1962) *Meth. Enzymol.* **5**, 339.

GALACTOSE OXIDASE

This copper-dependent enzyme [EC 1.1.3.9] catalyzes the reaction of D-galactose with dioxygen to produce D-*galacto*-hexodialdose and hydrogen peroxide.

A. Messerschmidt (1998) *Comprehensive Biological Catalysis: A Mechanistic Reference* **3**, 401.
N. Ito, P. F. Knowles & S. E. V. Phillips (1995) *Meth. Enzymol.* **258**, 235.
J. W. Whittaker (1995) *Meth. Enzymol.* **258**, 262.
E. T. Adman (1991) *Adv. Protein Chem.* **42**, 145.
B. G. Malmström, L.-E. Andréasson & B. Reinhammer (1975) *The Enzymes*, 3rd ed., **12**, 507.

GALACTOSE-1-PHOSPHATE URIDYLYLTRANSFERASE

This enzyme [EC 2.7.7.10], also known as UTP:hexose-1-phosphate uridylyltransferase, catalyzes the reaction of UTP with α-D-galactose 1-phosphate to produce UDP-galactose and pyrophosphate (or, diphosphate). α-D-Glucose 1-phosphate can also function as a substrate, albeit not as effectively.

L. S. Mullins & F. M. Raushel (1995) *Meth. Enzymol.* **249**, 398.
P. A. Frey (1992) *The Enzymes*, 3rd ed., **20**, 141.
P. A. Frey (1989) *Adv. Enzymol.* **62**, 119.
P. A. Frey, L.-J. Wong, K.-F. Sheu & S.-L. Yang (1982) *Meth. Enzymol.* **87**, 20.

α-GALACTOSIDASE

This enzyme [EC 3.2.1.22] catalyzes the hydrolysis of melibiose to yield galactose and glucose. It will also act on terminal, nonreducing α-D-galactose residues in α-D-galactosides. In addition, it will hydrolyze α-D-fucosides.

E. Conzelmann & K. Sandhoff (1987) *Adv. Enzymol.* **60**, 89.
P. M. Dey & E. del Campillo (1984) *Adv. Enzymol.* **56**, 141.
R. O. Brady (1983) *The Enzymes*, 3rd ed., **16**, 409.
H. M. Flowers & N. Sharon (1979) *Adv. Enzymol.* **48**, 29.

β-GALACTOSIDASE

This enzyme [EC 3.2.1.23], also known as LacZ, catalyzes the hydrolysis of terminal (nonreducing) β-D-galactose residues in β-D-galactosides.

M. W. Bauer, S. B. Halio & R. M. Kelly (1996) *Adv. Protein Chem.* **48**, 271.
G. Mooser (1992) *The Enzymes*, 3rd ed., **20**, 187.
M. A. Ator & P. R. Ortez de Montellano (1990) *The Enzymes*, 3rd ed. **19**, 213.
E. Conzelmann & K. Sandhoff (1987) *Adv. Enzymol.* **60**, 89.
H. Kresse & J. Glössl (1987) *Adv. Enzymol.* **60**, 217.

GALACTOSIDE FUCOSYLTRANSFERASES

Galactoside 2-L-fucosyltransferase [EC 2.4.1.69], also known as blood group H α-2-fucosyltransferase, catalyzes the reaction of GDP-L-fucose with β-D-galactosyl–R to produce GDP and α-L-fucosyl-1,2-β-D-galactosyl–R. Lactose can also act as the acceptor substrate.

Galactoside 3(4)-L-fucosyltransferase [EC 2.4.1.65], also known as blood group Lewis α-4-fucosyltransferase, catalyzes the reaction of GDP-L-fucose with 1,3-β-D-galactosyl-*N*-acetyl-D-glucosaminyl–R to produce GDP and 1,3-β-D-galactosyl-(α-1,4-L-fucosyl)-*N*-acetyl-D-glucosaminyl–R. It can also act on the corresponding 1,4-galactosyl derivative, forming 1,3-L-fucosyl links.

Galactoside 3-fucosyltransferase [EC 2.4.1.152], also known as Lewis-negative α-3-fucosyltransferase, catalyzes the reaction of GDP-L-fucose with 1,4-β-D-galactosyl-*N*-acetyl-D-glucosaminyl—R to produce GDP and 1,4-β-D-galactosyl-(α-1,3-L-fucosyl)-*N*-acetyl-D-glucosaminyl–R. Unlike the galactoside 3(4)-L-fucosyltransferase [EC 2.4.1.65] above, this enzyme exhibits no action on the corresponding 1,3-galactosyl derivative.

T. A. Beyer, J. E. Sadler, J. I. Rearick, J. C. Paulson & R. L. Hill (1981) *Adv. Enzymol.* **52**, 23.

GALACTOSIDE SIALYLTRANSFERASES

The following articles describe some of the molecular and physical properties of this enzyme.

T. A. Beyer, J. E. Sadler, J. I. Rearick, J. C. Paulson & R. L. Hill (1981) *Adv. Enzymol.* **52**, 23.
A. P. Grollman (1966) *Meth. Enzymol.* **8**, 351.

GAMMA (Γ or γ)

Γ 1. Symbol for surface concentration (SI units = moles per square meter). 2. The symbol for the limit of the second derivative of the reciprocal of the initial rate of an enzyme-catalyzed reaction vs. the reciprocal of the substrate concentration as the substrate concentration approaches infinity (thus, a measure of the degree of curvature in a double-reciprocal plot). Hence, $\Gamma = \lim_{[A]\to\infty}(\delta^2(\nu^{-1})/\delta([A]^{-1})^2)$. This function provides a means for estimating the degree of cooperativity in an enzyme-catalyzed reaction which does not have a linear double-reciprocal plot[1]. If Γ is positive, then the system exhibits positive cooperative; if negative, then negative cooperativity. This measurement of cooperativity is not commonly used since it is difficult to measure changes in slopes at high substrate concentrations[1]. **See** *Cooperativity; Hill Plot.* 3. Symbol for the gamma function, an important mathematical relationship that appears in a number of physical systems: for example, in immobilized enzyme kinetics[2]. It is defined by $\Gamma(1 + z) = \int_0^\infty e^{-x}x^z dx$ and has the property $\Gamma(z + 1) = z\ \Gamma(z)$. The gamma function also satisfies the relationship $\pi/(\sin(\pi z)) = \Gamma(z)\Gamma(1 - z)$ for all z whose absolute value is between zero and unity. Hence, $\Gamma(1/2)$ equals $\pi^{1/2}$.

γ 1. Symbol for activity coefficient. (γ_m on a molality basis, γ_c on a molarity basis, γ_x on a mole fraction basis). 2. Symbol for surface tension. 3. Symbol for photon. 4. Symbol for 10^{-4} gauss. 5. Symbol for shear strain. 6. Symbol for mass density. 7. Symbol for ratio of heat capacities, $\gamma = C_p/C_v$. 8. The third letter in the Greek alphabet; hence, used to denote the third item in a series (for example, the third methylene carbon in a fatty acid). $\gamma_\pm$, Symbol for mean ionic activity coefficient.

[1]K. E. Neet (1980) *Meth. Enzymol.* **64**, 139.
[2]K. J. Laidler & P. S. Bunting (1980) *Meth. Enzymol.* **64**, 227.

GAUSS

A unit, symbolized by G, for magnetic flux density (or, magnetic induction) equal to a maxwell per square centimeter. One gauss is equivalent to 10^{-4} tesla. If a conductor, one centimeter long, moves through a magnetic field at a velocity of one centimeter per second at right angles to the magnetic flux, then one gauss is the magnetic flux density that will induce an electromotive force of 10^{-8} volt (or, one abvolt).

GDP-MANNOSE 4,6-DEHYDRATASE

The following is a review on the molecular and physical properties of this enzyme [EC 4.2.1.47] which catalyzes the dehydration of GDP-mannose to yield GDP-4-dehydro-6-deoxy-D-mannose and water.

L. Glaser & H. Zarkowsky (1971) *The Enzymes*, 3rd ed., **5**, 465.

GDP-MANNOSE DEHYDROGENASE

This enzyme [EC 1.1.1.132] catalyzes the following reaction: GDP-mannose + 2 NAD$^+$ + H$_2$O to yield GDP-D-mannuronate + 2 NADH + 2H$^+$.

S. Shankar, R. W. Ye, D. Schlictman & A. M. Chakrabarty (1995) *Adv. Enzymol.* **70**, 221.

GDP-MANNOSE:DOLICHYL-PHOSPHATE MANNOSYLTRANSFERASE

This enzyme [EC 2.4.1.83] which catalyzes the following reaction: GDP-mannose + dolichyl phosphate to yield GDP + dolichyl D-mannosyl phosphate.

K. A. Presper & E. C. Heath (1983) *The Enzymes*, 3rd ed., **16**, 449.

GDP-MANNOSYLTRANSFERASES

This set of enzymes represent transferases that utilize GDP-mannose:

Glucomannan 4-β-mannosyltransferase [EC 2.4.1.32] catalyzes the reaction of GDP-mannose with [glucomannan]$_n$ to produce GDP and [glucomannan]$_{(n + 1)}$.

Heteroglycan α-mannosyltransferase [EC 2.4.1.48] catalyzes the reaction of GDP-mannose with a heteroglycan to produce GDP and a 1,2(or 1,3)-α-D-mannosylheteroglycan. In this reaction, in which the heteroglycan primer contains mannose, galactose, and xylose, 1,2- and 1,3-mannosyl bonds are formed.

Undecaprenyl-phosphate mannosyltransferase [EC 2.4.1.54] catalyzes the phosphatidylglycerol-dependent reaction of GDP-mannose with undecaprenyl phosphate to produce GDP and D-mannosyl-1-phosphoundecaprenol.

Phosphatidyl-*myo*-inositol α-mannosyltransferase [EC 2.4.1.57] catalyzes the transfer of one or more α-D-mannose units from GDP-mannose to positions 2, 6, and others in 1-phosphatidyl-*myo*-inositol.

GDP-mannose:dolichyl-phosphate mannosyltransferase [EC 2.4.1.83]: **See** *GDP-Mannose:Dolichyl-Phosphate Mannosyltransferase*

tRNA-queuosine β-mannosyltransferase [EC 2.4.1.110] catalyzes the reaction of GDP-mannose with tRNA^{Asp}-queuosine to produce GDP and tRNA^{Asp}-$O^{5''}$-β-D-mannosylqueuosine.

Glycolipid 2-α-mannosyltransferase [EC 2.4.1.131] catalyzes the transfer of an α-D-mannosyl residue from GDP-mannose into a lipid-linked oligosaccharide, forming an α-1,2-D-mannosyl-D-mannose linkage. The two 1,2-linked mannosyl residues in the mammalian lipid-linked oligosaccharide of the structure $Glc_3Man_9GlcNAc_2$ are produced by this enzyme.

Glycolipid 3-α-mannosyltransferase [EC 2.4.1.132], also known as mannosyltransferase II, catalyzes the transfer of an α-D-mannosyl residue from GDP-mannose into a lipid-linked oligosaccharide, forming an α-1,3-D-mannosyl-D-mannose linkage. The 1,3-linked mannosyl residue in the mammalian lipid-linked oligosaccharide of the structure $Glc_3Man_9GlcNAc_2$ is produced by this enzyme.

Chitobiosyldiphosphodolichol α-mannosyltransferase [EC 2.4.1.142] catalyzes the reaction of GDP-mannose with chitobiosyldiphosphodolichol to produce GDP and α-D-mannosylchitobiosyldiphosphodolichol.

GEAR ALGORITHM

A predictor-corrector algorithm for automatic computer-assisted integration of stiff ordinary differential equations. This procedure carries the name of its originator[1].

Numerical integration of rate equations can be challenging whenever rate constants of widely disparate magnitudes are involved. For example, if a two-reaction process has a very large rate constant for the first reaction, followed by a very small rate constant for the second reaction, then many numerical integration procedures require one to make very small steps to achieve sufficient accuracy for the first decay process, and this necessitates using many of the same small step time intervals for the second reaction. The simulation becomes extraordinarily slow if one is dealing with a series of reactions with very slow and very fast processes nested together. This condition is often referred to as "stiffness" or "stiffness instability". In the Gear Algorithm, the step size is chosen automatically so that once a fast reaction is simulated the step size can change to make the numerical integration of much slower steps far more tractable. *See Stiffness or Stiffness Instability*

[1]C. W. Gear (1971) *Numerical Initial-Value Problems in Ordinary Differential Equations*, Prentice-Hall, Englewood Cliffs, NJ.

GELSOLIN

A calcium-regulated actin regulatory protein that binds to and severs actin filaments, thereby controlling the gel-sol transitions of the actin cytoskeleton. After severing

an actin filament, gelsolin remains tightly bound on the ($+$)- or barbed-end of one fragmented filament.

P. J. McLaughlin & A. G. Weeds (1995) *Ann. Rev. Biophys. Biomol. Struct.* **24**, 643.

T. P. Stossel, C. Chaponnier, R. M. Ezzell, J. H. Harding, P. A. Janney, D. J. Kwiatkowski, S. E. Lind, D. B. Smith, F. S. Southwick, H. L. Yin & K. S. Zaner (1985) *Ann. Rev. Cell Biol.* **1**, 353.

GENERAL ACID CATALYSIS

Catalysis of a reaction by a Brønsted acid, including the protonated solvent S, forming the lyonium ion, SH^+. The rate of a general acid-catalyzed reaction is increased by an increase in the lyonium ion concentration and/or by an increase in the concentration of other acids, even when $[SH^+]$ is held constant.

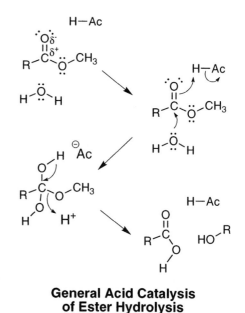

**General Acid Catalysis
of Ester Hydrolysis**

The rate of a general acid-catalyzed reaction is equal to $\Sigma_{HA}k_{HA}[HA]$ multiplied by some function of the substrate concentration(s) where [HA] is the general acid concentration and k_{HA} is the corresponding catalytic rate constant. Experimentally, general acid catalysis can be distinguished from specific acid catalysis by analysis of the effect of buffer concentration on the overall reaction rate. *See also Specific Acid Catalysis; Catalysis; General Base Catalysis*

GENERAL BASE CATALYSIS

Catalysis of a reaction by a Brønsted base, including the deprotonated solvent SH or SH^+, forming the lyate ion, S^- or S, respectively. The rate of such a catalyzed reaction is equal to $\Sigma_B k_B[B]$ multiplied by some function of the substrate concentration(s), where [B] is the general base concentration and k_B is the corresponding catalytic rate constant.

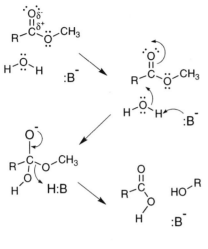

**General Base Catalysis
of Ester Hydrolysis**

Experimentally, general base catalysis can be distinguished from specific base catalysis by analysis of the effect of buffer concentration on the overall reaction rate. **See also** *Specific Base Catalysis; Catalysis; General Acid Catalysis*

GENETIC CODE

The following table indicates the standard codons that serve as the basis of the genetic code. Note: (a) Stop codons have no amino acids assigned to them; (b) Met* indicates the AUG start codon.

GENTAMICIN 2″-NUCLEOTIDYLTRANSFERASE

This enzyme [EC 2.7.7.46] catalyzes the reaction of a nucleoside triphosphate with gentamicin to produce a 2″-nucleotidylgentamicin and pyrophosphate (or, diphosphate). The nucleoside triphosphate can be ATP, dATP, CTP, ITP, or GTP. Kanamycin, tobramycin, and sisomicin can also act as the acceptor substrate. The nucleotidyl residue is transferred to the 2-hydroxyl group of the 3-amino-3-deoxy-D-glucose moiety in the acceptor substrate.

P. A. Frey (1989) *Adv. Enzymol.* **62**, 119.

GENTISATE 1,2-DIOXYGENASE

This iron-dependent enzyme [EC 1.13.11.4] catalyzes the following reaction: 2,5-dihydroxybenzoate + O_2 = maleylpyruvate.

M. R. Harpel & J. D. Lipscomb (1990) *Meth. Enzymol.* **188**, 101.

GEOMETRICAL ANALOGUE

Sequential Binding of Substrates A and B:

Binding of A-B Geometric Analogue:

	MIDDLE-BASE				
5′-BASE	U	C	A	G	3′-BASE
U	UUU = Phe	UCU = Ser	UAU = Tyr	UGU = Cys	U
U	UUC = Phe	UCC = Ser	UAC = Tyr	UGC = Cys	C
U	UUA = Leu	UCA = Ser	UAA = Stop	UGA = Stop	A
U	UUG = Leu	UCG = Ser	UAG = Stop	UGG = Trp	G
C	CUU = Leu	CCU = Pro	CAU = His	CGU = Arg	U
C	CUC = Leu	CCC = Pro	CAC = His	CGC = Arg	C
C	CUA = Leu	CCA = Pro	CAA = Gln	CGA = Arg	A
C	CUG = Leu	CCG = Pro	CAG = Gln	CGG = Arg	G
A	AUU = Ile	ACU = Thr	AAU = Asn	AGU = Ser	U
A	AUC = Ile	ACC = Thr	AAC = Asn	AGC = Ser	C
A	AUA = Ile	ACA = Thr	AAA = Lys	AGA = Arg	A
A	AUG = Met*	ACG = Thr	AAG = Lys	AGG = Arg	G
G	GUU = Val	GCU = Ala	GAU = Asp	GGU = Gly	U
G	GUC = Val	GCC = Ala	GAC = Asp	GGC = Gly	C
G	GUA = Val	GCA = Ala	GAA = Glu	GGA = Gly	A
G	GUG = Val	GCG = Ala	GAG = Glu	GGG = Gly	G

A multisubstrate inhibitor that achieves high-affinity because it interacts with binding determinants in more than a single substrate binding pocket on the enzyme. These inhibitors work with enzymes that have kinetic mechanisms that simultaneously require two or more of the substrates on the enzyme for catalysis. By binding to two ligand binding sites at the same time, the binding energy for ligand complexation is much greater than observed with competitive inhibitors acting only at a single substrate binding site. *See* *Adenylate Kinase Inhibitors*

GEOMETRIC MEAN, RULE OF THE

In the study of isotope effects, the isotope partitioning factor at a particular site is independent of the isotopic composition at any other site[1-3]. Thus, an isotope effect for a diisotopically substituted species should be the product of the kinetic isotope effects observed for each of the monoisotopically substituted compounds.

[1]V. Gold (1969) *Adv. Phys. Org. Chem.* **7**, 259.
[2]R. A. More O'Ferrall & A. J. Kresge (1980) *J. Chem. Soc., Perkin Trans.* **2**, 1840.
[3]K. B. Schowen & R. L. Schowen (1982) *Meth. Enzymol.* **87**, 551.

GERANOYL-CoA CARBOXYLASE

This biotin-dependent enzyme [EC 6.4.1.5] catalyzes the following reaction: ATP + geranoyl-CoA + HCO_3^- = ADP + P_i + 3-(4-methylpent-3-en-1-yl)pent-2-enedioyl-CoA.

H. G. Wood & R. E. Barden (1977) *Ann. Rev. Biochem.* **46**, 385.

GERANYL*TRANS*TRANSFERASE

This enzyme [EC 2.5.1.10], also known as farnesyl-diphosphate synthase, catalyzes the reaction of geranyl diphosphate with isopentenyl diphosphate to produce *trans,trans*-farnesyl diphosphate and pyrophosphate (or, diphosphate). Some forms of this enzyme will also utilize dimethylallyl diphosphate as a substrate. However, the enzyme will not accept larger prenyl diphosphates as an efficient substitute for geranyl diphosphate.

R. A. Gibbs (1998) *Comprehensive Biological Catalysis: A Mechanistic Reference* **1**, 31.
H. Kleinig (1989) *Ann. Rev. Plant Physiol. Plant Mol. Biol.* **40**, 39.
F. M. Raushel & J. J. Villafranca (1988) *Crit. Rev. Biochem.* **23**, 1.

GIBBERELLIN 2β-HYDROXYLASE

This enzyme [EC 1.14.11.13], also known as gibberellin 2β-dioxygenase, catalyzes the following reaction: gibberellin-1 + 2-oxoglutarate + O_2 = 2-β-hydroxygibberellin-1 + succinate + CO_2. The enzyme acts on other gibberellins as well.

A. G. Prescott & P. John (1996) *Ann. Rev. Plant Physiol. Plant Mol. Biol.* **47**, 245.

GIBBERELLIN 3β-HYDROXYLASE

This enzyme [EC 1.14.11.15], also referred to as gibberellin 3β-dioxygenase, uses iron ions and ascorbate and catalyzes the following reaction: gibberellin-20 + 2-oxoglutarate + dioxygen = gibberellin-1 + succinate + carbon dioxide. Other related enzymes (including gibberellin 12α-hydroxylase, gibberellin 13-hydroxylase, gibberellin 7-oxidase, and gibberellin 20-oxidase) are also reviewed in the articles below.

P. Hedden & Y. Kamiya (1997) *Ann. Rev. Plant Physiol. Plant Mol. Biol.* **48**, 431.
A. G. Prescott & P. John (1996) *Ann. Rev. Plant Physiol. Plant Mol. Biol.* **47**, 245.

GIBBS FREE ENERGY

A fundamental thermodynamic parameter (typically symbolized by *G*) used to describe the thermodynamic driving forces governing a closed system(s) at constant temperature and pressure. The Gibbs free energy is a state function that equates the enthalpy, *H*, minus the absolute temperature, *T*, multiplied by the entropy, *S*: thus, $G = H - TS$. The Helmholtz energy, *A*, and the Gibbs energy are related by the expression $G = A + PV$, where *P* and *V* are pressure and volume. In a closed system at constant temperature and pressure, wherein only *PV* work is permitted, the system at equilibrium will have a minimum value for the Gibbs free energy.

Because most biological systems are operative under conditions of constant pressure, changes in *G* are quite useful to describe driving forces for a wide variety of processes. The sign of the change in Gibbs free energy, ΔG, is directly associated with the direction in which a particular chemical reaction can proceed. If $\Delta G < 0$ for a given set of conditions of a particular reaction, then the reaction will proceed spontaneously in the indicated direction until equilibrium is reached. Conversely, if ΔG is positive, then energy will be needed to shift the reaction further from its equilibrium condition. *See* *Helmholtz Energy; Endergonic; Exergonic; Enthalpy; Entropy; Thermodynamics; Biochemical Thermodynamics*

GIBBS FREE ENERGY DIAGRAM

A graphical depiction (frequently called a reaction coordinate diagram) of the relative Gibbs free energies (or, more commonly, the standard Gibbs free energies) of the reactants, reaction intermediates, transition state(s), and products plotted in the same sequence in which they appear in the chemical reaction(s) under study. The horizontal coordinate expresses the ordered sequence, but is otherwise undefined in terms of time or any other parameter. The points representing the relative free energies are usually connected by a smooth curve (often referred to as the free energy profile). However, it should be pointed out that experimental observations typically can only provide information for the maxima and minima of these profiles and not on points in between.

The highest point on a Gibbs free energy diagram often, but not always, corresponds to the transition state of the rate-limiting step for the overall reaction(s). The concentration of a substrate not involved at that step may determine another step to be rate-limiting if the concentration of that substrate is sufficiently low[1]. *See also* Potential Energy Profile; Enzyme Energetics

[1]IUPAC (1979) *Pure and Appl. Chem.* **51**, 1725.

GIBBS FREE ENERGY OF ACTIVATION

The standard Gibbs free energy difference between the reactant(s) of a reaction and the transition state. It is commonly symbolized by either $\Delta G^{\ddagger}$ or $\Delta^{\ddagger} G^{\ominus}$ and is equal to $-RT \ln K^{\ddagger}$ where $K^{\ddagger}$ represents the equilibrium constant between the reactant(s) and the transition state, R is the molar gas constant, and T is the absolute temperature. (It is more proper to designate $\Delta G^{\ddagger}$ as equal to $-RT \ln[K^{\ddagger}(c^{\ominus})^{n-1}]$ where $c^{\ominus}$ is the standard-state concentration and n is the molecularity of the reaction. Thus, $K^{\ddagger} = \{\exp[-\Delta G^{\ddagger}/RT]\}(c^{\ominus})^{1-n}$. However, standard states are usually unit concentrations, the unit being used the same as that of the rate constant.) We should also recall the fact that $K^{\ddagger}$ differs from a true equilibrium constant due to the altered number of degrees of freedom in the transition state. The Gibbs free energy of activation can be determined from the bimolecular rate constant and the Eyring equation: $\Delta G^{\ddagger} = RT(\ln[k_B/h] - \ln[k/T])$ where k_B is Boltzmann's constant, h is Planck's constant, and k is the bimolecular rate constant. *See* Transition-State Theory (Thermodynamics)

GIBBS-HELMHOLTZ EQUATION

An important thermodynamic relationship for the temperature dependence of the Gibbs free energy, G:

$$(\partial(G/T)/\partial T)_P = -H/T^2$$

where T is the absolute temperature, P is the pressure, and H is the enthalpy. *See* Gibbs Free Energy

GLOBAL ANALYSIS

A systematic algorithm for data analysis that permits the experimenter to recover estimates of lifetimes (τ) from a multiexponential time-dependent process $F(t)$. The process is described by the generalized sum-of-exponentials expression, involving n steps:

$$F(t) = \Sigma a_i \exp[-t/\tau_i]$$

where the summation is taken from ($i = 1$) to ($i = n$), and a_i is the amplitude associated with each of the individual steps. This algorithm *simultaneously* analyzes data from multiple experiments. This is accomplished applying an internally consistent set of fitting parameters to nonlinear least squares software adapted to perform model-dependent summations of these nonlinear least-squares equations.

Selected entries from *Methods in Enzymology* [vol, page(s)]:
In kinetic analysis of complex reactions, **210**, 382; fluorescence decay rate distributions, **210**, 357; implementation in Laplace deconvolution noniterative method, **210**, 293; in multiexponential decays, **210**, 296; partial global analysis by simulated annealing methods, **210**, 365; spectral resolution, **210**, 299.

GLUCAN ENDO-1,3(4)-β-GLUCANASE

This enzyme [EC 3.2.1.6] (also known as endo-1,3(4)-β-glucanase, glucan endo-β1,3-β1,4-glucosidase, endo-1,4-β-glucanase, endo-1,3-β-glucanase, and laminarinase) catalyzes the endohydrolysis of 1,3- or 1,4-linkages in β-D-glucans when the glucose residue, whose reducing group is involved in the linkage to be hydrolyzed, is itself substituted at C3. The enzyme will also utilize laminarin, lichenin, and cereal D-glucans as substrates. This glucanase is different from that of glucan endo-1,3-β-glucosidase [EC 3.2.1.39].

G. Davies, M. L. Sinnott & S. G. Withers (1998) *Comprehensive Biological Catalysis: A Mechanistic Reference* **1**, 119.

GLUCAN ENDO-1,2-β-GLUCOSIDASE

This enzyme [EC 3.2.1.71], also known as endo-1,2-β-glucanase, catalyzes the random hydrolysis of 1,2-glucosidic linkages in 1,2-β-D-glucans.

E. T. Reese & M. Mandels (1966) *Meth. Enzymol.* **8**, 607.

GLUCAN ENDO-1,3-α-GLUCOSIDASE

This enzyme [EC 3.2.1.59], also known as endo-1,3-α-glucanase, catalyzes the endohydrolysis of 1,3-α-D-glucosidic linkages in pseudonigeran (the products of which are nigerose and glucose), isolichenin, and nigeran.

G. Davies, M. L. Sinnott & S. G. Withers (1998) *Comprehensive Biological Catalysis: A Mechanistic Reference* 1, 119.

GLUCAN ENDO-1,3-β-GLUCOSIDASE

This enzyme [EC 3.2.1.39] (also known as $(1 \rightarrow 3)$-β-glucan endohydrolase, endo-1,3-β-glucanase, and laminarinase) catalyzes the hydrolysis of 1,3-β-D-glucosidic linkages in 1,3-β-D-glucans. The enzyme exhibits very limited action on mixed-link (1,3/1,4)-β-D-glucans. However, it catalyzes the hydrolysis of laminarin, paramylon, and pachyman. The enzyme is not identical with endo-1,3(4)-β-glucanase [EC 3.2.1.6].

G. Davies, M. L. Sinnott & S. G. Withers (1998) *Comprehensive Biological Catalysis: A Mechanistic Reference* 1, 119.
E. T. Reese & M. Mandels (1966) *Meth. Enzymol.* 8, 607.

GLUCAN ENDO-1,6-β-GLUCOSIDASE

This enzyme [EC 3.2.1.75], also known as endo-1,6-β-glucanase, catalyzes the random hydrolysis of 1,6-linkages in 1,6-β-D-glucans. The enzyme will also act on lutean, pustulan, and 1,6-oligo-β-D-glucosides.

E. T. Reese & M. Mandels (1966) *Meth. Enzymol.* 8, 607.

GLUCAN 1,3-β-GLUCOSIDASE

This enzyme [EC 3.2.1.58], also known as exo-1,3-β-glucanase and exo-1,3-β-glucosidase, catalyzes the successive hydrolysis of β-D-glucose units from the nonreducing ends of 1,3-β-D-glucans, resulting in the release of glucose. The enzyme will also act on oligosaccharides. However, laminaribiose is a weak substrate.

E. T. Reese & M. Mandels (1966) *Meth. Enzymol.* 8, 607.

1,3-β-GLUCAN SYNTHASE

This enzyme [EC 2.4.1.34] (also known as 1,3-β-D-glucan-UDP glucosyltransferase, UDP-glucose-1,3-β-D-glucan glucosyltransferase, and callose synthetase) catalyzes the reaction of UDP-glucose with $[(1,3)$-β-D-glucosyl$]_n$ to produce UDP and $[(1,3)$-β-D-glucosyl$]_{(n+1)}$.

D. S. Feingold (1966) *Meth. Enzymol.* 8, 404.

GLUCARATE DEHYDRATASE

This enzyme [EC 4.2.1.40] catalyzes the following reaction: D-glucarate = 5-dehydro-4-deoxy-D-glucarate + H_2O.

W. A. Wood (1971) *The Enzymes*, 3rd ed., 5, 573.

GLUCOAMYLASE

This enzyme [EC 3.2.1.3] catalyzes the hydrolytic release of β-D-glucose from terminal 1,4-linked α-D-glucose residues successively from nonreducing ends of polysaccharide chains.

E. H. Van Beers, H. A. Büller, R. J. Grand, A. W. C. Einerhand & J. Dekker (1995) *Crit. Rev. Biochem. Mol. Biol.* 30, 197.
G. Mooser (1992) *The Enzymes*, 3rd ed., 20, 187.

GLUCOCEREBROSIDASE

This enzyme [EC 3.2.1.45] catalyzes the hydrolysis of D-glucosyl-*N*-acylsphingosine to yield D-glucose and *N*-acylsphingosine.

G. A. Grabowski, S. Gatt & M. Horowitz (1990) *Crit. Rev. Biochem. Mol. Biol.* 25, 385.
E. Conzelmann & K. Sandhoff (1987) *Adv. Enzymol.* 60, 89.
R. O. Brady (1983) *The Enzymes*, 3rd ed., 16, 409.

GLUCOKINASE

This enzyme [EC 2.7.1.2] catalyzes the ATP-dependent transphosphorylation of D-glucose to yield D-glucose 6-phosphate + ADP. The Michaelis constant for D-glucose is nearly 100 times greater than the corresponding value for hexokinase. Although originally thought to be specific for glucose, this enzyme also phosphorylates D-fructose, which binds some 30 to 50 times more weakly. **See** Hexokinase

J. H. Hurley (1996) *Ann. Rev. Biophys. & Biomol. Struct.* 25, 137.
A. Cornish-Bowden & M. L. Cárdenas (1991) *Trends Biochem. Sci.* 16, 281.
D. L. Purich, H. J. Fromm & F. B. Rudolph (1973) *Adv. Enzymol.* 39, 249.
S. P. Colowick (1973) *The Enzymes*, 3rd ed., 9, 1.

GLUCONATE DEHYDRATASE

This enzyme [EC 4.2.1.39] catalyzes the reaction: D-gluconate ⇌ 2-dehydro-3-deoxy-D-glucarate + H_2O.

W. A. Wood (1971) *The Enzymes*, 3rd ed., 5, 573.

GLUCOSAMINATE AMMONIA-LYASE

This enzyme [EC 4.3.1.9] catalyzes the following reaction involving pyridoxal phosphate as a cofactor: D-glucosaminate = 2-dehydro-3-deoxy-D-gluconate + NH_3.

W. A. Wood (1971) *The Enzymes*, 3rd ed., 5, 573.
L. Davis & D. E. Metzler (1972) *The Enzymes*, 3rd ed., 7, 33.

GLUCOSAMINE *N*-ACETYLTRANSFERASE

This enzyme [EC 2.3.1.3] catalyzes the reaction of acetyl-CoA with Δ-glucosamine to yield coenzyme A and *N*-acetyl-Δ-glucosamine.

H. Kresse & J. Glössl (1987) *Adv. Enzymol.* 60, 217.

GLUCOSAMINE-6-PHOSPHATE ISOMERASE

This enzyme [EC 5.3.1.10] catalyzes the following isomerization and deamination reaction: D-glucosamine 6-phosphate + H_2O ⇌ D-fructose 6-phosphate + NH_3.

I. A. Rose (1975) Adv. Enzymol. **43**, 491.
J. M. Buchanan (1973) Adv. Enzymol. **39**, 91.
E. A. Noltmann (1972) The Enzymes, 3rd ed., **6**, 271.

GLUCOSAMINE-6-PHOSPHATE SYNTHASE

This enzyme [EC 2.6.1.16], also referred to as glutamine–fructose-6-phosphate aminotransferase (isomerizing), catalyzes the reaction of L-glutamine and D-fructose 6-phosphate to yield L-glutamate and D-glucosamine 6-phosphate.

H. Zalkin (1993) Adv. Enzymol. **66**, 203.
H. Zalkin (1985) Meth. Enzymol. **113**, 278.
J. M. Buchanan (1973) Adv. Enzymol. **39**, 91.

GLUCOSE OXIDASE

This enzyme [EC 1.1.3.4] catalyzes the following reaction: β-D-glucose + O_2 ⇌ D-glucono-1,5-lactone + hydrogen peroxide.

H. J. Bright & D. J. T. Porter (1975) The Enzymes, 3rd ed., **12**, 421.

GLUCOSE-6-PHOSPHATASE

This enzyme [EC 3.1.3.9] catalyzes the hydrolysis of D-glucose 6-phosphate to yield D-glucose and orthophosphate. Some glucose phosphatases also catalyze transphosphorylation reactions from carbamoyl phosphate, hexose phosphates, pyrophosphate, phosphoenolpyruvate and nucleoside di- and triphosphates, using D-glucose, D-mannose, 3-methyl-D-glucose, or 2-deoxy-D-glucose as phosphoryl acceptors. **See** Isotope Exchange (Reactions Away from Equilibrium)

K. A. Sukalski & R. C. Nordlie (1989) Adv. Enzymol. **62**, 93.
R. C. Nordlie (1982) Meth. Enzymol. **87**, 319.
R. C. Nordlie (1971) The Enzymes, 3rd ed., **4**, 543.

GLUCOSE-6-PHOSPHATE DEHYDROGENASE

This enzyme [EC 1.1.1.49] catalyzes the $NADP^+$-dependent reduction of D-glucose 6-phosphate that yields 6-phospho-D-gluconolactone, NADPH, and a proton.

L. V. Kletsova, M. Y. Kiriukhin, A. Y. Chistoserdov & Y. D. Tsygankov (1990) Meth. Enzymol. **188**, 335.
A. P. Sokolov & Y. A. Trotsenko (1990) Meth. Enzymol. **188**, 339.
H. R. Levy (1979) Adv. Enzymol. **48**, 97.

GLUCOSE-1-PHOSPHATE URIDYLYLTRANSFERASE

This enzyme [EC 2.7.7.9] catalyzes the UTP-dependent uridylylation of α-D-glucose 1-phosphate that yields UDP-glucose and pyrophosphate.

L. S. Mullins & F. M. Raushel (1995) Meth. Enzymol. **249**, 398.
P. A. Frey (1992) The Enzymes, 3rd ed., **20**, 141.
P. A. Frey (1989) Adv. Enzymol. **62**, 119.
R. L. Turnquist & R. G. Hansen (1973) The Enzymes, 3rd ed., **8**, 51.

GLUCOSE TRACER KINETICS

Wolfe[1] provided an excellent account of the systematic application of stable and radioactive isotope tracers in determining the kinetics of glucose production, glycogenolysis, and gluconeogenesis in living systems.

[1]R. R. Wolfe (1992) Radioactive and Stable Isotope Tracers in Biomedicine, pp. 283-316, Wiley-Liss, New York.

GLUCURONATE-1-PHOSPHATE URIDYLYLTRANSFERASE

This enzyme [EC 2.7.7.44] catalyzes the reaction of UTP with 1-phospho-α-D-glucuronate to produce UDP-D-glucuronate and pyrophosphate. CTP can also serve as the nucleotide substrate, albeit not as effectively.

F. A. Loewus & M. W. Loewus (1983) Ann. Rev. Plant Physiol. **34**, 137.

GLUCURONATE-2-SULFATE SULFATASE

This enzyme [EC 3.1.6.18], also known as glucuronate-2-sulfatase and chondro-2-sulfatase, catalyzes the hydrolysis of the 2-sulfate groups of the 2-O-sulfo-D-glucuronate residues of chondroitin sulfate, heparin, and heparitin sulfate. The enzyme does not act on iduronate 2-sulfate residues.

H. Kresse & J. Glössl (1987) Adv. Enzymol. **60**, 217.

β-GLUCURONIDASE

This enzyme [EC 3.2.1.31] catalyzes the hydrolysis of a β-D-glucuronoside to yield an alcohol and D-glucuronate.

H. Kresse & J. Glössl (1987) Adv. Enzymol. **60**, 217.
P. M. Dey & E. del Campillo (1984) Adv. Enzymol. **56**, 141.

GLUCURONOKINASE

This enzyme [EC 2.7.1.43] catalyzes the reaction of ATP with D-glucuronate to produce ADP and 1-phospho-α-D-glucuronate.

F. A. Loewus & M. W. Loewus (1983) Ann. Rev. Plant Physiol. **34**, 137.

GLUCURONOLACTONE REDUCTASE

This enzyme [EC 1.1.1.20] catalyzes the reaction of L-gulono-1,4-lactone with $NADP^+$ to produce D-glucurono-3,6-lactone and NADPH.

Y. Mano & N. Shimazono (1970) *Meth. Enzymol.* **18A**, 55.

GLUCURONOSYLTRANSFERASES

This set of enzymes [EC 2.4.1.17], also known as UDP-glucuronosyltransferases, catalyzes the reaction of UDP-glucuronate with an acceptor to produce UDP and the β-D-glucuronoside of the acceptor. This family of proteins accepts a wide range of substrates, including phenols, alcohols, amines, and fatty acids.

A. Markovitz & A. Dorfman (1962) *Meth. Enzymol.* **5**, 155.

GLUTACONYL-CoA DECARBOXYLASE

This enzyme [EC 4.1.1.70] catalyzes the conversion of pent-2-enoyl-CoA to but-2-enoyl-CoA and carbon dioxide. The enzyme isolated from *Acidaminococcus fermentans* is a biotinyl-protein, requires sodium ions, and acts as a sodium pump.

J. N. Earnhardt & D. N. Silverman (1998) *Comprehensive Biological Catalysis: A Mechanistic Reference* **1**, 495.

GLUTAMATE *N*-ACETYLTRANSFERASE

This enzyme [EC 2.3.1.35], also known as ornithine acetyltransferase, and ornithine transacetylase, catalyzes the reversible reaction of N^2-acetyl-L-ornithine with L-glutamate to produce L-ornithine and *N*-acetyl-L-glutamate. This protein also exhibits a low hydrolysis activity (about 1% of that of the transferase activity) of N^2-acetyl-L-ornithine to yield acetate and L-ornithine. This enzyme is not identical with *N*-acetylglutamate synthase [EC 2.3.1.1].

G. Dénes (1970) *Meth. Enzymol.* **17A**, 273.

D-GLUTAMATE: D-AMINO ACID AMINOTRANSFERASE

This enzyme catalyzes the transamination of a wide spectrum of α-amino acids and α-keto (or 2-oxo) acids, demonstrating absolute specificity for their D-isomers. The most likely physiologic role is to provide D-amino acids for peptidoglycan synthesis in bacterial cell wall formation.

W. M. Jones, T. S. Soper, H. Ueno & J. M. Manning (1985) *Meth. Enzymol.* **113**, 108.

GLUTAMATE DECARBOXYLASE

This pyridoxal-phosphate-dependent enzyme [EC 4.1.1.15] catalyzes the conversion of L-glutamate to 4-aminobutanoate and carbon dioxide. The mammalian brain enzyme also acts on L-cysteate, 3-sulfino-L-alanine, and L-aspartate.

M. A. Ator & P. R. Ortez de Montellano (1990) *The Enzymes*, 3rd ed., **19**, 213.
D. J. Creighton & N. S. R. K. Murthy (1990) *The Enzymes*, 3rd ed., **19**, 323.
M. H. O'Leary (1989) *Ann. Rev. Biochem.* **58**, 377.
J.-Y. Wu, L. Denner, C.-T. Lin & G. Song (1985) *Meth. Enzymol.* **113**, 3.
M. L. Fonda (1985) *Meth. Enzymol.* **113**, 11.
A. E. Martell (1982) *Adv. Enzymol.* **53**, 163.
E. A. Boeker & E. E. Snell (1972) *The Enzymes*, 3rd ed., **6**, 217.

GLUTAMATE DEHYDROGENASE

Glutamate dehydrogenase [EC 1.4.1.2] catalyzes the reaction of L-glutamate with NAD^+ and water to produce α-ketoglutarate (or, 2-oxoglutarate), ammonia, and NADH.

Glutamate dehydrogenase ($NADP^+$) [EC 1.4.1.4] catalyzes the reaction of L-glutamate with $NADP^+$ and water to produce α-ketoglutarate (or, 2-oxoglutarate), ammonia, and NADPH.

Glutamate dehydrogenase ($NAD(P)^+$) [EC 1.4.1.3] catalyzes the reaction of L-glutamate with $NAD(P)^+$ and water to produce α-ketoglutarate (or, 2-oxoglutarate), ammonia, and NAD(P)H.

J. DiRuggiero & F. T. Robb (1996) *Adv. Protein Chem.* **48**, 311.
H.-M. Lam, K. T. Coschigamo, I. C. Oliveira, R. Melo-Oliveira & G. M. Coruzzi (1996) *Ann. Rev. Plant Physiol. Plant Mol. Biol.* **47**, 569.
N. M. W. Brunhuber & J. S. Blanchard (1994) *Crit. Rev. Biochem. Mol. Biol.* **29**, 415.
R. F. Colman (1991) *A Study of Enzymes* **2**, 173.
R. F. Colman (1990) *The Enzymes*, 3rd ed., **19**, 283.
H. F. Fisher (1973) *Adv. Enzymol.* **39**, 369.

GLUTAMATE FORMIMINOTRANSFERASE

This pyridoxal-phosphate-dependent enzyme [EC 2.1.2.5], also known as glutamate formyltransferase, catalyzes the reaction of 5-formiminotetrahydrofolate with L-glutamate to produce tetrahydrofolate and *N*-formimino-L-glutamate. The enzyme will additionally catalyze the transfer of the formyl moiety from 5-formyltetrahydrofolate to L-glutamate. This protein occurs in eukaryotes as a bifunctional enzyme also having a formiminotetrahydrofolate cyclodeaminase activity [EC 4.3.1.4].

V. Schirch (1998) *Comprehensive Biological Catalysis: A Mechanistic Reference* **1**, 211.

J. I. Rader & F. M. Huennekens (1973) *The Enzymes*, 3rd ed., **9**, 197.

GLUTAMATE FORMYLTRANSFERASE

This enzyme designation was formerly classified as EC 2.1.2.6. It is now classified under glutamate formiminotransferase [EC 2.1.2.5].

J. I. Rader & F. M. Huennekens (1973) *The Enzymes*, 3rd ed., **9**, 197.

GLUTAMATE RACEMASE

This pyridoxal-phosphate-dependent enzyme [EC 5.1.1.3] catalyzes the interconversion of L-glutamate and D-glutamate.

M. E. Tanner & G. L. Kenyon (1998) *Comprehensive Biological Catalysis: A Mechanistic Reference* **2**, 7.

E. Adams (1976) *Adv. Enzymol.* **44**, 69.

E. Adams (1972) *The Enzymes*, 3rd ed., **6**, 479.

GLUTAMATE γ-SEMIALDEHYDE DEHYDROGENASE

This enzyme [EC 1.2.1.41], also known as γ-glutamylphosphate reductase and glutamate-5-semialdehyde dehydrogenase, catalyzes the reaction of L-glutamate 5-semialdehyde with orthophosphate and NADP$^+$ to produce L-γ-glutamyl 5-phosphate and NADPH.

E. Adams & L. Frank (1980) *Ann. Rev. Biochem.* **49**, 1005.

GLUTAMATE SYNTHASE

Glutamate synthase (NADPH) [EC 1.4.1.13], an iron-sulfur flavoprotein, catalyzes the reaction of L-glutamine with α-ketoglutarate (or, 2-oxoglutarate) and NADPH to produce NADP$^+$ and two glutamate molecules. Ammonia can act as the nitrogen donor substrate instead of L-glutamine, albeit weaker.

Glutamate synthase (NADH) [EC 1.4.1.14], which uses FMN, catalyzes the reaction of L-glutamine with α-ketoglutarate (or, 2-oxoglutarate) and NADH to produce NAD$^+$ and two glutamate molecules.

Glutamate synthase (ferredoxin) [EC 1.4.7.1], also known as ferredoxin-dependent glutamate synthase, is an iron-sulfur flavoprotein, that catalyzes the reaction of L-glutamine with α-ketoglutarate (or, 2-oxoglutarate) and two reduced ferredoxin molecules to produce two oxidized ferredoxin molecules and two glutamate molecules.

H.-M. Lam, K. T. Coschigamo, I. C. Oliveira, R. Melo-Oliveira & G. M. Coruzzi (1996) *Ann. Rev. Plant Physiol. Plant Mol. Biol.* **47**, 569.

L. A. Kleczkowski (1994) *Ann. Rev. Plant Physiol. Plant Mol. Biol.* **45**, 339.

H. Zalkin (1993) *Adv. Enzymol.* **66**, 203.

R. F. Colman (1990) *The Enzymes*, 3rd ed., **19**, 283.

A. Meister (1985) *Meth. Enzymol.* **113**, 327.

J. M. Buchanan (1973) *Adv. Enzymol.* **39**, 91.

GLUTAMINASE

This enzyme [EC 3.5.1.2], also known as L-glutamine amidohydrolase, catalyzes the hydrolysis of L-glutamine to produce L-glutamate and ammonia.

N. P. Curthoys & M. Watford (1995) *Ann. Rev. Nutr.* **15**, 133.

E. Kvamme, I. AA. Torgner & G. Svenneby (1985) *Meth. Enzymol.* **113**, 241.

H. G. Windmueller (1982) *Adv. Enzymol.* **53**, 201.

S. C. Hartman (1971) *The Enzymes*, 3rd ed., **4**, 79.

GLUTAMINE AMINOTRANSFERASE

This pyridoxal-phosphate-dependent enzyme [EC 2.6.1.15] (also known as glutamine:pyruvate aminotransferase, glutaminase II, glutamine:oxo-acid transaminase, and glutamine transaminase L) catalyzes the reaction of L-glutamine with pyruvate to produce 2-oxoglutaramate and L-alanine. The enzyme will also catalyze the transfer of an amino group from L-methionine and glyoxylate can substitute for pyruvate.

A. J. L. Cooper & A. Meister (1985) *Meth. Enzymol.* **113**, 338.

A. E. Braunstein (1973) *The Enzymes*, 3rd ed., **9**, 379.

GLUTAMINE SYNTHETASE

This enzyme [EC 6.3.1.2], officially known as glutamate:ammonia ligase, catalyzes the reaction of L-glutamate with ammonia and MgATP to produce L-glutamine, MgADP and orthophosphate. The enzyme will also act on 4-methylene-L-glutamate (however, glutamine synthetase activity is not identical with 4-methyleneglutamine synthetase [EC 6.3.1.7]) and β-glutamate. *See* Arsenolysis; Bridge-to-Nonbridge Oxygen Scrambling; Isotope Exchange at Equilibrium; Wedler-Boyer Technique; Cumulative Inhibition; Unconsumed Substrate; Cryptic Catalysis; Borohydride Reduction; Enzyme Cascade Kinetics

D. L. Purich (1998) *Adv. Enzymol.* **72**, 9.

J. DiRuggiero & F. T. Robb (1996) *Adv. Protein Chem.* **48**, 311.

H.-M. Lam, K. T. Coschigamo, I. C. Oliveira, R. Melo-Oliveira & G. M. Coruzzi (1996) *Ann. Rev. Plant Physiol. Plant Mol. Biol.* **47**, 569.

S.-H. Liaw & D. Eisenberg (1994) *Biochemistry* **33**, 675.

J. J. Villafranca & T. Nowak (1992) *The Enzymes*, 3rd ed., **20**, 63.

P. A. Frey (1992) *The Enzymes*, 3rd ed., **20**, 141.

A. Oaks & B. Hirel (1985) *Ann. Rev. Plant Physiol.* **36**, 345.

I. A. Rose (1979) *Adv. Enzymol.* **50**, 361.

A. Meister (1974) *The Enzymes*, 3rd ed., **10**, 699.

E. R. Stadtman & A. Ginsburg (1974) *The Enzymes*, 3rd ed., **10**, 755.

GLUTAMINE SYNTHETASE ADENYLYLTRANSFERASE

This enzyme [EC 2.7.7.42], also known as glutamate:ammonia-ligase adenylyltransferase, plays a key role in controlling bacterial nitrogen metabolism by controlling the state of adenylylation of glutamine synthetase.

E. R. Stadtman (1973) *The Enzymes*, 3rd ed., **8**, 1.

E. R. Stadtman & A. Ginsberg (1974) *The Enzymes*, 3rd ed., **10**, 755.

S. G. Rhee, P. B. Chock & E. R. Stadtman (1989) *Adv. Enzymol.* **62**, 37.

GLUTAMINYL-tRNA SYNTHETASE

This enzyme [EC 6.1.1.18], also known as glutamine:tRNA ligase, catalyzes the reaction of ATP with L-glutamine and tRNA^{Gln} to produce L-glutaminyl-tRNA^{Gln}, AMP, and pyrophosphate (or, diphosphate).

E. A. First (1998) *Comprehensive Biological Catalysis: A Mechanistic Reference* **1**, 573.

J. J. Perona, M. A. Rould & T. A. Steitz (1993) *Biochemistry* **32**, 8758.

P. Hoben & D. Söll (1985) *Meth. Enzymol.* **113**, 55.

γ-GLUTAMYL CYCLOTRANSFERASE

This enzyme [EC 2.3.2.4] catalyzes the conversion of a (5-L-glutamyl)-L-amino acid to 5-oxoproline and an L-amino acid.

A. Meister (1985) *Meth. Enzymol.* **113**, 438.

P. Van Der Werf & A. Meister (1975) *Adv. Enzymol.* **43**, 519.

M. Orlowski & A. Meister (1971) *The Enzymes*, 3rd ed., **4**, 123.

γ-GLUTAMYLCYSTEINE SYNTHETASE

This enzyme [EC 6.3.2.2], also known as glutamate:cysteine ligase, catalyzes the reaction of ATP with L-glutamate and L-cysteine to produce ADP, orthophosphate, and γ-L-glutamyl-L-cysteine. L-Aminohexanoate can act as a substrate in place of glutamate. Certain thiol-containing molecules, including glutathione, will inhibit the mammalian enzyme.

R. C. Fahey & A. R. Sundquist (1991) *Adv. Enzymol.* **64**, 1-53.

G. F. Seelig & A. Meister (1985) *Meth. Enzymol.* **113**, 379 and 390.

A. Meister (1974) *The Enzymes*, 3rd ed., **10**, 671.

GLUTAMYL ENDOPEPTIDASES

This is a set of enzymes that act on peptide bonds containing a glutamyl residue. They include:

Glutamyl endopeptidase [EC 3.4.21.19] (also known as staphylococcal serine proteinase, V8 proteinase, protease V8, and endoproteinase Glu-C), a member of the peptidase family S2B, catalyzes the hydrolysis of Asp–Xaa and Glu–Xaa peptide bonds. In appropriate buffers, the specificity of the bond cleavage is restricted to Glu–Xaa. Peptide bonds involving bulky side chains of hydrophobic aminoacyl residues are hydrolyzed at a lower rate.

Glutamyl endopeptidase II [EC 3.4.21.82], also known as glutamic acid-specific protease, catalyzes the hydrolysis of peptide bonds, exhibiting a preference for Glu–Xaa bonds much more than for Asp–Xaa bonds. The enzyme has a preference for prolyl or leucyl residues at P2 and phenylalanyl at P3. Hydrolysis of Glu–Pro and Asp–Pro bonds is slow. This endopeptidase is a member of the peptidase family S2A.

J. J. Birktoft & K. Breddam (1994) *Meth. Enzymol.* **244**, 114.

γ-GLUTAMYL KINASE

This enzyme [EC 2.7.2.11], also known as glutamate 5-kinase, catalyzes the reaction of ATP with L-glutamate to produce ADP and L-glutamate 5-phosphate, which is rapidly converted in aqueous solutions to 5-oxoproline and orthophosphate.

E. Adams & L. Frank (1980) *Ann. Rev. Biochem.* **49**, 1005.

γ-GLUTAMYL TRANSPEPTIDASE

This enzyme [EC 2.3.2.2], also known as γ-glutamyltransferase, catalyzes the reaction of a γ-glutamyl compound (such as glutathione, γ-glutamyl amino acid, glutamine, glutathione disulfide, leukotriene C$_4$, *etc.*) with an acceptor substrate to produce the γ-glutamylated acceptor derivative and the de-γ-glutamylated substrate. Thus, when the acceptor substrate is water, the enzyme catalyzes a glutathionase activity (that is hydrolysis of glutathione to produce L-glutamate and L-cysteinylglycine) and a glutaminase activity (L-glutamine + water → L-glutamate and ammonia). When the acceptor substrate is an amino acid (particularly L-cystine, L-methionine, L-glutamine, and L-alanine), the product is the corresponding isopeptide derivative (that is, γ-glutamyl-L-cystine, γ-glutamyl-L-methionine, γ-glutamyl-L-glutamine, and γ-glutamyl-L-alanine, respectively). Dipeptides and even glutathione can also serve as acceptor substrates. Elevated levels of amino acids can result in substrate inhibition of all the activities. The plant and yeast enzymes have a different range of specificities than that of the mammalian enzyme mentioned above.

R. D. Allison (1985) *Meth. Enzymol.* **113**, 419.
S. S. Tate & A. Meister (1985) *Meth. Enzymol.* **113**, 400.
S. C. Hartman (1971) *The Enzymes*, 3rd ed., **4**, 79.

GLUTAMYL-tRNA^Gln AMIDOTRANSFERASE

This enzyme catalyzes the ATP-dependent reaction of glutamyl-tRNA^Gln with glutamine to produce glutaminyl-tRNA^Gln, ADP, orthophosphate, and glutamate.

H. Zalkin (1993) *Adv. Enzymol.* **66**, 203.
H. Zalkin (1985) *Meth. Enzymol.* **113**, 303.
J. M. Buchanan (1973) *Adv. Enzymol.* **39**, 91.

GLUTAMYL-tRNA SYNTHETASE

This enzyme [EC 6.1.1.17], also known as glutamate:tRNA ligase, catalyzes the reaction of L-glutamate with tRNA^Glu and ATP to produce L-glutamyl-tRNA^Glu, AMP, and pyrophosphate.

P. Schimmel (1987) *Ann. Rev. Biochem.* **56**, 125.
J. Lapointe, S. Levasseur & D. Kern (1985) *Meth. Enzymol.* **113**, 42.
M. Proulx & J. Lapointe (1985) *Meth. Enzymol.* **113**, 50.

GLUTATHIONE

A naturally occurring intracellular redox metabolite (γ-L-glutamyl-L-cysteinylglycine; symbolized GSH) that is reversibly oxidized to glutathione disulfide [*i.e.*, 2 G–SH − 2 H⁻ = G–S–S–G; $E°$(pH 7) = −0.25 V]. Cellular glutathione concentrations are typically at 1–5 mM concentrations, and glutathione plays major roles in cellular metabolism: for example, in the detoxification of certain arenes, in the mercapturic acid pathway, in the reduction of the Fe(III) in methemoglobin, in leukotriene and prostaglandin biosynthesis, in L-cystine translocation (as well as that of certain other amino acids), in the maintenance of thiol groups, in the destruction of hydrogen peroxide, other peroxides, and free radicals, in disulfide exchange reactions, *etc.*

Selected entries from *Methods in Enzymology* [vol, page(s)]:
Biosynthesis, **77**, 59-63 assay, **77**, 373-382; depletion *in vitro*, **77**, 59-63 [by acrylamide, **77**, 54; by acrylic esters, **77**, 54; by acrylonitrile, **77**, 54; by aspirin, **77**, 57; by *tert*-butyl hydroperoxide, **77**, 56, 57]; depletion in cells [by buthionine sulfoximine, **77**, 62, 63; by cumene hydroperoxide, **77**, 56, 57; by diamide, **77**, 55, 56; by diazenecarboxylate derivatives, **77**, 55; by diethyl maleate, **77**, 52-54, 58; by diisopropylidene acetone, **77**, 54, 58; by ethylmorphine, **77**, 57; by formaldehyde, **77**, 57; by glycylglycine, **77**, 57; by products of cytochrome P-450, **77**, 54; by tetrathionate, **77**, 56; by transferase substrates, **77**, 54; by α,β-unsaturated carbonyls, **77**, 52-54; *in vivo*, **77**, 50-59]; distribution in tissue samples, **77**, 376; factors affecting levels, **77**, 51; S-substituted derivatives, **77**, 237 [synthesis, **77**, 137; turnover in kidney cells, **77**, 144, 145; turnover rate, **77**, 63; glutathione disulfide [assay, **77**, 373-382; N-ethylmaleimide, **77**, 375; with glutathione reductase, **77**, 381, 382; o-phthaldialdehyde, **77**, 377; sample preparation, **77**, 375]; distribution in tissue samples, **77**, 376; efflux, **77**, 20; formation, **77**, 20].

GLUTATHIONE DEHYDROGENASE (ASCORBATE)

This enzyme [EC 1.8.5.1], also known as glutathione:dehydroascorbate oxidoreductase, catalyzes the reaction of two molecules of glutathione with dehydroascorbate to produce glutathione disulfide and ascorbate.

W. W. Wells, D. P. Xu & M. P. Washburn (1995) *Meth. Enzymol.* **252**, 30.

GLUTATHIONE PEROXIDASE

This selenium-dependent enzyme [EC 1.11.1.9] catalyzes the reaction of two molecules of glutathione with hydrogen peroxide to produce glutathione disulfide and two water molecules. Hydrogen peroxide can be replaced by steroid and lipid hydroperoxides, albeit not as effectively (nevertheless, this enzyme is not identical with phospholipid-hydroperoxide glutathione peroxidase [EC 1.11.1.12]). However, the hydroperoxy products formed by the action of lipoxygenase [EC 1.13.11.12] are not substrates.

F. Ursini, M. Maiorino, R. Brigelius-Flohé, K. D. Aumann, A. Roveri, D. Schomburg & L. Flohé (1995) *Meth. Enzymol.* **252**, 38.
R. C. Fahey & A. R. Sundquist (1991) *Adv. Enzymol.* **64**, 1.
K. T. Douglas (1987) *Adv. Enzymol.* **59**, 103.
B. Mannervik (1985) *Meth. Enzymol.* **113**, 490.
T. C. Stadtman (1979) *Adv. Enzymol.* **48**, 1.

GLUTATHIONE REDUCTASE

This FAD-dependent enzyme [EC 1.6.4.2] catalyzes the reaction of NADPH with glutathione disulfide to produce NADP⁺ and two glutathione molecules. The enzyme activity is dependent on a redox-active disulfide group in each of the active sites.

R. C. Fahey & A. R. Sundquist (1991) *Adv. Enzymol.* **64**, 1.
P. A. Karplus & G. E. Schulz (1989) *J. Mol. Biol.* **210**, 163.
K. T. Douglas (1987) *Adv. Enzymol.* **59**, 103.
I. Carlberg & B. Mannervik (1985) *Meth. Enzymol.* **113**, 484.
G. Schulz & E. F. Pai (1983) *J. Biol. Chem.* **258**, 1752.
C. H. Williams, Jr. (1976) *The Enzymes*, 3rd ed., **13**, 89.

GLUTATHIONE SYNTHETASE

This enzyme [EC 6.3.2.3] catalyzes the reaction of ATP with γ-L-glutamyl-L-cysteine with glycine to produce ADP, orthophosphate, and glutathione.

R. C. Fahey & A. R. Sundquist (1991) *Adv. Enzymol.* **64**, 1.
A. Meister (1985) *Meth. Enzymol.* **113**, 393.
A. Meister (1974) *The Enzymes*, 3rd ed., **10**, 671.

GLUTATHIONE *S*-TRANSFERASE

This set of enzymes [EC 2.5.1.18] (also known as glutathione *S*-alkyltransferase, glutathione *S*-aryltransferase, *S*-(hydroxyalkyl)glutathione lyase, and glutathione *S*-aralkyltransferase) catalyzes the reaction of an RX with glutathione to produce HX and an *S*-substituted (with R) glutathione (thus, R–S–glutathione). R may be an aliphatic, aromatic, or heterocyclic group whereas X may be a sulfate, nitrate, or halide. In addition, some members of this set of enzymes, which typically have a broad specificity, will also catalyze the *S*-addition of glutathione to aliphatic epoxides and arene oxides. Others in this classification will catalyze the reduction of polyol nitrate by glutathione to polyol and nitrite as well as certain isomerization reactions and disulfide interchanges.

K. A. Marrs (1996) *Ann. Rev. Plant Physiol. Plant Mol. Biol.* **47**, 127.
D. J. Meyer & B. Ketterer (1995) *Meth. Enzymol.* **252**, 53.
R. N. Armstrong (1994) *Adv. Enzymol.* **69**, 1.
T. H. Rushmore & C. B. Pickett (1993) *J. Biol. Chem.* **268**, 11475.
R. C. Fahey & A. R. Sundquist (1991) *Adv. Enzymol.* **64**, 1.
B. Mannervik & U. H. Danielson (1988) *Crit. Rev. Biochem.* **23**, 283.
K. T. Douglas (1987) *Adv. Enzymol.* **59**, 103.
R. N. Armstrong (1987) *Crit. Rev. Biochem.* **22**, 39.
B. Mannervik (1985) *Adv. Enzymol.* **57**, 357.
W. B. Jakoby & T. A. Fjellstedt (1972) *The Enzymes*, 3rd ed., **7**, 199.

GLYCERALDEHYDE-3-PHOSPHATE DEHYDROGENASE

Glyceraldehyde-3-phosphate dehydrogenase (phosphorylating) [EC 1.2.1.12], catalyzes the reaction of D-glyceraldehyde 3-phosphate with orthophosphate and NAD^+ to produce 3-phospho-D-glyceroyl phosphate and NADH. The enzyme utilizes an active site thiol group to form a thiohemiacetal intermediate which upon oxidation is converted to a covalently bound thiolester. Based on his work with this enzyme, the metabolist Ephraim Racker was among the first to propose that enzyme catalysis can take advantage of the transient formation of covalently bound intermediates. Others, including Otto Warburg himself, initially surmised *incorrectly* that formation of covalently bound intermediates

would unnecessarily increase the number of bond-making/breaking steps and would decrease catalytic efficiency. D-Glyceraldehyde and several other aldehydes will also act as weaker substrates. A number of thiols can replace orthophosphate, thereby releasing the corresponding thiolester.

Glyceraldehyde-3-phosphate dehydrogenase ($NADP^+$) (nonphosphorylating) [EC 1.2.1.9], also referred to as triose-phosphate dehydrogenase, catalyzes the reaction of D-glyceraldehyde 3-phosphate with $NADP^+$ and water to produce 3-phospho-D-glycerate and NADPH.

Glyceraldehyde-3-phosphate dehydrogenase ($NADP^+$) (phosphorylating) [EC 1.2.1.13], also known as triose-phosphate dehydrogenase ($NADP^+$), catalyzes the reaction of D-glyceraldehyde 3-phosphate with orthophosphate and $NADP^+$ to produce 3-phospho-D-glyceroyl phosphate and NADPH.

M. W. W. Adams & A. Kletzin (1996) *Adv. Protein Chem.* **48**, 101.
R. Jaenicke, H. Schurig, N. Beaucamp & R. Ostendorp (1996) *Adv. Protein Chem.* **48**, 181.
H. F. Fisher (1988) *Adv. Enzymol.* **61**, 1.
K. E. Neet (1980) *Meth. Enzymol.* **64**, 139.
J. I. Harris & M. Waters (1976) *The Enzymes*, 3rd ed., **13**, 1.

GLYCERATE DEHYDROGENASE

This enzyme [EC 1.1.1.29], also known as hydroxypyruvate reductase and hydroxypyruvate dehydrogenase, catalyzes the reaction of (R)-glycerate with NAD^+ to produce hydroxypyruvate and NADH.

P. M. Goodwin (1990) *Meth. Enzymol.* **188**, 361.
J. W. Thorner & H. Paulus (1973) *The Enzymes*, 3rd ed., **8**, 487.

GLYCERATE KINASE

This enzyme [EC 2.7.1.31] catalyzes the reaction of ATP with (R)-glycerate to produce ADP and 3-phospho-(R)-glycerate.

P. M. Goodwin (1990) *Meth. Enzymol.* **188**, 361.
J. W. Thorner & H. Paulus (1973) *The Enzymes*, 3rd ed., **8**, 487.
S. Black (1962) *Meth. Enzymol.* **5**, 352.

GLYCEROL

A substance (1,2,3-propanetriol; $HOCH_2CH(OH)CH_2OH$) that is a component and metabolite of many lipids and is commonly used in enzyme kinetic studies, serving as a cryoprotectant and stabilizing agent for proteins[1]. Glycerol has also been utilized as a cryosolvent in a number of cryoenzymology studies: *e.g.*, catalase[2].

The freezing points of 10%, 30%, 50%, and 66.7% aqueous solutions (w/w) of glycerol are $-1.6°C$, $-9.5°C$, $-23.0°C$ and $-46.5°C$, respectively. The melting point of pure glycerol is $+17.8°C$. Because the vicinal hydroxyl groups form complexes with borates and boronates, borate-based buffers should be avoided if glycerol is present in the buffer. Likewise, glycerol interferes with periodate oxidation. The high viscosity of glycerol (at 30°C it is about 785 times that of pure water; at 25°C, this factor exceeds 1000) can influence NMR experiments[3].

[1] C. S. Milner & N. N. Dalton (1953) *Glycerol*, Reinhold, New York.
[2] G. K. Strother & E. Ackerman (1961) *Biochim. Biophys. Acta* **47**, 317.
[3] M.-D. Tsai (1982) *Meth. Enzymol.* **87**, 235.

GLYCEROL DEHYDRATASE

This cobalamin-dependent enzyme [EC 4.2.1.30] catalyzes the conversion of glycerol to 3-hydroxypropanal and water.

B. T. Golding & W. Buckel (1998) *Comprehensive Biological Catalysis: A Mechanistic Reference* **3**, 239.

GLYCEROL KINASE

This enzyme [EC 2.7.1.30], also known as glycerokinase and ATP:glycerol 3-phosphotransferase, catalyzes the reaction of ATP with glycerol to produce ADP and glycerol 3-phosphate. Both glycerone (or, dihydroxyacetone) and L-glyceraldehyde can serve as substrates. The nucleoside triphosphate can be substituted by UTP and, in the case of the yeast enzyme, ITP and GTP.

J. H. Hurley (1996) *Ann. Rev. Biophys. & Biomol. Struct.* **25**, 137.
J. D. Esko & C. R. H. Raetz (1983) *The Enzymes*, 3rd ed., **16**, 207.
E. C. C. Lin (1977) *Ann. Rev. Biochem.* **46**, 765.
J. W. Thorner & H. Paulus (1973) *The Enzymes*, 3rd ed., **8**, 487.

GLYCEROL-3-PHOSPHATE O-ACYLTRANSFERASE

This enzyme [EC 2.3.1.15] catalyzes the reaction of an acyl-CoA with sn-glycerol 3-phosphate to produce coenzyme A and a 1-acyl-sn-glycerol 3-phosphate. The acyl-CoA derivatives contain an acyl group with a chain length of at least ten carbon atoms. In addition, the acyl-CoA can be substituted by an acyl-[acyl-carrier protein] derivative.

I. Nishida & N. Murata (1996) *Ann. Rev. Plant Physiol. Plant Mol. Biol.* **47**, 541.
C. Kent (1995) *Ann. Rev. Biochem.* **64**, 315.
M. A. Scheideler & R. M. Bell (1992) *Meth. Enzymol.* **209**, 55.

D. Haldar & A. Vancura (1992) *Meth. Enzymol.* **209**, 64.
R. M. Bell & R. A. Coleman (1983) *The Enzymes*, 3rd ed., **16**, 87.
J. D. Esko & C. R. H. Raetz (1983) *The Enzymes*, 3rd ed., **16**, 207.
R. A. Pieringer (1983) *The Enzymes*, 3rd ed., **16**, 255.

GLYCEROL-3-PHOSPHATE DEHYDROGENASE

Glycerol-3-phosphate dehydrogenase (NAD+) [EC 1.1.1.8] catalyzes the reaction of sn-glycerol 3-phosphate with NAD+ to produce glycerone phosphate (or, dihydroxyacetone phosphate) and NADH. 1,2-Propanediol phosphate and glycerone sulfate can also act as substrates, having a weaker affinity.

Glycerol-3-phosphate dehydrogenase (NAD(P)+) [EC 1.1.1.94], also known as NAD(P)H-dependent dihydroxyacetone-phosphate reductase, catalyzes the reaction of sn-glycerol 3-phosphate with NAD(P)+ to produce glycerone phosphate (or, dihydroxyacetone phosphate) and NAD(P)H.

Glycerol-3-phosphate dehydrogenase [EC 1.1.99.5] is a flavoprotein that catalyzes the reaction of sn-glycerol 3-phosphate with an acceptor substrate to produce glycerone phosphate (dihydroxyacetone phosphate) and the reduced acceptor.

B. F. Cooper & F. B. Rudolph (1995) *Meth. Enzymol.* **249**, 188.
J. D. Esko & C. R. H. Raetz (1983) *The Enzymes*, 3rd ed., **16**, 207.
Y. Hatefi & D. L. Stiggall (1976) *The Enzymes*, 3rd ed., **13**, 175.

GLYCERONE KINASE

This enzyme [EC 2.7.1.29], also known as triokinase and dihydroxyacetone kinase, catalyzes the reaction of ATP with glycerone (or, dihydroxyacetone) to produce ADP and glycerone phosphate (or, dihydroxyacetone phosphate). *See also* Triokinase

L. V. Bystrykh, W. de Koning & W. Harder (1990) *Meth. Enzymol.* **188**, 445.
K. H. Hofmann & W. Babel (1990) *Meth. Enzymol.* **188**, 451.

GLYCINE ACETYLTRANSFERASE

This pyridoxal-phosphate-dependent enzyme [EC 2.3.1.29], also known as glycine C-acetyltransferase and 2-amino-3-ketobutyrate:coenzyme A ligase, catalyzes the reaction of acetyl-CoA with glycine to produce coenzyme A and 2-amino-3-oxobutanoate.

D. McGilvray & J. G. Morris (1971) *Meth. Enzymol.* **17B**, 585.
I. Lieberman & H. A. Barker (1955) *Meth. Enzymol.* **1**, 616.

GLYCINE ACYLTRANSFERASES

This classification is for those enzymes that transfer acyl moieties to glycine. They include:

Glycine *N*-acyltransferase [EC 2.3.1.13] catalyzes the reaction of an acyl-CoA derivative with glycine to produce coenzyme A and an *N*-acylglycine. The acyl-CoA derivative can be one of a number of aliphatic and aromatic acids. However, neither phenylacetyl-CoA nor indole-3-acetyl-CoA can act as substrates.

Glycine *N*-benzoyltransferase [EC 2.3.1.71] catalyzes the reaction of benzoyl-CoA with glycine to produce coenzyme A and *N*-benzoylglycine. This enzyme is not identical with glycine *N*-acyltransferase or glutamine *N*-acyltransferase [EC 2.3.1.68].

H. Chantrenne (1955) *Meth. Enzymol.* **2**, 346.

GLYCINE AMIDINOTRANSFERASE

This enzyme [EC 2.1.4.1], also known as L-arginine:glycine amidinotransferase, catalyzes the reaction of L-arginine with glycine to produce L-ornithine and guanidinoacetate. Canavanine can serve as the substrate instead of arginine.

J. B. Walker (1979) *Adv. Enzymol.* **50**, 177.
J. B. Walker (1973) *The Enzymes*, 3rd ed., **9**, 497.

GLYCINE AMINOTRANSFERASE

This pyridoxal-phosphate-dependent enzyme [EC 2.6.1.4] catalyzes the reaction of glycine with α-ketoglutarate (or, 2-oxoglutarate) to produce glyoxylate and L-glutamate. ***See also*** *Glycine:Oxaloacetate Aminotransferase; Glyoxylate Aminotransferase*

A. E. Braunstein (1973) *The Enzymes*, 3rd ed., **9**, 379.

GLYCINE DECARBOXYLASE

This pyridoxal-phosphate-dependent enzyme [EC 1.4.4.2], also known as glycine dehydrogenase (decarboxylating) and the glycine cleavage system P-protein, catalyzes the reaction of glycine with a lipoylprotein to produce *S*-aminomethyldihydrolipoylprotein and carbon dioxide. Lipoamide can also act as acceptor substrate. This enzyme activity is a component, with aminomethyltransferase [EC 2.1.2.10], of glycine synthase. ***See also*** *Glycine Synthase*

D. J. Oliver (1994) *Ann. Rev. Plant Physiol. Plant Mol. Biol.* **45**, 323.
E. A . Boeker & E. E. Snell (1972) *The Enzymes*, 3rd ed., **6**, 217.

GLYCINE FORMIMINOTRANSFERASE

This enzyme [EC 2.1.2.4] catalyzes the reaction of 5-formiminotetrahydrofolate with glycine to produce tetrahydrofolate and *N*-formiminoglycine.

J. I. Rader & F. M. Huennekens (1973) *The Enzymes*, 3rd ed., **9**, 197.

GLYCINE METHYLTRANSFERASE

This enzyme [EC 2.1.1.20] catalyzes the reaction of *S*-adenosyl-L-methionine with glycine to produce *S*-adenosyl-L-homocysteine and sarcosine.

F. Takusagawa, M. Fujioka, A. Spies & R. L. Schowen (1998) *Comprehensive Biological Catalysis: A Mechanistic Reference* **1**, 1.
M. H. Stipanuk (1986) *Ann. Rev. Nutr.* **6**, 179.

GLYCINE:OXALOACETATE AMINOTRANSFERASE

This pyridoxal-phosphate-dependent enzyme [EC 2.6.1.35] catalyzes the reaction of glycine with oxaloacetate to produce glyoxylate and L-aspartate. ***See also*** *Glycine Aminotransferase; Glyoxylate Aminotransferase*

A. E. Braunstein (1973) *The Enzymes*, 3rd ed., **9**, 379.

GLYCINE REDUCTASE

This enzyme catalyzes the reaction of glycine with orthophosphate to produce acetyl phosphate and ammonium ion. The oxidized form of the enzyme is subsequently reduced to its dithiol form in a NADH-linked step.

T. C. Stadtman (1996) *Ann. Rev. Biochem.* **65**, 83.
P. D. van Poelje & E. E. Snell (1990) *Ann. Rev. Biochem.* **59**, 29.
T. C. Stadtman (1990) *Ann. Rev. Biochem.* **59**, 111.
T. C. Stadtman (1980) *Ann. Rev. Biochem.* **49**, 93.
T. C. Stadtman (1979) *Adv. Enzymol.* **48**, 1.

GLYCINE SYNTHASE

This enzyme [EC 1.4.4.2], also known as glycine cleavage complex, consists of the components glycine dehydrogenase (decarboxylating) [EC 1.4.4.2], which catalyzes the pyridoxal-phosphate-dependent reaction of glycine with a lipoylprotein to produce *S*-aminomethyldihydrolipoylprotein and carbon dioxide, an aminomethyltransferase [EC 2.1.2.10], also known as glycine-cleavage system T-protein, which catalyzes the transfer of the group to tetrahydrofolate to generate (6*R*)-5,10-methylenetetrahydrofolate, ammonia, and dihydrolipoylprotein, and the L-protein which is a flavoenzyme that utilizes NAD^+ to oxidize the reduced lipoate.

V. Schirch (1998) *Comprehensive Biological Catalysis: A Mechanistic Reference* **1**, 211.

D. J. Oliver (1994) *Ann. Rev. Plant Physiol. Plant Mol. Biol.* **45**, 323.

J. I. Rader & F. M. Huennekens (1973) *The Enzymes*, 3rd ed., **9**, 197.

E. A. Boeker & E. E. Snell (1972) *The Enzymes*, 3rd ed., **6**, 217.

GLYCOGEN PHOSPHORYLASE

This enzyme [EC 2.4.1.1], also called phosphorylase, catalyzes the reaction of $[(1,4)\text{-}\alpha\text{-D-glucosyl}]_n$ with orthophosphate to produce $[(1,4)\text{-}\alpha\text{-D-glucosyl}]_{n-1}$ and $\alpha\text{-D-}$glucose 1-phosphate. The name to be used with this enzyme is dependent on the naturally occurring substrate: for example, glycogen phosphorylase, starch phosphorylase, maltodextrin phosphorylase, *etc.*

N. B. Madsen (1991) *A Study of Enzymes* **2**, 139.

N. D. Madsen (1986) *The Enzymes*, 3rd ed., **17**, 365.

D. J. Graves & J. H. Wang (1972) *The Enzymes*, 3rd ed., **7**, 435.

GLYCOGEN SYNTHASE

Glycogen synthase [EC 2.4.1.11], also known as starch synthase and UDP-glucose-glycogen glucosyltransferase, catalyzes the reaction of UDP-glucose with $[(1,4)\text{-}\alpha\text{-D-glucosyl}]_n$ to produce UDP and $[(1,4)\text{-}\alpha\text{-D-glucosyl}]_{(n+1)}$. The name to be used with this enzyme is dependent on the nature of the product that is being formed: for example, glycogen synthase and starch synthase. Glycogen synthase from mammalian tissues is a multiprotein complex containing a catalytic subunit and the protein glycogenin. The enzyme requires glucosylated glycogenin as a primer (the reaction product of glycogenin glucosyltransferase [EC 2.4.1.186]). Glycogen synthase [EC 2.4.1.21], also known as starch (bacterial glycogen) synthase and ADP-glucose:starch glucosyltransferase, catalyzes the reaction of ADP-glucose with $[(1,4)\text{-}\alpha\text{-D-glucosyl}]_n$ to produce ADP and $[(1,4)\text{-}\alpha\text{-D-glucosyl}]_{(n+1)}$. The name to be used with this enzyme is dependent on the nature of the product that is being formed: for example, bacterial glycogen synthase and starch synthase. This enzyme is similar to the glycogen synthase [EC 2.4.1.11] above except that the nucleotide substrate is ADP-glucose.

N. B. Madsen (1991) *A Study of Enzymes* **2**, 139.

J. Larner (1990) *Adv. Enzymol.* **63**, 173.

R. F. Colman (1990) *The Enzymes*, 3rd ed., **19**, 283.

P. Cohen (1986) *The Enzymes*, 3rd ed., **17**, 461.

P. J. Roach (1986) *The Enzymes*, 3rd ed., **17**, 499.

W. Stalmans & H. G. Hers (1973) *The Enzymes*, 3rd ed., **9**, 309.

GLYCOLATE OXIDASE

This FMN-dependent enzyme [EC 1.1.3.15], also known as (*S*)-2-hydroxy-acid oxidase, catalyzes the reaction of a (*S*)-2-hydroxy acid with dioxygen to produce a 2-oxo acid and hydrogen peroxide. The enzyme exists as two major isoenzymes. The A form of the protein preferentially oxidizes short-chain aliphatic hydroxy acids. The B form preferentially oxidizes long-chain and aromatic hydroxy acids. The rat isoenzyme B form also acts as an L-amino-acid oxidase.

B. A. Palfey & V. Massey (1998) *Comprehensive Biological Catalysis: A Mechanistic Reference* **3**, 83.

L. A. Kleczkowski (1994) *Ann. Rev. Plant Physiol. Plant Mol. Biol.* **45**, 339.

G. A. Hamilton (1985) *Adv. Enzymol.* **57**, 85.

GLYCOLYTIC OSCILLATION

The periodic and regular changes in glycolytic pathway kinetics, as first observed in yeast[1]. In their recent consideration of three control models for glycolytic oscillations in yeast, Teusink *et al.*[2] applied metabolic control analysis and used operational definitions to quantify the control properties of enzymes with regard to glycolytic oscillations. Control of both frequency and amplitudes of the metabolite fluctuations were found to be distributed among the enzymes. They found little evidence supporting the role of an oscillophore (*i.e.*, the enzyme primarily held responsible for the generation of the oscillation of the system).

[1] A. Ghosh & B. Chance (1964) *Biochem. Biophys. Res. Commun.* **16**, 174.

[2] B. Teusink, B. M. Bakker & H. V. Westerhoff (1996) *Biochim. Biophys. Acta* **1275**, 204.

GLYCOSIDASES

The following are recent reviews on the molecular and physical properties of this class of enzymes. Among the glycosidases are: oligoxyloglucan β-glycosidase [EC 3.2.1.120]; DNA-deoxyinosine glycosidase, also known as DNA(hypoxanthine) glycohydrolase [EC 3.2.2.15]; DNA-3-methyladenine glycosidase I, also known as DNA glycosidase I [EC 3.2.2.20]; DNA-3-methyladenine glycosidase II, also known as DNA glycosidase II [EC 3.2.2.21]; rRNA *N*-glycosidase [EC 3.2.2.22]; peptide-N^4-(*N*-acetyl-β-glucosaminyl)asparagine amidase; and glycopeptide *N*-glycosidase [EC 3.5.1.52]. **See** *specific enzyme*

G. Mooser (1992) *The Enzymes*, 3rd ed., **20**, 187.

C. T. Walsh (1984) *Ann. Rev. Biochem.* **53**, 493.

H. M. Flowers & N. Sharon (1979) *Adv. Enzymol.* **48**, 29.

GLYCYL ENDOPEPTIDASE

This enzyme [EC 3.4.22.25] catalyzes the hydrolysis of peptide bonds with a preference for Gly–Xaa in proteins and small molecule substrates. The enzyme, a member of the peptidase family C1, is isolated from the papaya plant, *Carica papaya*. It is not inhibited by chicken cystatin, unlike most other homologs of papain.

D. J. Buttle (1994) *Meth. Enzymol.* **244**, 539.

GLYCYL-tRNA SYNTHETASE

This enzyme [EC 6.1.1.14], also known as glycine:tRNA ligase, catalyzes the reaction of glycine with tRNA^{Gly} and ATP to produce glycyl-tRNA^{Gly}, AMP, and pyrophosphate (or, diphosphate).

P. Schimmel (1987) *Ann. Rev. Biochem.* **56**, 125.

GLYOXALASE I

This enzyme [EC 4.4.1.5], also known as lactoylglutathione lyase and methylglyoxalase, catalyzes the conversion of (*R*)-*S*-lactoylglutathione to produce glutathione and methylglyoxal. 3-Phosphoglycerol-glutathione will also serve as a substrate.

D. J. Creighton & N. S. R. K. Murthy (1990) *The Enzymes*, 3rd ed., **19**, 323.
K. T. Douglas (1987) *Adv. Enzymol.* **59**, 103.
A. S. Mildvan & D. C. Fry (1987) *Adv. Enzymol.* **59**, 241.

GLYOXALASE II

This enzyme [EC 3.1.2.6], also known as *S*-hydroxyacylglutathione hydrolase, catalyzes the hydrolysis of an (*S*)-(2-hydroxyacyl)glutathione to yield glutathione and a 2-hydroxy acid anion. *S*-Acetoacetylglutathione will act as a substrate as well, but not as efficiently.

K. T. Douglas (1987) *Adv. Enzymol.* **59**, 103.

GLYOXYLATE AMINOTRANSFERASE

This pyridoxal 5-phosphate-dependent enzyme [EC 2.6.1.4] catalyzes the transamination of glyoxylate from L-glutamate to produce glycine and α-ketoglutarate.

N. E. Tolbert (1981) *Ann. Rev. Biochem.* **50**, 133.

GLYOXYLATE REDUCTASE

Glyoxylate reductase [EC 1.1.1.26] catalyzes the reversible reaction of glycolate with NAD$^+$ to produce glyoxylate and NADH. The enzyme will also catalyze the NADH-dependent interconversion of hydroxypyruvate to D-glycerate. Glyoxylate reductase (NADPH) [EC 1.1.1.79] catalyzes the reversible reaction of glycolate with NADP$^+$ to produce glyoxylate and NADPH (as well as the hydroxypyruvate to D-glycerate conversion). The enzyme can use NAD$^+$ as a substrate, although not as effectively as NADP$^+$.

L. A. Kleczkowski (1994) *Ann. Rev. Plant Physiol. Plant Mol. Biol.* **45**, 339.

GMP SYNTHETASE

GMP synthetase [EC 6.3.4.1], also known as xanthosine-5′-phosphate:ammonia ligase, catalyzes the reaction of ATP with xanthosine 5′-phosphate and ammonia to produce GMP, AMP, and pyrophosphate (or, diphosphate). GMP synthetase (glutamine-utilizing) [EC 6.3.5.2] catalyzes the reaction of ATP with xanthosine 5′-phosphate, L-glutamine, and water to produce GMP, AMP, L-glutamate, and pyrophosphate.

H. Zalkin (1993) *Adv. Enzymol.* **66**, 203.
H. Zalkin (1985) *Meth. Enzymol.* **113**, 273.
D. E. Koshland, Jr., & A. Levitzki (1974) *The Enzymes*, 3rd ed., **10**, 539.
J. M. Buchanan (1973) *Adv. Enzymol.* **39**, 91.

GOLDMAN EQUATION

An equation (also referred to as the constant field equation, the Goldman-Hodgkin-Katz equation, and the GHK equation) which relates the membrane potential ($\Delta\psi$) to the individual permeabilities of the ions (and their concentrations) on both sides of the membrane. Thus,

$$\Delta\psi = \frac{RT}{F} \ln \left[\frac{\Sigma_+ P_i [M_i^+]_{out} + \Sigma_- P_j [X_j^-]_{in}}{\Sigma_+ P_i [M_i^+]_{in} + \Sigma_- P_j [X_j^-]_{out}} \right]$$

where R is the universal gas constant, T is the absolute temperature, F is the Faraday constant, the sums (Σ) are taken over all cations (Σ_+) and anions (Σ_-) having significant permeabilities, P refers to the relative membrane permeability of these ions, and [M$^+$] and [X$^-$] are their concentrations.

If any one ion has a significantly greater permeability than any of the other ions, then the Goldman equation reduces to the Nernst equation for membrane potential.

If the permeability of one ion is altered, while all concentrations and other permeabilities remain unchanged, the Goldman equation permits one to predict the changes in the membrane potential. For example, consider the case in which the permeability for a cation (the cation that is in higher concentration inside the cell) has increased. In such a case, that cation flows out of the cell and more positive charges now exist on the outer surface of the membrane and more negative charges on the inner surface. As can be seen from the Goldman equation, the membrane potential becomes more negative: the membrane is hyperpolarized.

GOODNESS-OF-FIT CRITERIA

Mannervik[1] has provided criteria for assessing the fitness of a particular kinetic model to the experimental data (utilizing regression analysis and statistical methods) and to aid in the discrimination between different models. He offers the following criteria: (1) a good model is expected to give convergence in the regression analysis; (2) a good model should give meaningful parameter values with low standard deviations; (3) a good model should give residuals showing random distribution about the zero level and lacking correlation with any of the dependent or independent variables; (4) a good model should give a low residual sum of squares that is compatible with the experimental variance. **See also** Global Analysis; Statistics (A Primer)

[1]B. Mannervik (1982) Meth. Enzymol. **87**, 370.

GOOD'S BUFFERS

A collection of buffers[1–3] used in biochemical studies exhibiting minimal buffer salt interaction with proteins and/or metal ions. These buffers span the pH ranges normally used in biological experiments. While they tend not to bind metal ions strongly, this should be verified when metal ion-requiring substrates are involved[4]. The list below includes commercially available Good's buffers, along with their reported pK_a values (at 25°C, unless otherwise noted) and their corresponding dpK_a/dT values when known[5]. Those marked with an asterisk indicate the buffers which were originally reported by Good et al.[1]

2-(N-Morpholino)ethanesulfonic acid* (abbreviated MES)	pK_a = 6.1, dpK_a/dT = −0.011
Bis(2-hydroxyethyl)iminotris(hydroxymethyl)methane (BIS-TRIS)	pK_a = 6.5
N-(2-Acetamido)-2-iminodiacetic acid* (ADA)	pK_a = 6.6, dpK_a/dT = −0.011
Piperazine-N,N'-bis(2-ethanesulfonic acid* (PIPES)	pK_a = 6.8, dpK_a/dT = −0.0085
N-(2-Acetamido)-2-aminoethanesulfonic acid* (ACES*)	pK_a = 6.8, dpK_a/dT = −0.020
1,3-Bis[tris(hydroxymethyl)-methylamino]propane (BIS-TRIS PROPANE)	$pK_{a,1}$ = 6.8, $pK_{a,2}$ = 9.0
3-(N-Morpholino)-2-hydroxy-propanesulfonic acid (MOPSO)	pK_a = 6.9
N,N-Bis(2-hydroxyethyl)-2-amino-ethanesulfonic acid* (BES)	pK_a = 7.1, dpK_a/dT = −0.016
3-(N-Morpholino)propanesulfonic acid (MOPS)	pK_a = 7.2
N-Tris(hydroxymethyl)methyl-2-aminoethanesulfonic acid* (TES)	pK_a = 7.45 dpK_a/dT = −0.020
N-2-Hydroxyethylpiperazine-N'-2-ethanesulfonic acid* (HEPES)	pK_a = 7.5, dpK_a/dT = −0.014
3-[N,N-Bis(2-hydroxyethyl)-amino]-2-hydroxypropane-sulfonic acid (DIPSO)	pK_a = 7.6
3-[N-Tris(hydroxymethyl)-methylamino-2-hydroxy-propanesulfonic acid (TAPSO)	pK_a = 7.6
N-(2-Hydroxyethyl)piperazine-N'-(2-hydroxypropanesulfonic acid) (HEPPSO)	pK_a = 7.8
Piperazine-N,N'-bis(2-hydroxy-propanesulfonic acid) (POPSO)	pK_a = 7.8
N-(2-Hydroxyethyl)piperazine-N'-(3-propanesulfonic acid) (EPPS, also called HEPPS)	pK_a = 8.0
N-Tris(hydroxymethyl)methyl-glycine* (Tricine)	pK_a = 8.1, dpK_a/dT = −0.021
N,N-Bis(2-hydroxyethyl)glycine* (Bicine)	pK_a = 8.3, dpK_a/dT = −0.018
Tris(hydroxymethyl)methylamino-propanesulfonic acid (TAPS)	pK_a = 8.4
3-[(1,1-Dimethyl-2-hydroxyethyl)-amino]-2-hydroxypropane-sulfonic acid (AMPSO)	pK_a = 9.0
3-(N-Cyclohexylamino)-ethanesulfonic acid (CHES)	pK_a = 9.3 (9.55 at 20°C)
3-(Cyclohexylamino)-2-hydroxy-1-propanesulfonic acid (CAPSO)	pK_a = 9.6
3-(Cyclohexylamino)-1-propanesulfonic acid (CAPS)	pK_a = 10.4

[1]N. E. Good, G. D. Winget, W. Winter, T. N. Connolly, S. Izawa & R. M. M. Singh (1966) Biochemistry **5**, 467.
[2]N. E. Good & S. Izawa (1972) Meth. Enzymol. **24B**, 53.
[3]W. J. Ferguson & N. E. Good (1980) Anal. Biochem. **104**, 300.
[4]J. F. Morrison (1979) Meth. Enzymol. **63**, 257.
[5]D. D. Perrin & B. Dempsey (1974) Buffers for pH and Metal Ion Control, Chapman and Hall, London.

G PROTEINS

A superfamily of GTPases (often referred to as GTP- or G-regulatory proteins) whose cardinal feature is a 200-residue domain comprising a central six-stranded β-pleated sheet surrounded by α-helices[1,2]. Five polypeptide loops constituting the guanine nucleotide binding site are highly conserved among members of this protein superfamily. Examples of G proteins are (a) transducin, which in its bound-GTP state stimulates cyclic GMP phosphodiesterase, (b) elongation factor Tu which in its EF-Tu·GTP complex moves an aminoacyl-tRNA into the ribosome-mRNA complex A site for aminoacylation of a growing polypeptide, and (c) Ras which as Ras·GTP activates its downstream tyrosine kinase targets. Cleavage of the terminal phosphoanhydride bond of GTP allows each of these regulatory proteins to transit from its stimulatory conformational state to a quiescent state.

G proteins exhibit the ability to modulate both metabolic activation or inhibition based on the phosphorylation state of the bound guanine nucleotide. In this model, the activated receptor *R plays a catalytic role by activating more than one G-protein. This is evident in the following cycle of reactions:

1. Agonist-A Starts Process by Activating Receptor-R:
   R + A = *R

2. Activated Receptor *R Binds to GDP form of G Protein:

   *R + G$_{GDP}$ = [*RG$_{GDP}$]

3. GDP Dissociates from [*RG$_{GDP}$] Complex:

   [*RG$_{GDP}$ = [*RG___] + GDP

4. GDP-free Form [*RG___] Binds GTP and Activates G Protein:

   [*RG___] + GTP = [*RG*$_{GTP}$]

5. Activated Receptor *R Departs, Leaving Behind Activated G*$_{GTP}$:

   [*RG*$_{GTP}$] = *R + [G*$_{GTP}$]

6. G*$_{GTP}$ Stimulates "Downstream" Metabolic Processes":

   [G*$_{GTP}$] + Downstream Process → Downstream Processes*

7. G*$_{GTP}$ Undergoes GTP Hydrolysis to Deactivated G$_{GDP}$ Form:

   G*$_{GTP}$ → G$_{GDP}$ + P$_i$

8. Activated Metabolic Response* is Terminated:
   Downstream Process* → Downstream Process

A G-protein has two states, each characterized by the presence of bound GTP or GDP, and their interconversion is accomplished by an intrinsic GTPase as well as the participation of nucleotide-exchange factors. Bound GTP is converted to bound GDP by the GTPase, whereas bound GDP is converted to bound GTP by release of GDP and binding/exchange of GTP. The rate of interconversion between active and inactive states is limited by GTPase hydrolysis, and GTPase-activating proteins (GAPs) can accelerate hydrolysis, as can G protein association with a conformational state of its effector. In this respect, certain G proteins are "clocks", others are "switches" or "adapters", while still others are "sensors".

These GTPases operate by in-line or S$_N$2 mechanisms, where the phosphoryl undergoes inversion of configuration as it is directly transferred to water. The kinetics of the GTPase and the guanine nucleotide exchange activities determine the time dependence of activated or inhibited states of the metabolic process under their direct control. G$_{i\alpha1}$ has a turnover number for GTP hydrolysis of around 3 min^{-1}; Ras has a k_{cat} of 0.3 min^{-1}; and EF-Tu has a value of 0.003 min^{-1}. Tetracoordinate AlF$_4^-$ ion activates G$_\alpha$ subunits by mimicking GTP's γ-phosphoryl group as it undergoes hydrolysis and assumes a pentavalent geometry. GTPγS binding sustains the activated state of G proteins.

[1]S. R. Sprang (1997) *Ann. Rev. Biochem.* **66**, 639.
[2]D. E. Coleman, A. M. Berghuis, E. Lee, M. E. Linder, A. G. Gilman & D. R. Sprang (1994) *Science* **265**, 1405.

Entire MIE Volumes Dedicated to This Topic: Vol. **237** (R. Iyengar, ed.); vol. **195** (R. A. Johnson & J. D. Corbin, eds.).

GRAPHICAL METHODS

See *Double-Reciprocal Plot; Hanes Plot; Direct Linear Plot; Dixon Plot; Dixon-Webb Plot; Eadie-Hofstee Plot; Substrate Concentration Range; Frieden Protocol; Fromm Protocol; Point-of-Convergence Method; Dalziel Phi Relationships; Scatchard Plots; Hill Plots*

GRAPH THEORY

A branch of mathematics concerned with the study and applications of graphs and their uses in kinetic behavior of linked processes. Graph theory, especially when applied to the dynamic nature of networks, has provided useful protocols for deriving enzyme rate expressions. **See** *Fromm's Method for Deriving Enzyme Rate Equations*

GREEN FLUORESCENT PROTEIN

An intensely fluorescent protein[1] isolated from the jellyfish *Aequorea victoria* and other *Aequorea* species. This protein produces its fluorescence from an intrinsic chromophore generated by a series of steps involving residues 65–67 of the polypeptide chain. As a result, GFP has broad application, especially when expressed as a fused or chimeric protein containing a second polypeptide region whose intracellular location is of interest[2].

The steps leading to the formation of the intrinsic chromophore have recently been investigated[3] kinetically with S65T-GFP. The process of chromophore formation is an ordered sequence of three distinct steps: (1) slow protein folding ($k_f = 2.44 \times 10^{-3}$ s^{-1}) that precedes chromophore modification; (2) an intermediate step occurs that includes, but may not be necessarily limited to, cyclization of the tripeptide chromophore motif ($k_c = 3.8 \times 10^{-3}$ s^{-1}); and (3) rate-limiting oxidation of the cyclized chromophore ($k_{ox} = 1.51 \times 10^{-4}$ s^{-1}). Reid and Flynn[3] also reasoned that because chromophore forms *de novo* from purified denatured protein and is a first-order process, GFP chromophore formation is likely to be an autocatalytic process.

[1]M. Chalfie, Y. Tu, G. Euskirchen, W. W. Ward & D. C. Prasher (1994) *Science* **263**, 802.
[2]S. R. Kain, M. Adams, A. Kondepudi, T. T. Yang, W. W. Ward & P. Kitts (1995) *Biotechniques* **19**, 650.
[3]B. G. Reid & G. C. Flynn (1997) *Biochemistry* **36**, 6786.

GROTTHUSS CHAIN

Extended chains of hydrogen-bonded water molecules proposed by Christian von Grotthuss in the early 19th century to explain electrolyte conductance in water[1].

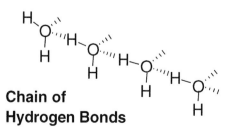

**Chain of
Hydrogen Bonds**

While these extended chains failed to explain the ionic mobilities of most inorganic ions, they remain a viable model for proton and hydroxide ion mobility. **See** *Grotthuss Mechanism; Water*

[1]The molecular structure of water was not known in Grotthuss' time, and he proposed that water molecules, designated as − +, were arranged between the positive (*p*) and negative (*n*) electrodes in the following manner: *p* − + − + − + − *n*. (J. R. Partington (1964) *A History of Chemistry*, vol. 4, pp. 26-27, Macmillan, London.)

GROTTHUSS-DRAPER LAW

The so-called first law of photochemistry stating that only the radiation absorbed by a molecular entity or substance is effective in producing a photochemical change.

GROTTHUSS MECHANISM

A mechanism that explains the high *apparent* ionic mobilities[1] of protons and hydroxide ions in terms of hydrogen bond-making/-breaking steps along extended chains of water molecules. Solvation of a proton can be represented simply[2] as H_3O^+, and this hydronium ion strongly interacts with other molecules in a hydrogen-bond network.

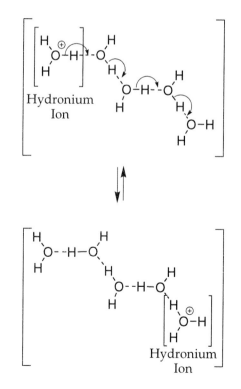

Shown in the figure above is a hypothetical scheme that illustrates how an organized series of protonation-deprotonation steps can have the net effect of rapidly "transferring" a proton from a hydronium ion (*i.e.*,

H_3O^+) in one region of an aqueous solution to produce a hydronium ion at a distant site. Note that the proton released locally from the initial H_3O^+ remains in its vicinity, and is not the same as the proton forming the hydronium ion at the distant site. For this reason, the ionic mobility appears to be much greater than would be expected on the basis of diffusion alone. Facilitated proton transfer along rigidly and accurately positioned hydrogen bonds could be of fundamental importance in enzyme catalysis. **See** Water

[1] Ionic mobility refers to the velocity of an ion moving toward an oppositely charged electrode when a 1-volt potential is applied across a 1-centimeter electrochemical cell.
[2] A strongly hydrated molecular cluster, such as $[H \cdot (OH_2)_4]^+$, is probably a more realistic representation (M. Eigen (1964) Angew. Chem. (Int. Eng. Edn.) **3**, 1).

GROUND STATE

The state of a molecular entity (or system) with the lowest Gibbs free energy. The context of the study usually makes clear the type of ground state under consideration (e.g., electronic, vibrational, rotational, etc.). **See also** Excited State; Fluorescence

GROUP TRANSFER REACTIONS

Enzyme-catalyzed reactions in which a functional group or moiety is transferred from one molecular entity to another.

GTP CYCLOHYDROLASE I

This enzyme [EC 3.5.4.16] catalyzes the reaction of GTP with two water molecules to produce formate and 2-amino-4-hydroxy-6-(erythro-1,2,3-trihydroxypropyl)-dihydropteridine triphosphate. The reaction involves hydrolysis of two C–N bonds and isomerization of the pentose unit. The recyclization step may be nonenzymatic.

E. R. Werner, H. Wachter & G. Werner-Felmayer (1997) Meth. Enzymol. **281**, 53.
C. A. Nichol, G. K. Smith & D. S. Duch (1985) Ann. Rev. Biochem. **54**, 729.
G. M. Brown & J. M. Williamson (1982) Adv. Biochem. **53**, 345.

GTP CYCLOHYDROLASE II

This enzyme [EC 3.5.4.25] catalyzes the reaction of GTP with three water molecules to produce formate, 2,5-diamino-6-hydroxy-4-(5-phosphoribosylamino)pyrimidine, and pyrophosphate (or, diphosphate). In this reaction, two C–N bonds are hydrolyzed, releasing formate, with

the simultaneous hydrolysis of the terminal pyrophosphate.

A. Bacher, G. Richter, H. Ritz, S. Eberhardt, M. Fischer & C. Krieger (1997) Meth. Enzymol. **280**, 382.
G. M. Brown & J. M. Williamson (1982) Adv. Biochem. **53**, 345.

GTP DEPLETION or REGENERATION

See ATP and GTP Depletion; Nucleoside 5'-Triphosphate Regeneration

GUANINE DEAMINASE

This enzyme [EC 3.5.4.3], also known as guanine aminohydrolase and guanase, catalyzes the hydrolysis of guanine to produce xanthine and ammonia.

C. L. Zielke & C. H. Suelter (1971) The Enzymes, 3rd ed., **4**, 47.

GUANOSINE DEAMINASE

This enzyme [EC 3.5.4.15], also known as guanosine aminohydrolase, catalyzes the hydrolysis of guanosine to yield xanthosine and ammonia.

C. L. Zielke & C. H. Suelter (1971) The Enzymes, 3rd ed., **4**, 47.

GUANYLATE CYCLASE

This enzyme [EC 4.6.1.2], also known as guanylyl cyclase and guanyl cyclase, catalyzes the conversion of GTP to 3',5'-cyclic GMP and pyrophosphate (or, diphosphate). Both ITP and dGTP can act as substrates.

Entire volume of MIE that covers this topic: R. A. Johnson & J. D. Corbin, eds. (1991) Meth. Enzymol., vol. **195**.

GUANYLATE KINASE

This enzyme [EC 2.7.4.8] catalyzes the reaction of ATP with GMP to produce ADP and GDP. dGMP can substitute for GMP and dATP can substitute for ATP.

E. P. Anderson (1973) The Enzymes, 3rd ed., **9**, 49.

GUEST

An inorganic or organic ion that binds to a host molecule. **See** Host Molecule; Inclusion Complexes

GUGGENHEIM METHOD (FIRST-ORDER RATE CONSTANT)

A systematic procedure[1,2] for determining the value of a first-order rate constant whenever the initial reactant concentration(s) or the final extent of reaction is unknown.

[1] E. A. Guggenheim (1926) Philos. Mag. **2**, 538.
[2] J. W. Moore & R. G. Pearson (1981) Kinetics and Mechanism, 3rd ed., Wiley, New York.

H

HABER-WEISS REACTION

A chemical reaction cycle involving hydrogen peroxide, hydroxide radical, molecular oxygen, and hydroxide ion:

$$H_2O_2 + \cdot OH \rightarrow H_2O + O_2^- + H^+ \text{ and}$$

$$H_2O_2 + O_2^- \rightarrow O_2 + OH^- + \cdot OH$$

The second reaction achieved notoriety as a possible source of hydroxyl radicals, but the reaction proceeds extremely slowly. Iron(III) complexes may catalyze this reaction; in this case, Fe(III) would first be reduced by the superoxide, followed by oxidation by hydrogen peroxide. **See also** Fenton Reaction

HALDANE RELATION

A mathematical equation[1] indicating how the equilibrium constant of an enzyme-catalyzed reaction (or half-reaction in the case of so-called ping pong reaction mechanisms) is related to the various kinetic parameters for the reaction mechanism. In the Briggs-Haldane steady-state treatment of a Uni Uni reaction mechanism, the Haldane relation can be written as follows:

$$K_{eq} = \frac{V_{max,forward} K_{product}}{V_{max,reverse} K_{substrate}} = \frac{k_{cat,forward}}{K_{substrate}} \times \frac{K_{product}}{k_{cat,reverse}}$$

where $V_{max,forward}$ (equal to $k_{cat,forward}[\text{Enzyme}_{total}]$) is the maximal reaction velocity in the forward direction, $K_{product}$ is the Michaelis constant for the reverse reaction, $V_{max,reverse}$ (equal to $k_{cat,reverse}[\text{Enzyme}_{total}]$) is the maximal reaction velocity in the reverse direction, and

$K_{substrate}$ is the Michaelis constant for the forward direction.

Rigorous adherence of enzymes to the Haldane relation is well illustrated by the case of wild-type and Glu[165]-to-Asp triose-phosphate isomerases[2]. These enzymes differ only with respect to a single methylene in the side-chain carboxyl group of residue 165. The steady-state parameters for the wild-type enzyme are: $k_{cat,forward}$, 430 s^{-1}; K_m, 0.97 mM for dihydroxyacetone phosphate; $k_{cat,reverse}$, 4300 s^{-1}; and K_m, 0.47 mM for D-glyceraldehyde 3-phosphate. Substitution of these values into the above equation gives K_{eq} equal to 21 (for the reaction written in the direction of dihydroxyacetone phosphate formation), a value that is in excellent agreement with independent determinations of the overall equilibrium constant ($K_{eq} = 22$)[3]. The steady-state parameters[2] for the Glu[165]-to-Asp enzyme are: $k_{cat,forward}$, 1.8 s^{-1}; K_m, 1.8 mM for dihydroxyacetone phosphate; $k_{cat,reverse}$, 2.8 s^{-1}; and K_m, 0.13 mM for D-glyceraldehyde 3-phosphate. Substitution of these values into the above equation again gives a value of 21.5 for K_{eq}. While there is every reason to expect that the wild-type and Glu[165]-to-Asp forms of triose-phosphate isomerase operate by virtually identical catalytic mechanisms, strict adherence to the Haldane relation must occur, even for enzymes using different catalytic pathways as long as one is dealing with the same overall reaction.

The use of Haldane relationships to verify the magnitude of the equilibrium constant or, conversely, to determine (or verify) one of the kinetic parameters requires that all constants be measured under the same experimental conditions (e.g., temperature, pH, buffer species, ionic strength, free metal ion concentrations, etc.) If not, the Haldane relationship has no meaning. In addition, kinetic data are often limited in precision, unlike equilibrium measurements. For multisubstrate reactions, there are at least two different Haldane relationships. Thus, there are a number of independent opportunities to compare the magnitude of derived equilibrium constant to that of the observed equilibrium constant. Such a procedure was used for nucleoside diphosphate kinase[4] and for galactose-1-phosphate uridylyltransferase[5].

Fromm[6] and Cleland[7] provide valuable discussions of the utility of Haldane relations in excluding certain kinetic reaction mechanisms based on a numerical evaluation of the constants on each side of the equal sign in the Haldane relation. If the equality is maintained, the candidate mechanism is consistent with the observed rate parameter data. Obviously, one must be concerned about the quality of experimentally derived estimates of rate parameters, because chemists have frequently observed that thermodynamic data (such as equilibrium constants) are often more accurate and precise than kinetically derived parameters. **See** *Haldane Relations for Multisubstrate Enzymes*

[1]J. B. S. Haldane (1930) *Enzymes*, Longmans, Green and Co., London. (republished in 1965 by M.I.T. Press, Cambridge, MA).
[2]R. T. Raines & J. R. Knowles (1986) *Ann. N. Y. Acad. Sci.* **471**, 266.
[3]I. A. Rose & Z. B. Rose (1969) *Comprehensive Biochemistry* **17**, 126.
[4]Graces & W. W. Cleland (1969) *Biochemistry* **8**, 633.
[5]P. A. Frey, L.-J. Wong, K.-F. Sheu & S.-L. Yang (1982) *Meth. Enzymol.* **87**, 20.
[6]H. J. Fromm (1975) *Initial Rate Enzyme Kinetics*, Springer-Verlag, Berlin.
[7]W. W. Cleland (1982) *Meth. Enzymol.* **87**, 366.

HALDANE RELATIONS FOR MULTISUBSTRATE ENZYMES

Alberty[1] first proposed the use of Haldane relations to distinguish among the ordered Bi Bi, the ordered Bi Bi Theorell-Chance, and the rapid equilibrium random Bi Bi mechanisms. Nordlie and Fromm[2] used Haldane relationships to rule out certain mechanisms for ribitol dehydrogenase.

Cleland[3] pointed out that most Haldane relations have the general form $K_{eq} = \{(V_{max,f})^n K_{(p)} K_{(q)} K_{(r)} \ldots\}/ \{(V_{max,r})^n K_{(a)} K_{(b)} K_{(c)} \ldots\}$ where $V_{max,f}$ and $V_{max,r}$ are the maximum velocities in the forward and reverse directions, respectively, n is an integer (ranging between -1, $0, 1, 2, \ldots$), $K_{(a)}, K_{(b)}, K_{(c)}$, *etc.*, are either the Michaelis constants or the dissociation constants for the substrates (A, B, C, *etc.*) and $K_{(p)}, K_{(q)}, K_{(r)}$, *etc.*, are either the Michaelis constants or dissociation constants for the products (P, Q, R, *etc.*). Cleland[4] distinguished two types

of Haldane relations: thermodynamic and kinetic. Thermodynamic Haldane relations contain only the equilibrium constants associated with the individual steps of the reaction mechanism. An example of the thermodynamic type can be given for the ping pong Bi Bi reaction scheme (E + A $\rightleftharpoons$ EX $\rightleftharpoons$ F + P followed by F + B $\rightleftharpoons$ FB $\rightleftharpoons$ E + Q). One Haldane is $K_{eq} = K_{ip} K_{iq}/(K_{ia} K_{ib})$ where K_{ip}, K_{iq}, K_{ia}, and K_{ib} refer to the dissociation constants of the products and reactants, respectively.

For a number of enzyme kinetic mechanisms, a thermodynamic Haldane cannot explicitly be defined in terms of K_{ix} parameters. An example is the ordered Bi Bi reaction:

$$E + A \underset{k_2}{\overset{k_1}{\rightleftharpoons}} EA$$

$$EA + B \underset{k_4}{\overset{k_3}{\rightleftharpoons}} EXY \underset{k_6}{\overset{k_5}{\rightleftharpoons}} EQ + P$$

$$EQ \underset{k_8}{\overset{k_7}{\rightleftharpoons}} E + Q$$

In order to derive a thermodynamic Haldane relation for this mechanism, the true dissociation constants for B (*i.e.*, k_4/k_3) and for P (*i.e.*, k_5/k_6) are needed. From the definitions for the various kinetic parameters (**See** *Ordered Bi Bi Mechanism*) it is readily seen that $k_4/k_3 = K_{ib} K_a V_{max,r}/(K_{ia} V_{max,f})$ and $k_5/k_6 = K_{ip} K_q V_{max,f}/ (K_{iq} V_{max,r})$. Thus, a thermodynamic Haldane relationship for the ordered Bi Bi mechanism can be written as $K_{eq} = V_{max,f}^2 K_{ip} K_q/(V_{max,r}^2 K_a K_{ib})$, where K_q and K_a are Michaelis constants.

Kinetic Haldane relations[4] use a ratio of apparent rate constants in the forward and reverse directions, if the substrate concentrations are very low. For an ordered Bi Bi reaction, the apparent rate constant for the second step is $V_{max,f}/K_b$ (where K_b is the Michaelis constant for B) and, in the reverse reaction, $V_{max,r}/K_p$. Each of these is multiplied by the reciprocal of the dissociation constant of A and Q, respectively. The "forward" product is then divided by the "reverse" product. Hence, the kinetic Haldane relationship for the ordered Bi Bi reaction is $K_{eq} = (K_{ia}^{-1} V_{max,f}/K_b)/(K_{iq}^{-1} V_{max,r}/K_p) = V_{max,f} K_p K_{iq}/ (V_{max,r} K_{ia} K_b)$. For completely random mechanisms, thermodynamic and kinetic Haldane relationships are equivalent.

In ping pong reactions, Haldane relations can also be written for the individual "half-reactions." In such cases, Haldane expressions assist in analyzing isotope exchange studies involving these partial reactions.

Haldane relationships can also be useful in characterizing isozymes or the same enzyme isolated from a different source. Reactions catalyzed by isozymes must have identical equilibrium constants, but the magnitudes of their kinetic parameters are usually different (*e.g.*, the case of yeast and mammalian brain hexokinase[5]). Note that the Haldane relationship for the ordered Bi Bi mechanism is $K_{eq} = V_{max,f}K_p K_{iq}/(V_{max,r}K_{ia}K_b)$. This same Haldane is also valid for the ordered Bi Bi Theorell-Chance mechanism and the rapid equilibrium random Bi Bi mechanism. The reverse reaction of the yeast enzyme is easily studied: an observation not true for the brain enzyme, even though both enzymes catalyze the exact same reaction. A crucial difference between the two enzymes is the dissociation constant (K_{iq}) for Q (in this case, glucose 6-phosphate). For the yeast enzyme, this value is about 5 mM whereas for the brain enzyme the value is 1 μM. Hence, in order for K_{eq} to remain constant (and assuming K_p, K_{ia}, and K_b are all approximately the same for both enzymes) the $V_{max,f}/V_{max,r}$ ratio for the brain enzyme must be considerably larger than the corresponding ratio for the yeast enzyme. In fact, the differences between the two ratios is more than a thousandfold. Hence, the Haldane relationship helps to explain how one enzyme appears to be more kinetically reversible than another catalyzing the same reaction.

[1]R. A. Alberty (1953) *J. Amer. Chem. Soc.* **75**, 1928.
[2]R. C. Nordlie & H. J. Fromm (1959) *J. Biol. Chem.* **234**, 2523.
[3]W. W. Cleland (1963) *Biochim. Biophys. Acta* **67**, 104.
[4]W. W. Cleland (1982) *Meth. Enzymol.* **87**, 366.
[5]D. L. Purich & H. J. Fromm (1972) *Curr. Top Cell. Reg.* **6**, 131.

HALF-LIFE

The period of time required for the decay of one-half of a substance via a first-order process. In a first-order process

$$[A] = [A_o]e^{-kt}$$

where [A] is the concentration of reactant at time t and [A$_o$] is the concentration at time zero. When [A] = $0.5[A_o]$, then $2 = e^{kt_{0.5}}$ (or, $\ln 2 = kt_{0.5}$). Hence, $t_{0.5} = (0.693147)/k$, where k is the first-order rate constant for the reaction.

The cardinal feature of all first-order processes is that the period required for reducing the reactant by a factor of 2 is independent of the amount of substance present.

CHEMICAL PROCESSES. In many chemical reactions and in *all* exchange reactions, the rate of decrease in the concentration of a reactant is directly proportional to the concentration of that reactant (*i.e.*, $v = k[A]$). For a first-order process, the relaxation period (corresponding to the period of time required to reach e^{-1}, or about 0.368, of the original amount) is given as:

$$t = 1/k = t_{0.5}/\ln 2$$

Many bimolecular rate processes can behave as though they are first-order processes, especially if one reactant is present at much greater concentration than the other. Such processes are called pseudo-first-order reactions.

RADIOACTIVE DECAY. Many atomic nuclei have unstable neutron-to-proton ratios and undergo spontaneous first-order decay through the emission of α, β^-, or β^+ particles or gamma rays.

The half-lives of radioisotopes typically used in biochemical experiments are:

Radionuclide	$t_{0.5}$*	*k
^{3}H	12.32 y	1.783×10^{-9} s^{-1}
^{11}C	20.3 m	5.69×10^{-4} s^{-1}
^{14}C	5.715×10^3 y	3.843×10^{-12} s^{-1}
^{13}N	9.97 m	1.16×10^{-3} s^{-1}
^{22}Na	2.605 y	8.432×10^{-9} s^{-1}
^{32}P	14.28 d	5.618×10^{-7} s^{-1}
^{33}P	25.3 d	3.17×10^{-7} s^{-1}
^{35}S	87.2 d	9.20×10^{-8} s^{-1}
^{36}Cl	3.01×10^5 y	7.30×10^{-14} s^{-1}
^{40}K	1.26×10^9 y	1.74×10^{-17} s^{-1}
^{42}K	12.36 h	1.558×10^{-5} s^{-1}
^{43}K	22.3 h	8.63×10^{-6} s^{-1}
^{45}Ca	162.7 d	4.931×10^{-8} s^{-1}
^{47}Ca	4.536 d	1.769×10^{-6} s^{-1}
^{51}Cr	27.70 d	2.896×10^{-7} s^{-1}
^{52}Fe	8.28 h	2.33×10^{-5} s^{-1}
^{55}Fe	2.73 y	8.05×10^{-9} s^{-1}
^{59}Fe	44.51 d	1.802×10^{-7} s^{-1}
^{57}Co	271.8 d	2.952×10^{-8} s^{-1}
^{58}Co	70.88 d	1.132×10^{-7} s^{-1}
^{60}Co	5.271 y	4.167×10^{-9} s^{-1}
^{123}I	13.2 h	1.46×10^{-5} s^{-1}
^{125}I	59.4 d	1.35×10^{-7} s^{-1}
^{131}I	8.040 d	9.978×10^{-7} s^{-1}

*Abbreviations are y = years, d = days, h = hours, m = minutes, and s = seconds.

For a radioisotope decay, the rate is given as

$$v = kN$$

where k is the first-order rate constant and N is the number of atoms (N = (number of moles)(N_A) = (number of moles)(6.0221367×10^{23} atoms·mol^{-1})). [In converting years to seconds, it is useful to recall that there are about 3.15×10^7 seconds/year, or roughly $\pi \times 10^7$ seconds/year].

HALF-REACTIONS

1. A balanced hypothetical chemical equation indicating the transfer of electrons between two different oxidation states of the same element of chemical species. 2. One segment of a ping-pong (or double-displacement) enzyme mechanism.

REDOX HALF-REACTIONS. Electron transfer reactions involve oxidation (or loss of electrons) of one component and reduction (or gain of electrons) by a second component. Therefore, a complete redox reaction can be treated as the sum of two half-reactions such that the stoichiometry and electric charge is balanced across a chemical equilibrium. For each such half-reaction, there is an associated standard potential E^o. The hydrogen ion-hydrogen gas couple is:

$$2H^+ + 2e = H_2$$

where by convention the charge on the electron is understood but not written. Because one cannot establish the absolute change in the electromotive force (emf) for any half-reaction, the E^o value for the above reaction is arbitrarily set a 0.0000 volts. The standard potential for any reaction can then be defined as the emf change [(+) or (−) as well as magnitude] of an electrode consisting of the half-reaction measure against the hydrogen ion-hydrogen gas half-reaction under standard conditions:

Reduction:
$$2H^+ + 2e^- = H_2 \qquad E^o = 0.000 \text{ volts}$$

Oxidation:
$$2Hg = Hg_2^{2+} + 2e^- \qquad E^o = -0.789 \text{ volts}$$

Overall Reaction:
$$2H^+ + 2Hg = Hg_2^{2+} + H_2 \qquad E^o = -0.789 \text{ volts}$$

Likewise, any two half-reactions may be summed to establish the standard potential for an overall reaction; in so doing, care must be exercised to reverse the sign of the emf associated with a half-reaction written in the opposite direction.

PING PONG HALF-REACTIONS. Many enzymes operate by double-displacement mechanisms involving covalent enzyme-substrate intermediates as shown in the following scheme:

First Half-Reaction $\qquad$ E + A $\rightleftharpoons$ EA $\rightleftharpoons$ FP $\rightleftharpoons$ F + P

Second Half-Reaction $\qquad$ F + B $\rightleftharpoons$ FB $\rightleftharpoons$ EQ $\rightleftharpoons$ E + Q

Overall Reaction $\qquad$ A + B $\rightleftharpoons$ P + Q

E and F represent free enzyme and the modified enzyme, respectively, and the free energy change overall is the sum of the changes in free energy for the individual half-reactions. Examples of ping-pong enzymes are nucleoside-5′-diphosphate kinase (NDPK) and the amino acid-keto acid aminotransferases (such as glutamate:oxaloacetate transaminase), and F's in these respective cases are phosphoryl-enzyme and enzyme-bound pyridoxamine compounds. Early investigators were persuaded that the formation of a covalent intermediate would unnecessarily add more bond making/breaking steps, thereby lowering an enzyme's catalytic effectiveness. We now recognize that enzymes frequently experience no difficulty in bond making/breaking; instead, product release often is rate-determining. The formation of covalent intermediates actually facilitates catalysis in several ways: (a) by conserving the group-transfer potential of a reactive grouping of atoms (example: a phosphoryl group) to promote subsequent transfer with the second substrate acting as an acceptor molecule; (b) by using a single binding site that recognizes A or Q (in the case of E) and B or P (in the case of F) in the scheme shown above; (c) by repetitive use of the same catalytic components for acid/base and electrostatic interactions in each half-reaction. The reader should note that each such half-reaction may be accompanied by changes in stereochemistry of the transferred group, if one can detect such inversions. For example, by using $[\gamma\text{-}(S)$ or $(R)\text{-}{}^{18}O,{}^{17}O,{}^{16}O]$ATP one may determine the stereochemistry of phosphoryl transfer reactions.

Because each half-reaction can behave like an independent chemical process, ping pong enzymes catalyze exchange reactions. For example, yeast NDPK catalyzes an ADP $\leftrightarrow$ ATP exchange reaction:

$$[^{14}C]ATP + E \rightleftharpoons E \cdot [^{14}C]ATP \rightleftharpoons F \cdot [^{14}C]ADP \rightleftharpoons F + [^{14}C]ADP$$

Therefore, by mixing enzyme, unlabeled ADP, and [^{14}C]ATP together, one can observe the synthesis of labeled ADP; this reaction can also be run in reverse by starting with enzyme, unlabeled ATP, and [^{14}C]ADP. (This is called a partial exchange reaction, because the total amount of ADP and ATP remains constant; only the isotopic forms change with time until they reach isotopic equilibrium.) *See Isotope Exchange; Substrate Synergism; Ping-Pong Mechanism; Covalent Intermediates; Phosphoryl Transfer Stereochemistry*

HALF-OF-THE-SITES REACTIVITY

A type of enzyme behavior [1-3] observed for some oligomeric proteins exhibiting cooperative behavior. For example, a negative, allosteric effector or covalent modifier (or even a substrate for an enzyme exhibiting negative homotropic cooperativity) may bind to one subunit, thereby inducing a conformational change that prevents the effector or covalent modifier from binding to the other subunit; hence the term, half-site reactivity. Lazdunski has also noted that half-site reactivity need not always exhibit cooperative curves in initial-rate kinetics[2,3].

[1]M. Lazdunski, C. Petitclerc, D. Chappelet & C. Lazdunski (1971) *Eur. J. Biochem.* **20**, 124.
[2]M. Lazdunski (1972) *Curr. Top. Cell. Regul.* **6**, 267.
[3]K. E. Neet (1980) *Meth. Enzymol.* **64**, 139.

HAMMERHEAD RIBOZYME

The smallest known catalytic RNA motif (containing around 30 nucleotides) first detected in certain small satellite and viroid RNAs. The descriptor "hammerhead" refers to the motif's shape when represented as a two-dimensional structure. Three Watson-Crick base-paired helices (designated I, II, and III) are linked by conserved sequences within the cleavage site located between helix I and III. When these two helices close with a hairpin loop, the resulting structure is the intramolecular (or *in cis*)-cleaving ribozyme. Closing only one of the helices leads to the intermolecular (or *in trans*)-cleaving form. Michaelis-Menten parameters were first established with the *in cis*-cleaving form[1], and pre-steady state kinetic studies have permitted the determination of individual rate constants for the Uni Bi random mechanism of the *in trans*-cleaving hammerhead ribozyme[2,3]. *See Catalytic RNA*

[1]J. Haseloff & W. L. Gerlach (1988) *Nature* **334**, 585.
[2]M. J. Fedor & O. C. Uhlenbeck (1992) *Biochemistry* **31**, 12042.
[3]K. J. Hertel, D. Herschlag & O. C. Uhlenbeck (1994) *Biochemistry* **33**, 3374.

HAMMETT EQUATION

An equation used to assess the effect of structure on reactivity: $\log(k_X/k_H) = \sigma_X \rho$ where k_H is the corresponding rate constant when the position in the reactant under study contains a hydrogen as a substituent at that position, k_X is the rate constant when the substituent is X, σ_X is a constant characteristic of the group X, and ρ (the rho-value) is a constant for a given reaction under a given set of conditions. The Hammett equation is a linear free-energy relationship that has been applied to many reactions and functional groups. *See ρ-Value; Linear Free-Energy Relationship; σ-Constant; Taft Equation; Extended Hammett Equation; Yukama-Tsuno Equation*

L. P. Hammett (1970) *Physical Organic Chemistry*, 2nd ed., McGraw-Hill, New York.
C. D. Johnson (1973) *The Hammett Equation*, Cambridge Univ. Press, Cambridge.
J. Shorter (1982) *Correlation Analysis of Organic Reactivity*, Wiley, New York.

HAMMOND PRINCIPLE/POSTULATE

The postulate stating that if a chemical mechanism involves two consecutive steps (or states) of nearly the same energy, then interconversion between such states will require only a minor reorganization of the molecular structure. Hence, for a highly exergonic reaction, the transition state more closely resembles the reactant(s) or the immediately preceding intermediate, rather than the product(s) or immediately succeeding intermediate. This same idea is expressed in Leffler's assumption, which is more commonly found in textbooks. In fact, the principle is sometimes referred to by the acronym "Bemahapothle" in recognition of the principle's major proponents (*i.e.*, <u>Be</u>ll, <u>Ma</u>rcus, <u>Ha</u>mmond, <u>Po</u>lanyi, <u>Th</u>ornton, and <u>Le</u>ffler).

Note that any factor that stabilizes a reaction intermediate will also stabilize the transition state leading to that intermediate. *See also Leffler's Assumption; Reacting Bond Rules*

J. E. Leffler (1953) *Science* **117**, 340.
G. S. Hammond (1955) *J. Amer. Chem. Soc.* **77**, 334.
D. Fárcasiu (1975) *J. Chem. Educ.* **52**, 76.
A. Williams (1984) *Acc. Chem. Res.* **17**, 425.

HANES PLOT

A linear graphical method for analyzing the initial rate kinetics of enzyme-catalyzed reactions. In the Hanes plot, $[A]/v$ is plotted as a function of $[A]$, where v is the initial rate and $[A]$ is the substrate concentration[1-3].

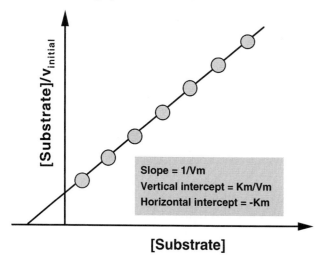

Slope = 1/Vm
Vertical intercept = Km/Vm
Horizontal intercept = -Km

[Substrate]

Also referred to as the Hanes-Hultin plot and the Hanes-Woolf (or, Woolf-Hanes) plot, the method is based on a transformation of the Michaelis-Menten equation (*i.e.*, the expression for the Uni Uni mechanism): $[A]/v = (K_a/V_{max}) + ([A]/V_{max})$ where V_{max} is the maximum forward velocity and K_a is the Michaelis constant for A. In the Hanes plot, the slope of the line is numerically equal to V_{max}, the vertical intercept is equivalent to K_a/V_{max}, and the horizontal intercept to $-K_a$. **See** *Double-Reciprocal Plot*

[1]B. Woolf (1932) quoted in *Allgemeine Chemie der Enzyme* (J. B. S. Haldane & K. G. Stern), p. 119, Steinkopff, Dresden-Leipzig.
[2]C. S. Hanes (1932) *Biochem. J.* **26**, 1406.
[3]E. Hultin (1967) *Acta Chem. Scand.* **21**, 1575.

HANSCH CONSTANT

A measure of the capability of a solute for hydrophobic interactions, based on the partition coefficient P for the distribution of the solute between 1-octanol and water.

C. Hansch & A. J. Leo (1979) *Substituent Constants for Correlation Analysis in Chemistry and Biology*, Wiley, New York.

HANSCH EQUATION

A quantitative relationship describing how the hydrophobicity of a parent compound is the result of additive contributions of substituents:

$$\pi = \log(P/P_o)$$

where P is the ratio of the solubility of a substituted molecule in *n*-octanol relative to its solubility in water, and P_o is the corresponding ratio for the unsubstituted parent compound. This relationship probably reflects the differences in the substituent's surface area, because the amount of water released in making a hydrophobic interaction is related to the water-accessible surface area.

C. Hansch & A. J. Leo (1979) *Substituent Constants for Correlation Analysis in Chemistry and Biology*, Wiley, New York.

HANSCH'S CORRELATION ANALYSIS

A procedure that utilizes quantitative structure-activity relationships to assist in the characterization and design of enzyme alternative substrates, inhibitors, and effectors. In this procedure, the inhibition (or activity) results of a series of structurally related compounds are analyzed to determine the coefficients in an equation of the form:

$$\log (1/C) = \Sigma a_n\pi_n + \Sigma b_n F_n + \Sigma c_n R_n + \Sigma d_n MR_n$$
$$+ \Sigma e_n E_{sn} + \ldots$$

where C is the molar concentration of the inhibitor (or effector, *etc.*) producing 50% inhibition of enzyme (or, some other rate-related constant), π is a hydrophobic constant (often obtained from partition coefficients of organic compounds serving as references for hydrophobic interactions of enzymes) in which the subscript n represents different values for π at different positions (*e.g.*, *ortho* and *meta* positions on an aromatic ring), F and R are the Swain and Lupton factors for electrostatic interactions, MR is the molar refractivity which models nonspecific binding in nonhydrophobic space, and E_s is Taft's steric parameter. Regression analysis of a set of closely related molecules permits one to determine the magnitude of the coefficients. The final expression can be quite useful in the design of other effectors as well as in the characterization of the relative contributions to binding of hydrophobic interactions, polarizability and dispersion forces, electronic effects, steric interactions, *etc.*

C. Hansch (1975) *Adv. Pharmacol. Chemother.* **13**, 45.
C. Silipo & C. Hansch (1976) *J. Med. Chem.* **19**, 62.
M. Yoshimoto & C. Hansch (1976) *J. Med. Chem.* **19**, 71.

HARMONIC BEHAVIOR

A regularly periodic motion characteristic of a pendulum, spring, or many chemical bonds. **See** *Hooke's Law*

HARTREE-FOCK MOLECULAR-ORBITAL CALCULATIONS

A method of molecular-orbital calculations, also referred to as the self-consistent field method (SCF), to characterize the bonding in unsaturated and aromatic molecules while neglecting electron-electron repulsion. The method has been extended to all valence electrons.

M. J. S. Dewar (1969) *The Molecular Orbital Theory of Organic Chemistry*, McGraw Hill, New York.
H. H. Jaffe (1969) *Acc. Chem. Res.* **2**, 136.

HEAD-TO-TAIL POLYMERIZATION

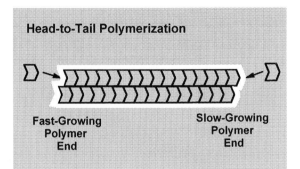

Head-to-Tail Polymerization

Fast-Growing Polymer End　　　**Slow-Growing Polymer End**

Addition reactions that are characteristic of head-to-tail addition of subunits to an elongating polymer. Note that the rates of subunit addition to each of the two ends usually differ as a consequence of differences in the geometry of the associated transition states.

A term used in describing the geometric disposition of subunits during indefinite polymerization of protein subunits into supramolecular assemblies. This is illustrated by the ATP hydrolysis-dependent polymerization of actin onto a trimeric nucleus (marked here with gray shading). Because most protein subunits or protomers exhibit some degree of polarity, one would anticipate that such molecules can form linear oligomeric and polymeric structures reflecting their ability to bind to each other in a so-called "head-to-tail" arrangement. Tubulin and tobacco mosaic virus coat protein also undergo head-to-tail polymerization. *See* Actin Assembly Kinetics; Microtubule Assembly Kinetics

HEAT CAPACITY

A parameter (measured at constant volume; C_v) equal to dq_v/dT where q_v is the heat absorbed at constant volume and T is the temperature. Heat capacity is also equal to $(\partial U/\partial T)_v$ where U is the internal energy. The heat capacity measured at constant pressure (C_p) of a system is equal to dq_p/dT where q_p is the heat absorbed

at constant pressure. This value of C_p is also equal to $(\partial H/\partial T)_p$ where H is the enthalpy at constant pressure. In general, the magnitude of C_p is typically larger than that of C_v although the numbers are very close for liquids and solids. An exception is the case of water between 1°C and 4°C.

Thus, $q_v = \int_{T_1}^{T_2} C_v \, dT$ at constant volume. If C_v is independent of temperature, then $q_v = C_v(T_2 - T_1)$. This is an expression for ΔU. Similarly, $q_p = \int_{T_1}^{T_2} C_p \, dT$. If C_p is independent of temperature, $q_p = C_p(T_2 - T_1)$, an expression for ΔH. It should also be recalled that heat capacities are not state functions.

$C_p - C_v$ equals $[P + (\partial U/\partial V)_T](\partial V/\partial T)_p$. The $(\partial U/\partial V)_T$ term is often referred to as the internal pressure and is large for liquids and solids (*See* Internal Pressure). Since ideal gases do not have internal pressure, $C_p - C_v = nR$ for ideal gases. The ratio of the heat capacities, C_p/C_v, is commonly symbolized by γ.

The SI unit for heat capacity is $J \cdot K^{-1}$. Molar heat capacities (C_m) are expressed as the ratio of heat supplied per unit amount of substance resulting in a change in temperature and have SI units of $J \cdot K^{-1} \cdot mol^{-1}$ (at either constant volume or pressure). Specific heat capacities (C_v or C_p) are expressed as the ratio of heat supplied per unit mass resulting in a change in temperature (at constant volume or pressure, respectively) and have SI units of $J \cdot K^{-1} \cdot kg^{-1}$. Debye's theory of specific heat capacities applies quantum theory in the evaluation of certain heat capacities.

It should also be recalled that $dS = (C_p/T)dT$ at constant pressure and $dS = (C_v/T)dT$ at constant volume where S is the entropy.

HEAVY ATOM ISOTOPE EFFECT

An isotope effect (on either kinetic or equilibrium process) resulting from substitution by isotopes other than those of hydrogen. *See* Kinetic Isotope Effects

HELIX-COIL TRANSITIONS OF NUCLEIC ACIDS

As discussed by Pörschke[1], one can distinguish several basic features of helix-coil transitions in polynucleotides. Formation of a helix from separated complementary strands most typically obeys second-order rate behavior[2,3], and forming a helix nucleus is considered to be the rate-determining step. With much longer chains, the so-called "zipping-up" phase that follows nucleation can be rate contributing. In such cases, one does not observe simple second-order kinetics. Whenever a mismatch occurs within a sequence of hydrogen-bonding base-pair partners, the annealing of the region to attain correct matching probably requires loss of helical structure, followed by a readjustment of interactions between partners. Another feature of helix-coil transitions is the dissociation behavior of long double helices, wherein the complexity of the available pathways for these transitions is reflected in the number of overlapping relaxation events. Spatz and Crothers[4] offered the following scheme for processes taking place during DNA denaturation: With the occurrence of a perturbation, there is attendent rupture of structure in AT-rich regions. Further perturbation causes the loss of base pairing in the regions adjacent to these loops; the helical winding of the double helix leads to coiling of the polynucleotide within adjacent regions of coiling. More extensive unwinding follows, resulting in the formation of larger loop structures.

See *DNA Unwinding Kinetics*

[1]D. Pörschke (1977) in *Chemical Relaxation in Molecular Biology* (I. Pecht & R. Rigler, eds.), pp. 190-218, Springer-Verlag, Berlin.
[2]P. D. Ross & J. M. Sturtevant (1960) *Proc. Natl. Acad. Sci. U.S.A.* **46**, 1360.
[3]J. G. Wetmur & N. Davidson (1968) *J. Mol. Biol.* **31**, 349.
[4]H. C. Spatz & D. M. Crothers (1969) *J. Mol. Biol.* **42**, 191.

HEME OXYGENASE

This enzyme [EC 1.14.99.3] catalyzes the reaction of heme with dioxygen and three AH_2 to produce biliverdin, carbon monoxide, Fe^{2+}, three A, and three water molecules. This enzyme uses NAD(P)H as a cofactor and requires the presence of NADPH:ferrihemoprotein reductase [EC 1.6.2.4].

H. B. Dunford (1998) *Comprehensive Biological Catalysis: A Mechanistic Reference* **3**, 195.
S. Yamamoto & Y. Ishimura (1991) *A Study of Enzymes* **2**, 315.

HEMOGLOBIN

Within the $\alpha_2\beta_2$ hemoglobin tetramer, the α and β subunits (made up of 141 and 146 amino acid residues, respectively) can be thought of as two $\alpha\beta$-dimeric units that are symmetrically positioned about a central cavity[1].

Hemoglobin binds oxygen cooperatively, but a simple all-or-nothing Hill-type binding model does not fit the observed saturation behavior; indeed, the observed Hill coefficient of 2.8 indicates that the system behaves as though the protein had 2.8 subunits that were infinitely cooperative. Because we know that four O_2 molecules are bound at saturation, a better model was proposed by Adair who could fit the entire binding equation by assuming that the intrinsic dissociation constants changed as a function of oxygen binding site occupancy.

A well-established fact is that deoxygenated and oxygenated forms differ with respect to each other by a slight 15° rotation and 1 Å translation between two $\alpha\beta$ dimers. The deoxygenated form is stabilized by protons as well as 2,3-bisphosphoglycerate, and these effector molecules respectively ensure higher than normal oxygen release during hypoxia and adaptation to high altitude. Perutz[2] first proposed that "the oxygenation of Hb is accompanied by structural changes in the subunits, triggered by shifts of the iron atoms relative to the porphyrin and, in the β-subunits, also by the steric effects of oxygen itself". Under physiological conditions, the oxygen saturation curve and the kinetics of hemoglobin's ligand binding reactions appear to fit the two-state Monod-Wyman-Changeux cooperativity model[3]. The relative abundance of the two states depend on the magnitude of the allosteric constant $\mathscr{L}$ in the Monod-Wyman-Changeux model (*i.e.*, $\mathscr{L} = [T_0]/[R_0]$).

[1]M. F. Perutz, A. J. Wilkinson, M. Paoli & G. G. Dodson (1998) *Annu. Rev. Biophys. Biomol. Struct.* **27**, 1.
[2]M. F. Perutz (1972) *Nature* **237**, 495.
[3]E. R. Henry, C. M. Jones, J. Hofrichter & W. R. Eaton. (1997) *Biochemistry* **36**, 6511.

Selected entries from *Methods in Enzymology* [vol, page(s)]:
Equilibrium Ligand Binding to Hemoglobin: Spectrophotometric procedures, **76**, 417; measurement of oxygen binding by means of a thin-layer optical cell, **76**, 427; measurement of accurate oxygen equilibrium curves by an automatic oxygenation apparatus, **76**, 438; thin-layer methods for determination of O_2 binding curves of hemoglobin solutions and blood, **76**, 449; analysis of ligand binding equilibria, **76**, 470; measurement of CO_2 equilibria: the chemical-chromatographic methods, **76**, 487; measurement of CO_2 binding: The ^{13}C NMR method, **76**, 496; continuous determination of the oxygen dissociation curve of whole blood, **76**, 511; measurement of the Bohr effect: dependence on pH of oxygen affinity of hemoglobin, **76**, 523; measurement of binding of nonheme ligands to hemoglobins, **76**, 533; use of NMR spectroscopy in studies in ion binding to hemoglobin, **76**, 552; rapid-rate equilibrium analysis of the interactions between organic phosphates and hemoglobins, **76**, 559; measurement of the oxidation-reduction equilibria of hemoglobin and myoglobin, **76**, 577; photochemistry of hemoproteins, **76**, 582; measurement and analysis of ligand-linked subunit dissociation equilibria in human

hemoglobins, **76**, 596; ligand binding and conformational changes measured by time-resolved absorption spectroscopy, **232**, 387; hemoglobin-liganded intermediates, **232**, 445; hemoglobin-oxygen equilibrium binding: rapid-scanning spectrophotometry and singular value decomposition, **232**, 460; oxygen equilibrium curve of concentrated hemoglobin, **232**, 486; Adair fitting to oxygen equilibrium curves of hemoglobin, **232**, 559; weighted nonlinear regression analysis of highly cooperative oxygen equilibrium curves, **232**, 576; dimer-tetramer equilibrium in Adair fitting, **232**, 597; effects of wavelength on fitting Adair constants for binding of oxygen to human hemoglobin, **232**, 606; Adair equation: rederiving oxygenation parameters, **232**, 632; linkage thermodynamics, **232**, 655.

Kinetics of Ligand Reaction with Hemoglobin: Stopped-flow, rapid mixing measurements of ligand binding to hemoglobin and red cells, **76**, 631; numerical analysis of kinetic ligand binding data, **76**, 652; flash photolysis of hemoglobin, **76**, 667; temperature jump of hemoglobin, **76**, 681; assignment of rate constants for O_2 and CO binding to α and β subunits within R- and T-state human hemoglobin, **232**, 363; femtosecond measurements of geminate recombination in heme proteins, **232**, 416; double mixing methods for kinetic studies of ligand binding in partially liganded intermediates of hemoglobin, **232**, 430; simulation of hemoglobin kinetics using finite element numerical methods, **232**, 517.

HEMOGLOBIN-S POLYMERIZATION

Deoxygenated sickle cell hemoglobin (deoxyHbS), the β-Glu-6-Val point mutant form[1] of adult hemoglobin, appears to obey the following empirical rate law for nucleation of polymerization:

$$\text{Rate} = k[\text{deoxyHbS}]^n$$

The exponent n has been taken to be the number of protomers (in this case tetramers) in the smallest stable nucleus on which further self-assembly of fibers takes place[2]. If nucleation is the rate-determining step in HbS polymerization, then one can estimate the rate of nucleation by determining the delay time t_d observed in lag-phase kinetic measurements of polymerization. A value of $n = 32$ was obtained from the slope of the plot shown below. One may question whether the apparent 32nd power dependence of polymerization rate on the concentration of HbS is physically realistic, especially when one recognizes that each tetramer within an assembled HbS fiber must make numerous contacts even in forming the so-called double strands. Moreover, the validity of the above kinetic analysis is valid only if implicit assumptions justifying the use of the above empirical rate law are satisfied. Among these assumptions are: that concentration and chemical activity are equivalent; that solute-solvent interactions are ideal; and that nucleation is the rate-determining step.

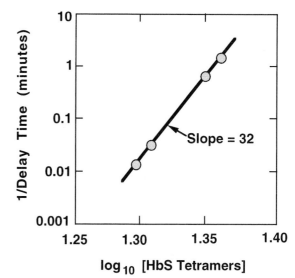

Plot of the reciprocal of the delay time for HbS fiber formation versus the logarithm of HbS concentration. From Hofrichter et al.[2] with permission of the authors and publisher.

This extraordinary result[1] suggests that polymerization is extremely sensitive to changes in HbS concentration, such that successful therapies for preventing sickling might be achievable by lowering the effective concentration of HbS. Modest changes in concentration should minimize the propensity for gelation long enough for the deoxygenated hemoglobin S molecules to rebind O_2 at the higher partial pressures of oxygen found in the lung. One should note that such high-order nucleation phenomena suggest that ligands such as carbon monoxide should paradoxically protect individuals from sickling, because even low concentrations of $HbS(CO)_4$ should interefere with the nucleation event. The late Eraldo Antonini and associates[3] used microspectrophotometry with single red blood cells by taking advantage of the photosensitivity of the carbon monoxide complex with Hb; this method allows ligand release under strong illumination, followed by reforming of the complex when the light is switched off. Briefly, they found that the time necessary for complete sickling, as evidenced by full deformation of the cell, ranged in all examined cells from 3 to 5 seconds. Interestingly, the extent of photodissociation was only about 60%, suggesting that carboxy HbS tetramers were not effectively inhibiting the nucleation of polymers, if nucleation remains rate-limiting *in vivo*. By lowering the intensity of the photo-dissociating light beam after sickling was observed, they discovered that the deformation could be maintained by a light intensity several times lower than required to induce sickling. Furthermore, the sickling process behaved as though it was fully reversible, an inference supported by the finding that recombination of hemoglobin and carbon monoxide after switching off the illuminating source was promptly attended by a return of the erythrocytes to their typical biconcave shape. The lack of any delay in these red cell sickling transitions leads one to question the physiologic relevance of the findings reported by Hofrichter and co-workers[2].

[1]R. E. Dickerson & I. Geis (1983) *Hemoglobin: Structure, Function, Evolution, and Pathology*, pp. 125-145, Benjamin-Cummings, Menlo Park.

[2]J. Hofrichter, P. Ross & W. Eaton (1974) *Proc. Natl. Acad. Sci. U.S.A.* **71**, 4864.

[3]E. Antonini, M. Brunori, B. Giardina, P. A. Benedetti, G. Bianchini & S. Grassi (1978) in *Biophysical Discussions: Fast Biochemical Reactions in Solutions, Membranes and Cells* (V. A. Parsegian, ed.), pp. 187-189, Rockefeller University Press, New York.

HENDERSON-HASSELBALCH EQUATION

An equation[1,2] relating the pH value of a solution and the concentrations of the basic and acidic species of an acid/base equilibria:

$$pH = pK_a' + \log([\text{basic species}]/[\text{acidic species}])$$

Note that K_a' is not the true dissociation constant for the weak acid (or weak base). For the reaction $HA \rightleftharpoons H^+ + A^-$, $K_a = (H^+)(A^-)/(HA)$ where (H^+), (A^-), and (HA), are the corresponding activities. However, K_a' is equal to $K_a f_{HA}/f_{A^-}$ where f_{HA} and f_{A^-} are the activity coefficients for HA and A^-, respectively. Under most practical conditions, particularly under conditions of low constant ionic strength, $K_a \cong K_a'$.

One should always be careful when using the Henderson-Hasselbalch equation: $pH = pK_a' + \log([A^-]/[HA])$. The value of [HA] is not always moles of the weak acid per unit volume, c_{HA}, that had been added to the moles of the conjugate base per unit volume, c_{A^-}. There are five species for this acid/base equilibria: HA, A^-, H^+, OH^-, and Na^+ (or, whatever cation portion of the salt of the conjugate base is present). These species are all interrelated by five expressions:

$$(H^+)(A^-)/(HA) = K_a$$

$$(H^+)(OH^-) = K_w$$

$$[HA] + [A^-] = c_{HA} + c_{A^-}$$

$$[Na^+] = c_{A^-}$$

$$[H^+] + [Na^+] = [OH^-] + [A^-]$$

where parentheses represent activities and brackets represent concentrations. Thus, $[A^-] = c_{A^-} + ((H^+)/f_{H^+}) - K_w f_{OH^-}/(H^+)$ and $[HA] = c_{HA} - ((H^+)/f_{H^+}) + K_w f_{OH^-}/(H^+)$. Hence, the equation relating pH to pK_a' now is $pH = pK_a' + \log[(c_{A^-} + (H^+)/f_{H^+} - K_w f_{OH^-}/(H^+))/(c_{HA} - (H^+)/f_{H^+} + K_w f_{OH^-}/(H^+))]$. This equation reduces to the standard Henderson-Hasselbalch equation if $[H^+]$ and $[OH^-]$ are much smaller than c_{HA} and/or c_{A^-}. If the $[H^+]$ or $[OH^-]$ values are less than 0.01 of the concentrations of the acidic and basic species, the error in neglecting the $[H^+]$ and $[OH^-]$ values is less than 0.009 pH unit. Perrin and Dempsey[3] provide a simple table on when best to use the simple Henderson-Hasselbalch equation:

c_{HA} or c_{A^-}	pH Range
5 mM	4.3–9.7
10 mM	4.0–10
50 mM	3.3–10.7
100 mM	3.0–11

Outside this range, the full expression should be used.

[1]L. J. Henderson (1908) *Amer. J. Physiol.* **21**, 169.

[2]K. Hasselbalch (1916) *Biochem. Zeit.* **78**, 112.

[3]D. D. Perrin & B. Dempsey (1974) *Buffers for pH and Metal Ion Control*, Chapman and Hall, London.

HENDERSON PLOT

A graphical protocol for the analysis of tight-binding inhibitors of enzyme-catalyzed reactions (see figure)[1,2].

This linearization of the tight-binding scheme allows the investigator the opportunity to calculate values for $[E_{total}]$ and K_i, the dissociation constant for the inhibitor. In the Henderson plot, $[I_{total}]/(1 - v/v_o)$ is plotted as a function of v_o/v where v_o is the steady-state velocity of the reaction in the absence of the inhibitor. The slope of the line is the apparent dissociation constant for the inhibitor. Secondary plots (from repeating the inhibition experiment at different substrate concentrations) will yield the K_i value. The vertical intercept is equal to $[E_{total}]$. Hence, repeating the experiment at a different concentration of enzyme will produce a parallel line.

[1]P. J. F. Henderson (1972) *Biochem. J.* **127**, 321.

[2]J. W. Williams & J. F. Morrison (1979) *Meth. Enzymol.* **63**, 437.

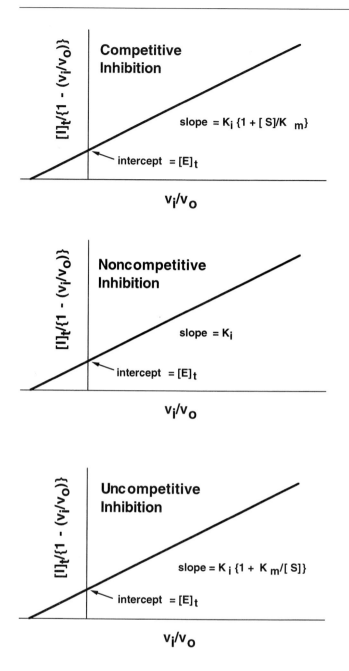

Competitive Inhibition

slope $= K_i \{1 + [S]/K_m\}$

intercept $= [E]_t$

$[I]_t/\{1 - (v_i/v_o)\}$

v_i/v_o

Noncompetitive Inhibition

slope $= K_i$

intercept $= [E]_t$

$[I]_t/\{1 - (v_i/v_o)\}$

v_i/v_o

Uncompetitive Inhibition

slope $= K_i \{1 + K_m/[S]\}$

intercept $= [E]_t$

$[I]_t/\{1 - (v_i/v_o)\}$

v_i/v_o

HENRY

The SI unit (symbolized by H) for both self-inductance and mutual inductance. One henry represents the inductance produced when an induced electromotive force of one volt is generated as the inducing current is altered at a rate of one ampere per second. (Hence, one henry is equivalent to one volt-second per ampere or one weber per ampere). **See** *Magnetic Susceptibility; Permeability*

HERTZ

The SI unit for frequency (symbolized by Hz) equal to reciprocal seconds. The term hertz should only be used with respect to cycles per second and not for radial (circular) frequency or angular velocity.

HESS' LAW

The amount of heat generated by a reaction is the same whether the reaction takes place in one step or in several steps. Hence, ΔH values (and, thus, ΔG values) are additive. This law, also known as the Law of Constant Heat Summation, was the earliest example of the Additivity Principle, which also states that ΔG values are additive. In biochemical processes, especially when dealing with multiple steps as in protein folding, application of the Additivity Principle can give spurious results if the accuracy and precision of the thermodynamic parameters is insufficient.

HETEROGENEOUS CATALYSIS

Catalysis of a reaction wherein one or more reactants is (are) in one phase and the catalytic process takes place at an interface between the phases. This phenomenon is also termed contact catalysis and surface catalysis. **See** *Homogeneous Catalysis*

HETEROGENEOUS REACTION

A reaction occurring at an interface of two phases. Some heterogeneous processes also display homogeneous aspects; for example, a particular reaction may occur in only one of the system's phases (*e.g.*, the rapid dissolution of a gas into a liquid followed by a particular reaction). **See** *Fractal Reaction Kinetics*

HETEROTROPIC EFFECT

Cooperative ligand binding in which a bound ligand molecule influences how a different type of ligand binds to its respective sites on a multisubunit macromolecule[1-3]. The influence of the ligand can result in either weakened binding (hence, negative heterotropic cooperativity) or in enhanced binding (hence, positive heterotropic cooperativity). An example of negative heterotropic cooperativity is the action of 2,3-bisphosphoglycerate or hydrogen ion to reduce the extent of oxygen binding to hemoglobin A.

In the context of the Monod-Wyman-Changeux concerted-transition model for allosteric effects, one usually considers the effects of specific site occupancy on the behavior of other binding sites. Thus, a more correct operational definition is that heterotropic effects represent the interactions on cooperative ligand binding exerted by molecules of unlike action. Applying this definition, we see that 2,3-bisphosphoglycerate or hydrogen ion act heterotropically with respect to oxygenation; yet, in terms of their mutual interactions with each other, 2,3-bisphosphoglycerate or hydrogen ion can be regarded as exerting homotropic effects. It is also true that while oxygen and carbon monoxide compete with respect to each other for the same sites (*i.e.*, the four heme groups on hemoglobin), both act on hemoglobin in the same manner—namely by increasing the fraction of macromolecules in the R-state. Thus, O_2 and CO can be said to act homotropically with respect to each other. ***See*** *Monod-Wyman-Changeux Model; Hemoglobin; Allosterism*

[1]J. Monad, J. Wyman & J.-P. Changeaux (1965) *J. Mol. Biol.* **12**, 88.
[2]K. E. Neet (1980) *Meth. Enzymol.* **64**, 139.
[3]K. E. Neet (1995) *Meth. Enzymol.* **249**, 519.

HEXA UNI PING PONG ENZYME MECHANISM

A ping pong enzyme-catalyzed reaction mechanism in which three substrates react to form three products[1-3]:

$$E + A \underset{k_2}{\overset{k_1}{\rightleftharpoons}} EX \underset{k_4}{\overset{k_3}{\rightleftharpoons}} F + P$$

$$F + B \underset{k_6}{\overset{k_5}{\rightleftharpoons}} FY \underset{k_8}{\overset{k_7}{\rightleftharpoons}} G + Q$$

$$G + C \underset{k_{10}}{\overset{k_9}{\rightleftharpoons}} GZ \underset{k_{12}}{\overset{k_{11}}{\rightleftharpoons}} E + R$$

where F and G are modified forms of the enzyme E, the substrates are represented by A, B, and C, and the products are P, Q, and R.

The steady-state determinants for each enzyme form (in the absence of abortive complexes) are:

$$[E] = k_5k_7k_9k_{11}(k_2 + k_3)[B][C]$$
$$+ k_2k_4k_9k_{11}(k_6 + k_7)[C][P]$$
$$+ k_2k_4k_6k_8(k_{10} + k_{11})[P][Q]$$

$$[EX] = k_1k_5k_7k_9k_{11}[A][B][C]$$
$$+ k_1k_4k_9k_{11}(k_6 + k_7)[A][C][P]$$
$$+ k_1k_4k_6k_8(k_{10} + k_{11})[A][P][Q]$$
$$+ k_4k_6kk_{10}k_{12}[P][Q][R]$$

$$[F] = k_1k_3k_9k_{11}(k_6 + k_7)[A][C]$$
$$+ k_1k_3k_6k_8(k_{10} + k_{11})[A][Q]$$
$$+ k_6k_8k_{10}k_{12}(k_2 + k_3)[Q][R]$$

$$[FY] = k_1k_3k_5k_9k_{11}[A][B][C]$$
$$+ k_1k_3k_5k_8(k_{10} + k_{11})[A][B][Q]$$
$$+ k_5k_8k_{10}k_{12}(k_2 + k_3)[B][Q][R]$$
$$+ k_2k_4k_8k_{10}k_{12}[P][Q][R]$$

$$[G] = k_1k_3k_5k_7(k_{10} + k_{11})[A][B]$$
$$+ k_5k_7k_{10}k_{12}(k_2 + k_3)[B][R]$$
$$+ k_2k_4k_{10}k_{12}(k_6 + k_7)[P][R]$$

$$[GZ] = k_1k_3k_5k_7k_9[A][B][C]$$
$$+ k_5k_7k_9k_{12}(k_2 + k_3)[B][C][R]$$
$$+ k_2k_4k_9k_{12}(k_6 + k_7)[C][P][R]$$
$$+ k_2k_4k_6k_8k_{12}k[P][Q][R]$$

Under initial rate conditions, where the product concentrations are effectively zero (*i.e.*, [P] $\approx$ 0, [Q] $\approx$ 0, and [R] $\approx$ 0), the steady-state expression for this reaction mechanism is

$$v = \frac{V_{\max}[A][B][C]}{\{K_c[A][B] + K_b[A][C] + K_a[B][C] + [A][B][C]\}}$$

where $V_{\max}$ is the maximum velocity in the forward direction (equal to $k_3k_7k_{11}[E_{total}]/(k_3k_7 + k_3k_{11} + k_7k_{11})$), K_a is the Michaelis constant for A (equal to $k_7k_{11}(k_2 + k_3)/\{k_1(k_3k_7 + k_3k_{11} + k_7k_{11})\}$), K_b is the Michaelis constant for B (equal to $k_3k_{11}(k_6 + k_7)/\{k_5(k_3k_7 + k_3k_{11} + k_7k_{11})\}$), and K_c is the Michaelis constant for C (equal to $k_3k_7(k_{10} + k_{11})/\{k_9(k_3k_7 + k_3k_{11} + k_7k_{11})\}$). This expression predicts that double-reciprocal plots ($1/v$ *vs.* the reciprocal concentration of one of the substrates in which one of the other substrate concentration is varied and the third substrate is kept at a constant, nonsaturating level) will all exhibit parallel lines. Secondary replots of the data will provide values for the kinetic constants.

Since "half-reactions" are present in this reaction mechanism (actually, "third-reactions"), partial exchange experiments are also available for investigation. ***See*** *Multisubstrate Mechanisms; Isotope Exchange at Equilibrium*

[1]Y. Milner, G. Michaels & H. G. Wood (1978) *J. Biol. Chem.* **253**, 878.
[2]N. H. Goss & H. G. Wood (1982) *Meth. Enzymol.* **87**, 51.
[3]C. S. Tsai, M. W. Burgett & L. J. Reed (1973) *J. Biol. Chem.* **248**, 8348.

HEXOKINASE

This phosphotransferase [EC 2.7.1.1] catalyzes phosphorylation of D-glucose as well as many other aldo- and keto-hexoses (with the notable exception of D-galactose)[1]. $MgATP^{2-}$ is the active phosphoryl donor, and the overall reaction is highly favorable (*i.e.*, K_{eq} = [ADP] [D-Glc-6-P]/[D-Glc][ATP] = 490 at pH 6.5; K_{eq} = 1500 at pH 7). The kinetic reaction for the yeast enzyme involves random sequential binding of hexose and $MgATP^{2-}$ to form a ternary complex. This was first demonstrated by the use of competitive inhibitors and isotope exchange at equilibrium. The rate of nucleotide exchange at equilibrium is about 1.5 times faster than the sugar ↔ sugar-phosphate exchange, suggesting that the dissociation of sugar or sugar-phosphate (or both) must be rate-limiting. Proteolyzed forms of the yeast enzyme appear to favor glucose binding. The mammalian enzyme also operates by a random ternary complex mechanism. Contrary to earlier reports asserting that glucose 6-phosphate is an allosteric inhibitor of the mammalian erythrocyte and brain enzymes, the potency of this sugar phosphate's action appears to arise through its tight binding (K_i = 12 μM) at the γ-phosphoryl subsite of the ATP binding site. Based on the stereochemistry[2] of phosphoryl transfer, the hexokinase reaction is best described as a single in-line or S_N2 reaction involving facilitated attack of the 6-hydroxy group on $MgATP^{2-}$.

See ATP and GTP Depletion; Deinhibition; Induced Fit Hypothesis

[1]D. L. Purich, F. B. Rudolph & H. J. Fromm (1973) *Adv. Enzymol.* **39**, 249.

[2]G. Lowe, P. M. Cullis, R. L. Jarvest, B. V. L. Potter & B. S. Sproat (1981) *Phil. Trans. R. Soc. London B* **293**, 75.

Selected entries from *Methods in Enzymology* [vol, page(s)]:
Activation, **63**, 258, 264, 292; assay, **63**, 32, 36, 37, 40; calorimetry, reactions catalyzed, **61**, 263, 268; competitive inhibition, **63**, 292, 476, 480; conformational change, **64**, 225, 226; cooperativity, **64**, 218, 221; hysteresis, **64**, 197, 215, 216, 218, 221, 222; inhibition, **63**, 5, 348; isomerization, **64**, 197; isotope exchange, **64**, 4, 7, 8, 25, 29-32, 37, 38; mechanism, **63**, 51; mnemonic mechanism, **64**, 221; nucleotide site, **64**, 180; product inhibition, **63**, 433, 434, 436; slow-binding inhibitor, **63**, 450; specificity, **63**, 378, 379; steady state assumption, **63**, 476; alternative substrate inhibition, **63**, 145, 494.

HEXOSAMINIDASES A AND B

This enzyme [EC 3.2.1.52] (also known as β-*N*-acetylhexosaminidase, β-hexosaminidase, hexosaminidase, and *N*-acetyl-β-glucosaminidase) catalyzes the hydrolysis of terminal nonreducing *N*-acetyl-D-hexosamine residues in *N*-acetyl-β-D-hexosaminides. The enzyme can also utilize *N*-acetylglucosides and *N*-acetylgalactosides as substrates. The associated human genetic disorders are known as Sandhoff's disease and Tay-Sachs disease.

G. Davies, M. L. Sinnott & S. G. Withers (1998) *Comprehensive Biological Catalysis: A Mechanistic Reference* **1**, 119.

E. Conzelmann & K. Sandhoff (1987) *Adv. Enzymol.* **60**, 89.

R. O. Brady (1983) *The Enzymes*, 3rd ed., **16**, 409.

HEXOSE-1-PHOSPHATE GUANYLYLTRANSFERASE

This enzyme [EC 2.7.7.29], also referred to as GDP-hexose pyrophosphorylase, catalyzes the reaction of GTP with α-D-hexose 1-phosphate to produce GDP-hexose and pyrophosphate (or, diphosphate).

R. G. Hansen, H. Verachtert, P. Rodriguez & S. T. Bass (1966) *Meth. Enzymol.* **8**, 269.

HIDDEN RETURN

A term used to designate any condition wherein a covalent molecule, R_1–R_2, is reformed without direct evidence of the prior formation of a tight ion pair (*i.e.*, $R_1^+R_2^-$).

HILDEBRAND PARAMETER

A measure of the cohesion of a solvent (*i.e.*, the energy required to create a cavity). This is an extension of earlier work by Thomson and others earlier in the twentieth century to treat solvent-solute interactions.

M. Chastrette, M. Rajzmann, M. Cannon & K. Purcell (1985) *J. Amer. Chem. Soc.* **107**, 1.

HILL COEFFICIENT

An all-or-none treatment of cooperative ligand binding, wherein a macromolecule with n sites for ligand exists in two states: (a) the uncomplexed form M; and (b) the fully ligated state ML_n.

To derive the Hill equation, we begin with the following equilibrium reaction:

$$M + nL \rightleftharpoons ML_n$$

Likewise, the following equilibrium constant can be written (in the form of a dissociation constant):

$$K = \frac{[M][L]^n}{[ML_n]} \quad \text{or} \quad [M] = K\frac{[ML_n]}{[L]^n}$$

Using the formalism of a saturation function, we get

$$Y_L = \frac{[ML_n]}{([M] + [ML_n])}$$

Y_L can vary over the interval from 0 to 1. Substituting for [M] in terms of $[ML_n]$ and rearranging yields:

$$Y_L = \frac{1}{\left(\dfrac{K}{[L]^n}\right) + 1} \qquad \text{or} \qquad Y_L = \frac{[L]^n}{K + [L]^n}$$

The Hill plot is obtained by first rearranging this expression to obtain

$$\frac{Y_L}{1 - Y_L} = \frac{[L]^n}{K}$$

then, by taking the decadic logarithm of both sides, we get

$$\log\left(\frac{Y_L}{1 - Y_L}\right) = n \log[L] - \log K$$

As shown in Fig. 1, the linear region of a plot of $\log\{Y_L/(1 - Y_L)\}$ versus $\log [L]$ yields a slope of n. When one obtains experimental data from a Hill Plot, the symbol for "n" is typically designated as n_H.

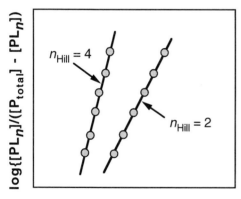

Fig. 1. Ligand binding behavior of a hypothetical protein exhibiting a Hill coefficient of 2 or 4.

Where n_H equals the number of sites determined by some other biophysical procedure, we say that the system shows "infinite cooperativity." No such behavior has been rigorously demonstrated for an enzyme or receptor. In the case of hemoglobin oxygenation (Fig. 2) under physiologic conditions, the Hill coefficient has a value of about 2.8. Of course, from X-ray structural information, we know that hemoglobin has four sites. Thus, we

can say that hemoglobin behaves as though it has 2.8 infinitely cooperative sites, or we can state that the four subunits of human hemoglobin fail to show complete cooperativity.

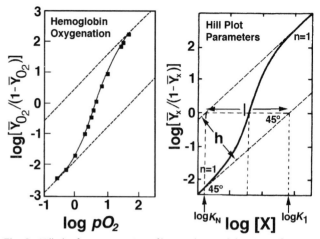

Fig. 2. Hill plot for oxygenation of human hemoglobin A as a function of the partial pressure (pO_2) of molecular oxygen. The diagram at the right shows that the Hill coefficient will reach a limiting value of one at both extremes of ligand concentration. For this reason this cooperativity index is best measured at ligand concentrations near half-maximal saturation.

Can the same enzyme yield a different Hill Plot under a different set of conditions? The answer is a most definite yes! The strength of interprotomer bonds determines the cooperativity of ligand binding, and agents that disrupt the quaternary structure of a protein can likewise change the nature of a Hill plot. The presence of an inhibitor or activator, treatment with a mercurial or some other alkylating agent, a change in pH or ionic strength, and changes in the temperature are all known to affect the cooperativity of ligand binding processes.

What would be the corresponding behavior of an enzyme obeying the Michaelis-Menten model? For the answer, let us restate the Michaelis-Menten model in the form of a saturation function:

$$M + L \rightleftharpoons ML$$

Likewise, the following equilibrium constant can be written (again in the form of a dissociation constant):

$$K = \frac{[M][L]}{[ML]} \qquad \text{or} \qquad [M] = K \frac{[ML]}{[L]}$$

Using the formalism of a saturation function, we get

$$Y_L = [ML]/\{[M] + [ML]\}$$

Because there is only one site for ligand binding, no "n" appears in either the numerator or denominator. Nevertheless, Y_L can still vary over the interval from 0 to 1. Substituting for [M] in terms of [ML] yields:

$$Y_L = [ML]/\{(K[ML]/[L]) + [ML])\}$$

or

$$Y_L = 1/\{(K/[L]) + 1\}$$

and multiplying the right-hand side of the equation by [L]/[L] gives the saturation function as

$$Y_L = [L]/\{K + [L]\}$$

This is rearranged to obtain:

$$Y_L/(1 - Y_L) = [L]/K$$

then, by taking the base 10 logarithm of both sides, we get

$$\log\{Y_L/(1 - Y_L)\} = \log[L] - \log K$$

The linear region of a plot of $\log\{Y_L/(1 - Y_L)\}$ versus $\log[L]$ yields a slope of one (see Fig. 3). Such ligand saturation behavior has been observed for myoglobin. Moreover, many oligomeric enzymes yield unit slopes, indicating that their binding sites operate independently (*i.e.*, there are no site-site interactions evident in their ligand binding).

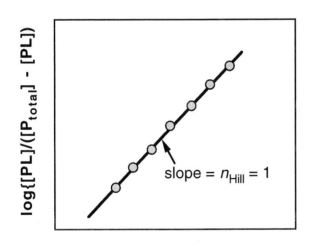

log [Ligand]

Fig. 3. Ligand binding behavior of a hypothetical protein exhibiting a unit Hill coefficient of ($n_H = 1$). Such behavior would be expected (a) for the case of a protein with only a single binding site or (b) for the case of an oligomeric protein containing n independent and noninteracting binding sites.

Should one use Hill plots to examine the initial velocity behavior of enzymes? Because infinite cooperativity is assumed to be the basis of the Hill treatment, only rapidly equilibrating systems are suitable for Hill analysis. In fact, no enzyme system displaying steady-state kinetic behavior will ever satisfy this requirement. Even the Michaelis-Menten equation only has an initial, rapid-equilibrium segment (namely substrate binding to free enzyme and substrate release from the EX complex); product release is considered to be a slow step. For this reason, one must avoid the use of kinetic data in any application of the Hill equation.

HiPIP

An abbreviation for <u>hi</u>gh-<u>p</u>otential <u>i</u>ron-sulfur <u>p</u>rotein, which is now regarded as a ferredoxin. In its role as a bacterial electron transfer component, this [4Fe–4S] cluster protein can undergo interconversion to the [4Fe–4S]$^{2+}$ and [4Fe–4S]$^{3+}$ states.

HISTAMINE *N*-METHYLTRANSFERASE

This enzyme [EC 2.1.1.8] catalyzes the reaction of *S*-adenosyl-L-methionine with histamine to produce *S*-adenosyl-L-homocysteine and N^τ-methylhistamine.

J. Axelrod (1971) *Meth. Enzymol.* **17B**, 766.

HISTIDINE AMMONIA-LYASE

This enzyme [EC 4.3.1.3] (also known as histidinase, histidase, and histidine α-deaminase) catalyzes the conversion of histidine to urocanate and ammonia. ***See*** *Dehydroalanine*

V. E. Anderson (1998) *Comprehensive Biological Catalysis: A Mechanistic Reference* **2**, 115.
K. R. Hanson & E. A. Havir (1972) *The Enzymes*, 3rd ed., **7**, 75.

HISTIDINE DECARBOXYLASE

This enzyme [EC 4.1.1.22] catalyzes the conversion of histidine to histamine and carbon dioxide. The enzyme requires either pyridoxal phosphate or pyruvate as a cofactor.

M. L. Hackert & A. E. Pegg (1998) *Comprehensive Biological Catalysis: A Mechanistic Reference* **2**, 201.
M. H. O'Leary (1992) *The Enzymes*, 3rd ed., **20**, 235.
H. Hayashi, H. Wada, T. Yoshimura, N. Esaki & K. Soda (1990) *Ann. Rev. Biochem.* **59**, 87.
D. J. Creighton & N. S. R. K. Murthy (1990) *The Enzymes*, 3rd ed., **19**, 323.
E. A. Boeker & E. E. Snell (1972) *The Enzymes*, 3rd ed., **6**, 217.

HISTIDINOL DEHYDROGENASE

This enzyme [EC 1.1.1.23] catalyzes the reaction of L-histidinol with two NAD^+ to produce L-histidine and two NADH. L-Histidinal will also serve as a substrate for this protein. The *Neurospora crassa* enzyme will also catalyze the reactions of phosphoribosyl-AMP cyclohydrolase [EC 3.5.4.19] and phosphoribosyl-ATP pyrophosphatase [EC 3.6.1.31].

R. G. Martin, M. A. Berberich, B. N. Ames, W. W. Davis, R. F. Goldberger & J. D. Yourno (1971) *Meth. Enzymol.* **17B**, 3.

HISTIDINOL-1-PHOSPHATE AMINOTRANSFERASE

This pyridoxal-phosphate-dependent enzyme [EC 2.6.1.9], also called imidazolylacetolphosphate aminotransferase and imidazole acetolphosphate transaminase, catalyzes the reversible reaction of histidinol phosphate with α-ketoglutarate (or, 2-oxoglutarate) to produce 3-(imidazol-4-yl)-2-oxopropyl phosphate and glutamate.

A. E. Braunstein (1973) *The Enzymes*, 3rd ed., **9**, 379.
R. G. Martin, M. A. Berberich, B. N. Ames, W. W. Davis, R. F. Goldberger & J. D. Yourno (1971) *Meth. Enzymol.* **17B**, 3.

HISTIDINOL-PHOSPHATASE

This enzyme [EC 3.1.3.15] catalyzes the hydrolysis of histidinol phosphate to generate histidinol and orthophosphate.

R. G. Martin, M. A. Berberich, B. N. Ames, W. W. Davis, R. F. Goldberger & J. D. Yourno (1971) *Meth. Enzymol.* **17B**, 3.

HISTIDYL-tRNA SYNTHETASE

This enzyme [EC 6.1.1.21], also referred to as histidine:tRNA ligase, catalyzes the reaction of histidine with $tRNA^{His}$ and ATP to produce histidyl-$tRNA^{His}$, AMP, and pyrophosphate. **See also** *Aminoacyl-tRNA Synthetases*

P. Schimmel (1987) *Ann. Rev. Biochem.* **56**, 125.

HISTONE KINASE

This enzyme [EC 2.7.1.70], also referred to as protamine kinase, catalyzes the reaction of ATP with protamine to produce ADP and an *O*-phosphoseryl residue in protamine. The enzyme will also phosphorylate histones and it requires cAMP.

A. M. Edelman, D.K . Blumenthal & E. G. Krebs (1987) *Ann. Rev. Biochem.* **56**, 567.

HOFMEISTER SERIES

An empirically derived relationship[1] that describes the systematic effects of different neutral salts on the solubility of proteins. Collins and Washabaugh[2] indicate that the order of ionic species eluding from a Sephadex G-10 column corresponds to the known order of effectiveness ions in the Hofmeister series on protein solubility:

$$SO_4^{2-} \simeq HPO_4^{2-} > F^- > Cl^- > I^- (\simeq ClO_4^-) > SCN^-$$

The order of elution indicates that ion size and charge density strongly influences the apparent molecular size of hydrated anions, hence the effect on the order of elution of these ions in gel permeation experiments. Those species eluding before chloride ion are termed polar kosmotropes to designate their ability to enhance water structure in their vicinity and to stabilize proteins. Those ions eluding after chloride ion are termed chaotropes, and they reduce local water structure and destabilize proteins. Chloride ion is regarded as having little effect on water structure or protein stability over a concentration range from 0.1–0.7 M^2. **See** *Chaotropic Agents; Kosmotropes*

[1]F. Hofmeister (1888) *Naunyn-Schmiedebergs Archiv für Experimentelle Pathologie und Pharmakologie (Leipzig)* **24**, 247.
[2]K. D. Collins & M. W. Washabaugh (1985) *Q. Rev. Biophys.* **18**, 323.

HOLO-[ACYL-CARRIER-PROTEIN] SYNTHASE

This enzyme [EC 2.7.8.7] catalyzes the reaction of coenzyme A with the apo-[acyl-carrier protein] to generate adenosine 3',5'-bisphosphate and the holo-[acyl-carrier protein].

R. H. Lambalot & C. T. Walsh (1997) *Meth. Enzymol.* **279**, 254.

HOMEOSTASIS

The self-regulated physiological/biochemical condition achieved through the action of built-in feedback loops that sense any departure from the system's set-point and then readjust input and/or output parameters to automatically return the system to that set-point. The term *homeostasis*, coined by American physiologist Walter Cannon to describe the resultant constancy of a living organism's internal state, has largely replaced the use of *milieu intérieur* (*i.e.*, internal environment) first proposed by the nineteenth century French physiologist Claude Bernard. The intellectual triumph of Bernard's recognition of this regulatory phenomenon is underscored by the fact that at just about the same time chemists were just realizing the occurrence of the dynamic equilibria in chemical reactions. Beyond the role of various organs in maintaining the internal environment, homeostasis is manifested at the subcellular level through feedback control of metabolic reactions and signal transduction processes.

Living organisms are self-stabilized by steady-state flows of metabolites, inorganic ions, and energy. As shown by Onsager and Prigogine, processes at steady state naturally resist change by outside forces. (*See* *Nonequilibrium Thermodynamics (A Primer)*) Homeostatic mechanisms endow cells with such remarkable resilience that early investigators mistook homeostasis as a persuasive indication that life is self-determining and beyond the laws of chemistry and physics. This tenet, known as vitalism, was based on Bergson's idea that living systems were maintained by a vital force (*élan vital*). To counter continued emphasis on vitalism, Bernard advanced the cause of the molecular life sciences by the following representative commentary[1]: "But if we consider it, we shall soon see that the spontaneity of living bodies is simply the appearance and the result of a certain mechanism in completely determined environments; so that it will be easy, after all, to prove that the behavior of living bodies, as well as the behavior of inorganic bodies, is dominated by a necessary determinism linking them with conditions of a purely physico-chemical order."

[1]C. Bernard (1865) *An Introduction to the Study of Experimental Medicine* (translated in 1927 by H. C. Greene) Dover edition, 1957, p. 61, Dover Publications, New York.

HOMOACONITATE HYDRATASE

This enzyme [EC 4.2.1.36], also referred to as homoaconitase, catalyzes the reversible conversion of 2-hydroxybutane 1,2,4-tricarboxylate to but-1-ene 1,2,4-tricarboxylate and water.

H. P. Broquist (1971) *Meth. Enzymol.* **17B**, 114.

HOMOCITRATE SYNTHASE

This enzyme [EC 4.1.3.21] catalyzes the reaction of 2-hydroxybutane 1,2,4-tricarboxylate with coenzyme A to produce acetyl-CoA, water, and α-ketoglutarate (or, 2-oxoglutarate).

M. J. P. Higgins, J. A. Kornblatt & H. Rudney (1972) *The Enzymes*, 3rd ed., **7**, 407.
H. P. Broquist (1971) *Meth. Enzymol.* **17B**, 113.

HOMOCYSTEINE DESULFHYDRASE

This pyridoxal-phosphate-dependent enzyme [EC 4.4.1.2] catalyzes the hydrolysis of homocysteine to produce ammonia, H_2S, and α-ketobutanoate (or, 2-oxobutanoate).

C. V. Smythe (1955) *Meth. Enzymol.* **2**, 315.

HOMOCYSTEINE S-METHYLTRANSFERASE

This enzyme [EC 2.1.1.10] catalyzes the reaction of *S*-adenosylmethionine with homocysteine to produce *S*-adenosylhomocysteine and methionine. With the bacterial enzyme, *S*-methylmethionine is a better substrate than *S*-adenosylmethionine.

S. K. Shapiro (1971) *Meth. Enzymol.* **17B**, 400.

HOMOGENEOUS CATALYSIS

Catalysis occurring in a single phase. Most enzymes are homogeneous catalysts; however if an enzyme embedded in a solid matrix acts on soluble substrates, aspects of heterogenous catalysis become important. *See* *Heterogeneous Catalysis*

HOMOGENEOUS REACTION

A reaction that occurs wholly within one phase. It should be pointed out that certain homogeneous systems may still exhibit nonhomogeneous effects. For example, a reaction in a solution may be influenced by the walls of the container. If such phenomena occur, then corrections would have to be made. In general, this is not a problem for most enzyme-catalyzed processes.

HOMOGENTISATE 1,2-DIOXYGENASE

This enzyme [EC 1.13.11.5], an iron-dependent system (also called homogentisicase and homogentisate oxygenase), catalyzes the reaction of homogentisate with dioxygen to produce 4-maleylacetoacetate.

O. Hayaishi, M. Nozaki & M. T. Abbott (1975) *The Enzymes*, 3rd ed., **12**, 119.
K. Adachi & Y. Takeda (1970) *Meth. Enzymol.* **17A**, 638.

HOMOISOCITRATE DEHYDROGENASE

This enzyme [EC 1.1.1.155] catalyzes the reaction of (−)-1-hydroxy-1,2,4-butane tricarboxylate with NAD^+ to produce α-ketoadipate (or, 2-oxoadipate), carbon dioxide, and NADH.

H. P. Broquist (1971) *Meth. Enzymol.* **17B**, 118.

HOMOPHILIC INTERACTION

Any set of geometrically and energetically equivalent binding interactions between two identical polypeptide chains or proteins. In a homophilic interaction, one polypeptide chain serves as a ligand for a receptor site on the other polypeptide chain, and *vice versa*.

HOMOSERINE *O*-ACETYLTRANSFERASE

This enzyme [EC 2.3.1.31] catalyzes the reaction of acetyl-CoA with homoserine to produce coenzyme A and *O*-acetylhomoserine.

S. Nagai & D. Kerr (1971) *Meth. Enzymol.* **17B**, 442.

HOMOSERINE DEHYDROGENASE

This enzyme [EC 1.1.1.3] catalyzes the reaction of homoserine with $NAD(P)^+$ to produce aspartate 4-semialdehyde and NAD(P)H. NAD^+ is the better of the two coenzymes with the yeast enzyme whereas $NADP^+$ is preferred by the *Neurospora* enzyme. ***See also*** *Aspartate Kinase*

I. Saint-Gerons, C. Parsot, M. M. Zakin, O. Bârzy & G. N. Cohen (1988) *Crit. Rev. Biochem.* **23**, S1.
P. Truffa-Bachi (1973) *The Enzymes*, 3rd ed., **8**, 509.

HOMOSERINE *O*-SUCCINYLTRANSFERASE

This enzyme [EC 2.3.1.46], also referred to as homoserine *O*-transsuccinylase, catalyzes the reaction of succinyl-CoA with homoserine to produce coenzyme A and *O*-succinylhomoserine.

I. Saint-Gerons, C. Parsot, M. M. Zakin, O. Bârzy & G. N. Cohen (1988) *Crit. Rev. Biochem.* **23**, S1.

HOMOTROPIC INTERACTION

A term used in describing the interactions among dissimilar ligands in cooperative saturation. If an enzyme binds two ligands, say X and Y, then the effect of the binding of molecule X on the binding of other molecules of X is termed a homotropic interaction. Likewise, the effect of the binding of molecule Y on the binding of other molecules of Y is termed a homotropic interaction. Effects arising from the influence of ligand X binding on ligand Y binding, and *vice versa,* are termed heterotropic interactions.

In the context of the Monod-Wyman-Changeux concerted-transition model for allosteric effects, one usually considers the effects of specific site occupancy on the behavior of other binding sites. Thus, the more correct operational definition states that homotropic effects represent the interactions on cooperative ligand binding exerted by molecules of *like action*. Applying this definition, we see that while oxygen and carbon monoxide compete with respect to each other for the same sites (*i.e.*, the four heme groups on hemoglobin), both act on hemoglobin in the same manner—namely by increasing the fraction of macromolecules in the R-state. Thus, O_2 and CO act homotropically with respect to each other. It is also true that 2,3-bisphosphoglycerate or hydrogen ion act heterotropically with respect to oxygenation; yet, in terms of their mutual interactions with each other, 2,3-bisphosphoglycerate or hydrogen ion can be regarded as exerting homotropic effects. ***See*** *Monod-Wyman-Changeux Model; Allosterism; Hemoglobin*

HOOKE'S SPRING LAW

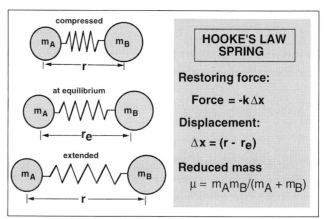

Fig. 1. Illustration of the forces produced during vibration of a Hooke's law spring.

The vibrational stretching mode of a chemical bond lies parallel to the direction of atom movement during bond scission. For this reason, by using Hooke's Law to estimate approximately the energy of bond stretching, one can infer aspects about the energetics that underlie kinetic isotope effects. Hooke's Law is a classical mechanical treatment of a massless spring joining two bodies of mass m_A and m_B at an equilibrium distance (r_e). Displacement of either body from this equilibrium spring length is taken as a mechanical analogue of a vibrating chemical bond joining two atoms or groups of atoms within a molecule (Fig. 1). Any such displacement is accompanied by an oppositely directed restoring force,

$$f = -k(r - r_e)$$

where $(r - r_e)$ is the displacement, and k is the linear proportionality constant. By applying Newton's Law ($f = ma = d^2r_A/dt^2 = d^2r_B/dt^2$) for mechanical force, one may describe the classical motion of the system as follows:

$$m_A \cdot d^2r_A/dt^2 = m_B \cdot d^2r_B/dt^2 = -k(r - r_e)$$

where r_A and r_B are distances of m_A and m_B from the molecule's center of gravity. Recalling that the center of mass is related to these displacements by $r_A = [m_A/(m_A + m_B)]r$ and $r_B = [m_B/(m_A + m_B)]r$, we can substitute either of these expressions into the above equation to obtain:

$$[m_A/(m_A + m_B)] \cdot d^2(r - r_e)/dt^2 = -k(r - r_e)$$

where $[m_A/(m_A + m_B)]$ equals the reduced mass (symbolized by μ_{AB}). Because r_e is itself a constant, we can rewrite the above equation as:

$$\mu_{AB} = d^2(\Delta x)/dt^2 = -k(\Delta x)$$

This second-order differential equation has a harmonic function as its solution:

$$\Delta x = A_{max}\sin[(k/\mu_{AB})^{1/2}t]$$

where A_{max} is the maximal amplitude (or displacement). Displacement is described by a sinusoidal function with respect to time—repeatedly achieving the identical displacement at any two times t_x and t_y differing by an integral multiple of the t_{period}, defined as the period of harmonicity (i.e., $A_x = A_y$ whenever $(t_x - t_y) = n \cdot t_{period}$, where $n = 1, 2, 3, \ldots$).

Differences in the potential energy U of a Hooke's Law spring can be calculated as the product of the restoring force ($-f$) and the distance dx over which the force acts. Thus, by substituting $-kx$ in place of f, we get

$$dU = -kx\, dx$$

Integration yields a parabolic function

$$U = (1/2)kx^2$$

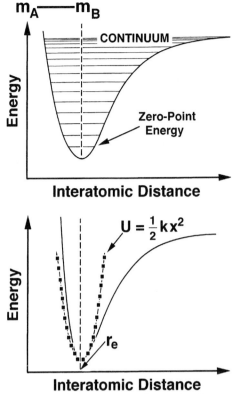

Fig. 2. *Top:* Energy well representation for a molecule m_A—m_B showing the various quantized states from the lowest-lying energy state (i.e., the zero-point energy state) to the classical mechanical continuum. *Bottom:* Same energy well with parabolic curve representing the potential energy of a Hooke's law spring. (Note that this approximation only correlates closely at the lowest energy level range. Because kinetic isotope effects are interpreted on the basis of changes in the zero-point energy, the Hooke's spring law approximation is useful in estimating the energy difference between two isotopomers.)

with non-negative energy values (hence, potential energy). This parabola is shown in Fig. 2 where there is good fit near the equilibrium position of the potential energy well for a chemical bond. This should not be surprising, because this classical mechanical model for a chemical bond assumed a linear restoring force; only the quantum mechanical treatment can fully account for the entire potential energy function, including the quantization of vibrational energies.

HORN-BÖRNIG PLOT

A graphical procedure used to determine, in cooperative systems, values for L (the ratio of the T to R state in the absence of any binding ligand in the Monod-Wyman-Changeux model) and n (the stoichiometry of binding) in exclusive binding systems ($c = 0$ where $c = K_R/K_T$; i.e., the ratio of the intrinsic dissociation constants for the two states [K_R and K_T are the dissociation constants of the ligand with respect to the R and T state, respectively]; thus, in this case, the T state has no significant affinity for the ligand)[1,2].

This plot is based on the equation

$$\log([\alpha V_{max,f}/v] - \alpha - 1) = (1 - n)\log(1 + \alpha) + \log L'$$

where α is the ratio of the ligand concentration to the dissociation constant of the ligand with respect to the R state, $V_{max,f}$ is the maximum velocity of the reaction in the forward direction, v is the initial velocity at the ligand concentration, and L' is the apparent allosteric constant equal to $(L(1 + \beta)^n/(1 + \gamma)^n$ where β is the concentration of an allosteric inhibitor divided by the dissociation constant for that inhibitor (to the T state) and γ is the concentration of an allosteric activator divided by the dissociation constant for that activator (to the R state); hence, if β and γ equal zero, then $L = L'$; or, if $\beta = \gamma$, then $L = L'$). Thus, for exclusive binding allosteric systems, a plot of $\log((\alpha V_{max,f}/v) - \alpha - 1)$ vs. $\log(1 + \alpha)$ will yield a straight line. The slope will be $1 - n$ and the vertical intercept will be $\log L'$. Values for L can be determined from data obtained in the absence of effectors. Thus, knowing the concentration of the effector can yield values for the dissociation constant for that effector. Note that this plot requires an accurate value for α. Hence, the dissociation constant of the ligand for the R state has to be known. Such a value can be obtained if an activator is available that completely changes the kinetics from nonhyperbolic to hyperbolic.

If the ligand can bind to either the R or T state (i.e., $c = 0$), the Horn-Börnig plot will yield nonlinear plots.

The chief limitation of this plotting method is that, unlike the Adair Equation, the plot requires that the system obey all of the conditions required by the MWC model. This assumption will often prove to be incorrect!

[1]A. Horn & H. Börnig (1969) FEBS Lett. **3**, 325.
[2]K. E. Neet (1980) Meth. Enzymol. **64**, 139.

HORSERADISH PEROXIDASE

This enzyme [EC 1.11.1.7] catalyzes the reaction of hydrogen peroxide with a donor to produce two water and the oxidized donor.

H. B. Dunford (1998) Comprehensive Biological Catalysis: A Mechanistic Reference **3**, 195.

O. Ryan, M. R. Smyth & C. Fágáin (1994) Essays in Biochem. **28**, 129.
M. A. Ator & P. R. Ortez de Montellano (1990) The Enzymes, 3rd ed., **19**, 213.

HOST-GUEST INTERACTIONS

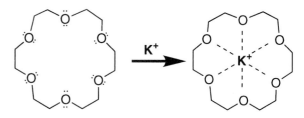

Binding of Potassium Ion to 18-Crown-6 Host Molecule

Structural recognition and binding interactions involving a naturally occurring or synthetic "host" (or template) and a smaller ligand known as a "guest"[1,2]. Examples of host molecules include the cyclodextrans and the so-called crown ethers. Host-guest interactions have served as useful systems for modeling enzymic catalysis, resolving optical isomers of complicated organic molecules, making analogues for micellar catalysis, and for understanding the selective cation transport properties of valinomycin-like antibiotics. For his studies on host-guest interactions, Professor Donald Cram received the Nobel Prize in Chemistry in 1989.

[1]D. W. Griffiths & M. L. Bender (1973) Advances in Catalysis **23**, 209.
[2]D. J. Cram (1976) Techniques of Chemistry, vol. **10**, pp. 815-873, Wiley, New York.

HOST MOLECULE

A molecule that contains one or more binding sites that can accommodate inorganic or organic ions referred to as guests. The binding site could even be a cavity within a crystal structure. Although enzymes clearly qualify as examples of host molecules, the term is usually restricted to structures such as crown ethers, macrocycles, and cyclodextrins. Nevertheless, these hosts do serve as models for molecular recognition. *See also* Crown Ethers; Macrocycles; Inclusion Complexes

HÜCKEL MOLECULAR-ORBITAL CALCULATIONS

A computational algorithm (also referred to as a HMO calculation) for characterizing delocalized chemical bonding patterns for unsaturated and aromatic molecules assumes that σ orbitals can be treated as local bonds[1,2]. The calculations involve only π electrons and either neglect or average out electron-electron repulsions. An extension of the Hückel method (often referred to as EHMO) has been used to treat other types of molecules[1,3]. **See also** *Hartree-Fock Molecular-Orbital Calculations; Ab Initio Molecular Orbital Calculations*

[1]K. Yates (1978) *Hückel Molecular Orbital Theory*, Academic Press, New York.
[2]C. A. Coulson, B. O'Leary & R. B. Mallion (1978) *Hückel Theory for Organic Chemists*, Academic Press, New York.
[3]R. Hoffman (1963) *J. Chem. Phys.* **39**, 1397.

HÜCKEL'S RULE

A rule for helping to predict if a monocyclic planar system of delocalized π electrons will exhibit aromatic character. The ring system will be aromatic if the number of delocalized π electrons equals $4n + 2$ where n is a positive integer or zero. For example, the ring system in phenylalanine obeys Hückel's rule. **See** *Aromaticity*

HUGHES-INGOLD THEORY FOR SOLVENT EFFECTS ON REACTIVITY

A principle applied in physical organic chemistry to account for the effect of solvation on the rates of nucleophilic reactions. This theory states that an increase in the ion-solvating power of the medium will tend to speed up the formation and concentration of charges, thereby inhibiting their breakdown or diffusion. This approach is predicated on the idea that a more polar solvent can potentially stabilize ionic intermediates and alter chemical reactivity.

Consider the attack of a nucleophile on the electrophilic carbon atom of a carbonyl group. Negative charge becomes concentrated on carbon and oxygen atoms of the carbonyl group as the reactants approach the the transition state. The Hughes-Ingold theory suggests that the reaction rate will be increased by any increase in the ion-solvating power of the medium.

To date, no single concept provides a quantitative means for estimating the magnitude of the contribution of reaction medium to rate processes. Nevertheless, the Hughes-Ingold principle offers some insight regarding how electrostatic interactions in enzyme active sites might finely tune the properties of nucleophilic steps in enzymic catalysis.

HUMMEL-DREYER TECHNIQUE

A gel filtration method[1] for measuring a ligand's equilibrium binding interactions with a macromolecule. In this technique, a gel filtration column is pre-equilibrated with a ligand (that can be measured by UV absorbance or by radioactivity), and then one adds a small sample of protein solution (itself containing ligand at the same total concentration as used to equilibrate the column). If the protein forms a complex with the ligand, then the protein peak will contain an excess of ligand depending on the protein concentration, the ligand concentration, and the strength of the equilibrium binding interaction. This method has proved very useful in determining dissociation constants and stoichiometries of ligand binding interactions[2,3].

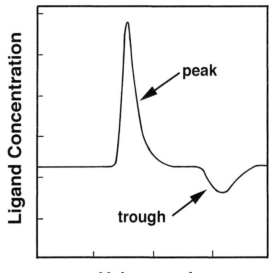

Volume, mL

Ligand binding to a macromolecule can be quantitatively analyzed by passage of the macromolecule through a gel filtration column that has been previously equilibrated at a chosen ligand concentration. Upon elution, the macromolecule with bound ligand will be excluded from the gel and will emerge as a peak followed by a trough which represents the depletion of ligand from the less mobile low-molecular-weight ligand. These experiments are typically carried out at various concentrations of ligand, and the area beneath the peak to the baseline should equal the area above the trough and up to the baseline. This baseline is constructed by drawing the best line through the horizontal segments of the elution profile.

[1]J. P. Hummel & W. J. Dreyer (1962) *Biochim. Biophys. Acta* **63**, 530.
[2]G. K. Ackers (1973) *Meth. Enzymology* **27**, 444.
[3]F. Sun, W. Kuo & R. A. Nash (1984) *J. Chromatog.* **288**, 377.

HYALURONATE LYASE

This enzyme [EC 4.2.2.1], also known as hyaluronidase, catalyzes the conversion of hyaluronate to n 3-(4-deoxy-β-D-gluc-4-enuronosyl)-*N*-acetyl-D-glucosamine. The enzyme is also capable of using chondroitin as a substrate.

A. Linker (1966) *Meth. Enzymol.* **8**, 650.

HYALURONIDASES

A number of enzymes have been referred to as hyaluronidases. Hyaluronate 4-glycanohydrolase [EC 3.2.1.35], also known as hyaluronoglucosaminidase and hyaluronidase, catalyzes the random hydrolysis of 1,4-linkages between *N*-acetyl-β-D-glucosamine and D-glucuronate residues in hyaluronate. It will also catalyze the hydrolysis of 1,4-β-D-glycosidic linkages between *N*-acetylgalactosamine or *N*-acetylgalactosamine sulfate and glucuronate in chondroitin, chondroitin 4- and 6-sulfates, and dermatan. Hyaluronate 3-glycanohydrolase [EC 3.2.1.36], also known as hyaluronoglucuronidase and hyaluronidase, catalyzes the random hydrolysis of 1,3-linkages between β-D-glucuronate and *N*-acetyl-D-glucosamine residues in hyaluronate. *See* Hyaluronate Lyase

W. L. Hynes & J. J. Ferretti (1994) *Meth. Enzymol.* **235**, 606.
H. Kresse & J. Glössl (1987) *Adv. Enzymol.* **60**, 217.
K. Meyer (1971) *The Enzymes*, 3rd ed., **5**, 307.

HYBRIDIZATION

1. A quantum mechanical phenomenon wherein certain orbitals (such as s, p, and d) combine into new orbitals (referred to as hybrid orbitals). For example, the electrons within the four C—H covalent bonds of methane occupy four hybrid orbitals (designated sp^3) that are formed from unhybridized *s* and *p* orbitals, originally present in the isolated carbon atom, upon combination of one carbon atom with four hydrogen atoms. Hybridization of these orbitals in carbon creates directed covalent bonds of equivalent energy and geometry. The sp^3 hybridization arises from only one of several approximate solutions to the Schrödinger equation, and it is worth noting that the four C–H bonds of methane are not always equivalent in their reactivity. 2. The formation of DNA–DNA and DNA–RNA duplexes from individual nucleic acid chains. Typically, a gel electropherogram containing a series of DNA bands is denatured under alkaline conditions and transferred to a nitrocellulose membrane to make a replica of the original electropherogram. The nitrocellulose-bound DNA species are brought into contact with a solution containing ^{32}P-labeled, single-stranded DNA (or RNA) molecules of interest, and the blot is incubated and subsequently extensively washed under conditions that form and maintain the structural integrity of DNA-[32P]DNA (or DNA-[32P]RNA) duplexes. Radioautography reveals the location(s) of the radiolabeled nucleic acid.

HYDRATION ATMOSPHERE

The short-range, semiordered water structure that surrounds ionic solutes and polar solutes as a consequence of dipolar and Coulombic interactions.

Anions and cations form strong coordination bonds with water molecules, and the polarizability of water factors into these interactions. Anions prefer to interact with the hydrogen atoms of a water molecule, whereas cations are attracted to the nonbonding electron pairs on the water oxygen. Because the nonbonded electron pairs tend to be localized and directional, cations make stronger bonds with water molecules than do anions of comparable ionic radius or electrical charge. This tightly bound water is said to electrostricted, a phenomenon associated with the ability of ions to attract and deform nearby water molecules, such that there is a slight increase in the density of bound water (*See* Electrostriction). Proceeding outwardly from this inner hydration sphere, there is an outer coordination shell containing a larger number of semiordered water molecules clustered together in extended hydrogen-bonded chains. The motions of water molecules in this cluster zone are more restricted than free water molecules in the bulk solvent; however, they are also less restricted in their motions than water molecules in the first hydration sphere. In this respect, water molecules form a two-zone hydration atmosphere that can profoundly affect the kinetics of ligand exchange reactions of metal ions in an aqueous environment (*See* Eigen-Tamm Mechanism). Note also that the solvated ion tends to create a centrosymmetric ordering of water molecules, and the degree of order is highest at short distances from that solute ion. This is to be contrasted with the hydration atmosphere surrounding a electrically charged membrane vesicle or a biomineral surface, where the solute ions and water molecules form planes or layers with predictable properties. *See* Biomineralization (3. The Double Layer)

HYDRATION NUMBER

The hydration number represents the number of water molecules directly associated with the solute ion. Some approximate hydration numbers for biologically significant metal ions are: Li^+, 3.7; Na^+ and K^+, 4; Ca^{2+}, 8; and Mg^{2+}, 12.

HYDRIDE

1. A negatively charged hydrogen ion (H:⁻). The hydride ion typically does not exist free in solution although it is an important species in a significant number of enzyme-catalyzed reactions. *See Hydride Ion Transfer*. 2. A molecular entity containing hydrogen in which one or more hydrogens has (have) a formal negative charge; for example, lithium aluminum hydride ($LiAlH_4$). 3. Compounds containing only hydrogen and one other element; examples of hydrides, based on this definition, are AsH_3, NH_3, H_2O, HCl, CH_4, H_2S, and PH_3.

HYDRIDE ION TRANSFER

The transfer of a hydride ion (H:⁻) in a reaction. Such transfers occur in many enzyme-catalyzed reactions; perhaps the most common being in pyridine nucleotide-linked dehydrogenases in which hydride transfer is stereospecific (*i.e.*, to the *re-* or *si*-face of the coenzyme)[1].

A number of methods have been utilized to probe hydride transfer reactions. These include the efforts to follow the course of the reaction in deuterated or tritiated water[2]. Perhaps one of the most effective methods is to study isotope effects[3-6].

[1]K.-S. You (1982) *Meth. Enzymol.* **87**, 101.
[2]I. A. Rose (1982) *Meth. Enzymol.* **87**, 84.
[3]K. B. Schowen & R. L. Schowen (1982) *Meth. Enzymol.* **87**, 551.
[4]D. B. Northrop (1982) *Meth. Enzymol.* **87**, 607.
[5]W. W. Cleland (1982) *Meth. Enzymol.* **87**, 625.
[6]W. W. Cleland (1995) *Meth. Enzymol.* **249**, 341.

HYDROCARBON MONOOXYGENASE

This enzyme system, also called ω-hydroxylase, catalyzes the NADH- and rubredoxin-dependent hydroxylation of terminal methyl groups of alkanes and fatty acids.

S. W. May & A. G. Katopodis (1990) *Meth. Enzymol.* **188**, 3.

HYDROGENASE

This enzyme [EC 1.18.99.1], also known as hydrogenlyase, catalyzes the reaction of H_2 with two oxidized ferredoxin to produce two H^+ and two reduced ferredoxin. This enzyme is an iron-sulfur protein and requires nickel ions. It can use molecular hydrogen to reduce a variety of substances. *See also Hydrogen Dehydrogenase; Cytochrome c₃ Hydrogenase*

M. A. Halcrow (1998) *Comprehensive Biological Catalysis: A Mechanistic Reference* **3**, 359.
R. J. Maier (1996) *Adv. Protein Chem.* **48**, 35.
M. W. W. Adams & A. Kletzin (1996) *Adv. Protein Chem.* **48**, 101.

R. Cammark, V. M. Fernandez & E. C. Hatchikian (1994) *Meth. Enzymol.* **243**, 43.
G. Palmer (1975) *The Enzymes*, 3rd ed., **12**, 1.

HYDROGEN BONDS

The primarily electrostatic bond formed by interaction of a hydrogen atom with a highly electronegative element (*e.g.*, oxygen, nitrogen, or fluorine) and a second electronegative atom (*e.g.*, oxygen, nitrogen, or fluorine).[1-3] The electrostatic bond is formed between a hydrogen donor (D-H) and an acceptor (A):

$$-D^{(\delta-)}-H^{(\delta+)} \cdots A^{(\delta-)}-$$

or simply as

$$-D-H \cdots A-$$

Hydrogen bonds (usually depicted with dotted or dashed lines) can be either intermolecular or intramolecular. In cases where oxygen or nitrogen acts as the electronegative element in the association, the oxygen or nitrogen atom can be singly or multiply bonded, albeit with different energies. The hydrogen bond is strongest (*i.e.*, most stable) when the three atoms of the bond are colinear (typically, within 15°). The distance between the donor and acceptor electronegative atoms is reduced from the calculated van der Waals radii as a result of the hydrogen bond interaction. Although they typically play only a compensatory role in stabilizing protein structure, hydrogen bonds can make significant contributions in both complexation and chemical reactions.

Jencks[4] has pointed out that it is conceptually useful to consider the hydrogen bond as an intermediate stage of proton transfer from an acid to a base. A sufficiently strong base will hold the proton closer, thereby stretching and weakening the proton's association with the acid. Complete transfer to the base breaks the hydrogen bond. Hence, a hydrogen bond strength increases with the acidity of the donor and the basicity of the acceptor.

As pointed out by Fersht[5,6], hydrogen bonds can also play significant roles in molecular recognition. An individual hydrogen bond can contribute 0.5–1.8 kcal mol⁻¹ (or, 2–7.5 kJ mol⁻¹) to the binding energy and this may increase specificity by a factor of 2–20. This was determined by assessing the effects of deleting a hydrogen bond to an uncharged hydrogen bond donor/acceptor pair. A larger magnitude (around 3–6 kcal mol⁻¹ or

12.5–25 kJ mol^{-1}) is typically observed if one deletes a hydrogen bond for a charged donor/acceptor. **See** Water; Low-Barrier Hydrogen Bonds

[1]L. Pauling (1960) *The Nature of the Chemical Bond*, Cornell Univ. Press, Ithaca, New York.

[2]M. D. Joesten & L. J. Schaad (1974) *Hydrogen Bonding*, Marcel Dekker, New York.

[3]G. C. Pimentel & A. L. McClellan (1960) *The Hydrogen Bond*, Freeman, San Francisco.

[4]W. P. Jencks (1969) *Catalysis in Chemistry and Enzymology*, pp. 338-350, McGraw-Hill, New York. [Also available as Jencks, W. P. (1986) *Catalysis in Chemistry and Enzymology*, Dover Publications, New York]

[5]A. R. Fersht (1987) *Trends Biochem. Sci.* **12**, 304.

[6]A. R. Fersht (1999) *Structure and Mechanism in Protein Science: A Guide to Enzyme Catalysis and Protein Folding*, pp. 650, W. H. Freeman, New York.

HYDROGEN DEHYDROGENASE

This enzyme [EC 1.12.1.2], also called hydrogenase on occasions, catalyzes the reaction of H_2 with NAD$^+$ to produce H^+ and NADH. This is a flavoprotein that requires iron and nickel. **See also** Hydrogenase; Cytochrome c_3 Hydrogenase

M. A. Halcrow (1998) *Comprehensive Biological Catalysis: A Mechanistic Reference* **3**, 359.

HYDROGEN EQUIVALENT

1. The number of replaceable hydrogen atoms per molecule of a particular substance. 2. The number of hydrogen atoms that will react with a molecule of a particular substance.

HYDROGEN EXCHANGE IN PROTEINS

The protons participating in hydrogen bonding in proteins are only released if their associated hydrogen bonds are broken and if there is direct access to additional hydrogens from the solvent. This is illustrated in the following scheme:

Dissociation:
$$[R-X \cdots H-Y-R'] \rightleftharpoons [R-X \quad H-Y-R']$$

Exchange:
$$[R-X \quad H-Y-R'] + D^+ \rightleftharpoons [R-X \quad D-Y-R'] + H^+$$

Therefore, those hydrogen atoms that are present in stable H-bonds will resist exchange with the solvent. H-bonds lying on the surface of many proteins readily undergo exchange. In addition to deuterium, one can also use tritium-labeled water to follow the kinetics of protein hydrogen exchange.

This technique was developed decades ago in the laboratory of Karl Linderstrøm-Lang. In his technique[1], the protein of interest is dissolved in D_2O where the protein is kept for long periods over which all protium atoms (*i.e.*, ^{1}H) attached to N and O are replaced by D (or, ^{2}H). After freeze drying and lyophilization to remove all solvent molecules, the protein is redissolved in H_2O and the time course of D $\rightleftharpoons$ H exchange is measured. Denatured proteins and random coiled proteins often exchange the hydrogens rapidly, whereas folded proteins typically show a spectrum of exchange rates.

This technique has been taken to new heights through the use of 2-D nuclear magnetic resonance[2-4] to assign many, if not all, exchangeable protons in terms of the residues to which they are attached. As a result, one can assess individual rate constants defining the kinetics of exchange.

While hydrogen exchange kinetics have been used to infer important features of protein folding pathways, Clarke *et al.*[5] argue that hydrogen exchange does not represent a short cut for studying protein-folding pathways because it is impossible to distinguish paths involving intermediates from side reactions. Their studies of barnase and chymotrypsin inhibitor-2 indicate no obvious relationship between hydrogen exchange at equilibrium and their folding pathways.

[1]K. Linderstrøm-Lang (1955) *Chem. Soc. (London) Spec. Publ.*, No. 2, p. 1.

[2]Y. Bai, T. R. Sosnick, L. Mayne & S. W. Englander (1995) *Science* **269**, 192.

[3]M. Jamin & R. L. Baldwin (1996) *Nature Struct. Biol.* **3**, 613.

[4]S. W. Englander, L. Mayne, Y. Bai & T.R. Sosnick (1997) *Protein Sci.* **6**, 1101.

[5]J. Clarke, L. S. Itzhaki & A. R. Fersht (1997) *Trends Biochem. Sci.* **22**, 284.

Selected entries from *Methods in Enzymology* [vol, page(s)]:
Chemistry, **232**, 31, 39-40; functional labeling method, **232**, 29, 32-35, 40-41; kinetic labeling method, **232**, 28-29; local unfolding model, **232**, 39-41; NMR analysis, **232**, 28; isotope dilution assays for cAMP, cGMP, and inositol 1,4,5-trisphosphate, **237**, 390-394; hydrogen-deuterium [amide I infrared band frequencies and, **232**, 169; with NMR analysis, **232**, 28, 29; hydrogen-tritium [functional labeling method, **232**, 29, 32-35; ^{3}H into normal water, nickel-iron hydrogenase assay, **243**, 57-59; kinetic labeling method, **232**, 28-29; tritium loss curve, **232**, 38]; amide groups, **259**, 345-348, 350; free energy change, **259**, 352; hemoglobin [cooperative free energy change loss, **259**, 355-356; tritium exchange, **259**, 353-355]; infrared spectroscopy, **259**, 406; locating energetic changes in proteins, **259**, 354-355; nuclear magnetic resonance, **259**, 349-350, 352; nucleotide bases, **259**, 346-347; in proteins, dependence on [denaturant concentration, **259**, 348-350; pH, **259**, 345-346, 352; protein unfolding, **259**, 344, 347; thermal stability, **259**, 350-352]; structural physics, **259**, 347-348; structural protection factor, **259**, 346.

HYDROGEN SULFIDE:FERRIC ION OXIDOREDUCTASE

This glutathione-dependent enzyme catalyzes the reaction of hydrogen sulfide with Fe^{3+} to produce sulfite and Fe^{2+}.

I. Suzuki (1994) *Meth. Enzymol.* **243**, 455.

HYDROGEN TUNNELING

The passage of one or more protons or hydride ions through a potential-energy barrier without requiring these reactants to possess sufficient energy to surmount the barrier. Hydrogen tunneling has been identified in a number of enzyme-catalyzed reactions[1], typically through isotope effect studies. Some evidence suggests that tunneling may be greater in the enzyme-catalyzed reaction as compared to the same reaction occurring in the absence of the enzyme. A number of factors have been suggested to contribute to this increase in the probability of tunneling. One is the internal thermodynamics of enzymic catalysis. When comparing different substrates and products for an enzyme that can act on a number of substrates, as the $\Delta H°$ value nears zero, the degree of tunneling increases[1-4]. In addition, the degree of tunneling is dependent on the ratio of the height and width of the reaction barrier[5]. Thus, compression of the reaction coordinate (*e.g.*, protein breathing motions bringing substrates close together thus increasing the probability of tunneling) could explain the increase seen in tunneling in enzyme-catalyzed reactions. Hence, both thermodynamic and geometric constraints seen in enzymes can potentially influence the occurrence of proton tunneling.

Large kinetic isotope effects are usually indicative of proton tunneling. However, if proton transfer is not rate-limiting or rate-contributing, proton tunneling may still be present, but its occurrence would be masked.

In studying both primary and secondary isotope effects, if the secondary effect is significantly reduced on the replacement of the primary H for D (*i.e.*, by the primary effect), then tunneling occurs. Note that this method assumes that the observed primary and secondary isotope effects occur in the same step of the reaction. A number of methods have been developed to characterize tunneling phenomena[1].

[1]B. J. Bahnson & J. P. Klinman (1995) *Meth. Enzymol.* **249**, 373.
[2]J. R. De La Vega (1982) *Acc. Chem. Res.* **15**, 185.
[3]Y. Cha, C. J. Murray & J. P. Klinman (1989) *Science* **243**, 1325.
[4]J. Rucker, Y. Cha, T. Jonsson, K. L. Grant & J. P. Klinman (1992) *Biochemistry* **31**, 11489.
[5]R. P. Bell (1980) *The Tunnel Effect in Chemistry*, Chapman and Hall, New York.

Selected entries from *Methods in Enzymology* [vol, page(s)]:
In enzyme catalysis, **249**, 373-397; [bovine serum amine oxidase, **249**, 393-394; coupled motion and, **249**, 386-388; demonstration, **249**, 374-386 (breakdown of rule of geometric mean in, **249**, 375-376, 388-389; by competitive comparison of k_H, k_D, and k_T, **249**, 382-383; exponential breakdown in, **249**, 376-378); factors affecting, **249**, 395-397; internal thermodynamics and, **249**, 396-397; monoamine oxidase B, **249**, 395; protein structure and, **249**, 397; verification, **249**, 386-395].

HYDROLASES

A major class of enzymes that catalyze hydrolytic cleavage reactions. Examples include esterases, phosphatases, sulfatases, nucleases, glycosidases, peptidases, proteinases, and amidases.

HYDROLYSIS

Any process of bond cleavage that is accompanied by the incorporation of the elements of a water molecule into the product(s):

$$R–X + H_2O \rightleftharpoons R—OH + H–X$$

Bonds typically hydrolyzed include carboxylic and phosphoric esters, amides, acetals, amidines, as well as metal ion complexes. (When a nucleophilic substitution reaction uses the solvent as the nucleophile, the reaction is often referred to as solvolysis.)

In the broader scope of this definition, we can also include the hydration of aldehydes or alkenes, where the $C{=}O$ or $C{=}C$ double bond is broken by incorporating the elements of water:

$$RHC{=}O + H_2O \rightleftharpoons RHC(OH)_2$$

$$RHC{=}CHR' + H_2O \rightleftharpoons RHC(OH)–CH_2R'$$

Jencks[1] considered the properties and mechanisms of hydrolysis and hydration. The interested reader may wish to consult his monograph (exact pages in parentheses) for information on the following: acetylacetone (p. 550); aldehydes (pp. 175–176; 211-215; 497–498); acylcyanides (p. 549); amides (pp. 523–526); amidines (pp. 471–477); anhydrides (pp. 508–517); anilides (pp. 224, 523–525; 553); aspirin (pp. 580–583); ATP (pp. 112–114); esters (pp. 399; 460; 483–487; 508–517; 570; 577–582);

ethyltrifluoroacetate (p. 515); formamidines (pp. 192, 234, 545–548; 553); hydration of carbonyl compounds (pp. 22, 33, 175–176, 198, 211–215, 497–498, 548-550, 608); imido esters (pp. 215–216, 487–489, 526–531, 536–542); ketones (pp. 548–550); *p*-nitrophenyl acetate (pp. 405, 408, 597–599); phenyl acetate (p. 314); phosphate esters (pp. 97, 169, 233, 305, 497); starch (p. 357); thiazolines (pp. 539–542, 552); and thiol esters (pp. 517–523, 534, 597).

[1]W. P. Jencks (1969) *Catalysis in Chemistry and Enzymology*, McGraw-Hill, New York. [Also available as Jencks, W. P. (1986) *Catalysis in Chemistry and Enzymology*, Dover Publications, New York]

HYDRON

The H$^+$ ion in its natural abundance, typically used when one wishes not to distinguish between isotopes. The term "proton" thus refers specifically to ^{1}H$^+$, "deuteron" to ^{2}H$^+$, and "triton" to ^{3}H$^+$.

HYDROPHILIC

Referring to the tendency or capacity of a molecular entity or one of its substituents to interact significantly with water or other polar solvents or polar groups.

HYDROPHOBIC EFFECT

The tendency[1–4] of apolar side chains of amino acids (or lipids) to reside in the interior nonaqueous environment of a protein (or membrane/micelle/vesicle). This process is accompanied by the release of water molecules from these apolar side-chain moieties. The effect is thermodynamically driven by the increased disorder (*i.e.*, $\Delta S > 0$) of the system, thereby overcoming the unfavorable enthalpy change (*i.e.*, $\Delta H < 0$) for water release from the apolar groups.

Cantor and Schimmel[5] provide a lucid description of the thermodynamics of the hydrophobic effect, and they stress the importance of considering both the unitary and cratic contributions to the partial molal entropy of solute-solvent interactions. Briefly, the partial molal entropy (S_A) is the sum of the *unitary* contribution (S_A') which takes into account the characteristics of solute A and its interactions with water) and the *cratic* term ($-R \ln \chi_{cA}$, where R is the universal gas constant and χ_A is the mole fraction of component A) which is a statistical term resulting from the mixing of component A with solvent molecules. The unitary change in entropy S_A'

for transfer of a hydrocarbon side chain from water to the nonaqueous interior of a protein is positive. Hence, its contribution ($-T\Delta S$) to the Gibbs free energy favors transfer of a side chain to the protein's interior. Note that in the presence of denaturing agent (such as 8 M urea), a side chain exhibits a reduced tendency to reside in the apolar interior[6]. About one-third to one-half of all amino acid residues in a protein have apolar side chains, thereby accounting for the cumulatively unfavorable effect of 8 M urea on protein folding.

Fersht and Serrano[7] recently described the systematic application of site-directed mutagenesis, together with calorimetry, to evaluate how specific side chains play roles in stabilizing protein structure. Many self-assembly processes are entropy driven, and Lauffer[8] proposed that water is released from apolar patches on the unpolymerized subunits as these patches interact with each other in the assembled state. In the case of microtubules, the warm-induced polymerization and cold-induced depolymerization are hallmarks of an entropy-driven process. Finally, although the common practices of enzyme purification are storage and processing at ice-water temperatures, some enzymes are cold-inactivated, and the experienced investigator will test the activity and structural properties of a newly characterized enzyme after storage at several different temperatures.

[1]W. Kauzmann (1959) *Adv. Protein Chem.* **14**, 1.
[2]C. Tanford (1962) *J. Amer. Chem. Soc.* **84**, 4240.
[3]C. Tanford (1980) *The Hydrophobic Effect: Formation of Micelles and Biological Membranes*, 2nd ed., Wiley, New York.
[4]F. M. Richards in *Protein Folding* (T. E. Creighton, ed.) pp. 1-58, Freeman, San Francisco.
[5]C. R. Cantor & P. R. Schimmel (1980) *Biophysical Chemistry*, vol. 1, pp. 281-288, Freeman, San Francisco.
[6]P. L. Whitley & C. Tanford (1962) *J. Biol. Chem.* **237**, 1735.
[7]A. R. Fersht & L. Serrano (1993) *Curr. Opin. Struct. Biol.* **3**, 75.
[8]M. A. Lauffer (1975) *Entropy-Driven Processes in Biology*, Springer-Verlag, Berlin.

D-2-HYDROXY-ACID DEHYDROGENASE

This enzyme [EC 1.1.99.6] catalyzes the reaction of (*R*)-lactate with an acceptor to produce pyruvate and a reduced acceptor. The enzyme, which requires FAD and a zinc ion, will utilize a number of (*R*)-2-hydroxy acids as substrates.

A. R. Clarke & T. R. Dafforn (1998) *Comprehensive Biological Catalysis: A Mechanistic Reference* **3**, 1.
Y. Hatefi & D. L. Stiggall (1976) *The Enzymes*, 3rd ed., **13**, 175.

(S)-2-HYDROXY-ACID OXIDASE

This enzyme [EC 1.1.3.15] (also referred to as glycolate oxidase, hydroxy-acid oxidase A, and hydroxy-acid oxidase B) catalyzes the reaction of an (S)-2-hydroxy acid with dioxygen to produce a 2-oxo acid and hydrogen peroxide. FMN is the cofactor for this enzyme. This oxidase exists as two major isoenzymes. The A form preferentially oxidizes short-chain aliphatic hydroxy acids whereas the B form preferentially oxidizes long-chain and aromatic hydroxy acids.

N. E. Tolbert (1981) *Ann. Rev. Biochem.* **50**, 133.

3-HYDROXYACYL-CoA DEHYDROGENASE

This enzyme [EC 1.1.1.35] catalyzes the reaction of an (S)-3-hydroxyacyl-CoA with NAD^+ to produce a 3-oxoacyl-CoA and NADH. The enzyme will also utilize S-3-hydroxyacyl-N-acylthioethanolamine and S-3-hydroxyacylhydrolipoate as substrates. The enzyme isolated from some sources can also utilize $NADP^+$ as the coenzyme, albeit as a weaker substrate. In addition, there is a broad specificity with respect to the acyl chain length (note that there is a long-chain-length 3-hydroxyacyl-CoA dehydrogenase [EC 1.1.1.211]).

F. Lynen & O. Wieland (1955) *Meth. Enzymol.* **1**, 566.

β-HYDROXYACYL THIOESTER (or, ACP) DEHYDRASE

This enzyme, officially known as 3-hydroxypalmitoyl-[acyl-carrier protein] dehydratase [EC 4.2.1.61], is the fatty-acid synthase component that catalyzes the conversion of (3R)-3-hydroxypalmitoyl-[acyl-carrier protein] to form 2-hexadecenoyl-[acyl-carrier protein] and water. This enzyme displays specificity toward 3-hydroxyacyl-[acyl-carrier protein] derivatives (with chain lengths from C_{12} to C_{16}), with highest activity on the palmitoyl derivative. *See also* Fatty Acid Synthetase

M. A. Ator & P. R. Ortez de Montellano (1990) *The Enzymes*, 3rd ed., **19**, 213.
S. J. Wakil & J. K. Stoops (1983) *The Enzymes*, 3rd ed., **16**, 3.
K. Bloch (1971) *The Enzymes*, 3rd ed., **5**, 441.

3-HYDROXYANTHRANILATE 3,4-DIOXYGENASE

This iron-dependent enzyme [EC 1.13.11.6] catalyzes the reaction of 3-hydroxyanthranilate and dioxygen to yield 2-amino-3-carboxymuconate semialdehyde. This product will spontaneously rearrange to quinolinic acid.

O. Hayaishi, M. Nozaki & M. T. Abbott (1975) *The Enzymes*, 3rd ed., **12**, 119.

3-HYDROXYANTHRANILATE OXIDASE

This enzyme [EC 1.10.3.5] catalyzes the reaction of 3-hydroxyanthranilate with dioxygen to produce 6-imino-5-oxocyclohexa-1,3-dienecarboxylate and hydrogen peroxide.

J. V. Schloss & M. S. Hixon (1998) *Comprehensive Biological Catalysis: A Mechanistic Reference* **2**, 43.

3-HYDROXYASPARTATE ALDOLASE

This enzyme [EC 4.1.3.14] reversibly converts *erythro*-3-hydroxy-L-aspartate to glycine and glyoxylate.

R. G. Gibbs & J. G. Morris (1970) *Meth. Enzymol.* **17A**, 981.

erythro-β-HYDROXYASPARTATE DEHYDRATASE

This pyridoxal-phosphate-dependent enzyme [EC 4.2.1.38] catalyzes the reaction of *erythro*-3-hydroxy-L-aspartate with water to produce oxaloacetate, ammonia, and water.

L. Davis & D. E. Metzler (1972) *The Enzymes*, 3rd ed., **7**, 33.
R. G. Gibbs & J. G. Morris (1970) *Meth. Enzymol.* **17A**, 981.

4-HYDROXYBENZOATE DECARBOXYLASE

This enzyme [EC 4.1.1.61] catalyzes the conversion of 4-hydroxybenzoate to phenol and carbon dioxide.

J. V. Schloss & M. S. Hixon (1998) *Comprehensive Biological Catalysis: A Mechanistic Reference* **2**, 43.

3-HYDROXYBENZOATE 4-MONOOXYGENASE

This enzyme [EC 1.14.13.23], also called 3-hydroxybenzoate 4-hydroxylase, catalyzes the reaction of 3-hydroxybenzoate with NADPH and dioxygen to produce 3,4-dihydroxybenzoate, $NADP^+$, and water. FAD is used as a cofactor and the enzyme can also use analogs of 3-hydroxybenzoate substituted in the 2, 4, 5, and 6 positions as substrates.

B. Entsch (1990) *Meth. Enzymol.* **188**, 138.
V. Massey & P. Hemmerich (1975) *The Enzymes*, 3rd ed., **12**, 191.

3-HYDROXYBENZOATE 6-MONOOXYGENASE

This enzyme [EC 1.14.13.24], also known as 3-hydroxybenzoate 6-hydroxylase, catalyzes the reaction of 3-hydroxybenzoate with NADH and dioxygen to produce 2,5-dihydroxybenzoate, NAD^+, and water. NADPH can also be used as a substrate, although not as effectively as NADH. FAD is used as a cofactor and the enzyme can utilize a number of analogs of 3-hydroxybenzoate substituted in the 2, 4, 5, and 6 positions as substrates.

B. Entsch (1990) *Meth. Enzymol.* **188**, 138.
V. Massey & P. Hemmerich (1975) *The Enzymes*, 3rd ed., **12**, 191.

4-HYDROXYBENZOATE 3-MONOOXYGENASE

This enzyme [EC 1.14.13.2], also known as *p*-hydroxybenzoate hydroxylase, catalyzes the reaction of 4-hydroxybenzoate, NADPH, and dioxygen to produce protocatechuate, $NADP^+$, and water. FAD is used as a cofactor by this enzyme. The enzyme from *Pseudomonas* is very specific for the coenzyme substrate whereas 4-hydroxybenzoate 3-monooxygenase (NAD(P)H) [EC 1.14.13.33] can utilize NADH and NADPH equally well. The enzyme isolated from *Corynebacterium cyclohexanicum* is highly specific for 4-hydroxybenzoate.

B. A. Palfey & V. Massey (1998) *Comprehensive Biological Catalysis: A Mechanistic Reference* **3**, 83.
S. Yamamoto & Y. Ishimura (1991) *A Study of Enzymes* **2**, 315.
B. Entsch (1990) *Meth. Enzymol.* **188**, 138.
V. Massey & P. Hemmerich (1975) *The Enzymes*, 3rd ed., **12**, 191.

3-HYDROXYBUTYRATE DEHYDROGENASE

This enzyme [EC 1.1.1.30] catalyzes the reversible reaction of (*R*)-3-hydroxybutanoate with NAD^+ to produce acetoacetate and NADH. Other 3-hydroxymonocarboxylic acids can act as substrates as well.

A. R. Clarke & T. R. Dafforn (1998) *Comprehensive Biological Catalysis: A Mechanistic Reference* **3**, 1.
J. D. McGarry & D. W. Foster (1980) *Ann. Rev. Biochem.* **49**, 395.

4-HYDROXYBUTYRATE DEHYDROGENASE

This enzyme [EC 1.1.1.61] catalyzes the reaction of 4-hydroxybutanoate with NAD^+ to produce succinate semialdehyde and NADH.

J. K. Hardman (1962) *Meth. Enzymol.* **5**, 778.

3-HYDROXYBUTYRYL-CoA EPIMERASE

This enzyme [EC 5.1.2.3] catalyzes the interconversion of (*S*)-3-hydroxybutanoyl-CoA and (*R*)-3-hydroxybutanoyl-CoA.

J. R. Stern (1962) *Meth. Enzymol.* **5**, 557.

HYDROXYINDOLE O-METHYLTRANSFERASE

This enzyme [EC 2.1.1.4], also referred to as acetylserotonin methyltransferase, catalyzes the reaction of *S*-adenosylmethionine with *N*-acetylserotonin to produce *S*-adenosylhomocysteine and *N*-acetyl-5-methoxytryptamine.

D. Sugden, V. Ceña & D. C. Klein (1987) *Meth. Enzymol.* **142**, 590.

4-HYDROXY-2-KETOGLUTARATE ALDOLASE

This enzyme [EC 4.1.3.16], also known as 2-keto-4-hydroxyglutarate aldolase and 2-oxo-4-hydroxyglutarate aldolase, catalyzes the reversible conversion of 4-hydroxy-2-oxoglutarate to pyruvate and glyoxylate. Interestingly, the enzyme is reported to be able to act on both stereoisomers.

K. N. Allen (1998) *Comprehensive Biological Catalysis: A Mechanistic Reference* **2**, 135.
D. J. Creighton & N. S. R. K. Murthy (1990) *The Enzymes*, 3rd ed., **19**, 323.
W. A. Wood (1972) *The Enzymes*, 3rd ed., **7**, 281.

HYDROXYLAMINE REDUCTASE

This enzyme [EC 1.6.6.11] catalyzes the reaction of hydroxylamine with NADH to produce ammonia, water, and NAD^+. A number of hydroxamates can also act as substrates.

M. Zucker & A. Nason (1955) *Meth. Enzymol.* **2**, 416.

4-HYDROXYMANDELONITRILE LYASE

This enzyme [EC 4.1.2.11], also known as hydroxymandelonitrile lyase or hydroxynitrile lyase, catalyzes the conversion of (*S*)-4-hydroxymandelonitrile to cyanide and 4-hydroxybenzaldehyde. Aliphatic hydroxynitriles do not serve as substrates for this enzyme, unlike that of (*S*)-hydroxynitrilase [EC 4.1.2.39].

M. K. Seely & E. E. Conn (1971) *Meth. Enzymol.* **17B**, 239.

β-HYDROXY-β-METHYLGLUTARYL-CoA LYASE

This enzyme [EC 4.1.3.4] catalyzes the conversion of (*S*)-3-hydroxy-3-methylglutaryl-CoA to acetyl-CoA and acetoacetate.

D. J. Creighton & N. S. R. K. Murthy (1990) *The Enzymes*, 3rd ed., **19**, 323.
K. M. Gibson (1988) *Meth. Enzymol.* **166**, 219.
M. J. P. Higgins, J. A. Kornblatt & H. Rudney (1972) *The Enzymes*, 3rd ed., **7**, 407.

β-HYDROXY-β-METHYLGLUTARYL-CoA REDUCTASE

β-Hydroxy-β-methylglutaryl-CoA reductase (NADPH) [EC 1.1.1.34] catalyzes the reaction of (S)-3-hydroxy-3-methylglutaryl-CoA and two NADPH to produce (R)-mevalonate, coenzyme A, and two NADP⁺. The enzyme is inactivated by β-hydroxy-β-methylglutaryl-CoA reductase (NADPH) kinase [EC 2.7.1.109] and reactivated by [hydroxymethylglutaryl-CoA reductase (NADPH)]-phosphatase [EC 3.1.3.47]. β-Hydroxy-β-methylglutaryl-CoA reductase (NADH) [EC 1.1.1.88] catalyzes the reaction of 3-hydroxy-3-methylglutaryl-CoA and two NADH to produce mevalonate, coenzyme A, and two NAD⁺.

D. M. Gibson & R. A. Parker (1987) *The Enzymes*, 3rd ed., **18**, 179.
T.-Y. Chang (1983) *The Enzymes*, 3rd ed., **16**, 491.
Z. H. Beg & H. B. Brewer, Jr. (1981) *Curr. Top. Cell. Reg.* **20**, 139.
V. W. Rodwell, D. J. McNamara & D. J. Shapiro (1973) *Adv. Enzymol.* **38**, 373.

β-HYDROXY-β-METHYLGLUTARYL-CoA REDUCTASE (NADPH) KINASE

This enzyme [EC 2.7.1.109], also known simply as reductase kinase, catalyzes the reaction of ATP with the enzyme 3-hydroxy-3-methylglutaryl-CoA reductase (NADPH) producing ADP and the phosphorylated form of the reductase. This phosphorylation inactivates the reductase. Histones can substitute for the reductase as substrates.

A. M. Edelman, D. K. Blumenthal & E. G. Krebs (1987) *Ann. Rev. Biochem.* **56**, 567.

β-HYDROXY-β-METHYLGLUTARYL-CoA SYNTHASE

This enzyme [EC 4.1.3.5] catalyzes the reversible reaction of acetyl-CoA with acetoacetyl-CoA and water to generate (S)-3-hydroxy-3-methylglutaryl-CoA and coenzyme A.

D. J. Creighton & N. S. R. K. Murthy (1990) *The Enzymes*, 3rd ed., **19**, 323.
M. J. P. Higgins, J. A. Kornblatt & H. Rudney (1972) *The Enzymes*, 3rd ed., **7**, 407.

3-HYDROXY-2-METHYLPYRIDINE-4,5-DICARBOXYLATE 4-DECARBOXYLASE

This enzyme [EC 4.1.1.51] catalyzes the conversion of 3-hydroxy-2-methylpyridine-4,5-dicarboxylate to 3-hydroxy-2-methylpyridine-5-carboxylate and carbon dioxide.

R. W. Burg (1970) *Meth. Enzymol.* **18A**, 634.

5-HYDROXYMETHYLPYRIMIDINE KINASE

This enzyme [EC 2.7.1.49] catalyzes the reaction of ATP with 4-amino-2-methyl-5-hydroxymethylpyrimidine to produce ADP and 4-amino-2-methyl-5-phosphomethylpyrimidine. In addition, ATP can be replaced as the substrate by CTP, UTP, and GTP.

E. P. Anderson (1973) *The Enzymes*, 3rd ed., **9**, 49.

6-HYDROXYNICOTINATE REDUCTASE

This enzyme [EC 1.3.7.1], also referred to as 6-oxotetrahydronicotinate dehydrogenase, catalyzes the reaction of 1,4,5,6-tetrahydro-6-oxonicotinate with oxidized ferredoxin to produce 6-hydroxynicotinate and the reduced ferredoxin.

L. Tsai & E. R. Stadtman (1971) *Meth. Enzymol.* **18B**, 233.

2-HYDROXY-3-OXOPROPIONATE REDUCTASE

This enzyme [EC 1.1.1.60], also referred to as tartronate semialdehyde reductase, catalyzes the reversible reaction of (R)-glycerate with NAD(P)⁺ to produce 2-hydroxy-3-oxopropanoate and NAD(P)H.

H. L. Kornberg & A. M. Gotto (1966) *Meth. Enzymol.* **9**, 240.

4-HYDROXYPHENYLACETATE 3-MONOOXYGENASE

This enzyme [EC 1.14.13.3], also called *p*-hydroxyphenylacetate 3-hydroxylase, catalyzes the reaction of 4-hydroxyphenylacetate with NADH and dioxygen to produce 3,4-dihydroxyphenylacetate, NAD⁺, and water. FAD is used as a cofactor.

B. A. Palfey & V. Massey (1998) *Comprehensive Biological Catalysis: A Mechanistic Reference* **3**, 83.

4-HYDROXYPHENYLPYRUVATE HYDROXYLASE

This enzyme [EC 1.13.11.27], also known as 4-hydroxyphenylpyruvate dioxygenase, catalyzes the reaction of 4-hydroxyphenylpyruvate with dioxygen to generate homogentisate and carbon dioxide. Iron ions are needed as cofactors.

A. G. Prescott & P. John (1996) *Ann. Rev. Plant Physiol. Plant Mol. Biol.* **47**, 245.

S. Lindstedt & B. Odelhög (1987) *Meth. Enzymol.* **142**, 139 and 143.

J. H. Fellman (1987) *Meth. Enzymol.* **142**, 148.

O. Hayaishi, M. Nozaki & M. T. Abbott (1975) *The Enzymes*, 3rd ed., **12**, 119.

4-HYDROXYPROLINE EPIMERASE

This enzyme [EC 5.1.1.8] catalyzes the interconversion of *trans*-4-hydroxy-L-proline to *cis*-4-hydroxy-D-proline. It also interconverts *trans*-4-hydroxy-D-proline and *cis*-4-hydroxy-L-proline.

M. E. Tanner & G. L. Kenyon (1998) *Comprehensive Biological Catalysis: A Mechanistic Reference* **2**, 7.

E. Adams & L. Frank (1980) *Ann. Rev. Biochem.* **49**, 1005.

E. Adams (1976) *Adv. Enzymol.* **44**, 69.

R. Kuttan & A. N. Radhakrishnan (1973) *Adv. Enzymol.* **37**, 273.

E. Adams (1972) *The Enzymes*, 3rd ed., **6**, 479.

β-HYDROXYPROPIONATE DEHYDROGENASE

This enzyme [EC 1.1.1.59] catalyzes the reaction of 3-hydroxypropanoate with NAD^+ to produce 3-oxopropanoate and NADH.

M. J. Coon & W. G. Robinson (1962) *Meth. Enzymol.* **5**, 451.

15-HYDROXYPROSTAGLANDIN DEHYDROGENASE

15-Hydroxyprostaglandin dehydrogenase (NAD^+) [EC 1.1.1.141] catalyzes the reaction of (5Z,13E)-(15S)-11α, 15-dihydroxy-9-oxoprost-13-enoate with NAD^+ to produce (5Z,13E)-11α-hydroxy-9,15-dioxoprost-13-enoate and NADH. This enzyme will also use prostaglandins E_2, $F_{2\alpha}$ and B_1 as substrates. However, prostaglandin D_2 will not serve as a substrate. In this case, 15-hydroxyprostaglandin D dehydrogenase ($NADP^+$) [EC 1.1.1.196] is used. 15-Hydroxyprostaglandin dehydrogenase ($NADP^+$) [EC 1.1.1.197] catalyzes the reaction of (5Z,13E)-(15S)-11α,15-dihydroxy-9-oxoprost-13-enoate with $NADP^+$ to produce (5Z,13E)-11α-hydroxy-9,15-dioxoprost-13-enoate and NADPH. This enzyme will also use prostaglandins E_2, $F_{2\alpha}$ and B_1 as substrates. However, prostaglandin D_2 will not serve as a substrate. In some systems, this enzyme may be identical with prostaglandin E_2 9-reductase [EC 1.1.1.189].

C. R. Pace-Asciak & W. L. Smith (1983) *The Enzymes*, 3rd ed., **16**, 543.

HYDROXYPYRUVATE REDUCTASE

This enzyme [EC 1.1.1.81], also referred to as D-glycerate dehydrogenase, catalyzes the reversible reaction of D-glycerate with $NAD(P)^+$ to produce hydroxypyruvate and NAD(P)H. *See also* Glycerate Dehydrogenase

L. A. Kleczkowski (1994) *Ann. Rev. Plant Physiol. Plant Mol. Biol.* **45**, 339.

C. Krema & M. E. Lidstrom (1990) *Meth. Enzymol.* **188**, 373.

3α-HYDROXYSTEROID DEHYDROGENASE

This enzyme [EC 1.1.1.50], also referred to as hydroxyprostaglandin dehydrogenase, catalyzes the reaction of androsterone with $NAD(P)^+$ to produce 5α-androstane-3,17-dione and NAD(P)H. Other 3α-hydroxysteroids can act as substrates as well as 9-, 11- and 15-hydroxyprostaglandins. The stereochemistry is B-specific with respect to the pyridine coenzymes.

O. Berséus, H. Danielsson & K. Einarsson (1969) *Meth. Enzymol.* **15**, 551.

S. S. Koide (1969) *Meth. Enzymol.* **15**, 651.

3(or 17)β-HYDROXYSTEROID DEHYDROGENASE

This enzyme [EC 1.1.1.51] catalyzes the reaction of testosterone with $NAD(P)^+$ to produce androst-4-ene-3,17-dione and NAD(P)H. The enzyme can also use other 3β- or 17β-hydroxysteroids as substrates.

J. Jarabak (1969) *Meth. Enzymol.* **15**, 746.

11β-HYDROXYSTEROID DEHYDROGENASE

This enzyme [EC 1.1.1.146] catalyzes the reaction of an 11β-hydroxysteroid with $NADP^+$ to produce an 11-oxosteroid and NADPH.

R. Benediktsson & C. R. W. Edwards (1996) *Essays in Biochem.* **31**, 23.

16α-HYDROXYSTEROID DEHYDROGENASE

This enzyme [EC 1.1.1.147] catalyzes the reaction of a 16α-hydroxysteroid with $NAD(P)^+$ to produce a 16-oxosteroid and NAD(P)H.

H. Breuer & R. Knuppen (1969) *Meth. Enzymol.* **15**, 691.

20-HYDROXYSTEROID DEHYDROGENASE

This enzyme [EC 1.1.1.53] catalyzes the reaction of androstane-3α,17β-diol with NAD^+ to produce 17β-hydroxyandrostan-3-one and NADH. The 3α-hydroxyl group or the 20β-hydroxyl group of pregnane and androstane steroids can act as the substrate.

W. G. Wiest (1969) *Meth. Enzymol.* **15**, 638.

21-HYDROXYSTEROID DEHYDROGENASE

21-Hydroxysteroid dehydrogenase (NAD⁺) [EC 1.1.1.150] catalyzes the reaction of pregnan-21-ol with NAD⁺ to produce pregnan-21-al and NADH. Other 21-hydroxycorticosteroids can also serve as substrates. 21-Hydroxysteroid dehydrogenase (NADP⁺) [EC 1.1.1.151] catalyzes the reaction of pregnan-21-ol with NADP⁺ to produce pregnan-21-al and NADPH. Other 21-hydroxycorticosteroids can also serve as substrates.

C. Monder & C. S. Furfine (1969) *Meth. Enzymol.* **15**, 667.

HYPERCONJUGATION

Delocalization of electrons in one or more σ-bonds that interact with the electrons in a π-electron network.

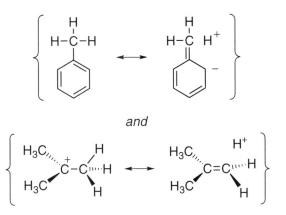

and

Hyperconjugation is less effective than conjugation in providing a mechanism for delocalizing electrons. This effect can often be detected spectrally by a slight shift to higher wavelength of UV light absorption, and the magnitude of the spectral shift can be predicted by the Woodward's rules.

N. Muller & R. S. Mulliken (1958) *J. Amer. Chem. Soc.* **80**, 3489.
M. J. S. Dewar (1962) *Hyperconjugation*, Ronald Press, New York.
T. Sorensen (1967) *J. Amer. Chem. Soc.* **89**, 3782.
E. M. Arnett & J. W. Larsen (1969) *J. Amer. Chem. Soc.* **91**, 34.
L. Radom (1982) *Prog. Theoret. Org. Chem.* **3**, 1.

HYPERCYCLE

A model for biological self-organization based on autocatalytic processes during early stages of biological evolution. The concept was introduced by Eigen[1] to characterize a functional entity which integrates information stored in individual self-replicating elements. These self-replicating elements compete, much like species do in Nature, such that a single species eventually is selected, attended by the disappearance of less efficient competitors. To suppress competition, Eigen and Schuster[2,3] introduced specific second-order coupling terms [that are proportional to the product of two population numbers (or concentration in chemical processes)] into their kinetic equations that form a closed catalytic cycle.

[1]M. Eigen (1971) *Quart. Rev. Biophys.* **4**, 149.
[2]M. Eigen & P. Schuster (1978) *Naturwissenschaften* **65**, 341.
[3]M. Eigen & P. Schuster (1979) *The Hypercycle- A Principle of Natural Self-Organization*, Springer-Verlag, Berlin.

HYPERPOLARIZATION

With respect to membranes, an increase in the polarization of a membrane. **See** *Depolarization; Action Potential*

HYPERREACTIVITY

Having an increased or elevated reactivity. This term has been used in reference to the relative activity of amino acyl residues at the active sites of enzyme. The immediate environment (*i.e.*, the microenvironment) may allow simple reagents to react faster with the amino acid than would normally be expected. Thus, in labeling of proteins with active site-directed reagents, an investigator should always consider the basis of increased reactivity: Is it due to facilitation of the reaction by increased affinity (*i.e.*, affinity labeling), or is it due to increased activity of the amino acyl side chain (*e.g.*, perhaps increased nucleophilicity due to the microenvironment).

Procedures[1-3] for distinguishing between the two means of facilitation include: (1) Use of a reagent that does not bind as an affinity label. If facilitation is due to hyperreactivity, this reagent should also be more reactive. (2) Use of transition-state analysis. A favorable change in the entropy of activation ($\Delta S^{\ddagger}$) would imply facilitation via affinity labeling whereas a more favorable change in the enthalpy of activation ($\Delta H^{\ddagger}$) implies hyperreactivity. However, a certain caution should always be exercised since other factors, *e.g.* differential solvation effects, can result in a certain degree of compensation between $\Delta H^{\ddagger}$ and $\Delta S^{\ddagger}$.

[1]E. P. Lennette & B.V. Plapp (1979) *Biochemistry* **18**, 3933.
[2]E. P. Lennette & B.V. Plapp (1979) *Biochemistry* **18**, 3938.
[3]B. V. Plapp (1982) *Meth. Enzymol.* **87**, 469.

HYPERTONIC

1. Referring to a larger osmotic pressure (or ionic strength) of a particular solution relative to some stan-

dard solution (*e.g.*, blood plasma). In a hypertonic solution, cells would tend to shrink. 2. Referring to states that have a larger degree of tension.

HYPOTAUROCYAMINE KINASE

This enzyme [EC 2.7.3.6] catalyzes the reaction of ATP with hypotaurocyamine to produce ADP and N^{ω}-phosphohypotaurocyamine. Taurocyamine will also act as a substrate, but not as efficiently.

N. v. Thoai, L.-A. Pradel & R. Kassab (1970) *Meth. Enzymol.* **17A**, 1002.

HYPOTHESIS

A trial idea, assumption, or conjecture regarding the behavior of a system. Such unproven assumptions help to establish organized thought and to develop experimental design. *See Statistics; Alternative Hypothesis; Null Hypothesis*

HYPOTHESIS TESTING

A statistical protocol for testing the likely validity of one hypothesis about the behavior of an experimental parameter against another, thereby affording a rational basis for accepting or rejecting a particular set of assumptions. *See Statistics*

HYPOXANTHINE (GUANINE) PHOSPHORIBOSYLTRANSFERASE

This enzyme [EC 2.4.2.8] (also known as hypoxanthine phosphoribosyltransferase, IMP pyrophosphorylase, transphosphoribosidase, and guanine phosphoribosyltransferase) catalyzes the purine salvage reaction of hypoxanthine with 5-phospho-α-D-ribose 1-diphosphate to produce IMP and pyrophosphate (or, diphosphate). The enzyme can also use guanine and 6-mercaptopurine as substrates.

W. N. Keley & J. B. Wyngaarden (1974) *Adv. Enzymol.* **41**, 1.

HYPSOCHROMIC SHIFT

An effect observed in the spectrum of a chemical species in which a substituent, solvent, change in environment, or other effect causes the electronic absorption spectrum to shift to shorter wavelengths. The opposite effect is referred to as a bathochromic shift. The hypsochromic shift is also known as the blue shift.

HYSTERESIS

Slow transitions produced by enzyme isomerizations. This behavior can lead to a type of cooperativity that is generally associated with ligand-induced conformational changes[1,2]. A number of enzymes are also known to undergo slow oligomerization reactions, and these enzymes may display unusual kinetic properties. If this is observed, it is advisable to determine the time course of enzyme activation or inactivation following enzyme dilution. *See Cooperativity; Bifurcation Theory; Lag Time*

[1]K. E. Neet & G. R. Ainslie, Jr., (1980) *Meth. Enzymol.* **64**, 192.
[2]K. E. Neet (1995) *Meth. Enzymol.* **249**, 519.

Selected entries from *Methods in Enzymology* [vol, page(s)]: Activation energetics, **64**, 225, 226; analysis, **64**, 196-199; artifact, **64**, 196-200; classification, **64**, 217-220 [cooperativity (monomeric, **64**, 221; oligomeric, **64**, 222); incidental to function, **64**, 217-220; physiological damping mechanism, **64**, 220, 221]; computer program, **64**, 203-205; deadend branch, **64**, 206-208; graphical analysis, **64**, 200-203; isomerization, **64**, 194; isotope trapping, **64**, 58; model, **64**, 205-213; oligomerization, **64**, 214-217; oxygen-18 exchange, **64**, 82; physiological role, **64**, 195, 196; statistical method, **64**, 203-205; structural basis, **64**, 223-225.

I

I$_{50}$ (or I$_{0.5}$)

The concentration of an inhibitor that reduces the velocity of an enzyme-catalyzed reaction or other metabolic process to 50% (or 0.5) the activity observed in the inhibitor's absence. For reversible inhibitors, the magnitude of I$_{50}$ or I$_{0.5}$ is directly related to the enzyme·inhibitor dissociation constant(s) defining competitive, noncompetitive, and uncompetitive inhibition. However, for very tight-binding inhibitors, the magnitude of I$_{0.5}$ can be substantially different than the dissociation constant. Likewise, if a metabolic inhibitor acts on more than one step (as effectors often do in metabolic pathways) or if one is dealing with cooperative inhibitor binding (as observed in allosteric control), then I$_{0.5}$ may again yield an inaccurate estimate of the microscopic inhibition constants.

ICEBERG EFFECT

A negative change in entropy associated with the decreased freedom of motion of water molecules in the vicinity of a nonpolar group. As pointed out by Jencks[1], one should not be too literal when interpreting this behavior, because the system is dynamic, and not frozen. The term flickering cluster implies some reduction in libration and rotation without complete loss.

[1]W. P. Jencks (1969) *Catalysis in Chemistry and Enzymology*, McGraw-Hill, New York. [Also available as Jencks, W. P. (1986) *Catalysis in Chemistry and Enzymology*, Dover Publications, New York].

IDURONATE-2-SULFATASE

This enzyme [EC 3.1.6.13], also known as iduronate-2-sulfate sulfatase and chondroitinsulfatase, catalyzes the hydrolysis of the 2-sulfate groups of the L-iduronate 2-sulfate units of dermatan sulfate, heparan sulfate, and heparin.

H. Kresse & J. Glössl (1987) *Adv. Enzymol.* **60**, 217.

α-L-IDURONIDASE

This enzyme [EC 3.2.1.76] catalyzes the hydrolysis of α-L-iduronosidic linkages in desulfated dermatan.

H. Kresse & J. Glössl (1987) *Adv. Enzymol.* **60**, 217.

IMBALANCE

A condition arising when reaction parameters for different bond-breaking or bond-making processes have developed synchronously as the transition state is approached. Imbalance is common in elimination and addition reactions as well as in proton transfer reactions. **See also** Synchronous; Synchronization

C. F. Bernasconi (1992) *Adv. Phys. Org. Chem.* **27**, 119.

IMIDAZOLEACETATE HYDROXYLASE

This enzyme [EC 1.14.13.5], also known as imidazoleacetate 4-monooxygenase, catalyzes the reaction of 4-imidazole acetate with NADH and dioxygen to produce 5-hydroxy-4-imidazole acetate, NAD$^+$, and water. FAD is the cofactor for this enzyme.

S. Yamamoto & Y. Ishimura (1991) *A Study of Enzymes* **2**, 315.

IMIDAZOLEGLYCEROL-PHOSPHATE DEHYDRATASE

This enzyme [EC 4.2.1.19] catalyzes the conversion of D-*erythro*-1-(imidazol-4-yl)glycerol 3-phosphate to 3-(imidazol-4-yl)-2-oxopropyl phosphate and water.

R. G. Martin, M. A. Berberich, B. N. Ames, W. W. Davis, R. F. Goldberger & J. D. Yourno (1971) *Meth. Enzymol.* **17B**, 3.

IMIDAZOLONEPROPIONASE

This enzyme [EC 3.5.2.7], also referred to as imidazolone-5-propionate hydrolase, catalyzes the hydrolysis of 4-imidazolone-5-propanoate to generate *N*-formimino-L-glutamate.

B. Magasanik, E. Kaminskas & J. Kimhi (1971) *Meth. Enzymol.* **17B**, 55.

IMINE INTERMEDIATE

An enzymatic reaction intermediate formed by nucleophilic attack by an amine on a carbonyl group of an aldehyde (forming an aldimine) or a ketone (forming a ketimine), followed by elimination of water.

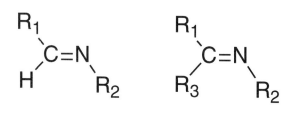

Such intermediates are known to form between substrate and enzyme in the aldolase reaction and between pyridoxal 5-phosphate and the amino group of enzyme or substrate in aminotransferase reactions. In the latter case, aldimine formation accounts for the high affinity of coenzyme binding to apotransaminases.

IMMOBILIZED ENZYME

Biological catalysts in the form of enzymes, cells, organelles, or synzymes that are tethered to a fixed bed, polymer, or other insoluble carrier or entrapped by a semi-impermeable membrane[1,2]. Immobilization often confers added stability, permits reuse of the biocatalyst, and allows the development of flow reactors. The mode of immobilization may produce distinct populations of biocatalyst, each exhibiting different activities within the same sample. The study of immobilized enzymes can also provide insights into the chemical basis of enzyme latency, a well-known phenomenon characterized by the limited availability of active enzyme as a consequence of immobilization and/or encapsulization.

The specialized kinetic properties[3-9] of matrix-bound enzymes are thought to result from at least four factors: (a) altered enzyme conformation and stability attending immobilization; (b) altered solvent and electrostatic en-

vironment for the immobilized enzyme; (c) changes in the effective substrate concentration accessible to the enzyme within the matrix; and (d) altered diffusional control on substrate binding.

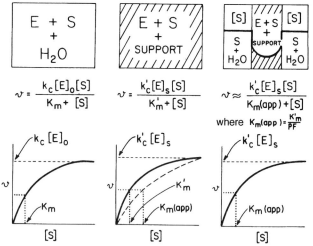

Effect of immobilization on the dynamic properties of enzymes. Examples of three different experimentally imposed conditions include: (a) enzyme and substrate are present in an aqueous solution; (b) enzyme and substrate are immobilized on or within a solid support; and (c) immobilized enzyme acts on a substrate present within an aqueous solution. The rate laws and curves that are presented below each diagram indicate the expected rate behavior. (From K. J. Laidler & P. S. Bunting (1982) *Meth. Enzymol.* **64**, 229.)

Although the kinetics of immobilized enzymes is especially relevant to industrial process chemistry, there is growing recognition that enzymes in crowded solutions may likewise experience diffusional control of the catalytic properties.

Use of immobilized catalysts requires thorough characterization of physical and chemical properties, and one cannot overemphasize the importance of describing a system adequately to ensure reproducibility in the hands of other experimenters. Buchholz and Klein[10] offered the following recommendations: (1) Provide a general description of the immobilized enzyme system, including the reaction scheme, type of carrier, exact source of the biocatalyst(s), known side reactions, and method of immobilization. (2) Describe biocatalyst(s) preparation in sufficient detail for others to replicate the preparation. Dry-weight yield of the preparation should be given as well as the amount of activity remaining in the supernatant after immobilization. The method for determining catalytic activity should always be provided as well. (3) Define physical and chemical properties, including the shape of the biocatalyst particles, their size distribution, typical swelling behavior, as well as the methods used for making these determinations. In column configurations, dimensions should be provided along with compression behavior and flow rates when pertinent. In stirred systems, effectiveness of suspension as well as any abrasion should also be assessed. It is worth noting that abrasion can also occur with fluidized beds. (4) Determine kinetic parameters that describe the system: (a) Compare the dependence of initial rate on substrate concentration for both immobilized and free biocatalyst(s). The time elapsed between the preparation of the immobilized system and the kinetic experiments should be noted as well as pertinent storage conditions. (b) Assess effects of pH as well as buffer type and concentration on the kinetics and stability of the system. (c) If temperature effects are studied, then those conditions should be thoroughly described. (d) When using intact cells as biocatalyst, it is advisable to describe effects of growth conditions on the immobilized system. (e) Determine the diffusion limitations of the system (including the effects of particle size or enzyme load). (f) Present information concerning the degree of conversion for different values of residence time. (g) Describe details concerning the storage stability of the immobilized system should be as actual data values rather than as extrapolated parameters. (h) Study the operational stability (*i.e.*, residual initial rates after different periods of operation).

Immobilized enzymes often behave differently than their soluble counterparts with respect to their susceptibility to inhibition[11].

[1]K. Mosbach, ed. (1977) *Meth. Enzymol.* vol. **44**, (1987) vols. **135** and **136**, and (1988) vol. **137**.
[2]M.-D. Legoy & D. Thomas, eds. (1995) *Ann. New York Acad. Sci.*, vol. **750**.
[3]K. J. Laidler & P. S. Bunting (1973) *The Chemical Kinetics of Enzyme Action*, Oxford Univ. Press, London.
[4]L. Goldstein (1976) *Meth. Enzymol.* **44**, 397.
[5]J. M. Engasser & C. Horvath (1976) in *Applied Biochemistry and Bioengineering* (L. B. Wingard, Jr., E. Katchalski-Katzir & L. Goldstein, eds.), Vol. **1**, p. 127, Academic Press, New York.
[6]K. J. Laidler & P. S. Bunting (1980) *Meth. Enzymol.* **64**, 227.
[7]D. Gabel & V. Kasche (1976) *Meth. Enzymol.* **44**, 526.
[8]S. K. Duggal & K. Buchholz (1982) *Eur. J. Appl. Microbiol. Biotech.* **16**, 81.
[9]D. S. Clark & J. E. Bailey (1984) *Biotechnol. Bioeng.* **16**, 231.
[10]K. Buchholz & J. Klein (1987) *Meth. Enzymol.* **135**, 3.
[11]V. V. Mozhaev & K. Martinek (1982) *Enzyme Microb. Technol.* **4**, 299.

Selected entries from *Methods in Enzymology* [**vol**, page(s)]:
Functional groups on enzymes suitable for binding to matrices, **44**, 11; a survey of enzyme coupling techniques, **135**, 30; immobilization of enzymes to agar, agarose, and Sephadex supports, **44**, 19; enzymes immobilized to cellulose, **44**, 46; immobilization of enzymes to various acrylic copolymers, **44**, 53; immobilization of enzymes on hydroxylalkyl methacrylate gels, **44**, 66; enzymes covalently bound to polyacrylic and polymethacrylic copolymers, **44**, 84; immobilization of enzymes on neutral and ionic carriers, **44**, 107; immobilization of enzymes on nylon, **44**, 118; covalent coupling methods for inorganic support materials, **44**, 134; tresyl chloride-activated supports for enzyme immobilization, **135**, 65; immobilization of proteins and ligands using chlorocarbonates, **135**, 84; 1,1'-carbonyldiimidazole-mediated immobilization of enzymes and affinity ligands, **135**, 102; covalent immobilization of proteins by techniques that permit subsequent release, **135**, 130; immobilization of glycoenzymes through carbohydrate chains, **135**, 141; immobilization of enzymes and microorganisms by radiation polymerization, **135**, 146; use of monoclonal antibodies for preparation of highly active immobilized enzymes, **135**, 160.

IMP CYCLOHYDROLASE

This enzyme [EC 3.5.4.10] catalyzes the hydrolysis of the purine ring of IMP to produce 5-formamido-1-(5-phosphoribosyl)imidazole-4-carboxamide.

J. G. Flaks & L. N. Lukens (1963) *Meth. Enzymol.* **6**, 52.

IMP DEHYDROGENASE

This oxidoreductase [EC 1.1.1.205], also known as inosine-5'-monophosphate dehydrogenase, catalyzes the reaction of IMP with NAD^+ and water to produce xanthosine 5'-phosphate and NADH.

B. F. Cooper & F. B. Rudolph (1995) *Meth. Enzymol.* **249**, 188.

INDEFINITE POLYMERIZATION

Any polymerization reaction in which the product of each elongation step can itself also undergo further polymerization. When the same types of bonds and/or conformational states that are present in the reactant(s) are generated within product(s) during elongation, the process is referred to as isodesmic polymerization. Such is the case for the indefinite polymerization of actin, tubulin, hemoglobin S, and tobacco mosaic virus coat protein. *See* Nucleation; Protein Polymerization; Actin Assembly Kinetics; Microtubule Assembly Kinetics; Microtubule Assembly Kinetics

INDOLEAMINE 2,3-DIOXYGENASE

This enzyme [EC 1.13.11.42], also referred to as indoleamine-pyrrole 2,3-dioxygenase, catalyzes the reaction of tryptophan with dioxygen to form *N*-formylkynurenine. Heme participates as a cofactor. Many substituted and unsubstituted indoleamines, including melatonin, act as substrates. In fact, among other roles, this enzyme participates in the degradation of melatonin.

H. B. Dunford (1998) *Comprehensive Biological Catalysis: A Mechanistic Reference* **3**, 195.

S. Kaufman (1993) *Ann. Rev. Nutr.* **13**, 261.

S. Yamamoto & Y. Ishimura (1991) *A Study of Enzymes* **2**, 315.

R. Yoshida & O. Hayaishi (1987) *Meth. Enzymol.* **142**, 188.

O. Hayaishi, M. Nozaki & M. T. Abbott (1975) *The Enzymes*, 3rd ed., **12**, 119.

INDOLEGLYCEROL PHOSPHATE SYNTHASE

This enzyme [EC 4.1.1.48] catalyzes the conversion of 1-(2-carboxyphenylamino)-1-deoxy-D-ribulose 5-phosphate to generate 1-(indol-3-yl)glycerol 3-phosphate, carbon dioxide, and water. In some organisms, this enzyme is part of a multifunctional protein together with one or more components of the system for the biosynthesis of tryptophan (*i.e.*, anthranilate synthase, anthranilate phosphoribosyltransferase, tryptophan synthase, and phosphoribosylanthranilate isomerase).

R. Bentley (1990) *Crit. Rev. Biochem. Mol. Biol.* **25**, 307.

K. Kirschner, H. Szadkowski, T. S. Jardetzky & V. Hager (1987) *Meth. Enzymol.* **142**, 386.

INDUCED FIT MODEL

A conceptual framework[1-3] for interpreting the action of substrate binding in reorienting otherwise inactive groups on the enzyme into their catalytically active configuration. This implies an essential isomerization step:

$$E + S = ES = (ES)' \rightarrow E + P$$

Yeast hexokinase provides good examples of induced fit in two ways: first, the enzyme undergoes a dramatic conformational change upon binding glucose; and second, the binding of certain pentoses reorganizes the catalytic site so that water can replace glucose's 6-hydroxymethyl group with a resulting 50,000 times increase in ATPase activity.

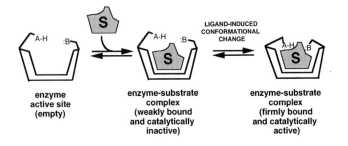

enzyme active site (empty)   enzyme-substrate complex (weakly bound and catalytically inactive)   LIGAND-INDUCED CONFORMATIONAL CHANGE   enzyme-substrate complex (firmly bound and catalytically active)

Jencks[4] has offered the view that the induced fit model provides a means for understanding the specificity and control of catalysis in terms of a substrate-induced conformational change, while not involving binding forces in reducing the reaction's activation energy. He indicates that induced fit has two consequences: (a) reduction in the observed binding constant relative to that which would have occurred were there perfect initial fit, and (b) the substrate that best utilizes specific binding forces to effect a conformational change should have the greatest maximal velocity. *See* Cooperativity; Koshland-Nemethy-Filmer Model

[1]D. E. Koshland, Jr. (1958) *Proc. Natl. Acad. Sci. U.S.A.* **44**, 98.

[2]D. E. Koshland, Jr. (1960) *Adv. Enzymol.* **22**, 45.

[3]D. E. Koshland, Jr. & K. E. Neet (1968) *Annu. Rev. Biochem.* **37**, 359.

[4]W. P. Jencks (1969) *Catalysis in Chemistry and Enzymology*, pp. 289-291, McGraw-Hill, New York. [Also available as Jencks, W. P. (1986) *Catalysis in Chemistry and Enzymology*, Dover Publications, New York].

INDUCED SUBSTRATE INHIBITION

A term first introduced by Cleland[1,2] to indicate that for ordered substrate binding mechanisms, addition of an inhibitor mimicking the first substrate may still permit binding of the second substrate. Hence, as long as the addition of the first substrate is not of the rapid equilibrium type, the presence of the inhibitor will "induce" substrate inhibition by the second substrate. An example of induced substrate inhibition is provided in the thymidylate synthase reaction[3] where the second substrate methylene tetrahydrofolate becomes an inhibitor, but only in the presence of the inhibitor bromodeoxyuridine 5'-monophosphate.

[1]W. W. Cleland (1990) *The Enzymes*, 3rd ed., **19**, 119.

[2]W. W. Cleland (1979) *Meth. Enzymol.* **63**, 500.

[3]P. V. Danenberg & K. D. Danenberg (1978) *Biochemistry* **17**, 4018.

INDUCED TRANSPORT ANALYSIS

An experimental protocol designed to detect and characterize isoenzyme mechanisms (that is, enzyme mechanisms requiring one or more isomerization steps of the free enzyme before the catalytic cycle is completed)[1-5]. In iso mechanisms, the substrate and product bind to two different conformational forms of the enzyme. The induced transport approach requires preincubation of both labeled substrate and labeled product with enzyme, allowing the system to reach equilibrium; a large amount of unlabeled substrate is then added. If an iso step is present, the labeled product will not compete with the unlabeled substrate, and the enzyme will catalyze the reverse synthesis of labeled substrate. Hence, there will be a transient decrease in labeled product concentration.

[1]H. G. Britton (1966) *Arch. Biochem. Biophys.* **117**, 167.
[2]H. G. Britton (1973) *Biochem. J.* **133**, 255.
[3]H. G. Britton (1965) *Nature* **205**, 1323.
[4]H. G. Britton & J. B. Clarke (1968) *Biochem. J.* **110**, 161.
[5]K. L. Rebholz & D. B. Northrop (1995) *Meth. Enzymol.* **249**, 211.

INDUCTION

1. Cellular: A process by which the amount of a protein is increased through the activation of gene expression. 2. Electromagnetic: A change in the intensity or direction of a magnetic field resulting in an electromotive force in any conductor within the field: $V = -d\Phi/dt$ where V is the induced potential difference, Φ is the magnetic flux, and t is time. 3. Electrostatic: The process by which an electron charge is produced on a conductor due to the influence of an electric field.

INDUCTION PERIOD

Any slow phase (of a chemical reaction or physical process) that is followed by a faster phase. Such behavior is observed for systems prior to attaining the steady-state condition. **See also** *Series Reactions*

INDUCTIVE EFFECT

The polarization of a chemical bond caused by the polarization of an adjacent or nearby bond. The inductive effect depends only on the nature of the bonds and is transmitted through a chain of atoms by electrostatic induction. **See also** *Field Effect*

INDUCTOMERIC EFFECT

A molecular polarizability effect occurring by the inductive mechanism of electron displacement[1-3]. Substituent polarizability is a factor governing reactivity.

[1]IUPAC (1994) *Pure and Appl. Chem.* **66**, 1077.
[2]C. K. Ingold (1953) *Structure and Mechanism in Organic Chemistry*, Cornell Univ. Press, Ithaca, New York.
[3]R. W. Taft & R. D. Topson (1987) *Progr. Phys. Org. Chem.* **61**, 1.

INFLECTION, POINT OF

A point on a curve at which the concavity of the curve changes (*e.g.*, up to down or *vice versa*). The second derivative of the equation of the curve will be zero at this point. A number of curves found in enzyme kinetic studies have inflection points: *e.g.*, cooperative systems, pH-rate studies, *etc.*

INHIBITION

The process by which the presence of a substance, known as the inhibitor, decreases reaction rate. **See** *Degree of Inhibition; Specific Type of Inhibition*

INHIBITOR

A substance, agent, or factor that lowers the rate of a reaction. An inhibitor that acts directly by binding to an enzyme is often referred to as an enzyme inhibitor. **See also** *Degree of Inhibition; Effector; Modifier; Inactivator; Competitive Inhibitor; Noncompetitive Inhibitor; Uncompetitive Inhibitor; Partial Inhibitor; Irreversible Inhibitor*

INHIBITORY INDEX

A value used to assess the effectiveness of an inhibitor acting on an enzyme or enzyme system derived from two different sources. It is a ratio equal to the concentration of the inhibitor X causing 50% inhibition of the host enzyme divided by the concentration of the inhibitor X causing 50% inhibition of the enzyme during some pathophysiologic condition or disease.

C. Silipo & C. Hansch (1976) *J. Med. Chem.* **19**, 62.

INITIAL RATE CONDITION

One of the basic assumptions in kinetic studies of an enzyme-catalyzed reaction is that true initial rates are being measured. In such cases, a plot of the product concentration *versus* time must yield a straight line. (This behavior is only observed when the substrate is at or near its initial (or, $t = 0$) concentration. As time increases, product accumulation and substrate depletion will result in a curvature of this progress curve; hence, the reaction velocity at these later times would be correspondingly lower.)

An oft-suggested check on the initial rate condition is that, over the duration of the assay, the initial substrate concentration (*i.e.*, [S]) should not decrease by more than 5% (and ideally, significantly less). However, the validity of this suggestion depends on a number of other factors. If the reaction equilibrium constant K_{eq} favors substrate(s), rather than product(s), then 10% conversion may amount to a significant fraction of that attainable at equilibrium, and the system would clearly not satisfy initial rate conditions. Initial rate conditions are apt to apply when the substrate concentration is less than 5–10% the $\Delta[S]$ value, where $\Delta[S]$ is $[S]_{initial} - [S]_{equilibrium}$. If the reaction rate is not linear over this change in substrate, the experimenter is well advised to adopt a more sensitive rate assay.

The initial-rate phase may quickly end if one or more products accumulate to concentrations approaching the respective inhibition constants. In these cases, one may seek to minimize this problem by including an auxiliary enzyme (a) to remove product(s) or (b) to regenerate the substrate concentration. Ideally, the auxiliary enzyme should have a lower K_m value for the product of the primary reaction than the corresponding K_i value for the primary enzyme. *See* *Chemical Kinetics; ATP/GTP regeneration*

R. D. Allison & D. L. Purich (1979) *Meth. Enzymol.* **63**, 3.

INITIAL RATE ENZYME DATA, REPORTING OF

The initial rate assumption is one of the most powerful and widely used assumptions in the kinetic characterization of enzyme action. The proper choice of reaction conditions that satisfy the initial rate assumption is itself a challenge, but once conditions are established for initial rate measurements, the kinetic treatment of an enzyme's rate behavior becomes much more tractable[1]. In reporting initial rate data, investigators would be well advised to provide the following information:

1. Provide a reference or direct experimental proof that the conditions chosen do provide initial rate measurements. At a minimum, the percentage of substrate consumed during the course of an initial rate determination should be specified. One should also show that under initial rate conditions a doubling of enzyme concentration should exactly produce a doubling in the observed initial rate. Likewise, if an auxiliary enzyme assay is used to monitor the primary enzyme's activity the observed rate at low or high substrate concentration should not depend on the concentration of additional enzyme(s), substrate(s), or factors used for the coupled assay. 2. Describe all assay conditions (*e.g.*, concentrations of substrates, products, inhibitors, and/or activators; enzyme concentration; temperature; pH and buffer composition;

other solutes; as well as the exact reaction volume and duration of assay, *etc.*). The specific activity and source of each enzyme should be explicitly indicated. Occasionally, the dilution of an enzyme stock solution to initiate the assay will result in a slow conformational change that effects catalytic activity. If this is the case, a complete description of assay conditions will also require an indication of the enzyme concentation in the stock solution and in the final assay itself. 3. All velocity data should be expressed as the molarity change per unit time, and not as micromoles per unit time. This is a requirement for all chemical kinetic studies, because the molarity change per unit time is an intensive variable. 4. Indicate the number of replicate kinetic runs as well as measures of the statistical variation in the rate data and in any derived kinetic parameters (*e.g.*, K_m, V_{max}, V_{max}/K_m, K_i, *etc.*). 5. Specify any unusual circumstances needed to obtain initial rate conditions. For example, the presence of saturating levels of an activator may be required to generate a linear double-reciprocal plot for a cooperative enzyme. Also indicate if any stabilizer is needed to reduce loss of enzyme activity and/or proteolysis. 6. Include adequate descriptions of any graphical treatment, computer modeling technique (*e.g.*, KINSIM, FITSIM, least squares analysis, ENZEQ, *etc.*), global analysis method, or any other data fitting technique.

[1]R. D. Allison & D. L. Purich (1979) *Meth. Enzymol.* **63**, 3.

INITIAL RATE PHASE

The period in a reaction curve that (a) is characterized by the full effect of initial substrate concentration and (b) for which any change in product concentration is minimal and has no measurable effect on the rate. There should be a linear relationship between product formation and time during the initial rate phase, and a doubling of the reaction time will result in a doubling of product formed. The relationship between reaction velocity and enzyme concentration (in the absence of self-association of the enzyme) should also be linear. Initial rates typically fail to be obtained when $[E_{total}]$ greater than or equal to $[substrate]_{initial}/100$. As a general rule, the substrate concentration will not have changed more than 5–10% from its initial value over the initial rate phase of the reaction. This oft-stated rule-of-thumb applies to reactions that are thermodynamically favorable, and investigators are well advised to work well below that 5–10% substrate depletion.

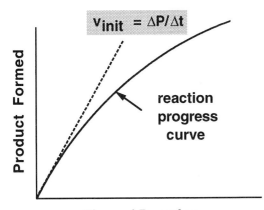

Figure 1. Plot of the change in product concentration as a function of time of reaction. The initial rate phase corresponds to the early linear region, and a tangent to this early region has a slope corresponding to the initial reaction velocity. (For a detailed description of how one obtains rate constants using the initial rate assumption, *See Chemical Kinetics*.)

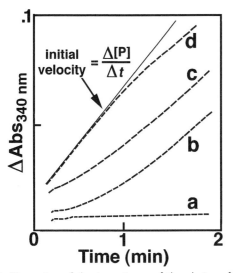

Figure 2. Illustration of the importance of the choice of reaction conditions on the determination of initial velocity. Shown are four conditions applied to examine the rate behavior of *Escherichia coli* NAD⁺-dependent coenzyme A-linked aldehyde dehydrogenase (Reaction: NAD⁺ + CoA-SH + Acetaldehyde = NADH + Acetyl-S-CoA + H⁺). All assay mixtures contained enzyme, 0.5 mM NAD⁺, 8 μM CoA-SH, 16 mM acetaldehyde, and 22.5 mM Tris buffer at pH 8.1. (a) Time-course observed when enzyme was added to the standard assay; (b) time-course observed when enzyme was added to standard assay augmented with 10 mM 2-mercaptoethanol; (c) time-course observed when enzyme was first preincubated for 15 min with 8 μM CoA-SH, 16 mM acetaldehyde, 10 mM 2-mercaptoethanol, and 22.5 mM Tris buffer at pH 8.1, and the reaction was initiated by addition of NAD⁺; (d) time-course observed when enzyme was preincubated with 10 mM 2-mercaptoethanol for 15 min and then added to standard assay augmented with 10 mM 2-mercaptoethanol. The data are most compatible with the idea that the enzyme has an active-site thiol group that must be reduced to express full catalytic activity during assay.

The development of an initial rate enzyme assay can also aid the discovery of special properties of the enzyme under investigation. Shown in Fig. 2, for example, are four conditions applied to examine the rate behavior of *Escherichia coli* NAD⁺-dependent Coenzyme A-linked aldehyde dehydrogenase (Reaction: NAD⁺ + CoA-SH + Acetaldehyde = NADH + Acetyl-S-CoA + H⁺). All assay mixtures contained enzyme, 0.4 mM NAD⁺, 8 μM CoA-SH, 16 mM acetaldehyde, and 22.5 mM Tris buffer (pH 8.1). The experiments demonstrated that the enzyme most probably contains an essential active-site thiol group.

When one or more product(s) exhibits a very small K_i value, the investigator must develop a more sensitive assay, such that true initial rates are actually obtained[1].

Occasionally, one can maintain initial rate conditions by using a coupled reaction system to regenerate one of the limiting substrates. For example, to regenerate ATP in a phosphotransferase reaction, one can use: creatine phosphate and creatine kinase; acetylphosphate and acetate kinase; or phosphoenolpyruvate and pyruvate kinase.

[1] R. D. Allison & D. L. Purich (1979) *Meth. Enzymol.* **63**, 3.

INNATE THERMODYNAMIC QUANTITIES

Structural and molecular biologists often study the temperature dependence of the equilibrium position of a reaction or process. The Gibbs free energy undoubtedly provides the correct thermodynamic criterion of equilibrium. An understanding of this parameter can be achieved from either a macroscopic level (classical thermodynamics) or a molecular level (statistical thermodynamics). Ultimately, one seeks to understand the factors influencing $\Delta G°$ for a specific reaction.

Innate Thermodynamic Quantities*. Certain components of the total change in $\Delta G°$ are *innate*, because such parameters have nonzero values, even when extrapolated to 0 K. Other components change with temperature (*e.g.*, at $T = 0$ K, $TS = 0$). Because $A = U - TS$ and $G = H - TS$[1-3], then $(H_o° = G_o°) \cong (A_o° = U_o°)$ at absolute zero. Except for entropy, the residual values of these quantities are the same at absolute zero, and they describe the innate thermodynamic behavior of the system.

Thermodynamic state functions change with temperature[1-3]; this will be true if values of the heat capacity of any component is nonzero (which is almost always true). Whenever the heat capacity is not a constant, the various thermodynamic state functions will show nonlinear dependencies on temperature.

Reaction enthalpies are frequently measured at or near room temperature, such that

$$\Delta H_{298}^{\circ} = \Delta H^{\circ}(T_{o}) + \int_{0}^{298} \Delta C_{p}^{\circ} \, dT$$

where this last term represents the thermal agitation energy (heat capacity integrals): a variation of the Kirchhoff equation. With reactions involving small molecules only, one can derive $\Delta H_{\text{reaction}}^{\circ}$ in terms of inherent bond energy ($\Delta H(T_{o})$ or ΔH_{0}° in some texts) by ignoring differences between this parameter and ΔH_{298}°. Cottrell[4] calculated that the effect on chemical bond energy is only $1–2$ kcal mol^{-1}, compared to the $80–100$ kcal mol^{-1} total. This is *not* true for systems involving biological macromolecules, and Chun[5-9] has shown that the heat capacity integrals may be $40–50\%$ of the magnitude of experimentally measured values of ΔH_{298}°.

Planck-Benzinger Thermal Work Function. In 1971, Benzinger[10] proposed a thermal work function to account for both Boltzmann statistical energy effects and quantum-mechanical bond energies. While the latter are not usually significantly altered in reactions involving small molecules, Benzinger's conjecture was that the large-scale and long-range changes of conformation upon protein folding and/or assembly might generate significant energy differences arising from cumulative alteration of their numerous covalent bond structures.

$$\Delta G - \Delta H(T_{o}) = -\left\{ T \int_{0}^{T} (\Delta C_{p}/T) dT - \int_{0}^{T} \Delta C_{p} \, dT \right\}$$
$$= \Delta W$$

ΔW represents the thermal components of any intra-/intermolecular bonding term in a system (*i.e.*, energy other than the inherent difference of the 0 K portion of the interaction energy). Thus, ΔW completely expresses the thermal energy difference of the process involved. Application of the thermal work function permits one to separate so-called 0 K energy differences from those

associated with the heat capacity integrals. This offers a fuller understanding of reaction energies[5-9]. From this expression, it is possible to determine the temperature-invariant enthalpy at equilibrium conditions,

$$\Delta H^{\circ}(T_{o}) = \Delta W^{\circ}(T_{s})_{\max} + \Delta G^{\circ}(T_{s})_{\min}$$

At $\langle T \rangle_{s}$, the temperature-invariant enthalpy is a primary source of the chemical bond energy essential for any reaction to proceed in an interacting system. The magnitude of $\Delta H^{\circ}(T_{o})$ is an indication of the type of macromolecular interaction taking place. This relationship has been designated as the Planck-Benzinger thermal work function. An analogous treatment developed by Giauque[11] for chemical systems has not been fruitfully applied to biological systems.

Application to Macromolecular Interactions. Chun[9] describes how one can analyze the thermodynamics of a particular biological system as well as the thermal transition taking place. Briefly, it is necessary to extrapolate thermodynamic parameters over a broad temperature range. Enthalpy, entropy, and heat capacity terms are evaluated as partial derivatives of the Gibbs free energy function defined by Helmholtz-Kelvin's expression, assuming that the heat capacities integral is a continuous function.

$$\partial \Delta G(T)/\partial T = -\Delta S(T)$$
$$\{\partial \Delta G(T)/T\}/\partial T = -\Delta H(T)/T^{2}$$
$$\partial \Delta H(T)/\partial T = \Delta C_{p}(T)$$
$$\partial \Delta S(T)/\partial T = C_{p}(T)/T$$

In this method, values of various thermodynamic properties are mathematically derived from the available data, based on thermodynamic principles. The method requires very accurate experimental data for the Gibbs free energy change showing the direction of the chemical change toward the minimum Gibbs function. Equilibrium sedimentation, calorimetric measurements, or spectroscopic data are useful in determinations of the temperature-invariant enthalpy. As to the question of whether the heat capacity integral is a continuous function, no phase change occurs over the accessible temperature range.

For self-associating protein systems, third-order polynomial functions provided a good fit over the accessible range[5–9]. The data on $\Delta G°$ must show the direction of the chemical change, toward the minimum in the Gibbs function. If this proves true, the equation can be applied in the standard or nonstandard state. For protein unfolding or DNA unwinding, nonlinear models are needed[6–9]. Consistent with Occam's razor, the simplest description is used to describe the system, and complexity is increased only if warranted by the experimental results.

The range of temperature over which any given biochemical system can be studied is quite limited: above some point, the system is destroyed; however, at too low a temperature, biological activity ceases. In favorable cases, data can be obtained from 283 to 333 K. Extrapolation to 0 K is a difficult, though manageable, problem.

This formalism may also be valuable in addressing protein unfolding, DNA unwinding, cold denaturation, and effects of site-direct mutagenesis on lysozyme unfolding[9].

[1]G. N. Lewis & M. Randall (1961) *Thermodynamics*, 2nd. ed. (revised by K. S. Pitzer & L. Brewer), McGraw-Hill, New York.
[2]J. W. Gibbs (1878) *Amer. J. Sci.* **16**, 441.
[3]M. Planck (1927) *Treatise on Thermodynamics*, 7th ed. (translated by A. Ogg, 3rd ed.), pp. 179-214 and 272-289, Longmans, Green and Co., London.
[4]T. L. Cottrell (1950) *The Strengths of Chemical Bonds*, pp. 21-70, Academic Press, London.
[5]P. W. Chun (1988) *Internat. J. Quantum Chem.* **15**, 247.
[6]P. W. Chun (1994) *J. Phys. Chem.* **86**, 6851.
[7]P. W. Chun (1995) *J. Biol. Chem.* **270**, 13925.
[8]P. W. Chun (1996) *J. Phys. Chem.* **100**, 7283.
[9]P. W. Chun (1997) *Meth. Enzymol.* **295**, 227.
[10]T. H. Benzinger (1971) *Nature* **229**, 100.
[11]W. F. Giauque (1930) *J. Am. Chem. Soc.* **52**, 4808 and 4816.

*The authors thank their colleague, Professor Paul W. Chun, who is the leading proponent of the Planck-Benzinger thermal work function (P. W. Chun (1997) *J. Phys. Chem.* **101**, 7885; (1998) *Meth. Enzymol.* **295**, 227). Those interested in additional information may wish to contact Professor Chun by at: pwchun@biochem.med.ufl.edu.

INNER COORDINATION SPHERE

The region of a metal ion complex where ligands make direct binding interactions with the central metal ion. When the ligands do not completely neutralize the positive ionic charge of the central ion, other ions or electron-rich substances will become loosely associated with the complex through so-called outer coordination sphere interactions.

INNER FILTER EFFECT

1. An apparent decrease in the emission quantum yield and/or distortion of the band shape due to the reabsorption of emitted radiation. If such an effect is not corrected or compensated for, results of an emission experiment may prove to be incorrect. This is especially true in fluorescence quenching experiments conducted to evaluate the stoichiometry and affinity of ligand binding. 2. In a light irradiation experiment, the absorption of incident radiation by a species or molecular entity other than the intended primary absorber. **See** *Fluorescence*

Comm. on Photochem. (1988) *Pure and Appl. Chem.* **60**, 1055.

INNER-SPHERE ELECTRON TRANSFER

1. Electron transfer between two metal centers sharing a ligand or atom in their respective coordination shells. **See** *Metal Ion Coordination (Basic Chemical Behavior)* 2. Electron transfer between two centers in which the interaction between the donor and acceptor centers in the transition state is significant (*i.e.*, greater than 20 kJ/mol). **See also** *Outer Sphere Electron Transfer*

INOSAMINE-PHOSPHATE AMIDINOTRANSFERASE

This enzyme [EC 2.1.4.2] catalyzes the reaction of arginine with 1-amino-1-deoxy-*scyllo*-inositol 4-phosphate to produce ornithine and 1-guanidino-1-deoxy-*scyllo*-inositol 4-phosphate. Other substrates include 1D-1-guanidino-3-amino-1,3-dideoxy-*scyllo*-inositol 6-phosphate, streptamine phosphate and 2-deoxystreptamine phosphate. Canavanine can substitute for arginine.

J. B. Walker (1973) *The Enzymes*, 3rd ed., **9**, 497.

INOSINE KINASE

This enzyme [EC 2.7.1.73] catalyzes the ATP-dependent phosphorylation of inosine to generate IMP and ADP.

E. P. Anderson (1973) *The Enzymes*, 3rd ed., **9**, 49.

INOSINE NUCLEOSIDASE

This enzyme [EC 3.2.2.2], also called inosinase, catalyzes the hydrolysis of inosine to generate hypoxanthine and D-ribose.

G. Davies, M. L. Sinnott & S. G. Withers (1998) *Comprehensive Biological Catalysis: A Mechanistic Reference* **1**, 119.

myo-INOSITOL 2-DEHYDROGENASE

This enzyme [EC 1.1.1.18] catalyzes the reaction of *myo*-inositol with NAD$^+$ to produce 2,4,6/3,5-pentahydroxy-cyclohexanone and NADH.

J. Larner (1962) *Meth. Enzymol.* **5**, 326.

myo-INOSITOL 1-KINASE

This enzyme [EC 2.7.1.64] catalyzes the reaction of ATP with *myo*-inositol to produce ADP and L-*myo*-inositol 1-phosphate.

F. A. Loewus & M. W. Loewus (1983) *Ann. Rev. Plant Physiol.* **34**, 137.

myo-INOSITOL 1-MONOPHOSPHATASE

This enzyme [EC 3.1.3.25], also referred to as *myo*-inositol-1(or 4)-monophosphatase and *myo*-inositol-1-phosphatase, catalyzes the hydrolysis of *myo*-inositol 1-monophosphate to generate *myo*-inositol and orthophosphate. Both enantiomers of *myo*-inositol 1-phosphate and *myo*-inositol 4-phosphate can act as substrates. However, the enzyme does not hydrolyze inositol bisphosphates, trisphosphates, or tetrakisphosphates.

P. W. Majerus (1992) *Ann. Rev. Biochem.* **61**, 225.
V. S. Bansal & P. W. Majerus (1990) *Ann. Rev. Cell Biol.* **6**, 41.

myo-INOSITOL OXYGENASE

This enzyme [EC 1.13.99.1] catalyzes the reaction of *myo*-inositol with dioxygen to produce D-glucuronate and water. The enzyme requires iron for activity.

F. A. Loewus & M. W. Loewus (1983) *Ann. Rev. Plant Physiol.* **34**, 137.

myo-INOSITOL-1-PHOSPHATE SYNTHASE

This enzyme [EC 5.5.1.4] catalyzes the conversion of D-glucose 6-phosphate to 1L-*myo*-inositol 1-phosphate. NAD$^+$ is required as a cofactor.

D. J. Creighton & N. S. R. K. Murthy (1990) *The Enzymes*, 3rd ed., **19**, 323.

INOSITOL POLYPHOSPHATE 1-PHOSPHATASE

Also known as inositol-1,4-bisphosphate 1-phosphatase [EC 3.1.3.57], this enzyme catalyzes the hydrolysis of D-*myo*-inositol 1,4-bisphosphate to generate D-*myo*-inositol 4-phosphate and orthophosphate. This enzyme can also act on inositol 1,3,4-trisphosphate (forming inositol 3,4-bisphosphate). Inositol-1,4,5-trisphosphate 1-phos-

phatase [EC 3.1.3.61] catalyzes the hydrolysis of D-*myo*-inositol 1,4,5-trisphosphate to produce D-*myo*-inositol 4,5-bisphosphate and orthophosphate.

P. W. Majerus (1992) *Ann. Rev. Biochem.* **61**, 225.
V. S. Bansal & P. W. Majerus (1990) *Ann. Rev. Cell Biol.* **6**, 41.

INOSITOL POLYPHOSPHATE 3-PHOSPHATASE

Inositol 1,3,4,5-tetrakisphosphate 3-phosphatase [EC 3.1.3.62] catalyzes the hydrolysis of D-*myo*-inositol 1,3,4,5-tetrakisphosphate to generate D-*myo*-inositol 1,4,5-trisphosphate and orthophosphate. Inositol 1,3-bisphosphate 3-phosphatase [EC 3.1.3.65] catalyzes the hydrolysis of D-*myo*-inositol 1,3-bisphosphate to produce D-*myo*-inositol 1-monophosphate and orthophosphate. This enzyme is magnesium-ion-independent and is different from inositol-1,4-bisphosphate 1-phosphatase [EC 3.1.3.57], inositol-1,4,5-trisphosphate 1-phosphatase [EC 3.1.3.61], and inositol 3,4-bisphosphate 4-phosphatase [EC 3.1.3.66].

P. W. Majerus (1992) *Ann. Rev. Biochem.* **61**, 225.
V. S. Bansal & P. W. Majerus (1990) *Ann. Rev. Cell Biol.* **6**, 41.

INOSITOL POLYPHOSPHATE 4-PHOSPHATASE

Inositol-3,4-bisphosphate 4-phosphatase [EC 3.1.3.66] catalyzes the hydrolysis of D-*myo*-inositol 3,4-bisphosphate to produce D-*myo*-inositol 3-monophosphate and orthophosphate. This enzyme is magnesium-ion-independent and is different from inositol-1,4-bisphosphate 1-phosphatase [EC 3.1.3.57], inositol-1,4,5-trisphosphate 1-phosphatase [EC 3.1.3.61], and inositol-1,3-bisphosphate 3-phosphatase [EC 3.1.3.65].

P. W. Majerus (1992) *Ann. Rev. Biochem.* **61**, 225.
V. S. Bansal & P. W. Majerus (1990) *Ann. Rev. Cell Biol.* **6**, 41.

INOSITOL POLYPHOSPHATE 5-PHOSPHATASE

Inositol-1,4,5-trisphosphate 5-phosphatase [EC 3.1.3.56], also known as inositol trisphosphate phosphomonoesterase and inositol polyphosphate 5-phosphatase, catalyzes the hydrolysis of D-*myo*-inositol 1,4,5-trisphosphate to produce D-*myo*-inositol 1,4-bisphosphate and orthophosphate. The type I enzyme (but not the type II enzyme) will also hydrolyze inositol 1,3,4,5-tetrakisphosphate at the 5-position. However, neither of the two types of the enzyme will utilize inositol 1,4-bisphosphate or inositol 1,3,4-trisphosphate as substrates.

P. W. Majerus (1992) *Ann. Rev. Biochem.* **61**, 225.
V. S. Bansal & P. W. Majerus (1990) *Ann. Rev. Cell Biol.* **6**, 41.

INOSITOL-1,3,4,6-TETRAPHOSPHATE 5-KINASE

This enzyme [EC 2.7.1.140], also referred to as 1D-*myo*-inositol-tetrakisphosphate 5-kinase, catalyzes the reaction of ATP with 1D-*myo*-inositol 1,3,4,6-tetrakisphosphate to produce ADP and 1D-*myo*-inositol 1,3,4,5,6-pentakisphosphate.

V. S. Bansal & P. W. Majerus (1990) *Ann. Rev. Cell Biol.* **6**, 41.

INOSITOL-3,4,5,6-TETRAPHOSPHATE 1-KINASE

This enzyme [EC 2.7.1.134], also referred to as 1D-*myo*-inositol-tetrakisphosphate 1-kinase, catalyzes the reaction of ATP with 1D-*myo*-inositol 3,4,5,6-tetrakisphosphate to produce ADP and 1D-*myo*-inositol 1,3,4,5,6-pentakisphosphate.

V. S. Bansal & P. W. Majerus (1990) *Ann. Rev. Cell Biol.* **6**, 41.

INOSITOL-1,3,4-TRISPHOSPHATE 6-KINASE

This enzyme [EC 2.7.1.133], also referred to as 1D-*myo*-inositol-trisphosphate 6-kinase, catalyzes the reaction of ATP with 1D-*myo*-inositol 1,3,4-trisphosphate to produce ADP and 1D-*myo*-inositol 1,3,4,6-tetrakisphosphate.

V. S. Bansal & P. W. Majerus (1990) *Ann. Rev. Cell Biol.* **6**, 41.

INOSITOL-1,4,5-TRISPHOSPHATE 3-KINASE

This enzyme [EC 2.7.1.127], also referred to as 1D-*myo*-inositol-trisphosphate 3-kinase, catalyzes the reaction of ATP with 1D-*myo*-inositol 1,4,5-trisphosphate to produce ADP and 1D-*myo*-inositol 1,3,4,5-tetrakisphosphate. The enzyme requires calcium ions as a cofactor.

P. W. Majerus (1992) *Ann. Rev. Biochem.* **61**, 225.
V. S. Bansal & P. W. Majerus (1990) *Ann. Rev. Cell Biol.* **6**, 41.

INSTANTANEOUS PROCESS

Any reaction or change that is immediate and occurs at the present, without any perceptible delay. In fact, the time-evolution or progress of many apparently instantaneous processes can be determined by using ultrafast methods for detecting and/or quantifying a change in the system. The descriptor "instantaneous" should not be confused with "spontaneous"; the latter describes a process that is energetically poised to progress without the intervention of any external force or treatment. Although a spontaneous chemical process may be thermodynamically favorable, kinetic constraints may prevent the system from progressing to its equilibrium position. If the energy of activation is appreciable, then attainment of the transition state is less likely, and an otherwise spontaneous process may fail to progress.

INSULYSIN

This zinc-dependent enzyme [EC 3.4.24.56], a member of the peptidase family M16, catalyzes the degradation of insulin, glucagon, and other polypeptides.

A. B. Becker & R. A. Roth (1995) *Meth. Enzymol.* **248**, 693.

INTENSITY

1. In reference to a spectral feature, the scalar or magnitude of that feature. 2. Chromatic purity. 3. The magnitude of a particular force or energy per unit (*e.g.*, surface, charge, mass, time, volume, *etc.*). 4. Synonym for photon irradiance. 5. Synonym for fluence rate. 6. Synonym for irradiance; illuminance. 7. Synonym for radiant power. 8. Symbolized by I, synonym for radiant intensity. 9. *See* *Magnetic Field Strength*. 10. *See* *Electric Field Strength*. 11. With respect to sound, symbolized by I, the rate of energy transfer per unit area normal to the direction of propagation. Thus, $I = (1/2)\rho_o ca^2 f^2$ where ρ_o is the mean density of the medium, a is the amplitude, f is the frequency, and c is the velocity of the sound.

INTERACTIVE

An adjective describing reciprocal action of two or more bodies, particles, or molecules upon each other. Such two-way action reflects the ability of interacting species to transfer energy. This descriptor is often used to describe the ability of ligand acting at one binding site on an enzyme or receptor to influence ligand binding at other sites.

INTERCEPT

1. The point at which a line, curve, or figure intersects a specified coordinate axis; this term also applies to the coordinates of that point. For example, in the double-reciprocal plot of the initial rates of the Uni Uni mecha-

nism ($1/v = 1/V_{\max} + (K_m/V_{\max})(1/[S])$ where v in the initial velocity, $V_{\max}$ is the maximum forward velocity, K_m is the Michaelis constant, and [S] is the substrate concentration) the vertical intercept (*i.e.*, the ordinate intercept) is $1/V_{\max}$ whereas the horizontal intercept (*i.e.*, the abscissa intercept) is $-1/K_m$. 2. A point of intersection for two lines, curves, or figures.

INTERFACE ACTIVATION IN LIPOLYSIS

The hydrolysis of lipids rarely occurs in a single homogeneous phase, and the behavior of lipases at membrane-solvent and micelle-solvent interfaces has been discussed in detail by Verger[1] and Jain *et al.*[2] ***See*** *Micellar Catalysis*

[1]R. Verger (1980) *Meth. Enzymol.* **64**, 340.
[2]M. K. Jain, M. H. Gelb, J. Rogers & O. G. Berg (1995) *Meth. Enzymol.* **249**, 570.

Selected entries from *Methods in Enzymology* [vol, page(s)]:
Adsorption effect, **64**, 345, 346; anchoring site, **64**, 390; bile salt interaction, **64**, 345; chain length effect, **64**, 344; charge density, **64**, 386; energy, **64**, 385; enzyme aggregation, **64**, 387, 388; enzyme inactivation, **64**, 347, 368, 372-375; free energy of adsorption, **64**, 365; hydration state, **64**, 385-386; insoluble lipid effect, **64**, 345-353; lipolytic enzymes, **64**, 341-345, 384-392; Michaelis constant, **64**, 356-358; orientation effect, **64**, 357, 385; insoluble lipid penetration, **64**, 345, 346; size effect, **64**, 356-358, 362; substrate concentration, **64**, 384, 385; surface dilution model, **64**, 355; surface pressure, **64**, 386, 387; organization and dynamics of phospholipids in aqueous dispersions, **249**, 570; structural features of phospholipase A$_2$, **249**, 572; neutral diluents, substrates, and transition-state mimics, **249**, 575; key considerations for interpreting interfacial catalysis, **249**, 578; kinetic variables at interface, **249**, 581; constraining variables for high processivity, **249**, 582; kinetic formalism for interfacial catalysis in scooting mode, **249**, 586; additional features of hydrolysis of vesicles in scooting mode, **249**, 586; determination of interfacial equilibrium parameters, **249**, 587; interfacial equilibrium constant determination by protection method, **249**, 588; table of kinetic and equilibrium constants, **249**, 591; binding of phospholipase A$_2$ to interface, **249**, 592; integrated Michaelis-Menten equation for interfacial lipolysis, **249**, 595; full activity by enzyme monomer, **249**, 600; competitive substrate specificity, **249**, 601; covalent modifiers of enzyme activity, **249**, 603; hydrolysis of zwitterionic vesicles, **249**, 605; phospholipid-detergent mixed micelles, **249**, 607; kinetic problems on monolayer interfaces, **249**, 610.

INTERMEDIACY

The state or condition of being intermediate, as in the middle place in a process. In most chemical reactions, intermediacy is characterized by a transition state barrier to reactivity.

INTERMEDIATE

1. A substance formed in the course of a chemical reaction that then proceeds further through the reaction. 2. A substance formed as a product of an enzyme-catalyzed reaction and is precursor to other compounds in a metabolic pathway.

INTERMEDIATES, COVALENT ENZYME-SUBSTRATE

Many enzyme-catalyzed reactions proceed via the formation of enzyme-substrate covalent intermediates. Formation of an enzyme-bound covalent compound need not require its participation in enzyme catalysis. Establishing the catalytic competence of covalent intermediates is a major objective of enzyme chemistry. This competence has to be evaluated by a number of different methods[1], including compatibility with initial rate data, isotope exchange behavior, results of thermal or chemical trapping, spectrophotometric detection, stereochemical course of the reaction to observe inversion of configuration and/or isotope scrambling, the behavior of analogs of the putative covalent intermediate, and site-directed mutagenesis studies.

[1]D. L. Purich (1982) *Meth. Enzymol.* **87**, 3.

INTERMOLECULAR

An adjective indicating an interaction between two or more distinct molecular entities. ***See also*** *Intramolecular*

INTERMOLECULAR ISOTOPE EFFECT

An isotope effect (either kinetic or equilibrium) resulting from a comparison made between isotopically different reactant molecular entities. ***See*** *Intramolecular Isotope Effect; Kinetic Isotope Effect; Equilibrium Isotope Effect*

IUPAC (1979) *Pure and Appl. Chem.* **51**, 1725.

INTERNAL EFFECTORS

Those metabolic intermediates whose concentrations are determined by the flux through the pathway in which they are formed and transformed. Consider the hypothetical pathway,

$$A \to B \to C \to D \to \cdots \to P$$

The internal intermediates are $B, C, D, \ldots$, and A and P are the respective initial substrate and final product. Within cells, the enzymes forming internal intermediates are rarely saturated with respect to the concentration of their individual substrates which are internal effectors of the pathway dynamics.

B. Crabtree & E. A. Newsholme (1987) *Trends Biochem. Sci.* **12**, 4.

INTERNAL ENERGY

The change of internal energy (symbolized ΔU) in a closed system is equal to the sum(s) of the transfer-of-heat (q) on the system plus the performance of work (w) on the system. Thus, $\Delta U = q + w$; or, in differential form, $dU = dq + dw$. The internal energy, U, can only be defined in terms of a difference in energy between two different states. In some instances, a convenient standard state may be defined to be zero (e.g., at absolute zero, etc.). The internal energy of a system is the sum of all the potential and kinetic energies of all the molecular entities within the system. This includes all energies associated with conformational changes of molecular entities as well as chemical reactions. If electromagnetic fields are present, then the internal energy includes electromagnetic energy. The energy function U is of importance in representing the first law of thermodynamics. The SI unit for ΔU is the joule.

INTERNAL ENTROPY

The internal freedom of motion of substrate molecules, intermediates, or reactive enzyme groups occurring during the binding and reaction of an enzyme-catalyzed reaction. When the bond hybridization changes during the course of a reaction (e.g., sp^3 to sp^2 and then back to sp^3 in the triose-phosphate isomerase reaction), loss of internal rotatory motion (in this example, about the C–C bond during the formation of the ene-diolate intermediate) leads to a change in the internal entropy of this reaction. Jencks[1] has discussed how restricted motions during catalysis are apt to influence reactivity.

[1]W. P. Jencks (1969) *Catalysis in Chemistry and Enzymology*, pp. 363-366, McGraw-Hill, New York. [Also available as Jencks, W. P. (1986) *Catalysis in Chemistry and Enzymology*, Dover Publications, New York].

INTERNAL EQUILIBRIA

Equilibria involving the productively bound substrates and the products formed during an enzyme-catalyzed reaction. These equilibria can be treated in terms of internal equilibrium constants (K_{int}) between these enzyme-bound species.

Burbaum *et al.*[1] considered how kinetic/thermodynamic features of present-day enzyme-catalyzed reactions suggest that enzyme evolution tends to maximize catalytic effectiveness. They analyzed Uni Uni enzymes in terms of reaction energetics. Catalytically optimized enzymes were proposed to have "internal" equilibrium constants that depend on how close to equilibrium they maintain their reactions *in vivo*. For enzymes operating near equilibrium, the optimum values of K_{int} are near unity. For those operating far away from equilibrium, catalytic efficiency is less sensitive to the magnitude of K_{int}. Accordingly, they suggested that the internal thermodynamics of enzyme-catalyzed reactions dynamically match the concentrations of substrate- and product-containing complexes to the steady state *in vivo*. Burbaum & Knowles[2] also determined K_{int} for enolase (which interconverts 2-phosphoglycerate and phosphoenolpyruvate) and creatine kinase (a bisubstrate enzyme forming $MgATP^{2-}$ and creatine from creatine phosphate and ADP). K_{int} values were determined by rapidly quenching equilibrium mixtures of enzyme and radiolabeled substrate and product, and these investigators chose conditions where virtually all of the marker substrate and product were bound. They also showed how published K_{int} values agree with their theory for kinetically optimized enzymes.

[1]J. J. Burbaum, R. T. Raines, W. J. Albery & J. R. Knowles (1989) *Biochemistry* **28**, 9293.
[2]J. J. Burbaum & J. R. Knowles (1989) *Biochemistry* **28**, 9306.

INTERNAL PRESSURE

The change in internal energy per change in volume at constant temperature (expressed in differential terms as $(\partial U/\partial V)_T$, where U is the internal energy, V is the volume, and T is the temperature). For gases, this value (usually symbolized by P_i) is small. However, for liquids and solids, the internal pressure is large. The magnitude of the internal pressure provides an estimate of the magnitude of the cohesive forces within the liquid or solid state (or even the gaseous state for nonideal gases). The internal pressure is also obtained from one of the thermodynamic equations of state: namely, $(\partial U/\partial V)_T = T(\partial P/\partial T)_V - P$ where P is the pressure. For an ideal gas, the internal pressure is zero, because there are, by definition, no cohesive or intermolecular forces that apply to an ideal gas. In general, dipolar liquids have larger values than nonpolar liquids. For example, at 1 atm of external pressure and a temperature of 298 K, diethyl ether and benzene have P_i values of 2.40×10^8 Pa (or 2.37×10^3 atm) and 3.69×10^8 Pa (or 3.64×10^3 atm), respectively. By virtue of its strong hydrogen bond interactions, water has a corresponding P_i value of 2.0×10^9 Pa (or 2.0×10^4 atm).

INTERNET

The Internet has become a powerful resource for molecular life scientists, and the only limitation appears to be that many URL's are not maintained as stable sites for long-term use. Nonetheless, the reader may find the following addresses as a useful initial point of departure in efforts to locate valuable software for analyzing protein and nucleic acid structure.

BLAST (NCBI Sequence Similarity Search of Nucleotide and Protein Databases)
http://www.ncbi.nlm.nih.gov/BLAST/

Biochemist On-Line
http://www.biochemist.com/

Brookhaven Protein Data Base (Structural):
http://www.pdb.bnl.gov/

dbEST (Database of Expressed Sequence Tags)
http://www.ncbi.nlm.nih.gov/dbEST/index.html

Biochemical Pathways (Searchable Index)
http://expasy.hcuge.ch/cgi-bin
/search-biochem-index

ENZYME (Enzyme Nomenclature Database)
http://expasy.hcuge.ch/sprot/enzyme.html

ExPASy Tools (A Menu of Sequence Analysis Tools)
http://expasy.hcuge.ch/www/tools.html

GenBank (Text and Similarity Searching)
http://www.ncbi.nlm.nih.gov/Web/Search
/index.html

MMDB (The Molecular Modelling Database)
http://www.ncbi.nlm.nih.gov/Structure/

MULTIIDENT (Proteins Identification Using pI, MW, Amino Acid Composition, Sequence Tag and Peptide Mass Fingerprinting Data)
http://expasy.hcuge.ch/sprot/multiident.html

ORF Finder (Open Reading Frame Finder)
http://www.ncbi.nlm.nih.gov/gorf/gorf.html

Pedro's BioMolecular Research Tools
http://www.public.iastate.edu/~pedro
/research_tools.html

Pfam (Comprehensive Collection of Protein Domain Families)
http://www.sanger.ac.uk/Pfam/

Principles of Protein Structure Using the Internet
http://www.cryst.bbk.ac.uk/PPS2/top.html

PROSITE (Dictionary of Protein Sites and Patterns)
http://expasy.hcuge.ch/sprot/prosite.html

PubMed (Med-Line Search Engine) :
http://www4.ncbi.nlm.nih.gov/htbin-post
/Entrez

Structural Classification of Proteins
http://scop.mrc-lmb.cam.ac.uk/scop/

SWISS-3DIMAGE (Database of Annotated 3D Images)
http://expasy.hcuge.ch/sw3d/sw3d-top.html

Swiss-Model (An Automated Knowledge-Based Protein Modelling Server)
http://expasy.hcuge.ch/swissmod
/SWISS-MODEL.html

SWISS-PROT (Annotated Protein Sequence Database)
http://expasy.hcuge.ch/sprot/sprot-top.html

Translate (Convert DNA or RNA Sequence to Protein Sequence)
http://expasy.hcuge.ch/www/dna.html

INTERPOLATION

Any mathematical algorithm or technique for estimating the magnitude of $f(x)$ for a continuous function of the variable x or for evaluating a parameter lying within an interval for which the first and last values of the interval are known.

Selected entries from *Methods in Enzymology* [vol, page(s)]: Graphical techniques, **210**, 306; polynomial methods, **210**, 307; with rational functions, **210**, 311; with spline functions, **210**, 312; with trigonometric functions, **210**, 312

INTERSYSTEM CROSSING

A term in photochemistry and photophysics describing an isoenergetic radiationless transition between two electronic states having different multiplicities. Such a process often results in the formation of a vibrationally excited molecular entity, at the lower electronic state, which then usually deactivates to its lowest vibrational energy level. **See also** Internal Conversion; Fluorescence

Comm. on Photochem. (1988) *Pure and Appl. Chem.* **60**, 1055.

INTRAMOLECULAR CATALYSIS

Acceleration of a chemical reaction (at a particular site on a molecular entity or complex) by one or more functional groups located at other site(s) within the same molecular entity. Intramolecular catalysis can often be expressed by comparison to the rate of the reaction when the functional group(s) is(are) removed from the same molecular entity or by determination of the effective molarity of the catalytic group.

INTRAMOLECULAR ISOTOPE EFFECT

An isotope effect (either kinetic or equilibrium) resulting from reactions in which the different isotopes occupy chemically equivalent alternative reactive sites within the same molecular entity. In such cases, isotopically distinct products are formed. **See** *Intermolecular Isotope Effect; Kinetic Isotope Effect; Equilibrium Isotope Effect*

IUPAC (1979) *Pure and Appl. Chem.* **51**, 1725.

INTRAMOLECULAR KINETIC ISOTOPE EFFECT

A kinetic isotope effect observed by a single reactant, having isotopic atoms at equivalent reactive positions, which reacts to produce isotopomeric products with a nonstatistical distribution. The pathway favored will be the one having lower force constants for the displacement of the isotopic nuclei in the transition state.

INTRINSIC BARRIER

The Gibbs energy of activation ($\Delta G^{\ddagger}$) when the thermodynamic driving force has been eliminated (*i.e.*, in the limiting case where $\Delta G^{\ddagger} = 0$). According to the Marcus equation, this barrier is directly related to the reorganization energy, λ, by the expression $\Delta G^{\ddagger} = \lambda/4$. **See** *Marcus Equation; Reorganization Energy*

R. D. Cannon (1980) *Electron Transfer Reactions*, Butterworths, London.
C. J. Schlesener, C. Amatore & J. K. Kochi (1986) *J. Phys. Chem.* **90**, 3747.

INTRINSIC BINDING ENERGY

The standard free energy change (ΔG_{int}) corresponding to the energy released when a protein binds to a compound (or a substituent group of that compound) without destabilization or energy losses attributable to translational, rotational, and internal entropy. As pointed out by Jencks[1], the intrinsic binding energy will never be directly observable, because binding of one molecule to another is almost always attended by a net loss of translational and rotational entropy. In other words, the observed binding energy of protein-ligand bond formation typically reflects

losses of translational and rotational degrees of freedom, often amounting to -40 entropy units. The intrinsic binding energy will also be underestimated by the net effect of the mutual freezing out of rotations and vibrations in the interacting components. Likewise, a protein or enzyme that destabilizes its ligand by distortion (*i.e.*, transition-state stabilization as opposed to ground-state stabilization) will also have an intrinsic binding energy that is better estimated through the binding of a so-called transition state analogue. Furthermore, Jencks[1] has stressed that the free energy of binding is affected by induced-fit interactions as well as nonproductive substrate complexation; these interactions are likely to exercise control over specificity without contributing significantly to catalytic rate enhancement.

[1]W. P. Jencks (1975) *Adv. Enzymol.* **43**, 358.

INTRINSIC KINETIC ISOTOPE EFFECT

An effect of isotopic substitution within a reactant or substrate on a specific step in an enzyme-catalyzed reaction. The magnitude of an intrinsic isotope effect may not equal the magnitude of an isotope effect on collective rate parameters such as V_{max} or V_{max}/K_m, unless the isotopically sensitive step is the rate-limiting or rate-contributing step. **See** *Kinetic Isotope Effect*

INULINASE

This enzyme [EC 3.2.1.7], also referred to as 2,1-β-D-fructan fructanohydrolase and inulase, catalyzes the endohydrolysis of 2,1-β-D-fructosidic linkages in inulin.

G. Avigad & S. Bauer (1966) *Meth. Enzymol.* **8**, 621.

INULOSUCRASE

This enzyme [EC 2.4.1.9], also referred to as sucrose 1-fructosyltransferase, catalyzes the reaction of sucrose with $[(2,1)\text{-}\beta\text{-}\text{D-fructosyl}]_n$ to produce D-glucose and $[(2,1)\text{-}\beta\text{-}\text{D-fructosyl}]_{(n+1)}$. Thus, this enzyme participates in the conversion of sucrose into inulin and glucose. Other fructosyl-containing sugars can serve as substrates as well instead of sucrose.

G. Mooser (1992) *The Enzymes*, 3rd ed., **20**, 187.

INVERSE ISOTOPE EFFECT

A kinetic isotope effect in which the ratio k_{light}/k_{heavy} is less than one. In such cases, there is an increase in the corresponding force constants on passing from the reactant to the transition state. **See** *Kinetic Isotope Effect*

INVERTASE

This enzyme [EC 3.2.1.26], also known as β-D-fructofuranosidase and saccharase, catalyzes the hydrolysis of terminal nonreducing β-D-fructofuranoside residues in β-D-fructofuranosides. Hence, the enzyme will also catalyze the hydrolysis of sucrose to glucose and fructose. In addition, the protein also exhibits a fructotransferase activity. **See also** Sucrase

G. Davies, M. L. Sinnott & S. G. Withers (1998) *Comprehensive Biological Catalysis: A Mechanistic Reference* **1**, 119.
P. M. Dey & E. del Campillo (1984) *Adv. Enzymol.* **56**, 141.
J. O. Lampen (1971) *The Enzymes*, 3rd ed., **5**, 291.

INVERTED/INVERSE MICELLE

A structure formed by the reversible association of amphiphiles in apolar solvents. In inverted micelles, the polar portion of the amphiphile is concentrated in the interior of the macrostructure. Such association usually occurs with aggregation and is not typically characterized by a definite nucleation stage. Thus, inverted micelles (also referred to as inverse or reverse micelles) often fail to exhibit critical micelle concentration behavior. **See** Micelle

IODIDE PEROXIDASE

This enzyme [EC 1.11.1.8], also known as iodotyrosine deiodase, iodinase, and thyroid peroxidase, catalyzes the reaction of iodide with hydrogen peroxide to produce iodine and two water. The cofactor for this enzyme is heme.

M. Morrison (1970) *Meth. Enzymol.* **17A**, 653 and 658.

IODINE-125 (^{125}I)

A radioactive iodine nuclide that has three principal radiation emissions γ (0.035 MeV), Kα X-ray (0.027 MeV), and Kβ X-ray (0.031 MeV). The decay half-life is 60.14 days. The following decay data indicate the time (in days) followed by the fraction of original amount at the specified time: 0, 1.000; 2, 0.977; 4, 0.955; 6, 0.933; 8, 0.912; 10, 0.891; 12, 0.871; 14, 0.851; 16, 0.831; 18, 0.812; 20, 0.794; 40, 0.630; 60, 0.500; 80, 0.397; 100, 0.315; 120, 0.250; 140, 0.198; 160, 0.157; 180, 0.125; 200, 0.099; 220, 0.079; 240, 0.063.

Special precautions: Store this radioisotope behind 3 mm thick lead shielding. The effective *biological* half-life in humans is around 140 days. The thyroid gland is the critical organ in terms of dose, and one would be well advised to adhere to institutional regulations regarding thyroid monitoring for localized uptake.

IODINE-131 (^{131}I)

A radioactive iodine nuclide that has three principal radiation emissions: β (0.606 MeV), γ (0.364 MeV), and γ (0.637 MeV). The decay half-life is 8.040 days. The following decay data indicate the time followed by the fraction of original amount at the specified time: 0, 1.000; 6 hours, 0.979; 12, 0.958; 18, 0.938; 24, 0.918; 30, 0.898; 36, 0.879; 42, 0.860; 48, 0.842; 54, 0.824; 60, 0.807; 66, 0.789; 3 days, 0.773; 6, 0.597; 9, 0.461; 12, 0.356; 15, 0.275; 18, 0.213; 21, 0.164; 24, 0.127; 27, 0.098; 30, 0.076; 33, 0.059; 36, 0.045.

Special precautions: The thyroid gland is the critical organ in terms of dose. The effective *biological* half-life in humans is around 140 days. Store this radioisotope behind lead shielding. Follow institutional regulations regarding (a) type and location of dosimeters; and (b) monitoring thyroid for localized uptake.

IODOACETATE

An alkylating agent (ICH_2COO^-) that acts as a potent irreversible inhibitor of enzymes containing reactive thiol, ε-amino, and/or imidazole side-chain groups within their active sites. The carboxymethylation reaction with enzymes is typically a bimolecular process that takes place without rate-saturation behavior[1].

[1]F. R. N. Gurd (1972) *Meth. Enzymol.* **25**, 424.

Selected entries from *Methods in Enzymology* [vol, page(s)]:
Principle of carboxymethylation, **25**, 424; precautions, **25**, 425; reaction products, **25**, 425-426; analysis of reaction time-course, **25**, 427-428; typical reaction conditions, **25**, 429-438; in affinity labeling, **87**, 469.

ION ATMOSPHERE

A term used in electrostatic descriptions of ions to denote the continuous electric charge density [$\rho(r)$] surrounding an ionic species. On average, an ion will be surrounded by a spherically symmetrical distribution of counterions that form its ion atmosphere. **See** Hydration Atmosphere

ION CURRENT

1. The flux of ions across a membrane barrier, usually in the downward direction with respect to ion concentra-

tion. 2. The rate of chemical ionization in a mass spectrometer.

Selected entries from *Methods in Enzymology* **[vol**, page(s)]: Analysis, software for, **207**, 717; barrier models, **207**, 818; closed and open time estimation, **207**, 755; data acquisition, **207**, 747; modal behavior analysis, **207**, 757; multiple channel problem, **207**, 756; single-channel [extraction of kinetic information, **207**, 765; measurement in tissue slices, **207**, 220]; synaptic, resolution improvement in patch clamp recording, **207**, 216; whole-cell recording in calcium channel, **207**, 181; fluctuation analysis, **207**, 192.

ION CYCLOTRON RESONANCE MASS SPECTROMETRY

An analytical mass spectrometric technique that traps an ionic species in a fluctuating magnetic field, such that the ion attains an angular or, in this case, a cyclotron frequency. This frequency ω_c equals the charge-to-mass ratio (z/m) multiplied by the product of the charge on a electron (e) and the magnetic field strength B. The ion cyclotron phenomenon is accomplished by using alternating current power source to create a fluctuating magnetic field. Coherent circular motion of resonating ions establishes a so-called image current that is used to determine the charge-to-mass ratio.

IONIC BOND

An electrostatic bond between two elements in which one or more electrons from one element have been donated to empty orbitals of the other element, thus imparting full charges to both elements; the bond between ions having opposite charges. The stable interaction of electropositive ions (*i.e.*, cations) and electronegative ions (*i.e.*, anions) resulting from a balancing of long-range ionic attraction and nearby repulsive forces. The bond length corresponds to the internuclear distance r between two charges (q_1 and q_2) in the following expression which includes terms for attractive potential ($U_{\text{attraction}} = -q_1q_2/r$) and the repulsive potential ($U_{\text{repulsion}} = be^{-r/a}$) where a and b are constants:

$$U = -q_1q_2/r + be^{-r/a}$$

This expression gives reasonable values for the equilibrium ionic bond length r and dipole moment μ, but the calculated curve for U versus r does not exactly match the experimental curve.

IONIC RADIUS

The calculated dimension (usually expressed in nm which is the preferred SI unit) of ions, based on X-ray crystallographic studies of ionic crystals. Pauling[1] describes attempts to obtain self-consistent estimations based on measurements with a variety of ionic crystals. The following tabulated values are based on his calculations.

Table I

Ionic Radii of Metal Ions of Specified Coordination Number (CN)

| Ion | CN[a] | | | | | |
	2	4	6	8	10	12
Li$^+$	—	0.73Å	0.90Å	1.06Å	—	—
Na$^+$	—	1.13Å	1.16Å	1.32Å	—	1.53Å
Mg^{2+}	—	0.71Å	0.86Å	1.03Å	—	—
Al^{3+}	—	0.53Å	0.68Å	—	—	—
K$^+$	—	1.51Å	1.52Å	1.65Å	1.73Å	—
Ca^{2+}	—	—	1.14Å	—	—	—
V^{2+}	—	—	0.93Å	—	—	—
V^{3+}	—	—	0.81Å	—	—	—
Cr^{2+}	—	0.72Å (HS)	0.79Å (LS) 0.89Å (HS)	—	—	—
Cr^{3+}	—	0.76Å	—	—	—	—
Mn^{2+}	—	0.80Å (HS)	0.81Å (LS) 0.97Å (HS)	—	—	—
Mn^{3+}	—	—	0.72Å (LS) 0.79Å (HS)	—	—	—
Fe^{2+}	—	0.77 (HS)	0.75 (LS) 0.92 (HS)	—	—	—
Fe^{3+}	—	0.63Å (HS)	0.69Å (LS) 0.79Å (HS)	—	—	—
Co^{2+}	—	0.72Å (HS)	0.79Å (LS) 0.89Å (HS)	—	—	—
Co^{3+}	—	—	0.69Å (LS) 0.75Å (HS)	—	—	—
Ni^{2+}	—	0.69Å 0.63Å sq	0.83Å	—	—	—
Ni^{3+}	—	—	0.70Å (LS) 0.74Å (HS)	—	—	—
Cu$^+$	0.60Å	0.74Å	0.91Å	—	—	—
Cu^{2+}	—	0.71Å	0.87Å	—	—	—
Zn^{2+}	—	0.74Å	0.88Å	1.04Å	—	—
Rb$^+$	—	—	1.66Å	1.75Å	1.80Å	1.86Å
Mo^{3+}	—	—	0.83Å	—	—	—
W^{4+}	—	—	0.80Å	—	—	—

[a]LS = low-spin state; HS = high-spin state; sq = square planar

[1]L. Pauling (1960) *The Nature of the Chemical Bond*, 3rd ed., p. 514, Cornell University Press, Ithaca.

IONIC STRENGTH

An ionic concentration function (symbolized by I) designed to account for electrostatic interactions among charged substances in solutions, such that ions of higher charge are given greater weight as indicated by the following equation

$$I = 0.5 \sum m_i z^2 I$$

where m_i is the molal concentration (*i.e.*, the number of moles of the *i*-th component per 1000 g of solvent) and z_i is the number corresponding to the ionic charge (*i.e.*, $z = -1$ or $+1$ for monovalent anions and monovalent cations, respectively; $z = -2$ or $+2$ for divalent anions and divalent cations; *etc.*). It should be noted that the ionic strength for a solution is the summation for *all* ions present in the solution.

For dilute solutions, the molar concentration, c, can be used in place of molality. Thus, $I \cong (1/(2\rho_o)) \sum c_i z^2_i$ where ρ_o is the density of the solvent. Hence, for aqueous solutions, $I \cong 0.5 \sum c_i z^2_i$. The following examples emphasize the importance of net charge in determining the magnitude of the ionic charge parameter: (a) for 10 mM NaCl, $I = 0.5[0.01 \times (-1)^2 + 0.01 \times (+1)^2] = 0.01$; (b) for 10 mM Na_2SO_4, $I = 0.5[0.01 \times (-2)^2 + 0.02 \times (+1)^2] = 0.03$; for 10 mM $MgSO_4$, $I = 0.5[0.01 \times (-2)^2 + 0.01 \times (+2)^2] = 0.04$; and (d) for 10 mM Na_3ATP, $I = 0.5[0.01 \times (-3)^2 + 0.03 \times (+1)^2] = 0.06$.

Rates of reactions involving ionic species, solubilities, activity coefficients, and a wide variety of other phenomena can be affected by the ionic strength. ***See also*** *Davies Equation; Debye-Hückel Treatment; Isotonic Buffers*

IONIZATION

Any process that generate one or more ions by (a) a unimolecular heterolysis of a neutral molecular entity into two or more ions or (b) a heterolytic substitution reaction involving neutral molecules. ***See also*** *Ionization Energy; Dissociation*

IONIZATION ENERGY

The minimum amount of energy or work (*i.e.*, enthalpy) required in the gaseous phase to remove an electron from an isolated molecular entity (typically in its vibrational ground state). Adiabatic ionization energy is that energy needed to produce a new molecular entity that is also in its vibrational ground state. If the product is not in this ground state and possesses the energy dictated by the Franck-Condon principle, the energy needed is referred to as the vertical ionization energy.

IONOPHORE

A compound that binds to an ion in a manner which greatly facilitates the bound ion's permeability across a membrane. Naturally occurring ionophores include both mobile carriers (*e.g.*, valinomycin and nigericin) and channel formers (*e.g.*, gramicidin A).

B. C. Pressman (1976) *Ann. Rev. Biochem.* **45**, 501.
M. Dobler (1981) *Ionophores and Their Structures*, Wiley, New York.

ION PAIR

1. An intimate pair of oppositely charged ions, mutually attracted to each other by noncovalent, electrostatic interactions. Ion pairs often behave experimentally as a single unit. According to Bjerrum, an ion pair has atomic centers that are held closer together than a distance corresponding to the expression: $q = (8.36 \times 10^6 (z_+ z_-/\varepsilon_r T))$, where q is expressed in picometers, z_+ and z_- are the charges of the ions, and ε_r is the relative permittivity of the medium. Ion pairs satisfying this requirement are often referred to as Bjerrum ion pairs. ***See*** *Tight Ion Pairs; Loose Ion Pairs; Solvent-Shared Ion Pair; Solvent-Separated Ion Pair*

2. The pair of oppositely charged ions produced in a single ionization reaction.

3. Occasionally, the monocation and electron generated in a redox half-reaction are also collectively referred to as an ion pair.

ION PUMP

A protein that brings about the active transport of ions by coupling the vectorial movenent of an ion to some source of Gibbs free energy, often directly from ATP hydrolysis. These pumps exhibit tight ion binding in one state (*i.e.*, on the low ion concentration side of a membrane) and weak binding in a second state (*i.e.*, the high ion concentration side). This change in specificity (*i.e.*, affinity) is driven by an ATP hydrolysis-dependent conformational change.

Jencks[1] has considered the mechanism of action of the sarcoplasmic reticulum calcium pump, which catalyzes the following reaction:

$$2\,Ca^{2+}_{cytoplasmic} + ATP \rightleftharpoons 2\,Ca^{2+}_{luminal} + ADP + P_i$$

where there is stoichiometric coupling between the ATPase reaction and the active transport of an ion. One can write a reaction cycle involving two principal states E_1 and E_2:

Sarcoplasmic Reticulum ATPase: The Ion Pump Cycle

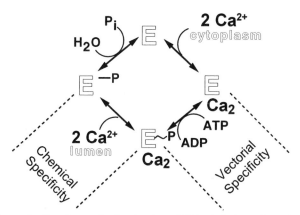

Figure 1. Reaction cycle showing how ATP hydrolysis and transient formation of a phosphoryl-enzyme intermediate are linked to the vectorial transport of calcium in the sarcoplasmic reticulum.

By analogy to the model proposed by Makinose[2], the cyclic transformations can also be depicted as follows:

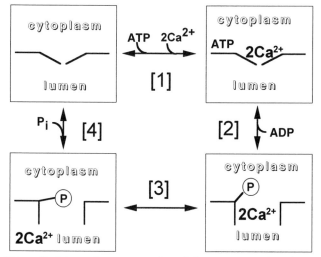

Figure 2. Schematic representation of the steps occurring during calcium ion pumping by the calcium-dependent ATPase.

In both schemes, the specificities of the pump for catalysis change in the two enzyme states. Jencks[1] points out that coupling is determined (a) by the chemical specificity achieved in catalyzing phosphoryl transfer to and from the enzyme (wherein $E \cdot Ca_2$ reversibly binds ATP, and E reacts reversibly with orthophosphate), and (b) by the vectorial specificity for ion binding and dissociation (wherein E reversibly binds/dissociates cytoplasmic calcium ion, and E–P reversibly binds/dissociates luminal calcium). There must be a single conformation change during the reaction cycle: between E_1 and E_2 in the free enzyme and from $E_1 \sim P \cdot Ca_2$ to $E_2 - P - Ca_2$ after enzyme phosphorylation.

Another indication that conformational changes are likely to account for changes in ion affinity was provided by Keillor and Jencks[3] who recently examined the kinetic properties of the sheep kidney Na^+,K^+-ATPase. This enzyme was preincubated with 120 mM sodium and 3 mM magnesium and then allowed to react with ATP (concentration range = 0.01–2.00 mM) to form a covalent phosphoenzyme intermediate (E–P). These investigators observed that the first-order rate constant for phosphorylation increases hyperbolically with ATP concentration to a maximum value of $4.6 \times 10^2\ s^{-1}$ and $K_{0.5} = 75\ \mu M$. If phosphoryl-transfer is assumed to be rate-limiting, then the approach to equilibrium that yields 50% E–P in the presence of ADP obeys the following rate law: $k_{observed} = k_f + k_r + 9.2 \times 10^2\ s^{-1}$. However, for E–P formation from $E \cdot Na_3$ with 1.0 mM ATP and 2.0 mM ADP, $k_{observed} = 4.2 \times 10^2\ s^{-1}$, indicating that phosphoryl transfer from bound ATP to the enzyme is not rate-limiting for E–P formation from $E \cdot Na_3$. Their results suggest that there is likely to be a rate-limiting conformational change of the $E \cdot Na_3 \cdot ATP$ intermediate, followed by rapid phosphoryl transfer, with $k_{cat} = 3000\ s^{-1}$. **See** *Membrane Transport; Binding Change Mechanism*

[1]W. P. Jencks (1980) *Adv. Enzymol.* **51**, 75.
[2]M. Makinose (1973) *FEBS Lett.* **37**, 140.
[3]J. W. Keillor & W. P. Jencks (1996) *Biochemistry* **35**, 2750.

IRON-SULFUR CLUSTER

A structural unit within iron-sulfur proteins, comprising two or more iron atoms and bridging sulfur atoms (also termed sulfide ligands or sulfido-bridges). These bridging sulfur atoms are acid-labile, generating H_2S in the presence of strong acid. The designation of a cluster is based on the iron and sulfide content, and is given in square brackets (*e.g.*, [2Fe−2S], [4Fe−4S], and [3Fe−4S]). The possible oxidation levels are indicated by the net charge excluding the ligands (*e.g.*, [4Fe−4S]$^{2+}$ cluster and [4Fe−4S]$^{1+}$ cluster). Note that the oxidized and reduced states of all such clusters differ by one formal charge, and this difference is independent of the number of iron atoms present. Individual iron atoms within a cluster can attain different oxidation states. **See** *Iron-Sulfur Proteins*

IRON-SULFUR PROTEINS

Iron-sulfur proteins contain nonheme iron that is liganded with sulfur atoms from cysteinyl residues and also usually contain inorganic sulfur (or, acid-labile sulfur) bridging two iron atoms. There are three major categories: (1) the rubredoxins, a group that lacks acid-labile sulfur, (2) simple iron-sulfur proteins containing only iron-sulfur clusters, and (3) complex iron-sulfur proteins with an additional redox center(s), such as a flavin, a heme, or another metal, often molybdenum. The clusters mainly function in electron transfer, but they can have other functions, including catalysis of hydratase/dehydratase reactions, maintenance of protein structure, and regulation of enzyme activity.

B. G. Fox (1998) *Comprehensive Biological Catalysis: A Mechanistic Reference* 3, 261.

H. Cheng & J. L. Markley (1995) *Ann. Rev. Biophys. Biomol. Struct.* **24**, 209.

T. E. Meyer (1994) *Meth. Enzymol.* **243**, 435.

J. B. Howard & D. C. Rees (1991) *Adv. Protein Chem.* **42**, 199.

G. Palmer (1975) *The Enzymes*, 3rd ed., **12**, 1.

Selected entries from *Methods in Enzymology* [vol, page(s)]: Detection, **69**, 238-249; electron paramagnetic resonance spectroscopy, **69**, 239-245 [dark chemical reduction with reducing agents, **69**, 241, 242; oxidation-reduction potentiometry, **69**, 242-245; photoreduction, **69**, 240, 241]; quantitative identification, **69**, 781-784 [materials, **69**, 781; methods, **69**, 781-783, 791, 792 (precautions, **69**, 789-791; variations, **69**, 784-789)]; membrane-bound [chromatophore fragments, from photosynthetic bacteria, **69**, 247-249; oxygen-evolving organisms, **69**, 245-248]; occurrence, **69**, 238-249, 779, 780; redox nature, core extrusion, **69**, 785; reduction, **69**, 218, 221; types, **69**, 779, 780.

IRREVERSIBLE PATHS, REACTIONS, AND PROCESSES

If one or all of the intermediate states of a path, process, or reaction are not equilibrium states, then the path is regarded as irreversible. **See** *Nonequilibrium Thermodynamics (A Primer)*

IRREVERSIBLE POLYMERIZATION

Cohen[1] presented a method for determining the rate constants for irreversible polymerization where the different rate constants apply to the initial step and the propagation reactions. Data are obtained in the usual manner at moderate concentrations of primer. The approach is equally valid for template-directed and template-independent polymerization. Cohen applied this method to obtain rate constants for polyadenylate [(poly(A)] polymerase (EC 2.7.7.19). The report also contains useful information about simulating the time course of irreversible polymerization.

[1]R. J. Cohen (1991) *J. Theoret. Biol.* **150**, 529.

ISOAMYLASE

This enzyme [EC 3.2.1.68], also called debranching enzyme, catalyzes the hydrolysis of α-(1,6)-D-glucosidic branch linkages in glycogen, amylopectin, and their β-limit dextrins. This particular enzyme is distinguished from α-dextrin endo-(1,6)-α-glucosidase [EC 3.2.1.41] by the inability of isoamylase to act on pullulan, and by the limited action on α-limit dextrins. Nevertheless, the action on glycogen is complete in contrast to the limited action by α-dextrin enzyme. Isoamylase will hydrolyze only 1,6-linkages at a branch point.

G. Mooser (1992) *The Enzymes*, 3rd ed., **20**, 187.

I. S. Pretorius, M. G. Lambrechts & J. Marmur (1991) *Crit. Rev. Biochem. Mol. Biol.* **26**, 53.

M. Vihinen & P. Mäntsälä (1989) *Crit. Rev. Biochem. Mol. Biol.* **24**, 329.

E. Y. C. Lee & W. J. Whelan (1971) *The Enzymes*, 3rd ed., **5**, 191.

ISOBUTYRYL-CoA MUTASE

This enzyme [EC 5.4.99.13] catalyzes the interconversion of 2-methylpropanoyl-CoA and butanoyl-CoA. The enzyme requires a cobalt ion for activity.

B. T. Golding & W. Buckel (1998) *Comprehensive Biological Catalysis: A Mechanistic Reference* 3, 239.

ISOCHORISMATE SYNTHASE

This enzyme [EC 5.4.99.6], also known as isochorismate synthase and isochorismate hydroxymutase, catalyzes the interconversion of chorismate and isochorismate.

R. Bentley (1990) *Crit. Rev. Biochem. Mol. Biol.* **25**, 307.

ISOCITRATE DEHYDROGENASE

Isocitrate dehydrogenase (NAD$^+$) [EC 1.1.1.41], also known as β-ketoglutaric : isocitric carboxylase, catalyzes the reaction of isocitrate (specifically, (1R,2S)-1-hydroxypropane 1,2,3-tricarboxylate, formerly termed *threo*-D(S)-isocitrate) with NAD$^+$ to produce α-ketoglutarate (or, 2-oxoglutarate), carbon dioxide, and NADH. The enzyme is reported not to decarboxylate added oxalosuccinate. Isocitrate dehydrogenase (NADP$^+$) [EC 1.1.1.42], also known as oxalosuccinate decarboxylase, catalyzes the reaction of isocitrate with NADP$^+$ to produce α-ketoglutarate, carbon dioxide, and NADPH. The enzyme is reported to be able to decarboxylate added oxalosuccinate.

A. R. Clarke & T. R. Dafforn (1998) *Comprehensive Biological Catalysis: A Mechanistic Reference* **3**, 1.
J. V. Schloss & M. S. Hixon (1998) *Comprehensive Biological Catalysis: A Mechanistic Reference* **2**, 43.
L. N. Johnson & D. Barford (1993) *Ann. Rev. Biophys. Biomol. Struct.* **22**, 199.
R. F. Colman (1990) *The Enzymes*, 3rd ed., **19**, 283.
D. J. Creighton & N. S. R. K. Murthy (1990) *The Enzymes*, 3rd ed., **19**, 323.
W. W. Cleland (1980) *Meth. Enzymol.* **64**, 104.
W. W. Cleland (1977) *Adv. Enzymol.* **45**, 373.
K. Dalziel (1975) *The Enzymes*, 3rd ed., **11**, 1.

ISOCITRATE LYASE

This enzyme [EC 4.1.3.1] (also referred to as isocitrase, isocitritase, and isocitratase) catalyzes the conversion of isocitrate (specifically, (1R,2S)-1-hydroxypropane 1,2,3-tricarboxylate) to succinate and glyoxylate.

J. V. Schloss & M. S. Hixon (1998) *Comprehensive Biological Catalysis: A Mechanistic Reference* **2**, 43.
D. J. Creighton & N. S. R. K. Murthy (1990) *The Enzymes*, 3rd ed., **19**, 323.
L. B. Spector (1972) *The Enzymes*, 3rd ed., **7**, 357.

ISODESMIC REACTION

A reaction in which the types of bonds formed in the product(s) are the same types of bonds broken in the reactant(s). **See** *Indefinite Polymerization*

ISOELECTRONIC

Referring to two or more molecular entities having the same number and connectivity of atoms (although not necessarily the same elements) as well as the same number of valence electrons. Thus, CO, N_2, and NO$^+$ are isoelectronic, as are ketene (CH$_2$=C=O) and diazomethane (CH$_2$=N=N). **See** *Electroneutrality Principle*

ISOENTROPIC

A term referring to a reaction series in which each of the individual reactions has the same entropy of activation.

ISOEQUILIBRIUM RELATIONSHIP

A linear relationship between the standard enthalpies and entropies of a series of structurally related molecular entities undergoing the same reaction: thus, $\Delta H^\circ - \beta \Delta S^\circ$ = constant or $\Delta \Delta H^\circ = \beta \Delta S^\circ$. When $\beta \geq 0$, this relationship is referred to as an isoequilibrium relationship. When the absolute temperature equals the factor β (often referred to as the isoequilibrium temperature), then all substituent effects on the reaction disappear (*i.e.*, $\Delta \Delta G^\circ = 0$). In other words, a reaction studied at $T = \beta$ will exhibit no substituent effects. This would suggest that, when one studies substituent effects on a reaction rate, the reaction should be studied at more than one temperature. Note also that the ρ factor in the Hammett equation changes sign at the isoequilibrium temperature. **See** *Isokinetic Relationship*

J. E. Leffler (1955) *J. Org. Chem.* **20**, 1202.

ISOERGONIC COOPERATIVITY

A novel form of allosteric protein-ligand binding behavior in which subsequent binding steps exhibit very different enthalpic properties, with only minor differences between their successive binding constants. One can detect isoergonic cooperativity by determinations of the change in the constant-pressure heat capacity (ΔC_p), but such behavior cannot be observed in experiments solely designed to determine the number of ligand molecules bound to a macromolecule.

H. F. Fisher & J. Tally (1998) *Meth. Enzymol.* **295**, 331.

ISOKINETIC RELATIONSHIP

A condition said to exist for a series of structurally related molecular entities undergoing the same general reaction, wherein the enthalpies and entropies of activation can be related by the expression $\Delta H^\ddagger - \beta \Delta S^\ddagger$ = constant (*i.e.*, $\Delta \Delta H^\ddagger = \beta \Delta \Delta S^\ddagger$), where the factor β is independent of temperature. When the absolute temperature equals this factor (*i.e.*, $T = \beta$), all reactants of the series react at the same rate, and the temperature is referred to as the isokinetic temperature. If errors in determinations of $\Delta H^\ddagger$ result in compensating errors in $\Delta S^\ddagger$, values for β may be spurious. Thus, one must employ appropriate methods in establishing any isokinetic relationship. **See also** *Isoequilibrium Relationship*

J. E. Leffler (1955) *J. Org. Chem.* **20**, 1202.

ISOLATED SYSTEM

Any system (chemical, biological, mechanical, *etc.*) that does not interact with its surroundings, such that matter and/or heat are not transferred into or out of the system. A change in environmental conditions does not result in a change in the system conditions.

An example of such a system is one in which the internal energy, U, and the volume, V, are constant. If these are the only two constraints on the system then, at thermodynamic equilibrium, the entropy, S, is at a maximum. On the other hand, if entropy and volume are constant for the isolated system then, at thermodynamic equilibrium, the internal energy is at a minimum. **See also** *Closed System; Open System*

ISOLEUCYL-tRNA SYNTHETASE

This enzyme [EC 6.1.1.5], also referred to as isoleucine:tRNA ligase, catalyzes the reaction of isoleucine with tRNA^{Ile} and ATP to produce isoleucyl-tRNA^{Ile}, AMP, and pyrophosphate.

W. Freist (1988) *Angew. Chem. Int. Ed. Engl.* **27**, 773.
P. Schimmel (1987) *Ann. Rev. Biochem.* 56, 125.
D. Söll & P. R. Schimmel (1974) *The Enzymes*, 3rd ed., **10**, 489.

ISOMALTASE

This enzyme [EC 3.2.1.10] (also referred to as oligo-1,6-glucosidase, sucrase-isomaltase, and limit dextrinase) catalyzes the hydrolysis of 1,6-α-D-glucosidic linkages in isomaltose and dextrin products generated from starch and glycogen *via* α-amylase. **See also** *Sucrase*

E. H. Van Beers, H. A. Büller, R. J. Grand, A. W. C. Einerhand & J. Dekker (1995) *Crit. Rev. Biochem. Mol. Biol.* **30**, 197.
G. Mooser (1992) *The Enzymes*, 3rd ed., **20**, 187.
M. Vihinen & P. Mäntsälä (1989) *Crit. Rev. Biochem. Mol. Biol.* **24**, 329.

ISO MECHANISMS

Enzyme mechanisms for which the enzyme, upon release of product(s), assumes a different form (*i.e.*, a different conformational structure, or one requiring rehydration, deprotonation; *etc.*) than the enzyme form originally present at the beginning of the catalytic cycle. Such mechanisms require reisomerization of the free enzyme back to the original form before another round of catalysis can occur. For example, the Iso Uni Uni mechanism is given by the following scheme: E + A $\rightleftharpoons$ EA $\rightleftharpoons$ FP $\rightleftharpoons$ F + P followed by F $\rightleftharpoons$ E. A number of tools are

available to aid in the identification and characterization of Iso mechanisms[1,2]. These include product inhibition studies, dead-end inhibition, isotope effect studies, and induced transport experiments. Other observations that may reflect Iso mechanisms. These include: biphasicity observed in Arrhenius plots as well as plots of pressure effects and unusual viscosity effects on kinetic parameters.

[1]F. B. Rudolph (1979) *Meth. Enzymol.* **63**, 411.
[2]K. L. Rebholz & D. B. Northrop (1995) *Meth. Enzymol.* **249**, 211.

ISOMER

One of a number of molecular entities having the same atomic composition, but different line formulas and/or different stereochemical configurations. A distinction is usually made between structural isomers (molecular entities have the same atomic composition albeit different line formuls), stereoisomers (entities having different arrangements, usually noninterconvertible and nonsuperimposable, in space), and *cis-trans* isomers (having different positions for substituents with respect to double bonds or sides of a ring structure).

ISOMERASES

A major class of enzymes that catalyze changes within one molecule. Examples include racemases, epimerases, mutases, and tautomerases.

ISOMERIZATION

A chemical reaction in which the principal product is isomeric with respect to the principal reactant. Isomerization can occur in the absence of molecular rearrangements (for example, between conformational isomers).

ISOPENTENYL-DIPHOSPHATE Δ-ISOMERASE

This enzyme [EC 5.3.3.2] catalyzes the interconversion of isopentenyl pyrophosphate and dimethylallyl pyrophosphate.

R. A. Gibbs (1998) *Comprehensive Biological Catalysis: A Mechanistic Reference* **1**, 31.
P. W. Holloway (1972) *The Enzymes*, 3rd ed., **6**, 565.

ISOPEPTIDE BOND

A substituted amide linkage between two amino acids in which the carbonyl group and/or the nitrogen is not directly bonded to an α-carbon of one of the amino acids. Perhaps the most common example of an isopeptide bond in biological systems is the peptide linkage between the glutamyl residue and the cysteinyl residue in glutathione.

ISOPRENE RULE

A rule proposed by Ruzicka[1] that monoterpenes and sesquiterpenes are derived by electrophilic additions and rearrangements from a common intermediate.

[1]L. Ruzicka (1953) *Experentia* **9**, 357.

3-ISOPROPYLMALATE DEHYDRATASE

This enzyme [EC 4.2.1.33], also referred to as isopropylmalate hydro-lyase, catalyzes the conversion of 3-isopropylmalate to 2-isopropylmaleate and water.

G. B. Kohlhaw (1988) *Meth. Enzymol.* **166**, 423.

3-ISOPROPYLMALATE DEHYDROGENASE

This enzyme [EC 1.1.1.85] catalyzes the reaction of 3-carboxy-2-hydroxy-4-methylpentanoate and NAD^+ to produce 3-carboxy-4-methyl-2-oxopentanoate and NADH. The product then decarboxylates, generating 4-methyl-2-oxopentanoate.

L. L. Searles & J. M. Calvo (1988) *Meth. Enzymol.* **166**, 225.
G. B. Kohlhaw (1988) *Meth. Enzymol.* **166**, 429.

2-ISOPROPYLMALATE SYNTHASE

This enzyme [EC 4.1.3.12] catalyzes the reaction of 3-carboxy-3-hydroxy-4-methylpentanoate with coenzyme A to produce acetyl-CoA, 3-methyl-2-oxobutanoate, and water. This enzyme has a potassium-ion requirement.

G. B. Kohlhaw (1988) *Meth. Enzymol.* **166**, 414.
M. J. P. Higgins, J. A. Kornblatt & H. Rudney (1972) *The Enzymes*, 3rd ed., **7**, 407.

ISOPULLULANASE

This enzyme [EC 3.2.1.57] catalyzes the hydrolysis of pullulan to produce isopanose (6-α-maltosylglucose). Panose (4-α-isomaltosylglucose) is a substrate and is hydrolyzed to isomaltose and glucose.

M. Vihinen & P. Mäntsälä (1989) *Crit. Rev. Biochem. Mol. Biol.* **24**, 329.

ISOSBESTIC POINT

A wavelength in the absorption spectra (measured as a wavelength, frequency, or wavenumber) in which absorption coefficients are equal. At such a point, the absorption of a sample does not change. For example the isosbestic point in the conversion of thionicotinamide adenine dinucleotide (thionicotinamide-NAD^+), an alternative substrate for many dehydrogenases, to the reduced form of the coenzyme, is 342 nm[1]. Hence, the absorption of the sample at this wavelength will not change over the course of the reaction. (An obsolete synonym is isoabsorption point.) *See Beer-Lambert Law; Absorption Spectroscopy*

[1]F. B. Rudolph & H. J. Fromm (1970) *Biochemistry* **9**, 4660.

ISOSTERIC AMINES AS CARBOCATION ANALOGUES

Positively charged amines that are structural analogues and are isosteric with putative carbocation intermediates in enzymic reactions. These compounds have proved their value in efforts to characterize enzyme mechanisms that proceed by the transient formation of carbocation intermediates.

ISOSTERY

A physiological mechanism of metabolic regulation via competitive inhibition by structural analogues of the substrate.

ISOTHERMAL PROCESS

Any process which occurs at constant temperature (thus, $dT = 0$). Such processes typically have a heat source or thermostat to control the temperature of the system. *See Adiabatic Process*

ISOTONIC

1. Referring or relating to or demonstrating equal tension. 2. Often referring to an equivalence of osmotic pressure (*i.e.*, isosmotic). Thus, "isotonic" often refers to solutions having identical ionic strength.

ISOTONIC BUFFERS

Buffer solutions that are isosmotic with respect to some standard, typically chosen such that suspended cells will neither shrink nor expand. Sodium chloride solutions (0.90% weight/volume or 0.155 M) at 37°C is often used to represent physiological conditions. These buffer systems are also important in studies of intact cells and membranal organelles; likewise, many pharmaceutical formulations must be prepared as isotonic solutions. Most enzyme-catalyzed reactions are affected by ionic strength, and kinetic studies are typically conducted in solutions at constant ionic strength, approximately in the physiologic range. Care must be exercised to achieve an appropriate balance of sodium and potassium ions (*i.e.*, cytosolic = 140 mM K^+ and 12 mM Na^+; extracellular fluid or blood plasma = 4 mM K^+ and 145 mM Na^+).

Sörenson[1] introduced the concept of pH and was among the first to consider the effects of pH on enzyme-catalyzed reactions. That same publication describes a phosphate buffer system, now widely known as a Sörenson buffer. In its original formulation these buffer solutions were hypotonic, varying slightly in osmotic properties throughout the pH range. Curie and Sciarrone[2] described a modified Sörenson buffer system that is isotonic throughout the pH range. These authors provided a table for preparing isotonic phosphate buffers having a tonicity of 0.92%. For example, 50 mL 0.0667 M Na_2HPO_4 (*i.e.*, 9.470 g/L) is added to 50 mL 0.0667 M NaH_2PO_4 (*i.e.*, 9.208 g/L of $NaH_2PO_4a \cdot H_2O$) together with 0.480 g of NaCl (for every 100 mL solution), and deionized water, yields a buffer of tonicity 0.92 and pH 6.72 at 25°C (6.71 at 37°C).

Krebs-Ringer physiologic buffer is prepared by mixing 100 mL solution A (0.9% w/v NaCl), 4 mL solution B (1.15% w/v KCl), 3 mL solution C (1.22% w/v $CaCl_2$), 1 mL solution D (2.11% w/v KH_2PO_4), and 1 mL solution E (3.8% w/v $MgSO_4 \cdot 7H_2O$). Krebs-Ringer-bicarbonate buffer is prepared in the same manner, except that 21 mL solution F (1.3% w/v $NaHCO_3$) is added. Krebs-improved-Ringer buffer solution is prepared by mixing 80 mL solution A, 4 mL solution B, 3 mL solution C, 1 mL solution D, 1 mL solution E, 21 mL solution F, 4 mL 0.16 M sodium pyruvate, 7 mL 0.1 M sodium fumarate, 4 mL 0.16 M sodium glutamate, and 5 mL 0.3 M glucose.

Palitzsch[3] described buffers utilizing sodium tetraborate and boric acid. Several authors[2,4] have presented tables for modified Palitzsch buffers which are isotonic (tonicity of 0.90 at 37°C). For example, 30 ml of 0.05 M sodium tetraborate ($Na_2B_4O_7 \cdot 10H_2O$ dissolved in deionized water) and 70 ml of 0.2 M boric acid together with 240 mg NaCl for every 100 ml will yield a solution having a pH of 8.12 at 25°C (and 8.11 at 37°C). However, care should always be exercised when using borate-based buffers since borate can form complexes with alcohols, polyols, and especially carbohydrates and ribonucleotides.

[1]S. P. L. Sörensen (1909) *Biochem. Zeit.* **21**, 131.
[2]A. P. Curie & B. J. Sciarrone (1969) *J. Pharm. Sci.* **58**, 990.
[3]S. Palitzsch (1915) *Biochem. Zeit.* **70**, 333.
[4]D. Perrin & B. Dempsey (1974) *Buffers for pH and Metal Ion Control*, Chapman and Hall, London.

ISOTOPE DILUTION

An analytical chemical technique that utilizes radioactive (or stable) isotopes for the quantitative analysis of the amount of substance. In the absence of a kinetic isotope effect, isotopic isomers react identically with respect to their unlabeled counterparts. The method offers the advantage that specific activity (or gram-atom excess in the case of stable isotopes) is an intensive variable. Therefore, one only needs to recover sufficient labeled metabolite to determine amount of substance and disintegrations per minute (or, gram-atom excess) to reach an accurate determination of specific activity. The technique is feasible so long as one can accurately determine the initial and final specific activities.

We can define the specific activity of a substance A as,

$$\text{Specific Activity} = SA_A = \text{mol A*}/(\text{mol A} + \text{mol A*})$$

When mol A* is much less than mol A (*i.e.*, if A* is truly a tracer species), then

$$SA_A = \text{mol A*}/\text{mol A} \propto \text{dpm A*}/\text{mol A}$$

Because specific activity is always an intensive variable, one need not recover all of the sample to determine the ratio mol A*/mol A. To evaluate SA_A, one need only obtain an amount that provides an accurate measurement of A* and total A. The former is easily achieved by liquid scintillation counting, and spectrophotometry or some other enzymatic assay usually allows accurate determination of mol A in a sample.

To illustrate how specific activity can be used to analyze the amount-of-substance, suppose that we combine isotopically labeled sample (SA_A as written above) with an unlabeled sample containing an unknown amount of the substance (call it A_x). Then, the new specific activity will be lower, such that

$$SA_A' = \text{mol A*}/(\text{mol A} + \text{mol A*} + \text{mol A}_x)$$

If the tracer contributes vanishingly little to the denominator, then

$$SA_A' = mol\ A^*/(mol\ A + mol\ A_x)$$

Rewriting each of the above equations in terms of mol A^*, and we obtain the following

$$SA_A(mol\ A) = mol\ A^*$$

$$SA_A'(mol\ A + mol\ A_x) = mol\ A^*$$

Setting these equations equal to each other, we get

$$SA_A(mol\ A) = SA_A'(mol\ A + mol\ A_x)$$

or

$$SA_A/SA_A' = (mol\ A + mol\ A_x)/(mol\ A)$$

Dividing the right-hand numerator and denominator by (mol A), we arrive at the following expression:

$$SA_A'/SA_A = 1 + (mol\ A_x/mol\ A)$$

Thus, accurate determination of the specific activities SA_A and SA_A' immediately provides the value of the unknown A_x. Note: Whenever $A_x < A$, the isotope dilution method suffers from any statistical variation in the measurements of quantities that are nearly identical. Should this prove to be true for your measurements, use a lesser amount of isotopic probe, so that $A_x \ll A$.

ISOTOPE EXCHANGE

1. A reaction in which the reactants and products exchange atoms or groups of atoms that are chemically identical and differ only in isotopic composition. For example, the exchange of deuterium in D_2O with the labile hydrogens of glucose in solution represents this type of isotope exchange. 2. The exchange of an isotope from one isotopically substituted substrate in a multisubstrate enzyme-catalyzed reaction, with at least one of the other substrates.

ISOTOPE EXCHANGE (Rate Law)

Duffield & Calvin[1] were among the first to derive the expression for an exchange reaction involving the transfer of isotope X^* between two reactants:

$$A-X + B-X^* = A-X^* + B-X$$

The gross rate of exchange R depends on the fraction of BX total that is present as BX^* and the fraction of AX_{total} present as AX.

$$-\ln (1 - F) = R[(a + b)/ab]\ t$$

where F is the fractional attainment of isotopic equilibrium (*i.e.*, $F = [A-X^*]_t/[A-X^*]_{eq} = [B-X^*]_t/[B-X^*]_{eq}$), a is the concentration of $A-X$, and b is the concentration of $B-X$. R has units of molarity per unit time (*i.e.*, M/s).

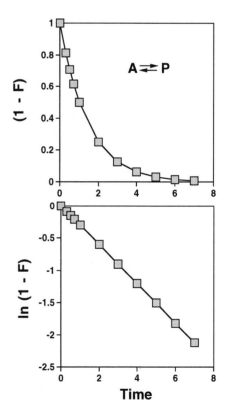

The expression[2] given above for R indicates that the exchange process is first order, and it will remain so provided tracer quantities are employed. It is also possible to roughly measure R by single-point determinations, but it is preferable to use several sampling times and to confirm the first-order nature of the measured exchange rate. In enzyme-catalyzed reactions, R depends upon the fraction of enzyme capable of reacting with the labeled substrate (or product), and all substrates and products affect R. This is true because the concentrations of substrate(s) and product(s) control the concentration of all Michaelis complexes as well as any "mixed" substrate-product-enzyme complexes resulting from abortive complex formation. In some cases, this equation may be further simplified if the level of one exchange partner is considerably greater than the other. If a is much greater than b, the rate equation for exchange will become:

$$R = -\frac{[\ln(1-F)][X][H_2O]}{([X]+[H_2O])\,t} = -\frac{[\ln(1-F)][X]}{t}$$

This rate law, for example, applies to the tritium exchange from labeled dihydroxyacetone phosphate to water in the aldolase reaction or the exchange from labeled malate to water in the fumarase reaction.

[1] R. B. Duffield & M. Calvin (1946) *J. Am. Chem. Soc.* **68**, 557.
[2] A. A. Frost & R. G. Pearson (1961) *Kinetics and Mechanism*, 2nd ed., Wiley, New York.

ISOTOPE EXCHANGE (Reactions At Equilibrium)

Even at dynamic equilibrium, uncatalyzed and catalyzed reactions relentlessly shuttle reactants and products forth and back, and the flux in each direction is essentially constant on the time scale used in most kinetic experiments. Inasmuch as the rate of enzymic catalysis depends upon the concentration of enzyme-substrate(s) and enzyme-product(s) complexes, the enzyme's behavior at equilibrium is a complex composite of the availability of ligand binding sites and the magnitude of rate constants governing steps in the forward and reverse directions. It is also true that the equilibrium flux in each direction depends upon enzyme concentration, and one may adjust the concentration of enzyme to suit the needs of the particular experiment. The cardinal feature of all exchange processes is that they obey simple first-order kinetics, and despite the complexity of enzyme interactions with substrate(s) and effector(s), the experimental determination of rate behavior is therefore relatively straightforward. Introduction of a labeled substrate or product allows one to trace quantitatively the course of exchange reactions. The most common applications of equilibrium exchange kinetics are the determination of substrate binding mechanisms (see Tables I and II) and the influ-

ence of regulatory molecules on enzymic catalysis. On the other hand, equilibrium exchange reactions can provide evidence for certain chemical intermediates as well as positional isotope exchange.

Depending on the position of the isotopic atom(s) in the labeled substrate or product, various exchanges may be examined[1-3]. The types of exchanges subject to measurement can be illustrated by considering a bisubstrate reaction:

$$A-X + B \rightleftharpoons A + B-X$$

For exchange to occur, atom(s) (or functional groupings of atoms, such as phosphoryl groups) must be transferred between two substrate-product partners. Thus, only three exchanges can occur in this example: $A-X$ with A; $A-X$ with $B-X$; B with $B-X$; but *not* B with A. In the hexokinase reaction, for example, one can observe exchange reactions between glucose and glucose 6-phosphate, ATP and ADP, and ATP and glucose 6-phosphate. But, there is no exchange of atoms between glucose and ADP. Likewise, NAD$^+$-dependent dehydrogenases never exhibit exchange between oxidized substrate and oxidized coenzyme. More complicated enzyme systems may have a number of exchange processes associated with catalysis, and the glutamine synthetase reaction provides an excellent example of the types of possible exchange reactions: [^{14}C]ATP with ADP; [^{14}C]glutamate with glutamine; [γ-^{32}P]ATP with orthophosphate; [^{15}N]ammonia with glutamine; [γ-^{18}O]ATP with orthophosphate; [^{18}O]glutamate with orthophosphate; [^{18}O]glutamine with orthophosphate; [^{18}O]ATP with glutamate; [^{18}O]ATP with glutamine; and [^{18}O]glutamate with glutamine. These do not include so-called positional isotope exchange (PIX) reactions which tell us about stereochemical properties and the reversibility of certain exchange reactions.

To maintain equilibrium, attention must be given to the mass-action ratio, which is the multiplicand of product concentrations (that is, [P], [Q], *etc.*) divided by the product of the substrate concentrations (that is, [A], [B], *etc.*):

$$K_{eq} = \{[P][Q]\ldots\}/\{[A][B]\ldots\}$$

Typically, the system can be treated in terms of pairwise concentrations of substrate and product (*e.g.*, let $\alpha = [P]/[A]$ and $\beta = [Q]/[B]$, where the product $\alpha\beta$ equals

Table I   Isotope Exchange-Rate Expressions for Selected Uni and Bi Substrate Systems[a]

1. Uni Uni

$$R_{A \leftrightarrow P} = \frac{V_1[A]}{K_a[1 + [A]/K_{ia} + [P]/K_{ip}]}$$

2. Ordered Bi Uni

$$R_{A \leftrightarrow P} = \frac{V_1[A][B]}{(K_aK_b + K_a[B])[1 + [A]/K_{ia} + [P]/K_{ip}]} \qquad R_{B \leftrightarrow P} = \frac{V_1[B]}{K_b[1 + K_{ia}/[A] + K_{ia} + [P]/K_{ip}[A]]}$$

3. Random Bi Uni (Rapid Equilibrium)

$$R_{all} = \frac{V_1}{1 + K_a/[A] + K_b/[B] + (K_{ia}K_b/[A][B])(1 + [P]/K_{ip})}$$

4. Random Bi Uni

$$R_{A \leftrightarrow P} = \frac{[k_1k_3k_9[A][B] + k_7k_9[A](k_2 + k_3[B])[B]/K_{ib}]E_0}{[k_2(k_4 + k_8 + k_9) + k_3[B](k_8 + k_9)][1 + [A]/K_{ia} + [B]/K_{ib} + [A][B]/K_{ia}K_b]}$$

$$R_{B \leftrightarrow P} = \frac{[k_5k_7k_9[A][B] + k_3k_9[B](k_6 + k_7[A])([A]/K_{ia})]E_0}{[k_6(k_4 + k_8 + k_9) + k_7[A](k_4 + k_9)][1 + [A]/K_{ia} + [B]/K_{ib} + [A][B]/K_{ia}K_b]}$$

5. Ordered Bi Bi

$$R_{A \leftrightarrow Q} = \frac{V_1[A][B]}{[K_{ia}K_b + K_a[B] + K_{ia}K_bK_q[P]/K_{iq}K_p][1 + ([A]/K_{ia})(1 + [B]/K_{eb}) + [Q]/K_{iq}]}$$

$$R_{A \leftrightarrow P} = \frac{V_1[A][B]}{[K_{ia}K_b + K_a[B]][1 + ([A]/K_{ia})(1 + [B]/K_{eb}) + [Q]/K_{iq}]} \qquad R_{B \leftrightarrow P} = \frac{V_1[B]}{K_b[1 + K_{ia}/[A] + [B]/K_{eb} + K_{ia}[Q]/K_{iq}[A]]}$$

$$R_{B \leftrightarrow Q} = \frac{V_1[B]}{[K_b + K_bK_q[P]/K_{iq}K_p][1 + K_{ia}/[A] + [B]/K_{eb} + K_{ia}[Q]/K_{iq}[A]]}$$

6. Theorell–Chance Bi Bi

$$R_{A \leftrightarrow Q} = \frac{V_1[A][B]}{[K_a[B] + K_{ia}K_b + K_{iq}[P]/K_{eq}][1 + [A]/K_{ia} + [Q]/K_{iq}]} \qquad R_{A \leftrightarrow P} = \frac{V_1[A][B]}{[K_{ia}K_b + K_a[B]][1 + [A]/K_{ia} + [Q]/K_{iq}]}$$

$$R_{B \leftrightarrow Q} = \frac{V_1[B]}{[K_b + K_{iq}[P]/K_{ia}K_{eq}][1 + K_{ia}/[A] + K_{ia}[Q]/K_{iq}[A]]} \qquad R_{B \leftrightarrow P} = \frac{V_1[B]}{K_b[1 + K_{ia}/[A] + K_{ia}[Q]/K_{iq}[A]]}$$

7. Random Bi Bi (Rapid Equilibrium)

$$R_{all} = \frac{V_1}{1 + K_a/[A] + K_b/[B] + K_{ia}K_b/[A][B][1 + [P]/K_{ip} + [Q]/K_{iq} + [P][Q]/K_{ip}K_q]}$$

8. Random Bi Bi

$$R_{A \leftrightarrow Q} = \frac{[k_9k_{11} + k_{13}(k_{10}[P] + k_{11})][k_1k_3[A][B] + k_7([A][B]/K_{1b})(k_2 + k_3[B])]E_0}{(Z)\{k_2[k_{10}[P](k_4 + k_8 + k_{13}) + k_{11}(k_4 + k_8 + k_9 + k_{13})] + k_3[B][k_{10}[P](k_8 + k_{13}) + k_{11}(k_8 + k_9 + k_{13})]\}}$$

$$R_{A \leftrightarrow P} = \frac{[k_{13}k_{15} + k_9(k_{14}[Q] + k_{15})][k_1k_3[A][B] + k_7([A][B]/K_{1b})(k_2 + k_3[B])]E_0}{(Z)\{k_2[k_{14}[Q](k_4 + k_8 + k_9) + k_{15}(k_4 + k_8 + k_9 + k_{13})] + k_3[B][k_{14}[Q](k_8 + k_9) + k_{15}(k_8 + k_9 + k_{13})]\}}$$

$$R_{B \leftrightarrow Q} = \frac{[k_9k_{11} + k_{13}(k_{10}[P] + k_{11})][k_5k_7[A][B] + k_3([A][B]/K_{ia})(k_6 + k_7[A])]E_0}{(Z)\{k_6[k_{10}[P](k_4 + k_8 + k_{13}) + k_{11}(k_4 + k_8 + k_9 + k_{13})] + k_7[A][k_{10}[P](k_4 + k_{13}) + k_{11}(k_4 + k_9 + k_{13})]\}}$$

$$R_{B \leftrightarrow P} = \frac{[k_{13}k_{15} + k_9(k_{14}[Q] + k_{15})][k_5k_7[A][B] + k_3([A][B]/K_{ia})(k_6 + k_7[A])]E_0}{(Z)\{k_6[k_{14}[Q](k_4 + k_8 + k_9) + k_{15}(k_4 + k_8 + k_9 + k_{13})] + k_7[A][k_{14}[Q](k_4 + k_9) + k_{15}(k_4 + k_9 + k_{13})]\}}$$

where $Z = \{1 + [A]/K_{ia}(1 + [B]/K_b) + [B]/K_{1b} + [P]/K_{ip} + [Q]/K_{iq}\}$

9. Random On–Ordered Off Bi Bi

$$R_{A \leftrightarrow Q} = \frac{k_9k_{11}[k_1k_3[A][B] + k_7([A][B]/K_{iab})(k_2 + k_3[B])]E_0}{\{k_2[k_{10}[P](k_4 + k_8) + k_{11}(k_4 + k_8 + k_9)] + k_3[B][k_8k_{10}[P] + k_{11}(k_8 + k_9)]\} (Y)}$$

$$R_{A \leftrightarrow P} = \frac{k_9[k_1k_3[A][B] + k_7([A][B]/K_{ib})(k_2 + k_3[B])]E_0}{\{k_3[B](k_8 + k_9) + k_2(k_4 + k_8 + k_9)\} (Y)}$$

$$R_{B \leftrightarrow Q} = \frac{k_9k_{11}[k_5k_7[A][B] + k_3([A][B]/K_{ia})(k_6 + k_7[A])]E_0}{\{k_6[k_{10}[P](k_4 + k_8) + k_{11}(k_4 + k_8 + k_9)] + k_7[A][k_4k_{10}[P] + k_{11}(k_4 + k_9)]\} (Y)}$$

$$R_{B \leftrightarrow P} = \frac{k_9[k_5k_7[A][B] + k_3([A][B]/K_{ia})(k_6 + k_7[A])]E_0}{\{k_7[A](k_4 + k_9) + k_6(k_4 + k_8 + k_9)\} (Y)}$$

where $Y = \{1 + ([A]/K_{ia})(1 + [B]/K_b) + [B]/K_{ib} + [Q]/K_{iq}\}$

10. Ping Pong Bi Bi

$$R_{A \leftrightarrow P} = \frac{V_1[A]}{K_a[1 + [A]/K_{ia} + K_{ip}[A]/K_{ia}[P] + [Q]/K_{iq}]} \qquad R_{B \leftrightarrow Q} = \frac{V_1[B]}{K_b[1 + [B]/K_{ib} + [P]/K_{ip} + K_{iq}[B]/K_{ib}[Q]]}$$

$$R_{A \leftrightarrow Q} = \frac{V_1[A][B]}{[K_a[B] + (K_{ia}K_b[P]/K_{ip})][1 + [A]/K_{ia} + [Q]/K_{iq} + K_{ip}[A]/K_{ia}[P]]}$$

---

[a] The expressions are derived for systems at chemical equilibrium and in the absence of abortive complexes. All expressions, except for the Uni Uni and Rapid Equilibrium cases, were derived assuming only one central catalytic complex.

Table II

Exchange Profiles for Enzyme Kinetic Mechanisms[a]

Mechanism	Exchange	Varied substrate-product pair(s)				
		A-P	B-P	A-Q	B-Q	A-B-P-Q[b]
1. Uni Uni	$A \leftrightarrow P$					
2. Ordered Bi Uni	$A \leftrightarrow P$					
	$B \leftrightarrow P$					
3. Random Bi Uni (Rapid Equilibrium)	Any					
4. Random Bi Uni	$A \leftrightarrow P$					
	$B \leftrightarrow P$					
5. Ordered Bi Bi	$A \leftrightarrow P$					
	$B \leftrightarrow P$					
	$A \leftrightarrow Q$					
	$B \leftrightarrow Q$					
6. Theorell-Chance Bi Bi	$A \leftrightarrow P$					
	$B \leftrightarrow P$					
	$A \leftrightarrow Q$					
	$B \leftrightarrow Q$					
7. Random Bi Bi (Rapid Equilibrium)	Any					
8. Random Bi Bi	Any					

[a]See Table I for a parallel listing of rate equations.
[b]A-B-P-Q refers to the Wedler-Boyer protocol of varying all substrates and products in a constant ratio.

the mass-action ratio). In this way, each substrate-product pair can be held constant with respect to the ratio of the absolute concentrations of substrate and product comprising that pair. In the case of the hexokinase reaction, for which the apparent equilibrium constant is about 500 at pH 6.5, we could chose α = [glucose 6-phosphate]/[glucose] = 25 and β = [ADP]/[ATP] = 20, such that $\alpha\beta$ equals 500. The sugar pair could be held at any set of absolute concentrations, say for instance 10 mM and 0.4 mM, while one can establish a number of nucleotide pairs (e.g., β_1 = ([ADP] = 20 mM)/([ATP] = 1 mM); β_2 = ([ADP] = 16 mM)/([ATP] = 0.8 mM); β_3 = ([ADP] = 12 mM)/([ATP] = 0.6 mM); β_4 = ([ADP] = 8 mM)/([ATP] = 0.4 mM); β_5 = ([ADP] = 4 mM)/([ATP] = 0.2 mM). Any $\alpha\beta$ product still satisfies the mass-action ratio, but this procedure allows one to vary the degree of saturation of hexokinase, in this case, with respect to the nucleotide pair. The symmetrical alteration of the sugar pair affords the opportunity to characterize the effects of increasing saturation of hexokinase by glucose and glucose 6-phosphate. Moreover, the nonvaried substrate-product pair can be prepared to contain buffer, metal ions, salt, and modifiers (if desired), and enzyme can be added to completely equilibrate the system. Because care has been exercised in choosing the above conditions, the reaction should be at or very near equilibrium, even before addition of enzyme. At this point, one may now introduce a small aliquot of stable or radioactive isotopically labeled reactant (where reactant is taken either to be substrate or product); the progress of the equilibrium exchange reaction may then be monitored by quenching an aliquot of the complete reaction sample, followed by chromatographic separation of labeled substrate and product. A useful check on systematic and/or consistent errors is that the progress curves for all exchange reactions are strictly first-order; moreover, the maximal rate of exchange should be linearly proportional to enzyme concentration.

Let us now examine the behavior of enzymes operating by way of ordered and random kinetic bisubstrate mechanisms:

1. Ordered Bi Bi Ternary Complex Mechanism

$$E \underset{k_2}{\overset{k_1[A]}{\rightleftharpoons}} EA \underset{k_4}{\overset{k_3[B]}{\rightleftharpoons}} EAB \underset{k_6}{\overset{k_5}{\rightleftharpoons}} EPQ \underset{k_8[P]}{\overset{k_7}{\rightleftharpoons}} EQ \underset{k_{10}[Q]}{\overset{k_9}{\rightleftharpoons}} E$$

The A↔Q exchange velocity as a function of the absolute concentrations of the [B]/[P] pair is the diagnostic test for compulsory ordered mechanisms wherein one substrate A must precede the binding of the other substrate, here B, and one product P must precede the release of the second product, here designated as Q. As the concentrations of [B] and [P] are raised, while maintaining the [B]/[P] ratio, a great deal of the enzyme will be drawn into the central ternary complexes, and the fraction of total enzyme present as uncomplexed E will become progressively lower. For A↔Q exchange to occur, labeled A* or Q* must bind to free enzyme because the rate of all exchange processes reflects the fraction of reactant X that can combine with the fraction of another reactant Y. Since uncomplexed enzyme is required to form EA* (or EQ* in the reverse reaction), the A↔Q exchange will increase as we proceed from low to moderate absolute concentrations of the [B]/[P] pair. However, the A↔Q exchange will not be favored as we continue to raise the absolute concentrations of the [B]/[P] pair. Note that the A↔Q exchange will not be depressed at high [A]/[Q] absolute levels, because increasing the A and Q concentrations actually favors the binding of A* (or Q*). Interestingly, the B↔P exchange requires sufficient concentrations of EA and EQ for combination with B* or P*, respectively. Thus, raising the absolute concentrations of components of the [A]/[Q] or [B]/[P] pair will increase the B↔P exchange rate, and a plot of the exchange rate *versus* absolute concentrations of either pair will be hyperbolic. This behavior is fully consistent with saturation kinetics. In summary, the depression of the A↔Q exchange at high absolute levels of B and P serves to categorize an enzyme as a ordered ternary complex type.

The now classical example is lactate dehydrogenase. Silverstein and Boyer[4] were the first to determine the rates of exchange between cognate pairs of reactants (i.e., lactate and pyruvate as well as NAD+ and NADH). Convenient [NADH]/[NAD+] and [pyruvate]/[lactate] ratios were chosen such that when combined they satisfied the apparent equilibrium constant for the LDH reaction. These investigators first established that each exchange rate was directly proportional to the duration of exchange and likewise directly proportional to enzyme concentration. As an additional control, they also demonstrated the equality of the pyruvate ↔ lactate exchange rates to within ten percent by starting with either [1-^{14}C]pyruvate or [1-^{14}C]lactate. Then, to examine the order of substrate binding/release, Silverstein and Boyer examined the effect of raising the absolute concentrations of lactate and pyruvate on the rate of equilibrium exchange between pyruvate and lactate and between NAD$^+$ and NADH. As shown in Fig. 1, the exchange between pyruvate and lactate rises to what is an essentially stable plateau, whereas the exchange between co-enzymes is markedly depressed over this concentration range of pyruvate and lactate. These observations exclude a random mechanism of substrate addition (see next paragraph) and were consistent with an ordered mechanism in which the coenzymes form the "outer" Michaelis complexes and each combines with its respective three-carbon acid to form "inner" ternary complexes.

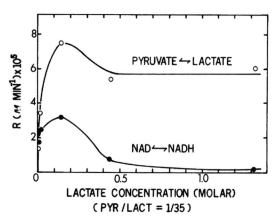

Figure 1. Effect of lactate and pyruvate concentrations on the equilibrium exchange rates of rabbit muscle lactate dehydrogenase at pH 7.9.[4] Reaction mixtures contained 1.68-1.70 mM NAD$^+$, 32.2-45.5 mM NADH, and the indicated concentrations of pyruvate and lactate.

2. Random Bi Bi Ternary Complex Mechanism

The random or noncompulsory ordered mechanism is noticeably symmetrical, with two different paths for producing the EAB complex from free enzyme E and its substrates A and B, as well as two different pathways for producing the EPQ complex from free enzyme E and its products P and Q:

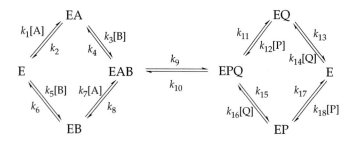

For this reason, these alternative routes for isotope combination with enzyme-substrate and/or enzyme-product complexes ensures that raising the [A]/[Q] or [B]/[P] pair will not depress either the A$\leftrightarrow$Q or the B$\leftrightarrow$P exchanges. Fromm, Silverstein, and Boyer[5] conducted a thorough analysis of the equilibrium exchange kinetic behavior of yeast hexokinase, and the data shown in Fig. 2 indicate that there is a random mechanism of substrate addition and product release.

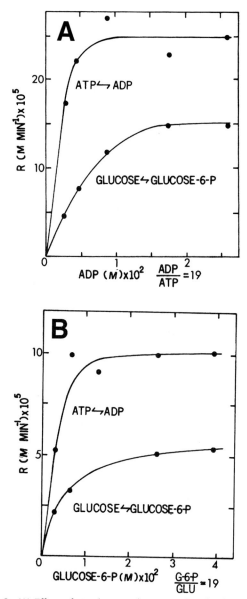

Figure 2. (A) Effect of simultaneously raising the absolute concentrations of ATP and ADP on the indicated equilibrium exchange reactions catalyzed by yeast hexokinase. (B) Effect of simultaneously raising the absolute concentrations of glucose and glucose 6-phosphate on the indicated equilibrium exchange reactions of yeast hexokinase.

These qualitative conclusions can be justified in terms of rate equations derived for each exchange catalyzed by ordered and random ternary complex mechanisms[1,3].

[1]P. D. Boyer (1959) *Arch. Biochem. Biophys.* **82**, 387.
[2]P. D. Boyer (1978) *Acc. Chem. Res.* **11**, 218.
[3]D. L. Purich & R. D. Allison (1982) *Meth. Enzymol.* **64**, 3.
[4]E. Silverstein & P. D. Boyer (1964) *J. Biol. Chem.* **239**, 3901 and 3908.
[5]H. J. Fromm, E. Silverstein & P. D. Boyer (1964) *J. Biol. Chem.* **239**, 3645.

Selected entries from *Methods in Enzymology* [vol, page(s)]: Abortive complex interaction, **64**, 39-45; enzyme stability and, **64**, 23, 24; equilibrium exchange [equations, **63**, 76-79; **64**, 7-18; experimental protocol, **64**, 19-27, 31; initial rate method, **64**, 32, 33; ionic strength effect, **64**, 27, 32; kinetic parameter, **64**, 17, 18, 25, 31, 32; pH effect, **64**, 23; profile, **64**, 13, 17; quenching exchange reactions, **64**, 25; rapid equilibrium model, **64**, 6, 7, 17, 18; side reaction effect, **64**, 24; substrate inhibition, **64**, 34-36; substrate-product ratio, **64**, 22, 23; substrate stability, **64**, 24; theory, **64**, 7-18, 30, 32, 33, 44; types, **64**, 4, 5]; intermediate, **64**, 25, 26, 34, 36-38; irreversible enzyme reactions, **64**, 36-38; modifier effects, **87**, 647; substrate synergism, **64**, 38, 39; equations for isotope exchange, **64**, 7-18 [derivation, **63**, 76-83; Cleland method, **63**, 80-83; equilibrium assumption, **63**, 76-79; steady state assumption, **63**, 79-83].

ISOTOPE EXCHANGE (Reactions Away From Equilibrium)

Although most enzyme exchange studies have been investigated at equilibrium, the back exchange of labeled product while the reaction is proceeding in the forward direction can provide valuable information about enzymic catalysis. Under favorable conditions, one may utilize such isotope exchange data to learn about the order of product release and the presence of covalent enzyme-substrate compounds. One of the first systems to be characterized in this way was glucose-6-phosphatase[1,2].

During the course of glucose 6-phosphate hydrolysis, radiolabeled glucose was added and the back exchange to form labeled glucose 6-phosphate was examined. (Likewise, the possibility of incorporation of labeled phosphate into glucose 6-phosphate was examined under identical conditions, but none was observed.) The rate of the glucose ↔ glucose 6-phosphate exchange correlated rather well with the amount of glucose inhibition of the phosphatase. These findings indicated that glucose is the first product to leave and that the negative free energy of hydrolysis is preserved as a phosphoryl-enzyme compound. It is also interesting to note that Zatman, Kaplan, and Colowick[3] used such exchange phenomena away from equilibrium to synthesize [^{14}C]NAD$^+$ from [^{14}C]-

nicotinamide and NAD$^+$ with bovine spleen NADase. They also observed that labeled adenosine diphosphoribosyl pyrophosphate (ADPR) failed to exchange in a similar fashion.

These mechanisms involve ordered release of products, and they are formally of the ping pong Bi Bi type with H_2O presumably entering the catalytic process after release of the first product. Labeled product, P, can exchange back to substrate A in the absence of the other product Q only when there is a significant level of EQ in the steady state and P* is sufficiently high (*i.e.*, equal or greater than the respective K_i value for P). An exchange from Q* back to A may only occur when there is a significant level of P present. If Q* does exchange without P present, one must conclude that the order is noncompulsory (random) with an adequate level of the EP complex in the steady state to support significant exchange. In this respect, observation of a P* ↔ A exchange is not strict proof of ordered release, and one may argue that a random mechanism with slow EQ breakdown to enzyme and Q is operative[4].

[1]L. F. Hass & W. L. Byrne (1960) *J. Am. Chem. Soc.* **82**, 947.
[2]R. C. Nordlie (1982) *Meth. Enzymol.* **87**, 319.
[3]L. J. Zatman, N. O. Kaplan & S. P. Colowick (1953) *J. Biol. Chem.* **200**, 197.
[4]D. L. Purich & R. D. Allison (1982) *Meth. Enzymol.* **64**, 1.

ISOTOPE RATIO MASS SPECTROMETER

An instrument that measures the isotopic mass ratio of a gas by bombarding the sample in an electron beam, such that the molecular ions generated can be deflected in their trajectories through a magnetic field in accordance to their charge/mass ratios. These devices are extremely accurate and reliable, and many stable isotope experiments can be analyzed by converting the isotopically substituted metabolite into carbon dioxide, water, or molecular nitrogen prior to IRMS measurements.

ISOTOPE SCRAMBLING

A change in the distribution of isotopes within a specified set of atoms of a particular molecular entity or entities. *See* Positional Isotope Exchange; Bridge-to-Nonbridge Oxygen Exchange

ISOTOPE TRAPPING

A technique for evaluating the rate (and rate constant) for the dissociation of an enzyme-substrate covalent compound or noncovalent complex, as compared to its rate (and rate constant) for interconversion into product. While the method was first applied by Meister's group[1] in 1962, its general utility remained unappreciated until the studies of Rose and co-workers[2]. (Among many practicing kineticists, the experimental strategy is now frequently called the Rose experiment.)

Rose and co-workers[2] first demonstrated that a proteolyzed form of hexokinase forms a "sticky" (or sluggishly dissociable) complex with glucose. The generalized application of this approach to the kinetic characterization of multisubstrate enzymes has been treated in detail[3]. **See also** *Partition Coefficient; Radiospecific Activity; Stickiness*

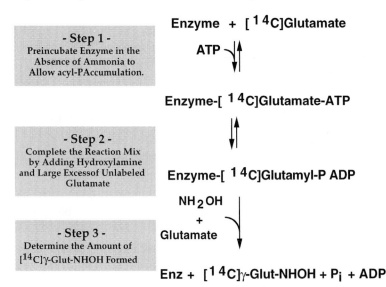

To adduce evidence for a γ-glutamyl phosphate intermediate in the sheep brain glutamine synthetase reaction, Meister's group incubated that enzyme with [14C]glutamate and ATP [in strict absence of the third substrate ammonia (or hydroxylamine)]. This solution was subsequently mixed with a second solution containing a great molar excess of unlabeled glutamate and the third substrate to initiate the overall reaction to form glutamine. The basic idea was that the formation of the γ-glutamyl phosphate intermediate in the absence of the third substrate would prime the enzymatic reaction to favor radioactive glutamine formation *versus* desorption of substrates. The reasoning went that if no acyl phosphate intermediate was formed, the radiolabeled glutamate would probably dissociate and become diluted with the excess of unlabeled glutamate, such that the amount of [14C]glutamine formed enzymatically would be miniscule. Rose offered the more rigorous interpretation that whether or not a covalent intermediary is in fact formed, the amount of [14C]glutamine depended only on the partition coefficient $[k_{\text{onward}}/(k_{\text{onward}} + k_{\text{dissoc}})]$, where k_{onward} is the onward rate constant for product formation and k_{dissoc} is the rate constant for dissociation of the complex of enzyme and radioactive substrate.

[1]P. R. Krishnaswamy, V. Pamiljans & A. Meister (1962) *J. Biol. Chem.* **237**, 2932.

[2]I. A. Rose, E. L. O'Connell, S. Litwin & J. Bar-Tana (1974) *J. Biol. Chem.* **249**, 5163.

[3]I. A. Rose (1980) *Meth. Enzymol.* **64**, 47.

Selected entries from *Methods in Enzymology* [vol, page(s)]: Enzyme-substrate complex formation, **64**, 53; data analysis, **64**, 56, 57; extensions of technique, **64**, 57-59; as evidence for occurrence of intermediate, **64**, 47-59; kinetic equation, **64**, 49-52; limitations, **64**, 57-59; mixing procedure, **64**, 53-56; reaction condition, **64**, 56, 57; termination, **64**, 56, 57.

ISOTOPICALLY SENSITIVE STEP

An elementary reaction or step (in a chemical process) for which the rate constants are altered by an isotopic substitution in substrate, product, or solvent. *See* Kinetic Isotope Effect; Solvent Isotope Effect

ISOTOPIC PERTURBATION

Any observable effect of isotopic substitution on the rate or extent of a chemical/physical process. Equilibrium isotopic perturbation measurements can provide valuable information about kinetic isotope effects on enzymic catalysis[1]. NMR shift difference measurements are also useful in detecting the effects of isotopic substitution on a fast (degenerate) equilibrium between two species differing only in their specific isotopic substitution[2,3]. The latter technique allows one to differentiate a rapidly equilibrating mixture with time-averaged symmetry from that of a single structure with higher symmetry[2,3].

[1]W. W. Cleland (1980) *Meth. Enzymol.* **64**, 104.
[2]IUPAC (1994) *Pure and Appl. Chem.* **66**, 1077.
[3]H. U. Siehl (1987) *Adv. Phys. Org. Chem.* **23**, 63.

ISOTOPOLOGUES

Molecules having the same type and number of isotopic atoms, albeit in different positions. CH_2DCH $(NH_3^+)COO^-$ and $CH_3CD(NH_3^+)COO^-$ are examples of isotopomers. Certain stereoisomers and *cis-trans* isomers can also be regarded as isotopomers: for example, (*R*)- and (*S*)-1-deuteroethanol, CH_3CHDOH.

ISOTROPIC

A term used with respect to a medium or substance or body having physical properties which are independent of direction. **See also** Anisotropic

ISOTROPIC MOTION

In the standard mathematical expressions for the contribution of quadrupolar relaxation to the relaxation rates of the quadrupolar nucleus (in nuclear magnetic resonance), rapid isotropic motion is assumed to occur. This behavior, in most cases, will not be true in the solid or liquid crystalline state[1-4].

[1]A. Abragam (1961) *The Principles of Nuclear Magnetism*, Oxford Univ. Press, London.
[2]T. L. James (1975) *Nuclear Magnetic Resonance in Biochemistry*, Academic Press, New York.
[3]K. Wuthrich (1976) *NMR In Biological Reactions*, Academic Press, New York.
[4]H. H. Mantsch, H. Saito & I. C. P. Smith (1977) *Prog. Nucl. Magn. Res. Spect.* **11**, 211.

ISO UNI UNI MECHANISM

The general reaction scheme for this mechanism can be drawn as shown below using Cleland's notation.

The complete steady-state rate expression for this scheme is

$$v = \frac{V_{max,f}V_{max,r}([A] - \{[P]/K_{eq}\})}{V_{max,r}K_a + V_{max,r}[A] + (V_{max,f}[P]/K_{eq}) + (V_{max,r}[A][P]/K_{iip})}$$

where $V_{max,f}$ and $V_{max,r}$ are the forward and reverse maximum velocities, respectively (in which $V_{max,f} = k_3k_5[E_{total}]/(k_3 + k_5)$ and $V_{max,r} = k_2k_6[E_{total}]/(k_2 + k_6)$), K_a is the Michaelis constant for A (equal to $(k_2 + k_3)(k_5 + k_6)/\{k_1(k_3 + k_5)\}$), K_{eq} is the equilibrium constant (equal to $V_{max,f}K_p/V_{max,r}K_a$ in which K_p is the Michaelis constant for P equal to $(k_2 + k_3)(k_5 + k_6)/\{k_4(k_2 + k_6)\}$), and K_{iip} is equal to $(k_3 + k_5)/k_4$. Note that this general expression reduces to the same equation as is seen for the Uni Uni mechanism in the absence of the product $P^{1,2}$.

[1]H. J. Fromm (1979) *Initial Rate Enzyme Kinetics*, Springer-Verlag, New York.
[2]K. L. Rebholz & D. B. Northrop (1995) *Meth. Enzymol.* **249**, 211.

ISOVALERYL-CoA DEHYDROGENASE

This FAD-dependent enzyme [EC 1.3.99.10] catalyzes the reaction of 3-methylbutanoyl-CoA with an electron-transferring flavoprotein to produce 3-methylbut-2-enoyl-CoA and the reduced electron-transferring flavoprotein. Together with the electron-transferring flavoprotein and the electron-transferring-flavoprotein dehydrogenase [EC 1.5.5.1], this enzyme forms a multienzyme system that can reduce ubiquinone and other acceptors. Isovaleryl-CoA dehydrogenase is not identical with butyryl-CoA dehydrogenase, acyl-CoA dehydrogenase, or 2-methylacyl-CoA dehydrogenase.

Y. Ikeda & K. Tanaka (1988) *Meth. Enzymol.* **166**, 155 and 374.

ISOZYMES

Enzymes catalyzing the same reaction, but differing from each other with respect to their primary structure. Such amino acid substitutions are often detected as differences in one or more physical and/or chemical properties, most notably electrophoretic mobility. Isoenzymes, although they catalyze the same reaction, may have different kinetic parameters, different regulatory characteristics, and/or different stabilities. Enzymes that catalyze the same reaction and have the same primary structure, while still differing by some measurable property, are often called pseudoisozymes (or, pseudoisoenzymes). Examples would be proteins that have undergone different degrees or types of covalent modification (including the cases of different enzymes derived from the same zymogen via varying levels of proteolysis). Other examples of pseudoisozymes are oligomeric proteins that form different degrees of aggregation.

J

JABLONSKI DIAGRAM

A diagram (plotting potential energy on the vertical axis *versus* interatomic distance for a bond on the horizontal axis) for illustrating how the excited state of a molecular entity undergoes an allowed transition between ground and excited states. The energy wells for the ground and excited states (often drawn as a Leonard-Jones potential energy diagram) have different interatomic distances representing the greater equilibrium bond length in the excited vibrational and electronic states. Vibrational states are depicted as a series of horizontal lines starting with the zero-point energy level, and the spacing between the vibrational state reflects the quantized nature of energy differences between successive vibrational states. In the electronic transition, one of the two electrons from a chemical bond is promoted to a higher energy state, and this will be a singlet excited state if the excited electron has a spin that is opposite that of the other electron in the electron-pair bond.

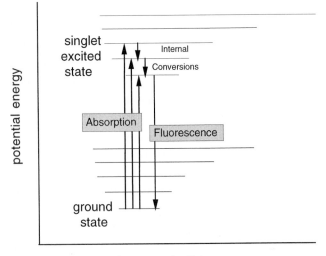

Electronic transitions occur much more rapidly than do the motions of nuclei (Franck-Condon principle), and this results in a vertical transition (*i.e.*, changes in the bond length cannot immediately attend photon absorption or emission).

If the promoted electron in the excited singlet state undergoes a change to a parallel spin relative to its partner, then it is said to have undergone an intersystem transfer from a singlet excited state to a triplet excited state. Longer-lived phosphorescent states are depicted as metastable states positioned at lower energy relative to the initial excited state, and the first excited triplet state is reached via radiationless transfer that delays photon emission in phosphorescence relative to the promptness of fluorescence emissions. Moreover, phosphorescence occurs with a lifetime ($\tau \approx 10^{-3}$ seconds or greater) that is much slower than the time scale of nuclear motions; hence this transition to the singlet ground state can be regarded as a nonvertical transition.

The name of this diagram (or plot) honors the pioneering contributions of Jablonski, who recognized that phosphorescence involves emission from a metastable triplet state lying at lower energy relative to the lowest lying excited singlet state.

J. R. Lakowicz (1983) *Principles of Fluorescence Spectroscopy*, Chapter 1, Plenum, New York.

JENCKS' CLOCK

A semiquantitative procedure used to estimate the lifetimes of carbocations and oxocarbenium ions by using diffusion-controlled trapping of the cations by nucleophiles[1,2]. Ions of intermediate stability react with azide ions at a constant, diffusion-controlled rate and react with water by an activated process. The ratio of the products obtained from the azide path and the water path is dependent on the electronic characteristics of the cation.

The Jencks' clock, as well as pulse radiolysis experiments, indicate that oxocarbenium cations often can be quite stabilized, with a half-life on the order of milliseconds. For example, at room temperature, the structures shown here (left to right) have respective lifetimes of

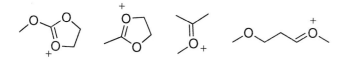

0.25 millisecond, 0.7 microsecond, 1 nanosecond, and 0.5 nanosecond[2,3]. The estimated lifetime of glucosyl cation in aqueous solution is about 1 picosecond[4]. While longer than the period required for a bond vibration, the magnitude of this value suggests that a discrete cationic intermediate is probably not meaningful in enzyme-catalyzed glycosyl transfer reactions, because the intermediates in those reactions typically persist for periods on the order of milliseconds. Nonetheless, a longer lived resonance-stabilized oxocarbenium ion may still form in the transition state of enzymic reactions.

[1]P. R. Young & W. P. Jencks (1977) *J. Amer. Chem. Soc.* **99**, 8238.

[2]J. P. Richard, M. E. Rothenberg & W. P. Jencks (1984) *J. Amer. Chem. Soc.* **106**, 1361.

[3]M. Steenken & R. A. McClelland (1989) *J. Amer. Chem. Soc.* **111**, 4967.

[4]T. L. Amyes & W. P. Jencks (1989) *J. Amer. Chem. Soc.* **111**, 7888.

JENKINS MECHANISM

A hypothetical reaction scheme for enzyme action and cooperativity[1]. In this ligand-substitution model, dissociation steps (for example, EP ⇌ E + P) are not treated as unimolecular steps; rather, dissociation is bimolecular, requiring that an incoming ligand displace the bound ligand (thus, EP + X ⇌ EX + P). A ligand X, referred to as the surrogate, is displaced by the incoming substrate or displaces the outgoing product. Once the enzyme-product binary complex (EP) is formed in the scheme above, EP can only undergo ligand displacement by a new S molecule or by the surrogate X. In other words, the catalytic conversion of ES to EP is unidirectional (*i.e.*, there is no k_{-3} term or direction; such a rate constant is present in the Briggs-Haldane type model). This would suggest an untenable outcome, namely that in the S to P conversion the reaction essentially proceeds unidirectionally through the transition state. Derivation of an expression for a scheme containing a k_{-3} step yields a rate equation containing both $[X]^2$ and $[S]^2$ terms (note that the scheme above should also yield an expression having $[S]^2$ terms since S can bind to two different forms of the enzyme (namely EX and EP)). At high substrate concentrations, nonlinearity should be observed in double-reciprocal plots.

[1]W. T. Jenkins (1982) *Adv. Enzymol.* **53**, 307.

JOB PLOT

A plotting protocol (also known as the method of continuous variation) that provides useful information about protein-ligand and protein-protein interactions. Mole fractions X_A and X_B of two interacting substances, say A and B, are varied such that the total molarity remains constant. Note that $X_A = \{[A]/([A] + [B])\}$ and $X_B = \{[B]/([A] + [B])\}$, such that $(X_A + X_B) = 1$. If an enzyme prefers to bind AB as a one-to-one complex, then the enzymatic activity will be maximal at a mole fraction X_A of 0.5 (*i.e.*, the point at which A and B are present in a one-to-one stoichiometry). Similarly, if AB_2 is the active species, then the enzyme will be most active at a X_A value of 0.33. In this manner, the stoichiometry of binding can be readily determined, and the technique can yield information concerning the affinity of the enzyme for the active species.

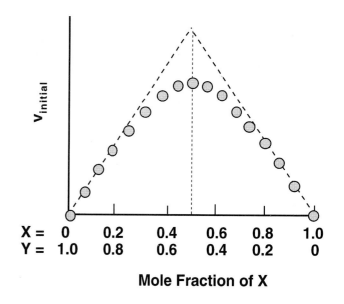

| X = | 0 | 0.2 | 0.4 | 0.6 | 0.8 | 1.0 |
| Y = | 1.0 | 0.8 | 0.6 | 0.4 | 0.2 | 0 |

Mole Fraction of X

Huang[1] has provided an excellent account of the application of this method of continuous variation in studies of enzymic catalysis. The technique can also assist in distinguishing between different cooperative models.

[1]C. Y. Huang (1982) *Meth. Enzymol.* **87**, 509.

JOULE

A unit of energy (symbolized by J) equal to 10^7 ergs. In the SI system, the joule replaces the calorie, which is equivalent to 4.184 J. Formally, the joule equals the heat generated by an ampere flowing through a 1 ohm resistor over a one second interval.

JUGLONE MONOOXYGENASE

This enzyme [EC 1.14.99.27] catalyzes the reaction of 5-hydroxy-1,4-naphthoquinone with AH_2 and dioxygen to produce 3,5-dihydroxy-1,4-naphthoquinone, A, and water. The enzyme can also use 1,4-naphthoquinone, naphthazarin, and 2-chloro-1,4-naphthoquinone as substrates, but not other related compounds.

H. Rettenmaier & F. Lingens (1985) *Biol. Chem. Hoppe Seyler* **366**, 637.

K

K_a

1. Symbol for the association constant. Thus, for the reaction $A \rightleftharpoons P + Q$, $K_a = [A_{eq}]/([P_{eq}][Q_{eq}])$. This constant is the reciprocal of the dissociation constant and its negative logarithm is pK_a. **See** *Association Constant*. 2. The symbol for the dissociation constant for an acid. Thus, for the reaction $HA \rightleftharpoons H^+ + A^-$, $K_a = [H^+][A^-]/[HA]$. The negative logarithm of this constant is pK_a. 3. Symbol (also K_A) for the Michaelis constant for substrate A in an enzyme-catalyzed reaction.

KALLIKREIN

Plasma kallikrein [EC 3.4.21.34], also known as kininogenin and serum kallikrein, catalyzes the hydrolysis of Arg—Xaa and Lys—Xaa bonds in polypeptides. This includes the Lys—Arg and Arg—Ser bonds in human kininogen, thus producing bradykinin. Tissue kallikrein [EC 3.4.21.35] catalyzes the hydrolysis of peptide bonds, preferentially Arg—Xaa, in small-molecule substrates. It catalyzes the breaking of the appropriate bonds in kininogen resulting in the release of lysyl-bradykinin.

C. W. Wharton (1998) *Comprehensive Biological Catalysis: A Mechanistic Reference* **1**, 345.
J. S. Bond & P. E. Butler (1987) *Ann. Rev. Biochem.* **56**, 333.

KAMLET-TAFT SOLVENT PARAMETERS

Parameters of the Kamlet-Taft solvatochromic relationship. These parameters measure the contributions to overall solvent polarity of the hydrogen bond donor, the hydrogen bond acceptor, and the dipolarity/polarizability properties of solvents.

M. J. Kamlet, J. L. M. Abboud & R. W. Taft (1981) *Progr. Phys. Org. Chem.* **13**, 485.

KAPPA (κ)

1. Symbol for conductivity. 2. Symbol for compressibility (κ_T for isothermal compressibility $[-(1/V)(\partial V/\partial P)_T]$ and

κ_s for isentropic compressibility $[-(1/V)(\partial V/\partial P)_s]$). 3. Symbol for reciprocal radius of ionic atmosphere.

KATAL

A unit of enzyme catalytic activity equal to the conversion of one mole of substrate per second in a specified assay system. The katal (kat) is more commonly used in clinical enzymology. One unit of enzyme activity (*i.e.*, one micromole per minute) corresponds to 16.67 nkat.

K_b

1. Symbol for the dissociation constant of a base. Thus, for the reaction $BOH \rightleftharpoons B^+ + OH^-$, $K_b = [B^+][OH^-]/[BOH]$. 2. Symbol (also, K_B) for the Michaelis constant for substrate B in an enzyme-catalyzed reaction.

k_{cat}

See *Turnover Number*

k_{cat}/K_m

The ratio of the turnover number (*i.e.*, $V_{max}/[E_{total}]$) to the K_m value of a substrate in a particular enzyme-catalyzed reaction. When k_{cat} and K_m are the true steady-state parameters, this ratio (or the ratio V_{max}/K_m) is an excellent gauge of the specificity of the enzyme for that substrate. The larger the ratio, the more effective that substrate is used by the enzyme under study. In addition, the effects of a number of mechanistic probes of enzyme action on this ratio (for example, pH effects, isotope effects, temperature effects, the influence of various modifiers, *etc.*) can provide much information on the catalytic and binding mechanism. **See** V_{max}/K_m

K_d

Symbol for dissociation constant, equal to the reciprocal of the association constant. The negative logarithm of this constant is the pK_d. **See** *Dissociation Constant;* K_a

KEKULÉ STRUCTURES

Canonical forms of benzene with alternating double bonds (*i.e.*, "cyclohexatriene"), structures which contribute equally and predominantly to benzene's resonance hybrid structure.

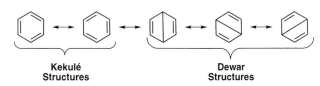

Kekulé Structures   **Dewar Structures**

They are estimated by calculation to account for about 75% of the resonance stabilization of benzene with Dewar structures accounting for most of the remaining stabilization. However, such resonance structures have no separate physical reality or independent existence. A dotted circle is often used to indicate the resonance-stabilized bonding of benzene. Nonetheless, the most frequently appearing structures of benzene are the two Kekulé structures.

KELVIN

A unit of thermodynamic temperature (symbolized by K) that is one of the seven SI base units. The kelvin is equal to $1/(273.16)$ of the thermodynamic temperature of the triple point of water. (Note: Absolute zero is 0 K, and not 0°K.)

K_{eq}

Symbol for the thermodynamic equilibrium constant for a reaction.

3-KETOACID CoA-TRANSFERASE

This enzyme [EC 2.8.3.5], also known as succinyl-CoA:3-ketoacid CoA-transferase and 3-oxoacid CoA-transferase, catalyzes the reversible reaction of succinyl-CoA with a 3-oxo acid to produce succinate and a 3-oxo-acyl-CoA derivative.

W. P. Jencks (1973) *The Enzymes*, 3rd ed., **9**, 483.

β-KETOACYL-ACP REDUCTASE

This enzyme [EC 1.1.1.100], also known as 3-oxoacyl-[acyl-carrier protein] reductase, catalyzes the reaction of a (3*R*)-3-hydroxyacyl-[acyl-carrier protein] with NADP$^+$ to produce 3-oxoacyl-[acyl-carrier protein] and NADPH. The enzyme prefers acyl-carrier protein derivatives over simple CoA derivatives as substrates.

J. Browse & C. Somerville (1991) *Ann. Rev. Plant Physiol. Plant Mol. Biol.* **42**, 467.
S. J. Wakil & J. K. Stoops (1983) *The Enzymes*, 3rd ed., **16**, 3.

β-KETOACYL-ACP SYNTHASE

This enzyme [EC 2.3.1.41], also known as 3-oxoacyl-[acyl-carrier protein] synthase, catalyzes the reaction of an acyl-[acyl-carrier protein] with malonyl-[acyl-carrier protein] to produce a 3-oxoacyl-[acyl-carrier protein], carbon dioxide, and the [acyl-carrier protein]. ***See also*** *Fatty Acid Synthetase*

J. V. Schloss & M. S. Hixon (1998) *Comprehensive Biological Catalysis: A Mechanistic Reference* **2**, 43.
J. Browse & C. Somerville (1991) *Ann. Rev. Plant Physiol. Plant Mol. Biol.* **42**, 467.
S. J. Wakil & J. K. Stoops (1983) *The Enzymes*, 3rd ed., **16**, 3.
P. R. Vagelos (1973) *The Enzymes*, 3rd ed., **8**, 155.

α-KETOBUTYRATE SYNTHASE

This enzyme [EC 1.2.7.2], also known as 2-oxobutyrate synthase, catalyzes the reaction of α-ketobutyrate (or, 2-oxobutanoate) with coenzyme A, and oxidized ferredoxin to produce propanoyl-CoA, carbon dioxide, and reduced ferredoxin.

B. B. Buchanan (1972) *The Enzymes*, 3rd ed., **6**, 193.

2-KETO-3-DEOXY-L-ARABINONATE DEHYDRATASE

This enzyme [EC 4.2.1.43], also referred to as 2-dehydro-3-deoxy-L-arabinonate dehydratase, catalyzes the conversion of 2-dehydro-3-deoxy-L-arabinonate to 2,5-dioxopentanoate and water.

K. N. Allen (1998) *Comprehensive Biological Catalysis: A Mechanistic Reference* **2**, 135.
W. A. Wood (1971) *The Enzymes*, 3rd ed., **5**, 573.

2-KETO-3-DEOXY-D-GLUCARATE ALDOLASE

This enzyme [EC 4.1.2.20], also referred to as 2-dehydro-3-deoxyglucarate aldolase, catalyzes the reversible reaction of 2-dehydro-3-deoxy-D-glucarate to pyruvate and tartronate semialdehyde.

W. A. Wood (1972) *The Enzymes*, 3rd ed., **7**, 281.

5-KETO-4-DEOXYGLUCARATE DEHYDRATASE

This enzyme [EC 4.2.1.41], also referred to as 5-dehydro-4-deoxyglucarate dehydratase, catalyzes the conversion of 5-dehydro-4-deoxy-D-glucarate to 2,5-dioxopentanoate, water, and carbon dioxide.

W. A. Wood (1971) *The Enzymes*, 3rd ed., **5**, 573.

2-KETO-3-DEOXY-L-PENTONATE ALDOLASE

This enzyme [EC 4.1.2.18], also referred to as 2-dehydro-3-deoxy-L-pentonate aldolase or 2-keto-3-deoxyarabonate aldolase, catalyzes the reversible conversion of 2-dehydro-3-deoxy-L-pentonate to pyruvate and glycolaldehyde.

W. A. Wood (1972) *The Enzymes*, 3rd ed., **7**, 281.

2-KETO-3-DEOXY-6-PHOSPHOGALACTONATE ALDOLASE

This enzyme [EC 4.1.2.21], also referred to as 6-phospho-2-dehydro-3-deoxygalactonate aldolase, catalyzes the reversible conversion of 2-dehydro-3-deoxy-D-galactonate 6-phosphate to pyruvate and D-glyceraldehyde 3-phosphate.

D. J. Creighton & N. S. R. K. Murthy (1990) *The Enzymes*, 3rd ed., **19**, 323.
W. A. Wood (1972) *The Enzymes*, 3rd ed., 7, 281.

2-KETO-3-DEOXY-6-PHOSPHOGLUCONATE ALDOLASE

This enzyme [EC 4.1.2.14], also referred to as phospho-2-dehydro-3-deoxygluconate aldolase, catalyzes the reversible conversion of 2-dehydro-3-deoxy-D-gluconate 6-phosphate to pyruvate and D-glyceraldehyde 3-phosphate. 2-Oxobutanoate can also serve as a substrate for this enzyme.

K. N. Allen (1998) *Comprehensive Biological Catalysis: A Mechanistic Reference* **2**, 135.
D. J. Creighton & N. S. R. K. Murthy (1990) *The Enzymes*, 3rd ed., **19**, 323.
W. A. Wood (1972) *The Enzymes*, 3rd ed., **7**, 281.

α-KETOGLUTARATE DEHYDROGENASE COMPLEX

This multiprotein complex contains as its central enzyme α-ketoglutarate dehydrogenase (lipoamide) [EC 1.2.4.2], also referred to as oxoglutarate dehydrogenase (lipoamide) and oxoglutarate decarboxylase. This enzyme catalyzes the reaction of α-ketoglutarate (or, 2-oxoglutarate) with lipoamide to produce *S*-succinyldihydrolipoamide and carbon dioxide, using thiamin pyrophosphate as a cofactor.

R. N. Perlman (1991) *Biochemistry* **30**, 8501.
M. Hamada & H. Takenaka (1991) *A Study of Enzymes* **2**, 227.
S. J. Yeaman (1986) *Trends Biochem. Sci.* **11**, 293.

α-KETOGLUTARATE SEMIALDEHYDE DEHYDROGENASE

This enzyme [EC 1.2.1.26], also known as 2,5-dioxovalerate dehydrogenase, catalyzes the reaction of 2,5-dioxopentanoate with $NADP^+$ and water to produce α-ketoglutarate (or, 2-oxoglutarate) and NADPH.

α-KETOGLUTARATE SYNTHASE

This enzyme [EC 1.2.7.3], also called 2-oxoglutarate synthase, catalyzes the reversible reaction of α-ketoglutarate (or, 2-oxoglutarate) with coenzyme A and oxidized ferredoxin to produce succinyl-CoA, carbon dioxide, and reduced ferredoxin.

B. B. Buchanan (1972) *The Enzymes*, 3rd ed., **6**, 193.

KETONIZATION

The process of converting an enol to a ketone. Pyruvate kinase catalyzes a ketonization reaction in the conversion of the enolpyruvate intermediate to pyruvate. **See** *Tautomerization*

I. A. Rose (1982) *Meth. Enzymol.* **87**, 84.

3-KETO-5β-STEROID Δ⁴-DEHYDROGENASE

This enzyme [EC 1.3.99.6], also known as 3-oxo-5β-steroid 4-dehydrogenase and Δ⁴-3-ketosteroid 5β-reductase, catalyzes the reaction of a 3-oxo-5β-steroid and an acceptor to produce a 3-oxo-Δ⁴-steroid and the reduced acceptor.

S. J. Davidson (1969) *Meth. Enzymol.* **15**, 656.

3-KETOSTEROID Δ⁵-ISOMERASE

This enzyme [EC 5.3.3.1], also called steroid Δ-isomerase and Δ⁵-3-ketosteroid isomerase, catalyzes the interconversion of a 3-oxo-Δ⁵-steroid and a 3-oxo-Δ⁴-steroid.

J. V. Schloss & M. S. Hixon (1998) *Comprehensive Biological Catalysis: A Mechanistic Reference* **2**, 43.
M. A. Ator & P. R. Ortez de Montellano (1990) *The Enzymes*, 3rd ed., **19**, 213.
D. J. Creighton & N. S. R. K. Murthy (1990) *The Enzymes*, 3rd ed., **19**, 323.
P. Talalay & A. M. Benson (1972) *The Enzymes*, 3rd ed., **6**, 591.

KEXIN

This calcium-ion-activated enzyme [EC 3.4.21.61] catalyzes the hydrolysis of peptide bonds at LysArg—Xaa and ArgArg—Xaa to process yeast α-factor pheromone and killer toxin precursors.

C. Brenner, A. Bevan & R. S. Fuller (1994) *Meth. Enzymol.* **244**, 152.

K_i (or, K_I)

Symbol for the dissociation constant of an inhibitor.

K_{ia}, K_{ib}, and K_{ic}

Symbol for the dissociation constant of substrate A, B, and C.

K_{ii}

Symbol for the dissociation constant of an inhibitor with respect to a particular form of the enzyme. This dissociation constant is associated with the intercept term in the double-reciprocal form of the initial-rate equation. For example, consider an inhibitor that can bind to either the free enzyme, E, or the binary central complex, EX, of a Uni Uni mechanism. K_{ii} would be the dissociation constant for the EX + I $\rightleftharpoons$ EXI step and is equal to [EX][I]/[EXI]. The binding of I to the free enzyme (*i.e.*, E + I $\rightleftharpoons$ EI) is governed by K_{is} (equal to [E][I]/[EI]).

KINEMATIC VISCOSITY

The ratio (symbolized by ν) of the dynamic viscosity, or simply viscosity η, to the fluid density, ρ. Kinematic viscosity has SI units of meters squared per second.

KINESIN

The following are recent reviews dealing with this ATP-dependent motor protein that is responsible for vesicle locomotion along microtubules in eukaryotes. *See* Force Effects on Molecular Motors

S. Khan & M. Sheetz (1997) *Ann. Rev. Biochem.* **66**, 785.
R. A. Walker & M. P. Sheetz (1993) *Ann. Rev. Biochem.* **62**, 429.
J. R. McIntosh & M. E. Porter (1989) *J. Biol. Chem.* **264**, 6001.

KINETIC AMBIGUITY

Any condition of kinetic equivalence characterized by the absence of unique properties permitting one to chose among rival mechanisms. *See* Kinetic Equivalence; Black Box

KINETIC CONTROL OF PRODUCT COMPOSITION

When one or more molecular entity(ies) participates in two or more parallel and irreversible reactions in which different products are formed, the faster-forming product will accumulate by the reaction having the lowest activation energy. Thus, kinetically controlled processes are those whose proportion of products is governed by the relative rates of the competing reactions. If the reac-tions are reversible and the process is stopped prior to the establishment of equilibrium, the overall process will also be kinetically controlled since the faster-formed product is still present in greatest abundance. *See* Thermodynamic Control

G. W. Klumpp (1982) *Reactivity in Organic Chemistry*, pp. 36-89, Wiley, New York.

KINETIC ELECTROLYTE EFFECT

The effect of added electrolyte and/or ionic strength on the rate constant(s) of a reaction in solution, exclusive of any role of that the electrolyte as a reactant or catalyst.

Recall from transition state theory that the rate of a reaction depends on k_o (the catalytic rate constant at infinite dilution in the given solvent), the activity of the reactants, and the activity of the activated complex. If one or more of the reactants is a charged species, then the activity coefficient of any ion can be expressed in terms of the Debye-Hückel theory. The latter treats the behavior of dilute solutions of ions in terms of electrical charge, the distance of closest approach of another ion, ionic strength, absolute temperature, as well as other constants that are characteristic of each solvent. If any other factor alters the effect of ionic strength on reaction rates, then one must look beyond Debye-Hückel theory for an appropriate treatment.

Ionic strength also has a great influence on protein structure, thus influencing enzyme-catalyzed rates above and beyond the effects on individual rate constants.

Although these effects are often collectively referred to as salt effects, IUPAC regards that term as too restrictive. If the effect observed is due solely to the influence of ionic strength on the activity coefficients of reactants and transition states, then the effect is referred to as a primary kinetic electrolyte effect or a primary salt effect. If the observed effect arises from the influence of ionic strength on pre-equilibrium concentrations of ionic species prior to any rate-determining step, then the effect is termed a secondary kinetic electrolyte effect or a secondary salt effect. An example of such a phenomenon would be the influence of ionic strength on the dissociation of weak acids and bases. *See* Ionic Strength

J. N. Brønsted (1922) *Z. Phys. Chem.* **102**, 169.
N. Bjerrum (1924) *Z. Phys. Chem.* **108**, 82.
IUPAC (1979) *Pure and Appl. Chem.* **51**, 1725.

KINETIC EQUIVALENCE

Any two reaction schemes are considered to be "kinetically equivalent" if they imply the same orders of reaction for all chemical species involved. For example, assume that the rate expression for a particular scheme for a chemical reaction is given by $d[P]/dt = v = k_1 k_3 [A][B]/(k_2 + k_3[B])$. For a second scheme, the expression is found to be $k_1' k_3' [A][B]/(k_2' + k_3'[B])$. Both expressions have the same general form and have the same dependence on reactant concentrations. Thus, the two schemes are kinetically equivalent. **See** Kinetic Ambiguity

KINETIC HALDANE RELATIONSHIPS

Haldane relationships based on the ratio of apparent rate constants for both the forward and reverse directions. Every enzyme mechanism has at least one kinetic Haldane relationship. For a Uni Uni mechanism[1], the kinetic Haldane relationship is: $K_{eq} = (V_{max,f}/K_{m,A})/(V_{max,r}/K_{m,P})$. **See** Haldane Relationships; Thermodynamic Haldane Relationships

[1]W. W. Cleland (1982) Meth. Enzymol. **87**, 366.

KINETIC ISOTOPE EFFECTS

GENERAL COMMENTS. Isotopically substituted molecules having identical geometrical, optical, and conformational properties are called isotopic isomers (or isotopomers), and depending on the position of isotopic substitution, two isotopic isomers may differ in their reactivity (*i.e.*, their rate of reaction). Any observed rate difference is called a kinetic isotope effect, and the occurrence and magnitude of the isotope effect can provide valuable information and insight regarding reaction mechanisms. One can also employ isotope effect studies to ascertain whether the bond-breaking step is either a rate-limiting, or at least a rate-contributing, feature of catalysis. In this section, we discuss the essential properties of isotopic isomers that yield kinetic isotope effects, and we briefly analyze their utility in mechanistic studies of enzymic catalysis. The interested reader will obtain more advanced treatments of isotope effects elsewhere[1-4].

A kinetic isotope effect (or KIE) is usually written as a ratio (*e.g.*, v_H/v_D or k_H/k_D) to indicate the magnitude of the rate (or rate constant) obtained for one isotopomer (in this case H, or ^1H) divided by the observed rate or rate constant achieved with the heavier isotopically labeled reactant (in this case D, or ^2H). For simplicity, some kineticists use ^Dk as the shorthand version of k_H/k_D. In enzyme-catalyzed kinetics, one must necessarily deal with the behavior of a multistep reaction scheme. For initial rate enzyme processes, one typically deals with collections of rate constants which appear in the form of the maximal velocity V_m (shortened to V) or the specificity constant V_m/K_m (shortened to V/K). Accordingly, enzyme kineticists will use ^DV and $^DV/K$ as an easy way to indicate the respective isotope effects $[(V_m)_H/(V_m)_D]$ and $[(V_m/K_m)_H/(V_m/K_m)_D]$, respectively.

TYPES OF ISOTOPE EFFECTS. Kinetic isotope effects provide information regarding the configuration of reactants as they approach the transition state of a reaction[1]. These inferences are accessible, because the transition state, in large measure, determines the type and magnitude of observed changes in chemical reactivity. Early work on reaction mechanisms relied on the introduction of various substituents (*e.g.*, methyl, ethyl, propyl, *etc.*) near the reaction center, but such "homologues" often introduced confounding variables, such as differences in their solvation, acid/base chemistry, and conformation. Thus, on the basis of studies on structural homologues, one could not unambiguously assign differences in reaction rate to changes in the configuration of atoms within the transition state. Moreover, for enzymes, there is always the problem that substrate selectivity/specificity limits the extent to which one may introduce substituents into reactant's structure without sacrificing the ability to bind at the active site. Isotopic substitution, by contrast, leads to more subtle and often far more predictable changes in active site geometry[2-4].

When the reaction center (*i.e.*, the site of bond-breaking/making) is itself the site of isotopic substitution, any observable rate change is termed a primary isotope effect. Thus, substitution of deuterium or tritium for the transferred hydrogen atom of NADH can give rise to a primary kinetic isotope effect, and when observed, one can infer that hydrogen transfer is at least a rate-contributing step in a particular dehydrogenase reaction. Isotopic substitution outside the reaction center most often will have considerably less influence on reaction rate; such rate changes are termed secondary kinetic isotope effects. These isotope effects typically decrease in magnitude as the point of substitution lies farther from the reaction center, and α, β, and γ are used to designate the positions of isotopic substitution:

$$-C_\gamma-C_\beta-C_\alpha-X$$

For the purposes of this introductory treatment, we will confine our discussion to primary kinetic isotope effects as well as secondary α isotope effects.

PRIMARY ISOTOPE EFFECT. Without engaging a detailed statistical mechanical argument[1] about the origin of primary isotope effects, we can list the basic assumptions of the Bigeleisen-Mayer treatment of isotope effects: (a) the Born-Oppenheimer approximation holds, such that the electronic potential energy does not change upon isotopic substitution; (b) one can parse the partition functions into the independent contributions attributable to translational, vibrational, and rotational motions; (c) classical mechanics is sufficient to describe translational and rotational motion, and vibrational motion is harmonic; (d) isotopic substitution is without effect on κ, the transition-state transmission coefficient; and (e) no quantum mechanical tunneling occurs. These approximations are adequate to describe most kinetic isotope effects, with the notable exception of hydrogen transfer, for which tunneling can make a significant contribution.

Let us now examine an approximate method[5] for comprehending their nature and magnitude. Consider the diagram in Fig. 1 where A—B and A—B' bonds of two

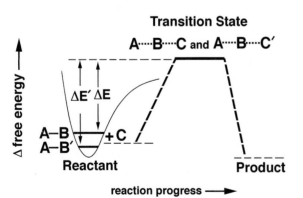

Figure 1. Reaction coordinate diagram for the reaction of A—B and A—B' with another agent, designated here as C. Note that the heavier isotopomer will lie at lower energy, as would be expected on the basis of its shorter equilibrium bond length and its consequentially lower zero point energy. Because the energy levels of the transition states for both reactions will be approximately the same, the slight difference in ground state zero point energy means that a greater fraction of A—B will reach the transition state, compared to A—B'.

isotopic isomers are cleaved. During reaction, the A—B and A—B' stretching frequencies differ only in the mass of the nucleus and are identical with respect to their essential electronic interactions.

During reaction, the A—B and A—B' stretching frequency is lost in the transition state, where B and B' are located at essentially identical positions. (In fact, vibrational frequencies other than the A—B and A—B' stretching mode may be likewise partly lost, but this represents what may be regarded as a minor refinement in most cases.) In the diagram above, the reader will recognize that there will always be a difference in the vibrational energies of the isotopic isomers A—B and A—B' reflecting the energy differences in the lowest lying vibrational level which has a finite zero-point energy (Z.P.E.). That a molecule has a finite Z.P.E. is a requirement of the Uncertainty Principle; any quantized harmonic oscillator will still experience unavoidable disturbances (hence, energy), even when that oscillator is dampened to its lowest energy state. The vibrational frequency of a bond stretching mode can be estimated by using an equation that accounts for the simple harmonic motion of a spring (**See** Hooke's Law Spring):

$$\nu_{A-B} = (1/2\,\pi)\,[k/\mathrm{mass_{reduced}}]^{1/2}$$

where k is the so-called force constant (in this case, a property of the electron density of the bond undergoing scission) and $\mathrm{mass_{reduced}}$ is the reduced mass of the A—B molecule [in this case, $m_A m_B/(m_A + m_B)$]. We already indicated that isotopic substitution has no effect on the electron density of the bond, and for this reason, the value of k is the same for A—B and A—B'. Thus, for two isotopic isomers having essentially identical vibrational force constants, we can divide the equation for A—B above by its counterpart for A—B', and this yields:

$$\frac{\nu_{A-B}}{\nu_{A-B'}} = \left\{\frac{m_B(m_A + m_{B'})}{m_{B'}(m_A + m_B)}\right\}^{1/2} = M$$

By using this equation, one can estimate how the change in isotopic substitution can result in a small but experimentally significant change in the zero-point energy:

$$\Delta E = E_{A-B} - E_{A-B'} = 0.5\,h\nu_{A-B}(1 - M)$$

where h is Planck's constant. The lighter isotopic isomer actually lies closer to the transition state by the amount given in this last equation.

This energy difference ΔE can be estimated reasonably accurately by using infrared absorption data to determine the magnitude of ν_{A-B}. If the A—B vibration is completely lost in going to the transition state during a chemical reaction, then the difference in zero-point energies for A—B and A—B′ stretching will lead to a corresponding rate decrease for A—B′ relative to A—B. This can be expressed as follows:

$$k_{A-B}/k_{A-B'} = \exp\{(-h\nu_{A-B}/2RT)[1-M]\}$$

where R is the gas constant, h is Planck's constant, and T is the absolute temperature. For example, a C—H bond has a value of 3000 cm^{-1} for $\nu_{C-H'}$, whereas ν_{C-H} for a C—D bond is 3000 cm^{-1}. In this case, ΔE is about 1.2 kcal/mol, and if this difference is fully expressed in the primary kinetic isotope effect, k_H/k_D will be about seven at room temperature. In some cases, although the primary kinetic isotope probably occurs in the isotopically sensitive step within a reaction mechanism, the experimenter is left with little indication thereof, because the particular step has no effect on reaction rate (*i.e.*, it is faster than the rate-determining step). Nonetheless, one may use the above expression to estimate the following primary isotope effects, which are taken from Hine[5]:

k_{C-H}/k_{C-D}	8.2 (at 0°C)
k_{C-H}/k_{C-D}	6.9 (at 25°C)
k_{N-H}/k_{N-D}	9.2 (at 25°C)
k_{C-H}/k_{C-T}	16 (at 25°C)
$k_{^{12}C-^{12}C}/k_{^{12}C-^{13}C}$	1.03 (at 0°C)
$k_{^{12}C-^{12}C}/k_{^{12}C-^{14}C}$	1.09 (at 0°C)

These estimations engage the assumption that any additional effects due to bending vibrations will tend to contribute relatively weakly to the isotope effect. In fact, however, the above estimates are probably only "ball park" estimates, in that coupling of molecular vibrations of multiatomic molecules is ignored[1]. Another key point is that a covalent bond need not be completely broken in the transition state, and two isotopic isomers may behave slightly differently in the transition state. Jencks[6] discusses this matter as well as the problem of nonlinear transition states, a condition that takes into account the fact that bending frequencies often lessen the developed magnitude of isotope effects.

Examination of one real-life case may benefit the reader's understanding. Strittmatter[7] studied the primary kinetic isotope effects arising in the NADH-dependent cytochrome b_5 reductase (EC 1.6.2.2). The oxidation of NADH and subsequent reduction of cytochrome b_5 is facilitated by the enzyme-bound FAD group, and the kinetics of the direct transfer of a hydrogen from the A-face (or *pro-R*) of NADH to the flavin can be monitored by the loss of the 340 nm absorbance of the NADH's dihydropyridine ring. Using deuterated isotopic isomers of NADH and several related compounds, Strittmatter[7] obtained the primary kinetic isotope effect data compiled in the table below.

Rates of Oxidation of NADH and Several Pyridine Ring Analogues by Cytochrome b_5 Reductase[a]

Nucleotide	Observed Turnover Number (mol product/mol enzyme/min)	k_H/k_D
NADH	29,600	—
NADD (*pro-R* Deuterium)	8,100	3.66
NADD (*pro-S* Deuterium)	29,700	1.00
AcPyADH[b]	3,560	—
AcPyADD (*pro-R* Deuterium)	340	10.45
AcPyADD (*pro-S* Deuterium)	3,530	0.99
PyAlADH[c]	414	—
PyAlADD (*pro-R* Deuterium)	48	8.7

[a]These kinetic experiments were carried out in microcuvettes at 25°C, which contained 25 nmol reduced pyridine nucleotide, 50 nmol potassium ferricyanide (added at zero time as the electron acceptor), 1 nmol reductase in 0.2 mL 0.1 M Tris-acetate (pH 8.1) containing 1 mM EDTA.
[b]AcPyADH = 3-Acetylpyridine analogue of NADH.
[c]PyAlADH = 3-Pyridinealdehyde analogue of NADH.

The values of k_H/k_D were quite large for coenzymes deuterated in the *pro-R* position, and the authors carefully conducted control experiments with coenzymes labeled at the *pro-S* position. The experiments clearly demonstrated that the overall transfer of electrons to ferricyanide involves NAD—H_A bond-breaking in the rate-limiting step of the catalyzed reaction. Moreover, the isotope effect is greatest with alternative substrates, suggesting that rate constants defining the reaction with NADD/H may mask the full extent of the kinetic isotope effect. The reader will also note that substitution at the *pro-S* position was without effect on the rates of hydrogen transfer. Had there been such an effect, we would call it a secondary kinetic isotope effect, because the isotope is not at an atom directly undergoing bond scission.

Classical mechanical treatments of primary kinetic isotope effects require the reactants to surmount an energy barrier in order for them to form products. By contrast, quantum mechanics allows for the phenomenon of tunneling, wherein reactants of insufficient energy to pass over the top of an energy barrier can still be converted to product, provided that the width of the barrier is small. Only particles having de Broglie wavelengths ($\lambda = h/mv$) that are on the same scale as the barrier thickness can "penetrate" the barrier by quantum mechanical tunneling. Protons and hydride ions have 1—2 Å de Broglie wavelengths, and several enzymic reactions have been demonstrated to exhibit tunneling. Tell-tale features of tunneling are isotope effects that exceed the magnitude predicted from barrier heights obtained by applying the Bigeleisen-Mayer treatment as well as the temperature dependence of the observed isotope effect. Tunneling is favored by (a) high, thin potential energy barriers; (b) low temperature; and (c) steric hindrance which appears to influence solvent exclusion.

SECONDARY ISOTOPE EFFECTS. Changes in reaction may also result from isotopic substitutions at positions that are immediately adjacent to the reaction center (*i.e.*, the bond broken/made in the chemical reaction under investigation). We deal here only with so-called secondary α-isotope effects, and we will limit the scope further by considering only α deuterium and α tritium isotope effects on carbon. Isotopic substitution by heavier nuclides will also give rise to α isotope effects, but they are quite small. The magnitudes of the α isotope effects for C—^{1}H$_\alpha$ compared to C—^{2}H$_\alpha$ as well as for C—^{1}H$_\alpha$ compared to C—^{3}H$_\alpha$ are also relatively small, frequently necessitating the use of special techniques.

One of the principal applications of secondary isotope effects is to distinguish S_N1 and S_N2 reactions. The α kinetic isotope effect for S_N1 reactions is typically significantly larger than that observed in S_N2 reactions. Such behavior reflects changes in the geometry of the carbon atom undergoing nucleophilic substitution (Fig. 2). If a nucleophilic substitution reaction is first-order (S_N1), the carbon atom changes from a tetrahedral *sp*^{3} arrangement to a flattened *sp*^{2} configuration prior to nucleophile attack.

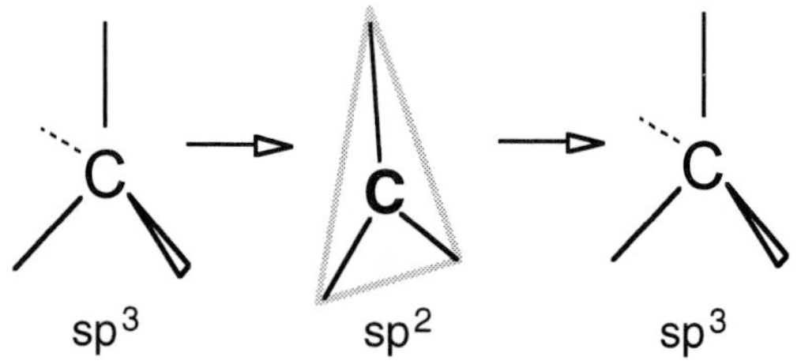

Figure 2. Rearrangement in molecular orbital hybridization and geometry of the carbon atom undergoing a first-order nucleophilic substitution (or S_N1) reaction.

In a second-order (S_N2)nucleophilic substitution reaction, the carbon atom simultaneously experiences the effects of the attacking nucleophile (N) and leaving group or exiphile (E). These mutual effects (shown in Fig. 3) serve to diminish energy diferences between each isotopic substrate and its corresponding transition state.

Figure 3. Structural rearrangements during a second-order nucleophilic substitution or S_N2 reaction.

Why does an S_N1 reaction have a significant secondary isotope effect? Detailed analysis[1,2,5] of secondary kinetic isotope effects suggests that it is difficult to describe their origin explicitly in quantitative terms. One qualitative approach is to separate the electronic, vibrational, rotational, and translational energies of a molecule, and then to deal with each to explore any differences, say between C—^{1}H$_\alpha$ and C—^{2}H$_\alpha$. Considerable theoretical treatment has led to the conclusion that the potential energy surfaces of hydrogen- and deuterium-containing isotopic isomers are essentially identical[8], suggesting that the secondary isotope effect results from different energies of nuclear motions (*i.e.*, vibrations, rotations, translations) and not from electronic differences. Of those involving nuclear motion, the vibrational energy dominates most, and the deviation from the equilibrium bond length (r_e) of the C—H$_\alpha$ and C—D$_\alpha$ can be taken as the dominant factor in the isotope effect. Because D is heavier, its mean-square amplitude will be closer to r_e which is the potential energy minimum. When steric repulsions between H$_\alpha$ (or D$_\alpha$) and the other groups are relieved, then the H$_\alpha$ isotopic isomer will undergo a more favorable transition by dropping through a greater change of potential energy than will its D_α isotopic isomer. The phenomenon can be qualitatively described by stating that a deuteration decreases the steric repulsion, and thus the formation of a flat (sp^2-like) geometry from tetrahedral (or sp^3) bonding is less favorable for the heavier isotopic isomer[9]. Another way of dealing with the origin of α-KIE's is to view the primary contributing factors as the out-of-plane vibrations in the ground state of the reactant. The α effect is dependent on the reduction in the size of the hydrogen-carbon-exiphile bending force constant in passing from reactant to its activated intermediate state (*i.e.*, $f_{HCX} \to f^\ddagger_{HCX}$), where as usual we use "$\ddagger$" to denote the transition state. An independent method of correlating the H—C—X bending constant with geometry is needed to permit a fuller analysis of geometry inferred on the basis of an observed kinetic α isotope effect.

Let us consider the classical secondary kinetic isotope experiments[10] on lysozyme-catalyzed hydrolysis of glycosides. Earlier X-ray work[11] indicated that the enzyme

binds oligosaccharides, such that the enzyme interacts with five monosaccharide units (designated alphabetically as subsites A, B, C, D, and E) stretching across the enzyme's adsorption pocket. Cleavage occurs between sites D and E, and the enzyme appeared to distort substrate analogues into a half-chair conformation, suggestive of a flattened oxocarbonium ion intermediate as illustrated in Fig. 4.

Chemical studies also demonstrated that lysozyme hydrolysis of the glycosidic bond proceeds with retention of configuration to at least 99.7%. Retention of configuration provides strong evidence against an in-line S_N2-type mechanism, but the authors were particularly interested in determining the extent to which a secondary kinetic isotope effect is expressed during lysozyme catalysis. Using a phenyl glycoside with the following structure (Fig. 5),

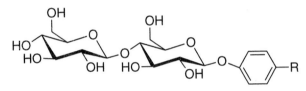

Figure 5. Phenylglycoside substrate substituted at C1 with either ^{1}H and ^{2}H.

the effect of deuteration at the C1 carbon of the glucosyl residue on the rate of phenol formation was measured. Actually, these investigators used either [^{3}H-phenyl]- and [^{14}C-phenyl]substrate to prepare the respective C1-^{1}H and C1-^{2}H isotopic isomers. The two isotopic isomers could then be combined with each other prior to enzyme addition, and the isotope effect (k_H/k_D) could be measured by determining changes in the ^{3}H/^{14}C ratio for the isolated phenol products[12]. This strategy permits one to attain high sensitivity and precision, something that is difficult to achieve with separate kinetic runs using isotopic isomers separately. Their experimental findings are given below. The first tabulated value agrees with the oxocarbonium ion mechanism for acid-catalyzed glycoside hydrolysis[13]. The second value for base-catalyzed hydrolysis likewise agrees with that expected for an in-line S_N2-type mechanism. Finally, the k_H/k_D of 1.11 at all three pH values fits best with the idea of an oxocarbonium mechanism, and the observed retention of stereochemical configuration at C1 suggests that the reaction involves restricted entry of the incoming nucleophile

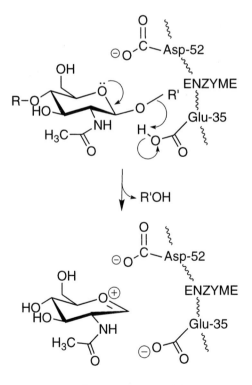

**Lysozyme Oxocarbonium Ion:
Formation and Stabilization**

Figure 4. Essential features of the hydrolytic reaction catalyzed by hen egg white lysozyme. Notice that two nearby carboxyl groups should be able to stabilize the oppositely charged oxocarbonium ion intermediate.

(*i.e.*, water). If a fully developed oxocarbonium ion in 2M HCl yields a k_H/k_D of 1.13, then the value of 1.11 is really quite close to this limit. However, as noted below, the magnitude of an observed kinetic isotope effect may be observed by other kinetic considerations.

Observed Isotope Effects for Hydrolysis of Phenyl-4-O-(2-acetamido-2-deoxy-β-D-glucopyranosyl)-β-D-glucopyranoside[a]

Conditions	Observed k_H/k_D
Acid (2 M HCl)	1.13
Base (3 M NaOCH₃)	1.03
Lysozyme (pH 5.5)	1.11
Lysozyme (pH 3.1)	1.11
Lysozyme (pH 8.3)	1.11

[a]From Dahlquist *et al.*[10] with permission of the authors and the American Chemical Society. Reaction was permitted to proceed to 2–5% completion and was then adjusted with acid or base to pH 5.6. The mixture was extracted with ether, and the extract was washed twice with saturated NaCl, then dried with anhydrous sodium sulfate, and added to scintillation fluid for counting of radiolabeled phenol.

Why do secondary isotope effects in S_N2 reactions vary from about 0.95 to 1.06 upon deuterium substitution in the α position? The most basic explanation for this behavior relates to the fact that the repulsion lost as the leaving group departs is partly balanced by the repulsions developed by the incoming nucleophile. The presence of both nucleophile and exiphile within the transition state maintains the zero point energy difference throughout the course of the S_N2 reaction. The magnitude of k_H/k_D depends upon differences between the corresponding force constants (specifically $f_{HCX} \rightarrow f^{\ddagger}_{HCX}$ and $f_{HCX} \rightarrow f^{\ddagger}_{HCN}$, where the subscripted X and N indicate the exiphile and nucleophile, respectively). This shows that bonding by both the nucleophile and exiphile to the carbon atom undergoing substitution is important in S_N2 reactions. The degree to which the entry and removal of the nucleophile and exiphile are well "choreographed" will affect the magnitude of the observed secondary isotope effects, which tend to be 1.00 ± 0.06. Thus, if an enzyme-catalyzed reaction yields a k_H/k_D isotope effect that is less than 1.07, the observed kinetic isotope effect is of dubious value in diagnosing the actual type of mechanism.

Are S_N1 and S_N2 mechanisms the whole story? For solution-phase nucleophilic substitution reactions, S_N1 and S_N2 mechanisms are limiting cases that lie along a continuum of possible interactions[14]:

$$
\begin{array}{cccccccc}
R{-}X & \rightleftharpoons & R^+X^- & \rightleftharpoons & R^+\|X^- & \rightleftharpoons & R^+ \text{ and } X^- \\
\downarrow {}^-Nu\colon & & \downarrow {}^-Nu\colon & & \downarrow {}^-Nu\colon & & \downarrow {}^-Nu\colon \\
Nu{-}R{-}X & & {}^{\delta-}Nu{-}R^+{-}X^{\delta-} & & {}^-Nu{-}R^+\|X^- & & Nu{-}R + X^- \\
\downarrow & & \downarrow & & \downarrow & & \downarrow \\
NuR + X^- & & NuR + X^- & & NuR + X^- & & RNu \text{ or } NuR + X^-
\end{array}
$$

where [R—X] is the covalently bonded form, [R⁺X⁻] is the tight ion pair, [R⁺‖X⁻] is the loosely associated ion pair, [R⁺ + X⁻] represents the fully dissociated ions, ⁻N: is the nucleophile, and RN and NR are the two optical isomers formed by attack on either face of the electrophilic center. These intermediate types of mechanisms obviously should have geometries that lie on a continuum; so also must the secondary kinetic isotope effects associated with these hybrid mechanisms. Other factors (*e.g.*, nucleophilic strength, solvent nucleophilicity, other solvation effects, the detailed nature of the leaving group, and temperature) can greatly influence the mechanistic course of a reaction. Even examples from physical organic chemistry (where reactant adsorption and release from an enzyme active site is not a factor) often fail to allow one to discriminate between these possible mechanisms. In this sense, it is not difficult to imagine that isotope effects often serve only as a guide for limiting the number of most likely mechanisms. In fact, the formation of tight-ion pairs, or even covalent intermediates, can introduce further complexity in enzymic reactions. Taking the case of lysozyme as an example, there still remains the possibility for transient formation of an acylal intermediate; this species could form in a post-rate-determining step, wherein the oxocarbonium ion and the active site carboxylate collapse to form a covalent bond. Such a pathway would lead to two Walden inversions: one forming the acylal, and a second inversion during hydrolysis of the acylal. This mechanistic route would therefore give the same apparent overall retention of stereochemical configuration predicted for the S_N1 case.

INFLUENCE OF OTHER STEPS ON THE MAGNITUDE OF OBSERVED ISOTOPE EFFECTS. As noted earlier, nonenzymatic reaction mechanisms do not involve those complexities imposed by substrate binding order, rates of substrate binding/release, as well as conformational changes that attend enzyme catalysis. As a result, the opportunity for detecting isotope effects is greater with organic chemical reactions than with their enzymic counterparts. If one considers the fact that bond breaking/making steps occur in the central complexes for enzyme processes, it is easy to recognize that the speed of substrate binding and/or product release can greatly influence the magnitude of observed kinetic isotope effects for enzymes. This well-known limitation on certain steady-state and equilibrium processes means that one obtains collections of rate constants in the form of the Michaelis constant (K_m), the maximal velocity (V_m), or the specificity constant (V_m/K_m). The diagram shown in Fig. 6 illustrates that one may link an observed kinetic isotope effect to a certain segment of an enzyme-catalyzed reaction. This depiction should be regarded as an idealized situation, because the magnitude of the isotope effect can be greatly influenced by the relative magnitude of the rate constants in certain steps.

$$E + S \underset{k_2}{\overset{k_1}{\rightleftharpoons}} ES \overset{k_3}{\rightarrow} EP \overset{k_5}{\rightarrow} E + Product$$

where k_3 is the bond-breaking step. [Note that k_4 and k_6 are not included because ES-to-EP and the EP-to-E steps are treated as irreversible reactions.] The operant steady-state rate equation is:

$$\nu = k_1 k_3 k_5 [S][E_{total}]/\{k_5(k_2 + k_3) + k_1(k_3 + k_5)[S]\}$$

where:

$$V_{max} = k_3 k_5 [E_{total}]/(k_3 + k_5)$$

$$K_S = k_5(k_2 + k_3)/\{k_1(k_3 + k_5)\}$$

$$V_{max}/K_S = k_1 k_3 [E_{total}]/(k_2 + k_3)$$

By assuming that the primary isotope effect (here noted as a deuterium isotope effect) only influences the bond-

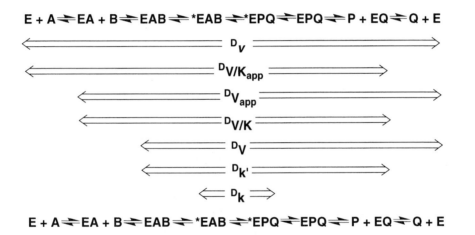

Steady-state Dependency of a Family of Deuterium Isotope Effects

Figure 6. The steady-state dependence of various isotope effects for a hypothetical enzyme catalyzing a reaction described as an ordered ternary complex kinetic mechanism.

Such considerations raise the concept of the *intrinsic kinetic isotope effect*—the effect of isotopic substitution on a specific step in an enzyme-catalyzed reaction. The magnitude of an intrinsic isotope effect may not equal the magnitude of an isotope effect on collective rate parameters such as V_{max} or V_{max}/K_m, unless the isotopically sensitive step is the rate-limiting or rate-contributing step. To tackle this problem, Northrop[15] extended the kinetic theory for primary isotope effects in enzyme-catalyzed reactions. His approach can be illustrated with the following example of a one-substrate/two-intermediate enzyme-catalyzed reaction:

breaking step (*i.e.*, ES → EP), Northrop rewrote the maximal velocity expression in terms of the single isotopically sensitive step:

$$V_H = k_{3H} k_{5H}[E_{total}]/(k_{3H} + k_{5H}) = k_{5H}[E_{total}]/(1 + k_{5H}/k_{3H})$$

$$V_D = k_{3D} k_{5H}[E_{total}]/(k_{3D} + k_{5H}) = k_{5H}[E_{total}]/(1 + k_{5H}/k_{3D})$$

And by dividing the first by the second, he obtained:

$$(V_H/V_D) = \{(k_{3H}/k_{3D}) + (k_{3H}/k_{5H})\}/(1 + k_{3H}/k_{5H})$$

Taking the limit of this equation as (k_{3H}/k_{5H}) approaches zero (*i.e.*, letting k_3 become very small so that the isotopically sensitive step is rate-limiting), we get

$$(V_H/V_D) \approx k_{3H}/k_{3D}$$

(See Footnote[16] for corresponding V/K expression.) We can see that in this case the isotope effect will be fully expressed in terms of (V_H/V_D) when the isotopically sensitive step is the rate-limiting step. One can now ask what happens if k_{5H} is not much larger than k_{3H}? Let us assess the hypothetical case where the primary isotope is assumed to be 10:

(1) If $k_{3H}/k_{5H} = 1$, then (V_H/V_D)
$$= \{10 + 1\}/(1 + 1) = 5.5$$

(2) If $k_{3H}/k_{5H} = 10$, then (V_H/V_D)
$$= \{10 + 10\}/(1 + 10) = 1.8$$

These simple calculations indicate how the other rate constants in a kinetic equation collectively act to greatly suppress the magnitude of the observed primary isotope effect[17].

With this background, let us examine how Northrop[15] was able to develop a way to extract the true (or as we more commonly say, *intrinsic*) isotope effect. Recognizing that an underdetermined system can often be solved if an additional quantitative relation is considered, Northrop proposed to measure two isotope effects, namely $(V/K)_H/(V/K)_D$ and $(V/K)_H/(V/K)_T$ and to take advantage of the following relationship:

$$(k_{3H}/k_{3T}) = (k_{3H}/k_{3D})^{1.442}$$

This exponential relationship is called the Swain relationship in honor of the senior author of a report[18] on how one can calculate the magnitude of primary isotope effects. Use of this relationship allowed Northrop to obtain the following expression[19]:

$$\frac{[(V/K)_H/(V/K)_D] - 1}{[(V/K)_H/(V/K)_T] - 1} = \frac{[(k_{3H}/k_{3D}) - 1]}{[(k_{3H}/k_{3D})^{1.442} - 1]}$$

OTHER EXAMPLES OF KINETIC ISOTOPE EFFECTS. The power of kinetic isotope effects in enzymology is well illustrated in the work of Rose[20] and Knowles[21] dealing with hydrogen effects in proton transfer to and from carbon. Abstraction of a proton from a tetrahedral carbon is a fundamental step in many enzyme-catalyzed reactions. Intramolecular proton transfer as well as partial loss (wash-out) migrating protons have provided important clues in mechanistic investigations. Enol and enediolate formation constitute several

of the best examples of how isotope effects have become a major tool for unraveling subtle properties of enzyme catalysis and control. The prototypical enzyme system is the triose-phosphate isomerase reaction where a *cis*-enediol intermediate is formed upon removal of the *pro-R* proton from C-1 of dihydroxyacetone phosphate or the lone proton of C-2 in D-glyceraldehyde 3-phosphate. This proton abstraction reaction is facilitated by the γ-carboxylate of E165 in the isomerase[22,23]. Rose's group demonstrated that a proton must also be transferred between the oxygen of C-1 and the oxygen of C-2 to achieve isomerization of dihydroxyacetone to D-glyceraldehyde 3-phosphate. Evidence for these important steps in the isomerase mechanism were provided by a combination of tritium isotope-exchange experiments and kinetic isotope-effect measurements[24].

Isotope effects have also been applied extensively to studies of $NAD^+/NADP^+$-linked dehydrogenases. We typically treat these enzymes as systems whose catalytic rates are limited by product release. Nonetheless, Palm[25] clearly demonstrated a primary tritium kinetic isotope effect on lactate dehydrogenase catalysis, a finding that indicated that the hydride transfer step is rate-contributing. Plapp's laboratory[26] later demonstrated that liver alcohol dehydrogenase has an intrinsic k_H/k_D isotope effect of 5.2 with ethanol and an intrinsic k_H/k_D isotope effect of 3-6-4.3 with benzyl alcohol. Moreover, Klinman[27] reported the following intrinsic isotope effects in the reduction of *p*-substituted benzaldehydes by yeast alcohol dehydrogenase: k_H/k_D for *p*-Br-benzaldehyde = 3.5; k_H/k_D for *p*-Cl-benzaldehyde = 3.3; k_H/k_D for *p*-H-benzaldehyde = 3.0; k_H/k_D for *p*-CH$_3$-benzaldehyde = 5.4; and k_H/k_D for *p*-CH$_3$O-benzaldehyde = 3.4.

Bell[28] described how quantum mechanical tunneling can be an important factor in the rate of chemical reactions, and Bahnson and Klinman[29] have extended these considerations to define how isotope effects can be utilized to infer important features of hydrogen transfer reactions. They point out that the tunneling probability depends on the degree of solvent reorganization during catalysis[30], thermodynamic relationships between substrates and products[24,31], and the height and width of the energy barrier along the reaction coordinate[32]. Bahnson and Klinman[29] describe how the temperature dependence of enzyme kinetic isotope effects is an important criterion for demonstrating tunneling. They provide examples from studies on alcohol dehydrogenase, bovine serum amine oxidase, and monamine oxidase B.

Although there are literally hundreds of reports (see reviews[3,4,33-35]) dealing with kinetic isotope effects, the recent study of Scheuring & Schramm[36] on the ADP-ribosylase activity of *Bordetella pertussis* toxin provides an excellent example of the combined power of primary and secondary kinetic isotope effect experiments as well as solvent isotope effects. For the ribosyl-acceptor substrate, these investigators used a synthetic cysteine-containing peptide corresponding to the last twenty amino acids of the α-subunit of the G-protein G_i3. A family of kinetic isotope effects (KIE) were evaluated by using [3]H-, [14]C-, and ^{15}N-labeled NAD^+ as substrates. Primary kinetic isotope effects were 1.050 ± 0.006 for [1'N—[14]C] and 1.021 ± 0.002 for [1N—[15]N]. The double primary [1'N—[14]C,1N—[15]N] effect was found to be 1.064 ± 0.002. Secondary kinetic isotope effects were also evaluated: [1'N—[3]H]secondary-KIE $= 1.208 \pm 0.014$; [2'N–[3]H] secondary-KIE $= 1.104 \pm 0.010$; [4'N–[3]H]secondary-KIE $= 0.989 \pm 0.001$; and [5'N–[3]H]secondary-KIE $= 1.014 \pm 0.002$. The measured commitment factor of 0.01, based on isotope trapping experiments, indicated that the observed isotope effects are intrinsic. The observed inverse solvent D_2O kinetic isotope effect suggested possible deprotonation of the attacking cysteinyl residue prior to reaching the transition state. The transition state structure was determined by a normal mode bond vibrational analysis. The transition state is characterized by a nicotinamide leaving group bond order of 0.14 (corresponding to a bond length of 2.06 Å). Their analysis also suggests a bond order of 0.11 for the incoming thiolate nucleophile, corresponding to a distance of 2.47 Å. The ribose ring appears to have strong oxocarbenium ion character. By comparing the likely transition states for the ADP-ribosylase activity and the inherent NAD^+ hydrolase, Scheuring & Schramm[36] inferred that the cysteinyl sulfur is likely to be a more nucleophilic in the transferase reaction than is H_2O in the hydrolytic reaction.

See also *Born-Oppenheimer Approximation; Equilibrium Isotope Effect; Hydrogen Tunneling; Intramolecular Kinetic Isotope Effect; Inverse Kinetic Isotope Effect; Quantum Mechanical Tunneling; Swain Relationship; Enzyme Energetics (The Case of Proline Racemase); Isotopic Perturbation*

[1]L. Melander (1960) *Isotope Effects on Reaction Rates*, p.181, Raven Press, New York.

[2]E. K. Thornton & E. R. Thornton (1978) in *Transition States of Biochemical Processes* (R. D. Gandour & R. L. Schowen, eds.), pp. 3-76, Plenum Press, New York.

[3]W. W. Cleland, M. H. O'Leary & D. B. Northrop, eds. (1977) *Isotope Effects on Enzyme-Catalyzed Reactions*, University Park Press, Baltimore.

[4]R. D. Gandour & R. L. Schowen, eds. (1978) *Transition States of Biochemical Processes*, Plenum Press, New York.

[5]J. Hine (1962) *Physical Organic Chemistry*, McGraw-Hill, New York.

[6]W. P. Jencks (1969) *Catalysis in Chemistry and Enzymology*, McGraw-Hill, New York. [Also available as Jencks, W. P. (1986) *Catalysis in Chemistry and Enzymology*, Dover Publications, New York]

[7]P. Strittmatter (1964) *J. Biol. Chem.* **240**, 4481.

[8]This is sometimes confused since C—H$_a$ and C—H$_a$ have slightly different dipole moments. As noted by Thornton & Thornton[2], the different dipoles arise because of different average geometries due to the stretching vibration. Thus, the dipole moment is not strictly an electronic property.

[9]A corollary of this argument is that a change from sp^2 to sp^3 hybridization should result in an inverse secondary isotope effect (*i.e.*, $k_H/k_D < 1$).

[10]F. W. Dahlquist, T. Rand-Meir & M. A. Raftery (1969) *Biochemistry* **8**, 4214.

[11]D. C. Phillips (1967) *Sci. Am.* **215** (Nov.), 78; (1964) *Proc. Natl. Acad. Sci. U.S.A.* **57**, 484.

[12]Dahlquist *et al.* [10] also provided a useful appendix for converting radioactivity counting data into observed isotope effects.

[13]C. A. Bunton (1969) *Nucleophilic Substitution at the Saturated Carbon Atom*, McGraw-Hill, New York.

[14]S. Winstein, E. Clippinger, A. H. Fainberg, R. Heck & G. C. Robinson (1956) *J. Amer. Chem. Soc.* **78**, 328.

[15]D. B. Northrop (1977) *Biochemistry* **14**, 2644.

[16]We can also rewrite the expression for V/K (from the list above) in terms of k_{3H} and k_{3D} to obtain:

$$(V/K)_H = k_{1H}k_{3H}[E_{total}]/(k_{2H} + k_{3H})$$

$$(V/K)_D = k_{1H}k_{3D}[E_{total}]/(k_{2H} + k_{3D})$$

Upon dividing the first equation by the second, we arrive at:

$$(V/K)_H/(V/K)_D = \{k_{3H}/(k_{2H} + k_{3H})\}/\{k_{3D}/(k_{2H} + k_{3D})\}$$

which upon rearrangement becomes

$$[(V/K)_H/(V/K)_D] - 1 = \{(k_{3H}/k_{3D}) - 1\}/\{(k_{3H}/k_{2H}) + 1\}$$

If $k_{3H} \ll k_{2H}$, then we get the desired final expression:

$$[(V/K)_H/(V/K)_D] - 1 = \{(k_{3H}/k_{3D}) - 1\}$$

[17]The situation would, of course, be far worse for secondary isotope effects. For example, if we were to take 1.22 as the limiting 2° isotope effect for a S_N1 substitution reaction and 1.06 as the largest secondary isotope effect for a S_N2 substitution reaction, then we could ask what value of k_{3H}/k_{4H} would yield a value of 1.06 (*i.e.*, lower k_{3H}/k_{3D} to the point where one could not distinguish the S_N1 and S_N2 cases.) In this case, $1.06 = \{1.22 + k_{3H}/k_{4H}\}/(1 + k_{3H}/k_{4H})$, or $k_{3H}/k_{4H} = 2.66$. The result reminds us that secondary isotope effects are exquisitely sensitive to the relative magnitudes of rate constants.

[18]C. G. Swain, E. C. Stivers, J. F. Reuwer, Jr., & L. J. Schaad (1958) *J. Am. Chem. Soc.* **80**, 5885.

[19]This treatment is based on the following assumptions[18]: (a) all differences, except for zero point vibrational energy are negligible; (b) the hydrogen isotope is attached to a heavy polyatomic molecule, such that isotopic substitution affects only the vibrations of

the bond holding the hydrogen; (c) the bond may be treated as a harmonic oscillator; and (d) no tunneling through the potential energy barrier occurs. This limits the value of k_H/k_T and k_H/k_D to mass effects only, such that

$$\frac{k_H}{k_T} = \left(\frac{k_H}{k_T}\right)^{\left[\frac{1/\sqrt{m_H} - 1/\sqrt{m_D}}{1/\sqrt{m_H} - 1/\sqrt{m_T}}\right]} = \left(\frac{k_H}{k_T}\right)^{1.442}$$

where m_H is 1.0081, m_D is 2.0147, and m_T is 3.0170. Although beyond the scope of this description, some dispute exists regarding the best value for the exponent, which may range from 1.33 to 1.58. In this respect, use of four significant figures may not be justified.

[20]I. A. Rose (1982) *Meth. Enzymol.* **87**, 84.

[21]J. R. Knowles (1980) *Ann. Rev. Biochem.* **49**, 877.

[22]S. G. Waley, J. C. Mill, I. A. Rose & E. L. O'Connell (1970) *Nature* **227**, 181.

[23]E. A. Komives, L. CX. Chang, E. Lolis, R. F. Tilton, G. A. Petsko & J. R. Knowles (1991) *Biochemistry* **30**, 3011.

[24]W. J. Albery & J. R. Knowles (1976) *Biochemistry* **15**, 5631.

[25]D. Palm (1968) *Eur. J. Biochem.* **5**, 270.

[26]B. V. Plapp (1970) *J. Biol. Chem.* **245**, 1725; (1973) *J. Biol. Chem.* **248**, 3470; *Meth. Enzymol.* **249**, 91.

[27]J. P. Klinman (1977) in *Isotope Effects on Enzyme-Catalyzed Reactions* (W. W. Cleland, M. H. O'Leary & D. B. Northrop, eds.), p. 176, Univ. Park Press, Baltimore.

[28]R. P. Bell (1980) *The Tunnel Effect in Chemistry*, Chapman & Hall, New York.

[29]B. J. Bahnson & J. P. Klinman (1995) *Meth. Enzymol.* **249**, 373.

[30]P. A. Bartlett & C. K. Marlowe (1987) *Biochemistry* **26**, 8553.

[31]K. P. Nambiar, D. M. Stauffer, P. A. Kolodziej & S. A. Benner (1983) *J. Amer. Chem. Soc.* **105**, 5886.

[32]J. Rodgers, D. A. Femec & R. L. Schowen (1982) *J. Amer. Chem. Soc.* **104**, 3263.

[33]D. L. Purich, ed. (1980) *Meth. Enzymol.* vol. **64**; (1982) *Meth. Enzymol.* vol. **87**; and (1995) *Meth. Enzymol.* vol. **249**, Academic Press, New York.

[34]P. F. Cook, ed. (1991) *Enzyme Mechanism from Isotope Effects*, CRC Press, Boca Raton, Florida.

[35]D. Devault (1984) *Quantum Mechanical Tunneling in Biological Processes*, 2nd ed., Cambridge University Press, Cambridge.

[36]J. Scheuring & V. I. Schramm (1997) *Biochemistry* **36**, 8215.

KINETIC PARAMETERS

1. The K_m, V_{max} (or, k_{cat}), K_{is}, K_{ii}, *etc.*, values of an enzyme-catalyzed reaction. 2. The value(s) of the individual rate constant(s) of a reaction or process.

KINETIC RESOLUTION

Partial or complete separation of enantiomers within a racemic mixture as a result of unequal rates of reaction with another agent. The latter reagent, catalyst, solvent, or micelle must itself exhibit chirality, resulting in its stereoselective or stereospecific action on the racemic compound.

KING-ALTMAN METHOD

A method to assist the investigator in the derivation of enzyme-rate expressions[1]. This method is schematic in nature, involving the drawing of all possible patterns of a geometric figure minus a loop. It is a very convenient

method for straightforward enzyme schemes. However, for reactions which contain three or more substrates or products, or a number of abortive complexes, or isomerization steps, *etc.*, other methods are available that are easier to follow and aren't as prone to mistakes (*e.g.*, it is a common mistake in the King-Altman method to miss one or more of the geometrical forms with complex mechanisms)[2]. ***See*** *Enzyme Rate Equations (2. Derivation of Initial Velocity Equations); Fromm's Method for Deriving Enzyme Rate Equations; Net Rate Constant Method (applies only to unbranched mechanisms)*

[1]E. L. King & C. Altman (1956) *J. Chem. Phys.* **60**, 1375.

[2]C. Y. Huang (1979) *Meth. Enzymol.* **63**, 54.

KINSIM/FITSIM

A computer software package for numerical integration of rate equations, first developed for use on VAX computers by Barshop, Wrenn and Frieden[1] under the name KINSIM to utilize kinetic information acquired during the full time course of a reaction to analyze the properties of candidate mechanisms. The second program, designated FITSIM, was developed by Zimmerle and Frieden[2] to carry out statistical fitting of rate data to kinetic simulations. These programs are now supported on the following platforms[3]: VMS, Silicon Graphics Iris (IRIX), IBM PC (in MS-DOS, Windows, and OS/2 versions), and Apple Macintosh.

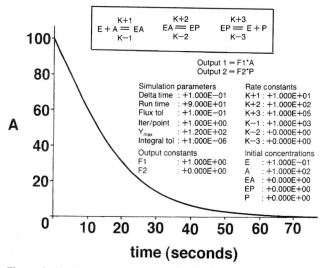

Figure 1. Uni Uni enzyme reaction scheme along with rate constants and a typical kinetic simulation.

Frieden[4,5] has presented an overview of the KINSIM program, and a typical printed display for the two-intermediate Uni Uni enzyme kinetic mechanism is shown in Fig. 1.

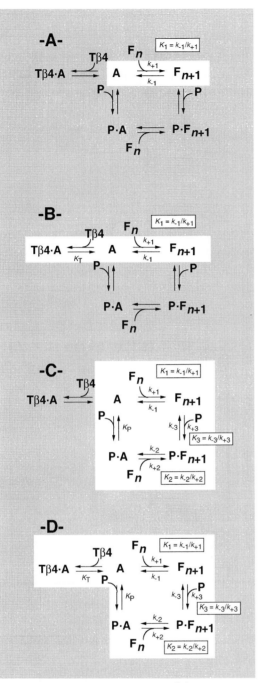

Figure 2. A diagrammatic representation of those segments of an overall reaction mechanism for barbed-end actin polymerization in the absence and presence of profilin and/or thymosin-β4. For clarity the overall scheme is indicated. Experimental data were obtained for individual segments (indicated in black type), and those parts of the scheme that were not evaluated in a particular experiment are shown in gray.

Kang et al.[6] recently used the KINSIM/FITSIM software to model barbed-end actin polymerization (*See Actin Assembly Kinetics*) in the absence and presence of profilin and/or thymosin-β4. This data analysis protocol permitted them to derive rate constants from a series of

experiments where the rates and extents of polymerization were first determined by right-angle light scattering. Fig. 2 illustrates the segments of the overall scheme investigated in any single set of experiments, and KINSIM/FITSIM was used to achieve global data analysis on a series of runs at various protein concentrations (see one illustrative set of polymerization runs in Fig. 3).

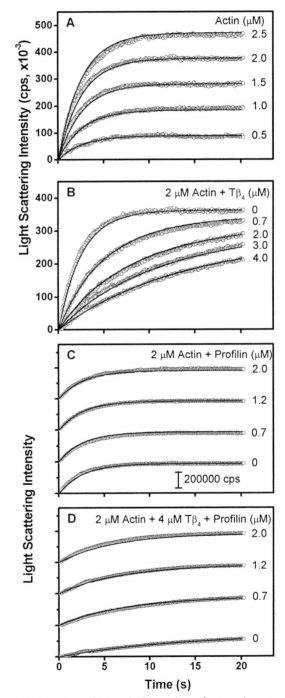

Figure 3. KINSIM/FITSIM applied to analysis of actin polymerization, showing actual light scattering data points and kinetic simulations (solid lines).

These experiments indicated (a) that profilin·actin·ATP can participate in actin filament, (b) that the derived rate and equilibrium constants define a model which satisfies a closed thermodynamic cycle, and (c) that contrary to earlier suggestions, there is no absolutely need to invoke any special property of profilin in promoting irreversible ATP hydrolysis during actin polymerization.

[1]B. A. Barshop, R. F. Wrenn & C. Frieden (1983) *Anal. Biochem.* **130**, 134.

[2]C. T. Zimmerle & C. Frieden (1989) *Biochem J.* **258**, 381.

[3]Those interested in obtaining information about the KINSIM/FIT-SIM software (KFSIM40.EXE), should consult the kinsim archive over the internet at: "http://wuarchive.wustl.edu."

[4]C. Frieden (1993) *Trends in Biochem. Sci.* **18**, 58.

[5]C. Frieden (1997) *Trends in Biochem. Sci.* **22**, 317.

[6]F. Kang, F.S. Southwick & D. L. Purich, unpublished findings.

Selected entries from *Methods in Enzymology* [vol, page(s)]: Data analysis flow chart, **240**, 314-315; data point number requirements, **240**, 314; determination of enzyme kinetic parameters multisubstrate, **240**, 316-319; single substrate, **240**, 314-316; enzyme mechanism testing, **240**, 322; evaluation of binding processes, **240**, 319-321; instructions for use, **240**, 312-313.

K_{ip}, K_{eq}, K_{ir}, . . .

Symbol for the dissociation constant for product P, Q, R, . . .

K_{is}

Symbol for the dissociation constant of an inhibitor with respect to the free enzyme, E. In the double-reciprocal form of the initial-rate equation, this term is associated with the slope portion of the expression. **See** K_{ii}

KITZ AND WILSON REPLOT

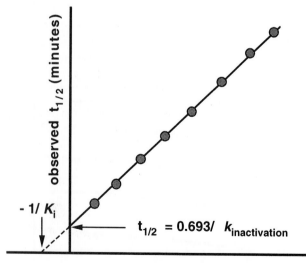

A graphical procedure[1] used in the study of irreversible inhibitors, particularly in mechanism-based inactiva-

tion[2], and with affinity labels[3]. In this method, the half-life ($t_{1/2}$) is measured at a number of different inhibitor concentrations (at constant enzyme concentration). The Kitz and Wilson plot is a replot of these $t_{1/2}$ values as a function of 1/[inactivator]. From this plot, k_{inact} and K_I can be obtained.

[1]R. Kitz & I. B. Wilson (1962) *J. Biol. Chem.* **237**, 3245.

[2]R. B. Silverman (1995) *Meth. Enzymol.* **249**, 240.

[3]B. V. Plapp (1982) *Meth. Enzymol.* **87**, 469.

KLOTZ PLOT

A ligand binding plot (1/n *versus* 1/[A]) that permits one to determine the number of ligand binding sites and the dissociation constant for the protein-ligand complex. The typical binding equation for *n*-independent sites is

$$\nu = (n[\text{Ligand}]/K_d)/\{1 + [\text{Ligand}]/K_d\}$$

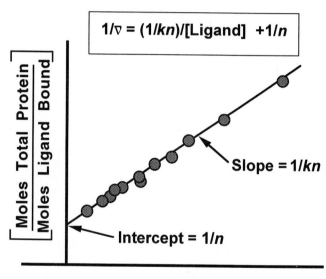

For the Klotz plot, this expression is transformed to yield:

$$1/\nu = \{(K_d/n)/[\text{Ligand}]\} + 1/n$$

where *n* is the fractional saturation of *n* sites with equivalent binding. A plot of $1/\nu$ *versus* 1/[A] gives a straight line with slope equal to K_d/n and an intercept of $1/n$. In the case of equilibrium dialysis or Hummel-Dreyer ligand binding experiments, the net accrual of ligand as macromolecule-ligand complex is obtained directly. In the case of an observed fluorescence change upon ligand binding, then the fraction of ligand bound at any given concentration of ligand is given as

$$X = (F_{observed} - F_{ligand})/(F_{maximum} - F_{ligand})$$

where $F_{maximum}$ is the fluorescence at saturating ligand concentration. Then a plot of the reciprocal of ν, which is again the number of moles of bound ligand per mole protein, *versus* the reciprocal of the free ligand complex relies on the following equation:

$$1/\nu = K_d/\{n\,(1-X)[\text{Ligand}]_{\text{total}}\} + 1/n$$

I. M. Klotz, F. M. Walker & R. G. Bivan (1946) *J. Am. Chem. Soc.* **68**, 1486.

K_m (or, K_M)

Symbol for the Michaelis constant (or, Michaelis-Menten constant). **See** *Michaelis Constant*

KOSHLAND-NÉMETHY-FILMER COOPERATIVITY MODEL

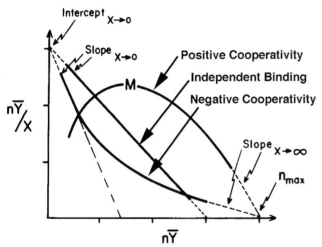

Scatchard plots for ligand saturation behavior of an oligomeric protein exhibiting positive cooperativity, negative cooperativity, or independent binding behavior. For addition details, **See** *Monod-Wyman-Changeux Cooperativity Model; Scatchard Plot*.

A model used to explain cooperativity on the basis of ligand-induced changes in conformation that may or may not alter the subunit-subunit interfaces of oligomeric enzymes and receptors[1-3]. This model has also been referred to as the Adair-Koshland-Némethy-Filmer model (AKNF model), the induced-fit model, and the sequential model.

The model assumes: (1) that the protein is oligomeric (or that the protein contains several topologically distinct ligand binding sites on a single polypeptide chain); (2) that in the absence of the binding ligand (or substrate), the protein has only one conformational form; (3) that upon binding, the ligand induces a conformational

change in the subunit to which it is bound (hence, there is no symmetry requirement as is essential for the Monod-Wyman-Changeux model), (4) that this conformational change affects the interface to one or more of the adjacent subunits and this conformational change alters the affinity of the new ligand binding to that adjacent subunit, and (5) that binding of the ligand(s) is a rapid equilibrium step(s).

The Koshland model can account for positive and negative cooperativity as well as independent ligand binding. The key is how much binding energy is released as each ligand interacts with protein as saturation of sites is achieved. For energy diagrams explaining these features, **See** *Monod–Wyman Changeux Model*.

The KNF model is an appealing model that builds on the well-known ability of ligands to disturb protein structure beyond the binding site *per se*. Moreover, many proteins exhibit negative cooperativity, and the Monod model simply cannot accommodate the possibility of negative cooperativity. **See** *Induced Fit Model; Cooperativity; Allosterism; Monod-Wyman-Changeux Model; Negative Cooperativity; Positive Cooperativity; Independent Binding; Linked Functions; Adair Equation*

[1] D. E. Koshland, Jr., G. Némethy & D. Filmer (1966) *Biochemistry* **5**, 365.
[2] K. E. Neet (1980) *Meth. Enzymol.* **64**, 139.
[3] K. E. Neet (1995) *Meth. Enzymol.* **249**, 519.

KOSMOTROPES

Small ions of high charge density that bind water molecules more strongly than the strength of water-water interactions in bulk solvent[1]. The major intracellular anions (*e.g.*, phosphates and carboxylates) are kosmotropes. These ions differ from chaotropes, which are large ions of low charge density and only weakly bind water molecules. **See** *Chaotropic Agents; Hofmeister Series*

[1] K. D. Collins (1997) *Biophys. J.* **72**, 65.

K_p, K_q, K_r, $\cdots$

Symbol for the Michaelis constant for product P, Q, R, . . .

K_S

1. Symbol for the dissociation constant of a substrate, S, for a particular enzyme. Note that, when this symbol is

used, the investigator is assuming that the substrate binds in a rapid equilibrium fashion. The K_S value is not necessarily equal to the Michaelis constant. 2. Symbol for the Michaelis constant for product S. **See** *Michaelis-Menten Equation; Uni Uni Mechanism*

K SYSTEMS

A subset in allosteric models of cooperativity. If an allosteric effector, upon binding to a cooperative enzyme, alters the Michaelis or dissociation constants (or $[S_{0.5}]$ value) for the substrate(s) (but not the V_{max} values), then that protein is a *K* system enzyme. **See** *Monod-Wyman-Changeux Model*

K. E. Neet (1980) *Meth. Enzymol.* **64**, 139.

KUHN-ROTH OXIDATION

A method for the determination of carbon-linked methyl groups (and, by extension, used in the analysis of chiral methyl groups in stereochemical studies of enzyme-catalyzed reactions acting on methyl groups). In this procedure, the methyl groups are converted to acetic acid by oxidation of the metabolite with chromic and sulfuric acids. Milder versions are also available.

H. Simon & H. G. Floss (1967) *Bestimmung der Isotopenverteilung in markierten Verbindungen*, Springer-Verlag, Berlin.
J. W. Cornforth, R. H. Cornforth, A. Pelter, M. G. Horning & G. Popjak (1959) *Tetrahedron* **5**, 311.

KURGANOV METHOD FOR ANALYZING COOPERATIVITY

A procedure[1,2] for estimating the cooperativity of a protein by using the initial rates at only three different concentrations of substrate to estimate the shape of the cooperative curve.

[1]B. I. Kurganov (1973) *Acta Biol. Med. Germ.* **31**, 181.
[2]B. I. Kurganov (1982) *Allosteric Enzymes: Kinetic Behavior*, Wiley, New York.

K_w

Symbol for the product of the H^+ concentration (or, H_3O^+ concentration) and the OH^- concentration of an aqueous solution; the autoprotolysis constant. **See** *Water, Temperature Effects of pK_w of*

KYNURENINASE

This pyridoxal-phosphate-dependent enzyme [EC 3.7.1.3] catalyzes the hydrolysis of kynurenine to produce anthranilate and alanine. 3′-Hydroxykynurenine and some other (3-arylcarbonyl)alanines can also be acted upon by this enzyme.

A. E. Martell (1982) *Adv. Enzymol.* **53**, 163.
K. Soda & K. Tanizawa (1979) *Adv. Enzymol.* **49**, 1.

KYNURENINE AMINOTRANSFERASE

Kynurenine:oxoglutarate aminotransferase [EC 2.6.1.7] catalyzes the pyridoxal-phosphate-dependent reaction of kynurenine and α-ketoglutarate (or, 2-oxoglutarate) to reversibly produce 4-(2-aminophenyl)-2,4-dioxobutanoate and glutamate. Kynurenine:glyoxylate aminotransferase [EC 2.6.1.63] catalyzes the reversible reaction of kynurenine with glyoxylate to produce 4-(2-aminophenyl)-2,4-dioxobutanoate and glycine.

M. C. Tobes (1987) *Meth. Enzymol.* **142**, 217.
K. Tanizawa, Y. Asada & K. Soda (1985) *Meth. Enzymol.* **113**, 90.
R. A. Hartline (1985) *Meth. Enzymol.* **113**, 664.
A. E. Braunstein (1973) *The Enzymes*, 3rd ed., **9**, 379.

KYNURENINE 3-HYDROXYLASE

This FAD-dependent enzyme [EC 1.14.13.9], also called kynurenine 3-monooxygenase, catalyzes the reaction of kynurenine with NADPH and dioxygen to produce 3-hydroxykynurenine, $NADP^+$, and water.

V. Massey & P. Hemmerich (1975) *The Enzymes*, 3rd ed., **12**, 191.

L

LABILE

1. Referring to an unstable and/or transient chemical species. 2. Referring to a substituent, atom, or group of a molecular entity that is easily removable; *e.g.*, the labile proton of the carboxyl group in trichloroacetic acid. 3. In reference to coordinator complexes, referring to ligands that can be readily replaced by other ligands. 4. On occasions, the term is also used with respect to stable, yet reactive, chemical species.

LACCASE

This copper-dependent enzyme [EC 1.10.3.2] catalyzes the reaction of four benzenediol and dioxygen to produce four benzosemiquinone and two water.

A. Messerschmidt (1998) *Comprehensive Biological Catalysis: A Mechanistic Reference* 3, 401.
J. V. Schloss & M. S. Hixon (1998) *Comprehensive Biological Catalysis: A Mechanistic Reference* 2, 43.
B. G. Malmström, L.-E. Andréasson & B. Reinhammer (1975) *The Enzymes*, 3rd ed., 12, 507.

β-LACTAMASES

Zinc-dependent enzymes [EC 3.5.2.6], including penicillinase and cephalosporinase, with varying specificity in their catalysis of β-lactam hydrolysis. Some act more readily on penicillins, whereas the catalysis of others is more efficient with cephalosporins.

C. W. Wharton (1998) *Comprehensive Biological Catalysis: A Mechanistic Reference* 1, 345.
M. Jamin, J.-M. Wilkin & J.-M. Frère (1995) *Essays in Biochem.* 29, 1.
M. A. Ator & P. R. Ortez de Montellano (1990) *The Enzymes*, 3rd ed., 19, 213.
N. Citri (1971) *The Enzymes*, 3rd ed., 4, 23.

LACTASE

β-Galactosidase [EC 3.2.1.23], also called lactase, catalyzes the hydrolysis of terminal, nonreducing β-D-galac-tose residues in β-D-galactosides. Lactase [EC 3.2.1.108] catalyzes the hydrolysis of lactose to produce glucose and galactose.

M. W. Bauer, S. B. Halio & R. M. Kelly (1996) *Adv. Protein Chem.* 48, 271.
E. H. Van Beers, H. A. Büller, R. J. Grand, A. W. C. Einerhand & J. Dekker (1995) *Crit. Rev. Biochem. Mol. Biol.* 30, 197.

LACTATE DEHYDROGENASE

This NAD$^+$-dependent oxidoreductase [EC 1.1.1.27] catalyzes the following reversible reaction:

$$\text{Lactate} + \text{NAD}^+ \rightleftharpoons \text{Pyruvate} + \text{NADH} + \text{H}^+$$

with hydride transfer occurring to or from the A-side of the pyridine ring. The kinetic reaction mechanism is best described as an ordered Bi Bi ternary complex scheme, and formation of Enz·NAD$^+$·Pyruvate or Enz·NADH·Lactate abortive complexes can lead to substrate inhibition under favorable conditions. ***See also*** *Abortive Complexes; Isotope Exchange at Equilibrium*

R. Jaenicke, H. Schurig, N. Beaucamp & R. Ostendorp (1996) *Adv. Protein Chem.* 48, 181.
N. J. Oppenheimer & A. L. Handlon (1992) *The Enzymes*, 3rd ed., 20, 453.
M. A. Ator & P. R. Ortez de Montellano (1990) *The Enzymes*, 3rd ed., 19, 213.
H. F. Fisher (1988) *Adv. Enzymol.* 61, 1.
J. J. Holbrook, A. Liljas, S. J. Steindel & M. J. Rossman (1975) *The Enzymes*, 3rd ed., 11, 191.
K. Dalziel (1975) *The Enzymes*, 3rd ed., 11, 1.
M. G. Rossman, A. Liljas, C.-I. Brändén & L. J. Banaszak (1975) *The Enzymes*, 3rd ed., 11, 61.
J. Everse & N. O. Kaplan (1973) *Adv. Enzymol.* 37, 61.

Selected entries from *Methods in Enzymology* [vol, page(s)]:
Enzyme-substrate complex formation, 64, 53; data analysis, 64, 57; extensions of technique, 64, 57-59; as evidence for occurrence of intermediate, 64, 47-59; kinetic equation, 64, 49-52; limitations, 64, 57-59; mixing procedure, 64, 53-56; reaction condition, 64, 56, 57; termination, 64, 56, 57.

D-LACTATE DEHYDROGENASE (CYTOCHROME)

This FAD-dependent enzyme [EC 1.1.2.4] catalyzes the reaction of (*R*)-lactate with two ferricytochrome *c* to produce pyruvate and two ferrocytochrome *c*.

T. A. Hansen (1994) *Meth. Enzymol.* **243**, 21.

Y. Hatefi & D. L. Stiggall (1976) *The Enzymes*, 3rd ed., **13**, 175.

L-LACTATE DEHYDROGENASE (CYTOCHROME)

This FMN- and protoheme IX-dependent enzyme [EC 1.1.2.3] catalyzes the reaction of (S)-lactate with two ferricytochrome *c* to produce pyruvate and two ferrocytochrome *c*.

B. A. Palfey & V. Massey (1998) *Comprehensive Biological Catalysis: A Mechanistic Reference*, **3**, 83.

T. A. Hansen (1994) *Meth. Enzymol.* **243**, 21.

Y. Hatefi & D. L. Stiggall (1976) *The Enzymes*, 3rd ed., **13**, 175.

LACTATE 2-MONOOXYGENASE

This FMN-dependent enzyme [EC 1.13.12.4] catalyzes the reaction of (S)-lactate with dioxygen to produce acetate, carbon dioxide, and water.

B. A. Palfey & V. Massey (1998) *Comprehensive Biological Catalysis: A Mechanistic Reference* **3**, 83.

M. A. Ator & P. R. Ortez de Montellano (1990) *The Enzymes*, 3rd ed., **19**, 213.

V. Massey & P. Hemmerich (1975) *The Enzymes*, 3rd ed., **12**, 191.

LACTATE RACEMASE

This enzyme [EC 5.1.2.1], also referred to as lacticoracemase and hydroxyacid racemase, catalyzes the interconversion of (S)-lactate and (R)-lactate.

M. E. Tanner & G. L. Kenyon (1998) *Comprehensive Biological Catalysis: A Mechanistic Reference* **2**, 7.

E. Adams (1976) *Adv. Enzymol.* **44**, 69.

L. Glaser (1972) *The Enzymes*, 3rd ed., **6**, 355.

LACTOPEROXIDASE

This enzyme [EC 1.11.1.7] catalyzes the reaction of a donor with hydrogen peroxide to produced the oxidized donor and two water. **See also** Peroxidase

H. B. Dunford (1998) *Comprehensive Biological Catalysis: A Mechanistic Reference* **3**, 195.

E. Cadenas (1989) *Ann. Rev. Biochem.* **58**, 79.

A. Naqui, B. Chance & E. Cadenas (1986) *Ann. Rev. Biochem.* **55**, 137.

LACTOSE SYNTHASE

This enzyme [EC 2.4.1.22] is a protein complex of two proteins (designated A and B) and catalyzes the reaction of UDP-galactose with D-glucose to generate UDP and lactose. In the absence of the α-lactalbumin (protein B), the enzyme catalyzes the transfer of galactose from UDP-galactose to *N*-acetylglucosamine (*i.e.*, the activity of *N*-acetyllactosamine synthase, EC 2.4.1.90).

G. Davies, M. L. Sinnott & S. G. Withers (1998) *Comprehensive Biological Catalysis: A Mechanistic Reference* **1**, 119.

R. L. Hill & K. Brew (1975) *Adv. Enzymol.* **43**, 411.

K. E. Ebner (1973) *The Enzymes*, 3rd ed., **9**, 363.

K. Brew (1970) *Essays in Biochem.* **6**, 93.

LAG TIME

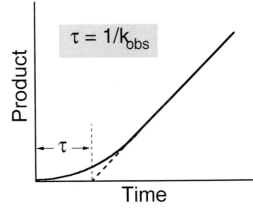

1. The pre-steady-state portion of progress curve, during which there is an exponential increase in the slope of the product concentration *versus* time plot. 2. The slow increase in the slope of the product concentration *vs.* time progress curve often observed when using coupled-enzyme assay protocols. Investigators strive to minimize such lag times[1]. 3. A slow increase in the slope of the product concentration *versus* time or progress curve. Such systems respond slowly to rapid changes in substrate concentrations and are often associated with reaction schemes containing a slow isomerization step[2,3]. **See also** Hysteresis; Pre-Steady-State; Autocatalysis; Microtubule Assembly Kinetics

[1]F. B. Rudolph, B. W. Baugher & R. S. Beissner (1979) *Meth. Enzymol.* **63**, 22.

[2]K. E. Neet & G. R. Ainslie (1980) *Meth. Enzymol.* **64**, 192.

[3]K. E. Neet (1995) *Meth. Enzymol.* **249**, 519.

LAMBDA (Λ, λ)

Λ 1. Symbol for molar conductivity of an electrolyte. 2. Symbol for Ostwald solubility coefficient.

λ 1. Symbol for Avogadro's number (rarely). 2. Symbol for wavelength. 3. Symbol for radioactive constant (decay rate constant). 4. Symbol for absolute chemical activity. 5. Symbol for thermal conductivity. 6. Symbol for mean free path.

LAMBDA (or, Λ-) ISOMERS OF METAL ION-NUCLEOTIDE COMPLEXES

Coordination isomers of metal ion-nucleotide complexes. For most metal ions, for example Mg^{2+}, the metal

ion exchange times are fast, and it is difficult to separate and characterize particular isomers. However, when the metal ion is Cr^{3+} or Co^{3+}, exchange times are very slow and the corresponding nucleotides can be used as inhibitors, chirality probes, and in distance measurements in NMR experiments[1]. If the terminal phosphate of a nucleotide triphosphate or a nucleotide diphosphate is chelated to the metal ion, two possible isomers can exist, Λ and Δ, which are defined by the screw sense of the coordination. If the Cr^{3+} or Co^{3+} is bound to both the β and γ phosphate groups of NTP, there are a total of four coordination isomers, two Λ and two Δ. These subsets of isomers are designated δ (or, pseudoaxial) and λ (or, pseudoequatorial). All of these isomers have proved to be quite useful in probes of an enzyme's active site[1].
See *Exchange-Inert Metal-Nucleotide Complexes*

[1]W. W. Cleland (1982) *Meth. Enzymol.* **87**, 159.

LANGMUIR ISOTHERM

A theoretical curve that accounts for the adsorption of gases onto a surface. Because the strength of adsorption depends on the absolute temperature, binding must be measured at a single temperature (hence the term "isotherm"). The treatment assumes (a) that each binding site can accommodate a single adsorbed molecule; (b) that the affinity is the same at all sites; (c) that binding at each site is completely independent of binding at any other site; and (d) that the total number of sites per unit area of the surface is a constant. The fraction of occupied sites is represented by θ, and the fraction of unoccupied sites is given as $(1 - \theta)$. At equilibrium, the Langmuir isotherm can be derived from the balancing of the rate of adsorption ($k_{adsorption}(1 - \theta)P$, where P is the partial pressure of the adsorbing gas) and the rate of desorption ($k_{desorption}\theta$). The resulting equation can be rearranged to yield: $\theta = bP/[1 + bP]$, where b is the adsorption coefficient, the ratio $k_{adsorption}/k_{desorption}$. Because both ligand-receptor and substrate-enzyme saturation curves have the same mathematical form, they are sometimes said to be Langmuir isotherms. **See** *Biomineralization; Micelle*

LANGMUIR TROUGH AND BALANCE

A device for direct measurement of the surface pressure exerted by a surface film on liquids. The basic features of the trough include a horizontally positioned tray with a mechanically positioned compressing barrier that retains the surface film on one end of the trough and a second float typically constructed of a mica strip to which is attached a torsion spring balance.

Langmuir Trough

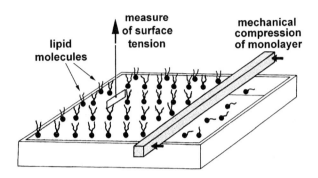

In a typical experiment, a small volume of an insoluble surface-active material (dissolved in a water-insoluble solvent such as benzene) is placed atop a clean water surface. As the solvent evaporates away, a film remains and the moving barrier can be adjusted so that the surface film exerts pressure on the mica float. A calibrated torsion balance is used to measure the force that the film exerts on the float. That force divided by the length of the float is the force per unit length or the surface pressure. For studies of lipolysis kinetics[1,2], a Langmuir trough can be constructed so that one can measure lipase action under first-order and zero-order conditions.

[1]R. Verger & G. H. de Haas (1973) *Chem. Phys. Lipids* **10**, 127.
[2]R. Verger (1980) *Meth. Enzymol.* **64**, 372.

LAPLACE TRANSFORM

A linear operator that permits one to solve the initial value problem

$$a\frac{d^2y}{dt^2} + b\frac{dy}{dt} + cy = f(t)$$

[where $y(0) = y_o$ and $y'(0) = y'_o$ for the continuous (or piecewise continuous) function $f(t)$, itself defined over the interval $0 \leq t < \infty$] in a manner analogous to that required for solving algebraic equations. The Laplace transform $\mathscr{L}\{f(t)\}$ is written as:

$$\mathscr{L}\{f(t)\} = \int_0^\infty e^{-st}f(t)dt = \lim_{A\to\infty}\int_0^A e^{-st}f(t)dt$$

where s is of sufficiently large magnitude and the parameter A lies in another interval. This transform is a linear operator, such that

$$\mathcal{L}\{r_1 f_1(t) + r_2 f_2(t)\} = \int_0^\infty e^{-st}[r_1 f_1(t) + r_2 f_2(t)]\mathrm{d}t$$
$$= \mathcal{L}\{r_1 f_1(t)\} + \mathcal{L}\{r_2 f_2(t)\}$$

The method of Laplace transformation represents a powerful tool for practicing kineticists seeking to solve linear differential equations. The three-step process includes: (a) transforming a differential equation (or group of differential equations) into a subsidiary algebraic equation(s); (b) solving this(these) subsidiary equation(s) using algebraic manipulations; and (c) retransforming the resultant equation(s) to obtain the sought after solution to the original equation(s). Many third-semester calculus textbooks, especially those applying differential equations to engineering problems, provide lucid accounts of the formalism of using Laplace transform methods.

Laplace transformation is particularly useful in pharmacokinetics where a number of series first-order reactions are used to model the kinetics of drug absorption, distribution, metabolism, and excretion. Likewise, the relaxation kinetics of certain multistep chemical and physical processes are well suited for the use of Laplace transforms.

Selected entries from *Methods in Enzymology* [vol, page(s)]:
Claverie approach to Padé-Laplace algorithm for sums of exponentials, **210**, 59; Taylor series expansion in analysis of sums of exponentials, **210**, 56.

LARMOR PRECESSION

The precession observed in a plane of an electron orbit of an atom when that atom lies within a uniform magnetic field. The plane will precess about the direction of the field such that a cone is traced out having an axis normal to the direction of the field. The frequency of this precession is usually symbolized by ν_L and is equal to $eB/(4\pi mc)$ where e is the electron charge, B is the field strength, m is the mass, and c is the velocity of the electron. Larmor precession is also important in NMR studies of enzymes.

LASER-FLASH KINETIC ANALYSIS

A rapid reaction technique for analyzing the kinetics of photo-sensitive or photo-responding biochemical reactions. One application deals with the determination[1] of the rate constants and likely intermediate steps involved in the fast electron transfer from metalloproteins such as plastocyanin or cytochrome c_6 to photosystem I (PSI). P700 in PSI can be rapidly photo-oxidized by a saturating laser flash, thereby inducing almost instantaneously an 820 nm absorbing species that can then react with reductant (in this case the metalloproteins above). The time course of this reaction can be monitored by loss of the 820 nm absorbance, and the oscilloscope tracings of the reaction's time course can be analyzed to obtain valuable kinetic information about the mechanism.

[1]M. Hervas, J. Navarro, A. Diaz, H. Bottin & M. De la Rosa (1995) *Biochemistry* **34**, 11321.

LATENT ACTIVITY

The amount of total enzymatic activity that becomes manifest only after disruption of membranous barriers between enzyme and substrate or upon removal of some otherwise inhibitory factor. Membrane disruption is often achieved by treatment with detergent to solubilize the enzyme. One example is the so-called *microsomal* glucose-6-phosphatase[1], an enzymatic activity that is located in the lumen of the endoplasmic reticulum but becomes trapped as a latent activity in microsome vesicles upon mechanical disruption of cells.

[1]R. C. Nordlie (1982) *Meth. Enzymol.* **87**, 319.

LATENT HEAT

Heat released or absorbed in a process other than a change in temperature. For example, latent heat of fusion and latent heat of vaporization refer to the heat associated with the change of state with no temperature change.

LAW OF CONSTANT HEAT SUMMATION

See Hess' Law; Additivity Principle

LEAST-CONDUCTIVE STEP

A kinetic term introduced by Ray[1] to designate the bottleneck in a steady-state enzyme kinetic pathway as that step in which the enzyme form accumulates in highest concentration at saturating substrate concentration. Highest accumulation of this enzyme reflects the fact that this species faces the highest barrier that precedes an irreversible step in the forward direction.

[1]W. J. Ray, Jr. (1983) *Biochemistry* **22**, 4452.

LEAST MOTION, PRINCIPLE OF

The hypothesis that, for a given set of reactants, those elementary reactions having the least change in nuclear positions will also have the lowest energies of activation.

J. Hine (1966) *J. Org. Chem.* **31**, 1236.
J. Hine (1977) *Adv. Phys. Org. Chem.* **51**, 1.

LEAST SQUARES METHOD

A statistical algorithm, also known as linear regression analysis, for systems where Y (the random variable) is linearly dependent on another quantity X (the ordinary or controlled variable). The procedure allows one to fit a straight line through points (x_1,y_1), (x_2,y_2), (x_3,y_3), . . . , (x_n,y_n) where the values x_i are defined before the experiment and y_i values are obtained experimentally and are subject to random error. The best fit line through such a series of points is called a "least squares fit", and the protocol provides measures of the reliability of the data and quality of the fit.

For such a series of ordered pairs, the vertical displacement of a sample value, say (x_j,y_j), from the straight line $y = a + bx$ is $|y_j - a - bx_j|$, and the sum of these squares is given as:

$$Q = \sum_{j=1}^{n} (y_j - a - bx_j)^2$$

The values of a and b are then chosen to minimize Q which is "the sum of the squares of $(y_j - a - b\,x_j)$", hence the name *Least Squares*. Accordingly, the equation for the best fit least squares line is:

$$[y = a + bx]_{Q\,min}$$

This line is frequently referred to as a regression line, and its slope b is the regression coefficient. This procedure can be extended to polynomial functions wherein segments can be treated as linear or nearly linear functions of Y on X.

Cleland[1] and Mannervik[2] have described least squares programs and procedures for treating enzyme kinetic data. The interested reader will also wish to consult numerous articles in vols. 210 and 259 in *Methods in Enzymology* (L. Brand & M. L. Johnson, eds.) dealing with numerical computer methods for statistical treatment of kinetic and equilibrium data.

[1]W. W. Cleland (1979) *Meth. Enzymol.* **63**, 103.
[2]B. Mannervik (1982) *Meth. Enzymol.* **87**, 370.

Selected entries from *Methods in Enzymology* [vol, page(s)]:
Computer program, **63**, 111-138; evaluation, **63**, 109-111; evaluation of oxygen equilibria data, **76**, 478-481, 483; iterative fitting, **63**, 106; for numerical analysis of ligand binding kinetics, **76**, 661-662; nonlinear, **63**, 104, 105; residual, **63**, 110-112; surface, **63**, 106-109; weighting, **63**, 105; χ^2 method [goodness-of-fit, **210**, 102; parabolic extrapolation, **210**, 10; residual value frequencies, **210**, 93]; nonlinear [global analysis packages, construction, **210**, 44; in iterative reconvolution program, **210**, 511]; parameter estimation from experimental data, **210**, 1{assumptions, **210**, 4; χ^2 method, parabolic extrapolation, **210**, 10; confidence intervals, **210**, 20 [asymptotic standard errors, **210**, 24; grid search method, **210**, 22; linear joint confidence intervals, **210**, 25; Monte Carlo method, **210**, 21; propagation, **210**, 28]; Gauss-Newton method, **210**, 11; Marquardt method, **210**, 16; Nelder-Mead simplex method, **210**, 18; performance methods, **210**, 9; sample analysis, **210**, 29; steepest descent method, **210**, 15}; simultaneous [free energy of site-specific DNA-protein interactions, **210**, 471; for model testing, **210**, 463; for parameter estimation, **210**, 463; separate analysis of individual experiments, **210**, 475; for testing linear extrapolation model for protein unfolding, **210**, 465.

LEAVING GROUP

An atom, group, or moiety (either charged or neutral) that detaches from the main portion of the substrate or reactant in a given reaction. If the leaving group detaches and carries away the electron pair that formerly constituted the bond to the central atom, the leaving group is said to be nucleofugal. If the electron pair is not removed, the leaving group is said to be electrofugal. A leaving group may also be termed an exiphile. *See also* Electrofuge; Nucleofuge; Entering Group; Nucleophilic Substitution Reaction; Nucleophile

LE CHATELIER'S PRINCIPLE

The tendency for a system at equilibrium to adjust in a manner that minimizes the effect of a change imposed by external factors (such as changes in temperature, pressure or electric field strength).

All homeostatic mechanisms can be regarded as biological feedback mechanisms designed to take maximal advantage of the self-compensating nature described as Le Chatelier's Principle[1]. An example is the linked binding functions associated with oxygen binding in the Monod-Wyman-Changeux model for hemoglobin; an increase in oxygen partial pressure is compensated by a change in the distribution of hemoglobin from the T-state to the R-state.

[1]H. L. Le Chatelier (1884) *Compt. Rend.* **99**, 786; (1888) *Compt. Rend.* **106**, 355, 598, 687, and 1008.

LECITHIN-CHOLESTEROL ACYLTRANSFERASE

This enzyme [EC 2.3.1.43] catalyzes the reaction of a phosphatidylcholine with a sterol to produce a sterol ester and a 1-acylglycerophosphocholine. The acyl moiety transferred can be a palmitoyl, oleoyl, or linoleoyl group. A number of sterols, including cholesterol, can act as the acyl acceptor.

C. J. Fielding & X. Collet (1991) *Meth. Enzymol.* **197**, 420.

LEE-WILSON TREATMENT

A treatment suggested by Lee and Wilson[1] for analyzing enzyme kinetic data in which a significant proportion of the initial substrate concentration has undergone conversion (thus, not under true initial rate conditions). It was suggested that the $[\bar{S}]$, that is, the arithmetic mean substrate concentration $(\{[S_0] + [S]\}/2)$ be used in the double-reciprocal form of the normal kinetic expression. Note that this approach is improved if the product(s) is(are) continuously removed from the reaction system (for example, by a coupling system), that substrate inhibition is absent, that all other cosubstrates are saturating, and that the reverse reaction is poor. If products cannot be removed, then the procedure should not be used if product inhibition is significant.

[1]H.-J. Lee & I. B. Wison (1971) *Biochim. Biophys. Acta* **242**, 519.

LEFFLER'S ASSUMPTION

The transition state of a reaction bears greater resemblance to the less stable species (of reactant or reaction intermediate or product). *See* Hammond Postulate

J. E. Leffler (1953) *Science* **117**, 340.

LEGENDRE TRANSFORMATION

A mathematical tool of proven utility in dealing with certain differential equations. Consider a function f (a dependent variable) having $x_1, x_2, x_3, \ldots, x_n$ as independent variables. Then $df = f_1 dx_1 + f_2 dx_2 + \ldots + f_n dx_n$ where $f_1 = (\partial f/\partial x_1)_{x_2 \ldots x_n}$, $f_2 = (\partial f/\partial x_2)_{x_1 x_3 \ldots x_n}$, *etc.* Let the new function g be equal to $f - f_1 x_1$. It can then be shown that $dg = -x_1 df_1 + f_2 dx_2 + \ldots + f_n dx_n$. This tool has been utilized to generate expressions for enthalpy, Helmholtz energy, and Gibbs free energy from the differential expression for internal energy.

LEGUMAIN

This enzyme [EC 3.4.22.34] catalyzes the hydrolysis of peptide bonds in proteins (for example, in azocasein) and polypeptides at Asn–Xaa, the preferential bond.

S.-i. Ishii (1994) *Meth. Enzymol.* **244**, 604.

LEISHMANOLYSIN

This zinc- and calcium-dependent enzyme [EC 3.4.24.36] catalyzes the hydrolysis of peptide bonds, exhibiting a preference for hydrophobic amino acid residues at positions P1 and P1′ , and basic amino acid residues at P2 and P3′.

J. Bouvier, P. Schneider & R. Etges (1995) *Meth. Enzymol.* **248**, 614.

LEUCINE 2,3-AMINOMUTASE

This cobalamin-dependent enzyme [EC 5.4.3.7] catalyzes the interconversion of $(2S)$-α-leucine and $(3R)$-β-leucine.

J. M. Poston (1988) *Meth. Enzymol.* **166**, 130.
O. W. Griffith (1986) *Ann. Rev. Biochem.* **55**, 855.
J. M. Poston (1986) *Adv. Enzymol.* **58**, 173.
B. M. Babior & J. S. Krouwer (1979) *Crit. Rev. Biochem.* **6**, 35.

LEUCINE AMINOPEPTIDASE

This zinc-dependent enzyme [EC 3.4.11.1], also referred to as cytosol aminopeptidase, leucyl aminopeptidase, and peptidase S, catalyzes the hydrolysis of a terminal peptide bond such that there is a release of an N-terminal amino acid, Xaa–Xbb-, in which Xaa is preferably a leucyl residue, but may be other aminoacyl residues including prolyl (although not arginyl or lysyl). Xbb may be prolyl. In addition, amino acid amides and methyl esters are also readily hydrolyzed, but the rates with arylamides are exceedingly slow. The enzyme is activated by heavy metal ions.

W. L. Mock (1998) *Comprehensive Biological Catalysis: A Mechanistic Reference* **1**, 425.
H. Kim & W. N. Lipscomb (1994) *Adv. Enzymol.* **68**, 153.
B. L. Vallee & A. Galdes (1984) *Adv. Enzymol.* **56**, 283.
R. J. DeLange & E. L. Smith (1971) *The Enzymes*, 3rd ed., **3**, 81.

LEUCINE AMINOTRANSFERASE

This pyridoxal-phosphate-dependent enzyme [EC 2.6.1.6] catalyzes the reversible reaction of leucine with α-ketoglutarate (or, 2-oxoglutarate) to produce 4-methyl-2-oxopentanoate and glutamate. *See also* Branched-Chain Amino Acid Aminotransferase

A. E. Braunstein (1973) *The Enzymes*, 3rd ed., **9**, 379.

LEUCINE DEHYDROGENASE

This enzyme [EC 1.4.1.9] catalyzes the reaction of leucine with NAD$^+$ and water to produce 4-methyl-2-oxopentanoate, ammonia, and NADH. Isoleucine, valine, norvaline, and norleucine can serve as substrates for this enzyme.

A. R. Clarke & T. R. Dafforn (1998) *Comprehensive Biological Catalysis: A Mechanistic Reference* **3**, 1.
N. M. W. Brunhuber & J. S. Blanchard (1994) *Crit. Rev. Biochem. Mol. Biol.* **29**, 415.
G. Livesey & P. Lund (1988) *Meth. Enzymol.* **166**, 282.

LEUCINE KINETICS

The biosynthetic incorporation of amino acids into proteins makes these metabolites valuable endogenous tracers for the characterization of protein turnover[1]. Of the naturally occurring amino acids, administration of a bolus dose of [^{3}H]leucine is widely used as a tracer in kinetic investigations of protein synthesis and secretion.

Leucine as a Tracer in Protein Turnover

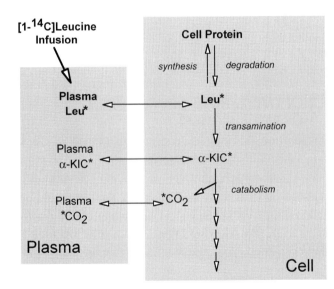

After intravenous injection, the specific radioactivity of plasma leucine decays over several orders of magnitude within the first 12 hours, followed by a slow decay lasting a number of weeks. This "tail" in the decay curve is the consequence of recycling of the leucine tracer as proteins are degraded, allowing the liberated [^{3}H]leucine to reenter the plasma pool. Kinetic data can be analyzed by mathematical compartmental modeling (**See** *SAAM*), and plasma leucine activity is used as a forcing function to drive the input of [^{3}H]leucine into the protein synthesis pathway. Leucine has many advantages as a protein

tracer: (a) it is an essential amino acid, and (b) the by-products of its degradation are not reincorporated into other amino acids. The initial step in leucine catabolism is transamination and formation of α-ketoisocaproic acid. This reaction is readily reversible. A subsequent series of oxidative reactions generate hydroxymethylglutaryl coenzyme A and acetate. In the oxidation of one of the intermediates, isovaleryl-coenzyme A, the tritium on position C-4 is lost, and the C-5 tritium can appear in steroids or fatty acids. Typically, any such tracer that becomes incorporated into lipid is greatly diluted. **See** *Protein Turnover Kinetics; SAAM*

[1]W. R. Fisher, L. A. Zech, L. L. Kilgore & P. W. Stacpoole (1991) *J. Lipid Res.* **32**, 1823.

LEUCOLYSIN

This zinc- and calcium-dependent venom endopeptidase [EC 3.4.24.6] hydrolyzes the following peptide bonds in the insulin B chain: Phe^1–Val^2, His^5–Leu^6, Ala^{14}–Leu^{15}, Gly^{20}–Glu^{21}, Gly^{23}–Phe^{24}, and Phe^{24}–Phe^{25}. It will also act on N-blocked dipeptides.

J. B. Bjarnason & J. W. Fox (1995) *Meth. Enzymol.* **248**, 345.

LEUCYL-tRNA SYNTHETASE

This enzyme [EC 6.1.1.4], also called leucine:tRNA ligase, catalyzes the reaction of leucine with tRNA^{Leu} and ATP to produce leucyl-tRNA^{Leu}, AMP, and pyrophosphate.

E. A. First (1998) *Comprehensive Biological Catalysis: A Mechanistic Reference* **1**, 573.
A. Hampel & R. Tritz (1988) *Meth. Enzymol.* **166**, 260.

LEUKOTRIENE A$_4$ HYDROLASE

This enzyme [EC 3.3.2.6] catalyzes the hydrolysis of (7E,9E,11Z,14Z) - (5S,6S) - 5,6 - epoxyicosa - 7,9,11,14 - tetraenoate to form (6Z,8E,10E,14Z)-(5S,12R)-5,12-dihydroxyicosa-6,8,10,14-tetraenoate. This enzyme is highly specific yet it also converts 4,5-leukotriene A$_4$ into leukotriene B$_4$. Note also that epoxide hydrolase [EC 3.3.2.3] is not identical with this enzyme.

N. Ohishi, T. Izumi, Y. Seyama & T. Shimizu (1990) *Meth. Enzymol.* **187**, 286.
J. Z. Haeggström (1990) *Meth. Enzymol.* **190**, 324.

LEUKOTRIENE C$_4$ SYNTHASE

This enzyme [EC 2.5.1.37] catalyzes the reaction of leukotriene A$_4$ with glutathione to generate leukotriene C$_4$ and water. This enzyme is not identical with glutathione transferase [EC 2.5.1.18]. **See also** *Glutathione S-Transferase*

M. Söderström, B. Mannervik & S. Hammarström (1990) *Meth. Enzymol.* **187**, 306.

R. J. Soberman (1990) *Meth. Enzymol.* **187**, 335.

LEVANASE

This enzyme [EC 3.2.1.65], also known as 2,6-β-D-fructan fructanohydrolase, catalyzes the random hydrolysis of 2,6-β-D-fructofuranosidic linkages in 2,6-β-D-fructans (levans) containing more than three fructosyl units.

G. Avigad & S. Bauer (1966) *Meth. Enzymol.* **8**, 621.

LEVANSUCRASE

This enzyme [EC 2.4.1.10], also known as sucrose 6-fructosyltransferase, catalyzes the reaction of sucrose with [(2,6)-β-D-fructosyl]$_n$ to produce glucose and [(2,6)-β-D-fructosyl]$_{(n+1)}$. A few other fructosyl-containing sugars can act as substrates.

G. Davies, M. L. Sinnott & S. G. Withers (1998) *Comprehensive Biological Catalysis: A Mechanistic Reference* **1**, 119.

G. Mooser (1992) *The Enzymes*, 3rd ed., **20**, 187.

LEVELING EFFECT

A property of solvents affecting the apparent acidity of Brønsted acids. This effect reflects the tendency of a solvent to make all Brønsted acids with acidities exceeding a certain value appear equally acidic. In such cases, there is a transfer of a proton to the protophilic solvent from dissolved acid (which is a stronger acid than the conjugate acid of the solvent). Hence, the only acid present to any significant extent is the lyonium ion. For example, water has a leveling effect on the acidities of $HClO_4$, HCl, and HI. In this respect, the apparent strength of an acid depends on the basicity of the solvent: the more basic the solvent, the more the reactants will become ionized. Hence, to compare relative acidities, the solvent must be selected to prevent complete ionization of the acids being studied. A similar phenomenon occurs with strong bases except, in these cases, the solvent is protogenic.

LEWIS ACID

Any chemical species having a vacant orbital and thus acting as an electron-pair acceptor from a Lewis base. Examples of Lewis acids include: BF_3, Fe^{3+}, Na^+, Ca^{2+}, and SO_3. **See** *Lewis Base; Acidity; Lewis Acidity*

LEWIS ACIDITY

The thermodynamic tendency of a substance to act as a Lewis acid. The strength of a Lewis acid depends on the nature of the base with which the Lewis acid forms a Lewis adduct. Hence, comparative measures of Lewis acidities are given by equilibrium constants for the formation of the adducts by a common reference base. **See** *Lewis Acid; Electrophilicity; Hard Acids; Soft Acids; Acceptor Number*

LEWIS BASE

Any chemical species that can donate a pair of electrons to a Lewis acid to form a Lewis adduct. Examples of Lewis bases include: NH_3, RNH_2, ROH, CO, CO_3^{2-}, and RSH. **See** *Lewis Acid*

LEWIS BASICITY

The thermodynamic tendency of a substance to act as a Lewis base. The strength of a Lewis base depends on the nature of the acid with which the Lewis base forms a Lewis adduct. Hence, comparative measures of Lewis basicities are given by equilibrium constants for the formation of the adducts by a common reference acid. **See** *Lewis Base; Nucleophilicity; Hard Bases; Soft Bases; Donor Number*

L'HÔPITAL'S RULE (or, L'HOSPITAL'S RULE)

A differential technique[1] in mathematics used to obtain information about the composite function $f(x)/g(x)$ where $f(x)$ and $g(x)$ are functions of x. One version of the method requires that both functions are differentiable, symbolized by $f'(x)$ and $g'(x)$, at every point x greater than a certain fixed $M > 0$ and that $g'(x) \neq 0$. If $\lim_{x \to \infty} (f'(x)/g'(x))$ exists and has a certain value, then $\lim_{x \to \infty} (f(x)/g(x))$ also exists and has that same value.

This rule has found a number of uses in enzymology. A simple example applies to the standard Michaelis-Menten equation, $v = V_{max}[S]/(K_m + [S])$. Here, $f(x) = V_{max}[S]$ and $g(x) = K_m + [S]$. Hence, $\lim_{[S] \to \infty} (f[S]/g([S])) = \lim_{[S] \to \infty} (f'([S])/g'[S]) = \lim_{[S] \to \infty} (V_{max}/1) = V_{max}$. Thus, at $[S] = \infty$, $v = V_{max}$. L'Hôpital's rule has also proved useful in determining expressions for the maximum rate of exchange in equilibrium exchange studies[2].

[1]G. F. A. de l'Hôpital (1696) *Analyse des infiniment petits*, Paris.
[2]D. L. Purich & R. D. Allison (1980) *Meth. Enzymol.* **64**, 3.

LIFETIME

The time (symbolized by τ) needed for a concentration of a molecular entity to decrease, in a first-order decay process to e^{-1} of its initial value. In this case, the lifetime (sometimes called mean lifetime) is equal to the reciprocal of the sum of rate constants for all concurrent first-order decompositions. If the process is not first-order, the term *apparent lifetime* should be used, and the initial concentration of the molecular entity should be provided. The terms lifetime and half-life should not be confused. *See Half-Life; Fluorescence*

LIGAND

[from the Latin, *ligare*, to bind] 1. A chemical group, ion, or molecule coordinately bound to the central atom in a complex; for example, the cyano moiety bound to the cobalt in cyanocobalamin or the porphyrin portion of heme. 2. An ion or molecule that binds noncovalently to a recognition site on a macromolecule (for example, D-glucose binding to hexokinase or a hormone binding to a receptor protein). 3. The analyte in a competitive binding assay (*e.g.*, in radioimmunoassays). *See also Ligand Diffusion to Receptor; Metal Ion Coordination Reactions; Linked Functions; Scatchard Plot; Hill Equation and Plot*

LIGAND BINDING ANALYSIS

Ligand binding experiments are essential for determining the number of binding sites, the dissociation constant(s) for ligand binding site(s), and the most likely mechanism of binding interactions. Winzor and Sawyer[1] and Klotz[2] present an excellent account of the repertoire of available techniques, including equilibrium dialysis, ultrafiltration, partition equilibrium methods, gel chromatography, UV/visible and fluorescence methods, NMR and ESR measurements, ion-specific/selective electrodes, biosensor methods as well as other competitive binding assays. The choice of technique obviously depends on the ability to detect and quantify binding site occupancy, and one must consider (a) the amount of ligand and macromolecule required for a given technique, (b) the concentration range over which saturation of binding sites is to be assessed, (c) linearity between a technique's signal and the extent of site occupancy,

and (d) stability of the components under the conditions required to make a specific type of measurement.

There are four basic methods for graphically representing ligand binding data: (a) fractional saturation versus ligand concentration; (b) Scatchard plots of $\nu/[\mathrm{L}]$ versus ν; (c) double-reciprocal plots involving (fractional saturation)$^{-1}$ versus $[\mathrm{L}]^{-1}$; and the Klotz plot. These are illustrated here by using different symbols for low and high ligand ranges to permit the reader the opportunity to assess the nature of each plot's representation of binding data. *See also Adair Equation; Linked Functions; Scatchard Plot; Hill Equation and Plot; Womack-Colowick Dialysis Method*

[1]D. J. Winzor & W. H. Sawyer (1995) *Quantitative Characterization of Ligand Binding*, pp. 168, Wiley-Liss, New York.
[2]I. M. Klotz (1997) *Ligand-Receptor Energetics: A Guide For the Perplexed*, Wiley, New York.

Selected entries from *Methods in Enzymology* [vol, page(s)]:
Ligand Binding Measurements: By equilibrium gel penetration, **117**, 342; by gel chromatography, **117**, 346; by ultrafiltration, **117**, 354; using hydroxyapatite columns, **117**, 370; by electrophoretic and spectroscopic techniques, **117**, 381; by fluorescence spectroscopy, **117**, 400; by sedimentation velocity to investigate ligand-induced protein self-association, **117**, 459; linked to protein self-associations, **117**, 496; the meaning of Scatchard and Hill plots, **48**, 270; ligand-binding by associating systems, **48**, 299; measurement of protein dissociation constants by tritium exchange, **48**, 321; fluorescence methods for measuring reaction equilibria and kinetics, **48**, 380; theoretical calculations of relative affinities of binding, **202**, 497; quantitative structure-activity relationships and molecular graphics in evaluation of enzyme-ligand interactions, **202**, 512; theoretical calculations of relative affinities of binding, **202**, 497; quantitative structure-activity relationships and molecular graphics in evaluation of enzyme-ligand interactions, **202**, 512; ligand binding density functions in analysis of multiple ligand binding, **208**, 270; nucleic acid binding, **212**, 427, 447; theoretical behavior for data collection by reverse titration, **208**, 272; determination of binding constants using linear dichroism, **226**, 243; equilibrium ligand-nucleic acid interactions [analysis, **212**, 402; dependence on salt concentration, **212**, 411]; lanthanide ions, spectroscopic determination, **226**, 506; protein-nucleic acid interaction, measurement, **208**, 226; site-specific protein-DNA interactions [electrolyte effects, **208**, 308; equilibria, **208**, 295; solute effects, **208**, 306; temperature effects, **208**, 301; thermodynamics, **208**, 295]; numerical analysis of binding data, **210**, 481.

Hemoglobin: Assignment of rate constants for O_2 and CO binding to α and β subunits within R- and T-state human hemoglobin, **232**, 363; hemoglobin ligand binding and conformational changes measured by time-resolved absorption spectroscopy, **232**, 387; femtosecond measurements of geminate recombination in heme proteins, **232**, 416; double mixing methods for kinetic studies of ligand binding in partially liganded intermediates of hemoglobin, **232**, 430; hemoglobin-liganded intermediates, **232**, 445; hemoglobin-oxygen equilibrium binding: rapid-scanning spectrophotometry and singular value decomposition, **232**, 460; oxygen equilibrium curve of concen-

LIGAND DIFFUSION TO RECEPTOR

The random motion process by which a ligand approaches and penetrates a tissue surface: first, by passing down a concentration gradient by three-dimensional Brownian motion in the bulk solution; then, by reaching the surface where the dimensionality of random excursions reduces to a two-dimensional process; and finally, by docking at the receptor site[1-3]. Once even weakly associated within the tissue surface, receptor electrostatics and mobility probably work together to guide the ligand to its binding site. Moreover, the kinetics of ligand diffusion to its receptor(s) is likely to greatly influence the time dependence of metabolite and drug action.

[1]H. C. Berg & E. M. Purcell (1977) *Biophys. J.* **20**, 193.
[2]J. Crank (1975) *The Mathematics of Diffusion*, 2nd ed., Clarendon Press, Oxford.
[3]T. Kenakin (1993) *Pharmacologic Analysis of Drug-Receptor Interaction*, 2nd ed., pp. 20-24, Raven Press, New York.

LIGAND EXCLUSION MODEL

A multiligand binding mechanism observed in some multisubstrate enzyme-catalyzed reactions[1,2]. Consider the case of a two-substrate enzyme in which substrate A binds deep within a cleft of the protein's active site. The second substrate, B, then binds over A to form the productive ternary complex, EAB. This is a specific type of ordered binding reaction. Binding of substrate A may also induce a conformational change in the enzyme, such that B binds more easily. In some ligand exclusion mechanisms, a competitive inhibitor of substrate B will also prevent the binding of A, resulting in nonhyperbolic behavior.

[1]D. G. Cross & H. F. Fisher (1970) *J. Biol. Chem.* **245**, 2612.
[2]H. F. Fisher, R. E. Gates & D. G. Cross (1970) *Nature* **228**, 247.

LIGAND FIELD SPLITTING

The loss of degeneracy of atomic or molecular levels in a molecular entity with a given symmetry by the attachment or removal of ligands to produce reduced symmetries. Ligand field theory treats metal ligand complexation as a consequence of molecular orbital formation, whereas crystal field splitting considers ligands as point

charges that interact electrostatically with a central electropositive metal ion.

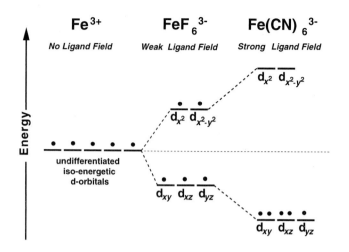

An example of ligand field theory is shown for the case of uncomplexed and complexed forms of Fe(III). Uncomplexed Fe^{3+} has 5 electrons occupying the $3d$ electron shell, one in each of the five isoenergetic orbitals. Upon complexation, each of the six negatively charged or electron-rich ligands are drawn toward the overall positively charge Fe(III) metal ion, leading to repulsion between the metal ion's $3d$ electrons and the electrons of the ligand. Recall that two of the $3d$ orbitals are geometrically disposed so that electrons in these orbitals will experience greater repulsion by ligands occupying their respective positions in a hexacoordinate complex. If the electron density of the ligands can repulsively interact with electrons in Fe(III), the $3d$ orbitals will split into high- and low-lying orbitals of different potential energy. For the FeF_6^{3-} complex, the extent to which the d orbitals split is relatively small, and the electrons remain unpaired by occupying each of the five d orbitals. With the $Fe(CN)_6^{3-}$ complex, however, the degree of splitting is considerably greater. The interpretation is that the energy difference is of sufficient magnitude that all five d-orbital electrons now occupy the three lower energy orbitals. With an uneven number of d-orbital electrons, one electron remains unpaired. **See** *Crystal Field Splitting*

LIGAND-INDUCED CONFORMATIONAL CHANGE

Although Fischer's Lock-and-Key hypothesis for ligand binding strongly influenced scientific thought about enzyme-substrate and ligand-receptor interactions, there is abundant evidence that many ligands induce conformational changes, however large or subtle, within their macromolecular host. The mechanism of ligand binding can generally be written as a two-step process, although additional isomerizations may intervene:

$$\text{Ligand} + \text{Protein} \rightleftharpoons X_1 \rightleftharpoons X_2 \rightleftharpoons X_3 \rightleftharpoons \ldots \rightleftharpoons X_{n-1} \rightleftharpoons X_n$$

The intermediate states X_1, X_2, X_3, ..., X_{n-1}, and X_n can often be directly observed by fast reaction kinetic approaches, or their existence can be inferred from the magnitude of the observed equilibrium constants and reasonable limits on rate constants for ligand binding and release. Koshland[1] first proposed the "induced-fit" model to account for substrate-induced conformational changes that bring catalyzing groups into closer contact with reacting groups on the substrate. As originally propounded, the model did not consider that such conformational changes may serve either to introduce strain within the substrate or to stabilize the transition state configuration of a reactant. However, one can reasonably argue that the concept of a ligand-induced conformational change logically anticipates these related features of enzymic catalysis. **See also** *Linked Functions; Lock-and-Key Model of Enzyme Action; Mapping Substrate Interactions Using Substrate Data*

[1]D. E. Koshland, Jr., (1958) *Proc. Natl. Acad. Sci. U.S.A.* **44**, 98.

LIGAND-INDUCED PROTEIN SELF-ASSOCIATION

When a binding protein exists in various oligomeric states linked by reversible reactions, differences in the strength of ligand-binding affinity by monomer and the higher oligomer species are often manifested as changes in the degree of oligomerization as the ligand concentration is altered. Such behavior constitutes an example of how the favorable Gibbs free energy of ligand binding ($\Delta G_{\text{binding}}$) creates a thermodynamic "pull" that alters the poise of the pre-existing the monomer $\rightleftharpoons$ oligomer equilibrium in a manner that leads to the accumulation of the latter.

Nichol and Winzor[1] described the binding equations that apply to such ligand-induced changes in receptor oligomerization. They also presented the following equation to describe the joint operation of allosteric ligand binding cooperativity and receptor self-association.

$$r = \frac{\sum_j \left\{ m_j \sum_{i=1}^{i=\tau_j} i \left\{ \prod_{l=1}^{l=i} K_{j,l} \right\} m_S^i \right\}}{\sum_j \left\{ j m_j \left(1 + \sum_{i=1}^{i=\tau_j} i \left\{ \prod_{l=1}^{l=i} K_{j,l} \right\} m_S^i \right) \right\}}$$

In this expression, j denotes the acceptor states (monomer, dimer, etc.), τ_j values describe the numbers of binding sites per molecule of the acceptor in each of its oligomeric states, the K's are ligand binding equilibrium constants, and S is the ligand.

An excellent example is the binding of 5'-AMP to glycogen phosphorylase *b*, an event that is attended by conversion of inactive dimeric enzyme into a catalytically active tetramer resembling phosphorylase *a*. In what by now should be regarded as a classic paper in enzymology, Huang and Graves[2] used specific activity measurements and light scattering experiments to determine how AMP binding is coupled to tetramer formation and catalytic activity.

Another example is the ability of taxol to promote tubulin polymerization: taxol exhibits no detectable interaction with the $\alpha\beta$-tubulin dimer; however taxol binds extremely tightly to microtubules. Howard and Timasheff[3] examined tubulin polymerization into microtubules as well as tubulin-colchicine aggregation. The linkage free energy provided by taxol binding is approximately -3.0 kcal/mol $\alpha\beta$-tubulin dimer, whereas this quantity is reduced to approximately -0.5 kcal/mol tubulin-colchicine, indicating that polymerization of the latter requires the expenditure of much more of the $\Delta G_{\text{binding}}$ to overcome likely steric hindrance and geometric strain.

[1]L. W. Nichol & D. J. Winzor (1981) in *Protein-Protein Interactions*, (C. Frieden and L. W. Nichol, eds.) Wiley-Interscience, New York.
[2]C. Y. Huang & D. J. Graves (1970) *Biochemistry* **9**, 660.
[3]W. D. Howard & S. N. Timasheff (1988) *J. Biol. Chem.* **263**, 1342.

LIGASES

A major class of enzymes, also referred to as synthetases, that catalyze the joining of two entities with the concomitant hydrolysis of molecules such as ATP or GTP.

LIGHT POLARIZATION

A condition that is said to occur when the end point of the electric vector of the light moves along a certain path. Light is *linearly* polarized when the end point of the electric vector moves along a straight line in the direction of light propagation. Light is (a) *circularly* polarized if the end point traces a circle or (b) *elliptically* polarized if the electric vector moves in an elliptical path as the light propagates.

This method was the first accurate spectroscopic method for determining chemical reaction rates. In the mid-eighteenth century, kinetic measurements of changes in the rotation of plane polarized light upon acid-catalyzed hydrolysis of sucrose led to the concept of a dynamic equilibrium.

LIGHT SCATTERING

Classical elastic light scattering[1,2] arises when light impinges on a molecule (the dimensions of which are much smaller than the wavelength of the light) and induces an oscillating dipole μ equal to the electric field strength **E** multiplied by the polarizability α. This oscillating dipole radiates light, and Rayleigh showed that for N identical particles the intensity I of the scattered light is given by:

$$I = I_o N\{(8\pi^2\alpha^2/\lambda^4 r^2)(1 + \cos^2 2\theta)$$

where I_o is the intensity of the light beam per square centimeter, λ is the wavelength of light, r is the distance in centimeters from the sample to detector, and 2θ is the angle between the incident and the scattered light rays. The inverse fourth power dependence of I on wavelength (or in other words, its fourth power dependence on frequency) explains why shorter wavelength light is scattered far more effectively than light of longer wavelength.

In practice, depending on the angle between the incident and scattered light paths, one can make several types of light scattering measurements:

(a) Turbidity is a parameter relating the intensity I of the light passing through the sample to a photomultiplier tube at $\theta = 0$ (*i.e.*, along its original path between the light source and the light trap) to the incident light intensity I_o. The following equation for turbidity is mathematically analogous to that describing absorbance:

$$I/I_o = \exp(-\tau x)$$

where τ is the turbidity and x is the path length of the sample cuvette. Turbidity is related to the so-called Rayleigh ratio ($R_\theta' = I_\theta r^2/I_o$) at 90° as follows:

$$R_\theta' = (16\pi/3)R_{90}'$$

Turbidity has proven to be especially useful in studies of protein polymerization, where one can demonstrate that the turbidity is directly proportional to the polymer mass concentration. This is illustrated in the following plot (Fig. 1) obtained for assembled microtubules.

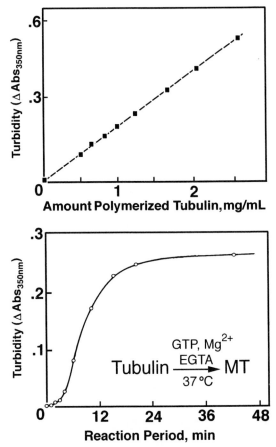

Figure 1. *Top:* Turbidity, measured at 350 nm, as a function of microtubule polymer mass concentration (expressed as mg/mL polymerized tubulin). Tubulin solutions of varying concentrations were polymerized until they reached stable plateau values in a Cary 118C spectrophotometer. Each sample was then transferred to an ultracentrifuge tube, and microtubules were pelleted, separated from the unpolymerized tubulin in the supernatant fraction, and then resuspended for protein concentration determination. The corresponding turbidity and polymer mass concentrations are plotted here. *Bottom:* Time-course of tubulin polymerization assayed by turbidity. Reproduced from MacNeal and Purich[3] with permission from the American Society for Biochemistry and Molecular Biology.

(b) One can also measure the intensity of light scattering when the photomultiplier is positioned at an angle other than 0°, and the detector is set at 90° as shown in Fig. 2. The angular dependence of scattered light intensity can provide valuable information about the hydrodynamic and thermodynamic properties of macromolecules.

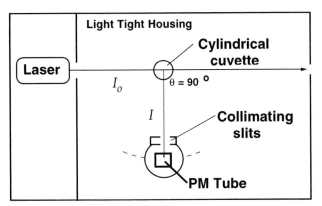

Figure 2. General layout of a light scattering apparatus. Light emitted from a light source impinges on a sample located in a cylindrical cell, and one can adjust the position of the photomultiplier (here shown at 90° relative to the transmitted light beam) along an arc centered at the midpoint of the sample cell.

Zimm[4] demonstrated that when the scattering particle dimensions are not small relative to the wavelength of the incident light, one can use light scattering to infer important information about the size and shape of a macromolecule as well as the interactions of macromolecules. Accurate determination of molecular weight using light scattering requires the use of a double-extrapolation method known as a Zimm plot. Because this Handbook focuses on kinetic behavior, interested readers will wish to consult excellent discussions by Tanford[1] and Cantor and Schimmel[2]. *See Actin Assembly Assays; Microtubule Assembly Kinetics; Fluorescence*

[1]C. Tanford (1961) *Physical Chemistry of Macromolecules*, pp. 275-316, Wiley, New York.
[2]C. R. Cantor & P. R. Schimmel (1980) *Biophysical Chemistry*, Part II, pp. 838-842, Freeman, San Francisco.
[3]R. K. MacNeal & D. L. Purich (1978) *J. Biol. Chem.* **253**, 4683.
[4]B. H. Zimm (1948) *J. Chem. Phys.* **16**, 1093.

LIGHT SCATTERING (DYNAMIC)

Time-dependent fluctuations in the intensity and spectrum of light scattered as a consequence of fluctuations in the number of scattering molecules within a small aperture. Doppler shifts resulting from molecular motions also alter the wavelength of scattered radiation. The technique for measuring various dynamic light scattering is widely known as quasi-elastic light scattering (or QELS), because the motions of the scattering centers are not totally elastic. The scattered light spectrum produced by

Brownian motion of a single molecular species results in a Lorentzian spectrum defined by the relationship:

$$I(S,n) = \langle N \rangle \, \pi^{-1} \{ S^2 D / (4\pi^2 \nu^2 + (S^2 D)^2 \}$$

where I is the intensity, S is $2 \sin\theta/\lambda$, ν is the frequency of incident light, $\langle N \rangle$ is the number average of molecules in the sampled volume, and D is the diffusion coefficient. The half-width of the Lorentzian-shaped frequency distribution (or spectrum) of the scattered light equals $S^2 D$, and the diffusion constant can be used to determine molecular weight. Considerable effort has been devoted to understanding how macromolecules in real solutions affect the intensity and spectrum of scattered light.

C. R. Cantor & P. R. Schimmel (1980) *Biophysical Chemistry*, Part II, pp. 839-843, Freeman, San Francisco.

LIGNAN PEROXIDASE

This heme-dependent enzyme [EC 1.11.1.14], also known as diarylpropane peroxidase, diarylpropane oxygenase, and ligninase I, catalyzes the reaction of 1,2-bis(3,4-dimethoxyphenyl)propane-1,3-diol with hydrogen peroxide to produce veratraldehyde, 1-(3,4-dimethylphenyl)ethane-1,2-diol, and four water molecules. The enzyme brings about the oxidative cleavage of C—C bonds in a number of model compounds and also oxidizes benzyl alcohols to aldehydes or ketones.

H. B. Dunford (1998) *Comprehensive Biological Catalysis: A Mechanistic Reference* **3**, 195.
F. P. Guengerich (1990) *Crit. Rev. Biochem. Mol. Biol.* **25**, 97.
T. K. Kirk, M. Tien, P. J. Kersten, B. Kalyanaraman, K. E. Hammel & R. L. Farrell (1990) *Meth. Enzymol.* **188**, 154.

LIMIT DEXTRAN

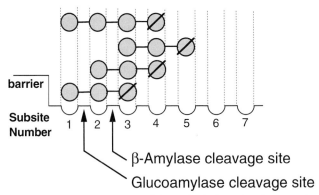

Diagrammatic representation of the individual saccharide unit subsites and cleavage sites in β-amylase and glucoamylase.

A oligosaccharide that cannot be degraded further by a polysaccharide hydrolase or polysaccharide phosphorylase. The types of limit dextrans formed by various amylases and phosphorylases can provide useful information about ligand binding at the active site, and the structure of a limit dextran is also useful for recognizing and identifying new enzyme activities.

LIMITING ISOTOPE EFFECT

A term used in the study of isotope effects in enzyme-catalyzed reactions, namely $^{D}V_{max}/K_m$ (equivalent to $^{H}V_{max}/^{H}K_m)/(^{D}V_{max}/^{D}K_m)$) and $^{D}V_{max}$ (equivalent to $^{H}V_{max}/^{D}V_{max}$).

D. B. Northrop (1982) *Meth. Enzymol.* **87**, 607.

LINEAR COMBINATION OF ATOMIC ORBITALS

A theoretical approach in molecular orbital studies to formulate an expression for the wave function of a molecular orbital (both for bonding and antibonding orbitals) by linear combinations of the overlapping atomic orbitals with appropriate weighting factors.

LINEAR FREE-ENERGY CORRELATIONS

A plot of the logarithm of a rate constant (or an equilibrium constant) for one series of reactions *versus* the logarithm of the rate constant (or the equilibrium constant) for a related series of reactions. (Recall that at constant temperature and pressure the logarithm of an equilibrium constant is proportional to $\Delta G°$, and the logarithm of a rate constant is proportional to $\Delta G^{\ddagger}$). An example of a linear free energy relationship is provided by the Hammett $\sigma\rho$-equation. With equilibrium constants, this relationship is given by the expression:

$$\log(K_x/K_H) = \sigma_X\rho$$

where K_H is the equilibrium constant where the substituent is H, K_x is the equilibrium constant where the substituent is X, σ_X is a constant characteristic of the substituent X, and ρ is a constant for a given reaction under a set of conditions (**See** *Rho Value*). Recall that $\Delta G°_x = -RT \ln K_x$ and $\Delta G°_H = -RT \ln K_H$. Thus, the Hammett equation can be rewritten as $[-\Delta G_x/(2.3RT)] + [\Delta G_H/(2.3RT)] = \sigma\rho$; or, $-\Delta G_x = \sigma\rho 2.3RT -\Delta G_H$. Hence, for a given set of conditions (*i.e.*, ΔG_H, ρ, R, T are all constant) σ is linear with respect to ΔG_x.

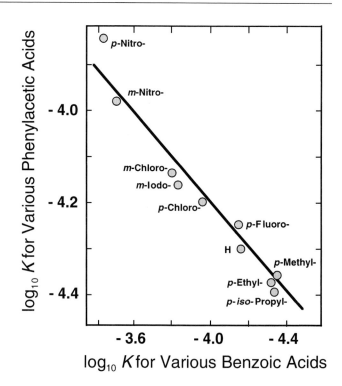

Other examples of linear free energy relationships include the Brønsted relation, the Grunwald-Winstein relationship, and the Taft equation. **See** *Rho Value*

LINEAR SOLVATION ENERGY RELATIONSHIPS

Equations containing a number of solvent parameters in linear or multiple linear regression and expressing the effect of the solvent on the rate of the reaction or the thermodynamic equilibrium constant. **See** E_T Values; Kamlet-Taft Solvent Parameter; Koppel-Palm Solvent Parameter; Z Value

LINEAR STRAIN

A dimensionless quantity, symbolized by either ε or e, for the change in length (l) due to some force or interaction per unit length: thus, $\varepsilon = \Delta l/l$. Linear strain is also known as longitudinal strain. Relative elongation is also measured by $\Delta l/l$.

LINE-SHAPE ANALYSIS

Determination of chemical exchange rate constants based on the shapes of spectroscopic lines (frequently NMR resonances) of dynamic processes.

LINEWEAVER-BURK PLOT

See Double-Reciprocal Plot

LINKED FUNCTION

If a macromolecule engages in two different association reactions with ligands X and Y at nearby (or structurally linked) binding sites, there arises the likelihood that the binding functions for these ligands will exhibit an interdependence that stems from the interaction between these sites. Such sites are said to be "linked," and the saturation function describing their binding interactions is called a linked function.

The following treatment of linked function behavior closely follows Wyman's approach[1,2] where one considers a macromolecule P having q sites, each able to combine with one molecule of ligand X. One can denote X as the total concentration of ligand bound by the macromolecule:

$$X = [P_0] \sum_{i=0}^{q} i K_i x^i$$

where $[P_0]$ is the concentration of the uncomplexed macromolecule, x is the activity of the ligand, and K_i is the apparent macroscopic association constant for the reaction:

$$P + iX \rightleftharpoons PX_i$$

Note also that $K_0 = 1$. The total concentration of macromolecule in all forms is:

$$P = [P_0] + [P_1] + [P_2] + \cdots + [P_q]$$

$$P = [P_0] \sum_{i=0}^{q} K_i x^i$$

Using $\overline{X}$ to designate the amount of ligand bound per mole of macromolecule, we get:

$$\overline{X} = \sum_i i K_i x^i / \sum_i K_i x^i$$

If the K_i's are independent of the total extent of the association reaction, then the above expression can be written in the following form:

$$\overline{X} = \frac{\partial \ln \sum K_i x^i}{\partial \ln x}$$

The fractional saturation of the macromolecule (written here as $\overline{x}$) is then:

$$\overline{x} = \overline{X}/q = \{d \ln \sum K_i x^i / d \ln x\}/q$$

A second ligand binding to another nearby site, then the total concentration of the macromolecule in all its forms will be:

$$P = [P_0] \sum_{i=0}^{q} \sum_{i=0}^{r} K_{ij} x^i y^i$$

where K_{ij} is the apparent equilibrium constant for the reaction:

$$P + iX + jY \rightleftharpoons PX_iY_j$$

This means that we can now write separate expressions for the amounts of ligands X and Y bound per mole of macromolecule:

$$\overline{X} = \frac{\partial \ln \sum_{i=0}^{q} \sum_{i=0}^{r} K_{ij} x^i y^j}{\partial \ln x} \quad \text{and} \quad \overline{Y} = \frac{\partial \ln \sum_{i=0}^{q} \sum_{i=0}^{r} K_{ij} x^i y^j}{\partial \ln y}$$

Letting $\sum\sum$ represent the double sum, we take advantage of the fact that $d \ln \sum\sum$ is a perfect differential that yields

$$\left(\frac{\partial \overline{X}}{\partial \ln y} \right)_x = \left(\frac{\partial \overline{Y}}{\partial \ln x} \right)_y$$

This expression is called a linked function and indicates how the binding of ligands at nearby sites can influence each other. *See also* Basic Regulatory Kinetics; Cooperativity; Allosterism; Feedback Effectors; Bohr Effect; Hemoglobin; Le Chatelier's Principle; Adair Equation

[1] J. T. Edsall & J. Wyman (1958) *Biophysical Chemistry*, pp. 654-660, Academic Press, New York.
[2] J. Wyman & S. J. Gill (1990) *Binding and Linkage*, Univ. Sci. Books, Mill Valley, CA.

LINOLEATE ISOMERASE

This enzyme [EC 5.2.1.5] catalyzes the interconversion of 9-*cis*,12-*cis*-octadecadienoate and 9-*cis*,11-*trans*-octadecadienoate.

S. Seltzer (1972) *The Enzymes*, 3rd ed., **6**, 381.
C. R. Kepler & S. B. Tove (1969) *Meth. Enzymol.* **14**, 105.

LIPASES

Triacylglycerol lipases [EC 3.1.1.3] (also known as triglyceride lipases, tributyrases, or simply as lipases) catalyze the hydrolysis of a triacylglycerol to produce a diacylglycerol and a fatty acid anion. The pancreatic enzyme acts only on an ester-water interface; the outer ester links in the substrate are the ones which are preferentially hydrolyzed. *See* *Lipoprotein Lipase; Micelles; and specific lipase.*

D. M. Quinn & S. R. Feaster (1998) *Comprehensive Biological Catalysis: A Mechanistic Reference* **1**, 455.

M. E. Lowe (1997) *Ann. Rev. Nutr.* **17**, 141.

Z. S. Derewenda (1994) *Adv. Protein Chem.* **45**, 1.

L. C. Smith, F. Faustinella & L. Chan (1992) *Curr. Opin. Struct. Biol.* **2**, 490.

H. Okuda (1991) in *A Study of Enzymes* (S. A. Kuby, ed.) **2**, 579.

P. Strålfars, H. Olsson & P. Belfrage (1987) *The Enzymes*, 3rd ed., **18**, 147.

M. A. Wells & N. A. DiRenzo (1983) *The Enzymes*, 3rd ed., **16**, 113.

R. L. Jackson (1983) *The Enzymes*, 3rd ed., **16**, 141.

J. C. Khoo & D. Steinberg (1983) *The Enzymes*, 3rd ed., **16**, 183.

R. Verger (1980) *Meth. Enzymol.* **64**, 340.

M. Sémériva & P. Desnuelle (1979) *Adv. Enzymol.* **48**, 319.

P. Desnuelle (1972) *The Enzymes*, 3rd ed., **7**, 575.

LIPID ACTIVATION

Stimulation in the activity of enzymes that are normally associated with biomembranes by general or specific lipids within that membrane. If this activation phenomenon is relatively nonspecific, the term most often applied is *interface activation*[1].

[1] R. Verger (1980) *Meth. Enzymol.* **64**, 340.

LIPID PHASE TRANSITION KINETICS

An emergent field dealing with the physical/chemical processes that underlie the changes of lipid phase state during such cellular events as membrane fusion, vesicle trafficking, and cell disjunction.

M. Caffrey (1989) *Ann. Rev. Biophys. Biophys. Chem.* **18**, 159.

LIPID TRACER KINETICS

Wolfe[1] has presented an excellent description of the systematic application of stable and radioactive isotope tracers in determining the kinetics of intestinal fat absorption, hepatic triglyceride synthesis, lipid mobilization, triglyceride-fatty acid recycling, and cholesterol turnover.

[1] R. R. Wolfe (1992) *Radioactive and Stable Isotope Tracers in Biomedicine*, pp. 317-340, Wiley-Liss, New York.

LIPOIC ACID AND DERIVATIVES

An acyl-transfer and redox coenzyme containing two sulfhydryl groups that form a dithiolane ring in the oxidized (disulfide) form. The redox potential at pH 7 is −0.29 volts. Lipoic acid is attached to the ε-amino group of lysyl residues of transacetylases (subunit of α-ketoacid dehydrogenase complexes), thereby permitting acyl

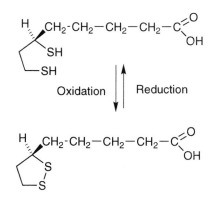

transfer as a swinging arm akin to that of carboxybiocytin.

Selected entries from *Methods in Enzymology* [vol, page(s)]:
Turbidimetric and polarographic assays for lipoic acid using mutants of *Escherichia coli*, **18A**, 269; chemical syntheses of ^{14}C-α-lipoic acid, turbidimetric and polarographic assays for lipoic acid using mutants of *Escherichia coli*, **18A**, 269; lipoic acid transport, **18A**, 276; lipoic acid activation, **18A**, 282; purification and properties of lipoamidase, **18A**, 292; purification and properties of lipoamide dehydrogenases from pig heart α-keto acid dehydrogenase complexes, **18A**, 298; chromatographic and spectral properties of lipoic acid and its metabolites, **62**, 129; thin-layer chromatography of lipoic acid, lipoamide, and their persulfides, **62**, 135; preparation and separation of *S*-oxides of α-lipoic acid, **62**, 137; photochemical preparation of acyl lipoic acids, **62**, 145; lipoyl disulfide reducing polymers, **62**, 147; biosynthesis of lipoic acid via unsaturated fatty acids, **62**, 152; turnover of protein-bound lipoic acid, **62**, 159; the radial diffusion assay of lipoamide dehydrogenase, **62**, 166; asparagusate dehydrogenase and lipoyl dehydrogenase from asparagus, **62**, 172; preparation and determination of asparagusic acid, **62**, 18; methodology employed for anaerobic spectrophotometric titrations and for computer-assisted data analysis, **62**, 185.

LIPOPHILIC

Referring to bodies, substances, molecular entities, or portions of molecular entities that have a tendency to be soluble in apolar solvents. *See* *Hydrophobic Interaction*

LIPOPOLYSACCHARIDE N-ACETYLGLUCOSAMINYLTRANSFERASE

This enzyme [EC 2.4.1.56] catalyzes the reaction of UDP-*N*-acetyl-D-glucosamine with a lipopolysaccharide to produce UDP and an *N*-acetyl-D-glucosaminyl-lipopolysaccharide. Thus, this enzyme transfers *N*-acetylglucosaminyl residues to a D-galactose residue in the partially completed lipopolysaccharide core.

M. J. Osborn & L. I. Rothfield (1966) *Meth. Enzymol.* **8**, 456.

LIPOPOLYSACCHARIDE GALACTOSYLTRANSFERASE

This enzyme [EC 2.4.1.44] catalyzes the reaction of UDP-galactose with a lipopolysaccharide to produce UDP and a D-galactosyl-lipopolysaccharide. Thus, the enzyme transfers D-galactosyl residues to D-glucose in the partially completed core of a lipopolysaccharide.

M. J. Osborn & L. I. Rothfield (1966) *Meth. Enzymol.* **8**, 456.

LIPOPOLYSACCHARIDE GLUCOSYLTRANSFERASE

Lipopolysaccharide glucosyltransferase I [EC 2.4.1.58] catalyzes the reaction of UDP-glucose with a lipopolysaccharide to produce UDP and a D-glucosyl-lipopolysaccharide. Thus, this enzyme transfers glucosyl residues to the backbone portion of a lipopolysaccharide. Lipopolysaccharide glucosyltransferase II [EC 2.4.1.73] catalyzes the reaction of UDP-glucose with a lipopolysaccharide to produce UDP and a D-glucosyl-lipopolysaccharide. Thus, this enzyme transfers glucosyl residues to the D-galactosyl-D-glucosyl side chains in the partially completed core of a lipopolysaccharide.

M. J. Osborn & L. I. Rothfield (1966) *Meth. Enzymol.* **8**, 456.

LIPOPROTEIN LIPASE

This enzyme [EC 3.1.1.34] (also called clearing factor lipase, diglyceride lipase, and diacylglycerol lipase) catalyzes the hydrolysis of a triacylglycerol to produce a diacylglycerol and a fatty acid anion. This enzyme hydrolyzes triacylglycerols in chylomicrons and in low-density lipoproteins and also acts on diacylglycerols. *See also Lipases*

R. Potenz, J.-Y. Lo, E. Zsigmond, L. C. Smith & L. Chan (1996) *Meth. Enzymol.* **263**, 319.
H. Okuda (1991) *A Study of Enzymes* **2**, 579.
A. Bensadoun (1991) *Ann. Rev. Nutr.* **11**, 217.
R. L. Jackson (1983) *The Enzymes*, 3rd ed., **16**, 141.
P. Desnuelle (1972) *The Enzymes*, 3rd ed., **7**, 575.

LIPOXYGENASE

Lipoxygenase [EC 1.13.11.12] catalyzes the reaction of linoleate with dioxygen to produce (9Z,11E)-(13S)-13-hydroperoxyoctadeca-9,11-dienoate. This iron-dependent enzyme can also oxidize other methylene-interrupted polyunsaturated fatty acids. *See also specific enzyme*

B. G. Fox (1998) *Comprehensive Biological Catalysis: A Mechanistic Reference* **3**, 261.
B. J. Gaffney (1996) *Ann. Rev. Biophys. Biomol. Struct.* **25**, 431.
A. G. Prescott & P. John (1996) *Ann. Rev. Plant Physiol. Plant Mol. Biol.* **47**, 245.
J. N. Siedow (1991) *Ann. Rev. Plant Physiol. Plant Mol. Biol.* **42**, 145.
M. A. Ator & P. R. Ortez de Montellano (1990) *The Enzymes*, 3rd ed., **19**, 213.

T. Schewe, S. M. Rapoport & H. Kühn (1986) *Adv. Enzymol.* **58**, 191.
H. Kühn, T. Schewe & S. M. Rapoport (1986) *Adv. Enzymol.* **58**, 273.
O. Hayaishi, M. Nozaki & M.T. Abbott (1975) *The Enzymes*, 3rd ed., **12**, 119.

5-LIPOXYGENASE

This iron-dependent enzyme [EC 1.13.11.34], better known as arachidonate 5-lipoxygenase and occasionally referred to as leukotriene A_4 synthase, catalyzes the reaction of arachidonate with dioxygen to produce (6E,8Z,11Z,14Z)-(5S)-5-hydroperoxyicosa-6,8,11,14-tetraenoate, which rapidly converts to leukotriene A_4.

B. G. Fox (1998) *Comprehensive Biological Catalysis: A Mechanistic Reference* **3**, 261.
A. W. Ford-Hutchinson, M. Gresser & R. N. Young (1994) *Ann. Rev. Biochem.* **63**, 383.
S. Yamamoto & Y. Ishimura (1991) *A Study of Enzymes* **2**, 315.
M. A. Ator & P. R. Ortez de Montellano (1990) *The Enzymes*, 3rd ed., **19**, 213.
T. Schewe, S. M. Rapoport & H. Kühn (1986) *Adv. Enzymol.* **58**, 191.
H. Kühn, T. Schewe & S. M. Rapoport (1986) *Adv. Enzymol.* **58**, 273.
C. R. Pace-Asciak & W. L. Smith (1983) *The Enzymes*, 3rd ed., **16**, 543.

8-LIPOXYGENASE

This iron-dependent enzyme [EC 1.13.11.40], better known as arachidonate 8-lipoxygenase, catalyzes the reaction of arachidonate with dioxygen to produce (5Z,9E,11Z,14Z)-(8R)-8-hydroperoxyicosa-5,9,11,14-tetraenoate.

C. R. Pace-Asciak & W. L. Smith (1983) *The Enzymes*, 3rd ed., **16**, 543.

12-LIPOXYGENASE

This iron-dependent enzyme [EC 1.13.11.31], better known as arachidonate 12-lipoxygenase, catalyzes the reaction of arachidonate with dioxygen to produce (5Z,8Z,10E,14Z)-(12S)-12-hydroperoxyicosa-5,8,10,14-tetraenoate, which then converts rapidly to the corresponding 12S-hydroxy compound.

B. G. Fox (1998) *Comprehensive Biological Catalysis: A Mechanistic Reference* **3**, 261.
S. Yamamoto & Y. Ishimura (1991) *A Study of Enzymes* **2**, 315.
T. Schewe, S. M. Rapoport & H. Kühn (1986) *Adv. Enzymol.* **58**, 191.
H. Kühn, T. Schewe & S. M. Rapoport (1986) *Adv. Enzymol.* **58**, 273.
C. R. Pace-Asciak & W. L. Smith (1983) *The Enzymes*, 3rd ed., **16**, 543.

15-LIPOXYGENASE

This iron-dependent enzyme [EC 1.13.11.33], better known as arachidonate 15-lipoxygenase or arachidonate ω-6-lipoxygenase, catalyzes the reaction of arachidonate with dioxygen to produce (5Z,8Z,11Z,13E)-(15S)-15-hydroperoxyicosa-5,8,11,13-tetraenoate, which rapidly converts to the corresponding 15S-hydroxy compound.

B. G. Fox (1998) *Comprehensive Biological Catalysis: A Mechanistic Reference* **3**, 261.
S. Yamamoto & Y. Ishimura (1991) *A Study of Enzymes* **2**, 315.
M. A. Ator & P. R. Ortez de Montellano (1990) *The Enzymes*, 3rd ed., **19**, 213.
C. R. Pace-Asciak & W. L. Smith (1983) *The Enzymes*, 3rd ed., **16**, 543.

LOCK-AND-KEY MODEL OF ENZYME ACTION

The original concept[1] offered to explain why enzymes exhibit such a high degree of substrate specificity. The active site of an enzyme was viewed as a topological template for a particular reactant; hence, there is an enzyme-substrate complementarity[2,3]. **See** *Induced-Fit Model; Ligand-Induced Conformational Change*

[1]E. Fischer (1894) *Ber.* **27**, 2985.
[2]A. Fersht (1977) *Enzyme Structure and Mechanism*, 2nd. ed., Freeman, San Francisco.
[3]A. R. Fersht (1999) *Structure and Mechanism in Protein Science: A Guide to Enzyme Catalysis and Protein Folding*, pp. 650, W. H. Freeman, New York.

LOG-DOSE RESPONSE CURVE FOR ENZYME COOPERATIVITY

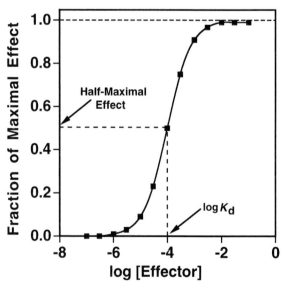

A method for graphically representing cooperativity in which the velocity (or percent response) is plotted as a function of the logarithm of the substrate concentration (or concentration of the binding ligand)[1,2]. Such curves are very commonly used in pharmacological studies.

A log dose-response curve affords an investigator an easy method for assessing cooperativity. By comparing the concentration of the ligand that yields a 90% response (or 90% V_{max}) to the concentration that generates a 10% response, the ratio obtained (the R_s value) will provide information on the degree of cooperativity ($R_s < 81$ is positive cooperativity and $R_s > 81$ for negative cooperativity). The R_s value and the Hill coefficient are mathematically related: $R_s = 81^{1/n_H}$.

[1]K. E. Neet (1980) *Meth. Enzymol.* **64**, 139.
[2]K. E. Neet (1995) *Meth. Enzymol.* **249**, 519.

LOG VELOCITY *VERSUS* LOG[SUBSTRATE] PLOT

A plot that permits the characterization of enzyme rate behavior over a large range of substrate concentrations: the logarithm of the initial steady-state velocity is plotted as a function of the logarithm of the substrate concentration. If the enzyme obeys Michaelis-Menten kinetics, then a single S-shaped curve will be obtained , and the slope at the mid-point will be equal to one. If a sample contains two enzymes displaying markedly different Michaelis constants (*i.e.*, differing by a factor of seven or more), and if there are comparable amounts of each enzymic activity, then a log velocity-substrate plot may exhibit two S-shaped curves. Enzymes that display negative or positive cooperativity will give slopes other than unity.

LONDON FORCES

Attractive forces (sometimes also referred to as dispersion forces) between apolar molecules arising from the mutual polarizability of the interacting molecules. London forces also contribute to the interactive forces between polar molecules. **See also** *van der Waals Forces; Noncovalent Interactions*

LONDON-STECK PLOT

A graphical method[1] for analyzing enzyme activation (a metal ion for the case shown above). Initial velocities are plotted in Fig. 1 as a function of the total substrate (S) concentration at different, constant concentrations of the activator (A). An analogous plot (Fig. 2) is made

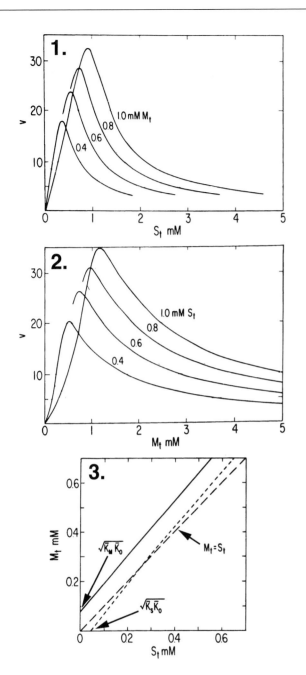

complex or the ES complex. This analysis assumes rapid equilibrium binding.

[1]W. P. London & T. L. Steck (1969) *Biochemistry* **8**, 1767.

LONE PAIR

A pair of nonbonding electrons localized in the valence shell on a single atom. Examples are the lone electron pair on the nitrogen atom in ammonia and the two lone pairs on the oxygen atom in water. In these cases, the lone pairs participate in hydrogen bonding interactions.

LONG-CHAIN ACYL-CoA DEHYDROGENASE

This enzyme [EC 1.3.99.13] catalyzes the reaction of a long-chain acyl-CoA with an electron-transferring flavoprotein to produce a 2,3-dehydroacyl-CoA and the reduced electron-transferring flavoprotein.

B. A. Palfey & V. Massey (1998) *Comprehensive Biological Catalysis: A Mechanistic Reference* **3**, 83.

LONG-CHAIN ALDEHYDE DEHYDROGENASE

This enzyme [EC 1.2.1.48] catalyzes the reaction of a long-chain aldehyde with NAD^+ to produce a long-chain acid anion and NADH. The best substrate is reported to be dodecylaldehyde.

M. Ueda & A. Tanaka (1990) *Meth. Enzymol.* **188**, 176.

LONG-CHAIN FATTY ACYL-CoA SYNTHETASE

Long-chain fatty acyl-CoA synthetase [EC 6.2.1.3] catalyzes the reaction of ATP with a long-chain carboxylic acid and coenzyme A to produce an acyl-CoA, AMP, and pyrophosphate. While utilizing a wide range of long-chain saturated and unsaturated fatty acids as substrates, enzymes from different tissues vary in their specificity.

J. C. Londesborough & L. T. Webster, Jr. (1974) *The Enzymes*, 3rd ed., **10**, 469.

LONG-RANGE LOOPS

Initial protein folding intermediates formed as a consequence of nonlocal interactions on the reaction path from unfolded to partially folded states. These loops are likely to determine the trajectory of later folding steps. The character of nonlocal interactions is likely to be much different than structures inferred from the results

versus the activator concentration at different fixed concentrations of the substrate. If the ES complex cannot proceed to form products and the true substrate is the SA complex (for example, the binding of a required metal ion to a nucleotide), then the two plots will consist of curves in which the velocity increases to a maximum value, then decreases. If these maximum velocities are replotted (Fig. 3) as a function of the corresponding substrate concentration (or activator concentration), a new curve is generated which has a horizontal intercept that is a function of the dissociation constant for the SA complex and the dissociation constant for either the EA of studies on unfolded states formed by perturbing an already well-folded protein. In the latter case, one implicitly assumes that the structure and reactivity of the species formed by the action of a pertubant on a folded protein are in fact related to structures formed in reactions that commence from a globally unfolded state. Ittah and Haas[1] provide an excellent account of how nonlocal interactions may act to stabilize long-range loops in the initial folding of reduced bovine pancreatic trypsin inhibitor.

[1]V. Ittah & E. Haas (1995) *Biochemistry* **34**, 4493.

LOOSE ION PAIR

An ion pair in which the constituent ions are separated by one or more solvent (or other neutral) molecules. If X^+ and Y^- represent the constituent ions, a loose ion pair is usually symbolized by $X^+\|Y^-$. The constituent ions of a loose ion pair can readily exchange with other ions in solution; this provides an experimental means for distinguishing loose ion pairs from tight ion pairs. In addition, there are at least two types of loose ion pairs: solvent-shared and solvent-separated. *See Ion Pair; Tight Ion Pair; Solvent-Shared Ion Pair; Solvent-Separated Ion Pair*

LOSSEN REARRANGEMENT

Originally, the Lossen rearrangement specifically referred to the conversion of a hydroxamic acid to an amine with the loss of carbon dioxide[1,2]:

$$R-C(=O)-NHOH \xrightarrow{} RNCO \xrightarrow{H_2O} RNHCOOH \xrightarrow{}$$
$RNH_2 + CO_2$. A thio-Lossen rearrangement has been proposed to occur spontaneously for the product generated by the action of myrosinase on sinigrin[3].

[1]W. Lossen (1872) *Ann.* **161**, 347.
[2]W. Lossen (1874) *Ann.* **175**, 271 and 313.
[3]G. Davies, M. L. Sinnott & S. G. Withers (1998) *Comprehensive Biological Catalysis: A Mechanistic Reference* **1**, 119.

LOW-BARRIER HYDROGEN BONDS
(Potential Role In Catalysis)

Low-barrier hydrogen bonds (abbreviated: LBHBs) are short, very strong hydrogen bonds formed when the electron-pair donors (most commonly oxygen or nitrogen atoms in proteins) sharing the hydrogen have comparable pK_a values.

Hydrogen bonds involve the sharing of a hydrogen atom by two electronegative partners by way of electrostatic and dispersion energy interactions. Pauling identified the properties of typical hydrogen bonds in polypeptides: (a) they are formed between electronegative atoms (most often amide hydrogens and amide carbonyls); (b) the strongest hydrogen bonds are linear, and any bending can severely lower their bond strength; and (c) they have bond lengths between 2.8 and 3.4 Å. When the bond lengths are compared to the sum of the van der Waals radii, one often observes a 10–25% reduction in distance upon hydrogen bond formation. There is an activation barrier for hydrogen "transfer" from one of the electronegative bonding partners to the other within a hydrogen bond. This barrier decreases as the difference in the pK_a values of the electron pair donors become comparable, hence the designation low-barrier hydrogen bonds. LBHBs have bond lengths of less than 2.5 Å. By contrast, weaker hydrogen bonds are formed when the pK_a values do not match.

The presence of LBHBs within enzyme active sites may partly explain the origin of enzymic rate enhancements[1,2]. Frey *et al.*[1] obtained evidence with chymotrypsin and model compounds suggesting that a low-barrier hydrogen bond may be a mechanistic feature in serine protease catalysis. Cleland & Kreevoy[3] have suggested that formation of low-barrier hydrogen bonds may supply 10–20 kilocalories per mole (or, 40–84 kJ/mol), thus facilitating difficult reactions such as enolization of carboxylate groups. The idea is that strong, low-barrier hydrogen bonds may greatly influence the transition state configuration. Induced conformational changes that more closely match the pK_a values may give rise to a strong, low-barrier bond: this is especially important when the pK_a values refer to those of a transition state or activated enzyme intermediate.

More recently, Cassidy *et al.*[4] conducted additional ^{1}H NMR experiments to evaluate the basicities of the dyad H57-D102 in the tetrahedral complexes of chymotrypsin with the peptidyl trifluoromethyl ketones (TFKs): *N*-acetyl-L-Leu-DL-Phe-CF$_3$ and *N*-acetyl-DL-Phe-CF$_3$. The proton bridging His-57 and Asp-102 is part of a low-barrier hydrogen bond (LBHBs). In ^{1}H NMR spectra at pH 7.0, these protons appear at δ 18.9 and 18.6 ppm, and the pK_a values of the dyads are 12.0 ± 0.2 and 10.8 ± 0.1, respectively. The available structural data on chymotrypsin and other serine proteases, along with the high pK_a values of the hemiketals formed with TFKs, suggest LBHB formation in catalysis arises through a substrate-induced conformational transition leading to steric compression between His-57 and Asp-102. The LBHB's N-to-O distance is shorter than the van der Waals contact distance, and the LBHB may stabilize the tetrahedral intermediate. In their mechanistic proposal, substrate-induced steric compression within the dyad increases the basicity of $N(\varepsilon)2$ in His-57, rendering it more effective as a base to abstract a proton from Ser-195 in tetrahedral intermediate formation. In this model, the energy for conformational compression may be provided by binding interactions remote from the scissile bond.

Warshel *et al.*[5] offer the argument that energy considerations suggest that low-barrier hydrogen bonds do not offer any special catalytic advantage as compared to ordinary hydrogen bonds. They evaluated the extent of transition state (TS) stabilization that might be achieved by low-barrier hydrogen bonds within enzyme active sites. They focused on three issues: (a) low-barrier hydrogen bonds are less stable than regular hydrogen bonds in water; (b) transition states are more stable in the enzyme active sites than in water, and (c) a nonpolar active site would destabilize the TS relative to its energy in water. By considering these points and related observations in a physically consistent frame-work, Warshel *et al.*[5] are drawn to conclude that the occurrence of low-barrier hydrogen bonds cannot stabilize the TS more effectively than ordinary hydrogen bonds. ***See also*** *Hydrogen Bond; Water*

[1] P. A. Frey, S. A. Whitt & J. B. Tobin (1994) *Science* **264**, 1927.
[2] J. A. Gerlt, M. M. Kreevoy, W. W. Cleland & P. A. Frey (1997) *Chem. Biol.* **4**, 259.
[3] W. W. Cleland & M. M. Kreevoy (1994) *Science* **264**, 1887.
[4] C. S. Cassidy, J. Lin & P. A. Frey (1997) *Biochemistry* **36**, 4576.
[5] A. Warshel & A. Papazyan (1996) *Proc. Natl. Acad. Sci. U.S.A.* **93**, 13665.

LUCIFERASE

The firefly enzyme (EC 1.13.12.7) catalyzes the intermediate formation of D($-$)-luciferyl adenylate and pyrophosphate from D($-$)-luciferin and ATP, followed by the oxidative reaction of the acyl adenylate with molecular oxygen to form an enzyme-bound product in the excited state[1–3]. Decay of this species to the ground state is attended by photon emission at a wavelength of 562 nm. This emission maximum is for *Photinus pyralis*, and other fireflies have slightly different emission properties.

The enzyme shows total specificity for ATP, but other nucleotides are poor competitive inhibitors of ATP. Those attempting to measure ATP exclusively must exercise care to block transphosphorylases (*e.g.*, nucleoside 5′-diphosphate kinase) that may indirectly synthesize ATP from ADP and another nucleoside 5′-triphosphate. Luciferase assays[1,2] can be carried out using a kit available from Sigma Chemical Co. (St. Louis, MO; product numbers L1759, L5256, and L9009). Although the enzyme has a low turnover number, sensitive ATP assays down to the femtomole range can be conveniently achieved using a luminometer.

The bacterial luciferase[4,5] [EC 1.14.14.3], also called alkanal monooxygenase (FMN-linked) and aldehyde monooxygenase, catalyzes the reaction of an aldehyde (RCHO) with dioxygen and $FMNH_2$ to produce a carboxylic acid (RCOOH), FMN, water, and light. The enzyme reaction mechanism involves incorporation of a molecule of dioxygen into reduced FMN and the subsequent reaction with the aldehyde to form an activated $FMN \cdot H_2O$ complex which breaks down with the emission of light. The enzyme is highly specific for reduced FMN and for long-chain aliphatic aldehydes with eight carbons or more.

[1] F. R. Leach & J. J. Webster (1986) *Meth. Enzymol.* **133**, 51.
[2] S. Lin & H. P. Cohen (1968) *Anal. Biochem.* **24**, 531.
[3] M. DeLuca (1976) *Adv. Enzymol.* **44**, 37.
[4] J. W. Hastings (1978) *Crit. Rev. Biochem.* **5**, 163.
[5] V. Massey & P. Hemmerich (1975) *The Enzymes*, 3rd ed., **12**, 191.

LUMEN

1. The SI unit of luminous flux, abbreviated lm. One lumen is the amount of luminous flux emitted in a unit solid angle of one steradian by a uniform point source having a luminous intensity of one candela. 2. The interior space of a tubular structure such as an intestine or artery.

LUMINESCENCE

An emission of radiation occurring spontaneously from an electronically or vibrationally excited species that is not in thermal equilibrium with its environment.

LYASES

A major class of enzymes that catalyze the cleavage of bonds by means other than hydrolysis or oxidation. When acting on a single substrate, at least one of the products formed by lyases is unsaturated.

LYATE ION

An anion produced as a result of the action of a Brønsted base on a protogenic solvent or of the action of autoprotolysis (loss of a hydron) of a solvent. Thus, OH^- is the lyate ion of water.

LYONIUM ION

A cation produced by the protonation of a solvent molecule. **See also** Leveling Effect

LYSINE 2,3-AMINOMUTASE

This enzyme [EC 5.4.3.2] catalyzes the interconversion of lysine and (3S)-3,6-diaminohexanoate. The enzyme is stimulated by S-adenosylmethionine and pyridoxal phosphate.

G. H. Reed & M. D. Ballinger (1995) *Meth. Enzymol.* **258**, 362.
P. A. Frey & G. H. Reed (1993) *Adv. Enzymol.* **66**, 1.
E. J. Brush & J. W. Kozarich (1992) *The Enzymes*, 3rd ed., **20**, 317.
T. C. Stadtman (1973) *Adv. Enzymol.* **38**, 413.

β-LYSINE 5,6-AMINOMUTASE

This cobalamin-dependent enzyme [EC 5.4.3.3] catalyzes the interconversion of (3S)-3,6-diaminohexanoate and (3S,5S)-3,5-diaminohexanoate.

B. T. Golding & W. Buckel (1998) *Comprehensive Biological Catalysis: A Mechanistic Reference* **3**, 239.
P. A. Frey & G. H. Reed (1993) *Adv. Enzymol.* **66**, 1.
T. C. Stadtman (1973) *Adv. Enzymol.* **38**, 413.
T. C. Stadtman (1972) *The Enzymes*, 3rd ed., **6**, 539.

D-LYSINE 5,6-AMINOMUTASE

This cobalamin-dependent enzyme [EC 5.4.3.4], also called D-α-lysine mutase, catalyzes the interconversion of D-lysine and 2,5-diaminohexanoate.

P. A. Frey & G. H. Reed (1993) *Adv. Enzymol.* **66**, 1.
T. C. Stadtman (1973) *Adv. Enzymol.* **38**, 413.
T. C. Stadtman (1972) *The Enzymes*, 3rd ed., **6**, 539.

LYSINE 6-AMINOTRANSFERASE

This enzyme [EC 2.6.1.36], also called lysine ε-aminotransferase, catalyzes the pyridoxal-phosphate-dependent reaction of L-lysine with α-ketoglutarate (or, 2-oxoglutarate) to produce 2-aminoadipate 6-semialdehyde and L-glutamate. The aldehyde product, also known as allysine, then undergoes dehydration to form 1-piperidine 6-carboxylate.

K. Tanizawa, S. Toyama & K. Soda (1985) *Meth. Enzymol.* **113**, 96.
A. E. Braunstein (1973) *The Enzymes*, 3rd ed., **9**, 379.

LYSINE DECARBOXYLASE

This pyridoxal-phosphate-dependent enzyme [EC 4.1.1.18] catalyzes the conversion of L-lysine to form cadaverine and carbon dioxide. 5-Hydroxy-L-lysine can also be decarboxylated by this enzyme.

E. A. Boeker & E. E. Snell (1972) *The Enzymes*, 3rd ed., **6**, 217.

LYSINE DEHYDROGENASE (α-DEAMINATING)

This enzyme [EC 1.4.1.15] catalyzes the reaction of L-lysine with NAD^+ to produce 1,2-didehydropiperidine 2-carboxylate, ammonia, and NADH.

N. M. W. Brunhuber & J. S. Blanchard (1994) *Crit. Rev. Biochem. Mol. Biol.* **29**, 415.

LYSINE 2-MONOOXYGENASE

This FAD-dependent enzyme [EC 1.13.12.2] catalyzes the reaction of L-lysine with dioxygen to produce 5-aminopentanamide, carbon dioxide, and water. Other diamino acids can serve as substrates as well.

S. Yamamoto & Y. Ishimura (1991) *A Study of Enzymes* **2**, 315.
V. Massey & P. Hemmerich (1975) *The Enzymes*, 3rd ed., **12**, 191.

LYSOPHOSPHOLIPASE

This enzyme [EC 3.1.1.5] (also referred to as lecithinase B, lysolecithinase, and phospholipase B) catalyzes the hydrolysis of a 2-lysophosphatidylcholine to produce glycerophosphocholine and a fatty acid anion.

E. A. Dennis, ed. (1991) *Meth. Enzymol.* **197**.
J. D. Esko & C. R. H. Raetz (1983) *The Enzymes*, 3rd ed., **16**, 207.
E. A. Dennis (1983) *The Enzymes*, 3rd ed., **16**, 307.

LYSOPHOSPHOLIPASE D

This enzyme [EC 3.1.4.39], also known as alkylglycerophosphoethanolamine phosphodiesterase, catalyzes the hydrolysis of 1-alkyl-sn-glycero-3-phosphoethanolamine to produce 1-alkyl-sn-glycerol 3-phosphate and ethanolamine. The enzyme will also act on the acyl and choline analogs of the lipid.

R. L. Wykle & J. C. Strum (1991) *Meth. Enzymol.* **197**, 583.

LYSOPINE DEHYDROGENASE

This enzyme [EC 1.5.1.16], also referred to as D-lysopine synthase, catalyzes the reaction of N^2-(D-1-carboxy-ethyl)-L-lysine with $NADP^+$ and water to produce L-lysine, pyruvate, and NADPH. In the reverse reaction, a number of L-amino acids can serve as substrates instead of L-lysine.

J. Thompson & S. P. F. Miller (1991) *Adv. Enzymol.* **64**, 317.

LYSOZYME

This enzyme [EC 3.2.1.17], also called muramidase, catalyzes the hydrolysis of the 1,4-β-linkages between *N*-acetyl-D-glucosamine and *N*-acetylmuramic acid in peptidoglycan heteropolymers of prokaryotic cell walls. Some chitinases [EC 3.2.1.14] also exhibit this activity.
See Kinetic Isotope Effect

G. Davies, M. L. Sinnott & S. G. Withers (1998) *Comprehensive Biological Catalysis: A Mechanistic Reference* **1**, 119.
G. Mooser (1992) *The Enzymes*, 3rd ed., **20**, 187.
H. A. McKenzie & F. H. White, Jr. (1991) *Adv. Protein Chem.* **41**, 174.
L. N. Johnson, J. Cheetham, P. J. McLaughlin, K. R. Acharya, D. Barford & D. C. Phillips (1988) *Curr. Top. Microbiol. Immunol.* **139**, 81.
A. J. Kirby (1987) *Crit. Rev. Biochem.* **22**, 283.
T. Imoto, L. N. Johnson, A. C. T. North, D. C. Phillips & J. A. Rupley (1972) *The Enzymes*, 3rd ed., **7**, 665.
A. Tsugita (1971) *The Enzymes*, 3rd ed., **5**, 343.

Selected entries from *Methods in Enzymology* [vol, page(s)]:
N-Acetylglucosamine interaction, **64**, 275; antigenic sites, **70**, 46, 47; binding sites, **61**, 515-522; use in cell lysis, **65**, 119, 168, 186, 861; chemical modification, **63**, 213; clefts, **61**, 445-448 [¹³C NMR, **61**, 469, 471-474, 477-479, 491-510, 531, 533; carbon count, **61**, 482, 488; chemical modification, **61**, 515, 526, 538-541; linewidths, **61**, 486; proton-coupling conditions, **61**, 492, 497; self-association, **61**, 542-544; specific assignments, **61**, 531]; conformation studies, **74**, 257; cryoenzymology, **63**, 338; calorimetry of denaturation, **61**, 266; difference spectral titration, **63**, 332, 333; effect of structure on antibody specificity, **70**, 42-46; frictional coefficient, **61**, 111; inhibitor, **63**, 401; interaction parameters, **61**, 48; low-frequency dynamics, **61**, 443-445; microcalorimetric studies of binding, **61**, 302; near-UV circular dichroism, **61**, 349, 350, 377; partial specific volumes, **61**, 48; pH study, **63**, 231-234; perturbed p*K* value, **63**, 210, 233; spin immunoassay, **74**, 151; structure, **70**, 43; subsite binding energy, **64**, 260, 275.

LYSYL ENDOPEPTIDASE

This enzyme [EC 3.4.21.50] catalyzes the hydrolysis of peptide bonds with a preference for Lys—Xaa (including Lys—Pro).

F. Sakiyama & T. Masaki (1994) *Meth. Enzymol.* **244**, 126.

LYSYL HYDROXYLASE

This enzyme [EC 1.14.11.4] catalyzes the reaction of a procollagen L-lysyl residue with α-ketoglutarate (or, 2-oxoglutarate) and dioxygen to produce a procollagen 5-hydroxy-L-lysyl residue, succinate, and carbon dioxide. Iron and ascorbate are essential cofactors.

B. G. Fox (1998) *Comprehensive Biological Catalysis: A Mechanistic Reference* **3**, 261.
O. Hayaishi, M. Nozaki & M. T. Abbott (1975) *The Enzymes*, 3rd ed., **12**, 119.
V. Ullrich & W. Duppel (1975) *The Enzymes*, 3rd ed., **12**, 253.

LYSYL OXIDASE

This enzyme [EC 1.4.3.13], also referred to as protein–lysine 6-oxidase, catalyzes the reaction of a peptidyl-L-lysyl-peptide with dioxygen and water to produce a peptidyl-allysyl-peptide, ammonia, and hydrogen peroxide. The enzyme will also act on 5-hydroxylysyl residues in proteins.

C. Anthony (1998) *Comprehensive Biological Catalysis: A Mechanistic Reference* **3**, 155.
A. Messerschmidt (1998) *Comprehensive Biological Catalysis: A Mechanistic Reference* **3**, 401.
C. Hartmann & W. S. McIntire (1997) *Meth. Enzymol.* **280**, 98.
J. P. Klinman (1996) *J. Biol. Chem.* **271**, 27189.
H. M. Kagan & P. Cai (1995) *Meth. Enzymol.* **258**, 122.
M. A. Ator & P. R. Ortez de Montellano (1990) *The Enzymes*, 3rd ed., **19**, 213.
G. J. Cardinale & S. Udenfriend (1974) *Adv. Enzymol.* **41**, 245.

LYSYLTRANSFERASE

This enzyme [EC 2.3.2.3], also referred to as lysyl-tRNA:phosphatidylglycerol transferase, catalyzes the reaction of L-lysyl-tRNA with phosphatidylglycerol to produce tRNA and 3-phosphatidyl-1'-(3'-*O*-L-lysyl) glycerol.

R. L. Soffer (1974) *Adv. Enzymol.* **40**, 91.

α-LYTIC PROTEINASE

This enzyme [EC 3.4.21.12] catalyzes the hydrolysis of peptide bonds in proteins, especially at peptide bonds adjacent to alanyl and valyl residues of bacterial cell walls, elastin, and other proteins. The enzyme is a member of the peptidase family S2A. *See also* Chymotrypsin; Catalytic Triad

K. Morihara (1974) *Adv. Enzymol.* **41**, 179.

LYXOSE ISOMERASE

This enzyme [EC 5.3.1.15], also referred to as lyxose ketol-isomerase, catalyzes the interconversion of D-lyxose and D-xylulose.

E. A. Noltmann (1972) *The Enzymes*, 3rd ed., **6**, 271.

M

MACROMOLECULAR SEQUENCE ANALYSIS (Computer Methods)

There are numerous computer software programs that permit one to investigate sequence alignment and phylogenetic relationships among (a) various proteins, domains, motifs, modules, *etc.*, and (b) nucleic acid sequences in DNA and RNA. In addition to those presented below, various Internet-based algorithms afford rapid and convenient analysis of macromolecular sequences.

Selected entries from *Methods in Enzymology* [vol, page(s)]:

Databases and Resources: Information services of European Bioinformatics Institute, **266**, 3; TDB: new databases for biological discovery, **266**, 27; PIR-international protein sequence database, **266**, 41; superfamily classification in PIR-international protein sequence database, **266**, 59; gene classification artificial neural system, **266**, 71; blocks database and its applications, **266**, 88; indexing and using sequence databases, **266**, 105; SRS: information retrieval system for molecular biology data banks, **266**, 114.

Searching through Databases: Applications of network BLAST server, **266**, 131; Entrez: molecular biology database and retrieval system, **266**, 141; applying motif and profile searches, **266**, 162; consensus approaches in detection of distant homologies, **266**, 184; identification of sequence patterns with profile analysis, **266**, 198; effective large-scale sequence similarity searches, **266**, 212; effective protein sequence comparison, **266**, 227; discovering and understanding genes in human DNA sequence using GRAIL, **266**, 259; linguistic analysis of nucleotide sequences: algorithms for pattern recognition and analysis of codon strategy, **266**, 281; protein sequence comparison at genome scale, **266**, 295; iterative template refinement: protein-fold prediction using iterative search and hybrid sequence/structure templates, **266**, 322.

Multiple Alignment and Phylogenetic Trees: Multiple protein sequence alignment: algorithms and gap insertion, **266**, 343; progressive alignment of amino acid sequences and construction of phylogenetic trees from them, **266**, 368; using CLUSTAL for multiple sequence alignments, **266**, 383; combined DNA and protein alignment, **266**, 402; inferring phylogenies from protein sequences by parsimony, distance, and likelihood methods, **266**, 418; reconstruction of gene trees from sequence data, **266**, 427; estimating evolutionary distances between DNA sequences, **266**, 449; local alignment statistics, **266**, 460; parametric and inverse-parametric sequence alignment with XPARAL, **266**, 481.

Secondary Structure Considerations: Identification of functional residues and secondary structure from protein multiple sequence alignment, **266**, 497; prediction and analysis of coiled-coil structures, **266**, 513; PHD: predicting one-dimensional protein structure by profile-based neural networks, **266**, 525; GOR method for predicting protein secondary structure from amino acid sequence, **266**, 540; analysis of compositionally biased regions in sequence databases, **266**, 554.

Three-Dimensional Considerations: Discrimination of common protein folds: application of protein structure to sequence/structure comparisons, **266**, 575; three-dimensional profiles for measuring compatability of amino acid sequence with three-dimensional structure, **266**, 598; SSAP: sequential structure alignment program for protein structure comparison, **266**, 617; understanding protein structure: using scop for fold interpretation, **266**, 635; detecting structural similarities: a user's guide, **266**, 643; alignment of three-dimensional protein structures: network server for database searching, **266**, 653; converting sequence block alignments into structural insights, **266**, 662.

MACROSCOPIC CONSTANTS

Observable constants, usually thermodynamic dissociation constants, for the association of a particular ligand with two or more sites on a larger molecular entity (*e.g.*, a macromolecule). Macroscopic constants are composites of microscopic (*i.e.*, intrinsic) constants.

The larger molecular entity need not be a macromolecule. Consider the proton dissociation steps of the simplest amino acid, glycine. The dissociation of the first proton (*i.e.*, the most acidic proton) can be represented by

$$\text{GlycineH}_2{}^+ \rightleftharpoons \text{H}^+ + \text{GlycineH}$$

and has an observable, macroscopic dissociation constant ($K_1 = 4.45 \times 10^{-3}$, $pK_1 = 2.35$). This macroscopic constant is a composite of two different dissociations:

$$1)\ {}^+\text{H}_3\text{NCH}_2\text{COOH} \rightleftharpoons \text{H}^+ + {}^+\text{H}_3\text{NCH}_2\text{COO}^-$$

$$2)\ {}^+\text{H}_3\text{NCH}_2\text{COOH} \rightleftharpoons \text{H}^+ + \text{H}_2\text{NCH}_2\text{COOH}$$

having the microscopic dissociation constants k_1 and k_2, respectively.

As can be seen in the entry on zwitterions, $K_1 = k_1 + k_2$. The observable macroscopic constant is indeed a composite of the intrinsic microscopic constants. **See** *Microscopic Constant; Zwitterion*

MACROSCOPIC DIFFUSION CONTROL

The limiting reaction rate achieved by diffusion occurring after mixing solutions containing two or more reactants. This form of control, also called mixing control, is typically more relevant for reactions involving heterogeneous systems (*e.g.*, solid/liquid or liquid/gas). **See also** *Microscopic Diffusion Control*

MAGNESIUM ION (Intracellular)

Beyond the effect of magnesium ion concentration on the equilibrium hydrolysis of adenosine triphosphate to adenosine diphosphate[1,2], there is ample evidence that $MgATP^{2-}$ is generally the most widespread substrate in kinase-type phosphotransferase reactions as well as other ATP-dependent processes. The extent to which $MgATP^{2-}$ is formed in solution depends on the free (or uncomplexed) magnesium ion concentration, as shown by the following equilibrium constant:

$$K_{formation} = [MgATP^{2-}]/[Mg_{free}^{2+}][ATP^{4-}]$$

For example, Bachelard[3] used $[Mg_{total}]/[ATP_{total}^{4-}] = 1$ in his rate studies, and he obtained a slightly sigmoidal plot of initial velocity versus "substrate ATP" concentration. This culminated in the erroneous proposal that brain hexokinase was allosterically activated by magnesium ions and by magnesium ion-adenosine triphosphate complex. Purich and Fromm[4] demonstrated that failure to achieve adequate experimental control over the free magnesium ion concentration can wreak havoc on the examination of enzyme kinetic behavior. Indeed, these investigators were able to account fully for the effects obtained in the previous hexokinase study[3].

Nonetheless, there remained some uncertainty regarding the actual intracellular concentration of uncomplexed magnesium ion, which controls the fraction of total ATP complexed with Mg^{2+}:

$$Fraction_{MgATP} = [MgATP^{2-}]/\{[ATP^{4-}] + [MgATP^{2-}]\}$$

or

$$Fraction_{MgATP} = 1/\{[ATP^{4-}]/[MgATP^{2-}] + 1\}$$

By substituting from the first equilibrium expression (*i.e.*, $[ATP^{4-}]/[MgATP^{2-}] = [Mg_{free}^{2+}]/K_{formation}$), this fraction becomes:

$$Fraction_{MgATP} = 1/\{1 + [Mg_{free}^{2+}]/K_{formation}\}$$

This underscores the need to know two parameters, namely $[Mg_{free}^{2+}]$ and $K_{formation}$ to assess $Fraction_{MgATP}$.

Accurate determination of the free intracellular magnesium ion concentration has been challenging because the available methods, including dye injection, microelectrodes, and nuclear magnetic resonance measurements, are either invasive or indirect. On the basis of substantially identical ^{31}P NMR measurements of the resonances in the spectrum of $MgATP^{2-}$, Wu *et al.*[5] put $[Mg_{free}^{2+}]$ at 2.5 mM in guinea pig heart muscle, whereas other investigators[6,7] arrived at a value of 0.6 mM. Later assessment[8] of the experimental evidence regarding the free intracellular magnesium concentration (as well as the limitations of the analytical methods) suggested that intracellular concentration of free magnesium is low, probably around 0.4 mM, and varies with time and conditions. Clarke *et al.*[9] recently introduced the β/α phosphoryl peak height ratio of ATP as a means for measuring free $[Mg^{2+}]$ using ^{31}P NMR. **See** *Energy Charge; Metal Ions in Nucleotide-Dependent Reactions; Biochemical Thermodynamics*

[1]R. A. Alberty (1968) *J. Biol. Chem.* **243**, 1337.
[2]R. C. Phillips, P. George & R. J. Rutman (1969) *J. Biol. Chem.* **244**, 3330.
[3]H. S. Bachelard (1971) *Biochem. J.* **125**, 249.
[4]D. L. Purich & H. J. Fromm (1972) *Biochem. J.* **130**, 63.
[5]S. T. Wu, G. M. Pieper, J. M. Salhany & R. S. Eliot (1981) *Biochemistry* **20**, 7399.
[6]R. K. Gupta & R. D. Moore (1980) *J. Biol. Chem.* **255**, 3987.
[7]R. K. Gupta, J. L. Benkovic & Z. B. Rose (1978) *J. Biol. Chem.* **253**, 6165.
[8]L. Garfinkel, R. A. Altschuld & D. Garfinkel (1986) *J. Mol. Cell. Cardiol.* **18**, 1003.
[9]K. Clarke, Y. Kashiwaya, M. T. King, D. Gates, C. A. Keon, H. R. Cross, G. K. Radda & R. L. Veech (1996) *J. Biol. Chem.* **271**, 21142.

MAGNESIUM-PROTOPORPHYRIN *O*-METHYLTRANSFERASE

This enzyme [EC 2.1.1.11] catalyzes the reaction of *S*-adenosylmethionine with magnesium protoporphyrin to produce *S*-adenosylhomocysteine and magnesium protoporphyrin monomethyl ester.

S. Granick & S. I. Beale (1978) *Adv. Enzymol.* **46**, 33.

MAGNETICALLY ANISOTROPIC GROUP

A functional moiety in a molecule that does not behave identically along all three axes when placed in a magnetic field. Examples are aromatic rings and triple bonds.

MAGNETIC SUSCEPTIBILITY

The ratio of the magnetization M to the magnetic field strength $\mathbf{H}$, multiplied by $[(\mu/\mu_0) - 1]$, where the permeability μ is the magnetic counterpart of the electrical permittivity, and μ_0 is the permeability of a vacuum. Magnetic susceptibility is measured with a magnetic balance, in which a sample suspended from a weighing balance is situated between the poles of a strong electromagnet. A diamagnetic substance will be pushed out of the field, whereas ferromagnetic and paramagnetic substances will be drawn into the field. Using a magnifying viewer, the experimentalist measures the force necessary to restore the sample to the same position as when the field is turned off. Paramagnetism is related to orbital angular momenta and electron spins within a chemical substance, and measurements of magnetic susceptibility can provide valuable mechanistic information related to ligand binding interactions with certain metal ions. *See Hemoglobin; Diamagnetic; Paramagnetic*

Selected entries from *Methods in Enzymology* [vol, page(s)]: Theoretical aspects, **76**, 354-356; diagmagnetic contribution, **76**, 358-359; experimental methods, **76**, 356-360; Faraday balance technique, **76**, 360-361; Gouy technique, **76**, 357, 360; instrumentation, **76**, 360-369; oxygen contribution, **76**, 360, 368; sources of experimental errors, **76**, 359-360; SQUID magnetometer use, **76**, 364-365; thermal equilibria, **76**, 358, 370; thermal expansion, **76**, 358.

MAGNETIZATION TRANSFER

An NMR method for determining the kinetics of chemical exchange by perturbing (*e.g.*, by saturation or inversion) the magnetization of nuclei in a particular site or sites and following the rate at which the magnetic equilibrium is restored.

MAGNETOTACTIC

Referring to the ability to become oriented with respect to a magnetic field. In addition to describing the properties of certain chemical substances, the term also applies to those organisms that can both sense the presence of, and orient relative to, the earth's magnetic field.

MALAPRADE REACTION
(Periodate Oxidation)

If a compound contains two hydroxyl groups, each attached to adjacent carbon atoms (*i.e.*, 1,2-glycols) or contains one hydroxyl group and one primary or secondary amino group (*i.e.*, β-amino alcohols), each attached to adjacent carbon atoms, then periodic acid (HIO_4) will cleave that carbon-carbon bond. If neither of those carbon atoms were terminal carbons, then each will be converted to the corresponding aldehyde or ketone, depending on what other groups are attached to those carbon atoms: *e.g.*, $RCH(OH)CH(OH)R' + HIO_4 \rightarrow RCHO + R'CHO + H_2O + HIO_3$ and $RCH(OH)CH(NH_2)R' + HIO_4 \rightarrow RCHO + R'CHO + NH_3 + HIO_3$. If there are three or more hydroxyl groups on three or more adjacent carbon atoms, the middle carbon atoms are converted to formic acid.

Periodic acid oxidation will also act on α-hydroxy aldehydes and ketones, α-diketones, α-keto aldehydes, and glyoxal. If the two neighboring hydroxyl groups are on an aromatic ring, the carbon-carbon bond is not cleave but the reactant is still oxidized. Thus, catechol is oxidized to the corresponding quinone.

Periodic acid reacts well in aqueous solution. Usually, if the reactant has to be run in organic solvents, lead tetraacetate is used as the reagent. Interestingly, periodic acid will not act on α-keto acids or α-hydroxy acids whereas lead tetraacetate will. The corresponding reactions are actually oxidative decarboxylations.

Periodic acid oxidation has proved to be a very useful tool in enzymology since a wide variety of biochemicals contain hydroxyl groups on adjacent carbon atoms. For example, periodate-oxidized ATP (also called adenosine 5'-triphosphate 2',3'-dialdehyde) has often been used as an alternative substrate or an irreversible inhibitor for a wide variety of ATP-utilizing enzymes. This compound, and many others, are now commercially available, even though they are readily synthesized: *e.g.*, periodic acid oxidized ADP, AMP, adenosine, P^1,P^5-di(adenosine-5')pentaphosphate, P^1,P^4-di(adenosine-5')tetraphosphate, GTP, GDP, GMP, guanosine, CTP, CDP, CMP, *etc.* In the case of the nucleosides, commercial sources also can supply the dialcohol form of the nucleoside: *i.e.*, the nucleoside has first been oxidized with periodic acid and then reduced to the dialcohol with borohydride.

L. Malaprade (1928) *Bull Soc. Chim. France* [4] **43**, 683.

L. Malaprade (1928) *Compte Rend.* **186**, 382.

K. W. Bentley (1973) in *Elucidation of Organic Structures by Physical and Chemical Methods*, pt. 2 (K. W. Bentley & G. W. Kirby, eds.), pp. 137-254 (vol. 4 of A. Weissberger, *Techniques in Chemistry*), Wiley, New York.

H. O. House (1972) *Modern Synthetic Reactions*, 2nd ed., Benjamin, New York

B. Sklarz (1967) *Q. Rev., Chem. Soc.* **21**, 3.

MALATE DEHYDROGENASE

1. Malate dehydrogenase [EC 1.1.1.37], also called malic dehydrogenase, catalyzes the reaction of (S)-malate with NAD^+ to produce oxaloacetate and NADH. Other 2-hydroxydicarboxylic acids can act as substrates as well. 2. Malate dehydrogenase ($NADP^+$) [EC 1.1.1.82] catalyzes the reaction of (S)-malate with $NADP^+$ to produce oxaloacetate and NADPH. This enzyme is activated by light. 3. Malate dehydrogenase (acceptor) [EC 1.1.99.16] catalyzes the FAD-dependent reaction of (S)-malate with an acceptor to produce oxaloacetate and the reduced acceptor. 4. Malate dehydrogenase (oxaloacetate-decarboxylating) [EC 1.1.1.38], also referred to as malic enzyme and pyruvic-malic carboxylase, catalyzes the reaction of (S)-malate with NAD^+ to produce pyruvate, carbon dioxide, and NADH. The enzyme will also decarboxylate added oxaloacetate. 5. Malate dehydrogenase (decarboxylating) [EC 1.1.1.39], also referred to as the malic enzyme, catalyzes the reaction of (S)-malate with NAD^+ to produce pyruvate, carbon dioxide, and NADH. This enzyme will not decarboxylate added oxaloacetate. 6. Malate dehydrogenase (oxaloacetate-decarboxylating, $NADP^+$) [EC 1.1.1.40], also referred to as malic enzyme and pyruvic-malic carboxylase, catalyzes the reaction of (S)-malate with $NADP^+$ to produce pyruvate, carbon dioxide, and NADPH. The enzyme will also decarboxylate added oxaloacetate. 7. D-Malate dehydrogenase (decarboxylating) [EC 1.1.1.83] catalyzes the reaction of (R)-malate with NAD^+ to produce pyruvate, carbon dioxide, and NADH. *See* Abortive Complexes

A. R. Clarke & T. R. Dafforn (1998) *Comprehensive Biological Catalysis: A Mechanistic Reference* **3**, 1.

J. V. Schloss & M. S. Hixon (1998) *Comprehensive Biological Catalysis: A Mechanistic Reference* **2**, 43.

H. Eisenberg, M. Mevarech & G. Zaccai (1992) *Adv. Protein Chem.* **43**, 1.

W. W. Cleland (1980) *Meth. Enzymol.* **64**, 104.

L. J. Banaszak & R. A. Bradshaw (1975) *The Enzymes*, 3rd ed., **11**, 369.

K. Dalziel (1975) *The Enzymes*, 3rd ed., **11**, 1.

M. G. Rossman, A. Liljas, C.-I. Brändén & L. J. Banaszak (1975) *The Enzymes*, 3rd ed., **11**, 61.

MALATE SYNTHASE

This enzyme [EC 4.1.3.2], also known as malate condensing enzyme and glyoxylate transacetylase, catalyzes the reaction of L-malate with coenzyme A to produce acetyl-CoA, water, and glyoxylate.

R. Kluger (1992) *The Enzymes*, 3rd ed., **20**, 271.

D. J. Creighton & N. S. R. K. Murthy (1990) *The Enzymes*, 3rd ed., **19**, 323.

M. J. P. Higgins, J. A. Kornblatt & H. Rudney (1972) *The Enzymes*, 3rd ed., **7**, 407.

MALEATE ISOMERASE

This enzyme [EC 5.2.1.1] catalyzes the interconversion of maleate and fumarate.

S. Seltzer (1972) *The Enzymes*, 3rd ed., **6**, 381.

MALEYLACETOACETATE ISOMERASE

This enzyme [EC 5.2.1.2] catalyzes the interconversion of 4-maleylacetoacetate and 4-fumarylacetoacetate. Maleylpyruvate can also act as a substrate.

J. V. Schloss & M. S. Hixon (1998) *Comprehensive Biological Catalysis: A Mechanistic Reference* **2**, 43.

K. T. Douglas (1987) *Adv. Enzymol.* **59**, 103.

S. Seltzer (1972) *The Enzymes*, 3rd ed., **6**, 381.

MALEYLPYRUVATE ISOMERASE

This enzyme [EC 5.2.1.4] catalyzes the interconversion of 3-maleylpyruvate and 3-fumarylpyruvate.

K. T. Douglas (1987) *Adv. Enzymol.* **59**, 103.

S. Seltzer (1972) *The Enzymes*, 3rd ed., **6**, 381.

MALONYL-CoA:[ACYL-CARRIER-PROTEIN] TRANSACYLASE

This enzyme [EC 2.3.1.39], also known as acyl-carrier protein malonyltransferase, catalyzes the reaction of malonyl-CoA with an acyl-carrier-protein to produce coenzyme A and a malonyl-[acyl-carrier-protein].

S. J. Wakil & J. K. Stoops (1983) *The Enzymes*, 3rd ed., **16**, 3.

P. R. Vagelos (1973) *The Enzymes*, 3rd ed., **8**, 155.

MALTASE

This enzyme [EC 3.2.1.20] catalyzes the hydrolysis of terminal, nonreducing 1,4-linked D-glucose residues with resulting release of D-glucose. This classification represents a set of enzymes with specificity directed mainly toward the exohydrolysis of 1,4-α-glucosidic linkages, as well as enzymes that hydrolyze oligosaccharides rapidly, relative to polysaccharides. The intestinal enzyme also hydrolyzes polysaccharides, thereby exhibiting a glucan

1,4-α-glucosidase [EC 3.2.1.3] activity. This enzyme will also catalyze the hydrolysis of 1,6-α-D-glucose linkages, albeit less effectively. **See also** α-*Glucosidase*

E. H. Van Beers, H. A. Büller, R. J. Grand, A. W. C. Einerhand & J. Dekker (1995) *Crit. Rev. Biochem. Mol. Biol.* **30**, 197.

G. Mooser (1992) *The Enzymes*, 3rd ed., **20**, 187.

M. Vihinen & P. Mäntsälä (1989) *Crit. Rev. Biochem. Mol. Biol.* **24**, 329.

MALTOSE PHOSPHORYLASE

This enzyme [EC 2.4.1.8] catalyzes the reaction of maltose with orthophosphate to produce D-glucose and β-D-glucose 1-phosphate.

J. J. Mieyal & R. H. Abeles (1972) *The Enzymes*, 3rd ed., **7**, 515.

MALYL-CoA LYASE

This enzyme [EC 4.1.3.24] catalyzes the conversion of (3S)-3-carboxy-3-hydroxypropanoyl-CoA to acetyl-CoA and glyoxylate.

P. M. Goodwin (1990) *Meth. Enzymol.* **188**, 361.

A. J. Hacking & J. R. Quayle (1990) *Meth. Enzymol.* **188**, 379.

P. J. Arps (1990) *Meth. Enzymol.* **188**, 386.

MAMILLARY MODEL

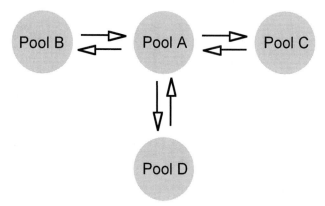

One of several general models for metabolite compartmentation in which a central compartment is directly linked to or feeds from (hence the name) other compartments that do not communicate with each other aside from their connection to the central pool. **See** *Catenary Model; Compartmental Analysis*

MANDELATE 4-MONOOXYGENASE

This iron-dependent enzyme [EC 1.14.16.6], also known as L-mandelate 4-hydroxylase, catalyzes the reaction of (S)-2-hydroxy-2-phenylacetate with tetrahydropteridine and dioxygen to produce (S)-4-hydroxymandelate, dihydropteridine, and water.

R. Y. Stanier (1955) *Meth. Enzymol.* **2**, 273.

MANDELATE RACEMASE

This enzyme [EC 5.1.2.2] catalyzes the interconversion of (S)-mandelate and (R)-mandelate.

M. E. Tanner & G. L. Kenyon (1998) *Comprehensive Biological Catalysis: A Mechanistic Reference* **2**, 7.

J. V. Schloss & M. S. Hixon (1998) *Comprehensive Biological Catalysis: A Mechanistic Reference* **2**, 43.

G. L. Kenyon & G. D. Hegeman (1979) *Adv. Enzymol.* **50**, 325.

E. Adams (1976) *Adv. Enzymol.* **44**, 69.

L. Glaser (1972) *The Enzymes*, 3rd ed., **6**, 355.

MANDELONITRILE LYASE

This enzyme [EC 4.1.2.10], also known as hydroxynitrile lyase and (R)-oxynitrilase, catalyzes the conversion of mandelonitrile to cyanide and benzaldehyde.

B. A. Palfey & V. Massey (1998) *Comprehensive Biological Catalysis: A Mechanistic Reference* **3**, 83.

MANGANESE PEROXIDASE

This heme-dependent enzyme [EC 1.11.1.13] catalyzes the reaction of two Mn^{2+} with two H^+ and hydrogen peroxide to generate two Mn^{3+} and two water. The enzyme participates in the oxidative degradation of lignin in white rot basidiomycetes.

H. B. Dunford (1998) *Comprehensive Biological Catalysis: A Mechanistic Reference* **3**, 195.

J. E. Penner-Hahn (1998) *Comprehensive Biological Catalysis: A Mechanistic Reference* **3**, 439.

β-MANNANASE

This enzyme [EC 3.2.1.78], also known as mannan endo-1,4-β-mannosidase and endo-1,4-mannanase, catalyzes the random hydrolysis of 1,4-β-D-mannosidic linkages in mannans, galactomannans, glucomannans, and galactoglucomannans.

M. W. Bauer, S. B. Halio & R. M. Kelly (1996) *Adv. Protein Chem.* **48**, 271.

MANNONATE DEHYDRATASE

This enzyme [EC 4.2.1.8], also known as D-mannonate hydro-lyase, catalyzes the conversion of D-mannonate to 2-dehydro-3-deoxy-D-gluconate and water.

J. Robert-Baudouy, J. Jimeno-Abendano & F. Stoeber (1982) *Meth. Enzymol.* **90**, 288.

W. A. Wood (1971) *The Enzymes*, 3rd ed., **5**, 573.

MANNOSE ISOMERASE

This enzyme [EC 5.3.1.7] catalyzes the interconversion of D-mannose and D-fructose. Both D-lyxose and D-rhamnose can act as substrates as well.

E. A. Noltmann (1972) *The Enzymes*, 3rd ed., **6**, 271.

MANNOSE-1-PHOSPHATE GUANYLYLTRANSFERASE

This enzyme [EC 2.7.7.13], also known as GTP-mannose-1-phosphate guanylyltransferase, catalyzes the reaction of GTP with α-D-mannose 1-phosphate to produce GDP-mannose and pyrophosphate (or, diphosphate). The bacterial enzyme can also use ITP and dGTP as the phosphoryl donors.

J. Preiss (1966) *Meth. Enzymol.* **8**, 271.

MANNOSE-6-PHOSPHATE ISOMERASE

This zinc-dependent enzyme [EC 5.3.1.8] (also known as phosphomannose isomerase, phosphohexoisomerase, and phosphohexomutase) catalyzes the interconversion of D-mannose 6-phosphate and D-fructose 6-phosphate.

K. U. Yüksel & R. W. Gracy (1991) *A Study of Enzymes* **2**, 457.
D. J. Creighton & N. S. R. K. Murthy (1990) *The Enzymes*, 3rd ed., **19**, 323.
I. A. Rose (1975) *Adv. Enzymol.* **43**, 491.
E. A. Noltmann (1972) *The Enzymes*, 3rd ed., **6**, 271.

MANNOSIDASES

α-Mannosidase [EC 3.2.1.24] catalyzes the hydrolysis of terminal, nonreducing α-D-mannose residues in α-D-mannosides. In addition, α-D-lyxosides and heptopyranosides (with the same configuration at C2, C3, and C4 as mannose) will serve as substrates as well.

β-Mannosidase [EC 3.2.1.25], also known as mannanase and mannase, catalyzes the hydrolysis of terminal, nonreducing β-D-mannose residues in β-D-mannosides.

M. W. Bauer, S. B. Halio & R. M. Kelly (1996) *Adv. Protein Chem.* **48**, 271.
S. C. Fry (1995) *Ann. Rev. Plant Physiol. Plant Mol. Biol.* **46**, 497.
E. Conzelmann & K. Sandhoff (1987) *Adv. Enzymol.* **60**, 89.
P. M. Dey & E. del Campillo (1984) *Adv. Enzymol.* **56**, 141.
K. A. Presper & E. C. Heath (1983) *The Enzymes*, 3rd ed., **16**, 449.
H. M. Flowers & N. Sharon (1979) *Adv. Enzymol.* **48**, 29.

MANOMETRIC ASSAY METHODS

For reactions in which one or more reactants or products is a gas, manometry (the measurement of pressure differences) can provide a convenient means for monitoring the course and kinetics of the reaction[1]. Thus, enzymes that can be assayed with this method include oxidases, urease, carbonic anhydrase, hydrogenase, and decarboxylases. For example, bacterial glutamate decarboxylase is readily assayed by utilizing a Warburg flask and measuring the volume of gas evolved at different times using a constant-pressure respirometer[2].

Note that manometric methods can also be used with reactions that generate acids[3]. If bicarbonate is a component of the reaction mixture, any acid formed will result in the generation of CO_2. Thus, even peptidases, proteinases, and esterases can be assayed manometrically[4,5]. Even some phosphotransferases have been assayed with this procedure[6]. ***See also*** *Eudiometer*

[1] M. Dixon (1951) *Manometric Methods*, Cambridge Univ. Press, Cambridge.
[2] M. L. Fonda (1985) *Meth. Enzymol.* **113**, 11.
[3] D. E. Green, D. M. Needham & J. G. Dewan (1937) *Biochem. J.* **31**, 2327.
[4] H. A. Krebs (1930) *Biochem. Z.* **220**, 283.
[5] R. Ammon (1934) *Pflug. Arch. Ges. Physiol.* **233**, 486.
[6] S. P. Colowick & H. M. Kalcker (1941) *J. Biol. Chem.* **137**, 789.

MAPPING SUBSTRATE INTERACTIONS USING KINETIC DATA

Although X-ray crystallography provides the most definitive information on the stereochemical considerations governing substrate binding, atomic structures of enzyme complexes often fail to reveal the nature of ligand-induced conformational changes that strongly influence the kinetics and thermodynamics of enzyme-ligand interactions. Nonetheless, one can frequently infer stereochemical relationships by comparing observed binding constants, inhibition constants, and in some cases Michaelis constants, for interactions of inhibitors and substrates with enzymes.

In the case of sheep brain glutamine synthetase, the goal was to discover why both D- and L-glutamate serve as substrates, and Gass and Meister[1] studied these substrates along with ten of their monomethyl derivatives. Of the latter, only three were active as substrates, and by making the assumption that amino acid substrates bind to glutamine synthetase in their most extended conformations, these investigators found that those hydrogens, whose methyl substituents yielded enzymatic activity, lie on one side of the molecule (*i.e.*, the side that resides behind the plane for the structures of L-glutamate and D-glutamate as shown in Fig. 1). This led to the hypothesis that the enzyme approaches its amino acid substrate from above the plane of the paper, thereby permitting methyl groups of the three active methyl derivatives to project backward and away from the enzyme's surface. To test this proposal, these investigators prepared L-*cis*-1-amino-1,3-dicarboxycyclohexane as a conformationally restricted L-glutamate analogue wherein the two backward-extending hydrogens of D- and L-glutamate are connected by the three additional methylene groups within the cyclohexane ring (see Fig. 1). This analogue proved to be an effective substrate for the sheep brain enzyme, thereby fixing the stereochemistry of enzyme-substrate complexation without any atomic-level information on the enzyme's structure. In retrospect, this conformationally restricted derivative would have also served as a valuable mechanistic probe of the mechanism of carboxyl group activation, because reaction of the analogue with ATP in an ammonia-free solution should form the corresponding acyl phosphate intermediate, and any intramolecular attack of the α-amino group to a pyroglutamate-like product (resembling a pyrrolidone carboxylate) should be far less likely. A similar argument applies to the use of β-glutamate which should form an acyl phosphate compound that resists intramolecular attack by the amino group.

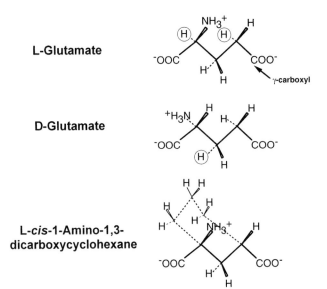

L-Glutamate

D-Glutamate

L-cis-1-Amino-1,3-dicarboxycyclohexane

Figure 1. Stereochemical representation of L-glutamate, D-glutamate, and L-cis-1-amino-1,3-dicarboxycyclohexane. Both D- and L-glutamate as well as α-aminoadipic acid are substrates for the enzyme, but only three of ten monomethyl derivatives of D- and L-glutamate are substrates. Assuming that the substrates are completely extended when bound to the enzyme, active substrates were obtained when the circled hydrogens are substituted by a methyl group. These hydrogens lie behind the plane of the paper, and they most probably face outward from the enzyme-bound substrate toward the bulk solvent. L-cis-1-Amino-1,3-dicarboxycyclohexane is an alternative substrate for glutamate in the glutamine synthetase reaction. Note that the cyclohexane ring is connected by three successive methylenes that bridge the two replaceable (circled) hydrogen atoms in L-glutamate.

Hexokinase provides another example of the stereochemical inferences that can be drawn on the basis of kinetic studies with alternative substrates and competi-

tive inhibitors[2]. Although the hexokinases show broad specificity with regard to their hexose substrates, they often differ with respect to the relative affinities and relative rates of phosphorylation of the sugar substrate. To rationalize the different affinities and relative rates of hexose phosphorylation, Robert Crane presented an elaborate model for the interaction of hexokinase with its sugar substrates. He proposed that (a) β-D-glucopyranose as the C1 conformer (chair equatorial) combines with hexokinase through the interaction of specific binding sites on the phosphotransferase with the hydroxyl groups located on carbon atoms 1, 3, and 6 of the sugar molecule; (b) this enzyme-substrate Michaelis complex then undergoes a contraction (or induced fit) to yield the C1 conformer (chair axial); and (c) only with this latter enzyme-substrate complex does phosphorylation occur. The chief reason for his postulating a substrate-induced conformational change was to rationalize the inability of hexokinase to act on compounds like N-acetylglucosamine which do act as potent competitive inhibitors with respect to the sugar substrate. The presence of such bulky substituents as the N-acetyl group, Crane reasoned, should stabilize the hexose in the C1 conformation, which is inactive.

While this model explained the action of the brain enzyme on a number of hexose substrates and nonsubstrate inhibitory analogs, the mode had its weaknesses. It assumed that the other conformations of a hexose that are in equilibrium with the "active" conformer act as competitive inhibitors relative to this conformer. One cannot evaluate the effect of a competitive inhibitor which is present in a constant proportion relative to the active substrate by initial velocity measurements. Moreover, the use of apparent Michaelis constants may not provide accurate estimates of affinity, which is more directly related to a dissociation constant. The chief limitation of the model, however, is that an equally great number of experimental facts can be satisfactorily explained in terms of a simpler scheme involving the binding and phosphorylation of the C1 conformer. Furthermore, one can understand more directly how the enzyme can phosphorylate glucopyranose and fructofuranose equally well.

Purich et al.[2] offered the following scheme which illustrates the structural similarities between C1 β-D-glucopyranose and the corresponding conformer of β-D-fructofuranose (Fig. 2):

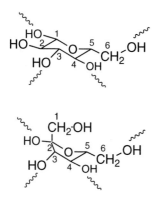

Hexokinase Interactions with Glucose (top) and Fructose (bottom)

Figure 2. Hypothetical scheme for the selectivity-conferring interactions (indicated as wavy lines) of hexokinase with keto- and aldhexoses. This model accounts for all known specificity and selectivity of yeast hexokinase toward hexose substrates and pentose ligands.

They noted that the hydroxyl substituents on carbon atoms 1, 3, 4, and 6 of glucose are oriented in approximately the same way as the hydroxyl groups located on carbon atoms 2, 3, 4, and 6 of fructose. By hypothesizing that these positions are the specificity-imparting groups on the hexose (as indicated by the wavy lines), they could rationalize the relative unimportance of the hydroxyl substituent at carbon atom 2 of glucose, mannose, and 2-deoxyglucose. The relative importance of the hydroxyl substituent at carbon 1 of glucose can be further appreciated by the fact that 1,5-anhydro-D-glucitol and 1,5-anhydro-D-mannitol are bound less tightly than their respective hexoses. That these anhydro sugar alcohols can be phosphorylated, however, indicates that the presence of a hydroxyl group in this position is not required. On the other hand, the 4 position of glucose appears essential in that galactose is bound very poorly and is not phosphorylated. It would be of interest to know whether 4-deoxyglucose can be bound and phosphorylated; this information would permit one to determine whether the inactivity of galactose is due to the axial orientation of the hydroxyl group or to the lack of a hydroxyl group in the equitorial orientation for binding. Finally, the inability of brain hexokinase to act on *N*-acetylglucosamine can be attributed to interaction of the bulky *N*-acetyl substituent with the groups on the enzyme responsible for recognizing the hydroxyl group at carbon atom 1 or 3 of glucose.

[1]J. D. Gass & A. Meister (1970) *Biochemistry* **9**, 1380.
[2]D. L. Purich, F. B. Rudolph & H. J. Fromm (1973) *Advan. Enzymol.* **39**, 249.

MARCUS EQUATION

An equation relating the Gibbs energy of activation ($\Delta G^{\ddagger}$) with the driving force of the reaction ($\Delta_r G^{\ominus\prime}$): $\Delta G^{\ddagger} = (\lambda/4)(1 + \Delta_r G^{\ominus\prime}/\lambda)^2$ where λ is the reorganization energy, $\lambda/4$ is the intrinsic barrier of the reaction, and $\Delta_r G^{\ominus\prime}$ is the standard Gibbs free energy for the reaction corrected for the electrostatic work required to bring the reactants together. This equation has been applied to electron transfer reactions as well as atom and group transfer reactions.

R. A. Marcus (1964) *Ann. Rev. Phys. Chem.* **15**, 155.
W. J. Albery (1980) *Ann. Rev. Phys. Chem.* **31**, 227.

MARCUS RATE THEORY

The Marcus rate theory[1,2] provides a means for addressing issues of proton transfer reactions, particularly in the analysis of Brønsted plots. The slope β in a Brønsted plot is often used to assist in characterizing a reaction mechanism and to estimate the degree of charge transfer between reactant(s) and product(s) in the transition state. In the Marcus rate theory, this slope is $d\Delta G^{\ddagger}/d\Delta G_R^{\ominus}$ where $\Delta G^{\ddagger}$ is the Gibbs activation energy for the proton transfer and $\Delta G_R^{\ominus}$ is the standard Gibbs free energy of the reaction with the required active site conformation. The Marcus rate theory has been applied to carbonic anhydrase[3-5].

[1]R. A. Marcus (1968) *J. Phys. Chem.* **72**, 891.
[2]A. J. Kresge (1975) *Acc. Chem Res.* **8**, 354.
[3]R. S. Rowlett & D. N. Silverman (1982) *J. Amer. Chem. Soc.* **104**, 6737.
[4]D. N. Silverman, C. Tu, X. Chen, S. M. Tanhauser, A. J. Kresge & P. J. Laipis (1993) *Biochemistry* **32**, 10757.
[5]D. N. Silverman (1995) *Meth. Enzymol.* **249**, 479.

MARKOV CHAIN

Any process comprising a sequence of events (or states), such that the probability of any one state only depends on the immediately preceding event. Markov (or Markoff) chains are frequently used to describe stochastic processes, such as Brownian motion, the random opening/closing of ion channels, and population genetics. This treatment affords the opportunity to treat chemical events, such as self-assembly, as discrete events rather than that achieved using continuous functions which might result in the absurdity of loss or gain of a fraction of one subunit in a depolymerization or polymerization event.

W. R. Gilks (1996) *Markov Chain Monto Carlo in Practice*, Chapman & Hall, New York.

A. Borodin & P. Salminen (1996) *Handbook of Brownian Motion — Facts and Formulae*, Springer, Berlin.

MARQUARDT ALGORITHM

An algorithm used in the statistical fitting of cooperative binding data via a nonlinear least-squares analysis. This algorithm is an efficient method for analyzing nonlinear binding or nonlinear kinetic functions. It is applicable for any type of saturation function[1-3]. In this process, a parameter, λ, is defined that controls "the interpolation between the extremes of the analytical or the gradient search by increasing the diagonal of the curvature matrix[3]." **See** *Computer Algorithms & Software*

[1]P. R. Bevington (1969) *Data Reduction and Error Analysis for the Physical Sciences*, McGraw-Hill, New York.

[2]P. J. Kasvinsky, N. B. Madsen, R. J. Fletterick & J. Sygusch (1978) *J. Biol. Chem.* **253**, 1290.

[3]K. E. Neet (1980) *Meth. Enzymol.* **64**, 139.

MASS-ACTION RATIO

The ratio of the product of the concentrations (or activities) of all the products of a reaction to the product of the concentration (or activities) of all the reactants (or substrates). This ratio, often symbolized by Γ, will change as the reaction progresses until equilibrium is reached (*i.e.*, at $t = \infty$), at which point $\Gamma = K_{eq}$.

MASSIEU FUNCTION

A thermodynamic function, symbolized by J, equal to the negative of the Helmholtz energy divided by the absolute temperature: thus, $J = -A/T$. The SI units are joules per kelvin. **See also** *Planck Function; Helmholtz Energy*

MASS SPECTROMETRY

An analytical technique that exploits magnetic fields to analyze molecules on the basis of their mass and electrical properties to determine (a) qualitative separation of mixtures of inorganic and organic species, (b) quantitative determination of the amount of substance, (c) isotopic abundance of atoms in simple and complex molecules, and (d) structures of biological and other organic molecules by use of special fragmentation methods.

General Features of Mass Spectrometers. To separate molecules into discrete chemical species, mass spectrometers utilize a sample inlet device, an ion source for imparting charge by bombardment with electrons, chemical ions, molecules, or photons. These systems are maintained at high vacuum so that the input sample is volatilized to a gaseous state and ported into the mass analyzer and detector. Mass analyzers differentiate molecules on the basis of their mass-to-charge (m/z) ratio by using magnetic sector analyzers, double-focusing analyzers, quadrupole mass filters, ion trapped analyzers, or time-of-flight analyzers. The power of mass spectrometry to separate chemical species is expressed in terms of resolution, which can be defined as $R = m/\Delta m$, where m is the mass of the first peak and Δm is the difference in mass between two successive peaks. Typical commercial spectrometers have resolutions of up to 500,000.

Types of Ion Sources. The versatility of mass spectrometry relates to the ability to generate ionized species using a variety of ion sources. Most frequently one utilizes electron ionization (EI) in which molecules in the gas phase are ionized by energetic electrons typically emitted from a tungsten or rhenium and accelerated by an electric potential of around 70 volts. The second most popular method for generating ionized species is chemical ionization (CI), a technique that relies on collision of analyte molecules with ions that have been previously produced by bombarding a reagent gas with electrons. Here, a common reagent gas is methane which upon electron bombardment can generate the following reactive species: CH_4^+, CH_3^+, and CH_2^+. High-molecular-weight analytes and polyelectrolytes cannot be readily introduced in the gaseous state, and in such cases, one can use fast atom bombardment (FAB) to produce positive and negative analyte ions that are desorbed from the surface of the source. In the case of proteins, matrix-assisted laser desorption ionization (MALDI) is utilized to introduce the macromolecule into a mass analyzer, typically of the time-of-flight type. Another ion source technique is thermospray (or electrospray) ionization (ES) which imparts positive charges to a nebularized analyte solution.

Selected entries from *Methods in Enzymology* [vol, page(s)]:
General Techniques: Methods of ionization, **193**, 3; mass analyzer, **193**, 37; detectors, **193**, 61; selected-ion measurements, **193**, 86; liquid chromatography-mass spectrometry, **193**, 107; tandem MS: multisector magnetic instruments, **193**, 131; tandem MS:

MATRIX ISOLATION

The isolation of a reactive or unstable molecular entity by dilution in an inert matrix.

MAXIMUM LIKELIHOOD METHOD

Maximum (maximized) likelihood is a statistical term that refers to the probability of randomly drawing a particular sample from a population, maximized over the possible values of the population parameters.

MAXWELL

A unit, symbolized by Mx, for magnetic flux. It is the flux through a square centimeter normal to a field of one gauss. One maxwell is equal to 10^{-8} weber.

MAXWELL'S THERMODYNAMIC RELATIONS

Fundamental relationships involving first partial differentials of absolute temperature (T), pressure (P), volume (V), and entropy (S) for a homogeneous system with a constant, given mass.

$$(\partial T/\partial V)_s = -(\partial P/\partial S)_v$$

$$(\partial T/\partial P)_s = (\partial V/\partial S)_p$$

$$(\partial P/\partial T)_v = (\partial S/\partial V)_T$$

$$(\partial V/\partial T)_p = -(\partial S/\partial P)_T$$

In this set of four fundamental thermodynamic expressions, one should note that only entropy cannot be directly determined experimentally, whereas P, V, and T are all measurable properties.

MEAN TRANSIT TIME

The average period of time during which a metabolite (or drug) remains in the volume of distribution; also referred to as the mean residence time. *See* Compartmental Analysis

MECHANISM-BASED INHIBITOR

A molecular entity that (a) resembles the substrate (or product) of an enzymic reaction, (b) is chemically unreactive in the absence of the enzyme, (c) binds to the enzyme; (d) undergoes activation by the enzyme to yield a chemically reactive species, (e) can be converted to product, and (f) occasionally inactivates the enzyme prior to its release from that protein.

Suicide inhibition follows the general scheme[1,2]:

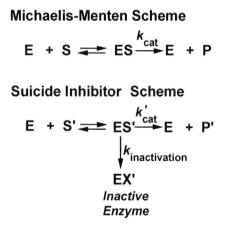

Michaelis-Menten Scheme

$$E + S \rightleftharpoons ES \xrightarrow{k_{cat}} E + P$$

Suicide Inhibitor Scheme

$$E + S' \rightleftharpoons ES' \xrightarrow{k'_{cat}} E + P'$$
$$\downarrow k_{inactivation}$$
$$EX'$$
Inactive Enzyme

where enzyme and inhibitor combine to form an EI complex which is interconverted to EX, the chemically activated species that either forms product P or covalently inactivates the enzyme. The kinetic behavior of a suicide inhibitor is shown below. The amount of product formed before complete inactivation is a direct measure of the relative magnitudes of rate constants for the product-forming and enzyme-inactivating steps.

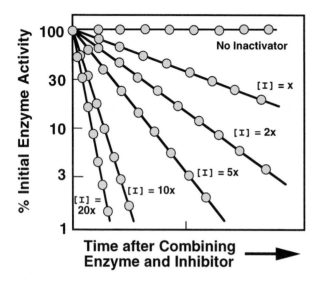

The synthesis of suicide inhibitors and the kinetics of enzyme inactivation have greatly advanced the study of enzyme mechanisms as well as the design of drugs and antimetabolites.

[1]R. B. Silverman (1988) *Mechanism-Based Enzyme Activation: Chemistry and Enzymology*, vols. **1** and **2**, CRC Press, Boca Raton.
[2]R. B. Silverman (1995) *Meth. Enzymol.* **249**, 240.

Selected entries from *Methods in Enzymology* [vol, page(s)]: Applications, **249**, 263-283; basic kinetics, **249**, 243-245; biphasic kinetics, **249**, 253; competitive inhibitor protection, **249**, 247, 256; criteria for, **249**, 247-249; definition, **249**, 240-242; in drug design, **249**, 246-247; experimental protocols for, **249**, 249-263; inactivation prior to release of active species, **249**, 248-249, 259-260; involvement of catalytic step, **249**, 248, 259; irreversibility, **249**, 248, 256-257; kinetic constants, determination, **249**, 254-256; non-pseudo-first-order kinetics, **249**, 251-253; partition ratio, **249**, 251; partition ratio determination, **249**, 260-263; pseudo-first-order kinetics, **249**, 250-251; saturation, **249**, 247, 254-256; stoichiometry, **249**, 248, 257-259; in study of enzyme mechanisms, **249**, 245-246; substrate protection, **249**, 247, 256; time dependence of inactivation, **249**, 247, 249-253; uses, **249**, 245-247.

MEDIATED AMPEROMETRIC BIOSENSOR

Any electrochemical device using a low molecular weight redox couple to shuttle electrons from the redox center of an enzyme to the surface of an indicator electrode, thereby increasing the effectiveness of amperometry in the detection of a substrate for the particular enzyme. The internal cavities of six-, seven-, and eight-membered cyclodextrins are trapezoids of revolution with larger "open mouths" dimensions (*i.e.*, respective diameters of 0.49, 0.62, and 0.79 nm and depths typically of 0.8 nm) suitable for the binding of many redox-active substances as host-guest complexes. *See Biosensor*

J. H. T. Luong, R. S. Brown, K. B. Male, M. V. Cattaneo & S. Zhao (1995) *Trends in Biotechnology* **13**, 457.

MEDIUM

1. The phase or environment in which one or more molecular entities and/or their reactions are being studied. 2. The solvent holding one or more solutes in solution or suspension. 3. Culture medium; a substance or mixture used for cultivation, growth, storage, *etc.*, of microorganisms.

MEERWEIN-PONDORFF-VERLEY REDUCTION

The reversible reduction of aldehydes or ketones to the corresponding alcohols using aluminum alkoxides such as aluminum isopropoxide ($Al[OCH(CH_3)_2]_3$), the reducing alcohol usually being isopropyl alcohol (*i.e.*, $RCOR' + CH_3CH(OH)CH_3 \rightarrow RCH(OH)R' + CH_3COCH_3$)[1-3]. The reaction is facilitated by the removal of acetone by distillation[4-6]. This reaction has served as a model for metal-ion-dependent hydride transfer reactions catalyzed by certain enzymes. The reverse reaction is known as the Oppenauer oxidation[7,8].

[1]H. Meerwein & R. Schmidt (1925) *Ann.* **444**, 221.
[2]W. Ponndorf (1926) *Angew. Chem.* **39**, 138.
[3]A. Verley (1925) *Bull. Soc. Chim. France* **37**, 537 and 871.
[4]J. Hutton (1979) *Synth. Commun.* **9**, 483.
[5]J. L. Namy, J. Souppe, J. Collin & H. B. Kagen (1984) *J. Org. Chem.* **49**, 2045.
[6]T. Okano, M. Matsuoka, H. Konishi & J. Kiji (1987) *Chem. Lett.* **1987**, 181.
[7]R. V. Oppenauer (1937) *Rec. Trav. Chim.* **56**, 137.
[8]C. Djerassi (1951) *Org. React.* **6**, 207.

MEGA-

A prefix, symbolized by M, used in multiples of units and corresponding to a value of 10^6.

MEISENHEIMER ADDUCT

A cyclohexadienyl Lewis adduct or salt formed by the reaction of a Lewis base with an aromatic compound[1]. Such an adduct is apparently formed from the reaction of OH^- with 4-(*N*-2-aminoethyl-2'-pyridyl disulfide)-7-nitrobenzo-2-oxa-1,3-diazole (2PROD)[2]. 2PROD is a two-protonic-state electrophile used as a probe for enzyme active site nucleophiles and as a fluorescent reporter-group delivery vehicle[3]. **See** *Two-Protonic State Electrophiles*

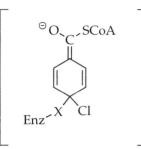

Yang *et al.*[4] recently reported on the mechanism of 4-chlorobenzoyl coenzyme A dehalogenase, an enzyme that catalyzes the hydrolytic dehalogenation of 4-chlorobenzoyl coenzyme A (4-CBA-CoA) to form 4-hydroxybenzoyl coenzyme A (4-HBA-CoA). The mechanism involves attack of an active site carboxylate at C4 of the substrate benzoyl ring to form a Meisenheimer complex (shown above). Loss of chloride ion from this intermediate then forms an arylated enzyme intermediate that is hydrolyzed to free enzyme plus 4-HBA-CoA by the addition of water at the acyl carbon. In later work, Taylor *et al.*[5] examined the activation of the 4-CBA-CoA toward nucleophilic attack by the active site carboxylate group.

[1]J. Meisenheimer (1902) *Liebig's Ann. Chem.* **323**, 205.
[2]B. S. Baines, G. Allen & K. Brocklehurst (1977) *Biochem. J.* **163**, 189.
[3]K. Brocklehurst (1982) *Meth. Enzymol.* **87**, 427.
[4]G. Yang, P.-H. Liang & D. Dunaway-Mariano (1994) *Biochemistry* **33**, 8527.
[5]K. L. Taylor, R. Q. Liu, P. H. Liang, J. Price, D. Dunaway-Mariano, P. J. Tonge, J. Clarkson & P. R. Carey (1995) *Biochemistry* **34**, 13881.

MELILOTATE HYDROXYLASE

This FAD-dependent enzyme [EC 1.14.13.4], also known as melilotate 3-monooxygenase and 2-hydroxyphenyl-propionate hydroxylase, catalyzes the reaction of 3-(2-hydroxyphenyl)propanoate with NADH and dioxygen to produce 3-(2,3-dihydroxyphenyl)propanoate, NAD[+], and water.

B. A. Palfey & V. Massey (1998) *Comprehensive Biological Catalysis: A Mechanistic Reference* **3**, 83.
V. Massey & P. Hemmerich (1975) *The Enzymes*, 3rd ed., **12**, 191.

MELTING

1. The process by which a solid converts into a liquid, usually as a result of an increase in temperature. [The melting point is the temperature at which the solid and liquid phases are in equilibrium with each other at a defined pressure (usually one atmosphere)]. While the melting point should have the same value as the freezing point of a substance, "freezing" is defined as the conversion of a liquid to a solid. 2. A term typically used in biochemistry to describe loss of structure attending thermal denaturation (most frequently applied to polynucleotides such as DNA). This denaturation is a cooperative process and the kinetics of the process depends on a number of factors: for example, the GC content, the ionic strength, the nature of the solvent, the pH, *etc.* The melting temperature, T_m, is the midpoint of a melting curve. Crothers *et al.*[1] have studied the kinetics and mechanism of the thermal melting of *E. coli* tRNA$_f^{Met}$.

[1]D. M. Crothers, P. E. Cole, C. W. Hilbers & R. G. Shulman (1974) *J. Mol. Biol.* **87**, 63.

MEMBRANE POTENTIAL

The voltage difference, sometimes referred to as the transmembrane potential or the membrane electric potential, that is due to the difference of concentrations of ions on either side of the membrane. Typically, such potentials are measured in millivolts (mV) and usually have values between −30 and −70 mV. In most cells, the interior of a cell is negative with respect to the exterior.

For the simplified case of only a single permeant ion, the equilibrium or Nernst potential defines the condition at which there is no net change in ion concentration. Under this condition of equilibrium, the electrochemical potential μ_j for moving the ion j^{z+} across the membrane will be zero, and

$$\mu_j = RT \ln[j]_{\text{exterior}}/[j]_{\text{interior}} - z\mathscr{F}E$$

where R is the gas constant, T is the absolute temperature, $\mathscr{F}$ is Faraday's constant (equal to 96,500 coulombs/mol), z is the ionic charge, and E is the transmembrane potential. Thus,

$$E_j = (RT/z\mathscr{F})\ln[j]_{\text{exterior}}/[j]_{\text{interior}}$$
$$= (58\,\text{mV}/z)\log_{10}[j]_{\text{exterior}}/[j]_{\text{interior}}$$

Thus, a 10:1 transmembrane gradient of a single monovalent ion, say potassium, will generate a membrane potential of 58 mV. **See** *Resting Potential; Action Potential; Depolarization; Threshold Potential; Nernst Equation; Goldman Equation; Patch-Clamp Technique*

MEMBRANE TRANSPORT

Transporters, particularly those carrying nonlipophilic species across biomembranes or model membranes, can be regarded as vectorial catalysts[1] (and are also called carriers, translocators, permeases, pumps, and ports [*e.g.*, symports and antiports]). Many specialized approaches and techniques have been developed to characterize such systems. This is reflected by the fact that there are currently twenty-three volumes in the *Methods in Enzymology* series (vols. 21, 22, 52-56, 81, 88, 96-98, 125-127, 156-157, 171-174, and 191-192) devoted to biomembranes and their constituent proteins. Chapters in each of these volumes will be of interest to those investigating transport kinetics. Other volumes are devoted to ion channels (207), membrane fusion techniques (220 and 221), lipids (14, 35, 71, and 72), plant cell membranes (148), and a volume on the reconstitution of intracellular transport (219). ***See*** *Ion Pumps*

BASIC TRANSPORTER KINETICS. In the kinetic identification and characterization of a transport system [including a demonstration of the ion dependency (*e.g.*, Na⁺-dependency) or lack of dependency], it is essential to demonstrate that saturation of the transport of a particular substrate occurs[2-4]. Simple diffusion of a substrate across a membrane is nonsaturable. (Note that for many substrates, a slow rate of diffusion across a membrane may take place simultaneously with the carrier-linked transport system, and appropriate corrections for basal rates of diffusion are necessary.) In addition, mediated-transport systems are typically very specific with respect to the nature and structure of the transporting molecular entity. An excellent probe of carrier-mediated transport is the stimulation observed of substrate uptake by the presence of a "countersubstrate" on the other side of the membrane. This acceleration by an internal substrate represents a facultative counterexchange and is a stringent criteria of carrier-mediated transport[4].

The rate equation for a single class of transporters acting on a metabolite is analogous to the Michaelis-Menten equation, except that one must also account for the effects of passive diffusion:

$$\nu = (\nu_{\text{carrier}} + \nu_{\text{diffusion}}) = \left\{ \frac{V_{\text{max}}[\text{S}]}{[\text{S}] + K_{\text{m}}} + k_{\text{D}}[\text{S}] \right\}$$

The contribution of diffusion to the rate does not follow rate saturation. Therefore, as one proceeds toward carrier saturation with substrate S, the rate will not simply reach a plateau, as is most often the case for one-substrate enzymes operating on extremely slow uncatalyzed reactions. Instead, a plot of *v versus* [S] will continue to rise with a constant slope, equal to k_{D}. It is usually a simple matter to extrapolate this noncarrier-mediated contribution back to each time-point for which a measurement was obtained and to obtain the corrected carrier-mediated rate by subtraction of the diffusional contribution.

Net carrier transport can be examined whenever a transportable substrate is present on both sides of a membrane. Again for the case of a single class of carriers, the inward transport rate and outward transport rates show rate-saturation behavior:

$$\nu_{\text{inward}} = \frac{V_{\text{max}}^{\text{out}}[\text{S}_{\text{outside}}]}{[\text{S}_{\text{outside}}] + K_{\text{m}}^{\text{out}}} \quad \text{and} \quad \nu_{\text{outward}} = \frac{V_{\text{max}}^{\text{in}}[\text{S}_{\text{inside}}]}{[\text{S}_{\text{inside}}] + K_{\text{m}}^{\text{in}}}$$

and the expression for net transfer, $\nu_{\text{net}} = (\nu_{\text{outside}} - \nu_{\text{inside}})_{\text{outward}}$, including the contribution from passive diffusion, can be stated as follows:

$$v_{net} = \left\{ \frac{V_{max}^{out}[S_{outside}]}{[S_{outside}] + K_m^{out}} \right\} - \left\{ \frac{V_{max}^{in}[S_{inside}]}{[S_{inside}] + K_m^{in}} \right\}$$
$$+ k_D([S_{outside}] - [S_{inside}])$$

Inhibition of carrier-mediated transport usually obeys the same rate laws describing competitive inhibition and noncompetitive inhbition for one-substrate enzymes.

If more than one class of carriers is mediating the transport of a common substrate S, then the rate law for transport becomes:

$$v = (v_1 + v_2) = \left\{ \frac{V_{max,1}[S]}{[S] + K_{m,1}} + \frac{V_{max,2}[S]}{[S] + K_{m,2}} \right\}$$

KINETIC CHARACTERIZATION OF TRANSPORTERS. Stein[3,5] has provided a useful outline of kinetic approaches to characterize transport systems. In each of the approaches, the investigator must accurately determine the amount of substrate transported at a defined point of time after the substrate and the membrane system have contacted each other. The kinetic approaches include:

(1) Equilibrium Exchange Methods: The membrane-enclosed container (*e.g.*, a cell, an organelle, or a vesicle) is preloaded with substrate at a particular concentration and then placed in a medium containing the substrate at the same concentration. (Note: One of the two "pools" of substrate is isotopically labeled.) The investigator then measures the amount of radiolabel transferred from vesicle to medium or *vice versa* as a function of time. A plot of the logarithm of the fraction of label exchange will be linear provided the system is homogeneous. It should also be noted that the same rate constant for equilibrium exchange should be obtained for both efflux and influx experiments. This should be demonstrated for each transport system studied. Finally, the rate constant for equilibrium exchange should be determined over a wide range of substrate concentrations. The dependency of the rate constant (or lack of) on the substrate concentration will reflect the type(s) of the transport system(s) present.

(2) Zero-Trans Procedures: These methods involve measurements of substrate transport into or out of a container (*i.e.*, a cell, organelle, vesicle, *etc.*) in which one side of the membrane (the trans side) is free of the

presence of the substrate. The other side of the membrane (termed the cis side) is preloaded with different concentrations of the substrate and the rate of transport is measured. This technique has a number of experimental difficulties. For example: measurements must be made rapidly; otherwise, integrated equations are required to account for the time evolution of the process. Thus, measurements must commence as soon as the cells are loaded. Because transport begins immediately, efflux and influx studies often yield different results as a consequence of the inherent asymmetry of the transporter. Transport of the substrate to the trans side is also often accompanied by osmotic and/or pH changes. When such experiments are done and these technical difficulties are properly addressed, the velocity of zero-trans transport can be plotted as a function of substrate concentration to obtain valid kinetic parameters.

(3) Infinite Trans Procedures: In this experimental methodology, the membrane-enclosed container (*e.g.*, cells, organelles, vesicles, *etc.*) is preloaded with a high concentration of unlabeled substrate. They are then placed in a medium containing the labeled substrate and the rate of influx is measured. The experiment is then repeated at different concentrations of the substrate in the surrounding medium. Note that the concentration of the substrate within the cell must be very high (in effect, infinite). This requires some awareness of the approximate magnitude of the K_m value of the substrate, and preliminary experiments should be undertaken. An investigator may also wish to demonstrate that an increase in the concentration of the internalized unlabeled substrate does not result in changes in the value of the observed kinetic parameters. While one could also do an infinite trans experiment for efflux transport (in which the medium contains a high concentration of substrate), this is difficult on technical grounds.

(4) Infinite *Cis* Methods: With this approach, membrane-enclosed containers (once again, cells, organelles, vesicles, *etc.*) are preloaded with high concentrations of unlabeled substrate. The cells are then placed in a medium containing various concentrations of unlabeled substrate (including no substrate and a saturating level of substrate). The investigator then measures the net loss of substrate from the container.

(5) Counterflow Experiments: Cells are once again preloaded with a high concentration of unlabeled substrate. The cells are then placed in a medium containing a low concentration of labeled substrate or a low concentration of an alternative substrate which utilizes the same transport system. The investigator then measures the rate of entry of the labeled substrate or of the alternative substrate.

EFFECT OF ELECTRICAL POTENTIAL ON TRANSPORT. The Gibbs free energy associated with membrane transport can be expressed as $\Delta G = RT \ln(C_{internal}/C_{external})$. Nonetheless, the actual concentrations achieved depend on the generation of transmembrane electrical potential, itself written as $\Delta G = z\mathscr{F} E$, where z is the valence, $\mathscr{F}$ is Faraday constant (23.5 kcal/$V \cdot$mol). E is the transmembrane potential difference, expressed in volts. Thus the minimal amount of free energy change associated with the movement of 1 mol solute can be written as:

$$\Delta G = RT \ln(C_{internal}/C_{external}) + z\mathscr{F} E$$

For a more detailed discussion of this topic, **See** *Membrane Potential*

[1]G. Sachs & S. Fleischer (1989) *Meth. Enzymol.* **171**, 3.
[2]T. Rosenberg & W. Wilbrandt (1957) *J. Gen. Physiol.* **41**, 289.
[3]W. D. Stein (1986) *Transport and Diffusion Across Cell Membranes*, Academic Press, Orlando, Florida.
[4]M. Klingenberg (1989) *Meth. Enzymol.* **171**, 12.
[5]W. D. Stein (1989) *Meth. Enzymol.* **171**, 23.

Selected entries from *Methods in Enzymology* [vol, page(s)]:
Transport Theory: An overview of transport machinery, **171**, 3; survey of carrier methodology: strategy for identification, isolation, and characterization of transport systems, **171**, 12; kinetics of transport: analyzing, testing, and characterizing models using kinetic approaches, **171**, 23; kinetics of ion movement mediated by carriers and channels, **171**, 62; inhibition kinetics of carrier systems, **171**, 113; design of simple devices to measure solute fluxes and binding in monolayer cell cultures, **171**, 133; utilization of binding energy and coupling rules for active transport and other coupled vectorial processes, **171**, 145; generation of steady-state rate equations for enzyme and carrier-transport mechanisms: a microcomputer program, **171**, 164.

Model Membranes and Their Characteristics: Liposome preparation and size characterization, **171**, 193; preparation of microcapsules from human erythrocytes: use in transport experiments of glutathione and its S-conjugate, **171**, 217; planar lipid-protein membranes: strategies of formation and of detecting dependencies of ion transport functions on membrane conditions, **171**, 225; spontaneous insertion of integral membrane proteins into preformed unilamellar vesicles, **171**, 253; gentle and fast transmembrane reconstitution of membrane proteins, **171**, 265; ion carriers in planar bilayers: relaxation techniques and noise analysis, **171**, 274; ion interactions at membranous polypeptide sites using nuclear magnetic resonance: determining rate and binding

constants and site locations, **171**, 286; measuring electrostatic potentials adjacent to membranes, **171**, 342; determination of surface potential of biological membranes, **171**, 364; lipid coumarin dye as a probe of interfacial electrical potential in biomembranes, **171**, 376; modulation of membrane protein function by surface potential, **171**, 387.

Isolation of Cells for Transport Studies: Use of nonequilibrium thermodynamics in the analysis of transport: general flow-force relationships and the linear domain, **171**, 397; cell isolation techniques: use of enzymes and chelators, **171**, 444; cell separation by gradient centrifugation methods, **171**, 462; cell separation by elutriation: major and minor cell types from complex tissues, **171**, 482; cell separation using velocity sedimentation at unit gravity and buoyant density centrifugation, **171**, 497; separation of cell populations by free-flow electrophoresis, **171**, 513; separation of cells and cell organelles by partition in aqueous polymer two-phase systems, **171**, 532; flow cytometry: rapid isolation and analysis of single cells, **171**, 549; use of cell-specific monoclonal antibodies to isolate renal epithelia, **171**, 581; pancreatic acini as second messenger models in exocrine secretion, **171**, 590; ascites cell preparation: strains, caveats, **171**, 593.

Polar Cell Systems for Membrane Transport Studies: Direct current electrical measurement in epithelia: steady-state and transient analysis, **171**, 607; impedance analysis in tight epithelia, **171**, 628; electrical impedance analysis of leaky epithelia: theory, techniques, and leak artifact problems, **171**, 642; patch-clamp experiments in epithelia: activation by hormones or neurotransmitters, **171**, 663; ionic permeation mechanisms in epithelia: bi-ionic potentials, dilution potentials, conductances, and streaming potentials, **171**, 678; use of ionophores in epithelia: characterizing membrane properties, **171**, 715; cultures as epithelial models: porous-bottom culture dishes for studying transport and differentiation, **171**, 736; volume regulation in epithelia: experimental approaches, **171**, 744; scanning electrode localization of transport pathways in epithelial tissues, **171**, 792.

Modification of Cells for Transport Experiments: Experimental control of intracellular environment, **171**, 817; implantation of isolated carriers and receptors into living cells by Sendai virus envelope-mediated fusion, **171**, 829; resonance energy transfer microscopy: visual colocalization of fluorescent lipid probes in liposomes, **171**, 850.

Red Blood Cells: Methods and analysis of erythrocyte anion fluxes, **173**, 54; Cation fluxes in the red blood cell: Na^+,K^+ pump, **173**, 80; recording single-channel currents from human red cells, **173**, 112; identification of amino acid transporters in the red blood cell, **173**, 122; transport measurement of anions, nonelectrolytes, and water in red blood cell and ghost systems, **173**, 160; kinetic properties of Na^+/H^+ exchange and Li^+/Na^+, Na^+/Na^+, and Na^+/Li^+ exchanges of human red cells, **173**, 176; water channels across the red blood cell and other biological membranes, **173**, 192; sugar transport in red blood cells, **173**, 231; nucleoside transport across red cell membranes, **173**, 250; transport studies in red blood cells by measuring light scattering, **173**, 263; cation-anion cotransport, **173**, 280; sodium-calcium and sodium-proton exchangers in red blood cells, **173**, 292; monocarboxylate transport in red blood cells: kinetics and chemical modification, **173**, 300; alkali metal/proton exchange, **173**, 330; preparation and properties of one-step inside-out vesicles from red cell membranes, **173**, 368; Na^+,K^+-pump stoichiometry and coupling in inside-out vesicles from red blood cell membranes, **173**, 377; isolation, reconstitution, and assessment of transmembrane orientation of the anion-exchange protein, **173**, 410.

Transport Studies with Intact Cells: Transport of glutathione, glutathione disulfide, and glutathione conjugates across the hepatocyte plasma membrane, **173**, 523; Ca^{2+} fluxes and phosphoinositides in hepatocytes,**173**, 534; magnesium transport in eukaryotic and prokaryotic cells using magnesium-28 ion, **173**, 546; measurement of amino acid transport by hepatocytes in suspension or monolayer culture, **173**, 564; distinguishing amino acid transport systems of a given cell or tissue, **173**, 576; measuring hexose transport in suspended cells, **173**, 616; preparation and culture of embryonic and neonatal heart muscle cells: modification of transport activity, **173**, 634; isolation of calcium-tolerant atrial and ventricular myocytes from adult rat heart, **173**, 662; measurement of Na$^+$ pump in isolated cells, **173**, 676; measurement of Na$^+$-K$^+$ pump in muscle, **173**, 695; measurement of transport versus metabolism in cultured cells, **173**, 714; purification and reconstitution of the phosphate transporter from rat liver mitochondria, **173**, 732; measurement of vacuolar pH and cytoplasmic calcium in living cells using fluorescence microscopy, **173**, 745; transport in mouse ascites tumor cells: symport of Na$^+$ with amino acids, **173**, 771; measurements of cytoplasmic pH and cellular volume for detection of Na$^+$/H$^+$ exchange in lymphocytes, **173**, 777.

Plasma Membranes and Derived Transporters: Measurement of transported calcium in synaptosomes, **174**, 3; glutamate accumulation into synaptic vesicles, **174**, 9; identification of bile acid transport protein in hepatocyte sinusoidal plasma membranes, **174**, 25; transport of alanine across hepatocyte plasma membranes, **174**, 31; purification and reconstitution of glucose transporter from human erythrocytes, **174**, 39; isolation of bilitranslocase, the anion transporter from liver plasma membrane for bilirubin and other organic anions, **174**, 50.

Intracellular Organelles: Measurement of intactness of rat liver endoplasmic reticulum, **174**, 58; calcium ion transport in mitochondria, **174**, 68; measurement of proton leakage across mitochondrial inner membranes and its relation to protonmotive force, **174**, 85; influence of calcium ions on mammalian intramitochondrial dehydrogenases, **174**, 95; effects of hormones on mitochondrial processes, **174**, 118; use of fluorescein isothiocyanate-dextran to measure proton pumping in lysosomes and related organelles, **174**, 131; cystine exodus from lysosomes: cystinosis, **174**, 154; isolation of physiologically responsive secretory granules from exocrine tissues, **174**, 162; transport of nucleotides in the golgi complex, **174**, 173.

Transport in Plants: Water flow in plants and its coupling to other processes: an overview, **174**, 183; plasmolysis and deplasmolysis, **174**, 225; passive permeability, **174**, 246; compartmentation in root cells and tissues: X-ray microanalysis of specific ions, **174**, 267; xylem transport, **174**, 277; phloem transport, **174**, 288; patch clamp measurements on isolated guard cell protoplasts and vacuoles, **174**, 312; intracellular and intercellular pH measurement with microelectrodes, **174**, 331; water relations of plant cells: pressure probe technique, **174**, 338; transport systems in algae and bryophytes: an overview, **174**, 366; uptake of sugars and amino acids by *Chlorella*, **174**, 390; electrophysiology of giant algal cells, **174**, 403; ion transport in *Chara* cells, **174**, 443; light-induced hyperpolarization in *Nitella*, **174**, 479; ATP-driven chloride pump in giant alga *Acetabularia*, **174**, 490.

Membrane Transport in Organelles: Metabolite levels in specific cells and subcellular compartments of plant leaves, **174**, 518; isolation of plant vacuoles and measurement of transport, **174**, 552; transport in fungal cells; kinetic studies of transport in yeast, **174**, 567; proton extrusion in yeast, **174**, 592; transport in isolated yeast vacuoles: characterization of arginine permease, **174**, 504; ion transport in yeast including lipophilic ions, **174**, 603; sugar transport in normal and mutant yeast cells, **174**, 617; transport of amino acids and selected anions in yeast, **174**, 623; accumulation of electroneutral and charged carbohydrates by proton cotransport in *Rhodotorula*, **174**, 629; proton-potassium symport

in walled eukaryotes: *Neurospora*, **174**, 654; isolation of everted plasma membrane vesicles from *Neurospora crassa* and measurement of transport function, **174**, 667.

Epithelial Membrane Transport: An introduction, **191**, 1; determination of paracellular shunt conductance in epithelia, **191**, 4.

KIDNEY: Isolated perfused and nonfiltering kidney, **191**, 31; multiple indicator dilution and the kidney: kinetics, permeation, and transport *in vivo*, **191**, 34; micropuncture techniques in renal research, **191**, 72; microperfusion-double-perfused tubule *in situ*, **191**, 98; transcapillary fluid transport in the glomerulus, **191**, 107; preparation and study of isolated glomeruli, **191**, 130; isolation and study of glomerular cells, **191**, 141; isolation and culture of juxtaglomerular and renomedullary interstitial cells, **191**, 152; microdissection of kidney tubule segments, **191**, 226; measurement of transmural water flow in isolated perfused tubule segments, **191**, 232; identification and study of specific cell types in isolated nephron segments using fluorescent dyes, **191**, 253; functional morphology of kidney tubules and cells *in situ*, **191**, 265; an electrophysiological approach to the study of isolated perfused tubules, **191**, 289; hormonal receptors in the isolated tubule, **191**, 303; metabolism of isolated kidney tubule segments, **191**, 325; endocytosis and lysosomal hydrolysis of proteins in proximal tubules, **191**, 340; flux measurements in isolated perfused tubules, **191**, 354; measurements of volume and shape changes in isolated tubules, **191**, 371; isolated renal cells; transport in isolated cells from defined nephron segments, **191**, 380; primary culture of isolated tubule cells of defined segmental origin, **191**, 409; transport studies by optical methods, **191**, 469; stoichiometry of coupled transport systems in vesicles, **191**, 479; phosphate transport in established renal epithelial cell lines, **191**, 494; ATP-driven proton transport in vesicles from the kidney cortex, **191**, 505; reconstitution and fractionation of renal brush border transport proteins, **191**, 583.

STIMULUS SECRETION COUPLING IN EPITHELIA: Receptor identification, **191**, 609; camp technologies, functional correlates in gastric parietal cells, **191**, 640; stimulus-secretion coupling: general models and specific aspects in epithelial cells, **191**, 661; metabolism and function of phosphatidylinositol-derived arachidonic acid, **191**, 676; measurement of intracellular free calcium to investigate receptor-mediated calcium signaling, **191**, 691; two-stage analysis of radiolabeled inositol phosphate isomers, **191**, 707.

PHARMACOLOGICAL AGENTS IN EPITHELIAL TRANSPORT: Pharmacological agents of gastric acid secretion: receptor antagonists and pump inhibitors, **191**, 721; cation transport probes: the amiloride series, **191**, 739; photoaffinity-labeling analogues of phlorizin and phloretin: synthesis and effects on cell membranes, **191**, 755; diuretic compounds structurally related to furosemide,**191**, 781; chloride channel blockers, **191**, 793.

STOMACH: Intracellular ion activities and membrane transport in parietal cells measured with fluorescent probes, **192**, 38; electrophysiological techniques in the analysis of ion transport across gastric mucosa, **192**, 82; gastric glands and cells: preparation and *in vitro* methods, **192**, 93; permeabilizing parietal cells,**192**, 108; pepsinogen secretion *in vitro*, **192**, 124; secretion: stomach, duodenum, **192**, 139; isolation of H$^+$,K$^+$-ATPase-containing membranes from the gastric oxyntic cell, **192**, 151; peptic granules from the stomach, **192**, 165; isolation and primary culture of endocrine cells from canine gastric mucosa, **192**, 176.

PANCREAS: Membrane potential measurements in pancreatic β cells with intracellular microelectrodes, **192**, 235; stimulation of secretion by secretagogues, **192**, 247; pancreatic secretion: *in vivo*, perfused gland, and isolated duct studies, **192**, 256; dispersed pancreatic acinar cells and pancreatic acini, **192**, 271; permeabilizing cells: some methods and applications for the study of intracellular processes, **192**, 280; electrophysiology of pancreatic acinar cells, **192**, 300.

INTESTINE: Characterization of a membrane potassium ion conductance in intestinal secretory cells using whole cell patch-clamp and calcium-sensitive dye techniques, **192**, 309; isolation of intestinal epithelial cells and evaluation of transport functions, **192**, 324; isolation of enterocyte membranes, **192**, 341; established intestinal cell lines as model systems for electrolyte transport studies, **192**, 354; sodium chloride transport pathways in intestinal membrane vesicles, **192**, 389; advantages and limitations of vesicles for the characterization and the kinetic analysis of transport systems, **192**, 409; isolation and reconstitution of the sodium-dependent glucose transporter, **192**, 438; calcium transport by intestinal epithelial cell basolateral membrane, **192**, 448; electrical measurements in large intestine (including cecum, colon, rectum), **192**, 459

LIVER: Use of isolated perfused liver in studies of biological transport processes, **192**, 485; measurement of unidirectional calcium ion fluxes in liver, **192**, 495; preparation and specific applications of isolated hepatocyte couplets, **192**, 501; characterizing mechanisms of hepatic bile acid transport utilizing isolated membrane vesicles, **192**, 517; preparation of basolateral (sinusoidal) and canalicular plasma membrane vesicles for the study of hepatic transport processes, **192**, 534.

OTHER EPITHELIA: Electrogenic and electroneutral ion transporters and their regulation in tracheal epithelium, **192**, 549; transformation of airway epithelial cells with persistence of cystic fibrosis or normal ion transport phenotypes, **192**, 565; cell culture of bovine corneal endothelial cells and its application to transport studies, **192**, 571; methods for studying eccrine sweat gland function *in vivo* and *in vitro*, **192**, 583.

MEPRIN

These zinc-dependent endopeptidases (meprin A [EC 3.4.24.18] and meprin B [EC 3.4.24.63]) are members of the peptidase family M12A. They catalyze the hydrolysis of peptide bonds in proteins and peptide substrates. Meprin A, a membrane-bound enzyme that has been isolated from mouse and rat kidney and intestinal brush borders as well as salivary ducts, acts preferentially on carboxyl side of hydrophobic amino acyl residues. Meprin A and B are insensitive to inhibition by phosphoramidon and thiorphan.

R. L. Wolz & J. S. Bond (1995) *Meth. Enzymol.* **248**, 325.
J. S. Bond & P. E. Butler (1987) *Ann. Rev. Biochem.* **56**, 333.

3-MERCAPTOPYRUVATE SULFURTRANSFERASE

This zinc-dependent enzyme [EC 2.8.1.2] catalyzes the reaction of 3-mercaptopyruvate with cyanide to produce pyruvate and thiocyanate. Other substrates include sulfite, sulfinates, mercaptoethanol, and mercaptopyruvate.

J. V. Schloss & M. S. Hixon (1998) *Comprehensive Biological Catalysis: A Mechanistic Reference* **2**, 43.

MERCURY(II) REDUCTASE

This enzyme [EC 1.16.1.1], also known as mercuric reductase, catalyzes the reaction of Hg with NADP$^+$ and H$^+$ to produce Hg^{2+} and NADPH.

A. R. Clarke & T. R. Dafforn (1998) *Comprehensive Biological Catalysis: A Mechanistic Reference* **3**, 1.
B. A. Palfey & V. Massey (1998) *Comprehensive Biological Catalysis: A Mechanistic Reference* **3**, 83.

MESO FORM (or, *MESO* ISOMER)

A stereoisomer having more than one chiral center and having an internal plane of symmetry. Hence, *meso* compounds do not exhibit optical activity.

Meso isomers can prove to be quite useful at times, particularly in assessing the topology of the active site of a protein. γ-Glutamyl transpeptidase can catalyze the transfer of the γ-glutamyl moiety of glutathione to an amino acid, forming a γ-glutamyl amino acid isopeptide[1]. The enzyme is stereospecific for L-amino acids and L-cystine serves as an excellent substrate. As one might expect, *meso*-cystine can also act as a substrate although there can only be one productive binding mode.

[1] R. D. Allison (1985) *Meth. Enzymol.* **113**, 419.

MESOLYTIC CLEAVAGE

Cleavage of a bond in a radical ion such that a separate radical product and ion product are formed.

P. Maslak & J. N. Narvaez (1990) *Angew. Chem. Int. Ed. Engl.* **29**, 283.

METABOLIC CONTROL ANALYSIS

An approach for interpreting changes in metabolic flux (symbolized by J) for a pathway made up of multiple steps, each catalyzed by an enzyme. The basic tenets of this approach are that no single enzyme-catalyzed step controls flux through a pathway and that the relative importance of individual steps in determining the overall flux can be altered by changes in metabolic state or by externally applied conditions[1]. The sensitivity of a pathway to changes in the activity of a particular enzyme-catalyzed step is typically evaluated by using inhibitors or molecular genetic approaches to alter the activity of the particular step.

The formalized application of metabolic control analysis deals with several parameters[2]: (a) The *flux control coefficient* is defined as the fractional change in pathway flux that attends an infinitesimal fractional change in activity of one particular enzyme in the pathway. A value of zero indicates that the metabolic flux is completely insensitive to a change in the activity of a particular enzyme, and a value of one would mean that the pathway is completely dominated by the activity of one particular step. (b) The *summation theorem* requires that all flux control coefficients sum to unity. (c) *Elasticity coefficients* represent measures of the fraction change in metabolic flux caused by an infinitesimal change in the concentration of a positive or negative effector. (d) The *connectivity theorem* indicates that the control of a system can be solved by determining the flux control coefficients and elasticity coefficients for each component step in a pathway. **See** *Top-Down Control Analysis*

[1]H. Kacser & J. A. Burns (1987) *Trends in Biochem. Sci.* **12**, 5.
[2]P. A. Quant (1993) *Trends in Biochem. Sci.* **18**, 26.

METABOLONS

Organized, multienzyme structures of particular biochemical pathways such that they provide efficient flux through the sequential reactions of those pathways[1]. Examples of pathways that have been suggested to act through metabolons include the tricarboxylic acid cycle[2], β-oxidation[3,4], the biosynthesis of specific classes of isoprenoids[5,6], and nucleotide biosynthesis[7–9]. There are now excellent kinetic criteria for judging whether a system involves substrate channeling. **See** *Substrate Channeling*

[1]P. A. Srere (1987) *Ann. Rev. Biochem.* **56**, 89.
[2]B. Sumegi, A. D. Sherry, C. R. Malloy, C. Evans & P. A. Srere (1991) *Biochem. Soc. Trans.* **4**, 1002.
[3]P. A. Srere & B. Sumegi (1994) *Biochem. Soc. Trans.* **22**, 446.
[4]M. A. Nada, W. J. Rhead, H. Sprecher, H. Schulz & C. R. Roe (1995) *J. Biol. Chem.* **270**, 530.
[5]J. Chappell (1995) *Plant Physiol.* **107**, 1.
[6]J. Chappell (1995) *Ann. Rev. Plant Physiol. Plant Mol. Biol.* **46**, 521.
[7]R. J. Zeleznikar, R. A. Heyman, R. M. Graeff, T. F. Walseth, S. M. Davis, E. A. Butz & N. D. Goldberg (1990) *J. Biol. Chem.* **265**, 300.
[8]F. N. Gellerich, Z. A. Khuchua & A. V. Kuznetsov (1993) *Biochem. Biophys. Acta* **1140**, 327.
[9]F. Savabi (1994) *Mol. Cell Biochem.* **134**, 145.

METAL ION CATALYSIS

Acceleration of reaction rate as a result of the presence of one or more metal ions. In many cases, the general acid properties of the metal ion contribute to catalysis. Metal ions play organizing roles as templates for reactants. The metal ion may also stabilize the transition state. Enzyme-bound metal ions (many enzymes are metalloproteins with tightly bound ions) can transiently undergo changes in their oxidation state to facilitate catalysis.

METAL ION COMPLEXES (Determining Stability Constants)

An essential aspect to understanding the influence of metal ions on enzyme-catalyzed reactions is the knowledge of how tight different metal ions bind to a wide variety of substrates (particularly nucleotides and other phosphoryl-containing molecular entities), products, and effectors and that binding phenomena are altered by the experimental conditions (*e.g.*, the effects of pH, temperature, ionic strength, *etc.*). This necessitates the experimental determination of the stability constant (an association constant) for the metal ion-ligand complex. O'Sullivan and Smithers[1] have reviewed the theory and the various techniques for such determinations and have provided values for many of the more common, biochemically relevant complexes.

[1]W. J. O'Sullivan & G. W. Smithers (1979) *Meth. Enzymol.* **63**, 294.

METAL ION COORDINATION REACTIONS

Metal ions interact with many substances, and the term "ligand" (from the Latin verb *ligare*: to join together) refers to a substance that interacts ionically to form an electrovalent bond or donates electrons to form coordinate covalent bonds. In biological systems, metal ions are typically aquated (*i.e.*, complexed to water) as well as combined with other ligands. Because water is so plentiful, aqua complexes of metal ions must first lose one or more H_2O molecules before other ligands can take their place. Water is neutral, and its reaction with a complex, say $[L_5MX]$ where X is an anionic or electron-rich substance, is usually relatively slow. **See** *Water; Water Exchange Rates of Aqua Metal Ions*. By contrast, the displacement of a neutral water molecule by a negatively charged ligand (*e.g.*, $[L_5M(H_2O)]^n +$ $X^- \rightarrow [L_5MX]^{n-1} + H_2O$) generally proceeds rapidly.

Acid hydrolysis of an octahedral metal ion complex is typically a dissociative or S_N1-type reaction. In the case of base hydrolysis, reactions tend to display S_N2-type reaction mechanisms, although others take place by what is termed an S_N1-conjugate base mechanism. The latter involves attack by an electrophile to abstract a proton from a bound ligand to yield its conjugate base form, and this route is limited to ligands that contain protons that can dissociate.

Oxidation-reduction reactions can occur if two redox-active atoms or groups come sufficiently close to each other for electron transfer to occur. With metal ion complexes, a change in the oxidation state of the metal ion may lead to a rearrangement of ligand bonding geometry. The term "outer-sphere" mechanism describes the situation where the ligands within the metal ion complex are so firmly held that the electron from the oxidant must transfer through this fully occupied coordination shell. By contrast, "inner-sphere" electron transfer reactions occur if the oxidizing agent binds to or shares a ligand with the metal ion complex. *See also Water; Water Exchange Rates of Aqua Metal Ions*

F. A. Cotton & G. Wilkinson (1988) *Advanced Inorganic Chemistry*, 5th ed., Wiley, New York.

METAL IONS IN NUCLEOTIDE-DEPENDENT REACTIONS

The actual substrate in most nucleotide-dependent reactions is not the free nucleotide (*i.e.*, NTP^{4-} [or $NTPH^{3-}$] or NDP^{3-} [or $NDPH^{2-}$]); rather, the actual phosphoryl donor species is the metal ion-nucleotide complex. Examples include the hexokinase reaction ($MgATP^{2-}$ + glucose $\rightleftharpoons$ $MgADP^{1-}$ + glucose 6-phosphate) and the adenylate kinase reaction ($MgADP^{1-}$ + ADP^{3-} $\rightleftharpoons$ $MgATP^{2-}$ + AMP^{2-}). In fact, the uncomplexed nucleotide may even act as a potent inhibitor. Such is the case with the brain hexokinase reaction in which the true phosphoryl donor is $MgATP^{2-}$ complex, whereas uncomplexed ATP^{4-} is a competitive inhibitor[1]. Interested readers should consult Morrison[2] for a detailed account of recommended procedures for studying enzymes that utilize metal-nucleotide substrates.

A surprisingly common, but ill-advised practice in studies of ATP-dependent kinases, synthetases, ATPases, and related enzymes is the use a single, high concentration of divalent metal ion throughout the entire kinetic investigation. The resultant problem is that elevated concentrations of metal ions often prove to be inhibitory; even more significant is the fact that the free metal ion concentration will not remain constant as the nucleotide concentration is increased. The other substrates in a multisub-

strate reaction may also nonspecifically bind the metal ions that are present at elevated concentrations. Another erroneous practice is to vary the absolute concentrations of metal ion and nucleotide simultaneously, say by maintaining both at constant ratio relative to each other. In such cases, the fraction of nucleotide present as the metal-nucleotide complex will change with absolute concentration, a phenomenon that can result in false co-operativity[1] observed as sigmoidal plots of v_o *versus* [substrate]. These effects are illustrated in Fig. 1 for a hypothetical ATP-dependent enzyme utilizing $MgATP^{2-}$ complex as the substrate.

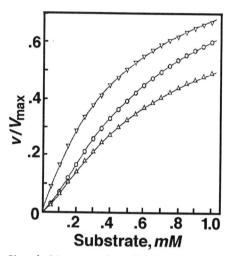

Figure 1. Plot of v/V_{max} *versus* the millimolar concentration of total substrate for a model enzyme displaying Michaelis-Menten kinetics with respect to its substrate MA (*i.e.*, metal ion M complexed to otherwise inactive ligand A). The concentrations of free A and MA were calculated assuming a stability constant of 10,000 M^{-1}. The Michaelis constant for MA and the inhibition constant for free A acting as a competitive inhibitor were both assumed to be 0.5 mM. The ratio v/V_{max} was calculated from the Michaelis-Menten equation, taking into account the action of a competitive inhibitor (when present). The upper curve represents the case where the substrate is both A and MA. The middle curve deals with the case where MA is the substrate and where A is not inhibitory. The bottom curve describes the case where MA is the substrate and where A is inhibitory. In this example, $[M]_{total} = [A]_{total}$ at each concentration of A plotted on the abscissa. Note that the bottom two curves are reminiscent of allosteric enzymes, but this false cooperativity arises from changes in the fraction of total "substrate A" that has metal ion bound. For a real example of how brain hexokinase cooperatively was debunked, consult D. L. Purich & H. J. Fromm (1972) *Biochem. J.* **130**, 63.

For kinetic experiments that are designed to determine the rate dependence of an enzyme-catalyzed reaction on nucleotide concentration, care must be exercized to maintain a constant free metal ion concentration under all experimental conditions. Too low a metal ion concentration allows uncomplexed nucleotides to inhibit an enzyme, and too high a metal ion concentration can lead to nonspecific binding and inhibition. In cases where the cosubstrate also binds metal ions, one may need to consider these binding interactions as well. Likewise, one may also want to use a buffer that does not bind metal ions and does not interfere with the strength of metal-nucleotide complexation[2].

Preliminary rate measurements should allow one to make a plot of initial velocity v_o *versus* [metal ion], and this should provide information on the optimal metal ion concentration. (For many MgATP^{2-}-dependent enzymes, the optimum is frequently 1–3 mM uncomplexed magnesium ion.) Then, by utilizing published values for formation constants (also known as stability constants) defining metal ion-nucleotide complexation, one can readily design experiments to keep free metal ion concentration at a fixed level. To compensate properly for metal ion complexation in ATP-dependent reactions, one must chose a buffer for which a stability constant is known. For example, in 25 mM Tris-HCl (pH 7.5), the stability constant for MgATP^{2-} is approximately 20,000 M^{-1}. Thus, one can write the following equation:

$$K = 20,000 \text{ M}^{-1} = [\text{MgATP}^{2-}]/\{[\text{Mg}^{2+}\text{free}][\text{ATP}^{4-}]\}$$

and if, for example, one wishes to work at 1 mM free (or uncomplexed) magnesium ion, the following holds:

$$(20,000 \text{ M}^{-1})[\text{Mg}^{2+}\text{free}] = 20 = [\text{MgATP}^{2-}]/[\text{ATP}^{4-}]$$

The ratio [MgATP^{2-}]/[ATP^{4-}] can be treated as x/y, where x represents [MgATP^{2-}], y represents [ATP^{4-}], and $(x + y)$ equals [ATP$_{total}$], the total nucleotide concentration[3]. For the above case, we see that $x = (20/21)$ [ATP$_{total}$] and $y = (1/21)$[ATP$_{total}$], such that $(20/21)$ [ATP$_{total}$] $+ (1/21)$[ATP$_{total}$] $=$ [ATP$_{total}$]. If one choses to work with 25 mM HEPES (pH 7.5) as the buffer, the stability constant for MgATP^{2-} is 100,000 M^{-1}; in this case, at 1 mM free magnesium ion, $x = (100/101)$ [ATP$_{total}$] and $y = (1/101)$[ATP$_{total}$].

To prepare a series of assay samples with different concentrations of phosphoryl acceptor substrate (designated here as "phosphoryl acceptor") and ATP, while still maintaining the desired [Mg^{+2}]$_{free}$, the following procedure is recommended. First, one prepares four different component solutions:

0.30 mL Buffer–Mg$^{2+}_{\text{"free"}}$ solution

0.30 mL Phosphoryl-Acceptor solution (at highest concentration)

0.30 mL ATP–Mg$^{2+}_{\text{"bound"}}$ solution (at highest concentration)

<u>0.10 mL Enzyme solution</u>

1.00 mL Reaction sample when combined

(The total volume here is chosen for most UV/Vis spectrophotometric assays, but the total reaction volume can be whatever the experimenter desires.) Note also that the magnesium ion is present in *two* solutions, and only when they are combined in the overall 1.00 mL sample will the final free concentration be what one desires.

Let us, for example, prepare 10 mL of the component solution that is 0.025 M buffer, 1 mM magnesium ion, and 5 mM of a nonvaried substance Y from stock solutions of buffer (say 1 M), magnesium ion (say 0.1 M), and Y (say 20 mM). The following equation allows us to calculate the volume of each stock solution while accounting for the dilution factor (in this case, 3.33 or 1.0 divided by 0.30):

$$\frac{(1.00/0.30) \times 0.025 \text{ M} \times 10 \text{ mL}}{1 \text{ M Buffer Stock Solution}} = \frac{0.83 \text{ mL}}{\text{Stock Buffer}}$$

$$\frac{(1.00/0.30) \times 1 \text{ mM} \times 10 \text{ mL}}{100 \text{ mM Magnesium Stock Solution}} = \frac{0.33 \text{ mL Stock}}{\text{Magnesium}}$$

$$\frac{(1.00/0.30) \times 5 \text{ mM} \times 10 \text{ mL}}{25 \text{ mM "Y" Stock Solution}} = \frac{6.67 \text{ mL}}{\text{Stock "Y"}}$$

The solutions are combined and then brought to 10 mL final volume with the addition of 2.17 mL deionized water. [Note: The (1.00/0.30)-factor corrects for the final combination of all four component solutions into the final reaction sample.]

To prepare say 10 mL of the 2 mM Phosphoryl-Acceptor (or Acc) solution from a 10 mM stock solution, we use the same procedure:

$$\frac{(1.00/0.30) \times 2 \text{ mM} \times 10 \text{ mL}}{\begin{array}{c}10 \text{ mM Phosphoryl–Acc.} \\ \text{Stock Solution}\end{array}} = \frac{6.67 \text{ mL Stock}}{\text{Phosphoryl–Acc.}}$$

The solutions are combined and then brought to 10 mL final volume with the addition of deionized water. Further dilutions of this well-mixed solution can be used to generate working solutions, such as $[Acc]_1$, $[Acc]_3$, $[Acc]_5$, $[Acc]_7$, and $[Acc]_9$, by combining one volume of solution at $[Acc]$ with 0, 2, 4, 6, and 8 volumes of deionized water.

Next, we can prepare 10 mL ATP–$Mg^{2+}_{\text{"bound"}}$ solution in a similar manner:

$$\frac{(1.00/0.30) \times 10\,\text{mM} \times 10\,\text{mL}}{100\,\text{mM ATP Stock Solution}} = \frac{3.33\,\text{mL}}{\text{Stock ATP}}$$

$$\frac{(1.00/0.30) \times (20/21) \times 10\,\text{mM} \times 10\,\text{mL}}{1.00\,\text{M Magnesium Stock Solution}} = \frac{0.79\,\text{mL Stock}}{\text{Magnesium}}$$

These latter two aliquots are combined and brought to 10 mL final volume with the addition of 5.88 mL deionized water. [Note that the complexation factor has a value of 20/21 in the second entry. This value will depend on the stability constant and the desired free magnesium ion concentration. For example, in 25 mM Tris-HCl (pH 7.5), the stability constant is 20,000 M^{-1}; thus, the complexation factor will be (20/21) at 1 mM uncomplexed magnesium ion. With 25 mM HEPES (pH 7.5), the stability constant is 100,000 M^{-1}, and the complexation factor value would be (100/101) at 1 mM free magnesium ion. If another free magnesium ion concentration is desired or if another buffer is used, then one must recalculate the fraction of bound metal ion.] Further dilutions of this well mixed solution can be used to generate $[ATP]_1$, $[ATP]_3$, $[ATP]_5$, . . ., working solutions as described above. When finally combined with the three other component solutions to make up a reaction sample, the free metal ion will always be 1 mM for the example given.

Lastly, to prepare the enzyme solution, say 10 nM from a 200 nM stock enzyme solution, we use the same procedure, except that the dilution factor in this case is (1.00/0.10) in terms of the 0.1 mL enzyme solution volume cited above.

$$\frac{(1.00/0.10) \times 10\,\text{nM} \times 1\,\text{mL}}{200\,\text{nM Enzyme Stock Solution}} = 0.50\,\text{mL Stock Enzyme}$$

This volume is then brought to the specified 1 mL final volume with the addition of 0.5 mL deionized water (or buffer and any other factors needed to maintain enzyme stability).

In this example, one can construct the following matrix of samples for determining the kinetic parameters of a two-substrate enzymic reaction:

$[Acc]_1+[ATP]_1$ $[Acc]_3+[ATP]_1$ $[Acc]_5+[ATP]_1$ $[Acc]_7+[ATP]_1$ $[Acc]_9+[ATP]_1$
$[Acc]_1+[ATP]_3$ $[Acc]_3+[ATP]_3$ $[Acc]_5+[ATP]_3$ $[Acc]_7+[ATP]_3$ $[Acc]_9+[ATP]_3$
$[Acc]_1+[ATP]_5$ $[Acc]_3+[ATP]_5$ $[Acc]_5+[ATP]_5$ $[Acc]_7+[ATP]_5$ $[Acc]_9+[ATP]_5$
$[Acc]_1+[ATP]_7$ $[Acc]_3+[ATP]_7$ $[Acc]_5+[ATP]_7$ $[Acc]_7+[ATP]_7$ $[Acc]_9+[ATP]_7$
$[Acc]_1+[ATP]_9$ $[Acc]_3+[ATP]_9$ $[Acc]_5+[ATP]_9$ $[Acc]_7+[ATP]_9$ $[Acc]_9+[ATP]_9$

The most important point is that this procedure provides a reliable and effective way for dealing with metal ion interactions with nucleotides. By maintaining the free magnesium ion at a single fixed concentration, the experimenter avoids any chance of uncontrolled variation in the $[MgATP^{2-}]/[ATP^{4-}]$ ratio.

The following table lists several frequently used stability constants for metal ion-nucleotide complexes. Because complexation is driven by neutralization of electric charge on the components, one should immediately appreciate that the values apply only for the specified solution conditions. A more complete list is provided elsewhere[4–5].

Stability Constants for Complexes of ATP[a]

Complex Formed	$K_{\text{stability}}$ (M^{-1})
MgH_2ATP	20
$MgHATP^{1-}$	500
$MgATP^{2-}$	73,000
CaH_2ATP	20
$CaHATP^{1-}$	400
$CaATP^{2-}$	35,000
MnH_2ATP	~100
$MgHATP^{1-}$	~1000
$MgATP^{2-}$	100,000
$KATP^{3-}$	15
$NaATP^{3-}$	14

[a]From reference 4.

Finally, because nucleotides can also act as allosteric effectors, an investigator seeking to characterize the kinetic properties of allosteric enzymes should consider using the methods described above to determine whether the true effector is the free nucleotide or the form complexed with metal ion.

[1]D. L. Purich & H. J. Fromm (1972) *Biochem. J.* **130**, 63.
[2]J. F. Morrison (1979) *Meth. Enzymol.* **63**, 257. (This reference provides additional advice for those seeking to characterize metal ion-dependent enzymes.)

[3]Recall that if one knows the ratio, r, of any two quantities a and b, the relative amounts of those two quantities is also known, because $r = a/b = x/(1 - x)$.

[4]D. D. Perrin & B. Dempsey (1974) *Buffers for pH and Metal Ion Control*, Wiley, New York. (This handy reference provides detailed procedure that allow one to compensate for pH and temperature effects on metal-ligand interactions as well as buffers.)

[5]W. J. O'Sullivan & G. W. Smithers (1979) *Meth. Enzymol.* **63**, 295. (This reference lists many stability constants and also provides a detailed account of the procedures for determining stability constants.)

METALLO-BIOCHEMICAL METHODS

Because metal ions bind to and modify the reactivity and structure of enzymes and substrates, a wide spectrum of techniques has been developed to examine the nature of metal ions which serve as templates, redox-active cofactors, Lewis acids/bases, ion-complexing agents, *etc.*

Selected entries from *Methods in Enzymology* [vol, page(s)]:

General Methods: Preparation of metal-free water, **158**, 3; elimination of adventitious metals, **158**, 6; metal-free dialysis tubing, **158**, 13; metal-free chromatographic media, **158**, 15; preparation of metal-free enzymes, **158**, 21; metal-buffered systems, **158**, 33; standards for metal analysis, **158**, 56; methods for metal substitution, **158**, 71; preparation of metal-hybrid enzymes, **158**, 79; introduction of exchange-inert metal ions into enzymes, **158**, 95; use of chelating agents to inhibit enzymes, **158**, 110.

Analytical Techniques: Atomic absorption spectrometry, **158**, 117; multielement atomic absorption methods of analysis, **158**, 145; ion microscopy in biology and medicine, **158**, 157; flame atomic emission spectrometry, **158**, 180; inductively coupled plasma-emission spectrometry, **158**, 190; inductively coupled plasma-mass spectrometry, **158**, 205; atomic fluorescence spectrometry, **158**, 222; electrochemical methods of analysis, **158**, 243; neutron activation analysis, **158**, 267.

Analysis of Specific Metals: Aluminum, **158**, 289; total calcium, **158**, 302; ionized calcium, **158**, 320; chromium, **158**, 334; cobalt, **158**, 344; copper, **158**, 351; complexed iron, **158**, 357; magnesium, **158**, 365; molybdenum, **158**, 371; nickel, **158**, 382; selenium, **158**, 391; vanadium, **158**, 402; zinc, **158**, 422.

Probing Metalloproteins: Electronic absorption spectroscopy of copper proteins, **226**, 1; electronic absorption spectroscopy of nonheme iron proteins, **226**, 33; cobalt as probe and label of proteins, **226**, 52; biochemical and spectroscopic probes of mercury(II) coordination environments in proteins, **226**, 71; low-temperature optical spectroscopy: metalloprotein structure and dynamics, **226**, 97; nanosecond transient absorption spectroscopy, **226**, 119; nanosecond time-resolved absorption and polarization dichroism spectroscopies, **226**, 147; real-time spectroscopic techniques for probing conformational dynamics of heme proteins, **226**, 177; variable-temperature magnetic circular dichroism, **226**, 199; linear dichroism, **226**, 232; infrared spectroscopy, **226**, 259; Fourier transform infrared spectroscopy, **226**, 289; infrared circular dichroism, **226**, 306; Raman and resonance Raman spectroscopy, **226**, 319; protein structure from ultraviolet resonance Raman spectroscopy, **226**, 374; single-crystal micro-Raman spectroscopy, **226**, 397; nanosecond time-resolved resonance Raman spectroscopy, **226**, 409; techniques for obtaining resonance Raman spectra of metalloproteins, **226**, 431; Raman optical activity, **226**, 470; surface-enhanced resonance Raman scattering, **226**, 482; luminescence spectroscopy, **226**, 495; circularly polarized luminescence, **226**, 539; low-temperature stopped-flow rapid-scanning spectroscopy: performance tests and use of aqueous salt cryosolvents, **226**, 553; transition-metal complexes as spectroscopic probes for selective covalent labeling and cross-linking of proteins, **226**, 565; ruthenium complexes as luminescent reporters of DNA, **226**, 576; detecting metal-metal interactions and measuring distances between metal centers in metalloproteins, **226**, 594; two-dimensional nuclear magnetic resonance of paramagnetic metalloproteins, **227**, 1; cadmium-113 nuclear magnetic resonance applied to metalloproteins, **227**, 16; lanthanide shift reagents, **227**, 43; alkali metal nuclear magnetic resonance, **227**, 78; calcium nuclear magnetic resonance, **227**, 107; pulsed electron nuclear multiple resonance spectroscopic methods for metalloproteins and metalloenzymes, **227**, 118; continuous wave electron nuclear double resonance spectroscopy, **227**, 190; vanadyl(IV) electron nuclear double resonance/electron spin echo envelope modulation spin probes, **227**, 232; multidimensional nuclear magnetic resonance methods to probe metal environments in proteins, **227**, 244; optically detected magnetic resonance of triplet states in proteins, **227**, 290; electron paramagnetic resonance, **227**, 330; electron paramagnetic resonance spectroscopy of iron complexes and iron-containing proteins, **227**, 353; intrinsic and extrinsic paramagnets as probes of metal clusters, **227**, 384; electron paramagnetic resonance spectroelectrochemical titration, **227**, 396; magnetic susceptibility, **227**, 412; multifield saturation magnetization of metalloproteins, **227**, 437; combining Mössbauer spectroscopy with integer spin electron paramagnetic resonance, **227**, 463; voltammetric studies of redox-active centers in metalloproteins adsorbed on electrodes, **227**, 479; direct and indirect electrochemical investigations of metalloenzymes, **227**, 501; pulse radiolysis, **227**, 522; physical methods to locate metal atoms in biological systems, **227**, 535.

METALLOTHIONEINS

Thioneins are apoproteins that are exceptionally sulfur-rich (composed of greater 30 mol% cysteine). These proteins are found in high abundance in liver and kidney cytoplasm where they form metallothioneins (the holoprotein forms) upon complexation with metal ions. Thionein synthesis is induced by the presence of metals, especially zinc, copper, mercury, and cadmium.

J. F. Riordan & B. L. Vallee, eds. (1991) *Meth. Enzymol.*, vol. **205**, Academic Press, San Diego.

META-MODEL

A computer program[1] for calculating steady-state fluxes and metabolite concentrations of metabolic systems on the IBM-PC and compatible computers. The software provides a simple means for calculating the control structure of a pathway.

[1]A. Cornish-Bowden & J. H. Hofmeyr (1991) *Comput. Appl. Biosci.* **7**, 89.

METAPHOSPHATE

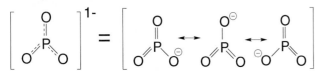

The unstable PO_3^- ion proposed to be an intermediate in a number of enzyme-catalyzed phosphotransfer reactions. **See** *Acyl-Phosphate*

METASTABLE STATE

An unstable equilibrium state (or configuration) that is at maximal potential energy. A metastable state is at equilibrium, and its potential energy [written here as $U(x)$] is such that any displacement (dx) from $x_{equilibrium}$ will result in the loss of potential energy.

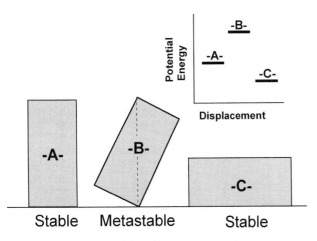

Although similar principles hold for transition states of chemical reactions, use of the term "metastable state" to refer to the transition state is actively discouraged by scientific convention. **See** *Transient Chemical Species; Jablonski Diagram*

METHANE MONOOXYGENASE

This enzyme [EC 1.14.13.25] catalyzes the reaction of methane with NAD(P)H and dioxygen to produce methanol, NAD(P)⁺, and water. This enzyme is reported to exhibit a broad specificity. Many alkanes can be hydroxylated and alkenes are converted into the corresponding epoxides. Carbon monoxide is oxidized to carbon dioxide, ammonia is oxidized to hydroxylamine, and some aromatic compounds and cyclic alkanes can also be hydroxylated, albeit not as efficiently.

A. R. Clarke & T. R. Dafforn (1998) *Comprehensive Biological Catalysis: A Mechanistic Reference* **3**, 1.

B. G. Fox (1998) *Comprehensive Biological Catalysis: A Mechanistic Reference* **3**, 261.
A. Messerschmidt (1998) *Comprehensive Biological Catalysis: A Mechanistic Reference* **3**, 401.
J. B. Howard & D. C. Rees (1991) *Adv. Protein Chem.* **42**, 199.
S. J. Pilkington & H. Dalton (1990) *Meth. Enzymol.* **188**, 181.
P. G. Fox, W. A. Froland, D. R. Jollie & J. D. Lipscomb (1990) *Meth. Enzymol.* **188**, 191.

METHANOL DEHYDROGENASE

This enzyme [EC 1.1.99.8], also known as alcohol dehydrogenase (acceptor), catalyzes the reaction of a primary alcohol with an acceptor to produce an aldehyde and the reduced acceptor. A wide range of primary alcohols (*e.g.*, methanol) can serve as substrates.

C. Anthony (1998) *Comprehensive Biological Catalysis: A Mechanistic Reference* **3**, 155.
J. P. Klinman & D. Mu (1994) *Ann. Rev. Biochem.* **63**, 299.
M. A. Ator & P. R. Ortez de Montellano (1990) *The Enzymes*, 3rd ed., **19**, 213.
J. Frank & J. A. Duine (1990) *Meth. Enzymol.* **188**, 202.
D. J. Day & C. Anthony (1990) *Meth. Enzymol.* **188**, 210.
A. R. Long & C. Anthony (1990) *Meth. Enzymol.* **188**, 216.
N. Arfman & L. Dijkhuizen (1990) *Meth. Enzymol.* **188**, 223.
J. A. Duine, J. Frank, & J. A. Jongejan (1987) *Adv. Enzymol.* **59**, 169.

5,10-METHENYLTETRAHYDROFOLATE CYCLOHYDROLASE

This enzyme [EC 3.5.4.9] catalyzes the hydrolysis of 5,10-methenyltetrahydrofolate to produce 10-formyltetrahydrofolate.

V. Schirch (1998) *Comprehensive Biological Catalysis: A Mechanistic Reference* **1**, 211.
V. Schirch (1997) *Meth. Enzymol.* **281**, 146.
R. E. MacKenzie (1997) *Meth. Enzymol.* **281**, 171.
S. W. Ragsdale (1991) *Crit. Rev. Biochem. Mol. Biol.* **26**, 261.
J. I. Rader & F. M. Huennekens (1973) *The Enzymes*, 3rd ed., **9**, 197.

5,10-METHENYLTETRAHYDROFOLATE SYNTHETASE

This enzyme [EC 6.3.3.2], also known as 5-formyltetrahydrofolate cyclo-ligase, catalyzes the reaction of ATP with 5-formyltetrahydrofolate to produce ADP, orthophosphate, and 5,10-methenyltetrahydrofolate.

V. Schirch (1998) *Comprehensive Biological Catalysis: A Mechanistic Reference* **1**, 211.
V. Schirch (1997) *Meth. Enzymol.* **281**, 146.
J. Jolivet (1997) *Meth. Enzymol.* **281**, 162.
J. I. Rader & F. M. Huennekens (1973) *The Enzymes*, 3rd ed., **9**, 197.

METHIONINE ADENOSYLTRANSFERASE

This enzyme [EC 2.5.1.6], also known as *S*-adenosylmethionine synthetase, catalyzes the reaction of ATP with L-methionine and water to produce S-adenosyl-L-methionine, orthophosphate, and pyrophosphate (or, diphosphate).

I. Saint-Gerons, C. Parsot, M. M. Zakin, O. Bârzy & G. N. Cohen (1988) *Crit. Rev. Biochem.* **23**, S1.
M. H. Stipanuk (1986) *Ann. Rev. Nutr.* **6**, 179.
C. W. Tabor & H. Tabor (1984) *Adv. Enzymol.* **56**, 251.
S. H. Mudd (1973)*The Enzymes*, 3rd ed., **8**, 121.

METHIONINE γ-LYASE

This pyridoxal-phosphate-dependent enzyme [EC 4.4.1.11], also known as L-methioninase, catalyzes the conversion of L-methionine to methanethiol, ammonia, and α-ketobutyrate (or, 2-oxobutanoate).

M. A. Ator & P. R. Ortez de Montellano (1990) *The Enzymes*, 3rd ed., **19**, 213.
N. Esaki & K. Soda (1987) *Meth. Enzymol.* **143**, 459.

METHIONINE SULFOXIDE REDUCTASE

This enzyme [EC 1.8.4.5], also known as methionine S-oxide reductase, catalyzes the reaction of L-methionine with oxidized thioredoxin to produce L-methionine S-oxide and reduced thioredoxin. Dithiothreitol can substitute for reduced thioredoxin in the reverse reaction. In addition, other methyl sulfoxides can replace methionine sulfoxide in the reverse reaction.

S. Black (1962) *Meth. Enzymol.* **5**, 992.

METHIONYL-tRNA FORMYLTRANSFERASE

This enzyme [EC 2.1.2.9] catalyzes the reaction of 10-formyltetrahydrofolate with L-methionyl-tRNA and water to produce tetrahydrofolate and N-formylmethionyl-tRNA.

V. Schirch (1998) *Comprehensive Biological Catalysis: A Mechanistic Reference* **1**, 211.
J. I. Rader & F. M. Huennekens (1973) *The Enzymes*, 3rd ed., **9**, 197.

METHIONYL-tRNA SYNTHETASE

This enzyme [EC 6.1.1.10], also known as methionine:tRNA ligase, catalyzes the reaction of methionine with tRNA^{Met} and ATP to produce methionyl-tRNA^{Met}, AMP, and pyrophosphate (or, diphosphate). **See also** *Aminoacyl-tRNA Synthetases*

E. A. First (1998) *Comprehensive Biological Catalysis: A Mechanistic Reference* **1**, 573.
P. Schimmel (1987) *Ann. Rev. Biochem.* **56**, 125.

4-METHOXYBENZOATE MONOOXYGENASE (O-Demethylating)

This enzyme [EC 1.14.99.15] catalyzes the reaction of 4-methoxybenzoate with AH$_2$ and dioxygen to produce 4-hydroxybenzoate, formaldehyde, A, and water. The bacterial enzyme consists of a ferredoxin-type protein and an iron-sulfur flavoprotein (FMN). 4-Ethoxybenzoate, N-methyl-4-aminobenzoate, and toluate can serve as substrates as well. The fungal enzyme acts best on veratrate.

B. G. Fox (1998) *Comprehensive Biological Catalysis: A Mechanistic Reference* **3**, 261.
V. Ullrich & W. Duppel (1975) *The Enzymes*, 3rd ed., **12**, 253.

3-METHYLADENINE-DNA GLYCOSYLASE

3-Methyladenine–DNA glycosylase I [EC 3.2.2.20], also known as DNA-3-methyladenine glycosylase I and DNA glycosidase I, catalyzes the hydrolysis of alkylated DNA, releasing 3-methyladenine. 3-Methyladenine-DNA glycosylase II [EC 3.2.2.21], also known as DNA-3-methyladenine glycosylase II and DNA glycosidase II, catalyzes the hydrolysis of alkylated DNA, releasing 3-methyladenine, 3-methylguanine, 7-methylguanine, and 7-methyladenine.

A. B. Britt (1996) *Ann. Rev. Plant Physiol. Plant Mol. Biol.* **47**, 75.
A. Sancar & G. B. Sancar (1988) *Ann. Rev. Biochem.* **57**, 29.
B. K. Duncan (1981) *The Enzymes*, 3rd ed., **14**, 565.

N-METHYL-L-ALANINE DEHYDROGENASE

This enzyme [EC 1.4.1.17] catalyzes the reaction of N-methyl-L-alanine with NADP$^+$ and water to produce pyruvate, methylamine, and NADPH.

N. M. W. Brunhuber & J. S. Blanchard (1994) *Crit. Rev. Biochem. Mol. Biol.* **29**, 415.

METHYLAMINE DEHYDROGENASE

This enzyme [EC 1.4.99.3], also known as amine dehydrogenase and primary-amine dehydrogenase, catalyzes the reaction of R–CH$_2$–NH$_2$ with water and an acceptor to produce R–CHO, ammonia, and the reduced acceptor. Tryptophan tryptophylquinone (TTQ) is the cofactor for this enzyme. **See** *Resonance Raman Spectroscopy; Topaquinone*

C. Anthony (1998) *Comprehensive Biological Catalysis: A Mechanistic Reference* **3**, 155.
C. Hartmann & W. S. McIntire (1997) *Meth. Enzymol.* **280**, 98.
W. S. McIntire (1995) *Meth. Enzymol.* **258**, 149.
V. L. Davidson, H. B. Brooks, M. E. Graichen, L. H. Jones & Y.-L. Hyun (1995) *Meth. Enzymol.* **258**, 176.

J. P. Klinman & D. Mu (1994) *Ann. Rev. Biochem.* **63**, 299.
V. L. Davidson (1990) *Meth. Enzymol.* **188**, 241.
J. A. Duine, J. Frank, & J. A. Jongejan (1987) *Adv. Enzymol.* **59**, 169.

3-METHYLASPARTATE AMMONIA-LYASE

This cobalamin-dependent enzyme [EC 4.3.1.2], also known as β-methylaspartase, catalyzes the conversion of L-*threo*-3-methylaspartate to mesaconate and ammonia.

K. R. Hanson & E. A. Havir (1972) *The Enzymes*, 3rd ed., **7**, 75.

β-METHYLASPARTATE-GLUTAMATE MUTASE

This cobalamin-dependent enzyme [EC 5.4.99.1], also known as glutamate mutase or methylaspartate mutase, catalyzes the interconversion of L-*threo*-3-methylaspartate and L-glutamate.

B. T. Golding & W. Buckel (1998) *Comprehensive Biological Catalysis: A Mechanistic Reference* **3**, 239.
H. A. Barker (1985) *Meth. Enzymol.* **113**, 121.
B. M. Babior & J. S. Krouwer (1979) *Crit. Rev. Biochem.* **6**, 35.
H. A. Barker (1972) *The Enzymes*, 3rd ed., **6**, 509.

METHYLATION (As a Mechanistic Probe)

Certain active site residues and/or reaction intermediates can be detected by direct methylation using diazomethane or nitromethane. Among the numerous examples are the treatment of the *N*-1'-carboxybiotin in carboxylases with diazomethane to yield the corresponding stable dimethyl ester[1] and the reaction of nitromethane at the active-site dehydroalanyl residue of phenylalanine ammonia lyase[2]. Another example is the use of diazomethane to identify the amino acid sequence of a peptide bearing the activator CO_2 in the ribulose-bisphosphate carboxylase reaction[3]. This enzyme was first activated by reaction of an activator CO_2, forming an ε-amino carbamate that was converted to the methoxycarbonyl derivative by diazomethane treatment, followed by trypsinization and sequencing of the resultant labeled peptides. The lysyl residue bearing the activator CO_2 is 26 residues removed from a lysyl residue identified by using *N*-bromoacetylethanolamine phosphate to react with the enzyme's active-site.

[1]F. Lynen, J. Knappe, E. Lorch, G. Jutting & E. Ringelman (1959) *Angew. Chem. Int. Ed. Engl.* **71**, 481.
[2]E. A. Havir & K. R. Hanson (1975) *Biochemistry* **14**, 1620.
[3]G. H. Lorimer (1981) *Biochemistry* **20**, 1236.

2-METHYL-BRANCHED-CHAIN ACYL-CoA DEHYDROGENASE

The acyl-CoA dehydrogenases are a family of mitochondrial flavoenzymes involved in fatty acid and branched chain amino-acid metabolism. In addition to long chain acyl-CoA dehydrogenases (LCADs), there are short/branched chain acyl-CoA dehydrogenase (SBCAD) that act on 2-methyl branched chain acyl-CoA substrates of varying chain lengths.

Y. Ikeda & K. Tanaka (1988) *Meth. Enzymol.* **166**, 360.

β-METHYLCROTONYL-CoA CARBOXYLASE

This biotin-dependent enzyme [EC 6.4.1.4] catalyzes the reaction of ATP with 3-methylcrotonyl-CoA and bicarbonate to produce ADP, orthophosphate, and 3-methylglutaconyl-CoA.

J. N. Earnhardt & D. N. Silverman (1998) *Comprehensive Biological Catalysis: A Mechanistic Reference* **1**, 495.
F. Lynen (1979) *Crit. Rev. Biochem.* **7**, 103.
A. W. Alberts & P.R. Vagelos (1972) *The Enzymes*, 3rd ed., **6**, 37.

2-METHYLENEGLUTARATE MUTASE

This cobalamin-dependent enzyme [EC 5.4.99.4] catalyzes the interconversion of 2-methyleneglutarate and 2-methylene-3-methylsuccinate.

B. T. Golding & W. Buckel (1998) *Comprehensive Biological Catalysis: A Mechanistic Reference* **3**, 239.
B. M. Babior & J. S. Krouwer (1979) *Crit. Rev. Biochem.* **6**, 35.
H. A. Barker (1972) *The Enzymes*, 3rd ed., **6**, 509.

5,10-METHYLENETETRAHYDROFOLATE DEHYDROGENASE

5,10-Methylenetetrahydrofolate dehydrogenase (NADP⁺) [EC 1.5.1.5] catalyzes the reaction of 5,10-methylenetetrahydrofolate with $NADP^+$ to produce 5,10-methenyltetrahydrofolate and NADPH. 5,10-Methylenetetrahydrofolate dehydrogenase (NAD⁺) [EC 1.5.1.15] catalyzes the reaction of 5,10-methylenetetrahydrofolate with NAD^+ to produce 5,10-methenyltetrahydrofolate and NADH.

V. Schirch (1998) *Comprehensive Biological Catalysis: A Mechanistic Reference* **1**, 211.
V. Schirch (1997) *Meth. Enzymol.* **281**, 146.
R. E. MacKenzie (1997) *Meth. Enzymol.* **281**, 171.
D. R. Appling & M. G. West (1997) *Meth. Enzymol.* **281**, 178.
S. W. Ragsdale (1991) *Crit. Rev. Biochem. Mol. Biol.* **26**, 261.

5,10-METHYLENETETRAHYDROFOLATE REDUCTASE

5,10-Methylenetetrahydrofolate reductase (NADPH) [EC 1.5.1.20] is an FAD-dependent enzyme that catalyzes the reaction of 5-methyltetrahydrofolate with $NADP^+$ to produce 5,10-methylenetetrahydrofolate and NADPH. 5,10-Methylenetetrahydrofolate reductase ($FADH_2$) [EC 1.7.99.5] is an FAD-dependent enzyme that catalyzes the reaction of 5-methyltetrahydrofolate with an acceptor to produce 5,10-methylenetetrahydrofolate and the reduced acceptor.

V. Schirch (1998) *Comprehensive Biological Catalysis: A Mechanistic Reference* **1**, 211.
S. W. Ragsdale (1991) *Crit. Rev. Biochem. Mol. Biol.* **26**, 261.

5,10-METHYLENETETRAHYDROFOLATE: tRNA (URACIL-5-)-METHYLTRANSFERASE ($FADH_2$ OXIDIZING)

This enzyme [EC 2.1.1.74], also known as folate-dependent ribothymidyl synthase, catalyzes the reaction of 5,10-methylenetetrahydrofolate with tRNA (with UψC) and $FADH_2$ to produce tetrahydrofolate, tRNA (with TψC) and FAD.

D. Söll & L. K. Kline (1982) *The Enzymes*, 3rd ed., **15**, 557.

3-METHYLGLUTACONYL-CoA HYDRATASE

This enzyme [EC 4.2.1.18] catalyzes the conversion of (S)-3-hydroxy-3-methylglutaryl-CoA to *trans*-3-methylglutaconyl-CoA and water.

K. M. Gibson (1988) *Meth. Enzymol.* **166**, 214.

N-METHYLGLUTAMATE SYNTHASE

This enzyme [EC 2.1.1.21], also known as methylamine–glutamate methyltransferase, catalyzes the reaction of methylamine and L-glutamate to produce ammonia and N-methyl-L-glutamate.

B. A. Palfey & V. Massey (1998) *Comprehensive Biological Catalysis: A Mechanistic Reference* **3**, 83.
L. B. Hersh (1985) *Meth. Enzymol.* **113**, 36.

γ-METHYL-γ-HYDROXY-α-KETOGLUTARATE ALDOLASE

This enzyme [EC 4.1.3.17], also known as 4-hydroxy-4-methyl-2-oxoglutarate aldolase and 4-hydroxy-4-methyl-2-oxoglutarate pyruvate-lyase, catalyzes the reversible conversion of 4-hydroxy-4-methyl-2-oxoglutarate to produce two pyruvate. 4-Hydroxy-4-methyl-2-oxoadipate and 4-carboxy-4-hydroxy-2-oxohexadioate can serve as substrates as well.

W. A. Wood (1972) *The Enzymes*, 3rd ed., **7**, 281.

METHYLHYDROXYPYRIDINECARBOXYLATE DIOXYGENASE

This FAD-dependent enzyme [EC 1.14.12.4], also known as 3-hydroxy-2-methylpyridinecarboxylate dioxygenase, catalyzes the reaction of 3-hydroxy-2-methylpyridine-5-carboxylate with NAD(P)H and dioxygen to produce 2-(acetamidomethylene)succinate and $NAD(P)^+$.

B. A. Palfey & V. Massey (1998) *Comprehensive Biological Catalysis: A Mechanistic Reference* **3**, 83.
O. Hayaishi, M. Nozaki & M. T. Abbott (1975) *The Enzymes*, 3rd ed., **12**, 119.

METHYLITACONATE Δ-ISOMERASE

This enzyme [EC 5.3.3.6] catalyzes the interconversion of methylitaconate and dimethylmaleate.

J. V. Schloss & M. S. Hixon (1998) *Comprehensive Biological Catalysis: A Mechanistic Reference* **2**, 43.

(R)-2-METHYLMALATE DEHYDRATASE

This iron-dependent enzyme [EC 4.2.1.35], also known as citraconase and citraconate hydratase, catalyzes the conversion of (R)-2-methylmalate to 2-methylmaleate and water.

H. A. Lardy (1969) *Meth. Enzymol.* **13**, 314.

(S)-2-METHYLMALATE DEHYDRATASE

This enzyme [EC 4.2.1.34], also known as mesaconase and mesaconate hydratase, catalyzes the conversion of (S)-2-methylmalate to 2-methylfumarate and water. The enzyme will also catalyze the hydration of fumarate to (S)-malate.

H. A. Lardy (1969) *Meth. Enzymol.* **13**, 314.
C. C. Wang & H. A. Barker (1969) *Meth. Enzymol.* **13**, 331.

METHYLMALONATE-SEMIALDEHYDE DEHYDROGENASE (Acylating)

This enzyme [EC 1.2.1.27] catalyzes the reaction of 2-methyl-3-oxopropanoate with coenzyme A and NAD^+ to produce propanoyl-CoA, carbon dioxide, and NADH. The enzyme will also catalyze the conversion of propanal to propanoyl-CoA.

K. Hatter & J. R. Sokatch (1988) *Meth. Enzymol.* **166**, 389.

METHYLMALONYL-CoA DECARBOXYLASE

This enzyme [EC 4.1.1.41], also known as propionyl-CoA carboxylase, catalyzes the conversion of (S)-2-methyl-3-oxopropanoyl-CoA to propanoyl-CoA and carbon dioxide. The enzyme from *Veillonella alcalescens* is a biotinyl-protein, requires sodium ions, and acts as a sodium pump.

J. N. Earnhardt & D. N. Silverman (1998) *Comprehensive Biological Catalysis: A Mechanistic Reference* **1**, 495.

METHYLMALONYL-CoA EPIMERASE

This enzyme [EC 5.1.99.1] interconverts (R)-2-methyl-3-oxopropanoyl-CoA (or, (R)-methylmalonyl-CoA) to (S)-2-methyl-3-oxopropanoyl-CoA (or, (S)-methylmalonyl-CoA).

M. E. Tanner & G. L. Kenyon (1998) *Comprehensive Biological Catalysis: A Mechanistic Reference* **2**, 7.
S. P. Stabler & R. H. Allen (1988) *Meth. Enzymol.* **166**, 400.
E. Adams (1976) *Adv. Enzymol.* **44**, 69.
L. Glaser (1972) *The Enzymes*, 3rd ed., **6**, 355.

D-METHYLMALONYL-CoA HYDROLASE

This enzyme, officially known as (S)-methylmalonyl-CoA hydrolase [EC 3.1.2.17], catalyzes the hydrolysis of (S)-methylmalonyl-CoA to form methylmalonate and coenzyme A.

R. J. Kovachy, S. P. Stabler & R. H. Allen (1988) *Meth. Enzymol.* **166**, 393.

L-METHYLMALONYL-CoA MUTASE

This cobalamin-dependent enzyme [EC 5.4.99.2] catalyzes the conversion of (R)-2-methyl-3-oxopropanoyl-CoA (or, (R)-methylmalonyl-CoA) to succinyl-CoA.

B. T. Golding & W. Buckel (1998) *Comprehensive Biological Catalysis: A Mechanistic Reference* **3**, 239.
M. L. Ludwig & R. G. Matthews (1997) *Ann. Rev. Biochem.* **66**, 269.
J. F. Kolhouse, S. P. Stabler & R. H. Allen (1988) *Meth. Enzymol.* **166**, 407.
H. A. Barker (1972) *The Enzymes*, 3rd ed., **6**, 509.

N^5-METHYLTETRAHYDROFOLATE: HOMOCYSTEINE METHYLTRANSFERASE

This cobalamin-dependent enzyme [EC 2.1.1.13], also known as methionine synthase and tetrahydropteroyl-glutamate methyltransferase, catalyzes the reaction of 5-methyltetrahydrofolate with L-homocysteine to produce tetrahydrofolate and L-methionine. Interestingly, the bacterial enzyme is reported to require S-adenosyl-L-methionine and $FADH_2$. **See also** *Tetrahydropteroyl-triglutamate Methyltransferase*

M. L. Ludwig & R. G. Matthews (1997) *Ann. Rev. Biochem.* **66**, 269.
R. Banerjee, Z. Chen & S. Gulati (1997) *Meth. Enzymol.* **281**, 189.
F. Takusagawa, M. Fujioka, A. Spies & R. L. Schowen (1998) *Comprehensive Biological Catalysis: A Mechanistic Reference* **1**, 1.
V. Schirch (1998) *Comprehensive Biological Catalysis: A Mechanistic Reference* **1**, 211.
B. T. Golding & W. Buckel (1998) *Comprehensive Biological Catalysis: A Mechanistic Reference* **3**, 239.
J. T. Jarrett, C. W. Goulding, K. Fluhr, S. Huang & R. G. Matthews (1997) *Meth. Enzymol.* **281**, 196.
R. T. Taylor & H. Weissbach (1973) *The Enzymes*, 3rd ed., **9**, 121.

METHYLTETRAHYDROMETHANOPTERIN: COENZYME M METHYLTRANSFERASE

This cobalamin-dependent enzyme catalyzes the reaction of methyltetrahydromethanopterin with coenzyme M to produce methyl-coenzyme M and tetrahydromethanopterin.

M. L. Ludwig & R. G. Matthews (1997) *Ann. Rev. Biochem.* **66**, 269.
K. M. Noll (1995) *Meth. Enzymol.* **251**, 470.

METHYLTHIOADENOSINE NUCLEOSIDE HYDROLASE

This enzyme [EC 3.2.2.16], also known as methylthioadenosine nucleosidase, catalyzes the hydrolysis of methylthioadenosine to produce adenine and 5-methylthio-D-ribose. S-Adenosylhomocysteine is not a substrate for this enzyme. **See also** *S-Adenosylhomocysteine Nucleosidase*

F. Schlenk (1983) *Adv. Enzymol.* **54**, 195.

METHYL TRANSFER REACTIONS USING S-ADENOSYLMETHIONINE

Methyltransferases that utilize S-adenosyl-L-methionine as the methyl donor (and thus generating S-adenosyl-L-homocysteine) catalyze: (a) N-methylation (*e.g.*, norepinephrine methyltransferase, histamine methyltransferase, glycine methyltransferase, and DNA-(adenine-N^6) methyltransferase), (b) O-methylation (*e.g.*, acetylserotonin methyltransferase, catechol methyltransferase, and tRNA-(guanosine-$O^{2'}$) methyltransferase), (c) S-methylation (*e.g.*, thiopurine methyltransferase and methionine S-methyltransferase), (d) C-methylation (*e.g.*, DNA-(cytosine-5) methyltransferase and indolepyruvate methyltransferase), and even (e) Co(II)-methylation during the course of the reaction catalyzed by methionine synthase[1-3].

Crystal structures of methyltransferases suggest that S-adenosyl-L-methionine binds in an extended conformation. The stereochemistry of many SAM-dependent methyltransferases has been determined by using chiral methyl groups containing all three isotopes of hydrogen. In the enzymes studied, the methyl group undergoes inversion of configuration, suggesting that the transfer of the methyl group is a direct S_N2 reaction with the methyl acceptor.

Either hydroxyl group of catechol can be methylated by catechol O-methyltransferase, albeit at different rates (*i.e.*, the enzyme does not exhibit absolute regiospecificity). The k_{cat} value for the 3-hydroxyl group is about 1 s^{-1} whereas that at the 4-position is about 0.1 or 0.2 s^{-1}. The mechanism has been reported to be ordered with SAM binding first, followed by magnesium ion, and then catechol. Interestingly, it appears that the rate-limiting step is the actual catalytic event.

Glycine N-methyltransferase is also reported to have an ordered binding mechanism with SAM binding first to the enzyme, there being no metal-ion dependency. Cooperative behavior is observed with SAM binding. The cooperative nature can be eliminated by the tryptic hydrolysis of the N-terminal eight amino acid residues.

With DNA-(cytosine-C^5) methyltransferases, the binding mechanism also appears to be ordered with an active-site cysteinyl residue required in the enzyme reaction path. However, in this case SAM is the second substrate. The binding mechanism appears to be random with DNA-(adenine-N^6) methyltransferase.

[1]R. T. Borchardt, Y. S. Wu, J. A. Huber & A. F. Wycpalek (1976) *J. Med. Chem.* **19**, 1104.
[2]J. Olsen, Y. S. Wu, R. T. Borchardt & R. L. Schowen (1979) *Transmethylation* (R. T. Borchardt, C. R. Creveling & E. Esdin, eds.), pp. 127-133, Elsevier, New York.
[3]F. Takusagawa, M. Fujioka, A. Spies & R. L. Schowen (1998) *Comprehensive Biological Catalysis: A Mechanistic Reference* **1**, 1.

MEVALDATE REDUCTASE

Mevaldate reductase [EC 1.1.1.32] catalyzes the reaction of (R)-mevalonate with NAD$^+$ to produce mevaldate and NADH. Mevaldate reductase (NADPH) [EC 1.1.1.33] catalyzes the reaction of (R)-mevalonate with NADP$^+$ to produce mevaldate and NADPH. There are reports that this enzyme may be identical with alcohol dehydrogenase (NADP$^+$) [EC 1.1.1.2].

G. Popják (1969) *Meth. Enzymol.* **15**, 393.

MEVALONATE KINASE

This enzyme [EC 2.7.1.36] catalyzes the reaction of ATP with (R)-mevalonate to produce ADP and (R)-5-phosphomevalonate. The nucleotide substrate can also be provided with CTP, GTP, or UTP.

R. A. Dixon, P. M. Dey & C. J. Lamb (1983) *Adv. Enzymol.* **55**, 1.
G. J. Schroepfer, Jr. (1981) *Ann. Rev. Biochem.* **50**, 585.
E. D. Beytía & J. W. Porter (1976) *Ann. Rev. Biochem.* **45**, 113.

MEYERHOF OXIDATION QUOTIENT

A quantitative index describing the effect of the absence or presence of molecular oxygen on the rate of glycolysis or fermentation:

$$Q_{\text{Meyerhof}} = \frac{Rate_{\text{Anaerobic Fermentation}} - Rate_{\text{Aerobic Fermentation}}}{Rate_{\text{Molecular Oxygen Uptake}}}$$

The quotient is usually about two for most cells, meaning that the consumption of one O$_2$ molecule usually diverts two molecules of either lactate or ethanol by forming their oxidative counterparts.

MICELLES

Organized supramolecular aggregates formed by surface active substances in aqueous solutions (Fig. 1). Micellar monolayers or bilayers are formed by burying apolar groups toward the interior while exposing hydrophilic head groups at the water-surfactant interface. Two types of surfactants form micelles: ionic surfactants (*i.e.*, those containing positively or negatively charged head groups) or nonionic surfactants (*i.e.*, those with nonionic hydrophilic head groups).

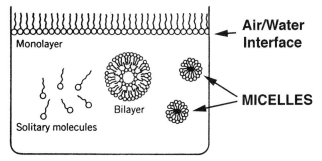

Figure 1. Various physical states of phospholipids in aqueous solution. Note the following features: (a) phospholipids residing at the air/water interface are arranged such that their polar head groups maximize contact with the aqueous environment, whereas apolar side chains extend outward toward the air; (b) solitary phospholipid molecules remain in equilibrium with various monolayer and bilayer structures; (c) bilayer vesicles and micelles remain in equilibrium with solitary phospholipid molecules, provided that the total lipid content exceeds the critical micelle concentration.

The term "mixed micelle" refers to those micelles composed of two or more surface active agents. The sizes of micelles in a solution obey a distribution function that is characteristic of their chemical composition and the ionic nature of the solution in which they reside.

FORMATION. Aqueous solutions of highly surface-active substances spontaneously tend to reduce interfacial energy of solute-solvent interactions by forming micelles. The critical micelle concentration (or, c.m.c.) is the threshold surfactant concentration, above which micelle formation (also known as micellization) is highly favorable. For sodium dodecyl sulfate, the c.m.c. is 5.6 mM at 0.01 M NaCl or about 3.1 mM at 0.03 M NaCl. The lower c.m.c. observed at higher salt concentration results from a reduction in repulsive forces among the ionic head groups on the surface of micelles made up of ionic surfactants. As would be expected for any entropy-driven process, micelle formation is less favorable as the temperature is lowered.

At their critical micelle concentrations, surface active agents (such as sodium dodecyl sulfate, Triton X-100, lysolecithin, and bile salts) self-associate into spherical or rod-shaped structures. Because dilution to below the c.m.c. results in rapid disassembly or dissolution of these detergent micelles, micelles are in dynamic equilibrium with other dissolved detergent molecules in the bulk solution.

ENERGETICS. The standard free energy change for micelle formation from 1.0 mole surfactant is given by the relationship,

$$\Delta G^\circ = -(RT/m)\ln K$$

where K is the equilibrium constant and m is number of monomeric units per micelle. If c is taken to be the stoichiometric concentration of the surfactant solution, and if x designates the fraction of monomers in the aggregated state, then:

$$K = (cx/m)/[c(1-x)]^m$$

and the standard state Gibbs free energy can also be restated as:

$$\Delta G^\circ = -(RT/m)\ln[cx/m] + RT\ln[c(1-x)]$$

At the critical micelle concentration (or c.m.c.), x must be zero because all the monomers are disaggregated; thus,

$$\Delta G^\circ = RT\ln[\text{c.m.c.}]$$

(Note the change of sign.) Likewise, we can define the standard entropy and enthalpy for micellization as follows:

$$\Delta S^\circ = d(\Delta G^\circ)/dT = -RT\,d\{\ln[\text{c.m.c.}]\}/dT - RT\ln[\text{c.m.c.}]$$

$$\Delta H^\circ = -RT^2\,d\{\ln[\text{c.m.c.}]\}/dT$$

In general, the standard enthalpy of micellization is large and negative, and an increase in temperature results in an increase in the c.m.c.; the positive entropy of micellization relates to the increased mobility of hydrocarbon side chains deep within the micelle as well as the hydrophobic effect. Hoffmann and Ulbricht[1] have provided a detailed account of the thermodynamics of micellization, and the interested reader will find that their tabulated thermodynamic values and treatment of models for micellar aggregation processes are especially worthwhile.

MICELLAR CATALYSIS. Chemical reactions can be accelerated by concentrating reactants on a micelle surface or by creating a favorable interfacial electrostatic environment that increases reactivity. This phenomenon is generally referred to as micellar catalysis. As pointed out by Bunton[2], the term "micellar catalysis" is used loosely because enhancement of reactivity may actually result from a change in the equilibrium constant for a reversible reaction. Because catalysis is strictly viewed as an enhancement of rate without change in a reaction's thermodynamic parameters, one must exercise special care to distinguish between kinetic and equilibrium effects. This is particularly warranted when there is evidence of differential interactions of substrate and product with the micelle. Micelles composed of optically active detergent molecules can also display stereochemical action on substrates[3].

Catalysis can occur when micellized surfactants are themselves chemically inert, and this effect relies on electrostatic and hydrophobic interactions that alter the susceptibility of reactants to nucleophilic attack or electron transfer. Catalysis can also take place with the micelle providing the acidic/basic as well as nucleophilic/electrophilic groups needed for reaction. In either case, the rate enhancement is designated as k_m/k_w where the subscripts indicate rate constants applying to the micellar and water phases:

Micellar Phase $D_n + S = SD_n \rightarrow P$

Water Phase $S \rightarrow P$

where D_n is made up of n detergent molecules, and S and P are substrate and product, respectively. (Note the analogy between the micellar path and the Michaelis-Menten kinetic mechanism for enzyme action. Both display saturation kinetics, the former dependent on substrate binding at the interface of the micelle instead of at a discrete enzyme active site.) In Bunton's treatment[2], the concentration of micelles C_M is equal to $(C_D -$ c.m.c.$)/N$ where C_D is the total concentration of surfactant (detergent) and N is the number of detergent molecules per micelle. The rate constant for the above mechanism can be obtained as,

$$k_\psi = (k_w + k_m K C_m)/(1 + K C_m)$$

This equation indicates that the observed rate will be equal to the water phase reaction rate until the added surfactant is present at sufficient concentrations to form micelles. Added complexity arises if the substrate selectively interacts with a fraction of the micelles having a particular size and/or shape.

MICELLAR SUBSTRATES. Phospholipids in micelles are frequently found to be more active substrates in lipolysis than those phospholipids residing in a lipid bilayer[4]. Dennis[4] first described the use of Triton X-100 to manipulate the amount of phospholipid per unit surface area of a micelle in a systematic analysis of the interfacial interactions of lipases with lipid micelles. Verger[5] and Jain et al.[6] have presented cogent accounts of the kinetics of interfacial catalysis by phospholipases. The complexity of the problem is illustrated in the diagram shown in Fig. 2 showing how the enzyme in the aqueous phase can bind to the interface (designated by the asterisk) and then become activated. Once this is achieved, E* catalyzes conversion of S* to release P*.

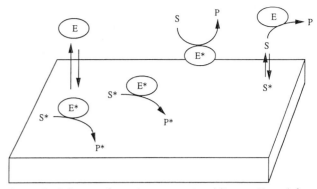

Figure 2. Behavior of membrane-associated lipases. From left to right: (a) catalytic action of an enzyme that first requires attachment to the substrate at the water-membrane interface; (b) action of an integral membrane enzyme that remains attached to the membrane where the enzyme finds its substrate; (c) action of a membrane-bound enzyme on substrates in the aqueous medium; and (d) action of an enzyme in the aqueous phase on a substrate that must first desorb from the membrane before it can interact with enzyme. From Jain et al.[6] with permission of the authors.

[1]H. Hoffmann & W. Ulbricht (1986) *Thermodynamic Data for Biochemistry and Biotechnology* (H.-J. Hinz, ed.), pp. 297-348, Springer-Verlag, Berlin.
[2]C. A. Bunton (1976) in *Techniques of Chemistry*, **10**, pp. 731-814, Wiley, New York.
[3]R. A. Moss, C. J. Talkowski, D. W. Reger & C. E. Powell (1973) *J. Amer. Chem. Soc.* **95**, 3215.
[4]E. A. Dennis (1973) *J. Lipid Res.* **14**, 152.
[5]R. Verger (1980) *Meth. Enzymol.* **64**, 340.
[6]M. K. Jain, M. H. Gelb, J. Rogers & O. G. Berg. (1995) *Meth. Enzymol.* **249**, 567.

Selected entries from *Methods in Enzymology* [**vol**, page(s)]: Activation of lipolytic enzymes by interfaces, **64**, 341; model for lipase action on insoluble lipids, **64**, 345; interfacial enzyme inactivation, **64**, 347; reversibility of the adsorption step, **64**, 347; monolayer substrates, **64**, 349; kinetic models applicable to partly soluble amphiphilic lipids, **64**, 353; surface dilution model, **64**, 355 and 364; practical aspects, **64**, 357.

MICHAELIS COMPLEX (Distal Motions Affecting E·S Formation)

Two-dimensional heteronuclear (^{1}H-^{15}N) nuclear magnetic relaxation studies indicate that the dihydrofolate reductase·folate complex exhibits a diverse range of backbone fluctuations on the time-scale of picoseconds to nanoseconds[1]. To assess whether these dynamical features influence Michaelis complex formation, Miller et al.[2] used mutagenesis and kinetic measurements to assess the role of a strictly conserved residue, namely Gly-121, which displays large-amplitude backbone motions on the nanosecond time scale. Deletion of Gly-121 dramatically reduces the hydride transfer rate by 550 times; there is also a 20-times decrease in NADPH cofactor binding affinity and a 7-fold decrease for NADP$^+$ relative to wild-type. Insertion mutations significantly decreased both substrate and cofactor binding. Their results suggest that distant residues, such as Gly-121 in DHFR, can strongly influence the formation of liganded complexes as well as the proper orientation of substrate and cofactor during the catalytic cycle.

[1]D. M. Epstein, S. J. Benkovic & P. E. Wright (1995) *Biochemistry* **34**, 11037.
[2]G. P. Miller & S. J. Benkovic (1998) *Biochemistry* **37**, 6327.

MICHAELIS CONSTANT

1. The quotient of rate constants obtained in steady-state treatments of enzyme behavior to define a substrate's interaction with an enzyme. While the Michaelis constant (with overall units of molarity) is a rate parameter, it is not itself a rate constant. Likewise, the Michaelis constant often is only a rough gauge of an enzyme's affinity for a substrate. 2. Historically, the term "Michaelis constant" referred to the true dissociation constant for the enzyme-substrate binary complex, and this parameter was obtained in the Michaelis-Menten rapid-equilibrium treatment of a one-substrate enzyme-catalyzed reaction. In this case, the Michaelis constant is usually symbolized by K_s. 3. The value equal to the concentration of substrate at which the initial rate, v, is one-half the maximum velocity (V_{max}) of the enzyme-catalyzed reaction under steady state conditions.

In the Briggs-Haldane steady-state treatment of a one-substrate enzyme system, the Michaelis constant, usually symbolized by K_m, is $(k_2 + k_3)/k_1$. For more complex reactions (*e.g.*, with several substrates and/or isomerization steps), the Michaelis constant for a given substrate is a more complex collection of rate constants. For a multisubstrate enzyme having substrates A and B, the Michaelis constants are usually symbolized by K_a and K_b, by K_{mA} and K_{mB}, or by K_m^A and K_m^B, respectively.

MICHAELIS CONSTANT (Apparent)

1. A parameter that depends on the concentration of another substrate in a multisubstrate reaction or on one or more cofactors or substances that influence reaction rate. It is an approximation of the true Michaelis constant. 2. A parameter obtained under conditions that do not rigorously satisfy the requirements of initial rate measurements.

MICHAELIS-MENTEN EQUATION

A model first advanced by Victor Henri[1] and later by Leonor Michaelis and Maud Menten[2] to account for the kinetic properties of a one-substrate, one-product enzyme-catalyzed reaction.

ASSUMPTIONS. Although one does not frequently state all of the requirements for deriving the Michaelis-Menten equation, the model assumes (a) that all experimental conditions (*e.g.*, pH, temperature, ionic strength, *etc.*) remain constant over the course of the experiment; (b) that the enzyme, substrate, and product are all stable under the assay conditions over the time course of the experiment; (c) that at least one binary central complex does, in fact, form (*e.g.*, ES); (d) that E and S react rapidly and reversibly to form this complex; (e) that the binding of S to E can be described by an equilibrium dissociation constant, $K_s = k_2/k_1$; (f) that the conversion of ES to E + P is slow and irreversible (or, if reversible, the concentration of [P] over the time course of the experiment is so low that the reverse reaction is negligible); (g) that the total enzyme concentration [E_{total}] is the sum of uncomplexed enzyme concentration, [E], and the enzyme-substrate central complex (*i.e.*, [E_{total}] = [E] + [ES]; (h) that the velocity, v, of the reaction, equal to d[P]/dt, equals [ES] multiplied by the unimolecular rate constant k_3 (thus, only considering the forward reaction [assumption (f) above]); (i) that the concentration of substrate is essentially unchanged during the initial velocity period (*i.e.*, $\Delta[S] \ll [S_{initial}]$); (j) that the total substrate concentration, [S_{total}], far exceeds the total enzyme concentration, [E_{total}], such that [S_{total}] $\approx$ [S], where [S] is the uncomplexed substrate concentration.

DERIVATION OF RAPID EQUILIBRIUM EXPRESSION. We can begin with two equations:

$$K_s = [E][S]/[ES] \quad \text{and} \quad [E_{total}] = [E] + [ES]$$

The first equation can be rewritten as [E] = K_s[ES]/[S] and substituted for [E] in the second equation to yield

$$[E_{total}] = (K_s[ES]/[S]) + [ES]$$

Dividing both sides by [ES], we get

$$[E_{total}]/[ES] = (K_s/[S]) + 1$$

Then, by multiplying the left side by k_3/k_3 (noting that $v = k_3$[ES] and defining the maximum velocity, V_{max}, as being k_3[E_{total}]), we get:

$$v = \frac{V_{max}[S]}{K_s + [S]}$$

The derivation of the steady-state enzyme rate equation for the single substrate enzyme-catalyzed reaction is provided in the entry entitled *Enzyme Rate Equations (I. The Basics)*.

GRAPHICAL REPRESENTATION. The above expression represents the equation of a hyperbola (*i.e.*, $f(x) = ax/(b + x)$ where a and b are constants) and a plot of the initial velocity as a function of [S] will result in a rectangular hyperbola. Another way for representing the Michaelis-Menten equation is by using the double-reciprocal (or, Lineweaver-Burk[3]) transformation:

$$\frac{1}{v} = \frac{1}{V_{max}} + \left(\frac{K_m}{V_{max}}\right)\frac{1}{[S]}$$

This method is widely used because it is a linear transformation of the hyperbolic function describing the rate saturation process. Both plots are shown below for the same data set.

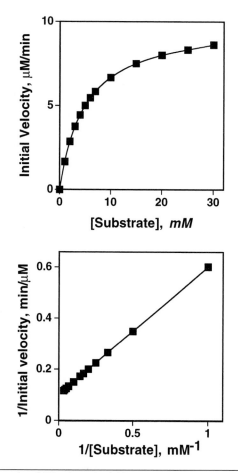

Double-reciprocal plots can be reasonably accurate if rate data can be obtained over a reasonable range of saturation, say from $0.3V_{max}$ to $0.8V_{max}$. Any indication of a curved double-reciprocal plot necessitates consideration of substrate activation (downward curvature as one proceeds to higher (1/[S]) values), substrate inhibition (usually evident by upward curvature as one approaches the $1/v$-axis), multiple binding of S, failure to measure true initial rates, or cooperativity.

INTEGRATED MICHAELIS-MENTEN EQUATION. Integrated rate equations for the treatment of enzyme kinetics can be traced back to the studies of Henri[4]. In principle, the integrated rate equation approach offers the advantage that a single reaction progress curve can provide all of the rate parameters typically obtained in initial rate studies carried out in the absence and presence of product (***See*** *Progress Curve Analysis*). Furthermore, the method avoids mixing errors that can influence the outcome of a series of initial rate measurements made separately at different substrate and product concentrations. Stayton and Fromm[5] have written an illuminating discussion regarding the validity of the integrated Michaelis-Menten equation, based on their detailed computer simulation studies.

[1]V. Henri (1903) *Lois Generales de l'Action des Diastases* (translated: "General Laws of Diastase Action"), Hermann, Paris.
[2] L. Michaelis & M. L. Menten (1913) *Biochem. Zeit.* **49**, 333.
[3]H. Lineweaver & D. Burk (1934) *J. Amer. Chem. Soc.* **56**, 658.
[4]V. Henri (1904) *Arch. Fisiol.* **1**, 299.
[5]M. M. Stayton & H. J. Fromm (1979) *Meth. Enzymol.* **78**, 309.

MICHAELIS-MENTEN KINETICS

An enzyme is said to "obey" Michaelis-Menten kinetics, if a plot of the initial reaction rate (in which the substrate concentration is in great excess over the total enzyme concentration) *versus* substrate concentration(s) produces a hyperbolic curve. There should be no cooperativity apparent in the rate-saturation process, and the initial rate behavior should comply with the Michaelis-Menten equation, $v = V_{max}[A]/(K_a + [A])$, where v is the initial velocity, [A] is the initial substrate concentration, V_{max} is the maximum velocity, and K_a is the dissociation constant for the substrate, A, binding to the free enzyme. The original formulation of the Michaelis-Menten treatment assumed a rapid pre-equilibrium of E and S with the central complex EX. However, the steady-state or Briggs-Haldane derivation yields an equation that is isomorphic: $v = V_{max}[A]/(K_m + [A])$ where K_m is a composite of a number of rate constants and is numerically equivalent to the substrate concentration at which the initial velocity is $0.5V_{max}$. Thus, the general rubric of "Michaelis-Menten kinetics" extends to systems exhibiting hyperbolic kinetics, irrespective of whether one relies on the original rapid equilibrium condition or the later steady-state assumption.

Some investigators also unnecessarily apply the further restriction that Michaelis-Menten kinetics refers only to enzymes catalyzing the conversion of a single substrate to a single product. Were this taken to its extreme, only isomerases would qualify, because most one-substrate systems utilize water as a second substrate or product. *See* Michaelis-Menten Equation; Uni Uni Mechanism; Enzyme Rate Equations (1. The Basics)

MICRO-

A prefix, symbolized by μ, used in submultiples of units and corresponding to a value of 10^{-6}.

MICROENVIRONMENT

The ionic and solvent environment of a specific localized region (*e.g.*, the near environment surrounding a substrate or residue residing within an enzyme's active site).

MICROSCOPIC DIFFUSION CONTROL

Any limitation on reaction rates in a homogeneous medium by the frequency of forming encounter complexes from reacting molecular entities (also called "encounter control"). For any instantaneously occurring bimolecular reaction in a homogeneous medium (*i.e.*, a reaction proceeding significantly more quickly than the encounter rate), the overall rate of the reaction is governed solely by the diffusion rates of the reactants. The alternative term is "total microscopic diffusion control". The term "partial microscopic diffusion control" is often used to describe cases where diffusion and chemical conversion occur on comparable time-scales. *See also* Macroscopic Diffusion Control

MICROSCOPIC REVERSIBILITY

A principle requiring that, under equilibrium conditions, any molecular process (or reaction) and the reverse of that process (or reaction) will occur with the same frequency. This principle was verified through the use of certain quantum mechanical expressions for transition probabilities.

The Principle of Microscopic Reversibility and its large-scale consequence, known as the Principle of Detailed Balancing enable investigators to understand the mechanism of the reverse reaction to the same level of accuracy as that achieved for the forward reaction.

This principle has only limited application to reactions that are not at equilibrium. Furthermore, the Principle of Microscopic Reversibility does not apply to reactions commencing with photochemical excitation. *See also* Detailed Balancing, Principle of; Chemical Reaction; Chemical Kinetics

MICROTUBULE ASSEMBLY KINETICS

The $\alpha\beta$ tubulin dimer contains all of the structural information required for the spontaneous formation of microtubules[1,2]. Tubulin polymerization readily occurs *in vitro* when the concentration of the $\alpha\beta$ tubulin dimer is sufficient to support spontaneous nucleation and elongation (see below). The following sections describe *in vitro* microtubule (MT) self-assembly. Intracellular MT assembly and diassembly uses similar elongation, dynamic instability, and treadmilling processes. However, intracellular MTs are stabilized through the association of one MT end with microtubule-organizing centers (*e.g.*, centrioles and ciliary/flagellar axonemes). *See also* Biochemical Self-Assembly; Actin Assembly Kinetics; Hemoglobin S Polymerization; Indefinite Polymerization; Isodesmic Polymerization; Microtubule Treadmilling

1. NUCLEATION. The polymer self-assembly theory of Oosawa and Kasai[3] treats nucleation as a highly cooperative and unfavorable event. Their kinetic theory for nucleation permits one to obtain information about the size of the polymerization nuclei, provided that two basic assumptions can be satisfied experimentally. First, the rate of nuclei formation is assumed to be proportional to the i_0th power of the protomer concentration, with i_0 representing the number of protomers required to create the nucleus. Second, the treatment deals only with that period during which the polymerization rate greatly exceeds the rate of protomer loss from the polymers (*i.e.*, the initial stage of polymerization when the protomer concentration is the highest).

1a. Theoretical Considerations. Oosawa and Kasai[3] also assumed nucleation involved formation of a higher order "closed" structure from an "open" structure with the same number of protomers. The nucleation step involved the conversion of an unstable linear trimer to a closed helical trimer; the greater stability of the closed structure arises from additional interprotomer bonding in the closed structure. Assuming that the rate of nuclei formation is proportional to the third power of the protomer concentration, c_1^3, the following equation results:

$$\frac{dc_p}{dt} = (k_+ c_1 - k_-) \int_0^t A c_1^3 dt = -\frac{dc_1}{dt}$$

where c_p is the total concentration of protomer in the polymer form, k_+ and k_- are the assembly and disassembly rate constants, and A is yet another constant which incorporates the rate constants for nucleation. The integral in this equation calculates the polymer number concentration at time t, and the net rate of polymer formation, dc_p/dt, is simply the number concentration times the rate of polymer formation for a single polymer (i.e., $k_+ c_1 - k_-$). The last part of the equality reflects the definition that polymer formation parallels the rate of loss of protomer concentration.

The above equation can be solved only if one adds a further restriction that $k_+ c_1$ greatly exceeds k_-. Under this condition, we arrive at the following expression:

$$\ln \left[\frac{1 + [1 - (c_1/c_0)^3]^{1/2}}{1 - [1 - (c_1/c_0)^3]^{1/2}} \right] = 3[(2/3)k_+ A]^{1/2} c_0^{3/2} t$$

Here, c_0 is the total protein concentration, and a plot of polymer formed *versus* reaction time can be made by substituting decreasing values of c_1 (recalling that c_p equals c_0 minus c_1) into this second expression and solving for t. Oosawa and Asakura[4] presented a more general solution for a nucleus composed of i_0 protomers, and the observed cooperativity depends on the number of protomers engaging in nucleation, as shown in the following expression:

$$\ln \left[\frac{1 + [1 - (c_1/c_0)^{i_0}]^{1/2}}{1 - [1 - (c_1/c_0)^{i_0}]^{1/2}} \right] = 2i_0^{1/2}[k_+ k_+{}^*]^{1/2} c_0^{i_0/2} t$$

Note that $k_+{}^*$ is the forward rate constant for nucleation. If one chooses a particular value of c_1/c_0 (e.g., one-half), then the left-hand side of the equation will remain constant, and one may measure the time, t, required to

obtain this particular c_1/c_0 ratio for different values of c_0. Under these conditions, this expression can be rearranged into the following forms:

$$t = (\text{constant})c_0^{-i_0/2}$$

$$\log t = -(i_0/2)\log c_0 + \log (\text{constant})$$

A plot of $\log t$ *versus* $\log c_0$, where t is the time required to achieve a final c_1/c_0 ratio, yields the number of protomers in the nucleus by determining the slope. Although the Oosawa treatment originally used a c_1/c_0 value equal to 0.5, detailed computer simulation work shows that the $k_+ c_1 \gg k_-$ assumption introduces significant error in the estimation of i_0 by the time that the reaction proceeds to the point that c_1/c_0 is at the value of one-half[1]. Thus, one must actually work under conditions where a smaller c_1/c_0 value is obtained. The above theory resembles the Hill Equation for cooperative ligand binding, because the nucleation step is assumed to be of high reaction order and no intermediate states of fewer than i_0 protomers accumulate. A difficulty with this theory is the need to establish that there is only protomer in solution. In the case of cold-depolymerized microtubule (MT) protein, for example, there are many oligomeric species that complicate the experimental outcome.

Wegner and Engel[5] extended the Oosawa analysis by dropping the assumption of irreversibility and by applying a steady-state approximation. They focused on a dimeric nucleus, and there is no readily available way for extending their treatment to more cooperative systems to yield tractable rate expressions. Nonetheless, their steady-state solutions agreed well with those obtained by direct numerical integration of the rate laws for two different sets of dimerization rate constants, and their analysis provides a rather satisfying view for actin polymerization.

1b. Experimental Considerations. Weisenberg[6] first showed MT reassembly *in vitro* is a spontaneous process, one that does not require addition of nucleating centers. Borisy and Olmsted[7] reported that microtubule assembly from porcine brain homogenates was clearly stimulated by the presence of high-molecular-weight structures and not just tubulin monomers. Cell extracts obtained by high-speed centrifugation formed few tubules, even though they were rich in tubulin. By electron microscopy, they observed characteristically disk-shaped structures in the low-speed supernatants that disappeared as MT reassembly progressed. That these ring structures disappear during tubule formation does not establish a direct causal relationship between ring structures and the true nucleating species for reassembly. In the strictest sense, one must reserve the designation "nucleation center" for the simplest "embryonic" oligomeric structures that permit elongation reactions. Nonetheless, their discovery of a means for achieving seeded assembly from the ring structures stimulated widespread interest in oligomeric rings as microtubule precursors.

There are several potential fates of ring structures during MT assembly[1]. First, they might serve as direct intermediates in MT assembly in much the same way that tobacco mosaic virus (TMV) protein disks stack to form rodlike particles. Second, rings may act as templates controlling the geometrical arrangement of protomers during assembly. Third, rings could break open into linear structures resembling protofilaments or they might open slightly to form "lock-washer" structures with their own incipient helical configuration. Such protofilament-like structures could interact with each other laterally to form structures that close to form a MT-like arrangement. Fourth, the ring structure *per se* might not be a true nucleating center; yet, oligomeric structures issuing from rings could isomerize into nucleation centers. Finally, ring structures could merely be a cold-stable tubulin polymorph, lacking function in reassembly formation. In this case, the ring structures could still be a depot or reservoir of tubulin. Timasheff and Grisham[8] also pointed out: (a) that the various mechanistic proposals need not be mutually exclusive; (b) that solvent composition could influence the pathway of reassembly; and (c) that ring structures, while interesting structural isomers of oligomeric species, may have no direct role in assembly.

Johnson and Borisy[9] first showed that the lag phase in the plot of turbidity (*i.e.*, polymer weight concentration) *versus* time accounted for only 5—10% of the entire amplitude obtained upon completion of the polymerization process. By fitting the elongation phase to a single exponential process, these investigators arrived at the correct conclusion that microtubule number concentration becomes relatively stable within the first minutes

after warming of cold-depolymerized microtubule protein. The constancy of microtubule number concentration during elongation has also been directly observed with bovine brain microtubule assembly by electron microscopy using samples fixed by dilution into microtubule stabilizing buffer under conditions which disfavor further nucleation[10]. Thus, nucleation does not occur to any appreciable extent once the lag phase ends.

Consider the case where the protein consists of $\alpha\beta$ dimers exclusively at the very beginning of an assembly experiment. Suppose further that spontaneous nucleation is sufficiently infrequent as the polymerization reaction reaches 5–10% of its maximal amplitude achieved over the remaining course of elongation. In this case, a reduction of the protomer concentration from about 20 to 18 μM would reduce the apparent extent of nucleation by a factor of about 10–20, such that the polymer number concentration remains fixed throughout the ensuing elongation phase. If nucleation were viewed as a one-step cooperative event, then the rate of nucleation would be proportional to the ith power of the protomer concentration if I protomers cooperatively form the polymerization nucleus:

$$i(\text{Protomers}) = (\text{Nucleus})$$

The equilibrium constant for this process would be given as follows.

$$K = [\text{Protomer}]^i/[\text{Nucleus}]$$

One would conclude that I must approximately equal 28 for this process! Hofrichter *et al.*[11] found a similar behavior in nucleation of human hemoglobin S (HbS); the apparent reaction order for the nucleation of HbS aggregation was about 32 (**See** *Hemoglogin S Polymerization*). Of course, such analyses are not fully justifiable, because one cannot assume ideality in the solution properties at high protein concentrations (**See** *Molecular Crowding*).

Engelbroughs *et al.*[12] carried out a kinetic analysis with the porcine brain protein utilizing exposure of rings to 0.8 *M* without reformation of ring structures to any major extent, as inferred from the much slower rate of assembly of such diluted protein as compared with diluted protein not receiving the high-salt treatment. They concluded that nucleation starts by the association of rings that are probably broken open upon warming. Because salt treatment obviates the need to fractionate the microtubule-associated proteins (MAPs) from the tubulin, their studies provided a means for analyzing the system at constant MAPs/tubulin ratios.

Karr and Purich[13] concluded that ring structures are themselves not microtubule assembly intermediates. They estimated that a 1–2 mg/mL sample of whole microtubule protein would contain around 10^{-7} M ring structures, if nearly 40% of the total cold-depolymerized protein elutes with the ring fraction. Yet, the MT number concentration during self-assembly is much lower, especially because the free tubulin concentration would be in considerable excess over the critical MT protein concentration. Karr and Purich[13] used 1 mM GDP to inhibit microtubule elongation and followed the time-course of ring disintegration after warming to 30°C. By assuming the change in turbidity at 350 nm is a linear measure of the protein weight concentration, the loss in ring structure concentration followed an exponential decay ($t_{0.5} = 73$ seconds). Ring dissociation was sufficiently long-lived to permit production of smaller oligomers that might serve as nucleation centers. Terry and Purich[14] also carried out pulse-chase measurements that demonstrated nearly complete equilibration of radiolabeled tubulin (Tb) E-site nucleotide warmed in the presence of unlabeled nucleotide. The MT protein concentrations used in these experiments were in the range of 1–2 mg/mL; at higher concentrations, elongation may be sufficiently rapid so as to incorporate more tubulin oligomers into microtubule polymer.

Mandelkow et al.[15] provided direct kinetic studies of nucleation by using 1 Å synchrotron radiation to obtain time-resolved scattering data during cycles of assembly and disassembly after temperature shifts between 4 and 36°C. Small-angle scattering theory requires independent scattering from all particles in the solution, and the theory relates the intensity of scattering to other parameters as follows:

$$I(h, t) \propto \Sigma \, x_k(t)^* p_k(t)^* i_k(h)$$

where h is $(4\pi \sin \theta)/\lambda$ with θ the angle of scattering and λ the wavelength of X-radiation. The sum, $\Sigma x_k(t)$ is taken over all aggregates k; $x_k(t)$ is the fraction of Tb promoters in an aggregate k; $p_k(t)$ represents the degree of polymer-ization (i.e., the number of protomers in a given aggregate); and $i_k(h)$ is the scattering of function characteristic of an aggregate k normalized to unity at $h = 0$. Because $x_k(t)$ is the fraction of subunits in a particular aggregate, $\Sigma x_k(t)$ represents a conservation term. The $i_k(h)$ term contains structural information about each aggregate. Mandelkow et al.[15] treated the rings as having apparent diameters of 375 Å and the MTs as having a mean diameter of about 350 Å. They recorded sufficient intensity data over time frames of 15 seconds during the assembly and disassembly events, and the angular dependence was evaluated from 0 to 30 mrad. They confirmed that ring structures breakdown to form small oligomers within 1 minute following warming to 36°C. Loss of rings leads to a condition where the main contribution to the scattering intensity can be interpreted in terms of a solution containing tubulin protomers and oligomers for the most part.

2. ELONGATION KINETICS. Elongation is the repetitive addition of protomers to the growing points at the ends of the polymer[1]. The foregoing discussion indicates that the life-span of nucleation should be relatively short. In this respect, nucleation and elongation phases are temporally separate and can be treated as only slightly overlapping kinetic phases. Elongation can also be studied in the absence of spontaneous nucleation by using pre-existing microtubules or microtubule organizing centers (MTOCs).

2a. Theoretical Considerations. To treat elongation kinetics[1], one typically assumes: (1) that all of the growing points are identical and well described by a single bimolecular rate constant for addition and a second rate constant for protomer loss; (2) that the addition of a protomer to a growing point creates a new growing point of identical reactivity, such that the rate for addition or loss does not depend on polymer length; (3) that continuous nucleation does not occur and the polymer number concentration remains constant; and (4) that other pathways for elongation, including addition of oligomers or polymers, are negligible. If these assumptions hold, elongation reactions can be written simply as follows:

$$MT_i + Tb \underset{k_-}{\overset{k_+}{\rightleftharpoons}} MT_{i+1}$$

where MT_i is a microtubule containing i protomers, Tb is a tubulin protomer, and k_+ and k_- are the corresponding rate constants for protomer addition and loss, respectively. The rate equation for this process is

$$dc_1/dt = k_-m - k_+mc_1 = (k_- - k_+c_1)m$$

where c_1 is the concentration of free protomers, and m is the constant concentration of polymer ends. Assuming that only a small fraction of the initial protein concentration (c_0) is depleted to create the nuclei during nucleation, then c_1 at the onset of elongation equals c_0. This also requires that any preexisting oligomers (*e.g.*, ring structures and protofilaments) comprise only a small part of the total protein at the start of the elongation phase. This rate equation describes a single exponential process, and the following integrated rate law describes the time evolution of the elongation process:

$$c_1(t) = k_-/k_+ + [c_0 - (k_-/k_+)]\exp(-k_+mt)$$

The equilibrium value of c_1 is k_-/k_+ or K_D, and the equilibrium constant should be independent of the total protein concentration in the polymerization reaction. Because all of the protein is considered to be in either the protomer or polymer forms, we may use the relation that $c_p = c_0 - c_1$, where c_p is the total concentration of protomers in the polymeric state. This relationship can be substituted into the above equation to yield a result for the time course of polymer formation:

$$c_p(t) = [c_0 - [k_-/k_+)][1 - \exp(-k_+mt)]$$

The quantity $c_p(t)$ can be measured experimentally by several techniques, but turbidimetric procedures are generally the preferred means. If the polymer is to form, c_0 must be greater than k_-/k_+. Thus, Oosawa's model predicts that there is a critical protomer concentration of k_-/k_+ (or K_D) below which polymer cannot exist.

The above theory can be extended to deal with other more complex cases. For example, the two ends of a biopolymer need not behave identically, and, as noted earlier, MTs are helical polymers of asymmetric protomer units. Thus, two sets of on- and off-constants might be necessary. In other cases, such as in the polymerization of tubulin in the presence of tubulin–colchicine complex (Sternlicht *et al.*[16]), one may need to consider copolymerization. The kinetics of microtubule depolymerization are the reverse of elongation, and are gener-

ally unaffected by processes as complicated as nucleation. Thus, depolymerization can often provide elongation rate constants more directly.

2b. Experimental Considerations. There are several methods for obtaining elongation rate constants. The rate constant for polymerization, k_+, can be determined from assembly kinetics data and a knowledge of the polymer number concentration, m at time t and equilibrium (eq), by plotting $\ln\{[c_p(t) - c_p(eq)]/c_p(eq)]\}$ *versus* time. The slope of this plot is $(-k_+m)$, as can be readily determined using a value of m obtained by microtubule length distribution measurements from dark-field[16] or electron[17] microscopy. In both cases, the average polymer length is determined and subsequently divided into the total polymer length. This quantity is estimated from the amount of protein in the polymer form and the number of protomers per unit length of polymer. Once k_+ is evaluated, one may obtain k_- with a knowledge of the critical concentration, K_D. A variation on this theme is provided by utilizing needed assembly with a known number of microtubule seeds prepared by mechanical shearing and measuring the initial rate of assembly at various concentrations of protomer. Alternatively, one may resort to depolymerization experiments to get k_- under conditions where the polymers are induced to disassemble irreversibly. This may be accomplished by rapidly dropping the temperature of the polymer solution or by rapidly reducing the protomer concentration to a value far below the prevailing critical concentration. The latter has the advantage that the rate constants refer to the temperature at which other tubule elongation reactions are most typically examined[1]. The kinetics of irreversible disassembly are presented elsewhere. **See** *Dilution-Induced Depolymerization*

2c. Role of GTP Hydrolysis. Enzymologists typically seek to identify those mechanistic features that explain catalytic rate enhancement; however, cell biologists focus their attention on regulation as the key feature of biological catalysis[18]. Tubulin may be considered to be a low-k_{cat} GTPase that acts as a highly regulated switch for controlling cell shape and motility. An even larger family of GTPases, commonly referred to as GTP- (or G-) regulatory systems, links the phosphorylation state of the bound guanine nucleotide to their inhibitor or activator potency. Each G-protein has two states, each characterized by the binding of GTP or GDP, with their interconversion accomplished directly by GTPase or indirectly by nucleotide exchange. In much the same way, tubulin exists in two principal states: the assembly-competent $Tb_E \cdot GTP$ complex and the assembly-incompetent $Tb_E \cdot GDP$ complex. Signal processing arises from the conformationally restricted properties of $Tb_E \cdot GTP$ and $Tb_E \cdot GDP$ in unpolymerized and polymerized forms, and although the microtubule literature is burgeoning, the scope of the present review is limited to this issue.

MacNeal and Purich[19] first demonstrated that hydrolysis of exchangeable site GTP is tightly coupled to assembly. These observations, along with changes in the critical concentration in the presence of GDP, led Karr et al.[20] to propose their boundary stabilization model for microtubule growth and stabilization. Here, $Tb_E \cdot GTP$ participates in tight-binding interactions that result in tubule stabilization. In this model, $Tb_E \cdot GDP$ (itself arising from assembly-induced hydrolysis) occurs within the interior microtubule lattice. Each internal $Tb_E \cdot GDP$ is stabilized by its surrounding $Tb_E \cdot GDP$ neighbors or by $Tb_E \cdot GTP$ at the microtubule ends which are referred to as boundaries. Loss of the single terminal layer (or boundary) would raise the critical concentration as a consequence of the much weaker binding of $Tb_E \cdot GDP$ molecules at the tubule ends. Such a model is also fully consistent with GDP inhibition of assembly as well as with the ability of enzymatic conversion of GTP to GDP resulting in microtubule disassembly. Delayed GTP hydrolysis is a model where $Tb_E \cdot GTP$ molecules persist randomly throughout the microtubule lattice for a period following polymerization[18]. There is no evidence supporting such a model, and isotope exchange data obtained by Angelastro and Purich[21] would suggest that this is an unlikely mechanism. Indeed, the microtubule lattice may exclude tubulin-bound GTP except at the terminal boundary. A third model involves formation of a multilayer "cap" comprised of many $Tb_E \cdot GTP$ molecules; this possibility suggested by the results of experiments conducted in the presence of the microtubule-stabilizing drug taxol[22]. Assembly under what may be regarded as "forcing conditions" probably has little or no relevance to microtubule stabilization in vivo, especially when one recognizes that $Tb_E \cdot GTP$, $Tb_E \cdot GDP$, and even E-site nucleotide-free Tb readily polymerize in the presence of taxol.

There is, in fact, no evidence supporting the existence of a "cap" containing many $Tb_E \cdot GTP$ molecules, even at very high tubulin concentrations where rapid polymerization is favored. Over an extraordinarily large range of tubulin concentration, O'Brien and Erickson[23] observed no accumulation of GTP beyond that anticipated on the basis of the boundary stabilization model of Karr et al.[20] This was confirmed independently by Jordan and Wilson[24].

Based on the use of aluminum tetrafluoride anion (AlF_4^-) as a phosphate analogue, Carlier et al.[25] suggested that orthophosphate may bind within the γ-phosphoryl pocket of the exchangeable nucleotide site in a manner that confers at least some of the stabilization displayed by the bound-GTP conformation. In particular, they proposed that microtubule-bound tubulin-GDP-P_i is a stable intermediate in the GTPase reaction, and they suggested that the dynamic instability of microtubules could be governed by loss of such "caps" or by P_i release into the medium. Formation of such a stable intermediate creates a paradox; if release of P_i increases the instability of microtubule ends (i.e., $\Delta G_{assembly}$ is more positive), this would require a correspondingly less favorable dissociation of P_i. Were $Tb_E \cdot GDP \cdot P_i$ and $Tb_E \cdot GTP$ isoenergetic, or even nearly so, the stable oxygen-18 exchange studies[21] would have readily indicated this through the loss of oxygen-18 atoms during such reversals.

Microtubules contain tubulin protomers that are regularly arranged in a head-to-tail manner[1]. Biased growth by addition of tubulin to one end is a consequence of the head-to-tail arrangement, and the transition states for tubulin addition reactions need not be identical for both polymer ends. Activation energies for the reactions at the two ends will in general be different, and the corresponding bimolecular rate constants will then be different. Microtubule polarity also accounts for the maintenance of different critical concentrations in the presence of GTP, and each end has different equilibrium constants for tubulin addition/release reactions. Such differential affinity arises from unequal retention of the Gibbs free energy of GTP hydrolysis; the less stable end (i.e., the (−)-end) does not release as much free energy of hydrolysis during each step in microtubule elongation.

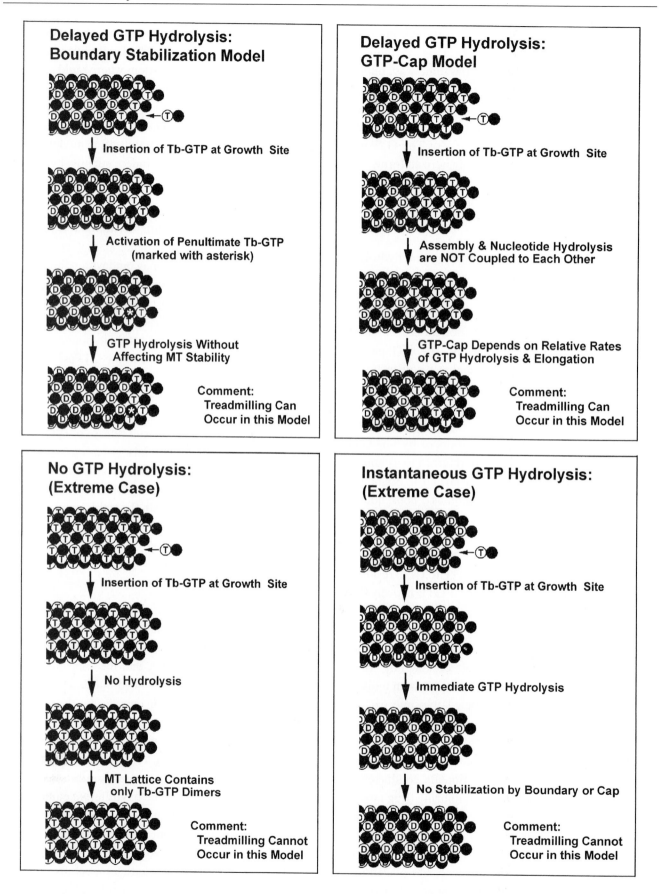

Delayed GTP Hydrolysis: Boundary Stabilization Model

Insertion of Tb-GTP at Growth Site

Activation of Penultimate Tb-GTP (marked with asterisk)

GTP Hydrolysis Without Affecting MT Stability

Comment: Treadmilling Can Occur in this Model

Delayed GTP Hydrolysis: GTP-Cap Model

Insertion of Tb-GTP at Growth Site

Assembly & Nucleotide Hydrolysis are NOT Coupled to Each Other

GTP-Cap Depends on Relative Rates of GTP Hydrolysis & Elongation

Comment: Treadmilling Can Occur in this Model

No GTP Hydrolysis: (Extreme Case)

Insertion of Tb-GTP at Growth Site

No Hydrolysis

MT Lattice Contains only Tb-GTP Dimers

Comment: Treadmilling Cannot Occur in this Model

Instantaneous GTP Hydrolysis: (Extreme Case)

Insertion of Tb-GTP at Growth Site

Immediate GTP Hydrolysis

No Stabilization by Boundary or Cap

Comment: Treadmilling Cannot Occur in this Model

3. MICROTUBULE ASSEMBLY/DISASSEMBLY KINETICS.

Cellular microtubules must undergo turnover, and nucleotide hydrolysis appears to play a central role in priming microtubules for their eventual disassembly. Two fundamentally different assembly/disassembly mechanisms persist during what has been termed steady-state polymerization; both rely on GTP hydrolysis to provide a source of Gibbs free energy to sustain the steady-state condition[1,18].

3a. Treadmilling. In the original formulation for actin dynamics[26], the concept of cytoskeletal polymer "treadmilling" was shown to arise from ATP hydrolysis-induced differences in the critical subunit concentrations in equilibrium with each filament end. Such a process implies that the macroscopic critical concentration [*i.e.*, K_∞ equals the algebraic sum of the off-rate constants for both ends divided by the algebraic sum of the on-rate constants for both ends] lies between the microscopic critical concentrations for the more stable and less stable filament ends; at a steady state of assembly/disassembly, this will consequentially lead to loss of subunits from the less stable filament end and accumulation of subunits at the more stable end. In the context of microtubule dynamics, Margolis and Wilson[27] proposed a microtubule treadmilling model that operates *via* the exclusive addition and release of tubulin dimers from opposite ends of microtubules. Because rapid dilution of microtubule protein to below the critical concentration resulted in prompt depolymerization with a rate constant of around 0.1 min^{-1} for complete loss of polymer mass, Karr and Purich[28] noted that this disassembly rate exceeded the steady-state treadmilling rate by a factor of 1000. They therefore proposed a minimal model wherein the possibility of reversibility at the so-called primary disassembly end was also indicated by use of a broken arrow. Caplow[29] correctly indicated that *all* four rate constants must be considered to account for the assembly/disassembly kinetics of tubulin dimer interactions with microtubule ends. ***See also*** *Actin Assembly Kinetics*

3b. Dynamic Instability. Mitchison and Kirschner[30] presented another explanation for microtubule steady-state dynamics in terms of GTP hydrolysis and the stochastics of losing the stabilizing cap (or boundary) of tubulin molecules containing unhydrolyzed GTP at their E-sites. In the "dynamic instability" model, length changes in microtubules at steady-state are thought to arise from the overall balance of two phases: the first involving slow growth of the majority of microtubule polymers; and the second arising from the rapid disassembly of a smaller fraction of polymers. They proposed that microtubules at steady-state contain $Tb_E \cdot GTP$ protomers at the polymer ends, forming a cap of stably bound protomers. In contrast, the microtubule interior lattice contains largely $Tb_E \cdot GDP$ protomers that are lost by endwise depolymerization whenever microtubules lose $Tb_E \cdot GTP$ protomers stabilizing their ends. Taken with the mounting evidence (*vide supra*) that there is no cap, the dynamic instability model most probably arises from the presence or loss of the stabilizing boundary formed by a few terminally bound $Tb_E \cdot GTP$ protomers.

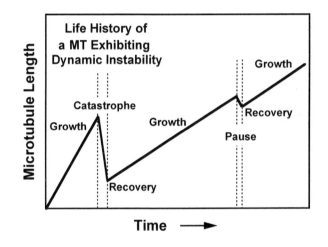

Tubules grow and remain stable as a consequence of $Tb_E \cdot GTP$ bound at the growth sites. They undergo stochastic disassembly in the improbable case that all growth points lose their $Tb_E \cdot GTP$ molecules. Under the influence of as yet unidentified intracellular signal(s), terminal $Tb_E \cdot GTP$ molecules are probably released from growth points. Then, disassembly promptly takes place in what can be regarded as a catastrophe. Depending on the circumstance, disassembling tubules may completely depolymerize or may recover such that regrowth begins the cycle anew. Kristofferson *et al.*[31] used biotinylated tubulin and antibody methods to analyze the time evolution of microtubule length redistribution which is a measure of steady-state microtubule dynamics. They were able to confirm the basic tenets of the dynamic instability model, and they clearly demonstrated that some tubules exhibit catastrophic depolymerization. Microtubules showed no evidence of treadmilling.

Video microscopy has permitted direct observation of microtubule assembly/disassembly dynamics *in vitro*. Horio and Hotani[32] first used dark-field optics to observe the growth and shrinkage phases, but so-called Allen video-enhanced contrast microscopy has become most convenient.

4. INTRACELLULAR DYNAMICS. Evaluating microtubule assembly/disassembly dynamics directly in living cells has proven to be a challenging task requiring great technical ingenuity. No single method provides both temporal and spatial resolution for obtaining (a) accurate estimates of kinetic constants and (b) the size and intracellular location of stable and dynamic microtubule pools. Direct microscopic observation of microtubules, for example, offers an attractive means for analyzing net rates of tubulin gain or loss[32–35], but the resolving power of light and fluorescence microscopy cannot distinguish single tubules from bundled microtubules or even a pair of microtubules that run closely parallel to each other. Thus, even as image reconstruction techniques are improved electronically, the physics of light refraction still will limit the technique to cells or cell regions containing only a few microtubules. In the case of microinjection, proteins can be introduced into unfertilized oocytes[36], thereby allowing the investigator to observe details of cytoskeletal assembly and disassembly after the subsequently fertilized egg proceeds through embryogenesis. Such an approach works best with large oocytes, as in the case of *Xenopus*, so that the embryo's development does not lead to significant dilution of the original microinjected reporter protein. Recent studies in *Xenopus* exemplify how microinjection of rhodamine-labeled or bis-"caged" fluorescein-labeled tubulin during the first cleavage division can be used to examine microtubule movement. At the appropriate stage of embryonic development, the "caged" fluorophore can be photoactivated by brief exposure to a focused light beam, and intracellular tubule dynamics can be recorded with a silicon-intensified target video camera. In experiments of this sort one must contend with problems of photo-induced oxidation of the fluorescent tag itself as well as damage to light-sensitive subcellular components. Because dioxygen is essential for photooxidation, efforts to eliminate O_2 by redox scavengers or by physically excluding this gas helps to minimize photo-damage. However, oxygen is an essential nutrient for maintaining

the [ATP]/[ADP] and [GTP]/[GDP] concentration ratios *via* oxidative phosphorylation. Likewise, the [ATP]/[ADP] ratio is important in modulating the action of microtubule-based motors [*i.e.*, dynein and kinesin], and various regulatory protein kinases. While the metabolic labeling technique can provide estimates of rates and extents of tubulin-microtubule exchange, results obtained with this approach must be interpreted using other data to gain insight about the cellular locations of stable and dynamic tubules. For example, there is mounting evidence for the occurrence of stable microtubules located in neurite outgrowths of PC12 cells grown in the presence of nerve growth factor.

The transition from G to M stages of the cell cycle is accompanied by complete microtubule cytoskeleton disassembly. The interphase microtubule cytoskeleton must be fully dismantled to allow mitosis to commence. Thus, upon transition from a growing to a shrinking tubule in the dynamic instability model, totality of depolymerization occurs, and this may be a very desirable outcome. During microtubule disassembly, Tba·GDP molecules rapidly issue from microtubule ends and probably crowd the vicinity of each unwinding microtubule end, and any entry and binding of Tb_E·GTP molecules would probably be most unlikely. Moreover, restoration of a Tb_E·GTP boundary layer and restabilization of a microtubule end would be especially improbable if Tb_E·GTP molecules must coexist at several growth sites simultaneously before stabilization can be restored. There must be multiple hierarchies of regulatory interactions that trigger global microtubule disassembly. Modulation of microtubule dynamics may be elicited by microtubule-associated proteins (MAPs) and enzymes, metabolic signals, as well as other low-molecular-weight factors. Protein kinase-mediated MAP phosphorylation releases this constraint of the latter *in vitro*[37]. Even more intriguing is the possibility of enhanced instability that may be achieved through the action of a *Xenopus* oocyte protein found to sever microtubules after mitotic activation[38]. Microtubules have their (+)-ends near the cell margin and distal to the centrosomes which appear to bind and stabilize the (−)-ends; loss of subunits only from the (+)-ends may be insufficient to permit rapid microtubule disassembly during transitions in the cell cycle. If both ends were free to disassemble, tubulin dimer release could proceed with rate constants of 200–500 s^{-1}, corresponding to length changes of about 6–12 μm/min. Severing long tubules into several shorter fragments could thus increase both the number and kind of disassembling ends, thereby allowing for rates upward of 50 μm/min. Vale's experiments on the severing of taxol-stabilized tubules are consistent with extensive, multiple fragmentation. In terms of control mechanisms, one should also not discount the opportunity for tubule disassembly in response to local surges of calcium ion, and any factor causing a drop in the cellular [GTP]/[GDP] poise could likewise tip the balance to favor disassembly. Post-translational modification[2] of tubulin [*e.g.*, acetylation, tyrosination, glutamylation, glycylation, and ADP-ribosylation] also may control the stability and dynamics of microtubules.

[1] D. L. Purich & D. Kristofferson (1984) *Adv. Prot. Chem.* **36**, 133.

[2] B. J. Terry & D. L. Purich (1982) *Adv. Enzymol.* **53**, 113.

[3] F. Oosawa & S. Asakura (1975) *Thermodynamics of the Polymerization of Protein*, Academic Press, New York.

[4] F. Oosawa & M. Kasai (1962) *J. Mol. Biol.* **4**, 10.

[5] A. Wegner & J. Engel (1975) *Biophys. Chem.* **3**, 215.

[6] R. Weisenberg (1972) *Science* **177**, 1104.

[7] G. G. Borisy & J. B. Olmsted (1972) *Science* **177**, 1196.

[8] S. N. Timasheff & L. M. Grisham (1980) *Annu. Rev. Biochem.* **49**, 565.

[9] K. A. Johnson & G. G. Borisy (1977) *J. Mol. Bio.* **117**, 1.

[10] D. Kristofferson & D. L. Purich (1981) *Arch. Biochem. Biophys.* **211**, 222.

[11] J. Hofrichter, P. D. Ross & W. A. Eaton (1974) *Proc. Natl. Acad. Sci. U.S.A.* **71**, 4864.

[12] Y. Engelbroughs, L. C. M. DeMaeyer & N. Overbergh (1977) *FEBS Lett.* **80**, 81.

[13] T. L. Karr & D. L. Purich (1980) *Biochem. Biophys. Res. Commun.* **95**, 1885.

[14] B. J. Terry & D. L. Purich (1980) *J. Biol. Chem.* **255**, 10532.

[15] E.-M. Mandelkow, A. Harmsen, E. Mandelkow & J. Bordas (1980) *Nature* **287**, 595.

[16] H. Sternlicht & I. Ringel (1979) *J. Biol. Chem.* **254**, 10540.

[17] T. L. Karr, D. Kristofferson & D. L. Purich (1980) *J. Biol. Chem.* **255**, 8560 and 8567.

[18] D. L. Purich & J. M. Angelastro (1994) *Adv. Enzymol.* **69**, 121.

[19] R. K. MacNeal & D. L. Purich (1978) *J. Biol. Chem.* **253**, 4683.

[20] T. L. Karr, A. E. Podrasky & D. L. Purich (1979) *Proc. Natl. Acad. Sci. U.S.A.* **76**, 5475.

[21] J. M. Angelastro & D. L. Purich (1990) *Eur. J. Biochem.* **191**, 507.

[22] M.-F. Carlier & D. Pantaloni (1981) *Biochemistry* **20**, 1918.

[23] E. T. O'Brien & H. P. Erickson (1989) *Biochemistry* **28**, 1413.

[24] M. A. Jordan & L. Wilson (1990) *Biochemistry* **29**, 2730.

[25] M.-F. Carlier, D. Didry, R. Melki, M. Chabre & D. Pantaloni (1988) *Biochemistry* **27**, 3555.

[26] A. Wegner (1976) *J. Mol. Biol.* **108**, 139.

[27] R. L. Margolis & L. Wilson (1978) *Cell* **13**, 1.

[28] T. L. Karr & D. L. Purich (1979) *J. Biol Chem.* **254**, 10885.

[29] M. Caplow (1992) *Curr. Opin. Cell Biol.* **4**, 58.

[30] T. Mitchison & M. W. Kirschner (1984) *Nature* **312**, 237.

[31] D. Kristofferson, T. Mitchison & M. W. Kirschner (1986) *J. Cell. Biol.* **102**, 1007.

[32] T. Horio & H. Hotani (1986) *Nature* **321**, 605.

[33] M. Caplow, R. Ruhlen, J. Shanks, R. A Walker, & E. D. Salmon (1989) *Biochemistry* **28**, 8136.

[34] L. Cassimeris, N. K. Pryer & E. D. Salmon (1988) *J. Cell Biol.* **107**, 2223.

[35] D. G. Drubin, S. C. Feinstein, E. M. Shooter & M. W. Kirschner (1985) *J. Cell Biol.* **101**, 1799.

[36] E. M. Tanaka & M. W. Kirschner (1991) *J. Cell. Biol.* **115**, 345.

[37] N. Raffaelli, P. S. Yamauchi & D. L. Purich (1992) *FEBS Letters* **296**, 21.

[38] R. D. Vale (1991) *Cell* **64**, 827.

MICROTUBULE CYTOSKELETON AS A "SENSISOME"

The "sensisome" is a hypothetical intracellular sensory organelle that may exploit the internal channels of microtubules (MTs) to acquire and conduct regulatory signals from the cell's periphery to its deepmost interior via the centrosome[1]. Athough the structural and mechanical properties of microtubules remain a central focus of modern cell biology, one inescapable feature of the microtubule—namely its lumen or internal channel—has attracted surprisingly little attention. When viewed in terms of MT cytoskeletal organization and assembly/disassembly dynamics, the MT lumen appears to be uniquely suited to serve as a sensory organelle.

Relative Dimensions of Microtubule Channels and Other Cell Components. The microtubule channel is surprisingly large. If filled only with water, the channel of a one micrometer-long microtubule would contain upwards of 1–2×10^6 water molecules. The channel's 15–16 nm diameter has an approximate cross-sectional area of about 180 nm^2, whereas the approximate cross-sectional areas of 100 kDa globular proteins fall in the range of 35-40 nm^2, suggesting that the microtubule channel is of sufficient dimension to conduct most biological molecules. Particles the size of the ribosome (molecular mass = 2.5×10^6 Da; long-axis = 25 nm) obviously cannot enter these channels, but even smaller molecules may experience limited access. It is well known that molecular diffusion through large pores depends on the pore's cross-sectional area as well as the size and shape of the diffusing molecule[2-4]. By analogy, as the dimension of a molecule approaches that of the MT channel diameter, movement into and through the channel will become hindered. This is to be expected, because molecules must enter without hitting the rim of the channel. One can most readily deal with the case where (a) no electrostatic interactions occur, (b) bulk solvent flow is inconsequential, and (c) the motion of a rigid spherical protein through the channel's orifice occurs by pure diffusion. In this limiting case, the channel's effective cross-sectional area for entry, $A_{effective}$, will be reduced by the radius of the entering molecule. For a sphere entering a cylindrical orifice, $A_{effective} = A_o(r - a)2/K_1 r^2$, where A_o is the actual cross-sectional area, a is the sphere's radius, and K_1 is the drag coefficient; the latter of which is a complex function of a/r. These considerations indicate that molecules with molecular masses above 300 kDa or with one dimension exceeding 10 nm may be effectively excluded from entry into the channel. Another factor that can alter the rate of diffusion within narrow-bore tubes is hindered motion. This phenomenon relates to friction imparted by the inner MT wall, and the effect is likely to be significant for larger protein molecules (molecular masses > 200 kDa) that might be present in microtubule channel. Even 100 kDa molecules might experience a 2-3 times reduction in diffusional mobility as a consequence of hindered motion, but experiments are needed to settle this issue definitively. Even if hindered motion does in fact occur, this need not detract from the potential suitability of the MT channels acting collectively as a "sensisome."

Nature of One-Dimensional Diffusion. The 15-16 nm diameter of the microtubule channel is comparatively small relative to the 5-50 micrometer microtubule lengths observed in most living cells. This feature allows one to treat diffusion within a microtubule virtually as a one-dimensional process, and we may begin by considering a long channel, or tube, running parallel with respect to the x-axis. Suppose further that the system is initially configured, such that the concentration to the left of the origin is c_1 and that to the right of the origin is c_2. For the case where $c_1 < c_2$, the mean square displacement $(\overline{x^2})$ of a particle can be described by the following function: $\overline{x^2} = (2k_B T/f)\tau$, where k_B is the Boltzmann constant, f is the frictional coefficient (Note: $f = -F/v$, for friction force F and particle velocity v), T is the absolute temperature, and τ is time over which a particle has a certain kinetic energy. During this small period τ, the diffusing molecules will possess the above-defined mean-square displacement; because the molecules can move in either direction, 50% of the molecules within a distance of $\overline{x^2}$ will pass the origin. Accordingly, leftward and rightward displacements will correspond to $(c_1 A/2)(\overline{x^2})^{1/2}$ and $(c_2 A/2)(\overline{x^2})^{1/2}$, respectively, and the net flow in the $(+)$-x direction is: $dn/dt = n/\tau = (c_1 - $

$c_2)A/2\tau$. The average change in the concentration gradient over time τ is: $dc/dn = (c_1 - c_2)A/2(\overline{x^2})^{1/2}$, and upon substitution into the prior equation, one arrives at: $dn/dt = -(A\overline{x^2}/2\tau)(\partial c/\partial x)$. [Note: This treatment describes motion along the x-axis; were one dealing with three-dimensional diffusion, the mean squared displacement $\overline{r^2}$ would be equal to $\overline{x^2} + \overline{y^2} + \overline{z^2}$, and $\overline{r^2} = (6kT/f)\tau$.] Based on the magnitude of diffusion constants for globular (*i.e.*, spherical) protein molecules, one can estimate that proteins like hemoglobin (mass = 68 kDa; diffusivity = 6.2 m^2/s $\times 10^{11}$) can readily diffuse 20–30 micrometers within 10–20 seconds. By contrast, low-molecular-weight metabolites, such as glycine (mass = 75; diffusivity = 95 m^2/s $\times 10^{11}$) and arginine (mass = 174; diffusivity = 58 m^2/s $\times 10^{11}$), will rapidly traverse these distances in a few seconds or less.

Advantages of Reducing the Dimensionality of Diffusion. Although molecular diffusion within bulk solutions is often best treated in a manner akin to three-dimensional Brownian motion, many biological processes are constrained geometrically, and the timing and efficiency of certain events can be strongly influenced by the ability of living systems to limit or reduce the dimensionality of diffusion. For example, two-dimensional surface diffusion increases the probability that a component will arrive at a particular locus within the cell. While transport along membrane surfaces represents an obvious example, movement within the microtubule channel represents a form of one-dimensional diffusion. Some of the virtues of one-dimensional diffusion have been indicated by Eigen who first proposed that the rapid action of repressors and other gene regulatory proteins might arise from a reduction in diffusion resulting in one-dimensional movement along DNA. The major advantages are (a) speed, (b) directionality, (c) processivity, and (d) the certainty of signal arrival at a desired destination.

Adam and Delbrück[5] pointed out how the ability to manage the dimensionality of diffusion represents a selective advantage in evolution, suggesting that fully developed, internally compartmentalized cells enjoy a great range of options for controlling metabolic processes. Briefly, the mean time τ_i for diffusion to a target can be described by the following two-component equation:

$$\tau^{(i)} = \frac{b^2}{D^{(i)}} f^{(i)}(a/b)$$

The first component $b2/D^{(1)}$ is the period of time required to traverse a distance b in any direction, whereas the second term $f^{(i)}(a/b)$ strongly depends on the dimensionality. Adam and Delbrück[6] define appropriate boundary conditions and equations describing the concentration of molecules in the diffusion space in terms of space coordinates and time. They treated four cases: (1) one-dimensional diffusion in the linear interval $a \leq x \leq b$; (2) two-dimensional diffusion on the circular ring $a \leq r \leq b$; (3) three-dimensional diffusion in a spherical shell $a \leq r \leq b$; and (4) combined three-dimensional and surface diffusion. They provide a useful account of how reduced dimensionality of diffusion can (a) lower the time required for a metabolite or particle originating at point P to reach point Q, and (b) improve the likelihood for capture (or catch) of regulatory molecules by other molecules localized in the immediate vicinity of some target point Q.

Properties of Microtubule Cytoskeleton that Commend its Use as a "Sensisome." The intriguing notion that the internal channels of cytoskeletal microtubules may act as a sensory organelle rests on several properties. First, microtubule channels should, on the basis of the above discussion, readily conduct small- and medium-sized molecules within seconds over cellular dimensions. Second, microtubules adopt and maintain a well defined cytoskeletal organization; during interphase, their $(-)$-ends are attached to the centrosome which resides in a perinuclear locale near the Golgi apparatus, and during mitosis they assemble into astral, pole-to-pole, and pole-to-kinetochore arrays. The $(+)$-ends of the microtubule in nondividing cells are localized in the cell's periphery, nearest to the plasma membrane where metabolic transport and signal transduction are most active. Third, the $(+)$-ends of microtubules in living cells are known to undergo stochastic disassembly and regrowth as a consequence of dynamic instability. This property allows MTs to act as self-filling "syringes" that can quickly capture an aliquot of the peripheral cytoplasm, thereby allowing the centrosome to be informed about recent changes in the solute composition of the peripheral cytoplasm. A second outcome of dynamic instability is that microtubules should be self-clearing, and larger molecules that become accidentally lodged in the inner MT channels can be released during catastrophic depolymerization. Fourth, the continuity of the microtubule channels from the cell periphery to the centrosome makes certain that metabolic signals will arrive at the centrosome. Finally, in some cells the total MT channel volume is likely to occupy a significant fraction of cellular space. In the case of the axon, microtubules appear as well spaced colinear arrays, and the MT-to-MT distances observed in EM cross-sections suggests that the inner channels may contain upwards of 2% of the axon's total volume.

If one accepts the view that Nature is extraordinarily opportunistic—never wasting even a gesture, then the microtubule lumen presents the opportunity for use as an elegant sensory organelle that may be present in all eukaryotic cells.

[1] One of the authors (D.L.P.) wishes to thank Ms. Dagny Ulrich for her continuing work on molecular diffusion within microtubule channels. Professor Robert J. Cohen also provided helpful insights about molecular diffusion as well as references to Smoluchowski's early work.

[2] C. M. Feldherr, R. J. Cohen & J. A. Ogburn (1983) *J. Cell Biol.* **96**, 1486.

[3] W. L. Haberman & R. M. Sayre (1958) Motion of Rigid and Fluid Spheres in Stationary and Moving Liquids Inside Cylindrical Tubes, *David Taylor Model Basin Report No. 1143*, U. S. Navy, Washington, D. C.

[4] P. L. Paine & P. Scherr (1975) *Biophys. J.* **15**, 1087.

[5] G. Adam & M. Delbrück (1968) in *Structural Chemistry and Biology* (A. Rich & N. Davidson, eds.) pp. 198-215, Freeman, San Francisco.

MICROTUBULE TREADMILLING

The biased opposite-end assembly and disassembly behavior of tubulin dimer interactions with assembled microtubules resulting from the differentially greater stability of the $(+)$-end, a consequence of that end's greater release of free energy attending GTP hydrolysis. The treadmilling phenomenon was originally observed *in vitro* with actin filaments[1,2]. Margolis and Wilson[3] first observed microtubule treadmilling and suggested that the microtubule assembly-disassembly "equilibrium" is the summation of two different steady-state reactions occurring at the opposite ends of the microtubule. They proposed that assembly and disassembly occur predominantly (if not exclusively) at the opposite ends under steady-state conditions *in vitro*. Terry and Purich[4] demonstrated that GTP hydrolysis was absolutely essential for the treadmilling process and that nonhydrolyzable analogues such as Gpp(NH)p (or, 5'-guanylylimidodiphosphate) and Gpp(CH₂)p (or, β,γ-methyleneguanosine 5'-triphosphate) promoted microtubule assembly, but not treadmilling.

The significance of microtubule treadmilling in living cells has remained a matter of dispute, especially because cytoplasmic formation and growth of microtubules are thought to be template-driven, using so-called microtubule-organizing centers such as the centrosome. With the minus ends anchored at the centrosome, microtubule turnover is believed to be dominated by dynamic instability[5] which reflects the disassembly/reassembly properties of the plus ends of microtubules. Nonetheless, Rodionov and Borisy[6] recently observed that microtubules in cytoplasmic fragments of fish melanophores detach from their nucleation site and depolymerize from their minus ends. Free microtubules moved toward the periphery by treadmilling growth at one end and shortening from the opposite end. Rodionov and Borisy[6] suggested that frequent release from nucleation sites may be a general centrosome property that permits minus-end release and microtubule turnover by treadmilling. *See Microtubule Assembly Kinetics*

[1]A. Wegner (1975) *Biophys. Chem.* **3**, 215.
[2]A. Wegner (1976) *J. Mol. Biol.* **108**, 139.
[3]R. L. Margolis & L. Wilson (1978) *Cell* **13**, 1.
[4]B. J. Terry & D. L. Purich (1980) *J. Biol. Chem.* **255**, 10532.
[5]T. Mitchison & M. W. Kirschner (1984) *Nature* **312**, 237.
[6]V. I. Rodionov & G. G. Borisy (1997) *Science* **275**, 215.

MIGRATION

1. The movement, relocation, or transfer of an atom, moiety, or group in a rearrangement reaction. Such transfer is usually intramolecular, and the migrating group can move with its electron pair (often regarded as a nucleophile), without its electron pair (thus regarded as an electrophile), or with a single electron (as in free-radical rearrangements). 2. The movement of a bond (within a given molecular entity) to a new position; often referred to as bond migration. Such a bond migration occurs in the conversion of zymosterol to $\Delta^{7,24}$-cholestadienol during cholesterol biosynthesis. 3. The movement of ions while under the influence of an electric field. Because cations and anions need not always move at the same velocity, they may transport different fractions of the current. *See also Molecular Rearrangement*

MINICHAPERONES

A polypeptide consisting of residues 191-345 of GroEL that, when immobilized on agarose, acts as a very efficient chaperone with proteins that resist renaturation by conventional refolding methods. Immobilized minichaperones may be useful in column chromatography procedures or batchwise to renature proteins in high yield and with biological activity.

M. M. Altamirano, R. Golbik, R. Zahn, A. M. Buckle & A. R. Fersht (1997) *Proc. Natl. Acad. Sci. U.S.A.* **94**, 3576.
A. R. Fersht (1999) *Structure and Mechanism in Protein Science: A Guide to Enzyme Catalysis and Protein Folding*, pp. 650, W. H. Freeman, New York.

MINIMAL LETHAL DOSE

The minimum dose of a toxic substance that is lethal when assayed with various experimental animals. It is usually followed by a subscript number equal to the percentage of the animals that died under the assay conditions (*e.g.*, LD_{50}).

MINIMAL STRUCTURAL CHANGE, POSTULATE OF

A prediction that only the minimum number of bond-making and bond-breaking steps occurs in the conversion of reactant(s) to product(s). Thus, during the course of a substitution reaction, a new substituent enters the exact same position previously occupied by the displaced moiety. *See Molecular Rearrangement; Rearrangement Stage*

MITOCHONDRIAL INTERMEDIATE PEPTIDASE

This zinc-dependent enzyme [EC 3.4.24.59] of the peptidase M3 family catalyzes the hydrolysis of a peptide bond such that there is a release of an N-terminal octapeptide at the second stage of processing of some proteins imported into the mitochondrion. The natural substrates are precursor proteins that already have been processed by the mitochondrial processing peptidase.

G. Isaya & F. Kalousek (1995) *Meth. Enzymol.* **248**, 556.

MITOGEN-ACTIVATED PROTEIN KINASES

As integral elements of the signal transduction pathway connecting the binding of a protein growth factor on the extracellular surface to specific gene expression, these protein kinases migrate from cytosol to nucleus and phosphorylate various transcription factors, including Jun/AP-1, Fos, and Myc. The following are recent reviews on the molecular and physical properties of this phosphotransferase (or, MAP kinase).

C. J. Marshall & S. J. Leevers (1995) *Meth. Enzymol.* **255**, 273.

D. R. Alessi, P. Cohen, A. Ashworth, S. Crowley, S. L. Leevers & C. J. Marshall (1995) *Meth. Enzymol.* **255**, 279.

C. A. Lange-Carter & G. L. Johnson (1995) *Meth. Enzymol.* **255**, 290.

MITOGEN-ACTIVATED PROTEIN KINASE KINASE

One of the downstream elements of the signal transduction pathway, MAPKK (or, MEK) is phosphorylated by a Ser/Thr protein kinase, known as Raf, and phosphorylated MAPKK then catalyzes phosphoryl transfer to MAP kinase. The following is a recent review on the molecular and physical properties of this phosphotransferase (or, MAP kinase kinase).

D. R. Alessi, P. Cohen, A. Ashworth, S. Crowley, S. L. Leevers & C. J. Marshall (1995) *Meth. Enzymol.* **255**, 279.

MIXED-FUNCTION OXIDASE

The following is review on the molecular and physical properties of this class of monooxygenases, which are also known as hydroxylases. A typical monooxygenase reaction is the hydroxylation of an alkane to an alcohol which involves a reduced cosubstrate that reduces a second atom within the O_2 molecule to form water. Flavin-containing monooxygenases include lysine oxygenase and 4-hydroxybenzoate hydroxylase. Reduced pteridines are involved in the phenylalanine hydroxylase and tryptophan hydroxylase reactions. ***See also*** *Cytochrome P-450*

C. H. Williams, Jr. (1976) *The Enzymes*, 3rd ed., **13**, 89.

MIXED-TYPE INHIBITION

A limiting case of noncompetitive inhibition, characterized by *third*-quadrant convergence of double-reciprocal plots of $1/v$ *versus* $1/[S]$ in the absence and presence of several constant levels of the inhibitor.

Consider the standard Uni Uni mechanism, $E + S \rightleftarrows EX \rightleftarrows E + P$, in which an inhibitor can bind reversibly to either the free enzyme, E, or to the central complex, EX, albeit with different affinities. Using the Cleland nomenclature, let K_{is} represent the dissociation constant of the inhibitor with respect to the free enzyme (thus, $K_{is} = [E][I]/[EI]$ and K_{ii} is the corresponding dissociation constant with respect to the central complex ($K_{ii} = [EX][I]/[EXI]$). Thus, $1/v = \{(1/V_{max})(1 + ([I]/K_{ii}))\} + \{(K_m/(V_{max}[S]))(1 + ([I]/K_{is}))\}$. If $K_{is} < K_{ii}$, the intersection will occur in the second quadrant of a double-reciprocal plot and if $K_{is} > K_{ii}$, the intersection point will be in the third quadrant. Thus, note that mixed-type inhibition is really a subset of noncompetitive inhibition.

MIXED VALENCY

A condition where metal ions within a coordination complex or cluster are present in more than one oxidation state. In such systems, there is often complete delocalization of the valence electrons over the entire complex or cluster, and this is thought to facilitate electron-transfer reactions. Mixed valency has been observed in iron-sulfur proteins. Other terms for this behavior include mixed oxidation state and nonintegral oxidation state.

MIXING TIME

The period that elapses before two or more solutions are thoroughly mixed in a chemical kinetic experiment. In most manually controlled chemical kinetic studies, the mixing time is rarely a factor affecting accurate data acquisition; however, the mixing time can be significant in rapid kinetic processes studied by continuous and stopped-flow kinetic techniques.

MODELS

Schematic representations serving as tools for assessing current knowledge and for designing better experiments to reveal the integrated nature of a process.

In chemistry and biology, mechanistic inquiry requires models that offer the virtues of precision, simplicity, and generativity. Simplicity arises from the symbolic representation of the interactions involving a minimal, yet sufficient, number of components to account for all observed properties of the system under investigation. Precision emerges from the consideration of how rival models lead to nonisomorphic features (*i.e.*, testable differences) that distinguish each candidate model. Generativity results from the act of recombining the constituent elements from one or more models to arrive at a revised model and to achieve added clarity that stimulates a new round of experimental tests. In this recursive manner, effective use of models permits one to determine the fitness of one or more models in describing the system. In biochemistry and molecular biology, this includes efforts to account fully for the effects of concentrations of each component, the time evolution of biochemical events, the accumulation of transient intermediates, changes in such positional properties as conformation, configuration, and/or location, as well as the thermodynamics or bioenergetics of all of the molecular interactions.

MODIFIER

A term applied to substances or agents that either depress or elevate the action of a particular catalyst. While often used synonymously with effector, the International Union of Biochemistry has suggested that the term modifier is more appropriate for those substances that are artificially added to *in vitro* systems. Other investigators restrict the use of the term modifier to those agents that covalently modify the structure of the catalyst, thereby altering the catalyst's activity. **See** *Effector; Activator; Inhibitor*

MODULATION

1. Any alteration in the kinetics or flux of an enzyme-catalyzed reaction or biochemical pathway. 2. The regulation of the rate of translation of mRNA by a modulating codon. 3. The fluctuation of cells, either functionally or morphologically or both, in response to one or more changes in the environment of the cells.

MOLAR ABSORPTION COEFFICIENT

A fundamental unit in spectroscopy (synonymous with molar extinction coefficient) and symbolized by ε. It is equal to the absorbance of light per unit path length and per unit concentration. **See** *Absorption Spectroscopy; Beer-Lambert Law*

MOLAR GAS CONSTANT

A constant (symbolized by R) and used throughout the disciplines of kinetics and thermodynamics, equal to $8.314510 \ \mathrm{J \cdot mol^{-1} \cdot K^{-1}}$ or $1.987216 \ \mathrm{cal \cdot mol^{-1} \cdot K^{-1}}$. [The term was originally developed as the proportionality constant for the ideal gas law ($PV = nRT$ where P, V, n, and T are pressure, volume, number of moles, and absolute temperature of the ideal gas, respectively), hence the appearance of "gas" in its name.] **See** *Boltzmann Constant*

MOLECULAR CROWDING

A term referring to volume-exclusion effects of added solutes on the activity and reactivity of chemical species. Although most biochemical binding interactions take place in low reactant concentrations, the cumulative effect of a cell's numerous solutes is that nearly every cellular compartment is highly concentrated[1-3]. In fact, one cannot escape the conclusion that cellular components are packed within cells to such an extent that certain nonideal behavior occurs. This is analogous to the behavior of a real gas whose molecules occupy a finite volume, rather than being infinitesimally small as in the case of ideal gas theory. Molecular crowding is probably of great significance in intracellular ligand binding and polymerization reactions.

Excluded Volume:

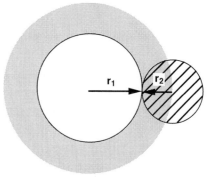

Volume from which solute molecules (of radii $r_1 + r_2$) are excluded by virtue of their mutual impenetrability.

$$V_{excluded}/molecule = (4/3) \pi [r_1 + r_2]^2$$

$$V_{excluded}/mole = (4/3) \pi N_o [r_1 + r_2]^2$$

Altered thermodynamic activity of proteins in solution arises when unreactive (or inert) macromolecules are added to a solution and occupy more than a few percent of total solution volume. Terms such as "unreactive", "background," or "inert" are used to emphasize that the added protein need not exhibit and direct binding interaction with the protein of interest. Instead, the consequences have more to do with molecular crowding, and approximate theoretical models show that this effect depends on the shapes and sizes of the macromolecules. Thus, biological fluids are anything but ideal or dilute solutions.

Muramatsu and Minton[4] studied the diffusion of tracer proteins at low concentration in solutions containing "background" proteins at concentrations of up to 200 g/liter. The fractional reduction of the diffusion coefficient of tracer in the presence of a given weight per volume concentration of background species generally increases with increasing size of tracer species and with decreasing size of background species. The diffusion constants of three out of four tracer species depended upon the concentrations of four background species in a way that can be treated semiquantitatively by a simple hard particle model. They concluded that the diffusive transport of larger proteins and aggregates may be slower by several orders of magnitude than that occurring in dilute solution. These investigators applied scaled particle theory to evaluate the probability that the target volume for a step in a random walk is free of any macromolecules.

Minton et al.[5] suggested that large excluded volume effects in cytoplasm (which they term "macromolecular crowding") make otherwise soluble macromolecules associate with membrane proteins to a greater extent than might be expected on the basis of simple considerations of mass action. Han et al.[6] extended this approach by choosing a different target volume which also allows the calculation to be extended to the diffusion of a hard sphere among hard spherocylinders. They concluded that, to the extent that proteins can be approximated as hard particles, the hindrance of globular proteins by other proteins is reduced when the background proteins aggregate. Greater aggregation has the effect of further decreasing particle surface area. Moreover, they indicate that the hindrance arising from the presence of rod-shaped background particles is reduced slightly when such rod-like particles are aligned.

Minsky et al.[7,8] recently suggested that the emergence of eukaryotes led to a highly crowded environment that may have promoted DNA self-assembly, leading to extremely condensed and thermodynamically stable DNA aggregates such as nucleosomes and solenoid structures of chromatin.

[1] A. P. Minton (1983) Mol. Cell. Biochem. 55, 119.
[2] S. B. Zimmerman & A. P. Minton (1993) Annu. Rev. Biophys. Biomol. Struct. 22, 27.
[3] A. P. Minton (1998) Meth. Enzymol. 295, 127.
[4] N. Muramatsu & A. P. Minton (1988) Proc. Natl. Acad. Sci. U.S.A. 85, 2984.
[5] A. P. Minton, G. C. Colclasure & J. C. Parker (1992) Proc. Natl. Acad. Sci. U.S.A. 89, 10504.
[6] J. Han & J. Herzfeld (1993) Biophys. J. 65, 1155.
[7] A. Minsky, R. Ghirlando & Z. Reich (1997) J. Theor. Biol. 188, 379.
[8] A. P. Minton (1998) Meth. Enzymol. 295, 127.

MOLECULAR DYNAMICS

Any of a series of computational methods for deducing molecular structure and reactivity by treating molecules or a system atoms comprising a segment of a macromolecule in terms of quantum and molecular mechanics. The basic goal is to explain molecular structure on the basis of the forces associated with and between bonded and nonbonded atoms. Quantum mechanical treatments become cumbersome when the number of atoms within the system is increased. For this reason one tends to use molecular mechanics, which is an empirical method, also called force-field calculations. These approximate methods allow one to estimate the relative energy differences among various molecular conformations by taking into account natural bond lengths and angles, influence of strain, torsional interactions, and attractive and/or repulsive electronic, dipolar, and van der Waals forces.

Selected entries from *Methods in Enzymology* [vol, page(s)]: Application to [free energy calculations, 243, 599-601; protein force field, 243, 595-598; anti-hapten monoclonal antibodies, 203, 31; and distance geometry, in NMR analysis of protein structure, 202, 268; DNA structure modeling, 211, 449; peptide modeling, 202, 427; RATTLE algorithm, 243, 597; restrained, 239, 416, 429-433, 437-438; types of runs, 243, 598; SHAKE algorithm, 243, 597; simplification for computational speed, 243, 596-598; simulations [applications, 241, 178-182; in drug design, 241, 382-384; in four dimensions, 239, 645-646; historical background, 241, 178-182; HIV-1 protease, 241, 371-373, 376-381]; theoretical model, 241, 187-189; methodology, 241, 186-187; searching configurational space, 239, 640; refinement [associated lowering of interval energy of distance geometry structures, 202, 290; physical reliability, dependence on force field accuracy, 202, 279; solvent inclusion during, 202, 297]; simulations [in free energy evaluation, 202, 499; metal force fields in proteins, 227, 285; applications, 241, 178-182; in drug design, 241, 382-384; in four dimensions, 239, 645-646; historical background, 241, 178-182; HIV-1 protease, 241, 371-373, 376-381 [Gln-88 mutant, 241, 189-192; structure-function analysis, 241, 178, 182-195; theoretical model, 241, 187-189]; HIV-1 protease inhibitor design method, 241, 370-384 [symmetric structures, 241, 340]; HIV reverse transcriptase, 241, 381; methodology, 241, 186-187; modified, 239, 641; multiple copy simultaneous search method, 241, 383-384; searching configurational space, 239, 640]; with time-averaged NOE restraints, 202, 299.

MOLECULAR ENTITY

Any chemically or isotopically distinct atom, molecule, ion, radical, complex, etc., capable of existing as a separately distinguishable unit or entity. The term generally refers to single entities, irrespective of their nature. The term chemical species stands for a set or ensemble of molecular entities.

MOLECULAR "HITCH-HIKING"

Certain biological molecules exhibiting even relatively weak binding to cytoskeletal motors may "hitch-hike" for certain periods of time; when summed over time, this behavior can represent a physiologically significant form of vectoral conveyance. Depending on the persistence of their interactions with cytoskeletal motors, the effective diffusional dimensionality of these hitch-hiking molecules can be substantially reduced. During the course of their studies on the delivery of cytoplasmic proteins to the presynaptic terminals of CNS neurons, Garner et al.[1] found unique characteristics of one protein (designated p118) conveyed in slow component b (or SCb) of axonal transport, the large group of proteins representing the cytoplasmic matrix. Alone among the SCb group, p118 coisolated with the synaptic junctional complex, and sequencing of this protein revealed it is type I or brain hexokinase (HK, EC 2.7.1.1). Brain HK associates primarily with the mitochondrial outer membrane, but Garner et al.[1] found that most of type I HK is transported at a rate at least 10 times slower than mitochondria. Accordingly, they suggested that the enzyme's transient interactions with the more rapidly moving mitochondria may account for the kinetics of axonal transport of HK toward the presynaptic terminal.

[1]J. A. Garner, K. D. Linse & R. K. Polk (1996) *J. Neurochem.* **67**, 845.

MOLECULARITY

An integer indicating the molecular stoichiometry of an elementary reaction. A fundamental assumption in chemical kinetics is that the kinetic form of a *one-step* reaction will be identical to its stoichiometric form. In terms of transition-state theory, molecularity equals the number of molecules (or entities) that are used to form the activated complex. For reactions in solution, solvent molecules are counted in the molecularity only if they enter into the overall process, not if they only exert an environmental or solvent effect[1-3].

The molecularity of a reaction is always an integer and only applies to elementary reactions. That is not always so for the order of a reaction, thus emphasizing the difference between molecularity and order. Molecularity

is the theoretical description of an elementary process, whereas order refers to the empirically derived rate expression. Usually, a bimolecular reaction is second order; however, the converse may not be true. Thus, unimolecular (sometimes called monomolecular), bimolecular, and ter- or trimolecular reactions refer to elementary reactions that involve one, two, or three entities, respectively, that combine to form an activated complex. The rate constant (or, specific reaction rate) for a reaction is commonly designated by k. The unit of time in the International System of units (SI), adopted in 1960 by the 11th General Conference on Weights and Measures, is the second (symbolized by s). Thus, unimolecular rate constants are typically expressed in s^{-1}, the value being concentration independent. A slight difficulty arises with regard to SI units and bi- and trimolecular rate constants. Concentrations in the SI system would be $mol \cdot m^{-3}$. Nevertheless, concentrations are typically expressed in $mol \cdot dm^{-3}$ (or, $mol \cdot L^{-1}$ or M^{-1}). Thus, a bimolecular rate constant typically has units of $M^{-1}s^{-1}$ whereas a trimolecular rate constant is expressed in units of $M^{-2}s^{-1}$. When gas-phase reactions are studied, (or, even with gases in solution), "concentrations" are often reported in terms of partial pressures rather than molarity. Nevertheless, it is usually preferable, when able, to convert such units to true concentration units.

Both unimolecular and bimolecular reactions are common throughout chemistry and biochemistry. Binding of a hormone to a reactor is a bimolecular process as is a substrate binding to an enzyme. Radioactive decay is often used as an example of a unimolecular reaction. However, this is a nuclear reaction rather than a chemical reaction. Examples of chemical unimolecular reactions would include isomerizations, decompositions, and dissociations. ***See also*** *Chemical Kinetics; Elementary Reaction; Unimolecular; Bimolecular; Transition-State Theory; Elementary Reaction*

[1]Nomenclature Committee of the International Union of Biochemistry (1982) *Eur. J. Biochem.* **128**, 281.
[2]J. W. Moore & R. G. Pearson (1981) *Kinetics and Mechanism*, Wiley-Interscience, New York.
[3]W. P. Jencks (1969) *Catalysts in Chemistry and Enzymology*, McGraw-Hill, New York. [Also available as Jencks, W. P. (1986) *Catalysis in Chemistry and Enzymology*, Dover Publications, New York.]

MOLECULAR ORBITALS

Geometrically defined regions of electron density surrounding two or more atomic nuclei. These orbitals are considered a result of an overlap of atomic orbitals. Two types of molecular orbitals are generated: bonding and antibonding orbitals. Bonding obitals have electron density positioned in a manner that favors bond formation, whereas the opposite is true for antibonding orbitals.

MOLECULAR REARRANGEMENT

A chemical process involving intramolecular repositioning or another alteration in the connectivity of atoms within a molecule, such that the product is isomeric with the reactant. Such reactions violate the principle of minimum structural change. Any reaction that includes migration of an atom, group, or bond falls under this class of molecular rearrangement reactions. If the entering group in a chemical reaction bonds to a different atom or position than was occupied by the leaving group, the reaction is also classified as a rearrangement.

In intermolecular rearrangement reactions (also referred to as "apparent rearrangement reactions"), the migrating group has been completely separated from the parent species and is then attached at a new position in a subsequent step. In intramolecular rearrangements the migrating group is never separated from the parent compound.

MOLECULAR SIMILARITY

1. A systematic approach[1,2] for the rational design of extremely high-affinity transition-state analogues based on the use of detailed reaction mechanism information to deduce the most likely transition-state configuration, followed by molecular mechanical modeling of the electrostatic surface potential of that transition-state configuration as well as numerous candidate analogues to obtain the best matching analogue. Subsequent testing of the inhibitory action of the most promising analogues provides an additional route for refining the investigator's structural model of the transition state. 2. A numerical index for evaluating the similarity of geometric and electronic properties of an analogue and the most likely transition-state configuration for an enzyme-catalyzed reaction.

The "molecular similarity" strategy is based on Pauling's long-standing postulate that catalytic rate enhancement arises from the catalyst's ability to lower the energy input required to reach the transition state by selectively binding (and hence stabilizing) reactants in their transition-state configuration. Vibrational modeling of a data set for experimentally determined kinetic isotope effects allows one to obtain a geometric model of the ground state and transition state for the nucleoside hydrolase reaction. From these data, one creates an electrostatic potential surface for free inosine, the above-mentioned transition state, as well as the reaction products.

[1]B. A. Horenstein & V. L. Schramm (1993) *Biochemistry* **32**, 7089.
[2]B. A. Horenstein, D. W. Parkin, B. Estupinan & V. L. Schramm (1991) *Biochemistry* **30**, 10788.

MOLE FRACTION

A measure of the fractional composition of a solution containing one or more substances in amounts n_A, n_B, n_C, . . . (where n_i in the amount-of-substance i expressed in mol), such that for any component, say B:

$$\text{mole fraction} = X_B = n_B/[n_A + n_B + n_C + \ldots]$$

Technically, one includes solvent as one of the components when expressing mole fraction in chemical thermodynamics, but in describing dilute biological solutions, $n_{solvent}$ is often omitted. With multicomponent solutions, one may chose to analyze the fractional composition of any two (or more) substances while not including others held constant in the experiment (*e.g.*, concentrations of buffer components, proton, supporting electrolyte(s), enzyme, *etc.*). For two components,

$$X_A = n_A/[n_A + n_B] \quad \text{and} \quad X_B = n_B/[n_A + n_B]$$

In this case, $X_A + X_B = 1$, and this formulation allows one to analyze binding interactions between two such components by the method of continuous variation. **See** *Concentration; Method of Continuous Variation*

MOLYBDENUM COFACTOR (MoCo)

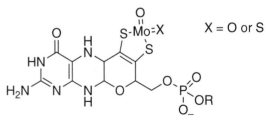

The molybdenum complex of the molybdopterin prosthetic group (ligand). In the molybdenum cofactor the minimal coordination of the Mo atom is thought to be provided by the chelating dithiolenato group of the molybdopterin and either two oxo or one oxo and one sulfido ligands.

MOLYBDOPTERIN

The prosthetic group associated with the molybdenum atom of the molybdenum cofactor found in most molybdenum-containing enzymes except nitrogenase (*See Molybdenum Cofactor*). Many of these enzymes catalyze two-electron redox reactions involving the net exchange of an oxygen atom between the substrate and water. In bacterial enzymes a nucleotide is linked to the phosphoryl group.

MONOD-WYMAN-CHANGEUX MODEL

A ligand binding model[1] that explains the positive cooperativity of many allosteric proteins and enzymes. Contrary to what is often stated in general biochemistry textbooks, the MWC cooperativity does not increase the affinity of ligand binding as one progresses toward higher ligand concentration; indeed, the MWC model states that the intrinsic binding constants are the same for each site. Cooperativity in this model is actually the result of the allosteric transitions that increase sites available in a conformation permitting ligand binding.

The MWC model (also known as the all-or-nothing model, the concerted transition model, and symmetry-conserving model[2,3]) relies on the following definitions and properties of allosteric systems: (1) Homotropic effects are interactions between binding sites for the same ligands. For example, the oxygenation of the four heme groups on hemoglobin are described as homotropic interactions. (2) Heterotropic interactions are interactions between binding sites for different ligands. For example, the effects of 2,3-bisphosphoglycerate on oxygenation of hemoglobin can be described as heterotropic interactions. (3) Most allosteric proteins are oligomeric, and changes in protein quaternary structure are factors in ligand binding cooperativity. (4) Homotropic interactions are most often cooperative for allosteric proteins, and few if any systems display only heterotropic interactions. (5) Cooperativity in the action of heterotropic ligands most likely stems from cooperative homotropic binding interactions. (6) Conditions, treatments, or mutations that alter heterotropic interactions also simultaneously alter the homotropic interactions.

The derivation of equilibrium ligand binding equations for this model involves the following assumptions: First, the protein is oligomeric and contains a finite number of identical subunits (called protomers). Second, the protein exists in two different symmetrical states (historically referred to as the R- and T-states), even in the absence of the binding ligand. Third, R- and T-states are in equilibrium with each other, even in the absence of bound ligand. Fourth, each state has a different affinity for the binding ligand or substrate (and the less abundant of the two states has the tighter affinity for the ligand); thus, the equilibrium that exists between the two forms of the protein can be altered by the presence of the ligand. Fifth, the symmetry of the oligomeric protein is preserved in the conformational transition between the two forms of the protein. (All binding sites on each state of the oligomeric protein are initially equivalent in the absence of bound ligand, and remain equivalent at all degrees of saturation.) And sixth, in the case of allosteric enzymes, binding of the substrate is treated as a rapid equilibrium phenomenon (that is, binding is rapid relative to any catalytic event(s)).

MONOD-WYMAN-CHANGEUX COOPERATIVITY MODEL

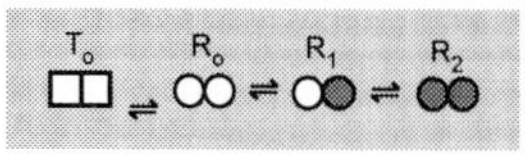

There is a pre-existing equilibrium between T_0 and R_0, even in the absence of ligand F.

$$\mathscr{L} = [\square\square] / [OO]$$

One can use α to represent the so-called reduced (or normalized) concentration of ligand F..

$$\alpha = [F] / K_F$$

Ligand F can bind to either of the two open sites on R_0, hence the "2" in the numerator.

$$K_F = 2[OO][F] / [\bullet O] \quad or \quad [\bullet O] = 2[OO][F] / K_F$$
$$[\bullet O] = 2\alpha[OO]$$

Ligand F can bind to only one open site, but the "2" in the denominator indicates two ways for the ligand to be released from the fully bound dimer.

$$K_F = [\bullet O][F] / 2[\bullet\bullet] \quad or \quad [\bullet\bullet] = [\bullet O][F] / 2K_F$$
$$[\bullet\bullet] = \alpha[\bullet O] / 2$$
$$[\bullet\bullet] = \alpha^2 [OO]$$

The ligand saturation function accounts for the moles of bound ligand divided by the total moles of binding sites (hence the "n" in the denominator.

$$\overline{Y}_F = \{[\bullet O] + 2[\bullet\bullet]\} / n\{[\square\square] + [OO] + [\bullet O] + [\bullet\bullet]\}$$

The ligand saturation function can then be re-expressed as:

$$\overline{Y}_F = \{2\alpha[OO] + 2\alpha^2[OO]\} / 2\{\mathscr{L}[OO] + [OO] + 2\alpha[OO] + \alpha^2[OO]\}$$

$$\overline{Y}_F = \{2\alpha + 2\alpha^2\} / 2\{\mathscr{L} + (1 + 2\alpha + \alpha^2)\}$$

$$= \alpha(1 + \alpha) / \{\mathscr{L} + (1 + \alpha)^2\}$$

Figure 1. Derivation of the MWC ligand saturation equation for a dimeric protein.

In Fig. 1, we show the full derivation of the saturation function for the case of ligand F binding to a dimer (open circles = unbound R-state; open squares = unbound T-state). Moreover, to demonstrate how the allosteric constant L influences the sigmoidicity of the ligand saturation curve, we present the dimer case again in Figure 2. At higher values of L (i.e., the greater the $[T_o]/[R_o]$ ratio), more of the ligand is present in the T-state, thereby requiring higher concentrations of ligand F (expressed here as the reduced concentration $\alpha = [F]/K_R$) to achieve significant saturation. The left-hand panel of Fig. 2 shows that the saturation curve becomes less S-shaped as the value of L approaches unity, a condition where the R_o and T_o states would be equally stable (i.e., $\Delta G_{R\text{-to-T-transition}} \approx 0$). The right-hand panel shows how the concentrations T_o, R_o, R_1, and R_2 change as a function of α for the case where $L = 1000$.

One can use free energy relationships to analyze the effect of ligand binding on the equilibrium between R and T states (see schematic diagrams shown in Fig. 3). Note that the energy difference between R_o and R_1 is exactly the same as that for R_1 and R_2. This is a required feature of the MWC model, because the strength of ligand F binding is exactly the same for each and every site in the R-state. Also note that the relative stability of T_o with respect to R_o will alter the fraction of total protein that is available for binding to ligand F.

Monod-Wyman-Changeux Model: Free Energy Changes

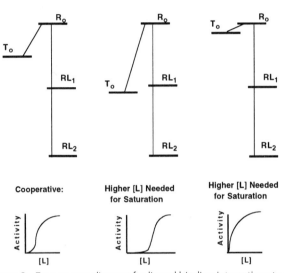

Figure 3. Free energy diagram for ligand binding interactions to an MWC dimer. Note that the relative stability of the T_0 to R_0 alters the ligand saturation curve as shown in each graph. The energy changes between each R-state are equivalent because each ligand binding interaction has the same equilibrium constant.

Shown in Fig. 4 are related energy diagrams showing (a) how inhibitor stabilizes the T-state and further reduces the ease of ligand F binding, and (b) how activator stabilizes the R state, increasing the fraction of total sites in the R state and thereby enhancing the binding of ligand F.

The energy diagrams presented in Fig. 5 allow one to compare the MWC model with the Koshland-Némethy-Filmer[2-4] model for ligand-induced conformation changes in ligand affinity. While describing key aspects of positive cooperativity, the Monod-Wyman-Changeux model cannot account for negative cooperativity. In this respect, any experimental evidence demonstrating nega-

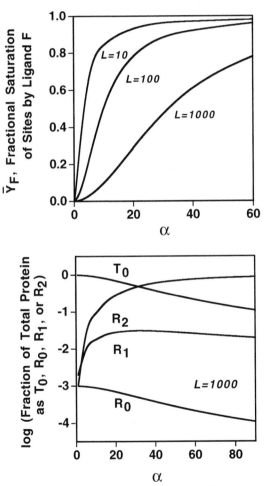

Figure 2. *Top:* Effect of the magnitude of the allosteric constant on ligand saturation behavior of a dimer obeying the MWC model. *Bottom:* Fraction of total dimeric protein present in various T- and R-species.

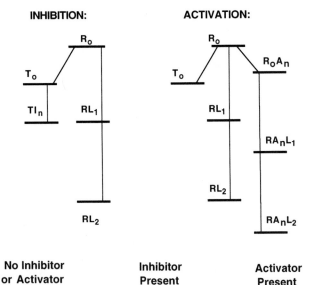

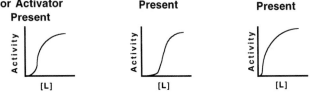

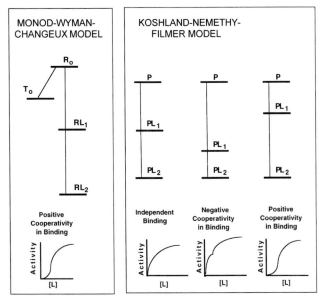

Figure 4. Free energy diagrams illustrating that binding of an inhibitor stabilizes the T-state and that binding of an activator stabilizes the R-state.

tive cooperativity immediately excludes the MWC model as a candidate mechanism.

Figure 5. Free energy diagrams showing the salient differences between the Monod and Koshland models. The MWC model is a two-state model with equivalent ligand binding interactions (indicated here by the equal spacing between R_0 and RL_1 states and between RL_1 and RL_2 states). In the KNF model, the amount of energy released determines whether binding will be independent or show negative or positive cooperativity.

[1]J. Monod, J. Wyman & J.-P. Changeux (1965) *J. Mol. Biol.* **12**, 88.
[2]K. E. Neet (1980) *Meth. Enzymol.* **64**, 139.
[3]K. E. Neet (1995) *Meth. Enzymol.* **249**, 519.
[4]D. E. Koshland, G. Némethy & D. Filmer (1966) *Biochemistry* **5**, 365.

MONOGLYCERIDE LIPASE

This enzyme [EC 3.1.1.23], also known as monoacylglycerol lipase, catalyzes the hydrolysis of glycerol monoesters of long-chain fatty acids. **See also** *Lipases*

J. C. Khoo & D. Steinberg (1983) *The Enzymes*, 3rd ed., **16**, 183.
P. Desnuelle (1972) *The Enzymes*, 3rd ed., **7**, 575.

MONO-ISO MECHANISMS

Enzyme reaction mechanisms that contain one ligand-independent enzyme isomerization step. **See** *Iso Mechanisms.*

MONO-ISO PING PONG BI BI REACTION SCHEME

A reaction scheme for a ping pong reaction in which there is a ligand-independent enzyme isomerization step following the second half-reaction. Thus, the first half-reaction of the scheme is represented by E + A $\rightleftharpoons$ EA $\rightleftharpoons$ FP $\rightleftharpoons$ F + P, the second half-reaction by F + B $\rightleftharpoons$ FB $\rightleftharpoons$ GQ $\rightleftharpoons$ G + Q, and the isomerization step by G $\rightleftharpoons$ E. In the absence of product, the steady-state rate expression predicts parallel lines in the double-reciprocal plots, identical to the lines one would observe in a normal ping pong Bi Bi reaction scheme. However, the complete rate equation predicts different product inhibition patterns[1,2]. This rate expression is identical to that of the straight-forward ping pong bi bi mechanism except that it includes additional terms in the denominator containing [A][Q], [A][B][Q], and [A][P][Q].

[1]F. B. Rudolph (1979) *Meth. Enzymol.* **63**, 411.
[2]K. L. Rebholz & D. B. Northrop (1995) *Meth. Enzymol.* **249**, 211.

MONOMER—POLYMER EXCHANGE KINETICS

Oosawa[1] originally described the phases of protein polymerization reactions as nucleation, growth, and attainment of monomer-polymer equilibrium. In the final phase, redistribution of polymer lengths causes the polymer length distribution to resemble a decaying exponential shape. Following the end of the rapid growth phase, no significant change occurs in the concentration of monomers. We use the term "steady state" to describe this period of constant monomer concentration. Wegner[2] showed that head-to-tail polymerization may occur during steady state, but only if monomer addition to polymers is accompanied by nucleotide hydrolysis. He postulated this mechanism to account for the unusually rapid exchange of radioactive monomers into F-actin at steady state. Margolis & Wilson[3] observed the same phenomenon for microtubules which use GTP. These investigators also recognized that radioactive label would be incorporated into polymers through the normal process of equilibrium exchange at the polymer ends. Polymers undergoing head-to-tail polymerization should incorporate label more quickly than by equilibrium exchange. The only theory describing equilibrium exchange rates was that of Kasai & Oosawa[4] and was based on an approximation. Wegner used this approximation to compare the label uptake rates resulting from the two mechanisms.

Kristofferson & Purich[5] discovered a simple solution that avoids Oosawa's approximation. Derivation of the equilibrium exchange rate of labeled monomer incorporation into biopolymers involves the following assumptions that are easily realized in practice. (1) The polymers are very long so there is no significant change in the number of polymers. (2) Exchange at one end will not interfere with the exchange dynamics at the opposite end of the polymer. (3) With equilibrium exchange, the probabilities of gaining or losing a monomer at a polymer end are the same. (4) All monomers free in solution are radiolabeled and dilution of label is negligible. One can also define the following quantities: N is the number of polymer ends; n is the number of "steps," where a "step" is either the addition or loss of a monomer at a polymer end. The average duration of a step is the reciprocal of the depolymerization rate constant at the polymer end. Karr, Kristofferson & Purich[6] determined a value of 113 s^{-1} for the net microtubule rate constant, or an average of 56.5 s^{-1}/end. Thus, the average step duration is about 18 ms. One should note that the theory does not consider dilution of the specific radioactivity of label which would result in lower label uptake values after long periods of time.

[1]F. Oosawa (1970) *J. Theor. Biol.* **27**, 69.
[2]A. Wegner (1976) *J. Mol. Biol.* **108**, 139.
[3]R. L. Margolis & L. Wilson (1978) *Cell* **13**, 1.
[4]M. Kasai & F. Oosawa (1969) *Biochim. Biophys. Acta.* **172**, 300.
[5]D. Kristofferson & D. L. Purich (1981) *J. Theor. Biol.* **92**, 85.
[6]T. L. Karr, D. Kristofferson & D. L. Purich (1980) *J. Biol. Chem.* **255**, 8560.

MONOMOLECULARITY

A synonym for "unimolecularity", referring to a reaction in which a single molecular entity is involved in the chemical event of an elementary reaction.

MONOPHENOL MONOOXYGENASE

This copper-dependent enzyme [EC 1.14.18.1] (also known as tyrosinase, phenolase, monophenol oxidase, and cresolase) catalyzes the reaction of L-tyrosine with L-dopa and dioxygen to produce L-dopa, dopaquinone, and water. This classification actually represents a set of copper proteins that also catalyze the reaction of catechol oxidase [EC 1.10.3.1] if only 1,2-benzenediols are available as substrates.

A. Messerschmidt (1998) *Comprehensive Biological Catalysis: A Mechanistic Reference* **3**, 401.
J. V. Schloss & M. S. Hixon (1998) *Comprehensive Biological Catalysis: A Mechanistic Reference* **2**, 43.
V. J. Hearing, Jr. (1987) *Meth. Enzymol.* **142**, 154.
K. Lerch (1987) *Meth. Enzymol.* **142**, 165.

MONTE CARLO METHODS

A computer-based, computationally intensive technique for analyzing chemical dynamics by using simulations based on the integration of simultaneous equations describing the time evolution of motions, usually classical in nature, of colliding atoms or molecules. The method requires a mathematical algorithm for randomly assigning initial conditions, including positional coordinates, energies, velocities, and related impact parameters for each atom or molecule. The overall distribution of energies is typically restricted to those permitted on the basis of the Maxwell-Boltzmann equilibrium distribution. The randomness of assignment was initially likened to the haphazard behavior of roulette wheels and other games of chance practiced in Monte Carlo, hence the cognomen. The method has also been used extensively to obtain approximate solutions to otherwise mathematically intractable problems by defining a range of values, each having a calculable probability of being the problem's solution.

Selected entries from *Methods in Enzymology* [vol, page(s)]: Algorithms, **259**, 579-580; computational time, **240**, 430, 436-437; membrane modeling [attainment of equilibrium, **240**, 577-578, 591-592; biased sampling, **240**, 590; cluster statistics, **240**, 583-585; data collection, **240**, 579-580; decision making, **240**, 576-577, 591-592; defining Monte Carlo cycle, **240**, 575; end detection, **240**, 577; equilibrium fluctuation, **240**, 590-592; Gibbs energy change, **240**, 576; parameter fitting, **240**, 580; phase transition of small unilamellar dipalmitoylphosphatidylcholine vesicles, **240**, 569-570; random selection of lattice point, **240**, 575-

MORE O'FERRALL-JENCKS DIAGRAM

A depiction of a hypothetical potential energy surface for a reacting system as a function of two chosen coordinates (*e.g.*, the lengths of two bonds being broken)[1-5]. Such diagrams are useful in assessing structural effects on transition states for stepwise or concerted pathways. An example of More O'Ferrall-Jencks diagrams for β-elimination reactions is shown below.

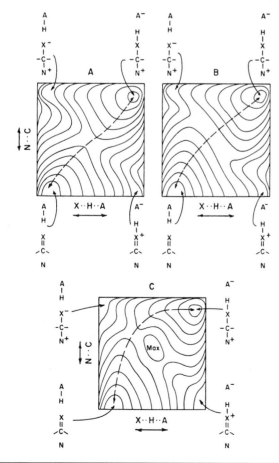

Structural changes influencing vibrational modes of the transition states alter transition state geometry. Changes in the direction of the reaction coordinate (reactant or product stabilizing or destabilizing factors) cause changes in accordance with the Hammond principle. Structural changes perpendicular to the reaction coordinate (so-called "anti-Hammond" effects or perpendicular effects) cause changes that run counter to the Hammond behavior (*i.e.*, the easier the process related to the structural change, the more advanced it will be at the transition state)[1].

[1]IUPAC (1994) *Pure and Appl. Chem.* **66**, 1077.
[2]R. A. More O'Ferrall (1970) *J. Chem. Soc. (B)*, 274.
[3]W. P. Jencks (1972) *Chem. Rev.* **72**, 705.
[4]W. P. Jencks (1980) *Acc. Chem. Res.* **13**, 161.
[5]D. A. Winey & E. R. Thornton (1975) *J. Amer. Chem. Soc.* **97**, 3102.

MÖSSBAUER SPECTROSCOPY

The Mössbauer effect refers to the absorption and re-radiation of a gamma ray by a nonrecoiling nucleus. Gamma rays of extremely narrowly defined wavelength exhibit this effect with certain nuclei (*e.g.*, the ^{57}Fe nucleus), and Mössbauer spectroscopy can detect even very subtle changes in electron configuration, chemical bonding, and oxidation state during substrate binding and enzymic catalysis. A lucid description of the theory and application of Mössbauer spectroscopy to the characterization of an enzyme mechanism appears in vol. **4** of *Comprehensive Biological Catalysis: A Mechanistic Reference* (M. Sinnott, ed.).

Selected entries from *Methods in Enzymology* [**vol**, page(s)]: Characteristic features of spectra, **76**, 334-335 [of Fe proteins, **69**, 775-777; of Mo-Fe proteins, **69**, 771-773]; cryogenic temperatures and external magnetic fields, **76**, 338-339; experimental running time, **76**, 342-344; experimental techniques, **76**, 340-347; of hemoglobins, **76**, 347-354 [by photodissociation, **76**, 349-351; by selective enrichment, **76**, 349; in strong magnetic fields, **76**, 351-354]; information sources, **76**, 345-347; Mössbauer effect as basis, **76**, 329-332; obtaining transmission spectra, **76**, 340-342; phase determination in X-ray structure analysis, **76**, 354; pulse processing equipment, **76**, 337-338; radiation shielding, **76**, 339-340; selective excitation double technique, **76**, 352-354; sources and detectors, **76**, 336-337; spectrometer components, **76**, 335-340; spectrum analysis, **76**, 344-345; superoperator approach, **76**, 345; transmission apparatus, **76**, 340; transmission geometry, **76**, 332-333.

mRNA (GUANINE-N^7-)-METHYLTRANSFERASE

This enzyme [EC 2.1.1.56] catalyzes the addition of the N^7-methyl group to the cap of mRNA: thus, *S*-adenosyl-L-methionine reacts with G(5')pppR-RNA to produce

S-adenosyl-L-homocysteine and m[7]G(5')pppR-RNA (the R group may be guanosine or adenosine).

P. Narayan & F. M. Rottman (1992) *Adv. Enzymol.* **65**, 255.

MU (μ)

1. Symbol for micro-. 2. Symbol for micron. 3. Symbol for dynamic viscosity. 4. Symbol for magnetic or electric dipole moment of a molecule. 5. Symbol for chemical potential. 6. Symbol for permeability; μ_0 for permeability of vacuum; μ_r for relative permeability. 7. Symbol for reduced mass.

$\tilde{\mu}$ Symbol for electrochemical potential.

mμ Symbol for millimicro- or millimicron.

MUCONATE CYCLOISOMERASE

This manganese-dependent enzyme [EC 5.5.1.1] catalyzes the interconversion of 2,5-dihydro-5-oxofuran-2-acetate and *cis,cis*-hexadienedioate. In the reverse reaction, 3-methyl-*cis,cis*-hexadienedioate can serve as a substrate and *cis,trans*-hexadienedioate can act as a weak substrate. The enzyme differs from chloromuconate cycloisomerase [EC 5.5.1.7] and dichloromuconate cycloisomerase [EC 5.5.1.11].

R. B. Meagher, K.-L. Ngai & L. N. Ornston (1990) *Meth. Enzymol.* **188**, 126.

MUCONOLACTONE Δ-ISOMERASE

This enzyme [EC 5.3.3.4] catalyzes the interconversion of 2,5-dihydro-5-oxofuran-2-acetate and 3,4-dihydro-5-oxofuran-2-acetate.

R. B. Meagher, K.-L. Ngai & L. N. Ornston (1990) *Meth. Enzymol.* **188**, 130.

MUCROLYSIN

This zinc-dependent endopeptidase [EC 3.4.24.54] from Chinese habu snake venom catalyzes the hydrolysis of the Ser[9]—His[10], His[10]—Leu[11], Ala[14]—Leu[15], Leu[15]—Tyr[16], and the Tyr[16]—Leu[17] peptide bonds in the insulin B chain.

J. B. Bjarnason & J. W. Fox (1995) *Meth. Enzymol.* **248**, 345.

MULTIENZYME

Any protein exhibiting more than one catalytic function as a result of distinct parts of a polypeptide chain ('domains') or distinct subunits, or both.

IUB Nomenclature Committee (1989) *Eur. J. Biochem.* **185**, 485.

MULTIPLE DEAD-END INHIBITION

Depressed catalytic activity occurring when an inhibitor binds more than once to a single enzyme form (or forms). While standard double-reciprocal plots are usually linear, secondary replots of the data (*i.e.*, plots of slopes and/or intercepts *vs.* [I], the concentration of the inhibitor) will be nonlinear depending on the relative magnitude of the $[I]^2$, $[I]^3$, . . ., and $[I]^n$ terms in the rate expression.

Cleland[1] presented more complex examples of multiple dead-end inhibition in which the kinetic expression for the slope or intercept is a function containing polynomials of the inhibitor concentration in both the denominator and numerator. For example, if the slope is equal to a function having the form $(a + b[I] + c[I]^2)/(d + e[I])$ in which *a*, *b*, *c*, *d*, and *e* are constants or collections of constants, then the slope is said to be a 2/1 function (the numbers representing the highest power of [I] in the numerator and denominator, respectively). The nonlinearity of the slope replot, in this case, is dependent on the relative magnitudes of the constants in the expression.

[1]W. W. Cleland (1963) *Biochim. Biophys. Acta* **67**, 173.

MULTIPLE-TURNOVER CONDITIONS

Reaction conditions permitting a catalyst to pass through many catalytic rounds. Multiple-turnover conditions are usually obtained by maintaining the substrate concentration in excess over the concentration of active catalyst. This technique usually allows one the opportunity to evaluate the catalytic rate constant k_{cat}, which is the first-order decay rate constant for the rate-determining step for each cycle of catalysis, and one can evaluate the magnitude of other parameters such as the substrate's dissociation constant or Michaelis constant.

MULTIPLICATIVE MODEL

Any mechanism in which the observed effect of two or more modifiers equals the product of effects observed when those modifiers act alone. Two agents acting competitively (*i.e.*, by binding at the same site) cannot act multiplicatively.

MULTIPLICITY

For a given total electron spin quantum number (*S*), the multiplicity is the number of possible orientations of the spin angular momentum for the same spatial electronic wavefunction. Thus, the multiplicity equals $2S + 1$. For a state of singlet multiplicity, $S = 0$ and $2S + 1 = 1$. If $S > L$, where L is the total electron orbital angular momentum quantum number, there are only $2L + 1$ possible orientations.

Comm. on Photochem. (1988), *Pure and Appl. Chem.* **60**, 1055.

MULTISITE PING PONG REACTION SCHEMES

Certain enzymes catalyze their reactions by way of a multisite mechanism in which the covalently linked intermediate is attached to a long arm that "swings" from one subsite to another subsite within the enzyme. In some cases, the covalently tethered intermediate can actually be transferred between subunits that form the active site. An example is *Propionibacterium shermanii* transcarboxylase[1], an enzyme that catalyzes the biotin-dependent conversion of methylmalonyl-CoA and pyruvate to propionyl-CoA and oxaloacetate. Carboxylated biotin allows the two catalytic subsites to operate on the same reaction intermediate.

While the initial rate and exchange studies indicated a standard ping pong mechanism for this enzyme, the product inhibition results were not consistent with the traditional single-site mechanism[1,2]. The two-site ping-pong scheme accounts for the product inhibition results observed with the transcarboxylase. A more complex multisite ping pong mechanism has also been presented for pyruvate dehydrogenase[3].

[1]D. B. Northrop (1969) *J. Biol. Chem.* **244**, 5808.
[2]D. B. Northrop & H. G. Wood (1969) *J. Biol. Chem.* **244**, 5820.
[3]W. W. Cleland (1973) *J. Biol. Chem.* **248**, 8353.

MULTISTAGE REACTIONS

Chemical reactions with one or more intermediates separated by transition states (usually indicated by the superscript symbol ‡):

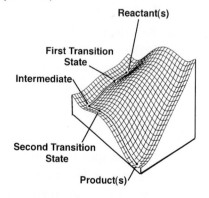

Potential Energy Surface for a Multistage Reaction

The overall rate of multistage reactions depends on the activation energy, the minimal energy needed by a reactant to reach the highest lying transition state. Any reactant(s) reaching this transition state can either re-form reactant(s) or proceed to form products. In so doing, energy is released, usually by momentum exchanges with colliding molecules. If reactants can achieve equilibrium with products, then the Law of Microscopic Reversibility requires that the product-to-reactant path will be identical to the reactant-to-product path.

MULTISUBSTRATE ANALOGUE

A molecule that possesses the capacity to span two or more substrate binding sites within the active site of an enzyme. These analogues are to be distinguished from transition-state analogues which typically mimic the geometry or bond order of the transition state. Early examples of multisubstrate analogues were P^1,P^4-di(adenosine-5')tetraphosphate[1] and P^1,P^5-di(adenosine-5')pentaphosphate[2] which were found to be potent inhibitors of adenylate kinase. **See** *Adenylate Kinase Inhibitors; Geometrical Analogue*

[1]D. L. Purich & H. J. Fromm (1972) *Biochim. Biophys. Acta* **276**, 563.
[2]G. E. Lienhard & I. I. Secemski (1973) *J. Biol. Chem.* **248**, 1121.

MULTISUBSTRATE MECHANISMS

The following diagrams indicate the binding interactions for enzyme kinetic mechanisms. To conserve space, the notation used in this Handbook is a compact version of the diagrams first introduced by Cleland[1]. His diagram for the Ordered Uni Bi Mechanism is as follows:

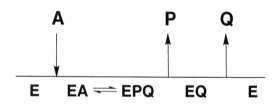

We use the following single-line compact form:

$$_E\underline{A \downarrow (EA \leftrightarrow EPQ) \downarrow P(EQ) \downarrow Q}_E$$

The enzyme surface is represented by the underline, in this case preceded by a subscript E to indicate the unbound (or free) enzyme before any substrate addition and after all products desorb. An arrow pointing to the line indicates binding, and the reader should understand that reversible binding is taken for granted. Moreover, unlike the original Cleland diagram, product release always is written by a downward arrow, because this systematic usage emphasizes the symmetry of certain mechanisms. The symbol "**A ↓**" indicates that substrate A adds; the symbol "**A or B ↓ ↓**", "**A or B or C ↓ ↓ ↓**", etc., indicates *random* addition of two or three substrates, respectively. Interconversions of enzyme-bound reactants can be reversible (↔) or irreversible (→).

Taking the Ordered Uni Bi mechanism as an example, we can consider several additional possibilities:

$$\underset{E}{\text{A ↓ (EA↔EX↔EPQ) ↓ P(EQ) ↓ Q}}_{E}$$

$$\underset{E}{\text{A ↓ (EA↔EX→EPQ) ↓ P(EQ) ↓ Q}}_{E}$$

$$\underset{E}{\text{A ↓ (EA→EX→EPQ) ↓ P(EQ) ↓ Q}}_{E}$$

For the case of so-called iso mechanisms, the compact diagrams are as follows:

$$\underset{E}{\text{A ↓ (EA↔EX→EPQ) ↓ P ↓ Q}}_{F↔E}$$

where F ⇌ E represents the isomerization step.

The following mechanisms are typically encountered in studies of enzymic catalysis:

Ordered Uni Bi Mechanism

$$\underset{E}{\text{A ↓ (EA↔EPQ) ↓ P(EQ) ↓ Q}}_{E}$$

Random Uni Bi Mechanism

$$\underset{E}{\text{A ↓ (EA↔EPQ) ↓ ↓ P or Q}}_{E}$$

Random Bi Uni Mechanism

$$\underset{E}{\text{A or B ↓ ↓ (EAB↔EP) ↓ P}}_{E}$$

Ordered Bi Uni Mechanism

$$\underset{E}{\text{A ↓ (EA)B ↓ (EAB↔EP) ↓ P}}_{E}$$

Ordered Bi Bi Mechanism

$$\underset{E}{\text{A ↓ (EA)B ↓ (EAB↔EPQ) ↓ P(EQ) ↓ Q}}_{E}$$

Ordered Bi Bi - Subsite Mechanism

$$\underset{E}{\text{A ↓ B ↓ (EAB ↓ EPQ) ↓ P(EQ) ↓ Q}}_{E}$$

$$\text{B ↑ (EB)} \qquad \text{(EP) ↑ P}$$

Ordered Bi Bi-Theorell Chance Mechanism

$$\underset{E}{\text{A ↓ (EA)B ↓ ↓ P(EQ) ↓ Q}}_{E}$$

Ping Pong Bi Bi Mechanism

$$\underset{E}{\text{A ↓ (EA↔FP) ↓ P(F)B ↓ (FB↔EQ) ↓ Q}}_{E}$$

Random Bi Bi Mechanism

$$\underset{E}{\text{A or B ↓ ↓ (EAB↔EPQ) ↓ ↓ P or Q}}_{E}$$

Ordered Ter Ter-Theorell Chance Mechanism

$$\underset{E}{\text{A ↓ (EA)B ↓ (EAB)C ↓ ↓ P(EQR) ↓ (EQ)Q ↓ R}}_{E}$$

Ordered Ter Ter Mechanism

$$\underset{E}{\text{A ↓ (EA)B ↓ (EAB)C ↓ (EABC↔EPQR) ↓ P(EQR) ↓ Q(ER) ↓ R}}_{E}$$

Random Ter Ter Mechanism

$$\underset{E}{\text{A or B or C ↓ ↓ ↓ (EACB↔EPQR) ↓ ↓ ↓ P or Q or R}}_{E}$$

Random AB, Random QR Ter Ter Mechanism

$$\underset{E}{\text{A or B ↓ ↓ (EAB)C ↓ (EABC↔EPQR) ↓ P ↓ (EQR) ↓ Q or R}}_{E}$$

Random BC, Random PQ Ter Ter Mechanism

$$\underset{E}{\text{A ↓ (EA)B or C ↓ ↓ (EABC↔EPQR) ↓ ↓ P or Q(ER) ↓ R}}_{E}$$

MUSCLE MASS DETERMINATION BY ISOTOPIC TRACERS

The isotope tracer method[1] for estimating the total muscle mass of an organism by determining the amount of creatinine clearance. The method relies on the fact that muscle creatine phosphate is converted to creatine, and the latter is metabolized into creatinine.

[1]R. R. Wolfe (1992) *Radioactive and Stable Isotope Tracers in Biomedicine*, pp. 193, Wiley-Liss, New York.

MUTAROTASES

Enzymes that catalyze the interconversion of the anomeric carbon of saccharides[1-3]. The name is derived from the ability of these enzymes to change or "mutate" the optical rotatory properties of sugars.

These enzymes catalyze mutarotation via acid/base catalysis, and the open-chain form of the sugar is an enzyme-bound intermediate in the reaction pathway. In principle, any protein that binds the ring-open form of a sugar will be capable of catalyzing mutarotation, and such an activity may have little to do with the physiologic function of that protein. These enzymes catalyze epimerization reactions at the hemiacetal or hemiketal carbon. Aldose 1-epimerase [EC 5.1.3.3], also known as mutarotase and aldose mutarotase, catalyzes the interconversion of α-D-glucose and β-D-glucose. The enzyme is reported to also utilize L-arabinose, D-xylose, D-galactose, maltose, and lactose as substrates. *See also* Anomers; *Fischer Projection; Aldose 1-Epimerase*

[1]J. M. Bailey, P. G. Pentchev, P. H. Fishman, S. A. Mulhern & J. W. Kusiak (1973) *Adv. Chem. Ser.* **117**, 264.
[2]E. Adams (1976) *Adv. Enzymol.* **44**, 69.
[3]M. E. Tanner (1998) *Comprehensive Biological Catalysis: A Mechanistic Reference* **1**, 208.

MUTAROTATION KINETICS

Anomeric interconversion of sugars and their derivatives is typically relatively slow, but anomerization of phosphorylated sugars is usually faster. Values for the rate constants can be obtained by nuclear magnetic resonance techniques. For example, the conversion of β-fructose 1,6-bisphosphate to the straight-chain form, at 25°C and pH 7.2, has a rate constant of 35 s^{-1}, and the rate constant for the reverse reaction is 1450 s^{-1}. Conversion of the α-anomer to the straight-chain form has a rate constant of 8.5 s^{-1} and a 70 s^{-1} value for the reverse reaction[2].

[1]S. J. Benkovic (1979) *Meth. Enzymol.* **63**, 370.
[2]C. F. Midelfort, R. K. Gupta & I. A. Rose (1976) *Biochemistry* **15**, 2178.

MYELOBLASTIN

This enzyme [EC 3.4.21.76] catalyzes the hydrolysis of peptide bonds in proteins, including elastin.

J. R. Hoidal, N. V. Rao & B. Gray (1994) *Meth. Enzymol.* **244**, 61.

MYELOPEROXIDASE

This enzyme [EC 1.11.1.7] catalyzes the reaction of a donor substrate with hydrogen peroxide to produce an oxidized donor and two water.

H. B. Dunford (1998) *Comprehensive Biological Catalysis: A Mechanistic Reference* **3**, 195.

E. Cadenas (1989) *Ann. Rev. Biochem.* **58**, 79.
J. K. Hurst & W. C. Barrette, Jr. (1989) *Crit. Rev. Biochem. Mol. Biol.* **24**, 271.
A. Naqui, B. Chance & E. Cadenas (1986) *Ann. Rev. Biochem.* **55**, 137.

MYOGLOBIN OXYGENATION

Although myoglobin was the first protein for which the atomic structure was determined, the detailed molecular mechanism of reversible oxygenation remains an area of active research[1]. The binding of oxygen is a multistep process, involving displacement of a noncoordinated water molecule at a distal pocket, iron atom movement into the plane of the porphyrin ring to create a hexacoordinate complex, and formation of a hydrogen bond between His-64's N$^\varepsilon$ and the second bound oxygen atom[2]. This can be described schematically as follows:

$$X + A \underset{ms}{\rightleftharpoons} [C_1, \ldots, C_n] \underset{ns}{\rightleftharpoons}$$
$$[B_1, \ldots, B_n] \underset{ps}{\rightleftharpoons} A \text{ or } MbX$$

where X is the free ligand, $[C_1, \ldots, C_n]$ is the water molecule displacement step, $[B_1, \ldots, B_n]$ is the formation of a contact pair, and A or MbX is the ground state complex. The overall association rate constants are 0.5 $\mu M^{-1}s^{-1}$ for CO, 17 $\mu M^{-1}s^{-1}$ for O$_2$, and 22 $\mu M^{-1}s^{-1}$ for NO·. Because photodissociation of the heme-bound ligand can be achieved by using an intense exitation pulse, the kinetics of these events can be recorded on the time scales shown beneath each species in the above scheme.

[1]Q. H. Gibson (1989) *J. Biol. Chem.* **264**, 20155.
[2]J. S. Olson & G. N. Phillips, Jr. (1996) *J. Biol. Chem.* **257**, 17593.

The following are other recent reviews on the molecular and physical properties of this oxygen binding protein.

H. B. Gray & J. R. Winkler (1996) *Ann. Rev. Biochem.* **65**, 537.
J. N. Onuchic, D. N. Beratan, J. R. Winkler & H. B. Gray (1992) *Ann. Rev. Biophys. Biomol. Struct.* **21**, 349.
J. B. Wittenberg & B. A. Wittenberg (1990) *Ann. Rev. Biophys. Biophys. Chem.* **19**, 217.
R. E. Dickerson & I. Geis (1983) *Hemoglobin*, Benjamin/Cummings, Menlo Park, CA.

MYOSIN ATPase

This ATPase activity [EC 3.6.1.32] is directly responsible for muscle contraction. In the absence of actin filaments, myosin is a feeble ATPase with a k_{cat} of only 0.05 s^{-1}, because product release is much slower than the rapid release of a proton in the P—O—P bond-cleavage step forming ADP and P$_i$ from bound ATP. Interaction with actin filaments, however, strongly stimulates the ATPase activity, increasing k_{cat} by a factor of about 200. This acceleration occurs as a consequence of the expedited release of P_i and then ADP.

In the so-called Lymm-Taylor model, the actomyosin ATPase reaction proceeds in a stepwise manner:
(1) Myosin binds ATP and releases the actin filament.

Filament·Myosin + ATP → Filament + Myosin·ATP

(2) Myosin·ATP complex is hydrolyzed rapidly to produce an conformationally energized state (often designated by an asterisk*).

Myosin·ATP + HOH → Myosin*·ADP·P_i

(3) Myosin*·ADP·P_i then binds to an actin filament, presumably at another position along the filament.

Filament + Myosin*·ADP·P_i →
Filament·Myosin*·ADP·P_i

(4) The filament-bound complex relaxes through the release of stored conformational energy and returns to the resting state, myosin·ADP·P_i.

Filament·Myosin*·ADP·P_i → Filament·Myosin·ADP·P_i

(5) This conformational relaxation is followed by release of P_i.

Filament·Myosin·ADP·P_i →
Filament·Myosin·ADP + P_i

(6) Finally, ADP is released to regenerate filament-bound myosin.

Filament·Myosin*·ADP → Filament·Myosin + ADP

When appropriately arranged in the context of the thick and thin filaments of the myofibril, repeated cycles of reaction steps *1-6* result in contraction. Other forms of myosin are responsible for contractile steps that are associated with the crawling of nonmuscle cells. **See also** *Force Effects on Molecular Motors*

S. Khan & M. Sheetz (1997) *Ann. Rev. Biochem.* **66**, 785.
I. Rayment (1996) *J. Biol. Chem.* **271**, 15850.
K. M. Ruppel & J. A. Spudich (1996) *Ann. Rev. Cell Dev. Biol.* **12**, 543.
M. S. Mooseker & R. E. Cheney (1995) *Ann. Rev. Cell Dev. Biol.* **11**, 633.
I. Rayment & H. M. Holden (1994) *Trends Biochem Sci.* **19**, 129.
K. A. Johnson (1992) *The Enzymes*, 3rd ed., **20**, 1.
J. R. Sellers & R. S. Adelstein (1987) *The Enzymes*, 3rd ed., **18**, 381.
Y. Tonomura & F. Oosawa (1972) *Ann. Rev. Biophys. Bioeng.* **1**, 159.

N

n→π* TRANSITION

An electronic transition in which an electron in a nonbonding (*e.g.*, lone pair) orbital (called an n-orbital) is promoted to a π-antibonding orbital. The excited state arising from such a promotion is often referred to as an n-π* state. An electron in an n-orbital typically interacts strongly with a polar solvent; this is less likely to be the case for an electron in a π* orbital. Therefore, the energy difference between n and π* orbital electrons will increase when a substance is placed in a more polar solvent; this is manifested as a shift to shorter wavelength (often called a blue shift) for light absorption. **See** *Absorption Spectroscopy*

n→σ* DELOCALIZATION

Delocalization of a free electron pair into an antibonding σ-orbital. **See** *Hyperconjugation; Resonance; Absorption Spectroscopy*

n→σ* TRANSITION

An electronic transition in which an electron in a nonbonding (*e.g.*, lone-pair) orbital is promoted to an antibonding σ orbital. **See** *Absorption Spectroscopy*

NAD⁺ GLYCOHYDROLASE

This enzyme [EC 3.2.2.5] (also known as NAD⁺ nucleosidase, NADase, DPNase, DPN hydrolase, and ADP-ribosylcyclase) catalyzes the hydrolysis of NAD⁺ to nicotinamide and ADP-ribose. The enzyme from some mammalian sources also catalyzes the transfer of ADP-ribose residues.

G. Davies, M. L. Sinnott & S. G. Withers (1998) *Comprehensive Biological Catalysis: A Mechanistic Reference* 1, 119.
N. J. Oppenheimer & A. L. Handlon (1992) *The Enzymes*, 3rd ed., 20, 453.

NADH-CYTOCHROME *b*₅ REDUCTASE

This FAD-dependent enzyme [EC 1.6.2.2], also known as cytochrome *b*₅ reductase, catalyzes the reaction of NADH with two molecules of ferricytochrome *b*₅ to produce NAD⁺ and two ferrocytochrome *b*₅.

P. W. Holloway (1983) *The Enzymes*, 3rd ed., 16, 63.
C. H. Williams, Jr. (1976) *The Enzymes*, 3rd ed., 13, 89.

NADH DEHYDROGENASE

NADH dehydrogenase (ubiquinone) [EC 1.6.5.3] (also called ubiquinone reductase, type I dehydrogenase, and complex I dehydrogenase) catalyzes the reaction of NADH with ubiquinone to produce NAD⁺ and ubiquinol. The complex, which uses FAD and iron-sulfur proteins as cofactors, is found in mitochondrial membranes and can be degraded to form NADH dehydrogenase [EC 1.6.99.3]. NADH dehydrogenase [EC 1.6.99.3] catalyzes the reaction of NADH with an acceptor to produce NAD⁺ and the reduced acceptor. Iron-sulfur and flavoproteins are still being used as cofactors with this component of EC 1.6.5.3. Interestingly, after certain preparations have been followed, cytochrome *c* may serve as the acceptor substrate.

U. Schulte & H. Weiss (1995) *Meth. Enzymol.* 260, 3.
J. E. Walker, J. M. Skehel & S. K. Buchanan (1995) *Meth. Enzymol.* 260, 14.
Y. Hatefi & D. L. Stiggall (1976) *The Enzymes*, 3rd ed., 13, 175.
G. Palmer (1975) *The Enzymes*, 3rd ed., 12, 1.

NADH DEHYDROGENASE (Other Than Complex I)

Other NADH dehydrogenases include NADH dehydrogenase (quinone) [EC 1.6.99.5] which catalyzes the reaction of NADH with an acceptor to produce NAD⁺ and the reduced acceptor. Menaquinone can serve as the acceptor substrate. This dehydrogenase is inhibited by AMP and 2,4-dinitrophenol but not by dicoumarol or folic acid derivatives.

Y. Hatefi & D. L. Stiggall (1976) *The Enzymes*, 3rd ed., 13, 175.

NADH/NADPH SURFACE FLUORESCENCE IN LIVING TISSUES

Surface fluorescence of NADH/NADPH can be recorded continuously with a DC fluorimeter and correlated with changes in experimental conditions. A mercury arc lamp (with a 340-375 nm filter in front) is used as a light source for fluorescence excitation. The fluorescence response of reduced NADH/NADPH was measured at 450-510 nm. The DC fluorimeter and the Hg arc lamp are connected to the kidney by a trifurcated fiber optics light guide. NADH/NADPH fluorescence emission can be corrected for changes in tissue opacity by a 1:1 subtraction of reflectance changes at 340-375 nm from the fluorescence. To determine NADH/NADPH redox state of the total surface area of kidney cortex and to evaluate whether certain areas were insufficiently perfused, fluorescence photographs of the total surface area were taken. The study demonstrated that the surface fluorescence method is simple and provides specific information about the mitochondrial oxidation-reduction state.

H. Franke, C. H. Barlow & B. Chance (1980) *Contrib. Nephrol.* **19**, 240.

NADH PEROXIDASE

This FAD-dependent enzyme [EC 1.11.1.1] catalyzes the reaction of NADH with hydrogen peroxide to produce NAD^+ and two water. Ferricyanide, quinones, and other compounds can replace hydrogen peroxide in this enzyme-catalyzed reaction.

B. A. Palfey & V. Massey (1998) *Comprehensive Biological Catalysis: A Mechanistic Reference* **3**, 83.

NAD⁺ KINASE

This enzyme [EC 2.7.1.23] catalyzes the reaction of ATP with NAD^+ to produce ADP and $NADP^+$. This reaction has the potential for altering the intracellular $[NADP^+]/[NADPH]$ redox potential by converting NAD^+ to $NADP^+$, but details on the regulation of this enzyme have not been worked out.

E. P. Anderson (1973) *The Enzymes*, 3rd ed., **9**, 49.
S. Pinder, J. B. Clark & A. L. Greenbaum (1971) *Meth. Enzymol.* **18B**, 20.

NADPH DEHYDROGENASE

This enzyme [EC 1.6.99.1] (also known as old yellow enzyme and NADPH diaphorase) catalyzes the reaction of NADPH with an acceptor to produce $NADP^+$ and the reduced acceptor. The yeast enzyme uses FMN whereas the plant enzyme uses FAD.

B. A. Palfey & V. Massey (1998) *Comprehensive Biological Catalysis: A Mechanistic Reference* **3**, 83.
L. M. Schopfer & V. Massey (1991) *A Study of Enzymes* **2**, pp. 247.
H. J. Bright & D. J. T. Porter (1975) *The Enzymes*, 3rd ed., **12**, 421.

NAD(P)H DEHYDROGENASE (Quinone)

This enzyme [EC 1.6.99.2] (also known as NAD(P)H:quinone reductase, DT diaphorase, quinone reductase, azoreductase, phylloquinone reductase, and menadione reductase) catalyzes the reaction of NAD(P)H with an acceptor to produce $NAD(P)^+$ and the reduced acceptor. This FAD-dependent enzyme is inhibited by dicoumarol.

B. A. Palfey & V. Massey (1998) *Comprehensive Biological Catalysis: A Mechanistic Reference* **3**, 83.
E. Cadenas, P. Hochstein & L. Ernster (1992) *Adv. Enzymol.* **65**, 97.
C. Lind, E. Cadenas, P. Hochstein & L. Ernster (1990) *Meth. Enzymol.* **186**, 287.

NAD(P)⁺ TRANSHYDROGENASE

This enzyme [EC 1.6.1.1] (also known as NAD(P)⁺ transhydrogenase (B-specific), pyridine nucleotide transhydrogenase, and nicotinamide nucleotide transhydrogenase) catalyzes the reversible reaction of NADPH with NAD^+ to produce $NADP^+$ and NADH. This FAD-dependent enzyme is B-specific with respect to both pyridine coenzymes. In addition, deamino coenzymes will also serve as substrates.

A. R. Clarke & T. R. Dafforn (1998) *Comprehensive Biological Catalysis: A Mechanistic Reference* **3**, 1.
J. Rydström, J. B. Hoek & L. Ernster (1976) *The Enzymes*, 3rd ed., **13**, 51.

NAD⁺ SYNTHETASE

NAD^+ synthetase (ammonia-utilizing) [EC 6.3.1.5] catalyzes the reaction of ATP with deamido-NAD^+ and ammonia to produce NAD^+, AMP, and pyrophosphate (or, diphosphate). NAD^+ synthetase (glutamine-utilizing) [EC 6.3.5.1] catalyzes the reaction of ATP with deamido-NAD^+, glutamine, and water to produce NAD^+, glutamate, AMP, and pyrophosphate.

V. Micheli & S. Sestini (1997) *Meth. Enzymol.* **280**, 211.
H. Zalkin (1993) *Adv. Enzymol.* **66**, 203.
H. Zalkin (1985) *Meth. Enzymol.* **113**, 297.
J. M. Buchanan (1973) *Adv. Enzymol.* **39**, 91.

NANO-

A prefix, symbolized by n, used in submultiples of units and corresponding to a value of 10^{-9}.

NANOSECOND ABSORPTION SPECTROSCOPY

Selected entries from *Methods in Enzymology* [vol, page(s)]:
Overview, **226**, 119, 147; absorption apparatus, **226**, 131; apparatus, **226**, 152; detectors, **226**, 126; detector systems, **226**, 125; excitation source, **226**, 121; global analysis, **226**, 146, 155; kinetic applications, **226**, 134; heme proteins, **226**, 142; multiphoton effects, **226**, 141; nanosecond time-resolved recombination, **226**, 141; quantum yields, **226**, 139; singular value decomposition, **226**, 146, 155; spectral dynamics, **226**, 136; time delay generators, **226**, 130.

NARCISSISTIC REACTION

A chemical reaction in which the reactant is converted into its mirror image, hence the name alluding to the self-admiring nature of the mythic Narcissus. In a narcissistic process, a chiral reactant can be converted into its enantiomer. Racemases catalyze the most obvious narcissistic biochemical reactions. Another example of a narcissistic reaction is degenerate rearrangement.

L. Salem (1971) *Acc. Chem. Res.* **4**, 322.

NATURAL ABUNDANCE

A quantitative measure of isotope composition relative to the abundance of all isotopic forms found in nature. Values for those stable isotopes most commonly employed in biological tracer experiments are: ^1H, 99.985%; ^2H, 0.015%; ^{12}C, 98.90%; ^{13}C, 1.10%; ^{14}N 99.63%; ^{15}N, 0.37%; ^{16}O, 99.76%; ^{17}O, 0.04%; and ^{18}O, 0.20%. ***See*** *Isotope Exchange Kinetics; Compartmental Analysis*

NEGATIVE COOPERATIVITY

Any set of ligand interactions with oligomeric or polymeric macromolecules where binding of the the first (or preceding) ligand molecule decreases the likelihood for binding of the next (or subsequent) ligand molecule. In the Koshland-Némethy-Filmer model, negative cooperativity occurs when the dissociation constant for ligand binding to the $(i + 1)$-site is greater than the dissociation constant for ligand binding to the ith site.

Note that negative cooperativity cannot occur in the Monod-Wyman-Changeux allosteric transition model, because the dissociation constant is equivalent for all sites. Thus, positive cooperativity can only result in this binding mechanism as a consequence of the "recruitment" of binding sites from the T-state in an all-or-none transition to the R-state. Any occurrence of negative cooperativity can be regarded as *prima facie* evidence

against the applicability of the Monod-Wyman-Changeux model to the system under investigation. ***See*** *Ligand Binding Cooperativity; Koshland-Némethy-Filmer Model; Scatchard Equation; Ligand-Induced Conformational Change*

NEIGHBORING GROUP MECHANISM

The condition observed with aliphatic nucleophilic substitution reactions displaying either a greater than expected rate of reaction or an unexpected retention of configuration at a chiral carbon atom. March[1] indicates that such effects are usually attributable to the presence of a group with an unshared electron pair located one or more carbon atoms away from the reaction center. A neighboring group mechanism involves two S_N2 reactions, with apparent overall retention of configuration occurring by two successive inversions. First, the neighboring group behaves as the nucleophile and facilitates expulsion of the leaving group; then the external nucleophile attacks from an angle of 180° with respect to the neighboring group, thereby expelling the neighboring group and resulting in a chirality that matches that of the original reactant. Because the neighboring group is contained as part of the reactant, its local concentration is high, thereby accounting for the higher observed reactivity. ***See also*** *Anchimeric Assistance*

[1] J. March (1992) *Advanced Organic Chemistry*, 4th ed., Wiley, New York.

NEIGHBORING-GROUP PARTICIPATION

The direct interaction of the reaction center of a molecular entity with a lone pair of electrons of an atom within that same molecular entity that is not associated with the reaction center; or, interaction of the reaction center with the σ- or π-electrons of a bond that is neither a part of the reaction center nor conjugated with the reaction center. Rate acceleration by such a process is referred to as anchimeric assistance. ***See*** *Intramolecular Catalysis; Synartetic Acceleration.*

N-END RULE

A model[1] that accounts for the selective degradation of proteins based on the amino acid that is present on the amino- or N-end of nascent proteins. Intracellular processing of nascent, noncompartmentalized proteins generates the mature protein via the action of amino-terminal peptidases. In model studies using β-galactosidase with different N-terminal aminoacyl residues, the time scale of degradation can be as short as 3 min or as long as a day. Commitment to degrade is encoded by classes of amino acids whose presence at the N-terminus places them into three general classes of turnover kinetics (*e.g.*, fast, intermediate, and slow), depending on whether they are strongly destabilizing, mildly destabilizing, or stabilizing with respect to turnover. The most stable contain N-terminal methionyl, seryl, alanyl, glycyl, threonyl, and valyl residues. Those of intermediate stability contain isoleucyl, glutaminyl, glutamyl, and tyrosyl residues, whereas the fast degrading proteins are strongly destabilized by the presence of leucyl, phenylalanyl, aspartyl, lysyl, or arginyl residues at the amino terminus. **See** *Protein Turnover Kinetics*

[1]A. Bachmair, D. Finley & A. Varshavsky (1986) *Science* **234**, 179.

NEPRILYSIN

This zinc-dependent endopeptidase [EC 3.4.24.11] catalyzes the hydrolysis of peptide bonds and exhibits preferential cleavage at the amino group of hydrophobic residues in proteins and polypeptides. Neprilysin is a membrane-bound glycoprotein that is inhibited by phosphoramidon and thiorphan.

C. Li & L. B. Hersh (1995) *Meth. Enzymol.* **248**, 253.
B. P. Roques, F. Noble, P. Crine & M.-C. Fournié-Zaluski (1995) *Meth. Enzymol.* **248**, 263.

NERNST EQUATION

The thermodynamic equation that expresses the electromotive force E in terms of the standard potential E^o and the concentrations of reactants and products in a redox reaction.

Basic Principles of the Nernst Equation. Consider the following generalized oxidation/reduction reaction

$$a\text{A} + b\text{B} \rightleftharpoons c\text{C} + d\text{D}$$

where the lower-case letters indicate the stoichiometry. In this case, the Gibbs free energy is given in the usual way as:

$$\Delta G = \Delta G^o + RT \ln\{(\text{C})^c(\text{D})^d/((\text{A})^a(\text{B})^b)\}$$

in terms of activities where R is the universal gas constant and T is the absolute temperature or in terms of reactant and product concentrations:

$$\Delta G = \Delta G^o + RT \ln\{[\text{C}]^c[\text{D}]^d/([\text{A}]^a[\text{B}]^b)\}$$

The expressions $\{\Delta G = -n\mathscr{F}E$ and $\Delta G^o = -n\mathscr{F}E^o$ where $\mathscr{F}$ is the Faraday constant$\}$ allow us to rewrite this expression in the following form:

$$-n\mathscr{F}E = -n\mathscr{F}E^o + RT \ln\{[\text{C}]^c[\text{D}]^d/([\text{A}]^a[\text{B}]^b)\}$$

Dividing each side of the expression by $(-nF)$, we obtain the Nernst equation:

$$E = E^o - (RT/n\mathscr{F}) \ln\{[\text{C}]^c[\text{D}]^d/([\text{A}]^a[\text{B}]^b)\}$$

At a temperature of 298.15 K, with $R = 8.314510$ J/mol^{-1}K^{-1}, the faraday $\mathscr{F}$ equals 96485.309 coulombs/mol, and converting to decadic logarithms by multiplying by the reciprocal of $\log_{10}e$ (or, 2.302585), the Nernst equation has the following abbreviated form (*useful only at 25°C*):

$$E = E^o - (0.05916/n) \log\{[\text{C}]^c[\text{D}]^d/([\text{A}]^a[\text{B}]^b)\}$$

Using the Nernst Equation. One may wish to evaluate $E_{\text{Cu}^{2+},\text{Cu}}$ for the following half-reaction, written here as the reduction (*i.e.*, gain of electrons) by Cu^{2+} to form elemental copper:

$$\text{Cu}^{2+} + 2\,\text{e}^- \rightleftharpoons \text{Cu}$$

Because elemental copper is insoluble, its activity or concentration is by convention set to a value of 1, and the number of equivalents of transferred electrons is two (from the equation immediately above). Therefore the Nernst equation is:

$$E_{\text{Cu}^{2+},\text{Cu}} = E_{\text{Cu}^{2+},\text{Cu}}{}^o - (0.05916/2) \log\{1/[\text{Cu}^{2+}]\}$$

or by inserting $+0.3419$ as the tabulated value for E^o and changing the sign on the logarithm upon inverting the Cu^{2+} concentration, we get

$$E_{\text{Cu}^{2+},\text{Cu}} = +0.3419 + 0.02958 \log[\text{Cu}^{2+}]$$

If one knows the concentration of cupric ions in solution, the value of $E_{\text{Cu}^{2+},\text{Cu}}$ can be evaluated quantitatively.

Calculation of Electromotive Force for an Overall Redox Reaction. In this case, there are two half-reactions:

$$\text{A}_{\text{oxidized}} + \text{e}^- \rightleftharpoons \text{A}_{\text{reduced}} \quad E_{\text{A}}{}^o$$
$$\text{B}_{\text{oxidized}} + \text{e}^- \rightleftharpoons \text{B}_{\text{reduced}} \quad E_{\text{B}}{}^o$$

And these must be written to obtain an electronically and stoichiometrically balanced equation:

$$A_{oxidized} + e^- \;\rightleftharpoons\; A_{reduced} \qquad\qquad E_A^{\,o}$$
$$B_{reduced} \;\rightleftharpoons\; B_{oxidized} + e^- \qquad\qquad -E_B^{\,o}$$

$$A_{oxidized} + B_{reduced} \;\rightleftharpoons\; A_{reduced} + B_{oxidized} \quad \Delta E^o = (E_A^{\,o} - E_B^{\,o})$$

Rees & Farrelly[1] give an excellent account of how solution conditions affect the electromotive force of biological electron transfer reactions. Moreover, they describe how the protein can influence the value of E^o: (a) by controlling the chemical nature of the ligands; (b) by imposing geometrical constraints on the redox center; (c) by altering the electrostatics by such noncovalent interactions as steric effects, solvent effects, placement of hydrophobic groups, and ionic effects (experienced directly as a result of ion binding to the redox protein or indirectly by altering the solution environment). **See also** *Electron Transfer Kinetics; Half-Reactions*

[1]D. C. Rees & D. Farrelly (1990) *The Enzymes* **19**, 37.

NET RATE CONSTANT METHOD FOR DERIVING ENZYME EQUATIONS

Cleland[1] introduced the net rate constant method to simplify the treatment of enzyme kinetic mechanisms that do *not* involve branched pathways. This method can be applied to obtain rate laws for isotope exchange, isotope partitioning, and positional isotope exchange. Since the net-rate constant method allows one to obtain V_{max}/K_m and K_m in terms of the individual rate constants, this method has greatest value for the characterization of isotope effects on V_{max}/K_m and K_m. Because only nonbranched kinetic schemes can be treated, most individuals will find Fromm's method[2,3] or Huang's modification[4] of Fromm's method to be much more generally useful, while still avoiding the inconvenience of the King-Altman method.

EXAMPLE 1. ONE-SUBSTRATE STEADY-STATE RATE EQUATION. We begin with the simplest two-intermediate example of a one-substrate mechanism:

$$E + S \underset{k_2}{\overset{k_1}{\rightleftharpoons}} ES \overset{k_3}{\rightarrow} ES' \overset{k_4}{\rightarrow} E + P$$

One then converts the scheme to a series of unidirectional or net rate constants (indicated by primes). The steady-state flux through each step as well as the distribution of enzyme species remains the same, and we get

$$E + S \overset{k_1'}{\rightarrow} ES \overset{k_3'}{\rightarrow} ES' \overset{k_4'}{\rightarrow} E + P$$

where each step has an associated net rate constant k_i'. Because the mechanism is a linear series, the rate for the overall reaction is the sum of the inverse of the net-rate constants through the individual steps. Thus, the rate equation must be:

$$(v/[E_{total}]) = 1/\{1/k_1' + 1/k_3' + 1/k_4'\}$$

where these net rate constants are expressed in terms of the actual rate constants as follows. One begins with any irreversible step in the original reaction scheme. In this case, there are two such steps, that converting ES into ES' and the other converting ES' to E + P. For an irreversible step, the net rate constant and the real rate constant will be identical:

$$k_3' = k_3 \quad \text{and} \quad k_2' = k_2$$

Moving leftward, we express k_1' as the real forward rate constant (in this case k_1) multiplied by a factor that relates the fraction of ES reacting in the forward direction as opposed to the fraction that returns to E. [Note: All of these constants are unimolecular rate constants, and all bimolecular rate constants must be multiplied by the concentration of substrate or product. This will become clearer in the later steps of this algorithm for deriving rate equations.] In this case, we get:

$$k_1' = k_1[S]\{k_3/(k_2 + k_3)\}$$

Now returning to the first equation and substituting the k_i' terms above

$$v = [E_{total}]/\{1/k_1' + 1/k_3' + 1/k_4'\}$$
$$v = k_1 k_3 k_4[S][E_{total}]/\{k_4(k_2 + k_3) + k_1[A](k_3 + k_4)\}$$
$$v = \{k_3 k_4/(k_3 + k_4)\}[E_{total}]/\{1 + \{k_4(k_2 + k_3)/k_1(k_3 + k_4)\}/[S]\}$$

Because this equation is now in the form

$$v = V_{max}/\{1 + K_m/[S]\}$$

then,

$$V_{max} = \{k_3 k_4/(k_3 + k_4)\}[E_{total}]$$
$$K_m = k_4(k_2 + k_3)/\{k_1(k_3 + k_4)\}$$

When the goal is to obtain V_{max} or V_{max}/K_m, the following procedure avoids the need to derive the entire rate equation:

(a) V_m is obtained at saturating S concentrations, such that k_1' will be infinite and the $(1/k_1')$ term drops out, such that for the above mechanism

$$V_{max} = [E_{total}]/\{1/k_3' + 1/k_4'\} = \{k_3k_4/(k_3 + k_4)\}[E_{total}]$$

(b) V_{max}/K_m is obtained as [S] is extrapolated to near zero. In this case, the k_1' term becomes the smallest net rate constant, and

$$V_{max}/K_m = k_1'[E_{total}] = k_1k_3[S][E_{total}]/(k_2 + k_3)$$

EXAMPLE 2. ORDERED UNI BI MECHANISM.

We now repeat the process for a slightly more complicated case.

$$E \underset{k_2}{\overset{k_1[A]}{\rightleftharpoons}} EA \underset{k_4}{\overset{k_3}{\rightleftharpoons}} EPQ \underset{k_6[P]}{\overset{k_5}{\rightleftharpoons}} EQ \overset{k_7}{\rightarrow} E + Q$$

One converts the scheme to a series of unidirectional or net rate constants. The steady-state flux through each step as well the distribution of enzyme species remains the same, and we get

$$E \overset{k_1'}{\rightarrow} EA \overset{k_3'}{\rightarrow} EPQ \overset{k_5'}{\rightarrow} EQ \overset{k_7'}{\rightarrow} E + Q$$

Because the mechanism is a linear series, the rate for the overall reaction is the sum of the inverse of the net rate constants through the individual steps:

$$(v/[E_{total}]) = 1/\{1/k_1' + 1/k_3' + 1/k_5' + 1/k_7'\}$$

where these net rate constants are expressed in terms of the actual rate constants below. [Recall again that these constants are unimolecular rate constants, and all bimolecular rate constants must be multiplied by the concentration of substrate or product.]

One begins with any irreversible step in the original reaction scheme. In this case, there is only one such irreversible step, that of converting EQ into E + Q. For an irreversible step, the net rate constant and the real rate constant are identical:

$$k_7' = k_7$$

Moving leftward, we express k_5' as the real forward rate constant multiplied by a factor that relates the fraction of EQ reacting in the forward direction as opposed to the fraction that returns to EPQ. In this case, because $k_7' = k_7$, we get:

$$k_5' = k_5\{k_7/(k_7 + k_6[P])\} = k_5k_7/(k_7 + k_6[P])$$

Again moving leftward, we get

$$k_3' = k_3\{k_5'/(k_4 + k_5')\}$$

then replacing k_5' by $k_5k_7/(k_7 + k_6[P])$, one gets

$$k_3' = \{k_3k_5k_7/(k_7 + k_6[P])\}/\{k_4 + \{k_5k_7/(k_7 + k_6[P])\}\}$$

$$k_3' = k_3k_5k_7/\{k_5k_7 + k_4k_6[P] + k_4k_7\}$$

And then we deal with the leftwardmost term k_1':

$$k_1' = k_1[A]k_3'/(k_2 + k_3')$$

which upon substituting k_3' by $k_3k_5k_7/\{k_5k_7 + k_4k_6[P] + k_4k_7\}$, gives

$$k_1' = k_1k_3k_5k_7[A]/\{k_2k_4k_6[P] + k_2k_4k_7 + k_2k_5k_7 + k_3k_5k_7\}$$

Now returning to the first equation and substituting the k_i' terms above

$$v = [E_{total}]/\{1/k_1' + 1/k_3' + 1/k_5' + 1/k_7'\}$$

$$v = [E_{total}]/\{\{(k_2k_4k_6[P] + k_2k_4k_7 + k_2k_5k_7 + k_3k_5k_7)/\{k_1k_3k_5k_7[A]\}\} + \{(k_5k_7 + k_4k_6[P] + k_4k_7)/(k_3k_5k_7)\} + \{(k_7 + k_6[P])/(k_5k_7)\} + (1/k_7)\}$$

$$= k_1k_3k_5k_7[E_{total}][A]/\{k_2k_4k_7 + k_2k_5k_7 + k_3k_5k_7 + k_2k_4k_6[P] + k_1(k_7(k_3 + k_4 + k_5) + k_3k_5)[A] + k_1k_6(k_3 + k_4)[A][P]\}$$

[1] W. W. Cleland (1975) *Biochemistry* **14**, 3220.
[2] H. J. Fromm (1970) *Biochem. Biophys. Res. Commun.* **40**, 692.
[3] H. J. Fromm (1975) *Initial Rate Enzyme Kinetics*, Springer-Verlag, New York.
[4] C. Y. Huang (1979) *Meth. Enzymol.* **63**, 54.

NEUROLYSIN

This endopeptidase [EC 3.4.24.16] catalyzes the hydrolysis of peptide bonds, preferentially cleaving neurotensin between Pro-10 and Tyr-11. However, there is no absolute requirement for a prolyl bond.

A. J. Barrett, M. A. Brown, P. M. Dando, C. G. Knight, N. McKie, N. D. Rawlings & A. Serizawa (1995) *Meth. Enzymol.* **248**, 529.
F. Checler, H. Barelli, P. Danch, V. Dive, B. Vincent & J. P. Vincent (1995) *Meth. Enzymol.* **248**, 593.

NEWTON

The SI unit (symbolized by N) for force; equal to kilogram-meter-second^2. The name of this classical mechanical term stems from Isaac Newton's work showing that $\mathbf{F} = m\mathbf{a}$, where $\mathbf{a} = d\mathbf{v}/dt = d^2\mathbf{r}/dt^2$.

NEWTON-RAPHSON METHOD

An algorithm for solving integrated rate equations, thereby facilitating the systematic analysis of progress curves of enzyme-catalyzed reactions.

I. A. Nimmo & G. L. Atkins (1974) *Biochem. J.* **141**, 913.
R. G. Duggleby & J. F. Morrison (1977) *Biochim. Biophys. Acta* **481**, 297.
R. G. Duggleby (1995) *Meth. Enzymol.* **249**, 61.

N–H EXCHANGE

Any technique[1,2] designed to probe protein structure and/or stability in terms of the rate and extent of deuterium-protium (^{2}H–^{1}H) or tritium-protium (^{3}H–^{1}H) exchange with peptide N–H groups. These methods are based on the resistance to exchange by N–H groups participating in hydrogen bonding with O=C–groups within α-helices. These hydrogen bonds must be broken for exchange to occur, and this fact forms the basis of the technique which can be applied (a) grossly by using one-dimensional NMR to measure exchange of N–H groups in a protein or (b) specifically by using two-dimensional NMR spectroscopy and site-directed mutagenesis to assign each resonance unambiguously to a certain residue in the protein. The observed rate constant for exchange by a single N–H group of residue i is of the product of the intrinsic rate of exchange times the relative fraction of groups in the nonbonded state (i.e.,

$$k_{observed} = k_{intrinsic}[1 - f\mathrm{B}(i)]).$$

[1]C. A Rohl & R. L. Baldwin (1998) *Meth. Enzymol.* **295**, 1.
[2]S. W. Englander & N. R. Kallenbach (1984) *Quart. Rev. Biophys.* **16**, 521.

NICKEL-DEPENDENT ENZYMES

Perhaps the best known nickel-dependent enzyme is urease (EC 3.5.1.5) which catalyzes the hydrolysis of urea to ammonium ion and carbamate; carbamate then undergoes nonenzymatic hydrolysis to bicarbonate and ammonia. Jack bean urease, the first ever enzyme crystallized, contains two Ni(II) atoms. Why urease utilizes Ni(II), perhaps the least Lewis-acidic of the transition metals, instead of another cation remains unclear[1]. Other nickel-dependent enzymes include hydrogenase (EC 1.18.99.1), carbon monoxide dehydrogenase (EC 1.2.99.2), hydrogen dehydrogenase (EC 1.12.1.2), and coenzyme F420 hydrogenase (EC 1.12.99.1).

[1]M. A. Halcrow (1998) *Comprehensive Biological Catalysis: A Mechanistic Reference* **1**, 506.

NICOTINAMIDASE

This enzyme [EC 3.5.1.19] catalyzes the hydrolysis of nicotinamide to produce nicotinate and ammonia.

S. Pinder, J. B. Clark & A. L. Greenbaum (1971) *Meth. Enzymol.* **18B**, 20.

NICOTINAMIDE COENZYMES

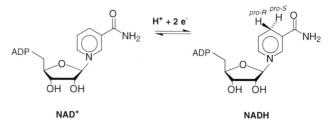

Reduced forms of these coenzymes absorb ultraviolet light near 340 nm, whereas the oxidized forms do not. For NAD$^+$ at neutral pH, the maximal absorbance band ($\varepsilon = 18000$ M^{-1}cm^{-1}) occurs at 260 nm; another absorbance band ($\varepsilon = 8000$ M^{-1}cm^{-1}) occurs at 230 nm. For NADH, the maximal absorbance band ($\varepsilon = 16900$ M^{-1}cm^{-1}) occurs at 259 nm; a second absorbance band ($\varepsilon = 6220$ M^{-1}cm^{-1})[1] occurs at 339 nm; two weaker bands occur at 234 and 290 nm with respective ε values of 6600 M^{-1}cm^{-1} and 1300 M^{-1}cm^{-1}. The same is true for NADP$^+$ and NADPH at 339 nm. Occasionally, investigators have used thio-NADH which has a much stronger absorbance around 366 nm. *See* Absorption Spectroscopy (Fig. 4, pg. 5); Nicotinic Acid: Analogs and Coenzymes

[1]The International Federation of Clinical Chemistry recommends a value of 6300 M^{-1}cm^{-1} at 339 nm and 30°C.

NICOTINAMIDE *N*-METHYLTRANSFERASE

This enzyme [EC 2.1.1.1] catalyzes the reaction of *S*-adenosyl-L-methionine with nicotinamide to produce *S*-adenosyl-L-homocysteine and 1-methylnicotinamide.

G. L. Cantoni (1955) *Meth. Enzymol.* **2**, 257.

NICOTINAMIDE MONONUCLEOTIDE ADENYLYLTRANSFERASE

This enzyme [EC 2.7.7.1], also known as NAD$^+$ pyrophosphorylase, catalyzes the reaction of ATP with nicotinamide ribonucleotide to produce NAD$^+$ and pyrophosphate (or, diphosphate). Nicotinate nucleotide is also a substrate. Nicotinate-nucleotide adenylyltransferase [EC 2.7.7.18] uses nicotinate ribonucleotide as the pyridine substrate, thereby producing deamido-NAD$^+$ and pyrophosphate.

G. Magni, N. Raffaelli, M. Emanuelli, A. Amici, P. Natalini & S. Ruggieri (1998) *Adv. Enzymol.* **73**, 240.

V. Micheli & S. Sestini (1997) *Meth. Enzymol.* **280**, 211.

G. Magni, M. Emanuelli, A. Amici, N. Raffaelli & S. Ruggieri (1997) *Meth. Enzymol.* **280**, 241 and 248.

NICOTINAMIDE PHOSPHORIBOSYLTRANSFERASE

This enzyme [EC 2.4.2.12], also known as NMN pyrophosphorylase, catalyzes the reaction of nicotinamide with 5-phospho-α-D-ribose 1-diphosphate to produce nicotinamide D-ribonucleotide and pyrophosphate.

V. Micheli & S. Sestini (1997) *Meth. Enzymol.* **280**, 211.

W. D. L. Musick (1981) *Crit. Rev. Biochem.* **11**, 1.

NICOTINATE DEHYDROGENASE

This enzyme [EC 1.5.1.13], also known as nicotinate hydroxylase, catalyzes the reaction of nicotinate with $NADP^+$ and water to produce 6-hydroxynicotinate and NADPH. This iron-dependent flavoprotein will also oxidize NADPH.

T. C. Stadtman (1980) *Ann. Rev. Biochem.* **49**, 93.

L. Tsai & E. R. Stadtman (1971) *Meth. Enzymol.* **18B**, 233.

NICOTINATE MONONUCLEOTIDE ADENYLYLTRANSFERASE

This enzyme [EC 2.7.7.18], also known as nicotinate-nucleotide adenylyltransferase and deamido-NAD^+ pyrophosphorylase, catalyzes the reaction of ATP with nicotinate ribonucleotide to produce deamido-NAD^+ and pyrophosphate.

V. Micheli & S. Sestini (1997) *Meth. Enzymol.* **280**, 211.

S. Pinder, J. B. Clark & A. L. Greenbaum (1971) *Meth. Enzymol.* **18B**, 20.

NICOTINATE-NUCLEOTIDE DIMETHYLBENZIMIDAZOLE PHOSPHORIBOSYLTRANSFERASE

This enzyme [EC 2.4.2.21] catalyzes the reaction of β-nicotinate D-ribonucleotide with dimethylbenzimidazole to produce nicotinate and N^1-(5-phospho-α-D-ribosyl)-5,6-dimethylbenzimidazole. Benzimidazole also serves as a substrate. The clostridial enzyme acts on adenine to produce 7-α-D-ribosyladenine 5'-phosphate.

J. A. Fyfe & H. C. Friedmann (1971) *Meth. Enzymol.* **18B**, 197.

NICOTINATE PHOSPHORIBOSYLTRANSFERASE

This enzyme [EC 2.4.2.11] catalyzes the reaction of nicotinate with 5-phospho-α-D-ribose 1-diphosphate to produce nicotinate D-ribonucleotide and pyrophosphate.

V. Micheli & S. Sestini (1997) *Meth. Enzymol.* **280**, 211.

W. D. L. Musick (1981) *Crit. Rev. Biochem.* **11**, 1.

NICOTINIC ACID: ANALOGUES AND COENZYMES

Selected entries from *Methods in Enzymology* [vol, page(s)]: Determination of nicotinamide, **66**, 3; fluorometric quantitation of picomole amounts of 1-methylnicotinamide and nicotinamide in serum, **66**, 5; temperature dependence of the spectroscopic properties of NADH, **66**, 8; purification of commercial NADH, **66**, 11; isolation and analysis of pyridine nucleotides and related compounds by liquid chromatography, **66**, 23; affinity chromatography of NAD^+ on immobilized dehydrogenase columns, **66**, 39; preparation of stereospecific tritium-labeled reduced nicotinamide adenine dinucleotide phosphate, **66**, 51; kinetic methods for detecting inhibitors in NADH for NADH-dependent enzymes, **66**, 55; simple methods for preparing nicotinamide mononucleotide and related analogs, **66**, 62; preparation and purification of nicotinamide mononucleotide analogs, **66**, 71; preparation of 3-aminopyridine adenine dinucleotide and 3-aminopyridine adenine dinucleotide phosphate, **66**, 81; preparation of α-$NADP^+$, **66**, 87; an improved method for measuring quinolinic acid in biological specimens, **66**, 91; NAD^+ kinase from sea urchin eggs, **66**, 101; nicotinic acid adenine dinucleotide phosphate ($NAADP^+$), **66**, 105; preparation of 2'P-ADP, **66**, 112; convenient method for enzymic synthesis of [^{14}C]nicotinamide riboside, **66**, 120; formation of nicotinamide ribose diphosphate ribose, a new metabolite of the NAD^+ pathway, by *Aspergillus niger*, **66**, 123; NAD^+ glycohydrolase from bovine seminal plasma, **66**, 144; NAD^+ glycohydrolases from rat liver nuclei, **66**, 151; poly(ADP-ribose) synthetase from rat liver nuclei, **66**, 154; poly(ADP-ribose) synthetase from calf thymus, **66**, 159; extraction and quantitative determination of larger than tetrameric endogenous polyadenosine diphosphoribose from animal tissues, **66**, 165; covalent modification of proteins by metabolites of NAD^+, **66**, 168; coenzyme activity of NAD^+ bound to polymer supports through the adenine moiety, **66**, 176; use of differently immobilized nucleotides for binding NAD^+-dependent dehydrogenases, **66**, 192.

NIH SHIFT

An intramolecular hydrogen migration observed in the hydroxylation of aromatic rings in certain enzyme-catalyzed reactions as well as some chemical reactions. The rearrangement was first observed[1-3] at the National Institutes of Health (hence the name "NIH") in studies of the synthesis of L-tyrosine from L-phenylalanine via phenylalanine hydroxylase. Observation of this shift requires appropriate deuteration of the aromatic reactant.

Other enzymes exhibiting this migration include cytochrome P450 and *p*-hydroxyphenylpyruvate dioxygenase

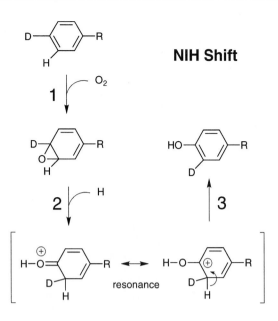

NIH Shift

resonance

(which involves an alkyl group migration rather than a hydride ion). In general, the NIH shift is attributed to the formation of an arene oxide intermediate, although other pathways have been suggested. Nonenzymatic routes that appear to involve NIH shifts have been reported[4].

[1] G. Guroff, D. Jerina, J. Rensen, S. Udenfriend & B. Witkop (1967) *Science* **157**, 1524.
[2] D. Jerina (1973) *Chem. Technol.* **4**, 120.
[3] R. P. Hanzlik, K. Hogberg & C. M. Judson (1984) *Biochemistry* **23**, 3048.
[4] T. C. Bruice & P. Bruice (1976) *Acc. Chem. Res.* **9**, 378.

NITRATE REDUCTASE

(1) Nitrate reductase (NADH) [EC 1.6.6.1], also known as assimilatory nitrate reductase, catalyzes the reaction of NADH with nitrate to produce NAD$^+$, nitrite, and water. This enzyme uses FAD or FMN, heme, and a molybdenum ion as cofactors. (2) Nitrate reductase (NAD(P)H) [EC 1.6.6.2], also known as assimilatory nitrate reductase, catalyzes the reaction of NAD(P)H with nitrate to produce NAD(P)$^+$, nitrite, and water. This enzyme uses FAD or FMN, heme, and a molybdenum ion as cofactors. (3) Nitrate reductase (NADPH) [EC 1.6.6.3] catalyzes the reaction of NADPH with nitrate to produce NADP$^+$, nitrite, and water. This enzyme uses FAD, heme, and a molybdenum ion as cofactors. (4) Nitrate reductase (cytochrome) [EC 1.9.6.1] catalyzes the reaction of nitrate with ferrocytochrome to produce nitrite and ferricytochrome. (5) Nitrate reductase (acceptor) [EC 1.7.99.4], also known as respiratory nitrate

reductase, catalyzes the reaction of nitrate with a reduced acceptor to produce nitrite and the oxidized acceptor. The *Pseudomonas* enzyme is a cytochrome-dependent system, but the enzyme from *Micrococcus halodenitrificans* is an iron protein containing molybdenum.

A. R. Clarke & T. R. Dafforn (1998) *Comprehensive Biological Catalysis: A Mechanistic Reference* **3**, 1.
P. E. Baugh, D. Collison, C. D. Garner & J. A. Joule (1998) *Comprehensive Biological Catalysis: A Mechanistic Reference* **3**, 377.
H. C. Huppe & D. H. Turpin (1994) *Ann. Rev. Plant Physiol. Plant Mol. Biol.* **45**, 577.
L. P. Solomonson & M. J. Barber (1990) *Ann. Rev. Plant Physiol. Plant Mol. Biol.* **41**, 225.
M. G. Guerrero, J. M. Vega & M. Losada (1981) *Ann. Rev. Plant Physiol.* **32**, 169.
R. C. Bray (1980) *Adv. Enzymol.* **51**, 107.
G. Palmer (1975) *The Enzymes*, 3rd ed., **12**, 1.

NITRENE

A molecule containing a neutral nitrogen atom with four nonbonding electrons (R–N̈:). While two of these nonbonding electrons are paired, the other two may have parallel spins (thus, triplet state) or antiparallel spins (singlet state). Nitrenes are the nitrogen analogues of carbenes.

W. Lwowski (1970) *Nitrenes*, Interscience, New York.
E. F. V. Scriven (1982) *React. Intermed.* **2**, 1.

NITRENIUM ION

A cationic species exhibiting significant positive charge residing on a dicoordinated nitrogen atom. Nitrenium ions are the nitrogen analogs of carbenium ions and may exist as intermediates in certain reactions. There are at least two types of nitrenium ion, R–N$^+$–R′ (note that if R is a proton, this species is simply a protonated nitrene) and RR′C=N$^+$. These ions can exist in either triplet or singlet states.

P. G. Gassman (1970) *Acc. Chem. Res.* **3**, 26.
P. T. Lansbury (1970) in *Nitrenes* (W. Lwowski, ed.), pp. 405-419, Interscience, New York, pp. 405-419.

NITRIC OXIDE SYNTHASE

This enzyme [EC 1.14.13.39] catalyzes the reaction of L-arginine with 1.5 NADPH, 0.5 H$^+$, and two dioxygen to produce L-citrulline, nitric oxide, 1.5 NADP$^+$, and two water. The enzyme in brain, but not that induced in lung or liver by endotoxin, requires calcium ions.

H. B. Dunford (1998) *Comprehensive Biological Catalysis: A Mechanistic Reference* **3**, 195.
L. Packer, ed. (1996) *Meth. Enzymol.*, vols. **268** and **269**.
S. R. Jaffrey & S. H. Snyder (1995) *Ann. Rev. Cell Dev. Biol.* **11**, 417.

D. S. Bredt & S. H. Snyder (1994) *Ann. Rev. Biochem.* **63**, 175.

M. Marletta (1993) *J. Biol. Chem.* **268**, 12231.

D. J. Stuehr & O. W. Griffith (1992) *Adv. Enzymol.* **65**, 287.

Selected entries from *Methods in Enzymology* [vol, page(s)]:
Assays, **233**, 250-258, 265 [citrulline method, **233**, 252, 255-256; hemoglobin method, **233**, 252-255; nitrite/nitrate method, **233**, 252, 256-258]; biopterin content determination, **233**, 262-264; co-factor requirements, **233**, 251, 259; distribution, **233**, 250, 258; flavin content determination, **233**, 262; isoforms [activity, **233**, 258-259; assay considerations, **233**, 251-252; cDNA, isolation and cloning, **233**, 266-269; properties, **233**, 258, 263; purification (from bovine aortic endothelial cells, **233**, 261-262; from induced RAW 264.7 macrophages, **233**, 260-261; from neuronal tissue, **233**, 259-260, 265-266); isolation, **233**, 250-251]; pH optimum, **233**, 251-252; properties, **233**, 250; reaction catalyzed by, **233**, 250-251, 258, 264-265; stability, **233**, 252; storage, **233**, 252; structure, **233**, 250.

NITRILE HYDRATASE

Nitrilase [EC 3.5.5.1], also known as nitrile aminohydrolase and nitrile hydratase, catalyzes the hydrolysis of a nitrile to produce a carboxylate and ammonia. The enzyme acts on a wide range of aromatic nitriles. Nitrile hydratase [EC 4.2.1.84], also known as nitrilase, catalyzes the hydrolysis of a nitrile to produce an aliphatic amide. The enzyme acts on short-chain aliphatic nitriles, converting them into the corresponding acid amides. However, this particular enzyme does not further hydrolyze these amide products; nor does the enzyme act on aromatic nitriles.

B. G. Fox (1998) *Comprehensive Biological Catalysis: A Mechanistic Reference* **3**, 261.

NITRITE REDUCTASE

(1) Nitrite reductase (NAD(P)H) [EC 1.6.6.4] catalyzes the reaction of three NAD(P)H with nitrite to yield three NAD(P)$^+$, NH$_4$OH, and water. Cofactors for this enzyme include FAD, non-heme iron, and siroheme. (2) Nitrite reductase (cytochrome) [EC 1.7.2.1] is a copper-dependent system that catalyzes the reaction of nitric oxide with two ferricytochrome c and water to produce nitrite and two ferrocytochrome c. (3) Ferredoxin–nitrite reductase [EC 1.7.7.1], a heme- and iron-dependent enzyme, catalyzes the reaction of ammonia with three oxidized ferredoxin to produce nitrite and three reduced ferredoxin. (4) Nitrite reductase [EC 1.7.99.3] is a copper- and FAD-dependent enzyme that catalyzes the reaction of two nitric oxide with an acceptor substrate and two water to produce two nitrite and the reduced acceptor.

A. R. Clarke & T. R. Dafforn (1998) *Comprehensive Biological Catalysis: A Mechanistic Reference* **3**, 1.

A. Messerschmidt (1998) *Comprehensive Biological Catalysis: A Mechanistic Reference* **3**, 401.

M.-C. Liu, C. Costa & I. Moura (1994) *Meth. Enzymol.* **243**, 303.

E. T. Adman (1991) *Adv. Protein Chem.* **42**, 145.

T. E. Meyer & M. D. Kamen (1982) *Adv. Protein Chem.* **35**, 105.

Y. Hatefi & D. L. Stiggall (1976) *The Enzymes*, 3rd ed., **13**, 175.

NITROGENASE

(1) Nitrogenase (ferredoxin) [EC 1.18.6.1] catalyzes the reaction of three reduced ferredoxin molecules with protons, N$_2$, and n ATP molecules to produce three oxidized ferredoxin molecules, two ammonia molecules, n ADP molecules, and n orthophosphate molecules where n is between 12 and 18. This iron-sulfur system also uses either molybdenum or vanadium ions. (2) Nitrogenase (flavodoxin) [EC 1.19.6.1] catalyzes the reaction of six reduced flavodoxin molecules with protons, N$_2$, and n ATP molecules to produce six oxidized flavodoxin molecules, two ammonia molecules, n ADP molecules, and n orthophosphate molecules. This system uses iron-sulfur and molybdenum ions.

G. J. Leigh (1998) *Comprehensive Biological Catalysis: A Mechanistic Reference* **3**, 349.

P. E. Baugh, D. Collison, C. D. Garner & J. A. Joule (1998) *Comprehensive Biological Catalysis: A Mechanistic Reference* **3**, 377.

J. B. Howard & D. C. Rees (1994) *Ann. Rev. Biochem.* **63**, 235.

L. E. Mortenson, L. C. Seefeldt, T. V. Morgan & J. T. Bolin (1993) *Adv. Enzymol.* **67**, 299.

H. Deng & R. Hoffman (1993) *Angew. Chem. Int. Ed. Engl.* **32**, 1062.

J. T. Bolin, N. Campobasso, S. W. Muchmore, T. V. Morgan & L. E. Mortenson (1993) in *Molybdenum Enzymes, Cofactors & Model Systems* (E. I. Stiefel, D. Coucouvanis & W. E. Newton, eds.) pp. 186-195, Amer. Chem. Soc., Washington, D.C.

S. Hunt & D. B. Layzell (1993) *Ann. Rev. Plant Physiol. Plant Mol. Biol.* **44**, 483.

R. H. Burris (1991) *J. Biol. Chem.* **266**, 9339.

R. C. Bray (1980) *Adv. Enzymol.* **51**, 107.

G. Palmer (1975) *The Enzymes*, 3rd ed., **12**, 1.

NITROGEN FIXATION

The process that assimilates molecular nitrogen, N$_2$, into ammonia and other nitrogenous metabolites. The first reaction is catalyzed by nitrogenase [EC 1.18.2.1] which is composed of two different protein components–one containing iron, molybdenum, and acid-labile sulfur (variously named as Mo-Fe protein, Component I, or molybdoferrodoxin) and a second component containing iron and sulfur only (called Fe protein, Component II, or azoferrodoxin). In bacteria, nitrogen fixing polypeptides are encoded in *nif* operon, and for *E. coli*, *Klebsiella*, and *Salmonella*, the *nif* operon is tightly coupled to the adenylylation/deadenylylation cascade that controls the synthesis of glutamine. In fact, transcription of the *nif* operon is promoted by glutamine synthetase (unadenyly-lated), the enzyme form that persists when ammonia stores are low.

Selected entries from *Methods in Enzymology* [vol, page(s)]:
Nitrogen fixation-assay methods and techniques, **24**, 415; preparation and properties of clostridial ferredoxins, **24**, 431; purification of nitrogenase from *Clostridium pasteurianum*, **24**, 446; nitrogenase complex and its components, **24**, 456; preparation of nitrogenase from nodules and separation into components, **24**, 470; clostridial rubredoxin, **24**, 477; purification of nitrogenase and crystallization of its Mo-Fe protein, **24**, 480; N_2-fixing plant-bacterial symbiosis in tissue culture, **24**, 497.

NOISE

Any unwanted, extraneous, and random disturbance that degrades accuracy and/or precision in the recording or measuring of a signal. The occurrence of noise is manifested in experimental science as a threshold that defines the sensitivity of measurement.

One usually evaluates the quality of an analytical measuring device on the basis of its signal-to-noise (or, S/N) ratio. The average signal strength or output of an instrument is proportional to the concentration of an analyte, whereas the amplitude (*i.e.*, strength) of noise is not. For this reason, the signal strength is proportional to some mean value $\bar{x}$, and the noise is best described by a standard deviation *sd*. The relative standard deviation $sd/\bar{x}$ is therefore the reciprocal of S/N. In most cases, accuracy and precision will be tolerant to S/N values that do not drop below 2-3. Noise can be categorized according to its source. Instrument noise involves thermal noise, shot noise, environmental noise, input power fluctuations, and a less well understood phenomenon known as flicker. Any impurities in chemical reagents or any inhomogeneously mixed sample can be a source of noise that is more related to the chemistry or physics of the sample whose properties are being measured.

Thermal noise (also known as Johnson noise) relates to the thermal motions or agitations of electrons in electrical devices. For example, a photomultiplier tube is set at a high voltage so that impinging photons can release electrons from the photocathode surface. Proper voltage adjustments must be made for optimal response to absorbed photons: (a) too low a voltage will result in no observable signal, and (b) too high a voltage will result in the spurious release of too many electrons, even in the absence of photoexcitation. Thus, thermal noise is an unavoidable or intrinsic property. In instruments relying on photomultiplier tubes, thermal noise is measured as a dark current: this simply means that there is a measurable electrical output even in the absence of light. At 298 K, the dark current in an RCA 928 photomultiplier tube is often around 400-500 counts/second; however, by reducing the temperature to around 240 K, the dark current drops into the 12-20 cpm range. This means that a system that is thermal noise-limited will experience a 25-40 times better signal-to-noise ratio at $-33°C$ than at $25°C$. For this reason, photomultiplier tubes in many spectrophotometers and fluorimeters are often housed within Peltier cooling devices.

Shot noise relates to randomly fluctuating electron movements across the anode-to-cathode junction. While this behavior does not alter the shape of a visible or UV spectrum, the presence of shot noise can reduce sensitivity.

Environmental noise refers to the effects of fluctuations in the surrounding electromagnetic fields generated by lightning, the presence of nearby high-voltage electrical power transmission lines, radiofrequency noise, and arcing from nearby high-current electrical motors or switches.

NONCOMPETITIVE INHIBITION

Reversible inhibition of an enzyme by an inhibitor binding to the enzyme whether or not the substrate is bound at the active site[1,2]. A noncompetitive inhibitor binds at a site that is distinct from the active site, and saturating levels of substrate cannot completely remove the inhibitory influence of this agent.

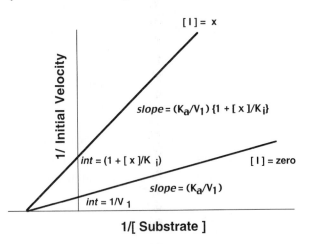

Consider the standard Uni Uni mechanism (E + A $\rightleftharpoons$ EX $\rightarrow$ E + P). A noncompetitive inhibitor, I, can bind reversibly to either the free enzyme (E) to form an EI complex (having a dissociation constant K_{is}), or to the central complex (EX) to form the EXI ternary complex (having a dissociation constant K_{ii}). Both the slope and vertical intercept of the standard double-reciprocal plot ($1/v$ *vs.* $1/[A]$) are affected by the presence of the inhibitor. If the secondary replots of the slopes and the intercepts (thus, slopes or vertical intercepts *vs.* [I]) are linear (**See** *Nonlinear Inhibition*), then the values of those dissociation constants can be obtained from these replots. If $K_{is} = K_{ii}$, then a plot of $1/v$ *vs.* $1/[A]$ at different constant concentrations of the inhibitor will have a common intersection point on the horizontal axis (if not, **See** *Mixed-Type Inhibition*). Note that the above analysis assumes that the inhibitor binds in a rapid equilibrium fashion. If steady-state binding conditions are present, then nonlinearity may occur, depending on the magnitude of the $[I]^2$ and $[A]^2$ terms in the rate expression.

See also *Mixed Type Inhibition*

[1]J. A. Todhunter (1979) *Meth. Enzymol.* **63**, 383.
[2]H. J. Fromm (1995) *Meth. Enzymol.* **249**, 123.

Selected entries from *Methods in Enzymology* [vol, page(s)]:
Computer program, **63**, 115, 128, 129, 137; coupled enzyme assay, **63**, 40; experimental design, **63**, 394; graph, **63**, 389; metal-ion, **63**, 292; one-substrate, **63**, 470-472; partial, **63**, 385; pH effect, **63**, 214; product, **63**, 412-414; progress curve, **63**, 164, 170, 171; rate expression, **63**, 393; tight-binding, **63**, 470-472.

NONCOVALENT INTERACTIONS

Any interaction between ions, molecules, or functional groups that does not involve covalency (*i.e.*, the sharing of a pair of electrons within a well defined molecular orbital). While such interactions are considerably weaker than those involved in the making/breaking of covalent bonds, significant binding energy can nonetheless arise owing to the great number of such interactions within structurally organized biological macromolecules. The same can be said for ligand binding interactions with macromolecules. The most common examples of noncovalent interactions are: (a) ion-ion interactions[1] (typically involving interaction energies in the range of 10 to 70 kcal/mol (or, 40 to 335 kJ/mol); (b) dipole-dipole interactions[2] (frequently characterized by interaction energies of less than 1 kcal/mol); (c) ion-induced dipole interactions[3] (with interaction energies also less than 1 kcal/mol); (d) dipole-induced dipole interactions[4] (with

interaction energies of 1 to 10 kcal/mol); and (e) London dispersion interactions[5] (also in the 1–10 kcal/mol range). Hydrophobic interactions may also be included along with this general group of noncovalent interactions; however, hydration involves weak coordinate covalent interactions, and hydrogen bonding is most often treated as a unique three-center interaction.

[1]The energy for ion-ion interactions is proportional to the product of the charges, q_1 and q_2, divided by the product of the dielectric constant D and the distance between the charges. If q_1 and q_2 are both of the same sign, the energy will be positive (*i.e.*, unfavorable), as would be expected for repulsion. If q_1 and q_2 are of opposite sign, the noncovalent interaction will be negative (*i.e.*, favorable).
[2]Dipole-dipole interaction energies are proportional to the product of the dipole moments, μ_1 and μ_2, divided by the product of the dielectric constant D and the sixth power of the distance between the dipoles.
[3]Ion-induced dipole interaction energies are proportional to the product of the square of the charge and the polarizability α of the atom/group with which the ion interacts, divided by the product of the dielectric constant D and the fourth power of the distance between the dipoles.
[4]Dipole-induced dipole interaction energies are proportional to the product of the square of the dipole moment and the polarizability α of the atom/group with which the ion interacts, divided by the product of the square of dielectric constant D and the sixth power of the distance between the dipole and the polarizable group.
[5]Dispersion interactions have energies that are proportional to the product of polarizabilities α_1 and α_2, divided by the sixth power of the distance between two polarizable atoms (or groups of atoms).

NONEQUILIBRIUM THERMODYNAMICS (A Primer)

Although many chemical and physical processes readily attain thermodynamic equilibrium, living systems characteristically maintain steady-state flows of energy and/or matter. Heat flow from a region of high temperature to one at low temperature and a gas expanding through a nozzle into a vacuum are examples of irreversible processes. So too are electric current flow and Fick's law of diffusion for neutral particles. In living systems, other examples include the steady-state rates of metabolism as well as the downward flow of protons along transmembrane gradients that drive ATP synthesis. Indeed, the cardinal feature of living systems is homeostasis—the compensatory adjustment from one steady state to another in response to an external stimulus or to changes in the availability of energy and/or matter. Any detailed thermodynamic description of nonequilibrium processes is necessarily more complicated, requiring additional parameters beyond those needed to specify the thermodynamic properties of systems at equilibrium. Even for the simplest nonequilibrium processes (such as the unidirectional transfer of only a single component), one must account for spatial constraints as well as the rates of change in the entropy of the system[1-4].

Steady-state processes are unidirectional and irreversible. Their special thermodynamic behavior stems from a lack of balance in the system. In all irreversible processes, ΔS_{total} will be a positive quantity. In the above mentioned case of the transfer of a quantity q of heat from a hot body to a cold body, we can explicitly express the ΔS_{system} as follows:

$$\Delta S_{\text{system}} = -q/T_{\text{hot}} + q/T_{\text{cold}}$$

where T_{hot} and T_{cold} are the temperatures of the hot and cold bodies. ΔS_{total} is the sum of ΔS_{system} and $\Delta S_{\text{surroundings}}$, and in the absence of any exchange with the surrounding environment, the total entropy change will equal the system's entropy change:

$$\Delta S_{\text{total}} = -q/T_{\text{hot}} + q/T_{\text{cold}} > 0$$

such that $q/T_{\text{cold}} > q/T_{\text{hot}}$.

Any directional flow of energy and/or matter between two interconnected systems, say A and B, can be treated as occurring across a metastable zone joining A and B. One can assume that systems A and B are each at equilibrium internally (*i.e.*, there are no differences within A or B in terms of temperature, pressure, composition, thermodynamic activities of each component, or even chemical reactivity). Nonetheless, systems A and B are not at equilibrium with respect to each other, thereby allowing for the net transfer of energy and/or matter from the system of higher energy to the other. Within the zone of transfer, one can apply the so-called method of local isolation to examine the thermodynamic properties describing a point or region. This point or region will experience the flow of energy/matter from A to B, and if the rate into that point or region is exactly balanced by its rate of loss of energy/matter, then that point or region is said to satisfy the steady-state approximation. This virtually time-independent rate of transfer can be described by writing out expressions of JX_i, the nonequilibrium flux in terms of each component X_i:

$$JX_i = -dX_{iA}/dt = dX_{iB}/dt$$

where the subscripts indicate component X_i present in system A and B.

This flux through the transfer zone leads to a net increase in entropy. While it is certainly true that the earlier definition of entropy requires that we consider reversible phenomena (*i.e.*, $dS = dq_{\text{reversible}}/T$), this limitation can be overcome, in principle, by treating the transition or transfer zone as a series of layers running perpendicular to the direction of flux. This approach offers the virtue that one can assume that T and P are well defined and constant within each of the layers or lamina. Onsager succeeded in achieving a rigorous theoretical framework that allows one to specify the thermodynamic properties of irreversible systems, including those at steady state. While a detailed consideration of the Onsager method lies beyond the scope of this brief description, the following statements comprise its essence. One uses the Principle of Microscopic Reversibility which clearly states that under equilibrium conditions, any process and its reverse follow the same path and take place at the same rate on a time-averaged basis. The thermodynamic equations of motion for each transport process maybe written with the flux equal to the sum of terms accounting for the transfer to and from systems A and B in a manner that is proportional to the thermodynamic force. Note that the forces characterizing irreversible thermodynamics should not be confused with those of classical mechanics, and no accelerative effects are implied. Rather each flux has a conjugate force associated with it, and the multiplicative product of a flux and its conjugate force will always have units corresponding to the rate of entropy production. For a chemical reaction, the generalized flux J equals $(1/V)d\chi/dt$ (where ξ is the extent of reaction), and the conjugate force is the affinity $\mathscr{A}$, equal to $-\Sigma \nu_i \mu_i$ (where ν_i and μ_i are the stoichiometric coefficient and the chemical potential of the i-th component).

For a system in which a chemical reaction occurs at constant T and P, d_eS is the flow of entropy into the system from the exterior, such that

$$Td_eS = dq = dU + PdV$$

The rate of change in d_iS (defined as the time-dependent change in the entropy change within the system) can then be written as:

$$d_iS/dt = -(1/T)(d\xi/dt) \Sigma \nu_i\mu_i = (d\xi/dt) \mathscr{A}V = \mathbf{v} \mathscr{A}$$

where the flux J is the reaction rate $\mathbf{v}$ (or, $(d\xi/dt)$ per unit volume V) and the generalized force $\mathscr{A}$ is as defined earlier. For systems undergoing steady state transfer, it can be further demonstrated that the rate of entropy production is at a minimum. Prigogine later demonstrated that the steady state can be quite stable, and small perturbations will not substantially alter steady state flux.

[1]B. C. Eu (1994) *Kinetic Theory and Irreversible Thermodynamics*, Wiley, New York.
[2]D. K. Kondepudi & I. Prigogine (1998) *Modern Thermodynamics: From Heat Engines to Dissipative Structures*, Wiley, New York.
[3]D. Jou & J. E. Llebot (1990) *Introduction to the Thermodynamics Processes of Biological Processes*, Prentice Hall, New York.
[4]H. J. Kreuzer (1981) *Nonequilibrium Thermodynamics and Its Statistical Foundations*, Oxford University Press, London.

NONHYPERBOLIC KINETICS

Referring to an enzyme whose kinetic properties do not yield hyperbolic saturation curves in plots of the initial rate as a function of the substrate concentration.

While the term is most typically applied to cooperative enzymes, nonhyperbolic behavior may indicate: (1) that the steady-state condition is not a valid description of an enzyme's behavior; (2) that one or more of the experimental conditions (*e.g.*, pH, temperature, ionic strength, *etc.*) changes over the course of the experiment or varies with the substrate concentration; (3) that the total enzyme concentration is not appreciably lower than the substrate concentration or the concentration of an effector; (4) that nonlinear substrate inhibition occurs; (5) that more than one molecule of a substrate is consumed during one catalytic "round" (*e.g.*, MgATP in carbamoyl phosphate synthetase or glutathione as both the γ-glutamyl donor and acceptor in γ-glutamyltranspeptidase); (6) that the product(s) has (have) a relatively low K_i value(s), thereby resulting in significant product inhibition; and (7) that the enzyme obeys a steady-state random multisubstrate reaction mechanism, such that squared terms in substrate concentration are evident.

NONLINEAR INHIBITION

This term usually applies to reversible inhibition of an enzyme-catalyzed reaction in which nonlinearity is detected (a) in a double-reciprocal plot (*i.e.*, $1/v$ *versus* $1/[S]$) in the presence of different, constant concentrations of inhibitor or (b) in replots of slope or intercept values obtained from primary plots of $1/v$ *versus* $1/[S]$). Nonline-

arity replots can be attributable to a number of factors: cooperativity; multiple inhibition (for example, the formation of an EI_2 complex); partial inhibition; substrate inhibition; inhibitor-induced substrate inhibition; tight-binding enzyme inhibition; or in certain cases of product inhibition. Before proceeding to consider these more complicated explanations, one should first take measures to exclude trivial reasons for nonlinearity, such as lack of constant pH, temperature variability, instability of substrate and protein, *etc.* **See also** *Cleland Rules for Dead-End Inhibition; Abortive Complex*

NONLINEAR LEAST SQUARES ANALYSIS

A class of statistical methods frequently used to analyze kinetic and thermodynamic data. Most of these methods require a preliminary estimate of the constants followed by cycles of iterative calculations that converge on a final value(s). Cleland[1] has presented a protocol for the statistical estimation of kinetic data. A nonlinear analysis has also been applied to progress curves[2].

[1]W. W. Cleland (1979) *Meth. Enzymol.* **63**, 103.
[2]R. G. Duggleby (1995) *Meth. Enzymol.* **249**, 61.

Selected entries from *Methods in Enzymology* [vol, page(s)]: Accuracy of computer programs, **240**, 21-22; accuracy of method, **240**, 6; algorithms, **240**, 5, 13; assumptions, **240**, 7-14, 55; curve fitting, **240**, 3-5, 10; data points [minimal number, **240**, 14; smoothing, **240**, 14]; determination of confidence intervals, **240**, 15-21, 28, 462; determination of statistical significance, **240**, 18-20; deviations, **240**, 3; effect of instrument delay, **240**, 13; fitting of heat capacity function, **240**, 510, 543; weighted sum, **240**, 3.

NONRADIATIVE DECAY

Disappearance of an excited molecular species arising from a radiationless transition. The energy of a nonradiative decay is dissipated vibrationally as thermal energy.

D-NOPALINE DEHYDROGENASE

This enzyme [EC 1.5.1.19], also known as D-nopaline synthase, catalyzes the reaction of N^2-(D-1,3-dicarboxypropyl)-L-arginine with $NADP^+$ and water to produce L-arginine, NADPH, and α-ketoglutarate (or, 2-oxoglutarate). In the reverse direction, the enzyme catalyzes the formation of D-nopaline from L-arginine as well as D-ornaline from L-ornithine.

J. Thompson & J. A. Donkersloot (1992) *Ann. Rev. Biochem.* **61**, 517.
J. Thompson & S. P. F. Miller (1991) *Adv. Enzymol.* **64**, 317.

NOREPINEPHRINE *N*-METHYLTRANSFERASE

This enzyme [EC 2.1.1.28], also known as phenylethanolamine *N*-methyltransferase, catalyzes the reaction of *S*-adenosyl-L-methionine with phenylethanolamine to produce *S*-adenosyl-L-homocysteine and *N*-methylphenylethanolamine. The enzyme will act on a number of phenylethanolamines and will catalyze the conversion of noradrenalin (or norepinephrine) into adrenalin (or epinephrine).

R. W. Fuller (1987) *Meth. Enzymol.* **142**, 655.

NORMAL DISTRIBUTION

A symmetrical bell-shaped curve corresponding to a graph of the frequency $f(x)$ of a value x, when μ is the mean and σ^2 is the variance, such that

$$f(x) = \{\exp[-(x - \mu)^2/2\sigma^2]\}/\sigma(2\pi)^{1/2}$$

When this function appears to satisfy the observed distribution of quantitative measurements, we say that such measurements are normally distributed. **See** *Statistics (A Primer)*

NORMAL ERROR CURVE

A statistical method for plotting the relative frequency (dN/N) of a probable error in a single measured value x *versus* the deviation (z) from μ, the mean of the data, in units of standard deviation (σ), such that $z = (x - \mu)/\sigma$. The standard error curve (shown below) does not depend on either the magnitude of the mean or the standard deviation of the data set. The maximum of the normal error curve is poised at zero, indicating that the mean is the most frequently observed value.

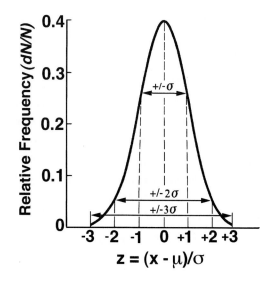

The area beneath the standard error curve can be obtained by integrating the following equation:

$$dN/N = (1/2\pi) \exp[-z^2/2]dz$$

The area under the entire curve equals a probability of unity. Over the interval from -1σ to $+1\sigma$, the integrated area equals 68.3% of the total area under the curve; thus, the probability of finding a value within that interval is 0.683. For $\pm 2\sigma$, the probability rises to 0.955, and for $\pm 3\sigma$, the probability is 0.997. Note that for a set of N data values, the estimated mean will progressively deviate less and less from the true mean as the number of data points increases. This so-called standard error of the mean (s_m) is inversely related to $N^{1/2}$. **See** *Statistics (A Primer)*

NORMALITY (or, NORMAL CONCENTRATION)

A unit of concentration (symbolized by N) equal to the number of equivalents of a chemical substance (*i.e.*, the grams of the substance divided by the equivalent weight) per liter of solution. The term ''normal solution'' refers to a solution of 1.0 N. For example, the gram-molecular-weight and the gram-equivalent-weight for a monoprotic acid are the same; hence, a 1.0 M HNO_3 is also 1.0 N. For a diprotic acid, the gram-molecular-weight is exactly twice the gram-equivalent-weight, and a 1.0 M H_2SO_4 solution is therefore 2.0 N. **See also** *Weight Normality*

NORTHROP ISOTOPE EFFECT METHOD

A procedure using both deuterium and tritium isotope effects on V_{max}/K_m to obtain *intrinsic* isotope effects for enzyme-catalyzed reactions[1-3]. **See** *Kinetic Isotope Effects*

[1]D. B. Northrop (1975) *Biochemistry* **14**, 2644.
[2]W. W. Cleland, M. H. O'Leary & D. B. Northrop, eds. (1977) *Isotope Effects on Enzyme-Catalyzed Reactions*, University Park Press, Baltimore.
[3]W. W. Cleland (1995) *Meth. Enzymol.* **249**, 341.

NOTEBOOK ENTRIES

Although every experimentalist develops her/his own habits for recording research observations, maintenance of adequate and timely documentation is still insufficiently stressed. Exceptions to this statement may be cited, especially in industry where laboratory notebooks are read and endorsed on a regular schedule for proprietary motives. Next to the writing of monthly or quarterly progress reports, nothing seems at first glance more inefficient than the humdrum recording of day-to-day observations. Nonetheless, laboratory notebooks and other supporting documentation are the stuff of science, and while there are no rules for keeping a laboratory notebook, the following are a few suggestions that can be used to good advantage.

(1) Chose a sturdy notebook, one that will survive your need to open and close it on innumerable occasions and one that will not burst when you have taped or pasted in chart paper tracings, scintillation counter output, chemical manufacturer's product information slips, graphs, *etc*. (Even superbly constructed laboratory notebooks can be damaged through normal use; so protect your notebook out of respect for its archival value.)

(2) Lab notebooks should be permanently bound, with all pages numbered. Reserve the first 3-4 pages for a fully annotated Table of Contents. This will speed up data retrieval when you're rushing to complete a required report or to beat the competition in submitting a manuscript for publication.

(3) Use a pen with permanent ink, and NOT a pencil, to make your entries. Date each entry, and be sure to put a date next to comments that you find necessary to add on a later occasion. (Such careful chronicling will allow you to recollect why you made certain changes in protocol or observations, especially those based on some later realization.)

(4) While you should keep your notebooks out of harm's way during an experiment, the lab notebook rightfully belongs at the bench, and not the desk. Record data directly into the book, rather than commiting data to memory or temporarily using a sheet of paper (or worse, a paper towel) in place of the notebook itself.

(5) Always opt for recording primary data, because errors and assumptions can influence your reduction of primary data into their final report-ready form. In kinetic experiments, you should always try to avoid the use of unitless parameters, such as "relative activity", because the ratio of two activity measurements, if recorded in the absence of the originally measured rate values, obscures later analysis of your experiments.

(6) Describe your experiments in sufficient detail to ensure that others can replicate your findings. Avoid codes or your own jargon, unless you have clearly defined the term in writing. Many biochemical experiments involve many operations that, while combined in different ways, should be clearly described in their order of execution for each experiment.

(7) Get in the habit of using any hiatus, however brief, during your experiments to keep your notebook current. (Reward yourself by combining this activity with time-out for some refreshment or relaxing music.)

(8) Use flow charts and "time lines" as a convenient means for recording an experiment. You can frequently write out a better diagram (especially as your experiment proceeds long into the night) than you can record any paragraph of sentences. The following is an example of a time line describing the assay of coenzyme-A-linked acetaldehyde dehydrogenase (abbreviated AldDH). The horizontal axis is the time axis and vertical downward arrows are used to indicate the sequence of various operations and/or treatments.

MIX: (at 4°C in cuvette)
0.5 mL standard buffer
(for composition, see
p. 25 in Notebook)
including:
10 mM DTT
23 μM CoASH
0.5 mM NADH
                    +
0.5 mL AldDH sample   Start reaction with
(dialyzed overnight   1.0 mL 50 mM
versus 100 vols   acetaldehyde   make duplicate
standard buffer)   in std. buffer at 28°   assay

         15 min at 28°C   assay immediately
        to activate enzyme   in SPEX Fluorolog
                               excite at 340 nm; read at 460 nm

(9) Note any unusual occurrences, even if they indicate some laziness or inattentiveness on your part. Despite all earnest attempts to keep an experiment from going awry, the complexity of most experiments conspires to introduce error or accidents into all observations. (It's far better to acknowledge an erroneous or suspicious measurement than to send later readers (including yourself) off on a series of experiments to follow up on a misleading observation that appears in your notebook without comment.)

(10) Record the source of the kinetic treatment (or some new derivation) directly into your notebook. Because kinetic measurements usually involve use of standard or newly derived rate equations, you should specify a complete list of assumptions and comments on the reasonableness of each assumption in your particular experiments.

(11) The source data for all published graphs and tables also should be marked in the notebook's Table of Contents. Likewise, after submitting the corrected version of your doctoral dissertation, the notebook's Table of Contents should be marked to indicate all figures and tables that appear in the dissertation. This will help your successors, should they wish to reanalyze a figure in your publications or dissertation.

(12) Finally, keep in mind that scientists must maintain high standards in conducting and reporting their findings. The following statements are excerpted from the 1989 report, entitled *On Being a Scientist*, published by the National Academy of Sciences' Committee on the Conduct of Science:

"Error caused by inherent limits on scientific theories can be discovered only through the gradual advancement of science, but error of a more human kind also occurs in science. Scientists are not infallible; nor do they have limitless working time or access to unlimited resources. Even the most responsible scientist can make an honest mistake. When such errors are discovered, they should be acknowledged, preferably in writing in the same journal in which the mistaken information was published. Scientists who make such acknowledgements promptly and graciously are not usually condemned by colleagues. Others can imagine making similar mistakes."

"Mistakes made while trying to do one's best are tolerated in science; mistakes made through negligent work are not. Haste, carelessness, inattention—any number of faults can lead to work that does not meet the standards demanded in science. In violating the methodological standards required by a discipline, a scientist damages not only his or her own work but the work of others. Furthermore, because the source of error may be hard to identify, sloppiness can cost years of effort, both for the scientist who makes the error and for others who try to build on that work."

The most recent version of *On Being a Scientist* is available over the internet at http://www.nas.edu/nap/online/obas/

NOYES EQUATION

An equation allowing an investigator to determine the chemical reaction order of a non-enzyme-catalyzed reaction and the rate expression for a non-first-order process by noting that half-lives for non-first-order reactions are dependent on the initial reactant concentration.

NU (ν)

1. Symbol for frequency. 2. Symbol for stoichiometric number. 3. Symbol for kinematic viscosity. 4. Symbol (with overbar) for the number of moles of ligand bound per mole of macromolecule.

NUCLEARITY

The number of central atoms joined in a single coordination entity or cluster by bridging ligands or by metal-metal bonds. Such complexes are referred to as being dinuclear, trinuclear, tetranuclear, polynuclear, *etc.*

NUCLEAR MAGNETIC RESONANCE

Space constraints permit us only to consider the essence of the phenomenon known as nuclear magnetic resonance. An applied magnetic field H_o interacts with the magnetic field generated by a nucleus (here, for simplicity, a proton with nuclear spin states of $+1/2$ and $-1/2$) having angular momentum as a consequence of its spinning motion. The spinning nucleus has a magnetic field strength, indicated by its magnetic moment μ which becomes directionally oriented with the applied magnetic field. Because the nucleus precesses, μ is oriented at angles θ and $(180° - \theta)$ relative to the lines of force in an applied field (Fig. 1).

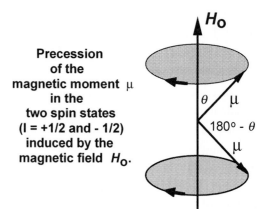

Precession of the magnetic moment μ in the two spin states (I = +1/2 and -1/2) induced by the magnetic field H_o.

At angle θ, μ is aligned with the direction of the applied field, whereas at angle $(180° - \theta)$, the opposite is true. For this reason, the respective potential energy of these two states is $-\mu H_o \sin \theta$ and $+\mu H_o \sin \theta$, resulting in a quantized energy difference ΔE given by the expression:

$$\Delta E = 2\mu H_o \sin \theta = h\nu$$

where h is Planck's constant. Thus, electromagnetic radiation with a frequency ν can be absorbed, thereby promoting the system to the higher energy state. Such energies correspond to photons in the radiofrequency range, and a radiofrequency generating coil is positioned around the sample within the cavity of the NMR spectrometer. NMR provides valuable information regarding molecular structure, and we can briefly consider two phenomena, namely chemical shift and spin-spin coupling.

A chemical shift is defined as a displacement in the magnetic resonance frequency of a nucleus as a consequence of the electronic environment in which the nucleus resides. Because moving electrons generate their own magnetic fields, a nucleus surrounded by these electrons experiences an effective field, H_{eff}, which is defined by $(1 - \alpha)H_o$, where α is the so-called screening constant and H_o is the applied magnetic field. A chemical shift is typically reported as a dimensionless displacement (units = parts per million, or simply ppm) from a reference standard. If the magnetic field is varied while the radiofrequency ν is held constant, then the chemical shift (ppm) equals $\{[H_{sample} - H_{reference}]/H_{reference}\} \times 10^6$. If the radiofrequency is varied at constant magnetic field, the chemical shift (ppm) equals $\{[\nu_{sample} - \nu_{reference}]/\nu_{reference}\} \times 10^6$. Tetramethylsilane is frequently used as a proton chemical shift reference compound in organic chemistry, because it yields a single sharp NMR signal in a region well removed from the resonance lines of most other protons.

One begins to sense the great versatility and power of NMR as a structural probe by considering so-called splitting patterns (or, more formally, absorption band multiplicities). These splitting patterns define the positions of nuclei within a molecule. Splitting is a consequence of reciprocal magnetic interactions between spinning nuclei, and the interaction likewise depends on electronic structure. Spin-spin coupling creates multiplets from

what would otherwise appear as sharp resonance lines, and the coupling strength J depends on the particular nuclei under examination. For example, in the case of n spin-coupled protons, the multiplet will have $n + 1$ lines. A single nearby proton will split the observed proton resonance into a doublet of equal or (1:1) intensity, two nearby electrons will split the observed proton resonance to a triplet with intensities of 1:2:1, and so on. The intensities are always coefficients of a binomial expansion. *See* Pascal's Triangle

So-called multidimensional NMR techniques can provide important information about macromolecular conformation. In these cases, the sequence of a protein is already known, and establishing covalent connectivity between atoms is not the goal. Rather, one seeks through-space information that can reveal the solution conformation of a protein or other macromolecule. Two- or three-dimensional techniques use pulses of radiation at different nuclear frequencies, and the response of the spin system is then recorded as a free-induction decay (FID). Techniques like COSY and NOESY allow one to deduce the structure of proteins with molecular weights less than 20,000–25,000.

Finally, as tabulated below, many elements having useful natural abundance and nuclear spin are naturally present in biomolecules. Still others can be substituted into biological molecules to provide a diverse range of opportunities. For example, fluorine can be substituted for hydrogen in many cases.

Table I

Several Biologically Important Nuclei for NMR Experiments

Nuclide	Natural Abundance	Spin
^1H	99.985%	$I = 1/2$
^2H	0.015%	$I = 1$
^{13}C	1.10%	$I = 1/2$
^{15}N	0.366%	$I = 1/2$
^{17}O	0.003%	$I = 5/2$
^{19}F	100%	$I = 1/2$
^{31}P	100%	$I = 1/2$

Selected entries from *Methods in Enzymology* [vol, page(s)]:
Techniques: Proton nuclear magnetic resonance in aqueous solutions, **49**, 253; fluorine nuclear magnetic resonance studies of proteins, **49**, 270; stopped-flow nuclear magnetic resonance spectroscopy, **49**, 295; nuclear relaxation measurements of the geometry of enzyme-bound substrates and analogs, **49**, 322; nuclear magnetic resonance kinetics viewed as enzyme kinetics, **49**,

NUCLEASE S1

This zinc-dependent endonuclease [EC 3.1.30.1] can act on either ribo- or deoxyribonucleic acids, with a preference for single-stranded substrates. It catalyzes the endonucleolytic cleavage of those nucleic acids to 5′-phosphomononucleotide and 5′-phosphooligonucleotide endproducts.

I. R. Lehman (1981) *The Enzymes*, 3rd ed., **14**, 193.

NUCLEATION

Any highly cooperative chemical or physical process that serves to generate structures that then grow by linear accretion or elongation. Nucleation is an important phase in indefinite polymerization processes (*e.g.*, actin filament and microtubule assembly). Nucleation is typically of a high kinetic order,

$$\text{Nucleation Rate} = k[\text{X}]^n$$

where n is the number of X subunits that must simultaneously collide to form the critical nucleus for further growth. In the case of actin assembly, n is around three. Once formed, these nuclei are frequently unstable and dissociate unless additional growth occurs. *See Protein Polymerization; Actin Assembly Kinetics; Microtubule Assembly Kinetics*

NUCLEOFUGE

A leaving group that departs with the bonding electron pair. The tendency of such a moiety to leave is referred to as nucleofugicity. *See also Electrofuge; Nucleophile*

NUCLEOPHILE

A molecular entity preferentially attracted to an electrophilic region or some other site of low electron density in another molecule. A nucleophile typically brings a pair of electrons to the second molecular entity, thereby qualifying it as an electron-pair donor (or Lewis base). *See also Electrophile*

NUCLEOPHILIC CATALYSIS

Catalysis by a Lewis base in which a Lewis adduct is a reaction intermediate.

NUCLEOPHILIC COMPETITION

A mechanistic strategy for detecting the presence of an electrophilic intermediate in a two-step enzyme-catalyzed hydrolysis reaction by introducing a second nucleophile to compete with water and thereby create a new reaction product. An illustrative case is catalysis of the *Escherichia coli* β-galactosidase reaction in the absence of the small subunit[1]. Rate-determining hydrolysis of the glycosyl-enzyme intermediate was demonstrated by pre-steady-state measurements and nucleophilic competition, the latter accomplished by using methanol as the competitive nucleophile to form the corresponding methylglycoside.

[1]S. V. Calugaru, B. G. Hall & M. L. Sinnott (1995) *Biochem. J.* **312**, 281.

NUCLEOPHILICITY

1. The relative reactivity of a nucleophilic reagent. Nucleophilicity of a Lewis base is typically measured by relative rate constants of different reagents toward a common substrate. 2. The property of being nucleophilic. *See Electrophilicity*

NUCLEOPHILIC SUBSTITUTION REACTION

A heterolytic reaction in which a nucleophilic reagent brings (*i.e.*, attacks with) a free pair of electrons on a molecular entity.

$$R_1: + R_2–R_3 \rightarrow R_1–R_2 + :R_3$$

This pair of electrons forms a new bond, and the leaving group (the nucleofuge) exits with an electron pair.

Types of Nucleophilic Substitution Reactions. A reaction described as S_N1, short for substitution, nucleophilic (unimolecular), comprises two steps:

$$\text{Step-1: } R - X \overset{\text{slow}}{\rightleftharpoons} R^{\oplus} + X^{\ominus}$$

$$\text{Step-2: } R^{\oplus} + Y^{\ominus} \overset{\text{fast}}{\rightleftharpoons} R - Y$$

The first reaction is the slow, or rate-determining, ionization of the substrate to form a carbocation intermediate. The products of this first step will tend to be stabilized best in polar solvents. The S_N1 type of reaction can also lead to the formation of ion pair intermediates, as shown in the following reaction scheme:

$$R - X \rightleftharpoons R^{\oplus} X^{\ominus} \rightleftharpoons R^{\oplus} \| X^{\ominus} \rightleftharpoons R^{\oplus} + X^{\ominus}$$

where the first of the two intermediates is called an intimate ion pair, and the second is termed a loose ion pair. The intimate ion pair is apt to undergo recombination to form the original substrate, and this process is frequently termed internal return.

A reaction described as S_N2, abbreviation for substitution, nucleophilic (bimolecular), is a one-step process, and no intermediate is formed. This reaction involves the so-called backside attack of a nucleophile Y on an electrophilic center RX, such that the reaction center (*i.e.*, the carbon or other atom attacked by the nucleophile) undergoes inversion of stereochemical configuration. In the transition-state nucleophile and exiphile (leaving group) reside at the reaction center. Aside from stereochemical issues, other evidence can be used to identify S_N2 reactions. First, because both nucleophile and substrate are involved in the rate-determining step, the reaction is second order overall: rate = k[RX][Y]. Moreover, one can use kinetic isotope effects to distinguish S_N1 and S_N2 cases (**See** Kinetic Isotope Effects).

A less frequent case is a reaction described as $S_N i$, short for substitution, nucleophilic (internal). This reaction type involves a unimolecular process in which the substrate dissociates to form an intimate ion pair; however, no inversion of configuration occurs, presumably because steric hindrance forces the nucleophile to enter at the same side from which the leaving group departed.

NUCLEOSIDE DEOXYRIBOSYLTRANSFERASE

This enzyme [EC 2.4.2.6] catalyzes the reaction of 2-deoxy-D-ribosyl-base[1] with base[2] to produce 2-deoxy-D-ribosyl-base[2] and base[1] in which base[1] and base[2] represent various purines and pyrimidines.

W. S. McNutt (1955) *Meth. Enzymol.* **2**, 464.

NUCLEOSIDE DIPHOSPHATASE

This enzyme [EC 3.6.1.6] catalyzes the hydrolysis of a nucleoside diphosphate to produce a nucleotide (that is, a nucleoside monophosphate) and orthophosphate. NDP substrates include IDP, GDP, UDP, as well as D-ribose 5-diphosphate.

G. W. E. Plaut (1963) *Meth. Enzymol.* **6**, 231.

NUCLEOSIDE DIPHOSPHATE KINASE

This enzyme [EC 2.7.4.6], also known as nucleoside diphosphokinase and nucleoside 5'-diphosphate phosphotransferase, catalyzes the reaction of ATP with a nucleoside diphosphate to produce ADP and a nucleoside triphosphate. ATP can be substituted by a number of nucleoside triphosphate and deoxynucleoside triphosphate compounds.

N. B. Ray & C. K. Mathews (1992) *Curr. Top. Cell. Reg.* **33**, 343.
P. A. Frey (1989) *Adv. Enzymol.* **62**, 119.
D. L. Purich & R. D. Allison (1980) *Meth. Enzymol.* **64**, 3.
R. E. Parks, Jr., & R. P. Agarwal (1973) *The Enzymes*, 3rd ed., **8**, 307.

NUCLEOSIDE PHOSPHOTRANSFERASE

This enzyme [EC 2.7.1.77] catalyzes the reaction of a nucleotide with a 2'-deoxynucleoside to produce a nucleoside and a 2'-deoxynucleoside 5'-monophosphate. The nucleotide substrate can be substituted with phenyl phosphate and nucleoside 3'-phosphates, although they are not as effective.

P. A. Frey (1992) *The Enzymes*, 3rd ed., **20**, 141.
P. A. Frey (1989) *Adv. Enzymol.* **62**, 119.

NUCLEOSIDE TRIPHOSPHATE:HEXOSE-1-PHOSPHATE NUCLEOTIDYLTRANSFERASES

This enzyme [EC 2.7.7.28], also known as NDP-hexose pyrophosphorylase, catalyzes the reaction of a nucleoside triphosphate with a hexose 1-phosphate to produce a NDP-hexose and pyrophosphate (or, diphosphate). In the reverse reaction the NDP-hexose can be, in decreasing order of activity, guanosine, inosine, and adenosine diphosphate hexoses in which the sugar is either glucose or mannose.

K. A. Presper & E. C. Heath (1983) *The Enzymes*, 3rd ed., **16**, 449.

NUCLEOSIDE 5′-TRIPHOSPHATE REGENERATION

The free energy of hydrolysis ($\Delta G_{hydrolysis}$) of nucleoside 5′-triphosphate (NTP) is thermodynamically coupled to many biosynthetic reactions as well as polymerization processes, and nucleoside 5′-diphosphate (NDP) accumulates during the course of experiments. Proper design of kinetic studies on these systems requires maintenance of NTP concentration throughout the course of any measurements, and a NTP-regenerating reaction should be incorporated into all such reaction samples. The late Ephriam Racker first introduced the practice of supplementing a reaction sample with one or more auxiliary enzymes and metabolites to regenerate any NTP that had become converted to NDP.

Three enzymes are typically employed for this purpose: (a) creatine kinase, (b) acetate kinase, and (c) pyruvate kinase. Their NDP specificity, the nature of the co-substrate, and both the kinetic and thermodynamic properties of each of these enzyme-catalyzed reactions commends their use in NTP regeneration. Likewise, their commercial availability at high purity also factors into their widespread usage. These three enzyme systems are not interchangeable, and judicious choice of regenerating enzyme is to be strongly advised. In particular, the experimenter will wish to confirm that the regenerating enzyme and the cosubstrate do not possess any contaminating enzymatic activity or inhibitory/activating property that interferes with the primary reaction under investigation.

CREATINE KINASE. The creatine kinase reaction[1] can be written as follows:

$$\text{phosphocreatine} + \text{MgADP}^- \rightleftharpoons \text{creatine} + \text{MgATP}^{2-}$$

The enzyme preferentially acts on adenine nucleotide, and the equilibrium constant for the reaction in the direction written above is approximately 40, depending on magnesium ion and proton concentrations. The enzyme is inhibited by 20 mM chloride ion. Typically, one uses 1–2 international units of the rabbit muscle enzyme along with 10 mM phosphocreatine.

ACETATE KINASE. MacNeal et al.[2] first proposed the use of Escherichia coli acetate kinase as a nearly ideal system for ATP (or GTP) resynthesis from ADP (or GDP). The reaction catalyzed is:

$$\text{acetyl phosphate} + \text{MgADP}^- \rightleftharpoons \text{acetate} + \text{MgATP}^{2-}$$

or

$$\text{acetyl phosphate} + \text{MgGDP}^- \rightleftharpoons \text{acetate} + \text{MgGTP}^{2-}$$

Earlier work[3,4] indicated that ADP and GDP were efficient phosphoryl-acceptor substrates. The equilibrium constant for these reactions is highly favorable ($K_{eq} = 3000$); thus, even when 90% of the initial acetyl phosphate concentration is converted to acetate, the [GTP]/[GDP] or [ATP]/[ADP] concentration ratio will still exceed 200-300, provided sufficient enzyme is present to maintain the equilibrium. Because acetate kinase and acetyl phosphate are not present in eukaryotes, one gains an added advantage over the above-mentioned regenerating systems: there are no contaminating levels of either even in crude animal tissue extracts. Use of acetate kinase and acetyl phosphate is highly recommended. The lithium salt of acetyl phosphate is stable for long-term storage in a desiccator at freezer temperatures.

For most reactions, one can use 1-2 mM acetyl phosphate and 1-2 international units of acetate kinase. The enzyme is usually supplied as a crystalline suspension in 3 M ammonium sulfate, and 10 mM ammonium ion is inhibitory. Therefore, a useful practice is to snip off 0.5 cm from a disposable Eppendorf micropipette tip to facilitate removal of 10-20 microliters of the crystalline suspension; then spin down the enzyme in a 1.5 ml disposable conical plastic centrifuge tube, and remove the ammonium sulfate solution with a wick of twisted Kimwipe. The enzyme precipitate can now be taken up directly into your working buffer. Note: Acetate kinase is inactivated by cold exposure, but incubation with 10^{-4} M ATP or GTP reactivates the enzyme if warmed to room temperature for 5-10 min.

PYRUVATE KINASE. This enzyme catalyzes phosphoryl transfer from the glycolytic intermediate phosphoenolpyruvate:

$$\text{P-enolpyruvate} + \text{MgADP}^- \rightleftharpoons \text{acetate} + \text{MgATP}^{2-}$$

The pyruvate kinase reaction is virtually irreversible under typical intracellular conditions[5], and the enzyme requires Mg^{2+} or Mn^{2+} in addition to the monovalent cation

K^+. Pyruvate kinase is inhibited by 3-5 mM ATP, as well as by acetyl-S-CoA and long-chain fatty acids. Rabbit muscle pyruvate kinase is typically used in ATP regeneration.

REGENERATION OF NTP'S OTHER THAN ATP or GTP. Although most experiments require regeneration of either ATP or GTP, occasionally one may wish to maintain the phosphorylation state of other nucleotides. This is readily accomplished by using acetate kinase in combination with nucleoside-5′-diphosphate kinase (NDPK) in the presence of ATP and acetyl phosphate. For the sake of exemplifying this method, let us consider the regeneration of UTP from UDP. The overall equilibrium constant for rephosphorylation of UDP to UTP is the same as the acetate kinase equilibrium constant, because the equilibrium constant for the NDPK reaction is approximately one. Typically, one can include 2-4 international units of commercially available yeast NDPK along with 0.5 mM ATP in the reaction mixture.

[1] F. B. Rudolph, B. W. Baugher & R. S. Beissner (1979) *Meth. Enzymol.* **63**, 22.

[2] R. K. MacNeal, B. C. Webb & D. L. Purich (1977) *Biochem. Biophys. Res. Commun.* **74**, 440.

[3] R. S. Anthony & L. B. Spector (1972) *J. Biol. Chem.* **247**, 2120.

[4] B. C. Webb, J. A. Todhunter & D. L. Purich (1976) *Arch. Biochem. Biophys.* **173**, 282.

[5] D. E. Metzler (1976) Biochemistry: The Chemical Reactions of Living Cells, p. 401, Academic Press, New York.

NUCLEOSOME CORE PARTICLE SELF-ASSEMBLY

The self-assembly of this fundamental building block of chromatin is a topic of enduring interest. DNase I digestion experiments as well as spectroscopic studies indicate that nucleosome core particles can be reconstituted by salt-jump (*i.e.*, diluting NaCl concentration from 2.0 to 0.2 M) or by direct mixing of histones and DNA at the lower salt concentration. Daban and Cantor[1] used the increase in eximer fluorescence to investigate the reassembly process in terms of a two-state model:

$$N' \rightleftharpoons N$$

and

$$K_{obs} = [N]/[N'] = (F_e - F_{N'})/(F_N - F_e)$$

where K_{obs} is the equilibrium constant between the N and N′ states, F_e is the fluorescence intensity of the equilibrium mixture, $F_{N'}$ is the fluorescence intensity of

N′, and F_N is the fluorescence intensity of the N state. Their studies actually produced evidence favoring a slightly more complicated scheme:

$$DNA + Histone \rightleftharpoons N'' \rightleftharpoons N' \rightleftharpoons N$$

where DNA and histone bind rapidly and a two-exponential rate equation gave values of 10 s^{-1} and $7 \times 10^{-3} \text{ s}^{-1}$ for the N″-to-N′ and N′-to-N steps, respectively. From the concentration dependence of the equilibrium constant, Daban and Cantor[1] suggest that about 20 NaCl molecules are involved in the process.

In a companion study, Daban and Cantor[2] proposed a role for histone pairs H2A,H2B and H3,H4 in the self-assembly of nucleosome core particles. A modified mechanism for *in vitro* nucleosome core particle formation was offered later by Aragay *et al.*[3] who probed the interaction of different histone oligomers with nucleosomes by using nondenaturing gel electrophoresis. Their experimental results indicate that H2A,H2B, H3,H4, and the four core histones can migrate spontaneously from the aggregated nucleosomes containing excess histones to free core DNA. Moreover, on the basis of electron microscopy, solubility, and supercoiling assays, transfer of excess histones from oligonucleosomes to free circular DNA can be demonstrated. Their results suggest that nucleosome core particles can be formed in 0.2 M NaCl by the following mechanisms: (1) by transfer of excess core histones from oligonucleosomes of free DNA; (2a) by transfer to excess H2A,H2B and H3,H4 associated separately with oligonucleosomes to free DNA or (2b) by transfer to excess H2A,H2B initially associated with oligonucleosomes to DNA, followed by the reaction of the resulting DNA-(H2A,H2B) complex with oligonucleosomes containing excess H3,H4, and (3) a two-step transfer reaction similar to that indicated in (2b), in which excess histones H3,H4 are transferred to DNA before the reaction with oligonucleosomes containing excess H2A,H2B. Samso and Danban[4] recently described synchrotron X-ray scattering data on core DNA-(H2A,H2B) complexes, using different histone to DNA weight ratios and ionic conditions ranging from very low ionic strength to 0.2 M NaCl to show that histones H2A,H2B are unable to fold core DNA. Reconstituted complexes prepared at a physiological salt concentration either with the four core histones or with histones H3,H4 without H2A,H2B were found to be completely folded particles with a radius of gyration that is similar to that of native nucleosome core. They found that the DNA of the extended complexes containing histones H2A,H2B becomes completely folded after the histone pair exchange reaction which occurs spontaneously between preformed DNA-(H2A,H2B) and DNA-(H3,H4) complexes.

[1] J.-R. Daban & C. R. Cantor (1982) *J. Mol. Biol.* **156**, 749.
[2] J.-R. Daban & C. R. Cantor (1982) *J. Mol. Biol.* **156**, 771.
[3] A. M. Aragay, X. Fernandez-Busquets & J. R. Daban (1991) *Biochemistry* **30**, 5022.
[4] M. Samso & J. R. Daban (1993) *Biochemistry* **32**, 4609.

3'-NUCLEOTIDASE

This enzyme [EC 3.1.3.6] catalyzes the hydrolysis of a 3'-ribonucleotide to produce a ribonucleoside and orthophosphate. The enzyme exhibits a wide specificity for the 3'-nucleotides.

G. I. Drummond & Y. Yamamoto (1971) *The Enzymes*, 3rd ed., **4**, 337.

5'-NUCLEOTIDASE

This enzyme [EC 3.1.3.5] catalyzes the hydrolysis of a 5'-ribonucleotide to produce a ribonucleoside and orthophosphate. The enzyme exhibits a wide specificity for the 5'-nucleotides.

G. I. Drummond & Y. Yamamoto (1971) *The Enzymes*, 3rd ed., **4**, 337.

NUCLEOTIDE EXCHANGE FACTORS

Proteins that catalyze the exchange of protein-bound nucleotide with nucleotide free in solution. These exchange factors represent an unclassified type of isomerase, but there has been no Enzyme Commission of this class.

Several well-known regulatory[1-3] and cytoskeletal[4] proteins carry out the metabolic functions as "two-state" proteins. Binding of a nucleoside 5'-triphosphate (NTP) induces one conformational state and binding the corresponding nucleoside 5'-diphosphate (NDP) produces a second state. In such models, only one form is active, and the free energy of NTP hydrolysis drives the formation of the second NDP-bound state. The rate of hydrolysis serves as a metabolic clock that sets the delay in the conversion from NTP-bound to NDP-bound forms. Moreover, rebinding of NTP is needed to reload the regulatory protein, and this requires nucleotide exchange which can occur spontaneously, albeit slowly, even in the absence of a nucleotide exchange factor. Cells achieve a more robust range of responses through the biosynthesis of nucleotide exchange factors that accelerate exchange in response to other metabolic signals. The GTP-binding proteins in signal transduction cascades operate through the use of low-molecular-weight proteins that act as exchange factors. Likewise, the high rates of intracellular actin filament assembly/disassembly may lead to rate-limiting exchange of actin-bound ADP with cytosolic ATP to regenerate actin-ATP complex, the most active polymerizing form of this cytoskeletal protein. The 15 kDa protein profilin may fulfill this role in promoting actin-based motility. **See** *Actin-Based Motility*

How then might one represent the facilitated exchange process? The facilitation of nucleotide exchange is in all respects analogous to an enzyme-catalyzed reaction (*i.e.*, the rate of a reaction is accelerated without altering the equilibrium position of the reaction). In this way, one can represent the facilitated nucleotide exchange reaction as one would any enzymic process:

$$F + P_{NDP} \rightleftharpoons F \cdot P_{NDP}$$

$$[NDP \cdot F \cdot P\_\_\_ \rightleftharpoons [F \cdot P\_\_\_] + NDP$$

$$[F \cdot P\_\_\_] + NTP \rightleftharpoons [NTP \cdot F \cdot P\_\_\_]$$

$$[NTP \cdot F \cdot P\_\_\_] \rightleftharpoons F \cdot P_{NTP}$$

$$F \cdot P_{NTP} \rightleftharpoons F + P_{NTP}$$

where F is the factor, and P is the protein in the various nucleotide-bound and nucleotide-free forms.

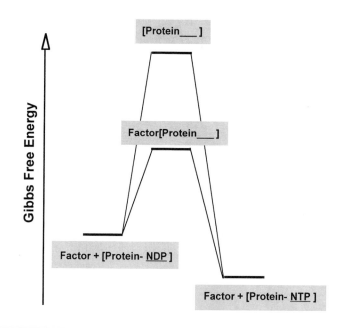

The diagram above also illustrates how the nucleotide-free form (indicated as [Protein___]) is stabilized by binding of the protein to the exchange factor. The nucleotide-free form is analogous to the transition state, and it lies at higher free energy (*i.e.*, it is less stable) than the bound NTP and NDP forms. Note also that the concentrations of NTP and/or NDP in the solution serve as a driving force favoring binding, as opposed to unbound forms, of the regulatory protein. Note also that the energy change from the P_{NDP} state to the P_{NTP} state is unaffected by the factor in this model.

[1] R. D. Vale (1996) *J. Cell Biol.* **135**, 291.
[2] A. G. Gilman (1987) *Ann. Rev. Biochem.* **56**, 615.
[3] L. Stryer (1991) *J. Biol. Chem.* **266**, 10711.
[4] D. L. Purich & J. M. Angelastro (1994) *Adv. Enzymol.* **69**, 121.

NUMERICAL COMPUTER METHODS

Enzyme kinetic data has been routinely presented in the form of variously named plots (*e.g.*, Lineweaver-Burk plot, Eadie-Hofstee plot, *etc.*), and the same is true for ligand binding (*e.g.*, Scatchard plot, Hill plot, *etc.*). With the advent of inexpensive digital computers, one can make strong arguments that direct linear plots (such as Initial Rate *versus* [Substrate] or Degree of Saturation *versus* [Ligand]) are more reliable and easier to treat statistically. The entries presented below illustrate how numerical computer methods can provide better ways for analyzing rate and equilibrium binding data.

Selected entries from *Methods in Enzymology* [vol, page(s)]:
Parameter estimation by least-squares methods, **210**, 1; global analysis of biochemical and biophysical data, **210**, 37; use of weighting functions in data fitting, **210**, 68; analysis of residuals: criteria for determining goodness-of-fit, **210**, 87; analysis of ligand-binding data with experimental uncertainties in independent variables, **210**, 106; Monte Carlo method for determining complete confidence probability distributions of estimated model parameters, **210**, 117; method of moments and treatment of nonrandom error, **210**, 237; interpolation methods, **210**, 305; practical aspects of kinetic analysis, **210**, 374; compartmental analysis of enzyme-catalyzed reactions, **210**, 391; numerical analysis of binding data: advantages, practical aspects, and implications, **210**, 481; oxygen equilibrium curve of concentrated hemoglobin, **232**, 486; Adair fitting to oxygen equilibrium curves of hemoglobin, **232**, 559; weighted nonlinear regression analysis of highly cooperative oxygen equilibrium curves, **232**, 576; dimer-tetramer equilibrium in Adair fitting, **232**, 597; effects of wavelength on fitting Adair constants for binding of oxygen to human hemoglobin, **232**, 606; Adair equation: rederiving oxygenation parameters, **232**, 632; linkage thermodynamics, **232**, 655; simulation of hemoglobin kinetics using finite element numerical meth-

ods, **232**, 517; least-squares techniques in biochemistry, **240**, 1; parameter estimates from nonlinear models, **240**, 23; sequential versus simultaneous analysis of data: differences in reliability of derived quantitative conclusions, **240**, 89; model-independent quantification of measurement error: empirical estimation of discrete variance function profiles based on standard curves, **240**, 121; impact of variance function estimation in regression and calibration, **240**, 150; application of Kalman filter to computational problems in statistics, **240**, 171; modeling chemical reactions: Jacobian paradigm and related issues, **240**, 181; the mathematics of biological oscillators, **240**, 198; analysis of diffusion-modulated energy transfer and quenching by numerical integration of diffusion equation in Laplace space, **240**, 216; applications of KinSim and FitSim computer simulation and fitting programs, **240**, 311; determination of rate and equilibrium binding constants for macromolecular interactions by surface plasmon resonance, **240**, 323; characterization of enzyme-complex formation by analysis of nuclear magnetic resonance line shapes, **240**, 438; analysis of drug-DNA binding isotherms: a Monte Carlo approach, **240**, 593; estimating binding constants for site-specific interactions between monovalent ions and proteins, **240**, 645; applying bifurcation theory to enzyme kinetics, **240**, 781.

NUMERICAL INTEGRATION

Any mathematical transformation that maps a function $(t = 0, q_0) \rightarrow (t = x, q_x)$ and is then iterated to map $(t = x, q_x) \rightarrow (t = x + y, q_x + y)$, and so forth to solve for the time evolution of an initial value problem.

Although one cannot obtain a general solution for solving the initial value problem

$$dy/dt = f(t,y) \qquad \text{where} \quad y(t_o) = y_o$$

there are a number of numerical procedures. These methods allow one to obtain reaction progress curves for kinetic systems that are characterized by one or more first-order differential equations. While a detailed consideration of these methods lies beyond the scope of this Handbook, many engineering mathematics textbooks describe the subject. A most useful reference by Braun[1] describes some of the most widely used algorithms, including the Euler method, the Runge-Kutta procedure, the three-term Taylor series method, and an improved Euler method. Also presented are short-statement Fortran and Pascal programs for these techniques, along with some practical advice. ***See*** *Gear Algorithm; Numerical Computer Methods; Stiffness*

[1] M. Braun (1991) *Differential Equations and Their Applications: An Introduction to Applied Mathematics*, pp. 96-126, Springer-Verlag, New York.

O

OCCAM'S RAZOR

The concept (known also as the Principle of Parsimony) grounded in scientific logic and stating that, in the absence of evidence to the contrary, one should always adopt the simplest explanation or model[1]. The metaphor "razor" (probably introduced by the French logician Condillac) is used to indicate a sharp and swift device that one can use to cut away and dispose of unneeded complexity. First advanced by William of Occam (1285–1349), this philosophic strategy tacitly assumes that Nature is straightforward, rarely wasting even a gesture[2,3]. This principle is of particular use in biochemical kinetics where the temptation to adopt complexity unnecessarily is to be resisted. **See** Model

[1]Stated in the Latin, "*Entia non sunt multiplicanda præter necessitate.*," or in English, "Entities should not be multiplied unnecessarily."
[2]Such being evident from numerous examples, one should not assume that Nature has achieved perfection in the elimination of inefficiency. Evolution works through accumulated advantage, the product of which may only be sufficiently adapted, while still appearing to be perfectly suited for a particular purpose.
[3]E. A. Moody (1967) in *The Encyclopedia of Philosophy* (P. Edwards, ed.) vol. **8**, pp. 306-317, Macmillan, New York.

OCCUPANCY THEORY OF DRUG ACTION

A mass action treatment of drug binding at a binding site on a receptor, such that the degree of biological response (either positive or negative) to drug binding is proportional to the drug's fractional occupancy of the receptor. Consider the binding of drug A to receptor R, with the formation of an AR complex:

$$A + R \underset{k_2}{\overset{k_1}{\rightleftarrows}} AR, \quad \text{where } K_A = k_2/k_1$$

$$= [A_{free}][R_{free}]/[AR]$$

in which k_1 and k_2 are the respective bimolecular and unimolecular rate constants for complex formation and dissociation, respectively. At equilibrium, we can use K_A to define the right-hand expression, where $[A_{free}]$ is the concentration of uncomplexed drug, $[R_{free}]$ is the concentration of uncomplexed receptor, and $[AR]$ is the concentration of drug-receptor complex. If the total drug concentration, $[A_t]$, greatly exceeds the total receptor concentration, $[R_{tot}]$, then $[A_{tot}] \simeq [A_{free}]$, whereas $[R_{tot}] = [R_{free}] + [AR]$. From these relationships, we get:

$$[AR]/[R_{tot}] = [A_{free}]/\{[A_{free}] + K_A\}$$

If the biological response elicited by drug binding is directly proportional to occupancy, then

$$E_a/E_m = [A_{free}]/\{[A_{free}] + K_A\}$$

where E_a is the effect observed as AR is formed, and E_m is the maximal effect occurring at full occupancy of available sites R_t. Typical dose-response (or more correctly, concentration-response) curves show hyperbolic functionality. This behavior only applies if $[A_t] >> [R_t]$ and the drug displays either rapid equilibrium or steady-state binding to the receptor. Very tight-binding or slow-binding drugs will not give the concentration-response curves shown above. Furthermore, drugs may not elicit just one effect, and side effects are common for almost all. If more than one receptor binds a particular drug, then the drug's concentration-response curve is likely to be more complicated than that shown above. Other confounding variables include: (a) nonuniform access of the drug to its receptor (**See** Diffusion of Ligand to Receptor); (b) cooperativity in drug binding, such that a single equilibrium cannot accurately account for the binding process; and (c) failure of the system to obey equilibrium or steady-state conditions required in deriving the mass action equilibrium.

OCTAHEDRAL COORDINATION

The eight-sided geometric solid corresponding to a hexacoordinate metal-ligand complex , as observed with ferricyanide $[Fe(III)CN_6]^{3-}$ or calcium ion complex with EGTA.

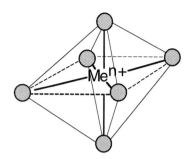

The six coordination sites are generated by hybridization of two $3d$, one $4s$, and three $4p$ orbitals to yield six equivalent d^2sp^3 orbitals, each situated at the corners of a regular octahedron.

OCTANOL DEHYDROGENASE

This enzyme [EC 1.1.1.73] will catalyze the reaction of 1-octanol with NAD^+ to produce 1-octanal and NADH. The enzyme can act on other long-chain alcohols, albeit not as effectively.

M. Ueda & A. Tanaka (1990) *Meth. Enzymol.* **188**, 171.

trans-OCTAPRENYLTRANSTRANSFERASE

This enzyme [EC 2.5.1.11], also known as nonaprenyl-diphosphate synthase and solanesyl-diphosphate synthase, catalyzes the reaction of all-*trans*-octaprenyl diphosphate with isopentenyl diphosphate to produce all-*trans*-nonaprenyl diphosphate and pyrophosphate. The enzyme will also utilize geranyl diphosphate and all-*trans*-prenyl diphosphates of intermediate size as substrates, but it will not use dimethylallyl diphosphate.

H. Kleinig (1989) *Ann. Rev. Plant Physiol. Plant Mol. Biol.* **40**, 39.
E. D. Beytía & J. W. Porter (1976) *Ann. Rev. Biochem.* **45**, 113.

OCTOPINE DEHYDROGENASE

This enzyme [EC 1.5.1.11], also known as D-octopine synthase, catalyzes the reversible reaction of N^2-(D-1-carboxyethyl)-L-arginine with NAD^+ and water to produce L-arginine, pyruvate, and NADH. The enzyme, in the reverse direction, will also act on L-ornithine, L-lysine, and L-histidine.

J. Thompson & J. A. Donkersloot (1992) *Ann. Rev. Biochem.* **61**, 517.
J. Thompson & S. P. F. Miller (1991) *Adv. Enzymol.* **64**, 317.
H. F. Fisher (1988) *Adv. Enzymol.* **61**, 1.

OFFORD PLOT

A plot of the logarithm of electrophoretic mobility of peptides as a function of molecular weight. Parallel lines are obtained with peptides having different net charges.

R. E. Offord (1977) *Meth. Enzymol.* **47**, 51.

OHM'S LAW

$I = E/R$ where I is the current, E is the electromotive force, and R is the resistance. The SI units for each of these is amperes, volts, and ohms, respectively. Ohm's law is also expressed as $I = \Delta F/R$ where ΔF is the difference in electric potential. The resistance is dependent upon the dimensions of the conductor.

2',5'-OLIGOADENYLATE SYNTHETASE

The following are recent reviews on the molecular and physical properties of this ligase (or, 2',5'-oligo(A) polymerase or oligo-isoadenylate synthetase E).

A. V. Itkes & E. S. Severin (1987) *Adv. Enzymol.* **59**, 213.
P. Lengyel (1982) *Ann. Rev. Biochem.* **51**, 251.
L. A. Ball (1982) *The Enzymes*, 3rd ed., **15**, 281.

OLIGOMERIZATION

The formation of multisubunit complexes (for example, protein complexes). Such complex formation is usually a cooperative phenomenon.

OMEGA (Ω or ω)

Ω Symbol for ohm, the SI unit of electrical resistance. ω 1. Symbol for solid angle. 2. Symbol for circular frequency. 3. Symbol for angular velocity.

OMPTIN

This serine-protease [EC 3.4.21.87] catalyzes the hydrolysis of peptide bonds at Xaa—Yaa in which there is a preference for arginyl or lysyl residues at Xaa and Yaa.

W. F. Mangel, D. L. Toleda, M. T. Brown, K. Worzalla, M. Lee & J. J. Dunn (1994) *Meth. Enzymol.* **244**, 384.

OPEN SYSTEM

Any chemical, biological, or mechanical system in which both heat and matter within the system can be transferred to and from its surrounding environment. Thus, a change in the immediate environment can result in a change in the system under consideration. **See also** *Closed System; Isolated System*

OPTICAL ACTIVITY

The property of a substance, molecular entity, or material to rotate the plane of polarized light as that light passes through a crystal, liquid, or solution containing that substance. The degree of rotation is dependent on a number of factors (*e.g.*, concentration, temperature, pressure, *etc.*) as well as on the wavelength of the light.

OPTICAL DENSITY

A now archaic synonym, abbreviated by O.D., for absorbance or turbidity. **See** *Absorbance; Turbidity; Absorption Spectroscopy*

OPTICAL ROTATION

The change, measured in terms of an angle α, of the plane of polarized light of a given wavelength passing through an optically active substance. The specific optical rotation, symbolized by $[\alpha]$, is the angle that the plane of polarized light has changed upon passing through a solution of one gram of an optically active substance per milliliter of solution having a path length of one decimeter, using light of wavelength equal to the D line of sodium. Optical rotation is a useful property in structure studies and has often proved valuable in kinetics as well.

OPTICAL TWEEZERS

A force-measuring technique that uses the photic energy of a tightly focused laser beam to "grab" hold of small particles through momentum transfers from impinging photons, the wavelength of which must be in nonabsorbing region of the spectrum. No mechanical contact with the object is required. In most biological applications, manipulated objects are large compared to the wavelength of light, and the resultant forces are explained in terms of principles of momentum conservation arising as a consequence of the wave-particle duality of light. The beam of light is a stream of photons that exerts what has been termed "radiation pressure" which can stop organelle movement when appropriately oriented relative to direction of force exerted by molecular motors.

Ashkin and Dziedzic[1] first reported the optical trapping and manipulation of viruses and bacteria by laser radiation pressure with single-beam gradient traps. Individual tobacco mosaic viruses and dense oriented arrays of viruses were trapped in aqueous solution with no apparent damage, even at ~120 milliwatts of argon laser power. Trapping and manipulation of individual motile bacteria were also demonstrated in a high-resolution microscope at powers of a few milliwatts. Ashkin *et al.*[2] also introduced the use of infrared (IR) light to greatly improve laser traps with much less optical consequential damage to living cells. Ashkin *et al.*[3] later examined organelle transport along microtubules and estimated the force generated by a single molecular motor to be roughly

$2-3 \times 10^{-7}$ dynes. **See** *Force Effects on Molecular Motors*

[1] A. Ashkin & J. M. Dziedzic (1987) *Science* **235**, 1517.
[2] A. Ashkin A, J. M. Dziedzic & T. Yamane (1987) *Nature* **330**, 769.
[3] A. Ashkin, K. Schutze, J. M. Dziedzic, U. Euteneuer & M. Schliwa (1990) *Nature* **348**, 346.

OPTICAL YIELD

The ratio of the optical purity of the product of a reaction (involving chiral reagents and products) to the optical purity of the product's precursor. The optical yield of a reaction is not related to the chemical yield.

OPTIMAL pH

The pH at which a particular reaction or process is most effective; often symbolized by pH_{opt}. In most instances, the optimum pH value for an enzyme-catalyzed reaction or biological process will fall within normal physiological pH ranges. However, when the pH_{opt} value is not within physiological ranges, then a number of possibilities are worth considering:

(1) The specific assay conditions have altered or modified the protein(s) or system such that the experimentally observed pH_{opt} is not the physiological pH_{opt}. For example, some component may be missing from the reaction mixture, a component that is present in the reaction mixture (and not present in physiological contexts) has altered the system, a nonphysiological ionic strength or temperature has altered the pH_{opt}, *etc.* **See** *Anion Effects on pH-Rate Data*

(2) The physiological pH value is not the pH value experienced by the *in vivo* system. The *in vivo* protein is optimally active in a pH range present in a localized volume (or, compartment) that is different than the pH range of the larger physiological environment.

(3) The protein(s) is relatively unstable at its true pH_{opt}, and this lack of stability has not been corrected in the pH-activity plot. Thus, the observed pH_{opt} is a compromise of the effect of pH on both catalytic activity (under the assay conditions) and protein denaturation and/or conformation.

(4) The pH_{opt} may not reflect the effect of pH on protein activity (or on biological processes) but may be a result of the effect of pH on the ionization(s) of the substrate(s). This may be particularly true when utilizing artificial substrates having pK_a value(s) significantly dif-

ferent that the naturally occurring substrate's pK_a value(s). Note that it is not uncommon that use of different substrates will result in different pH-activity curves.

(5) The observed pH_{opt} is identical to the physiological pH_{opt} even though it lies outside the normal pH range. If such a situation is true, then solution acidity becomes an important regulatory control of the protein's biological activity.

ORBITAL STEERING

A model attempting to account for the remarkable rate enhancements observed in enzyme catalysis in terms of an apparent reduction in reaction order and the optimal alignment of molecular orbitals to promote reactivity[1,2]. This model was introduced at a time when Woodward & Hoffmann[3] formalized their ideas for predicting the nature of organochemical reactions using molecular orbital calculations. Proponents of orbital steering argued that the angular orientation of each reactant could be restricted to sufficiently small angles, such that rate enhancements of eight orders of magnitude would be accessible. The model drew almost immediate criticism[4] on the grounds that the proposed angular orientation would lead to unfavorable thermodynamics (i.e., too much energy would be required to reduce the rotational degrees of freedom to give the needed angular orientation). Page and Jencks[5,6] also estimated the magnitude of rate enhancements achieved by bringing two reactants from the freedom of the bulk solution to form what amounts to a unimolecular complex on the enzyme's surface. Hackney[7] has presented a useful account of the orbital steering model in his review on the role of binding energy in promoting enzyme catalysis. We should note that while the orbital steering hypothesis remains out of favor, the merit of this proposal will only be known when it is possible to assess quantitatively how much binding energy plays a role in distorting the ground-state ES complex to reach its transition-state configuration. There is still much work needed to establish the energetics of complementarity of the enzyme and its transition state.

[1]D. R. Storm & D. E. Koshland, Jr. (1970) Proc. Natl. Acad. Sci. U.S.A. **66**, 445; (1972) J. Am. Chem. Soc. **94**, 5805.
[2]A. Dafforn & D. E. Koshland, Jr. (1971) Bioorg. Chem. **1**, 129; Proc. Natl. Acad. Sci. U.S.A. **66**, 445.
[3]R. B. Woodward & R. Hoffmann (1969) Angew. Chem. Int. Ed. Engl. **8**, 781.
[4]T. C. Bruice, A. Brown & D. O. Harris (1971) Proc. Natl. Acad. Sci. U.S.A. **68**, 658.
[5]M. I. Page & W. P. Jencks (1971) Proc. Natl. Acad. Sci. U.S.A. **68**, 1678.
[6]M. Page (1973) Chem. Soc. Revs. **2**, 295.
[7]D. Hackney (1990) The Enzymes, 3rd ed., **19**, 1.

ORBITAL SYMMETRY CONSIDERATIONS

The interpretation of chemical reactivity in terms of molecular orbital symmetry. The central principle is that orbital symmetry is conserved in concerted reactions. An orbital must retain a certain symmetry element (for example, a reflection plane) during the course of a molecular reorganization in concerted reactions. It should be emphasized that orbital-symmetry rules (also referred to as Woodward-Hoffmann rules) apply only to concerted reactions. The rules are very useful in characterizing which types of reactions are likely to occur under thermal or photochemical conditions. Examples of reactions governed by orbital symmetry restrictions include cycloaddition reactions and pericyclic reactions.

R. B. Woodward & R. Hoffmann (1969) Angew. Chem. Int. Ed. (Engl.) **8**, 781.

ORDERED BI BI ENZYME MECHANISM

A sequential enzyme-catalyzed reaction mechanism in which two substrates react to form two products and in which there is a preferred order in the binding of substrates and release of products. Several enzymes have been reported to have this type of binding mechanism, including alcohol dehydrogenase[1], carbamate kinase[2], lactate dehydrogenase[3–5], and ribitol dehydrogenase[6].

The Rapid Equilibrium Mechanism: In this scheme, the rate-determining step is the interconversion of the central complex, such that all binding steps can be described via dissociation constants.

$$E + A \underset{}{\overset{K_{ia}}{\rightleftharpoons}} EA$$

$$EA + B \overset{K_b}{\rightleftharpoons} EAB \underset{k_6}{\overset{k_5}{\rightleftharpoons}} EPQ \overset{K_p}{\rightleftharpoons} EQ + P$$

$$EQ \overset{K_{iq}}{\rightleftharpoons} E + Q$$

Under initial rate conditions (i.e., [P] and [Q] $\cong 0$) and no abortive complexes being formed, the initial-rate expression is

$$v = V_{max}[A][B]/\{K_{ia}K_b + K_b[A] + [A][B]\}$$

where $V_{max} = k_5[E_{total}]$, $K_{ia} = [E][A]/[EA]$, and $K_b = [EA][B]/[EAB]$. When $1/v$ is plotted as a function of

1/[A] at different constant concentrations of B, a series of straight lines will be obtained having a common intersection point. The intersection point will always be in the second quadrant. A double-reciprocal plot of $1/v$ vs. $1/[B]$ at different constant concentrations of A will consist of a series of straight lines all intersecting on the vertical axis. This observation is characteristic of the rapid equilibrium ordered system but not of the steady-state scheme.

The Steady-State Mechanism: If the interconversion step is not the sole rate-determining step and binding steps are not in rapid equilibrium, then a steady-state description of the reaction is applicable. If there is a single ternary complex, EXY, (thus, [EXY] = [EAB] + [EPQ]), the scheme can be depicted as

$$E \underset{k_2}{\overset{k_1[A]}{\rightleftharpoons}} EA \underset{k_4}{\overset{k_3[B]}{\rightleftharpoons}} EXY \underset{k_6[P]}{\overset{k_5}{\rightleftharpoons}} EQ \underset{k_8[Q]}{\overset{k_7}{\rightleftharpoons}} E$$

The steady-state determinants are:

$$[E] = k_2 k_7 (k_4 + k_5) + k_3 k_5 k_7 [B] + k_2 k_4 k_6 [P]$$

$$[EA] = k_1 k_7 (k_4 + k_5)[A] + k_1 k_4 k_6 [A][P] + k_4 k_6 k_8 [P][Q]$$

$$[EXY] = k_1 k_3 k_7 [A][B] + k_2 k_6 k_8 [P][Q] + k_1 k_3 k_6 [A][B][P] + k_3 k_6 k_8 [B][P][Q]$$

$$[EQ] = k_2 k_8 (k_4 + k_5)[Q] + k_1 k_3 k_5 [A][B] + k_3 k_5 k_8 [B][Q]$$

A double-reciprocal plot of $1/v$ vs. $1/[A]$ at varying constant concentrations of B will be a series of straight lines having a common intersection point. A double-reciprocal plot of $1/v$ vs. $1/[B]$ at varying constant concentrations of A will also be a series of straight lines having a common intersection point. Secondary replots of the initial-rate data will yield values for the kinetic parameters. The overall steady-state rate expression is

$v =$

$$\cfrac{V_{max,f} V_{max,r}([A][B] - [P][Q]/K_{eq})}{\left\{ \begin{array}{c} V_{max,r} K_{ia} K_b + V_{max,r} K_b[A] + V_{max,r} K_a[B] + V_{max,r}[A][B] \\ + \dfrac{V_{max,f} K_q[P]}{K_{eq}} + \dfrac{V_{max,f} K_p[Q]}{K_{eq}} + \dfrac{V_{max,f}[P][Q]}{K_{eq}} \\ + \dfrac{V_{max,f} K_q[A][P]}{K_{ia} K_{eq}} + \dfrac{V_{max,r} K_a[B][Q]}{K_{iq}} \\ + \dfrac{V_{max,r}[A][B][P]}{K_{ip}} + \dfrac{V_{max,f}[B][P][Q]}{K_{ip} K_{eq}} \end{array} \right\}}$$

where

$$V_{max,f} = k_5 k_7 [E_{total}]/(k_5 + k_7)$$

$$V_{max,r} = k_2 k_4 [E_{total}]/(k_2 + k_4)$$

$$K_{ia} = k_2/k_1$$

$$K_{ib} = (k_2 + k_4)/k_3$$

$$K_{ip} = (k_5 + k_7)/k_6$$

$$K_{iq} = k_7/k_8$$

$$K_a = k_5 k_7/\{k_1(k_5 + k_7)\}$$

$$K_b = k_7(k_4 + k_5)/\{k_3(k_5 + k_7)\}$$

$$K_p = k_2(k_4 + k_5)/\{k_6(k_2 + k_4)\}$$

$$K_q = k_2 k_4/\{k_8(k_2 + k_4)\}$$

$$K_{eq} = k_1 k_3 k_5 k_7/\{k_2 k_4 k_6 k_8\} = V_{max,f} K_p K_{iq}/\{V_{max,r} K_{ia} K_b\}$$

See also Random Bi Bi Mechanism; Ordered Bi Bi Theorell-Chance Mechanism; Multisubstrate Mechanisms

[1]C. C. Wratten & W. W. Cleland (1963) Biochemistry **2**, 935.
[2]M. Marshall & P. P. Cohen (1966) J. Biol. Chem. **241**, 4197.
[3]M. T. Hakala, A. J. Glaid & G. W. Schuert (1956) J. Biol. Chem. **221**, 191.
[4]V. Zewe & H. J. Fromm (1965) Biochemistry **4**, 782.
[5]S. R. Anderson, J. R. Florini & C. S. Vestling (1964) J. Biol. Chem. **239**, 2991.
[6]R. C. Nordlie & H. J. Fromm (1959) J. Biol. Chem. **234**, 2523.

ORDERED BI BI THEORELL-CHANCE MECHANISM

A sequential enzyme-catalyzed reaction mechanism in which two substrates react to form two products, in which the substrates bind and products are released in an ordered fashion and in which the ternary complex intermediate is not kinetically relevant under the reaction conditions:

$$E \underset{k_2}{\overset{k_1[A]}{\rightleftharpoons}} EA \underset{k_4[P]}{\overset{k_3[B]}{\rightleftharpoons}} EQ \underset{k_6[Q]}{\overset{k_5}{\rightleftharpoons}} E$$

The steady-state initial rate expression, in the absence of products and abortive complexes, is:

$$v = V_{max}[A][B]/\{K_{ia} K_b + K_b[A] + K_a[B] + [A][B]\}$$

where $V_{max} = k_5[E_{total}]$, $K_{ia} = k_2/k_1$, $K_a = k_5/k_1$, and $K_b = k_5/k_3$. The complete steady-state determinants for the individual enzyme forms are: $[E] = k_2 k_5 + k_3 k_5[B] +$

$k_2k_4[P]$; $[EA] = k_1k_5[A] + k_1k_4[A][P] + k_4k_6[P][Q]$; and $[EQ] = k_2k_6[Q] + k_1k_3[A][B] + k_3k_6[B][Q]$. A double-reciprocal plot of $1/v$ vs. $1/[A]$ at varying constant concentrations of $[B]$ will produce a series of straight lines having a common intersection point. Similarly, a double-reciprocal plot of $1/v$ vs. $1/[B]$ at varying constant concentrations of A will also yield a series of straight lines having a common intersection point. Replots of the slopes and intercepts as a function of the reciprocal of the concentration of the other substrate will yield values for the kinetic parameters.

ORDERED BI TER MECHANISM

An enzyme reaction scheme in which there are two substrates (A and B) and three products (P, Q, and R) and in which the substrates bind and the products are released in an ordered fashion. This reaction scheme is exemplified by the malic enzyme[1]. The initial rate expression, in the absence of abortive complexes and products, is identical to the corresponding equation for the ordered Bi Bi mechanism. **See** Multisubstrate Mechanisms; Ordered Bi Bi Mechanism

[1] R. Y. Hsu, H. A. Lardy & W. W. Cleland (1967) J. Biol. Chem. **242**, 5315.

ORDERED BI UNI MECHANISM

A sequential enzyme-catalyzed reaction scheme in which two substrates (A and B) react and form a single product and in which the substrates bind to the enzyme in a distinct order (*i.e.*, only A and the product P can bind to the free enzyme). The reverse scheme of this mechanism is the ordered Uni Bi system. (**See also** Ordered Uni Bi Mechanism)

The Rapid Equilibrium Mechanism: The following scheme depicts this mechanism:

$$E + A \underset{}{\overset{K_{ia}}{\rightleftharpoons}} EA$$

$$EA + B \underset{}{\overset{K_{ib}}{\rightleftharpoons}} EAB \underset{k_6}{\overset{k_5}{\rightleftharpoons}} EP \overset{K_{ip}}{\rightleftharpoons} E + P$$

In this rapid equilibrium mechanism, with all binding steps expressed with dissociation constants, then under initial rate conditions (*i.e.*, $[P] \cong 0$) the initial rate expres-

sion for this scheme (in the absence of abortive complexes and isomerization steps) is

$$v = V_{max}[A][B]/\{K_{ia}K_{ib} + K_{ib}[A] + [A][B]\}$$

where $V_{max} = k_5[E_{total}]$, $K_{ia} = [E][A]/[EA]$, and $K_{ib} = [EA][B]/[EAB]$. A double-reciprocal plot of $1/v$ vs. $1/[A]$ at different constant concentrations of B will result in a series of straight lines. A replot of the slopes vs. $1/[B]$ yields a line having a slope of $K_{ia}K_{ib}/V_{max}$. A replot of the vertical intercepts vs. $1/[B]$ will yield a horizontal intercept of $-1/K_{ib}$. Correspondingly, a double-reciprocal plot of $1/v$ vs. $1/[B]$ at different constant concentrations of A will produce a series of straight lines having a common intersection point on the vertical axis (at $1/v = 1/V_{max}$).

The Steady-State Mechanism: If we consider the ordered Bi Uni mechanism with only one central complex,

$$E + A \underset{k_2}{\overset{k_1}{\rightleftharpoons}} EA$$

$$EA + B \underset{k_4}{\overset{k_3}{\rightleftharpoons}} EXY \underset{k_6}{\overset{k_5}{\rightleftharpoons}} E + P$$

then the steady-state enzyme determinants are: $[E] = k_2(k_4 + k_5) + k_3k_5[B]$; $[EA] = k_1(k_4 + k_5)[A] + k_4k_6[P]$; and $[EXY] = k_1k_3[A][B] + k_3k_6[B][P] + k_2k_6[P]$.

The initial rate expression (in which $[P] = 0$) is

$$v = V_{max}[A][B]/\{K_{ia}K_b + K_b[A] + K_a[B] + [A][B]\}$$

where $V_{max} = k_5[E_{total}]$, $K_{ia} = k_2/k_1$, $K_b = (k_4 + k_5)/k_3$, and $K_a = k_5/k_1$. A double-reciprocal plot of $1/v$ vs. $1/[A]$ at different constant concentrations of B yields a set of straight lines. Correspondingly, a double-reciprocal plot of $1/v$ vs. $1/[B]$ at different constant concentrations of A will be a series of straight lines. Secondary replots of the initial rate data will yield values for the kinetic parameters.

ORDERED BI UNI UNI BI PING PONG MECHANISM

Melilotate hydrolase[1] and threonyl-tRNA synthetase[2] have been reported to use this reaction scheme. The reaction scheme for this three-substrate (A, B, and C), three-product (P, Q, and R) mechanism can be depicted as

$$E + A \underset{k_2}{\overset{k_1}{\rightleftharpoons}} EA$$

$$EA + B \underset{k_4}{\overset{k_3}{\rightleftharpoons}} EXY \underset{k_6}{\overset{k_5}{\rightleftharpoons}} F + P$$

$$F + C \underset{k_8}{\overset{k_7}{\rightleftharpoons}} FZ \underset{k_{10}}{\overset{k_9}{\rightleftharpoons}} ER + Q$$

$$ER \underset{k_{12}}{\overset{k_{11}}{\rightleftharpoons}} E + R$$

The initial rate expression for this scheme, in the absence of products or abortive complexes, is

$$v = \frac{V_{max}[A][B][C]}{\{K_{ia}K_b[C] + K_c[A][B] + K_b[A][C] + K_a[B][C] + [A][B][C]\}}$$

where $V_{max} = k_5 k_9 k_{11}[E_{total}]/\{k_5 k_9 + k_5 k_{11} + k_9 k_{11}\}$, $K_{ia} = k_2/k_1$, $K_a = k_5 k_9 k_{11}/\{k_1(k_5 k_9 + k_5 k_{11} + k_9 k_{11})\}$, $K_b = k_9 k_{11}(k_4 + k_5)/\{k_3(k_5 k_9 + k_5 k_{11} + k_9 k_{11})\}$, and $K_c = k_5 k_{11}(k_8 + k_9)/\{k_7(k_5 k_9 + k_5 k_{11} + k_9 k_{11})\}$.

See *Multisubstrate Mechanisms*

[1]S. Strickland & V. Massey (1973) *J. Biol. Chem.* **248**, 2953.
[2]C. C. Allende, H. Chaimovich, M. Gatica & J. E. Allende (1970) *J. Biol. Chem.* **245**, 93.

ORDERED BI UNI UNI TER PING PONG MECHANISM

A three-substrate, four-product enzyme reaction scheme in which the first two substrates bind in an ordered fashion followed by the release of the first product. The third substrate then binds and the remaining three products are released in order (thus, E + A ⇌ EA is followed by EA + B ⇌ (EAB ⇌ FP) ⇌ F + P and then F + C ⇌ (FC ⇌ EQRS) ⇌ ERS + Q followed by ERS ⇌ ES + R and ES ⇌ E + S). There are indications that asparagine synthetase from certain sources may have this reaction scheme.

ORDERED ON-RANDOM OFF BI BI DUAL-THEORELL-CHANCE MECHANISM

A sequential enzyme-catalyzed binding mechanism for a two substrate-two product system in which substrates A and B have to bind in a certain order but either P or Q can be released in a Theorell-Chance step upon the binding of B. Following this step, the other substrate is released (thus, E + A ⇌ EA followed by either EA + B ⇌ EQ + P or EA + B ⇌ EP + Q; then EQ ⇌ E + Q and EP ⇌ E + P).

ORDERED ON-RANDOM OFF BI BI MECHANISM

An enzyme reaction mechanism involving A binding before B and followed with the random release of products. In the absence of products and abortive complexes, the steady-state rate expression is identical to the rate expression for the ordered Bi Bi mechanism[1,2]. A random on-ordered off Bi Bi mechanism has been proposed for a mutant form of alcohol dehydrogenase[3].

[1]C. Y. Huang (1979) *Meth. Enzymol.* **63**, 54.
[2]S. Cha (1968) *J. Biol. Chem.* **243**, 820.
[3]A. J. Ganzhorn & B. V. Plapp (1988) *J. Biol. Chem.* **263**, 5446.

ORDERED TER BI MECHANISM

A three-substrate (A, B, and C), two-product (P and Q) enzyme reaction scheme in which all substrates and products bind and are released in an ordered fashion. Glyceraldehyde-3-phosphate dehydrogenase[1] has been reported to have this reaction scheme. The steady-state and rapid equilibrium expressions, in the absence of products and abortive complexes, are identical to the ordered Ter Ter mechanism. **See** *Ordered Ter Ter Mechanism*

[1]B. A. Orsi & W. W. Cleland (1972) *Biochemistry* **11**, 102.

ORDERED TER BI THEORELL-CHANCE MECHANISM

A three-substrate, two-product enzyme reaction scheme in which the substrates bind and products are released in an ordered pattern and the third substrate on and first product off participate in a Theorell-Chance step. Thus, E + A ⇌ EA is followed by EA + B ⇌ EAB and EAB + C ⇌ EQ + P after which comes EQ ⇌ E + Q.

ORDERED TER TER MECHANISM

A three-substrate, three-product enzyme reaction scheme in which all substrates bind and all products are released in an ordered sequence. Glutamate dehydrogenase[1] at a pH 8.8 has been reported to have this reaction scheme.

The Rapid Equilibrium Reaction: If the rate-determining step for the overall reaction is the interconversion step of the central complexes, then all of the binding steps can be described in terms of dissociation constants:

$$E + A \overset{K_{ia}}{\rightleftharpoons} EA$$

$$EA + B \overset{K_b}{\rightleftharpoons} EAB$$

$$EAB + C \overset{K_c}{\rightleftharpoons} EABC \underset{k_8}{\overset{k_7}{\rightleftharpoons}} EPQR \overset{K_p}{\rightleftharpoons} EQR + P$$

$$EQR \overset{K_q}{\rightleftharpoons} ER + Q$$

$$ER \overset{K_{ir}}{\rightleftharpoons} E + R$$

The rate expression, in the absence of products or abortive complexes, is:

$$v = \frac{V_{max}[A][B][C]}{\{K_{ia}K_bK_c + K_bK_c[A] + K_c[A][B] + [A][B][C]\}}$$

where $V_{max} = k_7[E_{total}]$, $K_{ia} = [E][A]/[EA]$, $K_b = [EA][B]/[EAB]$, and $K_c = [EAB][C]/[EABC]$.

The Steady-State Mechanism: If substrate binding and product release are not rapid, then a steady-state description of the reaction is applicable. If one considers the reaction scheme with a single central complex (that is, EXYZ), the reaction can be depicted as

$$E \underset{k_2}{\overset{k_1[A]}{\rightleftharpoons}} EA \underset{k_4}{\overset{k_3[B]}{\rightleftharpoons}} EAB \underset{k_6}{\overset{k_5[C]}{\rightleftharpoons}} EXYZ \underset{k_8[P]}{\overset{k_7}{\rightleftharpoons}}$$

$$EQR \underset{k_{10}[Q]}{\overset{k_9}{\rightleftharpoons}} ER \underset{k_{12}[R]}{\overset{k_{11}}{\rightleftharpoons}} E$$

The steady-state expression for this reaction scheme, in the absence of any products and abortive complexes, is

$$v = \frac{V_{max}[A][B][C]}{\{K_{ia}K_{ib}K_c + K_{ib}K_c[A] + K_{ia}K_b[C] + K_c[A][B] + K_b[A][C] + K_a[B][C] + [A][B][C]\}}$$

where

$$V_{max} = k_7k_9k_{11}[E_{total}]/\{k_7k_9 + k_7k_{11} + k_9k_{11}\}$$

$$K_{ia} = k_2/k_1$$

$$K_{ib} = k_4/k_3$$

$$K_a = k_7k_9k_{11}/\{k_1(k_7k_9 + k_7k_{11} + k_9k_{11})\}$$

$$K_b = k_7k_9k_{11}/\{k_3(k_7k_9 + k_7k_{11} + k_9k_{11})\}$$

$$K_c = k_9k_{11}(k_6 + k_7)/\{k_5(k_7k_9 + k_7k_{11} + k_9k_{11})\}$$

[1]E. Silverstein (1974) *Biochemistry* **13**, 3750.

ORDERED UNI BI MECHANISM

An enzyme reaction involving the conversion of a single substrate (A) to two products (P and Q) which are released in a preferred order [1,2].

The Rapid Equilibrium Reaction: If the interconversion step between the central complexes is rate-determining (*i.e.*, rapid equilibrium), then for the reaction $E + A \underset{k_2}{\overset{k_1}{\rightleftharpoons}} EA \overset{k_3}{\rightarrow}$ products, in the absence of products and abortive complexes, the rate expression is identical to the rapid equilibrium Michaelis-Menten equation,

$$v = V_{max}[A]/(K_{ia} + [A])$$

where $K_{ia} = k_2/k_1 = [E][A]/[EA]$ and $V_{max} = k_3[E_{total}]$.

The Steady-State Reaction: If central complex interconversion is not the sole rate-determining step (or, some other step is rate-limiting) then,

$$E \underset{k_2}{\overset{k_1[A]}{\rightleftharpoons}} EX \underset{k_4[P]}{\overset{k_3}{\rightleftharpoons}} EQ \underset{k_6[Q]}{\overset{k_5}{\rightleftharpoons}} E$$

The steady-state determinants are: $[E] = k_5(k_2 + k_3) + k_2k_4[P]$; $[EA + EPQ] = [EX] = k_1k_5[A] + k_1k_4[A][P] + k_4k_6[P][Q]$; and $[EQ] = k_6(k_2 + k_3)[Q] + k_1k_3[A]$. In the absence of products P and Q, with no abortive complexes and without isomerization steps,

$$v = V_{max,f}[A]/(K_a + [A])$$

where $V_{max,f} = k_3k_5[E_{total}]/(k_3 + k_5)$ and the Michaelis constant for $A = K_a = k_5(k_2 + k_3)/\{k_1(k_3 + k_5)\}$. The complete rate expression[3,4] is

$$v =$$

$$\frac{V_{max,f}V_{max,r}([A] - [P][Q]/K_{eq})}{\left\{ V_{max,r}K_a + V_{max,r}[A] + \dfrac{V_{max,f}K_q[P]}{K_{eq}} + \dfrac{V_{max,f}K_p[Q]}{K_{eq}} + \dfrac{V_{max,f}[P][Q]}{K_{eq}} + \dfrac{V_{max,r}[A][P]}{K_{ip}} \right\}}$$

where $V_{max,r} = k_2[E_{total}]$, $K_p = (k_2 + k_3)/k_4$, $K_q = k_2/k_6$, $K_{ip} = (k_3 + k_5)/k_4$, $K_{ia} = k_2/k_1$, $K_{iq} = k_5/k_6$, and $K_{eq} = k_1k_3k_5/\{k_2k_4k_6\} = V_{max,f}K_{ip}K_q/\{V_{max,r}K_{ia}\} = V_{max,f}K_pK_{iq}/\{V_{max,r}K_a\}$.

Adenylosuccinate lyase[5], malyl coenzyme A lyase[6], and NADase[7] are reported to have this mechanism. It should also be recognized that the termed "ordered Uni Bi" is often used to refer to enzyme-catalyzed reactions which are actually "ping pong Bi Bi" mechanisms in which water is the second substrate. *See* Multisubstrate Mechanisms

[1] W. W. Cleland (1963) *Biochem. Biophys. Acta* **67**, 104.
[2] H. J. Fromm (1979) *Meth. Enzymol.* **63**, 42.
[3] I. H. Segel (1975) *Enzyme Kinetics* , Wiley, New York.
[4] H. J. Fromm (1975) *Initial Rate Enzyme Kinetics*, Springer-Verlag, New York.
[5] W. A. Bridger & L.H. Cohen (1968) *J. Biol. Chem.* **243**, 644.
[6] L. B. Hersh (1974) *J. Biol. Chem.* **249**, 5208.
[7] L. J. Zatman, N. O. Kaplan & S. P. Colowick (1953) *J. Biol. Chem.* **200**, 197.

ORDERED UNI UNI BI BI PING PONG MECHANISM

A three-substrate, three-product enzyme reaction scheme in which the first substrate binds and the first product is release before the ordered binding of the second and third substrates:

$$E + A \underset{k_2}{\overset{k_1}{\rightleftharpoons}} (EA \rightleftharpoons FP) \underset{k_4}{\overset{k_3}{\rightleftharpoons}} F + P$$

$$F + B \underset{k_6}{\overset{k_5}{\rightleftharpoons}} FB$$

$$FB + C \underset{k_8}{\overset{k_7}{\rightleftharpoons}} (FBC \rightleftharpoons EQR) \underset{k_{10}}{\overset{k_9}{\rightleftharpoons}} ER + Q$$

$$ER \underset{k_{12}}{\overset{k_{11}}{\rightleftharpoons}} E + R$$

The steady-state rate expression, in the absence of products, abortive complexes, and any isomerization steps is

$$v = \frac{V_{max}[A][B][C]}{\{K_cK_{ib} + K_c[A][B] + K_b[A][C] + K_a[B][C] + [A][B][C]\}}$$

where V_{max} is the maximum forward velocity equal to $k_3k_9k_{11}[E_{total}]/\{k_3k_9 + k_3k_{11} + k_9k_{11}\}$, K_{ib} is equal to k_6/k_5, $K_a = k_9k_{11}(k_2 + k_3)/\{k_1(k_3k_9 + k_3k_{11} + k_9k_{11})\}$, $K_b = k_3k_9k_{11}/\{k_5(k_3k_9 + k_3k_{11} + k_9k_{11})\}$, and $K_c = k_3k_{11}(k_8 + k_9)/\{k_7(k_3k_9 + k_3k_{11} + k_9k_{11})\}$.

ORDERED UNI UNI, UNI BI PING PONG MECHANISM

A two-substrate, three-product enzyme reaction scheme in which the first substrate (A) binds to the enzyme, producing the first product (P) in a ping pong half-reaction. The second substrate (B) and the remaining two products then follow an ordered Uni Bi scheme.

ORDER OF REACTION

The overall order of a chemical reaction is equal to the sum of the exponents of all the concentration terms in the differential rate expression for a reaction considered in one direction only. For example, if the empirical rate law for a particular chemical reaction was

$$v = k[A]^{\alpha}[B]^{\beta}[C]^{\gamma}$$

then the overall order of the reaction would be $\alpha + \beta + \gamma$. In addition to the overall order, there are also partial reaction orders with respect to individual reactants. For the rate expression above, the reaction is of partial order α with respect to reactant A, partial order β with respect to B, and partial order γ with respect to C. If the concentration of one or more of the reactants is held constant, or nearly constant, over the time frame of experiment, then the reaction will be pseudo-nth order in which n is the sum of the exponents of the concentration terms in the rate expression that are not held constant. This is precisely the case with an enzyme, since the catalyst concentration remains constant.

With chemical reactions, the exponents in a rate expression are usually integers. However, the exponents can be fractions or even negative depending on the complexity of the reaction. Reaction order should not be confused with molecularity. Order is an empirical concept whereas molecularity refers to the actual molecular process. However, for elementary reactions, the reaction order equals the molecularity. *See* Chemical Kinetics; Molecularity; First-Order Reactions; Rate Constants

D-ORNITHINE 4,5-AMINOMUTASE

This enzyme [EC 5.4.3.5] catalyzes the interconversion of D-ornithine and D-*threo*-2,4-diaminopentanoate. The enzyme requires cobalamin, dithiothreitol, and pyridoxal phosphate.

P. A. Frey & G. H. Reed (1993) *Adv. Enzymol.* **66**, 1.
T. C. Stadtman (1972) *The Enzymes*, 3rd ed., **6**, 539.

ORNITHINE AMINOTRANSFERASE

This pyridoxal-phosphate-dependent enzyme [EC 2.6.1.13] (also known as ornithine:oxo-acid aminotransferase and ornithine keto-acid aminotransferase) catalyzes the reversible reaction of L-ornithine with an α-keto acid (or, a 2-oxo acid) to produce L-glutamate 5-semialdehyde and an L-amino acid.

H. Hayashi, H. Wada, T. Yoshimura, N. Esaki & K. Soda (1990) *Ann. Rev. Biochem.* **59**, 87.
A. J. L. Cooper (1985) *Meth. Enzymol.* **113**, 76.
E. Adams & L. Frank (1980) *Ann. Rev. Biochem.* **49**, 1005.
A. E. Braunstein (1973) *The Enzymes*, 3rd ed., **9**, 379.

ORNITHINE DECARBOXYLASE

This pyridoxal-phosphate-dependent enzyme [EC 4.1.1.17] catalyzes the conversion of L-ornithine to putrescine and carbon dioxide.

R. A. John (1998) *Comprehensive Biological Catalysis: A Mechanistic Reference* **2**, 173.
S.-I. Hayashi (1995) *Essays in Biochem.* **30**, 37.
T. Hashimoto & Y. Yamada (1994) *Ann. Rev. Plant Physiol. Plant Mol. Biol.* **45**, 257.
E. A. Boeker & E. E. Snell (1972) *The Enzymes*, 3rd ed., **6**, 217.

ORNITHINE TRANSCARBAMOYLASE

This enzyme [EC 2.1.3.3], also known as ornithine carbamoyltransferase and citrulline phosphorylase, catalyzes the reaction of carbamoyl phosphate with L-ornithine to produce L-citrulline and orthophosphate.

P. P. Cohen (1981) *Curr. Top. Cell. Reg.* **18**, 1.
H. J. Vogel & R. H. Vogel (1974) *Adv. Enzymol.* **40**, 65.
S. Ratner (1973) *Adv. Enzymol.* **39**, 1.

OROTATE PHOSPHORIBOSYL-TRANSFERASE

This enzyme [EC 2.4.2.10], also known as orotidylic acid phosphorylase and orotidine-5'-phosphate pyrophosphorylase, catalyzes the reaction of orotate with 5-phospho-α-D-ribose 1-diphosphate to produce orotidine 5'-phosphate and pyrophosphate (or, diphosphate).

W. D. L. Musick (1981) *Crit. Rev. Biochem.* **11**, 1.
M. E. Jones (1980) *Ann. Rev. Biochem.* **49**, 253.

OROTIDYLATE DECARBOXYLASE

This enzyme [EC 4.1.1.23], also known as orotidine-5'-phosphate decarboxylase, catalyzes the conversion of orotidine 5'-phosphate to UMP and carbon dioxide.

M. E. Jones (1992) *Curr. Top. Cell. Reg.* **33**, 331.

ORTHOPHOSPHATE CONTINUOUS ASSAY

Because ATP and GTP hydrolysis play such central roles in biochemical processes, kinetic assays of orthophos-phate (or, P_i) formation are widely utilized in mechanistic investigations. Traditionally, this has involved stopped-time assays in which samples must be removed from a reaction mixture and quenched prior to colorimetric determination of phosphate content. The discovery of a phosphate-binding protein that becomes fluorescent upon phosphate binding now permits continuous fluorimetric assay[1] in the same way that one measures NADH or NADPH production. The technique utilizes the A197C mutant of *Escherichia coli* phosphate binding protein containing the fluorophore *N*-[2-(1-maleimidyl) ethyl]-7-(diethylamino)coumarin-3-carboxamide attached to the introduced thiol[1]. The modified protein has an excitation maximum at 425 nm and emission maximum at 474 nm in the absence of P_i, which shifts to 464 nm with a 5.2-times enhancement of fluorescence upon complexation with P_i. The binding reaction occurs at pH 7.0, low ionic strength, 22°C, and similar spectral properties are observed at pH 8 or at ionic strength up to 1 M. The binding of orthophosphate is characterized by a K_d of about 0.1 μM, a bimolecular rate constant of 1.36×10^8 $M^{-1}s^{-1}$, and a dissociation rate constant of 21 s^{-1}. ATP only weakly inhibits P_i-induced fluorescence enhancement. In view of the presence of phosphate in many biological samples, Brune *et al.*[1] depleted endogenous orthophosphate to concentrations below 0.1 μM using 7-methylguanosine and purine nucleoside phosphorylase, with conversion of free P_i to ribose 1-phosphate. Using this probe to measure single turnover P_i release rates for actomyosin subfragment 1 hydrolysis of ATP, these investigators observed a small lag prior to rapid P_i release at both high and low [ATP].

White *et al.*[2] measured the rates of phosphate release during a single turnover of actomyosin nucleoside triphosphate (NTP) hydrolysis using a double-mixing stopped-flow spectrofluorometer, at very low ionic strength to increase the affinity of myosin-ATP and myosin-ADP-P_i to actin. Myosin subfragment 1 and a series of nucleoside triphosphates were mixed and incubated for approximately 1–10 s to allow NTP to bind to myosin and generate a steady-state mixture of myosin-NTP and myosin-NDP-P_i. The steady-state intermediates were then mixed with actin.

[1]M. Brune, J. L. Hunter, J. E. Corrie & M. R. Webb (1994) *Biochemistry* **33**, 8262.
[2]H. D. White, B. Belknap & M. R. Webb (1997) *Biochemistry* **36**, 11828.

OSTWALD RIPENING

A process by which a two-phase crystalline suspension (experiencing a positive interfacial tension) proceeds to decrease in free energy by minimizing the interfacial surface area at constant pressure and temperature[1]. This process of "ripening" reaches equilibrium only when the surface area reaches a minimum. As a result, a population of small crystals or a noncrystalline suspension will spontaneously dissolve, allowing its component subunits to add to other larger crystals. Some of these larger crystals then dissolve away and their subunits add to still fewer crystals, and the process ultimately culminates in the survival of a single crystal. *See* Biomineralization

[1]A. E. Neilsen (1964) *Kinetics of Precipitation*, pp. 108-117, Pergamon Press, Oxford.

OSTWALD SOLUBILITY COEFFICIENT

The volume in milliliters of gas dissolved per milliliter of liquid at one atmosphere of partial pressure of the gas at any given temperature. Thus, the Ostwald coefficient (Λ) differs from the Bunsen solubility coefficient (α) which is based on standard temperature and pressure. The two coefficients are related by the equation $\Lambda = \alpha(1 + 0.00367t)$ where t is the temperature in degrees Celsius.

OUTER SPHERE ELECTRON TRANSFER REACTION

A well-established electron transfer mechanism in inorganic chemistry, typically observed when two reacting species take longer to exchange ligands than undergo electron transfer. Each complex maintains its full complement of bound ligands during electron transfer, and the transferred electron must traverse both coordination shells. *See* Metal Ion Coordination (Basic Chemical Behavior)

OVERSATURATION

A phenomenon associated with noncompetitive product inhibition, wherein the half-times for approach to equilibrium divided by the initial substrate concentration are observed to increase with increasing substrate concentrations. As pointed out by Cleland[1], this would not be the case for such time courses (normalized with respect to substrate concentration) if the product inhibition were competitive. In the case of proline racemase[2], the observation of oversaturation suggests that the enzyme oper-

ates by an Iso mechanism (*i.e.*, the form of free enzyme reacting with the L-isomer is different from the enzyme form reacting with the D-isomer. Another case of oversaturation was described by Silverman *et al.*[3] for carbonic anhydrase.

[1]W. W. Cleland (1990) *The Enzymes* **19**, 99.
[2]L. M. Fisher, W. J. Albery & J. R. Knowles (1986) *Biochemistry* **25**, 2529.
[3]D. N. Silverman & S. H. Vincent (1983) *Crit. Rev. Biochem.* **14**, 207.

OXALOACETATE DECARBOXYLASE

This biotinylated enzyme [EC 4.1.1.3] catalyzes the decarboxylation of oxaloacetate to pyruvate and carbon dioxide.

M. H. O'Leary (1992) *The Enzymes*, 3rd ed., **20**, 235.
D. J. Creighton & N. S. R. K. Murthy (1990) *The Enzymes*, 3rd ed., **19**, 323.

OXALYL-CoA DECARBOXYLASE

This thiamin pyrophosphate-dependent enzyme [EC 4.1.1.8] catalyzes the conversion of oxalyl-CoA to formyl-CoA and carbon dioxide.

J. R. Quayle (1969) *Meth. Enzymol.* **13**, 369.

OXIDOREDUCTASES

A major class of enzymes that catalyze oxidation-reduction reactions. This class includes dehydrogenases, reductases, oxygenases, peroxidases, and a few synthases. Examples include alcohol dehydrogenase (EC 1.1.1.1), aldehyde oxidase (EC 1.2.3.1), orotate reductase (EC 1.3.1.14), glutamate synthase (EC 1.4.1.14), NAD(P)$^+$ transhydrogenase (EC 1.6.1.1), and glutathione peroxidase (EC 1.11.1.9).

5-OXOPROLINASE

This enzyme [EC 3.5.2.9], also known as pyroglutamase (ATP-hydrolyzing), catalyzes the reaction of ATP with 5-oxo-L-proline and two water to produce L-glutamate, ADP, and orthophosphate.

A. Meister, O. W. Griffith & J. M. Williamson (1985) *Meth. Enzymol.* **113**, 445.
A. P. Seddon, L. Li & A. Meister (1985) *Meth. Enzymol.* **113**, 451.
P. Van Der Werf & A. Meister (1975) *Adv. Enzymol.* **43**, 519.

OXYGEN ELECTRODE

An elecrochemical device for detecting dissolved oxygen content. The well known Clark electrode[1] consists of a platinum wire tip surrounded by a thin film of electrolyte solution that is shrouded by a plastic membrane. The membrane is permeable to oxygen, but impermeable to water and other solutes. When an appropriate potential is applied, oxygen is electrochemically destroyed on the metal's surface, and the flow of electrons from the tip to the infusing oxygen can be used to measure dissolved oxygen *in vitro* as well as in physiologic fluids (such as plasma). Although manometry was first applied to investigate mitochondrial oxidation, the O_2 electrode offers the advantage of rapid detection and continuous recording of changes in oxygen concentration[2]. Briefly, when ADP is added to a suspension of tightly coupled mitochondria, one can measure the P/O ratio by measuring ATP formation and molecular oxygen uptake.

If one assumes that the reaction system is at thermodynamic equilibrium, the oxygen electrode method also allows one to infer the partial pressure of oxygen in equilibrium with the test solution.

[1]L. C. Clark (1962) *Ann. N. Y. Acad. Sci.* **102**, 29.
[2]R. W. Estabrook (1967) *Meth. Enzymol.* **10**, 41.

OXYGEN ELECTRODE FOR MONITORING HEMOGLOBIN OXYGENATION

Polarography[1] can be used to measure the hemoglobin O_2 saturation in whole blood, employing up to 10 μL sample in an anaerobic stainless-steel cuvette (1 mL). Three oxygen tension values are determined: (a) that of an air-equilibrated buffer before the addition of the sample; (b) that after the addition of the sample; and (c) that after the addition of an oxidant. The oxygen saturation is then calculated from the three oxygen tension values, the volume of the reagents, and the solubility coefficient of oxygen. This simple, inexpensive, and accurate method correlates well with other standard methods.

[1]M. Samaja & E. Rovida (1983) *J. Biochem. Biophys. Methods* **7**, 143.

OXYGEN, OXIDES, & OXO ANIONS

DIOXYGEN or O_2. The most common allotrope of this element, dioxygen is a colorless gas (mp = $-218.4°C$ and bp = $-182.692°C$) that reacts with all elements, except halogens, noble gases, and a few metals. At saturation, O_2 is about 1.4 mM in water, and is readily soluble in organic solvents. Indeed, the act of pouring an organic solvent in the presence of air will often result in saturating the solvent with O_2. This observation regarding or-

ganic solvents should always be a consideration in handling air-sensitive materials.

Elemental oxygen ($Z = 8$ and $A_r = 15.9994$) occurs as 13 isotopes, but only four are of particular importance. Oxygen-16 (^{16}O) is the most common of the oxygen isotopes, representing 99.76% of all naturally occurring oxygen. Oxygen-17 (^{17}O) is a stable isotope representing only 0.04% of naturally occurring oxygen. With a nuclear spin of 5/2, ^{17}O is utilized in resonance studies. The remaining 0.20% of natural oxygen is oxygen-18 (^{18}O), a stable isotope used extensively in mass spectroscopy studies and in NMR experiments on enzymic systems. Oxygen-15 is the longest half-life (122.2 seconds; positron-emitting) radioisotope. ^{15}O is utilized in respiration studies, with detection achieved by positron emission tomography.

The structure of dioxygen is predicted by valence bond theory to be $O{=}O$, having one σ and one π bond, with each oxygen having two pairs of unshared electrons and a bond length of 0.1208 nm. Such a structure does not account for the observation that O_2 is paramagnetic in all phases, and molecular orbital theory correctly predicts the two unpaired electrons in dioxygen both in π^* orbitals. The singlet state of oxygen (the $^1\Delta_g$ state), in which the two π electrons are paired, is actually an excited state and is about 92 kJ·mol^{-1} above the triplet ground state (the $^3\Sigma_g^-$ state). Singlet dioxygen can react with many unsaturated compounds and can be readily generated by numerous photochemical reactions. Singlet oxygen is generated and/or utilized in many biochemical processes, particularly in the presence of light.

OZONE or O_3. The other molecular oxygen allotrope is ozone, a colorless gas (mp = $-192.7°C$ and bp = $-111.9°C$) that is also a powerful oxidant. The molecule is symmetrical and bent, having a bond angle of 116.8° and an oxygen—oxygen bond length of 0.1278 nm (another report lists 117.47° and 0.12716 nm). There appears to be considerable double-bond character present which can be accounted by resonance: $O{-}O{=}O \leftrightarrow O{=}O{-}O$. The conversion of ozone to dioxygen ($O_3 \rightleftharpoons (3/2)\ O_2$) is substantially endothermic ($\Delta H = -142$ kJ·mol^{-1}). This reaction is slow in the absence of catalysts and of light. One centrally important biological role of ozone is its absorption of ultraviolet light. Approximately 27% of this planet's atmosphere between the altitudes of 15 and 25 km (or, about 9 to 16 miles) is ozone, and this layer provides a UV-protective shield for many living organisms.

SUPEROXIDE or $O_2^{\cdot-}$. This paramagnetic radical anion is generated in a number of biochemical processes. For example, $O_2^{\cdot-}$ is the product of the reaction catalyzed by xanthine oxidase. Superoxide is a strong oxidizing agent and can have significant effects on a cell. Several superoxide dismutases (EC 1.15.1.1) catalyze the dismutation of superoxide: $2 O_2^{\cdot-} + 2 H^+ \rightarrow O_2 + H_2O_2$. In aqueous solutions, $O_2^{\cdot-}$ is highly hydrated. The oxygen–oxygen bond length is slightly longer (0.133 nm) in superoxide than in dioxygen. Stable ionic superoxides are formed by certain of the alkali metals. Potassium superoxide (KO_2) is the most common of these. It is a yellow crystal (mp = 380°C) that is soluble (and decomposes) in water. Transient levels of superoxide in aqueous solution can be generated a number of ways.

OXIDES. These are compounds consisting of oxygen and another element or radical. Nonmetal oxides are usually either acidic or neutral. Metal oxides are usually ionic and tend to generate a base upon reaction with water. When insoluble in water, metal oxides usually dissolve in dilute acid (e.g., $MgO_{(s)} + 2H^+ \rightarrow Mg_{(aq)}^{2+} + H_2O$). Amphoteric oxides act as a base with respect to a strong acid and as an acid with respect to a strong base. An example is zinc oxide ($ZnO + 2H_{(aq)}^+ \rightarrow Zn^{2+} + H_2O$ and $ZnO + 2OH^- + H_2O \rightarrow Zn(OH)_4^{2-}$). Such oxides can often be found with the metalloids (e.g., boron, arsenic, germanium). If a particular element forms several different oxides, then that with the element in its highest oxidation state is the most acidic oxide. Many oxides have important roles in biological processes:

(1) **Water or H_2O.** The $O-H$ bond lengths in water are 0.0958 nm (or 0.958 Å) and the bond angle is 104.45° (another report has 0.09575 nm and 104.51°). **See** Water

(2) **Hydroxide anion or OH^-.** This anion is the conjugate base of water. For electropositive elements such as sodium, potassium, and barium, discrete OH^- ions exist. For such hydroxides, the substance will act as a strong base when dissolved in water: thus, $M^+OH_{(s)}^- + nH_2O$

$\rightarrow M_{(aq)}^+ + OH_{(aq)}^-$. However, when there is a strong covalent bond between M and the oxygen, then the substance can act as an acid: that is, $MOH + nH_2O = MO_{(aq)}^- + H_3O_{(aq)}^+$. Amphoteric hydroxides will act as a base in the presence of a strong acid and as an acid in the presence of a strong base. Hydroxide ions can also form bridges between metal ions. **See** Water; pK_w; Bases

(3) **Hydronium ion or H_3O^+. See** Water; pK_w; Acids

(4) **Hydroxide radical or $HO\cdot$.** The hydroxide radical is an extremely reactive species that can damage cells. It can be generated by activated macrophages via superoxide and nitric oxide (especially peroxynitrite).

(5) **Hydrogen peroxide or HOOH.** Hydrogen peroxide is a relatively unstable compound and readily breaks down to water and dioxygen, a reaction catalyzed by powdered metals as well as the enzyme catalase ($2H_2O_2 = O_2 + 2H_2O$). The $O-O$ bond length is 0.1475 nm and the $O-H$ length is 0.097 nm. The $O-O-H$ bond angle is 96.9°; the two hydrogens do not lie in the same plane, and the dihedral angle is 93.9°. Many physical properties of hydrogen peroxide resemble those of water. There is extensive hydrogen bonding, and the density of neat hydrogen peroxide is greater than that of water. The pure, colorless liquid (melting point of −0.41°C; boiling point of 150.2°C) has a high dielectric constant that is slightly larger than water (e.g., a 65% aqueous hydrogen peroxide solution has a dielectric constant of 120). H_2O_2 is also more acidic than water when in dilute aqueous solutions ($H_2O_2 = H^+ + HO_2^-$, $K_a = 1.5 \times 10^{-12}$ at 20°C). Dilute (e.g., 30%) solutions of H_2O_2 are commonly used as oxidants, oxidation proceeding more rapidly in basic conditions than in acid solutions (thus, excess H_2O_2 can be best destroyed by heating in a basic solution). Hydrogen peroxide can be generated naturally (e.g., via amino acid oxidases, amine oxidases, glycolate oxidase, glycerol-3-phosphate oxidase, etc.) or by the autoxidation of a number of pharmaceuticals. Elevated levels of H_2O_2 can be very damaging to cells. For example, H_2O_2 (a) can oxidize iron(II) present in a number of heme proteins to Fe(III), (b) can act on double bonds of unsaturated fatty acyl groups of phospholipids of cell membranes (which can ultimately lead to rupture of the membrane), (c) can participate in the nonenzymatic oxidative decarboxylation of α-keto acids, and (d) can oxi-

dize methionyl and cysteinyl residues in proteins. Hydrogen peroxide is also acted upon by several other enzymes (EC 1.11.1.x): for example, the seleno-protein glutathione peroxidase [2 glutathione + H_2O_2 = 2 H_2O + glutathione disulfide] has an important role in regulating cellular H_2O_2 levels.

(6) Hydroperoxy radical or HOO·. This is a very reactive free radical that can be generated from hydrogen peroxide, particularly via metal ion-catalyzed decomposition. This radical can generate superoxide.

(7) Carbon monoxide or CO. Carbon monoxide is a colorless and practically odorless gas (mp = $-199°C$ and bp = $-191.5°C$) The C-to-O bond length is 0.11283 nm. The linear carbon monoxide is able to use vacant p orbitals to form end-on bonds with metal ions. Thus, CO can inhibit a wide range of metalloproteins, particularly heme proteins such as hemoglobin and cytochrome c oxidase (the fact that carbon monoxide complexed with heme proteins can dissociate by the action of light was an important tool in the early history of the characterization of heme proteins and electron transport). At least two enzymes catalyze reactions producing CO: the mammalian heme oxygenase (EC 1.14.99.3) and 3,4-dihydroxyquinoline 2,4-dioxygenase (EC 1.13.99.5). Other enzymes utilize CO as a substrate: *e.g.*, carbon monoxide oxygenase (or, cytochrome b-561) (EC 1.2.2.4), carbon monoxide oxidase (EC 1.2.3.10), and carbon monoxide dehydrogenase (EC 1.2.99.2). CO has been reported to be an alternative activator of guanylate cyclase. CO may play a role as a cellular messenger in mammals.

(8) Carbon dioxide or CO_2. Carbon dioxide is a linear molecule, O=C=O with a mean carbon-oxygen bond length of 0.116 nm. *See Carbon Dioxide*

(9) Nitric oxide or ·N=O. This oxide is a colorless gas (mp = $-163.6°C$ and bp = $-151.8°C$) that is very soluble in water. The N—O bond length is 0.11506 nm. The molecule is paramagnetic, and the electron configuration is $(\sigma_1)^2(\sigma_2)^2(\sigma_3)^2(\pi)^4(\pi^*)$. Note that the unpaired electron is in an antibonding orbital and is relatively easily lost. Hence, the nitrogen-oxygen bond in the nitrosonium ion, NO^+, is actually stronger and shorter (0.1062 nm) than the corresponding bond in nitric oxide. The bond order in nitric oxide is 2.5. In physiological solu-

tions, NO has a short half-life ($\sim$ five seconds), reacting rapidly with dissolved dioxygen to form the nitrogen dioxide radical (2·NO + O_2 = 2·NO_2). The latter, a free radical, then forms dinitrogen tetroxide which, in aqueous solutions, forms the stable anions nitrite and nitrate or will react with nitric oxide to form nitrous anhydride.

(10) Phosphate (or, Orthophosphate) or PO_4^{3-}. The conjugate base of HPO_4^{2-}, one of the oxo anions of phosphoric acid (the pK_a values of H_3PO_4 are, at 25°C, 2.15, 7.20, and 12.33). The structure of the resonance-stabilized PO_4^{3-} is tetrahedral and can form complexes with metal ions and other cations.

(11) Pyrophosphate (also known as diphosphate) or $P_2O_7^{4-}$. The conjugate base of pyrophosphoric acid (or, diphosphoric acid), $H_4P_2O_7$ (having pK_a values, at 25°C, of 0.85, 1.96, 6.60, and 9.41). Crystals of tetrasodium pyrophosphate (mp = 988°C) are very soluble in water. The pyrophosphate anion hydrolyzes to orthophosphate; however, the rate of hydrolysis is much slower than that of pyrophosphoric acid. The hydrated form of tetrasodium pyrophosphate exhibits a slight degree of efflorescence.

(12) Metaphosphate or PO_3^-. This ion, which is stable in the gas phase, is quickly converted to $H_2PO_4^-$ in aqueous solutions (note that the rate of conversion is slower at low temperatures). Metaphosphate has been postulated to be an intermediate in a number of enzyme-catalyzed reactions. *See Metaphosphate*

(13) Vanadium oxides. There are several oxides of vanadium, primarily depending on the oxidation state of this transition metal: vanadium(V) oxide (V_2O_5), vanadium(IV) oxide (VO_2), vanadium(III) oxide (V_2O_3), and vanadium(II) oxide (VO). In VO, the V—O bond length is 0.15893 nm. The vanadates consist of VO_4^{3-} (the conjugate base of $VO_3(OH)^{2-}$ which has a pK_a value of about 1.0) and a series of aggregated structures (*e.g.*, $V_{10}O_{26}^{6-}$). Sodium orthovanadate (Na_3VO_4) is commercially available. Orthovanadate is an analog of phosphate and has proved very useful in its action as an inhibitor of a wide variety of phosphate-dependent reactions. In the presence of orthovanadate, (Na^+/K^+)-ATPase will form VO_5^{5-}, which has a trigonal bipyramidal structure.

Metavanadate (VO_3^-) can form indefinite chains as well as cyclic structures. Sodium metavanadate is also commercially available. Hence, the investigator should be aware of which vanadate is being utilized in the studies.

(14) Arsenate. The conjugate base (AsO_4^{3-}) of $HAsO_4^{2-}$, an oxo anion of arsenic acid, H_3AsO_4 (pK_a values, at 18°C, of 2.25, 6.77, and 11.60). *See* Arsenolysis

Selected entries from *Methods in Enzymology* [vol, page(s)]:
Chemistry of Oxygen and Intermediate States of Its Reduction: Dioxygen, **105**, 3; oxygen radical species, **105**, 22; characterization of singlet oxygen, **105**, 36; role of iron in oxygen radical reactions, **105**, 47.

Isolation and Assays of Enzymes or Substances Resulting in Formation or Removal of Oxygen Radicals: Biological sources of O_2^-, **105**, 59; overview of superoxygenase, **105**, 61; methods for the study of superoxide chemistry in nonaqueous solutions, **105**, 71; generation of superoxide radicals in aqueous and ethanolic solutions by vacuum-UV photolysis, **105**, 81.

Isolation, Purification, Characterization, and Assay of Antioxygenic Enzymes: Isolation and characterization of superoxide dismutase, **105**, 88; superoxide dismutase assays, **105**, 93; assays of glutathione peroxidase, **105**, 114; catalase *in vitro*, **105**, 121; assays of lipoxygenase, **105**, 126.

Detection and Characterization of Oxygen Radicals: Pulse radiolysis methodology, **105**, 167; electron and hydrogen atom transfer reactions: determination of free radical redox potentials by pulse radiolysis, **105**, 179; spin trapping, **105**, 188; spin trapping of superoxide and hydroxyl radicals, **105**, 198; reaction of ·OH, **105**, 209; electron spin resonance spin destruction methods for radical detection, **105**, 215; low-level chemiluminescence as an indicator of singlet molecular oxygen in biological systems, **105**, 221; hplc-electrochemical detection of oxygen free radicals, **105**, 231; survey of the methodology for evaluating negative air ions: relevance to biological studies, **105**, 238; pulse radiolysis, **233**, 3; pulse radiolysis for investigation of nitric oxide-related reactions, **233**, 20; sulfhydryl free radical formation enzymatically by sonolysis, by radiolysis, and thermally: vitamin A, curcumin, muconic acid, and related conjugated olefins as references, **233**, 34; suppression of hydroxyl radical reactions in biological systems: considerations based on competition kinetics, **233**, 47; deoxyribose assay for detecting hydroxyl radicals, **233**, 57; detection of hydroxyl radicals by aromatic hydroxylation, **233**, 67;

measurement of iron and copper in biological systems: bleomycin and copper-phenanthroline assays, **233**, 82; oxygen radicals generation and DNA scission by anticancer and synthetic quinones, **233**, 92; spin trapping of hydroxyl radicals in biological systems, **233**, 105; *in vivo* detection of radical adducts by electron spin resonance, **233**, 112; measurement of superoxide reaction by chemiluminescence, **233**, 154.

Lipid Peroxidation: Chemistry, **105**, 273; overview of methods used for detecting lipid peroxidation, **105**, 283; chemical methods for the detection of lipid hydroperoxides, **105**, 293; comparative studies on different methods of malonaldehyde determination, **105**, 299; concentrating ethane from breath to monitor lipid peroxidation *in vivo*, **105**, 305.

Production, Detection, and Characterization of Oxygen Radicals and Other Species: Pulse radiolysis in study of oxygen radicals, **186**, 89; free radical initiators as source of water- or lipid-soluble peroxyl radicals, **186**, 100; generation of iron(IV) and iron(V) complexes in aqueous solutions, **186**, 108; sources of compilations of rate constants for oxygen radicals in solution, **186**, 113; use of polyaminocarboxylates as metal chelators, **186**, 116; preparation of metal-free solutions for studies of active oxygen species, **186**, 121; use of ascorbate as test for catalytic metals in simple buffers, **186**, 125; spin-trapping methods for detecting superoxide and hydroxyl free radicals *in vitro* and *in vivo*, **186**, 127; detection of singlet molecular oxygen during chloride peroxidase-catalyzed decomposition of ethyl hydroperoxide, **186**, 133; detection and quantitation of hydroxyl radical using dimethyl sulfoxide as molecular probe, **186**, 137; distinction between hydroxyl radical and ferryl species, **186**, 148; determination of hydroperoxides with fluorometric reagent diphenyl-1-pyrenylphosphine, **186**, 157; phycoerythrin fluorescence-based assay for reactive oxygen species, **186**, 161; sulfur-centered free radicals, **186**, 168; quinoid compounds: high-performance liquid chromatography with electrochemical detection, **186**, 180; preparation of tocopheroxyl radicals for detection by electron spin resonance, **186**, 197.

Enzymes Involved in Forming or Removing Oxygen Radicals and Derived Products: Determination of superoxide dismutase activity by purely chemical system based on NAD(P)H oxidation, **186**, 209; assays for superoxide dismutase autoxidation of hematoxylin, **186**, 220; assay for superoxide dismutase based on chemiluminescence of luciferin analog, **186**, 227; automated assay of superoxide dismutase in blood, **186**, 232; oxidative reactions of hemoglobin, **186**, 265; assays for cytochrome P-450 peroxygenase activity, **186**, 273; reductive cleavage of hydroperoxides by cytochrome P-450, **186**, 278; prostaglandin H synthase, **186**, 283; DT-diaphorase: purification, properties, and function, **186**, 287.

P

PACEMAKER REACTION

The slowest enzyme-catalyzed step in a metabolic pathway; sometimes referred to as the rate-limiting step (although that term is also used to describe a step in the reaction pathway of a single reaction). The pacemaker step may change, depending on availability of substrates and/or effectors as well as reaction conditions. For example, in red blood cells, hexokinase is thought to be the pacemaker for glycolysis, but hexokinase and phosphofructokinase both act as glycolytic pacemakers in many other cell types. **See also** Rate-Determining Step

H. A. Krebs (1957) Endeavour **16**, 125.

PACKING DENSITY

The fraction of total volume occupied by atoms in a molecule or solid. A van der Waals envelope for a molecule or segment of molecules can be readily determined from atomic models based on X-ray diffraction studies of crystals. The packing density of closest packed spheres is 0.74, that of closest packed infinite cylinders is 0.91, and that of a continuous solid is taken to be unity. The average packing density of ribonuclease S[1] is nearly 0.75, with several regions reaching 0.82–0.84. From the work of Monaco et al.[2], it is clear that the packing density at the subunit interfaces of aspartate transcarbamylase is also nearly 0.75. These values attest to the close-packed nature of polypeptide chains in globular proteins.

[1]F. M. Richards (1974) J. Mol. Biol. **82**, 1.
[2]H. L. Monaco, J. L. Crawford & W. N. Lipscomb (1978) Proc. Natl. Acad. Sci. U.S.A. **75**, 5276.

PANTOTHENIC ACID, COENZYME A, AND DERIVATIVES

Metabolic derivatives of pantothenic acid are of fundamental importance in acyl transfer reactions and in condensation reactions requiring an acidic α-proton. The discovery that coenzyme A was the so-called "acetylation coenzyme" resulted in the award of the 1953 Nobel Prize in Medicine and Physiology to Fritz Lipmann. Panthothenate is also found in acyl-carrier protein (ACP) which functions in a manner akin to CoA.

Selected entries from Methods in Enzymology [vol, page(s)]:
Gas chromatography of pantothenic acid, **18A**, 311; pantothenic acid and coenzyme A: determination of CoA by phosphotransacetylase from Escherichia coli B, **18A**, 314; preparation and assay of CoA-SS-glutathione, **18A**, 318; pantothenic acid and coenzyme A: preparation of CoA analogs, **18A**, 322; synthesis of coenzyme A analogues, **18A**, 338; pantothenic acid and coenzyme A: phosphopantothenoylcysteine synthetase from rat liver (pantothenate 4'-phosphate:L-cysteine ligase, EC 6.3.2.5), **18A**, 350; pantothenic acid and coenzyme A: phosphopantothenoylcysteine decarboxylase from rat liver [4'-phospho-N-(D-pantothenoyl)-L-cysteine carboxy-lyase], **18A**, 354; preparation of pantothenate-14C labeled pigeon liver fatty acid synthetase: release of the 4'-phosphopantetheine 14C-labeled prosthetic group, **18A**, 364; enzymatic synthesis of 14C-labeled coenzyme A, **18A**, 371; microbiological assay of pantothenic acid, **62**, 201; purification and properties of ketopantoate hydroxymethyl-transferase, **62**, 204; ketopantoyl lactone reductases, **62**, 209; pantothenate synthetase from Escherichia coli [D-pantoate:β-alanine ligase (AMP-forming)], **62**, 215; microbial synthesis of sugar derivatives of D-pantothenic acid, **62**, 220; syntheses of glycosyl derivatives of pantothenic acid and pantetheine, **62**, 227; synthesis of coenzyme A and its biosynthetic intermediates by microbial processes, **62**, 236; phosphopantothenoylcysteine decarboxylase from horse liver [4'-phospho-N-(D-pantothenoyl)-L-cysteine carboxy-lyase], **62**, 245; interconversion of apo- and holofatty acid synthetases of rat and pigeon liver, **62**, 249; purification and properties of pantetheinase from horse kidney, **62**, 262; pantothenase from Pseudomonas fluorescens, **62**, 267.

PAPAIN

This endopeptidase [EC 3.4.22.2], a member of the C1 peptidase family hydrolyzes peptide bonds in proteins, exhibiting a broad specificity for those bonds. There is a preference for an amino acyl residue bearing a large hydrophobic side chain at the P_2 position and the enzyme does not accept a valyl residue at P_1.

A. C. Storer & R. Ménard (1994) Meth. Enzymol. **244**, 486.
D. J. Buttle (1994) Meth. Enzymol. **244**, 539.
E. Shaw (1990) Adv. Enzymol. **63**, 271.
J. Drenth, J. N. Jansonius, R. Koekoek & B. G. Wolthers (1971) The Enzymes, 3rd ed., **3**, 484.
A. N. Glazer & E. L. Smith (1971) The Enzymes, 3rd ed., **3**, 502.

PARALLEL FIRST-ORDER REACTIONS

Two Different Reactants. A set of two or more first-order reactions, the reactants of which are independent of each other, generating at least one product that is common for each chemical reaction. For example, consider the following set of parallel reactions:

$$A \xrightarrow{k_1} P$$

$$B \xrightarrow{k_2} Q$$

If the reactions are followed by the rates of disappearance of A and B (*i.e.*, $-d[A]/dt$ and $-d[B]/dt$), then the first-order rate constants k_1 and k_2 are directly obtained via standard analysis. However, if the rate of formation of the common product concentration or the rate of disappearance of the sum of A and B is measured, the evaluation of the individual rate constants is more complex.

One Common Reactant. A set of two or more chemical reactions for a given reactant, each of which is first-order.

$$A \xrightarrow{k_1} P$$

$$A \xrightarrow{k_2} Q$$

(An example would be the reaction or decomposition of a substance by two pathways.) The complete rate law must include all of the alternative pathways. If $[A_o]$ represents the concentration of A at time $t = 0$, and $[A]$, $[P]$, and $[Q]$ represent the concentrations of the corresponding substances at $t = t$, then $-d[A]/dt = k_1[A] + k_2[A] = (k_1 + k_2)[A]$. Integration yields $\ln([A_o]/[A]) = (k_1 + k_2)t$ or $[A] = [A_o]e^{-(k_1+k_2)t}$. Note that the overall reaction is first order with respect to A.

PARALLEL REACTIONS

1. A set of reactions in which a specific reactant can proceed to react by two or more pathways, each independent of the others. 2. A set of reactions in which a common product is produced from different reactants, each reaction being independent of the others. 3. A less satisfactory usage of the term "parallel reactions" is as a synonym for "side reactions." **See** *Specific type of parallel reaction; Parallel First-Order Reactions*

PARAMAGNETIC

A physical property of substances having a positive, nonzero magnetic susceptibility (albeit small). All substances with unpaired electrons will exhibit paramagnetism, as will metals (as a consequence of the magnetic moments associated with the spins of conducting electrons). **See also** *Diamagnetic; Permeability; Magnetic Susceptibility; Electron Spin Resonance; Diamagnetic*

PARTIAL COMPETITIVE INHIBITION

This type of inhibition differs from that exhibited by classical competitive inhibitors, because the substrate can still bind to the EI complex and the EIS complex can go on to form product (albeit at a slower rate) without the inhibitor being released from the binding site. While standard double-reciprocal plots of partial competitive inhibitors will be linear (except for some steady-state, *i.e.*, non-rapid-equilibrium, cases), secondary slope replots will be nonlinear. **See** *Nonlinear Inhibition*

Consider the simplest case on a rapid equilibrium Uni Uni enzyme (*i.e.*, an enzyme in which all of the binding steps are fast relative to the catalytic step and all binding steps can be characterized by dissociation constants). In the absence of the inhibitor, $E + A \rightleftharpoons EX \rightarrow$ product where $K_{ia} = [E][A]/[EX]$ and the conversion step has the unimolecular rate constant k_3. The inhibitor I can bind to the free enzyme to form the EI complex ($E + I \rightleftharpoons EI$) in which $K_i = [E][I]/[EI]$ and the inhibitor can bind to the EX complex to form EXI (where $K_{ii} = [EX][I]/[EXI]$). The EI complex can bind substrate and also form EXI (in which $K_{iii} = [EI][A]/[EXI]$). The ternary EXI complex can now proceed to form product, $EXI \rightarrow$ product. In this simplest case, let the rate constant for the formation of product from EXI also be equal to k_3. Hence, $d[\text{product}]/dt = v = k_3([EX] + [EXI])$. The derivation of the rate expression is fairly straightforward, yielding $v = V_{max}[A](1 + ([I]/K_{ii}))/\{K_{ia}(1 + ([I]/K_i)) + [A](1 + ([I]/K_{ii}))\}$ where $V_{max} = k_3[E_{total}]$. A double-reciprocal plot ($1/v$ *vs.* $1/[A]$ at different constant concentrations of I) will consist of a series of straight lines, all intersecting at a common point on the vertical axis (at $1/v = 1/V_{max}$), as would be the case for a normal, that is, linear, competitive inhibitor. A secondary replot of the slopes, however, will produce a curve that is either hyperbolic up or down, depending on the relative sizes of the dissociation constants for the inhibitor. If $K_i = K_{ii}$, no inhibition will be detected. Numerical analysis of the rate data can provide good estimates of the dissociation constants.

For non-rapid-equilibrium cases (*i.e.*, steady-state cases) the enzyme rate expression is much more complex, containing terms with $[A]^2$ and with $[I]^2$. Depending on the relative magnitude of those terms in the initial rate expression, there may be nonlinearity in the standard double-reciprocal plot. In such cases, computer-based numerical analysis may be the only means for obtaining estimates of the magnitude of the kinetic parameters involving the partial inhibition. **See** *Competitive Inhibition*

PARTIAL DERIVATIVES (Basic Properties)

Because partial derivatives are used so prominently in thermodynamics (**See** *Maxwell's Relationships*), we briefly consider the properties of partial derivatives for systems having three variables x, y, and z, of which two are independent. In this case, $z = z(x,y)$, where x and y are treated as independent variables. If one deals with infinitesimal changes in x and y, the corresponding changes in z are described by the partial derivatives:

$$dz = \left(\frac{\partial z}{\partial x}\right)_y dx + \left(\frac{\partial z}{\partial y}\right)_x dy$$

The subscripts indicate the variables that are held constant when taking the partial derivatives. Likewise, one can chose to treat y and z as independent variables, thereby writing x in terms of these variables, such that $x = x(y,z)$. In this case, we can obtain the following expression for infinitesimal changes in these independent variables:

$$dx = \left(\frac{\partial x}{\partial y}\right)_z dy + \left(\frac{\partial x}{\partial z}\right)_y dz$$

If z is kept constant, we can evaluate $(\partial x/\partial y)_z$ as follows, noting that $dz = 0$:

$$0 = \left(\frac{\partial z}{\partial x}\right)_y dx + \left(\frac{\partial z}{\partial y}\right)_x dy$$

and

$$\frac{dx}{dy} = -\frac{(\partial z/\partial y)_x}{(\partial z/\partial x)_y}$$

PARTIAL INHIBITION

A form of reversible inhibition in which the inhibitor-substrate(s)-enzyme complex can still generate product(s), without the inhibitor dissociating (albeit at a slower rate when compared to the inhibitor-free system). **See** *Partial Competitive Inhibition; Inhibition*

PARTIAL MOLAR VOLUMES

A thermodynamic parameter $(\partial V/\partial n_B)_{T,P,n_{j\neq B}}$ which describes how the volume of component B in a multicomponent system depends on the change in its amount expressed in mol. Høiland[1] recently summarized the partial molar volumes of numerous biochemical compounds in aqueous solution. **See** *Dalton's Law of Partial Pressures; Concentrations; Molecular Crowding*

[1]H. Høiland (1986) in *Thermodynamic Data for Biochemistry and Biotechnology* (H.-J. Hinz, ed.), pp. 1-44, Springer-Verlag, Berlin.

PARTIAL PRESSURE

For a mixture of gases in a specified volume, the partial pressure of a component of the mixture is the pressure exerted by that compound if it were the sole occupant of the volume. For example, the partial pressure of oxygen is around 152 mm of mercury or about 0.21 atmospheres at sea level and 298 K. **See** *Dalton's Law of Partial Pressures; Concentration*

PARTIAL RATE FACTOR

The rate of substitution for a given group and a given reaction at a single position in an aromatic compound relative to the rate of substitution at one position in benzene[1].

Partial rate factors have proven quite useful in predicting isomer distributions from related aromatic reactants under the same reaction conditions. If two or more substituents are present on the aromatic ring, the effect of those substituents are usually considered to be additive. However, the procedure is less effective in the presence of steric effects, *ipso* attack, and resonance interactions between substituents.

The term has also been applied to Hammett equations (**See** *Selectivity Factor*) and for other substituted reactants undergoing parallel reactions at different sites with the same rate law.

[1]L. M. Stock & H. C. Brown (1963) *Adv. Phys. Org. Chem.* **1**, 35.

PARTITION FACTOR

A parameter (often symbolized as R) used in the analysis of kinetic isotope effects as a measure of the rate of a particular step relative to the rate of all preceding steps in the reaction sequence. This factor indicates the fate of the intermediate that immediately precedes the step in which the change in bonding to the isotopic atom occurs[1]. If the key step is described by the constant k_i (and its isotopically altered value k_i^*), then

$$(k/k^*)_{observed} = (k_i/k_i^* + R)/(1 + R)$$

where the value of R depends on all rate constants up to and including k_{i-1}. For example, in the one-intermediate, one-substrate mechanism:

$$E + S \underset{k_2}{\overset{k_1}{\rightleftharpoons}} ES \overset{k_3}{\rightarrow} E + P$$

where k_3 is the isotopically sensitive step. $R = k_3/k_2$.

[1]M. H. O'Leary (1976) in *Isotope Effects on Enzyme-Catalyzed Reactions* (W. W. Cleland, M. H. O'Leary & D. B. Northrop, eds) pp. 236-238, University Park Press, Baltimore.

PARVULIN

The smallest member of a new family of prolyl isomerases (unrelated to the cyclophilins or the FK-506 binding proteins) that catalyzes the proline-limited folding of a variant of ribonuclease T1 with a k_{cat}/K_m value of 30,000 M^{-1} s^{-1}. With the tetrapeptide succinyl-Ala-Leu-Pro-Phe-4-nitroanilide as a substrate in parvulin-catalyzed prolyl isomerization, this parameter is 1.1×10^7 M^{-1} s^{-1}. Parvulin also accelerates its own refolding in an autocatalytic fashion.

C. Scholz, J. Rahfeld, G. Fischer & F. X. Schmid (1997) *J. Mol. Biol.* **273**, 752.

PASCAL

The SI unit (symbolized by Pa), equal to one newton per square meter, for pressure or stress. One pascal equals ten baryes, 10^{-5} bar, and 9.869233×10^{-6} atmosphere.

PASCAL'S TRIANGLE

A systematic scheme for generating the coefficients of terms in a polynomial formed by raising a binomial $(a + b)$ to the exponent n, such that: $(a + b)^0 = 1$; $(a + b)^1 = 1a + 1b$; $(a + b)^2 = 1a + 2ab + 1b$; $(a + b)^3 = 1a + 3a^2b + 3ab^2 + 1b^3$; $(a + b)^4 = 1a^4 + 4a^3b + 6a^2b^2 + 4ab^3 + 1a^4$; *etc.* Pascal's triangle offers a useful check on coefficients used in derivations of multiple equilibria models of ligand-receptor interactions.

```
1  1   1   1   1   1  1  1
1  2   3   4   5   6  7
1  3   6  10  15  21
1  4  10  20  35
1  5  15  35
1  6  21
1  7
1
```

The triangle can be generated by remembering three properties: (a) the first row and first column are filled at every step with 1's; (b) the second row and second column are sets of counting numbers (*i.e.*, 1, 2, 3, . . .); and (3) each of the remaining numbers is the sum of numbers to its left and above it. Shown here is Pascal's triangle for the first seven steps.

PATCH CLAMP

An electrophysiologic ion current-measuring device[1] that comprises (a) an excised membrane patch that has become sealed to the orifice of a glass micropipette and (b) a voltage clamp electronic circuit for measuring ion channel currents. The patch clamp relies on formation of a gigaohm seal between the recording pipette and the cell membrane. This approach allows the investigator to prepare (a) inside-out patches with their cytoplasmic membrane face exposed to the bath solution, and (b) outside-out patches with their extracellular membrane face exposed to the bath solution (see diagram). The technique offers the advantage that solution composition on each side of the membrane patch can be readily manipulated. Electrophysiologic measurements are both accurate and precise.

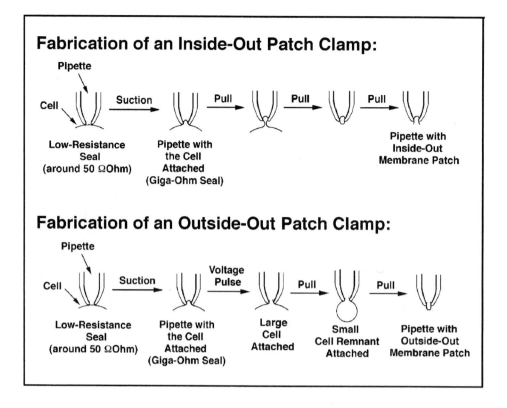

Fabrication of an Inside-Out Patch Clamp:

Pipette

Cell — Suction → Pull → Pull → Pull → Pipette with Inside-Out Membrane Patch

Low-Resistance Seal (around 50 ΩOhm) — Pipette with the Cell Attached (Giga-Ohm Seal)

Fabrication of an Outside-Out Patch Clamp:

Pipette

Cell — Suction → Voltage Pulse → Pull → Pull → Pipette with Outside-Out Membrane Patch

Low-Resistance Seal (around 50 ΩOhm) — Pipette with the Cell Attached (Giga-Ohm Seal) — Large Cell Attached — Small Cell Remnant Attached

Sakman and Neher[2] should be consulted regarding refinements in the extracellular patch clamp method that enable higher current resolution, direct membrane patch potential control, and physical isolation of membrane patches. They describe a convenient method for the fabrication of patch recording pipettes and procedures for achieving so-called "giga-seals"—namely pipette-membrane seals with resistances of 10^9–10^{11} Ω. The basic patch clamp recording circuit and properties of single acetylcholine-activated channels in muscle membrane illustrate the improved current and time-resolution that can be achieved with giga-seals. This monograph also considers the various ways for physically isolating patches of membrane from cells, thereby enabling one to record single channel currents with well-defined solutions on both sides of the membrane.

[1]E. Neher & J. H. Steinbach (1978) *J. Physiol.* **277**, 153.
[2]B. Sakman & E. Neher, eds. (1983) *Single-Channel Recording*, Plenum, New York.

Selected entries from *Methods in Enzymology* [vol, page(s)]: Technique and sample preparation, **207**, 159; outside-out configuration, **207**, 10; amplifiers, **238**, 153, 310; scrape loading, **238**, 363-364; trituration method, **238**, 374; data acquisition with fluorescence measurements, **238**, 312-313, 318-320; exocytosis monitoring, **238**, 320; pipette, **238**, 314-315; recording, **238**, 142, 152-153; stimulator, **238**, 311; temperature control, **238**, 314; tip-dip method, **238**, 344-348; ion channels from [*Esche-richia coli*, giant cells, and giant spheroplasts, **207**, 687; microbes, **207**, 681; organelles, **207**, 673; yeast (mitochondria, **207**, 686; spheroplasts, **207**, 685; vacuoles, **207**, 686).

pCa

The negative decadic (*i.e.*, base 10) logarithm of the calcium ion concentration.

pD

The negative decadic (*i.e.*, base 10) logarithm of the D+ (deuteron) concentration. If a pH electrode has been standardized with aqueous buffers, it can be used with D_2O systems with the following relationship: pD = (meter reading) + 0.4.

PECTINESTERASE

This enzyme [EC 3.1.1.11] also known as pectin methylesterase, pectin demethoxylase, and pectin methoxylase, catalyzes the reaction of pectin with *n* water to produce pectate and *n* methanol.

R. L. Fischer & A. B. Bennett (1991) *Ann. Rev. Plant Physiol. Plant Mol. Biol.* **42**, 675.

PENICILLOPEPSIN

This endopeptidase [EC 3.4.23.20], a member of the peptidase family A1, hydrolyzes proteins with a broad specificity similar to that of pepsin A.

D. R. Davies (1990) *Ann. Rev. Biophys. Biophys. Chem.* **19**, 189.

PEPSINS AND PEPSINOGENS

(1) Pepsin A [EC 3.4.23.1], the predominant endopeptidase in the gastric juice of vertebrates, is formed from pepsinogen A by limited proteolysis. Occurring in five molecular forms, this member of the A1 peptidase family preferentially cleaves peptide bonds in which hydrophobic, preferably aromatic, amino acyl residues are in the P_1 and $P_{1'}$ positions. (2) Pepsin B [EC 3.4.23.2] is an endopeptidase that catalyzes the degradation of gelatin. It exhibits very little activity with hemoglobin. The enzyme is formed from pepsinogen B and is a member of the peptidase family A1. (3) Pepsic C [EC 3.4.23.3], better known as gastricsin, is formed from progastricsin. Gastricsin exhibits an even more restricted specificity than pepsin A, with preferential cleavage at Tyr—Xaa bonds.

T. D. Meek (1998) *Comprehensive Biological Catalysis: A Mechanistic Reference* **1**, 327.

J. S. Fruton (1982) *Adv. Enzymol.* **53**, 239.

Selected entries from *Methods in Enzymology* [vol, page(s)]:
Pepsin: Activation energy, **63**, 243-245; isotope exchange, **64**, 10; kinetic constants, **63**, 244; kinin-releasing enzyme, **80**, 174; porcine, homology to cathepsin D, **80**, 578; spectrokinetic probe, **61**, 323; tight-binding inhibition, **63**, 462, 466; active site, structure, **241**, 214-215; amino-terminal lobe, pep.pep homodimer, **241**, 198, 201-203; assay, **241**, 213-214; crystal structure, **241**, 214-216; kinetic parameters, **241**, 222; pK_a values, **241**, 222
Pepstatin: As aspartyl protease inhibitor, **241**, 160, 183; inhibition of aspartic peptidases, **244**, 10; pepstatin-protease complex, **241**, 170-171.

PEPTIDE METHIONINE SULFOXIDE REDUCTASE

This reductase, officially known as protein-methionine-*S*-oxide reductase [EC 1.8.4.6], catalyzes the reaction of protein L-methionine with oxidized thioredoxin to yield protein L-methionine *S*-oxide and reduced thioredoxin. Dithiothreitol can replace reduced thioredoxin in the reverse reaction. Free methionine is not a substrate; however free methionine is similarly acted upon by an analogous reductase [EC 1.8.4.5].

N. Brot, M. A. Rahman, J. Moskovitz & H. Weissbach (1995) *Meth. Enzymol.* **251**, 462.

PEPTIDOMIMETIC COMPOUND

Analogues in which one or more of the naturally occurring peptide bonds in a substrate, effector, or ligand has been replaced by some other chemical structure. Such compounds are useful in characterizing binding specificity (particularly conformational specificity), and

by virtue of omitting the peptide bond(s), they often exhibit much longer *in vivo* half-life. Peptidomimetics can also occur naturally. Although morphine's structure has been known from Knorr's work nearly a century ago, the peptidomimetic nature of this analgesic agent wasn't recognized until about seventy years later.

PEPTIDYL-ASPARTATE ENDOPEPTIDASE

This metalloendoproteinase [EC 3.4.24.33] catalyzes the hydrolysis of Xaa—Asp, Xaa—Glu, and Xaa—cysteic acid bonds. Because of its limited specificity, this enzyme is useful in protein sequencing applications.

M.-L. Hagmann, V. Geuss, S. Fischer & G.-B. Kresse (1995) *Meth. Enzymol.* **248**, 782.

PEPTIDYL GLYCINE α-AMIDATING MONOOXYGENASE

Peptidylglycine monooxygenase [EC 1.14.17.3], also known as peptidyl α-amidating enzyme and peptidylglycine 2-hydroxylase, catalyzes the reaction of a peptidylglycine with ascorbate and dioxygen to produce a peptidyl(2-hydroxyglycine), dehydroascorbate, and water.

A. S. Kolhekar, R. E. Mains & B. A. Eipper (1997) *Meth. Enzymol.* **279**, 35.

PEPTIDYLTRANSFERASE

This ribosomal enzyme [EC 2.3.2.12] catalyzes the acyl transfer from a peptidyl-tRNA[1] with an aminoacyl-tRNA[2] to produce tRNA[1] and a peptidylaminoacyl-tRNA[2].

H. F. Noller (1993) *J. Bacteriol.* **175**, 5297.
H. F. Noller, V. Hoffarth & L. Zimniak (1992) *Science* **256**, 1416.
R. L. Soffer (1974) *Adv. Enzymol.* **40**, 91.

PERIODATE CLEAVAGE

The oxidative C—C bond-breaking reaction that occurs when periodate ion (IO_4^-) reacts with an organic compound having one or more vicinal hydroxyl groups. The specificity of the reaction results from the formation of an initial complex of the 1,2-diol with periodate prior to the oxidative cleavage step. This reaction is useful in determining the cyclic structures of carbohydrates as well as in the preparation of electrophilic substrate analogues that can be immobilized for affinity chromatography used in enzyme purification. *See Malaprade Reaction*

PERMEABILITY

1. The scalar, symbolized by μ, that relates the magnetic flux density, *B*, in a body or medium with the external

magnetic field strength, H, inducing it. Thus, $B = \mu H$ where μ has SI units of newtons per square ampere (which are equivalent to henrys per meter). The permeability of a vacuum, symbolized by μ_o and synonymous with the term magnetic constant, is set equal to $4\pi \times 10^{-7}$ N A^{-2} (or, $12.56637\ldots \times 10^{-7}$ N A^{-2}). The relative permeability, μ_r, is equal to μ/μ_o. For diamagnetic substances, $\mu_r < 1$. If the substance is paramagnetic, $\mu_r > 1$. The μ_r values for ferromagnetic substances are large and vary with the field strength. **See also** *Magnetic Susceptibility*

2. The ability of a substance or ion to pass through a membrane; typically measured by permeability constants. **See** *Permeability Constant*

PERMEABILITY COEFFICIENT

A parameter (usually symbolized by P, and often containing a subscript to indicate the specific ion) that is a measure of the ease with which an ion can cross a unit area of membrane by simple (or passive) diffusion through a membrane experiencing a 1.0 M concentration gradient. For a particular biological membrane, the permeabilities are dependent on the concentration and activity of various channel or transporter proteins. In an electrically active cell (*e.g.*, a neuron), increasing the permeability of K$^+$ or Cl$^-$ will usually result in hyperpolarization of the membrane. Increasing P_{Na} will cause depolarization.

The permeability coefficient (units = cm/s) can be written as follows:

$$P = -\frac{(c_2 - c_1)}{J} = \frac{KD}{l}$$

where $(c_2 - c_1)$ represents the concentration difference across a semipermeable membrane, J is the net transport rate (usually expressed as moles transferred per square centimeter per second), K is the lipid-water partition coefficient, D is the diffusion coefficient, and l is the thickness of the membrane. For the human erythrocyte membrane, the approximate permeability coefficients (cm/s) for K$^+$, Na$^+$, Cl$^-$, glucose, and water are 2×10^{-10}, 10^{-10}, 10^{-4}, 2×10^{-5}, and 5×10^{-3}. For the peripheral membrane of giant squid axon, the approximate permeability coefficients (cm/s) for K$^+$, Na$^+$, and Cl$^-$ are: (un-

der resting conditions) 6×10^{-8}, 2×10^{-8}, and 10^{-8}; and (upon excitation) 2×10^{-4}, 5×10^{-6}, and 10^{-8}.

M. K. Jain & R. C. Wagner (1980) *Introduction to Biological Membranes*, Wiley, New York.

PERMISSIVE ACTIVE SITE

A descriptor for an enzyme active site that permits binding of a family of related compounds (*e.g.*, mimics of the reaction intermediate) that can be derived from the initial binding and conformational changes in the substrate. This concept arose from the observation that a number of monoterpene cyclases were incapable of discriminating between enantiomers of the reaction intermediate, even though the enzyme catalyzes the synthesis of an enantiomerically pure product from an achiral substrate. An example is trichodiene synthase[1] which catalyzes the cyclization of farnesyl diphosphate to trichodiene.

[1]D. E. Cane, G. Yang, R. M. Coates, H.-J. Pyun & T. M. Hohn (1992) *J. Org. Chem.* **57**, 3454.

PERMUTATIONS AND COMBINATIONS

A permutation is a change that is primarily based on a rearrangement of existing elements or things. A combination of elements or things is any selection of one or more, taken without concern regarding their order.

Permutations. There are several important statements about permutations and their properties. First, the number of permutations of n different things in a row, taken all at a time, is given by the expression:

$$n! = (1)(2)(3) \ldots (n - 1)(n)$$

where $n!$ is "n factorial". This rule is reasonable because the first row can have n possible things, the second row can have $(n - 1)$ things, the third row $(n - 2)$ things, and so forth until the nth row contains only a single thing. Second, if n things are divided into c classes, where the same class contains things that are alike, and where those within different classes are different, the number of permutations of these things, taken all at a time, is given by the expression:

$$\frac{n!}{(n_1!)(n_2!)(n_3!) \ldots (n_c!)}$$

where the subscripts indicate all the classes from 1 to c, and $n = n_1 + n_2 + n_3 + \ldots + n_c$. Third, the number of different permutations of n different things taken k at a time without repetitions is given by:

$$n(n-1)(n-2)\ldots(n-k+1) = \frac{n!}{(n-k)!}$$

Combinations. The number of different combinations of n different things, taken k at a time, without repetitions, is given by the expression:

$$\binom{n}{k} = \frac{n!}{k!(n-k)!} = \frac{n(n-1)\ldots(n-k+1)}{k!(n-k)!}$$

Stirling's Approximation. When n gets large, evaluation of $n!$ becomes cumbersome, even with modern hand-held calculators. In such cases, one can use an approximation first recognized by the Scottish mathematician James Stirling:

$$n \simeq (2\pi n)^{1/2}(n/e)^n$$

See also Statistics (A Primer); Linked Functions

PEROXIDASE

This heme-dependent enzyme [EC 1.11.1.7] catalyzes the reaction of a donor with hydrogen peroxide to produce an oxidized donor and two water. **See also** Horseradish Peroxidase; Ovoperoxidase; Myeloperoxidase

H. B. Dunford (1998) Comprehensive Biological Catalysis: A Mechanistic Reference 3, 195.
M. A. Ator & P. R. Ortez de Montellano (1990) The Enzymes, 3rd ed., 19, 213.
F. P. Guengerich (1990) Crit. Rev. Biochem. Mol. Biol. 25, 97.

PERSISTENCE TIME

1. The period over which a system sustains a particular chemical behavior or biological response. 2. The time of continual progress of a chemotactic microorganism locomoting in one direction. **See** Twiddling

PERSISTENT

Referring to a reaction intermediate or free radical that has a lifetime longer than that of a transient species, typically on the time-scale of at least several minutes in dilute solution in inert solvents. Persistence is therefore a kinetic property related to reactivity. The *stability* of an intermediate or free radical is a thermodynamic property, often expressed in terms of the appropriate bond strengths. **See** Transient Chemical Species

D. Griller and K. U. Ingold (1976) Acc. Chem. Res. 9, 13.

PERTUSSIS TOXIN

An ADPribosyltransferase transferase produced by the *Bordetella pertussis* that catalyzes the transfer of an ADPR group onto the G-protein $G_{i\alpha}$, thereby blocking exchange of bound GDP for GTP.

D. J. Carty (1994) Meth. Enzymol. 237, 63.
L. Passador & W. Iglewski (1994) Meth. Enzymol. 235, 617.
N. J. Oppenheimer & A. L. Handlon (1992) The Enzymes, 3rd ed., 20, 453.
C.-Y. Lai (1986) Adv. Enzymol. 58, 99.

PHARMACOKINETICS

A special branch of drug metabolism/physiology dealing with the development of quantitative models for treating the time evolution of processes involved in drug absorption by the body, drug distribution throughout the various compartments (plasma, organs, tissues, populations of single cells, as well as defined regions within cells), drug conversion into various metabolized forms, and finally, drug excretion. These models are absolutely essential in developing an understanding of how various individuals dispose of a drug, and the information can be of paramount importance in efforts to improve the efficacy of a drug.

Many factors can influence drug response, including but not limited to the following: (a) physiologic factors such as gastrointestinal function, cardiovascular function, liver function, immune responsiveness, renal function, and stress (additionally for females: luteal cycle, pregnancy, lactation, and menopause); (b) physical factors such as season, time of day, duration of daylight, altitude, and ambient temperature; (c) individual factors such as genetics, age, sex, weight (obesity), state of health, exercise, cigarette smoking, occupational hazards, chemical exposure, alcohol consumption, and immunization; (d) site(s) and mode(s) of drug administration; (e) the various derivatized and/or formulated forms of the drug, including factors influencing drug stability and dissolution kinetics; (f) interactions with other drugs taken by the subject; and (e) biochemical and physiologic individuality (such as blood albumin levels, mutations affecting receptors, transporters, acid/base balance, *etc.*).

Detailed pharmacokinetic investigations are a necessary prelude to the formal approval protocols of the Federal Drug Administration and its counterparts worldwide. Although this topic lies beyond the scope of this Handbook, those interested will want to consult an excellent treatment, entitled "Basics Pharmacokinetics". This internet textbook developed by M. C. Makoid, P. J. Vuchetich, and U. V. Banakar at Creighton University (http://kiwi.creighton.edu) includes a mathematical review (through Laplace transforms) as well as chapters on pharmacological response, kinetics of intravenous bolus and intravenous infusion, biopharmaceutic factors, oral dosing, bioavailability, clearance, dosage, multicompartment modeling, protein binding, and nonlinear kinetics behavior.

PHASE

1. Any homogeneous part of a system that itself is heterogeneous (*i.e.*, separated from the other parts of the system by some physical boundary). For example, there are three phases present in a glass half-full of water with ice floating in the water: ice, liquid water, and air (containing water vapor). 2. A particular state of some periodic function or operation. 3. A point in time or fraction of a period that has elapsed in which the item undergoing change is of harmonic motion, uniform circular motion, or any periodic change (such as rotation, oscillation, *etc.*) that follows simple harmonic laws. The phase of such a harmonic quantity usually refers to some standard position or direction such as at the instant of starting. Particles in periodic motion are in phase if they are moving in the same direction as the relative displacement. Waves having the same frequency and waveform are in phase if they reach the same values simultaneously.

PHASE MODULATION FLUORESCENCE

A kinetic technique for determining a fluorophore's excited state lifetime by using a light source whose intensity is modulated sinusoidally at a certain frequency, such that the intensity of the fluorescence emission likewise varies sinusoidally but with an added delay from the finite relaxation constant for fluorescence decay. The period of the sinusoidal modulation is chosen to be in the neighborhood of the magnitude of the fluorescence lifetime.

J. R. Lakowicz (1983) *Principles of Fluorescence Spectroscopy*, pp. 53-56, Plenum, New York.

PHASE-TRANSFER CATALYSIS

The rate enhancement observed in a reaction occurring between chemical species present in different phases (for example, in two immiscible solvents) by the addition of a small amount of some agent (often referred to as the phase-transfer catalyst). In many cases in biological systems, this agent is a surfactant that extracts one of the reactants from one phase to the other phase, thus permitting the reaction to proceed.

pH EFFECTS ON ENZYMIC CATALYSIS

Because enzymes and/or substrates often contain one or more catalytically functional acid/base groups, it is not at all surprising that pH is a major factor influencing enzymic catalysis. That enzymes often display such interpretable and quantitatively analyzable behavior is quite remarkable, considering the great number of ionizable groups that they contain. This fact suggests (a) that the active center of an enzyme is reasonably well isolated from the effects of other ionizable groups outside the active site and (b) that an enzyme employs relatively few acid or base groups to achieve its great rate enhancement.

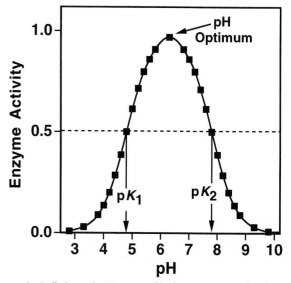

Figure 1. Bell-shaped pH-rate profile for an enzyme that has two ionizable groups affecting enzyme activity. The rate is directly proportional to the degree of ionization of the first group (determined by pK_1) and inversely proportional to the degree of ionization of the second group (governed by pK_2).

In characterizing the pH dependence of enzyme activity, one often observes (a) a bell-shaped curve in plots of activity *versus* pH (Fig. 1), or (b) "S"-shaped activity *versus* pH curves (either falling from optimal activity or rising to optimal activity) with an inflection point at some pH value. Michaelis and Davidsohn[1] were among the first to offer a kinetic scheme for explaining the nature of pH-rate profiles for enzyme-catalyzed reactions. Von Euler[2] extended these ideas by writing the following generalized scheme involving three enzyme states:

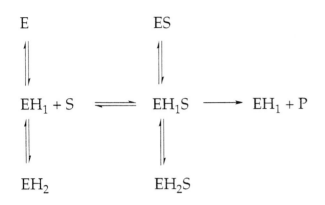

where the pertinent equilibrium constants can be written for the acidic (designated by subscript "a") and basic (designated by subscript "b") equilibria.

$$K_S = [E][S]/[ES]$$

$$K_{Eb} = [E][H^+]/[EH_1]$$

$$K_{ESb} = [ES][H^+]/[EH_1S]$$

$$K_{Ea} = [EH_1][H^+]/[EH_2]$$

$$K_{ESa} = [EH_1S][H^+]/[EH_2S]$$

If EH_1 and S combine rapidly, and if the conversion of EH_1S to $EH_1 + P$ is the slow step, one can write a rate law similar to the Michaelis-Menten equation. In this case, however, the enzyme is distributed into additional forms (including four nonproductive forms E, ES, EH_2 and EH_2S):

$$[E_{total}] = [ES] + [EHS] + [EH_2S] + [E] + [EH] + [EH_2]$$

The resulting rate law can be written as:

$$v = V_{max}/\{[1 + [H^+]/K_{ESa} + K_{ESb}/[H^+]] + (K_S/[S])[1 + [H^+]/K_{ESa} + K_{ESb}/[H^+]]\}$$

If $pK_{Eb} \approx pK_{ESb}$ and $pK_{Ea} \approx pK_{ESa}$, and if the two sets of pK's are separated sufficiently on the pH scale, then one can demonstrate that a plot of velocity *versus* pH will rise to a maximum at $(pK_{Eb} + pK_{Ea})/2$. At a pH value

below this point, the increased hydrogen ion "drives" the enzyme into the inactive EH_2 and EH_2S forms; similarly, as one moves toward higher pH, the inactive E and ES forms accumulate (Fig. 2).

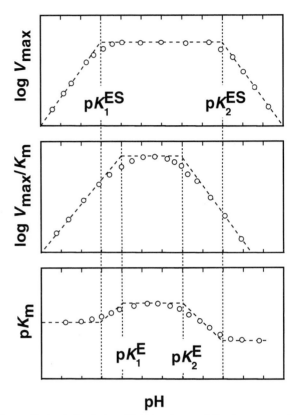

Figure 2. Analysis of the pH-dependence of the key parameters influencing the initial rate behavior of an enzyme containing two ionizable groups that determine catalytic activity.

The reader should note that ionizable groups on the substrate could also result in pH-rate profiles that pass through a maximum. For example, one can write the following scheme:

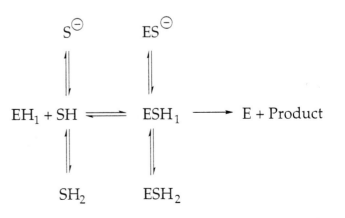

pH kinetic studies can help one to identify the type of amino acid side-chain group that is participating in catalysis. When combined with studies of temperature effects, solvent effects, isotope effects, *etc.*, pH kinetics can be a powerful tool. A thorough treatment of pH kinetics for enzymes lies beyond the scope of this Handbook, and the interested reader is encouraged to consult Fromm[3], Tipton & Dixon[4], and Cleland[5] for more details on the effects of pH on enzymic activity.

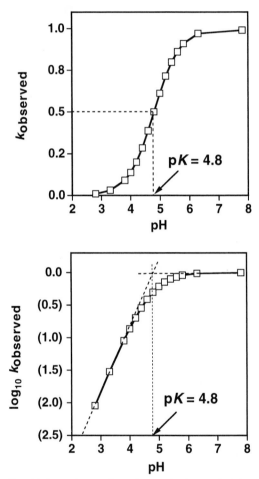

Figure 3. pH-dependence of an observed rate constant k_{obs}.

When it is feasible to identify and characterize a single rate process, the effect of pH is more straightforward. This is frequently the case for nonenzymatic reactions obeying simple rate equations, but fast reaction techniques sometimes allow one to examine a single elementary reaction in an enzymic process. In either case, the behavior shown in Fig. 3 typifies the manner in which an observed reaction rate constant depends on the pK_a of an ionizable group.

[1]L. Michaelis & H. Davidsohn (1911) *Biochem Z.* **35**, 386.
[2]H. v. Euler, K. Josephson & K. Myrbäck (1924) *Hoppe-Seyler's Z. Physiol. Chem.* **134**, 39.
[3]H. J. Fromm (1975) *Initial Rate Enzyme Kinetics*, pp. 201-224, Springer-Verlag, New York.
[4]K. F. Tipton & H. B. F. Dixon (1979) *Meth. Enzymol.* **63**, 183.
[5]W. W. Cleland (1982) *Meth. Enzymol.* **87**, 390.

pH EFFECTS ON NONENZYMATIC HYDROLYSIS

In his famous treatise on acid and base catalysis, Bell[1] consolidated modern views on acids and bases, particularly with respect to their effects on the rates of hydrolysis of various substances. Though not the originator of the concepts of general acid and general base catalysis, he promoted broader recognition that catalysis does not always occur simply through the action of hydrogen and hydroxyl ions. Bell also advanced the acidity and basicity concepts of Brønsted and Lewis, while investigating the mechanistic origin for the fact that hydrolysis of certain substances has a "signature" profile of log(reaction rate) *versus* pH.

Typical Profiles for Rate *versus* pH

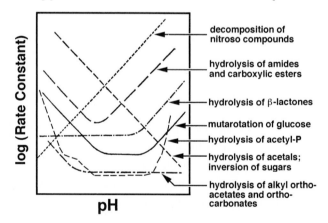

The diagram shown above is based, in part, on Bell's compilation of pH-rate profiles. The interested reader will wish to consult his classic presentation as well as specific entries on the properties and actions of acids and bases throughout this Handbook.

[1]R. P. Bell (1941) *Acid-Base Catalysis*, Clarendon Press, Oxford.

pH ELECTRODE

A glass electrode which takes advantage of rapid and reproducible ion exchange across a glass membrane located as a thin-walled bulb at the tip of the electrode. This membrane has the following liquid-membrane junction:

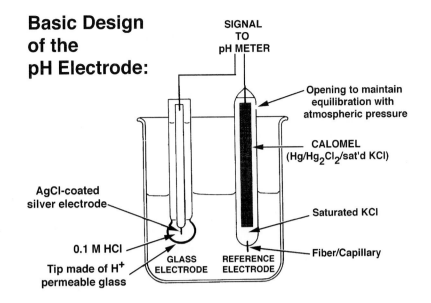

Basic Design of the pH Electrode:

By its chemical nature, the very thin dry layer of glass permits ion exchange without any possibility of electrochemical redox reactions. If the membrane is a sufficiently thin layer, one can achieve millisecond response times to changes in pH, and these properties can be put to full advantage with rapid mixing devices to study chemical reactions with half-lives in the 5–10 msec range.

PHENOL HYDROXYLASE

This FAD-dependent enzyme [EC 1.14.13.7], also known as phenol 2-monooxygenase, catalyzes the reaction of phenol with NADPH and dioxygen to produce catechol, NADP⁺, and water. Other substrates include resorcinol and *o*-cresol.

B. A. Palfey & V. Massey (1998) *Comprehensive Biological Catalysis: A Mechanistic Reference* **3**, 83.
B. G. Fox (1998) *Comprehensive Biological Catalysis: A Mechanistic Reference* **3**, 261.
J. V. Schloss & M. S. Hixon (1998) *Comprehensive Biological Catalysis: A Mechanistic Reference* **2**, 43.
S. Yamamoto & Y. Ishimura (1991) in *A Study of Enzymes* (S. A. Kuby, ed.), vol. **2**, pp. 315-344, CRC Press, Boca Raton.
J. H. Law, P. E. Dunn & K. J. Kramer (1977) *Adv. Enzymol.* **45**, 389.
V. Massey & P. Hemmerich (1975) *The Enzymes*, 3rd ed., **12**, 191.
V. Ullrich & W. Duppel (1975) *The Enzymes*, 3rd ed., **12**, 253.

PHENYLALANINE AMINOTRANSFERASE

This enzyme [EC 2.6.1.58], also referred to as phenylalanine(histidine) aminotransferase, catalyzes the reversible reaction of L-phenylalanine with pyruvate to produce phenylpyruvate and L-alanine. This enzyme will also use L-histidine and L-tyrosine as substrates. In the reverse reaction, L-methionine, L-serine, and L-gluta-mine can replace L-alanine. ***See also*** *Aromatic Amino Acid Aminotransferase*

M. Fujioka, Y. Morino & H. Wada (1970) *Meth. Enzymol.* **17A**, 585.

PHENYLALANINE AMMONIA-LYASE

This enzyme [EC 4.3.1.5] catalyzes the conversion of L-phenylalanine to *trans*-cinnamate and ammonia. The enzyme may also act on L-tyrosine. ***See also*** *Dehydroalanine; Borohydride Reduction*

V. E. Anderson (1998) *Comprehensive Biological Catalysis: A Mechanistic Reference* **2**, 115.
C. W. Abell & R.-S. Shen (1987) *Meth. Enzymol.* **142**, 242.
E. A. Havir (1987) *Meth. Enzymol.* **142**, 248.
R. A. Dixon, P. M. Dey & C. J. Lamb (1983) *Adv. Enzymol.* **55**, 1.
K. R. Hanson & E. A. Havir (1972) *The Enzymes*, 3rd ed., **7**, 75.

PHENYLALANINE DECARBOXYLASE

This pyridoxal-phosphate-dependent enzyme [EC 4.1.1.53] catalyzes the conversion of L-phenylalanine to phenethylamine and carbon dioxide. Tyrosine and other aromatic amino acids can also serve as substrates.

M. Fujioka, Y. Morino & H. Wada (1970) *Meth. Enzymol.* **17A**, 585.

PHENYLALANINE DEHYDROGENASE

This enzyme [EC 1.4.1.20] catalyzes the reaction of L-phenylalanine with NAD⁺ and water to produce phenylpyruvate, ammonia, and NADH. The enzymes isolated from *Bacillus badius* and *Sporosarcina ureae* are highly specific for L-phenylalanine, whereas that isolated from *Bacillus sphaericus* also acts on L-tyrosine.

A. R. Clarke & T. R. Dafforn (1998) *Comprehensive Biological Catalysis: A Mechanistic Reference* **3**, 1.

N. M. W. Brunhuber & J. S. Blanchard (1994) *Crit. Rev. Biochem. Mol. Biol.* **29**, 415.

PHENYLALANINE HYDROXYLASE (Phenylalanine 4-Monooxygenase)

This iron-dependent enzyme [EC 1.14.16.1], more widely known as phenylalanine hydroxylase and phenylalaninase, catalyzes the reaction of L-phenylalanine with tetrahydrobiopterin and dioxygen to produce L-tyrosine, dihydrobiopterin, and water.

P. F. Fitzpatrick (1998) *Comprehensive Biological Catalysis: A Mechanistic Reference* **3**, 181.

S. Kaufman (1993) *Adv. Enzymol.* **67**, 77.

S. Kaufman (1987) *The Enzymes*, 3rd ed., **18**, 217.

V. Massey & P. Hemmerich (1975) *The Enzymes*, 3rd ed., **12**, 191.

V. Ullrich & W. Duppel (1975) *The Enzymes*, 3rd ed., **12**, 253.

PHENYLALANINE RACEMASE

This enzyme [EC 5.1.1.11], also known as phenylalanine racemase (ATP-hydrolyzing), catalyzes the reaction of ATP with L-phenylalanine to produce D-phenylalanine, AMP, and pyrophosphate. In this unusual racemase reaction, a thiol group of an enzyme-bound pantotheine forms a thiolester from an initial aminoacyl-AMP intermediate; then, as is typical of acyl thioesters, the α-proton becomes labile, thereby permitting reversible inversion of configuration to produce an equilibrated mixture of thiolester-bound enantiomers. Hydrolysis of the thiolester yields the product.

M. E. Tanner & G. L. Kenyon (1998) *Comprehensive Biological Catalysis: A Mechanistic Reference* **2**, 7.

E. Adams (1972) *The Enzymes*, 3rd ed., **6**, 479.

PHENYLMETHYLSULFONYL FLUORIDE

A sulfonylating agent (abbreviated PMSF) that irreversibly inhibits many "serine esterases" and "serine proteases". Target enzymes usually react with PMSF and related alkylating agents through the activated imidazole group of a histidyl residue that is part of the catalytic triad.

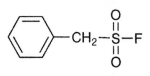

Gold[1] suggests that the reaction can be written analogously to the Michaelis-Menten equation:

$$E-H + RSO_2F \rightleftharpoons E \cdot RSO_2F \rightarrow E-SO_2R + HF$$

Because PMSF fails to inactivate acetylcholinesterase, this reagent is much less toxic than diisopropylfluorophosphate, and is also recommended as an alternative to the neurotoxic fluorophosphates and fluorophosphonates. PMSF is freshly prepared as a 1–3 mM solution in water (higher concentrations will precipitate spontaneously). A better procedure is to first prepare a 20 mM PMSF solution in 2-propanol or dioxane; this solution can then be added to the biological fluid with vortex mixing to achieve a 1–3 mM final concentration as a homogeneous solution. One should confirm that the alcohol or dioxane has little or no undesirable effect on enzymes or proteins of interest. ***See*** *Chymotrypsin; Protease Inhibitor "Cocktails"*

[1]A. M. Gold (1967) *Meth. Enzymol.* **11**, 706.

Selected entries from *Methods in Enzymology* [vol, page(s)]:
Sulfonylation reaction, **11**, 706; reaction kinetics, **11**, 707; second-order rate constants for inactivation of chymotrypsin, trypsin, and acetylcholine esterase by PMSF and related sulfonylating agents, **11**, 707; reactivation of PMS-chymotrypsin, **11**, 710; as inhibitor [of calcium-activated factor, **80**, 674; of cathepsin G, **80**, 565; of crayfish trypsin, **80**, 639; of elastase, **80**, 587; of prolylcarboxypeptidase, **80**, 465; of protease Re, **80**, 691; of protease So, **80**, 695; of protein C_a, **80**, 329]; proteolysis, **76**, 7.

PHENYLPYRUVATE SYNTHASE

This enzyme [EC 1.3.1.12], also known as hydroxyphenylpyruvate synthase or prephenate dehydrogenase, catalyzes the NAD^+-dependent oxidation of prephenate to form 4-hydroxyphenylpyruvate, CO_2 and NADH. The following is a recent review on the molecular and physical properties of this synthase.

B. B. Buchanan (1972) *The Enzymes*, 3rd ed., **6**, 193.

PHI (Φ or ϕ)

Φ 1. Symbol for potential energy. 2. Symbol for magnetic flux. 3. Symbol for radiant energy per time (radiant power).

ϕ 1. Symbol for quantum yield. 2. Symbol for one of the space coordinates in the three-dimensional, spherical polar coordination system. 3. Symbol for electric potential. 4. Symbol for volume fraction. 5. With a subscript designation, symbol for a Dalziel coefficient. 6. Symbol for fugacity coefficient. 7. Symbol for osmotic coefficient. 8. Symbol for heat flow rate.

PHI VALUES IN BISUBSTRATE MECHANISMS

Initial rate enzyme kinetics are useful in defining the order of substrate binding interactions in multisubstrate kinetic mechanisms. Even in the absence of corroborating evidence (from competitive inhibition studies, product inhibition, isotope exchange at equilibrium, *etc.*), sequential mechanisms should be discernible from the ping pong case. Indeed, as shown below for bisubstrate cases, sequential and ping pong mechanisms have two different functional dependencies on changes in the concentrations of substrates:

Sequential Mechanisms: $E_o/v = \phi_o + \phi_1/[A] + \phi_2/[B]$
$+ \phi_{12}/[A][B]$

Ping-Pong Mechanism: $E_o/v = \phi_o + \phi_1/[A] + \phi_2/[B]$

where the phi values represent collections of rate constants. Concentration ranges for substrates A and B can be chosen to achieve adequate saturation of the enzyme, thereby yielding a suitable range of experimentally determined reaction velocities (**See** *Initial Rate Enzyme Assays*). Associated with each A_xB_y pair will be an experimentally determined initial reaction velocity v_{xy}, and these can be plotted (1) as $1/v$ *versus* $1/[A]$, yielding a family of five straight-line plots corresponding to the five constant levels of B, and (2) as $1/v$ *versus* $1/[B]$, yielding a family of five straight-line plots corresponding to the five constant levels of A. Because sequential mechanisms have a $\phi_{12}/[A][B]$ term, each family of curves will converge to the left of the $(1/v)$-axis. This was observed to be the case in the yeast hexokinase reaction[1]. The absence of the $\phi_{12}/[A][B]$ term in ping pong mechanisms means that one should obtain two families of parallel line data, as shown for yeast nucleoside-5'-diphosphate kinase[2].

While such considerations suggest that initial rate studies alone can be a useful diagnostic of mechanism type, parallel-line data may result whenever the $\phi_{12}/[A][B]$ is sufficiently small so as to prevent the experimenter from observing convergent-line data characteristic of sequential mechanisms. This was recognized as a possibility in the case of the brain hexokinases[3–6] which yield parallel-line (or nearly so) data in initial rate studies with D-glucose and MgATP^{2-}. Because this class of enzymes exhibits activity with other hexoses, such as D-fructose or D-mannose, the initial rate results conducted with the former monosaccharide were informative regarding the kinetic mechanism of the brain hexokinases. The double-reciprocal plots below indicate the initial rate behavior of rat brain hexokinase with D-glucose and D-fructose and MgATP^{2-}.

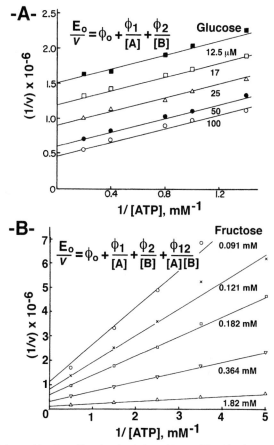

Initial rate kinetics of bovine brain hexokinase (A) with glucose and ATP as substrates, and (B) with fructose and ATP. Note that the apparent parallel-line kinetics observed with glucose conform to a rate equation lacking a ϕ_{12} term.

The results indicate that brain hexokinase does indeed operate by way of a sequential kinetic mechanism, and subsequent kinetic studies with reversible inhibitors support this conclusion. These comments reinforce the wisdom of remaining wary of mechanistic inferences drawn solely on the basis of initial rate studies alone. **See also** *Initial Rate Enzyme Assays*

[1] H. J. Fromm & V. Zewe (1962) *J. Biol. Chem.* **237**, 3027.
[2] E. Garces & W. W. Cleland (1967) *Biochemistry* **8**, 633.
[3] M. Copley & H. J. Fromm (1967) *Biochemistry* **6**, 3503.
[4] H. J. Fromm & J. Ning (1968) *Biochem. Biophys. Res. Commun.* **32**, 672.
[5] J. Ning, D. L. Purich & D. L. Fromm (1969) *J. Biol. Chem.* **244**, 3840.
[6] D. L. Purich & H. J. Fromm (1971) *J. Biol. Chem.* **246**, 3456.

PHOSPHATE ACETYLTRANSFERASE
(Phosphotransacetylase)

This enzyme [EC 2.3.1.8], perhaps more widely known as phosphotransacetylase and less so as phosphoacylase, catalyzes the reaction of acetyl-CoA with orthophosphate to produce coenzyme A and acetyl phosphate.

Other short-chain acyl-CoA derivatives can act as substrates.

L. L. Barton (1994) *Meth. Enzymol.* **243**, 94.

PHOSPHATE CONTINUOUS ASSAY
See *Orthophosphate Continuous Assay*

PHOSPHATE DIESTER HYDROLYSIS

Phosphate diesters resist acid hydrolysis, and they are only inefficiently hydrolyzed in alkaline solutions. The exception is any five-membered ring containing a phosphodiester, such as that formed upon alkaline treatment of RNA. Such five-membered cyclic phosphodiesters are rapidly hydrolyzed in alkaline solutions.

PHOSPHATE MONOESTER HYDROLYSIS

Phosphate monoesters [$R-OPO(OH)_2$] resist alkaline hydrolysis by converting to unreactive dianions [$R-OPO_3^{2-}$]. Phosphate monoester monoanions [$R-OPO(OH)O^{1-}$] are, however, susceptible to facile hydrolysis under slightly acidic conditions.

PHOSPHATIDATE CYTIDYLYLTRANSFERASE

This enzyme [EC 2.7.7.41], also known as CTP:phosphatidate cytidylyltransferase and CDP-diacylglycerol synthase, catalyzes the reaction of CTP with phosphatidate to produce CDP-diacylglycerol and pyrophosphate.

C. Kent (1995) *Ann. Rev. Biochem.* **64**, 315.
C. P. Sparrow (1992) *Meth. Enzymol.* **209**, 237.
J. D. Esko & C. R. H. Raetz (1983) *The Enzymes*, 3rd ed., **16**, 207.
R. A. Pieringer (1983) *The Enzymes*, 3rd ed., **16**, 255.

PHOSPHATIDATE PHOSPHATASE

This enzyme [EC 3.1.3.4] catalyzes the hydrolysis of a 3-*sn*-phosphatidate to generate a 1,2-diacyl-*sn*-glycerol and orthophosphate.

C. Kent (1995) *Ann. Rev. Biochem.* **64**, 315.
G. M. Carman & J. J. Quinlan (1992) *Meth. Enzymol.* **209**, 219.
G. M. Carman & Y.-P. Lin (1991) *Meth. Enzymol.* **197**, 548.
R. M. Bell & R. A. Coleman (1983) *The Enzymes*, 3rd ed., **16**, 87.
J. D. Esko & C. R. H. Raetz (1983) *The Enzymes*, 3rd ed., **16**, 207.
R. A. Pieringer (1983) *The Enzymes*, 3rd ed., **16**, 255.
Selected entries from *Methods in Enzymology* [vol, page(s)]:
  Phosphatidate phosphatase from yeast, **197**, 548; characterization and assay of phosphatidate phosphatase, **197**, 553.

PHOSPHATIDYLCHOLINE:SERINE O-PHOSPHATIDYLTRANSFERASE

This transferase, officially known as CDP-diacylglycerol—serine *O*-phosphatidyltransferase [EC 2.7.8.8], catalyzes the reaction of CDP-diacylglycerol with L-serine to form CMP and *O-sn*-phosphatidyl-L-serine.

W. R. Bishop & R. M. Bell (1988) *Ann. Rev. Cell Biol.* **4**, 579.

PHOSPHATIDYLETHANOLAMINE N-METHYLTRANSFERASE

This enzyme [EC 2.1.1.17] catalyzes the reaction of *S*-adenosyl-L-methionine with phosphatidylethanolamine to produce *S*-adenosyl-L-homocysteine and phosphatidyl-*N*-methylethanolamine.

D. Rhodes & A. D. Hanson (1993) *Ann. Rev. Plant Physiol. Plant Mol. Biol.* **44**, 357.
N. D. Ridgway & D. E. Vance (1992) *Meth. Enzymol.* **209**, 366.
J. D. Esko & C. R. H. Raetz (1983) *The Enzymes*, 3rd ed., **16**, 207.

PHOSPHATIDYLGLYCEROL PHOSPHATE PHOSPHOHYDROLASE

This enzyme [EC 3.1.3.27], also known as phosphatidylglycerophosphatase, catalyzes the hydrolysis of phosphatidylglycerophosphate to produce phosphatidylglycerol and orthophosphate.

W. Dowhan & C. R. Funk (1992) *Meth. Enzymol.* **209**, 224.
J. D. Esko & C. R. H. Raetz (1983) *The Enzymes*, 3rd ed., **16**, 207.
R. A. Pieringer (1983) *The Enzymes*, 3rd ed., **16**, 255.

1-PHOSPHATIDYLINOSITOL 4-KINASE

This enzyme [EC 2.7.1.67] catalyzes the reaction of ATP with 1-phosphatidyl-1D-*myo*-inositol to produce ADP and 1-phosphatidyl-1D-*myo*-inositol 4-phosphate.

C. Kent (1995) *Ann. Rev. Biochem.* **64**, 315.
G. M. Carman, C. J. Belunis & J. T. Nickels, Jr. (1992) *Meth. Enzymol.* **209**, 183.
A. Moritz, J. Westerman, P. N. E. de Graan & K. W. A. Wirtz (1992) *Meth. Enzymol.* **209**, 202.

PHOSPHATIDYLINOSITOL-4-PHOSPHATE 5-KINASE

This enzyme [EC 2.7.1.68], also known as diphosphoinositide kinase, catalyzes the reaction of ATP with 1-phosphatidyl-1D-*myo*-inositol 4-monophosphate to produce ADP and 1-phosphatidyl-1D-*myo*-inositol 4,5-bisphosphate.

C. Kent (1995) *Ann. Rev. Biochem.* **64**, 315.
C. E. Bazenet & R. A. Anderson (1992) *Meth. Enzymol.* **209**, 189.
A. Moritz, J. Westerman, P. N. E. de Graan & K. W. A. Wirtz (1992) *Meth. Enzymol.* **209**, 202.

1-PHOSPHATIDYLINOSITOL PHOSPHODIESTERASE

This enzyme [EC 3.1.4.10], also known as monophosphatidylinositol phosphodiesterase and phosphatidylinositol phospholipase C, catalyzes the conversion of 1-phosphatidyl-1D-*myo*-inositol to D-*myo*-inositol 1,2-cyclic phosphate and a diacylglycerol. The animal enzyme, but not that from bacteria, also hydrolyzes the cyclic phosphate, forming inositol 1-phosphate as the final product.

J. E. Bleasdale, J. C. McGuire & G. A. Bala (1990) *Meth. Enzymol.* **187**, 226.

PHOSPHATIDYLINOSITOL SYNTHASE

This enzyme [EC 2.7.8.11], also known as CDP-diacylglycerol—inositol 3-phosphatidyltransferase, catalyzes the reaction of CDP-diacylglycerol with *myo*-inositol to produce CMP and phosphatidyl-1D-*myo*-inositol.

C. Kent (1995) *Ann. Rev. Biochem.* **64**, 315.
G. M. Carman & A. S. Fischl (1992) *Meth. Enzymol.* **209**, 305.
J. D. Esko & C. R. H. Raetz (1983) *The Enzymes*, 3rd ed., **16**, 207.

PHOSPHATIDYL-*N*-METHYLETHANOLAMINE *N*-METHYLTRANSFERASE

This enzyme [EC 2.1.1.71] catalyzes the reaction of *S*-adenosyl-L-methionine with phosphatidyl-*N*-methylethanolamine to produce *S*-adenosyl-L-homocysteine and phosphatidyl-*N*-dimethylethanolamine. The enzyme will catalyze the transfer of an additional methyl group, thus producing phosphatidylcholine.

J. Bremer (1969) *Meth. Enzymol.* **14**, 125.

PHOSPHATIDYLSERINE DECARBOXYLASE

These pyridoxal-phosphate-dependent (or pyruvate-dependent) enzymes [EC 4.1.1.65] catalyze the decarboxylation of phosphatidyl-L-serine to produce phosphatidylethanolamine and carbon dioxide.

M. L. Hackert & A. E. Pegg (1998) *Comprehensive Biological Catalysis: A Mechanistic Reference* **2**, 201.
W. Dowhan (1997) *Meth. Enzymol.* **280**, 81.
W. Dowhan & Q.-X. Li (1992) *Meth. Enzymol.* **209**, 348.
D. R. Voelker & E. B. Gordon (1992) *Meth. Enzymol.* **209**, 360.
J. D. Esko & C. R. H. Raetz (1983) *The Enzymes*, 3rd ed., **16**, 207.
R. A. Pieringer (1983) *The Enzymes*, 3rd ed., **16**, 255.

PHOSPHATIDYLSERINE SYNTHASE

This enzyme [EC 2.7.8.8], also known as CDP-diacylglycerol:L-serine *O*-phosphatidyltransferase, catalyzes the reaction of CDP-diacylglycerol with L-serine to produce CMP and *O-sn*-phosphatidyl-L-serine.

C. Kent (1995) *Ann. Rev. Biochem.* **64**, 315.
W. Dowhan (1992) *Meth. Enzymol.* **209**, 287.
G. M. Carman & M. Bae-Lee (1992) *Meth. Enzymol.* **209**, 298.
R. A. Pieringer (1983) *The Enzymes*, 3rd ed., **16**, 255.

PHOSPHOACETYLGLUCOSAMINE MUTASE

This enzyme [EC 5.4.2.3], also known as acetylglucosamine phosphomutase and *N*-acetylglucosamine-phosphate mutase, catalyzes the interconversion of *N*-acetyl-D-glucosamine 1-phosphate and *N*-acetyl-D-glucosamine 6-phosphate. The enzyme is activated by *N*-acetyl-D-glucosamine 1,6-bisphosphate.

J. L. Reissig & L. F. Leloir (1966) *Meth. Enzymol.* **8**, 175.
D. M. Carlson (1966) *Meth. Enzymol.* **8**, 179.

3′-PHOSPHOADENYLYL SULFATE SULFOHYDROLASE

This manganese-dependent enzyme [EC 3.6.2.2], also known as phosphoadenylylsulfatase, catalyzes the hydrolysis of 3′-phosphoadenylyl sulfate to yield adenosine 3′,5′-bisphosphate and sulfate.

A. B. Roy (1971) *The Enzymes*, 3rd ed., **5**, 1.

PHOSPHODIESTERASES

Venom exonuclease [EC 3.1.15.1], also known as venom phosphodiesterase, catalyzes the exonucleolytic cleavage of RNA or DNA (preferring single-stranded substrates) in the 3′ to 5′ direction to yield 5′-phosphomononucleotides. Similar enzymes include hog kidney phosphodiesterase and the *Lactobacillus* exonuclease. ***See also*** *specific phosphodiesterase*

J. A. Gerlt (1992) *The Enzymes*, 3rd ed., **20**, 95.

PHOSPHOENOLPYRUVATE CARBOXYKINASE

Phosphoenolpyruvate carboxykinase (GTP) [EC 4.1.1.32], also known as phosphoenolpyruvate carboxylase and phosphopyruvate carboxylase, catalyzes the reaction of GTP with oxaloacetate to produce GDP, phosphoenolpyruvate, and carbon dioxide. ITP can replace GTP as the phosphorylating substrate.

Phosphoenolpyruvate carboxykinase (ATP) [EC 4.1.1.49], also known as phosphopyruvate carboxylase (ATP) and phosphoenolpyruvate carboxylase, catalyzes the reaction of ATP with oxaloacetate to produce ADP, phosphoenolpyruvate, and carbon dioxide.

J. V. Schloss & M. S. Hixon (1998) *Comprehensive Biological Catalysis: A Mechanistic Reference* **2**, 43.
A. Matte, L. W. Tari, H. Goldir & L. T. J. Delbaere (1997) *J. Biol. Chem.* **272**, 18105.

R. W. Hanson & Y. M. Patel (1994) *Adv. Enzymol.* **69**, 203.

J. J. Villafranca & T. Nowak (1992) *The Enzymes*, 3rd ed., **20**, 63.

M. H. O'Leary (1992) *The Enzymes*, 3rd ed., **20**, 235.

M. F. Utter & H. M. Kolenbrander (1972) *The Enzymes*, 3rd ed., **6**, 117.

PHOSPHOENOLPYRUVATE CARBOXYKINASE (Pyrophosphate)

This enzyme [EC 4.1.1.38] (also known as phosphoenolpyruvate carboxytransphosphorylase, phosphopyruvate carboxylase, and phosphoenolpyruvate carboxylase) catalyzes the reaction of phosphoenolpyruvate with orthophosphate and carbon dioxide to produce oxaloacetate and pyrophosphate (or diphosphate). The enzyme also catalyzes the reaction of phosphoenolpyruvate with orthophosphate to produce pyruvate and pyrophosphate.

H. G. Wood, W. E. O'Brien & G. Michaels (1977) *Adv. Enzymol.* **45**, 85.

M. F. Utter & H. M. Kolenbrander (1972) *The Enzymes*, 3rd ed., **6**, 117.

PHOSPHOENOLPYRUVATE CARBOXYLASE

This enzyme [EC 4.1.1.31] catalyzes the reaction of orthophosphate with oxaloacetate to produce phosphoenolpyruvate, carbon dioxide, and water.

J. V. Schloss & M. S. Hixon (1998) *Comprehensive Biological Catalysis: A Mechanistic Reference* **2**, 43.

R. Chollet, J. Vidal & M. H. O'Leary (1996) *Ann. Rev. Plant Physiol. Plant Mol. Biol.* **47**, 273.

M. H. O'Leary (1992) *The Enzymes*, 3rd ed., **20**, 235.

P. M. Goodwin (1990) *Meth. Enzymol.* **188**, 361.

M. F. Utter & H. M. Kolenbrander (1972) *The Enzymes*, 3rd ed., **6**, 117.

PHOSPHOENOLPYRUVATE MUTASE

This enzyme [EC 5.4.2.9], also known as phosphoenolpyruvate phosphomutase, catalyzes the interconversion of phosphoenolpyruvate and 3-phosphonopyruvate. This enzyme participates in the biosynthesis of the $C-P$ bond. However, the equilibrium greatly favors phosphoenolpyruvate.

J. V. Schloss & M. S. Hixon (1998) *Comprehensive Biological Catalysis: A Mechanistic Reference* **2**, 43.

A. C. Hengge (1998) *Comprehensive Biological Catalysis: A Mechanistic Reference* **1**, 517.

PHOSPHOENOLPYRUVATE PHOSPHATASE

This phosphatase [EC 3.1.3.60] catalyzes the hydrolysis of phosphoenolpyruvate to form pyruvate and phosphate. The enzyme also acts less efficiently on a wide range of other monophosphates.

W. C. Plaxton (1996) *Ann. Rev. Plant Physiol. Plant Mol. Biol.* **47**, 185.

PHOSPHOFRUCTOKINASE

This enzyme [EC 2.7.1.11], also known as phosphohexokinase and phosphofructokinase I, catalyzes the reaction of ATP with D-fructose 6-phosphate to produce ADP and D-fructose 1,6-bisphosphate. Both D-tagatose 6-phosphate and sedoheptulose 7-phosphate can act as the sugar substrate. UTP, CTP, GTP, and ITP all can act as the nucleotide substrate. This enzyme is distinct from that of 6-phosphofructo-2-kinase. **See also** *ATP & GTP Depletion*

H. C. Huppe & D. H. Turpin (1994) *Ann. Rev. Plant Physiol. Plant Mol. Biol.* **45**, 577.

T. Schirmer & P. R. Evans (1990) *Nature* **343**, 140.

S. J. Pilkis, T. H. Claus, P. D. Kountz & M. R. El-Maghrabi (1987) *The Enzymes*, 3rd ed., **18**, 3.

E. Van Schaffingen (1987) *Adv. Enzymol.* **59**, 315.

K. Uyeda (1979) *Adv. Enzymol.* **48**, 193.

D. P. Bloxham & H. A. Lardy (1973) *The Enzymes*, 3rd ed., **8**, 239.

Selected entries from *Methods in Enzymology* [vol, page(s)]:
Activation, **64**, 221; anomeric specificity, **63**, 373, 374; assay, **63**, 376; burst analysis, **64**, 197; cascade control, **64**, 325; contamination, **63**, 7; equilibrium shift, **64**, 220; hysteresis, **64**, 195; inactivation, **64**, 221; mechanism, **63**, 51; physiological damping, **64**, 217, 220, 221; product inhibition, **63**, 433; rate properties, initial, **63**, 156; substrate analog, **63**, 373, 374; substrate inhibition, **63**, 501.

6-PHOSPHOFRUCTO-2-KINASE

This enzyme [EC 2.7.1.105], also known as phosphofructokinase 2, catalyzes the reaction of ATP with D-fructose 6-phosphate to produce ADP and D-fructose 2,6-bisphosphate. This enzyme is distinct from that of 6-phosphofructokinase. It copurifies with fructose-2,6-bisphosphatase.

S. J. Pilkis, T. H. Claus, I. J. Kurland & A. J. Lange (1995) *Ann. Rev. Biochem.* **64**, 799.

S. J. Pilkis, T. H. Claus, P. D. Kountz & M. R. El-Maghrabi (1987) *The Enzymes*, 3rd ed., **18**, 3.

E. Van Schaffingen (1987) *Adv. Enzymol.* **59**, 315.

6-PHOSPHO-β-GALACTOSIDASE

This enzyme [EC 3.2.1.85], also known as β-D-phosphogalactoside galactohydrolase, catalyzes the hydrolysis of a 6-phospho-β-D-galactoside to produce an alcohol and 6-phospho-D-galactose.

M. W. Bauer, S. B. Halio & R. M. Kelly (1996) *Adv. Protein Chem.* **48**, 271.

PHOSPHOGLUCOMUTASE

This enzyme [EC 5.4.2.2], also known as glucose phosphomutase, catalyzes the interconversion of α-D-glucose 1-phosphate and α-D-glucose 6-phosphate. For this reaction to occur, α-D-glucose 1,6-bisphosphate must be present. This bisphosphate is an intermediate in the reaction, being formed by the transfer of a phosphate residue from the enzyme to the substrate. However, the dissociation of bisphosphate from the enzyme complex is much slower than the overall isomerization.

S. Shankar, R. W. Ye, D. Schlictman & A. M. Chakrabarty (1995) Adv. Enzymol. **70**, 221.
J.-B. Dai, Y. Liu, W. J. Ray, Jr. & M. Konno (1992) J. Biol. Chem. **267**, 6322.
I. A. Rose (1975) Adv. Enzymol. **43**, 491.
W. J. Ray, Jr., & E. J. Peck, Jr. (1972) The Enzymes, 3rd ed., **6**, 407.

6-PHOSPHOGLUCONATE DEHYDROGENASE

This enzyme [EC 1.1.1.43] catalyzes the reaction of 6-phospho-D-gluconate with $NAD(P)^+$ to produce 6-phospho-2-dehydro-D-gluconate and $NAD(P)H$.

M. J. Adams, G. H. Ellis, S. Gover, C. E. Naylor & C. Phillips (1994) Structure **2**, 651.
M. H. O'Leary (1992) The Enzymes, 3rd ed., **20**, 235.
R. F. Colman (1990) The Enzymes, 3rd ed., **19**, 283.
D. J. Creighton & N. S. R. K. Murthy (1990) The Enzymes, 3rd ed., **19**, 323.
L. V. Kletsova, M. Y. Kiriukhin, A. Y. Chistoserdov & Y. D. Tsygankov (1990) Meth. Enzymol. **188**, 335.
A. P. Sokolov & Y. A. Trotsenko (1990) Meth. Enzymol. **188**, 339.
K. Dalziel (1975) The Enzymes, 3rd ed., **11**, 1.

PHOSPHOGLUCOSE ISOMERASE

This enzyme [EC 5.3.1.9] (also known as glucose-6-phosphate isomerase, phosphohexose isomerase, phosphohexomutase, oxoisomerase, hexosephosphate isomerase, phosphosaccharomutase, phosphoglucoisomerase, and phosphohexoisomerase) catalyzes the interconversion of D-glucose 6-phosphate and D-fructose 6-phosphate. The enzyme also catalyzes the anomerization (or, mutarotation) of D-glucose 6-phosphate.

S. H. Seeholzer (1993) Proc. Natl. Acad. Sci. U.S.A. **90**, 1237.
D. J. Creighton & N. S. R. K. Murthy (1990) The Enzymes, 3rd ed., **9**, 323.
E. A. Noltmann (1972) The Enzymes, 3rd ed., **6**, 271.

6-PHOSPHO-β-GLUCOSIDASE

This enzyme [EC 3.2.1.86] catalyzes the hydrolysis of 6-phospho-β-D-glucoside-(1,4)-D-glucose to produce D-glucose 6-phosphate and glucose. The enzyme will also catalyze the hydrolysis of several other phospho-β-D-glucosides; but not their nonphosphorylated forms.

M. W. Bauer, S. B. Halio & R. M. Kelly (1996) Adv. Protein Chem. **48**, 271.

3-PHOSPHOGLYCERATE DEHYDROGENASE

This enzyme [EC 1.1.1.95] catalyzes the reaction of 3-phosphoglycerate with NAD^+ to produce 3-phosphohydroxypyruvate and NADH.

A. R. Clarke & T. R. Dafforn (1998) Comprehensive Biological Catalysis: A Mechanistic Reference **3**, 1.
L. I. Pizer & E. Sugimoto (1971) Meth. Enzymol. **17B**, 325.
H. J. Sallach (1966) Meth. Enzymol. **9**, 216.

3-PHOSPHOGLYCERATE KINASE

This enzyme [EC 2.7.2.3] catalyzes the reaction of ATP with 3-phospho-D-glycerate to produce ADP and 3-phospho-D-glyceroyl phosphate.

R. Jaenicke, H. Schurig, N. Beaucamp & R. Ostendorp (1996) Adv. Protein Chem. **48**, 181.
P. A. Frey (1992) The Enzymes, 3rd ed., **20**, 141.

PHOSPHOGLYCERATE MUTASE

This enzyme [EC 5.4.2.1] (also known as phosphoglycerate phosphomutase and phosphoglyceromutase) catalyzes the reaction of 2-phospho-D-glycerate with 2,3-diphosphoglycerate to produce 3-phospho-D-glycerate and 2,3-diphosphoglycerate. The enzymes isolated from mammals and from yeast are phosphorylated by $(2R)$-2,3-bisphosphoglycerate, which is also an intermediate. With the rabbit muscle enzyme, the dissociation of bisphosphate from the enzyme is much slower that the overall isomerization.

A. C. Hengge (1998) Comprehensive Biological Catalysis: A Mechanistic Reference **1**, 517.
L. A. Fothergill-Gilmore & H. C. Watson (1989) Adv. Enzymol. **62**, 227.
Z. B. Rose (1982) Meth. Enzymol. **87**, 42.
Z. B. Rose (1980) Adv. Enzymol. **51**, 211.
W. J. Ray, Jr., & E. J. Peck, Jr. (1972) The Enzymes, 3rd ed., **6**, 407.

PHOSPHOGLYCERIDE: LYSOPHOSPHATIDYLGLYCEROL ACYLTRANSFERASE

The following is a recent review on the molecular and physical properties of this transferase.

K. Y. Hostetler, J. Huterer & J. R. Wherrett (1992) Meth. Enzymol. **209**, 104.

PHOSPHOKETOLASE

This thiamin pyrophosphate-dependent enzyme [EC 4.1.2.9] catalyzes the reaction of D-xylulose 5-phosphate with orthophosphate to produce acetyl phosphate, D-glyceraldehyde 3-phosphate, and water.

M. Goldberg, J. M. Fessenden & E. Racker (1966) Meth. Enzymol. **9**, 515.
B. L. Horecker (1962) Meth. Enzymol. **5**, 261.

PHOSPHOLIPASE

This broad class of hydrolases constitutes a special category of enzymes which bind to and conduct their catalytic functions at the interface between the aqueous solution and the surface of membranes, vesicles, or emulsions. In order to explain the kinetics of lipolysis, one must determine the rates and affinities that govern enzyme adsorption to the interface of insoluble lipid structures[1,2]. One must also account for the special properties of the lipid surface as well as for the ability of enzymes to scoot[2] along the lipid surface. **See** *specific enzyme; Micelle; Interfacial Catalysis*

[1]R. Verger (1980) Meth. Enzymol. **64**, 340.
[2]M. K. Jain, M. H. Gelb, J. Rogers & O. G. Berg (1995) Meth. Enzymol. **249**, 567.

Selected entries from Methods in Enzymology [vol, page(s)]:
Assays & Kinetics: Assay strategies and methods for phospholipases, **197**, 3; utilization of labeled *Escherichia coli* as phospholipase substrate, **197**, 24; nuclear magnetic resonance spectroscopy to follow phospholipase kinetics and products, **197**, 31; monolayer techniques for studying phospholipase kinetics, **197**, 49; thio-based phospholipase assay, **197**, 65; chromogenic substrates and assay of phospholipases A$_2$, **197**, 75; coupled enzyme assays for phospholipase activities with plasmalogen substrates, **197**, 79; phospholipase A$_2$ assays with fluorophore-labeled lipid substrates, **197**, 90; assays of phospholipases on short-chain phospholipids, **197**, 95; phospholipase A$_2$-catalyzed hydrolysis of vesicles: uses of interfacial catalysis in the scooting mode, **197**, 112; assay of phospholipases C and D in presence of other lipid hydrolases, **197**, 125; preparation of alkyl ether and vinyl ether substrates for phospholipases, **197**, 134; inositol phospholipids and phosphates for investigation of intact cell phospholipase C substrates and products, **197**, 149; chromatographic analysis of phospholipase reaction products, **197**, 158; assays for measuring arachidonic acid release from phospholipids, **197**, 166; phosphoinositide-specific phospholipase C activation in brain cortical membranes, **197**, 183; quantitative analysis of water-soluble products of cell-associated phospholipase C- and phospholipase D-catalyzed hydrolysis of phosphatidylcholine, **197**, 191.
Structure-Function: Dissection and sequence analysis of phospholipases A$_2$, **197**, 201; cloning, expression, and purification of porcine pancreatic phospholipase A$_2$ and mutants, **197**, 214; preparation of antibodies to phospholipases A$_2$, **197**, 223; thermodynamics of phospholipase A$_2$-ligand interactions, **197**, 234; activation of phospholipase A$_2$ on lipid bilayers, **197**, 249;

phospholipase stereospecificity at phosphorus, **197**, 258; phospholipase A$_2$: microinjection and cell localization techniques, **197**, 269; analysis of lipases by radiation inactivation, **197**, 280; phosphorylation of phospholipase C *in vivo* and *in vitro*, **197**, 288.

PHOSPHOLIPASE A$_1$

This calcium-dependent enzyme [EC 3.1.1.32] catalyzes the hydrolysis of a phosphatidylcholine to generate 2-acylglycerophosphocholine and a fatty acid anion. This enzyme has a much broader specificity than phospholipase A$_2$. **See** *Micelle*

J. D. Esko & C. R. H. Raetz (1983) The Enzymes, 3rd ed., **16**, 207.
E. A. Dennis (1983) The Enzymes, 3rd ed., **16**, 307.

Selected entries from Methods in Enzymology [vol, page(s)]:
Detergent-resistant phospholipase A$_1$ from *Escherichia coli* membranes, **197**, 309; phospholipase A$_1$ activity of guinea pig pancreatic lipase, **197**, 316; purification of rat kidney lysosomal phospholipase A$_1$, **197**, 325; purification and substrate specificity of rat hepatic lipase, **197**, 331; human postheparin plasma lipoprotein lipase and hepatic triglyceride lipase, **197**, 339; phospholipase activity of milk lipoprotein lipase, **197**, 345.

PHOSPHOLIPASE A$_2$

This calcium-dependent enzyme [EC 3.1.1.4] (also known as phosphatidylcholine 2-acylhydrolase, lecithinase A, phosphatidase, and phosphatidolipase) catalyzes the hydrolysis of phosphatidylcholine to produce 1-acyl-glycerophosphocholine and a fatty acid anion. The enzyme can also hydrolyze phosphatidylethanolamine, choline plasmalogen, and phosphatides, removing the fatty acyl group attached to the C2 position of the glycerol moiety. **See** *Micelle*

D. M. Quinn & S. R. Feaster (1998) Comprehensive Biological Catalysis: A Mechanistic Reference **1**, 455.
M. H. Gelb, M. K. Jain, A. M. Hanel & O. G. Berg (1995) Ann. Rev. Biochem. **64**, 653.
M. K. Jain, M. H. Gelb, J. Rogers & O. G. Berg (1995) Meth. Enzymol. **249**, 567.
J. D. Esko & C. R. H. Raetz (1983) The Enzymes, 3rd ed., **16**, 207.
E. A. Dennis (1983) The Enzymes, 3rd ed., **16**, 307.
R. Verger (1980) Meth. Enzymol. **64**, 340.
D. J. Hanahan (1971) The Enzymes, 3rd ed., **5**, 71.

Selected entries from Methods in Enzymology [vol, page(s)]:
Cobra venom phospholipase A$_2$: *Naja naja naja*, **197**, 359; phospholipase A$_2$ from rat liver mitochondria, **197**, 365; assay and purification of phospholipase A$_2$ from human synovial fluid in rheumatoid arthritis, **197**, 373; purification of mammalian nonpancreatic extracellular phospholipases A$_2$, **197**, 381; spleen phospholipases A$_2$, **197**, 390; purification and characterization of cytosolic phospholipase A$_2$ activities from canine myocardium and sheep platelets, **197**, 400.

PHOSPHOLIPASE B

This enzyme, known officially as lysophospholipase [EC 3.1.1.5] as well as by other common names: lecithinase B, lysolecithinase, and phospholipase B, catalyzes the hydrolysis of 2-lysophosphatidylcholine to yield glycerophosphocholine and a fatty acid. *See* Micelle

K. Saito, J. Sugatani & T. Okumura (1991) *Meth. Enzymol.* **197**, 446.

PHOSPHOLIPASE C

This zinc-dependent enzyme [EC 3.1.4.3] (also known as lipophosphodiesterase I, lecithinase C, *Clostridium welchii* α-toxin, and *Clostridium oedematiens* β- and γ-toxins) catalyzes the hydrolysis of a phosphatidylcholine to produce 1,2-diacylglycerol and choline phosphate. The enzyme isolated from bacterial sources also acts on sphingomyelin and phosphatidylinositol; however, the enzyme isolated from seminal plasma does not act on phosphatidylinositol. *See* Micelle

N. H. Williams (1998) *Comprehensive Biological Catalysis: A Mechanistic Reference* **1**, 543.
W. D. Singer, H. A. Brown & P. C. Sternweis (1997) *Ann. Rev. Biochem.* **66**, 475.
J. D. Esko & C. R. H. Raetz (1983) *The Enzymes*, 3rd ed., **16**, 207.
E. A. Dennis (1983) *The Enzymes*, 3rd ed., **16**, 307.

Selected entries from *Methods in Enzymology* [vol, page(s)]:
Phosphatidylinositol-specific phospholipases C from *Bacillus cereus* and *Bacillus thuringiensis*, **197**, 493; assays for phosphoinositide-specific phospholipase C and purification of isozymes from bovine brain, **197**, 502; properties of phospholipase C isozymes, **197**, 511; phosphatidylinositol-specific phospholipase C from human platelets, **197**, 518; purification of guinea pig uterus phosphoinositide-specific phospholipase C, **197**, 526.

PHOSPHOLIPASE D

This enzyme [EC 3.1.4.4] (also known as lipophosphodiesterase II, lecithinase D, and choline phosphatase) catalyzes the hydrolysis of a phosphatidylcholine to produce choline and a phosphatidate. The enzyme will also act on other phosphatidyl esters. *See* Micelle

J. H. Exton (1997) *J. Biol. Chem.* **272**, 15579.
H. A. Brown & P. C. Sternweis (1995) *Meth. Enzymol.* **257**, 313.
E. P. Bowman, D. J. Uhlinger & J. D. Lambeth (1995) *Meth. Enzymol.* **256**, 246.
E. A. Dennis (1983) *The Enzymes*, 3rd ed., **16**, 307.

Selected entries from *Methods in Enzymology* [vol, page(s)]:
Glycosylphosphatidylinositol-specific phospholipase D, **197**, 567; solubilization and purification of rat tissue phospholipase D, **197**, 575; lysophospholipase D, **197**, 583.

PHOSPHOLIPID FLIP-FLOP

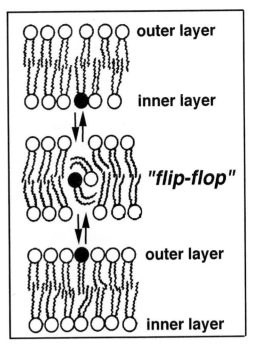

The pair-wise exchange of phospholipid molecules across a membrane bilayer. The polar head groups of phospholipids resist entry into the apolar core of the membrane bilayer, and phospholipid flip-flop is both improbable and sluggish (*i.e.*, $k_{exchange} \approx 10^{-5}\,s^{-1}$). Kornberg and McConnell[1] used nitroxide spin-labeled phospholipid vesicles to follow the time course of flip-flop. In their clever experiments, ascorbic acid was added to the bulk solvent to reduce any nitroxide spin label on the exterior surface of the lipid. Then, the time course of flip-flop could be determined by the time course of the diminution in the amplitude of the ESR spectrum that measured the amount of spin-label remaining on the inner, and thus solvent-inaccessible membrane surface.

[1]R. Kornberg & H. M. McConnell (1971) *Biochemistry* **10**, 1111.

PHOSPHOMANNOMUTASE

This enzyme [EC 5.4.2.8] catalyzes the interconversion of D-mannose 1-phosphate and D-mannose 6-phosphate. Either D-mannose 1,6-bisphosphate or D-glucose 1,6-bisphosphate can act as the cofactor.

S. Shankar, R. W. Ye, D. Schlictman & A. M. Chakrabarty (1995) *Adv. Enzymol.* **70**, 221.

5-PHOSPHOMEVALONATE KINASE

This enzyme [EC 2.7.4.2] catalyzes the reaction of ATP with (*R*)-5-phosphomevalonate to produce ADP and (*R*)-5-diphosphomevalonate.

PHOSPHOPANTOTHENOYLCYSTEINE SYNTHETASE

This enzyme [EC 6.3.2.5], also known as phosphopantothenoylcysteine ligase, catalyzes the reaction of CTP with (R)-4′-phosphopantothenate and L-cysteine to produce as yet unidentified products of CTP breakdown and (R)-4′-phosphopantothenoyl-L-cysteine. Cysteine can be replaced in this reaction by some of its derivatives.

P. D. van Poelje & E. E. Snell (1990) *Ann. Rev. Biochem.* **59**, 29.

PHOSPHOPENTOMUTASE

This enzyme [EC 5.4.2.7], also known as phosphodeoxyribomutase, catalyzes the interconversion of D-ribose 1-phosphate and D-ribose 5-phosphate. This enzyme will also catalyze the interconversion of 2-deoxy-D-ribose 1-phosphate and 2-deoxy-D-ribose 5-phosphate. The bisphosphate compound is the cofactor: thus, D-ribose 1,5-bisphosphate, 2-deoxy-D-ribose 1,5-bisphosphate, or even α-D-glucose 1,6-bisphosphate.

W. J. Ray, Jr., & E. J. Peck, Jr. (1972) *The Enzymes*, 3rd ed., **6**, 407.

PHOSPHOPROTEIN PHOSPHATASES

This set of enzymes [EC 3.1.3.16], also known as serine/threonine-specific protein phosphatases, catalyzes the hydrolysis of a seryl- or threonyl-bound phosphate group from a wide range of phosphoproteins, including a number of enzymes which have been phosphorylated under the action of a kinase. Thus, phosphoprotein + water yields protein + orthophosphate.

A. C. Hengge (1998) *Comprehensive Biological Catalysis: A Mechanistic Reference* **1**, 517.
R. D. Smith & J. C. Walker (1996) *Ann. Rev. Plant Physiol. Plant Mol. Biol.* **47**, 101.
S. I. Walaas & P. Greengard (1987) *The Enzymes*, 3rd ed., **18**, 285.
L. M. Ballou & E. H. Fischer (1986) *The Enzymes*, 3rd ed., **17**, 311.

PHOSPHORESCENCE

1. Long-lived luminescence. 2. Luminescence involving a change in spin multiplicity (*e.g.*, triplet-to-singlet, singlet-to-triplet, quartet state-to-doublet state, *etc.*). **See** *Fluorescence; Jablonski Diagram*

Comm. on Photochem. (1988) *Pure and Appl. Chem.* **60**, 1055.

PHOSPHORIBOSYLAMINOIMIDAZOLE-CARBOXAMIDE FORMYLTRANSFERASE

This enzyme [EC 2.1.2.3], also known as AICAR transformylase, catalyzes the reaction of 10-formyltetrahydrofolate with 5′-phosphoribosyl-5-amino-4-imidazole-carboxamide to produce tetrahydrofolate and 5′-phosphoribosyl-5-formamido-4-imidazolecarboxamide.

V. Schirch (1998) *Comprehensive Biological Catalysis: A Mechanistic Reference* **1**, 211.

PHOSPHORIBOSYL-AMP CYCLOHYDROLASE

This enzyme [EC 3.5.4.19] catalyzes the hydrolysis of 5-phosphoribosyl-AMP to produce 5-(5-phospho-D-ribosylaminoformimino)-1-(5-phosphoribosyl)imidazole-4-carboxamide. The enzyme isolated from *Neurospora crassa* can also catalyze the reactions of histidinol dehydrogenase and phosphoribosyl-ATP pyrophosphatase.

R. G. Martin, M. A. Berberich, B. N. Ames, W. W. Davis, R. F. Goldberger & J. D. Yourno (1971) *Meth. Enzymol.* **17B**, 3.

PHOSPHORIBOSYLANTHRANILATE ISOMERASE

This enzyme [EC 5.3.1.24], also known as *N*-(5′-phosphoribosyl)anthranilate isomerase, catalyzes the interconversion of *N*-(5-phospho-β-D-ribosyl)anthranilate and 1-(2-carboxyphenylamino)-1-deoxy-D-ribulose 5-phosphate. In some organisms, this enzyme is part of a multifunctional protein, together with one or more components of the system for the biosynthesis of tryptophan (anthranilate phosphoribosyltransferase, indole-3-glycerol-phosphate synthase, anthranilate synthase, and tryptophan synthase).

R. Bentley (1990) *Crit. Rev. Biochem. Mol. Biol.* **25**, 307.
K. Kirschner, H. Szadkowski, T. S. Jardetzky & V. Hager (1987) *Meth. Enzymol.* **142**, 386.

PHOSPHORIBOSYL-ATP PYROPHOSPHATASE

This enzyme [EC 3.6.1.31] catalyzes the hydrolysis of 5-phosphoribosyl-ATP to produce 5-phosphoribosyl-AMP and pyrophosphate (or, diphosphate). The *Neurospora crassa* enzyme also catalyzes the reactions of histidinol dehydrogenase and phosphoribosyl-AMP cyclohydrolase.

R. G. Martin, M. A. Berberich, B. N. Ames, W. W. Davis, R. F. Goldberger & J. D. Yourno (1971) *Meth. Enzymol.* **17B**, 3.

N-(5′-PHOSPHORIBOSYLFORMIMINO)-5-AMINO-1-(5″-PHOSPHORIBOSYL)-4-IMIDAZOLECARBOXAMIDE ISOMERASE

This enzyme [EC 5.3.1.16] catalyzes the interconversion of *N*-(5′-phospho-D-ribosylformimino)-5-amino-1-(5″-

phosphoribosyl)-4-imidazolecarboxamide and *N*-(5'-phospho-D-1'-ribulosylformimino)-5-amino-1-(5"-phosphoribosyl)-4-imidazolecarboxamide.

R. G. Martin, M. A. Berberich, B. N. Ames, W. W. Davis, R. F. Goldberger & J. D. Yourno (1971) *Meth. Enzymol.* **17B**, 3.

PHOSPHORIBOSYLGLYCINAMIDE FORMYLTRANSFERASE

This enzyme [EC 2.1.2.2], also known as 5'-phosphoribosylglycinamide transformylase, catalyzes the reaction of 10-formyltetrahydrofolate with 5'-phosphoribosylglycinamide to produce tetrahydrofolate and 5'-phosphoribosyl-*N*-formylglycinamide.

V. Schirch (1998) *Comprehensive Biological Catalysis: A Mechanistic Reference* **1**, 211.
J. I. Rader & F. M. Huennekens (1973) *The Enzymes*, 3rd ed., **9**, 197.

PHOSPHORUS-32 (^{32}P)

A radioactive β-emitting phosphorus nuclide with a half-life of 14.28 days, and a beta energy of 1.71 MeV. The following decay data indicate the time (in days) followed by the fraction of original amount at the specified time: 0.0, 1.000; 0.5, 0.976; 1.0, 0.953; 2.0, 0.908; 3.0, 0.865; 4.0, 0.824; 5.0, 0.785; 10.0, 0.616; 15, 0.483; 20, 0.379; 25, 0.297; 30, 0.233; 35, 0.183; 40, 0.144; 45, 0.113; 50, 0.088; 55, 0.069; 60, 0.054.

Special precautions: The maximum range in air is 6 meters, and the maximum range in water is 8 mm. Because bone concentrates phosphate, ^{32}P is the critical organ in terms of dose. Store this radioisotope behind 1–3 cm Lucite™ shielding. Avoid working over open containers, and exercise institutional guidelines on the type and location of dosimeters.

PHOSPHORYLASE KINASE

This enzyme [EC 2.7.1.38], also known dephosphophosphorylase kinase, catalyzes the reaction of four ATP with two phosphorylase *b* to produce four ADP and phosphorylase *a*.

E. G. Krebs (1986) *The Enzymes*, 3rd ed., **17**, 3.
G. M. Carlson, P. J. Bechtel & D. J. Graves (1979) *Adv. Enzymol.* **50**, 41.
D. A. Walsh & E. G. Krebs (1973) *The Enzymes*, 3rd ed., **8**, 555.

PHOSPHORYLASE PHOSPHATASE

This enzyme [EC 3.1.3.17] catalyzes the hydrolysis of phosphorylase *a* to produce two phosphorylase *b* and four orthophosphate.

L. M. Ballou & E. H. Fischer (1986) *The Enzymes*, 3rd ed., **17**, 311.
N. D. Madsen (1986) *The Enzymes*, 3rd ed., **17**, 365.
D. J. Graves & T. M. Martensen (1980) *Meth. Enzymol.* **64**, 325.

PHOSPHORYLATION STATE RATIO

A metabolic status parameter (abbreviated R_p) that indicates the phosphorylating power of ATP in cells as well as the availability of ADP and orthophosphate for oxidative phosphorylation: $R_p = [ATP]/\{[ADP][Orthophosphate]\}$. This ratio is buffered on the immediate time scale by substrate-level phosphorylation of ADP and on the intermediate time scale by mitochondrial phosphorylation of ATP which is then translocated to the cytoplasm and other compartments.

PHOSPHOSERINE AMINOTRANSFERASE

This pyridoxal-phosphate-dependent enzyme [EC 2.6.1.52] catalyzes the reaction of *O*-phospho-L-serine with α-ketoglutarate (or, 2-oxoglutarate) to produce 3-phosphonooxypyruvate and L-glutamate.

A. E. Braunstein (1973) *The Enzymes*, 3rd ed., **9**, 379.

3-PHOSPHOSHIKIMATE 1-CARBOXYVINYLTRANSFERASE

This enzyme [EC 2.5.1.19], also known as 5-enolpyruvylshikimate-3-phosphate synthase and 3-enolpyruvoylshikimate-5-phosphate synthase, catalyzes the reaction of phosphoenolpyruvate with 3-phosphoshikimate to produce orthophosphate and 5-*O*-(1-carboxyvinyl)-3-phosphoshikimate.

K. J. Gruys & J. A. Sikorski (1998) *Comprehensive Biological Catalysis: A Mechanistic Reference* **1**, 273.
A. Lewendon & J. R. Coggins (1987) *Meth. Enzymol.* **142**, 342.
D. M. Mousdale & J. R. Coggins (1987) *Meth. Enzymol.* **142**, 348.

PHOSPHOTRANSFERASE SYSTEM

The following are recent reviews on the molecular and physical properties of this transferase involved in group translocation.

M. Feese, D. W. Pettigrew, N. D. Meadow, S. Roseman & S. J. Remington (1994) *Proc. Natl. Acad. Sci. U.S.A.* **91**, 3544.
B. Erni (1992) *Int. Rev. Cytology* **137A**, 127.
N. D. Meadow, D. K. Fox & S. Roseman (1990) *Ann. Rev. Biochem.* **59**, 497.

PHOSPHOTYROSINE PROTEIN PHOSPHATASE

This enzyme [EC 3.1.3.48], also known as protein-tyrosine-phosphatase, catalyzes the hydrolysis of a phosphorylated tyrosyl residue in a phosphoprotein (the result of the action of a protein-tyrosine kinase) to produce a dephosphorylated tyrosyl residue and orthophosphate.

See also *Phosphoprotein Phosphatases*

A. C. Hengge (1998) *Comprehensive Biological Catalysis: A Mechanistic Reference* **1**, 517.
Z-Y. Zhang & J. E. Dixon (1994) *Adv. Enzymol.* **68**, 1.
R. L. Stone & J. E. Dixon (1994) *J. Biol. Chem.* **269**, 31323.
L. M. Ballou & E. H. Fischer (1986) *The Enzymes*, 3rd ed., **17**, 311.

PHOTOACOUSTIC SPECTROSCOPY

A spectroscopic technique based on the photoacoustic effect—a phenomenon occurring when a sample is irradiated with monochromatic light passing through a beam chopper at a frequency in the acoustical frequency range, such that light absorption causes periodic pressure fluctuations that are detected by a sensitive microphone. This technique allows one to investigate the electronic and infrared spectral properties of molecules in solid and semi-solid samples. For example, one can detect the oxyhemoglobin spectrum in whole blood or red blood cells directly. Furthermore, spectra can be obtained from samples present as spots on thin-layer chromatograms.

See *Fourier Transform IR/Photoacoustic Spectroscopy to Assess Secondary Structure*

J. W. Lin & L. P. Dubek (1979) *Anal. Chem.* **51**, 1627.
A. Rosencwaig (1980) *Photoacoustics and Photoacoustic Spectroscopy*, Wiley, New York.

PHOTOAFFINITY LABELING

A technique and mechanistic tool, pioneered by American chemist Frank Westheimer, in which a photochemically reactive substance is photoexcited in order to covalently attach a molecular entity to another chemical species, often a biomolecule. The advantage of this technique is that photochemically generated species are highly reactive and tend to be promptly captured by electrophiles in the near vicinity of the ligand, such that ligand release typically does not compete with the photo-cross-linking event. A disadvantage is that unbound photoaffinity ligands also absorb light and tend to react with solutes and/or solvent, thereby reducing the amount of useful reagent.

Chowdhry & Westheimer[1] described many of the strategies and applications of photoaffinity labeling to the investigation of biological systems. (***See*** *Azido Photoaffinity Reagents*) To provide an illustrative case, we describe

the recent studies of Bukhtiyarov *et al.*[2] who used photoreactive analogues of prenyl diphosphates with different carbon chain lengths as probes of recombinant human protein prenyltransferases. A geranylgeranyl diphosphate analogue, 2-diazo-3,3,3-trifluoropropionyloxy-farnesyl diphosphate (DATFP-FPP), was found to be the most effective inhibitor of both protein farnesyltransferase (PFT) and protein geranylgeranyltransferase-I (PFFT-I). Shorter photoreactive isoprenyl diphosphate analogues with geranyl and dimethylallyl moieties and the DATFP derivative of farnesyl monophosphate were much poorer inhibitors. DATFP-FPP was a competitive inhibitor of both PFT and PGGT-I with K_i values of 100 and 18 nM, respectively. [^{32}P]DATFP-FPP specifically photoradiolabeled the β-subunits of both PFT and PGGT-I. Photoradiolabeling of PGGT-I was inhibited more effectively by geranylgeranyl diphosphate than farnesyl diphosphate, whereas photoradiolabeling of PFT was inhibited better by farnesyl diphosphate than geranylgeranyl diphosphate. These findings led the authors to conclude that the β-subunits of protein prenyltransferases must contribute significantly to the recognition and binding of the isoprenoid substrate. ***See also*** *Azido Photoaffinity Reagents; Affinity Labeling*

[1]V. Chowdhry & F. H. Westheimer (1979) *Annu. Rev. Biochem.* **48**, 293.
[2]Y. E. Bukhtiyarov, C. A. Omer & C. M. Allen (1995) *J. Biol. Chem.* **270**, 19035.

PHOTOCHEMICAL REACTION

A chemical reaction induced by the absorption of a photon having energy corresponding to electromagnetic radiation in the ultraviolet or visible range.

PHOTOCHROMISM

A photo-induced transformation of a molecular entity in which there is a spectral change (usually, but not always, in the visible region of the spectrum). Such processes are photochemically or thermally reversible.

PHOTO-CROSS-LINKING

The photochemically induced formation of a covalent linkage between two different molecules (usually biomacromolecules) or between different moieties in the same molecule. ***See*** *Photoaffinity Labeling*

PHOTODIMERIZATION

A photochemical reaction in which the generation of free radicals results in dimerization of the starting molecule. The cross-linking of pyrimidine bases in DNA represents a fundamentally important example of photodimerization in biology.

PHOTOISOMERIZATION

A photochemical reaction in which the reactant is converted into an isomer. The most common example of such a reaction is photochemical *cis-trans* isomerization. Another example of photoisomerization is the effect of light on cholesta-3,5-diene.

PHOTOLYSIS

A bond-cleavage reaction that has been light-induced. The term has also been used to describe irradiation of a sample by light; however, such usage is discouraged. **See** *Flash Photolysis; Photoreactive Caged Compounds*

PHOTON COUNTING

The recording of sequential photon pulses for measurements of low levels of electromagnetic radiation as well as the recording of emission decays. The pulses are recorded from electron emission events from some photosensitive layer in conjunction with a photomultiplier system. **See also** *Time-Correlated Single Photon Counting; Fluorescence*

PHOTOOXIDATION

An oxidation reaction induced by light. For example, the loss of an electron as a result of photoexcitation or the reaction with O_2 due to the influence of light. **See** *Photooxygenation; Photoreduction*

PHOTOOXYGENATION

A photooxidative reaction in which molecular oxygen is incorporated into the reaction products(s). Three mechanisms appear to be common for such processes: (a) reaction of triplet O_2 with free radicals that have been generated photochemically; (b) reaction of photochemically produced singlet oxygen with a molecular species; and (c) the production of superoxide anion which then acts as the reactive species. **See also** *Photooxidation*

PHOTOPHYSICAL PROCESS

A photoexcited process that, while exhibiting no chemical change, leads to different states of a molecular entity by way of radiation and radiationless transitions. Light absorption is a photophysical process, as is fluorescence.

PHOTOREACTIVATION

The repair of macromolecules and biological systems by electromagnetic irradiation. This process is wavelength dependent. An example is DNA repair by certain enzymes.

PHOTOREACTIVE CAGED COMPOUNDS

Photosensitive precursors that generate chemically reactive species upon irradiation with a laser or flash-lamp. Suitably caged derivatives of ATP and other nucleotides are obtained by reaction of the metabolite with diazo compounds generated from 2-nitrobenzyldiazomethane. As pointed out in the excellent review of this topic by McCray & Trentham[1], 20 mJ from a near-UV light source (wavelength = 347 nm) can promptly liberate 2 mM ATP from a 5 mM initial concentration of caged P^3-1-(2-nitrophenyl)ethyl ester of ATP over an area of 10 mm^2 and at a depth of 0.1 mm. If the photolysis is sufficiently rapid relative to the later utilization of the liberated metabolite, one can synchronize the kinetics of the subsequent processes. Caged compounds have been used to investigate the kinetics of such diverse processes as muscle contraction, ion-pump ATPases, ion channels, receptors, bacterial flagella motility. **See** *Photolysis*

[1]J. A. McCray & D. R. Trentham (1989) *Ann. Rev. Biophys. Biophys. Chem.* **18**, 239.

PHOTOREDUCTION

A reduction reaction induced by the absorption of a photon in the UV or visible wavelength range of light; for example, the addition of one or more electrons to a photoexcited species and the photochemical hydrogenation of a substance.

PHOTOSYNTHESIS

The photoreductive synthetic process that promotes the assimilation of carbon dioxide into carbohydrates, other reduced metabolites, as well as ATP (synthesis of the latter is termed photophosphorylation). Photosynthesis is the primary mechanism for transducing solar energy into biomass, and green plants utilize chlorophyll *a* to capture a broad spectrum of solar radiant energy reaching the Earth's surface. Photosynthetic bacteria typically produce NADPH, the reductive energy of which is converted to ATP.

PHOTOSYNTHETIC REACTION CENTER

The following are recent reviews on the molecular and physical properties of this plant redox system.

H. B. Gray & J. R. Winkler (1996) *Ann. Rev. Biochem.* **65**, 537.
J. Deisenhofer & J. R. Norris, eds. (1993) *The Photosynthetic Reaction Center*, vols. I & II, Academic Press, Orlando.
M. Y. Okamura & G. Feher (1992) *Ann. Rev. Biochem.* **61**, 861.
J. B. Howard & D. C. Rees (1991) *Adv. Protein Chem.* **42**, 199.
D. C. Rees & D. Farrelly (1990) *The Enzymes*, 3rd ed., **19**, 37.

PHOTOSYNTHETIC WATER OXIDATION

The oxidation of water to dioxygen occurs as the consequence of Photosystem II-dependent generation of a very strong oxidant. Protons liberated by the water-oxidation reaction then contribute to the thylakoid transmembrane electrochemical gradient that drives ATP synthesis. Brudvig *et al.*[1] describe how flash-induced proton-release measurements have resolved key steps that provide insights on how the O$_2$-evolving center of PSII mediates this four-electron oxidation of water.

[1]G. W. Brudvig, W. F. Beck & J. C. de Paula (1989) *Ann. Rev. Biophys. Chem.* **18**, 25.

PHOTOSYSTEM I

A macromolecular complex that allows plants to harvest the sun's photic energy by absorbing photons and using their energy to catalyze photooxidation of plastocyanin, the copper protein situated in the lumen of thylakoid membranes, which undergoes subsequent electron transfer reactions. These reactions are illustrated in Fig. 1.

The following are recent reviews on the molecular and physical properties of Photosystems I and II.

H. Levanan & K. Möbius (1997) *Ann. Rev. Biophys. & Biomol. Struct.* **26**, 495.
B. Hankamer, J. Barber & E. J. Boekema (1997) *Ann. Rev. Plant Physiol. Plant Mol. Biol.* **48**, 641.
P. Horton, A. V. Ruban & R. G. Walters (1996) *Ann. Rev. Plant Physiol. Plant Mol. Biol.* **47**, 655.
B. R. Green & D. G. Durnford (1996) *Ann. Rev. Plant Physiol. Plant Mol. Biol.* **47**, 685.
S. Krömer (1995) *Ann. Rev Plant Physiol. Plant Mol. Biol.* **46**, 45.
B. A. Barry (1995) *Meth. Enzymol.* **258**, 303.
J. Barber, ed. (1992) *The Photosystems: Structure, Function & Molecular Biology*, Elsevier, New York.
G. W. Brudvig (1991) *Acc. Chem. Res.* **24**, 311.
L. Bogorad & I. K. Vasil (eds.) (1991) *The Photosynthetic Apparatus*, Academic Press, Orlando.

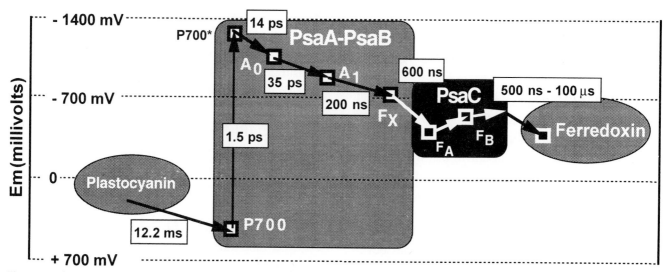

Figure 1. Electron transfer in Photosystem I. The values plotted in the vertical direction are the mid-point potentials, and the numbers next to each arrow are the half-lives for electron transfer. P700 is the primary reaction center; PsaA and PsaB are transmembrane proteins; and PsaC is a peripheral cytoplasmic component. Courtesy of Professor Parag Chitnis, Iowa State University.

pH-RATE PROFILE

A plot of the rate constant, the initial velocity, the maximum velocity of a catalyzed reaction, or the V_{max}/K_m ratio of an enzyme-catalyzed reaction (or the decadic logarithm of any of these quantities) as a function of the pH value of the solution, all other variables being held constant. *See also* pH Effects

pH STAT

Any electromechanical device that utilizes an automated feedback servomotor to regulate the addition of titrant (a standardized solution of acid or base within a syringe) into a reaction vessel or sample to maintain pH. The rate at which the syringe expels its contents allows one to determine the rate of a chemical reaction producing or consuming protons. There are many such enzyme-catalyzed reactions whose kinetics can be examined with a pH Stat. For maximal sensitivity, one must use weakly buffered solutions. In his classical kinetic investigation of DNA bond scission by DNase, Thomas[1] measured the rate of base addition in a pH Stat. The number of bonds cleaved was linear with time, and this was indicative of random scission.

Another means for following the kinetics of reactions that produce or consume protons is through the use indicator dyes and spectrophotometry, and an excellent example is the examination of hexokinase kinetics[2]. In such cases, the pK of the indicator dye must be carefully chosen to assure maximal sensitivity.

Although the glass electrode displays sufficient sensitivity and time response for many chemical reactions, special circumstances may require an ultrathin glass electrode to obtain millisecond response times. Less sensitive types of the pH Stat are also used in fermentation chambers to determine the rate/extent of fermentation or to maintain the viability and efficiency of microbes during fermentation.

[1]C. A. Thomas, Jr. (1956) *J. Am. Chem. Soc.* **78**, 1861.
[2]G. G. Hammes & D. Kochavi (1962) *J. Am. Chem. Soc.* **84**, 2069, 2073, and 2076.

pH VALUE

The symbol pH for $-\log[H^+]$, where $[H^+]$ is the hydrogen ion concentration (or, alternatively, pH $= -\log a_{H^+}$ where a_{H^+} is the hydrogen ion activity.) The two letters "p" and "H" stand for: "potency of H". Typically, pH values are measured with glass electrodes in which the electromotive force (e.m.f.) of an unknown solution is converted to a pH measurement. *See* pH Measurement

S. P. L. Sörensen (1909) *Biochem. Zeit.* **21**, 131.
A. Albert & E. P. Serjeant (1984) *The Determination of Ionization Constants*, 3rd ed., Chapman and Hall, London.

pH* VALUE

The apparent pH value of an aqueous-organic solvent system; an important factor in cryoenzymology experiments[1]. In such cases, pH* $= -\log a_H^{+*}$ where a_H^{+*} is the proton activity in the mixed solvent.

[1]A. L. Fink & M. A. Geeves (1979) *Meth. Enzymol.* **63**, 336.

PHYSICAL ORGANIC CHEMISTRY TERMINOLOGY (IUPAC Recommendations)

If the word you are searching for in the Word Finder directed you here, then it is a term that lies beyond the scope of this Handbook. The listed word will be found in an authoritative internet-accessible glossary dealing with physical organic chemistry: http://www.chem.qmw.ac.uk/iupac/gtpoc/. The glossary was developed under the auspices of the IUPAC Organic Chemistry Division's Commission on Nomenclature of Organic Chemistry and Commission on Physical Organic Chemistry. All definitions used in the internet glossary are identical to those in the published document[1], which the reader should cite if the definition is used verbatim or nearly so.

[1]P. Müller (1994) *Pure and Appl. Chem.* **66**, 1077.

6-PHYTASE

This enzyme [EC 3.1.3.26], also known as phytase, phytate 6-phosphatase, and *myo*-inositol-hexaphosphate 6-phosphohydrolase, catalyzes the hydrolysis of *myo*-inositol hexakisphosphate to produce 1-*myo*-inositol 1,2,3,4,5-pentakisphosphate and orthophosphate.

F. A. Loewus & M. W. Loewus (1983) *Ann. Rev. Plant Physiol.* **34**, 137.

PHYTOENE SYNTHASE COMPLEX

Phytoene synthase [EC 2.5.1.32] (also known as geranylgeranyl-diphosphate geranylgeranyltransferase and prephytoene-diphosphate synthase) catalyzes the reaction of two geranylgeranyl diphosphate to produce pyrophosphate (or, diphosphate) and prephytoene diphosphate. Isopentenyl pyrophosphate isomerase [EC 5.3.3.2] catalyzes the interconversion of isopentenyl diphosphate and dimethylallyl diphosphate. **See also** *Geranylgeranyl Diphosphate Geranylgeranyltransferase*

R. A. Gibbs (1998) *Comprehensive Biological Catalysis: A Mechanistic Reference* **1**, 31.
J. Chappel (1995) *Ann. Rev. Plant Physiol. Plant Mol. Biol.* **46**, 521.
B. Camara (1993) *Meth. Enzymol.* **214**, 352.
O. Dogbo, A. Laferrière, A. d'Harlingue & B. Camara (1988) *Proc. Natl. Acad. Sci. U.S.A.* **85**, 7054.

PI (II, π)

Π 1. Symbol for osmotic pressure. 2. Symbol for the mathematical product of a series. π 1. Symbol for surface pressure. 2. Symbol for the ratio of the circumference of a circle to its diameter (3.14159...).

π-BOND (PI-Bond)

A bond (a π molecular orbital) between adjacent atoms that has a nodel plane which includes the internuclear bond axis.

$\pi \rightarrow \pi^*$ TRANSITION

An electronic transition in which a π orbital electron is promoted to an antibonding orbital. The excited state arising from such a promotion is often referred to as a $\pi-\pi^*$ state. **See** *Absorption Spectroscopy*

$\pi \rightarrow \sigma^*$ TRANSITION

An electronic transition in which an π orbital electron is promoted to a σ antibonding orbital. **See** *Absorption Spectroscopy*

PICO-

A prefix, symbolized by p, used in submultiples of units and corresponding to a value of 10^{-12}. **See also** *Nano-; Femto-*

PICORNAIN

Picornain 2A [EC 3.4.22.29] catalyzes the hydrolysis of peptide bonds including the selective cleavage of a particular peptide bond in the picornavirus polyprotein. Picornain 3C [EC 3.4.22.28] catalyzes the selective hydrolysis of the Gln—Gly bond in the poliovirus polyprotein.

T. Skern & H.-D. Liebig (1994) *Meth. Enzymol.* **244**, 583.

PING PONG BI BI MECHANISM

An enzyme-catalyzed "double-displacement" reaction mechanism in which two substrates react to form two products, but one product must be released before the second substrate binds to the enzyme.

$$\mathrm{E} + \mathrm{A} \underset{k_2}{\overset{k_1}{\rightleftharpoons}} \mathrm{EX} \underset{k_4}{\overset{k_3}{\rightleftharpoons}} \mathrm{F} + \mathrm{P}$$

$$\mathrm{F} + \mathrm{B} \underset{k_6}{\overset{k_5}{\rightleftharpoons}} \mathrm{FY} \underset{k_8}{\overset{k_7}{\rightleftharpoons}} \mathrm{E} + \mathrm{Q}$$

In the scheme above, E represents free enzyme, A and B are the substrates, P and Q are the products, EX represents the binary complexes for the first substrate-product pair (EA $\rightleftharpoons$ ... $\rightleftharpoons$ FP), F is a modified form of the enzyme, and FY represents the binary complexes of the second substrate-product pair (FB $\rightleftharpoons$ ... $\rightleftharpoons$ EQ). A large number of enzymes have been reported to have this type of binding mechanism, including the aminotransferases, adenine phosphoribosyltransferase, glucose oxidase, coenzyme A transferases, *etc.*

The steady-state derivation of the ping pong Bi Bi mechanism provides the following determinants for each of the enzyme forms:

$$[E] = k_5 k_7 (k_2 + k_3)[B] + k_2 k_4 (k_6 + k_7)[P]$$

$$[EX] = k_1 k_5 k_7 [A][B] + k_1 k_4 (k_6 + k_7)[A][P] + k_4 k_6 k_8 [P][Q]$$

$$[F] = k_1 k_3 (k_6 + k_7)[A] + k_6 k_8 (k_2 + k_3)[Q]$$

$$[FY] = k_1 k_3 k_5 [A][B] + k_5 k_8 (k_2 + k_3)[B][Q] + k_2 k_4 k_8 [P][Q]$$

Under initial rate conditions where [P] and [Q] are effectively zero, the steady-state expression for this reaction scheme is $v = V_{max}[A][B]/\{K_b[A] + K_a[B] + [A][B]\}$ where $V_{max} = k_3 k_7 [E_{total}]/(k_3 + k_7)$, $K_a = k_7(k_2 + k_3)/\{k_1(k_3 + k_7)\}$, and $K_b = k_3(k_6 + k_7)/\{k_5(k_3 + k_7)\}$. Note that the denominator does not contain a substrate-independent term (*i.e.*, a term not multiplied by either [A] or [B] or both). This characteristic is distinctive from ordered or random mechanisms.

A double-reciprocal plot of $1/v$ *vs.* $1/[A]$ at varying levels of [B] will yield a series of parallel lines for this reaction scheme (having no abortives or isomerization steps), the slope equal to K_a/V_{max}, the vertical intercepts equal to $(1/V_{max})(1 + K_b/[B])$, and the horizontal intercepts equal to $1/[A] = -(1/K_a)(1 + K_b/[B])$. Thus, a replot of the vertical intercepts *vs.* $1/[B]$ will produce a straight line having a slope of K_b/V_{max}, a vertical intercept of $1/V_{max}$, and a horizontal intercept of $1/[B] = -1/K_b$. Similarly, a secondary replot of $1/v$ *vs.* $1/[B]$ data will provide a value for K_a.

The complete steady-state rate expression for the ping pong Bi Bi reaction is

$$v =$$

$$\frac{V_{max,f}V_{max,r}([A][B] - [P][Q]/K_{eq})}{\left\{ \begin{array}{l} V_{max,r}K_b[A] + V_{max,r}K_a[B] + V_{max,r}[A][B] + \dfrac{V_{max,f}K_q[P]}{K_{eq}} \\[2mm] + \dfrac{V_{max,f}K_p[Q]}{K_{eq}} + \dfrac{V_{max,f}[P][Q]}{K_{eq}} + \dfrac{V_{max,f}K_q[A][P]}{K_{ia}K_{eq}} + \dfrac{V_{max,r}K_a[B][Q]}{K_{iq}} \end{array} \right\}}$$

where $V_{max,f}$ is identical to V_{max} in the previous expression, $V_{max,r} = k_2 k_6 [E_{total}]/(k_2 + k_6)$, $K_q = k_2(k_6 + k_7)/\{k_8(k_2 + k_6)\}$, $K_p = k_8(k_2 + k_3)/\{k_4(k_2 + k_6)\}$, $K_{ia} = k_2/k_1$, and $K_{iq} = k_7/k_8$. Then the Haldane relationships are $K_{eq} = k_1 k_3 k_5 k_7/\{k_2 k_4 k_6 k_8\} = K_{ip}K_{iq}/(K_{ia}K_{ib}) = V_{max,f}K_{ip}K_q/(V_{max,r}K_{ia}K_b) = V_{max,f}K_pK_{iq}/(V_{max,r}K_aK_{ib}) = V_{max,f}^2 K_P K_q/(V_{max,r}^2 K_a K_b)$.

Another important characteristic of ping pong reaction is the presence of exchange reactions in the absence of other substrates. **See** *Isotope Exchange at Equilibrium; Multisite Ping Pong Bi Bi Mechanism; Half-Reaction; Substrate Synergism*

PLANCK FUNCTION

A thermodynamic function, symbolized by Y, equal to the negative of the Gibbs free energy divided by the absolute temperature: $Y = -G/T$. The SI units are joules per kelvin. **See also** *Gibbs Free Energy; Massieu Function*

PLANCK'S CONSTANT

A proportionality constant, symbolized by h, that relates the frequency of electromagnetic radiation to its quanta of energy (*i.e.*, $E = h\nu$): $h = 6.6260755 \times 10^{-34}$ J·s (or $6.6260755 \times 10^{-27}$ erg·seconds). The quantity $\hbar$ equals $h/2\pi$.

PLASMALOGEN SYNTHASE

This enzyme [EC 2.3.1.25], also known as 1-alkenyl-glycerophosphocholine acyltransferase, catalyzes the reaction of an acyl-CoA with 1-*O*-alk-1-enylglycero-3-phosphocholine to produce coenzyme A and plasmenyl-choline.

P. C. Choy & C. R. McMaster (1992) *Meth. Enzymol.* **209**, 86.

PLASMANYLETHANOLAMINE Δ^1-DESATURASE

This magnesium-dependent enzyme [EC 1.14.99.19], also known as alkylacylglycerophosphoethanolamine desaturase, catalyzes the reaction of an O-1-alkyl-2-acyl-*sn*-glycero-3-phosphoethanolamine with AH_2 and dioxygen to produce O-1-alk-1-enyl-2-acyl-*sn*-glycero-3-phosphoethanolamine, A, and two water. Either NADPH or NADH can serve as the AH_2 substrate. The reaction may require the participation of cytochrome b_5 and it has been reported that the enzyme requires ATP.

M. L. Blank & F. Snyder (1992) *Meth. Enzymol.* **209**, 390.

PLASMIN

Plasmin [EC 3.4.21.7], also known as fibrinase and fibrinolysin, is a peptidase (a member of the peptidase family S1) that exhibits preferential cleavage at Lys—Xaa > Arg–Xaa (there is actually greater selectivity than displayed by trypsin). Plasmin converts fibrin into soluble products. It is formed from plasminogen by proteolysis, resulting in multiple forms of the active plasmin.

F. B. Ablondi & J. J. Hagan (1960) *The Enzymes*, 2nd ed., **4**, 176.

[31]P MAGNETIC RESONANCE OF CONTRACTILE SYSTEMS

Because muscle cells are especially rich in terms of phosphorus-containing metabolites (*e.g.*, ATP, ADP, phosphocreatine, and orthophosphate), nuclear magnetic resonance[1-3] has proved to be a valuable noninvasive probe of metabolic changes attending muscle activity. The spectral sensitivity of ^{31}P is especially high relative to other nuclei, and one can detect cellular concentrations as low as 0.5 mM as well as utilize chemical shift data to define intracellular pH and free magnesium ion concentrations.

See also *Nuclear Magnetic Resonance; Chemical Shift*

[1]D. I. Hoult, S. J. W. Busby, D. G. Gadian, G. K. Radda, R. E. Richards & P. J. Seeley (1974) *Nature* **252**, 285.
[2]M. Barany & T. Glondek (1982) *Meth. Enzymol.* **85**, 624.
[3]M. J. Dawson, D. G. Gadian & D. R. Wilkie (1977) *J. Physiol. (London)* **267**, 703.

pMg

The negative base$_{10}$ logarithm of the magnesium ion concentration.

POINT-OF-CONVERGENCE METHOD

A procedure that assists in the characterization of binding mechanisms for sequential (*i.e.*, non-ping pong) reactions[1,2]. The same general initial rate expression applies to the steady-state ordered Bi Bi reaction, the rapid-equilibrium random Bi Bi reaction, and the Theorell-

Chance Bi Bi mechanism. (This expression also applies to isomerization mechanisms for the ordered and Theorell-Chance schemes in which a stable enzyme form undergoes isomerization.) The points of intersection for each different mechanism can be used to exclude certain possibilities, depending on the quality of the rate data.

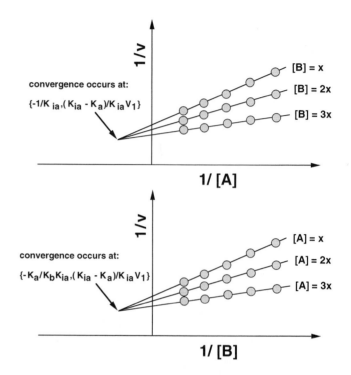

Another growing application for the technique is in evaluating mechanistic differences between wild-type enzymes and enzymes obtained from site-directed mutagenesis, as well as between different isozymes, between enzymes from different sources, or between the same enzyme acting on alternative substrates.

[1]J. D. Lueck, W. R. Ellison & H. J. Fromm (1973) *FEBS Lett.* **30**, 321.
[2]F. B. Rudolph & H. J. Fromm (1979) *Meth. Enzymol.* **63**, 138.

POINT OF INFLECTION

A location on a curve identifiable when progress from a position prior to that point to a position beyond that point requires crossing of a tangent line to that point. In calculus, the point of inflection occurs wherever the second derivative is zero and changes sign of $f''(x)$ on either side of that point.

POISE

A unit (symbolized by P) of dynamic viscosity. One poise is equivalent to one dyne-second per square centimeter, to one gram per centimeter-second, and to 0.1 pascal-second. This is the tangential force in dynes needed per unit area (in cm^2) to maintain a difference in velocity of one cm per second between two parallel plates separated by one cm of fluid.

POISSON DISTRIBUTION

The probability distribution for the number of events (occurring randomly at a rate μ) corresponding to the function

$$f(x) = (\mu^x/x!) \exp[-\mu]$$

where x equals $0, 1, 2, \ldots$. This probability density function is actually a limiting case of the binomial distribution. **See** Statistics (A Primer)

POLAR EFFECT

1. The effect of the relative electronegativity of a substituent (R) and/or the delocalization of electrons on a chemical property of a substance. Thus, for a molecule R$-$R', the polar effect refers to all nonsteric influences and modifications of electrostatic forces operating at the reaction center (R'), relative to some standard molecule, R$_o$$-$R'. Hence, the term is synonymous with the electronic effect.

2. The nonsteric influence exerted by nonconjugated substituents on reaction rates. In this case, electron delocalization between the substituent and the rest of the molecule is not considered, and the polar effect is not synonymous with the electronic effect. **See** Inductive Effect; Field Effect

POLARITY

1. With respect to a solvent, the overall solvation capability for solutes. 2. A property of bodies or systems that have a distinct direction; i.e., that have different or opposing physical properties or characteristics at different points. For example, an amino acid sequence in a polypeptide has polarity in that there is an amino end and a carboxyl end of the sequence. Similarly, microtubules and actin filaments have plus (+)-ends and minus (−)-ends that establish directionality for cellular and intracellular locomotion. 3. The state in which there is either a positive or negative aspect relative to the two poles of a magnet or to electrification. 4. Attraction toward an object or attraction in a specific direction. 5. In mathematics, the positive or negative sign of numbers.

POLARIZABILITY

A proportionality factor measuring the ease of distortion of the electron cloud of a chemical species as that species is subjected to an electric field (such as that artificially applied or due to the proximity of a charged reagent). The term "electric polarizability" has also been used. Polarizability (symbolized by the Greek letter "α") is determined by measuring the magnitude of the induced electric dipole moment of the molecule, $\mu_{induced}$, and the size of the local electric field strength, E, that is the source of the induced effect. Thus, $\mu_{induced} = \alpha E$. (In the more general case in which the direction of the induced dipole moment is not the same as the direction of the field, $\mu_{induced} = \tilde{\alpha} E$ where $\tilde{\alpha}$ is a tensor). The polarizability is the induced moment per unit field strength. In most cases, α is actually a mean polarizability, being an average over the three Cartesian coordinates of the molecule, and can be related to the index of refraction. With some molecules (e.g., diatomic species such as O$_2$) distinctions can be made between longitudinal polarizability (i.e., polarizability along the bond) and transverse polarizability (i.e., perpendicular to the bond).

C. K. Ingold (1953) Structure and Mechanism in Organic Chemistry, Cornell Univ. Press, New York.
R. W. Taft, Jr., & R .D. Topson (1987) Progr. Phys. Org. Chem. **16**, 1.

POLARIZATION

1. Any action or phenomenon leading to a separation of electrical charge and/or magnetization, including: the linear and circular polarization of light, the unequal electron transfer across an electrode or battery pole (such that $i_\rightarrow \neq i_\leftarrow$), and separation of ionic or electric charge within molecules. 2. Any positional separation of biological components into morphologically differentiated areas at opposite ends of a living cell. Examples of the latter include: polarized distribution of membrane molecules that depend on tight junctions; antigen presentation on cell surfaces; localization of glucose transporters; epithelial polarity that is essential for vectorial reabsorption and secretion of ions; as well as the differential insertion of channels, transporters, and related proteins into apical or basolateral membranes. 3. Any separation of ions across a selectively permeable membrane, including those achieved biologically through the action of ion pumps as well as the separation of electrolyte solutions by reverse osmosis.

POLY(ADP-RIBOSE) SYNTHETASE

This enzyme [EC 2.4.2.30] (also referred to as NAD^+ ADP-ribosyltransferase, poly(ADP) polymerase, poly-(adenosine diphosphate ribose) polymerase, and ADP-ribosyltransferase (polymerizing)) catalyzes the reaction of NAD^+ with $[ADP\text{-}D\text{-}ribosyl]_n$ to produce nicotinamide and $[ADP\text{-}D\text{-}ribosyl]_{(n+1)}$. The ADP-D-ribosyl group of NAD^+ is transferred to an acceptor carboxyl group on a histone or on the enzyme itself, and further ADP-ribosyl groups are transferred to the $2'$-position of the terminal adenosine moiety, building up a polymer with an average chain length of twenty to thirty units.

N. J. Oppenheimer & A. L. Handlon (1992) *The Enzymes*, 3rd ed., **20**, 453.

POLYGALACTURONASE

This enzyme [EC 3.2.1.15], also known as pectin depolymerase and pectinase, catalyzes the random hydrolysis of 1,4-α-D-galactosiduronic linkages in pectate and other galacturonans.

H. J. Phaff (1966) *Meth. Enzymol.* **8**, 636.

POLYMER FLEXING AND EXTENDING

Polymers spontaneously flex (or bend), and they can also extend. As pointed out by Oosawa and Asakura[1], any polymer behaving like a uniform rod will have an elastic (*i.e.*, reversible) modulus for flexing that is given by the following relationship:

$$\varepsilon = YI$$

where Y is the elastic modulus for stretching (often called the Young's modulus) and I is the moment of inertia ($I = \pi \alpha^4/4$ for a cylinder of uniform radius a). When an assembled polymer composed of protein subunits flexes, there develops a restoring potential energy that is related to the energy required to stretch such bonds from their equilibrium bond length to their extended length. Oosawa and Asakura[1] demonstrate how one can obtain the change in the free energy of intersubunit bonding within the polymer's lattice when the degree of stretching is relatively small. They also show that deformation by bending is far more pronounced than any deformation attributable to stretching.

[1] F. Oosawa & S. Asakura (1975) *Thermodynamics of Protein Polymerization*, pp. 162-168, Academic Press, New York.

POLYMERIZATION

1. The spontaneous self-assembly or template-directed assembly of component monomeric units into polymeric biological macromolecules. 2. The enzyme-catalyzed joining of monomeric units (such as amino acids, sugars, nucleotides) into covalently linked oligomeric or polymeric forms.

Biopolymers are typically formed from subunit components through covalent and/or noncovalent bonds. Examples of the covalently linked biopolymers include proteins, nucleic acids, and polysaccharides. Examples of biopolymers formed by noncovalent forces include hemoglobin ($\alpha_2\beta_2$), tubulin ($\alpha\beta$), and lactate dehydrogenase (M_4, M_3H_1, M_3H_1, M_2H_2, M_1H_3, and H_4 tetramers formed from muscle M and heart H polypeptide chains) or polymeric structures (such as microtubules, filamentous actin, and intermediate filaments). These polymerize spontaneously in response to changes in pH, temperature, or in the presence/absence of other solutes (metal ions, metabolites, or other inorganic anions). **See also** *Actin Assembly Kinetics; Microtubule Assembly Kinetics; Quasi-equivalence; Actin-Based Motility; Critical Concentration; Self-Assembly; Processivity*

POLYMERIZATION KINETICS (Irreversible)

Consider a series of irreversible polymerization reactions for which the same rate constant k: applies to the initial and all subsequent steps.

$$A_1 + A_1 \rightarrow A_2$$
$$A_2 + A_1 \rightarrow A_3$$
$$A_3 + A_1 \rightarrow A_4$$

etc.

The polymerization kinetics will be first-order with respect to $[A_{t=0}]$, and the polymerization reaction will exhibit zero-order kinetics for any set of rate measurements conducted at one fixed concentration of monomer.

If only monomer A_1 is initially present, the integrated rate law for this process can be written as:

$$[A_{poly,total}] = ([A_1] + [A_2] + [A_3] + \ldots + [A_n])$$

$$= [A_{t=0}]kt[(4 + kt)/(2 + kt)^2]$$

where $[A_{poly,total}]$ is the total amount of polymeric species, and $[A_{t=0}]$ is the initial monomer concentration. The polymerization rate, $d[A_{poly,total}]/dt$ is obtained by taking the derivative of the quotient[1], yielding the result:

$$d[A_{poly,total}]/dt = 8[A_t=0]k/(2 + kt)^3$$

[1]Recall that $d(u/v)/dx = \{(vdu/dx) - (udv/dx)\}/v^2 = \{(1/v)(du/dx)\} - \{(u/v^2)(dv/dx)\}$.

POLYNUCLEOTIDE 5'-HYDROXYL-KINASE

This enzyme [EC 2.7.1.78] catalyzes the reaction of ATP with 5'-dephospho-DNA to produce ADP and 5'-phospho-DNA. The enzyme can also act on 5'-dephospho-RNA 3'-mononucleotides. *See also* DNA Kinases

C. C. Richardson (1981) *The Enzymes*, 3rd ed., **14**, 299.
K. Kleppe & J. R. Lillehaug (1979) *Adv. Enzymol.* **48**, 245.

POLYNUCLEOTIDE PHOSPHORYLASE

This enzyme [EC 2.7.7.8], also known as polyribonucleotide nucleotidyltransferase, catalyzes the reaction of $[RNA]_{(n + 1)}$ with orthophosphate to produce $[RNA]_n$ and a nucleoside diphosphate. In the reverse reaction, ADP, IDP, GDP, UDP, and CDP can act as substrates.

U. Z. Littauer & H. Soreq (1982) *The Enzymes*, 3rd ed., **15**, 517.
T. Godefroy-Colburn & M. Grunberg-Manago (1972) *The Enzymes*, 3rd ed., **7**, 533.

POLYPHOSPHATE GLUCOKINASE

This enzyme [EC 2.7.1.63], also known as polyphosphate-glucose phosphotransferase, catalyzes the reaction of $[phosphate]_n$ with D-glucose to yield $[phosphate]_{(n-1)}$ and D-glucose 6-phosphate. The enzyme requires the presence of a neutral salt (such as potassium chloride) for maximum activity. Glucosamine can also serve as a substrate.

H. G. Wood & J. E. Clark (1988) *Ann. Rev. Biochem.* **57**, 235.

POLYPHOSPHATE KINASE

This enzyme [EC 2.7.4.1] catalyzes the reaction of ATP with $[phosphate]_n$ to produce $[phosphate]_{(n + 1)}$ and ADP.

H. G. Wood & J. E. Clark (1988) *Ann. Rev. Biochem.* **57**, 235.

POLYSACCHARIDE DEPOLYMERASE

This enzyme [EC 3.2.1.87], also called capsular-polysaccharide endo-1,3-α-galactosidase, catalyzes the random hydrolysis of 1,3-α-D-galactosidic linkages in the *Aerobacter aerogenes* capsular polysaccharide. It hydrolyzes the galactosyl-α-1,3-D-galactose linkages only in the complex substrate, bringing about depolymerization. *See also* Amylases; Lysozyme

A. Tsugita (1971) *The Enzymes*, 3rd ed., **5**, 343.

POOL SIZE

The total amount or mass of a tracee in a pool, corresponding to the volume of distribution (V) multiplied by the tracee concentration in the same pool; frequently abbreviated Q_T. *See* Compartmental Analysis

P/O RATIO

A measure of oxidative phosphorylation, equal to the ratio of the number phosphate groups esterified (*i.e.*, ATP formation from ADP and phosphate) relative to the atoms of oxygen consumed by the mitochondria.

PORPHOBILINOGEN DEAMINASE

This enzyme [EC 4.3.1.8], also known as hydroxymethylbilane synthase, catalyzes the dipyrromethane-dependent reaction of four porphobilinogen molecules with water to produce hydroxymethylbilane and four molecules of ammonia. In the presence of a second enzyme, uroporphyrinogen-III synthase [EC 4.2.1.75], the product is cyclized to form uroporphyrinogen-III.

G. V. Louie, P. D. Brownlie, R. Lambert, J. B. Cooper, T. L. Blundell, S. P. Wood, M. J. Warren, S. C. Woodstock & P. M. Jordan (1992) *Nature* **359**, 33.
S. Granick & S. I. Beale (1978) *Adv. Enzymol.* **46**, 33.

POSITIONAL ISOTOPE EXCHANGE ("PIX")

Positional isotope exchange ("PIX") is a very valuable technique in determining enzyme mechanisms, particularly those utilizing ATP, GTP, or another NTP substrate[1,2]. For example, the nucleotide substrate can be labeled with ^{18}O in bridging (*i.e.*, P—O—P or phosphoanhydride oxygen) and/or its nonbridging positions.

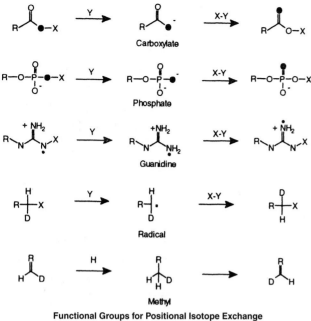

Functional Groups for Positional Isotope Exchange

By using ^{31}P NMR in the presence of the enzyme, an investigator can identify bridged-to-nonbridge (and the reverse) isotope exchanges and thereby identify probable intermediates on the reaction pathway. The procedure[2] is useful for any enzyme-catalyzed reaction in which the individual atoms of a functional group within a substrate, intermediate, or product become torsionally equivalent during the course of a reaction.

[1]I. A. Rose (1979) *Adv. Enzymol.* **50**, 361.
[2]L. S. Mullins & F. M. Raushel (1995) *Meth. Enzymol.* **249**, 398.

Selected entries from *Methods in Enzymology* [vol, page(s)]:
Adenylosuccinate synthetase, **249**, 423; D-alanine-D-alanine ligase, **249**, 417-418; aminoacyl-tRNA synthetases, **249**, 423; argininosuccinate lyase, **249**, 413-414; carbamoyl-phosphate synthetase, **249**, 418-423; CTP synthase, **249**, 423-424; functional groups for, **249**, 400-401; galactose-1-phosphate uridylyltransferase, **249**, 416-417; glutamine synthetase, **249**, 453; as mechanistic probe, **249**, 398; phosphoenolpyruvate carboxykinase, **249**, 424; Ping Pong mechanism analysis, **249**, 409-413; pyruvate-phosphate dikinase, **249**, 424; qualitative and quantitative approaches, **249**, 401-404; sequential mechanism analysis, **249**, 404-409 [enhancement, **249**, 404-407; inhibition, **249**, 407-409]; UDPglucose pyrophosphorylase, **249**, 414-416; variation of nonlabeled substrates and products, **249**, 404-413.

POSITIVE COOPERATIVITY

Any set of ligand interactions with oligomeric or polymeric macromolecules, such that binding of the first (or preceding) ligand molecule increases the likelihood for binding of the next (or subsequent) ligand molecule. In the Monod-Wyman-Changeux allosteric transition model, the dissociation constant for ligand interactions is equivalent for all sites, but cooperativity results from the disproportionate "recruitment" of binding sites from the so-called T-state in an all-or-none transition to the R-state. In the Koshland-Némethy-Filmer model, the dissociation constant for ligand binding to the $PL_{(i)}$ state is lower than the dissociation constant for ligand binding to the $PL_{(i-1)}$ state. **See** *Ligand Binding Cooperativity; Monod-Wyman-Changeux Model; Koshland-Némethy-Filmer Model; Hill Equation & Plot; Adair Equation; Allosterism; Linked Function; Hemoglobin; Aspartate Carbamoyltransferase; Scatchard Plot; Cooperativity; $\Delta\Delta G$ as Index of Cooperativity; Negative Cooperativity*

POSTTRANSLATIONAL MODIFICATION OF PROTEINS

Although allostery and protein turnover are effective regulatory mechanisms for enzymes, posttranslational modification represents another important means for achieving metabolic regulation. The range of enzyme-catalyzed posttranslational modification reactions is immense (*e.g.*, phosphorylation, adenylylation, ADPribosylation, hydroxylation, glycosylation, prenylation, fatty acylation, sulfation, proteolysis, acetylation, arginylation, glycylation, glutaminylation, glutamylation, *etc.*). Graves *et al.*[1] have provided a lucid account of the enzymology and kinetics of these modification reactions, including considerations of the structural consequences of such covalent modifications. Those interested in the stability and analysis of modified amino acids in proteins will wish to consult Table 4.2 in Graves *et al.*[1]

[1]D. J. Graves, B. L. Martin & J. H. Wang (1994) *Co- and Post-Translational Modification of Proteins: Chemical Principles and Biological Effects*, pp. 348ff, Oxford University Press, Oxford.

POTENTIAL ENERGY

The energy associated with position of a body or entity with respect to another body or entity or with respect to relative positions of parts within that same body. For an entity or body at position r associated with a force F, the potential energy, E_p or U, is $-\int F \cdot dr$. The strictly classical mechanical definition of potential energy typically leads one to consider a system in terms of stored energy and position, and in that case, one deals with the magnitude of force and distance between interacting bodies or particles. In this case, a chemical bond is treated by the so-called Hooke's Law model: a pair of atoms or substituents of mass m_A and m_B are attached at opposite ends of a vibrating spring, and the potential energy of the bond at any instant is determined by the reduced mass, $m_A m_B/(m_A + m_B)$, the force constant k of the spring, and a distance indicating whether the spring is compressed, extended, or exactly at its equilibrium position. However, as was clearly demonstrated a century ago by the physicist James Clerk Maxwell, the concept of potential energy also embodies the contributions of electric field effects and magnetic field effects, integrated over all space. Quantum mechanical treatments can rigorously specify the potential energy, but this is most frequently accomplished for very simple systems having only a few atoms. *See* *Energy; Kinetic Energy*

POTENTIAL ENERGY PROFILE

A depiction of the variation of the potential energy of the system of atoms or molecules (*e.g.*, the reactants, intermediates, transition states, and products of a particular reaction) as a function of a reaction coordinate. For an elementary reaction, this horizontal reaction coordinate represents the path of the energetically easiest passage of reactant(s) to product(s). For a series of elementary reactions (*e.g.*, a set of successive reaction steps), the horizontal coordinate is equivalent to a set of successive reaction coordinates. *See also* *Reaction Coordinate Diagram; Potential Energy Surface; Gibbs Free Energy Diagram*

POTENTIAL ENERGY SURFACE

A diagram depicting the variation of the potential energy associated with the reactants and products of an elementary reaction in which the energetically easiest progress (for movement of a representative point of the reaction system) is plotted as a function of two or three coordinates (usually representing molecular geometries) of a multidimensional, potential energy hypersurface. Typically, the diagram is a depiction having isoenergetic contour lines and resembles a topographic map. In most two-coordinate systems, for elementary reactions, the coordinates chosen are related to two significant variables that change during the course of a reaction (*e.g.*, two distinct interatomic distances).

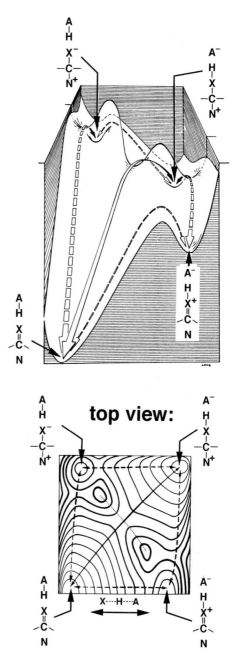

For more complicated reactions involving a number of elementary reactions, care should be exercised in selecting the coordinates in such systems. In some publications a third coordinate represents the standard Gibbs free energy instead of potential energy. The energetically easiest route on the contour map is referred to as the potential energy profile. *See* *Reaction Coordinate; Gibbs Free Energy Diagram; Potential Energy Profile; More-O'Ferrell-Jencks Diagram; Enzyme Energetics; Kinetic Isotope Effects*

J. W. Moore & R. G. Pearson (1981) *Kinetics and Mechanism*, 3rd ed., Wiley, New York.
IUPAC (1994) *Pure and Appl. Chem.* **66**, 1077.

PRECISION

The variability achieved in replicate measurements of the same quantity. The term high precision implies low variation within a set of determinations. **See also** Accuracy; Statistics (A Primer)

PRECURSOR

A metabolite, molecular entity, or some other event/process that precedes another component in a longer sequence of events or conversions. For example, the isoprenoid metabolite squalene is a precursor of cholesterol and glucose 6-phosphate is a precursor of glycogen, ribose, and pyruvate. **See** Series First Order Reaction; Pulse-Chase Experiments

PREDISSOCIATION

1. Dissociation occurring by tunneling from a bound to an unbound rovibronic state (*i.e.*, a state corresponding to a particular rotational energy level of a vibrational level of an electronic state). 2. The appearance of a diffuse band region within a series of sharp bands of an absorption spectrum.

Comm. on Photochem. (1988) *Pure and Appl. Chem.* **60**, 1055.

PREEQUILIBRIUM

A term (also referred to as "prior equilibrium") denoting any reversible step that precedes an irreversible step or the rate-limiting step in a multistage reaction mechanism. The so-described reaction step must rapidly establish an equilibrium between its reactants and products. The first association/dissociation equilibrium leading to the formation of EX complex from E and S in the Michaelis-Menten treatment is an example of a preequilibrium. **See** Michaelis-Menten Kinetics

PREEXPONENTIAL FACTOR

A constant in the Arrhenius equation and commonly symbolized by A: $k = A e^{-E_a/RT}$. It is also referred to as the A-factor. **See** Arrhenius Equation; Collision Theory; Transition-State Theory; Entropy of Activation

PRENYL-PROTEIN-SPECIFIC ENDOPEPTIDASE

A rat liver microsomal protease that cleaves an intact tripeptide, VIS, from S-farnesylated-CVIS tetrapeptide.

L. Liu, G.-F. Jang, C. C. Farnsworth, K. Yokoyama, J. A. Glomset & M. H. Gelb (1995) *Meth. Enzymol.* **250**, 189.

B. A. Gilbert, Y.-T. Ma & R. R. Rando (1995) *Meth. Enzymol.* **250**, 206.

PREORGANIZATION

1. The initial step(s) in a chemical reaction mechanism, as treated by Marcus Theory. (**See** Marcus Theory) 2. A term also applied to hosts, such as spherands, that have binding sites for ligands fully organized prior to any complexation. Studies of spherands and podands demonstrate that preorganization is a central determinant of binding strength[1,2]. The principle of preorganization states that the more highly organized and desolvated are hosts and guests prior to their complexation, the more stable their complexes will be. Preorganization is governed by both enthalpic and entropic terms. **See also** Podand; Spherand; Host; Guest; Crown Ether

[1] D. J. Cram (1988) *Angew. Chem. Int. Ed. Engl.* **27**, 1009.

[2] D. J. Cram, M. P. de Grandpro, C. B. Knobler & K. N. Trueblood (1984) *J. Amer. Chem. Soc.* **106**, 3286.

PREPHENATE DEHYDRATASE

This enzyme [EC 4.2.1.51] catalyzes the conversion of prephenate to phenylpyruvate, water, and carbon dioxide. This enzyme in enteric bacteria also possesses a chorismate mutase activity and converts chorismate into prephenate.

J. V. Schloss & M. S. Hixon (1998) *Comprehensive Biological Catalysis: A Mechanistic Reference* **2**, 43.

R. Bentley (1990) *Crit. Rev. Biochem. Mol. Biol.* **25**, 307.

B. E. Davidson (1987) *Meth. Enzymol.* **142**, 432.

B. E. Davidson & G. S. Hudson (1987) *Meth. Enzymol.* **142**, 440.

R. Fischer & R. Jensen (1987) *Meth. Enzymol.* **142**, 507.

PREPHENATE DEHYDROGENASE

Prephenate dehydrogenase [EC 1.3.1.12] catalyzes the reaction of prephenate with NAD^+ to produce 4-hydroxyphenylpyruvate, carbon dioxide, and NADH. This enzyme in enteric bacteria also possesses chorismate mutase activity and converts chorismate into prephenate. Prephenate dehydrogenase ($NADP^+$) [EC 1.3.1.13] catalyzes the reaction of prephenate with $NADP^+$ to produce 4-hydroxyphenylpyruvate, carbon dioxide, and NADPH.

M. H. O'Leary (1992) *The Enzymes*, 3rd ed., **20**, 235.

M. H. O'Leary (1989) *Ann. Rev. Biochem.* **58**, 377.

R. Fischer & R. Jensen (1987) *Meth. Enzymol.* **142**, 503.

PRESSURE-JUMP METHOD

A fast-reaction kinetic technique used to achieve a rapid change in external pressure that results in a sudden change in the equilibrium constant for a particular system. The investigator then analyzes the rate of approach of the system to the new equilibrium position. *See Chemical Kinetics*

PRE-STEADY-STATE PHASE

Refers to that initial period of nonlinear product formation, commencing with the initiation of the reaction and ending when the system is at steady state. Typically, the pre-steady-state phase lasts from milliseconds to a few seconds after mixing reactants. The time course of pre-steady-state rate processes often can be evaluated using stopped-flow, temperature-jump, and mix-quench methods.

PRIMARY PHOTOCHEMICAL PROCESS

Any elementary chemical process (also referred to as a primary photoreaction) in which an electronically excited molecular entity yields a primary photoproduct. *See Primary Photoproduct*

Comm. on Photochem. (1988) *Pure and Appl. Chem.* **60**, 1055.

PRIMARY PHOTOPRODUCT

The first observable entity, with distinctly different chemical properties compared to those of the reactant, produced in a primary photochemical process.

Comm. on Photochem. (1988) *Pure and Appl. Chem.* **60**, 1055.

PRIMITIVE CHANGE

A simple molecular change into which an elementary reaction can conceptually be further dissected. Primitive changes would include bond rupture, bond rotation, change in bond length, a redistribution of charge, *etc.* The concept of primitive change is useful in describing elementary reactions, but a primitive change does not itself represent a discrete chemical process. *See Elementary Reaction*

PRION PLAQUE FORMATION

Protein aggregation is a hallmark of scrapie, and scrapie protein can be induced to aggregate *in vitro* into forms that are indistinguishable from pathological brain-derived fibrils. Prusiner proposed that prion disease involves a mechanism for autocatalytic conversion of a host protein into a genetically identical, but conformationally different, prion state. Lundberg *et al.*[1] examined the kinetics and mechanism of amyloid formation by the prion protein H1 peptide as determined by time-dependent ESR. Their analysis suggests a model whereby the amorphous aggregate has a previously unsuspected dual role of releasing monomer into solution and also providing initiation sites for fibril growth. These findings suggest that the β-sheet-rich scrapie prion protein may be stabilized by aggregation. Harper and Lansbury[2] have also discussed simple mechanistic models involving nucleation-dependent polymerization, suggesting (a) that with such a mechanism, aggregation would be dependent on protein concentration and time and (b) that amyloid formation can be seeded by preformed fibrils. Eigen[3] has compared kinetic aspects of mechanisms of prion diseases based on the "protein only" hypothesis. He found that Prusiner's model almost surely also involves cooperativity, given realistic values of rate parameters. Eigen further states that were cooperativity occurring, then the model becomes phenomenologically indistinguishable from the Lansbury mechanism of plaque formation which is also a form of (passive) autocatalysis. Although both mechanisms differ with respect to relative stability of the two monomeric protein conformations, both require a thermodynamically favored aggregated state. While these considerations allow for a critical comparison of the mechanisms, they do not reveal the actual mechanism of infection. Nonetheless, Eigen argues that aggregation is likely to be a necessary, but possibly not sufficient, prerequisite of infection.

[1] K. M. Lundberg, C. J. Stenland, F. E. Cohen, S. B. Prusiner & G. L. Millhauser (1997) *Chem. Biol.* **4**, 345.
[2] J. D. Harper & P. T. Lansbury, Jr. (1997) *Annu. Rev. Biochem.* **66**, 385.
[3] M. Eigen (1996) *Biophys. Chem.* **63**, A1.

PROBABILITY

A statistical measure of the frequency or regularity of occurrence of a measured outcome (or measurement) in an experiment subject to random fluctuations in the measurement. The fundamental axioms of mathematical probability are: (a) if E is any event in the sample space S, then the probability of E's occurrence [written $P(E)$] ranges over the interval from zero to one; (b) over the entire sample space, $P(E)$ equals one; and (c) if events A and B are mutually exclusive, then the probability of their combined occurrence is the sum of the probabilities of their occurring individually. **See** *Statistics (A Primer); Permutations and Combinations*

PROBABILITY DENSITY FUNCTION

A function applied in statistics to predict the relative distribution if the frequency of occurrence of a continuous random variable (*i.e.,* a quantity that may have a range of values which cannot be individually predicted with certainty but can be described probabilistically) from which the mean and variance can be estimated.

PROBABLE ERROR

A statistical term for the deviation from the true value within which lies an experimentally measured value with a probability of 0.50. This corresponds to 0.674 σ (*i.e.,* 0.674 times the standard deviation). **See** *Statistics (A Primer); Normal Error Curve*

PROCESSIVITY

A measure of the tendency of an enzyme catalyzing a polymerization (or depolymerization) reaction to proceed through multiple catalytic rounds of bond making/breaking without dissociating from its polymeric substrate or template. The reader should note that processivity is technically a partition factor indicating the propensity of an enzyme to choose among several competing kinetic pathways. This partition constant has the general form: $k_{catalysis}/\{k_{catalysis} + k_{dissociation}\}$, and the greater its magnitude, the greater is the persistence of the enzyme on the polymer. One may also consider the processivity of molecular motors, such as dynein and myosin, in terms of their ability to proceed in the translocating/locomotory function by remaining attached to their cytoskeletal scaffold. Because polymerases, depolymerases, and cytoskeletal motors may be regulated by the concentration(s) of cosubstrate(s) and/or regulatory effector(s). Their processivity may well depend on the presence of these agents.

McClure and Chow[1] presented a useful protocol for determining the processivity of DNA polymerases. That approach relies on the ability to resolve discrete product lengths on acrylamide gels in 7 M urea and to analyze the resulting length distribution data using the Kuhn distribution law. The template challenge method[2] provides yet another means for observing the frequency of enzyme dissociation from its polymeric substrate, thereby allowing its binding to a second competing substrate/inhibitor. One measures cycling-time perturbation[2], wherein steady-state velocity measurements allow one to estimate the average time that an enzyme spends on its template. In the case of DNA polymerases, processivity may be linked to fidelity of nucleotide incorporation, because the binding energy of enzyme-polymer interactions are apt to factor into so-called editing steps where incorrectly inserted nucleotides are hydrolyzed out of the polymer before the next round of phosphodiester synthesis. **See also** *Random Scission Kinetics*

[1]W. R. McClure & Y. Chow (1980) *Meth. Enzymol.* **64**, 277.
[2]R. A. Bambara, D. Uyemura & T. Choi (1978) *J. Biol. Chem.* **253**, 413.

Selected entries from *Methods in Enzymology* [vol, page(s)]: Assay of DNA polymerases [autoradiogram interpretation, **262**, 277; DNA traps, **262**, 272-273; enzyme concentration, **262**, 272-273; gel electrophoresis, **262**, 271, 276-277; primer-template concentration, **262**, 273-274; radiolabeling of substrate [end labeling of primer, **262**, 274-275; internal labeling, **262**, 274, 277-278]; three nucleotide assay, **262**, 270-271; assay of nucleases, **262**, 279-280; bacteriophage 29 DNA polymerase, **262**, 44, 49; DNA polymerase β, **262**, 117; DNA polymerase ε, **262**, 98; DNA polymerase II, **262**, 14; DNA polymerase III, **262**, 22; HIV-1 reverse transcriptase [ribonuclease H activity, **262**, 279-280; transcriptase activity, **262**, 143-144]; Klenow fragment, **262**, 264-265.

PROCESSIVITY OF KINESIN ATPase

Kinesin and dynein are two microtubule-associated ATP-dependent motors responsible for intracellular motility. Gilbert *et al.*[1] made direct measurements of the dissociation kinetics of kinesin from microtubules (MTs), the release of orthophosphate and ADP, and the rebinding of this motor to MTs. They observed processivity in ATP hydrolysis amounting to 10 molecules ATP per site at low salt concentration and 1 molecule of ATP per site at a higher concentration of salt. After hydrolysis, the dissociation of kinesin from the MT is rate-limiting, and rebinding of kinesin·ADP to MTs is fast. The authors discuss how this behavior differs from that of skeletal myosin.

[1]S. P. Gilbert, M. R. Webb, M. Brune & K. A. Johnson (1995) *Nature* **373**, 671.

PROCHIRALITY

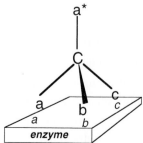

A stereochemical property of compounds arising from the ability of an enzyme's active site to distinguish between two chemically identical substituents covalently bound to a tetrahedral center (usually carbon and, in some cases, phosphorus). Prochirality is also termed pro-stereoisomerism. The classical example is citrate with its two carboxymethyl group substituents. Likewise, the C1 carbon atom of ethanol has two prochiral hydrogens. *See* Chirality; Chirality Probes

PROCOLLAGEN *N*-ENDOPEPTIDASE

This enzyme [EC 3.4.24.14], also known as procollagen *N*-proteinase, catalyzes the hydrolysis of the *N*-propeptide of the collagen chain α-1(I) at Pro—Gln and of α-2(II) chain at Ala—Gln. As a result, *N*-terminal propeptides of type I and II collagens are released prior to fibril assembly. However, it does not act on type III procollagen.

K. E. Kadler, S. J. Lightfoot & R. B. Watson (1995) *Meth. Enzymol.* **248**, 756.

PRODUCT

The substance or unstable species formed or generated by any chemical or photochemical reaction. The reaction product is usually symbolized by P; in reactions resulting in more than one product, P_1, P_2, . . . , or P, Q, R, . . . are commonly used.

PRODUCT-DETERMINING STEP

The step within a multistep reaction or process that determines the final product distribution. This step may be identical to or may even occur after the rate-determining step in the process.

PRODUCT DEVELOPMENT CONTROL

Referring to a kinetically controlled reaction in which selectivity parallels the relative thermodynamic stabilities of the products. Often in such reactions, the transition state occurs late along the reaction coordinate.

PRODUCT INHIBITION STUDIES

Experiments designed to reach conclusions about an enzyme-catalyzed reaction by examining how one or more products of the reaction alter the kinetic behavior of the enzyme. The diagnostic value of these approaches can be limited by formation of E·substrate·product abortive complexes in multisubstrate mechanisms.

(1) Evaluating the magnitude of dissociation constants of products can provide useful information on enzyme interactions of products. In most cases, these values are true dissociation constants, not Michaelis constants. Initial rate studies of the reverse reaction will provide K_m values and enable one to compare K_d and K_m values. This can aid in identifying key rate constants or steps in the reaction scheme.

(2) Under favorable circumstances, product inhibition studies can offer insights for distinguishing between rival kinetic schemes[1-3]. For example, consider a two-substrate two-product enzyme. A series of initial rate studies are undertaken in which double-reciprocal plots are made with each substrate (*i.e.*, $1/v$ *vs.* $1/[A]$ and $1/v$ *vs.* $1/[B]$ in which the nonvaried cosubstrate is held at a high, albeit nonsaturating, level) at different concentrations of the inhibitor. Identification of what type of inhibitor the product is with respect to each substrate (*i.e.*, competitive, noncompetitive, uncompetitive) and which, if any, of the slope or intercept replots are nonlinear can be of great use in identifying possible mechanisms. Further information can be obtained by repeating the series of experiments at saturating levels of the nonvaried cosubstrate. Cleland[4] has presented a set of rules to predict the type of inhibition one might expect for certain enzyme schemes.

Fromm[1] and Rudolph[2] have discussed the practical limitations on interpreting product inhibition experiments. The table below illustrates the distinctive kinetic patterns observed with bisubstrate enzymes in the absence or presence of abortive complex formation. It should also be noted that the random mechanisms in this table (and in similar tables in other texts) are usually for rapid equilibrium random mechanism schemes. Steady-state random mechanisms will contain squared terms in the product concentrations in the overall rate expression. The presence of these terms would predict nonlinearity in product inhibition studies. This nonlinearity might not be obvious under standard initial rate protocols, but products that would be competitive in rapid equilibrium systems might appear to be noncompetitive in steady-state random schemes[1,4,5], depending on the relative magnitude of those squared terms. *See* Abortive Complex

(3) Product inhibition studies are also very useful in identifying and characterizing the mechanisms of enzymes involving isomerization steps[1,3,4].

Use of Product Inhibitors to Distinguish Certain Bireactant Enzyme Systems[a]

Mechanism	Product	1/v versus 1/[A]	1/v versus 1/[A]
Theorell-Chance	P	N	C
	Q	C	N
Theorell-Chance (with abortives)	P	N	C
	Q	C	N
Ordered Bi Bi	P	N	N
	Q	C	C
Ordered Bi Bi (with abortives)	P	N[b]	N[c]
	Q	C	N
Random Bi Bi	P	C	C
	Q	C	C
Random Bi Bi (with abortives)	P	N	C
	Q	C	N
Ping Pong Bi Bi	P	N	C
	Q	C	N
Ping Pong Bi Bi (with abortives)	P	N[c]	NL[d]
	Q	C	NL[d]

[a]C = Competitive inhibition pattern; N = Noncompetitive inhibition pattern, and U = Uncompetitive inhibition pattern.
[b]Intercept replots versus concentration will be parabolic (concave up).
[c]Slope replots versus inhibitor concentration will be parabolic (concave up).
[d]Hyperbolic (concave up).

(5) With respect to enzymes having three or more products in the enzyme-catalyzed reaction, it is also possible to vary the concentration of two products (while keeping their ratio at a constant value) at the same time. If the rate expression for one reaction scheme has terms containing both those products, potentially that scheme can be distinguished from alternative reaction mechanisms that don't have those same terms in the rate expression.

[1]H. J. Fromm (1975) Initial Rate Enzyme Kinetics, Springer-Verlag, New York.
[2]F. B. Rudolph (1979) Meth. Enzymol. **63**, 411.
[3]K. L. Rebholz & D. B. Northrop (1995) Meth. Enzymol. **249**, 211.
[4]W. W. Cleland (1963) Biochim. Biophys. Acta **67**, 188.
[5]F. B. Rudolph & H. J. Fromm (1971) J. Biol. Chem. **246**, 6611.

PRODUCTION

A term used in compartmental analysis to designate the appearance of a tracee in its pool (or, more properly, its volume of distribution), the result from tracee biosynthesis and/or tracee transit from another pool. *See* Compartmental Analysis

PRODUCT RULE

A procedure in calculus for obtaining the derivative of the product of two functions, $f(x)$ times $g(x)$,

$$d[f(x)g(x)]/dx = f(x)[dg(x)/dx] + g(x)[df(x)/dx]$$

where each function must be differentiable.

PROGRESS CURVE ANALYSIS

While traditional steady-state enzyme kinetic experiments have focused on initial velocity treatments[1,2], the complete time course or reaction progress curve is actually a more robust source of kinetic information. Initial rates are estimated from the slope of a tangent line drawn to the earliest portion of a reaction progress curve, allowing one to obtain a velocity measurement as the change in product (or substrate) concentration per unit time. Initial velocity measurements offer the advantage that the instantaneous substrate concentrations, as well as those of products and/or other effector molecules, can be directly manipulated independently of stoichiometric relationships that hold during the course of the complete reaction process. Moreover, the measurements are usually accomplished under conditions where the state of the enzyme is generally well characterized and is unlikely to change by any irreversible process during the few moments over which a measurement is made. On the other hand, the complete reaction progress curve offers the opportunity to characterize the rate behavior at various extents of substrate depletion and product accumulation. Duggleby[3] has recently presented a useful detailed description of the progress curve analysis method, including nonlinear regression methods for evaluating the experimental data. He also discusses general approaches for deriving the integrated rate equations that describe the progress of one- and two-substrate enzyme systems. *See* KINSIM/FITSIM; Chemical Kinetics

[1]H. J. Fromm (1975) Initial Velocity Enzyme Kinetics, Springer-Verlag, New York.
[2]W. W. Cleland (1990) The Enzymes **19**, 99.
[3]R. G. Duggleby (1995) Meth. Enzymol. **249**, 61.

Selected entries from *Methods in Enzymology* [vol, page(s)]: Theory, **63**, 159-162; activation effect, **63**, 174, 175; analysis, **63**, 140, 159-183; burst, **64**, 20, 203, 215; enzyme concentration, **63**, 175-177; hysteresis, **64**, 197, 200-204; limitations, **63**, 181-183; plotting, **63**, 177-180; practical methods, **63**, 175-177; reversible inhibitor action, **63**, 163-175; reversible reaction, **63**, 171-175; simulation of, **63**, 180; advantages and disadvantages, **249**, 61-62; analysis, in kinetic models of inhibition, **249**, 168-169; concave-down, **249**, 156; concave-up, **249**, 156; with enzyme-product complex instability, **249**, 88; with enzyme-substrate instability, **249**, 88; equations [for classical inhibition, **249**, 176; for tight-binding inhibition, **249**, 174-178, 180]; inhibition rate constant determination, **244**, 448-449; in kinetic models of inhibition, **249**, 168; with noncompetitive product inhibition [Foster-Niemann plots, **249**, 226-228; nonlinear regression analysis, **249**, 225-226; oversaturation plots, **249**, 228]; nonlinear regression analysis, **249**, 61-90 [AGIRE computer program for, **249**, 79-81, 225-226; comparison to analysis based on rates, **249**, 61-63; complex reactions, **249**, 75-78; experimental design, **249**, 84-85; inhibitor effects, **249**, 71-75; potato acid phosphatase product inhibition, **249**, 73-74; preliminary fitting, **249**, 82-84; prephenate dehydratase product inhibition, **249**, 72-73; product inhibition effects, **249**, 72-73; prostate acid phosphatase phenyl phosphate hydrolysis, **249**, 70; reactions with two substrates, **249**, 75-77; reversible reactions, **249**, 77-78; with simple Michaelian enzyme, **249**, 63-71 [fitting equations, **249**, 63]; with slow-binding inhibitors, **249**, 88; with unstable enzymes, for kinetic characterization, **249**, 85-89.

PROLINE DEHYDROGENASE

This FAD-dependent enzyme [EC 1.5.99.8] catalyzes the reaction of L-proline with an acceptor and water to produce (S)-1-pyrroline 5-carboxylate and the reduced acceptor.

R. C. Scarpulla & R. L. Soffer (1978) *J. Biol. Chem.* **253**, 5997.
E. Adams & L. Frank (1980) *Ann. Rev. Biochem.* **49**, 1005.

PROLINE DIPEPTIDASE

This manganese-dependent enzyme [EC 3.4.13.9] (also known as Xaa—Pro dipeptidase, X—Pro dipeptidase, imidodipeptidase, prolidase, peptidase D, and γ-peptidase) catalyzes the hydrolysis of Xaa—Pro dipeptides (except for prolylproline). The dipeptidase also acts on aminoacylhydroxyproline derivatives. This cytosolic enzyme, a member of the peptidase family M24A, is found in most animal tissues.

N. C. Davies & E. L. Smith (1957) *J. Biol. Chem.* **224**, 261.
E. L. Smith (1955) *Meth. Enzymol.* **2**, 93.

PROLINE IMINOPEPTIDASE

This manganese-dependent enzyme [EC 3.4.11.5] catalyzes the release of an N-terminal prolyl residue from a peptide. The mammalian enzyme, which is not specific for prolyl bonds, is possibly identical with cytosol aminopeptidase [EC 3.4.11.1].

R. J. DeLange & E. L. Smith (1971) *The Enzymes*, 3rd ed., **3**, 81.
S. Sarid, A. Berger & E. Katchalski (1962) *J. Biol. Chem.* **237**, 2207.

PROLINE RACEMASE

This enzyme [EC 5.1.1.4] catalyzes the interconversion of L-proline and D-proline. ***See*** *Enzyme Energetics (The Case of Proline Racemase)*

M. E. Tanner & G. L. Kenyon (1998) *Comprehensive Biological Catalysis: A Mechanistic Reference* **2**, 7.
E. Adams (1976) *Adv. Enzymol.* **44**, 69.
E. Adams (1972) *The Enzymes*, 3rd ed., **6**, 479.

PROLINE REDUCTASE

This pyruvate-dependent enzyme [EC 1.4.4.1], also referred to as D-proline reductase (dithiol), catalyzes the reaction of 5-aminopentanoate with lipoate to produce D-proline and dihydrolipoate. Other dithiols can function as reducing agents.

P. D. van Poelje & E. E. Snell (1990) *Ann. Rev. Biochem.* **59**, 29.
P. A. Recsei & E. E. Snell (1984) *Ann. Rev. Biochem.* **53**, 357.

PROLIPOPROTEIN SIGNAL PEPTIDASE

This enzyme [EC 3.4.99.35] (better known as signal peptidase II and also known as bacterial leader peptidase I) catalyzes the cleavage of N-terminal leader sequences from membrane prolipoproteins.

K. Sankaran & H. C. Wu (1995) *Meth. Enzymol.* **248**, 169.

PROLYL 3-HYDROXYLASE

This enzyme [EC 1.14.11.7], also known as procollagen:proline, 2-oxoglutarate 3-dioxygenase, catalyzes the reaction of a prolyl residue in procollagen with dioxygen and α-ketoglutarate (or, 2-oxoglutarate) to produce a *trans*-3-hydroxyprolyl residue in procollagen, succinate, and carbon dioxide. This reaction also requires iron ions and ascorbate.

K. I. Kivirikko & T. Pihlajaniemi (1998) *Adv. Enzymol.* **72**, 325.
S. Englard & S. Seifter (1986) *Ann. Rev. Nutr.* **6**, 365.
E. Adams & L. Frank (1980) *Ann. Rev. Biochem.* **49**, 1005.

PROLYL 4-HYDROXYLASE

This enzyme [EC 1.14.11.2], also known as procollagen:proline, 2-oxoglutarate 4-dioxygenase, catalyzes the reaction of an L-prolyl residue in procollagen with dioxygen and α-ketoglutarate (or, 2-oxoglutarate) to produce a *trans*-4-hydroxyprolyl residue in procollagen, succinate, and carbon dioxide. This reaction also requires iron ions and ascorbate.

K. I. Kivirikko & T. Pihlajaniemi (1998) *Adv. Enzymol.* **72**, 325.
B. G. Fox (1998) *Comprehensive Biological Catalysis: A Mechanistic Reference* **3**, 261.
A. G. Prescott & P. John (1996) *Ann. Rev. Plant Physiol. Plant Mol. Biol.* **47**, 245.
O. Hayaishi, M. Nozaki & M. T. Abbott (1975) *The Enzymes*, 3rd ed., **12**, 119.

PROLYL ISOMERASE

This enzyme [EC 5.2.1.8] (also known as peptidylprolyl *cis-trans* isomerase, peptidylprolyl isomerase, and cyclophilin) catalyzes the interconversion of a peptidylprolyl derivative (in which $\omega = 180°$) and a peptidylprolyl derivative (in which $\omega = 0°$).

F. X. Schmid (1993) *Ann. Rev. Biophys. Biomol. Chem.* **22**, 123.
R. L. Stein (1993) *Adv. Protein Chem.* **44**, 1.
F. X. Schmid, L. M. Mayr, M. Mücke & E. R. Schönbrunner (1993) *Adv. Protein Chem.* **44**, 25.

PROPAGATION

Those steps or reactions within chain reactions that generate one or more reactive intermediates (often free radicals). In such propagation reactions, the newly formed reactive intermediate (*e.g.*, free radical) can then react with another molecule to produce another reactive intermediate. **See** *Chain Reaction; Chain Transfer; Initiation; Termination*

PROPANEDIOL DEHYDRATASE

This cobalamin-dependent enzyme [EC 4.2.1.28] catalyzes the conversion of propane-1,2-diol to propanal and water. The enzyme also dehydrates ethylene glycol to acetaldehyde.

B. T. Golding & W. Buckel (1998) *Comprehensive Biological Catalysis: A Mechanistic Reference* **3**, 239.
J. C. Murrell & W. Ashraf (1990) *Meth. Enzymol.* **188**, 26.

PROPIONYL-CoA CARBOXYLASE

This biotin-dependent enzyme [EC 6.4.1.3] catalyzes the reaction of ATP with propanoyl-CoA and HCO_3^- to produce ADP, orthophosphate, and (*S*)-methylmalonyl-CoA. Butanoyl-CoA will also serve as a substrate and the enzyme will also catalyze transcarboxylations.

J. N. Earnhardt & D. N. Silverman (1998) *Comprehensive Biological Catalysis: A Mechanistic Reference* **1**, 495.
J. V. Schloss & M. S. Hixon (1998) *Comprehensive Biological Catalysis: A Mechanistic Reference* **2**, 43.
H. G. Wood & R. E. Barden (1977) *Ann. Rev. Biochem.* **46**, 385.
A. W. Alberts & P. R. Vagelos (1972) *The Enzymes*, 3rd ed., **6**, 37.

PROPIONYL-CoA SYNTHETASE

This enzyme [EC 6.2.1.17], also known as propionate:CoA ligase, catalyzes the reaction of ATP with propanoate and coenzyme A to produce propanoyl-CoA, AMP, and pyrophosphate (or, diphosphate). Propenoate can also act as the substrate. This enzyme is not identical with acetyl-CoA synthetase or with butyryl-CoA synthetase.

J. C. Murrell & W. Ashraf (1990) *Meth. Enzymol.* **188**, 26.

PRO-PROCHIRAL

Referring to a compound containing an atom such that, replacement of an atom or group bonded to that particular atom with a different atom or group, would generate a prochiral compound. A simple example of a pro-prochiral molecule is provided by acetic acid. Replacement of one of the methyl hydrogens with another substituent (*e.g.*, Cl) will generate a prochiral compound (in this case, chloroacetic acid). The remaining two hydrogens are enantiotopic since compounds such as BrClCHCOOH are chiral. **See** *Prochirality; Chirality*

pro-R Group

A stereochemical condition that describes a stereoheterotopic group C in the compound $XABC_2$ containing a prochiral tetrahedral atom X having substituents A, B, C, and C.

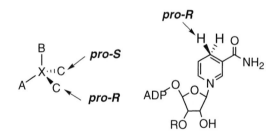

To make the assignment, *arbitrarily* assign a higher Cahn-Ingold-Prelog priority over the other stereoheterotopic group C, as though one had mentally substituted the first with a heavier isotope atom. Now, the "higher priority" group C is said to be *pro-R* if the configuration of the mentally generated chiral center is now *R*.

pro-S Group

A stereochemical condition that describes a stereoheterotopic group C in the compound $XABC_2$ containing a prochiral tetrahedral atom X having substituents A, B, C, and C.

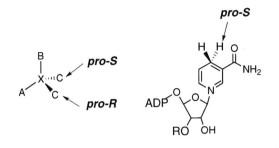

To make the assignment, *arbitrarily* assign a higher Cahn-Ingold-Prelog priority over the other stereoheterotopic group C, as though one had mentally substituted the first with a heavier isotope atom. Now, the "higher priority" group C is said to be *pro-S* if the configuration of the mentally generated chiral center is now *S*.

PROSTACYCLIN SYNTHASE

This heme thiolate-dependent enzyme [EC 5.3.99.4] catalyzes the interconversion of prostaglandin H_2 and prostaglandin I_2 (prostacyclin).

C. R. Pace-Asciak & W. L. Smith (1983) *The Enzymes*, 3rd ed., **16**, 543.

PROSTAGLANDIN E: 9-KETO (α)-REDUCTASE

This enzyme [EC 1.1.1.189], also known as prostaglandin E_2 9-reductase, catalyzes the NADPH-dependent reduction of prostaglandin E_2 to prostaglandin $F_{2\alpha}$ and $NADP^+$.

C. R. Pace-Asciak & W. L. Smith (1983) *The Enzymes*, 3rd ed., **16**, 543.

PROSTAGLANDIN H, PROSTAGLANDIN D ISOMERASE

This glutathione-dependent enzyme [EC 5.3.99.2], also known as prostaglandin D synthase and prostaglandin H_2:D isomerase, catalyzes the opening of the epidioxy bridge in (5*Z*,13*E*)-(15*S*)-9α,11α-epidioxy-15-hydroxyprosta-5,13-dienoate to yield (5*Z*,13*E*)-(15*S*)-9α,15-dihydroxy-11-oxoprosta-5,13-dienoate.

C. R. Pace-Asciak & W. L. Smith (1983) *The Enzymes*, 3rd ed., **16**, 543.

PROSTAGLANDIN H, PROSTAGLANDIN E ISOMERASE

This glutathione-dependent enzyme [EC 5.3.99.3], also known as prostaglandin E synthase, prostaglandin H_2 E isomerase, and endoperoxide isomerase, catalyzes the

opening of the epidioxy bridge, thereby converting (5*Z*,13*E*)-(15*S*)-9α,11α-epidioxy-15-hydroxyprosta-5,13-dienoate into (5*Z*,13*E*)-(15*S*)-11α,15-dihydroxy-9-oxoprosta-5,13-dienoate.

C. R. Pace-Asciak & W. L. Smith (1983) *The Enzymes*, 3rd ed., **16**, 543.

PROSTAGLANDIN H₂ SYNTHASE

By acting both as a dioxygenase and as a peroxidase, this enzyme [EC 1.14.99.1] catalyzes the reaction of arachidonate with reduced acceptor AH_2 and two dioxygen molecules to produce prostaglandin H_2, oxidized acceptor A, and water.

H. B. Dunford (1998) *Comprehensive Biological Catalysis: A Mechanistic Reference* **3**, 195.
W. L. Smith, R. M. Garavito & D. L. DeWitt (1996) *J. Biol. Chem.* **271**, 33157.
D. Picot, P. J. Loll & M. Garavito (1994) *Nature* **367**, 243.
J. A. Boyd (1990) *Meth. Enzymol.* **186**, 283.
C. R. Pace-Asciak & W. L. Smith (1983) *The Enzymes*, 3rd ed., **16**, 543.
O. Hayaishi, M. Nozaki & M. T. Abbott (1975) *The Enzymes*, 3rd ed., **12**, 119.

PROSTHETIC GROUP

A very tightly bound coenzyme. Examples of prosthetic groups include the hemes, cobalamin coenzymes, pyridoxal 5-phosphate, lipoic acid, and biotin. FAD and FMN are also very tightly held by most flavin-dependent oxidoreductases. In many instances, distinctions are made purely on the basis of affinity and do not apply consistently. For example, while NAD^+ and $NADP^+$ binding affinities are relatively high, dialysis is frequently sufficient for resolving (removing) these coenzymes from their binding sites on NAD(P)$^+$-dependent oxidoreductases.

PROTEASE ACCESSIBILITY (as Probe of Protein Structure)

A method for analyzing protein structure based on limited proteolysis. This method is especially useful in investigations of membrane proteins whose membrane association limits the repertoire of techniques that can be gainfully applied to infer structural features. For example, Davis *et al.*[1] used four proteases to assess the topology of yeast H$^+$-ATPase reconstituted into phosphatidylserine vesicles. Limited proteolysis by trypsin and α-chymotrypsin inactivates the enzyme and produces stable, membrane-bound fragments. Sequence analyses of these peptides localized these cleavage sites to opposite sides of a central hydrophilic domain containing consensus sequences for the site of phosphorylation and fluorescein isothiocyanate binding of several related ATPases. Limited proteolysis of the ATPase by elastase cuts approximately fifty amino acids from the C terminus, and the remaining membrane-bound fragment is active. Proteolysis by carboxypeptidase Y in the presence and absence of detergent suggests that the C terminus is on the inside of the vesicle in this reconstitution. The authors presented a model for the transmembrane arrangement of the polypeptide, with the C terminus on the inside of the vesicle and the N terminus on the outside. They suggested that the ATP binding region is likewise on the outside, and that the polypeptide passes through the membrane a minimum of five times.

[1]C. B. Davis & G. G. Hammes (1989) *J. Biol. Chem.* **264**, 370.

PROTEASE INHIBITOR "COCKTAILS"

Proteolysis can be arrested or minimized through the use of so-called cocktails of inhibitors of serine, cysteine, aspartic proteases and metalloproteases. Among the commonly used inhibitors are the sulfonylating agents acting on serine proteases, phenylmethanesulfonyl chloride (PMSF) and 4-(2-aminoethyl)benzenesulfonyl fluoride (AEBSF). Antipain is [(S)-carboxy-2-phenylethyl] carbamoyl-L-arginyl-L-valyl-L-arginal which has an I_{50} (*i.e.*, inhibitor concentration that yields 50% inhibition) of 0.15 μg/mL for papain, 0.25 μg/mL for trypsin, 1.2 μg/mL for cathepsin A, and is relatively ineffective with chymotrypsin. Leupeptin is *N*-propionyl-L-leucyl-L-leucyl-L-arginal which has a I_{50} of 2 μg/mL with trypsin, 8 μg/mL with plasmin, 0.5 μg/mL with papain, 0.5 μg/mL with cathepsin B and is relatively ineffective with chymotrypsin. Chymostatin (or, *N*-[((S)-carboxy-2-phenylethyl)carbamoyl]-α-[2-imidohexahydro-4(S)-pyrimidyl]-L-glycyl-L-leucyl-L-phenylalaninal) is a potent chymotrypsin inhibitor (with a I_{50} of 0.1–0.2 μg/mL). Many of these cocktails are commercially available.

General Protease Cocktails. These should contain the following inhibitors: AEBSF, *trans*-epoxysuccinyl-L-leucylamido(4-guanidino)butane (E-64), bestatin, leupeptin, aprotinin (and sodium EDTA if the protein or enzyme being purified does not require a divalent metal cofactor for stability or activity).

Fungal and Yeast Protease Cocktail. These should contain the following inhibitors: AEBSF, pepstatin A, E-64, and 1,10-phenanthroline.

Mammalian Cell Protease Inhibitor Cocktail. These should contain AEBSF, pepstatin A, E-64, bestatin, leupeptin, and aprotinin. (Metal chelators can be added to suppress the activity of calcium ion-dependent proteases such as calpain. Again, one must determine whether the protein or enzyme being purified does not require a divalent metal cofactor for stability or activity.)

Bacterial Extract Protease Cocktail. Inhibitors present are pepstatin A, 4-(2-amino-ethyl)benzenesulfonyl fluoride (AEBSF), *trans*-epoxysuccinyl-L-leucylamido(4-guanidino)butane (E-64), bestatin, and sodium EDTA.

PROTEASE La

This enzyme [EC 3.4.21.53], also known as endopeptidase La, ATP-dependent serine proteinase, and ATP-dependent protease La, catalyzes the hydrolysis of peptide bonds in large proteins (for example, globin, casein, and denatured serum albumin) in the presence of ATP (which is hydrolyzed to ADP and orthophosphate). Vanadate ion inhibits both reactions. A similar enzyme occurs in animal mitochondria. Protease La belongs to the peptidase family S16.

A. L. Goldberg, R. P. Moerschell, C. H. Chung & M. R. Maurizi (1994) *Meth. Enzymol.* **244**, 350.
S. Kuzela & A. L. Goldberg (1994) *Meth. Enzymol.* **244**, 376.

PROTEINASE A

This enzyme [EC 3.4.23.19], also known as aspergillopepsin II, proctase A, and *Aspergillus niger* var. *macrosporus* aspartic proteinase, catalyzes the hydrolysis of peptide bonds in proteins. It has been isolated from *Aspergillus niger* var. *macrosporus* and is distinct from aspergillopepsin I in specificity and in insensitivity to pepstatin.

K. Takahashi (1995) *Meth. Enzymol.* **248**, 146.

PROTEIN COMPLEMENTATION

A process that partially or fully restores the biological activity of two proteins, each lacking a functional domain or motif, through the formation of an oligomeric species. *See* Domain Swapping

I. Zabin & M. R. Villarejo (1975) *Annu. Rev. Biochem.* **44**, 295.

PROTEIN DISULFIDE ISOMERASE

This enzyme [EC 5.3.4.1], also known as "S—S rearrangase," catalyzes the rearrangement of intrachain or interchain disulfide bonds within proteins to form their native structures. The enzyme requires reducing agents or the partly reduced enzyme. The reaction operates by sulfhydryl-disulfide interchange.

H. F. Gilbert (1998) *Comprehensive Biological Catalysis: A Mechanistic Reference* **1**, 609.
R. B. Freedman, H. C. Hawkins & S. H. McLaughlin (1995) *Meth. Enzymol.* **251**, 397.
W. W. Wells, Y. Yang & T. L. Deits (1993) *Adv. Enzymol.* **66**, 149.
N. J. Bulleid (1993) *Adv. Protein Chem.* **44**, 125.
R. C. Fahey & A. R. Sundquist (1991) *Adv. Enzymol.* **64**, 1.

PROTEIN-DISULFIDE REDUCTASE (Glutathione)

This enzyme [EC 1.8.4.2], also known as glutathione:insulin transhydrogenase and insulin reductase, catalyzes the reaction of two glutathione with a disulfide bond in a protein to produce glutathione disulfide and a protein with two new thiol groups. The enzyme can reduce insulin and a number of other proteins.

H. H. Tomizawa (1971) *Meth. Enzymol.* **17B**, 515.

PROTEIN FARNESYLTRANSFERASES

This class of farnesyltransferases catalyzes the posttranslational modification of proteins by the cholesterol precursor, farnesylpyrophosphate. One of the substrates of this enzyme is Ras.

R. A. Gibbs (1998) *Comprehensive Biological Catalysis: A Mechanistic Reference* **1**, 31.
P. J. Casey & M. G. Seabra (1996) *J. Biol. Chem.* **271**, 5289.
F. L. Zhang & P. J. Casey (1996) *Ann. Rev. Biochem.* **65**, 241.
S. Clarke (1992) *Ann. Rev. Biochem.* **61**, 355.

PROTEIN GERANYLGERANYL-TRANSFERASES

The following are recent reviews on the molecular and physical properties of these transferases which transfer geranylgeranyl groups to defined sites on protein substrates.

R. A. Gibbs (1998) *Comprehensive Biological Catalysis: A Mechanistic Reference* **1**, 31.
P. J. Casey & M. G. Seabra (1996) *J. Biol. Chem.* **271**, 5289.
F.L. Zhang & P.J. Casey (1996) *Ann. Rev. Biochem.* **65**, 241.
Y. Jiang, G. Rossi & S. Ferro-Novick (1995) *Meth. Enzymol.* **257**, 21.
S. A. Armstrong, M. S. Brown, J. L. Goldstein & M. C. Seabra (1995) *Meth. Enzymol.* **257**, 30.
S. Clarke (1992) *Ann. Rev. Biochem.* **61**, 355.

PROTEIN KINASE

Any of a broad class of phosphoryl-transfer enzymes [EC 2.7.1.x] that catalyze the ATP-dependent phosphorylation of proteins, most often occurring at seryl, threonyl, and tyrosyl residues. These enzymes are central participants in cellular signal transduction pathways, and their discovery and recognition as primary control components of the cell culminated in the award of the 1992 Nobel Prize in Medicine and Physiology to American enzymologists Edwin Krebs and Edward Fischer. There is reason to believe that approximately 2% of the coding sequences in the human genome specify some 2000 different kinases that phosphorylate protein substrates. The prototypical enzyme is known as 3′,5′-cAMP-stimulated protein kinase (or, protein kinase A). *See specific protein kinase*

S. S. Taylor, D. R. Knighton, J. Zheng, L. F. Ten Eyck & J. M. Sowadski (1992) *Ann. Rev. Cell Biol.* **8**, 429.
J. Larner (1990) *Adv. Enzymol.* **63**, 173.
A. M. Edelman, D. K. Blumenthal & E. G. Krebs (1987) *Ann. Rev. Biochem.* **56**, 567.
S. I. Walaas & P. Greengard (1987) *The Enzymes*, 3rd ed., **18**, 285.
H. V. Rickenberg & B. H. Leichtling (1987) *The Enzymes*, 3rd ed., **18**, 419.
E. G. Krebs (1986) *The Enzymes*, 3rd ed., **17**, 3.
D. A. Walsh & E. G. Krebs (1973)*The Enzymes*, 3rd ed., **8**, 555.

Selected entries from *Methods in Enzymology* [vol, page(s)]:
General: Protein kinase classification, **200**, 3; protein kinase catalytic domain sequence database: identification of conserved features of primary structure and classification of family members, **200**, 38; protein kinase phosphorylation site sequences and consensus specificity motifs: tabulations, **200**, 62.
Assay Methods: Enzyme activity dot blots for assaying protein kinases, **200**, 85; solid-phase protein-tyrosine kinase assays, **200**, 90; nonradioactive assays of protein-tyrosine kinase activity using anti-phosphotyrosine antibodies, **200**, 98; use of synthetic amino acid polymers for assay of protein-tyrosine and protein-serine kinases, **200**, 107; assay of phosphorylation of small substrates and of synthetic random polymers that interact chemically with adenosine 5′-triphosphate, **200**, 112; assay of protein kinases using peptides with basic residues for phosphocellulose binding, **200**, 115; design and use of peptide substrates for protein kinases, **200**, 121; synthetic peptide substrates for casein kinase II, **200**, 134; nucleotide photoaffinity labeling of protein kinase subunits, **200**, 477; separation of phosphotyrosine, phosphoserine, and phosphothreonine by high-performance liquid chromatography, **201**, 3; chemical properties and separation of phosphoamino acids by thin-layer chromatography and/or electrophoresis, **201**, 10; determination of phosphoamino acid composition by acid hydrolysis of protein blotted to Immobilon, **201**, 21; comparison of three methods for detecting tyrosine-phosphorylated proteins, **201**, 28; generation and use of antibodies to phosphothreonine, **201**, 44; generation and use of anti-phosphotyrosine antibodies raised against bacterially expressed Abl protein, **201**, 53; preparation and use of anti-phosphotyrosine antibodies to study structure and function of insulin receptor, **201**, 65; generation of monoclonal antibodies against phosphotyrosine and their use for affinity purification of phosphotyrosine-containing proteins, **201**, 79; isolation of tyro-

sine-phosphorylated proteins and generation of monoclonal antibodies, **201**, 92; generation and use of anti-phosphotyrosine antibodies for immunoblotting, **201**, 101.

Determination of Phosphorylation Stoichiometry: By biosynthetic labeling, **201**, 245; by separation of phosphorylated isoforms, **201**, 251; by microchemical determination of phosphate in proteins isolated from polyacrylamide gels, **201**, 261; production of phosphorylation state-specific antibodies, **201**, 264.

Protein Kinase Inhibitors: Pseudosubstrate-based peptide inhibitors, **201**, 287; utilization of the inhibitor protein of adenosine cyclic monophosphate-dependent protein kinase, and peptides derived from it, as tools to study adenosine cyclic monophosphate-mediated cellular processes, **201**, 304; use of sphingosine as inhibitor of protein kinase C, **201**, 316; properties and use of H-series compounds as protein kinase inhibitors, **201**, 328; use and specificity of staurosporine, UCN-01, and calphostin C as protein kinase inhibitors, **201**, 340; inhibition of protein-tyrosine kinases by tyrphostins, **201**, 347; use and specificity of genistein as inhibitor of protein-tyrosine kinases, **201**, 362; use and selectivity of herbimycin a as inhibitor of protein-tyrosine kinases, **201**, 370; use of erbstatin as protein-tyrosine kinase inhibitor, **201**, 379.

PROTEIN KINASE C

Members of the protein kinase C family promote signal transduction by catalyzing ATP-dependent protein phosphorylation [general EC number 2.7.1.37] in response to various signals that promote lipid hydrolysis. The primary activator is diacylglycerol. **See also** *Calcium/Calmodulin-Dependent Protein Kinase*

A. C. Newton (1993) *Ann. Rev. Biophys. Biomol. Struct.* **22**, 1.
S. S. Taylor, D. R. Knighton, J. Zheng, L. F. Ten Eyck & J. M. Sowadski (1992) *Ann. Rev. Cell Biol.* **8**, 429.
J. Larner (1990) *Adv. Enzymol.* **63**, 173.
U. Kikkawa, A. Kishimoto & Y. Nishizuka (1989) *Ann. Rev. Biochem.* **58**, 31.

PROTEIN OXIDATION (In Aging & Disease)

Berlett and Stadtman[1] recently discussed the pathways by which oxidatively modified forms of proteins accumulate during aging, oxidative stress, and in some pathological conditions. Reactive oxygen species (ROS), generated by physiologic as well as nonphysiologic means, result in amino acid residue side-chain oxidation, protein-protein cross-linkages, and protein backbone oxidation. Henle and Linn[2] have reviewed related issues regarding the formation, prevention, and repair of DNA damage by iron/hydrogen peroxide (so-called Fenton oxidants).

[1]B. S. Berlett & E. R. Stadtman (1997) *J. Biol. Chem.* **272**, 20313.
[2]E. S. Henle & S. Linn (1997) *J. Biol. Chem.* **272**, 19095.

PROTEIN PRENYLTRANSFERASES

The following is a recent review on the molecular and physical properties of these transferases which prenylate suitable protein substrates.

Entire Volume on this Topic: P. J. Casey & J. E. Buss, eds. (1995) *Meth. Enzymol.*, vol. **250**.

PROTEIN-PROTEIN RECOGNITION KINETICS

Janin[1] considered the kinetics of protein-protein association based on a simple kinetic model for association that assumes proteins are relatively rigid bodies (*i.e.*, no large conformation change occurs) in the reaction. Association commences with the random collision at the rate $k_{collision}$ predicted by the Einstein-Smoluchowski equation, thereby creating an encounter pair whose evolution into a stable complex occurs provided that the two molecules are correctly oriented and positioned. This is treated by assigning a probability $p(r)$. In the absence of long-range interactions, the bimolecular association rate would be $p(r)k_{collision}$. He also noted that long-range electrostatic interactions should affect both $k_{collision}$ and $p(r)$. The collision rate is multiplied by $q(t)$, a factor larger than 1 when the molecules are oppositely charged; thus, coulombic attraction increases the frequency of collision. The parameter $q(t)$ is less than 1 for repulsive interactions. The product $p(r)q(t)$ represents the "steering effect" of electric dipoles acting to preorient the molecules even before collision. Janin[1] first applied this model on the kinetics of barnase-barstar association[2]. He found that, when long-range electrostatic interactions are fully screened or abolished by site-directed mutagenesis (such that $p(r)q(t) \sim 1$), the rate of productive collision is approximately 10^5 M^{-1}s^{-1}. Under such conditions, $p(r)$ is approximately 1.5×10^{-5} and is determined by the loss of rotational freedom. This value is in agreement with computer simulations of protein-protein docking, suggesting that a rotational entropy loss ΔS_{rot} is about 22 entropy units in the transition state for protein-protein recognition. As might be expected, low ionic strength enhances long-range electrostatic interactions and accelerates barnase-barstar association by a factor $q(t)q(r)$ of up to 10^5, suggesting further that favorable charge-charge interactions work together with charge-dipole interactions to make the association rate much faster than otherwise expected on the basis of free diffusion.

Schreiber and Fersht[2] noted earlier that the association of oppositely charged proteins proceeds through the rate-determining formation of an early, weakly specific complex. The formation of such a complex is thought to be dominated by long-range electrostatic interactions[3], and by precise docking to form the high affinity complex occurring afterward. They suggest that this mode of binding may be widespread in nature.

[1] J. Janin (1997) *Proteins* **28**, 153.
[2] G. Schreiber & A. Fersht (1996) *Nature Struct. Biol.* **3**, 427.
[3] A. R. Fersht (1999) *Structure and Mechanism in Protein Science: A Guide to Enzyme Catalysis and Protein Folding*, pp. 650, W. H. Freeman, New York.

PROTEIN SYNTHESIS INHIBITORS

Agents that retard or arrest protein synthesis by interacting with essential reactions in the multistep process. Cycloheximide effectively inhibits eukaryotic protein synthesis on cytoplasmic ribosomes, and 1 μg/mL results in 70–75% inhibition for cell-free translation, whereas 30 μg/mL almost totally blocks intracellular protein synthesis. Chloramphenicol blocks the peptidyltransferase reaction, but DNA synthesis can also be reduced by this inhibitor. Other protein synthesis inhibitors include: α-amino-β-butyric acid (an analogue of valine); aurintricarboxylic acid (an inhibitor of initiation at 50 μM); 7-azatryptophan (a tryptophan analogue that blocks bacteriophage synthesis); azetidine-2-carboxylic acid (a four-membered ring analogue of proline that can be incorporated into bacterial proteins); canavanine (an arginine analogue that inhibits bacterial protein synthesis) emetine (binds to the 40S eukaryotic ribosome and inhibits translocation); erythromycin (bacterial protein synthetase inhibitor); ethionine (methionine analogue that inhibits bacterial protein synthesis); 5-fluorotryptophan; 6-fluorotryptophan; methyltryptophan; fusidic acid; kasugamycin; 7-methylguanosine 5′-phosphate; ω-methyllysine; *O*-methylthreonine; norleucine; pactamycin; puromycin (potent inhibitor of prokaryotic and eukaryotic protein synthesis); selenomethionine; sparsomycin; streptomycin; tetracycline antibiotics; and tryptazan.

PROTEIN TURNOVER

Cyclical synthesis/degradation of proteins in cells. Schimke[1] treated the rate of protein synthesis as a zero-order process and the rate of protein degradation as a first-order reaction:

$$d[P_{total}]/dt = k_s - k_d[P_{total}]$$

with $[P_{total}]$ the concentration of a particular protein with k_s and k_d as the rate constants for synthesis and degradation. The zero-rate rate constant k_s (the units being molarity·time^{-1}) depends on many factors, such as a cell's mRNA content, ribosome content, pool size(s) for any potentially limiting amino acid(s) or aminoacyl-tRNA(s), and other factors affecting the efficiency of translation. In principle, the rate, extent, and location of mRNA synthesis/degradation can also strongly influence the magnitude of k_s. Although degradation rate is treated here as a first-order process, the suitability of the first-order approximation should always be verified for each individual protein. At a steady state,

$$k_s = k_d[P_{total}]_{Steady-State}$$

Moreover, with changes in cellular metabolism in response to changes in nutrients and metabolic effectors, the concentration of a specific protein can undergo a transition from one steady-state to another.

Experiments[2] on the incorporation of [^{14}C]leucine into rat liver tryptophan oxygenase in response to treatment with tryptophan and/or hydrocortisone illustrate how the judicious application of enzyme activity and specific radioactivity measurements can provide powerful insights regarding the mechanisms controlling enzyme synthesis and degradation. (**See also** *Leucine Kinetics*) Yamamoto[3] has discussed the related topic of steroid receptor-regulated transcription of specific genes and gene networks.

As shown in the table below, the half-lives of protein degradation are of widely disparate magnitudes.

Approximate Biological Half-Lives for Selected Proteins

Protein	Tissue Source	Half-life
Ornithine decarboxylase	Liver	12 min
Tyrosine aminotransferase	Liver	120 min
Glucokinase	Liver	12 h
Hexokinase	Liver	48–72 h
Fructose 1,6-bis-P aldolase	Muscle	>100 h
Lactate dehydrogenase	Muscle	>140 h
Cytochrome *c*	Muscle	>150 h

There are indications that the rates of degradation depend on the thermostability and oxidation of proteins, findings that suggest that the integrity of protein ternary structure may in part determine the overall intracellular lifetime of a protein. Dice and Goldberg[4] have carried out a statistical analysis of the relationship between protein degradative rate and protein molecular weight. Likewise, the amino acid at the N terminus of a protein appears to influence the protein's lifetime in cells (*See N-End Rule*). Goldberg and St. John[5] presented a very readable and informative account of investigations of intracellular protein degradation in mammalian and bacterial cells.

[1]R. T. Schimke (1969) *Current Topics in Cellular Regulation* **1**, 77.
[2]R. T. Schimke, E. W. Sweeney & C. M. Berlin (1965) *J. Biol. Chem.* **240**, 322 and 4609.
[3]K. R. Yamamoto (1985) *Ann. Rev. Genet.* **19**, 209.
[4]J. F. Dice & A. L. Goldberg (1975) *Arch. Biochem. Biophys.* **170**, 213.
[5]A. L. Goldberg & A. C. St. John (1976) *Ann. Rev. Biochem.* **55**, 1091.

PROTOCATECHUATE 3,4-DIOXYGENASE

This iron-dependent enzyme [EC 1.13.11.3], also known as protocatechuate oxygenase, catalyzes the reaction of 3,4-dihydroxybenzoate with dioxygen to produce 3-carboxy-*cis,cis*-muconate.

B. G. Fox (1998) *Comprehensive Biological Catalysis: A Mechanistic Reference* **3**, 261.
J. B. Howard & D. C. Rees (1991) *Adv. Protein Chem.* **42**, 199.
J. W. Whittaker, A. M. Orville & J. D. Lipscomb (1990) *Meth. Enzymol.* **188**, 82.
O. Hayaishi, M. Nozaki & M. T. Abbott (1975) *The Enzymes*, 3rd ed., **12**, 119.

PROTOCATECHUATE 4,5-DIOXYGENASE

This iron-dependent enzyme [EC 1.13.11.8], also known as protocatechuate 4,5-oxygenase, catalyzes the reaction of protocatechuate with dioxygen to produce 4-carboxy-2-hydroxymuconate semialdehyde.

B. G. Fox (1998) *Comprehensive Biological Catalysis: A Mechanistic Reference* **3**, 261.
J. V. Schloss & M. S. Hixon (1998) *Comprehensive Biological Catalysis: A Mechanistic Reference* **2**, 43.
D. M. Arciero, A. M. Orville & J. D. Lipscomb (1990) *Meth. Enzymol.* **188**, 89.
O. Hayaishi, M. Nozaki & M. T. Abbott (1975) *The Enzymes*, 3rd ed., **12**, 119.

PROTOCHLOROPHYLLIDE REDUCTASE

This enzyme [EC 1.3.1.33], also known as NADPH-protochlorophyllide oxidoreductase, catalyzes the reaction of chlorophyllide a with $NADP^+$ to produce protochlorophyllide, having the (7*S*,8*S*)-configuration, and NADPH. It catalyzes a light-dependent *trans*-reduction of the D-ring of protochlorophyllide.

PROTOGENIC

A term applied to a solvent that can act as a proton donor. Water and ethanol are both protogenic and protophilic. ***See*** *Protophilic; Amphiprotic*

PROTON AFFINITY

The energy released when a proton (or, hydron) reacts with a molecular entity (in a gas-phase reaction, either real or hypothetical) to produce the corresponding conjugate acid; *i.e.*, the negative of the enthalpy change.

S. G. Lias, R. P. Liebman & R. P. Levin (1984) *J. Phys. Chem. Ref. Data* **13**, 695.

PROTON INVENTORY TECHNIQUE

An isotope effect technique applied in enzyme-catalyzed isotope exchange experiments measuring reaction rate as a function of the $[D_2O]/[H_2O]$ ratio. Such studies often provide information concerning the effect of the environment of the transition state on exchangeable protons.

K. B. Schowen & R. L. Schowen (1982) *Meth. Enzymol.* **87**, 551.

PROTON TRANSFER IN AQUEOUS SOLUTION

The transfer of protons in the vapor phase is strongly influenced by proton affinities as well as factors affecting the stability of hydrogen-bonded molecular complexes (among them are the linearity of hydrogen bonds, the distance between the two atoms sharing the bonding hydrogen, and the electrostatic potentials of those two atoms). Eigen and co-workers[1–3] treated proton transfer as the summation of three discernibly distinct steps:

Step 1 $A{-}H + :B = A{-}H:B$

Step 2 $A{-}H:B = A^- \cdots H:B^+$

Step 3 $A^- \cdots H:B^+ = A^- + H:B^+$

The first step is a bimolecular reaction leading to the formation of a hydrogen bond; the second step is the breaking of the hydrogen bond such that the protonated species $H:B^+$ is formed; the third step is the dissociation reaction to form the products. In aqueous solutions, the bimolecular reaction proceeds much faster than would be predicted from gas phase kinetic studies, and this underscores the complexity of proton transfer in solvents with extensive hydrogen-bonding networks capable of creating parallel pathways for the first step. In their authoritative consideration of proton transfer in aqueous solutions, Schuster *et al.*[4] presented several summary statements about the nature of this process. First, transfer of a proton to solvent can occur equally well by protolysis:

$$A-H (+ H_2O) = A^- \cdots H^+(H_2O)$$

and by hydrolysis

$$A-H + OH^- = A^- (+ H_2O)$$

Second, rate constants for intermolecular proton transfer obey the Brønsted relationship, and the rates can be strongly influenced by electrostatics as well as the difference in the pK_a's for the two acids. Third, proton transfer can proceed at slower than diffusion-limited values with compounds forming weak hydrogen bonds in water.

Carbonic anhydrase presents an instructive case where the catalytic efficiency is so great ($k_{cat} > 10^6$ s^{-1}) that proton transfer becomes rate-limiting. The rate was found to depend on the concentration of the protonated form of buffers in the solution. Indeed, Silverman and Tu[5] adduced the first convincing evidence for the role of buffer in carbonic anhydrase catalysis through their observation of an imidazole buffer-dependent enhancement in equilibrium exchanges of oxygen isotope between carbon dioxide and water. The effect is strictly on k_{cat}, and k_{cat}/V_{max} is unaffected because the latter is described by a rate law that does not include proton transfer steps.

Other reviews on proton transfer deal with the following topics: hydrogen bonding and H$^+$ transfer[6-8], general acid/base catalysis[9-11], and proton transfers in biological systems involving proteins[12,13].

[1]M. Eigen, G. G. Hammes & K. Kustin (1960) *J. Am. Chem. Soc.* **82**, 3482.
[2]M. Eigen, W. Kruse, G. Maass & L. C. M. DeMaeyer (1964) *Prog. React. Kin.* **2**, 287.
[3]M. Eigen & G. G. Hammes (1963) *Adv. Enzymol.* **25**, 1.
[4]P. Schuster, W. Wolschann & K. Tortschanoff (1977) in *Chemical Relaxation in Molecular Biology* (I. Pecht & R. Rigler, eds.) pp. 107-190, Springer-Verlag, Berlin.
[5]D. N. Silverman & C. K. Tu (1975) *J. Amer. Chem. Soc.* **97**, 2263.
[6]R. P. Bell (1973) *The Proton in Chemistry*, 2nd ed., Cornell Univ. Press, Ithaca.
[7]A. J. Kresge (1975) *Accts. Chem. Res.* **8**, 354.
[8]J. Crossley (1970) *Advan. Mol. Relax. Proc.* **2**, 69.
[9]E. F. Caldin & V. Gold (1975) *Proton Transfer Reactions*, Chapman-Hall, London.
[10]W. J. Albery (1967) *Progr. React. Kin.* **4**, 353.
[11]E. Grunwald & E. K. Ralph (1971) *Accts. Chem. Res.* **4**, 107.
[12]G. G. Hammes & P. R. Schimmel (1970) *The Enzymes* **2**, 67.
[13]H. Gutfreund (1971) *Ann. Rev. Biochem.* **40**, 315.

PROTOPHILIC

A term applied to a solvent exhibiting affinity for protons and therefore acting as a proton acceptor. Water and ethanol are both protophilic; moreover, both are protogenic (*i.e.*, they can release protons in ionization reactions). **See** *Protogenic; Amphiprotic*

PROTOPORPHYRINOGEN OXIDASE

This FMN-dependent enzyme [EC 1.3.3.4] catalyzes the reaction of protoporphyrinogen-IX with dioxygen to produce protoporphyrin-IX and water. The enzyme will also use mesoporphyrinogen-IX as a substrate, albeit not as effectively.

T. A. Dailey & H. A. Dailey (1997) *Meth. Enzymol.* **281**, 340.
S. Granick & S. I. Beale (1978) *Adv. Enzymol.* **46**, 33.

PSEUDO BUFFERS

Solutions in which the buffering action is due to the solvent rather than any added solute[1]. Strongly acidic or basic aqueous solutions will show little change in pH when additional increments of acid or base are added (recall that the pK_a value for H_3O^+ is −1.74, and that for H_2O is 15.74)[1,2]. Because the solvent is in such high concentration, the buffering capacity for pseudo buffers is larger than for conventional buffers. **See** *Buffer Capacity*

[1]D. D. Perrin & B. Dempsey (1974) *Buffers for pH and Metal Ion Control*, Chapman and Hall, London.
[2]Z. H. S. Harned & R. A. Robinson (1940) *Trans. Faraday Soc.* **36**, 973.

PSEUDO CATALYSIS

The acceleration of a reaction by a substance that is also *consumed* during the process. An example of such a phenomenon is the acceleration of a reaction by a Brønsted acid present in large excess or maintained by a nearly constant concentration by a buffer. The acceleration of the hydrolysis of an amide by a certain Brønsted acid is actually general acid promotion rather than general acid catalysis. The term "promotion" has been used as a synonym for pseudo catalysis.

PSEUDOHALIDES

Anions such as thiocyanate, azide, and cyanide that can act in place of halides as alternative substrates or inhibitors in certain enzymatic reactions. The standard reduction potentials of Cl⁻, Br⁻, and I⁻ are 1.36 V, 1.05 V, and 0.53 V, respectively. The corresponding values for NCS⁻ and CN⁻ are 0.77 V and 0.375 V. Thus, haloperoxidase action on these anions is not at all surprising.

PSEUDOROTATION

A process observed in some pentacovalent phosphorane compounds (or intermediates) in which there is an interchange of apical and equatorial ligands.

In many nucleophilic substitution reactions on phosphorylated compounds, in the initial attack of the nucleophile to form a pentavalent intermediate, the entering nucleophile is in a so-called apical position, and the leaving group likewise resides in an apical position. However, in the case of an adjacent mechanism, the nucleophile enters on the same side that has the group destined to be the leaving group. Accordingly, Westheimer[1] described how the pentavalent phosphorus atom must undergo pseudorotation to place the eventual leaving group in an apical position. Only then will the leaving group depart from the reaction center. This mechanism was unambiguously demonstrated through the use of chiral thiophosphorus-containing substrates for the ribonuclease reaction in a classical set of experiments[2,3] and heralded the widespread and systematic application[4–8] of chiral [¹⁶O,¹⁷O,¹⁸O]thiophosphate esters to define the stereochemical course of ATP/GTP-dependent reactions catalyzed by phosphatases, AMP- and ADP-forming synthetases, and phosphotransferases.

[1]F. H. Westheimer (1968) *Acc. Chem. Res.* **1**, 70.
[2]D. A. Usher, D. I. Richardson & F. Eckstein (1970) *Nature (London)* **228**, 663.
[3]D. A. Usher, E. S. Erenrich & F. Eckstein (1972) *Proc. Natl. Acad. Sci. U.S.A.* **69**, 115.
[4]F. Eckstein, P. J. Romaniuk & B. A. Connoly (1982) *Meth. Enzymol.* **87**, 197.
[5]P. A. Frey, J. P. Richard, H. T. Ho, R. S. Brody, R. D. Sammons & K.-F. Sheu (1982) *Meth. Enzymol.* **87**, 213.
[6]M.-D. Tsai (1982) *Meth. Enzymol.* **87**, 235.
[7]S. L. Buchwald, D. E. Hansen, A. Hassett & J. R. Knowles (1982) *Meth. Enzymol.* **87**, 279.
[8]M. R. Webb (1982) *Meth. Enzymol.* **87**, 301.

PSI (Ψ)

1. Symbol for electric flux. 2. Symbol for wave function.

PTERIDINES, ANALOGUES & FOLATE COENZYMES

Metabolites that contain a ring system formally defined by IUPAC as pyrimidino[4,5-*b*]pyrazine or pyrazine [2,3-*d*]pyrimidine. These metabolites participate in various redox reactions, and the vitamin folic acid is the principal metabolic form from which these derivatives are synthesized. Folic acid is biosynthesized in microorganisms through the cleavage of the imidazole ring of GMP with loss of carbon-8 and subsequent ring closure to form the pteridine ring system.

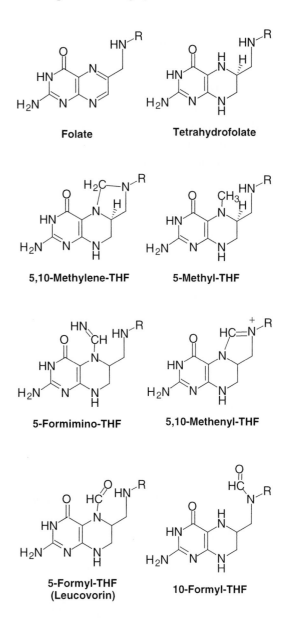

Folate   **Tetrahydrofolate**

5,10-Methylene-THF   **5-Methyl-THF**

5-Formimino-THF   **5,10-Methenyl-THF**

5-Formyl-THF (Leucovorin)   **10-Formyl-THF**

Selected entries from *Methods in Enzymology* [vol, page(s)]: Chromatographic analysis of pteridines, **66**, 429; thin-layer chromatography of pteroylmonoglutamates and related compounds, **66**, 437; chromatography of folates on Sephadex G-10, **66**, 443; separation of folic acid derivatives and pterins by high-performance liquid chromatography, **66**, 452; determination of folate by use of radioactive folate and binding proteins, **66**, 468; PMR characteristics of folic acid and analogs, **66**, 483; identification of the intracellular folate coenzymes of different cell types, **66**, 501;

isolation and preparation of pterins from biological materials, **66**, 508; methods for reduction, stabilization, and analyses of folates, **66**, 517; solid-phase synthesis of pteroylpolyglutamates, **66**, 523; preparation of tritiated dihydrofolic acid of high specific activity, **66**, 529; preparation of folic acid specifically labeled with carbon-13 in the benzoyl carbonyl, **66**, 533; preparation of aminopterin and p-aminobenzoylglutamic acid specifically labeled with carbon-13 in the benzoyl carbonyl, **66**, 536; folate and its reduced forms labeled with deuterium at carbon-7, **66**, 538; N-5-formyltetrahydrofolate (citrovorum factor) labeled stereospecifically with deuterium or tritium at carbon-6, **66**, 541; enzymic synthesis of (d)-L-5-methyltetrahydropteroylglutamate of high specific radioactivity, **66**, 545; synthesis of 2-amino-4-hydroxy-6-hydroxymethylpteridine pyrophosphate, **66**, 553; dihydroneopterin aldolase from *Escherichia coli*, **66**, 556; radioassay for dihydropteroate-synthesizing enzyme activity, **66**, 560; a simple radioassay for dihydrofolate synthetase activity in *Escherichia coli* and its application to an inhibition study of new pteroate analogs, **66**, 576; intracellular distribution, purification, and properties of dihydrofolate synthetase from pea seedlings, **66**, 581; synthetase, **66**, 585; formiminotransferase-cyclodeaminase: a bifunctional enzyme from porcine liver, **66**, 626; pteroylpolyglutamate synthase assay, **66**, 630; pteroylmonoglutamate conversion of into pteroylpolyglutamates, **66**, 638; conversion of pteroylmonoglutamates to pteroylpolyglutamates, **66**, 642; enzymic synthesis of 10-formyl-H₄-pteroyl-γ-glutamylglutamic acid from glutamic acid and 10-formyl-H₄-pteroylglutamic acid, **66**, 648; bacterial degradation of folic acid, **66**, 652; preparation and purification of pteroic acid from pteroylglutamic acid (folic acid), **66**, 657; a radioactive assay of pteroylpolyglutamate hydrolases (conjugases), **66**, 660; assay of folylpolyglutamate hydrolase using pteroyl-labeled substrates and selective short-term bacterial uptake for product determination, **66**, 663; purification of folate binding factors, **66**, 678; preparation and use of affinity columns with bovine milk folate-binding protein (FBP) covalently linked to Sepharose 4B, **66**, 686; purification and properties of the folate-binding protein, **66**, 690; kinetics of folate-protein binding, **66**, 694; purification of thymidylate synthetase from enzyme-poor sources by affinity chromatography, **66**, 709; preparation of an antiserum to sheep liver dihydropteridine reductase, **66**, 723.

PULLULANASE

This enzyme [EC 3.2.1.41] (also known as α-dextrin endo-1,6-α-glucosidase, pullulan 6-glucanohydrolase, limit dextrinase, debranching enzyme, and amylopectin 6-glucanohydrolase) catalyzes the hydrolysis of (1,6)-α-glucosidic linkages in pullulan and starch to form maltotriose. It is the starch debranching enzyme.

M. W. Bauer, S. B. Halio & R. M. Kelly (1996) *Adv. Protein Chem.* **48**, 271.

G. Mooser (1992) *The Enzymes*, 3rd ed., **20**, 187.

I. S. Pretorius, M. G. Lambrechts & J. Marmur (1991) *Crit. Rev. Biochem. Mol. Biol.* **26**, 53.

E. Y. C. Lee & W. J. Whelan (1971) *The Enzymes*, 3rd ed., **5**, 191.

PULSE-CHASE EXPERIMENTS

A radiometric technique for estimating the rates of intracellular synthesis and/or turnover of a metabolite. To measure the rates of synthesis, cells are exposed to a radioisotopically labeled precursor (such as [³H]-, [¹⁴C]-, [¹⁵N]-, [¹³C]-, [²H]leucine or [³⁵S]methionine in the case of proteins) for various periods, after which the cells are lysed and the protein of interest is purified (often by immunoprecipitation with specific antibodies). The time course of isotope incorporation gives information about the rate (slope of curve) and extent (amplitude of curve) of the proteins' synthesis. To measure degradation, cells are first pulse-labeled (*i.e.*, exposed to radiolabeled precursor for a fixed period, after which sufficient nonlabeled precursor is added to reduce the radiospecific activity of the precursor). Then, the cells are further incubated, and the radiospecific activity of a particular protein of interest is determined (again usually after immunoprecipitation or some other means for achieving its isolation from other cellular proteins). The key point is that the chase allows one to stop radiolabel uptake almost instantaneously, thereby permitting the kinetic

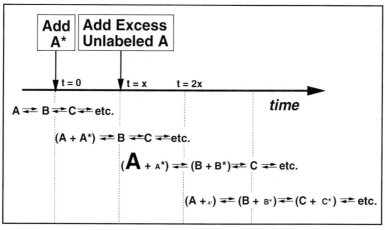

Diagram showing the addition of radiotracer A* at $t = 0$ (termed the "pulse") and the subsequent addition of excess unlabeled A (termed the "chase"). The relative size of the lettering serves to indicate the time-course of radiolabeled entry and exit from each metabolic pool.

analysis of subsequent steps in a reaction or series of reactions (see diagram above).

BASIC PULSE-CHASE KINETICS. For the case of a single pool of A that becomes promptly labeled with tracer $A*$, bolus administration of tracer will initially increase the tracer/tracee ratio (indicated by the initial isotopic enrichment E_o), which will subsequently suffer exponential decay[1]. As is the case for all first-order processes, the rate law is:

$$E_t = E_o \, e^{-kt}$$

where E_t is the isotopic enrichment at time t, and k is the first-order rate constant. Accordingly, a plot of ln (E_t/E_o) versus t should be linear, with a slope of $-k$. [In studies with whole organs or organisms, the kinetics of tracer incorporation into a large tracee pool may be sufficiently long so as to require the use of what is called a priming tracer dose[2].] When a tracee enters two different compartments, each with its own rate constant, the kinetics will be described as a sum of exponentials. If such a process exhibits a multiexponential form, the semilog plot will be nonlinear, often showing discrete linear segments if each successive exponential process is well resolved. Nonlinear least-squares fitting methods or software programs such as KINSIM and FITSIM can help one to simulate the time-evolution of E_t/E_o.

APPLICATIONS OF THE PULSE-CHASE TECHNIQUE. Consideration of a few illustrative cases should serve to underscore the utility of pulse-chase experiments in the characterization of biological systems.

Alice *et al.*[3] studied the turnover kinetics of *Listeria monocytogenes*-secreted p60 protein (a murein hydrolase) by host cell cytosolic proteasomes. J774 cells, seeded in flasks and incubated overnight in culture medium, were infected with log-phase cultures of *L. monocytogenes* for 30 min, washed, and incubated in culture medium for 3 h, with gentamicin (50 μg/ml) added after the first 30 min to inhibit extracellular bacterial growth. Cells then were washed and placed in methionine-free medium with spectinomycin, gentamicin, the eukaryotic protein synthesis inhibitors [cycloheximide (50 μg/mL) and anisomycin (30 μg/ml),] and 25 μM calpain inhibitor I. After 30 min, [^{35}S]methionine was added, and the cells were pulse-labeled for periods of 20 to 60 min. Cells

were washed and chased for the time intervals in culture medium with 20 μG/mL tetracycline to inhibit any further intracellular bacterial growth. Cells were harvested in 1% Triton X-100 lysis buffer containing protease inhibitors, and the detergent-containing lysates were clarified by centrifugation at 4°C. Then, p60 was immunoprecipitated with 25 μL protein A-Sepharose (50% slurry) and 5 μL anti-p60 antiserum for 1 h at 4°C. Beads were washed four times and resuspended in sample buffer, and the samples were electrophoresed on SDS-10% polyacrylamide gels under reducing conditions. Gels were enhanced with 0.5 M salicylate, 3% glycerol, dried, and exposed for autoradiography. Radioactive signals were quantified using a Bio-Rad GS-250 Molecular Imager. Half-lives were determined by plotting percentages of remaining p60 against the chase time intervals. The authors demonstrated that p60 is an N-end rule substrate. Because a protein's N-terminal amino acid can determine its rate of degradation, they mutagenized this residue in p60 into known stabilizing and destabilizing residues. N-terminal valine dramatically stabilized cytosolic p60, whereas aspartic acid substitution resulted in rapid degradation. **See** N-End Rule

Bai *et al.*[4] recently studied how the pattern of sulfation of heparan affects its turnover by examining heparan sulfate catabolism in wild-type Chinese hamster ovary (CHO) cells and mutant cells that are defective in 2-O-sulfation of uronic acid residues. Pulse-chase experiments were used to demonstrate that both mutant and wild-type cells transport newly synthesized heparan sulfate proteoglycans to the plasma membrane, where they shed into the medium or move into the cell through endocytosis. Internalization of the cell-associated molecules leads to sequential endoglycosidase (heparanase) fragmentation of the chains and eventual lysosomal degradation. Both wild-type CHO and the mutant cells were grown to near confluence and pulse-labeled for 1 h with 100 μCi/mL Na$_2$^{35}SO$_4$ (25–40 Ci/mg) in sulfate-free culture medium. After removing the labeled medium, the cell layers were washed three times, and fresh F-12 medium supplemented with 1 mM Na$_2$SO$_4$ was added as a chase. At various times, the medium was collected, and the cells were washed three times with phosphate-buffered saline. Cell surface proteoglycans were released by treating the cells with 0.125% (*w/v*) trypsin for 5–10 min at 37°C. The cells were centrifuged at 800g for 3 min, and the released proteoglycans were recovered in the supernatant. The pellets were designated as the intracellular pool. [For steady-state labeling of heparan sulfate, wild-type CHO and mutant pgsF-17 cells were labeled to constant radiospecific activity after 24 h with $Na_2{}^{35}SO_4$ (50 μCi/mL) in sulfate-free medium. The lack of sulfate in the medium did not cause undersulfation of the chains, because CHO cells derive adequate sulfate from the catabolism of cysteine and methionine.] Radiolabeled glycosaminoglycan chains were isolated from various fractions, and specific activities were determined. Their findings suggest that 2-O-sulfated iduronic acid residues in heparan sulfate are important for cleavage by endogenous heparanases but not for the overall catabolism of the chains.

To characterize further the cellular function of sterol carrier protein-2 (SCP-2), Baum et al.[5] transfected McA-RH7777 rat hepatoma cells with a pre-SCP-2 cDNA expression construct. In stable transfectants, pre-SCP-2 processing resulted in an 8-fold increase in peroxisomal levels of SCP-2. SCP-2 overexpression increased the rates of newly synthesized cholesterol transfer to the plasma membrane and plasma membrane cholesterol internalization by 4-fold. To make these measurements, pulse-chase analysis of cholesterol biosynthesis was accomplished using cells (5 × 10^6) incubated in 10 mL medium containing 5% lipoprotein-deficient serum (LPDS) and [^{3}H]acetate (100 μCi) at 37°C. After 1 h, the medium was removed and fresh chase medium containing 10 mM sodium acetate (unlabeled) was added. At the indicated times, cells were extensively rinsed with phosphate-buffered saline, dissociated with trypsin, extracted, and analyzed by HPLC for the radioactivity in cholesterol and sterol intermediates. Radiolabeled low-density lipoprotein was used in pulse-chase studies to estimate the transfer rate of LDL-derived cholesterol to the plasma membrane as described previously with minor modifications. Cells were seeded into 60-mm dishes at 10^6 cells/dish. After 48 h, the medium was changed to culture medium with 2% LPDS, and cells were incubated for 24 h. Cells were then cooled by rinsing with PBS at 20°C, followed by the addition of fresh DMEM medium containing 2% LPDS, radiolabeled LDL (100 μg/mL), and compound 58-035 (2 μg/mL). After 2 h at 20°C, monolayers were extensively rinsed with PBS at 4°C followed by addition of 5 ml of DMEM medium with

2% LPDS, LDL (100 μg/mL), and compound 58-035 (2 μg/mL) at 37°C. At various times, cells were harvested, treated with cholesterol oxidase, and lipid extracts analyzed by HPLC.

Pulse-chase experiments are also powerful tools in enzyme kinetics. For example, Esteban et al.[6] investigated the relationship between RNA folding and ribozyme catalysis through a detailed kinetic analysis of structural derivatives of the hairpin ribozyme. They used pulse-chase experiments to evaluate substrate dissociation. A small amount of 5-^{32}P-labeled substrate (less than 1 nM) was incubated with a saturating excess of ribozyme (200 nM) in the reaction buffer for 2 min (wild type (wt) ribozymes) or 30 s (modified SV5 ribozymes) at 25°C. These incubation times allow essentially complete formation of the ribozyme·substrate complex since the binding half-times under these conditions are about 30 and 1 s, for the wt and the modified ribozymes, respectively. The chase step was initiated by adding a large excess (5 μM final concentration) of either a DNA oligonucleotide that was fully complementary to the wt substrate or, alternatively, a nonradiolabeled SV5 substrate for reactions carried out with wt or SV5 ribozymes, respectively. A complementary DNA oligonucleotide was used instead of the unlabeled wt RNA substrate because this molecule forms stable dimers at high concentrations. During the chase period, aliquots were removed and quenched with an equal volume of loading buffer (15 mM EDTA, 97% formamide). Samples were analyzed and quantified as described above. Parallel control reactions were carried out in the absence of the chase molecule. The efficiency of the chase step was evaluated by mixing the labeled substrate with the chase molecule (either the complementary DNA oligonucleotide or the unlabeled RNA substrate) prior to the addition of ribozyme. No significant cleavage of the labeled substrate was observed under these conditions, indicating that there is no rebinding of the labeled substrate during the chase step. Typically, time courses carried out in the presence of the chase molecule displayed monophasic behavior and, therefore, were fitted to single-exponential equations. Control reactions in the absence of chase showed biphasic kinetics, and hence double-exponential equations were used. Estimation of the kinetic parameters (amplitudes and rates) was carried out as described above.

These investigators[6] also used pulse-chase experiments to measure the kinetics of substrate association. Several ribozyme concentrations, ranging from 12.5 nM to 200 nM for wt ribozymes or from 1 nM to 10 nM SV5 ribozymes, were combined with a trace amount (less than 0.1 nM) of the corresponding 5-^{32}P-labeled substrate in reaction buffer at 25°C. For each ribozyme concentration, several chase reactions were initiated at different times, ranging from 10 s to 4 min. The chase molecule was a complementary DNA oligonucleotide, in the case of the wt substrate, or unlabeled RNA substrate, in the case of the SV5 substrate. The final concentration of the chase molecule was 5 μM or 1 μM for reactions carried out with wt or SV5 substrates, respectively. Reactions were incubated for 1 h at 25°C after addition of the chase molecule. This time is sufficient to ensure a quantitative cleavage of the 5-^{32}P-labeled substrate·ribozyme complexes (the half-time for the cleavage reaction was about 5 min). Time courses of the cleavage reaction were fitted to single-exponential equations, and the observed rates were plotted versus ribozyme concentration to obtain k_{on} and k_{off}.

[1] R. R. Wolfe (1992) *Radiative and Stable Isotope Tracers in Biomedicine: Principles and Practice of Kinetic Analysis*, pp. 471, Wiley-Liss, New York.

[2] H. L. Searle, E. H. Strisower & I. L. Chaikoff (1954) *Am. J. Physiol.* **176**, 190.

[3] A. J. A. M. Sijts, I. Pilip & E. G. Pamer (1997) *J. Biol. Chem.* **272**, 19261.

[4] X. Bai, K. J. Bame, H. Habuchi, K. Kimata & J. D. Esko (1997) *J. Biol. Chem.* **272**, 23172.

[5] C. L. Baum, E. J. Reschly, A. K. Gayen, M. E. Groh & K. Schadick (1997) *J. Biol. Chem.* **272**, 6490.

[6] J. A. Esteban, A. R. Banerjee & J. M. Burke (1997) *J. Biol. Chem.* **272**, 13629.

PULSE RADIOLYSIS

A rapid reaction kinetic technique (time scale = 10–1000 ps) that typically uses a Van de Graff accelerator or a microwave linear electron accelerator to promptly generate a pulse of electrons at sufficient power levels for excitation and ionization of target substances by electron impact. The technique is the direct radiation chemical analog of flash photolysis[1], and the ensuing kinetic measurements are accomplished optically by IR/visible/UV adsorption spectroscopy or by fluorescence spectroscopy.

A good example is the examination[2] of the time course of superoxide depletion catalyzed by human manganese superoxide dismutase (MnSOD). The reaction was observed spectrophotometrically by measuring the absorbance of superoxide at 250–280 nm following pulse radiolysis to rapidly generate 6 μM superoxide anion. Catalysis showed an initial burst of activity lasting approximately 1 ms, followed by the rapid emergence of a greatly inhibited catalysis of zero-order rate. These catalytic properties of human MnSOD are qualitatively similar to those reported[3] for MnSOD from *Thermus thermophilus*. However, generation of the inhibited form is approximately 30-fold more rapid for human MnSOD. The turnover number for human MnSOD at pH 9.4 and 20°C was $k_{cat} = 4 \times 10^4$ s^{-1} and $k_{cat}/K_m = 8 \times 10^8$ M^{-1} s^{-1}, determined by KINSIM simulation and fitting pulse radiolysis data to the model of Bull *et al.*[3]

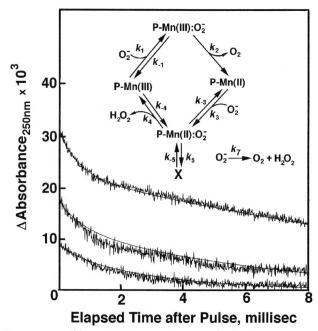

Time-course of human manganese superoxide dismutase reaction after generation of superoxide by pulse radiolysis[2]. The experiment demonstrates decrease in absorbance at 250 nm (ε = 2000 M^{-1} cm^{-1}) for a solution containing 0.5 μM enzyme, 50 μM EDTA, 10 mM sodium formate, and 2 mM sodium pyrophosphate at pH 9.4 and 20°C. The starting concentrations of superoxide were 11.6 μM (upper curve), 6.5 μM (middle curve), and 3.4 μM (lower curve). The calculated progress curves shown as solid lines were obtained by using the KINSIM software and the model of Bull *et al.*[3] Reproduced here with the permission of the authors and the American Chemical Society.

Other examples using pulse radiolysis include: (a) studies[4] on cytochrome c-d_1 nitrate reductase from *Thiosphaera pantotropha* to provide evidence for a fast intramolecular electron transfer from c-heme to d_1-heme; (b) determination[5] of the rate constant for the reaction of hydroxyl and oxide radicals with cysteine in aqueous solution; (c) direct demonstration[6] of the catalytic action of monodehydroascorbate reductase; (d) investigation[7] of the generation and reactions of the disulfide radical anion derived from metallothionein; (e) the kinetics[8] of NO· reactions with O_2^{-} and HO_2· ; and the enhanced intramolecular electron transfer[9] in the ascorbate oxidase reaction in the presence of oxygen.

[1]L. M. Dorfman (1974) in *Investigation of Rates and Mechanisms of Reactions*, Part II, 3rd ed. (G. G. Hammes, ed.) vol. VI, pp. 463*ff*, Wiley-Interscience, New York.

[2]J. L. Hsu, Y. Hsieh, C. Tu, D. O'Connor, H. S. Nick & D. N. Silverman (1996) *J. Biol. Chem.* **271**, 17687.

[3]C. Bull, E. C. Niederhoffer, T. Yoshida & J. A. Fee (1991) *J. Am. Chem. Soc.* **113**, 4069.

[4]K. Kobayashi, A. Koppenhofer, S. J. Ferguson & S. Tagawa (1997) *Biochem.* **36**, 13611.

[5]S. P. Mezyk (1996) *Radiat. Res.* **145**, 102.

[6]K. Kobayashi, S. Tagawa, S. Sano & K. Asada (1995) *J. Biol. Chem.* **270**, 27551.

[7]X. Fang, J. Wu, G. Wei, H. P. Schuchmann & C. von Sonntag (1995) *Int. J. Radiat. Biol.* **68**, 459.

[8]S. Goldstein & G. Czapski (1995) *Free Radic. Biol. Med.* **19**, 505.

[9]O. Farver, S. Wherland & I. Pecht (1994) *J. Biol. Chem.* **269**, 22933.

PURINE NUCLEOSIDASE

Thie enzyme [EC 3.2.2.1], also known as inosine-uridine preferring nucleoside hydrolase and IU-nucleoside hydrolase, catalyzes the hydrolysis of an *N*-D-ribosylpurine to produce a purine and D-ribose.

T. P. Wang (1955) *Meth. Enzymol.* **2**, 456.

PURINE NUCLEOSIDE PHOSPHORYLASE

The enzyme [EC 2.4.2.1], also known as inosine phosphorylase, catalyzes the reaction of a purine nucleoside with orthophosphate to produce a purine and α-D-ribose 1-phosphate. The enzyme will also catalyze the activity of nucleoside ribosyltransferase.

G. Davies, M. L. Sinnott & S. G. Withers (1998) *Comprehensive Biological Catalysis: A Mechanistic Reference* **1**, 119.

R. E. Parks, Jr., & R. P. Agarwal (1972) *The Enzymes*, 3rd ed., **7**, 483.

PUTRESCINE CARBAMOYLTRANSFERASE

This enzyme [EC 2.1.3.6] catalyzes the reaction of carbamoyl phosphate with putrescine to produce orthophosphate and *N*-carbamoylputrescine. The plant enzyme will also catalyze the reactions of ornithine carbamoyltransferase, carbamate kinase, and agmatine deiminase.

C. W. Tabor & H. Tabor (1984) *Ann. Rev. Biochem.* **53**, 749.

PUTRESCINE *N*-METHYLTRANSFERASE

This enzyme [EC 2.1.1.53] catalyzes the reaction of *S*-adenosyl-L-methionine with putrescine to produce *S*-adenosyl-L-homocysteine and *N*-methylputrescine.

T. Hashimoto & Y. Yamada (1994) *Ann. Rev. Plant Physiol. Plant Mol. Biol.* **45**, 257.

PUTRESCINE OXIDASE

This FAD-dependent enzyme [EC 1.4.3.10] catalyzes the reaction of putrescine with dioxygen and water to produce 4-aminobutanal, ammonia, and hydrogen peroxide. 4-Aminobutanal then condenses nonenzymically to 1-pyrroline.

H. Yamada (1971) *Meth. Enzymol.* **17B**, 726.

P-VALUE

1. The lowest level of significance for which an obtained test statistic is significant. 2. The probability that a variate (often called the random variable would strictly by chance assume a value greater than or equal to the observed value.

PYRAZOLE

A potent five-membered heterocyclic inhibitor (the chemical systematic name is 1,2-diazole) that strongly inhibits liver alcohol dehydrogenase. **See** *Alcohol Dehydrogenase*

PYRIDOXAL 4-DEHYDROGENASE

This enzyme [EC 1.1.1.107] catalyzes the reaction of pyridoxal with NAD^+ to produce 4-pyridoxolactone and NADH. The enzyme acts on the hemiacetal form of the substrate.

R. W. Burg (1970) *Meth. Enzymol.* **18A**, 634.

PYRIDOXAL KINASE

This enzyme [EC 2.7.1.35] (also known as pyridoxine kinase, pyridoxamine kinase, and vitamin B$_6$ kinase) catalyzes the reaction of ATP with pyridoxal to produce ADP and pyridoxal 5′-phosphate. Pyridoxine, pyridoxamine, and various other derivatives can also act as substrates.

S. L. Ink & L. M. Henderson (1984) *Ann. Rev. Nutr.* **4**, 455.

PYRIDOXAMINE:PYRUVATE AMINOTRANSFERASE

This pyridoxal-phosphate-dependent enzyme [EC 2.6.1.30] catalyzes the reaction of pyridoxamine with pyruvate to produce pyridoxal and L-alanine.

A. E. Braunstein (1973) *The Enzymes*, 3rd ed., **9**, 379.

PYRIDOXINE & PYRIDOXAL (Analogues And Derivatives)

Metabolites that contain pyridoxol (2-methyl-3-hydroxy-4,5-di[hydroxymethyl]pyridine), the water-soluble vitamin B_6. Phosphorylated forms of pyridoxal (the aldehyde form) and pyridoxamine (the amino form), known respectively as pyridoxal 5-phosphate (PLP) and pyridoxamine 5-phosphate (PMP), are tightly-bound coenzymes.

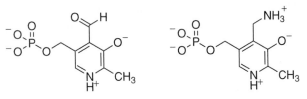

Pyridoxal 5'-phosphate　　　**Pyridoxamine 5'-phosphate**

PLP-dependent enzymes catalyze the following types of reactions: (1) loss of the α-hydrogen as a proton, resulting in racemization (example: alanine racemase), cyclization (example: aminocyclopropane carboxylate synthase), or β-elimation/replacement (example: serine dehydratase); (2) loss of the α-carboxylate as carbon dioxide (example: glutamate decarboxylase); (3) removal/replacement of a group by aldol cleavage (example: threonine aldolase; and (4) action via ketimine intermediates (example: selenocysteine lyase).

Selected entries from *Methods in Enzymology* [vol, page(s)]:
Analyzing spectra of vitamin B_6 derivatives, **18A**, 433; fluorometric determination of pyridoxal and pyridoxal 5'-phosphate in biological materials by the reaction with cyanide, **18A**, 471; nuclear magnetic resonance spectroscopy of vitamin B_6, **18A**, 475; mass spectrometry of vitamin B_6, **18A**, 483; fluorometric assay of pyridoxal, **62**, 405; fluorometric determination of pyridoxal phosphate in enzymes, **62**, 407; high-voltage electrophoresis and thin-layer chromatographic separation of vitamin B_6 compounds, **62**, 410; high-performance chromatography of vitamin B_6, **62**, 415; carbon-13 nuclear magnetic resonance spectroscopy of the vitamin B_6 group, **62**, 422; ^{13}C NMR spectroscopy of the vitamin B_6 group, **62**, 436; synthesis and biological activity of vitamin B_6 analogues, **62**, 454; vitamin B_6 antagonists of natural origin, **62**, 483; spin-labeled vitamin B_6 derivatives: synthesis and interaction with aspartate aminotransferase, **62**, 495; the use of 4'-N-(2,4-dinitro-5-fluorophenyl)pyridoxamine 5'-phosphate, **62**, 510; immobilization of pyridoxal 5'-phosphate and pyridoxal 5'-phosphate-dependent enzymes on Sepharose, **62**, 517; pyridoxal 5'-phosphate and analogs as probes of coenzyme-protein interac-

tion, **62**, 528; crystalline enzyme-substrate complexes of pyridoxal phosphate-dependent enzymes, **62**, 551; dithionite-induced changes in the spectra of free and enzyme-bound pyridoxal phosphate, **62**, 558; assay of pyridoxal phosphate and pyridoxamine phosphate, employing S-o-nitrophenyl-L-cysteine, a chromogenic substrate of tryptophanase, **62**, 561; pyridoxamine (pyridoxine) 5'-phosphate oxidase from rabbit liver, **62**, 568; pyridox(al, amine) 5'-phosphate hydrolase from rat liver, **62**, 574; plasma binding of B_6 compounds, **62**, 582.

PYRIMIDINE DIMER DNA GLYCOSYLASE

This enzyme [EC 3.2.2.17], also known as deoxyribodipyrimidine endonucleosidase, catalyzes the hydrolysis of the N-glycosidic bond between the 5'-pyrimidine residue in cyclobutadipyrimidine (in the DNA) and the corresponding deoxy-D-ribose residue.

B. K. Duncan (1981) *The Enzymes*, 3rd ed., **14**, 565.

PYROGLUTAMYL PEPTIDE HYDROLASE

This enzyme [EC 3.4.19.3], a member of the C15 peptidase family, is also known as pyroglutamyl-peptidase I, 5-oxoprolyl-peptidase, pyrrolidone-carboxylate peptidase, and pyroglutamyl aminopeptidase. This hydrolase catalyzes the conversion of a 5-oxoprolyl-peptide to produce 5-oxoproline and a peptide. The enzyme will not act on the 5-oxoprolyl peptide if the adjacent amino acid is L-proline. Enzyme activity is inhibited by thiol-blocking reagents.

E. Shaw (1990) *Adv. Enzymol.* **63**, 271.
R. J. DeLange & E. L. Smith (1971) *The Enzymes*, 3rd ed., **3**, 81.

PYROLYSIS

Decomposition, involving the uncatalyzed cleavage of one or more covalent bonds, due to exposure to elevated temperatures; sometimes referred to as thermolysis.

PYROPHOSPHATASE, INORGANIC

Inorganic pyrophosphatase [EC 3.6.1.1] plays a central role in phosphorus metabolism by catalyzing the hydrolysis of the phosphoanhydride bond of inorganic pyrophosphate (or, diphosphate; PP_i). This cleavage reaction acts in conjunction with pyrophosphate-forming ligases to provide an additional thermodynamic impetus for certain biosynthetic reactions. For example:

$$X + Y + MgATP^{2-} = X - Y + AMP + PP_i\ (\Delta G \approx 0)$$

$$PP_i + H_2O = 2P_i\ (\Delta G << 0)$$

Catalysis requires divalent metal ions which affect the apparent pK_a values of the essential general acid and base on the enzyme, and the pK_a of the substrate[1-3]. Three to five metal ions are required for maximal activity, depending on pH and enzyme source.

[1]B. S. Cooperman (1982) *Meth. Enzymol.* **87**, 526.
[2]P. Heikinheimo, J. Lehtonen, A. Baykov, R. Lahti, B. S. Cooperman & A. Goldman (1996) *Structure* **4**, 1491.
[3]A. A. Baykov, T. Hyytia, S. E. Volk, V. N. Kasho, A. V. Vener, A. Goldman, R. Lahti & B. S. Cooperman (1996) *Biochemistry* **35**, 4655.

PYROPHOSPHATE FRUCTOSE-6-PHOSPHATE 1-PHOSPHOTRANSFERASE

This enzyme [EC 2.7.1.90] (also known as diphosphate:fructose-6-phosphate 1-phosphotransferase, 6-phosphofructokinase (pyrophosphate), and pyrophosphate-dependent 6-phosphofructose-1-kinase) catalyzes the reaction of pyrophosphate (or, diphosphate) with D-fructose 6-phosphate to produce orthophosphate and D-fructose 1,6-bisphosphate.

H. C. Huppe & D. H. Turpin (1994) *Ann. Rev. Plant Physiol. Plant Mol. Biol.* **45**, 577.
E. Van Schaffingen (1987) *Adv. Enzymol.* **59**, 315.
H. G. Wood, W. E. O'Brien & G. Michaels (1977) *Adv. Enzymol.* **45**, 85.

PYROPHOSPHOMEVALONATE DECARBOXYLASE

This enzyme [EC 4.1.1.33] (also known as mevalonic acid 5-pyrophosphate anhydrodecarboxylase, mevalonate-5-diphosphate decarboxylase, and diphosphomevalonate decarboxylase) catalyzes the reaction of ATP with (R)-5-diphosphomevalonate to produce ADP, orthophosphate, isopentenyl diphosphate, and carbon dioxide.

R. A. Gibbs (1998) *Comprehensive Biological Catalysis: A Mechanistic Reference* **1**, 31.
R. A. Dixon, P. M. Dey & C. J. Lamb (1983) *Adv. Enzymol.* **55**, 1.

Δ¹-PYRROLINE-5-CARBOXYLATE DEHYDROGENASE

This enzyme [EC 1.5.1.12] catalyzes the reaction of 1-pyrroline-5-carboxylate with NAD^+ and water to produce L-glutamate and NADH. The enzyme will also act on 3-hydroxy-1-pyrroline-5-carboxylate to generate 4-hydroxyglutamate.

E. Adams & L. Frank (1980) *Ann. Rev. Biochem.* **49**, 1005.

Δ¹-PYRROLINE-5-CARBOXYLATE REDUCTASE

This enzyme [EC 1.5.1.2] catalyzes the reversible reaction of 1-pyrroline-5-carboxylate with NAD(P)H to produce L-proline and $NAD(P)^+$. In addition, the enzyme will also act on 1-pyrroline-3-hydroxy-5-carboxylate, reducing it to L-hydroxyproline.

E. Adams & L. Frank (1980) *Ann. Rev. Biochem.* **49**, 1005.
D. M. Greenberg (1962) *Meth. Enzymol.* **5**, 959.

1-PYRROLINE-4-HYDROXY-2-CARBOXYLATE DEAMINASE

This enzyme [EC 3.5.4.22] catalyzes the hydrolysis of 1-pyrroline-4-hydroxy-2-carboxylate to generate 2,5-dioxopentanoate and ammonia.

E. Adams (1971) *Meth. Enzymol.* **17B**, 266.

PYRUVATE CARBOXYLASE

This biotin-dependent enzyme [EC 6.4.1.1] catalyzes the reaction of ATP with pyruvate and HCO_3^- to produce ADP, orthophosphate, and oxaloacetate. The enzyme from yeast requires zinc whereas the enzyme isolated from animal tissues needs manganese as well as acetyl-CoA.

J. N. Earnhardt & D. N. Silverman (1998) *Comprehensive Biological Catalysis: A Mechanistic Reference* **1**, 495.
P. V. Attwood (1995) *Int. J. Biochem. Cell Biol.* **27**, 231.
M. H. O'Leary (1992) *The Enzymes*, 3rd ed., **20**, 235.
M. F. Utter, R. E. Barden & B. L. Taylor (1975) *Adv. Enzymol.* **42**, 1.
M. C. Strutton & M. R. Young (1972) *The Enzymes*, 3rd ed., **6**, 1.

PYRUVATE DECARBOXYLASE

This thiamin pyrophosphate-dependent enzyme [EC 4.1.1.1] catalyzes the conversion of an α-keto acid (or, a 2-oxo acid) to an aldehyde and carbon dioxide. This enzyme will also catalyze acyloin formation.

R. L. Schowen (1998) *Comprehensive Biological Catalysis: A Mechanistic Reference* **2**, 217.
J. V. Schloss & M. S. Hixon (1998) *Comprehensive Biological Catalysis: A Mechanistic Reference* **2**, 43.
A. Schellenberger, G. Hübner & H. Neef (1997) *Meth. Enzymol.* **279**, 131.
F. Dyda, W. Furey, S. Swaminathan, M. Sax, B. Farrenkopf & F. Jordan (1993) *Biochemistry* **32**, 6165.
Y. Lindqvist & G. Schneider (1993) *Curr. Opin. Struct. Biol.* **3**, 896.
M. H. O'Leary (1992) *The Enzymes*, 3rd ed., **20**, 235.

PYRUVATE DEHYDROGENASE

Pyruvate dehydrogenase (lipoamide) [EC 1.2.4.1], which requires thiamin pyrophosphate, catalyzes the reaction of pyruvate with lipoamide to produce S-acetyldihydrolipoamide and carbon dioxide. It is a component of the pyruvate dehydrogenase complex (which also includes dihydrolipoamide dehydrogenase [EC 1.8.1.4] and dihydrolipoamide acetyltransferase [EC 2.3.1.12]). Pyruvate dehydrogenase (cytochrome) [EC 1.2.2.2] catalyzes the reaction of pyruvate with ferricytochrome b_1 and water to produce acetate, carbon dioxide, and ferrocytochrome b_1. It also requires thiamin pyrophosphate.

J. V. Schloss & M. S. Hixon (1998) *Comprehensive Biological Catalysis: A Mechanistic Reference* **2**, 43.
A. Schellenberger, G. Hübner & H. Neef (1997) *Meth. Enzymol.* **279**, 131.
R. N. Perham (1995) *Meth. Enzymol.* **251**, 436.
R. N. Perham & P. N. Lowe (1988) *Meth. Enzymol.* **166**, 330.
L. J. Reed & S. J. Yeaman (1987) *The Enzymes*, 3rd ed., **18**, 77.
G. Palmer (1975) *The Enzymes*, 3rd ed., **12**, 1.

PYRUVATE DEHYDROGENASE KINASE

This enzyme [EC 2.7.1.99] catalyzes the reaction of ATP with [pyruvate dehydrogenase (lipoamide)] to produce ADP and [pyruvate dehydrogenase (lipoamide)] phosphate. This is an enzyme that is associated with the pyruvate dehydrogenase complex. Phosphorylation of pyruvate dehydrogenase (lipoamide) [EC 1.2.4.1] inactivates that enzyme.

L. J. Reed & S. J. Yeaman (1987) *The Enzymes*, 3rd ed., **18**, 77.
D. A. Flockhart & J. D. Corbin (1982) *Crit. Rev. Biochem.* **12**, 133.
D. A. Walsh & E. G. Krebs (1973) *The Enzymes*, 3rd ed., **8**, 555.

PYRUVATE DEHYDROGENASE PHOSPHATASE

This enzyme [EC 3.1.3.43] catalyzes the hydrolysis of the phosphorylated form of pyruvate dehydrogenase (lipoamide) to generate the dephosphorylated dehydrogenase and orthophosphate. The enzyme is associated with the pyruvate dehydrogenase complex.

L. J. Reed & S. J. Yeaman (1987) *The Enzymes*, 3rd ed., **18**, 77.

PYRUVATE FORMATE LYASE

This enzyme [EC 2.3.1.54], also known as formate *C*-acetyltransferase, catalyzes the reaction of acetyl-CoA with formate to produce coenzyme A and pyruvate.

J. Knappe & A. F. V. Wagner (1995) *Meth. Enzymol.* **258**, 343.
E. J. Bruch & J. W. Kozarich (1992) *The Enzymes*, 3rd ed., **20**, 317.
J. A. Stubbe (1989) *Ann. Rev. Biochem.* **58**, 257.

PYRUVATE KINASE

This enzyme [EC 2.7.1.40] catalyzes the reaction of ADP with phosphoenolypyruvate to produce ATP and pyruvate. Other nucleotides that can be used as substrates include UDP, GDP, CDP, IDP, and dADP. The enzyme will also phosphorylate hydroxylamine and fluoride in the presence of carbon dioxide. *See Nucleoside 5'-Triphosphate Regeneration*

J. V. Schloss & M. S. Hixon (1998) *Comprehensive Biological Catalysis: A Mechanistic Reference* **2**, 43.
H. C. Huppe & D. H. Turpin (1994) *Ann. Rev. Plant Physiol. Plant Mol. Biol.* **45**, 577.
J. J. Villafranca & T. Nowak (1992) *The Enzymes*, 3rd ed., **20**, 63.
R. F. Colman (1990) *The Enzymes*, 3rd ed., **19**, 283.
L. Engström, P. Ekman, E. Humble & Ö. Zetterqvist (1987) *The Enzymes*, 3rd ed., **18**, 47.

PYRUVATE, ORTHOPHOSPHATE DIKINASE

This enzyme [EC 2.7.9.1], also referred to as pyruvate, phosphate dikinase, catalyzes the reaction of ATP with pyruvate and orthophosphate to produce AMP, phosphoenolpyruvate, and pyrophosphate (or, diphosphate).

P. A. Frey (1992) *The Enzymes*, 3rd ed., **20**, 141.
F. M. Raushel & J. J. Villafranca (1988) *Crit. Rev. Biochem.* **23**, 1.
N. H. Goss & H. G. Wood (1982) *Meth. Enzymol.* **87**, 51.
H. G. Wood, W. E. O'Brien & G. Michaels (1977) *Adv. Enzymol.* **45**, 85.
R. A. Cooper & H. L. Kornberg (1974) *The Enzymes*, 3rd ed., **10**, 631.

PYRUVATE OXIDASE

This enzyme [EC 1.2.3.3], which requires thiamin pyrophosphate and FAD, catalyzes the reaction of pyruvate with orthophosphate, dioxygen, and water to produce acetyl phosphate, carbon dioxide, and hydrogen peroxide.

R. L. Schowen (1998) *Comprehensive Biological Catalysis: A Mechanistic Reference* **2**, 217.
A. Schellenberger, G. Hübner & H. Neef (1997) *Meth. Enzymol.* **279**, 131.
Y. A. Muller, Y. Lindqvist, W. Furey, G. E. Schulz, F. Jordan & G. Schneider (1993) *Structure* **1**, 95.

PYRUVATE SYNTHASE

This enzyme [EC 1.2.7.1], also known as pyruvate:ferredoxin 2-oxidoreductase, catalyzes the reaction of pyruvate with coenzyme A and oxidized ferredoxin to produce acetyl-CoA, carbon dioxide, and reduced ferredoxin.

L. L. Barton (1994) *Meth. Enzymol.* **243**, 94.
E. J. Brush & J. W. Kozarich (1992) *The Enzymes*, 3rd ed., **20**, 317.
B. B. Buchanan (1972) *The Enzymes*, 3rd ed., **6**, 193.

PYRUVATE, WATER DIKINASE

This manganese-dependent enzyme [EC 2.7.9.2], also known as phosphoenolypyruvate synthase, catalyzes the reaction of ATP with pyruvate and water to produce AMP, phosphoenolpyruvate, and orthophosphate.

P. A. Frey (1992) *The Enzymes*, 3rd ed., **20**, 141.
R. A. Cooper & H. L. Kornberg (1974) *The Enzymes*, 3rd ed., **10**, 631.

Q

Q_{10}

Symbol for the temperature coefficient, a quotient equal to v_{T+10}/v_T, where v_{T+10} and v_T are the rates of a process (*e.g.*, an enzyme-catalyzed reaction) at two temperatures differing by 10°C. This parameter is usually evaluated at saturating concentrations of substrate(s), so that temperature-dependent changes in Michaelis constant(s) are inconsequential. The Q_{10} value is a characteristic property of a particular enzyme from a specific organism and cell type. For example, one cannot use the Q_{10} value for one hexokinase from yeast to infer the temperature dependence of another hexokinase, say from rat brain. Likewise, the Q_{10} value need not remain the same for a mutant form and a wild-type enzyme.

Because the energy of activation, E_a, for a reaction having a rate constant k, equals $RT^2(\mathrm{d} \ln k)/\mathrm{d}T$ (where R is the universal gas constant and T is the absolute temperature), then, for a temperature increase of 10°, $E_a = \{RT(T + 10)\ln Q_{10}\}/10$. Since 10 is significantly less than T, this relationship reduces to $E_a = \{RT^2 \ln Q_{10}\}/10$. Q_{10} is temperature-dependent, and the above analysis assumes that all other reaction conditions are constant and that the enzyme is stable over this temperature range.

Fumarase[1] is reported to have a Q_{10} of 1.994 at 30°C and, at temperatures below 22°C, rat liver mitochondrial ATPase has a value of 29[2].

In addition, recall that $E_a = \Delta H^{\ddagger} + RT$. Thus, there is a minimal value for Q_{10} at a given temperature. This value is given by $Q_{10,\mathrm{min}} = e^{10/(T + 10)}$. Hence, at 25°C, $Q_{10,\mathrm{min}}$ is 1.034.

[1]V. Massey (1953) *Biochem. J.* **53**, 72.
[2]J. K. Raison (1973) *Symposia of the Soc. for Exper. Biol.* **27**, 485.

Q_{CO_2}

Symbol for the amount (in microliters) of CO_2 given off (under standard conditions of pressure and temperature) per milligram of tissue per hour.

Q TEST

A means for deciding whether to keep or reject a particular measurement within a set of measurements.[1] For example, suppose an investigator weighed the same sample of a particular substance five separate times and obtained weights of 10.18 g, 9.98 g, 10.07 g, 10.20 g, and 10.38 g. The experimental Q value, Q_{exp}, is equal to the difference between the questionable measurement and its nearest neighbor [in the case of the 10.38 value, this is (10.38 − 10.20) = 0.18] divided by the overall spread of the entire set of measurements (thus, 10.38 − 9.98 = 0.40). Hence, $Q_{\mathrm{exp}} = 0.18/0.40 = 0.45$. If this value exceeds a certain critical value, termed Q_{crit}, then the measurement is rejected. For a 90% confidence level with five measurements, $Q_{\mathrm{crit}} = 0.64$. Thus, the measurement would be retained in the example above. Tables of Q_{crit} are available[1] [for 90% confidence levels: for 3 samples or observations ($Q_{\mathrm{crit}} = 0.94$), for 4 (0.76), for 5 (0.64), for 6 (0.56), for 7 (0.51), for 8 (0.47), for 9 (0.44), and for 10 (0.41)]. However, one should always be wary of rejecting certain data, particularly if the set of measurements is small. In addition, if the investigator is aware that there were errors or difficulties with a particular measurement, then he/she has a basis for rejecting that value, even if the value satisfies the Q test. One should always exercise good judgment. It is often a good practice to report median values rather than means. Finally, Q tests should not be applied to derived values (for example, K_m values). There are other statistical means for presenting such data.

[1]R. B. Dean & W. J. Dixon (1951) *Anal. Chem.* **23**, 636.

QUADREACTANT SYSTEMS

Enzymes utilizing four substrates in a single reaction. An example is carbamoyl phosphate synthetase which has four substrates (L-glutamine, CO_2, 2 ATP) in addition to the nonvaried substrate H_2O. Note that multienzyme complexes (*e.g.*, pyruvate dehydrogenase complex) frequently use four substrates; however, these complexes generally involve a number of enzyme-catalyzed steps with fewer than four substrates in a single step. These would not be considered as quadreactant systems.

QUADRUPOLE

Two magnetic dipoles associated together as a single unit or four electric charges organized as a single unit. One of the isotopes of oxygen (^{17}O) has a large quadrupole moment, and because quadrupolar nuclei have significant magnetic resonance properties, investigators have been able to use these properties in probes of enzyme mechanisms: for example, with ^{31}P NMR[1,2].

[1] M.-D. Tsai (1979) *Biochemistry* **18**, 1468
[2] M.-D. Tsai (1982) *Meth. Enzymol.* **87**, 235.

QUALITY INDEX

Symbolized by Q, the quality index is a parameter used in studying the stereochemical nature of a particular enzyme-catalyzed reaction using chiral [$^{16}O,^{17}O,^{18}O$] phosphoric monoesters[1]. This parameter, which is dependent on the fractional ^{17}O content, allows the investigator a means to assess the reliability of the study(ies).

[1] S. L. Buchwald, D. E. Hansen, A. Hassett & J. R. Knowles (1982) *Meth. Enzymol.* **87**, 279.

QUANTASOME

Small granular structural unit that serves as the basic building block of the thylakoid membrane and contains multiple chlorophyll molecules as well as cytochromes.

QUANTUM

1. A discrete unit or parcel of energy (electric, photic, vibrational, rotatory, *etc.*) absorbed by or released from a quantized state of matter, usually at the molecular, atomic, or subatomic level. Within the context of the wave-particle duality of electromagnetic radiation, the quantum and photon are the basic units of energy change. 2. The minimum amount by which a certain property (for example, angular momentum, energy, *etc.*) can change.

QUANTUM MECHANICAL TUNNELING

A kinetic condition wherein a small particle, such as a proton, electron, or a hydride ion may, as a consequence of its de Broglie wavelength ($\lambda = h/mv$), "pass" or "tunnel" through a potential energy barrier rather than be required to be of adequate energy to surmount the activation energy barrier. This is a requirement in quantum mechanics: the Heisenberg Uncertainty Principle allows a small particle having a large enough de Broglie wavelength to exist on either side of a sufficiently narrow energy barrier. Protons have a de Broglie wavelength of 1-2 Å, just about the expected width for some proton transfer reactions. More massive particles will have correspondingly smaller de Broglie wavelengths, and quantum mechanical tunneling is less likely, even for deuterium. The net effect would be that tunneling should increase the magnitude of k_H/k_D. Tunneling is most likely in reactions possessing symmetrical transition states, a condition where the proton's motion makes the greatest contribution to the reaction coordinate. Interestingly, steric hindrance is also believed to favor tunneling. The remaining mystery is how an enzyme might finely tune the configuration of catalytic acid/base groups near the substrate in order to take maximal advantage of quantum mechanical tunneling. *See* Chemical Kinetics; Hydrogen Tunneling; Kinetic Isotope Effects

QUANTUM REQUIREMENT

A stoichiometric number indicating the number of photons required per molecule of O_2 produced in biosynthesis. *See* Quantum Yield

QUANTUM YIELD

A value (symbolized by ϕ) equal to the number of molecules transformed via a reaction per quantum of light absorbed. It is synonymous to quantum efficiency and is the reciprocal of the quantum requirement.

A primary quantum yield is the fraction of light-absorbing molecules that are converted in a particular process. Unfortunately, such values are often very difficult to measure. A product quantum yield is the ratio of the number of molecules of product formed per number of quanta absorbed by the reactant. That product formation is the rate-determining step is a strong possibility if the observed ϕ value does not vary with experimental conditions.

The differential quantum yield is given by $\phi = \{d[x]/dt\}/n$ where $d[x]/dt$ is the rate of change of the concentration or of the amount of some measurable quantity and n is the amount (*e.g.*, moles) of photons absorbed per unit time. The integral quantum yield is simply defined as the number of events per number of photons absorbed[1]. *See Actinometer; Fluorescence*

[1]IUPAC (1988) *Pure and Appl. Chem.* **60**, 1055.

QUASI-EQUILIBRIUM ASSUMPTION

Another term used to indicate a rapid equilibrium assumption in a kinetic process. The most prominent example in biochemistry is the assumption that enzyme and substrate rapidly form a preequilibrium enzyme·substrate complex in the Michaelis-Menten treatment.

QUASI-EQUIVALENCE

The concept[1] used in biological self-assembly to account for the ability of otherwise chemically identical subunits to adopt different conformational states, such that alternative geometrical or stereochemical bonding interactions (or valences) within the final assembled structure can be satisfied by the same polymerization subunit. The classical case is the assembly of the coat protein of icosahedral viruses, wherein pentavalent and hexavalent interactions of the identical protein subunit generate a structure that has an overall minimum in free energy. Such behavior requires (a) that there be only small changes in the free energies of conformations capable of pentavalent and hexavalent bonding interactions, and/or (b) that the ability to form closed geometrical structures overcomes any difference in the free energy of these two conformational states. Because the transition states are likely to be different in forming pentavalent or hexavalent structures, the kinetics of individual steps during self-assembly may proceed at different rates.

[1]D. L. D. Caspar (1980) *Biophys. J.* **32**, 103.

QUATERNARY COMPLEXES

Complexes formed from the ordered or random combination of four distinct entities. Many three-substrate enzymes proceed via the formation of a reversible E·A·B·C quaternary complex.

QUENCHER

1. In a photochemical process, a molecular entity that deactivates an excited state of another molecular entity.
2. A substance, complex, body, system, or parameter that

stops, or slows down, a chemical or enzyme-catalyzed reaction. *See Quenching*

QUENCH-FLOW EXPERIMENTS

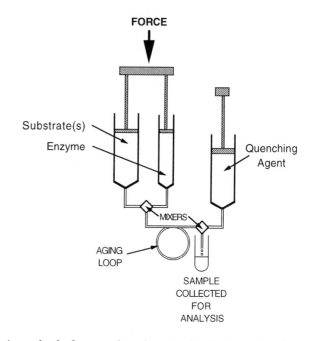

A method often used to characterize the transient kinetics of chemical and enzymic processes. The reaction is initiated by pushing two or more separate solutions through a single mixing chamber. The reaction mixture then flows through a tube for a known period of time (determined by the flow rate and the length of the tube) until a quenching solution is rapidly combined with the reacting mixture. When the substrate concentration is larger than that of the enzyme, one can often detect a burst of product formation that transiently exceeds the steady-state rate. When the concentration of substrate is the limiting factor, so-called single-turnover kinetics can be measured in rapid quench experiments.

K. A. Johnson (1995) *Meth. Enzymol.* **249**, 38.

QUENCHING

1. The process or procedure for arresting the progress of a chemical or enzyme-catalyzed reaction. The kineticist must ensure that the quenching procedure does, in fact, completely stop the reaction at the time of sampling (when measuring product formed or substrate remaining). In addition, it is necessary to demonstrate that the substrate and product (particularly labeled substrate and product) are stable throughout the quenching procedure. Another goal is to chose conditions for sufficiently

rapid quenching. An excellent test of quenching efficiency is to subject a typical reaction mixture (which includes all reaction components including the enzyme(s)) to the quenching protocol, *after which* a radiolabeled substrate is added. If labeled product is formed, then quenching was most probably incomplete. Methods for quenching include: (a) addition of specific agents (*e.g.*, organic solvents, EDTA, EGTA, thiol agents, acids, bases, *etc.*), (b) rapid changes in temperature (both heating and cooling have been used), and (c) dilution of the reaction mix to halt intermolecular processes.

2. A term often used in liquid scintillation counting in which the energy spectrum has been shifted to a lower energy. This often results from the presence of some interfering substance in the solution (*e.g.*, a colored agent or other foreign chemical). This phenomenon is probably responsible for many errors in rate studies, particularly when biomacromolecules or high salt concentrations cause phase changes in the scintillation cocktail. Three methods can be used to correct for the effects of quenching[1]. First, most modern liquid scintillation counters are equipped with external sources that serve to detect quenching and to correct for its effects. After the radioactivity of a sample is determined, the external source is moved into place, and a standard dose ($X_{standard}$) from this source should result in a certain number of disintegrations per unit time. The observed value ($X_{observed}$) is a direct measure of the counting efficiency (*i.e.*, efficiency = $X_{observed}/X_{standard}$. Second, a known amount of a reference standard of the same radionuclide as is utilized in the tracer studies can be added to an experimental sample, and the counting efficiency can be determined by the ratio of the observed increase in radioactivity divided by the expected increase in radioactivity. The third method takes advantage of the shift in a radionuclide's energy spectrum to lower energy in the presence of a quenching agent. The "channel ratios" method allows one to use the pulse-height spectrum to divulge any evidence of quenching.

3. In Geiger counters, the process of inhibiting a continuous discharge. This process is usually accomplished by the presence of a gas such as methane mixed with argon or neon.

[1]Y. Kobayashi & D. V. Marcdsley (1974) *Biological Applications of Liquid Scintillation Counting*, Academic Press, New York.

QUERCETIN 2,3-DIOXYGENASE

This enzyme [EC 1.13.11.24] catalyzes the reaction of quercetin, a flavonol, with dioxygen to produce 2-protocatechoylphloroglucinolcarboxylate and carbon monoxide. Copper ions are needed as a cofactor.

A. Messerschmidt (1998) *Comprehensive Biological Catalysis: A Mechanistic Reference*, **3**, 401.
O. Hayaishi, M. Nozaki & M. T. Abbott (1975) *The Enzymes*, 3rd ed., **12**, 119.

QUINOLINATE PHOSPHORIBOSYLTRANSFERASE

This enzyme [EC 2.4.2.19], also referred to as nicotinate mononucleotide pyrophosphorylase (carboxylating), catalyzes the reversible reaction of pyridine 2,3-dicarboxylate and 5-phospho-α-D-ribose 1-diphosphate to produce nicotinate D-ribonucleotide, carbon dioxide, and pyrophosphate.

L. M. Henderson (1983) *Ann. Rev. Nutr.* **3**, 289.
W. D. L. Musick (1981) *Crit. Rev. Biochem.* **11**, 1.

QUIN-2 TETRAACETOXY ESTER

A membrane-permeant form of Quin-2 (systematic name: 2-[(2-amino-5-methylphenoxy)methyl]-6-methoxy-8-aminoquinoline-N,N,N',N'-tetraacetate) which upon binding of calcium ion results in fluorescence. **See** *Fura-2*

QUOTIENT FUNCTION

Any ratio of parameters, such as V_{max}/K_m or the fraction of the protein in the R state to that fraction in the T state in the Monod-Wyman-Changeux allosterism model.

QUOTIENT, RESPIRATORY

A ratio of the amount of CO_2 produced by tissue metabolism to the amount of oxygen consumed.

QUOTIENT, SYNERGISM (Q_{syn})

A parameter used in isotope exchange experiments of enzyme-catalyzed reactions purported to have a ping pong mechanism. This parameter assesses whether substrate synergism is present and to what degree. For a ping pong Bi Bi mechanism, $Q_{syn} = \{(R_{max, A-P})^{-1} +$

$(R_{\mathrm{max, \ B\text{-}Q}})^{-1}\}/(V_{\mathrm{max, \ f}}^{-1} + V_{\mathrm{max, \ r}}^{-1})$ where $V_{\mathrm{max, \ f}}$ and $V_{\mathrm{max, \ r}}$ are the maximum initial-rate velocities for the forward and reverse reaction, respectively, and $R_{\mathrm{max, \ A\text{-}P}}$ and $R_{\mathrm{max, \ B\text{-}Q}}$ are the maximum rates of exchange seen in the half-reactions. A synergism quotient of unity indicates a ping pong mechanism with no substrate synergism. For values significantly larger than one, either the mechanism is not a simple ping pong scheme or substrate synergism is present[1].

[1]D. L. Purich & R. D. Allison (1980) *Meth. Enzymol.* **64**, 3.

R

RACEMASES

A class of enzymes that catalyze the interconversion of one enantiomer with its mirror image. Care must be exercised in applying this term. For example, the enzyme that interconverts D-methylmalonyl-CoA to L-methylmalonyl-CoA is not a racemase, but is instead an epimerase; the two coenzyme A derivatives are diastereoisomeric, and not enantiomeric, with respect to each other.

RACEMIC MIXTURE (or Racemate)

A mixture of equal amounts of two enantiomers, mirror images of each other. Such a mixture will not rotate the plane of polarized light. These mixtures can be designated by the prefix *RS*-, ($\pm$)-, or DL-. The prefix *dl*- should be avoided.

RACEMIZATION

Any reaction, either chemical or biochemical, which results in the conversion of an optically active molecular entity to a racemic mixture.

RAD

1. Unit for the dose absorbed from ionizing radiation. One rad equals 100 ergs of radiation absorbed per gram of tissue. One gray equals 100 rad. 2. Abbreviation for radian.

RADIAN

A supplementary SI unit (symbolized by rad) for plane angle: 2π radians equals 360°. Thus, one radian is equivalent to 57.2957795 . . . ° (or, 57° 17′ 44.806 . . .″).

RADIANCE

A measure of the radiant power (P) for a parallel beam of electromagnetic radiation leaving or passing through an infinitesimal element of surface in a given direction from the source divided by the orthogonally projected area of the element in a plane normal to the given direction of the beam (θ). It is symbolized by L and has SI units of watts per square meter. Thus, $L = (dP/dS)/\cos \theta$. If the radiant power is constant over the surface area, then $L = P/(S \cos \theta)$.

If the beam is divergent from an elementary cone of the solid angle ($d\Omega$) containing the given direction (θ), then $L = d^2P/(d\Omega dS \cos \theta)$. In this instance, the radiance has units of watts per square meter per steradian.

Comm. on Photochem. (1988) *Pure and Appl. Chem.* **60**, 1055.

RADIANT ENERGY

The radiant energy (Q) is equal to the total energy emitted, transferred, or received as radiation in a given period of time. It has SI units of joules. If the radiant power (P) is constant over the time period considered, then $Q = Pt$.

Comm. on Photochem. (1988) *Pure and Appl. Chem.* **60**, 1055.

RADIANT (or, ENERGY) FLUX

Symbolized by ϕ, the radiant flux is equivalent to the radiant power P. When the radiant energy Q is constant over the time frame considered, $\phi = Q/t$. Care should be exercised when using this term and its symbol in photochemical and related studies since ϕ is more commonly reserved for quantum yield.

Comm. on Photochem. (1988) *Pure and Appl. Chem.* **60**, 1055.

RADIATIONLESS TRANSITION

The transition that takes place without photon emission or absorption, between two states of a system. ***See*** *Fluorescence*

RADIATIVE LIFETIME

The lifetime of an excited state of a molecular entity (in the absence of any radiationless transition). Radiative lifetime is symbolized by τ_0 and equals the reciprocal of the first-order rate constant (or, sum of first-order rate constants).

Comm. on Photochem. (1988) *Pure and Appl. Chem.* **60**, 1055.

RADICAL (or FREE RADICAL)

1. Any molecular entity possessing an unpaired electron. The modifier "unpaired" is preferred over "free" in this context. The term "free radical" is to be restricted to those radicals which do not form parts of radical pairs. Further distinctions are often made, either by the nature of the central atom having the unpaired electron (or atom of highest electron spin density) such as a carbon radical (*e.g.*, $\cdot CH_3$) or whether the unpaired electron is in an orbital having more *s* character (thus, σ radicals) or more *p* character (hence, π radicals). Whenever presenting the radical molecular entity in a manuscript, the structure should always be written with a superscript dot or, preferably, a center-spaced bullet (*e.g.*, $\cdot OH$, $\cdot CH_3$, $Cl\cdot$).

2. Any substituent or moiety bound to a molecular entity. IUPAC suggests that this older term should be abandoned, preferring instead usage of "groups," "moieties," or "substituents."

IUPAC (1994) *Pure and Appl. Chem.* **66**, 1077.

RADICAL CENTER

The atom within a molecule or ion on which the unpaired electron is mainly localized. This center could actually be a group of atoms of a moiety of the molecule.

RADICAL ION

An electrically charged radical. Hence, a radical cation carries a positive charge and a radical anion carries a negative charge. The charge and the odd electron are often localized with the same atoms of the molecular entity. According to IUPAC recommendations, if the unpaired electron and the charge cannot be associated with specific atoms of the molecular entity (or they are both localized to the same atom), then a superscript (or center space) dot or bullet should precede the superscript charge (+ or −). Thus, a benzene radical cation would be $C_6H_6^{\cdot+}$. This is opposite of the convention used in mass spectroscopy.

IUPAC (1994) *Pure and Appl. Chem.* **66**, 1077.

RADIOACTIVE SUBSTANCES (General Precautions)

The following are general recommendations, which may be superseded by institutional requirements[1].

(1) Make certain that areas for radionuclide use are well marked, so that infrequent visitors to the laboratory will be immediately aware of areas to avoid. Ideally, these rooms should be locked and inaccessible to all but authorized personnel. A wise practice is to assume that these rooms will be entered by individuals lacking training in safe practices, and this mandates in favor of safe and secure storage procedures.

(2) Prohibit eating, drinking, and all mouth-pipetting in rooms containing designated radioactivity areas.

(3) Be properly dressed; *e.g.*, safety eyeglasses, long pants, labcoat, *closed-toe* shoes, gloves (these must be appropriate for the chemical properties of radiolabeled substances), lead apron (if recommended), and wrist guards. Dosimeters must be worn in recommended locations (wrist, finger, lapel, *etc.*).

(4) All radioactive material in storage and refuse containers should be clearly marked, preferably indicating the type of nuclide in use.

(5) Work with absorbent (preferably plastic-backed) benchtop coverings. Always use a spill tray containing an absorbent liner.

(6) Know the physical properties of the substances with which you are working. Keep in mind that some compounds (such as acetaldehyde and tritiated water) have low boiling points. Again, keep in mind that some gloves do not offer an adequate barrier to certain chemicals. Some compounds enter the body with such facility that special care must be exercised when they are in use. One example is dimethyl sulfoxide, which as a solvent facilitates the entry of many solutes into the body. There are many known cases where radiolabeled compounds contaminated individuals who failed to consider this power of DMSO as a solute vehicle.

(7) Handle any volatile substances in an approved and suitably marked fume hood or glove box. Chemical traps

may prove useful in preventing loss of certain volatile substances.

(8) Use a Geiger-Müller counter (and periodically use paper discs to swipe surfaces for scintillation counting) to determine the radioisotope as well as the sites and amounts of contaminating radioactivity.

(9) Keep accurate records on each compound, indicating radionuclide, amount remaining, location, date, and name of responsible user (usually the principal investigator's name).

(10) Maintain good housekeeping practices (*e.g.*, on ending an experiment, be sure to return all radionuclides to their safe storage locations, dispose of all contaminated items, and decontaminate surfaces). Remember that any unsuitably shielded waste can be hazardous to lab workers and custodial personnel.

(11) Again, one should always remember to WASH hands.

(12) Use of certain isotopes requires local or full-body scintillation counting to confirm the extent and location of any contaminant. Regulations concerning such monitoring vary according to institutional policy.

[1]For certification by the U. S. Nuclear Regulatory Commission, each institution establishes its own standards of safe practice that must meet or exceed N. R. C. Standards. Practices of individual investigators must adhere to their own institutional policies and regulations.

RADIOCARBON DATING

A cosmogenic method for dating carbonaceous objects based on the generation of carbon-14 from atmospheric nitrogen by cosmic radiation in the form of thermal neutrons; newly formed ^{14}C is then oxidized to $^{14}CO_2$ which enters the biosphere/geosphere where active exchange occurs between living objects and nutrient carbon dioxide. When such objects die, they end active exchange with the biosphere, and their radiocarbon content decreases by radioactive decay. This technique was pioneered by Professor Willard F. Libby (University of California, Los Angeles) who calibrated the ^{14}C content using carbonaceous objects from Egyptian tombs well-dated by dynastic hieroglyphic inscriptions and/or other archeologic documentation. Because the half-life for carbon-

14 is 5715 years, and because the precision of measuring carbon-14 content ranges from 4–6 percent, the technique is limited to dating objects less than 50,000 years. One should recognize that the intensity of cosmic radiation impinging on earth's atmosphere is not constant. Volcanic eruptions can likewise release carbon dioxide in amounts sufficient to dilute the ^{14}C content in the atmosphere. Thus, radiocarbon dating can be inaccurate. *See* Cosmogenic Dating

RADIOLYSIS

1. The cleavage or scission of a bond as a result of high-energy radiation. 2. The generation of any chemical process induced by high-energy radiation. *See* Pulsed Radiolysis

Comm. on Photochem. (1988) *Pure and Appl. Chem.* **60**, 1055.

RAMAN SCATTERING

A dispersive phenomenon occurring when the wavelength of scattered electromagnetic radiation in the mid-infrared spectral region is shifted relative to that of the incident beam of exciting radiation. Spectral excitation is typically measured at a nonabsorbing wavelength, and the Raman effect occurs when the polarizability of a bond varies with the internuclear distance, as specified by the equation:

$$\alpha = \alpha_o + (r - r_{eq})(\partial\alpha/\partial r)$$

where α_o is the bond's polarizability at an equilibrium distance r_{eq}, and r is the instantaneous internuclear distance. The internuclear distance depends on the vibrational frequency ν_{vib}, such that $(r - r_{eq}) = r_{max}\cos(2\,\pi\nu_{vib}t)$, and substitution into the first expression yields:

$$\alpha = \alpha_o + r_{maximal}\cos(2\,\pi\nu_{vibrational}t)(\partial\alpha/\partial r)$$

For the dipole moment μ, one can write a standard expression (*i.e.*, $\mu = \alpha E = \alpha E_o \cos(2\pi\nu_{ex}t)$, where ν_{ex} is the frequency of excitation). Upon substitution for α, we get

$$\mu = \alpha_o E_o \cos(2\,\pi\nu_{ex}t)$$
$$+ E_o r_{maximal}(\partial\alpha/\partial r)\cos(b\,2\,\pi\nu_{vib}t)\cos(2\,\pi\nu_{ex}t)$$

The last term contains the product of the two cosine functions[1] and can be re-expressed as:

$$\mu = \alpha_o E_o \cos(2\pi\nu_{ex}t)$$
$$+ 0.5 E_o r_{max}(\partial\alpha/\partial r)\{\cos[2\pi(\nu_{ex} - \nu_{vib})t]$$
$$+ \cos[2\pi(\nu_{ex} + \nu_{vib})t]\}$$

If only the $\alpha_o E_o \cos(2\pi\nu_{ex}t)$ term is present, there is no energy loss by the elastic collision of a photon and the molecule; in this case, we have the classical Rayleigh light scattering equation [i.e., $\mu = \alpha_o E_o \cos(2\pi\nu_{ex}t)$]. The term $0.5 E_o r_{max}(\partial\alpha/\partial r)\cos[2\pi(\nu_{ex} - \nu_{vib})t]$ represents the so-called Stokes frequency; and the term $0.5 E_o r_{max}(\partial\alpha/\partial r) \cos[2\pi(\nu_{ex} + \nu_{vib})t]$ is the so-called anti-Stokes frequency. One can say that the excitation wavelength has been changed or modulated when it experiences Raman scattering. We must further note that the energies represented by these inelastic collisions are too small to promote an electron to even the next electronic state; as such, they are not quantized, and they are treated as virtual states lying on a continuum between the ground and excited states. In resonance Raman spectroscopy, one takes advantage of the fact that Raman line intensities are enhanced greatly (often by a factor of 10^2 to 10^6) by excitation wavelengths approaching that of a electronic absorption in a molecule of interest. This increases the sensitivity of the technique, thereby lowering the concentration needed for spectral characterization. The availability of tunable dye lasers has stimulated the development of resonance Raman spectroscopy, and a fuller description of the technique and applications to enzyme studies are described elsewhere[2].

The Raman technique is particularly useful for investigating the nature of metal-ligand bonds, for which the vibrational energies typically range from 100–700 cm^{-1}. Spectra can be directly observed in aqueous solutions, offering a major advantage over infrared spectroscopy for the study of biological molecules owing to the opacity of water with respect to IR radiation.

[1]Note: The product of cos a and cos b can be rewritten as 0.5[cos (a + b) + cos(a + b)])

[2]M. Sinnott, C. D. Garner, E. First & G. Davies (1998) *Comprehensive Biological Catalysis: A Mechanistic Reference*, vol. **4**, p. 86, Academic Press, San Diego.

RANDOM AB/QR, ORDERED C/P TER TER MECHANISM

A three-substrate, three-product enzyme-catalyzed reaction scheme in which two substrates (A and B) can bind in any order but the third substrate (C) can only bind after the other two have already bound. Then, following the catalytic event, a particular product (P) is released followed by the random release of the remaining two products (Q and R). Guanylate cyclase has been reported[1] to have this mechanism.

[1]D. L. Garbers, J. G. Hardman & F. B. Rudolph (1974) *J. Biol. Chem.* **13**, 4166.

RANDOM AC/PR, ORDERED B/Q TER TER MECHANISM

A three-substrate, three-product enzyme-catalyzed reaction scheme in which a particular substrate (B) has to bind second, but the other two substrates (A and C) can either bind first or third. Then, following the catalytic event, a particular product (Q) has to be the second product released, but the other two products (P and R) can be either the first or third product released. *See Multisubstrate Mechanisms*

RANDOM BC/PQ, ORDERED A/R TER TER MECHANISM

A three-substrate, three-product enzyme-catalyzed reaction scheme in which one particular substrate (A) is the only substrate that can bind to the free enzyme. After the EA binary complex has formed, the other two substrates (B and C) can bind in any order. Following the catalytic event, two of the products (P and Q) can be released in a random sequence, but the third product (R) has to be the last product released. Citrate cleavage enzyme[1] and γ-glutamylcysteine synthetase[2] are reported to operate by this mechanism. *See Multisubstrate Mechanisms*

[1]K. M. Plowman & W. W. Cleland (1967) *J. Biol. Chem.* **242**, 4239.
[2]B. P. Yip & F. B. Rudolph (1976) *J. Biol. Chem.* **251**, 3563.

RANDOM BI BI MECHANISM

A two-substrate, two-product enzyme-catalyzed reaction scheme in which both the substrates (A and B) and the products (P and Q) bind and are released in any order. Note that this definition does not imply that there is an equal preference for each order (that is, it is not a requirement that the flux of the reaction sequence in which A binds first has to equal the flux of the reaction sequence in which B binds first). In fact, except for rapid equilibrium schemes, this is rarely true. There usually is a distinct preference for a particular pathway in a random mechanism. A number of kinetic tools and protocols (*e.g.*, isotope exchange at equilibrium experiments) can assist the investigator in identifying those reaction pathways.

Quite a few enzymes have been reported to have this reaction scheme: for example, hexokinase[1], adenylate kinase[2], and phosphofructokinase[3].

The rate expression for the rapid equilibrium random Bi Bi mechanism, in the absence of products, is $v = V_{\max}[A][B]/\{K_{ia}K_b + K_b[A] + K_a[B] + [A][B]\}$ where $V_{\max}$ is the maximum velocity in the forward direction (equal to $k_9[E_{total}]$ where k_9 is the rate constant for the EAB → EPQ conversion), $K_{ia} = [E][A]/[EA]$, $K_a = [EB][A]/[EAB]$, and $K_b = [EA][B]/[EAB]$. The steady-state random Bi Bi rate expression is a more complex equation containing additional terms of $[A]^2[B]$ and $[A][B]^2$ in the numerator and $[A]^2$, $[B]^2$, $[A]^2[B]$, and $[A][B]^2$ in the denominator. Rudolph and Fromm[4] have looked at the effect of the magnitude of these other terms on initial rate and product inhibition studies. **See** *Multisubstrate Mechanisms*

[1]J. Ning, D. L. Purich & H. J. Fromm (1969) *J. Biol. Chem.* **244**, 3840.
[2]D. G. Rhoads & J. M. Lowenstein (1968) *J. Biol. Chem.* **243**, 3963.
[3]R. L. Hanson, F. B. Rudolph & H. A. Lardy (1973) *J. Biol. Chem.* **248**, 7852.
[4]F. B. Rudolph & H. J. Fromm (1971) *J. Biol. Chem.* **246**, 6611.

RANDOM BI BI DUAL THEORELL-CHANCE MECHANISM

A two-substrate, two-product enzyme-catalyzed reaction scheme in which either substrate, A or B, can bind first. However, the second substrate participates in a Theorell-Chance step in which either one of the two products is released. **See** *Multisubstrate Mechanisms*

RANDOM BI UNI MECHANISM

An enzyme-catalyzed reaction scheme in which the two substrates (A and B) can bind in any order, resulting in the formation of a single product of the enzyme-catalyzed reaction (hence, this reaction is the reverse of the random Uni Bi mechanism). Usually the mechanism is distinguished as to being rapid equilibrium (*i.e.*, the rate-determining step is the central complex interconversion, EAB ↔ EP) or steady-state (in which the substrate addition and/or product release steps are rate-contributing). **See** *Multisubstrate Mechanisms*

Rapid Equilibrium Case. In the absence of significant amounts of product (*i.e.*, initial rate conditions; thus, [P] 0), the rate expression for the rapid equilibrium random Bi Uni mechanism is $v = V_{\max}[A][B]/(K_{ia}K_b + K_b[A] + K_a[B] + [A][B])$ where K_{ia} is the dissociation constant for the EA complex, K_a and K_b are the dissociation constants for the EAB complex with regard to ligands A and B, respectively, and $V_{\max} = k_9[E_{total}]$ where k_9 is the forward unimolecular rate constant for the conversion of EAB to EP. Double-reciprocal plots ($1/v$ *vs.* $1/[A]$ at different constant concentrations of B and $1/v$ *vs.* $1/[B]$ at different constant concentrations of A) will be intersecting lines. Slope and intercept replots will provide values for the kinetic parameters.

Steady-State Expression. In the absence of significant amounts of product, P (thus, initial rate conditions in which $[P] \approx 0$), the steady-state expression for the random Bi Uni mechanism having two central complexes

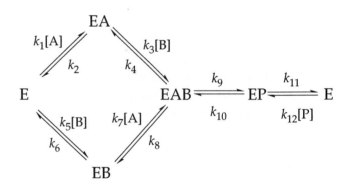

is $v = \{\Phi_{NAB}[A][B] + \Phi_{NA^2B}[A]^2[B] + \Phi_{NAB^2}[A][B]^2\}/ \{\Phi_{D0} + \Phi_{DA}[A] + \Phi_{DB}[B] + \Phi_{DAB}[A][B] + \Phi_{DA^2}[A]^2 + \Phi_{DB^2}[B]^2 + \Phi_{DA^2B}[A]^2[B] + \Phi_{DAB^2}[A][B]^2\}$ where the Φ values are collections of rate constants (not to be confused with Dalziel Φ coefficients) and are designated in subscripts by N or D (representing numerator or denominator, respectively) and the corresponding letter(s) for substrate(s) (with exponent when needed) to which that Φ value is a coefficient. For the above scheme with two central complexes,

$$\Phi_{NAB} = k_9 k_{11}(k_1 k_3 k_6 + k_2 k_5 k_7)[E_{total}]$$

$$\Phi_{NA^2B} = k_1 k_3 k_7 k_9 k_{11}[E_{total}]$$

$$\Phi_{NAB^2} = k_3 k_5 k_7 k_9 k_{11}[E_{total}]$$

$$\Phi_{D0} = k_2 k_6((k_4 + k_8)(k_{10} + k_{11}) + k_9 k_{11})$$

$$\Phi_{DA} = k_2 k_7 (k_4(k_{10} + k_{11}) + k_9 k_{11})$$
$$+ k_1 k_6 ((k_4 + k_8)(k_{10} + k_{11}) + k_9 k_{11})$$

$$\Phi_{DB} = k_3 k_6 (k_8(k_{10} + k_{11})$$
$$+ k_9 k_{11}) + k_2 k_5 ((k_4 + k_8)(k_{10} + k_{11}) + k_9 k_{11})$$

$$\Phi_{DAB} = (k_1 k_3 (k_6 + k_8) + k_5 k_7 (k_2 + k_4))(k_{10} + k_{11})$$
$$+ k_9 (k_3 (k_1 k_6 + k_7 k_{11}) + k_2 k_5 k_7)$$

$$\Phi_{DA^2} = k_1 k_7 (k_4(k_{10} + k_{11}) + k_9 k_{11})$$

$$\Phi_{DB^2} = k_3 k_5 (k_8(k_{10} + k_{11}) + k_9 k_{11})$$

$$\Phi_{DA^2B} = k_1 k_3 k_7 (k_9 + k_{10} + k_{11})$$

$$\Phi_{DAB^2} = k_3 k_5 k_7 (k_9 + k_{10} + k_{11})$$

If the terms that contain $[A]^2$ or $[B]^2$ are negligible relative to the other terms, then one will still observe straight lines in double-reciprocal plots, $V_{max} = \Phi_{NAB}/\Phi_{DAB}$, $K_a = \Phi_{DB}/\Phi_{DAB}$, and $K_b = \Phi_{DA}/\Phi_{DAB}$.

If one considers the simpler random Bi Uni scheme in which the two central complexes are grouped together as EXY (thus, [EAB] + [EP] = [EXY] and k_{11} in the above scheme would become k_9 while k_{12} becomes k_{10}), the individual enzyme determinants are:

$$[E] = k_2 k_6 (k_4 + k_8 + k_9) + k_2 k_7 (k_4 + k_9)[A]$$
$$+ k_3 k_6 (k_8 + k_9)[B] + k_3 k_7 k_9 [A][B]$$

$$[EA] = k_1 k_6 (k_4 + k_8 + k_9)[A] + k_1 k_7 (k_4 + k_9)[A]^2$$
$$+ k_4 k_5 k_7 [A][B] + k_4 k_6 k_{10}[P] + k_4 k_7 k_{10}[A][P]$$

$$[EB] = k_2 k_5 (k_4 + k_8 + k_9)[B] + k_3 k_5 (k_8 + k_9)[B]^2$$
$$+ k_1 k_3 k_8 [A][B] + k_2 k_8 k_{10}[P] + k_3 k_8 k_{10}[B][P]$$

$$[EXY] = (k_1 k_3 k_6 + k_2 k_5 k_7)[A][B] + k_1 k_3 k_7 [A]^2[B]$$
$$+ k_3 k_5 k_7 [A][B]^2 + k_3 k_6 k_{10}[B][P]$$
$$+ k_2 k_7 k_{10}[A][P] + k_3 k_7 k_{10}[A][B][P]$$
$$+ k_2 k_6 k_{10}[P]$$

RANDOM BI UNI UNI BI PING PONG MECHANISM

A three-substrate, three-product enzyme-catalyzed reaction scheme in which two substrates (A and B) can bind to the enzyme in any order. A catalytic event then occurs and a product (P) is released. After that desorption, the third substrate (C) binds, another catalytic event occurs, and the remaining two products (Q and R) are released in any order. Leucyl-tRNA synthetase has been re-

ported[1] to operate by this mechanism. *See* *Multisubstrate Mechanisms*

[1]C.-S. Linn, R. Irwin & J. G. Chirikjian (1975) *J. Biol. Chem.* **250**, 9299.

RANDOM ERROR

Deviations that arise probabilistically and have two characteristics: (a) the magnitude of these errors is more typically small, and (b) positive and negative deviations of the same magnitude tend to occur with the same frequency. Random error is normally distributed, and the bell-shaped curve for frequency of occurrence *versus* magnitude of error is centered at the true value of the measured parameter. *See* *Statistics (A Primer)*

RANDOM EXPERIMENT

An experiment that can result in different outcomes, even though the experiment is repeated by exactly the same procedure every time. A random variable is a numerical outcome of a random experiment. *See* *Statistics (A Primer)*

RANDOMIZATION

A strategy that attempts to obtain a statistically unbiased sample or data set for a series of experimental measurements by simulating a chance distribution or chance sequence. *See* *Statistics (A Primer)*

RANDOM MECHANISMS

Multisubstrate or multiproduct enzyme-catalyzed reaction mechanisms in which one or more substrates and/or products bind and/or are released in a random fashion. Note that this definition does not imply that there has to be an equal preference for any particular binding sequence. The flux through the different binding sequences could very easily be different. However, in rapid equilibrium random mechanisms, the flux rates are equivalent. *See* *Multisubstrate Mechanisms*

RANDOM-ON/ORDERED-OFF BI BI MECHANISM

A two-substrate, two-product enzyme-catalyzed reaction scheme in which the two substrates (A and B) can bind to the enzyme in any order but the two products are released in a distinct order (P being released before Q). *See* *Multisubstrate Mechanisms*

RANDOM SCISSION KINETICS

Although some depolymerases act processively in cleaving their polymeric substrates, others act by what can be described as multiple attack which results in nonselective scission or random scission. The analysis of cleavage products during the course of enzyme-catalyzed depolymerization can provide important clues about the nature of the reaction. With random scission, the rate of bond scission must be proportional to the total number of unbroken bonds present in the solution. Thomas[1] measured the rate of base addition in a pH-Stat (a device with an automated feedback servomotor that expels titrant from a syringe to maintain pH) to follow the kinetics of DNA bond scission by DNase. The number of bonds cleaved was linear with time, and this was indicative of random scission. In other cases, one may apply the template challenge method to assess the processivity of nucleic acid polymerases[2]. **See** Processivity

Tanford[3] presented a cogent kinetic treatment of random scission of a polymer, and the complete analysis is beyond the scope of this Handbook. The basic idea is that a molecule M_y can yield a molecule M_x (where x and y indicate the degree of polymerization, and where $y > x$) in two different ways. The x-mer formation rate from y-mers is twice the rate of bond scission at the concentration of the y-mer; thus,

$$\frac{d[M_x]}{dt} = 2k \sum_{y=x+1}^{\infty} [M_y]$$

from y equal to $(x + 1)$ to infinity. The rate of conversion of x-mers into shorter polymer chains occurs by rupture of any of their $(x - 1)$ bonds, and this effectively amplifies the magnitude of the rate constant k:

$$-\frac{d[M_x]}{dt} = (x - 1)k[M_x]$$

The rate of change of the x-mer concentration is

$$-\frac{d[M_x]}{dt} = (x - 1)k[M_x] + 2k \sum_{y=x+1}^{\infty} [M_y]$$

again where from y equal to $(x + 1)$ to infinity. If $x = n$, then upon integration, this expression takes on the following first-order rate form:

$$[M_n] = [M_n]_0 e^{-k(n-1)t}$$

Likewise, we can write corresponding equations for $[M_{n-1}]$, $[M_{n-2}]$, $[M_{n-3}]$, etc.:

$$[M_{n-1}] = [M_n]_0 e^{-k(n-2)t}(1 - e^{-kt})(2)$$

$$[M_{n-2}] = [M_n]_0 e^{-k(n-3)t}(1 - e^{-kt})(3 - e^{-kt})$$

$$[M_{n-3}] = [M_n]_0 e^{-k(n-4)t}(1 - e^{-kt})(4 - 2e^{-kt})$$

except for when it is undefined (*i.e.*, $x = n$). Then, by using N_{total} as the total number of subunits per volume, and by expressing $[M_x]$ as the number concentration of x-mers, this treatment can account for the time evolution of random polymer scission.

[1] C. A. Thomas, Jr. (1956) *J. Amer. Chem. Soc.* **78**, 1861.
[2] W. R. McClure & Y. Chow (1980) *Meth. Enzymol.* **64**, 277.
[3] C. Tanford (1961) *Physical Chemistry of Macromolecules*, p. 615, Wiley, New York.

RANDOM TER TER MECHANISM

A three-substrate, three-product enzyme-catalyzed reaction scheme in which the three substrates (A, B, and C) and three products (P, Q, and R) can bind to and be released in any order. A number of enzymes have been reported to have this mechanism: for example, adenylosuccinate synthetase[1,2], glutamate dehydrogenase[3], glutamine synthetase[4,5], formyltetrahydrofolate synthetase[6], and tubulin:tyrosine ligase[7]. **See** Multisubstrate Mechanisms

[1] F. B. Rudolph & H. J. Fromm (1969) *J. Biol. Chem.* **244**, 3832.
[2] G. D. Markham & G. H. Reed (1977) *Arch. Biochem. Biophys.* **184**, 24.
[3] P. C. Engel & K. Dalziel (1970) *Biochem. J.* **118**, 409.
[4] F. C. Wedler & P. D. Boyer (1972) *J. Biol. Chem.* **247**, 984.
[5] R. D. Allison, J. A. Todhunter & D. L. Purich (1977) *J. Biol. Chem.* **252**, 6046.
[6] B. K. Joyce & R. H. Hines (1966) *J. Biol. Chem.* **241**, 5716.
[7] N. L. Deans, R. D. Allison & D. L. Purich (1992) *Biochem. J.* **286**, 243.

RANDOM UNI BI MECHANISM

A single-substrate, two-product enzyme-catalyzed reaction scheme in which a substrate (A) binds to the free enzyme and is converted to two products (P and Q) which can be released in any order. Many enzymes described with this mechanism are actually random pseudo-Uni Bi schemes in which the second substrate is water.

Rapid Equilibrium Mechanism. If the rate-determining step is the catalytic step and all binding steps can be described by dissociation constants (*e.g.*, $K_{ia} = [E][A]/[EA]$), then, in the absense of products (*i.e.*, $[P]$ and $[Q]$ ≈ 0), the initial rate equation for the rapid equilibrium Uni Bi mechanism is identical to that of the Uni Uni mechanism: $v = V_{max,f}[A]/\{K_{ia} + [A]\}$ where $V_{max,f}$ is the maximum velocity in the forward direction and is equal to $k_3[E_{total}]$ where k_3 is the unimolecular forward rate constant for the EA ↔ EPQ conversion. If product P (but not Q) is present, the rate expression (in the absence of abortive complexes and ligand-independent isomerization steps) becomes $v = V_{max,f}[A]/\{K_{ia} + [A] + K_{ia}[P]/K_{ip}\}$ where $K_{ip} = [E][P]/[EP]$. As can readily be seen from the double-reciprocal form of this expression, P will be a competitive inhibitor with respect to A. A similar conclusion can be obtained for Q.

Steady-State Mechanism. Consider the reaction scheme with only one central complex (thus, EA + EPQ = EXY):

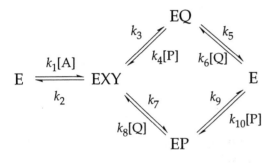

the individual steady-state enzyme determinants are:

$$[E] = k_5k_9(k_2 + k_3 + k_7) + k_4k_9(k_2 + k_7)[P] \\ + k_5k_8(k_2 + k_3)[Q] + k_2k_4k_8[P][Q]$$

$$[EXY] = k_1k_5k_9[A] + k_1k_4k_9[A][P] + k_1k_5k_8[A][Q] \\ + k_1k_4k_8[A][P][Q] + k_4k_6k_8[P][Q]^2 \\ + k_4k_8k_{10}[P]^2[Q] + (k_4k_6k_9 + k_5k_8k_{10})[P][Q]$$

$$[EQ] = k_1k_3k_9[A] + k_6k_9(k_2 + k_3 + k_7)[Q] \\ + k_6k_8(k_2 + k_3)[Q]^2 + k_1k_3k_8[A][Q] \\ + k_3k_8k_{10}[P][Q]$$

$$[EP] = k_1k_5k_7[A] + k_5k_{10}(k_2 + k_3 + k_7)[P] \\ + k_4k_{10}(k_2 + k_7)[P]^2 + k_1k_4k_7[A][P] \\ + k_4k_6k_7[P][Q]$$

In the absence of products, the initial rate equation is

$$v = V_{max,f}[A]/\{K_a + [A]\}$$

where

$$V_{max,f} = k_5k_9(k_3 + k_7)[E_{total}]/\{k_9(k_3 + k_5) + k_5k_7\}$$

$$K_a = k_5k_9(k_2 + k_3 + k_7)/\{k_1(k_9(k_3 + k_5) + k_5k_7)\}$$

If the $[P]^2$ and $[Q]^2$ terms are negligible relative to the other terms in the rate expression, the overall rate expression suggests that both P and Q may act like noncompetitive inhibitors (as opposed to the case in the rapid equilibrium scheme).

RANDOM UNI UNI BI BI PING PONG MECHANISM

A three-substrate, three-product enzyme-catalyzed reaction scheme in which a particular substrate (A) is the only substrate that can bind to the free, unmodified form of the enzyme. A chemical event then occurs, the first product (P) is released, and the remaining two substrates (B and C) can then bind in any order. Another catalytic event occurs, and the last two products (Q and R) are be released in any order. **See** *Multisubstrate Mechanisms*

RANDOM VARIABLES

In statistics, a variable, often denoted by a capital letter, that associates a number with the outcome of a random experiment. More formally, a random variable is a function that assigns a real number to each outcome in the sample space of a random experiment. **See** *Statistics (A Primer)*

RANDOM WALK

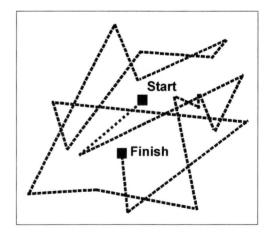

An aimless path resulting from a set of successive steps (or excursions) whose individual lengths and directions are only randomly related to each other. As with any stochastic process, the preceding step has no influence at all on any of the subsequent steps. **See** *Diffusion; Persistence Time; Twiddling*

RAPID BUFFER EXCHANGE

Many biological molecules are stored in a stabilizing buffer containing solutes that interfere with kinetic experiments. This is especially true of commercial enzyme products, and the "cocktail" of stabilizing agents for maintaining a long shelf-life is sometimes a trade secret that a company may refuse to divulge. (This is particularly true for enzymes used for special purposes, such as in clinical autoanalyzers. For this reason, even though the same company offers two or more preparations of the same enzyme at the same specific activity, one cannot be absolutely certain that both are stabilized in an identical manner.) Commercial enzymes are often kept active by the presence of 1–3 M ammonium sulfate or 1–2 M glycerol. Other stabilizing substances include metal ion chelating agents (*e.g.*, EDTA, EGTA, or *o*-phenanthroline), antifungals and antibacterials, Bentonite clay, talcum powder, boric acid, and various protease inhibitors. In some cases, a reversibly bound ligand (including one or more of the substrates, an activator, or an inhibitor) is added to maintain the enzyme in its catalytically active conformation. Such agents must be removed before conducting kinetic experiments, especially detailed mechanistic investigations, and membrane dialysis or gel filtration on Sephadex G-25 is most frequently employed for this purpose. (Dialysis is known in some instances to introduce trace elements, plasticizers, and even minor levels of enzymatic activities.) The problem with dialysis and gel filtration is that large samples (> 1 mL) are required, and samples suffer dilution of up to a factor of three. Use of much smaller protein and enzyme aliquots (in the 25–150 μL range) are not manageable with dialysis or column gel filtration.

Penefsky[1] introduced the centrifuged-column procedure (see diagram below) that can be used for small aliquots of enzymes and other proteins (a) to exchange buffers and/or other solutes, (b) to measure ligand binding, and (c) to concentrate protein by as much as 10 times. His two-step protocol for rapid buffer exchange can be summarized as follows:

Step 1. Load a disposable mini-chromatography column[2,3] with Sephadex G-50 (fine) dextran that was previously permitted to equilibrate with 5 volumes of the desired "new" buffer. Allow all the liquid to drain under the earth's gravitation to leave a 1 mL bed volume. Transfer this spin-column to an appropriately sized centrifuge tube and spin at $100 \times g$ (calculated for the radial distance to the tip of the column itself, and *not* the bottom of the centrifuge tube) for 2 min in a swinging bucket rotor[4]. The volume of the now semi-dry gel bed will be reduced by 25–40%, and the media will appear as a gel mass that may even break away from the walls of the chromatography tube. (This is normal and will not affect performance.)

Step 2. Transfer the disposable spin-column to a clean centrifuge tube and then add 25–150 mL protein in "old" buffer. Centrifuge the tube exactly as before, and the filtrate (volume usually 90–110% the volume applied to the semi-dry gel) will contain 95–100% of the protein applied to the column.

By reducing the duration or speed of the second centrifugation step, Penefsky[1] found that the entire protein sample will become concentrated into a smaller volume;

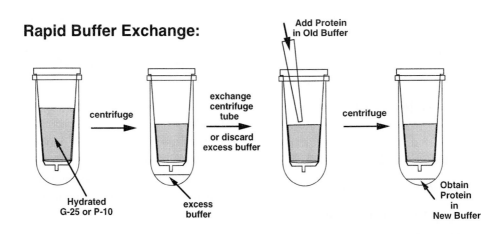

Rapid Buffer Exchange:

Add Protein in Old Buffer

centrifuge

exchange centrifuge tube

or discard excess buffer

centrifuge

Hydrated G-25 or P-10

excess buffer

Obtain Protein in New Buffer

hence, one can achieve up to 10x concentration of the protein with only a modest loss of enzyme.

[1]H. H. Penefsky (1979) *Meth. Enzymol.* **56**, 527.

[2]BioRad produces Bio-Spin disposable polypropylene chromatography columns (Catalog No. 732-6008) that can be packed with up to 1.2 mL gel filtration media. Their micro Bio-Spin columns hold 0.8 mL media. Both are autoclavable columns that have snap-off column tips, polyethylene media bed supports, and fit standard centrifuge tubes.

[3]In his work, Penefsky[1] simply used 1 mL Plastipak tuberculin syringes fitted with porous disks cut with a cork borer (0.193 inch inside diameter) from polyethylene sheets (1.5 mm thick, 70 micrometer pore size) available from Bolab, Inc. (Deery, NH). Note: The rubberized plungers of most syringes frequently contain lubricants, and these agents may bind to enzymes or they may dissolve and contaminate some enzyme preparations.

[4]Because various swinging bucket centrifuges behave differently with respect to acceleration and braking, some optimization of centrifugation conditions may be advisable. This can be readily accomplished by UV and visible spectrophotometry if one uses BSA (or some other inexpensive protein sample) along with a small amount of dichromate as a yellow colored metal ion species.

RAPID EQUILIBRIUM MECHANISMS

Reaction mechanisms (and binding mechanisms for macromolecules not catalyzing reactions) in which the binding of substrates and products is rapid relative to the interconversion and isomerization steps. Hence, those binding steps can be easily characterized by dissociation constants. The derivations of enzyme rate expressions using the rapid equilibrium assumption are very easy and it is common to find such equations in publications addressing the kinetics of a particular enzyme. Nevertheless, it is relatively rare to find papers in which the authors have actually demonstrated that the enzyme under study is truly a rapid equilibrium case. The rapid equilibrium rate equation may not be the appropriate expression to describe that specific protein. Rapid equilibrium enzymes are not as common as one may have assumed. Product desorption and/or substrate adsorption are often rate-contributing or rate-limiting. In those cases, the steady-state rate expression may describe the initial rate results better.

It should also be noted that the common models of allosterism (*e.g.*, the Monod-Wyman-Changeux model and the Koshland-Némethy-Filmer model) assume rapid equilibrium binding.

RAPID MIXING

As mentioned in an earlier entry (**See** *Mixing Time*), the time needed to mix all the components of a reaction together should be as short as possible. In transient kinetic studies (*i.e.*, rapid reaction kinetics or pre-steady-state kinetics) this factor is of even greater importance. In stopped-flow experiments, the mixing time should not be longer than 2 msec. **See** *Quench-Flow Experiments*

C. A. Fierke & G. G. Hammes (1995) *Meth. Enzymol.* **249**, 3.

RAPID SCAN SPECTROSCOPY OF FAST REACTIONS

Any advanced absorbance/fluorescence spectrophotometer designed for routine acquisition of absorption or emission on the subsecond time scale. The basic goal is to obtain a series of complete UV/visible or fluorescence spectra as a function of time, usually after samples are mixed in a stopped-flow device. Such data help the investigator to infer the most likely structures of transient intermediates whose electronic spectra or fluorescence spectra can be determined by deconvoluting the spectra with appropriate reaction kinetic simulation software or by some other global analysis method (Fig. 1).

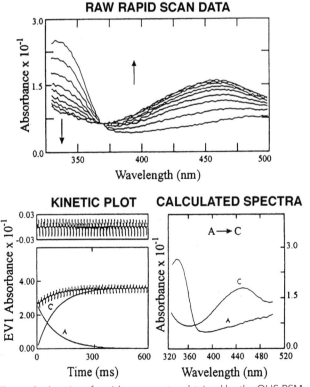

Figure 1. A series of rapid scan spectra obtained by the OLIS RSM-1000 Rapid-Scanning Monochromator (Courtesy of Olis Instruments, Inc., www.olisweb.com).

Rapid scan devices may employ a spinning prism, a diode array detector, or the newest spinning-disk technology. The OLIS spectrophotometers rely on the DeSa monochromator, a 0.25-meter double-grating monochromator. In place of a stationary central slit found in other monochromators is the "ScanDisk Module", a spinning disk with 16 slits driven by a synchronous motor at a rotational speed of 62.5 Hz to achieve millisecond spectral scans (or "snap shots") of a reaction. Because the DeSa monochromator is "subtractive", the entire output beam is spectrally homogeneous, regardless of the chosen resolution. Photometric accuracy exceeds that achievable with CCD diode array detectors, and the spinning disk is mechanically rugged. The instrument produces spectra (*i.e.*, signal strength information *versus* wavelength) as a function of time, and data analysis proceeds in three steps: (a) by determining the number of species by factor analysis; (b) by determining the rate constants using kinetic modeling software; and (c) by reconstructing time-invariant spectra of intermediates.

RAPID-START COMPLEX

The complex that RNA polymerase forms at the promoter site just prior to initiation. Some bacterial promoters require high NTP concentrations to initiate efficient transcription, because this represents a "status report" on the stores of ATP, UTP, GTP, and CTP needed for RNA synthesis. Nature has evolved a kinetic control device: high initiating ATP and GTP concentrations must be present to stabilize an otherwise short-lived polymerase-promoter complex. The reader may also recall that bacterial translation is also tightly controlled, and amino acid starvation leads to ppGpp synthesis, the so-called stringent-response agent that also potently inhibits RNA polymerase. Such kinetic control ensures that NTP and amino acid concentrations are adequate before transcription and translation occur.

RATE CONSTANT

The proportionality constant allowing one to equate reaction velocity (having units of $M \cdot s^{-1}$) to the mo-

larity of reactant(s) involved in the reaction. For unimolecular processes, $v = k[X]$, where k has units of s^{-1}; for bimolecular processes, $v = k[X]^2$ or $k[X][Y]$, where k has units of $M^{-1}s^{-1}$; and for trimolecular processes, $v = k[X]^3$ or $k[X]^2[Y]$ or $k[X][Y]^2$ or $k[X][Y][Z]$, where k has units of $M^{-2}s^{-1}$. **See** *Chemical Kinetics*

RATE-CONTRIBUTING STEP

Any reaction step having a rate constant whose magnitude is nearly that of the rate constant in the slowest step in a reaction mechanism. Such a step is said to be a contributing factor in slowing down the reaction rate.

RATE-CONTROLLING STEP

Synonym for rate-determining step and for rate-limiting step. If a rate constant for an elementary reaction has a stronger influence on the overall rate than any other rate constant in the mechanism, then the step associated with that rate constant is referred to as the rate-controlling step. Note that the rate-controlling step may change with a corresponding change in reaction conditions.

The rate-controlling step is the elementary reaction that has the largest control factor (CF) of all the steps. The control factor for any rate constant in a sequence of reactions is the partial derivative of $\ln v$ (where v is the overall velocity) with respect to $\ln k_i$ in which all other rate constants (k_j) and equilibrium constants (K_j) are held constant. Thus, $CF = (\partial \ln v / \partial \ln k_i)_{K_j, k_j}$. This definition is useful in interpreting kinetic isotope effects. **See** *Rate-Determining Step; Kinetic Isotope Effects*

RATE-DETERMINING STEP OR RATE-LIMITING STEP

Designations of the slowest step(s) in a rate process. These terms should be regarded as synonymous to the preferred term "rate-controlling" step. However, all three are frequently used to describe a slow step within a series of otherwise more rapid steps. **See** *Rate-Controlling Step*

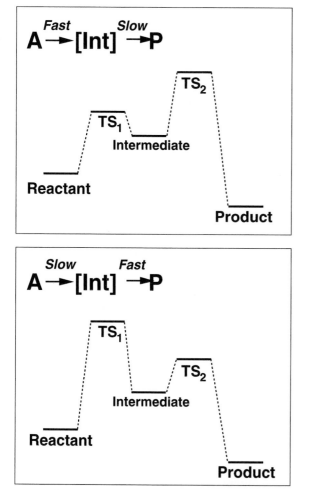

Energetics of Rate-Determining Steps. Although the initial and final states are the same in this reaction coordinate diagram, the rate of intermediate formation will be faster for the first example, where the barrier to intermediate formation is lower. Conversion of intermediate back to reactant or onward to product will also depend on the relative heights of the two barriers (marked here as TS$_1$ and TS$_2$). If the free energy barrier for conversion of a reaction intermediate in one direction is significantly lower than for conversion in the opposite direction, then the reaction will proceed many times faster by the lower energy route.

RATE LAW

A mathematical expression (usually in the form of a differential rate equation) showing how the rate of the reaction depends on the concentrations of chemical entities and rate constants (and/or equilibrium constants) and any partial orders of reactions. **See** Chemical Kinetics

RATE OF APPEARANCE

The rate of a reaction in which an increase in the concentration of one (or more) product(s) is actually what is being measured. It is symbolized by v and sometimes has a subscript for the species being measured (e.g., v_P or v_{pyruvate}). **See** Chemical Kinetics

RATE OF DISAPPEARANCE

The rate of a reaction in which a decrease in the concentration of substrate or reactant (or several reactants) is actually what is being measured. It is symbolized by v and may have a subscript for the species being measured (e.g., v_A or v_{GTP}). **See** Chemical Kinetics

RATE OF REACTION

For a chemical reaction,

$$aA + bB + \ldots \rightarrow pP + qQ + \ldots$$

where A, B, . . . , are the reactants, P, Q, . . . , are the products, and the corresponding lower-case letters are the stoichiometry numbers. The rate of the reaction, v, is defined as

$$v = -(1/a)(d[A]/dt) = -(1/b)(d[B]/dt) = \ldots$$
$$= (1/p)(d[P]/dt) = (1/q)(d[Q]/dt) = \ldots$$

in which t is time (usually seconds) and the entities in brackets are concentrations.

The above definition applies only if there is no accumulation of intermediates or formation of side products. IUPAC strongly suggests that this term be applied only when those conditions have been experimentally established. IUPAC also recommends that the terms "rate of appearance" or "rate of disappearance" (or "rate of consumption") be used if these conditions are not met.

It should also be recalled that the term "rate of conversion" (or, rate of extent of reaction), symbolized by dx/dt and equal to $-(1/a)(dn_A/dt) = -(1/b)(dn_B/dt) = \ldots = (1/p)(dn_P/dt) = (1/q)(dn_Q/dt) = \ldots$ where n, with the appropriate subscript, designates the amount of substance (e.g., moles or multiple thereof) also has this restriction. **See** Chemical Kinetics

IUPAC (1994) Pure and Appl. Chem. **66**, 1077.

RATE SATURATION BEHAVIOR

A term used to refer to any chemical process whose rate depends upon saturation of a binding site. Rate saturation is observed in enzyme kinetics, metabolic transport, host-guest reactions, and even heterogenous catalysis such as hydrogenation on metallic surfaces of platinum and nickel.

RAY-ROSCELLI TREATMENT (Isomerization)

A method[1] used to assess enzyme mechanisms which may have one or more enzyme isomerization steps[2]. In this procedure, $[P_t]/[P]$ is plotted as a function of time at different initial concentrations of substrate, where $[P_t]$ is the concentration of the product at time t and $[P_\infty]$ is the concentration of P once the system has reached equilibrium. These data can be used to calculate the K_{iip} term, associated with the isomerization step, in the rate expression.

[1]W. J. Ray, Jr., & G. A. Roscelli (1964) *J. Biol. Chem.* **239**, 3935.
[2]K. L. Rebholz & D. B. Northrop (1995) *Meth. Enzymol.* **249**, 211.

RE-

The *re* Face of Acetaldehyde

This stereochemical term (short for the Latin word *rectus*) is the designation for a stereoheterotopic face of a trigonal atom whose ligands appear in a clockwise sense in the order of the Cahn-Ingold-Prelog scheme when viewed from that side of the face.

G. P. Moss (1996) *Pure Appl. Chem.* **68**, 2193.

REACTANCY

The number of reactants partaking in an enzyme-catalyzed reaction. Because most enzyme reactions have an unequal number of substrates and products, one must specify the reactancy for a specified direction of the reaction. As an example, a multisubstrate reaction having two substrates and three products has a reactancy of two in the forward direction and three in the reverse direction. Cleland introduced the prefixes "Uni", "Bi", "Ter", and "Quad" to indicate reactancies of one, two, three, and four, respectively. Thus, the example given above can be called a "Bi Ter" reaction. Water molecules and protons are not usually considered when specifying reactancy.

REACTING BOND RULES

A set of rules describing how motions can affect a molecule's transition-state structure and reactivity.

1. When considering a molecule's internal motion corresponding to progress along the reaction coordinate and over a transition state, any change making that motion more difficult will result in a new geometry at the energy maximum, in which the motion has proceeded further.

2. For an internal motion corresponding to a vibration, any change altering the equilibrium point of the vibration in a particular direction will serve to shift the equilibrium in that direction.

3. During the course of reaction, effects on bonds that are either made or broken are the most significant. Therefore, bonds nearest sites of structural change are apt to be most strongly affected.

IUPAC (1994) *Pure Appl. Chem.* **66**, 1077.
E. R. Thornton (1967) *J. Amer. Chem. Soc.* **89**, 2915.

REACTING ENZYME CENTRIFUGATION

A centrifugation technique used to determine the oligomeric or monomeric status of an enzyme during catalysis. For example, this method can address such questions as: Is the catalytically active form of the enzyme a dimer or monomer? Is there cooperativity in the interconversion between the forms?

The reacting enzyme centrifugation technique has been used with yeast hexokinase[1] and with phosphoenolpyruvate carboxykinase[2].

[1]J. P. Shill & K. E. Neet (1975) *J. Biol. Chem.* **250**, 2259.
[2]Y.-B. Chaio (1975) Doctoral Dissertation, Case-Western Res. Univ., Cleveland, Ohio.

REACTION COORDINATE DIAGRAM

A graphical representation of changes in reactant structure and/or free energy along the reaction coordinate, which indicates the reaction's progress (*i.e.*, as the reactant(s) is(are) converted to product(s) in a process). For elementary reactions, a geometric parameter (such as bond length and/or angle) can represent a measure of the progress of the reaction. (Occasionally, the bond order is used to indicate progress along a reaction coordinate.) In any case, the horizontal axis must not be confused with time; rather, the reaction coordinate only specifies that a certain stage of the reaction has been achieved.

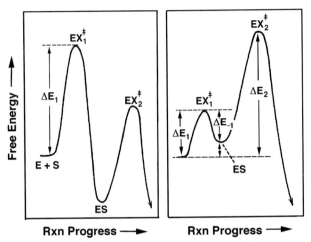

Figure 1. A hypothetical reaction coordinate diagram for an enzyme-catalyzed chemical reaction.

Reaction coordinate diagrams (Fig. 1) are important tools for analyzing the structure and energy relationships that describe a reaction mechanism. The following statements may prove helpful for constructing or interpreting reaction coordinate diagrams.

(a) The preferred unit for the vertical coordinate is ΔG/mol. In the absence of such information, one may use a more arbitrary representation (such as the change in a system's potential energy) to construct a hypothetical reaction coordinate diagram.

(b) The baseline in reaction coordinate diagrams is arbitrary, and two different reactants (each with their own reaction coordinate diagram) may be placed at the same vertical position at the start of the reaction. This is true even if the reactants have different stability, because reactivity is only a function of the differences in energy of a reactant's initial, transition-state, and final positions.

(c) The vertical positions of each chemical species (*e.g.*, reactants, products, intermediates, and enzyme-bound species) are defined by the $\Delta G°$ values for equilibria existing between them. Using the Additivity Principle, the differences in energy between successive steps are additive sums of their free energy changes, taking care to keep in mind that the signs of each energy change

must be written to agree with the direction of the reaction coordinate.

(d) Changes in free energy are estimated from experimentally determined equilibrium constants using the Gibbs equation ($\Delta G° = -nRT \ln K_{eq}$). (Although one may occasionally rely on equilibrium data dealing with cognate nonenzymatic reactions, the internal equilibria between enzyme-bound species are frequently strongly influenced by the enzyme. This must be an anticipated feature of all catalyzed processes, because transition state stabilization is the dominant factor leading to catalytic rate enhancement.

(e) The heights of barriers between successive chemical species can, under favorable conditions, be determined from information obtained from studies of temperature dependence or through the use of kinetic isotope effects.

(f) The rate-determining step is the highest point on the diagram. Any change in reaction conditions that lowers the energy difference from reactants to transition state will result in faster reaction rate (*i.e.*, greater reactivity). *See* Saddle-Point

(g) If the substrate (or reactant) is less stable than the product, the transition state is apt to appear early in the process and should resemble the starting material. And if the product is less stable, then the transition state will appear later and be more product-like in its structure.

(h) One can also infer that an unstable intermediate in a reaction should resemble the structure of the transition-state structure more closely than either substrate or product.

(i) Note that reversible reactions must pass through the same transition state, irrespective of whether the reaction is proceeding from reactant(s)-to-product(s) or *vice versa*.

(j) If a reactant has two pathways to product, and if the rates through paths A and B toward product are v_A and v_B, respectively, then the rates in the opposite directions along each pathway (and through their respective transition states) must have the same fractional contribution. This is a consequence of microscopic reversibility or the law of detailed balance, and $v_A/(v_A + v_B)$ will be the same in both directions; so also will this hold true for $v_B/(v_A + v_B)$.

In what is now a classical study in enzyme kinetics, W. J. Albery and J. R. Knowles[1] developed a strategy for establishing a reaction coordinate diagram (shown in Fig. 2) for triose-phosphate isomerase catalysis using solvent exchange and kinetic isotope effect data.

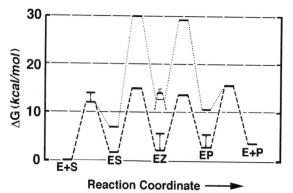

Figure 2. Free energy profile for converting dihydroxyacetone phosphate, the substrate (abbreviated S) and glyceraldehyde 3-phosphate, the product (abbreviated P), with intermediate formation of the enediolate (abbreviated Z). Catalysis occurs either by a free carboxyl group (levels connected by dotted lines) or by triose-phosphate isomerase (levels connected by dashed lines). The vertical arrows show the limits of those states that are less well defined as a result of uncertainty in the experimental data. The transition state marked "e" refers to the exchange of protons between the solvent and the enzyme-bound enediol intermediate (EZ). Reproduced with permission of the authors and the American Chemical Society.

In Fig. 2, the starting and ending positions for both reactions are the same, but the transition state lies a lower energy for the enzyme-catalyzed process. Their analysis led to the efficiency function, a quantitative expression that describes the effectiveness of a catalyst in accelerating a chemical reaction. The function, which depends on magnitude of the rate constants describing individual steps in the reaction, reaches a limiting value of unity when the reaction rate is controlled by diffusion. For the interconversion of dihydroxyacetone phosphate and glyceraldehyde 3-phosphate, the efficiency function equals 2.5×10^{-11} for a simple carboxylate catalyst and 0.6 for the enzyme-catalyzed process. Albery and Knowles[1] suggest that evolution has produced a nearly perfect catalyst. ***See*** *Enzyme Energetics (The Case of Proline Racemase); Potential-Energy Profile*

[1]W. J. Albery & J. R. Knowles (1977) *Biochemistry* **15**, 5631.

REACTION CYCLE

A cyclical depiction of an enzyme-mediated reaction written to account for the regenerative nature of catalytic processes. The reaction cycle for a typical one-substrate one-product enzyme mechanism may be written as follows:

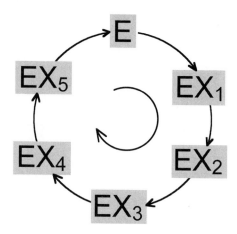

This reaction cycle has more steps than the simple Michaelis-Menten scheme. Nonetheless, the steady-state rate equations describing these reaction cycles have indistinguishable functions, and one cannot determine the number of intermediary steps by steady-state kinetics alone.

REACTION MECHANISM

A representation of all of the elementary reactions that lead to the overall chemical change being investigated. This representation would include a detailed analysis of the kinetics, thermodynamics, stereochemistry, solvent and electrostatic effects, and, when possible, the quantum mechanical considerations of the system under study. Among many items, this representation should be consistent with the reaction rate's dependence on concentration, the overall stoichiometry, the stereochemical course, presence and structure of intermediate, the structure of the transition state, effect of temperature and other variables, *etc.* ***See*** *Chemical Kinetics*

REACTION PROGRESS CURVES FOR UNSTABLE ENZYMES

Duggleby[1] provides a lucid account of how one can extract useful kinetic information from reaction progress curves. The nonlinear regression methods allow one to treat many cases, and they account for the fact that, after an enzyme is mixed with its substrate, the catalyzed rate declines over time as a consequence of substrate utilization and product accumulation. In an earlier report[2], he also considered the effects of enzyme instability, occurring either spontaneously or as a result of some added agent. Under most circumstances, catalysis practically ceases before the substrate is totally exhausted, such that the amount of substrate remaining is related to the inactivation rate constants for various intermediates on the catalytic pathway. A graphical method for estimating these inactivation rate constants is presented, and expressions for the reaction half-time are given for some certain cases. **See** *Progress Curve Analysis*

[1]R. G. Duggleby (1995) *Meth. Enzymol.* **249**, 61.
[2]R. G. Duggleby (1995) *J. Theoret. Biol.* **123**, 67.

REACTION RATE/VELOCITY

The rate of a chemical reaction, typically expressed as: $v = \Delta[X]/\Delta t$, the change in the molarity of a reactant per unit time. **See** *Chemical Kinetics*

Reaction velocity is related to the molar concentrations of reactant(s) by the rate constant:

(A) For zero-order processes (where k has units of $M\,s^{-1}$):

$$v = k$$

(B) For first-order processes (where k has units of s^{-1}):

$$v = k[X]$$

(C) For second-order processes (where k has units of $M^{-1}s^{-1}$):

$$v = k[X]^2 \text{ or } k[X][Y]$$

(D) For third-order processes (where k has units of $M^{-2}s^{-1}$):

$$v = k[X]^3 \text{ or } k[X]^2[Y] \text{ or } k[X][Y]^2 \text{ or } k[X][Y][Z]$$

REACTIVATION KINETICS
(Temperature Effects)

The study of temperature effects on the reactivation of an enzyme that has been completely unfolded allows one to distinguish between "reactivation" (referring to kinetic analysis exclusively) and "renaturation," the latter of which would reflect both the refolding transition and the formation of misfolded or aggregated by-products.

For example, the temperature effects on the reconstitution of D-glyceraldehyde-3-phosphate dehydrogenase from the hyperthermophilic eubacterium *Thermotoga maritima* were recently investigated[1]. The rate and yield of reactivation exhibited a complex behavior. At low temperatures, no significant regain of activity is seen. Between 35°C and 80°C reactivation exceeds 80%. At higher temperatures, thermal denaturation competes with refolding. Shifting the temperature from 0°C to a temperature equal to or greater than 30°C appears to release a trapped intermediate. High speed sedimentation equilibria of this trapped intermediate at 0°C reveals a nonlinear plot of ln c *versus* r^2. The concentration dependency of reactivation shows that renaturation of the enzyme involves consecutive folding and association steps. At high enzyme concentrations, association reaches its diffusion-controlled limit so that the overall reaction is fully determined by the structure formation at the monomer level. Assuming that the tetrameric enzyme is the only active species, the authors were able to determine the corresponding first-order and second-order rate constants ($k_1 = 2.3 \times 10^{-4}s^{-1}$ for folding and $k_2 = 8.3 \times 10^3\ M^{-1}s^{-1}$ for association).

[1]V. Rehaber & R. Jaenicke (1992) *J. Biol. Chem.* **267**, 10999.

REAGENT PURITY

Because enzymes are often inhibited/activated by low concentrations of metal ions, organic impurities, substrate or product analogs, and other effectors, it is incumbent that the investigator always to be concerned with the purity of all reagents used, including water. For most of the common reagents used in a laboratory, purity standards based on the latest ACS specifications are normally sufficient. Researchers should always be aware of particular commercial lot specifications. Even at low concentrations, heavy metals are known to significantly inhibit some enzymes, and for some systems, further purification steps are mandatory. The investigator should always exercise caution with regards to the specification claims provided by the commercial supplier. It might prove useful to incubate the enzyme with a mixture containing all reaction components (minus the substrate(s)) for different times, followed by a standard assay protocol, to detect any significant time-dependent change in enzyme activity. **See also** *Substrate Purity; Water Purity; Enzyme Purity*

REDOX-ACTIVE AMINO ACIDS

Hydoxylated amino acid residues that participate in redox-active enzyme reactions. *See* *Topaquinone*

Selected entries from *Methods in Enzymology* [vol, page(s)]:
Precursors of quinone-tanning: Dopa-containing proteins, **258**, 1; isolation of 2,4,5-trihydroxyphenylalanine quinone (Topa quinone) from copper amine oxidases, **258**, 20; spectrophotometric detection of Topa quinone, **258**, 34; model studies of Topa quinone: synthesis and characterization of Topa quinone derivatives, **258**, 39; catalytic aerobic deamination of activated primary amines by a model for the quinone cofactor of mammalian copper amine oxidases, **258**, 53; detection of reaction intermediates in Topa quinone enzymes, **258**, 69; mass spectrometric studies of the primary sequence and structure of bovine liver and serum amine oxidase, **258**, 90; cloning of mammalian Topa quinone-containing enzymes, **258**, 114; isolation of active site peptides of lysyl oxidase, **258**, 122; resonance Raman spectroscopy of quinoproteins, **258**, 132; redox-cycling detection of dialyzable pyrroloquinoline quinone (PQQ) and quinoproteins, **258**, 140; tryptophan tryptophylquinone (TTQ) in bacterial amine dehydrogenases, **258**, 149; model studies of cofactor tryptophan tryptophylquinone (TTQ), **258**, 164; detection of intermediates in tryptophan tryptophylquinone (TTQ) enzymes, **258**, 176; biogenesis of pyrroloquinone quinone from ^{13}C-labeled tyrosine, **258**, 227; X-ray crystallographic studies of cofactors in galactose oxidase, **258**, 235; spectroscopic studies of galactose oxidase, **258**, 262; use of rapid kinetics methods to study the assembly of the diferric-tyrosyl radical cofactor of *E. coli* ribonucleotide reductase, **258**, 278; tyrosyl radicals in photosystem II, **258**, 303; role of tryptophans in substrate binding and catalysis by DNA photolyase, **258**, 319; glycyl free radical in pyruvate formate lyase: synthesis, structure characteristics, and involvement in catalysis, **258**, 343; characterization of a radical intermediate in the lysine 2,3-aminomutase reaction, **258**, 362; role of oxidized amino acids in protein breakdown and stability, **258**, 379.

REDOX POTENTIAL

The reduction-oxidation potential (typically expressed in volts) of a compound or molecular entity measured with an inert metallic electrode under standard conditions against a standard reference half-cell. Any oxidation-reduction reaction, or redox reaction, can be divided into two half-reactions, one in which a chemical species undergoes oxidation and one in which another chemical species undergoes reduction. In biological systems the standard redox potential is defined at pH 7.0 versus the hydrogen electrode and partial pressure of dihydrogen of 1 bar.

REDUCED CONCENTRATION

A normalization parameter used in treating ligand binding equilibria to convert two extensive variables, $K_{\text{dissociation}}$ and substrate concentration, into a parameter whose value is related to the fractional saturation of ligand binding sites. For the simple Michaelis-Menten treatment, $v = V_{\text{max}}/\{1 + K_{\text{m}}/[S]\}$, if R is the reduced

concentration parameter ($R = [S]/K_{\text{m}}$), then $v = V_{\text{max}}/\{1 + R^{-1}\}$. The extent of substrate saturation of two Michaelis-Menten enzymes with different K_{m} values will be identical if their R values are the same. For example, if $R = 1$, then $v/V_{\text{max}} = 0.5$ for both enzymes. Likewise, R values of 5 and 10 yield respective velocities of $0.833V_{\text{max}}$ and $0.909V_{\text{max}}$.

Reduced concentrations are also useful in the Monod-Wyman-Changeux cooperativity model, where $\alpha = [F]/K_{\text{r}}$ and $c\alpha = [F]/K_{\text{t}}$. This makes polynomial functions simpler to handle. For example, if ligand F binds exclusively to the R-state, then the ligand F saturation function, Y_{F}, for an n-site protein equals $(1 + \alpha)^{n-1}/\{\mathscr{L} + (1 + \alpha)^n\}$, where $\mathscr{L}$ is the allosteric constant. Similarly, $\beta = [I]/K_{\text{i}}$ and $\gamma = [A]/K_{\text{a}}$, where [I] and [A] are the inhibitor and activator concentrations, respectively. In this case, the effects of I and/or A on the ligand F saturation function Y_{F} are also easily expressed as: $\mathbf{Y}_{\text{F}}$ equals $(1 + \alpha)^{n-1}/\{\mathscr{L}(1 + \gamma)^n/(1 + \beta)^n + (1 + \alpha)^n\}$. Reduced concentrations have less utility for the Koshland-Némethy-Filmer ligand-induced conformational change model, because all the equilibrium constants for ligand binding need not be identical.

REDUCED MASS

A quantity, μ, used in collision theory for the collision of two molecules having masses m_{A} and m_{B}. It is equal to $m_{\text{A}}m_{\text{B}}/(m_{\text{A}} + m_{\text{B}})$. *See* *Hooke's Law Spring; Kinetic Isotope Effects*

REDUCING AGENT (or Reductant)

An agent that brings about a reduction of some other molecular entity. Hence, the reductant becomes oxidized during the course of the reaction. *See* *Reduction*

REDUCTIO AD ABSURDUM

A systematic method in logic for disproving a proposition by demonstrating how the application of the proposition results in absurd, self-contradictory conclusions (hence the Latin expression "reduction to absurdity"). It is also referred to as the rule of indirect proof.

REDUCTION

1. The transfer of one or more electrons to a molecular entity. The term is synonymous to electronation. 2. A less general (and earlier, historically) definition was the

loss of one or more oxygen atoms from a molecular entity. **See** *Oxidation; Electrode Kinetics*

REFERENCE REACTION

A nonenzymic reaction used to evaluate the magnitude of the catalytic rate enhancement achieved by a corresponding enzyme-catalyzed reaction. For example, Radzicka and Wolfenden[1] reported that orotic acid is decarboxylated with a $t_{1/2}$ of 78 million years (or, about 2.5×10^{15} seconds) at room temperature in neutral aqueous solution, as suggested by the kinetics determined in sealed quartz tubes maintained at elevated temperatures. Thus, based on the maximal rate of the extremely proficient enzyme orotidine-5′-phosphate decarboxylase, the rate enhancement is estimated to be somewhere around 10^{17}. Based on a transition-state binding model, Radzicka and Wolfenden[1] estimated the intrinsic binding energy of the altered substrate in its transition state corresponds to a dissociation constant lower than 5×10^{-24} M! There are limitations in the use of reference reactions for this purpose[2], and the chief concern relates to the possibility that two related reactions (or even the same reaction) may proceed by different mechanisms. Moreover, there can be changes in the molecularity of the nonenzymic and enzymic processes. If a covalent intermediate forms in the enzymic reaction, there may simply be no appropriate cognate nonenzymic reaction. Finally, the order of a multistep process may not be the same for both reactions, even though the reactions are otherwise fundamentally similar. **See** *Catalytic Proficiency*

[1]A. Radzicka & R. Wolfenden (1995) *Science* **267**, 90.
[2]W. P. Jencks (1975) *Adv. Enzymol.* **43**, 373.

REGIOSELECTIVITY

1. Any reaction in which a bond-making or bond-breaking step exhibits a preference in a direction over all other directions. Regioselectivity[1,2] is described as being complete or partial if the product exhibits 100% or *x*%, respectively, preference of one site *versus* another site. 2. The ability of an enzyme to direct its catalysis toward a specific functional group in a substrate containing other identical or nearly identical group(s)[3,4]. For example, catechol *O*-methyltransferase will catalyze the methylation of both hydroxyl groups of catechol, albeit at different rates. The effect of pH on the isotope effects of 3-methylation *vs.* 4-methylation suggests that the rate-limiting step for the more rapid 3-methylation changes with more alkaline pH values whereas methyl transfer is still rate-determining for 4-methylation.

Regioselectivity can also have stereochemical issues associated with the catalytic event. For example, glutamine synthetase will catalyze amide bond formation using β-glutamate as an alternative substrate to produce each of the stereoisomeric β-glutamine products, albeit at different rates.

[1]A. Hassner (1968) *J. Org. Chem.* **33**, 2684.
[2]D. L. Adams (1992) *J. Chem. Ed.* **69**, 451.
[3]S. E. Wu, W. P. Huskey, R. T. Borchardt & R. L. Schowen (1984) *J. Amer. Chem. Soc.* **106**, 5762.
[4]F. Takusagawa, M. Fujioka, A. Spies & R. L. Schowen (1998) *Comprehensive Biological Catalysis: A Mechanistic Reference* **1**, 1.

REGULATION OF ENZYME ACTIVITY

The control via activation or inhibition of the rate(s) of an enzyme-catalyzed reaction(s). This control includes the increase or decrease in the stability or half-life of the enzyme(s). There are many different means by which control can be achieved. These include: 1. Substrate availability and reaction conditions (*e.g.*, pH, temperature, ionic strength, lipid interface activation); 2. Magnitude of V_{max} and K_m values; 3. Activation (particularly, feedforward activation); 4. Isozyme formation; 5. Compartmentalization and channeling; 6. Oligomerization/polymerization; 7. Feedback inhibition and cooperativity (particularly, allosterism and/or hysteresis); 8. Covalent modification; and 9. Gene regulation (induction & repression)

RELATIVE BIOLOGICAL EFFECTIVENESS

A ratio that assesses the biological effectiveness of absorbed radiation doses with respect to different types and energies of ionizing radiation. It is equal to the absorbed dose of a particular radiation divided by the absorbed dose of a standard radiation required to produce identical biological effects in a given organ, tissue, or organism.

RELATIVE DENSITY

The relative density, symbolized by *d*, is a unitless parameter equal to the density, *r*, of a substance at a given temperature, divided by the density of a standard, $\rho^{\ominus}$. Thus, $d = \rho/\rho^{\ominus}$. The standard density, for liquids and solids, is usually the density of pure water at the temperature of water's maximum density.

RELATIVE SUPERSATURATION (RS or RSS)

The ratio of the instantaneous solute concentration c to the solute's solubility s, where the latter is the solute concentration in equilibrium with its crystalline or precipitated phase. Hence, RS = c/s, and a supersaturated solution experiences a thermodynamic driving force ($\Delta G = -RT \ln[\text{RS}]$). A supersaturated solution will remain as a metastable state, because crystallization or precipitation requires a mechanism for relieving the supersaturated condition (*e.g.*, nucleation or addition of crystallite/precipitate). *See Biomineralization*

RELAXATION

The adjustment of a system of linked chemical reactions to a new state of equilibrium in response to the abrupt addition/removal of energy to/from the system. In temperature-jump relaxation methods, energy is added by electrical discharge of a capacitor which results in ohmic heating of ions suddenly moving toward their respective electrodes. In pressure-jump methods, the sudden rupture of a foil diaphragm causes depressurization, and the system responds to a loss of energy. In concentration correlation analysis, the chaotic movement of molecules leads to an increase or decrease in the statistically average number of molecules in a volume element, and the system relaxes in response to the perturbation from thermodynamic equilibrium. *See Chemical Kinetics; Fast Reaction Techniques*

RELAXATION PERIOD FOR PRECIPITATION

The time t that must elapse before a supersaturated solution reaches a quasi-stationary concentration[1]. In dilute aqueous solutions, t is typically around 10^{-8} seconds, but for some solutes this period has proven to be much greater. For example, Dunning[2] found that the relaxation period for concentrated sucrose solutions is about 100 hours. *See Biomineralization*

[1]A. E. Nielsen (1964) *Kinetics of Precipitation*, pp. 20-23, Pergamon Press, Oxford.
[2]W. J. Dunning (1949) *Disc. Faraday Soc.* **5**, 79.

RELAXATION TIME

The period of time that must elapse for an exponentially decreasing variable to fall from its original value X_{initial} to X_{initial}/e (or, approximately $0.3679 X_{\text{initial}}$). In 1866, J. C. Maxwell was the first to use the term "time of relaxation" to represent the period needed for the elastic force of a fluid to decay to X/e, where X is the initial value of the imposed force. *See Fast Reaction Techniques*

RENATURATION

The return of a biomacromolecule back to its natural and active conformation. The kinetics of such processes often exhibit significant aspects of cooperativity, as well as being aided, physiologically, by other macromolecules.

RENIN

This enzyme [EC 3.4.23.15], also known as angiotensin-forming enzyme and angiotensinogenase, catalyzes the hydrolysis of the Leu—Leu bond in angiotensinogen to generate angiotensin I. It belongs to the peptidase family A1.

T. D. Meek (1998) *Comprehensive Biological Catalysis: A Mechanistic Reference* **1**, 327.
M. A. Ondetti & D. W. Cushman (1982) *Ann. Rev. Biochem.* **51**, 283.

Selected entries from *Methods in Enzymology* [vol, page(s)]: Angiotensin I radioimmunoassay, **80**, 436-438; assay, **80**, 435-439; distribution, **80**, 427-429; fluorometric assay, **80**, 436, 438, 439; hog [amino acid composition, **80**, 441; isoelectric point, **80**, 440; kinetics, **80**, 439; molecular weight, **80**, 440; optimum pH, **80**, 440; properties, **80**, 440; purification, **80**, 431-433]; human [amino acid composition, **80**, 441; isoelectric point, **80**, 440; kinetics of activation, **80**, 439; molecular weight, **80**, 440; optimum pH, **80**, 440; properties, **80**, 440; purification, **80**, 433-435; stability, **80**, 435]; physiological function, **80**, 427; substrate specificity, **80**, 439-441; active site, **241**, 255-256; amino acid sequence, **241**, 216; pH of optimum human renin, **241**, 214-215; inhibitors, screening for HIV-1 protease inhibitors, **241**, 318-321; substrate specificity, **241**, 279.

REORGANIZATION ENERGY

Symbolized by λ, the reorganization energy of a one-electron transfer reaction is that energy needed for all structural adjustments, not only in the two reactants but in the neighboring solvent molecules as well, required for the two reactants to assume the correct configuration needed to transfer the sole electron. *See Intrinsic Barrier; Marcus Equation*

REPLOTS

Secondary plots of kinetic data are used to obtain various rate constants and other kinetic parameters such as K_m and V_{max}. To simply the analysis, one choses a algebraic transform of the rate equation that allows the observed data to be graphed in a linear format.

For example, consider the double-reciprocal plot of a two-substrate enzyme ($1/v$ vs. $1/[A]$ at different, constant concentrations of the other substrate B). A replot of the primary plot would be either a plot of the slope or the vertical intercept as a function of $1/[B]$.

Intercept Replot:

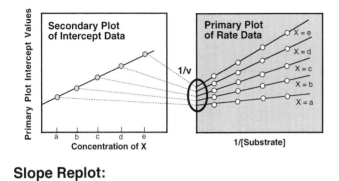

Slope Replot:

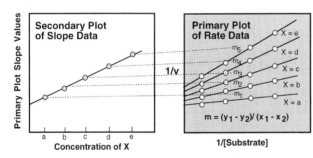

In the examples above, the secondary plots utilized the slopes or intercepts of the original plot. However, replots are secondary plots for any functional dependency using data obtained from a primary graphing procedure. Secondary replots can also be used with inhibition studies. In these cases, the slope or intercept of a double-reciprocal plot is graphed as a function of the inhibitor concentration.

RESOLUTION

The separation (as well as the corresponding procedure) of a racemic mixture into its two enantiomeric components[1,2]. There are several methods available, including:

1. Mechanical Separation of Crystals. The first instance of resolution was by L. Pasteur who was able to resolve crystals of sodium ammonium tartrate (which recrystallizes in two distinct, nonsuperimposable forms below 27°C). Although this procedure is rarely used, one might be able to seed a racemic solution resulting in only one enantiomer crystallizing[3]. In a related fashion, the addition of a chiral reagent can result in enantioselective recrystallization[4].

2. Formation of Diastereoisomers. The most common method of resolving a racemic mixture, also discovered by Pasteur, is the generation of diastereoisomers by reacting the mixture with an enantiomerically-pure reagent. The resulting two diastereoisomers will have different physical properties that permit separation (e.g., via solubility, fractional recrystallization, gas chromatography, HPLC, etc.). The purified diastereoisomer will now be used to regenerate the optically pure enantiomer (hence, the first step of reacting with the chiral reagent has to be reversible). If the racemic mixture has one or more carboxyl groups, it is a common procedure to react the mixture with an optically pure base (for example, S-brucine). Other reagents are available for other functional moieties. If the structure of the racemic mixture does not contain an available functional group, it may be possible to generate such a group and then reverse that process after the separation.

3. Chromatographic Techniques. If a chromatographic column consists of a chiral substance, it may be possible to resolve a racemic mixture. This is a very useful protocol and several chiral materials are now commercially available.

4. Enzymatic Reactions. Most enzymes are quite enantiomerically selective. Hence, reaction of a racemic mixture with the appropriate enzyme will destroy one enantiomer, leaving the other optically pure.

5. Reactivity Techniques. Enantiomers will react with chiral substances at different rates. Hence, it is possible to achieve at least partial resolution of the mixture when using chiral reagents[5].

[1]J. Jacques, A. Collett & S. H. Wilen (1981) *Enantiomers, Racemates, and Resolutions*, Wiley, New York.
[2]S. H. Wilen, A. Collett & J. Jacques (1977) *Tetrahedron* **33**, 2725.
[3]R. M. Secor (1963) *Chem. Rev.* **63**, 297.
[4]I. Weissbuch, L. Addadi, Z. Berkovitch-Yellin, E. Gati, S. Weinstein, M. Lahav & L. Leiserowitz (1983) *J. Amer. Chem. Soc.* **105**, 6615.

[5]L. Addadi, Z. Berkovitch-Yellin, I. Weissbuch, M. Lahav, & L. Leiserowitz (1986) *Top. Stereochem.* **16**, 1.

RESONANCE RAMAN SPECTROSCOPY

A spectroscopic technique that is associated with the enhancement of Raman line intensities upon photon absorption in the electronic spectral range corresponding to an absorption peak. *See* *Raman Spectroscopy*

Selected entries from *Methods in Enzymology* [vol, page(s)]:
Charge transfer transitions, **246**, 433, 438; data collection, **258**, 137-138; detectors for, **232**, 207; ferredoxin, **246**, 450-454; hemerythrin, **246**, 456-457; hemocyanin, **246**, 457-460; hemoglobin, **246**, 455-456; hemoglobin quaternary conformational changes, **232**, 58; high-potential iron protein, **246**, 453-454; instrumentation [excitation sources, **246**, 426-429; Fourier transform spectrometers, **246**, 425-426; grating spectrometers, **246**, 424-425; multichannel spectrometers, **246**, 424-425; sampling devices (flow cell, **246**, 430; low temperature, **246**, 430-432; NMR tube, **246**, 429-430; wavelength selection, **246**, 424-426)]; intensity of bands, **246**, 423, 438; intensity of scattered radiation, **246**, 420; isotopic substitution and band assignment, **258**, 139; laser-induced changes in sample monitoring, **258**, 137; laser selection, **258**, 136; low-temperature spectroscopy, **246**, 432; methylamine dehydrogenase, **258**, 138-139; multiple chromophore excitation, **246**, 439; picosecond, kinetic hole-burning experiment, **232**, 221-225; polarizability tensor component, **246**, 421-422; porphyrins, **246**, 435-437; as probe of geminate recombination, **232**, 212-228; protein concentration requirements, **246**, 387, 389; quinoprotein sample preparation [concentration, **258**, 135-136; derivatization, **258**, 133-135]; resonance enhancement, **246**, 382-383, 391, 417, 423; rubredoxin, **246**, 447-449; sample requirements, **246**, 391, 417; scattering mechanism, **246**, 418-423; sensitivity, **258**, 132, 137; Topa quinone peptides, **258**, 31-32, 34, 139; ultraviolet spectra of proteins, **246**, 386-387.

RESONANCE STABILIZATION ENERGY

The energy released by a molecular entity as a consequence of resonance, the magnitude of which equals the difference between the actual molecule's energy and the energy of the least energetic canonical form. Because the isolated canonical form does not exist factually, the resonance stabilization energy can only be estimated.

The following constitutes a brief description of how one makes effective use of the resonance method to assign the most likely resonance-stabilized structure(s): (1) Resonance structures must have the same number of paired electrons, and those that typically have identical locations for each pair are said to be the strongest contributors to the overall hybrid. (2) Ionic structures can be significant contributors if they contain elements of significantly different electronegativity. (3) The relative energies of each resonance structure can be estimated from bond energies, the presence of steric distortion,

and the presence of electron-attracting ionic structures. (4) When there are two or more contributing structures of identical or nearly equal energy, then resonance stabilization is greatest, and the overall hybrid will resemble these major contributors.

RESPIRATORY BURST OXIDASE

This FAD- and heme-containing oxidase catalyzes the NADPH-dependent conversion of two molecules of molecular oxygen to produce two molecules of superoxide. The enzyme plays a major role in destroying bacteria that are phagocytosed by neutrophils, eosinophils, and monocytes. *See also* *NADPH Oxidase*

C. Lamb & R. A. Dixon (1997) *Ann. Rev. Plant Physiol. & Mol. Biol.* **48**, 251.

S. J. Chanock, J. E. Benna, R. M. Smith & B. M. Babior (1994) *J. Biol. Chem.* **269**, 24519.

B. M. Babior (1992) *Adv. Enzymol.* **65**, 49.

B. M. Shapiro & P. B. Hopkins (1991) *Adv. Enzymol.* **64**, 291.

RESPIRATORY EXCHANGE RATIO

A ratio of the net output (in moles or volume at standard temperature and pressure per unit time) of CO_2 divided by the net uptake (in identical units) of O_2 at a given site.

RESTRICTION ENZYMES

Type I site-specific deoxyribonucleases [EC 3.1.21.3], also known as type I restriction enzymes, catalyze the ATP-dependent endonucleolytic cleavage of DNA to give random double-stranded fragments with terminal 5′-phosphates, also producing ADP and orthophosphate. These enzymes have an absolute requirement for ATP (or dATP) and *S*-adenosyl-L-methionine and recognize specific short DNA sequences, cleaving at sites remote from the recognition sequence. Type II site-specific deoxyribonucleases [EC 3.1.21.4], also known as type II restriction enzymes, catalyze the magnesium ion-dependent endonucleolytic cleavage of DNA to give specific double-stranded fragments with terminal 5′-phosphates. These enzymes recognize specific short DNA sequences and cleave either within, or at a short specific distance from, the recognition site. Type III site-specific deoxyribonucleases [EC 3.1.21.5], also known as Type III restriction enzymes, catalyze the endonucleolytic cleavage of DNA to give specific double-stranded fragments with terminal 5′-phosphates. There is an absolute requirement for ATP; however, the ATP is not hydrolyzed. *S*-Adenosyl-L-methionine stimulates the reaction, but is not absolutely required. These enzymes recognize specific short DNA sequences and cleave a short distance away from the recognition sequence. **See also** *Deoxyribonucleases*

N. H. Williams (1998) *Comprehensive Biological Catalysis: A Mechanistic Reference* **1**, 543.
A. S. Bhagwat (1992) *Meth. Enzymol.* **216**, 199.
P. Modrich (1982) *Crit. Rev. Biochem.* **13**, 287.
B. Endlich & S. Linn (1981) *The Enzymes*, 3rd ed., **14**, 137.
R. D. Wells, R. D. Klein & C. K. Singleton (1981) *The Enzymes*, 3rd ed., **14**, 157.
R. J. Roberts (1976) *Crit. Rev. Biochem.* **4**, 123.

REVERSE TRANSCRIPTASE

This enzyme [EC 2.7.7.49], also known as RNA-directed DNA polymerase, DNA nucleotidyltransferase (RNA-directed), and revertase, catalyzes the RNA-template-directed extension of the 3'-end of a DNA strand by one deoxynucleotide at a time: n deoxynucleoside triphosphate to produce n pyrophosphate (or, diphosphate) and DNA_n. The enzyme cannot initiate a DNA chain *de novo* and requires a DNA or RNA primer. **See also** *Viral Polymerases*

R. A. Katz & A. M. Skalka (1994) *Ann. Rev. Biochem.* **63**, 133.
C. M. Joyce & T. A. Steitz (1994) *Ann. Rev. Biochem.* **63**, 777.
A. M. Skalka & S. P. Goff (eds.) (1993) *Reverse Transcriptase*, Cold Spring Harbor Press, Cold Spring Harbor.
J. A. Peliska & S. J. Benkovic (1992) *Science* **258**, 1112.
I. M. Verma (1981) *The Enzymes*, 3rd ed., **14**, 87.

RHAMNOSE ISOMERASE

This enzyme [EC 5.3.1.14] catalyzes the interconversion of L-rhamnose and L-rhamnulose.

E. A. Noltmann (1972) *The Enzymes*, 3rd ed., **6**, 271.

α-L-RHAMNOSIDASE

This enzyme [EC 3.2.1.40] catalyzes the hydrolysis of terminal nonreducing α-L-rhamnoase residues in α-L-rhamnosides.

P. M. Dey & E. del Campillo (1984) *Adv. Enzymol.* **56**, 141.

RHAMNULOSE-1-PHOSPHATE ALDOLASE

This enzyme [EC 4.1.2.19] catalyzes the reversible conversion of L-rhamnulose 1-phosphate to dihydroxyacetone phosphate (or, glycerone phosphate) and (*S*)-lactaldehyde.

J. V. Schloss & M. S. Hixon (1998) *Comprehensive Biological Catalysis: A Mechanistic Reference* **2**, 43.
D. J. Creighton & N. S. R. K. Murthy (1990) *The Enzymes*, 3rd ed., **19**, 323.
D. S. Feingold & P. A. Hoffee (1972) *The Enzymes*, 3rd ed., **7**, 303.

RHIZOPUSPEPSIN

This enzyme [EC 3.4.23.21], also known as *Rhizopus* aspartic proteinase, catalyzes the hydrolysis of peptide bonds in proteins with a broad specificity similar to that of pepsin A, preferring hydrophobic residues at P_1 and $P_{1'}$. It clots milk and activates trypsinogen. It does not hydrolyze the Gln4—His5 bond, but it does hydrolyze the His10—Leu11 and the Val12—Glu13 peptide bonds in the B chain of insulin. The enzyme, a member of the peptidase family A1, is isolated from the zygomycete fungus *Rhizopus chinensis*. A similar endopeptidase is found in *R. niveus*. **See** *Pepsin & Pepsinogen*

T. D. Meek (1998) *Comprehensive Biological Catalysis: A Mechanistic Reference* **1**, 327.
D. R. Davies (1990) *Ann. Rev. Biophys. Biophys. Chem.* **19**, 189.

RHO (ρ)

1. Symbol for density. (*i.e.*, mass density). 2. Symbol for charge density. 3. Symbol for radiant energy density.

RHODANESE

This enzyme [EC 2.8.1.1] catalyzes the reaction of thiosulfate with cyanide to produce sulfite and thiocyanate. A number of other sulfur-containing compounds can act as substrates.

D. P. Kelly & A. P. Wood (1994) *Meth. Enzymol.* **243**, 501.
J. Westley (1973) *Adv. Enzymol.* **39**, 327.

RHODOPSIN

The following are recent reviews on the molecular and physical properties of this centrally important visual protein. A light-sensitive chromoprotein that serves as the photon acceptor for rod cells, consisting of a protein (opsin) and an aldehyde derivative of vitamin A (11-*cis*-retinal). This cofactor forms a imine with the ε-amino group of a specific opsin lysyl residue. The primary photo event involves conversion of 11-*cis*-retinal to all-*trans*-retinal. Conformational changes in opsin upon *cis*-to-*trans* isomerization trigger signal transduction events culminating in a nerve impulse.

W. D. Hoff, K.-H. Jung & J. L. Spudich (1997) *Ann. Rev. Biophys. Biomol. Struct.* **26**, 223.
V. R. Rao & D. D. Oprian (1996) *Ann. Rev. Biophys. Biomol. Struct.* **25**, 287.
B. J. Litman & D. C. Mitchell (1996) in *Rhodopsin and G-Protein Linked Receptors* (A. G. Lee, ed.), p. 1, JAI Press, Stamford, CT.
H. G. Khorana (1992) *J. Biol. Chem.* **267**, 1.
J. Nathans (1992) *Biochemistry* **31**, 4923.
P. A. Hargrave & J. H. McDowell (1992) *FASEB J.* **6**, 2323.
R. R. Birge (1990) *Ann. Rev. Phys. Chem.* **41**, 683.

RHODOPSIN KINASE

This enzyme [EC 2.7.1.125] catalyzes the reaction of ATP with rhodopsin to produce ADP and phosphorylated rhodopsin. The kinase acts on the bleached or activated form of rhodopsin.

X. Y. Zhao, K. Palczewski & H. Ohguro (1995) *Biophys. Chem.* **56**, 183.

A. M. Edelman, D. K. Blumenthal & E. G. Krebs (1987) *Ann. Rev. Biochem.* **56**, 567.

RHO-VALUE (ρ-Value)

A measure of the susceptibility to any substitution effect of reaction series for *families* of organic compounds as modified by σ-constants in an empirical $\rho\sigma$-equation. The ρ-value, usually with subscripts and/or superscripts designating particular systems. For example, for comparing the effect of substituents at the *meta-* and *para-*positions of certain reactions of aromatic compounds, $\log(k_x/k_H) = \rho\sigma_x$ where σ_x is a constant characteristic of substituent X and its position.

Reactions with positive ρ-values are accelerated, provided that the σ-constant is also positive. ***See also*** *Hammett Equation*

RIBITOL-5-PHOSPHATE CYTIDYLYLTRANSFERASE

This enzyme [EC 2.7.7.40], also known as CDP-ribitol pyrophosphorylase, catalyzes the reaction of CTP with D-ribitol 5-phosphate to produce CDP-ribitol and pyrophosphate (or, diphosphate).

D. R. D. Shaw (1966) *Meth. Enzymol.* **8**, 244.

RIBITOL-5-PHOSPHATE 2-DEHYDROGENASE

This enzyme [EC 1.1.1.137] catalyzes the reaction of D-ribitol 5-phosphate with NAD(P)$^+$ to produce D-ribulose 5-phosphate and NAD(P)H.

L. Glaser (1966) *Meth. Enzymol.* **8**, 240.

RIBOFLAVINASE

This enzyme [EC 3.5.99.1] catalyzes the hydrolysis of riboflavin to produce ribitol and lumichrome.

C. S. Yang & D. B. McCormick (1971) *Meth. Enzymol.* **18(B)**, 571.

RIBOFLAVIN KINASE

This enzyme [EC 2.7.1.26], also known as flavokinase, catalyzes the reaction of ATP with riboflavin to produce ADP and FMN.

E. P. Anderson (1973) *The Enzymes*, 3rd ed. **9**, 49.

D. B. McCormick (1971) *Meth. Enzymol.* **18(B)**, 544.

RIBOFLAVIN SYNTHASE

This enzyme [EC 2.5.1.9] catalyzes the conversion of two 6,7-dimethyl-8-(1-D-ribityl)lumazine to produce riboflavin and 4-(1-D-ribitylamino)-5-amino-2,6-dihydroxypyrimidine.

A. Bacher, S. Eberhardt, M. Fischer, S. Mörtl, K. Kis, K. Kugelbrey, J. Scheuring & K. Schott (1997) *Meth. Enzymol.* **280**, 389.

RIBONUCLEASE

This enzyme [often abbreviated by RNase; EC 3.1.27.5], also known as pancreatic ribonuclease, ribonuclease A, and ribonuclease I, catalyzes the endonucleolytic cleavage to 3'-phosphomononucleotides and 3'-phosphooligonucleotides ending in Cp or Up with 2',3'-cyclic phosphate intermediates. RNA hydrolysis occurs by general acid-base catalysis[1]. During the enzyme-catalyzed reaction, nucleoside-2',3'-cyclic phosphodiesters accumulate, indicating a two-step reaction: first, facilitated attack of the 2'-hydroxyl of RNA which leads to the formation of the cyclic intermediate; second, hydrolysis of the cyclic phosphodiester to yield the 3'-monoester. Now classical studies, chiefly from the laboratory of Moore & Stein, established that His-12 and His-119 play critically important roles, the former acting in the first step as a general base, and the latter serving as a general acid to protonate the leaving group. The second step is the reverse of the first, and the enzyme returns to its original protonation state upon completing the reaction cycle. *Aspergillus oryzae* RNase T1 partially digests RNA into short fragments by preferential cleavage after guanine residues, and pancreatic RNase catalyzes bond scission after pyrimidine residues. ***See*** *Pseudorotation; Diethylpyrocarbonate*

N. H. Williams (1998) *Comprehensive Biological Catalysis: A Mechanistic Reference* **1**, 543.

X. Parés, M. V. Nogués, R. de Llorens & C. M. Cuchillo (1991) *Essays in Biochem.* **26**, 89.

V. Shen & D. Schlessinger (1982) *The Enzymes*, 3rd ed. **15**, 501.

W. W. Cleland (1977) *Adv. Enzymol.* **45**, 373.

F. M. Richards & H. W. Wyckoff (1971) *The Enzymes*, 3rd ed., **4**, 647.

---

[1] In general acid catalysis, the reaction rate increases because the transition state for the reaction is lowered by proton transfer from a Brønsted acid; in general base catalysis, the reaction rate increases by virtue of proton abstraction by a Brønsted base.

RIBONUCLEASE F

This enzyme [EC 3.1.27.7] catalyzes the endonucleolytic cleavage of an RNA precursor into two. The cleavage takes place between a cytosine and an adenine residue, leaving 5′-hydroxyl and 3′-phosphate groups.

N. H. Williams (1998) *Comprehensive Biological Catalysis: A Mechanistic Reference* **1**, 543.
R. A. Katz & A. M. Skalka (1994) *Ann. Rev. Biochem.* **63**, 133.
C. M. Joyce & T. A. Steitz (1994) *Ann. Rev. Biochem.* **63**, 777.
J. M. Whitcomb & S. H. Hughes (1992) *Ann. Rev Cell Biol.* **8**, 275.

RIBONUCLEASE II

Ribonuclease II [EC 3.1.13.1], also called exoribonuclease II, catalyzes the exonucleolytic cleavage of the polynucleic acid, preferring single-stranded RNA, in the 3′- to 5′-direction to yield 5′-phosphomononucleotides. The enzyme processes 3′-terminal extra-nucleotides of monomeric tRNA precursors, following the action of ribonuclease P. Similar enzymes include RNase Q, RNase BN, RNase PIII, and RNase Y. Ribonuclease T_2 [EC 3.1.27.1] is also known as ribonuclease II.

V. Shen & D. Schlessinger (1982) *The Enzymes*, 3rd ed., **15**, 501.

RIBONUCLEASE III

This enzyme [EC 3.1.26.3], also known as RNase O and RNase D, catalyzes the endonucleolytic cleavage of RNA to 5′-phosphomonoesters. The enzyme cleaves multimeric tRNA precursors at the spacer region and is also involved in the processing of precursor rRNA, hnRNA, and early T7-mRNA. This enzyme can also act on double-stranded DNA.

R. Kole & S. Altman (1982) *The Enzymes*, 3rd ed., **15**, 469.
J. J. Dunn (1982) *The Enzymes*, 3rd ed., **15**, 485.

RIBONUCLEASE IV

This enzyme [EC 3.1.26.6], also known as endoribonuclease IV and poly(A)-specific ribonuclease, catalyzes the endonucleolytic cleavage of poly(A) to fragments terminated by 3′-hydroxyl and 5′-phosphate groups. Oligonucleotides are formed with an average chain length of ten.

V. Shen & D. Schlessinger (1982) *The Enzymes*, 3rd ed., **15**, 501.

RIBONUCLEASE P

This system [EC 3.1.26.5] catalyzes the endonucleolytic cleavage of RNA, removing the 5′-extra-nucleotide from the tRNA precursor. This step is essential for tRNA processing. It generates the 5′-termini of mature tRNA molecules. A similar activity is observed with RNase P3. *See* Catalytic RNA

J. A. Grasby (1998) *Comprehensive Biological Catalysis: A Mechanistic Reference* **1**, 563.
K. R. Groom. Y. L. Dang, G.-J. Gao, Y. C. Lou, N. C. Martin, C. A. Wise & M. J. Morales (1996) *Meth. Enzymol.* **264**, 86.
S. Altman (1989) *Adv. Enzymol.* **62**, 1.
R. Kole & S. Altman (1982) *The Enzymes*, 3rd ed., **15**, 469.

RIBONUCLEOTIDE REDUCTASE

Ribonucleotide reductase (diphosphate) [EC 1.17.4.1], also known as ribonucleoside-diphosphate reductase, catalyzes the reaction of a 2′-deoxyribonucleoside diphosphate with oxidized thioredoxin and water to produce a ribonucleoside diphosphate and reduced thioredoxin. This system requires the presence of iron ions and ATP. Ribonucleotide reductase (triphosphate) [EC 1.17.4.2], also known as ribonucleoside-triphosphate reductase, catalyzes the reaction of a 2′-deoxyribonucleoside triphosphate with oxidized thioredoxin and water to produce a ribonucleoside triphosphate and reduced thioredoxin. In this case, cobalt and ATP are cofactors.

B. T. Golding & W. Buckel (1998) *Comprehensive Biological Catalysis: A Mechanistic Reference* **3**, 239.
B. G. Fox (1998) *Comprehensive Biological Catalysis: A Mechanistic Reference* **3**, 261.
J. E. Penner-Hahn (1998) *Comprehensive Biological Catalysis: A Mechanistic Reference* **3**, 439.
J. M. Bollinger, Jr., W. H. Tong, N. Ravi, B. H. Huynh, D. E. Edmondson & J. Stubbe (1995) *Meth. Enzymol.* **258**, 278.
P. Reichard (1993) *J. Biol. Chem.* **268**, 8383.
W. W. Wells, Y. Yang & T. L. Deits (1993) *Adv. Enzymol.* **66**, 149.
M. Fontecave, P. Nordlund, H. Eklund & P. Reichard (1992) *Adv. Enzymol.* **65**, 147.
E. J. Brush & J. W. Kozarich (1992) *The Enzymes*, 3rd ed., **20**, 317.
J. B. Howard & D. C. Rees (1991) *Adv. Protein Chem.* **42**, 199.
J. Stubbe (1990) *Adv. Enzymol.* **63**, 349.
M. A. Ator & P. R. Ortez de Montellano (1990) *The Enzymes*, 3rd ed., **19**, 213.

RIBOSE-5-PHOSPHATE ADENYLYLTRANSFERASE

This enzyme [EC 2.7.7.35], also known as ADP-ribose phosphorylase, catalyzes the reaction of ADP with D-ribose 5-phosphate to produce orthophosphate and ADP-ribose.

W. R. Evans (1971) *Meth. Enzymol.* **23**, 566.

RIBOSE-5-PHOSPHATE EPIMERASE

This enzyme [EC 5.3.1.6] (also known as phosphopentosisomerase, phosphoriboisomerase, and ribose 5-phosphate isomerase) catalyzes the interconversion of D-ribose 5-phosphate and D-ribulose 5-phosphate. Both D-ribose 5-diphosphate and D-ribose 5-triphosphate can also function as substrates.

E. A. Noltmann (1972) *The Enzymes*, 3rd ed., **6**, 271.

RIBOSEPHOSPHATE PYROPHOSPHOKINASE

This enzyme [EC 2.7.6.1], also known as phosphoribosyl pyrophosphate synthetase, catalyzes the reaction of ATP with D-ribose 5-phosphate to produce AMP and 5-phospho-α-D-ribose 1-diphosphate. dATP can also function as a substrate.

J. J. Villafranca & F. M. Raushel (1980) *Ann. Rev. Biophys. Bioengin.* **9**, 363.
M. A. Becker, K. O. Raivio & J. E. Seegmiller (1979) *Adv. Enzymol.* **49**, 281.
A. S. Mildvan (1979) *Adv. Enzymol.* **49**, 103.
W. N. Keley & J. B. Wyngaarden (1974) *Adv. Enzymol.* **41**, 1.
R. L. Switzer (1974) *The Enzymes*, 3rd ed., **10**, 607.

D-RIBULOSE-1,5-BISPHOSPHATE CARBOXYLASE/OXYGENASE

Probably the most abundant naturally occurring catalyst, this enzyme [EC 4.1.1.39], also known as "rubisco", catalyzes the reaction of D-ribulose 1,5-bisphosphate with carbon dioxide to produce two 3-phospho-D-glycerate. The enzyme can also use dioxygen as a substrate instead of carbon dioxide, producing 3-phospho-D-glycerate and 2-phosphoglycolate.

J. V. Schloss & M. S. Hixon (1998) *Comprehensive Biological Catalysis: A Mechanistic Reference* **2**, 43.
M. Stitt & U. Sonnewald (1995) *Ann. Rev. Plant Physiol. Plant Mol. Biol.* **46**, 341.
S. Krömer (1995) *Ann. Rev. Plant Physiol. Plant Mol. Biol.* **46**, 45.
F. C. Hartman & M. R. Harpel (1994) *Ann. Rev. Biochem.* **63**, 197.
D. R. Geiger & J. C. Servaites (1994) *Ann. Rev. Plant Physiol. Plant Mol. Biol.* **45**, 235.
L. A. Kleczkowski (1994) *Ann. Rev. Plant Physiol. Plant Mol. Biol.* **45**, 339.
F. C. Hartman & M. R. Harpel (1993) *Adv. Enzymol.* **67**, 1.
M. H. O'Leary (1992)*The Enzymes*, 3rd ed., **20**, 235.
M. I. Siegel, M. Wishnick & M. D. Lane (1972) *The Enzymes*, 3rd ed., **6**, 169.

RIBULOSE-5-PHOSPHATE 3-EPIMERASE

This enzyme [EC 5.1.3.1] (also known as phosphoribulose epimerase, erythrose-4-phosphate epimerase, and pentose-5-phosphate 3-epimerase) catalyzes the interconversion of D-ribulose 5-phosphate and D-xylulose 5-phosphate. The enzyme can also act on D-erythrose 4-phosphate.

M. E. Tanner & G. L. Kenyon (1998) *Comprehensive Biological Catalysis: A Mechanistic Reference* **2**, 7.
E. Adams (1976) *Adv. Enzymol.* **44**, 69.
L. Glaser (1972) *The Enzymes*, 3rd ed., **6**, 355.

RIBULOSE-5-PHOSPHATE 4-EPIMERASE

This enzyme [EC 5.1.3.4] catalyzes the interconversion of L-ribulose 5-phosphate and D-xylulose 5-phosphate.

E. Adams (1976) *Adv. Enzymol.* **44**, 69.
L. Glaser (1972) *The Enzymes*, 3rd ed., **6**, 355.

RIESKE IRON-SULFUR PROTEIN

A mitochondrial respiratory chain iron-sulfur protein containing a [2Fe–2S] cluster with two coordinated cysteinyl sulfur ligands and two coordinated histidyl imidazole ligands. The term is also applied to similar proteins isolated from photosynthetic organisms and microorganisms that contain similarly coordinated [2Fe–2S] clusters.

RITCHIE EQUATION

A linear free energy equation[1,2] for reaction between nucleophiles and certain large organic cations (*e.g.*, triarylmethyl cations, aryltropylium cations, *etc.*). Thus, log k_N = log k_o + N_+ where k_N is the rate constant for the reaction of a particular cation with a given nucleophile in a given solvent, k_o is the corresponding rate constant for the same cation with water in water (*i.e.*, the solvent is water), and N_+ is the cation-independent parameter that is characteristic for that nucleophile system.

[1]C. D. Ritchie (1972) *Acc. Chem. Res.* **5**, 348.
[2]C. D. Ritchie (1986) *Can. J. Chem.* **64**, 2239.

RNA LIGASE

This enzyme [EC 6.5.1.3], also known as polyribonucleotide synthase (ATP), catalyzes the reaction of ATP with [ribonucleotide]$_n$ and [ribonucleotide]$_m$ to produce [ribonucleotide]$_{(n + m)}$, AMP, and pyrophosphate (or, diphosphate). This enzyme converts linear RNA to a circular form by the transfer of the 5'-phosphate to the 3'-hydroxyl terminus.

P. A. Frey (1992) *The Enzymes*, 3rd ed., **20**, 141.
O. C. Uhlenbeck & R. I. Gumport (1982) *The Enzymes*, 3rd ed., **15**, 31.

RNA POLYMERASES

RNA-directed RNA polymerase [EC 2.7.7.48] catalyzes the RNA-template-directed extension of the 3'-end of an RNA strand by one nucleotide at a time: thus, *n* nucleoside triphosphate produces n pyrophosphate and RNA_n. This enzyme can initiate a chain *de novo*.

DNA-directed RNA polymerase [EC 2.7.7.6] catalyzes the DNA-template-directed extension of the 3′-end of an RNA strand by one nucleotide at a time: thus, n nucleoside triphosphate generate RNA_n and n pyrophosphate. The enzyme can initiate a chain *de novo*. Three forms of the enzyme have been distinguished in eukaryotes on the basis of sensitivity of α-amanitin and the type of RNA synthesized. ***See also*** *Replicase*

S. M. Uptain, C. M. Kane & M. J. Chamberlain (1997) *Ann. Rev. Biochem.* **66**, 117.
S. Adhya, ed. (1996) *Meth. Enzymol.*, vols. **273** and **274**.
M. E. Kang & M. E. Dahmus (1995) *Adv. Enzymol.* **71**, 41.
S. S. Sastry, H. P. Spielmann & J. E. Hearst (1993) *Adv. Enzymol.* **66**, 85.
J. E. Coleman (1992) *Ann. Rev. Biochem.* **61**, 897.
D. A. Erie, T. D. Yager & P. H. von Hippel (1992) *Ann. Rev. Biophys. Biomol. Struct.* **21**, 379.
P. A. Frey (1992) *The Enzymes*, 3rd ed., **20**, 141.
M. J. Chamberlin (1982) *The Enzymes*, 3rd ed., **15**, 61.
M. Chamberlin & T. Ryan (1982) *The Enzymes*, 3rd ed., **15**, 87.
M. K. Lewis & R. R. Burgess (1982) *The Enzymes*, 3rd ed., **15**, 110.
R. Losick & J. Pero (1976) *Adv. Enzymol.* **44**, 165.
P. Chambon (1974) *The Enzymes*, 3rd ed., **10**, 261.
M. J. Chamberlin (1974) *The Enzymes*, 3rd ed., **10**, 333.

ROENTGEN

Symbolized by R, the roentgen is the international unit of exposure dose. It is equal to that quantity of radiation that will produce a charge of 2.58×10^{-4} coulombs on all the ions of one sign, when all the electrons released in a volume of air (at STP) of a mass of one kilogram are completely stopped.

ROENTGEN-EQUIVALENT-MAN (rem)

A unit of dose equal to the amount of ionizing radiation that produces in humans the same biological effect as one rad of X-rays or gamma rays. One rem is equal to 0.01 sievert.

ROTATIONAL CORRELATION TIME

The period required for reduction of the fraction of some rotationally correlated property to $1/e$ (or 0.367) of its initial value. ***See*** *Correlation Function*

Selected entries from *Methods in Enzymology* [vol, page(s)]:
Electron paramagnetic resonance [effect on line width, **246**, 596-598; motional narrowing spin label spectra, **246**, 595-598; slow motion spin label spectra, **246**, 598-601]; helix-forming peptides, **246**, 602-605; proteins, **246**, 595; Stokes-Einstein relationship, **246**, 594-595; temperature dependence, **246**, 602, 604.

ROTATIONAL DIFFUSION

The haphazard rotational motions of molecules or one or more segments of a molecule. This diffusional process strongly influences the mutual orientation of molecules (particularly large ones) as they encounter each other and proceed to form complexes. Rotational diffusion can be characterized by one or more relaxation times, τ_i, describing the motion of a molecule or segment of volume, V, in a medium of viscosity, η, as shown in the following equation:

$$\tau_i = \eta V/(RT)$$

where R and T are the universal gas constant and the absolute temperature, respectively. Because the rotating unit is rarely spherically symmetric in size and electronic charge, its rotational diffusional behavior is likely to be a composite of several tau values. Rod-like structures are especially likely to exhibit such behavior. The same is true for molecules having regions or segments that rotate relatively independently of each other. Under favorable conditions where the rotating unit is characterized by a fluorescence excited state lifetime of suitable magnitude, one can use time-resolved fluorescence anisotropy measurements to infer the time scale of rotational motions[1–3]. For example, the rotational dynamics of the purified dicyclohexylcarbodiimide-sensitive H^+-ATPase reconstituted into phospholipid vesicles and of the ATPase coreconstituted with the proton pump bacteriorhodopsin have been probed by time-resolved phosphorescence emission anisotropy[4]. These investigators used the phosphorescent probe erythrosin isothiocyanate to covalently label the γ-polypeptide of the ATPase before reconstitution. The rotational correlation time was independent of the viscosity of the external medium, but increased significantly as the microviscosity of the membrane increased. Thus, the rotational correlation times are a measure of the enzyme's motion in the membrane. At 4°C, the 100–180 μs correlation time was unaffected by the addition of substrates and the presence of a transmembrane pH gradient, suggesting further that molecular rotation does not appear to play an important role either in enzyme catalysis or ion pumping. ***See*** *Fluorescence*

[1]G. Weber (1971) *J. Chem. Phys.* **55**, 2399.
[2]J. Ygueribide, H. Epstein & L. Stryer (1970) *J. Mol. Biol.* **51**, 573.
[3]J. R. Lakowicz & G. Weber (1980) *Biophys. J.* **32**, 591.
[4]K. M. Musier-Forsyth & G. G. Hammes (1990) *Biochemistry* **29**, 3236.

rRNA (ADENINE-6-)-METHYLTRANSFERASE

This enzyme [EC 2.1.1.48] catalyzes the reaction of *S*-adenosyl-L-methionine with rRNA to produce *S*-adenosyl-L-homocysteine and rRNA containing N^6-methyladenine. The enzyme will also methylate 2-aminoadenosine to 2-methylaminoadenosine.

S. J. Kerr & E. Borek (1973) *The Enzymes*, 3rd ed., **9**, 167.

rRNA (GUANINE-N^1-)-METHYLTRANSFERASE

This enzyme [EC 2.1.1.51] catalyzes the reaction of *S*-adenosyl-L-methionine with rRNA to produce *S*-adenosyl-L-homocysteine and rRNA containing N^1-methylguanine.

S. J. Kerr & E. Borek (1973) *The Enzymes*, 3rd ed., **9**, 167.

R_s

A parameter used to assess the degree of cooperativity exhibited by an enzyme[1-3]. R_s equals the ratio of $[S]_{0.9}/[S]_{0.09}$; that is, the ratio of the substrate concentration needed for 90% saturation divided by the substrate concentration needed for 10% saturation. For a normal, hyperbolic, noncooperative curve, R_s equals 81. Thus, positively cooperative systems will have an R_s ratio less than 81, whereas negatively cooperative systems will have values larger than 81. The ratio is insensitive to the shape of the curve and does not address any questions concerning the substrate concentration range between the 10% and 90% points.

[1]D. E. Koshland, Jr., G. Némethy & D. Filmer (1966) *Biochemistry* **5**, 365.
[2]M. E. Kirtley & D. E. Koshland, Jr. (1967) *J. Biol. Chem.* **242**, 4192.
[3]K. E. Neet (1980) *Meth. Enzymol.* **64**, 139.

RUBBER
cis-POLYPRENYL*CIS*TRANSFERASE

This enzyme [EC 2.5.1.20], also known as rubber allyltransferase and rubber transferase, catalyzes the reaction of poly-*cis*-polyprenyl diphosphate (or, rubber particles) with isopentenyl diphosphate to produce a poly-*cis*-polyprenyl diphosphate longer by one C_5 unit and pyrophosphate (or, diphosphate).

R. A. Gibbs (1998) *Comprehensive Biological Catalysis: A Mechanistic Reference* **1**, 31.
E. D. Beytía & J. W. Porter (1976) *Ann. Rev. Biochem.* **45**, 113.
B. L. Archer & E. G. Cockbain (1969) *Meth. Enzymol.* **15**, 476.

RUBREDOXIN

A class of iron-sulfur proteins lacking acid-labile sulfur, but similar in function to the ferredoxins. The iron center is coordinated by four sulfur-containing ligands, usually from cysteinyl residues. Where known, these proteins function as electron carriers.

RUBREDOXIN:NAD⁺ REDUCTASE

This iron-dependent enzyme [EC 1.18.1.1], also known as rubredoxin reductase, catalyzes the reaction of reduced rubredoxin and NAD^+ to produce oxidized rubredoxin and NADH.

G. Palmer (1975) *The Enzymes*, 3rd ed., **12**, 1.

RUNGE-KUTTA ALGORITHM

A systematic stepwise method for numerical integration of a rate expression [indeed, of any differential equation $y' = f(x,y)$ with an initial value $y(x_o) = y_o$] to determine the time evolution of the rate process. ***See also*** *Numerical Computer Methods; Numerical Integration; Stiffness; Gear Algorithm*

R_v

A ratio used to assess the degree of cooperativity exhibited by an enzyme. It is equal to the true V_{max} value (typically extrapolated from the high-substrate-concentration end of a double-reciprocal plot) divided by the apparent V_{max} value obtained from extrapolating the asymptote in the low-substrate-concentration portion of the double-reciprocal plot. For a noncooperative system, R_v will equal one; positively cooperative systems will have values greater than one; and negatively cooperative systems will have values less than one[1,2]. This method requires good estimates of the asymptotes.

[1]G. R. Ainslie, Jr., J. P. Shill & K. E. Neet (1972) *J. Biol. Chem.* **247**, 7088.
[2]K. E. Neet (1980) *Meth. Enzymol.* **64**, 139.

S

$S_{0.5}$

A useful parameter in the investigation of cooperative enzymes. It is the concentration of substrate needed for one-half of the maximal saturation of the cooperative enzyme.

$S_{0.9}/S_{0.1}$ (or, R_S)

A kinetic parameter, introduced by Koshland[1], to indicate the ratio of substrate concentrations needed to achieve reaction velocities equal to $0.1 V_{max}$ and $0.9 V_{max}$. For an enzyme obeying the Michaelis-Menten equation, $S_{0.9}/S_{0.1}$ equals 81, indicating that such enzymes exhibit modest sensitivity of reaction rate relative to changes in the substrate concentration. Many positively cooperative enzymes have $S_{0.9}/S_{0.1}$ values between five and ten, indicating that they can be turned on or off over a relatively narrow substrate concentration range.

[1]D. E. Koshland, Jr. (1969) *Curr. Top. Cell. Reg.* **1**, 1.

SAAM

An abbreviation commonly used for "*s*imulation, *a*nalysis, *a*nd *m*odeling" program, a set of digital computer algorithms based on the compartmental analysis approach introduced by Berman[1-3]. Kinetic data are initially analyzed to determine the temporal and mass transport relationships between compartments by employing an interactive computer program for simulating kinetic models along with fitting procedures to evaluate the goodness of fit of the model to the experimental data. Each compartment can be related to one or more reacting chemical species. Hensley *et al.*[4] provides an excellent example of the application of compartmental analysis in the characterization of *Bam*HI endonuclease action.

[1]M. Berman (1971) in *Advances in Medical Physics* (J. S. Laughlin &

E. W. Webster, eds.), p. 279, Second International Conference on Medical Physics, Inc., Boston.
[2]M. Berman & M. F. Weiss (1978) *SAAM-27 Manual*, DHEW Publ. No. (NIH) 78-180.
[3]SAAM/CONSAAM software and manuals are available free of charge from the Laboratory of Mathematical Biology, Room 4B56/Building 10, National Cancer Institute, National Institutes of Health, Bethesda, MD 20892. For a listing of computer requirements as well as the down-loadable SAAM/CONSAAM software, the internet address is: http://www.saam.nci.nih.gov/
[4]P. Hensley, G. Nardone & M. E. Wastney (1992) *Meth. Enzymol.* **210**, 391.

SADDLE POINT

The point of minimal potential energy in the trajectory of reactants to products in a chemical reaction. A reaction's saddle point (or col[1]) indicates the geometry and energy of reactants as they approach and pass the transition state of a reaction.

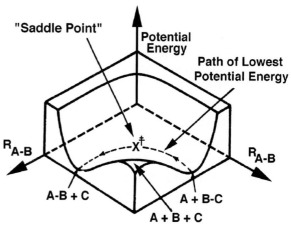

Graphical representation of the saddle point (here marked with an X) for the transfer of atom B as the substance A–B reacts with another species, C. Potential energy is plotted in the vertical direction. Note also that the surface resembles a horse saddle, with the horn of the saddle closest to the observer. As drawn here, the dissociation to form three discrete species (A + B + C) requires much more energy than that needed to surmount the path that includes the saddle point. A two-dimensional "slice" through a saddle point diagram is typically called a reaction-coordinate diagram or potential-energy profile.

The saddle point best represents the behavior of so-called elementary reactions occurring directly between small molecules; an example is the collision of gases and ions in crossed-beam, or translational kinetic experiments. Computer programs allow investigators to determine reaction path and transition state properties, to perform a normal mode analysis at a specified geometry, to calculate a trajectory based on a data set of reactant coordinates and momenta, to ascertain the minimum energy geometry for a specified potential energy surface, to calculate a trajectory beginning with one or two reactants, and to calculate a trajectory starting at a saddle point and proceeding toward reactants or products.

Multistep or multistage reactions in aqueous solutions are far more complicated than the process depicted above. Nonetheless, the conceptual picture of a saddle point allows one to comprehend the spatial organization and energy constraints influencing chemical reactivity.

[1]A col is a high pass through a mountain range or a saddle-like depression in the crest of a ridge.

SALICYLATE 1-MONOOXYGENASE

This FAD-dependent enzyme [EC 1.14.13.1], also known as salicylate hydroxylase, catalyzes the reaction of salicylate with NADH and dioxygen to produce catechol, NAD^+, carbon dioxide, and water.

B. A. Palfey & V. Massey (1998) Comprehensive Biological Catalysis: A Mechanistic Reference **3**, 83.

J. V. Schloss & M. S. Hixon (1998) Comprehensive Biological Catalysis: A Mechanistic Reference **2**, 43.

S. Yamamoto & Y. Ishimura (1991) A Study of Enzymes **2**, 315.

B. Entsch (1990) Meth. Enzymol. **188**, 138.

V. Massey & P. Hemmerich (1975) The Enzymes, 3rd ed., **12**, 191.

SAMPLE SPACE

A statistical term for the list or set (denoted as S) of all possible outcomes of a random experiment. In the case of a perfectly balanced, infinitely thin coin, there are only two outcomes, Heads and Tails, in the sample space.
See Random Experiment; Statistics (A Primer)

SAPONIFICATION

The reaction of an ester (most often a fatty acid ester) with a base to produce a carboxylate anion and an alcohol.

SARCOSINE DEHYDROGENASE

This FMN-dependent enzyme [EC 1.5.99.1] catalyzes the reaction of sarcosine with an acceptor and water to produce glycine, formaldehyde, and the reduced acceptor.

V. Schirch (1998) Comprehensive Biological Catalysis: A Mechanistic Reference **1**, 211.

M. H. Stipanuk (1986) Ann. Rev. Nutr. **6**, 179.

SARCOSINE OXIDASE

This FAD-dependent enzyme [EC 1.5.3.1] catalyzes the reaction of sarcosine with dioxygen and water to produce glycine, formaldehyde, and hydrogen peroxide.

V. Schirch (1998) Comprehensive Biological Catalysis: A Mechanistic Reference **1**, 211.

SATURATION

1. The condition wherein the concentration of a reactant, ligand, product, or effector is significantly greater than the dissociation constant of that same entity. 2. A solution becomes saturated with a particular substance when that substance has been dissolved to its maximum extent. 3. The process (or degree of) by which the number of π-bonds in a molecular entity decrease (that is, the removal of double and triple bonds and increase of σ-bonds).

SATURATION KINETICS

A term used to refer to any chemical process whose rate depends upon saturation of a binding site. Rate saturation is observed in enzyme kinetics, metabolic transport, host-guest reactions, and even heterogenous catalysis such as hydrogenation on metallic surfaces of platinum and nickel.

SCALAR COUPLING RELAXATION

A relaxation (and relaxation time) that contributes to the observed relaxation times (T_1 and T_2) in NMR experiments. Scalar coupling relaxation is the dominant relaxation process in $^{31}P(^{17}O)$ NMR.

SCATCHARD PLOT

A graphical procedure first used by Scatchard[1] to determine the stoichiometry and dissociation constant(s) for ligand binding to a macromolecule. Winzor and Sawyer[2] have pointed out that the Scatchard method[1,3] can be misleading with respect to the derived parameters. In fact, there are many computer programs[4,5] that permit one to analyze binding data without using graphical transformations like the Scatchard treatment.

Because the Scatchard equation is so widely used in the molecular life sciences, we show two different ways to derive the equation and we shall consider several special types of binding interactions.

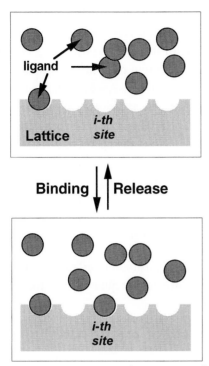

Figure 1. Ligand binding to a lattice of noninteracting binding sites.

Lattice Binding Model. Consider the diagram shown in Fig. 1 for a hypothetical array of binding sites on a protein, nucleic acid, organelle, or cell surface. Binding of small molecules (or ligands) can take place at any of the sites on the lattice. We assume: (a) that all sites are independent, such that occupancy at one site (marked here as the i-th site) in no way affects occupancy of other sites, even those immediately adjacent to the i-th site; (b) that all sites are capable of undergoing saturation by the ligand; (c) that the concentration of occupied sites can be experimentally determined using chemical analysis, but frequently relying on the use of radioisotopically labeled ligands or an appropriate spectroscopic technique that yields a change in signal to ligand occupancy. We can now define an equilibrium relationship:

$$L + S = LS$$

where L signifies the ligand and S is the site on the lattice. Because all sites are equivalent and noninteracting, the total concentration of sites can be denoted as C_S, the total concentration of sites with ligand bound can be written as C_B, and the concentration of free ligand as

C_F. The equilibrium dissociation constant can then be expressed as follows:

$$K = C_F(C_S - C_B)/C_B$$

which can be rearranged to yield:

$$C_B/C_F = (C_S - C_B)/K = (C_S/K) - (C_B/K)$$

and a plot of C_B/C_F *versus* C_B will be linear with the following parameters: (a) the slope equals $-(1/K)$; (b) the vertical-axis intercept equals C_S/K; and (c) the horizontal-axis intercept is C_S. The above expression is called the Scatchard equation, and the corresponding graphical representation is called a Scatchard plot (Fig. 2).

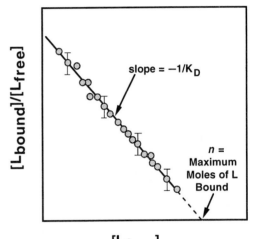

Figure 2. Scatchard plot for independent ligand binding.

How can the experimenter compensate for nonspecific binding of ligand? In addition to these high-affinity ligand binding sites, many macromolecules also bind ligands nonspecifically at one or more other sites, albeit with low affinity. This is often observed for drug binding at membrane receptor sites, but other cases, such as the binding of divalent cations to proteins, are also well documented. What is important to recognize about the Scatchard treatment is that one only considers high-affinity or so-called biospecific ligand binding. If there is an indication of weak secondary site interactions, one can typically correct for this by conducting binding studies with a nonphysiologic analogue. For drugs that act as antagonists or agonists of some physiologic process, this is often accomplished by using a pharmacologically ineffective enantiomer of an optically active drug. Such an analogue would not be expected to occupy the primary ligand sites on a receptor; otherwise, some pharmacologic effect would be manifested. Weak binding of ligand frequently shows a linear relation between amount bound and ligand concentration (see lower line in the upper panel of Fig. 3).

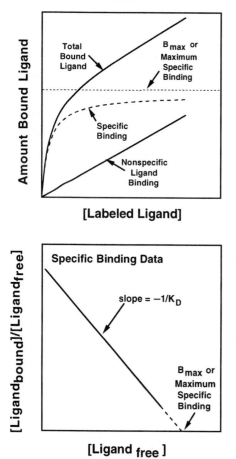

Figure 3. Treatment of specific and nonspecific binding of a ligand in equilibrium with a finite concentration of receptor sites.

This can be subtracted from the total amount of bound ligand (top curve) to arrive at the specific binding value for each ligand concentration. (Alternatively, in the absence of a pharmacologically ineffective ligand, one can extrapolate the slope of the total bound ligand curve to zero concentration, and then use this curve as the nonspecific binding curve. Of course, this requires one to first construct a linear plot starting at coordinates (0,0) with a slope matching that of the extrapolated line). Only then may one use the specific binding data to construct a Scatchard plot (shown in the lower panel of Fig. 3).

How may one extend the Scatchard treatment to deal with the case of steric hindrance among neighboring sites? This realistic situation occurs often, especially when the distances between binding sites is not much greater than the molecular size of the ligand or when ligand binding disturbs the structure of its macromolecular binding partner. One such case is the binding of acridine dyes by intercalation between DNA base pairs. Another might be the binding of the dynein motor or a bulky microtubule-associated protein onto the tubulin lattice within a microtubule. In such a case, we can define the concentration of potential sites at theoretical saturation to be C_o; we can define the fraction of sites actually bound at saturation as b_{ap}; we define the concentration of free ligand as C_F; and we can define $b_{ap}C_o$ as the concentration of total sites bound, or C_S. We then write the equilibrium expression:

$$K_{ap} = C_F(C_S - C_B)/C_B = C_F(b_{ap}C_o - C_B)/C_B$$

If we divide the right-hand numerator and denominator by C_o, and if we let r equal C_B/C_o, the above equation becomes

$$K_{ap} = C_F(b_{ap} - r)/r$$

Rearranging this expression yields a modified version of the Scatchard equation:

$$r/C_F = (b_{ap}/K_{ap}) - (r/K_{ap})$$

and a plot of r/C_F *versus* r will have a slope of $-(1/K_{ap})$ and a horizontal-axis intercept of b_{ap}. If such a plot is carried out for acridine binding to DNA, the value of b_{ap} is about 0.25, indicating that only about one in every four base pairs can accommodate a bound acridine molecule at saturation.

Standard Binding Model. The more commonly recognized form of the Scatchard equation can be written as follows:

$$\nu/[L] = K(n - \nu)$$

where ν is the fractional saturation of n binding sites by ligand L, and K is the dissociation constant. Consider the following sequence of ligand binding interactions for a macromolecule, where M_i represents the ith site on the macromolecule: $M_0 + L \rightleftharpoons M_1$; $M_1 + L \rightleftharpoons M_2$;...; $M_{n-1} + L \rightleftharpoons M_n$. The binding expression described above can be derived by defining ν as follows:

$$\nu = \frac{\sum\limits_{i=0}^{n} i[M_i]}{\sum\limits_{i=0}^{n} [M_i]}$$

where the concentration of species are shown in brackets. We can write the equilibrium for L binding to the i-th site (S_i) as follows:

$$S_{(i-1)} + L = S_i$$

Now let ν_i be defined as the saturation function for the i-th site:

$$\nu_i = [\text{Ligand bound at Site}_i]/\{[\text{Site}_{(i-1)}] \\ + [\text{Ligand bound at Site}_i]\}$$

$$\nu_i = [\text{Site}_i]/\{[\text{Site}_{(i-1)}] + [\text{Site}_i]\}$$

and (a) dividing the right-hand numerator and denominator by $[\text{Site}_{(i-1)}]$, and (b) using the expression ($[S_i]/[S_{(i-1)}] = [L]/K_i$) from the dissociation constant expression (*i.e.*, $K_i = [S_{(i-1)}][L]/[S_i]$), we get

$$\nu_i = ([L]/K_i)/\{1 + ([L]/K_i)\}$$

To convert the binding at the i-th site to the binding at all n sites, we use the expression:

$$\nu = \Sigma n \nu_i = n([L]/K_i)/\{1 + ([L]/K_i)\}$$

which directly rearranges to the familiar form of the Scatchard equation:

$$\nu = \frac{n[L]/K_i}{1 + ([L]/K_i)}$$

Upon rearrangement, we get the familiar Scatchard equation:

$$\nu/[L] = (n/K_i) - (\nu/K_i)$$

where a plot of ($\nu/[L]$) *versus* ν has a slope of $-(1/K_i)$ and a horizontal-axis intercept is the number of saturable sites n.

What about the likely case that there are two classes of binding interactions—one strong and one weak? To treat this case, we write two separate saturation functions, one for each class of sites:

$$\text{Class 1:} \quad r_1/[L] = (b_1 - r_1)/K_1$$

$$\text{Class 2:} \quad r_2/[L] = (b_2 - r_2)/K_2$$

where r_1 and r_2 are defined as the fraction of ligand bound to sites 1 and 2, respectively, and b_1 and b_2 are the fractions of bound sites that are type-1 and type-2. We can define r_T to be the sum of r_1 and r_2, and the analytical technique generally allows us to measure total ligand bound, or $r_T[L_{total}]$. Thus, we can write $r_T/[L]$ as follows:

$$r_T/[L] = \{(b_1 - r_1)/K_1\} + \{(b_2 - r_2)/K_2\}$$

or, if we substitute ($r_T - r_2$) for r_1, we obtain

$$r_T/[L] = \{[b_1 - (r_T - r_2)]/K_1\} + \{(b_2 - r_2)/K_2\}$$

A plot of ($r_T/[L]$) *versus* r_T begins on the vertical axis at the intercept (equal to $b_1/K_1 + b_2/K_2$) and has an initial slope (*i.e.*, at r_T values ≈ 0) equal to $-(1/K_1)$; this initial linear region extrapolates to a horizontal-axis value of b_1. As one progresses to even higher ligand concentrations, the curve bows to the right and eventually reaches the horizontal intercept r_T. Thus, on the basis of these experimentally derived values of K_1 and b_1, one can iteratively fit the entire experimental binding curve.

This behavior is illustrated in Fig. 4 for the case where the first ligand binds with a dissociation constant that is 30 times lower than the equilibrium constant for dissociation of ligand from the second site. The dotted line shows the behavior of a Scatchard plot in the absence of second-site binding; in this case, the dotted line extrapolates to a value of 1.0 on the horizontal axis.

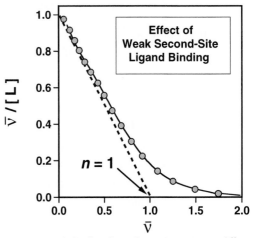

Figure 4. Scatchard plot for a ligand interacting at two different sites on the same protein. See text for details.

Scatchard *et al.*[3] found that thiocyanate binding to human serum albumin gave 10 sites with a 1–2 mM dissociation constant, and another 30 with a dissociation constant that was nearly 40 mM.

Finally, with homodimeric proteins, both sites are identical in the absence of any ligand. If a homodimeric protein exhibits a Scatchard plot shaped like that shown in Fig. 4, such an observation is diagnostic for negative cooperativity in ligand binding. The Koshland-Neméthy-Filmer model for cooperativity treats cases where ligand binding at the first site on a homodimer induces a conformational change that reduces the affinity of ligand binding at a second site. Negative cooperativity is not a possibility in the Monod-Wyman-Changeux model for cooperative ligand binding.

[1]G. Scatchard (1949) *New York Acad. Sci.* **51**, 660.
[2]D. J. Winzor & W. H. Sawyer (1995) *Quantitative Characterization of Ligand Binding*, p. 45, Wiley-Liss, New York.
[3]G. Scatchard, J. S. Coleman & A. L. Shen (1957) *J. Am. Chem. Soc.* **79**, 12.
[4]L. Brand & M. L. Johnson, eds., (1992) *Meth. Enzymol.*, vol. **210**, Academic Press, San Diego.
[5]M. L. Johnson & L. Brand, eds., (1994) *Meth. Enzymol.*, vol. **240**, Academic Press, San Diego.

SCAVENGER

Any compound, substance, or material that reacts with (or binds, chelates, or removes from a system) a trace component or reaction intermediate.

SCHIFF BASE (or Schiff's Base)

The product formed by the reaction of an aldehyde with an amine.

$$R_1 - NH_2 + R_2 - CHO \rightarrow R_1 - N = CHR_2 + H_2O$$

Historically, the amine was an aromatic amine but is now generalized to any amine. A Schiff base, also called an aldimine, is formed in the pyridoxal 5-phosphate-dependent aminotransferase reactions.

SCHRÖDINGER EQUATION

A class of partial differential equations first proposed by Erwin Schrödinger in 1926 to account for the so-called quantized wave behavior of molecules, atoms, nuclei, and electrons. Solutions to the Schrödinger equation are wave functions based on Louis de Broglie's proposal in 1924 that all matter has a dual nature, having properties of both particles and waves. These solutions are of a probabilistic character, and they describe how the structure of atoms, molecules, and closely packed assemblies is altered discontinuously in discrete increments of energy change (called a quantum). While Schrödinger equations offer the opportunity to attain remarkable accuracy, the number of atoms within an enzyme (or even its active site complexes with substrates) greatly limits the accuracy needed to predict changes occurring during catalysis. For this reason, transition state theory of enzymic reactions is typically analyzed by classical mechanics. Nonetheless, the Schrödinger equation predicts that electron hydride as well as proton transfer can occur without surmounting an energy barrier. **See** *Quantum Mechanical Tunneling*

The Schrödinger equation for a one-electron system is:

$$\frac{\partial^2 \Psi}{\partial x^2} + \frac{\partial^2 \Psi}{\partial y^2} + \frac{\partial^2 \Psi}{\partial z^2} + \frac{8\pi^2 m}{h^2}(E - V)\Psi = 0$$

where m is the mass of the electron, E and V are the electron's total and potential energy respectively, h is Planck's constant, and Ψ is the wave function for the electron. Chemical properties are determined by the energies of electrons as well as the relative positions of electrons with respect to the nucleus and other electrons within a molecule, ion, radical, or atom. Solutions to the Schrödinger equation can only be obtained for certain energy values, known as eigenvalues. They represent the allowable quantized states of the system, and for each eigenvalue, there is at least one eigenfunction Ψ.

P. S. C. Matthews (1986) *Quantum Chemistry of Atoms and Molecules*, Cambridge Univ. Press, Cambridge.

SCINTILLATORS

Compounds and materials, including mixtures, which can undergo radioluminescence and are used in the measurement of radioactivity. Characteristics of these substances include a high fluorescence quantum efficiency and a short fluorescence lifetime.

SCISSILE BOND

A covalent bond that is broken during the course of a chemical reaction.

SCREW SENSE OF METAL-NUCLEOTIDE COMPLEXES

The conformation and helical sense of the structure interconnecting the metal-coordinated oxygens and the phosphorus atoms of the nucleotide.

In the Δ series of CrATP, there are two screw-sense isomers, δ (pseudoequatorial) and λ (pseudoaxial). A number of nucleotide-dependent enzymes have been shown to be specific for one of the screw-sense isomers. **See** *Exchange-Inert Metal-Nucleotide Complexes*

D. Dunaway-Mariano & W. W. Cleland (1980) *Biochemistry* **19**, 1496.

D. Dunaway-Mariano & W. W. Cleland (1980) *Biochemistry* **19**, 1506.

W. W. Cleland (1982) *Meth. Enzymol.* **87**, 159.

SECOND

A unit of time (symbolized by s) that serves as one of the seven base SI units, equal to the duration corresponding to 9,192,631,770 times the period for transition between the two hyperfine levels of the ground state of an atom of ^{133}Cs.

SECONDARY ISOTOPE EFFECT

A change in reaction rate that results from isotopic substitution adjacent to the site of bond breaking/making. The classical example in enzymology is the hydrolysis of phenylglucosides by lysozyme[1,2], for which the so-called α-deuterium isotope effect (symbolized as k_H/k_D) was found to be 1.14. This result was consistent with an expected change in electronic hybridization at positions C-1 and oxygen (within the glucopyranose ring) if a carbonium ion (or carboxonium) intermediate formed during the reaction cycle. **See** *Kinetic Isotope Effect*

[1]F. W. Dahlquist, T. Rand-Meir & M. A. Raftery (1968) *Proc. Natl. Acad. Sci. U.S.A.* **61**, 1194.

[2]L. E. H. Smith, L. H. Mohr & M. A. Raftery (1973) *J. Am. Chem. Soc.* **95**, 7497.

SECOND-ORDER REACTION

With a Single Reaction Component. A reaction of a substance A is said to be second order if the following rate expression applies:

$$\text{Reaction Rate} = v = -d[A]/dt = k[A]^2$$

where k is the second-order rate constant (having units of reciprocal concentration multiplied by reciprocal time) and t is time. Integration of this expression yields $[A]^{-1} = kt + [A_o]^{-1}$ where $[A_o]$ and $[A]$ are the concentrations of the reactant at $t = 0$ and $t = t$, respectively. Thus, a plot of $[A]^{-1}$ *vs.* t will result in a straight line having a slope[1] of k. If the reaction indeed follows the

above rate expression with only a single reaction component, the second-order rate constant can also be obtained by determining the reaction half-life, $t_{1/2}$. For a second-order reaction, $t_{1/2} = (k[A_o])^{-1}$. Notice that the half-life for a second-order process varies with the initial reactant concentration.

With Two Different Reaction Components. A reaction occurring between two substances, A and B, is said to be second order if the reaction rate agrees with the rate expression[1]:

$$\text{Reaction Rate} = v = -d[A]/dt = k[A][B]$$

where t is time and k is the second-order rate constant. As mentioned in the entries on order and molecularity, this second-order rate constant is the bimolecular rate constant for an elementary reaction. Letting $[A_o]$ and $[B_o]$ represent the initial concentrations of the two reactants at $t = 0$, letting $[A]$ and $[B]$ represent the concentrations at $t = t$, and recalling that $[A_o] - [A] = [B_o] - [B]$, integration yields

$$\frac{1}{[B_o] - [A_o]} \ln\left(\frac{[A_o][B]}{[B_o][A]}\right) = kt$$

Plotting $([B_o] - [A_o])^{-1} \ln[[A_o][B]/([B_o][A])]$ vs. t or plotting $\ln([B]/[A])$ vs. t will yield a straight line having slope k. A value for k can also be obtain by substituting successive known values of $[A]$, $[B]$, and t into the above equation, calculate several values for k, and obtain a mean (assuming these values display some constancy). If x represents the decrease in reactant concentration at time t, the equation above becomes

$$\frac{1}{[B_o] - [A_o]} \ln\left(\frac{[A_o]([B_o] - x)}{[B_o]([A_o] - x)}\right) = kt$$

As before, values of k can be calculated using the equation or by graphical analysis (*e.g.*, plotting $\ln[([B_o] - x)/([A_o] - x)]$ vs. t). If the initial concentrations of A and B are equal (*i.e.*, $[A_o] = [B_o]$), the above equations have zero in the denominator, and are undefined.

If the concentration of the two reacting molecules remain identical throughout the course of reaction, then the rate expression becomes analogous to a second-order reaction with one component. Let $[D_o] = [A_o] + [B_o]$ and let $[D]$ represent the concentrations of A and B at $t = t$. Then, the integrated rate expression becomes $[D]^{-1} = kt + [D_o]^{-1}$. Once again, plotting $[D]^{-1}$ vs. t will yield a straight line having a slope of k.

There are many examples of second-order reactions: for example, associations, addition reactions, *etc*. The decomposition of gaseous nitrogen dioxide ($2NO_2 \rightarrow 2NO + O_2$) follows the rate law, $-d[NO_2]/dt = k[NO_2]^2$. The binding of a hormone to a receptor protein is also second-order.

If the concentration of one of the reactants is much larger than that of the other (for example $[A_o] >> [B_o]$), then that higher concentration can be treated as a constant as its value will not change much over the time course of the reaction. The rate expression then becomes

$$\ln\left(\frac{[B_o]}{[B_o] - x}\right) = [A_o]kt = k_{obs}t$$

where k_{obs} is the pseudo-first-order rate constant equal to $k[A_o]$. The second-order rate constant can be obtained by determining k_{obs} at different concentrations of A_o and plotting k_{obs} vs. $[A_o]$. As Jencks has pointed out, pseudo-rate constants are very sensitive to impurities[2]. It may be necessary to repurify the reactant. Nevertheless, pseudo-rate constants are of value as they provide the investigator with a means of separating variables that affect a chemical reaction mechanism[2]. An example of a pseudo-first-order reaction is the hydrolysis (or, inversion) of

sucrose catalyzed by strong acids (H_2O + sucrose $\xrightarrow{H^+}$ D-glucose + D-fructose). This reaction is of historical interest as it is the first reaction studied kinetically. Wilhelmy (1812–1864) used a polarimeter to follow the course of this reaction. The rate expression he reported, in his terminology, was $-dZ/dt = MZS$ where t was time, Z was the sucrose concentration, S was the acid concentration, and M was a constant[3]. This equation, in modern terminology, becomes $-d[\text{sucrose}]/dt = k[\text{sucrose}][H_2O][H^+]$. Wilhelmy is justly regarded by many as the father of chemical kinetics. *See Chemical Kinetics; Order of Reaction; Noyes Equation; Molecularity; Autocatalysis; First-Order Reaction.*

[1]J. W. Moore & R. G. Pearson (1981) *Kinetics and Mechanism*, Wiley, New York.
[2]W. P. Jencks (1969) *Catalysis in Chemistry and Enzymology*, McGraw-Hill, New York. [Also available as Jencks, W. P. (1986) *Catalysis in Chemistry and Enzymology*, Dover Publications, New York]
[3]L. Wilhelmy (1850) *Ann. der Physik und Chemie* **81**, 413 and 526.

SEDIMENTATION EQUILIBRIUM ULTRACENTRIFUGATION

An analytical ultracentrifugation method for determining the molecular mass, diffusion coefficient, and/or state of oligomerization of a macromolecule by conducting sedimentation conditions to establish an equilibrium distribution of the macromolecule from the meniscus to the bottom of the observation cell.

Selected entries from *Methods in Enzymology* [vol, page(s)]: Association constant determination, **259**, 444-445; buoyant mass determination, **259**, 432-433, 438, 441, 443, 444; cell handling, **259**, 436-437; centerpiece selection, **259**, 433-434, 436; centrifuge operation, **259**, 437-438; concentration distribution, **259**, 431; equilibration time, estimation, **259**, 438-439; molecular weight calculation, **259**, 431-432, 444; nonlinear least-squares analysis of primary data, **259**, 449-451; oligomerization state of proteins [determination, **259**, 439-441, 443; heterogeneous association, **259**, 447-448; reversibility of association, **259**, 445-447]; optical systems, **259**, 434-435; protein denaturants, **259**, 439-440; retroviral protease, analysis, **241**, 123-124; sample preparation, **259**, 435-436; second virial coefficient [determination, **259**, 443, 448-449; nonideality contribution, **259**, 448-449]; sensitivity, **259**, 427; stoichiometry of reaction, determination, **259**, 444-445; terms and symbols, **259**, 429-431; thermodynamic parameter determination, **259**, 427, 443-444, 449-451.

SEDIMENTATION VELOCITY ULTRACENTRIFUGATION

An analytical ultracentrifugation method for determining the molecular mass, diffusion coefficient, and/or state of oligomerization of a macromolecule by conducting sedimentation conditions that allow the macromolecule to migrate at constant velocity. The sedimentation coefficient s equals the sedimentation rate divided by the relative centrifugal field: $s = dr/dt/\omega^2 r$, where s is the sedimentation coefficient (expressed in svedberg units) coefficient, ω is the angular velocity in radians per second, and dr/dt is the sedimentation rate, and r is the radial displacement.

Selected entries from *Methods in Enzymology* [vol, page(s)]: Boundary analysis [baseline correction, **240**, 479, 485-486, 492, 501; second moment, **240**, 482-483; time derivative, **240**, 479, 485-486, 492, 501; transport method, **240**, 483-486]; computation of sedimentation coefficient distribution functions, **240**, 492-497; diffusion effects, correction [differential distribution functions, **240**, 500-501; integral distribution functions, **240**, 501]; weight average sedimentation coefficient estimation, **240**, 497, 499-500.

SELECTION RULE

A rule in quantum chemistry which states when a particular transition is allowed or forbidden.

L-SELENOCYSTEINE

A naturally occurring amino acid that is identical in structure to cysteine, but contains the micronutrient selenium in place of sulfur. The pK_a of the $R-SeH$ moeity is around 5.2, contrasting markedly from the corresponding 8.3 value for cysteine. Hence, at neutral pH selenocysteine is largely $R-Se^-$, whereas cysteine is largely $R-SH$. Selenocysteine is an essential component of glutathione peroxidase, an enzyme which protects cellular membranes from peroxidative damage. This amino acid is also present in bacterial formate dehydrogenase and glycine reductase. **See also** *L-Seryl-tRNA(SeC) Selenium Transferase*

T. C. Stadtman (1979) *Adv. Enzymol.* **48**, 1.

L-SELENOCYSTEINE β-LYASE

This pyridoxal-phosphate-dependent enzyme [EC 4.4.1.16], also known as selenocysteine reductase, catalyzes the reaction of L-selenocysteine with a reduced acceptor to produce hydrogen selenide, L-alanine, and the acceptor. The enzyme can use dithiothreitol or 2-mercaptoethanol as the reducing agent. Cysteine, serine, or chloroalanine are not alternative substrates for this enzyme.

T. C. Stadtman (1990) *Ann. Rev. Biochem.* **59**, 111.
N. Esaki & K. Soda (1987) *Meth. Enzymol.* **143**, 415.
N. Esaki & K. Soda (1987) *Meth. Enzymol.* **143**, 493.

SELENOPHOSPHATE SYNTHETASE

This magnesium-dependent enzyme [EC 2.7.9.3], also known as selenide,water dikinase and selenium donor protein, catalyzes the reaction of ATP with selenide and water to produce AMP, selenophosphate, and orthophosphate.

T. C. Stadtman (1996) *Ann. Rev. Biochem.* **65**, 83.

SELF-ABSORPTION

Self-absorption occurs when the fluorescence due to some excited molecular entity(ies) is absorbed by other molecular entities of the same species which are in the ground state.

SELF-PROTECTION MECHANISM

A mechanism seen on occasion in affinity labeling experiments when an active site-directed reagent binds to the active site of an enzyme and protects that active site against a bimolecular reaction with a second molecule

of the reagent. Appropriate design of the affinity labeling experiment will curtail this mechanism from occurring[1].

[1]B. V. Plapp (1982) *Meth. Enzymol.* **87**, 469.

SELF-QUENCHING

The event which occurs when an excited atom or molecular entity is quenched by an interaction with another atom or molecular entity of the species, which is in the ground state.

SENSITIVITY

1. The capacity to respond to a change in the environment or the influence of some signal.

2. A relationship used in the analysis of enzyme cascade systems. It is symbolized by S (and also referred to as the sensitivity index) and is equal to $8.89[e_{0.5}]/([e_{0.9}] - [e_{0.1}])$ where $[e_{0.5}]$ is the concentration of effector required to attain 50% maximal amplitude, $[e_{0.9}]$ is the concentration of effector required for 90% maximal amplitude, $[e_{0.1}]$ is the concentration of effector required for 10% maximal amplitude, and 8.89 is a normalizing constant used to yield a reference value of 1.00 for a pure hyperbolic binding isotherm[1].

[1]P. B. Chock & E. R. Stadtman (1980) *Meth. Enzymol.* **64**, 297.

SEPARATION OF VARIABLES

The process of solving for the integrated form of a differential equation, achieved by rearranging the equation into a form where each side of the resulting expression can be integrated with respect to one of the variables.

SEQUENTIAL MECHANISMS

Enzyme-catalyzed reaction schemes involving two or more substrates and/or two or more products in which there are no ping pong half-reaction steps. **See** *Multisubstrate Mechanisms*

SERIES REACTIONS

A set of reactions in which the product(s) of at least one reaction serves as a reactant for a subsequent reaction; also referred to as consecutive reactions.

The simplest set of series reactions involves two first-order reactions:

$$A \xrightarrow{k_1} B$$
$$B \xrightarrow{k_2} C$$

Harcourt and Esson[1] were the first to study such reactions (in their case, the reaction of potassium permanganate with oxalic acid). Esson[2] was the first to integrate the differential equation. This scheme can be treated exactly by considering the differential equations:

$$d[A]/dt = -k_1[A]$$

$$d[B]/dt = k_1[A] - k_2[B]$$

$$d[C]/dt = k_2[B]$$

If $[A_o]$ and $[A]$ are the concentrations of A at $t = 0$ and $t = t$, respectively, the integration of the first equation yields $[A] = [A_o]e^{-k_1t}$. Note that the concentration of A decreases with time exponentially, as would be expected for any first-order process. Substituting this expression for [A] into the second differential equation and assuming $[B_o] = 0$ at $t = 0$, integration yields

$$[B] = [[A_o](k_1/k_2 - k_1)](e^{-k_1t} - e^{-k_2t})$$

To derive an integrated expression for C, two relationships must be used. The three differential equations above tell us that $d[A]/dt + d[B]/dt + d[C]/dt = 0$. And, the conservation relationship provides $[A] + [B] + [C] = [A_o]$ at any time t. Assuming $[B_o]$ and $[C_o]$ are zero at $t = 0$, the $[C] = \{[A_o] - [A] - [B]\}$ or

$$[C] = [A_o]\{1 + (k_1 - k_2)^{-1}(k_2e^{-k_1t} - k_1e^{-k_2t})\}$$

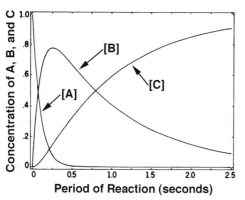

Time evolution of a series first-order process wherein $k_1 = 0.1$ min^{-1} and $k_2 = 0.05$ min^{-1}.

As can be seen in the plot, the concentration of A decreases exponentially with time. The concentration of B increases to a maximum value and then decreases. The concentration of C increases gradually until, at $t = \infty$,

$[C_\infty] = [A_o]$. Formation of [C] is slowest in the early stage of the progress curve (i.e. the induction period) which persists from $t = 0$ to the time at which the inflection point of the progress curve for [C] is reached. This inflection point occurs at the same time that [B] reaches its maximum concentration (since $d^2[C]/dt^2 = 0$ when $d[B]/dt = 0$). This is a characteristic of any series first-order sequence, however many steps are included, provided that only A is initially present. If $k_2 << k_1$ in the above scheme, the expression for [A] remains the same, the expression for [B] becomes $[B] = [A_o]e^{-k_2t}$, and [C] becomes $[A_o](1 - e^{-k_2t})$. In this case, notice the concentration of the final product at any time t depends only on the smallest rate constant and the initial concentration of the first reactant. Note that when $k_2 >> k_1$, [C] will equal $[A_o](1 - e^{-k_1t})$.

Series first-order processes are especially important in understanding metabolic pathway kinetics. Likewise, one should recognize that unassisted protein folding[3,4] is unlikely to be a single-step reaction (say U → N), where U represents unfolded protein, and N is the native conformation. At a minimum, realistic schemes for protein folding are apt to occur as a series of isomerizations:

$$U \rightarrow I_1 \rightarrow I_2 \rightarrow I_3 \rightarrow N$$

where I_1, I_2, and I_3 are intermediate states of folding. The greatest challenge is to develop spectroscopic tools that detect the intermediates with sufficient accuracy and precision so as to allow quantifiable kinetic measurements. If the number of steps exceeds two, one may utilize various software programs for numerical integration.

[1] A. V. Harcourt & W. Esson (1865) *Proc. Roy. Soc.* **14**, 470.
[2] W. Esson (1866) *Phil. Trans. Roy. Soc. London Ser. A.* **156**, 220.
[3] P. S. Kim & R. L. Baldwin (1990) *Ann. Rev. Biochem.* **59**, 631.
[4] T. E. Creighton (1990) *Biochem. J.* **270**, 1.

SERINE DEHYDRATASE

L-Serine dehydratase [EC 4.2.1.13], also known as serine deaminase and L-hydroxyaminoacid dehydratase, catalyzes the pyridoxal-phosphate-dependent hydrolysis of L-serine to produce pyruvate, ammonia, and water. In a number of organisms, this reaction is also catalyzed by threonine dehydratase.

D-Serine dehydratase [EC 4.2.1.14], also known as D-hydroxyaminoacid dehydratase, catalyzes the pyridoxal-phosphate-dependent hydrolysis of D-serine to produce pyruvate, ammonia, and water. The enzyme will also act on D-threonine, albeit not as effectively.

A. E. Braunstein & E. V. Goryachenkova (1984) *Adv. Enzymol.* **56**, 1.
L. Davis & D. E. Metzler (1972) *The Enzymes*, 3rd ed., **7**, 33.

SERINE DEHYDROGENASE

This enzyme [EC 1.4.1.7] catalyzes the reaction of L-serine with NAD^+ and water to produce 3-hydroxypyruvate, ammonia, and NADH.

N. M. W. Brunhuber & J. S. Blanchard (1994) *Crit. Rev. Biochem. Mol. Biol.* **29**, 415.

SERINE HYDROXYMETHYLTRANSFERASE

This pyridoxal-phosphate-dependent enzyme [EC 2.1.2.1], which has a recommended EC name of glycine hydroxymethyltransferase, catalyzes the reversible reaction of 5,10-methylenetetrahydrofolate with glycine and water to produce tetrahydrofolate and L-serine. The enzyme will also catalyze the reaction of glycine with acetaldehyde to form L-threonine as well as with 4-trimethylammoniobutanal to form 3-hydroxy-N^6,N^6,N^6-trimethyl-L-lysine.

V. Schirch (1997) *Meth. Enzymol.* **281**, 146.
D. J. Oliver (1994) *Ann. Rev. Plant Physiol. Plant Mol. Biol.* **45**, 323.
M. E. Lidstrom (1990) *Meth. Enzymol.* **188**, 365.
L. Schirch (1982) *Adv. Enzymol.* **53**, 83.
J. I. Rader & F. M. Huennekens (1973) *The Enzymes*, 3rd ed., **9**, 197.

SERINE PALMITOYLTRANSFERASE

This pyridoxal-phosphate-dependent enzyme [EC 2.3.1.50] catalyzes the reaction of palmitoyl-CoA with L-serine to produce coenzyme A, 3-dehydro-D-sphinganine, and carbon dioxide.

A. H. Merrill, Jr., & E. Wang (1992) *Meth. Enzymol.* **209**, 427.

SERPINS (Inhibitory Mechanism)

A serpin is a *serine proteinase inhibitor* that forms catalytically inactive complexes which, after cleavage of the $P^1 - P^{1\prime}$ linkage, releases the inhibitor very slowly. O'Malley *et al.*[1] recently demonstrated that antichymotrypsin (I) binds to chymotrypsin (E) to form an E*I* complex *via* a three-step mechanism:

$$E + I \rightleftharpoons E{\cdot}I \rightleftharpoons EI' \rightleftharpoons E*I*$$
$$\downarrow$$
$$E + I$$

In this scheme, EI' retains the $P^1 - P^{1\prime}$ linkage and is formed in a partly, or largely, rate-determining step.

Using a rapid quench-flow kinetic assay for post-complex fragment formation, Nair and Cooperman[2] showed that the E·I encounter complex of serpin and enzyme forms both E*I* and the post-complex fragment with the same rate constant, indicating that both species arise from EI' conversion to E*I*. These results support the conclusions (a) that the peptide bond remains intact within the EI' complex, and (b) that E*I* is likely to be either the acyl-enzyme or the tetrahedral intermediate formed after water attack on acyl-enzyme.

[1]K. H. O'Malley, S. A. Nair, H. Rubin & B. S. Cooperman (1997) *J. Biol. Chem.* **272**, 5354.
[2]S. A. Nair & B. S. Cooperman (1998) *J. Biol. Chem.* **273**, 17459.

L-SERYL-tRNA(SeC) SELENIUM TRANSFERASE

This enzyme [EC 2.9.1.1], also known as selenocysteine (or SeC) synthase, L-selenocysteinyl-tRNA(SeC) synthase, L-selenocysteinyl-tRNA(Sel) synthase, cysteinyl-tRNA(SeC)-selenium transferase, and cysteinyl-tRNA-(Sel)-selenium transferase, catalyzes the reaction of L-seryl-tRNA and selenophosphate to form L-selenocysteinyl-tRNA(SeC), orthophosphate, and water. The enzyme is responsible for the insertion of selenocysteine into selenoproteins. ***See*** *L-Selenocysteine*

SHIELDING

Interelectronic interactions that alter how any particular electron in a multi-electronic atom interacts with the nucleus and *vice versa*. These effects lead to so-called chemical shifts in NMR experiments, thereby providing valuable structural information concerning a molecule's bonding and conformation.

Atoms have a nuclear charge equal to the number of protons (*i.e.*, the atomic number), and an equivalent number of electrons are arranged in electronic orbitals, designated $s, p, d, f, g, \ldots$. These orbitals are not equienergetic, and their shape and directionality creates a noncentrosymmetric electric field surrounding the nucleus. This results in shielding: the different degrees of electron-neutron attraction created by the electron occupancy of different orbitals. In an applied magnetic field (H_o), electrons move in a way that induces their own magnetic field (H'). The magnetic moment of this induced field opposes that of the applied field, making the net magnetic field (H) experienced by the nucleus (say a proton in an organic compound) less than that of the applied field: $H = H_o - H'$. Any nucleus experiencing a field that is smaller than the applied magnetic field is said to be shielded. Electromagnetic radiation at radiofrequency wavelengths is used to excite nuclei residing in the applied magnetic field of an NMR spectrometer, causing them to "spin flip" (*i.e.*, transit to a higher energy state), and *diamagnetic shielding* alters the radiofrequency at which this transition occurs. **See** *Chemical Shift*

SHIFTED BINDING

A phenomenon that has been observed with enzymes acting on a polymeric substrate. The active site of these enzymes consists of a number of subsites. Shifted binding is "the binding of a second substrate molecule in a nonproductive mode within the subsites so as to sterically shift the otherwise more energetically favorable binding of a productively bound substrate."[1] In such cases, it may be necessary to keep the concentration of the polymeric substrate low enough to avoid such problems. Nevertheless, the investigator should always examine the bond cleavage frequencies over a wide range of substrate concentrations. If shifted binding is present, the frequencies should change with increasing substrate concentrations.

[1]J. D. Allen (1980) *Meth. Enzymol.* **64**, 248.

SHIFTING SPECIFICITY MODEL FOR ENZYME CATALYSIS

A new model[1,2] for enzyme catalysis that challenges the long-standing concept of transition state complementarity as the sole source of enzymatic catalytic efficacy. This shifting model states that: (a) enzymes evolved to bind substrates; (b) enzyme·substrate complexes have evolved to bind transition states; and (c) stronger interactions of substrate with the enzyme facilitate rapid conversion to product. This model questions the concept that strong interactions of enzyme and substrate reduce catalytic efficiency.

[1]B. M. Britt (1993) *J. Theor. Biol.* **164**, 181.
[2]B. M. Britt (1997) *Biophys. Chem.* **69**, 63.

SHIKIMATE DEHYDROGENASE

This enzyme [EC 1.1.1.25] catalyzes the reaction of shikimate with NADP$^+$ to produce 5-dehydroshikimate and NADPH.

R. Bentley (1990) *Crit. Rev. Biochem. Mol. Biol.* **25**, 307.
S. Chaudhuri, I. A. Anton & J. R. Coggins (1987) *Meth. Enzymol.* **142**, 315.

SHIKIMATE KINASE

This enzyme [EC 2.7.1.71] catalyzes the reaction of ATP with shikimate to produce ADP and shikimate 3-phosphate.

R. Bentley (1990) *Crit. Rev. Biochem. Mol. Biol.* **25**, 307.
R. De Feyter (1987) *Meth. Enzymol.* **142**, 355.

SHUTTLE GROUP

Any residue that facilitates entry or departure of a proton from an enzyme active site or to another region within an active site. Occasionally, proton shuttle groups can be removed by site-directed mutagenesis, and chemical rescue can be demonstrated by providing buffer salts to replace the shuttle group. **See** *Chemical Rescue*

An example is the hydration of CO_2, as catalyzed by carbonic anhydrase[1]. The catalytic reaction requires proton transfer from the zinc-bound water at the active site to solution to regenerate $Zn \cdot OH^-$ in each catalytic cycle. The most efficient isozyme forms use His-64 as a nearby proton shuttle group; other forms contain residues that are less effective in proton transfer and limit overall catalytic efficiency.

[1]J. N. Earnhardt, M. Qian, C.-K. Tu, P. J. Laipis & D. N. Silverman (1998) *Biochemistry* **37**, 7649.

SI-

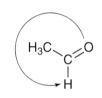

The *si* Face of Acetaldehyde

This stereochemical term (short for the Latin word *sinestrus*) is the designation for a stereoheterotopic face of a trigonal atom whose ligands appear in a counterclockwise sense in the order of the Cahn-Ingold-Prelog scheme when viewed from that side of the face.

G. P. Moss (1996) *Pure Appl. Chem.* **68**, 2193.

SIALIDASES (or Neuraminidases)

These enzymes [EC 3.2.1.18], better known as exo-α-sialidases and neuraminidases, catalyze the hydrolysis of 2,3-, 2,6-, and 2,8-glycosidic linkages joining terminal nonreducing *N*- or *O*-acylneuraminyl residues to galactose, *N*-acetylhexosamine, or *N*- or *O*-acylated neuraminyl residues in oligosaccharides, glycoproteins, glycolipids, or colominic acid. The enzymes do not act on 4-*O*-acetylated sialic acids. Endo-α-sialidase [EC 3.2.1.129] catalyzes the endohydrolysis of α-2,8-sialosyl linkages in oligo- or poly(sialic) acids. Although the name endo-*N*-acetylneuraminidase has also been used for this enzyme, this is misleading since its activity is not restricted to acetyl-substituted substrates. Anhydrosialidase [EC 3.2.1.138] catalyzes the hydrolysis of α-sialosyl linkages in *N*-acetylneuraminic acid glycosides, releasing 2,7-anhydro-α-*N*-acetylneuraminic acid.

E. Conzelmann & K. Sandhoff (1987) *Adv. Enzymol.* **60**, 89.
R. O. Brady (1983) *The Enzymes*, 3rd ed., **16**, 409.
H. M. Flowers & N. Sharon (1979) *Adv. Enzymol.* **48**, 29.
A. Gottschalk & A. S. Bhargava (1971) *The Enzymes*, 3rd ed., **5**, 321.

O-SIALOGLYCOPROTEASE

This enzyme [EC 3.4.24.57], also known as *O*-sialoglycoprotein endopeptidase and glycoprotease, catalyzes the hydrolysis of *O*-sialoglycoproteins; for example, it cleaves the Arg[31] — Asp[32] peptide bond in glycophorin A.

A. Mellors & R. Y. C. Lo (1995) *Meth. Enzymol.* **248**, 728.

SIALYLTRANSFERASES

This set of enzymes consists of those proteins that catalyze the transfer of sialyl moieties. (1) CMP-*N*-acetyl-neuraminate–β-galactoside α-2,6-sialyltransferase [EC 2.4.99.1] catalyzes the reaction of CMP-*N*-acetylneuraminate with β-D-galactosyl-1,4-acetyl-β-D-glucosamine to produce CMP and α-*N*-acetylneuraminyl-2,6-β-D-galactosyl-1,4-*N*-acetyl-β-D-glucosamine. (2) Monosialoganglioside sialyltransferase [EC 2.4.99.2] catalyzes the reaction of CMP-*N*-acetylneuraminate with D-galactosyl-*N*-acetyl-D-galactosaminyl-(*N*-acetylneuraminyl)-D-galactosyl-D-glucosylceramide to produce CMP and *N*-acetylneuraminyl-D-galactosyl-*N*-acetyl-D-galactosaminyl-(*N*-acetylneuraminyl)-D-galactosyl-D-glucosylceramide. This enzyme may be identical with β-galactoside α-2,3-sialyltransferase [EC 2.4.99.4]. (3) Lacto-sylceramide α-2,6-*N*-sialyltransferase [EC 2.4.99.11] catalyzes the reaction of CMP-*N*-acetylneuraminate with β-D-galactosyl-1,4-β-D-glucosylceramide to produce CMP and α-*N*-acetylneuraminyl-2,6-β-galactosyl-1,4-β-D-glucosylceramide.

Y. Kishimoto (1983) *The Enzymes*, 3rd ed., **16**, 357.
K. A. Presper & E. C. Heath (1983) *The Enzymes*, 3rd ed., **16**, 449.
R. J. Staneloni & L. F. Leloir (1982) *Crit. Rev. Biochem.* **12**, 289.
T. A. Beyer, J. E. Sadler, J. I. Rearick, J. C. Paulson & R. L. Hill (1981) *Adv. Enzymol.* **52**, 23.

SIDE REACTIONS

A reaction other than the physiologically important reaction catalyzed by an enzyme. There are numerous exam-

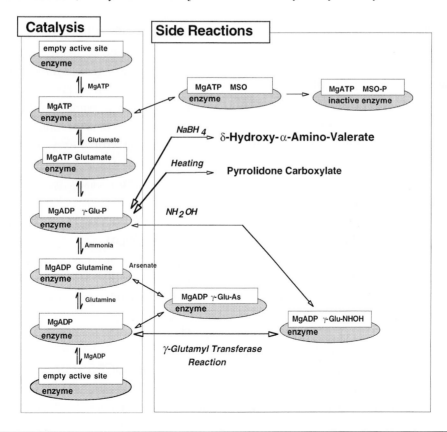

ples of side reactions in enzymic catalysis: (1) hexokinase displays an ATPase activity that is markedly enhanced when lyxose, a pentose that binds within the glucose binding site; (2) carbonic anhydrase can be an effective esterase, depending on reaction conditions; and (3) glutamine synthetase undergoes ATP-dependent inactivation by the metabolic inhibitor methionine sulfoximine.

Enzymes can also undergo other side reactions under conditions that divert a chemically reactive intermediate from its usual catalytic function. Again, glutamine synthetase is an excellent example (see figure above), because its side reactions include acyl-phosphate reduction by borohydride, pyroglutamate formation, and the formation of γ-glutamyl hydroxamate in the presence of hydroxylamine and arsenate.

SIGMA (σ)

1. The symbol for a symmetry designation of a molecular orbital. Such an orbital is without a nodal plane including the internuclear axis (as is the case with a π orbital). 2. The localized two-center bond associated with an *s* orbital. 3. The symbol for standard deviation. 4. The symbol for wavenumber. 5. The symbol for surface tension.

σ-ADDUCT (Sigma Adduct)

1. The product formed by the covalent attachment of some "entering" entity (an electrophile, nucleophile, or radical) to a ring carbon of a aromatic molecular entity, thus forming a new σ-bond and disrupting the original conjugation. 2. The product formed (as in definition 1) by the reaction of an "entering" entity to any unsaturated and conjugated system.

σ-CONSTANT (Sigma Constant)

1. The substituent constant for *meta*- and *para*-substituents in benzene derivatives, defined by Hammett on the basis of the K_a value of a substituted benzoic acid derivative in water at 25°C. 2. Referring to a set of related electronic substituent constants based on other standard reaction series (*e.g.*, σ^+, σ^-, σ°) or dissected on the basis of electronic effects (*e.g.*, σ_I, and σ_R). A large negative value for a sigma constant suggests that the substituent has a high electron-releasing power relative to hydrogen (as the substituent). Correspondingly, a large positive value for a sigma constant suggests a high electron-with-drawing power relative to hydrogen, either by inductive and/or resonance effects.

IUPAC (1994) *Pure and Appl. Chem.* **66**, 1077.

N. B. Chapman & J. Shorter (eds.) (1972) *Advances in Linear Free Energy Relationships*, Plenum, New York.

C. D. Johnson (1973) *The Hammett Equation*, Cambridge Univ. Press, Cambridge.

J. Shorter (1973) *Correlation Analysis in Organic Chemistry. An Introduction to Linear Free Energy Relationships*, Oxford Univ. Press, Oxford.

N. B. Chapman & J. Shorter (eds.) (1978) *Correlation Analysis in Chemistry: "Recent Advances"*, Plenum, New York.

σ → σ* TRANSITION

An electronic transition in which an electron is promoted from a bonding σ orbital to an antibonding σ* orbital. The transition energies are often close to those of Rydberg transitions. **See** *Absorption Spectroscopy*

Comm. on Photochem. (1988) *Pure and Appl. Chem.* **60**, 1055.

SIGMATROPIC REARRANGEMENT

A molecular rearrangement in which a σ-bond is formed between two atoms of the chemical species not previously linked. There may also be a relocation of π-bonds. Nevertheless, the total number of σ- and π-bonds will remain unchanged. Some authors restrict this term to intramolecular pericyclic reactions.

SIGNAL AMPLIFICATION

A parameter used to assist in the characterization of enzyme cascade systems. Symbolized by SA, it is equal to $[e_{0.5E}]/[e_{0.5I}]$ where $[e_{0.5E}]$ is the concentration of effector required for 0.5 activation of the converter enzyme E and $[e_{0.5I}]$ is the concentration of effector at which 50% of the interconvertible enzyme (I) has been modified[1]. **See** *Enzyme Cascade Kinetics*

[1]P. B. Chock & E. R. Stadtman (1980) *Meth. Enzymol.* **64**, 297.

SIGNAL PEPTIDASE, EUKARYOTIC MICROSOMAL

This proteolytic enzyme catalyzes the removal of N-terminal signal peptide extensions of secretory proteins and certain intracellular proteins.

M. O. Lively, A. L. Newsome & M. Nusier (1994) *Meth. Enzymol.* **244**, 301.

SIGNAL PROCESSOR

Any device that modifies the transduced signal from a detector in a manner that increases the signal strength (a process known as amplification) or converts an alternating current signal to a direct current output, or *vice versa*.

SIGNIFICANT FIGURES

The digits required to express the numerical results of a measurement to the precision with which that measurement is made. Suppose, for example, that the results of rate measurements, each made by two different methods (*e.g.*, absorbance and mass spectrometry in this case) gave values of 10.6 and 10.715 mM/min. Then the first measurement requires no fewer than three digits to convey the precision of the absorbance measurement, and the latter would not be fairly represented by any fewer than five in the case of the mass spectrometric determination. At the same time, any more than 3 or 5 digits for these respective cases would not be significant relative to the precision of the determinations. Note that these values could be equally well expressed as 1.06×10^{-2} and 1.0715×10^{-2} M/min, with the value to the left of the decimal point typically chosen to fall in the interval greater than one but less than 10. In this use of scientific notation, the exponent is utilized to locate the decimal place and does not reflect the number of significant figures. **See also** Precision

SINGLE-POOL MODEL FOR METABOLITE COMPARTMENTATION

A model based on the assumption that a metabolite is present within a single compartment with defined rate constants for absorption and elimination of the metabolite. The rate of appearance of a tracee and the infusion of tracer are assumed to take place in a single pool that is instantly well-mixed. Wolfe[1] has described in detail how the constant tracer infusion method allows one to calculate half-life, pool size, turnover time, mean residence time, and clearance time.

[1]R. R. Wolfe (1992) *Radioactive and Stable Isotope Tracers in Biomedicine*, Wiley-Liss, New York.

SINGLE-STEP REACTION

1. A reaction that cannot be reduced further, as in the case of *elementary reactions*. 2. Any reaction characterized by a single transition state.

SINGLET MOLECULAR OXYGEN

The excited singlet state of dioxygen (O_2). The term "singlet oxygen" should not be used since it is unclear whether the writer is referring to dioxygen or the excited singlet oxygen atom (in a ^1S or ^1D state).

SINGLET STATE

Any state which has a total electron spin quantum number of zero.

SINGLET-TRIPLET ENERGY TRANSFER

The transfer of excitation energy present in an atom or molecular species in an excited singlet state to another atom or molecular species, generating an electronically excited species in a triplet state.

SINGLE-TURNOVER CONDITIONS

Reaction conditions that only permit a catalyst to pass through a single round of catalysis. Single-turnover conditions are usually obtained by limiting the substrate concentration relative to the concentration of active catalyst. Occasionally, single-turnover conditions can also be achieved by limiting the period of reaction.

SITE-SPECIFIC ISOTOPE FRACTIONATION IN BIOCHEMICAL MECHANISMS

Zhang-Yunianta and Martin[1] recently discussed how studies of site-specific natural isotope fractionation by nuclear magnetic resonance can provide information and insight concerning chemical or biochemical mechanisms. The natural abundance of isotopic distribution in fermentation ethanol has been shown to reflect the physiological and environmental conditions of its carbohydrate precursors. Redistribution coefficients relating the natural site-specific isotope contents in end products to those of their precursors characterize the traceability of specific hydrogens in end products to their parent hydrogens in the starting materials. Average hydrogen isotope effects associated with the transfer of hydrogen from the water pool to the methyl or methylene site of product can also be estimated.

[1]B. L. Zhang-Yunianta & M. L. Martin (1995) *J. Biol. Chem.* **270**, 16023.

SI UNITS

The International System of Units (SI) built on seven base units: the unit of mass is the kilogram, the unit of time is the second, the unit of length is the meter, the unit of electric current is the ampere, the unit of temperature is kelvin, the unit of amount-of-substance is the mole, and the unit of luminous intensity is the candela.

SLOPE

1. The ratio of the rise to the run of any two points on a straight line or line segment. On a line drawn in a Cartesian coordinate system, it is also equal to the tangent of the angle that line makes with the horizontal axis (*i.e.*, the *x*-axis). 2. The first derivative of the equation of a curve at a given point on that curve.

SLOPE-INTERCEPT EQUATION

The equation of a line in the Cartesian coordinate system: $y = mx + b$ where x and y are the two coordinates of a point on the line, m is the slope of the line, and b is the vertical intercept.

SLOPE REPLOTS (or, Slope Secondary Replots)

Except for very simple systems, initial rate experiments of enzyme-catalyzed reactions are typically run in which the initial velocity is measured at a number of substrate concentrations while keeping all of the other components of the reaction mixture constant. The set of experiments is run again a number of times (typically, at least five) in which the concentration of one of those other components of the reaction mixture has been changed. When the initial rate data is plotted in a linear format (for example, in a double-reciprocal plot, $1/v$ *vs.* $1/[S]$), a series of lines are obtained, each associated with a different concentration of the other component (for example, another substrate in a multisubstrate reaction, one of the products, an inhibitor or other effector, *etc.*). The slopes of each of these lines are replotted as a function of the concentration of the other component (*e.g.*, slope *vs.* [other substrate] in a multisubstrate reaction; slope *vs.* $1/[inhibitor]$ in an inhibition study; *etc.*). Similar replots may be made with the vertical intercepts of the primary plots. The new slopes, vertical intercepts, and horizontal intercepts of these replots can provide estimates of the kinetic parameters for the system under study. In addition, linearity (or lack of) is a good check on whether the experimental protocols have valid steady-state conditions. Nonlinearity in replot data can often indicate cooperative events, slow binding steps, multiple binding, *etc.*

It is always good procedure to include the slope and intercept replots with the primary plots when reporting the results (perhaps best accomplished by having the replots as an inset of the primary graphs). They enable a reader to better evaluate the data.

SLOW-BINDING INHIBITOR

If there is a relatively long period of time before the interaction between a particular inhibitor and an enzyme has reached equilibrium (typically, on the order of 0.5–2 minutes, or longer), the inhibitor is considered to be a slow-binding inhibitor, and special procedures are used to characterize the kinetics[1-3]. Note that slow-binding inhibition does not necessarily imply tight-binding inhibition. An important characteristic of slow-binding inhibitors is that the progress curve will change depending on whether the reaction is initiated with addition of enzyme or with addition of substrate.

There are two basic kinetic schemes for slow-binding enzyme inhibition: Mechanism A involves the slow interaction of an inhibitor with enzyme and Mechanism B involves the rapid formation of an E·I complex that subsequently undergoes a slow isomerization to E·I′. The initial interaction of enzyme and inhibitor may not be sufficiently fast to allow free enzyme, E, and E·I and E·I′ to reach steady-state equilibrium. This would generate a more general mechanism. Sculley *et al.*[4] attempted to ascertain whether it might be possible to distinguish between the three possible inhibition mechanisms by steady-state kinetic techniques. By using three different ratios for the two rate constants that determine which mechanism applies, these investigators modeled the kinetic behavior by creating theoretical data sets for the most general mechanism. These reaction progress curve data were then "fitted" to the rate equations for the other two mechanisms. Their findings show how the values for the kinetic parameters, as determined from fits of the data to the equations for Mechanisms A and B, can be considerably in error.

[1] J. W. Williams & J. F. Morrison (1979) *Meth. Enzymol.* **63**, 437.
[2] J. F. Morrison & C. T. Walsh (1988) *Adv. Enzymol.* **59**, 201.
[3] S. E. Szedlacsek & R. G. Duggleby (1995) *Meth. Enzymol.* **249**, 144.
[4] M. J. Sculley, J. F. Morrison & W. W. Cleland (1996) *Biochim. Biophys. Acta* **1298**, 78.

SLOW TEMPERATURE-JUMP METHODS

Biochemical processes such as protein unfolding/refolding and supramolecular assembly/disassembly take place on a time scale of seconds to minutes after readjusting the temperature of a system. Most commercially available glass-jacketed cuvettes are not suitable for temperature jumps on this time scale, as a result of the slow kinetics of heat transfer across substances with characteristically high dielectric constants, and their use can convolute the time scale of the temperature change onto the time scale

In all cases, degassing of solutions by stirring *in vacuo* or by centrifugation (*i.e.*, pressurized degassing) is advised, because gas bubbles can accumulate on cuvette optical faces. Experience indicates that bubble clearance frequently is first-order and surprisingly slow!

[1] K. A. Johnson & G. G. Borisy (1977) *J. Mol. Biol.* **117**, 1.
[2] T. L. Karr, D. Kristofferson & D. L. Purich (1980) *J. Biol. Chem.* **255**, 8560 and 8567.
[3] F. M. Pohl (1968) *Eur. J. Biochem.* **4**, 373.
[4] F. M. Pohl (1977) in *Chemical Relaxation in Molecular Biology* (I. Pecht & R. Rigler, eds.) p. 289, Springer-Verlag, Berlin.

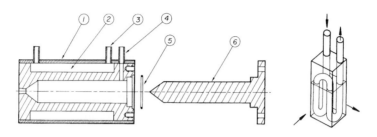

Rapid Heat Exchanger (left) and Heating/Cooling Cuvette (right)

Left: Diagram of a rapid heat exchanger. The overall block length is 10.2 cm with other dimensions drawn to scale. The heat exchanger block (1) was fabricated to accommodate a syringe with a Luer taper. Other features include the water jacket (2), water inlet and outlet ports (3), sample exit port (4), O-ring seal (5), and solid copper cold finger insert (6). This particular exchanger was built to minimize shear forces that might alter microtubule polymer length distributions, but one can also fabricate exchangers for dealing with smaller volumes to conserve sample. All copper parts are gold-plated to reduce chemical reactivity with biological molecules and buffer salts. *Right*: Temperature-regulated cuvette for examining temperature-dependent properties of biochemical reactions.

of the biochemical process *per se*. For example, early work on the kinetics of cold-induced microtubule depolymerization suggested that the process was essentially a zero-order process for over half of the disassembly process[1]. Nonetheless, later studies[2] utilizing a gold-plated, copper-block flow device for rapid heat exchange demonstrated that the depolymerization process was clearly not zero-order. **See** End-wise Depolymerization Pohl[3–4] has described a number of useful methods for achieving reasonably prompt changes in temperature. One approach (shown on the right in the above diagram) uses a fluorescence microcell fitted with an in-dwelling heating/cooling coil that does not obstruct the right-angle light path. The coil can be made of very thin-walled glass or, better still, of copper or aluminum with an appropriately inert, yet heat-conducting, coating. Pohl[3] also describes the modification of a commercial flow cell from Zeiss, Inc., to achieve fast heat exchange (half-time of 0.3 seconds) with a 0.05 mL sample volume. He also suggests that one can use a stopped-flow mixer to combine two solutions, each held at different temperatures, to achieve faster temperature changes.

SLOW, TIGHT-BINDING INHIBITOR

Inhibition of enzyme activity by a chemical species that binds slowly and is tight-binding as well (*i.e.*, has a low dissociation constant). Such inhibitors require special kinetic analysis[1,2]. The most common method of obtaining the inhibition parameters is by nonlinear regression analysis of the progress curves.

[1] J. W. Williams & J. F. Morrison (1979) *Meth. Enzymol.* **63**, 437.
[2] S. E. Szedlacsek & R. G. Duggleby (1995) *Meth. Enzymol.* **249**, 144.

SMOLUCHOWSKI TREATMENT FOR AGGREGATION

A probabilistic kinetic model describing the rapid coagulation or aggregation of small spheres that make contact with each other as a consequence of Brownian motion. Smoluchowski[1] recognized that the likelihood of a particle (radius = r_1) hitting another particle (radius = r_2; concentration = c_2) within a time interval (dt) equals the diffusional flux $(\partial c_2/\partial \rho)_{\rho=R_{12}}$ into a sphere of radius R_{12}, equal to $(r_1 + r_2)$. The effective diffusion coefficient D_{12} was taken to be the sum of the diffusion coefficients for the two particles, and the probability of a collision over the time interval may be stated as follows:

$$P dt = 4\pi R_{12}^2 D_{12}(\partial c_2/\partial \rho)_{\rho = R_{12}} dt = 4\pi R_{12}^2 D_{12} c_2 dt$$

This result implicitly requires: (a) that no strongly attractive/repulsive electrostatic interaction occurs as the particles approach each other, and (b) that a stationary diffusional concentration field surrounds the particle. [Note: Einstein's result[2] for spherical particle diffusivity (*i.e.*, $D = k_B T/6\pi\eta r$, where k_B is the Boltzmann constant, T is temperature in kelvin, η is the viscosity of the medium, and r is the radius) indicates that $R_{12}D$ will be approximately $4k_B T/6\pi\eta$.]

[1]M. von Smoluchowski (1918) *Z. Physik Chem.* **92**, 129.
[2]A. Einstein (1906) *Annalen der Physik* **19**, 371.

S$_N$ DISPLACEMENT REACTIONS

Substitution (type: nucleophilic) reactions in which an electron-rich agent (or nucleophile) attacks and forms a transient or stable bond with an electron-poor center (often carbon or phosphorus in biological systems) which has undergone prior or simultaneous expulsion of the leaving group (or exiphile).

The two limiting cases of nucleophilic displacement reactions are designated as S$_N$1 or S$_N$2 to indicate those mechanisms that respectively display overall first-order or second-order kinetics. These mechanisms are illustrated by the classical case of the reaction of hydroxide ion with chloromethane, and they differ with respect to the timing of the bond-breaking step relative to the bond-making step.

S$_N$1 REACTIONS. This mechanism can be represented as an initial rate-determining step, followed by the rapid reaction between the intermediate carbocation and the nucleophile (in this case, hydroxide ion):

$$CH_3-Cl \overset{slow}{\rightleftharpoons} CH_3^+ + Cl^-$$

In this first reaction, unless the process is occurring at very high temperatures, solvation of the ionic products is needed to provide the energy needed in the bond-breaking step. Because the carbocation has planar sp^2 hybridization, the second reaction can occur on either face. This second step may just as well be written in either of the following ways:

$$CH_3^+ + OH^- \overset{fast}{\longrightarrow} CH_3-OH$$

where the hydroxide ion attacks directly, or as the deprotonation of a hydrated carbocation whose positive charge is delocalized onto the oxygen atom of one solvent molecule:

$$CH_3^+ + HOH \rightarrow CH_3-OH_2^+ \overset{OH^-}{\longrightarrow} CH_3-OH + HOH$$

The reaction rate (v) for an S$_N$1 reaction can be written as:

$$v = k[CH_3-Cl]^1[OH^-]^0 = k[CH_3-Cl]$$

In the above expression, the quantity to the right of the first equal sign indicates that the order of the reaction with respect to the specific reactants is one and zero, respectively; and because the order with respect to hydroxide is zero, then $[OH^-]^0 = 1$, and the overall reaction order is the sum of the orders of the respective reactants. Hence, a S$_N$1 reaction is first-order overall.

S$_N$2 REACTIONS. This bimolecular reaction mechanism can be represented as a backside attack, in which the nucleophile approaches at an angle of 180° with respect to the leaving group (hence, the frequently used descriptor: "in-line" mechanism). This geometrical constraint results in an inversion of configuration about the central atom. (Because such inversions were first described in the 1890's by the carbohydrate chemist Walden, they are often called Walden inversions.) Any energy for bond scission is provided by the simultaneous making of an incipient bond between the nucleophile and the reaction center. As the reaction reaches its transition state, the central atom is pseudo-pentavalent, and the central atom has progressed from its sp^3 hybridization in the ground state to an sp^2 hybridization in the transition state. (The bond order remains four, because the central atom has three nonreacting substituents as well as another bond that is the sum of the entering and leaving groups.) The reaction occurs in one step, and no intermediate is formed:

$$HO^- \overset{\frown}{\ } CH_3 \overset{\frown}{Cl} \overset{slow}{\longrightarrow} CH_3-OH + Cl^-$$

where the curved arrows indicate the pairwise movement of electrons.

The reaction rate (v) for an S_N2 reaction can then be written as:

$$v = k[CH_3-Cl]^1[OH^-]^1$$

For this expression, the quantity to the right of the first equal sign indicates that the order of the reaction with respect to either reactant is one, and the overall reaction order is the sum of the orders of the respective reactants. Hence, a S_N2 reaction is second-order overall. In certain cases, such as solvolysis, the rate expression takes on pseudo-first-order kinetic behavior.

Technically, for solution-phase nucleophilic substitution reactions, S_N1 and S_N2 mechanisms are limiting cases on a continuum of possible interactions:

$$
\begin{array}{ccccccc}
R-X & \rightleftharpoons & R^+X^- & \rightleftharpoons & R^+\|X^- & \rightleftharpoons & R^+ \text{ and } X^- \\
\downarrow {}^-Nu: & & \downarrow {}^-Nu: & & \downarrow {}^-Nu: & & \downarrow {}^-Nu: \\
Nu-R-X & & {}^{\delta-}Nu-R^+-X^{\delta-} & & {}^-Nu-R^+\|X^- & & Nu-R+X^- \\
\downarrow & & \downarrow & & \downarrow & & \downarrow \\
NuR+X^- & & NuR+X^- & & NuR+X^- & & RNu \text{ or } NuR+X^-
\end{array}
$$

where $R-X$ is the covalently bonded form, R^+X^- is the tight (or intimate ion) ion pair, $R^+\|X^-$ is the loosely associated or solvent-separated ion pair, R^+ and X^- represent the fully dissociated ions, ${}^-Nu:$ is the nucleophile, X^- is the counterion (or, gegenion), and RNu and NuR are the two optical isomers formed by attack on either face of the electrophilic center. **See** Solvolysis

SODIUM–22 (^{22}Na)

A radioactive sodium nuclide that has three principal radiation emissions β^+ (0.546 MeV), γ (1.275 MeV), and the typical annihilation peak at 0.512 MeV. The maximum range in air is 1.4 meters. The decay half-life is 2.605 years. The following decay data indicate the time (in days) followed by the fraction of the original amount at the specified time: 0.0, 1.000; 30, 0.978; 60, 0.957; 90, 0.936; 120, 0.916; 150, 0.896; 180, 0.877; 210, 0.858; 240, 0.839; 270, 0.821; 300, 0.804; 600, 0.646; 900, 0.519; 1200, 0.417; 1500, 0.335; 1800, 0.269; 2100, 0.216; 2400, 0.174; 2700, 0.140; 3000, 0.112.

Special precautions: Store behind lead shielding. Follow institutional regulations regarding type and location of dosimeter. Avoid working immediately above open containers. Because no organ selectively concentrates sodium metabolites, the entire body is considered to be the critical organ in terms of dose. The effective *biological* half-life in humans is around 11 days.

SOLUBILITY PRODUCT

The product of the concentration of the ions in a saturated solution of a chemical species, typically symbolized by K_{sp}. This expression is valid for sparingly soluble salts.

If the product of the concentrations of the ionic species exceeds the K_{sp} value then either precipitation occurs or a metastable condition of supersaturation develops. When reporting solubility products, the temperature should always be provided. Examples of a few relevant K_{sp} values (at 25°C) are given below.

Barium carbonate ($BaCO_3$)	2.58×10^{-9}
Barium hydroxide octahydrate ($Ba(OH)_2 \cdot 8H_2O$)	2.55×10^{-4}
Calcium carbonate ($CaCO_3$)	4.96×10^{-9}
Calcium hydroxide ($Ca(OH)_2$)	4.68×10^{-6}
Calcium oxalate monohydrate ($CaC_2O_4 \cdot H_2O$)	2.34×10^{-9}
Calcium phosphate ($Ca_3(PO_4)_2$)	2.07×10^{-33}
Iron(II) carbonate ($FeCO_3$)	3.07×10^{-11}
Iron(II) hydroxide ($Fe(OH)_2$)	4.87×10^{-17}
Iron(III) hydroxide ($Fe(OH)_3$)	2.64×10^{-39}
Magnesium carbonate ($MgCO_3$)	6.82×10^{-6}
Magnesium hydroxide ($Mg(OH)_2$)	5.61×10^{-12}
Magnesium oxalate dihydrate ($MgC_2O_4 \cdot 2H_2O$)	4.83×10^{-6}
Magnesium phosphate ($Mg_3(PO_4)_2$)	9.86×10^{-25}
Manganese(II) carbonate ($MnCO_3$)	2.24×10^{-11}
Manganese(II) hydroxide ($Mn(OH)_2$)	2.06×10^{-13}
Manganese(II) oxalate dihydrate ($MnC_2O_4 \cdot 2H_2O$)	1.70×10^{-7}
Zinc carbonate ($ZnCO_3$)	1.19×10^{-10}
Zinc carbonate monohydrate ($ZnCO_3 \cdot H_2O$)	5.41×10^{-11}

SOLUTION

A homogeneous mixture of a liquid, known as the solvent, with one or more solids or gases (the solute) dissolved in that liquid. Solutions can also be formed by two or more miscible liquids. In such cases, the solvent is that liquid with the largest mole fraction.

SOLVATION

A stabilizing interaction between a solute and the solvent, or between groups on an insoluble substance and the surrounding solvent. **See** *Hydration Atmosphere*

SOLVENT-ACCESSIBLE SURFACE AREA

The surface[1] traced out by the center of a solvent probe molecule as it is rolled over the surface of a protein whose three-dimensional structure has been determined at the atomic level. Solvent-accessible surface areas can be calculated by various computer algorithms[2], and differences in solvent-accessible areas can be used to characterize the energetics of surface hydration as a function of changes in protein conformation, oligomerization, and complexation.

[1]B. Lee & F. M. Richards (1971) *J. Mol. Biol.* **55**, 379.
[2]B. M. Baker & K. P. Murphy (1998) *Meth. Enzymol.* **295**, 294.

SOLVENT ISOTOPE EFFECT

A change in chemical reactivity occurring as a consequence of changes in the isotopic composition of the solvent. **See** *Proton Inventory; Kinetic Isotope Effects*

SOLVENT TRAPPING OF ENZYME INTERMEDIATES

An X-ray crystallographic technique[1,2] resulting in the accumulation of transient enzyme-bound intermediates (in reactions using water as a substrate) by appropriately manipulating polarity through the addition of organic solvents. After enzyme crystals are grown normally to an acceptable size, they are placed in a capillary that permits flow of solvent. One can use protein cross-linking agents to stabilize a crystalline enzyme prior to solvent transfer[1]. The crystals are dried by capillarity using filter paper wicks, and organic solvent is then introduced. (Included with the organic solvent are activated molecular sieves, and these zeolites contain pores that trap and remove all water from the sample.) This is followed by introduction of substrate(s) to allow enzyme-bound intermediates to accumulate. Under favorable conditions, crystals containing such solvent-trapped intermediates are suitable for structure determination, and the electron density map so obtained is a weighted average favoring the most abundant enzyme form.

[1]P. A. Fitzpatrick, A. C. Steinmetz, D. Ringe & A. M. Klibanov (1993) *Proc. Natl. Acad. Sci. U.S.A.* **90**, 8653.
[2]N. H. Yennawar, H. P. Yennawar & G. K. Farber (1994) *Biochemistry* **33**, 7326.

SOLVOLYSIS

Whenever a nucleophilic substitution reaction uses the solvent as the nucleophile, the reaction can be referred to as a solvolysis reaction (or simply as solvolysis). Examples of solvolysis include hydrolysis (reaction with water), ethanolysis (reaction with ethanol), acetolysis (reaction with acetic acid), and formolysis (reaction with formic acid). All solvolysis reactions exhibit first-order kinetics, because the concentration of the solvent does not change to any appreciable extent during the solvolytic reaction. This fact is often misinterpreted as an indication that the reaction type is S_N1. To be sure, polar solvents such as water, alcohols, and amines should be regarded as good nucleophilic agents. A traditional method for distinguishing S_N1 and S_N2 reactions in physical organic chemistry is to add a low concentration of a substance that is a better nucleophile than the solvent. If the rate of substitution remains unchanged, then one can infer that the reaction is of the S_N1 type; however, should the nucleophilic substitution rate increase, an S_N2 type reaction is likely. Of course, the most direct means for distinguishing S_N1 and S_N2 reactions is by stereochemistry, when practicable.

SOMO

Abbreviation for *s*ingly *o*ccupied *m*olecular *o*rbital, a term that applies to, among other cases, the unpaired electrons of radicals.

SONICATION

The irradiation of a system with sound waves (usually ultrasound). Often used to disrupt cell membranes and in early steps in protein purification, it should also be noted that sonication can increase rates of reaction as well as assist in the preparation of vesicles.

SORBITOL DEHYDROGENASE

This enzyme [EC 1.1.1.14], also known as L-iditol 2-dehydrogenase and polyol dehydrogenase, catalyzes the reaction of L-iditol with NAD^+ to produce L-sorbose and NADH. The enzyme will also convert D-glucitol to D-fructose.

J. Jeffery & H. Jornvall (1988) *Adv. Enzymol.* **61**, 47.

SPECIFIC ACTIVITY

1. The intrinsic variable expressed usually as international units (U) of enzyme activity per milligram protein

(or nucleic acid in the case of ribozymes). One U corresponds to the conversion of 1 micromole substrate into product per minute. A katal corresponds to the conversion of one mole substrate per second. Hence, 1 U corresponds to 16.67 nkat. As an experimental parameter, specific activity is especially useful during enzyme purification when one must develop strategies for maximizing specific activity while ensuring that adequate yield is maintained. Likewise, one may use specific activity as a measure of purity of any protein or other biomolecule that possesses some assayable response (such as drug binding or metabolite transport).

2. The intrinsic variable expressed as units of radioactivity (in becquerels or, more traditionally, curies) per mole of a substance. One Bq corresponds to 1 disintegration per second (dps) and one Ci to 3.70×10^{10} Bq. This parameter is especially useful in quantifying the amount of substance in biological samples. For example, if SA_s is the standard specific radioactivity (say, x dps/y mol) of a standard, and if SA_e is the experimental specific activity (say, x dps/$(y + z)$ mol), then the content z in a sample can be determined from the expression $(SA_s/SA_e) = (y + z)/y$ or $z = y ([SA_s/SA_e] - 1)$. This intrinsic variable can also be expressed as the gram-atom excess of a stable isotope per mole of a substance. The numerator is typically determined using a ratio mass spectrometer, and the denominator can be estimated by chemical and/or spectroscopic techniques.

SPECIFIC CATALYSIS

The acceleration of a reaction by a unique catalyst rather than a family of related substances or materials. The term is commonly used with respect to H^+ or OH^- catalysis (thus, specific acid and specific base catalysis). **See** General Acid Catalysis; General Base Catalysis

SPECIFICITY

A measure and/or description of how specific an enzyme is toward a substrate or class of substrates or toward an effector or class of effectors. For effectors (or for ligands binding to macromolecules that are not enzymes), this specificity is readily measured by dissociation (or, association) constants. For enzymes, specificity is best quantitated by the V_{max}/K_m ratio. **See** Specificity Constant. It is crucial, in the complete characterization of an enzyme, that the specificity of the enzyme be known in detail.

This includes stereospecificity and anomeric specificity when appropriate. It should also be recalled that biomolecules often exist in different stable conformations and in different states of ionization and/or metal chelation. For example, for ATP-dependent enzymes, it may be necessary to know the relative specificity of $MgATP^{2-}$ and $MgHATP^-$ or with $MnATP^{2-}$ and $MnHATP^-$. If accurate values are known for the V_{max}/K_m ratios of a large set of substrates and alternative substrates, it may be possible to determine the topology of the enzyme's active site and to design specific active site inhibitors. **See** Mapping Substrate Interactions Using Kinetic Data

SPECIFICITY CONSTANT

The ratio of maximum velocity, V_{max}, or k_{cat} value to the true K_m value for a particular substrate, with units of a bimolecular rate constant $M^{-1}s^{-1}$. In comparing substances that act as substrates for a single enzyme, a higher V_{max}/K_m or k_{cat}/K_m ratio reflects a higher apparent rate of enzyme-substrate complexation.

SPIN LABEL

A stable paramagnetic group that has been attached to a molecular entity. Experiments with ESR measurements can reveal aspects of the microenvironment of the spin label. If the paramagnetic group is not covalently attached to the molecular entity of interest, the term "spin probe" is usually applied. **See also** Electron Spin Resonance

SPONTANEOUS PROCESS

Any reaction or change that is energetically poised to progress without the intervention of any external force or treatment. Although a spontaneous chemical process may be thermodynamically favorable, kinetic constraints may prevent the system from progressing to its equilibrium position. If the energy of activation is appreciable, then attainment of the transition state is less likely, and an otherwise spontaneous process may fail to progress. The descriptor "spontaneous" should not be confused with "instantaneous"; the latter describes a process that is immediate and occurs at the present, without any perceptible delay. In fact, the time-evolution or progress of many apparently instantaneous processes can be determined by using ultrafast methods for detecting and/or quantifying a change in the system.

SQUALENE SYNTHASE

This enzyme [EC 2.5.1.21], officially known as farnesyl-diphosphate farnesyltransferase (and also referred to as farnesyltransferase and presqualene-diphosphate synthase), catalyzes the conversion of two molecules of farnesyl diphosphate to yield presqualene diphosphate and pyrophosphate (or, diphosphate). In its polymeric form, the enzyme then catalyzes the NADPH-dependent reduction of presqualene diphosphate to form squalene. *See also* Farnesyl Diphosphate Farnesyltransferase

J. Chappel (1995) *Ann. Rev. Plant Physiol. Plant Mol. Biol.* **46**, 521.
R. Kluger (1992) *The Enzymes*, 3rd ed., **20**, 271.
G. J. Schroepfer, Jr. (1982) *Ann. Rev. Biochem.* **51**, 555.
G. J. Schroepfer, Jr. (1981) *Ann. Rev. Biochem.* **50**, 585.
E. D. Beytía & J. W. Porter (1976) *Ann. Rev. Biochem.* **45**, 113.

STABILITY CONSTANT

An equilibrium association or formation constant (units expressed as molarity^{-1} or M^{-1}) for ligand (L) binding to a metal (Me) ion:

$$Me^n + L^m = MeL^{n+m}; \text{ where } K = [MeL^{n+m}]/[Me^n][L^m]$$

where m and n are integers. *See* Metal Ion Complexation

STABILITY CONSTANT DETERMINATION BY A LINEAR PLOT METHOD

Li and Wang[1] described a convenient linear plotting protocol that permits one to evaluate stability constants for the formation of reversible metal·ligand complexes. Added accuracy is achieved over classical approaches, because their generalized equations quantitatively account for side-reactions of metal ion and/or ligand as well as the difference between the total and free ligand concentration.

[1]H. Li & J. Wang (1989) *Analyt. Chim. Acta.* **226**, 323.

STABILITY, PRODUCT

A product of an enzyme-catalyzed reaction should be stable enough over the time course of the experiment, including isolation, identification, and measurement of that product. If not, it may be necessary to alter the reaction conditions to increase the stability of a product. A number of enzymes have been studied at lower temperatures for this reason. The comments made in the substrate stability entry are also pertinent here. *See* Substrate Stability

Identification of an alternative substrate may likewise allow one to reinvestigate a particular protein, especially if the resulting product is more stable than the earlier studied product.

STANDARD CURVE

Any calibrated curve showing the dependent variable (frequently ΔAbsorbance) on the vertical axis and the independent variable (frequently concentration or time) on the horizontal axis. Error bars should be included to indicate the precision of experimental determinations. *See* Beer-Lambert Law; Absorption Spectroscopy

Selected entries from *Methods in Enzymology* [vol, page(s)]:
Generation, **240**, 122-123; confidence limits, **240**, 129-130; discrete variance profile, **240**, 124-126, 128-129, 131-133, 146, 149; error response, **240**, 125-126, 149-150; Monte Carlo validation, **240**, 139, 141, 146, 148-149; parameter estimation, **240**, 126-129; radioimmunoassay, **240**, 122-123, 125-127, 131-139; standard errors of mean, **240**, 135; unknown sample evaluation, **240**, 130-131; zero concentration response, **240**, 138, 150.

STANDARD DEVIATION

1. A measure, symbolized by σ, of the dispersion of a distribution and equal to the square root of the variance. Hence, it is used to describe the distribution about a mean value. For a normal distribution curve centered on some mean value, μ, multiples of the standard deviation provides information on what percentage of the values lie within $\pm n\sigma$ of that mean. Thus, 68.3% of the values lie within $\pm$ one standard deviation of the mean, 95.5% within $\pm 2\sigma$, and 99.7% within $\pm 3\sigma$. 2. The corresponding statistic, s, used to estimate the true standard deviation: $\sigma^2 = (\Sigma(x_i - x)^2)/(n - 1)$. *See* Statistics (A Primer)

STAPHYLOCOCCAL NUCLEASE

This hydrolase, officially known as micrococcal nuclease [EC 3.1.31.1 (formerly 3.1.4.7)], catalyzes endonucleolytic cleavage of single-/double-stranded DNA (as well as RNA) to form 3'-phospho-mononucleotide and 3'-phospho-oligonucleotide end-products. The enzyme also displays exonucleolytic activity.

J. A. Gerlt (1992) *The Enzymes*, 3rd ed., **20**, 95.
A. S. Mildvan & D. C. Fry (1987) *Adv. Enzymol.* **59**, 241.
F. A. Cotton & E. E. Hazen, Jr. (1971) *The Enzymes*, 3rd ed., **4**, 153.
C. B. Anfinsen, P. Cuatrecasas & H. Taniuchi (1971) *The Enzymes*, 3rd ed., **4**, 177.

STARCH PHOSPHORYLASE

This enzyme [EC 2.4.1.1] catalyzes the reaction of starch (that is, [(1,4)-α-D-glucosyl]$_n$) with orthophosphate to produce $[(1,4)\text{-}\alpha\text{-D-glucosyl}]_{(n-1)}$ and α-D-glucose 1-phosphate. **See** *Glycogen Phosphorylase*

STARCH SYNTHASE

Starch synthase [EC 2.4.1.11] catalyzes the reaction of UDP-glucose with starch (that is, $[(1,4)\text{-}\alpha\text{-D-glucosyl}]_n$) to generate UDP and $[(1,4)\text{-}\alpha\text{-D-glucosyl}]_{(n+1)}$. Starch (bacterial glycogen) synthase [EC 2.4.1.21], also referred to as glycogen synthase and ADP-glucose-starch glucosyltransferase, catalyzes the reaction of ADP-glucose and $[(1,4)\text{-}\alpha\text{-D-glucosyl}]_n$ to produce ADP and $[(1,4)\text{-}\alpha\text{-D-glucosyl}]_{(n+1)}$. This enzyme is similar to glycogen synthase and starch synthase except that it utilizes ADP-glucose. The recommended name various according to the source of the enzyme and the nature of its synthetic product; for example, starch synthase, bacterial glycogen synthase, *etc.*

A. M. Smith, K. Denyer & C. Martin (1997) *Ann. Rev. Plant Physiol. Plant Mol. Biol.* **48**, 67.

O. Nelson & D. Pan (1995) *Ann. Rev. Plant Physiol. Plant Mol. Biol.* **46**, 475.

STATE FUNCTION

A function (such as energy) that is dependent only on the state of the system and not on the pathway as to how that state was reached. Examples of state functions include internal energy, enthalpy, entropy, Gibbs free energy, Helmholtz free energy, pressure, volume, temperature, *etc.* Work and heat are not state functions.

STATIONARY STATE

1. A synonym for steady state. 2. In quantum mechanics, a state which does not change or evolve with time. **See** *Steady State*

STATISTICAL ANALYSIS OF ENZYME RATE DATA

The proper evaluation and assessment of the calculated or graphically determined values of the kinetic parameters requires the application of statistical analysis[1-5]. This is also true when looking for possible patterns in the various plots (*e.g.*, parallel lines *vs.* intersecting lines). When reporting kinetic values, the error limits should always be provided. Programs are available that statistically evaluates kinetic data[3]. **See** *Statistics (A Primer)*

[1]G. N. Wilkinson (1961) *Biochem. J.* **80**, 324.
[2]W. W. Cleland (1967) *Adv. Enzymol.* **29**, 1.
[3]W. W. Cleland (1979) *Meth. Enzymol.* **63**, 103.
[4]B. Mannervik (1982) *Meth. Enzymol.* **87**, 370.
[5]R. G. Duggleby (1995) *Meth. Enzymol.* **249**, 61.

STATISTICAL FACTORS

Terms appearing in binding equations describing the linked equilibria for the binding of a ligand (or proton)

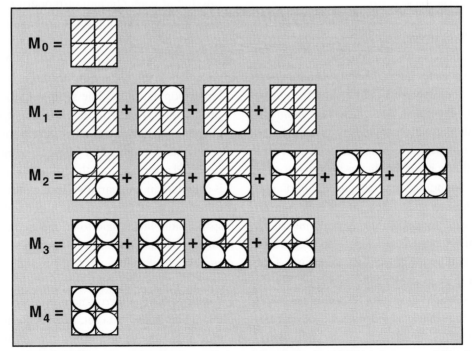

Figure 1. A schematic diagram indicating the multiple path by which a ligand (○) binds to an available site (▨).

to polyvalent receptors. Statistical factors permit one to account for the multiple microscopically distinct species formed in the course of filling multiple sites on a polyvalent acid, a multisite protein, or any lattice of sites available for ligand binding. As discussed in detail by Edsall and Wyman[1] and Cantor and Schimmel[2], these terms are required to account for the various ways for adding a ligand from each state of ligand occupancy. Failure to include statistical factors, will result in an inability to obtain the polynomial function known as the ligand saturation function.

Taking ligand binding to a tetramer P_4 as an example (see Fig. 1), there are four ways that such a tetramer can bind the first molecule of ligand L: one for each of the four equivalent sites. In the general equation for binding of a ligand to a protein containing n sites that are equivalent (and therefore independent), the intrinsic microscopic dissociation constant k is the same for each ligand-receptor interaction. For the second ligand molecule, there are six different ways to convert a singly occupied tetramer P_4L_1 to generate a doubly occupied tetramer P_4L_2. Generally, to bind i ligands on n-sites, there are $\Omega_{n,i} = n!/\{(n - i)!i!\}$ ways. Thus, for i equal to 1, $\Omega_{4,1} = 4$, and for i equal to 1, $\Omega_{4,2} = 6$.

[1]J. T. Edsall & J. Wyman (1958) *Biophysical Chemistry*, pp. 623-626, Academic Press, New York.
[2]C. R. Cantor & P. R. Schimmel (1980) *Biophysical Chemistry*, Part III, pp. 852-853, Freeman, San Francisco.

STATISTICS (A Primer)

The scientific method has several elemental stages: study, hypothesis, and experimentation. We rely on knowledge and insight to advance new conjectures about the nature of a process being examined, and one cannot minimize the importance of a good command of the associated literature. (Library research is time well spent.) A well considered hypothesis is the predicate for effective experimentation. A sound understanding of the particular experimental technique and all relevant theory is indispensable. Successful investigators almost invariably adopt research plans that allow a larger problem to be broken down into a series of easily tested parts. One should also heed the engineer's exhortation—"Keep it simple, stupid!"—when designing experiments. This adage also reminds us that straightforward experimental design often minimizes the chance introduction of unforeseen variables.

We usually seek to distinguish between two possibilities: (a) the *null hypothesis*—a conjecture that the observed set of results arises simply from the random effects of uncontrolled variables; and (b) the *alternative hypothesis* (or *research hypothesis*)—a trial idea about how certain factors determine the outcome of an experiment. We often begin by considering theoretical arguments that can help us decide how two rival models yield nonisomorphic (*i.e.*, characteristically different) features that may be observable under a certain set of imposed experimental conditions. In the latter case, the null hypothesis is that the observed differences are again haphazard outcomes of random behavior, and the alternative hypothesis is that the nonisomorphic feature(s) is (are) useful in discriminating between the two models.

The logic and rigor of statistics also reside at the heart of modern science. Statistical evaluation is the first step in considering whether a body of measurements allows one to discriminate among rival models for a biochemical process. Beyond their value in experimental sciences, probabilistic considerations also help us to formulate theories about the behavior of molecules and particles and to conceptualize stochastic and chaotic events.

We cannot possibly provide a detailed account of the full range of statistical methods and strategies that constitute the armamentarium of modern science. Yet, there is value in considering some basic statistical principles.

RANDOM SAMPLING AND RANDOM VARIABLES. A sample n is a representative subset of the larger population N. The sampling space, designated S, that is associated with an experiment is the set whose elements correspond to all possible sampling points. Experiments rarely yield the identical numerical value upon repeated sampling, and the value of the numerical event is called a random variable. In fact, the random variable is a real-valued function that is defined over the sample space. Sampling is said to be random if each of possible samples n_i, taken from N has an equal probability of occurring.

A second kind or error, namely systematic error, introduces bias in the data set, because systematic errors are most often the same for each observation made with a given apparatus or measuring device. This will tend to displace the frequency distribution curve describing a random variable in one direction. Moreover, certain threshold phenomenon, such a quenching in liquid scintillation counting, may favor detection of higher energy β particles. Behavior of this sort may also have a nonrandom distribution of data about a population mean. Other problems arise from personal (or operator) error, and the occurrence of mistakes (honest errors) is unavoidable in most circumstances. Individual bias can also wreak havoc on experiments. While the effects of this type of error can be mitigated somewhat by double-blinded strategies, within the molecular life sciences, the detailed plan for most experiments is almost always known by the researcher during the course of the experiment. Two

related matters deserve consideration: first, a contemporaneous laboratory notebook record can prove invaluable as one justifiably discounts data points known or suspected to be erroneous; second, keeping short the time elapsed between data acquisition and analysis is an imperative. Except for Sherlock Holmes, most sleuths abhor a cold stiff body.

Randomization is a strategy that helps the experimenter to obtain a statistically unbiased sample or data set for a series of experimental measurements by simulating a chance distribution or chance sequence. Randomization often eliminates or minimizes interference by otherwise irrelevant variables. There are, in fact, many instances

Table I

A collection of "random" numbers, based on the first 1392 integers found in the numerical value of π

31	41	59	26	53	58	97	93	23	84	62	64	33	83	27	95	02	88	41	97	16	93	99	75	10	58	20	97	49
44	59	23	07	81	64	06	28	62	08	99	86	28	03	48	25	34	21	17	06	79	86	20	89	98	62	80	34	82
53	42	11	70	67	98	21	48	08	65	13	28	23	06	64	70	93	84	46	09	55	05	82	23	17	25	35	94	08
12	84	81	11	74	50	28	41	02	70	19	38	52	11	05	55	96	44	62	29	48	95	49	30	38	19	64	42	88
10	97	56	65	93	34	46	12	84	75	64	82	33	78	67	83	16	52	71	20	19	09	14	56	48	56	69	23	46
03	48	61	04	54	32	66	48	21	33	93	60	72	60	24	91	41	27	37	24	58	70	06	60	63	15	58	81	74
88	15	20	92	09	62	82	92	54	09	17	15	36	43	67	89	25	90	36	00	11	33	05	30	54	88	20	46	65
21	38	41	46	95	19	41	51	16	09	43	30	57	27	03	65	75	95	91	95	30	92	18	61	17	38	19	32	61
17	93	10	51	18	54	80	74	46	23	79	96	27	49	56	73	51	88	57	52	72	48	91	22	79	38	18	30	11
94	91	29	83	36	73	36	24	40	65	66	43	08	60	21	39	49	46	39	52	24	73	71	90	70	21	79	86	09
43	70	27	70	53	92	17	17	62	93	17	67	52	38	46	74	81	84	67	66	94	05	13	20	00	56	81	27	14
52	63	56	08	27	78	57	71	34	27	57	78	96	09	17	36	37	17	87	21	46	84	40	90	12	24	95	34	30
14	65	49	58	53	71	05	07	92	27	96	89	25	89	23	54	20	19	95	61	12	12	90	21	96	08	64	03	44
18	15	98	13	62	97	74	77	13	09	96	05	18	70	72	11	34	99	99	99	83	72	97	80	49	95	10	59	73
17	32	81	60	96	31	85	95	02	44	59	45	53	46	90	83	02	64	25	22	30	82	53	34	46	85	03	52	61
93	11	88	17	10	10	00	31	37	83	87	52	88	65	87	53	32	08	38	14	20	61	71	77	66	91	47	30	35
98	25	34	90	42	87	55	46	87	31	15	95	62	86	38	82	35	37	87	59	37	51	95	77	81	85	77	80	53
21	71	22	68	06	61	30	01	92	78	76	61	11	95	90	92	16	42	01	98	93	80	95	25	72	01	06	54	85
86	32	78	86	59	36	15	33	81	82	79	68	23	03	01	95	20	35	30	18	52	96	89	95	77	36	22	59	94
13	89	12	49	72	17	75	28	34	79	13	15	15	57	48	57	24	24	54	15	06	95	95	08	29	53	31	16	86
17	27	85	58	89	07	50	98	38	17	54	63	74	64	93	93	19	25	50	60	40	09	27	70	16	71	13	90	09
84	88	24	01	28	58	36	16	03	56	37	07	66	01	04	71	01	81	94	29	55	59	61	98	94	67	67	83	74
49	44	82	55	37	97	74	72	68	47	10	40	47	53	46	46	20	80	46	68	42	59	06	94	91	29	33	13	67
70	28	98	91	52	10	47	52	16	20	56	96	02	40	58	03	81	50	19	35	11	25	33	82	43	00	35	58	74

when randomization can prevent the introduction of human bias in the choice of sampling schemes. The mere anticipation of a certain result has been known to influence the outcome of a measurement, particularly when a measurement is of a qualitative nature. With enzyme rate assays, a decision to order all of the measurements from low to high substrate/effector concentration, or vice versa, may lead to an incorrect data set if the stock solution of enzyme or substrate undergoes some time-dependent change that influences chemical reactivity. In fact, time is a variable that is difficult to hold constant, because one can rarely carry out all measurements simultaneously. Likewise, various measuring instruments can introduce bias when the data collected are uncorrected for time-dependent drift in the signal strength, line voltage, and thermal noise.

Perhaps the simplest algorithm for obtaining a randomized pairing of variables is the toss of a coin. However, to randomize the order of executing a larger series of measurements, one typically employs a random number generator or a random number table to assign random numbers to each of the measurements.

With a random number table (such as the one shown above), one can arrange a small number of experimental conditions into a random order. If one were to have ten samples, initially arranged from lowest to highest substrate concentration, then the experimenter can use the table shown above to randomize these sample with respect to the order of performing the measurements. To assign a random number to the first sample, an eraser on the back of a pencil can be allowed to contact some number in the table, while the experimenter keeps his/her eyes closed. By proceeding an upward or downward (or leftward or rightward) direction according to some predetermined sequence, say every other number in the table, each of the subsequent samples can be assigned a number from the table. The measurements are then executed in the order that corresponds to the increasing (or decreasing) magnitude of the numbers chosen from the random number table. Should two identical numbers be assigned from the table, a coin flip can readily determine the order (e.g., "heads" before "tails") of execution for these samples.

GENERAL DEFINITIONS. Effective use of statistical principles requires a clear understanding of certain fundamental terms.

The mean $\bar{x}$ (also known as the arithmetic mean or average) is given as:

$$\bar{x} = \frac{\sum\limits_{i=1}^{n} X_i}{n}$$

The variance V is estimated by the mean value of the square of an observation's deviation from the arithmetic mean:

$$V = \frac{(X_1 - \bar{x})^2 + (X_2 - \bar{x})^2 + \cdots + (X_n - \bar{x})^2}{n-1}$$

$$V = \frac{1}{n-1} \sum\limits_{i=1} n_i (X_1 - n)^2$$

The standard deviation equals the square root of the variance V. This parameter, symbolized variously as S, s, or σ, is easier to visualize than the variance. Statisticians prefer to analyze a system in terms of variance. However, chemists and biochemists favor standard deviation which allows one to present the value for a particular set of measurements and its standard deviation, all in the same units.

The coefficient of variance (CV) is 100 times the ratio of the standard deviation divided to the arithmetic mean.

The frequency f_i equals n_i/n, where n_i is the number of things having the same value or property X_i divided by the total population of things. It follows that the mean can also be calculated from knowledge of the frequency, such that

$$\bar{x} = \sum\limits_{i=1}^{n} (f_i X_i)$$

FREQUENCY DISTRIBUTIONS. All of the members of some class of things is referred to as a population, and the properties of such a population are inferred from the behavior of a smaller group of things called a sample. Any single property of an individual within a population may differ from the same property for another individual, and we usually measure a sample mean to obtain a mean for the entire population. One often wishes to observe how a given property is distributed for a population of individuals, and the frequency distribution of a variable is estimated on the basis of a frequency distribution within the sample.

1. Binomial (or Bernoulli) Distribution. This distribution applies when we are concerned with the number of times an event A occurs in n independent trials of an experiment, subject to two mutually exclusive outcomes A or B. (Note: The descriptor "independent" indicates that the outcome of one trial has no effect on the outcome of any other trial.) In each trial, we assume that outcome A has a probability $P(A) = p$, such that q, the probability of outcome A not occurring, equals $(1 - q)$. Assuming that the experiment is carried out n times, we can consider the random variable X as the number of times that outcome A takes place. X takes on values $1, 2, 3, \cdots, n$. Considering the event $X = x$ (meaning that A occurs in x of the n performances of the experiment), all of the outcomes A occur x times, whereas all the outcomes B occur $(n - x)$ times. The probability $P(X = x)$ of the event $X = x$ can be written as:

$$f(x) = \binom{n}{x} p^x q^{n-x}$$

where $x = 1, 2, 3, \cdots, n$. The frequency distribution determined by this probability function is the Binomial distribution.

2. Poisson Distribution. This distribution is a limiting case of the binomial distribution and is described by the probability function:

$$f(x) = \frac{\mu^x}{x!} e^{-\mu}$$

This result is obtained from the binomial distribution if we let p approach 0 and n approach infinity. In this case, the mean $\mu = p$ approaches a finite value. The variance of a Poisson distribution is given as: $\sigma^2 = \mu$.

For a λ of fixed value and an experiment with outcome y equal to $1, 2, 3, \ldots$, the Poisson distribution can also be expressed as:

$$p(y) = \frac{\lambda^y}{y!} e^{-\lambda}, \qquad \text{where } y = 0, 1, 2, \ldots$$

As an example, one may consider that a random process occurs where a protein may bind at a given location within the cell as often as $y = 1, 2, 3, \ldots$, times per second, but on average the protein binds once per second. If y is described by a Poisson probability function, estimate the probability that the protein will not bind at the site during a one-second interval. In this case, the time period is one second and the mean binding steps is λ equal to 1. Therefore,

$$p(y) = \frac{(1)^y e^{-1}}{y!} = \frac{e^{-1}}{y!}$$

The binding event that is missed in a one second period corresponds to $y \geq 0$, such that

$$p(0) = \frac{e^{-1}}{0!} = \frac{e^{-1}}{1} = 0.368$$

The probability that the protein will bind at least once is the event that $y \geq 1$; therefore:

$$P(y \geq 1) = \sum_{y=1}^{\infty} p(y) = 1 - p(0) = 1 - e^{-1} = 0.632$$

3. Gaussian (or Bell-Shaped) Distribution. One of the most frequently encountered frequency distributions is the Gaussian, or normal, distribution. This function has the density:

$$f(x) = (1/\sigma\sqrt{2\pi}) e^{-1/2((x-\mu)/\sigma)^2}$$

where μ is the mean and σ is the standard deviation of the distribution. The corresponding distribution curve is bell-shaped. A random variable having this distribution is said to be normally distributed, or just "normal." If the area under this curve is Q, then (a) over the interval $\mu - \sigma$ to $\mu + \sigma$, the area is about $0.68Q$; (b) over the interval $\mu - 2\sigma$ to $\mu + 2\sigma$, the area is about $0.955Q$; and (c) over the interval $\mu - 3\sigma$ to $\mu + 3\sigma$, the area is about $0.997Q$.

For data that do not conform to a Gaussian (or normal) distribution, one often can use a mathematical procedure that transforms the observed values to create a corresponding Gaussian distribution. When one knows population distribution, transforming data values to create a Gaussian distribution becomes a useful means for adapting the data set for analysis with statistical tests requiring a Gaussian distribution. If the data are Counts C (from a Poisson distribution), use the square root of C as the normalizing transformation; for Proportion P (from a binomial distribution), use the arc-sine of square root of P as the normalizing transformation; for a Measurement M (from a log-normal distribution), use the logarithm of M as the normalizing transformation; and for duration (D), use $1/D$ as the normalizing transformation.

CONFIDENCE LEVELS. The normal distribution curve can also be used to assign confidence limits. Recall that the population mean (μ) of a measurement is a constant that remains unknown unless one samples every individual within a population. In the absence of systematic errors, limits may be set within which the population mean is expected to lie with some given probability; these limits are called confidence limits. The population mean can be said to lie within $\mu \pm z\sigma$. For a confidence level of 50%, $z = 0.67$; for 68%, $z = 1.00$; for 80%, $z = 1.29$; for 90%, $z = 1.64$; for 95%, $z = 1.96$; for 96%, $z = 2.00$; for 99%, $z = 2.58$; for 99.7%, $z = 3.00$; and for 99.9%, $z = 3.29$. The equation for calculating a desired confidence level is:

$$\text{Confidence Level for } m = \bar{x} \pm \frac{z\sigma}{\sqrt{N}}$$

As an example, let's say that for a set of three determinations (*i.e.*, $N = 3$), the arithmetic mean for the rate measurements and its standard deviation are 1.63 μM/s and 0.1 μM/s. If we wanted to know the range of values over which there is a 95% confidence of obtaining the true mean, we would carry out the following short computation:

$$95\% \text{ Confidence Level} = 1.63 \pm \frac{1.96 \times 0.1}{\sqrt{3}}$$
$$= 1.63 \pm 0.11 \ \mu\text{M/s}$$

Note: We used the value of 1.96 for the value of z given above for the 95% confidence level.

One can demonstrate that the effect of random error may be reduced without limit by using an ever larger sample size. In fact, the standard deviation of the mean (σ_m) can be written as:

$$\sigma_m = \frac{\sigma}{\sqrt{N}}$$

Therefore, one would need to make 100 times as many observations to increase the accuracy of a measurement by 10, and 10,000 times as many to improve accuracy by a factor of 100. This indicates that one quickly obtains diminishing returns in accuracy for increasing numbers of observations.

Analysis of Variance (ANOVA). Keeping in mind that the total variance is the sum of squares of deviations from the grand mean, this mathematical operation allows one to partition variance. ANOVA is therefore a statistical procedure that helps one to learn whether sample means of various factors vary significantly from one another and whether they interact significantly with each other. One-way analysis of variance is used to test the null hypothesis that multiple population means are all equal.

One uses ANOVA when comparing differences between three or more means. For two samples, the one-way ANOVA is the equivalent of the two-sample (unpaired) t test. The basic assumptions are: (a) within each sample, the values are independent and identically normally distributed (*i.e.*, they have the same mean and variance); (b) samples are independent of each other; (c) the different samples are all assumed to come from populations having the same variance, thereby allowing for a pooled estimate of the variance; and (d) for a multiple comparisons test of the sample means to be meaningful, the populations are viewed as fixed, meaning that the populations in the experiment include all those of interest.

Several or many factors may contribute simultaneously to the variance observed in an experiment. Thus the total variance can be expressed as:

$$\sigma_{total}^2 = \sigma_1^2 + \sigma_2^2 + \sigma_3^2 + \cdots + \sigma_n^2$$

This fact leads to the concept of total variance, a parameter corresponding to the sum of variances arising from different components of the variation. Analysis of variance represents a statistical means for partitioning the total variance into individual components. This method attempts to analyze variation by assigning portions of the variance to each set of independent variables. In general, the experimenter will not or cannot include all the variables influencing the response in an experiment, and random variation in the response is observed even if all independent variables are kept constant. The variability of a set of n trial measurements is proportional to the sum of the deviations, $\sum_{i=1}^{n} (y_i - \bar{y})^2$. ANOVA separates the sum of the squares of deviations into parts that are individually attributable to one of the experiment's independent variables, such that

$$SS_{total} = \sum_{i=1}^{n} (y_i - \bar{y})^2$$
$$= SS\, V_1 + SS\, V_2 + \cdots + SS\, V_n + SS\, E$$

where $SS\, V_1$, $SS\, V_2$, and $SS\, V_n$ are the sums of squares for variables 1, 2 and n. The last term is the sum of the squares assigned to error. For independent random samples from two normally distributed populations having means of μ_1 and μ_2 and equal variances, $\sigma_1^2 = \sigma_2^2 = \sigma^2$, and the total variance of measurements around their respective means for the two samples can be written as:

$$SS_{total} = \sum_{i=1}^{2} \sum_{j=1}^{n_i} (y_{ij} - \bar{y})^2$$
$$= n_i \sum_{i=1}^{2} (\bar{y}_i - \bar{y})^2 + \sum_{i=1}^{2} \sum_{j=1}^{n_i} (y_{ij} - \bar{y})^2$$

where $\bar{y}_i$ is the average of the measurements in the ith sample. The first sum (on the rightmost side) is often called the sum-of-the-squares for treatments, and the second is the sum of squares of deviations that one uses to evaluate s^2. Without presenting a detailed description, suffice it to say that analysis of variance allows one to evaluate when the sum of the squares for treatments is sufficiently large to detect a statistically significant difference between means of μ_1 and μ_2.

LEAST SQUARES TREATMENT. Here, we deal briefly with the statistical algorithm, also known as linear regression analysis, for systems where Y (the random variable) is linearly dependent on another quantity X (the ordinary or controlled variable). The procedure allows one to fit a straight line through points (x_1,y_1), (x_2,y_2), (x_3,y_3), $\cdots$, (x_n,y_n) where the values x_i are defined before the experiment and y_i values are obtained experimentally and are subject to random error. The best fit line through such a series of points is called a "least squares fit," and the protocol provides measures of the reliability of the data and quality of the fit.

For such a series of ordered pairs, the vertical displacement of a sample value, say (x_j,y_j), from the straight line $y = a + bx$ is $|y_j - a - bx_j|$, and the sum of these squares is given as:

$$Q = \sum_{j=1}^{n} (y_j - a - bx_j)^n$$

Values of a and b are the chosen to minimize the parameter Q which is referred to "the sum of the squares of $(y_j - a - bx_j)$", hence the name "Least Squares". Accordingly, the equation for the best fit least squares line is:

$$[y = a + bx]_{Q\,min}$$

This line is frequently referred to as a regression line, and its slope b is the regression coefficient. This procedure can be extended to polynomial functions wherein segments can be treated as linear or nearly linear functions of Y on X.

COMPOUNDING OF ERRORS. Data collected in an experiment seldom involves a single operation, a single adjustment, or a single experimental determination. For example, in studies of an enzyme-catalyzed reaction, one must separately prepare stock solutions of enzyme and substrate, one must then mix these and other components to arrive at desired assay concentrations, followed by spectrophotometric determinations of reaction rates. A Lowry determination of protein or enzyme concentration has its own error, as does the spectrophotometric determination of ATP that is based on a known molar absorptivity. All operations are subject to error, and the error for the entire set of operations performed in the course of an experiment is said to involve the compounding of errors. In some circumstances, the experimenter may want to conduct an error analysis to assess the contributions of statistical uncertainties arising in component operations to the error of the entire set of operations. Knowledge of standard deviations from component operations can also be utilized to estimate the overall experimental error.

One begins by assuming that a final result, call it r, is related to a series of components x_i in some functional manner:

$$r = F(x_1, x_2, x_3, \cdots, x_n)$$

Small variations in x_i can be written as dx_i, and the incremental effect on result r is dr, such that:

$$dr = \frac{\partial F}{\partial x_1} dx_1 + \frac{\partial F}{\partial x_2} dx_2 + \frac{\partial F}{\partial x_3} dx_3 + \cdots + \frac{\partial F}{\partial x_i} dx_i$$
$$= \sum_{i=1}^{n} \frac{\partial F}{\partial x_i} dx_i$$

As noted earlier, the variance V is the square of the standard deviation s; thus, if each error is independent of the other, the overall variance for the above expression can be written as:

$$\sigma^2 = (dr)^2 = \sum_{i=1}^{n} \left(\frac{\partial F}{\partial x_i}\right)^2 dx_i^2$$

This confirms an earlier stated property of total variance:

$$\sigma^2 = \sigma_1^2 + \sigma_2^2 + \sigma_3^2 + \cdots + \sigma_i^2$$

Expressed in terms of the standard deviation,

$$\sigma = \sqrt{\sigma_1^2 + \sigma_2^2 + \sigma_3^2 + \cdots + \sigma_i^2}$$

For example, compounding the error of three operations, each with standard deviations of in concentration, say ± 2.0 mM, ± 1.1 mM, and ± 1.4 mM, would yield an overall standard deviation:

$$\sigma_{total} = \sqrt{(2.0)^2 + (1.1)^2 + (1.4)^2} \simeq 2.7$$

This exercise reminds us that the compounded error will always be larger than the single standard deviation for any single operation. If the standard deviations for component operations do not have the same units, they cannot be treated as described here. In many cases, the investigator can convert the uncertainty in one type of operation into units from another. For example, one might have a standard deviation of 0.01 mM in substrate concentration and a standard deviation of 0.6 absorbance units. If conversion of 0.01 mM substrate to product resulted in an absorbancy of 0.4 (*i.e.*, $A = \varepsilon[S] = 0.4$), then the overall standard deviation would be $[(0.4)^2 + (0.6)^2]^{1/2}$ or 0.72 absorbance units.

One may also consider the compound effect of error when the result r is related to a ratio of components x_1/x_2. In this case,

$$\left(\frac{\sigma}{r}\right) = \sqrt{\left(\frac{\sigma_1}{x_1}\right)^2 + \left(\frac{\sigma_2}{x_2}\right)^2}$$

CRITICAL t TEST FOR COMPARING AVERAGES.
Whenever two sets of analytical determinations have different arithmetic averages, but comparable standard deviations, one may apply the t test to assess the statistical significance of the difference in these averages.

For two averages, say $\bar{x}$ and $\bar{y}$, the parameter t is given by the following equation:

$$t = \frac{|\bar{x} - \bar{y}|}{s} \sqrt{nm(n + m)}$$

where the vertical bars indicate absolute value in order to make t greater than zero for two nonidentical averages. The coefficients m and n correspond to the number of values that were averaged to obtain $\bar{x}$ and $\bar{y}$.

Table II

Critical t Values for Comparing Averages

Degrees of Freedom $(n + m) - 2$	Confidence Level	
	95%	99%
1	12.7	63.7
2	4.3	9.9
3	3.2	5.8
4	2.8	4.6
5	2.6	4.0
6	2.5	3.7
8	2.3	3.4
10	2.3	3.2
15	2.1	2.9
20	2.1	2.8

If the calculated t value exceeds the tabulated value for the indicated degrees of freedom, then chances are 95-out-of-100 or 99-out-of-100 that the two averages are significantly different.

Before interpreting the results of a t test, one would do well to think carefully about whether an appropriate test has been chosen. First, one should not confuse t tests with other statistical concepts such as correlation and regression; the t test compares one variable between two groups, and one should use correlation and regression to learn how two different variables vary together. Second, do not confuse t tests with ANOVA; while t tests and related nonparametric tests exactly compare two groups, ANOVA compares three or more groups. Third, remember that t tests also must not be confused with analyses that use so-called contingency tables (*e.g.*, Fisher's or Chi Square (χ^2) tests); the t test is employed when one wishes to compare continuous variables, while a contingency table is used to analyze categorical variables.

W. Mendenhall, R. L. Schaeffer & D. D. Wackerly (1981) *Mathematical Statistics with Applications*, 2nd ed., Duxbury Press, Boston.

A. Papoulis (1991) *Probability, Random Variables, and Stochastic Processes*, 3rd ed., McGraw Hill, New York.

W. W. Daniel (1995) *Biostatistics*. 6th ed., Wiley, New York.

H. R. Lindman (1992) *Analysis of Variance in Experimental Design*. Springer-Verlag, New York.

R. G. Miller, Jr. (1996) *Beyond ANOVA, Basics of Applied Statistics*, 2nd ed., Chapman & Hall, London.

J. Neter, W. Wasserman, & M. H. Kutner (1990) *Applied Linear Statistical Models*, 3rd ed., Irwin Press, Homewood, IL.

B. Rosner (1995) *Fundamentals of Biostatistics*, 4th ed., Duxbury Press, Belmont, California.

B. J. Winer, D. R. Brown & K. M. Michels (1991) *Statistical Principles in Experimental Design*, 3rd ed., McGraw-Hill, New York.

J. H. Zar (1996) *Biostatistical Analysis*, 3rd ed., Prentice-Hall, Upper Saddle River, New Jersey.

STEADY STATE

1. A condition achieved in a multistep process when the rate of formation of unstable intermediates is time-invariant, or nearly so. 2. In stirred flow systems, the steady-state condition refers to the procedure in which the concentration of all components is kept constant, via additions to and withdrawals from the system.

STEADY-STATE ASSUMPTION

A mathematical simplification of rate behavior of a multistep chemical process assuming that over a period of time a system displays little or no change in the concentration(s) of intermediate species (*i.e.*, $d[\text{intermediate}]/dt \approx 0$). In enzyme kinetics, the steady-state assumption allows one to write and solve the differential equations defining the rates of interconversion of various enzyme species. This is especially useful in initial rate studies.

If the concentration of substrate is not at least 100 times the concentration of enzyme, the steady state will not persist over the time course of most experiments. In such cases, the resulting initial rate data cannot be analyzed by standard initial rate kinetic procedures. *See also* Enzyme Kinetics; Numerical Integration

STEPWISE REACTION

A sequence of at least two elementary reactions and having at least one reaction intermediate. *See Multistage Reaction*

STEREOCHEMICAL FIDELITY

Enzymatic catalysis that is stereochemically specific or highly selective, arising from the ability of the enzyme to orient the chiral center within the active site in a manner that leads to highly efficient stereochemical discrimination. An example is the hydride transfer from lactate's (*S*)-enantiomer to the *re*- or A-side of bound NAD$^+$. This reaction proceeds with stereochemical fidelity of over 99.999998%. The occurrence of such stereochemical fidelity has at least three explanations[1]: (a) that greater catalytic efficiency attends stereochemical fidelity; (b) that by chance evolution has inherited one type of transfer reaction; or (c) that its absence in some potentially stereoselective cases may reflect that the natural selective advantage for fidelity did not outweigh other advantages gained by its absence. In the case of NAD$^+$-dependent enzymes such as LDH, the NADH product is equally active metabolically, suggesting that some catalytic advantage is gained[2].

[1]D. J. Creighton & N. S. R. K. Murthy (1990) *The Enzymes*, 3rd ed., **19**, 325.

[2]V. E. Anderson & R. D. LaReau (1988) *J. Am. Chem. Soc.* **110**, 3695.

STEREOCHEMICAL TERMINOLOGY (IUPAC Recommendations)

If the word you are searching for in the Wordfinder directed you here, then it is a term that lies beyond the scope of this Handbook. The listed word will be found in an authoritative internet-accessible glossary dealing with stereochemistry [http://www.chem.qmw.ac.uk/iupac/stereo/]. The glossary was developed under the auspices of the IUPAC Organic Chemistry Division's Commission on Nomenclature of Organic Chemistry and Commission on Physical Organic Chemistry. All definitions used in the internet glossary are identical to those in the published document[1], which the reader should cite if the definition is used verbatim or nearly so.

[1]G. P. Moss (1996) *Pure and Appl. Chem.* **68**, 2193.

STEREOCHEMISTRY OF HYDRIDE TRANSFER TO/FROM NADH

NADH has two faces, designated the A (or *re*) face and the B (or *si*) face. Nearly fifty years ago, Vennesland, Westheimer, and their associates first developed a durable strategy for determining the stereochemical course of hydride transfer to and from the A or B face of

NADH. These experiments were pioneering with respect to contemporary enzymology, especially with regard to early recognition that coenzymes are held within enzyme active sites in stereochemically preferred ways. One typically utilizes NADH that contains a tritium or deuterium atom in the $4R$ or $4S$ position, and the success or failure of substrate deuteration/tritiation indicates the stereochemistry. Westheimer[1] has tabulated the known examples of dehydrogenases that exhibit specificity for a particular face of NADH. Creighton and Murthy[2] have reproduced this tabulation in their comprehensive review on the stereochemistry of enzyme-catalyzed reactions at carbon.

[1]F. H. Westheimer (1987) in *Pyridine Nucleotide Coenzymes*, (D. Dolphin, R. Poulson, & O. Avramoric, eds.) Part A, p. 253, Wiley, New York.
[2]D. J. Creighton & N. S. R. K. Murthy (1990) *The Enzymes*, 3rd ed., **19**, 325.

STEREOCHEMISTRY OF HYDROGEN TRANSFER TO/FROM FADH₂

Flavoenzymes use their tightly bound FAD or FMN coenzyme to promote oxidation-reduction reactions involving hydrogen transfer to the *re* or *si* face of their coenzyme. Fisher & Walsh[1] showed that the glutathione reductase reaction involves direct hydrogen transfer, and Pai and Schulz[2] advanced a general strategy for determining the stereochemical course of flavoenzyme-catalyzed reactions based on the stereochemistry of glutathione reductase. Creighton and Murthy[3] present the stereochemistry of sixteen flavoenzyme reactions in their comprehensive review on the stereochemistry of enzyme-catalyzed reactions at carbon.

[1]J. Fisher & C. T. Walsh (1974) *J. Am. Chem. Soc.* **96**, 4345.
[2]E. F. Pai & P. E. Schulz (1983) *J. Biol. Chem.* **258**, 1752.
[3]D. J. Creighton & N. S. R. K. Murthy (1990) *The Enzymes*, 3rd ed., **19**, 324.

STEREOCHEMISTRY OF ISOMERASES INVOLVING PROTON TRANSFER

Creighton and Murthy[1] recently reviewed the stereochemistry and related mechanistic issues associated with enzyme-catalyzed isomerizations that proceed by 1,2-hydrogen transfer or by 1,3-allylic hydrogen transfer. In the first case, the prototypical aldose-ketose isomerase is triose-phosphate isomerase (or TPI), an enzyme that uses the carboxylate of Glu-165 as a base for abstracting a proton from the substrate during catalysis. Δ^5-3-Keto-

steroid isomerase is likewise a leading example of a reaction involving 1,3-allylic proton transfer.

[1]D. J. Creighton & N. S. R. K. Murthy (1990) *The Enzymes*, 3rd ed., **19**, 344.

STEREOELECTRONIC

Referring to or pertaining to a dependence of the properties of a molecular entity or a transition-state species in a particular electronic state on the relative nuclear geometry. Such effects are usually due to different geometrical arrangements.

IUPAC (1994) *Pure and Appl. Chem.* **66**, 1077.

STEREOSELECTIVITY

The preferential formation of one stereoisomer over another in a chemical reaction. If the stereoisomers of concern are enantiomers then the term enantioselectivity is often used. The degree of enantioselectivity is usually expressed by the enantiomeric excess. If the stereoisomers of concern are diastereoisomers, then the term used is diastereoselectivity. It is quantitatively measured by the diastereoisomer excess.

E. L. Eliel (1962) *Stereochemistry of Carbon Compounds*, McGraw-Hill, New York.

STEREOSPECIFICITY

1. A stereospecific chemical reaction is one in which starting substrates or reactants, differing only in their configuration, are converted into stereoisomeric products. Note, with this definition a stereospecific reaction has to be stereoselective whereas the inverse statement (that is, with respect to a stereoselective reaction or process) is not necessarily true. 2. Referring to reactions that act on only one stereoisomer (or, have a preference for one stereoisomer). Thus, many enzyme-catalyzed reactions are stereospecific, and characterization of that stereospecificity is always an issue to be addressed for a particular enzyme.

E. L. Eliel (1962) *Stereochemistry of Carbon Compounds*, McGraw-Hill, New York.

STERIC-APPROACH CONTROL

The means of achieving stereoselectivity due to steric hindrance toward the attack of a reagent or cosubstrate. The less hindered face will be the more susceptible side. Many enzymes utilize this method of product control.

STERIC HINDRANCE

A steric effect on chemical reactivity resulting from a crowding of substituents around an otherwise reactive center.

STEROID 11β-HYDROXYLASE

This heme-thiolate-dependent enzyme [EC 1.14.15.4], also known as steroid 11β-monooxygenase, catalyzes the reaction of a steroid with reduced adrenal ferredoxin and dioxygen to produce an 11β-hydroxysteroid, oxidized adrenal ferredoxin, and water. The enzyme also catalyzes the hydroxylation of steroids at the 18-position and can catalyze the conversion of 18-hydroxycorticosterone into aldosterone.

M. J. Coon & D. R. Koop (1983) *The Enzymes*, 3rd ed., **16**, 645.
V. Ullrich & W. Duppel (1975) *The Enzymes*, 3rd ed., **12**, 253.

STEROID 17α-HYDROXYLASE

This heme-thiolate-dependent enzyme [EC 1.14.99.9], also known as steroid 17α-monooxygenase, catalyzes the reaction of a steroid with AH_2 and dioxygen to produce a 17α-hydroxysteroid, A, and water.

M. J. Coon & D. R. Koop (1983) *The Enzymes*, 3rd ed., **16**, 645.

STEROID 21-HYDROXYLASE

This heme-thiolate-dependent enzyme [EC 1.14.99.10], also known as steroid 21-monooxygenase, catalyzes the reaction of a steroid with AH_2 and dioxygen to produce a 21-hydroxysteroid, A, and water.

M. J. Coon & D. R. Koop (1983) *The Enzymes*, 3rd ed., **16**, 645.

STEROID SULFOTRANSFERASE

This enzyme [EC 2.8.2.15] catalyzes the reaction of 3'-phosphoadenylylsulfate with a phenolic steroid to produce adenosine 3',5'-bisphosphate and a steroid O-sulfate. The enzyme is very similar in its activity to alcohol sulfotransferase. However, steroid sulfotransferase can utilize estrone as a substrate.

S. G. Ramaswamy & W. B. Jakoby (1987) *Meth. Enzymol.* **143**, 201.

STICKY SUBSTRATES

A designation given to a substrate (or ligand) that displays a strong tendency to remain bound to its binding site, and, in the case of enzymes, to undergo enzymatic catalysis with greater ease than to dissociate. Cleland[1] states that the stickiness of a substrate is a measure of the ratio of the net constant for reaction of the first collision complex through the first irreversible step to the rate constant for dissociation of that collision complex. The strength of enzyme-substrate complexation can be altered by solution variables (*e.g.*, temperature, ionic strength, pH, solvent polarity, *etc.*) or factors such as proteolytic alteration of enzyme structure as in the case of yeast hexokinase's interaction with D-glucose[2]. *See Isotope Trapping*

[1]W. W. Cleland (1982) *Meth. Enzymol.* **87**, 631.
[2]I. A. Rose, E. L. O'Connell, S. Litwin & J. Bar-Tana (1974) *J. Biol. Chem.* **249**, 5163.

STIFFNESS INSTABILITY

A term used in numerical integration of rate processes to describe computational difficulties arising whenever reaction steps within a mechanism are each characterized by widely disparate time constants. In such a case, the most rapid step must be integrated repetitively over such small time intervals that these intervals are far too small for efficient integration of the slower reaction process. As a result, excessively long periods for computation are required for accurate integration of both steps in a mechanism. This problem was first noted in 1952[1], and Gear[2] has advanced a predictor-corrector algorithm to vary the size of the time interval to speed up numerical integration. *See Gear Algorithm*

[1]C. F. Curtiss & J. O. Hirschfelder (1952) *Proc. Natl. Acad. Sci. U.S.A.* **38**, 235.
[2]C. W. Gear (1971) *Numerical Initial-Value Problems in Ordinary Differential Equations*, Prentice-Hall, Englewood Cliffs, N. J.

STOICHIOMETRIC NUMBER

A number (usually symbolized by n followed by a subscript denoting the species) equal to the coefficient of that species in a particular reaction. For example, for the reaction A + 2B = 3P, the stoichiometric number for P is 3. By convention, stoichiometric numbers are positive for products and negative for reactants.

STOKES SHIFT

If an electronic transition results in both absorption and luminescence, then the Stokes shift is the difference (in either wavelength or frequency units) between the band maxima. If the luminescence occurs at a shorter wavelength, the difference is often referred to as an anti-Stokes shift.

STOPPED-FLOW MASS SPECTROMETRY IN ENZYME KINETICS

Northrop and Simpson[1] recently described how the interfacing of mass spectrometry with a stopped-flow mixing device promises to become a powerful approach for examining kinetic mechanisms as well as details of the chemical transformations attending catalysis. For example, one might apply this technique to an enzyme system suspected of forming a transient enzyme-substrate covalent intermediate. In this case, the covalently modified enzyme might be intercepted before it reacts with a second substrate or before it is transformed into a reaction product. Grieg et al.[2] already succeeded in determining dissociation constants for reversible protein·ligand complexes by using an electrospray injection port, and the technique may be especially useful for analyzing the properties of tight-binding inhibitors[1].

[1]D. B. Northrop & F. B. Simpson (1997) Bioorg. & Med. Chem. **5**, 641.
[2]M. J. Grieg, H. Gaus, L. L. Cummins, H. Sasmor & R. H. Griffey (1995) J. Am. Chem. Soc. **117**, 10765.

STOPPED-FLOW TECHNIQUES

Any of the rapid kinetic methods[1,2] relying on two syringes for passage of two solutions through a mixer and an observation cuvette into a third, so-called stopping-syringe. Filling of the latter mechanically trips an switch that starts the electronic acquisition of spectral data (usually accomplished by absorption or fluorescence spectroscopy). Such data provides kinetic information concerning the changes in reactant concentrations almost immediately after the chemical reaction is initiated by the mixing of the reactant-containing (or reaction-initiating) solutions.

The dead-time of stopped-flow devices is the sum of two parameters, namely the mixing time and the delay time. The mixing time is controlled by the efficiency of turbulent flow, usually achieved in a multi-jet arrangement that promotes intimate mixing of the fluids. Mixing time is strongly influenced by the need to choose mixing conditions that avoid or minimize the effect of cavitation, the mechanically induced formation of a vapor-filled cavity within a rapid-flow device arising as a consequence of fluid viscosity and high shear forces. Cavitation within the cuvette compartment creates an inhomogeneous medium that scatters the incident light beam used in the photo-electronic detection of absorbance or fluorescence signals. Delay time, the second contributor to the device's dead-time, is the elapsed time required for all electromechanical events associated with signal detection.

A derivative technique, known as stopped-flow/temperature-jump[3], allows one to take advantage of the stopped-flow strategy to establish a steady-state rate condition which can then be perturbed and analyzed by using the elements of a temperature-jump device.

[1]B. Chance (1964) in Rapid Mixing and Sampling Techniques in Biochemistry (B. Chance, R. Eisenhardt, Q. H. Gibson & K. Lonberg-Holm, eds) p. 125, Academic Press, New York.
[2]Q. H. Gibson (1969) Meth. Enzymol. **16**, 187.
[3]G. G. Hammes (1974) in "Investigations of Rates and Mechanisms of Reactions: Part II, Investigation of Elementary Steps in Solu-

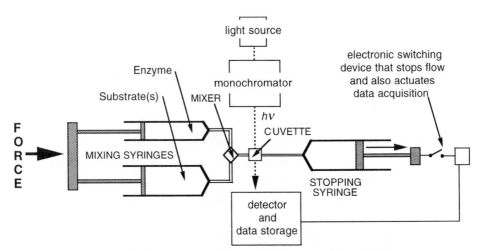

Figure 1. Construction and flow path of a stopped-flow device.

tion and Very Fast Reactions," *Techniques in Chemistry*, vol. VI (G. G. Hammes, ed.) p. 165, Wiley-Interscience, New York.

Selected entries from *Methods in Enzymology* [vol, page(s)]: Construction details, **16**, 189-194; Gibson-Milnes mixer, **16**, 192; temperature control, **16**, 194-197; light source, **16**, 197; mono-chromators, **16**, 198; photomultiplier and power supply, **16**, 198-200; noise levels, **16**, 205-207; problems (including air bubbles, leakage, sample contamination by diffusion from one working syringe to another, syringe breakage, anaerobic operations, sample poisoning by contact with stainless steel components), **16**, 207-210; Strittmatter microstopped-flow apparatus, **16**, 210-214; observation of rapid fluorescence changes, **16**, 214-218; pH detector for stopped flow, **16**, 218; polarographic recording, **16**, 218; dual-wavelength recording, **16**, 220-222; factors affecting dead time, **249**, 18; rapid scanning diode array, **249**, 19; pre-steady-state kinetics, **249**, 18; single turnover kinetics, **249**, 18; alcohol dehydrogenase, elucidation of mechanism, **246**, 16; circular dichroism, **246**, 61; copper amine oxidase [anaerobic conditions, **258**, 77; data analysis, **258**, 77-78, 81; deuterium isotope effects, **258**, 82-83; rapid-scanning analysis, **258**, 80-83; relaxation characteristics, **258**, 81-82]; dead time, **246**, 62, 172, 470, 472; high-pressure experiments, **259**, 373, 397; limitations, **249**, 18-19; methylamine dehydrogenase [amicyanin oxidation, **258**, 190; apparatus, **258**, 187, 189; methylamine reduction, **258**, 189-190]; ribonucleotide reductase diferric-tyrosyl radical cofactor assembly [absorbance changes, **258**, 282-283; apparatus, **258**, 281; mixing reactions, **258**, 281-282; rate constants, **258**, 283-285, 293; requirements, **258**, 280-281]; in thiyl radical reaction analysis, **251**, 50-51; time resolution, **246**, 16, 62; **251**, 50.

STRAIN

1. A property present in a molecular entity (or a transition state) if the energy of that entity or state is enhanced due to unfavorable bond lengths, bond angles, or dihedral angles relative to some appropriate standard. It is the standard enthalpy of a structure relative to a "strainless" structure (real or hypothetical). 2. The change of volume or shape of a body, or portion of a body, due to the influence of one or more applied forces.

STREPTOKINASE

A metallopeptidase isolated from hemolytic streptococci. It hydrolyzes peptide bonds in plasminogen, producing plasmin.

J. T. Radek, D. J. Davidson & F. J. Castellino (1993) *Meth. Enzymol.* **223**, 145.

STURGILL-BILTONEN METHOD

A procedure used in the characterization of cooperative proteins in which the second moment of the derivative of the saturation function is measured.

T. Sturgill & R. L. Biltonen (1976) *Biopolymers* **15**, 337.

SUBSITE MAPPING

A systematic kinetic analysis[1] of the binding site of an enzyme with the subunits of a polymeric substrate, in-

cluding specificity-imparting interactions at and surrounding the constellation of catalytically active groups participating in bond-making/breaking steps. The most rigorous approach relies on the additivity of subsite binding energies, based on the individual free energies of enzyme binding to the monomeric residues of the substrate. This approach requires thorough investigation of an enzyme's ability to bind polymeric substrates in various registrations, and this often can be inferred by studying the inhibitory properties of oligomeric competitive inhibitors.

In principle, the method affords the opportunity to predict the action pattern of an enzyme that is capable of binding a polymeric substrate in alternative ways, thereby leading to a preferred distribution of products. These methods are especially well suited for the investigation of polysaccharide hydrolyzing enzymes as well as proteinases. The binding of a maltotetraose to an enzyme possessing five glucosyl residue binding subsites illustrates the potential positional isomer interactions of the substrate: (a) only those substrate molecules spanning the cleavage site will undergo hydrolysis; (b) the action pattern of the hydrolase must be assessed by analysis of the limit dextran products; and (c) the other binding interactions, while not productive in terms of hydrolase activity, actually serve as inhibitors.

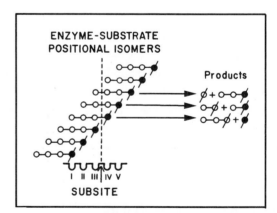

Hypothetical interactions of a five-subsite enzyme with a maltotetraose substrate shown initially with a terminally radiolabeled (filled circles) nonreducing end. The vertical arrow indicates the glycosyl bond cleavage region; the roman numerals indicate the subsites, and the 4 indicates the chain length. Only three binding interactions lead to hydrolysis: IV,4; V,4; and VI,4, and their different associated rate constants emphasize that the rates of hydrolysis need not be identical. *From Allen and Thoma[2] with permission of the Biochemical Society (London).*

[1]J. D. Allen (1980) *Meth. Enzymol.* **64**, 248.
[2]J. D. Allen & J. A. Thoma (1976) *Biochem. J.* **159**, 105.

SUBSTITUENT

An atom, group of atoms bonded together, or moiety that can be considered to replace a hydrogen atom in a parent molecular entity (real or hypothetical).

SUBSTITUTION REACTION

A reaction in which an atom, group, or moiety in a molecular entity is replaced by another. *See* S_N *Reactions*

SUBSTRATE

1. The substance or reactant being acted upon by a catalyst. The substrate is often symbolized by S in one-substrate reactions. In multisubstrate reactions, the substrates are commonly symbolized by A, B, C, *etc.* 2. The base or foundation upon which an organism lives or grows. 3. The substance or compound of particular interest, with which a reaction with some other chemical reagent is under study.

SUBSTRATE-ASSISTED CATALYSIS

The facilitation of enzymic catalysis achieved by the presence of a substituent located on the substrate at a site away from the position of bond making/breaking. Substrate-assisted catalysis by serine proteases is well documented[1]. Protease substrates are usually described as having sequences . . . P3—P2—P1—P1'—P2'—P3' . . . , where P1—P1' lie on each side of the scissile bond. Corey *et al.*[2] recently described how trypsin specificity is increased through substrate-assisted catalysis. They found that a strategically placed histidyl residue at position P2 or P1' in the peptide substrate was able to increase trypsin-catalyzed cleavage by 70–300 times. These investigators also used site-directed mutagenesis to explore this substrate-assisted catalysis.

[1]P. Carter & J. A. Wells (1987) *Science* **237**, 394.
[2]D. R. Corey, W. S. Willet, G. S. Coombs & C. S. Craik (1995) *Biochemistry* **34**, 11521.

SUBSTRATE CHANNELING

Any process by which there is direct transfer of noncovalently bound substrate(s) or coenzyme from the active site of the enzyme producing the metabolite to the active site of an enzyme catalyzing a succeeding step in a metabolic pathway or to another enzyme utilizing the coenzyme product.

An implied feature of substrate channeling is that there must be direct contact or complexation between the two enzymes participating in the direct transfer. The following account dealing with conflicting interpretations regarding substrate channeling underscores the need for caution in proposing a kinetic model based on limited experimental findings and in the absence of an adequate kinetic theory. Srivastava and Bernhard[1] initially reported that reduction of benzaldehyde and *p*-nitrobenzaldehyde by NADH using horse liver alcohol dehydrogenase (LADH) was faster when NADH was bound to glyceraldehyde-3-phosphate dehydrogenase (GPDH) than with free NADH. The rate of reduction of aldehyde substrate with GPDH-NADH followed a Michaelis-type concentration dependence on GPDH-NADH. The reaction velocity was independent of GPDH concentration when [GPDH] was greater than [NADH]$_{total}$. The K_m for GPDH-NADH was higher than that for free NADH. They suggested that transfer of NADH from GPDH to LADH proceeds through the initial formation of a GPDH-NADH-LADH complex! These investigators[2] also proposed that glycolytic enzymes form multienzyme complexes for the direct transfer of metabolites from the producing enzyme to the utilizing one.

Chock and Gutfreund[3] reexamined the kinetics of NADH transfer between its complexes with glycerol-3-phosphate dehydrogenase and with lactate dehydrogenase. Their work led them to conclude that coenzyme transfer between GPDH and LDH proceeds by a free-diffusion mechanism, *not* by direct transfer through a ternary complex. While their conclusions were disputed by Srivastava *et al.*[4], Wu *et al.*[5] carried out further studies on NADH binding to GPHD and LDH and on the displacement of enzyme-bound NADH by LDH or GPDH, adducing self-consistent evidence for a dissociative mechanism, as opposed to any direct transfer mechanism. Furthermore, while Brooks and Storey[6] reported that they could reproduce the results of Srivastava *et al.*[4], they also found that a mathematical solution of the direct-transfer-mechanism equations of Srivastava *et al.*[6] did not adequately describe the experimental properties of the reaction rate at increasing LDH concentrations.

Those believing they have obtained evidence for direct nondissociative substrate channeling are strongly encouraged to consult the illuminating report of Wu *et al.*[7], who developed a rigorous set of criteria for differentiating mechanisms for ligand exchange reactions. They provide analytical expressions for the kinetics of (a) the dissociative mechanism (*i.e.*, where the leaving ligand must first dissociate prior to the binding of the incoming ligand) and (b) the associative mechanism (where a ternary complex is formed between the incoming ligand and the complex containing the leaving ligand). Analysis of these equations showed that an associative mechanism only can yield an increasing kinetic pattern for the observed pseudo-first-order rate constants as a function of increasing concentration of the incoming ligand and plateaus, in most cases, at a value higher than the off-rate constant of the leaving ligand. By contrast, the dissociative mechanism can produce increasing or decreasing conditions, depending on the magnitudes of the individual rate constants.

[1]D. K. Srivastava & S. A. Bernhard (1984) *Biochemistry* **23**, 4538.
[2]D. K. Srivastava & S. A. Bernhard (1986) *Science* **234**, 1081.
[3]P. B. Chock & H. Gutfreund (1988) *Proc. Natl. Acad. Sci. U.S.A.* **85**, 8870.
[4]D. K. Srivastava, P. Smolen, G. F. Betts, T. Fukushima, H. O. Spivey & S. A. Bernhard (1989) *Proc. Natl. Acad. Sci. U.S.A.* **86**, 6464.
[5]X. M. Wu, H. Gutfreund, S. Lakatos & P. B. Chock (1991) *Proc. Natl. Acad. Sci. U.S.A.* **88**, 497.
[6]S. P. Brooks & K. B. Storey (1991) *Biochem. J.* **278**, 875.
[7]X. M. Wu, H. Gutfreund & P. B. Chock (1992) *Biochemistry* **31**, 2123.

SUBSTRATE INHIBITION

The reduction in enzymatic activity that results from the formation of nonproductive enzyme complexes at high substrate concentration. The most straightforward explanation for substrate inhibition is that a second set of lower affinity binding sites exists for a substrate[1], and occupancy of these sites ties up the enzyme in nonproductive or catalytically inefficient forms. Other explanations include (a) the removal of an essential active site metal ion or other cofactor from the enzyme by high concentrations of substrate, (b) an excess of unchelated substrate (such as ATP^{4-}, relative to the metal ion-substrate complex (such as $CaATP^{2-}$ or $MgATP^{2-}$) which is the true substrate[2,3]; and (c) the binding of a second molecule of substrate at a subsite of the normally occupied substrate binding pocket, such that neither substrate molecule can attain the catalytically active conformation[4]. For multisubstrate enzymes, nonproductive dead-end complexes can also result in substrate inhibition in the presence of one of the reaction products. In these cases, substrate inhibition can inform the experimentalist about the kinetic mechanism for substrate binding and/or product release[4–6]. **See** *Abortive Complex Formation*

Haldane[1] was probably the first to treat the case of substrate inhibition arising as a consequence of second site binding. His kinetic scheme was as follows:

$$E + S \underset{}{\overset{K_s}{\rightleftharpoons}} ES \overset{k_3}{\rightarrow} E + P \text{ and}$$

$$ES + S \underset{}{\overset{K_i}{\rightleftharpoons}} ES_2$$

The rate equation for this mechanism is:

$$V_{max}/v = 1 + K_s/[S] + [S]/K_i$$

where K_s and K_i are dissociation constants for the equilibria in the scheme above. The substrate inhibition behavior of an enzyme will depend on the relative magnitudes of K_s and K_i, and the figure presented below illustrates the general features of substrate inhibition.

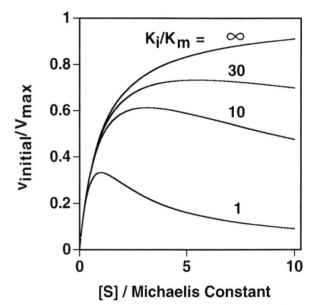

Plot of the fractional saturation of enzyme active sites (v/V_{max}) *versus* the reduced substrate concentration $[S]/K_s$ where $K_i/K_s = 1$, 10, 30, and infinity. Notice that there is an optimum in the rate dependence on substrate concentration.

[1]J. B. S. Haldane (1930) *Enzymes*, p. 84, Longmans, Green, & Co., London.
[2]D. L. Purich & H. J. Fromm (1972) *Biochem. J.* **130**, 63.
[3]R. D. Allison & D. L. Purich (1979) *Meth. Enzymol.* **63**, 1.
[4]W. W. Cleland (1971) *The Enzymes*, 3rd ed., **2**, 1.
[5]H. J. Fromm (1975) *Initial Rate Enzyme Kinetics*, Springer-Verlag, New York.
[6]W. W. Cleland (1979) *Meth. Enzymol.* **63**, 500.

SUBSTRATE OXIDATION KINETICS

Wolfe[1] has presented an excellent description of the systematic application of stable and radioactive isotope tracers in determining the kinetics of substrate oxidation, carbon dioxide formation (including $^{13}CO_2$ breath tests), glucose oxidation, and fat oxidation in normal and diseased states. Quantification of the rate and extent of substrate oxidation can be achieved by using a specific ^{13}C- or ^{14}C-substrate which upon oxidation releases radioactive carbon dioxide.

[1] R. R. Wolfe (1992) *Radioactive and Stable Isotope Tracers in Biomedicine*, p. 235, Wiley-Liss, New York.

SUBSTRATE PURITY

Substrate purity, standardization, and handling are issues of great significance with respect to enzyme kinetics. The presence of impurities (or alternative substrates) is not only a source of experimental error; they can easily mislead an investigator regarding the mechanism of enzymatic catalysis and/or control. Incorrect values for the true substrate concentrations can lead to substantial errors in the reported values for the kinetic parameters. Thus, efforts should be made to confirm substrate purity, including the stereochemical purity, and this should be accomplished enzymatically, chemically, chromatographically, or spectrally prior to any kinetic experiment. If pure stereoisomers are not available, then a statement with regard to the stereochemical purity must be specified in any report of experimental findings. Biochemical substances tend to be relatively unstable, and impurities can be readily generated. All reagent bottles should be marked with the date on which the item was received as well as the date when the container was first opened. Such information is often quite valuable to co-workers in the same laboratory. The information provided by the commercial vendor regarding purity should not be accepted as definite, and some further purification may be required. For example, some commercial sources of amino acids contain heavy metal ions which may inhibit enzyme activity. Both L-glutamate and L-glutamine are usually recrystallized from solutions containing EDTA by investigators before kinetic studies are carried out. Substrate purity is of prime importance in investigations of partial exchanges in ping pong mechanisms. For example, the ATP $\leftrightarrow$ AMP exchange observed with PRPP synthetase was reported to be due to impurities[1].

A good illustration of the influence of an impurity can be provided by noting the effect on initial-rate kinetics by the presence of a particular type of inhibitor in the substrate stock solution[2]. Substrates and competitive inhibitors are often structural analogs of each other. It would not be surprising if an unknown competitive inhibitor was present with the substrate. Hence, upon serial dilutions of the stock substrate solution, the inhibitor concentration would be in constant ratio relative to the substrate. Thus, the initial rate expression in the presence of a competitive inhibitor,

$$1/v = (1/V_{max}) + (K_m/V_{max})(1/[S])\{1 + [I]/K_i\}$$

is transformed to

$$1/v = (1/V_{max}) + (K_m/V_{max})(1/[S])\{1 + \alpha[S]/K_i\}$$

where $\alpha = [I]/[S]$ and K_i is the dissociation constant for the EI complex. This form of the equation is indistinguishable from the standard expression for a one-substrate, enzyme-catalyzed reaction. Graphical analysis of initial-rate data will thus produce wrong estimates of the kinetic parameters; for example, the observed Michaelis constant from the above expression would actually be $K_m K_i/(K_i + \alpha K_m)$.

If a noncompetitive or an uncompetitive inhibitor were present with the substrate at constant ratio, then graphical analysis would suggest that the phenomenon of substrate inhibition is present. If an investigator analyzed the apparent substrate inhibition via a Marmasse plot, wrong estimates of both the K_m and K_{is} values would be reported and the investigator would be mislead with respect to the kinetic mechanism. If partial inhibitors or alternative substrates are present in constant ratio, depending on the relative sizes of the V_{max} and K_m values, a variety of nonlinear graphical depictions can occur.

Kuby[3] also discusses the possibility of the presence of an impurity which forms a complex with the substrate. If the impure ligand is in constant ratio with the substrate (*i.e.*, $[L_{total}] = a[A_{total}]$), then a kinetic equation of the form $\alpha[A] + \beta[A]^2/(\gamma + \delta[A] + \varepsilon[A]^2)$ can be derived. Kuby points out that, in order for sigmoidal kinetics to be exhibited, $\beta\gamma > \alpha\delta$. However, this will not be true with this mechanism. Nevertheless, nonlinearity in double-reciprocal plots can still occur.

If an investigator remains concerned about the substrate purity following repurification procedures and subsequent analysis, then a number of other approaches may be considered. For example, different lots of substrate could be analyzed with the enzyme to learn if identical kinetic parameters are obtained. If one has some idea as to the identity of the possible contamination, then the impurity can be added to the stock substrate solution at a known amount and the accuracy of the kinetic parameters in the presence of the adduct can be assessed. Since the researcher already knows the degree of sensitivity of the various chemical, enzymatic, and/or spectral methods used to assess substrate purity, this known addition provides a means for determining the maximum amount of impurity present. Combined with the observations seen with the known addition of the impurity, such information will provide an idea on the level of accuracy of the kinetic parameters.

Closely related to issues of substrate purity are concerns related to the purity of the other components of the reaction mixture, especially water. For example, many enzymes that act on amine-containing molecules are inhibited by ammonium ions. As ammonia can be a common contaminant in many laboratories, doubly-distilled water can still contain enough ammonia to perturb certain enzyme activities. In such cases, it is important to test for ammonia and/or to determine the dissociation constant for ammonia with the enzyme under study.

Similarly, CO_2 and bicarbonate levels should be kept to a minimum by freshly preparing any alkaline buffer solutions. For similar reasons, stock solutions of NaOH, KOH, or other bases used to adjust pH values must always be freshly prepared.

When utilizing radiolabeled substrates, it is especially important to search for impurities resulting from radiolysis. This is particularly true if the stabilizing agent, often supplied by the vendor, has been removed during any laboratory procedure. It should be recalled that a radioactive compound can decompose chemically as well as via radioactive decay. In many cases, the rate of chemical decomposition for a radioactive compound is greater than the nonradioactive counterpart. This increased decomposition, which can result in a progressive increase in the apparent specific activity, may be a result of the formation of free radicals by the action of the radiation on the solvent. Thus, it is prudent to analyze for chemical and radiopurity prior to use. It may be necessary to develop procedures to stabilize the radiolabel following purification.

Selection of the location of the radiolabel is also of importance. The position of the label within the compound should be such that no primary or secondary isotope effects perturb the kinetics of the system (unless, of course, the investigator is studying isotope effects).

[1]R. L. Switzer & P. D. Simcox (1974) *J. Biol. Chem.* **249**, 5304.
[2]R. D. Allison & D. L. Purich (1979) *Meth. Enzymol.* **63**, 3.
[3]S. A. Kuby (1990) *A Study of Enzymes*, vol. 1, p. 285, CRC Press, Boca Raton.
[4]H. J. Fromm (1975) *Initial Rate Enzyme Kinetics*, Springer-Verlag, New York.

SUBSTRATE STABILITY & HANDLING

The inherent instability of many biological compounds (particularly substances that can serve as substrates, products, inhibitors, and effectors of enzyme-catalyzed reactions) can be a major factor in kinetic studies if decay occurs within the time course of the experiment. For example, equilibrium exchange studies of malate dehydrogenase were done at 1°C to avoid problems with oxaloacetate instability[1]. The degree of stability of each substrate and component of the assay mixture should be known under the specific assay conditions. For multisubstrate enzymes, it is a fairly straightforward procedure to examine each substrate for instability problems. For example, by using identical concentrations of two substrates containing different isotopic labels, one can verify the stoichiometry of multisubstrate reactions by observing the maintenance of the ratio of the two forms of radioactivity[2]. For example, when using L-[^{3}H]glutamate and [^{14}C]ATP with glutamine synthetase, an alteration in the ^{3}H:^{14}C ratio as the reaction progresses would indicate that either one or both of the substrates are inherently unstable under the assay conditions or that a contaminating enzyme is present and preferentially acting on one of the substrates. Another common method for demonstrating the stability of a substrate is to have the substrate present in a reaction mixture containing all components except one of the other substrates. If a side reaction is present for the substrate under consideration, then the assay protocol[3] for that substrate should demonstrate its existence.

Proper storage of substrates, stock solutions, and reagents is also an issue of importance in maintaining substrate, effector, and reagent stability as well as an issue of purity[4]. If stock solutions or reagents are to be kept frozen when not being used then, when needed, they should be thawed in a water bath and thoroughly mixed before use. If reagents are kept in a desiccator in the cold, the desiccator should be warmed to room temperature before opening. Due to the high instability of many biomolecules, efforts should also be made to keep the desiccant fresh. *See* Substrate Purity; Enzyme Purity; Water Purity

[1] E. Silverstein & G. Sulebele (1969) *Biochemistry* **8**, 2543.
[2] R. D. Allison & D. L. Purich (1979) *Meth. Enzymol.* **63**, 3.
[3] H. U. Bergmeyer (1983) *Methods of Enzymatic Analysis*, 3rd ed., Springer-Verlag, Berlin.
[4] H. J. Fromm (1975) *Initial Rate Enzyme Kinetics*, Springer-Verlag, New York.

SUBSTRATE SYNERGISM

The cardinal feature of Ping Pong mechanisms is the ability of the enzyme to catalyze partial exchange reactions as a result of the independence of the substrate's interactions with the enzyme. The second substrate is obliged to await the dissociation of the first product before it may bind to the enzyme. Multisubstrate enzymes frequently mediate such partial reactions which may be related to important steps in catalysis. One enzyme proposed to be of this sort is succinyl-CoA synthetase, but the partial reactions are relatively slow, and the participation of such reactions in catalysis becomes difficult to assess. Indeed, slow partial exchanges have been interpreted as proof of contamination of a particular enzyme with another enzyme or a small amount of the second substrate, an indication that the mechanism is not Ping Pong, or that the presence of other substrates may markedly increase the rates of elementary reactions giving rise to the partial exchange. Bridger *et al.*[1] proposed that the latter phenomenon be termed substrate synergism, and they examined this enzyme-substrate interaction by deriving appropriate rate laws for various exchanges. Their conclusion was that the rate of a partial exchange reaction must exceed the rate of the same exchange reaction in the presence of all the other substrates if the same catalytic steps and efficiencies are involved. If the opposite relation is observed, one must consider the possibility that synergism exists. Lueck and Fromm[2] examined the significance of partial exchange rate comparisons and focused on often misleading comparisons of exchange rates made with respect to initial velocity data.

[1] W. A. Bridger, W. A. Millen & P. D. Boyer (1968) *Biochemistry* **7**, 3608.
[2] J. D. Lueck & H. J. Fromm (1973) *FEBS Lett.* **32**, 184.

SUBTILISIN

This enzyme [EC 3.4.21.62], a serine endopeptidase that evolved independently of chymotrypsin, contains no cysteinyl residues. This enzyme catalyzes the hydrolysis of peptide bonds in proteins and has a broad specificity, with a preference for a large uncharged aminoacyl residue in the P1 subsite.

E. Lolis & G. A. Petsko (1990) *Ann. Rev. Biochem.* **59**, 597.
R. J. Coll, P. D. Compton & A. L. Fink (1982) *Meth. Enzymol.* **87**, 66.
J. S. Fruton (1982) *Adv. Enzymol.* **53**, 239.
J. Kraut (1971) *The Enzymes*, 3rd ed., **3**, 547.
F. S. Markland, Jr., & E. L. Smith (1971) *The Enzymes*, 3rd ed., **3**, 562.

Selected entries from *Methods in Enzymology* [vol, page(s)]: Cleavage site specificity, **244**, 114; catalytic triad, **244**, 37; subtilisin family members, **244**, 37-39; oxyanion hole, **244**, 37; sequence homology, **244**, 37, 39; thiol dependence, **244**, 37, 39; unit activity, **235**, 566; zymography in nondissociating gels with copolymerized substrates, **235**, 588-590.

SUCCESSOR COMPLEX

A radical ion pair (*e.g.*, $A^{\cdot-} B^{\cdot+}$) produced immediately upon transfer of an electron from one molecular entity, designated here as B, to another molecular entity, in this case, A.

SUCCINATE DEHYDROGENASE

Succinate dehydrogenase (ubiquinone) [EC 1.3.5.1], a multiprotein complex found in the mitochondria, catalyzes the reaction of succinate with ubiquinone to produce fumarate and ubiquinol. The enzyme requires FAD and iron-sulfur groups. It can be degraded to form succinate dehydrogenase [EC 1.3.99.1], a FAD-dependent system that catalyzes the reaction of succinate with an acceptor to produce fumarate and the reduced acceptor, but no longer reacts with ubiquinone.

K. M. Robinson & B. D. Lemire (1995) *Meth. Enzymol.* **260**, 34.
Y. Anraku (1988) *Ann. Rev. Biochem.* **57**, 101.
G. von Jagow & W. Sebald (1980) *Ann. Rev. Biochem.* **49**, 281.
Y. Hatefi & D. L. Stiggall (1976) *The Enzymes*, 3rd ed., **13**, 175.
G. Palmer (1975) *The Enzymes*, 3rd ed., **12**, 1.
T. P. Singer, E. B. Kearney & W. C. Kenney (1973) *Adv. Enzymol.* **37**, 189.

SUCCINYL-CoA SYNTHETASE

Succinyl-CoA synthetase (GDP) [EC 6.2.1.4], also known as succinate:CoA ligase, catalyzes the reversible reaction of GTP with succinate and coenzyme A to produce GDP, succinyl-CoA, and orthophosphate. The nucleotide substrate can be replaced with ITP and itaconate can substitute for succinate.

Succinyl-CoA synthetase (ADP) [EC 6.2.1.5] catalyzes the reversible reaction of ATP with succinate and coenzyme A to produce ADP, succinyl-CoA, and orthophosphate.

W. T. Wolodk, M. E. Fraser, M. N. G. James & W. A. Bridger (1994) *J. Biol. Chem.* **269**, 10883.
P. A. Frey (1992) *The Enzymes*, 3rd ed., **20**, 141.
J. S. Nishimura (1986) *Adv. Enzymol.* **58**, 141.
W. A. Bridger (1974) *The Enzymes*, 3rd ed., **10**, 581.

O-SUCCINYLHOMOSERINE (THIOL)-LYASE

This pyridoxal-phosphate-dependent enzyme [EC 4.2.99.9], also known as cystathionine γ-synthase, catalyzes the reaction of *O*-succinyl-L-homoserine with L-cysteine to produce cystathionine and succinate. The enzyme can also use hydrogen sulfide and methanethiol as substrates, producing homocysteine and methionine, respectively. In the absence of a thiol, the enzyme can also catalyze a β,γ-elimination reaction to form 2-oxobutanoate, succinate, and ammonia.

M. A. Ator & P. R. Ortez de Montellano (1990) *The Enzymes*, 3rd ed., **19**, 213.
D. J. Creighton & N. S. R. K. Murthy (1990) *The Enzymes*, 3rd ed., **19**, 323.
I. Saint-Gerons, C. Parsot, M. M. Zakin, O. Bârzy & G. N. Cohen (1988) *Crit. Rev. Biochem.* **23**, S1.
A. E. Martell (1982) *Adv. Enzymol.* **53**, 163.
L. Davis & D. E. Metzler (1972) *The Enzymes*, 3rd ed., **7**, 33.

SUCRASE

Sucrose α-D-glucohydrolase [EC 3.2.1.48], also known as sucrase and sucrase-isomaltase, catalyzes the hydrolysis of sucrose and maltose by an α-D-glucosidase-type action. This enzyme is isolated from intestinal mucosa as a single polypeptide chain that also exhibites activity toward isomaltose (an oligo-1,6-glucosidase activity, EC 3.2.1.10) by hydrolyzing 1,6-α-D-glucosidic linkages in isomaltose and dextrins. **See also** *Invertase; Isomaltase*

E. H. Van Beers, H. A. Büller, R. J. Grand, A. W. C. Einerhand & J. Dekker (1995) *Crit. Rev. Biochem. Mol. Biol.* **30**, 197.
H. Hauser & G. Semenza (1983) *Crit. Rev. Biochem.* **14**, 319.

SUCROSE PHOSPHORYLASE

This enzyme [EC 2.4.1.7], also known as sucrose glucosyltransferase, catalyzes the reaction of sucrose with orthophosphate to produce D-fructose and α-D-glucose 1-phosphate. In the forward reaction, arsenate may replace phosphate as the substrate. However, the resulting product is unstable in aqueous solutions. In the reverse reaction, various ketoses and L-arabinose may replace D-fructose. **See** *Arsenolysis*

G. Mooser (1992) *The Enzymes*, 3rd ed., **20**, 187.
J. J. Mieyal & R. H. Abeles (1972) *The Enzymes*, 3rd ed., **7**, 515.

SULFATE ADENYLYLTRANSFERASE

This enzyme [EC 2.7.7.4], also known as ATP sulfurylase, catalyzes the reaction of ATP with sulfate to produce adenylylsulfate and pyrophosphate.

C. Dahl, N. Speich & H. G. Trüper (1994) *Meth. Enzymol.* **243**, 331.
C. Dahl & H. G. Trüper (1994) *Meth. Enzymol.* **243**, 400.
J. A. Schiff & T. Saidha (1987) *Meth. Enzymol.* **143**, 329.
I. H. Segel, F. Renosto & P. A. Seubert (1987) *Meth. Enzymol.* **143**, 334.
H. D. Peck, Jr. (1974) *The Enzymes*, 3rd ed., **10**, 651.

SULFATE ADENYLYLTRANSFERASE (ADP)

This enzyme [EC 2.7.7.5], also known as ADP sulfurylase, catalyzes the reaction of ADP with sulfate to produce orthophosphate and adenylylsulfate.

A. Schmidt & K. Jäger (1992) *Ann. Rev. Plant Physiol. Plant Mol. Biol.* **43**, 325.
H. D. Peck, Jr. (1974) *The Enzymes*, 3rd ed., **10**, 651.

SULFIDES

1. Inorganic sulfur compounds containing another (usually more electropositive) element. When the other element is an alkali or alkaline earth, the sulfide is ionic in character. Metal sulfides often have unusual stoichiometries. Examples of sulfides include H_2S, Na_2S, FeS, and HgS. 2. Organic sulfides are also referred to as thioethers and have the general structure R—S—R′. Biochemical examples of sulfides include methionine, cystathionine, and djenkolic acid. If the two R groups are identical, the substance can be referred to as a symmetrical sulfide (biological examples of which are lanthionine and homolanthionine).

SULFITE DEHYDROGENASE

This enzyme [EC 1.8.2.1] catalyzes the reaction of sulfite with two ferricytochrome *c* and water to produce sulfate and two ferrocytochrome *c*. The enzyme is associated with cytochrome *c*-551.

D. P. Kelly & A. P. Wood (1994) *Meth. Enzymol.* **243**, 501.

SULFITE OXIDASE

This heme- and molybdenum-dependent enzyme [EC 1.8.3.1] catalyzes the reaction of sulfite with dioxygen and water to produce sulfate and hydrogen peroxide.

C. Kisker, H. Schindelin & D. C. Rees (1997) *Ann. Rev. Biochem.* **66**, 233.
I. Suzuki (1994) *Meth. Enzymol.* **243**, 447.
R. C. Bray (1980) *Adv. Enzymol.* **51**, 107.
R. C. Bray (1975) *The Enzymes*, 3rd ed., **12**, 299.

SULFITE REDUCTASE

Sulfite reductase (NADPH) [EC 1.8.1.2] catalyzes the reaction of H_2S with three $NADP^+$ and three water molecules to produce sulfite and three NADPH. The enzyme requires FAD, FMN, and heme. Sulfite reductase (ferredoxin) [EC 1.8.7.1] catalyzes the iron-dependent reaction of H_2S with three oxidized ferredoxin and three water to produce sulfite and three reduced ferredoxin. Sulfite reductase (acceptor) [EC 1.8.99.1] catalyzes the iron-dependent reaction of H_2S with an acceptor and three water to produce sulfite and the reduced acceptor. A stoichiometry of six molecules of reduced methyl viologen per molecule of sulfide formed was reported. *See also Desulfofuscidin; Desulforubidin*

I. Moura & A. R. Lino (1994) *Meth. Enzymol.* **243**, 296.
C. Dahl, N. Speich & H. G. Trüper (1994) *Meth. Enzymol.* **243**, 331.
C. Dahl & H. G. Trüper (1994) *Meth. Enzymol.* **243**, 400.
H. G. Trüper (1994) *Meth. Enzymol.* **243**, 422.
Y. Hatefi & D. L. Stiggall (1976) *The Enzymes*, 3rd ed., **13**, 175.
G. Palmer (1975) *The Enzymes*, 3rd ed., **12**, 1.

SULFONIUM SALTS

Compounds containing a pyramidally arranged (hence, chiral) sulfur to which are linked three alkyl or aryl groups, resulting in a net positive charge on the sulfur. A biologically important example is *S*-adenosyl-L-methionine chloride. Sulfonium salts can also be utilized as analogs or mimics of carbocation intermediates in enzyme-catalyzed reactions. For example, methyl-(4-methylpent-3-en-1-yl)vinylsulfonium perchlorate proved to be an excellent inhibitor ($K_i = 2.5$ μM) of the enzyme that catalyzes the formation of the bicyclic (+)-α-pinene[1].

[1]R. Croteau, C. J. Wheeler, R. Aksela & A. C. Oehlschlager (1986) *J. Biol. Chem.* **261**, 7257.

SULFOTRANSFERASES

This class constitutes a large set of enzymes that catalyze the transfer of sulfate groups [EC 2.8.2.x]. *See also* specific enzyme

G. Lowe (1998) *Comprehensive Biological Catalysis: A Mechanistic Reference* **1**, 627.
S. G. Ramaswamy & W. B. Jakoby (1987) *Meth. Enzymol.* **143**, 201.

SULFUR-35 (^{35}S)

A radioactive β-emitting sulfur nuclide with a decay half-life of 87.2 days and a beta energy of 0.167 MeV. The following decay data indicate the time (in days) followed by the fraction of original amount at the specified time: 0.0, 1.000; 3, 0.976; 6, 0.954; 9, 0.931; 12, 0.909; 15, 0.888; 18, 0.867; 21, 0.847; 24, 0.827; 27, 0.807; 30, 0.788; 60, 0.621; 90, 0.490; 120, 0.386; 150, 0.304; 180, 0.240; 210, 0.189; 240, 0.149; 270, 0.118; 300, 0.093.

Special precautions: Avoid sulfurous vapors by working in a chemical fume hood. Because no organ selectively concentrates sulfur metabolites, the entire body is the critical organ in terms of dose. The effective *biological half-life* in humans is around 90 days. The maximum range in air is 0.24 meters.

SULFUR DIOXYGENASE

This enzyme [EC 1.13.11.18] catalyzes the iron-dependent reaction of sulfur with dioxygen and water to produce sulfite.

D. P. Kelly & A. P. Wood (1994) *Meth. Enzymol.* **243**, 501.

SUPEROXIDE DETECTION/ MEASUREMENT

Fridovich[1] recently summarized important aspects concerning the accurate detection and measurement of superoxide. He indicates that univalent reduction of O_2 to superoxide is a facile process, but the instability of superoxide in aqueous solutions hinders its detection and measurement. To measure intracellular superoxide, he favors use of the rapid inactivation of [4Fe-4S]-containing dehydratases (such as aconitase) by oxidation of their iron-sulfur clusters. *See Oxygen, Oxides & Oxygen Radicals*

[1]I. Fridovich (1997) *J. Biol. Chem.* **272**, 18515.

SUPEROXIDE DISMUTASES

This enzyme [EC 1.15.1.1] catalyzes the reaction of two superoxide ($2 O_2^-$) and two H^+ to produce dioxygen and hydrogen peroxide. The enzyme uses copper and zinc ions; or iron ions; or manganese ions.

I. Fridovich (1995) *Ann. Rev. Biochem.* **64**, 97.
E. Cadenas, P. Hochstein & L. Ernster (1992) *Adv. Enzymol.* **65**, 97.
C. Bowler, M. Van Montagu & D. Inzé (1992) *Ann. Rev. Plant Physiol. Plant Mol. Biol.* **43**, 83.
E. T. Adman (1991) *Adv. Protein Chem.* **42**, 145.
I. Fridovich (1986) *Adv. Enzymol.* **58**, 61.
B. L. Vallee & A. Galdes (1984) *Adv. Enzymol.* **56**, 283.
B. G. Malmström, L.-E. Andréasson & B. Reinhammer (1975) *The Enzymes*, 3rd ed., **12**, 507.
I. Fridovich (1974) *Adv. Enzymol.* **41**, 35.

SUPERACID

A substance, mixture, or molecular entity having an unusually high degree of acidity, generally significantly higher than 100% sulfuric acid. An example is magic acid, an equimolar mixture of HSO_3F and SbF_5.

P. J. Gillespie (1968) *Acc. Chem. Res.* **1**, 202.
G. A. Olah (1985) *Superacids*, Wiley, New York.

SUPERACID CATALYSIS

A term used occasionally in enzymology with respect to catalysis by metal ions.

SUPERBASE

A substance, mixture, or molecular entity having a very high degree of basicity. An example is lithium diisopropylamide.

SUPERIMPOSABILITY (or Superposability)

The ability to use translation and rotation to bring two stereochemical structures into three-dimensional coincidence without any breaking and remaking of bonds that alter the stereochemical configuration around each atom in the two stereochemical structures.

SUPERRADIANCE

Spontaneous emission that has been amplified by a passage through a population-inverted medium.

Comm. on Photochem. (1988) *Pure and Appl. Chem.* **60**, 1055.

SUPERSATURATION

A metastable condition in which the amount of a dissolved substance is greater than the condition in which the dissolved molecular entity is in equilibrium with the undissolved solid. *See* Biomineralization

SURFACE CROSSING

The crossing of two potential energy surfaces of the electron energies of two states having different symmetry.

Comm. on Photochem. (1988) *Pure and Appl. Chem.* **60**, 1055.

SURFACE TENSION

The work required to increase the area of a surface reversibly and isothermally by a unit area. The surface tension can also be described as the force exerted perpendicularly to the liquid surface. Surface tension is typically expressed in units of ergs/cm^2 or as dynes/cm.

SWAIN-LUPTON RELATION

An approach to the correlation analysis of substituent effects in chemical reactions and processes. In this treatment, there are two parameters, a field constant (F) and a resonance constant (R)[1-3]. The approach has received some criticism[4-8].

[1]C. G. Swain & E. C. Lupton (1968) *J. Amer. Chem. Soc.* **90**, 4328.
[2]C. G. Swain, S. H. Unger, N. R. Rosenquist & M. S. Swain (1983) *J. Amer. Chem. Soc.* **105**, 492.
[3]IUPAC (1994) *Pure and Appl. Chem.* **66**, 1077.
[4]W. F. Reynolds & R. D. Topson (1984) *J. Org. Chem.* **49**, 1989.
[5]A. J. Hoefnagel, W. Oosterbeek & B. M. Wepster (1984) *J. Org. Chem.* **49**, 1993.
[6]M. Charton (1984) *J. Org. Chem.* **49**, 1997.
[7]C. G. Swain (1984) *J. Org. Chem.* **49**, 2005.
[8]C. Hanson, A. J. Leo & R. W. Taft (1991) *Chem. Rev.* **91**, 165.

SWAIN RELATIONSHIPS

Quantitative relationships[1-3] between magnitudes of deuterium and tritium primary kinetic isotope effects on chemical reactivity.

$$k_H/k_T = (k_H/k_D)^{1.442}$$

$$k_H/k_T = (k_D/k_T)^{3.26-3.34}$$

See Kinetic Isotope Effects

[1]C. G. Swain, E. C. Stivers, J. F. Reuwer, Jr., & L. J. Schaad (1958) *J. Amer. Chem. Soc.* **80**, 5885.
[2]D. B. Northrop (1982) *Meth. Enzymol.* **87**, 607.
[3]B. J. Bahnson & J. P. Klinman (1995) *Meth. Enzymol.* **249**, 373.

SYMBOLIC COMPUTING

The systematic application of artificial intelligence methods for the programmed manipulation of symbols in accordance with mathematical rules[1]. Symbolic computing allows one to obtain symbolic solutions to equilibrium binding expressions which otherwise require manual solution of simultaneous equations.

[1]H. R. Halvorson (1992) *Meth. Enzymol.* **210**, 601.

SYMPORT

A transport system that couples the transport of two or more molecular entities and/or ions across a biomembrane in the same direction.

SYNCHROTRON X-RAY FOOTPRINTING

All footprinting techniques are designed to characterize binding interactions by determining the accessibility of the backbone of macromolecules to chemical/enzymatic cleavage or modification reactions. Synchrotron X-ray footprinting uses synchrotron radiation to generate reactive chemical species (such as hydroxide radical ·OH) to detect and quantitate backbone accessibility.

B. Sclavi, S. Woodson, M. Sullivan, M. Chance & M. Brenowitz (1998) *Meth. Enzymol.* **295**, 379.

SYNERGISM

Any condition of enhanced potency achieved by two agents acting in concert, thereby exceeding the sum of each agent's separate effect. Synergism implies the cooperative or coordinated action of agents (frequently called agonists and effectors), and the molecular life sciences are replete with examples of synergism.

A classical case in pharmacology is the effect of cholinergic agents on the behavior of ion channels of the electric eel, *Electrophorus*. Measurements of membrane potentials using microelectrodes permitted Changeux and Podleski[1] to analyze the extent of depolarization as a function of the concentration of various agonists, either separately or in combination with each other. With phenyltrimethylammonium ion (or PTA), cooperativity was evident as a sigmoidal dose-response curve (Hill coefficient = 1.7). While antagonists had no effect on the cooperativity of PTA action, the presence of a second cholinergic agonist (decamethonium) completely abolished any evidence of PTA cooperativity. Thus, the two agonists acted synergistically, and their combined action was not simply the additive effect of their separate actions.

Another excellent example of synergism is the activation of a soluble adenylyl cyclase by forskolin and the α-

subunit of the heterotrimeric guanine nucleotide-binding protein (abbreviated: $G_s\alpha$)[1]. Shown in the figure below are plots of cyclase activity *versus* the concentration of forskolin (Panel A) or $G_s\alpha$ (Panel B). The additive sum of these agents acting separately as well as their concerted action are plotted in Panel C.

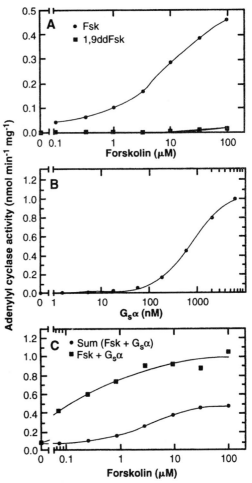

Plot of the enzymatic activity[2] of recombinant adenylyl cyclase (IC$_1$IIC$_2$-L$_3$). A, Activation by forskolin (Fsk), but not the inactive analogue 1,9-dideoxyforskolin (1,9-ddFsk). B, Activation by $G_s\alpha$ in the presence of GTPγS. C, Synergistic activation by Fsk and $G_s\alpha$ (again in the presence of GTPγS). *(Reproduced with permission of the authors and the American Association for the Advancement of Science.)*

Cooperative binding interactions of allosteric systems are also well-known cases of synergism. The archetype is hemoglobin which interacts with molecular oxygen, 2,3-bisphosphoglycerate, protons, and poisons such as carbon monoxide. In the Monod-Wyman-Changeux symmetry-conserving model for cooperativity, partial oxygenation displaces the so-called $R_o \rightleftharpoons T_o$ equilibrium to provide a high concentration of sites for ligand binding with *no* effect on the affinity of each site for O_2. In the

Koshland-Némethy-Filmer ligand-induced conformational change model, binding of oxygen at the first site results in a conformational change that alters the affinity of neighboring sites for oxygen. The MWC model affects the concentration of available binding sites, and the KNF model achieves synergism by altering the binding constants for ligands. *See* Allosterism; Cooperativity; Monod-Wyman-Changeux Model; Koshland-Némethy-Filmer Model

[1]J. P Changeux & T. R. Poleski (1968) *Proc. Natl. Acad. Sci. U.S.A.* **59**, 944.
[2]W. J. Tang & A. G. Gilman (1995) *Science* **268**, 1769.

SYNERGISTIC INHIBITION

A term used to describe inhibition induced by a second inhibitor. A particular inhibitor (I_1) may be a weak inhibitor for a certain enzyme; however, in the presence of a second inhibitor (I_2), inhibition is greatly enhanced. An example of such a system is the inhibitory effect of inorganic pyrophosphate on trichodiene synthase by certain aza analogues[1]. Both enantiomers of serine are weak inhibitors of γ-glutamyl transpeptidase (L-serine has a K_i value of 10.7 mM), but in the presence of borate ion, they are potent inhibitors[2–5].

Another example is the concerted inhibition model illustrated in the following scheme:

$$E + S \rightleftharpoons ES \rightarrow E + P$$
$$E + I \rightleftharpoons EI$$
$$EX + S \rightleftharpoons ESX \rightarrow EX + P$$
$$EX + Y \rightleftharpoons EXY$$

where X and Y are feedback inhibitors. Ligand X will have no effect on the reaction rate if the K_m and V_{max} remain the same as those obtained in the absence of X. Ligand Y is likewise ineffective alone, because the binding of ligand Y requires the presence of EX complex. However, in the presence of both X and Y, the enzyme is drawn into the catalytically inactive EXY complex.

[1]D. E. Cane, G. Yang, R. M. Coates, H.-J. Pyun & T. M. Hohn (1992) *J. Org. Chem.* **57**, 3454.
[2]J. P. Revel & E. G. Ball (1959) *J. Biol. Chem.* **234**, 577.
[3]S. S. Tate & A. Meister (1978) *Proc. Natl. Acad. Sci. U.S.A.* **75**, 4806.
[4]G. A. Thompson & A. Meister (1977) *J. Biol. Chem.* **252**, 6792.
[5]R. D. Allison (1985) *Meth. Enzymol.* **113**, 419.

SYSTEM

A term used in thermodynamics to designate a region separated from the rest of the universe by definite boundaries. The system is considered to be isolated if any change in the surroundings (*i.e.*, the portion of the universe outside of the boundaries of the system) does not cause any changes within the system. *See* Closed System; Isolated System; Open System

SYSTEMATIC ERRORS

Those errors arising either from a specific step in the experimental design or as a result of error introduced by an apparatus or instrument (*e.g.*, calibrated micropipette, pH meter or electrode, analytical balance, scintillation counter, *etc.*). Systematic errors are frequently present when one first attempts to institute a new experimental protocol, and thoughtful consideration of each step in a procedure helps to eliminate or minimize systematic errors. There are three types of error: (a) systematic error; (b) random error, itself related to random variation in sample quality, handling, or measurement; and (c) consistent error which is constant and independent of experimental conditions. As an example of systematic error, the failure to remove a yellow colored substance introduced in a specific step in an experiment may lead to quenching of radioactive samples during liquid scintillation counting. This would introduce a one-sided error by yielding lower than actual values for the observed counts per minute (cpm) for all radioactive samples measured in the experiment. Impurity can often prove to be the source of many systematic errors. Likewise, some pH electrodes do not give the correct pH in the presence of Tris buffer; failure to recognize this problem would introduce systematic error. ***Also see*** Random Error; Statistics (A Primer)

T

TANDEM MASS SPECTROMETRY

A high-resolution, double mass spectrometric (or MS/MS) technique that uses one mass spectrometer to separate mixtures of molecular ions into groups which upon entry into a second mass spectrometer are further fragmented and resolved into their characteristic mass spectra. Typically, the first mass spectrometer uses a chemical ionization method to impart kinetic energy with little or no fragmentation of each species; the second MS unit then relies on collisions of the parent ions with helium atoms to generate daughter ions which are characteristic and often predictable fragments of their respective parent ions.

TAU (τ)

1. Symbol for relaxation time. 2. Symbol for a time constant. 3. Symbol for mean life.

TAUTOMERISM

A rapid and usually reversible internal isomerization between two or more structurally distinct compounds. The intramolecular process entails a heterolytic cleavage of a chemical bond followed by a recombination of the fragments, usually accompanied by the migration of a double bond or by ring opening and ring closing. The most common form of tautomerism involves the transfer of a proton (*i.e.*, prototropy or prototropic rearrangement). An example of this form of tautomerism is keto-enol tautomerism between a carbonyl group having an α hydrogen and its enol form:

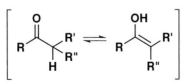

Keto-Enol Tautomerism

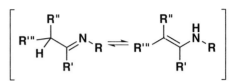

Imine-Enamine Tautomerism

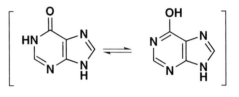

Lactam-LactimTautomerism

This reaction can be facilitated by both general acid and general base catalysis. In most cases, the equilibrium favors the keto form. For example, the percent enol content of acetaldehyde is only 1×10^{-3} while the enol percentage in acetone is 6×10^{-7}. Aqueous solutions of pyruvate and acetoacetate have less than 0.1% enol; however, the enol content of acetylacetone is 76.4%. The extent of enolization is greatly affected by the presence of double bonds or aromatic moieties that can be conjugated to the enol. In addition, the equilibria is affected by the nature of the solvent, by concentration, temperature, pH of the solution, and by the presence of metal ions. Thus, any perturbation of one or more of the factors can result in a displacement of the keto-enol equilibrium. This observation provides a tool in probing tautomeric specificity of particular proteins as well as the ability to assay enzymes that catalyze a keto-enol interconversion[1]; for example, phenylpyruvate tautomerase[2] [EC 5.3.2.1] and oxaloacetate tautomerase[3] [EC 5.3.2.2]. Following displacement of the equilibria, rates of ketonization or enolization can be followed by coupling the reaction to an enzyme that is specific for only one of the tautomeric species[3,4]. **See** *Rearrangements; Anomeric Specificity; Valence Tautomerism*

[1]I. A. Rose (1982) *Meth. Enzymol.* **87**, 84.
[2]F. Blasi, F. Fragomele & I. Covelli (1969) *J. Biol. Chem.* **244**, 4864.

[3]R. G. Arnett & G. W. Kosicki (1969) *J. Biol. Chem.* **244**, 2059.
[4]G. W. Kosicki (1962) *Can. J. Chem.* **40**, 1280.

TELOMERASE

An enzyme that uses an RNA template to add DNA to the ends of chromosomes. Telomerase is normally active only in stem cells and those cells giving rise to sperm and egg, but telomerase also undergoes activation when cells become cancerous. In the latter case, telomerase action allows transformed cells to replicate without a limit, a process termed "immortalization".

C. W. Greider (1996) *Ann. Rev. Biochem.* **65**, 337.
E. H. Blackburn (1992) *Ann. Rev. Biochem.* **61**, 113.

TEMPERATURE

A measure of the energy content or thermal content of a body or entity. A parameter that determines the directional transfer of heat to or from other bodies or entities. The SI temperature unit is the kelvin (abbreviated as K, and not °K).

TEMPERATURE CONTROL

Because enzymic reactions are so sensitive to variations in temperature, suitable care must be exercised such that the temperature of the system is constant and under adequate control. Ideally, the temperature-regulated water bath and/or circulator should reduce any variability to within less than $\pm 0.1°C$. Before initiating any reaction kinetic measurement, all samples should be thermally equilibrated. This process may take several minutes, depending on the nature of the sample container and the volume of the sample.

When assays are conducted with a instrument such as a spectrophotometer, the sample compartment in the instrument should also be equilibrated to the same temperature. Typically, the control of temperature inside the sample compartment is less than that of the water bath or circulator. Many instruments now provide a temperature readout for the sample compartment that allows an investigator the opportunity to directly assess temperature control issues in the instrument.

If the assay temperature is significantly different than room temperature (*e.g.*, more than 10°C above or below), temperature control becomes a more serious concern. Simply mixing a reaction sample may perturb the sample temperature. Hence, after thermal equilibrium, handling

should be minimized. A wide variety of protocols have been utilized to achieve this end[1].

Ideally, the enzyme solution to be added to the reaction solution should be likewise thermally equilibrated prior to initiating the reaction. However, the thermal instability of many enzymes may require their storage in an ice bucket. An aliquot of the cold enzyme can then be brought up to temperature just prior to the initiation of the reaction. If the enzyme solution cannot be pre-warmed, then care has to be exercised in the analysis of the experimental data. For example, assume that an enzyme had a temperature dependency (Q_{10}) of two (*i.e.*, for every 10°C change in temperature, there is a twofold change in reaction rate). Thus, if a 0.1 mL aliquot of enzyme at 0°C is added to a reaction mixture of 3.0 mL at 30°C, the reaction velocity may be perturbed by as much as 20%. In such cases, the initial velocity should only be measured after the ongoing reaction has reached thermal equilibration.

[1]R. D. Allison & D. L. Purich (1979) *Meth. Enzymol.* **63**, 3.

TEMPERATURE DEPENDENCE

Dixon and Webb[1] provided a useful list of various causes for the shape of the temperature effect seen in enzyme-catalyzed reactions: (1) effect on enzyme stability; (2) effect on the actual velocity of the reaction (especially on k_{cat}); (3) effect on affinity(ies) of the substrate(s) (*i.e.*, K_d values); (4) effect on the pH function of any or all components of the reactions (including the buffer); (5) effect on the affinity(ies) of enzyme effector(s); (6) an alteration in the rate determining (or rate contributing) step(s); (7) effect on the coupling enzymes of the assay; and (8) effect on physical properties (*e.g.*, solubility of substrates, particularly gas substrates such as O_2 and N_2, dielectric constant of the solvent, *etc.*).

Clearly, many of these effects can be addressed through preliminary studies. Incubation of the enzyme at one temperature while assaying at a different temperature will address issues concerned with protein stability as well as effects of temperature on any coupling enzymes in the assay protocol. Nevertheless, temperature studies on each enzyme in a multienzyme system should be addressed individually at an early point in the investigation. Issues related to the effects on affinities can usually be minimized by assaying under saturating conditions. Care should be exercised in this regard since, if the affinity of a substrate is altered significantly, a concentration that is saturating at one temperature may not be saturating at a different temperature. In a related issue, saturation may be difficult to achieve if the enzyme expresses substrate inhibition behavior at elevated substrate concentrations. Finally, since the pH may vary significantly with temperature and since pK_a values are temperature-dependent, a good temperature study of an enzyme is usually preceded by a pH study. At the very least, one should check the pH of the solution at different temperatures over the temperature range to be studied. *See Arrhenius Law; van't Hoff Relationship; Transition-State Theory*

[1]M. Dixon & E. C. Webb (1979) *Enzymes*, Academic Press, New York.

TEMPERATURE DEPENDENCE OF KINETIC ISOTOPE EFFECTS

Observation of the temperature dependence of k_H/k_T and k_D/k_T and k_H/k_D kinetic isotope effects *via* Arrhenius plot is very useful in detecting and characterizing proton tunneling[1].

[1]B. J. Bahnson & J. P. Klinman (1995) *Meth. Enzymol.* **249**, 373.

TEMPLATE CHALLENGE METHOD

A procedure used to assess the processivity of nucleic acid polymerases[1]. In this method, the investigator perturbs the polymerization reaction proceeding on one template by the addition of a new template. *See Processivity*

[1]W. R. McClure & Y. Chow (1980) *Meth. Enzymol.* **64**, 277.

TENSEGRITY

A model[1] for explaining how cells convert the mechanical signals (*i.e.*, the physical forces of gravity, hemodynamic stresses, and movement) into a chemical response. The model relies on recent experimental findings suggesting that cells use tensegrity architecture for their organization. Tensegrity predicts that cells are hardwired for immediate response to mechanical stresses transmitted over cell surface receptors coupling the cytoskeleton to extracellular matrix. Mechanical signals are thought to be integrated with other environmental signals and transduced into a biochemical response through force-dependent changes in scaffold geometry or molecular mechanics.

[1]D. E. Ingber (1997) *Annu. Rev. Physiol.* **59**, 575.
[1]D. E. Ingber (1998) *Sci. Amer.* **278** (1), 48.

TERMINATION

1. The step[1-3] or steps in a chain reaction in which reactive intermediates are stabilized and rendered inactive, thus ending the chain reaction. For example, a common termination reaction is the combination of two radicals: *i.e.*, $R\cdot + R'\cdot \rightarrow R-R'$. Another type of termination reaction is disproportionation: *e.g.*, $2CH_3-CH_2\cdot \rightarrow$ ethane + ethene. 2. The quenching of a reaction. 3. The halting of any process.

[1]S. J. Lapporte (1960) *Angew. Chem.* **72**, 759.
[2]M. J. Gibian & R. C. Corley (1973) *Chem. Rev.* **73**, 441.
[3]I. V. Khudyakov, P. P. Levin, & V. A. Kuzman (1980) *Russ. Chem. Rev.* **49**, 982.

TERNARY COMPLEX

A complex involving three species (such as the EAB complex formed in the ordered Bi Bi mechanism). *See Multisubstrate Mechanisms*

TER TER REACTION SYSTEMS

An enzyme-catalyzed reaction in which there are three substrates and three products. *See Multisubstrate Mechanisms*

TER UNI REACTION

An enzyme-catalyzed reaction in which there are three substrates and a single product. *See Multisubstrate Mechanisms*

TESLA

The SI unit for magnetic flux density or magnetic induction (symbolized by T) equal to a weber per square meter. One tesla is equivalent to 10^4 gauss.

TETRAHEDRAL INTERMEDIATE

An intermediate, often transient, appearing in a chemical or enzymatic reaction in which a carbon atom, which had been double-bonded (*i.e.*, in a trigonal structure) in a particular molecular entity, has been transformed to a carbon center having a tetrahedral arrangement of substituents. Tetrahedral intermediates of proteases have been stabilized with cryoenzymological techniques[1].

[1]R. J. Coll, P. D. Compton & A. L. Fink (1982) *Meth. Enzymol.* **87**, 66.

TETRAHYDROPTEROYLTRIGLUTAMATE METHYLTRANSFERASE

This enzyme [EC 2.1.1.14], also known as 5-methyltetrahydropteroyltriglutamate:homocysteine *S*-methyltransferase and methionine synthase, catalyzes the reaction of 5-methyltetrahydropteroyltri-L-glutamate with L-homocysteine to produce tetrahydropteroyltri-L-glutamate and L-methionine. The reaction requires the presence of phosphate. The enzyme isolated from *E. coli* also requires a reducing system. ***See*** *N^5-Methyltetrahydrofolate:Homocysteine Methyltransferase*

I. Saint-Gerons, C. Parsot, M. M. Zakin, O. Bârzy & G. N. Cohen (1988) *Crit. Rev. Biochem.* **23**, S1.

THERAPEUTIC RATIO

A ratio that characterizes the relative effectiveness of a pharmaceutical. It is equal to the maximally tolerated dose divided by the minimal curative or effective dose; or, to LD_{50}/ED_{50}.

THERMAL EQUILIBRATION

A necessary step for any initial rate study. Whenever practicable, the reaction solution should be preincubated at the final temperature. (This might not be possible for components that are unstable at that final temperature.) If enzyme is used to initiate the reaction, the enzyme stock solution should be briefly warmed. Occasionally a rapidly warmed enzyme may display hysteresis as a result of some temperature-dependent conformational change.

THERMAL LENSING

A method for determining the alteration of the refractive index of a medium as a result of the temperature rise in the path of a beam of coherent light absorbed by the medium. Thermal lensing can also occur with pigmented proteins, and this phenomenon can influence the accuracy of concentration gradient measurements in small aperture flow cuvettes as well as in ultracentrifugation.

Comm. of Photochem. (1988) *Pure and Appl. Chem.* **60**, 1055.

THERMAL TRAPPING

The isolation of an intermediate or unstable species by the lowering of the temperature[1,2]. The basic strategy is that one can "accumulate" otherwise unstable chemical species within potential energy wells along a reaction coordinate. ***See also*** *Cryoenzymology*

[1]A. L. Fink & M. A. Geeves (1979) *Meth. Enzymol.* **63**, 336.
[2]D. L. Purich (1982) *Meth. Enzymol.* **87**, 3.

THERMODYNAMIC CONTROL

If a particular molecular entity(ies) participates in two or more parallel reactions and the proportion of the resulting products is determined by the relative equilibrium constants for the interconversion of reaction intermediates on or after the rate-determining step(s), then the more prevalent product is said to be thermodynamically controlled (*i.e.*, the more stable product will be the one formed in highest amounts). If the reactions are reversible and the system is allowed to go to equilibrium, the favored product is the thermodynamically controlled species. A synonymous term is equilibrium control. ***See also*** *Kinetic Control*

G. W. Klumpp (1982) *Reactivity in Organic Chemistry*, p. 36, Wiley, New York.

THERMODYNAMIC CYCLE

A closed cycle of ligand binding interactions or chemical reactions that can be fruitfully analyzed by taking advantage of thermodynamic relationships holding for branched pathways.

Consider the case of glucose (Glc) and ATP interactions with hexokinase (HK) in the absence of catalysis (which can be accomplished by omitting magnesium ion).

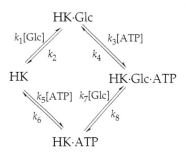

where $K_1 = k_2/k_1 = $ [HK][Glc]/[HK·Glc], $K_2 = k_4/k_3 = $ [HK·Glc][ATP]/[HK·Glc·ATP], $K_3 = k_6/k_5 = $ [HK][ATP]/[HK·ATP], and $K_4 = k_8/k_7 = $ [HK·ATP][Glc]/[HK·Glc·ATP]. The changes in free energy from one enzyme form to another are independent of the path taken. Thus, $K_1K_2 = K_3K_4$, and knowledge of any three of the equilibrium constants allows one to determine the fourth.

Another excellent example is δ-chymotrypsin, a protease whose active conformation is maintained by a salt bridge between the α-ammonium group on isoleucine-16 and the carboxylate anion of aspartate-194. Because pH-dependent changes in enzyme activity are linked to the maintenance or loss of this salt bridge, Fersht & Requena[1,2] set out to determine how pH affects the enzyme's conformational equilibria. In a clever set of stopped-flow experiments in which the enzyme and the dye proflavin were rapidly combined, they took advantage of the fact that there is a burst in the 465 nm absorbance when the dye binds to the active conformation. This burst is then followed by a first-order activation attended by additional proflavin binding. By analyzing the kinetics of proflavin binding as a function of pH, these investigators constructed the following thermodynamic diagram for the making/breaking of this essential salt bridge. For an example of a thermodynamic cycle in the analysis of actin polymerization, **See** *KINSIM/FITSIM.*

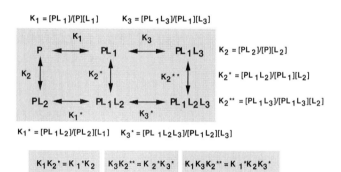

Finally, in order to deal with the complexity of some ligand binding systems, one requires the use of more than a single thermodynamic cycle. While such a system is necessarily more complicated, the equilibrium relationships among dissociation constants allows one to dissect important features of the process.

[1]A. R. Fersht & Y. Requena (1971) *J. Mol. Biol.* **60**, 279.
[2]A. R. Fersht (1971) *Cold Spring Harbor Symp.* **36**, 71.

THERMODYNAMIC EQUATIONS OF STATE

Expressions for internal energy, U, and enthalpy, H, in terms of pressure (P), volume (V), and absolute temperature (T):

$$(\partial U/\partial T)_T = -P + T(\partial P/\partial T)_V$$

$$(\partial H/\partial P)_T = V - T(\partial V/\partial T)_P$$

THERMODYNAMIC pK_a

The pK_a value of an ionizable group extrapolated to zero ionic strength.

THERMODYNAMICS, LAWS OF

Zeroth Law of Thermodynamics. If any two bodies are each in thermal equilibrium with a third body, then all three bodies are in thermal equilibrium with each other.

First Law of Thermodynamics. The total amount of energy within a closed system is constant (*i.e.*, the total energy of the system is conserved). Mathematically, this can be expressed as $\Delta U = q + w$ where ΔU is the change in internal energy, q is the heat transferred to the system, and w is the work done on the system (or, $dU = dq + dw$). Internal energy is a state function (*i.e.*, it is dependent only on the initial and final states and not on the path between those states). In addition, the validity of the first law means that perpetual motion machines are impossible. **See** *Conservation of Energy*

If nuclear reactions are to be considered, then the first law has to consider changes in mass associated with changes in energy. However, for normal chemical and biochemical processes, the change in mass is too small to be a factor.

Second Law of Thermodynamics. There have been numerous statements of the second law. To paraphrase Clausius: It is impossible to devise an engine or process which, working in a cycle, will produce no effect other than the transfer of heat from a colder to a warmer body. According to Caratheodory, the Second Law can be stated as follows[1]: "Arbitrarily close to any given state of any closed system, there exists an unlimited number of other states which it is impossible to reach from a given state as a result of any adiabatic process, whether reversible or not".

A corollary to this law is that all reversible Carnot cycles, operating between the same two temperatures, must have the same efficiency. For a perfectly reversible cycle in which the only pressure, P, is uniform and external, then $dU = TdS - PdU$ where dU is the differential change in internal energy, T is the absolute temperature, dS is the differential change in entropy, and dV is the differential change in volume. This relation, which is a combination of the first and second laws, is for systems of constant composition in which the only work done on or by the system is *PV* work. Another expression of the second law is that the entropy of a closed system increases with time. **See also** *Entropy; Caratheodory, Principle of; Carnot Cycle; Efficiency; Innate Thermodynamic Quantities*

Third Law of Thermodynamics. Also referred to as the Nernst heat theorem, this law states that it is impossible to reduce the temperature of any system, via a finite set of operations, to absolute zero. For any changes involving perfectly crystalline solids at absolute zero, the change in total entropy is zero (thus, $\Delta S_{0\,K} = 0$). A corollary to this statement is that every substance, at $T > 0$ K, must have a positive and finite entropy value. The entropy of that substance is zero only at absolute zero when that substance is in pure, perfect crystalline form. **See** *Entropy*

[1]J. T. Edsall & J. Wyman (1958) *Biophysical Chemistry*, p. 147, Academic Press, New York.

Selected entries from *Methods in Enzymology* [vol, page(s)]:
Pathway of allosteric control as revealed by intermediate states of hemoglobin, **259**, 1; probes of energy transduction in enzyme catalysis, **259**, 19; macromolecules and water: probing with osmotic stress, **259**, 43; linkage of protein assembly to protein-DNA binding, **259**, 95; linkage at steady state: allosteric transitions of thrombin, **259**, 127; thermal denaturation methods in the study of protein folding, **259**, 144; kinetics of lipid membrane phase transitions: a volume perturbation calorimeter study, **259**, 169; tight binding affinities determined from thermodynamic linkage to protons by titration calorimetry, **259**, 183; calorimetric methods for interpreting protein-ligand interactions, **259**, 194; extracting thermodynamic data from equilibrium melting curves for oligonucleotide order-disorder transitions, **259**, 221; predicting thermodynamic properties of RNA, **259**, 242; thermodynamics and mutations in RNA-protein interactions, **259**, 261; melting studies of RNA unfolding and RNA-ligand interactions, **259**, 281; structural-perturbation approaches to thermodynamics of site-specific protein-DNA interactions, **259**, 305; thermodynamic parameters from hydrogen exchange measurements, **259**, 344; application of pressure to biochemical equilibria: the other thermodynamic variable, **259**, 357; molecular volume, **259**, 377; hydrostatic and osmotic pressure as tools to study macromolecular recognition, **259**, 395; sedimentation equilibrium as thermodynamic tool, **259**, 427; footprint phenotypes: structural models of DNA-binding proteins from chemical modification analysis of DNA, **259**, 452; low-temperature electrophoresis methods, **259**, 468; use of multiple spectroscopic methods to monitor equilibrium unfolding of proteins, **259**, 487; probing structural and physical basis of protein energetics linked to protons and salt, **259**, 512; evaluating contribution of hydrogen bonding and hydrophobic bonding to protein folding, **259**, 538; analyzing solvent reorganization and hydrophobicity, **259**, 555; simple force field for study of peptide and protein conformational properties, **259**, 576; probes for analysis of stability of different variants of aspartate aminotransferase, **259**, 590;

thermodynamic approaches to understanding aspartate transcarbamylase, **259**, 608; on the interpretation of data from isothermal processes, **259**, 628.

THERMOLYSIN

This enzyme [EC 3.4.24.27], also known as *Bacillus thermoproteolyticus* neutral proteinase, is a thermostable extracellular metalloendopeptidase containing zinc and four calcium ions. A member of the peptidase family M4, this enzyme catalyzes the hydrolysis of peptide bonds with a preference for Xaa—Leu > Xaa—Phe.

J. E. Coleman (1992) *Ann. Rev. Biochem.* **61**, 897.
D. W. Christianson (1991) *Adv. Protein Chem.* **42**, 281.
E. Lolis & G. A. Petsko (1990) *Ann. Rev. Biochem.* **59**, 597.
B. L. Vallee & A. Galdes (1984) *Adv. Enzymol.* **56**, 283.
J. S. Fruton (1982) *Adv. Enzymol.* **53**, 239.

THETA (Θ)

1. Symbol (Θ) for characteristic temperature. 2. Symbol (θ) for degree of saturation of binding sites as defined in the Langmuir isotherm treatment for adsorption of a ligand onto a surface. **See** *Langmuir Isotherm*. 3. Symbol (θ) for plane angle. 4. Symbol (θ) for one of the space coordinates in the three-dimensional, spherical polar coordinate system. 5. Symbol (θ) for Celsius temperature.

THIAMIN: PHOSPHATES & ANALOGUES

A vitamin whose pyrophosphorylated form is an essential coenzyme in so-called α condensation and α cleavage reactions, whose cardinal feature is that the scissile bond always lies immediately adjacent to the carbonyl group.

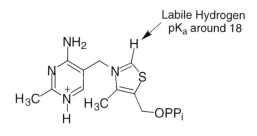

Selected entries from *Methods in Enzymology* [vol, page(s)]:
Acid dye method for the analysis of thiamin, **18A**, 73; electrophoretic separation and fluorometric determination of thiamin and its phosphate esters, **18A**, 91; catalytic polarography in the study of the reactions of thiamin and thiamin derivatives, **18A**, 93; preparation of thiamin derivatives and analogs, **18A**, 141; preparation of the mono- and pyrophosphate esters of 2-methyl-4-amino-5-hydroxymethylpyrimidine for thiamin biosynthesis, **18A**, 162; formation of the pyrophosphate ester of 2-methyl-4-amino-5-hydroxymethylpyrimidine by enzymes from brewers' yeast in thiamin biosynthesis, **18A**, 203; resolution, reconstitution, and other methods for the study of binding of thiamin pyrophos-

phate to enzymes, **18A**, 238; thiamin antagonists, **18A**, 245; enzymatic preparation, isolation, and identification of 2-α-hydroxyalkylthiamin pyrophosphates, **18A**, 259; an improved procedure for the determination of thiamin, **62**, 51; differential determination of thiamin and its phosphates and hydroxyethylthiamin and pyrithiamine, **62**, 54; fluorometric determination of thiamin and its mono-, di-, and triphosphate esters, **62**, 58; separation of thiamin phosphoric esters on Sephadex cation exchanger, **62**, 59; high-pressure liquid chromatography of thiamin, thiamin analogs, and their phosphate esters, **62**, 63; electrophoretic separation of thiamin and its mono-, di-, and triphosphate esters, **62**, 68; thiamin phosphate pyrophosphorylase, **62**, 69; synthesis of pyrimidine ^{14}C-labeled thiamin, **62**, 73; thiamin transport in *Escherichia coli*, **62**, 76; assay of thiamin-binding protein, **62**, 91; isolation and characterization of *Escherichia coli* mutants auxotrophic for thiamin phosphates, **62**, 94; a radiometric assay for thiamin pyrophosphokinase activity, **62**, 101; assay of thiamin pyrophosphokinase (ATP:thiamin pyrophosphotransferase) using anion-exchange paper disks, **62**, 103; affinity chromatography of thiamin pyrophosphokinase from rat brain on thiamin mono-phosphate-agarose, **62**, 105; enzymatic formation of thiamin pyrophosphate in plants, **62**, 107; preparation of thiamin triphosphate, **62**, 112; determination of thiaminase activity using thiazole-labeled thiamin, **62**, 113; membrane-bound thiamin triphosphatase, **62**, 118; transition-state analogs of thiamin pyrophosphate, **62**, 120.

THIOLESTER INTERMEDIATE

Thermodynamically unstable substances (structure below) that serve as enzymatic reaction intermediates and are typically formed by carboxylation (*i.e.*, carboxyl group transfer) of a sulfhydryl group on a cysteinyl residue of an enzyme.

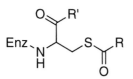

Thiolester intermediates are formed in the glyceraldehyde-3-phosphate dehydrogenase and many CoA-linked enzymatic reactions.

THIOLS

Sulfhydryl groups play important roles in enzymatic reactions, serving in some cases as active site nucleophiles and in other instances as polar groups that stabilize subunit interactions in multisubunit proteins.

Selected entries from *Methods in Enzymology* [vol, page(s)]: Assay with 2,2-dithiobisnitrobenzoic acid method, **233**, 381-382; pH effects, **233**, 384-385; formed by incubation with dithiothreitol, labeling, **233**, 409-410; protein thiol assay, **234**, 273-274; labeling, **233**, 414; reactivity with free radicals and reactive oxygen species, **233**, 405; reaction with ferrylmyoglobin, **233**, 196-197.

THREONINE DEHYDRATASE

This pyridoxal-phosphate-dependent enzyme [EC 4.2.1.16], also known as threonine deaminase and L-ser-

ine dehydratase, catalyzes the reaction of L-threonine with water to produce 2-oxobutanoate, ammonia, and water. The enzyme will also use L-serine as a substrate (thus, the activity of serine dehydratase).

H. E. Umbarger (1973) *Adv. Enzymol.* **37**, 349.
L. Davis & D. E. Metzler (1972) *The Enzymes*, 3rd ed., **7**, 33.

THRESHOLD POTENTIAL

The membrane potential at an axon hillock below which an action potential will be generated.

THROMBIN

Thrombin [EC 3.4.21.5], also known as fibrinogenase, catalyzes the hydrolysis of peptide bonds, exhibiting preferential cleavage for the Arg—Gly peptide bond. The enzyme, a member of the peptidase family S1, activates fibrinogen to fibrin and releases fibrinopeptide A and B. Thrombin, formed from prothrombin, is more selective in peptide hydrolysis than trypsin or plasmin.

E. Di Cera, Q. D. Dang, Y. Ayala & A. Vindigni (1995) *Meth. Enzymol.* **259**, 127.
M. T. Stubbs & W. Bode (1993) *Thrombosis Res.* **69**, 1.
L. Lorand & K. G. Mann, eds. (1993) *Meth. Enzymol.*, vols. **222** and **223**.
W. Bode, D. Turk & A. Karshikov (1992) *Prot. Sci.* **1**, 426.
L. J. Berliner (1992) *Thrombin, Structure & Function*, Plenum Press, New York.
M. F. Scully (1992) *Essays in Biochem.* **27**, 17.
E. W. Davie, K. Fujikawa, K. Kurachi & W. Kisiel (1979) *Adv. Enzymol.* **48**, 277.
S. Magnusson (1971) *The Enzymes*, 3rd ed., **3**, 278.

THROMBOXANE A SYNTHASE

This heme-thiolate-dependent enzyme [EC 5.3.99.5] catalyzes the conversion of (5Z,13E)-(15S)-9α,11α-epidioxy-15-hydroxyprosta-5,13-dienoate to (5Z,13E)-(15S)-9α,11α-epoxy-15-hydroxythromba-5,13-dienoate. Thus, it converts prostaglandin H_2 into thromboxane A_2.

H. B. Dunford (1998) *Comprehensive Biological Catalysis: A Mechanistic Reference* **3**, 195.
C. R. Pace-Asciak & W. L. Smith (1983) *The Enzymes*, 3rd ed., **16**, 543.

THYMIDINE 2'-HYDROXYLASE

This enzyme [EC 1.14.11.3], also known as thymidine:2-oxoglutarate dioxygenase, catalyzes the reaction of 2-deoxyuridine with α-ketoglutarate (or, 2-oxoglutarate) and dioxygen to produce uridine, succinate, and carbon dioxide. The enzyme, which also can act on thymidine, requires iron ions and ascorbate.

O. Hayaishi, M. Nozaki & M. T. Abbott (1975) *The Enzymes*, 3rd ed., **12**, 119.

THYMIDYLATE SYNTHASE

Thymidylate synthase [EC 2.1.1.45] reductively methylates 2'-deoxyuridine-5'-monophosphate to form 2'-deoxythymidine-5'-monophosphate in the following folate-dependent reaction: dUMP + N^5,N^{10}-methylene-tetrahydrofolate → dTMP + dihydrofolate.

Mechanistic and structural investigations, largely carried out by Santi's and Stroud's groups at the University of California, San Francisco, have provided important insights about the reaction mechanism of thymidylate synthase (TS), as well as the mode of inhibition of the antineoplastic agent fluorouracil[1]. The ordered binding sequence of the substrates deoxyuridine monophosphate (dUMP) and methylenetetrahydrofolate (CH$_2$H$_4$folate) can be understood from the structures, where each forms a large part of the binding site for the other. The enzyme forms a complex with 5-fluoro-2'-deoxyuridine 5'-monophosphate (FdUMP), and the TS-FdUMP-CH$_2$H$_4$folate complex shows a differential absorbance maximum at 326 nm, and is stable to SDS-PAGE. FdUMP is a slow, tight binding inhibitor of TS. The nature of covalently bound FdUMP has also been probed by ^{19}F nuclear magnetic resonance[2]. Thermodynamic analysis[3] of the binding of 5-FdUMP to thymidylate synthase over a range of temperatures supports the observation that this inhibitor binds to TS without producing profound conformational changes in the protein dimer. TS also catalyzes the dehalogenation of 5-bromo- and 5-iodo-2'-dUMP[4]. Wataya & Santi[5] observed a secondary kinetic isotope effect in the cysteine-promoted dehalogenation of 5-bromo-2'-deoxyuridine, providing evidence for the transient formation of a 5,6-dihydropyrimidine intermediate in the TS reaction.

Thymidylate synthase also catalyzes an exchange[6] of tritium of [5-^{3}H]dUMP for water protons in the absence of CH$_2$H$_4$folate. The turnover number for this exchange reaction is about 1/45,000 that of dTMP formation, and the K_m is 1.2×10^{-5} M, about the same as the K_d of the enzyme-dUMP complex estimated by equilibrium dialysis. The exchange reaction provided compelling evidence that the enzymic reaction involves attack of an enzyme nucleophile on the 6 position of dUMP to pro-vide a 5,6-dihydro-dUMP intermediate that becomes covalently bound to the enzyme.

Birdsall *et al.*[7] examined the X-ray structures of ternary complexes of the wild-type TS in attempts to discern the structural cues that choreograph three steps along the reaction pathway: (a) binding of substrates; (b) orientation; and (c) release of products. Each complex was formed by diffusion of either the CH$_2$H$_4$folate or its analog 10-propargyl-5,8-dideazafolate into crystals of thymidylate synthase with bound dUMP. A two-substrate/enzyme complex forms without further reaction. These investigators found that the imidazolidine ring is unopened, and the pterin ring of the CH$_2$H$_4$folate binds at an unproductive "alternate" site. They suggested that pterin binding at this site is an initial interaction that precedes all catalysis. The structure of the enzyme complex with dUMP and 10-propargyl-5,8-dideazafolate showed the substrate and cofactor analog bound in orientations favorable for initiating catalysis and may resemble the productive complex. Another complex involving dTMP and 7,8-dihydrofolate revealed the geometry after catalysis and just prior to product release.

[1]C. W. Carreras & D. V. Santi (1995) *Annu. Rev. Biochem.* **64**, 721.

[2]T. L. James, A. L. Pogolotti, Jr, K. M. Ivanetich, Y. Wataya, S. S. Lam & D. V. Santi (1976) *Biochem. Biophys. Res. Commun.* **72**, 404.

[3]L. Garcia-Fuentes, P. Reche, O. Lopez-Mayorga, D. V. Santi, D. Gonzalez-Pacanowska & C. Baron (1995) *Eur. J. Biochem.* **232**, 641.

[4]Y. Wataya & D. V. Santi (1975) *Biochem. Biophys. Res. Commun.* **67**, 818.

[5]Y. Wataya & D. V. Santi (1977) *J. Am. Chem. Soc.* **99**, 4534.

[6]A. L. Pogolotti, Jr., C. Weill, D. V. Santi (1979) *Biochemistry* **18**, 2794.

[7]D. L. Birdsall, J. Finer-Moore & R. M. Stroud (1996) *J. Mol. Biol.* **255**, 522.

The following are additional discussions of the molecular and physical properties of this enzyme.

D. R. Knighton, C.-C. Kan, E. Howland, C. A. Janson, Z. Hostomska, K. M. Welsh & D. A. Matthews (1994) *Nature Struct. Biol.* **1**, 186.

M. A. Ator & P. R. Ortez de Montellano (1990) *The Enzymes*, 3rd ed., **19**, 213.

P. Reichard (1988) *Ann. Rev. Biochem.* **57**, 349.

C. Heidelberger, P. V. Danenberg & R. G. Moran (1983) *Adv. Enzymol.* **54**, 57.

S. J. Benkovic (1980) *Ann. Rev. Biochem.* **49**, 227.

T. W. Bruice, C. Garrett, Y. Wataya & D. V. Santi (1980) *Meth. Enzymol.* **64**, 125.

M. Friedkin (1973) *Adv. Enzymol.* **38**, 235.

J. I. Rader & F. M. Huennekens (1973) *The Enzymes*, 3rd ed., **9**, 197.

TIGHT-BINDING INHIBITOR KINETICS

The study and characterization of tight-binding inhibitors requires special techniques and protocols[1]. Recall that an investigator typically varies the concentration of an inhibitor within a range bracketing the dissociation constant of that inhibitor. But, if an inhibitor has a very low K_d value, that concentration range could easily be close in magnitude to the concentration of the enzyme. In order for steady-state kinetics to be applicable to a system, not only must the substrate concentration be in great excess over the total enzyme concentration, but the enzyme concentration has to be much smaller than the concentration of any effector (in this case, an inhibitor). In such situations, different protocols have to be followed[1]. Not only will the investigator have to measure initial rates at different concentrations of substrate and fixed concentrations of inhibitor, but it will also be necessary to look at the steady-state velocities as a function of total enzyme concentration (at fixed levels of the inhibitor). **See** *Henderson Plot*

[1]J. W. Williams & J. F. Morrison (1979) *Meth. Enzymol.* **63**, 437.

TIGHT ION PAIR

An ion pair in which the constituent ions are not separated by a solvent or other intervening molecule. Tight ion pairs are also referred to as contact ion pairs. If X^+ and Y^- represent constituent ions, then a tight ion pair would be symbolized by X^+Y^-. An example of a tight ion pair would be the case in which an enzyme stabilizes a carbonium ion with juxtaposed negatively charged side-chain groups. **See** *Loose Ion Pair; Ion Pair; Solvent-Shared Ion Pair; Solvent-Separated Ion Pair.*

TIME

A reference frame that, while resisting unambiguous definition, serves to measure the progress/motion of chemical/physical processes. The SI unit for time is the second (symbolized as s). **See** *Time & Chronomals*

Time is readily sensed physiologically, and the so-called "flow of time"[1,2] is measured by the regularity of motions and changes occurring within and around us. Plato's interpretation of Heraclitus' analogy[3] of life to a river is that all things are in flux. We are immersed in a universe of processes that act as clocks, and even in solitude, we can still sense heartbeat, pulse, heat flow and the motions of other internal processes. Consciousness itself appears to depend on the pace of coordinated depolarization/repolarization of interneurons within the larger context of the brain's neural networks[4]. Channels opening, ions flowing down concentration gradients, channels closing, ion pump ATPases actively restoring transmembrane potentials—these are all subcellular processes serving as miniature clocks. Time can thus be described as the reciprocal of the frequency of motional change. The precision with which periodic change can be determined allows time to serve as a standard frame of reference for marking the progress of physicochemical processes. The *second* now corresponds to the duration of 9,192,631,770 periods of the radiation corresponding to the transition between two hyperfine levels of the cesium-133 atom in the ground state. This precision leads to the universal use of *dynamical time* as the independent variable in differential equations (*i.e.*, of the form dX/dt) to describe the time evolution of changes in location and energy. **See** *Circadian Rhythm*

[1]The term "flow of time" actually implies the absurd: that we can write an equation for the change in time per unit time (*i.e.*, dΔt/dt). In this regard, our intuitive sense of time passing quickly relates more to our having been distracted from an awareness of time's passage.

[2]"The besetting sin of philosophers, scientists, and, indeed all who reflect about time is describing it as if it were a dimension of space". C. W. K Mundle (1967) in *The Encyclopedia of Philosophy*, **8**, p. 138, Macmillan, New York.

[3]"Upon those stepping in the same river, ever different waters flow". This statement, attributed to Heraclitus of Ephesus, who was probably the first Grecian philosopher, survives only as a fragment of metrical verse torn from its original context.

[4]F. H. C. Crick (1994) *The Astonishing Hypothesis*, Scribner, New York. This fascinating book describes Crick's quest for an understanding of the chemical basis of human consciousness.

TIME & CHRONOMALS

A chronomal is a dimensionless parameter [symbolized by I or $I(x)$] that is proportional to time. Chronomals are especially useful in dealing with diffusion, chemical reactions, and other related processes. One can chose to express the properties of such systems as t equal to $K_D I_D$, where K_D contains all the physical constants and has overall units of time, whereas I_D is a chronomal expressed in terms of the extent of reaction ξ. In many respects, the chronomal can be regarded as dimensionless time.

In the following example dealing with the kinetics of precipitation, we will use the parameter ξ, the extent of reaction to designate the amount of substance that has undergone reaction to the amount of substance able to react. If the initial concentration is c_o at t_o, then

$$\xi = (c_o - c)/(c_o - s) \quad \text{or} \quad (c - s) = (c_o - s)(1 - \xi)$$

where s is the solubility of the substance. Now, if one treats the precipitate as a monodisperse set of spherical particles whose single radius r depends on time (*i.e.*, $r(t)$ changes as the reaction progresses), then

$$r(t) = r_1 \xi^{1/3}$$

Combining the last two equations with the rate law for precipitation [*i.e.*, $dr/dt = vD(c - s)/r$] and separating variables, we obtain an expression that can be written in form, $t = K_D I_D$, where $K_D = r_1^{2/3} vD(c_o - s)$ and I_D is the definite integral from $x = 0$ to ξ of $[x^{-1/3}(1 - x) - 1 \, dx]$. K_D contains all the physical constants and has units of time, whereas I_D is the chronomal. Nielsen presents a useful account of chronomals in the appendix to his treatment of crystallization and precipitation[1].

[1] A. E. Nielsen (1964) *Kinetics of Precipitation*, pp. 135-138, Pergamon, Oxford Univ. Press.

TIME-CORRELATED SINGLE PHOTON COUNTING

The measurement of the time histogram of a sequence of photons with respect to a periodic event. This technique requires a time-to-amplitude converter. **See also** *Photon Counting*

Comm. on Photochem. (1988) *Pure and Appl. Chem.* **60**, 1055.

TIME-OF-FLIGHT MASS SPECTROMETRY

A mass spectrometer employing brief exposure to a pulse of electrons to impart kinetic energy on sample molecules as well as the use of a drift tube to separate molecules on the basis of molecular mass. Because the imparted kinetic energies are virtually identical before entering the drift tube, and because the path length from entry port to detector is the same for all ionic species, the time-of-flight (or TOF) depends solely on the mass of a particular species. For biological macromolecules exhibiting little tendency to vaporize, one may use matrix-assisted laser-desorbed ion TOF mass spectrometry (or simply MALDI-TOF MS). In this technique, macromolecules are mixed with a matrix substance that is placed on an inert target and transferred to the sample compartment of the mass spectrometer. Immediately upon laser photo-excitation of the target, the matrix vaporizes so rapidly that the contained macromolecule(s) are likewise swept into the sample stream for electron bombardment and time-of-flight determination.

TIME-RESOLVED MACROMOLECULAR CRYSTALLOGRAPHY

A X-ray crystallographic method[1-3] for detecting the transient accumulation of intermediates in enzyme catalysis, protein folding, ligand-binding interactions, and other processes involving macromolecules. The approach is premised on the well documented retention of substantial reactivity of biological macromolecules, even in the crystalline state.

[1] H. Bartunik (1983) *Nucl. Instrum. Methods* **208**, 523.
[2] J. Hajdu, K. R. Acharya, D. I. Stuart, D. Barford & L. M. Johnson (1988) *Trends in Biochem. Sci.* **13**, 104.
[3] K. Moffat (1989) *Ann. Rev. Biophys. Biophys. Chem.* **18**, 309.

TIME-RESOLVED SPECTROSCOPY

Spectra recorded at a series of time intervals following excitation or other perturbation.

TIME-RESOLVED X-RAY DIFFRACTION

A method[1-3] for analyzing the kinetics of chemical/physical processes by combining the advantages of a focused, monochromatic synchrotron X-ray beam with a suitable live time-imaging device, thereby permitting preferably continuous monitoring of changes in the the diffracted X-rays. Caffrey[3] discusses how this technique can be gainfully applied to investigate lipid phase transitions.

[1] B. Batterman & N. W. Ashcroft (1979) *Science* **206**, 157.
[2] P. Eisenberger (1986) *Science* **231**, 687.
[3] M. Caffrey (1989) *Ann. Rev. Biophys. Biophys. Chem.* **18**, 159.

TIME-SCALE SEPARATION AND METABOLIC PATHWAYS CONTROL

Delgado *et al.*[1] recently demonstrated that time-scale separation is an effective way to localize metabolic control to only a few enzymes. They considered model pathways in which the eigenvalues of the Jacobian of the system are widely separated (*i.e.*, systems with time-scale separation). Their treatment assumes the system possesses a unique, asymptotically stable steady-state and that the reaction steps of the system under analysis are well represented by linear kinetics around the steady state. While cells may produce many enzymes at far greater concentrations than needed to maintain a certain steady-state, they can achieve time-scale separation by controlling expression of only a few enzymes. The over-expressed enzymes catalyze what are termed "fast" reactions and lead to small response times of the system. For these "fast" reactions, metabolite control coefficients are small compared with "slow" reactions and do not effectively control overall flux. Nonetheless, at pathway branch points, "fast" reactions may be mutually competitive and result in significant control coefficients for fluxes to the branches.

[1]J. Delgado & J. C. Liao (1995) *Biosystems* **36**, 55.

TLCK

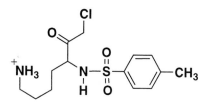

The inhibitor tosyl-lysine chloromethyl ketone which reacts irreversibly with the active site serine hydroxyl group of trypsin. In the initial step of the inactivation process, TLCK's ε-amino group directs the inhibitor into the negatively charged specificity-conferring pocket where lysyl or argininyl side-chains normally are bound within trypsin's active site. Then, nucleophilic attack by the active-site serine hydroxyl leads to the formation of an ether linkage, accompanied by expulsion of chloride ion. **See** *Chymotrypsin*

TOLUENE DIOXYGENASE

This iron-sulfur oxygenase [EC 1.14.12.11] catalyzes the reaction of molecular oxygen with toluene and NADH to produce (1S,2R)-3-methylcyclohexa-3,5-diene-1,2-diol and NAD+. This reductase is an iron-sulfur flavo-protein (FAD) that contains ferredoxin. Ethylbenzene, 4-xylene, and some halogenated toluenes can likewise undergo conversion to the corresponding *cis*-dihydro-diols.

L. P. Wackett (1990) *Meth. Enzymol.* **188**, 39.

TONICITY

1. The tension or osmotic pressure of a solution; also, ionic strength, usually measured as weight percentage. Often the tonicity of a solution is presented as relative to some physiological solution (*e.g.*, blood plasma). **See** *Hypertonic; Hypotonic; Isotonic; Isotonic Buffers*

2. An isotope tracer method[1] for estimating the total amount of water within an organism.

[1]R. R. Wolfe (1992) *Radioactive and Stable Isotope Tracers in Bio-medicine*, p. 190, Wiley-Liss, New York.

TOPAQUINONE

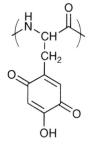

A trihydroxyphenylalanyl residue (symbolized TPQ) that plays an essential cofactor role in catalysis of amine oxidases that use molecular oxygen and copper ions.

D. Wemmer, A. J. Smith, S. Kaur, D. Maltby, A. L. Burlingame, and J. P. Klinman (1990) *Science* **248**, 981.
D. Mu, K. F. Medzihradsky, G. W. Adams, P. Mayer, W. M. Hines, A. L. Burlingame, A. J. Smith. D. Cai, and J. Klinman (1994) *J. Biol. Chem.* **269**, 9926.

TOTAL ENERGY EXPENDITURE

The energy intake by an organism plus or minus any change in body mass or composition. Lifson and McClin-tock[1] introduced the use of doubly labeled water (^2H_2O and $H_2^{18}O$) to estimate energy expenditure. In this strategy, ^{18}O is lost from the body as labeled water and carbon dioxide, while loss of the deuterium label allows one to measure turnover of the water pool. Wolfe[2] provides a lucid account of how the technique is applied, along with assumptions, strengths and weaknesses.

[1]N. Lifson, W. S. Little, D. G. Levitt & R. M. Henderson (1975) *J. Appl. Physiol.* **39**, 657.
[2]R. R. Wolfe (1992) *Radioactive and Stable Isotope Tracers in Bio-medicine*, pp. 207-233, Wiley-Liss, New York.

TPCK

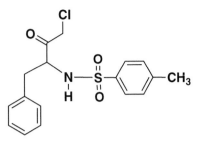

The active-site-directed inhibitor tosylphenylalanine chloromethyl ketone that specifically and irreversibly inhibits chymotrypsin. This chloroketone inhibitor relies on its toluene sulfonyl (or tosyl) group for binding into the aromatic binding pocket of chymotrypsin's active site. Inactivation occurs by alkylation of histidine-57 (pseudo-first order rate constant 0.2 min^{-1}). **See** *Chymotrypsin*

TRACEE

The unlabeled metabolite, drug, or other substance under study in a kinetic analysis of its biosynthesis/entry (appearance) and catabolism/clearance (disappearance). **See** *Compartmental Analysis*

TRACER

The labeled form of a metabolite, drug, or other substance used in a kinetic analysis of tracee biosynthesis/entry (appearance) and tracee catabolism/clearance (disappearance). **See** *Compartmental Analysis*

TRACER PERTURBATION METHOD

A enzyme kinetic technique, introduced by Britton and co-workers[1-4], that permits one to measure the equilibrium distribution of enzyme-bound substrate(s), intermediate(s), and product(s). In this procedure, radiolabeled substrate or product is initially permitted to react with enzyme for sufficient time to equilibrate. At thermodynamic equilibrium, all the different enzyme-bound species will be present at concentrations reflecting their stability relative to each other. One can then add a large excess of unlabeled substrate. Under this condition, any unbound or newly released radiolabel will mix with the large unlabeled pool of substrate or product, where it will undergo substantial reduction in its radiospecific activity. This dilution effectively reduces or eliminates any significant recycling of released radiolabel. One can then sample the system as a function of time, and the rates and amounts of labeled substrate and labeled product depend on the kinetics of synthesis and/or interconversion of bound radiolabel to substrate and product, which upon release is unlikely to rebind.

[1]H. G. Britton & J. B. Clarke (1972) *Biochem. J.* **110**, 161.
[2]H. G. Britton & J. B. Clarke (1972) *Biochem. J.* **130**, 397.
[3]H. G. Britton, J. Carreras & S. Grisolia (1971) *Biochemistry* **10**, 4522.
[4]H. G. Britton, J. Carreras & S. Grisolia (1971) *Biochemistry* **11**, 3008.

TRACER/TRACEE RATIO

A parameter used as a measure of isotope enrichment in stable isotope exchange and compartmental analysis experiments:

$$R = [r_{sample} - r_{reference}]/[r_{blank} - r_{reference}]$$

where r_{sample}, $r_{reference}$, and r_{blank} are the ion current ratios determined for an experimental sample, a reference sample, and a blank, respectively. The tracer/tracee ratio approximates specific activity (*i.e.*, [mol tracer]/[mol tracer + mol tracee]) when the tracer content is negligible relative to the tracee content. The tracer/tracee ratio R can be most directly determined using an ion ratio mass spectrometer to measure the ratio of ion currents for isotopomers such as $^{12}CO_2$ and $^{13}CO_2$ (at masses 44 and 45, respectively). **See** *Compartmental Analysis*

TRANSALDOLASE

This enzyme [EC 2.2.1.2], also known as dihydroxyacetone transferase and glycerone transferase, catalyzes the reversible reaction of sedoheptulose 7-phosphate with D-glyceraldehyde 3-phosphate to produce D-erythrose 4-phosphate and D-fructose 6-phosphate.

R. Kluger (1992) *The Enzymes*, 3rd ed., **20**, 271.
P. R. Levering & L. Dijkhuizen (1990) *Meth. Enzymol.* **188**, 405.
O. Tsolas & B. L. Horecker (1972) *The Enzymes*, 3rd ed., **7**, 259.

TRANSCARBOXYLASE

This enzyme [EC 2.1.3.1], also known as methylmalonyl-CoA carboxyltransferase, catalyzes the reaction of (*S*)-2-methyl-3-oxopropanoyl-CoA with pyruvate to produce propanoyl-CoA and oxaloacetate. The enzyme requires biotin, cobalt, and zinc as cofactors.

J. N. Earnhardt & D. N. Silverman (1998) *Comprehensive Biological Catalysis: A Mechanistic Reference* **1**, 495.
J. V. Schloss & M. S. Hixon (1998) *Comprehensive Biological Catalysis: A Mechanistic Reference* **2**, 43.

A. S. Mildvan, D. C. Fry & E. H. Serpersu (1991) in *A Study of Enzymes* (S.A. Kuby, ed.) vol. **2**, p. 105, CRC Press, Boca Raton.

J. R. Knowles (1989) *Ann. Rev. Biochem.* **58**, 195.

I. A. Rose (1982) *Meth. Enzymol.* **87**, 84.

H. G. Wood (1972) *The Enzymes*, 3rd ed., **6**, 83.

TRANSCRIPTION COMPLEX KINETICS

von Hippel[1] recently developed a model that integrates the structural, thermodynamic, and kinetic properties of the prokaryotic transcription complex that allows RNA polymerase to catalyze high-fidelity copying of a gene's DNA template strand into a cRNA transcript. This model allows one to understand how the transcription complex chooses between three competing pathways (*e.g.*, moving forward along the template with attending elongation of the RNA chain, moving backward along the template with or without transcript shortening, or dissociating from the template). The choice between elongation and termination appears to favor elongation during the polymerization reaction, because the $\Delta G_{\text{release}}^{\ddagger} > \Delta G_{\text{forward}}^{\ddagger}$: however, when the polymerase arrives at a termination position, $\Delta G_{\text{release}}^{\ddagger} \approx \Delta G_{\text{forward}}^{\ddagger}$. To render the polymerase pathway kinetically accessible, the change in $\Delta G_{\text{release}}^{\ddagger}$ may be related to a massive destabilization of the transcription complex as it approaches terminators along the DNA template.

[1]P. H. von Hippel (1998) *Science* **281**, 660.

TRANSCYTOSIS

The transport of a vesicle from one side of a cell to another. Typically, the vesicle is generated via endocytosis or phagocytosis and then exocytosed.

TRANSFERASES

A major class of enzymes that catalyze the transfer of a group or moiety from one compound to another. The groups being transferred can be one-carbon units such as methyl, hydroxylmethyl, carbamoyl, or amidino moieties. Enzymes transferring aldehyde or ketonic groups such as transketolase are members of this class. Other examples include acyltransferases, glycosyltransferases, aminotransferases, phosphotransferases, and sulfotransferases.

TRANSFORMATION

Any conversion of reactant (substrate) into a particular product, irrespective of reagents or mechanisms involved. Transformations are distinct from reactions, because a full description of the latter would state or imply all the reactants and all the products.

IUPAC (1994) *Pure and Appl. Chem.* **66**, 1077.

TRANSGLUTAMINASE

This enzyme [EC 2.3.2.13], also known as protein: glutamine γ-glutamyltransferase, catalyzes the calcium ion-dependent reaction of the side chain of a glutamyl residue in a protein with an alkylamine to generate an N^5-alkylglutaminyl residue in the protein and ammonia.

F. Tokunaga & S. Iwanaga (1993) *Meth. Enzymol.* **223**, 378.

J. E. Folk (1983) *Adv. Enzymol.* **54**, 1.

J. E. Folk & J. S. Finlayson (1977) *Adv. Protein Chem.* **31**, 1.

J. E. Folk & S. I. Chung (1973) *Adv. Enzymol.* **38**, 109.

TRANSIENT SPECIES (or, Transient Chemical Species)

A relative term referring to an intermediate in a reaction or series of reactions which is short-lived under the given experimental conditions and assay procedures. Such species have a shorter half-life than persistent chemical species.

The term "metastable" has often been used to describe transient species. However, care should be exercised in such cases. Metastability relates to the thermodynamic stability of a molecule whereas the term "transient" refers to a kinetic property. Nevertheless, most transients are unstable relative to reactants and products. **See also** Persistent

TRANSIENT-STATE KINETIC METHODS

Any experimental technique that discloses the time-resolved behavior of a chemical/physical process. These approaches allow one to surmount the inherent limitations of steady-state and/or equilibrium kinetic measurements in the detection and quantification of species that comprise the internal equilibria of enzymic catalysis.

Johnson[1,2] and Fierke & Hammes[3] have presented detailed accounts of how rapid reaction techniques allow one to analyze enzymic catalysis in terms of pre-steady-state events, single-turnover kinetics, substrate channeling, internal equilibria, and kinetic partitioning. **See** *Chemical Kinetics; Stopped-Flow Techniques*

[1]K. A. Johnson (1992) *The Enzymes*, 3rd ed., **20**, 1.

[2]K. A. Johnson (1995) *Meth. Enzymol.* **249**, 38.

[3]C. A. Fierke & G. G. Hammes (1995) *Meth. Enzymol.* **249**, 1.

TRANSITION COORDINATE

The normal reaction coordinate at the transition state corresponding to a vibration with an imaginary frequency. Motion along the transition coordinate will lead to either reactants or products[1]. **See** *Reaction Coordinate*

[1]IUPAC (1979) *Pure and Appl. Chem.* **51**, 1725.

TRANSITION DIPOLE MOMENT

The amplitude of the oscillating electric or magnetic moment that is induced in a molecular entity by an electromagnetic wave.

Comm. on Phorochem. (1988) *Pure and Appl. Chem.* **60**, 1055.

TRANSITION POLARIZATION

The direction of the transition moment in a molecular entity.

TRANSITION PROBABILITY FUNCTIONS

Relationships used in [^{18}O]phosphate analysis to determine the probability that an enzyme's reaction with a phosphate containing m labeled oxygens in the presence of unlabeled water will result in the release of a phosphate that contains n labeled oxygens.

D. D. Hackney, K. E. Stempel & P. D. Boyer (1980) *Meth. Enzymol.* **64**, 60.

TRANSITION STATE

In terms of elementary reactions, a transition state (the assembly of atoms at the transition state is often called an activated complex) is that species having a more positive Gibbs free energy than the products or reactants of that elementary reaction. This species is represented by that assembly(ies) of atoms which, if placed in these geometries, energies, and bonds (both complete and partial) would have an equal probability of forming either the reactants or products of that elementary reaction. Note that the transition state could be represented by a set of states, each having the ability to form products or reactants. This set of states is characterized by the partition function for the transition state. The transition state is characterized by one and only one imaginary frequency.

The terms "early" and "late" transition states are qualitative descriptions related to concepts such as those advanced by the Hammond principle. An early transition state is that assembly of atoms or moieties that closely resembles the reactant(s), such that only a relatively small reorganization will generate the reactant(s). Analogously, a late transition state more closely resembles the structure of the reaction product(s). **See** *Chemical Kinetics; Transition State Theory; Potential Energy Surface; Hammond Principle; Transition Structure*

TRANSITION-STATE ANALOGUES

A molecule designed to mimic the properties, structure, and/or geometries of the transition state of a particular reaction. Such compounds are often potent inhibitors of enzymes. **See** *Molecular Similarity*

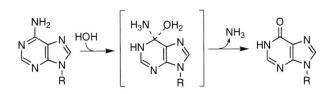

Likely Transition State in the Adenosine Deaminase Reaction

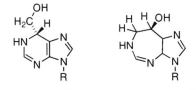

NDHPR (K$_i$ = 10^{-5} M)   Coformycin (K$_i$ = 10^{-11} M)

Figure 1. Adenosine deaminase reaction mechanism, indicating nucleophilic substitution reaction at position-6 of the purine ring. Note that the inhibitors NDHPR and coformycin also contain sp^3 hybridization in the position analogous to the carbon atom undergoing substitution in the adenosine deaminase reaction. Tight binding is thought to be a consequence of the complementarity of the enzyme for analogues mimicking the transition-state configuration.

As pointed out by Wolfenden[1,2], the classical Pauling-Haldane model of transition-state stabilization model for enzyme catalysis offers a means to rationalize the extraordinarily high affinities achieved by some substrate analogues. Using the example of the adenosine deaminase reaction (Fig. 1), we recognize that a key feature in this aromatic nucleophilic substitution reaction is the backside entry of a water molecule to attain loss of aromaticity as well as sp^3 hybridization at position-6 of the purine ring. Agents that can mimic the stereochemical and/or electrostatic arrangement achieved in the transition state are thought to possess a special binding advantage over typical substrates. This special advantage relates to the fact that such analogues are complementary to the transition-state configuration of the enzyme, whereas substrates in the ground state are not. Thus, while a substrate must necessarily suffer some loss in binding energy as it is contorted along its path to reach the transition state, the same is not true for transition-state analogues.

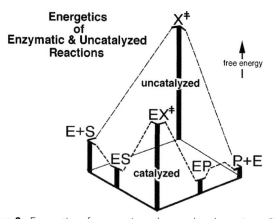

Figure 2. Energetics of enzymatic and uncatalyzed reactions. Conversion of substrate S directly to the transition state in the absence of enzyme is energetically more unfavorable than the corresponding conversion on the enzyme surface.

The energetics of enzymatic and their corresponding uncatalyzed reference reactions can be understood by the cyclic path that allows for substrate conversion to product by the uncatalyzed and enzymatic routes (Fig. 2). Note that the uncatalyzed reaction is characterized by a transition state that is far less stable than its enzymatic counterpart. Note also that the initial and final conditions are the same for either route, an absolute requirement for any catalyzed process (*i.e.*, no effect on the overall equilibrium constant).

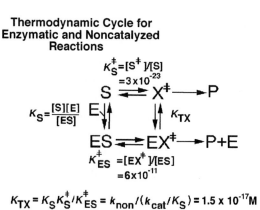

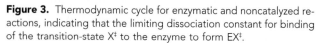

Figure 3. Thermodynamic cycle for enzymatic and noncatalyzed reactions, indicating that the limiting dissociation constant for binding of the transition-state $X^{\ddagger}$ to the enzyme to form $EX^{\ddagger}$.

In principle, one can never exactly duplicate the transition state, because transition state theory requires that such an intermediate species would disproportionate back to E·Substrate complex as well as proceed onward to E·Product complexes. However, the scheme shown in Fig. 3 permits one to estimate the maximal affinity that should be achievable if one were to approximate closely the electronic and stereochemical configuration of the enzyme and substrate in the transition state. An accurate estimation of K_{TX} requires detailed knowledge that the uncatalyzed reference reaction follows the same mechanism as the enzyme-catalyzed process. **See** *Enzyme Proficiency; Reference Reaction*

[1]R. Wolfenden (1969) *Nature (London)* **223**, 704.
[2]A. Radzicka & R. Wolfenden (1994) *Meth. Enzymol.* **249**, 284.

TRANSITION STATES IN BIOLOGICAL CATALYSIS

Rather than adopting a purely fundamentalist viewpoint centering on strong reactant(s) binding of a transition-state geometry, Cannon *et al.*[1] recently offered a different perspective on enzyme rate enhancements. By re-examining the reaction coordinate, their analysis suggested that the ground state of the enzyme-substrate complex and the transition state are related by the mean force acting along the reaction path. Their treatment would suggest that catalysis is not easily resolvable into the effects of ground state destabilization or transition-state stabilization. The ability of the enzyme to preorganize the reaction environment may instead be a dominant factor, and gating of solvent water molecules may eliminate or minimize the inhibitory effects of water molecules on chemical reactivity.

[1]W. R. Cannon, S. F. Singleton & S. J. Benkovic (1996) *Nature Struct. Biol.* **3**, 821.

TRANSITION-STATE THEORY

Theoretical descriptions of absolute reaction rates in terms of the rate-limiting formation of an activated complex during the course of a reaction. Transition-state theory (pioneered by Eyring[1–4], Pelzer and Wigner[5], and Evans and Polanyi[6]) has been enormously valuable, and beyond its application to chemical reactions, the theory applies to a wider spectrum of rate processes (*e.g.*, diffusion, flow of liquids, internal friction in large polymers, *etc.*). Transition state theory assumes: (1) that classical mechanics can be used to calculate trajectories over potential energy surfaces; (2) that reactants and activated complexes maintain equilibrium with each other and that the energetics of components follow Boltzmann distributions of energy; and (3) that reactions will occur once a certain point on the potential energy surface has been passed.

This theory assumes that the rate of a reaction at a given temperature is proportional to the concentration of an activated complex that is in equilibrium with the unactivated reactants. In proceeding from substrates to products, the reactants form an activated complex, also said to be in the transition state. As an example, consider the bimolecular reaction in the scheme below, in which the moiety X is being transferred.

$$A-X + B \rightleftharpoons A + B-X$$

In transition state theory, an activated complex occurs at an intermediary point prior to the formation of products:

$$A-X + B \rightleftharpoons [A\text{-}\text{-}\text{-}X\text{-}\text{-}\text{-}B]^{\ddagger} \rightleftharpoons A + B-X$$

The "double-dagger" symbol, $\ddagger$, is exclusively used to designate activated complexes or the transition state. This activated complex will only form when the reacting molecules have sufficient energy (i.e., $E \geq E_a$). The activated complex is that configuration of interacting molecules at the maximum point in the potential energy barrier for a reaction progressing up a potential energy "hill". In some reactions (e.g., proton or electron transfer), the proton or electron may leak through a potential energy barrier by a process called tunneling. Tunneling is clearly significant in electron-transfer reactions, but it can also be appreciable in the transfer of protons[7].

In transition-state theory, the absolute rate of a reaction is directly proportional to the concentration of the activated complex at a given temperature and pressure. The rate of the reaction is equal to the concentration of the activated complex times the average frequency with which a complex moves across the potential energy surface to the product side. If one assumes that the activated complex is in equilibrium with the unactivated reactants, the calculation of the concentration of this complex is greatly simplified. Except in the cases of extremely fast reactions, this equilibrium can be treated with standard thermodynamics or statistical mechanics[8]. The case of

extremely fast reactions introduces a small difference of about 10% between experimental rates and predicted rates. *See* Chemical Kinetics; Collision Theory; Temperature Dependence; Q_{10}; Unimolecular Reactions and Transition-State Theory; Transition-State Theory (Thermodynamics); Tunneling

[1]S. Glasstone, K. J. Laidler & H. Eyring (1941) *The Theory of Rate Processes*, McGraw-Hill, New York.
[2]H. Eyring (1935) *Chem. Rev.* **17**, 65.
[3]H. Eyring (1935) *J. Chem. Phys.* **3**, 107.
[4]W. F. K. Wynne-Jones & H. Eyring (1935) *J. Chem. Phys.* **3**, 492.
[5]H. Pelzer & E. Wigner (1932) *Zeit. Physik Chem.* **B15**, 445.
[6]M. G. Evans & M. Polanyi (1935) *Trans. Faraday Soc.* **31**, 875.
[7]R. P. Bell (1973) *The Proton in Chemistry*, 2nd ed., Cornell Univ. Press, Ithaca.
[8]R. D. Present (1959) *J. Chem. Phys.* **31**, 747.

TRANSITION-STATE THEORY (Thermodynamics)

Transition-state theory can also be expressed in thermodynamic terms[1]. Since $K^{\ddagger}$ in the Eyring equation ($k = k_B T K^{\ddagger}/h = RTK^{\ddagger}/Lh$ where k is the bimolecular rate constant, k_B is Boltzmann's constant, T is the absolute temperature, h is Planck's constant, L is Avogadro's constant, R is the universal gas constant, and $K^{\ddagger}$ is the equilibrium constant between the reactants and the transition state) is similar to a true equilibrium constant, let $\Delta G^{\ddagger}$ be the standard Gibbs free energy of activation between the reactants and the activated complex in the transition state. Strictly speaking, $\Delta G^{\ddagger} = -RT \ln [K^{\ddagger}(c^{\ominus})^{n-1}]$ where $c^{\ominus}$ is a standard-state concentration and n is the molecularity of the reaction. Thus, $K^{\ddagger} = e^{-\Delta G^{\ddagger}/RT}(c^{\ominus})^{1-n}$. However, standard states are, in general, unit concentrations, using whatever unit is being used to evaluate the rate constant. The bimolecular rate constant, in the Eyring equation ($k = (k_B T/h) K^{\ddagger}$), can be expressed as

$$k = \frac{k_B T}{h} e^{-\Delta G^{\ddagger}/RT}(c^{\ominus})^{1-n} = \frac{k_B T}{h} e^{\Delta S^{\ddagger}/R} e^{-\Delta H^{\ddagger}/RT}(c^{\ominus})^{1-n}$$

These parameters ($\Delta G^{\ddagger}$, $\Delta H^{\ddagger}$, and $\Delta S^{\ddagger}$) differ slightly from normal standard parameters in that the contribution of motion along the reaction coordinate toward the transition state is not included. The values are the difference in free energy, enthalpy, and entropy between 1 mole of activated complex and 1 mole of each reactant, all substances being at their standard-state concentrations (usually 1 M).

For a simple one-substrate, one-product enzyme-catalyzed reaction there are at least eighteen thermodynamic parameters that characterize the reaction coordinate profile. However, it should be noted that the Michaelis binary complex and the activated complex (*i.e.*, the Eyring complex) are <u>not</u> equivalent. An activated complex must form prior to the Michaelis complex, its formation governed by k_1, the first bimolecular rate constant. In addition, at least one activated complex occurs between the ES and the EP binary complexes. Knowing the $\Delta H^{\ominus}$ and $\Delta H^{\ddagger}$ values of each step in an enzyme-catalyzed reaction, allows the investigator to draw a reaction coordinate diagram. Similar diagrams can be plotted using $\Delta G^{\ominus}$ and $\Delta G^{\ddagger}$ values or $\Delta S^{\ominus}$ and $\Delta S^{\ddagger}$ values. Dixon and Webb[2] have provided a table reviewing the different procedures for the determination of the various thermodynamic quantities for a simple one-substrate, one-product enzyme-catalyzed reaction. A truly thorough analysis necessitates the ability to separate each rate constant.

If an investigator studies temperature effects on K_m and V_{max} values of an enzyme-catalyzed reaction, then interpretation of the results depends on the precise definitions of these kinetic parameters[3]. If the rate-determining step in the reaction changes with temperature then either one or both of the plots with respect to K_m or V_{max} may be curvilinear. When analyzing binding behavior ideally, true K_d values (rather than K_m values) should be studied. In addition, the ionization state of the protein and/or substrates and effectors (including the buffer) may change with temperature as well. Often temperature studies should be preceded by a pH study. *See* Arrhenius Equation; Transition-State Theory; Gibbs Free Energy of Activation; Enthalpy of Activation; Entropy of Activation; Volume of Activation; Transition-State Theory in Solutions; Diffusion Control; Temperature Dependence; Entropy and Enthalpy of Activation, Interpretations

[1]J. W. Moore & R. G. Pearson (1981) *Kinetics and Mechanism*, Wiley, New York.
[2]M. Dixon & E. C. Webb (1979) *Enzymes*, 3rd ed., Academic Press, New York.
[3]K. D. Gibson (1953) *Biochim. Biophys. Acta* **10**, 221.

TRANSITION STRUCTURE

The saddle point on a three-dimensional potential-energy surface, characterized by one negative force constant in the harmonic force constant matrix.

IUPAC (1994) *Pure and Appl. Chem.* **66**, 1077.

TRANSKETOLASE

This thiamin pyrophosphate-dependent enzyme [EC 2.2.1.1], also known as glycolaldehyde transferase, catalyzes the reversible reaction of sedoheptulose 7-phosphate with D-glyceraldehyde 3-phosphate to produce D-ribose 5-phosphate and D-xylulose 5-phosphate. The enzyme exhibits a wide specificity for both reactants. It also can catalyze the reaction of hydroxypyruvate with $R-CHO$ to produce carbon dioxide and $R-CH(OH)-C(=O)-CH_2OH$. Transketolase isolated from *Alkaligenes faecalis* shows high activity with D-erythrose as the acceptor substrate.

R. L. Schowen (1998) *Comprehensive Biological Catalysis: A Mechanistic Reference* **2**, 217.
J. V. Schloss & M. S. Hixon (1998) *Comprehensive Biological Catalysis: A Mechanistic Reference* **2**, 43.
A. Schellenberger, G. Hübner & H. Neef (1997) *Meth. Enzymol.* **279**, 131.
Y. Lindqvist & G. Schneider (1993) *Curr. Opin. Struct. Biol.* **3**, 896.
R. Kluger (1992) *The Enzymes*, 3rd ed., **20**, 271.
C. J. Gubler (1991) *A Study of Enzymes* **2**, 117.

TRANSMISSION COEFFICIENT

A ratio in transition state theory (symbolized by κ) that represents the probability that the activated complex will go on to form product(s) rather than return to reactants. In most cases, this value is approximately one; however, if reactants do not obey the Boltzmann law or if the temperature is very high, then the coefficient can be less than one. ***See*** Transition-State Theory

TRANSMISSION DENSITY

Symbolized by D, the transmission density is the negative decadic logarithm of the transmittance; thus, $D = -\log \tau$. ***See*** Transmittance

TRANSMITTANCE

A dimensionless quantity, symbolized by τ or T, equal to the transmitted radiant power, P_{tr}, divided by the radiant power incident on the sample, P_0; thus, $\tau = P_{tr}/P_0$. It is a measure of the ability of a body, solution, entity, *etc.*, to transmit electromagnetic radiation. It is synonymous with transmission factor. ***See also*** Internal Transmittance; Transmission Density; Total Transmittance; Beer-Lambert Law; Absorption Spectroscopy

TRANSPORT NUMBER

A dimensionless number, symbolized by t (or by t^+ or t_+ for a cation and t^- or t_- for an anion), that is equal to the fraction of the total current (of an electrolytic solution) carried by that ion. Thus, at infinite dilution $t_0^+ = \Lambda_0^+/\Lambda_0$ and $t_0^- = \Lambda_0^-/\Lambda_0$ where the subscripts refer to infinite dilution, Λ_0 is the molar conductance at infinite dilution and Λ_0^+ and Λ_0^- are the molar ionic conductances at infinite dilution. Transport numbers, also called transference numbers, can be measured by a number of means: *e.g.*, the Hittorf method or the moving boundary method. By knowing the molar conductance of an electrolytic solution, knowledge of the transport number enables one to determine the mobility of an ion.

TRAPPING

The binding, reaction, or interception of a reactive molecular entity or transitory intermediate in a reaction pathway to convert the substance to a more stable form and/or remove that substance from the system. Trapping may involve binding or reaction with another molecular entity or involve the alteration of some parameter (*e.g.*, thermal trapping)[1].

[1]D. L. Purich (1982) *Meth. Enzymol.* **87**, 3.

TRIMOLECULAR

Trimolecular reactions (also referred to as termolecular) involve elementary reactions where three distinct chemical entities combine to form an activated complex[1]. Trimolecular processes are usually third order, but the reverse relationship is not necessarily true. All truly tri- or termolecular reactions studied so far have been gas-phase processes. Even so, these reactions are very rare in the gas-phase. They should be very unlikely in solution due, in part, to the relatively slow-rate of diffusion in solutions. **See** *Molecularity; Order; Transition-State Theory; Collision Theory; Elementary Reactions*

[1]J. W. Moore & R. G. Pearson (1981) *Kinetics and Mechanism*, Wiley, New York.

TRIOSE-PHOSPHATE ISOMERASE

This enzyme [EC 5.3.1.1], also known as triosephosphate mutase (TIM) and phosphotriose isomerase, catalyzes the interconversion of D-glyceraldehyde 3-phosphate and dihydroxyacetone phosphate (IUPAC: glycerone phosphate). As pointed out by Rose[1], this enzyme is chiefly responsible for the largely symmetrical conversion of the two three-carbon segments of glucose to lactate and for the nearly uniform distribution of ^{14}C from pyruvate in the glucosyl units of liver glycogen. The reaction favors dihydroxyacetone phosphate, DHAP, with an equilibrium constant of 22. This enzyme is extraordinarily efficient, displaying a specific activity of 6700 mmol/min/mg at saturating G3P concentrations[2]. Like phosphoglucoisomerase, a *cis*-enediolate intermediate is a key catalytic feature[3], and Albery and Knowles[4] used kinetic isotope effect measurements to analyze the energetics of TIM catalysis. In fact, this enzyme and proline racemase are among the extensively characterized enzymes in terms of the energetics of the catalytic reaction cycle[5-7]. **See** *Haldane Relation; Enzyme Energetics (the Case of Proline Racemase); Kinetic Isotope Effect; Affinity Labeling; Reaction Coordinate Diagram*

[1]I. A. Rose (1969) in *Comprehensive Biochemistry* (M. Florkin & E. H. Stokes, eds.), vol. **17**, p. 126, Elsevier, Amsterdam.
[2]E. A. Noltmann (1972) *The Enzymes*, 3rd ed., **6**, 271.
[3]I. A. Rose (1982) *Meth. Enzymol.* **87**, 84.
[4]W. J. Albery & J. R. Knowles (1976) *Biochemistry* **15**, 5631.
[5]J. R. Knowles (1991) *Nature* **350**, 121.
[6]R. C. Davenport, P. A. Bash, B. A. Seaton, M. Karplus, G. A. Petsko & D. Ringe (1991) *Biochemistry* **30**, 5821.
[7]J. V. Schloss & M. S. Hixon (1998) *Comprehensive Biological Catalysis: A Mechanistic Reference* **2**, 43.

TRIPLE-COMPETITIVE METHOD

A method used to determine primary intrinsic isotope effects[1]. In this procedure, three differently labeled substrates are used to react with a labeled cosubstrate and the distribution of the labels in the products is measured.

[1]D. B. Northrop (1982) *Meth. Enzymol.* **87**, 607.

TRIPLET STATE

A state having a total electron spin quantum number of one. **See** *Fluorescence*

Comm. on Photochem. (1988) *Pure and Appl. Chem.* **60**, 1055.

TRIPLET-TRIPLET ANNIHILATION

A reaction between two species (*i.e.*, atoms or molecular entities), both of which are in triplet electronic states, resulting in at least two products, one of which is in its ground singlet state and the other in an excited singlet state. Delayed fluorescence often accompanies such processes. **See also** *Annihilation; Wigner Spin Conservation Rule; Delayed Luminescence*

TRIPLET-TRIPLET ENERGY TRANSFER

Energy transfer from a donor that is in an electronically excited triplet state to produce an electronically excited acceptor which is also in its triplet state.

TRIPLET-TRIPLET TRANSITIONS

Electron transitions in which both the initial and final states are triplet states.

Comm. on Photochem. (1988) *Pure and Appl. Chem.* **60**, 1055.

TRITIUM

A hydrogen isotope with a nucleus consisting of one proton and two neutrons (the nucleus is referred to as the triton). Tritium, a radioisotope symbolized by ^3_1H or T, decays by negative beta emission (0.01860 MeV) with a half-life of 12.32 years. The atomic weight of tritium is 3.01605 amu. It is frequently used in metabolic and kinetic experiments. The following decay data indicate the time followed by the fraction of original amount at the specified time: 0, 1.000; 1 month, 0.995; 2, 0.991; 3, 0.986; 4, 0.981; 5, 0.977; 6, 0.972; 7, 0.968; 8, 0.963; 9, 0.959; 10, 0.954; 11, 0.950; 1 year, 0.945; 2, 0.893; 3, 0.844; 4, 0.798; 5, 0.754; 6, 0.713; 7, 0.674; 8, 0.637; 9, 0.602; 10, 0.569; 11, 0.538; 12, 0.509.

Special precautions: The maximum range in air is 0.47 cm. Because no organ selectively concentrates tritiated metabolites, the entire body is considered to be the critical organ in terms of dose. The effective *biological* half-life in humans is around 10 days. This period can be shortened by increased water intake. Double gloving is sufficient, but changing the outer glove is recommended every 20 min when direct contact with gloves cannot be avoided.

TRITON

1. The name for a radioactive hydrogen ion. **See** *Tritium*
2. An abbreviation used for Triton X-100, a nonionic detergent that is used to solubilize membrane-bound proteins.

Triton X-100 has proved to be of great value in the surface dilution model[1,2] for lipolytic enzyme action. In this experimental strategy, the surface concentration of phospholipid in mixed micelles is reduced by the addition of Triton as a neutral diluent, thereby increasing the average distance between phospholipids. This allows one to draw mechanistic inferences about the binding interactions of lipases and phospholipases with their lipid substrates[1].

[1]T. G. Warner & E. A. Dennis (1975) *J. Biol. Chem.* **250**, 8044.
[2]R. A. Deems, B. R. Eaton & E. A. Dennis (1975) *J. Biol. Chem.* **250**, 9013.

TRYPSIN

This endopeptidase [EC 3.4.21.4] , a member of the peptidase family S1, hydrolyzes peptide bonds at Arg—Xaa and Lys—Xaa. **See** *Chymotrypsin; Catalytic Triad; Acyl-Serine Intermediate*

D. R. Corey & C. S. Craik (1992) *J. Amer. Chem. Soc.* **114**, 1784.
A. A. Kossiakoff (1987) in *Biological Macromolecules & Assemblies* (F. A. Jurnak & A. McPherson, eds.) **3**, p. 369, Wiley, New York.
R. J. Coll, P. D. Compton & A. L. Fink (1982) *Meth. Enzymol.* **87**, 66.
J. S. Fruton (1982) *Adv. Enzymol.* **53**, 239.
B. Keil (1971) *The Enzymes*, 3rd ed., **3**, 250.

TRYPTOPHANASE

This enzyme [EC 4.1.99.1], also known as L-tryptophan indole-lyase, catalyzes the hydrolysis of L-tryptophan to generate indole, pyruvate, and ammonia. The reaction requires pyridoxal phosphate and potassium ions. The enzyme can also catalyze the synthesis of tryptophan from indole and serine as well as catalyze 2,3-elimination and β-replacement reactions of some indole-substituted tryptophan analogs of L-cysteine, L-serine, and other 3-substituted amino acids.

I. Behbahani-Najad, J. L. Dye & C. H. Suelter (1987) *Meth. Enzymol.* **142**, 414.
A. E. Braunstein & E. V. Goryachenkova (1984) *Adv. Enzymol.* **56**, 1.
A. E. Martell (1982) *Adv. Enzymol.* **53**, 163.
E. E. Snell (1975) *Adv. Enzymol.* **42**, 287.
L. Davis & D. E. Metzler (1972) *The Enzymes*, 3rd ed., **7**, 33.

TRYPTOPHAN SYNTHASE

This pyridoxal-phosphate-dependent enzyme [EC 4.2.1.20] catalyzes the reaction of L-serine with 1-(indol-3-yl)glycerol 3-phosphate to produce L-tryptophan and glyceraldehyde 3-phosphate. The enzyme will also catalyze (a) the conversion of serine and indole into tryptophan and water and (b) conversion of indoleglycerol phosphate into indole and glyceraldehyde phosphate.

K. A. Johnson (1992) *The Enzymes*, 3rd ed., **20**, 1.
E. W. Miles (1991) *Adv. Enzymol.* **64**, 93.
C. C. Hyde & E. W. Miles (1990) *Biotechnology* **8**, 27.
E. W. Miles, R. Bauerle & S. A. Ahmed (1987) *Meth. Enzymol.* **142**, 398.
E. W. Miles (1979) *Adv. Enzymol.* **49**, 127.
C. Yanofsky & I. P. Crawford (1972) *The Enzymes*, 3rd ed., **7**, 1.

TUBULIN-GTP HYDROLYSIS
(Apparent Irreversibility)

The pathway of exchangeable-site GTP hydrolysis associated with bovine brain microtubule polymerization was investigated using an assay of intermediate ^{18}O-positional isotope exchange reactions[1]. Under a variety of conditions that influence the rate and extent of tubulin self-assembly, GTP hydrolysis proceeded without any evidence of multiple reversals that are characteristic of any reversible phosphoanhydride-bond cleavage. These results also accord with published findings[2] that ATP hydrolysis during actin polymerization fails to display intermediate exchange reactions. **See** *Microtubule Assembly Kinetics*

[1]J. M. Angelastro & D. L. Purich (1990) *Eur. J. Biochem.* **191**, 507.
[2]M. F. Carlier, D. Pantaloni, J. A. Evans, P. K. Lambooy, E. D. Korn & M. R. Webb (1988) *FEBS Lett.* **235**, 211.

TUBULIN POLYMERIZATION ASSAYS

Microtubule assembly and disassembly results from the repetitive endwise gain or loss of tubulin dimers[1,2], and a number of useful procedures have been developed to study the kinetics of tubulin polymerization and microtubule depolymerization. These include[3]: (a) Turbidity measurements which take advantage of the fact that the turbidity of microtubule protein solutions is a direct measure of the polymer weight concentration; (b) Electron microscopy which allows one to directly measure polymer length changes in samples fixed at various times after polymerization or depolymerization is induced; (c) Viscosity which is based on the use of low-shear conditions to observe changes in the apparent viscosity of solutions containing microtubules; (d) Sedimentation which is based on the fact that large protein aggregates such as microtubules can be separated from much more slowly sedimenting tubulin dimers; (e) Flow birefringence which allows one to orient long rod structures under the influence of an imposed flow shear force; (f) Dark-field microscopy which is based on Tyndall's classical 19th century design of a "ultramicroscope" for detecting particles by their light scattering properties; (g) X-ray scattering methods based on the small-angle scattering of synchrotron radiation; (h) Filtration[4,5] to separate polymeric and protomeric tubulin forms and to utilize the trapping of radioactively labeled GTP into the assembled microtubules; and (i) Biotinylated tubulin[6] to directly visualize the addition of tubulin molecules on each end of microtubules after subsequent streptavidin binding and immunocytochemical detection. **See** *Microtubule Assembly Kinetics*

[1]D. L. Purich & D. Kristofferson (1984) *Adv. Prot. Chem.* **36**, 133.
[2]D. L. Purich & J. M. Angelastro (1994) *Adv. Enzymol.* **69**, 121.
[3]F. Gaskin (1982) *Meth. Enzymol.* **85**, 433.
[4]R. B. Maccioni & N. W. Seeds (1978) *Arch. Biochem. Biophys.* **185**, 262.
[5]L. Wilson, H. P. Miller, K. W. Farrell, K. B. Snyder, W. C. Thompson & D. L. Purich (1985) *Biochemistry* **23**, 5254.
[6]D. Kristofferson, T. Mitchison & M. W. Kirschner (1986) *J. Cell Biol.* **102**, 1007.

TUBULIN:TYROSINE LIGASE

This enzyme [EC 6.3.2.25] catalyzes the ATP-dependent religation of L-tyrosine to the C-terminal L-glutamate residue in the detyrosinated $\alpha\beta$-tubulin dimer, yielding ADP, orthophosphate, and the religated tubulin dimer. L-Phenylalanine and 3,4-dehydroxy-L-phenylalanine are also substrates. This is the only known C-terminal peptide bond-forming reaction involving a protein substrate. **See also** *Arsenolysis*

N. L. Deans, R. D. Allison & D. L. Purich (1992) *Biochem. J.* **286**, 243.
J. Wehland, H. C. Schröder & K. Weber (1986) *Meth. Enzymol.* **134**, 170.
B. J. Terry & D. L. Purich (1982) *Adv. Enzymol.* **53**, 113.

TUNGSTOENZYMES

Enzymes that require tungsten as a metallic cofactor for catalytic activity. These enzymes include formate dehydrogenase (from *Clostridium thermoaceticum* and *Eubacterium acidaminophilum*), acetylene hydratase from *Peleobacter acetylenicus*, aldehyde ferredoxin oxidoreductase (from *Pyrococcus furiosus*, *P.* sp. ES-4, and *Thermococcus* sp. ES-1), formylmethanofuran dehydrogenase (from *Methanobacterium wolfei* and *M. thermoautotrophicum*), carboxylic acid reductase (*Clostridium thermoaceticum* and *C. formicoaceticum*), formaldehyde ferredoxin oxidoreductase (*Thermococcus litoralis* and *Pyrococcus furiosus*), aldehyde dehydrogenase (from *Desulfovibrio gigas*), and glyceraldehyde-3-phosphate ferredoxin oxidoreductase (from *Pyrococcus furiosus*).

A. Kletzin & M. W. W. Adams (1996) *FEMS Microbiol. Rev.* **18**, 5.
P. A. Bertram, R. A. Schmitz, D. Linder & R. K. Thauer (1994) *Arch. Microbiol.* **161**, 220.
F. da Silva & R. J. P. Williams (1991) *The Biological Chemistry of the Elements: the Inorganic Chemistry of Life*, Clarendon Press, Oxford.

TURBIDITY

The apparent absorption of light resulting from scattering of incident radiation out of the transmitted light path. The intensity of light scattered in various directions from a beam of incident light passing through a solution of macromolecules or other large particles is diminished, such that $I = I_o e^{-\tau\lambda}$ where turbidity τ is the measure of decreased intensity I of the incident beam intensity I_o per unit length, λ, of a solution. This expression is formally analogous to the Beer-Lambert Law for absorption of light. The increase in turbidity is an important kinetic tool in studying macromolecule polymerizations. Recalling that the Rayleigh ratio, R_θ, is an important parameter in light scattering experiments, the turbidity is related to the Rayleigh ratio, R_{90}', for N particles of uniform size per unit volume V at $\theta = 90°$, by the relationship $\tau = (16\pi/3)R_{90}'$. **See also** *Absorption Spectroscopy; Light Scattering; Actin Assembly Assays; Microtubule Assembly Kinetics; End-wise Depolymerization*

TURNOVER

1. The amount of material (*e.g.*, macromolecule or metabolite) metabolized or processed per unit time. 2. The balance between synthesis and degradation of a biomolecule. 3. One complete turn of a reaction cycle. **See** *Turnover Number; Protein Turnover; Pulse-Chase Experiments; Exponential Decay*

TURNOVER TIME

The period required for a pool of tracer to be replaced by newly biosynthesized substrate (also termed the tracee), corresponding to about five half-lives [*i.e.*, $1 - (0.5)^5$] or about 97% replacement of tracer by unlabeled substrate. **See** *Compartmental Analysis*

TWIDDLING

The random mechanochemical action of chemotactic bacteria, responsible for resetting their vectorial movement in gradients of chemotractant. Bacteria that experience a constant or increasing concentration of attractant twiddle infrequently, whereas those experiencing a drop in attractant concentration have a higher frequency of twiddling. Hence, those moving toward an attractant tend to continue in their migration, while those off-course use twiddling to reorient themselves. In their quantitative analysis, Dahlquist *et al.*[1] describe the importance of twiddling in making chemotaxis efficient.

Chen's laboratory[2,3] used quasi-elastic light scattering to track migrating chemotactic bands of *Escherichia coli* in a buffer solution. The temporal development of the bacterial density profile is observed by the intensity of scattered light as the band migrates through a fixed laser beam. They offered a model accounting for some aspects of the observed time evolution of the density profile. The microscopic motility characteristics of the *E. coli* in the band were simultaneously examined by photon correlation spectroscopy, and measured correlation functions yielded the spatial dependence of the half-width within the band. By analyzing the correlation function as a superposition of bacteria proceeding in a straight line and those engaging in twiddling motions, they obtain good agreement between their theoretical treatment and the measured angular dependence of the line shape. This effort helped them to estimate the fraction of twiddling bacteria in the center of the migrating band. **See** *Persistence*

[1]F. W. Dahlquist, P. Lovely & D. E. Koshland, Jr. (1972) *Nature New Biol.* **236**, 120.
[2]M. Holz & S. H. Chen (1978) *Biophys. J.* **31**, 15.
[3]P. C. Wang & S. H. Chen (1981) *Biophys. J.* **36**, 203.

TWO-PHOTON EXCITATION

The excitation of a molecular entity due to absorption, successively or simultaneously, of two photons. If successively, the species that absorbs the second photon still has some of the excitation energy imparted by the first photon.

Comm. on Photochem. (1988) *Pure and Appl. Chem.* **60**, 1055.

TWO-PHOTON PROCESS

A process triggered by a two-photon excitation.

TWO-PROTONIC-STATE ELECTROPHILES

Compounds exhibiting different electrophilic reactivities depending on the compound's protonic or ionization state[1]. Such molecules serve as excellent probes of active-site thiols and as reporter groups. **See also** *Affinity Labeling*

[1]K. Brocklehurst (1982) *Meth. Enzymol.* **87**, 427.

TYPE 1,2,3 COPPER BINDING SITES

The copper-binding sites in copper-containing proteins are characterized by three distinct classes. In Type-1, or blue copper centers, the copper is coordinated to at least two imidazole nitrogens from two histidyl residues and one sulfur from a cysteinyl residue. Type-1 coppers have small copper hyperfine couplings and a strong visible absorption in the Cu(II) state. Type-2 or non-blue copper sites contain copper that is bound mainly to imidazole nitrogens from histidyl residues. Type 3 copper centers are comprised of two spin-coupled copper ions, bound to imidazole nitrogens.

TYROSINE AMINOTRANSFERASE

This pyridoxal-phosphate-dependent enzyme [EC 2.6.1.5], also known as tyrosine transaminase, catalyzes the reaction of L-tyrosine with α-ketoglutarate (or, 2-oxoglutarate) to produce 4-hydroxyphenylpyruvate and L-glutamate. L-Phenylalanine can act as the substrate instead of tyrosine. In some systems, the mitochondrial enzyme may be identical with aspartate aminotransferase.

H. Hayashi, H. Wada, T. Yoshimura, N. Esaki & K. Soda (1990) *Ann. Rev. Biochem.* **59**, 87.

A. E. Braunstein (1973) *The Enzymes*, 3rd ed., **9**, 379.

TYROSINE KINASES, PROTEIN

The following are recent reviews on the molecular and physical properties of this enzyme [EC 2.7.1.112]. This enzyme [EC 2.7.1.112], also known as protein-tyrosine kinase and tyrosyl protein kinase, catalyzes the reaction of ATP with the hydroxyl group of a tyrosyl residue in a protein to produce ADP and the phosphotyrosyl residue.

B. M. Sefton & M.-A. Campbell (1991) *Ann. Rev. Cell Biol.* **7**, 257.
T. Hunter & B. M. Sefton, eds. (1991) *Meth. Enzymol.*, vols. **200** and **201**, Academic Press, San Diego.
J. Larner (1990) *Adv. Enzymol.* **63**, 173.
T. Hunter & J. A. Cooper (1986) *The Enzymes*, 3rd ed., **17**, 191.
M. F. White & C. R. Kahn (1986) *The Enzymes*, 3rd ed., **17**, 247.
T. Hunter & J. A. Cooper (1985) *Ann. Rev. Biochem.* **54**, 897.
D. A. Flockhart & J. D. Corbin (1982) *Crit. Rev. Biochem.* **12**, 133.

U

UBIQUINOL:CYTOCHROME *c* OXIDOREDUCTASE

This enzyme complex [EC 1.10.2.2], also known as cytochrome bc_1 and complex III, catalyzes the reaction of ubiquinol (QH_2) with two ferricytochrome *c* to produce ubiquinone (Q) and two ferrocytochrome *c*. The complex also contains cytochrome *b*-562, cytochrome *b*-566, cytochrome c_1, and a two-iron ferredoxin.

J. V. Schloss & M. S. Hixon (1998) *Comprehensive Biological Catalysis: A Mechanistic Reference* **2**, 43.
A. Tzagoloff (1995) *Meth. Enzymol.* **260**, 51.
U. Schulte & H. Weiss (1995) *Meth. Enzymol.* **260**, 63.
H.-P. Braun & U. K. Schmitz (1995) *Meth. Enzymol.* **260**, 70.
H. Schägger, U. Brandt, S. Gencic & G. von Jagow (1995) *Meth. Enzymol.* **260**, 82.
U. Brandt & B. Trumpower (1994) *Crit. Rev. Biochem. Mol. Biol.* **29**, 165.
B. L. Trumpower & R. B. Gennis (1994) *Ann. Rev. Biochem.* **63**, 675.
Y. Hatefi (1985) *Ann. Rev. Biochem.* **54**, 1015.
M. Wikström, K. Krab & M. Saraste (1981) *Ann. Rev. Biochem.* **50**, 623.
G. Palmer (1975) *The Enzymes*, 3rd ed., **12**, 1.

UBIQUITIN-PROTEIN LIGASES

These enzymes [EC 6.3.2.19] catalyze the reaction of ATP with ubiquitin and a lysyl residue in a protein to produce a protein containing an *N*-ubiquityllysyl residue, AMP, and pyrophosphate (or, diphosphate). Ubiquitin is coupled to the protein by an isopeptide bond between the C-terminal glycine of ubiquitin and ϵ-amino groups of lysyl residues in the protein. An intermediate in the reaction contains one ubiquitin residue bound as a thiolester to the enzyme, and a residue of ubiquitin adenylate noncovalently bound to the enzyme.

K. D. Wilkinson (1995) *Ann. Rev. Nutr.* **15**, 161.
A. Hershko & A. Ciechanover (1992) *Ann. Rev. Biochem.* **61**, 761.

UDENFRIEND'S REAGENT

A mixture of ascorbic acid, oxygen, EDTA, and a ferrous salt that is used in a number of oxidation reactions, including the hydroxylation of an aromatic carbon[1,2]. ***See also*** *Fenton Reaction*

[1]S. Udenfriend, C. T. Clark, J. Axelrod & B. B. Brodie (1954) *J. Biol. Chem.* **208**, 731.
[2]B. B. Brodie, P. A. Shore & S. Udenfriend (1954) *J. Biol. Chem.* **208**, 741.

UMBRELLA EFFECT

An effect observed with a number of compounds which have apparent chiral centers on elements other than carbon. For example, secondary and tertiary amines have a pyramidal structure in which the unshared pair of electrons is at the top of the pyramid. If the three substituents linked to the nitrogen are all different, one might suspect that the tertiary amine would give rise to optical activity and be resolvable. However, rapid oscillation of the unshared pair of electrons on one side of the nitrogen to the other (hence, pyramidal inversion) in effect causes interconversion of the two enantiomers and prevents resolution. If the nitrogen is at a bridgehead, this "umbrella effect" is inhibited and optical isomers can be isolated.

Other elements besides nitrogen can exhibit an umbrella effect: *e.g.*, phosphorus, sulfur, and arsenic. Fortunately, the effect is slow in most of these cases. Hence, one can prepare optically pure phosphoesters (such as, $^{16}O^{17}O^{18}OP-OR$). Thus, chiral analogs of ATP can be utilized to study enzyme stereochemistry. Similarly, chiral sulfones and sulfoxides have proved useful in assessing the topology of a protein's active site.

J. B. Lambert (1972) *Topics Stereochem.* **6**, 19.

UNCOMPETITIVE INHIBITION

Inhibition of an enzyme-catalyzed reaction in which the inhibitor does not bind to the free, uncomplexed enzyme and does not compete with the substrate for the enzyme's active site[1,2]. For a Uni Uni mechanism (E + A $\rightleftharpoons$ EX $\rightarrow$ E + P), an uncompetitive inhibitor would bind to the EX binary complex (thus, EX + I $\rightleftharpoons$ EXI), with a dissociation constant of K_{ii} (equal to [EX][I]/[EXI]). Both the steady-state and rapid equilibrium derivations yield an equation with the same general format:

$$v = \frac{V_{\max}[A]}{K_m + [A]\left(1 + \dfrac{[I]}{K_{ii}}\right)}$$

However, note that K_m is replaced with K_{ia} (the dissociation constant of A for the free enzyme) in the rapid-equilibrium equation. A standard double-reciprocal plot (1/v vs. 1/[A]) at different concentrations of inhibitor will yield a series of parallel lines. A vertical intercept vs. [I] secondary replot will provide a value for K_{ii} on the horizontal axis. If questions arise as to whether the lines are truly parallel, one possibility is to replot the data via a Hanes plot ([A]/v vs. [A]). In such a plot, the lines of an uncompetitive inhibitor intersect on the vertical axis.

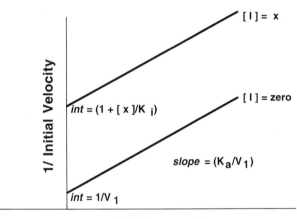

Uncompetitive inhibition can also be a possibility in multisubstrate reactions. For example, in an ordered Bi Bi reaction, a competitive inhibitor with respect to the second substrate B, will act as an uncompetitive inhibitor with respect to the first substrate, A.

[1]J. A. Todhunter (1979) Meth. Enzymol. **63**, 383.
[2]H. J. Fromm (1995) Meth. Enzymol. **249**, 123.

UNCONSUMED SUBSTRATE OR COFACTOR

Any reactant that remains unchanged (or is rapidly regenerated) before and after (but not necessarily during) an enzymic reaction. These could include a metal ion, an activator (e.g., calmodulin), a dissociable regulatory

subunit, a free-radical-containing subunit (e.g., the tyrosyl radical-containing subunit of ribonucleotide reductase), or any unconsumed components. Examples of the last case are ADP and arsenate in the γ-glutamyltransferase reaction catalyzed by glutamine synthetase:

$$\text{glutamine} + NH_2OH \underset{\text{ADP, arsenate}}{\overset{Mg^{2+} \text{ or } Mn^{2+}}{\rightleftharpoons}} \gamma\text{-glutamyl–}NHOH + NH_3$$

In this reaction, arsenate ester analogues of the γ-glutamylphosphate and ADP are likely to be formed in a manner analogous to that of the biosynthetic reaction; however, γ-glutamyl-arsenate and ADP-arsenate undergo rapid hydrolysis in each catalytic round, and the concentrations of ADP and arsenate remain unchanged. Polymeric substrates can also behave as unconsumed substrates. For example, glycogen contains n glucosyl units, and when serving as substrate for an exo-carbohydrase such as glycogen phosphorylase, the corresponding product with $(n-1)$ glucosyl units is also deemed unaltered when n is large.

Unconsumed substrates are treated as "substrates" or essential activators in deriving rate equations and studying detailed mechanisms. Nonetheless, one must indicate whether an unconsumed substrate (U) remains bound to the enzyme or not (in this case, U also becomes an unaltered product) in the reaction scheme. In practice, unconsumed substrates are likely to be involved in all the typical multisubstrate kinetic mechanisms[1]. Only one case is illustrated here, namely that the unconsumed substrate S_U activates catalysis when bound in a rapid-equilibrium ordered mechanism:

$$E + S_U \rightleftharpoons E{\cdot}S_U, \quad K_u = [E][S_U]/[E{\cdot}S_U]$$

$$E{\cdot}S_U + S \rightleftharpoons E{\cdot}S_U{\cdot}S, \quad K_s = [E{\cdot}S_U][S]/[E{\cdot}S_U{\cdot}S]$$

$$E{\cdot}S_U{\cdot}S \rightarrow E{\cdot}S_U + P, \quad v = k[E{\cdot}S_U{\cdot}S]$$

The initial rate law for this mechanism is:

$$\frac{1}{v} = \frac{1}{V_{\max}} + \frac{K_s}{V_{\max}[S]}\left\{1 + \frac{K_u}{[S_U]}\right\}$$

This rate equation is identical to that for a rapid equilibrium ordered addition bisubstrate mechanism (i.e., a scheme where substrate A rapidly binds prior to the addition of the second substrate B). Huang[1] has presented the theoretical basis for mechanisms giving rise to typical rapid-equilibrium ordered kinetic patterns, along with a helpful discussion on various sequential cases.

[1]C. Y. Huang (1990) *Biochemistry* **29**, 158.

UNDERSHOOT

A temporary or transient decrease below a final steady-state value.

UNIMOLECULAR

Elementary reactions (also termed monomolecular reactions) that involve only a single entity in the formation of an activated complex. Unimolecular rate constants, k, are concentration-independent and are typically expressed in units of s^{-1}. Unimolecular reactions are expected to be first order (*i.e.*, $-dc/dt = kc$ where c is the concentration and t is time). Examples of unimolecular processes include radioactive disintegrations, isomerizations, disassociations, and decompositions. Reactions in solution are unimolecular only if the solvent is not covalently incorporated into the product(s).

An entire mechanism can be unimolecular, provided that the rate-limiting step is unimolecular. *See also* Chemical Kinetics, Molecularity; Order of Reaction; Elementary Reactions

UNINTERESTING ENZYME

The null set. Just as there can be no uninteresting numbers in mathematics, each enzyme has the power to teach us fundamental *and* interesting facets of chemistry and biology.

UNIPORT

A system that transports one molecule or ion without concomitant transport of another molecule or ion. *See* Membrane Transport

UNI UNI ENZYME KINETIC MECHANISM

Many enzymes (especially isomerases, racemases, and epimerases) catalyze intramolecular reactions that can be described by a Uni Uni mechanism. These are typically written with one or two central complexes:

$$\text{Scheme 1: } E + A \underset{k_2}{\overset{k_1}{\rightleftharpoons}} EX \underset{k_4}{\overset{k_3}{\rightleftharpoons}} E + P$$

or

$$\text{Scheme 2: } E + A \underset{k_2}{\overset{k_1}{\rightleftharpoons}} EA \underset{k_4}{\overset{k_3}{\rightleftharpoons}} EP \underset{k_6}{\overset{k_5}{\rightleftharpoons}} E + P$$

Both schemes yield the same generalized steady-state initial rate equation:

$$v = \frac{V_{\text{max,f}}V_{\text{max,r}}([A] - [P]/K_{\text{eq}})}{(V_{\text{max,r}}K_a + V_{\text{max,r}}[A] + (V_{\text{max,f}}[P]/K_{\text{eq}}))}$$

In Scheme 1, the rate parameters $V_{\text{max,f}}$ and $V_{\text{max,r}}$ are the maximum velocities in the forward and reverse direction, respectively (such that $V_{\text{max,f}} = k_3[E_{\text{total}}]$ and $V_{\text{max,r}} = k_2[E_{\text{total}}]$), K_a is the Michaelis constant for substrate A ($K_a = (k_2 + k_3)/k_1$), and K_{eq} is the equilibrium constant (equal to k_1k_3/k_2k_4) and having the Haldane relationships $K_{\text{eq}} = V_{\text{max,f}}K_p/(V_{\text{max,r}}K_a) = K_{\text{ip}}/K_{\text{ia}}$ where K_p is the Michaelis constant for P (equal to $(k_2 + k_3)/k_4$) and K_{ia} and K_{ip} are the dissociation constants for A and P, respectively ($K_{\text{ia}} = k_2/k_1$ and $K_{\text{ip}} = k_3/k_4$)). The determinants for the two enzyme forms are $[E] = k_2 + k_3$ and $[EX] = k_1[A] + k_4[P]$. In the absence of the product P (*i.e.*, in initial rate studies where velocities are observed with less than 10% substrate conversion), the expression above reduces to $v = V_{\text{max,f}}[A]/(K_a + [A])$. For Scheme 2, the kinetic parameters are now: $V_{\text{max,f}} = k_3k_5[E_{\text{tot}}]/(k_3 + k_4 + k_5)$; $V_{\text{max,r}} = k_2k_4[E_{\text{tot}}]/(k_2 + k_3 + k_4)$; $K_a = (k_2k_4 + k_2k_5 + k_3k_5)/(k_1(k_3 + k_4 + k_5))$; $K_p = (k_2k_4 + k_2k_5 + k_3k_5)/(k_6(k_2 + k_3 + k_4))$; $K_{\text{ia}} = k_2/k_1$; $K_{\text{ip}} = k_5/k_6$; and $K_{\text{eq}} = k_1k_3k_5/(k_2k_4k_6) = V_{\text{max,f}}K_p/(V_{\text{max,r}}K_a)$. The determinants for the enzyme forms for Scheme 2 are: $[E] = k_2(k_4 + k_5) + k_3k_5$; $[EA] = k_1[A](k_4 + k_5) + k_4k_6[P]$; and $[EP] = k_6[P](k_2 + k_3) + k_1k_3[A]$. *See also* Michaelis-Menten Equation; Iso Uni Uni Mechanism; Rapid Equilibrium Assumption

H. J. Fromm (1975) *Initial Rate Enzyme Kinetics*, Springer-Verlag, New York.

UNREACTIVE

The opposite of reactive; hence, a body or molecular entity fails to react (or reacts more slowly) in some reaction, under specified conditions, as does some standard. The term is not synonymous with stable (however, a stable entity may be more reactive than some reference species).

IUPAC (1979) *Pure and Appl. Chem.* **51**,1725.

UNSTABLE

A body or molecular entity is said to be unstable with reference to another substance, if its molar Gibbs free energy is greater than the reference substance. This descriptor is not synonymous with reactive or transient. **See** *Metastable*

UNSTABLE AFFINITY LABELING REAGENT

Some affinity labeling reagents react with water to undergo deactivation, and depending on the kinetics of the reagent deactivation compared to the enzyme's alkylation by the active affinity labeling substance, the kinetics of enzyme inactivation can appear to be anomalous. For example, the inactivation of horse alcohol dehydrogenase by diethylpyrocarbonate (also known as ethoxyformic anhydride) does not proceed by pseudo-first-order kinetics, and this would be indicated by the nonlinearity of a plot of the logarithm of remaining enzyme activity *versus* the period of exposure to diethylpyrocarbonate. Because this alkylating reagent has a half-life of only 20 min at pH 8 at room temperature, one must take its hydrolysis into account in defining the kinetics of its action on alcohol dehydrogenase.

The following scheme describes reagent R reacting with an enzyme as well as the reagent inactivation process:

$$E + R \xrightarrow{k_1} E{-}X$$

$$R \xrightarrow{k_2} Q$$

where E, R, E$-$X, and Q are active enzyme, active reagent, inactivated enzyme, and inactivated reagent, respectively. The following differential equations describe the kinetics of the above reactions:

$$-dE/dt = k_1[E][R] \quad \text{and} \quad -dR/dt = k_2[R]$$

By integrating the second equation first, we get $[R] = [R]_{initial}e^{-k_2t}$; thus, we may substitute this expression for $[R]$ into the first differential equation to obtain the rate law that allows for enzyme inactivation by a reagent which itself is undergoing deactivation. After separating variables and integrating the combined expression, we obtain

$$\ln([E]_t/[E]_{initial}) = (k_1[R]_{initial}/k_2)[e^{-k_2t} - e^0]$$

or

$$[E]_t/[E]_{initial} = e - \{(k_1[R]_{initial}/k_2)(1 - e^{-k_2t}\}$$

The observed time course of an enzyme inactivation process can thus be fitted to this integrated rate law to obtain the value of k_1 and k_2, although in practice one can independently obtain a value for the latter in the absence of the enzyme.

UPCONVERSION

A nonlinear optical effect in which the frequency of light has increased.

UREA KINETICS

Wolfe[1] has presented an excellent description of the systematic application of stable and radioactive isotope tracers in determining the kinetics of urea production, urea recycling, and interorgan nitrogen transfer in living systems.

[1]R. R. Wolfe (1992) *Radioactive and Stable Isotope Tracers in Biomedicine*, pp. 341-355, Wiley-Liss, New York.

UREASE

Urease [EC 3.5.1.5], a protein containing two tightly bound nickel ions, catalyzes the hydrolysis of urea. Technically, the enzymatic phase proceeds up to the formation of carbamate:

Enzyme-Catalyzed Step:

$$H_2NC(=O)NH_2 + H_2O \rightarrow H_2NCO_2^- + NH_4^+$$

Nonenzymatic Step:

$$H_2NCO_2^- + H_2O \rightarrow HCO_3^- + NH_3$$

Carbamate then degrades nonenzymatically under aqueous conditions to yield carbon dioxide and a second molecule of ammonia. **See** *Nickel-Dependent Enzymes*

M. A. Halcrow (1998) *Comprehensive Biological Catalysis: A Mechanistic Reference* **1**, 506.
F. J. Reithel (1971) *The Enzymes*, 3rd ed., **4**, 1.
J. B. Sumner (1955) *Meth. Enzymol.* **2**, 378.
J. B. Sumner (1926) *J. Biol. Chem.* **69**, 435.

UV ENDONUCLEASES

Uvr(A)BC acts on a very broad spectrum of DNA damage, and in the damage recognition and incision steps, an intricate ATP hydrolysis-coupled process is employed rather than direct damage binding-incision utilized by most of the damage-specific repair enzymes.

A. B. Britt (1996) *Ann. Rev. Plant Physiol. Plant Mol. Biol.* **47**, 75.
L. Grossman & S. Thiagalingam (1993) *J. Biol. Chem.* **268**, 16871.
B. Van Houten (1990) *Microbiol. Rev.* **54**, 18.
B. Weiss & L. Grossman (1987) *Adv. Enzymol.* **60**, 1.

V

VALENCE-BOND METHOD

A procedure for obtaining an approximate wave function of a molecular orbital by treating the molecule as different canonical forms. The overall wave expression is thus the weighted sum of the wave equation for each of these canonical forms.

VALINE DEHYDROGENASE

This enzyme [EC 1.4.1.8] catalyzes the reaction of valine with NADP$^+$ and water to produce 3-methyl-2-oxobutanoate, ammonia, and NADPH.

A. R. Clarke & T. R. Dafforn (1998) *Comprehensive Biological Catalysis: A Mechanistic Reference* **3**, 1.

N. M. W. Brunhuber & J. S. Blanchard (1994) *Crit. Rev. Biochem. Mol. Biol.* **29**, 415.

VANADATE ION

The vanadium oxide VO_4^{3-} [conjugate base of the acid $VO_3(OH)^{2-}$] that adopts trigonal bipyramidal geometry and acts as a transition-state mimic for in-line phosphoryl transfer enzymes. **See** Oxygen, Oxides, Oxo Anions

Vanadate can be prepared as follows: dissolve vanadium(V) oxide in 2.1 molar equivalents of 1 N NaOH; stir 2-3 days, until the yellow color becomes faint; dilute with water to a final concentration of 100 mM.

Vanadate-inhibited enzymes often reactivate on addition of EDTA, which forms high-affinity vanadate:EDTA complexes; if divalent cations must be chelated, use EGTA. Vanadate interacts with many other organic compounds and buffer salts, HEPES being a useful exception. Some commercial preparations of nucleotide triphosphates and carbamoyl-P contain vanadate, and one should ascertain that reagents are vanadate-free before use in kinetic and mechanistic studies. Vanadate also complexes with peroxide anion to produce pervanadate, a reagent that can oxidatively inactivate enzymes containing active-site thiols.

Vanadium in Biological Systems (N. D. Chasteen, ed.) Kluwer Academic Publishers, Netherlands.

A. Butler, M. J. Clague & G. E. Meister (1994) *Chem. Rev.* **94**, 625.

A. Tracey, M. Gresser & S. Liu (1988) *J. Am. Chem. Soc.* **110**, 5869.

VANADIUM-DEPENDENT ENZYMES

Enzymes requiring vanadium for catalytic activity. Perhaps the best studied of these are the vanadium-dependent nitrogenases [EC 1.18.6.1]. Other vanadium-dependent enzymes include vanadium haloperoxidase, vanadium chloroperoxidase, and vanadium bromoperoxidase. In the vanadium chloroperoxidase and bromoperoxidase reactions, the vanadium(V) is coordinated in a trigonal bipyramidal site to a histidyl residue, three nonprotein oxygens, and, presumably, to a hydroxide.

G. J. Leigh (1998) *Comprehensive Biological Catalysis: A Mechanistic Reference* **3**, 349.

A. Butler (1998) *Comprehensive Biological Catalysis: A Mechanistic Reference* **3**, 427.

VAN DER WAALS FORCES

Attractive or repulsive forces between molecular entities or groups within the same molecular entity (*i.e.*, both intermolecular and intramolecular) not due to bond formation or to electrostatic interactions of ions or ionic groups with one another or with neutral molecules. The origin of van der Waals forces is in electric polarization of uncharged atoms, groups, or molecules and includes dipole-dipole interactions, dipole-induced dipole interactions, and London forces (induced dipole-induced dipole interactions).

VAN'T HOFF RELATIONSHIP

Thermodynamic considerations led van't Hoff to conclude[1] that the equilibrium constant, K_{eq}, for a reaction (scheme below) is equal to the ratio of the rate constants

$$A \underset{k_{-1}}{\overset{k_1}{\rightleftharpoons}} P$$

for that reaction, $K_{eq} = k_1/k_{-1}$. He also noted[2] that the effect of temperature on this equilibrium constant (at constant pressure) was described by the following equation,

$$d \ln K_{eq}/dT = \Delta U^{\ominus}/RT^2$$

where R is the molar gas constant (*i.e.*, 8.31431 $J \cdot K^{-1} \cdot mol^{-1}$ or 1.98722 $cal \cdot K^{-1} \cdot mol^{-1}$), T is the absolute temperature, and ΔU is the change in internal energy. The superscript symbol $^{\ominus}$ (or °) denotes standard conditions of temperature, pressure, and ionic strength of an ideal solution at 1 M[3-5]. In most biological systems, processes typically occur at constant pressure and usually involve only small changes in volume. Recall that the change in enthalpy, ΔH, is equal to the heat absorbed at constant pressure, q_p, when no work other than $P\Delta V$ work is being done: thus, $\Delta H = \Delta U + P\Delta V$ where P is the pressure and ΔV is the change in volume. For reactions in solutions, there usually is very little $P\Delta V$ changes when the pressure is low (*e.g.*, 1 atmosphere). Hence, $\Delta H^{\ominus}$ and $\Delta U^{\ominus}$ are very nearly equal. At substantially higher pressures, such as those found deep in the ocean, the $P\Delta V$ change can be a contributing factor. Substituting $\Delta H^{\ominus}$ into first equation and integrating yields $\ln K_{eq} = -(\Delta H^{\ominus}/RT) + $ constant. If the thermodynamic parameters are to be measured under standard conditions, then $\Delta H^{\ominus}$ can be determined from a plot of $\ln K_{eq}$ as a function of T^{-1} (a van't Hoff plot), the slope of the line being equal to $-\Delta H^{\ominus}/R$. The vertical intercept will be equal to $\Delta S^{\ominus}/R$ (since $-\ln K_{eq} = \Delta G^{\ominus}/RT = (\Delta H^{\ominus}/RT) - \Delta S^{\ominus}/R$). **See** *Collision Theory; Arrhenius Law; Transition-State Theory; Temperature Dependence*

[1]J. H. van't Hoff (1877) *Chem. Ber* **10**, 669.
[2]J. H. van't Hoff (English translation, 1896) *Studies in Chemical Dynamics*, Chem. Publ. Co., Easton, Penn.
[3]I. Mills (1988) *Quantities, Units and Symbols in Physical Chemistry*, Blackwell Sci. Publ., Oxford.
[4]IUPAC-IUPAB-IUB, Intercommission on Biothermodynamics (1985) *Eur. J. Biochem.* **153**, 429.
[5]Interunion Commission on Biothermodynamics (1976) *J. Biol. Chem.* **251**, 6879.

VAN'T HOFF'S LAWS

1. The osmotic pressure, at constant temperature, of a dilute solution is proportional to the concentrations of the dissolved substances; *i.e.*, $\Pi = RT \Sigma c_i$ where Π is the osmotic pressure, R is the molar gas constant, T is the absolute temperature and c_i is the molar concentration of species i. 2. In stereochemistry, all optically active molecules contain one or more multivalent atoms united with other atoms such that there is an unsymmetrical arrangement in space around those multivalent atoms.

VARIANCE

A statistical measure of the dispersion of the distribution of the random variable, typically obtained by taking the expected value of the square of the difference between the random variable and its mean. The variance is the square of the standard deviation. **See** *Statistics (A Primer)*

VARIANCE RATIO

The ratio of two independent estimates of the same variance from a Gaussian distribution based on sample sites $(n + 1)$ and $(m + 1)$, respectively. This ratio, often symbolized by F, can be used to test a null hypothesis. **See** *Statistics (A Primer)*

VELOCITY

A term usually synonymous to rate of reaction in the field of enzyme kinetics. Rate is the preferred terminology. **See** *Rate of Reaction; Rate of Conversion*

VELOCITY COEFFICIENT

The rate of transformation of a unit mass of substance in a reaction.

VELOCITY, MACROSCOPIC

The macroscopic velocity is the sum of all microscopic velocities of an enzyme system which acts on a polymer at several different positions, thus yielding a number of different products from the same substrate. One good example is polysaccharide depolymerase[1].

[1]J. D. Allen (1980) *Meth. Enzymol.* **64**, 248.

VENOM EXONUCLEASE

This enzyme [EC 3.1.15.1], also called venom phosphodiesterase, catalyzes the exonucleolytic cleavage of polynucleic acids (with a preference for a single-stranded substrate) in the 3′ to 5′ direction to yield 5′-phosphomononucleotides.

M. Laskowski, Sr. (1971) *The Enzymes*, 3rd ed., **4**, 313.

VESICLE TRANSPORT IN CELLS

The kinetic control of vesicle transport, from vesicle budding in the Golgi (and other origins) to fusion of the vesicle with the target membrane, is currently a vigorous area of research. The formation of COP-coated vesicles

(vesicles carrying newly synthesized proteins from Golgi cisternae) requires binding of a G-protein referred to as ARF (ADP-ribosylation factor). The ARF-GDP complex first binds to budding vesicle, followed by a GDP exchange. This event appears to trigger the binding of coatomer which causes the bud to "pinch" off from the Golgi. Hence, coatomer binding appears to drive budding. When GTP is hydrolyzed, ARF and coatomer are released and the vesicle is ready to fuse with the target membrane. Thus, the triggering of GTP hydrolysis allows the vesicle to fuse with its target. Just how all the steps are controlled and directed is a significant aspect of the study on this topic.

J. Ostermann, L. Orci, K. Tani, M. Amherdt, M. Ravazzola, Z. Elazar & J. E. Rothman (1993) *Cell* **75**, 1015.
J. E. Rothman (1994) *Nature* **372**, 55.
J. E. Rothman (1996) *Sci. Amer.* **274**, 70.

VIBRATIONAL COUPLING

Interaction between electronic and vibrational motions in a molecular entity.

Comm. on Photochem. (1988) *Pure and Appl. Chem.* **60**, 1055.

VIBRATIONAL REDISTRIBUTION

An intramolecular redistribution of energy among all the vibrational modes. For large molecules, this redistribution does not require collisions.

VIBRATIONAL RELAXATION

The loss of vibrational excitation energy by a molecular entity through energy transfer to the environment due to collisions.

Comm. on Photochem. (1988) *Pure and Appl. Chem.* **60**, 1055.

VIBRATIONAL SPECTROSCOPY

Any spectroscopic technique that directly provides information about the vibrational behavior of chemical substances.

Selected entries from *Methods in Enzymology* [vol, page(s)]:
Biomolecular vibrational spectroscopy, **246**, 377; Raman spectroscopy of DNA and proteins, **246**, 389; resonance Raman spectroscopy of metalloproteins, **246**, 416; structure and dynamics of transient species using time-resolved resonance Raman spectroscopy, **246**, 460; infrared spectroscopy applied to biochemical and biological problems, **246**, 501; resonance Raman spectroscopy of quinoproteins, **258**, 132.

VIBRONIC TRANSITION

A transition which is purely electronic or purely vibrational. The transition involves a change in both vibra-tional and electronic quantum numbers and a change in both vibrational and electronic energy.

VICINAL

A term used to designate the geometric arrangement of moieties attached to two adjacent atoms within a molecule. An example of a vicinal diol would be ethylene glycol ($HOCH_2CH_2OH$) and a vicinal dihalide would be 1-bromo-2-chloroethane.

VINYLACETYL-CoA ISOMERASE

This enzyme [EC 5.3.3.3] catalyzes the interconversion of vinylacetyl-CoA and crotonoyl-CoA. 3-Methylvi-nylacetyl-CoA can also be utilized as a substrate.

D. J. Creighton & N. S. R. K. Murthy (1990) *The Enzymes*, 3rd ed. **19**, 323.

VINYL CATION (or Vinylic Cation)

A very short-lived carbocation of the form $R_2C{=}C^+R$ that has been proposed as an intermediate in a number of chemical mechanisms.

VIRTUAL TRANSITION STATE

An approach and construct[1,2] used to understand isotope effects. The isotope effect observed in an enzyme-catalyzed reaction is a weighted average of several steps in the reaction. The transition state that one constructs from these studies is also a "weighted" average of several transition states: thus, the virtual transition state.

[1]R. L. Schowen (1978) in *Transition States of Biochemical Processes* (R. D. Gandour & R. L. Schowen, eds.), p. 77, Plenum, New York.
[2]K. B. Schowen & R. L. Schowen (1982) *Meth. Enzymol.* **87**, 551.

VISCOSITY

A measure of the frictional resistance that a fluid in motion offers to an applied shearing force. Let F be the frictional force, and let S be the area of the interface between a stationary surface plane and a fluid with velocity dv/dr. Then the resisting or frictional force is given as:

$$F = \eta S dv/dr$$

where the constant η is termed the coefficient of viscosity. The SI units are $kg \cdot m^{-1} \cdot s^{-1}$, with the cgs unit (called poise) equal to one-tenth of the SI unit. The following are variously defined terms based on the measured solvent viscosity η_o and the solution viscosity η, where c is the concentration in g/dL: (1) Relative Viscosity = η/η_o; (2) Specific Viscosity = $(\eta - \eta_o)/\eta_o$; (3) Reduced Viscosity $= c^{-1}(\eta - \eta_o)/\eta_o$; and (4) Intrinsic Viscosity $= \text{limit}_{c\to 0}$ $[c^{-1}(\eta - \eta_o)/\eta_o]$.

VISCOSITY COEFFICIENT

A parameter equal to the value of the force per unit area required to maintain a unit relative velocity between two parallel planes a unit distance apart.

VITAMIN A GROUP

Metabolic derivatives of the fat-soluble vitamin retinol, which was first described in 1913. These metabolites play a central role in the visual process where the primary photoevent is attended by conversion of 11-*cis*-retinal to all-*trans*-retinal. The oxidized metabolite retinoic acid is an agonist in cell growth and proliferation.

Selected entries from *Methods in Enzymology* [vol, page(s)]: Colorimetric estimation of vitamin A with trichloroacetic acid, **67**, 189; indirect spectrophotometry on vitamin A products: peak signal readout, **67**, 195; colorimetric analysis of vitamin A and carotene, **67**, 199; analysis of geometrically isomeric vitamin A compounds, **67**, 203; hplc of vitamin A metabolites and analogs, **67**, 220; gas chromatography, gas chromatography-mass spectrometry, and high-pressure liquid chromatography of carotenoids and retinoids, **67**, 233; separation of carotenoids on lipophilic Sephadex, **67**, 261; carotenoid biosynthesis by cultures and cell-free preparations of flavobacterium R1560, **67**, 264; synthesis and properties of glycosyl retinyl phosphates, **67**, 270; fluorescence assay of retinol-binding holoprotein, **67**, 282; purification of cellular retinol and retinoic acid-binding proteins from rat tissue, **67**, 288; cellular retinol-, retinal-, and retinoic acid-binding proteins from bovine retina, **67**, 296; isolation and purification of bovine rhodopsin, **67**, 301; quantitative analysis of retinal and 3-dehydroretinal by high-pressure liquid chromatography, **123**, 53; enzymatic synthesis and separation of retinyl phosphate mannose and dolichyl phosphate mannose by anion-exchange high-performance liquid chromatography, **123**, 61; separation and quantitation of retinyl esters and retinol by high-performance liquid chromatography, **123**, 68; separation of geometric isomers of retinol and retinoic acid in nonaqueous high-performance liquid chromatography, **123**, 75; use of enzyme-linked immunosorbent assay technique for quantitation of serum retinol-binding protein, **123**, 85; Fourier transform infrared spectroscopy of retinoids, **123**, 92; purification of interstitial retinol-binding protein from the eye, **123**, 102; quantification of physiological levels of retinoic acid, **123**, 112.

VITAMIN D GROUP

The vitamin and metabolic derivatives of a fat-soluble substance calciferol. These antirachitic vitamins are produced from $\Delta^{5,7}$-unsaturated sterols upon irradiation with ultraviolet light. Population biologists have adduced strong evidence for the hypothesis that lighter skin color, the primary characteristic of race, is related to the lower solar radiance in northern climes. Vitamin D plays a central role in calcium metabolism, both in terms of

absorption of this metal ion and its subsequent role in bone mineralization.

Selected entries from *Methods in Enzymology* [vol, page(s)]: Colorimetric determination of vitamin D_2 (calciferol), **67**, 323; analysis of vitamin D_2 isomers, **67**, 326; gas-liquid chromatography of vitamin D and analogs, **67**, 335; gas-liquid chromatographic determination of vitamin D in multiple vitamin tablets and their raw materials, **67**, 343; gas-liquid chromatographic determination of vitamin D, **67**, 347; thin layer-gas chromatographic determination of 25-hydroxycholecalciferol, **67**, 355; high-performance liquid chromatographic determination of vitamin D and metabolites, **67**, 357; HPLC of vitamin D metabolites and analogs, **67**, 370; mass fragmentographic assay of 25-hydroxyvitamin D_3, **67**, 385; determination of vitamin D and its metabolites in plasma, **67**, 393; assay of 24*R*,25-dihydroxycholecalciferol in human serum, **67**, 414; enzymic preparation of [^{3}H]1,25-dihydroxyvitamin D_3, **67**, 424; vitamin D metabolites: extraction from tissue and partial purification prior to chromatography, **67**, 426; purification and properties of vitamin D hydroxylases, **67**, 430; 25-hydroxylation of vitamin D_3 in liver microsomes and their smooth and rough subfractions, **67**, 441; measurement of the chicken kidney 25-hydroxyvitamin D_3 1-hydroxylase and 25-hydroxyvitamin D_3 24-hydroxylase, **67**, 445; purification, characterization, and quantitation of the human serum binding protein for vitamin D and its metabolites, **67**, 449; competitive binding assay for 25-hydroxyvitamin D using specific binding proteins, **67**, 459; competitive protein binding assay for plasma 25-hydroxycholecalciferol, **67**, 466; steroid competition assay for determination of 25-hydroxyvitamin D and 24,25-dihydroxyvitamin D, **67**, 473; features of the receptor proteins for the vitamin D metabolites, **67**, 479; measurement of kinetic rate constants for the binding of 1α,25-dihydroxyvitamin D_3 to its chick intestinal mucosa receptor using a hydroxyapatite batch assay, **67**, 488; structural aspects of the binding of 1α,25-dihydroxyvitamin D_3 to its receptor system in chick intestine, **67**, 494; radioimmunoassay for chick intestinal calcium-binding protein, **67**, 500; purification of chick intestinal calcium-binding protein, **67**, 504; characteristics and purification of the intestinal receptor for 1,25-dihydroxyvitamin D, **67**, 508; a sensitive radioreceptor assay for 1α,25-dihydroxyvitamin D in biological fluids, **67**, 522; use of chick kidney to enzymatically generate radiolabeled 1,25-dihydroxyvitamin D and other vitamin D metabolites, **67**, 529; isolation and identification of vitamin D metabolites, **123**, 127; a new pathway of 25-hydroxyvitamin D_3 metabolism, **123**, 141; enzyme-linked immunoabsorption assay for vitamin D-induced calcium-binding protein, **123**, 154; measurement of 25-hydroxyvitamin D 1α-hydroxylase activity in mammalian kidney, **123**, 159; quantitation of vitamin D_2, vitamin D_3, 25-hydroxyvitamin D_2, and 25-hydroxyvitamin D_3 in human milk, **123**, 167; 1,25-dihydroxyvitamin D microassay employing radioreceptor techniques, **123**, 176; assay for 1,25-dihydroxyvitamin D using rabbit intestinal cytosol-binding protein, **123**, 185; cytoreceptor assay for 1,25-dihydroxyvitamin D, **123**, 190; monoclonal antibodies as probes in the characterization of 1,25-dihydroxyvitamin D_3 receptors, **123**, 199.

VITAMIN D 25-HYDROXYLASE

The following are recent reviews on the molecular and physical properties of this liver enzyme which converts cholecalciferol (vitamin D_3) to 25-hydroxycholecalciferol.

H. F. DeLuca & C. Zierold (1998) *Nutr. Rev.* **56**, 54.
H. L. Henry & A. W. Norman (1984) *Ann. Rev. Nutr.* **4**, 493.

H. F. DeLuca & H. K. Schnoes (1983) *Ann. Rev. Biochem.* **52**, 411.

H. F. DeLuca & H. K. Schnoes (1976) *Ann. Rev. Biochem.* **45**, 631.

VITAMIN K-DEPENDENT γ-GLUTAMYL CARBOXYLASE

This enzyme catalyzes the vitamin K-dependent post-translational carboxylation of specific glutamic acid residues in prothrombin and other blood coagulation proteins to form 4-glutamylcarboxylate side chains.

P. Dowd, S.-W. Han, S. Naganathan & R. Hershline (1995) *Ann. Rev. Nutr.* **15**, 419.

J. W. Suttie (1993) *FASEB J.* **7**, 445.

K. J. Kotkow, D. A. Roth, T. J. Porter, B. C. Furie & B. Furie (1993) *Meth. Enzymol.* **222**, 435.

E. Cadenas, P. Hochstein & L. Ernster (1992) *Adv. Enzymol.* **65**, 97.

C. Vermeer (1990) *Biochem. J.* **266**, 625.

J. W. Suttie (1980) *Crit. Rev. Biochem.* **8**, 191.

J. Stenflo (1978) *Adv. Enzymol.* **46**, 1.

VITAMIN K EPOXIDASE

The following is a recent review on the molecular and physical properties of this enzyme.

R. E. Olson (1984) *Ann. Rev. Nutr.* **4**, 281.

VITAMIN K-2,3-EPOXIDE REDUCTASE

Vitamin-K-epoxide reductase (warfarin-sensitive) [EC 1.1.4.1] catalyzes the reaction of 2-methyl-3-phytyl-1,4-naphthoquinone with oxidized dithiothreitol and water to produce 2,3-epoxy-2,3-dihydro-2-methyl-3-phytyl-1,4-naphthoquinone and 1,4-dithiothreitol. In the reverse reaction, vitamin K 2,3-epoxide is reduced to vitamin K and possibly to vitamin K hydroquinone by 1,4-dithioerythritol (which is oxidized to the disulfide). Some other dithiols and butane-4-thiol can also act as substrates. This enzyme is strongly inhibited by warfarin.

Vitamin-K-epoxide reductase (warfarin-insensitive) [EC 1.1.4.2] catalyzes the reaction of 3-hydroxy-2-methyl-3-phytyl-2,3-dihydronaphthoquinone with oxidized dithiothreitol and water to produce 2,3-epoxy-2,3-dihydro-2-methyl-3-phytyl-1,4-naphthoquinone and 1,4-dithiothreitol. In the reverse reaction, vitamin K 2,3-epoxide is reduced to 3-hydroxy- (and 2-hydroxy-) vitamin K by 1,4-dithioerythritol (which is oxidized to the disulfide). The enzyme is not inhibited by warfarin.

R. E. Olson (1984) *Ann. Rev. Nutr.* **4**, 281.

VITAMIN K GROUP

This group of fat-soluble naphthoquinones contain isoprenoid side-chains of varying length and are important factors in coagulation (hence, K for the German term Koagulation). Vitamin K plays a central role as a cofactor in the synthesis of γ-carboxyglutamic acid (Gla), and deficiency in this vitamin leads to blood-clotting disorders characterized by a much greater tendency for hemorrhage. Dicoumarol is a potent antagonist of vitamin K, and this agent is used clinically to prevent blood clotting.

Selected entries from *Methods in Enzymology* [vol, page(s)]: Spectrophotometric determination of the K vitamins, **67**, 125; spectroscopic determination of vitamin K after reduction, **67**, 128; polarographic determination of vitamin K_5 in aqueous solution, **67**, 134; photochemical-fluorometric determination of the K vitamins, **67**, 140; determination of vitamin K_1 in photodegradation products by gas-liquid chromatography, **67**, 148; assay procedure for the vitamin K_1 2,3-epoxide-reducing system, **67**, 160; vitamin K-dependent carboxylase, **67**, 165; microsomal vitamin K-dependent carboxylase, **67**, 180; assay of coumarin antagonists of vitamin K in blood by high-performance liquid chromatography, **123**, 223; assay of K vitamins in tissues by high-performance liquid chromatography with special reference to ultraviolet detection, **123**, 235; analysis of bacterial menaquinone mixtures by reverse-phase HPLC, **123**, 251.

V_{max}

Symbol for maximal velocity of an enzyme-catalyzed reaction, usually expressed as the molarity change in product per unit time (usually, second). V_{max} must not be confused with k_{cat} or specific activity; the former has dimensions of time^{-1}, and the latter is usually expressed as micromol product per unit time per milligram of protein. *See Michaelis-Menten Equation; Enzyme Rate Equations (1. The Basics)*

V_{max}/K_m (As Measure of Substrate Capture and Product Release)

Although V_{max}/K_m is traditionally treated as a first-order rate constant for enzyme reactions at low substrate concentration, Northrop[1] recently pointed out that V_{max}/K_m actually provides a measure of the rate of *capture* of substrate by free enzyme into a productive complex or the complexes *destined* to go on to form products and complete a turnover at some later time. His analysis serves to underscore the concepts (a) that any catalytic cycle must be characterized by the efficiency of reactant capture and product release, and (b) the Michaelis constant takes on meaning beyond that typically associated with affinity for substrate. Consider the case of an enzyme and substrate operating by the following sequence of reactions:

$$E + S \rightleftharpoons ES \rightarrow EP \rightarrow E + P$$

where k_1 is the rate constant for ES formation, k_2 is the rate constant for ES dissociation to E and S, k_3 is the constant for the conversion from ES to EP, and k_4 is the rate constant for EP breakdown. For the case where ES formation is faster than EP breakdown (*i.e.*, the product is "sticky" and its release is slower than bond-breaking/making), the maximal velocity V_{max} and V_{max}/K_m are:

$$V_{max} = [ES]/(1/k_3 + 1/k_4)\, k_4[E_{total}]$$

$$V_{max}/K_m = k_1 k_3 [E]/(k_2 + k_3)$$

If the substrate is also "sticky" (*i.e.*, $k_2 \ll k_3$), we arrive at the diffusion-controlled parameter $V_{max}/K_m \approx k_1$ [E_{total}]. Northrop[1] defines V_{max}/K_m as the apparent rate constant for the *capture* of substrate into enzyme complexes that are *destined* to yield product(s) at some later time. Moreover, V_{max} is then defined as the apparent rate constant for the *release* of substrate from *captured complexes* in the form of free product(s). If one divides V_{max} by V_{max}/K_m, one obtains K_m:

$$K_m = V_{max}/(V_{max}/K_m) = k_5/k_1 = k_{release}/k_{capture}$$

and K_m is the ratio of the rate constants for release and capture. As pointed out by Northrop, this approach allows one to liken $1/K_m$ to the substrate affinity, even though K_m is not purely a dissociation constant. **See** *Specificity Constant*

[1]D. B. Northrop (1999) *Adv. Enzymol.* **72**, 12.

V_{max}-TYPE (or, V-Type) ALLOSTERIC SYSTEM

Most known allosteric enzymes rely on modulation of affinity, and are hence designated as *K*-type to signify changes in ligand binding affinity as reflected in changes in their apparent K_m values for substrates. Nonetheless, Monod *et al.*[1] anticipated that other enzymes would modulate their activity through changes in catalytic efficiency, and they used the descriptor "V_{max}-type or *V*-type" to describe this less frequently observed form of allosteric behavior.

One excellent example[1,3] of a V_{max}-type allosteric enzyme is *Escherichia coli* phosphoglycerate dehydrogenase (PGDH), a tetramer of identical subunits that catalyzes the formation of D-3-phosphohydroxypyruvate from D-3-phosphoglycerate in a reaction that uses NAD$^+$ as a redox cofactor. This regulatory enzyme is allosterically controlled by serine. All available information suggests that the effects on the K_m for substrate are minor and that enzyme activity depends on serine-induced changes in V_{max}. Inhibition studies with numerous amino acids and L-serine analogues indicate that all three functional groups of serine are required for optimal interaction. Kinetic studies indicate there are a minimum of two serine-binding sites, although the crystal structure of PGDH with bound serine as well as direct serine-binding studies revealed four serine-binding sites. The serine-binding sites reside at the interface between regulatory domains of adjacent subunits. Two serine molecules bind at each of the two regulatory domain interfaces in the enzyme. When all four serines are bound, one observes nearly total inhibition; however, with one bound serine at each interface 85% inhibition is observed. Tethering of the regulatory domains by multiple H-bonds from serine to each subunit appears to prevent structural reorientation that is required in the catalytic cycle. These investigators inferred that part of the conformational change may involve a hinge (formed where two antiparallel beta-sheets are joined within the regulatory domain). Serine binding may prevent the enzyme from closing the cleft between the substrate-binding domain and the nucleotide-binding domain. If this conformational change is necessary for catalysis, then a plausible model emerges for V_{max}-type allosteric control based on serine-regulated movement of rigid domains about flexible hinges.

[1]J. Monod, J. Wyman & J.-P. Changeux (1965) *J. Biol.* **12**, 88.
[2]G. A. Grant, D. J. Schuller & L. J. Banaszak (1996) *Protein Sci.* **5**, 34.
[3]D. J. Schuller, G. A. Grant & L. J. Banaszak (1995) *Nature Struct. Biol.* **2**, 69.

$V_{n,max}$

The initial velocity of a cooperative enzyme for which the Hill coefficient is maximal.

VOLATILE BUFFERS

Chemical species which can be used as buffers and, due to their volatile nature (*i.e.*, relatively low boiling points), can be readily removed from a system (*e.g.*, by evaporation). Such buffers have proved to be quite useful in procedures which later require ion-exchange chromatography or electrophoresis (particularly high voltage electrophoresis). Examples of some common components of volatile buffer systems (with their corresponding boiling points and pK_a values [at 25°C]) would be formic acid (100.5°C; 3.75), acetic acid (118°C; 4.76), pyridine (115.5°C; 5.23), triethanolamine (335.4°C; 7.76), ammonia ($-33.35°C$; 9.25), ethanolamine ($170.58°C$; 9.50), and trimethylamine ($2.9°C$; 9.80).

A few common volative buffer mixtures, along with their respective pH range are: pyridine-formic acid (2.3–3.5); trimethylamine-formic acid (or acetic acid) (3.0–6.0); triethanolamine-HCl (6.8–8.8); ammonia-formic acid (or acetic acid) (7.0–10.0). **See** *Buffers*

D. D. Perrin & B. Dempsey (1974) *Buffers for pH and Metal Ion Control*, Chapman and Hall, London.

VOLT

The SI unit for electric potential and for electromotive force (symbolized by V) equivalent to one joule per coulomb. It is the difference in electric potential needed for a one ampere current to flow through a resistance of one ohm.

VOLTAGE CLAMP

A circuit that uses a differential amplifier to maintain constant membrane potential by electronically balancing the ion channel current. This method allows the experimenter to analyze action potentials of excitable membranes resulting from an initial transient rise in sodium ion permeability followed by a transient rise in potassium ion permeability[1]. The technique is especially valuable for studying kinetic properties of voltage-gated channels as well as voltage-dependent channels. **See** *Membrane Potential; Patch Clamp Methods*

[1]A. L. Hodgkin & A. F. Huxley (1952) *J. Physiol.* **116**, 424; 472; and 497.

VOLUME OF ACTIVATION

A quantity (symbolized by $\Delta V^{\ddagger}$ or $\Delta^{\ddagger} V^{\ominus}$) derived from the pressure dependence of a reaction rate constant: $\Delta V^{\ddagger} = -RT(\partial \ln k/\partial P)_{T}$ where R is the molar gas constant, T is the absolute temperature, k is the reaction rate constant, and P is pressure[1]. For this equation, the rate constants of all non-first-order reactions are expressed in pressure-independent units (*e.g.*, molarity) at a fixed temperature and pressure.

In transition-state theory, this parameter represents the difference between the partial molar volume of the transition state ($V^{\ddagger}$) and the sums of the partial molar volumes of the reactants at the same temperature and pressure. Thus, $\Delta V^{\ddagger} = V^{\ddagger} - \Sigma(rV_{R})$ where r is the order of the reactant R and V_{R} is the partial molar volume. The net volume of activation for a bimolecular reaction is expected to be negative, whereas the volume of activation for a unimolecular reaction is expected to be zero or very small.

Moore[2] offers a concrete example in a problem asking the student to formulate an equation based on transition state theory to calculate the volume of activation per mole of a reaction that experiences a doubling in the bimolecular rate constant by an increasing pressure from 1 to 3000 atm at 298 K. Transition state theory requires that the activated complex remain in equilibrium with its reactants ($A + B = X^{\ddagger}$) where $\Delta G^{\ddagger} = -RT \ln K^{\ddagger}$. Recalling that $dG = VdP - SdT$, we can obtain the partial differential equation: $(\partial G^{\ddagger}/\partial P)_{T} = \Delta V^{\ddagger}$. Therefore, $(\partial \ln K^{\ddagger}/\partial P)_{T} = \Delta V^{\ddagger}/RT$. Again recalling that the reaction rate constant, $k_{\text{reaction}} = (k_{B}T/h)K^{\ddagger}$, we can obtain the expression $\ln(k_{\text{rxn (3000 atm)}}/k_{\text{rxn (1 atm)}}) = (P_{2} - P_{1})\Delta V^{\ddagger}/RT$. Thus, $\ln 2 = (3000 - 1)\Delta V^{\ddagger}/(82.05 \times 298) = 5.7 \text{ cm}^{3}$. This exercise indicates that reaction rate is relatively insensitive to pressure changes if $\Delta V^{\ddagger}$ is small. **See** *Transition-State Theory Expressed in Thermodynamic Terms; Gibbs Free Energy of Activation; Enthalpy of Activation; Entropy of Activation*

[1]IUPAC (1979) *Pure Appl. Chem.* **51**, 1725.
[2]W. J. Moore (1972) *Physical Chemistry*, 4th ed., p. 418, Wiley, New York.

VOLUME OF DISTRIBUTION

The volume or space over which a tracee is distributed within the body or structure. **See** *Compartmental Analysis*

VOLUME STRAIN

A dimensionless quantity for the change in volume, V, per unit volume. A synonymous term is bulk strain. An example of volume strain can be seen in bodies experiencing hydrostatic pressure. Volume strain is usually symbolized by θ: thus, $\theta = \Delta V/V_{0}$.

V SYSTEMS

Cooperative enzyme systems in which the presence of an allosteric effector results in an alteration of the $V_{\max}$ value of the system (as opposed to changes in the K_{m} value(s): K systems). **See** *$V_{\max}$-Type (or, V-Type) Allosteric System; Allosterism*

W

WALDEN INVERSION

The inversion of stereochemical configuration at a chiral center in a nucleophilic substitution reaction[1-4]. The uncatalyzed reaction is bimolecular: *i.e.*, it is an S_N2 reaction (in fact the Walden inversion was noted before the criteria for S_N2 reactions were presented). *See* S_N1 *and* S_N2 Reactions; Nucleophilic Substitution Reactions

[1]P. Walden (1895) *Ber.* **28**, 1287 and 2766.
[2]C. A. Bunton (1966) in *Studies of Chemical Structure and Reactivity* (J. H. Ridd, ed.), pp. 73-102, McGraw-Hill, New York.
[3]H. A. Bent (1968) *Chem. Rev.* **68**, 587.
[4]D. P. G. Harmon (1970) *J. Chem. Ed.* **47**, 398.

WALL EFFECT

Any detectable effect on the reaction or behavior of a particular system by the interior wall of the container or reaction vessel. Because proteins can form high-affinity complexes with glass and plastic surfaces, one must exercise caution in the choice of reaction kinetic conditions. Wall effects can be discerned if one determines catalytic activity under different conditions that minimize or maximize contact of the solution with the container. In principle, an enzyme-catalyzed reaction should proceed at the same rate if placed in a capillary or a culture tube; however, contact with the wall is maximized in a capillary, and wall effects should be more prominent. Some investigators add bovine serum albumin to prevent adsorption of their enzyme onto the container's walls.

WASHING LABWARE

Biochemical experiments require glassware that is scrupulously free of contaminants that can alter the structure or activities of proteins and other macromolecules. Thus, washing must be more rigorous than that used in a typical organic or inorganic laboratory. Simple visual inspection can often detect if any residual detergent remains; evidence of any remaining soap films can be readily detected with a pair of polarized sunglasses.

After cleaning with a phosphate-free detergent, labware should be rinsed extensively with deionized water and finally rinsed at least twice with doubly distilled water before drying. Avoid using paper towels, because lint and microscopic paper fibers often contain metal ions as well as acids/bases used in commercial treatment of paper products. Paper fibers also scatter light and interfere with spectrophotometric assays.

Once cleaned, glassware (and plasticware) should be covered or kept inverted. Even dust can be a troublesome contaminant.

Even scrupulously clean glass can become contaminated by fingerprints which often contain metal ions, amino acids, and oils. Heavy metal ions are present in the ink of permanent markers and can diffuse through the wall of certain plastic containers.

WASH-OUT

The loss of an isotopic label from one or more sites of a labeled compound in the absence or presence of an enzyme. Nonenzymatic tritium and deuterium wash-out occurs whenever a carbanion or ylide is sufficiently stable so as to promote proton release. Enzymatic tritium and deuterium wash-out also occurs when a carbanion is formed and the protonated base can undergo exchange with protons from the solution. The carbon-8 atom of many purines forms an ylide, and tritium or deuterium bound at this position can undergo facile exchange. For this reason, many tritium labeled nucleotides and nucleosides should be purified immediately prior to use. Another good example is proline racemase, which is thought to proceed by two-base catalysis through a transition state that has two thiols sandwiching a proline carbanion.

WATER

No single treatise can provide a sufficiently thorough account of the properties of water! Yet, the kinetics and thermodynamics of every biochemical process are linked to the molecular interactions of water with macromolecules, membranes, metabolites, anions, cations, protons and even electrons. For this reason, this handbook provides a brief overview of the structure and general properties of this most fascinating of all solvents. Where deemed appropriate, references are provided for further reading by those motivated to examine these topics at greater depth.

MOLECULAR STRUCTURE. The bond lengths and bond angle of water in its three forms (*e.g.*, H_2O, HDO, and D_2O) are well known, and Benedict *et al.*[1] gives the following bond lengths and angles: H_2O, 0.9572 Å and 104.52°; HDO, 0.9571 Å and 104.53°; and D_2O, 0.9575 Å and 104.47°. These are idealized values, and the bond lengths and angles are in constant flux. As pointed out by Eisenberg and Kauzmann[2], the near equivalence of bond lengths and angles for H_2O, HDO, and D_2O is consistent with the Born-Oppenheimer approximation that nucleus does not influence a molecule's electronic structure. (This fact is important in studies on kinetic isotope effects and solvent isotope effects.) In addition to its zero-point vibration, water molecules exhibit three vibrational modes: symmetric stretching along both $O-H$ bonds, asymmetric stretching along only one $O-H$, and bending. The dipole moment of water is 1.83 debye.

HYDROGEN BONDING. The stability of hydrogen bonds ($\Delta G_{formation} \approx -5$ kcal/mol or -21 kJ/mol) can be compared to the weaker van der Waals interactions ($\Delta G_{formation} \approx -0.3$ kcal/mol or -1.26 kJ/mol) or covalent bonds ($\Delta G_{formation} \approx -100$ kcal/mol or -420 kJ/mol). Because hydrogen bonding is so extensive in the case of liquid water[3], it is useful to consider some of the general features of hydrogen bonds. Three factors appear to be important in determining the strength of hydrogen bonds. First, the $O-H$ bond's polarization determines (a) the ease with which a hydrogen atom can be shared by its neighboring electronegative oxygen atoms, and (b) the net electrical charge on the oxygen atoms that share the hydrogen. Of course, acids form stronger hydrogen bonds than water and alcohols because their $O-H$ bonds are strongly polarized with greater net negative charge on the oxygen and more net positive charge on the hydrogen. This explains the propensity of acetic acid to form doubly hydrogen-bonded dimers in solid, liquid, and even vapor states. Second, the strongest hydrogen bonds are those most apt to have all three participating atoms arranged in a linear manner. Bent hydrogen bonds are considerably weaker. Third, the total interatomic distance between the two electronegative bonding partners also factors into bond strength, and the optimal interatomic distance for maximum bond overlap is around 3 Å.

Water is both a weak acid and a weak base, thereby promoting formation of good hydrogen bonding networks. Moreover, all of the above factors conspire to promote strong hydrogen bonding between water molecules. As discussed by Eisenberg and Kauzmann[2], the substantial effect of hydrogen bonding on the thermal energy of water is appreciated best when the thermodynamic properties of water are compared to those of other related liquids. They point out that the molal heats of vaporization of the related series of compounds (*i.e.*, H_2Te, H_2Se, and H_2S) decrease as their molecular weights increase. Water is the next lower molecular-weight homologue in this series; yet, water has a heat of vaporization that is more than twice that of hydrogen sulfide.

LIQUID WATER. Although there is still no universally accepted structure for liquid water, the ability to form strong hydrogen bonds is thought to dominate water structure in the liquid state. Indeed, there are also several models for the structure of liquid water that would suggest that the thermal energy of water relates to the hydrogen bonding equilibrium [written as: $O-H \cdots O = O-H + O$] for which a shift to the right (a) is favored at elevated temperatures of the liquid, and (b) represents an increased configurational potential energy. While ice has a reasonably ordered structure involving hydrogen bonds, the fact that ice has a lower density than liquid water reminds us that H_2O molecules in liquid water are somehow more densely packed.

X-ray diffraction studies yield radial distribution function data[4] which are dominated by the much greater scattering power of the more electron-rich oxygen atoms in water. These diffraction results tell us something about how closely packed are these oxygen atoms with respect to each other. Such experimental measurements must be analyzed using structural[5] or probabilistic[6] models to retrieve structural information from the convoluted radial distribution data, and this complexity arises from the lack of any regular arrangements of molecules within a fluid. These molecules may form short-lived clusters that relentlessly form and break up. Indeed, no straightforward model yet accounts for either the maximal density of water at 4°C or water's substantial expansion as one proceeds to cool the liquid below that point.

Narten *et al.*[4] arrived at 4.4 for the number of nearest neighboring H_2O molecules in solution. Furthermore, newer neutron diffraction studies[7] suggest that each water molecule has 4.5 to 5.0 nearest neighbors. Nèmethy and Scheraga[8] treated the radial distribution data as a composite of hydrogen bonding and non-hydrogen bonding populations. Bernal[9] suggested that water molecules may form irregular networks of two classes of ring structures: the first comprising four-membered rings, as in ice-IV; and those made up of two closely positioned six-membered rings with bent hydrogen bonds allowing closer packing of non-hydrogen-bonded molecules.

Rather than interpreting the nearly four-coordinate structure of liquid water in terms of ice-like regions within the liquid, Symons[10,11] envisages liquid water "as a three-dimensional network involving units fortuitously linked together, with four, five, six, and seven (membered) ring systems distinguishable but not of any particular significance because they must be haphazardly strung together". He suggests that water molecules form what he has termed as bifurcated hydrogen bonds where two OH groups share one lone pair or two lone pairs share an OH group. Furthermore, he suggests that the concentration of monomeric water molecules is likely to be quite low at room or physiologic temperatures.

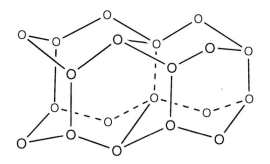

Positions of Water Oxygens in Hexagonal Ice

Stillinger[12,13] analyzed the structure of water by first considering hexagonal ice (illustrated above with solid lines indicating chemical bonds and dashed lines for hydrogen bonds). Starting with any oxygen atom, he points out that one can traverse many closed paths of contiguous hydrogen bonds by the simple rule that one always jumps to the nearest oxygen neighbor. Only polygonal paths with any even number of steps (or sides) offer a route that returns one to the origin, and this property is not observed with other hydrogen-bonded networks, such as that observed with HF or :NH_3. In the vapor phase, $(H_2O)_2$ dimers have linear hydrogen bonds with O-to-O interatomic distances of 2.98 Å, a value nearly 10% longer than the 2.74 Å for ice. Stillinger points out that quantum mechanical calculations support the linearity of hydrogen bonds, but they also suggest that rotations about those linear hydrogen bonds does not cost much in the way of lost stabilization energy. He finds appealing the ideas of Frank and Wen[14] who view liquid water as a cooperative hydrogen bonding network, where there are chains of hydrogen bonds that are shorter and more stable than observed with $(H_2O)_2$ dimers. Stillinger's own computer modeling[12,15] gives the following fractions of water molecules in liquid water (at 10°C) engaging in: zero H-bonds, ~4%; one H-bond ~18%; two H-bonds, ~36%; three H-bonds, ~32%; and 4 H-bonds, ~10%. This would suggest that the average water molecule in the liquid state makes 2.3 hydrogen bonds with its neighbors and that very few free water molecules (~4%) are present. By analyzing computer simulated molecular configurations of water molecules, Stillinger[13] argued in favor of the following model for liquid water: "This network has a local preference for tetrahedral geometry, but it contains a large proportion of strained and broken (hydrogen) bonds. These strained bonds appear to play a fundamental role in kinetic properties, because their presence enhances prospects for molecules to switch allegiances, trading a bond here for one there and thus altering the network topology".

IONIZATION AND SELF-DIFFUSION. Water molecules readily dissociate to form protons and hydroxide ions, and the dissociation constant is:

$$K_{water} = k_d/k_r = [H^+][OH^-]/[H_2O]$$

The rate constant k_d for dissociation is about 2×10^{-5} s^{-1}, and the reassociation rate constant k_r is about $1 \times$

10^{11} M^{-1}s^{-1} for the neutralization reaction[16]. The hydroxide ion and proton released in the dissociation process are strongly hydrated, and the hydrated proton can be represented in aqueous solution as $(H_2O)_4H^+$. The lifetime of any intermolecular complex, whether it is $(H_2O)_4H^+$ or other hydrogen-bonded clusters, is quite short, around 10^{-12} seconds. Moreover, a proton of a hydronium ion H_3O^+ or $(H_2O)_4H^+$ can jump from one water molecule to another by virtue of the extensive hydrogen bonding in the liquid state. Indeed, this proceeds with diffusion-limited rate constants of about 1.1×10^{10} M^{-1}s^{-1} for protons (*i.e.*, $v = k[H_2O][H^+]$, or about three times faster than proton transfer between water and hydroxide ion.

Eigen and DeMaeyer[16] summarized the following ionic properties of water (at 298 K): $K = [H^+][OH^-]/[HOH]$, 1.821×10^{-16} M; [H$^+$], 1.004×10^{-7} M; [OH$^-$], 1.004×10^{-7} M; k_d, 2.5×10^{-5} s^{-1}; k_a, 1.4×10^{11} M^{-1}s^{-1}; proton transfer rate (for $H_2O + H_3O^+ \rightarrow H_3O^+ + H_2O$), 10.6×10^9 M^{-1}s^{-1}; proton transfer rate (for $H_2O + HO^- \rightarrow HO^- + H_2O$), 3.8×10^9 M^{-1}s^{-1}.

Biological fluids are more complicated still, because their water structure is strongly influenced by the presence of many biological and inorganic solutes. Collins[17–19] has discussed how hydration of biomacromolecules is strongly influenced by the charge density and size of certain ions. He indicates that small ions of high charge density, known as kosmotropes, bind water molecules strongly relative to the strength of water-water interactions in the bulk solution. Another set of large monovalent ions, known as chaotropes, are of low charge density and bind water molecules weakly with respect to the strength of H_2O-H_2O in bulk solution. Phosphates and carboxylates are the major intracellular kosmotropes that form highly soluble, solvent-separated ion pairs with the intracellular chaotropes:potassium ion, arginine's guanidinium group, histidine's imidazolium group, and lysine's ϵ-ammonium group. Accordingly, Collins[17] has suggested that ion charge density may be regarded as a central determinant of structure and function in biological fluids. Lastly, the presence of macromolecular biological substances can also alter the fluid's viscosity and can restrict or promote certain interactions that are sensitive to the molecular crowding.

HYDROPHOBIC INTERACTIONS. These bonding interactions arise from the tendency[20–23] of nonpolar side chains of amino acids (or lipids) to reside in the interior, nonaqueous environment of a protein (or membrane/micelle/vesicle). This process is accompanied by the release of tightly bound water molecules from these apolar side-chain moieties. The hydrophobic effect is thermodynamically driven by the increased disorder (*i.e.*, $\Delta S > 0$) of the system, thereby overcoming the unfavorable enthalpy change (*i.e.*, $\Delta H < 0$) for water release from the apolar groups.

Cantor and Schimmel[24] provide a lucid description of the thermodynamics of the hydrophobic effect, and they stress the importance of considering both the unitary and cratic contributions to the partial molal entropy of solute-solvent interactions. Briefly, the partial molal entropy (S_A) is the sum of the *unitary* contribution (S_A') which takes into account the characteristics of solute A and its interactions with water) and the *cratic* term ($-R \ln \chi_A$) which is a statistical term resulting from the mixing of component A with solvent molecules. The unitary change in entropy for transfer of a hydrocarbon side chain from water to the nonaqueous interior of a protein is positive. Hence, its contribution ($-T\Delta S$) to the Gibbs free energy favors transfer of a side chain to the protein's interior. Note that in the presence of denaturing agent (such as 8 M urea), a side chain exhibits a reduced tendency to reside in the apolar interior[25]. About one-third to one-half of all amino acid residues in a protein have apolar side chains, thereby accounting for the cumulatively unfavorable effect of 8 M urea on protein folding.

Fersht and Serrano[26] recently described the systematic application of site-directed mutagenesis, together with calorimetry, to evaluate how specific side chains play roles in stabilizing protein structure. (***See also*** *Hydrogen Bond in Molecular Recognition*). Many self-assembly processes are entropy driven, and Lauffer[27] proposed the release of water from apolar patches on the unpolymerized subunits as these patches interact with each other in the assembled state. In the case of microtubules, the warm-induced polymerization and cold-induced depolymerization are hallmarks of an entropy-driven process. Finally, although the common practices of enzyme purification are storage and processing at ice-water temperature, some enzymes are cold-inactivated. The experienced investigator will test the activity and structural properties of a newly characterized enzyme after storage at several different temperatures.

[1]W. S. Benedict, H. H. Claasen & J. H. Shaw (1956) *J. Chem. Phys.* **24**, 1139.

[2]D. Eisenberg & W. Kauzmann (1969) *The Structure and Properties of Water*, Oxford Univ. Press, New York.

[3]L. Pauling (1960) *The Nature of the Chemical Bond*, 3rd ed., Cornell Univ. Press, Ithaca, N.Y.

[4]A. H. Narten & H. A. Levy (1969) *Science* **165**, 447.

[5]H. S. Frank (1972) in *Water: A Comprehensive Treatise* (F. Franks, ed.) pp. 515-543, Plenum Press, New York.

[6] A. Ben-Naim (1972) in *Water: A Comprehensive Treatise* (F. Franks, ed.) pp. 413-442, Plenum Press, New York.

[7]A. H. Narten (1972) *J. Chem. Phys.* **56**, 5681.

[8]G. Nèmethy & H. Scheraga (1962) *J. Chem. Phys.* **36**, 3382.

[9]J. D. Bernal (1964) *Proc. Royal Soc. London, Ser. A* **280**, 299.

[10]M. C. R. Symons (1975) *Philos. Trans. Royal Soc. London, Ser. B* **272**, 13.

[11]M. C. R. Symons (1981) *Acc. Chem. Res.* **14**, 179.

[12]F. H. Stillinger (1975) *Adv. Chem. Phys.* **31**, 1.

[13]F. H. Stillinger (1980) *Science* **209**, 451.

[14]H. S. Frank & W. Y. Wen (1957) *Discuss. Faraday Soc.* **24**, 133.

[15]F. H. Stillinger & A. Rahman (1974) *J. Chem. Phys.* **74**, 3677.

[16]M. Eigen & L. DeMaeyer (1958) *Proc. Roy. Soc., Series A* **247**, 505.

[17]K. D. Collins (1997) *Biophys. J.* **72**, 65.

[18]K. D. Collins (1995) *Proc. Natl. Acad. Sci. U.S.A.* **92**, 5553.

[19]K. D. Collins & M. W. Washabaugh (1985) *Q. Rev. Biophys.* **18**, 323.

[20]W. Kauzmann (1959) *Advances in Protein Chemistry* **14**, 1.

[21]C. Tanford (1962) *J. Amer. Chem. Soc.* **84**, 4240.

[22]C. Tanford (1980) *The Hydrophobic Effect: Formation of Micelles and Biological Membranes*, 2nd ed., Wiley, New York.

[23]F. M. Richards in *Protein Folding* (T. E. Creighton, ed.), pp. 1-58, Freeman, San Francisco.

[24]C. R. Cantor & P. R. Schimmel (1980) *Biophysical Chemistry*, vol. **1**, pp. 281-288, Freeman, San Francisco.

[25]P. L. Whitley & C. Tanford (1962) *J. Biol. Chem.* **237**, 1735.

[26]A. R. Fersht & L. Serrano (1993) *Curr. Opin. Struct. Biol.* **3**, 75.

[27]M. A. Lauffer (1975) *Entropy-Driven Processes in Biology*, Springer-Verlag, Berlin.

WATER, BOUND-

1. The water that is associated with a macromolecule, a colloid, or other substance such that it is not removable by simple filtration. 2. The water molecule(s) bound to the active site of an enzyme or to the surface of a protein.

WATER EXCHANGE RATES OF AQUA METAL IONS

Water molecules in the so-called coordination sphere of a metal ion complex can exchange with water molecules in the medium, and the rates depend largely on the nature of the metal ion and its electric charge, and to a lesser extent on the other coordinated substituents. The rate constants for substitutions of inner sphere water of various aqua ions were determined largely by Eigen's

group. Those with rate constants of $10^9–10^{10}$ s^{-1} are K^+, Rb^+, Cs^+, Ba^{2+}, and Hg^{2+}; $10^8–10^9$ s^{-1} are Li^+, Na^+, Ca^{2+}, Sr^{2+}, Cu^{2+}, Cr^{2+}, and Cd^{2+}; $10^7–10^8$ s^{-1} are Y^{3+}, Sc^{3+}, La^{3+}, Tb^{3+}, Gd^{3+}, and Zn^{2+}; and $10^6–10^7$ s^{-1} are In^{3+}, Fe^{2+}, and Mn^{2+}.

M. Eigen & K. Tamm (1962) *Z. Elektrochem.* **66**, 107.

F. Basolo & R. G. Pearson (1997) *Mechanism of inorganic reactions*, Wiley, New York.

WATER PURITY

The purity of the water used in kinetic and mechanistic studies should always be an issue of concern to an investigator. At a minimum level, all water used in a biochemical laboratory should be glass distilled. However, distillation is relatively ineffective in removing dissolved ionized gases and, although distillation will remove most dissolved ionized solids, not all are removed. It is best to combine glass distillation with one or more ion exchange resins. The American Society for Testing and Materials (ASTM) classifies ionically pure water as Type I water, if the maximum specific conductance is less than 0.06 micromho, less than 0.1 mg/L particulate matter, and pyrogen-free.

After purification, water should be stored in scrupulously cleaned glass containers to minimize metal ion contamination. If water or stock solutions are to be kept in plastic containers, then the containers should *not* be marked directly on the sides of the containers with marking pens commonly found in laboratories. Such commercial markers contain heavy metal ions which have been reported to permeate the walls of certain plastic containers.

As mentioned in the entry on *Substrate Purity*, ammonia is often a contaminant in many laboratories. It may be necessary to check for ammonia levels on a regular basis as ammonia can act as an inhibitor for many enzyme-catalyzed reactions acting on amine-containing substrates. **See** *Substrate Purity*

WATER (Temperature Effect on K_w)

If proton (or deuteron) concentrations and OH^- (or, OD^-) concentrations are measured in molalities, m_{H^+} (or, m_{D^+}) and m_{OH^-} (or, m_{OD^-}) respectively, then K_w is defined as the product of the molalities of the positively charged and negatively charged species. Since K_w is a thermodynamic constant, it is not surprising that it should vary with temperature. Defining $pK_w = -\log K_w$, the following indicates the pK_w values for pure H_2O (and D_2O) at the listed temperature under saturating vapor pressure: 0°C, 14.94 (5.97); 5°C, 14.73 (15.74); 10°C, 14.53 (15.53); 15°C, 14.34 (15.32); 20°C, 14.16 (15.13); 25°C, 13.99 (14.95); 30°C, 13.84 (14.78); 35°C, 13.69 (14.62); 40°C, 13.54 (14.46); 50°C, 13.27 (14.18); 75°C, 12.71 (13.57); and 100°C, 12.26 (13.10). Notice that, because of this temperature effect, neutrality of pure water is pH 7.0 at only 25°C. At lower temperatures, neutrality occurs at pH values greater than 7.0; at temperatures higher than 25°C, neutrality occurs below pH 7.0. At normal human adult body temperature (37°C), pK_w is about 13.6. Hence, at 37°C, neutrality will take place at pH 6.8. For a solution of D_2O at 37°C, neutrality occurs at a pD value of about 7.28.

W. L. Marshall & E. U. Franck (1981) *J. Phys. Chem. Ref. Data* **10**, 295.

R. E. Mesmer & D. L. Herting (1978) *J. Solution Chem.* **7**, 901.

WATT

The SI unit of power or radiant flux (symbolized by W) equal to one joule per second (or, meter^2 per kilogram per second^3). With respect to electric currents, the watt is the rate of work expressed by a current of one ampere and a potential difference of one volt. **See** *Power; Radiant Flux*

WEBER

The SI unit for magnetic flux (symbolized by Wb) equal to 10^8 maxwells. The weber is the magnetic flux that produces an electromotive force of one volt, in linking a circuit of one turn, when the flux is reduced to zero at a uniform rate of one ampere per second. One weber is equivalent to one volt-second.

WEDLER-BOYER ISOTOPE EXCHANGE PROTOCOL

A very useful strategy[1] applied in equilibrium exchange studies to obviate problems associated with the formation of abortive complexes. Rates of isotope exchange are measured as a function of the absolute concentration of any one of the substrates or products (see figure below) under conditions such that the concentrations of all substrates and products are being varied in a constant ratio (*i.e.*, K_{eq}). This technique also facilitates detailed exchange kinetic studies of the effects of regulatory effectors on the exchange rates of enzymes. **See** *Isotope Exchange at Equilibrium*

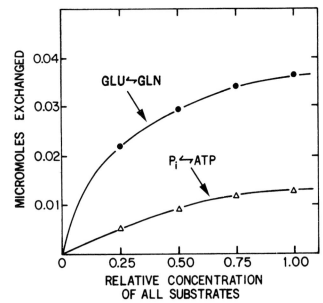

Experimental demonstration that the *Escherichia coli* glutamine synthetase reaction operates by a random order of addition of substrates and a random order of release of products.

[1]F. C. Wedler & P. D. Boyer (1972) J. Biol. Chem. **247**, 984.

WEIGHT

The force with which a body or entity is attracted to earth. The weight, W, is equal to the product of the mass, m, of the body and the acceleration of free fall in a vacuum (*i.e.*, $W = mg$). Since the earth is not a perfect sphere of uniform density, the weight of a body varies at various locales on the earth's surface. In practice, the mass of a body is determined by comparing its weight with a set of known standards: *i.e.*, $m_1/m_2 = W_1/W_2$.

WEIGHTING OF DATA

A statistical procedure that usually involves the multiplication of some quantity by a factor such that the result will be comparable to other quantities of the same type. For example, if a series of measurements of concentrations of a biomolecule are made, an investigator may wish to give greater "weight" to those values having the smallest associated error. **See** *Statistics (A Primer)*

Wilkinson[1] and Cleland[2] have pointed out the importance of proper weighting of initial rate data in the study of enzyme-catalyzed reactions. If the initial velocities vary by a factor of five or less, the weighting factor involves v_i^4 terms. If the velocities differ by a factor of ten or more, Cleland has suggested that the variance of the velocities (*i.e.*, the weighting factor) be proportional to v_i^2. **See** *Statistics (A Primer)*

[1]G. N. Wilkinson (1961) *Biochem. J.* **80**, 324.
[2]W. W. Cleland (1979) *Meth. Enzymol.* **63**, 103.

WIGNER SPIN-CONSERVATION RULE

A rule that affects energy transfers in photochemical reactions, particularly photosensitization processes. The total electron spin (*i.e.*, the vectorial overall spin angular momentum of the system) does not change after the electronic energy transfer between an excited molecular entity and another molecular entity.

WOMACK-COLOWICK DIALYSIS METHOD

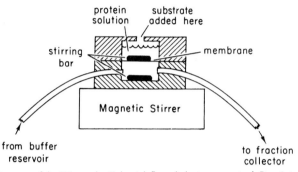

Diagram of the Womack–Colowick flow dialysis apparatus[1]. Reprinted with permission of the *Journal of Biological Chemistry*.

A ligand binding technique that relies on the ability to monitor the concentration of uncomplexed ligand in equilibrium with a macromolecule by determining the ligand's rate of transport across a dialysis membrane. The technique[1] allows one to make successive additions of ligand to the upper chamber (see diagram below) containing the protein, and one can typically obtain suf-ficient binding data within 20 min to construct a Scatchard plot for ligand binding at a single protein concentration. A miniaturized flow-dialysis cell has also been described[2].

[1]F. C. Womack & S. P. Colowick (1973) *Meth. Enzymol.* **17**, 464.
[2]N. C. Brown & P. Reichard (1969) *J. Mol. Biol.* **46**, 39.

WORK

A measure of the mechanical or nonmechanical energy change corresponding to the action of some force through a distance, expressed in joules (with force in newtons multiplied by distance in meters) or ergs (with force in dynes multiplied by distance in centimeters). Other examples of work include: the product of ΔG and the number of moles of reactant; the product of surface tension and area; the product of pressure and volume; the product of electromotive force and electrical charge; and the product of magnetic field and magnetization.

Reversible work is a conceptual measure of the energy required to apply a force through a distance along a reversible path connecting states that are all in equilibrium with each other. To ensure that all intermediate states can attain equilibrium with each other, transit along the interconnecting path must proceed slowly. By contrast, irreversible work is achieved when a force is applied along a path at a pace that precludes equilibration of intermediate states, thereby hindering the full transfer of energy stepwise along that path. Reversible work represents the maximal work that can be achieved in a system, and this conceptual measure forms the basis for thermodynamic and statistical mechanical treatments of energy changes.

X

XANTHINE DEHYDROGENASE

This molybdenum-dependent enzyme [EC 1.1.1.204] catalyzes the reaction of xanthine with NAD^+ and water to produce urate and NADH. A variety of purines and aldehydes can act as substrates. Mammalian xanthine dehydrogenase can be interconverted to a form that has xanthine oxidase [EC 1.1.3.22] activity. The liver enzyme exists primarily in the dehydrogenase form, but can be converted into xanthine oxidase: (a) on storage at $-20°C$, (b) on treatment with proteolytic agents or organic solvents, or (c) by thiol reagents. Xanthine dehydrogenase can also be converted into xanthine oxidase by the action of enzyme-thiol transhydrogenase (glutathione disulfide) in the presence glutathione disulfide. In other animal tissues, the enzyme exists almost entirely as xanthine oxidase. **See** Xanthine Oxidase

B. A. Palfey & V. Massey (1998) *Comprehensive Biological Catalysis: A Mechanistic Reference* **3**, 83.
P. E. Baugh, D. Collison, C. D. Garner & J. A. Joule (1998) *Comprehensive Biological Catalysis: A Mechanistic Reference* **3**, 377.
N. E. Tolbert (1981) *Ann. Rev. Biochem.* **50**, 133.
R. C. Bray (1980) *Adv. Enzymol.* **51**, 107.
R. C. Bray (1975) *The Enzymes*, 3rd ed., **12**, 299.

XANTHINE OXIDASE

This enzyme [EC 1.1.3.22] catalyzes the reaction of xanthine with dioxygen and water to produce urate and hydrogen peroxide. Enzymatic activity requires iron, FAD, and molybdenum. Hypoxanthine and some other purines and pterins can act as substrates. Under some conditions, the product is mainly superoxide rather than hydrogen peroxide: thus, $R-H$ reacts with two dioxygen and water to produce $R-OH$, two H^+, and two $O_2\cdot^-$ molecules. The *Micrococcus* enzyme can use ferredoxin as the acceptor substrate. The mammalian enzyme can be interconverted to xanthine dehydrogenase [EC 1.1.1.204]. **See** Xanthine Dehydrogenase

B. A. Palfey & V. Massey (1998) *Comprehensive Biological Catalysis: A Mechanistic Reference* **3**, 83.
P. E. Baugh, D. Collison, C. D. Garner & J. A. Joule (1998) *Comprehensive Biological Catalysis: A Mechanistic Reference* **3**, 377.
C. Kisker, H. Schindelin & D. C. Rees (1997) *Ann. Rev. Biochem.* **66**, 233.
S. C. Kundu & R. L. Willson (1995) *Meth. Enzymol.* **251**, 69.
R. C. Bray (1975) *The Enzymes*, 3rd ed., **12**, 299.
G. Palmer (1975) *The Enzymes*, 3rd ed., **12**, 1.

XI (x)

Symbol for extent or advancement (or extent of reaction) of a chemical reaction. ξ is the symbol for rate of conversion in which $\xi = d\ \xi/dt$; SI unit is $mol\cdot s^{-1}$ **See** *Extent of Reaction; Time & Chronomals*

XYLANASE

This enzyme [EC 3.2.1.8], also referred to as endo-1,4-β-xylanase and 1,4-β-D-xylan xylanohydrolase, catalyzes the endohydrolysis of 1,4-β-D-xylosidic linkages in xylans.

G. Davies, M. L. Sinnott & S. G. Withers (1998) *Comprehensive Biological Catalysis: A Mechanistic Reference* **1**, 119.
M. W. Bauer, S. B. Halio & R. M. Kelly (1996) *Adv. Protein Chem.* **48**, 271.
T. Hayashi (1989) *Ann. Rev. Plant Physiol. Plant Mol. Biol.* **40**, 139.

XYLOGLUCAN SYNTHASE

This enzyme, officially known as 1,4-β-D-xylan synthase [EC 2.4.2.24], catalyzes the reaction of UDP-xylose with $(1,4)$-β-D-xylan$_n$ to yield UDP and $(1,4)$-β-D-xylan$_{n+1}$. UDP-D-glucose is also a substrate.

T. Hayashi (1989) *Ann. Rev. Plant Physiol. Plant Mol. Biol.* **40**, 139.

D-XYLONATE DEHYDRATASE

This enzyme [EC 4.2.1.82], also referred to as D-*xylo*-aldonate dehydratase, catalyzes the conversion of D-xylonate to 2-dehydro-3-deoxy-D-xylonate and water.

W. A. Wood (1971) *The Enzymes*, 3rd ed., **5**, 573.

XYLOSE ISOMERASE

This enzyme [EC 5.3.1.5], which requires magnesium ions, catalyzes the interconversion of D-xylose to D-xylulose. This enzyme isolated from certain sources will also interconvert D-glucose and D-fructose.

C. I. F. Watt (1998) *Comprehensive Biological Catalysis: A Mechanistic Reference* **1**, 253.
I. A. Rose (1975) *Adv. Enzymol.* **43**, 491.
E. A. Noltmann (1972) *The Enzymes*, 3rd ed., **6**, 271.

α-XYLOSIDASE

The following is a recent review on the molecular and physical properties of this enzyme.

S. C. Fry (1995) *Ann. Rev. Plant Physiol. Plant Mol. Biol.* **46**, 497.

β-XYLOSIDASE

This enzyme [EC 3.2.1.37], also referred to as xylan 1,4-β-xylosidase and exo-1,4-β-xylosidase, catalyzes the hydrolysis of 1,4-β-D-xylans resulting in the removal of successive D-xylose residues from the nonreducing termini. Xylobiose can also be utilized as a substrate. The sheep liver enzyme has been reported to exhibit other exoglycosidase activities.

M. W. Bauer, S. B. Halio & R. M. Kelly (1996) *Adv. Protein Chem.* **48**, 271.
G. Mooser (1992) *The Enzymes*, 3rd ed., **20**, 187.
P. M. Dey & E. del Campillo (1984) *Adv. Enzymol.* **56**, 141.

D-XYLULOKINASE

This enzyme [EC 2.7.1.17], also known as xylulose kinase, catalyzes the reaction of ATP with D-xylulose to produce ADP and D-xylulose 5-phosphate.

W. L. Dills, Jr., P. D. Parsons, C. L. Westgate & N J. Komplin (1994) *Protein Expr. Purif.* **5**, 259.
M. S. Neuberger, B. S. Hartley & J. E. Walker (1981) *Biochem. J.* **193**, 513.

L-XYLULOKINASE

This enzyme [EC 2.7.1.53], also known as L-xylulose kinase, catalyzes the reaction of ATP with L-xylulose to produce ADP and L-xylulose 5-phosphate.

J. C. Sanchez, R. Gimenez, A. Schneider, W. D. Fessner, L. Baldoma, J. Aguilar & J. Badia (1994) *J. Biol. Chem.* **269**, 29665.
R. L. Anderson & W. A. Wood (1962) *J. Biol. Chem.* **237**, 1029.

Y

YAGA-OZAWA TREATMENT

A graphical procedure for analyzing effects of two competitive inhibitors at the active site of an enzyme: v_0/v_i is plotted as a function of ($[I_1] + [I_2]$) where v_0 is the initial-rate velocity in the absence of an inhibitor(s), v_i is the velocity in the presence of the inhibitor(s), and $[I_1]$ and $[I_2]$ are the two inhibitor concentrations[1,2]. If there are no interactions between the two inhibitors, then a straight line will be obtained; if there is an interaction at the active site, curvature will be observed. *See Yonetani-Theorell Treatment*

[1]K. Yagi & T. Ozawa (1960) *Biochim. Biophys. Acta* **42**, 381
[2]T. Yonetani (1982) *Meth. Enzymol.* **87**, 500.

YIELD COEFFICIENT (or, Coefficient of Yield)

A parameter used in the study of the growth and cultivation of microorganisms. It is equal to the mass of the amount of microorganism produced divided by the mass of the substrate consumed. The term "energy yield coefficient" is used if the ratio represents the yield of ATP produced per the theoretically possible yield.

YLIDES

A class of compounds in which a positively charged atom from group V or VI of the periodic table (*e.g.*, N, O, S, P, As, Se) is bonded to a carbon atom having an unshared pair of electrons. Whereas there is only one canonical form for nitrogen ylides ($R_3N^{\oplus}-\overset{..}{C}R_2^{\ominus}$), because of $p\pi$-$d\pi$ bonding, two canonical forms can be written for phosphorus (*i.e.*, $R_3P{=}CR_2 \leftrightarrow R_3P^{\oplus}-\overset{..}{C}R_2^{\ominus}$) and sulfur ylides ($R_2S{=}CR_2 \leftrightarrow R_3S^{\oplus}-\overset{..}{C}R_2^{\ominus}$). A number of enzyme-catalyzed reactions have been reported to utilize ylide-based chemistry. For example, the ylide form of the thiazolium ring on thiamin pyrophosphate is the nucleophile in pyruvate carboxylase. *See Carbanion*

YONETANI-THEORELL TREATMENT

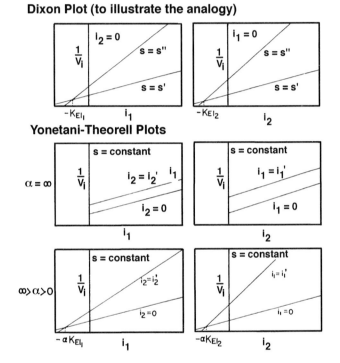

A graphical procedure for assessing the possible interaction of two different inhibitors binding at the active site of an enzyme. In this method, the investigator plots v_0/v_i as a function of the concentration of one of the two inhibitors at different, constant concentrations of the second substrate[1,2]. Such plots provide a means for the calculation of the inhibitor interaction constant which provides the investigator with a tool for assessing the degree of overlap of subsites at an enzyme's active center.

[1]T. Yonetani and H. Theorell (1964) *Arch. Biochem. Biophys.* **106**, 243.
[2]T. Yonetani (1982) *Meth. Enzymol.* **87**, 500.

Z

ZEATIN *CIS-TRANS* ISOMERASE

The following is a recent review on the molecular and physical properties of this enzyme.

A. N. Binns (1994) *Ann. Rev. Plant Physiol. Plant Mol. Biol.* **45**, 173.

ZEATIN REDUCTASE

This enzyme [EC 1.1.1.242] catalyzes the reversible reaction of dihydrozeatin with $NADP^+$ to produce zeatin and NADPH.

A. N. Binns (1994) *Ann. Rev. Plant Physiol. Plant Mol. Biol.* **45**, 173.

ZEOLITE

Highly crystalline, microporous substances comprised of an aluminosilicate framework wherein oxygen atoms join silicon and aluminum atoms located within tetrahedral units corresponding to $[SiO_4]^{4-}$ and $[AlO_4]^{5-}$. Zeolites were first discovered by Cronstedt[1] who recognized their ability to serve as boiling stones. Today, molecular seives have many applications, including water removal from organic solvents, ion-exchange applications, adsorption of odorant molecules, and chemical catalysis.

In aqueous solutions, the high negative surface charge of a zeolite must be neutralized by binding counterions, such as Na^+, K^+, and Ca^{2+}. The distribution of zeolite pore size can be modified, such that small molecules are included in the pores; those too large to diffuse into the pores are excluded. Structure types are named by a three-letter IUPAC code, based in part on the name of the zeolite first used to identify the specific type. They are also classified by pore size, framework density, and/or symmetry.

[1]A. F. Cronstedt (1756) *Kongl. Svenska Vetenskaps Acad. Handlinger* **17**, 120.

ZERO ORDER

Referring to reactions in which the reaction velocity is independent of the reactant under consideration. For example, for the reaction $A + B \rightarrow C$, if the empirical rate expression is $v = k[B]$, the reaction is first order with respect to B but zero order with respect to A. **See** *Chemical Kinetics; Rate Saturation; Michaelis-Menten Equation*

ZERO-ORDER REACTIONS

Reactions in which the velocity (v) of the process is independent of the reactant concentration, following the rate law $v = k$. Thus, the rate constant k has units of $M \cdot s^{-1}$. An example of a zero-order reaction is a Michaelis-Menten enzyme-catalyzed reaction in which the substrate concentration is much larger than the Michaelis constant. Under these conditions, if the substrate concentration is raised even further, no change in the velocity will be observed (since $v = V_{max}$). Thus, the reaction is zero-order with respect to the substrate. However, the reaction is still first-order with respect to total enzyme concentration. When the substrate concentration is not saturating then the reaction ceases to be zero order with respect to substrate. Reactions that are zero-order in each reactant are exceedingly rare. Thus, zero-order reactions address a fundamental difference between order and molecularity. Reaction order is an empirical relationship. Hence, the term pseudo-zero order is actually redundant. All zero-order reactions cease being so when no single reactant is in excess concentration with respect to other reactants in the system.

Another example of zero-order kinetics is the rate of dissolution of encapsulated solutes restricted in the egress by passage through a small orifice in the capsule. If a soluble salt is added in addition to the encapsulated solute, one obtains an osmotically driven solute release system. **See also** *Order of Reaction; Molecularity; Michaelis-Menten Equation; First-Order Reaction*

ZERO POINT ENERGY

The residual energy (designated E_o) of a harmonic oscillator in the ground state. The Heisenberg Uncertainty Principle does not permit any state of completely defined position *and* momentum. A one-dimensional harmonic oscillator has energy levels corresponding to:

$$E_\nu = (\nu + 0.5)h\nu$$

where $\nu = 0, 1, 2, 3, \cdots$, correspond to the vibrational energy levels, h is Planck's constant, and ν is the frequency. The zero-point energy occurs at the lowest energy state (*i.e.*, the ground state, where $\nu = 0$), and E_o equals $0.5h\nu$. **See** *Hooke's Law Spring; Kinetic Isotope Effects*

ZETA (ζ)

Symbol for zeta potential.

ZIPPER MECHANISMS FOR DNA UNFOLDING

The sequential and often cooperative disassembly of double-helical structure, occurring whenever the sample temperature exceeds the so-called melting temperature (T_m) for a given segment of DNA. Because of the low concentrations of intermediate states lying between helix and coil structures, the helix-coil transition can be approximated as a two-state, all-or-nothing process. **See** *DNA Unwinding: Kinetic Model for Small DNA*

H. C. Spatz & D. M. Carothers (1969) *J. Mol. Biol.* **42**, 191.
D. Pörschke & M. Eigen (1971) *J. Mol. Biol.* **62**, 361.

Z VALUE

A calculated transition energy used to assess the polarity of a solvent. The solvent ionizing capability directly affects the position of a peak, easily measured, in the ultraviolet region of the spectrum of the complex of an iodide ion with 2-methyl- or 1-ethyl-4-carbomethoxypyridinium ion. Water has a Z value of 94.6, ethanol has a value of 79.6, dimethyl sulfoxide's value is 71.1, and benzene has a value of 54. A similar polarity scale, known as $E_T(30)$ values, is related to the Z value scale: $Z = 1.41E_T(30) + 6.92$. **See** *Solvent Effects*

E. M. Kosower (1958) *J. Amer. Chem. Soc.* **80**, 3253, 3261, and 3267.
E. M. Kosower, G. Wu & T. S. Sorensen (1961) *J. Amer. Chem. Soc.* **83**, 3147.
M. H. Abraham, P. L. Grellier, J. M. Abboud & R. M. Doherty (1988) *Can. J. Chem.* **66**, 2673.

M. J. Kamlet, J. M. Abboud & R. W. W Taft (1981) *Prog. Phys. Org. Chem.* **13**, 485.
J. Shorter (1982) *Correlation Analysis of Organic Reactivity*, Wiley, New York .
C. Reichardt (1988) *Solvents and Solvent Effects in Organic Chemistry*, 2nd ed., VCH, New York.

ZWITTERGENTS

Zwitterionic detergents or surfactants. **See** *Detergents; Micelle*

ZWITTERION

A molecular entity containing oppositely charged acid-base groups. An example is $^+H_3NCH(CH_3)COO^-$, which is by far the predominant species of alanine at neutral pH. The term "zwitterion" is a German word designating the presence of an "inner salt" or an internal ionic species within a molecule. Most biochemistry textbooks indicate that zwitterions are electrically neutral, lacking any overall net charge. This is only true for a substance containing one positively charged and one negatively charged ionizable group. Polyvalent molecules may have zwitterionic forms with overall net charge. For glutamic acid, one zwitterionic species ($^+H_3NCH(CH_2CH_2COOH)COO^-$) is in fact neutral at low pH; the other is a monoanion present at higher pH ($^+H_3NCH(CH_2CH_2COO^-)COO^-$).

A zwitterion must not be confused with an ylide, which has opposite electric charges residing on adjacent atoms. The term "dipolar ion," while often used in place of zwitterion, is not considered by IUPAC to be an acceptable synonym, and should be avoided.

The relative concentrations of zwitterionic forms and the uncharged species is well illustrated in the case of glycine[1-4]. We begin by considering the relationship between macroscopic and microscopic constants. Titration experiments provide information on the proton dissociation constants; at 25°C and extrapolated to zero ionic strength, $K_1 = 4.45 \times 10^{-3}$ ($pK_1 = 2.35$) and $K_2 = 1.66 \times 10^{-10}$ ($pK_2 = 9.78$). The first macroscopic constant represents the dissociation of a proton from the fully protonated species ($^+H_3NCH_2COOH$) to form either the zwitterion ($^+H_3NCH_2COO^-$) or the neutral form (H_2NCH_2COOH). Similarly, the second macroscopic constant represents the dissociation of a proton from either the zwitterion or the uncharged species to form the anionic species, $H_2NCH_2COO^-$. To determine the relative abundance of the zwitterion and the uncharged species, the corresponding microscopic ionization equilibria must be evaluated. Note that these macroscopic constants are related to the microscopic events by the following expressions:

$$K_1 = \frac{([^+H_3NCH_2COO^-] + [H_2NCH_2COOH])[H^+]}{[^+H_3NCH_2COOH]}$$

$$K_2 = \frac{[H_2NCHCOO^-][H^+]}{[^+H_3NCH_2COO^-] + [H_2NCH_2COOH]}$$

Microscopic, or intrinsic, equilibrium constants are

$$k_1 = \frac{([^+H_3NCH_2COO^-][H^+]}{[^+H_3NCH_2COOH]}$$

$$k_2 = \frac{[H_2NCH_2COOH][H^+]}{[^+H_3NCH_2COOH]}$$

$$k_3 = \frac{[H_2NCH_2COO^-][H^+]}{[^+H_3NCH_2COO^-]}$$

$$k_4 = \frac{[H_2NCH_2COO^-][H^+]}{[H_2NCH_2COOH]}$$

Note that $K_1 = k_1 + k_2$ and $K_2 = [k_3^{-1} + k_4^{-1}]^{-1}$. The four microscopic constants are thermodynamically interdependent, such that: $k_1 k_3 = k_2 k_4$. The pK_a value for the amino group of glycine ethyl ester was experimentally determined to be 7.73 at 25°C; for glycine methyl ester, the value is 7.66[3,4]. If one assumes that this dissociation constant (an average value of 7.70) is identical to the dissociation constant of the amino proton from glycine, then $p k_2 = 7.70$ (*i.e.*, $k_2 = 2.00 \times 10^{-8}$). Thus, $k_1 = K_1 - k_2 \approx 4.45 \times 10^{-3}$. Combining the expression for K_2 above and the thermodynamic interrelationship of the microscopic constants, one finds that $k_3 = K_2(k_1 + k_2)/k_1 = K_1K_2/k_1 \approx 1.66 \times 10^{-10}$. Hence, k_4 equals $k_1 k_3/k_2$, or about 3.69×10^{-5}. From the definitions for the microscopic dissociation constants, $k_1/k_2 = k_4/k_3 = [^+H_3NCH_2COO^-]/[H_2NCH_2COOH] \approx 2.23 \times 10^5$. This ratio is pH-independent, and the zwitterion form predominates over the uncharged species at all pH's. Similar calculations can be made with other monoaminomonocarboxylic acids. Although there is some variability in the ratio for other common amino acids, the zwitterion is by far the major form of these amino acids.

For compounds with three pK_a values, a similar but somewhat more demanding analysis can be made[3]. The three pK_a values for aspartic acid are 1.99, 3.90, and 9.90, whereas the pK_a value for the diethyl ester is 6.5. The following compounds

$$^+H_3NCH(COO^-)CH_2COOH$$
$$^+H_3NCH(COOH)CH_2COO^-$$
$$H_2NCH(COOH)CH_2COOH$$

are present in solution in the approximate proportions 28,000:1600:1, respectively. For lysine's isoelectric structures

$$^+H_3NCH(COO^-)(CH_2)_4NH_2$$
$$H_2NCH(COO^-)(CH_2)_4NH_3^+$$
$$H_2NCH(COOH)(CH_2)_4NH_2$$

the relative proportions are $3.2 \times 10^5 : 1.8 \times 10^6 : 1$, respectively.

Because dissociation constants vary with temperature, the zwitterion/uncharged ratio may also vary with temperature[5]. For glycine[6], the temperature dependencies are $dpK_1/dT = -0.0020$ and $dpK_2/dT = -0.025$.

Finally, the zwitterion is not the major form for all "amino acids." For both *o*-aminobenzoic acid (anthranilic acid) and *p*-aminobenzoic acid, the uncharged species is the more prevalent form[3] whereas the zwitterion/uncharged ratio for *m*-aminobenzoic acid is about 2.31. This behavior arises primarily from the low pK_a values for the aromatic ring's amino group (*e.g.*, the pK_1 and pK_2 values for anthranilic acid are 2.05 and 4.95 and the pK_a value for the ester is only 2.09).

[1]L. Ebert (1926) *Z. Physik. Chem.* **121**, 385.

[2]J. T. Edsall & M. H. Blanchard (1933) *J. Amer. Chem. Soc.* **55**, 2337.

[3]J. T. Edsall (1943) in *Proteins, Amino Acids and Peptides as Ions and Bipolar Ions* (E. J. Cohn & J. T. Edsall, eds.), pp. 75-115, ACS Mongraph Series #1, Reinhold Publ. Corp., New York.

[4]J. T. Edsall & J. Wyman (1958) *Biophysical Chemistry*, vol. I, Academic Press, New York.

[5]R. M. C. Dawson, D. C. Elliott, W. H. Elliott & K. M. Jones (1986) *Data for Biochemical Research*, 3rd ed., Clarendon Press, Oxford.

[6]D. D. Perrin & B. Dempsey (1974) *Buffers for pH and Metal Ion Control*, Chapman and Hall, London.

ZYMOGEN

A precursor or storage form of an enzyme, sometimes used synonymously with proenzyme, in which the precursor is converted to the active (or more active) enzyme by proteolysis. This conversion can involve one or several hydrolytic steps. If more than one step occurs, the intermediate forms(s) may have partial activity. For example, prochymotrypsin (or, chymotrypsinogen) is converted to active α-chymotrypsin following three proteolytic cleavages and involves the formation of an active intermediate referred to as π-chymotrypsin. In some cases, the polypeptide products produced by proteolytic activation do separate from each other; in other instances, they remain associated in the final mature protein.

WORDFINDER

U se this Wordfinder to take fullest advantage of the text and reference material contained in this **HANDBOOK**.

This Wordfinder comprises all listed source words (CAPITALIZED) as well as other subheadings, keywords, or synonyms (written in lower case). Each of the latter is immediately followed by the recommended source entry (*CAPITALIZED & ITALICIZED*).

Note that many biochemical terms and enzymes (*e.g.*, ACAT, ATP, NADH, *etc.*) as well as computer software programs (*e.g.*, KINSIM, SAAM, FITSIM) are always capitalized. To avoid confusion, look for a comma following the entry: source entries stand alone and are not followed by a comma; the presence of a comma indicates that the entry is not a source word.

Many source words are also cross-referenced, and these source words (again *CAPITALIZED & ITALICIZED*) appear below each entry.

A

A, B, C, . . . P, Q, R, . . .
  CLELAND NOMENCLATURE
AB INITIO MOLECULAR-ORBITAL CALCULATIONS
ABM-1 & ABM-2 SEQUENCES IN ACTIN-BASED MOTORS
ABORTIVE COMPLEXES
  DEADEND COMPLEXES
  ENZYME REGULATION
  FRIEDEN DILEMMA
  INHIBITION
  ISOTOPE EXCHANGE AT EQUILIBRIUM
  ISOTOPE TRAPPING
  LACTATE DEHYDROGENASE
  LIGAND EXCLUSION MODEL
  NONPRODUCTIVE COMPLEXES
  PRODUCT INHIBITION
  SUBSTRATE INHIBITION
Abortive complexes in isotope exchange experiments,
  ABORTIVE COMPLEXES
Abortive complex interaction,
  ISOTOPE EXCHANGE AT EQUILIBRIUM
ABSCISSA
ABSOLUTE CONFIGURATION
  CONFIGURATION
  CAHN-INGOLD-PRELOG SYSTEM
  CORN RULE
  DIASTEREOMERS

  ENANTIOMERS
  (R/S)-CONVENTION
Absolute rate theory,
  TRANSITION-STATE THEORY
ABSOLUTE TEMPERATURE
ABSOLUTE UNCERTAINTY
  RELATIVE UNCERTAINTY
ABSOLUTE ZERO
ABSORBANCE
  BEER-LAMBERT LAW
  ABSORPTION COEFFICIENT
  ABSORPTION SPECTROSCOPY
ABSORBED DOSE
ABSORPTION
ABSORPTION COEFFICIENT
  BEER-LAMBERT LAW
  ABSORBANCE
  MOLAR ABSORPTION COEFFICIENT
  ABSORPTION SPECTROSCOPY
ABSORPTION SPECTROSCOPY
ABSORPTIVITY
  ABSORBANCE
  ABSORPTION COEFFICIENT
  BEER-LAMBERT LAW
  ABSORPTION SPECTROSCOPY
ABSTRACTION REACTION
Abzyme,
  See preferred descriptor CATALYTIC ANTIBODIES
ACAT,
  ACYL-CoA:CHOLESTEROL ACYLTRANSFERASE

ACCELERATION
  CATALYTIC RATE ENHANCEMENT
  CATALYTIC PROFICIENCY
  EFFICIENCY FUNCTION
Acceptor,
  FLUORESCENCE
  DONOR NUMBER
  LEWIS ACIDITY
Acceptor Number,
  DONOR NUMBER
  LEWIS ACIDITY
  PHYSICAL ORGANIC CHEMISTRY TERMINOLOGY
ACCRETION
ACCURACY
  PRECISION
Accuracy of measurement,
  NOISE
ACETALDEHYDE DEHYDROGENASE (ACETYLATING)
2-(Acetamidomethylene)succinate,
  METHYLHYDROXYPYRIDINECARBOXYLATE DIOXYGENASE
2-(ACETAMIDOMETHYLENE)SUCCINATE HYDROLASE
Acetate,
  LACTATE MONOOXYGENASE
  LEUCINE KINETICS
ACETATE KINASE
  ACYL-PHOSPHATE INTERMEDIATE
  NUCLEOSIDE-5'-TRIPHOSPHATE REGENERATION

N-ACYLNEURAMINATE 9-PHOSPHATASE

N-ACYLNEURAMINATE 9-PHOSPHATE SYN-
THASE

Acyloxyacyl hydrolase,

*LIPOPOLYSACCHARIDE DEACYLATING
ENZYME*

ACYL PHOSPHATASE

ACYL-PHOSPHATE-HEXOSE PHOSPHO-
TRANSFERASE

ACYL-PHOSPHATE INTERMEDIATE

D-ALANINE-D-ALANINE LIGASE

*MAPPING SUBSTRATE INTERACTIONS
USING KINETIC DATA*

Acyl phosphate reduction,

BOROHYDRIDE REDUCTION

ACYL-SERINE INTERMEDIATE

ACTIVE SITE TITRATION

Acyl transfer from a peptidyl-tRNA,

PEPTIDYL TRANSFERASE

ADAIR CONSTANTS

ADAIR EQUATION

ALLOSTERISM

COOPERATIVITY

KOSHLAND-NEMETHY-FILMER MODEL

Adair model,

COOPERATIVITY

Ada methyltransferase,

DNA REPAIR METHYLTRANSFERASES

β-Addition of water,

CROTONASE SUPERFAMILY

ADDITION REACTION

Additive feedback inhibition,

FEEDBACK INHIBITION

ADDITIVITY PRINCIPLE

ADDUCT

COMPLEX

LEWIS ADDUCT

MEISENHEIMER ADDUCT

Adenine,

*METHYLTHIOADENOSINE NUCLEOSIDE
HYDROLASE*

ADENINE DEAMINASE

ADENINE NUCLEOTIDE TRANSLOCASE

ADENINE PHOSPHORIBOSYLTRANS-
FERASE

Adenosine 3′,5′-bisphosphate synthesis,

*3′-PHOSPHOADENYLYL SULFATE SULFO-
HYDROLASE*

ADENOSINE DEAMINASE

Adenosine 5′-diphosphate,

ADP

Adenosine 5′-diphosphate-2′,3′-dialdehyde,

*MALAPRADE REACTION (PERIODATE
OXIDATION)*

ADENOSINE KINASE

Adenosine 5′-monophosphate,

AMP

Adenosine 5′-triphosphate,

ATP

Adenosine 5′-triphosphate-2′,3′-dialdehyde,

*MALAPRADE REACTION (PERIODATE
OXIDATION)*

S-Adenosylhomocysteine,

*MAGNESIUM-PROTOPORPHYRIN METH-
YLTRANSFERASE*

*METHYL TRANSFER REACTIONS USING
S-ADENOSYLMETHIONINE*

*NOREPINEPHRINE N-METHYLTRANS-
FERASE*

S-ADENOSYLHOMOCYSTEINE LYASE

S-ADENOSYLHOMOCYSTEINE NUCLEO-
SIDASE

*METHYLTHIOADENOSINE NUCLEOSIDE
HYDROLASE*

S-ADENOSYL-L-METHIONINE

METHYL TRANSFER REACTIONS

S-Adenosyl-L-methionine as a substrate,

S-ADENOSYLHOMOCYSTEINE LYASE

*S-ADENOSYLMETHIONINE DECARBOX-
YLASE*

S-ADENOSYLMETHIONINE HYDROLASE

*1-AMINOCYCLOPROPANE-1-
CARBOXYLATE SYNTHASE*

CATECHOL O-METHYLTRANSFERASE

COBALAMIN ADENOSYLTRANSFERASE

*COLUMBAMINE O-METHYLTRANS-
FERASE*

*DNA (CYTOSINE-5-)-METHYLTRANS-
FERASE*

GLYCINE METHYLTRANSFERASE

HISTAMINE N-METHYLTRANSFERASE

HOMOCYSTEINE METHYLTRANSFERASE

*HYDROXYINDOLE O-METHYLTRANS-
FERASE*

LYSINE 2,3-AMINOMUTASE

*MAGNESIUM-PROTOPORPHYRIN METH-
YLTRANSFERASE*

METHIONINE ADENOSYLTRANSFERASE

N^5*-METHYLTETRAHYDROFOLATE:
HOMOCYSTEINE METHYLTRANS-
FERASE*

*mRNA (GUANINE-7-)-METHYLTRANS-
FERASE*

*NICOTINAMIDE N-METHYLTRANS-
FERASE*

*NOREPINEPHRINE N-METHYLTRANS-
FERASE*

*PHOSPHATIDYLETHANOLAMINE N-
METHYLTRANSFERASE*

PUTRESCINE N-METHYLTRANSFERASE

*rRNA (ADENINE-6-)-METHYLTRANS-
FERASE*

*rRNA (GUANINE-1-)-METHYLTRANS-
FERASE*

S-ADENOSYLMETHIONINE DECARBOX-
YLASE

S-Adenosylmethionine decarboxylase reduction,

BOROHYDRIDE REDUCTION

S-Adenosylmethionine:guanidinoacetate,

*N-METHYLTRANSFERASE GUANIDI-
NOACETATE METHYLTRANS-
FERASE*

S-ADENOSYLMETHIONINE HYDROLASE

S-Adenosylmethionine:sterol methyltransferase,

Δ^{24}*-STEROL METHYLTRANSFERASE*

S-Adenosylmethionine synthetase,

METHIONINE ADENOSYLTRANSFERASE

ADENYLATE CYCLASE (or, ADENYLYL CY-
CLASE)

Adenylate energy charge,

ENERGY CHARGE

ADENYLATE ISOPENTENYLTRANSFERASE

ADENYLATE KINASE (or, MYOKINASE)

*METAL IONS IN NUCLEOTIDE-DEPEN-
DENT REACTIONS*

ENERGY CHARGE

5′-Adenylic acid aminohydrolase,

AMP AMINOHYDROLASE

ADENYLOSUCCINATE LYASE

ADENYLOSUCCINATE SYNTHETASE

Adenylyl cyclase,

ADENYLATE CYCLASE

ADENYLYLSULFATE–AMMONIA ADENYLYL-
TRANSFERASE

ADENYLYLSULFATE KINASE

ADENYLYLSULFATE REDUCTASE

ADENYLYLSULFATE SULFOHYDROLASE

Adhesion,

ACCRETION

BIOMINERALIZATION

Adiabatic compression and dilatation in a
sound wave,

CHEMICAL KINETICS

ADIABATIC PROCESS

ISOTHERMAL PROCESS

HEAT CAPACITY

1-ALKYL-2-LYSOGLYCERO-3-
    PHOSPHOCHOLINE *O*-ACETYLTRANS-
    FERASE

ALLANTOICASE

ALLANTOINASE

ALLENE OXIDE CYCLASE

ALLENE OXIDE SYNTHASE

ALLIIN LYASE

ALLOPHANATE HYDROLASE

All-or-none cooperativity,
    HILL EQUATION & PLOT

ALLOSE KINASE

ALLOSTERY
    ADAIR EQUATION
    COOPERATIVE LIGAND BINDING
    HILL EQUATION & PLOT
    KOSHLAND-NEMETHY-FILMER MODEL
    MONOD-WYMAN-CHANGEUX MODEL
    NEGATIVE COOPERATIVITY
    POSITIVE COOPERATIVITY
    HEMOGLOBIN

ALLOTOPIC EFFECT

Allowed electronic transitions,
    JABLONSKI DIAGRAM

ALLYLIC SUBSTITUTION REACTION

ALPHA (α)

ALPHA-ADDITION (α-ADDITION)

ALPHA CLEAVAGE (α-CLEAVAGE)

ALPHA EFFECT

ALPHA-ELIMINATION (α-ELIMINATION)

ALPHA EXPULSION

ALPHA PARTICLE

Alternant,
    *PHYSICAL ORGANIC CHEMISTRY NO-
        MENCLATURE*

ALTERNATIVE HYPOTHESIS
    STATISTICS (A PRIMER)

ALTERNATIVE PRODUCT INHIBITION
    ABORTIVE COMPLEXES

ALTERNATIVE SUBSTRATES
    COMPETITIVE INHIBITOR
    ABORTIVE COMPLEXES
    *MAPPING SUBSTRATE INTERACTIONS
        USING KINETIC DATA*
    MEMBRANE TRANSPORT
    ENERGY OF ACTIVATION
    Q_{10}
    ARRHENIUS EQUATION

ALTRONATE DEHYDRATASE

Aluminum,
    IONIC RADIUS

Aluminum alkoxides,

*MEERWEIN-PONDORFF-VERLEY RE-
    DUCTION*

Aluminum isopropoxide,
    *MEERWEIN-PONDORFF-VERLEY RE-
        DUCTION*

AMADORI REARRANGEMENT

"AMBER,"
    COMPUTER ALGORITHMS & SOFTWARE

"AMBER/OPLS,"
    COMPUTER ALGORITHMS & SOFTWARE

AMBIDENT NUCLEOPHILE

ω-AMIDASE

AMIDOPHOSPHORIBOSYLTRANSFERASE

Amine dehydrogenase,
    METHYLAMINE DEHYDROGENASE

AMINE OXIDASES

AMINE SULFOTRANSFERASE

Amino acid acetyltransferase,
    N-ACETYLGLUTAMATE SYNTHASE

D-AMINO ACID AMINOTRANSFERASE
    D-ALANINE AMINOTRANSFERASE

D-AMINO ACID DEHYDROGENASE

L-AMINO ACID DEHYDROGENASE

D-AMINO ACID OXIDASE

L-AMINO ACID OXIDASE

ω-AMINO ACID-PYRUVATE AMINOTRANS-
    FERASE

AMINO ACID RACEMASE

Amino acids, peptides & proteins,
    BIOCHEMICAL NOMENCLATURE

AMINO ACID TURNOVER KINETICS

AMINOACYLASE

AMINOACYL-tRNA HYDROLASE

Aminoacyl-tRNA synthetase,
    EDITING MECHANISMS

AMINOACYL-tRNA SYNTHETASES

2-AMINOADIPATE AMINOTRANSFERASE

2-AMINOADIPATE 6-SEMIALDEHYDE DEHY-
    DROGENASE

4-Aminobutanal, synthesis of,
    PUTRESCINE OXIDASE

4-AMINOBUTYRATE AMINOTRANSFERASE

α-AMINO-ε-CAPROLACTAM RACEMASE

1-AMINOCYCLOPROPANE-1-
    CARBOXYLATE SYNTHASE

2-AMINO-4-HYDROXY-6-
    HYDROXYMETHYLDIHYDROPTERI-
    DINE PYROPHOSPHOKINASE

5-Amino-4-imidazole carboxamide ribonucle-
    otide transformylase,
    *PHOSPHORIBOSYLAMINOIMIDAZOLE-
        CARBOXAMIDE FORMYLTRANS-
        FERASE*

β-AMINOISOBUTYRATE AMINOTRANS-
    FERASE

5-AMINOLEVULINATE AMINOTRANS-
    FERASE

5-AMINOLEVULINATE DEHYDRATASE

5-AMINOLEVULINATE SYNTHASE

AMINOMALONATE DECARBOXYLASE

(R)-3-Amino-2-methylpropionate–pyruvate
    aminotransferase,
    β-AMINOISOBUTYRATE AMINOTRANS-
        FERASE

5-Aminopentanoate,
    PROLINE REDUCTASE

AMINOPEPTIDASE
    See also specific aminopeptidases

AMINOPEPTIDASE P

Aminopropyltransferase,
    SPERMIDINE SYNTHASE

Ammonia as a substrate or product,
    ADENINE DEAMINASE
    ADENOSINE DEAMINASE
    ω-*AMIDASE*
    D-*AMINO ACID OXIDASE*
    L-*AMINO ACID OXIDASE*
    AMP DEAMINASE
    ASPARAGINASE
    ASPARAGINE SYNTHETASE
    CARBAMOYL PHOSPHATE SYNTHETASE
    GLUTAMATE DEHYDROGENASE
    GLUTAMINASE
    GLUTAMINE SYNTHETASE
    γ-*GLUTAMYL TRANSPEPTIDASE*
    GLYCINE CLEAVAGE COMPLEX
    LEUCINE DEHYDROGENASE
    *LYSINE DEHYDROGENASE (α-DEAMI-
        NATING)*
    LYSYL OXIDASE
    METHANE MONOOXYGENASE
    METHIONINE γ-LYASE
    METHYLAMINE DEHYDROGENASE
    3-METHYLASPARTATE AMMONIA-LYASE
    N-METHYLGLUTAMATE SYNTHASE
    UREASE

Ammonia, autoprotolysis constant,
    AUTOPROTOLYSIS

Ammonium ion,
    See Ammonia as a substrate or product

AMOUNT-OF-SUBSTANCE
    CONCENTRATION

AMP
    ADENYLATE KINASE
    AMP DEAMINASE

ASPARAGINE SYNTHETASE

CARNOSINE SYNTHETASE

LEUCYL-tRNA SYNTHETASE

LONG-CHAIN FATTY ACYL-CoA SYNTHETASE

METHIONYL-tRNA SYNTHETASE

NITROGENASE

PYRUVATE, ORTHOPHOSPHATE DIKINASE

PYRUVATE, WATER DIKINASE

AMP DEAMINASE

Amperometry

   MEDIATED AMPEROMETRIC BIO-SENSOR

AMPHIPHILIC

AMPHIPROTIC

   PROTOGENIC

Amphiprotic molecular entities,

   AUTOPROTOLYSIS

Amphiprotic solvent,

   AUTOPROTOLYSIS

AMPHOTERIC

AMPLIFICATION

   ENZYME CASCADE KINETICS

AMPLITUDE

AMP NUCLEOSIDASE

AMYLASES

   LIMIT DEXTRAN

γ-Amylases,

   GLUCOAMYLASE

Amylo-1,6-glucosidase,

   DEBRANCHING ENZYME

AMYLOSUCRASE

Anaerobic fermentation rate,

   MEYERHOF OXIDATION QUOTIENT

ANALOG COMPUTER SIMULATION

   COMPUTER SIMULATION; KINSIM/FITSIM

Analogues of naturally occurring peptide bonds,

   PEPTIDOMIMETIC COMPOUND

Ananain,

   BROMELAIN

ANAPLEROTIC

ANATION

ANCHIMERIC ASSISTANCE

   NEIGHBORING GROUP MECHANISM

ANCHOR PRINCIPLE

ANGIOTENSIN CONVERTING ENZYME

Angiotensin-forming enzyme,

   RENIN

Angiotensinogenase,

   RENIN

Angle strain,

   STEREOCHEMICAL TERMINOLOGY, IUPAC RECOMMENDATIONS

ANGULAR MOMENTUM

8-Anilino-1-naphthalene sulfonate,

   FLUORESCENCE

ANION EFFECTS ON pH-RATE DATA

   pH STUDIES

   ACTIVATION

   OPTIMUM pH

Anion radical,

   RADICAL ANION

ANISOTROPIC

   ISOTROPIC

Anisotropic diffusion coefficients,

   DIFFUSION WITHIN CELLULAR/EXTRA-CELLULAR SPACES

ANNULATION (and ANNELATION)

Annulene,

   PHYSICAL ORGANIC CHEMISTRY NOMENCLATURE

ANOMER,

   FISCHER PROJECTION

   EPIMERS

   DIASTEREOMERS

ANOMERIC EFFECT

ANS,

   FLUORESCENCE

ANTAGONIST

ANTARAFACIAL & SUPRAFACIAL MIGRA-TIONS

   ANTARAFACIAL & SUPRAFACIAL REAC-TIONS

ANTARAFACIAL & SUPRAFACIAL REAC-TIONS

ANTHOCYANIDIN SYNTHASE

ANTHRANILATE 1,2-DIOXYGENASE

ANTHRANILATE HYDROXYLASE

ANTHRANILATE PHOSPHORIBOSYLTRANS-FERASE

ANTHRANILATE SYNTHASE

ANTI

ANTIAROMATICITY

   AROMATIC

ANTI-ARRHENIUS BEHAVIOR

   ARRHENIUS EQUATION

   ARRHENIUS PLOT, NONLINEAR

Antibiosis,

   D-ALANINE-D-ALANINE LIGASE

ANTIBODY-HAPTEN INTERACTIONS

ANTIBONDING ORBITAL

   MOLECULAR ORBITAL

   BONDING ORBITAL

   ORBITAL

Antichymotrypsin,

   SERPINS (INHIBITORY MECHANISM)

Anticlinal,

   STEREOCHEMICAL TERMINOLOGY, IUPAC RECOMMENDATIONS

ANTICOOPERATIVITY

   LIGAND BINDING COOPERATIVITY

   NEGATIVE COOPERTIVITY

Antiperiplanar,

   STEREOCHEMICAL TERMINOLOGY, IUPAC RECOMMENDATIONS

ANTIPORT

   UNIPORT

   SYMPORT

   MEMBRANE TRANSPORT

Anti-Stokes shift,

   STOKES SHIFT

Anti-streptavidin antibody,

   BIOTIN AND DERIVATIVES

AO reductase,

   NAD(P)H DEHYDROGENASE (QUINONE)

ap,

   STEREOCHEMICAL TERMINOLOGY, IUPAC RECOMMENDATIONS

Ap_4A & Ap_5A,

   ADENYLATE KINASE

Apical,

   STEREOCHEMICAL TERMINOLOGY, IUPAC RECOMMENDATIONS

Apical and equatorial ligand interchange,

   PSEUDOROTATION

APICOPHILICITY

APOACTIVATOR

Apolar side-chain interactions,

   HYDROPHOBIC EFFECT

Apparent equilibrium constant,

   ATP, THERMODYNAMICS OF HYDRO-LYSIS

Apparent reaction order for microtubule nu-cleation,

   MICROTUBULE ASSEMBLY KINETICS

APROTIC

A PRIORI PROBABILITY

Aps kinase,

   ADENYLYLSULFATE KINASE

APYRASE

AQUACOBALAMIN REDUCTASE

AQUAMOLALITY

   CONCENTRATION

AQUATION
   HYDRATION: WATER
α-L-ARABINOFURANOSIDASE
D-ARABINOKINASE
ARABINONATE DEHYDRATASE
ARABINOSE ISOMERASE
ARABINOSE-5-PHOSPHATE ISOMERASE
Arachidonate 5-lipoxygenase,
   5-LIPOXYGENASE
ARACHIDONYL-CoA SYNTHETASE
Arene oxide intermediate,
   NIH SHIFT
ARGINASE
L-Arginine,
   D-NOPALINE SYNTHASE
   NITRIC OXIDE SYNTHASE
Arginine carboxypeptidase,
   CARBOXYPEPTIDASE N
ARGININE DECARBOXYLASE
ARGININE DEIMINASE
Arginine:glycine amidinotransferase,
   GLYCINE AMIDINOTRANSFERASE
ARGININE KINASE
ARGININE MONOOXYGENASE
ARGININE RACEMASE
L-Arginine synthesis,
   OCTOPINE DEHYDROGENASE
ARGININOSUCCINATE LYASE
ARGININOSUCCINATE SYNTHETASE
ARISTOLOCHENE SYNTHASE
Arogenate dehydrogenase,
   CYCLOHEXADIENYL DEHYDROGENASE
AROMATIC
   ANTIAROMATICITY
   HÜCKEL'S RULE
AROMATIC AMINO ACID AMINOTRANS-
   FERASE
AROMATIC AMINO ACID DECARBOX-
   YLASE
AROMATIC AMINO ACID–GLYOXYLATE
   AMINOTRANSFERASE
Aromatic amino acid hydroxylases,
   PHENYLALANINE HYDROXYLASE
   TYROSINE HYDROXYLASE
   TRYPTOPHAN HYDROXYLASE
ARRHENIUS CONSTANT
   ARRHENIUS EQUATION & PLOT
ARRHENIUS EQUATION & PLOT
   BOLTZMANN DISTRIBUTION
   COLLISION THEORY
   TEMPERATURE DEPENDENCY,
     TRANSITION-STATE THEORY

ENERGY OF ACTIVATION
   Q_{10}
ARRHENIUS PLOTS (NONLINEAR)
   ARRHENIUS EQUATION & PLOT
   Q_{10}
   TRANSITION-STATE THEORY
   VAN'T HOFF RELATIONSHIP
Arrhenius treatment,
   CHEMICAL KINETICS
Arsenate,
   ARSENOLYSIS
   OXYGEN, OXIDES & OXO ANIONS
ARSENOLYSIS
ARYLALKYLAMINE *N*-ACETYLTRANS-
   FERASE
ARYLAMINE *N*-ACETYLTRANSFERASE
Arylamine sulfotransferase,
   AMINE SULFOTRANSFERASE
ARYLFORMAMIDASE
ARYLSULFATASES
ARYLSULFOTRANSFERASE
ASCORBATE–CYTOCHROME b_5 RE-
   DUCTASE
ASCORBATE OXIDASE
ASCORBATE PEROXIDASE
ASPARAGINASE
ASPARAGINE AMINOTRANSFERASE
ASPARAGINE SYNTHETASE
1-Aspartamido-β-*N*-acetylglucosamine ami-
   dohydrolase,
   ASPARTYLGLUCOSAMINIDASE
ASPARTATE AMINOTRANSFERASE
ASPARTATE AMMONIA-LYASE
ASPARTATE CARBAMOYLTRANSFERASE
ASPARTATE α-DECARBOXYLASE
ASPARTATE β-DECARBOXYLASE
ASPARTATE KINASE
D-ASPARTATE OXIDASE
ASPARTATE RACEMASE
ASPARTATE-SEMIALDEHYDE DEHYDRO-
   GENASE
Aspartate transcarbamylase,
   ASPARTATE CARBAMOYLTRANSFERASE
ASPARTIC PROTEASES
β-ASPARTYL-*N*-ACETYLGLUCOSAM-
   INIDASE
β-ASPARTYLDIPEPTIDASE
ASPARTYL-tRNA SYNTHETASE

   AMINOACYL-RNA SYNTHETASES
Aspergillopepsin,
   PROTEINASE A

Assembly-induced GTP hydrolysis,
   MICROTUBULE ASSEMBLY KINETICS
Assimilatory nitrate reductase,
   NITRATE REDUCTASE
Assistance,
   NEIGHBORING GROUP MECHANISM
ASSOCIATION
ASSOCIATION CONSTANT
   DISSOCIATION CONSTANT
   STABILITY CONSTANT
Association/dissociation rate constants, de-
   termination,
   FLUORESCENCE
ASTACIN
ASYMMETRIC INDUCTION
ASYMMETRY PARAMETER
ASYMPTOTE
ATCase,
   ASPARTATE CARBAMOYLTRANSFERASE
ATMOSPHERE
ATOM PERCENT EXCESS
   TRACER/TRACEE RATIO
   COMPARTMENTAL ANALYSIS
   ISOTOPE EXCHANGE KINETICS
ATOMIC MASS UNIT
ATOMIC ORBITAL
ATOMIZATION
ATP
ATPase, Ca^{2+}
ATPase, F_1F_o,
   F_1F_o ATPase
   BINDING CHANGE MECHANISAM
   CF_1CF_O ATPase
ATPase, H^+/K^+
ATPase, Na^+/K^+
ATPase reaction,
   ION PUMPS
ATP CITRATE LYASE
ATP-dependent enzymes,
   ACETATE KINASE
   ACETYL-CoA CARBOXYLASE
   ACETYL-CoA SYNTHETASE
   N-ACETYLGLUCOSAMINE KINASE
   ACTIN ATPase
   ACTOMYOSIN ATPase
   N-ACYLMANNOSAMINE KINASE
   ADENINE NUCLEOTIDE TRANSLOCASE
   ADENOSINE KINASE
   ADENYLATE KINASE (MYOKINASE)
   ADENYLYLSULFATE KINASE
   D-ALANINE-D-ALANINE LIGASE

ALKYLGLYCEROL KINASE

ALLOPHANATE HYDROLASE

ALLOSE KINASE

AMINOACYL-tRNA SYNTHETASES

2-AMINO-4-HYDROXY-6-HYDROXY-
METHYLDIHYDROPTERIDINE PYRO-
PHOSPHOKINASE

APYRASE

D-ARABINOKINASE

ARACHIDONYL-CoA SYNTHETASE

ARGININE KINASE

ARGININOSUCCINATE SYNTHETASE

ASPARAGINE SYNTHETASE

ASPARTATE KINASE

ASPARTYL-tRNA SYNTHETASE

ATPase, Ca^{2+}

ATPase, F_1F_o

ATPase, H^+/K^+

ATPase, Na^+/K^+

ATP CITRATE LYASE

ATP:GLUCOSE-1-PHOSPHATE ADENYL-
YLTRANSFERASE

ATP PHOSPHORIBOSYLTRANSFERASE

BENZOATE-CoA LIGASE

BIOTIN HOLOCARBOXYLASE SYN-
THETASE

CARBAMOYL PHOSPHATE SYNTHETASE
I and II

CARNOSINE SYNTHETASE

CHAPERONES

CHOLINE KINASE

CHOLOYL-CoA SYNTHETASE

COBALAMIN ADENOSYLTRANSFERASE

4-COUMAROYL-CoA SYNTHETASE

CREATINE KINASE

CTP SYNTHETASE

CYTIDYLATE KINASE

2-DEHYDRO-3-DEOXYGLUCONOKINASE

DEHYDROGLUCONOKINASE

DEOXYADENOSINE KINASE

DEOXYADENYLATE KINASE

DEOXYCYTIDINE KINASE

(DEOXY)NUCLEOSIDE MONOPHOS-
PHATE KINASE

DEOXYTHYMIDINE KINASE

DEPHOSPHO-CoA KINASE

DETHIOBIOTIN SYNTHASE

DIACYLGLYCEROL KINASE

DIHYDROFOLATE SYNTHETASE

DNA GYRASES & DNA REVERSE GYRASE

ETHANOLAMINE KINASE

EXONUCLEASE V

FMN ADENYLYLTRANSFERASE

10-FORMYLTETRAHYDROFOLATE SYN-
THETASE

FRUCTOKINASE

GALACTOKINASE

GENTAMICIN 2″-NUCLEOTIDYLTRANS-
FERASE

GERANOYL-CoA CARBOXYLASE

GLUCOKINASE

GLUCURONOKINASE

GLUTAMINE SYNTHETASE

γ-GLUTAMYLCYSTEINE SYNTHETASE

GLUTATHIONE SYNTHETASE

GLYCERATE KINASE

GLYCEROL KINASE

GLYCERONE KINASE

GMP SYNTHETASE

GUANYLATE KINASE

HEXOKINASE

HISTONE KINASE

β-HYDROXY-β-METHYLGLUTARYL-CoA
REDUCTASE KINASE

5-HYDROXYMETHYLPYRIMIDINE KINASE

HYPOTAUROCYAMINE KINASE

INOSINE KINASE

INOSITOL-1,3,4,6-TETRAPHOSPHATE
5-KINASE

INOSITOL-3,4,5,6-TETRAPHOSPHATE
1-KINASE

INOSITOL-1,4,5-TRISPHOSPHATE
3-KINASE

LEUCYL-tRNA SYNTHETASE

LONG-CHAIN FATTY ACYL-CoA SYN-
THETASE

LUCIFERASE

5,10-METHENYLTETRAHYDROFOLATE
SYNTHETASE

METHIONINE ADENOSYLTRANSFERASE

METHIONYL-tRNA SYNTHETASE

β-METHYLCROTONYL-CoA CARBOX-
YLASE

MEVALONATE KINASE

MYOSIN and MYOSIN ATPase

NAD^+ KINASE

NAD^+ SYNTHETASE

NICOTINAMIDE MONONUCLEOTIDE
ADENYLYLTRANSFERASE

NUCLEOSIDE DIPHOSPHATE KINASE

5-OXOPROLINASE

PHENYLALANINE RACEMASE

1-PHOSPHATIDYLINOSITOL 4-KINASE

PHOSPHATIDYLINOSITOL-4-PHOSPHATE
5-KINASE

PHOSPHOFRUCTOKINASE

3-PHOSPHOGLYCERATE KINASE

5-PHOSPHOMEVALONATE KINASE

PHOSPHORYLASE KINASE

POLYNUCLEOTIDE 5′-HYDROXYL
KINASE

POLYPHOSPHATE KINASE

PROPIONYL-CoA CARBOXYLASE

PROPIONYL-CoA SYNTHETASE

PROTEASE La

PYRIDOXAL KINASE

PYROPHOSPHOMEVALONATE DECAR-
BOXYLASE

PYRUVATE CARBOXYLASE

PYRUVATE DEHYDROGENASE KINASE

PYRUVATE KINASE

PYRUVATE, WATER DIKINASE

RESTRICTION ENZYMES

RHODOPSIN KINASE

RIBOFLAVIN KINASE

RIBONUCLEOTIDE REDUCTASE

RIBOSEPHOSPHATE PYROPHOSPHO-
KINASE

RNA POLYMERASES

SELENOPHOSPHATE SYNTHETASE

SHIKIMATE KINASE

SULFATE ADENYLYLTRANSFERASE

TUBULIN:TYROSINE LIGASE

UBIQUITIN-PROTEIN LIGASES

D-XYLULOKINASE

ATP-dependent protease La,
PROTEASE La

ATP-dependent racemization,
PHENYLALANINE RACEMASE

ATP-dependent serine proteinase,
PROTEASE La

ATP depletion,
ATP & GTP DEPLETION

ATP:GLUCOSE-1-PHOSPHATE ADENYLYL-
TRANSFERASE

ATP & GTP DEPLETION

ATP hydrolysis-dependent conformational
change,
ION PUMPS

ATP PHOSPHORIBOSYLTRANSFERASE

ATP-regenerating enzymes,
ENERGY CHARGE

ATP regeneration,
NUCLEOSIDE 5′-TRIPHOSPHATE REGEN-
ERATION

ATP sulfurylase,
SULFATE ADENYLYLTRANSFERASE

B

TRANSITION-STATE THEORY

Bimolecular rate constant,
EYRING EQUATION

BINARY COMPLEX
MICHAELIS COMPLEX

BINDING CHANGE MECHANISM

BINDING INTERACTION
ALLOSTERIC INTERACTION
BINDING ISOTHERM
BIOSENSOR
COOPERATIVE LIGAND BINDING
EQUILIBRIUM CONSTANT
EQUILIBRIUM DIALYSIS
HUMMEL-DREYER METHOD
LINKED FUNCTION THEORY
KLOTZ PLOT
MACROSCOPIC CONSTANT
MICROSCOPIC CONSTANT
MOLECULAR CROWDING
SCATCHARD PLOT

Binding of small molecules,
SCATCHARD PLOT

BINDING SITE

BINOMIAL THEOREM
PASCAL'S TRIANGLE

BIOAVAILABILITY

BIOCATALYTIC ELECTRODE
BIOSENSOR

BIOCHEMICAL NOMENCLATURE: RECOM-
MENDATIONS

BIOCHEMICAL SELF-ASSEMBLY
ACTIN ASSEMBLY KINETICS
HEMOGLOBIN S POLYMERIZATION
MICROTUBULE ASSEMBLY KINETICS

BIOCONJUGATE

BIOCONVERSION

BIOLEACHING

Biological electron transfer reactions,
NERNST EQUATION
ELECTRON TRANSFER KINETICS
HALF-REACTIONS

Biologically important nuclei for NMR experiments,
NUCLEAR MAGNETIC RESONANCE

Biological response,
AGONIST

Biological response to drug binding,
*OCCUPANCY THEORY OF DRUG
ACTION*
DIFFUSION OF LIGAND TO RECEPTOR

Biological self-assembly,
QUASI-EQUIVALENCE

Biological self-organization,
HYPERCYCLE

BIOMASS

BIOMIMETIC

BIOMINERALIZATION
ACCRETION
IONIC STRENGTH
SOLUBILITY PRODUCT
LANGMUIR ISOTHERM

Bio-Rad chromatography columns,
RAPID BUFFER EXCHANGE

BIOREACTOR

BIOSENSOR
BIOCATALYTIC ELECTRODE

Bio-Spin columns,
RAPID BUFFER EXCHANGE

Biothermodynamics,
BIOCHEMICAL NOMENCLATURE

BIOTIN AND DERIVATIVES

Biotin-dependent enzymes,
ACETYL-CoA CARBOXYLASE
*BIOTIN HOLOCARBOXYLASE SYN-
THETASE*
DETHIOBIOTIN SYNTHASE
GERANOYL-CoA CARBOXYLASE
GLUTACONYL-CoA DECARBOXYLASE
*β-METHYLCROTONYL-CoA CARBOX-
YLASE*
*METHYLMALONYL-CoA DECARBOX-
YLASE*
OXALOACETATE DECARBOXYLASE
PROPIONYL-CoA CARBOXYLASE
PYRUVATE CARBOXYLASE
TRANSCARBOXYLASE

BIOTIN HOLOCARBOXYLASE SYNTHETASE

BIRADICAL
CARBENE
FREE RADICALS

BISPHOSPHOGLYCERATE MUTASE

Bi-substrate mechanism,
RANDOM BI BI MECHANISM
*RANDOM BI BI THEORELL-CHANCE
MECHANISM*
RANDOM BI UNI MECHANISM
*RANDOM BI UNI UNI BI PING PONG
MECHANISM*

BI TER REACTION

BI UNI REACTION

"BLACK BOX"

BLEACHING

BLEOMYCIN

Blood clotting cascade,
ENZYME CASCADE KINETICS

BLUE COPPER PROTEIN

Blue shift,
HYPSOCHROMIC

Bodenstein approximation,
*PHYSICAL ORGANIC CHEMISTRY
NOMENCLATURE*

BOHR

BOHR EFFECT
LINKED FUNCTIONS
HEMOGLOBIN
ADAIR EQUATION

BOLTZMANN DISTRIBUTION LAW
CRYOENZYMOLOGY

BOLUS

BOLUS TRACER INJECTION

Bond-cleavage, light-induced,
PHOTOLYSIS
FLASH PHOTOLYSIS

Bonding,
DELOCALIZATION

Bonding interactions in oligo-/polynucleo-
tides,
BASE STACKING

BONDING ORBITAL

BOND NUMBER

Bond opposition strain,
*STEREOCHEMICAL TERMINOLOGY,
IUPAC RECOMMENDATIONS*

BOND ORDER

BOND VALENCE SUM ANALYSIS

Bonkregic acid,
ADENINE NUCLEOTIDE TRANSLOCASE,

Borderline mechanism,
*PHYSICAL ORGANIC CHEMISTRY
NOMENCLATURE*

BORN-OPPENHEIMER APPROXIMATION
ADIABATIC PHOTOREACTION
DIABATIC PHOTOREACTION

BORNYL PYROPHOSPHATE SYNTHASE

BOROHYDRIDE REDUCTION
ACYL-PHOSPHATE INTERMEDIATE

BOSE-EINSTEIN CONDENSATE

Bound water,
WATER, BOUND-

Bowsprit,
*STEREOCHEMICAL TERMINOLOGY,
IUPAC RECOMMENDATIONS*

Boyer,
BINDING CHANGE MECHANISM

Carbohydrates,
  BIOCHEMICAL NOMENCLATURE
Carbohydrates, determining the cyclic structures of,
  PERIODATE CLEAVAGE
  MALAPRADE REACTION
Carbon-carbon bond formation,
  CROTONASE SUPERFAMILY
CARBON-14 (^{14}C)
Carbon-14 dating,
  RADIOCARBON DATING
CARBON DIOXIDE
  OXYGEN, OXIDES & OXO ANIONS
Carbon dioxide produced by tissue metabolism,
  QUOTIENT, RESPIRATORY
CARBONIC ANHYDRASE
  PROTON TRANSFER IN AQUEOUS SOLUTION
  MANOMETRIC ASSAY METHODS
  MARCUS RATE THEORY
CARBONIUM ION
  CARBOCATION
  CARBENE
CARBON MONOXIDE
  OXYGEN, OXIDES & OXO ANIONS
CARBON MONOXIDE DEHYDROGENASE
Carboxyatractylate,
  ADENINE NUCLEOTIDE TRANSLOCASE
N^2-(D-1-Carboxyethyl)-L-arginine,
  OCTOPINE DEHYDROGENASE
N^5-(CARBOXYETHYL)ORNITHINE SYNTHASE
4-Carboxy-2-hydroxymuconate semialdehyde,
  PROTOCATECHUATE 4,5-DIOXYGENASE
CARBOXYLESTERASES
Carboxyl group activation,
  ACYL-PHOSPHATE INTERMEDIATE
Carboxyl group transfer,
  ACYL ENZYME INTERMEDIATES
α-Carboxyl-phosphate intermediate,
  D-ALANINE-D-ALANINE LIGASE
3-Carboxy-*cis,cis*-muconate,
  PROTOCATECHUATE 3,4-DIOXYGENASE
CARBOXYPEPTIDASE A
CARBOXYPEPTIDASE B
CARBOXYPEPTIDASE C
CARBOXYPEPTIDASE D
Carboxypeptidase H (or, E),
  ENKEPHALIN CONVERTASE
CARBOXYPEPTIDASE M

CARBOXYPEPTIDASE N
CARBOXYPEPTIDASE T
5-*O*-(1-Carboxyvinyl)-3-phosphoshikimate, synthesis of,
  3-PHOSPHOSHIKIMATE 1-CARBOXYVINYLTRANSFERASE
CARDIOLIPIN SYNTHASE
CARNITINE
CARNITINE *O*-ACETYLTRANSFERASE
CARNITINE *O*-OCTANOYLTRANSFERASE
CARNITINE *O*-PALMITOYLTRANSFERASE
CARNOSINASE
CARNOSINE SYNTHETASE
CARNOT CYCLE
  CARNOT'S THEOREM
  EFFICIENCY
  THERMODYNAMICS
CARNOT'S THEOREM
  EFFICIENCY
  THERMODYNAMICS, LAWS OF
  CARNOT CYCLE
β-CAROTENE (or, β-CAROTENOID) 15,15'-DIOXYGENASE
Carotenoids,
  BIOCHEMICAL NOMENCLATURE
Carrier-mediated transport,
  FACILITATED TRANSPORT
CASBENE SYNTHASE
Cascade kinetics,
  AMPLIFICATION, ENZYME CASCADE KINETICS
CASPASES (or, Cysteinyl Aspartate-Specific Proteinases)
CATALASE
CATALYST
  ENZYME
CATALYST EFFECT ON EQUILIBRIUM CONSTANT
CATALYTIC ANTIBODIES
Catalytic center activity,
  TURNOVER NUMBER
CATALYTIC CONSTANT
  TURNOVER NUMBER
CATALYTIC DOMAIN
  MULTIENZYME POLYPEPTIDE
  AUTONOMOUS CATALYTIC DOMAIN
CATALYTIC ENHANCEMENT
CATALYTIC PROFICIENCY
  CATALYTIC EFFICIENCY
  REFERENCE REACTION
Catalytic rate acceleration,
  EFFICIENCY FUNCTION

Catalytic residues,
  AFFINITY LABELING
CATALYTIC RNA
Catalytic triad,
  CHARGE RELAY SYSTEM
CATASTROPHE THEORY
Catastrophic depolymerization of microtubules,
  MICROTUBULE ASSEMBLY KINETICS
Catechol,
  PHENOL HYDROXYLASE
CATECHOL 1,2-DIOXYGENASE
CATECHOL 2,3-DIOXYGENASE
CATECHOL *O*-METHYLTRANSFERASE
CATECHOL OXIDASE
  MONOPHENOL MONOOXYGENASE
  TYROSINE MONOOXYGENASE
CATENANE
CATENARY MODEL FOR COMPARTMENT ANALYSIS
CATHEPSIN B
CATHEPSIN D
CATHEPSIN E
CATHEPSIN G
CATHEPSIN H
CATHEPSIN L
CATHEPSIN S
CATHEPSIN T
CATIONIC INTERMEDIATE ANALOGUES
Cationotropic rearrangement (cationotropy),
  PHYSICAL ORGANIC CHEMISTRY NOMENCLATURE
CATION-π INTERACTIONS
Cation radical,
  RADICAL CATION
CDP-Abequose epimerase,
  CDP-PARATOSE EPIMERASE
CDP-Choline:1,2-diacylglycerol cholinephosphotransferase,
  CHOLINEPHOSPHOTRANSFERASE
CDP-4-DEHYDRO-6-DEOXYGLUCOSE REDUCTASE
CDP-Diacylglycerol,
  PHOSPHATIDATE CYTIDYLYLTRANSFERASE
CDP-DIACYLGLYCEROL:*sn*-GLYCEROL-3-PHOSPHATE PHOSPHATIDYLTRANSFERASE
CDP-Diacylglycerol–inositol 3-phosphatidyltransferase,
  PHOSPHATIDYLINOSITOL SYNTHASE
CDP-Diacylglycerol:L-serine *O*-phosphatidyltransferase,
  PHOSPHATIDYLSERINE SYNTHASE

CHOLERA TOXIN

See also ADP-RIBOSYLATION

Cholesterol acyltransferase,

ACYL-CoA:CHOLESTEROL O-ACYL-TRANSFERASE

Cholesterol turnover,

LIPID TRACER KINETICS.

CHOLESTERYL ESTER HYDROLASE

CHOLINE ACETYLTRANSFERASE

CHOLINE DEHYDROGENASE

CHOLINE KINASE

CHOLINE OXIDASE

Choline phosphatase,

PHOSPHOLIPASE D

CHOLINE-PHOSPHATE CYTIDYLYLTRANS-FERASE

CHOLINEPHOSPHOTRANSFERASE

Cholinesterase,

BUTYRYLCHOLINE ESTERASE

CHOLINE SULFATASE

CHOLINE SULFOTRANSFERASE

CHOLOYL-CoA SYNTHETASE

CHONDROITIN 4-SULFOTRANSFERASE

CHONDROSULFATASES

CHORISMATE MUTASE

CHORISMATE SYNTHASE

Chromatin self-assembly,

NUCLEOSOME CORE PARTICLE SELF-ASSEMBLY

CHROMIUM-51 (^{51}Cr)

Chromium ion,

IONIC RADIUS

Chromium(III)-ATP,

SCREW SENSE OF METAL-NUCLEOTIDE COMPLEXES

Chromium(III) nucleotide complex,

EXCHANGE-INERT METAL-NUCLEOTIDE COMPLEXES

CHROMIUM-NUCLEOTIDES

CHROMOGENIC SUBSTRATE

ABSORPTION SPECTROSCOPY

FLUORESCENCE SPECTROSCOPY

CHROMOPHORE

CHYMASE

CHYMOPAPAIN

CHYMOSIN (or, RENNIN)

CHYMOTRYPSIN

ACYL-SERINE INTERMEDIATE

AFFINITY LABELING

SERPINS (INHIBITORY MECHANISM)

CHYMOTRYPSIN C

Chymotrypsin hydrolysis of *p*-nitrophenyl ethylcarbonate or *p*-nitrophenyl ac-etate,

BURST KINETICS

CHYMOTRYPSIN-LIKE SERINE PROTEASE SUPERFAMILY

CIDNP (Chemically induced dynamic nuclear polarization),

PHYSICAL ORGANIC CHEMISTRY NO-MENCLATURE

Cine-substitution,

PHYSICAL ORGANIC CHEMISTRY NO-MENCLATURE

trans-Cinnamate,

PHENYLALANINE AMMONIA-LYASE

trans-CINNAMATE 4-MONOOXYGENASE

CIRCADIAN RHYTHM

CIRCE EFFECT

CIRCULAR DICHROISM SPECTROSCOPY

Cisoid conformation,

STEREOCHEMICAL TERMINOLOGY, IUPAC RECOMMENDATIONS

CITRAMALATE LYASE

CITRAMALYL-CoA LYASE

CITRATE LYASE

CITRATE SYNTHASE

L-Citrulline,

ORNITHINE TRANSCARBAMOYLASE

NITRIC OXIDE SYNTHASE

Citrulline phosphorylase,

ORNITHINE TRANSCARBAMOYLASE

CLAISEN CONDENSATION

Class (A) metal ion,

PHYSICAL ORGANIC CHEMISTRY NO-MENCLATURE

Class (B) metal ion,

PHYSICAL ORGANIC CHEMISTRY NO-MENCLATURE

CLATHRATE

CAGE COMPOUND

CLELAND NOMENCLATURE

Clinal,

STEREOCHEMICAL TERMINOLOGY, IUPAC RECOMMENDATIONS

Closed catalytic cycle,

HYPERCYCLE

CLOSED SYSTEM

Close-packing of polypeptide chains in glob-ular proteins,

PACKING DENSITY

Clostridium oedematiens β- and γ-toxins,

PHOSPHOLIPASE C

Clostridium welchii α-toxin,

PHOSPHOLIPASE C

CLOSTRIPAIN

C.M.C.

CMP-*N*-acetylneuraminate,

SIALYLTRANSFERASE

CMP-*N*-acetylneuraminate-β-galactoside α-2,6-sialyltransferase,

SIALYLTRANSFERASE

CMP-*N*-ACETYLNEURAMINATE-β-GALACTOSIDE SIALYLTRANSFERASES

COACTIVATOR

COBALAMIN ADENOSYLTRANSFERASE

Cobalamin-dependent enzymes,

AQUACOBALAMIN REDUCTASES

COBALAMIN ADENOSYLTRANSFERASE

ETHANOLAMINE AMMONIA-LYASE

GLYCEROL DEHYDRATASE

LEUCINE 2,3-AMINOMUTASE

β-LYSINE 5,6-AMINOMUTASE

3-METHYLASPARTATE AMMONIA-LYASE

2-METHYLENEGLUTARATE MUTASE

L-METHYLMALONYL-CoA MUTASE

D-ORNITHINE 4,5-AMINOMUTASE

PROPANEDIOL DEHYDRATASE

COBALAMINS AND COBAMIDES (B$_{12}$)

Cobalt(III) nucleotide complex,

EXCHANGE-INERT METAL-NUCLEOTIDE COMPLEXES

COBALT NUCLEOTIDES

COCOONASES

Coefficients in binomial expansion,

PASCAL'S TRIANGLE

Coenzyme A-dependent enzymes,

ACETALDEHYDE DEHYDROGENASE (ACETYLATING)

ACETYL-CoA ACETYLTRANSFERASE (or, THIOLASE)

ACETYL-CoA ACYLTRANSFERASE

ACETYL-CoA:ACP TRANSACYLASE

ACETYL-CoA CARBOXYLASE

ACETYL-CoA SYNTHETASE

N-ACETYLGLUTAMATE SYNTHASE

ACYL-CoA:CHOLESTEROL ACYLTRANS-FERASE

ACYL-CoA DEHYDROGENASES

ACYL-CoA DESATURASE

ACYL-CoA:RETINOL ACYLTRANSFERASE

1-ACYLGLYCEROL-3-PHOSPHATE ACYL-TRANSFERASE

1-ACYLGLYCEROL-3-PHOSPHO-CHOLINE ACYLTRANSFERASE

β-ALANYL-CoA AMMONIA LYASE

5-AMINOLEVULINATE SYNTHASE

ARACHIDONYL-CoA SYNTHETASE

CONCERTED FEEDBACK INHIBITION
  FEEDBACK INHIBITION
Concerted Monod-Wyman-Changeux model,
  COOPERATIVITY
CONCERTED PROCESS
  PRIMITIVE CHANGES
  ELEMENTARY REACTIONS
  SYNCHRONOUS
Condensation equilibrium,
  PROTEIN POLYMERIZATION
Condensation equilibrium model for nucleation,
  MICROTUBULE ASSEMBLY KINETICS
CONDENSATION REACTION
CONFIDENCE INTERVAL
CONFIDENCE LEVEL
CONFIGURATION, DETERMINATION OF
CONFORMATIONAL CHANGES IN PROTEINS
Conformational equilibria in chymotrypsin,
  THERMODYNAMIC CYCLES
CONFORMATIONAL EQUILIBRIA, MACROMOLECULAR
CONFORMATIONS
Conformers,
  CONFORMATIONS
Conjugated double bonds,
  DELOCALIZATION
Conjugate force,
  NONEQUILIBRIUM THERMODYNAMICS: A PRIMER
"CONNOLY MOLECULAR SURFACE" program,
  COMPUTER ALGORITHMS & SOFTWARE
Conrotatory,
  PHYSICAL ORGANIC CHEMISTRY NOMENCLATURE
Consecutive reactions,
  SERIES REACTIONS
CONSERVATION OF CHARGE, LAW OF
CONSERVATION OF ENERGY, LAW OF
CONSTANT RATIO METHODS
CONSTITUENT
Constitutional isomerism,
  STEREOCHEMICAL TERMINOLOGY, IUPAC RECOMMENDATIONS
Contact ion pair,
  TIGHT ION PAIR
Contamination
  SUBSTRATE PURITY
  REAGENT PURITY

  STEREOCHEMICAL PURITY
  WATER PURITY
Continued polymerization/depolymerization,
  PROCESSIVITY
Continuous variation method,
  JOB PLOT
CONTINUOUS *versus* STOPPED-TIME ASSAYS
CONTRADICTION
CONTRAST AGENT
CONTROL
CONTROL COEFFICIENT
Convergence-point method,
  POINT-OF-INTERSECTION
Conversion of pepsinogen to active pepsin,
  AUTOCATALYSIS
Cooperative oxygen binding,
  BOHR EFFECT
  COOPERATIVITY
Cooperative process,
  NUCLEATION
COOPERATIVITY
  ALLOSTERISM
  HYSTERESIS
  JENKINS MECHANISM
  JOB PLOT
  KOSHLAND-NEMETHY-FILMER MODEL
  LINKED FUNCTIONS
  LOG-RESPONSE CURVE FOR ENZYME COOPERATIVITY
  METAL IONS IN NUCLEOTIDE-DEPENDENT REACTIONS
  MICHAELIS-MENTEN EQUATION
  MICHAELIS-MENTEN KINETICS
  MONOD-WYMAN-CHANGEUX MODEL
  NEGATIVE COOPERATIVITY
  POSITIVE COOPERATIVITY
Cooperativity index,
  $S_{0.9}/S_{0.1}$
Cooperativity of ligand binding,
  MONOD-WYMAN-CHANGEUX MODEL
COORDINATE-COVALENT BOND
COORDINATION
  NUCLEARITY
COORDINATION NUMBER
Coordinatively unsaturated,
  PHYSICAL ORGANIC CHEMISTRY NOMENCLATURE
Copper amine oxidase,
  AMINE OXIDASE
  REDOX-ACTIVE AMINO ACIDS

Copper-dependent enzymes,
  ASCORBATE OXIDASE
  CATECHOL OXIDASE
  FERROXIDASE
  GALACTOSE OXIDASE
  LACCASE
  MONOPHENOL MONOOXYGENASE
  NITRITE REDUCTASE
  PHOTOSYSTEM I
  QUERCETIN 2,3-DIOXYGENASE
  SUPEROXIDE DISMUTASES
COPROPORPHYRINOGEN OXIDASE
CORNFORTH CHIRAL METHYL STEREOCHEMICAL METHOD
"CORN" RULE
Coronate,
  PHYSICAL ORGANIC CHEMISTRY NOMENCLATURE
CORPHIN
CORRELATION COEFFICIENT
CORRELATION DIAGRAM
CORRELATION FUNCTION
  CONCENTRATION CORRELATION ANALYSIS
CORRELATION SPECTROSCOPY
CORRELATION TIME
Corrinoids,
  BIOCHEMICAL NOMENCLATURE
CORTISONE β-REDUCTASE
COSMOGENIC DATING
  RADIOCARBON DATING
Cosphere,
  PHYSICAL ORGANIC CHEMISTRY NOMENCLATURE
COULOMB
COULOMB'S LAW
  DEBYE-HÜCKEL TREATMENT
COULOMETRIC TITRATION
4-COUMAROYL-CoA SYNTHETASE
COUNTERION
Counting efficiency,
  QUENCHING
Counting rate,
  COUNTING STATISTICS
COUNTING STATISTICS
COUPLED ENZYME ASSAYS
Coupling enzymes,
  ENZYME CONCENTRATION EFFECTS
COVALENT BOND
Covalent interconversion,
  ENZYME CASCADE KINETICS

Cytochrome-dependent reactions,
  ACYL-CoA DESATURASE
  ADRENODOXIN
  *ASCORBATE-CYTOCHROME b₅ RE-
    DUCTASE*
  CAMPHOR 5-MONOOXYGENASE
  CYTOCHROME c
  CYTOCHROME c₃ HYDROGENASE
  CYTOCHROME c PEROXIDASE
  CYTOCHROME P-450 REDUCTASE
  MIXED-FUNCTION OXIDASE
  NADH-CYTOCHROME b₅ REDUCTASE
  NADH DEHYDROGENASE
  NITRATE REDUCTASE
  *PLASMANYLETHANOLAMINE Δ¹-DESA-
    TURASE*
  PYRUVATE DEHYDROGENASE
  SULFITE DEHYDROGENASE
  *UBIQUINOL:CYTOCHROME c OXIDORE-
    DUCTASE*
CYTOCHROME P-450
CYTOCHROME P-450 REDUCTASE
β-(9-CYTOKININ)-ALANINE SYNTHASE
CYTOKININ 7β-GLUCOSYLTRANSFERASE
Cytoplasmic calcium ion,
  ION PUMPS
Cytoskeleton dynamics,
  ACTIN ASSEMBLY KINETICS
  ACTIN-BASED MOTILITY
  MICROTUBULE ASSEMBLY KINETICS
  TREADMILLING
  *FLUORESCENCE RECOVERY AFTER PHO-
    TOBLEACHING*

D

DAKIN-WEST REACTION
Dalziel relationships,
  PHI RELATIONSHIPS
Dark reaction,
  PHOTOSYNTHESIS
DARVEY PLOT
dATP,
  *RIBOSEPHOSPHATE PYROPHOSPHO-
    KINASE*
DAVIES EQUATION
  DEBYE-HÜCKEL TREATMENT
Day-night cycle,
  CIRCADIAN RHYTHM
dCMP DEAMINASE
Dead-end complexes,

ABORTIVE COMPLEXES
DEAD-END INHIBITION
DEAD-END INHIBITION (CLELAND'S
  RULES)
Deamido-NAD⁺ as substrate,
  NAD⁺ SYNTHETASE
  *NICOTINAMIDE MONONUCLEOTIDE
    ADENYLYLTRANSFERASE*
Deamido-NAD⁺ pyrophosphorylase,
  *NICOTINATE MONONUCLEOTIDE
    ADENYLYLTRANSFERASE*
DEBRANCHING ENZYME
DeBroglie wavelength,
  QUANTUM MECHANICAL TUNNELING
DEBYE
Debye-Hückel length,
  BIOMINERALIZATION
DEBYE-HÜCKEL TREATMENT
DEFINITE INTEGRAL
Degenerate rearrangement,
  NARCISSISTIC REACTION
  *PHYSICAL ORGANIC CHEMISTRY NO-
    MENCLATURE*
Degraded accuracy or sensitivity of mea-
  surement,
  NOISE
DEGREE OF ACTIVATION
  ACTIVATION
  BASAL RATE
DEGREE OF DISSOCIATION
  HENDERSON-HASSELBALCH EQUATION
DEGREE OF INHIBITION
  *INHIBITION (as well as specific type of in-
    hibition)*
DEGREES OF FREEDOM
DEHYDROALANINE
  BOROHYDRIDE REDUCTION
DEHYDROASCORBATE REDUCTASE
DEHYDROCHOLESTEROL REDUCTASE
2-DEHYDRO-3-DEOXY-D-GLUCONATE
  5-DEHYDROGENASE
2-DEHYDRO-3-DEOXY-D-GLUCONATE
  6-DEHYDROGENASE
2-DEHYDRO-3-DEOXYGLUCONOKINASE
2-Dehydro-3-deoxy-D-pentonate aldolase,
  *2-KETO-3-DEOXYPENTONATE
    ALDOLASE*
DEHYDROGLUCONOKINASE
3-DEHYDROQUINATE DEHYDRATASE
5-Dehydroquinate dehydrase reduction,
  BOROHYDRIDE REDUCTION
3-DEHYDROQUINATE SYNTHASE

5-Dehydroshikimate,
  SHIKIMATE DEHYDROGENASE
3-Dehydro-D-sphinganine,
  SERINE PALMITOYLTRANSFERASE
3-DEHYDROSPHINGANINE REDUCTASE
DEINHIBITION
Delayed fluorescence,
  JABLONSKI DIAGRAM
  PHOSPHORESCENCE
Delayed emission,
  FLUORESCENCE
Delay time in polymerization,
  HEMOGLOBIN-S POLYMERIZATION
DELOCALIZATION
Delocalization of a free electron pair,
  $n \rightarrow \sigma^*$ *DELOCALIZATION*
  HYPERCONJUGATION; RESONANCE
"DELPHI,"
  COMPUTER ALGORITHMS & SOFTWARE
DELTA (Δ or δ)
DEOXYADENOSINE KINASE
DEOXYADENYLATE KINASE
DEOXYCYTIDINE KINASE
DEOXYCYTIDYLATE HYDROXYMETHYL-
  TRANSFERASE
Deoxycytidylate kinase,
  CYTIDYLATE KINASE
3-DEOXY-D-manno-OCTULOSONATE
  ALDOLASE
3-DEOXYOCTULOSONATE 8-PHOSPHATE
  SYNTHASE
2'-Deoxynucleoside,
  NUCLEOSIDE PHOSPHOTRANSFERASE
2'-Deoxynucleoside 5'-monophosphate,
  NUCLEOSIDE PHOSPHOTRANSFERASE
(DEOXY)NUCLEOSIDE MONOPHOSPHATE
  KINASE
DEOXYRIBONUCLEASES
2'-Deoxyribonucleoside diphosphate,
  RIBONUCLEOTIDE REDUCTASE
2-DEOXYRIBOSE-5-PHOSPHATE ALDOLASE
DEOXYTHYMIDINE KINASE
DEP,
  DIETHYL PYROCARBONATE
DEPHOSPHO-CoA KINASE
Dephospho-CoA pyrophosphorylase,
  *PANTETHEINE-PHOSPHATE ADENYLYL-
    TRANSFERASE*
DEPOLARIZATION
  ACTION POTENTIAL
  HYPERPOLARIZATION
DEPOLYMERIZATION
  PROCESSIVITY

DIHYDROLIPOAMIDE DEHYDROGENASE

DIHYDRONEOPTERIN ALDOLASE

DIHYDRONEOPTERIN TRIPHOSPHATE PYROPHOSPHOHYDROLASE

DIHYDROOROTASE

DIHYDROOROTATE DEHYDROGENASE

DIHYDROOROTATE OXIDASE

DIHYDROPTERIDINE REDUCTASE

DIHYDROPTEROATE SYNTHASE

DIHYDROPYRIMIDINASE

DIHYDROURACIL DEHYDROGENASE

Dihydroxyacetone kinase,
   GLYCERONE KINASE

Dihydroxyacetone phosphate,
   *RHAMNULOSE-1-PHOSPHATE AL-
   DOLASE*

DIHYDROXYACETONE PHOSPHATE ACYL-
   TRANSFERASE

DIHYDROXYACETONE SYNTHASE

DIHYDROXY-ACID DEHYDRATASE

DIHYDROXYBIPHENYL 1,2-DIOXYGENASE

DIHYDROXYFUMARATE DECARBOXYLASE

2,3-DIHYDROXYINDOLE 2,3-DIOXYGENASE

3,4-DIHYDROXYPHENYLACETATE 2,3-DIOX-
   YGENASE

Dihydroxyphenylalanine decarboxylase,
   *AROMATIC AMINO ACID DECARBOX-
   YLASE*

DIHYDROXYPHTHALATE DECARBOXYLASE

2,5-DIHYDROXYPYRIDINE 5,6-DIOXY-
   GENASE

2,3-DIHYDROXY-9,10-SECOANDROSTA-
   1,3,5(10)-TRIENE-9,17-DIONE 4,5-
   DIOXYGENASE

DIIODOPHENYLPYRUVATE REDUCTASE

DIIODOTYROSINE AMINOTRANSFERASE

DIISOPROPYL FLUOROPHOSPHATE

Dilution below critical concentration,
   DEPOLYMERIZATION, END-WISE

DILUTION EFFECTS

Dilution effects on enzymes,
   ENZYME CONCENTRATION

Dilution-induced disassembly,
   DEPOLYMERIZATION, END-WISE

2-Dimensional NMR of exchangeable
   protons,
   HYDROGEN EXCHANGE IN PROTEINS

DIMERIZATION

DIMETHYLALLYL*TRANS*TRANSFERASE

DIMETHYLANILINE MONOOXYGENASE

Dimethylbenzimidazole,
   NICOTINATE-NUCLEOTIDE DIMETHYL-

*BENZIMIDAZOLE PHOSPHORIBO-
   SYLTRANSFERASE*

6,7-Dimethyl-8-(1-D-ribityl)lumazine,
   RIBOFLAVIN SYNTHASE

DIMINISHED ISOTOPE EFFECT

Dinitrogenase,
   NITROGENASE

Dinitrogenase reductase,
   NITROGENASE

1,2-Diol cleavage,
   PERIODATE CLEAVAGE
   MALAPRADE REACTION

DIOL DEHYDRASE

Dioxygen,
   OXYGEN, OXIDES & OXO ANIONS

DIPEPTIDASES

DIPEPTIDYL-PEPTIDASE I

DIPEPTIDYL-PEPTIDASE IV

Diphosphate,
   OXYGEN, OXIDES & OXO ANIONS

2,3-Diphosphoglycerate formation,
   PHOSPHOGLYCERATE MUTASE

Diphosphoglyceromutase,
   BISPHOSPHOGLYCEROMUTASE

Diphosphoinositide kinase,
   *PHOSPHATIDYLINOSITOL-4-PHOSPHATE
   5-KINASE*

(R)-5-Diphosphomevalonate, as substrate,
   *PYROPHOSPHOMEVALONATE DECAR-
   BOXYLASE*

(R)-5-Diphosphomevalonate, formation of,
   5-PHOSPHOMEVALONATE KINASE

DIPHOSPHOMEVALONATE DECARBOX-
   YLASE

DIPHTHERIA TOXIN

DIPOLAR BOND

DIPOLE-DIPOLE INTERACTION
   NONCOVALENT INTERACTIONS

Dipole-dipole resonance energy transfer,
   FLUORESCENCE

Dipole-induced dipole interactions,
   NONCOVALENT INTERACTIONS

DIPOLE MOMENT

Diradical,
   BIRADICAL

Directional flow of energy and/or matter,
   *NONEQUILIBRIUM THERMODYNAMICS:
   A PRIMER*

DIRECT-LINEAR PLOT

Direct microscopic observation of microtubules,
   MICROTUBULE ASSEMBLY KINETICS

"DISMAN,"
   COMPUTER ALGORITHMS & SOFTWARE

DISPROPORTIONATION

Disrotatory,
   *PHYSICAL ORGANIC CHEMISTRY NO-
   MENCLATURE*

DISSOCIATION

DISSOCIATION CONSTANT

DISSOCIATION CONSTANTS (Metal-Nucleotide Complexes)

DISSOCIATION CONSTANTS (Temperature
   Effects)

Dissociation constant(s) for ligand binding,
   SCATCHARD PLOT

DISSOCIATION KINETICS

Dissociation rate of an enzyme-substrate
   complex,
   ISOTOPE TRAPPING

Dissolution,
   FRACTAL REACTION KINETICS

Dissymmetry,
   *STEREOCHEMICAL TERMINOLOGY,
   IUPAC RECOMMENDATIONS*

Distonic radical cation,
   *PHYSICAL ORGANIC CHEMISTRY NO-
   MENCLATURE*

DISTORTION, FREE ENERGY OF

DISTRIBUTION

Disulfide bond reduction,
   DITHIOTHREITOL

Disulfide rearrangements in proteins,
   PROTEIN DISULFIDE ISOMERASE

Dithiolane ring formation,
   EFFECTIVE MOLARITY

DITHIONITE

DITHIOTHREITOL

DIXON PLOT

DIXON-WEBB PLOTS

DNA (ADENINE-N^6)-METHYLTRANSFERASE

DNA BREATHING

DNA (CYTOSINE-5-)-METHYLTRANSFERASE

DNA, exonucleolytic cleavage,
   PHOSPHODIESTERASES

DNA GLYCOSYLASES

DNA GYRASES & DNA REVERSE GYRASE

DNA LIGASES

DNA-3-Methyladenine glycosylase,
   3-METHYLADENINE-DNA GLYCOSYLASE

DNA METHYLTRANSFERASES

DNA nucleotidylexotransferase,

DNA OXIDATIVE DAMAGE AND REPAIR
   PHOTOREACTIVATION

DNA PHOTOLYASES

DNA POLYMERASES

EDITING MECHANISMS

DNA repair,

DNA OXIDATIVE DAMAGE AND REPAIR

PHOTOREACTIVATION

DNA Reverse gyrase,

DNA GYRASE & DNA REVERSE GYRASE

DNase,

DEOXYRIBONUCLEASES

DNase digestion,

NUCLEOSOME CORE PARTICLE SELF-
ASSEMBLY

DNase inhibition,

ACTIN ASSEMBLY ASSAYS

DNA structure,

NUCLEAR MAGNETIC RESONANCE

DNA TOPOISOMERASES

DNA Unwinding,

INNATE THERMODYNAMIC QUANTITIES

DNA UNWINDING (Kinetic Model for Small
DNA)

DOCKING METHODS

[D_2O]/[H_2O] ratio,

PROTON INVENTORY TECHNIQUE

Dolichyl-phosphate mannosyltransferase,

GDP-MANNOSE:DOLICHYL PHOSPHATE
MANNOSYLTRANSFERASE

DOMAIN

DOMAIN SWAPPING

DONNAN EQUILIBRIUM

Donor,

FLUORESCENCE

Donor number (DN),

PHYSICAL ORGANIC CHEMISTRY NO-
MENCLATURE

DOPA,

REDOX-ACTIVE AMINO ACIDS

DOPAMINE β-MONOOXYGENASE

DOSE

Dose, absorbed ionizing radiation,

RAD

DOSE, LETHAL

Dose-response curve,

CONCENTRATION-RESPONSE CURVE

Double displacement mechanism,

PING PONG BI BI MECHANISM

ISOTOPE EXCHANGE

MULTISITE PING PONG BI BI MECH-
ANISM

HALF-REACTIONS (Ping Pong Mech-
anism)

Double-layer region,

BIOMINERALIZATION

Double-layer theory,

BIOMINERALIZATION

DOUBLE-RECIPROCAL PLOT

Downfield,

PHYSICAL ORGANIC CHEMISTRY NO-
MENCLATURE

DPNase,

NAD+ GLYCOHYDROLASE

DPN hydrolase,

NAD+ GLYCOHYDROLASE

Drug,

AGONIST

DRUG ABSORPTION/ELIMINATION (FIRST-
ORDER MODEL)

Drug absorption, quantitative models for,

PHARMACOKINETICS

DRUG ACTION,

DIFFUSION OF LIGAND TO RECEPTOR

DRUG RECEPTOR

OCCUPANCY THEORY OF DRUG
ACTION

OPERATIONAL MODEL FOR DRUG
ACTION

CONCENTRATION-RESPONSE CURVE

DRUG ACTION (OPERATIONAL MODEL)

Drug binding to receptor,

OCCUPANCY THEORY OF DRUG
ACTION

DIFFUSION OF LIGAND TO RECEPTOR

Drug clearance, dosage,

PHARMACOKINETICS

Drug distribution kinetics,

PHARMACOKINETICS

Drug excretion rates,

PHARMACOKINETICS

Drug metabolism/physiology,

PHARMACOKINETICS

DRUG RECEPTOR

"DSPP,"

COMPUTER ALGORITHMS & SOFTWARE

dTMP,

THYMIDYLATE SYNTHASE

Dual substituent-parameter equation,

PHYSICAL ORGANIC CHEMISTRY NO-
MENCLATURE

dUMP,

THYMIDYLATE SYNTHASE

DYAD SYMMETRY

DYNAMIC EQUILIBRIUM

Dynamic instability,

PROTEIN POLYMERIZATION

Dynamic instability of microtubules,

MICROTUBULE ASSEMBLY KINETICS

Dynamic light scattering,

LIGHT SCATTERING (DYNAMIC)

DYNAMICS

DYNEIN

Dyotropic rearrangement,

PHYSICAL ORGANIC CHEMISTRY NO-
MENCLATURE

E

e

EADIE-HOFSTEE PLOT

Eclipsed,

STEREOCHEMICAL TERMINOLOGY,
IUPAC RECOMMENDATIONS

Eclipsing,

STEREOCHEMICAL TERMINOLOGY,
IUPAC RECOMMENDATIONS

Eclipsing Conformation,

STEREOCHEMICAL TERMINOLOGY,
IUPAC RECOMMENDATIONS

Eclipsing strain,

STEREOCHEMICAL TERMINOLOGY,
IUPAC RECOMMENDATIONS

EC_{50}

EC Nomenclature system

ENZYME COMMISSION NOMEN-
CLATURE

EC number,

ENZYME COMMISSION NOMEN-
CLATURE

EDITING

EDITING MECHANISMS

EDTA

Effective concentration,

EFFECTIVE MOLARITY

DEBYE-HÜCKEL TREATMENT

EFFECTIVE DOSE

EFFECTIVE MOLARITY,

INTRAMOLECULAR CATALYSIS

EFFECTOR

MODIFIER

ACTIVATOR

INHIBITOR

EFFICIENCY FUNCTION

EF HAND

Eigen,

CHEMICAL KINETICS

EIGENFUNCTION

EIGEN-TAMM MECHANISM FOR METAL
ION COMPLEXATION

EINSTEIN

EINSTEIN-SUTHERLAND EQUATION

DIFFUSION

Eisenthal-Cornish-Bowden plot,

DIRECT LINEAR PLOT

ELASTASE

ELASTICITY

ELASTICITY COEFFICIENT

ELECTRIC CHARGE

ELECTRIC POTENTIAL

ELECTROMOTIVE FORCE

ELECTROACTIVE SPECIES

ELECTROCATALYSIS

Electrocyclic reaction,

PHYSICAL ORGANIC CHEMISTRY NO-MENCLATURE

ELECTRODE KINETICS

ELECTROFUGE

ELECTROPHILE

NUCLEOFUGE

Electrolyte behavior,

DEBYE-HÜCKEL TREATMENT

Electromeric effect,

PHYSICAL ORGANIC CHEMISTRY NO-MENCLATURE

ELECTROMOTIVE FORCE

ELECTRIC POTENTIAL

NERNST EQUATION

ELECTRON

ELECTRON ACCEPTOR

ELECTRON AFFINITY

Electronation,

PHYSICAL ORGANIC CHEMISTRY NO-MENCLATURE

Electron attachment,

PHYSICAL ORGANIC CHEMISTRY NO-MENCLATURE

ELECTRON DENSITY

CHARGE DENSITY

ELECTRON DENSITY FUNCTION

ELECTRON DONOR

ELECTRONEGATIVITY

ELECTRONEUTRALITY PRINCIPLE

Electronic transitions,

JABLONSKI DIAGRAM

FLUORESCENCE

ABSORPTION SPECTROSCOPY

$n \rightarrow \pi^*$ *TRANSITION*

$n \rightarrow \sigma^*$ *TRANSITION*

$\pi^* \rightarrow \pi^*$ *TRANSITION*

Electron microscopy,

ACTIN ASSEMBLY ASSAYS

Electron movements across the anode-to-cathode junction,

NOISE

ELECTRON-NUCLEAR DOUBLE RESO-NANCE (ENDOR)

Electron-pair donor (or Lewis base),

NUCLEOPHILE

ELECTRON SINK

ELECTRON SPIN RESONANCE

ELECTRON TRANSFER

MARCUS EQUATION

ELECTRODE KINETICS

Electron transfer mechanism,

OUTER SPHERE ELECTRON TRANSFER REACTION

ELECTRON TRANSFER REACTIONS

ELECTRON TRANSPORT

ELECTROPHILE

ELECTROPHILICITY

NUCLEOPHILE

ELECTROPHILIC CATALYSIS

ELECTROPHILICITY

NUCLEOPHILICITY ELECTROPHILE

ELECTROPHILIC SUBSTITUTION REACTION

Electrophoretic mobility,

OFFORD PLOT

Electrostatic attraction/repulsion,

DEBYE-HÜCKEL TREATMENT

ELECTROSTATIC BOND

ELECTROSTATIC SURFACE POTENTIAL

ELECTROSTRICTION

ELECTROTAXIS

ELECTROVALENT BOND

ELEMENTARY CHARGE

ELEMENTARY REACTION

Elementary reaction stoichiometry,

MOLECULARITY

CHEMICAL KINETICS

UNIMOLECULAR

BIMOLECULAR

TRANSITION-STATE THEORY

ELEMENTARY REACTION

Element effect,

PHYSICAL ORGANIC CHEMISTRY NO-MENCLATURE

Element of chirality,

STEREOCHEMICAL TERMINOLOGY, IUPAC RECOMMENDATIONS

ELIMINAND

β-Elimination of water,

CROTONASE SUPERFAMILY

Elongation,

ACTIN ASSEMBLY KINETICS

MICROTUBULE ASSEMBLY KINETICS

ELONGATION (No Change in Polymer Number Concentration)

''EM'' algorithm,

COMPUTER ALGORITHMS & SOFTWARE

EMBRYO

EMISSION

EMISSION SPECTRUM

ACTION SPECTRUM

EXCITATION SPECTRUM

FLUORESCENCE

Empirical rate equations,

CHEMICAL KINETICS

Enantioconvergence,

STEREOCHEMICAL TERMINOLOGY, IUPAC RECOMMENDATIONS

Enantioenriched,

STEREOCHEMICAL TERMINOLOGY, IUPAC RECOMMENDATIONS

Enantiomer,

STEREOCHEMICAL TERMINOLOGY, IUPAC RECOMMENDATIONS

Enantiomer excess/Enantiomeric excess,

STEREOCHEMICAL TERMINOLOGY, IUPAC RECOMMENDATIONS

Enantiomeric ratio,

STEREOCHEMICAL TERMINOLOGY, IUPAC RECOMMENDATIONS

Enantiomerically enriched/Enantioenriched,

STEREOCHEMICAL TERMINOLOGY, IUPAC RECOMMENDATIONS

Enantiomerically pure/Enantiopure,

STEREOCHEMICAL TERMINOLOGY, IUPAC RECOMMENDATIONS

ENANTIOSELECTIVE REACTION

ASYMMETRIC INDUCTION

ENCOUNTER COMPLEX

ENCOUNTER-CONTROLLED RATE

DIFFUSION CONTROL FOR BIMOLECU-LAR COLLISIONS

ENDERGONIC PROCESS

ENDO-α (or β)-N-ACETYLGALACTOSAMI-NIDASE

ENDO-β-N-ACETYLGLUCOSAMINIDASE

ENDOENZYME

ENDO-β-GALACTOSIDASE

Endo-1,4-β-glucanase,

CELLULASE

Endo-1,4-β-mannanase,

MANNANASE

ENDONUCLEASE

DEOXYRIBONUCLEASES

RESTRICTION ENZYMES

ENDONUCLEASE III

ENDONUCLEASE IV

ENDONUCLEASE V (PYRIMIDINE DIMER)

ENDONUCLEASE VII

ENDO-OLIGOPEPTIDASE A

Endopeptidase,

NEUROLYSIN

ENDOPEPTIDASE Clp

Endopeptidase La,

PROTEASE La

ENDOPOLYGALACTURONIDASE

ENDOPOLYPHOSPHATASE

Endoskeleton and exoskeleton,

BIOMINERALIZATION

ENDOTHERMIC PROCESS

ENTHALPY

EXOTHERMIC

ENDOGONIC

ENDOERGIC

ENDOTHIAPEPSIN

Endo-1,4-β-xylanase,

XYLANASE

END PRODUCT

End-product regulation,

CUMULATIVE FEEDBACK INHIBITION

End rule,

N-END RULE

END-TO-END ANNEALING

End-wise depolymerization,

DEPOLYMERIZATION (END-WISE)

Eneamine,

ENOLIZATION

ENE REACTION

Energetics of biological macromolecules,

THERMODYNAMICS OF BIOLOGICAL MACROMOLECULES

ENERGY BARRIER

ENERGY CHARGE (or, ADENYLATE ENERGY CHARGE)

Energy level diagram,

FLUORESCENCE

ENERGY OF ACTIVATION

ARRHENIUS EQUATION

LEAST MOTION, PRINCIPLE OF

MARCUS EQUATION

MARCUS RATE THEORY

TRANSITION-STATE THEORY

Q_{10}

ENERGY TRANSFER

FÖRSTER RESONANCE EXCITATION TRANSFER (FRET)

RADIATIVE ENERGY TRANSFER

SPECTRAL OVERLAP

ENHANCER

ENOLASE

ENOLASE SUPERFAMILY

Enolate anion generated by nucleophilic aromatic addition,

CROTONASE SUPERFAMILY

Enolate anions,

ENOLIZATION

ENOLIZATION

2-Enolpyruvate-phosphate,

NUCLEOSIDE-5'-TRIPHOSPHATE REGENERATION

5-*ENOLPYRUVYLSHIKIMATE*-3-PHOSPHATE SYNTHASE

ENOYL-ACP REDUCTASE

FATTY ACID SYNTHETASE

ENOYL-CoA HYDRATASE (or, CROTONASE)

ENOYLPYRUVATE

Enrichment,

ISOTOPE ENRICHMENT

ATOM PERCENT EXCESS.

Ensemble average fluctuation with time,

CORRELATION FUNCTION

ENTATIC STATE

ENTERING GROUP

ENTEROPEPTIDASE

ENTEROTOXIN

ENTHALPY

ENTHALPY-DRIVEN REACTION

ENTHALPY OF ACTIVATION

TRANSITION-STATE THEORY (THERMODYNAMICS)

TRANSITION-STATE THEORY

GIBBS FREE ENERGY OF ACTIVATION

ENTROPY OF ACTIVATION

VOLUME OF ACTIVATION

ENTROPY

GIBBS FREE ENERGY

ENTHALPY

THERMODYNAMICS

NONEQUILIBRIUM THERMODYNAMICS: A PRIMER

ENTROPY AND ENTHALPY OF ACTIVATION (ENZYMATIC)

TRANSITION-STATE THEORY

TRANSITION-STATE THEORY (THERMODYNAMICS)

TRANSITION-STATE THEORY IN SOLUTIONS

ENTROPY OF ACTIVATION

VOLUME OF ACTIVATION

Entropy change for ligand binding,

CALORIMETRY

ENTROPY OF ACTIVATION

TRANSITION-STATE THEORY (THERMODYNAMICS)

GIBBS FREE ENERGY OF ACTIVATION

ENTHALPY OF ACTIVATION

VOLUME OF ACTIVATION

ENTROPY AND ENTHALPY OF ACTIVATION

Entropy, rate of change in,

NONEQUILIBRIUM THERMODYNAMICS: A PRIMER

ENTROPY TRAP MODEL

Envelope conformation,

STEREOCHEMICAL TERMINOLOGY, IUPAC RECOMMENDATIONS

Environmental noise,

NOISE

ENZYME

Enzyme activation,

BASAL RATE

Enzyme binding interaction with polymeric substrate,

PROCESSIVITY

ENZYME CASCADE KINETICS

Enzyme catalytic model,

ORBITAL STEERING

ENZYME CATALYSIS

ENZYME COMMISSION NOMENCLATURE

ENZYME CONCENTRATION EFFECTS

ENZYME DEPLETION (Apparent Noncompetitive Inhibition)

Enzyme electrodes,

BIOCATALYTIC ELECTRODE

ENZYME ENERGETICS (The Case of Proline Racemase)

ENZYME EQUIVALENT WEIGHT

ENZYME INHIBITION

Enzyme isomerization,

RAY-ROSCELLI PLOT

TRACER PERTURBATION METHOD

ENZYME-LINKED IMMUNOSORBENT ASSAYS (ELISA)

ENZYME MECHANISM

Enzyme purification,

ENZYME PURITY

ENZYME PURITY

ENZYME RATE EQUATIONS (1. The Basics)

FROMM'S METHOD FOR DERIVING ENZYME RATE EQUATIONS

UBIQUINOL:CYTOCHROME c OXIDORE-
DUCTASE

FERREDOXIN:NADP+ REDUCTASE

Ferric ion,
IONIC RADIUS

Ferricytochrome b_5 reduction,
NADH-CYTOCHROME b_5 REDUCTASE

FERROCHELATASE

FERROMAGNETIC

Ferromagnetic substances,
PERMEABILITY
MAGNETIC SUSCEPTIBILITY

Ferrous ion,
IONIC RADIUS

FERROXIDASE

[2Fe-2S] CLUSTER

[4Fe-4S] CLUSTER

F_1F_o ATPase and H+-TRANSPORTING
ATPase

F_1F_o ATPase mechanism,
BINDING CHANGE MECHANISM

Fibrils, pathological brain-derived,
PRION PLAQUE FORMATION

FICIN

FICK'S LAWS OF DIFFUSION
DIFFUSION
TRANSPORT PROCESSES

Fiducial group,
STEREOCHEMICAL TERMINOLOGY,
IUPAC RECOMMENDATIONS

FIELD EFFECT
INDUCTIVE EFFECT

"FIRM,"
COMPUTER ALGORITHMS & SOFTWARE

FIRST DERIVATIVE TEST

First law of thermodynamics,
CONSERVATION OF ENERGY
THERMODYNAMICS, LAWS OF

First-order kinetics,
ISOTOPE EXCHANGE AT EQUILIBRIUM

First-order process/reaction,
CHEMICAL KINETICS

First-order rate behavior,
AUTOPHOSPHORYLATION

FIRST-ORDER REACTION KINETICS
ORDER OF REACTION
HALF-LIFE
SECOND-ORDER REACTION
ZERO-ORDER REACTION
MOLECULARITY
MICHAELIS-MENTEN EQUATION

CHEMICAL KINETICS

First-order relaxation,
FLUORESCENCE

FISCHER PROJECTION

Fischer-Rosanoff Convention,
STEREOCHEMICAL TERMINOLOGY,
IUPAC RECOMMENDATIONS

"FITSIM" COMPUTER PROGRAM
KINSIM/FITSIM

FLAGELLAR ROTATION

Flagpole,
STEREOCHEMICAL TERMINOLOGY,
IUPAC RECOMMENDATIONS

FLASH PHOTOLYSIS
CHEMICAL KINETICS

Flash vacuum pyrolysis (FVP),
PHYSICAL ORGANIC CHEMISTRY
NOMENCLATURE

FLAVANONE 3β-HYDROXYLASE

FLAVINS AND DERIVATIVES

Flavokinase,
RIBOFLAVIN KINASE

FLAVONE SYNTHASE

FLAVONOL SYNTHASE

Flicker,
NOISE

Flip-flop of lipids across membrane bilayers,
PHOSPHOLIPID FLIP-FLOP

"FLOG,"
COMPUTER ALGORITHMS & SOFTWARE

Flow birefringence,
ACTIN ASSEMBLY ASSAYS

Fluctuating system,
CORRELATION FUNCTION

Fluctuation correlation,
FLUORESCENCE

FLUCTUATIONS

Fluorouracil,
THYMIDYLATE SYNTHASE

FLUORESCENCE (1. General Principles)

FLUORESCENCE (2. Time-resolved
Methods)

FLUORESCENCE (3. Resonance Energy
Transfer)

Fluorescence changes,
ACTIN ASSEMBLY ASSAYS

FLUORESCENCE CORRELATION SPEC-
TROSCOPY (FCS)

FLUORESCENCE DECAY
FLUORESCENCE

FLUORESCENCE DEPOLARIZATION

Fluorescence lifetime,

PHASE MODULATION FLUORESCENCE

FLUORESCENCE LIFETIME IMAGING
MICROSCOPY

Fluorescence phosphate assay,
ORTHOPHOSPHATE CONTINUOUS
ASSAY

FLUORESCENCE RECOVERY AFTER
PHOTOBLEACHING

Fluorimeter,
FLUORESCENCE

FLUOROACETATE/FLUOROCITRATE

Fluorocitrate
FLUOROACETATE/FLUOROCITRATE

Fluorometric assay for avidin and biotin,
BIOTIN AND DERIVATIVES

Fluorophore,
FLUORESCENCE

"FLUSYS,"
COMPUTER ALGORITHMS & SOFTWARE

FLUX
NONEQUILIBRIUM THERMODYNAMICS:
A PRIMER
RADIANT ENERGY FLUX
RADIANT POWER

FLUX CONTROL IN METABOLIC
PATHWAYS

FLUX-GENERATING STEPS

Fluxional,
PHYSICAL ORGANIC CHEMISTRY
NOMENCLATURE

FLUX RATIO

Flux through a pathway
METABOLIC CONTROL ANALYSIS
METABOLONS
META-MODEL

FMN,
L-LACTATE DEHYDROGENASE
(CYTOCHROME)
LACTATE MONOOXYGENASE
LUCIFERASE
4-METHOXYBENZOATE MONOOXY-
GENASE
RIBOFLAVIN KINASE

FMN ADENYLYLTRANSFERASE

FMN-dependent enzymes,
CYTOCHROME P-450 REDUCTASE
DIHYDROOROTATE OXIDASE
FMN ADENYLYLTRANSFERASE
GLUTAMATE SYNTHASE
GLYCOLATE OXIDASE
(S)-2-HYDROXY-ACID OXIDASE
L-LACTATE DEHYDROGENASE
(CYTOCHROME)

FUMARATE REDUCTASE

FURA-2

   CALCIUM ION INDICATOR DYES

   METAL ION COMPLEXATION

FURIN

G

ΔG of ATP hydrolysis,

   ATP, THERMODYNAMICS OF HYDRO-LYSIS

$\Delta G°$ (Relation to K_{eq})

   ACTIVITY COEFFICIENTS

   ADDITIVITY PRINCIPLE

   BIOCHEMICAL THERMODYNAMICS

   CHEMICAL POTENTIAL

   EQUILIBRIUM CONSTANTS

   HESS'S LAW

   INNATE THERMODYNAMIC QUANTITIES

   MOLECULAR CROWDING

   THERMODYNAMICS, LAWS OF

   THERMODYNAMIC CYCLE

   THERMODYNAMIC EQUATIONS OF STATE

G-actin,

   ACTIN ASSEMBLY KINETICS

$\Delta\Delta G$ AS AN INDEX OF COOPERATIVITY

GALACTARATE DEHYDRATASE

GALACTITOL 2-DEHYDROGENASE

Galactocerebrosidase,

   GALACTOSYLCERAMINIDASE

GALACTOKINASE

GALACTONATE DEHYDRATASE

GALACTOSE 1-DEHYDROGENASE

GALACTOSE OXIDASE

GALACTOSE-1-PHOSPHATE URIDYLYL-TRANSFERASE

Galactose-6-sulfate sulfatase,

   N-ACETYLGALACTOSAMINE-6-SULFATE SULFATASE

α-GALACTOSIDASE

β-GALACTOSIDASE

GALACTOSIDE FUCOSYLTRANSFERASES

GALACTOSIDE SIALYLTRANSFERASES

β-D-Galactosyl-1,4-β-D-glucosylceramide,

   SIALYLTRANSFERASE

"GAMESS,"

   COMPUTER ALGORITHMS & SOFTWARE

GAMMA (γ)

Gas-phase acidity,

   PHYSICAL ORGANIC CHEMISTRY NO-MENCLATURE

Gas-phase basicity,

   PHYSICAL ORGANIC CHEMISTRY NO-MENCLATURE

Gauche,

   STEREOCHEMICAL TERMINOLOGY, IUPAC RECOMMENDATIONS

Gauche effect,

   STEREOCHEMICAL TERMINOLOGY, IUPAC RECOMMENDATIONS

GAUSS

"GAUSSIAN",

   COMPUTER ALGORITHMS & SOFTWARE

GDP-Fucose synthase,

   FUCOSE-1-PHOSPHATE GUANYLYL-TRANSFERASE

GDP-MANNOSE 4,6-DEHYDRATASE

GDP-MANNOSE DEHYDROGENASE

GDP-MANNOSE:DOLICHYL PHOSPHATE MANNOSYLTRANSFERASE

GDP-MANNOSYLTRANSFERASES

GEAR ALGORITHM

   STIFFNESS OR STIFFNESS INSTABILITY

Gel-liquid crystalline phase transition,

   CALORIMETRY

GELSOLIN

Geminate recombination,

   PHYSICAL ORGANIC CHEMISTRY NO-MENCLATURE

GENERAL ACID CATALYSIS

   SPECIFIC ACID CATALYSIS

   CATALYSIS

   GENERAL BASE CATALYSIS

GENERAL BASE CATALYSIS

   SPECIFIC BASE CATALYSIS

   CATALYSIS

   GENERAL ACID CATALYSIS

GENETIC CODE

GENTAMICIN 2"-NUCLEOTIDYLTRANS-FERASE

GENTISATE 1,2-DIOXYGENASE

Geometric constraints on diffusion,

   FRACTAL REACTION KINETICS

Geometric inhibitor,

   D-ALANINE-D-ALANINE LIGASE

Geometric isomerism,

   STEREOCHEMICAL TERMINOLOGY, IUPAC RECOMMENDATIONS

GEOMETRIC MEAN, RULE OF THE

GERANOYL-CoA CARBOXYLASE

Geranyl-diphosphate cyclase,

   BORNYL PYROPHOSPHATE SYNTHASE

Geranyl diphosphate synthase,

   DIMETHYLALLYLTRANSTRANSFERASE

Geranylgeranyl diphosphate synthase,

   FARNESYLTRANSTRANSFERASE

Geranylgeranyltransferases,

   PROTEIN GERANYLGERANYLTRANS-FERASES

GERANYLTRANSTRANSFERASE

GIBBERELLIN 2β-HYDROXYLASE

GIBBERELLIN 3β-HYDROXYLASE

GIBBS FREE ENERGY

   BIOCHEMICAL THERMODYNAMICS

   ENDERGONIC

   EXERGONIC

   ENTHALPY

   ENTROPY

   HELMHOLTZ ENERGY

   THERMODYNAMICS

GIBBS FREE ENERGY DIAGRAM

   POTENTIAL ENERGY PROFILE

   ENZYME ENERGETICS

GIBBS FREE ENERGY OF ACTIVATION

   TRANSITION-STATE THEORY (THERMO-DYNAMICS)

GIBBS-HELMHOLTZ EQUATION

   GIBBS FREE ENERGY

GLOBAL ANALYSIS

1,4-α-Glucan branching enzyme,

   BRANCHING ENZYME

GLUCAN ENDO-1,3(4)-β-GLUCANASE

GLUCAN ENDO-1,2-β-GLUCOSIDASE

GLUCAN ENDO-1,3-β-GLUCOSIDASE

GLUCAN ENDO-1,6-β-GLUCOSIDASE

GLUCAN 1,3-β-GLUCOSIDASE

1,4-α-Glucanotransferase,

   DEBRANCHING ENZYME

1,4-α-Glucan:orthophosphate α-glucosyl-transferase,

   GLYCOGEN PHOSPHORYLASE

1,3-β-GLUCAN SYNTHASE

GLUCARATE DEHYDRATASE

D-Glucitol,

   SORBITOL DEHYDROGENASE

GLUCOAMYLASE

GLUCOCEREBROSIDASE

GLUCOKINASE

GLUCONATE DEHYDRATASE

Gluconeogenesis tracer kinetics,

   GLUCOSE TRACER KINETICS

GLUCOSAMINATE AMMONIA-LYASE

GLUCOSAMINE *N*-ACETYLTRANSFERASE

GLUCOSAMINE-6-PHOSPHATE ISOMERASE

GLUCOSAMINE-6-PHOSPHATE SYNTHASE

Glucose as a substrate or product,

   GLUCOKINASE

   HEXOKINASE

   LACTASE

   LACTOSE SYNTHASE

   LEVANSUCRASE

   MALTASE

   MALTOSE PHOSPHORYLASE

   MAPPING SUBSTRATE INTERACTIONS USING KINETIC DATA

   XYLOSE ISOMERASE

GLUCOSE OXIDASE

GLUCOSE-6-PHOSPHATASE

   ISOTOPE EXCHANGE (Reactions Away From Equilibrium)

Glucose-1-phosphate adenylyltransferase,

   ATP:GLUCOSE-1-PHOSPHATE ADENYL-YLTRANSFERASE

GLUCOSE-6-PHOSPHATE DEHYDRO-GENASE

α-D-Glucose 1-phosphate, interconversion with α-D-glucose 6-phosphate,

   PHOSPHOGLUCOMUTASE

D-Glucose 6-phosphate, interconversion with D-fructose 6-phosphate,

   PHOSPHOGLUCOSE ISOMERASE

Glucose-6-phosphate isomerase,

   PHOSPHOGLUCOISOMERASE

D-Glucose 6-phosphate synthesis,

   6-PHOSPHO-β-GLUCOSIDASE

GLUCOSE-1-PHOSPHATE URIDYLYLTRANS-FERASE

GLUCOSE TRACER KINETICS

Glucosylceramide β-D-glucosidase,

   GLUCOCEREBROSIDASE

GLUCURONATE-1-PHOSPHATE URIDYLYL-TRANSFERASE

GLUCURONATE-2-SULFATE SULFATASE

β-GLUCURONIDASE

GLUCURONOKINASE

GLUCURONOLACTONE REDUCTASE

GLUCURONOSYLTRANSFERASES

GLUTACONYL-CoA DECARBOXYLASE

Glutamate acetyltransferase,

   N-ACETYLGLUTAMATE SYNTHASE

   GLUTAMATE N-ACETYLTRANSFERASE

GLUTAMATE *N*-ACETYLTRANSFERASE

   N-ACETYLGLUTAMATE SYNTHASE

D-GLUTAMATE:D-AMINO ACID AMINO-TRANSFERASE

Glutamate-amino acid transaminases,

   See specific aminotransferase

Glutamate as a substrate or product,

   GLUTAMATE DECARBOXYLASE

   GLUTAMATE DEHYDROGENASE

   GLUTAMATE SYNTHASE

   GLUTAMINE SYNTHETASE

   γ-GLUTAMYL TRANSPEPTIDASE

   LEUCINE AMINOTRANSFERASE

   LYSINE 6-AMINOTRANSFERASE

   MAPPING SUBSTRATE INTERACTIONS USING KINETIC DATA

   β-METHYLASPARTATE-GLUTAMATE MUTASE

   N-METHYLGLUTAMATE SYNTHASE

   5-OXOPROLINASE

   ZWITTERION

GLUTAMATE DECARBOXYLASE

GLUTAMATE DEHYDROGENASE

GLUTAMATE FORMIMINOTRANSFERASE

GLUTAMATE FORMYLTRANSFERASE

Glutamate mutase,

   β-METHYLASPARTATE-GLUTAMATE MUTASE

GLUTAMATE RACEMASE

L-Glutamate 5-semialdehyde,

   ORNITHINE AMINOTRANSFERASE

GLUTAMATE γ-SEMIALDEHYDE DEHYDRO-GENASE

GLUTAMATE SYNTHASE

GLUTAMINASE

Glutamine amidotransferases,

   See specific amidotransferase

GLUTAMINE AMINOTRANSFERASE

Glutamine:PRPP amidotransferase,

   AMIDOPHOSPHORIBOSYLTRANSFERASE

GLUTAMINE SYNTHETASE

   ARSENOLYSIS

   BRIDGE-TO-NONBRIDGE OXYGEN SCRAMBLING

   ISOTOPE EXCHANGE AT EQUILIBRIUM

   WEDLER-BOYER TECHNIQUE

   CUMULATIVE INHIBITION

   UNCONSUMED SUBSTRATE

   CRYPTIC CATALYSIS

   BOROHYDRIDE REDUCTION

   ENZYME CASCADE KINETICS

GLUTAMINE SYNTHETASE ADENYLYL-TRANSFERASE

Glutamine synthetase reaction,

   ISOTOPE TRAPPING

Glutamine synthetase reaction intermediate,

   BOROHYDRIDE REDUCTION

Glutamine synthetase UTase/UR enzyme re-actions,

   BIFUNCTIONAL ENZYME

GLUTAMINYL-tRNA SYNTHETASE

   AMINOACYL-tRNA SYNTHETASES

γ-Glutamyl carboxylase,

   VITAMIN K-DEPENDENT γ-GLUTAMYL CARBOXYLASE

γ-GLUTAMYL CYCLOTRANSFERASE

γ-GLUTAMYLCYSTEINE SYNTHETASE

GLUTAMYL ENDOPEPTIDASES

γ-GLUTAMYL KINASE

γ-Glutamyl-phosphate,

   ISOTOPE TRAPPING

   BOROHYDRIDE REDUCTION

γ-Glutamyl phosphate reductase,

   GLUTAMATE γ-SEMIALDEHYDE DEHY-DROGENASE

γ-GLUTAMYL TRANSPEPTIDASE

GLUTAMYL-tRNA^{Gln} AMIDOTRANSFERASE

GLUTAMYL-tRNA SYNTHETASE

GLUTATHIONE

Glutathione as a substrate, product, or cofactor,

   γ-GLUTAMYLCYSTEINE SYNTHETASE

   γ-GLUTAMYL TRANSPEPTIDASE

   GLUTATHIONE PEROXIDASE

   GLUTATHIONE REDUCTASE

   GLUTATHIONE SYNTHETASE

   GLUTATHIONE S-TRANSFERASE

   LEUKOTRIENE C_4 SYNTHASE

GLUTATHIONE DEHYDROGENASE (ASCORBATE)

Glutathione disulfide (GSSG),

   γ-GLUTAMYL TRANSPEPTIDASE

   GLUTATHIONE PEROXIDASE

   GLUTATHIONE REDUCTASE

   XANTHINE DEHYDROGENASE

GLUTATHIONE PEROXIDASE

GLUTATHIONE REDUCTASE

GLUTATHIONE SYNTHETASE

GLUTATHIONE *S*-TRANSFERASE

Glutathionylspermidine synthetase/amidase,

   BIFUNCTIONAL ENZYME

GLYCERALDEHYDE-3-PHOSPHATE DEHY-DROGENASE

GLYCERATE DEHYDROGENASE

GLYCERATE KINASE

GLYCEROL

GLYCEROL DEHYDRATASE

GLYCEROL KINASE

GLYCEROL-3-PHOSPHATE O-ACYLTRANS-
FERASE

GLYCEROL-3-PHOSPHATE DEHYDRO-
GENASE

GLYCERONE KINASE

Glycerone phosphate,

*RHAMNULOSE-1-PHOSPHATE
ALDOLASE*

Glycinamide ribonucleotide transformylase,

*PHOSPHORIBOSYLGLYCINAMIDE
FORMYLTRANSFERASE*

Glycine,

GLYCINE SYNTHASE

SERINE HYDROXYMETHYLTRANSFERASE

GLYCINE ACETYLTRANSFERASE

GLYCINE ACYLTRANSFERASES

GLYCINE AMIDINOTRANSFERASE

GLYCINE AMINOTRANSFERASE

GLYCINE DECARBOXYLASE

GLYCINE FORMIMINOTRANSFERASE

GLYCINE ACETYLTRANSFERASE

GLYCINE ACYLTRANSFERASES

GLYCINE AMIDINOTRANSFERASE

GLYCINE AMINOTRANSFERASE

Glycine cleavage complex,

GLYCINE SYNTHASE

GLYCINE DECARBOXYLASE

GLYCINE FORMIMINOTRANSFERASE

[^{15}N]Glycine kinetics,

PROTEIN TURNOVER KINETICS

GLYCINE METHYLTRANSFERASE

GLYCINE:OXALOACETATE AMINOTRANS-
FERASE

GLYCINE REDUCTASE

GLYCINE SYNTHASE

Glycogen debranching enzyme,

DEBRANCHING ENZYME

Glycogenolysis tracer kinetics,

GLUCOSE TRACER KINETICS

GLYCOGEN PHOSPHORYLASE

GLYCOGEN SYNTHASE

GLYCOLATE OXIDASE

(S)-2-HYDROXY-ACID OXIDASE

Glycolysis,

MEYERHOF OXIDATION QUOTIENT

See also specific enzyme of glycolysis

GLYCOLYTIC OSCILLATION

GLYCOSIDASES,

See specific enzymes

Glycosyltransferases

See specific enzymes

GLYCYL ENDOPEPTIDASE

GLYCYL-tRNA SYNTHETASE

GLYOXALASE I

GLYOXALASE II

GLYOXYLATE AMINOTRANSFERASE

GLYOXYLATE REDUCTASE

GMP SYNTHETASE

GOLDMAN EQUATION

GOODNESS-OF-FIT CRITERIA

GLOBAL ANALYSIS

STATISTICS (A Primer)

GOOD'S BUFFERS

Gouy-Chapman layer,

BIOMINERALIZATION

G PROTEINS

''GRAFIT,''

COMPUTER ALGORITHMS & SOFTWARE

Gram atom excess

ATOM PERCENT EXCESS

OXYGEN ISOTOPE EXCHANGE

COMPARTMENTAL ANALYSIS

ISOTOPE EXCHANGE KINETICS

GRAPHICAL METHODS

DALZIEL PHI RELATIONSHIPS

DIRECT LINEAR PLOT

DIXON PLOT

DIXON-WEBB PLOT

DOUBLE-RECIPROCAL PLOT

EADIE-HOFSTEE PLOT

FRIEDEN PROTOCOL

FROMM PROTOCOL

HANES PLOT

HILL PLOT

INDIVIDUAL REACTION MECHANISMS

LINEWEAVER-BURKE PLOT

POINT-OF-CONVERGENCE METHOD

SCATCHARD PLOT

SUBSTRATE CONCENTRATION RANGE

GRAPH THEORY

*FROMM'S METHOD FOR DERIVING
RATE EQUATIONS*

Gray,

RAD

GREEN FLUORESCENT PROTEIN

''GROMOS,''

COMPUTER ALGORITHMS & SOFTWARE

GROTTHUSS CHAINS

GROTTHUSS-DRAPER LAW

GROTTHUSS MECHANISM

GROUND STATE

EXCITED STATE

FLUORESCENCE

GROUP TRANSFER REACTIONS

''GROW,''

COMPUTER ALGORITHMS & SOFTWARE

Growth processes,

AUTOCATALYSIS

GTPases,

See specific enzyme

GTP CYCLOHYDROLASE I

GTP CYCLOHYDROLASE II

GTP-dependent enzymes,

ADENYLOSUCCINATE SYNTHETASE

DEOXYTHYMIDINE KINASE

*FUCOSE-1-PHOSPHATE GUANYLYL-
TRANSFERASE*

*GENTAMICIN 2''-NUCLEOTIDYLTRANS-
FERASE*

GLYCEROL KINASE

G PROTEINS

GTP CYCLOHYDROLASES

GUANYLATE CYCLASE

*HEXOSE-1-PHOSPHATE GUANYLYL-
TRANSFERASE*

5-HYDROXYMETHYLPYRIMIDINE KINASE

*MANNOSE-1-PHOSPHATE GUANYLYL-
TRANSFERASE*

MEVALONATE KINASE

MICROTUBULE ASSEMBLY KINETICS

*PHOSPHOENOLPYRUVATE CARBOXY-
KINASE*

SUCCINYL-CoA SYNTHETASE

TUBULIN-GTP HYDROLYSIS

VESICLE TRANSPORT IN CELLS

GTP depletion,

ATP & GTP DEPLETION

GTP hydrolysis kinetics in tubulin polymerization,

MICROTUBULE ASSEMBLY KINETICS

GTP regeneration,

*NUCLEOSIDE 5'-TRIPHOSPHATE REGEN-
ERATION*

GUANINE DEAMINASE

GUANOSINE DEAMINASE

GUANYLATE CYCLASE

GUANYLATE KINASE

GUEST

HOST MOLECULE

INCLUSION COMPLEXES

GUGGENHEIM METHOD (FIRST-ORDER
RATE CONSTANT)

H

Homochiral,
STEREOCHEMICAL TERMINOLOGY, IUPAC RECOMMENDATIONS
HOMOCITRATE SYNTHASE
Homoconjugation,
PHYSICAL ORGANIC CHEMISTRY NOMENCLATURE
HOMOCYSTEINE DESULFHYDRASE
Homocysteine methylases,
N^5-METHYLTETRAHYDROFOLATE: HOMOCYSTEINE METHYLTRANS-FERASE TETRAHYDROPTEROYL-TRIGLUTAMATE METHYLTRANS-FERASE
HOMOCYSTEINE S-METHYLTRANSFERASE
HOMOGENEOUS CATALYSIS
HOMOGENEOUS REACTION
HOMOGENTISATE 1,2-DIOXYGENASE
HOMOISOCITRATE DEHYDROGENASE
Homoleptic,
PHYSICAL ORGANIC CHEMISTRY NO-MENCLATURE
Homomorphic,
STEREOCHEMICAL TERMINOLOGY, IUPAC RECOMMENDATIONS
HOMOPHILIC INTERACTION
HOMOSERINE O-ACETYLTRANSFERASE
Homoserine dehydratase,
CYSTATHIONINE γ-LYASE
HOMOSERINE DEHYDROGENASE
ASPARTATE KINASE
HOMOSERINE SUCCINYLTRANSFERASE
Homotopic,
STEREOCHEMICAL TERMINOLOGY, IUPAC RECOMMENDATIONS
HOMOTROPIC INTERACTION
MONOD-WYMAN-CHANGEUX MODEL
ALLOSTERISM
HEMOGLOBIN
"HONDO,"
COMPUTER ALGORITHMS & SOFTWARE
HOOKE'S LAW SPRING
Hormone,
AGONIST
HORN-BÖRNIG PLOT
HORSERADISH PEROXIDASE
HOST-GUEST INTERACTIONS
HOST MOLECULE
CROWN ETHERS
MACROCYCLES
INCLUSION COMPLEXES
HÜCKEL MOLECULAR-ORBITAL CALCULA-TIONS

HARTREE-FOCK MOLECULAR-ORBITAL CALCULATIONS
AB INITIO MOLECULAR ORBITAL CALCU-LATIONS.
HÜCKEL'S RULE
AROMATICITY
HUGHES-INGOLD THEORY FOR SOLVENT EFFECTS ON REACTIVITY
HUMMEL-DREYER TECHNIQUE
HYALURONATE LYASE
HYALURONIDASES
HYBRIDIZATION
HYDRATION ATMOSPHERE
HYDRATION NUMBER
Hydrazone reduction,
BOROHYDRIDE REDUCTION
HYDRIDE
HYDRIDE ION TRANSFER
Hydride transfer,
HYDROGEN TUNNELING
HYDROCARBON MONOOXYGENASE
HYDROGENASE
Hydrogen bonding,
WATER
Hydrogen-bonding interactions in DNA and RNA,
BASE PAIRING
NUCLEIC ACID RENATURATION KINETICS
HYDROGEN BONDS
HYDROGEN DEHYDROGENASE
HYDROGEN EQUIVALENT
HYDROGEN EXCHANGE IN PROTEINS
Hydrogen migration,
NIH SHIFT
Hydrogen peroxide,
OXYGEN, OXIDES & OXO ANIONS
CATALASE
HORSERADISH PEROXIDASE
LACTOPEROXIDASE
LIGNAN PEROXIDASE
LYSYL OXIDASE
MANGANESE PEROXIDASE
MYELOPEROXIDASE
OVOPEROXIDASE
PEROXIDASE
PYRUVATE OXIDASE
XANTHINE OXIDASE
Hydrogen selenide,
L-SELENOCYSTEINE β-LYASE
HYDROGEN SULFIDE:FERRIC ION OXIDO-REDUCTASE

Hydrogen transfer,
HYDROGEN TUNNELING
KINETIC ISOTOPE EFFECTS
HYDROGEN TUNNELING
HYDROLASES
ENZYME COMMISSION NOMEN-CLATURE
HYDROLYSIS
Hydrolysis pH profile of acyl-phosphate,
ACYL-PHOSPHATE INTERMEDIATE
Hydrolysis of NAD^+,
NAD^+ GLYCOHYDROLASE
Hydrolysis of nucleoside diphosphate,
NUCLEOSIDE DIPHOSPHATASE
Hydrolysis of 3'-ribonucleotide,
3'-NUCLEOTIDASE
Hydrolysis of 5'-ribonucleotide,
5'-NUCLEOTIDASE
Hydrolysis of ribo- or deoxyribonucleic acids,
NUCLEASE S1
HYDRON
Hydronium ion,
WATER
pK_a
pK_w
ACID-BASE RELATIONSHIPS
OXYGEN, OXIDES & OXO ANIONS
Hydroperoxy radical,
OXYGEN, OXIDES & OXO ANIONS
HYDROPHILIC
HYDROPHOBIC EFFECT
Hydroxide anion,
WATER
pK_a
pK_w
ACID-BASE RELATIONSHIPS
OXYGEN, OXIDES & OXO ANIONS
Hydroxide radical,
OXYGEN, OXIDES & OXO ANIONS
Hydroxonium ion,
WATER
pK_a
pK_w
ACID-BASE RELATIONSHIPS
OXYGEN, OXIDES & OXO ANIONS
D-2-HYDROXY-ACID DEHYDROGENASE
(S)-2-HYDROXY-ACID OXIDASE
3-HYDROXYACYL-CoA DEHYDROGENASE
β-HYDROXYACYL THIOESTER (or, ACP) DE-HYDRASE
FATTY ACID SYNTHETASE

L-Hydroxyamino acid dehydratase,
   SERINE DEHYDRATASE
3-HYDROXYANTHRANILATE 3,4-DIOXY-GENASE
3-HYDROXYANTHRANILATE OXIDASE
3-HYDROXYASPARTATE ALDOLASE
erythro-β-HYDROXYASPARTATE DEHY-DRATASE
4-HYDROXYBENZOATE DECARBOXYLASE
2-Hydroxybenzoate monooxygenase,
   SALICYLATE 1-MONOOXYGENASE
3-HYDROXYBENZOATE 4-MONOOXY-GENASE
3-HYDROXYBENZOATE 6-MONOOXY-GENASE
4-HYDROXYBENZOATE 3-MONOOXY-GENASE
D-3-HYDROXYBUTYRATE DEHYDRO-GENASE
4-HYDROXYBUTYRATE DEHYDROGENASE
4-HYDROXYBUTYRYL-CoA DEHYDRATASE
3-HYDROXYBUTYRYL-CoA EPIMERASE
Hydroxycinnamoyl-CoA ligase,
   4-COUMAROYL-CoA SYNTHETASE
β-Hydroxydecanoyl thioester dehydrase,
   β-HYDROXYACYL THIOESTER DEHY-DRASE
4-HYDROXY-2-KETOGLUTARATE AL-DOLASE
Hydroxylamine,
   ACYL-PHOSPHATE INTERMEDIATE
HYDROXYLAMINE REDUCTASE
Hydroxylation of aromatic rings,
   NIH SHIFT
4-HYDROXYMANDELONITRILE LYASE
β-HYDROXY-β-METHYLGLUTARYL-CoA LYASE
β-HYDROXY-β-METHYLGLUTARYL-CoA REDUCTASE
β-HYDROXY-β-METHYLGLUTARYL-CoA REDUCTASE (NADPH) KINASE
β-HYDROXY-β-METHYLGLUTARYL-CoA SYNTHASE
4-Hydroxy-4-methyl-2-oxoglutarate aldolase,
   γ-METHYL-γ-HYDROXY-α-KETOGLUTARATE ALDOLASE
3-Hydroxy-2-methylpyridinecarboxylate di-oxygenase,
   METHYLHYDROXYPYRIDINECARBOXY-LATE DIOXYGENASE
3-HYDROXY-2-METHYLPYRIDINE-4,5-DICARBOXYLATE 4-DECARBOXYLASE
5-HYDROXYMETHYLPYRIMIDINE KINASE
6-Hydroxynicotinate,
   NICOTINATE DEHYDROGENASE

6-HYDROXYNICOTINATE REDUCTASE
2-HYDROXY-3-OXOPROPIONATE RE-DUCTASE
4-HYDROXYPHENYLACETATE 3-MONO-OXYGENASE
4-HYDROXYPHENYLPYRUVATE HYDROX-YLASE
4-Hydroxyphenylpyruvate, synthesis of,
   PREPHENATE DEHYDROGENASE
4-HYDROXYPROLINE EPIMERASE
trans-3-Hydroxyprolyl residue in procollagen,
   PROLYL 3-HYDROXYLASE
trans-4-Hydroxyprolyl residue, formation in procollagen,
   PROLYL 4-HYDROXYLASE
β-HYDROXYPROPIONATE DEHYDRO-GENASE
15-HYDROXYPROSTAGLANDIN DEHYDRO-GENASE
3-Hydroxypyruvate,
   SERINE DEHYDROGENASE
HYDROXYPYRUVATE REDUCTASE
3α-HYDROXYSTEROID DEHYDROGENASE
3(or 17)β-HYDROXYSTEROID DEHYDRO-GENASE
11β-HYDROXYSTEROID DEHYDROGENASE
16α-HYDROXYSTEROID DEHYDROGENASE
20-HYDROXYSTEROID DEHYDROGENASE
21-HYDROXYSTEROID DEHYDROGENASE
HYPERCONJUGATION
HYPERCYCLE
HYPERPOLARIZATION
HYPERREACTIVITY
HYPERTONIC
HYPOTAUROCYAMINE KINASE
HYPOTHESIS
HYPOTHESIS TESTING
Hypothetical potential energy surface for a reacting system,
   MORE O'FERRALL-JENCKS DIAGRAM
HYPOXANTHINE (GUANINE) PHOSPHO-RIBOSYLTRANSFERASE
HYPSOCHROMIC SHIFT
HYSTERESIS
   BIFURCATION THEORY
   COOPERATIVITY
   LAG TIME

I

I_{50} (or $I_{0.5}$)
ICEBERG EFFECT

Icosahedral virus assembly,
   QUASI-EQUIVALENCE
IcsA,
   ACTIN-BASED BACTERIAL MOTILITY
Identity reaction,
   PHYSICAL ORGANIC CHEMISTRY NO-MENCLATURE
L-Iditol,
   SORBITOL DEHYDROGENASE
Iditol dehydrogenase,
   SORBITOL DEHYDROGENASE
IDURONATE-2-SULFATE SULFATASE
α-L-IDURONIDASE
IMBALANCE
   SYNCHRONOUS
   SYNCHRONIZATION
IMIDAZOLEACETATE HYDROXYLASE
IMIDAZOLE GLYCEROL PHOSPHATE DEHY-DRATASE
IMIDAZOLONEPROPIONASE
Imidodipeptidase,
   PROLINE DIPEPTIDASE
IMINE INTERMEDIATE
Imine reduction,
   BOROHYDRIDE REDUCTION
IMMOBILIZED ENZYMES
IMP CYCLOHYDROLASE
IMP DEHYDROGENASE
Imperfect synchronization,
   SYNCHRONIZATION
Impurities, presence of,
   SUBSTRATE PURITY
   ENZYME PURITY
Inclusion compound (or, inclusion complex),
   PHYSICAL ORGANIC CHEMISTRY NO-MENCLATURE
INDEFINITE POLYMERIZATION
   NUCLEATION
   PROTEIN POLYMERIZATION
   MICROTUBULE ASSEMBLY KINETICS
   ACTIN ASSEMBLY KINETICS
Index of refraction,
   REFRACTIVE INDEX
INDOLEAMINE 2,3-DIOXYGENASE
INDOLEGLYCEROL PHOSPHATE SYN-THASE
INDUCED FIT MODEL
   COOPERATIVITY
   KOSHLAND-NEMETHY-FILMER MODEL
INDUCED SUBSTRATE INHIBITION
INDUCED TRANSPORT ANALYSIS
INDUCTION

INDUCTION PERIOD
SERIES REACTIONS
INDUCTIVE EFFECT
FIELD EFFECT
INDUCTOMERIC EFFECT
INFLECTION, POINT OF
INHIBITION
INHIBITOR
INHIBITOR
DEGREE OF INHIBITION
EFFECTOR
MODIFIER
INACTIVATOR
COMPETITIVE INHIBITOR
NONCOMPETITIVE INHIBITOR
UNCOMPETITIVE INHIBITOR
PARTIAL INHIBITOR
IRREVERSIBLE INHIBITOR
INHIBITORY INDEX
Initial rate assumption,
CHEMICAL KINETICS
INITIAL RATE CONDITION
ATP/GTP REGENERATION
INITIAL RATE ENZYME DATA, REPORTING OF
INITIAL RATE PHASE
Initial reaction phase,
ENZYME KINETIC EQUATIONS (1. The Basics)
Initial value problem, solutions to,
NUMERICAL INTEGRATION
GEAR ALGORITHM
NUMERICAL COMPUTER METHODS
INNATE THERMODYNAMIC QUANTITIES
INNER COORDINATION SPHERE
INNER FILTER EFFECT
FLUORESCENCE
Inner-sphere coordination,
EIGEN-TAMM MECHANISM
INNER-SPHERE ELECTRON TRANSFER
OUTER SPHERE ELECTRON TRANSFER
METAL ION COORDINATION (Basic Chemical Behavior)
Inorganic pyrophosphatase,
PYROPHOSPHATASE, INORGANIC
INOSAMINE-PHOSPHATE AMIDINO-TRANSFERASE
INOSINE KINASE
INOSINE NUCLEOSIDASE
Inosine phosphorylase,
PURINE NUCLEOSIDE PHOSPHORYLASE

myo-INOSITOL 2-DEHYDROGENASE
myo-INOSITOL 1-KINASE
myo-INOSITOL 1-MONOPHOSPHATASE
myo-INOSITOL OXYGENASE
myo-INOSITOL-1-PHOSPHATE SYNTHASE
INOSITOL POLYPHOSPHATE 1-PHOSPHATASE
INOSITOL POLYPHOSPHATE 3-PHOSPHATASE
INOSITOL POLYPHOSPHATE 4-PHOSPHATASE
INOSITOL POLYPHOSPHATE 5-PHOSPHATASE
INOSITOL-1,3,4,6-TETRAPHOSPHATE 5-KINASE
INOSITOL-3,4,5,6-TETRAPHOSPHATE 1-KINASE
INOSITOL-1,3,4-TRISPHOSPHATE 6-KINASE
INOSITOL-1,4,5-TRISPHOSPHATE 3-KINASE
In-out isomerism,
STEREOCHEMICAL TERMINOLOGY, IUPAC RECOMMENDATIONS
Input power fluctuations,
NOISE
"INSIGHT/DISCOVER" Program,
COMPUTER ALGORITHMS & SOFTWARE
INSTANTANEOUS PROCESS
INSULYSIN
Integrated rate equations,
NEWTON-RAPHSON METHOD
Integrated rate expression for autocatalysis,
AUTOCATALYSIS
Intensive variable,
ISOTOPE DILUTION
INTENSITY
INTERACTIVE
INTERCEPT
Interconversion of enantiomers,
RACEMASES
Interconvertible enzymes,
ENZYME CASCADE KINETICS
Interelectronic interactions,
SHIELDING
INTERFACE ACTIVATION IN LIPOLYSIS
MICELLAR CATALYSIS
Interfaces as transition zones,
BIOMINERALIZATION
INTERMEDIACY
INTERMEDIATE
Intermediate detection,
CRYOENZYMOLOGY
Intermediate, lifetime of,

PERSISTENT
TRANSIENT CHEMICAL SPECIES
INTERMEDIATES, COVALENT ENZYME-SUBSTRATE
Intermediates in enzyme-catalyzed reactions,
ENOLIZATION
CRYOENZYMOLOGY
INTERMOLECULAR
INTRAMOLECULAR
Intermolecular effectors,
AUTOINHIBITION
INTERMOLECULAR ISOTOPE EFFECT
INTRAMOLECULAR ISOTOPE EFFECT
ISOTOPE EFFECT
EQUILIBRIUM ISOTOPE EFFECT
Internal compensation,
STEREOCHEMICAL TERMINOLOGY, IUPAC RECOMMENDATIONS
INTERNAL EFFECTORS
INTERNAL ENERGY
INTERNAL ENTROPY
INTERNAL EQUILIBRIA
INTERNAL PRESSURE
INTERNET (SOME USEFUL ADDRESSES)
INTERPOLATION
INTERSYSTEM CROSSING
INTERNAL CONVERSION
FLUORESCENCE
Intersystem transfer,
JABLONSKI DIAGRAM
Intimate ion pair,
ION PAIR
TIGHT ION PAIR
Intracellular behavior of regulatory enzymes,
ENERGY CHARGE
INTRAMOLECULAR CATALYSIS
Intramolecular hydrogen migration,
NIH SHIFT
INTRAMOLECULAR ISOTOPE EFFECT
INTERMOLECULAR ISOTOPE EFFECT
KINETIC ISOTOPE EFFECT
EQUILIBRIUM ISOTOPE EFFECT
INTRAMOLECULAR KINETIC ISOTOPE EFFECT
INTRINSIC BARRIER
MARCUS EQUATION
REORGANIZATION ENERGY
INTRINSIC BINDING ENERGY
Intrinsic fluorescence,
FLUORESCENCE

Intrinsic isotope effects on enzyme-catalyzed reactions,
NORTHROP ISOTOPE EFFECT METHOD
INTRINSIC KINETIC ISOTOPE EFFECT
KINETIC ISOTOPE EFFECT
Intrinsic viscosity,
VISCOSITY
INULINASE
INULOSUCRASE
INVERSE ISOTOPE EFFECT
KINETIC ISOTOPE EFFECT
Inversion of configuration,
NUCLEOPHILIC DISPLACEMENT RE-ACTION
IN-LINE MECHANISM
STEREOCHEMISTRY
WALDEN INVERSION
INVERTASE
SUCRASE
INVERTED/INVERSE MICELLE
MICELLE
IODIDE PEROXIDASE
IODINE-125 (^{125}I)
IODINE-131 (^{131}I)
IODOACETATE
ION ATMOSPHERE
ION CURRENT
ION CYCLOTRON RESONANCE MASS SPECTROMETRY
IONIC BOND
IONIC RADIUS
IONIC STRENGTH
ISOTONIC BUFFERS
DAVIES EQUATION
DEBYE-HÜCKEL THEORY
Ion-ion interactions,
NONCOVALENT INTERACTIONS
IONIZATION
IONIZATION ENERGY
DISSOCIATION
IONIZATION ENERGY
Ionization, of water,
AUTOPROTOLYSIS
IONOPHORE
ION PAIR,
TIGHT ION PAIRS
LOOSE ION PAIRS
SOLVENT-SHARED ION PAIR
SOLVENT-SEPARATED ION PAIR
Ion pair return,
INTERNAL ION-PAIR RETURN

EXTERNAL ION-PAIR RETURN
HIDDEN RETURN
Ion permeation across membrane,
PERMEABILITY CONSTANT
PERMEABILITY COEFFICIENT
ION PUMPS
MEMBRANE TRANSPORT
BINDING CHANGE MECHANISM
Ipso-attack,
PHYSICAL ORGANIC CHEMISTRY NO-MENCLATURE
IRMS,
ISOTOPE RATIO MASS SPECTROMETER
Iron(II) carbonate ($FeCO_3$),
SOLUBILITY PRODUCT
Iron-dependent enzymes,
ADENYLYLSULFATE REDUCTASE
ANTHRANILATE 1,2-DIOXYGENASE
BENZENE 1,2-DIOXYGENASE
BENZOATE 1,2-DIOXYGENASE
CARBON MONOXIDE DEHYDRO-GENASE
β-CAROTENE 15,15'-DIOXYGENASE
CATECHOL 1,2-DIOXYGENASE
COPROPORPHYRINOGEN OXIDASE
CYSTEAMINE DIOXYGENASE
CYSTEINE DIOXYGENASE
CYTOCHROME c_3 HYDROGENASE
DESATURASES, FATTY ACYL-CoA
DIHYDROOROTATE DEHYDROGENASE
3,4-DIHYDROXYPHENYLACETATE 2,3-DIOXYGENASE
2,3-DIHYDROXY-9,10-SECOANDROSTA-1,3,5(10)-TRIENE-9,17-DIONE 4,5-DIOXYGENASE
FLAVANONE 3β-HYDROXYLASE
GENTISATE 1,2-DIOXYGENASE
GLUTAMATE SYNTHASE
HOMOGENTISATE 1,2-DIOXYGENASE
HYDROGENASE
HYDROGEN DEHYDROGENASE
3-HYDROXYANTHRANILATE 3,4-DIOXY-GENASE
4-HYDROXYPHENYLPYRUVATE HYDROX-YLASE
myo-INOSITOL OXYGENASE
LIPOXYGENASE
8-LIPOXYGENASE
12-LIPOXYGENASE
15-LIPOXYGENASE
LYSYL HYDROXYLASE
MANDELATE 4-MONOOXYGENASE

(R)-2-METHYLMALATE DEHYDRATASE
NICOTINATE DEHYDROGENASE
NITRATE REDUCTASE
NITRITE REDUCTASE
PHENYLALANINE MONOOXYGENASE
PROLYL 3-HYDROXYLASE
PROLYL 4-HYDROXYLASE
PROTOCATECHUATE 3,4-DIOXYGENASE
PROTOCATECHUATE 4,5-DIOXYGENASE
RIESKE IRON-SULFUR PROTEIN
RUBREDOXIN
RUBREDOXIN:NAD$^+$ REDUCTASE
SUCCINATE DEHYDROGENASE
SULFITE REDUCTASE
SULFUR DIOXYGENASE
THYMIDINE 2'-HYDROXYLASE
XANTHINE OXIDASE
Iron(II) hydroxide ($Fe(OH)_2$),
SOLUBILITY PRODUCT
Iron(III) hydroxide ($Fe(OH)_3$),
SOLUBILITY PRODUCT
IRON-SULFUR CLUSTER
Iron-sulfur protein cofactors,
NADH DEHYDROGENASE
IRON-SULFUR PROTEINS
IRREVERSIBLE PATHS, REACTIONS, AND PROCESSES
NONEQUILIBRIUM THERMODYNAMICS (A Primer)
IRREVERSIBLE POLYMERIZATION
Irreversible process,
ENTROPY
NONEQUILIBRIUM THERMODYNAMICS (A Primer)
Isentropic process,
ADIABATIC PROCESS
ISOAMYLASE
"ISOBI,"
COMPUTER ALGORITHMS & SOFTWARE
"ISOBI-HS,"
COMPUTER ALGORITHMS & SOFTWARE
ISOBUTYRYL-CoA MUTASE
"ISOCALC4,"
COMPUTER ALGORITHMS & SOFTWARE
"ISOCALC5,"
COMPUTER ALGORITHMS & SOFTWARE
ISOCHORISMATE SYNTHASE
ISOCITRATE DEHYDROGENASE
ISOCITRATE LYASE
"ISO-COOP,"
COMPUTER ALGORITHMS & SOFTWARE

J

K

Kynurenine formamidase,
ARYLFORMAMIDASE

KYNURENINE 3-HYDROXYLASE

L

Labeled compounds,
BIOCHEMICAL NOMENCLATURE

LABILE

LACCASE

α-Lactalbumin molten globule state,
CALORIMETRY

(S)-Lactaldehyde,
*RHAMNULOSE-1-PHOSPHATE AL-
DOLASE*

β-LACTAMASES

LACTASE

LACTATE DEHYDROGENASE
ABORTIVE COMPLEXES
*D-LACTATE DEHYDROGENASE (CYTO-
CHROME)*
*L-LACTATE DEHYDROGENASE (CYTO-
CHROME)*
ISOTOPE EXCHANGE AT EQUILIBRIUM

D-LACTATE DEHYDROGENASE (CYTO-
CHROME)

L-LACTATE DEHYDROGENASE (CYTO-
CHROME)

LACTATE 2-MONOOXYGENASE

LACTATE RACEMASE

Lactate racemase reduction,
BOROHYDRIDE REDUCTION

LACTOPEROXIDASE
PEROXIDASE

Lactose
LACTASE
LACTOSE SYNTHASE

LACTOSE SYNTHASE

Lactosylceramide α-2,6-N-sialyltransferase,
SIALYLTRANSFERASE

Lag phase in microtubule assembly,
MICROTUBULE ASSEMBLY KINETICS

Lag phase in polymerization,
HEMOGLOBIN-S POLYMERIZATION

Lag-phase kinetics,
AUTOCATALYSIS

LAG TIME
HYSTERESIS
PRE-STEADY STATE
AUTOCATALYSIS
MICROTUBULE ASSEMBLY KINETICS

LAMBDA (Λ)

LAMBDA (or Λ-) ISOMERS OF METAL ION-
NUCLEOTIDE COMPLEXES

LANGMUIR ISOTHERM
BIOMINERALIZATION
MICELLE

LANGMUIR TROUGH AND BALANCE

LAPLACE TRANSFORM

LARMOR PRECESSION

LASER-FLASH KINETIC ANALYSIS

LATENT ACTIVITY

LATENT HEAT

Latent heat of fusion
LATENT HEAT

Latent heat of vaporization
LATENT HEAT

Lateral binding proteins,
ACTIN REGULATORY PROTEINS

Lateral diffusibility of molecules within mem-
branes,
*FLUORESCENCE RECOVERY AFTER PHO-
TOBLEACHING*

Lattice binding model,
SCATCHARD PLOT

Law of Constant Heat Summation,
HESS' LAW
ADDITIVITY PRINCIPLE

LCAO,
*LINEAR COMBINATION OF ATOMIC OR-
BITALS*

LCAT,
*LECITHIN-CHOLESTEROL ACYLTRANS-
FERASE*

Lead tetraacetate
*MALAPRADE REACTION (PERIODATE
OXIDATION)*

Leakage rate,
BASAL RATE

LEAST-CONDUCTIVE STEP

LEAST MOTION, PRINCIPLE OF

Least squares fit
LEAST SQUARES METHOD
MARQUARDT ALGORITHM

LEAST SQUARES METHOD

LEAVING GROUP
NUCLEOFUGE
ELECTROFUGE
NUCLEOPHILE
ENTERING GROUP
*NUCLEOPHILIC SUBSTITUTION RE-
ACTION*

LE CHATELIER'S PRINCIPLE

Lecithinase A,
PHOSPHOLIPASE A$_2$

Lecithinase B,
LYSOPHOSPHOLIPASE

Lecithinase C,
PHOSPHOLIPASE C

Lecithinase D,
PHOSPHOLIPASE D

LECITHIN-CHOLESTEROL ACYLTRANS-
FERASE

LEE-WILSON TREATMENT

LEFFLER'S ASSUMPTION
HAMMOND POSTULATE

Left-to-right convention,
*PHYSICAL ORGANIC CHEMISTRY NO-
MENCLATURE*

LEGENDRE TRANSFORMATION

LEGUMAIN

LEISHMANOLYSIN

Lennard-Jones potential energy diagram,
JABLONSKI DIAGRAM

Leucine,
LEUCINE 2,3-AMINOMUTASE
LEUCINE AMINOTRANSFERASE
LEUCINE DEHYDROGENASE
LEUCINE KINETICS
LEUCYL-tRNA SYNTHETASE
ZWITTERION

β-Leucine,
LEUCINE 2,3-AMINOMUTASE

LEUCINE 2,3-AMINOMUTASE

LEUCINE AMINOPEPTIDASE

LEUCINE AMINOTRANSFERASE
*BRANCHED-CHAIN AMINO ACID AMI-
NOTRANSFERASE*

LEUCINE DEHYDROGENASE

LEUCINE KINETICS
PROTEIN TURNOVER KINETICS
SAAM

Leucine:tRNA ligase,
LEUCYL-tRNA SYNTHETASE

LEUCOLYSIN

Leucyl aminopeptidase,
LEUCINE AMINOPEPTIDASE

Leucyl residue,
LEUCINE AMINOPEPTIDASE

Leucyl-tRNA^{Leu},
LEUCYL-tRNA SYNTHETASE

LEUCYL-tRNA SYNTHETASE

Leukotriene A$_4$,
LEUKOTRIENE A$_4$ HYDROLASE

Linear polymer disassembly,
DEPOLYMERIZATION, END-WISE
Linear regression analysis,
LEAST SQUARES METHOD
LINEAR SOLVATION ENERGY RELATION-
SHIPS
E_T *VALUES*
KAMLET-TAFT SOLVENT PARAMETER
KOPPEL-PALM SOLVENT PARAMETER
Z VALUE
LINEAR STRAIN
Line formula,
*PHYSICAL ORGANIC CHEMISTRY NO-
MENCLATURE*
LINE-SHAPE ANALYSIS
Lineweaver-Burk plot,
DOUBLE-RECIPROCAL PLOT
*ENZYME KINETIC EQUATIONS (1. The
Basics)*
LINKED FUNCTIONS
ALLOSTERISM
BASIC REGULATORY KINETICS
BOHR EFFECT
COOPERATIVITY
FEEDBACK EFFECTORS
LE CHATELIER'S PRINCIPLE
HEMOGLOBIN
ADAIR EQUATION
LINOLEATE ISOMERASE
Linoleoyl ester,
*LECITHIN-CHOLESTEROL ACYLTRANS-
FERASE*
LIPASES
LANGMUIR TROUGH AND BALANCE
LIPOPROTEIN LIPASE
MICELLES
LIPID ACTIVATION
Lipid mobilization kinetics,
LIPID TRACER KINETICS
Lipid phase state,
LIPID PHASE TRANSITION KINETICS
LIPID PHASE TRANSITION KINETICS
LIPID TRACER KINETICS
Lipid transfer across membrane bilayers,
PHOSPHOLIPID FLIP-FLOP
Lipoamide,
PYRUVATE DEHYDROGENASE
LIPOIC ACID AND DERIVATIVES
Lipolysis,
LANGMUIR TROUGH AND BALANCE
MICELLES

LIPOPHILIC
HYDROPHOBIC INTERACTION
Lipophosphodiesterase I,
PHOSPHOLIPASE C
Lipophosphodiesterase II,
PHOSPHOLIPASE D
Lipopolysaccharide,
*LIPOPOLYSACCHARIDE N-ACETYLGLU-
COSAMINYLTRANSFERASE*
*LIPOPOLYSACCHARIDE GALACTOSYL-
TRANSFERASE*
*LIPOPOLYSACCHARIDE GLUCOSYL-
TRANSFERASE*
LIPOPOLYSACCHARIDE N-ACETYLGLUCO-
SAMINYLTRANSFERASE
Lipopolysaccharide core,
*LIPOPOLYSACCHARIDE N-ACETYLGLU-
COSAMINYLTRANSFERASE*
*LIPOPOLYSACCHARIDE GALACTOSYL-
TRANSFERASE*
*LIPOPOLYSACCHARIDE GLUCOSYL-
TRANSFERASE*
LIPOPOLYSACCHARIDE GALACTOSYL-
TRANSFERASE
LIPOPOLYSACCHARIDE GLUCOSYLTRANS-
FERASE
LIPOPROTEIN LIPASE
LIPASES
LIPOXYGENASE
5-LIPOXYGENASE
8-LIPOXYGENASE
12-LIPOXYGENASE
15-LIPOXYGENASE
Listeria monocytogenes,
ACTIN-BASED BACTERIAL MOTILITY
*ACTIN FILAMENT GROWTH (Polymeriza-
tion Zone Model)*
Lithium,
IONIC RADIUS
Local isolation, method of,
*NONEQUILIBRIUM THERMODYNAMICS
(A Primer)*
Local maxima in a continuous and differenti-
able function,
FIRST DERIVATIVE TEST
Local minima in a continuous and differenti-
able function,
FIRST DERIVATIVE TEST
LOCK-AND-KEY MODEL OF ENZYME
ACTION
INDUCED-FIT MODEL
*LIGAND-INDUCED CONFORMATIONAL
CHANGE*
LOG-RESPONSE CURVE FOR ENZYME
COOPERATIVITY

LOG VELOCITY *VERSUS* LOG [SUBSTRATE]
PLOT
LONDON FORCES
VAN DER WAALS FORCES
NONCOVALENT INTERACTIONS
LONDON-STECK PLOT
LONE PAIR
Long-chain acid anion,
*LONG-CHAIN ALDEHYDE DEHYDRO-
GENASE*
*LONG-CHAIN FATTY ACYL-CoA SYN-
THETASE*
Long-chain acyl-CoA,
*LONG-CHAIN ACYL-CoA DEHYDRO-
GENASE*
*LONG-CHAIN FATTY ACYL-CoA SYN-
THETASE*
LONG-CHAIN ACYL-CoA DEHYDRO-
GENASE
Long-chain aldehyde,
*LONG-CHAIN ALDEHYDE DEHYDRO-
GENASE*
LONG-CHAIN ALDEHYDE DEHYDRO-
GENASE
LONG-CHAIN FATTY ACYL-CoA SYN-
THETASE
Longitudinal strain,
LINEAR STRAIN
Long-lived luminescence,
PHOSPHORESCENCE
JABLONSKI DIAGRAM
LONG-RANGE LOOPS
LOOSE ION PAIR
ION PAIR
SOLVENT-SEPARATED ION PAIR
SOLVENT-SHARED ION PAIR
TIGHT ION PAIR
LOSSEN REARRANGEMENT
LOW-BARRIER HYDROGEN BONDS (Poten-
tial Role in Catalysis)
HYDROGEN BONDS
WATER
Low-density lipoproteins,
LIPOPROTEIN LIPASE
Lowest-lying excited state,
FLUORESCENCE
Low-k_{cat} GTPase in tubulin polymerization,
MICROTUBULE ASSEMBLY KINETICS
LUCIFERASE
D(-)-Luciferin,
LUCIFERASE
D(-)-Luciferyl adenylate,
LUCIFERASE

LUMEN

Lumichrome,
  RIBOFLAVINASE

Luminal calcium ion,
  ION PUMPS

LUMINESCENCE
  FLUORESCENCE

Luminometer,
  LUCIFERASE

Luminous flux,
  LUMEN

LYASES
  ENZYME COMMISSION NOMEN-
    CLATURE

LYATE ION

LYONIUM ION
  LEVELING EFFECT

Lysine,
  LYSINE 2,3-AMINOMUTASE
  LYSINE 6-AMINOTRANSFERASE
  LYSINE DECARBOXYLASE
  LYSINE DEHYDROGENASE (α-DEAMI-
    NATING)
  LYSINE MONOOXYGENASE
  LYSOPINE DEHYDROGENASE
  ZWITTERION

D-Lysine,
  D-LYSINE 5,6-AMINOMUTASE

LYSINE 2,3-AMINOMUTASE

β-LYSINE 5,6-AMINOMUTASE

D-LYSINE 5,6-AMINOMUTASE

LYSINE 6-AMINOTRANSFERASE

Lysine ε-aminotransferase,
  LYSINE 6-AMINOTRANSFERASE

LYSINE DECARBOXYLASE

LYSINE DEHYDROGENASE (α-DEAMI-
  NATING)

LYSINE 2-MONOOXYGENASE

D-α-Lysine mutase,
  D-LYSINE 5,6-AMINOMUTASE

Lysolecithin,
  LYSOPHOSPHOLIPASE
  MICELLES

Lysolecithinase,
  LYSOPHOSPHOLIPASE

2-Lysophosphatidylcholine,
  LYSOPHOSPHOLIPASE

LYSOPHOSPHOLIPASE

LYSOPHOSPHOLIPASE D

Lysopine,
  LYSOPINE DEHYDROGENASE

LYSOPINE DEHYDROGENASE

D-Lysopine synthase,
  LYSOPINE DEHYDROGENASE

LYSOZYME
  KINETIC ISOTOPE EFFECT

LYSYL ENDOPEPTIDASE

LYSYL HYDROXYLASE

LYSYL OXIDASE

Lysyl residue,
  LYSYL HYDROXYLASE
  LYSYL OXIDASE

LYSYLTRANSFERASE

L-Lysyl-tRNA^Lys,
  LYSYLTRANSFERASE

Lysyl-tRNA:phosphatidylglycerol transferase,
  LYSYLTRANSFERASE

α-LYTIC PROTEINASE
  CHYMOTRYPSIN
  CATALYTIC TRIAD

D-Lyxose,
  LYXOSE ISOMERASE
  MANNOSE ISOMERASE

LYXOSE ISOMERASE

Lyxose ketol-isomerase,
  LYXOSE ISOMERASE

D-Lyxosides,
  MANNOSIDASES

M

MACROMOLECULAR SEQUENCE ANALY-
  SIS (Computer Methods)

MACROSCOPIC CONSTANTS
  MICROSCOPIC CONSTANTS
  ZWITTERION

MACROSCOPIC DIFFUSION CONTROL
  MICROSCOPIC DIFFUSION CONTROL

Magic acid,
  PHYSICAL ORGANIC CHEMISTRY
    NOMENCLATURE

Magnesium carbonate ($MgCO_3$),
  SOLUBILITY PRODUCT

Magnesium hydroxide ($Mg(OH)_2$),
  SOLUBILITY PRODUCT

Magnesium ion,
  ENERGY CHARGE
  IONIC RADIUS
  LAMBDA (or Λ-) ISOMERS OF METAL
    ION-NUCLEOTIDE COMPLEXES
  METAL IONS IN NUCLEOTIDE-
    DEPENDENT REACTIONS
  XYLOSE ISOMERASE

MAGNESIUM ION (Intracellular)
  ENERGY CHARGE
  METAL IONS IN NUCLEOTIDE-DEPEN-
    DENT REACTIONS
  BIOCHEMICAL THERMODYNAMICS

Magnesium oxalate dihydrate
  ($MgC_2O_4 \cdot 2H_2O$),
  SOLUBILITY PRODUCT

Magnesium phosphate ($Mg_3(PO_4)_2$),
  SOLUBILITY PRODUCT

Magnesium protoporphyrin,
  MAGNESIUM-PROTOPORPHYRIN
    O-METHYLTRANSFERASE

MAGNESIUM-PROTOPORPHYRIN O-
  METHYLTRANSFERASE

Magnesium protoporphyrin monomethyl
  ester,
  MAGNESIUM-PROTOPORPHYRIN
    O-METHYLTRANSFERASE

MAGNETICALLY ANISOTROPIC GROUP

Magnetic balance,
  MAGNETIC SUSCEPTIBILITY

Magnetic dipoles,
  QUADRUPOLE

Magnetic equivalence,
  PHYSICAL ORGANIC CHEMISTRY
    NOMENCLATURE

Magnetic field,
  LARMOR PRECESSION
  MAGNETICALLY ANISOTROPIC GROUP
  MAGNETOTACTIC

Magnetic field strength,
  MAGNETIC SUSCEPTIBILITY

Magnetic flux,
  MAXWELL

Magnetic flux density,
  PERMEABILITY
  MAGNETIC SUSCEPTIBILITY

MAGNETIC SUSCEPTIBILITY
  PARAMAGNETIC
  HEMOGLOBIN
  DIAMAGNETIC

Magnetization,
  MAGNETIC SUSCEPTIBILITY

MAGNETIZATION TRANSFER

MAGNETOTACTIC

Magnitude of k_H/k_D,
  QUANTUM MECHANICAL TUNNELING

MALAPRADE REACTION (Periodate Oxidation)

(R)-Malate,
  MALATE DEHYDROGENASE

(S)-Malate,
  ABORTIVE COMPLEXES
  MALATE DEHYDROGENASE
  MALATE SYNTHASE
  (S)-2-METHYLMALATE DEHYDRATASE
Malate condensing enzyme,
  MALATE SYNTHASE
MALATE DEHYDROGENASE
  ABORTIVE COMPLEXES
Malate dehydrogenase (acceptor),
  MALATE DEHYDROGENASE
Malate dehydrogenase (decarboxylating),
  MALATE DEHYDROGENASE
D-Malate dehydrogenase (decarboxylating),
  MALATE DEHYDROGENASE
Malate dehydrogenase (NADP$^+$),
  MALATE DEHYDROGENASE
Malate dehydrogenase (oxaloacetate-decar-
  boxylating),
  MALATE DEHYDROGENASE
Malate dehydrogenase (oxaloacetate-decar-
  boxylating, NADP$^+$),
  MALATE DEHYDROGENASE
MALATE SYNTHASE
MALDI-TOF MS,
  TIME-OF-FLIGHT MASS SPECTROMETRY
Maleate,
  MALEATE ISOMERASE
MALEATE ISOMERASE
4-Maleylacetoacetate,
  MALEYLACETOACETATE ISOMERASE
MALEYLACETOACETATE ISOMERASE
Maleylpyruvate,
  MALEYLACETOACETATE ISOMERASE
  MALEYLPYRUVATE ISOMERASE
MALEYLPYRUVATE ISOMERASE
Malic dehydrogenase,
  MALATE DEHYDROGENASE
Malic enzyme,
  MALATE DEHYDROGENASE
Malonyl-[acyl-carrier-protein],
  *MALONYL-CoA:ACYL-CARRIER-PROTEIN
    TRANSACYLASE*
MALONYL-CoA:ACYL-CARRIER-PROTEIN
  TRANSACYLASE
Malonyl-coenzyme A,
  *MALONYL-CoA:ACYL-CARRIER-PROTEIN
    TRANSACYLASE*
MALTASE
  α-GLUCOSIDASE

Maltose,
  MALTASE
  MALTOSE 1-EPIMERASE
  MALTOSE PHOSPHORYLASE
MALTOSE PHOSPHORYLASE
MALYL-CoA LYASE
MAMILLARY MODEL
  CATENARY MODEL
  COMPARTMENTAL ANALYSIS
Mandelate,
  MANDELATE 4-MONOOXYGENASE
  MANDELATE RACEMASE
L-Mandelate 4-hydroxylase,
  MANDELATE 4-MONOOXYGENASE
MANDELATE 4-MONOOXYGENASE
MANDELATE RACEMASE
Mandelonitrile,
  MANDELONITRILE LYASE
MANDELONITRILE LYASE
Manganese(II) carbonate (MnCO$_3$),
  SOLUBILITY PRODUCT
Manganese(II) hydroxide (Mn(OH)$_2$),
  SOLUBILITY PRODUCT
Manganese ion,
  GLUTAMINE SYNTHETASE
  MANGANESE PEROXIDASE
  *METAL IONS IN NUCLEOTIDE-DEPEN-
    DENT REACTIONS*
  PYRUVATE, WATER DIKINASE
  IONIC RADIUS
Manganese(II) oxalate dihydrate
  (MnC$_2$O$_4$·2H$_2$O),
  SOLUBILITY PRODUCT
MANGANESE PEROXIDASE
  PEROXIDASE
β-MANNANASE
  MANNOSIDASES
Mannan endo-1,4-β-mannosidase,
  β-MANNANASE
Mannans,
  β-MANNANASE
Mannase,
  MANNOSIDASES
D-Mannonate,
  MANNONATE DEHYDRATASE
MANNONATE DEHYDRATASE
D-Mannonate hydrolase,
  MANNONATE DEHYDRATASE
D-Mannose,
  MANNOSE ISOMERASE
  *MAPPING SUBSTRATE INTERACTIONS
    USING KINETIC DATA*

MANNOSE ISOMERASE
α-D-Mannose 1-phosphate,
  *MANNOSE-1-PHOSPHATE GUANYLYL-
    TRANSFERASE*
D-Mannose 6-phosphate,
  MANNOSE-6-PHOSPHATE ISOMERASE
MANNOSE-1-PHOSPHATE GUANYLYL-
  TRANSFERASE
D-Mannose 1-phosphate, interconversion to
  D-mannose 6-phosphate,
  PHOSPHOMANNOMUTASE
MANNOSE-6-PHOSPHATE ISOMERASE
D-Mannose residues,
  MANNOSIDASES
MANNOSIDASES
D-Mannosides,
  MANNOSIDASES
1,4-β-D-Mannosidic linkages,
  β-MANNANASE
MANOMETRIC ASSAY METHODS
  EUDIOMETER
Manometry,
  MANOMETRIC ASSAY METHODS
MAPPING SUBSTRATE INTERACTIONS
  USING KINETIC DATA
MARCUS EQUATION
MARCUS RATE THEORY
MARKOV CHAIN
Markovnikoff rule,
  *PHYSICAL ORGANIC CHEMISTRY NO-
    MENCLATURE*
MARQUARDT ALGORITHM
  COMPUTER ALGORITHMS & SOFTWARE
MASS-ACTION RATIO
Mass-action ratio, determination,
  ISOTOPE EXCHANGE AT EQUILIBRIUM
MASSIEU FUNCTION
  HELMHOLTZ ENERGY
  PLANCK FUNCTION
MASS SPECTROMETRY
Matrix of biominerals,
  BIOMINERALIZATION
MATRIX ISOLATION
MAXIMUM LIKELIHOOD METHOD
Maximum velocity,
  MICHAELIS-MENTEN EQUATION
MAXWELL
MAXWELL'S THERMODYNAMIC RELA-
  TIONS
Mean activity coefficient,
  DEBYE-HÜCKEL TREATMENT
Mean lifetime,
  LIFETIME
Mean residence time,
  MEAN TRANSIT TIME

MEAN TRANSIT TIME
  COMPARTMENTAL ANALYSIS

MECHANISM-BASED INHIBITOR

MEDIATED AMPEROMETRIC BIOSENSOR
  BIOSENSOR

MEDIUM

Medium effect,
  SOLVENT EFFECTS

MEERWEIN-PONDORFF-VERLEY RE-
DUCTION

MEGA-

MEISENHEIMER ADDUCT
  TWO-PROTONIC STATE ELECTROPHILES

Melilotate,
  MELILOTATE HYDROXYLASE

MELILOTATE HYDROXYLASE

Melilotate 3-monooxygenase,
  MELILOTATE HYDROXYLASE

MELTING

Melting curve,
  MELTING

Melting point,
  MELTING

Membrane,
  LIPID ACTIVATION
  MEMBRANE TRANSPORT

Membrane disruption,
  LATENT ACTIVITY

Membrane electric potential,
  MEMBRANE POTENTIAL

Membrane fusion,
  LIPID PHASE TRANSITION KINETICS

Membrane lipid asymmetry,
  PHOSPHOLIPID FLIP-FLOP

Membrane permeability,
  PERMEABILITY
  PERMEABILITY CONSTANT

MEMBRANE POTENTIAL
  ACTION POTENTIAL
  DEPOLARIZATION
  GOLDMAN EQUATION
  NERNST EQUATION
  RESTING POTENTIAL
  THRESHOLD POTENTIAL
  PATCH-CLAMP TECHNIQUE

Membrane protein dynamics,
  *FLUORESCENCE RECOVERY AFTER PHO-
  TOBLEACHING*

MEMBRANE TRANSPORT
  ION PUMPS

Membrane transport in organelles,
  MEMBRANE TRANSPORT

Menadione reductase,
  NAD(P)H DEHYDROGENASE (QUINONE)

MEPRIN

Mercaptoethanol,
  *3-MERCAPTOPYRUVATE SULFURTRANS-
  FERASE*

3-Mercaptopyruvate,
  *3-MERCAPTOPYRUVATE SULFURTRANS-
  FERASE*

3-MERCAPTOPYRUVATE SULFURTRANS-
FERASE

Mercuric ion,
  MERCURY(II) REDUCTASE

Mercuric reductase,
  MERCURY(II) REDUCTASE

Mercury,
  MERCURY(II) REDUCTASE

MERCURY(II) REDUCTASE

Mesaconase,
  (S)-2-METHYLMALATE DEHYDRATASE

Mesaconate,
  3-METHYLASPARTATE AMMONIA-LYASE

Mesaconate hydratase,
  (S)-2-METHYLMALATE DEHYDRATASE

MESO FORM (or, MESO ISOMER)

Meso isomer,
  MESO FORM (or, MESO ISOMER)

MESOLYTIC CLEAVAGE

Metabolic activity of an individual organism,
  BASAL RATE

METABOLIC CONTROL ANALYSIS
  TOP-DOWN CONTROL ANALYSIS

Metabolic flux,
  METABOLIC CONTROL ANALYSIS

Metabolite compartmentation,
  MAMILLARY MODEL

METABOLONS
  SUBSTRATE CHANNELING

Metal complexes,
  *METAL ION COORDINATION REAC-
  TIONS*

METAL ION CATALYSIS

METAL ION COMPLEXES (Determining Sta-
bility Constants)
  *METAL IONS IN NUCLEOTIDE-DEPEN-
  DENT REACTIONS*

METAL ION COORDINATION REACTIONS
  WATER
  *WATER EXCHANGE RATES OF AQUA
  METAL IONS*

Metal ion exchange times,
  *LAMBDA (or Λ-) ISOMERS OF METAL
  ION-NUCLEOTIDE COMPLEXES*

Metal ion-ligand complex,
  *METAL ION COMPLEXES (DETERMINING
  STABILITY CONSTANTS)*

Metal ion-nucleotide complexes,
  *LAMBDA (or Λ-) ISOMERS OF METAL
  ION-NUCLEOTIDE COMPLEXES*
  *METAL IONS IN NUCLEOTIDE-DEPEN-
  DENT REACTIONS*

Metal ion oxidation state,
  MIXED VALENCY

METAL IONS IN NUCLEOTIDE-DEPEN-
DENT REACTIONS

Metal ligand complexation,
  LIGAND FIELD SPLITTING

METALLO-BIOCHEMICAL METHODS

Metalloproteins,
  LASER-FLASH KINETIC ANALYSIS
  METAL ION CATALYSIS

METALLOTHIONEINS

Metal-nucleotide complex,
  *EXCHANGE-INERT METAL-NUCLEOTIDE
  COMPLEXES*

Metal oxides,
  OXYGEN, OXIDES & OXO ANIONS

META-MODEL

METAPHOSPHATE
  ACYL PHOSPHATES
  OXYGEN, OXIDES & OXO ANIONS

METASTABLE STATE
  JABLONSKI DIAGRAM
  TRANSIENT CHEMICAL SPECIES

Metastable triplet state,
  JABLONSKI DIAGRAM

Metavanadate,
  OXYGEN, OXIDES & OXO ANIONS

METHANE MONOOXYGENASE

Methanol, autoprotolysis constant,
  AUTOPROTOLYSIS

METHANOL DEHYDROGENASE

5,10-METHENYLTETRAHYDROFOLATE
CYCLOHYDROLASE

5,10-METHENYLTETRAHYDROFOLATE SYN-
THETASE

METHIONINE ADENOSYLTRANSFERASE

METHIONINE γ-LYASE

METHIONINE SULFOXIDE REDUCTASE

Methionine synthase,
  *N^5-METHYLTETRAHYDROFOLATE:
  HOMOCYSTEINE METHYLTRANS-
  FERASE*

METHYL TRANSFER REACTIONS USING S-ADENOSYLMETHIONINE

Methionine:tRNA ligase,
METHIONYL-tRNA SYNTHETASE

Methionyl-tRNA^{Met},
METHIONYL-tRNA FORMYLTRANS-FERASE
METHIONYL-tRNA SYNTHETASE

METHIONYL-tRNA FORMYLTRANSFERASE

METHIONYL-tRNA SYNTHETASE
AMINO-ACYL tRNA SYNTHETASES

Method of continuous variation,
JOB PLOT

4-Methoxybenzoate,
4-METHOXYBENZOATE MONOOXY-GENASE

4-METHOXYBENZOATE MONOOXYGEN-ASE (O-DEMETHYLATING)

3-Methyladenine,
3-METHYLADENINE-DNA GLYCOSYLASE

7-Methyladenine,
3-METHYLADENINE-DNA GLYCOSYLASE

3-METHYLADENINE-DNA GLYCOSYLASE

N-Methyl-L-alanine,
N-METHYL-L-ALANINE DEHYDRO-GENASE

N-METHYL-L-ALANINE DEHYDROGENASE

Methylamine,
N-METHYL-L-ALANINE DEHYDRO-GENASE
METHYLAMINE DEHYDROGENASE
N-METHYLGLUTAMATE SYNTHASE

METHYLAMINE DEHYDROGENASE
RESONANCE RAMAN SPECTROSCOPY
TOPAQUINONE

Methylamine-glutamate methyltransferase,
N-METHYLGLUTAMATE SYNTHASE

N-Methyl-4-aminobenzoate,
4-METHOXYBENZOATE MONOOXY-GENASE

β-Methylaspartase,
3-METHYLASPARTATE AMMONIA-LYASE

L-*threo*-3-Methylaspartate,
3-METHYLASPARTATE AMMONIA-LYASE
β-METHYLASPARTATE-GLUTAMATE MUTASE

3-METHYLASPARTATE AMMONIA-LYASE

β-METHYLASPARTATE-GLUTAMATE MUTASE

Methylaspartate mutase,
β-METHYLASPARTATE-GLUTAMATE MUTASE

METHYLATION (As a Mechanistic Probe)

2-METHYL-BRANCHED-CHAIN ACYL-CoA DEHYDROGENASE

2-Methyl-4-carbomethoxypyridinium ion,
Z VALUE

β-METHYLCROTONYL-CoA CARBOXYLASE,

3-Methylcrotonyl-coenzyme A
β-METHYLCROTONYL-CoA CARBOX-YLASE

2-Methyleneglutarate,
2-METHYLENEGLUTARATE MUTASE

2-METHYLENEGLUTARATE MUTASE

Methylene-interrupted polyunsaturated fatty acids,
LIPOXYGENASE

2-Methylene-3-methylsuccinate,
2-METHYLENEGLUTARATE MUTASE

5,10-Methylenetetrahydrofolate,
5,10-METHYLENETETRAHYDROFOLATE DEHYDROGENASE
5,10-METHYLENETETRAHYDROFOLATE REDUCTASE
5,10-METHYLENETETRAHYDROFOLATE: tRNA (URACIL-5-)-METHYLTRANS-FERASE (FADH$_2$ OXIDIZING)
SERINE HYDROXYMETHYLTRANSFERASE
THYMIDYLATE SYNTHASE

5,10-METHYLENETETRAHYDROFOLATE DE-HYDROGENASE

5,10-METHYLENETETRAHYDROFOLATE RE-DUCTASE

5,10-METHYLENETETRAHYDROFOLATE: tRNA (URACIL-5-)-METHYLTRANSFER-ASE (FADH$_2$ OXIDIZING)

2-Methylfumarate,
(S)-2-METHYLMALATE DEHYDRATASE

3-METHYLGLUTACONYL-CoA HYDRATASE

3-Methylglutaconyl-coenzyme A,
β-METHYLCROTONYL-CoA CARBOX-YLASE

3-METHYLGLUTACONYL-CoA HYDRATASE

N-Methyl-L-glutamate,
N-METHYLGLUTAMATE SYNTHASE

N-METHYLGLUTAMATE SYNTHASE

3-Methylguanine,
3-METHYLADENINE-DNA GLYCOSYLASE

7-Methylguanine,
3-METHYLADENINE-DNA GLYCOSYLASE

γ-Methyl-*γ*-hydroxy-*α*-ketoglutarate,
γ-METHYL-γ-HYDROXY-α-KETOGLUTARATE ALDOLASE

γ-METHYL-*γ*-HYDROXY-*α*-KETOGLU-TARATE ALDOLASE

METHYLHYDROXYPYRIDINECARBOXYLATE DIOXYGENASE

Methylitaconate,
METHYLITACONATE Δ-ISOMERASE

METHYLITACONATE Δ-ISOMERASE

(*R*)-2-Methylmalate,
(R)-2-METHYLMALATE DEHYDRATASE

(*S*)-2-Methylmalate,
(S)-2-METHYLMALATE DEHYDRATASE

(*R*)-2-METHYLMALATE DEHYDRATASE

(*S*)-2-METHYLMALATE DEHYDRATASE

2-Methylmaleate,
(R)-2-METHYLMALATE DEHYDRATASE

Methylmalonate semialdehyde,
METHYLMALONATE-SEMIALDEHYDE DE-HYDROGENASE

METHYLMALONATE-SEMIALDEHYDE DE-HYDROGENASE (ACYLATING)

(*R*)-Methylmalonyl-CoA,
METHYLMALONYL-CoA EPIMERASE
L-METHYLMALONYL-CoA MUTASE

(*S*)-Methylmalonyl-CoA,
METHYLMALONYL-CoA DECARBOX-YLASE
METHYLMALONYL-CoA EPIMERASE
D-METHYLMALONYL-CoA HYDROLASE
PROPIONYL-CoA CARBOXYLASE

METHYLMALONYL-CoA DECARBOXYLASE

METHYLMALONYL-CoA EPIMERASE

D-METHYLMALONYL-CoA HYDROLASE

L-METHYLMALONYL-CoA MUTASE

4-Methyl-2-oxopentanoate,
LEUCINE AMINOTRANSFERASE
LEUCINE DEHYDROGENASE
LEUCINE KINETICS

2-Methyl-3-oxopropanoate,
METHYLMALONATE-SEMIALDEHYDE DE-HYDROGENASE

(*R*)-2-Methyl-3-oxopropanoyl-coenzyme A,
METHYLMALONYL-CoA EPIMERASE
L-METHYLMALONYL-CoA MUTASE

(*S*)-2-Methyl-3-oxopropanoyl-coenzyme A,
METHYLMALONYL-CoA DECARBOX-YLASE
METHYLMALONYL-CoA EPIMERASE

N-Methylphenylethanolamine,
NOREPINEPHRINE N-METHYLTRANS-FERASE

N-Methylputrescine, synthesis of,
PUTRESCINE N-METHYLTRANSFERASE

5-Methyltetrahydrofolate,
5,10-METHYLENETETRAHYDROFOLATE REDUCTASE

N^5-METHYLTETRAHYDROFOLATE:

HYDROXYBUTYRATE DEHYDROGE-
NASES

β-HYDROXYPROPIONATE DEHYDRO-
GENASE

20-HYDROXYSTEROID DEHYDRO-
GENASE

IMIDAZOLEACETATE HYDROXYLASE

IMP DEHYDROGENASE

myo-INOSITOL 2-DEHYDROGENASE

myo-INOSITOL-1-PHOSPHATE SYN-
THASE

ISOCITRATE DEHYDROGENASE

3-ISOPROPYLMALATE DEHYDRO-
GENASE

LACTATE DEHYDROGENASE

LEUCINE DEHYDROGENASE

LONG-CHAIN ALDEHYDE DEHYDRO-
GENASE

LYSINE DEHYDROGENASE (α-DEAMI-
NATING)

MALATE DEHYDROGENASE

5,10-METHYLENETETRAHYDROFOLATE
DEHYDROGENASE METHYLMALO-
NATE-SEMIALDEHYDE DEHYDRO-
GENASE

NAD+ GLYCOHYDROLASE

NAD+ KINASE

OCTANOL DEHYDROGENASE

OCTOPINE DEHYDROGENASE

PHENYLALANINE DEHYDROGENASE

POLY(ADP-RIBOSE) SYNTHETASE (or,
SYNTHASE)

PREPHENATE DEHYDROGENASE

PYRIDOXAL 4-DEHYDROGENASE

Δ^1-PYRROLINE-5-CARBOXYLATE DEHY-
DROGENASE

RUBREDOXIN:NAD+ REDUCTASE

SERINE DEHYDROGENASE

SORBITOL DEHYDROGENASE

XANTHINE DEHYDROGENASE

NAD+ formation,
NADH PEROXIDASE

NAD+ GLYCOHYDROLASE

NADH,
NICOTINAMIDE COENZYMES (SPEC-
TRAL PROPERTIES)

NADH-dependent enzymes,
AQUACOBALAMIN REDUCTASES
BENZENE 1,2-DIOXYGENASE
DEHYDROASCORBATE REDUCTASE
GLUTAMATE SYNTHASE
GLYOXYLATE REDUCTASE
3-HYDROXYBENZOATE 6-MONOOXY-
GENASE

HYDROXYLAMINE REDUCTASE

4-HYDROXYPHENYLACETATE 3-MONO-
OXYGENASE

IMIDAZOLEACETATE HYDROXYLASE

NADH-CYTOCHROME b₅ REDUCTASE

NADH DEHYDROGENASE

NADH DEHYDROGENASE, OTHER THAN
COMPLEX I

NADH PEROXIDASE

NITRATE REDUCTASE

NADH-CYTOCHROME b₅ REDUCTASE

NADH DEHYDROGENASE

NADH DEHYDROGENASE (Other Than
Complex I)

NADH dehydrogenase (quinone),
NADH DEHYDROGENASE, OTHER THAN
COMPLEX I

NADH/NADPH SURFACE FLUORESCENCE
IN LIVING TISSUES

NADH oxidation,
NADH PEROXIDASE

NADH PEROXIDASE

NAD+ KINASE

NAD+ nucleosidase,
NAD+ GLYCOHYDROLASE

NADP+,
NICOTINAMIDE COENZYMES (SPEC-
TRAL PROPERTIES)

NADP+-dependent enzymes,
N-ACETYL-γ-GLUTAMYL-PHOSPHATE RE-
DUCTASE

ACYL-CoA DEHYDROGENASES

ACYLGLYCERONE-PHOSPHATE RE-
DUCTASE

ALDEHYDE DEHYDROGENASE

ALDOSE REDUCTASE

2-AMINOADIPATE 6-SEMIALDEHYDE DE-
HYDROGENASE

ASPARTATE-SEMIALDEHYDE DEHYDRO-
GENASE

N⁵-(CARBOXYETHYL)ORNITHINE SYN-
THASE

CORTISONE β-REDUCTASE

7-DEHYDROCHOLESTEROL REDUCTASE

2-DEHYDRO-3-DEOXY-D-GLUCONATE 6-
DEHYDROGENASE

3-DEHYDROSPHINGANINE REDUCTASE

DIAMINOPIMELATE DEHYDROGENASE

FERREDOXIN:NADP+ REDUCTASE

10-FORMYLTETRAHYDROFOLATE DEHY-
DROGENASE

GLUCOSE-6-PHOSPHATE DEHYDRO-
GENASE

GLUCURONOLACTONE REDUCTASE

GLUTAMATE DEHYDROGENASE

GLUTAMATE γ-SEMIALDEHYDE DEHY-
DROGENASE

GLYCERALDEHYDE-3-PHOSPHATE DEHY-
DROGENASE

HOMOSERINE DEHYDROGENASE

3-HYDROXYACYL-CoA DEHYDRO-
GENASE

15-HYDROXYPROSTAGLANDIN DEHY-
DROGENASE

HYDROXYSTEROID DEHYDROGENASES

ISOCITRATE DEHYDROGENASE

β-KETOACYL-ACP REDUCTASE

α-KETOGLUTARATE SEMIALDEHYDE DE-
HYDROGENASE

LYSOPINE DEHYDROGENASE

MALATE DEHYDROGENASE

MERCURY(II) REDUCTASE

N-METHYL-L-ALANINE DEHYDRO-
GENASE

5,10-METHYLENETETRAHYDROFOLATE
DEHYDROGENASE

NAD(P)+ TRANSHYDROGENASE

NICOTINATE DEHYDROGENASE

D-NOPALINE SYNTHASE

SHIKIMATE DEHYDROGENASE

SULFITE REDUCTASE

VALINE DEHYDROGENASE

ZEATIN CIS-TRANS ISOMERASE

NADPH DEHYDROGENASE

NAD(P)H DEHYDROGENASE (QUINONE)

NADPH-dependent enzymes,
ACYL-ACYL-CARRIER-PROTEIN Δ⁹-DESA-
TURASE

ACYL-CoA DEHYDROGENASES

ADRENODOXIN

ANTHRANILATE HYDROXYLASE

AQUACOBALAMIN REDUCTASES

N⁵-(CARBOXYETHYL)ORNITHINE SYN-
THASE

trans-CINNAMATE 4-MONOOXYGENASE

CYCLOHEXANONE MONOOXYGENASE

CYTOCHROME P-450 REDUCTASE

DIHYDROFOLATE REDUCTASE

DIMETHYLANILINE MONOOXYGENASE

FARNESYL-DIPHOSPHATE FARNESYL-
TRANSFERASE

FATTY ACID SYNTHASE

GLUTATHIONE REDUCTASE

HEME OXYGENASE

3-HYDROXYBENZOATE 6-MONOOXY-
GENASE

4-HYDROXYBENZOATE 3-MONOOXY-
GENASE

Nonadditivity of entropies and free energy for weak interactions,
ADDITIVITY PRINCIPLE

Nonaprenyl-diphosphate synthase,
trans-OCTAPRENYLTRANSTRANSFERASE

Nonbonded interactions,
STEREOCHEMICAL TERMINOLOGY, IUPAC RECOMMENDATIONS

NONCOMPETITIVE INHIBITION
MIXED-TYPE INHIBITION

Noncompetitive inhibition, limiting case of,
MIXED-TYPE INHIBITION

NONCOVALENT INTERACTIONS
ALLOSTERIC INTERACTION
BINDING INTERACTION
BINDING ISOTHERM
BIOSENSOR
COOPERATIVE LIGAND BINDING
EQUILIBRIUM CONSTANT
EQUILIBRIUM DIALYSIS
HUMMEL-DREYER METHOD
LINKED FUNCTION THEORY
KLOTZ PLOT
MACROSCOPIC CONSTANT
MICROSCOPIC CONSTANT
MOLECULAR CROWDING
SCATCHARD PLOT

NONEQUILIBRIUM THERMODYNAMICS (A Primer)

Non-first-order process,
NOYES EQUATION

NONHYPERBOLIC KINETICS

Noninteger reaction orders in elementary reactions,
FRACTAL REACTION KINETICS

NONLINEAR INHIBITION
CLELAND RULES FOR DEAD-END INHIBITION
ABORTIVE COMPLEX

Nonlinear kinetics of drug bioavailability,
PHARMACOKINETICS

NONLINEAR LEAST SQUARES ANALYSIS

Nonmetal oxides,
OXYGEN, OXIDES & OXO ANIONS

Nonprocessivity,
RANDOM SCISSION KINETICS

Nonproductive reversible complexes,
ABORTIVE COMPLEXES

Nonproductive substrate binding,
SHIFTED BINDING

NONRADIATIVE DECAY

D-NOPALINE SYNTHASE

NOREPINEPHRINE *N*-METHYLTRANS-FERASE

NORMAL DISTRIBUTION
STATISTICS (A Primer)

NORMAL ERROR CURVE
STATISTICS (A Primer)

NORMALITY (or, NORMAL CONCENTRATION)

NORTHROP ISOTOPE EFFECT METHOD
KINETIC ISOTOPE EFFECTS

NOTEBOOK ENTRIES

NOYES EQUATION

NU (ν)

NUCLEARITY

NUCLEAR MAGNETIC RESONANCE
LARMOR PRECESSION
LIGAND BINDING ANALYSIS
LINE-SHAPE ANALYSIS
LOW-BARRIER HYDROGEN BONDS: ROLE IN CATALYSIS
MAGNESIUM ION (INTRACELLULAR)
MAGNETIZATION TRANSFER

Nuclear pores,
DIFFUSION OF MOLECULES INTO A PORE

Nucleation,
ACTIN ASSEMBLY KINETICS
BIOCHEMICAL SELF-ASSEMBLY
BIOMINERALIZATION
PRION PLAQUE FORMATION

Nucleation as a highly cooperative process,
MICROTUBULE ASSEMBLY KINETICS

Nucleation center, in microtubule self-assembly,
MICROTUBULE ASSEMBLY KINETICS

Nucleation of polymerization,
HEMOGLOBIN-S POLYMERIZATION

Nucleic acid dynamics,
NUCLEAR MAGNETIC RESONANCE

NUCLEASE S1

NUCLEATION
ACTIN ASSEMBLY KINETICS
MICROTUBULE ASSEMBLY KINETICS
PROTEIN POLYMERIZATION KINETICS

NUCLEIC ACID RENATURATION KINETICS

Nucleic acid structure,
BASE STACKING

NUCLEOFUGE
ELECTROFUGE
NUCLEOPHILE

NUCLEOPHILE

RITCHIE EQUATION
ELECTROPHILE

NUCLEOPHILIC CATALYSIS

NUCLEOPHILIC COMPETITION

NUCLEOPHILICITY
ALPHA EFFECT
ELECTROPHILICITY

Nucleophilicity of hydrazines and hydroxylamines,
ALPHA EFFECT

NUCLEOPHILIC SUBSTITUTION REACTION

Nucleophilic substitution reactions on phosphorus,
CHIRAL ATP
PSEUDOROTATION

Nucleoside 2′,3′-cyclic phosphodiesters,
RIBONUCLEASE (RNase)

NUCLEOSIDE DEOXYRIBOSYLTRANS-FERASE

NUCLEOSIDE DIPHOSPHATASE

NUCLEOSIDE DIPHOSPHATE KINASE
HALF-REACTIONS

Nucleoside 5′-diphosphate phosphotransferase,
NUCLEOSIDE DIPHOSPHATE KINASE

Nucleoside diphosphokinase,
NUCLEOSIDE DIPHOSPHATE KINASE

Nucleoside monophosphate,
NUCLEOSIDE DIPHOSPHATASE

NUCLEOSIDE PHOSPHOTRANSFERASE

NUCLEOSIDE TRIPHOSPHATE: HEXOSE-1-PHOSPHATE NUCLEOTIDYLTRANS-FERASES

Nucleoside triphosphate reaction with a hexose 1-phosphate,
NUCLEOSIDE TRIPHOSPHATE HEXOSE-1-PHOSPHATE NUCLEOTIDYLTRANS-FERASES

NUCLEOSIDE-5′-TRIPHOSPHATE REGENERATION

NUCLEOSOME CORE PARTICLE SELF-ASSEMBLY

3′-NUCLEOTIDASE

5′-NUCLEOTIDASE

NUCLEOTIDE EXCHANGE FACTORS

Nucleotides and nucleic acids,
BIOCHEMICAL NOMENCLATURE

NUMERICAL COMPUTER METHODS

NUMERICAL INTEGRATION
STIFFNESS
GEAR ALGORITHM
NUMERICAL COMPUTER METHODS

O

PROTOPORPHYRINOGEN OXIDASE

PUTRESCINE OXIDASE

PYRUVATE OXIDASE

QUERCETIN 2,3-DIOXYGENASE

D-RIBULOSE-1,5-BISPHOSPHATE CAR-
BOXYLASE/OXYGENASE

SALICYLATE 1-MONOOXYGENASE

SARCOSINE OXIDASE

STEROID HYDROXYLASES

SULFITE OXIDASE

SULFUR DIOXYGENASE

THYMIDINE 2'-HYDROXYLASE

TOLUENE DIOXYGENASE

XANTHINE OXIDASE

OXYGEN ELECTRODE FOR MONITORING
HEMOGLOBIN OXYGENATION

OXYGEN, OXIDES & OXO ANIONS

Oxygen, photo-induced incorporation,
PHOTO-OXYGENATION

OXYGEN RADICALS

Oxygen transport in erythrocytes,
DIFFUSION OF OXYGEN WITHIN RED
CELLS

Oxygen uptake,
MEYERHOF OXIDATION QUOTIENT

Ozone,
OXYGEN, OXIDES & OXO ANIONS

P

PACEMAKER REACTION

PACKING DENSITY

Palitzsch buffers,
ISOTONIC BUFFERS

PANTOTHENIC ACID, COENZYME A, AND
DERIVATIVES

PAPAIN

Paracrystals,
BIOMINERALIZATION

PARALLEL FIRST-ORDER REACTIONS

Parallel line kinetics,
PHI VALUES IN BISUBSTRATE MECHA-
NISMS
INITIAL RATE ENZYME ASSAYS

PARALLEL REACTIONS
See specific type of parallel reaction

PARAMAGNETIC
PERMEABILITY
MAGNETIC SUSCEPTIBILITY
DIAMAGNETIC
ELECTRON SPIN RESONANCE

PARTIAL COMPETITIVE INHIBITION
COMPETITIVE INHIBITION

PARTIAL DERIVATIVES (Basic Properties)

Partial exchange reactions,
SUBSTRATE SYNERGISM

PARTIAL INHIBITION
PARTIAL COMPETITIVE INHIBITION
INHIBITION

PARTIAL MOLAR VOLUMES
DALTON'S LAW OF PARTIAL PRESSURES
CONCENTRATION
MOLECULAR CROWDING

PARTIAL PRESSURE
DALTON'S LAW OF PARTIAL PRESSURE
CONCENTRATION

PARTIAL RATE FACTOR
SELECTIVITY FACTOR

Partition and diffusion coefficients relative to
membrane thickness,
PERMEABILITY COEFFICIENT

PARTITIONING FACTOR

PARVULIN

PASCAL

PASCAL'S TRIANGLE

PATCH CLAMP

Pauling's postulate transition state stabili-
zation,
ENTROPY TRAP MODEL
MOLECULAR SIMILARITY
TRANSITION-STATE INHIBITORS

pCa

pD

Pectate,
PECTINESTERASE

Pectin demethoxylase,
PECTINESTERASE

PECTINESTERASE

Pectin methoxylase,
PECTINESTERASE

Pectin methylesterase,
PECTINESTERASE

Peltier cooling devices, noise reduction in
PM tubes,
NOISE

Penefsky centrifuged-column procedure,
RAPID BUFFER EXCHANGE

PENICILLOPEPSIN

Pentacovalent phosphorane intermediates,
PSEUDOROTATION

Pentavalent intermediate, in phosphotransfer reactions,
PSEUDOROTATION

Pentose-5-phosphate 3-epimerase,
RIBULOSE-5-PHOSPHATE 3-EPIMERASE

PEPSINS and PEPSINOGENS

Peptide bond(s),
PEPTIDOMIMETIC COMPOUND

PEPTIDE METHIONINE SULFOXIDE RE-
DUCTASE

PEPTIDOMIMETIC COMPOUND

PEPTIDYL-ASPARTATE ENDOPEPTIDASE

Peptidylglycine,
PEPTIDYL GLYCINE α-AMIDATING
MONOOXYGENASE

PEPTIDYL GLYCINE α-AMIDATING
MONOOXYGENASE

Peptidylglycine 2-hydroxylase,
PEPTIDYL GLYCINE α-AMIDATING
MONOOXYGENASE

Peptidylglycine monooxygenase,
PEPTIDYL GLYCINE α-AMIDATING
MONOOXYGENASE

Peptidylprolyl cis-trans isomerase,
PROLYL ISOMERASE

Peptidylprolyl isomerase,
PROLYL ISOMERASE

PEPTIDYLTRANSFERASE

Percent diastereoisomer excess,
STEREOCHEMICAL TERMINOLOGY,
IUPAC RECOMMENDATIONS

Percent enantiomer excess,
STEREOCHEMICAL TERMINOLOGY,
IUPAC RECOMMENDATIONS

Percolation clusters,
FRACTAL REACTION KINETICS

PERIODATE CLEAVAGE
MALAPRADE REACTION

Periodicity,
CIRCADIAN RHYTHM

Period of sustained chemical behavior or biological response,
PERSISTENCE TIME

Periplanar,
STEREOCHEMICAL TERMINOLOGY,
IUPAC RECOMMENDATIONS

Periselectivity,
PHYSICAL ORGANIC CHEMISTRY NO-
MENCLATURE

PERMEABILITY
MAGNETIC SUSCEPTIBILITY

PERMEABILITY COEFFICIENT

PERMISSIVE ACTIVE SITE

PERMUTATIONS AND COMBINATIONS
STATISTICS (A Primer)
LINKED FUNCTIONS

PEROXIDASE

HORSERADISH PEROXIDASE

OVOPEROXIDASE

MYELOPEROXIDASE

Perpendicular effect,

PHYSICAL ORGANIC CHEMISTRY NOMENCLATURE

PERSISTENCE TIME

TWIDDLING

PERSISTENT

PHYSICAL ORGANIC CHEMISTRY NOMENCLATURE

TRANSIENT CHEMICAL SPECIES

PERTUSSIS TOXIN

Phalloidin,

ACTIN ASSEMBLY KINETICS

PHARMACOKINETICS

PHASE

PHASE MODULATION FLUORESCENCE

PHASE-TRANSFER CATALYSIS

pH dependence in enzymic catalysis,

pH EFFECTS ON ENZYMIC CATALYSIS

pH EFFECTS ON ENZYMIC CATALYSIS

pH ELECTRODE

pH electrode correction for D_2O,

pD

Phenethylamine,

PHENYLALANINE DECARBOXYLASE

PHENOL HYDROXYLASE

Phenol 2-monooxygenase,

PHENOL HYDROXYLASE

Phenylalaninase,

PHENYLALANINE MONOOXYGENASE

Phenylalanine,

PHENYLALANINE AMINOTRANSFERASE

PHENYLALANINE AMMONIA-LYASE

PHENYLALANINE DECARBOXYLASE

PHENYLALANINE DEHYDROGENASE

PHENYLALANINE MONOOXYGENASE

PHENYLALANINE RACEMASE

PHENYLALANINE AMINOTRANSFERASE

AROMATIC AMINO ACID AMINOTRANSFERASE

PHENYLALANINE AMMONIA-LYASE

DEHYDROALANINE

BOROHYDRIDE REDUCTION

PHENYLALANINE DECARBOXYLASE

PHENYLALANINE DEHYDROGENASE

Phenylalanine(histidine) aminotransferase,

PHENYLALANINE AMINOTRANSFERASE

PHENYLALANINE HYDROXYLASE (Phenylalanine Monooxygenase)

PHENYLALANINE RACEMASE

PHENYLMETHYLSULFONYL FLUORIDE

CHYMOTRYPSIN

PROTEASE INHIBITOR "COCKTAILS"

Phenylpyruvate,

PHENYLALANINE DEHYDROGENASE

PREPHENATE DEHYDRATASE

PHENYLPYRUVATE SYNTHASE

PHI (ϕ or Φ)

PHI VALUES IN BISUBSTRATE MECHANISMS

INITIAL RATE ENZYME ASSAYS

Phosphate,

ORTHOPHOSPHATE CONTINUOUS ASSAY

OXYGEN, OXIDES & OXO ANIONS

PHOSPHATE ACETYLTRANSFERASE (Phosphotransacetylase)

PHOSPHATE CONTINUOUS ASSAY

ORTHOPHOSPHATE CONTINUOUS ASSAY

PHOSPHATE DIESTER HYDROLYSIS

Phosphate ester,

ARSENOLYSIS

PHOSPHATE MONOESTER HYDROLYSIS

Phosphatidase,

PHOSPHOLIPASE A_2

3-*sn*-Phosphatidate,

PHOSPHATIDATE PHOSPHATASE

PHOSPHATIDATE CYTIDYLYLTRANSFERASE

PHOSPHATIDATE PHOSPHATASE

Phosphatidolipase,

PHOSPHOLIPASE A_2

Phosphatidylcholine 2-acylhydrolase,

PHOSPHOLIPASE A_2

Phosphatidylcholine, hydrolysis of,

PHOSPHOLIPASE A_2

PHOSPHOLIPASE D

PHOSPHATIDYLCHOLINE:SERINE *O*-PHOSPHATIDYLTRANSFERASE

Phosphatidyl-*N*-dimethylethanolamine,

PHOSPHATIDYL-N-METHYLETHANOLAMINE N-METHYLTRANSFERASE

Phosphatidylethanolamine,

PHOSPHATIDYLETHANOLAMINE N-METHYLTRANSFERASE

PHOSPHATIDYLSERINE DECARBOXYLASE

PHOSPHATIDYLETHANOLAMINE N-METHYLTRANSFERASE

Phosphatidylglycerol,

PHOSPHATIDYLGLYCEROL PHOSPHATE PHOSPHOHYDROLASE

PHOSPHATIDYLGLYCEROL PHOSPHATE PHOSPHOHYDROLASE

Phosphatidylglycerophosphatase,

PHOSPHATIDYLGLYCEROL PHOSPHATE PHOSPHOHYDROLASE

Phosphatidylglycerophosphate,

PHOSPHATIDYLGLYCEROL PHOSPHATE PHOSPHOHYDROLASE

1-PHOSPHATIDYLINOSITOL 4-KINASE

PHOSPHATIDYLINOSITOL-4-PHOSPHATE 5-KINASE

Phosphatidyl-1D-*myo*-inositol,

PHOSPHATIDYLINOSITOL SYNTHASE

1-Phosphatidyl-1D-*myo*-inositol 4,5-bisphosphate synthesis,

PHOSPHATIDYLINOSITOL-4-PHOSPHATE 5-KINASE

1-Phosphatidyl-1D-*myo*-inositol 4-phosphate synthesis,

1-PHOSPHATIDYLINOSITOL 4-KINASE

1-PHOSPHATIDYLINOSITOL PHOSPHODIESTERASE

Phosphatidylinositol phospholipase C,

1-PHOSPHATIDYLINOSITOL PHOSPHODIESTERASE

PHOSPHATIDYLINOSITOL SYNTHASE

Phosphatidyl-*N*-methylethanolamine,

PHOSPHATIDYLETHANOLAMINE N-METHYLTRANSFERASE

PHOSPHATIDYL-N-METHYLETHANOLAMINE N-METHYLTRANSFERASE

O-*sn*-Phosphatidyl-L-serine,

PHOSPHATIDYLSERINE SYNTHASE

PHOSPHATIDYLSERINE DECARBOXYLASE

PHOSPHATIDYLSERINE SYNTHASE

Phosphinate inhibitor,

D-ALANINE-D-ALANINE LIGASE

PHOSPHOACETYLGLUCOSAMINE MUTASE

Phosphoacylase,

PHOSPHATE ACETYLTRANSFERASE

Phosphoadenylylsulfatase,

3'-PHOSPHOADENYLYL SULFATE SULFOHYDROLASE

3'-Phosphoadenylylsulfate hydrolysis,

3'-PHOSPHOADENYLYL SULFATE SULFOHYDROLASE

3'-PHOSPHOADENYLYL SULFATE SULFOHYDROLASE

Phosphoanhydride,

ACYL-PHOSPHATE INTERMEDIATE

ENERGY CHARGE MODEL

6-Phospho-2-dehydro-D-gluconate and NAD(P)H,

6-PHOSPHOGLUCONATE DEHYDROGENASE

Phosphodeoxyribomutase,

PHOSPHOPENTOMUTASE

PHOSPHODIESTERASES

See also specific enzyme

Phosphodiester hydrolysis,

EXONUCLEASES

Phosphoenolpyruvate,

PHOSPHOENOLPYRUVATE CARBOX-
YLASE

PYRUVATE KINASE

PYRUVATE, WATER DIKINASE

PYRUVATE, ORTHOPHOSPHATE DIK-
INASE

PHOSPHOENOLPYRUVATE CARBOXY-
KINASE

PHOSPHOENOLPYRUVATE CARBOXYKI-
NASE (Pyrophosphate)

PHOSPHOENOLPYRUVATE CARBOXYLASE

PHOSPHOENOLPYRUVATE CARBOXYKI-
NASE (Pyrophosphate)

PHOSPHOENOLPYRUVATE CARBOXY-
KINASE

Phosphoenolpyruvate carboxytransphosphorylase,

PHOSPHOENOLPYRUVATE CARBOXYKI-
NASE (PYROPHOSPHATE)

Phosphoenolpyruvate, interconversion with
3-phosphonopyruvate,

PHOSPHOENOLPYRUVATE MUTASE

PHOSPHOENOLPYRUVATE MUTASE

PHOSPHOENOLPYRUVATE PHOSPHATASE

Phosphoenolpyruvate phosphomutase,

PHOSPHOENOLPYRUVATE MUTASE

Phosphoenolpyruvate synthase,

PYRUVATE, WATER DIKINASE

Phosphoenolpyruvate synthesis,

PHOSPHOENOLPYRUVATE CARBOXY-
KINASE

PYRUVATE, WATER DIKINASE

PHOSPHOFRUCTOKINASE

ATP & GTP DEPLETION

6-PHOSPHOFRUCTO-2-KINASE

6-Phospho-D-galactose,

6-PHOSPHO-β-GALACTOSIDASE

6-PHOSPHO-β-GALACTOSIDASE

β-D-Phosphogalactoside galactohydrolase,

6-PHOSPHO-β-GALACTOSIDASE

PHOSPHOGLUCOMUTASE

6-PHOSPHOGLUCONATE DEHYDRO-
GENASE

PHOSPHOGLUCOSE ISOMERASE

6-PHOSPHO-β-GLUCOSIDASE

3-Phospho-D-glycerate,

D-RIBULOSE-1,5-BISPHOSPHATE CAR-
BOXYLASE/OXYGENASE

PHOSPHOGLYCERATE MUTASE

3-PHOSPHOGLYCERATE DEHYDRO-
GENASE

3-PHOSPHOGLYCERATE KINASE

PHOSPHOGLYCERATE MUTASE

Phosphoglycerate phosphomutase,

PHOSPHOGLYCERATE MUTASE

Phosphoglyceromutase,

PHOSPHOGLYCERATE MUTASE

3-Phospho-D-glyceroyl phosphate,

3-PHOSPHOGLYCERATE KINASE

PHOSPHOGLYCERIDE:LYSOPHOSPHATIDYL-
GLYCEROL ACYLTRANSFERASE

Phosphohexose isomerase,

PHOSPHOGLUCOSE ISOMERASE

Phosphohexomutase,

PHOSPHOGLUCOSE ISOMERASE

3-Phosphohydroxypyruvate formation,

3-PHOSPHOGLYCERATE DEHYDRO-
GENASE

PHOSPHOKETOLASE

PHOSPHOLIPASE

See specific enzyme

MICELLE

LIPASE

INTERFACIAL CATALYSIS

PHOSPHOLIPASE A$_1$

MICELLE

PHOSPHOLIPASE A$_2$

PHOSPHOLIPASE B

PHOSPHOLIPASE C

PHOSPHOLIPASE D

PHOSPHOLIPID FLIP-FLOP

PHOSPHOMANNOMUTASE

(R)-5-Phosphomevalonate, as substrate,

5-PHOSPHOMEVALONATE KINASE

5-PHOSPHOMEVALONATE KINASE

3-Phosphonopyruvate, interconversion with
phosphoenolpyruvate,

PHOSPHOENOLPYRUVATE MUTASE

(R)-4'-Phosphopantothenate, as substrate,

PHOSPHOPANTOTHENOYLCYSTEINE
SYNTHETASE

(R)-4'-Phosphopantothenoyl-L-cysteine,

PHOSPHOPANTOTHENOYLCYSTEINE
SYNTHETASE

PHOSPHOPANTOTHENOYLCYSTEINE SYN-
THETASE

PHOSPHOPENTOMUTASE

Phosphopentose isomerase,

RIBOSE-5-PHOSPHATE EPIMERASE

PHOSPHOPROTEIN PHOSPHATASES

Phosphopyruvate carboxylase,

PHOSPHOENOLPYRUVATE CARBOXY-
KINASE

PHOSPHOENOLPYRUVATE CARBOXYKI-
NASE (PYROPHOSPHATE)

PHOSPHORESCENCE

JABLONSKI DIAGRAM

FLUORESCENCE

5-Phospho-α-D-ribose 1-diphosphate,

RIBOSEPHOSPHATE PYROPHOSPHO-
KINASE

NICOTINAMIDE PHOSPHORIBOSYL-
TRANSFERASE

NICOTINATE PHOSPHORIBOSYLTRANS-
FERASE

Phosphoriboisomerase,

RIBOSE-5-PHOSPHATE EPIMERASE

5'-Phosphoribosyl-5-amino-4-
imidazolecarboxamide,

PHOSPHORIBOSYLAMINOIMIDAZOLE-
CARBOXAMIDE FORMYLTRANS-
FERASE

PHOSPHORIBOSYLAMINOIMIDAZOLECAR-
BOXAMIDE FORMYLTRANSFERASE

PHOSPHORIBOSYL-AMP CYCLOHY-
DROLASE

5-Phosphoribosyl-AMP, formation of,

PHOSPHORIBOSYL-ATP PYROPHOS-
PHATASE

5-Phosphoribosyl-AMP, hydrolysis of,

PHOSPHORIBOSYL-AMP CYCLOHY-
DROLASE

PHOSPHORIBOSYLANTHRANILATE ISO-
MERASE

5-Phosphoribosyl-ATP, hydrolysis of,

PHOSPHORIBOSYL-ATP PYROPHOS-
PHATASE

PHOSPHORIBOSYL-ATP PYROPHOS-
PHATASE

N^1-(5-Phospho-α-D-ribosyl)-5,6-
dimethylbenzimidazole,

NICOTINATE-NUCLEOTIDE DIMETHYL-
BENZIMIDAZOLE

PHOSPHORIBOSYLTRANSFERASE

5'-Phosphoribosyl-5-formamido-4-
imidazolecarboxamide,

PHOSPHORIBOSYLAMINOIMIDAZOLE-
CARBOXAMIDE FORMYLTRANS-
FERASE

N-(5'-PHOSPHORIBOSYLFORMIMINO)-5-
AMINO-1-(5''-PHOSPHORIBOSYL)-4-

PROTOPHILIC

PROTOGENIC

AMPHIPROTIC

PROTOPORPHYRINOGEN OXIDASE

Pseudo-asymmetric carbon atom,

STEREOCHEMICAL TERMINOLOGY, IUPAC RECOMMENDATIONS

Pseudo-axial,

STEREOCHEMICAL TERMINOLOGY, IUPAC RECOMMENDATIONS

PSEUDO BUFFERS

PSEUDOCATALYSIS

PSEUDOHALIDES

Pseudoisomeric forms,

BIOCHEMICAL THERMODYNAMICS

ATP, THERMODYNAMICS OF HYDROLYSIS

PSEUDOROTATION

PSI (ψ)

PTERIDINES, ANALOGUES & FOLATE COENZYMES

PULLULANASE

PULSE-CHASE EXPERIMENTS

N-END RULE

PULSE RADIOLYSIS

PURINE NUCLEOSIDASE

PURINE NUCLEOSIDE PHOSPHORYLASE

PUTRESCINE CARBAMOYLTRANSFERASE

PUTRESCINE *N*-METHYLTRANSFERASE

PUTRESCINE OXIDASE

Putrescine synthesis,

ORNITHINE DECARBOXYLASE

P-VALUE

Pyramidal inversion,

STEREOCHEMICAL TERMINOLOGY, IUPAC RECOMMENDATIONS

PYRAZOLE

ALCOHOL DEHYDROGENASE

Pyridine nucleotide transhydrogenase,

NAD(P)⁺ TRANSHYDROGENASE

PYRIDOXAL 4-DEHYDROGENASE

PYRIDOXAL KINASE

Pyridoxal-phosphate-dependent enzymes,

N²-ACETYLORNITHINE AMINOTRANSFERASE

O-ACETYLSERINE (THIOL)-LYASE

ALANINE AMINOTRANSFERASE

ALANINE:GLYOXYLATE AMINOTRANSFERASE

ALANINE RACEMASE

S-ALKYLCYSTEINE LYASE

ALLIIN LYASE

AMINO ACID RACEMASE

2-AMINOADIPATE AMINOTRANSFERASE

1-AMINOCYCLOPROPANE-1-CARBOXYLATE SYNTHASE

5-AMINOLEVULINATE AMINOTRANSFERASE

5-AMINOLEVULINATE DEHYDRATASE

5-AMINOLEVULINATE SYNTHASE

ARGININE DECARBOXYLASE

ARGININE RACEMASE

AROMATIC AMINO ACID AMINOTRANSFERASE

ASPARTATE AMINOTRANSFERASE

ASPARTATE β-DECARBOXYLASE

BRANCHED-CHAIN AMINO ACID AMINOTRANSFERASE

CYSTATHIONINE β-LYASE

CYSTATHIONINE β-SYNTHASE

CYSTEINE CONJUGATE β-LYASE

CYSTEINE LYASE

DIAMINOPIMELATE DECARBOXYLASE

DIIODOTYROSINE AMINOTRANSFERASE

GLUCOSAMINATE AMMONIA-LYASE

GLUTAMATE DECARBOXYLASE

GLUTAMATE FORMIMINOTRANSFERASE

GLUTAMINE AMINOTRANSFERASE

GLYCINE ACETYLTRANSFERASE

GLYCINE AMINOTRANSFERASE

GLYCINE DECARBOXYLASE

GLYCINE:OXALOACETATE AMINOTRANSFERASE

GLYCINE SYNTHASE

HISTIDINE DECARBOXYLASE

HISTIDINOL-1-PHOSPHATE AMINOTRANSFERASE

HOMOCYSTEINE DESULFHYDRASE

erythro-β-HYDROXYASPARTATE DEHYDRATASE

KYNURENINASE

KYNURENINE AMINOTRANSFERASE

LEUCINE AMINOTRANSFERASE

LYSINE 2,3-AMINOMUTASE

LYSINE 6-AMINOTRANSFERASE

LYSINE DECARBOXYLASE

METHIONINE γ-LYASE

ORNITHINE AMINOTRANSFERASE

PHENYLALANINE DECARBOXYLASE

PHOSPHATIDYLSERINE DECARBOXYLASE

PHOSPHOSERINE AMINOTRANSFERASE

PYRIDOXAMINE:PYRUVATE AMINOTRANSFERASE

L-SELENOCYSTEINE β-LYASE

SERINE DEHYDRATASE

SERINE PALMITOYLTRANSFERASE

O-SUCCINYLHOMOSERINE (THIOL)-LYASE

THREONINE DEHYDRATASE

TRYPTOPHANASE

TYROSINE AMINOTRANSFERASE

Pyridoxal 5'-phosphate, synthesis of,

PYRIDOXAL KINASE

PYRIDOXAMINE:PYRUVATE AMINOTRANSFERASE

PYRIDOXINE & PYRIDOXAL (Analogues and Derivatives)

4-Pyridoxolactone, synthesis of,

PYRIDOXAL 4-DEHYDROGENASE

PYRIMIDINE DIMER DNA GLYCOSYLASE

Pyroglutamase (ATP-hydrolyzing),

5-OXOPROLINASE

PYROGLUTAMYL PEPTIDE HYDROLASE

PYROLYSIS

PYROPHOSPHATASE, INORGANIC

Pyrophosphate,

OXYGEN, OXIDES & OXO ANIONS

Pyrophosphate as a substrate or product,

ACETATE KINASE (PYROPHOSPHATE)

ACYLNEURAMINATE CYTIDYLYLTRANSFERASE

ADENINE PHOSPHORIBOSYLTRANSFERASE

ADENYLATE CYCLASE

ALKALINE PHOSPHATASE

AMIDOPHOSPHORIBOSYLTRANSFERASE

AMINOACYL-tRNA SYNTHETASES

ANTHRANILATE PHOSPHORIBOSYLTRANSFERASE

ARACHIDONYL-CoA SYNTHETASE

ARGININOSUCCINATE SYNTHETASE

ASPARAGINE SYNTHETASE

ATP:GLUCOSE-1-PHOSPHATE ADENYLYLTRANSFERASE

ATP PHOSPHORIBOSYLTRANSFERASE

BENZOATE-CoA LIGASE

BIOTIN HOLOCARBOXYLASE SYNTHETASE

BORNYL PYROPHOSPHATE SYNTHASE

CARNOSINE SYNTHETASE

CHOLINE-PHOSPHATE CYTIDYLYLTRANSFERASE

CHOLOYL-CoA SYNTHETASE

COBALAMIN ADENOSYLTRANSFERASE

4-COUMAROYL-CoA SYNTHETASE

R

$R_{0.9}$,
  $S_{0.9}/S_{0.5}$
RACEMASES
Racemic conglomerate,
  *STEREOCHEMICAL TERMINOLOGY,
  IUPAC RECOMMENDATIONS*
RACEMIC MIXTURE (or Racemate)
RACEMIZATION
RAD
RADIAN
RADIANCE
RADIANT ENERGY
RADIANT (or, ENERGY) FLUX
Radiant power,
  RADIANCE
Radiationless transfer,
  JABLONSKI DIAGRAM
RADIATIONLESS TRANSITION
  FLUORESCENCE
RADIATIVE LIFETIME
RADICAL (or, FREE RADICAL)
RADICAL CENTER
RADICAL ION
Radioactive decay,
  CHEMICAL KINETICS
RADIOACTIVE SUBSTANCES (General Pre-
  cautions)
RADIOCARBON DATING
  COSMOGENIC DATING
RADIOLYSIS
  PULSE RADIOLYSIS
Radiometric technique,
  PULSE-CHASE EXPERIMENTS
RAMAN SCATTERING
RANDOM AB/QR, ORDERED C/P TER TER
  MECHANISM
RANDOM AC/PR, ORDERED B/Q TER TER
  MECHANISM
RANDOM BC/PQ, ORDERED A/R TER TER
  MECHANISM
RANDOM BI BI MECHANISM
  MULTISUBSTRATE MECHANISM
RANDOM BI BI DUAL-THEORELL-CHANCE
  MECHANISM
  MULTISUBSTRATE MECHANISM
Random Bi Bi ternary complex mechanism,
  ISOTOPE EXCHANGE AT EQUILIBRIUM
RANDOM BI UNI MECHANISM
RANDOM BI UNI UNI BI PING PONG
  MECHANISM

RANDOM ERROR
  STATISTICS (A Primer)
RANDOM EXPERIMENT
  STATISTICS (A Primer)
RANDOMIZATION
  STATISTICS (A Primer)
RANDOM MECHANISMS
  MULTISUBSTRATE MECHANISMS
Random number table,
  STATISTICS (A Primer)
RANDOM-ON/ORDERED-OFF BI BI MECH-
  ANISM
RANDOM SCISSION KINETICS
RANDOM TER TER MECHANISM
RANDOM UNI BI MECHANISM
RANDOM UNI UNI BI BI PING PONG
  MECHANISM
RANDOM VARIABLE
  STATISTICS (A Primer)
RANDOM WALK
  DIFFUSION
  PERSISTENCE TIME
  TWIDDLING
RAPID BUFFER EXCHANGE
Rapid equilibrium assumption,
  QUASI-EQUILIBRIUM ASSUMPTION
RAPID EQUILIBRIUM MECHANISMS
Rapid gel filtration of biomacromolecules,
  RAPID BUFFER EXCHANGE
RAPID MIXING
  QUENCHED FLOW EXPERIMENTS
Rapid quench experiments,
  QUENCH-FLOW EXPERIMENTS
Rapid reaction kinetics,
  PULSE RADIOLYSIS
Rapid reaction techniques,
  CRYOENZYMOLOGY
  PULSE RADIOLYSIS
Rapid removal of enzyme stabilizing agents,
  RAPID BUFFER EXCHANGE
RAPID SCAN SPECTROSCOPY OF FAST RE-
  ACTIONS
RAPID-START COMPLEX
Rate coefficients exhibiting temporal
  memory,
  FRACTAL REACTION KINETICS
RATE CONSTANT
  CHEMICAL KINETICS
RATE-CONTRIBUTING STEP
RATE-CONTROLLING STEP
  *RATE-DETERMINING STEP or RATE-LIM-
  ITING STEP*
  KINETIC ISOTOPE EFFECTS

RATE-DETERMINING STEP or RATE-LIM-
  ITING STEP
Rate enhancement,
  CATALYSIS
  CHEMICAL KINETICS
  ORBITAL STEERING
Rate enhancement observed in two-phase re-
  action,
  PHASE-TRANSFER CATALYSIS
RATE LAW
  CHEMICAL KINETICS
Rate-limiting step in inner-sphere coordi-
  nation,
  EIGEN-TAMM MECHANISM
Rate-limiting step in metabolic pathway,
  PACEMAKER REACTION
RATE OF APPEARANCE
  CHEMICAL KINETICS
RATE OF DISAPPEARANCE
  CHEMICAL KINETICS
Rate of intracellular synthesis and/or turn-
  over of a metabolite,
  PULSE-CHASE EXPERIMENTS
RATE OF REACTION
  CHEMICAL KINETICS
Rate processes in aqueous solutions,
  CHEMICAL KINETICS
RATE SATURATION BEHAVIOR
RAY-ROSCELLI TREATMENT
  ISOMERIZATION
  ISO MECHANISMS
RE-
REACTANCY
Reactant conversion into its mirror image,
  NARCISSISTIC REACTION
REACTING BOND RULES
REACTING ENZYME CENTRIFUGATION
REACTION COORDINATE DIAGRAM
  POTENTIAL ENERGY DIAGRAM
  SADDLE POINT
  *ENZYME ENERGETICS (The Case of Pro-
  line Racemase)*
REACTION CYCLE
Reaction intermediate,
  ACYL-PHOSPHATE INTERMEDIATE
  PERSISTENT
  TRANSIENT CHEMICAL SPECIES
REACTION MECHANISM
  CHEMICAL KINETICS
Reaction order of nonenzymic reaction,
  CHEMICAL KINETICS
  NOYES EQUATION

Riboflavin,
RIBOFLAVINASE
RIBOFLAVIN KINASE
RIBOFLAVIN SYNTHASE
RIBOFLAVINASE
RIBOFLAVIN KINASE
RIBOFLAVIN SYNTHASE
RIBONUCLEASE (or RNase)
PSEUDOROTATION
DIETHYL PYROCARBONATE
RIBONUCLEASE F
RIBONUCLEASE II
RIBONUCLEASE III
Ribonuclease inactivation,
DIETHYL PYROCARBONATE
RIBONUCLEASE IV
RIBONUCLEASE P
Ribonuclease, pancreatic,
RIBONUCLEASE (or RNase)
Ribonucleoside synthesis,
3'-NUCLEOTIDASE
5'-NUCLEOTIDASE
RIBONUCLEOTIDE REDUCTASE
D-Ribose 5-phosphate,
RIBOSE-5-PHOSPHATE EPIMERASE
*RIBOSEPHOSPHATE PYROPHOSPHO-
KINASE*
RIBOSE-5-PHOSPHATE ADENYLYLTRANS-
FERASE
RIBOSE-5-PHOSPHATE EPIMERASE
D-Ribose 1-phosphate, interconversion to
D-ribose 5-phosphate,
PHOSPHOPENTOMUTASE
Ribose 5-phosphate isomerase,
RIBOSE-5-PHOSPHATE EPIMERASE
RIBOSEPHOSPHATE PYROPHOSPHO-
KINASE
Ribozyme,
CATALYTIC RNA
ENZYME
D-Ribulose 1,5-bisphosphate,
*D-RIBULOSE-1,5-BISPHOSPHATE CAR-
BOXYLASE/OXYGENASE*
D-RIBULOSE-1,5-BISPHOSPHATECARBOXYL-
ASE/OXYGENASE
D-Ribulose 5-phosphate,
*RIBITOL-5-PHOSPHATE 2-DEHYDRO-
GENASE*
RIBULOSE-5-PHOSPHATE 3-EPIMERASE
L-Ribulose 5-phosphate,
RIBULOSE-5-PHOSPHATE 4-EPIMERASE
RIBULOSE-5-PHOSPHATE 3-EPIMERASE

RIBULOSE-5-PHOSPHATE 4-EPIMERASE
RIESKE IRON-SULFUR PROTEIN
Ring reversal (or ring inversion),
*STEREOCHEMICAL TERMINOLOGY,
IUPAC RECOMMENDATIONS*
Ring structures, role in microtubule as-
sembly,
MICROTUBULE ASSEMBLY KINETICS
RITCHIE EQUATION
RNA, exonucleolytic cleavage,
PHOSPHODIESTERASES
RNA LIGASE
RNA POLYMERASES
EDITING MECHANISMS
REPLICASE
RNA stability,
DIETHYL PYROCARBONATE
ROENTGEN
ROENTGEN-EQUIVALENT-MAN (rem)
Rosanoff convention,
*STEREOCHEMICAL TERMINOLOGY,
IUPAC RECOMMENDATIONS*
Rotamer,
*STEREOCHEMICAL TERMINOLOGY,
IUPAC RECOMMENDATIONS*
Rotary engine,
BINDING CHANGE MECHANISM
Rotational barrier,
*STEREOCHEMICAL TERMINOLOGY,
IUPAC RECOMMENDATIONS*
Rotational catalysis by ATP synthase,
BINDING CHANGE MECHANISM
ROTATIONAL CORRELATION TIME
CORRELATION FUNCTION
ROTATIONAL DIFFUSION
FLUORESCENCE
rRNA (ADENINE-6-)-METHYLTRANSFERASE
rRNA (GUANINE-1-)-METHYLTRANSFERASE
R_s
RUBBER *cis*-POLYPRENYL*CIS*TRANSFERASE
Rubidium ion,
IONIC RADIUS
Rubisco,
*D-RIBULOSE-1,5-BISPHOSPHATE CAR-
BOXYLASE/OXYGENASE*
RUBREDOXIN
RUBREDOXIN:NAD$^+$ REDUCTASE
RUNGE-KUTTA ALGORITHM
NUMERICAL COMPUTER METHODS
NUMERICAL INTEGRATION
STIFFNESS
GEAR ALGORITHM
R_v

S

$S_{0.5}$
$S_{0.9}/S_{0.1}$ (or, R_s)
SAAM
Sacrificial hyperconjugation,
*PHYSICAL ORGANIC CHEMISTRY NO-
MENCLATURE*
SADDLE POINT
SALICYLATE 1-MONOOXYGENASE
SAMPLE SPACE
RANDOM EXPERIMENT
STATISTICS (A Primer)
Sample turbidity,
FLUORESCENCE
SAPONIFICATION
Sarcoplasmic reticulum calcium pump,
ION PUMPS
SARCOSINE DEHYDROGENASE
SARCOSINE OXIDASE
SATURATION
Saturation function,
LIGAND BINDING
COOPERATIVE LIGAND BINDING
SATURATION KINETICS
Saturation ratio,
RELATIVE SUPERSATURATION
SCALAR COUPLING RELAXATION
Scatchard equation,
SCATCHARD PLOT
SCATCHARD PLOT
SCAVENGER
SCHIFF BASE (or, Schiff's Base)
Schiff base reduction,
BOROHYDRIDE REDUCTION
SCHRÖDINGER EQUATION
*QUANTUM MECHANICAL TUNNEL-
LING*
SCINTILLATORS
SCISSILE BOND
Scrambling,
ISOTOPIC SCRAMBLING
POSITIONAL ISOTOPE EXCHANGE
SCREW SENSE OF METAL-NUCLEOTIDE
COMPLEXES
Sea shell formation,
BIOMINERALIZATION
SECOND
SECONDARY ISOTOPE EFFECT
KINETIC ISOTOPE EFFECT

Secondary kinetic isotope effect,
  KINETIC ISOTOPE EFFECT
Second law of thermodynamics,
  THERMODYNAMICS, LAWS OF
  ENTROPY
  CARATHEODORY, PRINCIPLE OF
  CARNOT CYCLE
  EFFICIENCY
Second-order autocalytic reaction,
  AUTOCATALYSIS
Second-order rate constant,
  ENCOUNTER-CONTROLLED RATE
SECOND-ORDER REACTION
  CHEMICAL KINETICS
  ORDER OF REACTION
  NOYES EQUATION
  MOLECULARITY
  AUTOCATALYSIS
  FIRST-ORDER REACTION
SEDIMENTATION EQUILIBRIUM ULTRACEN-
  TRIFUGATION
SEDIMENTATION VELOCITY ULTRACEN-
  TRIFUGATION
Seeded assembly,
  POLYMERIZATION
  ELONGATION (No Change in Polymer
  Number Concentration)
SELECTION RULE
Selective protein degradation rates,
  N-END RULE
Selenium donor protein,
  SELENOPHOSPHATE SYNTHETASE
L-SELENOCYSTEINE
  L-SELENOCYSTEINE β-LYASE
L-SELENOCYSTEINE β-LYASE
  L-SELENOCYSTEINE
Selenophosphate,
  SELENOPHOSPHATE SYNTHETASE
SELENOPHOSPHATE SYNTHETASE
SELF-ABSORPTION
Self-activating process,
  AUTOCATALYSIS
Self-ordering reactants,
  FRACTAL REACTION KINETICS
Self-phosphorylation process,
  AUTOPHOSPHORYLATION
SELF-PROTECTION MECHANISM
SELF-QUENCHING
Self-replicating elements,
  HYPERCYCLE
Self-unmixing of reactants,
  FRACTAL REACTION KINETICS

SENSITIVITY
Sensitivity of measurement,
  NOISE
SEPARATION OF VARIABLES
SEQUENTIAL MECHANISMS
  MULTISUBSTRATE MECHANISMS
Sequential *versus* ping-pong mechanisms,
  PHI VALUES IN BISUBSTRATE MECHA-
  NISMS
  INITIAL RATE ENZYME ASSAYS
Series first-order processes,
  CHEMICAL KINETICS
SERIES REACTIONS
  CHEMICAL KINETICS
Serine
  SERINE DEHYDRATASE
  SERINE DEHYDROGENASE
  SERINE HYDROXYMETHYLTRANSFERASE
  SERINE PALMITOYLTRANSFERASE
Serine deaminase
  SERINE DEHYDRATASE
SERINE DEHYDRATASE
SERINE DEHYDROGENASE
Serine esterase inhibitor, irreversible,
  PHENYLMETHYLSULFONYL FLUORIDE
SERINE HYDROXYMETHYLTRANSFERASE
SERINE PALMITOYLTRANSFERASE
Serine protease inhibitor, irreversible,
  PHENYLMETHYLSULFONYL FLUORIDE
Serine proteinase inhibitor,
  SERPIN
Serine/threonine-specific protein phos-
  phatase,
  PHOSPHOPROTEIN PHOSPHATASES
SERPINS (INHIBITORY MECHANISM)
L-SERYL-tRNA ^Ser SELENIUM TRANSFERASE
  L-SELENOCYSTEINE
Severing protein,
  CYTOSKELETAL PROTEIN POLYMER-
  IZATION
SHIELDING
  CHEMICAL SHIFT
SHIFTED BINDING
SHIFTING SPECIFICITY MODEL FOR EN-
  ZYME CATALYSIS
Shigella,
  ACTIN-BASED BACTERIAL MOTILITY
  ACTIN FILAMENT GROWTH (Polymeriza-
  tion Zone Model)
Shikimate,
  SHIKIMATE DEHYDROGENASE
  SHIKIMATE KINASE

SHIKIMATE DEHYDROGENASE
SHIKIMATE KINASE
Shikimate 3-phosphate,
  SHIKIMATE KINASE
Shot noise,
  NOISE
SHUTTLE GROUP
SI-
SIALIDASES (or Neuraminidases)
O-SIALOGLYCOPROTEASE
SIALYLTRANSFERASE
SIDE REACTIONS
SIGMA (σ)
σ-ADDUCT (Sigma-Adduct)
σ-CONSTANT (Sigma-Constant)
$\sigma \rightarrow \sigma^*$ TRANSITION
  ABSORPTION SPECTROSCOPY
SIGMATROPIC REARRANGEMENT
Sigmoidal dose-response,
  ENZYME CASCADE KINETICS
SIGNAL AMPLIFICATION
  ENZYME CASCADE KINETICS
SIGNAL PEPTIDASE, EUKARYOTIC MICRO-
  SOMAL
SIGNAL PROCESSOR
Signal-to-noise ratio,
  NOISE
SIGNIFICANT FIGURES
  PRECISION
Simple diffusion through a membrane,
  PERMEABILITY COEFFICIENT
Simplest explanation or model,
  OCCAM'S RAZOR
Simplified treatment of enzyme kinetics,
  NET RATE CONSTANT METHOD
Simulation, analysis, and modeling program,
  SAAM
Single-electron transfer mechanism (SET),
  PHYSICAL ORGANIC CHEMISTRY NO-
  MENCLATURE
Single-ion activity coefficients,
  DEBYE-HÜCKEL TREATMENT
SINGLE-POOL MODEL FOR METABOLITE
  COMPARTMENTATION
SINGLE-STEP REACTION
Single-stranded RNA and DNA substrates,
  NUCLEASE S1
Singlet dioxygen,
  OXYGEN, OXIDES & OXO ANIONS
SINGLET MOLECULAR OXYGEN

Singlet oxygen,
OXYGEN, OXIDES & OXO ANIONS
SINGLET STATE
JABLONSKI DIAGRAM
SINGLET-TRIPLET ENERGY TRANSFER
SINGLE-TURNOVER CONDITIONS
Site-directed inhibitor,
AFFINITY LABELING
Site-site interaction,
COOPERATIVITY
Site-specific enzyme modification,
AFFINITY LABELING
SITE-SPECIFIC ISOTOPE FRACTIONATION
IN BIOCHEMICAL MECHANISMS
Site-specific labeling,
AFFINITY LABELING
SI UNITS
Skew,
STEREOCHEMICAL TERMINOLOGY,
IUPAC RECOMMENDATIONS
Slit widths,
FLUORESCENCE
SLOPE
SLOPE-INTERCEPT EQUATION
SLOPE REPLOTS (or, Slope Secondary Re-
plots)
SLOW-BINDING INHIBITOR
D-ALANINE-D-ALANINE LIGASE
Slowest enzymic step in metabolic pathway,
PACEMAKER ENZYME
Slow partial exchange reactions,
SUBSTRATE SYNERGISM
SLOW TEMPERATURE-JUMP METHODS
DEPOLYMERIZATION, END-WISE
SLOW, TIGHT-BINDING INHIBITOR
SMOLUCHOWSKI TREATMENT FOR AG-
GREGATION
S_N1 and S_N2 reactions,
KINETIC ISOTOPE EFFECTS
S_N DISPLACEMENT REACTIONS
SOLVOLYSIS
S/N ratio,
NOISE
SODIUM-22 (^{22}Na)
Sodium ion,
IONIC RADIUS
Sodium, potassium ATPase,
ATPases
Solanesyl-diphosphate synthase,
trans-OCTAPRENYLTRANSTRANSFERASE
Solid-state NMR of inhibited complex,
D-ALANINE-D-ALANINE LIGASE

Solubility,
BIOMINERALIZATION
SOLUBILITY PRODUCT
Solubility product constant,
BIOMINERALIZATION
Solute-solvent interactions,
HYDROPHOBIC EFFECT
SOLUTION
SOLVATION
HYDRATION ATMOSPHERE
CHEMICAL KINETICS
Solvatochromic relationship,
PHYSICAL ORGANIC CHEMISTRY NO-
MENCLATURE
Solvatochromism,
PHYSICAL ORGANIC CHEMISTRY NO-
MENCLATURE
SOLVENT-ACCESSIBLE SURFACE AREA
Solvent acidity function,
BUNNETT-OLSEN EQUATIONS
Solvent cage,
CHEMICAL KINETICS
SOLVENT-TRAPPED INTERMEDIATES
SOLVENT ISOTOPE EFFECT
KINETIC ISOTOPE EFFECT
Solvent parameter,
PHYSICAL ORGANIC CHEMISTRY NO-
MENCLATURE
SOLVENT TRAPPING OF ENZYME INTERME-
DIATES
SOLVOLYSIS
Solvophobicity parameter,
PHYSICAL ORGANIC CHEMISTRY NO-
MENCLATURE
SOMO
PHYSICAL ORGANIC CHEMISTRY NO-
MENCLATURE
SONICATION
SORBITOL DEHYDROGENASE
L-Sorbose,
SORBITOL DEHYDROGENASE
Speciation,
BIOMINERALIZATION
SPECIFIC ACTIVITY
ISOTOPE DILUTION
SPECIFIC CATALYSIS
GENERAL ACID CATALYSIS
GENERAL BASE CATALYSIS
SPECIFICITY
SPECIFICITY CONSTANT
MAPPING SUBSTRATE INTERACTIONS
USING KINETIC DATA

SPECIFICITY CONSTANT
Spectrophotofluorimeter design,
FLUORESCENCE
Spin adduct,
PHYSICAL ORGANIC CHEMISTRY NO-
MENCLATURE
Spin counting,
PHYSICAL ORGANIC CHEMISTRY NO-
MENCLATURE
Spin density,
PHYSICAL ORGANIC CHEMISTRY NO-
MENCLATURE
SPIN LABEL
ELECTRON SPIN RESONANCE
Spin multiplicity,
PHOSPHORESCENCE
JABLONSKI DIAGRAM
Spin-spin coupling,
NUCLEAR MAGNETIC RESONANCE
SPONTANEOUS PROCESS
Spontaneous resolution,
STEREOCHEMICAL TERMINOLOGY,
IUPAC RECOMMENDATIONS
SQUALENE SYNTHASE
"S—S rearrangase,"
PROTEIN DISULFIDE ISOMERASE
STABILITY CONSTANT
METAL ION COMPLEXATION
STABILITY CONSTANT DETERMINATION
BY A LINEAR PLOT METHOD
STABILITY CONSTANT DETERMINATION
BY A LINEAR PLOT METHOD
STABILITY, PRODUCT
SUBSTRATE STABILITY
Stabilization,
ENZYME STABILITY
STANDARD CURVE
BEER-LAMBERT LAW
ABSORPTION SPECTROSCOPY
STANDARD DEVIATION
STATISTICS (A Primer)
Standard deviation of the data set,
NORMAL ERROR CURVE
STATISTICS (A Primer)
Standard error curve,
NORMAL ERROR CURVE
STATISTICS (A Primer)
Standard error of the mean,
NORMAL ERROR CURVE
STATISTICS (A Primer}
STAPHYLOCOCCAL NUCLEASE
STARCH PHOSPHORYLASE

STARCH SYNTHASE

STATE FUNCTION

STATIONARY STATE

STEADY STATE

STATISTICAL ANALYSIS OF ENZYME RATE DATA

STATISTICS (A Primer)

STATISTICAL FACTOR

STATISTICS (A Primer)

Statistical mechanics,

ENTROPY

STATISTICS (A Primer)

STEADY STATE

Steady-state approximation for polymer nucleation,

MICROTUBULE ASSEMBLY KINETICS

STEADY-STATE ASSUMPTION

ENZYME KINETICS

NUMERICAL INTEGRATION

Steady-state processes,

NONEQUILIBRIUM THERMODYNAMICS (A Primer)

Stepwise depolymerization,

DEPOLYMERIZATION, END-WISE

STEPWISE REACTION

MULTISTAGE REACTION

STEREOCHEMICAL FIDELITY

STEREOCHEMICAL TERMINOLOGY (IUPAC Recommendations)

STEREOCHEMISTRY OF HYDRIDE TRANSFER TO/FROM NADH

STEREOCHEMISTRY OF HYDROGEN TRANSFER TO/FROM FADH$_2$

STEREOCHEMISTRY OF ISOMERASES INVOLVING PROTON TRANSFER

Stereoconvergence,

STEREOCHEMICAL TERMINOLOGY, IUPAC RECOMMENDATIONS

Stereodescriptor,

STEREOCHEMICAL TERMINOLOGY, IUPAC RECOMMENDATIONS

STEREOELECTRONIC

Stereogenic unit (or Stereogen),

STEREOCHEMICAL TERMINOLOGY, IUPAC RECOMMENDATIONS

Stereoheterotopic,

STEREOCHEMICAL TERMINOLOGY, IUPAC RECOMMENDATIONS

Stereoheterotopic face of a trigonal atom,

RE-

SI-

Stereoheterotopic groups in a prochiral atom,

pro-R GROUP

pro-S GROUP

Stereoisomerism,

STEREOCHEMICAL TERMINOLOGY, IUPAC RECOMMENDATIONS

Stereoisomers,

DIASTEREOMERS

STEREOSELECTIVITY

STEREOSPECIFICITY

STERIC-APPROACH CONTROL

STERIC HINDRANCE

Steric hindrance effect on ligand binding,

SCATCHARD PLOT

Stern layer,

BIOMINERALIZATION

STEROID 11β-HYDROXYLASE

STEROID 17α-HYDROXYLASE

STEROID 21-HYDROXYLASE

Steroids,

BIOCHEMICAL NOMENCLATURE

STEROID SULFOTRANSFERASE

STICKY SUBSTRATES

ISOTOPE TRAPPING

STIFFNESS (or Stiffness Instability)

GEAR ALGORITHM

Stoichiometric coupling,

ION PUMPS

STOICHIOMETRIC NUMBER

Stoichiometry of elementary reactions,

CHEMICAL KINETICS

MOLECULARITY

UNIMOLECULAR

BIMOLECULAR

TRANSITION-STATE THEORY

ELEMENTARY REACTION

STOKE'S SHIFT

STOPPED-FLOW MASS SPECTROMETRY IN ENZYME KINETICS

STOPPED-FLOW TECHNIQUES

Stopping a chemical reaction,

QUENCHING

STRAIN

Streak camera method,

FLUORESCENCE

Streptavidin,

BIOTIN AND DERIVATIVES

STREPTOKINASE

Stress,

PASCAL

Strong cooperativity,

BINDING CHANGE MECHANISM

STURGILL-BILTONEN METHOD

Suberyldicholine dissociation from acetylcholine receptor,

DISSOCIATION KINETICS

SUBSITE MAPPING

SUBSTITUENT

SUBSTITUTION REACTION

S$_N$ REACTIONS

Substitution reactions at a saturated carbon,

ALPHA EFFECT

SUBSTRATE

SUBSTRATE-ASSISTED CATALYSIS

SUBSTRATE CHANNELING

Substrate contaminated with the enzyme,

BASAL RATE

Substrate cycling,

BIFURCATION THEORY

Substrate depletion,

INITIAL RATE PHASE

Substrate dissociation rate,

ISOTOPE TRAPPING

STICKY SUBSTRATES

Substrate-induced conformational change,

INDUCED FIT MODEL

SUBSTRATE INHIBITION

ABORTIVE COMPLEX FORMATION

LACTATE DEHYDROGENASE

LEE-WILSON EQUATION

Substrate ionization, effect on enzyme catalysis,

pH EFFECTS ON ENZYME CATALYSIS

SUBSTRATE OXIDATION KINETICS

Substrate-product ratio,

ISOTOPE EXCHANGE AT EQUILIBRIUM

SUBSTRATE PURITY

SUBSTRATE STABILITY & HANDLING

SUBSTRATE SYNERGISM

SUBTILISIN

Substituent or moiety bound to a molecular entity,

RADICAL (or FREE RADICAL)

Subunit assembly in multisubunit proteins,

BIOCHEMICAL SELF-ASSEMBLY

Subzero enzymology,

CRYOENZYMOLOGY

SUCCESSOR COMPLEX

Succinate,

SUCCINYL-CoA SYNTHETASE

O-SUCCINYLHOMOSERINE (THIOL)-LYASE

SUCCINATE DEHYDROGENASE

Succinyl-CoA,
SUCCINYL-CoA SYNTHETASE
SUCCINYL-CoA SYNTHETASE
O-Succinyl-L-homoserine,
O-SUCCINYLHOMOSERINE (THIOL)-LYASE
O-SUCCINYLHOMOSERINE (THIOL)-LYASE
SUCRASE
SUCROSE PHOSPHORYLASE
ARSENOLYSIS
SULFATE ADENYLYLTRANSFERASE
SULFATE ADENYLYLTRANSFERASE (ADP)
Sulfhydryl-disulfide interchange,
PROTEIN DISULFIDE ISOMERASE
SULFIDES
Sulfite,
RHODANESE
SULFITE DEHYDROGENASE
SULFITE OXIDASE
SULFITE REDUCTASE
SULFONIUM SALTS
SULFOTRANSFERASES
SULFUR-35 (^{35}S)
SULFUR DIOXYGENASE
Sulfuric acid, autoprotolysis constant,
AUTOPROTOLYSIS
Superoxide,
OXYGEN, OXIDES & OXO ANIONS
PHOTOOXYGENATION
SUPEROXIDE DETECTION/MEASUREMENT
SUPEROXIDE DISMUTASES
SUPERACID
SUPERACID CATALYSIS
SUPERBASE
SUPERIMPOSABILITY (or Superposability)
SUPERRADIANCE
SUPERSATURATION
BIOMINERALIZATION
Suprafacial,
ANTARAFACIAL
ANTARAFACIAL AND SUPRAFACIAL MIGRATIONS
ANTARAFACIAL AND SUPRAFACIAL REACTIONS
Surface-adsorbed ions and molecules,
BIOMINERALIZATION
Surface charge,
BIOMINERALIZATION
SURFACE CROSSING
Surface diffusion,
BIOMINERALIZATION

Surface fluorescence,
NADH/NADPH SURFACE FLUORESCENCE IN LIVING TISSUES
Surface plasmon resonance (SPR),
BIOSENSOR
SURFACE TENSION
Sustained chemical behavior or biological response,
PERSISTENCE TIME
SWAIN-LUPTON RELATION
SWAIN RELATIONSHIP
KINETIC ISOTOPE EFFECTS
Swain-Schaad relationship,
EXPONENTIAL BREAKDOWN
SYMBOLIC COMPUTING
Symmetry-conserving allosteric model,
MONOD-WYMAN-CHANGEUX MODEL
SYMPORT
Symproportionation,
PHYSICAL ORGANIC CHEMISTRY NOMENCLATURE
Synartetic acceleration,
PHYSICAL ORGANIC CHEMISTRY NOMENCLATURE
SYNCHROTRON X-RAY FOOTPRINTING
Synclinal,
STEREOCHEMICAL TERMINOLOGY, IUPAC RECOMMENDATIONS
Synergistic feedback inhibition,
FEEDBACK INHIBITION
SYNERGISM
ALLOSTERISM
COOPERATIVITY
MONOD-WYMAN-CHANGEUX MODEL
KOSHLAND-NEMETHY-FILMER MODEL
SUBSTRATE SYNERGISM
SYNERGISTIC INHIBITION
Synperiplanar,
STEREOCHEMICAL TERMINOLOGY, IUPAC RECOMMENDATIONS
Synthesis of ATP,
BINDING CHANGE MECHANISM
Synthetases,
LIGASES
SYSTEM
SYSTEMATIC ERRORS
RANDOM ERROR
STATISTICS (A Primer)

T

TANDEM MASS SPECTROMETRY
TAU (τ)

TAUTOMERISM
REARRANGEMENTS
ANOMERIC SPECIFICITY
VALENCE TAUTOMERISM
Teeth,
BIOMINERALIZATION
Tele-substitution,
PHYSICAL ORGANIC CHEMISTRY NOMENCLATURE
TELOMERASE
TEMPERATURE
Temperature coefficient,
Q_{10}
TEMPERATURE CONTROL
TEMPERATURE DEPENDENCE
ARRHENIUS LAW
VAN'T HOFF RELATIONSHIP
TRANSITION-STATE THEORY
TEMPERATURE DEPENDENCE OF KINETIC ISOTOPE EFFECTS
Temperature dependence of rate constants,
ARRHENIUS EQUATION & PLOT
Temperature-jump method,
CHEMICAL KINETICS
TEMPLATE CHALLENGE METHOD
PROCESSIVITY
Template-directed assembly,
MICROTUBULE ASSEMBLY
Template-directed irreversible polymerization,
IRREVERSIBLE POLYMERIZATION
Template-directed self-assembly,
BIOCHEMICAL SELF-ASSEMBLY
Template-independent irreversible polymerization,
IRREVERSIBLE POLYMERIZATION
TENSEGRITY
N-Terminal aminoacyl residues,
N-END RULE
TERMINATION
Termolecular,
TRIMOLECULAR
TERNARY COMPLEX
MULTISUBSTRATE MECHANISMS
Ter quad reaction,
MULTISUBSTRATE MECHANISMS
Ter-substrate mechanism,
RANDOM AC/PR, ORDERED B/Q TER TER MECHANISM
RANDOM BC/PQ, ORDERED A/R TER TER MECHANISM
RANDOM TER TER MECHANISM

WEIGHT

WEIGHTING OF DATA
   STATISTICS (A Primer)

Wheland intermediate,
   *PHYSICAL ORGANIC CHEMISTRY NO-
   MENCLATURE*

Whole body turnover kinetics,
   PROTEIN TURNOVER KINETICS

WIGNER SPIN-CONSERVATION RULE

WOMACK-COLOWICK DIALYSIS METHOD

WORK

X

Xaa-Pro dipeptidase,
   PROLINE DIPEPTIDASE

XANTHINE DEHYDROGENASE
   XANTHINE OXIDASE

XANTHINE OXIDASE
   XANTHINE DEHYDROGENASE

XI (ξ)
   EXTENT OF REACTION
   TIME & CHRONOMALS

XYLANASE

XYLOGLUCAN SYNTHASE

D-XYLONATE DEHYDRATASE

XYLOSE ISOMERASE

D-Xylose 5-phosphate,
   RIBULOSE-5-PHOSPHATE 4-EPIMERASE

α-XYLOSIDASE

β-XYLOSIDASE

D-XYLULOKINASE

L-XYLULOKINASE

X-Pro dipeptidase
   PROLINE DIPEPTIDASE

X-ray crystallographic analysis of enzyme intermediates,
   SOLVENT-TRAPPED INTERMEDIATES

Y

YAGA-OZAWA TREATMENT
   YONETANI-THEORELL TREATMENT

Yeast hexokinase,
   INDUCED FIT MODEL

YIELD COEFFICIENT (or, Coefficient of
   Yield)

Yield, quantum,
   QUANTUM YIELD

YLIDES
   CARBANION

YONETANI-THEORELL TREATMENT

Z

Zaitsev rule,
   *PHYSICAL ORGANIC CHEMISTRY NO-
   MENCLATURE*

ZEATIN *CIS-TRANS* ISOMERASE

ZEATIN REDUCTASE

ZEOLITE

ZERO ORDER
   CHEMICAL KINETICS
   RATE SATURATION
   MICHAELIS-MENTEN EQUATION

ZERO-ORDER REACTIONS
   ORDER OF REACTION
   MOLECULARITY
   MICHAELIS-MENTEN EQUATION
   FIRST-ORDER REACTION

ZERO POINT ENERGY
   HOOKE'S LAW SPRING
   KINETIC ISOTOPE EFFECTS

Zeroth law of thermodynamics,
   THERMODYNAMICS, LAWS OF

ZETA (ζ)

Zinc carbonate ($ZnCO_3$),
   SOLUBILITY PRODUCT

Zinc carbonate monohydrate ($ZnCO_3 \cdot H_2O$),
   SOLUBILITY PRODUCT

Zinc-dependent enzymes,
   D-ALANYL-D-ALANINE PEPTIDASE
   ALCOHOL DEHYDROGENASE
   ALKALINE PHOSPHATASE
   5-AMINOLEVULINATE DEHYDRATASE
   ANGIOTENSIN CONVERTING ENZYME
   ASTACIN
   ATROXASE

   CARBONIC ANHYDRASE
   *CARBON MONOXIDE DEHYDRO-
   GENASE*
   CARBOXYPEPTIDASES
   CARNOSINASE
   DIHYDROOROTATE DEHYDROGENASE
   DIPEPTIDASES
   D-2-HYDROXY ACID DEHYDROGENASE
   INSULYSIN
   LEISHMANOLYSIN
   LEUCOLYSIN
   MANNOSE-6-PHOSPHATE ISOMERASE
   MEPRIN
   *MITOCHONDRIAL INTERMEDIATE PEP-
   TIDASE*
   MUCROLYSIN
   NEPRILYSIN
   NUCLEASE S1
   PHOSPHOLIPASE C
   PYRUVATE CARBOXYLASE
   SUPEROXIDE DISMUTASES
   THERMOLYSIN
   TRANSCARBOXYLASE

Zinc ion,
   IONIC RADIUS

ZIPPER MECHANISMS FOR DNA UN-
   FOLDING
   *DNA UNWINDING: Kinetic Model For
   Small DNA*

Zucker-Hammett hypothesis,
   *PHYSICAL ORGANIC CHEMISTRY NO-
   MENCLATURE*

Z VALUE
   SOLVENT EFFECTS

ZWITTERGENTS
   DETERGENTS
   MICELLE

ZWITTERION

ZYMOGEN

Zymogen activation,
   AUTOCATALYSIS

Zyxin,
   ACTIN-BASED MOTILITY

DATE DUE

This Library charges fines on overdues.			